Symbol	Description	Page
A'	Complement of event A	92
A, B	Events	82
$A \cup B$	Union of events A and B	90
AB	Intersection of events A and B	90
α (alpha)	Probability of rejecting H_0 when H_0 is true (Type I error)	279
b	Number of blocks in a randomized block design	428
B_i	Total of observations for block i in an analysis of variance	429
β (beta)	Probability of accepting H_0 when H_0 is false (Type II error)	280
β_0	y-intercept in regression models	584
$\hat{\beta}_0$	Least squares estimator of β_0	588
β_1	Slope of straight-line regression model	584
$\hat{\beta}_1$	Least squares estimator of β_1	588
β_i	Coefficient of independent variable x_i in a multiple regression model	638
$\hat{\beta}_i$	Least squares estimator of β_i	640
c_j	Total count in column j of a contingency table	556
CM	Correction for the mean in an analysis of variance	414
χ^2 (chi square)	Probability distribution of various test statistics	323
df	Degrees of freedom for the t-, χ^2-, and F-distributions	298
D_0	Hypothesized difference between population means or binomial proportions	338
$E(n_i)$	Expected number of outcomes of type i under the null hypothesis in a multinomial test	547
$E(n_{ij})$	Expected count in cell (i, j) of a contingency table	553
$\hat{E}(n_{ij})$	Estimated expected count in cell (i, j) of a contingency table	554
$E(x)$	Expected value of the random variable x	147
ε (epsilon)	Random error component in regression models	584
$f(x)$	Probability density of frequency function for the continuous random variable x	192
F	Test statistic used to compare two variances, to compare k population means, and to test several terms in a multiple regression	384
F_r	Test statistic for the Friedman nonparametric analysis of variance	511
F_α	Value of F-distribution with area α to its right	388
H	Test statistic for the Kruskal–Wallis nonparametric analysis of variance	505
H_a	Alternative (or research) hypothesis	278
H_0	Null hypothesis	278
IQR	Interquartile range	55
k	Number of treatments being compared in a completely randomized or randomized block design	409

Continued inside back cover

Third Edition

Statistics

Third Edition

Statistics

James T. McClave
University of Florida

Frank H. Dietrich, II
Northern Kentucky University

Dellen Publishing Company
San Francisco, California

Collier Macmillan Publishers
London

divisions of Macmillan, Inc.

© Copyright 1985 by Dellen Publishing Company,
a division of Macmillan, Inc.

Printed in the United States of America

Permissions: Dellen Publishing Company
               400 Pacific Avenue
               San Francisco, California 94133

Orders: Dellen Publishing Company
          c/o Macmillan Publishing Company
          Front and Brown Streets
          Riverside, New Jersey 08075

Collier Macmillan Canada, Inc.

Library of Congress Cataloging in Publication Data

McClave, James T.
  Statistics.

  Includes index.
  1. Statistics.   I. Dietrich, Frank H.   II. Title.
QA276.12.M4   1985   519.5   84-28645
ISBN 0-02-378760-0 (Macmillan)

Printing: 1 2 3 4 5 6 7 8   Year: 5 6 7 8 9 0

ISBN 0-02-378760-0

Contents

Preface

The third edition of *Statistics* maintains the same objectives as the earlier editions, namely to introduce students to the basic concepts of statistics and to show them how these concepts can be used in making inferences from experimental data and from sample surveys. The text is designed so that the early chapters can be used for a one-quarter (or a one-semester) introductory course for all undergraduates. The remainder of the text can be used as a second-quarter (semester) follow-up course, with emphasis on special applications such as analysis of variance, regression analysis, or one of the other methodologies included in the later chapters. As in the second edition, we have maintained a unified approach to the subject, attempting both to provide an overall picture of statistics and its role in business and the sciences, and to provide the student with some methodology that will be relevant and useful in other college courses and in subsequent fields of employment. We have also maintained the same level of presentation in this edition.

In addition to making important changes in wording that improve the readability of the text, the following major changes have been made:

Chapter 2: Methods for Describing Sets of Data   Two new sections have been added to this chapter. Section 2.2 explains how to construct and interpret stem and leaf displays. The data sets used to illustrate the methods of constructing stem and leaf displays and histograms are based on real data selected from newspapers and scientific journals. Section 2.8 explains how to use z-scores and box plots to identify outliers. Then this concept is used to introduce the notion of a rare event and a brief introduction to statistical inference in Section 2.9.

Chapter 5: Continuous Random Variables   In the second edition, all the examples in Section 5.2 were concerned with finding areas under the normal curve between two values of the standard normal random variable z or two values of a normal random variable x. We have added an example that deals with the reverse problem, finding a value z_0 corresponding to an area under the normal curve.

Chapter 6: Sampling Distributions   In Section 6.2, we have added a definition for a point estimator.

Chapter 7: Estimation and Tests of Hypotheses: Single Sample   This edition attempts to provide a better explanation of the role of the Type I and Type II errors in testing hypotheses. The new Case Study 7.2 explains the role of hypothesis testing in the formulation of computer security systems. The relevance of Type I and Type II errors is explained in terms of the probabilities of rejecting authorized users and accepting unauthorized users. In addition, new Example 7.3 provides a second example of the computation of the probability β of making a Type II error.

Chapter 9: Analysis of Variance: Comparing More Than Two Means This chapter has been greatly expanded by adding two new sections, many new examples, and corresponding exercises. New Section 9.3 explains how to perform an analysis of variance for a two-factor factorial experiment. Particular stress is placed on the importance of detecting factor interactions. New Section 9.4 presents Tukey's method for making multiple comparisons about a set of population means.

Applications Particular effort was made to add exercises that are based on real-life situations, data sets, and/or research results reported by the news media or in research journals. A total of 184 new conceptual exercises of this type have been added in this edition.

Case Studies Eight new case studies have been added in this edition to place greater emphasis on the relevance of statistics to problem solving in the real world.

The third edition of *Statistics* still retains the most important features of the second edition. The material is presented in a manner that permits flexibility in the amount of time devoted to particular topics. Sections that are not prerequisite to succeeding sections and chapters are marked "(Optional)." We have included several features in this book that will make it different from most introductory statistics texts currently available. These features, which assist the student in achieving an overview of statistics and an understanding of its relevance in the solution of real problems, are as follows:

Case Studies (See the list of 30 case studies on page xvii.) Many important concepts are emphasized by the inclusion of case studies, which consist of brief summaries of actual applications of statistical concepts and are often drawn directly from the research literature. These case studies allow the student to see applications of important statistical concepts immediately after their introduction. The case studies also help to answer by example the often-asked questions, "Why should I study statistics? Of what relevance is statistics to my program?" Finally, the case studies constantly remind the student that each concept is related to the dominant theme—statistical inference.

The Use of Examples as a Teaching Device We have introduced and illustrated almost all new ideas by examples. Our belief is that most students will better understand definitions, generalizations, and abstractions *after* seeing an application. In most sections, an introductory example is followed by a general discussion of the procedures and techniques, and then a second example is presented to solidify the understanding of the concepts.

A Simple, Clear Style We have tried to achieve a simple and clear writing style. Subjects that are tangential to our objective have been avoided, even though some may be of academic interest to those well-versed in statistics. We have not taken an encyclopedic approach in the presentation of material.

Many Exercises—Labeled by Type   The text has a large number (almost 1,200) of exercises illustrating applications in almost all areas of research. However, we believe that many students have trouble learning the mechanics of statistical techniques when problems are all couched in terms of realistic applications—the concept becomes lost in the words. Thus, the exercises at the ends of all sections are divided into two parts:

1. **Learning the mechanics**   These exercises are intended to be straightforward applications of the new concepts. They are introduced in a few words and are unhampered by a barrage of background information designed to make them "practical," but which often detracts from instructional objectives. Thus, with a minimum of labor, the student can recheck his or her ability to comprehend a concept or a definition.
2. **Applying the concepts**   The mechanical exercises described above are followed by realistic exercises that allow the student to see applications of statistics across a broad spectrum. Once the mechanics are mastered, these exercises develop the student's skills at comprehending realistic problems that describe situations to which the techniques may be applied.

On Your Own . . .   Each chapter ends with an exercise entitled "On Your Own. . . ." The intent of this exercise is to give the student some hands-on experience with an application of the statistical concepts introduced in the chapter. In most cases, the student is required to collect, analyze, and interpret data relating to some real application.

A Choice in Level of Coverage of Probability   One of the most troublesome aspects of an introductory statistics course is the study of probability. Probability is troublesome for instructors because they must decide on the level of presentation, and it is troublesome for students because they (often) find it difficult at any level. We believe that one cause for these problems is the mixture of probability and counting rules that occurs in most introductory texts. We have included the counting rules in a separate and optional section at the end of the chapter on probability. In addition, all exercises that require the use of counting rules are marked with an asterisk (*) to indicate this. Thus, the instructor can control the level of coverage of probability.

A word should be added about the length of the probability chapter. Although more space is devoted to probability than in many introductory texts, there are three simple explanations for this: more examples, more exercises, and the optional counting rule section mentioned above. We have included the usual die and coin examples to introduce concepts, but we also work many practical examples so that the connection between probability and statistics is clearly made. This same pattern is followed in the exercise sections. Thus, the 31 examples and 101 exercises account for the length of the chapter. We trust that these will make this troublesome subject easier to learn and to teach.

Where We've Been . . . Where We're Going . . .   The first page of each chapter is a "unification" page. Our purpose is to allow the student to see how the chapter fits into the scheme of statistical inference. First, we briefly show how the material presented in previous chapters helps us to achieve our goal (Where we've been). Then, we indicate what the next chapter (or chapters) contributes to the overall objective (Where we're going). This feature allows us to point out that we are constructing the foundation block by block, with each chapter an important component in the structure of statistical inference. Furthermore, this feature provides a series of brief resumés of the material covered as well as glimpses of future topics.

Chapter Table of Contents   Preceding each chapter is a chapter table of contents. This enables the student to see the sequence of topics to be covered in the chapter.

An Extensive Coverage of Multiple Regression Analysis and Model Building   This topic represents one of the most useful statistical tools for the solution of applied problems. Although an entire text could be devoted to regression modeling, we feel that we have presented a coverage that is understandable, usable, and much more comprehensive than the presentations in other introductory statistics texts. We devote three chapters to discussing the major types of inferences that can be derived from a regression analysis, showing how these results appear in computer printouts and, most important, selecting multiple regression models to be used in an analysis. Thus, the instructor has the choice of a one-chapter coverage of simple regression, a two-chapter treatment of simple and multiple regression, or a complete three-chapter coverage of simple regression, multiple regression, and model building. This extensive coverage of such useful statistical tools will provide added evidence to the student of the relevance of statistics to the solution of applied problems.

Footnotes   Although the text is designed for students with a noncalculus background, footnotes explain the role of calculus in various derivations. Footnotes are also used to inform the student about some of the theory underlying certain results. The footnotes allow additional flexibility in the mathematical and theoretical level at which the material is presented.

Supplementary Materials   A study guide, a solutions manual, and a 3,000 item test bank are available.

We have organized the text in what we believe is a logical sequence. For example, rather than place analysis of variance and nonparametric statistics at the end of the book, we have placed them immediately following the chapters on making inferences about one and two populations. The analysis of variance extends these methods to k populations, and nonparametric statistics gives procedures for making the inferences when the assumptions necessary for parametric methods are in doubt. The three regression and model building chapters, which we consider an essential and unique component, are the final three chapters, because regression represents a greater change in direction than the other inference-making techniques.

Because many introductory courses in statistics are of one-term duration, we show several possible sequences of coverage in the diagram below:

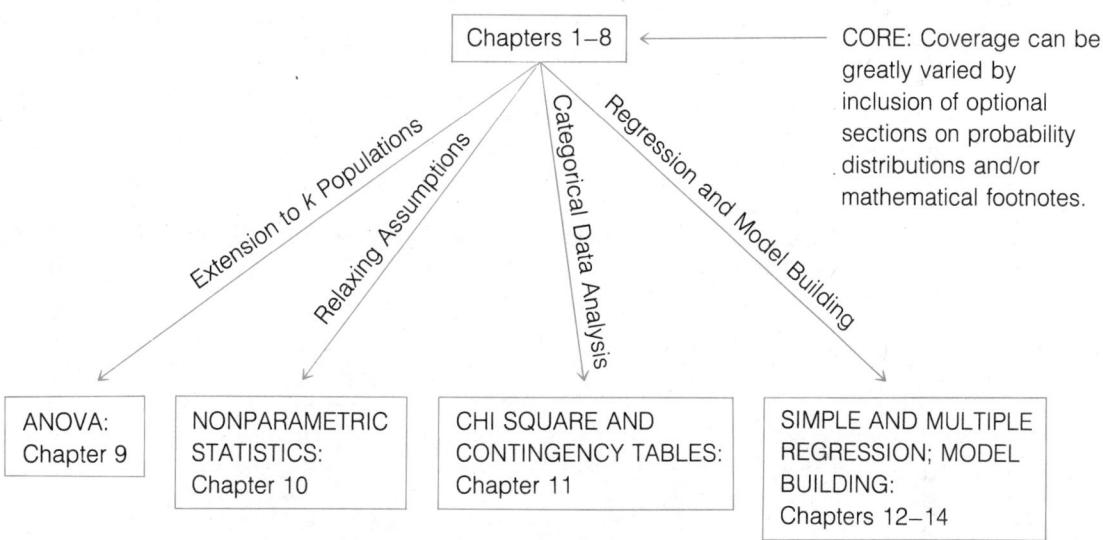

Thanks are due to many individuals who helped in the preparation of this text. Among them are William Beyer, University of Akron; Rudy Gideon, University of Montana; Jean L. Holton, Virginia Commonwealth University; John H. Kellermeier, Northern Illinois University; William G. Koellner, Montclair State University; Diane Lambert, Carnegie–Mellon University; James Lang, Valencia Junior College; Pi-erh Lin, Florida State University; William B. Owen, Central Washington University; Won J. Park, Wright State University; Charles W. Sinclair, Portland State University; Vasanth B. Solomon, Drake University; and Augustin Vukov, University of Toronto. Special thanks to John Dirksey, California State University at Bakersfield. Phyllis Niklas and Susan Reiland have our appreciation and admiration for editing and producing this book. Their work defies explanation; you have to see to believe the care and professionalism with which they work. Finally, we thank the thousands of students at the University of Florida who have helped us to form our ideas about teaching statistics. Their most common complaint seems to be that texts are written for the instructor rather than the student. We hope that this book is an exception.

Case Studies

Third Edition

Statistics

CHAPTER 1

What Is Statistics?

Where We're Going . . .

Statistics? Is it a field of study, a group of numbers that summarize the state of our national economy, the performance of a football team, the social conditions in a particular locale, or, as the title of a popular book (Tanur et al., 1978) suggests, "a guide to the unknown"? We will attempt to answer this question in Chapter 1. Throughout the remainder of the text, we will show you how statistics can be used to interpret experimental and sample survey data. Since many jobs in government, industry, medicine, and other fields require this facility, you will see how statistics can be beneficial to you.

Contents

1.1 Statistics: What Is It?

What does statistics mean to you? Does it bring to mind batting averages, Gallup polls, unemployment figures, numerical distortions of facts (lying with statistics!), or simply a college requirement you have to complete? We hope to convince you that statistics is a meaningful, useful science with a broad, almost limitless scope of application to business, government, and the sciences. We also want to show that statistics lie only when they are misapplied. Finally, our objective is to paint a unified picture of statistics to leave you with the impression that your time was well spent studying a subject that will prove useful to you in many ways.

Statistics means "numerical descriptions" to most people. Monthly unemployment figures, the failure rate of a particular type of steel-belted automobile tire, and the proportion of women who favor the Equal Rights Amendment all represent statistical descriptions of large sets of data collected on some phenomenon. Most often the purpose of calculating these numbers goes beyond the description of the set of data. Frequently, the data are regarded as a sample selected from some larger set of data. For example, a sampling of unpaid accounts for a large merchandiser would allow you to calculate an estimate of the average value of unpaid accounts. This estimate could be used as an audit check on the total value of all unpaid accounts held by the merchandiser. So, the applications of statistics can be divided into two broad areas: (1) describing large masses of data and (2) drawing conclusions (making estimates, decisions, predictions, etc.) about some set of data based on sampling. Let us examine some case studies that illustrate applications of statistics.

Case Study 1.1
A Survey: Where "Women's Work" Is Done by Men

The 1980 February/March issue of *Public Opinion* describes the results of a survey of several hundred married men from each of nine countries who responded to the following question:

> In the following list, which household jobs would you say it would be reasonable that the man would often take over from his wife: washing up (doing dishes), changing baby's napkin (diaper), cleaning house, ironing, organizing meal, staying at home with sick child, shopping, none of these?

The graphs in Figure 1.1 provide an effective summary of the thousands of opinions obtained and allow for an easy comparison of attitudes across countries. The area of statistics concerned with the summarization and description of data is called ***descriptive statistics.***

Case Study 1.2
An Experiment: Investigating an Effect of Smoking During Pregnancy

In an article in the *Journal of the American Medical Association,* M. Sexton and J. R. Hebel (1984) report on their research into the effects of maternal smoking on the birth weight of babies. Their experiment randomly assigned 935 pregnant women smokers into two groups: One continued smoking throughout pregnancy and the other, the **control group,** received smoking intervention (i.e., assistance to reduce to eliminate smoking). From the measured baby weights of these two groups of women, Sexton and Hebel inferred that "some fetal growth retardation can be overcome by the provision of antismoking assistance to pregnant women."

Figure 1.1
"Women's Work" Is
Rarely Done by Men
in Italy, Germany . . .
Note: The sample size for
each country exceeded 900
except Luxembourg, where
the sample was 334.
Source: Survey by the
European Economic
Community Commission.
"Women and Men of
Europe in 1978," October–
November 1977, as shown
in *Public Opinion*,
February–March 1980,
p. 37. © Copyright
American Enterprise
Institute.

Question: In the following list, which household jobs would you say it would be reasonable that the man would often take over from his wife: washing up (doing dishes), changing baby's napkin (diaper), cleaning house, ironing, organizing meal, staying at home with sick child, shopping, none of these?

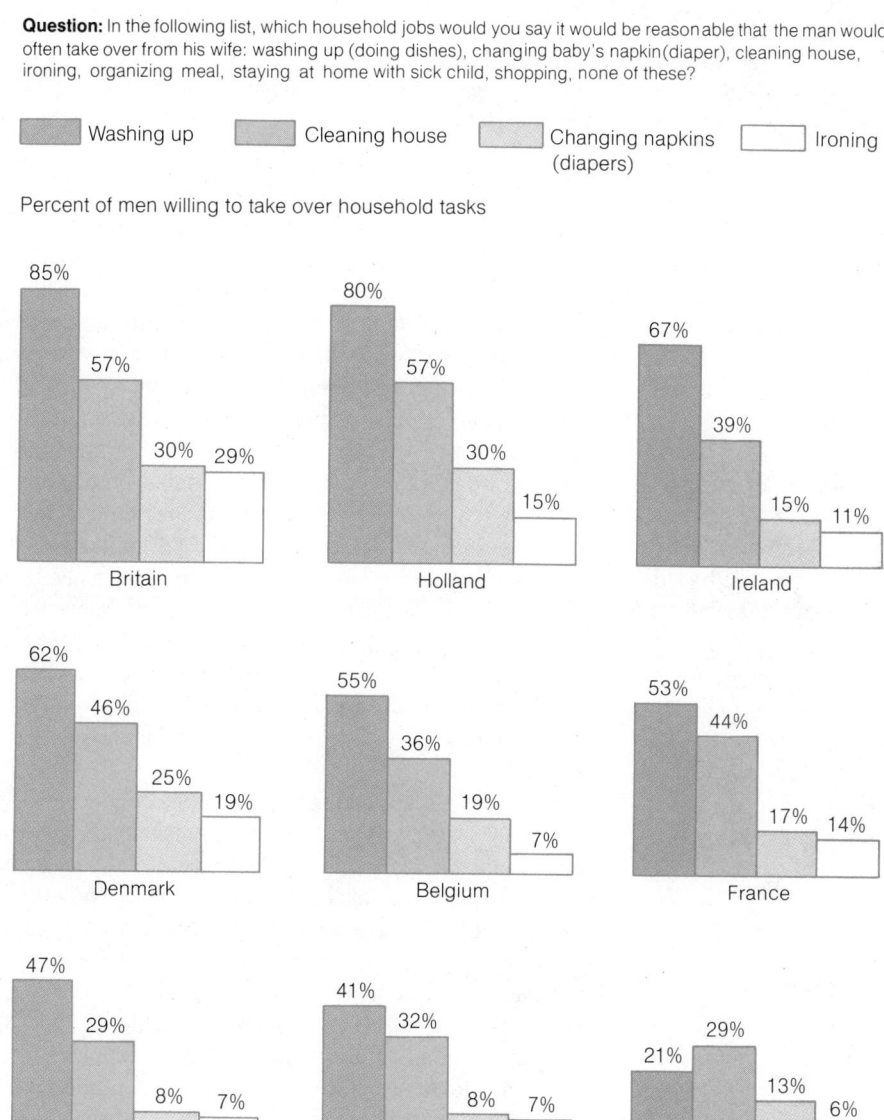

Washing up   Cleaning house   Changing napkins (diapers)   Ironing

Percent of men willing to take over household tasks

In making this statement, Sexton and Hebel made an inference about the impact of a smoking intervention program on all pregnant women who smoke, an inference based on the comparison of the samples of baby weights for the smoking and the control groups of women. The branch of statistics exemplified in this case study is known as *inferential statistics.*

Case Study 1.3
Does Judicial Action Affect the Probability of Conviction?

Defense attorneys often remind juries that accused criminals are innocent until proven guilty. But if the judge waits to deliver his or her own version of the reminder until after the testimony is in, jurors may be more likely to render a guilty verdict than if they had heard the reminder before the trial.

This quote is from an article in the June 1980 issue of *Psychology Today** that discusses a study conducted by two social psychologists at the University of Kansas.

Involved in the study were 107 student jurors, some of whom heard the judge's reminder at the start of a taped trial, some at the end, and some not at all. An analysis of the 107 student verdicts prompted the quote just given.

Note that a **sample** of 107 student jurors was observed and that their verdicts were used to **infer** that the jurors in any trial may be more likely to render a guilty verdict if they hear a reminder from the judge at the end of the trial. This case study, like Case Study 1.2, is an example of the use of inferential statistics.

Case Study 1.4
Taste Preference for Beer: Brand Image or Physical Characteristics of the Beer?

Two sets of data of interest to a firm's marketing department are (1) the set of taste-preference scores given by consumers to their product and their competitors' products when all brands are clearly labeled and (2) the taste-preference scores given by the same set of consumers when all brand labels have been removed and the consumer's only means of product identification is taste. With such information the marketing department should be able to determine whether taste preference arose because of perceived physical differences in the products or as a result of the consumer's image of the brand. (Brand image is, of course, largely a result of a firm's marketing efforts.) Such a determination should help the firm develop marketing strategies for their product.

A study using these two sets of data was conducted by Ralph Allison and Kenneth Uhl (1965) in an effort to determine whether beer drinkers could distinguish among major brands of unlabeled beer. A sample of 326 beer drinkers was randomly selected from the set of beer drinkers identified as males who drank beer at least three times a week. During the first week of the study each of the 326 participants was given a six-pack of unlabeled beer containing three major brands and was asked to taste-rate each beer on a scale from 1 (poor) to 10 (excellent). During the second week the same set of drinkers was given a six-pack containing six major brands. This time, however, each bottle carried its usual label. Again, the drinkers were asked to taste-rate each beer from 1 to 10. From a statistical analysis of the two sets of data yielded by the study, Allison and Uhl concluded that the 326 beer drinkers studied could not distinguish among brands by taste on an overall basis. This result enabled them to infer statistically that such was also the case for beer drinkers in general. Their results also indicated that brand labels and their associations did significantly influence the tasters' evaluations. These findings suggest that physical differences in the products have less to do with their success or failure in the marketplace than the image of the brand in the consumer's mind. As to the benefits of such a study, Allison and Uhl note, "to the extent that product images, and their changes, are believed to be a result of advertis-

* Reprinted from *Psychology Today* magazine. Copyright © 1980 by Ziff Davis Publishing Co.

ing . . . the ability of firms' advertising programs to influence product images can be more thoroughly examined.''

This case study, like Case Studies 1.2 and 1.3, is an example of the use of inferential statistics. Using data collected from a sample of 326 beer drinkers, Allison and Uhl make inferences about the ability of all beer drinkers to distinguish among major brands of unlabeled beer.

These case studies provide four examples of the uses of statistics. Note that each involves an analysis of data, either for the purpose of describing the data set (Case Study 1.1) or for making inferences about a much larger data set based on sampling (Case Studies 1.2, 1.3, and 1.4). They thus provide realistic examples of the two broad areas of statistical applications.

1.2 The Elements of Statistics

Although applications of statistics abound in almost every area of human endeavor, there are certain elements common to all statistical problems. The foundation of every statistical problem is a *population:*

Definition 1.1

The *population* is a set of data that characterizes some phenomenon.

Thus, we use the term *population* to represent a set of measurements (rather than a group of people) that characterizes some phenomenon of interest to us. For example, the employment status of every person in the United States is a set of data that characterizes a certain aspect of our national health and that, consequently, is of interest to government economists and sociologists. Similarly, the life spans of Florida spiny lobsters represent a data set that characterizes the life expectancy of this important source of seafood. These data would be relevant in evaluating the growth in numbers of lobsters over time and could be beneficial in establishing minimum lobster trapping weights. Other examples of populations are the number of errors contained in each of the many pages in an accountant's ledger and the lengths of survival, after onset of the disease, of children afflicted with leukemia. Thus, we think of a population as being a large—perhaps infinitely large—collection of measurements.

The second element of a statistical problem is the *sample:*

Definition 1.2

The *sample* is a subset of data selected from a population.

The sample is a subset (part of) the population. The employment status of 2,000 people selected from the total of all employable persons in the United States, the life spans of twenty Florida lobsters, and the number of errors per page on 10 pages of a 100-page ledger all represent samples of the respective populations. In everyday usage, the word *sample* implies a collection of objects—for example, a sample of 1,200 people from a city or a sample of ten transistors from a day's production. When selecting people or objects from some group, we sometimes utilize this terminology; i.e., we use *sample* to refer to the selected objects rather than to the collection of measurements made on the objects. Whether we are speaking of a collection of measurements (our definition) or the collection of objects on which the measurements are made (everyday usage) becomes clear from the context of the discussion.

The usefulness of the sample is clarified by considering the third element of a statistical problem—the **inference:**

Definition 1.3

A **statistical inference** is a decision, estimate, prediction, or generalization about the population based on information contained in a sample.

That is, we use the information in the smaller set of measurements (the sample) to make decisions, predictions, or generalizations about the large or whole set of measurements (the population). For example, we might use the number of accounting errors in a 10-page sample of a ledger to estimate the number of errors on all 100 pages of the ledger. Similarly, we could use a sample consisting of the employment status of 2,000 employable people to predict the national unemployment rate. Or we might try to infer the life characteristics of the Florida lobster based on a sample of observed life lengths of twenty lobsters. In each case, we are using the information in a sample to make inferences about the corresponding population.

The preceding definitions identify three of the four elements of a statistical problem. The fourth, and perhaps the most important, is the topic of the next section.

1.3 Statistics: Witchcraft or Science?

We have identified the primary objective of statistics as making inferences about a population based on information contained in a sample. However, inference making constitutes only part of our story. We also want to measure and report the **reliability** of each inference made—this is the fourth element of a statistical problem.

The measure of reliability that accompanies an inference separates the science of statistics from the art of fortune telling. A palm reader, like a statistician, may examine a sample (your hand) and make inferences about the population (your life). However, no measure of reliability can be attached to the reader's inferences. We, on the other hand, will always be sure to assess the reliability of our statistical inferences. For example, public opinion polls estimate the proportion of people favoring some issue or political candidate and give, along with their estimate, a "margin of error." What they are giving you is a *bound* on the *estimation error,* a number that the error of our estimation is not likely to exceed. Thus, the uncertainty of our estimation is measured by the size of the bound on the estimation errors. The reliability of our statistical inferences will be discussed at length throughout this text. For now, we simply want you to realize that an inference is incomplete without a measure of its reliability.

We conclude this section with a summary of the elements of an inferential statistical problem:

Four Elements Common to Inferential Statistical Problems

1. The population of interest, with a procedure for sampling the population
2. The sample and analysis of the information in the sample
3. The inference about the population, based on information contained in the sample
4. A measure of reliability for the inference

1.4 Why Study Statistics?

Why study statistics? The growth in data collection associated with scientific phenomena as well as the operations of business and government (quality control, statistical auditing, forecasting, etc.) has been truly remarkable over the past several decades. Published results of political, economic, and social surveys as well as increasing government emphasis on drug and product testing provide vivid evidence of the need to be able to evaluate data sets intelligently. Consequently, you will want to develop a discerning sense of rational thought that will enable you to evaluate numerical data. You may be called upon to use this ability to make intelligent decisions, inferences, and generalizations. For this reason, the study of statistics is an essential preparation for a role in modern society.

Exercises 1.1–1.9

1.1 To evaluate the current status of the dental health of schoolchildren, the American Dental Association conducted a survey to estimate the average number of cavities per child in grade school in the United States. One thousand schoolchildren from across the country were selected, and the number of cavities for each was recorded.
a. Describe the population of interest to the American Dental Association.
b. Describe the sample.

1.2 A first-year chemistry student conducts an experiment to determine the amount of hydrochloric acid necessary to neutralize 2 ounces of a basic solution. The student prepares five 2-ounce portions of the solution and adds a known concentration of hydrochloric acid to each. The amount of acid necessary to achieve neutrality of the solution is recorded for each of the five portions.
a. Describe the population of interest to the student.
b. Describe the sample.

1.3 A manufacturer of vacuum cleaners has decided that an assembly line is operating satisfactorily if less than 2% of the cleaners produced per day are defective. If 2% or more of the cleaners are defective, the line must be shut down and proper adjustments made. To check every cleaner as it comes off the line would be costly and time-consuming. The manufacturer decides to choose thirty cleaners at random from a specific day's production and test for defects.

a. Describe the population of interest to the manufacturer.

b. Describe the sample.

c. Give an example of an inference the manufacturer might make.

1.4 An insurance company would like to determine the proportion of all medical doctors who have been involved in one or more malpractice suits. The company selects 500 doctors at random and determines the number in the sample who have ever been involved in a malpractice suit.

c. Describe the population of interest to the insurance company.

b. Describe the sample.

c. Give an example of an inference the insurance company might make.

1.5 A new teaching method has been introduced in the third grade at a local elementary school. At the end of the first year, the school board wants to evaluate the new method and determine its effectiveness. A standardized test, traditionally given at the end of the third grade, will be used to determine whether the new method is effective.

a. Describe the population of interest to the school board.

b. Describe the sample.

c. Give an example of an inference the school board might make.

d. What should accompany any inference the school board makes?

1.6 The *Gainesville Sun* (March 14, 1984) reports on the topics that teenagers most want to discuss with their parents. The findings, the results of a Gallup Youth Poll, show that 46% would like more discussion about the family's financial situation, 37% would like to talk about school, and 30% would like to talk about religion. The survey was based on a national sampling of 505 teenagers.

a. Focus on a single question asked the 505 teenagers. Describe the sample observations.

b. Describe the population about which the pollsters wish to make an inference.

c. How is the inference expressed?

d. Newspaper accounts of most polls usually give a margin of error (a percentage) for the survey result. What is the purpose of the "margin of error" and what is its interpretation?

1.7 An article in the *Minneapolis Tribune* (October 24, 1982) reports on sampling of fish in the Great Lakes by the Michigan Department of Agriculture. The objective of the sampling was to determine the level of contaminants, DDT, PCB's, Dieldrin, etc., in lake fish.

Some 209 fish were randomly sampled from the lake, and the levels of the various contaminants were measured for each fish. Although the study found the level to be diminished, the Michigan Public Health Department advised the public not to eat more than one meal of lake fish per week and pregnant women were advised to eat none.

a. Describe the population of interest to the Michigan Department of Agriculture.

b. Describe the sample observations.

c. Describe the type of inference that the department wished to make about the population.

d. How might a measure of reliability of the inference be expressed?

1.8 According to *U.S. News and World Report* (May 10, 1982), a powerful new anti-biotic, piperacillin, was developed by the Lederle Laboratories and approved by the Food and Drug Administration. Tests on 600 patients showed the drug to be 92% effective in curing serious bacterial infection.

a. Do the 600 patients represent a population or a sample? Describe the population and sample of interest to the research physicians.

b. Describe the inference made by the researchers.

c. Do you have sufficient information to assess the reliability of their inference?

1.9 A *U.S. News and World Report* (June 21, 1982) article describes a new method of treating a major form of blindness in elderly people. The process, using laser beams to seal abnormal blood vessels in the eye, was tried on 224 patients. Of these, only 14% went blind in 1 year. In a control group of similar untreated patients, 42% went blind in 1 year. Therefore, to determine whether the laser beam treatment was effective, research physicians wished to compare the proportions of patients going blind in 1 year for *two* different populations.

a. Describe the two populations that the research physicians want to compare.

b. Identify the samples.

c. Describe the inference that the researchers will make.

d. Based on the description given here, do you have enough information to evaluate the reliability of their inference?

On Your Own . . .

Scan a recent issue of a daily newspaper and look for articles that contain numerical data. The data might be a summary of the results of a public opinion poll, the results of a vote by the United States Senate, crime rates, birth or death rates, an election result, etc. For each article containing data that you find, answer the following questions:

a. Do the data constitute a sample or an entire population? If a sample has been taken, clearly identify both the sample and the population; otherwise, identify the population.

b. If a sample has been observed, does the article present an explicit (or implied) inference about the population of interest? If so, state the inference made in the article.

c. If an inference has been made, has a measure of reliability been included? What is it?

References

Allison, R. I. & Uhl, K. P. "Influence of beer brand identification on taste perception." *Journal of Marketing Research,* August 1965, 36–39.

Careers in statistics. American Statistical Association and the Institute of Mathematical Statistics, 1974.

Rubenstein, C. "The presumption of innocence needs prompting." *Psychology Today,* June 1980, 30.

Sexton, M., & Hebel, J. R. "A clinical trial of change in maternal smoking and its effect on birth weight." *Journal of the American Medical Association,* 1984, *251.*

Tanur, J. M., Mosteller, F., Kruskal, W. H., Link, R. F., Pieters, R. S., & Rising, G. R. *Statistics: A guide to the unknown.* (E. L. Lehmann, special editor.) San Francisco: Holden-Day, 1978.

"'Women's work' is rarely done by men in Italy, Germany, ..." *Public Opinion,* February–March 1980, 37.

CHAPTER 2

Methods for Describing Sets of Data

Where We've Been . . .

By examining typical examples of the use of statistics, we identified four elements that are common to every inferential statistical problem: a population, a sample, an inference, and a measure of the reliability of the inference. The last two elements identify the goal of statistics—using sample data to make an inference (a decision, estimate, or prediction) about a population.

Where We're Going . . .

Before we make an inference, we must be able to describe a data set. Both graphic and numerical methods for describing sets of data are discussed in this chapter. As you will learn in Chapter 7, we will use some sample numerical descriptive measures to estimate the values of corresponding population descriptive measures. Therefore, our efforts in this chapter will ultimately lead to statistical inference.

Contents

Suppose we wish to evaluate the mathematical capabilities of a class of 1,000 college freshmen based on their quantitative Scholastic Aptitude Test (SAT) scores. How would you describe these 1,000 measurements? You can see that this is not an easy question to answer. The 1,000 scores provide too many bits of information for our minds to comprehend. It is clear that we need some method for summarizing the information in a data set. Methods for describing data sets are also essential for statistical inference. Most populations are large data sets. Consequently, if we are going to make descriptive statements (inferences) about a population based on information contained in a sample, we will once again need methods for describing a data set.

Two methods for describing data are presented in this chapter, one **graphic** and the other **numerical.** As you will subsequently see, both play an important role in statistics.

2.1
Types of Data

Although the number of phenomena that can be measured is almost limitless, data can generally be classified as one of two types: **quantitative** or **qualitative.**

Definition 2.1

Quantitative data are observations that are measured on a numerical scale.

The most common type of data is quantitative data, since many descriptive variables observed in nature are measured on numerical scales. Examples of quantitative data are:

1. The bacteria count in your drinking water

2. The monthly unemployment percentage

3. A person's blood pressure

4. The number of women executives in an industry

The measurements in these examples are all numerical.

All data that are not quantitative are qualitative:

Definition 2.2

If each measurement in a data set falls into one and only one of a set of categories, the data set is called *qualitative.*

Qualitative data are observations that are categorical rather than numerical. Examples of qualitative data are:

1. The political party affiliations of a group of people. Each person would have one and only one political party affiliation.

2. The brand of gasoline last purchased by each person in a sample of automobile owners. Again, each measurement would fall into one and only one category.

3. The state in which each firm in a sample of firms in the United States has its highest yearly sales.

Notice that each of the examples has nonnumerical, or qualitative, measurements.

As you would expect, the methods used for summarizing the information in a sample of measurements depend on the type of data being collected. Qualitative data can be summarized by giving the number of observations that fall in each of the classification categories, but the description of quantitative data is more complex. Consequently, this chapter presents methods for describing quantitative data sets.

Exercises 2.1–2.2

Learning the Mechanics

2.1 State whether the following types of data are qualitative or quantitative:

a. The bacteria count in the water at thirty city swimming pools

b. The occupation of each of 200 shoppers at a supermarket

c. The marital status of each person living on a city block

d. The number of months between auto maintenance for each of 100 sales representatives

2.2 State whether the following types of data are qualitative or quantitative:

a. The height of each student in your class

b. The length of time each of thirty patients must stay in a hospital

c. The political party of each United States senator

d. The religious affiliation of each patient of a psychologist

2.2 Graphic Methods for Describing Quantitative Data: Stem and Leaf Displays

Before we can use the information in a sample to make inferences about a population, we need methods to summarize, or describe, a set of data. For example, the Environmental Protection Agency (EPA) performs extensive tests on all new car models to determine their mileage rating. Suppose that the 100 measurements in Table 2.1 represent the results of such tests on a certain new car model. How can we summarize the information in this rather large sample?

Table 2.1

EPA Mileage Ratings on 100 Cars

36.3	41.0	36.9	37.1	44.9	36.8	30.0	37.2	42.1	36.7
32.7	37.3	41.2	36.6	32.9	36.5	33.2	37.4	37.5	33.6
40.5	36.5	37.6	33.9	40.2	36.4	37.7	37.7	40.0	34.2
36.2	37.9	36.0	37.9	35.9	38.2	38.3	35.7	35.6	35.1
38.5	39.0	35.5	34.8	38.6	39.4	35.3	34.4	38.8	39.7
36.3	36.8	32.5	36.4	40.5	36.6	36.1	38.2	38.4	39.3
41.0	31.8	37.3	33.1	37.0	37.6	37.0	38.7	39.0	35.8
37.0	37.2	40.7	37.4	37.1	37.8	35.9	35.6	36.7	34.5
37.1	40.3	36.7	37.0	33.9	40.1	38.0	35.2	34.8	39.5
39.9	36.9	32.9	33.8	39.8	34.0	36.8	35.0	38.1	36.9

A visual inspection of the data indicates some obvious facts. Most of the mileages are in the 30's, for example, with a smaller fraction in the 40's. But it is difficult to provide much additional information on the 100 mileage ratings without resorting to some method of summarizing the data. A graphic method of summarizing is provided by a **stem and leaf display.**

Figure 2.1 shows a stem and leaf display for the data. To construct this display, we first partition a typical observation into a **stem** and a **leaf.** For our display we chose

Figure 2.1 A Stem
and Leaf Display for the
EPA Mileage Ratings on
100 Cars

Stem	Leaf
30	.0
31	.8
32	.7 .5 .9 .9
33	.9 .1 .8 .9 .2 .6
34	.8 .0 .4 .8 .2 .5
35	.5 .9 .3 .9 .7 .6 .2 .0 .6 .1 .8
36	.3 .2 .3 .5 .8 .9 .9 .0 .7 .6 .4 .8 .5 .4 .6 .1 .8 .7 .7 .9
37	.0 .1 .3 .9 .2 .6 .3 .1 .9 .4 .0 .0 .1 .6 .8 .7 .0 .2 .4 .7 .5
38	.5 .6 .2 .3 .0 .2 .7 .8 .4 .1
39	.9 .0 .8 .4 .0 .7 .3 .5
40	.5 .3 .7 .2 .5 .1 .0
41	.0 .0 .2
42	.1
43	
44	.9

the stem portion of an observation as all digits at or to the left of the 1's digit place. The remaining portion of the observation to the right of the stem is the leaf. The stems and leaves for several mileage readings, 36.3, 32.7, and 40.5, are shown here:

Stem	Leaf		Stem	Leaf		Stem	Leaf
36	.3		32	.7		40	.5

After choosing the stem and leaf for an observation, we list the set of possible stems for the data set in a column from the smallest (30) to the largest (44). Then the leaf for each observation is recorded in the row of the display corresponding to the observation's stem. For example, the leaf (.3) of the first observation in Table 2.1 is written in the row corresponding to the stem (36). Similarly, the leaf (.7) for the second observation in Table 2.1 is recorded in the row of Figure 2.1 corresponding to the stem (32).

The stem and leaf display (Figure 2.1) presents a compact picture of the data set. You can see at a glance that the 100 mileage readings were distributed between 30.0 and 44.9 with most of them falling in stem rows 35 to 39. The 6 leaves in stem row 34 indicate that 6 of the 100 readings were at least 34.0 but less than 35.0. Similarly, the 11 leaves in stem row 35 indicate that 11 of the 100 readings were at least 35.0 but less than 36. Only five cars had readings equal to 41 or larger, and only one was as low as 30.

The stem and leaf display for a data set can be modified by changing the way we define the stem and the leaf for a measurement. In the stem and leaf display of Figure 2.1, for example, we defined the stem to be all digits to the left of the decimal place. Another definition for the stem would be all digits at or to the left of the 10's digit. Using this definition, the stems and leaves for 36.3 and 32.7 would be

Stem	Leaf		Stem	Leaf
3	6.3		3	2.7

If you look at the data, you can see why we did not define the stem in this way. Since all the mileage readings fall in the 30's and 40's, all the leaves would fall into two stem rows of the display. The resulting graphic picture of the data set would not be nearly as informative as Figure 2.1.

Example 2.1   The data in Table 2.2 give the percentages of the total number of college or university student loans by state that are in default. Construct a stem and leaf display for the data.

Table 2.2
Defaulted Student Loans
(in Millions of Dollars)

STATE	%	STATE	%	STATE	%	STATE	%
Ala.	12.0	Ill.	9.3	Mont.	6.4	R.I.	8.8
Alaska	19.7	Ind.	6.7	Nebr.	4.9	S.C.	14.1
Ariz.	12.1	Iowa	6.2	Nev.	10.1	S.Dak.	5.5
Ark.	12.9	Kans.	5.7	N.H.	7.9	Tenn.	12.3
Calif.	11.4	Ky.	10.3	N.J.	12.0	Tex.	15.2
Colo.	9.5	La.	13.5	N.Mex.	7.5	Utah	6.0
Conn.	8.8	Maine	9.7	N.Y.	11.3	Vt.	8.3
Del.	10.9	Md.	16.6	N.C.	15.5	Va.	14.4
D.C.	14.7	Mass.	8.3	N.Dak.	4.8	Wash.	8.4
Fla.	11.8	Mich.	11.4	Ohio	10.4	W.Va.	9.5
Ga.	14.8	Minn.	6.6	Okla.	11.2	Wis.	9.0
Hawaii	12.8	Miss.	15.6	Oreg.	7.9	Wyo.	2.7
Idaho	7.1	Mo.	8.8	Pa.	8.7		

Source: National Direct Student Loan Program.

Solution   The first step is to define the stem for an observation. Since the smallest observation in the data set is 2.7 and the largest is 19.7, a good choice for the stem would be all digits to the left of the decimal point. Thus, the stem and leaves for 2.7 and 19.7 would be

Stem	Leaf		Stem	Leaf
2	.7		19	.7

This stem and leaf display is shown in Figure 2.2.

Figure 2.2 Stem and Leaf Display for the Percentage of Student Loans (per State) in Default

Stem	Leaf						
2	.7						
3							
4	.9	.8					
5	.7	.5					
6	.7	.2	.6	.4	.0		
7	.1	.9	.5	.9			
8	.8	.3	.8	.7	.8	.3	.4
9	.5	.3	.7	.5	.0		
10	.9	.3	.1	.4			
11	.4	.8	.4	.3	.2		
12	.0	.1	.9	.8	.0	.3	
13	.5						
14	.7	.8	.1	.4			
15	.6	.5	.2				
16	.6						
17							
18							
19	.7						

You can see that Figure 2.2 provides a good graphic description of the percentage of student loans in default by state. Most of the observations fall in stem rows 6 through 12, indicating percentages of loans in default from 6.0% to 12.9%. All but two fall in stem rows 4 through 16, i.e., from 4% to 16.9%. ■

Apart from providing a good graphic picture of the data set, a stem and leaf display possesses two added advantages. If you want to recover the data from a stem and leaf display, you can readily reconstruct the values of the observations by recombining the leaves with the stems. Moreover, the construction of the display automatically arranges the observations in ordered sets. This feature makes it easy to arrange the observations from smallest to largest and, for example, to find the observation in the middle of this ordered arrangement.

One disadvantage of the stem and leaf display is that it is awkward (but possible) to control the number of stems. Thus, it is possible to decide on the approximate number of stems beforehand and then to define the stem in order to achieve this number. Some of the simplicity of constructing the stem and leaf display is lost in the process. A more obvious disadvantage is that the stem and leaf display is unsuitable when the number of observations in the data set is large. In this case the number of leaves in the stem rows becomes too large and the leaves run off the side of the display. To obtain a graphic description of large data sets, we display the data using a ***relative frequency histogram.*** This graphic method, which bears a similarity to the stem and leaf display, is discussed in the following section.

How to Construct a Stem and Leaf Display

1. Define the stem and leaf that you wish to use. You will probably wish to choose the stem so that the number of possible stems in the display is not too large.

2. Write the stems in a column from the smallest stem at the top to the largest at the bottom.

3. Record the leaf for each observation in the row corresponding to its stem.

Exercises 2.3–2.7

Learning the Mechanics

2.3 Consider the following data:

18.3	27.3	25.6	19.0	27.4
22.7	21.9	22.4	23.7	20.9
17.5	38.7	30.1	24.6	28.2
33.6	26.4	42.5	33.5	16.4
35.1	40.1	34.1	31.2	21.9

a. Construct a stem and leaf display with the stem defined to be all digits at or to the left of the 10's digit.

b. Construct a stem and leaf display with the stem defined to be all digits to the left of the decimal point.

c. Which stem and leaf display seems to give a better description of the data?

2.4 Construct a stem and leaf display for the following data. Define the stem to be all digits at or to the left of the 10's digit.

8.4	32.5	17.6	33.9	20.6
16.2	12.0	9.4	42.2	19.5
29.7	28.2	24.3	20.7	29.1
10.3	23.9	22.1	21.5	52.8

Explain how this stem and leaf display describes the data set.

Applying the Concepts

2.5 According to the National Center for Education Statistics only 72.1% of the nation's 3.8 million high school students, those scheduled to graduate in 1981, actually graduated (*Bakersfield Californian,* May 12, 1983). The percentages graduated by state, Minnesota with the highest percentage and the District of Columbia the lowest, are shown here:

Minn.	86.0	Hawaii	79.6	Vt.	75.5	Del.	72.7	Calif.	68.0
N.Dak.	84.9	Pa.	78.6	Ind.	75.3	Mich.	72.5	Ky.	67.3
Iowa	84.8	Ohio	78.4	Wash.	75.3	N.Mex.	71.9	Ala.	67.1
S.Dak.	82.8	N.H.	77.1	Md.	75.0	Maine	71.4	N.C.	67.1
Wis.	82.3	Okla.	76.9	Ark.	74.8	Conn.	71.1	Tenn.	66.7
Nebr.	81.3	Idaho	76.8	Mo.	74.0	Oreg.	70.3	N.Y.	65.9
Mont.	80.9	Mass.	76.8	W.Va.	74.0	Ariz.	69.0	Ga.	64.3
Kans.	80.5	Nev.	76.8	Va.	73.3	S.C.	68.8	Fla.	63.7
Utah	80.2	N.J.	76.7	Ill.	73.2	Tex.	68.8	La.	63.4
Wyo.	80.0	Colo.	76.5	R.I.	72.8	Alaska	68.7	Miss.	61.8
								D.C.	54.6

Describe this data set by using a stem and leaf display. After you have examined the display, describe the characteristics of the data set in words.

2.6 While producing many economic benefits to the state of Florida, gypsum and phosphate mines also produce a harmful byproduct: radiation. It has been known for a number of years that the mine tailings (waste) contain radioactive radon 222. In fact, new housing complexes built over the leveled piles of residue have shown disturbing radiation levels within the houses. The radiation levels in waste gypsum and phosphate mounds in Polk County, Florida, are regularly monitored by the Eastern Environmental Radiation Facility (EERF) and by the Polk County Health Department (PCHD), Winter Haven, Florida. Shown below are measurements of the exhalation rate (a measure of radiation) of soil samples taken on waste piles in Polk County, Florida. They represent part of the data contained in a report by Thomas R. Horton (1979) of EERF. Construct a stem and leaf display for the data set. Use the stem and leaf display to present a verbal description of the data set.

1,709.79	4,132.28	2,996.49	2,796.42	3,750.83	961.40	1,096.43	1,774.77
357.17	1,489.86	2,367.40	11,968.23	178.99	5,402.35	2,315.52	2,617.57
1,150.94	3,017.48	599.84	2,758.84	3,764.96	1,888.22	2,055.20	205.84
1,572.69	393.55	538.37	1,830.78	878.56	6,815.69	752.89	1,977.97
558.33	880.84	2,770.23	1,426.57	1,322.76	1,480.04	9,139.21	1,698.39

2.7 In an article on the relationship between worker strike activity and union activity, Neil A. Palomba (1969) gives the strike activity (measured as the percentage of the estimated total working time lost due to strikes) for each state (excluding Maryland) for 1964. Although dated, these percentages provide one interesting measure of a state's industrial climate in the 1960s. Describe them by using a stem and leaf display. Then locate the position of your state within the display and interpret its value in relation to the others.

STATE	%	STATE	%	STATE	%	STATE	%	STATE	%
Ala.	.14	Hawaii	.02	Mich.	.83	N.Y.	.11	Tenn.	.23
Alaska	.11	Idaho	.11	Minn.	.02	N.C.	.01	Tex.	.06
Ariz.	.09	Ill.	.18	Miss.	.14	N.Dak.	.03	Utah	.66
Ark.	.10	Ind.	.16	Mo.	.14	Ohio	.38	Vt.	.26
Calif.	.16	Iowa	.16	Mont.	.28	Okla.	.01	Va.	.04
Colo.	.04	Kans.	.11	Nebr.	.05	Oreg.	.12	Wash.	.16
Conn.	.08	Ky.	.17	Nev.	.36	Pa.	.14	W.Va.	.45
Del.	.41	La.	.10	N.H.	.03	R.I.	.09	Wis.	.21
Fla.	.20	Maine	.15	N.J.	.27	S.C.	.01	Wyo.	.01
Ga.	.13	Mass.	.07	N.Mex.	.09	S.Dak.	.16	Md.	—

2.3 Graphic Methods for Describing Quantitative Data: Relative Frequency Histograms

A relative frequency histogram for the 100 EPA mileage readings is shown in Figure 2.3. The horizontal axis of Figure 2.3, which gives the miles per gallon for a given automobile, is divided into intervals commencing with the interval from 29.95 to 31.45 and proceeding in intervals of equal size to 43.45 to 44.95 miles per gallon. The vertical axis gives the proportion (or **relative frequency**) of the 100 readings that fall in each interval. Thus,

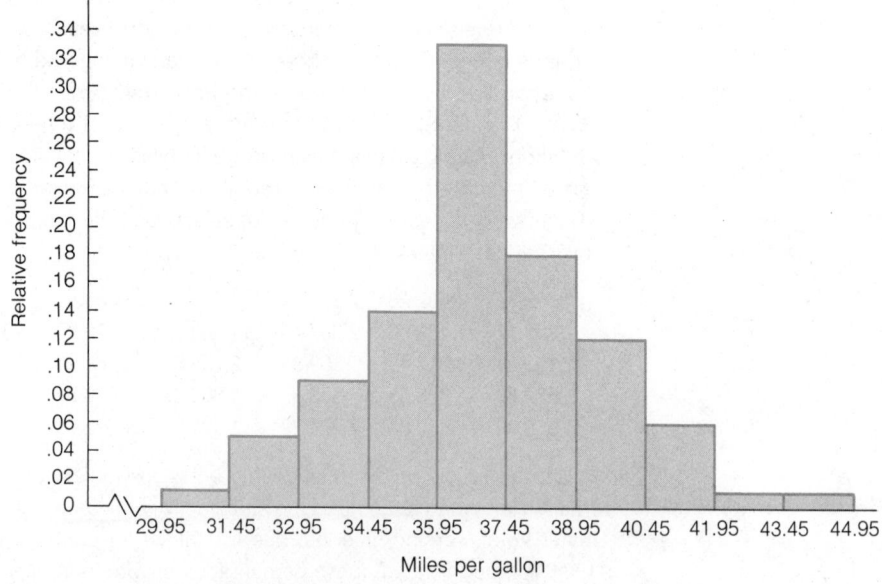

Figure 2.3   Histogram for EPA Mileage Data

you can see that 0.33, or 33%, of the owners obtained a mileage between 35.95 and 37.45. This interval contains the highest relative frequency, and the intervals tend to contain a smaller fraction of the measurements as the mileages get smaller or larger.

By summing the relative frequencies in the intervals 34.45–35.95, 35.95–37.45, and 37.45–38.95, you can see that 65% of the mileages are between 34.45 and 38.95. Similarly, only 2% of the cars obtained a mileage rating over 41.95. Many other summary statements can be made by further study of the histogram. When constructing a histogram, the general rules listed in the accompanying box should be followed.

How to Construct a Histogram

1. Identify the smallest and the largest measurements in the set of data.
2. Divide the interval between the smallest and the largest measurement into between five and twenty equal subintervals called *classes.* These classes should satisfy the following requirements:

   a. Each measurement falls into one and only one subinterval.
   b. No measurement falls on a boundary of a subinterval.

   Although the choice of the number of classes is arbitrary, you will obtain a better description of the data if you use a small number of subintervals when you have a small amount of data and use a large number of subintervals for a large amount of data.
3. Compute the proportion (relative frequency) of measurements in each subinterval.*
4. Using a vertical axis of about three-fourths the length of the horizontal axis, plot each relative frequency as a rectangle over the corresponding subinterval.

To construct the relative frequency histogram in Figure 2.3 we must first define the measurement classes. Since the number of measurements, $n = 100$, is moderately large, we will arbitrarily choose to construct ten measurement classes. The ten classes must span the distance between the smallest measurement, 30.0, and the largest measurement, 44.9. Thus, each class should have a width of

$$\text{Approximate class width} = \frac{\text{Largest measurement} - \text{Smallest measurement}}{\text{Number of classes}}$$

$$= \frac{44.9 - 30.0}{10} \approx 1.49$$

or, rounding upward to be certain of including all the observations, the class interval width is approximately equal to 1.50.

---

* Note that *frequencies* rather than relative frequencies may be used in constructing a histogram. The frequency is the actual number of measurements in each interval.

Locating the lower class boundary of the first class interval at 29.95 (slightly below the smallest measurement) and adding the class width of 1.5, we find the upper class boundary to be 31.45. Adding 1.5 again, we find the upper class boundary of the second class to be 32.95. Continuing this process, we obtain the ten class intervals shown in Table 2.3. Note that the points which locate the class boundaries are written with a 5 in the second decimal place. This makes it impossible for any of the miles per gallon observations to fall on a class boundary (since they were given only to the nearest tenth).

Table 2.3

Measurement Classes, Frequencies, and Relative Frequencies for the Car Mileage Data

MEASUREMENT CLASS	FREQUENCY	RELATIVE FREQUENCY
29.95–31.45	1	.01
31.45–32.95	5	.05
32.95–34.45	9	.09
34.45–35.95	14	.14
35.95–37.45	33	.33
37.45–38.95	18	.18
38.95–40.45	12	.12
40.45–41.95	6	.06
41.95–43.45	1	.01
43.45–44.95	1	.01
	100	1.00

The next step is to determine the class frequencies and calculate the class relative frequencies. These quantities are defined as follows:

Definition 2.3

The **class frequency** for a given class, say class i, is equal to the total number of measurements which fall in that class. The class frequency for class i is denoted by the symbol f_i.

Definition 2.4

The **class relative frequency** for a given class, say class i, is equal to the class frequency divided by the total number n of measurements, i.e.,

$$\text{Relative frequency for class } i = \frac{f_i}{n}$$

Scan the data and count the number of measurements in each class interval. This number, the **class frequency**, is entered in the second column of Table 2.3. Finally, calculate the **class relative frequency,** the proportion of the total number of measurements that fall in each interval. The class relative frequencies for the mileage data are obtained by dividing each of the class frequencies by the total number of measurements (100); these are listed in the third column of Table 2.3. As noted at the outset of this discussion, these relative frequencies were used to construct the relative frequency histogram of Figure 2.3.

By looking at a histogram (say, the relative frequency histogram in Figure 2.3), you can see two important facts. First, note the total area under the histogram and then note the proportion of the total area that falls over a particular interval of the x-axis. You will see that the proportion of the total area that falls above an interval is equal to the relative frequency of measurements falling in the interval. For example, the relative frequency for the class interval 35.95−37.45 is .33. Consequently, the rectangle above the interval contains .33 of the total area under the histogram.

Second, you can imagine the appearance of the relative frequency histogram for a very large set of data (say, a population). As the number of measurements in a data set is increased, you can obtain a better description of the data by decreasing the width of the class intervals. When the class intervals become small enough, a relative frequency histogram will (for all practical purposes) appear as a smooth curve (see Figure 2.4).

Figure 2.4   The Effect of the Size of a Data Set on the Outline of a Histogram

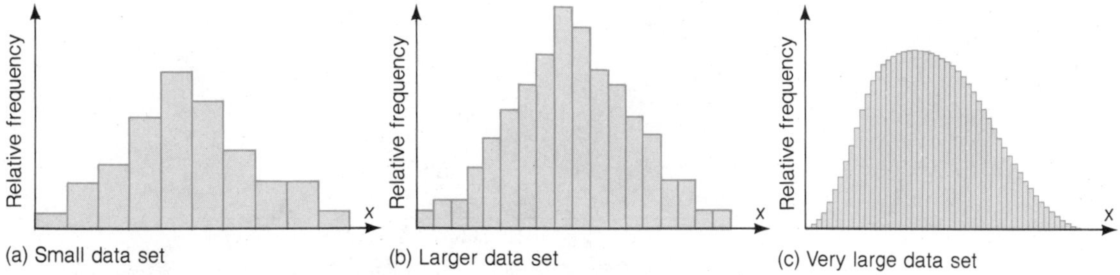

(a) Small data set   (b) Larger data set   (c) Very large data set

Case Study 2.1

Mercury Poisoning and the Dental Profession

The hazards to health traced to environmental pollution constitute an area of major national concern. Mercury has been identified as one source of environmental contamination. Recognition of mercury as a hazard can be traced to Theophrastus, a Greek scientist living about 400 B.C. The physiologic effects of mercury poisoning are a matter of record with death as the ultimate possibility.

This is the introductory paragraph to an article that appeared in the November 1974 issue of the *Journal of the American Dental Association* (Miller et al., 1974).* The article discusses mercury vapor contamination levels in the air of dental offices and the resulting threat to the health of dentists and dental assistants. Such contamination might result from spills in handling mercury, the unprotected storage of scrap amalgam, or aerosols

* Copyright by the American Dental Association. Reprinted by permission.

created by the use of high-speed rotary cutting instruments in removing old amalgam fillings.

The cumulative absorption of small quantities of mercury can result in serious medical problems. The constant daily exposure of dentists and their auxiliary personnel to possible mercury contamination is therefore an important concern. A level of .05 milligram of mercury per cubic meter of air is the largest amount considered safe for those working a 40-hour work week.

A determination of mercury vapor levels was made in sixty dental offices in San Antonio, Texas. A relative frequency histogram summarizing the information provided by these measurements is given in Figure 2.5. The histogram clearly shows that an alarming fraction ($\frac{6}{60}$, or $\frac{1}{10}$) were above the danger level of .05 milligram. You can see that these data have been effectively summarized and clearly indicate a need for strict policing of mercury vapor levels in dental offices.

Figure 2.5 Relative Frequency Histogram

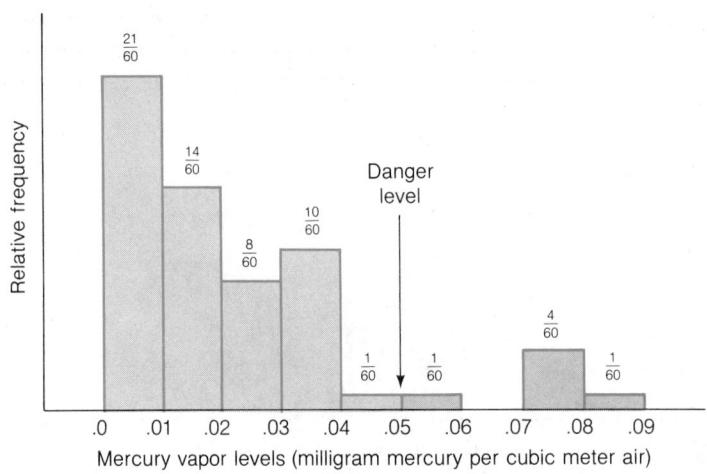

Case Study 2.1 illustrates a shortcoming of relative frequency histograms. Suppose you wish to use the relative frequency histogram to infer the nature of the population relative frequency distribution—that is, the distribution of the mercury vapor levels in the offices of all dentists practicing in the United States. It is true that the sample and population relative frequency distributions will be similar. But how similar? How can you explain how similar the two figures will be? Or, equivalently, how can you measure the reliability of the inference? In Section 2.4 we will explain how you can use one or more numbers (numerical descriptive measures) to describe a distribution of measurements. Further, you will see in Chapter 7 that we can use sample numerical descriptive measures to make inferences about their population counterparts and that we can measure the reliability of these inferences. In short, you will see that numerical descriptive measures are superior to graphic descriptive measures when you want to use the sample data to make inferences about the population from which the sample was selected.

Exercises 2.8–2.20

Learning the Mechanics

2.8 Construct a relative frequency histogram for the data summarized in the relative frequency table given in the margin.

2.9 A sample of twenty measurements is shown here:

26	34	21	32	32
36	28	38	17	29
22	12	26	39	25
31	30	23	27	19

a. Give the upper and lower boundaries for six measurement classes commencing the lower boundary of the first class at 10.5. Use a class width equal to 5. Determine the relative frequency for each of the six classes.

b. Construct a relative frequency histogram using the results of part a.

2.10 Use the following data to construct a relative frequency histogram with eight measurement classes:

5.9	5.3	1.6	7.4	9.8	1.7
8.6	1.2	2.1	4.0	6.5	7.2
7.3	8.4	8.9	6.7	9.2	2.8
4.5	6.3	7.6	9.7	9.4	8.8
3.5	1.1	4.3	3.3	3.1	1.3
8.4	1.6	8.2	6.5	4.1	3.1
1.1	5.0	9.4	6.4	7.7	2.7

2.11 Construct a relative frequency histogram for the following data:

23	12	82	12	67
52	24	17	15	60
32	49	37	55	81
39	99	88	24	30
12	19	53	50	18
16	40	51	61	35

2.12 Construct a relative frequency histogram for the data of Exercise 2.3. Compare your histogram with the stem and leaf display of Exercise 2.3. Which seems to describe the data set better?

2.13 Construct a relative frequency histogram for the data of Exercise 2.4. Compare your histogram with the stem and leaf display of Exercise 2.4. Which seems to describe the data set better?

Table in margin:

MEASUREMENT CLASS	RELATIVE FREQUENCY
0.5– 2.5	.10
2.5– 4.5	.15
4.5– 6.5	.25
6.5– 8.5	.20
8.5–10.5	.05
10.5–12.5	.10
12.5–14.5	.10
14.5–16.5	.05

Applying the Concepts

2.14 The graph below summarizes the scores obtained by 100 students on a questionnaire designed to measure aggressiveness. (Scores are integer values that range from 0 to 20. A high score indicates a high level of aggression.)

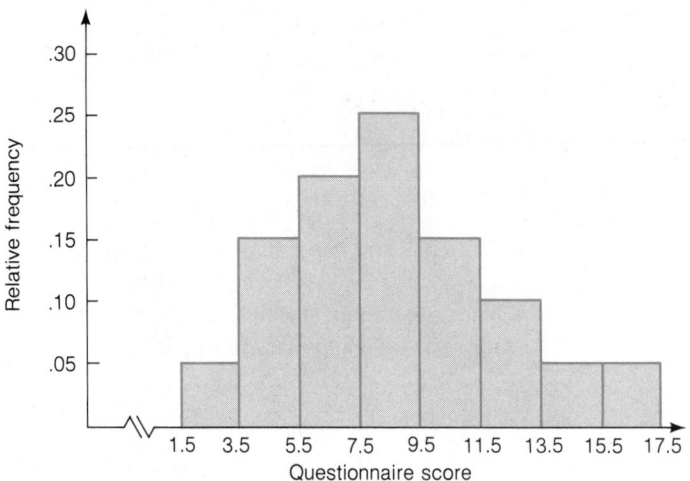

a. Which measurement class contains the highest proportion of test scores?
b. What proportion of the scores lie between 3.5 and 5.5?
c. What proportion of the scores are higher than 11.5?
d. How many students scored less than 5.5?

2.15 The numbers of years of experience (to the nearest half year) for the faculty of a certain statistics department are recorded as follows:

5.0	4.5	19.5	1.0	6.5
0.5	4.0	1.0	2.5	9.0
5.5	5.0	3.5	4.5	1.5
8.0	21.0	7.0	13.5	3.5

a. Construct a relative frequency histogram for these data. Use six measurement classes, commencing with .45 and ending with 21.45.
b. In order to be eligible for certain benefits, a faculty member must have more than 13 years of experience. What proportion of the statistics faculty are eligible for these benefits?

2.16 In the late 1970s and early 1980s gasoline prices rose at an incredible rate. The following is a sample of the prices charged (in cents per gallon) for regular grade gasoline at twenty-five gas stations in the greater Cincinnati area in January 1981:

128.9	121.9	133.9	119.9	115.9
115.9	127.9	119.8	116.8	122.8
135.9	118.8	115.9	121.9	126.9
117.9	115.8	121.9	124.9	129.9
122.8	131.9	132.8	120.8	124.9

a. Construct a relative frequency histogram for these data.

b. Suppose that you were to sample twenty-five gas stations from another area of the country, say New York City or rural Kansas. Do you think a histogram for the twenty-five measurements would look similar to the one you constructed in part a? Explain.

c. Explain the correspondence between areas under the relative frequency histogram and relative frequencies.

2.17 Considering the climate, is it economically feasible to start an orange grove in northern Florida? If the temperature falls below 32°F, oil-burning smudge pots must be lit to keep the orange trees from freezing. Suppose a prospective grower decides that a grove would be economically feasible if the pots have to be lit an average of 15 days or less each year. The grower selects 20 years since 1900 at random and obtains the total number of days per year that the temperature fell below 32°F:

20	16	13	12
9	25	16	6
15	10	18	11
14	12	17	13
13	28	14	15

a. Construct a relative frequency histogram for these data.

b. Based on these sample data, estimate the proportion of years in which the pots have to be lit 15 days or less. [*Note:* We will show you how to evaluate the reliability of this estimate in Chapter 7.]

2.18 Refer to the data (Exercise 2.5) on the percentage of high school students scheduled to graduate who actually graduated. These data, given for each state and the District of Columbia, are reproduced here:

Minn.	86.0	Hawaii	79.6	Vt.	75.5	Del.	72.7	Calif.	68.0
N.Dak.	84.9	Pa.	78.6	Ind.	75.3	Mich.	72.5	Ky.	67.3
Iowa	84.8	Ohio	78.4	Wash.	75.3	N.Mex.	71.9	Ala.	67.1
S.Dak.	82.8	N.H.	77.1	Md.	75.0	Maine	71.4	N.C.	67.1
Wis.	82.3	Okla.	76.9	Ark.	74.8	Conn.	71.1	Tenn.	66.7
Nebr.	81.3	Idaho	76.8	Mo.	74.0	Oreg.	70.3	N.Y.	65.9
Mont.	80.9	Mass.	76.8	W.Va.	74.0	Ariz.	69.0	Ga.	64.3
Kans.	80.5	Nev.	76.8	Va.	73.3	S.C.	68.8	Fla.	63.7
Utah	80.2	N.J.	76.7	Ill.	73.2	Tex.	68.8	La.	63.4
Wyo.	80.0	Colo.	76.5	R.I.	72.8	Alaska	68.7	Miss.	61.8
								D.C.	54.6

Construct a relative frequency histogram for the data. Compare the relative frequency distribution with the stem and leaf display of Exercise 2.5. Do these two graphic methods convey similar descriptions of the data set?

2.19 In Exercise 2.6 we constructed a stem and leaf display for the forty exhalation rate measurements on waste gypsum and phosphate mounds in Florida. Construct a relative frequency histogram for the data. Compare your relative frequency histogram with the stem and leaf display of Exercise 2.6. Do the two graphic displays convey essentially the same information about the data set?

2.20 Refer to the measures of strike activity in Exercise 2.7. (The data are reproduced below.) Construct a relative frequency histogram for the data. Describe your distribution in words.

STATE	%	STATE	%	STATE	%	STATE	%	STATE	%
Ala.	.14	Hawaii	.02	Mich.	.83	N.Y.	.11	Tenn.	.23
Alaska	.11	Idaho	.11	Minn.	.02	N.C.	.01	Tex.	.06
Ariz.	.09	Ill.	.18	Miss.	.14	N.Dak.	.03	Utah	.66
Ark.	.10	Ind.	.16	Mo.	.14	Ohio	.38	Vt.	.26
Calif.	.16	Iowa	.16	Mont.	.28	Okla.	.01	Va.	.04
Colo.	.04	Kans.	.11	Nebr.	.05	Oreg.	.12	Wash.	.16
Conn.	.08	Ky.	.17	Nev.	.36	Pa.	.14	W.Va.	.45
Del.	.41	La.	.10	N.H.	.03	R.I.	.09	Wis.	.21
Fla.	.20	Maine	.15	N.J.	.27	S.C.	.01	Wyo.	.01
Ga.	.13	Mass.	.07	N.Mex.	.09	S.Dak.	.16	Md.	—

2.4 Numerical Measures of Central Tendency

Now that we have presented some graphic techniques for summarizing and describing data sets, we turn to numerical methods for accomplishing this objective. When we speak of a data set, we refer to either a sample or a population. If statistical inference is our goal, we will wish ultimately to use sample numerical descriptive measures to make inferences about the corresponding measures for a population.

As you will see, there are a large number of numerical methods available to describe data sets. Most of these methods measure one of two data characteristics:

1. The **central tendency** of the set of measurements, i.e., the tendency of the data to cluster or to center about certain numerical values.
2. The **variability** of the set of measurements, i.e., the spread of the data.

In this section we concentrate on measures of central tendency. In the next section, we will discuss measures of variability.

The most popular and best understood measure of central tendency for a quantitative data set is the **arithmetic mean** (or simply the **mean**) of a data set:

Definition 2.5

The **mean** of a set of quantitative data is equal to the sum of the measurements divided by the number of measurements contained in the data set.

In everyday terms, the mean is the average value of the data set.

Example 2.2   Calculate the mean of the following five sample measurements: 5,   3,   8,   5,   6

Solution   The mean is the average of the five measurements, i.e.,

$$\frac{5+3+8+5+6}{5} = 5.4$$

Thus, the mean of this sample is 5.4.*   ∎

At this point, it will be advantageous to present some shorthand notation that will simplify calculation instructions for the mean as well as other more complicated numerical descriptive measures we will subsequently encounter. This notation is summarized in the accompanying box. Remember that such notation is used for one reason

---

x: The letter x is used to represent an arbitrary measurement of a sample.

n: A lowercase n is used to represent the number of measurements in a sample.

$\sum x$:   The Greek capital letter sigma followed by the letter x is used to mean "add all the measurements of a sample."

$\bar{x}$:   The letter x with a bar over it (read "x bar") is used to denote the sample mean.

Thus, using this notation, the sample mean is given by the formula**

$$\bar{x} = \frac{\sum x}{n} = \frac{\text{Sum of all the sample measurements}}{\text{Number of measurements in the sample}}$$

μ: The Greek letter mu is used to denote the mean of all the measurements in the population.

---

* In the examples given here, $\bar{x}$ is sometimes rounded to the nearest tenth, sometimes the nearest hundredth, sometimes the nearest thousandth. There is no specific rule for rounding when calculating $\bar{x}$ because $\bar{x}$ is specifically defined to be the sum of all measurements divided by n; i.e., it is a specific fraction. When $\bar{x}$ is used for descriptive purposes, it is often convenient to round the calculated value of $\bar{x}$ to the number of significant figures used for the original measurements. When $\bar{x}$ is to be used in other calculations, however, it may be necessary to retain more significant figures.

** We omit the formula for calculating the mean for grouped data (data presented in a relative frequency table). The reader interested in this special topic should consult the references at the end of this chapter.

only: to avoid having to repeat the same verbal descriptions over and over again. If you mentally substitute the verbal definition of a symbol each time you read it, you will soon become accustomed to its use.

Example 2.3

Calculate the sample mean for the 100 EPA mileages given in Table 2.1.

Solution

The mean gas mileage for the 100 cars is

$$\bar{x} = \frac{36.3 + 41.0 + \cdots + 38.1 + 36.9}{100} = \frac{3,699.4}{100} = 36.994$$

Given this information, you would be able to visualize a distribution of gas mileage readings centered in the vicinity of $\bar{x} = 37.0$. An examination of the relative frequency histogram (Figure 2.3) confirms that $\bar{x}$ does in fact fall near the center of the distribution.

In a practical situation, we rarely know the population mean μ, but we can use the sample mean $\bar{x}$ to estimate its value. Thus, we would infer that the mean gas mileage for all cars (of the model included in the sample) is near 37.0 miles per gallon. We will show you how to evaluate the reliability of this estimate in Chapter 7. ∎

Case Study 2.2

Hotels: A Rational
Method for
Overbooking

The most outstanding characteristic of the general hotel reservation system is the option of the prospective guest, without penalty, to change or cancel his reservation or even to "no-show" (fail to arrive without notice). Overbooking (taking reservations in excess of the hotel capacity) is practiced widely throughout the industry as a compensating economic measure. This has motivated our research into the problem of determining policies for overbooking which are based on some set of rational criteria.

So says Marvin Rothstein (1974) in an article that appeared in the journal for the American Institute for Decision Sciences. In this paper Rothstein introduces a method for scientifically determining hotel booking policies and applies it to the booking problems of the 133-room Sheraton Pocono Inn at Stroudsburg, Pennsylvania.

From the Sheraton Pocono Inn's records the number of reservations, walk-ins (people without reservations who expect to be accommodated), cancellations, and no-shows were tabulated for each day during the period August 1–28, 1971. The inn's records for this period included approximately 3,100 guest histories. From the tabulated data the mean or average number of room reservations per day for each of the 7 days of the week were computed. These appear in Table 2.4.

Table 2.4
Mean Number of
Room Reservations,
August 1–28, 1971,
133 Rooms

SUNDAY	MONDAY	TUESDAY	WEDNESDAY	THURSDAY	FRIDAY	SATURDAY
138	126	149	160	150	150	169

In applying his booking policy decision method to the Sheraton's data, Rothstein used the means listed in Table 2.4 to help portray the inn's demand for rooms.

The mean number of Saturday reservations during August 1–28, 1971, is 169. This may be interpreted as an estimate of μ, the mean number of rooms demanded via reservations (walk-ins also contribute to the demand for rooms) on a Saturday during 1971. If this reservation data for all Saturdays during 1971 had been tabulated, μ could have been computed. But since only August data are available, they were used to estimate μ. Can you think of some problems associated with using August's data to estimate the mean for the entire year?

Definition 2.6

The **median** is another important measure of central tendency. In general terms, the median is the middle number when the measurements in a data set are arranged in ascending (or descending) order.

Calculating a Median

Rank the n measurements in the data set from the smallest to the largest. The smallest will receive rank 1, the next largest rank 2, ..., and the largest rank n.

1. If the number n of measurements is odd, the median is the measurement in the middle of the ranking, i.e., the measurement with rank equal to $(n + 1)/2$.
2. If the number n of measurements is even, the median is the mean of the two middle measurements in the ranking, i.e., halfway between the measurement ranked $n/2$ and the measurement ranked $(n/2) + 1$.

The median is of most value in describing large data sets. If the data set is characterized by a relative frequency histogram (Figure 2.6), the median is the point on the x-axis such that half the area under the histogram lies above the median and half lies

Figure 2.6 Location of the Median

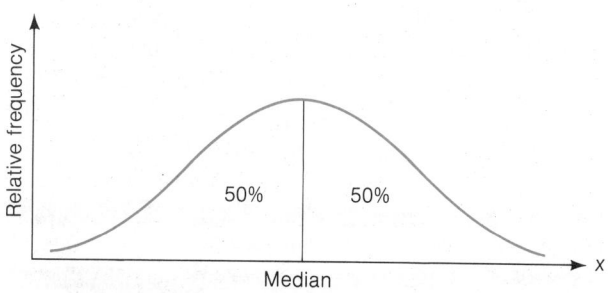

below. [*Note:*   In Section 2.3 we observed that the relative frequency associated with a particular interval on the *x*-axis is proportional to the amount of area under the histogram that lies above the interval.]

Example 2.4   Consider the following sample of $n = 7$ measurements: 5,   7,   4,   5,   20,   6,   2

a. Calculate the median of this sample.

b. Eliminate the last measurement (the 2) and calculate the median of the remaining $n = 6$ measurements.

Solution   **a.** The seven measurements in the sample are ranked in ascending order:

  2,   4,   5,   5,   6,   7,   20

Because the number of measurements is odd, the median is the middle measurement. Thus, the median of this sample is 5.

b. After removing the 2 from the set of measurements, we rank the sample measurements in ascending order as follows:

  4,   5,   5,   6,   7,   20

Now the number of measurements is even, so we average the middle two measurements. The median is $(5 + 6)/2 = 5.5$.   ■

In certain situations, the median may be a better measure of central tendency than the mean. In particular, the median is less sensitive than the mean to extremely large or small measurements. To illustrate, note that all but one of the measurements in part a of Example 2.4 center about $x = 5$. The single relatively large measurement, $x = 20$, does not affect the value of the median, 5, but it shifts the mean, $\bar{x} = 7$, to the right of most of the measurements. As another example, if you were interested in computing a measure of central tendency of the incomes of a company's employees, the mean might be misleading. If all blue-collar and white-collar employees' incomes are included in the data set, the high incomes of a few executives will influence the mean more than the median. Thus, the median will provide a more accurate picture of the typical income for employees of the company. Similarly, the median yearly sales for a sample of companies would locate the middle of the sales data. However, a few companies with very large yearly sales would greatly influence the mean, making it deceptively large. That is, the mean could exceed a vast majority of the sample measurements, making it a misleading measure of central tendency.

If you were to calculate the median for the 100 gas mileage readings in Table 2.1, you would find that the median, 37.0, and the mean, 36.994, are almost equal. This fact indicates that the data form an approximately **symmetric** distribution. (Compare the center figure in the accompanying box with Figure 2.3.) As indicated in the box, a comparison of the mean and median gives an indication of the **skewness** (non-symmetry) of a data set.

Comparing the Mean and the Median

If the median is less than the mean, the data set is skewed to the right:

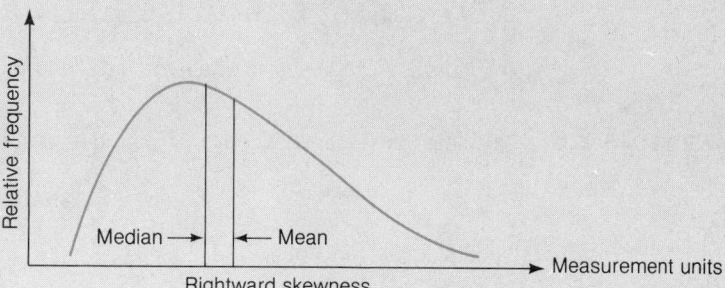

For symmetric data sets, the mean equals the median:

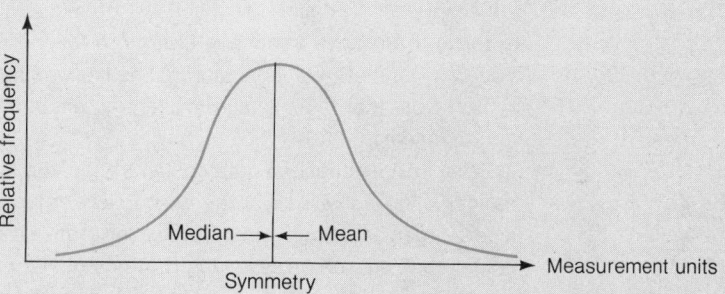

If the median is greater than the mean, the data set is skewed to the left:

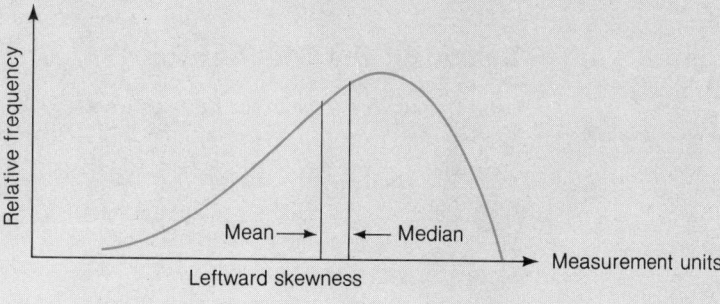

A third measure of central tendency is the **mode** of a set of measurements:

Definition 2.7

The **mode** is the measurement that occurs most frequently in the data set.

Therefore, the mode shows where the data tend to concentrate.

Example 2.5   Calculate the mode for the following ten quiz grades:

8, 7, 9, 6, 8, 10, 9, 9, 5, 7

Solution   Since 9 occurs most often, the mode is 9. ■

The mode is of primary value in describing large data sets. When a large data set has been described using a relative frequency histogram, the mode will be located in the class containing the largest relative frequency. This is called the **modal class.** Several definitions exist for locating the position of the mode within a modal class, but the simplest is to define the mode as the midpoint of the modal class. For example, examine the relative frequency histogram for the EPA mileage readings (Figure 2.3). You can see that the modal class is the interval 35.95–37.45. The mode (the midpoint) is 36.7. Note that this measure of central tendency is very close to the mean, 36.994, and the median, 37.0.

Because it emphasizes data concentration, the mode has applications in fields such as marketing as well as in the description of large data sets collected by state and federal agencies. For example, a retailer of men's clothing would be interested in the modal neck size and sleeve length of potential customers. A supermarket manager is interested in the cereal brand with the largest share of the market, i.e., the modal brand. The modal income class of the American worker is of interest to the Labor Department. Thus, the mode provides a useful measure of central tendency for various applications.

**Exercises
2.21–2.33**

Learning the Mechanics

2.21 Calculate the mean for samples where:
a. $n = 10$, $\sum x = 85$ **b.** $n = 16$, $\sum x = 400$
c. $n = 45$, $\sum x = 35$ **d.** $n = 18$, $\sum x = 242$

2.22 Calculate the mean, median, and mode for each of the following samples:
a. 7, -2, 3, 3, 0, 4
b. 2, 3, 5, 3, 2, 3, 4, 3, 5, 1, 2, 3, 4
c. 51, 50, 47, 50, 48, 41, 59, 68, 45, 37

2.23 Describe how the mean compares to the median for a distribution as follows:
a. Skewed to the left   **b.** Skewed to the right   **c.** Symmetric

2.24 Calculate the mean and median for each of the following samples:
a. 0, 2, 1, 3, 1, 2, 2, 1, 4
b. 1, 0, 4, 1, 1, 2, 3, 1, 20
c. −100, −99, −1, 0, 1, 99, 100
d. −100, −99, −1, 0, 1, 2, 3

Applying the Concepts

2.25 A psychologist has developed a new technique intended to improve rote memory. To test the method against other standard methods, twenty high school students are selected at random and each is taught the new technique. The students are then asked to memorize a list of 100 word phrases using the technique. The following are the number of word phrases memorized correctly by the students:

91	64	98	66	83
87	83	86	80	93
83	75	72	79	90
80	90	71	84	68

a. Define the terms *mean, median,* and *mode* in the context of this problem.
b. Construct a relative frequency histogram for the data.
c. Compute the mean, median, and mode for the data set and locate them on the histogram. Do these measures of central tendency appear to locate the center of the distribution of data?

2.26 Would you expect the data sets described below to possess relative frequency distributions that are symmetric, skewed to the right, or skewed to the left? Explain.
a. The salaries of all persons employed by a large university
b. The grades on an easy test
c. The grades on a difficult test
d. The amounts of time students in your class studied last week
e. The ages of automobiles on a used car lot
f. The amounts of time spent by students on a difficult examination (maximum time is 50 minutes)

2.27 One index used by social scientists to measure a person's socioeconomic status is personal income. To compile a breakdown of the percentage of drinkers and abstainers by socioeconomic status, a survey was taken. The yearly incomes (in dollars) of the respondents who claim to abstain from drinking are:

12,000	10,000	10,000	90,800	9,800
12,000	5,600	9,000	9,000	10,200
15,000	17,000	11,500	13,000	8,000
9,500	13,500	12,200	17,000	11,400

a. Compute the mean, median, and mode for this sample.

b. Now drop the highest value and repeat part a. What effect does dropping this large measurement have on the measures of central tendency computed in part a? Which measure of central tendency seems to be most sensitive to extremely high scores?

2.28 The scores for a statistics test are as follows:

87	76	96	77
94	92	88	85
66	89	79	95
50	91	83	88
82	58	18	69

a. Compute the mean, median, and mode for these data.

b. Which of the three measures of central tendency do you think would best represent the achievement of the class?

c. Eliminate the two lowest scores, and again compute the mean, median, and mode. Which measure of central tendency do you think is most affected by extremely low scores?

2.29 Ten presumably trained rats were released in a maze. Their times to escape (in seconds) are recorded below. The N's represent two rats that had still not escaped by the end of the experiment.

100	38	N	122	95
116	56	135	104	N

a. Can you calculate the mean for these data? Explain.

b. Is the median a meaningful measure of central tendency for these data? Explain. Calculate the median.

2.30 In the National Basketball Association (NBA) there are presently a handful of superstars with contracts worth millions of dollars annually. The majority of players, however, earn less than $100,000 per year. The players' association is presently negotiating with the owners for more fringe benefits. The owners want to show that the NBA players have very lucrative salaries and do not need additional benefits.

a. What measure of central tendency of NBA salaries should the owners use to support their position? Why?

b. What measure should the players' association use? Why?

2.31 Use the stem and leaf display of Exercise 2.5 to find the median of the data set. Find the mean of the data set and compare it with the median. Do the relative locations of the mean and median suggest skewness in the data set? Explain.

2.32 Use the stem and leaf display of Exercise 2.6 to find the median of the data set. Find the mean of the data set and compare it with the median. Do the relative locations of the mean and median suggest skewness in the data set? Explain.

2.33 Use the stem and leaf display of Exercise 2.7 to find the median of the data set. Find the mean of the data set and compare it with the median. Do the relative locations of the mean and median suggest skewness in the data set? Explain.

2.5
Numerical
Measures of
Variability

Measures of central tendency provide only a partial description of a quantitative data set. The description is incomplete without a measure of the variability, or spread, of the data set.

If you examine the two histograms in Figure 2.7, you will notice that both hypothetical data sets are symmetric with equal modes, medians, and means. However, data set 1 in Figure 2.7(a) has measurements spread with almost equal relative frequency over the measurement classes, while data set 2 in Figure 2.7(b) has most of its measurements clustered about its center. Thus, data set 2 is *less variable* than data set 1. Consequently, you can see that we need a measure of variability as well as a measure of central tendency to describe a data set.

Figure 2.7

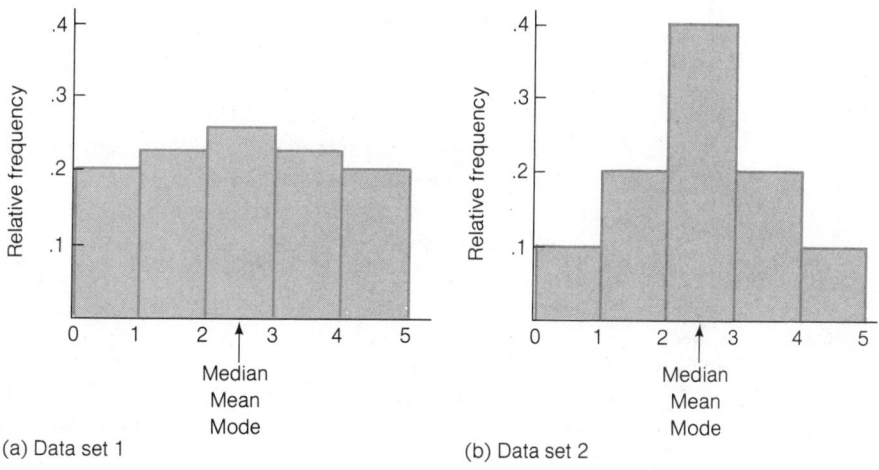

(a) Data set 1   (b) Data set 2

Perhaps the simplest measure of the variability of a quantitative data set is its *range.*

Definition 2.8

The *range* of a data set is equal to the largest measurement minus the smallest measurement.

The range is easy to compute and easy to understand, but it is a rather insensitive measure of data variation when the data sets are large. This is because two data sets can have the same range and be vastly different with respect to data variation. This phenomenon is demonstrated in Figure 2.7. Both distributions of data shown in the figure have the same range, but most of the measurements in data set 2 tend to concentrate near the center of the distribution. Consequently, the data are much less variable than the data in set 1. Thus, you can see that the range does not always detect differences in data variation for large data sets.

Table 2.5

	SAMPLE 1	SAMPLE 2
MEASUREMENTS	1, 2, 3, 4, 5	2, 3, 3, 3, 4
MEAN	$\bar{x} = \dfrac{1+2+3+4+5}{5} = \dfrac{15}{5} = 3$	$\bar{x} = \dfrac{2+3+3+3+4}{5} = \dfrac{15}{5} = 3$
DISTANCES OF MEASUREMENT VALUES FROM $\bar{x}$	$(1-3), (2-3), (3-3), (4-3), (5-3)$ or $-2, -1, 0, 1, 2$	$(2-3), (3-3), (3-3), (3-3), (4-3)$ or $-1, 0, 0, 0, 1$

Let us see if we can find a measure of data variation that is more sensitive than the range. Consider the two samples in Table 2.5; each has five measurements. (We have ordered the numbers for convenience.) Note that both samples have a mean of 3 and that we have also calculated the distance between each measurement and the mean. What information do these distances contain? If they tend to be large in magnitude, as in sample 1, the data are spread out, or highly variable. If the distances are mostly small, as in sample 2, the data are clustered around the mean, $\bar{x}$, and therefore do not exhibit much variability. You can see that these distances, displayed graphically in Figure 2.8, provide information about the variability of the sample measurements.

Figure 2.8 Dot Diagrams for Two Data Sets

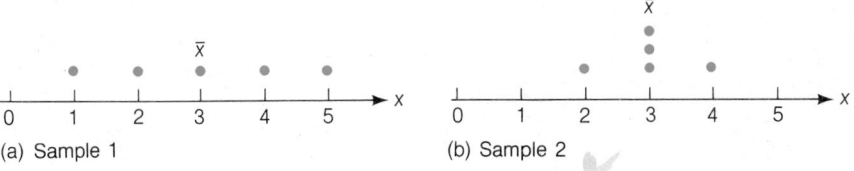

(a) Sample 1   (b) Sample 2

The next step is to condense the information in these distances into a single numerical measure of variability. Averaging the distances from $\bar{x}$ will not help because the negative and positive distances cancel; i.e., the sum of the deviations (and thus the average deviation) is always equal to zero.

Two methods come to mind for dealing with the fact that positive and negative distances from the mean cancel. The first is to treat all the distances as though they were positive, ignoring the sign of the negative distances. We will not pursue this line of thought because the resulting measure of variability (the mean of the absolute values of the distances) is difficult to interpret. A second method of eliminating the minus signs associated with the distances is to square them. The quantity we can calculate from the squared distances will provide a meaningful description of the variability of a data set.

To use the squared distances calculated from a data set, we first calculate the **sample variance:**

Definition 2.9

The **sample variance** for a sample of n measurements is equal to the sum of the squared distances from the mean divided by $(n-1)$. In symbols, using s^2 to represent the sample variance,

$$s^2 = \frac{\sum(x - \bar{x})^2}{n - 1}$$

Referring to the two samples in Table 2.5, you can calculate the variance for sample 1 as follows:

$$s^2 = \frac{(1 - 3)^2 + (2 - 3)^2 + (3 - 3)^2 + (4 - 3)^2 + (5 - 3)^2}{5 - 1}$$

$$= \frac{4 + 1 + 0 + 1 + 4}{4} = 2.5$$

The second step in finding a meaningful measure of data variability is to calculate the **standard deviation** of the data set:

Definition 2.10

The **sample standard deviation,** s, is defined as the positive square root of the sample variance, s^2. Thus,

$$s = \sqrt{s^2} = \sqrt{\frac{\sum(x - \bar{x})^2}{n - 1}}$$

The population variance, denoted by the symbol σ^2 (sigma squared) is the average of the squared distances of the observations from the mean μ, and σ (sigma) is the square root of this quantity. Since we never really compute σ^2 or σ from the population (the object of sampling is to avoid this procedure), we simply denote these two quantities by their respective symbols.

s^2 = Sample variance	s = Sample standard deviation
σ^2 = Population variance	σ = Population standard deviation

Notice that in contrast to the variance, the standard deviation is expressed in the original units of measurement. If the original measurements are in dollars, for example, the standard deviation will be expressed in dollars. Moreover, you may wonder why we use the divisor $(n - 1)$ instead of n when calculating the sample variance. This is because by using the divisor $(n - 1)$ you obtain a better estimate of σ^2 than you do by dividing the sum of the squared distances by n. Since we will ultimately want to use sample statistics to make inferences about the corresponding population parameters, $(n - 1)$ is preferred to n when defining the sample variance.

Example 2.6   Calculate the standard deviation of the following sample: 2,   3,   3,   3,   4

Solution   For this data set, $\bar{x} = 3$. Then

$$s = \sqrt{\frac{(2 - 3)^2 + (3 - 3)^2 + (3 - 3)^2 + (3 - 3)^2 + (4 - 3)^2}{5 - 1}}$$

$$= \sqrt{\frac{2}{4}} = \sqrt{.5} = 0.71$$ ■

Example 2.6 may have raised two thoughts in your mind. First, calculating s^2 and s can be very tedious if $\bar{x}$ is a number that contains a large number of significant figures or if there are a large number of measurements in a data set. Second, we have not explained how a sample standard deviation can be used to describe the variability of a data set. Fortunately, we have an easier method for calculating s^2 and s. This method is demonstrated in Example 2.7. The interpretation of s is the subject of Section 2.6.

Example 2.7   Calculate the sample variance s^2 and the sample standard deviation s for the 100 gas mileage readings given in Table 2.1.

Solution   Recall that $\bar{x} = 36.994$. Clearly it would be a formidable task to calculate $\sum(x - \bar{x})^2$ for all 100 readings. Instead, the calculation can be done in the following manner:

Shortcut Formula for Sample Variance

$$s^2 = \frac{\left(\begin{array}{c}\text{Sum of squares of}\\\text{sample measurements}\end{array}\right) - \dfrac{(\text{Sum of sample measurements})^2}{n}}{n - 1}$$

$$= \frac{\sum x^2 - \dfrac{(\sum x)^2}{n}}{n - 1}$$

Step 1. *Calculate the sum of the squared measurements.* With the aid of a calculator, we obtain

$$\sum x^2 = 137,434.38$$

Step 2. *Calculate the square of the sum of the measurements:*

$$(\textstyle\sum x)^2 = (3,699.4)^2 = 13,685,560.36$$

Step 3. *Find* $\sum(x - \bar{x})^2 = \sum x^2 - [(\sum x)^2/n]$, *the numerator of* s^2 *in the shortcut formula:*

$$\sum(x - \bar{x})^2 = \sum x^2 - \frac{(\sum x)^2}{n}$$

$$= 137,434.38 - \frac{13,685,560.36}{100} = 578.7764$$

We can now calculate s^2 by dividing this quantity by $(n - 1)$:

$$s^2 = \frac{578.7764}{99} = 5.85$$

$$s = +\sqrt{5.85} = 2.42$$

Note that this formula requires only the sum of the sample measurements, $\sum x$, and the sum of the squares of the sample measurements, $\sum x^2$. Be careful when you calculate these two sums. Rounding the values of x^2 that appear in $\sum x^2$ or rounding the quantity $(\sum x)^2/n$ can lead to substantial errors in the calculation of s^2. ■

One question occurs over and over again about the calculation of s^2. How many decimal places should you carry? The answer has *nothing* to do with the number of decimal places retained in the sample measurements. If the sample measurements are rounded to the nearest hundredth, tenth, or whatever, rounding off simply adds to the variability of the measurements. This added variation is reflected in the value of s^2, a number calculated precisely according to Definition 2.9.

Theoretically, s^2 is a number regardless of how many decimal places it takes to specify its value. In practice, we round the calculated value. There are no rules for the rounding procedure but, keeping in mind that we wish to calculate the standard deviation s, it is reasonable to retain twice as many decimal places in s^2 as you wish to have in s. If you wish to calculate s to the nearest hundredth, for example, you should calculate s^2 to the nearest ten-thousandth.

Exercises 2.34–2.39

Learning the Mechanics

2.34 Calculate the range, variance, and standard deviation for the following samples:
a. 4, 2, 1, 0, 1
b. 1, 6, 2, 2, 3, 0, 3
c. 8, −2, 1, 3, 5, 4, 4, 1, 3, 3
d. 0, 2, 0, 0, −1, 1, −2, 1, 0, −1, 1, −1, 0, −3, −2, −1, 0, 1

2.35 Calculate the variance and standard deviation for samples where:
a. $n = 10$, $\sum x^2 = 84$, $\sum x = 20$ **b.** $n = 40$, $\sum x^2 = 380$, $\sum x = 100$
c. $n = 20$, $\sum x^2 = 18$, $\sum x = 17$

2.36 Calculate the range, variance, and standard deviation for the following samples:
a. 39, 42, 40, 37, 41   **b.** 100, 4, 7, 96, 80, 3, 1, 10, 2
c. 100, 4, 7, 30, 80, 30, 42, 2

Applying the Concepts

2.37 Consider the following sample of five measurements: 2,   1,   1,   0,   3
a. Calculate the range, s^2, and s.
b. Add 3 to each measurement and repeat part a.
c. Subtract 4 from each measurement and repeat part a.
d. Considering your answers to parts a, b, and c, what seems to be the effect on the variability of a data set by adding the same number to or subtracting the same number from each measurement?

2.38 The final grades given by two professors in introductory statistics courses have been carefully examined. The students in the first professor's class had a grade-point average of 3.0 and a standard deviation of .2. Those in the second professor's class had grade points with an average of 3.0 and a standard deviation of 1.0. If you had a choice, which professor would you take for this course? Explain.

2.39 Consider the following two samples:

Sample 1:   10, 0, 1, 9, 10, 0, 8, 1, 1, 9
Sample 2:   0, 5, 10, 5, 5, 5, 6, 5, 6, 5

a. Examine both samples and identify the one that you believe has the greater variability.
b. Calculate the range for each sample. Does the result agree with your answer to part a? Explain.
c. Calculate the variance for each sample. Does the result agree with your answer to part a? Explain.
d. Which of the two, the range or the variance, provides a better measure of variability? Why?

2.6 Interpreting the Standard Deviation

As we have seen, if we are comparing the variability of two samples selected from a population, the sample with the larger standard deviation is the more variable of the two. Thus, we know how to interpret the standard deviation on a relative or comparative basis, but we have not explained how it provides a measure of variability for a single sample.

To understand how the standard deviation provides a measure of variability of a data set, consider a specific data set and answer the following questions: How many measurements are within 1 standard deviation of the mean? How many measurements are within 2 standard deviations? For example, consider the 100 mileage per gallon readings given in Table 2.1.

Recall that $\bar{x} = 36.994$ and $s = 2.42$. Then

$$\bar{x} - s = 34.574 \qquad \bar{x} - 2s = 32.154$$
$$\bar{x} + s = 39.414 \qquad \bar{x} + 2s = 41.834$$

If we examine the data, we find that 68 of the 100 measurements, or 68% of the measurements, are in the interval

$$\bar{x} - s \quad \text{to} \quad \bar{x} + s$$

Similarly, we find that 96, or 96%, of the 100 measurements are in the interval

$$\bar{x} - 2s \quad \text{to} \quad \bar{x} + 2s$$

These intervals are usually written $(\bar{x} - s, \bar{x} + s)$ and $(\bar{x} - 2s, \bar{x} + 2s)$.

These observations identify criteria for interpreting a standard deviation that apply to any set of data, whether a population or a sample. The criteria, expressed as a mathematical theorem and as a rule of thumb, are presented in Tables 2.6 and 2.7. In these tables we give two sets of answers to the questions of how many measurements fall within 1, 2, and 3 standard deviations of the mean. The first, which applies to *any* set of data, is derived from a theorem proved by the Russian mathematician Chebyshev. The second, which applies only to mound-shaped distributions of data, is based upon empirical evidence that has accumulated over the years. The frequency histogram of a mound-shaped sample is approximately symmetric, with a clustering of measurements about the midpoint of the distribution (the mean, median, and mode should all be about the same), tailing off rapidly as we move away from the center of the histogram. Thus, the histogram will have the appearance of a mound or bell, as shown in Figure 2.9 (page 44). The percentages given for the intervals in Table 2.7 provide remarkably good approximations even when the distribution of the data is slightly skewed or asymmetric.

Table 2.6
An Aid to Interpretation of a Standard Deviation: Chebyshev's Rule

Chebyshev's Rule

Chebyshev's rule applies to any sample of measurements, regardless of the shape of the frequency distribution:

a. It is possible that very few of the measurements will fall within 1 standard deviation of the mean $(\bar{x} - s, \bar{x} + s)$.

b. At least three-fourths of the measurements will fall within 2 standard deviations of the mean $(\bar{x} - 2s, \bar{x} + 2s)$.

c. At least eight-ninths of the measurements will fall within 3 standard deviations of the mean $(\bar{x} - 3s, \bar{x} + 3s)$.

Table 2.7
An Aid to Interpretation
of a Standard Deviation:
The Empirical Rule

The Empirical Rule

The Empirical Rule is a rule of thumb that applies to samples with frequency distributions that are mound-shaped:

a. Approximately 68% of the measurements will fall within 1 standard deviation of the mean $(\bar{x} - s, \bar{x} + s)$.

b. Approximately 95% of the measurements will fall within 2 standard deviations of the mean $(\bar{x} - 2s, \bar{x} + 2s)$.

c. Essentially all the measurements will fall within 3 standard deviations of the mean $(\bar{x} - 3s, \bar{x} + 3s)$.

Figure 2.9 Histogram of a Mound-Shaped Sample

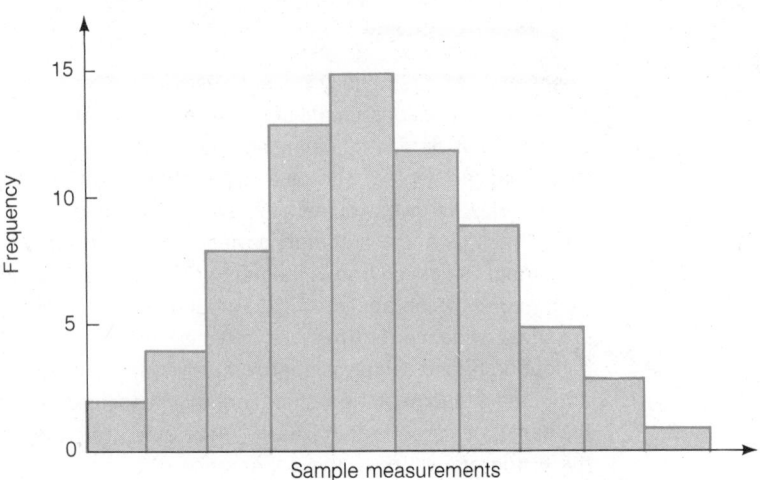

Example 2.8   Here is a sample of earnings per share data for thirty *Fortune* 500 companies:

1.97	.60	4.02	3.20	1.15	6.06
4.44	2.02	3.37	3.65	1.74	2.75
3.81	9.70	8.29	5.63	5.21	4.55
7.60	3.16	3.77	5.36	1.06	1.71
2.47	4.25	1.93	5.15	2.06	1.65

The mean and standard deviation of these data are 3.74 and 2.20, respectively. Calculate the fraction of the thirty measurements in the intervals $\bar{x} \pm s$, $\bar{x} \pm 2s$, and $\bar{x} \pm 3s$, and compare the results with those in Tables 2.6 and 2.7.

Solution   We first form the interval

$$(\bar{x} - s, \bar{x} + s) = (3.74 - 2.20, 3.74 + 2.20) = (1.54, 5.94)$$

A check of the measurements shows that twenty-three measurements are within this 1 standard deviation interval around the mean. This number represents $\frac{23}{30} \approx 77\%$ of the sample measurements.

The next interval of interest is

$$(\bar{x} - 2s,\ \bar{x} + 2s) = (3.74 - 4.40,\ 3.74 + 4.40) = (-0.66,\ 8.14)$$

All but two measurements are within this interval, so approximately 93% are within 2 standard deviations of $\bar{x}$.

Finally, the 3 standard deviation interval around $\bar{x}$ is

$$(\bar{x} - 3s,\ \bar{x} + 3s) = (3.74 - 6.60,\ 3.74 + 6.60) = (-2.86,\ 10.34)$$

All the measurements fall within 3 standard deviations of the mean.

These 1, 2, and 3 standard deviation percentages (77, 93, and 100) agree fairly well with the approximations of 68%, 95%, and 100% given by the Empirical Rule (Table 2.7) for mound-shaped distributions. If you look at the frequency histogram for this data set in Figure 2.10 you will note that the distribution is not really mound-shaped, nor is it extremely skewed. Thus, we get reasonably good results from the mound-shaped approximations. Of course, we know from Chebyshev's rule (Table 2.6) that no matter what the shape of the distribution, we would expect at least 75% and 89% (eight-ninths) of the measurements to lie within 2 and 3 standard deviations, respectively, of $\bar{x}$.

Figure 2.10 Histogram for Earnings per Share Data

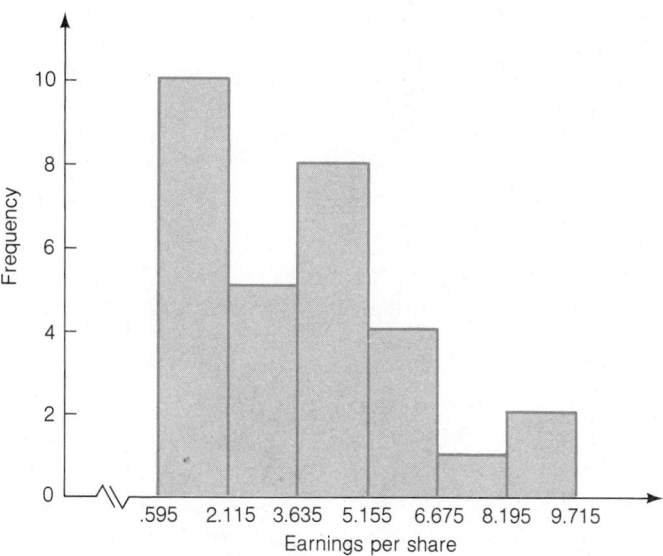

Example 2.9 Chebyshev's rule and the Empirical Rule (Tables 2.6 and 2.7) can be put to an immediate practical use as a check on the calculation of a standard deviation. Suppose you have a data set for which the smallest measurement is 20 and the largest is 80. You have

calculated the standard deviation of the data set to be

$$s = 190$$

How can you use the information in Tables 2.6 and 2.7 to provide a rough check on your calculated value of s?

Solution The larger the number of measurements in a data set, the greater will be the tendency for very large or very small measurements (extreme values) to appear in the data set. But from Tables 2.6 and 2.7 you know that most of the measurements (approximately 95% if the distribution is mound-shaped) will be within 2 standard deviations of the mean. And regardless of how many measurements are in the data set, almost all of them will fall within 3 standard deviations of the mean. Consequently, we would expect the range to be equal to somewhere between 4 and 6 standard deviations, i.e., between $4s$ and $6s$. For the given data set, the range is

$$\text{Range} = \text{Largest measurement} - \text{Smallest measurement}$$
$$= 80 - 20 = 60$$

Then, if we let the range equal $6s$, we obtain

$$\text{Range} = 6s$$
$$60 = 6s$$
$$s = 10$$

If we let the range equal $4s$ (see Figure 2.11), we obtain a larger (and more conservative) value for s, namely,

$$\text{Range} = 4s$$
$$60 = 4s$$
$$s = 15$$

Now you can see that it does not make much difference whether you let the range equal $4s$ (which is more realistic for most data sets) or $6s$ (which is reasonable for large

Figure 2.11 The Relationship Between the Range and the Standard Deviation

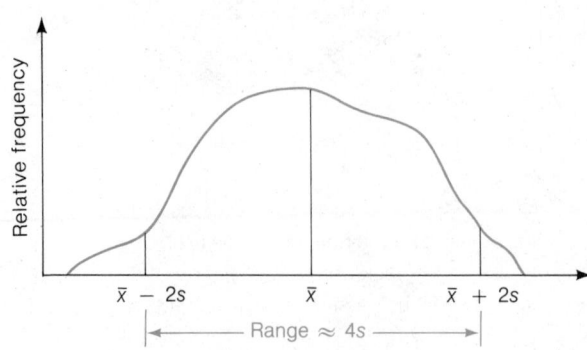

data sets). It is clear that your calculated value, $s = 190$, is too large, and you should check your calculations.* ■

Exercises
2.40–2.52

Learning the Mechanics

2.40 For any set of data, what can be said about the percentage of the measurements contained in each of the following intervals?

a. $\bar{x} - s$ to $\bar{x} + s$ **b.** $\bar{x} - 2s$ to $\bar{x} + 2s$ **c.** $\bar{x} - 3s$ to $\bar{x} + 3s$

2.41 For a set of data with a mound-shaped relative frequency distribution, what can be said about the percentage of the measurements contained in each of the following intervals?

a. $\bar{x} - s$ to $\bar{x} + s$ **b.** $\bar{x} - 2s$ to $\bar{x} + 2s$ **c.** $\bar{x} - 3s$ to $\bar{x} + 3s$

2.42 The following is a sample of twenty-five measurements:

7	6	6	11	8
9	11	9	10	8
7	7	5	9	10
7	7	7	7	9
12	10	10	8	6

a. Compute $\bar{x}$, s^2, and s for this sample.

b. Count the number of measurements in the intervals $\bar{x} \pm s$, $\bar{x} \pm 2s$, $\bar{x} \pm 3s$. Express the number of measurements in each interval as a percentage of the total number of measurements.

c. Compare the percentages found in part b to the percentages given by the Empirical Rule and Chebyshev's rule.

d. Calculate the range, and use it to obtain a rough approximation for s. Does the result compare favorably with the actual value for s found in part a?

Applying the Concepts

2.43 In Exercise 2.5 we gave, for each state, the percentage of high school students scheduled to graduate in 1981 who actually graduated. The data are reproduced here:

Minn.	86.0	Hawaii	79.6	Vt.	75.5	Del.	72.7	Calif.	68.0
N.Dak.	84.9	Pa.	78.6	Ind.	75.3	Mich.	72.5	Ky.	67.3
Iowa	84.8	Ohio	78.4	Wash.	75.3	N.Mex.	71.9	Ala.	67.1
S.Dak.	82.8	N.H.	77.1	Md.	75.0	Maine	71.4	N.C.	67.1
Wis.	82.3	Okla.	76.9	Ark.	74.8	Conn.	71.1	Tenn.	66.7
Nebr.	81.3	Idaho	76.8	Mo.	74.0	Oreg.	70.3	N.Y.	65.9
Mont.	80.9	Mass.	76.8	W.Va.	74.0	Ariz.	69.0	Ga.	64.3
Kans.	80.5	Nev.	76.8	Va.	73.3	S.C.	68.8	Fla.	63.7
Utah	80.2	N.J.	76.7	Ill.	73.2	Tex.	68.8	La.	63.4
Wyo.	80.0	Colo.	76.5	R.I.	72.8	Alaska	68.7	Miss.	61.8
								D.C.	54.6

* In the exercises and examples that follow, we will use range $\approx 4s$, or, equivalently, $s \approx$ range/4.

a. Calculate the mean and standard deviation of the data.

b. Locate your mean on the stem and leaf display (Exercise 2.5) and on the relative frequency distribution (Exercise 2.18). Does the mean appear to be located near the center of the distribution?

c. Calculate the percentages of measurements in the intervals $\bar{x} \pm s$, $\bar{x} \pm 2s$, and $\bar{x} \pm 3s$. Does the Empirical Rule, using $\bar{x}$ and s, provide an adequate description of the data set?

2.44 Refer to the measures of strike activity (Exercise 2.7). The data are reproduced below. Calculate $\bar{x}$, s^2, and s for the data. Then find the percentages of observations in the intervals $\bar{x} \pm s$, $\bar{x} \pm 2s$, and $\bar{x} \pm 3s$. How do these percentages compare with those given by the Empirical Rule?

STATE	%	STATE	%	STATE	%	STATE	%	STATE	%
Ala.	.14	Hawaii	.02	Mich.	.83	N.Y.	.11	Tenn.	.23
Alaska	.11	Idaho	.11	Minn.	.02	N.C.	.01	Tex.	.06
Ariz.	.09	Ill.	.18	Miss.	.14	N.Dak.	.03	Utah	.66
Ark.	.10	Ind.	.16	Mo.	.14	Ohio	.38	Vt.	.26
Calif.	.16	Iowa	.16	Mont.	.28	Okla.	.01	Va.	.04
Colo.	.04	Kans.	.11	Nebr.	.05	Oreg.	.12	Wash.	.16
Conn.	.08	Ky.	.17	Nev.	.36	Pa.	.14	W.Va.	.45
Del.	.41	La.	.10	N.H.	.03	R.I.	.09	Wis.	.21
Fla.	.20	Maine	.15	N.J.	.27	S.C.	.01	Wyo.	.01
Ga.	.13	Mass.	.07	N.Mex.	.09	S.Dak.	.16	Md.	

2.45 Refer to the measurements of radiation on soil samples collected from gypsum and phosphate waste mounds in Polk County, Florida. These data, discussed in Exercise 2.6, are reproduced here:

1,709.79	4,132.28	2,996.49	2,796.42	3,750.83	961.40	1,096.43	1,774.77
357.17	1,489.86	2,367.40	11,968.23	178.99	5,402.35	2,315.52	2,617.57
1,150.94	3,017.48	599.84	2,758.84	3,764.96	1,888.22	2,055.20	205.84
1,572.69	393.55	538.37	1,830.78	878.56	6,815.69	752.89	1,977.97
558.33	880.84	2,770.23	1,426.57	1,322.76	1,480.04	9,139.21	1,698.39

a. Calculate the sample mean and standard deviation.

b. Compare the values of $\bar{x}$ and s calculated in part a with the relative frequency histogram for the data (Exercise 2.19). Does $\bar{x}$ fall near the middle of the distribution? Use the range of the data to compute an approximate value for s; compare the result to the value you calculated in part a.

c. Calculate the percentages of observations in the intervals $\bar{x} \pm s$, $\bar{x} \pm 2s$, and $\bar{x} \pm 3s$. How do these percentages compare with those given by the Empirical Rule?

2.46 In Case Study 9.2, we will discuss a study designed to investigate the effects of two variables—(1) a student's level of mathematical anxiety and (2) teaching method—on a student's achievement in a course in mathematics. Students who had a low level of mathematical anxiety and were taught using the traditional expository method obtained a mean score of 183.43 and a standard deviation of 52.27. Use the Empirical Rule, along with this information, to give a verbal description of the data set.

2.47 For each day of last year, the number of vehicles passing through a certain intersection was recorded by a city engineer. One objective of this study was to determine the percentage of days that more than 425 vehicles used the intersection. If the mean for the data was 375 vehicles per day and the standard deviation was 25 vehicles:

a. What can be said about the percentage of days that more than 425 vehicles used the intersection? Assume that nothing is known about the shape of the relative frequency distribution for the data.

b. What is your answer to part a if you know that the relative frequency distribution for the data is mound-shaped?

2.48 A buyer for a lumber company must decide whether to buy a piece of land containing 5,000 pine trees. If 1,000 of the trees are at least 40 feet tall, the buyer will purchase the land; otherwise, he will not. The owner of the land reports that the height of the trees has a mean of 30 feet and a standard deviation of 3 feet. Based on this information, what is the buyer's decision?

2.49 A chemical company produces a substance composed of 98% cracked corn particles and 2% zinc phosphide for use in controlling rat populations in sugarcane fields. Production must be carefully controlled to maintain the 2% zinc phosphide because too much zinc phosphide will cause damage to the sugarcane and too little will be ineffective in controlling the rat population. Records from past production indicate that the distribution of the actual percentage of zinc phosphide present in the substance is approximately mound-shaped, with a mean of 2.0% and a standard deviation of .08%. If the production line is operating correctly, approximately what proportion of batches from a day's production will contain less than 1.84% zinc phosphide? Suppose one batch chosen randomly actually contains 1.80% zinc phosphide. Does this indicate that there is too little zinc phosphide in today's production? Explain your reasoning.

2.50 Solar energy is considered by many to be the energy of the future. A recent survey was taken to compare the cost of solar energy to the cost of gas or electric energy. Results of the survey revealed that the distribution of the amount of the monthly utility bill of a three-bedroom house using gas or electric energy had a mean of $125 and a standard deviation of $10.

a. If nothing is known about the distribution of the amounts of monthly utility bills, what can you say about the fraction of all three-bedroom homes using gas or electric energy having bills between $95 and $155?

b. If it is reasonable to assume that the distribution of the amounts of monthly utility bills is mound-shaped, approximately what proportion of three-bedroom homes would have monthly bills less than $135?

c. Suppose that three houses with solar energy units had the following monthly utility bills: $101, $98, $104. Does this suggest that solar energy units might result in lower utility bills? Explain. [*Note:* We will present a statistical method in Chapter 7 for testing this conjecture.]

2.51 The following data sets have been invented to demonstrate that the lower bounds given by Chebyshev's rule are appropriate. Notice that the data are contrived and would not be encountered in a real-life problem.

a. Consider a data set that contains ten 0's, two 1's, and ten 2's. Calculate $\bar{x}$, s^2, and s. What percentage of the measurements are in the interval $\bar{x} \pm s$? Compare this result to Chebyshev's rule.

b. Consider a data set that contains five 0's, thirty-two 1's, and five 2's. Calculate $\bar{x}$, s^2, and s. What percentage of the measurements are in the interval $\bar{x} \pm 2s$? Compare this result to Chebyshev's rule.

c. Consider a data set that contains three 0's, fifty 1's, and three 2's. Calculate $\bar{x}$, s^2, and s. What percentage of the measurements are in the interval $\bar{x} \pm 3s$? Compare this result to Chebyshev's rule.

d. Draw a histogram for each of the data sets in parts a, b, and c. What do you conclude from these graphs and the answers to parts a, b, and c?

2.52 In Exercise 9.26 (Chapter 9) we present the results of a study to compare the effects of fructose and glucose on the high endurance of women athletes. Six women athletes received 300 milliliters each of a certain drink (say water and glucose) and then ran until exhausted. Various measurements were then taken on each athlete. The mean and standard deviation of the performance times (in minutes) for the six athletes after receiving the glucose drink were 61.9 and 20.3, respectively. What do the mean and standard deviation tell you about this small data set?

2.7 Measures of Relative Standing

As we have seen, numerical measures of central tendency and variability describe the general nature of a data set (either a sample or a population). We may also be interested in describing the relative quantitative location of a particular measurement within a data set. Descriptive measures of the relationship of a measurement to the rest of the data are called ***measures of relative standing.***

One measure of the relative standing of a measurement is its ***percentile ranking.*** For example, suppose you scored an 80 on an examination and you want to know how you fared in comparison with others in your class. If the instructor tells you that you scored in the 90th percentile, it means that 90% of the examination grades were less than yours and 10% were greater. Thus, if the examination scores were described by the relative frequency histogram in Figure 2.12, the 90th percentile would be located at a point such that 90% of the total area under the relative frequency histogram lies below the 90th percentile and 10% lies above. If the instructor tells you that you scored in the 50th percentile (the median of the data set), 50% of the examination grades would be less than yours and 50% would be greater.

Percentile rankings are of practical value only for large data sets. Finding them involves a process similar to the one used in finding a median. The measurements are ranked in order and a rule is selected, similar to that used in locating a median, to de-

Figure 2.12 Location of 90th Percentile for Examination Grades

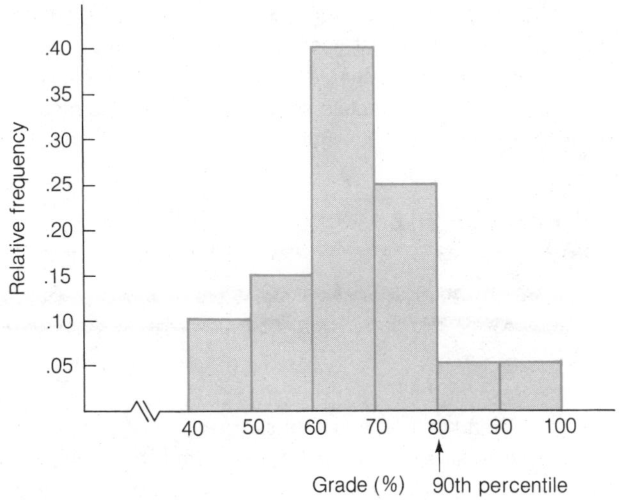

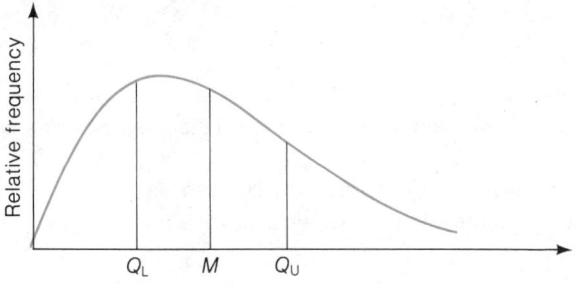

fine the location of each percentile. Since we are primarily interested in interpreting the percentile rankings of measurements (rather than finding particular percentiles for a data set), we will define the pth percentile of a data set as follows:

Definition 2.11

For any set of n measurements (arranged in ascending or descending order), the **pth percentile** is a number such that $p\%$ of the measurements fall below the pth percentile and $(100 - p)\%$ fall above it.

A less precise measure of relative location of an observation is based on three particular percentiles: the **quartiles** of a data set. The quartiles are values of x that partition the data set into four groups, each containing 25% of the measurements. The lower quartile Q_L is the 25th percentile, the middle quartile is the median M, and the upper quartile Q_U is the 75th percentile (see Figure 2.13). Knowing that you achieved

Figure 2.13 The Quartiles for a Data Set

a test score in the upper quartile gives you an approximate location of your score relative to the other test scores. You know that it is in the top 25% of the scores. Moreover, the quartiles enable you to construct a mental picture of the relative frequency distribution for the complete set of test scores. The median locates its center, and the lower and upper quartiles indicate its spread.

Definition 2.12

The *lower quartile* is the 25th percentile of a data set. The *middle quartile* is the median. The *upper quartile* is the 75th percentile.

There are a number of methods for estimating population quartiles based on sample data. One relatively easy method is to rank the sample data and choose Q_L as the observation with rank equal to $(n + 1)/4$. Since $(n + 1)/4$ may be a fraction, we round to the nearest integer. If $n = 25$, for example, $(n + 1)/4 = (25 + 1)/4 = 6.5$. Rounding upward, we would select Q_L as the measurement with rank equal to 7. Approximately 25%, actually $(\frac{7}{25})100 = 28\%$, of the measurements will be less than or equal to Q_L.

Similarly, to find Q_U we calculate $(n + 1)(\frac{3}{4})$ and round to the nearest integer. This integer is taken to be the rank of the measurement we select as Q_U. For example, $(n + 1)(\frac{3}{4}) = (25 + 1)(\frac{3}{4}) = 19.5$. Rounding downward, we would select Q_U as the measurement with rank 19. Approximately 25%, actually $(\frac{7}{25})(100) = 28\%$, of the measurements will be equal to or larger than Q_U.

Calculating Quartiles

Rank the n measurements in the data set from the smallest to the largest. The smallest will receive rank 1, the next largest rank 2, . . . , and the largest rank n.

1. To find Q_L, calculate $(n + 1)/4$ and round to the nearest integer. The measurement with this rank is Q_L. If $(n + 1)/4$ equals an integer plus $\frac{1}{2}$, round upward.

2. To find Q_U, calculate $(n + 1)(\frac{3}{4})$ and round to the nearest integer. The measurement with this rank is Q_U. If $(n + 1)(\frac{3}{4})$ equals an integer plus $\frac{1}{2}$, round downward.

Example 2.10   Find the lower and upper quartiles for the 100 EPA mileages of Table 2.1.

Solution   The easiest way to find Q_L and Q_U is to obtain their values from the stem and leaf display (Figure 2.1). If the observations are ranked from the smallest to the largest, then

Q_L is the observation with rank equal to

$$(n + 1)(\tfrac{1}{4}) = (100 + 1)(\tfrac{1}{4}) = 25.25 \text{ or } 25$$

Similarly, Q_U will be the observation with rank equal to

$$(n + 1)(\tfrac{3}{4}) = (100 + 1)(\tfrac{3}{4}) = 75.75 \text{ or } 76$$

Looking at the stem and leaf display of Figure 2.1, you can see that eighteen of the observations fall in stem row 34 or below. Therefore, Q_L is the observation with the seventh smallest leaf in stem row 35. Since this leaf is .6,

$$Q_L = 35.6$$

To find Q_U, we count the leaves moving downward from the largest stem (44) and note that twenty of the leaves fall in stem 39 or above. Therefore, Q_U is the observation with the fifth largest leaf in stem row 38. Since this leaf is .4,

$$Q_U = 38.4$$

Another measure of relative standing in popular use is the **z-score.** As you can see in Definition 2.13, the z-score makes use of the mean and standard deviation of the data set in order to specify the relative location of a measurement:

Definition 2.13

The **sample z-score** for a measurement x is

$$z = \frac{x - \bar{x}}{s}$$

The **population z-score** for a measurement x is

$$z = \frac{x - \mu}{\sigma}$$

Note that the z-score is calculated by subtracting $\bar{x}$ (or μ) from the measurement x and then dividing the result by s (or σ). The final result, the z-score, represents the distance between a given measurement x and the mean expressed in standard deviations.

Example 2.11   Suppose a sample of annual incomes for 200 steelworkers is taken. The mean and standard deviation are

$$\bar{x} = \$17{,}000 \qquad s = \$2{,}000$$

Suppose Joe Smith's annual income is $15,000. What is his sample z-score?

Solution   You can see that Joe Smith's annual income lies below the mean income of the 200 steelworkers:

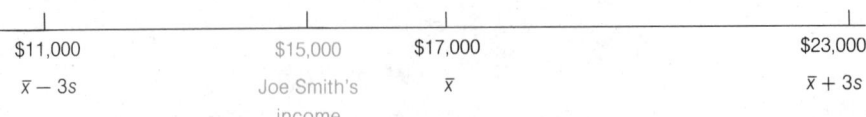

| $11,000 | | $15,000 | $17,000 | | $23,000 |

$\bar{x} - 3s$ Joe Smith's $\bar{x}$ $\bar{x} + 3s$
income

We compute

$$z = \frac{x - \bar{x}}{s} = \frac{\$15,000 - \$17,000}{\$2,000} = -1.0$$

which tells us that Joe Smith's annual income is 1.0 standard deviation *below* the sample mean, or, in short, his sample z-score is −1.0. ■

The numerical value of the z-score reflects the relative standing of the measurement. A large positive z-score implies that the measurement is larger than almost all other measurements, while a large negative z-score indicates that the measurement is smaller than almost every other measurement. If a z-score is zero or near zero, the measurement is located near the mean of the sample or population.

We can be more specific if we know that the frequency distribution of the measurements is mound-shaped. In this case, the following interpretation of the z-scores can be given:

Interpretation of z-Scores for Mound-Shaped Distributions of Data

---

1. Approximately 68% of the measurements will have a z-score between −1 and 1.
2. Approximately 95% of the measurements will have a z-score between −2 and 2.
3. All or almost all the measurements will have a z-score between −3 and 3.

Note that this interpretation of z-scores is identical to that given by the Empirical Rule for samples from mound-shaped distributions. The statement that a measurement falls in the interval $(\mu - \sigma)$ to $(\mu + \sigma)$ is equivalent to the statement that a measurement has a population z-score between −1 and 1, since all measurements between $(\mu - \sigma)$ and $(\mu + \sigma)$ are within 1 standard deviation of μ (see Figure 2.14).

We end this section with an example and a case study that indicate how z-scores may be used to accomplish our primary objective: the use of sample information to make inferences about the population.

Figure 2.14 Population *z*-scores for a Mound-Shaped Distribution

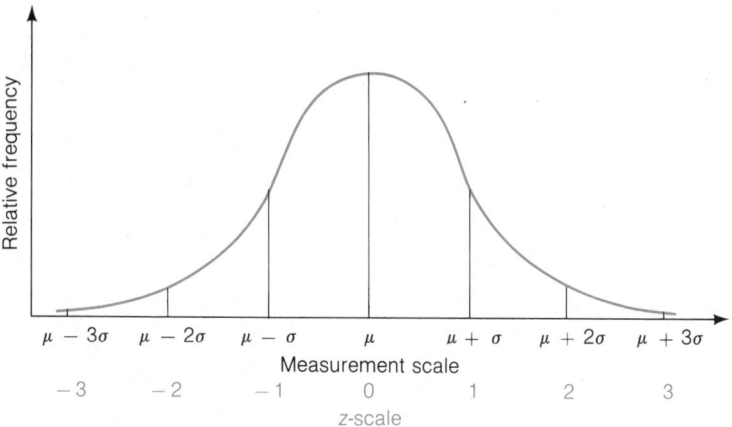

2.8
Detecting
Outliers

An **outlier** is an observation that falls far out in the tail of a distribution and may be a faulty observation. Suppose, for example, that you were to sample the weekly sales of fifty salespersons in a company and found that all but one of the sales ranged from $3,000 to $5,000. The sales for the single exception were $750. If the object of the sampling is to learn something about the weekly sales of full-time salespersons, then the $750 observation is suspect and merits investigation.

A further check of company records might indicate that the salesperson worked only a partial week because of sickness. In that case this observation is not from the population of interest to you and should be deleted from the sample. If investigation does not provide a reason for eliminating this observation from the sample, it should not be deleted. Even though extreme values are improbable, their occurrence is not impossible.

The most obvious test for an outlier is to calculate its *z*-score (Section 2.7). If the *z*-score for an observation is 4.2, for example, we know that it lies more than 4 standard deviations away from the sample mean. The Empirical Rule (Table 2.7) tells us that a *z*-score this large is highly improbable and points to the possibility of a faulty observation.

A second method for detecting outliers, known as a **box plot,** is based on the **interquartile range** (IQR), the distance between the upper and lower quartiles:

$$IQR = Q_U - Q_L$$

This rule of thumb procedure locates a box on the *x*-axis above the middle 50% of the observations, i.e., between the lower and upper quartiles. A box plot for the 100 EPA mileage ratings of Table 2.1 is shown in Figure 2.15 (page 56).

In Example 2.10, we found Q_L and Q_U for the 100 EPA mileage readings to be 35.6 and 38.4, respectively. Therefore, the interquartile range is

$$IQR = Q_U - Q_L = 38.4 - 35.6 = 2.8$$

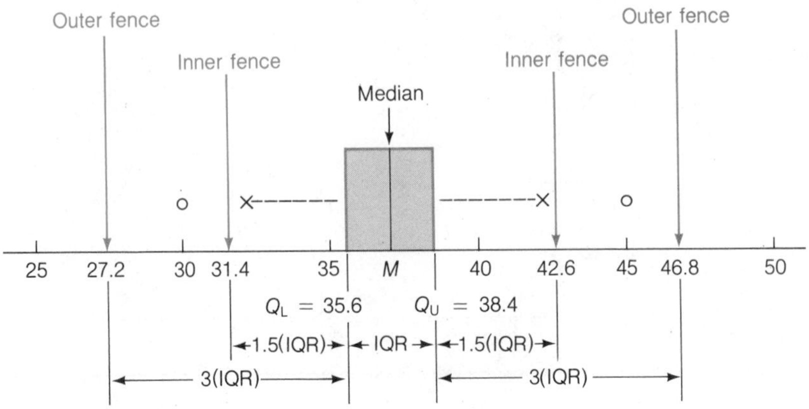

Figure 2.15 Box Plot for the 100 EPA Mileage Ratings of Table 2.1

The box plot (Figure 2.15) locates a box over the x-axis between Q_L and Q_U. Fifty percent of the observations lie between Q_L and Q_U, i.e., within the box. A vertical line is drawn inside the box to locate the median. If the data set is symmetric, the median will fall at the center of the box. If the data set is skewed to the left or right, the median will fall toward the right or left side of the box, respectively. To detect outliers, two sets of limits are constructed. ***Inner fences*** are located a distance of 1.5(IQR) = 4.2 below Q_L and above Q_U. Thus, the lower inner fence is located at 31.4 and the upper inner fence at 42.6 (see Figure 2.15). Any observations falling outside the inner fences are suspect outliers.

Similarly, ***outer fences*** are located a distance 3(IQR) = 8.4 below Q_L and above Q_U. The outer fences are located at 27.2 and 46.8 (see Figure 2.15). Any observations outside the outer fences are beyond the point of suspicion and are designated as outliers.

Suspect outliers—observations between the inner and outer fences—are located on the box plot by using small circles. Outliers—those that fall outside the outer fences—would be located with solid dots. To further highlight extreme values, "whiskers" are added to the box plot. The value in the region between Q_L and the lower inner fence that is closest to the inner fence is marked with an × and joined to the box with a dashed line—a whisker. In our example, the × is located at 31.8. Similarly, an × with attached whisker is located at 42.1, the most extreme value in the region between Q_U and the upper inner fence.

Further examination of the 100 mileage ratings (Table 2.1) shows that only two observations fall outside the inner fences. The smallest observation, 30.0, falls between the lower inner and outer fences. The largest observation, 44.9, falls between the upper inner and outer fences. These observations, shown as small circles on the box plot, are suspect outliers. (The corresponding z-scores are −2.89 and 3.26, respectively.) The circumstances concerning their collection should be investigated to determine whether these observations truly represent the sampled population.

The z-score and box plot methods both establish rule of thumb limits outside of which an observation is deemed to be an outlier. Since both methods locate the limits a specified distance from the center (relatively speaking) of the data set, it should be possible to compare the ability of the two methods to detect outliers. In fact, we might expect the two rules of thumb to produce similar results. We will lead you through this comparison for a specific type of data in Exercise 5.29.

2.9 Outliers, Rare Events, and Statistical Inference

The object of a method for detecting outliers is to identify faulty sample observations, i.e., observations that do not belong to the population you are attempting to sample. As the following example illustrates, the logic employed in detecting outliers is the basis for one of the two statistical methods for making statistical inferences about population parameters. We will introduce the logic here and then develop it further in Chapter 7.

Example 2.12

Suppose a female bank employee believes that her salary is low as a result of sex discrimination. To substantiate her belief, she collects information on the salaries of her male counterparts in the banking business. She finds that their salaries have a mean of $24,000 and a standard deviation of $2,000. Her salary is $17,000. Does this information support her claim of sex discrimination?

Solution

The analysis might proceed as follows: First, we calculate the z-score for the woman's salary with respect to those of her male counterparts. Thus,

$$z = \frac{\$17,000 - \$24,000}{\$2,000} = -3.5$$

The implication is that the woman's salary is 3.5 standard deviations *below* the mean of the male salary distribution. Furthermore, if a check of the male salary data shows that the frequency distribution is mound-shaped, we can infer that very few salaries in this distribution should have a z-score less than −3, as shown in Figure 2.16. Therefore, a z-score of −3.5 represents either a measurement from a distribution different from the male salary distribution or a very unusual (highly improbable) measurement for the male salary distribution.

Figure 2.16 Male Salary Distribution

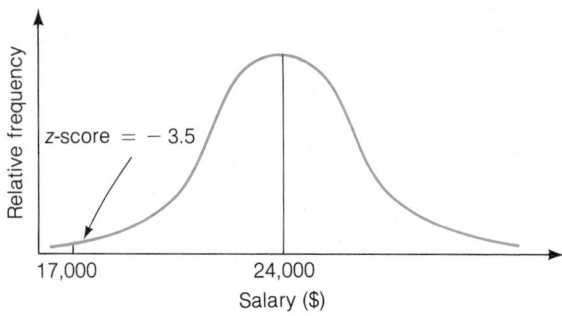

Which of the two situations do you think prevails? Do you think the woman's salary is simply unusually low in the distribution of salaries, or do you think her claim of salary discrimination is justified? Most people would probably conclude that her salary does not come from the male salary distribution. However, the careful investigator should require more information before inferring sex discrimination as the cause. We would want to know more about the data collection technique the woman used, and more about her competence at her job. Also, perhaps other factors such as length of employment should be considered in the analysis. ∎

Example 2.12 exemplifies an approach to statistical inference that might be called the **rare event approach.** An experimenter hypothesizes a specific frequency distribution to describe a population of measurements. Then a sample of measurements is drawn from the population. If the experimenter finds it unlikely that the sample came from the hypothesized distribution, the hypothesis is concluded to be false. Thus, in Example 2.12 the woman believes her salary reflects sex discrimination. She hypothesizes that her salary should be just another measurement in the distribution of her male counterparts' salaries if no discrimination exists. However, it is so unlikely that the sample (in this case, her salary) came from the male frequency distribution that she rejects that hypothesis, concluding that the distribution from which her salary was drawn is different from the distribution for the men.

This rare event approach to inference-making will be discussed further in later chapters. Proper application of the approach requires a knowledge of probability, the subject of our next chapter.

Case Study 2.3
Statistics and Air Quality Standards

H. E. Neustadter and S. M. Sidik (1974) discuss calculation procedures for obtaining an air quality standard (AQS) to use in judging whether a company is exceeding air pollution standards. A simplification of a method they discuss is outlined here:

1. Obtain a large set of daily measurements from a company known to be complying with established air pollution standards.

2. If necessary, transform the data so that they form a mound-shaped distribution to which the aids in Table 2.7 may be applied. For example, if the logarithm of each measurement is calculated for typical air quality data, the new set of data often forms a mound-shaped distribution.

3. Calculate the mean $\bar{x}$ and standard deviation s of the mound-shaped set of data and use this information to obtain the AQS. For example, if it is desired to obtain a standard that would be exceeded on approximatelly 2.5% of the industry's operating days, calculate

    $$AQS = \bar{x} + 2s$$

    The number of standard deviations to be added to $\bar{x}$ can be adjusted to reflect the percentage of days a company is to be permitted to exceed the standard. As Neustadter and Sidik point out, "a major objective of air quality monitoring is often to determine compliance with air quality standards which may in part consist of a 24-hour level not to be exceeded more than once a year." They show that one should calculate the mean plus approximately 2.7 standard deviations to meet this standard.

**Exercises
2.53–2.67**

Learning the Mechanics

2.53 Compute the z-score corresponding to each of the following values of x:

a. $x = 40$, $s = 5$, $\bar{x} = 30$ **b.** $x = 90$, $\mu = 89$, $\sigma = 2$

c. $\mu = 50$, $\sigma = 5$, $x = 50$ **d.** $s = 4$, $x = 20$, $\bar{x} = 30$

e. In parts a–d, state whether the z-score locates x within a sample or a population.

f. In parts a–d, state whether each value of x lies above or below the mean and by how many standard deviations.

2.54 For each of the following, use the z-score to determine whether the value of x is an outlier:

a. $x = 29$, $\bar{x} = 16$, $s = 7$ **b.** $x = 102$, $\bar{x} = 106$, $s = 1.4$

c. $x = 7.3$, $\bar{x} = 7.1$, $s = .79$ **d.** $x = 47$, $\bar{x} = 46$, $s = .25$

2.55 Give the percentage of measurements in a data set that are above and below each of the following percentiles:

a. 75th percentile   **b.** 50th percentile

c. 20th percentile   **d.** 84th percentile

2.56 Compare z-scores to decide which of the following x-values lie the greatest distance above the mean and the greatest distance below the mean.

a. $x = 100$, $\mu = 50$, $\sigma = 25$ **b.** $x = 1$, $\mu = 4$, $\sigma = 1$

c. $x = 0$, $\mu = 200$, $\sigma = 100$ **d.** $x = 10$, $\mu = 5$, $\sigma = 3$

2.57 Find the median, the upper and lower quartiles, and the interquartile range for the following data set:

17	15	13	18	15	30
16	19	16	18	13	15
10	12	14	14	23	13
12	16	10	12	14	14

2.58 Construct a box plot for the data of Exercise 2.57. Identify the outliers and suspect outliers and show them on your plot.

2.59 Find the median, the upper and lower quartiles, and the interquartile range for the following data set:

3.5	3.8	4.1	3.9	3.8
2.1	4.2	3.7	4.0	5.2
4.4	3.9	.8	4.5	3.7
3.9	4.0	3.5	3.8	4.1
3.8	4.0	3.9	3.9	3.8

2.60 Construct a box plot for the data of Exercise 2.59. Identify the outliers and suspect outliers and show them on your plot.

Applying the Concepts

2.61 The distribution of scores on a nationally administered college achievement test has a median of 520 and a mean of 540.

a. Explain why it is possible for the mean to exceed the median for this distribution of measurements.

b. Suppose you are told that the 90th percentile is 660. What does this mean?

c. Suppose you are told that you scored at the 94th percentile. Interpret this statement.

2.62 Many firms use on-the-job training to teach their employees computer programming. Suppose you work in the personnel department of a firm that just finished training a group of its employees to program and you have been requested to review the performance of one of the trainees on the final test that was given to all trainees. The mean and standard deviation of the test scores are 80 and 5, respectively, and the distribution of scores is mound-shaped.

a. The employee in question scored 65 on the final test. Compute the employee's z-score.

b. Approximately what percentage of the trainees will have z-scores equal to or less than the employee of part a?

c. If a trainee were arbitrarily selected from those who had taken the final test, is it more likely that he or she would score 90 or above, or 65 or below?

2.63 A city librarian claims that books have been checked out an average of 7 (or more) times in the last year. You suspect he has exaggerated the checkout rate (book usage) and that the mean number of checkouts per book per year is, in fact, less than 7. Using the card catalog, we randomly select one book and find that it has been checked out 4 times in the last year. Assume that we know from previous records that the standard deviation of the number of checkouts per book per year is 1.

a. If the mean number of checkouts per book per year really is 7, what is the z-score corresponding to 4?

b. Considering your answer to part a, is there evidence to indicate that the librarian's claim is incorrect?

c. If you knew that the distribution of the number of checkouts were mound-shaped, would your answer to part b change? Explain.

d. If the standard deviation of the number of checkouts per book per year were 2 (instead of 1), would your answers to parts b and c change? Explain.

2.64 Polychlorinated biphenyls (PCB's), considered to be extremely hazardous to humans, are often used in the insulation of large electrical transformers. On March 24, 1984, the *Gainesville Sun* reported on the discovery of a particularly high PCB count at a salvage company in Clay County, Florida. The company, which salvaged the copper in electrical transformers, allowed oil contaminated with PCB's to seep into the soil in and around the salvage site. One soil sample in the vicinity registered 200 parts per million (ppm) of PCB's, four times the safe limit established by the Florida Department of Environmental Regulation.

Suppose that the PCB count in samples of soil in the vicinity of the salvage operation has a distribution with mean equal to 25 ppm and standard deviation equal to 5 ppm of PCB's.* Would a soil sample showing 200 ppm be classified as an outlier? Explain.

* In Section 4.5 we explain why it is likely that σ is approximately equal to 5.

2.65 The data showing the percentage of high school students graduating in each state (Exercise 2.5) are reproduced here:

Minn.	86.0	Hawaii	79.6	Vt.	75.5	Del.	72.7	Calif.	68.0
N.Dak.	84.9	Pa.	78.6	Ind.	75.3	Mich.	72.5	Ky.	67.3
Iowa	84.8	Ohio	78.4	Wash.	75.3	N.Mex.	71.9	Ala.	67.1
S.Dak.	82.8	N.H.	77.1	Md.	75.0	Maine	71.4	N.C.	67.1
Wis.	82.3	Okla.	76.9	Ark.	74.8	Conn.	71.1	Tenn.	66.7
Nebr.	81.3	Idaho	76.8	Mo.	74.0	Oreg.	70.3	N.Y.	65.9
Mont.	80.9	Mass.	76.8	W.Va.	74.0	Ariz.	69.0	Ga.	64.3
Kans.	80.5	Nev.	76.8	Va.	73.3	S.C.	68.8	Fla.	63.7
Utah	80.2	N.J.	76.7	Ill.	73.2	Tex.	68.8	La.	63.4
Wyo.	80.0	Colo.	76.5	R.I.	72.8	Alaska	68.7	Miss.	61.8
								D.C.	54.6

a. Find the median, lower and upper quartiles, and the interquartile range for the data.
b. Construct a box plot for the data. Show all outliers and suspect outliers on the plot.

2.66 The radon 222 data set (Exercise 2.6) is reproduced here:

1,709.79	4,132.28	2,996.49	2,796.42	3,750.83	961.40	1,096.43	1,774.77
357.17	1,489.86	2,367.40	11,968.23	178.99	5,402.35	2,315.52	2,617.57
1,150.94	3,017.48	599.84	2,758.84	3,764.96	1,888.22	2,055.20	205.84
1,572.69	393.55	538.37	1,830.78	878.56	6,815.69	752.89	1,977.97
558.33	880.84	2,770.23	1,426.57	1,322.76	1,480.04	9,139.21	1,698.39

a. Find the median, lower and upper quartiles, and the interquartile range for the data.
b. Construct a box plot for the data. Show all outliers and suspect outliers on the plot.

2.67 The data set containing the percentages of total work-hours lost due to strikes for each state (excluding Maryland; see Exercise 2.7) is reproduced here:

STATE	%	STATE	%	STATE	%	STATE	%	STATE	%
Ala.	.14	Hawaii	.02	Mich.	.83	N.Y.	.11	Tenn.	.23
Alaska	.11	Idaho	.11	Minn.	.02	N.C.	.01	Tex.	.06
Ariz.	.09	Ill.	.18	Miss.	.14	N.Dak.	.03	Utah	.66
Ark.	.10	Ind.	.16	Mo.	.14	Ohio	.38	Vt.	.26
Calif.	.16	Iowa	.16	Mont.	.28	Okla.	.01	Va.	.04
Colo.	.04	Kans.	.11	Nebr.	.05	Oreg.	.12	Wash.	.16
Conn.	.08	Ky.	.17	Nev.	.36	Pa.	.14	W.Va.	.45
Del.	.41	La.	.10	N.H.	.03	R.I.	.09	Wis.	.21
Fla.	.20	Maine	.15	N.J.	.27	S.C.	.01	Wyo.	.01
Ga.	.13	Mass.	.07	N.Mex.	.09	S.Dak.	.16	Md.	—

a. Find the median, lower and upper quartiles, and the interquartile range for the data.

b. Construct a box plot for the data. Show all outliers and suspect outliers on the plot.

2.10 Distorting the Truth with Descriptive Techniques

While it may be true in telling a story that "a picture is worth a thousand words," it is also true that pictures can be used to convey a colored and distorted message to the viewer. So the old adage applies: "Let the buyer (reader) beware." Examine relative frequency histograms and, in general, all graphic descriptions with care.

We will mention a few of the pitfalls to watch for when interpreting a chart or graph. But first we should mention the **time series graph,** which is often the object of distortion. This type of graph records the behavior of some variable of interest recorded over time. Examples of variables commonly graphed as time series abound: economic indices, the United States food surplus, defense spending, presidential popularity index, etc. Since time series graphs often appear in newspapers or magazines, we will use some of them to demonstrate several ways in which pictures are commonly distorted.

One common way to change the impression conveyed by a graph is to change the scale on the vertical axis, the horizontal axis, or both. For example, Figure 2.17 is a **bar graph** that shows the market share of sales for a company for each of the years 1979 to 1984. If you want to show that the change in firm A's market share over time is moderate, you should pack in a large number of units per inch on the vertical axis. That is, make the distance between successive units on the vertical scale small, as shown in Figure 2.17. You can see that a change in the firm's market share over time is barely apparent.

Figure 2.17 Firm A's Market Share from 1979 to 1984—Packed Vertical Axis

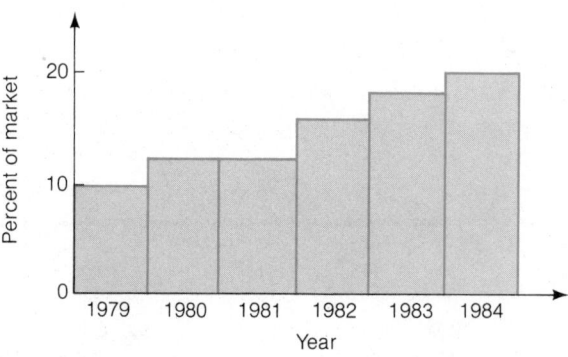

If you want to use the same data to make the changes in firm A's market share appear large, you should increase the distance between successive units on the vertical

Figure 2.18 Firm A's Market Share from 1979 to 1984—Stretched Vertical Axis

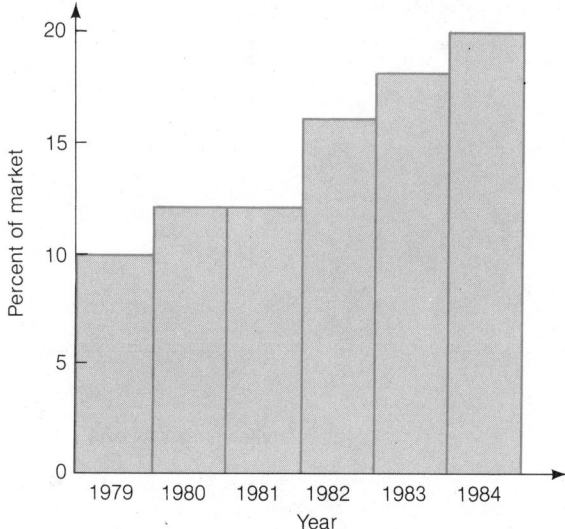

axis. That is, stretch the vertical axis by graphing only a few units per inch as in Figure 2.18. A telltale sign of stretching is a long vertical axis, but this is often hidden by starting the vertical axis at some point above zero, as shown in Figure 2.19(a). The same effect can be achieved by using a broken line for the vertical axis, as shown in Figure 2.19(b).

Figure 2.19 Changes in Money Supply from January to June 1980

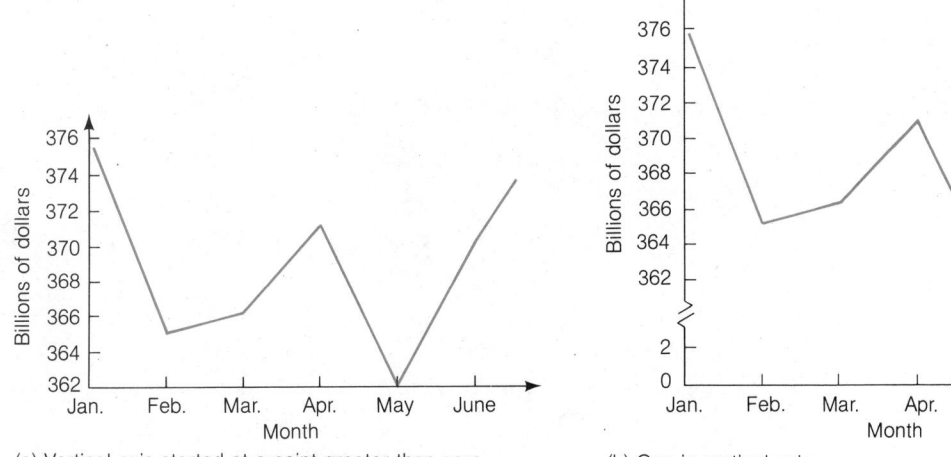

(a) Vertical axis started at a point greater than zero

(b) Gap in vertical axis

Figure 2.20   Gross National Product from 1979 to 1980

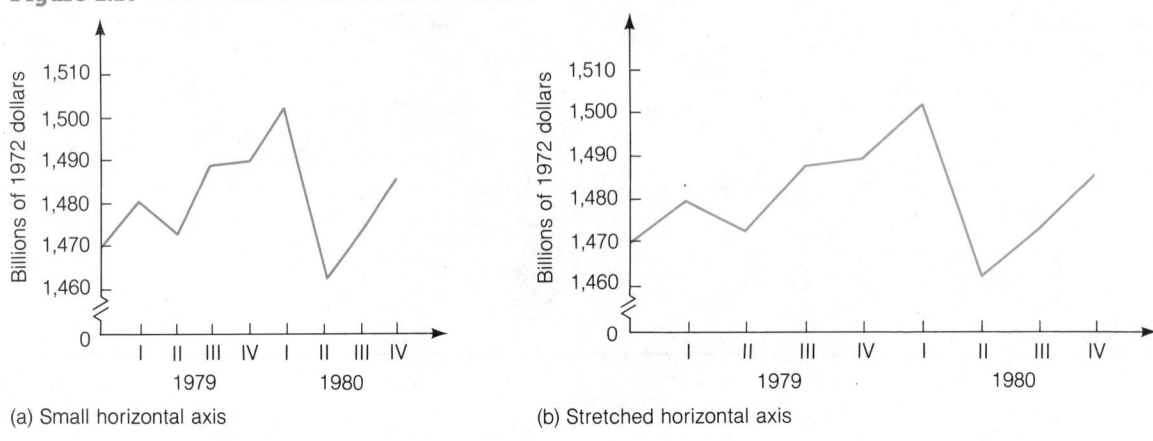

(a) Small horizontal axis   (b) Stretched horizontal axis

Stretching the horizontal axis (increasing the distance between successive units) may also lead you to incorrect conclusions. For example, Figure 2.20(a) depicts the change in the Gross National Product (GNP) from the first quarter of 1979 to the last quarter of 1980. If you increase the size of the horizontal axis, as in Figure 2.20(b), the change in the GNP over time seems less pronounced.

The changes in categories indicated by a bar graph can also be emphasized or deemphasized by stretching or shrinking the vertical axis. Another method of achieving visual distortion with bar graphs is by making the width of the bars proportional to their height. For example, look at the bar chart in Figure 2.21(a), which depicts the per-

Figure 2.21   Relative Share of the Automobile Market for Each of Four Major Manufacturers

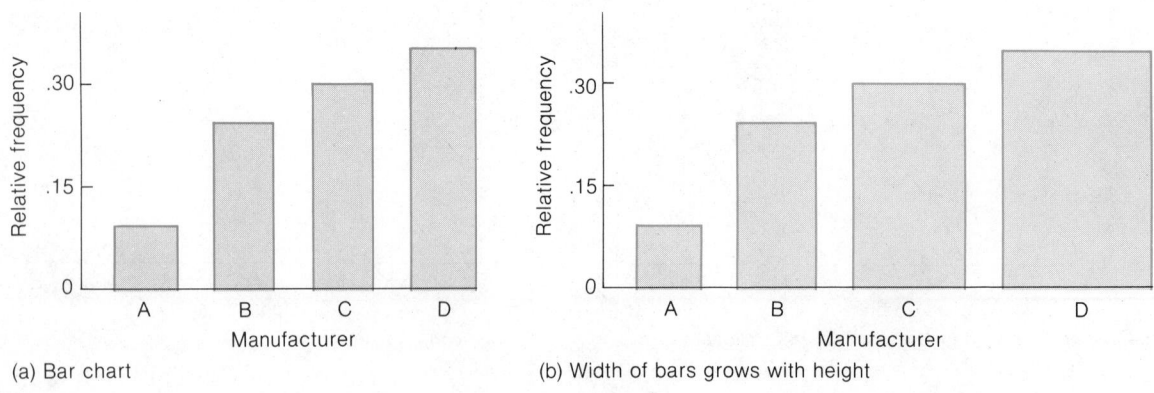

(a) Bar chart   (b) Width of bars grows with height

centage of a year's total automobile sales attributable to each of the four major manu-facturers. Now suppose we make the width as well as the height grow as the market share grows. This change is shown in Figure 2.21(b). The reader may tend to equate the *area* of the bars with the relative market share of each manufacturer. In fact, the true relative market share is proportional only to the height of the bars.

We have presented only a few of the ways that graphs can be used to convey mis-leading pictures of phenomena. However, the lesson is clear. Examine all graphic des-criptions of data with care. Particularly, check the axes and the size of the units on each axis. Ignore the visual changes and concentrate on the actual numerical changes indicated by the graph or chart. The information in a data set can also be distorted by using numerical descriptive measures, as Example 2.13 indicates.

Example 2.13

Suppose you are considering working for a small law firm that presently has a senior member and three junior members. You inquire about the salary you could expect to earn if you join the firm. Unfortunately, you receive two answers:

Answer A: The senior member tells you that an "average employee" earns $37,500.

Answer B: One of the junior members later tells you that an "average employee" earns $25,000.

Which answer can you believe? The confusion exists because the phrase "average employee" has not been clearly defined. Suppose the four salaries paid are $25,000 for each of the three junior members and $75,000 for the senior member. Thus,

$$\bar{x} = \frac{3(\$25,000) + \$75,000}{4} = \frac{\$150,000}{4} = \$37,500$$

Median = $25,000

You can now see how the two answers were obtained. The senior member reported the mean of the four salaries, and the junior member reported the median. The informa-tion you received was distorted because neither person stated which measure of central tendency was being used. ■

Another distortion of information in a sample occurs when *only* a measure of central tendency is reported. Both a measure of central tendency and a measure of variability are needed to obtain an accurate mental image of a data set.

Suppose you want to buy a new car and are trying to decide which of two models to purchase. Since energy and economy are both important issues, you decide to pur-chase model A because its EPA mileage rating is 32 miles per gallon in the city, while the mileage rating for model B is only 30 miles per gallon in the city.

However, you may have acted too quickly. How much variability is associated with the ratings? As an extreme example, suppose that further investigation reveals that the standard deviation for model A mileages is 5 miles per gallon, while that for model B is only 1 mile per gallon. If the mileages form a mound-shaped distribution, they might appear as shown in Figure 2.22. Note that the larger amount of variability associated

Figure 2.22 Mileage Distributions for Two Car Models

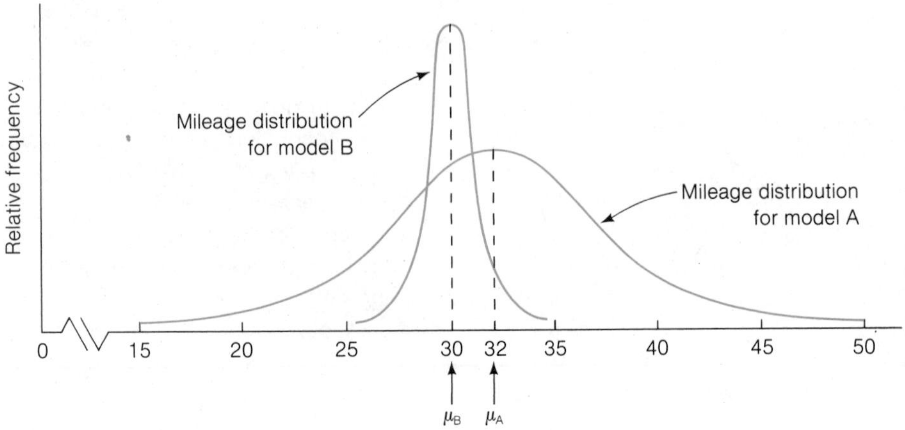

with model A implies that more risk is involved in purchasing model A. That is, the particular car you purchase is more likely to have a mileage rating that will greatly differ from the EPA rating of 32 miles per gallon if you purchase model A, while a model B car is not likely to vary from the 30 miles per gallon rating by more than 2 miles per gallon.

Case Study 2.4

Children Out of School in America: Making an Ugly Picture Look Worse

David L. Martin (1975) points out another method of distorting the truth with descriptive techniques in his article, "Firsthand report: How flawed statistics can make an ugly picture look even worse." In his critique of the Children Defense Fund's (CDF) 1973 report, *Children Out of School in America*, Martin quotes (italicized comments are Martin's):

25 percent of the 16 and 17 year olds in the Portland, Me., Bayside East Housing Project were out of school. *Only eight children were surveyed; two were found to be out of school.*

Of all the secondary school students who had been suspended more than once in census tract 22 in Columbia, S.C., 33 percent had been suspended two times and 67 percent had been suspended three or more times. *CDF found only three children in that entire census tract who had been suspended; one child was suspended twice and the other two children, three or more times.*

In the Portland Bayside East Housing Project, CDF says that 50 percent of all the secondary school children who had been suspended more than once had been suspended three or more times. *The survey found two secondary school children had been suspended in that area; one of them had been suspended three or more times.*

In each of these examples the reporting of percentages instead of the numbers themselves is misleading. Any inference one might draw from the cited examples would not be reliable. (We will see how to measure the reliability of estimated percentages in Chapter 7.) In short, either the numbers alone should be reported instead of percentages, or, better yet, the report should state that the numbers were too small to report by region. If several regions were combined, the numbers (and percentages) would be more meaningful.

Summary

Since we want to use sample data to make inferences about the population from which it is drawn, it is important for us to be able to describe the data. **Graphic methods** are important and useful tools for describing data sets. Our ultimate goal, however, is to use the sample to make inferences about the population. We are wary of using graphic techniques to accomplish this goal, since they do not lend themselves to a measure of the reliability for an inference. We therefore developed **numerical measures** to describe a data set.

These numerical methods for describing **quantitative** data sets can be grouped as follows:

1. Measures of central tendency
2. Measures of variability

The measures of central tendency we presented were the **mean, median,** and **mode.** The relationship between the mean and median provides information about the **skewness** of the frequency distribution. For making inferences about the population, the sample mean will usually be preferred to the other measures of central tendency. The **range, variance,** and **standard deviation** all represent numerical measures of variability. Of these, the variance and standard deviation are in most common use, especially when the ultimate objective is to make inferences about a population.

The mean and standard deviation may be used to make statements about the fraction of measurements in a given interval. For example, we know that at least 75% of the measurements in a data set will lie within 2 standard deviations of the mean. If the frequency distribution of the data set is mound-shaped, approximately 95% of the measurements will lie within 2 standard deviations of the mean.

Measures of relative standing provide still another dimension on which to describe a data set. The objective of these measures is to describe the location of a specific measurement relative to the rest of the data set. By doing so, you can construct a mental image of the relative frequency distribution. **Percentiles** and **z-scores** are important examples of measures of relative standing.

We found that the concept of an **outlier,** a rare or improbable sample outcome, is the basis for one type of statistical inference. The **rare event** concept of statistical inference means that if the chance that a particular sample came from a hypothetical population is very small, we can conclude either that the sample is extremely rare or that the hypothesized population is not the one from which the sample was drawn. The more unlikely it is that the sample came from the hypothesized population, the more strongly we favor the conclusion that the hypothesized population is not the true one. We need to be able to assess accurately the rarity of a sample, and this requires a knowledge of probability, the subject of our next chapter.

Finally, we gave some examples that demonstrated how descriptive statistics may be used to distort the truth. You should be very critical when interpreting graphic or numerical descriptions of data sets.

Supplementary Exercises 2.68–2.91

Learning the Mechanics

2.68 Consider the following data set:

50	38	56	57	48	64
41	51	52	46	51	50
60	46	41	53	65	59
42	37	50	45	52	42
56	47	63	58	49	55

a. Construct a stem and leaf display for the data. Use it to give a verbal description of the data set.
b. Construct a relative frequency histogram for this set of data.
c. Calculate $\bar{x}$, s^2, and s.
d. Calculate the intervals $\bar{x} \pm s$, $\bar{x} \pm 2s$, and $\bar{x} \pm 3s$. What percentage of the measurements would you expect to fall in each interval?
e. Calculate the actual percentage of the number of measurements falling in each interval. How does the result compare with your answers to part c?
f. Construct a box plot for the data and use it to identify any outliers.

2.69 Construct a relative frequency distribution for the data summarized in the following table:

MEASUREMENT CLASS	RELATIVE FREQUENCY	MEASUREMENT CLASS	RELATIVE FREQUENCY
0.00–0.75	.02	5.25–6.00	.15
0.75–1.50	.01	6.00–6.75	.12
1.50–2.25	.03	6.75–7.50	.09
2.25–3.00	.05	7.50–8.25	.05
3.00–3.75	.10	8.25–9.00	.04
3.75–4.50	.14	9.00–9.75	.01
4.50–5.25	.19		

2.70 Consider the following three measurements: 50, 70, 80. Find the z-score for each measurement if they are from a population with a mean and standard deviation equal to:
a. $\mu = 60$, $\sigma = 10$ **b.** $\mu = 60$, $\sigma = 5$
c. $\mu = 40$, $\sigma = 10$ **d.** $\mu = 40$, $\sigma = 100$

2.71 If the range of a set of data is 20, find a rough approximation to the standard deviation of the data set.

2.72 Calculate the mean, variance, and standard deviation of the following samples:
a. 0, −3, 4, 7, 2, 6   **b.** 5, 7, 4, 8, 6, 6, 3, 9, 5, 7
c. 12, 3, 22, 5, 19, 20, 9

2.73 Calculate the median, mode, and range for each of the following data sets:
a. 3, −4, 0, −6, 0, −2   **b.** 22, 31, 19, 25, 25, 35
c. 100, 27, 75, 80, 60, 52, 80, 83

2.74 Calculate the mean, variance, and standard deviation of the following samples:
a. 1.6, 2.4, 3.1, 1.8, 2.3, 4.3   **b.** .03, .09, .01, .04, .10, .07

2.75 Identify each of the following data sets as qualitative or quantitative:
a. The types of metal that are used to make automobiles
b. The amount of an anesthesia required for each of three surgical patients
c. The planting distances for certain types of garden vegetables
d. The nationality of each of twenty immigrants to the United States

Applying the Concepts

2.76 The owner of a service station decided to conduct a survey of service records to determine the length of time (in months) between customer oil changes. A random sample of the station records produced the following times between oil changes:

6	6	24	8	6
6	6	16	6	12
18	8	4	12	12

Compute the sample mean, variance, and standard deviation.

2.77 In Exercise 2.76, if the individual who changed the oil in his or her car every 24 months had instead changed it every 18 months, would the sample variance increase or decrease? Why? If instead of every 24 or 18 months, the person had changed the oil every 6 months, how would the resulting sample variance compare to the sample variance when the oil was changed every 24 months? Every 18 months?

2.78 As a result of government pressure, automobile manufacturers in the United States are deeply involved in research to improve their products' gas mileage. One manufacturer, hoping to achieve 30 miles per gallon on one of its full-size models, measured the mileage obtained by thirty test versions of the model with the following results (rounded to the nearest mile for convenience):

30	30	32	31	30	27
28	31	33	28	30	31
29	31	27	33	31	30
35	29	32	27	31	34
30	27	26	29	28	30

a. Construct a stem and leaf display for the data. Use it to give a verbal description of the data set.
b. Construct a relative frequency histogram to describe the data set.
c. If the manufacturer would be satisfied with a (population) mean of 30 miles per gallon, how would it react to the preceding test data?
d. Construct a box plot for the data. Do you have evidence to suggest outliers?
e. Compute $\bar{x}$, s^2, and s for the data set.

f. What percentage of the measurements would you expect to find in the intervals $\bar{x} \pm s$, $\bar{x} \pm 2s$, and $\bar{x} \pm 3s$?

g. Count the number of measurements that actually fall within the intervals of part f and express each interval count as a percentage of the total number of measurements. Compare these results with the results of part f.

2.79 A radio station claims that the amount of advertising per hour of broadcast time has an average of 3 minutes and a standard deviation equal to 2.1 minutes. You listen to the radio station for 1 hour, at a randomly selected time, and carefully observe that the amount of advertising time is equal to 7 minutes. Does this observation appear to disagree with the radio station's claim? Explain.

2.80 In a study of the generality of response to pain, subjects were exposed to two kinds of pain-producing stimuli. The object of the study was to see whether the subject showed consistency in pain response. The following data represent a sample of pain-responsivity scores from the study:

10	13	20	15	13
16	13	21	19	11
12	16	11	15	16

a. Calculate the range for the data and use the range to calculate a rough estimate of the sample standard deviation. Use this as a check on your arithmetic in part b.

b. Find the mean, variance, and standard deviation for the sample.

c. What proportion of the data falls within 2 standard deviations of the mean?

2.81 A severe drought affected several western states for 3 years. A Christmas tree farmer is worried about the drought's effect on the size of his trees. To decide whether the growth of the trees has been retarded, the farmer decides to take a sample of the heights of twenty-five trees and obtains the following results (recorded in inches):

60	57	62	69	46
54	64	60	59	58
75	51	49	67	65
44	58	55	48	62
63	73	52	55	50

a. Construct a stem and leaf display for the data. Use it to give a verbal description of the data set.

b. Construct a relative frequency histogram to describe the data set.

c. Construct a box plot for the data. Do you have evidence to suggest outliers?

d. Compute $\bar{x}$, s^2, and s for these data.

e. What percentage of the tree heights would you expect to find in the intervals $\bar{x} \pm s$, $\bar{x} \pm 2s$, $\bar{x} \pm 3s$?

f. Count the number of measurements that actually fall in each interval of part e, and express each interval count as a percentage of the total number of measurements. Compare these results to the results of part e.

2.82 The Community Attitude Assessment Scale (CAAS) measures citizens' attitudes toward fifteen life areas (e.g., education, employment, and health) on four dimensions—importance, influence, equality of opportunity, and satisfaction. In order to develop the CAAS, a number of households in each of twenty-five communities were randomly selected and sent questionnaires. Because relatively low response rates suggest that there could be a substantial but unknown opinion bias in the reported data, the percentage of the sample responding to the survey was determined in each community. The results are given here (in percent):

21	14	18	20	14
16	6	22	28	16
26	14	13	15	25
21	14	7	12	8
15	14	21	22	10

a. Construct a stem and leaf display for the data. Use it to give a verbal description of the data set.

b. Construct a relative frequency histogram for the data given, locating the mean, median, and mode.

c. Construct a box plot for the data as a check for outliers.

d. Find the range for the data and use it to calculate an approximate value for s. Use this value to check your answer to part e.

e. Calculate the variance and standard deviation for the data.

f. Find the proportion of the measurements that fall in the interval $\bar{x} \pm 2s$.

2.83 A small computing center has found that the number of jobs submitted per day to its computers has a distribution that is approximately mound-shaped, with a mean of 83 jobs and a standard deviation of 10.

a. On approximately what percentage of days will the number of jobs submitted be between 73 and 93?

b. On approximately what percentage of days will the number of jobs submitted be between 63 and 83?

c. On approximately what percentage of days will the number of jobs submitted be greater than 93?

2.84 A professor believes that if a class is allowed to work on an examination as long as desired, the times spent by the students would be approximately mound-shaped with mean 40 minutes and standard deviation 6 minutes. Approximately how long should be allotted for the examination if the professor wants almost all (say, 97.5%) of the class to finish?

2.85 A veterinarian was interested in determining how many animals were treated in the clinic each day. A random sample of 20 days' records produced the following results:

15	17	24	16
18	15	19	22
25	21	18	17
20	20	20	18
10	16	12	21

a. Calculate the sample mean, variance, and standard deviation.
b. Suppose the veterinarian had seen two additional animals on each of the 20 days. Again calculate the sample mean, variance, and standard deviation. Compare these numbers with your results in part a. What is the effect on the sample mean, variance, and standard deviation of adding a constant to each measurement in the sample?

2.86 A recently hired coach of distance runners was interested in knowing how many miles America's top distance runners usually run in a week. The coach surveyed fifteen of the best distance runners with the following results (in miles):

120	95	110	95	70
90	80	100	125	75
85	100	115 ·	130	90

a. Find the sample mean, variance, and standard deviation for these data.
b. Suppose each runner cut the mileage in half. Again calculate the mean, variance, and standard deviation for these data. Compare these results with those obtained in part a. What is the effect on the sample mean, variance, and standard deviation of multiplying each measurement in the sample by a constant?

2.87 By law a box of cereal labeled as containing 16 ounces must contain at least 16 ounces of cereal. It is known that the machine filling the boxes produces a distribution of fill weights that is mound-shaped, with mean equal to the setting on the machine and with a standard deviation equal to .03 ounce. To ensure that most of the boxes contain at least 16 ounces, the machine is set so that the mean fill per box is 16.09 ounces.

a. What percentage of the boxes will contain less than 16 ounces if the machine is set so that $\mu = 16.09$?
b. If the machine is set so that $\mu = 16.09$, is it likely that a randomly selected box would contain less than 16 ounces?
c. If the machine is set so that $\mu = 16.09$, is it likely that a randomly selected box of cereal would contain as little as 16.05 ounces? Explain.

2.88 A study was conducted to determine the effects of cigarette smoking on the lungs. The United States Surgeon General has warned that the two most dangerous ingredients contained in cigarettes are tar and nicotine. The amount of tar (in milligrams) present in ten selected brands is recorded below. Determine the mean, range, variance, and standard deviation for the data shown.

Winston	19	Viceroy	16
Marlboro	17	Raleigh	16
Vantage	11	Marlboro Lights	12
Kent Lights	8	Tareyton	17
Winston Lights	12	Parliament	10

2.89 Most people living in metropolitan areas receive impressions of what is happening in their area primarily through their major newspapers. A study was conducted to determine whether the *Uniform Crime Report*, compiled by the Federal Bureau of Investigation, and the daily newspaper gave consistent information about the trend and distribution of crime in a metropolitan area. An attention score, based on the amount of space devoted to a story, was calculated for each paper's coverage of murders, assaults, robberies, etc. Suppose μ, the average murder attention score of metropolitan newspapers across the country in 1980, was 60, with $\sigma = 4.5$. The *St. Louis Globe-Democrat* has a 1980 murder attention score of 69.

a. Approximately what percentage of the newspapers had a murder attention score higher than the *Globe-Democrat* in 1980? (Make no assumptions about the nature of the distribution of scores.)

b. Repeat part a, assuming attention scores were mound-shaped.

2.90 Audience sizes at concerts given by the Jacksonville Symphony over the past 2 years were recorded and found to have a sample mean of 3,125 and a sample standard deviation of 25. Calculate $\bar{x} \pm s, \bar{x} \pm 2s$, and $\bar{x} \pm 3s$. What fraction of the recorded audience sizes of the past 2 years would be expected to fall in each of these intervals?

2.91 A producer of alkaline batteries was interested in obtaining a statistical description of the shelf-life of the battery it manufactures. Twenty-five batteries were selected at random as they came off the assembly line and their shelf-lives were tested. The following are the lifetimes of the twenty-five batteries rounded to the nearest month:

24	21	24	20	19
25	27	24	30	21
23	24	24	19	24
26	22	25	24	24
25	23	28	23	25

a. Compute $\bar{x}$, s^2, and s for this data set.

b. What percentage of the measurements would you expect to find in the intervals $\bar{x} \pm s, \bar{x} \pm 2s$, and $\bar{x} \pm 3s$?

c. Count the number of measurements that actually fall within the intervals of part b and express each interval count as a percentage of the total number of measurements. Compare these results with the results of part b.

On Your Own . . .

We list below several sources of real-life data sets that have been obtained from Wasserman and Bernero's *Statistics Sources*. This index of data sources is very complete and is a useful reference for anyone interested in finding almost any type of data. First we list some almanacs:

CBS News Almanac

Information Please Almanac

World Almanac and Book of Facts

United States government publications are also rich sources of data:

Agricultural Statistics

Digest of Educational Statistics

Handbook of Labor Statistics

Housing and Urban Development Yearbook

Social Indicators

Uniform Crime Reports for the United States

Vital Statistics of the United States

Business Conditions Digest

Economic Indicators

Monthly Labor Review

Survey of Current Business

Bureau of the Census Catalog

Many data sources are published on an annual basis:

Commodity Yearbook

Facts and Figures on Government Finance

Municipal Yearbook

Standard and Poor's Corporation, Trade and Securities: Statistics

Some sources contain data that are international in scope:

Compendium of Social Statistics

Demographic Yearbook

United Nations Statistical Yearbook

World Handbook of Political and Social Indicators

Utilizing the data sources listed above, sources suggested by your instructor, or your own resourcefulness, find one real-life quantitative data set that stems from an area of particular interest to you.

a. Describe the data set by using a relative frequency histogram.

b. Find the mean, median, variance, standard deviation, and range of the data set.

c. Use Tables 2.6 and 2.7 to describe the distribution of this data set. Count the actual number of observations that fall within 1, 2, and 3 standard deviations of the mean of the data set and compare these counts with the description of the data set you developed in part b.

References   Horton, T. R. "A preliminary radiological assessment of radon exhalation from phosphate gypsum piles and inactive uranium mill tailings piles." EPA-520/5-79-004. Washington, D. C.: Environmental Protection Agency, 1979.

Huff, D. *How to lie with statistics.* New York: Norton, 1954.

Koopmans, L. H. *An introduction to contemporary statistics.* North Scituate, Mass.: Duxbury, 1981. Chapters 1 and 2.

Martin, D. L. "Children out of school: Firsthand report: how flawed statistics can make an ugly picture look even worse." *American School Board Journal,* 1975, *162,* 57–59.

Mendenhall, W. *Introduction to probability and statistics.* 6th ed. North Scituate, Mass.: Duxbury, 1983. Chapter 3.

Miller, S. L., Domey, R. G., Elston, S. F. A., & Milligan, G. "Mercury vapor levels in the dental office: a survey." *Journal of the American Dental Association,* November 1974, *89,* 1084–1091.

Neustadter, H. E., & Sidik, S. M. "On evaluating compliance with air pollution levels 'not to be exceeded more than once a year'." *Journal of Air Pollution Control Association,* 1974, *24*(6), 559–563.

Palomba, N. A. "Strike activity and union membership: an empirical approach." *University of Washington Business Review,* Winter 1969.

Rothstein, M. "Hotel overbooking as a Markovian sequential decision process." *Decision Sciences,* July 1974, *5,* 389–405.

Wasserman, P., & Bernero, J. *Statistics sources.* 5th ed. Detroit: Gale Research Company, 1978.

CHAPTER 3

Probability

Where We've Been . . .

We have identified inference, from a sample to a population, as the goal of statistics. To reach this goal, we must be able to describe a set of measurements. The use of graphic and numerical methods for describing data sets and for phrasing inferences was the topic of Chapter 2.

Where We're Going . . .

Now that we know how to phrase an inference about a population, we turn to the problem of making the inference. What is it that permits us to make the inferential jump from sample to population and then to give a measure of reliability for the inference? As you will subsequently see, the answer is *probability*. This chapter is devoted to a study of probability—what it is and some of the basic concepts of the theory behind it.

Contents

You will recall that statistics is concerned with decisions about a population based on sample information. Understanding how this is accomplished will be easier if you understand the relationship between population and sample. This understanding is enhanced by reversing the statistical procedure of making inferences from sample to population. In this chapter we assume the population *known* and calculate the chances of obtaining various samples from the population. Thus, probability is the reverse of statistics: In probability we use the population information to infer the probable nature of the sample.

Probability plays an important role in decision making. To illustrate, suppose you have an opportunity to invest in an oil exploration company. Past records show that out of ten previous oil drillings (a sample of the company's experiences), all ten resulted in dry wells. What do you conclude? Do you think the chances are better than 50–50 that the company will hit a producing well? Should you invest in this company? We think your answer to these questions will be an emphatic "No." If the company's exploratory prowess is sufficient to hit a producing well 50% of the time, a record of ten dry wells out of ten drilled is an event that is just too improbable. Do you agree?

As another illustration, suppose you are playing poker with what your opponents assure you is a well-shuffled deck of cards. In three consecutive 5-card hands, the person on your right is dealt 4 aces. Based on this sample of three deals, do you think the cards are being adequately shuffled? Again, we think your answer will be "No" and that you will reach this conclusion because dealing three hands of 4 aces is just too improbable assuming that the cards were properly shuffled.

Note that the decisions concerning the potential success of the oil drilling company and the decision concerning the card shuffling were both based on probabilities, namely the probabilities of certain sample results. Both situations were contrived so that you could easily conclude that the probabilities of the sample results were small. Unfortunately, the probabilities of many observed sample results are not so easy to evaluate intuitively. For these cases we will need the assistance of a theory of probability.

3.1 Events, Sample Spaces, and Probability

We begin our treatment of probability with simple examples that are easily described, thus eliminating any discussion that could be distracting. With the aid of simple examples, important definitions are introduced and the notion of probability is more easily developed.

Suppose a coin is tossed once and the up face is recorded. This is an ***observation,*** or ***measurement.*** Any process of making an observation is called an ***experiment.*** Our definition of experiment is broader than that used in the physical sciences, where you would picture test tubes, microscopes, etc. Other practical examples of statistical experiments are recording whether a customer prefers one of two brands of electronic calculators, recording a voter's opinion on an important political issue, measuring the amount of dissolved oxygen in a polluted river, observing the closing price of a stock, counting the number of errors in an inventory, and observing the fraction of insects killed by a new insecticide. This list of statistical experiments could be continued, but the point is that our definition of experiment is very broad.

Definition 3.1

An **experiment** is the process of making an observation or taking a measurement.

Consider another simple experiment consisting of tossing a die and observing the number on the up face. The six basic possible outcomes to this experiment are:

1. Observe a 1   **4.** Observe a 4
2. Observe a 2   **5.** Observe a 5
3. Observe a 3   **6.** Observe a 6

Note that if this experiment is conducted once, *you can observe one and only one of these six basic outcomes*. Also, these possibilities cannot be decomposed into more basic outcomes. These basic possible outcomes to an experiment are called **simple events.**

Definition 3.2

A **simple event** is the most basic outcome of an experiment.

Example 3.1   Two coins are tossed and their up faces are recorded. List all the simple events for this experiment.

Solution   Even for a seemingly trivial experiment, we must be careful when listing the simple events. At first glance the basic outcomes seem to be Observe two heads, Observe two tails, Observe one head and one tail. However, further reflection reveals that the last of these, Observe one head and one tail, can be decomposed into Head on coin 1, Tail on coin 2 and Tail on coin 1, Head on coin 2.* Thus, the simple events are as follows:

1. Observe *HH*   **3.** Observe *TH*
2. Observe *HT*   **4.** Observe *TT*

(where *H* in the first position means "Head on coin 1," *H* in the second position means "Head on coin 2," etc.).   ■

We will often wish to refer to the collection of all the simple events of an experiment. This collection will be called the **sample space** of the experiment. For example, there

---

* Even if the coins are identical in appearance, there are, in fact, two distinct coins. Thus, the designation of one coin as coin 1 and the other as coin 2 is legitimate in any case.

are six simple events in the sample space associated with the die toss experiment. The sample spaces for the experiments discussed thus far are shown in Table 3.1.

Definition 3.3

The **sample space** of an experiment is the collection of all its simple events.

Table 3.1
Experiments and Their
Sample Spaces

Experiment: Observe the up face on a coin.

Sample space: 1.   Observe a head
               2.   Observe a tail

This sample space can be represented in set notation as a set containing two simple events:

S: $\{H, T\}$

where H represents the simple event Observe a head and T represents the simple event Observe a tail.

Experiment: Observe the up face on a die.

Sample space: 1.   Observe a 1
               2.   Observe a 2
               3.   Observe a 3
               4.   Observe a 4
               5.   Observe a 5
               6.   Observe a 6

This sample space can be represented in set notation as a set of six simple events:

S: $\{1, 2, 3, 4, 5, 6\}$

Figure 3.1   Venn
Diagrams for the Three
Experiments from Table 3.1

(a) Experiment: Observe
the up face on a coin

(b) Experiment: Observe
the up face on a die

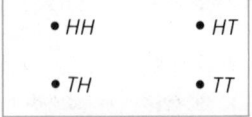

(c) Experiment: Observe
the up faces on two
coins

Experiment: Observe the up faces on two coins.

Sample space: 1.   Observe HH
               2.   Observe HT
               3.   Observe TH
               4.   Observe TT

This sample space can be represented in set notation as a set of four simple events:

S: $\{HH, HT, TH, TT\}$

Just as graphs are useful in describing sets of data, a pictorial method for presenting the sample space and its simple events will often be useful. Figure 3.1 shows such a representation for each of the experiments in Table 3.1. In each case, the sample space is shown as a closed figure, labeled S, containing a set of points, called **sample points,** with each point representing one simple event. Note that the number of sample points in a sample space S is equal to the number of simple events associated with the respective experiment: two for the coin toss, six for the die toss, and four for the two-coin toss. These graphic representations are called ***Venn diagrams.***

Now that we have defined simple events as the basic outcomes of the experiment and the sample space as the collection of all the simple events, we are prepared to discuss the probabilities of simple events. You have undoubtedly used the term *probability* and have some intuitive idea about its meaning. Probability is generally used synonymously with "chance," "odds," and similar concepts. We will begin our treatment of probability using these informal concepts and then solidify what we mean later. For example, if a fair coin is tossed, we might reason that both the simple events, Observe a head and Observe a tail, have the same chance of occurring. Thus, we might state that "the probability of observing a head is 50%," or "the odds of seeing a head are 50–50." Both these statements are based on an informal knowledge of probability.

The probability of a simple event is a number between 0 and 1 that measures the likelihood that the event will occur when the experiment is performed. This number is usually taken to be the relative frequency of the occurrence of a simple event in a very long series of repetitions of an experiment. When this information is not available, we select the number based on experience. For example, if we are assigning probabilities to the two simple events in the coin toss experiment (Observe a head and Observe a tail), we might reason that if we toss a balanced coin a very large number of times, the simple events Observe a head and Observe a tail will occur with the same relative frequency of .5. Thus, the probability of each simple event is .5.

In other cases we may choose the probability based on general information about the experiment. For example, if the experiment is observing whether a venture succeeds or fails (the simple events), we may assess the probability of success by considering the personnel managing the venture, the success of similar ventures, and any other information deemed pertinent. If we finally decide that the venture has an 80% chance of succeeding, we assign a probability of .8 to the simple event Success. We hope that .8 is a reasonably accurate measure of the likelihood of the occurrence of the simple event Success. If it is not, we may be misled on decisions based on this probability or based on calculations in which it appears.

No matter how you assign the probabilities to simple events, the probabilities assigned must obey two rules:

> **1.** All simple event probabilities *must* lie between 0 and 1.
> **2.** The probabilities of all the simple events within a sample space *must* sum to 1.

Although the probabilities of simple events are often of interest in their own right, it is usually probabilities of collections of simple events that are important. Example 3.2 demonstrates this point.

Example 3.2   A fair die is tossed and the up face is observed. If the face is even, you win $1. Otherwise, you lose $1. What is the probability that you win?

Solution   Recall that the sample space for this experiment contains six simple events:

$$S: \quad \{1, 2, 3, 4, 5, 6\}$$

Since the die is balanced, we assign a probability of $\frac{1}{6}$ to each of the simple events in this sample space. An even number will occur if one of the simple events, Observe a 2, Observe a 4, or Observe a 6, occurs. A collection of simple events such as this will be called an **event,** and we will denote this event by the letter A. Since the event A contains three simple events—all with probability $\frac{1}{6}$—and since no simple events can occur simultaneously, we reason that the probability of A is the sum of the probabilities of the simple events in A. Thus, the probability of A is $\frac{1}{6} + \frac{1}{6} + \frac{1}{6} = \frac{1}{2}$. This implies that *in the long run* you will win \$1 half the time and lose \$1 half the time.   ■

Figure 3.2 Die Toss Experiment with Event A: Observe an Even Number

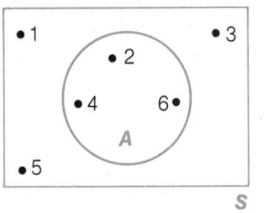

Figure 3.2 is a Venn diagram depicting the sample space associated with a die toss experiment and the event A, Observe an even number. The event A is represented by the closed figure inside the sample space S. This closed figure A contains all the simple events that comprise it.

How do you decide which simple events belong to the set associated with an event A? Test each simple event in the sample space S. If event A occurs when a particular simple event occurs, then that simple event is in the event A. For example, the event A, Observe an even number, in the die toss experiment will occur if the simple event Observe a 2 occurs. By the same reasoning, the simple events Observe a 4 and Observe a 6 are also in event A.

To summarize, we have demonstrated that an event can be defined in words or it can be defined as a specific set of simple events. This leads us to the following general definition of an event:

Definition 3.4

An **event** is a specific collection of simple events.

Example 3.3

SIMPLE EVENT	PROBABILITY
HH	$\frac{4}{9}$
HT	$\frac{2}{9}$
TH	$\frac{2}{9}$
TT	$\frac{1}{9}$

Consider the experiment of tossing two coins. Suppose the coins are *not* balanced and the correct probabilities associated with the simple events are given in the table. [*Note:* The necessary properties for assigning probabilities to simple events are satisfied.]

Consider the events

A: {Observe exactly one head}

B: {Observe at least one head}

Calculate the probability of A and the probability of B.

Solution   Event A contains the simple events HT and TH. Since two or more simple events cannot occur at the same time, we can easily calculate the probability of event A by summing the probabilities of the two simple events. Thus, the probability of observing exactly one head (event A), denoted by the symbol $P(A)$, would be

$$P(A) = P(\text{Observe } HT) + P(\text{Observe } TH)$$

$$= \tfrac{2}{9} + \tfrac{2}{9} = \tfrac{4}{9}$$

Similarly, since B contains the simple events HH, HT, and TH,

$$P(B) = \tfrac{4}{9} + \tfrac{2}{9} + \tfrac{2}{9} = \tfrac{8}{9}$$ ∎

The preceding example leads us to a general procedure for finding the probability of an event A:

> The probability of an event A is calculated by summing the probabilities of the simple events in A.

Thus, we can summarize the steps for calculating the probability of any event:*

Steps for Calculating Probabilities of Events

1. Define the experiment, i.e., describe the process used to make an observation and the type of observation that will be recorded.
2. List the simple events.
3. Assign probabilities to the simple events.
4. Determine the collection of simple events contained in the event of interest.
5. Sum the simple event probabilities to get the event probability.

Example 3.4   A bank wishes to divide its service windows into two groups corresponding to the type of customer account: commercial or personal. One problem facing the bank is deciding how to apportion the service windows to the categories of service. At this stage of our study, we do not have the tools to solve this problem, but we can say that one of the important factors affecting the solution is the proportion of the two types of customers that enter the bank at a particular time. To illustrate, what is the probability that an incoming customer will have a commercial account? What is the probability that the next two customers will have commercial accounts? What is the probability for the general case of k customers? Explain how you might attempt to solve this problem. [*Note:*   For this example we use the term *customer* to refer only to people who seek teller service.]

Solution   **Step 1.** Define the experiment. The experiment corresponding to the entrance of a single customer is identical in underlying structure to the coin toss experiment illustrated in Figure 3.1(a). A customer, who possesses either a commercial account (call this a

---

* A thorough treatment of this topic can be found in the classic text by W. Feller (1968).

head) or a personal account (call this a tail), is observed and the type of account is recorded.

Experiment: Observe the type of account a single customer possesses.

Step 2. List the simple events. There are only two possible outcomes of the experiment. These simple events are:

Simple events: 1.   *C*:   {Customer possesses a commercial account}
                 2.   *D*:   {Customer possesses a personal account}

Step 3. Assign probabilities to the simple events. The difference between this problem and the coin toss problem becomes apparent when we attempt to assign probabilities to the two simple events. What probability should we assign to the simple event *C*? Some people might say .5, as for the coin toss experiment, but you can see that finding $P(C)$, the probability of simple event *C*, is not so easy. Suppose that a check of the bank's records shows that 10% of their accounts are personal. Then, at first glance, it would appear that $P(C)$ is .10. But this may not be correct, because the probability will depend on how frequently the two types of customers use the accounts—i.e., the proportion of customers per week, on the average, that seek banking service. So the important point to note is that this is a case where equal probabilities are not assigned to the simple events. How can we find these probabilities? A good procedure might be to monitor the system for a period of time, and ask incoming customers which type of service they desire. Then the proportions of the two types of customers could be used to approximate the probabilities of the two simple events. We could then continue with steps 4 and 5 to calculate any probability of interest for this experiment with two simple events.

This experiment, assessing the service requirements of two customers, is identical to the experiment of Example 3.3, tossing two coins, except that the probabilities of the simple events are not the same. We will learn how to find the probabilities of the simple events for this experiment, or for the general case of *k* customers, in Section 3.5.   ■

For the experiments discussed thus far, listing the simple events has been easy. For more complex experiments, the number of simple events may be so large that listing them is impractical. In solving probability problems for experiments with many simple events we employ the same principles as for experiments with few simple events. The only difference is that we need ***counting rules*** for determining the number of simple events without actually enumerating all of them. In Section 3.8 (an optional section), we present several of the more useful counting rules.

Case Study 3.1

Comparing
Subjective
Probability
Assessments with
Relative
Frequencies

Preston and Baratta (1948) performed an experiment with the objective of comparing how an individual's subjective assessment of the probability of an event compares with the known probability (relative frequency of occurrence) of the event. The individuals selected for the experiment ranged from undergraduates with no training in probability theory to professors of mathematics and statistics with a "substantial acquaintance with probability theory." Each individual participated in a game in which he or she bet part of an initial stake on one of seven different outcomes of a combination card–dice

Figure 3.3 Observed Relationship Between True and Subjective Probabilities

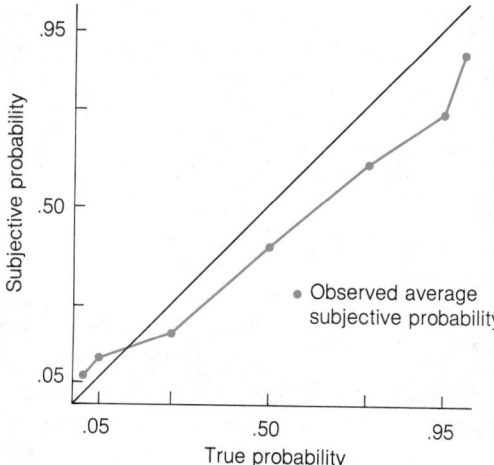

game. The probabilities of these seven different events, known only to the experimenters, were .01, .05, .25, .50, .75, .95, and .99. From the amount the subject is willing to bet, the individual's subjective probabilities can be determined and then compared with the actual probabilities of the events.

In Figure 3.3 we reproduce the authors' figure depicting the average subjective probability assessed by the experimental subjects compared to the true probabilities. Some of the conclusions reached were:

1. Events with probabilities less than .25 were subjectively overestimated.
2. Events with probabilities more than .25 were subjectively underestimated.
3. These conclusions were the same for both the probabilistically naive and sophisticated subjects.

For events with probabilities that are not clearly defined (such as the probability of rain tomorrow), it is important to have information on general tendencies in subjectively evaluating the probabilities. Of course, much more evidence has been collected on the subjective evaluation of probabilities since the Preston and Baratta article (not all of which corroborate their conclusions), but it remains an interesting evaluation of a person's ability to evaluate probabilities subjectively.

Exercises 3.1–3.21

Learning the Mechanics

3.1 An experiment results in one of the following simple events: E_1, E_2, E_3, E_4, and E_5.
a. Find $P(E_3)$ if $P(E_1) = .1$, $P(E_2) = .3$, $P(E_4) = .1$, and $P(E_5) = .1$.
b. Find $P(E_3)$ if $P(E_1) = P(E_3)$, $P(E_2) = .1$, $P(E_4) = .2$, and $P(E_5) = .2$.
c. Find $P(E_3)$ if $P(E_1) = P(E_2) = P(E_4) = P(E_5) = .1$.

3.2 An experiment results in one of the following simple events: E_1, E_2, E_3, and E_4.
a. It is known that $P(E_1) = P(E_2) = P(E_3) = P(E_4)$. Find $P(E_1)$.
b. It is known that $P(E_2) = P(E_3) = P(E_4) = .2$. Find $P(E_1)$.

3.3 The sample space for an experiment contains five simple events with probabilities as shown in the table. Find the probability of each of the following events:

SIMPLE EVENTS	PROBABILITIES
1	.05
2	.25
3	.30
4	.25
5	.15

A: {Either 1, 2, or 3 occurs}

B: {Either 1, 3, or 5 occurs}

C: {4 does not occur}

3.4 A nickel, dime, and quarter are tossed and the up faces are noted after each toss.
a. List the simple events in the sample space for this experiment.
b. Assign reasonable probabilities to the simple events.
c. Find the probability of each of the following events:

A: {At least one head appears}

B: {Exactly one head appears}

C: {The first toss is a head}

3.5 Repeat the experiment in Exercise 3.4 a total of 200 times and record the numbers of times events *A*, *B*, and *C* occur. Compare these observed relative frequencies with the probabilities calculated in Exercise 3.4. If you were to repeat the experiment many millions of times, how would the observed relative frequencies differ from those calculated from the 200 repetitions?

3.6 The Venn diagram depicts an experiment with six simple events. The events *A* and *B* are also shown. The probabilities of the simple events are as follows:

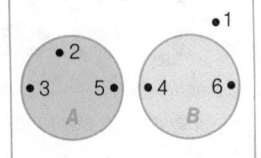

$$P(1) = P(2) = P(4) = \tfrac{2}{9} \qquad P(3) = P(5) = P(6) = \tfrac{1}{9}$$

a. Find $P(A)$.   **b.** Find $P(B)$.
c. Find the probability that the events *A* and *B* occur *simultaneously*.

3.7 Two fair dice are tossed, and the up face on each die is recorded.
a. List the thirty-six simple events contained in the sample space.
b. Find the probability of observing each of the following events:

A: {A 3 appears on each of the two dice}

B: {The sum of the numbers is even}

C: {The sum of the numbers is equal to 7}

D: {A 5 appears on at least one of the dice}

E: {The sum of the numbers is 10 or more}

3.8 Consider the experiment composed of one roll of a fair die followed by one toss of a fair coin. List the simple events. Assign a logical probability to each simple event. Determine the probability of observing each of the following events:

A: {6 on the die; *H* on the coin}

B: {Even number on the die; *T* on the coin}

C: {Even number on the die}

D: {*T* on the coin}

3.9 Two marbles are randomly drawn from a box containing two blue marbles and three red marbles. Determine the probability of observing each of the following events:

A: {Two blue marbles are drawn}

B: {A red and a blue marble are drawn}

C: {Two red marbles are drawn}

3.10 Simulate the experiment described in Exercise 3.9 using any five identically shaped objects, two of which are one color and three are another. Mix the objects, randomly draw two, record the results, and then replace the objects. Repeat the experiment a large number of times (at least 100). Calculate the proportion of times events *A*, *B*, and *C* occur. How do these proportions compare with the probabilities you calculated in Exercise 3.9? Should these proportions equal the probabilities? Explain.

Applying the Concepts

SIMPLE EVENTS	PROBABILITIES
SS	.81
SD	.09
DS	.09
DD	.01

3.11 A hospital reports that two patients have been admitted who have contracted Legionnaire's disease. Suppose our experiment consists of observing whether the patients survive or die as a result of the disease. The simple events and probabilities of their occurrence are shown in the table (where *S* in the first position means that patient 1 survives, *D* in the first position means that patient 1 dies, etc). Find the probabilities of each of the following events:

A: {Both patients survive the disease}

B: {At least one patient dies}

C: {Exactly one patient survives the disease}

3.12 In its "Intelligence Report," *Parade Magazine* (October 23, 1983) comments on a rare occurrence: the birth of a pair of twins, one white and one black, to an English couple. *Parade Magazine* notes that the odds that a pregnant woman will give birth to twins are 1 in 80. The probability that one of the twins is white and the other black is much smaller, even if you are given the fact that the children are the product of a mixed marriage.

Suppose that a child born of a mixed marriage has a 50–50 chance of being white or black and that the outcome for one of a pair of twins is independent of the outcome for the other. What is the probability that a woman in a mixed marriage who is pregnant with twins will:

a. Have twins, both black?   **b.** Have twins, one white and one black?

c. Have twins, both white?

3.13 The corporations in the highly competitive razor blade industry do a tremendous amount of advertising each year. Corporation G gave a supply of the three top name brands, G, S, and W, to a consumer and asked him to use them and rank them in order of preference. The corporation was, of course, hoping the consumer would prefer its brand and rank it first, thereby giving them some material for a consumer interview advertising campaign. If the consumer did not prefer one blade more than any other, but was still required to rank the blades, what is the probability that:

a. The consumer ranked brand G first?

b. The consumer ranked brand G last?

c. The consumer ranked brand G last and brand W second?

d. The consumer ranked brand W first, brand G second, and brand S third?

3.14 An individual's genetic makeup is determined by the genes obtained from each parent. For every genetic trait, each parent possesses a gene pair; and each contributes one-half of this gene pair, with equal probability, to their offspring, forming a new gene pair. The offspring's traits (eye color, baldness, etc.) come from this new gene pair, where each gene in this pair possesses some characteristic.

For the gene pair that determines eye color, each gene trait may be one of two types: dominant brown (*B*) or recessive blue (*b*). A person possessing the gene pair *BB* or *Bb* has brown eyes, while the gene pair *bb* produces blue eyes.

a. Suppose that both parents of an individual are brown-eyed, each with a gene pair of the type *Bb*. What is the probability that a randomly selected child of this couple will have blue eyes? [*Hint:* Construct the sample space for the experiment.]

b. If one parent has brown eyes, type *Bb*, and the other has blue eyes, what is the probability that a randomly selected child of this couple will have blue eyes?

c. Suppose that one parent is brown-eyed, type *BB*. What is the probability that a child has blue eyes?

3.15 Highway Traffic Safety Administration officials estimate that used car buyers are losing $2 billion a year because of falsified odometer readings (*Orlando Sentinel,* April 13, 1984)—a cost that, on the average, adds $750 to the price of a used car. Although it is against the law to falsify these readings, they estimate that nine out of ten heavily used cars, those owned by leasing companies, etc., have had the odometer mileage reduced. Suppose you are looking for two used cars and the dealer has four cars in stock that were formerly owned by an auto leasing company. If two of the four cars have had their odometers set back and you purchase two cars from among these four, list the simple events that can occur. Assign reasonable probabilities to the simple events and then find the probability that:

a. You select the two leasing cars that have had their odometer mileage set back.

b. You select the two with untampered odometers.

c. Exactly one of the two cars has had its odometer mileage set back.

3.16 Three people play a game called "Odd Man Out." In this game, each player flips a fair coin until the outcome (heads or tails) for one of the players is not the same as the other two players'. This player is then the odd man out and loses the game. Find the probability that the game ends (i.e., either exactly one of the coins will fall heads or exactly one of the coins will fall tails) after only one toss by each player. Suppose that one of the players, hoping to reduce his chances of being the odd man, uses a two-headed coin. Will this ploy be successful? Solve by listing the simple events in the sample space.

3.17 The breakdown of workers in a particular state according to their political affiliation and type of job held is as follows:

		POLITICAL AFFILIATION		
		Republican	Democrat	Independent
TYPE OF JOB	White-collar	12%	12%	6%
	Blue-collar	23%	43%	4%

Suppose a worker is selected at random within the state and the worker's political affiliation and type of job are noted.

a. List all simple events for this experiment.
b. What is the set of all simple events called?
c. Let A be the event that the worker is a white-collar worker. Find $P(A)$.
d. Let B be the event that the worker is a Republican. Find $P(B)$.
e. Let C be the event that the worker is a Democrat. Find $P(C)$.
f. Let D be the event that the worker is a white-collar worker and a Democrat. Find $P(D)$.

3.18 You are a lawyer for a client who has committed a felony, and there are seven judges who could hear your motion to set bail. Four judges are strict and the other three are lenient. As you walk into the courtroom, Judge A (a strict judge) is leaving to go home.

a. What is the probability of drawing a lenient judge for your client?
b. What is the probability of drawing a lenient judge for your client if the probability of getting Judge B (a strict judge) is .3 and the probability of getting Judge C (a strict judge) is .4 and the probabilities of getting any of the other four judges are equally likely?

3.19 According to David Dreman (*Forbes*, October 27, 1980, pp. 202–203) investment in new issues (the stock of newly formed companies) can be both suicidal and rewarding. Dreman based his comments on a Securities and Exchange Commission (SEC) study of 500 new issues that went public during the 1961–1962 stock boom. The SEC found that of the 500 companies, 43% went bankrupt, 25% were operating at losses, and only 20% showed a profit. Only 12 companies of the 500 appeared to have outstanding prospects. Suppose that back in 1961 you had randomly selected two of five new issues for investment and that, unknown to you, only two of the five would eventually show a profit. What is the probability that:

a. Both of the new issues in which you invested will eventually show a profit?
b. Neither of the two issues will eventually show a profit?
c. At least one of the two issues you selected will eventually show a profit?

3.20 High school students who participate in varsity track may enter at most three events during a meet. Suppose a star senior at a certain school can compete equally well in the following five events: high jump, triple jump, long jump, 220-yard dash, and 100-yard dash.

a. List the different ways this track star can choose three events in which to participate at the upcoming track meet.

b. Unknown to the senior, an opposing team will set state track records in the 100-yard dash and the triple jump. If the three events in which to participate are selected at random, what is the probability that the track star will compete in both events in which a record will be set?

3.21 Before placing a person in a highly skilled position, a company gives the applicants a series of three examinations. The first is a physical examination, and each applicant is classified as satisfactory or unsatisfactory. The other two are verbal and quantitative examinations, and the scores are used to classify each applicant as high, medium, or low in each area. Thus, each individual will receive a health score, a verbal score, and a quantitative score.

a. List the different sets of classifications that can result from this battery of examinations.

b. If all applicants who take the examinations are equally qualified, and all the variation in test scores is caused by random variation due to the types of test questions, what is the probability that an applicant receives the lowest classification on all three examinations?

c. If an applicant scores in the highest category on at least two of the three examinations, the applicant will get a position. What is the probability that a randomly selected applicant will get a position?

3.2 Compound Events

An event can often be viewed as a composition of two or more other events. Such events are called **compound events;** they can be formed (composed) in two ways.

Definition 3.5

The **union** of two events A and B is the event that occurs if either A or B or both occur on a single performance of the experiment. We will denote the union of events A and B by the symbol $A \cup B$.

Definition 3.6

The **intersection** of two events A and B is the event that occurs if both A and B occur on a single performance of the experiment. We will write AB for the intersection of events A and B.

Example 3.5 Consider the die toss experiment. Define the following events:

    A: {Toss an even number}

    B: {Toss a number less than or equal to 3}

a. Describe $A \cup B$ for this experiment.
b. Describe AB for this experiment.
c. Calculate $P(A \cup B)$ and $P(AB)$ assuming the die is fair.

Solution Draw the Venn diagram as shown in the margin.

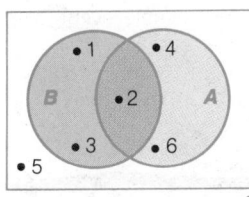

a. The union of *A* and *B* is the event that occurs if we observe either an even number, a number less than or equal to 3, or both on a single throw of the die. Consequently, the simple events in the event $A \cup B$ are those for which *A* occurs, *B* occurs, or both *A* and *B* occur. Testing the simple events in the entire sample space, we find that the collection of simple events in the union of *A* and *B* is

$$A \cup B = \{1, 2, 3, 4, 6\}$$

b. The intersection of *A* and *B* is the event that occurs if we observe *both* an even number and a number less than or equal to 3 on a single throw of the die. Testing the simple events to see which imply the occurrence of *both* events *A* and *B*, we see that the intersection contains only one simple event:

$$AB = \{2\}$$

In other words, the intersection of *A* and *B* is the simple event Observe a 2.

c. Recalling that the probability of an event is the sum of the probabilities of the simple events of which the event is composed, we have

$$P(A \cup B) = P(1) + P(2) + P(3) + P(4) + P(6)$$
$$= \tfrac{1}{6} + \tfrac{1}{6} + \tfrac{1}{6} + \tfrac{1}{6} + \tfrac{1}{6} = \tfrac{5}{6}$$

and

$$P(AB) = P(2) = \tfrac{1}{6}$$
                                                         ■

Unions and intersections can be defined for more than two events. For example, the event $A \cup B \cup C$ represents the union of three events, *A*, *B*, and *C*. This event, which includes the set of simple events in *A*, *B*, or *C*, will occur if any one or more of the events *A*, *B*, or *C* occurs. Similarly, the intersection *ABC* is the event that all three of the events *A*, *B*, and *C* occur. Therefore, *ABC* is the set of simple events that are in all three of the events *A*, *B*, and *C*.

Example 3.6 Refer to Example 3.5 and define the event

    C: {Toss a number greater than 1}

Find the simple events in:

a. $A \cup B \cup C$ **b.** *ABC*

where

> A: {Toss an even number}
>
> B: {Toss a number less than or equal to 3}

Solution   **a.** Event C contains the simple events corresponding to tossing a 2, 3, 4, 5, or 6, and event B contains the simple events 1, 2, and 3. Therefore, the event that either A, B, or C occurs contains all six simple events in S, that is, those corresponding to tossing a 1, 2, 3, 4, 5, or 6.

b. You can see that you will observe all of the events A, B, and C only if you observe a 2. Therefore, the intersection ABC contains the single simple event Toss a 2. ∎

3.3 Complementary Events

A very useful concept in the calculation of event probabilities is the notion of **complementary events:**

Definition 3.7

The **complement** of an event A is the event that A does not occur, i.e., the event consisting of all simple events that are not in event A. We will denote the complement of A by A'.

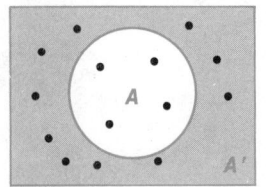

An event A is a collection of simple events, and the simple events included in A' are those that are not in A. Figure 3.4 demonstrates this idea. Note from the figure that all simple events in S are included in *either* A or A', and that *no* simple event is in both A and A'. This leads us to conclude that the probabilities of an event and its complement *must sum to 1:*

Figure 3.4   Venn Diagram of Complementary Events

The sum of the probabilities of complementary events equals 1, i.e.,

$$P(A) + P(A') = 1$$

In many probability problems it is easier to calculate the probability of the complement of the event of interest rather than the event itself. Then, since

$$P(A) + P(A') = 1$$

we can calculate P(A) by using the relationship

$$P(A) = 1 - P(A')$$

Example 3.7   Consider the experiment of tossing two fair coins. Calculate the probability of event A: {Observing at least one head} by using the complementary relationship.

Solution   We know that event A: {Observing at least one head} consists of the simple events

A: {HH, HT, TH}

The complement of A is defined as the event that occurs when A does not occur. Therefore,

A': {Observe no heads} = {TT}

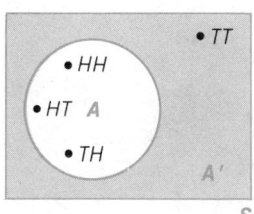

Figure 3.5   Complementary Events in the Toss of Two Coins

This complementary relationship is shown in Figure 3.5. Assuming the coins are balanced,

$$P(A') = P(TT) = \tfrac{1}{4}$$

and

$$P(A) = 1 - P(A') = 1 - \tfrac{1}{4} = \tfrac{3}{4}$$

■

Example 3.8   A fair coin is tossed ten times and the up face is recorded after each toss. What is the probability of event A: {Observe at least one head}?

Solution   We will solve this problem by following the five steps for calculating probabilities of events (see Section 3.1):

Step 1. Define the experiment. The experiment is to record the results of the ten tosses of the coin.

Step 2. List the simple events. A simple event consists of a particular sequence of ten heads and tails. Thus, one simple event is

HHTTTHTHTT

which denotes head on first toss, head on second toss, tail on third toss, etc. Others would be *HTHHHTTTTT* and *THHTHTHTTH*. There is obviously a very large number of simple events—too many to list. It can be shown (see Section 3.8) that there are $2^{10} = 1{,}024$ simple events for this experiment.

Step 3. Assign probabilities. Since the coin is fair, each sequence of heads and tails has the same chance of occurring and therefore all the simple events are equally likely. Then

$$P(\text{Each simple event}) = \frac{1}{1{,}024}$$

Step 4. Determine the simple events in event A. A simple event is in A if at least one H appears in the sequence of ten tosses. However, if we consider the complement of A, we find that

A': {No heads are observed in ten tosses}

Thus, A' contains only one simple event:

A': $\{TTTTTTTTTT\}$ and $P(A') = \dfrac{1}{1,024}$

Step 5. Since we know the probability of the complement of A, we use the relationship for complementary events:

$$P(A) = 1 - P(A') = 1 - \dfrac{1}{1,024} = \dfrac{1,023}{1,024} = .999$$

That is, we are virtually certain of observing at least one head in ten tosses of the coin.

■

**Exercises
3.22–3.30**

Learning the Mechanics

3.22 A fair coin is tossed three times and the events A and B are defined as follows:

A: {At least one head is observed}

B: {The number of heads observed is odd}

a. Identify the simple events in the events A, B, $A \cup B$, A', and AB.
b. Find $P(A)$, $P(B)$, $P(A \cup B)$, $P(A')$, and $P(AB)$ by summing the probabilities of the appropriate simple events.

3.23 A pair of fair dice is tossed. Define the following events:

A: {You will roll a 7}   (i.e., the sum of the numbers of dots on the upper faces of the two dice is equal to 7)

B: {At least one of the two dice is showing a 4}

a. Identify the simple events in the events A, B, AB, $A \cup B$, and A'.
b. Find $P(A)$, $P(B)$, $P(AB)$, $P(A \cup B)$, and $P(A')$ by summing the probabilities of the appropriate simple events.

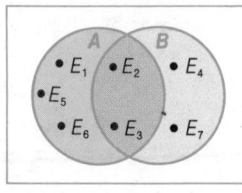

3.24 Consider the Venn diagram, where $P(E_1) = P(E_2) = P(E_3) = \frac{1}{5}$, $P(E_4) = P(E_5) = \frac{1}{20}$, $P(E_6) = \frac{1}{10}$, and $P(E_7) = \frac{1}{5}$. Find each of the following probabilities.

a. $P(A)$ **b.** $P(B)$ **c.** $P(A \cup B)$ **d.** $P(AB)$
e. $P(A')$ **f.** $P(B')$ **g.** $P(A \cup A')$ **h.** $P(A'B)$

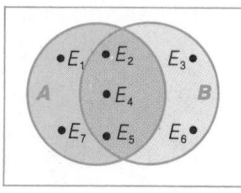

3.25 Consider the Venn diagram, where $P(E_1) = .13$, $P(E_2) = .05$, $P(E_3) = P(E_4) = .2$, $P(E_5) = .06$, $P(E_6) = .3$, and $P(E_7) = .06$. Find each of the following probabilities:

a. $P(A')$ **b.** $P(B')$ **c.** $P(A'B)$

d. $P(A \cup B)$ **e.** $P(AB)$ **f.** $P(A' \cup B')$

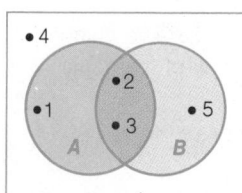

3.26 The Venn diagram portrays an experiment that contains five simple events. Two events, A and B, are also shown. The probabilities of the simple events are: $P(1) = \frac{1}{10}$, $P(2) = \frac{2}{10}$, $P(3) = \frac{3}{10}$, $P(4) = \frac{1}{10}$, $P(5) = \frac{3}{10}$. Find $P(A')$, $P(B')$, $P(A'B)$, $P(AB)$, $P(A \cup B)$, $P(A \cup B')$ $P[(AB)']$, and $P[(A \cup B)']$.

Applying the Concepts

3.27 A buyer for a large metropolitan department store must choose two firms from the four available to supply the store's fall line of men's slacks. The buyer has not dealt with any of the four firms before and considers their products equally attractive. Unknown to the buyer, two of the four firms are having serious financial problems that may result in their not being able to deliver the fall line of slacks as soon as promised. The four firms are identified as G_1 and G_2 (firms in good financial condition) and P_1 and P_2 (firms in poor financial condition). Simple events identify the pairs of firms selected. If the probability of the buyer selecting a particular firm from among the four is the same for each firm, the simple events and their probabilities for this buying experiment are those listed in the table. We will define the following events:

A: {At least one of the selected firms is in good financial condition}

B: {Firm P_1 is selected}

a. Define the event AB as a specific collection of simple events.
b. Define the event $A \cup B$ as a specific collection of simple events.
c. Define the event A' as a specific collection of simple events.
d. Find $P(A)$, $P(B)$, $P(AB)$, $P(A \cup B)$, and $P(A')$ by summing the probabilities of the appropriate simple events.

SIMPLE EVENTS	PROBABILITIES
$G_1 G_2$	$\frac{1}{6}$
$G_1 P_1$	$\frac{1}{6}$
$G_1 P_2$	$\frac{1}{6}$
$G_2 P_1$	$\frac{1}{6}$
$G_2 P_2$	$\frac{1}{6}$
$P_1 P_2$	$\frac{1}{6}$

3.28 A state energy agency mailed questionnaires on energy conservation to 1,000 homeowners in the state capital. Five hundred questionnaires were returned. Suppose that an experiment consists of randomly selecting one of the returned questionnaires. Consider the events:

A: {The home is constructed of brick}

B: {The home is more than 30 years old}

C: {The home is heated with oil}

Describe each of the following events in terms of unions, intersections, and complements ($A \cup B$, AB, A', etc.):

a. The home is more than 30 years old and is heated with oil.

b. The home is not constructed of brick.

c. The home is heated with oil or is more than 30 years old.

d. The home is constructed of brick and is not heated with oil.

3.29 One game that is very popular in many American casinos is roulette. Roulette is played by spinning a ball on a circular wheel that has been divided into thirty-eight arcs of equal length; these bear the numbers 00, 0, 1, 2, . . . , 35, 36. The number of the arc on which the ball comes to rest is the outcome of one play of the game. The numbers are also colored in the following manner:

Red: 1, 3, 5, 7, 9, 12, 14, 16, 18, 19, 21, 23, 25, 27, 30, 32, 34, 36

Black: 2, 4, 6, 8, 10, 11, 13, 15, 17, 20, 22, 24, 26, 28, 29, 31, 33, 35

Green: 00, 0

Players may place bets on the table in a variety of ways, including bets on odd, even, red, black, high, low, etc. Define the following events:

A: {Outcome is an odd number}   (00 and 0 are not considered odd or even)

B: {Outcome is a black number}

C: {Outcome is a low number (1–18)}

a. Define the event AB as a specific set of simple events.

b. Define the event $A \cup B$ as a specific set of simple events.

c. Find $P(A)$, $P(B)$, $P(AB)$, $P(A \cup B)$, and $P(C)$ by summing the probabilities of the appropriate simple events.

d. Define the event ABC as a specific set of simple events.

e. Find $P(ABC)$ by summing the probabilities of the simple events given in part d.

f. Define the event $(A \cup B \cup C)$ as a specific set of simple events.

g. Find $P(A \cup B \cup C)$ by summing the probabilities of the simple events given in part f.

3.30 Identifying managerial prospects who are both talented and motivated is difficult. A personnel manager constructed the following two-way table to define nine combinations of talent–motivation levels. The number in a cell is the manager's estimate of the probability that a managerial prospect will fall in that category.

		TALENT		
		High	*Medium*	*Low*
	High	.05	.16	.05
MOTIVATION	*Medium*	.19	.32	.05
	Low	.11	.05	.02

Suppose that the personnel manager has decided to hire a new manager. Define the following events:

A: {Prospect places in high motivation category}

B: {Prospect places in high talent category}

C: {Prospect is average or better in both categories}

D: {Prospect rates poor in at least one category}

E: {Prospect places highest in both categories}

a. Does the sum of the cell probabilities equal 1?

b. List the simple events in each of the events described above and find their probabilities.

c. Find $P(A \cup B)$, $P(AB)$, and $P(A \cup C)$.

d. Find $P(A')$ and explain what this means from a practical point of view.

3.4 Conditional Probability

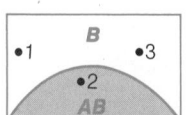

Figure 3.6 Reduced Sample Space for the Die Toss Experiment —Given That Event *B* Has Occurred

The event probabilities we have been discussing give the relative frequencies of the occurrences of the events when the experiment is repeated a very large number of times. They are called **unconditional probabilities** because no special conditions are assumed other than those that define the experiment.

Sometimes we may wish to alter the probability of an event when we have additional knowledge that might affect its outcome. This probability is called the **conditional probability** of the event. For example, we have shown that the probability of observing an even number (event *A*) on a toss of a fair die is $\frac{1}{2}$. However, suppose you are given the information that on a particular throw of the die the result was a number less than or equal to 3 (event *B*). Would you still believe that the probability of observing an even number on that throw of the die is equal to $\frac{1}{2}$? If you reason that making the assumption that *B* has occurred reduces the sample space from six simple events to three simple events (namely, those contained in event *B*), the reduced sample space is as shown in Figure 3.6.

Since the only even number of the three numbers in the reduced sample space *B* is the number 2 and since the die is fair, we conclude that the probability that *A* occurs *given that B occurs* is 1 in 3, or $\frac{1}{3}$. We will use the symbol $P(A|B)$ to represent the probability of event *A* given that event *B* occurs. For the die toss example

$$P(A|B) = \tfrac{1}{3}$$

To get the probability of event *A* given that event *B* occurs, we proceed as follows: We divide the probability of the part of *A* that falls within the reduced sample space *B*, namely $P(AB)$, by the total probability of the reduced sample space, namely $P(B)$. Thus, for the die toss example where event *A*: {Observe an even number} and event *B*: {Observe a number less than or equal to 3}, we find

$$P(A|B) = \frac{P(AB)}{P(B)} = \frac{P(2)}{P(1) + P(2) + P(3)} = \frac{\frac{1}{6}}{\frac{3}{6}} = \frac{1}{3}$$

This formula for $P(A|B)$ is true in general:

> To find the *conditional probability that event A occurs given that event B occurs*, divide the probability that *both A and B* occur by the probability that *B* occurs, that is,
>
> $$P(A|B) = \frac{P(AB)}{P(B)} \qquad \text{(We assume that } P(B) \neq 0.\text{)}$$

Example 3.9

SIMPLE EVENTS	PROBABILITIES
AC	.15
AC'	.25
A'C	.10
A'C'	.50

Many medical researchers have conducted experiments to examine the relationship between cigarette smoking and cancer. Let A represent the event that an individual smokes and let C represent the event that an individual develops cancer. Thus, AC is the simple event that an individual smokes and develops cancer, AC' is the simple event that an individual smokes and does not develop cancer, etc. Assume that the probabilities associated with the four simple events are as shown in the table for a certain section of the United States. How can these simple event probabilities be used to examine the relationship between smoking and cancer?

Solution

One method of determining whether these probabilities indicate that smoking and cancer are related is to compare the conditional probability that an individual acquires cancer given that he or she smokes with the conditional probability that an individual acquires cancer given that he or she does not smoke. That is, we will compare $P(C|A)$ with $P(C|A')$. The calculations are as follows:

$$P(C|A) = \frac{P(AC)}{P(A)}$$

where the event A, a person smokes, contains two simple events: AC (a person smokes and develops cancer) and AC' (a person smokes and does not develop cancer). Remembering that the probability of an event is the sum of the probabilities of its simple events, we obtain

$$P(A) = P(AC) + P(AC')$$
$$= .15 + .25 = .40$$

Then

$$P(C|A) = \frac{P(AC)}{P(A)} = \frac{.15}{.40} = .375$$

The conditional probability that a nonsmoking individual develops cancer is calculated in a similar manner:

$$P(C|A') = \frac{P(A'C)}{P(A')} = \frac{.10}{.60} = .167$$

where

$$P(A') = P(A'C) + P(A'C') = .10 + .50 = .60$$

or

$$P(A') = 1 - P(A) = 1 - .40 = .60$$

Either way we get the same answer.

Comparing the two conditional probabilities, we see that the probability that a smoker develops cancer (.375) is more than twice the probability that a nonsmoker develops cancer (.167). This does not imply that smoking *causes* cancer, but it does suggest a pronounced link between smoking and cancer. ■

Example 3.10   The investigation of consumer product complaints by the Federal Trade Commission (FTC) has generated much interest by manufacturers in the quality of their products. A manufacturer of an electromechanical kitchen aid conducted an analysis of a large number of consumer complaints and found that they fell into the six categories shown in Table 3.2. If a consumer complaint is received, what is the probability that the cause of the complaint is product appearance given that the complaint originated during the guarantee period?

Table 3.2
Distribution of Product
Complaints

	REASON FOR COMPLAINT			
	Electrical	*Mechanical*	*Appearance*	TOTALS
DURING GUARANTEE PERIOD	18%	13%	32%	63%
AFTER GUARANTEE PERIOD	12%	22%	3%	37%
TOTALS	30%	35%	35%	100%

Solution   Let *A* represent the event that the cause of a particular complaint is product appearance and let *B* represent the event that the complaint occurred during the guarantee period. Checking Table 3.2, you can see that $(18 + 13 + 32) = 63\%$ of the complaints occur during the guarantee period. Hence, $P(B) = .63$. The percentage of complaints that were caused by appearance and occurred during the guarantee period (the event *AB*) is 32%. Therefore, $P(AB) = .32$.

Using these probability values, we can calculate the conditional probability $P(A|B)$ that the cause of a complaint is appearance given that the complaint occurred during the guarantee time:

$$P(A|B) = \frac{P(AB)}{P(B)} = \frac{.32}{.63} = .51$$

Consequently, you can see that slightly more than half the complaints that occurred during the guarantee period were due to scratches, dents, or other imperfections in the surface of the kitchen devices.   ■

Case Study 3.2

Purchase Patterns and the Conditional Probability of Purchasing

In his doctoral dissertation, Alfred A. Kuehn (1958) examined sequential purchase data to gain some insight into consumer brand switching. He analyzed the frozen orange juice purchases of approximately 600 Chicago families during 1950–1952. The data were collected by the *Chicago Tribune* Consumer Panel. Kuehn was interested in determining the influence of a consumer's last four orange juice purchases on the next purchase. Thus, sequences of five purchases were analyzed.

Table 3.3 summarizes the data collected for Snow Crop brand orange juice and part of Kuehn's analysis of the data. In the column labeled "Previous Purchase Pattern" an S stands for the purchase of Snow Crop by a consumer and an O stands for the purchase of a brand other than Snow Crop. Thus, for example, SSSO is used to represent the purchase of Snow Crop three times in a row followed by the purchase of some other brand of frozen orange juice. The column labeled "Sample Size" lists the number of occurrences of the purchase sequences in the first column. The column labeled "Frequency" lists the number of times the associated purchase sequence in the first column led to the next purchase (i.e., the fifth purchase in the sequence) being Snow Crop.

The column labeled "Observed Approximate Conditional Probability of Purchase" contains the relative frequency with which each sequence of the first column led to the next purchase being Snow Crop. These relative frequencies, which give approximate conditional probabilities, are computed for each sequence of the first column by di-

Table 3.3

Observed Approximate Conditional Probability of Purchasing Snow Crop Given the Four Previous Brand Purchases

PREVIOUS PURCHASE PATTERN S = Snow Crop O = Other brand	SAMPLE SIZE	FREQUENCY	OBSERVED APPROXIMATE CONDITIONAL PROBABILITY OF PURCHASE P{Purchase|Previous purchase pattern}
SSSS	1,047	844	.806
OSSS	277	191	.690
SOSS	206	137	.665
SSOS	222	132	.595
SSSO	296	144	.486
OOSS	248	137	.552
SOOS	138	78	.565
OSOS	149	74	.497
SOSO	163	66	.405
OSSO	181	75	.414
SSOO	256	78	.305
OOOS	500	165	.330
OOSO	404	77	.191
OSOO	433	56	.129
SOOO	557	86	.154
OOOO	8,442	405	.048

viding the frequency of the sequence by the sample size of the sequence. For example, .806 is the approximate conditional probability that the next purchase will be Snow Crop given that the previous four purchases were also Snow Crop.

An examination of the approximate conditional probabilities in the fourth column indicates that both the most recent brand purchased and the number of times a brand is purchased have an effect on the next brand purchased. It appears that the influence on the next brand of orange juice purchased by the second most recent purchase is not as strong as the most recent purchase, but it is stronger than the third most recent purchase. In general, it appears that the probability of a particular consumer purchasing Snow Crop the next time he or she buys orange juice is inversely related to the number of consecutive purchases of another brand he or she made since last purchasing Snow Crop and is directly proportional to the number of Snow Crop purchases among the four purchases.

Kuehn conducts a more formal statistical analysis of these data, which we will not pursue here. We simply want you to see that probability is a basic tool for making inferences about populations using sample data.

Exercises 3.31–3.40

Learning the Mechanics

3.31 Given that $P(A) = .3$, $P(B) = .7$, and $P(AB) = .15$, find $P(A|B)$ and $P(B|A)$.

3.32 Two fair coins are tossed and the events A and B are defined as follows:

A: {At least one head appears}

B: {Exactly one head appears}

Find $P(A)$, $P(B)$, $P(AB)$, $P(A|B)$, and $P(B|A)$.

3.33 A sample space contains six simple events and events A, B, and C as shown in the Venn diagram. The probabilities of the simple events are $P(1) = .20$, $P(2) = .05$, $P(3) = .30$, $P(4) = .10$, $P(5) = .10$, $P(6) = .25$. Use the Venn diagram and the probabilities of the simple events to find:

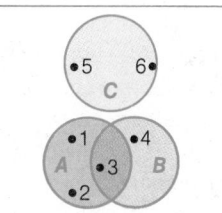

a. $P(A)$, $P(B)$, and $P(C)$

b. $P(AB)$, $P(AC)$, and $P(BC)$

c. $P(A|B)$, $P(B|C)$, $P(C|A)$, and $P(C|A')$

3.34 A box contains two white, two red, and two blue poker chips. Two chips are randomly chosen without replacement and their colors are noted. Define the following events:

A: {Both chips are of the same color}

B: {Both chips are red}

C: {At least one chip is red or white}

Find $P(B|A)$, $P(B|A')$, $P(B|C)$, $P(A|C)$, and $P(C|A')$.

Applying the Concepts

SIMPLE EVENTS	PROBABILITIES
SS	.09
SF	.21
FS	.21
FF	.49

3.35 A soap manufacturer has decided to market two new brands. An analysis of current market conditions and a review of the firm's past successes and failures with new brands have led the manufacturer to believe that the simple events and the probabilities of their occurrence in this marketing experiment are as shown in the table (where S means the brand succeeds and F means the brand fails in the first year). Define the following events:

A: {Both new brands are successful in the first year}

B: {At least one new brand is successful in the first year}

a. Find $P(A)$, $P(B)$, and $P(AB)$.   **b.** Find $P(A|B)$ and $P(B|A)$.

3.36 Six people apply for two identical positions in a company. Four are minority applicants and the remainder are nonminority. Define the following events:

A: {Both persons selected are nonminority candidates}

B: {Both persons selected are minority candidates}

C: {At least one of the persons selected is a minority candidate}

If all the applicants are equally qualified and the choice is essentially a random selection of two applicants from the six available, find:

a. $P(A)$ **b.** $P(B)$ **c.** $P(C)$ **d.** $P(B|C)$

e. Assume that the minority candidates are numbered 1, 2, 3, 4 for purposes of identification. Define the event

D: {Minority candidate 1 is selected}

Find $P(D|C)$.

3.37 An article in *Business Week* (September 12, 1983) reports on the problems that evolve from the failure to inform patients adequately of both the proper application of prescription drugs and the precautions to take in order to avoid potential side effects. This failure results in numerous cases of serious illness and, in some cases, even death. One study revealed that 300,000 U.S. hospital admissions each year are caused by adverse reactions to prescription drugs. Another study concluded that 7% of all hospital admissions are related to drug-induced problems resulting from imprudent prescriptions. One method of increasing patients' awareness of the problem is for physicians to provide Patient Medication Instruction (PMI) sheets. The American Medical Association, however, has found that only 20% of the doctors who prescribe drugs frequently distribute PMI sheets to their patients. Assume that 20% of all patients receive the PMI sheet with their prescriptions and that 12% receive the PMI sheet and are hospitalized because of a drug-related problem. What is the probability that a person will be hospitalized for a drug-related problem given that the person has received the PMI sheet?

3.38 A fast-food restaurant chain with 700 outlets in the United States describes the geographic location of its restaurants with the following table of percentages:

		REGION			
		NE	*SE*	*SW*	*NW*
	Under 10,000	5%	6%	3%	0%
POPULATION OF CITY	*10,000–100,000*	15%	15%	12%	5%
	Over 100,000	20%	4%	10%	5%

A restaurant is to be chosen at random from the 700 to test market a new style of chicken.

a. Given that the restaurant chosen is in a city with population over 100,000, what is the probability that it is located in the northeast?

b. Given that the restaurant chosen is in the southeast, what is the probability that it is located in a city with population under 10,000?

c. If the restaurant selected is located in the southwest, what is the probability that the city it is in has a population of 100,000 or less?

d. If the restaurant selected is located in the northwest, what is the probability that the city it is in has a population of 10,000 or more?

3.39 There are several methods of typing, or classifying, human blood. The most common procedure types blood into the general classifications of A, B, O, or AB. A method that is not as well known examines phosphoglucomutase (PGM) and classifies the blood into one of three main categories, 1-1, 2-1, or 2-2. Suppose that a certain geographic region of the United States has the PGM percentages shown in the following table:

		1-1	2-1	2-2
RACE	*White*	46.3%	39.2%	4.0%
	Black	6.7%	3.4%	0.4%

A person is to be chosen at random from this region.

a. What is the probability that a black person is chosen?

b. Given that a black is chosen, what is the probability he or she is PGM type 1-1?

c. Given that a white is chosen, what is the probability he or she is PGM type 1-1?

3.40 The following table of percentages describes the 1,000 apartment units in a large suburban apartment complex:

		APARTMENT SIZE		
		One bedroom	Two bedroom	Three bedroom
LOCATION WITHIN BUILDING	*First floor*	10%	30%	8%
	Second floor	18%	20%	14%

The manager of the complex is considering installing new carpets in all the apartments. Before doing so, he wants to wear-test the brand in which he is interested for 6 months in one of the 1,000 apartments. He plans to choose one apartment at random from the 1,000 and install a test carpet.

a. What is the probability that he will choose a first-floor, two-bedroom apartment?

b. Given that he chooses a second-floor apartment, what is the probability that the apartment has three bedrooms?

c. Given that he chooses a one-bedroom apartment, what is the probability that the apartment is on the second floor?

d. Given that the apartment selected is on the first floor, what is the probability that it has two or three bedrooms?

3.5 Probabilities of Unions and Intersections

Since unions and intersections of events are themselves events, we can always calculate their probabilities by adding the probabilities of the simple events that compose them. However, when the probabilities of certain events are known, it is easier to use two rules—the additive rule and the multiplicative rule—to calculate the probability of unions and intersections. How and why these rules work will be illustrated by example.

Example 3.11

A loaded (unbalanced) die is tossed and the up face is observed. The following two events are defined:

A: {Observe an even number}

B: {Observe a number less than 3}

Suppose that $P(A) = .4$, $P(B) = .2$, and $P(AB) = .1$. Find $P(A \cup B)$. [*Note:*   Assuming that we would know these probabilities in a practical situation is not very realistic, but the example will illustrate a point.]

Solution

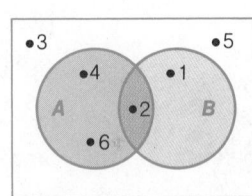

Figure 3.7   Venn Diagram for Die Toss

By studying the Venn diagram in Figure 3.7, we can obtain information that will help us find $P(A \cup B)$. We can see that

$$P(A \cup B) = P(1) + P(2) + P(4) + P(6)$$

Also, we know that

$$P(A) = P(2) + P(4) + P(6) = 4$$
$$P(B) = P(1) + P(2) = .2$$
$$P(AB) = P(2) = 1$$

If we add the probabilities of the simple events that comprise events A and B, we find

$$P(A) + P(B) = \overbrace{P(2) + P(4) + P(6)}^{P(A)} + \overbrace{P(1) + P(2)}^{P(B)}$$

$$= \overbrace{P(1) + P(2) + P(4) + P(6)}^{P(A \cup B)} + \overbrace{P(2)}^{P(AB)}$$

Thus, by subtraction,

$$P(A \cup B) = P(A) + P(B) - P(AB)$$
$$= .4 + .2 - .1 = .5$$

∎

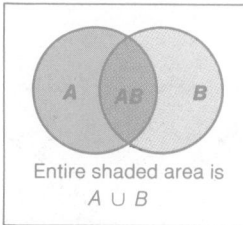

Entire shaded area is
$A \cup B$

S

Figure 3.8 Venn
Diagram of Union

By studying the Venn diagram in Figure 3.8, you can see that the method used in Example 3.11 may be generalized to find the union of two events for any experiment. The probability of the union of two events, A and B, can always be obtained by summing $P(A)$ and $P(B)$ and subtracting $P(AB)$. Note that we must subtract $P(AB)$ because the simple event probabilities in AB have been included twice—once in $P(A)$ and once in $P(B)$.

The formula for calculating the probability of the union of two events, often called the **additive rule of probability,** is given here:

Additive Rule of Probability

The probability of the union of events A and B is the sum of the probability of events A and B minus the probability of the intersection of events A and B, i.e.,

$$P(A \cup B) = P(A) + P(B) - P(AB)$$

Example 3.12

Hospital records show that 12% of all patients are admitted for surgical treatment, 16% are admitted for obstetrics, and 2% receive both obstetrics and surgical treatment. If a new patient is admitted to the hospital, what is the probability that the patient will be admitted either for surgery, obstetrics, or both?

Solution

Consider the following events:

A: {A patient admitted to the hospital receives surgical treatment}

B: {A patient admitted to the hospital receives obstetrics treatment}

Then, from the given information,

$$P(A) = .12 \qquad P(B) = .16$$

and the probability of the event that a patient receives both obstetrics and surgical treatments is

$$P(AB) = .02$$

The event that a patient admitted to the hospital receives either surgical treatment, obstetrics treatment, or both is the union $A \cup B$. The probability of $A \cup B$ is given by the additive rule of probability:

$$P(A \cup B) = P(A) + P(B) - P(AB)$$
$$= .12 + .16 - .02 = .26$$

Thus, 26% of all patients admitted to the hospital receive either surgical treatment, obstetrics treatment, or both. ■

A very special relationship exists between the events A and B when AB contains no simple events. In this case, we call the events A and B **mutually exclusive** events:

Definition 3.8

Events A and B are **mutually exclusive** if AB contains no simple events.

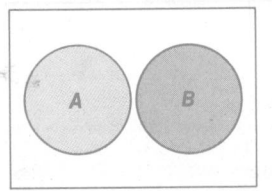

Figure 3.9 Venn Diagram of Mutually Exclusive Events

Figure 3.9 shows a Venn diagram of two mutually exclusive events. The events A and B have no simple events in common, i.e., A and B cannot occur simultaneously, and $P(AB) = 0$. Thus, we have the following important relationship:

If two events A and B are mutually exclusive, the probability of the union of A and B equals the sum of the probabilities of A and B; that is,

$$P(A \cup B) = P(A) + P(B)$$

Example 3.13

Consider the experiment of tossing two balanced coins. Find the probability of observing *at least* one head.

Solution

Define the events

  A: {Observe at least one head}
  B: {Observe exactly one head}
  C: {Observe exactly two heads}

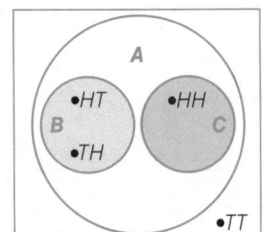

Figure 3.10 Venn Diagram for Coin Toss Experiment

Note that

$$A = B \cup C$$

and that BC contains no simple events (see Figure 3.10). Thus, B and C are mutually exclusive, so that

$$P(A) = P(B \cup C) = P(B) + P(C)$$
$$= \tfrac{1}{2} + \tfrac{1}{4} = \tfrac{3}{4}$$ ■

Although Example 3.13 is very simple, the concept of writing events with verbal descriptions that include the phrases "at least" or "at most" as unions of mutually exclusive events is a very useful one. This enables us to find the probability of the event by adding the probabilities of the mutually exclusive events.

The second rule of probability, which will help us find the probability of the intersection of two events, is illustrated by Example 3.14.

Example 3.14

An agriculturist, who is interested in planting wheat next year, is concerned with the following events:

A: {The production of wheat will be profitable}

B: {A serious drought will occur}

Based on available information, the agriculturist believes that the probability is .01 that production of wheat will be profitable *assuming* a serious drought will occur in the same year, and that the probability is .05 that a serious drought will occur. That is,

$$P(A|B) = .01$$
$$P(B) = .05$$

Based on the information provided, what is the probability that a serious drought will occur *and* that a profit will be made? That is, find $P(AB)$.

Solution

As you will see, we have already developed a formula for finding the probability of an intersection of two events. Recall that the conditional probability of A given B is

$$P(A|B) = \frac{P(AB)}{P(B)}$$

Multiplying both sides of this equation by $P(B)$, we obtain a formula for the probability of the intersection of events A and B. This is often called the ***multiplicative rule of probability*** and is given by

$$P(AB) = P(A|B)P(B)$$

Thus,

$$P(AB) = (.01)(.05) = .0005$$

The probability that a serious drought occurs *and* the production of wheat is profitable is only .0005. As we might expect, this intersection is a very rare event. ∎

Multiplicative Rule of Probability

$$P(AB) = P(A|B)P(B) = P(B|A)P(A)$$

Intersections often contain only a few simple events. In this case, the probability of an intersection is easy to calculate by summing the appropriate simple event probabilities. However, the formula for calculating intersection probabilities plays a very important role, particularly in an area of statistics known as ***Bayesian statistics.*** (More detailed discussions of Bayesian statistics are contained in the references at the end of the chapter.)

Example 3.15   Consider the experiment of tossing a fair coin twice and recording the up face on each toss. The following events are defined:

  A: {First toss is a head}

  B: {Second toss is a head}

Does *knowing* that event A has occurred affect the probability that B will occur?

Solution   Intuitively the answer should be no, since what occurs on the first toss should in no way affect what occurs on the second toss. Let us check our intuition. Recall the sample space for this experiment:

1. Observe *HH*   **3.** Observe *TH*
2. Observe *HT*   **4.** Observe *TT*

Each of these simple events has a probability of $\frac{1}{4}$. Thus,

$$P(B) = P(HH) + P(TH) \quad \text{and} \quad P(A) = P(HH) + P(HT)$$
$$= \tfrac{1}{4} + \tfrac{1}{4} = \tfrac{1}{2} \qquad\qquad\qquad = \tfrac{1}{4} + \tfrac{1}{4} = \tfrac{1}{2}$$

Now what is $P(B|A)$?

$$P(B|A) = \frac{P(AB)}{P(A)} = \frac{P(HH)}{P(A)}$$
$$= \frac{\frac{1}{4}}{\frac{1}{2}} = \frac{1}{2}$$

We can now see that $P(B) = \frac{1}{2}$ and $P(B|A) = \frac{1}{2}$. Knowing that the first toss resulted in a head does not affect the probability that the second toss will be a head. The probability is $\frac{1}{2}$ whether or not we know the result of the first toss. When this occurs, we say that the two events A and B are **independent.** ∎

Definition 3.9

Events A and B are **independent** if the occurrence of B does not alter the probability that A has occurred, i.e., events A and B are independent if

  $P(A|B) = P(A)$

When events A and B are **independent** it is also true that

  $P(B|A) = P(B)$

Events that are not independent are said to be **dependent.**

Example 3.16   Consider the experiment of tossing a fair die and let

A: {Observe an even number}

B: {Observe a number less than or equal to 4}

Are events A and B independent?

Solution   The Venn diagram for this experiment is shown in Figure 3.11. We first calculate

$$P(A) = P(2) + P(4) + P(6) = \tfrac{1}{2}$$
$$P(B) = P(1) + P(2) + P(3) + P(4) = \tfrac{4}{6} = \tfrac{2}{3}$$
$$P(AB) = P(2) + P(4) = \tfrac{2}{6} = \tfrac{1}{3}$$

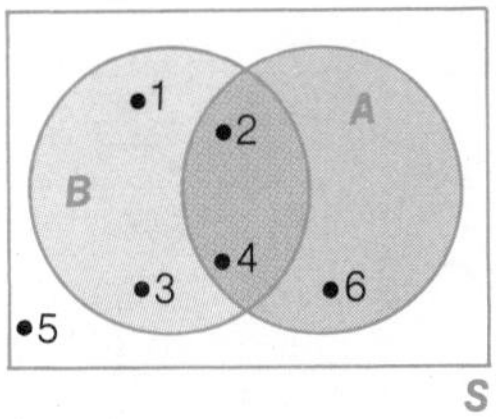

Now assuming that B has occurred, the conditional probability of A given B is

$$P(A|B) = \frac{P(AB)}{P(B)} = \frac{\tfrac{1}{3}}{\tfrac{2}{3}} = \frac{1}{2} = P(A)$$

Thus, assuming that event B occurs does not alter the probability of observing an even number—it remains $\tfrac{1}{2}$. Therefore, the events A and B are independent. Note that if we calculate the conditional probability of B given A, our conclusion is the same:

$$P(B|A) = \frac{P(AB)}{P(A)} = \frac{\tfrac{1}{3}}{\tfrac{1}{2}} = \frac{2}{3} = P(B)$$ ■

Figure 3.11   Venn Diagram for Die Toss Experiment

Example 3.17   Refer to the consumer product complaint study in Example 3.10. The percentages of complaints of various types during and after the guarantee period are shown in Table 3.2. Define the following events:

A: {Cause of complaint is product appearance}

B: {Complaint occurred during the guarantee term}

Are A and B independent events?

Solution   Events A and B are independent if $P(A|B) = P(A)$. We calculated $P(A|B)$ in Example 3.10 to be .51, and from Table 3.2 we can see that

$$P(A) = .32 + .03 = .35$$

Therefore, $P(A|B)$ is not equal to $P(A)$, and A and B are not independent events.   ■

We will make three final points about independence. The first is that the property of independence, unlike the mutually exclusive property, cannot be shown on or gleaned from a Venn diagram and you cannot trust your intuition. In general, the only way to check for independence is by performing the calculations of the probabilities in the definition.

The second point concerns the relationship between the mutually exclusive and independence properties. Suppose that events A and B are mutually exclusive, as shown in Figure 3.9. Are these events independent or dependent? That is, does the assumption that B occurs alter the probability of the occurrence of A? It certainly does, because if we assume that B has occurred, it is impossible for A to have occurred simultaneously. *Thus, mutually exclusive events are dependent events.*

The third point is that the probability of the intersection of independent events is very easy to calculate. Referring to the formula for calculating the probability of an intersection, we find

$$P(AB) = P(B)P(A|B)$$

Thus, since $P(A|B) = P(A)$ when A and B are independent, we have the following useful rule:

> **If events A and B are independent,** the probability of the intersection of A and B equals the product of the probabilities of A and B; that is,
>
> $$P(AB) = P(A)P(B)$$

In the die toss experiment, we showed in Example 3.16 that the events A: {Observe an even number} and B: {Observe a number less than or equal to 4} are independent if the die is fair. Thus,

$$P(AB) = P(A)P(B) = (\tfrac{1}{2})(\tfrac{2}{3}) = \tfrac{1}{3}$$

This agrees with the result

$$P(AB) = P(2) + P(4) = \tfrac{2}{6} = \tfrac{1}{3}$$

that we obtained in the example.

Example 3.18

In Example 3.4, a bank considered the problem of apportioning its service windows according to two types of accounts: personal or commercial. In the example, we discussed the problem of finding the probability that one, two, or, in general, k customers arriving at the bank possessed a commercial account. We are now ready to find the probability that both of two customers arriving at the bank possess commercial accounts. Suppose that a study of arriving customers shows that 20% of arriving customers intend to utilize their commercial accounts at the bank.

a. If two customers arrive at the bank, what is the probability that they will both utilize a commercial account?

b. If k customers arrive at the bank, what is the probability that all will utilize commercial accounts?

Solution

a. Let C_1 be the event that customer 1 will utilize a commercial account and let C_2 be a similar event for customer 2. The event that *both* customers will utilize commercial accounts is the intersection C_1C_2. Then, since it is not unreasonable to assume that the service requirements of the customers would be independent of one another, the probability that both will utilize commercial accounts is

$$P(C_1 C_2) = P(C_1)P(C_2)$$
$$= (.2)(.2) = (.2)^2 = .04$$

b. Let C_i represent the event that the ith customer will utilize a commercial account. Then the event that all three of three arriving customers will utilize a commercial account is the intersection of the event $C_1 C_2$ (from part a) with the event C_3. Assuming independence of the events, C_1, C_2, and C_3, we have

$$P(C_1 C_2 C_3) = P(C_1 C_2)P(C_3)$$
$$= (.2)^2(.2) = (.2)^3 = .008$$

Noting the pattern, you can see that the probability that all k out of k arriving customers will utilize a commercial account is the probability of $C_1 C_2 \cdots C_k$, or

$$P(C_1 C_2 \cdots C_k) = (.2)^k \qquad \text{for } k = 1, 2, 3, \ldots \qquad \blacksquare$$

Exercises 3.41–3.58

Learning the Mechanics

3.41 An experiment results in one of three mutually exclusive events, A, B, or C. It is known that $P(A) = .40$, $P(B) = .25$, and $P(C) = .35$.
a. Find $P(A \cup B)$.   **b.** Find $P(AC)$.
c. Find $P(A|B)$.   **d.** Find $P(B \cup C)$.
e. Are B and C independent events? Explain.

3.42 An experiment results in one of five simple events with the following probabilities: $P(E_1) = .22$, $P(E_2) = .31$, $P(E_3) = .15$, $P(E_4) = .22$, and $P(E_5) = .1$. The following events have been defined:

A: $\{E_1, E_3\}$
B: $\{E_2, E_3, E_4\}$
C: $\{E_1, E_5\}$

Find each of the following probabilities:
a. $P(A)$ **b.** $P(B)$ **c.** $P(AB)$
d. $P(A|B)$ **e.** $P(BC)$ **f.** $P(C|B)$

3.43 Three fair coins are tossed and the following events are defined:

A: {Observe at least one head}
B: {Observe exactly two heads}
C: {Observe exactly two tails}
D: {Observe at most one head}

a. Sum the probabilities of the appropriate simple events to find: $P(A)$, $P(B)$, $P(C)$, $P(D)$, $P(AB)$, $P(AD)$, $P(CB)$, and $P(BD)$.

b. Use the formulas of this section and your answers to part a to calculate: $P(A \cup B)$, $P(D \cup A)$, $P(B \cup C)$, $P(B|A)$, $P(A|D)$, and $P(C|B)$.

c. Are events A and B independent? Mutually exclusive?

d. Are events A and D independent? Mutually exclusive?

e. Are events B and C independent? Mutually exclusive?

3.44 Two fair dice are tossed and the following events are defined:

A: {Sum of the numbers showing is odd}

B: {Sum of the numbers showing is 9, 11, or 12}

a. Are events A and B independent? Why?

b. Are events A and B mutually exclusive? Why?

3.45 Two events, A and B, are mutually exclusive with $P(A) = .3$ and $P(B) = .4$.
a. Find $P(A \cup B)$.   **b.** Find $P(A|B)$.   **c.** Find $P(B|A)$.
d. Are A and B independent? Why?

3.46 For two events, A and B, $P(A) = .4$ and $P(B) = .2$.
a. If A and B are independent, find $P(AB)$, $P(A|B)$, and $P(A \cup B)$.
b. If A and B are dependent, with $P(A|B) = .6$, find $P(AB)$ and $P(B|A)$.

Applying the Concepts

3.47 Each of a random sample of fifty people was asked to name his or her favorite soft drink. The responses are shown below:

Pepsi-Cola	18	Sprite	4
Coca-Cola	16	Nehi Orange	1
Mr. Pibb	4	Dr Pepper	1
Seven-Up	6		

Suppose a person is selected at random from the survey. Let A be the event that the person preferred a soft drink bottled by the Coca-Cola Company (Coca-Cola, Mr. Pibb, or Sprite). Let B be the event that the person did *not* choose a cola (either Pepsi-Cola or Coca-Cola).
a. Find $P(A)$.   **b.** Find $P(B)$.   **c.** Find $P(A \cup B)$.
d. Find $P(AB)$.   **e.** Find $P(AB')$.   **f.** Find $P(A|B)$.
g. Are A and B independent events? Why?
h. Are A and B mutually exclusive events? Why?

3.48 The percentages of all the teenagers aged 14 to 19 in two small communities fell in the following six community–delinquency categories:

		COMMUNITY	
		A	B
	Nondelinquents	28%	42%
DELINQUENCY	First offenders	5%	15%
	Repeat offenders	7%	3%

Suppose a teenager was chosen at random from one of the two communities and the following events are defined:

A: {Teenager chosen is from community A}

B: {Teenager chosen is from community B}

C: {Teenager chosen is not delinquent}

D: {Teenager chosen has committed at least one crime}

Find the following probabilities:

a. $P(C')$ **b.** $P(A \cup D)$ **c.** $P(C|B)$ **d.** $P(D|A)$

3.49 The probability that a certain electronics component fails when first used is .10. If it does not fail immediately, the probability that it lasts for 1 year is .99. What is the probability that a new component will last 1 year?

3.50 Some strings of Christmas tree lights are wired in series; thus, if one bulb fails the entire string goes out. Suppose the probability of an individual bulb failing during a certain period of time is .05. What is the probability that a string of ten lights goes out during that period of time? What assumption did you make concerning the light bulbs? [*Hint:* The complement of the event, At least one of the ten lights fails, is the event, None of the ten lights fails.]

3.51 Refer to Exercise 3.15 and the problem of purchasing used automobiles that have had their odometer mileage set back. Suppose that you are shopping for a used car and the dealer has three good-looking cars in stock that were formerly owned by an auto leasing company. If, as stated by the Highway Traffic Safety Administration, nine out of ten of these cars have had their odometer mileage set back, what is the probability that:

a. All three cars have had their odometer mileage set back?

b. None have had their odometer mileage set back?

c. Exactly one has had its odometer mileage set back?

d. The car you select (from the three used cars available) has had its odometer mileage set back?

3.52 On April 25, 1980, the *Bakersfield Californian* reported on a record payout in the Pennsylvania lottery. The winning number in the state lottery that day, with options from 000 to 999, was 666. Harried state officials were denying the possibility of tampering with the lottery, but many bookies in Pittsburgh were refusing bets on any numbers involving 4's or 6's. Each digit of the winning number was determined by allowing a blast of air to propel one of ten ping pong balls, numbered 0, 1, 2, . . . , 9, into a basket. On this particular day, the TV host (eventually indicted and convicted) and some friends injected liquid into all the balls except those numbered 4 and 6. Thus, only numbers involving the 4 and the 6 digits could be winners.

a. If the balls had not been fixed, what is the probability that a number involving only 4's and 6's would win?

b. Given the conditions of part a, what is the probability that the number 666 would win?

c. Since the TV announcer and collaborators knew that the winning number would involve only 4's and 6's, what is the probability that the winning number would be 666?

3.53 A popular dice game, called "craps," is played in the following manner: A player starts by rolling two dice. If the result is a 7 or 11, the player wins. For any other sum appearing on the dice, the player continues to roll the dice until that outcome reoccurs (in which case the player wins) or until a 7 or 11 occurs (in which case the player loses). If on any roll the outcome is 2 (snake-eyes), the game is over and the player loses.
a. What is the probability that a player wins the game on the first roll of the dice? (Assume the dice are balanced.)
b. What is the probability that a player loses the game on the first roll of the dice?
c. If the player throws a total of 3 on the first roll, what is the probability that the game ends on the next roll?

3.54 Despite penicillin and other antibiotics, bacterial pneumonia still kills thousands of Americans every year. Just recently, the United States Food and Drug Administration (FDA) approved the use of a new antipneumonia vaccine called Pneumovax. It is designed especially for elderly or debilitated patients, who are usually the most vulnerable to bacterial pneumonia. Field trials proved the new vaccine to be 90% effective in stimulating the production of antibodies to pneumonia-producing bacteria (i.e., 90% successful in preventing a person exposed to pneumonia-producing bacteria from acquiring the disease). Suppose the probability of an elderly or debilitated person being exposed to these bacteria is .40 (whether inoculated or not), and after being exposed, the probability of each person contracting bacterial pneumonia if not inoculated with the vaccine is .95. Find the probability that an elderly or debilitated person inoculated with this new vaccine acquires pneumonia. What is the probability if this person has not been inoculated?

3.55 According to an article in the *Bakersfield Californian* (May 10, 1982), as many as 1 in every 500 blacks in the United States has sickle-cell anemia. One in 10 is a carrier of the trait and those who marry have a 1 in 4 chance of giving the disease to their children. Suppose that a carrier has three children and that the transmission of the disease from the carrier to any one child is independent of whether or not the carrier transmits to another. Find the probability that:
a. None of the children acquire the disease.
b. All three acquire the disease.
c. Exactly one acquires the disease.

3.56 One of the problems encountered in organ transplants is rejection by the body of the transplanted tissue, i.e., the tendency of the white blood cells to attack the foreign tissue. An article in *Newsweek* ("The New Era of Transplants," August 29, 1983) notes that the key to reception or rejection is the nature of the antigens attached to the tissue cells. If the antigens of the donor and receiver match, the body will accept the transplanted tissue. The article goes on to note that the antigens in twins always match. The probability of a match in siblings is .25 and that of a match in two people from the population at large is .001. Suppose that you need a kidney and you have two brothers and a sister.

a. If one of the three offers a kidney, what is the probability that the antigens will match?

b. If all three offer a kidney, what is the probability that all three will match?

c. If all three offer a kidney, what is the probability that none will match?

3.57 Refer to Exercise 3.56. Answer parts b and c but this time assume that the three donors were obtained from the population at large.

3.58 Refer to the kidney transplant exercise (Exercise 3.56). The *Newsweek* article (August 29, 1983) reports that Sandoz, a pharmaceutical firm, has developed a new drug, cyclosporine, that appears to retard a body's immune system from rejecting transplanted organs. Sandoz reports that kidney transplant patients who receive the drug have an 80% chance of living through the first year and it has also improved the survival rates for other types of transplants. Suppose a hospital performs four kidney transplants and all patients receive the drug cyclosporine. What is the probability that:

a. All four patients are alive at the end of 1 year?

b. None of the four patients is alive at the end of 1 year?

c. At least one of the patients is alive at the end of 1 year?

3.6 Probability and Statistics: An Example

We have introduced a number of new concepts in the preceding sections, and this makes the study of probability a particularly arduous task. It is therefore essential to establish clearly the connection between probability and statistics, which we will do in the remaining chapters. However, we present one brief example in this section so that you can begin to understand why some knowledge of probability is important in the study of statistics.

Suppose a psychologist is researching the hypothesis that rats which have been trained will pass on at least part of the training to their offspring. To test the hypothesis, three offspring (no two with the same parents) of trained rats are randomly selected and subjected to a training test. It is known from many previous experiments that the relative frequency distribution of the scores for untrained rats is mound-shaped with a mean of 60 and a standard deviation of 10. Suppose all three of the trained rats' offspring score more than 70 on the test. What can the research psychologist conclude?

The relative frequency distribution of the scores for untrained rats is shown in Figure 3.12 (page 116). If the distribution is mound-shaped and approximately symmetric about the mean, we can conclude that approximately 16% of untrained rats will score more than 70 on the test (see Table 2.7). Now define the events

A_1: {Offspring 1 scores more than 70}

A_2: {Offspring 2 scores more than 70}

A_3: {Offspring 3 scores more than 70}

We want to find $P(A_1 A_2 A_3)$, the probability that all three offspring score more than 70 on the training test.

Figure 3.12 Relative Frequency Distribution for Training Test Scores

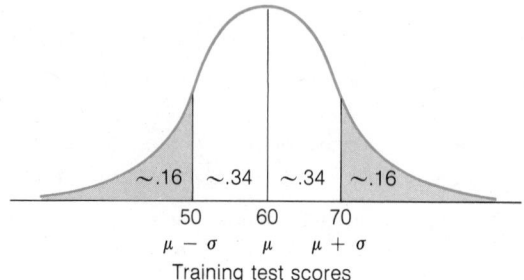

Since the offspring are selected so that they have different parents, it may be plausible to assume that the events A_1, A_2, and A_3 are independent. That is,

$$P(A_2|A_1) = P(A_2)$$

In words, knowing that the first offspring scores more than 70 on the test does not affect the probability that the second offspring scores more than 70. With the assumption of independence, we can calculate the probability of the intersection by multiplying the individual probabilities:

$$P(A_1A_2A_3) = P(A_1)P(A_2)P(A_3)$$
$$\approx (.16)(.16)(.16) = .004096$$

Thus, the probability that the research psychologist will observe all three offspring score more than 70 is only about .004 *if the offspring are untrained*. If this event were to occur, the psychologist might conclude that it lends credence to the theory that the offspring inherit some of the parents' training, *since it is so unlikely to occur if they are untrained*. Such a conclusion would be an application of the rare event approach to statistical inference, and you can see that the basic principles of probability play an important role.

3.7 Random Sampling

How a sample is selected from a population is of vital importance in statistical inference because the probability of an observed sample will be used to infer the characteristics of the sampled population. To illustrate, suppose you deal yourself 4 cards from a deck of 52 cards and all 4 cards are aces. Do you conclude that your deck is an ordinary bridge deck, containing only 4 aces, or do you conclude that the deck is stacked with more than 4 aces? It depends on how the cards were drawn. If the 4 aces were always placed at the top of a standard bridge deck, drawing 4 aces is not unusual—it is certain. On the other hand, if the cards are thoroughly mixed, drawing 4 aces in a sample of 4 cards is highly improbable. The point, of course, is that in order to use the observed sample of 4 cards to draw inferences about the population (the deck of 52 cards), you need to know how the sample was selected from the deck.

One of the simplest and most frequently employed sampling procedures is implied in the previous examples and exercises. It produces what is known as a ***random sample.***

> **Definition 3.10**
>
> If n elements are selected from a population in such a way that every set of n elements in the population has an equal probability of being selected, the n elements are said to be a ***random sample.****

Example 3.19   Suppose a lottery consists of ten tickets. (This number is small to simplify our example.) One ticket stub is to be chosen and the corresponding ticket holder will receive a generous prize. How would you select this ticket stub so that the prize will be awarded fairly?

Solution   If the prize is to be awarded fairly, it seems reasonable to require that each ticket stub has the same probability of being drawn. That is, each stub should have a probability of $\frac{1}{10}$ of being selected. A method to achieve the objective of equal selection probabilities is to *mix* the ten stubs thoroughly and *blindly* pick one of the stubs. If this procedure were repeatedly used, each time replacing the selected stub, a particular stub should be chosen approximately $\frac{1}{10}$ of the time in a long series of draws. This method of sampling is known as ***random sampling.***

   If a population is not too large and the elements can be numbered on slips of paper, poker chips, etc., you can physically mix the slips of paper or chips and remove n elements from the total. The numbers that appear on the chips selected would indicate the population elements to be included in the sample. Such a procedure will not guarantee a random sample, because it is often difficult to achieve a thorough mix, but it provides a reasonably good approximation to random sampling.   ■

Example 3.20   Suppose you wish to randomly sample 5 (we will keep the number in the sample small to simplify our example) from a population of 100,000 households. Give a procedure for selecting this random sample.

Solution   Since there are 100,000 households, it is not feasible to select a random sample by numbering slips of paper (or poker chips, etc.), mixing them, and choosing 5. Instead, we will enlist the aid of Table I in the Appendix.

   First, we number the households in the population from 1 to 100,000. Then, we turn to a page of Table I, say the first page. (A partial reproduction of the first page of Table I is shown in Figure 3.13, page 118.) Now, randomly select a starting number, say the random number appearing in the third row, second column. This number is 48360. Proceed down the second column to obtain the remaining four random numbers. The five selected random numbers are shaded in Figure 3.13. Using the first five digits to represent the households from 1 to 99,999 and the number 00000 to represent

---

* Strictly speaking, this is a ***simple random sample.*** There are many different types of random samples. The simple random sample is the most common.

Figure 3.13 Partial Reproduction of Table I in the Appendix

COLUMN ROW	1	2	3	4	5	6
1	10480	15011	01536	02011	81647	91646
2	22368	46573	25595	85393	30995	89198
3	24130	48360	22527	97265	76393	64809
4	42167	93093	06243	61680	07856	16376
5	37570	39975	81837	16656	06121	91782
6	77921	06907	11008	42751	27756	53498
7	99562	72905	56420	69994	98872	31016
8	96301	91977	05463	07972	18876	20922
9	89579	14342	63661	10281	17453	18103
10	85475	36857	53342	53988	53060	59533
11	28918	69578	88231	33276	70997	79936
12	63553	40961	48235	03427	49626	69445
13	09429	93969	52636	92737	88974	33488
14	10365	61129	87529	85689	48237	52267
15	07119	97336	71048	08178	77233	13916

household 100,000, you can see that the households numbered

48,360   93,093   39,975   6,907   72,905

should be included in your sample.   ■

Table I in the Appendix is just one example of a ***table of random numbers.*** Most samplers use such a table to obtain random samples. Random number tables are constructed in such a way that every number occurs with (approximately) equal probability. Further, the occurrence of any one number in a position is independent of any of the other numbers that appear in the table. To use a table of random numbers, number the N elements in the population from 1 to N. Then turn to Table I and select a starting number in the table. Proceeding from this number either across the rows or down the column, remove and record n numbers from the table. Use only the necessary number of digits in each random number to identify the element to be included in the sample.

3.8 Some Counting Rules (Optional)

In Section 3.1 we pointed out that experiments sometimes have so many simple events that it is impractical to list them all. However, many of these experiments possess simple events with identical characteristics. If you can develop a ***counting rule*** to count the number of simple events for such an experiment, it can be used to aid in the solution of the problems.

Example 3.21

A product can be shipped by four different airlines and each airline can ship via three different routes. How many distinct ways exist to ship the product?

Solution

A pictorial representation of the different ways to ship the product will aid in counting them. This representation, called a ***decision tree,*** is shown in Figure 3.14. At the start-

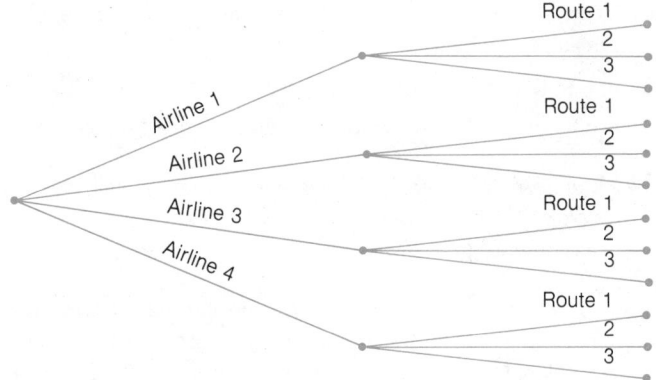

Figure 3.14 Decision Tree for Shipping Problem

ing point (stage 1), there are four choices—the different airlines—to begin the journey. Once we have chosen an airline (stage 2), there are three choices—the different routes—to complete the shipment and reach the final destination. Thus, the decision tree clearly shows that there are $(4)(3) = 12$ distinct ways to ship the product. ■

The method of solving this example can be generalized to any number of stages with sets of different elements. This method is given by the **multiplicative rule.**

The Multiplicative Rule

You have k sets of elements, n_1 in the first set, n_2 in the second set, ..., and n_k in the kth set. Suppose you wish to form a sample of k elements *by taking one element from each* of the k sets. The number of different samples that can be formed is the product

$$n_1 n_2 n_3 \cdot \cdots \cdot n_k$$

Example 3.22   There are twenty candidates for three different executive positions, E_1, E_2, and E_3. How many different ways could you fill the positions?

Solution   For this example, there are $k = 3$ sets of elements corresponding to:

> Set 1:   Candidates available to fill position E_1
>
> Set 2:   Candidates remaining (after filling E_1) that are available to fill E_2
>
> Set 3:   Candidates remaining (after filling E_1 and E_2) that are available to fill E_3

The numbers of elements in the sets are $n_1 = 20$, $n_2 = 19$, $n_3 = 18$. Therefore, the number of different ways of filling the three positions is $n_1 n_2 n_3 = (20)(19)(18) = 6,840$. ■

Example 3.23   Consider the experiment discussed in Example 3.8 of tossing a coin ten times. Show how we found that there were $2^{10} = 1,024$ simple events for this experiment.

Solution   There are $k = 10$ sets of elements for this experiment. Each set contains two elements: a head and a tail. Thus, there are

$$(2)(2)(2)(2)(2)(2)(2)(2)(2)(2) = 2^{10} = 1{,}024$$

different outcomes (simple events) of this experiment.   ∎

Example 3.24

Suppose there are five different dangerous military missions, each requiring one soldier. In how many different ways can five soldiers from a squadron of 100 be assigned to these five missions?

Solution   We can solve this problem by using the multiplicative rule. The entire set of 100 soldiers is available for the first mission, and after the selection of one soldier for that mission, 99 are available for the second mission, etc. Thus, the total number of different ways of choosing five soldiers for the five missions is

$$n_1 n_2 n_3 n_4 n_5 = (100)(99)(98)(97)(96) = 9{,}034{,}502{,}400$$ ∎

The arrangement of elements in a distinct order is called a **permutation.** Thus, in Example 3.24 there are more than 9 billion different *permutations* of five elements (soldiers) drawn from a set of 100 elements!

> ### Permutations Rule
>
> Given a *single set* of N distinctly different elements, you wish to select n elements from the N and *arrange* them within n positions. The number of different **permutations** of the N elements taken n at a time is denoted by P_n^N and is equal to
>
> $$P_n^N = N(N-1)(N-2) \cdots (N-n+1) = \frac{N!}{(N-n)!}$$
>
> where $n! = n(n-1)(n-2) \cdots (3)(2)(1)$ and is called **n factorial.** (Thus, for example, $5! = 5 \cdot 4 \cdot 3 \cdot 2 \cdot 1 = 120$.) The quantity $0!$ is defined to be equal to 1.

Example 3.25

Consider the following transportation problem: You wish to drive, in sequence, from a starting point to each of five cities, and you wish to compare the distances—and ultimately the costs—of the different routings. How many different routings would have to be compared?

Solution   Denote the cities as $C_1, C_2, \ldots, C_5$. Then a route moving from the starting point to C_2 to C_1 to C_3 to C_4 to C_5 would be represented as $C_2 C_1 C_3 C_4 C_5$. The total number of routings would equal the number of ways you could rearrange the $N = 5$ cities in $n = 5$ positions. This number is

$$P_n^N = P_5^5 = \frac{5!}{(5-5)!} = \frac{5!}{0!} = \frac{5 \cdot 4 \cdot 3 \cdot 2 \cdot 1}{1} = 120$$

(Recall that $0! = 1$.)   ∎

Example 3.26　There are four construction workers and you must assign three to job 1 and one to job 2. In how many different ways can you make this assignment?

Solution　To begin, suppose that each worker is to be assigned to a distinct job. Then, using the multiplicative rule, we obtain $(4)(3)(2)(1) = 24$ ways of assigning the workers to four distinct jobs. The 24 ways are listed in four groups in Table 3.4 (where ABCD signifies that worker A was assigned the first distinct job, worker B the second, etc.).

Table 3.4

GROUP 1	GROUP 2	GROUP 3	GROUP 4
ABCD	ABDC	ACDB	BCDA
ACBD	ADBC	ADCB	BDCA
BACD	BADC	CADB	CBDA
BCAD	BDAC	CDAB	CDBA
CABD	DABC	DACB	DBCA
CBAD	DBAC	DCAB	DCBA

Table 3.5

JOB 1	JOB 2
ABC	D
ABD	C
ACD	B
BCD	A

Now suppose that the first three positions represent job 1 and the last position represents job 2. We can now see that all the listings in group 1 represent the same outcome of the experiment of interest. That is, workers A, B, and C are assigned to job 1 and worker D to job 2. Similarly, group 2 listings are equivalent, as are group 3 and group 4 listings. Thus, there are only four different assignments of four workers to the two jobs. These are shown in Table 3.5.

To generalize this result, we point out that the final result can be found by

$$\frac{(4)(3)(2)(1)}{(3)(2)(1)(1)} = 4$$

The $(4)(3)(2)(1)$ is the number of different ways (*permutations*) the workers could be assigned four distinct jobs. The division by $(3)(2)(1)$ is to remove the duplicated permutations resulting from the fact that three workers are assigned the same jobs. And the division by 1 is associated with the worker assigned to job 2.　■

The Partitions Rule

There exists a *single set* of N distinctly different elements and you wish to partition them into k sets, the first set containing n_1 elements, the second containing n_2 elements, . . . , and the kth set containing n_k elements. The number of different partitions is

$$\frac{N!}{n_1! n_2! \cdot \cdots \cdot n_k!} \qquad \text{where } n_1 + n_2 + n_3 + \cdots + n_k = N$$

Example 3.27   You have twelve construction workers and you wish to assign three to job site 1, four to job site 2, and five to job site 3. In how many different ways can you make this assignment?

Solution   For this example, $k = 3$ (corresponding to the $k = 3$ different job sites), $N = 12$, and $n_1 = 3$, $n_2 = 4$, and $n_3 = 5$. Then the number of different ways to assign the workers to the job sites is

$$\frac{N!}{n_1!n_2!n_3!} = \frac{12!}{3!4!5!} = \frac{12 \cdot 11 \cdot 10 \cdots \cdot 3 \cdot 2 \cdot 1}{(3 \cdot 2 \cdot 1)(4 \cdot 3 \cdot 2 \cdot 1)(5 \cdot 4 \cdot 3 \cdot 2 \cdot 1)} = 27{,}720 \qquad \blacksquare$$

Example 3.28   How many samples of four courses can be selected from a list of twenty-five courses in a university catalog?

Solution   For this example, $k = 2$ (corresponding to the $n_1 = 4$ classes you *do* choose and the $n_2 = 21$ classes you *do not* choose) and $N = 25$. Then, the number of different ways to choose the four classes from twenty-five is

$$\frac{N!}{n_1!n_2!} = \frac{25!}{(4!)(21!)} = \frac{25 \cdot 24 \cdot 23 \cdots \cdot 3 \cdot 2 \cdot 1}{(4 \cdot 3 \cdot 2 \cdot 1)(21 \cdot 20 \cdots \cdot 2 \cdot 1)} = 12{,}650$$

(Perhaps you now understand better the confusion that surrounds registration time at many schools.) $\qquad \blacksquare$

This special application of the partitions rule—partitioning a set of N elements into $k = 2$ groups (the elements that appear in a sample and those that do not)—is very common. Therefore, we give a different name to the rule for counting the number of different ways of partitioning a set of elements into two parts: the **combinations rule.**

The Combinations Rule

A sample of n elements is to be chosen from a set of N elements. Then the number of different samples of n elements that can be selected from N is denoted by $\binom{N}{n}$ and is equal to

$$\binom{N}{n} = \frac{N!}{n!(N - n)!}$$

Note that the order in which the n elements are drawn is not important.

Example 3.29   Five soldiers are to be chosen for a dangerous mission from a squadron of 100 men. In how many ways can groups of five men be formed?

Solution   This problem is equivalent to sampling $n = 5$ elements from a set of $N = 100$ elements. Thus, the number of ways is the number of possible combinations of five soldiers selected from 100, or

$$\binom{100}{5} = \frac{100!}{(5!)(95!)} = \frac{100 \cdot 99 \cdot 98 \cdot 97 \cdot 96 \cdot 95 \cdot 94 \cdots 2 \cdot 1}{(5 \cdot 4 \cdot 3 \cdot 2 \cdot 1)(95 \cdot 94 \cdots 2 \cdot 1)}$$

$$= \frac{100 \cdot 99 \cdot 98 \cdot 97 \cdot 96}{5 \cdot 4 \cdot 3 \cdot 2 \cdot 1}$$

$$= 75,287,520$$

Compare this result with that of Example 3.24 where we found that the number of permutations of five elements drawn from 100 was more than 9 billion. Because the order of the elements does not affect combinations, there are *fewer* combinations than permutations.   ∎

Summary of Counting Rules

1. *Multiplicative rule:* If you are drawing *one element from each of k sets of elements,* with the sizes of the sets $n_1, n_2, \ldots, n_k$, the number of different results is

$$n_1 n_2 n_3 \cdots n_k$$

2. *Permutations rule:* If you are drawing *n elements from a set of N elements and arranging the n elements in a distinct order,* the number of different results is

$$P_n^N = \frac{N!}{(N - n)!}$$

3. *Partitions rule:* If you are partitioning the *elements of a set of N elements into k groups consisting of* $n_1, n_2, \ldots, n_k$ *elements* ($n_1 + n_2 + \cdots + n_k = N$), the number of different results is

$$\frac{N!}{n_1! n_2! \cdots n_k!}$$

4. *Combinations rule:* If you are drawing *n elements from a set of N elements without regard to the order of the n elements,* the number of different results is

$$\binom{N}{n} = \frac{N!}{n!(N - n)!}$$

[*Note:*   The combinations rule is a special case of the partitions rule when $k = 2$.]

When working a probability problem, you should carefully examine the experiment to see whether you can use one or more of the rules we have discussed in this section. We will illustrate how these rules can help solve a probability problem.

Example 3.30   A consumer testing service is commissioned to rank the top three brands of laundry detergent. A total of ten brands are to be included in the study.

a. In how many different ways can the consumer testing service arrive at the final ranking?

b. If the testing service can distinguish no difference among the brands and therefore arrives at the final ranking by random selection, what is the probability that company Z's brand is ranked first? In the top three?

Solution   **a.** Since the testing service is drawing three elements (brands) from a set of ten elements and arranging the three elements in a distinct order, we use the permutations rule to find the number of different results:

$$P^{10}_3 = \frac{10!}{(10-3)!} = 10 \cdot 9 \cdot 8 = 720$$

b. The steps for calculating the probability of interest are as follows:

Step 1. The experiment is to select and rank three brands of detergent from ten brands.

Step 2. There are too many simple events to list. However, we know from part a that there are 720 different outcomes (i.e., simple events) of this experiment.

Step 3. If we assume that the testing service determines the rankings at random, each of the 720 simple events should have an equal probability of occurrence. Thus,

$$P(\text{Each simple event}) = \tfrac{1}{720}$$

Step 4. One event of interest to company Z is that their brand receives top ranking. We will call this event A. The list of simple events that result in the occurrence of event A is long, but the *number* of simple events contained in event A is determined by breaking event A into two parts:

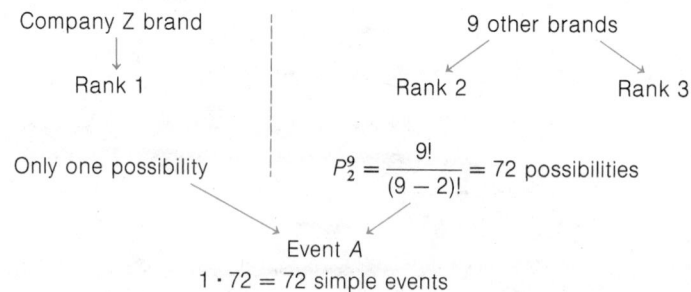

Thus, event A can occur in seventy-two different ways.

Now define B as the event that company Z's brand is ranked in the top three. Since event B only specifies that brand Z appear in the top three, we repeat the calculations above, fixing brand Z in position 2 and then in position 3. We conclude that the number of simple events contained in event B is $3(72) = 216$.

Step 5. The final step is to calculate the probabilities of events A and B. Since the 720 simple events are equally likely to occur, we find

$$P(A) = \frac{\text{Number of simple events in } A}{\text{Total number of simple events}} = \frac{72}{720} = \frac{1}{10}$$

Similarly,

$$P(B) = \frac{216}{720} = \frac{3}{10}$$

■

Example 3.31 Refer to Example 3.30. Suppose the consumer testing service is to choose the top three laundry detergents from the group of ten *but is not to rank the three.*

   a. In how many different ways can the testing service choose the three to be designated as top detergents?
   b. Assuming that the testing service makes its choice by a random selection and that company X has two brands in the group of ten, what is the probability that exactly one of the company X brands is selected in the top three? At least one?

Solution **a.** The testing service is selecting three elements (brands) from a set of ten elements *without regard to order*, so we can apply the combinations rule to determine the number of different results:

$$\binom{10}{3} = \frac{10!}{3!(10-3)!} = \frac{10 \cdot 9 \cdot 8}{3 \cdot 2 \cdot 1} = 120$$

   b. Step 1. The experiment is to select (*but not rank*) three brands from ten.
   Step 2. There are 120 simple events for this experiment.
   Step 3. Since the selection is made at random,

   $$P(\text{Each simple event}) = \tfrac{1}{120}$$

   Step 4. Define events A and B as follows:

   A: {Exactly one company X brand is selected}
   B: {At least one company X brand is selected}

Since each of the simple events is equally likely to occur, we need only know the number of simple events in A and B to determine their probabilities.

In order for event A to occur, exactly one company X brand must be selected, along with two of the remaining eight brands. We thus break A into two parts:

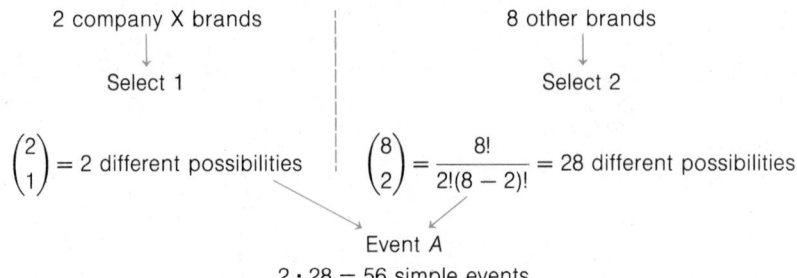

Event A

$2 \cdot 28 = 56$ simple events

Note that the one company X brand can be selected in two ways, while the two other brands can be selected in twenty-eight ways. (We use the combinations rule because the order of selection is not important.) Then, we use the multiplicative rule to combine one of the two ways to select a company X brand with one of the twenty-eight ways to select two other brands, yielding a total of fifty-six simple events for event A.

To count the simple events contained in B, we first note that

$$B = A \cup C$$

where A is defined as before and

C: {Both company X brands are selected}

We have determined that A contains fifty-six simple events. Using the same method for event C, we find:

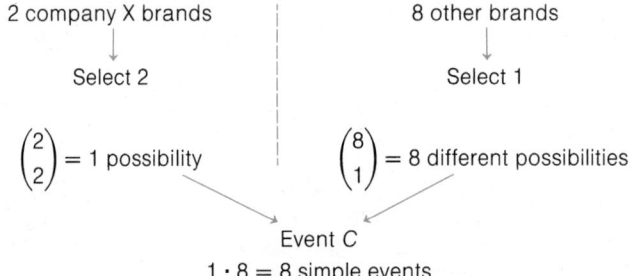

Event C

$1 \cdot 8 = 8$ simple events

Thus, event C contains eight simple events and $P(C) = \frac{8}{120}$.

Step 5. Since all the simple events are equally likely,

$$P(A) = \frac{\text{Number of simple events in } A}{\text{Total number of simple events}} = \frac{56}{120} = \frac{7}{15}$$

$$P(B) = P(A \cup C) = P(A) + P(C) - P(AC)$$

where $P(AC) = 0$ because the testing service cannot select exactly one *and* exactly two of the company X brands; that is, A and C are mutually exclusive events. Thus,

$$P(B) = \frac{56}{120} + \frac{8}{120} = \frac{64}{120} = \frac{8}{15}$$

Learning how to decide whether a particular counting rule applies to an experiment takes patience and practice. If you wish to develop this skill, attempt to use the rules to solve the following exercises and especially the optional exercises (marked with an asterisk) at the end of this chapter. Proofs of the counting rules can be found in the text by W. Feller (1968, Chapter 3).

Exercises 3.59–3.74

Learning the Mechanics

3.59 Determine the number of simple events contained in the sample space when you toss a coin the following number of times:
a. Twice   **b.** Three times   **c.** Five times   **d.** n times

3.60 Find the numerical values of:

a. $\binom{6}{3}$ **b.** P_2^5 **c.** P_2^4 **d.** $\binom{100}{98}$

e. $\binom{50}{50}$ **f.** $\binom{50}{0}$ **g.** P_3^5 **h.** P_0^{10}

3.61 An experiment consists of choosing objects without regard to order. How many simple events are there if you choose the following?
a. 3 objects from 7   **b.** 2 objects from 6
c. 2 objects from 30   **d.** 8 objects from 10
e. r objects from q

3.62 Determine the number of simple events contained in the sample space when you toss the following:
a. One die   **b.** Two dice   **c.** Four dice   **d.** n dice

Applying the Concepts

3.63 Flying into New York from a certain city, you can choose one of three airlines and can travel either first class or economy. How many travel options do you have?

3.64 How many different 5-card hands can be dealt from a 52-card bridge deck?

3.65 Suppose an automobile license plate is designed to show a number from 1 to 26, corresponding to the twenty-six counties in a state, and then is followed by a five-digit number. How many different license plates can the state issue?

3.66 A company sells five different styles of its intermediate model automobile. The buyer can get an automobile in one of eight colors with either standard or automatic transmission. Would it be reasonable to expect a dealer to stock at least one automobile in every style, color, transmission combination? At a minimum, how many automobiles would the dealer have to stock?

3.67 A salesperson living in city A wishes to visit four cities, B, C, D, and E.
a. If the cities are all interconnected by airlines, how many different travel plans could be constructed to visit each city exactly once and then return home?

b. Suppose that all cities are connected except that B and C are not directly connected. How many different flight plans would be available to the salesperson?

3.68 In an article titled "State's License Plates Will Soon Look Different," the *Gainesville Sun* (April 11, 1984) notes that the state of Florida is running out of combinations of letters and numbers for its license plates. The current license plate contains three letters of the alphabet followed by three digits selected from the ten digits 0, 1, 2, ..., 9. This system, according to the article, allows a maximum of 17,576,000 combinations of letters and numbers or, equivalently, 17,576,000 license plates. Is this number, 17,576,000, correct?

3.69 Refer to Exercise 3.68. According to the *Gainesville Sun* article, new Florida tags will reverse the existing procedure, starting with three digits followed by three letters.
a. How many new tags will the new system provide?
b. Since the new tag numbers will be added to the old numbers, what is the total number of licenses available for registration in Florida?

3.70 The five-digit zip code has become an integral part of the United States postal system. How many different zip codes are available for use by the postal service? (Assume any five-digit number can be used as a zip code.) The first three digits of a Chicago zip code are 606. If no other city in the United States has these first three digits as part of its zip, how many different zip codes may possibly exist in Chicago?

3.71 Suppose you are to choose a basketball team (five players) from eight available athletes.
a. How many ways can you choose a team (ignoring positions)?
b. How many ways can you choose a team composed of two guards, two forwards, and a center?
c. How many ways can you choose a team composed of tall guard, short guard, power forward, shooting forward, and center?

3.72 Intercollegiate volleyball rules require that after the opposing team has lost its serve, each of the six members of the serving team must rotate into new positions on the court. Hence, each player must be able to play all six different positions. How many different team combinations by player and position are possible during a volleyball game? If players are initially assigned to the positions in a random manner, find the probability that the best server on the team is in the serving position.

3.73 A college professor hands out a list of ten questions, five of which will appear on the final examination for the course. One of the students taking the course is pressed for time and can prepare for only seven of the ten questions. If the professor chooses the five questions at random from the ten, then:
a. What is the probability that the student will be prepared for all five questions that appear on the final examination?
b. What is the probability that the student will be prepared for less than three questions?
c. What is the probability that the student will be prepared for exactly four questions?

3.74 Consider 5-card poker hands dealt from a standard 52-card bridge deck. Two important events are:

A: {You draw a flush}

B: {You draw a straight}

a. Find $P(A)$. **b.** Find $P(B)$.

c. The event that both A and B occur, that is, AB, is called a straight flush. Find $P(AB)$.

[*Note:* A *flush* consists of any 5 cards of the same suit. A *straight* consists of any 5 cards with values in sequence. In a straight, the cards may be of any suit and an ace may be considered as having a value of 1 or a value higher than a king.]

Summary

We have developed some of the basic tools of probability to enable us to assess the probability of various sample outcomes given a specific population structure. Although many of the examples we presented were of no practical importance, they accomplished their purpose if you now understand the concepts and definitions necessary for a basic understanding of probability.

In the next several chapters we will present probability models that can be used to solve practical problems. You will see that for most applications, we will need to make inferences about unknown aspects of these probability models, i.e., we will need to apply inferential statistics to the problem.

Supplementary Exercises 3.75–3.101

[*Note:* Starred (*) exercises refer to the optional section in this chapter.]

Learning the Mechanics

3.75 A fair die is tossed and the up face is noted. If the number is even, the die is tossed again; if the number is odd, a fair coin is tossed. Define the events:

A: {A head appears on the coin}

B: {The die is tossed only one time}

a. List the simple events in the sample space.

b. Give the probability for each of the simple events.

c. Find $P(A)$ and $P(B)$.

d. Identify the simple events in A', B', AB, and $A \cup B$.

e. Find $P(A')$, $P(B')$, $P(AB)$, $P(A \cup B)$, $P(A|B)$, and $P(B|A)$.

f. Are A and B mutually exclusive events? Independent events? Why?

***3.76** Find the numerical value of:

a. $6!$ **b.** $\binom{10}{9}$ **c.** $\binom{10}{1}$ **d.** P_2^6

e. $\binom{6}{3}$ **f.** $0!$ **g.** P_4^{10} **h.** P_2^{50}

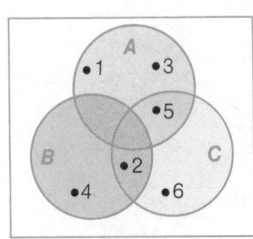

S

3.77 The Venn diagram shown in the margin illustrates a sample space containing six simple events and three events, A, B, and C. The probabilities of the simple events are

$$P(1) = .4 \qquad P(2) = .1 \qquad P(3) = .1$$
$$P(4) = .2 \qquad P(5) = .1 \qquad P(6) = .1$$

a. Find $P(AB)$, $P(BC)$, $P(A \cup C)$, $P(A \cup B \cup C)$, $P(B')$, $P(A'B)$, $P(B|C)$, and $P(B|A)$.
b. Are A and B independent? Mutually exclusive? Why?
c. Are B and C independent? Mutually exclusive? Why?

3.78 Psychologists tend to believe that there is a relationship between aggressiveness and order of birth. To test this belief, a psychologist chose 500 elementary school students at random and administered to each a test designed to measure the student's aggressiveness. Each student was classified according to one of four categories. The percentages of students falling in the four categories are shown here:

	FIRSTBORN	NOT FIRSTBORN
AGGRESSIVE	15%	15%
NOT AGGRESSIVE	25%	45%

a. If one student is chosen at random from the 500, what is the probability that the student is firstborn?
b. What is the probability that the student is aggressive?
c. What is the probability that the student is aggressive, given the student was firstborn?
d. If

  A: {Student chosen is aggressive}

  B: {Student chosen is firstborn}

  are A and B independent events? Explain.

3.79 Two events, A and B, are independent, with $P(A) = .3$ and $P(B) = .1$.
a. Are A and B mutually exclusive? Why?
b. Find $P(A|B)$ and $P(B|A)$.
c. Find $P(A \cup B)$.

3.80 Explain when it is appropriate to use each of the following counting rules to count the simple events in an experiment:
a. The multiplicative rule   **b.** The permutations rule
c. The partitions rule   **d.** The combinations rule

3.81 A balanced die is thrown once. If a 4 appears, a ball is drawn from urn 1; otherwise a ball is drawn from urn 2. Urn 1 contains four red, three white, and three black balls. Urn 2 contains six red and four white balls.

a. Find the probability that a red ball is drawn.

b. Find the probability that urn 1 was used given that a red ball was drawn.

3.82 Consider an experiment that consists of simultaneously flipping a nickel, dime, and quarter. List the simple events. Let A be the event that the up face on the dime is heads. Let B be the event of observing at least two heads. Let C be the event that the outcomes (on the up faces) on all three coins are the same.

a. Identify the simple events in BC.

b. Identify the simple events in $A \cup B$.

c. Find $P(BC)$ by summing the probabilities of the simple events in BC.

d. Find $P(A \cup B)$ by summing the probabilities of the simple events in $A \cup B$.

e. Find $P(B \cup C)$ by summing the probabilities of the simple events in $B \cup C$.

f. Find $P(B \cup C)$ using the additive rule of probability.

Applying the Concepts

3.83 In college basketball games a player may be afforded the opportunity to shoot two consecutive foul shots (free throws).

a. Suppose a player who scores on 80% of his foul shots has been awarded two free throws. If the two throws are considered independent, what is the probability that the player scores on both shots? Exactly one? Neither shot?

b. Suppose a player who scores on 80% of his first attempted foul shots has been awarded two free throws, and the outcome on the second shot is dependent on the outcome of the first shot. In fact, if this player makes the first shot he makes 90% of the second shots, and if he misses the first shot, he makes 70% of the second shots. In this case, what is the probability that the player scores on both shots? Exactly one? Neither shot?

c. In parts a and b, we considered two ways of *modeling* the probability a basketball player scores on two consecutive foul shots. Which model do you think is a more realistic attempt to explain the outcome of shooting foul shots, i.e., do you think two consecutive foul shots are independent or dependent? Explain.

3.84 Two companies, A and B, package and market a chemical substance and claim .15 of the total weight of the substance is sodium. However, a careful survey of 4,000 packages (half from each company) indicates the proportion varies around .15, with the following results:

		PROPORTION OF SODIUM			
		Less than .100	.100–.149	.150–.199	Over .200
CHEMICAL BRAND	A	25%	10%	10%	5%
	B	5%	5%	10%	30%

Suppose a package is chosen at random from the 4,000 packages. If

A: {Package chosen is brand A}

B: {Package chosen is brand B}

    C: {Package chosen contains less than .100 sodium}

    D: {Package chosen contains between .100 and .149 sodium}

    E: {Package chosen contains between .150 and .199 sodium}

    F: {Package chosen contains over .200 sodium}

then describe the characteristics of a package portrayed by the following events:

a. $A \cup B$ **b.** $B \cup F$ **c.** AD

d. EB **e.** $(AC) \cup (AD)$

3.85 Refer to Exercise 3.84. Find the probabilities of the following events by summing the probabilities of the appropriate simple events:

a. A, B, C, D, E, F **b.** $A \cup B$ **c.** BC

d. AF **e.** AB **f.** $C \cup D$

g. $(AC) \cup (AD)$

3.86 A survey of the faculty of a large university yielded the following breakdown according to sex and marital status:

		MARITAL STATUS			
		Married	Single	Divorced	Widowed
SEX	Male	60%	12%	8%	5%
	Female	10%	3%	2%	0%

Suppose a faculty member is chosen at random. If

    A: {Faculty member chosen is female}

    B: {Faculty member chosen is male}

    C: {Marital status of chosen faculty member is married}

    D: {Marital status of chosen faculty member is single}

    E: {Marital status of chosen faculty member is widowed}

    F: {Marital status of chosen faculty member is divorced}

then describe the characteristics of a faculty member portrayed by the following events:

a. $A \cup B$ **b.** BF **c.** $C \cup D$

d. EF **e.** $F \cup B$ **f.** $E \cup F$

3.87 Refer to Exercise 3.86. Find the probability of the following events by summing the probabilities of the appropriate simple events:

a. A, B, C, D, E, F **b.** $A \cup B$ **c.** BF

d. $C \cup D$ **e.** EF **f.** $F \cup B$

3.88 In the game of parcheesi each player rolls a pair of dice on each turn. In order to begin the game you must throw a 5 on at least one of the dice, or a total of 5 on the two dice. What is the probability that you can begin the game on your first turn? The second turn? The third turn? The *n*th turn?

3.89 Entomologists are often interested in studying the effect of chemical attractants (pheromones) on insects. One common technique is to release several insects equidistant from the pheromone being studied and from a control substance. If the pheromone has an effect, more insects will travel toward it rather than toward the control. Otherwise, the insects are equally likely to travel in either direction. Suppose the pheromone under study has no effect so that it is equally likely that an insect will move toward either the pheromone or the control.

a. If five insects are released, what is the probability that all five travel toward the pheromone?

b. Exactly four?

3.90 The probability of the union of three events, A, B, and C, is given by this expression (proof omitted):

$$P(A \cup B \cup C) = P(A) + P(B) + P(C) - P(AB) - P(AC) - P(BC) + P(ABC)$$

a. Sketch a Venn diagram showing the events A, B, and C and their intersections. Let the areas in the diagram corresponding to events A, B, C, AB, AC, BC, and ABC represent their probabilities. Then use the diagram to justify the formula for $P(A \cup B \cup C)$.

b. A national poll of biostatisticians was conducted to ascertain their professional responsibilities. An analysis of the responses gave the following distribution of professional responsibilities:

A:	{Research}	40%
B:	{Professional consultation}	64%
C:	{Data collection and analysis}	36%

Suppose 10% are involved in all three activities; 15% are involved in both A and C; and 17% are involved in both A and B. Using this information, find the percentage of all biostatisticians that are involved in both B and C (i.e., in professional consultation and in data collection and analysis). Assume that $P(A \cup B \cup C) = 1$.

3.91 A manufacturer of 35-mm cameras knows that a shipment of thirty cameras sent to a large discount store contains six defective cameras. The manufacturer also knows that the store will choose two of the cameras at random, test them, and accept the shipment if neither is defective.

a. What is the probability that the first camera chosen by the store will be defective?

b. Given that the first camera chosen passed inspection, what is the probability that the second camera chosen will fail inspection?

c. What is the probability that the shipment will be accepted?

3.92 The probability that a mini-computer salesperson sells a computer to a prospective customer on the first visit to the customer is .4. If the salesperson fails to make the sale on the first visit, the probability that the sale will be made on the second visit is .65. The salesperson never visits a prospective customer more than twice. What is the probability that the salesperson will make a sale to a particular customer?

3.93 Seventy-five percent of all women who submit to pregnancy tests are really pregnant. A certain pregnancy test gives a false positive result with probability .02 and a valid positive result with probability .99. If a particular woman's test is positive, what is the probability that she really is pregnant? [*Hint:* If A is the event that a woman is pregnant and B is the event that the pregnancy test is positive, then B is the union of the two mutually exclusive events AB and $A'B$. Also, the probability of a "false positive result" may be written as $P(B|A') = .02$.]

3.94 Blackjack, a favorite game of gamblers, is played by a dealer and at least one opponent and uses a 52-card bridge deck. At the outset of the game, 2 cards are dealt to the player and 2 cards to the dealer. Drawing an ace and a face card is called "blackjack." If the dealer draws it, he or she automatically wins. If the dealer does not draw a blackjack and the player does, the player wins.

a. What is the probability that the dealer will draw a blackjack?

b. What is the probability that the player wins with a blackjack?

3.95 A small brewery has two bottling machines. Machine A produces 75% of the bottles and machine B produces 25%. One out of every twenty bottles filled by A is rejected for some reason, while one out of every thirty bottles from B is rejected. What proportion of bottles is rejected? What is the probability that a bottle comes from machine A, given that it is accepted?

3.96 A clinical psychologist is asked to view tapes in which each of six experimental subjects is discussing his or her recent dreams. Three of the six subjects have previously been classified as "high-anxiety" individuals and the other three as "low-anxiety." The psychologist is told only that there are three of each type and is asked to select the three high-anxiety subjects.

a. List all possible outcomes (simple events) for this experiment.

b. Assuming that the psychologist guesses at the classifications of the subjects, assign probabilities to the simple events.

c. Find the probability that the psychologist guesses all classifications correctly.

d. Find the probability that the psychologist guesses at least two of the three high-anxiety subjects correctly.

*3.97 Refer to Exercise 3.96. Suppose that the clinical psychologist is given tapes of twenty subjects and is told only that ten are high-anxiety individuals and ten are low-anxiety individuals. The psychologist is asked to select the ten high-anxiety subjects.

a. Using the appropriate counting rule from Section 3.8, count the number of simple events for this experiment.

b. Assuming that the psychologist is guessing, assign probabilities to each of the simple events.

c. Find the probability that the psychologist guesses all classifications correctly.

d. Find the probability that the psychologist guesses at least nine of the ten high-anxiety subjects correctly.

*3.98 Suppose a manufacturer of television sets is prepared to send twenty new sets to three retail dealers. Dealer A is to get ten of the sets, dealer B six of the sets, and dealer C four of the sets.

a. In how many different ways can the twenty sets be divided among the dealers in the prescribed numbers?

b. Assuming that the sets are randomly divided among the dealers, and that there are three defective sets among the twenty, what is the probability that dealer A gets all three defective sets?

c. Making the same assumptions as in part b, what is the probability that dealer A gets two defective sets and dealer B the other defective set?

*3.99 Suppose there are 500 applicants for five equivalent positions at a factory and the company is able to narrow the field to thirty equally qualified applicants. Seven of the finalists are minority candidates. Assume that the five who are chosen are selected at random from this final group of thirty.

a. In how many different ways can the selection be made?

b. What is the probability that none of the minority candidates is hired?

c. What is the probability that no more than one minority candidate is hired?

*3.100 A poll is taken in an attempt to determine the top three prime-time television shows. Each person interviewed is asked to rank his or her first, second, and third favorite show from a list of forty-two shows (fourteen from each of the three major networks, ABC, CBS, NBC).

a. In how many different ways can the ranking be made by an interviewee?

b. If an interviewee ranks the top three shows by making a random selection, what is the probability that the first two programs selected are ABC programs while the third is an NBC program?

c. If the ranking is random, what is the probability that all three program selections are from the same network?

*3.101 According to the January 26, 1978, CBS morning news program, a very rare event recently occurred in Dubuque, Iowa. Each of four women playing bridge was astounded to note that she had been dealt a perfect bridge hand. That is, one woman was dealt all thirteen spades, another all thirteen hearts, another all the diamonds, and another all the clubs. What is the probability of this rare event?

On Your Own . . .

Obtain a standard deck of 52 playing cards (the kind commonly used for bridge, poker, or solitaire). An experiment will consist of drawing 1 card at random from the deck of cards and recording which card was observed. This will be simulated by shuffling the deck thoroughly and observing the top card. Consider the following two events:

   A: {Card observed is a heart}

   B: {Card observed is an ace, king, queen, or jack}

a. Find $P(A)$, $P(B)$, $P(AB)$, and $P(A \cup B)$.

b. Conduct the experiment ten times and record the observed card each time. Be sure to return the observed card each time and thoroughly shuffle the deck before making the draw. After 10 cards have been observed, calculate the proportion of observations that satisfy event A, event B, event AB, and event A ∪ B. Compare the observed proportions with the true probabilities calculated in part a.

c. Conduct the experiment forty more times to obtain a total of fifty observed cards. Now calculate the proportion of observations that satisfy event A, event B, event AB, and event A ∪ B. Compare these proportions with those found in part b and the true probabilities found in part a.

d. Conduct the experiment fifty more times to obtain a total of 100 observations. Compare the observed proportions for the 100 trials with those found previously. What comments do you have concerning the different proportions found in parts b, c, and d as compared to the true probabilities found in part a? How do you think the observed proportions and true probabilities would compare if the experiment were conducted 1,000 times? 1 million times?

References   Feller, W. *An introduction to probability theory and its applications.* 3d ed. Vol. I. New York: Wiley, 1968. Chapters 1, 3, 4, and 5.

Kuehn, A. "An analysis of the dynamics of consumer behavior and its implications for marketing management." Unpublished doctoral dissertation, Graduate School of Industrial Administration, Carnegie Institute of Technology, 1958.

Parzen, E. *Modern probability theory and its applications.* New York: Wiley, 1960. Chapters 1 and 2.

Preston, M. G., & Baratta, P. "An experimental study of the auction-value of an uncertain outcome." *American Journal of Psychology*, 1948, *61*, 183–193.

Scheaffer, R. L., & Mendenhall, W. *Introduction to probability: theory and applications.* North Scituate, Mass.: Duxbury, 1975. Chapters 1 and 2.

CHAPTER 4

Discrete Random Variables

Where We've Been . . .

By illustration we indicated in Chapter 3 how probability would be used to make an inference about a population from data contained in an observed sample. We also noted that probability would be used to measure the reliability of the inference.

Where We're Going . . .

Most experimental events in Chapter 3 were events described in words and denoted by capital letters. In real life, most sample observations are numerical—in other words, numerical data. In this chapter, we will learn that data are observed values of random variables. We will study several important random variables and learn how to find the probabilities of specific numerical outcomes.

Contents

You may have noticed that many of the examples of experiments in Chapter 3 generated quantitative (numerical) observations. The unemployment rate, the percentage of voters favoring a particular candidate, the cost of textbooks for a school term, and the amount of pesticide in the discharge waters of a chemical plant are all examples of numerical measurements of some phenomenon. Thus, most experiments have simple events that correspond to values of some numerical variable.

Definition 4.1

A ***random variable*** is a rule that assigns one (and only one) numerical value to each simple event of an experiment.*

Example 4.1

Three potential voters are asked whether they are in favor of a bond for building a fine arts center. Each response is recorded as Yes or No. Define a random variable that could be of interest in this experiment.

Solution

In this experiment, the simple events are *not* numerical in nature. In fact, the eight simple events listed in Table 4.1 are sequences of Yes's and No's. It is doubtful that anyone would be interested in the exact sequence of Yes's and No's, each corresponding to one of the voters polled. One random variable of interest is the *number of voters* in favor of the bond issue. Thus, the value 3 would be assigned to the first simple event, 2 to the second, etc., as shown in Table 4.1. Note that only one numerical value is assigned to each simple event, although several simple events have the same numerical value of the random variable. ∎

Table 4.1
Possible Outcomes
of Voter Sampling

| | VOTER | | | RANDOM VARIABLE |
	1	2	3	(NUMBER OF YES'S)
1	Yes	Yes	Yes	3
2	Yes	Yes	No	2
3	Yes	No	Yes	2
4	No	Yes	Yes	2
5	Yes	No	No	1
6	No	Yes	No	1
7	No	No	Yes	1
8	No	No	No	0

The term *random variable* is more meaningful than just the term *variable,* because the adjective *random* indicates that the experiment may result in one of the several possible values of the variable, according to the *random* outcome of the experiment. For example, if the experiment is to count the number of customers who use the drive-

---

* By *experiment,* we mean an experiment that yields random outcomes (as defined in Chapter 3).

up window of a bank each day, the random variable (the number of customers) will vary from day to day, partly because of the random phenomena that influence whether customers use the drive-up window. Thus, the possible values of this random variable range from zero to the maximum number of customers the window could possibly serve in a day.

We will define two different types of random variables, **discrete** and **continuous,** in Section 4.1. Then we will spend the remainder of this chapter discussing specific types of discrete random variables and the aspects that make them important to the statistician.

4.1
Two Types of Random Variables

Dividing one unit of probability among the simple events in a sample space and consequently assigning probabilities to the values of a random variable is not always as easy as the examples in Chapter 3 might lead you to believe. If the number of simple events is finite, that is, if they can be completely listed, the job is relatively easy. However, some experiments result in an infinite number of sample points, in which case assignment of probabilities will be more difficult. In fact, we will have to use different probability models depending on the number of values that a random variable can assume.

Example 4.2

A panel of ten wine experts is asked to taste a new white wine and assign a rating of 0, 1, 2, or 3. A score is then obtained by adding together the ratings of the ten experts. How many values can this random variable assume?

Solution

A simple event is a sequence of ten numbers associated with the rating of each expert. The random variable assigns a score to each one of these simple events by adding the ten numbers together. Thus, the smallest score is 0 (if all ten ratings are 0) and the largest score is 30 (if all ten ratings are 3). Since every integer between 0 and 30 is a possible score, the random variable x can assume thirty-one values.

This is an example of a **discrete random variable,** since there is a finite number of distinct possible values. Whenever all the possible values a random variable can assume can be listed (or *counted*), the random variable is *discrete*. ∎

Example 4.3

Suppose the Environmental Protection Agency (EPA) takes readings once a month on the amount of pesticide in the discharge water of a chemical company. If the amount of pesticide exceeds the maximum level set by the EPA, the company is forced to take corrective action and may be subject to penalty. Consider the random variable:

Number, x, of months before the company's discharge exceeds the EPA's maximum level

What values can x assume?

Solution

The company's discharge of pesticide may exceed the maximum allowable level on the first month of testing, the second month of testing, etc. It is possible that the company's discharge will *never* exceed the maximum level. Thus, the set of possible values for the number of months until the level is first exceeded is the set of all positive integers:

1, 2, 3, 4, . . .

If we can list the values of a random variable x, even though the list is never-ending, we call the list **countable** and the corresponding random variable *discrete*. Thus, the number of months until the company's discharge first exceeds the limit is a *discrete random variable.*   ▪

Example 4.4   Refer to Example 4.3. A second random variable of interest is the amount x of pesticide (in milligrams per liter) found in the monthly sample of discharge waters from the chemical company. What values can this random variable assume?

Solution   Unlike the *number* of months before the company's discharge exceeds the EPA's maximum level, the set of all possible values for the *amount* of discharge *cannot* be listed, i.e., is not countable. The possible values for the amount x of pesticide would correspond to the points on the line interval between zero and the largest possible value the amount of the discharge could attain, the maximum number of milligrams that could occupy 1 liter of volume. (Practically, the interval would be much smaller, say, between zero and 500 milligrams per liter.) When the values of a random variable are not countable, but instead correspond to the points on some line interval, we call it a **continuous random variable.** Thus, the *amount* of pesticide in the chemical plant's discharge waters is a *continuous random variable.*   ▪

Definition 4.2

Random variables that can assume a *countable* number of values are called **discrete.**

Definition 4.3

Random variables that can assume values corresponding to any of the points contained in one or more intervals on a line are called **continuous.**

Examples of discrete random variables are:

1. The number of sales made by a salesperson in a given week: $x = 0, 1, 2, \ldots$.
   [*Note:* Theoretically, x could become very large.]
2. The number of students in a sample of 500 who favor an increase in student activities and, correspondingly, an increase in student activity fees:
   $x = 0, 1, 2, \ldots, 500$.
3. The number of students applying to medical schools this year: $x = 0, 1, 2, \ldots$.
   [*Note:* Theoretically, x could become very large.]
4. The number of errors on a page of an accountant's ledger: $x = 0, 1, 2, \ldots$.

5. The number of customers waiting to be served in a restaurant at a particular time: $x = 0, 1, 2, \ldots$.

Note that each of the examples of discrete random variables begins with the words "The number of . . . ." This wording is very common, since the discrete random variables most frequently observed are counts. Examples of continuous random variables are:

1. The length of time between arrivals at a hospital clinic: $0 \le x < \infty$ (infinity).

2. For a new apartment complex, the length of time from completion until a specified number of apartments are rented: $0 \le x < \infty$.

3. The amount of carbonated beverage loaded into a 12-ounce can in a can filling operation: $0 \le x \le 12$.

4. The depth at which a successful oil drilling venture first strikes oil: $0 \le x \le c$, where c is the maximum depth obtainable.

5. The weight of a food item bought in a supermarket: $0 \le x \le 500$.
[*Note:*   Theoretically, there is no upper limit on x, but it is unlikely that it would exceed 500 pounds.]

Discrete random variables and their probability distributions are discussed in this chapter. Continuous random variables and their probability distributions are the topic of Chapter 5.

Exercises 4.1–4.3

Applying the Concepts

4.1 Classify the following random variables according to whether they are discrete or continuous:
a. The number of words spelled correctly by a student on a spelling test
b. The amount of liquid waste a plant purifies daily
c. The length of time an employee is late for work
d. The number of bacteria per cubic centimeter of drinking water
e. The amount of carbon monoxide produced per gallon of unleaded gas
f. Your weight

4.2 Identify the following random variables as discrete or continuous:
a. The number of patients entering a doctor's office during any given hour
b. The heart rate (number of beats per minute) of an American male
c. The time it takes a student to complete an examination
d. The barometric pressure at a given location
e. The number of persons in a given town who are disabled
f. Your heart rate

4.3 Identify the following random variables as discrete or continuous:
a. The length of time until recovery from a tonsillectomy
b. The number of violent crimes committed per month in your community
c. The number of commercial aircraft near-misses per month

d. The number of winners each week in a state lottery

e. Your blood pressure

f. Your height

4.2 Probability Distributions for Discrete Random Variables

A complete description of a discrete random variable requires that we *specify the possible values the random variable can assume and the probability associated with each value.* To illustrate, consider Example 4.5.

Example 4.5

Recall the experiment of tossing two coins (Chapter 3), and let *x* be the number of heads observed. Find the probability associated with each value of the random variable *x*, assuming the two coins are fair.

Solution

Recall from Chapter 3 that the sample space and simple events for this experiment are as shown in Figure 4.1, and the probability associated with each of the four simple events is $\frac{1}{4}$. The random variable *x* can assume values 0, 1, 2. Then, identifying the probabilities of the simple events associated with each of these values of *x*, we have

$$P(x = 0) = P(TT) = \frac{1}{4}$$

$$P(x = 1) = P(TH) + P(HT) = \frac{1}{4} + \frac{1}{4} = \frac{1}{2}$$

$$P(x = 2) = P(HH) = \frac{1}{4}$$

HH • *x* = 2	*HT* • *x* = 1
TH • *x* = 1	*TT* • *x* = 0

S

Figure 4.1   Venn Diagram for the Two-Coin Toss Experiment

Thus, we now know the values the random variable can assume, and we know how the probability is *distributed over* these values. This completely describes the random variable and will be referred to as the **probability distribution.** This probability distribution is given in the form of a table (Table 4.2) and as a graph (Figure 4.2). Denoting the probability of *x* by the symbol $p(x)$, we have $p(0) = \frac{1}{4}$, $p(1) = \frac{1}{2}$, and $p(2) = \frac{1}{4}$.

Table 4.2
Probability Distribution for Coin Toss Experiment: Tabular Form

x	p(x)
0	$\frac{1}{4}$
1	$\frac{1}{2}$
2	$\frac{1}{4}$

Figure 4.2   Probability Distribution for Coin Toss Experiment: Graphic Form

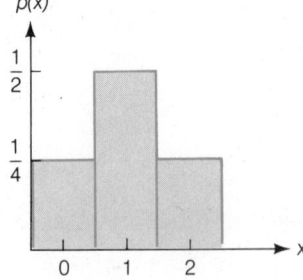

Despite the fact that the probabilities $p(0)$, $p(1)$, and $p(2)$ are concentrated at the points $x = 0$, 1, and 2, respectively, we represent the probabilities in Figure 4.2 as rectangles similar to the rectangles that are used in a relative frequency histogram. This emphasizes the relationship between probability distributions and relative frequency histograms, and we will use the same representation when we approximate certain types of probabilities.

We could also present the probability distribution for x as a formula, but this would unnecessarily complicate a very simple example. We will give the formulas for the probability distributions of some common discrete random variables later in this chapter.

Definition 4.4

The **probability distribution** of a discrete random variable is a graph, table, or formula that specifies the probability associated with each possible value the random variable can assume.

Two requirements must be satisfied by all probability distributions for discrete random variables:

Requirements for the Probability Distribution of a Discrete Random Variable x

$p(x) \geq 0$ for all values of x

$\sum p(x) = 1$

where the summation* of $p(x)$ is over all possible values of x.

Example 4.5 illustrates how the probability distribution for a discrete random variable can be derived, but for many practical situations the task is much more difficult. Fortunately, many experiments and associated discrete random variables observed in nature possess identical characteristics. Thus, you might observe a random variable in a psychology experiment that would possess the same probability distribution as a random variable observed in an engineering experiment or a social sample survey. We classify random variables according to type of experiment, derive the probability distribution for each of the different types, and then use the appropriate probability distribution when a particular type of random variable is observed in a practical situation. The probability distributions for most commonly occurring discrete random variables have already been derived. This fact simplifies the problem of finding the probability distributions for random variables.

---

* Unless otherwise indicated, summations will always be over all possible values of x.

Exercises
4.4–4.16

Learning the Mechanics

4.4 Consider the following probability distribution:

x	−4	0	1	3
$p(x)$	.1	.3	.4	.2

a. List the values that x may assume.
b. What value of x is most probable?
c. What is the probability that x is greater than zero?
d. What is the probability that $x = -2$?

4.5 A discrete random variable x can assume five possible values: 2, 3, 5, 8, and 10. Its probability distribution is shown here:

x	2	3	5	8	10
$p(x)$	.20	.10		.25	.15

a. What is $p(5)$?
b. What is the probability that x equals 2 or 10?
c. What is $P(x \le 8)$?

4.6 Explain why each of the following is or is not a valid probability distribution for a random variable x:

a.

x	0	1	2	3
$p(x)$	.1	.3	.3	.2

b.

x	−2	−1	0
$p(x)$	.25	.50	.25

c.

x	4	9	20
$p(x)$	−.3	.4	.3

d.

x	2	3	5	6
$p(x)$	.15	.15	.45	.35

4.7 The random variable x has the following discrete probability distribution:

x	0	1	2	3	4
$p(x)$	.10	.10	.25	.25	.30

a. Find $P(x \le 0)$.   **b.** Find $P(x < 0)$.   **c.** Find $P(x = 2)$.
d. Find $P(x \le 3)$.   **e.** Find $P(x > 0)$.   **f.** Find $P(2 \le x \le 4)$.

4.8 The random variable x has the following discrete probability distribution:

x	−2	−1	0	1	2
$p(x)$	.05	.20	.40	.20	.15

a. Find $P(x \le 0)$.   **b.** Find $P(x > -1)$.   **c.** Find $P(-1 \le x \le 1)$.
d. Find $P(x < 2)$.   **e.** Find $P(-1 < x < 2)$.   **f.** Find $P(x < 1)$.

4.9 Toss three fair coins and let x equal the number of heads observed.

a. Identify the simple events associated with this experiment and assign a value of x to each simple event.

b. Calculate $p(x)$ for each value of x.

c. Construct a probability histogram for $p(x)$.

d. What is $P(x = 2 \text{ or } x = 3)$?

4.10 A fair die is tossed twice and x, the sum of the up faces, is recorded.

a. Give the probability distribution for x in tabular form.

b. Find $P(x \geq 8)$.

c. Find $P(x < 8)$.

d. What is the probability that x is odd? Even?

e. What is $P(x = 7)$?

Applying the Concepts

4.11 In Exercise 3.55 we noted that 1 in every 500 blacks in the United States has sickle-cell anemia. One in 10 is a carrier of the trait, and those who marry have a 1 in 4 chance of giving the disease to their children. Suppose that a carrier has three children and let x equal the number who acquire the disease.

a. Find $p(x)$ for x = 0, 1, 2, 3.   **b.** Graph $p(x)$.   **c.** Find $P(x \geq 1)$.

4.12 Every human possesses two sex chromosomes. A copy of one or the other (equally likely) is contributed to an offspring. Males have one X chromosome and one Y chromosome. Females have two X chromosomes. If a couple has three children, what is the probability that they have at least one boy? [*Hint:* Define the random variable z as the number of male offspring, and find the probability distribution of z.]

4.13 Coach "Bear" Bryant of the University of Alabama, a legendary figure in college football, was known for his winning seasons. He consistently won nine or more games per season. Suppose that x equals the number of games won up to the halfway mark (six games) in a twelve game season. If Coach Bryant and his team had a probability $p = .70$ of winning any one game (and the winning or losing of one game was independent of another), then the probability distribution of the number x of winning games in a series of six games (we show how to calculate these probabilities in Section 4.4) is

x	0	1	2	3	4	5	6
$p(x)$	.001	.010	.060	.185	.324	.302	.118

Find the probability that the number of games won by Coach Bryant in the first half of a randomly selected season is:

a. 6   **b.** 5   **c.** Less than or equal to 4

4.14 Recent studies have found that 49% of all American households possess a gun of some kind, and of these gun owners, 61% favor gun control. Three households are chosen at random, and x, the number of households possessing a gun while also favoring gun control, is counted.

a. Calculate $p(x)$ for $x = 0, 1, 2, 3$.　　**b.** Graph $p(x)$.

c. Find the probability that at least one of the three households possesses a gun and favors gun control.

4.15 In Exercise 3.58 we noted that a new drug, cyclosporine, appears to retard a body's immune system from rejecting transplanted organs. The drug's developer, Sandoz, reports that kidney transplant patients who receive the drug have an 80% chance of living through the first year (*Newsweek,* August 29, 1983). Suppose that four kidney transplant patients are given cyclosporine, and let x equal the number living through the first year.

a. Calculate $p(x)$ for $x = 0, 1, 2, 3, 4$.　　**b.** Graph $p(x)$.　　**c.** Find $P(x < 2)$.

4.16 Suppose that two men and two women apply for two identical managerial positions. If all the candidates are equally qualified so that essentially two of the four are selected at random for the positions, find the probability distribution of x, the number of women who are hired. Graph $p(x)$.

4.3 Expected Values of Discrete Random Variables

If a discrete random variable x were observed a very large number of times and the data generated were arranged in a relative frequency distribution, the relative frequency distribution would be indistinguishable from the probability distribution for the random variable. Thus, the probability distribution for a random variable is a theoretical model for the relative frequency distribution of a population. To the extent that the two distributions are equivalent (and we will assume they are), the probability distribution for x possesses a mean μ and a variance σ^2 that are identical to the corresponding descriptive measures for the population. This section explains how you can find the mean value for a random variable. We will illustrate the procedure with an example.

Examine the probability distribution for x (the number of heads observed in the toss of two coins) in Figure 4.3. Try to locate the mean of the distribution intuitively. We may reason that the mean μ of this distribution is equal to 1 as follows: In a large number of experiments, $\frac{1}{4}$ should result in $x = 0$, $\frac{1}{2}$ in $x = 1$, and $\frac{1}{4}$ in $x = 2$ heads. Therefore, the average number of heads is

$$\mu = 0(\tfrac{1}{4}) + 1(\tfrac{1}{2}) + 2(\tfrac{1}{4})$$
$$= 0 \quad + \tfrac{1}{2} \quad + \tfrac{1}{2} = 1$$

Figure 4.3 Probability Distribution for a Two-Coin Toss

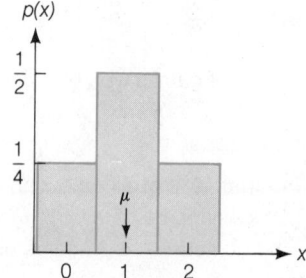

Note that to get the population mean of the random variable x, we multiply each possible value of x by its probability $p(x)$, and then we sum this product over all possible values of x. Another term often used as a substitute for the **mean of x** is the **expected value of x, denoted E(x).**

Definition 4.5

The **expected value** of a discrete random variable x is

$$\mu = E(x) = \sum xp(x)$$

Example 4.6

Suppose you work for an insurance company and you sell a $10,000 whole-life insurance policy at an annual premium of $290. Actuarial tables show that the probability of death during the next year for a person of your customer's age, sex, health, etc., is .001. What is the expected gain (amount of money made by the company) for a policy of this type?

Solution

The experiment is to observe whether the customer survives the upcoming year. The probabilities associated with the two simple events, Live and Die, are .999 and .001, respectively. The random variable you are interested in is the gain x, which can assume the following values:

GAIN x	SIMPLE EVENT	PROBABILITY
$290	Customer lives	.999
$290–$10,000	Customer dies	.001

If the customer lives, the company gains the $290 premium as profit. If the customer dies, the gain is negative because the company must pay $10,000, for a net "gain" of $(290 − 10,000)$. The expected gain is therefore

$$\mu = E(x) = \sum xp(x)$$
$$= (290)(.999) + (290 - 10{,}000)(.001)$$
$$= 290(.999 + .001) - 10{,}000(.001)$$
$$= 290 - 10 = \$280$$

In other words, if the company were to sell a very large number of 1-year $10,000 policies to customers possessing the characteristics described above, it would (on the average) net $280 per sale in the next year. ■

We want to measure the variability as well as the central tendency of a probability distribution. The **population variance** σ^2 is defined as the average of the squared distance of x from the population mean μ. Since x is a random variable, the squared distance, $(x - \mu)^2$, is also a random variable. Using the same logic used to find the mean value of x, we find the mean value of $(x - \mu)^2$ by multiplying all possible values of $(x - \mu)^2$ by $p(x)$ and then summing over all possible x-values.* This quantity,

$$E[(x - \mu)^2] = \sum_{\text{all } x} (x - \mu)^2 p(x)$$

is also called the **expected value of the squared distance from the mean;** that is, $\sigma^2 = E[(x - \mu)^2]$. The standard deviation of x is defined as the square root of the variance σ^2.

Definition 4.6

The **variance** of a discrete random variable x is

$$\sigma^2 = E[(x - \mu)^2] = \sum (x - \mu)^2 p(x)$$

Definition 4.7

The **standard deviation** of a discrete random variable is equal to the square root of the variance, i.e., to $\sigma = \sqrt{\sigma^2}$.

Example 4.7

Medical research has shown that a certain type of chemotherapy is successful 70% of the time when used to treat skin cancer. Suppose five skin cancer patients are treated with this type of chemotherapy and let x equal the number of successful cures out of the five. The probability distribution for the number x of successful cures out of five is given in the table:

x	0	1	2	3	4	5
$p(x)$	.002	.029	.132	.309	.360	.168

a. Find $\mu = E(x)$.   **b.** Find $\sigma = \sqrt{E[(x - \mu)^2]}$.

c. Graph $p(x)$. Locate μ and the interval $\mu \pm 2\sigma$ on the graph. Explain how μ and σ can be used to describe $p(x)$.

Solution   **a.** Applying the formula

$$\mu = E(x) = \sum xp(x)$$
$$= 0(.002) + 1(.029) + 2(.132) + 3(.309) + 4(.360) + 5(.168)$$
$$= 3.50$$

---

* It can be shown that $E[(x - \mu)^2] = E(x^2) - \mu^2$, where $E(x^2) = \sum x^2 p(x)$. Note the similarity between this expression and the shortcut formula $\sum (x - \bar{x})^2 = \sum x^2 - (\sum x)^2/n$ given in Chapter 2.

b. Now we calculate the variance of x:

$$\sigma^2 = E[(x - \mu)^2] = \sum(x - \mu)^2 p(x)$$
$$= (0 - 3.5)^2(.002) + (1 - 3.5)^2(.029) + (2 - 3.5)^2(.132)$$
$$+ (3 - 3.5)^2(.309) + (4 - 3.5)^2(.360) + (5 - 3.5)^2(.168)$$
$$= 1.05$$

Thus, the standard deviation is

$$\sigma = \sqrt{\sigma^2} = \sqrt{1.05} = 1.02$$

c. The graph of $p(x)$ is shown in Figure 4.4. Note that the mean μ and the interval $\mu \pm 2\sigma$ are shown on the graph. We can use μ and σ to describe the probability distribution $p(x)$ in the same way that we used $\bar{x}$ and s to describe a relative frequency distribution in Chapter 2. Note particularly that $\mu = 3.5$ locates the center of the probability distribution. If five skin cancer patients receive the chemotherapy treatment, we expect the number x that are cured to be near 3.5. Similarly, $\sigma = 1.02$ measures the spread of the probability distribution $p(x)$. Since this distribution is a theoretical relative frequency distribution that is moderately mound-shaped (see Figure 4.4), we expect (see Tables 2.6 and 2.7) at least 75% and, more likely, near 95% of observed x-values to fall in the interval $\mu \pm 2\sigma$, that is, between 1.46 and 5.54. Compare this result with the actual probability that x falls in the interval $\mu \pm 2\sigma$. From Figure 4.4 you can see that this probability includes the sum of

Figure 4.4 Graph of $p(x)$ for Example 4.7

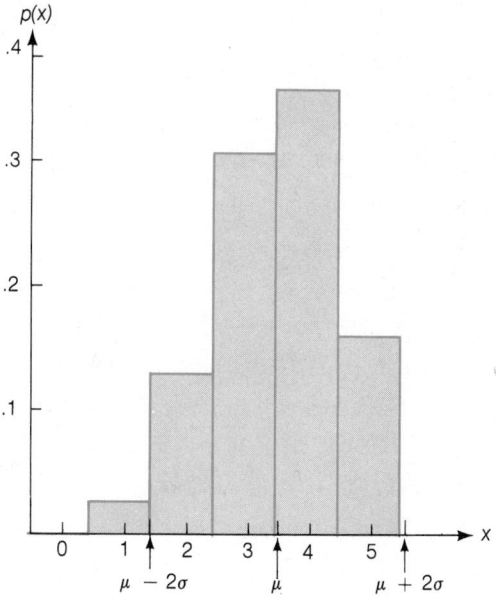

$p(x)$ for all values of x except $p(0) = .002$ and $p(1) = .029$. Therefore, 96.9% of the probability distribution lies within 2 standard deviations of the mean. This percentage is consistent with Tables 2.6 and 2.7. ■

Case Study 4.1

A Restaurant Chain
Fights Sales Tax
Claim

The June 1, 1977, business section of the Orlando, Florida, *Sentinel Star* featured the following headline: "Red Lobster to Fight Tax Claim." According to the *Sentinel Star*, the Red Lobster Inns of America, a national seafood chain, had decided to take the state of Florida to court. The dispute concerned the 4% sales tax levied on most purchases in the state and focused mainly on the state's "bracket collection system." According to the bracket system, a merchant must collect 1¢ for sales between 10¢ and 25¢, 2¢ for sales from 26¢ to 50¢, 3¢ for sales between 51¢ and 75¢, and 4¢ for sales between 76¢ and 99¢. Red Lobster contended that if this system is followed, merchants will always collect more than 4%. That is, if a sale were made for $10.41, 4% will be collected on the $10, but more than 4% will be collected on the 41¢. This, they contend, would amount to more than 4% on the total sale and therefore is not consistent with the 4% tax required by law.

Concrete evidence supplied by the state of Florida tax records does indeed support the contention that the amount of tax collected using the bracket system exceeds the 4% specified by law. It appears that the state sales tax receipts exceeded expected revenue (based on 4%) by $9.5 million in a single year.

What percent sales tax should the state expect to receive using the bracket system for computing the tax? (As noted, the tax on the whole dollar portion of the sale will be 4%.) Using the formula for calculating expected values, you can show (see Exercise 4.124) that the expected percent tax paid on the cents portion of a sale is 4.6%.

Exercises 4.17–4.29

Learning the Mechanics

4.17 Consider the following probability distribution for the random variable x:

x	10	20	30	40	50	60
$p(x)$	.10	.25	.30	.20	.10	.05

a. Find μ, σ^2, and σ.   **b.** Graph $p(x)$.
c. Locate μ and the interval $\mu \pm 2\sigma$ on your graph. What is the probability that x will fall within the interval $\mu \pm 2\sigma$?

4.18 Consider the following probability distribution for the random variable x:

x	1	2	3	4	5
$p(x)$	.05	.30	.35	.20	.10

a. Find μ, σ^2, and σ.   b. Graph $p(x)$.

c. Locate μ and the interval $\mu \pm \sigma$ on your graph. What is the probability that x will fall within the interval $\mu \pm \sigma$?

d. Locate the interval $\mu \pm 3\sigma$ on your graph. What is the probability that x falls within this interval?

4.19 Consider the following probability distribution:

x	1	2	4	10
$p(x)$	.3	.5	.1	.1

a. Find $\mu = E(x)$.   b. Find $\sigma^2 = E[(x - \mu)^2]$.   c. Find σ.

4.20 Consider the following probability distribution:

x	0	1	2	3	4
$p(x)$	.1	.4	.3	.1	.1

a. Find the mean of the distribution.   b. Find the variance.

c. Find the standard deviation.

4.21 Consider the following probability distribution:

x	-4	-3	-2	-1	0	1	2	3	4
$p(x)$	.02	.07	.10	.15	.30	.18	.10	.06	.02

a. Calculate μ, σ^2, and σ.

b. Graph $p(x)$. Locate μ, $\mu - 2\sigma$, and $\mu + 2\sigma$ on the graph.

c. What is the probability that x is in the interval $\mu \pm 2\sigma$?

4.22 Consider the following probability distributions:

x	0	1	2		y	0	1	2
$p(x)$	.3	.4	.3		$p(y)$	.1	.8	.1

a. Use your intuition to find the mean for each distribution. How did you arrive at your choice?

b. Which distribution appears to be more variable? Why?

c. Calculate μ and σ^2 for each distribution. Compare these answers to your answers in parts a and b.

Applying the Concepts

4.23 A rehabilitation officer at a county jail questioned each inmate to determine how many previous convictions, x, each had prior to the one for which he or she was now

serving. The relative frequencies corresponding to x are given in the following probability distribution:

x	0	1	2	3	4
p(x)	.16	.53	.20	.08	.03

If we can regard the relative frequencies as the approximate values for $p(x)$, find the
expected number of previous convictions for an inmate.

4.24 A hospital research laboratory purchases rats from a distributor for use in experiments. The distributor sells four different strains of rats, each at different prices, and
fills requests with one of the four strains depending on availability. The laboratory would
like to estimate how much it will have to spend on rats in the next year. The following
table lists the four strains, price per fifty rats for each strain, and the probability of the
purchase of each strain:

STRAIN	PRICE PER 50 RATS	PROBABILITY
A	$50.00	$\frac{1}{10}$
B	$75.00	$\frac{2}{5}$
C	$87.50	$\frac{3}{10}$
D	$100.00	$\frac{1}{5}$

a. If x is the price of a shipment of fifty rats, what is the expected price?
b. What is the variance of the price?
c. Graph $p(x)$. Locate μ and $\mu \pm 2\sigma$ on the graph. What proportion of the time does
x fall in this interval?

4.25 In Exercise 4.15, we found the probability distribution of the number x of four kidney
transplant patients who were treated with the drug cyclosporine and who survived at
least 1 year.
a. Find the expected value of x.   **b.** Find the variance of x.
c. Find the probability that x falls in the interval $\mu \pm 2\sigma$.

4.26 Exercise 4.13 gives a hypothetical probability distribution for the number x of football games that Coach Bear Bryant of Alabama might win in the first half of his season.
The probability distribution is reproduced here:

x	0	1	2	3	4	5	6
p(x)	.001	.010	.060	.185	.324	.302	.118

a. Find the expected number of games that Alabama would win in the first half of the
season.
b. Find σ.
c. Find the probability that x is in the interval $\mu \pm 2\sigma$.

Profit contribution	p(Profit contribution)
−$5,000[a]	.3
$10,000	.4
$30,000	.3

[a] A negative contribution is a loss.

4.27 A company's marketing and accounting departments have determined that if the company markets its newly developed line of party favors, the probability distribution in the table will describe the contribution of the new line to the firm's profit during the next 6 months. The company has decided it should market the new line of party favors if the expected contribution to profit for the next 6 months is over $10,000. Based on the probability distribution, will the company market the new line?

4.28 On a certain busy holiday weekend, a national airline has many requests for standby flights at half of the usual one-way air fare. However, past experience has shown that these passengers have only about a 1 in 5 chance of getting on the standby flight. When they fail to get on a flight as a standby, their only other choice is to fly first class on the next flight out. Suppose that the usual one-way air fare to a certain city is $70 and the cost of flying first class is $90. Should a passenger who wishes to fly to this city opt to fly as a standby? [*Hint:* Find the expected cost of the trip for a person flying standby.]

4.29 Odds makers try to predict which football teams will win and by how much (the *spread*). If the odds makers do this accurately, adding the spread to the underdog's score should make the final score a tie. Suppose a bookie will give you $6 for every $1 you risk if you pick the winners in three ballgames (adjusted by the spread). Thus, for every $1 bet you will either lose $1 or gain $5. What is the bookie's expected earnings per dollar wagered?

4.4
The Binomial Random Variable

Many experiments result in dichotomous responses, i.e., responses for which there exist two possible alternatives, such as Yes–No, Pass–Fail, Defective–Nondefective, Male–Female, etc. A simple example of such an experiment is the coin toss experiment. A coin is tossed a number of times, say ten. Each toss results in one of two outcomes, Head or Tail, and the probability of observing each of these two outcomes remains the same for each of the ten tosses. Ultimately, we are interested in the probability distribution of x, the number of heads observed. Many other experiments are equivalent to tossing a coin (either balanced or unbalanced) a fixed number n of times and observing the number x of times that one of the two possible outcomes occurs. Random variables that possess these characteristics are called **binomial random variables.**

Public opinion and consumer preference polls (e.g., the Gallup and Harris polls) frequently yield observations on binomial random variables. For example, suppose a sample of 100 students is selected from a large student body and each person is asked whether he or she favors (a Head) or opposes (a Tail) a certain campus issue. Ultimately, we are interested in x, the number of people in the sample who favor the issue. If each student is randomly selected from the student body and if the (unknown) proportion of students favoring the issue is p, then observing whether a student favors or is opposed to the issue is analogous to tossing an unbalanced coin. The chance that any randomly selected student favors the issue is p; the probability that he or she opposes the issue is $(1 - p)$. Sampling 100 students is analogous to tossing the coin

100 times. Thus, you can see that opinion polls which record the number of people who favor a certain issue are real-life equivalents of coin toss experiments.

The experiment we have been describing is called a **binomial experiment** and is identified by the following characteristics:

Characteristics of a Binomial Random Variable

1. The experiment consists of n identical trials.
2. There are only two possible outcomes on each trial. We will denote one outcome by S (for Success) and the other by F (for Failure).
3. The probability of S remains the same from trial to trial. This probability will be denoted by p, and the probability of F will be denoted by q. Note that $q = 1 - p$.
4. The trials are independent.
5. The binomial random variable x is the number of S's in n trials.

Example 4.8   For each of the following examples, decide whether x is a binomial random variable.

a. Suppose a university scholarship committee must select two students to receive a scholarship for the next academic year. The committee receives ten applications for the scholarships—six from male students and four from female students. Suppose the applicants are all equally qualified, so that the selections are randomly made. Let x be the number of female students who receive a scholarship.

b. Before marketing a new product on a large scale, many companies will conduct a consumer-preference survey to determine whether the product is likely to be successful. Suppose a company develops a new diet soda and then conducts a taste-preference survey with 100 randomly chosen consumers stating their preference among the new soda and the two leading sellers. Let x be the number of the 100 who choose the new brand over the two others.

c. Some surveys are conducted by using a method of sampling other than simple random sampling (defined in Chapter 3). For example, suppose a television cable company is trying to decide whether to establish a branch in a certain city. The company plans to conduct a survey to determine the fraction of households in the city that would use the cable television service. The sampling method is to choose a city block at random and then survey every household on that block. This sampling technique is called **cluster sampling.** Suppose ten blocks are so sampled, producing a total of 124 household responses. Let x be the number of the 124 households that would use the television cable service.

Solution   a. In checking the binomial characteristics, a problem arises with independence (characteristic 4 in the preceding box). Given that the first student selected is female, the probability that the second chosen is female is $\frac{3}{9}$. On the other hand, given that the first selection is a male student, the probability that the second is female is $\frac{4}{9}$. Thus, the conditional probability of a Success (choosing a female student to receive a scholarship) on the second trial (selection) depends on the outcome of the first trial, and the trials are therefore dependent. Since the trials are *not independent,* this is not a binomial random variable.

b. Surveys that produce dichotomous responses and use random sampling techniques are classic examples of binomial experiments. In our example, each randomly selected consumer either states a preference for the new diet soda or does not. The sample of 100 consumers is a very small proportion of the totality of potential consumers, so the response of one would be, for all practical purposes, independent of another. Thus, x is a binomial random variable.

c. This example is a survey with dichotomous responses (Yes or No to the cable service), but the sampling method is not simple random sampling. Again, the binomial characteristic of independent trials would very probably not be satisfied. The responses of households within a particular block almost surely would be dependent, since households within a block tend to be similar with respect to income, race, and general interests. Thus, the binomial model would not be satisfactory for x if the cluster sampling technique were employed. ∎

Example 4.9   The Heart Association claims that only 10% of adults over 30 years of age in the United States can pass the minimum fitness requirements established by the president's Physical Fitness Commission. Suppose that four adults are randomly selected and each is given the fitness test. Let x be the number of the four who pass the minimum requirements. Find the probability distribution for x, assuming that the Heart Association's claim is true.

Solution   Recall that a probability distribution describes a discrete random variable by assigning probabilities to each of its values. In this example, the possible values of x, the number of four adults who pass the minimum requirements, are 0, 1, 2, 3, 4. Furthermore, if you check the characteristics of this experiment, you can see that it is a binomial experiment.

Let us first consider the event $x = 0$, that is, the event that none of the four tested adults passes the test. You can see that the event $x = 0$ is equivalent to the simple event

FFFF

where F in the first position implies that adult 1 fails, F in the second position implies that adult 2 fails, etc. Since the trials are independent in this binomial experiment (knowing whether adult 1 passes should not affect the probability that adult 2 passes), we can find the probability of an intersection by multiplying the probabilities of the events. Thus,

$$P(x = 0) = P(FFFF) = P(F)P(F)P(F)P(F)$$
$$= (.9)(.9)(.9)(.9) = (.9)^4 = .6561$$

The event $x = 1$ implies that one of the four adults passes the physical fitness test and three fail it. The following list of simple events contains all simple events that imply

$x = 1$:

SFFF   *FSFF*   *FFSF*   *FFFS*

where S in the first position corresponds to adult 1 passing the test, S in the second position corresponds to adult 2 passing the test, etc. Note that each of these simple events will have the same probability, $(.1)(.9)^3$, where .1 corresponds to the one adult who passes the test and $(.9)^3$ corresponds to the three who fail it. Remembering from Chapter 3 that we obtain the probability of an event by summing the probabilities of the simple events of which it is composed, we get

$$P(x = 1) = 4[(.1)(.9)^3] = .2916$$

The event $x = 2$ implies that two adults pass the test and two fail it; this event consists of the following six simple events:

SSFF   *SFSF*   *SFFS*   *FSSF*   *FSFS*   *FFSS*

Each of these simple events has probability $(.1)^2(.9)^2$, so that

$$P(x = 2) = 6[(.1)^2(.9)^2] = .0486$$

Similarly,

$$P(x = 3) = 4[(.1)^3(.9)] = .0036$$
$$P(x = 4) = (.1)^4 = .0001$$

The complete probability distribution is given in Table 4.3 and shown in Figure 4.5.

Table 4.3
Probability Distribution for Physical Fitness Example: Tabular Form

x	$p(x)$
0	.6561
1	.2916
2	.0486
3	.0036
4	.0001

Figure 4.5 Probability Distribution for Physical Fitness Example: Graphic Form

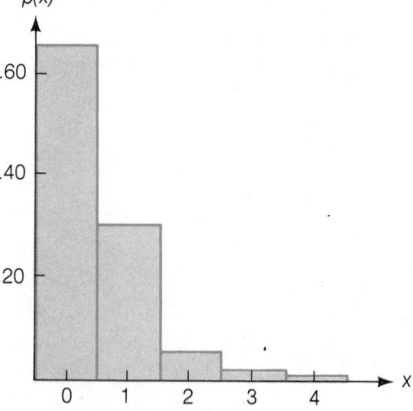

Before we give a formula for $p(x)$, we will refresh your memory on factorial notation. Particularly, the symbol $n!$ is to be read "n factorial" and is calculated by

$$n! = n(n - 1)(n - 2) \cdots \cdot 3 \cdot 2 \cdot 1$$

We define $0! = 1$. Thus, for example, $4! = 4 \cdot 3 \cdot 2 \cdot 1 = 24$.

Using factorial notation, we write the formula for a binomial probability distribution with $n = 4$ and $p = .2$:

$$p(x) = \frac{4!}{x!(4-x)!}(.1)^x(.9)^{4-x}$$

Then, for $x = 2$, we have

$$p(2) = \frac{4!}{2!(4-2)!}(.1)^2(.9)^{4-2} = \frac{4 \cdot 3 \cdot 2 \cdot 1}{(2 \cdot 1)(2 \cdot 1)}(.1)^2(.9)^2$$

$$= 6(.1)^2(.9)^2 = .0486$$

which agrees with our simple event calculation. Note that the first part of the formula, $4!/x!(4-x)!$, counts the number of simple events that result in x adults passing the physical fitness test. The second part of the formula, $(.1)^x(.9)^{4-x}$, is the probability assigned to each simple event that has x adults passing and $(4-x)$ failing. When we multiply the *number* of simple events by the *probability* assigned to each simple event, we get the probability that x adults pass the test. We will see that this formula can be generalized to give the probability distribution of any binomial random variable.

Note that $\binom{n}{x}$, shorthand for $n!/x!(n-x)!$, is the number of simple events that have x successes and $(n-x)$ failures (see Section 3.8 for a discussion of the combinations rule for counting simple events), and $p^x q^{n-x}$ is the probability assigned to each simple event that has x successes and $(n-x)$ failures. The product of these two quantities, $\binom{n}{x}p^x q^{n-x}$, is the probability that x successes and $(n-x)$ failures are observed.

The Binomial Probability Distribution

$$p(x) = \binom{n}{x}p^x q^{n-x} \qquad (x = 0, 1, 2, \ldots, n)$$

where

$p =$ Probability of a success on a single trial

$q = 1 - p$

$n =$ Number of trials

$x =$ Number of successes in n trials

$$\binom{n}{x} = \frac{n!}{x!(n-x)!}$$

The binomial probability distribution is so named because the probabilities, $p(x)$, $x = 0, 1, \ldots, n$, are terms of the binomial expansion, $(q + p)^n$.

Example 4.10   Refer to Example 4.9. Calculate μ and σ, the mean and standard deviation, respectively, of the number of the four adults who pass the test.

Solution   From Section 4.3 we know that the mean of a discrete probability distribution is

$$\mu = \sum x p(x)$$

Referring to Table 4.3, the probability distribution for the number x who pass the fitness test, we find

$$\mu = 0(.6561) + 1(.2916) + 2(.0486) + 3(.0036) + 4(.0001)$$
$$= .4 = 4(.1) = np$$

The relationship $\mu = np$ holds in general for a binomial random variable.
   The variance is

$$\sigma^2 = \sum(x - \mu)^2 p(x) = \sum(x - .4)^2 p(x)$$
$$= (0 - .4)^2(.6561) + (1 - .4)^2(.2916) + (2 - .4)^2(.0486)$$
$$+ (3 - .4)^2(.0036) + (4 - .4)^2(.0001)$$
$$= .104976 + .104976 + .124416 + .024336 + .001296$$
$$= .36 = 4(.1)(.9) = npq$$

The relationship $\sigma^2 = npq$ holds in general for a binomial random variable.
   Finally, the standard deviation of the number who pass the fitness test is

$$\sigma = \sqrt{\sigma^2} = \sqrt{.36} = .6$$

   We emphasize that you need not use the expectation summation rules to calculate μ and σ^2 for a binomial random variable. You can find them easily using the formulas $\mu = np$ and $\sigma^2 = npq$.

Mean, Variance, and Standard Deviation for a Binomial Random Variable

Mean: $\mu = np$

Variance: $\sigma^2 = npq$

Standard deviation: $\sigma = \sqrt{npq}$

   As we demonstrated in Chapter 2, the mean and standard deviation provide measures of the central tendency and variability, respectively, of a distribution. Thus, we can use μ and σ to obtain a rough visualization of the probability distribution for x when the calculation of the probabilities is too tedious. To illustrate the use of the binomial probability distribution, consider Example 4.11.

Example 4.11 A poll of twenty voters is taken in a large city. The purpose is to determine x, the number in favor of a certain candidate for mayor. Suppose that (unknown to us) 60% of all the city's voters favor this candidate.

a. Find the mean and standard deviation of x.
b. Find the probability that x is less than or equal to ten ($x \leq 10$).
c. Find the probability that x exceeds twelve ($x > 12$).
d. Find the probability that x equals eleven ($x = 11$).
e. Graph the probability distribution of x and locate the interval $\mu - 2\sigma$ to $\mu + 2\sigma$ on the graph.

Solution **a.** Given that the sample of twenty was randomly selected from a large number of voters, it is likely that x, the number of the twenty who favor the candidate, is a binomial random variable. The value of p is the fraction of the total voters who favor the candidate, that is, $p = .6$. Therefore, we calculate the mean and variance:

$$\mu = np = 20(.6) = 12 \qquad \sigma^2 = npq = 20(.6)(.4) = 4.8$$

The standard deviation is then

$$\sigma = \sqrt{4.8} = 2.2$$

b. Calculating binomial probabilities when n is large is a formidable task. For example, to find the probability that $x \leq 10$, we would calculate

$$P(x \leq 10) = p(0) + p(1) + p(2) + \cdots + p(10)$$

$$= \sum_{x=0}^{10} {}^{*}p(x) = \sum_{x=0}^{10} \binom{20}{x}(.6)^x(.4)^{20-x}$$

We can avoid these tedious calculations by making use of cumulative binomial probability tables (Table II in the Appendix). Part of Table II is shown in Figure 4.6 (page 160). The entries in Table II are the cumulative sums

$$P(x \leq k) = p(0) + p(1) + p(2) + \cdots + p(k)$$

for values of $k = 0, 1, 2, \ldots, (n - 1)$. Observe that the bottom row of the table, the one corresponding to $k = n$, is omitted. This is because the sum of $p(x)$ from $x = 0$ to $x = n$ is always equal to 1; that is, $P(x \leq n) = 1$ for any binomial random variable.

To find $P(x \leq 10)$ for $n = 20$ and $p = .6$, we first find the column corresponding to $p = .6$ and then the row corresponding to $k = 10$. The recorded value, shaded in Figure 4.6, is

$$P(x \leq 10) = .245$$

---

* The value of x below the $\sum$ symbol, $x = 0$, is the **first member,** or **lower limit,** of the summation. The value of x above the $\sum$ symbol, $x = 10$, is the **last member,** or **upper limit,** of the summation. Thus, $\sum_{x=0}^{10} p(x) = p(0) + p(1) + \cdots + p(9) + p(10)$. We will include these limits when the summation extends over only some of the possible values of x.

Figure 4.6   Partial Reproduction of Table II in the Appendix

h. $n = 20$

k \ p	0.01	0.05	0.10	0.20	0.30	0.40	0.50	0.60	0.70	0.80	0.90	0.95	0.99
0	.818	.358	.122	.012	.001	.000	.000	.000	.000	.000	.000	.000	.000
1	.983	.736	.392	.069	.008	.001	.000	.000	.000	.000	.000	.000	.000
2	.999	.925	.677	.206	.035	.004	.000	.000	.000	.000	.000	.000	.000
3	1.000	.984	.867	.411	.107	.016	.001	.000	.000	.000	.000	.000	.000
4	1.000	.997	.957	.630	.238	.051	.006	.000	.000	.000	.000	.000	.000
5	1.000	1.000	.989	.804	.416	.126	.021	.002	.000	.000	.000	.000	.000
6	1.000	1.000	.998	.913	.608	.250	.058	.006	.000	.000	.000	.000	.000
7	1.000	1.000	1.000	.968	.772	.416	.132	.021	.001	.000	.000	.000	.000
8	1.000	1.000	1.000	.990	.887	.596	.252	.057	.005	.000	.000	.000	.000
9	1.000	1.000	1.000	.997	.952	.755	.412	.128	.017	.001	.000	.000	.000
10	1.000	1.000	1.000	.999	.983	.872	.588	.245	.048	.003	.000	.000	.000
11	1.000	1.000	1.000	1.000	.995	.943	.748	.404	.113	.010	.000	.000	.000
12	1.000	1.000	1.000	1.000	.999	.979	.868	.584	.228	.032	.000	.000	.000
13	1.000	1.000	1.000	1.000	1.000	.994	.942	.750	.392	.087	.002	.000	.000
14	1.000	1.000	1.000	1.000	1.000	.998	.979	.874	.584	.196	.011	.000	.000
15	1.000	1.000	1.000	1.000	1.000	1.000	.994	.949	.762	.370	.043	.003	.000
16	1.000	1.000	1.000	1.000	1.000	1.000	.999	.984	.893	.589	.133	.016	.000
17	1.000	1.000	1.000	1.000	1.000	1.000	1.000	.996	.965	.794	.323	.075	.001
18	1.000	1.000	1.000	1.000	1.000	1.000	1.000	.999	.992	.931	.608	.264	.017
19	1.000	1.000	1.000	1.000	1.000	1.000	1.000	1.000	.999	.988	.878	.642	.182

c. To find the probability

$$P(x > 12) = p(13) + p(14) + \cdots + p(19) + p(20) = \sum_{x=13}^{20} p(x)$$

we use the fact that for all probability distributions, $\sum p(x) = 1$. Therefore,

$$P(x > 12) = 1 - [p(0) + p(1) + \cdots + p(12)]$$

$$= 1 - P(x \le 12) = 1 - \sum_{x=0}^{12} p(x)$$

Consulting Table II, we find the entry in row $k = 12$, column $p = .6$ to be .584. Thus,

$$P(x > 12) = 1 - .584 = .416$$

d. To find the probability that exactly eleven voters favor the candidate, recall that the entries in Table II are cumulative probabilities and use the relationship

$$P(x = 11) = [p(0) + p(1) + \cdots + p(10) + p(11)]$$

$$- [p(0) + p(1) + \cdots + p(9) + p(10)]$$

$$= P(x \le 11) - P(x \le 10)$$

Then

$$P(x = 11) = .404 - .245 = .159$$

Figure 4.7 The Binomial Probability Distribution for *x* in Example 4.11

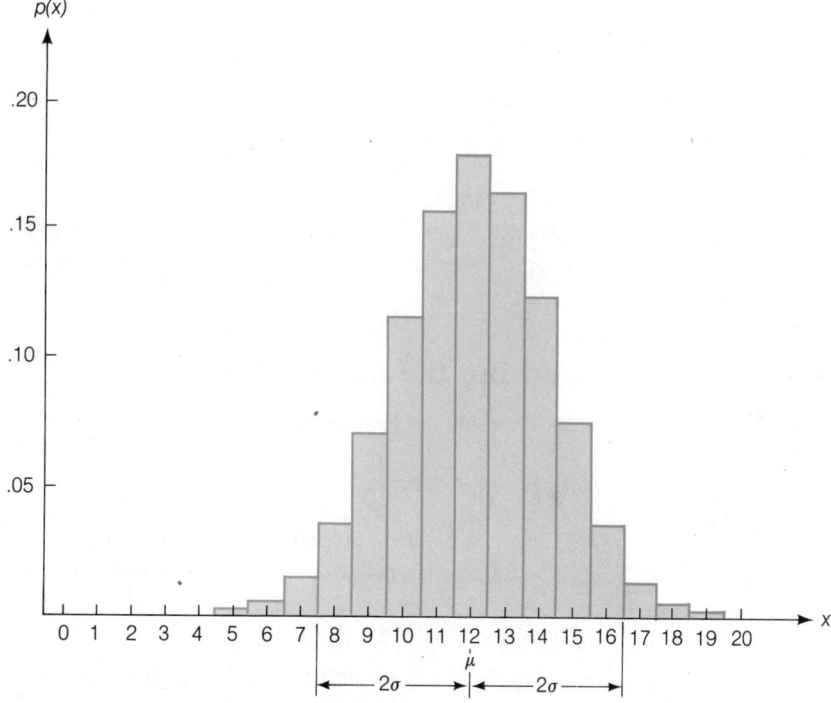

e. The probability distribution for *x* is shown in Figure 4.7. Note that

$$\mu - 2\sigma = 12 - 2(2.2) = 7.6$$
$$\mu + 2\sigma = 12 + 2(2.2) = 16.4$$

The interval $\mu - 2\sigma$ to $\mu + 2\sigma$ is shown in Figure 4.7. The probability that *x* falls in the interval, $\mu \pm 2\sigma$, that is, $P(x = 8, 9, 10, \ldots, 16) = P(x \le 16) - P(x \le 7) = .984 - .021 = .963$. Note that this probability is very close to the .95 given by the Empirical Rule. ∎

Case Study 4.2

A Survey of Children's Political Knowledge

Children's images of political leaders in Britain, France, and the United States were studied by Fred I. Greenstein (1975). Data were collected by means of interviews with small samples of children in the three countries "in order to examine various standard assumptions about political culture and socialization among the three nations, as well as black–white differences in the United States." During one phase of the study, twenty-five black children from the United States were asked to name the president of their country. This phase represents a binomial experiment with $n = 25$ trials and p equal to the proportion of all black children who could correctly name the president at that time (1969–1970). One objective of the experiment was to obtain an estimate of the value of p.

Of the sample of twenty-five black children, twenty-four correctly identified Richard Nixon as president. The implication of this result is that the proportion of all black children who could have made a correct identification must have been quite high. In fact, if the true proportion were equal to .8, the probability that at least twenty-four out of twenty-five would correctly identify the president is only .027. (You can verify this result by using Table II in the Appendix.) Thus, unless the observed outcome represents a rare event, the proportion of all black children who could have correctly identified the president was probably in excess of .8 at that time. In Chapter 7 we will develop a systematic approach for making inferences about proportions.

Exercises 4.30–4.53

Learning the Mechanics

4.30 Compute the following:

a. $\dfrac{5!}{3!(5-3)!}$ **b.** $\dbinom{6}{3}$ **c.** $\dbinom{8}{0}$ **d.** $\dbinom{5}{5}$ **e.** $\dbinom{6}{1}$

4.31 If x is a binomial random variable, compute $p(x)$ for each of the following cases:

a. $n=5,\quad x=1,\quad p=.3$ **b.** $n=4,\quad x=2,\quad q=.5$
c. $n=3,\quad x=0,\quad p=.2$ **d.** $n=5,\quad x=3,\quad p=.4$
e. $n=4,\quad x=2,\quad q=.6$ **f.** $n=3,\quad x=1,\quad p=.3$

4.32 Suppose x is a binomial random variable with $n=6$ and $p=.5$.
a. Display $p(x)$ in tabular form.
b. Compute the mean and variance of x.
c. Graph $p(x)$ and locate μ and the interval $\mu \pm 2\sigma$ on the graph.
d. What is the probability that x falls within the interval $\mu \pm 2\sigma$?

4.33 Suppose x is a binomial random variable with $n=4$ and $p=.2$.
a. Calculate the value of $p(x)$, $x=0,\ 1,\ 2,\ 3,\ 4,\ 5$, using the formula for a binomial probability distribution.
b. Using your answers to part a, give the probability distribution for x in tabular form.

4.34 If x is a binomial random variable, calculate μ, σ^2, and σ for each of the following:

a. $n=30,\quad p=.4$ **b.** $n=80,\quad p=.3$
c. $n=100,\quad p=.3$ **d.** $n=70,\quad p=.9$
e. $n=60,\quad p=.5$ **f.** $n=1,000,\quad p=.03$

4.35 If x is a binomial random variable, use Table II in the Appendix to find the following probabilities:

a. $P(x=2)$ for $n=10,\quad p=.3$
b. $P(x\le 5)$ for $n=15,\quad p=.5$
c. $P(x>1)$ for $n=5,\quad p=.2$
d. $P(x<10)$ for $n=25,\quad p=.8$
e. $P(x\ge 10)$ for $n=15,\quad p=.7$
f. $P(x=2)$ for $n=20,\quad p=.1$

4.36 Suppose x is a binomial random variable with $n = 6$ and $p = .4$.
a. Display $p(x)$ in tabular form.
b. Compute the mean and variance of x.
c. Graph $p(x)$ and locate $E(x)$ and the interval $\mu \pm 2\sigma$ on the graph.
d. What is the probability that x falls within the interval $\mu \pm 2\sigma$?

4.37 Suppose x is a binomial random variable with $n = 5$ and $p = .5$.
a. Display $p(x)$ in tabular form.
b. Compute the mean and variance of x.
c. Graph $p(x)$ and locate $E(x)$ and the interval $\mu \pm 2\sigma$ on the graph.
d. What is the probability that x falls within the interval $\mu \pm 2\sigma$?

4.38 Given that x is a binomial random variable, $n = 15$, and $p = .4$, use Table II to find the following probabilities:
a. $P(x \le 1)$ b. $P(x \ge 3)$ c. $P(x \le 5)$
d. $P(x < 10)$ e. $P(x > 10)$ f. $P(x = 6)$

4.39 The binomial probability distribution is a family of probability distributions with each single distribution depending on the values of n and p. Assume that x is a binomial random variable with $n = 4$.
a. Determine a value of p such that the probability distribution of x is symmetric.
b. Determine a value of p such that the probability distribution of x is skewed to the right.
c. Determine a value of p such that the probability distribution of x is skewed to the left.
d. Graph each of the binomial distributions you obtained in parts a, b, and c. Locate the mean for each distribution on its graph.
e. In general, for what values of p will a binomial distribution be symmetric? Skewed to the right? Skewed to the left?

Applying the Concepts

4.40 A large southern university has determined from past records that the probability a student who registers for fall classes will have his or her schedule rejected (due to overfilled classrooms, clerical error, etc.) is .2.
a. Suppose that 25,000 students register for fall classes, and x is the number of students who have their schedules rejected. Is x a binomial random variable? Explain.
b. Suppose that a random sample of 20 students is selected from the total of 25,000, and x is the number of these students who have their schedules rejected. Is x a binomial random variable? Explain.
c. Suppose that you sample the results of the first 1,000 students who register next fall and record x, the number of rejected registrations. Is x a binomial random variable? Explain.
d. For the random variables in parts a, b, and c that you identified as being binomial, find μ, σ^2, and σ.

4.41 Suppose that 60% of all people who are eligible for jury duty in a large Florida city are in favor of capital punishment. How does this finding affect the composition of a jury in a murder trial? Suppose that a jury of twelve is to be randomly selected from among all the prospective jurors in this city.

a. What is the expected number of jurors who favor capital punishment?

b. What is the probability that none of the twelve jurors selected favors capital punishment?

c. If the jury were really selected at random, would you be surprised if none of the jurors favored capital punishment? Explain.

4.42 A problem of considerable impact on the economy is the burgeoning cost of Medicare and other public-funded medical services. One aspect of this problem, reported in the "Behavior" section of *Time* (April 18, 1983), concerns the high percentage of people seeking medical treatment who, in fact, have no physical basis of their ailment. One conservative estimate is that the percentage of people who seek medical assistance but have no real physical ailment is 10%, and some doctors believe that it may be as high as 40%. Suppose that we randomly sample the records of a doctor and find that five of fifteen patients seeking medical assistance are physically healthy.

a. What is the probability of observing five or more physically healthy patients in a sample of fifteen if the proportion p that the doctor normally sees is 10%?

b. What is the probability of observing five or more physically healthy patients in a sample of fifteen if the proportion p that the doctor normally sees is 40%?

c. Why might your answer to part a make you believe that p is larger than .1?

4.43 The *Orlando Sentinel* (April 12, 1984) reports that "after a celebrity passes out of the limelight, the public quickly assumes that he has died." To substantiate this statement, they give the results of an experiment conducted by *Psychology Today*. The magazine asked its readers to guess whether twenty celebrities of the past were alive. On the average, the respondents guessed correctly on nine and did not even recognize six of the twenty. Suppose that a respondent possessed no knowledge about the state of health of any of the celebrities and simply guessed whether they were alive or dead. What is the probability that a respondent would guess correctly on nine or more of the twenty celebrities?

4.44 The *Wall Street Journal* (March 8, 1984) notes that "a blood clot dissolver that researchers hope can stop heart attacks passed its first test in humans but researchers said the substance must still undergo extensive trials before any life-saving potential can be determined." One aspect of the research by Frans Van de Werf, M.D., and colleagues (1984), reported in the *New England Journal of Medicine*, involved actual tests on seven humans aged 50 or more who suffered heart attacks and were treated with the new drug, *t*-PA. After treatment, the blood clots in six of the seven patients were dissolved. The blood clot did not dissolve in the seventh patient.

Assume that the drug is ineffective in dissolving blood clots and that without the drug the probability p that a heart attack patient's blood clot would dissolve of its own accord, in a short time, is very small, say less than .1. Let x equal the number of heart attack patients in the sample of seven whose blood clots dissolved in a short time.

a. If the drug is ineffective and $p = .1$, what is the probability that blood clots would have dissolved of their own accord in as many as six of seven heart attack patients?

b. If p really is equal to .1 and if the drug t-PA is ineffective in treating heart attacks, would you conclude that $x \geq 6$ is a rare event? Using the logic of Section 2.9, what do you think about the utility of t-PA in dissolving blood clots in heart attack patients?

4.45 Over the years, a physician has found that one out of every ten diabetics receiving insulin develops antibodies against the hormone, thus requiring a more costly form of medication.

a. Find the probability that, of the next five diabetic patients the physician treats, none will develop antibodies against insulin.

b. Find the probability that at least one will develop antibodies.

c. What assumptions are needed for solving this problem?

4.46 A problem of great concern to a manufacturer is the cost of repair and replacement required under a product's guarantee agreement. Assume it is known that 10% of all electronic pocket calculators purchased are returned for repair while their guarantee is still in effect. If a firm purchased twenty-five pocket calculators for its salespeople, what is the probability that five or more of these calculators will need repair while their guarantees are still in effect?

4.47 Experiments with animals are often conducted to determine whether certain chemicals are linked to cancer. Suppose that when a chemical is ingested in large doses by experimental rats, 24% of the rats contract thyroid tumors.

a. If each rat in a group of 200 ingests a large dose of the chemical, how many would be expected to be free of thyroid tumors?

b. Within what limits would we expect the number of rats free of thyroid tumors to fall? [*Hint:* Use Chebyshev's rule to assist in establishing the limits.]

4.48 An accountant believes that 90% of the company's invoices are free of errors. To check this theory the accountant randomly samples twenty-five invoices and finds that seven contain errors.

a. If the accountant's theory is correct, what is the probability that of the twenty-five invoices written, seven or more contain errors?

b. What assumptions do you have to make to solve this problem using the methodology of this section?

c. If these assumptions are satisfied and if the sample of twenty-five invoices produces seven that contain errors, do you think that more than 10% of the company's invoices contain errors? Explain.

4.49 Suppose you are a purchasing officer for a large company. You have purchased 5 million electrical switches and have been guaranteed by the supplier that the shipment will contain no more than .1% defectives. To check the shipment, you randomly sample 500 switches, test them, and find that four are defective. If the switches are as represented, calculate μ and σ for this sample of 500. Based on this evidence, do you think the supplier has complied with the guarantee? Explain. [*Hint:* Calculate μ and σ for this binomial random variable with $p = .001$ to see if a value of x as large as 4 is probable.]

4.50 According to the "January" theory, if the stock market is up in January it will be up for the whole year (and vice versa). Believe it or not, this indicator of stock market behavior has been correct for 29 of the last 34 years and 100% correct in odd-numbered years.* Suppose that there is no truth whatever in this theory and that stock prices are just as likely to move up or down in any given year, regardless of the direction of movement in January.

a. Find the probability of perfect agreement between the January and annual movements in stock prices over the 15 odd-numbered years.

b. What is the probability of perfect agreement between the January and annual movements in stock prices in at least 10 of the 15 years?

4.51 An experiment is to be conducted to see whether an acclaimed psychic has extrasensory perception (ESP). Five different cards are shuffled and one is chosen at random. The psychic will then try to identify which card was drawn without seeing it. The experiment is to be repeated twenty times and x, the number of correct decisions, is recorded. (Assume that the twenty trials are independent.)

a. If the psychic is guessing, i.e., if the psychic does not possess ESP, what is the value of p, the probability of a correct decision on each trial?

b. If the psychic is guessing, what is the expected number of correct decisions in twenty trials?

c. If the psychic is guessing, what is the probability of six or more correct decisions in twenty trials?

d. Suppose that the psychic makes six correct decisions in twenty trials. Is there evidence to indicate that the psychic is *not* guessing and actually has ESP? Explain.

4.52 A new drug has been synthesized that is designed to reduce a person's blood pressure. Twenty randomly selected hypertensive patients receive the new drug. Suppose eighteen or more of the patients' blood pressures drop.

a. Suppose that the probability that a hypertensive patient's blood pressure drops if he or she is *untreated* is .5. Then what is the probability of observing eighteen or more blood pressure drops in a random sample of twenty treated patients if the new drug is in fact ineffective in reducing blood pressure?

b. Considering this probability (part a), do you think you have observed a rare event or do you conclude that the drug is effective in reducing hypertension?

4.53 Most firms use sampling plans to control the quality of manufactured items ready for shipment or the quality of items that have been purchased. To illustrate the use of a sampling plan, suppose you are shipping electrical fuses in lots, each containing 10,000 fuses. The plan specifies that you will randomly sample twenty-five fuses from each lot and accept (or ship) the lot if x, the number of defective fuses in the sample, is less than 3. If $x \geq 3$, you will reject the lot and hold it for a complete reinspection.

a. What is the probability of accepting a lot ($x = 0, 1,$ or 2) if the actual fraction defective in the lot is:

(i) 1   **(ii)** .8   **(iii)** .5   **(iv)** .2   **(v)** .05   **(vi)** 0

---

* From the "Heard on the Street" section of the *Wall Street Journal* (February 1, 1984).

b. Construct a graph showing $P(A)$, the probability of lot acceptance, as a function of lot fraction defective, p. This graph is called the **operating characteristic curve** for the sampling plan.

c. Suppose the sampling plan calls for sampling $n = 25$ fuses and accepting a lot if $x \leq 3$. Calculate the quantities specified in part a and construct the operating characteristic curve for this sampling plan. Compare this curve with the curve obtained in part b. (Note how the curve characterizes the ability of the plan to screen bad lots from shipment.)

4.5
The Poisson Random Variable (Optional)

A type of probability distribution that is often useful in describing the number of events that will occur in a specific period of time or in a specific area or volume is the **Poisson distribution** (named after the eighteenth-century physicist and mathematician, Siméon Poisson). Typical examples of random variables for which the Poisson probability distribution provides a good model are:

1. The number of traffic accidents per month at a busy intersection.
2. The number of noticeable surface defects (scratches, dents, etc.) found by quality inspectors on a new automobile
3. The parts per million of some toxicant found in the water or air emission from a manufacturing plant
4. The number of diseased trees per acre of a certain woodland
5. The number of death claims received per day by an insurance company
6. The number of unscheduled admissions per day to a hospital

Characteristics of a Poisson Random Variable

1. The experiment consists of counting the number of times a certain event occurs during a given unit of time or in a given area or volume (or weight, distance, or any other unit of measurement).
2. The probability that an event occurs in a given unit of time, area, or volume is the same for all the units.
3. The number of events that occur in one unit of time, area, or volume is independent of the number that occur in other units.
4. The mean (or expected) number of events in each unit will be denoted by the Greek letter lambda, λ.

The characteristics of the Poisson random variable are usually difficult to verify for practical examples. The examples given above satisfy them well enough that the Poisson distribution provides a good model in many instances. As with all probability models, the real test of the adequacy of the Poisson model is in whether it provides a reasonable approximation to reality—that is, whether empirical data support it.

The Poisson probability distribution also provides a good approximation to a binomial probability distribution with mean

$$\lambda = np$$

when n is large and p is small (say $np \leq 7$). To illustrate with a relatively small value of n, if $n = 25$ and $p = .05$ then the exact value of the binomial probability $p(2)$ is .231. The Poisson approximation is .224. The approximations are better for $n \geq 100$ (see Exercise 4.58).

The probability distribution, mean, and variance for a Poisson random variable are shown in the accompanying box. Note that λ is the mean of the distribution and that e is the symbol for the irrational number 2.71828 . . . . To assist you in calculating $p(x)$, we provide values of $e^{-\lambda}$ for various values of λ in Table III of the Appendix. Example 4.12 illustrates the use of the Poisson probability distribution.

Probability Distribution, Mean, and Variance for a Poisson Random Variable

$$p(x) = \frac{\lambda^x e^{-\lambda}}{x!} \qquad (x = 0, 1, 2, \ldots)$$

$$\mu = \lambda \qquad \sigma^2 = \lambda$$

where

$\lambda =$ Mean number of events during given unit of time, area, volume, etc.

$e = 2.71828 \ldots$

Example 4.12

One way ecologists estimate population sizes of rare species of animals is to use the number of reported sightings of the animals during a fixed period of time. For example, suppose the number of reported sightings per week of blue whales is recorded. Denoting this random variable by x, assume that x has (approximately) a Poisson distribution with a mean of 2.5.

a. Find the mean and standard deviation of x, the number of blue whale sightings per week.

b. Find the probability that exactly five blue whale sightings will be reported during a 1-week period.

c. Find the probability that two or more sightings are made in a 1-week period.

Solution

a. The mean and variance of a Poisson random variable are both equal to λ. Thus, for this example,

$$\mu = \lambda = 2.5 \qquad \sigma^2 = \lambda = 2.5$$

Then the standard deviation is

$$\sigma = \sqrt{2.5} = 1.6$$

b. We want the probability that exactly five sightings are made in a week. The probability distribution for x is

$$p(x) = \frac{\lambda^x e^{-\lambda}}{x!}$$

Then, since $\lambda = 2.5$, $x = 5$, and $e^{-2.5} = .082085$ (from Table III),

$$p(5) = \frac{(2.5)^5 e^{-2.5}}{5!} = \frac{(2.5)^5 (.082085)}{5 \cdot 4 \cdot 3 \cdot 2 \cdot 1} = .067$$

c. To find the probability that two or more sightings are made, we need to find

$$P(x \geq 2) = p(2) + p(3) + p(4) + \cdots = \sum_{x=2}^{\infty} p(x)$$

Since we cannot compute this infinite summation, we use the complementary event:

$$P(x \geq 2) = 1 - P(x \leq 1) = 1 - [p(0) + p(1)]$$
$$= 1 - \frac{(2.5)^0 e^{-2.5}}{0!} - \frac{(2.5)^1 e^{-2.5}}{1!}$$
$$= 1 - \frac{1(.082085)}{1} - \frac{2.5(.082085)}{1}$$
$$= 1 - .287 = .713$$

According to our Poisson model, the probability that two or more sightings are made in a given week is .713.

The probability distribution for x in this example is shown in Figure 4.8 for x values between 0 and 9. The mean $\mu = 2.5$ and the interval $\mu \pm 2\sigma = (-.7, 5.7)$ are

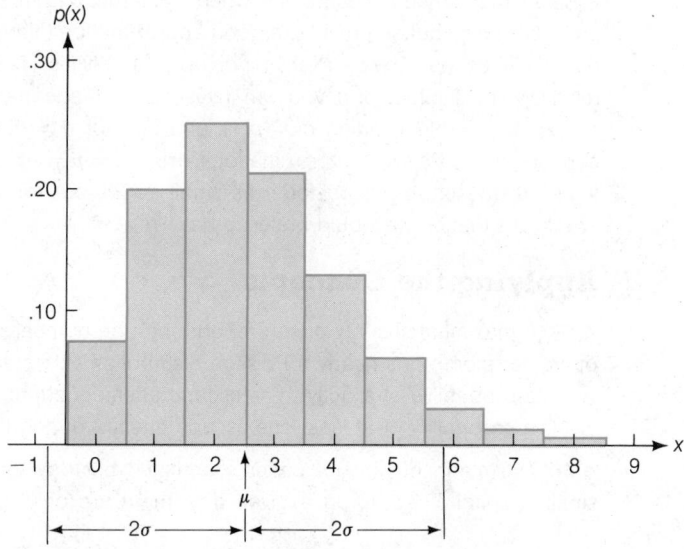

Figure 4.8 Poisson Probability Distribution for x in Example 4.12

indicated. The probability that x falls in the interval $\mu \pm 2\sigma$, that is, $P(x = 0, 1, 2, \ldots, 5)$, is .96. Note that this result is very close to the value .95 given by the Empirical Rule.   ■

Exercises 4.54–4.69

Learning the Mechanics

4.54 Use Table III in the Appendix to compute each of the following:

a. $e^{-3.0}$ **b.** $e^{-4.5}$ **c.** $e^{-1.7}$ **d.** $e^{-6.0}$

e. $\dfrac{(3)^2 e^{-3}}{2!}$ **f.** $\dfrac{(2.5)^3 e^{-2.5}}{3!}$ **g.** $\dfrac{(8.6)^4 e^{-8.6}}{4!}$

4.55 Given that x is a random variable for which a Poisson probability distribution provides a good characterization, compute the following:

a. $P(x \leq 2)$, when $\lambda = 1$ **b.** $P(x = 1)$, when $\lambda = 4$
c. $P(x \geq 1)$, when $\lambda = 2$ **d.** $P(x = 0)$, when $\lambda = 8$

4.56 Suppose that x is a random variable for which a Poisson probability distribution with $\lambda = 2$ provides a good characterization.

a. Graph $p(x)$ for $x = 0, 1, 2, \ldots, 9$.
b. Find μ and σ for x, and locate μ and the interval $\mu \pm 2\sigma$ on the graph.
c. What is the probability that x will fall within the interval $\mu \pm 2\sigma$?

4.57 Suppose that x is a random variable for which a Poisson probability distribution with $\lambda = 4$ provides a good characterization.

a. Graph $p(x)$ for $x = 0, 1, 2, \ldots, 9$.
b. Find μ and σ for x, and locate μ and the interval $\mu \pm 2\sigma$ on the graph.
c. What is the probability that x will fall within the interval $\mu \pm 2\sigma$?

4.58 As mentioned in Section 4.5, when n is large, p is small, and $np \leq 7$, the Poisson probability distribution provides a good approximation to the binomial probability distribution. Since we provide exact binomial probabilities (Table II in the Appendix) for relatively small values of n, you can investigate the adequacy of the approximation for $n = 25$. Use Table II to find $p(0)$, $p(1)$, and $p(2)$ for $n = 25$ and $p = .05$. Calculate the corresponding Poisson approximations using $\lambda = \mu = np$. [*Note:* These approximations are reasonably good for n as small as 25, but to use the approximation in a practical situation we would prefer to have $n \geq 100$.]

Applying the Concepts

4.59 A maximum security prison reports that the number of escape attempts by prisoners per month has nearly a Poisson distribution with mean equal to 1.5. Find:

a. The probability of exactly three escape attempts during the next month
b. The probability of at least one escape attempt during the next month

4.60 The mean number of patients admitted per day to the emergency room of a small hospital is 2.5. If, on a given day, there are only four beds available for new patients, what is the probability the hospital will not have enough beds to accommodate its newly admitted patients?

4.61 A can company reports that the number of breakdowns per 8-hour shift on its machine-operated assembly line follows a Poisson distribution with a mean of 1.5.
a. What is the probability of exactly two breakdowns on the midnight shift?
b. What is the probability of fewer than two breakdowns on the afternoon shift?
c. What is the probability of no breakdowns during three consecutive 8-hour shifts? (Assume that the machine operates independently across shifts.)

4.62 "Lawyers Glut Brings Call to Limit Numbers" headlines an article in the *Gainesville Sun* (April 13, 1984). The article states that there are approximately 650,000 lawyers in the nation today, as many as 1 for every 436 people in Florida, 1 for every 300 in California, and 1 for every 360 in Texas. Suppose you enter a Steak and Ale restaurant in Texas and the restaurant contains 400 patrons.
a. If the patrons represent a random selection of people from within the state of Texas, what is the probability that the restaurant contains at least one lawyer? [*Hint:* See Exercise 4.58.]
b. If you were to go to any Steak and Ale restaurant in Texas and the restaurant contained 400 patrons, the probability that the restaurant would contain at least one lawyer is probably larger than your answer to part a. Why?

4.63 The *Gainesville Sun* (March 19, 1984) notes that local health officials are taking a closer look at the high incidence of cancer in the Jacksonville area. The latest statistics (1981) show the county's incidence of cancer at 121.6 per 100,000 people compared with 76 per 100,000 for the whole of Florida. Although officials are not certain why the incidence of cancer in the Jacksonville area is so high, some think that it is due to a higher than usual incidence of lung cancer resulting from the use of asbestos in the area. Suppose that 1,000 ex-shipyard workers in the Jacksonville area are x-rayed to check for lung cancer. If the incidence of lung cancer in the former shipyard workers is the same as the overall incidence of cancer (say, approximately 120 per 100,000 people).
a. What is the probability that none of the persons examined in the screening process have lung cancer?
b. What is the probability that as many as three persons have lung cancer?

4.64 Refer to Exercise 4.63. Is it likely that as many as 8 among the 1,000 persons screened have lung cancer? If as many as 8 do have lung cancer, would you infer that the incidence of lung cancer in ex-shipyard workers exceeds the incidence of cancer in the general public?

4.65 The safety supervisor at a large manufacturing plant believes the expected number of industrial accidents per month is 3.4.
a. What is the probability of exactly two accidents occurring next month?
b. What is the probability of three or more accidents occurring next month?
c. What assumptions do you need to make to solve this problem using the methodology of this chapter?

4.66 As a check on the quality of the wooden doors produced by a company, its owner requested that each door undergo inspection for defects before leaving the plant. The plant's quality control inspector found that on the average 1 square foot of door surface contains .5 minor flaw. Subsequently, 1 square foot of each door's surface was examined for flaws. The owner decided to have all doors reworked that were found to have two or more minor flaws in the square foot of surface that was inspected.

a. What is the probability that a door will fail inspection and be sent back for reworking?

b. What is the probability that a door will pass inspection?

4.67 The number x of people who arrive at a cashier's counter in a bank during a specified period of time often possesses (approximately) a Poisson probability distribution. If we know the mean arrival rate λ, the Poisson probability distribution can be used to aid in the design of the customer service facility. Suppose you estimate that the mean number of arrivals per minute for cashier service at a bank is one person per minute.

a. What is the probability that in a given minute the number of arrivals will equal three or more?

b. Can you tell the bank manager that the number of arrivals will rarely exceed two per minute?

4.68 The Environmental Protection Agency (EPA) issues standards on air and water pollution that vitally affect the safety of consumers and the operations of industry. For example, the EPA states that manufacturers of vinyl chloride and similar compounds must limit the amount of these chemicals in plant air emissions to 10 parts per billion. Suppose you represent one of the manufacturers and you know that the mean emission of vinyl chloride for your plant is 4 parts per billion.

a. If the parts per billion of vinyl chloride in air follows a Poisson probability distribution and x is the parts per billion for a particular sample, what is the standard deviation of x for your plant?

b. If the mean parts per billion for your plant is really equal to 4, is it likely that a sample would yield a value of x that exceeds the EPA limits? Explain.

4.69 The National Transportation Safety Board is responsible for the investigation of aviation accidents. In 1979, some 208.9 billion passenger miles were flown by commercial airlines in the United States. During this period there were 279 fatalities. In 1980, some 200.1 billion passenger miles were flown and there were 13 fatalities. Based on data from 1975–1980, the average number of fatalities that occur per 100 million passenger miles flown is approximately .043 (*Statistical Abstract of the U.S.: 1981*, p. 641). Assuming that airlines fly approximately 17 billion passenger miles per month, the mean number of fatalities for a 1-month period is 7.31 (170 times the mean per 100 million miles). Suppose that the probability distribution for x, the number of fatalities per month, can be approximated by a Poisson probability distribution.

a. What is the probability that no fatalities will occur during any given month next year?

b. Find $E(x)$ and the standard deviation of x.

c. Refer to part b. Use your answers to part b to describe the probability that as many as twenty fatalities will occur in any given month.

4.6
The Hypergeometric Random Variable (Optional)

Consider the following examples of *hypergeometric random variables:*

1. An employer randomly selects and hires three applicants from a pool of ten equally qualified applicants, six of whom are men and four of whom are women. Let x be the number of women hired.

2. You randomly select three alkaline batteries from a collection of seven remaining in a department store's stock. Unknown to you, two of the seven batteries are defective. Let x be the number of defective batteries you select.

3. A city commission plans to select five individuals at random from a group of twenty volunteers to serve on the city's zoning committee. Suppose that twelve of the volunteers are Democrats and eight are Republicans. Let x be the number of Democrats chosen.

4. A certain professional football team must release three players from its squad before the season begins in order to meet league limits on the total number of players. The coaches have narrowed the choice of whom they will release to five players, and they plan to select at random the three to be released from these five. Suppose that two of the five players would eventually become starting players for the team if they were retained. Let x be the number of potential starting players who are released from the team.

The general characteristics of a hypergeometric random variable are given in the following box.

Characteristics of a Hypergeometric Random Variable

1. The experiment consists of randomly drawing n elements without replacement from a set of N elements, r of which are S's (for Success) and $(N - r)$ of which are F's (for Failure).
2. The hypergeometric random variable x is the number of S's in the draw of n elements.

Both the hypergeometric and binomial characteristics stipulate that each draw or trial results in one of two outcomes. The basic difference between these random variables is that the hypergeometric trials are dependent, while the binomial trials are independent. The draws are dependent because the probability of next drawing an S (or an F) depends on what occurred on preceding draws.

To illustrate, suppose that you select $n = 2$ elements from a set containing $N = 5$ elements, $r = 3$ S's and $N - r = 2$ F's. Although the probability of drawing an S on the first draw is $\frac{3}{5}$, the probability of drawing an S on the second draw depends on the outcome of the first. If the first draw results in an S, the probability of an S on the second draw is $\frac{2}{4}$ or $\frac{1}{2}$. However, if the first draw results in an F, the probability of an S on the second draw is $\frac{3}{4}$. Consequently, the results of the draws represent dependent events.

In general, it might seem that x can assume any one of the values, 0, 1, 2, . . . , n, but this is incorrect. The values that x can assume depend upon N, n, and r. For example, if $N = 20$, $n = 10$, and $r = 5$, then x can assume one of the values 0, 1, 2, . . . , 5. But if $N = 12$, $n = 10$, $r = 5$, you will always have at least three S's in the sample (because the largest number of F's that can fall in the sample is $N - r = 7$, leaving *at least* three of the $n = 10$ draws for S's). For this case, x can assume the values 3, 4, 5. The values that x can assume will be evident from the values of N, n, and r, and we give the probability distribution of x in the box.

Probability Distribution, Mean, and Variance of the Hypergeometric Random Variable

$$p(x) = \frac{\binom{r}{x}\binom{N-r}{n-x}}{\binom{N}{n}}$$

$$\mu = \frac{nr}{N} \qquad \sigma^2 = \frac{r(N-r)n(N-n)}{N^2(N-1)}$$

where

N = Total number of elements

r = Number of S's in the N elements

n = Number of elements drawn

x = Number of S's drawn in the n elements

Example 4.13   Suppose, as we mentioned earlier, an employer randomly selects three new employees from a total of ten applicants, six men and four women. Let x be the number of women who are hired.

a. Find the mean and standard deviation of x.

b. Find the probability that no women are hired.

Solution   **a.** Since x is a hypergeometric random variable with $N = 10$, $n = 3$, and $r = 4$, the mean and variance are

$$\mu = \frac{nr}{N} = \frac{(3)(4)}{10} = 1.2$$

$$\sigma^2 = \frac{r(N-r)n(N-n)}{N^2(N-1)} = \frac{4(10-4)3(10-3)}{(10)^2(10-1)} = \frac{(4)(6)(3)(7)}{(100)(9)} = .56$$

The standard deviation is

$$\sigma = \sqrt{.56} = .75$$

b. The probability that no women are selected by the employer, assuming the selection is truly random, is

$$P(x = 0) = p(0) = \frac{\binom{4}{0}\binom{10-4}{3-0}}{\binom{10}{3}} = \frac{\frac{4!}{0!(4-0)!}\frac{6!}{3!(6-3)!}}{\frac{10!}{3!(10-3)!}} = \frac{(1)(20)}{120} = \frac{1}{6}$$

The entire probability distribution for x in the example is shown in Figure 4.9. The mean $\mu = 1.2$ and the interval $\mu \pm 2\sigma = (-.3, 2.7)$ are indicated. You can see that if this random variable were observed over and over again a large number of times, most of the values of x (approximately 97%) would fall within the interval $\mu \pm 2\sigma$. ∎

Figure 4.9 Probability Distribution for x in Example 4.13

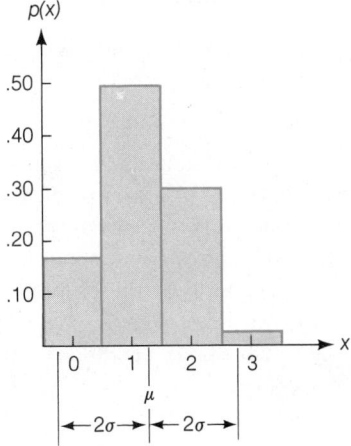

Exercises 4.70–4.85

Learning the Mechanics

4.70 Given that x is a hypergeometric random variable, $N = 8$, $n = 3$, and $r = 5$, compute the following:

a. $P(x = 1)$ **b.** $P(x = 0)$ **c.** $P(x = 3)$ **d.** $P(x \geq 4)$

4.71 Given that x is a hypergeometric random variable, compute $p(x)$ for each of the following cases:

a. $N = 5$, $n = 3$, $r = 3$, $x = 1$ **b.** $N = 9$, $n = 5$, $r = 3$, $x = 3$

c. $N = 4$, $n = 2$, $r = 2$, $x = 2$ **d.** $N = 3$, $n = 2$, $r = 2$, $x = 0$

4.72 Suppose that x is a hypergeometric random variable, $N = 10$, $n = 5$, and $r = 7$.

a. Display the probability distribution for x in tabular form.

b. Compute the mean and variance of x.

c. Graph $p(x)$ and locate μ and the interval $\mu \pm 2\sigma$ on the graph.

d. What is the probability that x will fall within the interval $\mu \pm 2\sigma$?

4.73 Suppose that x is a hypergeometric random variable, $N = 12$, $n = 8$, and $r = 6$.

a. Display the probability distribution for x in tabular form.

b. Compute μ and σ for x.

c. Graph $p(x)$ and locate μ and the interval $\mu \pm 2\sigma$ on the graph.

d. What is the probability that x will fall within the interval $\mu \pm 2\sigma$?

4.74 Suppose that x is a hypergeometric random variable with $N = 12$, $n = 8$, and $r = 7$.

a. Display the probability distribution for x in tabular form.

b. Compute μ and σ for x.

c. Graph $p(x)$ and locate μ and the interval $\mu \pm 2\sigma$ on the graph.

d. What is the probability that x will fall within the interval $\mu \pm 2\sigma$?

4.75 Use the results of Exercise 4.74 to find the following probabilities:

a. $P(x = 1)$ **b.** $P(x = 4)$ **c.** $P(x \leq 4)$

d. $P(x \geq 5)$ **e.** $P(x < 3)$ **f.** $P(x \geq 8)$

4.76 Imagine a population of $N = 25$ subjects of which 20% are of type A.

a. Compute the probability that 20% of a sample of $n = 5$ subjects are of type A if the sample is randomly selected *with replacement* (i.e., after a subject is drawn from the population, it is replaced before the next subject is drawn).

b. Calculate the probability specified in part a for the case where the sample is randomly selected *without replacement*.

c. What type of probability distribution is appropriate for $p(x)$ if sampling is with replacement? Without replacement?

d. Suppose that N is very large in relation to n, say $N = 100$ or larger. Compute the probabilities specified in parts a and b. How do the values of $p(x)$, sampling with and without replacement, compare when the number N of elements in the population is large relative to the sample size n?

Applying the Concepts

4.77 A social psychologist wishes to determine the effects of competition on performance. Each subject in the study is allowed to choose three of his or her own opponents in a particular game from a group of nine competitors. Suppose that in this group of nine, four are of equal or greater ability than the subject and the remaining five are of lesser ability. Assume that the subject in question chooses the opponents at random.

a. What is the probability that he or she will compete against three lesser opponents?

b. What is the probability that the subject will compete against at least one opponent of equal or greater ability?

4.78 Defective alternators have been mistakenly installed in three of the last six truck engines to emerge from an assembly line. It is not known which engines contain the defective alternators, but three of the six engines are randomly selected for inspection.

a. What is the probability that each contains a defective alternator?

b. What is the probability that none contain a defective alternator?

c. What is the probability that more than one contain a defective alternator?

4.79 The baseballs used at the major league level are individually hand-stitched and hence are not uniformly made. Most pitchers who throw curve balls prefer baseballs with high stitches due to the firmer grip they provide. From a box of twenty new baseballs, thirteen of which are high-stitched, an umpire chooses five at random to start the game.

a. Find the probability that exactly four of the five balls chosen to start the game are preferred by the starting curve-ball pitcher.

b. Find the probability that at least four of the five balls chosen to start the game are high-stitched.

4.80 Suppose you are purchasing cases of wine (twelve bottles per case) and that periodically you select a test case to determine the adequacy of the sealing process. To do this, you randomly select and test three bottles in the case. If a case contains one bottle of spoiled wine, what is the probability that it will appear in your sample?

4.81 Five individuals apply for two vacancies in the shipping department of your plant. Two of the five people have superior credentials. You have been instructed to choose randomly two of the five applicants to fill the open positions.

a. What is the probability that you will choose the two individuals with the superior credentials?

b. What is the probability that at least one of the individuals with the superior credentials will be selected?

4.82 If you are purchasing small lots of a manufactured product and it is very costly to test a single item, it may be desirable to test a sample of items from the lot rather than every item in the lot. Such a sampling plan would be based on a hypergeometric probability distribution. For example, suppose each lot contains ten items. You decide to sample four items per lot and reject the lot if you observe one or more defectives. If the lot contains one defective item, what is the probability that you will accept the lot? What is the probability that you will accept the lot if it contains two defective items? Three? Four?

4.83 Construct an operating characteristic curve for the sampling plan (Exercise 4.82) by plotting the probability of lot acceptance $P(A)$ versus the lot fraction defective. (Use the probabilities of lot acceptance calculated in Exercise 4.82.)

4.84 Suppose that we were to change the *acceptance number* for the sampling plan (Exercise 4.82) to 2; that is, we will accept the lot if the number x of defectives in the sample is $x \leq 2$. Construct an operating characteristic curve for this sampling plan and compare it with the operating characteristic curve of Exercise 4.83. Which sampling plan is more likely to detect lots containing defectives? Explain how the answer is apparent in the comparison of the operating characteristic curves.

4.85 A curious event is described in the *Minneapolis Star and Tribune* (May 27, 1983). The Minneapolis Community Development Agency (MCDA) makes home improvement grants each year to homeowners in depressed neighborhoods in the city. Of the $708,000 granted in 1983, some $233,000 was awarded by the city council using a "random selection" of 140 homeowners' applications from among a total of 743 applications, 601 from the north side and 142 from the south side of Minneapolis. Oddly, all 140 chosen were from the north side—clearly a highly improbable outcome if, in fact, the 140 winners were randomly selected from among the 743 applicants.

a. Suppose that the 140 winning applications were randomly selected from among the total of 743, and let x equal the number in the sample from the north side. Find the mean and standard deviation of x.

b. Use the results of part a to support a contention that the grant winners were not randomly selected.

4.7
The Geometric Random Variable (Optional)

Another common discrete random variable that has many applications is the **geometric random variable.** Like the binomial variable, it arises naturally from a discussion of a coin toss experiment (whether the coin is balanced or unbalanced). But instead of tossing the coin a fixed number of times and observing the number x of heads, we toss the coin and count the number x of tosses until the first head appears. As in the binomial experiment, we assume that the tosses are independent of each other. The geometric random variable has the characteristics listed in the box.

Characteristics of the Geometric Random Variable

1. The experiment consists of a sequence of independent trials.
2. Each trial results in one of two outcomes. We denote one of them by S and the other by F.
3. The probability of S remains the same from trial to trial. We will denote this probability by p.
4. The geometric random variable x is defined to be the number of trials until the first S is observed.

The probability distribution for the geometric random variable provides a good model for the length of time a customer must wait for some type of servicing. For this application, the time must be measured in whole units (minutes, hours, etc.). Then x is equal to the number of time units a customer must wait until he or she is served.

Probability Distribution, Mean, and Variance of a Geometric Random Variable

$$p(x) = q^{x-1}p \qquad (x = 1, 2, 3, \ldots)$$

$$\mu = \frac{1}{p} \qquad \sigma^2 = \frac{q}{p^2}$$

where

p = Probability of an S outcome

$q = 1 - p$

x = Number of trials until the first S is observed

The number of job applicants interviewed by an employer until the first suitable prospect is found is another discrete random variable that might be modeled by a geometric probability distribution. Or, for a sequence of independent interviews by a police detective on a murder case, x could represent the number of interviews until the first substantial lead is obtained.

Example 4.14

Let x be the number of prospective jurors who will be examined until one is admitted as a juror for a trial. Assume that x is a geometric random variable with p, the probability of a prospective juror being accepted, equal to .5.

a. Find the mean and standard deviation of x.

b. Find the probability that more than two prospective jurors must be examined before one is admitted to the jury.

Solution

a. The mean and variance for this geometric random variable are

$$\mu = \frac{1}{p} = \frac{1}{.5} = 2 \qquad \sigma^2 = \frac{q}{p^2} = \frac{.5}{(.5)(.5)} = 2$$

Then the standard deviation is

$$\sigma = \sqrt{\sigma^2} = \sqrt{2} = 1.41$$

b. To find the probability that more than two prospective jurors will be examined before one is admitted to the jury, we must find

$$P(x > 2) = p(3) + p(4) + p(5) + \cdots$$

Since the number of terms in this sum is infinitely large, we first find the probability of the complementary event, $P(x \leq 2)$. Then

$$P(x > 2) = 1 - P(x \leq 2)$$
$$= 1 - [p(1) + p(2)]$$

The next step is to find $p(1)$ and $p(2)$. Substituting into the formula for $p(x)$, we find

$$p(1) = q^{1-1}p = (.5)^0(.5) = .5$$
$$p(2) = q^{2-1}p = (.5)^1(.5) = .25$$

Then

$$P(x > 2) = 1 - P(x \leq 2)$$
$$= 1 - (.5 + .25) = .25$$

This result tells us that there is a .25 probability that more than two prospective jurors will be examined before one is admitted to the jury.

The probability distribution for x in this example is shown in Figure 4.10. The expected value of x and the interval $\mu \pm 2\sigma$ are indicated. The probability that x falls in the interval, $\mu \pm 2\sigma$, that is, $P(x = 1, 2, 3, \text{ or } 4)$, is .9375. Considering that the distribution (Figure 4.10) is highly skewed, it is interesting to observe that this probability is not too different from the .95 given for mound-shaped distributions by the Empirical Rule.

Figure 4.10 Probability Distribution for x in Example 4.14

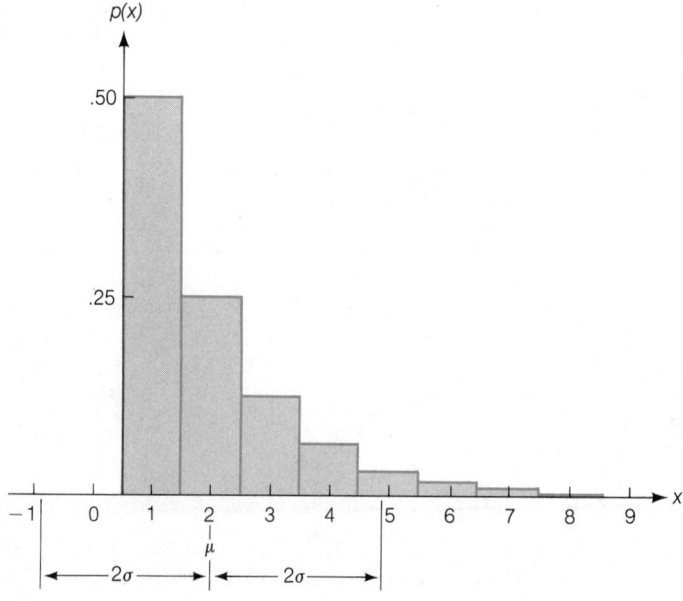

Note that the geometric probabilities always show $p(x)$ decreasing as x increases. Although Figure 4.10 shows only the probabilities up to $x = 8$, the geometric distribution assigns positive probabilities to all positive integers.   ▪

Exercises 4.86–4.96

Learning the Mechanics

4.86 Given that x is a geometric random variable with $p = .2$, compute the following:
a. $P(x = 2)$ **b.** $P(x = 3)$ **c.** $P(x \leq 3)$
d. $P(x \geq 1)$ **e.** $P(x = 4)$ **f.** $P(x > 2)$

4.87 Suppose that x is a geometric random variable. Calculate μ, σ^2, and σ when:
a. $p = .01$ **b.** $p = .1$ **c.** $p = .8$
d. $p = .3$ **e.** $p = .99$ **f.** $p = 1$

4.88 Suppose that x is a geometric random variable with $p = .7$.
a. Graph $p(x)$, $x = 1, 2, \ldots$.
b. Compute the mean and variance of x. Locate μ and the interval $\mu \pm 2\sigma$ on the graph.
c. What is the probability that x will fall within the interval $\mu \pm 2\sigma$?

4.89 Suppose that x is a geometric random variable with $p = .3$.
a. Graph $p(x)$, $x = 1, 2, \ldots$.
b. Compute the expected value and variance of x. Locate μ and the interval $\mu \pm \sigma$ on your graph.
c. What is the probability that x will fall within the interval $\mu \pm \sigma$?

4.90 Use the results of Exercise 4.89 to find the following probabilities:
a. $P(x \leq 2)$ **b.** $P(x > 3)$ **c.** $P(x > 0)$ **d.** $P(1 \leq x \leq 6)$

Applying the Concepts

4.91 A real estate broker estimates that when a client is shown available houses, the probability of clinching a sale is .20.
a. If the broker shows houses to a sequence of different clients, what is the probability of achieving the first sale before the fourth client is shown the houses?
b. On the average, how many clients will have to be shown houses before the broker makes the first sale?
c. Is it likely that the broker would have to show houses to as many as twenty clients before the first sale is made? Justify your answer.

4.92 An oil company has determined that the probability of striking oil on any particular drilling is .2. Accordingly, what is the probability that it would drill four dry wells before striking oil on the fifth drilling?

4.93 The personnel office of a large research and development firm has determined that the probability of any particular job applicant being qualified for the existing opening on the company's technical staff is .10.

a. What is the probability that the personnel office will not find a qualified applicant until the third person interviewed?

b. What is the probability that the first person interviewed will qualify?

4.94 Despite the fact that a manufacturer replaces all lights on his production floor periodically, the probability of at least one light failing on any given day is .05.

a. What is the probability that all lights on the production floor will survive the next 5 days?

b. What is the mean number of days until at least one light fails on the production floor?

c. Is it likely that the first failure will occur within a month of the periodic replacement? Justify your answer.

4.95 If the probability that a customer will be served by a clerk during any given minute he or she is in the store is .3, what is the probability that the customer will have to wait 4 or more minutes to be served?

4.96 Ten percent of the light bulbs produced by a company are defective.

a. If an inspector tests light bulbs randomly selected from the production line, what is the probability that the first defective bulb will be observed on or after the fourth test?

b. What is the expected number of tests the inspector will have to perform before the first defective bulb is observed?

Summary

Observations taken on discrete random variables (those that can assume a countable number of values) often have the characteristics of a **binomial, Poisson, hypergeometric,** or **geometric random variable.** In this chapter, we gave the identifying characteristics for each of these random variables, indicated some practical applications for which the probability models would be appropriate, and gave the formulas for their probability distributions, means, and variances.

Using the probability distribution for a random variable, we were able to calculate the probabilities of specific sample observations. When the probabilities were difficult to calculate, the means and standard deviations provided numerical descriptive measures that enabled us to visualize the probability distributions and thereby to make some very approximate probability statements about their behavior.

Supplementary Exercises 4.97–4.132

Learning the Mechanics

[*Note:* Starred (*) exercises refer to optional sections in this chapter.]

4.97 Suppose that x is a binomial random variable. Find $p(x)$ for each of the following combinations of x, n, and p.

a. $x = 1$, $\quad n = 3$, $\quad p = .1$ **b.** $x = 4$, $\quad n = 20$, $\quad p = .3$

c. $x = 0$, $\quad n = 2$, $\quad p = .4$ **d.** $x = 4$, $\quad n = 5$, $\quad p = .5$

e. $n = 15$, $\quad x = 12$, $\quad p = .9$ **f.** $n = 10$, $\quad x = 8$, $\quad p = .6$

4.98 Which of the following describe discrete random variables and which describe continuous random variables?
a. The number of damaged inventory items
b. The average monthly sales revenue generated by a salesperson over the past year
c. The number of square feet of warehouse space a company rents
d. The length of time a firm must wait before its copying machine is fixed

4.99 Consider the following discrete probability distribution:

x	10	12	18	20
$p(x)$	.2	.2	.1	.5

a. Calculate μ, σ^2, and σ.
b. What is the probability that $x < 15$?
c. Calculate $\mu \pm 2\sigma$.
d. What is the probability that x is in the interval $\mu \pm 2\sigma$?

4.100 Suppose that x is a binomial random variable with $n = 20$ and $p = .6$.
a. Find $P(x = 14)$.　　　**b.** Find $P(x \leq 10)$.　　　**c.** Find $P(x > 10)$.
d. Find $P(8 \leq x \leq 17)$.　　**e.** Find $P(8 < x < 17)$.　　**f.** Find μ, σ^2, and σ..
g. What is the probability that x is in the interval $\mu \pm 2\sigma$?

*__4.101__ Suppose that x is a hypergeometric random variable. Compute $p(x)$ for each of the following cases:
a. $N = 6$,　$n = 4$,　$r = 3$,　$x = 2$
b. $N = 5$,　$n = 2$,　$r = 2$,　$x = 2$
c. $N = 7$,　$n = 4$,　$r = 4$,　$x = 3$

*__4.102__ Suppose that x is a geometric random variable. Compute $p(x)$ for each of the following cases:
a. $p = .2$,　$x = 3$　　**b.** $p = .5$,　$x = 4$　　**c.** $p = .9$,　$x = 2$

*__4.103__ Suppose that x is a Poisson random variable. Compute $p(x)$ for each of the following cases:
a. $\lambda = 2$,　$x = 3$　　**b.** $\lambda = 1$,　$x = 4$　　**c.** $\lambda = .5$,　$x = 2$

*__4.104__ Suppose that x is a random variable for which a Poisson probability distribution with $\lambda = 3$ provides a good characterization. Compute the following:
a. $P(x = 1)$　　**b.** $P(x = 3)$　　**c.** $P(x = 0)$
d. $P(x = 5)$　　**e.** $P(x \leq 2)$　　**f.** $P(x \geq 2)$

*__4.105__ Suppose that x is a geometric random variable with $p = .2$. Compute:
a. $P(x = 2)$　　**b.** $P(x = 2, 3,$ or $4)$　　**c.** $P(x \leq 2)$
d. $P(x = 5)$　　**e.** $P(x > 4)$　　**f.** $P(x = 1, 2,$ or $5)$

*__4.106__ Suppose that x is a geometric random variable with $p = .5$.
a. Graph $p(x)$ for $x = 1, 2, \ldots, 7$.
b. Compute μ and σ for x. Locate μ and the interval $\mu \pm 2\sigma$ on the graph.
c. What is the probability that x will fall within the interval $\mu \pm 2\sigma$?

Applying the Concepts

4.107 Due to pollution, it is thought that 5% of the fish found in a certain river contain a level of mercury that is harmful to humans. To test this theory, environmentalists sample ten fish from the river and analyze each for the presence of a dangerous amount of mercury.

a. If the 5% contamination figure is correct, what is the probability that none of the ten fish contain a dangerous level of the substance?

b. What is the probability that no more than two of the fish contain a dangerous level of the substance?

4.108 Many minor operations at a hospital can be performed the same day the patient is admitted. A hospital serving a large metropolitan area has found that in the past, 20% of newly admitted patients needing an operation are scheduled for same-day surgery. Suppose that ten patients are randomly selected from those admitted to the hospital for surgery over the past year. If x, the number in the sample of ten who receive same-day surgery, possesses a binomial probability distribution:

a. What is the probability that exactly five of these patients have same-day surgery?

b. What is the probability that at most one has same-day surgery?

c. If the ten patients are selected from the admissions on a single given day, is it reasonable to expect x to possess the characteristics of a binomial random variable? That is, is this a binomial experiment? Explain.

4.109 The head of a large library claims that 60% of the books in the library have been published since 1960.

a. If twenty books are chosen at random from the library, what is the probability that at least fifteen were published since 1960, assuming the claim is true?

b. What is the probability that fewer than nine books chosen were published after 1960?

4.110 An advertisement for laundry soap claims that the soap is preferred over all others by 30% of American women.

a. Assuming that this claim is true, what is the probability that fewer than four women in a random sample of twenty-five prefer the advertiser's brand?

b. What is the probability that the number of women preferring the brand takes a value in the interval $3 \leq x \leq 13$?

c. If a sample of twenty-five American women was taken and only three preferred the brand, what conclusions would you draw? Why?

4.111 Suppose it is known that 5% of all radios produced by a manufacturer have defective tuning mechanisms. Your store receives a large shipment of the radios from which you choose ten to inspect. You have decided not to accept the shipment if you discover one or more defective radios. Before inspecting the ten radios, what is the probability that you will not accept the shipment?

4.112 An important function in any business is long-range planning. Additions to a firm's physical plant, for example, cannot be achieved overnight; their construction must be planned years in advance. Anticipating a substantial growth in sales over the next 5 years, a printing company is planning today for the warehouse space it will need 5 years hence. It obviously cannot be certain exactly how many square feet of storage space, x, it will need in 5 years, but the company can project its needs by using a probability distribution such as the following:

x	10,000	15,000	20,000	25,000	30,000	35,000
p(x)	.05	.15	.35	.25	.15	.05

What is the expected number of square feet of storage space the printing company will need in 5 years?

4.113 An employee of a firm has an option to invest $1,000 in the company's bonds. At the end of 1 year, the company will buy back the bonds at a price determined by its profits for the year. From past years, the company predicts it will buy the bonds back at the following prices with the associated probabilities (x = price paid for bonds):

x	$0	$500	$1,000	$1,500	$2,000
p(x)	.01	.22	.30	.22	.25

a. What is the probability the employees will receive $1,000 or less for the investment?
b. What is the expected price paid for the bonds?
c. What is the employee's expected profit?
d. Find σ^2 and σ for this probability distribution.

4.114 The probability that a person responds to a mailed questionnaire is .4.
a. What is the probability that of twenty questionnaires, more than twelve will be returned?
b. How many questionnaires should be mailed if you want to be reasonably certain that at least 100 will be returned?

4.115 The state highway patrol has determined that one out of every six calls for help originating from roadside call boxes is a hoax. Five calls for help have been received and five tow trucks dispatched.
a. What is the probability that none of the calls was a hoax?
b. What is the probability that only three of the callers really needed assistance?
c. What assumptions do you have to make in order to solve this problem?
d. If the highway patrol answers 10,000 calls for help next year and each call costs the patrol about $30 (labor, gas, etc.), approximately how much money will be wasted answering false alarms?

4.116 A sales manager has determined that a salesperson makes a sale to 70% of the retailers visited.
a. If the salesperson visits five retailers today and twenty tomorrow, what is the probability that he or she makes exactly four sales today *and* more than ten tomorrow?

b. If the salesperson visits five retailers today and twenty tomorrow, what is the probability that he or she makes more than fourteen sales out of the twenty-five visits?

c. If the salesperson visits four retailers today and five tomorrow, what is the probability that in these two days he or she will make exactly two sales?

COMPANY A		COMPANY B	
x	p(x)	x	p(x)
2	.05	2	.15
3	.15	3	.30
4	.20	4	.30
5	.35	5	.20
6	.25	6	.05

4.117 The owner of construction company A makes bids on jobs so that if awarded the job, company A will make a $10,000 profit. The owner of construction company B makes bids on jobs so that if awarded the job, company B will make a $15,000 profit. Each company describes the probability distribution of the number of jobs the company is awarded per year as shown in the table.

a. Find the expected number of jobs each will be awarded in a year.

b. What is the expected profit for each company?

c. Find the variance and standard deviation of the distribution of number of jobs awarded per year for each company.

d. Graph $p(x)$ for both companies A and B. For each company, what proportion of the time will x fall in the interval $\mu \pm 2\sigma$?

4.118 A large cigarette manufacturer has determined that the probability of a new brand of cigarettes obtaining a large enough market share to make production profitable is .3. Over the next 3 years, this manufacturer will introduce one new brand a year.

a. What is the probability that at least one new brand will obtain sufficient market share to make its production profitable?

b. What is the probability that all three new brands will obtain sufficient market share?

c. What assumptions do you have to make in order to solve this problem?

4.119 If it is known that 5% of the finished products coming off an assembly line are defective, what is the probability that exactly one of the next four products coming off the line is defective? What assumptions do you have to make to solve this problem using the methodology of this chapter?

4.120 Suppose you are an airport manager. In looking over your records for the past year, you note that 60% of the time the 8:10 PM flight from Atlanta is 20 or more minutes late.

a. If you assume that the probability of the 8:10 PM flight being 20 or more minutes late on each day during the upcoming 5-day period is .6, what is the probability that the plane will be 20 or more minutes late exactly three times in the next 5 days?

b. What is the probability that the plane will be late at least three times in the next 5 days?

4.121 In recent years, the use of the telephone as a data collection instrument for public opinion polls has been steadily increasing. However, one of the major factors bearing on the extent to which the telephone will become an acceptable data collection tool in the future is the refusal rate, i.e., the percentage of the eligible subjects actually contacted who refuse to take part in the poll. Suppose that past records indicate a refusal rate of 20% in a large city. A poll of twenty-five residents is to be taken and x is the number of residents contacted by telephone who refuse to take part in the poll.

a. Find the mean and variance of x.

b. Find $P(x \le 5)$.

c. Find $P(x > 10)$.

4.122 The efficacy of insecticides is often measured by the dose necessary to kill a certain percentage of insects. Suppose that a certain dose of a new insecticide is supposed to kill 80% of the exposed insects. To test the claim, twenty-five insects are put in contact with the insecticide.

a. If the insecticide really kills 80% of the exposed insects, what is the probability that fewer than fifteen die?

b. If you observed such a result, what would you conclude about the new insecticide? Explain your logic.

4.123 A physical fitness specialist claims that the probability is greater than .5 that an average adult male can improve his physical condition by spending 5 minutes per day on a certain exercise program. To test the claim, fifteen randomly selected adult males follow the program for a specified amount of time. Maximal oxygen uptake is measured before and after the program for each male and serves as the criterion for assessing physical condition. If the program is not really beneficial (i.e., the probability of improvement is only .5), what is the probability that eleven or more of the fifteen men have improved maximal oxygen uptake?

Suppose eleven or more showed increased maximal oxygen uptake. Assuming that the specialist's exercise program is ineffective, would you regard $x \ge 11$ as a rare event or would you conclude that, in actuality, the probability of improvement exceeds $p = \frac{1}{2}$ and that the program is effective?

Values of x	Tax (¢)
0, 1, . . . , 9	0
10, 11, . . . , 25	1
26, 27, . . . , 50	2
51, 52, . . . , 75	3
76, 77, . . . , 99	4

4.124 [*Warning:* This exercise is realistic, but the computations involved are tedious.] Refer to Case Study 4.1—the Red Lobster sales tax problem. Let $x = 0, 1, 2, \dots, 99$ be the number of cents (exceeding whole dollars) involved in a sale, and assume that the sales tax is assessed using the bracket system listed in the table. Suppose that x has a probability distribution $p(x) = .01$, $x = 0, 1, 2, \dots, 99$ (an assumption that might be fairly accurate for restaurant sales). Find the expected value of the percentage of tax paid on the cents portion of a sale. [*Hint:* You can write the percent tax—call it y—for each value of x. You also know the probabilities associated with each value of x and, consequently, each value of y. Then the expected percentage of tax paid is $E(y) = \sum y p(y)$.]

*****4.125** If 20% of the finished products coming off an assembly line are defective, what is the probability that more than three randomly selected finished products would have to be inspected before a defective product is found?

*****4.126** A small life insurance company has determined that on the average it receives five death claims per day.

a. What is the probability that the company will receive three claims or less on a particular day?

b. What is the probability that the company will receive exactly five claims on a particular day?

c. What assumptions must you make to find these probabilities?

*4.127 An emergency rescue vehicle is used an average of 1.3 times daily.
a. What is the probability that the vehicle will be used exactly twice tomorrow?
b. What is the probability that it will be used more than twice?
c. Exactly three times?

*4.128 By mistake a manufacturer of tape recorders includes three defective recorders in a shipment of ten going out to a small retailer. The retailer has decided to accept the shipment of recorders only if none are found to be defective. Upon receipt of the shipment the retailer examines only five of the recorders.
a. What is the probability that the shipment will be rejected?
b. If the retailer inspects six of the recorders, what is the probability that the shipment will be accepted?

*4.129 You have determined that one out of five customers who enter your furniture store makes a purchase. What is the probability that of the first four customers in a day the fourth is the initial customer to make a purchase?

*4.130 A wholesale office equipment outlet claims that on an average it sells 2.5 typewriters per day. If it has only five typewriters in stock at the close of business today and does not expect to receive a shipment of new typewriters until some time after the close of business tomorrow, what is the probability that the outlet's current supply of typewriters will not be sufficient to meet tomorrow's demand?

*4.131 Of the eight families in a particular neighborhood, six have incomes over $20,000 and two have incomes under $20,000.
a. If four of these eight families are randomly selected to participate in a taste test for a newly developed type of instant breakfast food, what is the probability that exactly three of the chosen families have an income over $20,000?
b. What is the probability that at least one family with an income under $20,000 is chosen?

*4.132 Large bakeries typically have fleets of delivery trucks. It was determined by one such bakery that the expected number of delivery truck breakdowns per day is 1.5. Assume that the number of breakdowns is independent from day to day.
a. What is the probability that there will be exactly two breakdowns today and exactly three tomorrow?
b. Fewer than two today and more than two tomorrow?

On Your Own . . .

Consider the following random variables:

1. The number x of people who recover from a certain disease out of a sample of five patients
2. The number x of voters in a sample of five who favor a method of tax reform
3. The number x of hits a baseball player gets in five official times at bat

In each case, x is a binomial random variable (or approximately so) with $n = 5$ trials. Assume that in each case the probability of Success is .3. (What is a Success in each of the examples?) If this were true, the probability distribution for x would be the same for each of the three examples. To obtain a relative frequency histogram for x, conduct the following experiment: Place ten poker chips (pennies, marbles, or any ten *identical* items) in a bowl and mark three of the ten Success—the remaining seven will represent Failure. Randomly select a chip from the ten, observing whether it was a Success or Failure. Then return the chip and randomly select a second chip from the ten available chips, and record this outcome. Repeat this process until a total of five trials has been conducted. Count the number x of Successes observed in the five trials. Repeat the entire process 100 times to obtain 100 observed values of x.

a. Use the 100 values of x obtained from the simulation to construct a relative frequency histogram for x. Note that this histogram is an approximation to $p(x)$.

b. Calculate the exact values of $p(x)$ for $n = 5$ and $p = .3$ and compare these values with the approximations found in part a.

c. If you were to repeat the simulation an extremely large number of times (say 100,000), how do you think the relative frequency histogram and true probability distribution would compare?

References   Greenstein, F. I. "The benevolent leader revisited: children's images of political leaders in three democracies." *American Political Science Review*, December 1975, *69*, 1371–1398.

Hogg, R. V., & Craig, A. T. *Introduction to mathematical statistics*. 4th ed. New York: Macmillan, 1978. Chapters 1 and 3.

Mendenhall, W., Scheaffer, R. L., & Wackerly, D. *Mathematical statistics with applications*. 2d ed. North Scituate, Mass.: Duxbury, 1980. Chapter 3.

Mood, A. M., Graybill, F. A., & Boes, D. C. *Introduction to the theory of statistics*. 3d ed. New York: McGraw-Hill, 1963. Chapter 3.

Van de Werf, F., et al. "Coronary thrombosis with tissue-type plasminogen activator in patients with evolving myocardial infarction." *New England Journal of Medicine,* 1984, *310*.

CHAPTER 5

Continuous Random Variables

Where We've Been...

Because numerical data represent observed values of random variables, we needed to find the probabilities associated with specific sample observations. As noted in Chapter 4, this task depends on whether a random variable is discrete or continuous. The probability theory of Chapter 3 provided the mechanism for finding the probabilities associated with discrete random variables. Finding and describing this set of probabilities—the probability distribution for a discrete random variable—was the subject of Chapter 4.

Where We're Going...

Since data may be derived from observations on continuous as well as discrete random variables, we need to know about probability distributions associated with continuous random variables and also how to use the mean and standard deviation to describe these distributions. Chapter 5 addresses this problem and, in particular, introduces the *normal probability distribution*. As you will subsequently see, the normal probability distribution is a useful distribution in many areas of research.

Contents

In this chapter we will consider some continuous random variables that are commonly encountered. Recall that a continuous random variable is one that can assume any value within some interval or intervals. For example, the length of time between a person's visits to a doctor, the thickness of sheets of steel produced in a rolling mill, and the yield of wheat per acre of farmland are all continuous random variables. The methodology we employ to describe continuous random variables will necessarily be somewhat different from that used to describe discrete random variables. We will first discuss the general form of **continuous probability distributions,** and then we will present three specific types that are used in making statistical decisions. The **normal probability distribution,** which plays a basic and important role in both the theory and application of statistics, is essential to the study of most of the subsequent chapters in this book. The other types have practical applications, but a study of these topics is optional.

5.1 Continuous Probability Distributions

The graphic form of the probability distribution for a continuous random variable x is a smooth curve that might appear as shown in Figure 5.1. This curve, a function of x, is denoted by the symbol $f(x)$ and is variously called a **probability density function,** a **frequency function,** or a **probability distribution.**

The areas under a probability distribution correspond to probabilities for x. For example, the area A beneath the curve between the two points a and b, as shown in Figure 5.1, is the probability that x assumes a value between a and b ($a < x < b$). Because there is no area over a point, say $x = a$, it follows that (according to our model) the probability associated with a particular value of x is equal to zero; that is, $P(x = a) = 0$ and hence $P(a < x < b) = P(a \leq x \leq b)$. In other words, the probability is the same regardless of whether you include the endpoints of the interval. Also, because areas over intervals represent probabilities, it follows that the total area under a probability distribution, the probability assigned to all values of x, should equal 1. Note that probability distributions for continuous random variables possess different shapes depending on the relative frequency distributions of real data that the probability distributions are supposed to model.

Figure 5.1   A Probability Distribution $f(x)$ for a Continuous Random Variable x

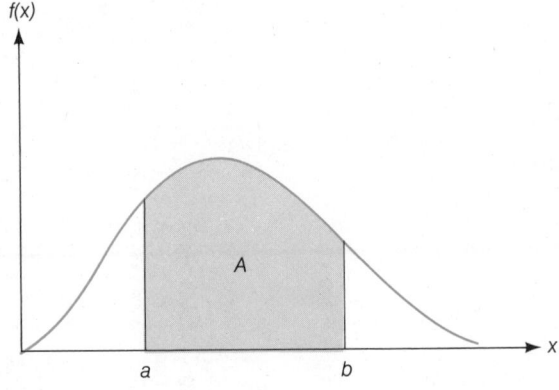

The areas under most probability distributions are obtained by use of the calculus or other numerical methods.* Because this is often a difficult procedure, we will give the areas for some of the most common probability distributions in tabular form in the Appendix. Then to find the area between two values of x, say $x = a$ and $x = b$, you simply have to consult the appropriate table.

For each of the continuous random variables presented in this chapter, we will give the formula for the probability distribution along with its mean and standard deviation. These two numbers, μ and σ, will enable you to make some approximate probability statements about a random variable even when you do not have access to a table of areas under the probability distribution.

5.2 The Normal Distribution

One of the most commonly observed continuous random variables has a **bell-shaped** probability distribution as shown in Figure 5.2. It is known as a **normal random variable** and its probability distribution is called a **normal distribution.**

You will see during the remainder of this text that the normal distribution plays a very important role in the science of statistical inference. Moreover, many phenomena generate random variables with probability distributions that are very well approximated by a normal distribution. For example, the error made in measuring a person's blood pressure may be a normal random variable, and the probability distribution for the yearly rainfall in a certain region might be approximated by a normal probability distribution. The normal distribution might also provide an accurate model for the probability distribution of the weights of loads of produce shipped to a supermarket. You can determine the adequacy of the normal approximation to an existing population of data by comparing the relative frequency distribution of a sample of the data (at least 200 measurements) to the normal probability distribution. Tests to detect disagreement between a set of data and the assumption of normality are available, but they are beyond the scope of this book.

Figure 5.2 A Normal Probability Distribution

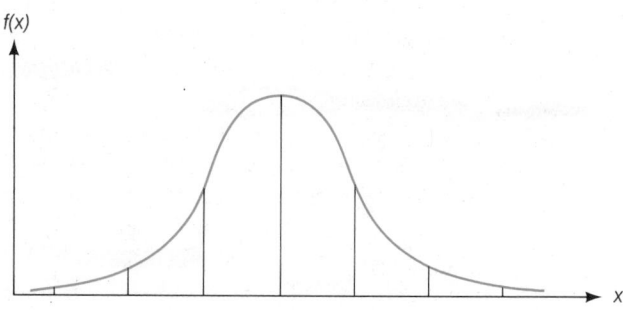

f(x)

x

---

* Students with a knowledge of calculus should note that the probability that x assumes a value in the interval $a < x < b$ is $P(a < x < b) = \int_a^b f(x)\,dx$, assuming the integral exists. Similar to the requirements for a discrete probability distribution, we require $f(x) \geq 0$ and $\int_{-\infty}^{\infty} f(x)\,dx = 1$.

Figure 5.3   Several
Normal Distributions with
Different Means and
Standard Deviations

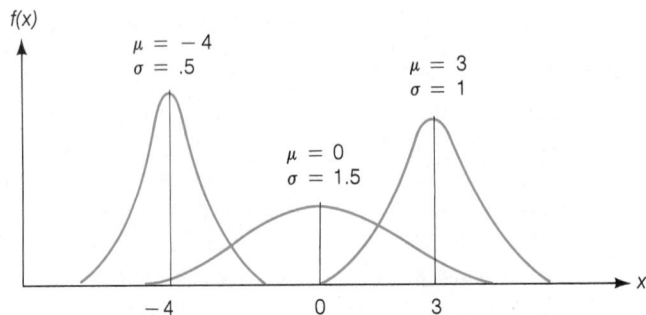

The normal distribution is perfectly symmetric about its mean μ, as can be seen in the examples in Figure 5.3. Its spread is determined by the value of its standard deviation σ.

The formula for the normal probability distribution is shown in the box.

Probability Distribution for a Normal Random Variable *x*

$$f(x) = \frac{1}{\sigma\sqrt{2\pi}}\, e^{-(1/2)[(x-\mu)/\sigma]^2}$$

where

μ = Mean of the normal random variable *x*

σ = Standard deviation

π = 3.1416...

e = 2.71828...

Note that the mean μ and standard deviation σ appear in this formula, so that no separate formulas for μ and σ are necessary. To graph the normal curve we have to know the numerical values of μ and σ.

Computing the area over intervals under the normal probability distribution is a difficult task.* Consequently, we will use the computed areas listed in Table IV of the Appendix. Since there is an infinitely large number of normal curves—one for each pair of values for μ and σ—we have formed a single table that will apply to any normal curve. This is done by constructing the table of areas as a function of the z-score (presented in Section 2.9). The population z-score for a measurement was defined as the

---

* The student with a knowledge of calculus should note that there is not a closed-form expression for $P(a < x < b) = \int_b^a f(x)\,dx$ for the normal probability distribution. The value of this definite integral can be obtained to any desired degree of accuracy by approximation procedures. For this reason, it is tabulated for the user.

distance between the measurement and the population mean, divided by the population standard deviation. Thus, the z-score gives the distance between a measurement and the mean in units equal to the standard deviation. In symbolic form, the z-score for the measurement *x* is

$$z = \frac{x - \mu}{\sigma}$$

Note that when $x = \mu$, we obtain $z = 0$.

To illustrate the use of Table IV, suppose we know that the length of time *x* between charges of a pocket calculator has a normal distribution with a mean of 50 hours and a standard deviation of 15 hours. If we were to observe the length of time that elapses before the need for the next charge, what is the probability that this measurement will assume a value between 50 and 70 hours? This probability is the area under the normal probability distribution between 50 and 70, as shown in the shaded area *A* of Figure 5.4.

Figure 5.4 Normal Distribution: $\mu = 50$, $\sigma = 15$

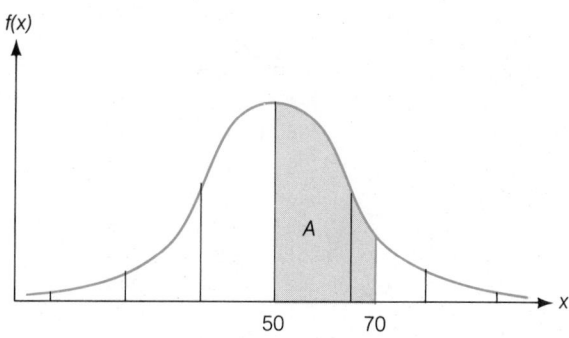

The first step in finding the area *A* is to calculate the z-score corresponding to the measurement 70. We calculate

$$z = \frac{x - \mu}{\sigma} = \frac{70 - 50}{15} = \frac{20}{15} = 1.33$$

Thus, the measurement 70 is 1.33 standard deviations above the mean, $\mu = 50$. The second step is to refer to Table IV (a partial reproduction of this table is shown in Figure 5.5, page 196). Note that z-scores are listed in the left-hand column of the table. To find the area corresponding to a z-score of 1.33, we first locate the value 1.3 in the left-hand column. Since this column lists *z* values to one decimal place only, we refer to the top row of the table to get the second decimal place, .03. Finally, we locate the number where the row labeled $z = 1.3$ and the column labeled .03 meet. This number represents the area between the mean μ and the measurement that has a z-score of 1.33:

$A = .4082$

Hence, the probability that the calculator operates between 50 and 70 hours before needing a charge is .4082.

Figure 5.5 Reproduction of Part of Table IV in the Appendix

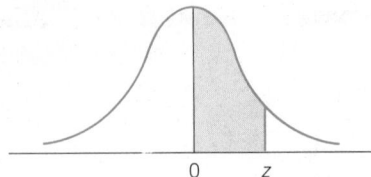

z	.00	.01	.02	.03	.04	.05	.06	.07	.08	.09
0.0	.0000	.0040	.0080	.0120	.0160	.0199	.0239	.0279	.0319	.0359
0.1	.0398	.0438	.0478	.0517	.0557	.0596	.0636	.0675	.0714	.0753
0.2	.0793	.0832	.0871	.0910	.0948	.0987	.1026	.1064	.1103	.1141
0.3	.1179	.1217	.1255	.1293	.1331	.1368	.1406	.1443	.1480	.1517
0.4	.1554	.1591	.1628	.1664	.1700	.1736	.1772	.1808	.1844	.1879
0.5	.1915	.1950	.1985	.2019	.2054	.2088	.2123	.2157	.2190	.2224
0.6	.2257	.2291	.2324	.2357	.2389	.2422	.2454	.2486	.2517	.2549
0.7	.2580	.2611	.2642	.2673	.2704	.2734	.2764	.2794	.2823	.2852
0.8	.2881	.2910	.2939	.2967	.2995	.3023	.3051	.3078	.3106	.3133
0.9	.3159	.3186	.3212	.3238	.3264	.3289	.3315	.3340	.3365	.3389
1.0	.3413	.3438	.3461	.3485	.3508	.3531	.3554	.3577	.3599	.3621
1.1	.3643	.3665	.3686	.3708	.3729	.3749	.3770	.3790	.3810	.3830
1.2	.3849	.3869	.3888	.3907	.3925	.3944	.3962	.3980	.3997	.4015
1.3	.4032	.4049	.4066	.4082	.4099	.4115	.4131	.4147	.4162	.4177
1.4	.4192	.4207	.4222	.4236	.4251	.4265	.4279	.4292	.4306	.4319
1.5	.4332	.4345	.4357	.4370	.4382	.4394	.4406	.4418	.4429	.4441

Example 5.1 Suppose you have a normal random variable x with $\mu = 50$ and $\sigma = 15$. Find the probability that x will fall within the interval $30 < x < 50$.

Solution The solution to this example can be seen from Figure 5.6. Note that both $x = 30$ and $x = 70$ lie the same distance from the mean, $\mu = 50$; $x = 30$ lies below the mean and $x = 70$ lies above it. Then, because the normal curve is symmetric about the mean, the

Figure 5.6 Normal Probability Distribution: $\mu = 50, \sigma = 15$

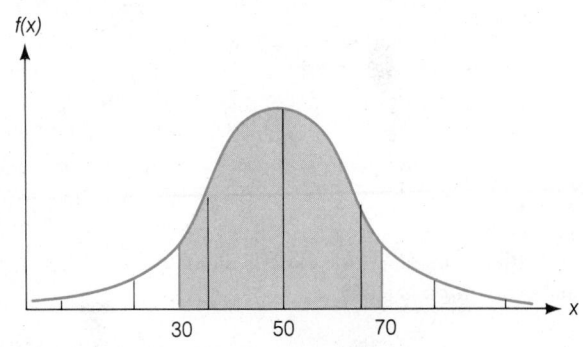

area representing the probability that x falls between $x = 30$ and $\mu = 50$ is equal to the area representing the probability that it falls between $\mu = 50$ and $x = 70$. The probability of observing a value between 50 and 70 (from Table IV) is .4082 (obtained in the previous discussion).

Because $x = 30$ lies to the left of the mean, the corresponding z-score should be negative and of the same magnitude as the z-score corresponding to $x = 70$. Checking, we obtain

$$z = \frac{x - \mu}{\sigma} = \frac{30 - 50}{15} = -1.33$$ ∎

In finding areas (probabilities) under the normal curve, it is easier to show the locations of the z-scores rather than the corresponding values of x. For example, the z-scores corresponding to $x = 30$ and $x = 70$ are located on the distribution of z-scores at the points shown in Figure 5.7. The distribution of z-scores, known as a **standard normal distribution**, always has a mean equal to zero and a standard deviation equal to 1.

Figure 5.7 A Distribution of z-Scores (a Standard Normal Distribution)

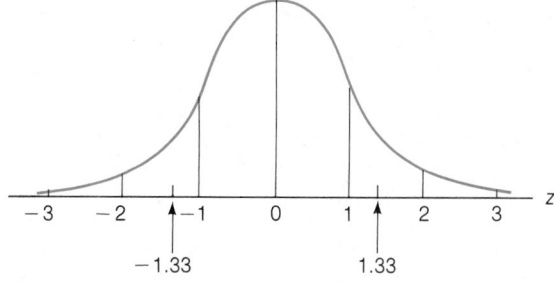

Example 5.2 Use Table IV to determine the area to the right of the z-score 1.64 for the standard normal distribution, i.e., find $P(z > 1.64)$.

Solution The probability that a normal random variable will fall more than 1.64 standard deviations to the right of its mean is indicated in Figure 5.8. Because the normal distribution is symmetric, half of the total probability (.5) lies to the right of the mean and half to the

Figure 5.8 Standard Normal Distribution: $\mu = 0, \sigma = 1$

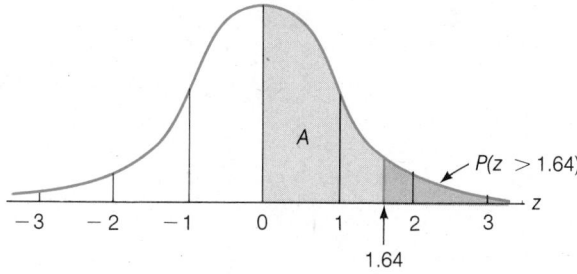

left. Therefore, the desired probability is

$$P(z > 1.64) = .5 - A$$

where A is the area between $\mu = 0$ and $z = 1.64$, as shown in the figure. Referring to Table IV, we find that the area A corresponding to $z = 1.64$ is .4495. So,

$$P(z > 1.64) = .5 - A = .5 - .4495 = .0505 \qquad \blacksquare$$

Example 5.3   Find the total area to the right of $z = -.74$ for the standard normal distribution. This area is $P(z > -.74)$.

Solution   The standard normal distribution is shown in Figure 5.9, with the area to the right of $-.74$, $P(z > -.74)$, shaded. Note that we have divided the total shaded area corresponding to $P(z > -.74)$ into two parts: the area to the left of $z = 0$ (A_1) and the area to the right of $z = 0$ (A_2). Whenever the desired area overlaps the mean, it is necessary to make this division and to find the areas separately in Table IV. The area A_2 is easy to find, since it is all the area to the right of the mean. Thus, $A_2 = .5$. The area A_1 is the area between $z = 0$ and $z = -.74$. This area, which is equivalent to the tabulated area (Table IV) between $z = 0$ and $z = .74$, is $A_1 = .2704$. Then the total area A to the right of $z = -.74$ is the sum of the areas A_1 and A_2:

$$P(z > -.74) = A_1 + A_2 = .2704 + .5 = .7704 \qquad \blacksquare$$

Figure 5.9   Standard Normal Distribution: $\mu = 0,\ \sigma = 1$

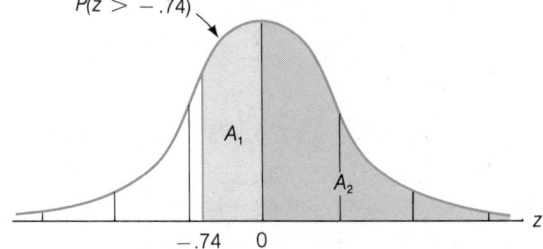

$P(z > -.74)$

A_1

A_2

$-.74 \quad 0$

z

Example 5.4   Find the total area to the right of $z = 1.96$ and to the left of $z = -1.96$. To put this in probabilistic terminology, find the probability that a normal random variable lies more than 1.96 standard deviations away from the mean.

Solution   The requested probability $P(z > 1.96$ or $z < -1.96)$ is the sum of the two areas A_1 and A_2 shown in Figure 5.10. Because the normal distribution is symmetric, the areas lying to the right of $z = 1.96$ and to the left of $z = -1.96$ must be equal. Therefore, $A_1 = A_2$.

Checking Table IV, we find the area corresponding to $z = 1.96$ to be .4750. This is the area between $z = 0$ and $z = 1.96$. Therefore,

$$A_2 = .5 - .4750 = .0250$$

Figure 5.10 Standard Normal Distribution: $\mu = 0$, $\sigma = 1$

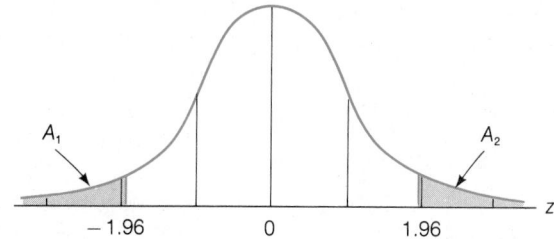

And, because of the symmetry of the normal distribution,

$$P(z > 1.96 \text{ or } z < -1.96) = A_1 + A_2 = .0250 + .0250 = .0500 \qquad \blacksquare$$

Example 5.5   Suppose an automobile manufacturer introduces a new model that has an advertised mean in-city mileage of 27 miles per gallon. Although such advertisements seldom report any measure of variability, suppose you write the manufacturer for the details of the tests, and you find that the standard deviation is 3 miles per gallon. This information leads you to formulate a probability model for the random variable x, the in-city mileage for this car model. You believe that the probability distribution of x can be approximated by a normal distribution with a mean of 27 and a standard deviation of 3.

a. If you were to buy this model of automobile, what is the probability that you would purchase one that averages less than 20 miles per gallon for in-city driving?

b. Suppose you purchase one of these new models and it does get less than 20 miles per gallon for in-city driving. Should you conclude that your probability model is incorrect?

Solution   **a.** The probability model proposed for x, the in-city mileage, is shown in Figure 5.11. We are interested in finding the area A to the left of 20, since this area corresponds to the probability that a measurement chosen from this distribution falls below 20. In other words, if this model is correct, the area A represents the fraction of cars that can be expected to get less than 20 miles per gallon for in-city driving. To find A, we first calculate the z-value corresponding to $x = 20$. That is,

$$z = \frac{x - \mu}{\sigma} = \frac{20 - 27}{3} = -\frac{7}{3} = -2.33$$

Figure 5.11 Normal Probability Distribution for x in Example 5.5: $\mu = 27$ Miles per Gallon, $\sigma = 3$ Miles per Gallon

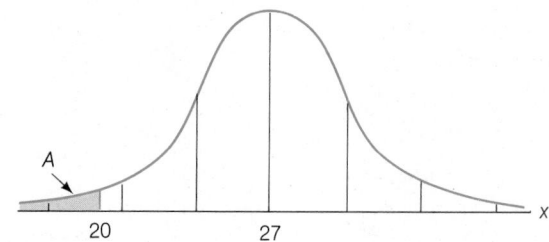

Since Table IV gives only areas to the right of the mean (and because the normal distribution is symmetric about its mean), we look up 2.33 in Table IV and find that the corresponding area is .4901. This is equal to the area between $z = 0$ and $z = -2.33$, so we find

$$A = .5 - .4901 = .0099 \approx .01$$

According to this probability model, you should have only about a 1% chance of purchasing a car of this make with an in-city mileage under 20 miles per gallon.

b. Now you are asked to make an inference based on a sample—the car you purchased. You are getting less than 20 miles per gallon for in-city driving. What do you infer? We think you will agree that one of two possibilities is true:

The probability model is correct. You simply were unfortunate to have purchased one of the cars in the 1% that get less than 20 miles per gallon in the city.

The probability model is incorrect. That is, if the manufacturer meant that the in-city mileage for the cars has a normal distribution with a mean equal to 27 and $\sigma = 3$, the claim is false.

You have no way of knowing with certainty which possibility is correct, but the evidence points to the second one. We are again relying on the rare event approach to statistical inference that we introduced earlier. The sample (one measurement in this case) was so unlikely to have been drawn from the proposed probability model that it casts serious doubt on the model. We would be inclined to believe that the model is somehow in error. Perhaps the assumption of a normal distribution is unwarranted, or the mean of 27 is an overestimate, or the standard deviation of 3 is an underestimate, or some combination of these errors was made. At any rate, the form of the actual probability model certainly merits further investigation. ■

Example 5.6

Find the value of z, call it z_0, such that $P(z \geq z_0) = .10$.

Solution

This example reverses the questions posed in Examples 5.1 to 5.5. Rather than finding an area under the normal curve over a specific interval on the z- or x-axis, we are given a specific area under the normal curve and asked to find the value of z that bounds the interval. Specifically, we want to find the value of z_0 that places an area of .10 in the upper tail of the z-distribution (see Figure 5.12).

Figure 5.12 Standard Normal Distribution for Example 5.6

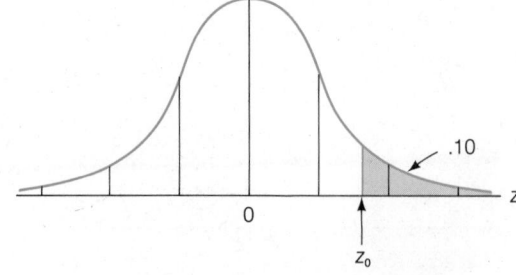

Since half the area under the normal curve lies to the right of the mean, $z = 0$, the area between $z = 0$ and z_0 is $(.5 - .1) = .4$. Thus, z_0 is the value of z in Table IV of the Appendix that corresponds to an area equal to .4000. Examining the body of Table IV, we find that the tabulated area nearest to .4000 is .3997. Since the value of z corresponding to this area is 1.28, the value of z_0 such that $P(z \geq z_0) = .10$ is $z_0 \approx 1.28$. ■

Aids in Determining Probabilities for Normal Random Variables

1. Sketch a normal curve and label the mean. Then shade the area that corresponds to the probability of interest.
2. Convert the x-values at the boundaries of the area to z-values using the formula

$$z = \frac{x - \mu}{\sigma}$$

3. In using Table IV of the Appendix, remember that the normal distribution is symmetric about its mean. Therefore, the area between the mean zero of the z-distribution and some value of z equals the area between the mean zero and $-z$. Also, the total area under the curve is 1, and this area is divided equally on each side of the mean.

Case Study 5.1
Grading on the Curve

How did teachers ever suppose that the statistician's bell-shaped curve ought somehow to be imposed on the results of their work? The famous curve was developed to describe the distribution of natural phenomena. If we weigh 10,000 grains of corn, or measure the heights of men and women, or their ability to learn nonsense syllables, the findings will cluster around some central value and taper at both ends. The bell-shaped curve is descriptive of raw or unselected phenomena, and then only if vast numbers of cases are used. When the teacher receives his pupils, their distribution with respect to some characteristics may follow the "normal curve." But, having received his charges, the skillful teacher sets out as fast as he can to destroy the "natural" state of affairs.

This statement is the introduction to Clyde W. Bresee's (1976) article about "grading on the curve." His main point is that many teachers who consistently grade on the curve are assuming that the measures of learning (test scores and the like) will always form a normal or bell-shaped distribution. These teachers may even use z-scores to determine grades and the corresponding areas under the normal curve (those in Table IV) to obtain the percentages of students who will receive each grade. The implication is that a student's performance will be measured only in relation to the other students in the class. Bresee relates the following anecdote:

After the completion of a particularly successful unit, a teacher was heard to say, "I don't know how I'll grade this thing because they all did so well." This is a sad and dangerous statement because this teacher is on the verge of undoing a month's or even a year's work. Here is a teacher so indoctrinated by an erroneous concept that he questions his own accomplishment even in the face of clear evidence. But question it he must, if he has been trained to grade "on the curve." In a class of 20, for example, the only possible explanation for 15 A's is that he is a weak teacher or a "soft grader," or both.

The caveat presented by Bresee can be more generally applied. The normal distribution is widely used and in many situations it provides an adequate approximation to reality. However, before each new application of the normal distribution, the situation should be carefully studied, and the user must be confident that all attendant assumptions are satisfied. When using the normal distribution to "grade on a curve," Bresee worries that the day may come when a supervisor admonishes a teacher, "Last year we sent you twenty children to teach. We spent a year's time and upwards of $20,000 on them, and all we have to show for it is the same old bell-shaped curve!"

Exercises 5.1–5.29

Learning the Mechanics

5.1 Find the area under the standard normal probability distribution between the following pairs of z-scores:
a. $z = 0$ and $z = 2.00$ **b.** $z = 0$ and $z = 1.00$
c. $z = 0$ and $z = 3.00$ **d.** $z = 0$ and $z = .58$
e. $z = -2.00$ and $z = 0$ **f.** $z = -1.00$ and $z = 0$
g. $z = -1.69$ and $z = 0$ **h.** $z = -.58$ and $z = 0$

5.2 Find each of the following:
a. $P(0 \leq z \leq 2.00)$ **b.** $P(-1.33 < z < 0)$
c. $P(0 \leq z < 3.06)$ **d.** $P(-.75 < z < 0)$

5.3 Find each of the following:
a. $P(-1 \leq z \leq 1)$ **b.** $P(-2 \leq z \leq 2)$ **c.** $P(-1.73 < z \leq .64)$
d. $P(-1.45 < z < .33)$ **e.** $P(z \geq -1.95)$ **f.** $P(z < 2.57)$

5.4 Find each of the following:
a. $P(z > 1.20)$ **b.** $P(z < -1.40)$ **c.** $P(.75 \leq z \leq 2.11)$
d. $P(-1.96 \leq z < -.61)$ **e.** $P(z \geq 0)$ **f.** $P(-2.33 < z < 1.10)$

5.5 Find a z-score, call it z_0, such that:
a. $P(z \geq z_0) = .5$ **b.** $P(z \geq z_0) = .025$ **c.** $P(z \leq z_0) = .025$
d. $P(z \geq z_0) = .0228$ **e.** $P(0 \leq z \leq z_0) = .4803$ **f.** $P(z < z_0) = .0401$

5.6 Find a z-score, call it z_0, such that:
a. $P(z > z_0) = .9808$ **b.** $P(z < z_0) = .9850$
c. $P(-z_0 \leq z \leq z_0) = .95$ **d.** $P(-z_0 \leq z \leq z_0) = .90$
e. $P(-z_0 \leq z \leq z_0) = .6826$ **f.** $P(-z_0 \leq z \leq z_0) = .9950$

5.7 Find each of the following:
a. $P(z \leq 1.86)$ **b.** $P(-1.68 \leq z \leq .85)$

c. $P(z \geq 2.77)$ **d.** $P(z \geq -1.62)$

e. $P(-1.96 \leq z \leq -1.12)$ **f.** $P(z \leq 3.00)$

g. $P(z \leq -3.05)$ **h.** $P(-1.45 \leq z \leq 1.45)$

5.8 Find a z-score, call it z_0, such that:

a. $P(z \leq z_0) = .0212$ **b.** $P(z \geq z_0) = .0885$

c. $P(z \leq z_0) = .7704$ **d.** $P(-z_0 \leq z \leq z_0) = .8414$

e. $P(-2 \leq z \leq z_0) = .6722$ **f.** $P(z_0 \leq z \leq 2.6) = .0312$

5.9 Give the z-score for a measurement from a normal distribution for the following:

a. One standard deviation above the mean

b. One standard deviation below the mean

c. Equal to the mean

d. Two and one-half standard deviations below the mean

e. Three standard deviations above the mean

5.10 Suppose that the random variable x is best described by a normal distribution with $\mu = 25$ and $\sigma = 5$. Find the z-score that corresponds to each of the following x values:

a. $x = 25$ **b.** $x = 30$ **c.** $x = 37.5$

d. $x = 10$ **e.** $x = 50$ **f.** $x = 32$

5.11 A random variable x is normally distributed with $\mu = 22$ and $\sigma = 5$. Determine the distance, in units of standard deviations, between each of the following values of x and the mean, $\mu = 22$:

a. $x = 10$ **b.** $x = 20$ **c.** $x = 25$

d. $x = 15$ **e.** $x = 0$ **f.** $x = 60$

5.12 Suppose that x is a normally distributed random variable with $\mu = 11$ and $\sigma = 2$. Find each of the following:

a. $P(10 \leq x \leq 12)$ **b.** $P(6 \leq x \leq 10)$ **c.** $P(13 \leq x \leq 16)$

d. $P(7.8 \leq x \leq 12.6)$ **e.** $P(x \geq 13.24)$ **f.** $P(x \geq 7.62)$

5.13 Suppose that x is a normally distributed random variable with $\mu = 45$ and $\sigma = 10$. Find each of the following:

a. $P(x \leq 50)$ **b.** $P(x \leq 35.6)$ **c.** $P(40.7 \leq x \leq 65.8)$

d. $P(22.9 \leq x \leq 33.2)$ **e.** $P(x \geq 25.3)$ **f.** $P(x \leq 25.3)$

5.14 Suppose that x is a normally distributed random variable with $\mu = 30$ and $\sigma = 8$. Find a value of the random variable, call it x_0, such that:

a. $P(x \geq x_0) = .5$ **b.** $P(x < x_0) = .025$

c. $P(x > x_0) = .10$ **d.** $P(x > x_0) = .95$

e. 10% of the values of x are less than x_0

f. 80% of the values of x are less than x_0

g. 1% of the values of x are greater than x_0

5.15 Suppose that x is a normally distributed random variable with mean 100 and standard deviation 8. Draw a rough graph of the distribution of x. Locate μ and the interval $\mu \pm 2\sigma$ on the graph. Find the following probabilities:

a. $P(\mu - 2\sigma \leq x \leq \mu + 2\sigma)$ **b.** $P(x \geq \mu + 2\sigma)$ **c.** $P(x \leq 92)$

d. $P(92 \leq x \leq 116)$ **e.** $P(92 \leq x \leq 96)$ **f.** $P(76 \leq x \leq 124)$

Applying the Concepts

5.16 In a survey of 2,000 long-distance telephone calls reported in the *Orlando Sentinel* (March 12, 1984), it was found that seven of eight long-distance phone companies were overcharging (charging for additional time) for a given call and that six were charging for unconnected calls. Suppose that the additional time being charged to a long-distance phone call has a normal distribution with a mean of 25 seconds and a standard deviation of 8 seconds.

a. Find the probability that a given long-distance call will be overcharged by at least 40 seconds.

b. By no more than 10 seconds.

5.17 The average salary for a major league baseball player has risen steadily from $19,000 per year in 1967 to $340,000 in 1984 (*U.S. News and World Report*, April 9, 1984).

a. If the 1984 distribution of salaries is normally distributed with a standard deviation equal to $90,000, what percentage of major league baseball players are making $500,000 per year or more?

b. Can you give a reason why it is unlikely that the distribution of major league baseball salaries is normally distributed?

5.18 When economic times are difficult, people keep their cars. During the recession of 1982–1983, the R. L. Polk statistics showed that the nation's automobile population was the oldest it had ever been, with a median age of 6.2 years in 1983—up from 4.9 years in 1970 (*Bakersfield Californian*, August 3, 1983). Suppose that the age of automobiles in 1983 was (approximately) normally distributed with standard deviation equal to 2.1 years.

a. Give (approximately) the mean age of the automobiles on the road.

b. Give the approximate percentage of automobiles 10 years old or older.

c. Give the approximate percentage of cars less than 3 years old.

5.19 The amount of oxygen dissolved in rivers and streams depends on the water temperature and on the amounts of decaying organic matter from natural processes or human disturbances that are present in the water. The Council on Environmental Quality (CEQ) considers a dissolved oxygen content of less than 5 milligrams per liter of water to be undesirable because it is unlikely to support aquatic life. Suppose that an industrial plant discharges its waste into a river and that the downstream daily oxygen content measurements are normally distributed with a mean equal to 6.3 milligrams per liter and a standard deviation of .6 milligram per liter.

a. What percentage of the days would the dissolved oxygen content in the river be considered undesirable by the CEQ?

b. Within what limits would we expect the dissolved oxygen content to fall?

5.20 The scores on a test designed to measure elementary school teachers' attitudes toward handicapped students are normally distributed with a mean score equal to 67 and a standard deviation equal to 10.8.

a. If an elementary school teacher is chosen at random, what is the probability that he or she would score above 95 on the test?

b. A program has been developed to improve teachers' attitudes toward handicapped students. One teacher who has completed the program is chosen at random and scores above 95 on the test. Would you conclude that the mean test score for teachers completing the program is higher than that of other teachers? Why?

5.21 The pulse rate per minute of the adult male population between 18 and 25 years of age in the United States is known to have a normal distribution with a mean of 72 beats per minute and a standard deviation of 9.7. If the requirements for military service state that anyone with a pulse rate over 100 is medically unsuitable for service, what proportion of the males between 18 and 25 years of age would be declared unfit because their pulse rates are too high?

5.22 If you are attracted to lotteries, one of the best is run by the U.S. government. Furthermore, with some research you can improve your chances of winning. We refer to the Interior Department's lottery for oil and gas leases. The *Wall Street Journal* (March 29, 1984) reports on abuses in the lottery—particularly the tendency of the Interior Department to include valuable oil leases, some worth millions of dollars, among the many included in the lottery. The Interior Department claims that only 5 to 10% of the leases included in the $75 per ticket lottery should be salable to an oil company. Yet 184 of 328 winners in the July 1980 lottery of Wyoming lands were able to sell their leases to oil companies. Assuming that the Interior Department's claim is correct, is this large number of salable leases a rare event? Answer the question by finding the z-score corresponding to $x = 184$ salable leases out of 328, given that as many as 10% of all leases included in the lottery are salable.

5.23 Ideally, a worker seeking a new job should acquire information about available wage rates in the industry in order to be able to compare wage rates offered by different firms. However, such a search could be time-consuming and costly. In particular, the longer an unemployed worker searches for a higher wage, the greater will be the loss in income. Therefore, workers may not find it worthwhile to search until they find the highest available wage rate, and managers may not have to pay top dollar to attract workers. These factors help explain the existing dispersion in wage rates. Suppose the distribution of wage rates nationwide that would be offered to a certain skilled worker can be approximated by a normal distribution with $\mu = \$10.50$ per hour and $\sigma = \$1.25$ per hour. In addition, assume that the worker is offered $12.00 per hour by the first firm contacted.

a. Suppose the worker were to undertake a nationwide job search. What proportion of the wage rates that would be offered to the worker would be greater than $12.00 per hour?

b. If the worker were to complete a nationwide job search and then randomly select one of the many job offers received, what is the probability that the wage rate would be more than $10.00 per hour?

c. The *median*, call it x_m, of a continuous random variable x is the value such that $P(x \geq x_m) = P(x \leq x_m) = .5$. That is, the median is the value x_m such that half the area under the probability distribution lies above x_m and half lies below it. Find the median of the random variable corresponding to the wage rate and compare it to the mean wage rate.

5.24 Do security analysts do a good job of forecasting corporate earnings growth and advising their clientele? David Dreman, a *Forbes* columnist, addresses this question in an article titled "Astrology Might Be Better" (*Forbes*, March 26, 1984). The basis of Dreman's article is a study by Professors Michael Sandretto of Harvard and Sudhir Milkrishnamurthi of MIT. The study surveys security analysts' forecasts of annual earnings for the (then) current year for more than 769 companies with five or more forecasts per company per year. The average forecast error for this large number of forecasts was plus or minus 31.3%. To apply this information to a practical situation, suppose the population of analysts' forecast errors is normally distributed with a mean of 31.3% and a standard deviation of 10%.

a. If you obtain a security analyst's forecast for a certain company, what is the probability that it will be in error by more than 50%?

b. If three analysts make the forecast, what is the probability that at least one of the analysts will err by more than 50%?

5.25 A machine used to regulate the amount of dye dispensed for mixing shades of paint can be set so that it discharges an average of μ milliliters of dye per can of paint. The amount of dye discharged is known to have a normal distribution with a standard deviation of .4 milliliter. If more than 6 milliliters of dye are discharged when making a certain shade of blue paint, the shade is unacceptable. Determine the setting for μ so that only 1% of the cans of paint will be unacceptable.

5.26 A physical fitness association is including the mile run in their secondary school fitness test for boys. The time for this event for boys in secondary school is approximately normally distributed with a mean of 450 seconds and a standard deviation of 40 seconds. If the association wants to designate the fastest 10% as "excellent," what time should the association set for this criterion?

5.27 The board of examiners that administers the real estate brokers' examination in a certain state found that the mean score on the test was 435 and the standard deviation was 72. If the board wants to set the passing score so that only the best 30% of all applicants pass, what is the passing score? Assume that the scores are normally distributed.

5.28 The distribution of the demand (in number of units per unit time) for a product can often be approximated by a normal probability distribution. For example, a bakery has determined that the number of loaves of its white bread demanded daily has a normal distribution with mean 7,200 loaves and standard deviation 300 loaves. Based on cost considerations, the company has decided that its best strategy is to produce a sufficient number of loaves so that it will fully supply demand on 94% of all days.

a. How many loaves of bread should the company produce?

b. Based on the production in part a, on what percentage of days will the company be left with more than 500 loaves of unsold bread?

5.29 What relationship exists between the z-score and the box plot methods (Section 2.8) for detecting outliers? Although the answer depends on the underlying distribution of the population being sampled, we will assume for this exercise that the population distribution is normal.

a. What are the z-scores corresponding to the lower and upper quartiles, Q_L and Q_U, of the population distribution?

b. In terms of standard deviations, specify the width of the interquartile range of the population distribution.

c. What are the z-scores corresponding to the inner fences?

d. What are the z-scores corresponding to the outer fences?

e. If an observation is randomly selected from the population, what is the probability that its z-score is less than -3 or greater than 3?

f. Refer to part e. What is the probability that the observation will fall outside the population inner fences?

g. Refer to parts e and f. What is the probability that the observation will fall outside the population outer fences?

h. Is there much difference between the ability of the z-score and the box plot methods for detecting outliers if the underlying population distribution is normal? Explain.

5.3
Approximating a Binomial Distribution with a Normal Distribution

When a binomial random variable can assume a large number of values, the calculation of its probabilities may become very tedious. To contend with this problem, we provide tables in the Appendix to give the probabilities for some values of n and p, but these tables are by necessity incomplete. In particular, the binomial probability table (Table II) can be used only for $n = 5, 10, 15, 20,$ or 25. To deal with this limitation, we seek approximation procedures for calculating the probabilities associated with a binomial probability distribution.

When n is large, a normal probability distribution may be used to provide a good approximation to the probability histogram of a binomial random variable. To show how this approximation works, we refer to Example 4.11, in which we used the binomial distribution to model the number x of twenty voters who favor a candidate. We assumed that 60% of all the eligible voters favored the candidate. The mean and standard deviation of x were found to be $\mu = 12$ and $\sigma = 2.2$. The binomial distribution for $n = 20$ and $p = .6$ is shown in Figure 5.13 (page 208), and the approximating normal distribution with mean $\mu = 12$ and standard deviation $\sigma = 2.2$ is superimposed.

As part of Example 4.11, we used Table II to find the probability that $x \leq 10$. This probability, which is equal to the sum of the areas contained in the rectangles (shown in Figure 5.13) that correspond to $p(0), p(1), p(2), \ldots, p(10)$, was found to equal .245. The portion of the approximating normal curve that would be used to approximate the

Figure 5.13
Binomial Distribution
for $n = 20$, $p = .6$
and Normal Distri-
bution with $\mu = 12$,
$\sigma = 2.2$

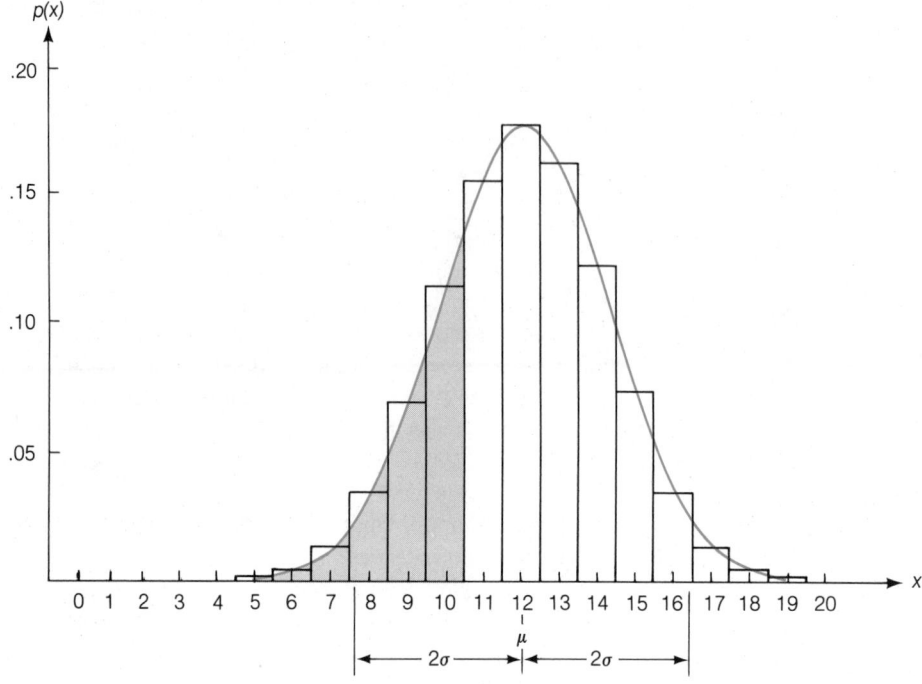

area $p(0) + p(1) + \cdots + p(10)$ is shaded in Figure 5.13. Note that this shaded area lies to the left of 10.5 (not 10), so we may include all of the probability in the rectangle corresponding to $p(10)$. The z-score corresponding to this value of x, 10.5, is

$$z = \frac{x - \mu}{\sigma} = \frac{10.5 - 12}{2.2} = -.68$$

From Table IV of the Appendix we find that the area between $z = 0$ and $z = -.68$ is .2517. Thus, the approximating normal probability is the area to the left of $x = 10.5$, or

$$P(x \le 10) \approx .5 - .2517 = .2483$$

You can see that this approximation yields a value that differs only slightly from the exact value, .245.

You may be wondering how large n should be in order for the normal distribution to provide an adequate approximation to the binomial. We will, as a rule of thumb, require that the interval $\mu \pm 3\sigma$ lie completely within the range of values for x, that is, within the interval from zero to n. When this condition is *not* satisfied, the binomial probability distribution will be skewed (to the right or left, depending upon the value of p) and the symmetric normal curve will provide a poor approximation to it. In the preceding example, $\mu \pm 3\sigma = 12 \pm 3(2.2) = 12 \pm 6.6 = (5.4, 18.6)$. This lies within the interval from zero to 20, so the normal approximation should be adequate.

> ### The Sample Size Necessary for the Normal Distribution to Provide a Good Approximation to the Binomial Probability Distribution
>
> The sample size should be large enough so that the interval $\mu \pm 3\sigma$ (where $\mu = np$ and $\sigma = \sqrt{npq}$) lies completely within the range of the values of x, that is, within the interval zero to n.

Example 5.7   An integral part of many modern electronic appliances (television sets, tape recorders, pocket calculators, etc.) is solid-state circuitry. Since this circuitry is mass-produced (stamped) by machine, many of these products have become relatively inexpensive. A problem with anything that is mass-produced is quality control, and the manufacturing process must somehow be monitored to make certain the fraction of defective items produced is kept at an acceptable level.

One method of dealing with this problem is **lot acceptance sampling,** in which a sample of the items produced is selected, and each item in the sample is carefully tested. The lot of items is then accepted (inferring there are few defectives in the entire lot) or rejected (inferring there is an unacceptable fraction of defectives in the entire lot), based on the number of defectives in the sample. For example, suppose a manufacturer of solid-state circuits for television sets chooses 200 stamped circuits from the day's production and determines x, the number of defective circuits in the sample. If the manufacturer is willing to accept the production of up to 6% defectives:

a. Find the mean and standard deviation of x, assuming that the true proportion of defectives is .06.

b. Use the normal approximation to determine the probability that twenty or more defectives are observed in the sample of 200 circuits, i.e., that $x \geq 20$.

Solution   **a.** The random variable x is binomial with $n = 200$ and the fraction defective $p = .06$. Thus,

$$\mu = np = 200(.06) = 12$$
$$\sigma = \sqrt{npq} = \sqrt{200(.06)(.94)} = \sqrt{11.28} = 3.36$$

Note that

$$\mu \pm 3\sigma = 12 \pm 3(3.36) = 12 \pm 10.08 = (1.92, 22.08)$$

lies completely within the range from zero to 200, so that the normal probability distribution should provide an adequate approximation to this binomial distribution.

b. To find the approximating area corresponding to $x \geq 20$, refer to Figure 5.14. Note that we want to include all of the binomial probability histogram from 20 to 200, inclusive. But in order to include the entire rectangle corresponding to $x = 20$, we

Figure 5.14   Normal
Approximation to the
Binomial Distribution with
$n = 200$, $p = .06$

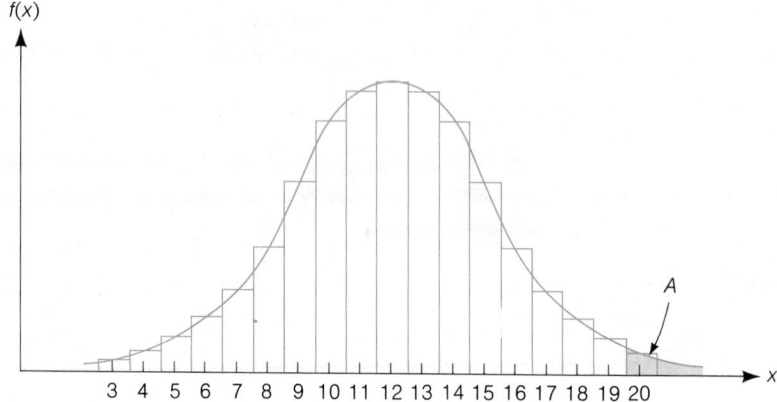

must begin the approximating area at $x = 19.5$. Thus, our z-value is

$$z = \frac{x - \mu}{\sigma} = \frac{19.5 - 12}{3.36} = \frac{7.5}{3.36} = 2.23$$

Referring to Table IV in the Appendix, we find that the area to the right of the mean
corresponding to $z = 2.23$ (see Figure 5.15) is .4871. So the area A is

$$A = .5 - .4871 = .0129$$

Thus, the normal approximation to the binomial probability is

$$P(x \geq 20) \approx .0129$$

In other words, the probability is extremely small that twenty or more defectives will
be observed in a sample of 200 circuits—*if in fact the true fraction defective is .06.*
If the manufacturer observes $x \geq 20$, the likely reason is that the process is producing more than the acceptable 6% defectives. The lot acceptance sampling procedure is another example of using the rare event approach to make inferences.

                                                                ■

Figure 5.15   Standard
Normal Distribution

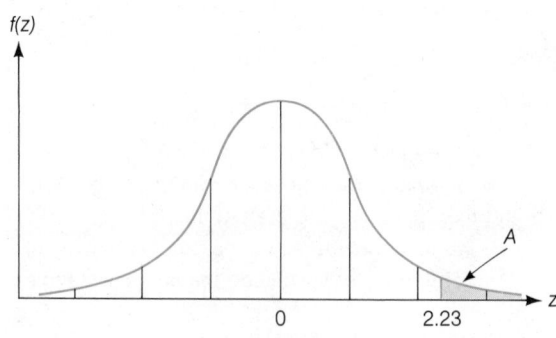

The following aids are helpful when using the normal probability distribution to approximate the binomial probability distribution:

Aids in Using Normal Probabilities to Approximate Binomial Probabilities

1. Find $\mu = np$ and $\sigma = \sqrt{npq}$. Check to make certain that the interval $\mu \pm 3\sigma$ lies completely within the interval from zero to n.

2. Sketch the probability rectangles you wish to approximate and sketch the approximating normal curve. Label the mean of the normal curve.

3. Shade the area of interest. Remember that when you compute z-values, the x-values that locate this area on the approximating normal curve always end in .5. You will be able to see this in your sketch (step 2).

4. Use the method discussed in Section 5.2 for finding normal areas to obtain the approximating normal probability.

Note: We do not recommend memorizing rules telling when to add or to subtract .5 in the numerator of the z-value. The best way to answer this question is to perform steps 2 and 3. Once you have decided which rectangles should be included, you will know whether .5 should be added to or subtracted from x.

Exercises 5.30–5.46

Learning the Mechanics

5.30 Why might you want to use a normal distribution to approximate a binomial distribution?

5.31 What conditions must be satisfied in order for the normal distribution to provide a good approximation to the binomial probability distribution?

5.32 Assume that x is a binomial random variable with $n = 25$ and $p = .4$. Use Table II in the Appendix and the normal approximation to find the exact and approximate values, respectively, for the following probabilities:
a. $P(x \leq 9)$ **b.** $P(x \leq 14)$ **c.** $P(x = 9)$ **d.** $P(8 \leq x \leq 15)$

5.33 Assume that x is a binomial random variable with $n = 50$ and $p = .4$. Use a normal approximation to find the following:
a. $P(x \leq 35)$ **b.** $P(22 \leq x \leq 29)$ **c.** $P(x \geq 24)$

5.34 Calculate the approximate probability that the binomial random variable x is larger than 50 for each of the following situations:
a. $n = 100$, $p = .5$ **b.** $n = 100$, $p = .6$ **c.** $n = 200$, $p = .15$
d. $n = 300$, $p = .2$ **e.** $n = 200$, $p = .2$

5.35 Assume that x is a binomial random variable with $n = 100$ and $p = .45$. Use a normal approximation to find the following:
a. $P(x \leq 45)$ **b.** $P(40 \leq x \leq 50)$ **c.** $P(x \geq 38)$

Applying the Concepts

5.36 The *Statistical Abstract of the U.S.: 1981* reports that 22.5% of the country's 79,108,000 households are inhabited by one person. If 1,000 randomly selected homes are to participate in a Nielsen survey to determine television ratings, find the approximate probability that no more than 250 of these homes are inhabited by one person.

5.37 It is against the law to discriminate against job applicants because of race, religion, sex, or age. Of the individuals who apply for an accountant's position in a large corporation, 40% are over 45 years of age. If the company decides to choose fifty of a very large number of applicants for closer credential screening, claiming that the selection will be random and not age-biased, what is the approximate probability that fewer than fifteen of those chosen are over 45 years of age? (Assume that the applicant pool is large enough so that x, the number in the sample over 45 years of age, has a binomial probability distribution.)

5.38 According to the "January" theory for investing in corporate stocks, the Dow Jones Industrial (stock) Average will show an increase (or a decrease) for the full year if it rises (or falls) in January. This barometer of stock market behavior has been correct for 29 of the past 34 years (*Wall Street Journal*, "Heard on the Street," February 1, 1984). The coincidence in the January and the full year movements in the Dow Jones Average may not be pure chance. If the Dow Jones Average is up (or down) in January, it certainly increases the probability that it will be up (or down) for the full year. Suppose that the probability of coincidence in any single year is .6. Use the normal curve approximation to the binomial probability distribution to find the probability that the January and the full year movements in the Dow Jones Average will coincide twenty-nine or more times in 34 years.

5.39 In Exercise 4.43 we presented some results of an *Orlando Sentinel* survey of its readers' knowledge of celebrities who had faded from the public's eye. In this survey, the respondent was asked to state whether each of the twenty celebrities was alive or dead. The average number of correct answers was nine. Suppose that a respondent guessed the correct answer for each of the celebrities so the probability of a correct answer was .5. Use the normal approximation to the binomial probability distribution to find the probability that:

a. The respondent would answer correctly for nine or more of the twenty celebrities. Compare your approximate value of the probability to the exact value obtained in Exercise 4.43.

b. The respondent would answer incorrectly for as many as twelve of the celebrities.

5.40 Melanoma, a malignant form of skin cancer, strikes more than 15,000 Americans per year and kills 45% of this number (*Time*, May 30, 1983).

a. What are the expected value and variance of x, the number of the 15,000 annual melanoma patients who die of the affliction?

b. Find the probability that x will exceed 6,900 patients per year.

c. Would you expect the number x dying of melanoma to exceed 7,000 in any single year? Explain.

5.41 In Exercise 5.22, we noted that the Interior Department claims that only 5 to 10% of all leases awarded in their $75 per ticket oil and gas lease lottery are salable to oil companies. In the July 1980 lottery of Wyoming government lands, 184 of 328 winners were salable to oil companies. Suppose that as many as 10% of all leases awarded by the Interior Department are salable to oil companies.

a. What is the probability that as many as fifty leases (i.e., fifty or more) would be salable to oil companies?

b. If 184 of the 328 leases awarded in the July 1980 Wyoming lottery were in fact salable to oil companies, would you regard this as a rare event? Explain.

c. Given the outcome of the Wyoming lottery and considering your answer to part b, do you believe the Interior Department's claim? Explain.

5.42 An advertising agency was hired to introduce a new product. It claimed that after its campaign, 30% of all consumers were familiar with the product. To check the claim, the manufacturer of the product surveyed 2,000 consumers. Of this number, 527 consumers had learned about the product through sources attributable to the campaign. What is the approximate probability that as few as 527 (i.e., 527 or less) would have learned about the product if the campaign was really 30% effective?

5.43 A recent study involving attrition rates at a major university has shown that 43% of all incoming freshmen do not graduate within 4 years of entrance.

a. If 200 freshmen are randomly sampled this year and their progress through college is followed, what is the approximate probability that no less than half will graduate within the next 4 years?

b. What is the approximate probability that the number of sampled freshmen graduating within 4 years will be between forty and eighty?

5.44 To check on the effectiveness of a new production process, 700 photoflash devices were randomly selected from a large number that had been produced. If the process actually produces 6% defectives, what is the approximate probability that:

a. More than fifty defectives appear in the sample of 700?

b. The number of defectives in the sample of 700 is forty-five or less?

5.45 The median time a patient waits to see a doctor in a large clinic is 20 minutes. On a day when 150 patients visit the clinic, what is the approximate probability that:

a. More than half will have to wait more than 20 minutes?

b. More than eighty-five will have to wait more than 20 minutes?

c. More than sixty but less than ninety will have to wait more than 20 minutes?

5.46 The percentage of fat in the bodies of American men is an approximate normal random variable with mean equal to 15% and standard deviation equal to 2%.

a. If these values were used to describe the body fat of men in the United States Army and if 20% or more body fat is characterized as obese, what is the approximate probability that a random sample of 10,000 soldiers will contain fewer than fifty who would be characterized as obese?

b. If the army actually were to check the percentage of body fat for a random sample of 10,000 men and if only thirty contained 20% (or higher) body fat, would you conclude that the army was successful in reducing the percentage of obese men below the percentage in the general population? Explain your reasoning.

5.4
The Uniform Distribution (Optional)

All the probability problems discussed in Chapter 3 had sample spaces that contained a finite number of simple events. In many of these problems, the simple events were assigned equal probabilities—for example, the die toss or the coin toss. For continuous random variables there is an infinite number of values in the sample space, but in some cases the values may appear to be equally likely. For example, if a short exists in a 5-meter stretch of electrical wire, it may have an equal probability of being in any particular 1-centimeter segment along the line. Or if a safety inspector plans to choose a time at random during the four afternoon work-hours to pay a surprise visit to a certain area of a plant, then each 1-minute time interval in this 4-work-hour period will have an equally likely chance of being selected for the visit.

Continuous random variables that appear to have equally likely outcomes over their range of possible values possess a **uniform probability distribution,** perhaps the simplest of all continuous probability distributions. Suppose that the random variable x can assume values only in an interval $c \leq x \leq d$. Then the uniform frequency function has a rectangular shape, as shown in Figure 5.16. Note that the possible values of x consist of all points in the interval between point c and point d. The height of $f(x)$ is constant in that interval and equals $1/(d - c)$. Therefore, the total area under $f(x)$ is given by

Total area of rectangle = (Base)(Height)

$$= (d - c)\left(\frac{1}{d - c}\right) = 1$$

The uniform probability distribution provides a model for continuous random variables that are **evenly distributed** over a certain interval. That is, a uniform random variable is one that is just as likely to assume a value in one interval as it is to assume a value in any other interval of equal size. There is no clustering of values around any value; instead, there is an even spread over the entire region of possible values.

The uniform distribution is sometimes referred to as the **randomness distribution** since one way of generating a uniform random variable is to perform an experiment in

Figure 5.16 The Uniform Probability Distribution

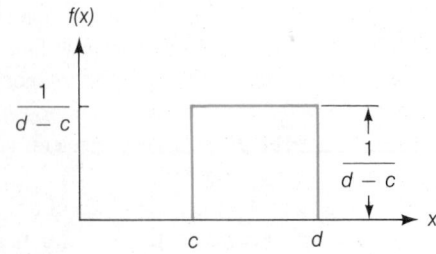

which a point is *randomly selected* on the horizontal axis between the points c and d. If we were to repeat this experiment infinitely often, we would create a uniform probability distribution like that shown in Figure 5.16. The random selection of points in an interval can also be used to generate random numbers such as those in Table I of the Appendix. Recall that random numbers are selected in such a way that every number would have an equal probability of selection. Therefore, random numbers are realizations of a uniform random variable. (Random numbers were used to draw random samples in Section 3.7.) The formulas for the uniform probability distribution, its mean, and standard deviation are shown in the box.

Probability Distribution, Mean, and Standard Deviation of a Uniform Random Variable x

$$f(x) = \frac{1}{d - c} \qquad (c \leq x \leq d)$$

$$\mu = \frac{c + d}{2} \qquad \sigma = \frac{d - c}{\sqrt{12}}$$

Suppose that the interval $a < x < b$ lies within the domain of x; that is, it falls within the larger interval $c \leq x \leq d$. Then the probability that x assumes a value within the interval $a < x < b$ is the area of the rectangle over the interval, namely $(b - a)/(d - c)$.*

Example 5.8

An unprincipled used car dealer sells a car to an unsuspecting buyer, even though the dealer knows that the car will have a major breakdown within the next 6 months. The dealer provides a warranty of 45 days on all cars sold. Let x represent the length of time until the breakdown occurs, and assume that x is a uniform random variable with values between zero and 6 months.

a. Calculate the mean and standard deviation of x. Graph the probability distribution of x, and show the mean on the horizontal axis. Also show 1 and 2 standard deviation intervals around the mean.

b. Calculate the probability that the breakdown occurs while the car is still under warranty.

Solution

a. To calculate the mean and standard deviation for x, we substitute zero and six months for c and d, respectively, in the formulas for uniform random variables. Thus,

$$\mu = \frac{c + d}{2} = \frac{0 + 6}{2} = 3 \text{ months}$$

---

* The student with a knowledge of calculus should note that

$$P(a < x < b) = \int_a^b f(x)\, dx = \int_a^b 1/(d - c)\, dx = (b - a)/(d - c)$$

and

$$\sigma = \frac{d - c}{\sqrt{12}} = \frac{6 - 0}{\sqrt{12}} = \frac{6}{3.464} = 1.73 \text{ months}$$

The uniform probability distribution is

$$f(x) = \frac{1}{d - c} = \frac{1}{6 - 0} = \frac{1}{6} \qquad (0 \leq x \leq 6)$$

The graph of this function is shown in Figure 5.17. The mean and 1 and 2 standard deviation intervals around the mean are shown on the horizontal axis.

Figure 5.17   Distribution for x in Example 5.8

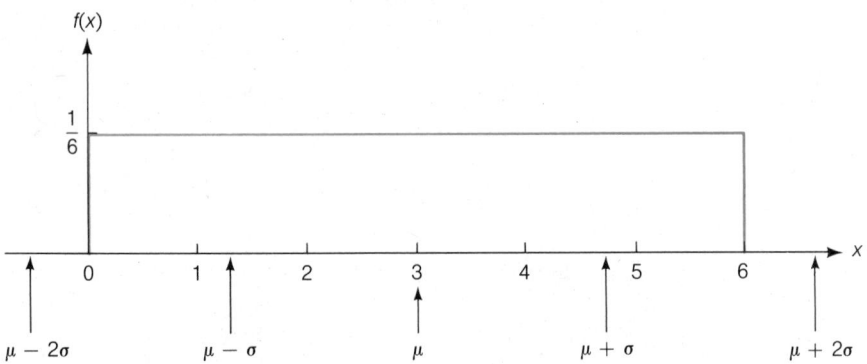

b. To find the probability that the car is still under warranty when it breaks down, we must find the probability that x, the time until the breakdown occurs, is less than 45 days or (about) 1.5 months. As indicated in Figure 5.18, we need to calculate the area under the frequency function $f(x)$ between the points $x = 0$ and $x = 1.5$. This is the area of a rectangle with base $1.5 - 0 = 1.5$ and height $\frac{1}{6}$. The probability that the unsuspecting buyer will be able to have the car repaired at the dealer's expense is then

$$P(0 < x < 1.5) = (\text{Base})(\text{Height}) = (1.5)(\tfrac{1}{6}) = .25 \qquad \blacksquare$$

Figure 5.18   Probability That Car Breaks Down Within 1.5 Months of Purchase

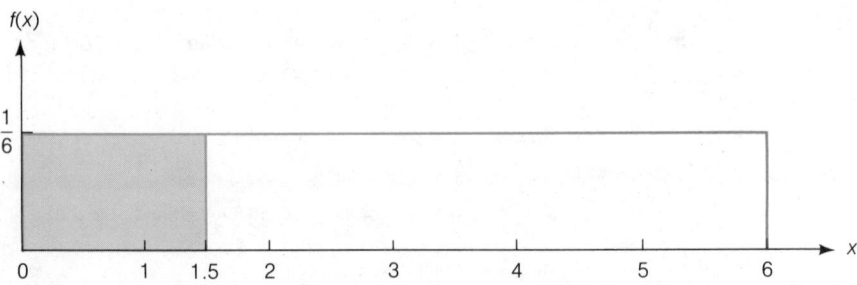

**Exercises
5.47–5.56**

Learning the Mechanics

5.47 Suppose that x is a random variable best described by a uniform probability distribution with $c = 10$ and $d = 15$.
a. Find $f(x)$.
b. Find the mean and standard deviation of x.
c. Graph $f(x)$ and locate μ and the interval $\mu \pm 2\sigma$ on the graph. Note that the probability that x assumes a value within the interval $\mu \pm 2\sigma$ is equal to 1.

5.48 Refer to Exercise 5.47. Find the following:
a. $P(10 \leq x \leq 12)$ **b.** $P(11 \leq x \leq 12.8)$ **c.** $P(x \geq 13.8)$
d. $P(x \leq 14)$ **e.** $P(x \leq 8)$ **f.** $P(0 \leq x \leq 12)$
g. $P(x \geq 10.5)$

5.49 Suppose that x is a random variable best described by a uniform probability distribution with $c = 2$ and $d = 4$.
a. Find $f(x)$.
b. Find the mean and standard deviation of x.
c. Find $P(\mu - \sigma \leq x \leq \mu + \sigma)$. **d.** Find $P(x > 2.78)$.
e. Find $P(2.4 \leq x \leq 3.7)$ **f.** Find $P(x < 2)$.

Applying the Concepts

5.50 The manager of a large department store with three floors reports that the time a customer on the second floor must wait for an elevator has a uniform distribution ranging from 0 to 4 minutes. Find the mean and standard deviation of x, the time a customer on the second floor waits for an elevator. If it takes the elevator 15 seconds to go from floor to floor, find the probability that a hurried customer can reach the first floor in less than 1.5 minutes after pushing the second-floor elevator button.

5.51 A bus is scheduled to stop at a certain bus stop every half hour on the hour and the half hour. At the end of the day, buses still stop about every 30 minutes, but due to delays that often occur earlier in the day, the bus is likely to be late. The director of the bus line claims that the length of time a bus is late is uniformly distributed and the maximum time that a bus is late is 20 minutes.
a. If the director's claim is true, what is the expected number of minutes a bus will be late?
b. If the director's claim is true, what is the probability that the last bus on a given day will be more than 19 minutes late?
c. If you arrive at the bus stop at the end of a day at exactly half-past the hour and must wait more than 19 minutes for the bus, what would you conclude about the director's claim? Why?

5.52 The manager of a local soft drink bottling company believes that when a new beverage-dispensing machine is set to dispense 7 ounces, it in fact dispenses an amount x at random anywhere between 6.5 and 7.5 ounces inclusive. Suppose that x has a uniform probability distribution.

a. Is the amount dispensed by the beverage machine a discrete or a continuous random variable? Explain.

b. Graph the frequency function for x, the amount of beverage the manager believes is dispensed by the new machine when it is set to dispense 7 ounces.

c. Find the mean and standard deviation for the distribution graphed in part b, and locate the mean and the interval $\mu \pm 2\sigma$ on the graph.

5.53 Refer to Exercise 5.52. Find the following probabilities:
a. $P(x \geq 7)$ **b.** $P(6.5 \leq x \leq 7.5)$ **c.** $P(x \leq 6.75)$
d. $P(x > 7.25)$ **e.** $P(x < 6)$ **f.** $P(6.5 \leq x \leq 7.25)$

5.54 Refer to Exercises 5.52 and 5.53. What is the probability that each of the next six bottles filled by the new machine will contain more than 7.25 ounces of beverage? Assume that the amount of beverage dispensed in one bottle is independent of the amount dispensed in another bottle.

5.55 The weather on a tropical island in January is fairly constant. Records indicate that the high temperatures for each day of the month tend to have a uniform distribution over the interval from 75°F to 90°F. A tourist arrives on the island on a randomly selected day in January.
a. What is the probability that the temperature will be above 80°F?
b. What is the probability that the temperature will be between 80°F and 85°F?
c. What is the expected temperature?

5.56 Rapid advances in technology in recent years have led to the development and manufacture of extremely complex equipment and, consequently, to the need to evaluate the equipment's reliability. The **reliability** of a piece of equipment is frequently defined to be the probability p that the equipment performs its intended function successfully for a given period of time under specific conditions (Martz & Waller, 1982). Because p varies from one point in time to another, some reliability analysts treat p as if it were a random variable with a uniform probability distribution over the interval from 0 to 1. This assumption implies, for example, that p is equally likely to be below .2 as it is to be above .8. Suppose an analyst characterizes the total uncertainty about the reliability of a particular robotic device used in an automobile assembly line using the following distribution:

$$f(p) = \begin{cases} 1 & 0 \leq p \leq 1 \\ 0 & \text{otherwise} \end{cases}$$

a. Graph the analyst's probability distribution for p.
b. Find the mean and variance of p.
c. According to the analyst's probability distribution for p, what is the probability that p is greater than .95? Less than .95?
d. Suppose that the analyst receives the additional information that p is definitely between .90 and .95, but that there is complete uncertainty about where it lies between these values. Describe the probability distribution the analyst should use to describe the available information about p.

5.5
The
Exponential
Distribution
(Optional)

The length of time between emergency arrivals at a hospital, the length of time between breakdowns of manufacturing equipment, the length of time between catastrophic events (floods, earthquakes, etc.), and the distance traveled by a wildlife ecologist between sightings of an endangered species are all random phenomena that we might want to describe probabilistically. The amount of time or distance between occurrences of random events like these can often be described by the **exponential probability distribution.** For this reason, the exponential distribution is sometimes called the **waiting time distribution.** The formula for the exponential probability distribution is shown in the box along with the mean and standard deviation of this frequency function.

Probability Distribution, Mean, and Standard Deviation for an Exponential Random Variable x

$$f(x) = \frac{1}{\theta} e^{-x/\theta} \qquad (x > 0)$$

$$\mu = \theta \qquad \sigma = \theta$$

Unlike the normal distribution, which has a shape and location determined by the values of the two quantities μ and σ, the shape of the exponential distribution is governed by a single quantity, θ. Further, it is a probability distribution with the property that its mean equals its standard deviation. Exponential distributions corresponding to $\theta = .5, 1,$ and 2 are shown in Figure 5.19.

Figure 5.19 Exponential Distributions

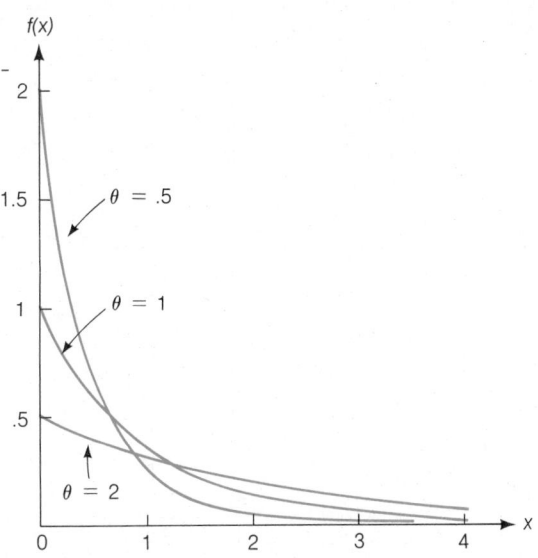

Figure 5.20 The Area A to the Right of a Number a for an Exponential Distribution

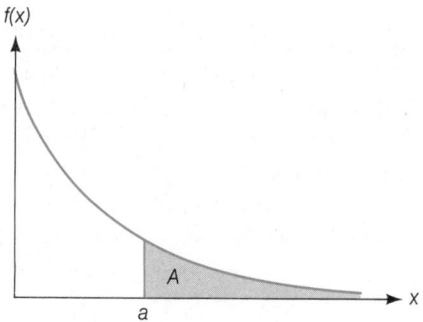

To calculate probabilities for exponential random variables, we need to be able to find areas under the exponential probability distribution. Suppose we want to find the area A, to the right of some number a, as shown in Figure 5.20. This area can be calculated by using the following formula:

> ### Finding the Area A to the Right of a Number a for an Exponential Distribution
>
> $$A = P(x \geq a) = e^{-a/\theta}$$

Use Table III in the Appendix to find the value of $e^{-a/\theta}$ after substituting the appropriate numerical values for θ and a.

Example 5.9

Suppose the length of time (in hours) between emergency arrivals at a certain hospital is modeled as an exponential distribution with $\theta = 2$. What is the probability that more than 5 hours pass without an emergency arrival?

Figure 5.21 Exponential Distribution for Example 5.9: $\theta = 2$

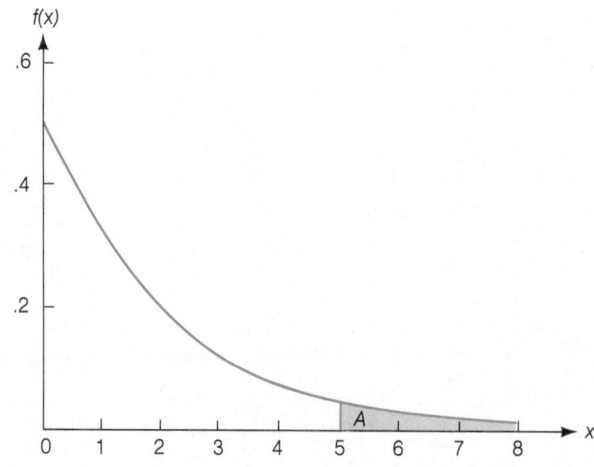

Solution   The probability we want is the area A to the right of $a = 5$ in Figure 5.21. To find this probability, use the formula given for area:

$$A = e^{-a/\theta} = e^{-(5/2)} = e^{-2.5}$$

Referring to Table III, we find

$$A = e^{-2.5} = .082085$$

Our exponential model indicates that the probability that more than 5 hours pass between emergency arrivals is about .08 for this hospital.   ◼

Example 5.10   A microwave oven manufacturer is trying to determine the length of warranty period it should attach to its magnetron tube, the most critical component in the oven. Preliminary testing has shown that the length of life (in years), x, of a magnetron tube has an exponential probability distribution with $\theta = 6.25$.

a. Find the mean and standard deviation of x.
b. Suppose a warranty period of 5 years is attached to the magnetron tube. What fraction of tubes must the manufacturer plan to replace, assuming that the exponential model with $\theta = 6.25$ is correct?
c. Find the probability that the length of life of a magnetron tube will fall within the interval $\mu - 2\sigma$ to $\mu + 2\sigma$.

Solution   a. Since $\theta = \mu = \sigma$, both μ and σ equal 6.25.
b. To find the fraction of tubes that will have to be replaced before the 5-year warranty period expires, we need to find the area between zero and 5 under the distribution. This area, A, is shown in Figure 5.22. To find the required probability, we recall the formula

$$P(x > a) = e^{-a/\theta}$$

Figure 5.22 Exponential Distribution for Example 5.10: $\theta = 6.25$

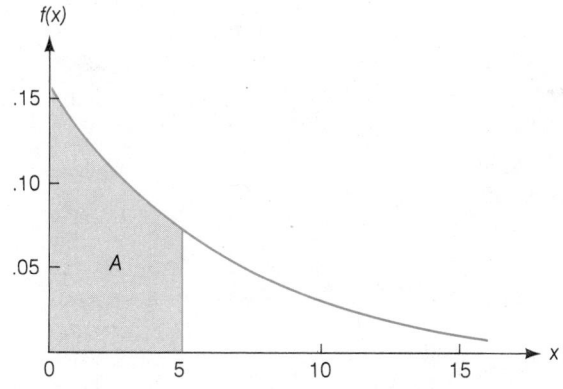

Using this formula, we can find

$$P(x > 5) = e^{-a/\theta} = e^{-5/6.25} = e^{-.80} = .449329$$

(see Table III). To find the area A, we use the complementary relationship:

$$P(x \leq 5) = 1 - P(x > 5) = 1 - .449329 = .550671$$

So approximately 55% of the magnetron tubes will have to be replaced during the 5-year warranty period.

c. We would expect the probability that the life of a magnetron tube, x, falls within the interval $\mu - 2\sigma$ to $\mu + 2\sigma$ to be quite large. A graph of the exponential distribution showing the interval $\mu - 2\sigma$ to $\mu + 2\sigma$ is given in Figure 5.23. Since the point $\mu - 2\sigma$ lies below $x = 0$, we only need to find the area between $x = 0$ and $x = \mu + 2\sigma = 6.25 + 2(6.25) = 18.75$. This area, P, which is shaded in Figure 5.23, is

$$P = 1 - P(x > 18.75)$$
$$= 1 - e^{-18.75/\theta} = 1 - e^{-18.75/6.25}$$
$$= 1 - e^{-3}$$

Checking Table III for the value of e^{-3}, we find $e^{-3} = .049787$. Therefore, the probability that the life x of a magnetron tube will fall within the interval $\mu - 2\sigma$ to $\mu + 2\sigma$ is

$$P = 1 - e^{-3}$$
$$= 1 - .049787 = .950213$$

You can see that this probability agrees very well with the interpretation of a standard deviation given by the Empirical Rule in Table 2.7 even though this probability distribution is not mound-shaped. (It is strongly skewed to the right.)   ■

Figure 5.23   Exponential Distribution for Example 5.10: $\theta = 6.25$

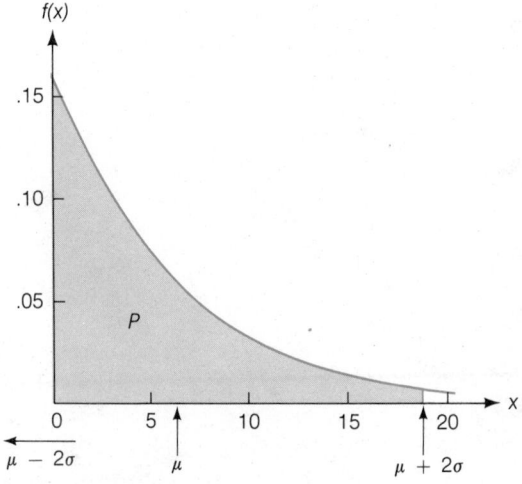

Case Study 5.2

Assessing the
Reliability of
Computer Software

In a discussion about the reliability of computer software, G. J. Schick (1974) says the following:

> Custom software . . . is expensive to develop and requires extensive testing—the goal being to certify that the software is in error-free condition, ready to support the mission for which it was designed. Similar economies should also be expected from an integrated statistical software test program. Traditionally, there are never enough time or resources to test all possible branches and data combinations in a computer program of reasonable size.
>
> Current practice is to design and develop a software system and then to test it to detect errors, until the amount of time and expense required to discover remaining errors is too great to justify further testing. . . . In principle, few large real-time computer programs ever have been tested completely and unequivocally in the sense that every logical data path has been successfully executed under every logical combination for the data at hand for all possible options. One management objective would be to test every logical path in the computer program at least once with some kind of numerical check. At the present state of the art, such a degree of testing is neither feasible nor realistic. In practice, the contractor must be willing to release and the customer willing to accept a level of risk associated with a program that has been less than completely checked.

In finding and correcting errors in a computer program (debugging) and determining the program's reliability, Schick and others have noted the importance of the distribution of the time until the next program error is found. If this distribution is assumed to be exponential, with

$$f(x) = \frac{1}{\theta} e^{-x/\theta} \qquad (x > 0, \theta > 0)$$

then, as Schick points out, its mean, θ, would be the average time required to find the next error.

In his article, Schick describes a method relative to software reliability for estimating the parameter θ of the exponential distribution. Using computer debugging data supplied by the United States Navy, Schick demonstrates how this estimation procedure and the exponential distribution can be used to estimate the reliability of a computer program. [*Note:* The model used by Schick to represent the distribution for the time until the next error is based on the exponential distribution, but it is slightly more complicated because he assumes that θ varies. For our purposes, however, nothing is lost by assuming the distribution he uses to be exponential.]

After twenty-six program errors were found, Schick estimated θ to be 23.8 days. This means that the average time it would take to find the next (twenty-seventh) error is approximately 24 days. Thus, using the estimated 23.8 days, the probability of it taking, say, 60 or more days to find the next error is approximately

$$P(x \geq 60) = e^{-60/23.8} = .08046$$

Over the next 290 days, five more errors were detected. Since this is a rate of about one error every 60 days and since $P(x \geq 60) \approx .08$, it seems unlikely that an exponential distribution with $\theta = 23.8$ is an appropriate representation of the distribution for the time until the next error. Based on the number of new errors found and the length of time it took to find them, Schick reestimated θ to be 278 days, meaning that on an average the next error (thirty-second) would not occur for 278 days. At this point, the length of time and, therefore, the cost required to find any remaining program errors may be prohibitive. Debugging should probably be discontinued.

Exercises 5.57–5.70

Learning the Mechanics

5.57 Find the value of e^{-c} for each of the following values of c:

a. $c = 6.25$ **b.** $c = 1.10$ **c.** $c = .75$

d. $c = 3.40$ **e.** $c = 2.75$ **f.** $c = 7.80$

5.58 Suppose that x has an exponential distribution with $\theta = .5$. Find the following probabilities:

a. $P(x > .5)$ **b.** $P(x \leq 3)$ **c.** $P(x > 1.5)$ **d.** $P(x \leq 5)$

5.59 Suppose that x has an exponential distribution with $\theta = 2.0$. Find the following probabilities:

a. $P(x \leq 4)$ **b.** $P(x > .5)$ **c.** $P(x \leq 2)$ **d.** $P(x > 3)$

5.60 Suppose the random variable x has an exponential probability distribution with $\theta = .5$. Find the mean and standard deviation of x. Find the probability that x will assume a value within the interval $\mu \pm 2\sigma$.

Applying the Concepts

5.61 After bacteria are subjected to a certain drug, the length of time until the bacteria die follows an exponential distribution with a mean of $\frac{1}{2}$ hour.

a. How long after the drug is administered will half the bacteria be dead?

b. What proportion of bacteria will die between .2 and 1 hour?

5.62 The length of time x it takes to produce a human reaction to tear gas has an exponential distribution with a mean of 3 minutes.

a. If tear gas is fired into a house to subdue a terrorist, how long should the police wait so that the probability is .99 that the tear gas has taken effect?

b. What proportion of people will be affected within 5 minutes?

5.63 An article in the Jacksonville, Florida, *Times Union* (March 11, 1984) reports on the unexplained crash of a small plane on takeoff and the resulting injury to its 24-year-old student pilot. The article notes that Shields Aviation, the renter of the plane, inspects (and presumably performs maintenance) on the aircraft every 100 flight hours. The plane had flown 30 hours since its last inspection. Suppose that x, the time between malfunctions for this particular plane, has an exponential distribution with mean equal to 300 hours. What is the probability that a plane of this type will malfunction within 30 hours after the last inspection?

5.64 The length of time between breakdowns of an essential piece of equipment is an important factor in deciding on the amount of auxiliary equipment needed to assure continuous service. A machine room supervisor believes the time between breakdowns of a certain electrical generator is best approximated by an exponential distribution with mean equal to 10 days.

a. What is the standard deviation of this exponential distribution?

b. Assuming the supervisor has correctly characterized the distribution for the time between breakdowns and that the generator broke down today, what is the probability that the generator will break down again within the next 14 days?

c. What is the probability that the generator will operate for more than 20 days without a breakdown?

5.65 The probability distribution of the length of service time is important in the design of service facilities, and it has a definite effect on sales. Suppose the time an individual has to wait in line to be served at a fast-food franchise is assumed to have an exponential distribution with mean equal to 1 minute.

a. What is the probability that an individual would have to wait more than 2 minutes before being served?

b. What is the probability that an individual will be served within 30 seconds after arriving in line?

5.66 An experienced mountain climber has found that the time between rock slides at a certain area of a mountain range is exponentially distributed with mean equal to 2.5 days. After observing a rock slide at this particular area, a team of climbers starts ascending the mountain. If it takes 6 hours to reach a cabin at the top of the mountain, find the probability that the team will be endangered by a rock fall before reaching the cabin.

5.67 An outbreak of a certain species of caterpillar, the spruce budworm, can cause extensive damage to the timberlands of the northern United States. It is known that an outbreak of this type of caterpillar occurs, on the average, every 30 years. Assuming that this phenomenon obeys an exponential probability law, what is the probability that catastrophic outbreaks of spruce budworm will occur within 6 years of each other?

5.68 The shelf-life of a perishable product is a random variable that is related to consumer acceptance and, ultimately, to sales and profit. Suppose that the shelf-life of bread is best approximated by an exponential distribution with mean equal to 2 days. What fraction of the loaves stocked today would you expect to still be salable (i.e., not stale) 3 days from now?

5.69 The length of time between arrivals at a hospital clinic and the length of clinical service time are two random variables that play important roles in designing a clinic and deciding how many physicians and nurses are needed for its operation. The probability distributions of both the length of time between arrivals and the length of service time are often approximately exponential. Suppose the mean time between arrivals for patients at a clinic is 4 minutes.

a. What is the probability that a particular interarrival time (the time between the arrival of two patients) is less than 1 minute?

b. What is the probability that the next four interarrival times are all less than 1 minute?

c. What is the probability that an interarrival time will exceed 10 minutes?

5.70 The following is a sample of the lengths of time between arrivals (rounded to the nearest minute) at the emergency room of a hospital:

1	10	3	3	9	7	1	4	3
14	6	4	5	7	9	5	1	9
13	4	2	6	12	5	1	10	5
8	3	23	11	14	18	15	12	1
4	22	26	37	3	8	19	5	8
18	16	28	31	21				

a. Draw a relative frequency histogram for these data. Does it appear that an exponential distribution could be used to characterize the length of time between arrivals? Explain.

b. If you were asked to model the length of time between arrivals for this emergency room, discuss how you would estimate θ.

Summary

Many **random variables** encountered in the real world have probability distributions that are well approximated by the **normal, uniform,** or **exponential probability distributions.** In this chapter we showed the graphic shape of these probability distributions, gave their means and standard deviations, and pointed out some practical applications for each of these probability models. Moreover, we showed that the normal probability distribution provides a good approximation for the binomial distribution when the number of trials, n, is sufficiently large.

Supplementary Exercises 5.71–5.104

[*Note: Starred (*) exercises refer to optional sections in this chapter.*]

Learning the Mechanics

5.71 Calculate the area under the standard normal probability distribution between the following pairs of z-scores:

a. 0 and 1.96   **b.** −1.96 and 1.96   **c.** −1.30 and 1.30

d. −2.14 and −1.07   **e.** .74 and 2.01   **f.** −1.43 and 1.95

5.72 Find the following probabilities:

a. $P(z \leq 1.1)$ **b.** $P(z \geq 1.1)$ **c.** $P(z \geq -1.86)$

d. $P(-2.75 \leq z \leq -.55)$ **e.** $P(-2.45 \leq z \leq .38)$ **f.** $P(z \leq -1.91)$

5.73 Find a z-score, call it z_0, such that:

a. $P(z \leq z_0) = .8023$ **b.** $P(z \geq z_0) = .0985$ **c.** $P(z \leq z_0) = .5$

d. $P(-z_0 \leq z \leq z_0) = .6212$ **e.** $P(z \geq z_0) = .8508$ **f.** $P(z \geq z_0) = .0055$

5.74 The random variable x has a normal distribution with $\mu = 70$ and $\sigma = 10$. Find the following probabilities:

a. $P(x \le 75)$ **b.** $P(x \ge 90)$ **c.** $P(60 \le x \le 75)$
d. $P(x > 75)$ **e.** $P(x = 75)$ **f.** $P(x \le 95)$

5.75 The random variable x has a normal distribution with $\mu = 60$ and $\sigma^2 = 64$. Find a value of x, call it x_0, such that:

a. $P(x \ge x_0) = .5$ **b.** $P(x \le x_0) = .9911$ **c.** $P(x \le x_0) = .0028$
d. $P(x \ge x_0) = .0228$ **e.** $P(x \le x_0) = .1003$ **f.** $P(x \ge x_0) = .7995$

5.76 Assume that x is a binomial random variable with $n = 100$ and $p = .6$. Use the normal probability distribution to approximate the following probabilities:

a. $P(x \le 48)$ **b.** $P(50 \le x \le 65)$ **c.** $P(x \ge 70)$
d. $P(55 \le x \le 58)$ **e.** $P(x = 62)$ **f.** $P(x \le 49 \text{ or } x \ge 72)$

*5.77 Assume that x is a random variable best described by a uniform distribution with $c = 30$ and $d = 70$.

a. Find $f(x)$.
b. Find the mean and standard deviation of x.
c. Graph the probability distribution for x and locate its mean and the interval $\mu \pm 2\sigma$ on the graph.
d. Find $P(x \le 45)$.   **e.** Find $P(x \ge 58)$.   **f.** Find $P(x \le 100)$.
g. Find $P(\mu - \sigma \le x \le \mu + \sigma)$.   **h.** Find $P(x > 60)$.

*5.78 Assume that x has an exponential distribution with $\theta = 2.0$. Find:

a. $P(x \le 1)$ **b.** $P(x > 1)$ **c.** $P(x = 1)$
d. $P(x \le 6)$ **e.** $P(2 \le x \le 10)$

Applying the Concepts

5.79 For a student to graduate, a high school requires that each student demonstrate competence in mathematics by scoring 70% or above on a mathematics achievement test. The scores of those students taking the test for the first time are normally distributed with a mean of 77% and a standard deviation of 7.3%. What percentage of students who take the test for the first time will pass the test?

5.80 Farmers often sell fruits and vegetables at roadside stands during the summer. One such roadside stand has a daily demand for tomatoes that is approximately normally distributed with a mean equal to 125 tomatoes per day and a standard deviation equal to 30 tomatoes per day.

a. If there are 90 tomatoes available to be sold at the roadside stand at the beginning of a day, what is the probability that they will all be sold?
b. If there are 200 tomatoes available to be sold, what is the probability that 50 or more will not be sold that day?
c. How many tomatoes must be available on any given day so that there will be only a 10% chance that all the tomatoes will be sold?

5.81 A firm believes the internal rate of return for its proposed investment can best be described by a normal distribution with mean 20% and standard deviation 3%. What is the probability that the internal rate of return for the investment will be the following?
a. Greater than 26% or less than 14%   **b.** At least 16%   **c.** More than 28.5%

5.82 The probability distribution of the number of people per month who open a savings account in a large banking system is approximately normally distributed with $\mu = 1,280$ and $\sigma = 265$.
a. If the bank gives a $10 gift to each new account holder, what portion of the time will the bank's payout for gifts exceed $15,000 per month?
b. What is the bank's mean monthly payout for gifts?

5.83 The length of time required to assemble a photoelectric cell is normally distributed with mean equal to 18.1 minutes and $\sigma = 1.3$ minutes. What is the probability that it will require more than 20 minutes to assemble a cell?

5.84 A local track club has decided to sponsor a 10,000-meter road race. From past results of races across the state, it is known that the length of time to complete the race has an approximately normal distribution with a mean of 49 minutes and a standard deviation of 8 minutes. The club has decided that everyone who completes the race will receive a T-shirt. Those who run between 34 and 60 minutes will also receive a medal, and those finishing under 34 minutes will receive a plaque.
a. What proportion of racers would you expect to receive a medal and a T-shirt?
b. What proportion of racers would you expect to receive a plaque and a T-shirt?

5.85 A woman has made a daily record of the length of time it takes to travel from her house to her place of work. The distribution is approximately normal with a mean of 20 minutes and a standard deviation of 2.1 minutes.
a. If the woman leaves for work 15 minutes before she is to start working, what is the probability that she will make it on time?
b. How long before she is to start working should she leave so that she is late only 10% of the time?

5.86 The amount of coal produced per day by a coal mine has a normal distribution with a mean of 60.8 tons and a standard deviation of 7.9 tons.
a. On any given day, what is the probability the mine produces less than 50 tons of coal?
b. Between 62 and 66 tons?

5.87 Suppose the present value of a risky investment is approximately normally distributed with mean $10,000 and standard deviation $4,000.
a. What is the probability that the present value of the investment is less than $1,000?
b. Greater than $20,000?

5.88 After extensive testing of a new low-tar cigarette, a tobacco company concludes that the number of milligrams of tar yielded by each cigarette has a probability distribution that is approximately normal with $\mu = 8$ and $\sigma = 1.9$ milligrams.
a. What is the probability that one of the new low-tar cigarettes will yield more than 10 milligrams of tar?

b. If two of the new low-tar cigarettes are chosen at random and tested, what is the probability that both will yield less than 6 milligrams of tar?

5.89 Eighty 10-pound bags of sugar are randomly sampled from the stock of a local sugar wholesaler and weighed. The following are the results rounded to the nearest tenth of a pound:

10.2	10.2	9.6	9.9	10.0	10.0	10.0	10.1
10.4	9.9	9.6	10.0	9.9	10.1	9.9	9.9
9.8	10.0	10.4	10.0	9.7	9.9	10.2	9.8
9.9	10.3	10.0	10.0	9.8	9.8	10.1	9.9
10.1	9.9	10.2	9.8	10.0	10.1	9.5	10.0
10.0	10.1	10.1	10.1	10.2	10.5	9.9	10.1
9.5	9.9	9.7	9.9	9.9	10.1	9.9	10.3
10.1	10.1	9.8	10.0	10.0	10.0	9.7	10.0
10.0	10.2	10.0	10.1	10.3	9.8	10.0	10.2
9.6	9.9	9.8	10.0	9.8	10.2	10.0	9.9

a. Construct a relative frequency histogram for these data and suggest a continuous probability distribution that could be used to approximate the weight distribution of the wholesaler's stock of 10-pound bags of sugar.

b. Using the histogram, estimate μ and σ for the weight distribution of the wholesaler's stock of 10-pound bags of sugar.

c. Calculate $\bar{x}$ and s for the sample of eighty weights. How do these estimates of μ and σ compare to your answer in part b?

5.90 A company has a lump-sum incentive plan for salespeople that is dependent upon their level of sales. If they sell less than $100,000 per year, they receive a $1,000 bonus; from $100,000 to $200,000, they receive $5,000; and above $200,000, they receive $10,000. Suppose the annual sales per salesperson has approximately a normal distribution with $\mu = \$180,000$ and $\sigma = \$50,000$.

a. Find p_1, the proportion of salespeople who receive a $1,000 bonus.

b. Find p_2, the proportion of salespeople who receive a $5,000 bonus.

c. Find p_3, the proportion of salespeople who receive a $10,000 bonus.

d. What is the mean value of the bonus payout for the company? [*Hint:* See the definition for the expected value of a random variable in Chapter 4.]

5.91 It is quite common for the standard deviation of a random variable to increase proportionally as the mean increases. When this occurs, the **coefficient of variation,**

$$CV = \frac{\sigma}{\mu}$$

which is the ratio of σ to μ, is the **proportionality constant.** To illustrate, the error (in dollars) in assessing the value of a house increases as the house increases in value. Suppose that long experience with assessors in a given region has shown that the coefficient of variation is .08 and that the probability distribution of assessed valuations on the same house by many different assessors is approximately normal with a mean we will call the "true value" of the house. Suppose the true value of your house is $50,000 and it is being assessed for taxation purposes. What is the probability the assessor will assess your house in excess of $55,000?

5.92 On the average, the main chute fails in one of every 1,000 parachutes. Suppose that during a lifetime a professional parachutist makes 4,000 jumps, and let x equal the number of times the main chute fails. What is the approximate probability that the parachutist's main chute fails on at least one jump? [*Note:* Because n is large and p is so small, the Poisson probability distribution will also provide a good approximation to this probability. (See Exercise 4.58.) If you covered Section 4.5, find the Poisson approximation to $P(x > 0)$.]

5.93 The net weight per package of a certain brand of corn chips is listed as 10 ounces. The weight actually delivered to each package by an automated machine is a normal random variable with mean 10.5 ounces and standard deviation .2 ounce. Suppose 100 packages are chosen at random and the net weights are ascertained. Let x be the number of the 100 selected packages that contain at least 10 ounces of corn chips. Then x is a binomial random variable with n = 100 and p = probability that a randomly selected package contains at least 10 ounces. What is the probability that they all contain at least 10 ounces of corn chips? What is the probability that at least 90% of the packages contain 10 ounces or more?

5.94 An admissions officer for a law school indicates that 35% of the applicants meet all ten requirements and 95% meet at least eight of the ten requirements.
a. If a random sample of 300 applicants is taken, what is the approximate probability that fewer than 250 will fail to meet all ten requirements?
b. What is the approximate probability that more than 280 will meet at least eight of the ten requirements?

5.95 Sixteen percent of the American black population is known to suffer from sickle-cell anemia. If 1,000 American black people are sampled at random, what is the approximate probability that:
a. More than 175 have the disease?
b. Fewer than 140 have the disease?
c. The number of people in the sample with the disease is between 130 and 180?

*5.96 Blending feeders are machines used to break up tobacco that has been aged in tightly packed hogsheads. One cigarette manufacturer determines that the time between breakdowns for each of its blending feeders is best represented by an exponential distribution with mean equal to 100 hours of operation. A particular feeder has just been repaired and put back into service.
a. What is the probability that it will not break down for at least 50 more hours?
b. What is the probability that it will break down within the next 100 hours?

5.97 A loan officer in a large bank has been assigned to screen sixty loan applications during the next week.

a. If her past record indicates that she turns down 20% of the applicants, what is the approximate probability that forty-one or more of the sixty applications will be approved?

b. What is the approximate probability that between forty-five and fifty of the applications will be approved?

*5.98 Assume that the length of the active life of baking yeast has an exponential distribution with mean equal to 5 months. Suppose the expiration date marked on a package of yeast is based on a life of 7 months. What is the probability that a package of the yeast will lose its potency before its expiration date?

5.99 Contrary to our intuition, very reliable decisions concerning the proportion of a large group of consumers favoring a certain product or a certain social issue can be based on relatively small samples. For example, suppose the target population of consumers contains 50 million people and we wish to decide whether the proportion of consumers, p, in the population that favor some product (or issue) is as large as some value, say .2. Suppose you randomly select a sample as small as 1,600 from the 50 million and you observe the number x of consumers in the sample who favor the new product. Assuming that $p = .2$, find the mean and standard deviation of x. Suppose that 400 (or 25%) of the sample of 1,600 consumers favor the new product. Why might this sample result lead you to conclude that p (the proportion of consumers favoring the product in the population of 50 million) is at least as large as .2? [*Hint:* Find the values of μ and σ for $p = .2$, and use them to decide whether the observed value of x is unusually large.]

5.100 Golf balls that do not meet a manufacturer's shape specifications are referred to as being "out of round" and may be sold as rejects. Assume that 10% of the balls produced by a certain machine are out of round. What is the approximate probability that of the next 200 balls produced by the machine, twenty-five or more are out of round?

*5.101 The lifetime (in hours) of a certain electronics component has an exponential distribution with $\theta = 2,000$.

a. What is the probability that a randomly chosen component lasts more than 2,500 hours?

b. Given that a component has lasted 1,000 hours, what is the probability it will last at least 2,500 hours longer?

[*Note:* Compare the answers to parts a and b. This is the result of a property of the exponential distribution. Items that possess an exponential lifetime distribution are never subject to fatigue. The probability that they will live for 100 hours is the same for an item that is 1,000 hours old as for a new item. Certainly, not too many objects in nature possess this lifetime distribution, but machines or equipment that are subject to periodic maintenance often do.]

5.102 The median income of residents in a certain community is $20,000. If 1,000 people are randomly sampled from the population of residents, let x equal the number

whose incomes are less than $20,000. Compute approximate values for the following:
a. $P(x > 500)$ **b.** $P(x \leq 480)$ **c.** $P(475 \leq x \leq 525)$

*5.103 The number of serious accidents in a manufacturing plant has (approximately) a Poisson probability distribution with a mean of two serious accidents per month. If x, the number of events per unit time, has a Poisson distribution with mean λ, then it can be shown that the time between two successive events has an exponential probability distribution with mean $\theta = 1/\lambda$.

a. If an accident occurs today, what is the probability that the next serious accident will not occur within the next month?

b. What is the probability that more than one serious accident will occur within the next month?

*5.104 The Poisson probability distribution, like the binomial, can be approximated by a normal probability distribution. This approximation, using $\mu = \lambda$ and $\sigma = \sqrt{\lambda}$, will be good when λ is large (large enough so that the distance between $x = 0$ and λ is at least $3\sigma = 3\sqrt{\lambda}$, or $\lambda \geq 9$). The number of union complaints per month at a certain manufacturing plant has a Poisson probability distribution with $\lambda = 40$ complaints per month. Use the normal approximation to the Poisson probability distribution.

a. Approximate the probability that the number of complaints in a given month will be less than thirty-five.

b. Approximate the probability that the number of complaints in a given month exceeds forty.

c. If the mean number of complaints per month remains constant and if the number of complaints in 1 month is independent of the number in any other, what is the probability that in each of 3 successive months, the number of complaints exceeds forty?

On Your Own...

For large values of n the computational effort involved in working with the binomial probability distribution is considerable. Fortunately, in many instances the normal distribution provides a good approximation to the binomial distribution. This exercise was designed to enable you to demonstrate to yourself how well the normal distribution approximates the binomial distribution.

a. Let the random variable x have a binomial probability with $n = 10$ and $p = .5$. Using the binomial distribution, find the probability that x takes on a value in each of the following intervals: $\mu \pm \sigma$, $\mu \pm 2\sigma$, and $\mu \pm 3\sigma$.

b. Find the probabilities requested in part a by using a normal approximation to the given binomial distribution.

c. Determine the magnitude of the difference between each of the three probabilities as determined by the binomial distribution and by the normal approximation.

d. Letting x have a binomial distribution with $n = 20$ and $p = .5$, repeat parts a, b, and c. Notice that the probability estimates provided by the normal distribution are more accurate for $n = 20$ than for $n = 10$.

e. Letting x have a binomial distribution with $n = 20$ and $p = .01$, repeat parts a, b, and c. Notice that the probability estimates provided by the normal distribution are very poor in this case. Explain why this occurs.

References

Bresee, C. W. "On 'grading on the curve.'" *Clearing House,* November 1976, *50*, 108–110.

Martz, H. F., & Waller, R. A. *Bayesian reliability analysis.* New York: Wiley, 1982. Pages 1 and 256.

Mendenhall, W., Scheaffer, R. L., & Wackerly, D. *Mathematical statistics with applications.* 2d ed. North Scituate, Mass.: Duxbury, 1980. Chapter 4.

Schick, G. J. "The search for a software reliability model." *Design Sciences,* October 1974, *5*, 529.

CHAPTER 6

Sampling Distributions

Where We've Been . . .

We have learned in earlier chapters that the objective of most statistical investigations is inference—that is, making decisions or predictions about a population based on information in a sample. To actually make the decision, we use the sample data to compute sample statistics, such as the sample mean or variance. The knowledge of random variables and their probability distributions enables us to construct theoretical models of populations.

Where We're Going . . .

Because sample measurements are observed values of random variables, the value that you compute for a sample statistic will vary in a random manner from sample to sample. In other words, since sample statistics are random variables, they therefore possess probability distributions that are either discrete or continuous, as discussed in Chapters 4 and 5. These probability distributions, called *sampling distributions* because they characterize the distribution of values of the various statistics over a very large number of samples, are the topic of this chapter. In particular, you will learn why many sampling distributions tend to be approximately normal, and you will see how sampling distributions can be used to evaluate the reliability of inferences made using the statistics.

Contents

In Chapters 4 and 5 we assumed that we knew the probability distribution of a random variable, and using this knowledge we were able to compute the mean, variance, and probabilities associated with the random variable. However, in most practical applications, this information is not available. To illustrate, in Example 4.11 we calculated the probability that the binomial random variable x, the number of twenty polled voters who favor a certain mayoral candidate, assumed specific values. To do this, it was necessary to assume some value for p, the proportion of all voters who favor the candidate. Thus, for the purposes of illustration, we assumed $p = .6$ when, in all likelihood, the exact value of p would be unknown. In fact, the probable purpose of taking the poll is to estimate p. Similarly, when we modeled the in-city gas mileage of a certain automobile model, we used the normal probability distribution with an *assumed* mean and standard deviation of 27 and 3 miles per gallon, respectively. In most situations, the true mean and standard deviation are unknown quantities that would have to be estimated. Numerical quantities that describe probability distributions are called ***parameters.*** Thus, p, the probability of a success in a binomial experiment, and μ and σ, the mean and standard deviation of a normal distribution, are examples of parameters.

Definition 6.1

A ***parameter*** is a numerical descriptive measure of a population.

We have also discussed the sample mean $\bar{x}$, sample variance s^2, sample standard deviation s, etc., which are numerical descriptive measures calculated from the sample. We will often use the information contained in these ***sample statistics*** to make inferences about the parameters of a population.

Definition 6.2

A ***sample statistic*** is a quantity calculated from the observations in a sample.

Note that the term *statistic* refers to a *sample* quantity and the term *parameter* refers to a *population* quantity.

Before we can show you how to use sample statistics to make inferences about population parameters, we need to be able to evaluate their properties. Does one sample statistic contain more information than another about a population parameter? On what basis should we choose the "best" statistic for making inferences about a parameter? The purpose of this chapter is to answer these questions.

6.1
What Is a Sampling Distribution?

If we want to estimate a parameter of a population—say, the population mean μ—three are a number of sample statistics that could be used for the estimate. Two possibilities are the sample mean $\bar{x}$ and the sample median m. Which of these do you think will provide a better estimate of μ?

Before answering this question, consider the following example: Toss a fair die and let x equal the number of dots showing on the up face. Suppose the die is tossed three times, producing sample measurements 2, 2, 6. The sample mean is $\bar{x} = 3.33$ and the sample median is $m = 2$. Since the population mean of x is $\mu = 3.5$, you can see that for this sample of three measurements, the sample mean $\bar{x}$ provides an estimate that falls closer to μ than does the sample median (see Figure 6.1(a)). Now suppose we toss the die three more times and obtain the sample measurements 3, 4, 6. The mean and median of this sample are $\bar{x} = 4.33$ and $m = 4$, respectively. This time m is closer to μ (see Figure 6.1(b)).

Figure 6.1 Comparing the Sample Mean ($\bar{x}$) and Sample Median (m) as Estimators of the Population Mean (μ)

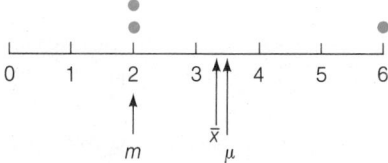 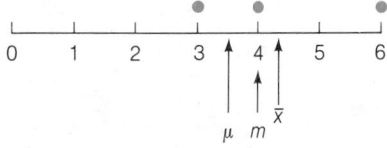

(a) Sample 1: $\bar{x}$ is closer than m to μ (b) Sample 2: m is closer than $\bar{x}$ to μ

This simple example illustrates an important point: Neither the sample mean nor the sample median will *always* fall closer to the population mean. Consequently, we cannot compare these two sample statistics, or, in general, any two sample statistics, on the basis of their performance for a single sample. Instead, we need to recognize that sample statistics are themselves random variables, because different samples can lead to different values for the sample statistics. As random variables, sample statistics must be judged and compared on the basis of their probability distributions, i.e., the *collection* of values and associated probabilities of each statistic that would be obtained if the sampling experiment were repeated a *very large number of times*. We will illustrate this concept with an example.

Suppose it is known that in a certain part of Canada the daily high temperature recorded for all past months of January has a mean of $\mu = 10°F$ and a standard deviation of $\sigma = 5°F$. Consider an experiment consisting of randomly selecting twenty-five daily high temperatures from the records of past months of January and calculating the sample mean $\bar{x}$. If this experiment were repeated a very large number of times, the value of $\bar{x}$ would vary from sample to sample. For example, the first sample of twenty-five temperature measurements might have a mean $\bar{x} = 9.8$, the second sample a mean $\bar{x} = 11.4$, the third sample a mean $\bar{x} = 10.5$, etc. If the sampling experiment were repeated a very large number of times, the resulting histogram of sample means would be approximately the probability distribution of $\bar{x}$. If $\bar{x}$ is a good estimator of μ, we would expect the values of $\bar{x}$ to cluster around μ as shown in Figure 6.2 (page 238). This probability distribution is called a **sampling distribution** because it is generated by repeating a sampling experiment a very large number of times.

Figure 6.2 Sampling
Distribution for $\bar{x}$
Based on a Sample
of $n = 25$ Measurements

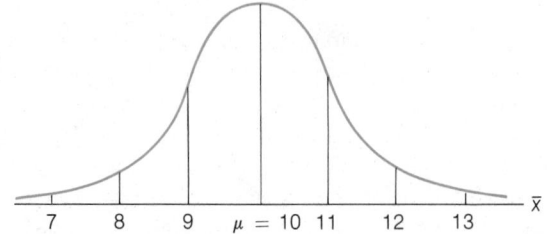

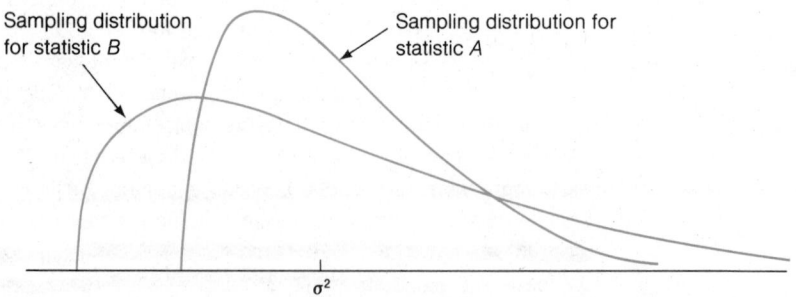

Definition 6.3

The **sampling distribution** of a sample statistic calculated from a sample of n measurements is the probability distribution of the statistic.

In actual practice, the sampling distribution of a statistic is obtained mathematically or (at least approximately) by simulating the sample on a computer using a procedure similar to that just described.

If $\bar{x}$ has been calculated from a sample of $n = 25$ measurements selected from a population with mean $\mu = 10$ and standard deviation $\sigma = 5$, the sampling distribution (Figure 6.2) provides all the information you may wish to know about its behavior. For example, the probability that you will draw a sample of twenty-five measurements and obtain a value of $\bar{x}$ in the interval $9 \leq \bar{x} \leq 10$ will be the area under the sampling distribution over that interval. Note that you need not know the actual value of μ in order to use the information that a sampling distribution provides about the distance between $\bar{x}$ and μ. All you need to know is the position of the sampling distribution relative to μ.

Since the properties of a statistic are typified by its sampling distribution, it follows that to compare two statistics you compare their sampling distributions. For example, if you have two statistics, A and B, for estimating the same parameter (for purposes of illustration, suppose the parameter is the population variance σ^2) and if their sampling distributions are as shown in Figure 6.3, you would choose statistic A in preference to statistic B. You would make this choice because the sampling distribution for statistic

Figure 6.3 Two
Sampling Distributions
for Estimating the
Population Variance, σ^2

Sampling distribution
for statistic B

Sampling distribution for
statistic A

A centers over σ^2 and has less spread (variation) than the sampling distribution for statistic B. When you draw a single sample in a practical sampling situation, the probability is higher that statistic A will fall nearer σ^2.

Remember that in practice we will not know the numerical value of the unknown parameter σ^2, so we will not know whether statistic A or statistic B is closer to σ^2 for a sample. We have to rely on our knowledge of the theoretical sampling distributions to choose the best sample statistic, and then use it sample after sample. The procedure for finding the sampling distribution for a statistic is demonstrated in Example 6.1.

Example 6.1

Consider a population consisting of the measurements 0, 3, and 12 and described by the following probability distribution:

x	0	3	12
$p(x)$	$\frac{1}{3}$	$\frac{1}{3}$	$\frac{1}{3}$

A random sample of $n = 3$ measurements is selected from the population.

a. Find the sampling distribution of the sample mean $\bar{x}$.
b. Find the sampling distribution of the sample median m.

Solution

Every possible sample of $n = 3$ measurements is listed in Table 6.1 along with the sample mean and median. Also, because any one sample is as likely to be selected as any other (random sampling), the probability of observing any particular sample is $\frac{1}{27}$.

Table 6.1

POSSIBLE SAMPLES	$\bar{x}$	m	PROBABILITY	POSSIBLE SAMPLES	$\bar{x}$	m	PROBABILITY
0, 0, 0	0	0	$\frac{1}{27}$	3, 3, 12	6	3	$\frac{1}{27}$
0, 0, 3	1	0	$\frac{1}{27}$	3, 12, 0	5	3	$\frac{1}{27}$
0, 0, 12	4	0	$\frac{1}{27}$	3, 12, 3	6	3	$\frac{1}{27}$
0, 3, 0	1	0	$\frac{1}{27}$	3, 12, 12	9	12	$\frac{1}{27}$
0, 3, 3	2	3	$\frac{1}{27}$	12, 0, 0	4	0	$\frac{1}{27}$
0, 3, 12	5	3	$\frac{1}{27}$	12, 0, 3	5	3	$\frac{1}{27}$
0, 12, 0	4	0	$\frac{1}{27}$	12, 0, 12	8	12	$\frac{1}{27}$
0, 12, 3	5	3	$\frac{1}{27}$	12, 3, 0	5	3	$\frac{1}{27}$
0, 12, 12	8	12	$\frac{1}{27}$	12, 3, 3	6	3	$\frac{1}{27}$
3, 0, 0	1	0	$\frac{1}{27}$	12, 3, 12	9	12	$\frac{1}{27}$
3, 0, '3	2	3	$\frac{1}{27}$	12, 12, 0	8	12	$\frac{1}{27}$
3, 0, 12	5	3	$\frac{1}{27}$	12, 12, 3	9	12	$\frac{1}{27}$
3, 3, 0	2	3	$\frac{1}{27}$	12, 12, 12	12	12	$\frac{1}{27}$
3, 3, 3	3	3	$\frac{1}{27}$				

a. From Table 6.1 you can see that $\bar{x}$ can assume the values 0, 1, 2, 3, 4, 5, 6, 8, 9, and 12. Because $\bar{x} = 0$ occurs in only one sample, $P(\bar{x} = 0) = \frac{1}{27}$. Similarly, $\bar{x} = 1$ occurs in three samples: $(0, 0, 3)$, $(0, 3, 0)$, and $(3, 0, 0)$. Therefore, $P(\bar{x} = 1) = \frac{3}{27} = \frac{1}{9}$. Calculating the probabilities of the remaining values of $\bar{x}$ and arranging them in a table, we obtain the following probability distribution:

$\bar{x}$	0	1	2	3	4	5	6	8	9	12
$p(\bar{x})$	$\frac{1}{27}$	$\frac{3}{27}$	$\frac{3}{27}$	$\frac{1}{27}$	$\frac{3}{27}$	$\frac{6}{27}$	$\frac{3}{27}$	$\frac{3}{27}$	$\frac{3}{27}$	$\frac{1}{27}$

b. In Table 6.1 you can see that the median m can assume one of the three values 0, 3, or 12. The value $m = 0$ occurs in seven different samples. Therefore, $P(m = 0) = \frac{7}{27}$. Similarly, $m = 3$ occurs in thirteen samples and $m = 12$ occurs in seven samples. Therefore, the probability distribution for the median m is as follows:

m	0	3	12
$p(m)$	$\frac{7}{27}$	$\frac{13}{27}$	$\frac{7}{27}$

■

Example 6.1 demonstrates the procedure for finding the exact sampling distribution of a statistic when the number of different samples that could be selected from the population is relatively small. In the real world, populations are often generated by random variables that can assume a very large number of values; the samples are difficult (or impossible) to enumerate. When this situation occurs, we may choose to obtain the approximate sampling distribution for a statistic by simulating the sampling over and over again and recording the proportion of times different values of the statistic occur. Example 6.2 illustrates this procedure.

Example 6.2 Suppose we perform the following experiment over and over again: Take a sample of eleven measurements from the uniform distribution shown in Figure 6.4. Calculate the two sample statistics

$$\bar{x} = \text{Sample mean} = \frac{\sum x}{11}$$

$m = $ Median = Sixth sample measurement when the eleven measurements are arranged in ascending order

Obtain approximations to the sampling distributions of $\bar{x}$ and m.

Figure 6.4 Uniform Distribution from 0 to 1

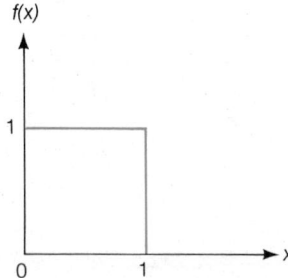

Solution We use a computer to generate 1,000 samples, each with $n = 11$ observations. Then, we compute $\bar{x}$ and m for each sample. Our goal is to obtain approximations to the

sampling distributions of $\bar{x}$ and m and to find out which sample statistic ($\bar{x}$ or m) contains more information about μ. (Note that, in this particular example, we *know* the population mean is $\mu = .5$.) The first ten of the 1,000 samples generated are presented in Table 6.2. For example, the first computer-generated sample from the uniform distribution (arranged in ascending order) contained the following measurements: .125, .138, .139, .217, .419, .506, .516, .757, .771, .786, .919. The sample mean $\bar{x}$ and median m computed for this sample are

$$\bar{x} = \frac{.125 + .138 + \cdots + .919}{11} = .481$$

m = Sixth ordered measurement = .506

Table 6.2

First Ten Samples of
$n = 11$ Measurements
from a Uniform
Distribution

SAMPLE	MEASUREMENTS										
1	.217	.786	.757	.125	.139	.919	.506	.771	.138	.516	.419
2	.303	.703	.812	.650	.848	.392	.988	.469	.632	.012	.065
3	.383	.547	.383	.584	.098	.676	.091	.535	.256	.163	.390
4	.218	.376	.248	.606	.610	.055	.095	.311	.086	.165	.665
5	.144	.069	.485	.739	.491	.054	.953	.179	.865	.429	.648
6	.426	.563	.186	.896	.628	.075	.283	.549	.295	.522	.674
7	.643	.828	.465	.672	.074	.300	.319	.254	.708	.384	.534
8	.616	.049	.324	.700	.803	.399	.557	.975	.569	.023	.072
9	.093	.835	.534	.212	.201	.041	.889	.728	.466	.142	.574
10	.957	.253	.983	.904	.696	.766	.880	.485	.035	.881	.732

The relative frequency histograms for $\bar{x}$ and m for the 1,000 samples of size $n = 11$ are shown in Figure 6.5.

Figure 6.5

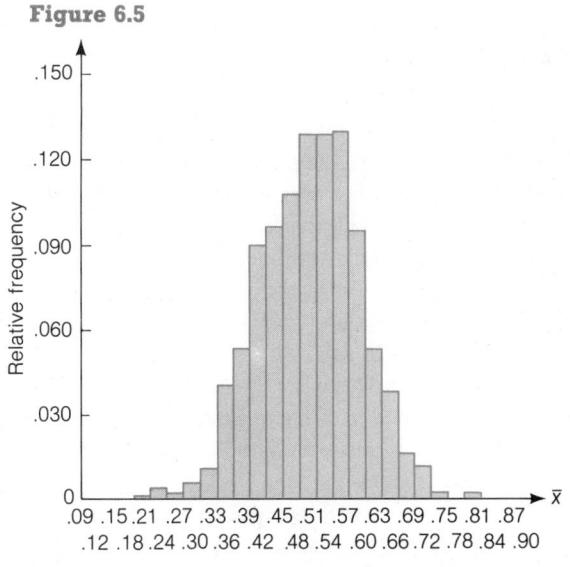

(a) Sampling distribution for $\bar{x}$ (based on 1,000 samples of $n = 11$ measurements)

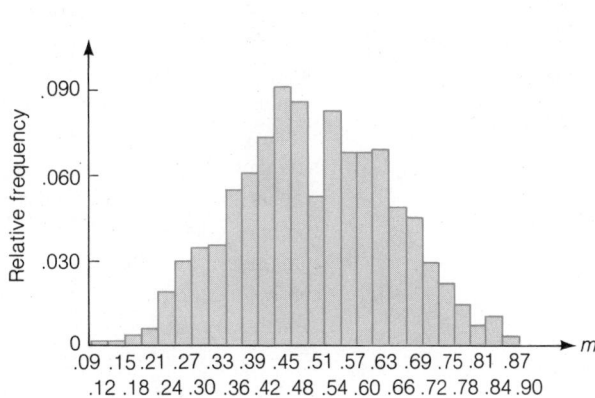

(b) Sampling distribution for m (based on 1,000 samples of $n = 11$ measurements)

You can see that the values of $\bar{x}$ tend to cluster around μ to a greater extent than do the values of m. Thus, on the basis of the observed sampling distributions, we conclude that $\bar{x}$ contains more information about μ than m does—at least for samples of $n = 11$ measurements from the uniform distribution.   ■

As noted earlier, many sampling distributions can be derived mathematically, but the theory necessary to do this is beyond the scope of this text. Consequently, when we need to know the properties of a statistic, we will present its sampling distribution and describe its properties. Several of the important properties we look for in sampling distributions are discussed in the next section.

Exercises 6.1–6.11

[*Note:* Starred (*) *exercises require the use of a computer.*]

6.1 The following probability distribution describes a population of measurements that can assume values of 0, 2, 4, and 6, each of which occurs with the same relative frequency:

x	0	2	4	6
$p(x)$	$\frac{1}{4}$	$\frac{1}{4}$	$\frac{1}{4}$	$\frac{1}{4}$

a. List all the different samples of $n = 2$ measurements that can be selected from this population.

b. Calculate the mean of each different sample listed in part a.

c. If a sample of $n = 2$ measurements is randomly selected from the population, what is the probability that a specific sample will be selected?

d. Assume that a random sample of $n = 2$ measurements is selected from the population. List the different values of $\bar{x}$ found in part b and find the probability of each. Then give the sampling distribution of the sample mean $\bar{x}$ in tabular form.

e. Construct a probability histogram for the sampling distribution of $\bar{x}$.

6.2 Simulate sampling from the population (Exercise 6.1) by marking the values of x, one on each of four identical coins (or poker chips, etc.). Place the coins (marked 0, 2, 4, and 6) into a bag, randomly select one, and observe its value. Replace this coin, draw a second coin, and observe its value. Finally, calculate the mean $\bar{x}$ for this sample of $n = 2$ observations randomly selected from the population (Exercise 6.1). Replace the coins, mix, and, using the same procedure, select a sample of $n = 2$ observations from the population. Record the numbers and calculate $\bar{x}$ for this sample. Repeat this sampling process until you acquire 100 values of $\bar{x}$. Construct a relative frequency distribution for these 100 sample means. This distribution will be an approximation to the exact sampling distribution of $\bar{x}$ found in part e of Exercise 6.1. Compare the two distributions. The distribution obtained in this exercise will not be exactly the same as the exact sampling distribution (Exercise 6.1, part e).

If you were to repeat the sampling procedure, drawing two coins not 100 times but 10,000 times, the relative frequency distribution for the 10,000 sample means would be almost identical to the sampling distribution of $\bar{x}$ found in Exercise 6.1, part e.

6.3 Consider the population described by the following probability distribution:

x	1	2	3	4	5
$p(x)$	.2	.3	.2	.2	.1

The random variable x is observed twice. If these observations are independent, verify that the different samples of size 2 and their probabilities are as follows:

SAMPLE	PROBABILITY	SAMPLE	PROBABILITY	SAMPLE	PROBABILITY
1, 1	.04	3, 1	.04	5, 1	.02
1, 2	.06	3, 2	.06	5, 2	.03
1, 3	.04	3, 3	.04	5, 3	.02
1, 4	.04	3, 4	.04	5, 4	.02
1, 5	.02	3, 5	.02	5, 5	.01
2, 1	.06	4, 1	.04		
2, 2	.09	4, 2	.06		
2, 3	.06	4, 3	.04		
2, 4	.06	4, 4	.04		
2, 5	.03	4, 5	.02		

a. Find the sampling distribution of the sample mean $\bar{x}$.
b. Construct a probability histogram for the sampling distribution of $\bar{x}$.
c. What is the probability that $\bar{x}$ is 4.5 or larger?
d. Would you expect to observe a value of $\bar{x}$ equal to 4.5 or larger? Explain.

6.4 Refer to Exercise 6.3 and find $E(x) = \mu$. Then use the sampling distribution of $\bar{x}$ found in Exercise 6.3 to find the expected value of $\bar{x}$. Note that $E(\bar{x}) = \mu$.

6.5 Refer to Exercise 6.3. Assume that a random sample of $n = 2$ measurements is randomly selected from the population.
a. List the different values that the sample median m may assume and find the probability of each. Then give the sampling distribution of the sample median.
b. Construct a probability histogram for the sampling distribution of the sample median and compare it with the probability histogram for the sample mean (Exercise 6.3, part b).

6.6 Consider the population described by the following probability distribution:

x	1	2	3
$p(x)$	.6	.3	.1

The random variable x is observed three times. Suppose the observations are independent.
a. List all possible values of $n = 3$ observations and give their probabilities.
b. List all possible values of the sample mean $\bar{x}$ and give their probabilities. Present the sampling distribution of $\bar{x}$ in tabular form.
c. Construct a probability histogram of the sampling distribution of $\bar{x}$.

6.7 Refer to Exercise 6.6.

a. Find the sampling distribution of the sample median m for samples of $n = 3$ observations.

b. Construct a probability histogram for the sampling distribution of m, and compare it with the probability histogram for the sampling distribution of the sample mean (Exercise 6.6, part c).

6.8 Refer to the probability distribution of Exercise 6.6.

a. List all possible samples of $n = 4$ observations and give their probabilities.

b. List all possible values of the sample mean $\bar{x}$ and give their probabilities. Present the sampling distribution of $\bar{x}$ in tabular form.

c. Construct a probability histogram for the sampling distribution of $\bar{x}$, and compare it with the corresponding sampling distribution based on samples of $n = 3$ observations (Exercise 6.6, part c). What is the effect of increasing the sample size on the sampling distribution of the sample mean $\bar{x}$?

6.9 Repeat the instructions of Exercise 6.8 for the sample median m. Compare the probability histogram of the sampling distribution of m based on $n = 4$ observations with the one based on $n = 3$ observations (Exercise 6.7, part b). What is the effect of increasing the sample size on the sampling distribution of the sample median m?

*6.10** In Example 6.2 we used the computer to generate 1,000 samples, each containing $n = 11$ observations, from a uniform distribution over the interval from 0 to 1. For this exercise, generate 500 samples, each containing $n = 15$ observations, from this population.

a. Calculate the sample mean for each sample. To approximate the sampling distribution of $\bar{x}$, construct a relative frequency histogram for the 500 values of $\bar{x}$.

b. Repeat part a for the sample median. Compare this approximate sampling distribution with the approximate sampling distribution of $\bar{x}$ found in part a.

*6.11** Consider a population that contains values of x equal to 00, 01, 02, 03, . . . , 96, 97, 98, 99. Assume that these values of x occur with equal probability. Generate 500 samples, each containing $n = 25$ measurements, from this population. Calculate the sample mean $\bar{x}$ and sample variance s^2 for each of the 500 samples.

a. To approximate the sampling distribution of $\bar{x}$, construct a relative frequency histogram for the 500 values of $\bar{x}$.

b. Repeat part a for the 500 values of s^2.

6.2 Properties of Sampling Distributions: Unbiasedness and Minimum Variance

The simplest type of statistic used to make inferences about a population parameter is a **point estimator**. A point estimator is a rule or formula that tells us how to use the sample data to calculate a single number that is intended to estimate the value of some population parameter. For example, the sample mean $\bar{x}$ is a point estimator of the population mean μ. Similarly, the sample variance s^2 is a point estimator of the population variance σ^2.

Definition 6.4

A *point estimator* of a population parameter is a rule or formula that tells us how to use the sample data to calculate a single number that can be used as an *estimate* of the population parameter.

Often, many different point estimators can be found to estimate the same parameter. Each will have a sampling distribution that provides information about the point estimator. By examining the sampling distribution, we can determine how large the difference between an estimate and the true value of the parameter (called the **error of estimation**) is likely to be. We can also tell whether an estimator is more likely to overestimate or to underestimate a parameter.

Example 6.3 Suppose two statistics, A and B, exist to estimate the same population parameter, θ (theta). (Note that θ could be any parameter, μ, σ^2, σ, etc.) Suppose the two statistics have sampling distributions as shown in Figure 6.6. Based on these sampling distributions, which statistic is more attractive as an estimator of θ?

Figure 6.6 Sampling Distributions of Unbiased and Biased Estimators

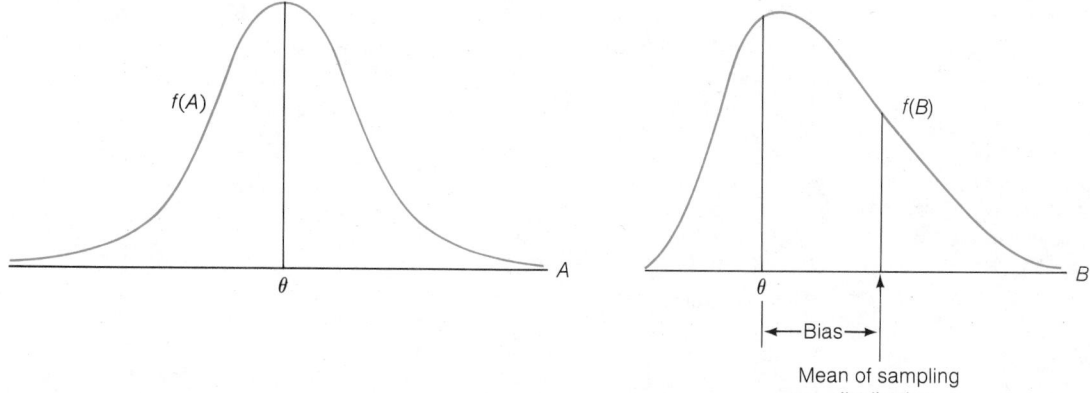

(a) Unbiased sample statistic for the parameter θ (b) Biased sample statistic for the parameter θ

Solution As a first consideration, we would like the sampling distribution to center over the parameter we wish to estimate. One way to characterize this property is in terms of the mean of the sampling distribution. Consequently, we say that a statistic is **unbiased** if the mean of the sampling distribution is equal to the parameter it is intended to estimate. This situation is shown in Figure 6.6(a) where the mean μ_A of statistic A is equal to θ. If the mean of a sampling distribution is not equal to the parameter it is intended to estimate, the statistic is said to be **biased.** The sampling distribution for a biased statistic is shown in Figure 6.6(b). The mean μ_B of the sampling distribution for statistic B is not equal to θ; in fact, it is shifted to the right of θ. ∎

You can see that biased statistics tend either to overestimate or to underestimate a parameter. Consequently, when other properties of statistics tend to be equivalent, we will choose an unbiased statistic to estimate a parameter of interest.*

Definition 6.5

If the sampling distribution of a sample statistic has a mean equal to the population parameter the statistic is intended to estimate, the statistic is said to be an **unbiased** estimate of the parameter.

If the mean of the sampling distribution is not equal to the parameter, the statistic is said to be a **biased** estimate of the parameter.

The standard deviation of a sampling distribution measures another important property of statistics—the spread of the estimates generated by repeated sampling. Suppose two statistics, A and B, are both unbiased estimators of the population parameter. Since the means of the two sampling distributions are the same, we turn to their standard deviations in order to decide which will provide estimates that fall closer to the unknown population parameter we are estimating. Naturally, we will choose the sample statistic that has the smaller standard deviation. Figure 6.7 depicts sampling distributions for A and B. Note that the standard deviation of the distribution of A is smaller than the standard deviation for B, indicating that over a large number of samples, the values of A cluster more closely around the unknown population parameter than do the values of B.

Figure 6.7 Sampling Distributions for Two Unbiased Estimators

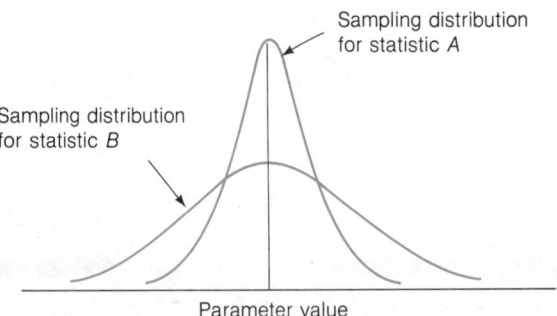

In summary, to make an inference about a population parameter, we will use the sample statistic with a sampling distribution that is unbiased and has a small standard deviation (usually smaller than the standard deviation of other unbiased sample statistics). The derivation of this sample statistic will not concern us, because the "best" statistic for estimating specific parameters is a matter of record. We will simply present an unbiased estimator with its standard deviation for each population parameter we

---

*Unbiased statistics do not exist for all parameters of interest.

consider. [*Note:* The standard deviation of the sampling distribution of a statistic is also called the **standard error of the statistic.**]

Example 6.4 In Example 6.1, we found the sampling distributions of the sample mean $\bar{x}$ and the sample median m for random samples of $n = 3$ measurements from a population defined by the probability distribution:

x	0	3	12
$p(x)$	$\frac{1}{3}$	$\frac{1}{3}$	$\frac{1}{3}$

The sampling distributions of $\bar{x}$ and m were found to be the following:

$\bar{x}$	0	1	2	3	4	5	6	8	9	12
$p(\bar{x})$	$\frac{1}{27}$	$\frac{3}{27}$	$\frac{3}{27}$	$\frac{1}{27}$	$\frac{3}{27}$	$\frac{6}{27}$	$\frac{3}{27}$	$\frac{3}{27}$	$\frac{3}{27}$	$\frac{1}{27}$

m	0	3	12
$p(m)$	$\frac{7}{27}$	$\frac{13}{27}$	$\frac{7}{27}$

a. Show that $\bar{x}$ is an unbiased estimator of μ in this situation.
b. Show that m is a biased estimator of μ in this situation.

Solution **a.** The expected value of a discrete random variable x (see Section 4.3) is defined to be $E(x) = \sum xp(x)$, where the summation is over all values of x. Then

$$E(x) = \mu = \sum xp(x) = (0)(\tfrac{1}{3}) + (3)(\tfrac{1}{3}) + (12)(\tfrac{1}{3}) = 5$$

The expected value of the discrete random variable $\bar{x}$ is

$$E(\bar{x}) = \sum (\bar{x})p(\bar{x})$$

summed over all values of $\bar{x}$. Or

$$E(\bar{x}) = (0)(\tfrac{1}{27}) + (1)(\tfrac{3}{27}) + 2(\tfrac{3}{27}) + \cdots + (12)(\tfrac{1}{27})$$
$$= 5$$

Since $E(\bar{x}) = \mu$, we see that $\bar{x}$ is an unbiased estimator of μ.
b. The expected value of the sample median m is

$$E(m) = \sum mp(m) = (0)(\tfrac{7}{27}) + (3)(\tfrac{13}{27}) + (12)(\tfrac{7}{27})$$

Since the expected value of m is not equal to μ ($\mu = 5$), the sample median m is a biased estimator of μ. ∎

Example 6.5 Refer to Example 6.4 and find the standard deviations of the sampling distributions of $\bar{x}$ and m. Which statistic would appear to be a better estimator for μ?

Solution The variance of the sampling distribution of $\bar{x}$ (we will denote it by the symbol $\sigma_{\bar{x}}^2$) is found to be

$$\sigma_{\bar{x}}^2 = E\{[\bar{x} - E(\bar{x})]^2\} = \sum (\bar{x} - \mu)^2 p(\bar{x})$$

where, from Example 6.4,

$$E(\bar{x}) = \mu = 5$$

Then

$$\sigma_{\bar{x}}^2 = (0 - 5)^2(\tfrac{1}{27}) + (1 - 5)^2(\tfrac{3}{27}) + (2 - 5)^2(\tfrac{3}{27}) + \cdots + (12 - 5)^2(\tfrac{1}{27})$$

$$= 8.6667$$

and

$$\sigma_{\bar{x}} = \sqrt{8.6667} = 2.94$$

Similarly, the variance of the sampling distribution of m (we will denote it by σ_m^2) is

$$\sigma_m^2 = E\{[m - E(m)]^2\}$$

where, from Example 6.4, the expected value of m is $E(m) = 4.56$. Then

$$\sigma_m^2 = E\{[m - E(m)]^2\} = \sum[m - E(m)]^2 p(m)$$

$$= (0 - 4.56)^2(\tfrac{7}{27}) + (3 - 4.56)^2(\tfrac{13}{27}) + (12 - 4.56)^2(\tfrac{7}{27})$$

$$= 20.9136$$

and

$$\sigma_m = \sqrt{20.9136} = 4.57$$

Which statistic appears to be the better estimator for the population mean μ: the sample mean $\bar{x}$ or the median m? To answer this question, we compare the sampling distributions of the two statistics. The sampling distribution of the sample median m is biased (i.e., it is shifted to the left of the mean μ) and its standard deviation $\sigma_m = 4.57$ is much larger than the standard deviation of the sampling distribution of $\bar{x}$, $\sigma_{\bar{x}} = 2.94$. Consequently, the sample mean $\bar{x}$ would be a better estimator of the population mean μ, for the population in question, than would the sample median m. ■

Exercises
6.12–6.18

[*Note:* Starred (*) exercises require the use of a computer.]

6.12 Consider the probability distribution:

x	0	1	5
$p(x)$	$\tfrac{1}{3}$	$\tfrac{1}{3}$	$\tfrac{1}{3}$

a. Find μ and σ^2.
b. Find the sampling distribution of the sample mean $\bar{x}$ for a random sample of $n = 2$ measurements from this distribution.
c. Show that $\bar{x}$ is an unbiased estimator for μ. [*Hint:* Show that $E(\bar{x}) = \sum \bar{x} p(\bar{x}) = \mu$.]
d. Find the sampling distribution of the sample variance s^2 for a random sample of $n = 2$ measurements from this distribution.
e. Show that s^2 is an unbiased estimator for σ^2.

6.13 Consider the following probability distribution:

x	2	4	9
$p(x)$	$\frac{1}{3}$	$\frac{1}{3}$	$\frac{1}{3}$

a. Calculate μ for this distribution.

b. Find the sampling distribution of the sample mean $\bar{x}$ for a random sample of $n = 3$ measurements from this distribution and show that $\bar{x}$ is an unbiased estimator of μ.

c. Find the sampling distribution of the sample median m for a random sample of $n = 3$ measurements from this distribution and show that the median is a biased estimator of μ.

d. If you wanted to estimate μ using a sample of three measurements from this population, which estimator would you use? Why?

6.14 Consider the following probability distribution:

x	0	1	2
$p(x)$	$\frac{1}{3}$	$\frac{1}{3}$	$\frac{1}{3}$

a. Find μ.

b. For a random sample of $n = 3$ observations from this distribution, find the sampling distribution of the sample mean.

c. Find the sampling distribution of the median of a sample of $n = 3$ observations from this population.

d. Refer to parts b and c and show that both the mean and median are unbiased estimators of μ for this population.

e. Find the variances of the sampling distributions of the sample mean and the sample median.

f. Which estimator would you use to estimate μ? Why?

*6.15 Generate 500 samples, each containing $n = 25$ measurements, from a population that contains values of x equal to 01, 02, . . . , 48, 49, 50. Assume that these values of x are equally likely. Calculate the sample mean $\bar{x}$ and median m for each sample. Construct relative frequency histograms for the 500 values of $\bar{x}$ and the 500 values of m. Use these approximations to the sampling distributions of $\bar{x}$ and m to answer the following questions:

a. Does it appear that $\bar{x}$ and m are unbiased estimators of the population mean? [*Note:* $\mu = 25.5$.]

b. Which sampling distribution displays greater variation?

6.16 Refer to Exercise 6.6.

a. Show that $\bar{x}$ is an unbiased estimator of μ.

b. Find $\sigma_{\bar{x}}^2$.

c. Find the probability that $\bar{x}$ will fall within $2\sigma_{\bar{x}}$ of μ.

6.17 Refer to Exercise 6.6.
a. Find the sampling distribution of s^2.
b. Find the population variance σ^2.
c. Show that s^2 is an unbiased estimator of σ^2.
d. Find the sampling distribution of the sample standard deviation s.
e. Show that s is a biased estimator of σ.

6.18 Refer to Exercise 6.7, where we found the sampling distribution of the sample median. Show that the median is or is not an unbiased estimator of the population mean μ.

6.3
The Central Limit Theorem

Estimating the mean useful life of automobiles, the mean number of crimes per month in a large city, and the mean yield per acre of a new soybean hybrid are practical problems with something in common. In each case we are interested in making an inference about the mean μ of some population. As we mentioned in Chapter 2, the sample mean $\bar{x}$ is, in general, a good estimator of μ. We will now develop pertinent information about the sampling distribution for this useful statistic.

Example 6.6

Suppose a population has the uniform probability distribution given in Figure 6.8. The mean and standard deviation of this probability distribution are $\mu = .5$ and $\sigma = .29$. (See Section 5.4 for the formulas for μ and σ.) Now suppose a sample of eleven measurements is selected from this population. Describe the sampling distribution of the sample mean $\bar{x}$ based on the 1,000 sampling experiments discussed in Example 6.2.

Figure 6.8 Sampled Uniform Population

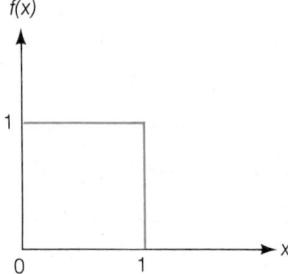

$f(x)$

Solution

You will recall that in Example 6.2 we generated 1,000 samples of $n = 11$ measurements each. The relative frequency histogram for the 1,000 sample means is shown in Figure 6.9 with a normal probability distribution superimposed. You can see that this normal probability distribution approximates the computer-generated sampling distribution very well.

To describe fully a normal probability distribution, it is necessary to know its mean and standard deviation. Inspection of Figure 6.9 indicates that the mean of the distribution of $\bar{x}$, $\mu_{\bar{x}}$, appears to be very close to .5, the mean of the sampled uniform population. Furthermore, for a mound-shaped distribution such as that shown in Figure 6.9, almost all the measurements should fall within 3 standard deviations of the mean. Since

Figure 6.9 Relative Frequency Histogram for $\bar{x}$ in 1,000 Samples of $n = 11$ Measurements with Normal Distribution Superimposed

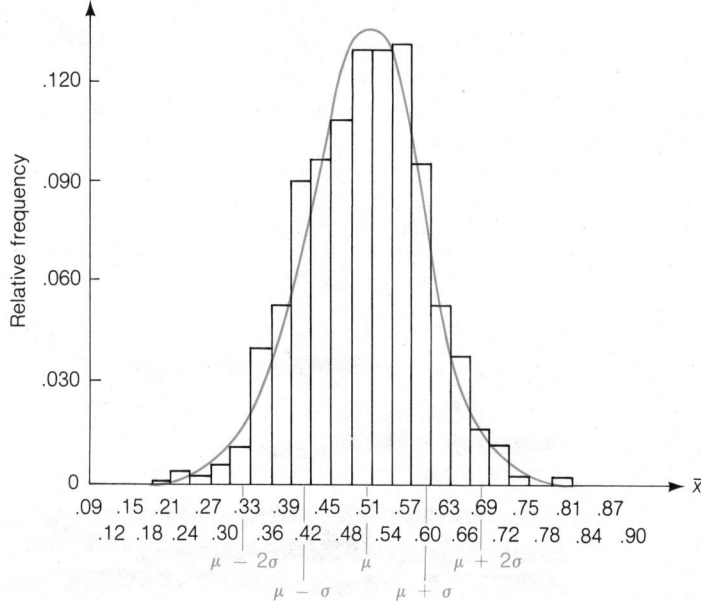

the number of values of $\bar{x}$ is very large (1,000), the range of the observed $\bar{x}$'s divided by 6 (rather than 4) should give a reasonable approximation to the standard deviation of the sample means, $\sigma_{\bar{x}}$. The values of $\bar{x}$ range from about .2 to .8, so we calculate

$$\sigma_{\bar{x}} \approx \frac{\text{Range of } \bar{x}\text{'s}}{6} = \frac{.8 - .2}{6} = .1$$

To summarize our findings based on 1,000 samples, each consisting of eleven measurements from a uniform population, the sampling distribution of $\bar{x}$ appears to be approximately normal with a mean of about .5 and a standard deviation of about .1.

It can be shown in general that the sampling distribution of $\bar{x}$ has the following properties:

Properties of the Sampling Distribution of $\bar{x}$

1. Mean of sampling distribution = Mean of sampled population
That is, $\mu_{\bar{x}} = E(\bar{x}) = \mu$.

2. Standard deviation of sampling distribution $= \dfrac{\left(\begin{array}{c}\text{Standard deviation of}\\ \text{sampled population}\end{array}\right)}{\text{Square root of sample size}}$

That is, $\sigma_{\bar{x}} = \sigma/\sqrt{n}$.

3. The sampling distribution of $\bar{x}$ is approximately normal for large sample sizes.

You can see that our approximation to $\mu_{\bar{x}}$ in Example 6.6 was precise, since property 1 assures us that the mean is the same as that of the sampled population: .5. Property 2 tells us how to calculate the standard deviation of the sampling distribution of $\bar{x}$. Substituting $\sigma = .29$, the standard deviation of the sampled uniform distribution, and the sample size $n = 11$ into the formula for $\sigma_{\bar{x}}$, we find

$$\sigma_{\bar{x}} = \frac{\sigma}{\sqrt{n}} = \frac{.29}{\sqrt{11}} = .09$$

Thus, the approximation we obtained in Example 6.6, $\sigma_{\bar{x}} \approx .1$, is very close to the exact value, $\sigma_{\bar{x}} = .09$.

The justification for property 3 is contained in one of the most important theoretical results in statistics, the **central limit theorem:**

Central Limit Theorem

For large sample sizes, the mean $\bar{x}$ of a sample from a population with mean μ and standard deviation σ possesses a sampling distribution that is approximately normal—*regardless of the probability distribution of the sampled population*. The larger the sample size, the better will be the normal approximation to the sampling distribution of $\bar{x}$.*

In summary, the sampling distribution of $\bar{x}$ is approximately normal with a mean equal to μ (that is, $E(\bar{x}) = \mu$) and a standard deviation equal to $\sigma/\sqrt{n}$, where n is the sample size.

Example 6.7

A manufacturer of automobile batteries claims that the distribution of the lengths of life of its best battery has a mean of 54 months and a standard deviation of 6 months. Suppose a consumer group decides to check the claim by purchasing a sample of fifty of these batteries and subjecting them to tests that determine their lives.

a. Assuming that the manufacturer's claim is true, describe the sampling distribution of the mean lifetime of a sample of fifty batteries.

b. Assuming that the manufacturer's claim is true, what is the probability the consumer group's sample has a mean life of 52 or fewer months?

Solution   **a.** Even though we have no information about the shape of the probability distribution of the lives of the batteries, we can use the central limit theorem to deduce

---

* Moreover, because of the central limit theorem the sum of the identically distributed independent random variables, $\sum x$, will possess a sampling distribution that is approximately normal for large samples. This distribution will have a mean equal to $n\mu$ and a variance equal to $n\sigma^2$. Proof of the central limit theorem is beyond the scope of this book, but it can be found in many mathematical statistics texts.

that the sampling distribution for a sample mean lifetime of fifty batteries is approximately normally distributed. Furthermore, the mean of this sampling distribution is the same as the mean of the sampled population, which is $\mu = 54$ months according to the manufacturer's claim. Finally, the standard deviation of the sampling distribution is given by

$$\sigma_{\bar{x}} = \frac{\sigma}{\sqrt{n}} = \frac{6}{\sqrt{50}} = .85 \text{ month}$$

Note that we used the claimed standard deviation of the sampled population, $\sigma = 6$ months. Thus, if we assume that the claim is true, the sampling distribution for the mean life of the fifty batteries sampled is as shown in Figure 6.10.

Figure 6.10 Sampling Distribution of $\bar{x}$ in Example 6.7 for $n = 50$

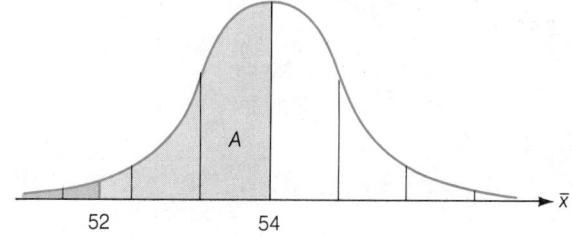

b. If the manufacturer's claim is true, the probability that the consumer group observes a mean battery life of 52 or fewer months for their sample of fifty batteries, $P(\bar{x} \le 52)$, is equivalent to the darker shaded area in Figure 6.10. Since the sampling distribution is approximately normal, we can find this area by computing the z-value:

$$z = \frac{\bar{x} - \mu_{\bar{x}}}{\sigma_{\bar{x}}} = \frac{\bar{x} - \mu}{\sigma_{\bar{x}}} = \frac{52 - 54}{.85} = -2.35$$

where $\mu_{\bar{x}}$, the mean of the sampling distribution of $\bar{x}$, is equal to μ, the mean of the lives of the sampled population, and $\sigma_{\bar{x}}$ is the standard deviation of the sampling distribution of $\bar{x}$. Note that the z is the familiar standardized distance (z-score) of Section 2.7 and, since $\bar{x}$ is approximately normally distributed, it will possess the standard normal distribution of Section 5.2.

The area A shown in Figure 6.10 between $\bar{x} = 52$ and $\bar{x} = 54$ (corresponding to $z = -2.35$) is found in Table IV of the Appendix to be .4906. Therefore, the area to the left of $\bar{x} = 52$ is

$$P(\bar{x} \le 52) = .5 - A = .5 - .4906 = .0094$$

Thus, the probability the consumer group will observe a sample mean of 52 or less is only .0094 if the manufacturer's claim is true. If the fifty tested batteries do result in a mean of 52 or fewer months, the consumer group will have strong evidence that the manufacturer's claim is untrue, because such an event is very unlikely to occur if the claim is true. (This is still another application of the *rare event approach* to statistical inference.)   ■

In addition to providing a very useful approximation for the sampling distribution of a sample mean, the central limit theorem offers an explanation for the fact that many relative frequency distributions of data possess mound-shaped distributions. Many of the macroscopic measurements we take in various areas of research are really means or sums of many microscopic phenomena. For example, a year's growth of a pine seedling is the total of the many individual components that affect the plant's growth. Similarly, the length of time a construction company takes to complete a house might be viewed as the total of the times taken to complete each of the distinct jobs necessary to build the house. The monthly demand for blood at a hospital may be viewed as the total of the individual patients' needs. Whether the observations entering into these sums satisfy the assumptions basic to the central limit theorem is questionable, but it is a fact that many distributions of data are mound-shaped and possess the appearance of normal distributions. Thus, the central limit theorem offers one explanation for the frequent occurrence of mound-shaped distributions in nature.

Exercises 6.19–6.31

Learning the Mechanics

[*Note:*   Starred (*) exercises require the use of a computer.]

6.19 Suppose that a random sample of n measurements is selected from a population with mean $\mu = 20$ and variance $\sigma^2 = 50$. For each of the following values of n, give the mean and standard deviation of the sampling distribution of the sample mean $\bar{x}$:

a. $n = 4$ **b.** $n = 25$ **c.** $n = 100$
d. $n = 50$ **e.** $n = 500$ **f.** $n = 1,000$

6.20 Suppose that a random sample of $n = 25$ measurements is selected from a population with mean μ and standard deviation σ. For each of the following values of μ and σ, give the values of $\mu_{\bar{x}}$ and $\sigma_{\bar{x}}$:

a. $\mu = 5$, $\sigma = 3$ **b.** $\mu = 100$, $\sigma = 2$
c. $\mu = 10$, $\sigma = 40$ **d.** $\mu = 10$, $\sigma = 200$

6.21 Consider the following probability distribution:

x	1	2	3	10
$p(x)$	.1	.4	.4	.1

a. Find μ, σ^2, and σ.
b. Find the sampling distribution of $\bar{x}$ for random samples of $n = 2$ measurements from this distribution.
c. Show that $\mu_{\bar{x}} = \mu$ and $\sigma_{\bar{x}} = \sigma/\sqrt{n} = \sigma/\sqrt{2}$.

6.22 Will the sampling distribution of $\bar{x}$ always be approximately normally distributed? Explain.

6.23 A random sample of $n = 100$ observations is selected from a population with $\mu = 30$ and $\sigma = 16$. Approximate the following probabilities:
a. $P(\bar{x} \geq 28)$ **b.** $P(22.1 \leq \bar{x} \leq 26.8)$ **c.** $P(\bar{x} \leq 28.2)$ **d.** $P(\bar{x} \geq 27.0)$

6.24 A random sample of $n = 60$ observations is selected from a population with $\mu = 1.5$ and $\sigma = 4$. Approximate the following probabilities:
a. $P(\bar{x} \leq 2.4)$ **b.** $P(\bar{x} \leq 2.5)$ **c.** $P(1.1 \leq \bar{x} \leq 1.4)$ **d.** $P(\bar{x} \geq 2.9)$

6.25 A random sample of $n = 900$ observations is selected from a population with $\mu = 100$ and $\sigma = 10$.
a. What are the largest and smallest values of $\bar{x}$ that you would expect to see?
b. How far, at the most, would you expect $\bar{x}$ to deviate from μ?
c. Did you have to know μ to answer part b? Explain.

*6.26 Consider a population that contains values of x equal to 00, 01, 02, ... , 97, 98, 99. Assume that the values of x are equally likely. For each of the following values of n, generate 500 random samples and calculate $\bar{x}$ for each sample. For each sample size, construct a relative frequency histogram of the 500 values of $\bar{x}$. What changes occur in the histograms as the value of n increases? What similarities exist? Use $n = 2$, $n = 5$, $n = 10$, $n = 30$, $n = 50$.

Applying the Concepts

6.27 The number of violent crimes per day in a certain city possesses a mean equal to 1.3 and a standard deviation equal to 1.7. A random sample of 50 days is observed, and the daily mean number of crimes for this sample, $\bar{x}$, is calculated.
a. Give the mean and standard deviation of the sampling distribution of $\bar{x}$.
b. Will the sampling distribution of $\bar{x}$ be approximately normal? Explain.
c. Find an approximate value of $P(\bar{x} < 1)$.
d. Find an approximate value of $P(\bar{x} > 1.9)$.

6.28 A manufacturer of steel bearings monitors samples of $n = 5$ bearing diameters every hour. If the cutting machine is working properly (is in control), the machine produces bearings with an inside diameter of 1.0 inch and standard deviation of .001 inch.
a. If the machine is in control, what is the probability that the mean $\bar{x}$ of the sample of $n = 5$ bearing diameters exceeds 1.001 inches?
b. If the machine is out of control and the mean inside diameter is 1.002 inches, what is the probability that $\bar{x}$ will be less than 1.001 inches?

6.29 An educational researcher has developed an IQ test that she claims is not biased against black children. It is known that scores of white children on the test have a mean equal to 100 and a standard deviation equal to 15. In order to test the claim of nonbias, a random sample of 200 black children are given the researcher's IQ test.
a. Assuming that the researcher's claim is true, describe the sampling distribution of the mean IQ score for a sample of 200 black children.
b. Assuming that the researcher's claim is true, what is the approximate probability that the sample mean IQ score is less than 97?

c. Suppose that the sample mean is actually 96.5. How would you interpret this value of $\bar{x}$ in view of the researcher's claim? Explain.

d. Suppose that the sample mean is actually 98.5. How would you interpret this value of $\bar{x}$ in view of the researcher's claim? Explain.

6.30 Last year a company initiated a program to compensate its employees for unused sick days, paying each employee a bonus of one-half the usual wage earned for each unused sick day. The question that naturally arises is: "Did this policy motivate employees to use fewer allotted sick days?" *Before* last year, the number of sick days used by employees had a distribution with a mean of 7 days and a standard deviation of 2 days.

a. Assuming that these parameters did not change last year, find the approximate probability that the sample mean number of sick days used by 100 employees chosen at random was less than or equal to 6.4 last year.

b. Suppose the sample mean for the 100 employees was, in fact, 6.4. How would you interpret this result?

6.31 Suppose a sample of $n = 50$ items is drawn from a population of manufactured products and the weight x of each item is recorded. Prior experience has shown that the weight has a probability distribution with $\mu = 6$ ounces and $\sigma = 2.5$ ounces. Then $\bar{x}$, the sample mean, will be approximately normally distributed (because of the central limit theorem).

a. Calculate $\mu_{\bar{x}}$ and $\sigma_{\bar{x}}$.

b. What is the approximate probability that the manufacturer's sample has a mean weight of between 5.75 and 6.25 ounces?

c. What is the approximate probability that the manufacturer's sample has a mean weight of less than 5.5 ounces?

d. How would the sampling distribution of $\bar{x}$ change if the sample size n were increased from 50 to, say, 100?

6.4
The Relation Between Sample Size and a Sampling Distribution

Suppose you draw two random samples, one containing $n = 5$ and the second $n = 10$ observations, from a population with mean μ and standard deviation σ. If you compute the mean $\bar{x}$ for each sample and obtain the results shown in the table, which estimate do you think contains more information about μ?

Intuitively, it would seem that the sample mean based on ten measurements contains more information than the mean based on five, but to answer the question correctly we need to compare the sampling distributions of these two statistics.

From Section 6.3 we know that the expected value of the sample means in repeated sampling is μ, regardless of the sample size. That is, both sample means are unbiased estimators of μ. The main difference in the sampling distributions lies in their standard deviations: The standard deviation of the mean based on $n = 5$ is $\sigma/\sqrt{5}$, while that based on $n = 10$ is $\sigma/\sqrt{10}$. Since the second standard deviation is smaller, we expect the sample means based on ten measurements to cluster more closely around μ in repeated

SAMPLE 1	SAMPLE 2
$n = 5$	$n = 10$
$\bar{x} = 12.6$	$\bar{x} = 13.1$

sampling than those based on five measurements. Thus, our intuitive feeling that $\bar{x}$ for $n = 10$ contains more information about μ is justified.

For the statistics you will encounter in this text, the variance of the sampling distribution is inversely proportional to the sample size.* Or you can say that the standard deviation of the sampling distribution is proportional to $1/\sqrt{n}$. (This relationship can be seen in the standard deviation of a sample mean $\bar{x}$: $\sigma_{\bar{x}} = \sigma/\sqrt{n}$.) To reduce the standard deviation of the sampling distribution of a statistic by $\frac{1}{2}$, you will therefore need four times as many observations in your sample. To reduce the standard deviation to one-third its original value, you will need nine times as many observations.

The sampling distributions for the sample mean $\bar{x}$, based on random samplings from a normally distributed population, are shown in Figure 6.11 for $n = 1$, 4, and 16 observations. The curve for $n = 1$ represents the probability distribution for the population. Those for $n = 4$ and $n = 16$ are sampling distributions for $\bar{x}$. Note how the distributions contract (variation decreases) for $n = 4$ and $n = 16$. The standard deviation for $\bar{x}$ based on $n = 16$ measurements is one-half the corresponding standard deviation for the distribution based on $n = 4$ measurements.

For most sampling distributions, the standard deviation of the distribution decreases as the sample size increases. We will use this result in Chapter 7 to help us determine the sample size needed to obtain a specified accuracy of estimation.

Figure 6.11 Three Sampling Distributions for $\bar{x}$

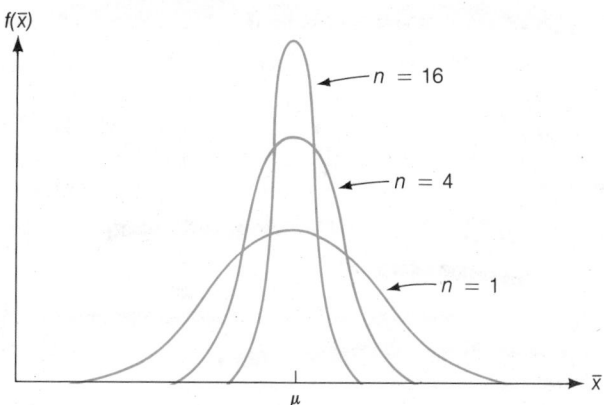

6.5
The Sampling Distribution for the Difference Between Two Statistics

Quite often we will wish to compare the proportion of people or objects in one group that possess some special attribute with the proportion in another. For example, we might wish to compare the proportion of Democrats in favor of certain legislation with the proportion of Republicans in favor of the same legislation, or we might wish to compare the proportions of consumers who prefer product 1 to those who prefer product 2. Or we might wish to compare the means of two populations, say the mean gas mileage for one compact car with the mean mileage of another. All three of these

* This is not true of all statistics, but it is true for most.

comparisons utilize a comparison of the corresponding sample statistics. For example, to estimate the difference in the means of two populations we use the difference between the means, $\bar{x}_1$ and $\bar{x}_2$, of independent random samples selected from the two populations. How close will the difference in the sample means $(\bar{x}_1 - \bar{x}_2)$ lie to the actual difference in the population means? To answer this question, we need to know something about the sampling distribution of the quantity $(\bar{x}_1 - \bar{x}_2)$.

We cannot completely specify the form of the sampling distribution for the difference between two statistics without considering particular cases, but we can say something about the mean, variance, and standard deviation of the sampling distribution.* We will give formulas for these quantities that will always apply and then demonstrate their use with an example.

Suppose you want to estimate the difference between two population parameters, θ_1 and θ_2. You have an unbiased statistic for estimating θ_1 (call it A) and another unbiased statistic for estimating θ_2 (call it B). Then, because these estimators are unbiased, it follows that $E(A) = \theta_1$ and $E(B) = \theta_2$ (i.e., the mean of the sampling distribution of A is θ_1 and the mean of the sampling distribution of B is θ_2). Further, assume that the variance of the sampling distribution of A is σ_A^2 and the variance of B is σ_B^2. Then it can be shown (the proof is omitted here) that the mean and variance of the sampling distribution of $(A - B)$, assuming A and B are independent, are

$$\mu_{(A-B)} = E(A - B) = \theta_1 - \theta_2 \qquad \sigma_{(A-B)}^2 = \sigma_A^2 + \sigma_B^2$$

We will illustrate the use of these formulas with Example 6.8.**

The Mean and Variance for the Difference Between Two Statistics A and B

$$\mu_{(A-B)} = E(A - B) = E(A) - E(B) = \theta_1 - \theta_2$$
$$\sigma_{(A-B)}^2 = \sigma_A^2 + \sigma_B^2$$

Note: The formula for $\sigma_{(A-B)}^2$ is based on the assumption that the two statistics, A and B, are independent.

In concluding, note that we have assumed that the two statistics, A and B, are independent. This assumption is often satisfied, but not always. Many types of comparisons are made between matched pairs of experimental units. For example, you might wish to compare gains in weight of pigs fed according to two different diets using pairs of pigs from the same litter. The weight gains for these pairs of pigs, one pig for each pair on diet A and the other on diet B, would be dependent because pigs within pairs

---

* The theory of statistics provides information on the form of the sampling distribution for the following class of statistics: The sums or differences of any number of normally distributed random variables will have a sampling distribution that is normally distributed. The random variables need not be independent of each other.

** Although it is not relevant to our discussion, it can also be shown that the variance of the sum of two independent statistics A and B is $\sigma_{(A+B)}^2 = \sigma_A^2 + \sigma_B^2$.

would possess the same genetic characteristics. The point is that you should carefully examine data to make certain that they satisfy the assumptions on which your methodology is based. You will learn more about matched-pair experiments in Section 8.3.

Example 6.8

Suppose you have two populations of investment returns that have means μ_1 and μ_2 and variances σ_1^2 and σ_2^2, respectively. Independent random samples of n_1 and n_2 observations are selected from the two populations: n_1 from population 1 and n_2 from population 2. The sample means $\bar{x}_1$ and $\bar{x}_2$ are computed from the samples. Find the expected value and standard deviation of the sampling distribution for the difference in two sample means, $(\bar{x}_1 - \bar{x}_2)$.

Solution

Since we know from Section 6.3 that $E(\bar{x}_1) = \mu_1$ and $E(\bar{x}_2) = \mu_2$, it follows that the mean of the sampling distribution of $(\bar{x}_1 - \bar{x}_2)$ is

$$E(\bar{x}_1 - \bar{x}_2) = E(\bar{x}_1) - E(\bar{x}_2) = \mu_1 - \mu_2$$

Recall that the standard deviation of the mean $\bar{x}$ of a random sample of n observations is $\sigma/\sqrt{n}$ (where σ is the standard deviation of the sampled population). Since the variance of a random variable is equal to the square of the standard deviation, it follows that the variances of the sample means $\bar{x}_1$ and $\bar{x}_2$ are

$$\sigma_{\bar{x}_1}^2 = \frac{\sigma_1^2}{n_1} \quad \text{and} \quad \sigma_{\bar{x}_2}^2 = \frac{\sigma_2^2}{n_2}$$

Then, using the formula for the variance of the difference of two independent statistics, we get

$$\sigma_{(\bar{x}_1 - \bar{x}_2)}^2 = \sigma_{\bar{x}_1}^2 + \sigma_{\bar{x}_2}^2 = \frac{\sigma_1^2}{n_1} + \frac{\sigma_2^2}{n_2}$$

and the standard deviation of the sampling distribution for $(\bar{x}_1 - \bar{x}_2)$ is

$$\sigma_{(\bar{x}_1 - \bar{x}_2)} = \sqrt{\frac{\sigma_1^2}{n_1} + \frac{\sigma_2^2}{n_2}}$$

It can be shown that the sampling distribution for $(\bar{x}_1 - \bar{x}_2)$ is approximately normal for large values of n_1 and n_2. Therefore, for large samples the sampling distribution for $(\bar{x}_1 - \bar{x}_2)$ appears as shown in Figure 6.12. ∎

Figure 6.12 Sampling Distribution for $(\bar{x}_1 - \bar{x}_2)$

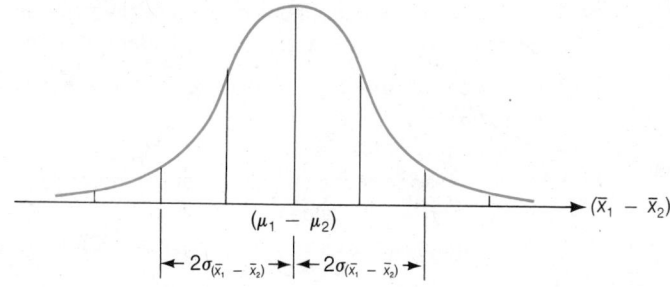

Summary

Many practical problems require that an inference be made about some population **parameter** (we call it θ). If we want to make this inference on the basis of information in a sample, we need to compute a **sample statistic** that contains information about θ. The amount of information a sample statistic contains about θ is reflected by its **sampling distribution,** the probability distribution of the sample statistic. In particular, we want a sample statistic that is an **unbiased** estimator of θ and has a smaller variance than any other unbiased sample statistic.

When the population parameter of interest is the mean μ, the sample mean provides an unbiased estimator with a standard deviation of $\sigma/\sqrt{n}$. Moreover, the **central limit theorem** assures us that the sampling distribution for the mean of a large sample is approximately normally distributed, no matter what the shape of the relative frequency distribution of the sampled population.

The amount of information in a sample that is relevant to some population parameter is related to the sample size. For example, the standard deviation of the sampling distribution of the sample mean $\bar{x}$ is inversely proportional to the square root of the sample size (that is, $1/\sqrt{n}$).

The sampling distributions for all the many statistics that can be computed from sample data could be discussed in detail, but this would delay discussion of the practical objective of this course—the role of statistical inference in decision-making. Consequently, we will comment further on the sampling distributions of statistics when we use them as estimators or decision-makers in the following chapters.

Supplementary Exercises 6.32–6.56

Learning the Mechanics

6.32 A random sample of $n = 3$ observations is selected from a population that is described by the following probability distribution:

x	0	2	3
$p(x)$	$\frac{1}{3}$	$\frac{1}{3}$	$\frac{1}{3}$

a. Calculate μ for this distribution.
b. List all the possible samples of $n = 3$ measurements from this population, and calculate the sample mean and median for each sample.
c. Find the sampling distributions of the sample mean and median.
d. Show that the mean is an unbiased estimator of μ and that the median is not.

6.33 A random sample of $n = 75$ observations is selected from a population with $\mu = 120$ and $\sigma^2 = 410$.
a. Find $\mu_{\bar{x}}$ and $\sigma_{\bar{x}}$.
b. What is the shape of the sampling distribution of $\bar{x}$?
c. Find the approximate value of $P(\bar{x} \le 118)$.
d. Find the approximate value of $P(115 \le \bar{x} \le 123)$.

e. Find the approximate value of $P(\bar{x} \leq 124.6)$.

f. Find the approximate value of $P(\bar{x} \geq 122.7)$.

6.34 A random sample of $n = 52$ observations is selected from a population with $\mu = 19.6$ and $\sigma = 2.5$. Approximate each of the following probabilities:

a. $P(\bar{x} \leq 19.6)$ **b.** $P(\bar{x} \leq 19)$ **c.** $P(\bar{x} \geq 20.1)$ **d.** $P(19.2 \leq \bar{x} \leq 20.6)$

6.35 Suppose that two fair dice are tossed. Let $\bar{x}$ be the mean of the two numbers that appear on the up faces of the dice.

a. Give the sampling distribution of $\bar{x}$.

b. Find the approximate value of $P(\bar{x} \leq 5)$.

6.36 Refer to Exercise 6.35. Simulate the sampling distribution of $\bar{x}$ by tossing a pair of balanced dice 100 times. Calculate $\bar{x}$ for each toss of the dice, and construct a relative frequency histogram for these 100 values of $\bar{x}$. Compare this graph with the exact sampling distribution for $\bar{x}$ of Exercise 6.35.

6.37 The following table contains fifty random samples of random digits, $x = 0, 1, 2, 3, \ldots, 9$, where $p(x) = \frac{1}{10}$. Each sample contains $n = 6$ measurements.

SAMPLE	SAMPLE	SAMPLE	SAMPLE
8, 1, 8, 0, 6, 6	7, 6, 7, 0, 4, 3	4, 4, 5, 2, 6, 6	0, 8, 4, 7, 6, 9
7, 2, 1, 7, 2, 9	1, 0, 5, 9, 9, 6	2, 9, 3, 7, 1, 3	5, 6, 9, 4, 4, 2
7, 4, 5, 7, 7, 1	2, 4, 4, 7, 5, 6	5, 1, 9, 6, 9, 2	4, 2, 3, 7, 6, 3
8, 3, 6, 1, 8, 1	4, 6, 6, 5, 5, 6	8, 5, 1, 2, 3, 4	1, 2, 0, 6, 3, 3
0, 9, 8, 6, 2, 9	1, 5, 0, 6, 6, 5	2, 4, 5, 3, 4, 8	1, 1, 9, 0, 3, 2
0, 6, 8, 8, 3, 5	3, 3, 0, 4, 9, 6	1, 5, 6, 7, 8, 2	7, 8, 9, 2, 7, 0
7, 9, 5, 7, 7, 9	9, 3, 0, 7, 4, 1	3, 3, 8, 6, 0, 1	1, 1, 5, 0, 5, 1
7, 7, 6, 4, 4, 7	5, 3, 6, 4, 2, 0	3, 1, 4, 4, 9, 0	7, 7, 8, 7, 7, 6
1, 6, 5, 6, 4, 2	7, 1, 5, 0, 5, 8	9, 7, 7, 9, 8, 1	4, 9, 3, 7, 3, 9
9, 8, 6, 8, 6, 0	4, 4, 6, 2, 6, 2	6, 9, 2, 9, 8, 7	5, 5, 1, 1, 4, 0
3, 1, 6, 0, 0, 9	3, 1, 8, 8, 2, 1	6, 6, 8, 9, 6, 0	4, 2, 5, 7, 7, 9
0, 6, 8, 5, 2, 8	8, 9, 0, 6, 1, 7	3, 3, 4, 6, 7, 0	8, 3, 0, 6, 9, 7
8, 2, 4, 9, 4, 6	1, 3, 7, 3, 4, 3		

a. Use the 300 random digits to construct a relative frequency histogram for the data. This relative frequency distribution should approximate $p(x)$.

b. Calculate the mean of the 300 digits. This will give an accurate estimate of μ (the mean of the population), which is 4.5.

c. Calculate s^2 for the 300 digits. This result should be close to the variance of x, $\sigma^2 = 8.25$.

d. Suppose you intend to make an inference about the mean μ using the median of a sample of $n = 6$ measurements. To see how well the sample median will estimate μ, calculate the median m for each of the fifty samples. Construct a relative frequency histogram for the sample medians to see how close they lie to the mean of $\mu = 4.5$. Calculate the mean and standard deviation of the fifty medians.

e. Calculate $\bar{x}$ for each of the fifty samples. Construct a relative frequency histogram for the sample means to see how close they lie to the mean of $\mu = 4.5$. Calculate the mean and standard deviation of the fifty means.

f. Which estimator, $\bar{x}$ or m, seems to be a better estimator of the population mean μ? Explain.

6.38 Suppose that x equals the number of heads observed when a single coin is tossed; that is, $x = 0$ or $x = 1$. The population corresponding to x is the set of 0's and 1's generated when the coin is tossed repeatedly a large number of times. Suppose that we select $n = 2$ observations from this population. (That is, we toss the coin twice and observe two values of x.)

a. List the three different samples (combinations of 0's and 1's) that could be obtained.

b. Calculate the value of $\bar{x}$ for each of the samples.

c. List the values that $\bar{x}$ can assume, and find the probabilities of observing these values.

d. Construct a graph of the sampling distribution of $\bar{x}$.

6.39 Suppose that x equals the number of heads observed when a single coin is tossed; that is, $x = 0$ or $x = 1$. The population corresponding to x is the set of 0's and 1's generated when the coin is tossed repeatedly a large number of times. Suppose that we select $n = 3$ observations from this population. (That is, we toss the coin three times and observe three values of x.)

a. List the four different samples (combinations of 0's and 1's) that could be obtained.

b. Calculate the value of $\bar{x}$ for each of the samples.

c. List the values that $\bar{x}$ can assume, and find the probabilities of observing these values.

d. Construct a graph of the sampling distribution of $\bar{x}$.

6.40 To see the effect of sample size on the standard deviation of the sampling distribution of a statistic, refer to Exercise 6.37 and combine pairs of samples (moving down the columns of the table) to obtain twenty-five samples of $n = 12$ measurements. Calculate the median for each sample.

a. Construct a relative frequency histogram for the twenty-five medians. Compare this with the histogram prepared for Exercise 6.37, part d, which is based on samples of $n = 6$ digits.

b. Calculate the mean and standard deviation of the twenty-five medians. Compare the standard deviation of this sampling distribution with the standard deviation of the sampling distribution in Exercise 6.37, part d. What relationship would you expect to exist between the two standard deviations?

6.41 Refer to Exercise 6.40. Repeat the exercise, but use the means of the samples rather than the medians. Compare the results with those of Exercise 6.37, part e.

6.42 A random sample of size n is to be drawn from a large population with mean 100 and standard deviation 10, and the sample mean $\bar{x}$ is to be calculated. To see the effect of different sample sizes on the standard deviation of the sampling distribution of $\bar{x}$, plot $\sigma/\sqrt{n}$ against n for $n = 1, 5, 10, 20, 30, 40,$ and 50.

Applying the Concepts

6.43 Over the last month a large supermarket chain has received many consumer complaints about the quantity of chips in 9-ounce bags of a particular brand of potato chips. Suspecting that the complaints are merely the result of the potato chips settling to the bottom of the bags during shipping, but wanting to be able to assure its customers they are getting their money's worth, the chain decides to examine the next shipment of chips received by their largest store. Thirty-five 9-ounce bags are randomly selected from the shipment, their contents weighed, and the sample mean weight computed. The chain's management decides that if the sample mean is less than 8.95 ounces, the shipment will be refused and a complaint registered with the potato chip company. Assume that the distribution of weights of the contents of all the potato chip bags in question has a mean of 8.9 ounces and a standard deviation of .13 ounce.

a. What is the approximate probability that the supermarket chain's investigation will lead to refusal of the shipment?

b. What assumptions did you have to make in order to answer part a? Justify the assumptions.

6.44 The distribution of the number of barrels of oil produced by a certain oil well each day for the past 3 years has a mean of 400 and a standard deviation of 75.

a. Describe the sampling distribution of the mean number of barrels produced per day for samples of 40 production days drawn from the past 3 years.

b. What is the approximate probability that the sample mean will be greater than 425?

c. What is the approximate probability that the sample mean will be less than 400?

6.45 Water availability is of prime importance in the life cycle of most reptiles. To determine the rate of evaporative water loss of a certain species of lizard at a particular desert site, thirty-four such lizards were randomly collected, weighed, and placed under the appropriate experimental conditions. After 24 hours, each lizard was removed, re-weighed, and its total water loss was calculated as the difference between initial body weight and body weight after treatment. Previous studies have shown that the distribution of water loss for the lizards has a mean of 3.1 grams and a standard deviation of .8 gram.

a. Find the approximate probability that the thirty-four lizards have a mean water loss of less than 3.0 grams.

b. Between 3.15 and 3.25 grams.

6.46 Electric power plants that use water for cooling their condensers sometimes discharge heated water into rivers, lakes, or oceans. It is known that water heated above certain temperatures has a detrimental effect on the plant and animal life in the water. Suppose it is known that the increased temperature of the heated water discharged by a certain power plant on any given day has a distribution with a mean of 5°C and a standard deviation of .5°C.

a. For 50 randomly selected days, what is the approximate probability that the average increase in temperature of the discharged water is greater than 5.0°C?

b. Less than 4.8°C?

c. What assumptions must be made for you to answer the questions?

6.47 ***Random number generators*** have many uses in statistics.* One type is designed to produce a sequence of numbers between 0 and 1. A number x can assume any value in the interval from 0 to 1 with equal probability, and any value of x is independent of the values of previous numbers that appear in the sequence. Furthermore, the probability distribution of x has a mean $\mu = .5$ and a standard deviation $\sigma = .29$. Let y be the average of n such random numbers.

a. Graph the probability distribution for x.

b. Give the mean and standard deviation of the sampling distribution of y.

c. What is the approximate form of the sampling distribution of y when n is large?

d. Sketch the sampling distribution of y and compare it with your graph from part a.

6.48 To determine whether a metal lathe that produces machine bearings is properly adjusted, a random sample of twenty-five bearings is collected and the diameter of each is measured. If the standard deviation of the diameters of the bearings measured over a long period of time is .001 inch, what is the approximate probability that the mean diameter $\bar{x}$ of the sample of twenty-five bearings will lie within .0001 inch of the population mean diameter of the bearings?

6.49 Refer to Exercise 6.48. The mean diameter of the bearings produced by the machine is supposed to be .5 inch. The company decides to use the sample mean (from Exercise 6.48) to decide whether the process is in control, i.e., whether it is producing bearings with a mean diameter of .5 inch. The machine will be considered out of control if the mean of the sample of $n = 25$ diameters is less than .4994 inch or larger than .5006 inch. If the true mean diameter of the bearings produced by the machine is .501 inch, what is the approximate probability that the test will imply that the process is out of control?

6.50 Suppose you wish to purchase a case of expensive wine. You plan to open two bottles for immediate use and you will keep the remaining bottles if the two are acceptable. Suppose there are ten bottles in the case, and, unknown to you, the condition of the wine in the bottles is as shown below (1 = good, 0 = bad):

Bottle	1	2	3	4	5	6	7	8	9	10
Condition	1	0	0	1	1	1	1	1	0	1

Since you are interested only in the ten bottles in the case, the collection of ten 0 or 1 responses is the population of interest to you.

a. If you randomly sample two bottles from the case, how many different samples (different pairs of bottles) could you select? List them.

b. Suppose you are going to accept the case only if both bottles in the sample are good. Identify all samples containing two good bottles. What is the probability

---

* Random number generators are used to produce random numbers such as those that appear in Table I of the Appendix.

that you will accept the case? [*Hint:* See the definition of a random sample in Section 3.7.]

c. Let *x* equal the number of good bottles in the sample of $n = 2$. Construct the sampling distribution of *x*.

6.51 The distribution of the number of characters printed per second by a particular kind of line printer at a computer terminal has the following parameters: $\mu = 45$ characters per second, $\sigma = 2$ characters per second.

a. Describe the sampling distribution of the mean number of characters printed per second for random samples of 1-minute intervals.

b. Find the approximate probability that the sample mean for a random sample of 60 seconds will be between 44.5 and 45.3 characters per second.

c. Find the approximate probability that the sample mean will be less than 44 characters per second.

6.52 This past year, an elementary school began using a new method to teach arithmetic to first graders. A standardized test, administered at the end of the year, was used to measure the effectiveness of the new method. The distribution of past scores on the standardized test produced a mean of 75 and a standard deviation of 10.

a. If the new method is no different from the old method, what is the approximate probability that the mean score $\bar{x}$ of a random sample of thirty-six students will be greater than 79?

b. What assumptions must be satisfied to make your answer valid?

6.53 As part of a company's quality control program, it is a common practice to monitor the quality characteristics of a product over time. For example, the amount of alkali in soap might be monitored by randomly selecting from the production process and analyzing $n = 5$ test quantities of soap each hour. The mean $\bar{x}$ of the sample would be plotted against time on a control chart as shown in Figure (a). If the process is in control, $\bar{x}$ should assume a distribution with a process mean μ and standard deviation σ. The control chart in Figure (b) shows a horizontal line to locate the process mean and two lines, located $3\sigma_{\bar{x}}$ above and below μ, which are called control limits. If $\bar{x}$ falls within

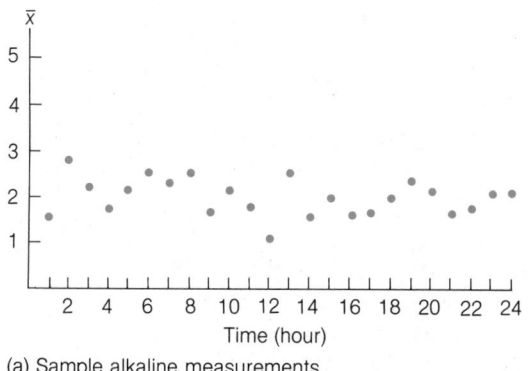

(a) Sample alkaline measurements

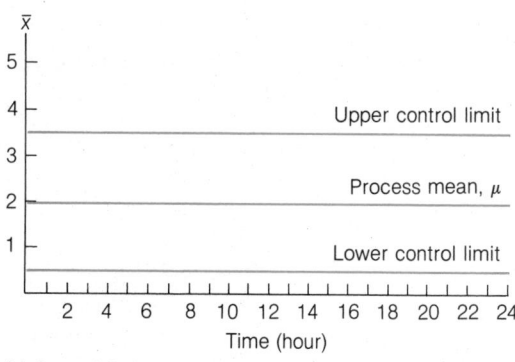

(b) Control limits

the control limits, the process is deemed to be in control. If $\bar{x}$ is outside the limits, the monitor flashes a warning and suggests that something is wrong with the process. Suppose for the soap process that experience has shown $\mu = 2\%$ and $\sigma = 1\%$.

a. If $n = 5$, how far away from μ should you locate the upper and lower control limits?

b. If the process is in control, what is the approximate probability that at any fixed point in time, $\bar{x}$ will fall outside the control limits? State any assumptions you must make in reaching a solution.

*6.54 [*Note:* This exercise refers to an optional section in Chapter 4.] A building contractor has decided to purchase a load of factory-reject aluminum siding as long as the average number of flaws per piece of siding in a sample of size 35 from the factory's reject pile is 2.1 or less. If it is known that the number of flaws per piece of siding in the factory's reject pile has a Poisson probability distribution with a mean of 2.5, find the approximate probability that the contractor will not purchase a load of siding. [*Hint:* If x is a Poisson random variable with mean λ, then σ_x^2 also equals λ.]

6.55 The distribution of the number of loaves of bread sold per day by a large grocery store over the past 5 years has a mean of 250 and a standard deviation of 45.

a. Describe the sampling distribution of the total number of loaves of bread sold per 30 randomly selected shopping days. [*Hint:* See the footnote in Section 6.3 that gives the application of the central limit theorem to the sum of the measurements in a sample.]

b. Give the approximate probability that the total number of loaves sold per 30 shopping days is between 7,000 and 8,000.

c. Give the approximate probability that the total is greater than 8,100 loaves.

*6.56 [*Note:* This exercise refers to an optional section in Chapter 5.] The exponential random variable (described in Section 5.5) is a continuous random variable with probability distribution $f(x) = (1/\theta)e^{-x/\theta}$. The mean and standard deviation of the exponential distribution are both equal to θ. Suppose n observations are randomly selected from such a distribution.

a. If n is large, what is the shape of the sampling distribution of the sample mean?

b. What are the mean and standard deviation of this sampling distribution?

On Your Own . . .

To understand the central limit theorem and sampling distribution, consider the following experiment: Toss four identical coins and record the number of heads observed. Then repeat this experiment four more times, so that you end up with a total of five observations for the random variable x, the number of heads when four coins are tossed.

Now derive and graph the probability distribution for x, assuming the coins are balanced. Note that the mean of this distribution is $\mu = 2$ and the standard deviation is $\sigma = 1$. This probability distribution represents the one from which you are drawing a random sample of five measurements.

Next, calculate the mean $\bar{x}$ of the five measurements, i.e., calculate the mean number of heads you observed in five repetitions of the experiment. Although you have repeated the basic experiment five times, you have only one observed value of $\bar{x}$. To derive the probability distribution or sampling distribution of $\bar{x}$ empirically, you have to repeat the entire process (of tossing four coins five times) many times. Do it 100 times.

The approximate sampling distribution of $\bar{x}$ can be derived theoretically by making use of the central limit theorem. We expect at least an approximate normal probability distribution with a mean $\mu = 2$ and a standard deviation

$$\sigma_{\bar{x}} = \frac{\sigma}{\sqrt{n}} = \frac{1}{\sqrt{5}} = .45$$

Count the number of your 100 $\bar{x}$'s that fall in each of the following intervals:

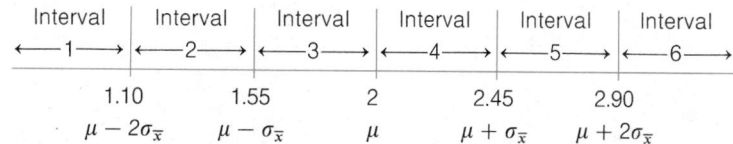

Use the normal probability distribution with $\mu = 2$ and $\sigma_{\bar{x}} = .45$ to calculate the expected number of the 100 $\bar{x}$'s in each of the intervals. How closely does the theory describe your experimental results?

References   Hogg, R. V., & Craig, A. T. *Introduction to mathematical statistics.* 4th ed. New York: Macmillan, 1978. Chapter 4.

Lindgren, B. W. *Statistical theory.* 3d. ed. New York: Macmillan, 1976. Chapter 2.

Mendenhall, W., Scheaffer, R. L., & Wackerly, D. *Mathematical statistics with applications.* 2d. ed. North Scituate, Mass.: Duxbury, 1980. Chapter 7.

CHAPTER 7

Estimation and Tests of Hypotheses: Single Sample

Where We've Been . . .

In the preceding chapters we learned that relevant information about populations can be characterized by numerical descriptive measures (called *parameters*) and that decisions about their values are based on sample statistics computed from sample data. Since statistics vary in a random manner from sample to sample, inferences based on sample statistics are subject to uncertainty. This property is reflected in the sampling (probability) distribution of a statistic.

Where We're Going . . .

This chapter begins to put some of the preceding material into practice. That is, we will estimate or make decisions about population means or proportions based on a single sample selected from a population and then use the sampling distribution of a sample statistic to assess the uncertainty associated with an inference.

Contents

The estimation of the mean gas mileage for a new car model, the testing of a claim that a certain brand of television tube has a mean life of 5 years, and the estimation of the mean potency of a drug are practical problems with a common element. In each case we are interested in making an inference about the mean of a population. This important problem constitutes one of the primary topics of this chapter.

We will concentrate on two types of inferences about a population parameter: **estimation of the parameter** and **tests of hypotheses, or claims, about the parameter.** You will see that different techniques are used for making inferences depending on whether a sample contains a large or small number of measurements. In either case, our objectives remain the same: We want to make the best use of information in the sample to make an inference and to assess its reliability.

In Sections 7.1 and 7.2 we consider large-sample methods for estimation and tests of hypotheses about population means. The small-sample analogs of these two topics are covered in Section 7.4. We consider large-sample inferences about a binomial population proportion in Section 7.5 and a method for determining the appropriate sample size in Section 7.6. Finally, in Section 7.7, we discuss estimation and tests of hypotheses about a population variance.

7.1 Large-Sample Estimation of a Population Mean

We will illustrate the **large-sample method** of estimating a population mean with an example. Suppose a large hospital wants to estimate the average length of time patients remain in the hospital. To accomplish this objective, the hospital administrators plan to sample 100 of all previous patients' records and to use the sample mean, $\bar{x}$, of the lengths of stay to estimate the mean stay, μ, of *all* patients' visits. Further, they plan to use the sampling distribution of the sample mean to assess the accuracy of their estimate. How will this be accomplished?

According to the central limit theorem, the sampling distribution of the sample mean is approximately normal for large samples, as shown in Figure 7.1. Let us calculate the interval

$$\bar{x} \pm 2\sigma_{\bar{x}} = \bar{x} \pm \frac{2\sigma}{\sqrt{n}}$$

Figure 7.1 Sampling Distribution of $\bar{x}$

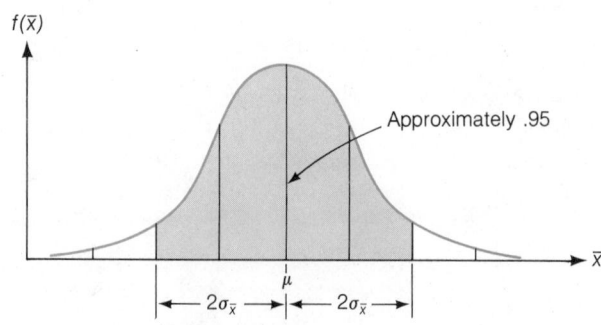

That is, we will form an interval with endpoints located 2 standard deviations above and below the sample mean. What are the chances (answer before we have drawn a sample) that this interval will enclose μ, the population mean?

To answer this question, refer to Figure 7.1. If the 100 measurements yield a value of $\bar{x}$ that falls between the two lines shown in color, i.e., within 2 standard deviations of μ, then the interval $\bar{x} \pm 2\sigma_{\bar{x}}$ will contain μ; if $\bar{x}$ falls outside these boundaries, the interval $\bar{x} \pm 2\sigma_{\bar{x}}$ will not contain μ. Since the area under the normal curve (the sampling distribution of $\bar{x}$) between these boundaries is about .95 (more precisely, from Table IV in the Appendix the area is .9544), we know that the interval $\bar{x} \pm 2\sigma_{\bar{x}}$ will contain μ with a probability approximately equal to .95.

To illustrate, suppose that the sum and the sum of squared deviations for the sample of 100 lengths of time spent in the hospital are

$$\sum x = 465 \text{ days} \qquad \text{and} \qquad \sum (x - \bar{x})^2 = 2{,}387$$

Then

$$\bar{x} = \frac{\sum x}{n} = \frac{465}{100} = 4.65$$

$$s^2 = \frac{\sum (x - \bar{x})^2}{n - 1} = \frac{2{,}387}{99} = 24.11 \qquad \text{and} \qquad s = 4.9$$

To form the interval of 2 standard deviations around $\bar{x}$, we calculate

$$\bar{x} \pm 2\sigma_{\bar{x}} = 4.65 \pm 2 \frac{\sigma}{\sqrt{100}}$$

But now we face a problem. You can see that without knowing the standard deviation σ of the original population, i.e., the standard deviation of the lengths of stay of *all* patients, we cannot calculate this interval. However, since we have a large sample ($n = 100$ measurements), we can approximate the interval by using the sample standard deviation s to approximate σ. Thus,

$$\bar{x} \pm 2 \frac{\sigma}{\sqrt{100}} \approx \bar{x} \pm 2 \frac{s}{\sqrt{100}} = 4.65 \pm 2 \left(\frac{4.9}{10} \right) = 4.56 \pm .98$$

That is, we estimate the mean length of stay in the hospital for all patients to fall in the interval 3.67 to 5.63 days.

Can we be sure that μ, the true mean, is in the interval 3.67 to 5.63? We cannot be certain, but we can be reasonably confident that it is. This confidence is derived from the knowledge that if we were to draw repeated random samples of 100 measurements from this population and form an interval of 2 standard deviations around $\bar{x}$ each time, approximately 95% of the intervals would contain μ. We have no way of knowing (without looking at all the patients' records) whether our sample interval is one of the 95% that contain μ or one of the 5% that do not, but the odds certainly favor its containing μ. Consequently, the interval 3.67 to 5.63 provides an estimate of the mean length of patient stay in the hospital. The formula that tells us how to calculate an interval estimate based on sample data is called an ***interval estimator.*** The probability, .95, that measures the confidence we can place in the interval estimate is called a ***confidence coefficient.*** The percentage, 95%, is called the **confidence level** for the interval estimate.

Definition 7.1

An ***interval estimator*** is a formula that tells us how to use sample data to calculate an interval that estimates a population parameter.

Definition 7.2

The ***confidence coefficient*** is the probability that an interval estimator encloses the population parameter if the estimator is used repeatedly a very large number of times.

The ***confidence level*** is the confidence coefficient expressed as a percentage.

Interval

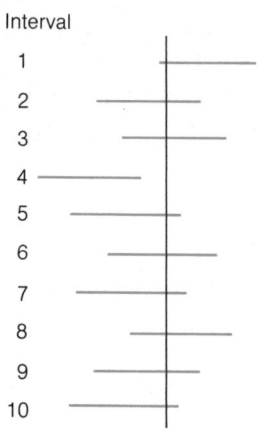

Figure 7.2 Interval Estimators for μ: Ten Samples

Now we have seen how an interval can be used to estimate a population parameter. This is a common statistical practice, because when we use an interval estimator we can usually assess the level of confidence we have that the interval actually contains the true value of the parameter. Figure 7.2 shows what happens when ten different samples are drawn from a population and a confidence interval for a parameter, say μ, is calculated from each. The true value of μ is located by the vertical line in the figure. Ten confidence intervals, corresponding to ten samples, are shown as horizontal line segments. Note that the confidence intervals move from sample to sample—sometimes containing μ and other times missing μ. If our confidence level is 95%, then in the long-run 95% of our sample confidence intervals will contain μ.

Suppose you wish to choose a confidence coefficient other than .95. Notice in Figure 7.1 that the confidence coefficient .95 is equal to the total area under the sampling distribution, less .05 of the area, which is divided equally between the two tails. Using this idea, we can construct a confidence interval with any desired confidence coefficient by increasing or decreasing the area (call it α) assigned to the tails of the sampling distribution (see Figure 7.3). For example, if we place area $\alpha/2$ in each tail and if $z_{\alpha/2}$

Figure 7.3 Locating $z_{\alpha/2}$ on the Standard Normal Curve

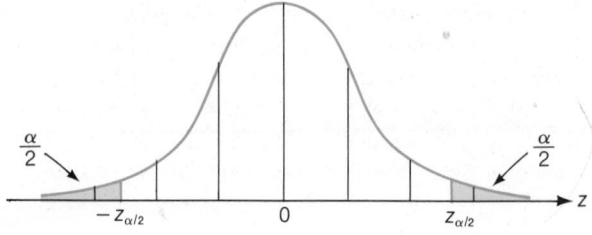

is the z-value such that the area $\alpha/2$ will lie to its right, then the confidence interval with confidence coefficient $(1 - \alpha)$ is

$$\bar{x} \pm z_{\alpha/2}\sigma_{\bar{x}}$$

To illustrate, for a confidence coefficient of .90 we have $(1 - \alpha) = .90$, $\alpha = .10$, $\alpha/2 = .05$, and $z_{.05}$ is the z-value that locates area .05 in the upper tail of the sampling distribution. Recall that Table IV in the Appendix gives the areas between the mean and a specified z-value. Since the total area to the right of the mean is .5, we find that $z_{.05}$ will be the z-value corresponding to an area of $.5 - .05 = .45$ to the right of the mean (see Figure 7.4). This z-value is $z_{.05} = 1.645$. Confidence coefficients used in practice (in published articles) range from .90 to .99. The most common confidence coefficients with corresponding values of α and $z_{\alpha/2}$ are shown in Table 7.1.

Figure 7.4 The z-Value ($z_{.05}$) Corresponding to an Area Equal to .05 in the Upper Tail of the z-Distribution

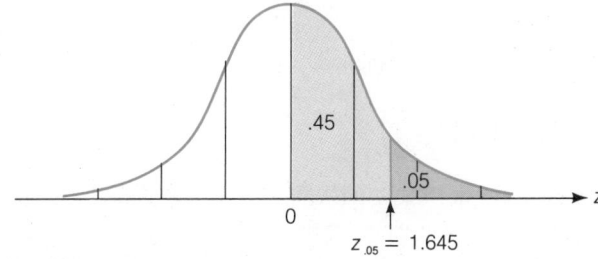

Table 7.1
Commonly Used Values of $z_{\alpha/2}$

CONFIDENCE LEVEL $100(1 - \alpha)$	α	$\alpha/2$	$z_{\alpha/2}$
90%	.10	.05	1.645
95%	.05	.025	1.96
99%	.01	.005	2.58

Large-Sample $100(1 - \alpha)$% Confidence Interval for μ

$$\bar{x} \pm z_{\alpha/2}\sigma_x = \bar{x} \pm z_{\alpha/2}\frac{\sigma}{\sqrt{n}}$$

where $z_{\alpha/2}$ is the z-value with an area $\alpha/2$ to its right (see Figure 7.3) and $\sigma_{\bar{x}} = \sigma/\sqrt{n}$. The parameter σ is the standard deviation of the sampled population and n is the sample size.

When n is equal to 30 or more, the confidence interval is approximately equal to

$$\bar{x} \pm z_{\alpha/2}\left(\frac{s}{\sqrt{n}}\right)$$

where s is the sample standard deviation.

Example 7.1

Unoccupied seats on flights cause the airlines to lose revenue. Suppose a large airline wants to estimate its average number of unoccupied seats per flight over the past year. To accomplish this, the records of 225 flights are randomly selected from the files, and the number of unoccupied seats is noted for each of the sampled flights. The sample mean and standard deviation are

$$\bar{x} = 11.6 \text{ seats} \qquad s = 4.1 \text{ seats}$$

Estimate μ, the mean number of unoccupied seats per flight during the past year, using a 90% confidence interval.

Solution

The general form of the 90% confidence interval for a population mean is

$$\bar{x} \pm z_{\alpha/2}\sigma_{\bar{x}} = \bar{x} \pm z_{.05}\sigma_{\bar{x}} = \bar{x} \pm 1.645\left(\frac{\sigma}{\sqrt{n}}\right)$$

For the 225 records sampled, we have

$$11.6 \pm 1.645\left(\frac{\sigma}{\sqrt{225}}\right)$$

Since we do not know the value of σ (the standard deviation of the number of unoccupied seats per flight for all flights of the year), we use our best approximation: the sample standard deviation s. Then the 90% confidence interval is, approximately,

$$11.6 \pm 1.645\left(\frac{4.1}{\sqrt{225}}\right) = 11.6 \pm .45$$

or from 11.15 to 12.05. That is, at the 90% confidence level, we estimate the mean number of unoccupied seats per flight to be between 11.15 and 12.05 during the sampled year. We stress that the confidence level refers to the procedure used. If we were to apply this procedure repeatedly to different samples, approximately 90% of the intervals would contain μ. ∎

Case Study 7.1

Dancing to the Customer's Tune: The Need to Assess Customer Preferences

The following quotations have been extracted from the December 13, 1976, issue of *Business Week:**

"We're dancing to the tune of the customer as never before," says J. Janvier Wetzel, vice-president for sales promotion at Los Angeles-based Broadway Department Stores. "With population growth down to a trickle compared with its previous level, we're no longer spoiled with instant success every time we open a new store. Traditional department stores are locked in the biggest competitive battle in their history."

The nation's retailers are becoming uncomfortably aware that today's operating environment is vastly different from that of the 1960s. Population growth is slowing, a growing singles market is emerging, family formations are coming at later ages, and more women are embarking on careers. Of the 71 million households in the U.S. today, the dominant consumer buying segment is families headed by persons over 45. But by 1980 this group will have lost its majority status to the 25 to 40-year-old group. Merchants must now reposition their stores to attract these new customers.

To do so retailers are using market research to ferret out new purchasing attitudes and life-styles and then translating this into customer buying segments. . . . Department stores are taking a hard look at some of the basics of their business by . . . spending heavily for far more elaborate market research. Data on demographics, psychographics (measurement of attitudes), and life-style are being fed into retailers' computers so they can make marketing decisions based on actual spending patterns and estimate their inventory needs with less risk.

In order to stock their various departments with the type and style of goods that appeal to their potential group of customers, a downtown department store should be interested in estimating the average age of downtown shoppers, not shoppers in general. Suppose a downtown department store questions forty-nine downtown shoppers concerning their age (the offer of a small gift certificate may help persuade shoppers to respond to such questions). The sample mean and standard deviation are found to be 40.1 and 8.6, respectively. The store could then estimate the mean age μ of all downtown shoppers with a 95% confidence interval as follows:

$$\bar{x} \pm 1.96\left(\frac{s}{\sqrt{n}}\right) = 40.1 \pm 1.96\left(\frac{8.6}{\sqrt{49}}\right) = 40.1 \pm 2.4$$

Thus, the department store should gear its sales to consumers with average age between 37.7 and 42.5.

**Exercises
7.1–7.16**

Learning the Mechanics

7.1 A random sample of 100 observations from a population produced the following summary statistics:

$$\sum x = 500 \qquad \sum x^2 = 6{,}066$$

a. Find a 95% confidence interval for μ.
b. Interpret the confidence interval you found in part a.

7.2 A random sample of sixty-four observations from a population produced the following summary statistics:

$$\sum x = 200 \qquad \sum x^2 = 639$$

a. Find a 95% confidence interval for μ.
b. Interpret the confidence interval you found in part a.

7.3 A random sample of sixty observations produced a mean $\bar{x} = 30.4$ and a standard deviation $s = 1.6$.
a. Find a 95% confidence interval for the population mean μ.
b. Find a 90% confidence interval for μ.
c. Find a 99% confidence interval for μ.

7.4 A random sample of n measurements was selected from a population with unknown mean μ and standard deviation σ. Calculate a 95% confidence interval for μ for each of the following situations:

a. $n = 50$, $\bar{x} = 28$, $s^2 = 12$ **b.** $n = 200$, $\bar{x} = 102$, $s^2 = 12$
c. $n = 100$, $\bar{x} = 17$, $s = .1$ **d.** $n = 100$, $\bar{x} = 1.02$, $s = 1.10$

7.5 A random sample of fifty observations from a population produced the following summary statistics:

$$\sum x = 506 \qquad \sum x^2 = 6{,}818$$

a. Find a 99% confidence interval for μ.
b. Find a 90% confidence interval for μ.

7.6 A random sample of 100 observations from a normally distributed population possesses a mean equal to 76.9 and a standard deviation equal to 4.8.
a. Find a 95% confidence interval for μ.
b. What is meant when you say that a confidence coefficient is .95?
c. Find a 99% confidence interval for μ.
d. What happens to the width of a confidence interval as the value of the confidence coefficient is increased while the sample size is held fixed?

7.7 Explain what is meant by the statement, "We are 95% confident that an interval estimate contains μ."

7.8 Will a large-sample confidence interval be valid if the population from which the sample is taken is not normally distributed? Explain.

7.9 The mean and standard deviation of a random sample of n measurements are equal to 27.4 and 2.5, respectively.
a. Find a 95% confidence interval for μ if $n = 100$.
b. Find a 95% confidence interval for μ if $n = 400$.
c. Find the widths of the confidence intervals found in parts a and b. What is the effect on the width of a confidence interval of quadrupling the sample size while holding the confidence coefficient fixed?

Applying the Concepts

7.10 A fact long known but little understood is that twins, in their early years, tend to have lower IQ's and pick up language more slowly than nontwins. Recently, psychologists have found that the slower intellectual growth of most twins may be caused by benign parental neglect. Suppose that it is desired to estimate the mean attention time given to twins per week by their parents. A sample of forty-six sets of $2\frac{1}{2}$-year-old twin boys is taken, and at the end of 1 week the attention time given to each pair is recorded. The results are as follows:

$\bar{x} = 22$ hours $s = 16$ hours

Using the data, find a 90% confidence interval for the mean attention time given to all twin boys by their parents. Interpret your confidence interval.

7.11 Suppose a large labor union wishes to estimate the mean number of hours per month a union member is absent from work. The union decides to sample 320 of its members at random and monitor their working time for 1 month. At the end of the month, the total number of hours absent from work is recorded for each employee. If the mean and standard deviation of the sample are $\bar{x} = 9.6$ hours and $s = 6.4$ hours, find a 95% confidence interval for the true mean number of hours absent per month per employee. Interpret your confidence interval.

7.12 An article in the *Wall Street Journal* (March 1, 1984) states that of forty-eight sources surveyed by blue chip economic indicators, Prudential Bache Securities gave the lowest estimate (3%) of the rise in the Consumer Price Index for 1984 and Prudential Insurance (Prudential Bache's owner) gave the highest (5.6%).

a. Find an approximate value for the sample standard deviation of the sample of forty-eight estimates. [*Hint:* Assume that the range is approximately equal to 4*s*.]

b. Assume that the forty-eight estimates represent a random sample of estimates from a large number of estimate sources. If the mean of the sample was 4.3%, find a 90% confidence interval for the mean estimated increase in the Consumer Price Index for this population of estimates associated with the estimate sources.

7.13 Automotive engineers are continually improving their products. Suppose a new type of brake light has been developed by General Motors. As part of a product safety evaluation program, General Motors' engineers wish to estimate the mean driver response time to the new brake light. (Response time is the length of time from the point that the brake is applied until the driver in the following car takes some corrective action.) Fifty drivers are selected at random and the response time (in seconds) for each driver is recorded, yielding the following results: $\bar{x} = .72$, $s^2 = .022$. Estimate the mean driver response time to the new brake light using a 99% confidence interval. Interpret your confidence interval.

7.14 As an aid in the establishment of personnel requirements, the director of a hospital wishes to estimate the mean number of people who are admitted to the emergency room during a 24-hour period. The director randomly selects sixty-four different 24-hour periods and determines the number of admissions for each. For this sample, $\bar{x} = 19.8$ and $s^2 = 25$. Estimate the mean number of admissions per 24-hour period with a 95% confidence interval. Interpret the result.

7.15 A sociologist wishes to estimate the average number of television viewing hours per American family per week. A random sample of 400 families yields a mean of 32.6 hours and a standard deviation of 9.9 hours. Estimate the mean viewing time with a 90% confidence interval. Interpret your confidence interval.

7.16 An article titled "Scientists Seek Upper Hand in Insect Wars" (*Wall Street Journal,* March 6, 1984) notes that the cockroach has had 300 million years to develop a resistance to destruction. That they are "reproductively brilliant" is evidenced by a study conducted by researchers for S. C. Johnson & Son, Inc. (manufacturers of Raid and Off). Five thousand roaches, the expected number in a roach-infested house, were released in the Raid test kitchen. One week later the kitchen was fumigated and 16,298 dead roaches were counted, a gain of 11,298 roaches for the 1-week period. Assume that none of the original roaches died during the 1-week period and that the standard deviation of x, the number of roaches produced per roach in a 1-week period, is 1.5. Use the number of roaches produced by the sample of 5,000 roaches to find a 95% confidence interval for the mean number of roaches produced per week for each roach in a typical roach-infested house.

7.2 A Large-Sample Test of an Hypothesis About a Population Mean

Suppose building specifications in a certain city require that the average breaking strength of residential sewer pipe be more than 2,400 pounds per foot of length (i.e., per lineal foot). Each manufacturer who wants to sell pipe in this city must demonstrate that its product meets the specification. Note that we are again interested in making an inference about the mean μ of a population. However, in this example we are less interested in estimating the value of μ than we are in testing an **hypothesis** about its value. That is, we want to decide whether or not the mean breaking strength of the pipe exceeds 2,400 pounds per lineal foot.

In general, the hypothesis that a researcher wishes to establish—the research hypothesis—is called the ***alternative hypothesis.*** To establish this alternative hypothesis, we first define a ***null hypothesis,*** a theory that directly opposes the research hypothesis. Then we attempt to gain support for the research (or alternative) hypothesis by producing evidence to show that the null hypothesis is false. This is accomplished by using our rare event approach of Chapter 2. That is, we attempt to show that the sample is inconsistent with the null hypothesis: If the null hypothesis is true, the sample represents a rare event.

In our example, the sewer pipe manufacturer wishes to show that the mean breaking strength of its pipe exceeds 2,400 pounds per lineal foot; that is, $\mu > 2,400$. To do this, the researcher will hypothesize that $\mu = 2,400$ and attempt to show that this hypothesis is false. Therefore, the null and research hypotheses are:

Null hypothesis (H_0): $\quad\quad\quad\quad\quad\quad\quad$ $\mu = 2,400$ (i.e., the manufacturer's pipe does not meet specifications)

Research (alternative) hypothesis (H_a): $\mu > 2,400$ (i.e., the manufacturer's pipe does meet specifications)

Next, we need a procedure for using the information in the sample to decide which hypothesis is true. Since we are testing hypotheses about a population mean μ, it is reasonable to use the sample mean $\bar{x}$ to decide between the two hypotheses. Specifically, we will reject the null hypothesis H_0 in favor of the research hypothesis H_a when the sample mean $\bar{x}$ strongly indicates that μ exceeds 2,400 pounds per lineal foot.

A convenient measure of the distance between $\bar{x}$ and the hypothesized mean value of 2,400 is the z-score:

$$z = \frac{\bar{x} - 2,400}{\sigma_{\bar{x}}} = \frac{\bar{x} - 2,400}{\sigma/\sqrt{n}}$$

Note that the z-score expresses the distance between $\bar{x}$ and the hypothesized mean μ in units of standard deviations of $\bar{x}$ (that is, $\sigma_{\bar{x}}$). How large a z-score will be required before you decide to reject the null hypothesis? If you examine Figure 7.5, you will note that the chance of observing $\bar{x}$ more than 1.645 standard deviations above 2,400 is only .05—*if in fact the true mean μ is 2,400*. Thus, if the sample mean is more than 1.645 standard deviations above 2,400, either H_0 is true and a relatively rare event has occurred (.05 probability) or H_a is true and the population mean exceeds 2,400. Deciding that the research hypothesis is true if in fact it is false is called a **Type I decision error.** As indicated in Figure 7.5, the risk of making a Type I error, that is, deciding in favor of the research hypothesis if in fact the null hypothesis is true, is denoted by the symbol α. That is,

$$\alpha = P(\text{Type I error})$$
$$= P(\text{Rejecting the null hypothesis if it is true})$$

In our example

$$\alpha = P(z > 1.645 \text{ if in fact } \mu = 2,400) = .05$$

Figure 7.5   The Sampling Distribution of $\bar{x}$, Assuming $\mu = 2,400$

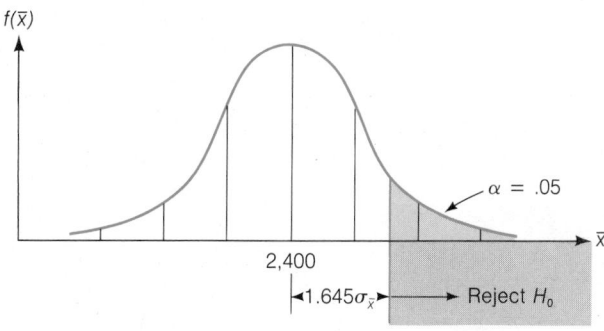

We now summarize the elements of the test:

H_0: $\mu = 2,400$ H_a: $\mu > 2,400$

Test statistic: $z = \dfrac{\bar{x} - 2,400}{\sigma_{\bar{x}}}$

Rejection region: $z > 1.645$, which corresponds to $\alpha = .05$

Note that the **rejection region** refers to the values of the test statistic for which we will **reject the null hypothesis.**

To illustrate the use of the test, suppose we test fifty sections of sewer pipe and find the mean and standard deviation for these fifty measurements to be

$\bar{x} = 2,460$ pounds per lineal foot $s = 200$ pounds per lineal foot

As in the case of estimation, we can use s to approximate σ when s is calculated from a large set of sample measurements.

The test statistic is

$$z = \frac{\bar{x} - 2{,}400}{\sigma_{\bar{x}}} = \frac{\bar{x} - 2{,}400}{\sigma/\sqrt{n}} \approx \frac{\bar{x} - 2{,}400}{s/\sqrt{n}}$$

Substituting $\bar{x} = 2{,}460$, $n = 50$, and $s = 200$, we have

$$z \approx \frac{2{,}460 - 2{,}400}{200/\sqrt{50}} = \frac{60}{28.28} = 2.12$$

Therefore, the sample mean lies $2.12\sigma_{\bar{x}}$ above the hypothesized value of μ, 2,400, as shown in Figure 7.6. Since this value of z exceeds 1.645, it falls in the rejection region. That is, we reject the null hypothesis that $\mu = 2{,}400$ and accept the research hypothesis, $\mu > 2{,}400$. Thus, it appears that the company's pipe has a mean strength that exceeds 2,400 pounds per lineal foot.

Figure 7.6   Location of the Test Statistic for a Test of the Hypothesis H_0: $\mu = 2{,}400$

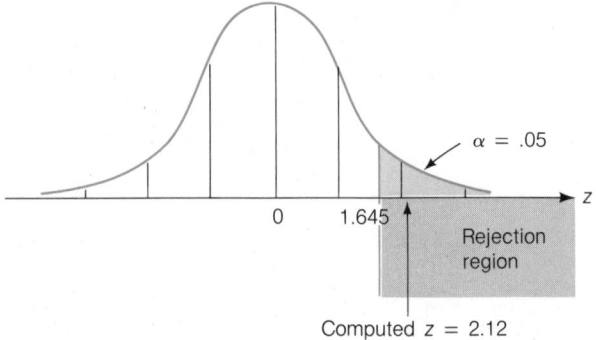

Computed $z = 2.12$

How much faith can be placed in this conclusion? What is the probability that our statistical test could lead us to reject the null hypothesis (and conclude that the company's pipe meets the city's specifications) if in fact the null hypothesis is true? The answer is "$\alpha = .05$." That is, we selected the level of risk, α, of making a Type I error when we constructed the test. Thus, the chances are only 1 in 20 that our test could lead us to conclude the manufacturer's pipe satisfies the city's specifications if in fact this conclusion is false.

Now suppose the sample data had not indicated that the sewer pipe met the city's specifications. What would we have concluded if we had computed a value of $z \le 1.645$? Failure to reject the null hypothesis indicates that the evidence in the sample was not sufficient to support the research hypothesis at the $\alpha = .05$ **level of significance.** Note that we carefully avoid stating that the null hypothesis is true, for then we would be risking a second type of error—concluding the null hypothesis is true (the pipe fails to meet specifications) if in fact the research hypothesis is true (the pipe does meet specifications). We call this a **Type II error.** The probability of committing a Type II error is usually denoted by the symbol β (beta).

For example, if μ is in fact equal to 2,475 pounds per lineal foot instead of the hypothesized 2,400 pounds, the sampling distribution of $\bar{x}$ would be shifted to the right

Figure 7.7 The Probability β of Failing to Reject H_0 If $\mu = 2{,}475$ Pounds

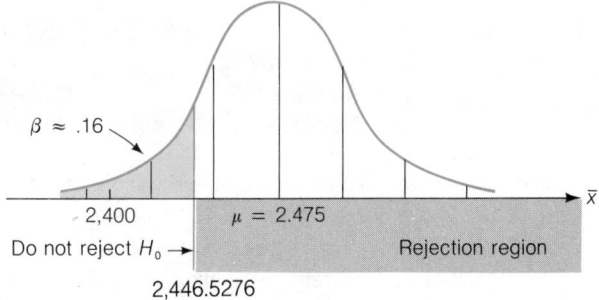

(see Figure 7.7). The probability β that $\bar{x}$ would fall in the "acceptance" region is the area under this sampling distribution (a normal curve) to the left of the value of $\bar{x}$ that locates the boundary of the rejection region. This boundary (see Figure 7.5) is located 1.645 $\sigma_{\bar{x}}$ to the right of the hypothesized mean, i.e., at

$$2{,}400 + 1.645\left(\frac{\sigma}{\sqrt{n}}\right) \approx 2{,}400 + 1.645\left(\frac{s}{\sqrt{n}}\right)$$

$$= 2{,}400 + 1.645\left(\frac{200}{\sqrt{50}}\right)$$

$$= 2{,}446.5276$$

The z-score corresponding to $\bar{x} = 2{,}446.5276$ when $\mu = 2{,}475$ is

$$z = \frac{\bar{x} - \mu}{\sigma_{\bar{x}}} = \frac{2{,}446.5276 - 2{,}475}{28.28} = -1.01$$

You can verify (see Table IV in the Appendix) that the lower tail area under the normal curve to the left of $z = -1.01$ is equal to .1562. Therefore, the probability β of failing to reject the null hypothesis if in fact $\mu = 2{,}475$ is approximately equal to .16. The risk of concluding that the pipe does not meet specifications if it is really 75 pounds stronger than the minimum 2,400 pounds per lineal foot is rather large ($\beta = .16$). This risk increases (see Figure 7.8) if the actual value of μ is closer to the hypothesized value, $\mu = 2{,}400$. The values of β, corresponding to the shaded areas in Figure 7.8, are given for $\mu = 2{,}425$, 2,450, and 2,475.

Figure 7.8 Values of β When $\mu = 2{,}425$, 2,450, and 2,475 Pounds per Lineal Foot

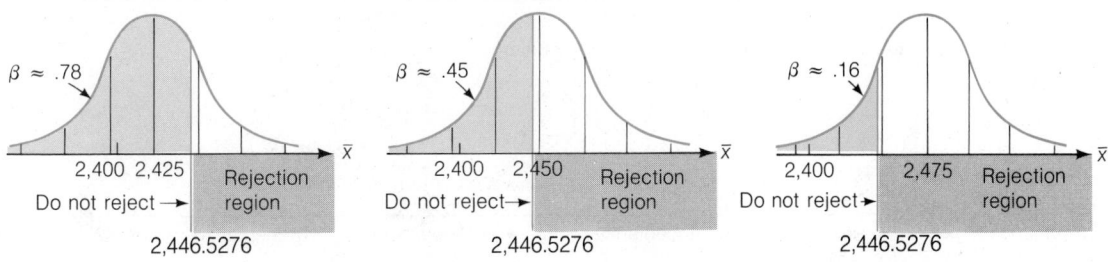

Table 7.2

Conclusions and
Consequences for a Test
of an Hypothesis

| | | TRUE STATE OF NATURE | |
		H_0 *true*	H_a *true*
CONCLUSION	H_0 *true*	Correct decision	Type II error (probability β)
	H_a *true*	Type I error (probability α)	Correct decision

Table 7.2 summarizes the four possible situations that might arise when an hypothesis is tested. The two possible states of nature correspond to the two columns of the table; that is, either H_0 is true or H_a is true. The two rows of the table indicate the two possible conclusions that can be reached; either H_0 is true or H_a is true. The two kinds of decisions are shown in the body of the table. Either the research hypothesis or the null hypothesis can be accepted. Associated with these two actions are two types of risk: the risk of making a Type I error, measured by α, and the risk of making a Type II error, measured by β. Note that a Type I error can be made *only* when the research hypothesis is accepted (which occurs when the null hypothesis is rejected) and a Type II error can be made *only* when the null hypothesis is accepted. Not too surprisingly, the measures of these two types of risk, α and β, are related.

You can see in Figure 7.9 that α is decreased by moving the rejection region farther out into the tail of the sampling distribution. By doing so, the rejection region becomes smaller and the acceptance region becomes larger. What happens to β as the acceptance region becomes larger?

Since β is the probability of accepting H_0 if in fact some alternative value of the parameter is true, β is the probability that the test statistic falls in the acceptance region. And the larger the acceptance region, the larger will be the value of β. So the relationship between α and β is what you might expect intuitively: As you decrease one type of risk (say, the risk α of incorrectly accepting H_a), you increase the other (the risk β of incorrectly accepting H_0). Fortunately, we can reduce both types of risk by increasing the sample size. The more information you have in the sample, the greater will be the ability of the test statistic to reach the correct decision.

In theory, we could consider the probabilities of the two types of risk, α and β, the possible losses attached to the Type I and Type II errors, and choose the rejection region to minimize the expected loss.

In practice, β is difficult to calculate for many tests, and it is impossible to specify a meaningful alternative to the null hypothesis for others. So as an introduction to tests of hypotheses, we suggest the following procedure: Select the null hypothesis as the theory opposing the research hypothesis (the one you want to support). Then if the null hypothesis is true, you will know the probability that the test will lead to an

Figure 7.9   Reducing α Reduces the Rejection Region and Enlarges the Acceptance Region

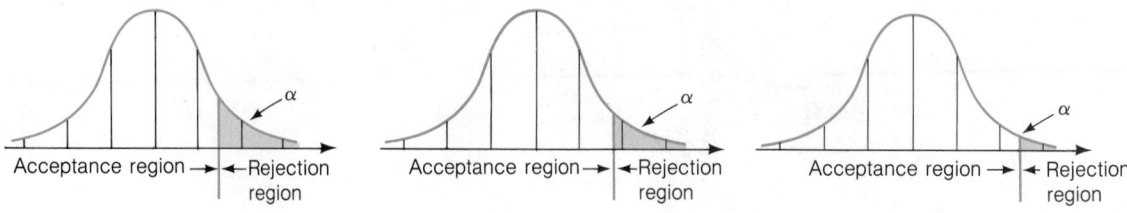

incorrect rejection of H_0. It will be α and you can choose this value as large or small as you wish prior to the selection of your sample. If the test statistic does not fall in the rejection region, *do not* accept the null hypothesis unless you know β. Withhold judgment and seek a larger sample size to lead you closer to a decision. Or estimate the parameter by using a confidence interval. This will give an interval estimate of its true value and give you a measure of the reliability of your inference.

The elements of a test of an hypothesis are summarized in the box.

Elements of a Test of an Hypothesis

1. *Null hypothesis* (H_0): A theory that is phrased in terms of the values of one or more population parameters. The theory is usually one that we wish to disprove.

2. *Alternative* (*research*) *hypothesis* (H_a): A theory that opposes the null hypothesis and that we wish to establish as true.

3. *Test statistic:* A sample statistic used to decide whether to reject the null hypothesis.

4. *Rejection region:* The numerical values of the test statistic for which the null hypothesis will be rejected. The rejection region is chosen so that the probability is α that it will contain the test statistic if the null hypothesis is true (thereby leading to an incorrect conclusion), where α is usually chosen to be small (say, .01, .05, or .10).

5. *Experiment and calculation of test statistic:* The sampling experiment is performed and the numerical value of the test statistic is determined.

6. *Conclusion:*

    a. If the numerical value of the test statistic falls in the rejection region, we conclude that the alternative hypothesis is true (i.e., reject the null hypothesis), and we know that the test procedure will lead to this conclusion incorrectly only $100\alpha\%$ of the time it is used.

    b. If the test statistic does not fall in the rejection region, we reserve judgment about which hypothesis is true. We do not accept the null hypothesis, because we do not (in general) know the probability β that our test procedure will lead us to accept H_0 incorrectly.

Example 7.2 A research psychologist plans to administer a test designed to measure self-confidence to a random sample of fifty professional athletes. The psychologist theorizes that professional athletes tend to be more self-confident than others. Since the national norm of the test is known to be 72, the theory may be partially validated if it can be shown that the mean score for all professional athletes, μ, exceeds 72.

Suppose the sample mean and standard deviation of the fifty scores are

$$\bar{x} = 74.1 \qquad s = 13.3$$

Do these data support the research hypothesis of the psychologist? Use $\alpha = .10$.

Solution   The elements of the test are

$$H_0: \quad \mu = 72 \qquad H_a: \quad \mu > 72$$

Test statistic: $z = \dfrac{\bar{x} - 72}{\sigma_{\bar{x}}} = \dfrac{\bar{x} - 72}{\sigma/\sqrt{n}} \approx \dfrac{\bar{x} - 72}{s/\sqrt{n}}$

Rejection region: $z > 1.28$ (the table value corresponding to $\alpha = .10$)

We now substitute the sample statistics into the test statistic to obtain

$$z \approx \frac{74.1 - 72}{13.3/\sqrt{50}} = 1.12$$

Thus, although the mean score for the sample of athletes exceeds the national norm by more than 2 points, the z-value of 1.12 does not fall in the rejection region (see Figure 7.10). Therefore, this sample does not provide sufficient evidence at the $\alpha = .10$ level to support the psychologist's theory.   ■

Figure 7.10   Location of the Test Statistic for Example 7.2

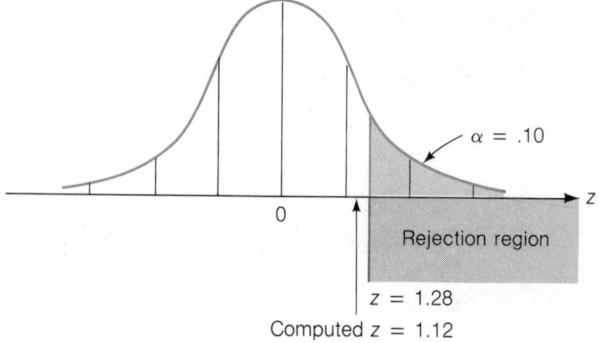

$z = 1.28$
Computed $z = 1.12$

Example 7.3   Calculate the probability β of a Type II error for the test in Example 7.2 if the mean score for professional athletes, μ, actually equals 75. This is the probability of failing to conclude that the professional athletes' mean score exceeds the national norm when, in fact, their mean score is as large as 75.

Solution   The rejection region for the test in Example 7.2 is $z > 1.28$ or, equivalently, values of $\bar{x}$ that lie more than $1.28\sigma_{\bar{x}}$ above the hypothesized mean, $\mu_0 = 72$. This point is shown on the $\bar{x}$-axis below the dotted distribution curve in Figure 7.11. Thus, the rejection region, expressed in terms of $\bar{x}$, is

$$\bar{x} > \mu_0 + 1.28\sigma_{\bar{x}}$$

or

$$\bar{x} > 72 + 1.28\left(\frac{13.3}{\sqrt{50}}\right) = 72 + 1.28(1.88)$$

or

$$\bar{x} > 74.41$$

The probability β of not rejecting H_0: $\mu = 72$ when μ is really equal to 75 is the probability that $\bar{x}$ falls in the nonrejection region, i.e.,

$\bar{x} < 74.41$

This is the shaded area under the normal curve with mean $\mu = 75$ in Figure 7.11. To find this area, we need to find the z-value corresponding to $\bar{x} = 74.41$ for $\mu = 75$, i.e.,

$$z = \frac{\bar{x} - \mu}{\sigma_{\bar{x}}} = \frac{74.41 - 75}{1.88} \approx -.31$$

The tabulated (Table IV in the Appendix) area corresponding to $z = .31$ is $A = .1217$; this gives the area under the normal curve between $\bar{x} = 74.41$ and $\mu = 75$. The area under the normal curve *below* $\bar{x} = 74.41$ is β, the probability that $\bar{x}$ will fall in the nonrejection region. Thus,

$$\beta = .5 - .1217 = .3783$$

Therefore, the probability that the test fails to indicate that the mean score for professional athletes exceeds 72 when in fact their true mean score equals 75 is approximately equal to .38. ∎

Figure 7.11 Calculating β When $\mu = 75$ (Example 7.3)

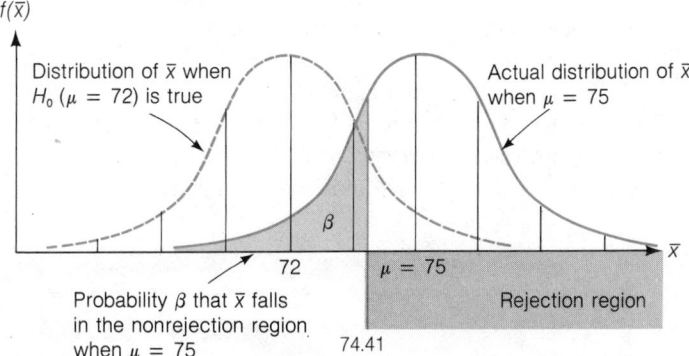

Although the value of the sample mean $\bar{x}$ could be used as a test statistic to test an hypothesis about a population mean μ, we will find it more convenient to use the z-statistic. In fact, for the preceding examples we based the decision either to reject or not to reject the null hypothesis on the computed value of z. For example, saying that you will reject H_0 in Example 7.2 if $\bar{x}$ lies more than $1.28\sigma_{\bar{x}}$ above $\mu = 72$ is the same as saying that you will reject H_0 if z is greater than 1.28.

The research hypothesis for Example 7.2, namely that $\mu > 72$, leads to a **one-tailed (upper tail)** statistical test, because we would reject the null hypothesis only for large values of z (values in the upper tail of the z-distribution). See Figure 7.10. Some statistical investigations seek to show that μ is *either* larger or smaller than some specified value. This type of research (alternative) hypothesis, for example,

H_a: $\mu > 72$ or $\mu < 72$

Figure 7.12 Rejection
Region for a Two-Tailed
Test of the Mean,
H_0: $\mu = \mu_0$

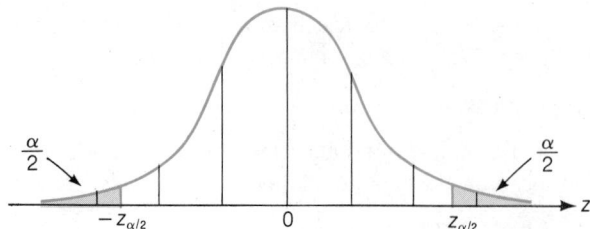

will be supported for large positive or negative values of z. Thus, the rejection region will be located in both tails of the z-distribution, splitting α between the two tails (see Figure 7.12). Such a statistical test is said to be **two-sided** or **two-tailed.** The value of z, denoted by the symbol $z_{\alpha/2}$, that places half of α in the upper tail of the z-distribution can be obtained from the table of areas under the normal curve (Table IV in the Appendix).

Notice that alternative hypotheses are always expressed as inequalities, i.e., you are always attempting to show that μ is larger than some value (upper one-tailed test), smaller than some value (lower one-tailed test), or not equal to some value (a two-tailed test). In contrast, the null hypothesis is always expressed as an equality. To illustrate, in Example 7.2 the alternative hypothesis is H_a: $\mu > 72$. Although the opposite of this alternative is $\mu \leq 72$, the null hypothesis is given as H_0: $\mu = 72$. This is because we need to determine from the sampling distribution of $\bar{x}$ those values that contradict the null hypothesis and support the alternative hypothesis, H_a: $\mu > 72$. Any value of $\bar{x}$ that leads to rejection of $\mu = 72$ in favor of $\mu > 72$ would certainly also lead to rejection of any value of μ less than 72. For this reason, the null hypothesis is given as the equality H_0: $\mu = 72$ rather than the inequality $\mu \leq 72$.

Steps for Selecting the Null and Alternative Hypotheses

1. Specify the hypothesis that you wish to support. Remember this will give a range of possible values for the parameter being tested and will be expressed as an inequality in the alternative hypothesis H_a.

Example: H_a: $\mu > 72$

2. Define the opposite of the alternative hypothesis. This will be the set of all possible values of the parameter that are not contained in H_a.

Example: $\mu \leq 72$

For the null hypothesis, H_0, choose the value of the parameter that is nearest in value to those specified in H_a.

Example: Of the values $\mu \leq 72$, the one nearest in value to those contained in $\mu > 72$ is $\mu = 72$. Thus, the null hypothesis is H_0: $\mu = 72$.

Example 7.4   A nutritionist believes that a 12-ounce box of breakfast cereal should contain an average of 1.2 ounces of bran. The nutritionist measures a random sample of sixty boxes of a popular cereal for bran content. Suppose the data yield

$$\bar{x} = 1.170 \text{ ounces of bran} \qquad s = .111 \text{ ounce of bran}$$

Do the data indicate that the mean bran content of all boxes of this brand of cereal differs from 1.2 ounces? Use $\alpha = .05$.

Solution   We wish to determine whether μ, the mean amount of bran, *differs* from 1.2 ounces; i.e., we wish to detect $\mu > 1.2$ or $\mu < 1.2$ if either of these situations exists. Therefore, we want to conduct a two-tailed test. The elements of the test are

$$H_0: \quad \mu = 1.2 \qquad H_a: \quad \mu \neq 1.2$$

Test statistic: $z = \dfrac{\bar{x} - 1.2}{\sigma_{\bar{x}}}$

Rejection region: $z > 1.96$ or $z < -1.96$ (see Figure 7.13)

Figure 7.13 Two-Tailed Rejection Region: $\alpha = .05$

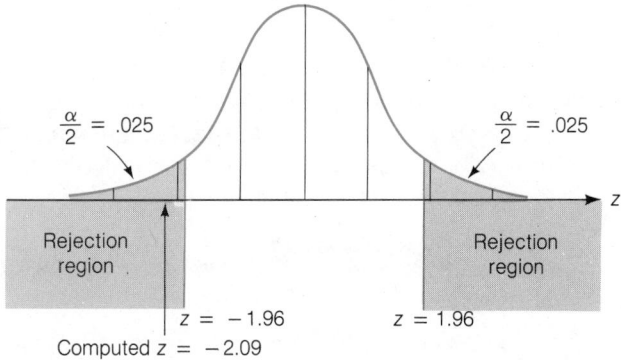

$\dfrac{\alpha}{2} = .025$ $\dfrac{\alpha}{2} = .025$

Rejection region   Rejection region

$z = -1.96$ $z = 1.96$

Computed $z = -2.09$

Note that $z = 1.96$ was chosen for the boundary of the rejection region because $P(z > 1.96) = .025$. This value is obtained from Table IV in the Appendix.

We now calculate

$$z = \frac{\bar{x} - 1.2}{\sigma_{\bar{x}}} = \frac{\bar{x} - 1.2}{\sigma/\sqrt{n}} = \frac{1.170 - 1.2}{\sigma/\sqrt{60}}$$

$$\approx \frac{1.170 - 1.2}{s/\sqrt{60}} = \frac{-.030}{.111/\sqrt{60}} = -2.09$$

You can see in Figure 7.13 that the calculated z-value, -2.09, is in the lower-tail rejection region, and there is evidence to indicate that the mean bran content, μ, differs from 1.2 ounces. It appears that, on average, the cereal boxes contain too little bran (as judged by the nutritionist). How reliable is this conclusion? We know that the test statistic will erroneously reject the null hypothesis only 5% of the time (because $\alpha = .05$).

Therefore, we are reasonably confident that this statistical test has led us to a correct conclusion. ∎

Large-Sample Test of an Hypothesis About μ

One-Tailed Test

H_0: $\mu = \mu_0$*

H_a: $\mu < \mu_0$
(or H_a: $\mu > \mu_0$)

Test statistic: $z = \dfrac{\bar{x} - \mu_0}{\sigma_{\bar{x}}}$

Rejection region: $z < -z_\alpha$
(or $z > z_\alpha$ when H_a: $\mu > \mu_0$)

where z_α is chosen so that
$$P(z > z_\alpha) = \alpha$$

Two-Tailed Test

H_0: $\mu = \mu_0$*

H_a: $\mu \neq \mu_0$

Test statistic: $z = \dfrac{\bar{x} - \mu_0}{\sigma_{\bar{x}}}$

Rejection region: $z < -z_{\alpha/2}$
or $z > z_{\alpha/2}$

where $z_{\alpha/2}$ is chosen so that
$$P(z > z_{\alpha/2}) = \alpha/2$$

As we have indicated by the preceding examples, a large-sample statistical test of an hypothesis concerning a population mean can be either one-tailed or two-tailed, depending on the nature of the research (alternative) hypothesis we wish to support. A summary of the test is given in the preceding box. The two possible conclusions resulting from the sample data are given below:

Possible Conclusions for a Test of an Hypothesis

1. If the calculated z-score falls in the rejection region, conclude that the research hypothesis is true. If this strategy is used repeatedly, then Type I errors are made approximately $100\alpha\%$ of the time if H_0 is true.
2. If the calculated z-score does not fall in the rejection region, state that the data do not provide evidence to support the research hypothesis. (The null hypothesis should not be accepted unless the probability β of making a Type II error is calculated. This is not easy to do for most hypothesis tests.)

Case Study 7.2
Hypothesis Tests in Computer Security Systems

Hackers are able to crack password codes and wander the phone lines, trespassing through data banks. Industrial spies raid secret data files, snatching new product designs. Computer criminals enrich themselves, plundering electronic fund transfer networks. The Defense Department is positive—almost—that its top-security computer systems are invulnerable to penetration by outsiders or, worse, insiders with evil intentions.

---

* The symbol μ_0 represents the numerical value assigned to μ under the null hypothesis.

This paragraph introduces an article by Daniel Kagan in the March 1984 issue of *Omni* on one of the most pressing problems in high-technology industries today: computer security. The movie *War Games* is a fictional account of a young man's successful attempt to crack the security codes of Defense Department computers, gaining access to highly confidential, sensitive, and, as it turns out in the movie, potentially dangerous information. Perhaps more disturbing, however, are the many factual reports of computer thieves obtaining access to bank accounts, government data bases, and university research computers. Several proposed solutions to the growing computer security problem are presented in Kagan's article "Locking Up Data."

The objective of computer security is to allow only authorized personnel access to the computer's data files. This goal is typically achieved by use of a *password*—a collection of symbols (usually letters and numbers) that must be supplied by the user before the computer permits access to the account. The problem is that persistent computer thieves can program their computers to generate millions of combinations of letters and numbers and to enter them into the computer into which access is desired until the correct password is found. Some accounts are doubly protected with an unlisted phone number providing a first level of protection, followed by the password. This measure simply means that gaining illegal access may take longer, since the thief must now wait until his computer cracks both levels of security. The problem is augmented by the more common, less intriguing, but equally damaging illegal access by former employees and friends of current employees who steal the necessary passwords.

Omni reports on several innovative new proposals for solving the computer security problem. One school of thought consists of "a move away from I.D. verification based on what the operator *has* (like a magnetic stripe card) and what he *knows* (like a password)—toward using what the operator *is*. . . . Authorized users are identified by unique body characteristics, 'You can't leave your body at home like a card or key; and no one can steal it,' quips Tom Catto of Palmguard, Inc., in Beaverton, Oregon." Palmguard's security system consists of computer identification of the user's palm before access is permitted. The system tests the hypothesis

H_0:   The proposed user is authorized

versus

H_a: The proposed user is unauthorized

by checking characteristics of the proposed user's palm against those stored in the authorized users' data bank. Palmguard reports that the Type I error rate is less than 1%, while the Type II error rate is .00025%.

Since the Type I error refers to the rejection of H_0 when H_0 is true, the implication is that an authorized user will be denied access by the computer less than 1% of the time. This is primarily an inconvenience, since the authorized user can try to gain access again with reasonable assurance of success. Of more importance is the Type II error rate, which refers to the acceptance of H_0 when H_0 is false, or the probability than an unauthorized user will be granted access to the account. Palmguard's claim implies that a Type II error will occur only 25 times in 10 million.

Another new security system, the EyeDentifyer, "spots authorized computer users by reading the one-of-a-kind patterns formed by the network of minute blood vessels across the retina at the back of the eye." The Type I and II error rates are reported to be .01% (1 in 10,000) and .005% (5 in 100,000), respectively.

As new security systems are developed, the Type I and Type II error rates provide objective standards with which to compare them. Of course, the ultimate security system would have a Type II error rate of zero, but as with tests of statistical hypotheses, this is probably an impractical (and unachievable) goal.

Exercises 7.17–7.35

Learning the Mechanics

7.17 Define each of the following:
a. Null hypothesis   **b.** Alternative hypothesis
c. Test statistic   **d.** Rejection region
e. Type I error   **f.** Type II error
g. α **h.** β

7.18 For each of the following rejection regions, sketch the sampling distribution for z and indicate the location of the rejection region:
a. $z > 1.96$ **b.** $z > 1.645$ **c.** $z > 2.58$ **d.** $z < -1.28$
e. $z < -1.645$ or $z > 1.645$ **f.** $z < -2.58$ or $z > 2.58$

7.19 If the rejection region is defined as in Exercise 7.18, what is the probability that a Type I error will be made in each case?

7.20 A random sample of n observations is taken from a population with unknown mean μ. Give the six elements of a test of hypothesis for each of the following situations:
a. H_0: $\mu = 12$; H_a: $\mu \neq 12$; $n = 70$; $\bar{x} = 13.1$; $s = 6.8$; $\alpha = .05$
b. H_0: $\mu = 27$; H_a: $\mu > 27$; $n = 50$; $\bar{x} = 35$; $s^2 = 502$; $\alpha = .10$
c. H_0: $\mu = .3$; H_a: $\mu < .3$; $n = 40$; $\bar{x} = .28$; $s = .02$; $\alpha = .01$

7.21 Suppose you are interested in conducting the following statistical test:

H_0: $\mu = 200$ H_a: $\mu > 200$

and you have decided to use the following decision rule: "Reject H_0 if the sample mean of a random sample of 100 items is more than 212." Assume that the standard deviation of the population is 80.
a. Express the decision rule in terms of z.
b. Find α, the probability of making a Type I error, by using this decision rule.

7.22 A random sample of n observations is selected from a population with unknown mean μ and variance σ^2. Give the six elements of a test of hypothesis for each of the following situations:
a. H_0: $\mu = 400$; H_a: $\mu > 400$; $n = 200$; $\bar{x} = 430$; $s = 280$; $\alpha = .05$
b. H_0: $\mu = 400$; H_a: $\mu > 400$; $n = 200$; $\bar{x} = 430$; $s = 125$; $\alpha = .01$
c. H_0: $\mu = 9.5$; H_a: $\mu \neq 9.5$; $n = 120$; $\bar{x} = 7.7$; $s^2 = 184.2$; $\alpha = .05$

7.23 A random sample of n observations is selected from a population with unknown mean μ and variance σ^2. Give the six elements of a test of hypothesis for each of the following situations:

a. H_0: $\mu = 0$; H_a: $\mu \neq 0$; $n = 175$; $\bar{x} = 6.5$; $s^2 = 496.3$; $\alpha = .01$
b. H_0: $\mu = 65$; H_a: $\mu < 65$; $n = 86$; $\bar{x} = 63.8$; $s = 10.3$; $\alpha = .10$
c. H_0: $\mu = 65$; H_a: $\mu < 65$; $n = 186$; $\bar{x} = 63.8$; $s = 10.3$; $\alpha = .10$

7.24 A random sample of sixty observations produced the following sums:

$$\sum x = 102.9 \qquad \sum x^2 = 202$$

a. Test the null hypothesis that $\mu = 1.86$ against the alternative hypothesis that $\mu < 1.86$. Use $\alpha = .05$.
b. Test the null hypothesis that $\mu = 1.86$ against the alternative hypothesis that $\mu \neq 1.86$. Use $\alpha = .05$.

7.25 In a test of hypothesis, who or what determines the size of the rejection region?

7.26 If you test an hypothesis and reject the null hypothesis in favor of your research hypothesis, does your test prove that the research hypothesis is correct? Explain.

7.27 When do you risk making a Type I error? A Type II error?

Applying the Concepts

7.28 An automobile manufacturer believes that the mean mileage per gallon of one of its new models exceeds the mean EPA (Environmental Protection Agency) rating of 43 miles per gallon. To gain evidence to support its belief, the manufacturer randomly selected forty of the cars and recorded the miles per gallon for each over a 100-mile course. The mean and standard deviation of the mileages per gallon for the sample of forty cars were $\bar{x} = 43.6$ and $s = 1.3$ miles per gallon.

a. Since the manufacturer wants to show that the mean miles per gallon for the cars exceeds 43, what should you choose for your alternative and null hypotheses?
b. Do the data provide sufficient evidence to support the manufacturer's belief? Use $\alpha = .05$.

7.29 A pain reliever currently being used in a hospital is known to bring relief to patients in a mean time of 3.5 minutes. To compare a new pain reliever with the one currently being used, the new drug is administered to a random sample of fifty patients. The mean time to relief for the sample of patients is 2.8 minutes and the standard deviation is 1.14 minutes. Do the data provide sufficient evidence to conclude that the new drug was effective in reducing the mean time until a patient receives relief from pain? Test using $\alpha = .10$.

7.30 Small increases in the mean level of bills for monthly long-distance telephone calls produce substantial increases in the profits for telephone companies. A telephone company's records indicate that the amounts paid by private customers per month for long-distance telephone calls have a distribution with mean $17.10 and standard deviation $21.21.

a. If a random sample of fifty bills is taken, what is the approximate probability that the sample mean is greater than $20?

b. If a random sample of 100 bills is taken, what is the approximate probability that the sample mean is greater than $20?

c. Suppose that a random sample of 100 customers' bills during a given month produced a sample mean of $22.10 expended for long-distance calls. Do these data suggest that the mean level of billing per private customer for long-distance calls is in excess of $17.10? Test using $\alpha = .10$.

7.31 In 1979, basic cable television service cost an average of $7.37 per month in the United States (*Statistical Abstract of the U.S.: 1981,* p. 565). In March 1983, the Federal Communications Commission (FCC) noted that the cost of basic cable service had risen only about 8% since 1979 and that it cost on average no more than $8.00 per month. A consumer advocacy group doubts the FCC's claim. In order to investigate the claim, the group randomly samples 33 of the more than 4,000 cable systems in the United States and asks each what its basic service charge was in early 1983. The following results were obtained:

$8.53	$8.41	$7.80	$8.20	$8.14	$7.89	$7.92
20.01	8.96	6.50	8.79	7.73	8.15	8.00
7.66	7.76	7.63	8.16	7.50	7.64	6.99
9.83	7.86	7.97	6.96	8.63	7.64	
8.13	8.35	7.83	7.88	7.65	7.75	

a. Specify the null and alternative hypotheses that should be used by the consumer advocacy group in investigating the FCC's claim.

b. With respect to the hypotheses you specified in part a, explain the practical implications of making a Type I error and a Type II error.

c. In conducting this hypothesis test, what is the probability of committing a Type I error?

d. Conduct the hypothesis test you described in part a and interpret the test's results in the context of this exercise. Use $\alpha = .10$.

7.32 The University of Minnesota uses thousands of fluorescent light bulbs each year. The brand of bulb it currently uses has a mean life of 900 hours. A manufacturer claims that its new brand of bulbs, which cost the same as the brand the university currently uses, has a mean life of more than 900 hours. The university has decided to purchase the new brand if, when tested, the test evidence supports the manufacturer's claim at the .05 significance level. Suppose sixty-four bulbs were tested with the following results:

$\bar{x} = 920$ hours $s = 80$ hours

Will the University of Minnesota purchase the new brand of fluorescent bulbs?

7.33 Refer to the strength test for sewer pipe (see Figure 7.5). Find the probability β of a Type II error when $\mu = 2,425$. Compare your answer with the one given in Figure 7.8.

7.34 Refer to the strength test for sewer pipe (see Figure 7.5). Find the probability β of a Type II error when $\mu = 2,450$. Compare your answer with the one given in Figure 7.8.

7.35 A machine is set to produce bolts with a mean length of 1 inch. Bolts that are too long or too short do not meet the customer's specifications and must be rejected. To avoid producing too many rejects, the bolts produced by the machine are sampled from time to time and tested to see whether the machine is still operating properly, i.e., producing bolts with a mean length of 1 inch. Suppose fifty bolts have been sampled, and $\bar{x} = 1.02$ inches and $s = .04$ inch.

a. At the $\alpha = .01$ significance level, does the sample evidence indicate that the machine is producing bolts with a mean length not equal to 1 inch; i.e., is the production process out of control?

b. What is the approximate probability β that the test will fail to reject H_0: $\mu = 1.0$ when, in fact, μ actually equals 1.005 inches?

7.3 Observed Significance Levels: *p*-Values

According to the statistical test procedure described in Section 7.2, the rejection region and, correspondingly, the value of α are selected prior to conducting the test, and the conclusions are stated in terms of rejecting or not rejecting the null hypothesis. A second method of presenting the results of a statistical test is one that reports the extent to which the test statistic disagrees with the null hypothesis and leaves to the reader the task of deciding whether to reject the null hypothesis. This measure of disagreement is called the ***observed significance level*** (or ***p-value***) for the test. For example, the value of the test statistic computed for the sample of $n = 50$ sections of sewer pipe was $z = 2.12$. Since the test is one-tailed, i.e., the alternative (research) hypothesis of interest is H_a: $\mu > 2,400$, values of the test statistic even more contradictory to H_0 than the one observed would be values larger than $z = 2.12$. Therefore, the observed significance level (*p*-value) for this test is

$$p = P(z \geq 2.12)$$

or, equivalently, the area under the standard normal curve to the right of $z = 2.12$ (see Figure 7.14).

The area A in Figure 7.14 is given in Table IV in the Appendix as .4830. Therefore, the upper-tail area corresponding to $z = 2.12$ is

$$p\text{-value} = .5 - .4830 = .0170$$

Figure 7.14 Finding the *p*-Value for an Upper-Tailed Test When $z = 2.12$

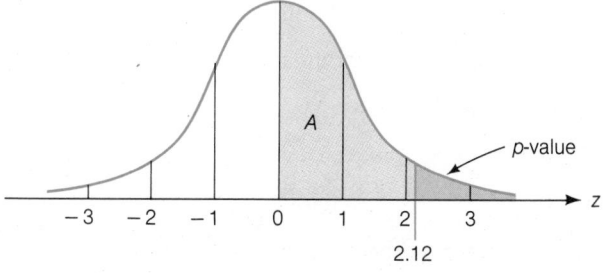

Consequently, we say that these test results are "very significant"; i.e., they disagree rather strongly with the null hypothesis, H_0: $\mu = 2{,}400$, and favor H_a: $\mu > 2{,}400$. The probability of observing a z-value as large as 2.12 is only .0170, if in fact the true value of μ is 2,400.

If you are inclined to select $\alpha = .05$ for this test, then you would reject the null hypothesis because the p-value for the test, .0170, is less than .05. In contrast, if you choose $\alpha = .01$ you would not reject the null hypothesis because the p-value for the test is larger than .01. Thus, the use of the observed significance level is identical to the test procedure described in the preceding sections except that the choice of α is left to you.

Definition 7.3

The **observed significance level**, or **p-value**, for a specific statistical test is the probability (assuming H_0 is true) of observing a value of the test statistic that is at least as contradictory to the null hypothesis, and supportive of the alternative hypothesis, as the one computed from the sample data.

Example 7.5   Find the observed significance level for the test of the mean weight of bran in cereal in Example 7.4.

Solution   Example 7.4 presented a two-tailed test of the hypothesis

H_0: $\mu = 1.2$ ounces

against the alternative hypothesis

H_a: $\mu \neq 1.2$ ounces

The observed value of the test statistic in Example 7.4 was $z = -2.09$, and any value of z less than -2.09 or larger than 2.09 (because this is a two-tailed test) would be even more contradictory to H_0. Therefore, the observed significance level for the test is

p-value $= P(z < -2.09$ or $z > 2.09)$

Consulting Table IV in the Appendix, we find

$P(z > 2.09) = .5 - .4817 = .0183$

Therefore, the p-value for the test is

$2(.0183) = .0366$

These test results would be called significant *in a statistical sense*. Whether the results are significant *in a practical sense* depends on how much the actual mean weight of bran differs from the desired weight of 1.2 ounces and whether the difference is large enough to be of significance from an economic and nutritional point of view.   ∎

When publishing the results of a statistical test of hypothesis in journals, case studies, reports, etc., many researchers make use of p-values. Instead of selecting α beforehand and then conducting a test as outlined in this chapter, the researcher computes and reports the value of the appropriate test statistic and its associated p-value. It is left to the reader of the report to judge the significance of the result, i.e., the reader must determine whether to reject the null hypothesis in favor of the alternative hypothesis, based on the reported p-value. This p-value is often referred to as the **attained significance level** of the test. Usually, the null hypothesis is rejected if the observed significance level is *less* than the fixed significance level, α, chosen by the reader. The inherent advantages of reporting test results in this manner are twofold: (1) Readers are permitted to select the maximum value of α that they would be willing to tolerate if they actually carried out a standard test of hypothesis in the manner outlined in this chapter, and (2) a measure of the degree of significance of the result (i.e., the p-value) is provided.

**Reporting Test Results as p-Values:
How to Decide Whether to Reject H_0**

1. Choose the maximum value of α that you are willing to tolerate.
2. If the observed significance level (p-value) of the test is less than the maximum value of α, reject the null hypothesis.

**Exercises
7.36–7.48**

Learning the Mechanics

7.36 In a test of the hypothesis H_0: $\mu = 50$ versus H_a: $\mu > 50$, a sample of $n = 100$ observations possesses mean and standard deviation $\bar{x} = 50.5$ and $s = 3.3$. Find and interpret the p-value for this test.

7.37 In a test of the hypothesis H_0: $\mu = 10$ versus H_a: $\mu \neq 10$, a sample of $n = 50$ observations possesses mean and standard deviation $\bar{x} = 9.5$ and $s = 2.1$. Find and interpret the p-value for this test.

7.38 Give the observed significance level for each of the following observed values of the z-statistic for testing H_0: $\mu = 21$ against H_a: $\mu > 21$.
a. $z = 1.45$ **b.** $z = 2.17$ **c.** $z = 2.63$
d. $z = 1.29$ **e.** $z = 1.16$ **f.** $z = 1.84$

7.39 Refer to Exercise 7.38. Interpret the observed significance levels that you acquired for parts a and b.

7.40 Give the observed significance level for each of the following observed values of the z-statistic for testing H_0: $\mu = .96$ against H_a: $\mu \neq .96$:
a. $z = 1.62$ **b.** $z = 2.81$ **c.** $z = -1.87$
d. $z = -2.01$ **e.** $z = 1.36$ **f.** $z = -1.55$

7.41 Refer to Exercise 7.40. Interpret the observed significance levels for parts a and f.

7.42 Suppose you were to test H_0: $\mu = 22.5$ against the alternative H_a: $\mu < 22.5$ when $n = 100$, $\bar{x} = 22.2$, and $s = 1.6$. Give the observed significance level for the test and interpret its value.

Applying the Concepts

7.43 *USA Today* reported recently that for the 1983–1984 academic year, 4-year private colleges charged students an average of $4,627 for tuition and fees, while at 4-year public colleges the average was $1,105 ("USA Snapshots," *USA Today*, August 17, 1983, p. 1). Suppose that for 1984–1985 a random sample of thirty colleges yields the following data on tuition and fees: $\bar{x} = \$5,000$ and $s = \$1,643$. Assume that $4,627 is the population mean for 1983–1984.

a. Specify the null and alternative hypotheses you would use to investigate whether the mean amount for tuition and fees in 1984–1985 was significantly larger (in the statistical sense) than it was in 1983–1984.

b. Calculate the *p*-value for the hypothesis test you described in part a, and explain what the *p*-value indicates about the statistical significance of the test results.

c. Explain the difference between statistical significance and practical significance in the context of this exercise.

7.44 According to advertisements, a strain of soybeans planted on soil prepared with a specified fertilizer treatment has a mean yield of 500 bushels per acre. Fifty farmers who belong to a cooperative plant the soybeans, each using a 40-acre plot, and each records the mean yield per acre. The mean and variance for the sample of fifty farms are $\bar{x} = 485$ and $s^2 = 10,045$.

a. Do the data provide sufficient evidence to indicate that the mean yield for the soybeans is different than advertised? Give the observed significance level for the test and interpret its value.

b. In reaching the conclusion in part a, would you have to qualify your conclusions because of the manner in which the sample was selected? Explain.

7.45 Refer to the automobile mileage test in Exercise 7.28. Find the observed significance level for the test and interpret its value.

7.46 A new blood pressure drug is advertised to reduce, after 1 week of medication, a patient's blood pressure an average of 10 units. Blood pressure reductions were recorded for thirty-seven patients after treatment with the drug. The mean and standard deviation for this sample were 8.7 and 6.8, respectively. Do the data appear to contradict the advertising claim? Explain. Find the observed significance level for the test.

7.47 Refer to the sampling of basic television services in Exercise 7.31. Find and interpret the *p*-value for the test.

7.48 Refer to the test of the length of life of the fluorescent light bulbs in Exercise 7.32. Find the observed significance level for this test and interpret its value.

7.4
Small-Sample Inferences About a Population Mean

Federal legislation requires pharmaceutical companies to perform extensive tests on new drugs before they can be marketed. Initially, a new drug is tested on animals. If the drug is deemed safe after this first phase of testing, the pharmaceutical company is then permitted to begin human testing on a limited basis. During this second phase, inferences must be made about the safety of the drug based on information in very small samples.

Suppose a pharmaceutical company must demonstrate that a prescribed dose of a certain new drug will result in an average increase in blood pressure of less than 3 points. Assume that only six patients can be used in the initial phase of human testing. The use of a *small sample* in making an inference about μ presents two immediate problems when we attempt to use the standard normal z as a test statistic.

Problem 1   The shape of the sampling distribution of the sample mean $\bar{x}$ (and the z-statistic) now depends on the shape of the population that is sampled. We can no longer assume that the sampling distribution of $\bar{x}$ is approximately normal, because the central limit theorem only assures normality for samples that are sufficiently large.

Solution   *The sampling distribution of $\bar{x}$ (and z) is exactly normal even for relatively small samples if the sampled population is normal.* It is approximately normal if the sampled population is approximately normal.   ■

Problem 2   Although it is still true that $\sigma_{\bar{x}} = \sigma/\sqrt{n}$, the sample standard deviation provides a poor estimate for σ when the sample size is small. This affects the sampling distribution of the z-statistic when we substitute s for σ.

Solution   Instead of using the statistic

$$z = \frac{\bar{x} - \mu}{\sigma_{\bar{x}}} = \frac{\bar{x} - \mu}{\sigma/\sqrt{n}}$$

which requires knowledge of, or a good approximation to, σ (in order that the sampling distribution of z be normally distributed), we use the statistic

$$t = \frac{\bar{x} - \mu}{s/\sqrt{n}}$$

(which replaces the population standard deviation σ by the sample standard deviation s) and determine its exact sampling distribution.   ■

The distribution of the *t-statistic* in repeated sampling was discovered by W. S. Gosset, a scientist in the Guinness brewery, who published his discovery in 1908 under the pen name of Student. The main result of Gosset's work is that if we are sampling from a normal distribution, the *t*-statistic has a sampling distribution very much like that of the *z*-statistic: mound-shaped, symmetric, with mean zero. The primary difference between the sampling distributions of t and z is that the t-distribution is more variable than the z, which follows intuitively when you realize that t contains two random quantities ($\bar{x}$ and s), while z contains only one ($\bar{x}$).

The actual amount of variability in the sampling distribution of t depends on the sample size n. A convenient way of expressing this dependence is to say that the t-statistic has $(n-1)$ **degrees of freedom.** Recall that the quantity $(n-1)$ is the divisor that appears in the formula for s^2. This number plays a key role in the sampling distribution of s^2 and will appear in discussions of other statistics in later chapters. Particularly, the smaller the number of degrees of freedom associated with the t-statistic, the more variable will be its sampling distribution.

In Figure 7.15 we show both the sampling distribution of z and the sampling distribution of a t-statistic with 4 degrees of freedom (df). You can see that the increased variability of the t-statistic means that the t-value, t_α, that locates an area α in the upper tail of the t-distribution is larger than the corresponding value z_α. For any given value of α, the t-value t_α increases as the number of degrees of freedom (df) decreases. Values of t that will be used in forming small-sample confidence intervals for μ and rejection regions for small-sample tests of hypotheses about μ are given in Table V of the Appendix. A partial reproduction of this table is shown in Figure 7.16.

Figure 7.15 Standard Normal (z) Distribution and t-Distribution with 4 df

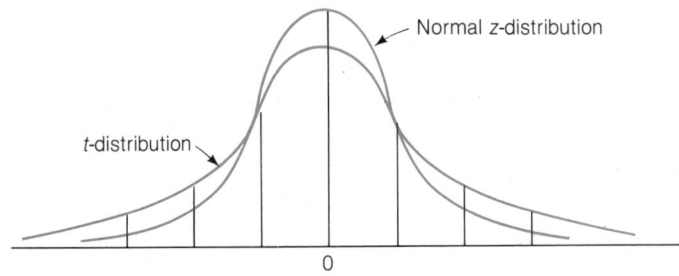

Note that t_α-values are listed for degrees of freedom from 1 to 29, where α refers to the tail area under the t-distribution to the right of t_α. For example, if we want the t-value with an area of .025 to its right and 4 df, we look in the table under the column $t_{.025}$ for the entry in the row corresponding to 4 df. This entry is $t_{.025} = 2.776$, as shown in Figure 7.17. The corresponding standard normal z-score is $z_{.025} = 1.96$.

Note that the last row of Table V, where df = infinity, contains the standard normal z-values. This follows from the fact that as the sample size n grows very large, s becomes closer to σ and thus t becomes closer in distribution to z. In fact, when df = 29, there is little difference between corresponding tabulated values of z and t. Thus, we choose the arbitrary cutoff of $n = 30$ (df = 29) to distinguish between the large-sample and small-sample inferential techniques.

Figure 7.16 Reproduction of Part of Table V in the Appendix

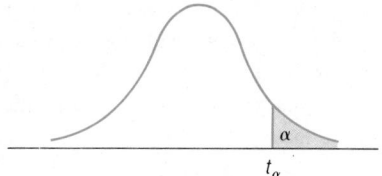

DEGREES OF FREEDOM	$t_{.100}$	$t_{.050}$	$t_{.025}$	$t_{.010}$	$t_{.005}$
1	3.078	6.314	12.706	31.821	63.657
2	1.886	2.920	4.303	6.965	9.925
3	1.638	2.353	3.182	4.541	5.841
4	1.533	2.132	2.776	3.747	4.604
5	1.476	2.015	2.571	3.365	4.032
6	1.440	1.943	2.447	3.143	3.707
7	1.415	1.895	2.365	2.998	3.499
8	1.397	1.860	2.306	2.896	3.355
9	1.383	1.833	2.262	2.821	3.250
10	1.372	1.812	2.228	2.764	3.169
11	1.363	1.796	2.201	2.718	3.106
12	1.356	1.782	2.179	2.681	3.055
13	1.350	1.771	2.160	2.650	3.012
14	1.345	1.761	2.145	2.624	2.977
15	1.341	1.753	2.131	2.602	2.947

Figure 7.17 The $t_{.025}$ Value in a t-Distribution with 4 df and the Corresponding $z_{.025}$ Value

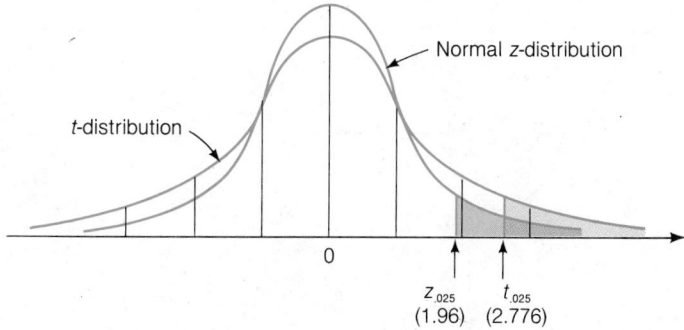

Returning to the example of testing a new drug, suppose that the six test patients have blood pressure increases of 1.7, 3.0, .8, 3.4, 2.7, and 2.1 points. We calculate

$$\bar{x} = \frac{\sum x}{n} = \frac{13.7}{6} = 2.28$$

$$s^2 = \frac{\sum(x - \bar{x})^2}{n - 1} = \frac{\sum x^2 - \frac{(\sum x)^2}{n}}{n - 1} = \frac{35.79 - \frac{(13.7)^2}{6}}{5} = .9017$$

$$s = \sqrt{s^2} = .950$$

We can now use these results to determine whether there is evidence that the new drug satisfies the requirement that the resulting increase in blood pressure averages less than 3 points. This can be accomplished by testing the null hypothesis that the true mean increase, μ, is equal to 3 against the alternative hypothesis that μ is less than 3. The elements of the test are

Null hypothesis H_0: $\mu = 3$

Alternative hypothesis H_a: $\mu < 3$

For the small sample we use the t-statistic.

Test statistic: $t = \dfrac{\bar{x} - \mu_0}{s/\sqrt{n}} = \dfrac{\bar{x} - 3}{s/\sqrt{n}}$

Assumption: The relative frequency distribution of the population of blood pressure increases associated with patients taking the drug is approximately normal.

If we want to test at the $\alpha = .05$ level, the rejection region is

Rejection region: $t < -t_{.05} = -2.015$, where df (degrees of freedom) is equal to $n - 1 = 5$

The rejection region is shown in Figure 7.18.

Figure 7.18   A t-Distribution with 5 df and the Rejection Region for the Drug Testing Example

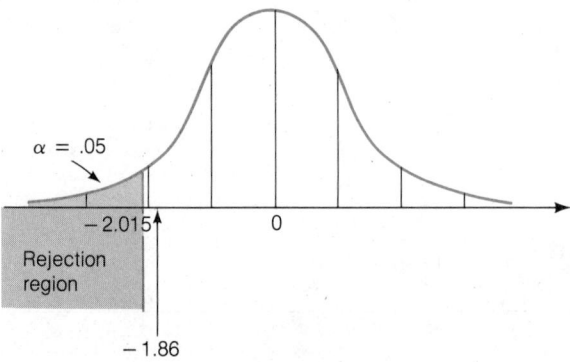

We now calculate

$$t = \frac{\bar{x} - 3}{s/\sqrt{n}} = \frac{2.28 - 3}{.95/\sqrt{6}} = -1.86$$

Since the value $t = -1.86$ that is calculated from the sample is not less than the tabulated value -2.015, we cannot conclude that the mean increase in blood pressure resulting from taking the drug is less than 3 points. The pharmaceutical company may decide to perform further tests on animals or to modify the composition of the drug before further testing.

It is interesting to note that the calculated t-value, -1.86, is *less than* the .05 level z-value, -1.645. The implication is that if we had *incorrectly* used a z-statistic for this test, we would have rejected the null hypothesis at the .05 level, concluding that the mean increase in blood pressure is less than 3. The important point is that the statistical procedure to be used must always be closely scrutinized and all the assumptions understood. Many statistical lies are the result of misapplications of otherwise valid procedures.

Small-Sample Test of an Hypothesis About μ

One-Tailed Test

$H_0: \quad \mu = \mu_0$

$H_a: \quad \mu < \mu_0$
    (or $H_a: \quad \mu > \mu_0$)

Test statistic: $\quad t = \dfrac{\bar{x} - \mu_0}{s/\sqrt{n}}$

Rejection region: $\quad t < -t_\alpha$
    (or $\quad t > t_\alpha$
    when $\quad H_a: \quad \mu > \mu_0$)

Two-Tailed Test

$H_0: \quad \mu = \mu_0$

$H_a: \quad \mu \neq \mu_0$

Test statistic: $\quad t = \dfrac{\bar{x} - \mu_0}{s/\sqrt{n}}$

Rejection region: $\quad t < -t_{\alpha/2}$
    or $\quad t > t_{\alpha/2}$

where t_α and $t_{\alpha/2}$ are based on $(n - 1)$ degrees of freedom

Assumption: A random sample is selected from a population with a relative frequency distribution that is approximately normal.

Example 7.6

A major car manufacturer wants to test a new engine to see whether it meets new air pollution standards. The mean emission μ of all engines of this type must be less than 20 parts per million of carbon. Ten engines are manufactured for testing purposes, and the mean and standard deviation of the emissions for this sample of engines are determined to be

$\bar{x} = 17.1$ parts per million $\qquad s = 3.0$ parts per million

Do the data supply sufficient evidence to allow the manufacturer to conclude that this type of engine meets the pollution standard? Assume that the manufacturer is willing to risk a Type I error with probability $\alpha = .01$.

Solution

The manufacturer wants to support the research hypothesis that the mean emission level μ for all engines of this type is less than 20 parts per million. The elements of this small-sample one-tailed test are

$H_0: \quad \mu = 20 \qquad H_a: \quad \mu < 20$

Test statistic: $\quad t = \dfrac{\bar{x} - 20}{s/\sqrt{n}}$

Figure 7.19   A
t-Distribution with 9 df and
the Rejection Region for
Example 7.6

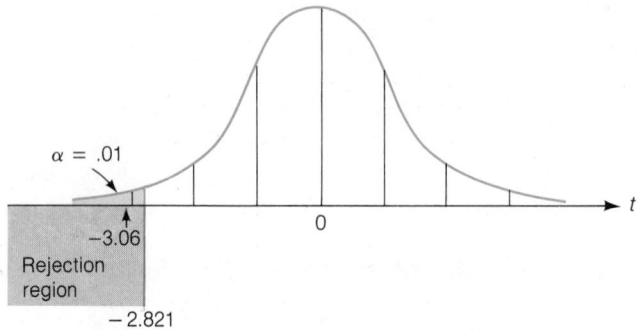

Assumption: The relative frequency distribution of the population of emission
level for all engines of this type is approximately normal.

Rejection region: For $\alpha = .01$ and df $= n - 1 = 9$, the one-tailed rejection
region (see Figure 7.19) is $t < -t_{.01} = -2.821$.

We now calculate the test statistic:

$$t = \frac{\bar{x} - 20}{s/\sqrt{n}} = \frac{17.1 - 20}{3.0/\sqrt{10}} = -3.06$$

Since the calculated *t* falls in the rejection region (see Figure 7.19), the manufacturer
concludes that $\mu < 20$ parts per million and the new engine type meets the pollution
standard. Are you satisfied with the reliability associated with this inference? The probability is only $\alpha = .01$ that the test would support the research hypothesis if in fact it
were false.   ■

Example 7.7   Find the observed significance level for the test described in Example 7.6.

Solution   The test of Example 7.6 was a lower-tail test: H_0: $\mu = 20$ versus H_a: $\mu < 20$. Since
the value of *t* computed from the sample data was $t = -3.06$, the observed significance
level (or *p*-value) for the test is equal to the probability that *t* would assume a value
less than or equal to -3.06 if in fact H_0 were true. This is equal to the area in the
lower tail of the *t*-distribution (shaded in Figure 7.20). To find this area, i.e., the *p*-value

Figure 7.20   The
Observed Significance
Level for the Test,
Example 7.7

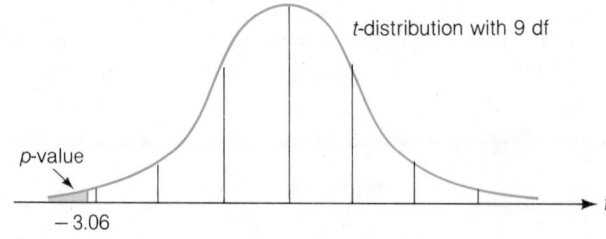

for the test, we consult the t-table (Table V in the Appendix). Unlike the table of areas under the normal curve, Table V gives only the t-values corresponding to the areas .100, .050, .025, .010, and .005. Therefore, we can only approximate the p-value for the test. Since the observed t-value was based on 9 degrees of freedom, we use the df $= 9$ row in Table V and move across the row until we reach the t-values that are closest to the observed $t = -3.06$. [*Note:* We ignore the minus sign.] The t-values corresponding to p-values of .010 and .005 are 2.821 and 3.250, respectively. Since the observed t-value falls between $t_{.010}$ and $t_{.005}$, the p-value for the test lies between .005 and .010. We could interpolate to locate the p-value for the test more accurately, but it is easier and adequate for our purposes to choose the larger area as the p-value and report it as .010. Thus, we would reject the null hypothesis, H_0: $\mu = 20$ parts per million, for any value of α larger than .01. ∎

We may also use the t-distribution to form a small-sample confidence interval for a population mean μ—*if the population is approximately normally distributed.* Recall that the large-sample confidence interval for μ is

$$\bar{x} \pm z_{\alpha/2}\sigma_{\bar{x}} = \bar{x} \pm z_{\alpha/2}\left(\frac{\sigma}{\sqrt{n}}\right)$$

where $100(1 - \alpha)\%$ is the desired confidence level. To form the small-sample confidence interval, replace σ by s and $z_{\alpha/2}$ by $t_{\alpha/2}$, where the number of degrees of freedom for the tabulated t-value is $(n - 1)$.

Small-Sample Confidence Interval for μ

$$\bar{x} \pm t_{\alpha/2}\left(\frac{s}{\sqrt{n}}\right)$$

where $t_{\alpha/2}$ is based on $(n - 1)$ degrees of freedom

Assumption: A random sample is selected from a population with a relative frequency distribution that is approximately normal.

Example 7.8 When food prices began their rapid increase in the early 1970s, some of the major television networks began periodically to purchase a grocery basket full of food at supermarkets around the country. They always bought the same items at each store so they could compare food prices. Suppose you want to estimate the mean price for a grocery basket in a specific geographical region of the country. You purchase the specified items at a random sample of twenty supermarkets in the region. The mean and standard deviation of the costs at the twenty supermarkets are

$$\bar{x} = \$26.84 \qquad s = \$2.63$$

Form a 95% confidence interval for the mean cost μ of a grocery basket for this region.

Solution   If we assume that the distribution of costs for the grocery basket at all supermarkets in the region is approximately normal, we can use the t-statistic to form the confidence interval. For a confidence level of 95%, we need the tabulated value of t with df $= n - 1 = 19$:

$$t_{\alpha/2} = t_{.025} = 2.093$$

Then the confidence interval is

$$\bar{x} \pm t_{.025}\left(\frac{s}{\sqrt{n}}\right) = 26.84 \pm 2.093\left(\frac{2.63}{\sqrt{20}}\right)$$

$$= 26.84 \pm 1.23 = (25.61, 28.07)$$

Thus, we are reasonably confident that the interval from \$25.61 to \$28.07 contains the true mean cost μ of the grocery basket. This is because, if we were to employ our interval estimator on repeated occasions, 95% of the intervals constructed would contain μ. ■

We have emphasized throughout this section that an assumption that the population is approximately normally distributed is necessary for making small-sample inferences about μ when using the t-statistic. While many phenomena do have approximately normal distributions, it is also true that many random phenomena have distributions that are not normal or even mound-shaped. Empirical evidence acquired over the years has shown that the t-distribution is rather insensitive to moderate departures from normality. That is, use of the t-statistic when sampling from mound-shaped populations generally produces credible results; however, for cases in which the distribution is distinctly nonnormal, either take a larger sample or use a **nonparametric method** (the topic of Chapter 10).

What Do You Do When the Population Relative Frequency Distribution Departs Greatly from Normality?

Answer: Use the nonparametric statistical methods of Chapter 10.

Exercises 7.49–7.69

Learning the Mechanics

7.49 Let t_0 be a specific value of t. Use Table V in the Appendix to find t_0-values such that the following statements are true:

a. $P(t \geq t_0) = .025$, where df $= 8$ **b.** $P(t \geq t_0) = .01$, where df $= 10$
c. $P(t \leq t_0) = .005$, where df $= 17$ **d.** $P(t \leq t_0) = .05$, where df $= 14$

7.50 A random sample of n observations is selected from a normal population to test the null hypothesis that $\mu = 10$. Specify the rejection region for each of the following combinations of H_a, α, and n:

a. H_a: $\mu \neq 10$; $\alpha = .05$; $n = 15$ **b.** H_a: $\mu > 10$; $\alpha = .01$; $n = 20$
c. H_a: $\mu > 10$; $\alpha = .10$; $n = 8$ **d.** H_a: $\mu < 10$; $\alpha = .01$; $n = 17$
e. H_a: $\mu \neq 10$; $\alpha = .10$; $n = 22$ **f.** H_a: $\mu < 10$; $\alpha = .05$; $n = 6$

7.51 A random sample of n observations is selected from a normal population with mean μ. For each of the following combinations of sample size and confidence level, specify the value of t needed to form a confidence interval for μ.

a. $n = 18$, confidence level of 90%   **b.** $n = 5$, confidence level of 99%
c. $n = 14$, confidence level of 99%   **d.** $n = 21$, confidence level of 95%
e. $n = 21$, confidence level of 90%

7.52 A random sample of six measurements from a normally distributed population yielded $\bar{x} = 4.9$ and $s = 1.4$.

a. Test the null hypothesis that the mean of the population is 6 against the alternative hypothesis, $\mu < 6$. Use $\alpha = .10$.
b. Test the null hypothesis that the mean of the population is 6 against the alternative hypothesis, $\mu \neq 6$. Use $\alpha = .10$.
c. Form a 95% confidence interval tor μ.
d. Form a 90% confidence interval for μ.

7.53 The following sample of five measurements was randomly selected from a normally distributed population: 4, 7, 3, 4, 6

a. Test the null hypothesis that the mean of the population is 6 against the alternative hypothesis, $\mu < 6$. Use $\alpha = .05$.
b. Test the null hypothesis that the mean of the population is 6 against the alternative hypothesis, $\mu \neq 6$. Use $\alpha = .05$.
c. Find the observed significance level for each test.
d. Find a 95% confidence interval for μ.

7.54 The following sample of six measurements was randomly selected from a normally distributed population: 1, 3, -1, 5, 1, 2

a. Test the null hypothesis that the mean of the population is 3 against the alternative hypothesis, $\mu < 3$. Use $\alpha = .05$.
b. Test the null hypothesis that the mean of the population is 3 against the alternative hypothesis, $\mu \neq 3$. Use $\alpha = .05$.
c. Find the observed significance level for each test.
d. Find a 95% confidence interval for μ.

7.55 A random sample of six measurements from a normally distributed population produced the following observations: 4, 6, 2, 7, 6, 5

a. Test the null hypothesis that $\mu = 2.4$ against the alternative hypothesis that $\mu > 2.4$. Use $\alpha = .01$.
b. Test the null hypothesis that $\mu = 2.4$ against the alternative hypothesis that $\mu \neq 2.4$. Use $\alpha = .01$.
c. Test the null hypothesis that $\mu = 2.4$ against the alternative hypothesis that $\mu \neq 2.4$. Use $\alpha = .05$.

d. Find a 95% confidence interval for μ.

e. Find a 99% confidence interval for μ.

Applying the Concepts

7.56 Pulse rate is an important measure of the fitness of a person's cardiovascular system. The mean pulse rate for American adult males is approximately 72 heart beats per minute. A random sample of twenty-one American adult males who jog at least 15 miles per week had a mean pulse rate of 52.6 beats per minute and a standard deviation of 3.22 beats per minute.

a. Find a 95% confidence interval for the mean pulse rate of all American adult males who jog at least 15 miles per week.

b. Interpret the interval found in part a.

c. What assumptions are required for the validity of the confidence interval?

7.57 A consumer protection group is concerned that a catsup manufacturer is filling its 20-ounce family-size containers with less than 20 ounces of catsup. The group purchases ten family-size bottles of this catsup, weighs the contents of each, and finds that the mean weight is equal to 19.86 ounces and the standard deviation is equal to .22 ounce.

a. Do the data provide sufficient evidence for the consumer group to conclude that the mean fill per family-size bottle is less than 20 ounces? Test using $\alpha = .05$.

b. If the test in part a were conducted on a periodic basis by the company's quality control department, is the consumer group more concerned about making a Type I error or a Type II error? (The probability of making this type of error is called the *consumer's risk*.)

c. The catsup company is also interested in the mean amount of catsup per bottle. It does not wish to overfill them. For the test conducted in part a, which type of error is more serious from the company's point of view—a Type I error or a Type II error? (The probability of making this type of error is called the *producer's risk*.)

7.58 Refer to Exercise 7.57. Find a 90% confidence interval for the mean number of ounces of catsup being dispensed.

7.59 Many investment advisory services provide individual investors with forecasts of the movement of the stock market and indicate stocks they believe will move upward or downward in the coming months. One advisory service lists one stock each week as a prime prospect for price appreciation. From their list of 1983 best prospects for price appreciation, we have randomly selected ten stocks and show the percentage change in price as of a date early in 1984. The data are shown in the table on page 307.

a. Suppose that the percentage changes in the ten stocks can be viewed as a random sample of changes in the stocks selected by the advisory service as prime prospects for investing. Do the data provide sufficient evidence to indicate that the mean percentage change in price in prime prospect stocks is positive? Test using $\alpha = .05$.

b. Find a 90% confidence interval for the mean percentage appreciation (or depreciation) in stock price for the advisory service's 1983 recommendations.

STOCK	PERCENTAGE CHANGE IN PRICE (1983–1984)
1	+16.2
2	−43.5
3	−18.8
4	−19.7
5	−37.0
6	−24.3
7	−45.8
8	+16.2
9	+4.8
10	−16.1

7.60 One of the most feared predators in the ocean is the great white shark. Although it is known that the white shark grows to a mean length of 21 feet, a marine biologist believes that the great white sharks off the Bermuda coast grow much longer due to unusual feeding habits. To test this claim, a number of full-grown great white sharks are captured off the Bermuda coast, measured, and then set free. However, because the capture of sharks is difficult, costly, and very dangerous, only three are sampled. Their lengths are 24, 20, and 22 feet.

a. Do the data provide sufficient evidence to support the marine biologist's claim? Use $\alpha = .10$.

b. Give the approximate observed significance level for the test in part a, and interpret its value.

c. What assumptions must be made in order to carry out the test?

d. Do you think these assumptions are likely to be satisfied in this sampling situation?

7.61 In an article titled "Huge Phone Bills Look Like Mobster Fraud," the *Orlando Sentinel* (March 15, 1984) comments on the rash of exorbitant telephone bills received by some AT&T customers in early 1984. Unexplained huge bills received during this brief period of time (often by individuals) possessed the following dollar values: $109,500, $61,180, $125,883, $35,236, $26,337, $93,315, and $36,063. Suppose these bills represent a random sample of the sizes of the thefts of telephone services that AT&T might expect in the future. Use the data to obtain an estimate of the mean size of a theft in the future. Use a 90% confidence interval.

7.62 A company purchases large quantities of naphtha in 50-gallon drums. Because the purchases are ongoing, small shortages in the drums can represent a sizable loss to the company. The weights of the drums vary slightly from drum to drum, so the weight of the naphtha is determined by removing it from the drums and measuring it. Suppose the company samples the contents of twenty drums, measures the naphtha in each, and calculates $\bar{x} = 49.70$ gallons and $s = .32$ gallon. Do the sample statistics provide sufficient evidence to indicate that the mean fill per 50-gallon drum is less than 50 gallons? Use $\alpha = .10$.

7.63 Refer to Exercise 7.62. Find a 90% confidence interval for the mean number of gallons of naphtha per drum.

7.64 An important problem facing strawberry growers is the control of nematodes. These organisms compete with the plants for nutrients in the soil, thereby reducing yield. For this reason, fumigation is normally a part of field preparation. In the past, the fumigants used yielded an average of 8 pounds of marketable fruit for a certain standard sized plot. Recently, a new fumigant has been developed. It is applied to six standard plots of strawberries, and the yield of marketable fruit (in pounds) for each plot is 9, 9, 13, 9, 10, 8.

a. Do the data indicate a significant increase in average yield at the .05 level of significance?

b. What assumptions are necessary for the procedure used to be valid?

7.65 A psychologist was interested in knowing whether male heroin addicts' assessments of self-worth differ from those of the general male population. On a test designed to measure assessment of self-worth, the mean score for males from the general population is 48.6. A random sample of twenty-five scores achieved by heroin addicts yielded a mean of 44.1 and a standard deviation of 6.2.

a. Do the data indicate a difference in assessment of self-worth between male heroin addicts and the general male population? Test using $\alpha = .01$.

b. Give the approximate observed significance level for the test and interpret its value.

7.66 In Exercise 4.44, we reported on research by Frans Van de Werf, M.D., and colleagues (*New England Journal of Medicine,* March 8, 1984) on a new drug, *t*-PA, which may prove to be effective in dissolving blood clots in heart attack patients. One aspect of their research involved measuring the length x of time for a heart attack patient's blood clot to be dissolved after treatment with *t*-PA. These times, recorded for $n = 7$ patients, were 50, 75, 0, 33, 57, 35, and 19 minutes.

a. Assume that the length x of time until a blood clot dissolves is normally distributed. Find a 90% confidence interval for μ, the mean length of time for a blood clot to be dissolved after treatment with *t*-PA.

b. Explain why the distribution of x might not be normally distributed.

c. If the distribution of x is not normally distributed, what effect would this information have on your confidence interval in part a?

7.67 The application of adrenaline is the prevailing treatment to reduce eye pressure in glaucoma patients. Theoretically, a new synthetic drug will cause the same mean drop in pressure (5.5 units) without the side effects caused by adrenaline. The new drug is given to five glaucoma patients and the reductions in eye pressure for the patients are 4.0, 3.8, 5.7, 5.3, and 4.6 units.

a. Look at the data. Using your intuition, do you think that the mean reduction in pressure for the new drug differs from that produced by adrenaline?

b. Now use a statistical test to answer the question in part a. Do the data provide sufficient evidence to indicate that the mean reduction in eye pressure due to the new drug is different from that produced by adrenaline? Test using $\alpha = .05$.

c. Give the approximate observed significance level for the test and interpret its value.

7.68 A problem that occurs with certain types of mining is that some byproducts tend to be mildly radioactive and these products sometimes get into our freshwater supply.

The EPA has issued regulations concerning a limit on the amount of radioactivity in supplies of drinking water. Particularly, the maximum level for naturally occurring radiation is 5 picocuries per liter of water. A random sample of twenty-four water specimens from a city's water supply produced the sample statistics $\bar{x} = 4.61$ picocuries per liter and $s = .87$ picocurie per liter.

a. Do these data provide sufficient evidence to indicate that the mean level of radiation is safe (below the maximum level set by the EPA)? Test using $\alpha = .01$.

b. Why should you want to use a small value of α for the test in part a?

7.69 What is an MBA degree really worth? *Forbes* (December 19, 1983) gives the median starting salaries for MBA's from fourteen business schools for spring 1983. Five of these median salaries, randomly selected from the group, are shown in the table. Suppose that the data could be viewed as a random sample of the median starting salaries in 1983 for all MBA programs. Estimate the mean of the population of median MBA starting salaries using a 95% confidence interval. Interpret your confidence interval.

SCHOOL	MEDIAN STARTING SALARY (SPRING 1983)
Dartmouth	$35,000
Northwestern	$31,100[a]
Columbia	$34,534
NYU	$30,600
Virginia	$33,000

[a] Estimate.

7.5 Large-Sample Inferences About a Binomial Population Proportion

In recent years the number of public opinion polls has grown at an astounding rate. Almost daily, the news media report the results of some poll. Pollsters regularly determine the percentage of people in favor of the president's energy program, the fraction of voters in favor of a certain candidate, the fraction of customers who favor a particular brand of wine, and the proportion of people who smoke cigarettes. In each case, we are interested in estimating the percentage (or proportion) of some group with a certain characteristic. In this section we will consider methods for making inferences about population proportions.

Example 7.9

The mid-1970s may well be remembered for political unrest across the country. Since the days of Watergate, public opinion polls have been conducted to estimate the fraction of Americans who trust the president. Suppose that 1,000 people are randomly chosen and 637 answer that they trust the president. How would you estimate the true fraction of *all* American people who trust the president?

Solution

What we have really asked is how would you estimate the probability p of success in a binomial experiment, where p is the probability that a person chosen trusts the president. One logical method of estimating p for the population is to use the proportion of successes in the sample. That is, we can estimate p by calculating

$$\hat{p} = \frac{\text{Number of people sampled who trust the president}}{\text{Number of people sampled}}$$

where $\hat{p}$ is read "p hat." Thus, in this case,

$$\hat{p} = \frac{637}{1,000} = .637$$ ∎

To determine the reliability of the estimator $\hat{p}$, we need to know its sampling distribution. That is, if we were to draw samples of 1,000 people over and over again, each time calculating a new estimate $\hat{p}$, what would the frequency distribution of all the $\hat{p}$'s be? The answer lies in viewing $\hat{p}$ as the average or mean number of successes per trial over the n trials. If each success is assigned a value equal to 1 and a failure is assigned a value of zero, then the sum of all n sample observations is x, the total number of successes, and $\hat{p} = x/n$ is the average or mean number of successes per trial in the n trials. The central limit theorem tells us that the relative frequency distribution of the sample mean for any population is approximately normal for large samples. Therefore, the sampling distribution of $\hat{p}$ has the characteristics listed in the accompanying box and shown in Figure 7.21.

Sampling Distribution of $\hat{p}$

1. The mean of the sampling distribution of $\hat{p}$ is p; that is, $\hat{p}$ is an unbiased estimator of p.
2. The standard deviation of the sampling distribution of $\hat{p}$ is $\sqrt{pq/n}$; that is, $\sigma_{\hat{p}} = \sqrt{pq/n}$, where $q = 1 - p$.
3. For large samples, the sampling distribution of $\hat{p}$ is approximately normal. A sample size is considered large if the interval $\hat{p} \pm 3\sigma_{\hat{p}}$ does not include 0 or 1. [*Note:* Usually p is unknown. You will have to approximate its value in order to apply this criterion.]

The fact that the sampling distribution of $\hat{p}$ is approximately normal for large samples allows us to form confidence intervals and test hypotheses about p in a manner which is completely analogous to that used for large-sample inferences about μ.

Figure 7.21 Sampling Distribution of $\hat{p}$

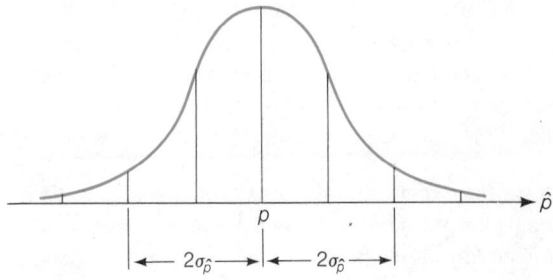

Thus, if 637 of 1,000 Americans say they trust the president, a 95% confidence interval for the proportion of *all* Americans who trust the president is

$$\hat{p} \pm z_{\alpha/2}\sigma_{\hat{p}} = .637 \pm 1.96 \sqrt{\frac{pq}{1,000}}$$

Table 7.3

Values of pq for Several Different Values of p

p	pq	$\sqrt{pq}$
.5	.25	.50
.6 or .4	.24	.49
.7 or .3	.21	.46
.8 or .2	.16	.40
.9 or .1	.09	.30

where $q = 1 - p$. Just as we needed an approximation for σ in calculating a large-sample confidence interval for μ, we now need an approximation for p. As Table 7.3 shows, the approximation for p does not have to be especially accurate, because the value of $\sqrt{pq}$ needed for the confidence interval is relatively insensitive to changes in p. Therefore, we can use $\hat{p}$ to approximate p. Keeping in mind that $\hat{q} = 1 - \hat{p}$, we substitute these values into the formula for the confidence interval:

$$\hat{p} \pm 1.96\sqrt{pq/1,000} \approx \hat{p} \pm 1.96\sqrt{\hat{p}\hat{q}/1,000}$$
$$= .637 \pm 1.96\sqrt{(.637)(.363)/1,000} = .637 \pm .030$$
$$= (.607, .667)$$

Then we can be 95% confident that the interval from 60.7% to 66.7% contains the true percentage of *all* Americans who trust the president. That is, in repeated construction of confidence intervals, approximately 95% of all samples would produce confidence intervals that enclose p.

The general form of a large-sample confidence interval for p is shown in the box.

Large-Sample Confidence Interval for *p*

$$\hat{p} \pm z_{\alpha/2}\sigma_{\hat{p}} \approx \hat{p} \pm z_{\alpha/2}\sqrt{\hat{p}\hat{q}/n} \quad \text{where } \hat{q} = 1 - \hat{p}$$

Tests of hypotheses concerning p are also analogous to those for population means (large samples).

Large-Sample Test of an Hypothesis About *p*

One-Tailed Test

H_0: $p = p_0$ (p_0 = hypothesized value of p)

H_a: $p < p_0$
(or H_a: $p > p_0$)

Test statistic: $z = \dfrac{\hat{p} - p_0}{\sigma_{\hat{p}}}$

Two-Tailed Test

H_0: $p = p_0$

H_a: $p \neq p_0$

Test statistic: $z = \dfrac{\hat{p} - p_0}{\sigma_{\hat{p}}}$

where, according to H_0, $\sigma_{\hat{p}} = \sqrt{p_0 q_0/n}$ and $q_0 = 1 - p_0$

Rejection region: $z < -z_\alpha$
(or $z > z_\alpha$
when H_a: $p > p_0$)

Rejection region: $z < -z_{\alpha/2}$
or $z > z_{\alpha/2}$

Example 7.10

The reputations (and hence sales) of many businesses can be severely damaged by shipments of manufactured items that contain an unusually large percentage of defectives. For example, a manufacturer of flashbulbs for cameras may want to be reasonably certain that fewer than 5% of its bulbs are defective. Suppose 300 bulbs are randomly selected from a very large shipment, each is tested, and ten defective bulbs are found. Does this provide sufficient evidence for the manufacturer to conclude that the fraction defective in the entire shipment is less than .05? Use $\alpha = .01$.

Solution

The objective of the sampling is to determine whether there is sufficient evidence to indicate that the fraction defective, p, is less than .05. Consequently, we will test the null hypothesis that $p = .05$ against the alternative hypothesis that $p < .05$. The elements of the test are

$$H_0: \quad p = .05 \qquad H_a: \quad p < .05$$

Test statistic: $\quad z = \dfrac{\hat{p} - .05}{\sigma_{\hat{p}}}$

Rejection region: $\quad z < -z_{.01} = -2.33 \quad$ (see Figure 7.22)

We now calculate the test statistic:

$$z = \frac{\hat{p} - .05}{\sigma_{\hat{p}}} = \frac{(10/300) - .05}{\sqrt{p_0 q_0 / n}} = \frac{.033 - .05}{\sqrt{p_0 q_0 / 300}}$$

Notice that we use p_0 to calculate $\sigma_{\hat{p}}$ because, in contrast to calculating $\sigma_{\hat{p}}$ for a confidence interval, the test statistic is computed on the assumption that the null hypothesis is true; that is, $p = p_0$. Therefore, substituting the values for $\hat{p}$ and p_0 into the z-statistic, we obtain

$$z \approx \frac{-.017}{\sqrt{(.05)(.95)/300}} = \frac{-.017}{.0126} = -1.35$$

As shown in Figure 7.22, the calculated z-value does not fall in the rejection region. Therefore, based on this test, there is insufficient evidence to indicate that the shipment contains fewer than 5% defective bulbs. ◼

Figure 7.22   Rejection Region for Example 7.10

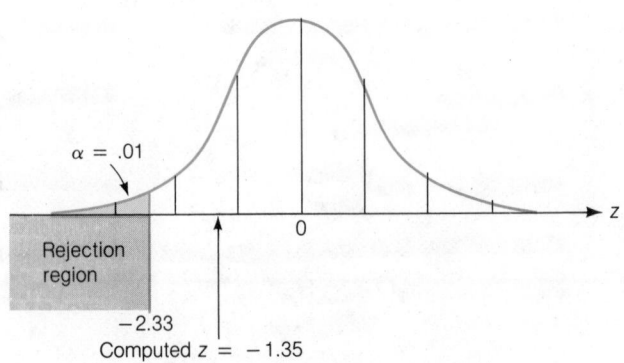

$\alpha = .01$

Rejection region

-2.33

Computed $z = -1.35$

0

z

Example 7.11   In Example 7.10 we found that we did not have sufficient·evidence, at the $\alpha = .01$ level of significance, to indicate that the fraction defective p of flashbulbs was less than $p = .05$. How strong was the weight of evidence favoring the alternative hypothesis (H_a: $p < .05$)? Find the observed significance level for the test.

Solution   The computed value of the test statistic z was $z = -1.35$. Therefore, for this lower-tail test, the observed significance level is

Observed·significance level $= P(z \le -1.35)$

This lower-tail area is shown in Figure 7.23. The area between $z = 0$ and $z = 1.35$ is given in Table IV in the Appendix as .4115. Therefore, the observed significance level is $.5 - .4115 = .0885$. Note that this probability is quite small. Although we did not reject H_0: $p = .05$ at $\alpha = .01$, the probability of observing a z-value as small or smaller than -1.35 is only .0885 if in fact H_0 is true. Therefore, we would reject H_0 if we choose $\alpha = .10$ (since the p-value is less than .10), and we would not reject H_0 (the conclusion of Example 7.10) if we choose $\alpha = .05$.   ∎

Figure 7.23   The Observed Significance Level for Example 7.11

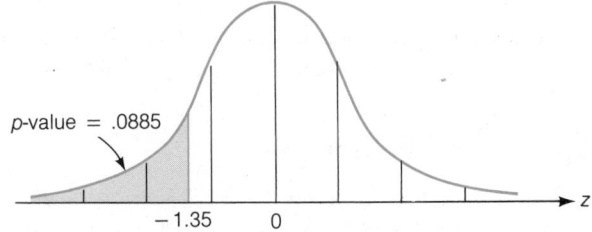

p-value $= .0885$

The confidence interval and test of an hypothesis for p in the previous examples are based on the assumption that the sample size n is large enough that $\hat{p}$ will have an approximately normal sampling distribution (according to the central limit theorem). As a rule of thumb, this condition is satisfied if the interval $\hat{p} \pm 3\sigma_{\hat{p}}$ does not contain 0 or 1.

Small-sample estimators and test procedures are also available for p. These are omitted from our discussion because most surveys use samples that are large enough to employ the large-sample estimators and tests presented in this section.

Exercises 7.70–7.82

Learning the Mechanics

7.70   A random sample of sixty-four observations is selected from a binomial population with unknown proportion of successes p. The computed value of $\hat{p}$ is equal to .56.
a.   Construct a 95% confidence interval for p.
b.   Construct a 90% confidence interval for p.

c. Test H_0: $p = .5$ against H_a: $p \neq .5$. Use $\alpha = .05$.

d. Test H_0: $p = .5$ against H_a: $p > .5$. Use $\alpha = .05$.

7.71 A random sample of 100 observations is selected from a binomial population with unknown probability of success p. The computed value of $\hat{p}$ is equal to .74.

a. Test H_0: $p = .65$ against H_a: $p > .65$. Use $\alpha = .01$.

b. Test H_0: $p = .65$ against H_a: $p > .65$. Use $\alpha = .10$.

c. Test H_0: $p = .90$ against H_a: $p \neq .90$. Use $\alpha = .05$.

d. Form a 95% confidence interval for p.

e. Form a 99% confidence interval for p.

7.72 Suppose that a random sample of 100 observations from a binomial population gives a value of $\hat{p} = .69$ and you wish to test the null hypothesis that the population parameter p is equal to .75 against the alternative hypothesis, $p < .75$.

a. Noting that $\hat{p} = .69$, what does your intuition tell you? Does the value of $\hat{p}$ appear to contradict the null hypothesis?

b. Use the large-sample z-test to test H_0: $p = .75$ against the alternative hypothesis, H_a: $p < .75$. Use $\alpha = .05$. How do the test results compare with your intuitive decision from part a?

7.73 What conditions must be satisfied for a large-sample confidence interval for p, or large-sample test of hypothesis for p, to be valid?

Applying the Concepts

7.74 According to a spokesperson for General Mills, the company's "cents-off" coupon offers are designed to get people to buy their products. Their refund offers (money returned with proof of repeated purchases) are designed to encourage people to continue buying their products. In a national survey conducted by the Nielsen Clearing House in 1975, some 65% of the respondents indicated that they used cents-off coupons when grocery shopping. In a 1980 survey, the Nielsen organization found that 76% of those surveyed used cents-off coupons ("A Penny Refunded Is a Penny Earned," *Minneapolis Star*, November 29, 1981, p. 7F).

a. Suppose the 1980 survey consisted of a random sample of 100 shoppers of which 76 indicated that they used cents-off coupons. Use this information to determine whether the percentage of shoppers using coupons in 1980 is significantly greater than 65%. Test using $\alpha = .05$.

b. Find the observed significance level for the test you conducted in part a and interpret its value.

7.75 The U.S. Commission on Crime wishes to estimate the fraction of crimes related to firearms in an area with one of the highest crime rates in the country. The commission randomly selects 600 files of recently committed crimes in the area and finds 380 in which a firearm was reportedly used. Find a 99% confidence interval for p, the true fraction of crimes in the area in which some type of firearm was reportedly used.

7.76 Following the examination of 209 fish that were randomly selected from the five Great Lakes, the Michigan Department of Agriculture reported that contamination of Great Lakes fish with toxic DDT, dieldrin, and PCB's is at its lowest level in years. Descriptions of the contaminated fish that were caught are given in the table. Even though contaminant levels are down, Michigan's Public Health Department still advises against eating more than one meal of Great Lakes fish a week—and none at all for pregnant women ("Level of Tainted Fish Falls in Great Lakes," *Minneapolis Tribune,* October 24, 1982, p. 15C).

TYPE OF FISH	NUMBER	LAKE	CONTAMINANT
Lake trout	1	Superior	DDT
Lake trout	3	Michigan	PCB's
Lake trout	7	Michigan	Dieldrin
Lake trout	1	Huron	Dieldrin
Whitefish	4	Michigan	Dieldrin
Chub	9	Michigan	Dieldrin

a. Is the sample of fish large enough to use the normal distribution to approximate the sampling distribution of $\hat{p}$, the proportion of contaminated fish in the sample? Explain.

b. Estimate the proportion of contaminated fish in the Great Lakes using an 80% confidence interval.

7.77 Shoplifting is an escalating problem for retailers. According to *U.S. News and World Report* (February 21, 1977), one New York City store randomly selected 500 shoppers and observed them while they were in the store. One in twelve was seen stealing. How accurate is this estimate? To help you answer this question, construct a 95% confidence interval for p, the proportion of all the store's customers who are shoplifters.

7.78 A method currently used by doctors to screen women for possible breast cancer fails to detect cancer in 15% of the women who actually have the disease. A new method has been developed that researchers hope will be able to detect cancer more accurately. A random sample of seventy women known to have breast cancer were screened using the new method. Of these, the new method failed to detect cancer in six.

a. Do the data provide sufficient evidence to indicate that the new screening method is better than the one currently in use? Test using $\alpha = .05$.

b. Find the observed significance level for the test and interpret its value.

7.79 In an article titled "Searching for a Forever Home," *Time* (May 2, 1983) reports on how television programs aid in the adoption of orphans who have physical or mental handicaps. One method of stimulating the adoption program is to present television profiles of the children, one or more per program. The article documents the success of these programs and notes that Oklahoma City's television station KOCO helped to place 92 of the 119 it profiled, New York's WCBS placed 21 of 35, and Atlanta's WXIA placed 79 of 177. How effective were these three television stations in promoting the adoption of the children?

a. Find a 95% confidence interval for each television station's placement success rate.
b. Suppose that the total 331 children profiled by the three stations could be regarded as a random sample from the population of all similar profiles that might be presented by television stations throughout the country. Do the data provide sufficient evidence to indicate that the national placement success rate exceeds .5? Test using $\alpha = .05$.

7.80 A *U.S. News and World Report* article (June 21, 1982) describes a new method of treating a major form of blindness in elderly people. The process, using laser beams to seal abnormal blood vessels in the eye, was tried on 224 patients and, of these, only 14% went blind in 1 year. In a control group of similar untreated patients, 42% went blind in 1 year. Suppose that the percentage of patients going blind in the population represented by the control group was actually 42%.

a. Do the data for the 224 treated patients provide sufficient evidence to indicate that the laser beam treatment was effective in reducing the probability that a patient will go blind after 1 year? Test using $\alpha = .05$.
b. Find the *p*-value for the test and interpret its value.
c. Find a 95% confidence interval for the probability of going blind after 1 year for a patient receiving the laser beam treatment.

7.81 According to *U.S. News and World Report* (May 10, 1982), major infections are the fifth leading cause of death in the United States and approximately 2 million of these cases occur in the nation's hospitals. A powerful new antibiotic, piperacillin, was developed by Lederle Laboratories and approved by the Food and Drug Administration in 1982. Tests on 600 patients showed the drug to be 92% effective in curing serious bacterial infections.

a. Find a 99% confidence interval for the probability p that a patient with a serious bacterial infection will be cured after treatment with piperacillin.
b. If all 2 million hospital patients with serious bacterial infections were treated with piperacillin, what is the expected value of the number x that would be cured?
c. Find the standard deviation of x.
d. Give an upper limit on the number x that would be cured.

7.82 An article in the *Gainesville Sun* (March 14, 1984) reports on the topics that teenagers most want to discuss with their parents. The findings, the results of a Gallup Youth Poll, showed that 46% would like more discussion about the family's financial situation, 37% would like to talk about school, and 30% would like to talk about religion. These and other percentages were based on a national sampling of 505 teenagers. What "margin of error" (using the language of the media) would you attach to these findings? Explain.

**7.6
Determining
the
Sample Size**

When you are planning an experiment with the intention of estimating a population parameter, say a mean μ or a binomial population proportion p, one of the first concerns is sample size. How can the appropriate sample size be determined? To answer this question you must decide how reliable you wish your estimate to be and then you will need to know how many measurements to include in your sample. Consider Example 7.12.

Example 7.12   In the hospital example in Section 7.1, we estimated the mean length of time, μ, that patients remain in a hospital. A sample of 100 patients' records produced an estimate, $\bar{x}$, that was within .98 day of μ with probability equal to .95. Suppose we wish to estimate the true mean to within .5 day with a probability equal to .95. How large a sample would be required?

Solution   For the sample size $n = 100$, we found an approximate 95% confidence interval to be

$$\bar{x} \pm 2\sigma_{\bar{x}} \approx 4.65 \pm .98$$

If we now want our estimator $\bar{x}$ to be within .5 day of μ, we must have

$$2\sigma_{\bar{x}} = .5$$

or

$$\frac{2\sigma}{\sqrt{n}} = .5$$

The necessary sample size is found by solving this equation for n.

To solve for n, we need an approximation for σ, which in this case comes from the original pilot sample of $n = 100$ records, $s = 4.9$ (see Section 7.1). Thus, we have

$$\frac{2\sigma}{\sqrt{n}} \approx \frac{2s}{\sqrt{n}} = .5$$

or

$$\frac{2(4.9)}{\sqrt{n}} = .5$$

$$\sqrt{n} = \frac{2(4.9)}{.5} = 19.6$$

$$n = (19.6)^2$$

$$= 384.16 \approx 384$$

The company will have to sample approximately 384 patients' records in order to estimate the mean length of stay, μ, to within .5 day with probability equal to .95. ■

A similar argument follows if we want to determine the sample size necessary for estimating a binomial population proportion to within a given bound B (often called *margin of error*) with a specified confidence level. The general equations for determining the sample size to estimate both μ and p are given in the box.

Sample Size Determination with $100(1 - \alpha)\%$ Confidence Desired

For estimating μ to within a bound B with probability $(1 - \alpha)$, solve the following equation for n:

$$z_{\alpha/2}\left(\frac{\sigma}{\sqrt{n}}\right) = B$$

The solution is

$$n = \frac{z_{\alpha/2}^2 \sigma^2}{B^2}$$

The value of σ substituted into these expressions is obtained from an estimate s of σ obtained from a prior sample (pilot study, etc.) or from an educated guess as to the value of the range R of the observations in the population. The value of σ is taken to be approximately $R/4$.

- -

For estimating p to within a bound B with probability $(1 - \alpha)$, solve the following equation for n:

$$z_{\alpha/2}\sqrt{\frac{pq}{n}} = B$$

The solution is

$$n = \frac{z_{\alpha/2}^2 pq}{B^2}$$

The value of p substituted into these expressions is obtained from an estimate based on a prior sample, obtained from an educated guess, or, most conservatively, chosen equal to .5. (The value $p = .5$ gives the largest value for $\sigma_{\hat{p}}$ and, consequently, results in a solution that is as large as or larger than the required sample size.)

Example 7.13   Refer to Example 7.10 in which a flashbulb manufacturer was making an inference about the fraction defective in a shipment. Suppose the manufacturer wants to estimate the true fraction p to within .01 (that is, $B = .01$) with a confidence coefficient equal to .90. How large a sample would be needed? (Assume the true value of p is near .05.)

Solution Since we want the error of estimation to be less than $B = .01$ with probability .90, we must have $\alpha = 1 - .90 = .10$. Then $z_{\alpha/2} = z_{.05} = 1.645$. Substituting these values into the formula for n, we get

$$n = \frac{z_{\alpha/2}^2 pq}{B^2} = \frac{(1.645)^2(.05)(.95)}{(.01)^2}$$

$$= 1{,}285.4 \approx 1{,}285$$

Thus, a fairly large sample—about 1,285 bulbs—must be tested if the manufacturer wants to be 90% certain that the estimate of the fraction defective will fall within .01 of the true value of p. ■

Note in Example 7.13 that the manufacturer had prior information concerning the approximate value of the fraction defective, p. Knowing that p would be small, the manufacturer assumed a value of p equal to .05 in order to obtain an approximate sample size necessary to estimate its true value. In many sampling situations, you will have no information on the approximate value of p. Then you may wish to substitute $p = .5$ into the formula for the sample size. The value of n you obtain will be the largest sample size needed to achieve the specified bound B on the error of estimation, regardless of the true value of p.

Exercises 7.83–7.96

Learning the Mechanics

7.83 Find the sample size needed to estimate μ for each of the following situations:
a. Bound is 3; $\sigma = 45$; 95% confidence is desired.
b. Bound is 1; $\sigma = 45$; 95% confidence is desired.
c. Bound is 4; $\sigma = 75$; 99% confidence is desired.
d. Bound is 6; $\sigma = 75$; 90% confidence is desired.

7.84 Find the sample size needed to estimate p for each of the following situations:
a. Bound is .02; p is near .9; 99% confidence is desired.
b. Bound is .04; p is near .9; 99% confidence is desired.
c. Bound is .03; p is near .6; 95% confidence is desired.
d. Bound is .02; p is near .4; 90% confidence is desired.
e. Bound is .05; no prior estimate of p is available; 95% confidence is desired.

7.85 If you wish to estimate a population mean correct to within .1 with probability .95 and you know, from prior sampling, that σ^2 is approximately equal to 1.5, how many observations have to be included in your sample?

7.86 Find the approximate sample size necessary to estimate a binomial proportion p correct to within .02 with probability equal to .90.
a. Assume that you know p is near .7.
b. Suppose that you have no knowledge of the value of p but you wish to be certain that your sample is large enough to achieve the specified accuracy for the estimate.

7.87 Suppose you wish to estimate a population mean correct to within .15 with probability equal to .90. You do not know σ^2, but you know that the observations will range in value between 39 and 42. Find the approximate sample size that will produce the desired accuracy of the estimate. You wish to be conservative to ensure that the sample size will be ample to achieve the desired accuracy of the estimate. [*Hint:* Using your knowledge of data variation from Section 2.6, assume that the range of the observations equals 4σ.]

7.88 Assume that you wish to estimate a population mean using a 95% confidence interval and that you know from prior information that $\sigma^2 \approx 1$.

a. To see the effect of the sample size on the width of the confidence interval, calculate the width of the confidence interval for $n = 16, 25, 49, 100, 400$.

b. Plot the widths as a function of sample size n on graph paper. Connect the points by a smooth curve and note how the width decreases as n increases.

Applying the Concepts

7.89 According to the *Minneapolis Star* ("Monitor: Vehicle Speeds," May 2, 1980), the federal government requires states to certify that they are enforcing the 55 mile per hour speed limit and that motorists are driving at that speed. A state is in jeopardy of losing millions of dollars in federal road funds if more than 60% of its vehicles on 55 mile per hour highways are exceeding the speed limit. The Minnesota Highway Patrol conducts seventy radar surveys each year at a total of fifty sites to estimate the proportion p of vehicles exceeding 55 miles per hour. Each sample survey involves at least 400 vehicles.

a. How large a sample should be selected at site 42 on Interstate 35W to estimate p to within 3% with 90% confidence? Last year approximately 60% of all vehicles exceeded 55 miles per hour.

b. The highway patrol also estimates μ, the average speed of vehicles on state highways. Accordingly, it wants to know whether the sample size determined in part a is large enough to also estimate μ to within .25 mile per hour with 90% confidence. Assume that the standard deviation of vehicle speeds is approximately 2 miles per hour. How large a sample should be taken at site 42 to estimate μ with the desired reliability?

7.90 According to a Food and Drug Administration (FDA) study, the average cup of coffee contains 115 milligrams of caffeine; the amount per cup ranges from 60 to 180 milligrams. In contrast, Sugar-Free Mr. Pibb tested at 58.8 milligrams per 12-ounce serving, Coca-Cola and Diet Coke at 45.6, and Pepsi at 38.4 (*Gainesville Sun,* March 15, 1984). Suppose that you wish to repeat the FDA experiment to obtain an estimate of the mean caffeine content in a cup of coffee correct to within 5 milligrams per cup. How many cups of coffee must you include in your sample?

7.91 The owner of a large turkey farm knows from previous years that the largest profit is made by selling turkeys when their average weight is 15 pounds. Approximately how many turkeys must be sampled in order to estimate their true mean weight to within .5 pound with probability equal to .95? Assume that prior knowledge indicates that the standard deviation of the weights of the turkeys is approximately 2 pounds.

7.92 To estimate the mean age (in months) at which an American Indian child learns to walk, how many Indian children must be sampled if the researcher desires an estimate to be within 1 month of the true mean with 99% confidence? Assume that the researcher knows only that the age at which these children begin walking ranges from 8 to 26 months.

7.93 A market researcher wants to select one sample to estimate both μ, the average age of people living within 5 miles of a proposed shopping mall site, and p, the proportion of people within that 5-mile radius that are between 20 and 40 years of age. He wants to estimate μ with a 95% confidence interval that is no more than 6 years wide and p with a 90% confidence interval of width no greater than .1. It is known from previous studies of this population that the standard deviation of the ages in the population is 10 years, and it is believed that p is near .4. How large a sample must the researcher draw in order to construct confidence intervals for both μ and p that satisfy the specifications?

7.94 If you want to estimate the proportion of operating automobiles that are equipped with air pollution devices, approximately how large a sample would be required to estimate p to within .02 with probability equal to .95?

7.95 Suppose a department store wants to estimate μ, the average age of the customers in its contemporary apparel department, correct to within 2 years with probability equal to .95. Approximately how large a sample would be required? [*Note:* The management does not know σ but guesses that the age of its customers ranges from 15 to 45. If you take this range to equal 4σ, you will have a conservative approximation to σ that can be used to calculate n.]

7.96 The EPA standards on the amount of suspended solids that can be discharged into rivers and streams is a maximum of 60 milligrams per liter daily, with a maximum monthly average of 30 milligrams per liter. Suppose you want to test a randomly selected sample of n water specimens in order to estimate the mean daily rate of pollution produced by a mining operation. If you want your estimate correct to within 1 milligram with probability equal to .90, how many water specimens must you include in your sample? Assume prior knowledge indicates that pollution readings in water samples taken during a single day have a standard deviation equal to 5 milligrams.

7.7 Inferences About a Population Variance (Optional)

Although many practical problems involve inferences about a population mean (or proportion), it is sometimes of interest to make an inference about a population variance, σ^2. To illustrate, a quality control supervisor in a cannery knows that the exact amount each can contains will vary, since there are certain uncontrollable factors that affect the amount of fill. The mean fill per can is important, but equally important is the variation of fill. If σ^2, the variance of the fill, is large, some cans will contain too little and others too much. Suppose regulatory agencies specify that the standard deviation

Figure 7.24 Several
χ^2 Probability Distributions

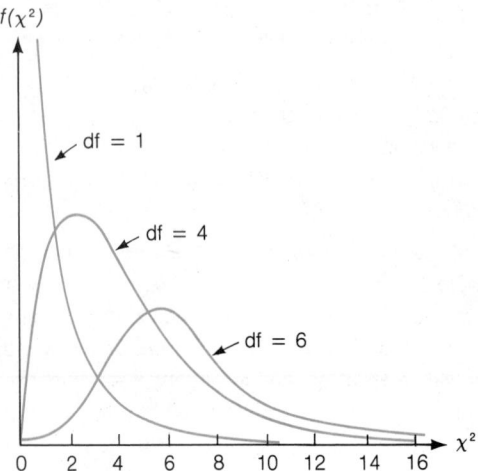

Figure 7.25 Repro-
duction of Part of Table VI
in the Appendix

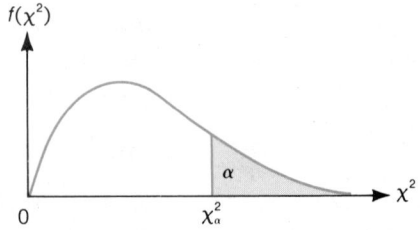

DEGREES OF FREEDOM	$\chi^2_{.100}$	$\chi^2_{.050}$	$\chi^2_{.025}$	$\chi^2_{.010}$	$\chi^2_{.005}$
1	2.70554	3.84146	5.02389	6.63490	7.87944
2	4.60517	5.99147	7.37776	9.21034	10.5966
3	6.25139	7.81473	9.34840	11.3449	12.8381
4	7.77944	9.48773	11.1433	13.2767	14.8602
5	9.23635	11.0705	12.8325	15.0863	16.7496
6	10.6446	12.5916	14.4494	16.8119	18.5476
7	12.0170	14.0671	16.0128	18.4753	20.2777
8	13.3616	15.5073	17.5346	20.0902	21.9550
9	14.6837	16.9190	19.0228	21.6660	23.5893
10	15.9871	18.3070	20.4831	23.2093	25.1882
11	17.2750	19.6751	21.9200	24.7250	26.7569
12	18.5494	21.0261	23.3367	26.2170	28.2995
13	19.8119	22.3621	24.7356	27.6883	29.8194
14	21.0642	23.6848	26.1190	29.1413	31.3193
15	22.3072	24.9958	27.4884	30.5779	32.8013
16	23.5418	26.2962	28.8454	31.9999	34.2672
17	24.7690	27.5871	30.1910	33.4087	35.7185
18	25.9894	28.8693	31.5264	34.8053	37.1564
19	27.2036	30.1435	32.8523	36.1908	38.5822

of the amount of fill should be less than .1 ounce. To determine whether the process is meeting this specification, the supervisor randomly selects ten cans, weighs the contents of each, and finds that the sample standard deviation of these measurements is .04. Do these data provide sufficient evidence to indicate that the variability is as small as desired? To answer this question, we need a procedure for testing an hypothesis about σ^2.

Intuitively it seems that we should compare the sample variance s^2 to the hypothesized value of σ^2 (or s to σ) in order to make a decision about the population's variability. The quantity

$$\frac{(n-1)s^2}{\sigma^2}$$

has been shown to have a sampling distribution called a **chi square** (χ^2) **distribution** when the population from which the sample is taken is *normally distributed*. Several chi square distributions are shown in Figure 7.24.

The upper-tail areas for this distribution have been tabulated and are given in Table VI of the Appendix, a portion of which is reproduced in Figure 7.25. The table gives the values of χ^2, denoted as χ_α^2, that locate an area of α in the upper tail of the chi square distribution; that is, $P(\chi^2 > \chi_\alpha^2) = \alpha$. In this case, as with the t-statistic, the shape of the chi square distribution depends on the degrees of freedom associated with s^2, namely $(n-1)$. Thus, for $n = 10$ and an upper-tail value $\alpha = .05$, you will have $n - 1 = 9$ df and $\chi_{.05}^2 = 16.9190$ (shaded area in Figure 7.25). To further illustrate the use of Table VI, we return to the can-filling example.

Example 7.14

According to the previous discussion, the quality control supervisor sampled $n = 10$ cans and calculated $s = .04$. Does this value of s provide sufficient evidence to indicate that the standard deviation σ of the fill measurements is less than .1 ounce?

Solution

Since the null and alternative hypotheses must be stated in terms of σ^2 (rather than σ), we want to test the null hypothesis that $\sigma^2 = .01$ against the alternative that $\sigma^2 < .01$. Therefore, the elements of the test are

H_0: $\sigma^2 = .01$ H_a: $\sigma^2 < .01$

Test statistic: $\chi^2 = \dfrac{(n-1)s^2}{\sigma^2}$

Assumption: The distribution of the amounts of fill is approximately normal.

Rejection region: The smaller the value of s^2 we observe, the stronger the evidence in favor of H_a. Thus, we reject H_0 for "small values" of the test statistic. With $\alpha = .05$ and 9 df, the χ^2 value for rejection is found in Table VI and pictured in Figure 7.26 (page 324). We will reject H_0 if $\chi^2 < 3.32511$.

Remember that the area given in Table VI is the area to the *right* of the numerical value in the table. Thus, to determine the lower-tail value, which has $\alpha = .05$ to its *left,* we used the $\chi_{.95}^2$ column in Table VI.

Figure 7.26   Rejection
Region for Example 7.14

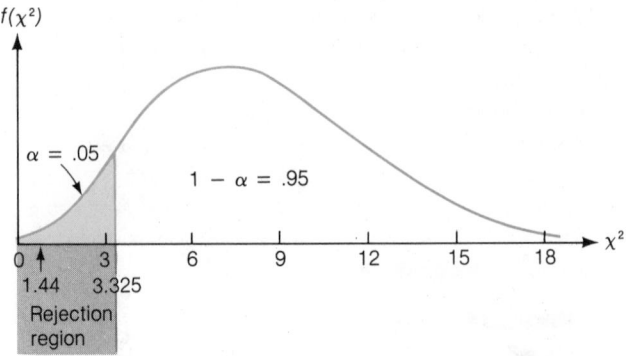

Since

$$\chi^2 = \frac{(n-1)s^2}{\sigma^2} = \frac{9(.04)^2}{.01} = 1.44$$

is less than 3.32511, the supervisor can conclude that the variance σ^2 of the population of all amounts of fill is less than .01 ($\sigma < .1$) with probability of a Type I error equal to $\alpha = .05$. If this procedure is repeatedly used, it will incorrectly reject H_0 only 5% of the time. Thus, the quality control supervisor is confident in the decision that the cannery is operating within the desired limits of variability.   ■

It is also possible to form a confidence interval for a population variance through the manipulation of the χ^2 distribution. This confidence interval, as well as the one-tailed and two-tailed tests of hypotheses for σ^2, are given in the accompanying boxes.

Test of an Hypothesis About σ^2

One-Tailed Test

H_0: $\sigma^2 = \sigma_0^2$ (σ_0^2 = hypothesized variance)

H_a: $\sigma^2 < \sigma_0^2$
   (or H_a: $\sigma^2 > \sigma_0^2$)

Test statistic: $\chi^2 = \dfrac{(n-1)s^2}{\sigma_0^2}$

Rejection region: $\chi^2 < \chi^2_{(1-\alpha)}$
   (or $\chi^2 > \chi^2_\alpha$
   when H_a: $\sigma^2 > \sigma_0^2$)

Two-Tailed Test

H_0: $\sigma^2 = \sigma_0^2$

H_a: $\sigma^2 \neq \sigma_0^2$

Test statistic: $\chi^2 = \dfrac{(n-1)s^2}{\sigma_0^2}$

Rejection region: $\chi^2 < \chi^2_{(1-\alpha/2)}$
   or $\chi^2 > \chi^2_{\alpha/2}$

where the distribution of χ^2 is based on $(n-1)$ degrees of freedom

Assumption: The population from which the sample is drawn is approximately normal.

Confidence Interval for σ^2

$$\frac{(n-1)s^2}{\chi^2_{\alpha/2}} < \sigma^2 < \frac{(n-1)s^2}{\chi^2_{(1-\alpha/2)}}$$

where the distribution of χ^2 is based on $(n-1)$ degrees of freedom

Assumption: The population from which the sample is drawn is approximately normal.

Example 7.15

Test scores are often used to discriminate among individuals applying for the same job, students applying to graduate school, attorneys trying to become members of the bar, etc. Suppose an employment agency uses a 500-point examination to help in determining which job applicants are best qualified for certain positions. The variability in these test scores should be considered when evaluating the test results. For example, if all applicants should somehow score exactly the same score on the test (no variability among the scores), the test would be of no value in deciding which applicants should be employed. A large amount of variability among the test scores would be desirable in order to differentiate the relative merits of the applicants. To evaluate the variability of the test scores, the employment agency randomly selects 100 test scores and calculates $s^2 = 127$. Use this information to form a 95% confidence interval for σ^2, the variability for *all* test scores.

Solution   The general form of the confidence interval is

$$\frac{(n-1)s^2}{\chi^2_{\alpha/2}} < \sigma^2 < \frac{(n-1)s^2}{\chi^2_{(1-\alpha/2)}}$$

where we assume the test scores are approximately normally distributed. Using Table VI in the Appendix, we obtain the tabulated values of chi square corresponding to $\alpha/2 = .025$ and $1 - \alpha/2 = .975$, based on $n - 1 = 100 - 1 = 99$ degrees of freedom:

$$\chi^2_{.025} \approx 129.56 \quad \text{and} \quad \chi^2_{.975} \approx 74.22$$

The interval of interest is thus

$$\frac{99(127)}{129.56} < \sigma^2 < \frac{99(127)}{74.22}$$

$$97.04 < \sigma^2 < 169.40$$

Thus, the employment agency can be 95% confident that the variance of all test scores is (approximately) between 97 and 170 or, equivalently, that the standard deviation is between 10 and 13, rounding to the nearest whole number.

This information can be used to determine whether the amount of variability is sufficient. If the standard deviation of 13 is deemed too small to allow the company to discriminate among the applicants, the structure of the test should be changed. ∎

**Exercises
7.97–7.109**

Learning the Mechanics

7.97 Let χ_0^2 be a particular value of χ^2. Find the value of χ_0^2 such that:
a. $P(\chi^2 > \chi_0^2) = .10$ for $n = 10$ **b.** $P(\chi^2 > \chi_0^2) = .05$ for $n = 13$
c. $P(\chi^2 > \chi_0^2) = .025$ for $n = 6$

7.98 A random sample of n observations is selected from a normal population to test the null hypothesis that $\sigma^2 = 25$. Specify the rejection region for each of the following combinations of H_a, α, and n:
a. H_a: $\sigma^2 \neq 25$; $\alpha = .05$; $n = 18$ **b.** H_a: $\sigma^2 > 25$; $\alpha = .01$; $n = 20$
c. H_a: $\sigma^2 > 25$; $\alpha = .10$; $n = 15$ **d.** H_a: $\sigma^2 < 25$; $\alpha = .01$; $n = 12$
e. H_a: $\sigma^2 \neq 25$; $\alpha = .10$; $n = 8$ **f.** H_a: $\sigma^2 < 25$; $\alpha = .05$; $n = 25$

7.99 A random sample of fifteen observations is selected from a normal population with variance σ^2. Give the values of $\chi_{\alpha/2}^2$ and $\chi_{(1-\alpha/2)}^2$ that would be used to form a confidence interval for σ^2 for each of the following levels of confidence:
a. 95%   **b.** 90%   **c.** 99%

7.100 A random sample of five measurements gave $\bar{x} = 9.4$ and $s^2 = 4.84$.
a. What assumptions must you make concerning the population in order to test an hypothesis about (or estimate) σ^2?
b. Suppose the assumptions in part a are satisfied. Test the null hypothesis, $\sigma^2 = 1$, against the alternative hypothesis, $\sigma^2 > 1$. Use $\alpha = .05$.
c. Test the null hypothesis that $\sigma^2 = 1$ against the alternative hypothesis that $\sigma^2 \neq 1$. Use $\alpha = .05$.
d. Find a 90% confidence interval for σ^2.

7.101 Refer to Exercise 7.100. Suppose we had $n = 100$, $\bar{x} = 9.4$, and $s^2 = 4.84$.
a. Test the null hypothesis, H_0: $\sigma^2 = 1$, against the alternative hypothesis, H_a: $\sigma^2 > 1$.
b. Compare your test results with those of Exercise 7.100.
c. Find a 90% confidence interval for σ^2. Compare this confidence interval with the confidence interval obtained in Exercise 7.100 and note the effect of an increase in sample size on the width of the interval.

7.102 A random sample of $n = 5$ observations from a normal population produced the following measurements: 5, 7, 3, 8, 1
a. Do the data provide sufficient evidence to indicate that $\sigma^2 > 20$? Test using $\alpha = .05$.
b. Find a 90% confidence interval for σ^2.

7.103 A random sample of $n = 7$ observations from a normal population produced the following measurements: 4, 0, 6, 3, 3, 5, 9
a. Do the data provide sufficient evidence to indicate that $\sigma^2 < 1$? Test using $\alpha = .05$.
b. Find a 90% confidence interval for σ^2.

Applying the Concepts

7.104 An educational testing service designed an achievement test so that the range in student scores would be at least 300 points. To see whether they achieved their objective, the testing service gave the test to a random sample of thirty students and found that the sample mean and variance were 759 and 1,943, respectively. Do the data provide sufficient evidence to indicate that the test does not achieve the desired dispersion in scores? Test using $\alpha = .05$. [*Hint:* Assume that range $= 6\sigma$.]

7.105 A marine biologist wishes to use male angelfish for experimental purposes due to the belief that their weight is fairly stable (i.e., the variability in weights among male angelfish is small). The biologist randomly samples sixteen male angelfish and finds that their mean weight is 4.1 pounds and the standard deviation is 1.73 pounds. Find a 95% confidence interval for the variability in weights of all male angelfish. What assumptions must you make in order to form the interval?

7.106 Refer to Exercise 7.105. It is suggested that the marine biologist use parrotfish instead of male angelfish in the experiment. Since these are more difficult to obtain, the biologist decides to use parrotfish only if there is evidence that the variance of their weights is less than 4. A random sample of ten parrotfish produces a mean of 4.3 pounds and a variance of 2. Is there sufficient evidence for the biologist to claim that the variability in weights among parrotfish is small enough to justify their use in the experiment? Test at $\alpha = .05$, and state any assumptions that are needed.

7.107 A new gun-like apparatus has been devised to replace the needle in administering vaccines. The apparatus, which is connected to a large supply of vaccine, can be set to inject different amounts of the serum, but the variance in the amount of serum injected to a given person must not be greater than .06 to ensure proper inoculation. A random sample of twenty-five injections resulted in a variance of .135. Do the data provide sufficient evidence to indicate the gun is not working properly? Use $\alpha = .10$.

7.108 It is essential in the manufacture of machinery to utilize parts that conform to specifications. In the past, diameters of the ball-bearings produced by a certain manufacturer had a variance of .00156. To cut costs, the manufacturer instituted a less expensive production method. The variance of the diameters of 100 randomly sampled bearings produced by the new process was .00211. Do the data provide sufficient evidence to indicate that diameters of ball-bearings produced by the new process are more variable than those produced by the old process?

7.109 To perform an experiment, a chemist has to use a substance that contains 50% sodium nitrate. The chemist suspects that a particular batch of the substance has not been mixed thoroughly, thus causing the amount of sodium nitrate to vary from one portion of the batch to another. The results of twenty randomly selected 10-ounce samples yield a sample standard deviation equal to .05 ounce. Estimate the true variance of the amount of sodium nitrate in 10-ounce samples selected from the batch. Use a 95% confidence interval.

Summary

The objective of statistics is to make inferences about a population based on information in a sample. In this chapter, we have presented several methods for accomplishing this objective.

The inference-making techniques we discussed are *estimation* and *hypothesis testing.* Estimation of a *population parameter* is accomplished by using an interval estimate with a probability of coverage (**confidence coefficient**) that is fixed by the experimenter at a high level (usually .90, .95, or .99). On the other hand, when a specific **research (alternative) hypothesis** about a parameter is tested, the probability α of incorrectly rejecting the **null hypothesis** and accepting the research hypothesis is chosen to be small. Thus, we try to control the chance of error in both these inference-making procedures.

One of the most important parameters about which inferences are made is the population mean μ. The sample mean $\bar{x}$ is used for making the inference, but the method depends on the **sample size.** When the sample size is large (we have specified $n > 30$ as large), the standard normal z-statistic is used. The **t-statistic** is employed when σ is unknown and a small sample is drawn from a normally (or approximately normally) distributed population.

Another important parameter in practical applications is the **binomial population proportion p.** This probability of success is estimated by the sample fraction of successes, $\hat{p}$, and the z-statistic is again used to form confidence intervals or to test an hypothesis. The sample size necessary for estimating a population mean μ or a binomial population proportion p can be determined by specifying the **confidence level** and the desired bound on the error of the estimate.

Finally, when inferences concerning a population variance are of interest, the **chi square (χ^2) distribution** enables us to conduct tests of hypotheses or to form confidence intervals for σ^2.

Supplementary Exercises 7.110–7.143

[*Note: List the assumptions necessary for the valid implementation of the statistical procedures you use in solving all these exercises. Starred (*) exercises refer to the optional section.*]

Learning the Mechanics

7.110 A random sample of twenty observations selected from a normal population produced $\bar{x} = 72.6$ and $s^2 = 19.4$.

a. Form a 90% confidence interval for the population mean.

b. Test H_0: $\mu = 80$ against H_a: $\mu < 80$. Use $\alpha = .05$.

c. Test H_0: $\mu = 80$ against H_a: $\mu \neq 80$. Use $\alpha = .01$.

d. Form a 99% confidence interval for μ.

e. How large a sample would be required to estimate μ to within 1 unit with 95% confidence?

7.111 A random sample of $n = 200$ observations from a binomial population yields $\hat{p} = .29$.

a. Test H_0: $p = .35$ against H_a: $p < .35$. Use $\alpha = .05$.

b. Test H_0: $p = .35$ against H_a: $p \neq .35$. Use $\alpha = .05$.

c. Form a 95% confidence interval for p.

d. Form a 99% confidence interval for p.

e. How large a sample would be required to estimate p to within .05 with 99% confidence?

7.112 A random sample of 175 measurements possessed a mean $\bar{x} = 8.2$ and a standard deviation $s = .79$.
a. Form a 95% confidence interval for μ.
b. Test H_0: $\mu = 8.3$ against H_a: $\mu \neq 8.3$. Use $\alpha = .05$.
c. Test H_0: $\mu = 8.4$ against H_a: $\mu \neq 8.4$. Use $\alpha = .05$.

*7.113 A random sample of forty-one observations from a normal population possessed a mean $\bar{x} = 88$ and a standard deviation $s = 6.9$.
a. Form a 90% confidence interval for σ^2.
b. Form a 99% confidence interval for σ^2.
c. Test H_0: $\sigma^2 = 30$ against H_a: $\sigma^2 > 30$. Use $\alpha = .05$.
d. Test H_0: $\sigma^2 = 30$ against H_a: $\sigma^2 \neq 30$. Use $\alpha = .05$.

Applying the Concepts

7.114 Failure to meet payments on student loans guaranteed by the United States government has been a major problem for both banks and the government. Approximately 50% of all student loans guaranteed by the government are in default. A random sample of 350 loans to college students in one region of the United States indicates that 147 loans are in default.
a. Do the data indicate that the proportion of student loans in default in this area of the country differs from the proportion of all student loans in the United States that are in default? Use $\alpha = .01$.
b. Find the observed significance level for the test and interpret its value.

7.115 In order to be effective, the mean length of life of a certain mechanical component used in a spacecraft must be larger than 1,100 hours. Due to the prohibitive cost of the components, only three can be tested under simulated space conditions. The lifetimes (hours) of the components were recorded and the following statistics were computed: $\bar{x} = 1{,}173.6$ and $s = 36.3$. Do the data provide sufficient evidence to conclude that the component will be effective? Use $\alpha = .01$.

7.116 The mean score on a Peace Corps application test, based on many tests conducted over a long period of time, is 80. Ten prospective applicants have taken a course designed to improve their scores on the test. The scores of the ten applicants who completed the course had a mean equal to 86.1 and a standard deviation equal to 12.4.
a. Do the data provide sufficient evidence to conclude that students taking the course will have a higher mean score than those who do not? Test using $\alpha = .10$.
b. Find the approximate observed significance level for the test and interpret its value.
c. What assumptions must be made in order for the procedure that you used in part a to be valid?

7.117 A sporting goods manufacturer who produces both white and yellow tennis balls claims that more than 75% of all tennis balls sold are yellow. A marketing study of the purchases of white and yellow tennis balls at a number of stores showed that of 470 cans sold, 410 were yellow and 60 were white.

a. Is there sufficient evidence to support the manufacturer's claim? Test using $\alpha = .01$.

b. Find the observed significance level for the test and interpret its value.

7.118 A discount store claims that its steel-belted radial tires last longer than those of a major tire company. The following experiment was performed to test this claim: On each of forty cars, one discount tire and one major company tire were mounted on the rear axle. After each car was driven 8,000 miles, the tires were inspected for wear. Suppose the tires of the discount store show less wear on thirty-two of the cars. Is there sufficient evidence to conclude that the discount store's claim is correct? Test using $\alpha = .10$. [*Hint:*   Let p equal the proportion of discount store tires that last longer than those of the major tire company. Then test H_0: $p = .5$ against H_a: $p > .5$.]

7.119 During past harvests, a farmer has averaged 68.2 bushels of corn per acre. A new fertilizer has been placed on the market, and after using the new fertilizer the farmer notes the yield of corn for four randomly selected fields of equal size. The mean yield is 72.4 bushels per acre and the standard deviation is 2.2 bushels.

a. If these data truly represent a random sample of corn yields that the farmer might expect (now and in the future) when using the new fertilizer, do they suggest that the mean yield of corn per acre has changed from past years? Test using $\alpha = .05$.

b. Note that the four yield measurements were selected from within the same year. Are these measurements a random sample selected from the population of interest to the farmer? If not, what information do the data provide the farmer?

7.120 The EPA sets a limit of 5 parts per million on PCB (a dangerous substance) in water. A major manufacturing firm producing PCB for electrical insulation discharges small amounts from the plant. The company management, attempting to control the PCB in its discharge, has given instructions to halt production if the mean amount of PCB in the effluent exceeds 3 parts per million. A random sample of fifty water specimens produced the following statistics: $\bar{x} = 3.1$ parts per million and $s = .5$ part per million.

a. Do these statistics provide sufficient evidence to halt the production process? Use $\alpha = .01$.

b. If you were the plant manager, would you want to use a large or a small value for α for the test in part a?

7.121 If rejection of the null hypothesis of a certain test would cause your firm to go out of business, would you want α to be small or large? Explain.

7.122 A company is interested in estimating μ, the mean number of days of sick leave taken by all its employees. The firm's statistician selects at random 100 personnel files and notes the number of sick days taken by each employee. The following sample statistics are computed:

$\bar{x} = 12.2$ days $s = 10$ days

a. Estimate μ using a 90% confidence interval.

b. How many personnel files would the statistician have to select in order to estimate μ to within 2 days with 99% confidence?

7.123 A large mail-order company has placed an order for 5,000 electric can openers with a supplier on condition that no more than 2% of the devices will be defective. To check the shipment, the company tests a random sample of 400 of the can openers and finds eleven are defective. Does this provide sufficient evidence to indicate that the proportion of defective can openers in the shipment exceeds 2%? Test using $\alpha = .05$.

7.124 Refer to Exercise 7.123. Suppose the company wants to estimate the proportion p of defective can openers in the shipment correct to within .04 with probability equal to .95. Approximately how large a sample would be required?

7.125 A university is considering a change in the way students pay for their education. Presently, the students pay $16 per credit hour. The university is contemplating charging each student a set fee of $240 per quarter, regardless of how many credit hours each takes. To see if this proposal would be economically feasible, the university would like to know how many credit hours, on the average, each student takes per quarter. A random sample of 250 students yields a mean of 14.1 credit hours and a standard deviation of 2.3 credit hours per quarter.
a. Estimate the mean credit hours per student per quarter using a 99% confidence interval.
b. Use this confidence interval to obtain an interval estimate for the mean tuition fee per student per quarter. Interpret the result.

7.126 In checking the reliability of a bank's records, auditing firms sometimes ask a sample of the bank's customers to confirm the accuracy of their savings account balances as reported by the bank. Suppose an auditing firm is interested in estimating p, the proportion of a bank's savings accounts on whose balances the bank and the customer disagree. Of 200 savings account customers questioned by the auditors, 15 said their balance disagreed with that reported by the bank.
a. Using a 95% confidence interval, estimate the actual proportion of the bank's savings accounts on whose balances the bank and customer disagree.
b. The bank claims that the true fraction of accounts on which there is disagreement is at most .05. You, as an auditor, doubt this claim. Test the bank's claim at the .10 significance level.

7.127 Refer to Exercise 7.126. How many savings account customers should the auditors question if they want to estimate p to within .02 with probability equal to .95?

7.128 The Chamber of Commerce of a small seaside resort would like to know the mean number of hours of labor required daily to clear litter from its public beach on weekends. A random sample of the labor expended on each of fifteen randomly selected Sunday mornings produced a mean of 3.6 hours and a standard deviation of .6 hour. Estimate the mean amount of labor required per Sunday morning to clear the beach of litter. Use a 95% confidence interval.

7.129 The mean grade-point average (GPA) at a certain university was 3.20 in 1974. To show that grade inflation has been reversed, a dean sets out to show that the mean GPA is now lower than 3.20. A random sample of 100 students yields a mean GPA of

3.05 and a standard deviation equal to .90. Do the data provide sufficient evidence to indicate that grade inflation has been reversed? (Test using $\alpha = .10$.)

7.130 The strength of a pesticide dosage is often measured by the proportion of pests that dosage will kill. To determine this proportion for a certain dosage of rat poison, 250 rats are fed the dosage of poison and 215 die. Use a 90% confidence interval to estimate the true proportion of rats that will succumb to the dosage. Interpret your result.

7.131 A meteorologist wishes to estimate the mean amount of snowfall per year in Spokane, Washington. A random sample of the recorded snowfalls for 20 years produces a sample mean equal to 54 inches and a standard deviation of 9.59 inches.

a. Estimate the true mean amount of snowfall in Spokane using a 99% confidence interval.

b. If you were purchasing snow-removal equipment for a city, what numerical descriptive measure of the distribution of depth of snowfall would be of most interest to you? Would it be the mean?

7.132 A leading cigarette manufacturer claims its cigarettes contain an average of less than 16 milligrams of tar. To check this claim, a random sample of cigarettes will be chosen and the mean amount of tar per cigarette will be estimated. If previous information indicates that the amount of tar per cigarette has a standard deviation of 2.5 milligrams, how many cigarettes should be sampled to estimate the true mean to within .45 milligram with probability .95?

7.133 A university dean is interested in determining the proportion of students who receive some sort of financial aid. Rather than examine the records for all students, the dean randomly selects 200 students and finds that 118 of them are receiving financial aid. Use a 95% confidence interval to estimate the true proportion of students who receive aid.

7.134 Before approval is given for the use of a new insecticide, the United States Department of Agriculture (USDA) requires that several tests be performed to see how the substance will affect wildlife. In particular, the USDA would like to know the proportion of starlings that will die after being exposed to the insecticide. A random sample of eighty starlings were caught and fed their regular food, which had been treated with the substance. After 10 days, ten starlings had died. Use a 99% confidence interval to estimate the true proportion of starlings that will be killed by the substance.

7.135 A recent poll of 200 college-age women from across the country indicated 130 approved of women seeking professional careers. Find a 95% confidence interval for the true proportion of college-age women who approve of women seeking professional careers.

7.136 Officials from a high school claim that at least 85% of the students who have graduated from the school have received a college degree or are enrolled in a college degree program. A random sample of sixty former graduates indicates that forty-seven have received or are enrolled in a program to receive a college degree. Do the data contradict the school officials' claim? (Use $\alpha = .05$.)

7.137 Many people think that a national lobby's successful fight against gun control legislation is reflecting the will of a minority of Americans. A random sample of 4,000 citizens yielded 2,250 who are in favor of gun control legislation. Use a 99% confidence interval to estimate the true proportion of Americans who favor gun control legislation. Interpret the result.

7.138 To help consumers assess the risks they are taking, the Food and Drug Administration (FDA) publishes the amount of nicotine found in all commercial brands of cigarettes. A new cigarette has recently been marketed. The FDA tests on this cigarette gave a mean nicotine content of 26.4 milligrams and standard deviation of 2.0 milligrams for a sample of $n = 9$ cigarettes. Using a 95% confidence interval, estimate the true mean nicotine content per cigarette for the brand. Interpret the results.

7.139 A health researcher wishes to estimate the mean number of cavities per child for children under the age of 12 who live in a specified environment. The number of cavities per child for a random sample of thirty-five children under the age of 12 has a mean of 2 and a standard deviation of 1.7. Construct a 90% confidence interval for the mean number of cavities per child under the age of 12 who lives in the sampled environment.

7.140 In the past, a chemical company produced 880 pounds of a certain type of plastic per day. Now, using a newly developed and less expensive process, the mean daily yield of plastic for the first 50 days of production is 871 pounds; the standard deviation is 21 pounds.
a. Do the data provide sufficient evidence to indicate that the mean daily yield for the new process is less than for the old procedure? (Test using $\alpha = .01$.)
b. What assumptions must you make in order to use the statistical test you employed?

*__7.141__ A machine used to fill beer cans must operate so that the amount of beer actually dispensed varies very little. If too much beer is released, the cans will overflow, causing waste. If too little beer is released, the cans will not contain enough beer, causing complaints from customers. A random sample of the fills for twenty cans yielded a standard deviation of .07 ounce. Using a 95% confidence interval, estimate the true variance of the fills.

*__7.142__ Ophthalmologists require an instrument that can rapidly measure interocular pressure for glaucoma patients. The device now in general use is known to yield readings of this pressure with a variance of 10.3. The variance of five pressure readings on the same eye by a newly developed instrument is equal to 9.8. Does this sample variance provide sufficient evidence to indicate that the new instrument is more reliable than the instrument currently in use? (Use $\alpha = .05$.)

*__7.143__ The standard deviation of the diameters of screw-top lids must not be larger than .6 millimeter to ensure that the lids will fit properly on glass jars. A random sample of fifteen lids yields a sample standard deviation of .78 millimeter. Do the data provide sufficient evidence to conclude the standard deviation of the lid diameters is larger than .6? (Use $\alpha = .01$.)

On Your Own . . .

Choose a population pertinent to your major area of interest that has an unknown mean (or, if the population is binomial, that has an unknown probability of success). For example, a marketing major may be interested in the proportion of consumers who prefer a certain product. A sociology major may be interested in estimating the proportion of people in a certain socioeconomic group or the mean income of people living in a certain part of a city. A political scientist may wish to estimate the proportion of an electorate in favor of a certain candidate, a certain amendment, or a certain presidential policy. A person interested in medicine might want to find the average length of time patients stay in the hospital or the average number of people treated daily in the emergency room. We could continue with examples, but the point should be clear—choose something of interest to you.

Define the parameter you want to estimate and conduct a *pilot study* to obtain an initial estimate of the parameter of interest and, more important, an estimate of the variability associated with the estimator. A pilot study is a small experiment (perhaps twenty to thirty observations) used to gain some information about the population of interest. The purpose is to help plan more elaborate future experiments. Using the results of your pilot study, determine the sample size necessary to estimate the parameter to within a reasonable bound (of your choice) with a 95% confidence interval.

References

Environmental Protection Agency. *Environment Midwest,* September–October 1976, Region V.

Freedman, D., Pisani, R., & Purves, R. *Statistics.* New York: W. W. Norton, 1978. Chapter 26.

Mendenhall, W. *Introduction to probability and statistics.* 6th ed. Boston: Duxbury, 1983. Chapters 8 and 9.

Snedecor, G. W., & Cochran, W. G. *Statistical methods.* 7th ed. Ames: Iowa State University Press, 1980. Chapters 4 and 5.

CHAPTER 8

Estimation and Tests of Hypotheses: Two Samples

Where We've Been . . .

Two methods for making statistical inferences, estimation and tests of hypotheses, were presented in Chapter 7. Confidence intervals and tests of hypotheses based on single samples were used to make inferences about the nature of sampled populations. Particularly, we gave confidence intervals and tests of hypotheses concerning a population mean μ and a binomial proportion p, and we learned how to select the sample size necessary to obtain a specified amount of information concerning a parameter. (Inferences about a population variance σ^2 were included in an optional section.)

Where We're Going . . .

Now that we have learned to make inferences about a single population, we will learn how to compare two populations. Such problems often arise in practice. We may wish to compare the mean gas mileages for two models of automobiles, the mean retirement ages of workers in the public and private sectors, or the mean reaction times of men and women to a visual stimulus. We may also wish to compare two population proportions, say the proportions of subjects in a psychological experiment that respond to two different stimuli. How to decide whether differences exist and how to estimate the differences between population means and proportions will be the subjects of this chapter.

Contents

Many experiments involve a comparison of two populations. For example, a sales manager for a steel company may want to estimate the difference in mean sales per customer between two different salespeople. A consumer group may want to test whether two major brands of food freezers differ in the mean amount of electricity they use. A political candidate may wish to estimate the difference in the proportions of voters in two districts who favor his or her candidacy. A professional golfer may wish to compare the variability in the distance that two competing brands of golf balls travel when struck with the same club. We will consider techniques for solving these two-sample inference problems in this chapter.

8.1 Large-Sample Inferences About the Difference Between Two Population Means: Independent Sampling

Many of the same procedures that are used to estimate and test hypotheses about a single parameter can be modified to make inferences about two parameters. Both the z- and t-statistics may be adapted to make inferences about the difference between two population means.

In this section we develop the large-sample z-statistic for comparing two population means. The t-statistic for making small-sample inferences about the difference between two population means is introduced in Section 8.2.

We will use Example 8.1 to introduce the procedures for making large-sample inferences about the difference between two population means.

Example 8.1

A dietitian has developed a diet that is low in fats, carbohydrates, and cholesterol. Although the diet was initially intended to be used by people with heart disease, the dietitian wishes to examine the effect this diet has on the weights of obese people. Two random samples of 100 obese people each are selected, and one group of 100 is placed on the low-fat diet. The other 100 are placed on a diet that contains approximately the same quantity of food but is not as low in fats, carbohydrates, and cholesterol. For each person, the amount of weight lost (or gained) in a 3-week period is recorded. Using the data given in the table, form a 95% confidence interval for the difference between the population mean weight losses for the two diets.

	LOW-FAT DIET	OTHER DIET
SAMPLE SIZE	100	100
SAMPLE MEAN WEIGHT LOSS	9.3 pounds	3.7 pounds
SAMPLE VARIANCE	22.4	16.3

Solution

Recall that the general form of a large-sample confidence interval for a single mean μ is $\bar{x} \pm z_{\alpha/2}\sigma_{\bar{x}}$. That is, we add and subtract $z_{\alpha/2}$ standard deviations of the sample estimate, $\bar{x}$, to the value of the estimate. We will employ a similar procedure to form the confidence interval for the difference between two population means.

Let μ_1 represent the mean of the conceptual population of weight losses for all obese people who could be placed on the low-fat diet. Let μ_2 be similarly defined for the other diet. We wish to form a confidence interval for $(\mu_1 - \mu_2)$. An intuitively appealing estimator for $(\mu_1 - \mu_2)$ is the difference between the sample means $(\bar{x}_1 - \bar{x}_2)$. Thus, we will form the confidence interval of interest by

$$(\bar{x}_1 - \bar{x}_2) \pm z_{\alpha/2}\sigma_{(\bar{x}_1 - \bar{x}_2)}$$

In Section 6.5, we noted that

$$\sigma_{(\bar{x}_1 - \bar{x}_2)} = \sqrt{\frac{\sigma_1^2}{n_1} + \frac{\sigma_2^2}{n_2}} \approx \sqrt{\frac{s_1^2}{n_1} + \frac{s_2^2}{n_2}}$$

Using the sample data and noting that $\alpha = .05$ and $z_{.025} = 1.96$, we find that the 95% confidence interval is, approximately,

$$(9.3 - 3.7) \pm 1.96 \sqrt{\frac{22.4}{100} + \frac{16.3}{100}} = 5.6 \pm (1.96)(.62)$$

$$= 5.6 \pm 1.22$$

or (4.38, 6.82). Using this estimation procedure over and over again for different samples, we know that approximately 95% of the confidence intervals formed in this manner will enclose the difference in population means $(\mu_1 - \mu_2)$. Therefore, we are reasonably confident that the mean weight loss for the low-fat diet is between 4.38 and 6.82 pounds more than the mean weight loss for the other diet. With this information, the dietitian better understands the potential of the low-fat diet as a weight-reducing diet. ■

The justification for the procedure used in Example 8.1 to estimate $(\mu_1 - \mu_2)$ relies on the properties of the sampling distribution of $(\bar{x}_1 - \bar{x}_2)$. These properties are summarized in the box, and the performance of the estimator in repeated sampling is pictured in Figure 8.1 (page 338).

Properties of the Sampling Distribution of $(\bar{x}_1 - \bar{x}_2)$

1. The sampling distribution of $(\bar{x}_1 - \bar{x}_2)$ is approximately normal for *large samples*.
2. The mean of the sampling distribution of $(\bar{x}_1 - \bar{x}_2)$ is $(\mu_1 - \mu_2)$.
3. If the two samples are independent, the standard deviation of the sampling distribution is

$$\sigma_{(\bar{x}_1 - \bar{x}_2)} = \sqrt{\frac{\sigma_1^2}{n_1} + \frac{\sigma_2^2}{n_2}}$$

where σ_1^2 and σ_2^2 are the variances of the two populations being sampled and n_1 and n_2 are the respective sample sizes.

In Example 8.1, we noted the similarity in the procedures for forming a large-sample confidence interval for one population mean and a large-sample confidence interval for the difference between two population means. When we are testing hypotheses, the

Figure 8.1 Sampling Distribution of $(\bar{x}_1 - \bar{x}_2)$

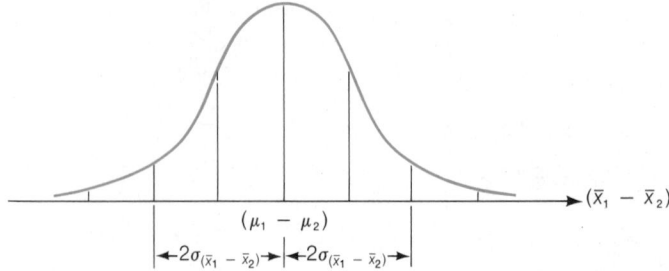

procedures are again very similar. The general large-sample procedures for forming confidence intervals and testing hypotheses about $(\mu_1 - \mu_2)$ are summarized in the following boxes.

Large-Sample Confidence Interval for $(\mu_1 - \mu_2)$

$$(\bar{x}_1 - \bar{x}_2) \pm z_{\alpha/2}\sigma_{(\bar{x}_1 - \bar{x}_2)} = (\bar{x}_1 - \bar{x}_2) \pm z_{\alpha/2}\sqrt{\frac{\sigma_1^2}{n_1} + \frac{\sigma_2^2}{n_2}}$$

Assumptions: The two samples are randomly selected in an independent manner from the two populations. The sample sizes, n_1 and n_2, are large enough so that $\bar{x}_1$ and $\bar{x}_2$ each have approximately normal sampling distributions and so that s_1^2 and s_2^2 provide good approximations to σ_1^2 and σ_2^2. This will be true if $n_1 \geq 30$ and $n_2 \geq 30$.

Large-Sample Test of an Hypothesis for $(\mu_1 - \mu_2)$

One-Tailed Test

H_0: $(\mu_1 - \mu_2) = D_0$

H_a: $(\mu_1 - \mu_2) < D_0$

   [or H_a: $(\mu_1 - \mu_2) > D_0$]

Two-Tailed Test

H_0: $(\mu_1 - \mu_2) = D_0$

H_a: $(\mu_1 - \mu_2) \neq D_0$

where $D_0 =$ Hypothesized difference between the means (this difference is often zero)

Test statistic: $z = \dfrac{(\bar{x}_1 - \bar{x}_2) - D_0}{\sigma_{(\bar{x}_1 - \bar{x}_2)}}$

where $\sigma_{(\bar{x}_1 - \bar{x}_2)} = \sqrt{\dfrac{\sigma_1^2}{n_1} + \dfrac{\sigma_2^2}{n_2}}$

Rejection region: $z < -z_\alpha$

   [or $z > z_\alpha$ when

   H_a: $(\mu_1 - \mu_2) > D_0$]

Test statistic: $z = \dfrac{(\bar{x}_1 - \bar{x}_2) - D_0}{\sigma_{(\bar{x}_1 - \bar{x}_2)}}$

Rejection region: $z < -z_{\alpha/2}$

   or $z > z_{\alpha/2}$

Assumptions: Same as for the large-sample confidence interval.

Example 8.2 The management of a restaurant wants to determine whether a new advertising campaign has increased its mean daily income (net). The incomes for each of 50 business days prior to the campaign's beginning are recorded. After conducting the advertising campaign and allowing a 20-day period for the advertising to take effect, the restaurant management records the income for 30 business days. These two samples will allow the management to make an inference about the effect of the advertising campaign on the restaurant's daily income. A summary of the results of the two samples is shown in the table. Do these samples provide sufficient evidence for the management to conclude that the mean income has been increased by the advertising campaign? Test using $\alpha = .05$.

BEFORE CAMPAIGN	AFTER CAMPAIGN
$n_1 = 50$	$n_2 = 30$
$\bar{x}_1 = \$1,255$	$\bar{x}_2 = \$1,330$
$s_1 = \$215$	$s_2 = \$238$

Solution We can best answer this question by performing a test of an hypothesis. Defining μ_1 as the mean daily income before the campaign and μ_2 as the mean daily income after the campaign, we will attempt to support the research (alternative) hypothesis that $\mu_2 > \mu_1$ (i.e., that $(\mu_1 - \mu_2) < 0$). Thus, we will test the null hypothesis, $(\mu_1 - \mu_2) = 0$, rejecting this hypothesis if $(\bar{x}_1 - \bar{x}_2)$ equals a large negative value. The elements of the test are as follows:

$$H_0: \quad (\mu_1 - \mu_2) = 0 \qquad (D_0 = 0)$$
$$H_a: \quad (\mu_1 - \mu_2) < 0 \qquad (\mu_1 < \mu_2)$$

Test statistic: $z = \dfrac{(\bar{x}_1 - \bar{x}_2) - D_0}{\sigma_{(\bar{x}_1 - \bar{x}_2)}} = \dfrac{(\bar{x}_1 - \bar{x}_2) - 0}{\sigma_{(\bar{x}_1 - \bar{x}_2)}}$

Rejection region: $z < -z_\alpha = -1.645$ (see Figure 8.2)

Figure 8.2 Rejection Region for Advertising Campaign Example

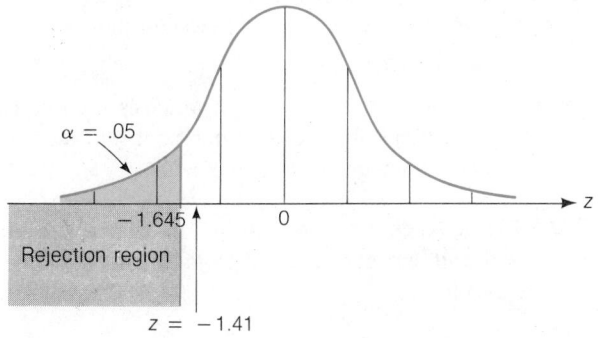

$\alpha = .05$

-1.645

Rejection region

0

z

$z = -1.41$

We now calculate

$$z = \frac{(\bar{x}_1 - \bar{x}_2) - 0}{\sigma_{(\bar{x}_1 - \bar{x}_2)}} = \frac{(1{,}255 - 1{,}330)}{\sqrt{\dfrac{\sigma_1^2}{n_1} + \dfrac{\sigma_2^2}{n_2}}}$$

$$\approx \frac{-75}{\sqrt{\dfrac{s_1^2}{n_1} + \dfrac{s_2^2}{n_2}}} = \frac{-75}{\sqrt{\dfrac{(215)^2}{50} + \dfrac{(238)^2}{30}}} = \frac{-75}{53.03} = -1.41$$

As you can see in Figure 8.2, the calculated z-value does not fall in the rejection region. The samples do not provide sufficient evidence, at the $\alpha = .05$ significance level, for the restaurant management to conclude that the advertising campaign has increased the mean daily income. ∎

Example 8.3   Find the observed significance level for the test in Example 8.2.

Solution   The alternative hypothesis in Example 8.2, H_a: $\mu_1 - \mu_2 < 0$, required a lower one-tailed test. Since the value of the test statistic calculated from the sample data was $z = -1.41$, the observed significance level (p-value) for the test is the probability of observing a value of z at least as contradictory to the null hypothesis as $z = -1.41$ if in fact H_0 is true, i.e.,

p-value $= P(z \le -1.41)$

This probability is equal to the shaded area shown in Figure 8.3.

Figure 8.3   The Observed Significance Level for the Test in Example 8.2

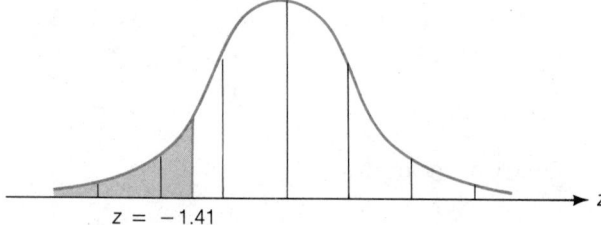

$z = -1.41$

The tabulated area corresponding to $z = 1.41$ (Table IV in the Appendix) is .4207. Therefore, the observed significance level for the test is

p-value $= .5 - .4207 = .0793$

Therefore, we would reject the null hypothesis (and conclude that $\mu_1 - \mu_2 < 0$) only for values of α larger than .0793. Consistent with the conclusion of Example 8.2, we cannot reject the null hypothesis for $\alpha = .05$. ∎

Example 8.4   Find a 95% confidence interval for the difference in mean daily incomes before and after the advertising campaign of Example 8.2, and discuss the implications of the confidence interval.

Solution The 95% confidence interval for $(\mu_1 - \mu_2)$ is

$$(\bar{x}_1 - \bar{x}_2) \pm z_{\alpha/2} \sqrt{\frac{\sigma_1^2}{n_1} + \frac{\sigma_2^2}{n_2}}$$

Once again, we will substitute s_1^2 and s_2^2 for σ_1^2 and σ_2^2, because these quantities will provide good approximations to σ_1^2 and σ_2^2 for samples as large as $n_1 = 50$ and $n_2 = 30$. Then the 95% confidence interval for $(\mu_1 - \mu_2)$ is

$$(1{,}255 - 1{,}330) \pm 1.96 \sqrt{\frac{(215)^2}{50} + \frac{(238)^2}{30}} = -75 \pm 103.94$$

Thus, we estimate the difference in mean daily income to fall in the interval $-\$178.94$ to $\$28.94$. In other words, we estimate that μ_2, the mean daily income *after* the advertising campaign, could be larger than μ_1, the mean daily income *before* the campaign, by as much as $\$178.94$ per day or it could be less than μ_1 by $\$28.94$ per day.

Now what should the restaurant management do? You can see that the sample sizes collected in the experiment are not large enough to detect a difference in the means. To be able to detect a difference (if in fact a difference exists), the management will have to repeat the experiment and increase the sample sizes. This strategy will reduce the width of the confidence interval for $(\mu_1 - \mu_2)$. The restaurant management's best estimate of $(\mu_1 - \mu_2)$ is the point estimate $(\bar{x}_1 - \bar{x}_2) = -\$75$. Thus, the management must decide whether the cost of conducting the advertising campaign is overshadowed by a possible gain in mean daily income estimated at $\$75$ (but which might be as large as $\$178.94$ or could be as low as $-\$28.94$). Based on this analysis, the management will decide whether to continue the experiment or reject the new advertising program as a poor investment. ■

Case Study 8.1
Productivity and Mobility

One theory regarding the mobility of college and university faculty members is that those who are most productive in the publishing of scholarly articles are also the most mobile. The logic behind this theory is that good researchers and publishers receive more job offers and are therefore more likely to move from one university to another. Since most people in the academic world are aware that a similar strong relationship does not exist for academics who are known to be fine teachers, we might wonder whether it holds for persons employed in industry. George F. Dreher (1982) considered this question by examining the personnel records of a large national oil company. Dreher obtained early career performance records for 529 of the company's employees. Of these, 174 were classified as *stayers,* those who stayed with the company; the other 355, who left the company at varying points during a 15-year period, were classified as *leavers.* Dreher made a number of comparisons between these two groups. Among these were a comparison of measures of the mean initial performances of the stayers and leavers, their rates of career advancement (number of promotions per year), and their final performance appraisals. The means and standard deviations for these sample comparisons are shown in Table 8.1.

Although a large-sample z-test would be appropriate, Dreher tests for differences between the means of stayers and leavers using a small-sample t-test. This test, to be discussed in the following section, is based on assumptions that are not necessary for the large-sample z-test. When the sample sizes are very large, as in Dreher's sampling, the numerical values of the t- and z-statistics are almost the same and lead to essentially the same results. Dreher's computed t-values are shown in Table 8.1.

We have not computed the values of the large-sample z-statistics for comparing the three pairs of population means, but it is clear (from examining Dreher's large t-values) that there is ample evidence to indicate a difference in means for all three comparisons. (The p-values for all three tests are extremely small.) In fact, it appears that stayers had higher mean measures of initial and final performance and a higher mean rate of career advancement than the leavers.

Table 8.1

Difference Between Leavers and Continuing Employees

VARIABLE	STAYERS $(n_1 = 174)$ $\bar{x}_1$	s_1	LEAVERS $(n_2 = 355)$ $\bar{x}_2$	s_2	t
Initial performance	3.51	.51	3.24	.52	5.23
Rate of career advancement	.43	.20	.31	.31	4.63
Final performance appraisal	3.78	.62	3.15	.68	9.76

Exercises 8.1–8.17

Learning the Mechanics

SAMPLE 1	SAMPLE 2
$n_1 = 36$	$n_2 = 40$
$\bar{x}_1 = 3.7$	$\bar{x}_2 = 4.2$
$s_1^2 = .25$	$s_2^2 = .36$

8.1 Select two independent random samples: forty observations from population 1 and fifty from population 2. The sample means and variances are shown in the table.
a. Form a 95% confidence interval for $(\mu_1 - \mu_2)$, the difference between the means of population 1 and population 2.
b. Test the null hypothesis H_0: $(\mu_1 - \mu_2) = 0$ against the alternative hypothesis H_a: $(\mu_1 - \mu_2) \neq 0$. Use $\alpha = .05$.
c. Find the observed significance level for the test and interpret its value.

SAMPLE 1	SAMPLE 2
$n_1 = 80$	$n_2 = 80$
$\bar{x}_1 = 17.6$	$\bar{x}_2 = 21.3$
$s_1 = 5.9$	$s_2 = 7.8$

8.2 Two independent random samples were selected from populations with means μ_1 and μ_2, respectively. The sample sizes, means, and standard deviations are shown in the table.
a. Form a 90% confidence interval for $(\mu_1 - \mu_2)$.
b. Form a 99% confidence interval for $(\mu_1 - \mu_2)$.
c. Test the null hypothesis H_0: $(\mu_1 - \mu_2) = 0$ against the alternative hypothesis H_a: $(\mu_1 - \mu_2) < 0$. Use $\alpha = .01$.
d. Find the observed significance level for the test and interpret its value.

SAMPLE 1	SAMPLE 2
$n_1 = 100$	$n_2 = 120$
$\bar{x}_1 = 25.3$	$\bar{x}_2 = 23.1$
$s_1^2 = 80.6$	$s_2^2 = 110.2$

8.3 Two independent random samples produced the results shown in the table.
a. Test H_0: $(\mu_1 - \mu_2) = 0$ against H_a: $(\mu_1 - \mu_2) > 0$. Use $\alpha = .10$.
b. Test H_0: $(\mu_1 - \mu_2) = 0$ against H_a: $(\mu_1 - \mu_2) \neq 0$. Use $\alpha = .05$.
c. Form a 95% confidence interval for $(\mu_1 - \mu_2)$.

8.4 Are the hypothesis test and confidence interval procedures given in this section valid if the sampled populations are not normally distributed? Explain.

Applying the Concepts

8.5 An experiment has been conducted at a university to compare the mean number of study hours expended per week by student athletes with the mean number of hours expended by nonathletes. A random sample of 55 athletes produced a mean equal to 20.6 hours studied per week and a standard deviation equal to 5.3 hours. A second random sample of 200 nonathletes produced a mean equal to 23.5 hours per week and a standard deviation equal to 4.1 hours.

a. Describe the two populations involved in the comparison.

b. Do the samples provide sufficient evidence to conclude that there is a difference in the mean number of hours of study per week between athletes and nonathletes? Test using $\alpha = .01$.

c. Construct a 99% confidence interval for $(\mu_1 - \mu_2)$.

d. Would a 95% confidence interval for $(\mu_1 - \mu_2)$ be narrower or wider than the one you found in part c? Why?

8.6 A new type of band has been developed by a dental laboratory for children who have to wear braces. The new bands are designed to be more comfortable, look better, and provide more rapid progress in realigning teeth. An experiment was conducted to compare the mean wearing time necessary to correct a specific type of misalignment between the old braces and the new bands. One hundred children were randomly assigned, fifty to each group. A summary of the data is shown in the table.

	OLD BRACES	NEW BANDS
$\bar{x}$	410 days	380 days
s	45 days	60 days

a. Is there sufficient evidence to conclude that the new bands do not have to be worn as long as the old braces? Use $\alpha = .01$.

b. Find a 95% confidence interval for the difference in mean wearing times for the two types of braces. Interpret the interval.

8.7 Suppose it is desired to compare two physical education training programs for preadolescent girls. A total of eighty girls are randomly selected, with forty assigned to each program. After three 6-week periods on the program, each girl is given a fitness test that yields a score between zero and 100. The means and variances of the scores for the two groups are shown in the table. Calculate a 99% confidence interval for the true difference in mean fitness scores for girls trained using these two programs.

	n	$\bar{x}$	s^2
PROGRAM 1	40	78.7	201.6
PROGRAM 2	40	75.3	259.2

8.8 An experiment was conducted to compare the yield of two varieties of tomatoes, A and B. Forty plants of each variety were randomly selected and planted in the same field. The yields, recorded in kilograms of tomatoes produced for each plant, possessed means of 10.5 kilograms per plant for variety A and 9.3 kilograms per plant for variety B. The variances for samples A and B were 2.1 and 2.8, respectively. Do the data provide sufficient evidence to conclude there is a difference between the mean weights of tomatoes produced per plant for the two varieties? Test using $\alpha = .05$.

8.9 It is often said that economic status is related to the commission of crimes. To test this theory, a sociologist selected a random sample of seventy people (who had no record of criminal conviction) from the census records of a certain city and recorded their annual incomes. Similarly, a random sample of sixty people, each of whom had committed their first crime, was selected from court records and the annual income (prior to arrest) was recorded for each. The means and variances of the annual incomes (in thousands of dollars) for the people in the two groups are shown in the table. Do the data provide sufficient evidence to indicate that the mean income of criminals, prior to committing their first offense, is lower than that for the noncriminal public? Test using $\alpha = .05$.

	$\bar{x}$	s^2
CRIMINALS	13.3	24.2
NONCRIMINALS	15.4	42.6

BRAND S	BRAND H
$\bar{x}_1 = 4.1$	$\bar{x}_2 = 5.2$
$s_1 = 1.2$	$s_2 = 1.4$

8.10 A large supermarket chain is interested in determining whether a difference exists between the mean shelf-life (in days) of brand S bread and brand H bread. Random samples of fifty freshly baked loaves of each brand were tested, with the results shown in the table.

a. Is there sufficient evidence to conclude that a difference does exist between the mean shelf-lives of brand S and brand H bread? Test at the $\alpha = .05$ level.

b. Find the observed significance level for the test and interpret its value.

c. Let μ_1 and μ_2 represent the mean shelf-lives for brands S and H, respectively. Construct a 90% confidence interval for $(\mu_1 - \mu_2)$. Give an interpretation of your confidence interval.

A	B
$\bar{x}_1 = 365$	$\bar{x}_2 = 352$
$s_1 = 23$	$s_2 = 41$

8.11 Two manufacturers of corrugated fiberboard both claim that the strength of their product tests on the average at more than 360 pounds per square inch. As a result of consumer complaints, a consumer products testing firm believes that firm A's product is stronger than firm B's. To test its belief, 100 fiberboards were chosen randomly from firm A's inventory and 100 were chosen from firm B's inventory. The results of tests run on the samples are shown in the table.

a. Does the sample information support the consumer products testing firm's belief? Test at the .05 significance level.

b. What assumptions did you make in conducting the test in part a? Do you think such assumptions could comfortably be made in practice? Why or why not?

c. Find the observed significance level for the test and interpret its value.

8.12 Give a practical example where each of the following hypotheses would be appropriate:

a. H_0: $(\mu_1 - \mu_2) = 0$ and H_a: $(\mu_1 - \mu_2) > 0$

b. H_0: $(\mu_1 - \mu_2) = 0$ and H_a: $(\mu_1 - \mu_2) < 0$
c. H_0: $(\mu_1 - \mu_2) = 0$ and H_a: $(\mu_1 - \mu_2) \neq 0$

8.13 The *Orlando Sentinel* (April 12, 1984) presented an article titled "Books Hold Their Own Against Television." The article reports on an hour-long survey of 1,961 persons selected from among the American public. The survey was conducted for the Book Industry Study Group, a nonprofit organization representing book publishers, manufacturers, wholesalers, etc. Among the statistics computed from the survey data, they found that the amount of time each person claimed to read per week averaged 11.7 hours versus 16.3 hours spent watching television. However, the article notes that the A. C. Nielsen Co., which measures television viewing for sponsors, claims that the average American watched $28\frac{1}{2}$ hours of television per week. Suppose that the standard deviations of the distributions of lengths of time spent per week reading and the lengths of time watching television are both equal to 5 hours. Assume also that the Nielsen estimate of mean television viewing time is based on an independent random sample of size 1,961.

a. Use the survey results to find a 95% confidence interval for the difference between the mean length of time per week reading and the mean length of time watching television. Interpret your results.

b. If the survey and A. C. Nielsen are both sampling the same population, what (approximately) is the probability that their sample estimates of mean television viewing time would differ by as much as $(28.5 - 16.3) = 12.2$ hours?

c. Can you explain why the "rare event" in part b might have occurred?

8.14 Refer to Case Study 8.1 and the comparison of mean performance characteristics between corporate stayers and leavers.

a. Use Dreher's statistics, shown in Table 8.1, to test the null hypothesis that there is no difference in mean initial performance measure between stayers and leavers. Test using $\alpha = .01$. Interpret your conclusion.

b. Give the approximate p-value for the test and interpret it.

8.15 Refer to Case Study 8.1 and the comparison of mean performance characteristics between corporate stayers and leavers.

a. Use Dreher's statistics, shown in Table 8.1, to test the null hypothesis that there is no difference in the rate of career advancement, i.e., the mean number of promotions per year, between stayers and leavers. Test using $\alpha = .01$. Interpret your conclusion.

b. Give the approximate p-value for the test and interpret it.

8.16 Refer to Case Study 8.1 and the comparison of mean performance characteristics between corporate stayers and leavers.

a. Use Dreher's statistics, shown in Table 8.1, to test the null hypothesis that there is no difference in mean final performance measure between stayers and leavers. Test using $\alpha = .01$. Interpret your conclusion.

b. Give the approximate p-value for the test and interpret it.

8.17 As part of a study in participative management, George H. Hines (1974) sampled workers from two types of New Zealand sociocultural backgrounds: those who believed in the existence of a class system and those who believed that they lived and worked in a classless society. Each worker in the sampling was selected from a work environment with participatory management. Do workers who consider themselves to be social equals with their management superiors possess different levels of job satisfaction than those workers who see themselves as socially different from management? Each worker in the independent random samples was asked to answer this question by rating his or her job satisfaction on a scale of 1 (poor) to 7 (excellent). Using the results of this study (shown in the table), what can you say about differences in job satisfaction for the two different sociocultural types of workers?

| | BELIEF IN EXISTENCE OF A CLASS SYSTEM | |
	Yes	No
SAMPLE SIZE	175	277
MEAN	5.42	5.19
STANDARD DEVIATION	1.24	1.17

8.2
Small-Sample Inferences About the Difference Between Two Population Means: Independent Sampling

Suppose a television network wants to determine whether major sports events or first-run movies attract more viewers in the prime-time hours. It selects twenty-eight prime-time evenings; of these, thirteen have programs devoted to major sports events and the remaining fifteen have first-run movies. The number of viewers (estimated by a television viewer rating firm) is recorded for each program. If μ_1 is the mean number of sports viewers per evening of sports programming and μ_2 is the mean number of movie viewers per evening, we want to detect a difference between μ_1 and μ_2—if such a difference exists. Therefore, we want to test the null hypothesis

$$H_0: \quad (\mu_1 - \mu_2) = 0$$

against the alternative hypothesis

$$H_a: \quad (\mu_1 - \mu_2) \neq 0 \qquad \text{(i.e., either } \mu_1 > \mu_2 \quad \text{or} \quad \mu_2 > \mu_1)$$

Since the sample sizes are small, estimates of σ_1^2 and σ_2^2 are unreliable and the z-test statistic is inappropriate for the test. But as in the case of a single mean (Section 7.4), we can construct a Student's t-statistic. This statistic (formula to be given subsequently) has the familiar t-distribution described in Chapter 7. To use the t-statistic, both sampled populations must be approximately normally distributed with equal population variances, and the random samples must be selected independently of each other. The normality and equal variances assumptions imply relative frequency distributions for the populations that would appear as shown in Figure 8.4.

Will these assumptions be satisfied for the television viewing problem? We think that both assumptions will be adequately satisfied in this sampling situation. Since we assume the two populations have equal variances ($\sigma_1^2 = \sigma_2^2 = \sigma^2$), it is reasonable to use the information contained in both samples to construct a **pooled sample estimator** of σ^2 for use in the t-statistic. Thus, if s_1^2 and s_2^2 are the two sample variances (both

Figure 8.4 Assumptions for the Two-Sample t: (1) Normal Populations, (2) Equal Variances

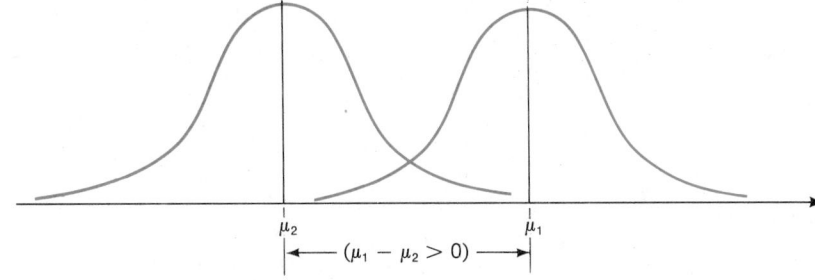

estimating the variance σ^2 common to both populations), the pooled estimator of σ^2, denoted as s_p^2, is

$$s_p^2 = \frac{(n_1 - 1)s_1^2 + (n_2 - 1)s_2^2}{(n_1 - 1) + (n_2 - 1)} = \frac{(n_1 - 1)s_1^2 + (n_2 - 1)s_2^2}{n_1 + n_2 - 2}$$

or

$$s_p^2 = \frac{\overbrace{\sum(x_1 - \bar{x}_1)^2}^{\substack{\text{From} \\ \text{sample 1}}} + \overbrace{\sum(x_2 - \bar{x}_2)^2}^{\substack{\text{From} \\ \text{sample 2}}}}{n_1 + n_2 - 2}$$

where x_1 represents a measurement from sample 1 and x_2 represents a measurement from sample 2. Recall that the term *degrees of freedom* was defined in Section 7.4 as 1 less than the sample size for each sample—that is, $(n_1 - 1)$ for sample 1 and $(n_2 - 1)$ for sample 2. Since we are pooling the information on σ^2 obtained from both samples, the degrees of freedom associated with the pooled variance s_p^2 is equal to the sum of the degrees of freedom for the two samples, namely the denominator of s_p^2; that is, $(n_1 - 1) + (n_2 - 1) = n_1 + n_2 - 2$.

To obtain the **small-sample test statistic** for testing H_0: $(\mu_1 - \mu_2) = D_0$, substitute the pooled estimate of σ^2 into the formula for the two-sample z-statistic (Section 8.1) to obtain

$$t = \frac{(\bar{x}_1 - \bar{x}_2) - D_0}{\sqrt{s_p^2 \left(\frac{1}{n_1} + \frac{1}{n_2} \right)}}$$

It can be shown that this statistic has a t-distribution with $(n_1 + n_2 - 2)$ degrees of freedom.

We will use the television viewer example to outline the final steps for this t-test: The hypothesized difference in mean number of viewers is $D_0 = 0$. The rejection region will be two-tailed and will be based on a t-distribution with $(n_1 + n_2 - 2)$ or $(13 + 15 - 2) = 26$ df. For $\alpha = .05$, the rejection region for the test would be

$$t < -t_{\alpha/2} \quad \text{or} \quad t > t_{\alpha/2}$$

The value for $t_{.025}$ given in Table V of the Appendix is 2.056. Thus, the rejection region for the television example is

$$t < -2.056 \quad \text{or} \quad t > 2.056$$

This rejection region is shown in Figure 8.5.

Figure 8.5 Rejection Region for a Two-Tailed t-Test: $\alpha = .05$, df $= 26$

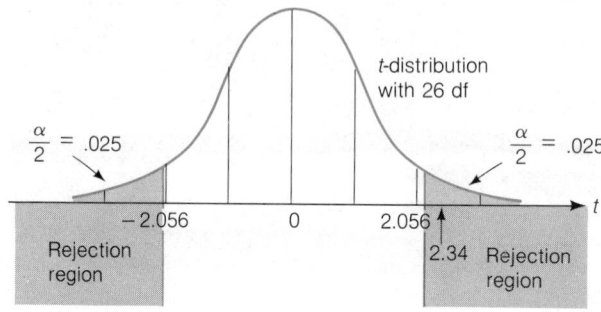

Example 8.5

Suppose the television network's samples produce the following results:

· SPORTS	MOVIE
$n_1 = 13$	$n_2 = 15$
$\bar{x}_1 = 6.8$ million	$\bar{x}_2 = 5.3$ million
$s_1 = 1.8$ million	$s_2 = 1.6$ million

Do the data provide sufficient evidence to indicate a difference between the mean numbers of viewers for major sports events and first-run movies shown in prime time? Test using $\alpha = .05$.

Solution   We first calculate

$$s_p^2 = \frac{(n_1 - 1)s_1^2 + (n_2 - 1)s_2^2}{n_1 + n_2 - 2} = \frac{(13 - 1)(1.8)^2 + (15 - 1)(1.6)^2}{13 + 15 - 2} = \frac{74.72}{26} = 2.87$$

Then

$$t = \frac{(\bar{x}_1 - \bar{x}_2) - D_0}{\sqrt{s_p^2 \left(\frac{1}{n_1} + \frac{1}{n_2}\right)}} = \frac{(6.8 - 5.3) - 0}{\sqrt{2.87 \left(\frac{1}{13} + \frac{1}{15}\right)}} = \frac{1.5}{.64} = 2.34$$

Since the observed value of t ($t = 2.34$) falls in the rejection region (see Figure 8.5), the samples provide sufficient evidence to indicate that the mean numbers of viewers differ for major sports events and first-run movies shown in prime time. Or we can say that the test results are statistically significant at the $\alpha = .05$ level of significance. Because the rejection was in the positive or upper tail of the t-distribution, it appears that the mean number of viewers for sports events exceeds that for movies. ■

Example 8.6   Find the approximate observed significance level for the test in Example 8.5.

Solution   The observed significance level (p-value) for the test is the probability of observing a value of the test statistic that is *at least* as contradictory to H_0: $\mu_1 = \mu_2$ as the value observed if in fact H_0 is true. Thus, since the test was two-sided, we have

$$p\text{-value} = P(t < -2.34 \quad \text{or} \quad t > 2.34)$$
$$= 2P(t > 2.34) \qquad \text{(because the } t\text{-distribution is symmetric about}$$
$$\text{its mean, zero)}$$

Turning to Table V in the Appendix for 26 degrees of freedom, you can see that

$$P(t < -2.479 \quad \text{or} \quad t > 2.479) = 2P(t > 2.479) = 2(.01) = .02$$

and

$$P(t < -2.056 \quad \text{or} \quad t > 2.056) = 2P(t > 2.056) = 2(.025) = .05$$

Since the observed value of t lies between 2.479 and 2.056, the p-value for the test lies between .02 and .05. We have agreed to choose the larger of these as the approximate observed significance level and would thus give the approximate p-value for the test as .05. A reader of this reported p-value would reject the null hypothesis (and conclude that a difference exists between the mean numbers of sports and movie viewers) for values of α larger than or equal to .05.   ∎

The same t-statistic can also be used to construct confidence intervals for the difference between population means. Both the confidence interval and the test of hypothesis procedures are summarized in the accompanying boxes.

Small-Sample Confidence Interval for ($\mu_1 - \mu_2$) (Independent Samples)

$$(\bar{x}_1 - \bar{x}_2) \pm t_{\alpha/2} \sqrt{s_p^2 \left(\frac{1}{n_1} + \frac{1}{n_2} \right)}$$

where

$$s_p^2 = \frac{(n_1 - 1)s_1^2 + (n_2 - 1)s_2^2}{n_1 + n_2 - 2}$$

and $t_{\alpha/2}$ is based on ($n_1 + n_2 - 2$) degrees of freedom.

Assumptions: 1.   Both sampled populations have relative frequency distributions that are approximately normal.
2.   The population variances are equal.
3.   The samples are randomly and independently selected from the populations.

Small-Sample Test of an Hypothesis for $(\mu_1 - \mu_2)$ (Independent Samples)

One-Tailed Test

H_0: $(\mu_1 - \mu_2) = D_0$

H_a: $(\mu_1 - \mu_2) < D_0$
    [or H_a: $(\mu_1 - \mu_2) > D_0$]

Test statistic:

$$t = \frac{(\bar{x}_1 - \bar{x}_2) - D_0}{\sqrt{s_p^2\left(\dfrac{1}{n_1} + \dfrac{1}{n_2}\right)}}$$

Rejection region:
  $t < -t_\alpha$
  [or $t > t_\alpha$ when
  H_a: $(\mu_1 - \mu_2) > D_0$]

Two-Tailed Test

H_0: $(\mu_1 - \mu_2) = D_0$

H_a: $(\mu_1 - \mu_2) \neq D_0$

Test statistic:

$$t = \frac{(\bar{x}_1 - \bar{x}_2) - D_0}{\sqrt{s_p^2\left(\dfrac{1}{n_1} + \dfrac{1}{n_2}\right)}}$$

Rejection region:
  $t < -t_{\alpha/2}$
  or $t > t_{\alpha/2}$

where t_α and $t_{\alpha/2}$ are based on $(n_1 + n_2 - 2)$ degrees of freedom.

Assumptions: Same as for the small-sample confidence interval for $(\mu_1 - \mu_2)$ in the previous box.

Example 8.7   Suppose you wish to compare a new method of teaching reading to "slow learners" to the current standard method. You decide to base this comparison on the results of a reading test given at the end of a learning period of 6 months. Of a random sample of twenty slow learners, eight are taught by the new method and twelve are taught by the standard method. All twenty children are taught by qualified instructors under similar conditions for a 6-month period. The results of the reading test at the end of this period are summarized in the table. Estimate the true mean difference $(\mu_1 - \mu_2)$ between the test scores for the new method and the standard method. Use a 90% confidence interval, and interpret the interval. What assumptions must be made in order that the estimate be valid?

NEW METHOD	STANDARD METHOD
$n_1 = 8$	$n_2 = 12$
$\bar{x}_1 = 76.9$	$\bar{x}_2 = 72.7$
$s_1 = 4.85$	$s_2 = 6.35$

Solution   The objective of this experiment is to obtain a 90% confidence interval for $(\mu_1 - \mu_2)$. To use the small-sample confidence interval for $(\mu_1 - \mu_2)$, the following assumptions must be satisfied:

1. We assume that the populations of test scores are normally distributed for both the new and the standard methods of instruction. Since a test score can be viewed as

a sum of the results on the various components of the test, the central limit theorem lends credence to this assumption.

2. The variance of the test scores is assumed to be the same for the two populations. Under the circumstances, we might expect the variation in test scores to be approximately the same for both methods.

3. The samples are randomly and independently selected from the two populations. We have randomly chosen twenty different slow learners for the two samples in such a way that the test score for one child is not dependent on the test score for any other child. Therefore, this assumption would probably be valid.

The first step in constructing the confidence interval is to calculate the pooled estimate of variance

$$s_p^2 = \frac{(n_1 - 1)s_1^2 + (n_2 - 1)s_2^2}{n_1 + n_2 - 2}$$

$$= \frac{(8 - 1)(4.85)^2 + (12 - 1)(6.35)^2}{8 + 12 - 2}$$

$$= 33.7892$$

where s_p^2 is based on $(n_1 + n_2 - 2) = (8 + 12 - 2) = 18$ degrees of freedom. Then the 90% confidence interval for $(\mu_1 - \mu_2)$, the difference between mean test scores for the two methods, is

$$(\bar{x}_1 - \bar{x}_2) \pm t_{\alpha/2} \sqrt{s_p^2 \left(\frac{1}{n_1} + \frac{1}{n_2}\right)} = (76.9 - 72.7) \pm t_{.05} \sqrt{33.7892 \left(\frac{1}{8} + \frac{1}{12}\right)}$$

$$= 4.20 \pm 1.734(2.653) = 4.20 \pm 4.60$$

or $(-.40, 8.80)$. This means that with a confidence coefficient equal to .90, we estimate the difference in mean test scores between using the new method of teaching and the standard method to fall in the interval $-.40$ to 8.80. In other words, we estimate the mean test score for the new method to be anywhere from .40 less than to 8.80 more than the mean test score for the standard method. Although the sample means seem to suggest that the new method is associated with a higher mean test score, there is insufficient evidence to indicate that $(\mu_1 - \mu_2)$ differs from zero because the interval includes zero as a possible value for $(\mu_1 - \mu_2)$. To show a difference in mean test scores (if it exists), you could increase the sample size and thereby narrow the width of the confidence interval for $(\mu_1 - \mu_2)$. An alternative is to design the experiment differently. This possibility will be discussed in the next section. ■

The two-sample t-statistic is a powerful tool for comparing population means when the assumptions are satisfied. It has also been shown to retain its usefulness when the sampled populations are only approximately normally distributed. And when the sample sizes are equal, the assumption of equal population variances can be relaxed. That is, if $n_1 = n_2$ then σ_1^2 and σ_2^2 can be quite different and the test statistic will still possess, approximately, a Student's t-distribution. When the experimental situation does not satisfy the assumptions, you can select larger samples from the populations or you can use other available statistical tests (nonparametric statistical tests, which are described in Chapter 10).

What Should You Do If the Assumptions Are Not Satisfied?

Answer: If you are concerned that the assumptions are not satisfied, use the Wilcoxon rank sum test for independent samples to test for a shift in population distributions. See Chapter 10.

Exercises 8.18–8.30

SAMPLE 1	SAMPLE 2
1.2	4.2
3.1	2.7
1.7	3.6
2.8	3.9
3.0	

SAMPLE 1	SAMPLE 2
$n_1 = 17$	$n_2 = 12$
$\bar{x}_1 = 5.4$	$\bar{x}_2 = 7.9$
$s_1 = 3.4$	$s_2 = 4.8$

Learning the Mechanics

8.18 Independent random samples from normal populations produced the results shown in the table.

a. Calculate the pooled estimate of σ^2.

b. Do the data provide sufficient evidence to indicate that $\mu_2 > \mu_1$? Test using $\alpha = .10$.

c. Find a 90% confidence interval for $(\mu_1 - \mu_2)$.

8.19 Independent random samples selected from two normal populations produced the sample means and standard deviations shown in the table.

a. Test H_0: $(\mu_1 - \mu_2) = 0$ against H_a: $(\mu_1 - \mu_2) \neq 0$. Use $\alpha = .05$.

b. Form a 95% confidence interval for $(\mu_1 - \mu_2)$.

8.20 To use the t-statistic to test for a difference between the means of two populations, what assumptions must be made about the two populations? About the two samples?

8.21 In the t-tests of this section, σ_1^2 and σ_2^2 are assumed to be equal. Thus, we say $\sigma_1^2 = \sigma_2^2 = \sigma^2$. Why is a pooled estimator of σ^2 used instead of either s_1^2 or s_2^2?

Applying the Concepts

8.22 The following data from *Sports Magazine* (March 1984) list the salaries of some of the highest-paid professional athletes in baseball and basketball. Suppose that these represent independent random samples of salaries from among the higher-paid players in both professions. Do the data provide sufficient evidence to indicate a difference in mean salaries between the professional basketball and baseball players? Test using $\alpha = .01$.

1. Larry Bird, $1.8 million
    forward, Boston Celtics

2. Mitch Kupchak, $1.6 million
    forward, Los Angeles Lakers

3. Moses Malone, $1.6 million
    center, Philadelphia 76ers

1. Steve Garvey, $1.85 million
    first baseman, San Diego Padres

2. Mike Schmidt, $1.652 million
    third baseman, Philadelphia Phillies

3. George Foster, $1.55 million
    outfielder, New York Mets

4. Kareem Abdul-Jabbar, $1.5 million center, Los Angeles Lakers	4. Dave Winfield, $1.531 million outfielder, New York Yankees
5. Ralph Sampson, $1.3 million center, Houston Rockets	5. Dale Murphy, $1.3 million outfielder, Atlanta Braves
6. Jack Sikma, $1.092 million center, Seattle SuperSonics	6. Fred Lynn, $1.25 million outfielder, California Angels
7. Otis Birdsong, $1 million guard, New Jersey Nets	7. Gary Carter, $1.242 million catcher, Montreal Expos
8. Kevin McHale, $1 million center-forward, Boston Celtics	8. Pete Rose, $1.2 million first baseman, Montreal Expos
9. Marques Johnson, $800,000 forward, Milwaukee Bucks	9. Rod Carew, $1.102 million first baseman, California Angels
10. Alex English, $773,000 forward, Denver Nuggets	10. Ozzie Smith, $1.1 million shortstop, St. Louis Cardinals

8.23 Amid concerns and protests about its potential dangers, nuclear-generated energy is reportedly on the rise and the growth rate of nuclear power facilities is expected to continue to increase. One reason is that nuclear power is cheaper to produce than conventional sources. The Atomic Industrial Forum, which represents the nuclear industry, reports that coal-produced energy costs 3.5¢ per kilowatt-hour to produce compared to 3.1¢ for nuclear power (*Time*, February 13, 1984).

Assume that these costs are averages and that the results were taken over nine nuclear power plants and eleven coal-powered plants. Assume also that the variances of the cost estimates for nuclear and coal-produced power were .05 and .04, respectively. Does this information provide sufficient evidence to say that nuclear-generated power is less expensive to produce than coal-produced power? Test using $\alpha = .05$. What assumptions are required for the test to be valid?

8.24 An industrial plant wants to determine which of two types of fuel—gas or electric—will produce more useful energy at the lower cost. One measure of economical energy production, called the *plant investment per delivered quad,* is calculated by taking the amount of money (in dollars) invested in the particular utility by the plant and dividing by the delivered amount of energy (in quadrillion British thermal units). The smaller this ratio, the less an industrial plant pays for its delivered energy.

Random samples of eleven plants using electrical utilities and sixteen plants using gas utilities were taken, and the plant investment per quad was calculated for each. The data produced the results shown in the table.

	ELECTRIC	GAS
SAMPLE SIZE	11	16
MEAN INVESTMENT/QUAD (BILLIONS)	$44.5	$34.5
VARIANCE	76.4	63.8

a. Do these data provide sufficient evidence at the $\alpha = .05$ level of significance to indicate a difference in the average investment per quad between the plants using gas and those using electrical utilities?

b. Find a 90% confidence interval for $(\mu_1 - \mu_2)$. Give a practical interpretation of this interval.

8.25 An experiment is conducted to investigate the effect of a drug on the time to complete a task. Twenty people are divided at random into two groups of ten each. One group is given a placebo, while the second experimental group is administered a drug thought to increase the ability to complete the task quickly. For the control group, the times required to complete the task had a mean of 14.8 minutes and a variance of 3.9; for the experimental group, the average was 12.3 minutes and the variance was 4.3.

a. Test the null hypothesis that the drug has no effect in reducing the mean length of time to complete the task against the alternative hypothesis that the mean time is less for those subjects who receive the drug. Conduct the test at the $\alpha = .10$ level of significance.

b. Find the approximate observed significance level for the test and interpret its value.

U.S. PLANTS	JAPANESE PLANTS
7.11%	3.52%
6.06%	2.02%
8.00%	4.91%
6.87%	3.22%
4.77%	1.92%

8.26 With the emergence of Japan as an industrial superpower, American businesses have begun taking a close look at Japanese management styles and philosophies. Some of the credit for the high quality of Japanese products has been attributed to the Japanese system of permanent employment for their workers. In the United States, high job turnover rates are common in many industries and are associated with high product defect rates. High turnover rates mean that U.S. plants are more highly populated with inexperienced workers who are unfamiliar with the company's product lines than is the case in Japan. In a recent study of the room air conditioner industry in Japan and the United States, David Garvin (1983) reported that the difference in the average annual turnover rate of workers between American plants and Japanese plants was 3.1%. In a different study, five Japanese and five American plants that manufacture room air conditioners were randomly sampled and their turnover rates were determined. (See the table.)

a. Do these data provide sufficient evidence to contradict the results of Garvin's study? Test using $\alpha = .10$.

b. Report the observed significance level of the test you conducted in part a.

c. List any assumptions you made in conducting the hypothesis test of part a.

8.27 Many armchair quarterbacks try to figure out what makes a good football team—offense or defense. Some even think one conference is better than another. Some data for the 1982 season are given in the table at the top of the next page (from *Sports Illustrated*, September 1, 1983).

a. Find the mean and variance of the two conferences for yards per game rushing.

b. Do the data provide sufficient evidence to indicate that the mean yards per game rushing differ for the two conferences? Test using $\alpha = .05$.

AFC	YARDS/GAME RUSHING	NFC	YARDS/GAME RUSHING
Bills	152	Cowboys	146
Patriots	150	Saints	140
Dolphins	149	Cardinals	134
Jets	146	Falcons	131
Steelers	132	Redskins	127
Chargers	125	Packers	120
Raiders	120	Rams	114
Colts	116	Lions	114
Broncos	113	Bears	110
Bengals	105	Bucs	106
Chiefs	105	Vikings	101
Browns	97	Giants	94
Oilers	89	Eagles	92
Seahawks	88	49ers	82

8.28 To compare two methods of teaching reading, randomly selected groups of elementary school children were assigned to each of the two teaching methods for a 6-month period. The criterion for measuring achievement was a reading comprehension test. The results are shown in the table. Do the data provide sufficient evidence to indicate a difference in mean scores on the comprehension test for the two teaching methods? Test using $\alpha = .05$.

	NUMBER OF CHILDREN PER GROUP	$\bar{x}$	s^2
METHOD 1	11	64	52
METHOD 2	14	69	71

8.29 Suppose you are the personnel manager for a company and you suspect a difference in the mean length of work time lost due to sickness for two types of employees: those who work at night versus those who work during the day. Particularly, you suspect that the mean time lost for the night shift exceeds the mean for the day shift. To check your theory, you randomly sample the records for ten employees for each shift category and record the number of days lost due to sickness within the past year. The data are shown in the table.

NIGHT SHIFT, 1		DAY SHIFT, 2	
21	2	13	18
10	19	5	17
14	6	16	3
33	4	0	24
7	12	7	1
$\bar{x}_1 = 12.8$		$\bar{x}_2 = 10.4$	
$\sum x_1^2 = 2{,}436$		$\sum x_2^2 = 1{,}698$	

a. Calculate s_1^2 and s_2^2.

b. Show that the pooled estimate of the common population standard deviation, σ, is 8.86. Look at the range of the observations within each of the two samples. Does it appear that the estimate, 8.86, is a reasonable value for σ?

c. If μ_1 and μ_2 represent the mean number of days per year lost due to sickness for the night and day shifts, respectively, test the null hypothesis $H_0: \mu_1 = \mu_2$ against the alternative $H_a: \mu_1 > \mu_2$. Use $\alpha = .05$. Do the data provide sufficient evidence to indicate that $\mu_1 > \mu_2$?

d. What assumptions must be satisfied so that the t-test from part c is valid?

PLANT DISCHARGE	UPSTREAM
30.1	29.7
36.2	30.3
33.4	26.4
28.2	27.3
29.8	31.7
34.9	32.3

8.30 Suppose your plant purifies its liquid waste and discharges the water into a local river. An EPA inspector has collected water specimens of the discharge of your plant and also water specimens in the river upstream from your plant. Each water specimen is divided into five parts, the bacteria count is read on each, and the mean count for each specimen is reported. The average bacteria counts for each of six specimens are reported in the table for the two locations.

a. Why might the bacteria counts shown here tend to be approximately normally distributed?

b. Do the data provide sufficient evidence to indicate that the mean of the bacteria count for the discharge exceeds the mean of the count upstream? Use $\alpha = .05$.

c. Find the approximate observed significance level for the test and interpret its value.

8.3 Inferences About the Difference Between Two Population Means: Paired Difference Experiments

In Example 8.7 we compared two methods of teaching reading to slow learners by means of a 90% confidence interval. Suppose it is possible to measure the slow learners' "reading IQ's" *before* they are subjected to a teaching method. Eight pairs of slow learners with similar reading IQ's are found, and one member of each pair is randomly assigned to the standard teaching method while the other is assigned to the new method. Do the data in Table 8.2 support the hypothesis that the population mean reading test score for slow learners taught by the new method is greater than the mean reading test score for those taught by the standard method?

We want to test

$$H_0: (\mu_1 - \mu_2) = 0 \qquad H_a: (\mu_1 - \mu_2) > 0$$

Table 8.2

Reading Test Scores for Eight Pairs of Slow Learners

PAIR	NEW METHOD	STANDARD METHOD
1	77	72
2	74	68
3	82	76
4	73	68
5	87	84
6	69	68
7	66	61
8	80	76
	$\bar{x}_1 = 76.0$	$\bar{x}_2 = 71.625$
	$s_1^2 = 48.0$	$s_2^2 = 49.1$

If we use the two-sample t-statistic (Section 8.2), we first calculate

$$s_p^2 = \frac{(n_1 - 1)s_1^2 + (n_2 - 1)s_2^2}{n_1 + n_2 - 2}$$

$$= \frac{(8 - 1)(48.0) + (8 - 1)(49.1)}{8 + 8 - 2}$$

$$= 48.55$$

and then the test statistic

$$t = \frac{(\bar{x}_1 - \bar{x}_2) - 0}{\sqrt{s_p^2 \left(\frac{1}{n_1} + \frac{1}{n_2}\right)}} = \frac{76.0 - 71.625}{\sqrt{48.55\left(\frac{1}{8} + \frac{1}{8}\right)}} = \frac{4.375}{3.485} = 1.26$$

This small t-value will not lead to rejection of H_0 when compared to the t-distribution with $n_1 + n_2 - 2 = 14$ df, even if α is chosen as large as .10 ($t_{.10} = 1.345$). Thus, from *this* analysis we might conclude that there is insufficient evidence to infer a difference in the mean test scores for the two methods.

If you carefully examine the data in Table 8.2, however, you will find this result difficult to accept. The test score of the new method is larger than the corresponding test score for the standard method *for every one of the eight pairs of slow learners.* This, in itself, seems to provide strong evidence to indicate that μ_1 exceeds μ_2. Why, then, did the t-test fail to detect this difference? The answer is: *The two-sample t is not a valid procedure to use with this set of data.*

The two-sample t is inappropriate because the assumption of independent samples is invalid. We have randomly chosen *pairs of test scores,* and thus, once we have chosen the sample for the new method, we have *not* independently chosen the sample for the standard method. The dependence between observations within pairs can be seen by examining the pairs of test scores, which tend to rise and fall together as we go from pair to pair. This pattern provides strong visual evidence of a violation of the assumption of independence required for the two-sample t-test of Section 8.2. In this situation, you will note the *large variation within samples* (reflected by the large value of s_p^2) in comparison to the relatively *small difference between the sample means.* Because s_p^2 is so large, the t-test of Section 8.2 is unable to detect a difference between μ_1 and μ_2.

We now consider a valid method of analyzing the data of Table 8.2. In Table 8.3 on page 358 we add the column of differences between the test scores of the pairs of slow learners. We can regard these differences in test scores as a random sample of differences for all pairs (matched on reading IQ) of slow learners, past and present. Then we can use this sample to make inferences about the mean of the population of differences, μ_D—which is equal to the difference ($\mu_1 - \mu_2$). That is, the mean of the population (and sample) of differences equals the difference between the population (and sample) means. Thus, our test becomes

$$H_0: \quad \mu_D = 0 \qquad (\mu_1 - \mu_2 = 0)$$
$$H_a: \quad \mu_D > 0 \qquad (\mu_1 - \mu_2 > 0)$$

Table 8.3

PAIR	NEW METHOD	STANDARD METHOD	DIFFERENCE (NEW METHOD — STANDARD METHOD)
1	77	72	5
2	74	68	6
3	82	76	6
4	73	68	5
5	87	84	3
6	69	68	1
7	66	61	5
8	80	76	4

$$\bar{x}_D = 4.375$$
$$s_D = 1.69$$

The test statistic is a one-sample t, since we are now analyzing a single sample of differences:

Test statistic: $t = \dfrac{\bar{x}_D - 0}{s_D / \sqrt{n_D}}$

where

$\bar{x}_D$ = Sample mean difference

s_D = Sample standard deviation of differences

n_D = Number of differences = Number of pairs

Assumptions: The population of differences in test scores is approximately normally distributed. The sample differences are randomly selected from the population of differences. [*Note:* We do not need to assume that $\sigma_1^2 = \sigma_2^2$.]

Rejection region: At significance level $\alpha = .05$, we will reject H_0 if $t > t_{.05}$, where $t_{.05}$ is based on $(n_D - 1)$ degrees of freedom.

Referring to Table V in the Appendix, we find the t-value corresponding to $\alpha = .05$ and $n_D - 1 = 8 - 1 = 7$ df to be $t_{.05} = 1.895$. Then we will reject the null hypothesis if $t > 1.895$. Note that the number of degrees of freedom has decreased from $n_1 + n_2 - 2 = 14$ to 7 by using the paired difference experiment rather than the two independent random samples design. Now calculate

$$t = \frac{\bar{x}_D - 0}{s_D / \sqrt{n_D}} = \frac{4.375}{1.69 / \sqrt{8}} = 7.32$$

Because this value of t falls in the rejection region, we conclude that the mean test score for slow learners taught by the new method exceeds the mean score for those taught by the standard method. We have confidence in this conclusion because the probability that this test would lead to the rejection of H_0, given that it is true, is only $\alpha = .05$.

This kind of experiment, in which observations are paired and the differences are analyzed, is called a **paired difference experiment.** In many cases, a paired difference experiment can provide more information about the difference between population means than an independent samples experiment. The idea is to compare population means by comparing the differences between pairs of experimental units (objects, people, etc.) that were very similar prior to the experiment. The differencing removes sources of variation that tend to inflate σ^2. For example, when two children are taught to read by two different methods, the observed difference in achievement may be due to a difference in the effectiveness of the two teaching methods *or* it may be due to differences in the initial reading levels and IQ's of the two children (random error). To reduce the effect of differences in the children on the observed differences in reading achievement, the two methods of reading are imposed on two children who are more likely to possess similar intellectual potentials, namely children with nearly equal IQ's. The effect of this pairing is to remove the larger source of variation that would be present if children with different abilities were randomly assigned to the two samples. Making comparisons within groups of similar experimental units is called **blocking,** and the paired difference experiment is a simple example of a **randomized block experiment.** In our example, pairs of children with matching IQ scores represent the blocks.

Some other examples for which the paired difference experiment might be appropriate are the following:

1. Suppose you want to estimate the difference $(\mu_1 - \mu_2)$ in mean price per gallon between two major brands of premium gasoline. If you choose two independent random samples of stations for each brand, the variability in price due to geographic location may be large. To eliminate this source of variability you could choose pairs of stations of similar size, one station for each brand, in close geographic proximity and use the sample of differences between the prices of the brands to make an inference about $(\mu_1 - \mu_2)$.

2. Suppose a college placement center wants to estimate the difference $(\mu_1 - \mu_2)$ in mean starting salaries for men and women graduates who seeks jobs through the center. If it independently samples men and women, the starting salaries may vary because of their different college majors and differences in grade-point averages. To eliminate these sources of variability, the placement center could match male and female job-seekers according to their majors and grade-point averages. Then the differences between the starting salaries of each pair in the sample could be used to make an inference about $(\mu_1 - \mu_2)$.

3. Suppose you wish to estimate the difference $(\mu_1 - \mu_2)$ in mean absorption rate into the bloodstream for two drugs that relieve pain. If you independently sample people, the absorption rates might vary because of age, weight, sex, blood pressure, etc. In fact, there are many possible sources of nuisance variability, and pairing individuals who are similar in all the possible sources would be quite difficult. However, it may be possible to obtain two measurements *on the same person*. First, we administer one of the two drugs and record the time until absorption. After a sufficient amount of time, the other drug is administered and a second measurement

on absorption time is obtained. The differences between the measurements for each person in the sample could then be used to estimate $(\mu_1 - \mu_2)$. This procedure would be advisable only if the amount of time allotted between drugs is sufficient to guarantee little or no carryover effect. Otherwise, it would be better to use different people matched as closely as possible on the factors thought to be most important.

The one-tailed and two-tailed hypothesis-testing procedures and the method of forming confidence intervals for the difference between two means using a paired difference experiment are summarized in the boxes.

Paired Difference Confidence Interval

$$\bar{x}_D \pm t_{\alpha/2} \frac{s_D}{\sqrt{n_D}}$$

where $t_{\alpha/2}$ is based on $(n_D - 1)$ degrees of freedom.

Assumptions: 1. The relative frequency distribution of the population of differences is normal.

   2. The sample differences are randomly selected from the population of differences.

Paired Difference Test of an Hypothesis

One-Tailed Test

H_0: $(\mu_1 - \mu_2) = D_0$;
   i.e., $(\mu_D = D_0)$

H_a: $(\mu_1 - \mu_2) < D_0$;
   i.e., $(\mu_D < D_0)$
   [or H_a: $(\mu_1 - \mu_2) > D_0$;
   i.e., $(\mu_D > D_0)$]

Test statistic: $t = \dfrac{\bar{x}_D - D_0}{s_D/\sqrt{n_D}}$

Rejection region:
   $t < -t_\alpha$
   [or $t > t_\alpha$
   when H_a: $(\mu_1 - \mu_2) > D_0$]

Two-Tailed Test

H_0: $(\mu_1 - \mu_2) = D_0$;
   i.e., $(\mu_D = D_0)$

H_a: $(\mu_1 - \mu_2) \neq D_0$;
   i.e., $(\mu_D \neq D_0)$

Test statistic: $t = \dfrac{\bar{x}_D - D_0}{s_D/\sqrt{n_D}}$

Rejection region:
   $t < -t_{\alpha/2}$
   or $t > t_{\alpha/2}$

where t_α and $t_{\alpha/2}$ are based on $(n_D - 1)$ degrees of freedom

Assumptions: 1. The relative frequency distribution of the population of differences is normal.

   2. The differences are randomly selected from the population of differences.

Example 8.8 A paired difference experiment is conducted to compare the starting salaries of male and female college graduates who find jobs. Pairs are formed by choosing a male and a female with the same major and similar grade-point averages. Suppose a random sample of ten pairs is formed in this manner and the starting annual salary of each person is recorded. The results are shown in Table 8.4. Test to see whether there is evidence that the mean starting salary, μ_1, for males exceeds the mean starting salary, μ_2, for females. Use $\alpha = .05$.

Table 8.4

PAIR	MALE	FEMALE	DIFFERENCE (MALE — FEMALE)
1	$14,300	$13,800	$ 500
2	16,500	16,600	−100
3	15,400	14,800	600
4	13,500	13,500	0
5	18,500	17,600	900
6	12,800	13,000	−200
7	14,500	14,200	300
8	16,200	15,100	1,100
9	13,400	13,200	200
10	14,200	13,500	700

Solution The elements of the paired difference test are

$$H_0: \quad \mu_D = 0 \qquad (\mu_1 - \mu_2 = 0)$$
$$H_a: \quad \mu_D > 0 \qquad (\mu_1 - \mu_2 > 0)$$

Note that we propose a one-sided research hypothesis, since we are interested in determining whether the data indicate that μ_1 exceeds μ_2, that is, male mean starting salary exceeds female mean starting salary.

Test statistic: $t = \dfrac{\bar{x}_D - 0}{s_D/\sqrt{n_D}}$

Assumptions: 1. The relative frequency distribution for the population of differences is normal.
2. The sample differences are randomly selected from the population.

Since the test is upper-tailed, we will reject H_0 if $t > t_\alpha$, where $t_{.05} = 1.833$ is based upon $(n_D - 1) = 9$ degrees of freedom. The rejection region is shown in Figure 8.6. We now calculate

$$\sum x_D = 500 + (-100) + \cdots + 700 = 4,000$$

and

$$\sum x_D^2 = 3,300,000$$

Figure 8.6 Rejection
Region for Example 8.8

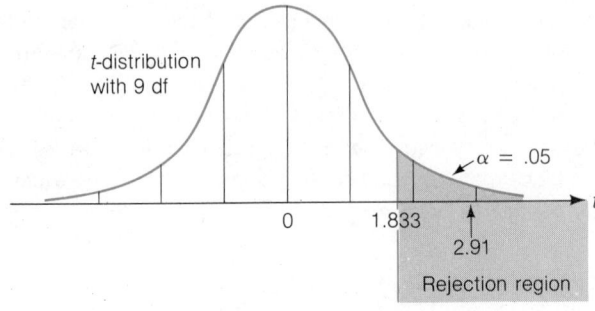

Then

$$\bar{x}_{\mathbf{D}} = \frac{\sum x_{\mathbf{D}}}{10} = \frac{4,000}{10} = 400$$

$$s_{\mathbf{D}}^2 = \frac{\sum (x_{\mathbf{D}} - \bar{x}_{\mathbf{D}})^2}{n_{\mathbf{D}} - 1} = \frac{\sum x_{\mathbf{D}}^2 - (\sum x_{\mathbf{D}})^2/10}{9}$$

$$= \frac{3,300,000 - (4,000)^2/10}{9} = 188,888.89$$

$$s_{\mathbf{D}} = \sqrt{s_{\mathbf{D}}^2} = 434.61$$

Substituting these values into the formula for the test statistic, we find that

$$t = \frac{\bar{x}_{\mathbf{D}} - 0}{s_{\mathbf{D}}/\sqrt{n_{\mathbf{D}}}} = \frac{400}{434.61/\sqrt{10}} = \frac{400}{137.44} = 2.91$$

As you can see in Figure 8.6, the calculated t falls in the rejection region. Thus, we conclude at the $\alpha = .05$ level of significance that the mean starting salary for males exceeds the mean starting salary for females. ∎

One measure of the amount of information about $(\mu_1 - \mu_2)$ gained by using a paired difference experiment rather than an independent samples experiment in Example 8.8 is the relative widths of the confidence intervals obtained by the two methods. A 95% confidence interval for $(\mu_1 - \mu_2)$ using the paired difference experiment is

$$\bar{x}_{\mathbf{D}} \pm t_{\alpha/2} \frac{s_{\mathbf{D}}}{\sqrt{n_{\mathbf{D}}}} = 400 \pm t_{.025} \frac{434.61}{\sqrt{10}}$$

$$= 400 \pm 2.262 \frac{434.61}{\sqrt{10}}$$

$$= 400 \pm 310.88$$

$$\approx 400 \pm 311 = (\$89, \$711)$$

If we analyzed the same data as though this were an independent samples experiment,* we would first calculate the following quantities:

MALES	FEMALES
$n_1 = 10$	$n_2 = 10$
$\bar{x}_1 = \$14{,}930$	$\bar{x}_2 = \$14{,}530$
$s_1^2 = 3{,}009{,}000$	$s_2^2 = 2{,}331{,}222.22$

Then

$$s_p^2 = \frac{(n_1 - 1)s_1^2 + (n_2 - 1)s_2^2}{n_1 + n_2 - 2} = \frac{9(3{,}009{,}000) + 9(2{,}331{,}222.22)}{10 + 10 - 2}$$

$$= 2{,}670{,}111.11$$

$$s_p = \sqrt{s_p^2} = 1{,}634.05$$

The 95% confidence interval is

$$(\bar{x}_1 - \bar{x}_2) \pm t_{.025} \sqrt{s_p^2 \left(\frac{1}{n_1} + \frac{1}{n_2}\right)} = 400 \pm (2.101) \sqrt{2{,}670{,}111.11 \left(\frac{1}{10} + \frac{1}{10}\right)}$$

$$= 400 \pm 1{,}535.35$$

$$\approx 400 \pm 1{,}535$$

$$= (-\$1{,}135, \$1{,}935)$$

The confidence interval for the independent sampling experiment is about five times wider than for the corresponding paired difference confidence interval. Blocking out the variability due to differences in majors and grade-point averages significantly increases the information about the difference in male and female mean starting salaries by providing a much more accurate (smaller confidence interval for the same confidence coefficient) estimate of $(\mu_1 - \mu_2)$.

You may wonder whether conducting a paired difference experiment is always superior to an independent samples experiment. The answer is: Most of the time, but not always. We sacrifice half the degrees of freedom in the t-statistic when a paired difference design is used instead of an independent samples design. This is a loss of information, and unless this loss is more than compensated for by the reduction in variability obtained by blocking (pairing), the paired difference experiment will result in a net loss of information about $(\mu_1 - \mu_2)$. Thus, we should be convinced that the pairing will significantly reduce variability before performing the paired difference experiment. Most of the time this will happen.

One final note: The pairing of the observations is determined *before* the experiment is performed (that is, by the *design* of the experiment). A paired difference experiment

---

* This is done only to provide a measure of the increase in the amount of information obtained by a paired design in comparison to an unpaired design. Actually, if an experiment is designed using pairing, an unpaired analysis would be invalid because the assumption of independent samples would not be satisfied.

is *never* obtained by pairing the sample observations after the measurements have been acquired.

> ### What Do You Do When the Assumption of a Normal Distribution for the Population of Differences Is Not Satisfied?
>
> Answer: Use the Wilcoxon signed rank test for the paired difference design (Chapter 10).

Case Study 8.2

Matched Pairing in Studying the Mentally Retarded

The statistical implications underlying matching procedures have frequently been overlooked in educational research with the mentally retarded population. It is, therefore, the purpose of this paper to point out some of the advantages and disadvantages of different matching procedures.

Stainback and Stainback (1973) describe a number of experimental situations in which blocking (matching) subjects before experimentation might reduce variability and thereby increase the amount of information obtained. One of the matching procedures discussed by the authors can be summarized as follows: Suppose it is desired to form two groups of mentally retarded subjects to compare two methods of educational therapy. Subjects could be randomly selected from an existing (large) group of subjects and ordered (from lowest to highest) on the scores of an appropriate matching variable. Several variables suggested by the authors are "pretest measures of the experimental criterion, measures of learning rate (mental age and Intelligence Quotients), chronological age, personal characteristics (sex, race), environmental conditions (socioeconomic level), or combinations of two or more variables." The two highest-ranking subjects on the matching variables would then form pair 1, the next two pair 2, etc., and one member from each pair would be randomly assigned to each therapy group. This experiment is a practical example of a paired difference experiment, and if the matching variables are correctly chosen, the responses of subjects will be more homogeneous within pairs than between pairs. The authors conclude:

> It is important that when comparing two groups on a criterion variable the two groups be as equal as possible on all relevant factors excepting only the independent variables. This consideration deserves particular emphasis in the area of mental retardation since the researchers are constantly dealing with diverse groups. It should be restated, therefore, that researchers in the area of mental retardation should become acutely aware of advantages and disadvantages of matching procedures and their alternatives.

Exercises 8.31–8.41

Learning the Mechanics

8.31 The data for a random sample of six paired observations are shown in the table.

a. Calculate the difference between each pair of observations by subtracting observation 2 from observation 1. Use the differences to calculate $\bar{x}_D$ and s_D^2.

PAIR	SAMPLE FROM POPULATION 1 Observation 1	SAMPLE FROM POPULATION 2 Observation 2
1	5	3
2	3	1
3	9	7
4	6	2
5	4	5
6	8	7

b. If μ_1 and μ_2 are the means of populations 1 and 2, respectively, express μ_D in terms of μ_1 and μ_2.

c. Form a 95% confidence interval for μ_D.

d. Test the null hypothesis H_0: $\mu_D = 0$ against the alternative hypothesis H_a: $\mu_D \neq 0$. Use $\alpha = .05$.

8.32 The data for a random sample of ten paired observations are shown in the table.

PAIR	SAMPLE FROM POPULATION 1	SAMPLE FROM POPULATION 2
1	19	24
2	25	27
3	31	36
4	52	53
5	49	55
6	34	34
7	59	66
8	47	51
9	17	20
10	51	55

a. Do the data provide sufficient evidence to indicate that μ_2, the mean of population 2, is larger than μ_1? (Test H_0: $\mu_D = 0$ against H_a: $\mu_D < 0$, where $\mu_D = \mu_1 - \mu_2$.) Test using $\alpha = .05$.

b. Form a 90% confidence interval for μ_D.

Applying the Concepts

PAIR	FEMALE	MALE
1	40.03	84.51
2	75.62	80.06
3	53.38	102.30
4	62.27	88.96

8.33 The data shown in the table are part of a study conducted to compare the abilities of men and women to perform the strenuous tasks of a firefighter (Phillips & Pepper, 1982). They represent the pulling force (in newtons) that a firefighter was able to exert in pulling the starter cord of a P-250 fire pump. Firefighters were matched in pairs according to weight, thus producing data for a matched pairs (or paired difference) experiment.

a. Do the data provide sufficient evidence to indicate a difference in mean pulling force between female and male firefighters? Test using $\alpha = .05$.

b. Find a 95% confidence interval for the difference in mean pulling force between female and male firefighters. Interpret the interval.

8.34 A new weight-reducing technique, consisting of a liquid protein diet, is currently undergoing tests by the Food and Drug Administration (FDA) before its introduction into the market. A typical test performed by the FDA is the following: The weights of a random sample of five people are recorded before they are introduced to the liquid protein diet. The five individuals are then instructed to follow the liquid protein diet for 3 weeks. At the end of this period, their weights (in pounds) are again recorded. The results are listed in the table. Let μ_1 be the true mean weight of individuals before starting the diet and let μ_2 be the true mean weight of individuals after 3 weeks on the diet. Construct a 95% confidence interval for the difference between the true mean weights before and after the diet is used. What assumptions are necessary to ensure the validity of the procedure you used?

PERSON	WEIGHT BEFORE DIET	WEIGHT AFTER DIET
1	150	143
2	195	190
3	188	185
4	197	191
5	204	200

8.35 A 1974 Supreme Court decision (*Milliken v. Bradley*) held that desegregation could not extend beyond the boundary of the school systems that were found to be segregated. One dissenting justice feared that the decision would trigger white flight, the migration of white families out of the inner cities to the suburbs. In discussing this

Racial Change and White Flight in Selected City School Systems, 1968–1972

CITY	PERCENTAGE MINORITY			PERCENTAGE CHANGE IN WHITE STUDENTS[a]	
	1968	1970	1972	1968–1970	1970–1972
Boston	31.5	35.9	40.4	−3.9	−7.4
Philadelphia	61.8	63.6	64.8	−5.7	−2.2
Baltimore	65.1	67.1	69.3	−5.6	−9.3
St. Louis	63.8	65.9	69.1	−9.4	−13.8
Chicago	62.3	65.4	69.2	−9.0	−14.7
Detroit	60.7	65.5	69.5	−15.6	−14.0
Atlanta	61.8	68.7	77.4	−22.3	−34.4
Charlotte	29.5	31.1	32.8	−3.1	−5.6
Jacksonville	28.2	29.4	32.6	−1.8	−11.4
Houston	46.7	50.6	56.4	−9.1	−17.5
San Francisco	58.8	63.1	68.2	−13.5	−22.4
San Diego	23.9	24.6	26.3	−1.1	−5.5

[a] Percentages based on beginning years.

Source: U.S. Dept. of Health, Education, and Welfare, Office for Civil Rights, *Directory of Public Elementary and Secondary Schools in Selected Districts*, fall 1968, fall 1970, and fall 1972.

decision, Charles T. Clotfelter (1976) presents data showing the percentages of minority students for 1968, 1970, and 1972 in the school systems of twelve large cities. Examine the last two columns of the table, which give the percentage change in white students over the two time periods, 1968–1970 and 1970–1972.

a. Was the shift in white families out of the inner cities larger in the time period 1970–1972 than for the comparable period, 1968–1970? Test using $\alpha = .05$.

b. What assumptions must be satisfied for the test in part a to be valid?

c. Explain why the data might or might not satisfy the assumptions of part b.

8.36 The data shown in the table provide information on the relationship between the mean daily air temperature and the cocoon temperature of woolly-bear caterpillars of the High Arctic (Kevan, Jensen, & Shorthouse, 1982). You can see from the table that the data indicate that the caterpillar's body temperature (inside the cocoon) is higher than the outside air temperature. Estimate the mean difference in temperature between the cocoon and the outside air. Use a 95% confidence interval.

| | TEMPERATURE (°C) | |
DAY	Air	Cocoon[a]
1	10.4	15.1
2	9.2	14.6
3	2.2	6.8
4	2.6	6.8
5	4.1	8.0
6	3.7	8.7
7	1.7	3.6
8	2.0	5.3
9	3.0	7.0
10	3.5	7.1
11	4.5	9.6
12	4.4	9.5

[a] Each cocoon temperature is the average of the temperatures of two cocoons.

8.37 In the past, many bodily functions were thought to be beyond conscious control. However, recent experimentation suggests that it may be possible for a person to control certain body functions if that person is trained in a program of *biofeedback* exercises. An experiment is conducted to show that blood pressure levels can be consciously reduced in people trained in this program. The blood pressure measurements (in millimeters of mercury) listed in the table at the top of the next page represent readings before and after the biofeedback training of six subjects.

a. Is there sufficient evidence to conclude that the mean blood pressure after people are trained in this program is less than the mean blood pressure before training? Use $\alpha = .05$.

SUBJECT	BEFORE	AFTER
1	136.9	130.2
2	201.4	180.7
3	166.8	149.6
4	150.0	153.2
5	173.2	162.6
6	169.3	160.1

b. Find the approximate observed significance level for the test and interpret its value.

8.38 While producing many economic benefits to the state of Florida, gypsum and phosphate mines also produce a harmful byproduct: *radiation*. It has been known for a number of years that the mine tailings (waste) contain radioactive radon 222. In fact, new housing complexes built on top of the leveled piles of residue have shown disturbing radiation levels within the houses. The radiation levels in waste gypsum and phosphate mounds in Polk County, Florida, are regularly monitored by the Eastern Environmental Radiation Facility (EERF) and by the Polk County Health Department (PCHD), Winter Haven, Florida. The table shows measurements of the exhalation rate (a measure of radiation) for fifteen soil samples obtained from waste mounds in Polk County, Florida. The exhalation rate was measured for each soil sample by both the PCHD and the EERF. The objective of selecting the paired measurements was to determine whether a bias exists, a difference in the mean readings, between PCHD and EERF. They represent part of the data contained in a report by Thomas R. Horton of EERF (Horton, 1979).

CHARCOAL CANISTER NO.	PCHD	EERF	CHARCOAL CANISTER NO.	PCHD	EERF
71	1,709.79	1,479.0	85	393.55	187.7
58	357.17	257.8	46	880.84	630.4
84	1,150.94	1,287.0	4	2,996.49	3,707.0
91	1,572.69	1,395.0	20	2,367.40	2,791.0
44	558.33	416.5	36	599.84	706.8
43	4,132.28	3,993.0	42	538.37	618.5
79	1,489.86	1,351.0	55	2,770.23	2,639.0
61	3,017.48	1,813.0			

a. Do the data provide sufficient evidence to indicate a difference in the mean exhalation rates between PCHD and EERF? Test using $\alpha = .05$.

b. Find a 95% confidence interval for the difference in mean measurements between PCHD and EERF. Interpret the interval.

8.39 A manufacturer of automobile shock absorbers was interested in comparing the durability of its shocks with that of the biggest competitor. To make the comparison, one of the manufacturer's and one of the competitor's shocks were randomly selected and installed on the rear wheels of six cars. After the cars had been driven 20,000 miles, the strength of each test shock was measured, coded, and recorded. The following are the results of the examination:

CAR NUMBER	MANUFACTURER'S	COMPETITOR'S
1	8.8	8.4
2	10.5	10.1
3	12.5	12.0
4	9.7	9.3
5	9.6	9.0
6	13.2	13.0

a. Do the data present sufficient evidence to conclude there is a difference in the mean strength of the two types of shocks after 20,000 miles of use? Let $\alpha = .05$.

b. What assumptions are necessary in order to apply a paired difference analysis to the data?

c. Construct a 95% confidence interval for $(\mu_1 - \mu_2)$. Interpret the meaning of your confidence interval.

8.40 Suppose the data in Exercise 8.39 are based on independent random samples.

a. Do the data provide sufficient evidence to indicate a difference between the mean strengths for the two types of shocks? Let $\alpha = .05$.

b. Construct a 95% confidence interval for $(\mu_1 - \mu_2)$. Interpret your result.

c. Compare the confidence intervals you obtained in Exercise 8.39 and part b of this exercise. Which is larger? To what do you attribute the difference in size? Assuming in each case that the appropriate assumptions are satisfied, which interval provides you with more information about $(\mu_1 - \mu_2)$? Explain.

d. Are the results of an unpaired analysis valid when the data have been collected from a paired experiment?

8.41 In "The Effects of Fructose and Glucose on High Endurance Performance," R. G. McMurray, J. R. Wilson, and B. S. Kitchell (1983) report on an experiment to measure the effects of fructose and glucose on high-endurance performance of athletes. Six trained female runners were used in the experiment. Each was given 300 milliliters of a liquid 45 minutes prior to running for 85 minutes or until they reached a state of exhaustion, whichever occurred first. Various measures of endurance, such as performance time (time to exhaustion), rating of perceived exertion, etc., were recorded at the end of each run. Four liquids (treatments) were used in the experiment. The first contained fructose, the second contained glucose, the third contained water sweetened with a calcium saccharine solution (a placebo designed to suggest the presence of fructose or glucose), and the fourth contained water alone. Each of the six subjects performed the run for each of the four liquids, which were arranged in random order. The table at the top of the next page gives the averages of the six runners' times (in minutes) to exhaustion for only two of the mixtures, glucose and the placebo. The table also gives the sample sizes and standard deviations for the two samples.

	GLUCOSE	PLACEBO
SAMPLE SIZE	6	6
MEAN	63.9	52.2
STANDARD DEVIATION	20.3	13.5

a. Describe the experiment and explain how and why it does or does not satisfy the assumptions of independent random sampling.

b. Suppose that the data were based on independent random samples. Would the data provide sufficient evidence to indicate a difference in the mean time to exhaustion between runners given the glucose mixture and those given the placebo? Test using $\alpha = .05$. Interpret your results.

c. Consider the manner in which the experiment was actually conducted. What do you gain or lose by analyzing the data using the method of part b?

8.4
Inferences About the Difference Between Population Proportions: Independent Binomial Experiments

Suppose a presidential candidate wants to compare the preference of registered voters in the northeastern United States (NE) to those in the southeastern United States (SE). Such a comparison would help determine where to concentrate campaign efforts. The candidate hires a professional pollster to randomly choose 1,000 registered voters in the northeast and 1,000 in the southeast and interview each to learn her or his voting preference. The objective is to use this sample information to make an inference about the difference $(p_1 - p_2)$ between the proportion p_1 of *all* registered voters in the northeast and the proportion p_2 of *all* registered voters in the southeast who plan to vote for the presidential candidate.

The two samples represent independent binomial experiments. (See Section 4.4 for the characteristics of binomial experiments.) The binomial random variables are the numbers x_1 and x_2 of the 1,000 sampled voters in each area who indicate they will vote for the candidate. The results are summarized in the table. We can now calculate the sample proportions $\hat{p}_1$ and $\hat{p}_2$ of the voters in favor of the candidate in the northeast and southeast, respectively:

NE	SE
$n_1 = 1,000$	$n_2 = 1,000$
$x_1 = 546$	$x_2 = 475$

$$\hat{p}_1 = \frac{x_1}{n_1} = \frac{546}{1,000} = .546 \qquad \hat{p}_2 = \frac{x_2}{n_2} = \frac{475}{1,000} = .475$$

The difference between the sample proportions $(\hat{p}_1 - \hat{p}_2)$ makes an intuitively appealing estimator of the difference between the population parameters $(p_1 - p_2)$. For our example, the estimate is

$$(\hat{p}_1 - \hat{p}_2) = .546 - .475 = .071$$

To judge the reliability of the estimator $(\hat{p}_1 - \hat{p}_2)$, we must observe its performance in repeated sampling from the two populations. That is, we need to know the sampling distribution of $(\hat{p}_1 - \hat{p}_2)$. The properties of the sampling distribution are given in the box. Remember that $\hat{p}_1$ and $\hat{p}_2$ can be viewed as means of the number of successes per trial in the respective samples, so the central limit theorem applies when the sample sizes are large.

Since the distribution of $(\hat{p}_1 - \hat{p}_2)$ in repeated sampling is approximately normal, we can use the z-statistic to derive confidence intervals for $(p_1 - p_2)$ or to test an hypothesis about $(p_1 - p_2)$.

Properties of the Sampling Distribution of $(\hat{p}_1 - \hat{p}_2)$

1. If the sample sizes n_1 and n_2 are large (see Section 5.3 for a clearer meaning of large), the sampling distribution of $(\hat{p}_1 - \hat{p}_2)$ is approximately normal.

2. The mean of the sampling distribution of $(\hat{p}_1 - \hat{p}_2)$ is $(p_1 - p_2)$,* that is,

$$E(\hat{p}_1 - \hat{p}_2) = p_1 - p_2$$

Thus, $(\hat{p}_1 - \hat{p}_2)$ is an unbiased estimator of $(p_1 - p_2)$.

3. The standard deviation of the sampling distribution of $(\hat{p}_1 - \hat{p}_2)$ is

$$\sigma_{(\hat{p}_1 - \hat{p}_2)} = \sqrt{\frac{p_1 q_1}{n_1} + \frac{p_2 q_2}{n_2}}$$

For the voter example, a 95% confidence interval for the difference $(p_1 - p_2)$ is

$$(\hat{p}_1 - \hat{p}_2) \pm 1.96\sigma_{(\hat{p}_1 - \hat{p}_2)}$$

or

$$(\hat{p}_1 - \hat{p}_2) \pm 1.96\sqrt{\frac{p_1 q_1}{n_1} + \frac{p_2 q_2}{n_2}}$$

The quantities $p_1 q_1$ and $p_2 q_2$ must be estimated in order to complete the calculation of the standard deviation, $\sigma_{(\hat{p}_1 - \hat{p}_2)}$, and hence the calculation of the confidence interval. In Section 7.5 we showed that the value of pq is relatively insensitive to the value chosen to approximate p. Therefore, $\hat{p}_1 \hat{q}_1$ and $\hat{p}_2 \hat{q}_2$ will provide satisfactory estimates to approximate $p_1 q_1$ and $p_2 q_2$, respectively. Then

$$\sqrt{\frac{p_1 q_1}{n_1} + \frac{p_2 q_2}{n_2}} \approx \sqrt{\frac{\hat{p}_1 \hat{q}_1}{n_1} + \frac{\hat{p}_2 \hat{q}_2}{n_2}}$$

and we will approximate the 95% confidence interval by

$$(\hat{p}_1 - \hat{p}_2) \pm 1.96\sqrt{\frac{\hat{p}_1 \hat{q}_1}{n_1} + \frac{\hat{p}_2 \hat{q}_2}{n_2}}$$

---

* The mean and variance of the sampling distribution of $(\hat{p}_1 - \hat{p}_2)$ can be derived by using the formulas given in Section 6.5.

Substituting the sample quantities yields

$$(.546 - .475) \pm 1.96 \sqrt{\frac{(.546)(.454)}{1,000} + \frac{(.475)(.525)}{1,000}}$$

or $.071 \pm .044$. Thus, we estimate the difference $(p_1 - p_2)$ to fall in the interval .027 to .115. We infer that there are between 2.7% and 11.5% more registered voters in the northeast than in the southeast who plan to vote for the presidential candidate. It seems that the candidate should direct a greater campaign effort in the southeast compared to the northeast. The confidence coefficient associated with our interval estimate is .95.

The general form of a confidence interval for the difference $(p_1 - p_2)$ between binomial proportions is given in the box.

Large-Sample $100(1 - \alpha)\%$ Confidence Interval for $(p_1 - p_2)$

$$(\hat{p}_1 - \hat{p}_2) \pm z_{\alpha/2} \sigma_{(\hat{p}_1 - \hat{p}_2)} = (\hat{p}_1 - \hat{p}_2) \pm z_{\alpha/2} \sqrt{\frac{p_1 q_1}{n_1} + \frac{p_2 q_2}{n_2}}$$

$$\approx (\hat{p}_1 - \hat{p}_2) \pm z_{\alpha/2} \sqrt{\frac{\hat{p}_1 \hat{q}_1}{n_1} + \frac{\hat{p}_2 \hat{q}_2}{n_2}}$$

The z-statistic,

$$z = \frac{(\hat{p}_1 - \hat{p}_2) - (p_1 - p_2)}{\sigma_{(\hat{p}_1 - \hat{p}_2)}}$$

is used to test the null hypothesis that $(p_1 - p_2)$ equals some specified difference, say D_0. For the special case where $D_0 = 0$, i.e., where we want to test the null hypothesis H_0: $(p_1 - p_2) = 0$ (or, equivalently, H_0: $p_1 = p_2$), the best estimate of $p_1 = p_2 = p$ is obtained by dividing the total number of successes $(x_1 + x_2)$ for the two samples by the total number of observations $(n_1 + n_2)$; that is,

$$\hat{p} = \frac{x_1 + x_2}{n_1 + n_2}$$

Then the best estimate of $\sigma_{(\hat{p}_1 - \hat{p}_2)}$ is

$$\sigma_{(\hat{p}_1 - \hat{p}_2)} = \sqrt{\frac{p_1 q_1}{n_1} + \frac{p_2 q_2}{n_2}} \approx \sqrt{\frac{\hat{p}\hat{q}}{n_1} + \frac{\hat{p}\hat{q}}{n_2}} = \sqrt{\hat{p}\hat{q}\left(\frac{1}{n_1} + \frac{1}{n_2}\right)}$$

The test is summarized in the box.

Large-Sample Test of an Hypothesis About $(p_1 - p_2)$

One-Tailed Test

H_0: $(p_1 - p_2) = D_0$

H_a: $(p_1 - p_2) < D_0$
  [or H_a: $(p_1 - p_2) > D_0$]

where $D_0 =$ Hypothesized value of $(p_1 - p_2)$

Test statistic: $z = \dfrac{(\hat{p}_1 - \hat{p}_2) - D_0}{\sigma_{(\hat{p}_1 - \hat{p}_2)}}$

Rejection region: $z < -z_\alpha$
  [or $z > z_\alpha$ when
  H_a: $(p_1 - p_2) > D_0$]

Two-Tailed Test

H_0: $(p_1 - p_2) = D_0$

H_a: $(p_1 - p_2) \neq D_0$

Test statistic: $z = \dfrac{(\hat{p}_1 - \hat{p}_2) - D_0}{\sigma_{(\hat{p}_1 - \hat{p}_2)}}$

Rejection region: $z < -z_{\alpha/2}$
  or $z > z_{\alpha/2}$

Note: $\sigma_{(\hat{p}_1 - \hat{p}_2)} = \sqrt{\dfrac{p_1 q_1}{n_1} + \dfrac{p_2 q_2}{n_2}}$

To calculate $\sigma_{(\hat{p}_1 - \hat{p}_2)}$, approximate p_1 and p_2 using $\hat{p}_1$ and $\hat{p}_2$ except for the special case when $D_0 = 0$. Then use

$$\hat{p}_1 = \hat{p}_2 = \hat{p} = \frac{x_1 + x_2}{n_1 + n_2}$$

For this special case,

$$\sigma_{(\hat{p}_1 - \hat{p}_2)} \approx \sqrt{\hat{p}\hat{q}\left(\frac{1}{n_1} + \frac{1}{n_2}\right)}$$

Example 8.9

In the past decade there have been intensive antismoking campaigns sponsored by both federal and private agencies. Suppose that the American Cancer Society randomly sampled 1,500 adults in 1982 and then sampled 2,000 adults in 1984 to determine whether there was evidence that the percentage of smokers had decreased. The results of the two sample surveys are shown in the table, where x_1 and x_2 represent the numbers of smokers in the 1982 and 1984 samples, respectively. Do these data indicate that the fraction of smokers decreased over this 2-year period? Use $\alpha = .05$.

1982	1984
$n_1 = 1,500$	$n_2 = 2,000$
$x_1 = 576$	$x_2 = 652$

Solution

If we define p_1 and p_2 as the true proportions of adult smokers in 1982 and 1984 the elements of our test are

$$H_0: \ (p_1 - p_2) = 0 \qquad H_a: \ (p_1 - p_2) > 0$$

(The test is one-tailed since we are interested only in determining whether the proportion of smokers *decreased*.)

Figure 8.7   Rejection
Region for Example 8.9

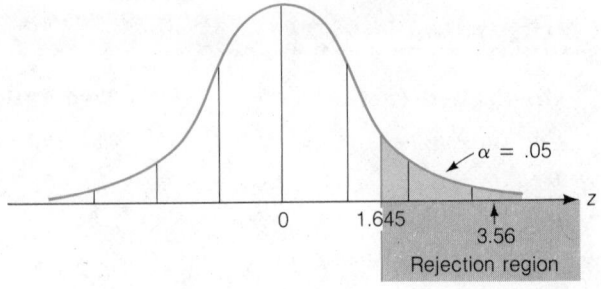

Test statistic: $z = \dfrac{(\hat{p}_1 - \hat{p}_2) - 0}{\sigma_{(\hat{p}_1 - \hat{p}_2)}}$

Rejection region: $\alpha = .05$

$z > z_\alpha = z_{.05} = 1.645$ (see Figure 8.7.)

We now calculate the sample proportions of smokers

$$\hat{p}_1 = \frac{576}{1,500} = .384 \qquad \hat{p}_2 = \frac{652}{2,000} = .326$$

Then

$$z = \frac{(\hat{p}_1 - \hat{p}_2) - 0}{\sigma_{(\hat{p}_1 - \hat{p}_2)}} \approx \frac{(\hat{p}_1 - \hat{p}_2)}{\sqrt{\hat{p}\hat{q}\left(\dfrac{1}{n_1} + \dfrac{1}{n_2}\right)}}$$

where

$$\hat{p} = \frac{x_1 + x_2}{n_1 + n_2} = \frac{576 + 652}{1,500 + 2,000} = .351$$

Thus,

$$z = \frac{.384 - .326}{\sqrt{(.351)(.649)\left(\dfrac{1}{1,500} + \dfrac{1}{2,000}\right)}} = \frac{.058}{.0164} = 3.56$$

There is sufficient evidence at the $\alpha = .05$ level to conclude that the proportion of all adults who smoke has decreased over the 1982–1984 period. We could place a confidence interval on $(p_1 - p_2)$ if we were interested in estimating the extent of the decrease.   ∎

**Exercises
8.42–8.60**

Learning the Mechanics

8.42   Construct a 95% confidence interval for $(p_1 - p_2)$ in each of the following situations:

a. $n_1 = 400,\quad \hat{p}_1 = .65;\quad n_2 = 400,\quad \hat{p}_2 = .58$

b. $n_1 = 180,\quad \hat{p}_1 = .31;\quad n_2 = 250,\quad \hat{p}_2 = .25$

c. $n_1 = 100,\quad \hat{p}_1 = .46;\quad n_2 = 120,\quad \hat{p}_2 = .61$

8.43 Independent random samples, each containing 800 observations, were selected from two binomial populations. The samples from populations 1 and 2 produced 320 and 400 successes, respectively.

a. Test the null hypothesis H_0: $(p_1 - p_2) = 0$ against the alternative hypothesis H_a: $(p_1 - p_2) \neq 0$. Use $\alpha = .05$.

b. Test H_0: $(p_1 - p_2) = 0$ against H_a: $(p_1 - p_2) \neq 0$. Use $\alpha = .01$.

c. Test H_0: $(p_1 - p_2) = 0$ against H_a: $(p_1 - p_2) < 0$. Use $\alpha = .01$.

d. Form a 90% confidence interval for $(p_1 - p_2)$.

8.44 The quantities $\hat{p}_1$ and $\hat{p}_2$ have been defined to be x_1/n_1 and x_2/n_2, respectively. What assumptions do we make about x_1 and x_2?

8.45 What are the characteristics of a binomial experiment?

8.46 Explain why the central limit theorem is important in finding an approximate distribution for $(\hat{p}_1 - \hat{p}_2)$. See Section 6.5.

8.47 Sketch the sampling distribution of $(\hat{p}_1 - \hat{p}_2)$ based on independent random samples of $n_1 = 100$ and $n_2 = 200$ observations from two binomial populations with success probabilities $p_1 = .1$ and $p_2 = .5$, respectively.

8.48 Explain how knowing the sampling distribution of $(\hat{p}_1 - \hat{p}_2)$ can help us measure the reliability of the estimator $(\hat{p}_1 - \hat{p}_2)$.

Applying the Concepts

8.49 Two surgical procedures are widely used to treat a certain type of cancer. To compare the success rates of the two procedures, random samples of the two types of surgical patients were obtained and the numbers of patients who showed no recurrence of the disease after a 1-year period were recorded. The data are shown in the table. Do the data provide sufficient evidence to indicate a difference in the success rates of the two procedures?

	n	NUMBER OF SUCCESSES
PROCEDURE A	100	78
PROCEDURE B	100	87

8.50 A new insect spray, type A, is to be compared with a spray, type B, that is currently in use. Two rooms of equal size are sprayed with the same amount of spray, one room with A, the other with B. Two hundred insects are released into each room, and after 1 hour the numbers of dead insects are counted. The results are given in the table.

	SPRAY A	SPRAY B
NUMBER OF INSECTS	200	200
NUMBER OF DEAD INSECTS	120	80

a. Do the data provide sufficient evidence to indicate that spray A is more effective than spray B in controlling the insects? Test using $\alpha = .05$.

b. Find a 90% confidence interval for $(p_1 - p_2)$, the difference in the rates of kill for the two sprays. Interpret this interval.

8.51 The *Orlando Sentinel* report (cited in Exercise 8.13) on a 1983 survey of the American public provides many other interesting statistics concerning our reading habits. One outcome of the sampling of 1,961 people showed that 56% claimed to read a book occasionally—a percentage, the article notes, that has "changed barely in 5 years." In 1978, the percentage was 55%. Despite the apparent small change in the sample percentages, small changes in the actual percentages could be very important for the publishing industry. Assume that the 1978 and 1983 surveys were based on independent random samples of 1,961 persons each.

a. Do the data provide sufficient evidence to indicate a change from 1978 to 1983 in the proportion of people in the literate American population who claim that they occasionally read a book? Test using $\alpha = .01$.

b. The survey also found that the percentage of heavy readers, those who read at least one book per week, increased from 18% in 1978 to 35% in 1983. Find a 99% confidence interval for the actual change in population proportions from 1978 to 1983. Interpret your results.

8.52 It is estimated that more than half the votes cast in upcoming elections will be cast by women. Of concern to the prevalent political parties is the possibility of a "gender gap" (*Parade Magazine*, March 4, 1984). This hypothesis states that there is a difference between male and female perceptions of which political party best addresses the nation's problems. Particularly, those who favor this hypothesis believe that the Democratic platform may appear to be more in line with philosophies of the majority of women. Suppose a Republican campaign manager wishes to test this theory. Assume that 300 registered voters (150 men, 150 women) were given the platforms of both the Republicans and the Democrats for five major issues and then asked to give their preference for one of the parties. Suppose 79 men and 71 women preferred the Republican views.

a. Does this result provide sufficient evidence to indicate that a gender gap exists? Test at $\alpha = .05$.

b. If there were no undecided responses, can we say that women prefer the Democrats' responses to issues? Test at $\alpha = .01$.

8.53 A *U.S. News and World Report* article (June 21, 1982) describes a new method of treating a major form of blindness in elderly people. The process, using laser beams to seal abnormal blood vessels in the eye, was tried on 224 patients, and of these only 14% went blind in 1 year. In a control group of similar untreated patients, 42% went blind in 1 year. Assume that the control group contained the same number of patients as the treated group.

a. Do the data provide sufficient evidence to indicate that the laser beam treatment was effective in reducing the probability that a patient will go blind after 1 year? Test using $\alpha = .05$.

b. Find the p-value for the test and interpret its value.

c. Find a 95% confidence interval for the reduction in the probability of going blind after 1 year for a patient receiving the laser beam treatment.

8.54 Suppose a firm switches its table salt container from a cylinder (expensive) to a rectangular box (inexpensive). The firm samples 1,000 households nationwide, both before and after the switch, to estimate the percentage of households that purchase its brand of salt. The following results are obtained:

	BEFORE	AFTER
SAMPLE SIZE	1,000	1,000
NUMBER OF HOUSEHOLDS USING FIRM'S BRAND	475	305

a. Estimate the difference between the true proportions of households using the firm's salt before and after the packaging switch. Use a 90% confidence interval.

b. Interpret the confidence interval of part a. Express the reliability of the interval.

8.55 The Reserve Mining Company of Minnesota commissioned a team of physicians to study the breathing patterns of its miners who were exposed to taconite dust. The physicians compared the breathing of 307 miners who had been employed in the Reserve's Babbit, Minnesota, mine for more than 20 years with thirty-five Duluth area men with no history of exposure to taconite dust. The physicians concluded that "there is no significant difference in respiratory symptoms or breathing ability between the group of men who have worked in the taconite industry for more than 20 years and a group of men of similar smoking habits but without exposure to taconite dust." Using the statistical procedures you have learned in this chapter, design an hypothesis test (give H_0, H_a, test statistic, etc.) that would have been appropriate for use in the physicians' study. [Source: Associated Press, *Minneapolis Tribune*, February 20, 1977.]

8.56 Refer to Exercise 8.55. Suppose the physicians determined that 61 of the 307 miners had breathing irregularities and that five of the thirty-five Duluth men had breathing irregularities.

a. Test to see whether these data indicate that a higher proportion of breathing irregularities exists among those who have been exposed to taconite dust than among those who have not been exposed.

b. Find the observed significance level for the test and interpret its value.

8.57 As part of their research on the enjoyment of work by full-time workers in the United States, N. D. Glenn and C. N. Weaver (1982) give a comparison of the 1955 and the 1980 percentages of full-time workers who claimed to enjoy their work so much that they had a difficult time putting it aside. The sample sizes and sample percentages, based on a national sampling, are shown in the table at the top of the next page for workers in three age groups. Test H_0: $p_1 = p_2$ against H_a: $p_1 \neq p_2$ for each of the age groups and calculate the p-values for your tests. State the practical

AGE	1955		1980	
	n	%	n	%
29 and under	136	44.1	387	25.6
30–49	413	50.8	577	32.1
50 and over	232	57.8	323	44.9

implications of your *p*-values.

8.58 The *Orlando Sentinel* (March 8, 1984) suggests that coffee drinking is dropping in favor of soda. The article notes that the "Winter Coffee Drinking Study" shows that 55.2% of all Americans now drink coffee as compared with 74.7% in 1962. No sample sizes are given, but let us assume that both surveys were based on samples of 1,000 adult Americans. Use this information and the results just cited to find a 95% confidence interval for the percentage drop in coffee-drinking adult Americans between 1962 and 1984.

8.59 The number of practicing attorneys has more than doubled in the decade 1973–1983, and the percentage of female attorneys has increased from 5% to 15% of the legal profession. A random sample of 400 female attorneys and 200 male attorneys, from among the approximately 606,000 total number of attorneys in the United States, revealed that 25% of the women finished in the top 10% of their classes versus 18% of the males. The survey also found that women tend to enter law school at a later age than men. (Nearly 33% of women began practicing after age 30 compared with 14% for men.) For older women and men beginning practice, women's starting salaries tend to be higher than men of the same age. Finally, the median salary for different age groups has increased with age, peaking for men at age 51–55 and leveling off thereafter (*American Bar Association Journal*, October 1983). Based on the independent random samples of 400 female and 200 male attorneys selected from all attorneys in the United States, do the data provide sufficient evidence to indicate that the probabilities of finishing in the upper 10% of their law school class differ between male and female lawyers?

8.60 According to an article in the *Bakersfield Californian* (November 18, 1982), beta blockers given to heart patients can reduce or increase the death rate, depending on when the patient receives the drug. A beta blocker is a drug that blocks the body's adrenaline from activating the heart. In a British study, 632 heart attack patients received a beta blocker while 471 others received a placebo (a pill that contains no drugs and is designed to make patients think they are taking a drug). According to the article, the medicine significantly improved the survival of people who took the drug within 4 months of their heart attack. Ninety-five percent were alive after 6 years versus 77% of those who took the placebos. For people who started taking the blocker between 1 and $7\frac{1}{2}$ years after a heart attack, the situation was reversed. Those who received the beta blocker had a 79% survival rate versus 92% who received the placebo. The *Bakersfield Californian* does not give the sample sizes involved in these two comparisons, nor does it answer a very curious question. Why is there an apparent difference in the percentages of survival among patients who received placebos?

Theoretically they should be identical. Let us suppose that both samples of patients who received placebos contained 471 people.

a. What is the probability that the percentages of survival for the two experiments, 77% and 92%, could differ by as much as 15%, given that the population survival rates were equal to the average of the two percentages, .845?

b. Suppose you were to test the null hypothesis that the percentages of survival were equal for the two populations of patients who received placebos. Find the p-value for the test and state your conclusion.

8.5 Determining the Sample Size

You can find the appropriate sample size to estimate the difference between a pair of parameters with a specified degree of reliability by using the method described in Section 7.6. That is, to estimate the difference between a pair of parameters correct to within B units with probability $(1 - \alpha)$, let $z_{\alpha/2}$ standard deviations of the sampling distribution of the estimator equal B. Then solve for the sample size. To do this, you have to solve the problem for a specific ratio between n_1 and n_2. Most often, you will want to have equal sample sizes, that is, $n_1 = n_2 = n$. We will illustrate the procedure with two examples.

Example 8.10

New fertilizer compounds are often advertised with the promise of increased yields. Suppose we want to compare the mean yield μ_1 of wheat when a new fertilizer is used to the mean yield μ_2 with a fertilizer in common use. The estimate of the difference in mean yield per acre is to be corrrect to within .25 bushel with a confidence coefficient of .95. If the sample sizes are to be equal, find $n_1 = n_2 = n$, the number of 1-acre plots of wheat assigned to each fertilizer.

Solution

To solve the problem, you need to know something about the variation in the bushels of yield per acre. Suppose that from past records you know the yields of wheat possess a range of approximately 10 bushels per acre. You could then approximate $\sigma_1 = \sigma_2 = \sigma$ by letting the range equal 4σ. Thus,

$$4\sigma \approx 10 \text{ bushels}$$

$$\sigma \approx 2.5 \text{ bushels}$$

The next step is to solve the equation

$$z_{\alpha/2}\sigma_{(\bar{x}_1 - \bar{x}_2)} = B$$

or

$$z_{\alpha/2}\sqrt{\frac{\sigma_1^2}{n_1} + \frac{\sigma_2^2}{n_2}} = B$$

for n, where $n = n_1 = n_2$. Since we want the estimate to lie within $B = .25$ of $(\mu_1 - \mu_2)$

with confidence coefficient equal to .95, we have $z_{\alpha/2} = z_{.025} = 1.96$. Then, letting $\sigma_1 = \sigma_2 = 2.5$ and solving for n, we have

$$1.96 \sqrt{\frac{(2.5)^2}{n} + \frac{(2.5)^2}{n}} = .25$$

$$1.96 \sqrt{\frac{2(2.5)^2}{n}} = .25$$

$$n = 768.32 \approx 768$$

Consequently, you will have to sample 768 acres of wheat for each fertilizer to estimate the difference in mean yield per acre to within .25 bushel. Since this would necessitate extensive and costly experimentation, you might decide to allow a larger bound (say, $B = .50$ or $B = 1$) in order to reduce the sample size, or you might decrease the confidence coefficient. The point is: We can obtain an idea of the experimental effort necessary to achieve a specified precision in our final estimate by determining the approximate sample size *before* the experiment is begun.   ■

Example 8.11   A production supervisor suspects a difference exists between the proportions p_1 and p_2 of defective items produced by two different machines. Experience has shown that the proportion defective for each of the two machines is in the neighborhood of .03. If the supervisor wants to estimate the difference in the proportions correct to within .005 with probability .95, how many items must be randomly sampled from the production of each machine? (Assume that you want $n_1 = n_2 = n$.)

Solution   For the specified level of reliability, $z_{\alpha/2} = z_{.025} = 1.96$. Then, letting $p_1 = p_2 = .03$ and $n_1 = n_2 = n$, we find the required sample size per machine by solving the following equation for n:

$$z_{\alpha/2} \sigma_{(\hat{p}_1 - \hat{p}_2)} = B$$

or

$$z_{\alpha/2} \sqrt{\frac{p_1 q_1}{n_1} + \frac{p_2 q_2}{n_2}} = B$$

$$1.96 \sqrt{\frac{(.03)(.97)}{n} + \frac{(.03)(.97)}{n}} = .005$$

$$1.96 \sqrt{\frac{2(.03)(.97)}{n}} = .005$$

$$n = 8{,}943.2$$

You can see that this may be a tedious sampling procedure. If the supervisor insists on estimating $(p_1 - p_2)$ correct to within .005 with probability equal to .95, approximately 9,000 items will have to be inspected for each machine.   ■

You can see from the calculations in Example 8.11 that $\sigma_{(\hat{p}_1 - \hat{p}_2)}$ (and hence the solution, $n_1 = n_2 = n$) depends on the actual (but unknown) values of p_1 and p_2. In fact, the required sample size $n_1 = n_2 = n$ is largest when $p_1 = p_2 = \frac{1}{2}$. Therefore, if you have no prior information on the approximate values of p_1 and p_2, use $p_1 = p_2 = \frac{1}{2}$ in the formula for $\sigma_{(\hat{p}_1 - \hat{p}_2)}$. If p_1 and p_2 are in fact close to $\frac{1}{2}$, then the values of n_1 and n_2 that you have calculated will be correct. If p_1 and p_2 differ substantially from $\frac{1}{2}$, then your solutions for n_1 and n_2 will be larger than needed. Consequently, using $p_1 = p_2 = \frac{1}{2}$ when solving for n_1 and n_2 is a conservative procedure because the sample sizes n_1 and n_2 will be at least as large as (and probably larger than) needed.

Sample Size Determination with $100(1 - \alpha)\%$ Confidence Desired

For estimating $(\mu_1 - \mu_2)$ to within a bound B with probability $(1 - \alpha)$ and with equal sample sizes, that is, $n_1 = n_2$, solve the following equation for n_1:

$$n_1 = \left(\frac{z_{\alpha/2}}{B}\right)^2 (\sigma_1^2 + \sigma_2^2)$$

The values of σ_1^2 and σ_2^2 substituted into this expression are obtained from estimates of σ_1^2 and σ_2^2—that is, values of s_1^2 and s_2^2 obtained from prior samples (pilot study, etc.) or from an educated guess based on the range R, that is, $s \approx R/4$.

For estimating $(p_1 - p_2)$ to within a bound B with probability $(1 - \alpha)$ and with equal sample sizes, that is, $n_1 = n_2$, solve the following equation for n_1:

$$n_1 = \left(\frac{z_{\alpha/2}}{B}\right)^2 (p_1 q_1 + p_2 q_2)$$

The values of p_1 and p_2 substituted into this expression can be obtained from estimates based on prior samples, obtained from educated guesses, or, most conservatively, chosen to equal $p_1 = p_2 = .5$.

Exercises
8.61–8.69

Learning the Mechanics

8.61 Find the appropriate values of n_1 and n_2 (assume $n_1 = n_2$) needed to estimate $(\mu_1 - \mu_2)$ to within:

a. A bound on the error of estimation equal to 2.8 with 95% confidence. From prior experience it is known that $\sigma_1 \approx 13$ and $\sigma_2 \approx 14$.

b. A bound on the error of estimation equal to 5 with 99% confidence. The range of each population is 40.

c. A bound on the error of estimation equal to .5 with 90% confidence. Assume that $\sigma_1^2 \approx 4.4$ and $\sigma_2^2 \approx 6.7$.

8.62 Assuming that $n_1 = n_2$, find the sample sizes needed to estimate $(p_1 - p_2)$ for each of the following situations:

a. Bound = .01 with 99% confidence. Assume that $p_1 \approx .2$ and $p_2 \approx .5$.

b. Bound = .05 with 90% confidence. Assume there is no prior information available to obtain approximate values of p_1 and p_2.

c. Bound = .05 with 90% confidence. Assume that $p_1 \approx .1$ and $p_2 \approx .3$.

Applying the Concepts

8.63 One reason high school seniors are encouraged to attend college is that the job opportunities are much better for those with college degrees than for those without. A high school counselor wants to estimate the difference in mean income per day between high school graduates who have a college education and those who have not gone on to college. Suppose it is decided to compare the daily incomes of 30-year-olds, and the range of daily incomes for both groups is approximately $200 per day. How many people from each group should be sampled in order to estimate the true difference between mean daily incomes correct to within $10 per day with probability .9? Assume that $n_1 = n_2$.

8.64 Rat damage creates a large financial loss in the production of sugarcane. One aspect of the problem that has been investigated by the U.S. Department of Agriculture concerns the optimal place to locate rat poison. To be most effective in reducing rat damage, should the poison be located in the middle of the field or on the outer perimeter? One way to answer this question is to determine where the greater amount of damage occurs. If damage is measured by the proportion of cane stalks that have been damaged by rats, how many stalks from each section of the field should be sampled in order to estimate the true difference between proportions of stalks damaged in the two sections to within .02 with probability .95?

8.65 Suppose you are interested in the growth rate of dividends. Consider investing $1,000 in a stock and suppose you want to estimate the dividend rate on your $1,000 investment at the end of 5 years. Particularly, you want to compare two types of stocks, electrical utilities and oil companies. To conduct your study, you plan to select n oil stocks and n electrical utility stocks at random. For each stock, you will check the records, calculate the number of shares of stock you could have purchased 5 years ago for $1,000, and then calculate the dividend rate (in percent) that the stock would be paying today on your $1,000 investment. Suppose you think the dividend rates will vary over a range of roughly 25%. How large should n be if you want to estimate the difference between the mean rates of dividend return correct to within 3% with probability equal to .95?

8.66 A television manufacturer wants to compare with a competitor the proportions of its best sets that need repair within 1 year. If it is desired to estimate the difference between proportions to within .05 with 90% confidence, and if the manufacturer plans to sample twice as many buyers (n_1) of its sets as buyers (n_2) of the competitor's sets, how many buyers of each brand must be sampled? Assume that the proportion of sets that need repair will be about .2 for both brands.

8.67 In seeking a good professional football running back, a coach is looking for a player with a large mean yards gained per carry and a small standard deviation. Suppose that you wish to compare the mean yards gained per carry for two major prospects based on independent random samples of their yards gained per carry in the early part of the coming pro football season. Suppose that data from last year indicate that $\sigma_1 = \sigma_2 \approx 5$ yards. If you wish to estimate the difference in means correct to within 1 yard with probability equal to .9, how many runs would have to be observed for each player? (Assume equal sample sizes.)

8.68 In Exercise 8.23, we compared the mean costs of electrical energy produced by nuclear and coal-powered electrical power plants. In that exercise, we were given the information that the variances in costs for the nuclear and coal-powered plants were $\sigma_1^2 = .05$ and $\sigma_2^2 = .04$, respectively. Suppose that you wish to estimate the difference in mean costs per kilowatt-hour correct to within .1¢ with probability .95. How many power plants of each type must you include in your samples?

8.69 In Exercise 8.59, we compared the proportions of female and male attorneys finishing in the upper 10% of their law school classes. Suppose that we wish to estimate this difference by using independent random samples of equal size selected from the two populations. If we wish our margin of error to be .03 and if we want our estimate to be within this margin with probability .95, how many attorneys must we include in each sample?

8.6 Comparing Two Population Variances: Independent Random Samples

Many times it is of practical interest to use the techniques developed previously in this chapter to compare the means or proportions of two populations. However, there are also important instances when it is desired to compare two population variances. For example, when two devices are available for producing precision measurements (scales, calipers, thermometers, etc.), we might want to compare the variability of the measurements of the devices before deciding which one to purchase. Or when two standardized tests can be used to rate applicants, the variability of the scores for both tests should be taken into consideration before deciding which test to use.

For problems like these we need to develop a statistical procedure to compare population variances. The common statistical procedure for comparing population variances, σ_1^2 and σ_2^2, makes an inference about the ratio σ_1^2/σ_2^2. In this section, we will show how to test the null hypothesis that the ratio σ_1^2/σ_2^2 equals 1 (the variances are equal) against the alternative hypothesis that the ratio differs from 1 (the variances differ):

$$H_0: \frac{\sigma_1^2}{\sigma_2^2} = 1 \quad (\sigma_1^2 = \sigma_2^2) \qquad H_a: \frac{\sigma_1^2}{\sigma_2^2} \neq 1 \quad (\sigma_1^2 \neq \sigma_2^2)$$

To make an inference about the ratio σ_1^2/σ_2^2 it seems reasonable to collect sample data and use the ratio of the sample variances, s_1^2/s_2^2. We will use the test statistic

$$F = \frac{s_1^2}{s_2^2}$$

To establish a rejection region for the test statistic, we need to know the sampling distribution of s_1^2/s_2^2. As you will subsequently see, the sampling distribution of s_1^2/s_2^2 is based on two of the assumptions already required for the t-test:

1. The two sampled populations are normally distributed.
2. The samples are randomly and independently selected from their respective populations.

When these assumptions are satisfied and when the null hypothesis is true (that is, $\sigma_1^2 = \sigma_2^2$), the sampling distribution of $F = s_1^2/s_2^2$ is the **F-distribution** with ($n_1 - 1$) numerator degrees of freedom and ($n_2 - 1$) denominator degrees of freedom, respectively. The shape of the F-distribution depends on the degrees of freedom associated with s_1^2 and s_2^2—that is, on ($n_1 - 1$) and ($n_2 - 1$). An F-distribution with 7 and 9 df is shown in Figure 8.8. As you can see, the distribution is skewed to the right, since s_1^2/s_2^2 cannot be less than zero but can increase without bound.

Figure 8.8 An
F-Distribution with
7 Numerator and
9 Denominator Degrees
of Freedom

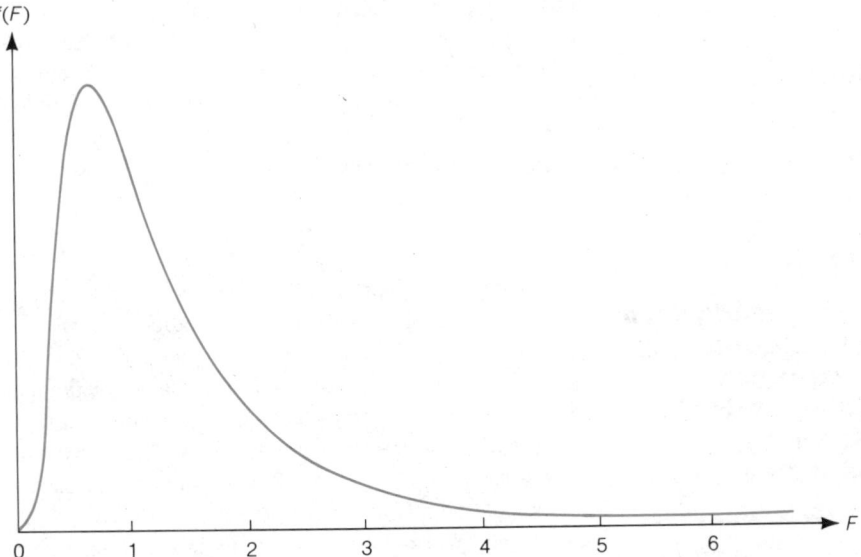

We need to be able to find F-values corresponding to the tail areas of this distribution in order to establish the rejection region for our test of hypothesis, because when the population variances are unequal, we expect the ratio F of the sample variances to be either very large or very small. The upper-tail F-values for $\alpha = .10$, .05, .025, and .01 can be found in Tables VII, VIII, IX, and X of the Appendix. Table VIII is partially reproduced in Figure 8.9. It gives F-values that correspond to $\alpha = .05$ upper-tail areas for different degrees of freedom. The columns of Tables VII, VIII, IX, and X correspond to the degrees of freedom v_1 for the numerator sample variance, s_1^2, while the rows correspond to the degrees of freedom v_2 for the denominator sample variance, s_2^2. Thus, if the numerator degrees of freedom is $v_1 = 7$ and the denominator degrees of freedom is $v_2 = 9$, we look in the seventh column and ninth row to find $F_{.05} = 3.29$. As shown

Figure 8.9 Reproduction of Part of Table VIII in the Appendix: $\alpha = .05$

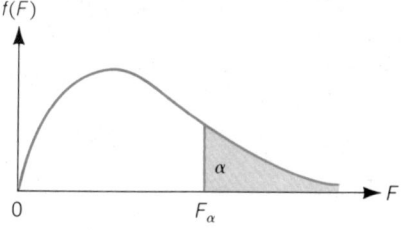

v_1 / v_2	NUMERATOR DEGREES OF FREEDOM								
	1	2	3	4	5	6	7	8	9
1	161.4	199.5	215.7	224.6	230.2	234.0	236.8	238.9	240.5
2	18.51	19.00	19.16	19.25	19.30	19.33	19.35	19.37	19.38
3	10.13	9.55	9.28	9.12	9.01	8.94	8.89	8.85	8.81
4	7.71	6.94	6.59	6.39	6.26	6.16	6.09	6.04	6.00
5	6.61	5.79	5.41	5.19	5.05	4.95	4.88	4.82	4.77
6	5.99	5.14	4.76	4.53	4.39	4.28	4.21	4.15	4.10
7	5.59	4.74	4.35	4.12	3.97	3.87	3.79	3.73	3.68
8	5.32	4.46	4.07	3.84	3.69	3.58	3.50	3.44	3.39
9	5.12	4.26	3.86	3.63	3.48	3.37	3.29	3.23	3.18
10	4.96	4.10	3.71	3.48	3.33	3.22	3.14	3.07	3.02
11	4.84	3.98	3.59	3.36	3.20	3.09	3.01	2.95	2.90
12	4.75	3.89	3.49	3.25	3.11	3.00	2.91	2.85	2.80
13	4.67	3.81	3.41	3.18	3.03	2.92	2.83	2.77	2.71
14	4.60	3.74	3.34	3.11	2.96	2.85	2.76	2.70	2.65

DENOMINATOR DEGREES OF FREEDOM

in Figure 8.10, $\alpha = .05$ is the tail area to the right of 3.29 in the F-distribution with 7 and 9 df. That is, if $\sigma_1^2 = \sigma_2^2$ then the probability that the F-statistic will exceed 3.29 is $\alpha = .05$.

Figure 8.10 An F-Distribution for $v_1 = 7$ and $v_2 = 9$ df: $\alpha = .05$

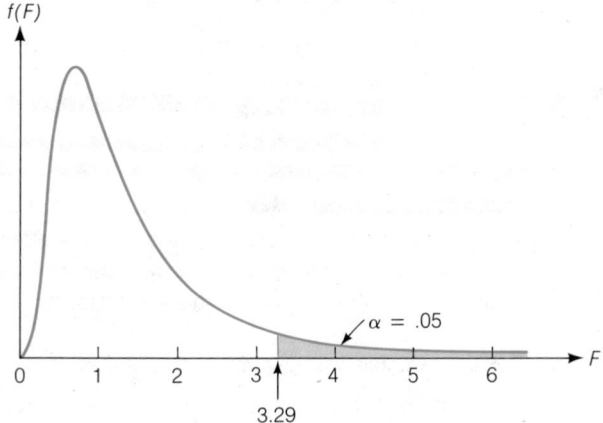

Example 8.12 An experimenter wants to compare the metabolic rates of white mice subjected to different drugs. The weights of the mice may affect their metabolic rates, and thus the experimenter wishes to obtain mice that are relatively homogeneous with regard to weight. Five hundred mice will be needed to complete the study. Presently, eighteen mice from supplier 1 and another thirteen mice from supplier 2 are available for comparison. The experimenter weighs these mice and obtains the following summary information:

SUPPLIER 1	SUPPLIER 2
$n_1 = 18$	$n_2 = 13$
$\bar{x}_1 = 4.21$ ounces	$\bar{x}_2 = 4.18$ ounces
$s_1^2 = .019$	$s_2^2 = .049$

Do these data provide sufficient evidence to indicate a difference in the variability of weights of mice obtained from the two suppliers? (Use $\alpha = .10$.) Using the results of this analysis, what would you suggest to the experimenter?

Solution Let

σ_1^2 = Population variance of weights of white mice from supplier 1

σ_2^2 = Population variance of weights of white mice from supplier 2

The hypotheses of interest are then

$$H_0: \quad \frac{\sigma_1^2}{\sigma_2^2} = 1 \quad (\sigma_1^2 = \sigma_2^2)$$

$$H_a: \quad \frac{\sigma_1^2}{\sigma_2^2} \neq 1 \quad (\sigma_1^2 \neq \sigma_2^2)$$

The nature of the F-tables given in the Appendix affects the form of the test statistic. To form the rejection region for a two-tailed F-test, we want to make certain that the upper tail is used, because only the upper-tail values of F are shown in Tables VII, VIII, IX, and X. To accomplish this, *we will always place the larger sample variance in the numerator of the F test statistic*. This has the effect of doubling the tabulated value for α, since we double the probability that the F-ratio will fall in the upper tail by always placing the larger sample variance in the numerator. That is, we make the test two-tailed by putting the larger variance in the numerator rather than establishing rejection regions in both tails.

Thus, for our example, we have a denominator s_1^2 with df $= v_2 = n_1 - 1 = 17$ and a numerator s_2^2 with df $= v_1 = n_2 - 1 = 12$. Therefore, the test statistic will be

$$F = \frac{\text{Larger sample variance}}{\text{Smaller sample variance}} = \frac{s_2^2}{s_1^2}$$

and we will reject H_0: $\sigma_1^2 = \sigma_2^2$ for $\alpha = .10$ when the calculated value of F exceeds the tabulated value:

$$F_{\alpha/2} = F_{.05} = 2.38 \quad \text{(see Figure 8.11)}$$

Figure 8.11 Rejection Region for Example 8.12

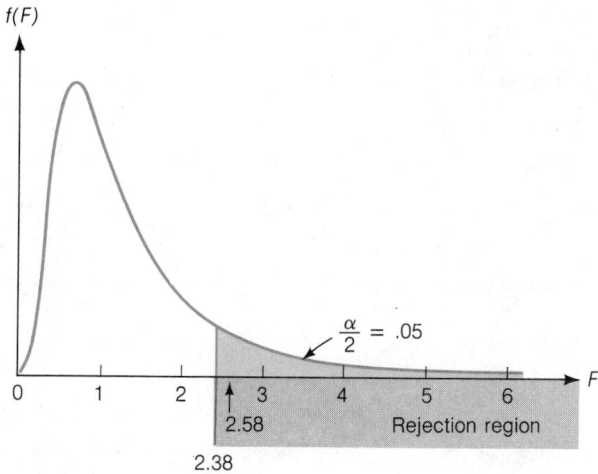

We can now calculate the value of the test statistic and complete the analysis:

$$F = \frac{s_2^2}{s_1^2} = \frac{.049}{.019} = 2.58$$

When we compare this result to the rejection region shown in Figure 8.11, we see that $F = 2.58$ falls in the rejection region. Therefore, the data provide sufficient evidence to indicate that the population variances differ. It appears that the weights of mice obtained from supplier 1 tend to be more homogeneous than the weights of those obtained from supplier 2. On the basis of this evidence, we would advise the experimenter to purchase the mice from supplier 1. ∎

What would you have concluded if the value of F calculated from the samples had not fallen in the rejection region? Would you conclude that the null hypothesis of equal variances is true? No, because then you risk the possibility of a Type II error (accepting H_0 if H_a is true) without knowing the value of β, the probability of accepting H_0: $\sigma_1^2 = \sigma_2^2$ if in fact it is false. Since we will not consider the calculation of β for specific alternatives in this text, when the F-statistic does not fall in the rejection region we simply conclude that insufficient sample evidence exists to refute the null hypothesis that $\sigma_1^2 = \sigma_2^2$.

The F-test for equal population variances is summarized in the box.

F-Test for Equal Population Variances

One-Tailed Test

H_0: $\sigma_1^2 = \sigma_2^2$

H_a: $\sigma_1^2 < \sigma_2^2$
   (or H_a: $\sigma_1^2 > \sigma_2^2$)

Test statistic:

$$F = \frac{s_2^2}{s_1^2}$$

$$\left(\text{or}\quad F = \frac{s_1^2}{s_2^2} \text{ when } H_a\text{: } \sigma_1^2 > \sigma_2^2\right)$$

Two-Tailed Test

H_0: $\sigma_1^2 = \sigma_2^2$

H_a: $\sigma_1^2 \neq \sigma_2^2$

Test statistic:

$$F = \frac{\text{Larger sample variance}}{\text{Smaller sample variance}}$$

$$= \frac{s_1^2}{s_2^2}\quad \text{when } s_1^2 > s_2^2$$

$$\left(\text{or}\quad \frac{s_2^2}{s_1^2}\quad \text{when } s_2^2 > s_1^2\right)$$

Rejection region:

$F > F_\alpha$
   (or $F > F_\alpha$ when H_a: $\sigma_1^2 > \sigma_2^2$)

Rejection region:

$F > F_{\alpha/2}$ when $s_1^2 > s_2^2$
   (or $F > F_{\alpha/2}$ when $s_2^2 > s_1^2$)

where F_α and $F_{\alpha/2}$ are based on v_1 = numerator degrees of freedom and v_2 = denominator degrees of freedom; v_1 and v_2 are the degrees of freedom for the numerator and denominator sample variances, respectively.

Assumptions: 1.   Both sampled populations are normally distributed.
2.   The samples are random and independent.

Example 8.13   Find the approximate p-value for the test in Example 8.12.

Solution   Since the observed value of the F-statistic in Example 8.12 was 2.58, the observed significance level of the test would equal the probability of observing a value of F at least as contradictory to H_0: $\sigma_1^2 = \sigma_2^2$ as $F = 2.58$, if in fact H_0 is true. Since we give the F-tables in the Appendix only for values of α equal to .10, .05, .025, and .01, we can only approximate the observed significance level. Checking Tables VII, VIII, IX, and X, we find $F_{.05} = 2.38$ and $F_{.025} = 2.82$. Since the observed value of F exceeds $F_{.05}$ but is less than $F_{.025}$, the observed significance level for the test is less than

Approximate p-value = 2(.05) = .10

Note that we double the α-value shown in Table VIII because this is a two-tailed test.

Example 8.14

An investor believes that although the price of stock 1 usually exceeds that of stock 2, stock 1 represents a riskier investment, where the risk of a given stock is measured by the variation in daily price changes. Suppose we obtain a random sample of twenty-five daily price changes for stock 1 and twenty-five for stock 2. The sample results are summarized in the table. Compare the risks associated with the two stocks by testing the null hypothesis that the variances of the price changes for the stocks are equal against the alternative that the price variance of stock 1 exceeds that of stock 2. Use $\alpha = .05$.

STOCK 1	STOCK 2
$n_1 = 25$	$n_2 = 25$
$\bar{x}_1 = .250$	$\bar{x}_2 = .125$
$s_1 = .76$	$s_2 = .46$

Solution

$H_0: \ \sigma_1^2 = \sigma_2^2 \qquad H_a: \ \sigma_1^2 > \sigma_2^2$

Test statistic: $F = \dfrac{s_1^2}{s_2^2}$

Assumptions: 1.   The changes in daily stock prices have relative frequency distributions that are approximately normal.
2.   The stock samples are randomly and independently selected from a set of daily stock reports.

Rejection region: $F > F_\alpha = F_{.05} = 1.98$, where $F_{.05}$ is based on $v_1 = 24$ df and $v_2 = 24$ df.

We calculate

$$F = \frac{s_1^2}{s_2^2} = \frac{(.76)^2}{(.46)^2} = 2.73$$

The calculated F exceeds the rejection value of 1.98. Therefore, we conclude that the variance of the daily price change for stock 1 exceeds that for stock 2. It appears that stock 1 is a riskier investment than stock 2. How much reliability can we place in this inference? Only one time in twenty (since $\alpha = .05$), on the average, would this statistical test lead us to conclude erroneously that σ_1^2 exceeds σ_2^2 if in fact they are equal.

Since this is a one-tailed test, the p-value is equal to the probability that F exceeds the computed value, 2.73; that is,

p-value $= P(F > 2.73)$

Checking Tables IX and X, we find that $F_{.025} = 2.27$ and $F_{.01} = 2.66$. Since the computed value of the F-statistic exceeds 2.66, the observed significance level for the test is slightly less:

Approximate p-value $= .01$ ∎

As a final example of an application, consider the comparison of population variances as a check of the assumption $\sigma_1^2 = \sigma_2^2$ needed for the two-sample t-test. Rejection of the null hypothesis $\sigma_1^2 = \sigma_2^2$ would indicate that the assumption is invalid. [*Note:* Nonrejection of the null hypothesis *does not* imply that the assumption is valid.] We will illustrate with an example.

Example 8.15

NEW METHOD	STANDARD METHOD
$n_1 = 8$	$n_2 = 12$
$\bar{x}_1 = 76.9$	$\bar{x}_2 = 72.7$
$s_1 = 4.85$	$s_2 = 6.35$

In Example 8.7 (Section 8.2) we used the two-sample t-statistic to compare the mean reading scores of two groups of slow learners who had been taught to read using two different methods. The data are repeated here for convenience. The use of the t-statistic was based on the assumption that the population variances of the test scores were equal for the two methods. Check this assumption by using $\alpha = .10$.

Solution We can test

$$H_0: \quad \sigma_1^2 = \sigma_2^2 \qquad H_a: \quad \sigma_1^2 \neq \sigma_2^2$$

using

Test statistic: $F = \dfrac{\text{Larger sample variance}}{\text{Smaller sample variance}} = \dfrac{s_2^2}{s_1^2}$

Rejection region: $\alpha = .10$

$$F > F_{\alpha/2} = F_{.05} = 3.6$$

where $F_{.05}$ is based on $v_1 = 11$ df and $v_2 = 7$ df. (We interpolate between 10 and 12 for the numerator degrees of freedom.) We calculate

$$F = \frac{(6.35)^2}{(4.85)^2} = 1.71$$

Since $F < 3.6$, we do not reject the null hypothesis that the population variances of the reading test scores are equal. It is here that the temptation to misuse the F-test is strongest. *We cannot conclude that the data justify the use of the t-statistic.* This is equivalent to accepting H_0, and we have repeatedly warned against this conclusion because the probability of a Type II error, β, is unknown. The α level of .10 protects us only against rejecting H_0 if it is true. This use of the F-test may prevent us from abusing the t-procedure when we obtain a value of F that leads to a rejection of the assumption that $\sigma_1^2 = \sigma_2^2$. But when the F-statistic does not fall in the rejection region, we know little more about the validity of the assumption than before we conducted the test. ∎

We have presented the F-test as a test of an hypothesis of equality of variances, that is, $\sigma_1^2 = \sigma_2^2$. Although this is the most common application of the test, it can also be used to test an hypothesis that the ratio between the population variances is equal to some specified value, $H_0: \quad \sigma_1^2/\sigma_2^2 = k$. The test would be conducted in exactly the same way as a test of an hypothesis concerning the equality of variances except that we would use the test statistic

$$F = \frac{s_1^2}{s_2^2}\left(\frac{1}{k}\right)$$

What Do You Do If the Assumption of Normal Population Distributions Is Not Satisfied?

Answer: The F-test is much less robust (i.e., much more sensitive) to departures from normality than the t-test for comparing population means (Section 8.2). If you have doubts about the normality of the population frequency distributions, use a nonparametric method for comparing the two population variances. A method can be found in the references listed at the end of this chapter.

Rather than test an hypothesis concerning σ_1^2/σ_2^2, you may wish to gain insight into the relative magnitudes of σ_1^2 and σ_2^2 by estimating their ratio. A confidence interval for the ratio of two population variances can be obtained by using the formula shown in the box.

Confidence Interval for σ_1^2/σ_2^2

$$\left(\frac{s_1^2}{s_2^2}\right)\left(\frac{1}{F_{L,\alpha/2}}\right) < \frac{\sigma_1^2}{\sigma_2^2} < \left(\frac{s_1^2}{s_2^2}\right)F_{U,\alpha/2}$$

where $F_{L,\alpha/2}$ is the tabulated value of F that places an area $\alpha/2$ in the upper tail of the F-distribution and that is based on $v_1 = (n_1 - 1)$ numerator and $v_2 = (n_2 - 1)$ denominator degrees of freedom, and $F_{U,\alpha/2}$ is the tabulated upper-tail value of F based on $v_1 = (n_2 - 1)$ numerator and $v_2 = (n_1 - 1)$ denominator degrees of freedom.

Assumptions: The samples were randomly and independently selected from populations that possess approximately normal distributions.

Example 8.16 Refer to Example 8.12 and find a 90% confidence interval for the ratio of the variances of the weights of mice obtained from the two suppliers.

Solution In Example 8.12, we were given the following information:

SUPPLIER 1	SUPPLIER 2
$n_1 = 18$	$n_2 = 13$
$\bar{x}_1 = 4.21$ ounces	$\bar{x}_2 = 4.18$ ounces
$s_1^2 = .019$	$s_2^2 = .049$

Then

$$n_1 - 1 = 18 - 1 = 17 \qquad n_2 - 1 = 13 - 1 = 12$$

and, from Table VIII of the Appendix,

$$F_{L,\alpha/2} = F_{L,.05} \approx 2.59 \quad \text{(where } v_1 = 17 \text{ and } v_2 = 12)$$

$$F_{U,\alpha/2} = F_{U,.05} = 2.38 \quad \text{(where } v_1 = 12 \text{ and } v_2 = 17)$$

Substituting these values into the formula for the confidence interval, we obtain

$$\left(\frac{s_1^2}{s_2^2}\right)\left(\frac{1}{F_{L,\alpha/2}}\right) < \frac{\sigma_1^2}{\sigma_2^2} < \left(\frac{s_1^2}{s_2^2}\right)F_{U,\alpha/2}$$

$$\left(\frac{.019}{.049}\right)\left(\frac{1}{2.59}\right) < \frac{\sigma_1^2}{\sigma_2^2} < \left(\frac{.019}{.049}\right)2.38$$

$$.150 < \frac{\sigma_1^2}{\sigma_2^2} < .923$$

According to this confidence interval, we estimate that σ_1^2, the variance in the weights of mice obtained from supplier 1, could be as small as .150 or as large as .923 times the size of σ_2^2, the variance in the weights of mice obtained from supplier 2. ∎

Exercises
8.70–8.79

Learning the Mechanics

8.70 Specify the appropriate rejection region for testing H_0: $\sigma_1^2 = \sigma_2^2$ in each of the following situations:

a. H_a: $\sigma_1^2 > \sigma_2^2$; $\alpha = .05$, $n_1 = 21$, $n_2 = 20$
b. H_a: $\sigma_1^2 < \sigma_2^2$; $\alpha = .05$, $n_1 = 10$, $n_2 = 21$
c. H_a: $\sigma_1^2 \neq \sigma_2^2$; $\alpha = .10$, $n_1 = 16$, $n_2 = 25$
d. H_a: $\sigma_1^2 < \sigma_2^2$; $\alpha = .01$, $n_1 = 31$, $n_2 = 41$
e. H_a: $\sigma_1^2 \neq \sigma_2^2$; $\alpha = .05$, $n_1 = 9$, $n_2 = 25$

8.71 Give the appropriate tabled F-values to form a 95% confidence interval for σ_1^2/σ_2^2 for each of the following combinations of n_1 and n_2:

a. $n_1 = 10$, $n_2 = 8$ **b.** $n_1 = 13$, $n_2 = 9$
c. $n_1 = 16$, $n_2 = 41$ **d.** $n_1 = 16$, $n_2 = 13$

SAMPLE 1	SAMPLE 2
$n_1 = 16$	$n_2 = 25$
$\bar{x}_1 = 22.5$	$\bar{x}_2 = 28.2$
$s_1^2 = 3.56$	$s_2^2 = 10.24$

8.72 Independent random samples were selected from each of two normally distributed populations, $n_1 = 15$ from population 1 and $n_2 = 27$ from population 2. The means and variances for the two samples are shown in the table.

a. Test the null hypothesis H_0: $\sigma_1^2 = \sigma_2^2$ against the alternative hypothesis H_a: $\sigma_1^2 \neq \sigma_2^2$. Use $\alpha = .05$.
b. Form a 95% confidence interval for σ_1^2/σ_2^2.
c. Test H_0: $\sigma_1^2 = \sigma_2^2$ against H_a: $\sigma_1^2 < \sigma_2^2$. Use $\alpha = .05$.

SAMPLE 1	SAMPLE 2
3.1	2.3
4.3	1.4
1.2	3.7
1.7	8.9
.6	
3.4	

8.73 Independent random samples were selected from each of two normally distributed populations, $n_1 = 6$ from population 1 and $n_2 = 4$ from population 2. The data are shown in the table.

a. Form a 90% confidence interval for σ_1^2/σ_2^2.
b. Test H_0: $\sigma_1^2 = \sigma_2^2$ against H_a: $\sigma_1^2 < \sigma_2^2$. Use $\alpha = .01$.
c. Test H_0: $\sigma_1^2 = \sigma_2^2$ against H_a: $\sigma_1^2 \neq \sigma_2^2$. Use $\alpha = .10$.

8.74 Under what conditions is the sampling distribution of s_1^2/s_2^2 an F-distribution?

Applying the Concepts

8.75 In Exercise 8.67, we planned to compare the mean yards gained per run for two professional football running backs. Suppose that a record of early season runs shows that the standard deviations of yards gained per run, based on samples of fifty runs per player, are $s_1 = 3.2$ and $s_2 = 5.7$ yards, respectively. Do these standard deviations indicate that the variation in the distributions of yards gained per carry differs for the two players? Test using $\alpha = .05$.

WOMEN	MEN
$n = 25$	$n = 20$
$\bar{x} = 42.7$	$\bar{x} = 41.8$
$s^2 = 18.3$	$s^2 = 8.5$

8.76 A series of experiments has been conducted to compare the quantity of hemoglobin in the blood of men and women who are between the ages of 20 and 30 years. One phase of the study deals with a comparison of the variability in the hemoglobin measurements between the two groups. In random samples of twenty-five women and twenty men, all between the ages of 20 and 30 years, the researcher recorded the amount of hemoglobin in the blood of each. (The quantity of hemoglobin is measured as a percentage of the total volume of blood.) The results are summarized in the table. Form a 90% confidence interval for the ratio of the variance of the amount of hemoglobin in women's blood to the corresponding variance for men.

8.77 The goalie is generally regarded as the most important player on a hockey team. One measure of a goalie's ability is the "goals against" average (GA average), i.e., the average number of goals the goalie gives up per game. However, most National Hockey League coaches agree that consistency in performance is just as important as the GA average. A consistent goalie is one whose number of goals given up per game varies only slightly from game to game. Two goalies with nearly equal GA averages are competing for the starting position on a hockey team. The coach will choose the starter on the basis of the better GA average only if there is no evidence of a difference in the consistency of the two goalies based on their performances in ten exhibition games (ten games per goalie). Otherwise, the more consistent goalie will win the starting position. The results of the exhibition games are given in the table. What decision does the coach make? Test at the $\alpha = .05$ level of significance.

	GOALIE A	GOALIE B
Number of games	10	10
$\bar{x}$ (GA average)	3.3	3.1
s^2	.68	2.77

8.78 Suppose a firm has been experimenting with two different physical arrangements of its assembly line. It has been determined that both arrangements yield approximately the same average number of finished units per day. To obtain an arrangement that produces greater process control, you suggest that the arrangement with the smaller variance in the number of finished units produced per day be per-

ASSEMBLY LINE 1	ASSEMBLY LINE 2
$n_1 = 21$ days	$n_2 = 21$ days
$s_1^2 = 1,432$	$s_2^2 = 3,761$

manently adopted. Two independent random samples yield the results shown in the table. Do the samples provide sufficient evidence at the .10 significance level to conclude that the variances of the two arrangements differ? If so, which arrangement would you choose? If not, what would you suggest the firm do?

8.79 The quality control department of a paper company measures the brightness (a measure of reflectance) of finished paper on a periodic basis throughout the day. Two instruments that are available to measure the paper specimens are subject to error, but they can be adjusted so that the mean readings for a control paper specimen are the same for both instruments. Suppose you are concerned about the precision of the two instruments and want to compare the variability in the readings of instrument 1 to those of instrument 2. Five brightness measurements were made on a single paper specimen using each of the two instruments. The data are shown in the table.

INSTRUMENT 1	INSTRUMENT 2
29	26
28	34
30	30
28	32
30	28

a. Form a 95% confidence interval for the ratio of the variance of the measurements obtained by instrument 1 to the variance of the measurements obtained using instrument 2.

b. Interpret the interval you found in part a in terms of the precision of the two instruments.

c. What assumptions must be satisfied for the confidence interval in part a to be valid?

Summary

We have presented various techniques for using the information in two samples to make inferences about the difference between population parameters. As you would expect, we are able to make reliable inferences with fewer assumptions about the sampled populations when the sample sizes are large. When we cannot take large samples from the populations, the **two-sample t-statistic** permits us to use the limited sample information to make inferences about the **difference between means** when the assumptions of normality and equal population variances are at least approximately true. The **paired difference experiment** offers the possibility of increasing the information about $(\mu_1 - \mu_2)$ by pairing similar observational units to control variability. In designing a paired difference experiment, we expect that the reduction in variability will more than compensate for the loss in degrees of freedom.

Two other inferential procedures for making comparisons between population parameters were presented in this chapter. A method for comparing two binomial parameters, p_1 and p_2, was presented. Practical examples of such comparisons are numerous and appear frequently in the analysis of surveys. (Applications of this and other techniques discussed in the chapter are suggested by the exercises that follow.)

The chapter concluded with a procedure for comparing two population variances, σ_1^2 and σ_2^2. We use an **F-statistic** to test the null hypothesis that two population variances are equal. This test is of practical importance because variances often represent measures of risk, error, or precision.

Supplementary Exercises 8.80–8.123

[Note: List the assumptions necessary to ensure the validity of the statistical procedures you use to work these exercises.]

Learning the Mechanics

SAMPLE 1	SAMPLE 2
$n_1 = 12$	$n_2 = 14$
$\bar{x}_1 = 17.8$	$\bar{x}_2 = 15.3$
$s_1^2 = 74.2$	$s_2^2 = 60.5$

8.80 Independent random samples were selected from two normally distributed populations with means μ_1 and μ_2, respectively. The sample sizes, means, and variances are shown in the table.
a. Test the null hypothesis H_0: $(\mu_1 - \mu_2) = 0$ against the alternative hypothesis H_a: $(\mu_1 - \mu_2) > 0$. Use $\alpha = .05$.
b. Form a 99% confidence interval for $(\mu_1 - \mu_2)$.
c. How large must n_1 and n_2 be if you wish to estimate $(\mu_1 - \mu_2)$ to within 2 units with 99% confidence? Assume that $n_1 = n_2$.

SAMPLE 1	SAMPLE 2
$n_1 = 20$	$n_2 = 15$
$\bar{x}_1 = 123$	$\bar{x}_2 = 116$
$s_1^2 = 31.3$	$s_2^2 = 120.1$

8.81 Two independent random samples were selected from normally distributed populations with means and variances (μ_1, σ_1^2) and (μ_2, σ_2^2), respectively. The sample sizes, means, and variances are shown in the table.
a. Form a 95% confidence interval for σ_1^2/σ_2^2.
b. Test H_0: $\sigma_1^2 = \sigma_2^2$ against H_a: $\sigma_1^2 \neq \sigma_2^2$. Use $\alpha = .05$.
c. Would you be willing to use a t-test to test the null hypothesis H_0: $(\mu_1 - \mu_2) = 0$ against the alternative hypothesis H_a: $(\mu_1 - \mu_2) \neq 0$? Why?

SAMPLE 1	SAMPLE 2
$n_1 = 135$	$n_2 = 148$
$\bar{x}_1 = 12.2$	$\bar{x}_2 = 8.3$
$s_1^2 = 2.1$	$s_2^2 = 3.0$

8.82 Two independent random samples are taken from two populations. The results of these samples are summarized in the table.
a. Form a 90% confidence interval for $(\mu_1 - \mu_2)$.
b. Test H_0: $(\mu_1 - \mu_2) = 0$ against H_a: $(\mu_1 - \mu_2) \neq 0$. Use $\alpha = .01$.
c. What sample sizes would be required if you wish to estimate $(\mu_1 - \mu_2)$ to within .2 with 90% confidence? Assume that $n_1 = n_2$.

SAMPLE 1	SAMPLE 2
$n_1 = 200$	$n_2 = 200$
$x_1 = 110$	$x_2 = 130$

8.83 Independent random samples were selected from two binomial populations. The sizes and number of observed successes for each sample are shown in the table.
a. Test the null hypothesis H_0: $(p_1 - p_2) = 0$ against the alternative hypothesis H_a: $(p_1 - p_2) < 0$. Use $\alpha = .10$.
b. Form a 95% confidence interval for $(p_1 - p_2)$.
c. What sample sizes would be required if we wish to estimate $(p_1 - p_2)$ to within .01 with 95% confidence? Assume that $n_1 = n_2$.

8.84 A random sample of five pairs of observations were selected, one of each pair from a population with mean μ_1, the other from a population with mean μ_2. The data are shown in the table at the top of the next page.

PAIR	VALUE FROM POPULATION 1	VALUE FROM POPULATION 2
1	28	22
2	31	27
3	24	20
4	30	27
5	22	20

a. Test the null hypothesis H_0: $\mu_D = 0$ against H_a: $\mu_D \neq 0$, where $\mu_D = \mu_1 - \mu_2$. Use $\alpha = .05$.

b. Form a 95% confidence interval for μ_D.

c. When are the procedures you used in parts a and b valid?

8.85 List the assumptions necessary for each of the following inferential techniques:

a. Large-sample inferences about the difference $(\mu_1 - \mu_2)$ between population means using a two-sample z-statistic

b. Small-sample inferences about $(\mu_1 - \mu_2)$ using an independent samples design and a two-sample t-statistic

c. Small-sample inferences about $(\mu_1 - \mu_2)$ using a paired difference design and a single-sample t-statistic to analyze the differences

d. Large-sample inferences about the difference $(p_1 - p_2)$ between binomial proportions using a two-sample z-statistic

Applying the Concepts

TAX-FREE BONDS	TAXABLE BONDS
$\bar{x}_1 = 9.8\%$	$\bar{x}_2 = 9.3\%$
$s_1 = 1.1\%$	$s_2 = 1.0\%$

8.86 To compare the rate of return an investor can expect on tax-free municipal bonds with the rate of return on taxable bonds, an investment advisory firm randomly samples ten bonds of each type and computes the annual rate of return over the past 3 years for each bond. The rate of return is then adjusted for taxes, assuming the investor is in a 30% tax bracket. The means and standard deviations for the adjusted returns are shown in the table.

a. Test to see whether there is a difference in the mean rates of return between tax-free and taxable bonds for investors in the 30% tax bracket. Use $\alpha = .05$.

b. What assumptions were necessary for the validity of the testing procedure you used in part a?

8.87 Refer to Exercise 8.86.

a. Test the hypothesis that the two population variances are equal. Use $\alpha = .10$.

b. Form a 90% confidence interval for σ_1^2/σ_2^2.

8.88 Management training programs are often instituted in order to teach supervisory skills and thereby increase productivity. Suppose a company psychologist administers a set of examinations to each of ten supervisors before such a training program begins and then administers similar examinations at the end of the program. The examinations are designed to measure supervisory skills, with higher scores indicating increased skill. The results of the tests are shown in the table. Test to see whether the data indicate that the training program is effective. Use $\alpha = .10$.

SUPERVISOR	BEFORE TRAINING PROGRAM	AFTER TRAINING PROGRAM
1	63	78
2	93	92
3	84	91
4	72	80
5	65	69
6	72	85
7	91	99
8	84	82
9	71	81
10	80	87

8.89 Lack of motivation is a problem of many students in inner-city schools. To cope with this problem, an experiment was conducted to determine whether motivation could be improved by allowing students greater choice in the structures of their curricula. Two schools with similar student populations were chosen, and fifty students were randomly selected from each to participate in the experiment. School A permitted its fifty students to choose only the courses they wanted to take. School B permitted its students to choose their courses and also to choose when and from which instructors to take the courses. The measure of student motivation was the number of times each student was absent from or late for a class during a 20-day period. The means and variances for the two samples are shown in the table. Do the data provide sufficient evidence to indicate that students from school B were late or absent less frequently than those from school A? (Use $\alpha = .1$.)

SCHOOL A	SCHOOL B
$\bar{x}_A = 20.5$	$\bar{x}_B = 19.6$
$s_A^2 = 26.2$	$s_B^2 = 24.1$

8.90 Will premium gasoline provide an increase in the mileage per gallon obtained by your automobile in comparison with the mileage for standard gasoline? Is the higher price of premium gasoline worth the increase (assuming an increase exists)? To assist in answering these questions, a government agency randomly selected 100 automobiles from its fleet and divided them into two equal groups. Each automobile was operated until it consumed 100 gallons of gasoline. One group used regular gasoline; the other used premium. At the conclusion of the experiment, the miles per gallon were calculated for each automobile. The means and standard deviations for the two samples are shown in the table. Estimate the difference in mean gasoline mileage between regular and premium gasolines using a 95% confidence interval. Interpret the interval estimate.

	NUMBER OF AUTOMOBILES	MEAN NUMBER OF MILES PER GALLON	s
REGULAR	50	19.3	2.1
PREMIUM	50	22.0	1.7

8.91 Refer to Exercise 8.90. Suppose the agency wishes to estimate the difference between mean mileages to within .5 mile per gallon with a probability of .95. How many automobiles would the agency have to include in each sample?

8.92 Advertising companies often try to characterize the average user of a client's product so that the advertisements can be targeted at specific segments of the buying community. Suppose a new movie is about to be released and an advertising company wants to determine whether to aim the advertisements at people under 25 years old or those over 25. They plan to arrange an advance showing of the movie to a number of individuals from each group and then to obtain an opinion about the movie from each individual. How many individuals should be included in each sample if the advertising company wants to estimate the difference in the proportions of those who like the movie to within .05 with 90% confidence? Assume that the sample size for each group will be the same and about half of each group will like the movie.

BANK 1	BANK 2
$\bar{x}_1 = 2.2$	$\bar{x}_2 = 1.8$
$s_1 = 1.15$	$s_2 = 1.10$

8.93 Two banks, bank 1 and bank 2, each independently sampled forty and fifty of their business accounts, respectively, and determined the number of the bank's services (loans, checking, savings, investment counseling, etc.) each sampled business was using. Both banks offer the same services. A summary of the data supplied by the samples is given in the table. Do the samples yield sufficient evidence to conclude that the average number of services used by bank 1's business customers is significantly greater (at the $\alpha = .10$ level) than the average number of services used by bank 2's business customers?

8.94 Find a 99% confidence interval for $(\mu_1 - \mu_2)$ in Exercise 8.93. Does the interval include zero? Interpret the confidence interval.

8.95 Radio stations sometimes conduct prize giveaways in an attempt to increase their share of the listening audience. Suppose a station manager calls 300 randomly selected households in a city and finds that 65 have members who regularly listen to the station. The station then conducts a 2-month promotional contest and follows it with a survey of 500 randomly chosen households. The survey shows that 154 households have members who regularly listen to the station.

a. Use a 90% confidence interval to estimate the difference between the proportions of those who regularly listen to the station before and after the promotional contest.

b. Construct a 95% confidence interval for the proportion of those who listen to the station after the promotion is over.

8.96 A consumer protection agency wants to compare the work of two electrical contractors in order to evaluate their safety records. The agency plans to inspect residences in which each of these contractors has done the wiring in order to estimate the difference in the proportions of residences that are electrically deficient. Suppose the proportions of residences with deficient work are expected to be about .10 for both contractors. How many homes should be inspected in order to estimate the difference in proportions to within .05 with 90% confidence?

8.97 The interocular pressure of glaucoma patients is often reduced by treatment with adrenaline. To compare a new synthetic drug with adrenaline, seven glaucoma patients were treated with both drugs, one eye with adrenaline and one with the synthetic drug. The reduction in pressure in each eye was then recorded, as shown in the table. Do the data provide sufficient evidence to indicate a difference in the mean reductions in eye pressure for the two drugs? Test using $\alpha = .10$.

PATIENT	ADRENALINE	SYNTHETIC
1	3.5	3.2
2	2.6	2.8
3	3.0	3.1
4	1.9	2.4
5	2.9	2.9
6	2.4	2.2
7	2.0	2.2

8.98 Suppose you have been offered similar jobs in two different locales. To help in deciding which job to accept, you would like to compare the cost of living in the two cities. One of your primary concerns is the cost of housing, so you obtain a copy of a newspaper from each locale and begin to study the housing prices in the classified advertisements. One convenient method for getting a general idea of prices is to compute the cost per square foot. This is done by dividing the price of the house by the heated area (in square feet) of the house. Random samples of sixty-three advertisements in locale 1 and seventy-eight in locale 2 produce the results given in the table. Is there evidence that the mean housing price per square foot differs in the two locales? Test using $\alpha = .05$.

LOCALE 1	LOCALE 2
$\bar{x}_1 = \$23.40$ per square foot	$\bar{x}_2 = \$25.20$ per square foot
$s_1 = \$2.50$ per square foot	$s_2 = \$2.80$ per square foot

8.99 Refer to Exercise 8.98. You also want to compare food prices in the two locales. You develop a list of fifteen food items of various types and obtain prices from the newspaper advertisements for a supermarket chain that has a store in each locale. The results are shown in the table.

FOOD ITEM	LOCALE 1	LOCALE 2	FOOD ITEM	LOCALE 1	LOCALE 2
1	$2.49	$2.55	9	$3.83	$3.75
2	.35	.37	10	2.93	3.11
3	5.12	5.05	11	1.03	1.25
4	1.33	1.52	12	2.13	2.05
5	.78	.85	13	6.25	6.25
6	3.03	2.98	14	2.14	2.30
7	.25	.35	15	1.98	1.97
8	4.16	4.29			

a. Do these data provide sufficient evidence to indicate a difference between the mean food prices in the two locales? Use $\alpha = .01$.

b. Can you think of a better way to design this experiment?

8.100 Some power plants are located near rivers or oceans so that the available water can be used for cooling the condensers. As part of an environmental impact study, suppose a power company wants to estimate the difference in mean water temperature between the discharge of its plant and the offshore waters. How many sample measurements must be taken at each site in order to estimate the true difference between means to within $.2°C$ with 95% confidence? Assume that the range in readings will be about $4°C$ at each site and the same number of readings will be taken at each site.

8.101 The use of preservatives by food processors has become a controversial issue. Suppose two preservatives are extensively tested and determined safe for use in meats. A processor wants to compare the preservatives for their effects on retarding spoilage. Suppose fifteen cuts of fresh meat are treated with preservative A and fifteen with B, and the number of hours until spoilage begins is recorded for each of the thirty cuts of meat. The results are summarized in the table.

PRESERVATIVE A	PRESERVATIVE B
$\bar{x}_1 = 106.4$ hours	$\bar{x}_2 = 96.5$ hours
$s_1 = 10.3$ hours	$s_2 = 13.4$ hours

a. Is there evidence of a difference in mean time until spoilage begins between the two preservatives at the $\alpha = .05$ level?

b. Can you recommend an experimental design that the processor could have used to reduce the variability in the data?

8.102 Refer to Exercise 8.101.

a. Construct a 95% confidence interval for the difference between the mean times until spoilage for the two preservatives.

b. Form a 95% confidence interval for σ_1^2/σ_2^2.

8.103 A physiologist wishes to study the effect of birth-control pills on exercise capacity. Five female subjects who have never taken the pill have their maximal oxygen uptake measured (in milliliters per kilogram of their body weight) during a treadmill session. The five subjects then take the pill for a specified length of time and their uptakes are measured again, as given in the table. Do the data provide sufficient

SUBJECT	MAXIMAL OXYGEN UPTAKE	
	Before	*After*
1	35.0	29.5
2	36.5	33.5
3	36.0	32.0
4	39.0	36.5
5	37.5	35.0

evidence to indicate that the mean maximal oxygen uptake after taking birth-control pills is less than the mean uptake before taking the pill? Use $\alpha = .01$.

8.104 An economist wants to investigate the difference in unemployment rates between an urban industrial community and a university community in the same state. She interviews 525 potential members of the work force in the industrial community and 375 in the university community. Of these, forty-seven and twenty-two, respectively, are unemployed. Use a 95% confidence interval to estimate the difference in unemployment rates in the two communities. Interpret this interval.

8.105 A careful auditing is essential to all businesses, large and small. Suppose a firm wants to compare the performances of two auditors it employs. One measure of auditing performance is error rate, so the firm decides to sample 200 pages at random from the work of each auditor and carefully examine each page for errors. Suppose the number of pages on which at least one error is found is seventeen for auditor A and twenty-five for auditor B. Test to see whether these data indicate a difference in the true error rates for the two auditors. Use $\alpha = .01$.

1979	1980
$\bar{x}_1 = \$676$	$\bar{x}_2 = \$853$
$s_1 = \$554$	$s_2 = \$715$
$n_1 = 325$	$n_2 = 375$

8.106 Since tourism is one of the largest industries in the state of Florida, the economy of the state depends heavily on the number of tourists who visit Florida annually and on the mean amount of money tourists spend while they are in the state. Suppose a study is conducted during two consecutive years, say 1979 and 1980, to compare the mean expenditure of tourists in Florida. Random samples of 325 and 375 tourists (a family is treated as one tourist) are selected in 1979 and 1980, respectively, and the total expenditure in the state is recorded for each. The results are given in the table.
a. Form a 90% confidence interval for the difference between the mean expenditures per tourist in 1979 and 1980.
b. What assumptions are necessary for the validity of the procedure in part a?

RABBIT	DRUG A	DRUG B
1	52	51
2	50	61
3	49	53
4	63	68
5	58	60

8.107 It is desired to compare two drugs that cause the pupil of the eye to dilate. Initially, five rabbits are to be used in studying the drugs. Each rabbit has one drug randomly assigned to the left eye, the other drug to the right eye, and the length of time in minutes until the pupil returns to normal is recorded for each eye. The results are listed in the table. Do these data provide sufficient evidence to indicate a difference in the true mean length of time until the pupil returns to normal for the two drugs? Test at the $\alpha = .10$ level of significance.

8.108 A large shipment of produce contains Valencia and navel oranges. To determine whether there is a difference in the proportions of nonmarketable fruit between the two varieties, random samples of 850 Valencia oranges and 1,500 navel oranges are independently selected and the number of nonmarketable oranges of each type is counted. It is found that thirty Valencia and ninety navel oranges from these samples are nonmarketable. Do these data provide sufficient evidence to indicate a difference between the proportions of nonmarketable Valencia and navel oranges? Test at the $\alpha = .05$ level of significance.

8.109 When new instruments are developed to perform chemical analyses of products (food, medicine, etc.), they are usually evaluated with respect to two criteria: accuracy and precision. *Accuracy* refers to the ability of the instrument to identify correctly the nature and amounts of a product's components. *Precision* refers to the consistency with which the instrument will identify the components of the same material. Thus, a large variability in the identification of a single sample of a product indicates a lack of precision. Suppose a pharmaceutical firm is considering two brands of an instrument designed to identify the components of certain drugs. As part of a comparison of precision, ten test-tube samples of a well-mixed batch of a drug are selected and then five are analyzed by instrument A and five by instrument B. The data shown in the table are the percentages of the primary component of the drug given by the instruments. Do these data provide evidence of a difference in the precision of the two machines? Use $\alpha = .10$.

INSTRUMENT A	INSTRUMENT B
43	46
48	49
37	43
52	41
45	48

8.110 A large department store plans to renovate one of its floors and increase the floor space for one department. The management has narrowed the decision to two departments: men's clothing and sporting goods. The final decision will be based on mean sales; the department having the greater mean will be enlarged. The last 12 months' sales data are shown in the table.

MONTH	MEN'S CLOTHING	SPORTING GOODS	MONTH	MEN'S CLOTHING	SPORTING GOODS
1	$15,726	$17,533	7	$15,525	$16,774
2	11,243	10,895	8	15,799	16,223
3	22,325	19,449	9	16,449	16,135
4	23,494	21,500	10	16,993	17,834
5	12,676	18,925	11	19,832	18,429
6	13,492	21,426	12	32,434	34,565

a. Use these data to form a 95% confidence interval for the difference between mean monthly sales for the two departments.

b. On the basis of the confidence interval formed in part a, can you make a recommendation to the store management as to which department should be enlarged?

c. What assumptions are necessary to make valid the procedure you used in part a?

8.111 Two basketball players engage in a foul-shooting contest in which each player takes 100 shots. It is desired to compare the percentages of shots made by each player.
a. What is the parameter of interest?
b. In this contest, player A made ninety-three shots and player B made eighty-six shots. Estimate the parameter of interest using a 95% confidence interval.

8.112 A political candidate conducted a sample survey to determine whether a television advertising campaign was worthwhile. Both before and after the advertising campaign, random samples were taken from the candidate's constituency and each person was asked his or her voter preference. The results of the survey are shown in the table. Using a 95% confidence interval, estimate the difference in the proportion of voters who favor the candidate before and after the campaign.

	SAMPLE SIZE	NUMBER WHO PREFER THE CANDIDATE
BEFORE ADVERTISING CAMPAIGN	200	85
AFTER ADVERTISING CAMPAIGN	300	139

8.113 Refer to Exercise 8.112. What size samples need to be taken in order to estimate the difference between proportions to within .04 with probability .95? Assume that $n_1 = n_2$.

8.114 The following experiment was conducted to compare two coatings designed to improve the durability of the soles of jogging shoes. A $\frac{1}{8}$-inch-thick layer of coating 1 was applied to one of a pair of shoes and a layer of equal thickness of coating 2 was applied to the other shoe. Ten joggers were given pairs of shoes treated in this manner and were instructed to record the number of miles covered in each shoe before the $\frac{1}{8}$-inch coating was worn through in any one place. The results are given in the table.

JOGGER	COATING 1	COATING 2	JOGGER	COATING 1	COATING 2
1	892	985	6	853	875
2	904	953	7	780	895
3	775	775	8	695	725
4	435	510	9	825	858
5	946	895	10	750	812

a. Do the data provide sufficient evidence to indicate a difference between the mean numbers of miles of wear that a runner might expect from the two coatings? Test using $\alpha = .05$.
b. Use a 90% confidence interval to estimate the true difference between the mean numbers of miles of wear for the two sole coatings. Interpret this interval.
c. Why is the design used for this experiment preferable to independent random sampling?

8.115 Many college and university professors have been accused of "grade promotion" over the past several years. This means they assign higher grades now than in the past, even though students' work is of the same caliber. If grade promotion has occurred, the mean grade-point average of today's students should exceed the mean of 10 years ago. To test the grade promotion theory at one university, a business professor randomly selects seventy-five business majors who are graduating with the present class and fifty who graduated 10 years ago. The results are shown in the table. Test to see whether the data indicate that the mean grade-point average of the present class exceeds the mean grade-point average of those who graduated 10 years ago. Use $\alpha = .05$.

10 YEARS AGO	PRESENT
$\bar{x}_1 = 2.82$	$\bar{x}_2 = 3.04$
$s_1 = .43$	$s_2 = .38$
$n_1 = 50$	$n_2 = 75$

8.116 Smoke detectors are highly recommended safety devices for early fire detection in homes and businesses. It is extremely important that the devices are not defective. Suppose that 100 brand A smoke detectors are tested and 12 fail to emit a warning signal. Subjected to the same test, 15 out of 90 brand B detectors fail to operate. Form a 90% confidence interval to estimate the difference between the fractions of defective smoke detectors produced by the two companies. Interpret this confidence interval.

8.117 The federal government is interested in determining whether salary discrimination exists between men and women in the private sector. Suppose random samples of fifteen women and twenty-two men are drawn from the population of first-level managers in the private sector. The information is summarized in the table.

WOMEN	MEN
$\bar{x}_1 = 18,400$	$\bar{x}_2 = 19,700$
$s_1 = 2,300$	$s_2 = 3,100$
$n_1 = 15$	$n_2 = 22$

a. Do these data provide sufficient evidence to indicate that the mean salary of male first-level managers exceeds the mean salary of females in that position? Use $\alpha = .10$.

b. What assumptions are necessary for the validity of the test used in part a?

8.118 Refer to Exercise 8.117. Conduct a test to determine whether the assumption of equal salary variances is reasonable. Use $\alpha = .10$.

8.119 An automobile manufacturer wants to estimate the difference between the mean miles per gallon ratings for two models of cars that the company produces. If the range of ratings is expected to be about 6 miles per gallon for each model, how many cars of each type must be tested in order to estimate the difference between means to within .5 mile per gallon with 95% confidence?

8.120 The state of Florida now requires all high school students to pass a literacy test before they receive a high school diploma. A student who fails the test can enroll in a refresher course and retake the test at a later date. To evaluate the effectiveness of the refresher course, eight students' test scores were compared, before and after, with the results shown in the table. Do the data provide sufficient evidence to conclude that the refresher course significantly improves the test scores? (Use $\alpha = .05$.)

STUDENT	BEFORE	AFTER
1	45	49
2	52	50
3	63	70
4	68	71
5	57	53
6	55	61
7	60	62
8	59	67

OLD DRUG	NEW DRUG
$n_1 = 12$	$n_2 = 12$
$s_1^2 = 14.3$	$s_2^2 = 8.2$

8.121 A drug currently being used to reduce the heart rate of patients before surgery works very well in some patients while having little effect in others. Researchers have developed a new drug that they hope will produce more consistent results. To test the new drug, twenty-four dogs were randomly chosen and divided into two groups of twelve dogs each. One group of dogs was injected with the new drug, while the other group was injected with the old drug. The reduction in heart rate for each dog was recorded, with the results summarized in the table.

a. Do the data provide sufficient evidence to conclude that the variation in the heart rate reductions is less for the new drug than the old? Use $\alpha = .05$.

b. Find the approximate observed significance level for the test and interpret its value.

ADDITIVE A	ADDITIVE B
$n_A = 8$	$n_B = 7$
$\bar{x}_A = 16.38$	$\bar{x}_B = 18.88$
$s_A^2 = 12$	$s_B^2 = 22$

8.122 An experiment was performed to evaluate the effect of two different additives on gasoline consumption. Fifteen cars with varying engine sizes, and thus varying demands for gasoline, were available for the test. The experimenter randomly assigned eight cars to receive additive A and seven cars to receive additive B. Each car was driven 100 miles over the same course, and the gasoline consumption was recorded for each. A summary of the data is given in the table (measurements are in miles per gallon). Find a 95% confidence interval for the true difference between the mean gasoline consumptions of the two additives. What assumptions are necessary for the procedure you use to be valid?

8.123 A football fan decides to place bets according to the predictions of one of two newspaper columnists. To decide which columnist to use, both columnists' predictions are randomly sampled over the preceding weeks and the number of correct predictions for each is determined. The results are shown in the table. Do the data provide sufficient evidence to indicate that one of the columnists is better at picking winners than the other?

COLUMNIST	TOTAL NUMBER OF PREDICTIONS	NUMBER OF CORRECT PREDICTIONS
1	60	48
2	50	42

On Your Own . . .

We have now discussed two methods of collecting data to compare two population means. In many experimental situations a decision must be made either to collect two independent samples or to conduct a paired difference experiment. The importance of this decision cannot be overemphasized, since the amount of information obtained and the cost of the experiment are both directly related to the method of experimentation that is chosen.

Choose two populations (pertinent to your major area) that have unknown means and for which you could both collect two independent samples and collect paired observations. Before conducting the experiment, state which method of sampling you think will provide more information (and why). To compare the two methods, first perform the independent sampling procedure by collecting ten observations from each population (a total of twenty measurements), and then perform the paired difference experiment by collecting ten pairs of observations.

Construct two 95% confidence intervals, one for each experiment you conduct. Which method provides the shorter confidence interval and thus more information on this performance of the experiment? Does this result agree with your preliminary expectations?

References   Clotfelter, C. T. "Detroit decision and white flight." *Journal of Legal Studies*, 1976, 5.

Dreher, G. F. "The role of performance in the turnover process." *Academy of Management Journal*, 1982, 25.

Freedman, D., Pisani, R., & Purves, R. *Statistics*. New York: W. W. Norton and Co., 1978.

Garvin, D. A. "Quality on the line." *Harvard Business Review*, September–October, 1983, 65–75.

Glenn, N. D., & Weaver, C. N. "Enjoyment of work by full-time workers in the U.S., 1955 and 1980." *Public Opinion Quarterly*, 1982, *46*.

Hines, G. H. "Influences on employee expectancy and participative management." *Academy of Management Journal*, 1974, *17*.

Horton, T. R. "Preliminary radiological assessment of radon exhalation from phosphate gypsum piles and inactive uranium mill tailings piles." EPA-520/5-79-004. Washington, D.C.: Environmental Protection Agency, 1979.

Kevan, P. G., Jensen, T. S., & Shorthouse, J. D. "Body temperatures and behavioral thermoregulation of High Arctic woolly-bear caterpillars and pupae (*Gynaephora rossii*, Lymantridae: Lepidoptera) and the importance of sunshine." *Arctic and Alpine Research*, 1982, *14*.

McMurray, R. G., Wilson, J. R., & Kitchell, B. S. "The effects of fructose and glucose on high-endurance performance." *Research Quarterly for Exercise and Sport*, 1983, *54*.

Mendenhall, W. *Introduction to probability and statistics*. 6th ed. Boston: Duxbury, 1983. Chapters 8 and 9.

Phillips, M. D., & Pepper, R. L. "Shipboard firefighting performance of females and males." *Human Factors*, 1982, *24*.

Snedecor, G. W., & Cochran, W. *Statistical methods*. 7th ed. Ames: Iowa State University Press, 1980.

Stainback, S., & Stainback, W. C. "Matched procedures and research in mental retardation." *Training School Bulletin*. May 1973, *70*, 33–37.

CHAPTER 9

Analysis of Variance: Comparing More Than Two Means

Where We've Been...

As we have seen in preceding chapters, the solutions of many practical problems are based on inferences about population means. Methods for estimating and testing hypotheses about a single mean and the comparison of two means were presented in Chapters 7 and 8.

Where We're Going...

This chapter extends the methods of Chapters 7 and 8 to the comparison of more than two means. We will use sampling procedures that are analogous to the independent sampling and paired difference designs of Chapter 8. Then we will investigate the effect of two independent variables on a response for data collected in a factorial experiment.

Contents

The independent sampling design and the paired difference experiment (sampling schemes employed in Chapter 8 for the comparison of two means) can be generalized to permit a comparison of more than two means. For example, a medical researcher might wish to compare the length of time for a patient to respond to treatment by three drugs, A, B, and C. Thirty patients are randomly selected and divided into three equal groups. Patients in the first group are treated with drug A, the second group with B, and the third group with C. Then the length of time for recovery is measured for each patient. This experiment, which requires a comparison of three means (the means of the populations of patient responses corresponding to the three drugs), is a generalization of the independent sampling design of Chapter 8. Because the patients employed in the experiment are randomly selected and randomly assigned to the three drug treatment categories, the sampling procedure is often called a **completely randomized design.**

Similarly, the paired difference experiment of Chapter 8 can be generalized to permit a comparison of more than two means. If we wanted to compare the mean density of cakes baked by three different mixes, A, B, and C, we would mix the batter for cakes prepared by each of the three mixes and bake them in the same oven at the same time. This way the cakes would be matched on oven environment (temperature, temperature variation, etc.) and would identify a block of three experimental units, one corresponding to cake mix A, the second to B, and the third to C. This procedure would be repeated a number of times, and each baking would produce a **block** of three cakes, one corresponding to each of the three mixes. Because the cakes would be randomly located within the oven during each baking to remove the possibility of a bias due to location (temperature may vary from one location in an oven to another), this method of sampling is called a **randomized block design.** Like the paired experiment, the strategy behind the design is to make comparisons of mixes under relatively stable experimental conditions.

Because the design of experiments first achieved importance in agricultural experimentation, the terminology of the subject has an agricultural flavor. Most agricultural experiments involve the *treatment* of experimental units in two or more different ways and then a comparison of the means of the populations of measurements corresponding to the different treatments. For example, they might compare the yield of plots of corn treated with four different types of fertilizer, say A, B, C, and D. The four fertilizer types would be called **treatments.** Similarly, they might compare the gain in the weight of pigs fed on four different diets, A, B, C, and D. Here too the diets are called treatments, and the objective of the experiment is to compare the means of the populations of measurements corresponding to the four treatments. As you will subsequently see, the design of an experiment involves selecting the treatments to be included in the experiment as well as deciding how to apply the treatments to the experimental units.

The advantages obtained by selecting data according to a designed experiment are twofold. You can often acquire more information than could be obtained from the same amount of data collected in an undesigned manner. Moreover, the data can be analyzed by using a simple procedure called an **analysis of variance.** In the sections that follow, we will show you how to compare two or more treatment means based on the completely randomized and the randomized block designs. We will then use an analysis of variance to analyze the data from designed two-variable experiments.

9.1 Comparing More Than Two Population Means: The Completely Randomized Design

The following example serves a double purpose. First, it will illustrate the basic features of a completely randomized design. Second, it will be used to introduce terminology in common use when experimental designs are discussed.

Suppose an experimenter wanted to compare the mean weight gains of young pigs fed on three different diets, A, B, and C. The experimenter randomly (and independently) chose ten pigs (each was 6 weeks old) to be fed each of the three diets. (A total of $3 \times 10 = 30$ pigs were employed in the experiment.) At the end of the 3-month period, the gain in weight for each animal was recorded.

The important item to note in this example is that the ten sample measurements (weight gains) for each diet were *randomly and independently chosen* from the three conceptual populations of all possible weight gains for 6-week-old pigs fed by the three diets. *Choosing the samples randomly and independently is the defining characteristic of a completely randomized design.*

Definition 9.1

A **completely randomized design** (or **independent sampling design**) is one in which independent random samples are drawn from each of the populations of interest.

The weight gain was recorded for each pig, or **experimental unit,** involved in the study. Generally, experimental units are the objects on which the sample measurements are obtained. As noted earlier, the three diets given to the pigs are referred to as treatments.

The notation necessary for the analysis of data generated by a completely randomized design is shown in Table 9.1 (page 410). The table applies to a comparison of the means of any number, say k, of populations, where k can equal 2, 3, 4, ...

To decide whether a difference exists among the treatment means $\mu_1, \mu_2, \ldots, \mu_k$, we examine the **spread** (or **variation**) among the sample means. The greater the variation, the greater will be the evidence to indicate differences among $\mu_1, \mu_2, \ldots, \mu_k$. This variation, measured by a weighted sum of squares of deviations of the sample means $\bar{x}_1, \bar{x}_2, \ldots, \bar{x}_k$ about the overall mean $\bar{\bar{x}}$, is called the **sum of squares for treatments** and is given by the expression

$$SST = n_1(\bar{x}_1 - \bar{\bar{x}})^2 + n_2(\bar{x}_2 - \bar{\bar{x}})^2 + \cdots + n_k(\bar{x}_k - \bar{\bar{x}})^2$$

Note that each squared distance between the sample mean and the overall mean is multiplied by a weight, the sample size n_i. Note also that the SST is large when the sample means are very different.

Table 9.1

Summary Notation for
a Completely Randomized
Design

		POPULATIONS (TREATMENTS)			
		1	2	3 ... k	
	MEAN	μ_1	μ_2	μ_3 ... μ_k	
	VARIANCE	σ_1^2	σ_2^2	σ_3^2 ... σ_k^2	
		INDEPENDENT RANDOM SAMPLES			
		1	2	3 ... k	
	SAMPLE SIZE	n_1	n_2	n_3 ... n_k	
	SAMPLE TOTALS	T_1	T_2	T_3 ... T_k	
	SAMPLE MEANS	$\bar{x}_1$	$\bar{x}_2$	$\bar{x}_3$ ... $\bar{x}_k$	

Total number of measurements $= n = n_1 + n_2 + n_3 + \cdots + n_k$
Sum of all n measurements $= \sum x$
Mean of all n measurements $= \bar{x}$
Sum of squares of all n measurements $= \sum x^2$

Now suppose we want to test the null hypothesis that the k treatment means are equal, i.e.,

$$H_0: \quad \mu_1 = \mu_2 = \cdots = \mu_k$$

against the alternative

H_a: At least two of the treatment means differ

Large values of the SST will show support for the alternative hypothesis. That is, if the sum of squared differences between the sample means and the overall mean is large, we will tend to believe that the population means differ.

How large must the SST be before we reject H_0 and accept H_a? We will compare this measure of variability among sample means to a measure of the variability of the experimental units themselves, i.e., the ***within-sample variability.***

You can see how this principle works by comparing the dot diagrams for two samples as shown in Figure 9.1. Five observations for sample 1 and five for sample 2 are shown. The locations of the sample means are indicated by the two arrows. Do you think the data provide sufficient evidence to indicate a difference in the corresponding population means μ_1 and μ_2? In our opinion, the difference between $\bar{x}_1$ and $\bar{x}_2$ is not large enough to indicate a difference between μ_1 and μ_2. *This is because the difference between sample means is small in relation to the variability within the sample observations.*

Figure 9.1 Dot
Diagrams for Two
Samples (No Evidence of
a Difference Between
Population Means)

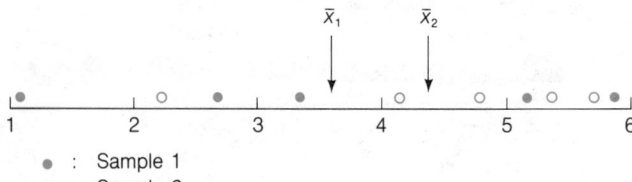

● : Sample 1
○ : Sample 2

Figure 9.2 Dot
Diagrams for Two Samples
(Evidence of a Difference
Between Population Means)

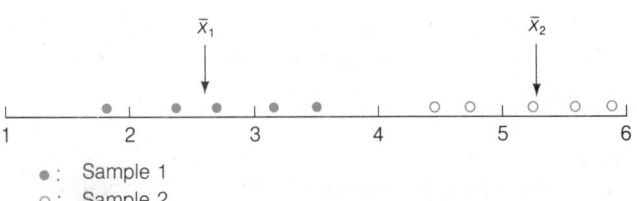

•: Sample 1
○: Sample 2

Now look at two other samples of $n_1 = n_2 = 5$ measurements, as shown in Figure 9.2. The data appear to give clear evidence of a difference between μ_1 and μ_2. *This is because the difference between the sample means, $\bar{x}_1$ and $\bar{x}_2$, is large in comparison with the variability within the sample observations.*

To measure the within-sample variability, we pool the within-sample sum of squared deviations about the sample means:

$$SSE = \sum(x_1 - \bar{x}_1)^2 + \sum(x_2 - \bar{x}_2)^2 + \cdots + \sum(x_k - \bar{x}_k)^2$$
$$= (n_1 - 1)s_1^2 + (n_2 - 1)s_2^2 + \cdots + (n_k - 1)s_k^2$$

where x_1 represents a measurement in sample 1, x_2 a measurement in sample 2, etc. We use **SSE** to denote the **sum of squared errors.** This quantity measures variability unexplained by the differences between the sample means. That is, the SSE is a pooled measure of the variability within the k samples.

We now want to compare the variability between treatment means (SST) to the within-sample variability (SSE). The first step is to divide each sum of squares by its degrees of freedom in order to obtain **mean squares.** We have $(k - 1)$ degrees of freedom for treatments—one for each of the k treatment means minus one for the estimation of the overall mean. Thus, we calculate

$$MST = \frac{SST}{k - 1}$$

where **MST** denotes **mean square for treatments.**

The degrees of freedom for error equals $(n - k)$, one for each of the n measurements minus one for each of the k treatment means we have estimated. Thus, we calculate

$$MSE = \frac{SSE}{n - k} = \frac{(n_1 - 1)s_1^2 + \cdots + (n_k - 1)s_k^2}{n_1 + n_2 + \cdots + n_k - k}$$

where **MSE** denotes **mean square for error.** For the special case where $k = 2$, MSE is equal to s_p^2, the pooled estimator of σ^2 used in the independent random samples t-test of Section 8.2.*

* For k greater than 2, MSE is an extension of the pooled estimator of σ^2 discussed in Section 8.2. For the two-sample case,

$$s_p^2 = \frac{\sum(x_1 - \bar{x}_1)^2 + \sum(x_2 - \bar{x}_2)^2}{n_1 + n_2 - 2}$$

The numerator of s_p^2 is SSE.

We now compare the two sources of variability—the source due to differences among the sample (treatment) means and the source due to within-sample differences among experimental units. Recall that in Section 8.6 we used the F-statistic to compare sample variances. For our application, MST and MSE both provide unbiased estimates of σ^2 when H_0 is true, that is, when $\mu_1 = \mu_2 = \cdots = \mu_k$. Therefore, as in Section 8.6, we use the F-statistic to compare these two sources of variability:

$$F = \frac{MST}{MSE}$$

Large values of the F-statistic indicate that the differences among the sample means are large and therefore support the alternative hypothesis that the population means differ. The test, with necessary assumptions, is summarized in the box. Because the F-statistic involves a comparison of two sources of variation, this procedure for comparing two or more population means is usually referred to as an **analysis of variance (ANOVA).**

Test to Compare k Treatment Means for a Completely Randomized Design

H_0: $\mu_1 = \mu_2 = \cdots = \mu_k$

H_a: At least two treatment means differ

Test statistic: $F = \dfrac{MST}{MSE}$

Assumptions: 1.   All k population probability distributions are normal.
2.   The k population variances are equal.
3.   Samples are randomly and independently selected from the respective populations.

Rejection region: $F > F_\alpha$, where F is based on $(k-1)$ numerator degrees of freedom (associated with the MST) and $(n-k)$ denominator degrees of freedom (associated with the MSE).

Example 9.1

Refer to our earlier discussion concerning a comparison of the mean weight gain of pigs for three different diets, A, B, and C. Independent random samples of ten 6-week-old pigs were chosen for each diet, and the weight gain for each pig was recorded. Set up the test to compare the population mean weight gains for the three diets. (We will give the data and perform the test in Example 9.2.)

Solution

We will test the null hypothesis that the mean weight gain for each diet is the same. Denoting the true mean gains for diets A, B, and C by μ_1, μ_2, and μ_3, respectively, we will test

H_0: $\mu_1 = \mu_2 = \mu_3$

H_a: At least two of the mean weight gains differ

The test statistic compares the variability between the sample mean weight gains to the within-sample variability:

Test statistic: $F = \dfrac{\text{MST}}{\text{MSE}}$

Assumptions: 1. The populations of weight gains for each of the three diets have (at least approximately) normal distributions.
2. The population variances of the distributions of weight gains are the same for the three diets.
3. The three samples are randomly and independently drawn from the respective populations.

Rejection region: Three treatments (the three diets) are being compared, so the number of treatment degrees of freedom is $(k - 1) = (3 - 1) = 2$. There are $n = 30$ measurements in the combined samples, so the error degrees of freedom is $(n - k) = (30 - 3) = 27$. Using $\alpha = .05$, we will reject the null hypothesis that the true means are the same if $F > F_{.05}$, where, from Table VIII in the Appendix, $F_{.05} = 3.35$.

This rejection is shown in Figure 9.3. Once the test is set up, we are prepared to collect samples, perform the calculations, and state a conclusion.

Figure 9.3 Rejection Region for Example 9.1: Numerator df $= 2$, Denominator df $= 27$

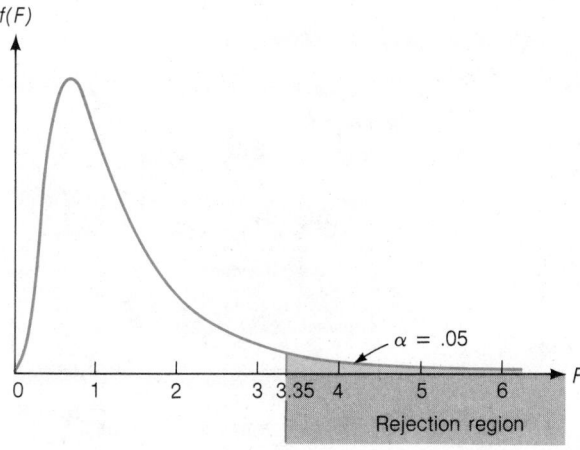

Although the sums of squares, SST and SSE, can be calculated by the formulas given earlier in this section, simpler computing formulas are available. Much of the theory and computation of analysis of variance rests on the concept of **_partitioning the sum of squares of deviations_** of all of the x-values about the overall mean:

$$\text{SS(Total)} = \sum (x - \bar{\bar{x}})^2$$

This is called the **_total sum of squares._** The partitioning is diagrammed in Figure 9.4.

Figure 9.4 Partitioning of the Total Sum of Squares for the Completely Randomized Design

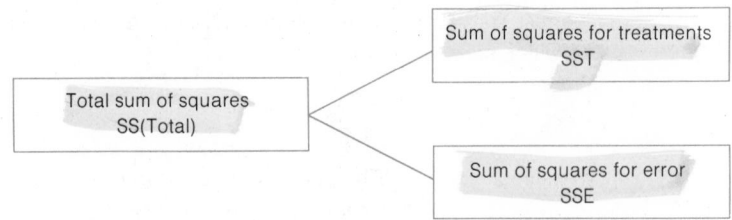

The computation of SSE by calculating and pooling the sums of squares of deviations for each of the samples is a tedious procedure. The computational simplification offered by the partitioning is that the SSE can be obtained by subtraction, i.e.,

$$SSE = SS(Total) - SST$$

We present the computational formulas that lead to the calculation of the F-statistic in the box.

Formulas for the Calculations in the Completely Randomized Design

$$CM = \text{Correction for mean} = \frac{(\text{Total of all observations})^2}{\text{Total number of observations}}$$

$$= \frac{(\sum x)^2}{n} = \frac{(T_1 + \cdots + T_k)^2}{n}$$

$SS(Total) = $ Total sum of squares

$$= (\text{Sum of squares of all observations}) - CM$$

$$= \sum x^2 - CM$$

$SST = $ Sum of squares for treatments

$$= \left(\begin{array}{c} \text{Sum of squares of treatment totals with} \\ \text{each square divided by number of} \\ \text{observations for that treatment} \end{array} \right) - CM$$

$$= \frac{T_1^2}{n_1} + \frac{T_2^2}{n_2} + \cdots + \frac{T_k^2}{n_k} - CM$$

$SSE = $ Sum of squares for error $= SS(Total) - SST$

$$MST = \text{Mean square for treatments} = \frac{SST}{k-1}$$

$$MSE = \text{Mean square for error} = \frac{SSE}{n-k}$$

$$F = \text{Test statistic} = \frac{MST}{MSE}$$

where k is the total number of treatments and n is the total number of observations.

Example 9.2 Refer to Example 9.1, where we set up a test to compare the mean weight gain of pigs fed on three different diets. The data for this experiment are given in Table 9.2. Perform the calculations required to obtain the F-statistic, and interpret the result.

Table 9.2

Diet: Weight Gains
(in Kilograms)

	DIET A	DIET B	DIET C
	30.1	34.5	28.8
	31.6	30.3	27.1
	25.4	31.7	28.3
	33.3	29.0	30.2
	29.0	30.5	31.5
	30.9	32.6	29.4
	28.5	33.1	30.4
	29.3	31.9	24.6
	31.2	32.2	32.3
	32.7	26.7	28.7
Totals	302.0	312.5	291.3

Solution From the table, the totals for the three samples are $T_1 = 302.0$, $T_2 = 312.5$, and $T_3 = 291.3$.

$$\sum x = T_1 + T_2 + T_3 = 905.8$$

$$\sum x^2 = (30.1)^2 + (31.6)^2 + \cdots + (28.7)^2$$
$$= 27,507.78$$

Then, following the order of calculations listed in the box, we find

$$CM = \frac{(\sum x)^2}{n} = \frac{(905.8)^2}{30} = 27,349.121$$

$$SS(Total) = \sum x^2 - CM$$
$$= 27,507.78 - 27,349.121$$
$$= 158.659$$

$$SST = \frac{T_1^2}{n_1} + \frac{T_2^2}{n_2} + \frac{T_3^2}{n_3} - CM$$
$$= \frac{(302.0)^2}{10} + \frac{(312.5)^2}{10} + \frac{(291.3)^2}{10} - 27,349.121$$
$$= 27,371.594 - 27,349.121$$
$$= 22.473$$

$$SSE = SS(Total) - SST$$
$$= 158.659 - 22.473$$
$$= 136.186$$

$$MST = \frac{SST}{k-1} = \frac{22.473}{2} = 11.237$$

$$MSE = \frac{SSE}{n-k} = \frac{136.186}{27} = 5.044$$

Finally, the value of the F-statistic is

$$F = \frac{MST}{MSE} = \frac{11.237}{5.044} = 2.23$$

This calculated F does not exceed the tabulated value $F_{.05} = 3.35$ (the value that locates the rejection region; see Example 9.1). Therefore, there is insufficient evidence to indicate a difference among the mean weight gains of the three diets. To reach a more definitive conclusion, the experimenter would have to collect more information on the three diets by conducting additional experiments. ∎

The results of an analysis of variance are often summarized in tabular form. The general form of an ***ANOVA table*** for a completely randomized design is shown in Table 9.3. *Source* refers to the source of variation, and for each source *df* refers to the degrees of freedom, *SS* to the sum of squares, *MS* to the mean square, and *F* to the *F*-statistic comparing the treatment mean square to the error mean square. Table 9.4 is the ANOVA summary table corresponding to the analysis of variance data for Examples 9.1 and 9.2.

Table 9.3
ANOVA Summary Table for a Completely Randomized Design

SOURCE	df	SS	MS	F
Treatments	$k-1$	SST	MST	MST/MSE
Error	$n-k$	SSE	MSE	
Total	$n-1$	SS(Total)		

Table 9.4
ANOVA Summary Table for Examples 9.1 and 9.2

SOURCE	df	SS	MS	F
Diet	2	22.473	11.237	2.23
Error	27	136.186	5.044	
Total	29	158.659		

Example 9.3

A sociologist conducted an experiment to compare the mean performances of college freshmen associated with four socioeconomic groups. Note that in this experiment, treatments (the socioeconomic conditions of the students) are not "applied" to the experimental units but instead identify the four populations of interest to the sociologist. Independent random samples of grade-point averages at the end of the freshman year were selected from each group. (The socioeconomic groups are in order from the lowest to the highest in terms of income.) The data are shown in Table 9.5. Do the data provide sufficient evidence to indicate a difference in mean grade-point average for the four socioeconomic groups? Test at the $\alpha = .05$ level of significance.

Table 9.5

	GROUP 1	GROUP 2	GROUP 3	GROUP 4
	2.87	3.23	2.61	2.25
	2.16	3.45	3.56	3.13
	3.14	3.67	2.97	2.44
	2.51	2.78	2.33	3.27
	1.80	3.77	3.64	2.81
	3.01		2.67	1.36
	2.16		3.31	2.70
			3.01	2.41
Totals	17.65	16.90	24.10	20.37

Solution

We wish to test the null hypothesis

$$H_0: \quad \mu_1 = \mu_2 = \mu_3 = \mu_4$$

against the alternative hypothesis

H_a: At least two of the means are not equal

where μ_1 = population mean grade-point average for group 1, etc.

Following the order of calculations listed earlier, we will construct an ANOVA table for this problem (Table 9.6).

$$\sum x = 17.65 + 16.90 + 24.10 + 20.37 = 79.02$$

$$\sum x^2 = (2.87)^2 + (2.16)^2 + \cdots + (2.41)^2 = 232.2534$$

$$CM = \frac{\left(\sum x\right)^2}{n} = \frac{(79.02)^2}{28} = 223.0057$$

$$SS(\text{Total}) = \sum x^2 - CM$$

$$= 232.2534 - 223.0057$$

$$= 9.2477$$

$$SST = \frac{T_1^2}{n_1} + \frac{T_2^2}{n_2} + \frac{T_3^2}{n_3} + \frac{T_4^2}{n_4} - CM$$

$$= \frac{(17.65)^2}{7} + \frac{(16.90)^2}{5} + \frac{(24.10)^2}{8} + \frac{(20.37)^2}{8} - 223.0057$$

$$= 3.0878$$

$$SSE = SS(\text{Total}) - SST$$

$$= 9.2477 - 3.0878 = 6.1599$$

Table 9.6

ANOVA Summary Table for Example 9.3

SOURCE	df	SS	MS	F
Socioeconomic group	3	3.0878	1.0293	4.01
Error	24	6.1599	.2567	
Total	27	9.2477		

$$MST = \frac{SST}{k-1} = \frac{3.0878}{3} = 1.0293$$

$$MSE = \frac{SSE}{n-k} = \frac{6.1599}{24} = .2567$$

$$F = \frac{MST}{MSE} = \frac{1.0293}{.2567} = 4.01$$

Since this observed value of the test statistic, $F = 4.01$, is larger than the tabulated F-value based on $v_1 = 3$ numerator and $v_2 = 24$ denominator degrees of freedom, $F_{.05} = 3.01$, we may conclude that there is a significant difference in the mean grade-point averages for at least two of the socioeconomic groups. We make this conclusion at the .05 level of significance. ■

What is the relationship between the F-statistic and the t-statistic? We compared two population means, based on independent random sampling, using a Student's t-statistic (Section 8.2), and now we find that this same test can be conducted (when k, the number of treatments, equals 2) using an F-statistic. The answer to this question is that for $k = 2$ (that is, $v_1 = 1$), it can be shown (proof omitted) that

$$F_\alpha = t^2_{\alpha/2}$$

and that the F-test is equivalent to a corresponding *two-tailed t-test*. For example, if $MSE = s^2$ is based on 20 degrees of freedom,

$t_{.025} = 2.086$ (from Table V in the Appendix)

$F_{.05} = 4.35$ (from Table VIII in the Appendix)

$F_{.05} = t^2_{.025} = (2.086)^2 = 4.35$

Thus, when $k = 2$ you can test the null hypothesis H_0: $\mu_1 = \mu_2$ against the alternative hypothesis H_a: $\mu_1 \neq \mu_2$ (i.e., a two-tailed test) using either an F-test or a t-test. However, if you wish to conduct a one-tailed test to detect, say, H_a: $\mu_1 < \mu_2$, you must use a Student's t-test.

For $k = 2$ treatments, that is, $v_1 = 1$,

$$F_\alpha = t^2_{\alpha/2}$$

For this particular case, the F-test for comparing $k = 2$ treatment means is equivalent to a two-tailed t-test.

Because the completely randomized design involves the selection of independent random samples, we can find a confidence interval for a single treatment mean by using the method of Section 7.4 or for the difference between two treatment means by using the method of Section 8.2. The estimate of σ^2 will be based on the pooled sum of squares within all k samples; that is,

$$MSE = s^2 = \frac{SSE}{n - k}$$

This is the same quantity that is used as the denominator for the analysis of variance F-test. The formulas for the confidence intervals of Chapters 7 and 8 are reproduced in the following box:

Confidence Intervals for Means

Single treatment mean (say, treatment i):

$$\bar{x}_i \pm t_{\alpha/2} \frac{s}{\sqrt{n_i}}$$

Difference between two treatment means (say, treatments i and j):

$$(\bar{x}_i - \bar{x}_j) \pm t_{\alpha/2} s \sqrt{\frac{1}{n_i} + \frac{1}{n_j}}$$

where $s = \sqrt{MSE}$ and $t_{\alpha/2}$ is the tabulated value of t (Table V in the Appendix) that locates $\alpha/2$ in the upper tail of the t-distribution and has $(n - k)$ degrees of freedom (the degrees of freedom associated with error in the ANOVA).

Example 9.4 Refer to Example 9.3 and find a 95% confidence interval for μ_1, the mean grade-point average of all possible freshmen belonging to the lowest socioeconomic group (group 1).

Solution From Table 9.6, MSE = .2567. Then

$$s = \sqrt{MSE} = \sqrt{.2567} = .5067$$

The sample mean grade-point average for group 1 is

$$\bar{x}_1 = \frac{T_1}{n_1} = \frac{17.65}{7} = 2.52$$

The tabulated value, $t_{.025}$ for 24 df (the same as for MSE), is 2.064. So a 95% confidence interval for μ_1 is

$$\bar{x}_1 \pm t_{.025} \frac{s}{\sqrt{n_1}} = 2.52 \pm (2.064)\left(\frac{.5067}{\sqrt{7}}\right)$$

or (2.12, 2.92). Thus, with 95% confidence the mean grade-point average for all freshmen in group 1 is between 2.12 and 2.92. ■

Example 9.5

Refer to Example 9.3. Suppose that prior to conducting the experiment, the sociologist particularly wished to estimate the difference in mean grade-point average between the freshmen in group 1 and those in group 2. Find a 95% confidence interval for this difference in mean grade-point averages.

Solution

The sample mean for group 2 is

$$\bar{x}_2 = \frac{T_2}{n_2} = \frac{16.90}{5} = 3.38$$

and, from Example 9.4, $\bar{x}_1 = 2.52$. The tabulated t-value with 24 df is $t_{.025} = 2.064$ (the same as for Example 9.4). Then the 95% confidence interval for $(\mu_1 - \mu_2)$, the difference in mean grade-point averages between the two groups, is

$$(\bar{x}_1 - \bar{x}_2) \pm t_{.025}s\sqrt{\frac{1}{n_1} + \frac{1}{n_2}} = (2.52 - 3.38) \pm (2.064)(.5067)\sqrt{\frac{1}{7} + \frac{1}{5}}$$

or $(-1.47, -.25)$. The fact that the interval contains only negative numbers means we can conclude, with 95% confidence, that the mean grade-point average of freshmen in group 2 exceeds the mean grade-point average of those in group 1.   ■

There is one important point to remember when using the method of Example 9.5. If you have decided to form confidence intervals for two or more treatment means, it is advisable to choose the pairs in advance (so that your choice will not be influenced by the data) and to keep the number of pairs of means as small as possible. The probability that a single confidence interval will enclose the difference between a pair of population means is the confidence coefficient $(1 - \alpha)$. But if you form a large number of confidence intervals, the probability that all the intervals will enclose the differences between population means is much smaller than $(1 - \alpha)$. Methods (called *multiple comparison procedures*) are available for forming confidence intervals for the difference between all pairs of population means or any group of them with a known confidence coefficient $(1 - \alpha)$. One of these procedures is presented in Section 9.4.

Case Study 9.1

Comparing the Strengths of Women Athletes

Muscle strength and endurance are two important aspects of neuromuscular performance.

This quote from Vivian Heyward and Leslie McCreary (1977) appeared in the introduction to an article in which they discussed the strength and endurance of women athletes.

A completely randomized design was used to compare the maximal grip strength of women athletes who engaged in eight different sports. The means and standard deviations for these data are shown in Table 9.7. The eight sports are the treatments, and the sample sizes for each sport are shown in parentheses.

The ANOVA table given in the article appears in Table 9.8. As can be seen, the response (maximal strength) is referred to as the *dependent variable* and the sources of variation, which we have called *treatment* and *error*, are referred to as *between sports* and *within sports*, respectively. (This terminology is often used in connection with completely randomized designs.) The F-statistic is equal to 3.15; it is significant at the .05 level of significance since the tabulated value of $F_{.05}$ corresponding to $v_1 = 7$ numerator

Table 9.7

Means and Standard
Deviations of
Neuromuscular
Performance Data for
Each Sport ($n = 50$)

SPORT	MAXIMAL STRENGTH (KILOGRAMS) $\bar{x}$	s
Basketball ($n_1 = 7$)	40.68	5.96
Field hockey ($n_2 = 7$)	39.16	4.84
Golf ($n_3 = 6$)	42.74	4.62
Gymnastics ($n_4 = 5$)	39.14	2.80
Swimming ($n_5 = 6$)	33.38	4.63
Tennis ($n_6 = 5$)	45.44	5.68
Track and field ($n_7 = 8$)	37.26	3.69
Volleyball ($n_8 = 6$)	38.37	5.24
Totals ($n = 50$)	39.29	5.51

Table 9.8

Summary of ANOVA
for Neuromuscular
Data ($n = 50$)

SOURCE OF VARIATION	df	SS	MS	F
Dependent variable, maximal strength				
Between sports (treatments)	7	511.50	73.07	3.15
Within sports (error)	42	974.88	23.21	

and $v_2 = 42$ denominator degrees of freedom is $F_{.05} \approx 2.25$. Thus, there is sufficient evidence to indicate that there is a difference in mean maximal grip strength among the eight sports.

According to Table 9.7, the largest (strongest) mean grip strengths were associated with women in tennis and golf, while the smallest was associated with women in swimming. Heyward and McCreary attribute this result to the extensive use of forearm and hand muscles in tennis and golf. Statistical comparison of specific pairs of means could be made by using a multiple comparison procedure. Most procedures for making pairwise comparisons among a group of means are based on the assumption that the samples are of equal size (see Section 9.4). A procedure applicable to unequal sample sizes, proposed by H. Scheffé, is discussed in Steele and Torrie (1980).

In concluding a discussion of the analysis of variance for a completely randomized design, we should comment on the assumptions that are required for its validity, i.e., the assumptions of equal population variances and normally distributed populations. Moderate departures from normality do not have much effect on the significance levels of tests or on the confidence coefficients associated with confidence intervals. Departures from the asumption of equal population variances can affect those measures of reliability, but the effect is less when the sample sizes are equal. When in doubt, use a nonparametric statistical method such as the Kruskal–Wallis H-test of Section 10.3.

What Do You Do When the Assumptions Are Not Satisfied for the Analysis of Variance for a Completely Randomized Design?

Answer: Use a nonparametric statistical method such as the Kruskal–Wallis H-test of Section 10.3.

**Exercises
9.1–9.13**

Learning the Mechanics

9.1 Independent random samples were selected from three normally distributed populations with common (but unknown) variance σ^2. The data are shown in the table.

SAMPLE 1	SAMPLE 2	SAMPLE 3
3.8	5.4	1.3
1.2	2.0	.7
2.9		3.1
3.3		

a. Compute the appropriate sums of squares and mean squares and fill in the appropriate entries in the ANOVA table:

SOURCE	df	SS	MS	F
Treatments				
Error				
Total				

b. Test the hypothesis that the population means are equal ($\mu_1 = \mu_2 = \mu_3$) against the alternative hypothesis that at least one mean is different from the other two. Test using $\alpha = .05$.

c. Find a 90% confidence interval for ($\mu_2 - \mu_3$). Interpret the interval.

d. What would happen to the width of the confidence interval in part c if you quadrupled the number of observations in the two samples?

e. Find a 95% confidence interval for μ_2.

f. Approximately how many observations would be required if you wished to be able to estimate a population mean correct to within .4 with probability equal to .95?

9.2 A partially completed ANOVA table for a completely randomized design is shown here:

SOURCE	df	SS	MS	F
Treatments	6	16.9		
Error				
Total	41	45.2		

a. Complete the ANOVA table.

b. How many treatments are involved in the experiment?

c. Do the data provide sufficient evidence to indicate a difference among the population means? Test using $\alpha = .10$.

d. Find the approximate observed significance level for the test in part c, and interpret it.

e. Suppose that $\bar{x}_1 = 3.7$ and $\bar{x}_2 = 4.1$. Do the data provide sufficient evidence to indicate a difference between μ_1 and μ_2? Assume that there are seven observations for each treatment. Test using $\alpha = .10$.

f. Refer to part e. Find a 90% confidence interval for $(\mu_1 - \mu_2)$.

g. Refer to part e. Find 90% confidence interval for μ_1.

9.3 Independent random samples were selected from three normally distributed populations with common (but unknown) variance σ^2. The data are shown in the table.

SAMPLE 1	SAMPLE 2	SAMPLE 3
3.1	5.4	1.1
4.3	3.6	.2
1.2	4.0	3.0
	2.9	

a. Compute the appropriate sums of squares and mean squares and fill in the appropriate entries in the ANOVA table:

SOURCE	df	SS	MS	F
Treatments				
Error				
Total				

b. Test the hypothesis that the population means are equal ($\mu_1 = \mu_2 = \mu_3$) against the alternative hypothesis that at least one mean is different from the other two. Test using $\alpha = .05$.

c. Find a 90% confidence interval for $(\mu_2 - \mu_3)$. Interpret the interval.

d. What would happen to the width of the confidence interval in part c if you quadrupled the number of observations in the two samples?

e. Find a 95% confidence interval for μ_2.

f. Approximately how many observations would be required if you wished to be able to estimate a population mean correct to within .4 with probability equal to .95?

9.4 A partially completed ANOVA table for a completely randomized design is shown here:

SOURCE	df	SS	MS	F
Treatments	4	24.7		
Error				
Total	34	62.4		

a. Complete the ANOVA table.

b. How many treatments are involved in the experiment?

c. Do the data provide sufficient evidence to indicate a difference among the population means? Test using $\alpha = .10$.

d. Suppose that $\bar{x}_1 = 3.7$ and $\bar{x}_2 = 4.1$. Do the data provide sufficient evidence to indicate a difference between μ_1 and μ_2? Assume that there are seven observations for each treatment. Test using $\alpha = .10$.

e. Refer to part d. Find a 90% confidence interval for $(\mu_1 - \mu_2)$.

f. Refer to part d. Find a 90% confidence interval for μ_1.

9.5 Describe in words the type of variability being measured by each of the following sums of squares:

a. SSE   **b.** SST   **c.** SS(Total)

9.6 Explain why the F-test used in the analysis of variance to test the null hypothesis H_0: $\mu_1 = \mu_2 = \cdots = \mu_k$ is a one-tailed, upper-tail test.

9.7 What assumptions are necessary for the validity of the F-test in a completely randomized design?

Applying the Concepts

9.8 Studies conducted at the University of Melbourne (Australia) indicate that there may be a difference between the pain thresholds of blonds and brunettes (*Family Weekly, Gainesville Sun,* Gainesville, Florida, February 5, 1978). Men and women of various ages were divided into four categories according to hair color: light blond, dark blond, light brunette, and dark brunette. The purpose of the experiment was to determine whether hair color is related to the amount of pain produced by common types of mishaps and assorted types of trauma. Each person in the experiment was given a pain threshold score based on his or her performance in a pain sensitivity test (the higher the score, the higher the person's pain tolerance). The results are given in the table.

HAIR COLOR			
Light blond	*Dark blond*	*Light brunette*	*Dark brunette*
62	63	42	32
60	57	50	39
71	52	41	51
55	41	37	30
48	43		35

a. Based on the given information, what type of experimental design appears to have been employed?

b. Is there evidence that mean pain thresholds differ among people possessing the four types of hair color? Use $\alpha = .01$.

c. Estimate the difference in mean pain threshold scores between light blonds and dark brunettes. Use a 95% confidence interval. Interpret the interval.

d. Estimate the difference in mean pain threshold scores between light brunettes and dark brunettes. Use a 95% confidence interval.

9.9 Some varieties of nematodes (roundworms that live in the soil and are frequently so small they are invisible to the naked eye) feed on the roots of lawn grasses and crops such as strawberries and tomatoes. This pest, which is particularly troublesome in warm climates, can be treated by the application of nematocides. However, because of the size of the worms, it is very difficult to measure the effectiveness of these pesticides directly. To compare four nematocides, the yields of equal-sized plots of one variety of tomatoes were collected. The data (yields in pounds per plot) are shown in the table.

NEMATOCIDE			
1	2	3	4
18.6	18.7	19.4	19.0
18.2	19.3	19.9	18.5
17.6	18.9	19.7	18.6
		19.1	

a. Do the data provide sufficient evidence to indicate a difference among the mean yields of tomatoes per plot for the four nematocides? Test using $\alpha = .05$.

b. Find the approximate observed significance level for the test in part a, and interpret its value.

c. Estimate the difference in mean yield between nematocide 2 and nematocide 3. Use a 90% confidence interval.

9.10 A research psychologist wishes to investigate the difference in maze test scores for a strain of laboratory mice trained under different laboratory conditions. The experiment is conducted using eighteen randomly selected mice of this strain, with six receiving no training at all (control group), six trained under condition 1, and six trained under condition 2. Then each of the mice is given a test score between zero and 100, depending on its performance in a test maze. The experiment produced the results in the table.

CONTROL	CONDITION 1	CONDITION 2
58	73	53
32	70	74
59	68	72
64	71	62
55	60	58
49	62	61

a. Is there sufficient evidence to indicate a difference among mean maze test scores for mice trained under the three different laboratory conditions? Use $\alpha = .05$.

b. Estimate the mean maze test score for the mice in the control group. Use a 90% confidence interval.

c. Estimate the difference in mean maze test scores between mice trained under condition 1 and those trained under condition 2. Use a 90% confidence interval.

9.11 How does flexitime—allowing workers to set their individual work schedules—affect a worker's job satisfaction? Researchers recently performed a study to compare a measure of job satisfaction for workers using three types of work scheduling: flexitime, staggered starting hours, and fixed hours. Workers in each group worked according to their specified scheduling system for 4 months. Although each worker filled out job satisfaction questionnaires both before and after the 4-month test period, we will examine only the posttest scores. The sample sizes, means, and standard deviations of the scores for the three groups are shown in the table at the top of the next page.

		GROUP	
	Flexitime	Staggered	Fixed
SAMPLE SIZE	27	59	24
MEAN	35.33	31.05	28.71
STANDARD DEVIATION	10.22	7.22	9.28

a. Assume that the data were collected according to a completely randomized design. Use the information in the table to calculate the treatment totals, CM, and SS(Treatments).

b. Use the values of the sample standard deviations to calculate the sum of squares of deviations *within* each of the three samples. Then calculate SSE, the sum of these quantities.

c. Construct an ANOVA table for the data.

d. Do the data provide sufficient evidence to indicate differences in mean job satisfaction scores among the three groups? Test using $\alpha = .05$.

e. Find a 90% confidence interval for the difference in mean job satisfaction scores between workers on flexitime and those on fixed schedules.

f. Do the data provide sufficient evidence to indicate a difference in mean scores between workers on flexitime and those using staggered starting hours? Test using $\alpha = .05$.

9.12 An experiment is conducted to determine whether there is a difference among the mean increases in growth produced by five inoculins of growth hormones for plants. The experimental material consists of twenty cuttings of a shrub (all of equal weight), with four cuttings randomly assigned to each of the five different inoculins. The results of the experiment are given in the table; all measurements represent an increase in weight (in grams).

		INOCULIN		
A	B	C	D	E
15	21	22	10	6
18	13	19	14	11
9	20	24	21	15
16	17	21	13	8

	BRAND	
A	B	C
310	261	233
235	219	289
279	263	301
306	247	264
237	288	273
284	197	208
259	207	245
273	221	271
219	244	298
301	228	276

a. With this information, can we conclude that there is a difference among the mean increases in weight for the five inoculins of growth hormone? Test at the $\alpha = .05$ level of significance.

b. What assumptions must the data satisfy to make the test in part a valid?

c. Find the approximate observed significance level for the test in part a, and interpret its value.

9.13 One of the selling points of golf balls is their durability. An independent testing laboratory is commissioned to compare the durability of three different brands of golf balls (brands A, B, and C). Balls of each type will be put into a machine that hits the balls with the same force that a golfer does on the course. The number of hits required until the outer covering is cracked is recorded for each ball. Ten balls from each manufacturer are randomly selected for testing. The results are given in the table.

a. Is there evidence that the mean durabilities of the three brands differ? Use $\alpha = .05$.

b. Estimate the difference between the mean durability of brands A and C. Use a 99% confidence interval.

9.2 Randomized Block Design

We mentioned in the introduction to this chapter that a randomized block design is an extension of a paired difference experiment. As noted in Chapter 8, a paired difference experiment for comparing two population means often provides more information because it makes a comparison of observations from two populations (call them A and B) between pairs of matched (or similar) experimental units. If you were comparing the miles per gallon ratings for two car models, for example, measurements would vary substantially for the same model because of the differences in the driving habits of the drivers. To remove this source of variation, you could pair an A and a B car and assign both cars to a *single* driver. The driver would drive both cars and obtain the ratings for each. By taking the difference between the pair of measurements for each driver, you would control the variation among drivers.

The extension of the paired difference design to a comparison of more than two means implies that larger groups of experimental units are matched. The following example demonstrates how a randomized block design is conducted. Suppose it is desired to compare the average yields of four varieties of wheat (A, B, C, and D). Five farms are available for planting the wheat that will eventually be harvested in order to compare the mean yields. It is probable that growing conditions (soil, moisture, etc.) vary from farm to farm, and consequently wheat yields would vary from one farm to another. To control this source of variability, each variety of wheat is grown on a 1-acre plot at each farm, as shown in Figure 9.5. The matched group of four experimental units (matched on farms) is called a **block** and the resulting design is called a **randomized block design.** Finally, the varieties of wheat are randomly assigned to the individual plots at each farm.

In general, a randomized block design consists of *blocks of experimental units that are relatively homogeneous.* The 1-acre plots (experimental units) within each farm

Figure 9.5 A Randomized Block Design for Comparing Four Varieties of Wheat

Blocks (Farms)

1	2	3	4	5
B	C	B	A	A
A	B	C	B	D
C	D	A	D	B
D	A	D	C	C

should have growing conditions that are more similar than plots from two different farms. The word *randomized* in the phrase *randomized block design* implies that the treatments (varieties of wheat) were randomly assigned to the experimental units (plots) within each block (farm). In general, we would define a randomized block design as follows:

Definition 9.2

A *randomized block design* is a design devised to compare the means for k treatments utilizing b matched blocks of k experimental units each. Each treatment appears once in every block.

We will now develop the procedures needed to analyze data obtained from a randomized block design. The design employs groups of homogeneous experimental units (matched as closely as possible) to compare the means of the populations associated with k treatments. Suppose there are b blocks of relatively homogeneous experimental units. Since each treatment must be represented in each block, the blocks will each contain k experimental units, which will be randomly assigned to the k treatments. The general format of the randomized block design is shown in Figure 9.6. Although we show the treatments in order within the blocks, in practice they would be assigned to the experimental units in a random order (thus the name *randomized block design*). The notation for the results of a randomized block experiment is summarized in Table 9.9.

We are interested in using the randomized block design to test the same null hypothesis that we tested using the completely randomized design, i.e.,

H_0: $\mu_1 = \mu_2 = \cdots = \mu_k$

H_a: At least two treatment means differ

The test statistic is also identical to that used for the completely randomized design:

Test statistic: $F = \dfrac{\text{MST}}{\text{MSE}}$

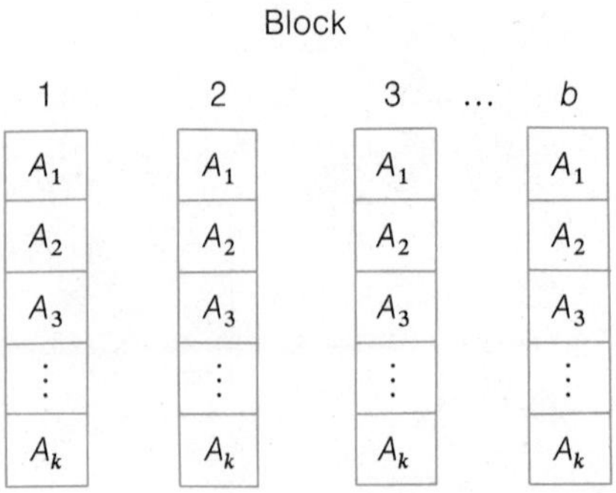

Figure 9.6 General Form of the Randomized Block Design (Treatment i Is Denoted by A_i)

Table 9.9

Notation for the Results of
a Randomized Block
Experiment

	TREATMENT 1	TREATMENT 2	...	TREATMENT k
SAMPLE SIZE	b	b	...	b
TOTALS	T_1	T_2	...	T_k
	BLOCK 1	BLOCK 2	...	BLOCK b
SAMPLE SIZE	k	k	...	k
TOTALS	B_1	B_2	...	B_k

Total number of observations $= bk = n$
Sum of all observations $= \sum x$
Sum of squares of all observations $= \sum x^2$

The numerator of the F-statistic, MST (mean square for treatments), is computed exactly as it was for the completely randomized design. However, the denominator, MSE (mean square for error), is computed differently. This is most easily seen from Figure 9.7, which illustrates the partitioning of the total sum of squares about the overall mean, SS(Total). Note that the SS(Total) is now partitioned into three parts. The SSE for the completely randomized experiment has been subdivided into two parts—the sum of squares for blocks, SSB, and the sum of squares for error, the SSE for the randomized block design. (We use the same symbol, SSE, for the sum of squares for error in both designs, although their computational formulas differ.)

Figure 9.7 may help you understand how a randomized block design works. Since

$$SS(Total) = SSB + SST + SSE$$

it follows that the SSE for this design is

$$SSE = SS(Total) - SST - SSB$$

Figure 9.7 Partitioning of the Total Sum of Squares for the Randomized Block Design

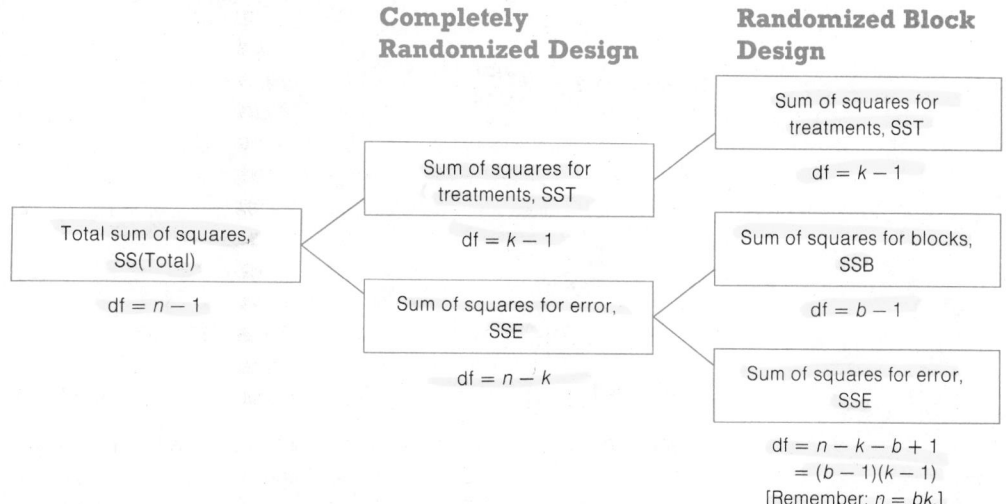

In other words, the SSE equals the sum of squares for error for the completely randomized design [SS(Total) − SST] *minus* SSB. Thus, the randomized block design permits us to remove the variation between blocks from within-sample variation and may even decrease the MSE. Remember: The smaller the value of the MSE (which appears in the denominator of the *F*-statistic), the more likely it is that we will detect a difference among the treatment means if such a difference exists. The test for treatment differences using the randomized block design is summarized in the accompanying box.

Test to Compare k Treatment Means: Randomized Block Design

H_0: $\mu_1 = \mu_2 = \cdots = \mu_k$

H_a: At least two treatment means differ

Test statistic: $F = \dfrac{\text{MST}}{\text{MSE}}$

Assumptions: 1.   The probability distributions of observations corresponding to all the block and treatment combinations are normal.
2.   The variances of all the probability distributions are equal.

Rejection region: $F > F_\alpha$, where F is based on $(k - 1)$ numerator degrees of freedom and $(n - k - b + 1)$ denominator degrees of freedom.

Example 9.6   A study was conducted in a large city to compare the mean supermarket prices of the four leading brands of coffee at the end of the year. Ten supermarkets in the city were selected, and the price per pound was recorded for each brand. Set up the test of the null hypothesis that the mean prices of the four brands sold in the city were the same at the end of the year. Use $\alpha = .05$.

Solution   We would expect prices to be more homogeneous within a store than between stores. In other words, the experimental units—the 1-pound cans of coffee—will be more homogeneous (with respect to price) within stores. The stores, then, act as blocks and the coffee brands act as treatments.

Denote the true mean prices of the four brands as μ_1, μ_2, μ_3, and μ_4. Then the elements of the test are

H_0: $\mu_1 = \mu_2 = \mu_3 = \mu_4$

H_a: At least two brands have different mean prices

Test statistic: $F = \dfrac{\text{MST}}{\text{MSE}}$

Assumptions: 1.   The probability distributions of coffee prices corresponding to all the supermarket and brand combinations are normal.
2.   The variances of the probability distributions are equal.

Rejection region: Since we have $k = 4$ treatments (brands) and $b = 10$ blocks (stores), we use a tabulated F-value with $k - 1 = 4 - 1 = 3$ df in the numerator and $n - k - b + 1 = (4)(10) - 4 - 10 + 1 = 27$ df in the denominator. Then, for $\alpha = .05$, we will reject H_0 if $F > F_{.05}$, where (from Table VIII in the Appendix) $F_{.05} = 2.96$.

The formulas needed for the analysis of a randomized block design are presented in the box on page 432. ∎

Example 9.7

The data (and totals) for the coffee price study described in Example 9.6 are given in Table 9.10. Calculate the F-statistic. Do the data provide sufficient evidence to indicate a difference in the mean prices for the four brands of coffee?

Table 9.10

Price per Pound of Coffee

SUPERMARKET	A	B	C	D	TOTALS
1	$ 2.43	$ 2.47	$ 2.47	$ 2.41	9.78
2	2.48	2.52	2.53	2.48	10.01
3	2.38	2.44	2.42	2.35	9.59
4	2.40	2.47	2.46	2.39	9.72
5	2.35	2.42	2.44	2.32	9.53
6	2.43	2.49	2.47	2.42	9.81
7	2.55	2.62	2.64	2.56	10.37
8	2.41	2.49	2.47	2.39	9.76
9	2.53	2.60	2.59	2.49	10.21
10	2.35	2.43	2.44	2.36	9.58
TOTALS	24.31	24.95	24.93	24.17	98.36

(Columns A–D grouped under heading: BRAND)

Solution   Following the order of calculations listed in the box on page 432, we have

$$\sum x^2 = (2.43)^2 + (2.48)^2 + \cdots + (2.36)^2$$
$$= 242.0966$$

$$CM = \frac{(\sum x)^2}{n} = \frac{(98.36)^2}{40}$$
$$= 241.86724$$

$$SS(Total) = \sum x^2 - CM$$
$$= 242.0966 - 241.86724 = .22936$$

$$SST = \frac{T_1^2}{10} + \frac{T_2^2}{10} + \frac{T_3^2}{10} + \frac{T_4^2}{10} - CM$$

$$= \frac{(24.31)^2}{10} + \frac{(24.95)^2}{10} + \frac{(24.93)^2}{10} + \frac{(24.17)^2}{10} - 241.86724$$

$$= 241.91724 - 241.86724 = .05000$$

Formulas for Calculations in the Randomized Block Design

CM = Correction for mean

$$= \frac{(\text{Total of all observations})^2}{\text{Total number of observations}}$$

$$= \frac{\left(\sum x\right)^2}{n}$$

SS(Total) = Total sum of squares

= (Sum of squares of all observations) − CM

$$= \sum x^2 - \text{CM}$$

SST = Sum of squares for treatments

$$= \left(\begin{array}{c}\text{Sum of squares of treatment totals with}\\ \text{each square divided by } b, \text{ the number of}\\ \text{observations for that treatment}\end{array}\right) - \text{CM}$$

$$= \frac{T_1^2}{b} + \frac{T_2^2}{b} + \cdots + \frac{T_k^2}{b} - \text{CM}$$

SSB = Sum of squares for blocks

$$= \left(\begin{array}{c}\text{Sum of squares for block totals with}\\ \text{each square divided by } k, \text{ the number}\\ \text{of observations in that block}\end{array}\right) - \text{CM}$$

$$= \frac{B_1^2}{k} + \frac{B_2^2}{k} + \cdots + \frac{B_b^2}{k} - \text{CM}$$

SSE = Sum of squares for error

= SS(Total) − SST − SSB

MST = Mean square for treatments

$$= \frac{\text{SST}}{k - 1}$$

MSB = Mean square for blocks

$$= \frac{\text{SSB}}{b - 1}$$

MSE = Mean square for error

$$= \frac{\text{SSE}}{n - k - b + 1} = \frac{\text{SSE}}{(b - 1)(k - 1)}$$

$$F = \frac{\text{MST}}{\text{MSE}}$$

where n = Total number of observations

b = Number of blocks

k = Number of treatments

$$SSB = \frac{B_1^2}{4} + \frac{B_2^2}{4} + \cdots + \frac{B_{10}^2}{4} - CM$$

$$= \frac{(9.78)^2}{4} + \frac{(10.01)^2}{4} + \cdots + \frac{(9.58)^2}{4} - 241.86724$$

$$= 242.04175 - 241.86724 = .17451$$

$$SSE = SS(Total) - SST - SSB$$

$$= .22936 - .05 - .17451 = .00485$$

$$MST = \frac{SST}{k - 1} = \frac{.05}{3}$$

$$= .016667$$

$$MSB = \frac{SSB}{b - 1} = \frac{.17451}{9}$$

$$= .019390$$

$$MSE = \frac{SSE}{n - k - b + 1} = \frac{.00485}{27}$$

$$= .00017963$$

$$F = \frac{MST}{MSE} = \frac{.016667}{.00017963}$$

$$= 92.8$$

Since the calculated $F = 92.8$ greatly exceeds the tabulated value of $F_{.05} = 2.96$, there is very strong evidence that at least two of the means for the populations of prices of the four coffee brands differ. ■

The ANOVA summary table for a randomized block analysis would appear as shown in Table 9.11. Table 9.12 is the ANOVA table for the data analysis in Example 9.7.

The formula for the confidence interval for the difference between a pair of treatment means is identical to the formula presented in Section 8.2 except for the estimate of σ^2,

Table 9.11
ANOVA Summary Table for a Randomized Block Design

SOURCE	df	SS	MS	F
Treatment	$k - 1$	SST	MST	MST/MSE
Blocks	$b - 1$	SSB	MSB	
Error	$n - k - b + 1$	SSE	MSE	
Total	$n - 1$	SS(Total)		

Table 9.12
ANOVA Table for Example 9.7

SOURCE	df	SS	MS	F
Treatment	3	.05000	.016667	92.8
Block	9	.17451	.019390	
Error	27	.00485	.00017963	
Total	39	.22936		

which is

$$s^2 = \text{MSE} = \frac{\text{SSE}}{n - k - b + 1}$$

This quantity appears in the ANOVA table. The formula for the confidence interval is shown in the accompanying box.

$100(1 - \alpha)\%$ Confidence Interval for the Difference Between a Pair of Treatment Means ($\mu_i - \mu_j$)

$$(\bar{x}_i - \bar{x}_j) \pm t_{\alpha/2} s \sqrt{\frac{1}{b} + \frac{1}{b}} \quad \text{or} \quad (\bar{x}_i - \bar{x}_j) \pm t_{\alpha/2} s \sqrt{\frac{2}{b}}$$

where b is the number of blocks and $t_{\alpha/2}$ is the tabulated value of t (Table V in the Appendix) that locates $\alpha/2$ in the upper tail of the t-distribution, based on $(n - k - b + 1)$ degrees of freedom.

Example 9.8   Refer to the coffee brand price data in Example 9.7 and find a 90% confidence interval for the difference between the mean prices of brand A and brand B.

Solution   From Example 9.7, the sample means for brands A and B (identified as 1 and 2, respectively) are

$$\bar{x}_1 = \frac{T_1}{b} = \frac{24.31}{10} = \$2.431 \qquad \bar{x}_2 = \frac{T_2}{b} = \frac{24.95}{10} = \$2.495$$

From Table V in the Appendix, the tabulated value of t based on 27 degrees of freedom is $t_{.05} = 1.703$, and from Table 9.12 we have

$$s^2 = \text{MSE} = .00017963 \qquad s = .0134$$

Then the 90% confidence interval for ($\mu_1 - \mu_2$) is

$$(\bar{x}_1 - \bar{x}_2) \pm t_{.05} s \sqrt{\frac{2}{b}}$$

Substituting into the formula, we get

$$(2.431 - 2.495) \pm (1.703)(.0134) \sqrt{\tfrac{2}{10}}$$

or $(-.074, -.054)$. Thus, we estimate the mean price of brand B to exceed the mean price for A by as little as $.054 or as much as $.074 per pound. ∎

We can also conduct a test of an hypothesis to determine whether differences exist among block means. This test will help us to decide whether blocking was successful in reducing the experimental error. That is, if the block means differ we know there is evidence that the experimental units are more homogeneous within blocks than between blocks and the use of the randomized block experiment is justified. Such information is useful if similar experiments are to be conducted.

The test procedure for block means is very similar to that for treatment means. We compare the variation among blocks, as measured by the mean square for blocks (MSB), to the variation due to error, as measured by the mean square for error (MSE). The test is summarized in the box.

Test to Compare b Block Means: Randomized Block Design

H_0: The b block means are equal

H_a: At least two block means differ

Test statistic: $F = \dfrac{\text{MSB}}{\text{MSE}}$

Assumptions: Same as for the test of treatment means.

Rejection region: The numerator degrees of freedom is $(b - 1)$, the number of blocks minus 1. The denominator degrees of freedom is $(n - k - b + 1)$. We will reject H_0 if $F > F_\alpha$, where F_α is based on $v_1 = (b - 1)$, the number of blocks minus 1, and $v_2 = (n - k - b + 1)$ degrees of freedom.

Example 9.9 Refer to Examples 9.6 and 9.7, in which we used a randomized block design to compare the mean prices of four brands of coffee. The blocks were supermarkets in a large city. Test the null hypothesis that the block means are the same, i.e., that the average price of coffee is the same for the ten supermarkets. Use $\alpha = .05$.

Solution The test for comparing the block means is

H_0: Mean coffee prices are the same for all ten supermarkets

H_a: Mean coffee prices differ for at least two supermarkets

Test statistic: $F = \dfrac{\text{MSB}}{\text{MSE}}$

Assumptions: Same as for the test comparing the mean prices of the four coffee brands (Example 9.6).

Rejection region: The numerator and denominator degrees of freedom for the F-statistic are $b - 1 = 10 - 1 = 9$ and $n - k - b + 1 = 27$, respectively. Then the rejection region is $F > F_{.05}$, where (from Table VIII in the Appendix) $F_{.05} = 2.25$.

We have previously calculated MSB and MSE in Example 9.7. Substituting these values into the F-statistic, we have

$$F = \frac{MSB}{MSE} = \frac{.019390}{.00017963} = 107.9$$

Since the calculated F greatly exceeds the tabulated F-value, we have strong evidence that the means of coffee prices differ among the ten supermarkets. The decision to use a randomized block design was wise. Blocking has the effect of reducing the MSE and increasing the amount of information in the experiment. Future price comparison studies might benefit from this information. Table 9.13 is the complete ANOVA summary table for this experiment. ▪

Table 9.13
Complete ANOVA
Summary Table for
Example 9.9

SOURCE	df	SS	MS	F
Treatment	9	.05000	.016667	92.8
Block	3	.17451	.019390	107.9
Error	27	.00485	.00017963	
Total	39	.22936		

We conclude this section with a caution: The result of the test for the equality of block means must be interpreted with care, especially when the calculated value of the F test statistic does not fall in the rejection region. This does not necessarily imply that the block means are the same, i.e., that blocking is unimportant. Reaching this conclusion would be equivalent to accepting the null hypothesis, a practice we have carefully avoided due to the unknown probability of committing a Type II error (that is, of accepting H_0 if H_a is true). In other words, even when a test for block differences is inconclusive, we may still want to use the randomized block design in similar future experiments. If the experimenter believes that the experimental units are more homogeneous within blocks than among blocks, he or she should use the randomized block design regardless of whether the test comparing the block means shows them to be different for one experiment.

If you think it likely that the assumptions are not satisfied, you may wish to use a nonparametric statistical method such as the Friedman F_r-test of Section 10.4. The assumptions of normality and equal population variances need not be satisfied when using this test.

What Do You Do When the Assumptions Are Not Satisfied for the Analysis of Variance for a Randomized Block Design?

Answer: Use a nonparametric statistical method such as the Friedman F_r-test of Section 10.4.

Exercises 9.14–9.26

Learning the Mechanics

9.14 A randomized block design was conducted to compare the mean responses for three treatments, A, B, and C, in four blocks. The data are shown in the table.

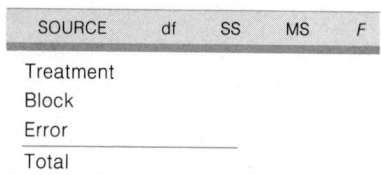

TREATMENT	BLOCK			
	1	2	3	4
A	4	5	3	3
B	2	7	1	5
C	5	4	2	2

a. Compute the appropriate sums of squares and mean squares, and fill in the entries in the ANOVA table:

SOURCE	df	SS	MS	F
Treatment				
Block				
Error				
Total				

b. Do the data provide sufficient evidence to indicate a difference among treatment means? Test using $\alpha = .05$.

c. Do the data provide sufficient evidence to indicate that blocking was effective in reducing the experimental error? Test using $\alpha = .05$.

d. Find a 90% confidence interval for $(\mu_A - \mu_B)$.

e. What assumptions must the data satisfy to make the F-tests in parts b and c valid?

9.15 The analysis of variance for a randomized block design produced the ANOVA table entries shown here:

SOURCE	df	SS	MS	F
Treatment	3	28.2		
Block	5		13.80	
Error		34.1		
Total				

a. Complete the ANOVA table.

b. Do the data provide sufficient evidence to indicate a difference among the treatment means? Test using $\alpha = .01$.

c. Do the data provide sufficient evidence to indicate that blocking was a useful design strategy for this experiment? Explain.

d. If the sample means for treatments A and B are $\bar{x}_A = 9.7$ and $\bar{x}_B = 12.1$, respectively, find a 90% confidence interval for $(\mu_A - \mu_B)$. Interpret the interval.

9.16 Explain the difference between a completely randomized design and a randomized block design.

9.17 Explain why the following statement is true: The smaller the value of MSE in relation to the value of MST, the more likely it is that we will detect a difference among the treatment means, if such a difference exists.

9.18 What assumptions are necessary for the validity of the F-test in a randomized block design?

Applying the Concepts

9.19 An evaluation of diffusion bonding of zircaloy components is performed. The main objective is to determine which of three elements—nickel, iron, or copper—is the best bonding agent. A series of zircaloy components are bonded with each of the possible bonding agents. Since there is a great deal of variation in components machined from different ingots, a randomized block design is used, blocking on the ingots. A pair of components from each ingot are bonded together using each of the three agents, and the pressure (in units of 1,000 pounds per square inch) required to separate the bonded components is measured. The data in the table are obtained.

| | BONDING AGENT | | |
INGOT	Nickel	Iron	Copper
1	67.0	71.9	72.2
2	67.5	68.8	66.4
3	76.0	82.6	74.5
4	72.7	78.1	67.3
5	73.1	74.2	73.2
6	65.8	70.8	68.7
7	75.6	84.9	69.0

a. Is there evidence of a difference in pressure required to separate the components among the three bonding agents? Use $\alpha = .05$.

b. Form a 95% confidence interval to estimate the difference in mean pressure between nickel and iron. Interpret this interval.

9.20 Two drugs, A and B, used for the treatment of glaucoma (an eye disease) were tested for effectiveness on ten diseased dogs. Drug A was administered to one eye (chosen randomly) of each dog and drug B to the other eye. Pressure measurements were taken 1 hour later on both eyeballs of each dog. The ten diseased dogs act as the blocks for comparing the two treatments, drugs A and B. Pressure measurements are given in the table. (The smaller the measurement, the less serious the eye disease.)

a. Perform an analysis of variance for these data. Do the data provide sufficient evidence to indicate a difference in mean pressure readings for the two treatments (i.e., is one of the glaucoma drugs better than the other)? Use $\alpha = .05$.

b. What is the purpose of using the dogs as blocks in this experiment?

c. Recall that a randomized block design with $k = 2$ treatments is a paired difference experiment (Chapter 8). Analyze the data as a paired difference experiment using a t-test to compare the treatment means. (Use $\alpha = .05$.)

	TREATMENT	
DOG	Drug A	Drug B
1	.17	.15
2	.20	.18
3	.14	.13
4	.18	.18
5	.23	.19
6	.19	.12
7	.12	.07
8	.10	.09
9	.16	.14
10	.13	.08

d. Compare the computed F- and t-values from parts a and c, and verify that $F = t^2$. Also verify that for the rejection region values of F and t, $F_\alpha = t_{\alpha/2}^2$.

e. Find the approximate observed significance level for the test in part a, and interpret its value.

9.21 A construction firm employs three cost estimators. Usually, only one estimator works on each potential job, but it is advantageous to the company if the estimators are consistent enough that it does not matter which of the three is assigned to a job. To check on the consistency of the estimators, several jobs are selected and all three estimators are asked to make estimates. The estimates for each job by each estimator are given in the table.

Estimates (in Thousands of Dollars)

	ESTIMATOR		
JOB	A	B	C
1	27.3	26.5	28.2
2	66.7	67.3	65.9
3	104.8	102.1	100.8
4	87.6	85.6	86.5
5	54.5	55.6	55.9
6	58.7	59.2	60.1

a. Do these estimates provide sufficient evidence that the means for at least two of the estimators differ? Use $\alpha = .05$.

b. Find the approximate observed significance level for the test in part a, and interpret its value.

c. Present the complete ANOVA summary table for this experiment.

d. Use a 90% confidence interval to estimate the difference between the mean responses given by estimators B and C.

9.22 A power plant, which uses water from the surrounding bay for cooling its condensers, is required by the EPA to determine whether discharging its heated water into the bay has a detrimental effect on the flora (plant life) in the water. The EPA requests that the power plant make its investigation at three strategically chosen locations, called *stations*. Stations 1 and 2 are located near the plant's discharge tubes; station 3 is located farther out in the bay. During one randomly selected day in each of 4 months, a diver descends to each of the stations, randomly samples a square meter area of the bottom, and counts the number of blades of the different types of grasses present. The results for one important grass type are given in the table.

	STATION		
MONTH	1	2	3
May	28	31	53
June	25	22	61
July	37	30	56
August	20	26	48

a. Is there sufficient evidence to indicate that the mean number of blades found per square meter per month differs for at least two of the three stations? Use $\alpha = .05$.

b. Is there sufficient evidence to indicate that the mean number of blades found per square meter differs among the 4 months? Use $\alpha = .05$.

c. Place a 90% confidence interval on the difference in means between stations 1 and 3.

9.23 The data shown in the table (*Gainesville Sun*, March 15, 1984) enable us to compare the flood levels (in feet) in March 1984 on the Florida Suwannee and Santa Fe rivers with similar levels in 1983, 1973, and 1948. (The crests of the flood levels for 1984 are expected to exceed the March 1984 levels shown in the table.) As you can see from the table, the water level depends on the river and the location on the river where the water height is measured. To compare flood levels for 1984 with the levels of 1983, 1973, and 1948, we examine the flood level measurements within each location.

STATION	FLOOD STAGE	1984	1983	1973	1948
Suwannee River					
White Springs	77.0	79.9	74.4	88.5	85.2
Ellaville	54.0	59.8	52.8	65.0	68.1
Branford	29.0	29.3	29.2	35.6	38.9
Wilcox	14.0	11.6	13.0	18.6	22.3
Santa Fe River					
Three Rivers	19.0	24.1	24.3	27.2	30.6
US 129 Bridge	21.0	23.8	23.8	30.7	34.2

a. What experimental design was used to collect the data shown in the table?

b. Perform an analysis of variance for the data.

c. Do the data provide sufficient evidence of differences in mean flood level among the years 1984, 1983, 1973, and 1948? Test using $\alpha = .05$.

d. Find a 95% confidence interval for the difference in mean flood levels between 1984 and 1948.

9.24 In Exercise 8.35, we discussed the Supreme Court decision (*Milliken* v. *Bradley*) that appears to have triggered the flight of white families out of the inner city to the suburbs. The data in Exercise 8.35 are reproduced here (Clotfelter, 1976). You can see that the data giving the percentage of minority students in a city for the years 1968, 1970, and 1972 are matched on cities. Therefore, to compare the mean percentage of minority students in large cities for the years 1968, 1970, and 1972 we analyze the data using the analysis of variance for a randomized block design.

Racial Change and White Flight in Selected City School Systems, 1968–1972

CITY	PERCENTAGE MINORITY			PERCENTAGE CHANGE IN WHITE STUDENTS[a]	
	1968	1970	1972	1968–1970	1970–1972
Boston	31.5	35.9	40.4	−3.9	−7.4
Philadelphia	61.8	63.6	64.8	−5.7	−2.2
Baltimore	65.1	67.1	69.3	−5.6	−9.3
St. Louis	63.8	65.9	69.1	−9.4	−13.8
Chicago	62.3	65.4	69.2	−9.0	−14.7
Detroit	60.7	65.5	69.5	−15.6	−14.0
Atlanta	61.8	68.7	77.4	−22.3	−34.4
Charlotte	29.5	31.1	32.8	−3.1	−5.6
Jacksonville	28.2	29.4	32.6	−1.8	−11.4
Houston	46.7	50.6	56.4	−9.1	−17.5
San Francisco	58.8	63.1	68.2	−13.5	−22.4
San Diego	23.9	24.6	26.3	−1.1	−5.5

[a] Percentages based on beginning years.
Source: U.S. Dept. of Health, Education, and Welfare, Office for Civil Rights, Directory of Public Elementary and Secondary Schools in Selected Districts, fall 1968, fall 1970, and fall 1972.

a. Perform an analysis of variance for the data, and display the results in an ANOVA table.
b. Do the data provide evidence to indicate differences in the mean percentage of minority students in cities among the years 1968, 1970, and 1972? Test using $\alpha = .05$.
c. Do the data provide sufficient evidence to indicate that the mean percentage of minority students in a city in 1972 is larger than in 1968? Test using $\alpha = .05$.

9.25 The Bell Telephone Company's long-distance phone charges appear to be exorbitant when compared with some of its competitors, but this is because a comparison of charges between competing companies is often analogous to comparing apples and eggs. Bell Telephone's charges for individuals are on a per call basis. In contrast, its competitors often charge a monthly minimum long-distance fee, reduce the charges as the usage rises, or both. The table on page 442 shows a sampling of long-distance charges from Orlando to twelve cities for three of the non-Bell companies offering long-distance service. The data were contained in an advertisement (*Orlando Sentinel*, March 19, 1984). A note in fine print below the advertisement states that the rates are based on "30 hours of usage" for each of the servicing companies. The data in the table are pertinent for companies making phone calls to large cities. Therefore, assume that we conducted the survey of charges and that the cities receiving the calls were randomly selected from among all large cities in the United States.
a. What type of design was used for the data collection?
b. Perform an analysis of variance for the data. Present the results in an ANOVA table.

FROM ORLANDO TO:	TIME	LENGTH OF CALL (MINUTES)	COMPANY 1	COMPANY 2	COMPANY 3
NEW YORK	Day	2	.77	.79	.66
CHICAGO	Evening	3	.69	.71	.59
LOS ANGELES	Day	2	.87	.88	.66
ATLANTA	Evening	1	.22	.23	.20
BOSTON	Day	3	1.15	1.19	.99
PHOENIX	Day	5	1.92	1.98	1.65
WEST PALM BEACH	Evening	2	.49	.42	.40
MIAMI	Day	3	1.12	1.05	.99
DENVER	Day	10	3.85	3.96	3.30
HOUSTON	Evening	1	.22	.23	.20
TAMPA	Day	3	1.06	1.00	.99
JACKSONVILLE	Day	3	1.06	1.00	.99

c. Do the data provide sufficient evidence to indicate differences in mean charges among the three companies? Test using $\alpha = .05$.

d. Company 3 placed the advertisement, so it might be more relevant to compare the charges for companies 1 and 2. Do the data provide sufficient evidence to indicate a difference in mean charges for these two companies? Test using $\alpha = .05$.

9.26 Refer to Exercise 8.41 (McMurray, Wilson, & Kitchell, 1983). The table presents the mean and the estimated standard deviation of the mean (also called standard error of the mean, SEM) for each of the four treatments and for seven response variables (performance time, rating of perceived exertion, etc.) observed in the experiment. Since each mean is based on six observations, the standard error of a sample mean is $s/\sqrt{6}$, that is, the sample standard deviation divided by $\sqrt{6}$. Therefore, the standard deviation for each of the samples can be calculated by multiplying the standard error by $\sqrt{6} = 2.45$. Focus your attention on any one of the response variables in the table, say performance time, and ignore the other response variables.

a. What experimental design was used to collect the data?

Metabolic and Performance Responses of Six Female Subjects Running Until Exhaustion After the Ingestion of Fructose, Glucose, Water, and a Placebo (Mean ± SEM)

	FRUCTOSE	GLUCOSE	PLACEBO	CONTROL (H_2O)
PERFORMANCE TIME (MIN)	61.9 ± 8.3	639 ± 8.5	52.2 ± 5.5[a]	65.6 ± 7.6[a]
RATING OF PERCEIVED EXERTION[b]	11.6 ± 1.3	13.1 ± 1.1	14.6 ± 1.3	11.9 ± 1.3
HEART RATE (BEATS/MIN)	173.4 ± 7.8	172.0 ± 5.9	175.4 ± 6.1	172.1 ± 6.5
RESTING OXYGEN UPTAKE (L/MIN)	.31 ± .06	.26 ± .08	.30 ± .03	.31 ± .05
EXERCISE OXYGEN UPTAKE (L/MIN)	3.15 ± .14	2.96 ± .07[a]	3.46 ± .11[a]	3.19 ± .14
RESTING R VALUE[c]	.94 ± .02	.99 ± .02	.73 ± .02	.71 ± .02
EXERCISE R VALUE[c]	.93 ± .02	.99 ± .02	.81 ± .03	.85 ± .03

[a] Significant difference ($p < .05$) from each other.
[b] Significant difference ($p < .05$) placebo vs. fructose or control trials.
[c] Significant difference ($p < .05$) fructose and glucose vs. control and placebo.

b. Exercise 9.11, at the end of Section 9.1, illustrates that it is possible to perform an analysis of variance for some data sets when given the sample means and standard deviations. Is it possible to perform an analysis of variance for the performance data based on the information in the table?

c. Suppose that the information on performance times given in the table is derived from a completely randomized design. Perform the analysis of variance, and present your results in an ANOVA table. Do the data provide sufficient evidence to indicate differences in mean performance times among the four treatments? Test using $\alpha = .05$ and explain the practical consequences of the test results.

d. Refer to part c. What is gained or lost by analyzing the data as if they had been collected according to a completely randomized design?

9.3
The Analysis of Variance for a Two-Way Classification of Data: Factorial Experiments

A randomized block design is often called a *two-way classification of data* because:

1. It involves two independent variables: one factor and one direction of blocking.

2. Each level of one independent variable occurs with every level of the other independent variable

A two-way classification of data always permits the display of the data in a two-way table, one containing r rows and c columns. For example, the data for the randomized block design displayed in Table 9.10, a two-way table, contains $r = 10$ rows and $c = 4$ columns. Each of the $rc = (10)(4) = 40$ cells of the table contains one observation.

The treatment selection for a two-factor experiment may also yield a two-way classification of data. Suppose that you wish to relate the mean number of defects on a finished item, say a new desk top, to two factors: type of nozzle for the varnish spray gun and length of spraying time. Three types of nozzles are to be used in the experiment (three levels) and two lengths of spraying times (two levels). If we choose the treatments for the experiment to include all combinations of the three levels of nozzle type with the two levels of spraying time, we will obtain a two-way classification of data. This selection of treatments is called a *complete 3 × 2 factorial experiment.*

Definition 9.3

A *factorial experiment* is a method for selecting the treatments (i.e., the factor–level combinations) to be included in an experiment. A complete factorial experiment is one in which observations are made for every combination of the factor levels.

If we include a third factor, say paint type, at three levels in the experiment, then a complete factorial experiment will include all of the $3 \times 2 \times 3 = 18$ combinations of nozzle type, spraying time, and paint type. The resulting collection of data would be called a *three-way classification of data.*

Factorial experiments are useful methods for selecting treatments because they permit the estimation and testing of hypotheses about factor interactions. In this section we will learn how to perform an analysis of variance for a two-way classification of data. In the process, you will learn that the computational procedure is the same for both the randomized block design and the two-factor experiment because both involve a two-way classification of data. You will also learn why the ANOVA F-tests differ and discover the practical interpretations that can be derived from them.

Suppose that a two-way classification represents a two-factor experiment with factor A at a levels and factor B at b levels. Further assume that the ab treatments of the factorial are replicated r times so that there are r observations for each of the ab treatment combinations. (That is, there are r observations in each of the ab cells of the two-way table.) Then the total number n of observations is $n = abr$ and the total sum of squares, SS(Total), can be partitioned into four parts: SSA, SSB, SS(AB), and SSE. The sources of variation and their respective degrees of freedom are shown in Figure 9.8.

Figure 9.8   Partitioning of the Total Sum of Squares for a Two-Factor Experiment

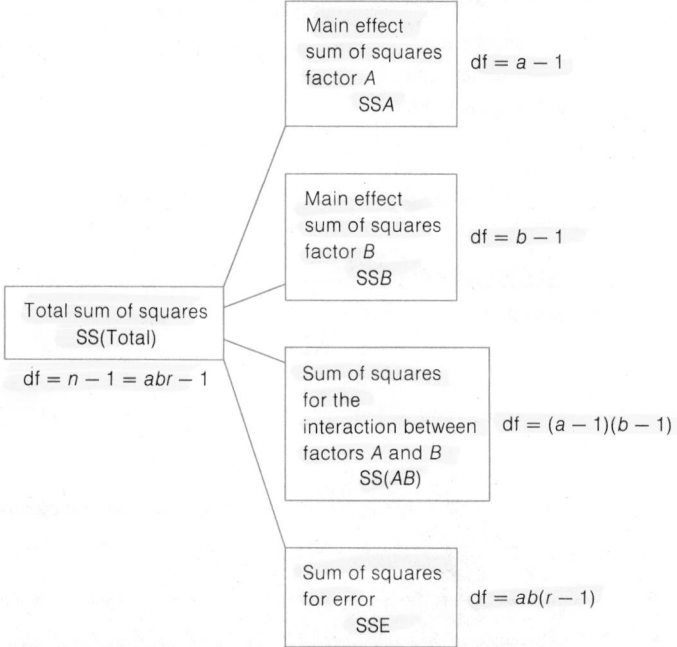

When the number of observations per cell for a two-way factorial experiment is the same—r observations per cell—then the degrees of freedom and sums of squares for the analysis of variance are additive, i.e.,

SS(Total) = SSA + SSB + SS(AB) + SSE

$$n - 1 = abr - 1 = (a - 1) + (b - 1) + (a - 1)(b - 1) + ab(r - 1)$$

and the ANOVA table would appear as shown in Table 9.14. Note that for a factorial experiment, the number r of observations per factor–level combination must always be two or more ($r \geq 2$). Otherwise, you will not have any degrees of freedom for SSE.

Table 9.14

ANOVA Table for a Two-Way Classification of Data with r Observations per Cell: Neither Variable Is a Direction of Blocking

SOURCE	df	SS	MS
Main effects A	$(a-1)$	SSA	SSA/$(a-1)$
Main effects B	$(b-1)$	SSB	SSB/$(b-1)$
AB interaction	$(a-1)(b-1)$	SS(AB)	SS(AB)/$(a-1)(b-1)$
Error	$ab(r-1)$	SSE	SSE/$ab(r-1)$
Total	$abr-1$	SS(Total)	

The notation and the formulas for calculating the sums of squares and mean squares for Table 9.14 are similar to those encountered in the analysis of variance for the completely randomized and the randomized block designs. They are shown in the boxes.

Notation for the Analysis of Variance for a Two-Way Classification of Data

a = Number of levels of independent variable 1

b = Number of levels of independent variable 2

r = Number of measurements in any pair of levels of independent variables 1 and 2

A_i = Total of all measurements of independent variable 1 at level i $(i = 1, 2, \ldots, a)$

$\bar{A}_i$ = Mean of all measurements of independent variable 1 at level i $(i = 1, 2, \ldots, a)$

$$= \frac{A_i}{br}$$

B_j = Total of all measurements of independent variable 2 at level j $(j = 1, 2, \ldots, b)$

$\bar{B}_j$ = Mean of all measurements of independent variable 2 at level j $(j = 1, 2, \ldots, b)$

$$= \frac{B_j}{ar}$$

AB_{ij} = Total of all measurements at the ith level of independent variable 1 and at the jth level of independent variable 2 $(i = 1, 2, \ldots, a;$ $j = 1, 2, \ldots, b)$

n = Total number of measurements = $a \times b \times r$

$\sum_{i=1}^{n} x_i^2$ = Sum of squares of all n measurements

$$CM = \frac{(\text{Total of all } n \text{ measurements})^2}{n} = \frac{\left(\sum_{i=1}^{n} x_i\right)^2}{n}$$

Formulas for the Calculations for a Two-Way Classification of Data

CM = Correction for the mean

$$= \frac{(\text{Total of all } n \text{ measurements})^2}{n}$$

$$= \frac{\left(\sum_{i=1}^{n} x_i\right)^2}{n}$$

$SS(\text{Total})$ = Total sum of squares

= Sum of squares of all n measurements − CM

$$= \sum_{i=1}^{n} x_i^2 - CM$$

SSA = Sum of squares for main effects, independent variable 1

$$= \left(\begin{array}{c}\text{Sum of squares of totals } A_1, A_2, \ldots, A_a \\ \text{divided by number of measurements} \\ \text{in a single total, namely } br\end{array}\right) - CM$$

$$= \frac{\sum_{i=1}^{a} A_i^2}{br} - CM$$

SSB = Sum of squares for main effects, independent variable 2

$$= \left(\begin{array}{c}\text{Sum of squares of totals } B_1, B_2, \ldots, B_b \\ \text{divided by number of measurements} \\ \text{in a single total, namely } ar\end{array}\right) - CM$$

$$= \frac{\sum_{j=1}^{b} B_j^2}{ar} - CM$$

$SS(AB)$ = Sum of squares for AB interaction

$$= \left(\begin{array}{c}\text{Sum of squares of cell totals} \\ AB_{11}, AB_{12}, \ldots, AB_{ab} \text{ divided by} \\ \text{number of measurements} \\ \text{in a single total, namely } r\end{array}\right) - SSA - SSB - CM$$

$$= \frac{\sum_{j=1}^{b} \sum_{i=1}^{a} AB_{ij}^2}{r} - SSA - SSB - CM$$

Before we actually work through an example, we need to know what we are going to do with these mean squares. Particularly, we need to know something about their practical significance.

To understand the practical significance of the four sources of variation in an ANOVA table (Table 9.14), visualize the data in the two-way table (Table 9.15). The ab cells of the two-way table correspond to the ab treatments in a completely randomized design. For each of the ab treatments, we have a random sample of r observations. Therefore, as in the case of the completely randomized design (Section 9.1), SSE is the pooled sum of squares *within* the ab samples and MSE is its corresponding mean square.

Table 9.15
Two-Way Table for a Two-Factor Factorial Experiment

r observations in each cell

	LEVEL	1	2	3	...	b	LEVEL MEANS FOR FACTOR A
	1						$\bar{A}_1$
	2						$\bar{A}_2$
FACTOR A AT a LEVELS	3						$\bar{A}_3$
	⋮						⋮
	a						$\bar{A}_a$
	LEVEL MEANS FOR FACTOR B	$\bar{B}_1$	$\bar{B}_2$	$\bar{B}_3$	...	$\bar{B}_b$	

Now suppose that we have calculated the row, column, and cell means in a table for a two-factor factorial experiment and the means appear as shown in Table 9.16. Notice that the difference in the means for levels 1 and 2 for factor A is $\bar{A}_1 - \bar{A}_2 = 4 - 7 = -3$. Now check the difference in the cell means for levels 1 and 2 of factor A when factor B is at level 1. This difference too is $7 - 10 = -3$. In fact, if you look at any level of factor B, the difference in the cell means for levels 1 and 2 of factor A is the same as the overall difference in the level means: $\bar{A}_1 - \bar{A}_2 = -3$. Similarly, if you compare the means of any pair of factor B levels, say levels 2 and 3, for a given level of factor A, the difference is constant and equal to the overall difference in B levels: $\bar{B}_2 - \bar{B}_3 = 2.33 - 3.33 = -1$. The numbers in Table 9.15 are fictitious, of course, but they illustrate a situation where the effect of one factor on the response is independent of the effect of the level of the other factor. When this situation exists, we say that factors A and B *do not interact.*

In contrast to Table 9.16, suppose that the cell means appeared as shown in Table 9.17. In this table you can see that the means of the observations for the different levels of factor A, 7.33, 6.33, and 6.66, differ very little but the cell means, for given levels of factor B, differ greatly. They even reverse themselves in sign. For example, the difference in the cell means for levels 1 and 2 of factor A for level 1 of factor B is $15 - 4 = 11$. The corresponding difference for level 2 of factor B is $5 - 10 = -5$. Clearly, the overall

Table 9.16
Cell, Row, and Column Means for a Fictitious 3 × 3 Factorial Experiment When No Interaction Is Present

	LEVEL	FACTOR B 1	2	3	MEANS
	1	7	2	3	$\bar{A}_1 = 4$
FACTOR A	2	10	5	6	$A_2 = 7$
	3	5	0	1	$\bar{A}_3 = 2$
	MEANS	$\bar{B}_1 = 7.33$	$\bar{B}_2 = 2.33$	$\bar{B}_3 = 3.33$	

Table 9.17

Cell, Row, and Column Means for a Fictitious 3 × 3 Factorial Experiment When Interaction Is Present

	LEVEL	FACTOR B			MEANS
		1	2	3	
FACTOR A	1	15	5	2	7.33
	2	4	10	5	6.33
	3	1	6	13	6.66
	MEANS	6.66	7	6.66	

difference in the observation means between levels 1 and 2 of factor A—$\bar{A}_1 - \bar{A}_2 = 7.33 - 6.33 = 1$—tells us very little about the difference in a pair of cell means for any specific level of factor B. Table 9.17 illustrates a situation where the mean response *depends* on the particular combination of levels of factors A and B. When this situation occurs, we say that factors A and B interact.

The interaction sum of squares, SS(AB), and its respective mean square, MS(AB), where

$$MS(AB) = \frac{SS(AB)}{(a-1)(b-1)}$$

provide a measure of the interaction between factors A and B. If MS(AB) is substantially larger than the mean square for error, MSE, we have evidence of factor interaction. To test the null hypothesis, H_0: There is *no interaction* between factors A and B, we use the familiar test statistic

$$F = \frac{MS(AB)}{MSE}$$

and reject H_0 if $F \geq F_\alpha$, where F_α is based on $v_1 = (a-1)(b-1)$ and $v_2 = ab(r-1)$ degrees of freedom.

The sum of squares, SSA, called the main effect sum of squares for factor A, measures the variation of the factor A level means, $\bar{A}_1, \bar{A}_2, \ldots, \bar{A}_a$, about the mean of all abr observations. This quantity is relevant only if there is no evidence of factor interaction. To test the null hypothesis, H_0: There are no differences among the mean values of the response for the levels of factor A, we compute the mean square for A,

$$MSA = \frac{SSA}{a-1}$$

and use the test statistic

$$F = \frac{MSA}{MSE}$$

and reject H_0 if $F \geq F_\alpha$, where F_α is based on $v_1 = a-1$ and $v_2 = ab(r-1)$ degrees of freedom. A similar test for evidence of the main effects for factor B uses the F-statistic,

$$F = \frac{MSB}{MSE}$$

and rejects H_0 if $F \geq F_\alpha$, where F_α is based on $v_1 = b-1$ and $v_2 = ab(r-1)$ degrees of freedom.

> ## Test Statistics and Rejection Regions for a Two-Factor Factorial Experiment
>
> ### For Factor Interaction
>
> Test statistic: $F = \dfrac{MS(AB)}{MSE}$
>
> Rejection region: $F \geq F_\alpha$, where F_α is based on $v_1 = (a-1)(b-1)$ and $v_2 = ab(r-1)$ degrees of freedom
>
> ### For Main Effects Factor A
>
> Test statistic: $F = \dfrac{MSA}{MSE}$
>
> Rejection region: $F \geq F_\alpha$, where F_α is based on $v_1 = a - 1$ and $v_2 = ab(r-1)$ degrees of freedom
>
> ### For Main Effects Factor B
>
> Test statistic: $F = \dfrac{MSB}{MSE}$
>
> Rejection region: $F \geq F_\alpha$, where F_α is based on $v_1 = b - 1$ and $v_2 = ab(r-1)$ degrees of freedom

If the test for factor interaction is statistically significant (i.e., there is evidence of factor interaction), there is no point in testing the main effects for factors A and B. The presence or absence of main effects is of no practical importance if factors A and B interact. When factor interaction is present, the cell means corresponding to the factor–level combinations (i.e., means of the cells of the two-way table) are of primary importance. The formulas for confidence intervals for a cell mean or the difference between a pair of cell means are shown in the next two boxes.

> ## $100(1 - \alpha)\%$ Confidence Interval for the Mean of a Single Cell of the Two-Way Table
>
> $$\bar{x}_{ij} \pm t_{\alpha/2} \frac{s}{\sqrt{r}}$$
>
> where $\bar{x}_{ij} =$ Cell mean for cell in ith row, jth column
>
> $r =$ Number of measurements per cell
>
> $s = \sqrt{MSE}$
>
> and $t_{\alpha/2}$ is based on $ab(r-1)$ df.

> ### $100(1 - \alpha)\%$ Confidence Interval for the Difference in a Pair of Cell Means
>
> Let
>
> $\bar{x}_1$ = Sample mean of r measurements in first cell
>
> $\bar{x}_2$ = Sample mean of r measurements in second cell
>
> Then the $100(1 - \alpha)\%$ confidence interval for the difference between the cell means is
>
> $$(\bar{x}_1 - \bar{x}_2) \pm t_{\alpha/2} s \sqrt{\frac{2}{r}}$$
>
> where $s = \sqrt{\text{MSE}}$ and $t_{\alpha/2}$ is based on $ab(r - 1)$ df.

Example 9.10   A manufacturer whose daily supply of raw materials is variable and limited can use the material to produce two different products in various proportions. The profit per unit of raw material obtained by producing each of the two products depends on the length of a product's manufacturing run and hence on the amount of raw material assigned to it. Other factors, such as worker productivity, machine breakdown, etc., affect the profit per unit as well, but their net effect on profit is random and uncontrollable. The manufacturer has conducted an experiment to investigate the effect of the level of supply of raw materials, S, and the ratio of its assignment, R, to the two product manufacturing lines on the profit per unit of raw material. The ultimate goal would be to be able to choose the best ratio R to match each day's supply S of raw materials. The levels of supply of the raw material chosen for the experiment were 15, 18, and 21 tons; the levels of the ratio of allocation to the two product lines were $\frac{1}{2}$, 1, and 2. The response was the profit (in cents) per unit of raw material supply obtained from a single day's production. Three replications of a complete 3×3 factorial experiment were conducted in a random sequence (i.e., a completely randomized design). The data for the 27 days are shown in Table 9.18.

Table 9.18

		RAW MATERIAL SUPPLY, TONS (S)		
		15	18	21
RATIO OF RAW MATERIAL ALLOCATION (R)	$\frac{1}{2}$	23, 20, 21	22, 19, 20	19, 18, 21
	1	22, 20, 19	24, 25, 22	20, 19, 22
	2	18, 18, 16	21, 23, 20	20, 22, 24

a. Calculate the appropriate sums of squares and construct an ANOVA table.

b. Do the data present sufficient evidence to indicate an S–R interaction?

c. Find a 95% confidence interval to estimate the mean profit per unit of raw materials when $S = 18$ tons and the ratio of production is $R = 1$.

d. Find a 95% confidence interval to estimate the difference in mean profit per unit of raw materials between $S = 18$, $R = \frac{1}{2}$ and $S = 18$, $R = 1$.

Solution a. The sums of squares for the ANOVA table are calculated as follows:

$$\text{CM} = \frac{(\text{Total of all } n \text{ measurements})^2}{n} = \frac{(558)^2}{27} = 11{,}532$$

$$\text{SS(Total)} = \sum_{i=1}^{n} x_i^2 - \text{CM} = 11{,}650 - 11{,}532 = 118$$

The next step is to construct a table showing the totals of x values for each combination of levels of supply S and ratio R and then the totals for each level of S and each level of R. These totals, computed from the raw data table, are shown in Table 9.19. Then

$$\text{SS(Supply)} = \frac{\sum_{i=1}^{3} S_i^2}{9} - \text{CM} = \frac{(177)^2 + (196)^2 + (185)^2}{9} - \text{CM}$$

$$= \frac{103{,}970}{9} - 11{,}532 = 20.22$$

$$\text{SS(Ratio)} = \frac{\sum_{i=1}^{3} R_i^2}{9} - \text{CM} = \frac{(183)^2 + (193)^2 + (182)^2}{9} - \text{CM}$$

$$= \frac{103{,}862}{9} - 11{,}532 = 8.22$$

$$\text{SS}(SR) = \sum_{j=1}^{3} \sum_{i=1}^{3} \frac{SR_{ij}^2}{3} - \text{SS(Supply)} - \text{SS(Ratio)} - \text{CM}$$

$$= \frac{(64)^2 + (61)^2 + (58)^2 + \cdots + (66)^2}{3} - 20.22 - 8.22 - 11{,}532$$

$$= \frac{34{,}820}{3} - 20.22 - 8.22 - 11{,}532$$

$$= 46.23$$

Table 9.19

		RAW MATERIAL SUPPLY, TONS (S)			
		15	18	21	
RATIO OF RAW MATERIAL ALLOCATION (R)	$\frac{1}{2}$	64	61	58	$R_1 = 183$
	1	61	71	61	$R_2 = 193$
	2	52	64	66	$R_3 = 182$
		$S_1 = 177$	$S_2 = 196$	$S_3 = 185$	Total $= 558$

$$SSE = SS(Total) - SS(Supply) - SS(Ratio) - SS(SR)$$
$$= 118.00 - 20.22 - 8.22 - 46.23$$
$$= 43.33$$

The ANOVA table is given in Table 9.20.

Table 9.20

SOURCE	df	SS	MS
Supply	2	20.22	10.11
Ratio	2	8.22	4.11
Supply–Ratio interaction	4	46.23	11.56
Error	18	43.33	2.41
Total	26	118.00	

b. To test the null hypothesis that supply S and ratio R do not interact, we use the F test statistic:

$$F = \frac{MS(SR)}{MSE} = \frac{11.56}{2.41} = 4.80$$

The degrees of freedom associated with $MS(SR)$ and s^2 are 4 and 18, respectively (given in Table 9.20). Therefore, we reject H_0 if $F \geq F_{.05}$, where $v_1 = 4$, $v_2 = 18$, and $F_{.05} = 2.93$. Since the computed value of F (4.80) exceeds $F_{.05}$, we reject H_0 and conclude that supply and ratio interact. The presence of interaction tells you that the mean profit depends on the combination of levels of supply S and ratio R. Consequently, there is little point in checking to see whether the means differ for the three levels of supply or whether they differ for the three levels of ratio. For example, the supply level that gave the highest mean profit (over all levels of R) might not be the same S–R level combination that produces the largest mean profit per unit of raw material.

c. A 95% confidence interval for the mean $E(x)$ when supply $S = 18$ and ratio $R = 1$ is

$$\bar{x}_{18.1} \pm t_{.025}\left(\frac{s}{\sqrt{r}}\right)$$

where $\bar{x}_{18.1}$ is the mean of the $r = 3$ values of x obtained for $S = 18$, $R = 1$, $s = \sqrt{MSE} = \sqrt{2.41} = 1.55$, and $t_{.025} = 2.101$ is based on 18 df. Substituting, we obtain

$$\frac{71}{3} \pm 2.101\left(\frac{1.55}{\sqrt{3}}\right)$$

$$23.67 \pm 1.88$$

Therefore, our interval estimate for the mean profit per unit of raw material where $S = 18$ and $R = 1$ is $21.79 to $25.55.

d. A 95% confidence interval for the difference in mean profit per unit of raw material for two different combinations of levels of S and R is

$$(\bar{x}_1 - \bar{x}_2) \pm t_{.025} s \sqrt{\frac{2}{r}}$$

where $\bar{x}_1$ and $\bar{x}_2$ represent the means of the $r = 3$ replications for the factor–level combinations $S = 18$, $R = \frac{1}{2}$ and $S = 18$, $R = 1$, respectively. Substituting, we obtain

$$\left(\frac{61}{3} - \frac{71}{3}\right) \pm (2.101)(1.55) \sqrt{\frac{2}{3}}$$

$$-3.33 \pm 2.66$$

Therefore, the interval estimate for the difference in mean profit per unit of raw material for the two factor–level combinations is $(-\$5.99, -\$.67)$. The negative values indicate that we estimate the mean for $S = 18$, $R = \frac{1}{2}$ to be less than the mean for $S = 18$, $R = 1$ by between \$.67 and \$5.99. ■

Case Study 9.2
Anxiety Levels and Mathematical Achievement

If you have a high level of anxiety when you take a mathematics examination, it may come as no surprise that there appears to be a relationship between the level of your anxiety and your achievement. Writing in the *Journal for Research in Mathematics Education*, Pamela S. Clute (1984) describes her doctoral research on this subject. Clute's research involved a study of the effect of three factors on a student's mathematics examination score taken after completing a survey course in mathematics. One factor was the college where the course was taken. Students took the course at both the University of California at Riverside and California State College, San Bernardino. Therefore, the factor College was at two levels. The second factor was the anxiety level of the student. Students were tested to determine their anxiety level and placed into one of three anxiety level groups. Therefore, the factor Anxiety level was at three levels. The third factor, at two levels, was the manner in which the instructional material was presented. Two methods were employed; a standard direct instruction method and a direct instruction discovery method. This latter method developed the subject by a sequence of questions, eventually leading the student to the mathematical principle that was the object of the lesson.

Since all the students were taught by the same instructor (P. S. Clute), one would not expect to see differences in research results between the two colleges and none were found in an analysis of the data. Consequently, we will combine the data for the two colleges and view Clute's experiment as a two-factor factorial experiment. The two factors are the mathematics anxiety level (three levels) and the method of instruction (two levels). The sample sizes, examination test score means, and standard deviations (in parentheses) for the six factor–level combinations are shown in Table 9.21. In the table, you can see that for both teaching methods the mean test score drops as the mathematics anxiety level increases. You may also note the unusually high mean score for students with a low anxiety level who were taught by the discovery method.

Table 9.21

Means and Standard Deviations on the Mathematics Achievement Test by Instructional Method and Level of Anxiety

METHOD		MATHEMATICS ANXIETY LEVEL						
		Low		Medium		High		
	n	M	n	M	n	M	TOTAL	
Discovery	14	352.50	14	299.71	13	225.38	294.17	
		(27.31)		(29.60)		(59.09)	(65.68)	
Expository	14	288.86	13	265.69	13	260.69	272.17	
		(79.94)		(81.24)		(54.17)	(72.26)	
Total		320.68		283.33		243.04	283.31	
		(66.98)		(61.52)		(58.38)	(69.46)	

Note: Maximum score = 400.

Clute's analysis of variance of the three-factor factorial experiment confirms our examination of the table of means (Table 9.21). Her analysis of variance is shown in Table 9.22. Although we have not explained how to compute the sums of squares for a three-factor factorial experiment, the interpretation of the ANOVA table is the same as the interpretation for a two-factor experiment. Sources of variation shown in Table 9.22 include the main effects for methods (M), anxiety (A), and college (C). They also include the three two-way interactions, $M \times A$, $M \times C$, and $A \times C$, and the three-way interaction $M \times A \times C$. Examining the computed values of the F-statistics, you can see that only one interaction, the $M \times A$ interaction, is statistically significant ($F = 4.96$) at the .01 level of significance. This result tells us that the teaching method to use for a group of students depends on their anxiety level. Note that none of the statistical tests involving colleges (interactions or main effects) were statistically significant. While this result does not imply that there is no difference in mean scores between colleges, it supports our initial thought that there is no reason to expect students at one school to score higher or lower than those at the other.

Clute concludes that "the most interesting finding of the study is the significant interaction between instructional method and the level of anxiety. The students with low mathematics anxiety tended to do better with the discovery method, whereas students with high mathematics anxiety tended to do better under the expository treatment."

Table 9.22

Analysis of Variance of Mathematics Achievement Scores

SOURCE	df	MS	F
Method (M)	1	7,945	2.30
Anxiety (A)	2	34,935	10.11[a]
College (C)	1	2,239	.65
$M \times A$	2	17,125	4.96[a]
$M \times C$	1	516	.15
$A \times C$	2	7,914	2.29
$M \times A \times C$	2	1,985	.58
Residual	69	3,455	

[a] $p < .01$.

Exercises 9.27–9.37

Learning the Mechanics

9.27 What do we mean when we say that two factors interact?

9.28 The partially completed table is for a 3 × 4 factorial experiment with two observations for each factor–level combination:

SOURCE	df	SS	MS	F
A		.8		
B		5.3		
AB		9.6		
Error			.5	
Total				

a. Describe the experiment, giving the number of factors, levels, etc.
b. Compute the ANOVA table.
c. What is meant by factor interaction, and what is its practical implication?
d. Do the data provide sufficient information to indicate an interaction between factors A and B? Test using $\alpha = .05$.
e. What are the practical implications of your test results in part d?

9.29 The partially completed table is for a two-factor factorial experiment:

SOURCE	df	SS	MS	F
A	3		.75	
B	1	.95		
AB			.30	
Error				
Total	23	6.5		

a. Give the number of levels for each factor.
b. How many observations were collected for each factor–level combination?
c. Complete the ANOVA table.
d. Test for factor interaction and factor main effects. (Use $\alpha = .05$.)
e. What are the practical implications of your test in part d?

9.30 The following two-way table gives data for a 2 × 3 factorial experiment with two observations for each factor–level combination:

	LEVEL	FACTOR B 1	FACTOR B 2	FACTOR B 3
FACTOR A	1	3.1, 4.0	4.6, 4.2	6.4, 7.1
	2	5.9, 5.3	2.9, 2.2	3.3, 2.5

a. Perform an analysis of variance for the data, and display the results in an ANOVA table.

b. Do the data provide evidence of factor interaction? Test using $\alpha = .05$.

c. What are the practical implications of the test results in part b?

d. Find a 90% confidence interval for the difference between the mean responses for factor–level combinations $A(1)$, $B(1)$ and $A(2)$, $B(1)$.

9.31 The following two-way table gives data for a 2×2 factorial experiment with two observations per factor–level combination:

		FACTOR B	
	LEVEL	1	2
FACTOR A	1	29.6	47.3
		35.2	42.1
	2	12.9	28.4
		17.6	22.7

a. Perform an analysis of variance for the data, and display the results in an ANOVA table.

b. Do the data provide evidence of factor interaction? Test using $\alpha = .05$.

c. Do the data provide sufficient evidence to indicate a main effect due to factor A? Test using $\alpha = .05$. What is the practical implication of this test result?

d. Do the data provide sufficient evidence to indicate a main effect due to factor B? Test using $\alpha = .05$. What is the practical implication of this test result?

e. Find a 95% confidence interval for the difference in mean response between levels 1 and 2 of factor B.

Applying the Concepts

9.32 A beverage distributor wishes to determine the combination of advertising agency (two) and advertising medium (three) that will produce the largest increase in sales per advertising dollar. Each of the advertising agencies prepares copy or film for each of the media: newspaper, radio, and television. Twelve small towns of roughly the same size are selected for the experiment, and two each are assigned to receive an advertisement prepared and transmitted by each of the six agency–medium combinations. The dollar increases in sales per advertising dollar, based on a 1-month sales period, are shown in the table.

		NEWSPAPER	RADIO	TELEVISION
AGENCY	1	15.3	20.1	12.7
		12.7	17.4	16.2
	2	18.9	24.3	12.5
		22.4	28.8	9.4

a. Perform an analysis of variance for the data, and display your results in an ANOVA table.

b. Do the data provide sufficient information to indicate an agency–medium interaction? Test using $\alpha = .05$.

c. What are the practical implications of the test in part b?

9.33 The sample means and standard deviations (in parentheses) shown in the table were presented in a paper by T. G. Kovacs and G. Leduc (1982). The data were the weight gains of rainbow trout (percentage of body weight) over a 20-day period for each of the four factor–level combinations of a 2×2 factorial experiment. One factor was the ration level (amount of food provided to the fish) and the second was the temperature in degrees Celsius ($^\circ$C) of the water. Note that the zero ration level produced a negative percentage weight gain, i.e., a loss in weight.

RATION LEVEL	WATER TEMPERATURE (°C)	
	6°	12°
0	−8.14 (3.27)	−10.33 (2.35)
2.5	29.12 (5.67)	38.05 (6.99)

a. Calculate the total weight gain (or loss) for the $n = 20$ weight gain measurements for each of the four categories of the 2×2 factorial experiment.

b. Calculate CM.

c. Use the results of parts a and b to calculate the sums of squares for ration, for temperature, and for the interaction ration $\times$ temperature.

d. Calculate each sample variance. Then calculate the sums of squares of deviations *within* each sample for each of the four samples.

e. Calculate SSE. [*Hint:* SSE is the pooled sum of squares of the deviations calculated in part b.]

f. Now that you know SS(Ration), SS(Temperature), SS(Ration × Temperature), and SSE, find SS(Total).

g. Arrange the calculations obtained in the previous parts in an ANOVA table.

h. Explain the practical significance of the presence (or absence) of ration × temperature interaction. Do the data provide sufficient evidence of a ration × temperature interaction?

i. Do the data provide sufficient evidence to indicate a difference in weight gain between fish at 6°C and those at 12°C when fed on the 2.5 ration level? Test using $\alpha = .05$.

9.34 How do women compare with men in their ability to perform laborious tasks requiring strength? Some information on this question is provided in a study by M. D. Phillips and R. L. Pepper (1982). Phillips and Pepper conducted a 2×2 factorial experiment to investigate the effect of the factor sex (male or female) and the factor weight (light or heavy) on the length of time for a person to perform a certain firefighting task. Eight persons were selected for each of the $2 \times 2 = 4$ sex–weight categories of the 2×2 factorial experiment, and the length of time (in seconds) needed to complete the task was recorded for each of the thirty-two persons. The means and standard deviations (SD) of the four samples are shown in the table at the top of the next page.

| | LIGHT WEIGHT | | HEAVY WEIGHT | |
	Mean	SD	Mean	SD
FEMALE	18.30	6.81	14.50	2.93
MALE	13.00	5.04	12.25	5.70

a. Calculate the total of the $n = 8$ time measurements for each of the four categories of the 2×2 factorial experiment.

b. Calculate CM.

c. Use the results of parts a and b to calculate the sums of squares for sex, for weight, and for the interaction sex $\times$ weight.

d. Calculate each sample variance. Then calculate the sums of squares of deviations *within* each sample for each of the four samples.

e. Calculate SSE. [*Hint:* SSE is the pooled sum of squares of the deviations calculated in part b.]

f. Now that you know SS(Sex), SS(Weight), SS(Sex $\times$ Weight), and SSE, find SS(Total).

g. Arrange the calculations obtained in the previous parts in an ANOVA table.

h. Explain the practical significance of the presence (or absence) of sex $\times$ weight interaction. Do the data provide sufficient evidence of a sex $\times$ weight interaction?

i. Do the data provide sufficient evidence to indicate a difference in mean time to complete the task between light men and women? Test using $\alpha = .05$.

j. Do the data provide sufficient evidence to indicate a difference in mean time to complete the task between heavy men and women? Test using $\alpha = .05$.

9.35 Refer to Exercise 9.34. Phillips and Pepper give data on another 2×2 factorial experiment utilizing twenty males and twenty females. The experiment involved the same treatments with ten persons assigned to each sex–weight category. The response measured for each person was the pulling force (in newtons, N) that the person was able to exert on the starter cord of a P-250 fire pump. The means and standard deviations (SD) of the four samples (corresponding to the $2 \times 2 = 4$ categories of the experiment) are shown in the table.

Force Exerted in the P-250 Task

| | GROUP 1 (LIGHT) | | GROUP 2 (HEAVY) | |
GENDER	Mean (N)	SD	Mean (N)	SD
FEMALES	46.26	14.23	62.72	13.97
MALES	88.07	8.32	86.29	12.45

a. Use the procedures outlined in Exercise 9.34 to perform an analysis of variance for the experiment. Display your sums of squares, etc., in an ANOVA table.

b. Explain the practical significance of the presence (or absence) of sex $\times$ weight interaction. Do the data provide sufficient evidence of a sex $\times$ weight interaction?

c. Do the data provide sufficient evidence to indicate a difference in mean force to complete the task between light men and women? Test using $\alpha = .05$.

d. Do the data provide sufficient evidence to indicate a difference in mean force to complete the task between heavy men and women? Test using $\alpha = .05$.

9.36 If you are overweight, how will your friends respond to a request for a favor? C. L. Steinberg and J. M. Birk (1983) sought an answer to this question by experimentation. Each subject in the experiment contacted a second person (a confederate) who was either male or female, overweight or normal weight. After following a standard procedure, the confederate requested a favor of the subject and a measure of the subject's compliance was recorded. The portion of the data summary that gives the measure of compliance of male subjects to the weight and sex of confederates is shown in the table.

		CONFEDERATE'S WEIGHT	
		Overweight	Normal
CONFEDERATE'S SEX	Male	32.0 (28.65)	34.67 (36.52)
	Female	24.27 (12.94)	36.67 (14.48)

As the table suggests, the study was conducted as a 2 × 2 factorial experiment. The two factors were the confederate's sex (two levels) and the confederate's weight (two levels). Fifteen male subjects were assigned to each of the confederate sex–weight groups. The means and standard deviations of each of the four samples are shown in the table (standard deviations are in parentheses).

a. Calculate the sample totals for each of the four samples.

b. Calculate the total for all $n = 60$ observations and calculate CM.

c. Calculate the three treatment sums of squares, SS(Sex), SS(Weight), and SS(Sex × Weight).

d. Use the sample standard deviations to calculate the sums of squares of the deviations within each of the four samples. *Hint:* If the variance of a sample of n observations is equal to s^2, then

$$s^2 = \frac{\sum_{i=1}^{n} (x_i - \bar{x})^2}{n - 1} \qquad \text{or} \qquad (n-1)s^2 = \sum_{i=1}^{n} (x_i - \bar{x})^2$$

e. Calculate SSE. [*Note:* SSE is the pooled sum of squares of deviations for the four samples.]

f. Arrange the results of parts a to e in an ANOVA table and complete the table.

g. Do the data provide sufficient evidence to indicate a sex × weight interaction? Test using $\alpha = .05$.

h. State the practical implications of your test results in part g. [*Note:* We will analyze these data further in Exercise 9.43.]

9.37 The data summary in Exercise 9.36, based on Steinberg and Birk's weight and compliance experiment, contained only half of the experimental results. The other half of the data was produced by the same experiment described in Exercise 9.36 except that sixty female subjects were used. Therefore, this data summary, shown in the table, enables us to determine the measure of compliance of female subjects to the four sex × weight factor–level combinations defining the characteristics of the confederates. (Standard deviations are in parentheses.)

| | | CONFEDERATE'S WEIGHT | |
		Overweight	Normal
CONFEDERATE'S SEX	Male	21.33 (15.86)	38.31 (20.04)
	Female	20.33 (16.42)	27.80 (16.21)

Perform an analysis of variance for these data following the question parts of Exercise 9.36. [*Note:* The combined sets of data for both male and female subjects constitute a three-factor factorial experiment and should be analyzed as such. Since the analysis of a three-factor experiment is beyond the scope of this text, we have performed a partial analysis of the data by partitioning the data into two two-factor factorial experiments.]

9.4 A Procedure for Making Multiple Comparisons

Many experiments are conducted to determine the largest (or the smallest) mean in a set. For example, a food technologist might wish to determine the preservative that produces the largest mean shelf-life for a cake mix. A research physician might wish to determine which of a number of drugs produces the largest mean drop in the blood pressure of heart patients. An industrial psychologist might wish to determine which of a set of working conditions produces the highest mean worker score on an attitudinal test.

Choosing the treatment with the largest mean from among five treatments might seem to be a simple matter. All that we need to do is to make a certain number, say $n_1 = n_2 = \cdots = n_5 = 10$ observations on each treatment, obtain the sample means, $\bar{x}_A, \bar{x}_B, \ldots, \bar{x}_E$, and compare them using Student's t-tests to determine whether differences exist among the pairs of means.

The problem with this procedure is that a Student's t-test, with its associated value of α, is valid only when the two treatments to be compared are selected prior to experimentation. After you have looked at the data, you cannot use a Student's t-statistic to compare the treatments for the largest and smallest sample means because they will always be farther apart, on the average, than the means for any pair of treatments selected at random. And if you conduct a series of t-tests, each with a chance α of indicating a difference between a pair of means when in fact no difference exists, then the risk of making at least one Type I error in a series of t-tests will be larger than the value of α specified for a single t-test.

There are a number of procedures for comparing and ranking a group of treatment means. The one that we present, known as **Tukey's method for multiple comparisons,** utilizes the **Studentized range**

$$q = \frac{\bar{x}_{\max} - \bar{x}_{\min}}{s/\sqrt{n}}$$

(where $\bar{x}_{max}$ and $\bar{x}_{min}$ are the largest and smallest sample means, respectively) to determine whether the difference in any pair of sample means implies a difference in the corresponding treatment means. The logic behind this multiple comparison procedure is that if we determine a critical value for the difference between the largest and smallest sample means, $|\bar{x}_{max} - \bar{x}_{min}|$, one that implies a difference in their respective treatment means, then any other pair of sample means that differ by as much or more than this critical value would also imply a difference in corresponding treatment means. Tukey's procedure selects this critical distance, ω, so that the probability of making one or more Type I errors (concluding that a difference exists between a pair of treatment means when in fact they are identical) is α. Therefore, the risk of making a Type I error applies to the whole procedure, i.e., to all of the comparisons of means, rather than to a single comparison.

Tukey's procedure assumes that the sample means, k in number, are based on independent random samples, each containing the same number n_t of observations. Then if $s = \sqrt{MSE}$ is the computed standard deviation for the analysis, the distance ω is

$$\omega = q_\alpha(k, v) \frac{s}{\sqrt{n_t}}$$

The tabulated statistic, $q_\alpha(k, v)$, is the critical value of the Studentized range, the value that locates α in the upper tail of the q-distribution. This critical value depends on

---

Tukey's Multiple Comparison Procedure

1. Calculate

$$\omega = q_\alpha(k, v) \frac{s}{\sqrt{n_t}}$$

where

k = Number of sample means

$s = \sqrt{MSE}$

v = Number of degrees of freedom associated with MSE

n_t = Number of observations in each of the k samples

$q_\alpha(k, v)$ = Critical value of the Studentized range (Tables XIV and XV in the Appendix)

2. Rank the k sample means. Any pair of sample means differing by more than ω implies a difference in the corresponding population means.

Assumptions: Samples were randomly and independently selected from normal populations with means $\mu_1, \mu_2, \ldots, \mu_k$ and common variance σ^2.

Table 9.23

A Portion of Table XIV: Tabulated Values of the Studentized Range, $q(k, v)$, Upper 5%

v \ k	2	3	4	5	6	7	8	9	10	11	12	13	14
1	17.97	26.98	32.82	37.08	40.41	43.12	45.40	47.36	49.07	50.59	51.96	53.20	54.33
2	6.08	8.33	9.80	10.88	11.74	12.44	13.03	13.54	13.99	14.39	14.75	15.08	15.38
3	4.50	5.91	6.82	7.50	8.04	8.48	8.85	9.18	9.46	9.72	9.95	10.15	10.35
4	3.93	5.04	5.76	6.29	6.71	7.05	7.35	7.60	7.83	8.03	8.21	8.37	8.52
5	3.64	4.60	5.22	5.67	6.03	6.33	6.58	6.80	6.99	7.17	7.32	7.47	7.60
6	3.46	4.34	4.90	5.30	5.63	5.90	6.12	6.32	6.49	6.65	6.79	6.92	7.03
7	3.34	4.16	4.68	5.06	5.36	5.61	5.82	6.00	6.16	6.30	6.43	6.55	6.66
8	3.26	4.04	4.53	4.89	5.17	5.40	5.60	5.77	5.92	6.05	6.18	6.29	6.39
9	3.20	3.95	4.41	4.76	5.02	5.24	5.43	5.59	5.74	5.87	5.98	6.09	6.19
10	3.15	3.88	4.33	4.65	4.91	5.12	5.30	5.46	5.60	5.72	5.83	5.93	6.03
11	3.11	3.82	4.26	4.57	4.82	5.03	5.20	5.35	5.49	5.61	5.71	5.81	5.90
12	3.08	3.77	4.20	4.51	4.75	4.95	5.12	5.27	5.39	5.51	5.61	5.71	5.80
13	3.06	3.73	4.15	4.45	4.69	4.88	5.05	5.19	5.32	5.43	5.53	5.63	5.71
14	3.03	3.70	4.11	4.41	4.64	4.83	4.99	5.13	5.25	5.36	5.46	5.55	5.64
15	3.01	3.67	4.08	4.37	4.60	4.78	4.94	5.08	5.20	5.31	5.40	5.49	5.57
16	3.00	3.65	4.05	4.33	4.56	4.74	4.90	5.03	5.15	5.26	5.35	5.44	5.52
17	2,98	3.63	4.02	4.30	4.52	4.70	4.86	4.99	5.11	5.21	5.31	5.39	5.47
18	2.97	3.61	4.00	4.28	4.49	4.67	4.82	4.96	5.07	5.17	5.27	5.35	5.43
19	2.96	3.59	3.98	4.25	4.47	4.65	4.79	4.92	5.04	5.14	5.23	5.31	5.39
20	2.95	3.58	3.96	4.23	4.45	4.62	4.77	4.90	5.01	5.11	5.20	5.28	5.36

α, the number of treatment means involved in the comparison, and v, the number of degrees of freedom associated with MSE. Values of $q_\alpha(k, v)$ for $\alpha = .05$ and .01 are given in Tables XIV and XV, respectively, in the Appendix. A portion of Table XIV is shown in Table 9.23. If you wish to make pairwise comparisons among $k = 5$ sample means and if MSE is based on $v = 12$ degrees of freedom, for example, then for $\alpha = .05$ we have $q_\alpha(5, 12) = 4.51$.

Example 9.11

Refer to the study of the effect of raw material supply and ratio of material allocation on a manufacturer's profit (Example 9.10). There we found evidence to indicate factor interaction and thus wish to examine the means of the nine treatments corresponding to the nine raw material supply (S) and ratio of raw material allocations (R) level combinations. Use Tukey's multiple comparison procedure, $\alpha = .05$, to rank the treatment means and to determine which of the means can be judged to be different.

Solution

The sample means for the nine factor–level combinations are shown in Table 9.24. Since the sample means in Table 9.24 are the mean profits per unit of raw material supply for a day's production, we would want to find the raw material supply S and ratio of allocation R that produces the maximum profit. The first step in the ranking procedure is to calculate ω for $k = 9$ treatment means, $n_t = 3$ observations per treatment, $\alpha = .05$, and $s = 1.55$ (calculated in part c of Example 9.10). Since MSE is based on

Table 9.24

Sample Means for $k = 9$ Treatments

| | | RAW MATERIAL SUPPLY, TONS (S) | | |
		15	18	21
RATIO OF RAW MATERIAL ALLOCATION (R)	$\frac{1}{2}$	21.33	20.33	19.33
	1	20.33	23.67	20.33
	2	17.33	21.33	22.00

$v = 18$ degrees of freedom, the value of $q_{.05}(k, v)$ is given in Table XIV of the Appendix:

$$q_{.05}(9, 18) = 4.96$$

and

$$\omega = q_{.05}(9, 18) \frac{s}{\sqrt{n_t}} = (4.96) \frac{1.55}{\sqrt{3}} = 4.44$$

Therefore, population means corresponding to pairs of sample means that differ by more than $\omega = 4.44$ will be judged to be different. The sample means for the nine S–R combinations are ranked as follows:

17.33, 19.33, 20.33, 20.33, 20.33, 21.33, 21.33, 22.00, 23.67

Using $\omega = 4.44$ as a yardstick to determine differences between pairs of treatments, you can see that there is no evidence to indicate a difference between any pairs of treatments that are adjacent in the ranking. There is evidence to indicate that the treatment means corresponding to sample means 22.00 and 23.67¢ are different from the treatment mean with sample mean equal to 17.33. There is no evidence to indicate a difference in the treatments corresponding to the eight largest sample means, 19.33 to 23.67 (indicated by the overbar). Further experimentation would be required to determine whether the observed differences among the eight largest sample means really imply differences among the corresponding population means. ■

Exercises 9.38–9.45

Learning the Mechanics

9.38 Give the values of $q_\alpha(k, v)$ for the following values of k, v, and α:

a. $k = 3$, $v = 9$, $\alpha = .05$ **b.** $k = 5$, $v = 15$, $\alpha = .01$

c. $k = 4$, $v = 8$, $\alpha = .01$ **d.** $k = 4$, $v = 12$, $\alpha = .05$

9.39 Independent random samples of four observations per sample were collected from five populations. The value of SSE for this completely randomized design was .72 and the sample means were $\bar{x}_1 = 4.3$, $\bar{x}_2 = 4.9$, $\bar{x}_3 = 3.9$, $\bar{x}_4 = 3.7$, and $\bar{x}_5 = 4.4$.

a. How many degrees of freedom are associated with SSE?

b. Calculate s^2.

c. Find ω for $\alpha = .05$ and explain what it is.

d. Order the sample means and determine which pairs, if any, appear to differ. Use $\alpha = .05$.

9.40 If you were to compare the means of two populations based on independent samples of equal size, you would test for a difference using Student's t-statistic of Section 8.2, i.e.,

$$t = \frac{\bar{x}_1 - \bar{x}_2}{s\sqrt{1/n_1 + 1/n_2}} = \frac{\bar{x}_1 - \bar{x}_2}{s\sqrt{2/n_1}} \qquad \text{when } n_1 = n_2$$

For $\alpha = .05$ the rejection region for the test is $t \geq t_{.025}$ or $t \leq -t_{.025}$. Suppose that $n_1 = n_2 = 4$ and $s^2 = 9$.

a. How many degrees of freedom would be associated with s^2?

b. How large could the difference be between $\bar{x}_1$ and $\bar{x}_2$ (either positive or negative) before you rejected the null hypothesis H_0: $\mu_1 = \mu_2$?

c. Find the value of ω for $k = 2$, $v = 6$, and compare it with your answer to part b.

d. Use the results of part b to explain the meaning of ω.

Applying the Concepts

9.41 Refer to Exercise 9.10. Use Tukey's multiple comparison procedure to determine which, if any, of the mean maze scores differ for the three conditions of the maze test. Use $\alpha = .05$ and interpret your results. The data are reproduced here:

CONTROL	CONDITION 1	CONDITION 2
58	73	53
32	70	74
59	68	72
64	71	62
55	60	58
49	62	61

9.42 In Exercise 9.12, we used an analysis of variance to compare the mean increases in growth for plants receiving five different types of growth hormones. The data are reproduced in the table. Use Tukey's multiple comparison procedure to determine whether differences exist among the mean growths for the five types of hormones. Use $\alpha = .05$ and interpret your results.

		INOCULIN		
A	B	C	D	E
15	21	22	10	6
18	13	19	14	11
9	20	24	21	15
16	17	21	13	8

9.43 Refer to the weight and compliance experimental results in Exercise 9.36. Use Tukey's method of paired comparisons to test for differences among the ranked means. Test using $\alpha = .05$. What are the practical implications of the test results? How do they add to the results of the test in Exercise 9.36?

9.44 Refer to the weight and compliance experimental results in Exercise 9.37. Use Tukey's method of paired comparisons to test for differences among the ranked means. Test using $\alpha = .05$. What are the practical implications of the test results? How do they add to the results of the test in Exercise 9.37?

9.45 Although we cannot conduct an analysis of variance for the complete three-factor weight and compliance experiment (Exercises 9.36 and 9.37), we can use Tukey's method of paired comparisons to test for differences among the eight factor–level combination means. The three factors in the combined $2 \times 2 \times 2$ factorial experiment are sex of subject (male or female), sex of confederate (male or female), and weight of confederate (overweight or normal). The means and standard deviations of the samples of fifteen subjects assigned to each of the eight treatment groups are shown in the table (Steinberg & Birk, 1983). (Standard deviations are shown in parentheses.) As far as this analysis is concerned, we can ignore the factorial aspects of the experiment and view the data as eight independent random samples of fifteen observations each.

| | CONFEDERATES | | | |
| | OVERWEIGHT | | NORMAL WEIGHT | |
SUBJECTS	Male	Female	Male	Female
Male	32.0	24.27	34.67	36.67
	(28.65)	(12.94)	(36.52)	(14.48)
Female	21.33	20.33	38.31	27.80
	(15.86)	(16.42)	(20.04)	(16.21)

Note: $n = 15$ for each group.

a. Use the sample standard deviations to calculate the sums of squares of deviations within each of the eight samples.
b. Calculate SSE, the pooled sum of squares of deviations for the eight samples.
c. Calculate s^2. [Note: s^2 is based on $n_1 + n_2 + \cdots + n_8 = 8$ df.]
d. Use Tukey's method of paired comparisons to test and determine which, if any, of the eight treatment means are different. Test using $\alpha = .05$. Are any of the mean measures of compliance unusually low or unusually high? Explain the practical significance of your test results.

Summary

This chapter presents an extension of the independent sampling and paired difference experiments to allow for the comparison of two or more means. The **independent sampling design** (or **completely randomized design**) uses independent samples from each of k populations to compare their means. The **randomized block design,** like the paired difference design, uses relatively **homogeneous blocks of experimental units,** with each **treatment** randomly assigned to one experimental unit in each block to compare the treatment means.

For both designs, the comparison of population (or treatment) means is made by comparing the sample **variation** among the treatment means, as measured by the **mean**

square for treatments (MST), to the variation attributable to differences among experimental units, as measured by the **mean square for error (MSE).** If the ratio of MST to MSE is large, we conclude that a difference exists between the means of at least two of the k populations.

The analysis of variance, used to analyze data from the completely randomized and randomized block design, was also used to analyze data obtained from a two-factor factorial experiment. In order to use an analysis of variance for this experiment, we must take an equal number r of observations ($r \geq 2$) for all factor–level combinations. The most important information to be derived from the analysis of variance for a factorial experiment is whether the factors appear to interact. If interaction is present, we know that we cannot examine the effects of the factors on the response independently of each other. Rather, we must focus attention on the means for the factor–level combinations. The ranking and comparison of a group of treatment means can be accomplished by using a multiple comparison procedure. Tukey's method for making multiple comparisons was explained in Section 9.4.

Supplementary Exercises 9.46–9.79

[*Note: List the assumptions necessary to ensure the validity of the procedure you use to solve these problems.*]

Learning the Mechanics

9.46 An experiment utilizing a randomized block design was conducted to compare the mean responses for four treatments, A, B, C, and D. The treatments were randomly assigned to the four experimental units in each of five blocks. The data are shown in the table.

TREATMENT	BLOCK 1	2	3	4	5
A	8.6	7.5	8.7	9.8	7.4
B	7.3	6.3	7.3	8.4	6.3
C	9.1	8.3	9.0	9.9	8.2
D	9.3	8.2	9.2	10.0	8.4

a. Do the data provide sufficient evidence to indicate a difference among treatment means? Test using $\alpha = .05$.

b. Do the data provide sufficient evidence to indicate a difference among block means? Test using $\alpha = .05$.

c. Find a 99% confidence interval for $(\mu_B - \mu_D)$.

9.47 Independent random samples were selected from four normally distributed populations with common (but unknown) variance σ^2. The data are shown in the table.

a. Do the data provide sufficient evidence to indicate a difference among population means? Test using $\alpha = .05$.

SAMPLE 1	SAMPLE 2	SAMPLE 3	SAMPLE 4
8	6	9	12
10	9	10	13
9	8	8	10
10	8	11	11
11	7	12	11

b. Form a 90% confidence interval for $(\mu_1 - \mu_2)$.

c. Form a 95% confidence interval for μ_4.

Applying the Concepts

SUPERMARKET			
A	B	C	D
4.2	4.5	5.0	3.8
4.0	4.7	4.0	4.0
4.3	4.4	4.3	3.7
4.5	4.4	4.7	3.7

9.48 Higher wholesale beef prices over the past few years have resulted in the sale of ground beef with higher fat content in an attempt to keep retail prices down. Four different supermarket chains were chosen, and four 1-pound packages of ground beef were randomly selected from each. The fat content (in ounces) was measured for each package with the results shown in the table.

a. What experimental design does this represent?

b. Do the data provide evidence at the $\alpha = .05$ level of a difference in the mean fat content among the four supermarket chains?

c. Use a 90% confidence interval to estimate the mean fat content per pound at supermarket C.

9.49 A company is planning to market a new cereal with one of three possible package designs. To determine which has the most appeal, five stores are supplied with all three package designs. All packages are priced the same, so that if any design outsells the others, it will be due primarily to visual attractiveness. The cereal is on the market for several months, and the number of sales for each design at each store is recorded in the table.

STORE	DESIGN		
	A	B	C
1	101	111	100
2	98	102	105
3	121	120	114
4	132	140	127
5	95	98	94

a. Test to see whether there is evidence of a difference among the mean numbers of sales for the three designs.

b. Give statistical justification why (or why not) blocking was necessary in this experiment.

c. What assumptions were necessary for the validity of the test you conducted in part a?

9.50 It has been hypothesized that treatment, after casting, of a plastic used in optic lenses will improve wear. Four different treatments are to be tested. To determine whether any differences in mean wear exist among treatments, twenty-eight castings from a single formulation of the plastic were made and seven castings were randomly assigned to each of the treatments. Wear was determined by measuring the increase in "haze" after 200 cycles of abrasion (better wear being indicated by small increases). The results are given in the table.

| | TREATMENT | | |
A	B	C	D
9.16	11.95	11.47	11.35
13.29	15.15	9.54	8.73
12.07	14.75	11.26	10.00
11.97	14.79	13.66	9.75
13.31	15.48	11.18	11.71
12.32	13.47	15.03	12.45
11.78	13.06	14.86	12.38

a. Is there evidence of a difference in mean wear among the four treatments? Use $\alpha = .05$.

b. Find the approximate observed significance level for the test in part a, and interpret its value.

c. Estimate the difference in mean haze increase between treatments B and C. Use a 99% confidence interval.

d. Find a 90% confidence interval for the mean wear for lenses receiving treatment A.

9.51 A professor wishes to determine whether performance in an introductory statistics course depends on the student's year in college. Random samples of the final grade-point averages of students from each of the freshman, sophomore, junior, and senior years produce the following data summary:

	FRESHMAN	SOPHOMORE	JUNIOR	SENIOR
$\sum x$	52.55	168.25	760.50	363.10
$\sum x^2$	108.25	360.19	1,630.75	1,045.33
n	25	75	350	150

a. Do the data provide sufficient evidence to indicate that students tend to perform differently depending on their year in college? Use $\alpha = .05$.

b. Find a 95% confidence interval for the difference in mean GPA between seniors and freshmen.

| | DRUG | |
1	2	3
1.70	1.73	1.67
1.72	1.79	1.63
1.81	1.76	1.67

9.52 A drug company synthesized three new drugs that should alleviate pain due to ulcers. To determine whether the drugs will be absorbed by the stomach (and hence have a possibility of being effective) nine pigs were randomly assigned, three to each drug, to receive oral doses. After a given amount of time, the concentration of the drug in the stomach lining of each pig was determined. The data are shown in the table.

a. Do the data provide sufficient evidence to indicate a difference in the mean concentration for the three drugs after the fixed period of time? Use $\alpha = .01$.

b. Use a 95% confidence interval to estimate the difference in the mean concentration for drugs 1 and 2. Interpret the interval.

c. Find a 95% confidence interval for the mean concentration for drug 3. Interpret the interval.

9.53 In a nutrition experiment, an investigator studied the effects of different rations on the growth of young rats. Forty rats from the same inbred strain were divided at random into four groups of ten and used for the experiment. A different ration was fed to each group and, after a specified length of time, the increase in growth of each rat was measured (in grams). The data are shown in the table.

RATION A		RATION B		RATION C		RATION D	
10	6	13	9	12	10	15	21
8	6	15	10	16	12	13	18
12	9	14	8	13	10	15	20
11	5	13	10	11	9	10	19
9	6	17	8	15	9	12	22

a. Do the data provide sufficient evidence to indicate a difference among mean growths for rats receiving the different rations? Test using $\alpha = .05$.

b. Estimate the difference between mean growths for rations A and D. Use a 90% confidence interval.

c. Suppose that prior to conducting the experiment, you expected the mean increase in weight for rats receiving ration B to exceed 10 grams. Do the data support this theory?

d. Find a 90% confidence interval for the mean gain for rats on ration D.

9.54 A farmer wants to determine the effect of five different concentrations of lime on the pH (acidity) of the soil on a farm. Fifteen soil samples are to be used in the experiment, five from each of three different locations. The five soil samples from each location are then randomly assigned to the five concentrations of lime, and 1 week after the lime is applied the pH of the soil is measured. The data are shown in the table.

| LOCATION | LIME CONCENTRATION | | | | |
	0	1	2	3	4
I	3.2	3.6	3.9	4.0	4.1
II	3.6	3.7	4.2	4.3	4.3
III	3.5	3.9	4.0	3.9	4.2

a. What experimental design was used here?

b. Do the data provide sufficient evidence to indicate that the five concentrations of lime have different mean soil pH levels? Use $\alpha = .05$.

c. Is there evidence of a difference in mean soil pH levels among locations? Test using $\alpha = .05$.

9.55 One important consideration in determining which location is best for a new retail business is the amount of traffic that passes the location each business day. Counters are placed at each of four locations on the five weekdays, and the number of cars passing each location is recorded in the table.

DAY	LOCATION			
	I	II	III	IV
1	453	482	444	395
2	500	605	505	490
3	392	400	383	390
4	441	450	429	405
5	427	431	440	430

a. What type of design does this represent?
b. Is there evidence of a difference in the mean number of cars per day at the four locations?
c. Estimate the difference between the mean numbers of cars that pass locations I and III each weekday.

9.56 Several companies are experimenting with the concept of paying production workers (generally paid by the hour) on a salary basis. It is believed that absenteeism and tardiness will increase under this plan, yet some companies feel that the working environment and overall productivity will improve. Fifty production workers under the salary plan are monitored at company A, and likewise fifty under the hourly plan at company B. The number of work-hours missed due to tardiness or absenteeism over a 1-year period is recorded for each worker. The results are partially summarized in the table.

SOURCE	df	SS	MS	F
Company		3,237.2		
Error		16,167.7		
Total	99			

a. Fill in the missing information.
b. Is there evidence at the $\alpha = .05$ level of significance that the mean number of hours missed differs for employees of the two companies?
c. Is there sufficient information given to form a confidence interval for the difference between the mean number of hours missed at the two companies?

9.57 Due to increased energy shortages and costs, utility companies are stressing ways in which home and apartment utility bills can be cut. One utility company reached an agreement with the owner of a new apartment complex to conduct a test of energy-saving plans for apartments. The tests were to be conducted before the apartments were rented. Four apartments were chosen that were identical in size, amount of shade, and direction faced. Four plans were to be tested, one on each apartment. The thermostat was set at 75°F in each apartment, and the monthly utility bill was recorded for each of the three summer months. The results are listed in the table.

MONTH	TREATMENT			
	1	2	3	4
June	$74.44	$68.75	$71.34	$65.47
July	86.96	73.47	83.62	72.33
August	82.00	71.23	79.98	70.87

Treatment 1: No insulation; no awnings
Treatment 2: Insulation in walls and ceilings; no awnings
Treatment 3: No insulation; awnings for windows
Treatment 4: Insulation; awnings for windows

a. Is there evidence of a difference among the mean monthly utility bills for the four treatments? Use $\alpha = .01$.

b. Is there evidence that blocking is important, i.e., that the mean bills differ for at least 2 of the 3 months? Use $\alpha = .05$.

c. To see whether awnings on the windows reduce costs, place a 95% confidence interval on the difference between the means of treatments 2 and 4.

9.58 A chemist has run an experiment to study the effect of four treatments on the glass transition temperature (in degrees Kelvin) of a certain polymer compound. Raw material used to make this polymer is bought in small batches. The material is thought to be fairly uniform within a batch but variable between batches. Therefore, each treatment was run on samples from each batch with the results listed in the table.

BATCH	TREATMENT			
	I	II	III	IV
1	576	584	562	543
2	515	563	522	536
3	562	555	550	530

a. Do the data provide sufficient evidence to indicate a difference in mean temperature among the four treatments? Use $\alpha = .05$.

b. Is there sufficient evidence to indicate a difference in mean temperature among the three batches? Use $\alpha = .05$.

c. If the experiment were to be conducted again in the future, would you recommend any changes in its design?

DIET		
I	II	III
500	505	825
620	765	870
685	730	695
440	570	740
645	760	850

9.59 An experimenter believes that weight gain of chickens can be increased by adding small amounts of thyroxin to their diet. To test this theory, fifteen chickens are divided into three groups of five chickens each. Each group is fed a regular diet to which differing amounts of thyroxin are added. The first diet has no thyroxin, the second diet contains 2 milligrams thyroxin per kilogram feed, and the third diet contains 5 milligrams thyroxin per kilogram feed. After 8 weeks on the rations, the gain in weight for each chicken is measured (in grams). The data are shown in the table.

a. Do the data provide sufficient evidence to conclude there is a difference in mean weight gain among the three feeds? Use $\alpha = .05$.

b. Estimate the difference in mean weight gain between diets II and III. Use a 95% confidence interval.

9.60 From time to time, one branch office of a company must make shipments to a branch office in another state. There are three package delivery services between the two cities where the branch offices are located. Since the price structures for the three delivery services are quite similar, the company wants to compare the delivery times. The company plans to make several different types of shipments to its branch office. To compare the carriers, each shipment will be sent in triplicate, one with each carrier. The results listed in the table are the delivery times in hours.

SHIPMENT	CARRIER I	II	III
1	15.2	16.9	17.1
2	14.3	16.4	16.1
3	14.7	15.9	15.7
4	15.1	16.7	17.0
5	14.0	15.6	15.5

a. Is there evidence of a difference in mean delivery times among the three companies? Use $\alpha = .05$.

b. Use a 99% confidence interval to estimate the difference between the mean delivery times for carriers I and II.

c. What assumptions are necessary for the validity of the procedures you used in parts a and b?

9.61 To compare the preferences of technicians for three brands of calculators, each technician was required to perform an identical series of calculations on each of the three calculators, A, B, and C. To avoid the possibility of fatigue, a suitable time period separated each set of calculations and the calculators were used in random order by each technician. A preference rating, based on a 0–100 scale, was recorded for each machine–technician combination. These data are shown in the table.

TECHNICIAN	CALCULATOR BRAND A	B	C
1	85	90	95
2	70	70	75
3	65	60	80

a. Do the data provide sufficient evidence to indicate a difference in technician preference among the three brands? Use $\alpha = .05$.

b. Why did the experimenter have each technician test all three calculators? Why not randomly assign three different technicians to each calculator?

9.62 An experiment was conducted to compare the yields of orange juice for six different juice extractors. Because of the possibility of variation in the amount of juice per orange from one truckload of oranges to another, equal weights of oranges from a single truckload were assigned to each extractor and this process was repeated for fifteen loads. The amount of juice recorded for each extractor for each truckload produced the following sums of squares:

SOURCE	df	SS	MS	F
Extractor		84.71		
Truckload		159.29		
Error		95.33		
Total		339.33		

a. Complete the ANOVA table.

b. Do the data provide sufficient evidence to indicate a difference in mean amount of juice extracted by the six extractors? Use $\alpha = .05$.

9.63 An experiment was conducted to compare corn yield per acre for four different fertilizer applications. Sixteen plots of equal size were prepared for planting and four each were assigned to receive the fertilizer treatments. Unfortunately, the crops in two of the plots were damaged during the growing season and were removed from the experiment. The yields of corn in bushels per plot are shown in the table.

FERTILIZER APPLICATION (POUNDS PER PLOT)			
5	10	15	20
57	62	65	43
51	59	72	55
45	75	67	49
	69	78	

a. Do the data suggest a difference in mean number of bushels of corn produced among the four fertilizer treatments?

b. Use a 95% confidence interval to estimate the difference between the mean number of bushels produced by plots that receive 5 and 20 pounds of fertilizer.

c. Use a 95% confidence interval to estimate the mean number of bushels produced by plots that receive 15 pounds of fertilizer.

9.64 A fast-food chain expects mean gross sales of $800,000 per year per franchise. A random sample of the chain's stores was selected in Los Angeles, Miami, and Chicago, and the results are given in the table.

Gross Sales
(in Units of $100,000)

MIAMI	LOS ANGELES	CHICAGO
8.7	8.7	7.8
7.4	8.0	7.6
7.9	9.0	6.9
8.0	8.3	5.7
8.5	9.0	
7.9		

a. Is there evidence of a difference in mean gross sales among these three cities? Use $\alpha = .01$.

b. Form a 90% confidence interval for the mean gross sales of the Miami stores in this fast-food chain. Interpret this interval.

9.65 Psychologists have studied the effect of the working environment or surroundings on the quality and quantity of work done. Many businesses have music piped into the work area to improve the environment. An experiment is performed to determine which type of music is best suited for a particular company. Three types of music—country, rock, and classical—are tried, each on four randomly selected workdays. The productivity is measured by recording the number of items produced on each of the days. The results are shown in the table. Using this information, can we conclude that the mean numbers of items produced differ for at least two of the three types of music?

COUNTRY	ROCK	CLASSICAL
857	791	824
801	753	847
795	781	881
842	776	865

DIVISION		
I	II	III
11.1	11.4	10.4
10.9	11.6	10.3
10.8	11.0	10.5

9.66 A corporation manages a very large number of stores, which it classifies into three geographic divisions. Three stores are randomly selected from each division, and a study is made to determine the mean inflation rate for the items in inventory. The inflation rate is recorded in the table as a percentage increase in price over a year's time.

a. Is there sufficient evidence to indicate a difference in the mean inflation rates among stores in different divisions?

b. In division I, the numbers 11.1, 10.9, and 10.8 represent a sample from what population?

9.67 One indicator of employee morale is the length of time employees stay with a company. A large corporation has three factories located in similar areas of the country. While the corporation management attempts to maintain uniformity in management, working conditions, employee relations, etc., at its various factories, it realizes that differences may exist between the various factories. To study this phenomenon, employee records are randomly selected at each of the three factories and the length of employee service with the company is recorded. The following is a summary of the data:

FACTORY	1	2	3
NUMBER IN SAMPLE	15	21	17
SST = 421.74;	SSE = 3574.06		

Is there evidence of a difference in mean length of service at the three factories? Use $\alpha = .05$.

9.68 England has experimented with different 40-hour work weeks to maximize production and minimize expenses. A factory tested a 5-day week (8 hours per day), a

4-day week (10 hours per day), and a $3\frac{1}{3}$-day week (12 hours per day). The weekly production results are shown in the table (in thousands of dollars worth of items produced).

8-HOUR DAY	10-HOUR DAY	12-HOUR DAY
87	75	95
96	82	76
75	90	87
90	80	82
72	73	65
86		

a. What experimental design was employed here?

b. Construct an ANOVA summary table for this experiment.

c. Is there evidence of a difference in the mean productivities for the three lengths of workdays?

d. Form a 90% confidence interval for the mean weekly productivity when 12-hour workdays are used.

9.69 Three anticoagulant drugs are studied to compare their effectiveness in dissolving blood clots. Each of five subjects receives the drugs at equally spaced time intervals and in random order. Time periods between drug applications permit a drug to be passed out of a subject's body before the subject receives the next drug. After each drug is in the bloodstream, the length of time required for a cut of specified size to stop bleeding is recorded. The results are shown in the table.

Clotting Time (Seconds)

PERSON	DRUG A	B	C
1	127.5	129.0	135.5
2	130.6	129.1	138.0
3	118.3	111.7	110.1
4	155.5	144.3	162.3
5	180.7	174.4	181.8

a. Do the data provide evidence to indicate a difference in mean clotting times among the three drugs? Test using $\alpha = .01$.

b. Find the approximate observed significance level for the test in part a, and interpret its value.

c. Was blocking effective in reducing the variation within the data? That is, do the data support the contention that mean clotting time varies from person to person?

d. Find a 90% confidence interval for the difference between mean clotting times for drugs B and C. Interpret this interval.

9.70 Mileage tests are performed to compare three different brands of regular gas. Four different automobiles are used in the experiment, and each brand of gas is used in each car until the mileage is determined. The results (in miles per gallon) are shown in the table.

BRAND	AUTOMOBILE 1	2	3	4
A	20.2	18.7	19.7	17.9
B	19.7	19.0	20.3	19.0
C	18.3	18.5	17.9	21.1

a. Is there evidence of a difference in the mean mileage rating among the three brands of gasoline? Use $\alpha = .05$.

b. Construct the ANOVA summary table for this experiment.

c. Is there evidence of a difference in the mean mileage for the four models, i.e., is blocking important in this type of experiment? Use $\alpha = .05$.

d. Form a 99% confidence interval for the difference between the mean mileage ratings of brands B and C.

METHOD A	B	C
8.2	7.9	7.1
7.1	8.1	7.4
7.8	8.3	6.9
8.9	8.5	6.8
8.8	7.6	
	8.5	

9.71 To reduce the time spent in transferring materials from one location to another, three methods have been devised. With no previous information available on the effectiveness of these three approaches, a study is performed. Each approach is tried several times, and the amount of time to completion (in hours) is recorded in the table.

a. What experimental design was used?

b. Is there evidence that the mean time to completion of the task differs for at least two of the three methods? Use $\alpha = .01$.

c. Form a 95% confidence interval for the mean time to completion for method B.

9.72 Methods of displaying goods can have an effect on their sales. The manager of a large produce market would like to try three different displays for a certain fruit. The locations for the three displays are chosen in a way that will make them equally accessible to customers. The three displays will be set up for five 1 week periods. Between each of these periods will be a 2-week period when a standard display is used. For each of the five experimental weeks, the sales (in dollars) of the fruit from each display are determined with the results shown in the table.

PERIOD	DISPLAY A	B	C
1	$125	$153	$108
2	137	135	113
3	110	122	105
4	119	133	112
5	141	144	136

a. Is there evidence of a difference in mean sales for the three types of display? Use $\alpha = .05$.

b. Do the data indicate that the use of weeks as blocks was necessary? Test at the $\alpha = .05$ level.

c. Construct the ANOVA summary table for this experiment.

d. Use a 95% confidence interval to estimate the difference between the mean sales for displays A and C.

PLAN			
1	2	3	4
27	25	34	30
25	28	29	33
29	30	32	31
26	27	31	
	24	36	

9.73 In hopes of attracting more riders, a city transit company plans to have express bus service from a suburban terminal to the downtown business district. These buses will travel along a major city street where there are numerous traffic lights that affect travel time. The city decides to perform a study of the effect of four different plans (a special bus lane, traffic signal progression, etc.) on the travel times for the buses. Travel times (in minutes) are measured for several weekdays during a morning rush-hour trip while each plan is in effect. The results are recorded in the table.

a. What experimental design was employed?

b. Is there evidence of a difference in the mean travel times for the four plans? Use $\alpha = .01$.

c. Form a 95% confidence interval for the difference between the mean travel times for plan 1 (express lane) and plan 3 (a control—no special travel arrangements).

9.74 An agricultural research unit wished to study the yields of three strains of wheat, A, B, and C. Because of the direction of drainage, it was suspected that the field used for the experiment varied substantially in fertility along the north–south line. Consequently, to reduce data variation the three strains of wheat were planted in plots laid out in east–west blocks. One plot of each type of wheat was assigned to each block, and the strains were assigned in random order within each block. The layout of the design and the data on wheat yields for each strain within each block (in bushels) are shown here:

A	B	C	A	B
53	62	47	56	58

B	C	A	B	C
60	44	49	55	45

C	A	B	C	A
49	57	61	43	51

a. Do the data provide sufficient evidence to indicate a difference in mean yields for the three strains of wheat? Test using $\alpha = .10$.

b. Find a 90% confidence interval for the difference in mean yields between strains A and B. Interpret this interval.

c. Can you use these data to obtain a valid confidence interval for the mean yield for a particular strain? Explain.

9.75 The concentration of a catalyst used in producing grouted sand is thought to affect its strength. An experiment designed to investigate the effects of three different concentrations of the catalyst utilized five test specimens of grout per concentration. The strength of a grouted sand was determined by placing the test specimen in a press and applying pressure until the specimen broke. The pressure required to break the specimens, expressed in pounds per square inch, is shown in the table.

CONCENTRATION OF CATALYST		
35%	40%	45%
5.9	6.8	9.9
8.1	7.9	9.0
5.6	8.4	8.6
6.3	9.3	7.9
7.7	8.2	8.7

a. Do the data provide sufficient evidence to indicate a difference in mean strength of the grouted sand among the three concentrations of catalyst? Test using $\alpha = .05$.

b. Find a 95% confidence interval for the difference in mean strength for specimens produced with a 35% concentration of catalyst versus those containing a 45% concentration of catalyst.

9.76 Refer to Case Study 9.2. Since there was no evidence to indicate any difference in mathematics scores between the two colleges, Clute combined the data for the two colleges and analyzed the portion of the mathematics examination scores that pertained to the low-level items on the test.

a. The sample sizes, mean scores, and standard deviations (in parentheses) for the six method–anxiety combinations are shown in the table. Examine the table and speculate on the effects of teaching method and anxiety level on the mean score. Does it appear that teaching method and anxiety level interact?

Means and Standard Deviations on the Low-Level Items by Instructional Method and Level of Anxiety

METHOD	n	MATHEMATICS ANXIETY LEVEL LOW	n	MEDIUM	n	HIGH	TOTAL
Discovery	14	218.79 (20.81)	14	183.43 (27.17)	13	146.85 (36.93)	183.90 (40.77)
Expository	14	183.43 (52.27)	13	182.46 (41.64)	13	185.92 (33.72)	183.92 (42.38)
Total		201.11 (43.99)		182.96 (34.20)		166.38 (39.97)	183.91 (41.31)

b. As you can see from the table in part a, the sample sizes for the six factor–level combinations are not equal. Therefore, you cannot use the formulas given in Section 9.3 to calculate the sums of squares of deviations for the various sources of variation. Clute's ANOVA table is shown at the top of the next page. Describe the practical implications of the test results, and explain how they agree or disagree with your speculations in part a.

Analysis of Variance for
Low-Level Items

SOURCE	df	MS	F
Method (M)	1	17	.01
Anxiety (A)	2	8,146	6.02[a]
M × A	2	9,341	6.90[a]
Residual	75	1,354	

[a] $p < .01$.

9.77 Refer to the analysis of the low-level test items in Exercise 9.76.

a. Use the mean squares and degrees of freedom in the ANOVA table of part b to calculate the missing sums of squares for each source of variation.

b. We explained in Exercise 9.76 that we cannot use the ANOVA formulas in Section 9.3 to calculate the sums of squares for Clute's ANOVA table. We can, however, check her value for SSE. Use the sample standard deviations and sample sizes for the six factor–level combinations to calculate the sums of squares of deviations within each of the six samples. Then pool the six sums of squares of deviations found in part b. This result will give you the value for SSE for the analysis of variance. Compare your answer with the value for SSE shown in Clute's ANOVA table (Exercise 9.76, part b).

c. Why can you not use the sums of squares formulas given in Section 9.3 to calculate the sums of squares for the analysis of variance of Clute's data?

9.78 Refer to Case Study 9.2. Since there was no evidence to indicate any difference in mathematics scores between the two colleges, Clute combined the data for the two colleges and analyzed the portion of the mathematics examination scores that pertained to the high-level items on the test.

a. The sample sizes, mean scores, and standard deviations (in parentheses) for the six method–anxiety combinations are shown in the table. Examine the table and speculate on the effects of teaching method and anxiety level on the mean score. Does it appear that teaching method and anxiety level interact?

Means and Standard
Deviations on the
High-Level Items by
Instructional Method and
Level of Anxiety

METHOD		MATHEMATICS ANXIETY LEVEL					TOTAL
	n	LOW	n	MEDIUM	n	HIGH	
Discovery	14	134.43	14	109.14	13	78.08	107.93
		(23.06)		(12.88)		(28.67)	(31.77)
Expository		104.71		91.69	13	74.77	90.75
		(31.42)		(47.92)		(37.39)	(40.26)
Total		119.57		100.74		76.42	99.44
		(30.99)		(34.95)		(32.69)	(37.01)

Note: Maximum score = 240.

b. As you can see from the table in part a, the sample sizes for the six factor–level combinations are not equal. Therefore, you cannot use the formulas given in Section 9.3 to calculate the sums of squares of deviations for the various sources of variation. Clute's ANOVA table is shown below. Describe the practical implications of the test results, and explain how they agree or disagree with your speculations in part a.

Analysis of Variance for
High-Level Items

SOURCE	df	MS	F
Method (M)	1	5,724	5.65[a]
Anxiety (A)	2	12,573	12.39[b]
M × A	2	1,176	1.16
Residual	75	1,015	

[a] $p < .05$.
[b] $p < .01$.

9.79 Refer to the analysis of the mathematics achievement test scores for low-level items in Exercise 9.76. There we found evidence of an interaction between teaching method and a student's anxiety level as they affect a student's achievement in a mathematics course. The next step in the analysis of the test scores might be a pairwise comparison of the mean scores for the six factor–level combinations. Could we use Tukey's method for paired comparisons? Explain.

On Your Own . . .

Due to ever-increasing food costs, consumers are becoming more discerning in their choice of supermarkets. It is usually more convenient to shop at only one market, as opposed to buying different items at different markets. Thus, it would be useful to compare the mean food expenditure for a market basket of food items from store to store. Since there is a great deal of variability in the prices of products sold at any supermarket, we will consider an experiment that blocks on products.

Choose three (or more) supermarkets in your area that you want to compare. Then choose approximately ten (or more) food products you typically purchase. For each food item, record the price each store charges in the following manner:

FOOD ITEM 1	FOOD ITEM 2	...	FOOD ITEM 10
Price store 1	Price store 1	...	Price store 1
Price store 2	Price store 2	...	Price store 2
Price store 3	Price store 3	....	Price store 3

Use the data you obtain to test

H_0:   Mean expenditures at the stores are the same

H_a: Mean expenditures for at least two of the stores are different

Also test to determine whether blocking on food items is advisable in this kind of experiment. Fully interpret the results of your analysis.

References   Clute, P. S. "Mathematics anxiety: instructional method and achievement in a survey course in college mathematics." *Journal for Research in Mathematics Education,* 1984, *15.*

Heyward, V., & McCreary, L. "Analysis of the static strength and relative endurance of women athletes." *Research Quarterly,* 1977, *48,* 703–709.

Kovacs, T. G., & Leduc, G. "Sublethal toxicity of cyanide to rainbow trout (*Salmo gairdneri*) at different temperatures." *Canadian Journal of Fisheries and Aquatic Sciences,* 1982, *39.*

McMurray, R. G., Wilson, J. R., & Kitchell, B. S. "The effects of fructose and glucose on high-endurance performance." *Research Quarterly for Exercise and Sport,* 1983, *54.*

Mendenhall, W. *Introduction to linear models and the design and analysis of experiments.* Belmont, Ca.: Wadsworth, 1968. Chapter 8.

Phillips, M. D., & Pepper, R. L. "Shipboard firefighting performance of females and males." *Human Factors,* 1982, *24.*

Snedecor, G. W., & Cochran, W. G. *Statistical methods.* 7th ed. Ames: Iowa State University Press, 1980.

Steele, R. G. D., & Torrie, J. H. *Principles and procedures of statistics: a biometrical approach.* 2nd ed. New York: McGraw-Hill, 1980.

Steinberg, C. L., & Birk, J. M. *Journal of General Psychology,* 1983, *109.*

Winer, B. J. *Statistical principles in experimental design.* 2nd ed. New York: McGraw-Hill, 1971.

CHAPTER 10

Nonparametric Statistics

Where We've Been . . .

Chapters 7, 8, and 9 presented techniques for making inferences about the mean of a single population and for comparing the means of two or more populations. Most of the techniques discussed in those chapters were based on the assumption that the sampled populations have probability distributions that are approximately normal with equal variances. But how can you analyze data that evolve from populations that do not satisfy these assumptions? How can you make comparisons between populations when you cannot assign specific numerical values to your observations?

Where We're Going . . .

In this chapter, we present statistical techniques for comparing two or more populations that are based on an ordering of the sample measurements according to their relative magnitudes. These techniques, which require fewer or less stringent assumptions concerning the nature of the probability distributions of the populations, are called *nonparametric statistical methods*. The statistical tests presented in this chapter apply to the same experimental designs as those covered in the introduction to an analysis of variance (Chapter 9). Thus, we will present nonparametric statistical techniques for comparing two or more populations using either a completely randomized or a randomized block design.

Contents

The *t*- and *F*-tests for comparing the means of two or more populations (Chapters 8 and 9) are unsuitable for analyzing some types of experimental data. For these tests to be appropriate, we assumed that the random variables being measured had normal probability distributions with equal variances. Yet in practice, the observations from one population may exhibit much greater variability than those from another, or the probability distributions may be decidedly nonnormal. For example, the distribution might be very flat, peaked, or strongly skewed to the right or left. When any of the assumptions required for the *t*- and *F*-tests are seriously violated, the computed *t*- and *F*-statistics may not follow the standard *t*- and *F*-distributions. In this case, the tabulated values of *t* and *F* (Tables V, VII, VIII, IX, and X in the Appendix) are not applicable, the correct value of α for the test is unknown, and the *t*- and *F*-tests are of dubious value.

Data that often occur in practical situations for which the *t*- and *F*-tests are inappropriate are responses that are not susceptible to a meaningful numerical measurement but can be *ranked in order of magnitude*. For example, if we want to compare the teaching ability of two college instructors based on subjective evaluations of students, we cannot assign a number that exactly measures the teaching ability of a single instructor. But we can compare two instructors by ranking them according to a subjective evaluation of their teaching abilities. If instructors A and B are evaluated by each of ten students, we have the standard problem of comparing the probability distributions for two populations of ratings—one for instructor A and one for B—but the *t*-test of Chapter 8 would be inappropriate, because the only data that can be recorded are the preference statements of the ten students, i.e., each student decides that either A is better than B or vice versa.

Consider another example of this type of data. Most firms that plan to market a new product nationally first test it in a few cities or regions to determine its acceptability. For a food product this may entail taste tests in which consumers rank the new product in order of preference with respect to one or more currently popular brands. A consumer probably has a preference for each product, but the strength of the preference is difficult, if not impossible, to measure. Consequently, the best we can do is to have each consumer examine the new product, along with a few established products, and rank them according to preference: 1 for the most preferred, 2 for second, etc.

The **nonparametric** counterparts of the *t*- and *F*-tests compare the probability distributions of the sampled populations rather than specific parameters of these populations (such as the means or variances). For example, nonparametric tests can be used to compare the probability distribution of the strengths of preferences for a new product to the probability distributions of the strengths of preferences for the currently popular brands. If it can be inferred that the distribution of the new product lies above (to the right of) the others (see Figure 10.1), the implication is that the new product tends to be more preferred than the currently popular products. Such an inference might lead to a decision to market the product nationally.

Many nonparametric methods use the **relative ranks** of the sample observations, rather than their actual numerical values. These tests are particularly valuable when we are unable to obtain numerical measurements of some phenomena but are able to rank them in comparison to each other. Statistics based on ranks of measurements

Figure 10.1 Probability Distributions of Strengths of Preference Measurements (New Product Is Preferred)

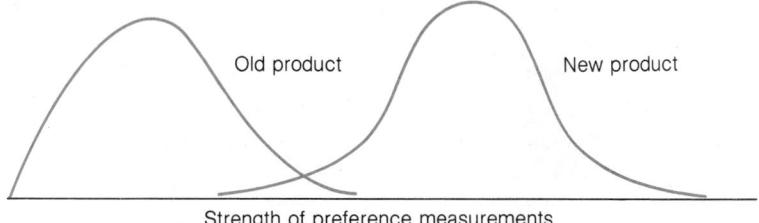

Old product   New product

Strength of preference measurements

are called **rank statistics.** In Sections 10.1 and 10.3, we will present rank statistics for comparing two probability distributions using independent samples. In Sections 10.2 and 10.4, the matched-pairs and randomized block designs are used to make nonparametric comparisons of populations. Finally, in Section 10.5, we present a nonparametric measure of correlation between two variables—**Spearman's rank correlation coefficient.**

10.1 Comparing Two Populations: Wilcoxon Rank Sum Test for Independent Samples

Suppose two independent random samples are to be used to compare two populations and the *t*-test of Chapter 8 is inappropriate for making the comparison. We may be unwilling to make assumptions about the form of the underlying population probability distributions or we may be unable to obtain exact values of the sample measurements. For either of these situations, if the data can be ranked in order of magnitude then the **Wilcoxon rank sum test** (developed by Frank Wilcoxon) can be used to test the hypothesis that the probability distributions associated with the two populations are equivalent.

Suppose an experimental psychologist wants to compare reaction times for adult males under the influence of drug A to those under the influence of drug B. Experience has shown that populations of reaction time measurements often possess probability distributions that are skewed to the right, as shown in Figure 10.2. Consequently, a *t*-test should not be used to compare the mean reaction times for the two drugs because the normality assumption that is required for the *t*-test may not be valid.

Suppose the psychologist randomly assigns seven subjects to each of two groups, one group to receive drug A and the other to receive drug B. The reaction time for

Figure 10.2 Typical Probability Distribution of Reaction Times

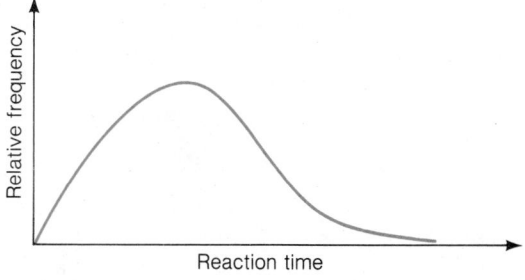

Reaction time

each subject is measured at the completion of the experiment. These data (with the exception of the measurement for one subject in group A who was eliminated from the experiment for personal reasons) are shown in Table 10.1.

Table 10.1

Reaction Times of Subjects Under the Influence of Drug A or B

| DRUG A | | DRUG B | |
Reaction time (seconds)	Rank	Reaction time (seconds)	Rank
1.96	4	2.11	6
2.24	.7	2.43	9
1.71	2	2.07	5
2.41	8	2.71	11
1.62	1	2.50	10
1.93	3	2.84	12
		2.88	13

The population of reaction times for either of the drugs, say drug A, is that which could conceptually be obtained by giving drug A to all adult males. To compare the probability distributions for populations A and B, *we first rank the sample observations as though they were all drawn from the same population.* That is, we pool the measurements from both samples and then rank the measurements from the smallest (a rank of 1) to the largest (a rank of 13). The results of this ranking process are shown in Table 10.1.

If the two populations were identical, we would expect the ranks to be *randomly mixed* between the two samples. If, on the other hand, one population tends to have longer reaction times than the other, we would expect the larger ranks to be mostly in one sample and the smaller ranks mostly in the other. Thus, the test statistic for the Wilcoxon test is based on the totals of the ranks for each of the two samples—that is, on the **rank sums.** When the sample sizes are equal, for example, the greater the difference in the rank sums, the greater will be the weight of evidence to indicate a difference between the probability distributions for populations A and B. In the reaction times example, we denote the rank sum for drug A by T_A and that for drug B by T_B. Then

$$T_A = 4 + 7 + 2 + 8 + 1 + 3 = 25$$
$$T_B = 6 + 9 + 5 + 11 + 10 + 12 + 13 = 66$$

The sum of T_A and T_B is always equal to $n(n + 1)/2$, where $n = n_1 + n_2$. So, for this example, $n_1 = 6$, $n_2 = 7$, and

$$T_A + T_B = \frac{13(13 + 1)}{2} = 91$$

Since $T_A + T_B$ is fixed, a small value for T_A implies a large value for T_B (and vice versa) and a large difference between T_A and T_B. Therefore, the smaller the value of one of the rank sums, the greater the evidence to indicate that the samples were selected from different populations.

Figure 10.3 Reproduction of Part of Table XI in the Appendix

a. $\alpha = .025$ one-tailed; $\alpha = .05$ two-tailed

n_2 \\ n_1	3		4		5		6		7		8		9		10	
	T_L	T_U	T_L	T_U	T_L	T_U	T_L	T_U	T_L	T_U	T_L	T_U	T_L	T_U	T_L	T_U
3	5	16	6	18	6	21	7	23	7	26	8	28	8	31	9	33
4	6	18	11	25	12	28	12	32	13	35	14	38	15	41	16	44
5	6	21	12	28	18	37	19	41	20	45	21	49	22	53	24	56
6	7	23	12	32	19	41	26	52	28	56	29	61	31	65	32	70
7	7	26	13	35	20	45	28	56	37	68	39	73	41	78	43	83
8	8	28	14	38	21	49	29	61	39	73	49	87	51	93	54	98
9	8	31	15	41	22	53	31	65	41	78	51	93	63	108	66	114
10	9	33	16	44	24	56	32	70	43	83	54	98	66	114	79	131

The test statistic for this test is the rank sum for the smaller sample or, in the case where $n_1 = n_2$, either rank sum can be used. Values that locate the rejection region for this rank sum are given in Table XI in the Appendix. A partial reproduction of this table is shown in Figure 10.3. The columns of the table represent n_1, the first sample size, and the rows represent n_2, the second sample size. *The T_L and T_U entries in the table are the boundaries of the lower and upper regions, respectively, for the rank sum associated with the sample that has fewer measurements.* If the sample sizes n_1 and n_2 are the same, either rank sum may be used as the test statistic. To illustrate, suppose $n_1 = 8$ and $n_2 = 10$. For a two-tailed test with $\alpha = .05$, we consult part a of the table and find that the null hypothesis will be rejected if the rank sum of sample 1 (the sample with fewer measurements), T, is less than or equal to $T_L = 54$ *or* greater than or equal to $T_U = 98$. The Wilcoxon rank sum test is summarized in the box on page 488.

Example 10.1

Do the data given in Table 10.1 provide sufficient evidence to indicate a shift in the probability distributions for drugs A and B, that is, that the probability distribution corresponding to drug A lies either to the right or left of the probability distribution corresponding to drug B? Test at the .05 level of significance.

Solution

H_0: The two populations of reaction times corresponding to drug A and drug B have the same probability distribution

H_a: The probability distribution for drug A is shifted to the right or left of the probability distribution corresponding to drug B

Test statistic: Since drug A has fewer subjects than drug B, the test statistic is T_A, the rank sum of drug A's reaction times.

Rejection region: Since the test is two-sided, we consult part a of Table XI for the rejection region corresponding to $\alpha = .05$. We will reject H_0 for $T_A \leq T_L$ or $T_A \geq T_U$. Thus, we will reject H_0 if $T_A \leq 28$ or $T_A \geq 56$.

Since T_A, the rank sum of drug A's reaction times in Table 10.1, is 25, it is in the rejection region (see Figure 10.4, page 489).[*]

[*] Figure 10.4 depicts only one side of the two-sided alternative hypothesis. The other would show distribution A shifted to the right of distribution B.

Wilcoxon Rank Sum Test: Independent Samples*

One-Tailed Test

H_0: Two sampled populations have identical probability distributions

H_a: The probability distribution for population A is shifted to the right of that for B

Test statistic:

The rank sum T associated with the sample with fewer measurements (if sample sizes are equal, either rank sum can be used)

Rejection region:

Assuming the smaller sample size is associated with distribution A (if sample sizes are equal, we use the rank sum T_A), we reject the null hypothesis if

$$T_A \geq T_U$$

where T_U is the upper value given by Table XI in the Appendix for the chosen *one-tailed* α-value.

Two-Tailed Test

H_0: Two sampled populations have identical probability distributions

H_a: The probability distribution for population A is shifted to the left *or* to the right of that for B

Test statistic:

The rank sum T associated with the sample with fewer measurements (if the sample sizes are equal, either rank sum can be used)

Rejection region:

$T \leq T_L$ or $T \geq T_U$, where T_L is the lower value given by Table XI in the Appendix for the chosen *two-tailed* α-value and T_U is the upper value from Table XI

[*Note:* If the one-sided alternative is that the probability distribution for A is shifted to the *left* of B (and T_A is the test statistic), we reject the null hypothesis if $T_A \leq T_L$.]

Assumptions: 1.   The two samples are random and independent.
2.   The observations obtained can be ranked in order of magnitude. [*Note:* No assumptions have to be made about the shape of the population probability distributions.]

* Another statistic used for comparing two populations based on independent random samples is the ***Mann–Whitney U-statistic.*** The U-statistic is a simple function of the rank sums. It can be shown that the Wilcoxon rank sum test and the Mann–Whitney U-test are equivalent.

Figure 10.4 Alternative Hypothesis and Rejection Region for Example 10.1

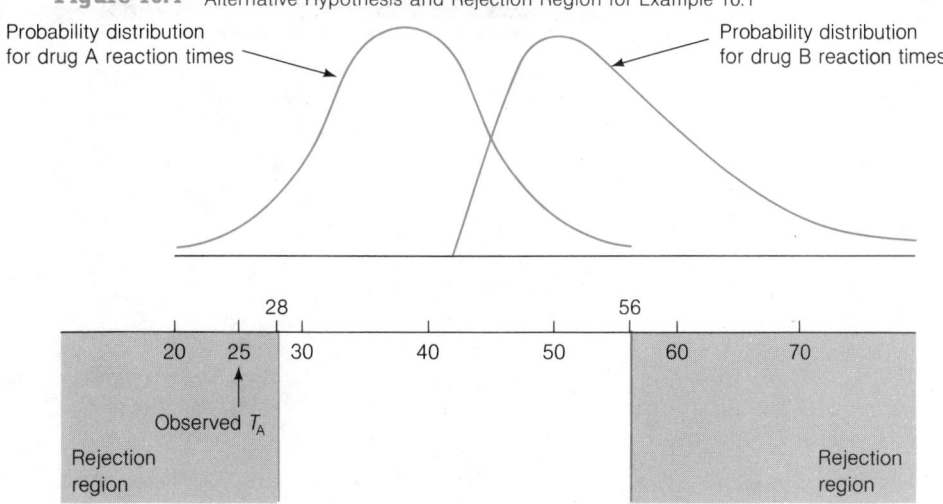

We can conclude that the probability distributions for drugs A and B are not identical. In fact, it appears that drug B tends to be associated with reaction times that are larger than those associated with drug A (because T_A fell in the lower tail of the rejection region).

■

When you apply the Wilcoxon rank sum test in a practical situation, you may encounter one or more ties in the observations. A tie occurs when two of the sample observations are equal. The Wilcoxon rank sum test is still valid if the number of ties is small in comparison with the number of sample measurements—*and if you assign to each tied observation the average of the ranks the two observations would have received if the observations had not been tied.* For example, suppose the fourth and fifth smallest observations were tied. Since these observations would have received the ranks 4 and 5, you should assign the average of these ranks, 4.5, to both of them and then proceed with the test in the usual manner.

Table XI in the Appendix gives values of T_L and T_U for values of n_1 and n_2 less than or equal to 10. When both sample sizes, n_1 and n_2, are 10 or larger, the sampling distribution of T_A can be approximated by a normal distribution with mean and variance

$$E(T_A) = \frac{n_1(n_1 + n_2 + 1)}{2} \qquad \text{and} \qquad \sigma^2_{T_A} = \frac{n_1 n_2(n_1 + n_2 + 1)}{12}$$

Therefore, for $n_1 \geq 10$ and $n_2 \geq 10$ we can conduct the Wilcoxon rank sum test using the familiar z-test of Chapters 8 and 9. The test is summarized at the top of the next page.

Wilcoxon Rank Sum Test: Large Independent Samples

One-Tailed Test

H_0:   Two sampled populations have identical probability distributions

H_a: The probability distribution for population A is shifted to the right of that for B

Test statistic: $z = \dfrac{T_A - \dfrac{n_1(n_1 + n_2 + 1)}{2}}{\sqrt{\dfrac{n_1 n_2(n_1 + n_2 + 1)}{12}}}$

Rejection region: $z > z_\alpha$

Assumptions: $n_1 \geq 10$ and $n_2 \geq 10$

Two-Tailed Test

H_0:   Two sampled populations have identical probability distributions

H_a: The probability distribution for population A is shifted to the left *or* to the right of that for B

Test statistic: $z = \dfrac{T_A - \dfrac{n_1(n_1 + n_2 + 1)}{2}}{\sqrt{\dfrac{n_1 n_2(n_1 + n_2 + 1)}{12}}}$

Rejection region: $z < -z_{\alpha/2}$
or $z > z_{\alpha/2}$

Exercises 10.1–10.13

Learning the Mechanics

10.1 Specify the test statistic and the rejection region for the Wilcoxon rank sum test for independent samples in each of the following situations:

a. H_0:   Two probability distributions, A and B, are identical

   H_a: Probability distribution for population A is shifted to the right or left of probability distribution for population B

   $n_A = 8$, $n_B = 6$, $\alpha = .10$

b. H_0:   Two probability distributions, A and B, are identical

   H_a: Probability distribution for population A is shifted to the right of probability distribution for population B

   $n_A = 5$, $n_B = 6$, $\alpha = .05$

c. H_0:   Two probability distributions, A and B, are identical

   H_a: Probability distribution for population A is shifted to the left of probability distribution for population B

   $n_A = 10$, $n_B = 8$, $\alpha = .025$

10.2 Suppose you wish to compare two treatments, A and B, and you want to determine whether the distribution of the population of B measurements is shifted to the right of that of A measurements. Assume $n_A = 9$, $n_B = 4$, and $\alpha = .05$.

a. Give the rejection region for the test.

b. Suppose you wish to detect a shift in the distributions, either A to the right of B or vice versa. Locate the rejection region for the test.

10.3 Refer to Exercise 10.2. Suppose you obtain the following independent random samples of observations on experimental units subjected to two treatments, A and B:

A: 36, 39, 33, 29, 42, 33, 35, 28, 34
B: 35, 48, 52, 66

Conduct a test of the hypotheses described in Exercise 10.2. Test using $\alpha = .05$.

10.4 Independent random samples are selected from two populations. The data are shown in the table.

SAMPLE FROM POPULATION 1	SAMPLE FROM POPULATION 2
15	6
16	13
13	8
14	9
12	7
17	5
	4
	10

a. Use the Wilcoxon rank sum test to determine whether the data provide sufficient evidence to indicate a shift in the locations of the probability distributions of the sampled populations. Test using $\alpha = .05$.

b. Do the data provide sufficient evidence to indicate that the probability distribution for population 1 is shifted to the right of the probability distribution for population 2? Use the Wilcoxon rank sum test with $\alpha = .05$.

10.5 Explain the difference between the one-tailed and two-tailed versions of the Wilcoxon rank sum test for independent random samples.

10.6 Suppose that you wish to compare two treatments, A and B, based on independent random samples of fifteen observations selected from each of the two populations. If $T_A = 173$, do the data provide sufficient evidence to indicate that distribution A is shifted to the left of distribution B? Test using $\alpha = .05$.

SUBDIVISION	
A	B
43	57
48	39
42	55
60	52
39	88
47	46
	41
	64

Applying the Concepts

10.7 A realtor wants to determine whether a difference exists between home prices in two subdivisions. Six homes from subdivision A and eight homes from subdivision B are sampled, and the prices (in thousands of dollars) are recorded in the table.

a. Use the two-sample t-test to compare the population mean price per house of the two subdivisions. What assumptions are necessary for the validity of this procedure? Do you think they are reasonable in this case?

b. Use the Wilcoxon rank sum test to see whether there is a difference (a shift in location) in the probability distributions of house prices in the two subdivisions.

U.S. PLANTS	JAPANESE PLANTS
7.11%	3.52%
6.06%	2.02%
8.00%	4.91%
6.87%	3.22%
4.77%	1.92%

10.8 Recall that the variance of a binomial sample proportion $\hat{p}$ depends on the value of the population parameter p. As a consequence, the variance of a sample percentage, $(100\hat{p})\%$, also depends on p. Thus, if you conduct an unpaired t-test (Section 8.2) to compare the means of two populations of percentages, you may be violating the assumption, $\sigma_1^2 = \sigma_2^2$, upon which the t-test is based. If the disparity in the variances is large, you will obtain more reliable test results using the Wilcoxon rank sum test for independent samples. In Exercise 8.26 we used a Student's t-test to compare the mean annual percentages of labor turnover between U.S. and Japanese manufacturers of air conditioners (Garvin, 1983). The annual percentage turnover rates for five U.S. and five Japanese plants are shown in the table. Do the data provide sufficient evidence to indicate that the mean annual percentage turnover for U.S. plants exceeds the corresponding mean for Japanese plants? Test using the Wilcoxon rank sum test with $\alpha = .05$.

10.9 An educational psychologist claims that the order in which test questions are asked affects a student's ability to answer correctly. To investigate this assertion, a professor randomly divides a class of thirteen students into two groups—seven in one group and six in the other. The professor prepares one set of test questions but arranges the questions in two different orders. On test A the questions are arranged in order of increasing difficulty (that is, from easiest to most difficult), while on test B the order is reversed. One group of students is given test A, the other test B, and the test score is recorded for each student. The results are as follows:

Test A: 90, 71, 83, 82, 75, 91, 65
Test B: 66, 78, 50, 68, 80, 60

TWIN BLADES	SINGLE BLADES
8	10
17	6
9	3
11	7
15	13
10	14
6	5
12	7

Do the data provide sufficient evidence to indicate a difference (a shift in location) in the probability distributions of student scores on the two tests? Test using $\alpha = .05$.

10.10 A major razor blade manufacturer advertises that its twin-blade disposable razor will "get you a lot more shaves" than any single-blade disposable razor on the market. A rival blade company, which has been very successful in selling single-blade razors, wishes to test this claim. Independent random samples of eight single-blade shavers and eight twin-blade shavers are taken, and the number of shaves that each gets before the razor is disposed of is recorded in the table.

a. Do the data support the twin-blade manufacturer's claim? Use $\alpha = .05$.

b. Do you think that this experiment was designed in the best possible way? If not, what design might have been better?

SCHEDULE A	SCHEDULE B
15	5
10	1
5	2
7	8
4	2
9	6
7	3

10.11 Fourteen rats were used in an experiment aimed at comparing two deprivation schedules, A and B, for their effect on hoarding behavior. An independent sampling design was used, with seven rats randomly assigned to each schedule. At the end of the deprivation period, the rats were permitted free access to food pellets and the number of pellets hoarded (taken but not eaten) during a given time period was recorded. The data are given in the table. Is there sufficient evidence to indicate that rats on one of the deprivation schedules have a greater tendency to hoard than those on the other schedule? Test using $\alpha = .05$.

DEAF CHILDREN	HEARING CHILDREN
2.75	1.15
3.14	1.65
3.23	1.43
2.30	1.83
2.64	1.75
1.95	1.23
2.17	2.03
2.45	1.64
1.83	1.96
2.23	1.37

10.12 In a comparison of visual acuity of deaf and hearing children, eye movement rates are taken on ten deaf and ten hearing children. (See the table.) A clinical psychologist believes that deaf children have greater visual acuity than hearing children. Test the psychologist's claim by using the data in the table. (The larger a child's eye movement rate, the more visual acuity the child possesses.) Use $\alpha = .05$.

10.13 Conduct the test in Exercise 10.12 by using the large-sample approximation for the Wilcoxon rank sum test. Compare the results with those found in Exercise 10.12.

10.2 Comparing Two Populations: Wilcoxon Signed Rank Test for the Paired Difference Experiment

Nonparametric techniques may also be employed to compare two probability distributions when a paired difference design is used. For example, consumer preferences for two competing products are often compared by having each of a sample of consumers rate both products. Thus, the ratings have been paired on each consumer. Here is an example of this type of experiment.

For some paper products, softness of the paper is an important consideration in determining consumer acceptance. One method of determining softness is to have judges give a sample of the products a softness rating. Suppose each of ten judges is given a sample of two products that a company wants to compare. Each judge rates the softness of each product on a scale from 1 to 10, with higher ratings implying a softer product. The results of the experiment are shown in Table 10.2.

Table 10.2
Softness Ratings of Paper

JUDGE	PRODUCT A	PRODUCT B	DIFFERENCE (A − B)	ABSOLUTE VALUE OF DIFFERENCE	RANK OF ABSOLUTE VALUE
1	6	4	2	2	5
2	8	5	3	3	7.5
3	4	5	−1	1	2
4	9	8	1	1	2
5	4	1	3	3	7.5
6	7	9	−2	2	5
7	6	2	4	4	9
8	5	3	2	2	5
9	6	7	−1	1	2
10	8	2	6	6	10

T_+ = Sum of positive ranks = 46
T_- = Sum of negative ranks = 9

Since this is a paired difference experiment, we analyze the differences between the measurements (see Section 8.3). However, the nonparametric approach requires that we calculate the ranks of the absolute values of the differences between the measurements, i.e., the ranks of the differences after removing any minus signs. *Note that tied absolute differences are assigned the average of the ranks they would receive if they were unequal but successive measurements.* After the absolute differences are ranked, the sum of the ranks of the positive differences of the original measurements, T_+, and the sum of the ranks of the negative differences of the original measurements, T_-, are computed.

We are now prepared to test the nonparametric hypothesis:

H_0: The probability distributions of the ratings for products A and B are identical.

H_a: The probability distributions of the ratings differ (in location) for the two products. (Note that this is a two-sided alternative and that therefore it implies a two-tailed test.)

Test statistic: $T =$ Smaller of the positive and negative rank sums T_+ and T_-

The smaller the value of T, the greater the evidence to indicate that the two probability distributions differ in location. The rejection region for T can be determined by consulting Table XII in the Appendix (part of the table is shown in Figure 10.5). This table gives

Figure 10.5 Reproduction of Part of Table XII in the Appendix

ONE-TAILED	TWO-TAILED	$n = 5$	$n = 6$	$n = 7$	$n = 8$	$n = 9$	$n = 10$
$\alpha = .05$	$\alpha = .10$	1	2	4	6	8	11
$\alpha = .025$	$\alpha = .05$		1	2	4	6	8
$\alpha = .01$	$\alpha = .02$			0	2	3	5
$\alpha = .005$	$\alpha = .01$				0	2	3
		$n = 11$	$n = 12$	$n = 13$	$n = 14$	$n = 15$	$n = 16$
$\alpha = .05$	$\alpha = .10$	14	17	21	26	30	36
$\alpha = .025$	$\alpha = .05$	11	14	17	21	25	30
$\alpha = .01$	$\alpha = .02$	7	10	13	16	20	24
$\alpha = .005$	$\alpha = .01$	5	7	10	13	16	19
		$n = 17$	$n = 18$	$n = 19$	$n = 20$	$n = 21$	$n = 22$
$\alpha = .05$	$\alpha = .10$	41	47	54	60	68	75
$\alpha = .025$	$\alpha = .05$	35	40	46	52	59	66
$\alpha = .01$	$\alpha = .02$	28	33	38	43	49	56
$\alpha = .005$	$\alpha = .01$	23	28	32	37	43	49
		$n = 23$	$n = 24$	$n = 25$	$n = 26$	$n = 27$	$n = 28$
$\alpha = .05$	$\alpha = .10$	83	92	101	110	120	130
$\alpha = .025$	$\alpha = .05$	73	81	90	98	107	117
$\alpha = .01$	$\alpha = .02$	62	69	77	85	93	102
$\alpha = .005$	$\alpha = .01$	55	61	68	76	84	92

a value T_0 for both one-tailed and two-tailed tests for each value of n, the number of matched pairs. For a two-tailed test with $\alpha = .05$, we will reject H_0 if $T \le T_0$. You can see in Figure 10.5 that the value of T_0 that locates the boundary of the rejection region for the judges' ratings for $\alpha = .05$ and $n = 10$ pairs of observations is 8. Therefore, the rejection region for the test (see Figure 10.6) is

Rejection region: $T \le 8$ for $\alpha = .05$

Figure 10.6 Rejection Region for Paired Difference Experiment

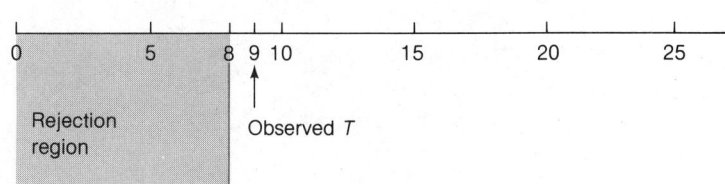

Since the smaller rank sum for the paper data, $T_- = 9$, does not fall within the rejection region, the experiment has not provided sufficient evidence to indicate that the two paper products differ with respect to their softness ratings at the $\alpha = .05$ level.

Note that if a significance level of $\alpha = .10$ had been used, the rejection region would have been $T \le 11$ and we would have rejected H_0. In other words, the samples do provide evidence that the probability distributions of the softness ratings differ at the $\alpha = .10$ significance level.

When the Wilcoxon signed rank test is applied to a set of data, it is possible that one (or more) of the paired differences may equal zero. The Wilcoxon test continues to be valid if the number of zeros is small in comparison to the number of pairs, but to perform the test you must delete the zeros and reduce the number of differences accordingly. For example, if you have $n = 12$ pairs and two of the differences equal zero, you should delete these pairs, rank the remaining $n = 10$ differences, and use $n = 10$ when locating T_0 in Table XII. *Ties in ranks are treated in the same manner as for the Wilcoxon rank sum test for independent samples: Assign to each of the tied ranks the average of the ranks the two observations would have received if the observations had not been tied.* The Wilcoxon signed rank test for a paired difference experiment is summarized in the box on page 496.

Example 10.2   Suppose the police commissioner in a small community must choose between two plans for patrolling the town's streets. Plan A, the less expensive plan, uses voluntary citizen groups to patrol certain high-risk neighborhoods. In contrast, plan B would utilize police patrols. As an aid in reaching a decision, both plans are examined by ten trained criminologists, each of whom is asked to rate the plans on a scale from 1 to 10. (High ratings imply a more effective crime prevention plan.) The city will adopt plan B (and hire extra police) only if the data provide evidence that criminologists tend to rate plan B more effective than plan A.

Wilcoxon Signed Rank Test for a Paired Difference Experiment

One-Tailed Test	Two-Tailed Test
H_0: Two sampled populations have identical probability distributions	H_0: Two sampled populations have identical probability distributions
H_a: The probability distribution for population A is shifted to the right of that for population B	H_a: The probability distribution for population A is shifted to the right or to the left of that for population B
Test statistic: T_-, the negative rank sum (we assume the differences are computed by subtracting each paired B measurement from the corresponding A measurement)	Test statistic: T, the smaller of the positive and negative rank sums, T_+ and T_-
Rejection region: $T_- \leq T_0$, where T_0 is found in Table XII (in the Appendix) for the one-tailed significance level α and the number of untied pairs, n	Rejection region: $T \leq T_0$, where T_0 is found in Table XII (in the Appendix) for the two-tailed significance level α and the number of untied pairs, n

[*Note:* If the alternative hypothesis is that the probability distribution for A is shifted to the left of B, we use T_+ as the test statistic and reject H_0 if $T_+ \leq T_0$.]

Assumptions: 1. A random sample of pairs of observations has been taken.
2. The absolute differences in the paired observations can be ranked. [*Note:* No assumptions have to be made about the form of the population probability distributions.]

The results of the survey are shown in Table 10.3. Do the data provide evidence at the $\alpha = .05$ level that the distribution of ratings for plan B lies above that for plan A?

Solution   The null and alternative hypotheses are

H_0: The two probability distributions of effectiveness ratings are identical

H_a: The effectiveness ratings of the more expensive plan (B) tend to exceed those of plan A

Observe that the alternative hypothesis is one-sided (i.e., we only wish to detect a shift in the distribution of the B ratings to the right of the distribution of A ratings) and there-

Table 10.3
Effectiveness Ratings by
Ten Qualified Crime
Prevention Experts

CRIME PREVENTION EXPERT
1
2
3
4
5
6
7
8
9
10

fore it impl e Figure 10.7). If the alternative
hypothesis an their paired A ratings, more
negative d and T_+ will be small. Because
Table XII is e will use T_+ as the test statistic
and reject

The diff shown in Table 10.3. Note that
one of the minate this pair from the ranking
and reduc alue to enter Table XII, you see
that for a ave $T_0 = 8$. Therefore, the test
statistic ar

Test sta

Rejectio

Summir Table 10.3, we find $T_+ = 15.5$.
Since this conclude that this sample pro-
vides insu port the alternative hypothesis.
The comm zing police patrols tends to be

Figure 10.7 The
Alternative Hypothesis for
Example 10.2: We Expect
T_+ to Be Small

Probability

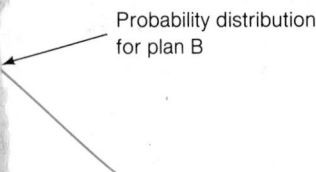

Probability distribution
for plan B

rated higher than the plan using citizen volunteers. That is, on the basis of this study, extra police will not be hired.   ■

As is the case for the rank sum test for independent samples, the sampling distribution of the signed rank statistic can be approximated by a normal distribution when the number n of paired observations is large (say $n \geq 25$). The large-sample z-test is summarized in the box.

Wilcoxon Signed Rank Test for a Paired Difference Experiment: Large Sample

One-Tailed Test

H_0: Two sampled populations have identical probability distributions

H_a: The probability distribution for population A is shifted to the right of that for population B

Test statistic: $z = \dfrac{T_+ - \dfrac{n(n+1)}{4}}{\sqrt{\dfrac{n(n+1)(2n+1)}{24}}}$

Rejection region: $z > z_\alpha$

Assumptions: $n \geq 25$

Two-Tailed Test

H_0: Two sampled populations have identical probability distributions

H_a: The probability distribution for population A is shifted to the right or to the left of that for population B

Test statistic: $z = \dfrac{T_+ - \dfrac{n(n+1)}{4}}{\sqrt{\dfrac{n(n+1)(2n+1)}{24}}}$

Rejection region: $z < -z_{\alpha/2}$
  or $z > z_{\alpha/2}$

Exercises 10.14–10.22

Learning the Mechanics

10.14 Specify the test statistic and the rejection region for the Wilcoxon signed rank test for the paired difference design in each of the following situations:

a. H_0: Two probability distributions, A and B, are identical

   H_a: Probability distribution for population A is shifted to the right or left of probability distribution for population B

   $n = 25$, $\alpha = .10$

b. H_0: Two probability distributions, A and B, are identical

   H_a: Probability distribution for population A is shifted to the right of probability distribution for population B

   $n = 41$, $\alpha = .05$

c. H_0: Two probability distributions, A and B, are identical

$\quad H_a$: Probability distribution for population A is shifted to the left of probability distribution for population B

$\quad n = 8, \quad \alpha = .005$

10.15 Suppose you wish to test an hypothesis that two treatments, A and B, are equivalent against the alternative that the responses for A tend to be larger than those for B.

a. If $n = 8$ and $\alpha = .01$, give the rejection region for a Wilcoxon signed rank test.

b. Suppose you wish to detect a difference in the locations of the distributions of the responses for A and B if such a difference exists. If $n = 7$ and $\alpha = .10$, give the rejection region for the Wilcoxon signed rank test.

10.16 A random sample of nine pairs of measurements is shown in the table.

PAIR	SAMPLE DATA FROM POPULATION 1	SAMPLE DATA FROM POPULATION 2
1	8	7
2	10	1
3	6	4
4	10	10
5	7	4
6	8	3
7	4	6
8	9	2
9	8	4

a. Use the Wilcoxon signed rank test to determine whether the data provide sufficient evidence to indicate that the probability distribution for population 1 is shifted to the right of the probability distribution for population 2. Test using $\alpha = .05$.

b. Use the Wilcoxon signed rank test to determine whether the data provide sufficient evidence to indicate that the probability distribution for population 1 is shifted either to the right or to the left of the probability distribution for population 2. Test using $\alpha = .05$.

10.17 Suppose you wish to test an hypothesis that two treatments, A and B, are equivalent against the alternative that the responses for A tend to be larger than those for B.

a. If the number of pairs equals 25, give the rejection region for the large-sample Wilcoxon signed rank test for $\alpha = .05$.

b. Suppose that $T_+ = 273$. State your test conclusions.

c. Find the p-value for the test and interpret it.

Applying the Concepts

10.18 In Exercise 8.35, we discussed the Supreme Court decision (*Milliken* v. *Bradley*) that appears to have triggered the flight of white families out of the inner city to the suburbs. The data from Exercise 8.35 are reproduced here in the table (Clotfelter, 1976). Examine the last two columns of the table, which give the percentage change in white students over two time periods, 1968–1970 and 1970–1972. Was the shift of white families out of the inner cities larger in the time period 1970–1972 than for the comparable period, 1968–1970? Test using a Wilcoxon signed rank test with $\alpha = .05$.

Racial Change and White Flight in Selected City School Systems, 1968–1972

CITY	PERCENTAGE MINORITY			PERCENTAGE CHANGE IN WHITE STUDENTS[a]	
	1968	1970	1972	1968–1970	1970–1972
Boston	31.5	35.9	40.4	−3.9	−7.4
Philadelphia	61.8	63.6	64.8	−5.7	−2.2
Baltimore	65.1	67.1	69.3	−5.6	−9.3
St. Louis	63.8	65.9	69.1	−9.4	−13.8
Chicago	62.3	65.4	69.2	−9.0	−14.7
Detroit	60.7	65.5	69.5	−15.6	−14.0
Atlanta	61.8	68.7	77.4	−22.3	−34.4
Charlotte	29.5	31.1	32.8	−3.1	−5.6
Jacksonville	28.2	29.4	32.6	−1.8	−11.4
Houston	46.7	50.6	56.4	−9.1	−17.5
San Francisco	58.8	63.1	68.2	−13.5	−22.4
San Diego	23.9	24.6	26.3	−1.1	−5.5

[a] Percentages based on beginning years.
Source: U.S. Dept. of Health, Education, and Welfare, Office for Civil Rights, Directory of Public Elementary and Secondary Schools in Selected Districts, fall 1968, fall 1970, and fall 1972.

MOUSE	COMPOUND	
	X	Y
1	4.7	5.1
2	3.3	4.6
3	8.5	8.7
4	3.9	3.6
5	7.0	6.1
6	4.7	4.1
7	5.2	5.1

10.19 Hypoglycemia is a condition in which blood sugar is below normal limits. To compare two compounds, X and Y, for treating hypoglycemia, each compound is applied to half the diaphragms of each of seven white mice. Blood glucose uptake in milligrams per gram of tissue is measured for each half, producing the results listed in the table. Do the data provide sufficient evidence to indicate that one of the compounds tends to produce higher blood sugar uptake readings than the other? Test using $\alpha = .10$.

TWIN PAIR	SCHOOL	
	A	B
1	65	69
2	72	72
3	86	74
4	50	52
5	60	47
6	81	72

10.20 Children completing the sixth grade at a school located in a large city have the choice of going to one of two junior high schools, A or B. Members of the school board want to compare the academic effectiveness of the two schools. The parents of six sets of identical twins agree to send one child to school A and the other to school B. Since each set of twins is in the same class at each grade level through the sixth grade, a paired difference design could be employed. Near the end of the ninth grade, an achievement test is given to each child in the experiment. The results are given in the table. Test to determine whether there is evidence of a difference (shift in location) in the probability distributions of achievement test scores at the two schools. Use $\alpha = .10$.

10.21 Twelve sets of identical twins are given psychological tests to determine whether the firstborn of the twins tends to be more aggressive than the secondborn. The results are shown in the table, where the higher score indicates greater aggressiveness. Do the data provide sufficient evidence to indicate that the firstborn of a pair of twins is more aggressive than the other? Test using $\alpha = .05$.

SET	FIRSTBORN	SECONDBORN	SET	FIRSTBORN	SECONDBORN
1	86	88	7	77	65
2	71	77	8	91	90
3	77	76	9	70	65
4	68	64	10	71	80
5	91	96	11	88	81
6	72	72	12	87	72

10.22 In Exercise 8.38, we compared matched pairs of measurements on the exhalation rate (a measure of radiation) of fifteen soil samples from waste gypsum and phosphate mounds in Polk County, Florida. Each soil sample was measured for exhalation rate by the Polk County Health Department (PCHD) and the Eastern Environmental Radiation Facility (EERF). The data are reproduced in the table. Do the data provide sufficient evidence to indicate that one of the measuring facilities, PCHD or EERF, tends to read higher or lower than the other? Test using the Wilcoxon signed rank test with $\alpha = .05$.

CHARCOAL CANISTER NO.	PCHD	EERF	CHARCOAL CANISTER NO.	PCHD	EERF
71	1,709.79	1,479.0	85	393.55	187.7
58	357.17	257.8	46	880.84	630.4
84	1,150.94	1,287.0	4	2,996.49	3,707.0
91	1,572.69	1,395.0	20	2,367.40	2,791.0
44	558.33	416.5	36	599.84	706.8
43	4,132.28	3,993.0	42	538.37	618.5
79	1,489.86	1,351.0	55	2,770.23	2,639.0
61	3,017.48	1,813.0			

10.3
Kruskal–Wallis
H-Test for a
Completely
Randomized
Design

In Chapter 9 we used an analysis of variance and the *F*-test to compare the means of *k* populations based on random sampling from populations that were normally distributed with a common variance σ^2. We now present a nonparametric technique for comparing the populations that requires no assumptions concerning the population probability distributions.

Suppose a health administrator wants to compare the unoccupied bed space for three hospitals located in the same city. She randomly selects ten different days from the records of each hospital and lists the number of unoccupied beds for each day (see Table 10.4). Because the number of unoccupied beds per day may occasionally

Table 10.4

Number of Available Beds

HOSPITAL 1		HOSPITAL 2		HOSPITAL 3	
Beds	Rank	Beds	Rank	Beds	Rank
6	5	34	25	13	9.5
38	27	28	19	35	26
3	2	42	30	19	15
17	13	13	9.5	4	3
11	8	40	29	29	20
30	21	31	22	0	1
15	11	9	7	7	6
16	12	32	23	33	24
25	17	39	28	18	14
5	4	27	18	24	16
$R_1 = 120$		$R_2 = 210.5$		$R_3 = 134.5$	

be quite large, it is conceivable that the population distributions of data may be skewed to the right and that this type of data may not satisfy the assumptions necessary for a parametric comparison of the population means. We therefore use a nonparametric analysis and base our comparison on the rank sums for the three sets of sample data. Just as with two independent samples (Section 10.1), the ranks are computed for each observation according to the relative magnitude of the measurements *when the data for all the samples are combined* (see Table 10.4). Ties are treated as they were for the Wilcoxon rank sum and signed rank tests by assigning the average value of the ranks to each of the tied observations.

We test

H_0: The probability distributions of the number of unoccupied beds are the same for all three hospitals

If we denote the three sample rank sums by R_1, R_2, and R_3, the test statistic is given by

$$H = \frac{12}{n(n + 1)} \sum \frac{R_j^2}{n_j} - 3(n + 1)$$

where n_j is the number of measurements in the jth sample and n is the total sample size ($n = n_1 + n_2 + \cdots + n_k$). For the data in Table 10.4, we have $n_1 = n_2 = n_3 = 10$ and $n = 30$. The rank sums are $R_1 = 120$, $R_2 = 210.5$, and $R_3 = 134.5$. Thus,

$$H = \frac{12}{30(31)} \left[\frac{(120)^2}{10} + \frac{(210.5)^2}{10} + \frac{(134.5)^2}{10} \right] - 3(31)$$

$$= 99.097 - 93 = 6.097$$

Figure 10.8 Several χ^2 Probability Distributions

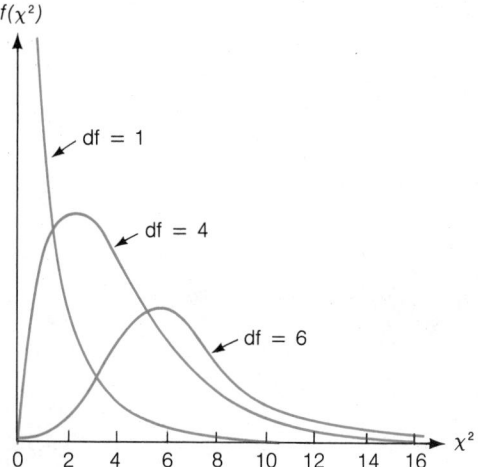

If the null hypothesis is true, the distribution of *H* in repeated sampling is approximately a χ^2 (chi square) distribution. This approximation for the sampling distribution of *H* is adequate as long as each of the *k* sample sizes exceeds 5. (See the references for more detail.) The χ^2 probability distribution is characterized by a single parameter called the *degrees of freedom associated with the distribution*. Several χ^2 probability distributions with different degrees of freedom are shown in Figure 10.8. The degrees of freedom corresponding to the approximate sampling distribution of *H* will always be $(k-1)$, one less than the number of probability distributions being compared. Because large values of *H* support the alternative hypothesis that the populations have different probability distributions, the rejection region for the test is located in the upper tail of the χ^2 distribution, as shown in Figure 10.9 (page 504).

For the data of Table 10.4, the approximate distribution of the test statistic *H* is χ^2 with $(k-1) = 2$ df. To determine how large *H* must be before we will reject the null hypothesis, we consult Table VI in the Appendix. (Part of this table is shown in Figure 10.9.) Entries in the table give an upper-tail value of χ^2, call it χ^2_α, such that $P(\chi^2 > \chi^2_\alpha) = \alpha$. The columns of the table identify the value of α associated with the tabulated value of χ^2_α, and the rows correspond to the degrees of freedom. Thus, for $\alpha = .05$ and df $= 2$, we can reject the null hypothesis that the three probability distributions are the same if

$$H > \chi^2_{.05} \qquad \text{where } \chi^2_{.05} = 5.99147$$

The rejection region is pictured in Figure 10.10. Since the calculated $H = 6.097$ exceeds the critical value of 5.99147, we conclude that at least one of the three hospitals tends to have a larger number of unoccupied beds than the others.

Note that prior to conducting the experiment we might have decided to compare the daily occupancy rates for a specific pair of hospitals. The Wilcoxon rank sum test

Figure 10.9 Reproduction of Part of Table VI in the Appendix

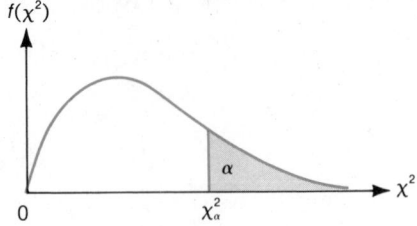

DEGREES OF FREEDOM	$\chi^2_{.100}$	$\chi^2_{.050}$	$\chi^2_{.025}$	$\chi^2_{.010}$	$\chi^2_{.005}$
1	2.70554	3.84146	5.02389	6.63490	7.87944
2	4.60517	5.99147	7.37776	9.21034	10.5966
3	6.25139	7.81473	9.34840	11.3449	12.8381
4	7.77944	9.48773	11.1433	13.2767	14.8602
5	9.23635	11.0705	12.8325	15.0863	16.7496
6	10.6446	12.5916	14.4494	16.8119	18.5476
7	12.0170	14.0671	16.0128	18.4753	20.2777
8	13.3616	15.5073	17.5346	20.0902	21.9550
9	14.6837	16.9190	19.0228	21.6660	23.5893
10	15.9871	18.3070	20.4831	23.2093	25.1882
11	17.2750	19.6751	21.9200	24.7250	26.7569
12	18.5494	21.0261	23.3367	26.2170	28.2995
13	19.8119	22.3621	24.7356	27.6883	29.8194
14	21.0642	23.6848	26.1190	29.1413	31.3193
15	22.3072	24.9958	27.4884	30.5779	32.8013
16	23.5418	26.2962	28.8454	31.9999	34.2672
17	24.7690	27.5871	30.1910	33.4087	35.7185
18	25.9894	28.8693	31.5264	34.8053	37.1564
19	27.2036	30.1435	32.8523	36.1908	38.5822

Figure 10.10 Rejection Region for the Comparison of Three Probability Distributions

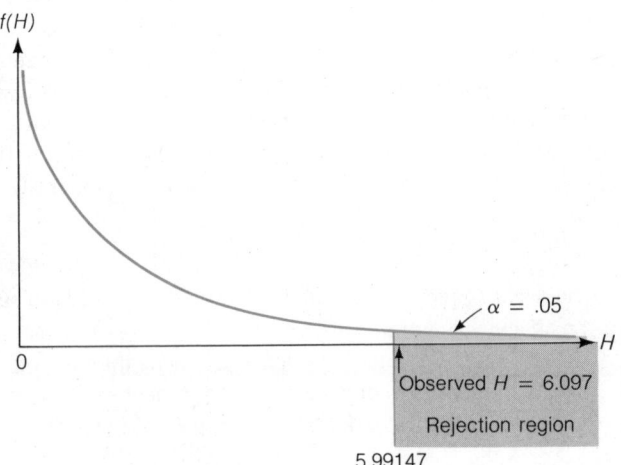

presented in Section 10.1 could be used for this purpose. The Kruskal–Wallis *H*-test for comparing more than two probability distributions is summarized in the box.

Kruskal–Wallis *H*-Test for Comparing *k* Probability Distributions

H_0:   The *k* probability distributions are identical

H_a: At least two of the *k* probability distributions differ in location

Test statistic: $H = \dfrac{12}{n(n + 1)} \sum \dfrac{R_j^2}{n_j} - 3(n + 1)$

where

n_j = Number of measurements in sample *j*

R_j = Rank sum for sample *j*, where the rank of each measurement is computed according to its relative magnitude in the totality of data for the *k* samples

n = Total sample size = $n_1 + n_2 + \cdots + n_k$

Assumptions: 1.   The *k* samples are random and independent.
                        2.   There are five or more measurements in each sample.
                        3.   The observations can be ranked.

[*Note:* No assumptions have to be made about the shape of the population probability distributions.]

Rejection region: $H > \chi_\alpha^2$ with $(k - 1)$ degrees of freedom

---

Example 10.3   Suppose a dairy farmer wants to compare the amount of milk produced by dairy cattle fed on four different diets. Five cattle are randomly assigned and maintained on each of the diets for a 3-month period. At the end of the 3 months, each cow's milk production is recorded for a 1-week period. The numbers given in Table 10.5 are the ranks of the productivity data. Do the data provide sufficient evidence to indicate that at least one of the diets tends to achieve greater milk production than the others? Test at the .05 level of significance.

Table 10.5
Ranks for Milk
Production Data

DIET 1	DIET 2	DIET 3	DIET 4
1	12	8	14
5	2	9	15
6	17	3	16
7	19	11	4
10	20	13	18
$R_1 = 29$	$R_2 = 70$	$R_3 = 44$	$R_4 = 67$

Solution   The elements of the test are as follows:

H_0:   The population probability distributions of milk production for the four diets are identical

H_a: At least two of the diets have probability distributions with different locations

Test statistic: $H = \dfrac{12}{n(n+1)} \sum \dfrac{R_j^2}{n_j} - 3(n+1)$

$$= \dfrac{12}{20(21)} \left[\dfrac{(29)^2}{5} + \dfrac{(70)^2}{5} + \dfrac{(44)^2}{5} + \dfrac{(67)^2}{5} \right] - 3(21)$$

$$= 69.5 - 63 = 6.5$$

Rejection region: Since we are comparing four probability distributions, there are $k - 1 = (4 - 1) = 3$ df associated with the test statistic. Thus, we will reject H_0 if $H > \chi_{.05}^2 = 7.81473$.

Since 6.5 is less than 7.81473, we have insufficient evidence to indicate that at least one of the diets tends to achieve greater milk productivity than the others.   ■

Exercises 10.23–10.34

Learning the Mechanics

10.23 Use Table VI in the Appendix to find each of the following χ^2 values:
a. $\chi_{.05}^2$,   df = 20   **b.** $\chi_{.025}^2$,   df = 15   **c.** $\chi_{.01}^2$,   df = 30
d. $\chi_{.10}^2$,   df = 80   **e.** $\chi_{.05}^2$,   df = 2   **f.** $\chi_{.005}^2$,   df = 10

10.24 Use Table VI in the Appendix to find each of the following probabilities:
a. $P(\chi^2 \geq 3.07382)$,   where df = 12   **b.** $P(\chi^2 \leq 24.4331)$,   where df = 40
c. $P(\chi^2 \geq 14.6837)$,   where df = 9   **d.** $P(\chi^2 < 34.1696)$,   where df = 20
e. $P(\chi^2 < 6.26214)$,   where df = 15   **f.** $P(\chi^2 \leq .584375)$,   where df = 3

10.25 Suppose you wish to use the Kruskal–Wallis H-test to compare the probability distributions of three populations. The following are independent random samples selected from the three populations:

I:	66	33	55	88	58	62	69	49
II:	22	31	16	25	30	33	40	
III:	75	96	102	75	88	78		

a. What experimental design was used?
b. Specify the null and alternative hypotheses you would test.
c. Specify the rejection region you would use for your hypothesis test at $\alpha = .01$.
d. Conduct the test at $\alpha = .01$.

10.26 Independent random samples were selected from four populations. The data are shown in the table.

SAMPLE 1	SAMPLE 2	SAMPLE 3	SAMPLE 4
8.1	7.8	2.1	4.8
2.3	3.5	3.3	3.9
3.4	5.2	1.0	2.6
5.1	5.0	4.5	6.7
3.6	4.6	.8	5.2
	7.7		

a. Do the data provide sufficient evidence to indicate a difference in location between at least two of the four probability distributions? Use the Kruskal–Wallis *H*-test with $\alpha = .05$.

b. What assumptions must be satisfied for the test results in part a to be valid?

10.27 Under what circumstances does the χ^2-distribution provide an appropriate characterization of the sampling distribution of the Kruskal–Wallis *H*-statistic?

Applying the Concepts

10.28 Sixth graders in an elementary school are taught reading by three different methods. Students were randomly assigned to three classes. One class used programmed instruction, a second used standard memorization techniques, and the third used an open classroom approach. The increases in reading levels attained by five students randomly selected from each of the three classes are shown in the table. Do the data provide sufficient evidence to indicate that the probability distributions of increases in reading level differ for at least two of the methods? Use $\alpha = .05$.

PROGRAMMED	STANDARD	OPEN
.9	1.0	1.7
1.5	.8	.5
.7	.9	1.6
1.1	1.2	1.4
.5	1.4	1.0

10.29 Random samples of six senior computer systems analysts were selected from each of three industries: banking, federal government, and retail sales. Their salaries

BANKING	FEDERAL GOVERNMENT	RETAIL SALES
$30,000	$28,000	$20,100
24,500	34,000	19,200
27,100	39,000	20,500
26,000	35,000	20,600
23,800	34,100	21,100
25,800	36,200	19,300

Source: Based on *The American Almanac of Jobs and Salaries,* 1982, p. 420.

are recorded in the table. You have been hired to determine whether differences exist among the salary distributions for senior systems analysts in the three industries.

a. Under what circumstances would it be appropriate to use the F-test for a completely randomized design to perform the required analysis?

b. Which assumption(s) required by the F-test may be violated in this problem? Explain.

c. Use the Kruskal–Wallis H-test to determine whether the salary distributions differ among the three industries. Specify your null and alternative hypotheses, and state your conclusions in the context of the problem. Use $\alpha = .05$.

LIST 1	LIST 2	LIST 3
48	41	18
43	36	42
39	29	28
57	40	38
21	35	15
47	45	33
58	32	31

10.30 Three lists of words, representing three levels of abstractness, are randomly assigned to twenty-one experimental subjects so that seven subjects receive each list. The subjects are asked to respond to each word on their list with as many associated words as possible within a given period of time. A subject's score is the total number of associates, summing over all words in the list. Scores for each list are given in the table. Do the data provide sufficient evidence to indicate a difference (shift in location) between at least two of the probability distributions of the numbers of word associates that subjects can name for the three lists? Use $\alpha = .05$.

10.31 An experiment was conducted to compare the length of time it takes a human to recover from each of the three types of influenza—Victoria A, Texas, and Russian. Twenty-one human subjects were selected at random from a group of volunteers and divided into three groups of seven each. Each group was randomly assigned a strain of the virus, and the influenza was induced in the subjects. All the subjects were then cared for under identical conditions, and the recovery time (in days) was recorded. The results are shown in the table.

VICTORIA A	TEXAS	RUSSIAN
12	9	7
6	10	3
13	5	7
10	4	5
8	9	6
11	8	4
7	11	8

a. Do the data provide sufficient evidence to indicate that the recovery times for one or more types of influenza tend to be longer than for the other types? Test using $\alpha = .05$.

b. Do the data provide sufficient evidence to indicate a difference in locations of the distributions of recovery times for the Victoria A and Russian types? Test using $\alpha = .05$.

10.32 The EPA wants to determine whether temperature changes in the ocean's water caused by a nuclear power plant will have a significant effect on the animal life in the region. Recently hatched specimens of a certain species of fish are randomly divided into four groups. The groups are placed in separate simulated ocean environments that are identical in every way except for water temperature. Six months later, the specimens are weighed. The results (in ounces) are given in the table. Do the data provide sufficient evidence to indicate that one or more of the temperatures tend to produce larger weight increases than the other temperatures? Test using $\alpha = .10$.

38°F	42°F	46°F	50°F
22	15	14	17
24	21	28	18
16	26	21	13
18	16	19	20
19	25	24	21
	17	23	

10.33 An experiment was conducted to determine whether a test designed to identify a certain form of mental illness could be easily interpreted with little psychological training. Thirty judges were selected to review the results of 100 tests, half of which were given to disturbed patients and half to normal people. Of the thirty judges chosen, ten were staff members of a mental hospital, ten were trainees at the hospital, and ten were undergraduate psychology majors. The results in the table give the number of the 100 tests correctly classified by each judge. Do the data provide sufficient evidence of a difference in the probability distributions of the number of correct identifications among the three types of judges? Use $\alpha = .05$.

STAFF		TRAINEES		UNDERGRADUATES	
78	76	80	69	65	74
79	86	75	81	70	80
85	88	72	76	74	73
93	84	68	72	78	75
90	81	75	76	68	73

BRAND		
A	B	C
36	49	71
48	33	31
5	60	140
67	2	59
53	55	42

10.34 Three different brands of magnetron tubes (the key components in microwave ovens) were subjected to stressful testing, and the number of hours each operated without repair was recorded. (See the table.) Although these times do not represent typical life lengths, they do indicate how well the tubes can withstand extreme stress.

a. Use the *F*-test for a completely randomized design (Chapter 9) to test the hypothesis that the mean length of life under stress is the same for the three brands. Use $\alpha = .05$. What assumptions are necessary for the validity of this procedure? Is there any reason to doubt these assumptions?

b. Use the Kruskal–Wallis *H*-test to determine whether evidence exists to conclude that the brands of magnetron tubes tend to differ in length of life under stress. Test using $\alpha = .05$.

10.4
The Friedman
F_r-Test for a
Randomized
Block Design

In Section 9.2 we employed an analysis of variance to compare k population means when the data were collected using a randomized block design. The **Friedman F_r-test** provides another method for testing to detect a shift in location of a set of k populations.* Like other nonparametric tests, it requires no assumptions concerning the nature of the populations other than that you be able to rank the individual observations.

In Section 10.1, we gave an example where a completely randomized design was used to compare the reaction times of subjects under the influence of one of two drugs. When the effect of a drug is short-lived (there is no carryover effect) and when the drug effect varies greatly from person to person, it may be beneficial to employ a *randomized block design*. Using the subjects as blocks, we would hope to eliminate the variability among subjects and thereby increase the amount of information in the experiment. Suppose that three drugs, A, B, and C, are to be compared using a randomized block design. Each of the three drugs is administered to the *same subject* with suitable time lags between the three doses. The order in which the drugs are administered is randomly determined for each subject. Thus, one drug would be administered to a subject, a reaction time noted, and after a sufficient length of time the second drug administered, etc.

Suppose that six subjects are chosen and that the reaction times for each drug are as shown in Table 10.6. To compare the three drugs, we rank the observations within each subject (block) and then compute the rank sums for each of the drugs (treatments). Tied observations within blocks are handled in the usual manner by assigning the average value of the ranks to each of the tied observations.

Table 10.6
Reaction Times
for Three Drugs

SUBJECT	DRUG A	RANK	DRUG B	RANK	DRUG C	RANK
1	1.21	1	1.48	2	1.56	3
2	1.63	1	1.85	2	2.01	3
3	1.42	1	2.06	3	1.70	2
4	2.43	2	1.98	1	2.64	3
5	1.16	1	1.27	2	1.48	3
6	1.94	1	2.44	2	2.81	3
		$R_1 = \overline{7}$		$R_2 = \overline{12}$		$R_3 = \overline{17}$

The null and alternative hypotheses are

H_0: The populations of reaction times are identically distributed for all three drugs

H_a: At least two of the drugs have probability distributions of reaction times that differ in location

* The Friedman F_r-test is the product of the Nobel prize winning economist Milton Friedman.

The Friedman F_r test statistic, which is based on the rank sums for each treatment, is

$$F_r = \frac{12}{bk(k+1)} \sum R_j^2 - 3b(k+1)$$

where b is the number of blocks, k is the number of treatments, and R_j is the jth rank sum. For the data in Table 10.6,

$$F_r = \frac{12}{(6)(3)(4)}[(7)^2 + (12)^2 + (17)^2] - 3(6)(4)$$

$$= 80.33 - 72 = 8.33$$

Like the Kruskal–Wallis H-statistic, the Friedman F_r-statistic has approximately a χ^2 sampling distribution with $(k-1)$ degrees of freedom. Empirical results show the approximation to be adequate if either b (the number of blocks) or k (the number of treatments) exceeds 5. The Friedman F_r-test for a randomized block design is summarized in the box.

Friedman F_r-Test for a Randomized Block Design

H_0: The probability distributions for the k treatments are identical

H_a: At least two of the probability distributions differ in location

Test statistic: $\quad F_r = \dfrac{12}{bk(k+1)} \sum R_j^2 - 3b(k+1)$

where

b = Number of blocks

k = Number of treatments

R_j = Rank sum of the jth treatment, where the rank of each measurement is computed relative to its position *within its own block*

Assumptions: 1. The treatments are randomly assigned to experimental units within the blocks.

2. The measurements can be ranked within blocks.

3. Either the number of blocks (b) or the number of treatments (k) should exceed 5 for the χ^2 approximation to be adequate.

[*Note:* No assumptions have to be made about the shape of the population probability distributions.]

Rejection region: $\quad F_r > \chi_\alpha^2$ with $(k-1)$ degrees of freedom

For the drug example, we will use $\alpha = .05$ to form the rejection region:

$$F_r > \chi_{.05}^2 = 5.99147$$

where $\chi_{.05}^2$ is based on $(k - 1) = 2$ degrees of freedom. Consequently, because the observed value, $F_r = 8.33$, exceeds 5.99147, we conclude that at least two of the three drugs have probability distributions of reaction times that differ in location.

Example 10.4

Suppose a marketing firm wants to compare the relative effectiveness of three different modes of advertising: direct-mail, newspaper ads, and magazine ads. For fifteen clients, all three modes are used over a 1-year period and the marketing firm records the year's percentage response to each type of advertising. That is, the firm divides the number of responses to a certain type of advertising by the total number of potential customers reached by the advertisements of that type. The results are shown in Table 10.7. Do these data provide sufficient evidence to indicate a difference in the locations of the probability distributions of response rates? Use the Friedman F_r-test at the $\alpha = .10$ level of significance.

Table 10.7

Percentage Response to Three Types of Advertising for Fifteen Different Companies

COMPANY	DIRECT-MAIL	RANK	NEWSPAPER	RANK	MAGAZINE	RANK
1	7.3	1	15.7	3	10.1	2
2	9.4	2	18.3	3	8.2	1
3	4.3	1	11.2	3	5.1	2
4	11.3	2	19.1	3	6.5	1
5	3.3	1	9.2	3	8.7	2
6	4.2	1	10.5	3	6.0	2
7	5.9	1	8.7	2	12.3	3
8	6.2	1	14.3	3	11.1	2
9	4.3	2	3.1	1	6.0	3
10	10.0	1	18.8	3	12.1	2
11	2.2	1	5.7	2	6.3	3
12	6.3	2	20.2	3	4.3	1
13	8.0	1	14.1	3	9.1	2
14	7.4	2	6.2	1	18.1	3
15	3.2	1	8.9	3	5.0	2
		$R_1 = 20$		$R_2 = 39$		$R_3 = 31$

Solution   The fifteen companies act as blocks in this experiment; thus, we rank the observations within each company (block) and then compute the rank sums for each of the three types of advertising (treatments). The null and alternative hypotheses are

H_0: The probability distributions for the response rates are the same for all three types of advertising

H_a: At least two of the probability distributions of response rates differ in location

Friedman F_r test statistic:

$$F_r = \frac{12}{bk(k+1)} \sum R_j^2 - 3b(k+1)$$

$$= \frac{12}{(15)(3)(4)} (R_1^2 + R_2^2 + R_3^2) - (3)(15)(4)$$

$$= \frac{12}{(15)(3)(4)} [(20)^2 + (39)^2 + (31)^2] - (3)(15)(4)$$

$$= 192.13 - 180 = 12.13$$

Rejection region: $F_r > \chi_{.10}^2 = 4.60517$ where $\chi_{.10}^2$ is based on $(k-1) = 2$ degrees of freedom

Since the calculated $F_r = 12.13$ exceeds the critical value of 4.60517 (see Figure 10.11), we conclude that the probability distributions of response rates differ in location for at least two of the three types of advertising.

Figure 10.11 Rejection Region for Example 10.4

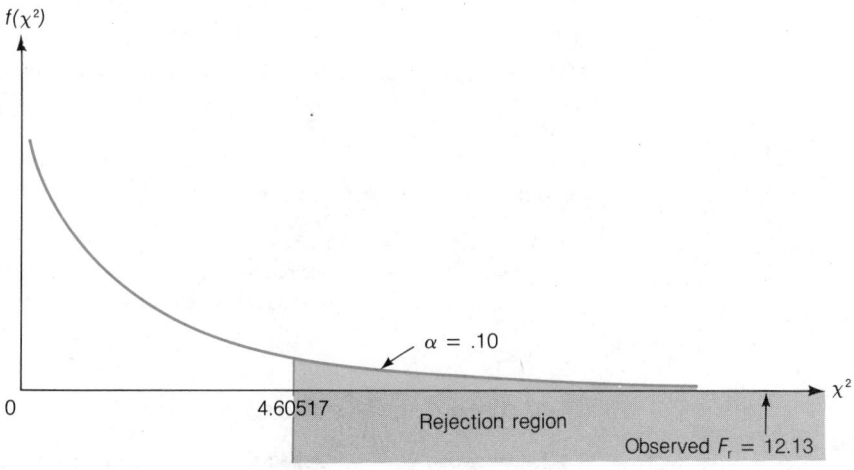

Clearly, the assumptions that the measurements are ranked within blocks and that the number of blocks (companies) is greater than 5 are satisfied. However, the experimenter must be sure that the treatments are randomly assigned to blocks. For the procedure to be valid, we assume that the three modes of advertising are used in a random order by each company. If this were not true, the difference in the response rates for the three advertising modes might be due to the order in which the modes are used.

■

Case Study 10.1
Consumer Rankings of Products

Since consumers' images reflect, to some extent, actions taken by marketers in dealing with many marketing variables, it is frequently desirable to determine if these images have patterns.

McClure (1971) studied the pattern of consumers' images of three appliances. Five attributes—price, looks, need for repair, ease of use, and familiarity—were examined for each of two brands. For example, a consumer was asked to rank brand A's refrigerators, ranges, and automatic clothes washers in terms of the attribute "ease of use." McClure states:

> Conceptually, it was an investigation of the "halo effect" . . . which refers to the individual's supposed tendency to imbue his evaluations of specific characteristics of an appliance with the same direction of general feeling expressed about the brand. The halo effect would be considered operative to the extent that the individual's general image of Brand X influences his rating of an individual appliance of that brand when asked to evaluate it.

Responses of 282 female heads of households were obtained in a large midwestern city. For each of the ten responses (five attributes for two brands) a Friedman analysis of variance was conducted, with the three types of appliances representing the treatments and the 282 consumers acting as blocks. A significant value of the test statistic F_r would indicate consistent ranking of the three appliances by the 282 subjects for the attribute, i.e., that the probability distributions of ranks given the three appliances differ. A small value of F_r would lend credence to the hypothesis that the rankings are randomly performed, i.e., that they have approximately the same probability distributions. McClure found that the rank probability distributions differ for all attributes except price for brand A (at the $\alpha = .10$ level); brand A refrigerators consistently tend to obtain the highest ranking. However, brand B has no such clear pattern, with only the familiarity rankings showing significant differences. The subjects seemed to be most familiar with the brand B clothes washer, but they could not agree on the ranking of the other attributes. McClure concluded his article:

> Many marketing researchers have not been aware of the Friedman two-way analysis of variance by ranks. However, it has potential for use in situations common to many consumer surveys in which sets of ordinal [rank] data are generated by each respondent.

Exercises 10.35–10.42

Learning the Mechanics

10.35 Suppose you have used a randomized block design to help you compare the effectiveness of three different treatments, A, B, and C. You obtained the data listed in the table and plan to conduct a Friedman F_r-test.

a. Specify the null and alternative hypotheses you would test.

BLOCK	TREATMENT A	B	C
1	9	11	18
2	13	13	13
3	11	12	12
4	10	15	16
5	9	8	10
6	14	12	16
7	10	12	15

b. Specify the rejection region for the test. Use $\alpha = .10$.

c. Conduct the test and interpret the results.

10.36 An experiment was conducted using a randomized block design with four treatments and six blocks. The ranks of the measurements within each block are shown in the table. Use the Friedman F_r-test for a randomized block design to determine whether the data provide sufficient evidence to indicate that at least two of the treatment probability distributions differ in location. Test using $\alpha = .05$.

TREATMENT	BLOCK 1	2	3	4	5	6
1	3	3	2	3	2	3
2	1	1	1	2	1	1
3	4	4	3	4	4	4
4	2	2	4	1	3	2

Applying the Concepts

10.37 A sociologist conducted an experiment to investigate the general public's perception of certain occupations. Each of a random sample of fifteen people was asked to rank, in order of prestige, five leading professions: lawyer, politician, physician, corporate president, and college professor. Do the data shown in the table provide sufficient evidence to indicate a difference in the amount of prestige the public attaches to these five professions? Test using $\alpha = .025$.

PERSON	LAWYER	POLITICIAN	PHYSICIAN	CORPORATE PRESIDENT	COLLEGE PROFESSOR
1	3	5	1	4	2
2	4	5	1	2	3
3	1	4	2	5	3
4	3	5	2	4	1
5	4	5	1	3	2
6	5	4	3	2	1
7	1	5	3	4	2
8	4	5	1	3	2
9	3	5	1	4	2
10	4	5	2	3	1
11	5	4	2	3	1
12	3	5	1	4	2
13	4	5	2	3	1
14	3	4	1	5	2
15	4	5	1	2	3

10.38 In Exercise 9.23, we compared the mean flood crests on the Suwannee and Santa Fe rivers for the flood level years 1984, 1983, 1973, and 1948. The data (in feet), matched by year at 6 locations on the rivers, are shown in the table (*Gainesville Sun*, March 15, 1984). Do the data indicate that the flood crests on the rivers were higher or lower in any one of the flood level years than any other? Test using the Friedman F_r-test with $\alpha = .05$.

STATION	FLOOD STAGE	1984	1983	1973	1948
Suwannee River					
White Springs	77.0	79.9	74.4	88.5	85.2
Ellaville	54.0	59.8	52.8	65.0	68.1
Branford	29.0	29.3	29.2	35.6	38.9
Wilcox	14.0	11.6	13.0	18.6	22.3
Santa Fe River					
Three Rivers	19.0	24.1	24.3	27.2	30.6
US 129 Bridge	21.0	23.8	23.8	30.7	34.2

METAL	SEALER I	II	III
1	4.6	4.2	4.9
2	7.2	6.4	7.0
3	3.4	3.5	3.4
4	6.2	5.3	5.9
5	8.4	6.8	7.8
6	5.6	4.8	5.7
7	3.7	3.7	4.1
8	6.1	6.2	6.4
9	4.9	4.1	4.2
10	5.2	5.0	5.1

EAR	SPRAY A	B	C
1	21	23	15
2	29	30	21
3	16	19	18
4	20	19	18
5	13	10	14
6	5	12	6
7	18	18	12
8	26	32	21
9	17	20	9
10	4	10	2

10.39 Corrosion of different metals is a problem in many mechanical devices. Three sealers used to retard the corrosion of metals were tested to see whether there were any differences among them. Samples of ten different metal compositions were treated with each of the three sealers, and the amount of corrosion was measured after exposure to the same environmental conditions for 1 month. The data are given in the table. Is there any evidence of a difference in the probability distributions of the amounts of corrosion among the three types of sealer? Use $\alpha = .05$.

10.40 A serious drought-related problem for farmers is the spread of aflatoxin, a highly toxic substance caused by mold, which contaminates field corn. In higher levels of contamination, aflatoxin is potentially hazardous to animal and possibly human health. (Officials of the FDA have set a maximum limit of 20 parts per billion aflatoxin as safe for interstate marketing.) Three sprays, A, B, and C, have been developed to control aflatoxin in field corn. To determine whether differences exist among the sprays, ten ears of corn are randomly chosen from a contaminated corn field and each is divided into three pieces of equal size. The sprays are then randomly assigned to the pieces for each ear of corn, thus setting up a randomized block design. The table gives the amount (in parts per billion) of aflatoxin present in the corn samples after spraying. Use the Friedman test to determine whether there are differences among the probability distributions of the amounts of aflatoxin present for the three sprays. Test at the $\alpha = .05$ level of significance.

10.41 In recent years, domestic car manufacturers have devoted more attention to the small car market. To compare the popularity of four domestic small cars within a city, a local trade organization obtained the information given in the table from four car dealers—one dealer for each of the four car makes. Is there evidence of a difference in location among the probability distributions of the number of cars sold for each type? Use $\alpha = .10$.

Number of Small Cars Sold

MONTH	MAKE OF CAR			
	A	B	C	D
1	9	17	14	8
2	10	20	16	9
3	13	15	19	12
4	11	12	19	11
5	7	18	13	8

RAT	CHEMICAL		
	A	B	C
1	6	5	3
2	9	8	4
3	6	9	3
4	5	8	6
5	7	8	9
6	5	7	6
7	6	7	5
8	6	5	7

10.42 An experiment is conducted to investigate the toxic effect of three chemicals, A, B, and C, on the skin of rats. Three adjacent 1-inch squares are marked on the backs of eight rats, and each of the three chemicals is applied to each rat. The squares of skin are then scored from 0 to 10, depending on the degree of irritation. The data are given in the table. Is there sufficient evidence to support the alternative hypothesis that at least two of the probability distributions of skin irritation scores corresponding to the three chemicals differ in location? Use $\alpha = .01$.

10.5 Spearman's Rank Correlation Coefficient

Suppose ten new paintings are shown to two art critics and each critic ranks the paintings from 1 (best) to 10 (worst). We want to determine whether the critics' ranks are related. Does a correspondence exist between their ratings? If a painting is rated high by critic 1, is it likely to be rated high by critic 2? Or do high rankings by one critic correspond to low rankings by the other? That is, we wish to determine whether the rankings of the critics are **correlated.**

If the rankings are as shown in the "Perfect Agreement" columns of Table 10.8, we immediately notice that the critics agree on the rank of every painting. High ranks correspond to high ranks and low ranks to low ranks. This is an example of perfect

Table 10.8

Rankings of Ten Paintings by Two Critics

PAINTING	PERFECT AGREEMENT		PERFECT DISAGREEMENT	
	Critic 1	Critic 2	Critic 1	Critic 2
1	4	4	9	2
2	1	1	3	8
3	7	7	5	6
4	5	5	1	10
5	2	2	2	9
6	6	6	10	1
7	8	8	6	5
8	3	3	4	7
9	10	10	8	3
10	9	9	7	4

positive correlation between the ranks. In contrast, if the rankings appear as shown in the "Perfect Disagreement" columns of Table 10.8, high ranks for one critic correspond to low ranks for the other. This is an example of perfect *negative correlation.*

In practice, you will rarely see perfect positive or negative correlation between the ranks. In fact, it is quite possible for the critics' ranks to appear as shown in Table 10.9. You will note that these rankings indicate some agreement between the critics, but not perfect agreement, thus indicating a need for a measure of rank correlation.

Table 10.9
Rankings of Paintings:
Less Than Perfect
Agreement

PAINTINGS	CRITIC 1	CRITIC 2	DIFFERENCE BETWEEN RANK 1 AND RANK 2 d	d^2
1	4	5	−1	1
2	1	2	−1	1
3	9	10	−1	1
4	5	6	−1	1
5	2	1	1	1
6	10	9	1	1
7	7	7	0	0
8	3	3	0	0
9	6	4	2	4
10	8	8	0	0
				$\sum d^2 = 10$

Spearman's rank correlation coefficient, r_s, provides a measure of correlation between ranks. The formula for this measure of correlation is given in the box. We also give a formula that is identical to r_s when there are no ties in rankings; this provides a good approximation to r_s when the number of ties is small relative to the number of pairs.

Note that if the ranks for the two critics are identical, as in the second and third columns of Table 10.8, the differences between the ranks, d, will all be zero. Thus,

$$r_s = 1 - \frac{6\sum d^2}{n(n^2 - 1)} = 1 - \frac{6(0)}{10(99)} = 1$$

That is, *perfect positive correlation* between the pairs of ranks is characterized by a Spearman correlation coefficient of $r_s = 1$. When the ranks indicate perfect disagreement, as in the fourth and fifth columns of Table 10.8, $\sum d_i^2 = 330$ and

$$r_s = 1 - \frac{6(330)}{10(99)} = -1$$

Thus, *perfect negative correlation* is indicated by $r_s = -1$.

For the data of Table 10.9,

$$r_s = 1 - \frac{6\sum d^2}{n(n^2 - 1)} = 1 - \frac{6(10)}{10(99)} = 1 - \frac{6}{99} = .94$$

Spearman's Rank Correlation Coefficient

$$r_s = \frac{SS_{uv}}{\sqrt{SS_{uu}SS_{vv}}}$$

where

$$SS_{uv} = \sum(u_i - \bar{u})(v_i - \bar{v}) = \sum u_i v_i - \frac{(\sum u_i)(\sum v_i)}{n}$$

$$SS_{uu} = \sum(u_i - \bar{u})^2 = \sum u_i^2 - \frac{(\sum u_i)^2}{n}$$

$$SS_{vv} = \sum(v_i - \bar{v})^2 = \sum v_i^2 - \frac{(\sum v_i)^2}{n}$$

u_i = Rank of the ith observation in sample 1

v_i = Rank of the ith observation in sample 2

n = Number of pairs of observations (number of observations in each sample)

Shortcut Formula for r_s*

$$r_s = 1 - \frac{6\sum d_i^2}{n(n^2 - 1)}$$

where

$d_i = u_i - v_i$ (difference in the ranks of the ith observations for samples 1 and 2)

The fact that r_s is *close* to 1 indicates that the critics tend to agree, but the agreement is not perfect.

The value of r_s always falls between -1 and $+1$, with $+1$ indicating perfect positive correlation and -1 indicating perfect negative correlation. The closer r_s falls to $+1$ or -1, the greater the correlation between the ranks. Conversely, the nearer r_s is to zero, the less the correlation. We summarize the properties of r_s in the box on page 520.

Note that the concept of correlation implies that two responses are obtained for each experimental unit. In the art critics example, each painting received two ranks (one for each critic) and the objective of the study was to determine the degree of positive correlation between the two rankings. Rank correlation methods can be used to measure the correlation between any pair of variables. If two variables are measured

---

* The shortcut formula is not exact when there are tied measurements, but it is a good approximation when the total number of ties is not large relative to n.

Properties of Spearman's Rank Correlation Coefficient

1. The value of r_s is always between -1 and $+1$.

2. r_s positive: The ranks of the pairs of sample observations tend to increase together.

3. $r_s = 0$:   The ranks are not correlated.

4. r_s negative: The ranks of one variable tend to decrease as the other variable's ranks increase.

---

on each of n experimental units, we rank the measurements associated with each variable separately. Ties receive the average of the ranks of the tied observations. Then we calculate the value of r_s for the two rankings. This value measures the rank correlation between the two variables. We illustrate the procedure in Example 10.5.

Example 10.5 A study is conducted to investigate the relationship between cigarette smoking during pregnancy and the weights of newborn infants. A sample of fifteen women smokers kept accurate records of the number of cigarettes smoked during their pregnancies, and the weights of their children were recorded at birth. The data are given in Table 10.10. Calculate and interpret Spearman's rank correlation coefficient for the data.

Table 10.10
Data and Calculations
for Example 10.5

WOMAN	CIGARETTES PER DAY	RANK	BABY'S WEIGHT (Pounds)	RANK	d	d^2
1	12	1	7.7	5	-4	16
2	15	2	8.1	9	-7	49
3	35	13	6.9	4	9	81
4	21	7	8.2	10	-3	9
5	20	5.5	8.6	13.5	-8	64
6	17	3	8.3	11.5	-8.5	72.25
7	19	4	9.4	15	-11	121
8	46	15	7.8	6	9	81
9	20	5.5	8.3	11.5	-6	36
10	25	8.5	5.2	1	7.5	56.25
11	39	14	6.4	3	11	121
12	25	8.5	7.9	7	1.5	2.25
13	30	12	8.0	8	4	16
14	27	10	6.1	2	8	64
15	29	11	8.6	13.5	-2.5	6.25
					Total $=$	795

Solution We first rank the number of cigarettes smoked per day, assigning a 1 to the smallest number (12) and a 15 to the largest (46). Note that the two ties receive the averages of their respective ranks. Similarly, we assign ranks to the fifteen babies' weights. Since the number of ties is relatively small, we will use the shortcut formula to calculate r_s. The differences between the ranks of the babies' weights and the ranks of the number of cigarettes smoked per day are shown in Table 10.10. The squares of the differences are also given. Thus,

$$r_s = 1 - \frac{6\sum d_i^2}{n(n^2 - 1)} = 1 - \frac{6(795)}{15(15^2 - 1)} = 1 - 1.42 = -.42$$

This negative correlation coefficient indicates that in this sample an increase in the number of cigarettes smoked per day is associated with (but is not necessarily the *cause* of) a decrease in the weight of the newborn infant. Can this conclusion be generalized from the sample to the population? That is, can we conclude that weights of newborns and the number of cigarettes smoked per day are negatively correlated for the populations of observations for *all* smoking mothers?

If we define ρ_s as the **population Spearman rank correlation coefficient** (i.e., the rank correlation coefficient that could be calculated from all (x, y) values in the population), this question can be answered by conducting the test

H_0: $\rho_s = 0$ (no population correlation between ranks)

H_a: $\rho_s < 0$ (negative population correlation between ranks)

Test statistic: r_s, the *sample* Spearman rank correlation coefficient

To determine a rejection region, we consult Table XIII in the Appendix, which is partially reproduced in Figure 10.12. Note that the left-hand column gives values of n, the number of pairs of observations. The entries in the table are values for an upper-tail rejection region, since only positive values are given. Thus, for $n = 15$ and $\alpha = .05$, the value .441 is the boundary of the upper-tail rejection region, so that $P(r_s > .441) = .05$

Figure 10.12 Reproduction of Part of Table XIII in the Appendix

n	$\alpha = .05$	$\alpha = .025$	$\alpha = .01$	$\alpha = .005$
5	.900	—	—	—
6	.829	.886	.943	—
7	.714	.786	.893	—
8	.643	.738	.833	.881
9	.600	.683	.783	.833
10	.564	.648	.745	.794
11	.523	.623	.736	.818
12	.497	.591	.703	.780
13	.475	.566	.673	.745
14	.457	.545	.646	.716
15	.441	.525	.623	.689
16	.425	.507	.601	.666
17	.412	.490	.582	.645
18	.399	.476	.564	.625
19	.388	.462	.549	.608
20	.377	.450	.534	.591

if H_0: $\rho_s = 0$ is true. Similarly, for negative values of r_s, we have $P(r_s < -.441) = .05$ if $\rho_s = 0$. That is, we expect to see $r_s < -.441$ only 5% of the time if there is really no relationship between the ranks of the variables. The lower-tail rejection region is therefore

Rejection region: $\alpha = .05$ $r_s < -.441$

Since the calculated $r_s = -.42$ is not less than $-.441$, we cannot reject H_0 at the $\alpha = .05$ level of significance. That is, this sample of fifteen smoking mothers provides insufficient evidence to conclude that a negative correlation exists between number of cigarettes smoked and the weight of newborns for the populations of measurements corresponding to all smoking mothers. This does not, of course, mean that no relationship exists. A study using a larger sample of smokers and taking other factors into account (father's weight, sex of newborn child, etc.) would be more likely to discover whether smoking and the weight of a newborn child are related. ∎

A summary of Spearman's nonparametric test for correlation is given in the box.

Spearman's Nonparametric Test for Rank Correlation

One-Tailed Test	**Two-Tailed Test**
H_0: $\rho_s = 0$	H_0: $\rho_s = 0$
H_a: $\rho_s > 0$ (or H_a: $\rho_s < 0$)	H_a: $\rho_s \neq 0$
Test statistic: r_s, the sample rank correlation (see the formulas for calculating r_s)	Test statistic: r_s, the sample rank correlation (see the formulas for calculating r_s)
Rejection region: $r_s > r_{s,\alpha}$ (or $r_s < -r_{s,\alpha}$ when H_a: $\rho_s < 0$)	Rejection region: $r_s < -r_{s,\alpha/2}$ or $r_s > r_{s,\alpha/2}$
where $r_{s,\alpha}$ is the value from Table XIII corresponding to the upper-tail area α and n pairs of observations	where $r_{s,\alpha/2}$ is the value from Table XIII corresponding to the upper-tail area $\alpha/2$ and n pairs of observations

---

Example 10.6   Manufacturers of perishable foods often use preservatives to retard spoilage. One concern is that using too much preservative will change the flavor of the food. Suppose an experiment is conducted using samples of a food product with varying amounts of preservative added. The length of time until the food shows signs of spoiling and a taste rating are recorded for each sample. The taste rating is the average rating for three tasters, each of whom rates each sample on a scale from 1 (good) to 5 (bad). Twelve sample measurements are shown in Table 10.11. Use a nonparametric test to find whether the spoilage times and taste ratings are correlated. Use $\alpha = .05$.

Table 10.11

Data for Example 10.6

SAMPLE	TIME UNTIL SPOILAGE (Days)	RANK	TASTE RATING	RANK
1	30	2	4.3	11
2	47	5	3.6	7.5
3	26	1	4.5	12
4	94	11	2.8	3
5	67	7	3.3	6
6	83	10	2.7	2
7	36	3	4.2	10
8	77	9	3.9	9
9	43	4	3.6	7.5
10	109	12	2.2	1
11	56	6	3.1	5
12	70	8	2.9	4

Note: Tied measurements are assigned the average of the ranks that would be given the measurements if they were different but consecutive.

Solution The test is two-tailed with

$$H_0: \quad \rho_s = 0 \qquad H_a: \quad \rho_s \neq 0$$

Test statistic: $r_s = 1 - \dfrac{6\sum d_i^2}{n(n^2 - 1)}$

Rejection region: Since the test is two-tailed, we need to halve the α-value before consulting Table XIII. For $\alpha = .05$, we calculate $\alpha/2 = .025$ and look up the value of $r_{s,.025}$ corresponding to $n = 12$ pairs of observations:

$$r_{s,\alpha/2} = r_{s,.025} = .591$$

We will reject H_0 if

$$r_s < -.591 \qquad \text{or} \qquad r_s > .591$$

The first step in the computation of r_s is to sum the squares of the differences between ranks:

$$\sum d_i^2 = (2 - 11)^2 + (5 - 7.5)^2 + \cdots + (8 - 4)^2 = 536.5$$

Then

$$r_s = 1 - \frac{6(536.5)}{12(144 - 1)} = -.876$$

Since $-.876 < -.591$, we reject H_0 and conclude that the amount of preservative is associated with the taste ratings of the food. The fact that r_s is negative suggests that the preservative has an adverse effect on the taste.   ∎

Case Study 10.2

Does Perceived Prestige of Occupation Depend on Titles of Jobs?

Wayne J. Villemez and Burton B. Silver (1976) argue that the prestige of occupations may depend on the title used to describe the occupation. This suggestion has important implications because of "the preoccupation of American sociology with status."

Table 10.12 shows ten occupational categories described in three slightly different ways. Each set of descriptions is accompanied by a ranking of those categories obtained from a random sample of 144 college students.* Villemez and Silver report Spearman's rank correlation coefficients between the different occupational descriptions. For description 1 versus description 2, $r_s = .59$; for description 1 versus description 3, $r_s = .49$; and for description 2 versus description 3, $r_s = .33$. The authors state:

These moderate (and nonsignificant) correlations resulting from using different words suggest at least that situs (category) ranking is not entirely consistent.... That is, when rating broad

Table 10.12

Occupational Descriptions and Ranks

OCCUPATIONAL CATEGORY	DESCRIPTION 1	RANK	DESCRIPTION 2	RANK	DESCRIPTION 3	RANK
1	Building and maintenance	9	Constructing and maintaining	7.5	Construction	6.5
2	Arts and entertainment	3	Culture and information	9	Aesthetics and information	3
3	Transportation	8	Movement of goods and people	7.5	Movement of materials and people	8
4	Extraction	10	Removing materials from the earth	10	Farming and mining	9
5	Health and welfare	6.5	Citizen well-being	5.5	Social, physical, and moral well-being	1
6	Commerce	4	Business	2	Trade	5
7	Legal authority	1	Law and control	5.5	Law enforcement	6.5
8	Finance and records	5	Monetary affairs	3	Records and accounts	4
9	Manufacturing	6.5	Production	4	Production and assembly	10
10	Education and research	2	Knowledge	1	Training and education	2

* In this study the lowest rank refers to the occupation rated *highest*. Although this method of assigning ranks is opposite to the method we have been using, the numerical value of r_s, the rank correlation, is the same for both methods.

categories of type of work, raters probably do so with sets of specific familiar occupations in mind, which biases their ratings. The changing titles in this study may have brought to mind different occupational sets.

Case Study 10.3

The Problem of Nonresponse Bias in Mail Surveys

Researchers who collect their sample data via mail questionnaires always run the risk that their respondents will not be a representative sample of the entire population. The reason for this, according to Rosenthal and Rosnow (1975), is the tendency for the respondents to be (1) better educated, (2) of higher social status, (3) more intelligent, (4) in need of social approval, (5) more social, and (6) more interested in the research topic than nonrespondents.

Researchers sometimes attempt to verify the representativeness of their sample by obtaining information about the demographic characteristics of the nonrespondents and comparing them to those of the sample of respondents. Finding similarities gives researchers confidence in the representativeness of their sample. Another approach is to contact a small sample of the nonrespondents and obtain responses to the questionnaire. These responses can then be compared to those of the original respondents. The more similar the patterns of responses, the more confident the researcher can be that the original sample of returned questionnaires is representative of the population sampled from.

In a recent marketing research study by David W. Finn, Chih-Kang Wang, and Charles W. Lamb (1983), a random sample of twenty nonrespondents to their mail questionnaire were contacted by telephone and asked to respond to one of the questions on the questionnaire. The question was: "In general, what is your willingness to buy products made in each of the following countries?" They were asked to indicate their willingness by responding on a 5-point scale that ranged from extremely willing to extremely unwilling. The mean willingness score was computed for each country, and the countries were ranked accordingly. Similarly, a rank ordering was developed for the 273 respondents to their mail questionnaire. These rankings are displayed in Table 10.13.

Table 10.13

Ranking of Consumer Willingness to Buy Products Made in Indicated Countries

COUNTRY	RANK ORDER Respondent	RANK ORDER Nonrespondent
UK	1	1
Japan	2	3
France	3	2
Taiwan	4	4
Brazil	5	5
India	6	6
Iran	7	7
Angola	8	8
USSR	9	9
Cuba	10	10

Note: Data were collected in the spring of 1977.

Finn, Wang, and Lamb compared the rank orderings for the respondents and non-respondents by using Spearman's rank correlation coefficient. They obtained $r_s = .9879$ (p-value < 0.01) and concluded "that respondents and nonrespondents in this study did not differ attitudinally" (p. 336). Further, they noted that even if respondents and nonrespondents to a mail survey differ demographically, as was the case in their study, the Spearman rank correlation result indicates that such differences should not automatically be interpreted as signaling the existence of nonresponse bias. The sample of opinions obtained from the respondents may, in fact, be representative of the population of opinions even though respondents and nonrespondents differ demographically.

Exercises 10.43–10.53

Learning the Mechanics

10.43 Specify the rejection region for Spearman's nonparametric test for rank correlation in each of the following situations:

a. H_0: $\rho_s = 0$; H_a: $\rho_s \neq 0$, $n = 9$, $\alpha = .05$
b. H_0: $\rho_s = 0$; H_a: $\rho_s > 0$, $n = 25$, $\alpha = .025$
c. H_0: $\rho_s = 0$; H_a: $\rho_s < 0$, $n = 30$, $\alpha = .005$

10.44 Compute Spearman's rank correlation coefficient for each of the following pairs of sample observations:

a.
x	30	55	60	19	40
y	26	36	65	25	35

b.
x	90	100	120	137	41
y	81	95	75	52	136

c.
x	1	15	4	10
y	11	26	15	21

d.
x	5	20	15	10	3
y	80	83	91	82	87

PAIR	VALUE OF x	VALUE OF y
1	65	58
2	57	61
3	55	58
4	38	23
5	29	34
6	43	38
7	49	37

10.45 A random sample of seven pairs of observations are recorded on two variables, x and y. The data are shown in the table. Use Spearman's nonparametric test for rank correlation to answer the questions.

a. Do the data provide sufficient evidence to conclude that ρ_s, the rank correlation between x and y, is greater than zero? Test using $\alpha = .05$.

b. Do the data provide sufficient evidence to conclude that $\rho_s \neq 0$? Test using $\alpha = .05$.

10.46 Compute r_s for the data in Case Study 10.3, and compare the result with the value given in the case study.

Applying the Concepts

10.47 Two expert wine tasters were asked to rank six brands of wine. Their rankings are shown in the table. Do the data present sufficient evidence to indicate a positive correlation in the rankings of the two experts?

BRAND	EXPERT 1	EXPERT 2
A	6	5
B	5	6
C	1	2
D	3	1
E	2	4
F	4	3

10.48 An experiment was designed to study whether eye pupil size is related to a person's attempt at deception. Eight students were asked to respond verbally to a series of questions. Before the questioning began, the pupil size of each student was noted and the students were instructed to answer some of the questions dishonestly. (The number of questions answered dishonestly was left to individual choice.) During questioning, the percentage increase in pupil size was recorded. Each student was then given a deception score based on the proportion of questions answered dishonestly. (High scores indicate a large number of deceptive responses.) The results are given in the table. Can you conclude that the percentage increase in eye pupil size is positively correlated with deception score? Use $\alpha = .05$.

STUDENT	DECEPTION SCORE	PERCENTAGE INCREASE IN EYE PUPIL SIZE
1	87	10
2	63	6
3	95	11
4	50	7
5	43	0
6	89	15
7	33	4
8	55	5

10.49 The decision to build a new plant or move an existing plant to a new location involves the long-term commitment of both human and monetary resources. Accordingly, such decisions should be made only after carefully considering the relevant factors associated with numerous alternative plant sites. G. M. Epping (1982) recently examined the relationship between the location factors deemed important by businesses that located in Arkansas and those that considered Arkansas but located elsewhere. A questionnaire that asked manufacturers to rate the importance of thirteen general location factors on a 9-point scale was completed by 118 firms that had moved a plant to Arkansas in the period 1955–1977 and by 73 firms that had recently considered Arkansas but located elsewhere. Epping averaged the importance ratings and arrived at the rankings shown in the table at the top of the next page. Calculate Spearman's rank correlation coefficient for these data. Do the data present significant evidence to indicate a rank correlation between rankings of factor importance for manufacturers locating in Arkansas and similar rankings by manufacturers deciding to locate elsewhere? Test using $\alpha = .05$.

FACTOR	RANK	
	Manufacturers locating in Arkansas	Manufacturers not locating in Arkansas
Labor	1	1
Taxes	2	2
Industrial Site	3	4
Information Sources and Special Inducements	4	5
Legislative Laws and Structure	5	3
Utilities and Resources	6	7
Transportation Facilities	7	8
Raw Material Supplies	8	10
Community	9	6
Industrial Financing	10	9
Markets	11	12
Business Services	12	11
Personal Preferences	13	13

10.50 It is frequently conjectured that income is one of the primary determinants of social status for an adult male. To investigate this theory, fifteen adult males are chosen at random from a community inhabited primarily by professional people, and their annual gross incomes are noted. Each subject is then asked to complete a questionnaire designed to measure social status within the community. The social status scores (higher scores correspond to higher social status) and gross incomes are given in the table for the fifteen adult males.

SUBJECT	SOCIAL STATUS	INCOME (Thousands of dollars)
1	92	29.9
2	51	18.7
3	88	32.0
4	65	15.0
5	80	26.0
6	31	9.0
7	38	11.3
8	75	22.1
9	45	16.0
10	72	25.0
11	53	17.2
12	43	9.7
13	87	20.1
14	30	15.5
15	74	16.5

a. Compute Spearman's rank correlation coefficient for these data.

b. Is there evidence that social status and income are positively correlated? Use $\alpha = .05$.

10.51 Many large businesses send representatives to college campuses to conduct job interviews. To aid the interviewer, one company decides to study the correlation between the strength of an applicant's references (the company requires three references) and the performance of the applicant on the job. Eight recently hired employees are sampled, and independent evaluations of both references and job performance are made on a scale from 1 to 20. The scores are given in the table.

EMPLOYEE	REFERENCES	JOB PERFORMANCE
1	18	20
2	14	13
3	19	16
4	13	9
5	16	14
6	11	18
7	20	15
8	9	12

a. Compute Spearman's rank correlation coefficient for these data.

b. Is there evidence that strength of references and job performance are positively correlated? Use $\alpha = .05$.

10.52 Recreation therapy is currently being used for the treatment of the mentally retarded. Particularly, psychologists theorize that certain types of recreation have a tranquilizing effect on highly excitable patients. In one experiment, nine mentally retarded patients were monitored for 1 week, and the total number of hours each spent in a specially designed recreation room was recorded. The patients were permitted to spend as much time as they wanted in the room. Also recorded was the number of times during the week that some type of tranquilizing medication had to be given to a patient in a highly excited state. Using the data in the table, determine whether there is sufficient

PATIENT	RECREATION TIME (Hours)	TRANQUILIZERS
1	16	3
2	22	1
3	10	4
4	8	9
5	14	5
6	34	2
7	26	0
8	13	10
9	5	9

evidence to indicate that recreation is negatively correlated with the number of times a tranquilizer has to be given. Test using $\alpha = .05$.

10.53 A large manufacturing firm wants to determine whether a relationship exists between the number of work-hours an employee misses per year and the employee's annual wages. A sample of fifteen employees produced the data in the table. Do these data provide evidence that the work-hours missed are related to annual wages? Use $\alpha = .05$.

EMPLOYEE	WORK-HOURS MISSED	ANNUAL WAGES (Thousands of dollars)
1	49	15.8
2	36	17.5
3	127	11.3
4	91	13.2
5	72	13.0
6	34	14.5
7	155	11.8
8	11	20.2
9	191	10.8
10	6	18.8
11	63	13.8
12	79	12.7
13	43	15.1
14	57	24.2
15	82	13.9

Summary

We have presented several useful **nonparametric techniques** for comparing two or more populations. Nonparametric techniques are useful when the underlying assumptions for their parametric counterparts are not justified or when it is impossible to assign specific values to the observations. **Rank sums** are the primary tools of nonparametric statistics. The **Wilcoxon rank sum statistic** and the **Wilcoxon signed rank statistic** can be used to compare two populations for either an **independent sampling experiment** or a **paired difference experiment**. The **Kruskal–Wallis H-test** is applied when comparing k populations using a **completely randomized design.** The **Friedman F_r-test** is used to compare k populations when a **randomized block design** is conducted.

The strength of nonparametric statistics lies in their general applicability. Few restrictive assumptions are required, and they may be used for observations that can be ranked but cannot be exactly measured. Therefore, nonparametric tests, in conjunction with the parametric tests of Chapters 7 to 9, provide a useful set of statistical methods for comparing the locations of two or more population frequency distributions.

Supplementary Exercises 10.54–10.90

Learning the Mechanics

10.54 The data for three independent random samples are shown in the table. It is known that the sampled populations are not normally distributed. Use an appropriate test to determine whether the data provide sufficient evidence to indicate that at least two of the populations differ in location. Test using $\alpha = .05$.

SAMPLE FROM POPULATION 1	SAMPLE FROM POPULATION 2	SAMPLE FROM POPULATION 3
18	12	87
32	33	53
43	10	65
15	34	50
63	18	64
		77

10.55 A random sample of nine pairs of observations are recorded on two variables, x and y. The data are shown in the table.

a. Do the data provide sufficient evidence to indicate that ρ_s, the Spearman rank correlation between x and y, differs from zero? Test using $\alpha = .05$.

b. Do the data provide sufficient evidence to indicate that the probability distribution for x is shifted to the right of that for y? Test using $\alpha = .05$.

PAIR	VALUE OF x	VALUE OF y
1	19	12
2	27	19
3	15	7
4	35	25
5	13	11
6	29	10
7	16	16
8	22	10
9	16	18

10.56 Two independent random samples produced the measurements listed in the table. Do the data provide sufficient evidence to conclude that there is a difference between the locations of the probability distributions for the sampled populations? Test using $\alpha = .05$.

SAMPLE FROM POPULATION 1	SAMPLE FROM POPULATION 2
1.2	1.5
1.9	1.3
.7	2.9
2.5	1.9
1.0	2.7
1.8	3.5
1.1	

10.57 An experiment was conducted using a randomized block design with five treatments and four blocks. The data are shown in the table on page 532. Do the data provide sufficient evidence to conclude that at least two of the treatment probability distributions differ in location? Test using $\alpha = .05$.

TREATMENT	BLOCK 1	2	3	4
1	75	77	70	80
2	65	69	63	69
3	74	78	69	80
4	80	80	75	86
5	69	72	63	77

Applying the Concepts

STORE 1	STORE 2
23.8	16.3
32.9	23.2
17.2	10.3
90.3	28.5
18.1	15.0
34.5	14.3
21.3	19.6
27.1	13.8
20.7	12.6
29.0	23.3

10.58 A national clothing store franchise operates two stores in one city—one urban and one suburban. To stock the stores with clothing suited to the customers' needs, a survey is conducted to determine the incomes of the customers. Ten customers in each store are offered significant discounts if they will reveal the annual income of their household. The results are listed in the table (in thousands of dollars). Is there evidence that customers of one store tend to have higher incomes than customers of the other store? Use $\alpha = .05$.

10.59 An experiment was conducted to compare two print types, A and B, to determine whether type A was easier to read. Ten subjects were randomly divided into two groups of five. Each subject was given the same material to read, one group receiving the material printed with type A, the other group receiving print type B. The time necessary for each subject to read the material (in seconds) is shown below:

Type A: 95, 122, 101, 99, 108
Type B: 110, 102, 115, 112, 120

Do the data provide sufficient evidence to indicate that print type A is easier to read? Test using $\alpha = .05$.

BEFORE	AFTER
12	4
5	2
10	7
9	3
14	8
6	

10.60 A study was conducted to determine whether the installation of a traffic light was effective in reducing the number of accidents at a busy intersection. Samples of 6 months prior to installation and 5 months after installation of the light yielded the number of accidents per month shown in the table.

a. Is there sufficient evidence to conclude that the probability distribution of the number of monthly accidents before installation of the traffic light is shifted to the right of the probability distribution of the number of accidents after installation of the light? Test using $\alpha = .025$.

b. Explain why this type of data might or might not be suitable for analysis using the *t*-test of Chapter 8.

10.61 The length of time required for a human to respond to a new painkiller was tested in the following manner. Seven randomly selected subjects were assigned to receive both aspirin and the new drug. The two treatments were spaced in time and assigned in random order. The length of time (in minutes) required for a subject to indicate that he or she could physically feel pain relief was recorded for both the aspirin and the drug. The data are shown in the table. Do the data provide sufficient evidence to indicate that the probability distribution of the times required to obtain relief with aspirin is shifted to the right of the probability distribution of the times required to obtain relief with the drug? Test using $\alpha = .05$.

SUBJECT	ASPIRIN	DRUG
1	15	7
2	20	14
3	12	13
4	20	11
5	17	10
6	14	16
7	17	11

10.62 A new diet that has recently appeared on the market is supposed to be effective in reducing weight. To determine whether the diet will produce a weight loss within a 1-week period, ten people were randomly chosen and placed on the new diet. Their weights (in pounds) were recorded at the beginning and end of 1 week, as listed in the table. Do the data provide sufficient evidence to indicate that the probability distribution of weights before the diet is shifted to the right of the probability distribution of weights after the diet? Test using $\alpha = .05$.

SUBJECT	BEFORE	AFTER
1	115	112
2	123	124
3	155	153
4	220	219
5	215	216
6	166	166
7	185	180
8	172	172
9	245	241
10	184	182

LOCATION	A	B
1	879	1,085
2	445	325
3	692	848
4	1,565	1,421
5	2,326	2,778
6	857	992
7	1,250	1,303
8	773	1,215

10.63 A manufacturer of household appliances is considering one of two chains of department stores to be the sales merchandiser for its product in the southwest. Before choosing one chain, the manufacturer wants to make a comparison of the product exposure that might be expected for the two chains. Eight locations are selected where both chains have stores and, on a specific day, the number of shoppers entering each store is recorded. The data are shown in the table. Do the data provide sufficient evidence to indicate that one of the chains tends to have more customers per day than the other? Test using $\alpha = .05$.

10.64 A state highway patrol was interested in knowing whether frequent patrolling of highways substantially reduces the number of speeders. Two similar interstate highways were selected for the study—one heavily patrolled and the other only occasionally patrolled. After 1 month, random samples of 100 cars were chosen on each highway and the number of cars exceeding the speed limit was recorded. This process was repeated on 5 randomly selected days. The data are shown in the table.

DAY	HIGHWAY 1 (Heavily patrolled)	HIGHWAY 2 (Occasionally patrolled)
1	35	60
2	40	36
3	25	48
4	38	54
5	47	63

a. Do the data provide evidence to indicate that the heavily patrolled highway tends to have fewer speeders per 100 cars than the occasionally patrolled highway? Test using $\alpha = .05$.

b. Use the paired t-test with $\alpha = .05$ to compare the population mean number of speeders per 100 cars for the two highways. What assumptions are necessary for the validity of this procedure?

10.65 A drug company has synthesized two new compounds to be used in sleeping pills. The data in the table represent the additional hours of sleep gained by ten patients through the use of the two drugs. Do the data present sufficient evidence to indicate that the probability distributions of additional hours of sleep differ for the two drugs? Test using $\alpha = .10$.

PATIENT	DRUG A	DRUG B	PATIENT	DRUG A	DRUG B
1	.4	.7	6	2.9	3.4
2	−.7	−1.6	7	4.0	3.7
3	−.4	−.2	8	.1	.8
4	−1.4	−1.4	9	3.1	.0
5	−1.6	−.2	10	1.9	2.0

10.66 Two fluoride toothpastes and one nonfluoride toothpaste were compared for their effectiveness in preventing cavities. Three randomly selected groups of subjects

FLUORIDE A	FLUORIDE B	NONFLUORIDE
0	2	4
1	0	3
3	3	5
1	3	4
2	0	4
0	1	3
1	2	4
3	1	5
2		
2		

used the toothpastes for 6 months, and each subject was examined before and after the study to determine the number of new cavities that developed. The data are shown in the table. Do the data provide sufficient evidence to indicate that the probability distributions of the number of new cavities differ for at least two of the toothpastes? Use $\alpha = .05$.

A	B	C
10.8	22.3	9.8
15.6	19.5	12.3
19.2	18.6	16.2
17.9	24.3	14.1
18.3	19.9	15.3
9.8	20.4	10.8
16.7	23.6	12.2
19.0	21.2	17.3
20.3	19.8	15.1
19.4	22.6	11.3

10.67 Weevils cause millions of dollars worth of damage each year to cotton crops. Three chemicals (A, B, and C) designed to control weevil populations were applied, one to each of three fields of cotton. After 3 months, ten plots of equal size were randomly selected within each field and the percentage of cotton plants with weevil damage was recorded for each. Do the data in the table provide sufficient evidence to indicate a difference in location among the distributions of damage rates corresponding to the three treatments? Use $\alpha = .05$.

10.68 An economist is interested in knowing whether property tax rates differ among three types of school districts—urban, suburban, and rural. A random sample of several districts of each type produced the data in the table. (The rate is in mills, where 1 mill = $\$1/1{,}000$.) Do the data indicate a difference in the level of property taxes among the three types of school districts? Use $\alpha = .05$.

URBAN	SUBURBAN	RURAL
4.3	5.9	5.1
5.2	6.7	4.8
6.2	7.6	3.9
5.6	4.9	6.2
3.8	5.2	4.2
5.8	6.8	4.3
4.7		

SUPERVISOR			
I	II	III	IV
20	17	16	8
19	11	15	12
20	13	13	10
18	15	18	14
17	14	11	9
	16		10

10.69 Suppose a company wants to study how personality relates to leadership. Four supervisors with different types of personalities are selected. Several employees are then selected from the group supervised by each, and these employees are asked to rate the leader of their group on a scale from 1 to 20 (where 20 signifies highly favorable). The resulting data are shown in the table.

a. Is there sufficient evidence to indicate that at least one of the supervisors tends to receive higher ratings than the others? Use $\alpha = .05$.

b. Suppose the company is particularly interested in comparing the ratings of the personality types represented by supervisors I and III. Make this comparison using $\alpha = .05$.

10.70 Do burn accident rates vary from season to season? To answer this question, twelve hospitals were randomly selected and the number of burn patients admitted during each of the past four seasons was recorded for each. The data are shown in the table on page 536. Is there sufficient evidence to suggest that the frequency of burn accidents varies for at least two seasons? Test using $\alpha = .10$.

HOSPITAL	SUMMER	FALL	WINTER	SPRING
1	20	14	25	16
2	5	4	7	4
3	15	10	14	8
4	18	11	17	12
5	35	28	32	30
6	10	8	10	7
7	27	20	25	19
8	32	16	31	20
9	15	10	12	8
10	7	8	10	6
11	17	7	14	8
12	23	10	21	9

10.71 Three new traps were tested to compare their ability to trap mosquitoes. Each of three traps, A, B, and C, were placed side-by-side at each of four different locations. After a specified length of time, the number of mosquitoes in each trap was recorded. Do the data in the table provide sufficient evidence to indicate a difference among the probability distributions of the number of mosquitoes trapped by the three devices?

	TRAP		
LOCATION	A	B	C
1	3	5	0
2	23	17	15
3	11	5	7
4	19	11	5

10.72 An experiment was conducted to compare three calculators, A, B, and C, according to ease of operation. To make the comparison, six randomly selected students were assigned to perform the same sequence of arithmetic operations on each of the three calculators. The order of use of the calculators varied in a random manner from student to student. The times necessary for the completion of the sequence of tasks (in seconds) are recorded in the table. Do the data provide sufficient evidence to indicate that the task completion times tend to be lower for at least one of the calculator types?

	CALCULATOR TYPE		
STUDENT	A	B	C
1	306	330	300
2	260	265	285
3	281	290	277
4	288	301	305
5	301	309	319
6	262	245	240

10.73 A union wants to determine the preferences of its members before negotiating with management. Ten union members are randomly selected, and an extensive questionnaire is completed by each member. The responses to the various aspects of the questionnaire will enable the union to rank in order of importance the items to be negotiated. The rankings are shown in the table. Is there sufficient evidence to indicate that one or more of the items are preferred to the others? Test using $\alpha = .05$.

PERSON	MORE PAY	JOB STABILITY	FRINGE BENEFITS	SHORTER HOURS
1	2	1	3	4
2	1	2	3	4
3	4	3	2	1
4	1	4	2	3
5	1	2	3	4
6	1	3	4	2
7	2.5	1	2.5	4
8	3	1	4	2
9	1.5	1.5	3	4
10	2	3	1	4

10.74 An insurance company wants to determine whether a relationship exists between the number of claims filed by owners of family policies and the annual incomes of the families. A random sample of ten policies is selected with the results listed in the table. Is there evidence of a relationship between the number of claims filed and the incomes of the family policyholders? Use $\alpha = .10$.

FAMILY	CLAIMS (3-year period)	ANNUAL INCOME (Thousands of dollars, averaged over 3 years)	FAMILY	CLAIMS (3-year period)	ANNUAL INCOME (Thousands of dollars, averaged over 3 years)
1	5	16.5	6	7	17.6
2	1	12.6	7	0	9.3
3	9	62.5	8	2	21.6
4	0	25.6	9	6	20.1
5	4	15.3	10	3	14.5

Number of Patients
Treated per Day

BEFORE	AFTER
26	28
25	30
27	27
26	31
24	29
20	23
22	28
21	29

10.75 The trend among doctors in some areas of the country is to form a group practice. By combining treatment resources, doctors hope to be more efficient. One doctor who recently joined a group family practice wants to compare the distribution of the number of patients he treated during a day when he was an individual practitioner with the distribution after joining the group practice. Records were checked for 8 days before and 8 days after, with the results listed in the table. Use the Wilcoxon rank sum test to determine whether these samples indicate that the doctor tends to treat more patients per day now than before. Use $\alpha = .05$.

10.76 Each applicant to a certain university is judged by his or her high school grade-point average and a score on a standard aptitude test. The GPA's and aptitude scores for eight randomly selected applicants are shown in the table. Do these data provide sufficient evidence to indicate a positive correlation between high school GPA and aptitude test score? Use $\alpha = .05$.

APPLICANT	GPA	APTITUDE TEST
1	3.25	1,200
2	2.85	890
3	3.01	980
4	4.00	1,150
5	3.10	1,510
6	2.90	950
7	2.75	1,010
8	3.35	1,080

10.77 Twelve samples of variously priced carpeting are selected and tested for wearability. The cost per square yard and the number of months of wear for each of the twelve samples are listed in the table. Do the data provide sufficient evidence to indicate that wearability tends to increase as the price increases? Test using $\alpha = .05$.

SAMPLE	COST	MONTHS OF WEAR	SAMPLE	COST	MONTHS OF WEAR
1	$ 9.95	32.5	7	$10.45	25.2
2	8.50	24.8	8	17.95	35.3
3	10.85	25.6	9	12.95	34.6
4	7.99	18.4	10	9.99	29.7
5	15.25	28.3	11	14.85	29.9
6	20.50	20.4	12	9.75	26.3

10.78 The coach of a mediocre basketball team has one all-star player who attempts the vast majority of the team's shots. For each of the last ten games, the star's number of shots and the team's winning margin (negative number implies a loss) are recorded:

Game:	1	2	3	4	5	6	7	8	9	10
Star's shots:	19	15	21	17	25	22	14	11	18	18
Winning margin:	−3	13	−8	−5	−2	16	4	1	−5	−12

Compute Spearman's rank correlation coefficient. Is there evidence of a relationship between the team's performance and the star's number of shots?

10.79 In recent years, many magazines have been forced to raise their prices because of increased postage, printing, and paper costs. Because magazines are now more expensive, some households may be subscribing to fewer magazines than they did 3 years ago. Ten households are selected at random, and the number of magazines subscribed to 3 years ago and now is determined. The results are listed in the table. Does this sample provide sufficient evidence to indicate that households tend to subscribe to fewer magazines now than they did 3 years ago? Use $\alpha = .05$.

Number of
Magazine Subscriptions

HOUSEHOLD	3 YEARS AGO	NOW	HOUSEHOLD	3 YEARS AGO	NOW
1	8	4	6	6	5
2	3	5	7	4	3
3	6	4	8	2	2
4	3	3	9	9	6
5	10	5	10	8	2

Number of No-Shows

BEFORE	AFTER
10	4
5	3
3	8
6	5
7	6
11	4
8	2
9	5
6	7
5	1

10.80 A hotel has had a problem with people reserving rooms for a weekend and then not honoring their reservations (no-shows). As a result, the hotel has developed a new reservation and deposit plan that it hopes will reduce the number of no-shows. One year after the new policy is initiated, the management evaluates its effect in comparison with the old policy. Compare the records given in the table for the ten nonholiday weekends preceding the institution of the new policy and the ten nonholiday weekends preceding the evaluation time. Has the situation improved under the new policy? Test at $\alpha = .05$.

10.81 A clothing manufacturer employs five inspectors who provide quality control of workmanship. Every item of clothing produced carries with it the number of the inspector who checked it. Thus, the company can evaluate an inspector by keeping records of the number of complaints received about products bearing his or her inspection number. Records for 6 months are given in the table.

Number of Returns

MONTH	INSPECTOR				
	I	II	III	IV	V
1	8	10	7	6	9
2	5	7	4	12	12
3	5	8	6	10	6
4	9	6	8	10	13
5	4	13	3	7	15
6	4	8	2	6	9

a. Do these data provide sufficient evidence to indicate a tendency to receive more complaints for at least one of the inspectors' products than for the others? Use $\alpha = .10$.

b. Use the Wilcoxon signed rank test to determine whether evidence exists to indicate that the performances of inspectors I and IV differ. Use $\alpha = .05$.

10.82 A savings and loan association is considering three locations in a large city as potential office sites. The company has hired a marketing firm to compare the incomes of people living in the area surrounding each location. The market researchers interview ten households chosen at random in each area to determine the type of job, length of employment, etc., of those in the households who work. This information will enable them to estimate the annual income of each household. The results are shown in the table.

a. Is there evidence that incomes in at least one of the locations tend to be higher than in the other locations? Use $\alpha = .05$.

Estimated Annual Income
(Thousands of Dollars)

LOCATION		
1	2	3
19.3	24.3	19.5
20.5	30.5	14.3
17.1	35.2	22.2
13.3	57.1	18.2
25.5	33.6	17.6
21.2	27.2	23.3
28.5	88.5	28.3
19.7	32.9	21.7
23.0	26.2	25.0
20.1	29.0	20.2

b. Use the Wilcoxon rank sum test to compare the probability distributions of incomes in locations 1 and 2. Use $\alpha = .05$.

10.83 A manufacturer wants to determine whether the number of defectives produced by its employees tends to increase as the day progresses. Unknown to the employees, a complete inspection is made of every item that is produced on one day, and the hourly fraction defective is recorded. The resulting data are given in the table. Do they provide evidence that the fraction defective increases as the day progresses? Test at the $\alpha = .05$ level.

HOUR	FRACTION DEFECTIVE
1	.02
2	.05
3	.03
4	.08
5	.06
6	.09
7	.11
8	.10

10.84 A businesswoman who is looking for a new investment considers a certain suburban community to be a good location for a new restaurant. She decides to survey some residents in the area to see what type of restaurant would be preferred. Ten people are chosen at random and each is asked to estimate how many times in the past 6 months he or she has eaten in each of three types of restaurants—fast-food, family menu, and smorgasbord. The businesswoman then ranks the numbers for each person to obtain a preference ranking. The results are shown in the table. Is there evidence of differences in locations among the probability distributions of preferences for the three restaurant types? Use $\alpha = .05$.

PERSON	PREFERENCE RANKING OF RESTAURANT TYPE Fast-food	Family menu	Smorgasbord
1	1	2.5	2.5
2	2	1	3
3	3	2	1
4	3	1	2
5	3	1	2
6	2	3	1
7	1.5	1.5	3
8	1	2	3
9	3	1	2
10	3	2	1

10.85 For many years, the Girl Scouts of America have sold cookies using various sales techniques. One troop experimented with several techniques and reported the number of boxes of cookies sold per scout, as listed in the table.

DOOR-TO-DOOR	TELEPHONE	GROCERY STORE STAND	DEPARTMENT STORE STAND
47	63	113	25
93	19	50	36
58	29	68	21
37	24	37	27
62	33	39	18
		77	31

a. Is there sufficient evidence to indicate that at least one sales method tends to produce a larger number of sales per scout than the others? Use $\alpha = .10$.

b. Does the door-to-door technique tend to produce a different number of sales per scout than the grocery store stand? Test using $\alpha = .05$.

10.86 A psychologist ranked a random sample of ten children on two subjective scales according to the amount of paranoid behavior they exhibit and the amount of aggressiveness they exhibit. The rankings according to these two criteria are:

Child:	1	2	3	4	5	6	7	8	9	10
Paranoia:	7	3	6	1	2	4	10	8	5	9
Aggression:	5	1	4	2	8	7	9	3	6	10

Compute Spearman's rank correlation coefficient. Is there evidence of a relationship between aggression and paranoia in children as judged by this psychologist?

10.87 Performance in a personal interview often determines whether a candidate is offered a job. Suppose the personnel director of a company interviews six potential job applicants without knowing anything about their backgrounds and then rates them on a scale from 1 to 10. Independently, the director's supervisor evaluates the background qualifications of each candidate on the same scale. The results are shown in the table. Is there evidence that a candidate's qualification score is positively correlated with interview performance score? Use $\alpha = .05$.

CANDIDATE	QUALIFICATIONS	INTERVIEW
1	10	8
2	8	9
3	9	10
4	4	5
5	5	3
6	6	6

10.88 Two car-rental companies have long waged an advertising war. An independent testing agency is hired to compare the number of rentals at one major airport. After 10 days, the agency has the data listed in the table. At this point, can either car-rental company claim to be number one at this airport? Use $\alpha = .05$.

DAY	RENTAL COMPANY A	RENTAL COMPANY B	DAY	RENTAL COMPANY A	RENTAL COMPANY B
1	29	22	6	16	20
2	26	29	7	35	30
3	19	30	8	43	45
4	28	25	9	29	38
5	27	26	10	32	40

10.89 The health food business has boomed in recent years. Suppose three different health food diets are compared by placing eight overweight individuals on each diet for 6 weeks. The values in the table on the next page represent the weight losses (in pounds) of the twenty-four individuals.

	DIET	
A	B	C
11	0	3
19	4	7
23	19	8
7	15	11
2	8	9
13	11	10
20	14	16
22	17	5

a. Can we conclude that the probability distributions of the weight losses differ for the three diets? Use $\alpha = .05$.

b. Compare the weight loss distributions for diets A and B. Use $\alpha = .10$.

10.90 A taste test conducted to compare three brands of beer utilized ten randomly selected beer drinkers. Each person was given three unmarked glasses of beer—one containing each brand—and was asked to rate each on a scale from 1 to 10. (A higher score indicates a better taste.) Do the data given in the table indicate that at least one of the brands of beer is preferred to the others? Test using $\alpha = .05$.

PERSON	A	B	C	PERSON	A	B	C
1	5	7	3	6	10	9	8
2	8	8	5	7	6	8	7
3	6	7	7	8	5	5	4
4	9	6	7	9	6	8	5
5	9	8	5	10	7	6	4

On Your Own . . .

In Chapters 9 and 10 we have discussed two methods of analyzing a randomized block design. When the populations have normal probability distributions and their variances are equal, we can employ the analysis of variance described in Chapter 9. Otherwise, we can use the Friedman F_r-test.

In the "On Your Own" section of Chapter 9, we asked you to conduct a randomized block design to compare supermarket prices, and to use an analysis of variance to interpret the data. Now use the Friedman F_r-test to compare the supermarket prices.

How do the results of the two analyses compare? How can you explain the similarity (or lack of similarity) between the two results?

References
Agresti, A., & Agresti, B. F. *Statistical methods for the social sciences.* San Francisco: Dellen, 1979.

Clotfelter, C. T. "Detroit decision and white flight." *Journal of Legal Studies,* 1976, 5.

Epping, G. M. "Importance factors in plant location in 1980." *Growth and Change,* April 1982, *13,* 47–51.

Finn, D. W., Wang, C.-K., & Lamb, C. W. "An examination of the effects of sample composition bias in a mail survey." *Journal of the Market Research Society,* October 1983, 25, 331–338.

Garvin, D. "Quality on the line." *Harvard Business Review,* September–October 1983, 65–75.

Gibbons, J. D. *Nonparametric statistical inference.* New York: McGraw-Hill, 1971.

Hollander, M., & Wolfe, D. A. *Nonparametric statistical methods.* New York: Wiley, 1973.

McClure, P. "Analyzing consumer image data using Friedman two-way analysis of variance by ranks." *Journal of Marketing Research,* 1971, 8, 370–371.

Noether, G. E. *Elements of nonparametric statistics.* New York: Wiley, 1967.

Rosenthal, R., & Rosnow, R. L. *The volunteer subject.* New York: Wiley, 1975.

Siegel, S. *Nonparametric statistics for the behavioral sciences.* New York: McGraw-Hill, 1956.

Villemez, W. J., & Silver, B. B. "Occupational situs as horizontal social position: a reconsideration." *Sociology and Social Research,* 1976, *61,* 320–335.

CHAPTER 11

The Chi Square Test and the Analysis of Contingency Tables

Where We've Been . . .

The preceding chapters have presented statistical methods for analyzing many types of data. Most of the techniques presented in Chapters 7 to 9, except those that pertained to binomial proportions, were appropriate for populations of data on quantitative random variables that were independent and, for small samples, had (at least approximately) normal probability distributions with a common variance. Nonparametric statistical procedures were presented in Chapter 10 to compare two or more populations when the assumptions of normality or common variance were likely to be violated or when the responses could only be ranked according to their relative magnitudes.

Where We're Going . . .

The methods of this chapter are appropriate for a type of data that is common in many disciplines, a type exemplified by the binomial data of Chapters 7 and 8. We refer to *count or enumerative data*. Recall that the binomial experiment (Chapter 4) consists of n trials, each of which results in one of *two* outcomes. Thus, we have two classifications into which all the data fall. In this chapter, we will consider the analysis of data that fall into *two or more* categories.

Contents

Many useful experiments consists of *enumerating* the number of occurrences of some event. For example, we may *count* the number of people who recover from leukemia when receiving a newly developed treatment, or the number of consumers who choose each of three brands of coffee, or the number of students who major in each of five different academic disciplines. In some instances, the objective of collecting the **count data** is to analyze the distribution of the counts in the various **classes** or **cells.** For example, we may want to estimate the proportion of smokers who prefer each of three different brands of cigarettes by counting the number in a sample of smokers who buy each brand. We will say that count data classified on a single scale have a **one-dimensional classification.** The analysis of one-dimensional count data is discussed in Section 11.1.

In other instances, the objective of collecting the count data is to study the relationship between two different categorical factors. For example, we may be interested in investigating whether the style of package purchased is related to the sex of the buyer. Or the relationship between socioeconomic status and political party affiliation could be of interest. When count data are classified in a **two-dimensional table,** we call the result a **contingency table.** The analyses of two types of contingency tables are discussed in Sections 11.2 and 11.3.

11.1 One-Dimensional Count Data: The Multinomial Distribution

We first consider experiments that result in classification according to a single criterion, i.e., classification on a one-dimensional scale. Suppose we wish to compare the percentage of voters favoring each of three political candidates running for the same elective position. The voting preferences of a random sample of 150 eligible voters are obtained, and the resulting count data are classified according to a single criterion: candidate preference. The data are shown in Table 11.1. Do you think these data indicate a voter preference for any of the candidates?

Table 11.1
Voter-Preference Survey

CANDIDATE 1	CANDIDATE 2	CANDIDATE 3
61	53	36

To answer this question with a valid statistical analysis, we need to know the underlying probability distribution of these count data. This distribution, called the **multinomial probability distribution,** is an extension of the binomial distribution (Section 4.4). The properties of a multinomial experiment are given in the box at the top of the next page.

Note that our voter-preference survey satisfies the properties of a multinomial experiment. The experiment consists of randomly sampling $n = 150$ voters from a large population of voters containing an unknown proportion p_1 who favor candidate 1, a proportion p_2 who favor candidate 2, and a proportion p_3 who favor candidate 3. Each voter sampled represents a single trial that can result in

---

Properties of the Multinomial Experiment

1. The experiment consists of n identical trials.
2. There are k possible outcomes to each trial.
3. The probabilities of the k outcomes, denoted by $p_1, p_2, \ldots, p_k$, remain the same from trial to trial, where $p_1 + p_2 + \cdots + p_k = 1$.
4. The trials are independent.
5. The random variables of interest are the counts $n_1, n_2, \ldots, n_k$ in each of the k cells.

---

one of three outcomes: The voter will favor either candidate 1, 2, or 3 with probabilities p_1, p_2, and p_3, respectively. (Assume that all voters will have a preference.) The voting preference of any single voter in the sample does not affect the preference of another; consequently, the trials are independent. And, finally, you can see that the recorded data are the numbers of voters in each of the three voter-preference categories. Thus, the voter-preference survey satisfies the five properties of a multinomial experiment. You can see that the properties of the multinomial experiment closely resemble those of the binomial experiment and that, in fact, a binomial experiment is a multinomial experiment for the special case where $k = 2$.

In the voter-preference survey, and in most practical applications of the multinomial experiment, the k outcome probabilities $p_1, p_2, \ldots, p_k$ are unknown and we want to use the survey data to make inferences about their values. The unknown probabilities in the voter-preference survey are

p_1 = Proportion of all voters who favor candidate 1

p_2 = Proportion of all voters who favor candidate 2

p_3 = Proportion of all voters who favor candidate 3

To decide whether the voters have a preference for any of the candidates, we will want to test the null hypothesis that the candidates are equally preferred (that is, $p_1 = p_2 = p_3 = \frac{1}{3}$) against the alternative hypothesis that one candidate is preferred (that is, at least one of the probabilities p_1, p_2, and p_3 exceeds $\frac{1}{3}$). Thus, we want to test

H_0: $p_1 = p_2 = p_3 = \frac{1}{3}$ (no preference)

H_a: At least one of the proportions exceeds $\frac{1}{3}$ (a preference exists)

If the null hypothesis is true and $p_1 = p_2 = p_3 = \frac{1}{3}$, the expected value (mean value) of the number of voters who prefer candidate 1 is given by

$E(n_1) = np_1 = (n)\frac{1}{3} = (150)\frac{1}{3} = 50$

Similarly, $E(n_2) = E(n_3) = 50$ if the null hypothesis is true and no preference exists.

Figure 11.1 Rejection Region for Voter-Preference Survey

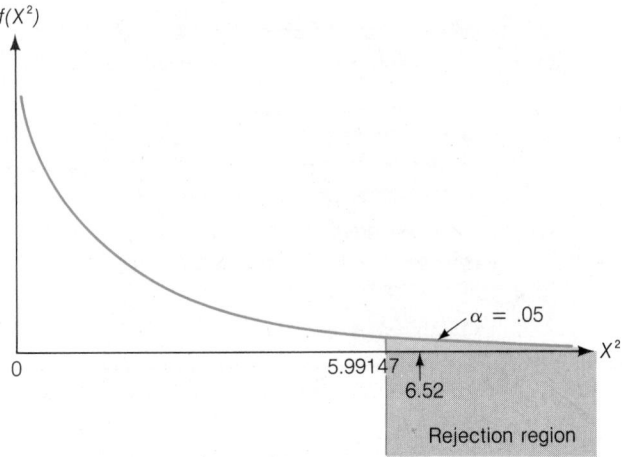

The following test statistic measures the degree of disagreement between the data and the null hypothesis:

$$X^2 = \frac{[n_1 - E(n_1)]^2}{E(n_1)} + \frac{[n_2 - E(n_2)]^2}{E(n_2)} + \frac{[n_3 - E(n_3)]^2}{E(n_3)}$$

$$= \frac{(n_1 - 50)^2}{50} + \frac{(n_2 - 50)^2}{50} + \frac{(n_3 - 50)^2}{50}$$

Note that the farther the observed numbers n_1, n_2, and n_3 are from their expected value (50), the larger X^2 will become. That is, large values of X^2 imply that the null hypothesis is false.

We have to know the distribution of X^2 in repeated sampling before we can decide whether the data indicate that a preference exists. When H_0 is true, X^2 can be shown to have approximately a χ^2-distribution with $(k - 1)$ degrees of freedom.* The properties of the χ^2-distribution and the use of Table VI in the Appendix, which gives the critical values of χ^2, were presented in optional Section 7.7 and again in Section 10.3. The rejection region for the voter-preference survey for $\alpha = .05$ and $k - 1 = 3 - 1 = 2$ df is

Rejection region: $X^2 > \chi^2_{.05}$

This value of $\chi^2_{.05}$ (found in Table VI) is 5.99147. (See Figure 11.1.) The computed value of the test statistic is

$$X^2 = \frac{(n_1 - 50)^2}{50} + \frac{(n_2 - 50)^2}{50} + \frac{(n_3 - 50)^2}{50}$$

$$= \frac{(61 - 50)^2}{50} + \frac{(53 - 50)^2}{50} + \frac{(36 - 50)^2}{50} = 6.52$$

* The derivation of the degrees of freedom for X^2 involves the number of linear restrictions imposed on the count data. In the present case, the only constraint is that $\Sigma n_i = n$, where n (the sample size) is fixed in advance. Therefore, df $= k - 1$. For other cases, we will give the degrees of freedom for each usage of X^2 and refer the interested reader to the references at the end of the chapter for more detail.

Since the computed $X^2 = 6.52$ exceeds the critical value of 5.99147, we conclude at the $\alpha = .05$ level of significance that there does exist a voter preference for one or more of the candidates.

Now that we have evidence to indicate that the proportions p_1, p_2, and p_3 are unequal, we can make inferences concerning their individual values using the methods of Section 7.5. [*Note:* We cannot use the methods of Section 8.4 to compare two proportions because the cell counts are dependent random variables.] The general form for a test of an hypothesis concerning multinomial probabilities is shown in the box.

A Test of an Hypothesis About Multinomial Probabilities

H_0: $p_1 = p_{1,0}$, $p_2 = p_{2,0}$, ..., $p_k = p_{k,0}$,
where $p_{1,0}, p_{2,0}, \ldots, p_{k,0}$ represent the hypothesized values of the multinomial probabilities

H_a: At least one of the multinomial probabilities does not equal its hypothesized value

Test statistic: $X^2 = \sum \dfrac{[n_i - E(n_i)]^2}{E(n_i)}$

where $E(n_i) = np_{i,0}$, the expected number of outcomes of type i assuming that H_0 is true. The total sample size is n.

Rejection region: $X^2 > \chi_\alpha^2$, where χ_α^2 has $(k - 1)$ df

Assumptions: 1. A multinomial experiment has been conducted. This is generally satisfied by taking a random sample from the population of interest.
2. $E(n_i) \geq 5$ for all n_i.*

Example 11.1 Suppose an educational television station has broadcast a series of programs on the physiological and psychological effects of smoking marijuana. Now that the series is finished, the station wants to see whether the citizens within the viewing area have changed their minds about how possession of marijuana should be considered legally. Before the series was shown, it was determined that 7% of the citizens favored legalization, 18% favored decriminalization, 65% favored the existing law (a person could be fined or imprisoned), and 10% had no opinion.

A summary of the opinions (after the series was shown) of a random sample of 500 people in the viewing area is given in Table 11.2 (next page). Test at the $\alpha = .01$ level to see whether these data indicate that the distribution of opinions differs significantly from the proportions that existed before the educational series was aired.

* The assumption that all expected cell counts are larger than 5 is necessary in order to ensure that the χ^2 approximation is appropriate. Exact methods for conducting the test of an hypothesis exist and may be used for small expected cell counts, but these methods are beyond the scope of this text.

Table 11.2

Distribution of Opinions About Marijuana Possession

LEGALIZATION	DECRIMINALIZATION	EXISTING LAWS	NO OPINION
39	99	336	26

Solution

Define the proportions after the airing to be

p_1 = Proportion of citizens favoring legalization

p_2 = Proportion of citizens favoring decriminalization

p_3 = Proportion of citizens favoring existing laws

p_4 = Proportion of citizens with no opinion

Then the null hypothesis representing no change in the distribution of percentages is

H_0: $p_1 = .07$, $p_2 = .18$, $p_3 = .65$, $p_4 = .10$

and the alternative is

H_a: At least one of the proportions differs from its null hypothesized value

Test statistic: $X^2 = \sum \dfrac{[n_i - E(n_i)]^2}{E(n_i)}$

where

$E(n_1) = np_{1,0} = 500(.07) = 35$

$E(n_2) = np_{2,0} = 500(.18) = 90$

$E(n_3) = np_{3,0} = 500(.65) = 325$

$E(n_4) = np_{4,0} = 500(.10) = 50$

Since all these values are larger than 5, the χ^2 approximation is appropriate. Also, if the citizens in the sample were randomly selected, the properties of the multinomial probability distribution are satisfied.

Rejection region: For $\alpha = .01$ and df $= k - 1 = 3$, reject H_0 if $X^2 > \chi^2_{.01}$, where (from Table VI in the Appendix) $\chi^2_{.01} = 11.3449$.

We now calculate the test statistic:

$$X^2 = \frac{(39 - 35)^2}{35} + \frac{(99 - 90)^2}{90} + \frac{(336 - 325)^2}{325} + \frac{(26 - 50)^2}{50} = 13.249$$

Since this value exceeds the table value of χ^2 (11.3449), the data provide sufficient evidence ($\alpha = .01$) that the opinions on legalization of marijuana have changed since the series was aired. ∎

If we focus on one particular outcome of a multinomial experiment, we can use the methods developed in Section 7.5 for a binomial proportion to establish a confidence interval for any one of the multinomial probabilities.* For example, if we want a 95% confidence interval for the proportion of citizens in the viewing area who have no opinion about the issue, we calculate

$$\hat{p}_4 \pm 1.96\sigma_{\hat{p}_4}$$

where

$$\hat{p}_4 = \frac{n_4}{n} = \frac{26}{500} = .052 \quad \text{and} \quad \sigma_{\hat{p}_4} \approx \sqrt{\frac{\hat{p}_4(1 - \hat{p}_4)}{n}}$$

Thus, we get

$$.052 \pm 1.96 \sqrt{\frac{(.052)(.948)}{500}} = .052 \pm .019$$

or (.033, .071). Thus, we estimate that between 3.3% and 7.1% of the citizens now have no opinion on the issue of marijuana legalization. The series of programs may have helped citizens who formerly had no opinion on the issue to form an opinion, since it appears that the proportion of "no opinions" is now less than 10%.

Exercises 11.1–11.14

Learning the Mechanics

11.1 Use Table VI of the Appendix to find each of the following χ^2-values:
a. $\chi^2_{.05}$ for df = 15 **b.** $\chi^2_{.990}$ for df = 100
c. $\chi^2_{.10}$ for df = 12 **d.** $\chi^2_{.005}$ for df = 2

11.2 Use Table VI of the Appendix to find the following probabilities:
a. $P(\chi^2 \leq .872085)$ for df = 6 **b.** $P(\chi^2 > 30.5779)$ for df = 15
c. $P(\chi^2 \geq 82.3581)$ for df = 100 **d.** $P(\chi^2 < 13.7867)$ for df = 30

11.3 Find the rejection region for a one-dimensional χ^2-test of a null hypothesis concerning $p_1, p_2, \ldots, p_k$ if:
a. $k = 3$; $\alpha = .10$ **b.** $k = 5$; $\alpha = .01$ **c.** $k = 4$; $\alpha = .05$

11.4 What conditions must n satisfy to make the χ^2-test valid?

11.5 A multinomial experiment with $k = 3$ cells and $n = 300$ produced the data shown in the table. Do these data provide sufficient evidence to contradict the null hypothesis that $p_1 = .25$, $p_2 = .25$, and $p_3 = .50$? Test using $\alpha = .05$.

	CELL	
1	2	3
n_i 70	66	164

11.6 A multinomial experiment with $k = 4$ cells and $n = 400$ produced the data shown in the table at the top of the next page. Do these data provide sufficient evidence to contradict the null hypothesis that $p_1 = .2$, $p_2 = .4$, $p_3 = .1$, and $p_4 = .3$? Test using $\alpha = .05$.

* Note that focusing on one outcome has the effect of lumping the other $(k - 1)$ outcomes into a single group. Thus, we obtain, in effect, two outcomes—or a binomial experiment.

	CELL			
	1	2	3	4
n_i	71	194	46	89

11.7 A multinomial experiment with four possible outcomes and 100 trials produced the data shown in the table.

	OUTCOME			
	1	2	3	4
n_i	29	16	32	23

a. Do these data provide sufficient evidence to indicate that the four outcomes are not equally likely? Test using $\alpha = .10$.

b. Form a 95% confidence interval for p_3, the probability that outcome 3 will occur on a particular trial.

Applying the Concepts

11.8 Overweight trucks are responsible for much of the damage sustained by our local, state, and federal highway systems. Though illegal, overweight trucks proliferate. Truckers have learned to avoid weigh stations run by enforcement officers by taking back roads when weigh stations are open and traveling during periods of the week when weigh stations are likely to be closed. A state highway planning agency recently monitored the movements of overweight trucks on an interstate highway using an unmanned, computerized scale that is built into the highway. Unknown to the truckers, the scale weighs their vehicles as they pass over it. For a particular week, each day's proportion of the week's total truck traffic (five-axle semitrailers) was as follows:

MONDAY	TUESDAY	WEDNESDAY	THURSDAY	FRIDAY	SATURDAY	SUNDAY
.191	.198	.187	.180	.155	.043	.046

Source: Dahlin and Owen (1984), p. 10.

During the same week, the number of overweight trucks per day was:

MONDAY	TUESDAY	WEDNESDAY	THURSDAY	FRIDAY	SATURDAY	SUNDAY
90	82	72	70	51	18	31

a. The planning agency would like to know whether the number of overweight trucks per week is distributed over the 7 days of the week in direct proportion to the volume of truck traffic. Test using $\alpha = .05$.

b. Find the approximate p-value for the test of part a.

11.9 A pharmacologist studied a random sample of 168 pregnant women from the time they were found to be pregnant through labor and delivery, and she recorded the number of different drug products that each used. A summary of the number of

Drug Products Used

| | | 5 OR |
NONE	1–4	MORE
4	8	156

women for each of three categories is shown in the table. Suppose the researcher theorizes that the proportion of women using no drug products during pregnancy is .05 and the proportion using five or more drug products is .80.

a. Do these data indicate that the proportions of pregnant women in the three categories differ significantly from the proportions represented by the pharmacologist's hypothesis? Test at the $\alpha = .01$ level of significance.

b. Find a 95% confidence interval for the proportion of women using five or more drug products during pregnancy. Interpret this interval.

11.10 After purchasing a policy from one life insurance company, a person has a certain period of time in which the policy can be canceled without financial obligation. An insurance company is interested in seeing whether those who cancel a policy during this time period are as likely to be in one policy-size category as another. Records for 250 people who canceled policies during this period are selected at random from company files with the results shown in the table. Is there sufficient evidence to conclude that the canceled policies are not distributed equally among the five policy-size categories? Use $\alpha = .05$.

Number of People Canceling per Policy-Size Category

| SIZE OF POLICY (THOUSANDS OF DOLLARS) | | | | |
10	15	20	25	30
31	39	67	54	59

11.11 In the game of chess, the first few moves play a very important role in determining the final outcome. Five different opening strategies are highly favored by chess experts. To determine whether one or more of these strategies is most preferred by grand masters in international competition, a random sample of 100 grand masters is taken, and each is asked which of the strategies he or she would prefer to employ. A summary of their responses is shown here:

Strategy: A   B   C   D   E
Frequency: 17   27   22   15   19

Do these data present sufficient evidence to indicate a preference for one or more of the strategies? Use $\alpha = .05$.

11.12 There are four standard surgical techniques, A, B, C, and D, presently used in abdominal surgery. To find out whether one method is preferred over any other, each of a random sample of 200 surgeons was asked which technique he or she preferred. A summary of the data is shown here:

A   B   C   D
48   68   45   39

Do the data provide sufficient evidence to indicate differences in preferences for the four techniques? Test using $\alpha = .05$.

11.13 According to a geneticist's theory, a crossing of red and white snapdragons should produce offspring that are 25% red, 50% pink, and 25% white. An experiment conducted to test the theory produces 30 red, 78 pink, and 36 white offspring in 144 crossings. Do the data provide sufficient evidence to contradict the geneticist's theory?

BRAND			
A	B	C	D
39	57	55	49

11.14 Supermarket chains often carry products with their own brand labels and usually price them lower than the nationally known brands. A supermarket conducted a taste test to determine whether there was a difference in taste among the four brands of ice cream it carried: a local brand (A) and three national brands (B, C, D). A sample of 200 people participated, and they indicated the preferences shown in the table. Is there evidence of a difference in preference for the four brands? Test at $\alpha = .05$.

11.2 Contingency Tables

In Section 11.1, we introduced the multinomial probability distribution and considered data classified according to a single criterion. We now consider multinomial experiments in which the data are classified according to two criteria, i.e., *classification with respect to two factors.*

For example, the energy shortage has made many consumers more aware of the size of the automobiles they purchase. Suppose an automobile manufacturer is interested in determining the relationship between the size and manufacturer of newly purchased automobiles. One thousand recent buyers of American-made cars are randomly sampled, and each purchase is classified with respect to the size and manufacturer of the automobile. The data are summarized in the **two-way table** shown in Table 11.3. This table is called a **contingency table;** it presents multinomial count data classified on two scales, or **dimensions,** of classification—namely automobile size and manufacturer.

Table 11.3

Contingency Table for Automobile Size Example

	MANUFACTURER				
	A	B	C	D	TOTALS
SMALL	157	65	181	10	413
INTERMEDIATE	126	82	142	46	396
LARGE	58	45	60	28	191
TOTALS	341	192	383	84	1,000

The symbols representing the cell counts for the multinomial experiment in Table 11.3 are shown in Table 11.4, part A, and the corresponding cell, row, and column probabilities are shown in Table 11.4, part B. Thus, n_{11} represents the number of buyers who purchase a small car of manufacturer A and p_{11} represents the corresponding cell probability. Note the symbols for the row and column totals and also the symbols for the probability totals. The latter are called **marginal probabilities** for each row and column. The marginal probability p_1 is the probability that a small car is purchased; the marginal probability p_A is the probability that a car by manufacturer A is purchased. Thus, $p_1 = p_{11} + p_{12} + p_{13} + p_{14}$ and $p_A = p_{11} + p_{21} + p_{31}$.

Table 11.4

(A) Observed Counts for Contingency Table 11.3

| | MANUFACTURER | | | | |
	A	B	C	D	TOTALS
SMALL	n_{11}	n_{12}	n_{13}	n_{14}	n_1
INTERMEDIATE	n_{21}	n_{22}	n_{23}	n_{24}	n_2
LARGE	n_{31}	n_{32}	n_{33}	n_{34}	n_3
TOTALS	n_A	n_B	n_C	n_D	n

(B) Probabilities for Contingency Table 11.3

| | MANUFACTURER | | | | |
	A	B	C	D	TOTALS
SMALL	p_{11}	p_{12}	p_{13}	p_{14}	p_1
INTERMEDIATE	p_{21}	p_{22}	p_{23}	p_{24}	p_2
LARGE	p_{31}	p_{32}	p_{33}	p_{34}	p_3
TOTALS	p_A	p_B	p_C	p_D	1

Thus, we can see that this really is a multinomial experiment with a total of 1,000 trials, $(3)(4) = 12$ cells or possible outcomes, and probabilities for each cell as shown in Table 11.4, part B. If the 1,000 recent buyers are randomly chosen, the trials are considered independent and the probabilities are viewed as remaining constant from trial to trial.

Suppose we want to know whether the two classifications, manufacturer and size, are dependent. That is, if we know which size car a buyer will choose, does that information give us a clue about the manufacturer of the car the buyer will choose? In a probabilistic sense we know (Chapter 3) that independence of events A and B implies $P(AB) = P(A)P(B)$. Similarly, in the contingency table analysis, if the two classifications are independent, the probability that an item is classified in any particular cell of the table is the product of the corresponding marginal probabilities. Thus, under the hypothesis of independence, in Table 11.4, part B, we must have

$$p_{11} = p_1 p_A \qquad p_{12} = p_1 p_B$$

and so forth.

To test the hypothesis of independence, we use the same reasoning employed in the one-dimensional tests of Section 11.1. First, we calculate the expected, or mean, count in each cell assuming that the null hypothesis of independence is true. We do this by noting that the expected count in a cell of the table is just the total number of multinomial trials, n, times the cell probability. Recall that n_{ij} represents the observed count in the cell located in the ith row and jth column. Then the expected cell count for the upper-left-hand cell (first row, first column) is

$$E(n_{11}) = np_{11}$$

or, when the null hypothesis (the classifications are independent) is true,

$$E(n_{11}) = np_1 p_A$$

Since these true probabilities are not known, we estimate p_1 and p_A by the sample proportions $\hat{p}_1 = n_1/n$ and $\hat{p}_A = n_A/n$. Thus, the estimate of the expected value $E(n_{11})$ is

$$\hat{E}(n_{11}) = n\left(\frac{n_1}{n}\right)\left(\frac{n_A}{n}\right) = \frac{n_1 n_A}{n}$$

Similarly, for each i, j,

$$\hat{E}(n_{ij}) = \frac{(\text{Row total})(\text{Column total})}{\text{Total sample size}}$$

Thus,

$$\hat{E}(n_{12}) = \frac{n_1 n_B}{n}$$

$$\vdots \qquad \vdots$$

$$\hat{E}(n_{34}) = \frac{n_3 n_D}{n}$$

Using the data in Table 11.3, we find

$$\hat{E}(n_{11}) = \frac{n_1 n_A}{n} = \frac{(413)(341)}{1,000} = 140.833$$

$$\hat{E}(n_{12}) = \frac{n_1 n_B}{n} = \frac{(413)(192)}{1,000} = 79.296$$

$$\vdots \qquad \vdots \qquad \vdots \qquad \vdots$$

$$\hat{E}(n_{34}) = \frac{n_3 n_D}{n} = \frac{(191)(84)}{1,000} = 16.044$$

The observed data and the estimated expected values (in parentheses) are shown in Table 11.5.

Table 11.5

Observed and Estimated Expected (in Parentheses) Counts

	MANUFACTURER			
	A	*B*	*C*	*D*
SMALL	157 (140.833)	65 (79.296)	181 (158.179)	10 (34.692)
INTERMEDIATE	126 (135.036)	82 (76.032)	142 (151.668)	46 (33.264)
LARGE	58 (65.131)	45 (36.672)	60 (73.153)	28 (16.044)

We now use the X^2-statistic to compare the observed and expected (estimated) counts in each cell of the contingency table:

$$X^2 = \frac{[n_{11} - \hat{E}(n_{11})]^2}{\hat{E}(n_{11})} + \frac{[n_{12} - \hat{E}(n_{12})]^2}{\hat{E}(n_{12})} + \cdots + \frac{[n_{34} - \hat{E}(n_{34})]^2}{\hat{E}(n_{34})}$$

$$= \sum \frac{[n_{ij} - \hat{E}(n_{ij})]^2}{\hat{E}(n_{ij})} \,^*$$

* The use of $\sum$ in the context of a contingency table analysis refers to a sum over all cells in the table.

Substituting the data of Table 11.5 into this expression, we get

$$X^2 = \frac{(157 - 140.833)^2}{140.833} + \frac{(65 - 79.296)^2}{79.296} + \cdots + \frac{(28 - 16.044)^2}{16.044} = 45.81$$

Large values of X^2 imply that the observed and expected counts do not closely agree and therefore imply that the hypothesis of independence is false. To determine how large X^2 must be before it is too large to be attributed to chance, we make use of the fact that the sampling distribution of X^2 is approximately a χ^2 probability distribution when the classifications are independent.

When testing the null hypothesis of independence in a two-way contingency table, the appropriate degrees of freedom will be $(r - 1)(c - 1)$, where r is the number of rows and c is the number of columns in the table.

For the size and make of automobiles example, the degrees of freedom for χ^2 is $(r - 1)(c - 1) = (3 - 1)(4 - 1) = 6$. Then, for $\alpha = .05$, we reject the hypothesis of independence when

$$X^2 > \chi^2_{.05} = 12.5916$$

Since the computed $X^2 = 45.81$ exceeds the value 12.5916, we conclude that the size and manufacturer of a car selected by a purchaser are dependent events.

The general form of a two-way contingency table containing r rows and c columns (called an $r \times c$ contingency table) is shown in Table 11.6. Note that the observed count in the (ij) cell is denoted by n_{ij}, the ith row total is r_i, the jth column total is c_j, and the total sample size is n. Using this notation, we give the general form of the contingency table test for independent classifications in the box.

General Form of a Contingency Table Analysis: A Test for Independence

H_0: The two classifications are independent

H_a: The two classifications are dependent

Test statistic: $X^2 = \sum \dfrac{[n_{ij} - \hat{E}(n_{ij})]^2}{\hat{E}(n_{ij})}$

where

$$\hat{E}(n_{ij}) = \frac{r_i c_j}{n}$$

Rejection region: $X^2 > \chi^2_\alpha$, where χ^2_α has $(r - 1)(c - 1)$ df.

Assumptions: 1. The n observed counts are a random sample from the population of interest. We may then consider this to be a multinomial experiment with $r \times c$ possible outcomes.
2. For the χ^2 approximation to be valid, we require that the estimated expected counts exceed 5 in all cells.

Table 11.6
General $r \times c$
Contingency Table

		COLUMN				ROW
		1	2	...	c	TOTALS
ROW	1	n_{11}	n_{12}	...	n_{1c}	r_1
	2	n_{21}	n_{22}	...	n_{2c}	r_2
	⋮	⋮	⋮		⋮	⋮
	r	n_{r1}	n_{r2}	...	n_{rc}	r_r
COLUMN TOTALS		c_1	c_2	...	c_c	n

Example 11.2

A social scientist wants to determine whether the marital status (divorced or not divorced) of American men is independent of their religious affiliation (or lack thereof). A sample of 500 American men is surveyed and the results are tabulated as shown in Table 11.7. Test to see whether there is sufficient evidence to indicate that the marital status of men who have been or are currently married is dependent upon religious affiliation. Test using $\alpha = .01$.

Table 11.7
Observed and Estimated
(in Parentheses) Expected
Counts, Example 11.2

	RELIGIOUS AFFILIATION					
	A	B	C	D	None	TOTALS
DIVORCED	39 (48.952)	19 (18.560)	12 (12.992)	28 (22.736)	18 (12.760)	116
NEVER DIVORCED	172 (162.048)	61 (61.440)	44 (43.008)	70 (75.264)	37 (42.240)	384
TOTALS	211	80	56	98	55	500

Solution   The first step is to calculate estimated expected cell frequencies under the assumption that the classifications are independent. Thus,

$$\hat{E}(n_{11}) = \frac{r_1 c_1}{n} = \frac{(116)(211)}{500} = 48.952$$

$$\hat{E}(n_{12}) = \frac{r_1 c_2}{n} = \frac{(116)(80)}{500} = 18.560$$

and so forth. All the estimated expected cell counts are shown in Table 11.7.

We are now ready to conduct the test for independence:

H_0:   The marital status of American men and their religious affiliation are independent

H_a: The marital status of American men and their religious affiliation are dependent

Test statistic: $X^2 = \sum \dfrac{[n_{ij} - \hat{E}(n_{ij})]^2}{\hat{E}(n_{ij})}$

Since all the estimated expected cell frequencies are greater than 5, the χ^2 approximation is appropriate. Assuming the men chosen were randomly selected from all married or previously married American men, the characteristics of the multinomial probability distribution are satisfied.

Rejection region: For $\alpha = .01$ and $(r-1)(c-1) = (1)(4) = 4$ df, reject H_0 if $X^2 > \chi^2_{.01}$, where $\chi^2_{.01} = 13.2767$

The calculated value of the test statistic is

$$X^2 = \frac{(39 - 48.952)^2}{48.952} + \frac{(19 - 18.560)^2}{18.560} + \cdots + \frac{(37 - 42.240)^2}{42.240}$$

$$= 7.135$$

Since $X^2 = 7.135$ is less than $\chi^2_{.01} = 13.2767$, we cannot conclude that the marital status of American men depends on their religious affiliation. (Note that we could not reject H_0 even with $\alpha = .10$, since $\chi^2_{.10} = 7.77944$.) ∎

Case Study 11.1

Evaluating a New Method for Treating Cancer

The typical method of treating moderately advanced cancer of the larynx (voice box) is removal by surgery. This method achieves initial control of the cancer in approximately 75% of the cases; if the cancer recurs, salvage treatment results in approximately 87% of the cancers ultimately controlled. Recent attempts have been made to treat cancer of the larynx by radiation therapy alone, thereby saving the patient's larynx. W. M. Mendenhall et al. (1984) present data on a group of patients treated by this method at the University of Florida's Shands Hospital. Eighteen patients with cancer of the larynx were treated by radiation alone and twenty-three were treated by surgery alone. Of those treated by radiation, eleven cancers were controlled at the primary site (the larynx). In seven patients the cancer recurred in the larynx; six were treated by surgery and one refused further treatment. Of these, four were controlled so that the number ultimately controlled by radiation therapy alone (eleven) or by surgical salvage (four) was fifteen of eighteen patients. Initial control of the cancer was achieved for eighteen of the twenty-three patients treated by surgery alone. Salvage treatment was successful for three of the remaining five. Thus, removal of the larynx by surgery alone achieved ultimate control in twenty-one of the twenty-three cases. These data are summarized in Table 11.8.

Table 11.8

Comparison of Two Methods for Treating Cancer of the Larynx

	SURGERY	RADIATION THERAPY
NUMBER OF CANCERS ULTIMATELY CONTROLLED	21	15
NUMBER NOT ULTIMATELY CONTROLLED	2	3
TOTALS	23	18

The data of Table 11.8 cannot be analyzed using the chi square test to detect differences in rates of control for the two methods of treatment. The expected numbers in the "no ultimate control" cells are too small. However, the same test can be conducted by using a small-sample method known as *Fisher's exact test*. This method calculates the exact probability (*p*-value) of observing sample results at least as contradictory to the hypothesis of independence as those observed for the researchers' data. Mendenhall and coworkers report the *p*-value for this test as .187, a value that indicates little evidence of a difference in the rates of achieving ultimate control for the two methods of treatment.

Mendenhall and colleagues clearly intend their research to be a first step in a study that will require the treatment of many more cancer patients using radiation alone. Small-sample methods are available for constructing a confidence interval for the rate of achieving ultimate control using radiation treatment alone, but the interval width would be quite large. As we learned in Chapter 7, it requires a fairly large sample to estimate a binomial proportion p with a small margin of error. This larger sample will be obtained by combining treatment results from other cancer clinics and by collecting data on patients who will be treated in the future.

Exercises 11.15–11.27

Learning the Mechanics

11.15 Find the rejection region for a test of independence of two classifications if the contingency table contains r rows and c columns and:

a. $r = 5$, $c = 5$, $\alpha = .05$ **b.** $r = 3$, $c = 6$, $\alpha = .10$
c. $r = 2$, $c = 3$, $\alpha = .01$

11.16 Test the null hypothesis of independence of the two classifications, A and B, of the 3×3 (that is, $r = 3$, $c = 3$) contingency table shown here. Test using $\alpha = .05$.

		B		
		B_1	B_2	B_3
	A_1	39	75	42
A	A_2	63	51	70
	A_3	30	38	29

11.17 a. Test the null hypothesis that the rows and columns for the 4×3 contingency table shown here are independent. Test using $\alpha = .01$.

		COLUMN		
		1	2	3
	1	20	30	50
	2	50	30	40
ROW	3	100	50	50
	4	40	0	60

b. Form a 90% confidence interval for p_{11}, the probability of observing a response in the first row and first column of the table.

Applying the Concepts

11.18 Many scientists believe that alcoholism is linked to social isolation. One measure of social isolation is marital status, i.e., whether a person is married or not. To test the notion that alcoholics are socially isolated, 280 adults were randomly selected and each was classified as a diagnosed alcoholic, undiagnosed alcoholic, or nonalcoholic and categorized according to his or her marital status. A summary of the responses is shown in the table. Can you conclude that there is a relationship between the marital status and alcoholic classifications? Test using $\alpha = .05$.

| | | ALCOHOLIC CLASSIFICATION | | |
		Diagnosed	Undiagnosed	Nonalcoholic
MARITAL STATUS	Married	21	37	58
	Not married	59	63	42

11.19 David Behar and Mark A. Stewart (1984) conducted a study of fifty-eight children admitted to the child psychiatry ward of the University of Iowa for aggressive conduct. The purpose of the study was to determine the influence of social class, sex, and age on the clinical characteristics of the children. A small portion of their data is shown in the table, which lists the numbers of children exhibiting antisocial behavior for two categories of social class. Classes I to III represent children from middle-class families. Those in Classes IV and V were from poor families.

	CLASSES I–III	CLASSES IV–V
Numbers exhibiting antisocial behavior	24	17
Sample size	28	30

a. Do the data provide sufficient evidence to indicate a dependence between antisocial behavior and social class? Test using $\alpha = .05$.
b. The authors give the approximate p-value for the test as .015. Calculate the approximate p-value for the test and compare it with the authors' value. Interpret the p-value.

11.20 Refer to Exercise 11.19 and the study by Behar and Stewart (1984) on aggressive behavior in children. Another comparison given in their paper was the numbers of male and female children for whom physical aggressiveness was a presenting complaint. The data are shown in the table.

	FEMALE	MALE
Number for which physical aggressiveness is a presenting complaint	7	40
Sample size	12	46

a. Do the data provide sufficient evidence to indicate a dependence between sex of the child and the proportion for which physical aggressiveness is a presenting complaint? Test using $\alpha = .05$.
b. The authors give the p-value for the test as .025. Find the p-value for your test and compare it with the authors' value. Interpret the p-value.

11.21 One criterion used to evaluate employees in the assembly section of a large factory is the number of defective pieces per 1,000 parts produced. The quality control department wants to find out whether there is a relationship between years of experience and defect rate. Since the job is repetitious, after the initial training period any improvement due to a learning effect might be offset by a decrease in the motivation of a worker. A defect rate is calculated for each worker for a yearly evaluation. The results for 100 workers are given in the table. Is there evidence of a relationship between defect rate and years of experience? Use $\alpha = .05$.

| | | YEARS OF EXPERIENCE (AFTER TRAINING PERIOD) | | |
		1	2–5	6–10
	High	6	9	9
DEFECT RATE	Average	9	19	23
	Low	7	8	10

11.22 An experimenter wishes to determine whether there is a relationship between hair color and eye color. One hundred people are randomly sampled and the eyes and hair of each person are judged to be light or dark. A summary of the number of people in each of the four categories is shown in the table. Do the data provide sufficient evidence to indicate a relationship between eye and hair color? Test using $\alpha = .10$.

	LIGHT HAIR	DARK HAIR	TOTALS
LIGHT EYES	31	21	52
DARK EYES	14	34	48
TOTALS	45	55	100

11.23 An insurance company that sells hospitalization policies wants to know whether there is a relationship between the amount of hospitalization coverage a person has and the length of stay in the hospital. Records are selected at random at a large hospital by hospital personnel, and the information on length of stay and hospitalization coverage is given to the insurance company. The results are summarized in the table. Can you conclude that there is a relationship between length of stay and hospitalization coverage? Use $\alpha = .01$.

| | | LENGTH OF STAY IN HOSPITAL (DAYS) | | | |
		5 or under	6–10	11–15	Over 15
	Under 25%	26	30	6	5
HOSPITALIZATION COVERAGE OF COSTS	25 to <50%[a]	21	30	11	7
	50 to <75%	25	25	45	9
	75% and over	11	32	17	11

[a] The symbol < is read as "less than."

11.24 Professors Leigh Lawton and A. Parasuraman (1980) have investigated the impact of marketing on the role of the consumer in influencing the design and development of new products. A total of 107 companies were involved in the study and, based on an analysis, were classified according to the extent to which they were perceived to have adopted the marketing concept (low, medium, or high). New products for these firms were then classified according to whether the new product idea was derived from a consumer-oriented source. The data are shown in the table. Do the data provide sufficient evidence to indicate that the proportion of new product ideas based on consumer sources depends on the extent to which a company is perceived to have adopted the marketing concept? Test using $\alpha = .05$.

| | | CONSUMER-ORIENTED SOURCE FOR IDEA | |
		Yes	No
	Low	6	17
EXTENT OF ADOPTION OF MARKETING CONCEPT	Medium	10	45
	High	6	23

11.25 A team of market researchers conducted a study involving 200 inhabitants of the United States to determine what people fear the most. The sex of each person polled was noted, and then each was asked which of the following was his or her greatest fear: speaking before a group, heights, bugs/insects, financial problems, sickness/death, and other. The results of the poll are given in the table. Do the data provide sufficient information to indicate a relationship between sex and greatest fear? Test at the $\alpha = .05$ level.

| | | GREATEST FEAR | | | | | |
		Speaking before a group	Heights	Bugs/insects	Financial problems	Sickness/ death	Other
SEX	Male	21	10	7	23	15	21
	Female	16	22	15	9	18	23

11.26 As noted earlier in our discussion, the X^2-statistic possesses approximately a chi square distribution when the sample size n is sufficiently large. The actual sample size n needed to achieve a satisfactory approximation depends on the application. To be safe, we suggested that n be large enough so that all expected cell counts exceed 5. Case Study 11.1 produced data that did not satisfy this criterion. Calculate the X^2-statistic for the data of Case Study 11.1 and find its p-value. Compare this approximate p-value with the exact p-value given by Mendenhall et al. This comparison will enable you to see how well the results for the two tests agree.

11.27 Over the years pollsters have found that the public's confidence in big business has been closely tied to the economic climate of the country. When businesses are growing and employment is increasing, public confidence is high. When the opposite occurs, public confidence is low. In a recent study, Harvey Kahalas (1981) explored the relationship between confidence in business and job satisfaction. He hypothesized that there is a relationship between level of confidence and job satisfaction and that this relationship holds true for both union and nonunion workers. To test his hypothesis he used the sample data given in the tables, which were collected by the National Opinion Research Center.

Sample of Union Members

| | | JOB SATISFACTION | | | |
		Very satisfied	*Moderately satisfied*	*Little dissatisfied*	*Very dissatisfied*
CONFIDENCE IN MAJOR CORPORATIONS	*A great deal*	26	15	2	1
	Only some	95	73	16	5
	Hardly any	34	28	10	9

Sample of Nonunion Workers

| | | JOB SATISFACTION | | | |
		Very satisfied	*Moderately satisfied*	*Little dissatisfied*	*Very dissatisfied*
CONFIDENCE IN MAJOR CORPORATIONS	*A great deal*	111	52	13	4
	Only some	246	142	37	18
	Hardly any	73	51	19	9

a. Kahalas concluded that his hypothesis was not supported by the data. Do you agree? Conduct the appropriate hypothesis tests using $\alpha = .05$. Be sure to specify the null and alternative hypotheses of your tests.

b. Find and interpret the approximate p-values of the tests you conducted in part a.

11.3 Contingency Tables with Fixed Marginal Totals

In many contingency table experiments the number of observations for each row (or column) is set at a fixed number before the data are collected. In the marital status–religion example (Example 11.2), the experimenter might decide to sample 100 men from each of the religious classifications. Thus, the observed data do not represent one random sample from the population of all American men, but represent *five individual random samples of men from each religious classification.*

The details of the analysis are the same when marginal row (or column) totals are fixed, but the assumptions are slightly different, as shown in the box on page 563.

Example 11.3

Suppose three different techniques are to be investigated for teaching elementary calculus. Technique 1 involves the use of computer-assisted instruction (CAI) in conjunction with lectures, technique 2 involves CAI only, and technique 3 involves lectures only.

Random samples of 100 students are assigned to each of the three teaching techniques, and the final grades of the students are to be used to compare the methods. Do the data in Table 11.9 provide sufficient evidence to indicate that the distribution

---

General Form of a Contingency Table Analysis:
A Test for Independence with Row Totals Fixed

If *row* totals are fixed:

H_0: The row proportions in each cell do not depend on the row;
that is, the distributions of observations in the column categories are
the same for each row

H_a: The row proportions in some (or all) of the cells depend on the row;
that is, the distributions of observations in the column categories differ
for at least two of the rows

Test statistic: $X^2 = \sum \dfrac{[n_{ij} - \hat{E}(n_{ij})]^2}{\hat{E}(n_{ij})}$

where

$\hat{E}(n_{ij}) = \dfrac{r_i c_j}{n}$

Rejection region: $X^2 > \chi_\alpha^2$, where χ_α^2 has $(r - 1)(c - 1)$ df

Assumptions: 1.   A random sample is selected from each of the row
category populations. These sample sizes, which are the row
totals, are specified prior to sampling.
2.   The samples are independently selected.
3.   We require the estimated expected value of each cell to
exceed 5 in order to use the χ^2 approximation.

To obtain the directions for conducting a χ^2-analysis for fixed column totals, all
you have to do is interchange the words *column* and *row* in these instructions.

Table 11.9

Final Grades for
Teaching Techniques

| | | TEACHING TECHNIQUE | | | |
		1	2	3	TOTALS
	A	15 (13.333)	13 (13.333)	12 (13.333)	40
	B	34 (32.333)	28 (32.333)	35 (32.333)	97
FINAL GRADE	C	40 (38.000)	36 (38.000)	38 (38.000)	114
	D	3 (9.333)	19 (9.333)	6 (9.333)	28
	F	8 (7.000)	4 (7.000)	9 (7.000)	21
TOTALS		100	100	100	300

of final grades depends on the teaching technique employed? Test at the .10 level
of significance.

Solution   We want to test

H_0: The proportions of final grades in the grade categories do not depend on the teaching technique

H_a: The proportions of final grades in the grade categories depend on the teaching technique

Test statistic: $X^2 = \sum \dfrac{[n_{ij} - \hat{E}(n_{ij})]^2}{\hat{E}(n_{ij})}$

Rejection region: For $\alpha = .10$ and $(r - 1)(c - 1) = (4)(2) = 8$ df, we will reject H_0 if $X^2 > \chi^2_{.10}$, where $\chi^2_{.10} = 13.3616$.

We calculate the estimated expected counts exactly as in Section 11.2:

$$\hat{E}(n_{11}) = \frac{r_1 c_1}{n} = \frac{(40)(100)}{300} = 13.333$$

$$\hat{E}(n_{12}) = \frac{r_1 c_2}{n} = \frac{(40)(100)}{300} = 13.333$$

and so forth. The estimated expected counts have been entered in parentheses in Table 11.9. Since all these values are greater than 5, the χ^2 approximation is appropriate. Then

$$X^2 = \frac{(15 - 13.333)^2}{13.333} + \frac{(13 - 13.333)^2}{13.333} + \cdots + \frac{(9 - 7.000)^2}{7.000} = 18.948$$

Since $X^2 = 18.948$ is greater than the critical value of χ^2 (13.3616), we conclude that the distribution of final grades does depend on the teaching technique used. We could now use the methods of Chapters 7 and 8 to make inferences about the individual cell probabilities or to compare cell probabilities between two columns.

Our procedure and conclusion are valid as long as the three samples are random and independent. We emphasize that we can make inferences only about the population from which the students were chosen. If all students were selected from the same university, for example, then strictly speaking we can make our inference only with respect to that university.   ■

Case Study 11.2
Does Aid to a Crime Victim Depend on Commitment of a Bystander?

Stewart and Cannon (1977) investigated the effect of the extent of a bystander's commitment to a victim on the bystander's response to a crime perpetrated on the victim. The investigators simulated a crime in which a shopper's bag was stolen. The extent of the experimental subject's commitment to the victim was determined as follows:

Commitment. The victim would look at the subject and say: "Would you watch this bag for me, please? I'll be back in a minute." Without waiting for a reply, the victim would then walk into a store and out of the subject's sight.

Noncommitment. The victim would sigh heavily, begin to walk away, turn back to the bag for a second, and then keep going into the store and out of sight.

After the victim had been out of sight for approximately 2 minutes, the thief would walk over near the bag, look at it, take a quick glance around, pick up the bag, and walk away, at a normal pace, in a direction opposite the victim. In each case, it was noted whether the subject intervened when the crime was committed. A total of seventy-nine subjects were involved, with thirty-nine committed and forty noncommitted. The resulting data appear in Table 11.10.

Table 11.10
Commitment Versus
Viewer Response

		COMMITMENT	
		Committed	Noncommitted
RESPONSE	Intervened	26	6
	Did not intervene	13	34

A χ^2-test was conducted and it was concluded ($X^2 = 21.88$) that the percentage of people who intervened depended on whether the subject was committed or noncommitted. This conclusion can be made with α as small as .001. With respect to the experimental variable, commitment, the authors conclude:

> The present findings seem to add a bit of clarity to the conceptualization of this variable, in suggesting that the simple act of identifying one's property as one's own and asking another to watch it makes the perpetrator's act less ambiguous. In fact, evidence showed that committed bystanders indicated more confidence than did noncommitted ones that the perpetrator's intentions were criminal.

When the column totals (or row totals) are fixed in a 2 × 2 contingency table, you are comparing two binomial proportions and could test an hypothesis of no difference between the proportions (equivalent to the chi square test of independence) or estimate their difference using the z-statistic and the methods of Section 8.4. For example, in Case Study 11.2 the experimenters selected samples from two binomial populations: $n_1 = 39$ subjects from a binomial population with parameter p_1, the proportion of committed subjects who would intervene; and $n_2 = 40$ subjects from a binomial population with parameter p_2, the proportion of noncommitted subjects who would intervene. Thus, they could have tested H_0: $p_1 = p_2$ by using either a chi square test or the z-test of Section 8.4. The chi square test for a 2 × 2 contingency table is equivalent to a two-tailed z-test. If you wish to conduct a one-tailed test of hypothesis using H_a: $p_1 < p_2$ or H_a: $p_1 > p_2$, the chi square test would be inappropriate; you should use the z-test of Section 8.4.

**Exercises
11.28–11.35**

Learning the Mechanics

11.28 Four independent random samples of 100 observations each were classified according to how they fall in one of three categories. The contingency table, with columns corresponding to the four samples and rows corresponding to the three categories, is shown at the top of the next page.

		COLUMN (SAMPLES)			
		1	2	3	4
ROW (CATEGORIES)	1	20	20	25	10
	2	30	40	40	50
	3	50	40	35	40

a. Do the data provide sufficient evidence to indicate that the distributions of observations within the rows depend on the columns? Test using $\alpha = .05$.

b. Form a 95% confidence interval for the probability of observing a response in the second row and third column.

c. Form a 95% confidence interval for the difference between the probability of observing a response in the first row, first column, and the probability of observing a response in the first row, fourth column.

Applying the Concepts

11.29 Refer to Exercise 7.79. In an article titled "Searching for a Forever Home," *Time* (May 2, 1983) reports on how television programs aid in the adoption of orphans who have physical or mental handicaps. One method of stimulating the adoption program is to present television profiles of the children, one or more per program. The article documents the success of these programs and notes that Oklahoma City's television station KOCO helped to place 92 of the 119 it profiled, New York's WCBS placed 21 of 35, and Atlanta's WXIA placed 79 of 177. How effective were these three television stations in promoting the adoption of the children?

a. Do the data provide sufficient evidence to indicate differences in the success rates for the three television stations? Construct a contingency table for the data and test using $\alpha = .05$.

b. Find a 95% confidence interval for the success ratio for television station WXIA.

c. Find a 95% confidence interval for the difference in success rates between stations KOCO and WXIA.

11.30 Is there a difference in vocational preference between firstborn children and those that follow? Robert M. Lynch and Janet Lynch (1981) sought an answer to this question by sampling 244 twelve to fourteen-year-old New York public school students, 122 firstborn and 122 second, third, or fourthborn. To measure vocational preference, each student took a self-administered vocational counseling test. The test attempts to identify a person's occupational preference by classifying the individual into one of six groups. The distributions of the 122 firstborns and of the 122 second, third, or fourthborns into the six vocational classes are shown in the table.

a. Do the data present sufficient evidence to indicate a difference between the distribution of students among the six vocational classes for firstborns and the corresponding distribution for second, third, or fourthborns? Test using $\alpha = .05$.

Bivariate Frequencies of
Vocational Class by
Ordinal Position

VOCATIONAL CLASS	FIRSTBORN	LATER BORN
Conventional	38 (31%)	9 (7%)
Realistic	26 (21%)	19 (16%)
Enterprising	24 (20%)	15 (12%)
Social	12 (10%)	15 (12%)
Artistic	12 (10%)	21 (17%)
Investigative	10 (9%)	43 (35%)
	122 (100%)	122 (100%)

$$\chi^2 = 44.39; \quad df = 5; \quad p < .001$$

b. What are the practical implications of the test results as they pertain to the question the Lynches sought to answer?

11.31 Nausea is a common symptom among postoperative patients. A group of physicians is interested in comparing two new drugs, A and B, for their effectiveness in preventing postoperative nausea. One hundred eighty patients scheduled for surgery at a large hospital are used in the study, with sixty assigned to receive drug A and sixty to receive drug B after their operations. The remaining sixty patients are given a placebo (no drug). A short time after the operation, each patient is classified according to the degree of nausea felt. The results are given in the table. Is there evidence of a difference among the drugs and the placebo with respect to their abilities to reduce postoperative nausea? Test using $\alpha = .05$.

	DEGREE OF NAUSEA				TOTALS
	None	*Slight*	*Moderate*	*Severe*	
DRUG A	40	10	6	4	60
DRUG B	36	12	4	8	60
PLACEBO	30	16	8	6	60
TOTALS	106	38	18	18	180

11.32 A study was conducted to determine whether a relationship exists between obesity in children and obesity in their parents. Random samples of fifty obese and fifty nonobese children were obtained. Then, for each child, it was determined whether one or both parents were obese. A summary of the data is shown in the table. Do the data provide sufficient evidence to indicate that child obesity is dependent on parental obesity?

		CHILD		TOTALS
		Obese	*Nonobese*	
PARENT	*Obese*	34	29	63
	Nonobese	16	21	37
TOTALS		50	50	100

11.33 An experiment was conducted to compare two methods for operating a group family medical practice. Four hundred patients were randomly assigned to two groups: One group received the conventional direct contact with physicians, while the other group made first contact with a nurse–practitioner and were then referred to a physician if a physician's services were deemed necessary. At the conclusion of the experiment the quality of each person's medical care was rated as satisfactory or unsatisfactory by an impartial medical observer in consultation with the patient. The results of the experiment are shown in the table. Do the data present sufficient evidence to indicate a difference in the proportions of satisfactory ratings for the two methods of patient care? Test using $\alpha = .05$.

	CONVENTIONAL	NURSE–PRACTITIONER	TOTALS
SATISFACTORY	148	161	309
UNSATISFACTORY	52	39	91
TOTALS	200	200	400

11.34 A study was conducted to determine whether the treatment psychiatric patients receive is determined by their social status. Generally, the treatment of a psychiatric patient is classified as follows: psychotherapy, organic treatment (physical–chemical), or no treatment (a patient receiving custodial care in an institution, but neither of the other two types of treatment). One hundred psychiatric patients were randomly sampled from each of four social classes, and each was classified according to the type of treatment received. Determine whether the data given in the table support the theory that type of psychiatric treatment and social status are dependent. Test at the $\alpha = .10$ level.

		SOCIAL CLASS			
		Upper	Upper-middle	Middle-lower	Lower
	Psychotherapy	78	53	31	17
TREATMENT	Organic	13	27	38	33
	No treatment	9	20	31	50

11.35 It is commonly assumed that the more experience a job applicant has, the better that person will perform the necessary duties. Other factors such as whether the person has a college degree or is male or female also may be indicative of future performance. H. M. Greenberg and J. Greenberg (1980) argue that for sales jobs the most important factor is the matching of the particular job requirement with an applicant's personal characteristics. This, they claim, will result in better retention of employees and produce higher levels of job performance. To validate this claim they studied two groups of recently hired sales personnel. Of these, 1,980 were job-matched and 3,961 were not. After 6 months they were evaluated. The aggregate data are shown in the table, where 1 represents the highest level of performance and 4 represents the lowest. The tabulated values are the percentages of total sales personnel contained in the respective samples.

	PERFORMANCE				QUIT OR	TOTALS
	1	2	3	4	FIRED	
JOB-MATCHED	9%	40%	32%	14%	5%	100%
NOT JOB-MATCHED	2%	17%	25%	31%	25%	100%

a. Use the percentages given in the table and the sample sizes to construct a contingency table that shows the numbers of sales personnel falling in each category of the table.

b. Do the data provide sufficient evidence to indicate that the proportions of sales personnel falling in the performance categories depend on whether the people are job-matched? Test using $\alpha = .05$.

c. Do the data provide sufficient evidence to indicate that the proportion of sales personnel receiving the highest rating (1) is larger if job-matched than if not? Test using $\alpha = .05$. [*Note:* This problem requires a one-sided test.]

11.4 Caution

Because the X^2-statistic for testing hypotheses about multinomial probabilities is one of the most widely applied statistical tools, it is also one of the most abused statistical procedures. The user should always be certain that the experiment satisfies the assumptions given with each procedure. Furthermore, the user should be certain that the sample is drawn from the correct population—that is, from the population about which the inference is to be made. If in Example 11.3 the experimenter had chosen 100 students from each of three different colleges, no valid inference could be made about the teaching methods. We would be comparing the three colleges, as well as teaching methods.

The use of the χ^2 probability distribution as an approximation to the sampling distribution for X^2 should be avoided when the expected counts are very small. The approximation can become very poor when these expected counts are small, and thus the true α level may be quite different from the tabled value. As a rule of thumb, an expected cell count of at least 5 means that the χ^2 probability distribution can be used to determine an approximate critical value.

If the X^2-value does not exceed the established critical value of χ^2, *do not accept* the hypothesis of independence. You would be risking a Type II error (accepting H_0 if it is false), and the probability β of committing such an error is unknown. The usual alternative hypothesis is that the classifications are dependent. Because there is literally an infinite number of ways two classifications can be dependent, it is difficult to calculate one or even several values of β to represent such a broad research hypothesis. Therefore, we avoid concluding that two classifications are independent, even when X^2 is small.

Finally, if a contingency table X^2-value does exceed the critical value, we must be careful to avoid inferring that a causal relationship exists between the classifications. Our alternative hypothesis states that the two classifications are statistically dependent—and statistical dependence does not imply causality. *Therefore, the existence of a causal relationship cannot be established by a contingency table analysis.*

Summary

The use of **count data** to test hypotheses about **multinomial probabilities** represents a useful statistical technique. In a **one-dimensional table** we can use count data to test the hypothesis that the multinomial probabilities are equal to specified values. In the **two-dimensional contingency table,** we can test the independence of the two classifications. And these by no means exhaust the uses of the X^2-statistic. Many other applications can be found in the references at the end of this chapter.

Caution should be exercised to avoid misuse of the χ^2-procedure. The experiment must be multinomial,* and the expected counts should not be too small if the χ^2 critical value is used. Moreover, the X^2-statistic should not always be viewed as the final answer. If two classifications are found to be dependent, many measures of association exist for quantifying the nature and strength of their dependence (see the references).

Supplementary Exercises 11.36–11.58

Learning the Mechanics

11.36 A random sample of 250 observations was classified according to the row and column categories shown in the table.

		COLUMN		
		1	2	3
	1	20	20	10
ROW	2	10	20	70
	3	20	50	30

a. Do the data provide sufficient evidence to conclude that the rows and columns are dependent? Test using $\alpha = .05$.
b. Would the analysis change if the row totals were fixed before the data were collected?
c. Do the assumptions required for the analysis to be valid differ according to whether the row (or column) totals are fixed? Explain.

11.37 A random sample of 150 observations was classified into the categories shown in the table.

		CATEGORY			
	1	2	3	4	5
n_1	28	35	33	25	29

a. Do the data provide sufficient evidence to indicate that the categories are not equally likely? Use $\alpha = .10$.
b. Form a 90% confidence interval for p_2, the probability that an observation will fall in category 2.

* When the row (or column) totals are fixed, each row (or column) represents a separate multinomial experiment.

Applying the Concepts

11.38 A computer used by a 24-hour banking service is supposed to assign each transaction to one of five memory locations at random. A check at the end of a day's transactions gives the following counts to each of the five memory locations:

1	2	3	4	5
90	78	100	72	85

Is there evidence to indicate a difference among the proportions of transactions assigned to the five memory locations? Test using $\alpha = .025$.

11.39 In a recent poll, 656 people were randomly selected and classified according to whether they were government employees and whether they believed that the quality of life had changed since 1970. A summary of the responses is shown in the table. Do the data provide sufficient evidence to indicate that government employees perceive the change in quality of life differently from others? Test using $\alpha = .05$.

	QUALITY OF LIFE	
	Worse	Better
GOVERNMENT EMPLOYEES	17	31
OTHER	317	291

11.40 A restaurateur who own restaurants in four cities is considering the possibility of building separate dining rooms for nonsmokers to accommodate customers who wish to dine in a smokefree environment. Since this change would involve significant expense, the restaurateur plans to survey the customers at each restaurant and ask them the following question: "Would you be more comfortable dining here if there were a separate dining room for nonsmokers only?" Suppose seventy-five people were randomly selected and surveyed at each restaurant with the results shown in the table. Is there sufficient evidence to indicate a difference among customer preferences for the four restaurants? Use $\alpha = .10$.

		ANSWERS TO QUESTION		
		Yes	No	It makes no difference
RESTAURANT	1	38	32	5
	2	42	26	7
	3	35	34	6
	4	37	30	8

11.41 A study was done on the accuracy of newspaper advertisements by the five types of food stores in a southeastern city. On each of 4 days, items were randomly selected from the advertisements for each type of store and the actual price was compared to the advertised price. Each of the stores in the city was classified as one of the following types: national, regional chain A, regional chain B, regional chain C, or independent. Values in the table represent the number of items that were correctly and incorrectly priced.

TYPE OF STORE	NUMBER CORRECTLY PRICED	NUMBER INCORRECTLY PRICED
National chain	89	10
Regional chain A	53	14
Regional chain B	43	12
Regional chain C	32	13
Independent	41	7

a. Determine whether these data provide sufficient evidence to conclude that the proportion of correctly priced items differs for at least two types of stores. Use $\alpha = .10$.

b. Use a 95% confidence interval to estimate the proportion of correctly priced items in the stores in the national chain category.

11.42 Despite a good winning percentage, a certain major league baseball team has not drawn as many fans as one would expect. In hopes of finding ways to increase attendance, the management plans to interview fans who come to the games to find out why they come. One thing that the management might want to know is whether there are differences in support for the team among various age groups. Suppose the information in the table was collected during interviews with fans selected at random. Can you conclude that there is a relationship between age and number of games attended per year? Use $\alpha = .05$.

		NUMBER OF GAMES ATTENDED PER YEAR		
		1 or 2	3–5	Over 5
	Under 20	78	107	17
	21–30	147	87	13
AGE OF FAN	31–40	129	86	19
	41–55	55	103	40
	Over 55	23	74	22

11.43 If a company can identify times of day when accidents are most likely to occur, extra precautions can be instituted during those times. A random sample of the accident reports over the last year at a plant gives the frequency of occurrence of accidents at the different hours of the workday. Can it be concluded from the data in the table that the proportions of accidents are different for at least two of the four time periods?

HOURS	1–2	3–4	5–6	7–8
NUMBER OF ACCIDENTS	31	28	45	47

11.44 A sociologist was interested in knowing whether sons have a tendency to choose the same occupation as their fathers. To investigate this question, 500 males were polled and questioned concerning their occupation and the occupation of their father. A summary of the numbers of father–son pairs falling in each occupational category is shown in the table. Do the data provide sufficient evidence to indicate a dependence between a son's choice of occupation and his father's occupation? Test using $\alpha = .05$.

		SON			
		Professional or business	Skilled	Unskilled	Farmer
FATHER	Professional or business	55	38	7	0
	Skilled	79	71	25	0
	Unskilled	22	75	38	10
	Farmer	15	23	10	32

11.45 Teenage alcoholism is a big problem in the United States. To discover why teenagers are turning to alcohol, a survey was conducted to find out whether a teenager's family status has any relationship to the amount of alcohol he or she consumes. A random sample of 200 teenagers between the ages of 15 and 19 were questioned concerning their use of alcohol. A summary of the responses is shown in the table. Do the data provide sufficient evidence to indicate a relationship between family status and the use of alcohol? Test using $\alpha = .05$.

		ALCOHOL		
		None	Occasional	Frequent
FAMILY STATUS	Upper class	4	16	10
	Upper middle class	11	40	24
	Lower middle class	9	47	9
	Lower class	6	17	7

11.46 An appliance store is having a sale and wants to determine which modes of advertising are effective. A survey of the customers who know about the sale reveals the breakdown given in the table. The mode reported is that from which each customer first learned of the sale.

TELEVISION	NEWSPAPER	RADIO	WORD OF MOUTH
53	36	32	48

a. Is there evidence that the proportions of customers who learned about the sale differ for at least two of the four modes of advertising? Use $\alpha = .05$.

b. Estimate the proportion who learn about the sale by word of mouth. Use a 90% confidence interval.

11.47 Employee integrity is important to the success of all businesses. Suppose an industrial security firm wants to conduct a study of criminal cases involving stolen company money in which employees have been found guilty. Among the data they record are the employee's salary (wages) and the amount of money stolen from the company for 400 recent cases. Does this information provide evidence of a relationship between employee income and amount stolen? Use $\alpha = .05$.

		AMOUNT STOLEN (THOUSANDS OF DOLLARS)			
		Under 5	5 to <10	10 to <20	20 and over
INCOME OF EMPLOYEE (THOUSANDS OF DOLLARS)	Under 15	46	39	17	5
	15–25	78	79	61	19
	Over 25	5	14	25	12

11.48 Five candidates have just entered the race for mayor of a large city. To determine whether any of the candidates has an early lead in popularity, a poll is conducted. Some 2,000 voters are asked to indicate the candidate they prefer. A summary of the responses is shown in the table. Do the data provide sufficient evidence to indicate a preference for at least one of the five candidates? Test using $\alpha = .01$.

			CANDIDATE		
	I	II	III	IV	V
RESPONSES	385	493	628	235	259

11.49 A national survey was conducted to determine how the general public views government involvement in domestic projects. Two hundred people from each of three income levels were asked if they thought the government was too involved, not involved enough, or involved just enough. A summary of their responses is shown in the table. Do the data provide sufficient information to indicate a relationship between income and view on government involvement in domestic projects? Test using $\alpha = .05$.

		INVOLVEMENT			
		Too little	Just enough	Too much	TOTALS
INCOME	Low	125	48	27	200
	Medium	103	58	39	200
	High	72	69	59	200
TOTALS		300	175	125	600

11.50 A local bank plans to offer a special service to its young customers. To determine their economic interests, a survey of 100 people under 30 years of age is conducted. Each person is asked to identify his or her top two financial priorities from the six choices listed in the table. Use a χ^2 test to determine whether the proportions of responses differ for at least two of the six pairs of priorities. Test at $\alpha = .10$.

FIRST PRIORITY	SECOND PRIORITY	NUMBER OF RESPONSES
Buy a car	Go on a trip	15
Buy a car	Save money	14
Save money	Buy a car	22
Save money	Go on a trip	23
Go on a trip	Buy a car	10
Go on a trip	Save money	16

11.51 Along with the technological age comes the problem of workers being replaced by machines. A labor management organization wants to study the problem of workers displaced by automation within three industries. Case reports for 100 workers whose loss of job is directly attributable to technological advances are selected within each industry. For each worker selected it is determined whether he or she was given another job within the same company, found a job with another company in the same industry, found a job in a new industry, or has been unemployed for longer than 6 months. The results are given in the table.

		SAME COMPANY	NEW COMPANY *(Same industry)*	NEW INDUSTRY	UNEMPLOYED
	A	62	11	20	7
INDUSTRY	B	45	8	38	9
	C	68	19	8	5

a. Does the plight of automation-displaced workers depend on the industry? Use $\alpha = .01$.

b. Estimate the difference between the proportions of displaced workers who find work in another industry for industries A and C. Use a 95% confidence interval.

11.52 A corporation owns several convenience stores that are open 24 hours a day. It is interested in knowing whether there is a relationship between time of day and size of purchase. One of its stores is selected at random to be involved in a study. Store records are collected over a period of several weeks and then 300 purchases are randomly selected. Since the register also prints the time of the purchase, this random selection procedure yields both amount and time of purchase. The information is summarized in the table.

		SIZE OF PURCHASE		
		$2 or less	$2.01–$7	Over $7
	8 AM *to* < 4 PM	65	38	14
TIME OF PURCHASE	4 PM *to* < 12 *midnight*	61	49	10
	12 *midnight to* < 8 AM	29	27	7

a. Is there a relationship between time and size of purchase? Use $\alpha = .05$.

b. Use a 90% confidence interval to estimate the proportion of customers who spend $2 or less during the period 8 AM–4 PM.

11.53 A political scientist wants to determine whether there is a relationship between a person's income and his or her political affiliation. The researcher randomly samples 265 registered voters and determines the income and political affiliation of each. A summary of the data is shown in the table. Do the data provide sufficient evidence to indicate a relationship between political affiliation and annual income? Test using $\alpha = .10$.

| | ANNUAL INCOME (THOUSANDS OF DOLLARS) | | | |
	25 or over	16 < 25	8 < 16	Below 8
REPUBLICAN	50	28	20	12
DEMOCRAT	14	35	35	41
OTHER	6	7	10	7

11.54 A city has three television stations. Each station has its own evening news program from 6:00 to 6:30 PM every weekday. An advertising firm wants to know whether there is an unequal breakdown of the evening news audience among the three stations. One hundred people are selected at random from those who watch the evening news on one of these three stations. Each is asked to specify which news program he or she watches. Do the results in the table provide sufficient evidence to indicate that the three stations do not have equal shares of the evening news audience? Use $\alpha = .05$.

STATION	1	2	3
NUMBER OF VIEWERS	35	43	22

11.55 Several life insurance firms have policies geared to college students. To get more information about this group, a major insurance firm interviews college students to find out the type of life insurance they prefer, if any. The following table is produced after surveying 1,600 students.

| | PREFERENCE | | |
	Term insurance	Whole-life insurance	No preference
FEMALES	116	27	676
MALES	215	33	533

a. Is there evidence that the life insurance preference of students depends on sex?

b. Estimate the proportion of female students who have no preference in life insurance.

11.56 In late 1977, many farmers across the United States went on strike, protesting that the prices of farm products, chiefly grains, were less than the cost of production. Although the main strike goal was to receive 100% of parity prices for all farm products, a second controversial goal was to induce farmers to reduce production, thereby reducing surpluses and boosting prices. A sample survey of 100 farmers was conducted to determine whether a relationship exists between a farmer's decision to participate in the strike and the farmer's opinion concerning the necessity for a cutback in production. The results are shown in the table. Is there evidence of a relationship between a farmer's strike position and the stand on a cutback in production? Use $\alpha = .05$.

		ON STRIKE	
		Yes	No
50% CUTBACK IN PRODUCTION	Favor	21	7
	Undecided	37	2
	Opposed	22	11

11.57 A statistical analysis is to be done on a set of data consisting of 1,000 monthly salaries. The analysis requires the assumption that the sample has been drawn from a normal distribution. A preliminary test, called the χ^2 *goodness-of-fit test,* can be used to determine whether it is plausible to assume that the sample is from a normal distribution. Suppose the mean and standard deviation of the 1,000 salaries are hypothesized to be $900 and $50, respectively. Using the standard normal tables, we can approximate the probabilities of being in the intervals listed in the table. The third column represents the expected number of the 1,000 salaries to be found in each interval if the sample was drawn from a normal distribution with $\mu = 900$ and $\sigma = 50$. Suppose the last column contains the actual observed frequencies in the sample. Large differences between the observed and expected frequencies cast doubt on the normality assumption.

INTERVAL	PROBABILITY	EXPECTED FREQUENCY	OBSERVED FREQUENCY
Less than $800	.023	23	26
$800 < $850	.136	136	146
$850 < $900	.341	341	361
$900 < $950	.341	341	311
$950 < $1,000	.136	136	143
$1,000 or above	.023	23	13

a. Compute the X^2-statistic based on the observed and expected frequencies—just as you would in a contingency table analysis.
b. Find the tabulated χ^2-value when $\alpha = .05$ and there are 5 df. (There are 5 df associated with this X^2-statistic.)
c. Based on the X^2-statistic and the tabulated χ^2-value, is there evidence that the salary distribution is nonnormal?*

---

* If we want to test the null hypothesis that a population's relative frequency distribution is normal with unspecified mean and variance, we will need to estimate μ and σ in order to estimate the k cell probabilities. We lose 2 df corresponding to these estimates, so that the χ^2 rejection region is based on $(k - 3)$ df.

11.58 Suppose a random variable is hypothesized to be normally distributed with mean zero and standard deviation 1. A random sample of 200 observations on the variable yields the frequencies in the intervals listed in the table. Do the data provide sufficient evidence to contradict the hypothesis that x is normally distributed with $\mu = 0$ and $\sigma = 1$? Use the technique developed in Exercise 11.57.

INTERVAL	FREQUENCY
$x < -2$	7
$-2 \leq x < -1$	20
$-1 \leq x < 0$	61
$0 \leq x < 1$	77
$1 \leq x < 2$	26
$x \geq 2$	9

On Your Own . . .

Many researchers rely on surveys to estimate the proportions of experimental units in populations that possess certain specified characteristics. A political scientist may want to estimate the proportion of an electorate in favor of a certain legislative bill. A social scientist may be interested in the proportions of people in a geographical region who fall in certain socioeconomic classifications. A psychologist might want to compare the proportions of patients who have different psychological disorders.

Choose a specific topic, similar to those described above, that interests you. Clearly define the population of interest, identify data categories of specific interest, and identify the proportions associated with them. Now *guesstimate* the proportions of the population that you think fall in each of the categories. For example, you might guess that all the proportions are equal, or that the first proportion is twice as large as the second but equal to the third, etc.

You are now ready to collect the actual data by obtaining a random sample from your population of interest. Select a sample size so that all expected cell counts are at least 5 (preferably larger), and collect the data.

Use the count data you have obtained to test the null hypothesis that the true proportions in the population equal your presampling guesstimates of these actual proportions. Would failure to reject this null hypothesis imply that your guesstimates are correct?

References   Agresti, A., & Agresti, B. F. *Statistical methods for the social sciences.* San Francisco: Dellen, 1979. Chapter 8.

Behar, D., & Stewart, M. A. "Aggressive conduct disorder: the influence of social class, sex, and age on the clinical picture." *Journal of Child Psychology and Psychiatry,* 1984, *25.*

Cochran, W. G. "The χ^2 test of goodness of fit." *Annals of Mathematical Statistics,* 1952, *23,* 315–345.

Dahlin, C., & Owen, F. "An analysis of data collected at the I-494 weighing-in-motion site." St. Paul: Minnesota Dept. of Transportation, 1984.

Greenberg, H. M., & Greenberg, J. "Job-matching for better sales performance." *Harvard Business Review,* September–October 1980.

Kahalas, H. "The relationship between confidence in business and job satisfaction for union and nonunion members." *Baylor Business Studies,* February–April 1981, *127,* 45–53.

Lawton, L., & Parasuraman, A. "The impact of the new marketing concept on new product planning." *Journal of Marketing,* 1980, *44,* 19–25.

Lynch, R. M., & Lynch, J. "Birth order and vocational preference." *Journal of Experimental Education,* 1981, *49.*

Mendenhall, W. M., Million, R. R., Sharkey, D. E., & Cassisi, N. J. "Stage T3 squamous cell carcinoma of the glottic larynx treated with surgery and/or radiation therapy." *International Journal of Radiation Oncology and Biological Physics,* 1984, *10.*

Savage, I. R. "Bibliography of nonparametric statistics and related topics." *Journal of the American Statistical Association,* 1953, *48,* 844–906.

Siegel, S. *Nonparametric statistics for the behavioral sciences.* New York: McGraw-Hill, 1956. Chapter 9.

Stewart, J. E., & Cannon, D. A. "Effects of perpetrator status and bystander commitment on responses to a simulated crime." *Journal of Police Science and Administration,* 1977, *5*(3), 318–323.

CHAPTER 12

Simple Linear Regression

Where We've Been . . .

We have learned how to estimate and test hypotheses about population parameters based on a random sample of observations from the population and have extended these methods to allow for a comparison of parameters from two or more populations.

Where We're Going . . .

For many sampling situations, we have much more information available on a random variable (and the population it generates) than that contained in a single random sample. For example, if we wanted to predict the rainfall at a given location on a given day, we could select a single random sample of n daily rainfalls, use the methods of Chapter 7 to estimate the mean daily rainfall μ, and then use this quantity to predict the rainfall for any given day. But this method fails to utilize scientific information that is available to any forecaster. We know the daily rainfall is related to barometric pressure, cloud cover, etc. By measuring barometric pressure and cloud cover at the same time we sample the daily rainfall, we hope to establish the relationship between these variables and to utilize them for prediction.

This chapter is devoted to the most elementary situation—relating two variables. The more complex problem of relating more than two variables is the topic of Chapter 13.

Contents

Much practical research is devoted to the topic of **modeling**—that is, trying to describe how variables are related. For example, an econometrician might be interested in modeling the relationship between the Gross National Product (GNP) and the current rate of unemployment. A behavioral psychologist might want to model the relationship between a child's motor activity and the concentration of a certain enzyme in the bloodstream. A sociologist might be interested in relating the rate of juvenile delinquency in a neighborhood to the percentage of children in broken homes in that neighborhood.

One method of modeling the relationship between variables is called **regression analysis**—this important topic is the subject of Chapters 12 to 14. In this chapter, we will discuss a simple **linear (straight-line) model** for the relationship between two variables, the **least squares method** of fitting regression models using sample data, how to make inferences about the model, and how to use the model for prediction.

12.1 Probabilistic Models

An important consideration when taking a drug is how it may affect one's perception or general awareness. Suppose you want to model the length of time it takes to respond to a stimulus (a measure of awareness) as a function of the percentage of a certain drug in the bloodstream. The first question to be answered is this: "Do you think an exact relationship exists between these two variables?" That is, do you think it is possible to state the exact length of time it takes an individual (subject) to respond if the amount of the drug in the bloodstream is known? We think you will agree with us that this is not possible for several reasons. The reaction time depends on many variables other than the percentage of the drug in the bloodstream; for example, the time of day, the amount of sleep the subject had the night before, the subject's visual acuity, the subject's general reaction time without the drug, and the subject's age would all probably affect reaction time. Even if many variables are included in a model (the topic of Chapter 13), it is still unlikely that we would be able to predict *exactly* the subject's reaction time. There will almost certainly be some variation in response times due strictly to *random phenomena* that cannot be modeled or explained.

If we were to construct a model that hypothesized an exact relationship between variables, it would be called a **deterministic model.** For example, if we believe that y, the reaction time (in seconds), will be exactly one and one-half times x, the amount of drug in the blood, we write

$$y = 1.5x$$

This represents a **deterministic relationship** between the variables y and x. It implies that y can always be determined exactly when the value of x is known. *There is no allowance for error in this prediction.*

If, on the other hand, we believe there will be unexplained variation in reaction times—perhaps caused by important but unincluded variables or by random phenomena—we will discard the deterministic model and use a model that accounts for this **random error.** This **probabilistic model** includes both a deterministic component and a random error component. For example, if we hypothesize that the response time y is related to

Figure 12.1 Possible Reaction Times, y, for Five Different Drug Percentages, x

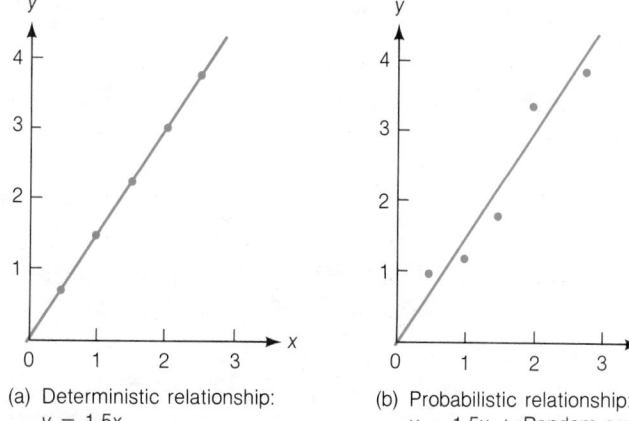

(a) Deterministic relationship: $y = 1.5x$

(b) Probabilistic relationship: $y = 1.5x +$ Random error

the percentage of drug x by

$$y = 1.5x + \text{Random error}$$

we are hypothesizing a **probabilistic relationship** between y and x. Note that the deterministic component of this probabilistic model is 1.5x.

Figure 12.1(a) shows the possible responses for five different values of x, the percentage of drug in the blood, when the model is deterministic. All the responses must fall exactly on the line because a deterministic model leaves no room for error.

Figure 12.1(b) shows a possible set of responses for the same values of x when we are using a probabilistic model. Note that the deterministic part of the model (the straight line itself) is the same. Now, however, the inclusion of a random error component allows the response times to vary from this line. Since we know that the response time does vary randomly for a given value of x, the probabilistic model provides a more realistic model for y than does the deterministic model.

General Form of Probabilistic Models

$y = $ Deterministic component $+$ Random error

where y is the variable to be predicted. We will always assume that the mean value of the random error equals zero. This is equivalent to assuming that the mean value of y, $E(y)$, equals the deterministic component of the model, i.e.,

$E(y) = $ Deterministic component

We begin with the simplest of probabilistic models—the **straight-line model**—which derives its name from the fact that the deterministic portion of the model graphs as a straight line. Fitting this model to a set of data is an example of **regression analysis** or *regression modeling.* The elements of the straight-line model are summarized in the box.

A First-Order (Straight-Line) Probabilistic Model

$$y = \beta_0 + \beta_1 x + \varepsilon$$

where

$y = $ *Dependent* or *response* variable (variable to be modeled)

$x = $ *Independent* or *predictor* variable (variable used as a predictor of y)*

ε (epsilon) $= $ Random error component

β_0 (beta zero) $= $ y-intercept of the line, i.e., point at which the line intercepts or cuts through the y-axis (see Figure 12.2)

β_1 (beta one) $= $ Slope of the line, i.e., amount of increase (or decrease) in the deterministic component of y for every 1 unit increase in x. You can see (Figure 12.2) that $E(y)$ increases by the amount β_1 as x increases from 2 to 3.

Figure 12.2   The Straight-Line Model

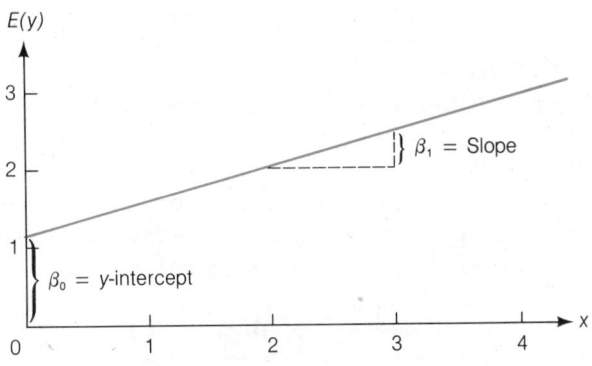

Note that we use Greek symbols, β_0 and β_1, to represent the y-intercept and slope of the model. They are population parameters with numerical values that will be known only if we have access to the entire population of (x, y) measurements.

It is helpful to think of regression modeling as a five-step procedure:

Step 1. Hypothesize the deterministic component of the probabilistic model.

Step 2. Use sample data to estimate unknown parameters in the model.

---

* The word *independent* should not be interpreted in a probabilistic sense, as defined in Chapter 3. The phrase *independent variable* is used in regression analysis to refer to a predictor variable for the response y.

Step 3. Specify the probability distribution of the random error term, and estimate any unknown parameters of this distribution.

Step 4. Statistically check the usefulness of the model.

Step 5. When you are satisfied that the model is useful, use it for prediction, estimation, etc.

In this chapter we will skip step 1 (which is difficult) and deal only with the straight-line model. Chapters 13 and 14 explain how to build more complex models.

Exercises 12.1–12.7

Learning the Mechanics

12.1 In each case graph the line that passes through the points:

a. (1, 0) and (2, 4)   **b.** (0, 3) and (3, 7)

c. (−1, −2) and (3, 1)   **d.** (2, 3) and (4, −5)

12.2 Give the slope and y-intercept for each of the lines graphed in Exercise 12.1.

12.3 Give the equations of the lines graphed in Exercise 12.1.

12.4 Graph the following lines:

a. $y = 1 + 1.5x$ **b.** $y = .5 - x$ **c.** $y = -1 + 4x$

d. $y = 3 - 2x$ **e.** $y = 2x$

12.5 Give the slope and y-intercept for each of the lines defined in Exercise 12.4.

Applying the Concepts

12.6 When is it appropriate to use a deterministic model to describe the relationship between two variables? Give an example of two variables that you think have a deterministic relationship.

12.7 When is it appropriate to use a probabilistic model to describe the relationship between two variables? Give an example of two variables that you think should be modeled by a probabilistic relationship.

12.2 Fitting the Model: Least Squares Approach

Suppose an experiment involving five subjects is conducted to determine the relationship between the percentage of a certain drug in the bloodstream and the length of time it takes to react to a stimulus. The results are shown in Table 12.1. (The number of measurements and the measurements themselves are unrealistically simple in order to avoid arithmetic confusion in this introductory example.) This set of data will be used to demonstrate the five-step procedure of regression modeling given in Section 12.1. In this section we will hypothesize the deterministic component of the model and estimate its unknown parameters (steps 1 and 2). Discussion of the model assumptions and the random error component (step 3) are the subjects of Sections 12.3 and 12.4, while Sections 12.5 to 12.7 assess the utility of the model (step 4). Finally, using the model for prediction and estimation (step 5) is the subject of Section 12.8.

Table 12.1

Reaction Time Versus
Drug Percentage

SUBJECT	AMOUNT OF DRUG x (%)	REACTION TIME y (seconds)
1	1	1
2	2	1
3	3	2
4	4	2
5	5	4

Step 1. *Hypothesize the deterministic component of the probabilistic model.* As stated before, we will consider only straight-line models in this chapter, and thus the complete model to relate mean response time $E(y)$ to drug percentage x is given by

$$E(y) = \beta_0 + \beta_1 x$$

Step 2. *Use sample data to estimate unknown parameters in the model.* This step is the subject of this section—namely, how can we best use the information in the sample of five observations in Table 12.1 to estimate the unknown y-intercept β_0 and slope β_1?

To determine whether a linear relationship between y and x is plausible, it is helpful to plot the sample data. Such a plot, called a **scattergram,** locates each of the five data points on a graph, as shown in Figure 12.3. Note that the scattergram suggests a general tendency for y to increase as x increases. If you place a ruler on the scattergram, you will see that a line may be drawn through three of the five points, as shown in Figure 12.4. To obtain the equation of this visually fitted line, note that the line intersects the y-axis at $y = -1$, so the y-intercept is -1. Also, y increases exactly 1 unit for every 1 unit increase in x, indicating that the slope is $+1$. Therefore, the equation is

$$\tilde{y} = -1 + 1(x) = -1 + x$$

where $\tilde{y}$ is used to denote the predicted y from the visual model.

Figure 12.3 Scatter-
gram for Data in Table 12.1

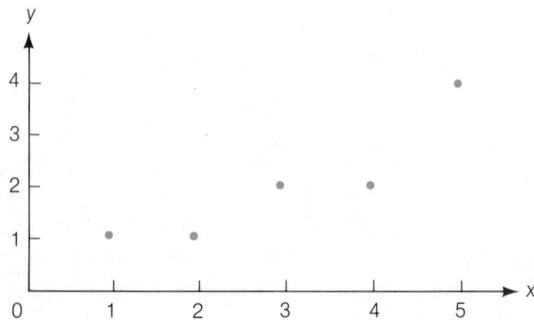

One way to decide quantitatively how well a straight line fits a set of data is to note the extent to which the data points deviate from the line. For example, to evaluate the model in Figure 12.4, we calculate the magnitude of the **deviations,** i.e., the differences between the observed and the predicted values of y. These deviations, or **errors,** are the vertical distances between observed and predicted values (see Figure 12.4). The

Figure 12.4 Visual Straight-Line Fit to the Data

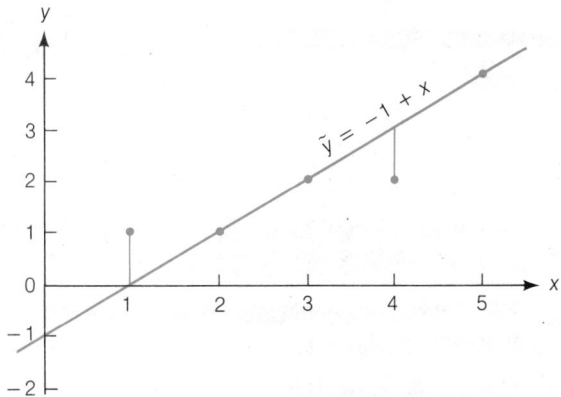

observed and predicted values of y, their differences, and their squared differences are shown in Table 12.2. Note that the *sum of errors* equals zero and the *sum of squares of the errors* (SSE), which gives greater emphasis to large deviations of the points from the line, is equal to 2.

Table 12.2

Comparing Observed and Predicted Values for the Visual Model

x	y	$\tilde{y} = -1 + x$	$(y - \tilde{y})$		$(y - \tilde{y})^2$
1	1	0	$(1 - 0) =$	1	1
2	1	1	$(1 - 1) =$	0	0
3	2	2	$(2 - 2) =$	0	0
4	2	3	$(2 - 3) =$	-1	1
5	4	4	$(4 - 4) =$	0	0
			Sum of errors =	0	Sum of squared errors (SSE) = 2

You can see by shifting the ruler around the graph that it is possible to find many lines for which the sum of the errors is equal to zero, but it can be shown that there is one (and only one) line for which the SSE is a *minimum*. This line is called the **least squares line,** the **regression line,** or **least squares prediction equation.**

To find the least squares line for a set of data, assume that we have a sample of n data points that can be identified by corresponding values of x and y, say (x_1, y_1), $(x_2, y_2), \ldots, (x_n, y_n)$. For example, the $n = 5$ data points shown in Table 12.2 are $(1, 1)$, $(2, 1)$, $(3, 2)$, $(4, 2)$, and $(5, 4)$. The straight-line model for the response y in terms of x is

$$y = \beta_0 + \beta_1 x + \varepsilon$$

The line that gives the mean (or expected) value of y for a given value of x is

$$E(y) = \beta_0 + \beta_1 x$$

and the fitted line, which we will determine, is written as

$$\hat{y} = \hat{\beta}_0 + \hat{\beta}_1 x$$

The "hats" can be read as "estimator of." Thus, $\hat{y}$ is an estimator of the mean value of y, $E(y)$, and a predictor of some future value of y; and $\hat{\beta}_0$ and $\hat{\beta}_1$ are estimators of β_0 and β_1, respectively.

For a given data point, say the point (x_i, y_i), the observed value of y is y_i and the predicted value of y would be obtained by substituting x_i into the prediction equation:

$$\hat{y}_i = \hat{\beta}_0 + \hat{\beta}_1 x_i$$

And the deviation of the ith value of y from its predicted value is

$$(y_i - \hat{y}_i) = [y_i - (\hat{\beta}_0 + \hat{\beta}_1 x_i)]$$

Then the sum of squares of the deviations of the y-values about their predicted values for all of the n data points is

$$SSE = \sum [y_i - (\hat{\beta}_0 + \hat{\beta}_1 x_i)]^2$$

The quantities $\hat{\beta}_0$ and $\hat{\beta}_1$ that make the SSE a minimum are called the **least squares estimates** of the population parameters β_0 and β_1, and the prediction equation $\hat{y} = \hat{\beta}_0 + \hat{\beta}_1 x$ is called the **least squares line.**

Definition 12.1

The **least squares line** is one that has a smaller SSE than any other straight-line model.

The values of $\hat{\beta}_0$ and $\hat{\beta}_1$ that minimize the SSE are (proof omitted) given by the formulas in the accompanying box.*

Formulas for the Least Squares Estimates

Slope: $\hat{\beta}_1 = \dfrac{SS_{xy}}{SS_{xx}}$

y-intercept: $\hat{\beta}_0 = \bar{y} - \hat{\beta}_1 \bar{x}$

where

$$SS_{xy} = \sum (x_i - \bar{x})(y_i - \bar{y}) = \sum x_i y_i - \frac{(\sum x_i)(\sum y_i)}{n}$$

$$SS_{xx} = \sum (x_i - \bar{x})^2 = \sum x_i^2 - \frac{(\sum x_i)^2}{n}$$

n = Sample size

---

* Students who are familiar with the calculus should note that the values of β_0 and β_1 that minimize $SSE = \sum (y_i - \hat{y}_i)^2$ are obtained by setting the two partial derivatives $\partial SSE/\partial \beta_0$ and $\partial SSE/\partial \beta_1$ equal to zero. The solutions to these two equations yield the formulas shown in the box. Furthermore, we denote the *sample* solutions to the equations by $\hat{\beta}_0$ and $\hat{\beta}_1$, where the "hat" denotes that these are sample estimates of the true population intercept β_0 and slope β_1.

Preliminary computations for finding the least squares line for the drug reaction example are presented in Table 12.3. We can now calculate

$$SS_{xy} = \sum x_i y_i - \frac{(\sum x_i)(\sum y_i)}{5} = 37 - \frac{(15)(10)}{5}$$

$$= 37 - 30 = 7$$

$$SS_{xx} = \sum x_i^2 - \frac{(\sum x_i)^2}{5} = 55 - \frac{(15)^2}{5}$$

$$= 55 - 45 = 10$$

Table 12.3

Preliminary Computations for the Drug Reaction Example

	x_i	y_i	x_i^2	$x_i y_i$
	1	1	1	1
	2	1	4	2
	3	2	9	6
	4	2	16	8
	5	4	25	20
TOTALS	$\sum x_i = 15$	$\sum y_i = 10$	$\sum x_i^2 = 55$	$\sum x_i y_i = 37$

Then the slope of the least squares line is

$$\hat{\beta}_1 = \frac{SS_{xy}}{SS_{xx}} = \frac{7}{10} = .7$$

and the y-intercept is

$$\hat{\beta}_0 = \bar{y} - \hat{\beta}_1 \bar{x} = \frac{\sum y_i}{5} - \hat{\beta}_1 \frac{(\sum x_i)}{5} = \frac{10}{5} - (.7)\frac{(15)}{5}$$

$$= 2 - (.7)(3) = 2 - 2.1 = -.1$$

The least squares line is thus

$$\hat{y} = \hat{\beta}_0 + \hat{\beta}_1 x = -.1 + .7x$$

The graph of this line is shown in Figure 12.5.

Figure 12.5 The Line $\hat{y} = -.1 + .7x$ Fit to the Data

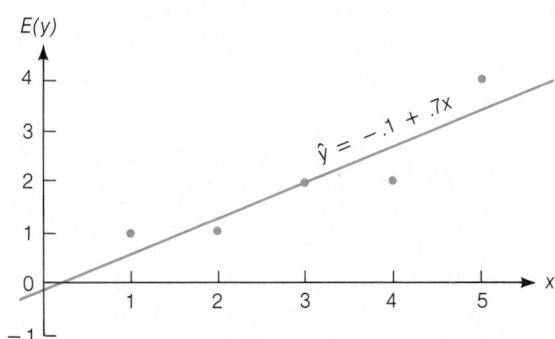

Table 12.4

Comparing Observed and Predicted Values for the Least Squares Prediction Equation

x	y	$\hat{y} = -.1 + .7x$	$(y - \hat{y})$	$(y - \hat{y})^2$
1	1	.6	$(1 - .6) = $　.4	.16
2	1	1.3	$(1 - 1.3) = -.3$	.09
3	2	2.0	$(2 - 2.0) = $　0	.00
4	2	2.7	$(2 - 2.7) = -.7$	.49
5	4	3.4	$(4 - 3.4) = $　.6	.36
			Sum of errors $= $　0	SSE $= 1.10$

The predicted value of y for a given value of x can be obtained by substituting into the formula for the least squares line. Thus, when $x = 2$ we predict y to be

$$\hat{y} = -.1 + .7x = -.1 + .7(2) = 1.3$$

We will show how to find a prediction interval for y in Section 12.8.

The observed and predicted values of y, the deviations of the y-values about their predicted values, and the squares of these deviations are shown in Table 12.4. Note that the sum of squares of the deviations, SSE, is 1.10, and (as we would expect) this is less than the SSE $= 2.0$ obtained in Table 12.2 for the visually fitted line.

To summarize, we have defined the best-fitting straight line to be the one that satisfies the least squares criterion; that is, the sum of the squared errors will be smaller than for any other straight-line model. This line is called the **least squares line,** and its equation is called the **least squares prediction equation.**

Exercises 12.8–12.17

Learning the Mechanics

12.8 Consider the four data points given in the margin.
a. Plot the points and visually fit a straight line through them.
b. Find the slope and the y-intercept of the line drawn in part a.
c. Fit a line to the data using the method of least squares, i.e., find the least squares estimates of β_0 and β_1, and give the equation of the least squares line.
d. Compare the answers to parts b and c. Are corresponding estimates of β_0 and β_1 equal? Should they be?

x	0	1	2	3
y	2.0	3.5	4.5	6.0

12.9 Consider the six data points given in the margin.
a. Plot the points and visually fit a straight line through them.
b. Find the slope and the y-intercept of the line drawn in part a.
c. Fit a line to the data using the method of least squares, i.e., find the least squares estimates of β_0 and β_1.
d. Graph the least squares line and compare it to the line drawn in part a. Comment.

x	1	2	3	4	5	6
y	-1	2	3	4	6	6

12.10 Use the method of least squares to fit a straight line to the five given data points.
a. What are the least squares estimates of β_0 and β_1?
b. Plot the data points and graph the least squares line. Does the line pass through the data points?
c. Predict the value of y for $x = 1$.
d. Predict the value of y for $x = -1.5$.

x	-2	-1	0	1	2
y	6	4	3	1	-1

Applying the Concepts

12.11 In recent years, physicians have used the "dividing reflex" to reduce abnormally rapid heartbeats in humans by briefly submerging the patient's face in cold water. The reflex, triggered by cold water temperatures, is an involuntary neural response that shuts off circulation to the skin, muscles, and internal organs and diverts extra oxygen-carrying blood to the heart, lungs, and brain. A research physician conducted an experiment to investigate the effects of various cold water temperatures on the pulse rate of small children. The data for seven 6-year-old children are shown in the table.

CHILD	TEMPERATURE OF WATER x (°F)	DECREASE IN PULSE RATE y (beats/minute)
1	68	2
2	65	5
3	70	1
4	62	10
5	60	9
6	55	13
7	58	10

a. Find the least squares line for the data.
b. Construct a scattergram for the data; then graph the least squares line as a check on your calculations.
c. If the water temperature is 60°F, predict the drop in pulse rate for a 6-year-old child. [*Note:* A measure of the reliability of these predictions is discussed in Section 12.8.]

12.12 Is the percentage of games won by a major league baseball team related to the team's batting average? The table shows the percentage of games won and the

TEAM	PERCENTAGE GAMES WON	TEAM BATTING AVERAGE	TEAM EARNED RUN AVERAGE
Detroit	64.8	.270	3.61
Toronto	56.9	.278	3.75
Baltimore	52.5	.247	3.59
New York	52.0	.275	3.71
Boston	51.6	.278	4.19
Cleveland	44.0	.261	4.28
Milwaukee	41.9	.259	3.89
Minnesota	52.8	.274	3.70
California	50.0	.252	3.83
Kansas City	49.2	.269	4.08
Chicago	48.4	.248	4.10
Oakland	48.0	.256	4.17
Seattle	48.0	.257	4.36
Texas	43.1	.254	3.94

batting averages for fourteen American League teams at the fourteenth week into the 1983 season.

a. If you were to model the relationship between a major league team's percentage of games won y and the team's batting average using a straight line, would you expect the slope of the line to be positive or negative?

b. Fit a simple linear regression model to the data. [*Hint*: To simplify the calculations, let $x = 1{,}000$ (Team batting average $-.265$).]

c. Construct a scattergram for the data and graph the least squares line. Does the slope of the least squares line seem to agree with the points on your scattergram?

d. Can you explain why the percentage of games won does not appear to be strongly related to a team's batting average?

12.13 Refer to Exercise 12.12. The earned run average of a baseball team is a measure of its pitching ability. It gives the average number of runs scored per game by opposing teams. Consider the relationship between a baseball team's percentage y games won and the team's earned run average.

a. If you were to model the relationship between a major league team's percentage of games won y and the team's earned run average using a straight line, would you expect the slope of the line to be positive or negative?

b. Fit a simple linear regression model to the data given in the table accompanying Exercise 12.12.

c. Construct a scattergram for the data and graph the least squares line. Does the slope of the least squares line seem to agree with the points on your scattergram?

d. Can you explain why the percentage of games won does not appear to be strongly related to a team's earned run average?

12.14 To investigate the relationship between yield of potatoes, y, and level of fertilizer application, x, an experimenter divides a field into eight plots of equal size and applies differing amounts of fertilizer to each. The yield of potatoes (in pounds) and the fertilizer application (in pounds) are recorded for each plot. The data are as follows:

x	1	1.5	2	2.5	3	3.5	4	4.5
y	25	31	27	28	36	35	32	34

a. Construct a scattergram for the data.

b. Find the least squares estimates for β_0 and β_1.

c. According to your least squares line, approximately how many pounds of potatoes would you expect from a plot to which 3.75 pounds of fertilizer has been applied? [*Note:* A measure of the reliability of these predictions is discussed in Section 12.8.]

12.15 In Exercise 8.36, we gave data on the mean daily air temperature and corresponding cocoon temperature of woolly-bear caterpillars of the High Arctic (Kevan, Jensen, & Shorthouse, 1982). The data, collected over 12 days, are reproduced in the table.

a. Fit a least squares line to relate the cocoon temperature y to the outside air temperature x.

DAY	TEMPERATURES (°C) Air	TEMPERATURES (°C) Cocoon[a]
1	10.4	15.1
2	9.2	14.6
3	2.2	6.8
4	2.6	6.8
5	4.1	8.0
6	3.7	8.7
7	1.7	3.6
8	2.0	5.3
9	3.0	7.0
10	3.5	7.1
11	4.5	9.6
12	4.4	9.5

[a] Each cocoon temperature is the average of the temperatures of two cocoons.

b. Plot the data points and graph your least squares line on the same sheet of graph paper. Does your line provide a good fit to the data?

12.16 The sand lance is a small fish that can be found in the Northwest Atlantic from Cape Hatteras to Greenland. In a study of the biological and demographic characteristics of the sand lance, G. H. Winters (1983) includes data on the mean length of specimens collected each year for the period 1969 through 1979 for sand lance ages 2 through 8 years:

MEAN LENGTH (MILLIMETERS)	176	194	212	226	236	244	254
AGE (YEARS)	2	3	4	5	6	7	8

a. Find the least squares prediction equation relating the mean length y of the sand lance to its age x.
b. Plot the data points and graph your least squares line on the same sheet of graph paper. Does your line appear to provide a good fit to the data?

12.17 The data shown in the table on the next page are part of a series of experiments conducted to investigate the effect of water temperature on the absorption by rainbow trout of sub-lethal levels of cyanide (Kovacs & Leduc, 1982). This specific set of data is the result of a preliminary experiment to determine the relationship between mean weight gain (percentage of body weight) over a 20-day period as a function of the ration fed the fish (percentage of body weight). Twenty fish were included in the experiment for each ration level and water temperature combination. The mean percentage weight gain for each sample of twenty fish is shown in the table. (Note that, as expected, a zero level of rations produced a negative weight gain, i.e., a loss.) Since all the sample sizes are equal, we can fit a simple linear regression model to the data for any water temperature level by fitting the model to the four sample means.

| RATION | MEAN WET WEIGHT GAIN (%) | | |
(% body weight per day)	6°C	12°C	18°C
.0	−8.14	−10.33	−13.21
.8	12.31		
1.2		14.29	
1.5	28.19		15.51
2.5	29.12	38.05	
3.5			51.13
4.0		60.13	
4.5			65.49
Maintenance ration	.32	.46	.69

a. Find the least squares line relating mean weight gain to ration level for each of the three water temperature levels.

b. Plot the data points and graph the least squares lines on the same sheet of graph paper. Do the least squares lines appear to provide good fits to their respective data sets? Does the relationship between mean weight gain and ration appear to depend on the water temperature? (In Chapter 14 we will test to determine whether differences exist.)

12.3 Model Assumptions

In Section 12.2 we assumed that the probabilistic model relating drug reaction time y to the percentage of drug x in the bloodstream is

$$y = \beta_0 + \beta_1 x + \varepsilon$$

and recall that the least squares estimate of the deterministic component of the model, $\beta_0 + \beta_1 x$, is

$$\hat{y} = \hat{\beta}_0 + \hat{\beta}_1 x = -.1 + .7x$$

Now we turn our attention to the random component ε of the probabilistic model and its relation to the errors of estimating β_0 and β_1. In particular, we will see how the probability distribution of ε determines how well the model describes the true relationship between the dependent variable y and the independent variable x.

Step 3 in a regression analysis requires us to specify the probability distribution of the random error ε. We will make four basic assumptions about the general form of this probability distribution:

Assumption 1. The mean of the probability distribution of ε is zero. That is, the average of the errors over an infinitely long series of experiments is zero for each setting of the independent variable x. This assumption implies that the mean value of y, $E(y)$, for a given value of x is $E(y) = \beta_0 + \beta_1 x$.

Assumption 2. The variance of the probability distribution of ε is constant for all settings of the independent variable x. For our straight-line model, this assumption means that the variance of ε is equal to a constant, say σ^2, for all values of x.

Figure 12.6 The Probability Distribution of ε

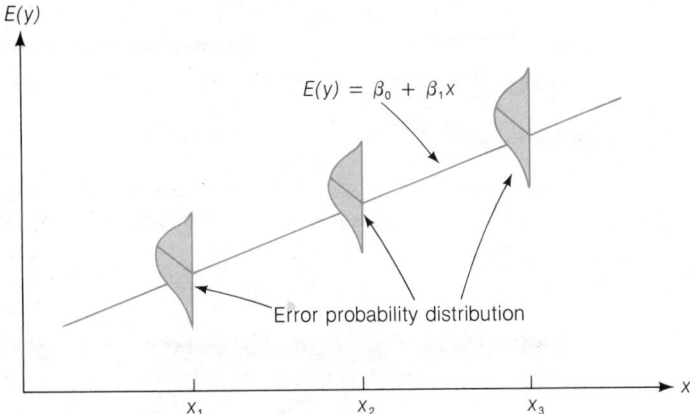

The probability distribution of ε is normal.

Assumption 4. The errors associated with any two different observations are independent. That is, the error associated with one value of y has no effect on the errors associated with other y-values.

The implications of the first three assumptions can be seen in Figure 12.6, which shows distributions of errors for three values of x, namely x_1, x_2, and x_3. Note that the relative frequency distributions of the errors are normal with a mean of zero and a constant variance σ^2. (All the distributions shown have the same amount of spread or variability.) The straight line shown in Figure 12.6 is the mean value of y for a given value of x. We will denote this mean value as $E(y)$. Then, the line of means is given by the equation

$$E(y) = \beta_0 + \beta_1 x$$

Various techniques exist for checking the validity of these assumptions, and there are remedies to be applied when they appear to be invalid. These topics are beyond the scope of this text, but they are discussed in some of the references listed at the end of the chapter. Fortunately, the assumptions need not hold exactly in order for least squares estimators and test statistics (to be described subsequently) to possess the measures of reliability that we would expect from a regression analysis. The assumptions will be satisfied adequately for many applications encountered in practice.

12.4 An Estimator of σ^2

It seems reasonable to assume that the greater the variability of the random error ε (which is measured by its variance σ^2), the greater will be the errors in the estimation of the model parameters β_0 and β_1 and in the error of prediction when $\hat{y}$ is used to predict y for some value of x. Consequently, you should not be surprised, as we proceed through this chapter, to find that σ^2 appears in the formulas for all confidence intervals and test statistics that we will be using.

In most practical situations, σ^2 is unknown and we must use our data to estimate its value. The best (proof omitted) estimate s^2 of σ^2 is obtained by dividing the sum of squares of the deviations of the y-values from the prediction line,

$$SSE = \sum(y_i - \hat{y}_i)^2$$

by the number of degrees of freedom associated with this quantity. We use 2 df to estimate the two parameters β_0 and β_1 in the straight-line model, leaving $(n-2)$ df for the error variance estimation.

Estimation of σ^2 for a (First-Order) Straight-Line Model

$$s^2 = \frac{SSE}{\text{Degrees of freedom for error}} = \frac{SSE}{n-2}$$

where

$$SSE = \sum(y_i - \hat{y}_i)^2 = SS_{yy} - \hat{\beta}_1 SS_{xy}$$

$$SS_{yy} = \sum(y_i - \bar{y})^2 = \sum y_i^2 - \frac{\left(\sum y_i\right)^2}{n}$$

Warning: When performing these calculations, you may be tempted to round the calculated values of SS_{yy}, $\hat{\beta}_1$, and SS_{xy}. Be certain to carry at least six significant figures for each of these quantities to avoid substantial errors in calculation of the SSE.

In the drug reaction example, we previously calculated $SSE = 1.10$ for the least squares line $\hat{y} = -.1 + .7x$. Recalling that there were $n = 5$ data points, we have $n - 2 = 5 - 2 = 3$ df for estimating σ^2. Thus,

$$s^2 = \frac{SSE}{n-2} = \frac{1.10}{3} = .367$$

is the estimated variance, and

$$s = \sqrt{.367} = .61$$

is the estimated standard deviation of ε.

You may be able to obtain an intuitive feeling for s by recalling the interpretation given to a standard deviation in Chapter 2 and remembering that the least squares line estimates the mean value of y for a given value of x. Since s measures the spread of the distribution of y-values about the least squares line, we should not be surprised to find that most of the observations lie within $2s$ or $2(.61) = 1.22$ of the least squares line. For this simple example (only five data points), all five data points fall within $2s$ of the least squares line. In Section 12.8, we will use s to evaluate the error of prediction when the least squares line is used to predict a value of y to be observed for a given value of x.

Exercises
12.18–12.21

Learning the Mechanics

12.18 Suppose that you fit a least squares line to nine data points and the calculated value of SSE is .429.
   a. Find s^2, the estimator of σ^2 (the variance of the random error ε).
   b. What is the largest deviation that you might expect between any one of the nine points and the least squares line?

12.19 Calculate SSE and s^2 for the least squares lines found in:
   a. Exercise 12.8   **b.** Exercise 12.9   **c.** Exercise 12.10

Applying the Concepts

12.20 An electronics dealer believes that there is a linear relationship between the number of hours of quadraphonic programming on a city's FM stations and sales of quadraphonic systems. Records for the dealer's sales during the last 6 months and the amount of quadraphonic programming for the corresponding months are:

MONTH	AVERAGE AMOUNT OF QUADRAPHONIC PROGRAMMING x (hours)	NUMBER OF QUADRAPHONIC SYSTEMS SOLD y
1	33.6	7
2	36.3	10
3	38.7	13
4	36.6	11
5	39.0	14
6	38.4	18

   a. Fit a least squares line to the data.
   b. Plot the data and graph the least squares line as a check.
   c. Calculate the SSE, s^2, and s.
   d. The deviations between the observed values of y and their corresponding predicted values (used in calculating SSE) are called *residuals*. Calculate the residuals for the six data points. Within what limits would you expect most of the residuals to fall?

12.21 A company keeps extensive records on its new salespeople on the premise that sales should increase with experience. A random sample of seven new salespeople

MONTHS ON JOB x	MONTHLY SALES y (thousands of dollars)
2	2.4
4	7.0
8	11.3
12	15.0
1	.8
5	3.7
9	12.0

produced the data on experience and sales shown in the table.

a. Fit a least squares line to the data.

b. Plot the data and graph the least squares line.

c. Predict the sales that a new salesperson would be expected to generate after 6 months on the job. After 9 months.

d. Calculate the SSE, s^2, and s.

e. The deviations between the observed values of y and their corresponding predicted values (used in calculating SSE) are the residuals. Calculate the residuals for the six data points. Within what limits would you expect most of the residuals to fall?

12.5 Assessing the Utility of the Model: Making Inferences About the Slope β_1

Now that we have specified the probability distribution of ε and found an estimate of the variance σ^2, we are ready to make statistical inferences about the model's utility for predicting the response y. This is step 4 in our regression modeling procedure.

Refer again to the data of Table 12.1 and suppose that the reaction times are *completely unrelated* to the percentage of drug in the bloodstream. What could be said about the values of β_0 and β_1 in the hypothesized probabilistic model

$$y = \beta_0 + \beta_1 x + \varepsilon$$

if x contributes no information for the prediction of y? The implication is that the mean of y, that is, the deterministic part of the model $E(y) = \beta_0 + \beta_1 x$, does not change as x changes. In the straight-line model, this means that the true slope, β_1, is equal to zero (see Figure 12.7). Therefore, to test the null hypothesis that x contributes no information for the prediction of y against the alternative hypothesis that these variables are linearly related with a slope differing from zero, we test

$$H_0: \quad \beta_1 = 0 \qquad H_a: \quad \beta_1 \neq 0$$

If the data support the alternative hypothesis, we will conclude that x does contribute information for the prediction of y using the straight-line model (although the true

Figure 12.7   Graphing the Model $y = \beta_0 + \varepsilon$ $(\beta_1 = 0)$

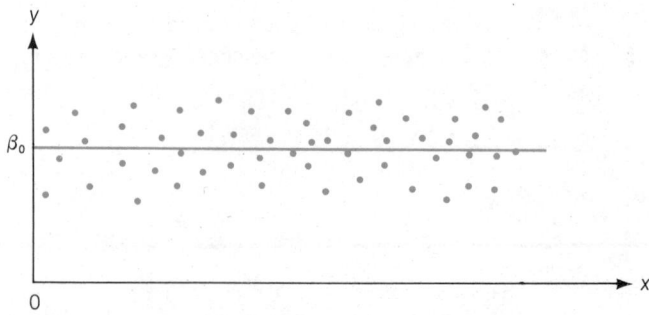

relationship between $E(y)$ and x could be more complex than a straight line). Thus, to some extent, this is a test of the utility of the hypothesized model.

The appropriate test statistic is found by considering the sampling distribution of $\hat{\beta}_1$, the least squares estimator of the slope β_1:

Sampling Distribution of $\hat{\beta}_1$

If we make the four assumptions about ε (see Section 12.3), the sampling distribution of the least squares estimator $\hat{\beta}_1$ of the slope will be normal with mean β_1 (the true slope) and standard deviation

$$\sigma_{\hat{\beta}_1} = \frac{\sigma}{\sqrt{SS_{xx}}} \qquad \text{(See Figure 12.8.)}$$

Figure 12.8 Sampling Distribution of $\hat{\beta}_1$

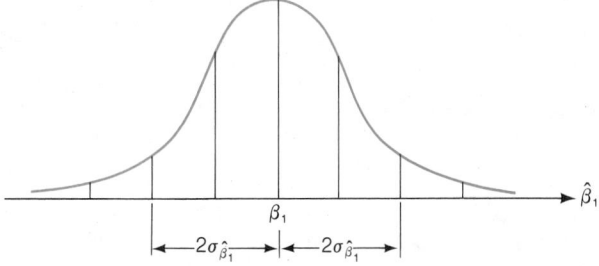

Since σ is usually unknown, the appropriate test statistic is a Student's t-statistic formed as follows:

$$t = \frac{\hat{\beta}_1 - \text{Hypothesized value of } \beta_1}{s_{\hat{\beta}_1}} \qquad \text{where } s_{\hat{\beta}_1} = \frac{s}{\sqrt{SS_{xx}}}$$

Thus,

$$t = \frac{\hat{\beta}_1 - 0}{s/\sqrt{SS_{xx}}}$$

Note that we have substituted the estimator s for σ and then formed $s_{\hat{\beta}_1}$ by dividing s by $\sqrt{SS_{xx}}$. The number of degrees of freedom associated with this t-statistic is the same as the number of degrees of freedom associated with s. Recall that this number is $(n-2)$ df when the hypothesized model is a straight line (see Section 12.4). The setup of our test of the utility of the model is summarized in the box on page 600.

For the drug reaction example, we will choose $\alpha = .05$ and, since $n = 5$, t will be based on $n - 2 = 3$ df and the rejection region will be

$$t < -t_{.025} = -3.182 \qquad \text{or} \qquad t > t_{.025} = 3.182$$

A Test of Model Utility

One-Tailed Test	Two-Tailed Test

One-Tailed Test

H_0: $\beta_1 = 0$

H_a: $\beta_1 < 0$

   (or H_a: $\beta_1 > 0$)

Test statistic: $t = \dfrac{\hat{\beta}_1}{s_{\hat{\beta}_1}} = \dfrac{\hat{\beta}_1}{s/\sqrt{SS_{xx}}}$

Rejection region: $t < -t_\alpha$

   (or $t > t_\alpha$

   when H_a: $\beta_1 > 0$)

Two-Tailed Test

H_0: $\beta_1 = 0$

H_a: $\beta_1 \neq 0$

Test statistic: $t = \dfrac{\hat{\beta}_1}{s_{\hat{\beta}_1}} = \dfrac{\hat{\beta}_1}{s/\sqrt{SS_{xx}}}$

Rejection region: $t < -t_{\alpha/2}$

   or $t > t_{\alpha/2}$

where t_α and $t_{\alpha/2}$ are based on $(n - 2)$ degrees of freedom

Assumptions: The four assumptions about ε listed in Section 12.3.

We previously calculated $\hat{\beta}_1 = .7$, $s = .61$, and $SS_{xx} = 10$. Thus,

$$t = \frac{\hat{\beta}_1}{s/\sqrt{SS_{xx}}} = \frac{.7}{.61/\sqrt{10}} = \frac{.7}{.19} = 3.7$$

Since this calculated t-value falls in the upper-tail rejection region (see Figure 12.9), we reject the null hypothesis and conclude that the slope β_1 is not zero. The sample evidence indicates that x contributes information for the prediction of y using a linear model for the relationship between reaction time and the amount of the drug in the bloodstream.

Figure 12.9 Rejection Region and Calculated t-Value for Testing H_0: $\beta_1 = 0$ Versus H_a: $\beta_1 \neq 0$

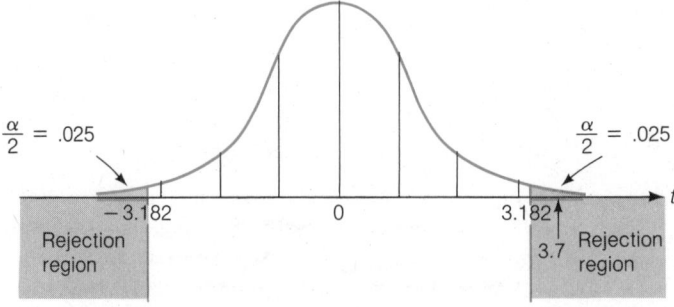

What conclusion can be drawn if the calculated t-value does not fall in the rejection region? We know from previous discussions of the philosophy of hypothesis-testing that such a t-value does *not* lead us to accept the null hypothesis. That is, we do not conclude that $\beta_1 = 0$. Additional data might indicate that β_1 differs from zero, or a more complex relationship may exist between x and y, requiring the fitting of a model other than the straight-line model. We will discuss several such models in Chapter 13.

Another way to make inferences about the slope β_1 is to estimate it using a confidence interval. This interval is formed as shown in the box.

A 100(1 − α)% Confidence Interval for the Slope β_1

$$\hat{\beta}_1 \pm t_{\alpha/2}\, s_{\hat{\beta}_1}$$

where

$$s_{\hat{\beta}_1} = \frac{s}{\sqrt{SS_{xx}}}$$

and $t_{\alpha/2}$ is based on $(n - 2)$ degrees of freedom.

Assumptions: The four assumptions about ε listed in Section 12.3.

For the drug reaction example, $t_{\alpha/2}$ is based on $(n - 2) = 3$ degrees of freedom. Therefore, a 95% confidence interval for the slope β_1, the expected change in reaction time for a 1% increase in the amount of drug in the bloodstream, is

$$\hat{\beta}_1 \pm t_{.025} s_{\hat{\beta}_1} = .7 \pm 3.182\left(\frac{s}{\sqrt{SS_{xx}}}\right)$$

$$= .7 \pm 3.182\left(\frac{.61}{\sqrt{10}}\right) = .7 \pm .61$$

Thus, the interval estimate of the slope parameter β_1 is .09 to 1.31.

Since all the values in this interval are positive, it appears that β_1 is positive and that the mean of y, $E(y)$, increases as x increases. However, the rather large width of the confidence interval reflects the small number of data points (and, consequently, a lack of information) in the experiment. We would expect a narrower interval if the sample size were increased.

**Exercises
12.22–12.31**

Learning the Mechanics

12.22 Give the values of $\hat{\beta}_1$, s^2, SS_{xx}, and the degrees of freedom associated with s^2 for each of the following exercises:
a. Exercise 12.8   **b.** Exercise 12.9   **c.** Exercise 12.10

12.23 For each of the following exercises, determine whether there is sufficient evidence to indicate that β_1 is greater than zero. Test using $\alpha = .05$.
a. Exercise 12.8   **b.** Exercise 12.9

12.24 For each of the following exercises, determine whether there is sufficient evidence to indicate that β_1 is less than zero. Test using $\alpha = .05$.
a. Exercise 12.10   **b.** Exercise 12.11

12.25 For each of the following exercises, determine whether there is sufficient evidence to indicate that β_1 differs from zero. Test using $\alpha = .05$.
a. Exercise 12.13   **b.** Exercise 12.14

Applying the Concepts

12.26 A breeder of thoroughbred horses wishes to model the relationship between the gestation period and the life span of a horse. The breeder believes that the two variables may follow a linear trend. The information in the table was supplied to the breeder from various thoroughbred stables across the state. (Note that the horse has the greatest variation of gestation period of any species, due to seasonal and nutritional factors.)

HORSE	GESTATION PERIOD x (days)	LIFE SPAN y (years)
1	416	24
2	279	25.5
3	298	20
4	307	21.5
5	356	22
6	403	23.5
7	265	21

a. Fit a least squares line to the data. Plot the data points and graph the least squares line as a check on your calculations.
b. Do the data provide sufficient evidence to support the horse breeder's hypothesis? Test using $\alpha = .05$.
c. Find a 90% confidence interval for β_1. Interpret this interval.

12.27 A group of children, ranging from 10 to 12 years of age, were administered a verbal test in order to study the relationship between the number of words used and the silence interval before response. The tester believes that a linear relationship exists between the two variables. Each subject was asked a series of questions, and the total number of words used in answering was recorded. The time (in seconds) before the subject responded to each question was also recorded. The data for eight children are given in the table.
a. Write a simple linear probabilistic model relating total words to total silence time, and use the least squares method to estimate the deterministic part of the model.
b. Does x contribute information for the prediction of y? Test the null hypothesis, slope $\beta_1 = 0$, against the alternative hypothesis, $\beta_1 \neq 0$. Use $\alpha = .05$. Interpret the results of the test.

SUBJECT	TOTAL WORDS y	TOTAL SILENCE TIME x (seconds)
1	61	23
2	70	37
3	42	38
4	52	25
5	91	17
6	63	21
7	71	42
8	55	16

12.28 A local brewery is interested in determining whether a linear relationship exists between the amount it spends on television advertising and total sales. The relevant data are listed in the table.

MONTH	SALES (Thousands of dollars)	TELEVISION ADVERTISING EXPENDITURES (Thousands of dollars)
January	50	.5
February	90	.9
March	30	.4
April	90	.7
May	91	1.1
June	95	.75
July	95	.8

a. Find the least squares line for the given data. Plot the data on a scattergram, and graph the line as a check on your calculations.

b. Letting $\alpha = .05$, test the null hypothesis that $\beta_1 = 0$. What alternative hypothesis would you select for this test? Draw appropriate conclusions concerning the adequacy of a linear model to describe the relationship between sales and television advertising expenditures.

c. Construct a 90% confidence interval for the slope parameter in the hypothesized linear model.

d. Interpret the confidence interval and explain what it tells you about the relationship between sales and television advertising expenditure.

12.29 In Exercise 12.15, we fit a least squares line to relate the cocoon temperature y of a woolly-bear caterpillar to the outside air temperature x. The data are reproduced in the table on page 604 (Kevan, Jensen, & Shorthouse, 1982).

a. Calculate SSE and s^2.

b. Do the data provide sufficient evidence to indicate that the outside air temperature provides information for predicting the woolly-bear caterpillar cocoon temperature? Test using $\alpha = .05$.

c. Find the p-value for the test and interpret it.

| DAY | TEMPERATURES (°C) | |
	Air	Cocoon[a]
1	10.4	15.1
2	9.2	14.6
3	2.2	6.8
4	2.6	6.8
5	4.1	8.0
6	3.7	8.7
7	1.7	3.6
8	2.0	5.3
9	3.0	7.0
10	3.5	7.1
11	4.5	9.6
12	4.4	9.5

[a] Each cocoon temperature is the average of the temperatures of two cocoons.

12.30 In Exercise 12.16, we fit a least squares line to relate the length of a sand lance to its age. The data are reproduced in the table (Winters, 1983).

MEAN LENGTH (MILLIMETERS)	176	194	212	226	236	244	254
AGE (YEARS)	2	3	4	5	6	7	8

a. Do the data provide sufficient evidence to indicate that age x contributes information for the prediction of the length y of a sand lance? Test using $\alpha = .05$.

b. Find the p-value for the test and interpret it.

12.31 In Exercise 12.17, we discussed an experiment conducted to determine the effect of ration level on the growth rate of rainbow trout (Kovacs & Leduc, 1982). Twenty fish were fed at each of four ration levels at 6°C water temperature. The experiment was repeated for water temperatures of 12°C and 18°C. The table gives the means and

| RATION (% body weight per day) | MEAN WET WEIGHT GAIN (%) | | |
	6°C	12°C	18°C
.0	−8.14 (3.27)	−10.33 (2.35)	−13.21 (1.96)
.8	12.31 (3.37)		
1.2		14.29 (5.32)	
1.5	28.19 (5.39)		15.51 (4.48)
2.5	29.12 (5.67)	38.05 (6.99)	
3.5			51.13 (7.14)
4.0		60.13 (12.71)	
4.5			65.49 (13.04)
Maintenance ration	.32	.48	.69

standard deviations (in parentheses) for each sample of twenty fish. Refer only to the data collected on fish fed in water maintained at 12°C.

a. Find SSE and s^2 for the $n = 4$ data points.

b. Find a 95% confidence interval for the mean gain (percentage of body weight) for a 1% increase in ration level.

c. State the assumptions required for your inference in part c to be valid.

d. Which assumption may not be satisfied? Explain why.

12.6 Correlation: A Measure of the Utility of the Model

The claim is often made that the number of cigarettes smoked and the incidence of lung cancer are "highly correlated." Another popular belief is that the crime rate and the unemployment rate are "correlated." Some people even believe that the Dow Jones Industrial Average and the lengths of fashionable skirts are "correlated." In this section we will discuss the concept of **correlation.**[*] A numerical descriptive measure of the correlation between two variables x and y is provided by the **Pearson product moment coefficient of correlation, r.**

Definition 12.2

The sample **Pearson product moment coefficient of correlation, r,** is defined as

$$r = \frac{SS_{xy}}{\sqrt{SS_{xx}SS_{yy}}}$$

It is a measure of the strength of the linear relationship between two random variables x and y.

Note that the computational formula for the correlation coefficient r given in Definition 12.2 involves the same quantities that were used in computing the least squares prediction equation. In fact, since the numerators of the expressions for $\hat{\beta}_1$ and r are identical, you can see that $r = 0$ when $\hat{\beta}_1 = 0$ (the case where x contributes no information for the prediction of y) and that r is positive when the slope is positive and negative when the slope is negative. Unlike $\hat{\beta}_1$, the correlation coefficient r is *scaleless* and assumes a value between -1 and $+1$, regardless of the units of x and y.

A value of r near or equal to zero implies little or no linear relationship between y and x. In contrast, the closer r comes to 1 or -1, the stronger the linear relationship between y and x. And if $r = 1$ or $r = -1$, all the sample points fall exactly on the least squares line. Positive values of r imply a positive linear relationship between y and x; that is, y increases as x increases. Negative values of r imply a negative linear

---

[*] Recall that we first discussed the concept of correlation in Section 10.5, where we presented Spearman's rank correlation coefficient.

Figure 12.10 Values of *r* and Their Implications

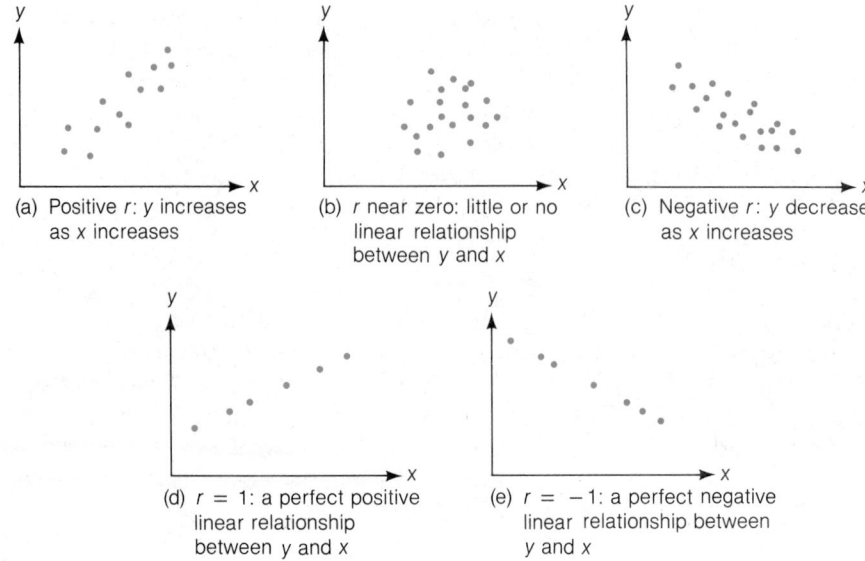

(a) Positive *r*: *y* increases as *x* increases

(b) *r* near zero: little or no linear relationship between *y* and *x*

(c) Negative *r*: *y* decreases as *x* increases

(d) *r* = 1: a perfect positive linear relationship between *y* and *x*

(e) *r* = −1: a perfect negative linear relationship between *y* and *x*

relationship between *y* and *x*; that is, *y* decreases as *x* increases. Each of these situations is portrayed in Figure 12.10.

We will demonstrate how to calculate the coefficient of correlation *r* using the data in Table 12.1 for the drug reaction example. The quantities needed to calculate *r* are SS_{xy}, SS_{xx}, and SS_{yy}. The first two quantities have been calculated previously and are repeated here for convenience:

$$SS_{xy} = 7 \qquad SS_{xx} = 10$$

$$SS_{yy} = \sum y^2 - \frac{\left(\sum y\right)^2}{n} = 26 - \frac{(10)^2}{5} = 26 - 20 = 6$$

We now find the coefficient of correlation:

$$r = \frac{SS_{xy}}{\sqrt{SS_{xx}SS_{yy}}} = \frac{7}{\sqrt{(10)(6)}} = \frac{7}{\sqrt{60}}$$

$$= .904$$

The fact that *r* is positive and near 1 in value indicates that the reaction time tends to increase as the amount of drug in the bloodstream increases—*for this sample of five subjects*. This is the same conclusion we reached when we found the calculated value of the least squares slope to be positive.

Example 12.1   A firm wants to know the correlation between the size of its sales force and its yearly sales revenue. The records for the past 10 years are examined, and the results listed in Table 12.5 are obtained. Calculate the coefficient of correlation *r* for the data.

Table 12.5
Sales Versus
Number of Salespeople

YEAR	NUMBER OF SALESPEOPLE x	SALES y (hundred thousand dollars)
1975	15	1.35
1976	18	1.63
1977	24	2.33
1978	22	2.41
1979	25	2.63
1980	29	2.93
1981	30	3.41
1982	32	3.26
1983	35	3.63
1984	38	4.15

Solution We need to calculate SS_{xy}, SS_{xx}, and SS_{yy}.

$$SS_{xy} = \sum xy - \frac{(\sum x)(\sum y)}{n} = 800.62 - \frac{(268)(27.73)}{10} = 57.456$$

$$SS_{xx} = \sum x^2 - \frac{(\sum x)^2}{n} = 7{,}668 - \frac{(268)^2}{10} = 485.6$$

$$SS_{yy} = \sum y^2 - \frac{(\sum y)^2}{n} = 83.8733 - \frac{(27.73)^2}{10} = 6.97801$$

Then the coefficient of correlation is

$$r = \frac{SS_{xy}}{\sqrt{SS_{xx}SS_{yy}}} = \frac{57.456}{\sqrt{(485.6)(6.97801)}} = \frac{57.456}{58.211} = .99$$

Thus, the size of the sales force and sales revenue are very highly correlated—at least over the past 10 years. The implication is that a strong positive linear relationship exists between these variables (see Figure 12.11). We must be careful, however, not to

Figure 12.11 Scatter-gram for Example 12.1

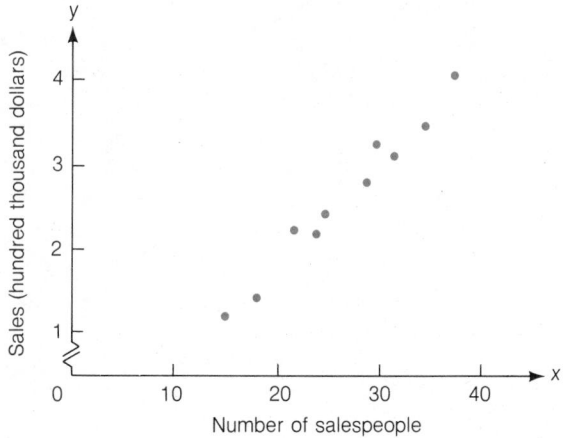

jump to any unwarranted conclusions. For instance, the firm may be tempted to conclude that the best thing it can do to increase sales is to hire a large number of new salespeople—that is, that there is a *causal relationship* between the two variables. However, high correlation does not imply causality. The fact is, many things have probably contributed both to the increase in the size of the sales force and to the increase in sales revenue. The firm's expertise has undoubtedly grown, the economy has inflated (so that 1984 dollars are not worth as much as 1975 dollars), and perhaps the scope of products and services sold by the firm has widened. *We must be careful not to infer a causal relationship on the basis of high sample correlation. The only safe conclusion when a high correlation is observed in the sample data is that a linear trend may exist between x and y.*   ■

Keep in mind that the correlation coefficient r measures the linear correlation between x-values and y-values in the sample and that a similar linear coefficient of correlation exists for the population from which the data points were selected. The **population correlation coefficient** is denoted by the symbol ρ (rho). As you might expect, ρ is estimated by the corresponding sample statistic, r. Or, rather than estimating ρ, we might want to test the hypothesis H_0: $\rho = 0$ against H_a: $\rho \neq 0$, i.e., test the hypothesis that x contributes no information for the prediction of y by using the straight-line model against the alternative that the two variables are at least linearly related.

However, we have already performed this *identical* test in Section 12.5 when we tested H_0: $\beta_1 = 0$ against H_a: $\beta_1 \neq 0$. When we tested the null hypothesis H_0: $\beta_1 = 0$ in connection with the drug reaction example, the data led to a rejection of the null hypothesis at the $\alpha = .05$ level. This rejection implies that the null hypothesis of a zero linear correlation between the two variables (drug and reaction time) can also be rejected at the $\alpha = .05$ level. The only real difference between the least squares slope $\hat{\beta}_1$ and the coefficient of correlation r is the measurement scale. Therefore, the information they provide about the utility of the least squares model is to some extent redundant. For this reason, we will use the slope to make inferences about the existence of a positive or negative linear relationship between two variables.

12.7 Coefficient of Determination

Another way to measure the contribution of x in predicting y is to consider how much the errors of prediction of y were reduced by using the information provided by x. If you do not use x, the best prediction for any value of y would be $\bar{y}$, and the sum of squares of the deviations of the y-values about $\bar{y}$ is the familiar

$$SS_{yy} = \sum(y - \bar{y})^2$$

On the other hand, if we use x to predict y, the sum of squares of the deviations of the y-values about the least squares line is

$$SSE = \sum(y - \hat{y})^2$$

Then the reduction in the sum of squares of deviations that can be attributed to x, expressed as a proportion of SS_{yy}, is

$$\frac{SS_{yy} - SSE}{SS_{yy}}$$

It can be shown that this quantity is equal to the square of the simple linear coefficient of correlation.

Definition 12.3

The square of the coefficient of correlation is called the **coefficient of determination.** It represents the proportion of the sum of squares of deviations of the y-values about their mean that can be attributed to a linear relationship between y and x.

$$r^2 = \frac{SS_{yy} - SSE}{SS_{yy}} = 1 - \frac{SSE}{SS_{yy}}$$

Note that r^2 is always between zero and 1 because r is between -1 and $+1$. Thus, $r^2 = .60$ means that 60% of the sum of squares of deviations of the y-values about their mean is attributable to the linear relation between y and x.

Example 12.2 Calculate the coefficient of determination for the drug reaction example using the formula given with Definition 12.3. The data are repeated in Table 12.6 for convenience.

Table 12.6

AMOUNT OF DRUG x (%)	REACTION TIME y (seconds)
1	1
2	1
3	2
4	2
5	4

Solution From previous calculations,

$$SS_{yy} = 6$$

and

$$SSE = \sum (y - \hat{y})^2 = 1.10$$

Then the coefficient of determination is given by

$$r^2 = \frac{SS_{yy} - SSE}{SS_{yy}} = \frac{6.0 - 1.1}{6.0} = \frac{4.9}{6.0}$$

$$= .82$$

(In Section 12.6, we calculated $r = .904$. Now we have $r^2 = (.904)^2 = .82$.) So we know that using the amount of drug in the blood, x, to predict y with the least squares line

$$\hat{y} = -.1 + .7x$$

accounts for 82% of the total sum of squares of deviations of the five sample y-values about their mean. ■

Case Study 12.1
Predicting the United States Crime Index

Reporting and analyzing crime rates is an important function of many law enforcement agencies. David Heaukulani (1975) comments: "The simple linear regression analysis remains one of the most useful tools for crime prediction." He demonstrates this point by fitting a straight-line model to predict the annual value of the United States crime index as a function of the United States population for each year. The data for the years 1963–1973 are given in Table 12.7, and the least squares line is shown in Figure 12.12.

Table 12.7

United States Population and Crime Index

YEAR	POPULATION	ACTUAL CRIME INDEX
1963	188,531,000	2,259,081
1964	191,334,000	2,604,426
1965	193,818,000	2,780,015
1966	195,857,000	3,243,400
1967	197,864,000	3,802,273
1968	199,861,000	4,466,573
1969	201,921,000	4,989,747
1970	203,184,772	5,568,197
1971	206,256,000	5,995,211
1972	208,232,000	5,891,924
1973	209,851,000	8,638,375

Heaukulani reports that the correlation is $r = .94$ ($r^2 = .88$), which provides sufficient evidence at $\alpha = .01$ to indicate that the size of the population is useful for predicting the crime index using the straight-line model. Heaukulani concludes that

most agencies are using too narrow a span of statistical measurement to evaluate the overall crime picture. Year-by-year comparisons and percent analyses do not take average fluctuations into consideration. A lack of knowledge about statistical principles on the part of laymen, especially those in the news media, leads to a distortion or an invalid representation of the crime figures. Any increase is reported either as "bucking a trend" or "in line with the rise in crime."

If we expect to make any progress in solving the crime problem, we must evaluate the crime index objectively. We have to accept the fact that the crime index will probably increase from year to year. We should be willing to accept an average amount of increase and a set maximum

Figure 12.12   Least Squares Line Relating Crime Index to Population Size (1963–1973)

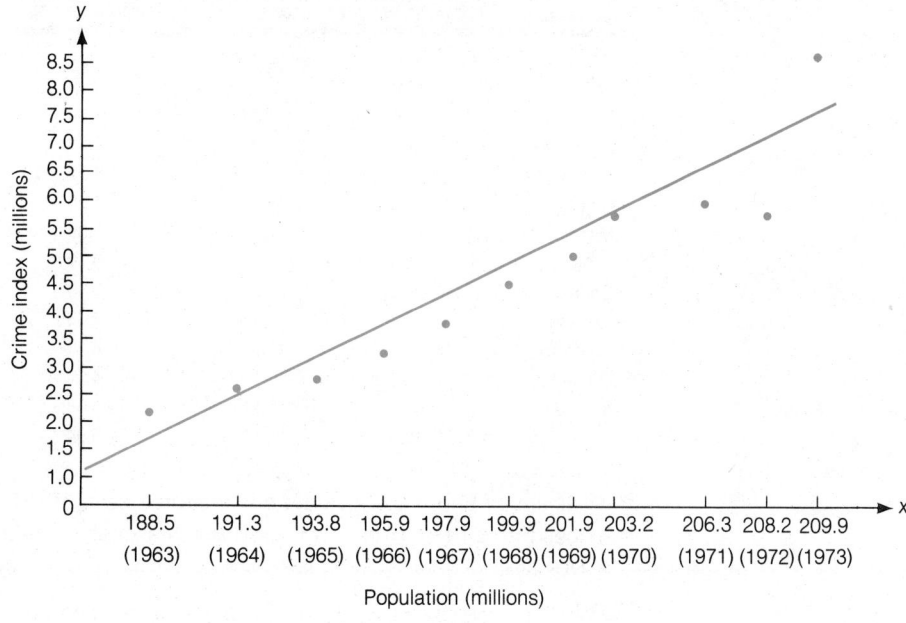

limit. If the crime index exceeds the limit, only then should we become concerned and attempt to examine the factors that may have been responsible for the excess deviation.

Note that the 1973 crime rate of 8.6 million greatly exceeded the predicted value of less than 7 million. Perhaps a special effort should be made to determine the causes of this "excess deviation."

Exercises 12.32–12.41

Learning the Mechanics

12.32 Calculate the values of r and r^2 for each of the following exercises:

a. Exercise 12.8   **b.** Exercise 12.9   **c.** Exercise 12.10

Applying the Concepts

12.33 In Exercise 12.12, we gave data on the percentage y of games won and the team batting average x for the fourteen American League baseball teams. The data are reproduced in the table on the next page.

a. Would you expect the correlation coefficient between percentage games won and team batting average to be positive or negative? Explain.

TEAM	PERCENTAGE GAMES WON	TEAM BATTING AVERAGE	TEAM EARNED RUN AVERAGE
Detroit	64.8	.270	3.61
Toronto	56.9	.278	3.75
Baltimore	52.5	.247	3.59
New York	52.0	.275	3.71
Boston	51.6	.278	4.19
Cleveland	44.0	.261	4.28
Milwaukee	41.9	.259	3.89
Minnesota	52.8	.274	3.70
California	50.0	.252	3.83
Kansas City	49.2	.269	4.08
Chicago	48.4	.248	4.10
Oakland	48.0	.256	4.17
Seattle	48.0	.257	4.36
Texas	43.1	.254	3.94

b. Calculate r and r^2 for the data and interpret them.

c. Does the value of r^2 in part b provide a reasonable characterization of the goodness of fit of your least squares line (Exercise 12.12) to the data points? Explain.

12.34 In Exercise 12.33, we gave data on the percentage y of games won and the team earned run average x for the fourteen American League baseball teams.

a. Would you expect the correlation coefficient between percentage games won and team earned run average to be positive or negative? Explain.

b. Calculate r and r^2 for the data and interpret them.

c. Does the value of r^2 in part b provide a reasonable characterization of the goodness of fit of your least squares line (Exercise 12.13) to the data points? Explain.

12.35 Calculate the coefficient of determination and the simple coefficient of correlation for the woolly-bear caterpillar data in Exercise 12.15. Interpret them.

12.36 When student seating in a large lecture course is by personal choice, is seat location correlated to a student's grade? In particular, do students who choose seats in the front of the classroom tend to obtain better grades than those who choose the back? The grades of ten randomly selected statistics students are given here along with their row location (number from the front of the classroom):

GRADE	93	68	89	98	66	86	73	55	80	71
ROW	7	1	4	3	25	12	9	30	13	27

a. Calculate r and r^2.

b. Do the data provide sufficient evidence to indicate that students sitting in the front of the classroom tend to receive better grades? Test using $\alpha = .05$.*

* Recall (Section 12.6) that the test of H_0: $\rho = 0$ is equivalent to the test of H_0: $\beta_1 = 0$.

12.37 Find the correlation coefficient and the coefficient of determination for the sample data listed in the table and interpret your results.

YEAR	NUMBER OF EIGHTEEN HOLE AND LARGER GOLF COURSES IN THE UNITED STATES	UNITED STATES DIVORCE RATE PER 1,000 POPULATION
1960	2,725	2.2
1965	3,769	2.5
1970	4,845	3.5
1972	5,385	4.1
1975	6,282	4.8

Source: U.S. Bureau of the Census, *Statistical Abstract of the U.S.:* 1976.

12.38 Is the maximal oxygen uptake, a measure often used by physiologists to indicate an individual's state of cardiovascular fitness, related to the performance of distance runners? Six long-distance runners submitted to treadmill tests for determination of their maximal oxygen uptake. The results, along with each runner's best mile time (in seconds), are shown in the table.

ATHLETE	MAXIMAL OXYGEN UPTAKE (Milliliters/kilogram)	MILE TIME (Seconds)
1	63.3	241.5
2	60.1	249.8
3	53.6	246.1
4	58.8	232.4
5	67.5	237.2
6	62.5	238.4

a. Calculate r and r^2.

b. Do the data provide sufficient evidence to indicate that mile time is negatively correlated with maximal oxygen uptake? Test H_0: $\rho = 0$ against the alternative hypothesis, H_a: $\rho < 0$, using $\alpha = .05$.*

12.39 Is there a correlation between the amount of education received by people living in urban centers and that received by people living in the urban fringes? To determine whether a correlation exists, a sociologist compared the percentages of people with 4 years of high school education or more for the two groups. The data are given in the table at the top of the next page.

a. Find the correlation coefficient for the data.

b. Find the coefficient of determination and explain its meaning in terms of this problem.

c. Is there sufficient evidence to indicate a nonzero correlation between x and y? Test using $\alpha = .05$.*

* Recall (Section 12.6) that the test of H_0: $\rho = 0$ is equivalent to the test of H_0: $\beta_1 = 0$.

CITY	URBAN CENTER y	URBAN FRINGE x	CITY	URBAN CENTER y	URBAN FRINGE x
Baltimore	28.2	42.3	Milwaukee	39.7	54.4
Boston	44.6	55.8	New Orleans	33.3	44.6
Chicago	35.3	53.9	New York	36.4	48.7
Cleveland	30.1	55.5	Philadelphia	30.7	48.0
Dallas	48.9	56.4	St. Louis	26.3	43.3
Detroit	34.4	47.5	San Francisco	49.4	57.9
Houston	45.2	50.1	Washington	47.8	67.5
Los Angeles	53.4	53.4			

Source: Computed from U.S. Bureau of the Census, *U.S. Census of Population: 1960, General Social and Economic Characteristics,* and *U.S. Census of Population and Housing: 1960, Census Tracts* (Washington, D.C.: Government Printing Office, 1961).

12.40 The data shown in the table give the monthly sales y (in dollars) of room air conditioners for a furniture store in Baton Rouge, Louisiana, during a 12-month period in 1981. Along with each monthly sales figure is the mean temperature for the month.

MONTH	SALES	TEMPERATURE	MONTH	SALES	TEMPERATURE
1	5,027	46.2	7	87,513	83.6
2	0	52.8	8	36,270	82.6
3	2,708	59.1	9	7,963	76.3
4	27,521	72.0	10	1,749	68.4
5	41,316	71.9	11	4,155	62.2
6	93,537	81.7	12	2,220	51.8

a. Would you expect mean monthly sales of room air conditioners to be positively correlated, negatively correlated, or not correlated at all with the mean monthly temperature? Explain.

b. Calculate r and r^2 and interpret their values.

c. Do the data provide sufficient evidence to indicate that the mean monthly temperature provides information for predicting monthly sales? Test using $\alpha = .05$.

d. List other independent variables that might be better predictors of mean monthly sales.

12.41 The data shown in the table give the estimated miles per gallon (MPG) and weights (in pounds) for sixteen 1983 American-made automobiles.

MPG	WEIGHT	MPG	WEIGHT	MPG	WEIGHT
29	2,156	16	2,940	28	2,327
21	2,588	22	2,455	22	2,493
22	3,695	20	2,684	34	2,175
18	3,091	28	2,098	37	1,980
31	2,098	16	3,760	22	2,455
28	2,317				

a. Would you expect miles per gallon y to be positively or negatively correlated with weight x? Explain.

b. Calculate r and r^2 and interpret them.

c. Fit a simple linear regression model to the data. Give your prediction equation.

d. Do the data provide sufficient evidence to indicate that miles per gallon decreases as the weight of the automobile increases? Test using $\alpha = .05$.

12.8
Using the Model for Estimation and Prediction

If we are satisfied that a useful model has been found to describe the relationship between reaction time and amount of drug in the bloodstream, we are ready for step 5 in our regression modeling procedure: using the model for estimation and prediction.

The most common uses of a probabilistic model for making inferences can be divided into two categories. The first is the use of the model for estimating the mean value of y, $E(y)$, for a specific value of x. For our drug reaction example, we may want to estimate the mean response time for all people whose blood contains 4% of the drug. The second use of the model entails predicting a particular y-value for a given x. That is, we may want to predict the reaction time for a specific person who possesses 4% of the drug in the bloodstream.

In the first case, we are attempting to estimate the mean value of y for a very large number of experiments at the given x-value. In the second case, we are trying to predict the outcome of a single experiment at the given x-value. Which of these model uses—estimating the mean value of y or predicting an individual value of y (for the same value of x)—can be accomplished with the greater accuracy?

Before answering this question, we first consider the problem of choosing an estimator (or predictor) of the mean (or individual) y-value. We will use the least squares prediction equation

$$\hat{y} = \hat{\beta}_0 + \hat{\beta}_1 x$$

both to estimate the mean value of y and to predict a specific value of y for a given value of x. For our example, we found

$$\hat{y} = -.1 + .7x$$

so that the estimated mean reaction time for all people when $x = 4$ (drug is 4% of blood content) is

$$\hat{y} = -.1 + .7(4) = 2.7 \text{ seconds}$$

The identical value is used to predict the y-value when $x = 4$. That is, both the estimated mean and the predicted value of y are $\hat{y} = 2.7$ when $x = 4$, as shown in Figure 12.13.

The difference between these two model uses lies in the relative accuracy of the estimate and the prediction. These accuracies are best measured by the repeated sampling errors of the least squares line when it is used as an estimator and as a predictor, respectively. These errors are given in the box on page 616.

Figure 12.13 Estimated Mean Value and Predicted Individual Value of Reaction Time y for $x = 4$

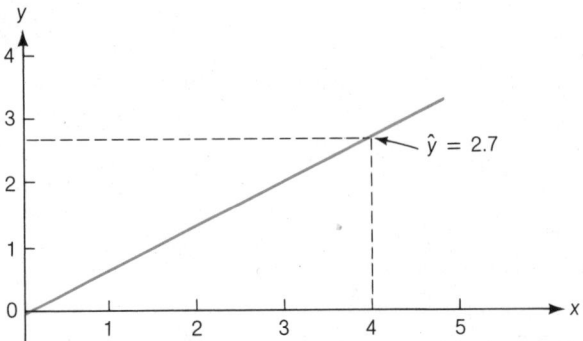

Sampling Errors for the Estimator of the Mean of y and the Predictor of an Individual y

1. The standard deviation of the sampling distribution of the estimator $\hat{y}$ of the mean value of y at a specific value of x, say x_p, is

$$\sigma_{\hat{y}} = \sigma \sqrt{\frac{1}{n} + \frac{(x_p - \bar{x})^2}{SS_{xx}}}$$

where σ is the standard deviation of the random error ε.

2. The standard deviation of the prediction error for the predictor $\hat{y}$ of an individual y-value at a specific value of x is

$$\sigma_{(y - \hat{y})} = \sigma \sqrt{1 + \frac{1}{n} + \frac{(x_p - \bar{x})^2}{SS_{xx}}}$$

where σ is the standard deviation of the random error ε.

The true value of σ is rarely known, so we estimate σ by s and calculate the estimation and prediction intervals as shown in the next two boxes.

A $100(1 - \alpha)\%$ Confidence Interval for the Mean Value of y at $x = x_p$

$\hat{y} \pm t_{\alpha/2}$(Estimated standard deviation of $\hat{y}$)

or

$$\hat{y} \pm t_{\alpha/2}s \sqrt{\frac{1}{n} + \frac{(x_p - \bar{x})^2}{SS_{xx}}}$$

where $t_{\alpha/2}$ is based on $(n - 2)$ degrees of freedom.

A 100(1 − α)% Prediction Interval* for an Individual **y** at **x = x_p**

$\hat{y} \pm t_{\alpha/2}[\text{Estimated standard deviation of } (y - \hat{y})]$

or

$$\hat{y} \pm t_{\alpha/2}s\sqrt{1 + \frac{1}{n} + \frac{(x_p - \bar{x})^2}{SS_{xx}}}$$

where $t_{\alpha/2}$ is based on $(n - 2)$ degrees of freedom.

Example 12.3

Find a 95% confidence interval for the mean reaction time when the concentration of the drug in the bloodstream is 4%.

Solution

For a 4% concentration, $x = 4$ and the confidence interval for the mean value of y is

$$\hat{y} \pm t_{\alpha/2}s\sqrt{\frac{1}{n} + \frac{(x_p - \bar{x})^2}{SS_{xx}}} = \hat{y} \pm t_{.025}s\sqrt{\frac{1}{5} + \frac{(4 - \bar{x})^2}{SS_{xx}}}$$

where $t_{.025}$ is based on $n - 2 = 5 - 2 = 3$ degrees of freedom. Recall that $\hat{y} = 2.7$, $s = .61$, $\bar{x} = 3$, and $SS_{xx} = 10$. From Table V in the Appendix, $t_{.025} = 3.182$. Thus, we have

$$2.7 \pm (3.182)(.61)\sqrt{\frac{1}{5} + \frac{(4 - 3)^2}{10}} = 2.7 \pm (3.182)(.61)(.55)$$

$$= 2.7 \pm 1.1$$

Therefore, when the percentage of drug in the bloodstream is 4% the 95% confidence interval for the mean reaction time for all possible subjects is 1.6 to 3.8 seconds. Note that we used a small amount of data (small sample size) for purposes of illustration in fitting the least squares line. The interval would probably be narrower if more information had been obtained from a larger sample. ∎

Example 12.4

Predict the reaction time for the next performance of the experiment for a subject with a drug concentration of 4%. Use a 95% prediction interval.

Solution

To predict the response time for an individual subject for whom $x = 4$, we calculate the 95% prediction interval as

$$\hat{y} \pm t_{\alpha/2}s\sqrt{1 + \frac{1}{n} + \frac{(x_p - \bar{x})^2}{SS_{xx}}} = 2.7 \pm (3.182)(.61)\sqrt{1 + \frac{1}{5} + \frac{(4 - 3)^2}{10}}$$

$$= 2.7 \pm (3.182)(.61)(1.14)$$

$$= 2.7 \pm 2.2$$

* The term *prediction interval* is used when the interval formed is intended to enclose the value of a random variable. The term *confidence interval* is reserved for estimation of population parameters (such as the mean).

Therefore, we predict that the reaction time for this individual will fall in the interval from .5 to 4.9 seconds. Like the confidence interval for the mean value of y, the prediction interval for y is quite large. This is because we have chosen a simple example (only five data points) to fit the least squares line. The width of the prediction interval could be reduced by using a larger number of data points.

A comparison of the confidence interval for the mean value of y and the prediction interval for a future value of y for 4% drug concentration ($x = 4$) is illustrated in Figure 12.14. Note that the prediction interval for an individual value of y is always wider than the corresponding confidence interval for the mean value of y. You can see this by examining the formulas for the two intervals and by studying Figure 12.14. ■

Figure 12.14 A 95% Confidence Interval for Mean Reaction Time and a Prediction Interval for Reaction Time When $x = 4$

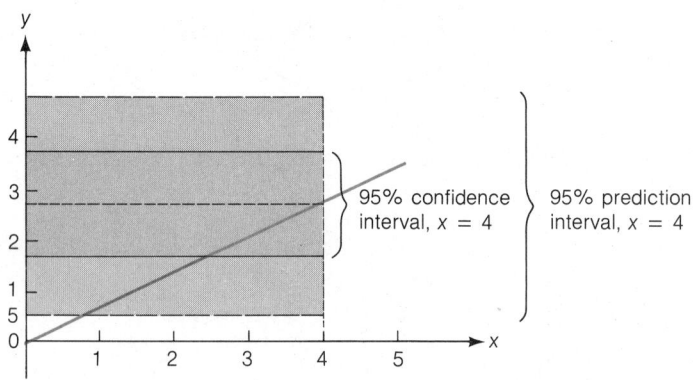

The error in estimating the mean value of y, $E(y)$, for a given value of x, say x_p, is the vertical distance between the least squares line and the true line of means, $E(y) = \beta_0 + \beta_1 x$. This error, shown in Figure 12.15, takes its smallest value when $x = \bar{x}$. The farther x lies from $\bar{x}$, the larger the error of estimation. In contrast, the error in predicting some future value of y is the sum of the two errors—the error of estimating the mean of y, $E(y)$, shown in Figure 12.15, plus the random error ε that is a component of the value of y to be predicted (see Figure 12.16). Consequently, the error of predicting a particular value of y is usually larger than the error of estimating the mean value of y for a given value of x. As the sample size n is increased, both the confidence interval and the prediction interval decrease. The least squares line comes closer and closer to the line of means, $E(y) = \beta_0 + \beta_1 x$, and the $100(1 - \alpha)\%$ prediction interval for y decreases to $\hat{y} \pm z_{\alpha/2} s$.

Be careful not to use the least squares prediction equation to estimate the mean value of y or to predict a particular value of y for values of x that fall outside the *range* of the values of x contained in your sample data. The model might provide a

Figure 12.15 Error of Estimating the Mean Value of y for a Given Value of x

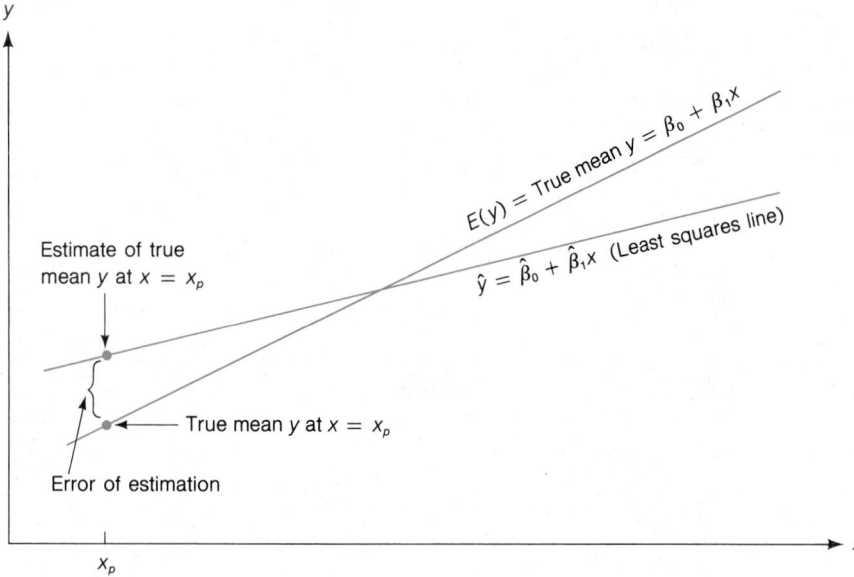

Figure 12.16 Error of Predicting a Future Value of y for a Given Value of x

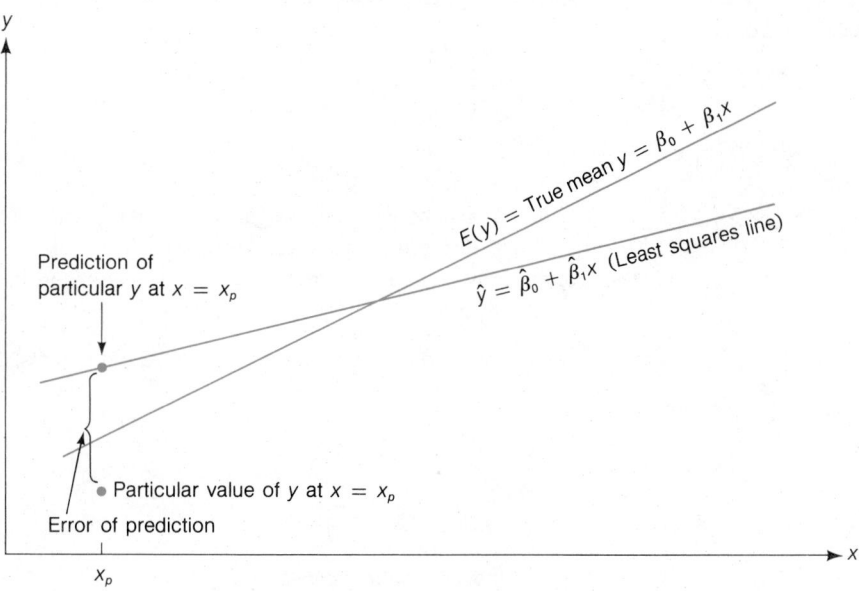

good fit to the data over the range of x-values contained in the sample but a very poor fit for values of x outside this region (see Figure 12.17). Failure to heed this warning may lead to errors of estimation and prediction that are much larger than expected.

Figure 12.17 The Danger of Using a Model to Predict Outside the Range of the Sample Values of x

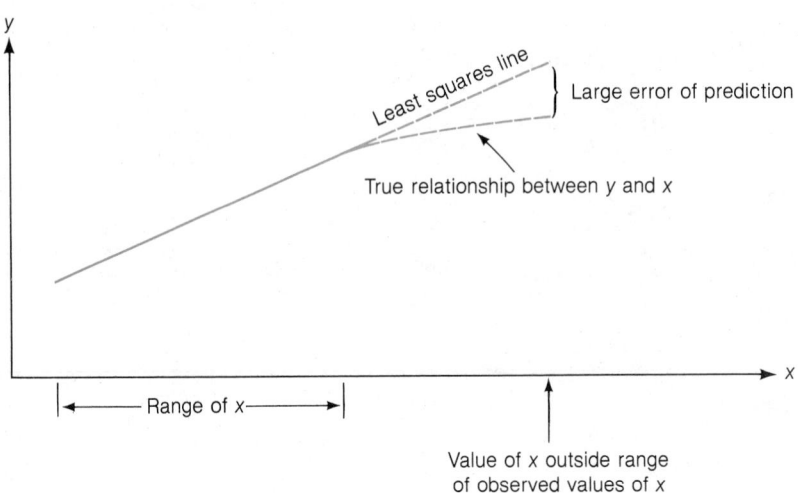

Exercises 12.42–12.49

Learning the Mechanics

12.42 The data from Exercise 12.10 are repeated here:

x	-2	-1	0	1	2
y	6	4	3	1	-1

a. Find a 90% confidence interval for the mean value of y when $x = -1$.
b. Find a 90% prediction interval for y when $x = -1$.
[*Note:* Both intervals in parts a and b are wide because of the small number of data points.]

12.43 In fitting a least squares line to $n = 22$ data points, the following quantities were computed:

$$SS_{xx} = 20 \quad \bar{x} = 3$$
$$SS_{yy} = 32 \quad \bar{y} = 4$$
$$SS_{xy} = 25$$

a. Find the least squares line.
b. Graph the least squares line.
c. Calculate SSE.
d. Calculate s^2.
e. Find a 95% confidence interval for the mean value of y when $x = 1$.
f. Find a 95% prediction interval for y when $x = 1.5$.
g. Find a 95% confidence interval for the mean value of y when $x = 0$.

Applying the Concepts

12.44 Certain dosages of a new drug developed to reduce a smoker's reliance on tobacco may reduce one's pulse rate to dangerously low levels. To investigate the drug's effect on pulse rate, different dosages of the drug were administered to six randomly selected patients, and 30 minutes later the decrease in each patient's pulse rate was recorded.

PATIENT	DOSAGE x (cubic centimeters)	DECREASE IN PULSE RATE y (beats/minute)
1	2.0	15
2	1.5	9
3	3.0	18
4	2.5	16
5	4.0	23
6	3.0	20

a. Is there evidence of a linear relationship between drug dosage and change in pulse rate? Test at $\alpha = .10$.

b. Find a 95% confidence interval for the slope β_1.

c. Find a 99% confidence interval for the mean decrease in pulse rate corresponding to a dosage of 3.5 cubic centimeters.

d. Find a 99% prediction interval for the decrease in pulse rate corresponding to a dosage of 3.5 cubic centimeters.

12.45 Refer to the woolly-bear caterpillar data in Exercises 12.15 and 12.29.

a. Find a 95% confidence interval for the mean cocoon temperature when the air temperature is 7°C. Interpret the interval.

b. Suppose that you were to place a single woolly-bear caterpillar cocoon in a controlled environment of 7°C. Find a 95% prediction interval for the cocoon temperature. Interpret the interval.

MPH	MPG
50	34.8, 33.6
55	34.6, 34.1
60	32.8, 31.9
65	32.6, 30.0
70	31.6, 31.8
75	30.9, 31.7

12.46 Will the national 55 mile per hour (MPH) highway speed limit provide a substantial savings in fuel? To investigate the relationship between automobile gasoline consumption and driving speed, a small economy car was driven twice over the same stretch of interstate freeway at each of six different speeds. The numbers of miles per gallon (MPG) measured for each of the twelve trips are shown in the table.

a. Fit a least squares line to the data.

b. Is there sufficient evidence to conclude there is a linear relationship between speed and gasoline consumption?

c. Construct a 90% confidence interval for β_1.

d. Construct a 95% confidence interval for miles per gallon when the speed is 72 miles per hour.

e. Construct a 95% prediction interval for miles per gallon when the speed is 58 miles per hour.

12.47 In planning for an orientation meeting with new accounting majors, the chairperson of the Accounting Department wants to emphasize the importance of doing well in the major courses in order to get better-paying jobs after graduation. To support this point, the chairperson plans to show that there is a strong positive correlation between starting salaries for recent accounting graduates and their grade-point averages in the major courses. Records for seven of last year's accounting graduates are selected at random and are given in the table.

GRADE-POINT AVERAGE IN MAJOR COURSES x	STARTING SALARY y (thousands of dollars)
2.58	11.5
3.27	13.8
3.85	14.5
3.50	14.2
3.33	13.5
2.89	11.6
2.23	10.6

a. Find the least squares prediction equation.
b. Plot the data and graph the line as a check on your calculations.
c. Find the values of r and r^2 and interpret these values.
d. Find a 95% prediction interval for a graduate whose grade-point average is 3.2.
e. What is the mean starting salary for graduates with grade-point averages equal to 3.0? Use a 95% confidence interval.

12.48 In Exercise 12.16, we found the least squares line relating the age x of a sand lance to its length y.
a. Find a 95% confidence interval for the mean length of sand lances that are 4 years old. Interpret the interval.
b. Find a 95% prediction interval for the length of a sand lance that is 4 years old. Interpret the interval. Explain the difference between this interval and the interval obtained in part a.

12.49 Refer to Exercise 12.31. Find a 95% confidence interval for the mean weight gain of rainbow trout (percentage of body weight) over a 20-day period when the ration level is 2.5% of body weight and the water temperature is 12°C. Interpret the confidence interval.

12.9
Simple Linear Regression: An Example

In the preceding sections we have presented the basic elements necessary to fit and use a straight-line regression model. In this final section we will assemble these elements by applying them in an example.

Suppose a fire insurance company wants to relate the amount of fire damage in major residential fires to the distance between the residence and the nearest fire station. The study is to be conducted in a large suburb of a major city; a sample of fifteen recent fires in this suburb is selected. The amount of damage, y, and the distance, x,

Table 12.8

Fire Damage Data

DISTANCE FROM FIRE STATION x (miles)	FIRE DAMAGE y (thousands of dollars)	DISTANCE FROM FIRE STATION x (miles)	FIRE DAMAGE y (thousands of dollars)
3.4	26.2	2.6	19.6
1.8	17.8	4.3	31.3
4.6	31.3	2.1	24.0
2.3	23.1	1.1	17.3
3.1	27.5	6.1	43.2
5.5	36.0	4.8	36.4
.7	14.1	3.8	26.1
3.0	22.3		

between the fire and the nearest fire station are recorded for each fire. The results are given in Table 12.8.

Step 1. First, we hypothesize a model to relate fire damage, y, to the distance from the nearest fire station, x. We will hypothesize a straight-line probabilistic model:

$$y = \beta_0 + \beta_1 x + \varepsilon$$

Step 2. Next, we use the data to estimate the unknown parameters in the deterministic component of the hypothesized model. We make some preliminary calculations:

$$SS_{xx} = \sum x^2 - \frac{\left(\sum x\right)^2}{n} = 196.16 - \frac{(49.2)^2}{15}$$

$$= 196.160 - 161.376 = 34.784$$

$$SS_{yy} = \sum y^2 - \frac{\left(\sum y\right)^2}{n} = 11{,}376.48 - \frac{(396.2)^2}{15}$$

$$= 11{,}376.480 - 10{,}464.963 = 911.517$$

$$SS_{xy} = \sum xy - \frac{\left(\sum x\right)\left(\sum y\right)}{n} = 1{,}470.65 - \frac{(49.2)(396.2)}{15}$$

$$= 1{,}470.650 - 1{,}299.536 = 171.114$$

Then the least squares estimates of the slope β_1 and intercept β_0 are

$$\hat{\beta}_1 = \frac{SS_{xy}}{SS_{xx}} = \frac{171.114}{34.784} = 4.919$$

$$\hat{\beta}_0 = \bar{y} - \hat{\beta}_1 \bar{x} = \frac{396.2}{15} - 4.919 \left(\frac{49.2}{15}\right)$$

$$= 26.413 - (4.919)(3.28) = 26.413 - 16.134 = 10.279$$

and the least squares equation is

$$\hat{y} = 10.279 + 4.919x$$

This prediction equation is graphed in Figure 12.18 along with a plot of the data points.

Figure 12.18 Least Squares Model for the Fire Damage Data

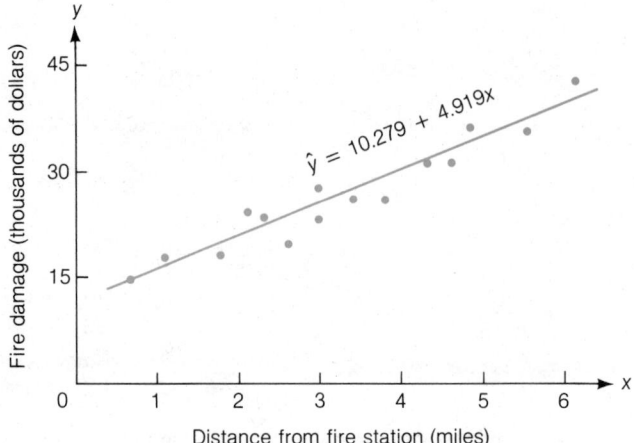

Step 3. Now we specify the probability distribution of the random error component ε. The assumptions about the distribution are identical to those listed in Section 12.3. Although we know that these assumptions are not completely satisfied (they rarely are for practical problems), we are willing to assume they are approximately satisfied for this example. We have to estimate the variance σ^2 of ε, so we calculate

$$\text{SSE} = \sum(y - \hat{y})^2 = \text{SS}_{yy} - \hat{\beta}_1 \text{SS}_{xy}$$

where the last expression represents a shortcut formula for SSE. Thus,

$$\text{SSE} = 911.517 - (4.919)(171.114)$$
$$= 911.517 - 841.709766 = 69.807234^*$$

To estimate σ^2, we divide SSE by the degrees of freedom available for error, $n - 2$. Thus,

$$s^2 = \frac{\text{SSE}}{n - 2} = \frac{69.807234}{15 - 2} = 5.3698 \qquad s = \sqrt{5.3698} = 2.32$$

Step 4. We can now check the utility of the hypothesized model—that is, whether x really contributes information for the prediction of y using the straight-line model. First test the null hypothesis that the slope β_1 is zero, i.e., that there is no linear relationship between fire damage and the distance from the nearest fire station, against the alternative hypothesis that fire damage increases as the distance increases, that is, H_a: $\beta_1 > 0$. We test

$$H_0: \ \beta_1 = 0 \qquad H_a: \ \beta_1 > 0$$

Test statistic: $t = \dfrac{\hat{\beta}_1 - 0}{s_{\hat{\beta}_1}} = \dfrac{\hat{\beta}_1}{s / \sqrt{\text{SS}_{xx}}}$

Assumptions: The same assumptions made about ε in Section 12.3

* The values for SS_{yy}, $\hat{\beta}_1$, and SS_{xy} used to calculate SSE are exact for this example. For other problems where rounding is necessary, at least six significant figures should be carried for these quantities. Otherwise, the calculated value of SSE may be substantially in error.

Rejection region: For $\alpha = .05$, we will reject H_0 if $t > t_\alpha$, where, for $n - 2 = 13$ df, we find $t_{.05} = 1.771$.

We then calculate the t-statistic:

$$t = \frac{\hat{\beta}_1}{s_{\hat{\beta}_1}} = \frac{\hat{\beta}_1}{s/\sqrt{SS_{xx}}} = \frac{4.919}{2.32/\sqrt{34.784}} = \frac{4.919}{.393} = 12.5$$

This large t-value leaves little doubt that mean fire damage and distance between the fire and the station are at least linearly related, with mean fire damage increasing as the distance increases.

We gain additional information about the relationship by forming a confidence interval for the slope β_1. A 95% confidence interval is

$$\hat{\beta}_1 \pm t_{.025}s_{\hat{\beta}_1} = 4.919 \pm (2.160)(.393)$$
$$= 4.919 \pm .849 = (4.070, 5.768)$$

We estimate that the interval from \$4,070 to \$5,768 encloses the mean increase (β_1) in fire damage per additional mile distance from the fire station.

Another measure of the utility of the model is the coefficient of correlation r. We have

$$r = \frac{SS_{xy}}{\sqrt{SS_{xx}SS_{yy}}} = \frac{171.114}{\sqrt{(34.784)(911.517)}}$$
$$= \frac{171.114}{178.062} = .96$$

The high correlation confirms our conclusion that β_1 is greater than zero; it appears that fire damage and distance from the fire station are positively correlated.

The coefficient of determination is

$$r^2 = (.96)^2 = .92$$

which implies that 92% of the sum of squares of deviations of the y-values about $\bar{y}$ is explained by the distance x between the fire and the fire station. All signs point to a strong relationship between x and y.

Step 5. We are now prepared to use the least squares model. Suppose the insurance company wants to predict the fire damage if a major residential fire were to occur 3.5 miles from the nearest fire station. The predicted value is

$$\hat{y} = \hat{\beta}_0 + \hat{\beta}_1 x = 10.279 + (4.919)(3.5)$$
$$= 10.279 + 17.216 = 27.5$$

(We round to the nearest tenth to be consistent with the units of the original data in Table 12.8.) If we want a 95% prediction interval, we calculate

$$\hat{y} \pm t_{.025}s\sqrt{1 + \frac{1}{n} + \frac{(x - \bar{x})^2}{SS_{xx}}} = 27.5 \pm (2.16)(2.32)\sqrt{1 + \frac{1}{15} + \frac{(3.5 - 3.28)^2}{34.784}}$$
$$= 27.5 \pm (2.16)(2.32)\sqrt{1.0681}$$
$$= 27.5 \pm 5.2 = (22.3, 32.7)$$

The model yields a 95% prediction interval of \$22,300 to \$32,700 for fire damage in a major residential fire 3.5 miles from the nearest station.

One caution before closing: We would not use this prediction model to make predictions for homes less than .7 mile or more than 6.1 miles from the nearest fire station. A look at the data in Table 12.8 reveals that all the x-values fall between .7 and 6.1. It is dangerous to use the model to make predictions outside the region in which the sample data fall. A straight line might not provide a good model for the relationship between the mean value of y and the value of x when stretched over a wider range of x-values.

Summary

We have introduced an extremely useful tool in this chapter—the **method of least squares** for fitting a prediction equation to a set of data. This procedure, along with associated statistical tests and estimations, is called a **regression analysis.** In five steps we showed how to use sample data to build a model relating a dependent variable y to a single independent variable x:

1. The first step is to hypothesize a **probabilistic model.** In this chapter, we confined our attention to the straight-line model, $y = \beta_0 + \beta_1 x + \varepsilon$.
2. The second step is to use the method of least squares to estimate the unknown parameters in the **deterministic component,** $\beta_0 + \beta_1 x$. The least squares estimates yield a model $\hat{y} = \hat{\beta}_0 + \hat{\beta}_1 x$ with a sum of squared errors (SSE) that is smaller than that produced by any other straight-line model.
3. The third step is to specify the probability distribution of the **random error component, ε.**
4. The fourth step is to assess the utility of the hypothesized model by making inferences about the slope β_1, calculating the **coefficient of correlation r,** and calculating the **coefficient of determination r^2.**
5. Finally, if we are satisfied with the model we are prepared to use it. We used the model to estimate the mean y-value, $E(y)$, for a given x-value and to predict an individual y-value for a specific value of x.

The following two chapters develop the concepts introduced in this chapter.

Supplementary Exercises 12.50–12.65

Learning the Mechanics

12.50 In fitting a least squares line to $n = 15$ data points, the following quantities were computed:

$$SS_{xx} = 50 \qquad \bar{x} = 1.3$$
$$SS_{yy} = 25 \qquad \bar{y} = 27$$
$$SS_{xy} = -30$$

a. Find the least squares line.
b. Graph the least squares line.

c. Calculate SSE.

d. Calculate s^2.

e. Find a 90% confidence interval for β_1. Interpret this estimate.

f. Find a 90% confidence interval for the mean value of y when $x = 1.8$.

g. Find a 90% prediction interval for y when $x = 1.8$.

12.51 Consider the following ten data points:

x	3	5	6	4	3	7	6	5	4	7
y	4	3	2	1	2	3	3	5	4	2

a. Plot the data on a scattergram.

b. Calculate the values of r and r^2.

c. Is there sufficient evidence to indicate that x and y are linearly correlated? Test at the $\alpha = .10$ level of significance.

Applying the Concepts

12.52 A study was conducted to determine whether the final grade in an introductory sociology course was related to a student's performance on a verbal ability test administered before college entrance. The verbal test scores and final grades for a random sample of ten students are shown in the table.

STUDENT	VERBAL ABILITY TEST SCORE x	FINAL SOCIOLOGY GRADE y
1	39	65
2	43	78
3	21	52
4	64	82
5	57	92
6	47	89
7	28	73
8	75	98
9	34	56
10	52	75

a. Assuming that a linear relationship exists between verbal scores and final grades, find the least squares line relating y to x.

b. Plot the data points and graph the least squares line.

c. Do the data provide sufficient evidence to indicate that a positive correlation exists between verbal score and final grade? (Use $\alpha = .01$.)

d. Find a 95% confidence interval for the slope β_1.

e. Predict a student's final grade in the introductory course when his or her verbal test score is 50. (Use a 90% prediction interval.)

f. Find a 95% confidence interval for the mean final grade for students scoring 35 on the college entrance verbal exam.

12.53 A large supermarket chain has its own store brand for many grocery items. These tend to be priced lower than other brands. For a particular item, the chain wants to study the effect of varying the price for the major competing brand on the sales of the store brand item, while the prices for the store brand and all other brands are held fixed. The experiment is conducted at one of the chain's stores over a 7-week period, and the results are shown in the table.

WEEK	COMPETITOR'S PRICE x (cents)	STORE BRAND SALES y
1	37	122
2	32	107
3	29	99
4	35	110
5	33	113
6	31	104
7	35	116

a. Find the least squares line relating store brand sales y to major competitor's price x.
b. Plot the data and graph the line as a check on your calculations.
c. Does x contribute information for the prediction of y?
d. Calculate r and r^2 and interpret their values.
e. Find a 90% confidence interval for mean store brand sales when the competitor's price is 33¢.
f. Suppose you were to set the competitor's price at 33¢. Find a 90% prediction interval for next week's sales.

12.54 As part of the first-year evaluation for new salespeople, a large food-processing firm projects the second-year sales for each salesperson based on his or her sales for the first year. The sales data for eight salespeople are given in the table.

FIRST-YEAR SALES x (thousands of dollars)	SECOND-YEAR SALES y (thousands of dollars)
75.2	99.3
91.7	125.7
100.3	136.1
64.2	108.6
81.8	102.0
110.2	153.7
77.3	108.8
80.1	105.4

a. Use the data in the table to fit a simple linear prediction model for second-year sales based on the first year's sales. Assume the data have been adjusted in terms of a base year to discount inflation effects.

b. Plot the data and graph the line as a check on your calculations.

c. Do the data provide sufficient information to indicate that x contributes information for the prediction of y?

d. Calculate r^2 and interpret its value.

e. If a salesperson has first-year sales of $90,000, find a 90% prediction interval for the second year's sales.

12.55 The data shown in the table give the consumer purchases of oranges, y, and the price per box for an 8-month period.

MONTH	U.S. CONSUMER PURCHASES OF ORANGES y (thousands of boxes)	PRICE PER BOX x (dollars)
October	6,100	6.20
November	6,800	6.05
December	8,400	6.20
January	6,000	8.10
February	5,800	8.70
March	5,500	8.75
April	4,400	9.40
May	4,000	9.75

a. Construct a scattergram of the data.

b. Find the least squares line for the data and plot it on your scattergram. The least squares line may be viewed as an estimate of the short-run demand function for oranges.

c. Define β_1 in the context of this problem.

d. Test the hypothesis that the price per box of oranges contributes no information for the prediction of the number of boxes consumed when a linear model of short-run demand is used (let $\alpha = .05$). Draw the appropriate conclusions.

e. Find a 90% confidence interval for β_1. Interpret your results.

f. Find the coefficient of correlation for the given data.

g. Find the coefficient of determination for the linear model of demand you constructed in part b. Interpret your result.

h. Find a 90% prediction interval for the number of boxes of oranges that will be consumed if the price per box is $8.

i. Find a 95% confidence interval for the expected number of boxes that will be consumed at a price of $8 per box.

12.56 In placing a weekly order, a concessionaire who provides services at a baseball stadium must know what size crowd is expected during the coming week in order to know how much food, etc., to order. Since advanced ticket sales give an indication of expected attendance, food needs might be predicted on the basis of the advanced sales. Data from seven previous weeks of games are given in the table on page 630.

a. Use the data in the table to develop a simple linear model for hot dogs purchased as a function of advanced ticket sales.

HOT DOGS PURCHASED DURING WEEK y (thousands)	ADVANCED TICKET SALES FOR WEEK x (thousands)
39.1	54.0
35.9	48.1
20.8	28.8
42.4	62.4
46.0	64.4
40.7	59.5
29.9	42.3

b. Plot the data and graph the line as a check on your calculations.

c. Do the data provide sufficient information to indicate that advanced ticket sales provide information for the prediction of hot dog demand?

d. Calculate r^2 and interpret its value.

e. Find a 90% confidence interval for the mean number of hot dogs purchased when the advanced ticket sales equal 50,000.

f. If the advanced ticket sales this week equal 55,000, find a 90% prediction interval for the number of hot dogs that will be purchased this week at the game.

12.57 At temperatures approaching absolute zero (273° below 0°C), helium exhibits traits that defy many laws of conventional physics. An experiment has been conducted with helium in solid form at various temperatures near absolute zero. The solid helium is placed in a dilution refrigerator along with a solid impure substance, and the fraction (in weight) of the impurity passing through the solid helium is recorded. (This phenomenon of solids passing directly through solids is known as *quantum tunneling*.) The data are given in the table.

TEMPERATURE x (°C)	PROPORTION OF IMPURITY PASSING THROUGH HELIUM y	TEMPERATURE x (°C)	PROPORTION OF IMPURITY PASSING THROUGH HELIUM y
−262.0	.315	−272.0	.935
−265.0	.202	−272.4	.957
−256.0	.204	−272.7	.906
−267.0	.620	−272.8	.985
−270.0	.715	−272.9	.987

a. Fit a least squares line to the data.

b. Test the null hypothesis, H_0: $\beta_1 = 0$, against the alternative hypothesis, H_a: $\beta_1 < 0$, at the $\alpha = .01$ level of significance.

c. Compute r^2 and interpret your results.

d. Find a 95% prediction interval for the percentage of the solid impurity passing through solid helium at −273°C. (Note that this value of x is outside the experimental region, where use of the model for prediction may be dangerous.)

12.58 A certain manufacturer evaluates the sales potential for a product in a new marketing area by selecting several stores in the area to sell the product on a trial basis for a 1-month period. The sales figures for the trial period are then used to project sales for the entire area. [*Note:* The same number of trial stores are used each time.]

TOTAL SALES DURING TRIAL PERIOD x (hundreds of dollars)	TOTAL SALES FOR FIRST MONTH FOR ENTIRE AREA y (hundreds of dollars)
16.8	48.2
14.0	46.8
18.3	54.3
22.1	59.7
14.9	48.3
23.2	67.5

a. Use the data in the table to develop a simple linear model for predicting first-month sales for the entire area based on sales during the trial period.

b. Plot the data and graph the line as a check on your calculations.

c. Do the data provide sufficient evidence to indicate that total sales during the trial period contribute information for predicting total sales during the first month?

d. Use a 90% prediction interval to predict total sales for the first month for the entire area if the trial sales equal $2,000.

12.59 The management of a manufacturing firm is considering the possibility of setting up its own market research department rather than continuing to use the services of a market research firm. The management wants to know what salary should be paid to a market researcher, based on years of experience. An independent consultant checks with several other firms in the area and obtains the information shown in the table on market researchers.

ANNUAL SALARY y (thousands of dollars)	EXPERIENCE x (years)
21.3	2
21.2	1.5
30.0	11
34.1	15
30.4	9
26.9	6

a. Fit a least squares line to the data.

b. Plot the data and graph the line as a check on your calculations.

c. Calculate r and r^2. Explain how these values measure the utility of the model.

d. Estimate the mean annual salary of a market researcher with 8 years of experience. Use a 90% confidence interval.

e. Predict the salary of a market researcher with 7 years of experience. Use a 90% prediction interval.

12.60 A study was conducted to determine whether there is a linear relationship between the breaking strength, y, of wooden beams and the specific gravity, x, of the wood. Ten randomly selected beams of the same cross-sectional dimensions were stressed until they broke. The breaking strengths and the density of the wood are shown in the table for each of the ten beams.

BEAM	SPECIFIC GRAVITY x	STRENGTH y	BEAM	SPECIFIC GRAVITY x	STRENGTH y
1	.499	11.14	6	.528	12.60
2	.558	12.74	7	.418	11.13
3	.604	13.13	8	.480	11.70
4	.441	11.51	9	.406	11.02
5	.550	12.38	10	.467	11.41

a. Fit the model $y = \beta_0 + \beta_1 x + \varepsilon$.
b. Test H_0: $\beta_1 = 0$ against the alternative hypothesis H_a: $\beta_1 \neq 0$.
c. Estimate the mean strength for beams with specific gravity .590. Use a 90% confidence interval.

12.61 Although the income tax system is structured so that people with higher incomes should pay a higher percentage of their incomes in taxes, there are many loopholes and tax shelters available for people with higher incomes. A sample of seven individual 1984 tax returns gave the data listed in the table on income and taxes paid.

INDIVIDUAL	GROSS INCOME x (thousands of dollars)	TAXES PAID y (percentage of total income)
1	35.8	16.7
2	80.2	21.4
3	14.9	15.2
4	7.3	10.1
5	9.1	12.2
6	150.7	19.6
7	25.9	17.3

a. Fit a least squares line to the data.
b. Plot the data and graph the line as a check on your calculations.
c. Calculate r and r^2 and interpret each.
d. Find a 90% confidence interval for the mean percent taxes paid for individuals with gross incomes of $70,000.

12.62 The data in the table were collected to calibrate a new instrument for measuring interocular pressure. The interocular pressure for each of ten glaucoma patients was measured by the new instrument and by a standard, reliable, but more time-consuming method.
a. Fit a least squares line to the data.
b. Calculate r and r^2. Interpret each of these quantities.

PATIENT	RELIABLE METHOD x	NEW INSTRUMENT y
1	20.2	20.0
2	16.7	17.1
3	17.1	17.2
4	26.3	25.1
5	22.2	22.0
6	21.8	22.1
7	19.1	18.9
8	22.9	22.2
9	23.5	24.0
10	17.0	18.1

c. Predict the pressure measured by the new instrument when the reliable method gives a reading of 20.0. Use a 90% prediction interval.

12.63 In Exercise 12.17, we discussed an experiment conducted to determine the effect of ration level on the growth rate of rainbow trout (Kovacs & Leduc, 1982). Twenty fish were fed at each of four ration levels at 6°C water temperature. The experiment was repeated for water temperatures of 12°C and 18°C. The table gives the means and standard deviations (in parentheses) for each sample of twenty fish. Refer only to the data collected on fish fed in water maintained at 18°C.

RATION (% body weight per day)	MEAN WET WEIGHT GAIN (%) 6°C	12°C	18°C
.0	−8.14 (3.27)	−10.33 (2.35)	−13.21 (1.96)
.8	12.31 (3.37)		
1.2		14.29 (5.32)	
1.5	28.19 (5.39)		15.51 (4.48)
2.5	29.12 (5.67)	38.05 (6.99)	
3.5			51.13 (7.14)
4.0		60.13 (12.71)	
4.5			65.49 (13.04)
Maintenance ration	.32	.4ʹ	.69

a. Find SSE and s^2 for the $n = 4$ data points.

b. Find a 95% confidence interval for the mean gain (percentage of body weight) for a 1% increase in ration level.

c. State the assumptions required for your inference in part c to be valid.

d. Find a 95% confidence interval for the mean weight gain of rainbow trout (percentage of body weight) over a 20-day period when the ration level is 4% of body weight.

12.64 Explain why for a given x-value the prediction interval for an individual y-value is always wider than the confidence interval for a mean value of y.

12.65 Explain why the confidence interval for the mean value of y for a given x-value gets wider the farther x is from $\bar{x}$. What are the implications of this phenomenon for estimation and prediction?

On Your Own . . .

There are many dependent variables in all areas of research that are the subject of regression modeling efforts. We list five such variables here:

1. Crime rate in various communities
2. Daily maximum temperature in your town
3. Grade-point average of students who have completed one academic year at your college
4. Gross National Product of the United States
5. Points scored by your favorite football team in a single game

Choose one of these dependent variables that is of particular interest to you or choose some other dependent variable for which you want to construct a prediction model. There may be a large number of independent variables that should be included in a prediction equation for the dependent variable you choose. List three potentially important independent variables, x_1, x_2, and x_3, that you think might be (individually) strongly related to your dependent variable. Next, obtain ten data values, each of which consists of a measure of your dependent variable y and the corresponding values of x_1, x_2, and x_3.

a. Use the least squares formulas given in this chapter to fit three straight-line models—one for each independent variable—for predicting y.
b. Interpret the sign of the estimated slope coefficient $\hat{\beta}_1$ in each case, and test the utility of the model by testing H_0: $\beta_1 = 0$ against H_a: $\beta_1 \neq 0$. What assumptions must be satisfied to assure the validity of these tests?
c. Calculate the coefficient of determination r^2 for each model. Which of the independent variables predicts y best for the ten sampled sets of data? Is this variable necessarily best in general (i.e., for the entire population)? Explain.

Be sure to keep the data and the results of your calculations, since you will need them for the "On Your Own" sections in Chapters 13 and 14.

References   Graybill, F. *Theory and application of the linear model.* North Scituate, Mass.: Duxbury, 1976. Chapter 5.

Heaukulani, D. "The normal distribution of crime." *Journal of Police Science and Administration,* 1975, 3(3), 312–318.

Kovacs, T. G., & Leduc, G. "Sublethal toxicity to rainbow trout (*Salmo gairdneri*) at different temperatures." *Canadian Journal of Fisheries and Aquatic Sciences,* 1982, 39.

Mendenhall, W. *Introduction to linear models and the design and analysis of experiments.* Belmont, Ca.: Wadsworth, 1968.

Mendenhall, W., & McClave, J. T. *A second course in business statistics: regression analysis.* San Francisco: Dellen, 1981.

Neter, J., & Wasserman, W. *Applied linear statistical models.* Homewood, Ill.: Richard Irwin, 1974. Chapters 2–6.

Winters, G. H. "Analysis of the biological and demographic parameters of northern sand lance, *Ammodytes dubins,* from the Newfoundland Grand Bank." *Canadian Journal of Fisheries and Aquatic Sciences,* 1983, *40.*

Younger, M. S. *A handbook for linear regression.* North Scituate, Mass.: Duxbury, 1979.

CHAPTER 13

Multiple Regression

Where We've Been . . .

In Chapter 12 we demonstrated how to model the relationship between a dependent variable y and an independent variable x using a straight line. We fit the straight line to the data points, used r and r^2 to measure the strength of the relationship between y and x, and then used the resulting prediction equation to estimate the mean value of y or to predict some future value of y for a given value of x.

Where We're Going . . .

This chapter will convert the basic concept of Chapter 12 into a powerful estimation and prediction device by modeling the mean value of y as a function of two or more independent variables. The techniques developed will enable you to model a response, y, as a function of both quantitative and qualitative variables. As in the case of a simple linear regression, a multiple regression analysis involves fitting the model to a data set, testing the utility of the model, and using it for the estimation of the mean value of y for given values of the independent variables. We will also use the model to predict a value of y to be observed in the future.

Contents

Most practical applications of regression analysis utilize models that are more complex than the simple straight-line model. For example, a realistic probabilistic model for reaction time to a stimulus would include more than just the amount of a particular drug in the bloodstream. Factors such as age, a measure of visual perception, and sex of the subject are a few of the many variables that might be related to reaction time. Thus, we would want to incorporate these and other potentially important independent variables into the model in order to make accurate predictions.

Probabilistic models that include terms involving x^2, x^3 (or higher-order terms), or more than one independent variable are called **multiple regression models.** The general form of these models is

$$y = \beta_0 + \beta_1 x_1 + \beta_2 x_2 + \cdots + \beta_k x_k + \varepsilon$$

The dependent variable y is now written as a function of k independent variables, $x_1, x_2, \ldots, x_k$. The random error term is added to make the model probabilistic rather than deterministic. The value of the coefficient β_i determines the contribution of the independent variable x_i, and β_0 is the y-intercept. The coefficients $\beta_0, \beta_1, \ldots, \beta_k$ are usually unknown, since they represent population parameters.

At first glance it might appear that the regression model shown above would not allow for anything other than straight-line relationships between y and the independent variables, but this is not true. Actually, $x_1, x_2, \ldots, x_k$ can be functions of variables as long as the functions do not contain unknown parameters. For example, the dollar sales, y, in new housing in a region could be a function of the independent variables

x_1 = Mortgage interest rate

x_2 = (Mortgage interest rate)^2 = x_1^2

x_3 = Unemployment rate in the region

and so on. You could even insert a cyclical term (if it would be useful) of the form $x_4 = \sin t$, where t is a time variable. The multiple regression model is quite versatile and can be made to model many different types of response variables.

The same steps we followed in developing a straight-line model are applicable to the multiple regression model:

Step 1. First, hypothesize the form of the model. This step involves the choice of the independent variables to be included in the model.

Step 2. Next, estimate the unknown parameters $\beta_0, \beta_1, \ldots, \beta_k$.

Step 3. Then specify the probability distribution of the random error component ε and estimate its variance σ^2.

Step 4. Now check the utility of the model.

Step 5. Finally, use the fitted model to estimate the mean value of y or to predict a particular value of y for given values of the independent variables.

The initial step—hypothesizing the form of the model—is the subject of Chapter 14. In this chapter we will assume that the form of the model is known, and we will discuss steps 2 to 5 for a given model.

Case Study 13.1
Predicting
Corporate
Executive
Compensation

Towers, Perrin, Forster & Crosby (TPF&C), an international management consulting firm, has developed a unique application of multiple regression analysis. Many firms are interested in evaluating their management salary structure, and TPF&C uses multiple regression models to accomplish this evaluation. The Compensation Management Service, as TPF&C calls it, measures both the internal and external consistency of a company's pay policies to determine whether they reflect the management's intent.

The dependent variable y used to measure executive compensation is annual salary. The independent variables used to explain salary structure include the executive's age, education, rank, and bonus eligibility; number of employees under the executive's direct supervision; as well as variables that describe the company for which the executive works, such as annual sales, profit, and total assets.

The initial step in developing models for executive compensation is to obtain a sample of executives from various client firms, which TPF&C calls the Compensation Data Bank. The data for these executives are used to estimate the model coefficients (the β-parameters), and these estimates are then substituted into the linear model to form a prediction equation. To predict an executive's compensation, TPF&C substitutes into the prediction equation the values of the independent variables that pertain to the executive (age, rank, etc.). This application of multiple regression analysis is developed more fully in Section 13.7.

13.1
Model
Assumptions

We noted in our introduction that the multiple regression model is of the form

$$y = \beta_0 + \beta_1 x_1 + \beta_2 x_2 + \cdots + \beta_k x_k + \varepsilon$$

where y is the response variable that you wish to predict; $\beta_0, \beta_1, \ldots, \beta_k$ are parameters with unknown values; $x_1, x_2, \ldots, x_k$ are independent information-contributing variables, and ε is a random error component. Since $\beta_0, \beta_1, \ldots, \beta_k$ and $x_1, x_2, \ldots, x_k$ are non-random, the quantity

$$\beta_0 + \beta_1 x_1 + \beta_2 x_2 + \cdots + \beta_k x_k$$

represents the deterministic portion of the model. Therefore, y is composed of two components—one fixed and one random—and, consequently, y is a random variable.

$$y = \overbrace{\beta_0 + \beta_1 x_1 + \cdots + \beta_k x_k}^{\substack{\text{Deterministic} \\ \text{portion of model}}} + \overbrace{\varepsilon}^{\substack{\text{Random} \\ \text{error}}}$$

We will assume (as in Chapter 12) that the random error can be positive or negative and that for any setting of the x-values, $x_1, x_2, \ldots, x_k$, the random error ε possesses a normal probability distribution with mean equal to zero and variance equal to σ^2. Further, we assume that the random errors associated with any (and every) pair of y-values are probabilistically independent. That is, the error ε associated with any one y-value is independent of the error associated with any other y-value. These assumptions are summarized as follows:

Assumptions for Random Error ε

1. For any given set of values of $x_1, x_2, \ldots, x_k$, the random error ε has a normal probability distribution with mean equal to zero and variance equal to σ^2.

2. The random errors are independent (in a probabilistic sense).

The assumptions that we have described for a multiple regression model imply that the mean value $E(y)$ for a given set of values of $x_1, x_2, \ldots, x_k$ is equal to

$$E(y) = \beta_0 + \beta_1 x_1 + \beta_2 x_2 + \cdots + \beta_k x_k$$

Models of this type are called **linear** statistical models because $E(y)$ is a **linear function** of the unknown parameters $\beta_0, \beta_1, \ldots, \beta_k$.

All the estimation and statistical test procedures described in this chapter depend on the data satisfying the assumptions described in this section. Since we will rarely, if ever, know for certain whether this situation occurs, we will want to know how well a regression analysis works and how much faith we can place in our inferences when certain assumptions are not satisfied. We will have more to say on this topic after we discuss the methods of a regression analysis more thoroughly and have shown how they are used in a practical situation.

13.2 Fitting the Model: Least Squares Approach

The method of fitting multiple regression models is identical to that of fitting the simple straight-line model: the method of least squares. That is, we choose the estimated model

$$\hat{y} = \hat{\beta}_0 + \hat{\beta}_1 x_1 + \cdots + \hat{\beta}_k x_k$$

that minimizes

$$\text{SSE} = \sum (y - \hat{y})^2$$

As in the case of the simple linear model, the sample estimates $\hat{\beta}_0, \hat{\beta}_1, \ldots, \hat{\beta}_k$ are obtained as a solution of a set of simultaneous linear equations.*

The primary difference between fitting the simple and multiple regression models is computational difficulty. The $(k + 1)$ simultaneous linear equations that must be solved to find the $(k + 1)$ estimated coefficients $\hat{\beta}_0, \hat{\beta}_1, \ldots, \hat{\beta}_k$ are difficult (sometimes nearly impossible) to solve with a pocket or desk calculator. Consequently, we resort to the use of computers. Many computer packages have been developed to fit a multiple regression model using the method of least squares. We will present output from the SAS System** computer package instead of presenting the tedious hand calculations

---

* Students who are familiar with the calculus should note that $\hat{\beta}_0, \hat{\beta}_1, \ldots, \hat{\beta}_k$ are the solutions to the set of equations $\partial \text{SSE}/\partial \hat{\beta}_0 = 0$, $\partial \text{SSE}/\partial \hat{\beta}_1 = 0, \ldots, \partial \text{SSE}/\partial \hat{\beta}_k = 0$. The solution is usually given in matrix form, but we will not present the details here. See Mendenhall (1968) for details.

** SAS is the registered trademark of SAS Institute Inc., Cary, N.C., U.S.A.

required to fit the models. The regression output of the SAS System is similar to that of most other package regression programs. We will compare the SAS output with two other regression program outputs in Section 13.8. We demonstrate the SAS regression procedure with the following example.

There is a growing concern about the amount of energy consumed by American homeowners. In all-electric homes the amount of electricity expended is of interest to consumers, builders, and groups involved with energy conservation. Suppose we wish to investigate the monthly electrical usage, y, in all-electric homes and its relationship to the size, x, of the home. Moreover, suppose we think that monthly electrical usage in all-electric homes is related to the size of the home by the model

$$y = \beta_0 + \beta_1 x + \beta_2 x^2 + \varepsilon$$

To estimate the unknown parameters β_0, β_1, and β_2, values of y and x are collected for ten homes during a particular month. The data are shown in Table 13.1.

Table 13.1
Home Size–
Electrical Usage
Data

SIZE OF HOME x (square feet)	MONTHLY USAGE y (kilowatt-hours)	SIZE OF HOME x (square feet)	MONTHLY USAGE y (kilowatt-hours)
1,290	1,182	1,840	1,711
1,350	1,172	1,980	1,804
1,470	1,264	2,230	1,840
1,600	1,493	2,400	1,956
1,710	1,571	2,930	1,954

Notice that we include a term involving x^2 in this model because we expect curvature in the graph of the response model relating y to x. The term involving x^2 is called a **quadratic term.** Figure 13.1 illustrates that the electrical usage appears to increase in a curvilinear manner with the size of the home. This provides some support for inclusion of the quadratic term x^2 in the model.

Figure 13.1 Scatter-gram of the Home Size–Electrical Usage Data

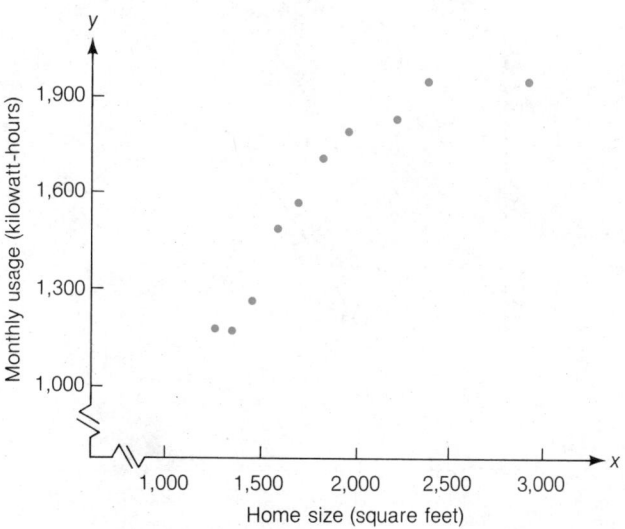

Figure 13.2   SAS Output for the Home Size–Electrical Usage Data

SOURCE	DF	SUM OF SQUARES	MEAN SQUARE	F VALUE	PR > F
MODEL	2	831069.54637065	415534.77318533	189.71	0.0001
ERROR	7	15332.55362935	2190.36480419		STD DEV
				R-SQUARE	46.8013333
CORRECTED TOTAL	9	846402.10000000		0.981885	

PARAMETER	ESTIMATE	T FOR H0: PARAMETER = 0	PR > |T|	STD ERROR OF ESTIMATE
INTERCEPT	−1216.14388700	−5.01	0.0016	242.80636850
X	2.39893018	9.76	0.0001	0.24583560
X ∗ X	−0.00045004	−7.62	0.0001	0.00005908

Note: Computer printers represent products such as $x_1 x_2$ as X1 ∗ X2. Consequently, x^2 is shown at the lower left-hand corner of the computer printout as X ∗ X.

Part of the output from the SAS multiple regression routine for the data in Table 13.1 is reproduced in Figure 13.2. The least squares estimates of the β-parameters appear in the column labeled ESTIMATE. You can see that $\hat{\beta}_0 = -1{,}216.1$, $\hat{\beta}_1 = 2.3989$, and $\hat{\beta}_2 = -.00045$. Therefore, the equation that minimizes the SSE for the data is

$$\hat{y} = -1{,}216.1 + 2.3989x - .00045x^2$$

The minimum value of the SSE, $15{,}332.6$, also appears in the printout. [*Note:* Throughout this chapter we will shade the aspects of the printout that are under discussion.]

Note that the graph of the multiple regression model (Figure 13.3, a response curve) provides a good fit to the data of Table 13.1. Furthermore, the small value of $\hat{\beta}_2$ does

Figure 13.3 Least Squares Model for the Home Size–Electrical Usage Data

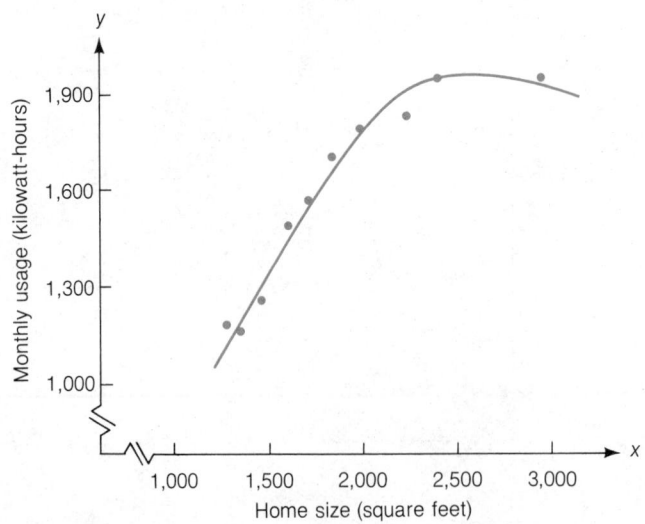

not imply that the curvature is insignificant, since the numerical value of $\hat{\beta}_2$ depends on the scale of the measurements. We will test the contribution of the quadratic coefficient $\hat{\beta}_2$ in Section 13.4.

The ultimate goal of this multiple regression analysis is to use the fitted model to predict electrical usage y for a home of a specific size (area) x. And, of course, we will want to give a prediction interval for y so that we will know how much faith we can place in the prediction. That is, if the prediction model is used to predict electrical usage y for a given size of home, x, what will be the error of prediction? To answer this question, we need to estimate σ^2, the variance of ε.

13.3 Estimation of σ^2, the Variance of ε

The specification of the probability distribution of the random error component ε of the multiple regression model follows the same general outline as for the straight-line model. We assume that ε is normally distributed with mean zero and constant variance σ^2 for any set of values for the independent variables $x_1, x_2, \ldots, x_k$. Furthermore, the errors are assumed to be independent. Given these assumptions, the remaining task in describing the probability distribution of ε is to estimate σ^2.

For example, in the quadratic model describing electrical usage as a function of home size, we found a minimum SSE = 15,332.6. We now want to use this quantity to estimate the variance of ε. Recall that the estimator for the straight-line model was $s^2 = \text{SSE}/(n-2)$ and note that the denominator is n minus the number of estimated β-parameters, which is $n - 2$ in the straight-line model. Since we must estimate one more parameter, β_2, for the quadratic model $y = \beta_0 + \beta_1 x + \beta_2 x^2 + \varepsilon$, the estimator of σ^2 is

$$s^2 = \frac{\text{SSE}}{n-3}$$

That is, the denominator becomes $(n-3)$ because there are now three β-parameters in the model.

The numerical estimate of σ^2 for this example is

$$s^2 = \frac{\text{SSE}}{10-3} = \frac{15,332.6}{7} = 2,190.36$$

where s^2 is called the **mean square for error,** or **MSE.** This estimate of σ^2 is shown in Figure 13.2 in the column titled MEAN SQUARE and the row titled ERROR.

For the general multiple regression model

$$y = \beta_0 + \beta_1 x_1 + \beta_2 x_2 + \cdots + \beta_k x_k + \varepsilon$$

we must estimate the $(k + 1)$ parameters $\beta_0, \beta_1, \beta_2, \ldots, \beta_k$. Thus, the estimator of σ^2 is the SSE divided by the quantity n minus the number of estimated β-parameters, or $s^2 = \text{SSE}/[n - (k + 1)]$.

We will use the estimator of σ^2 both to check the utility of the model (Sections 13.4 and 13.5) and to provide a measure of the reliability of predictions and estimates when the model is used for those purposes (Section 13.6). Thus, you can see that the estimation of σ^2 plays an important part in the development of a regression model.

Estimator of σ^2 for Multiple Regression Model with k Independent Variables

$$MSE = \frac{SSE}{n - (\text{Number of estimated } \beta\text{-parameters})}$$

$$= \frac{SSE}{n - (k + 1)}$$

13.4 Estimating and Testing Hypotheses About the β-Parameters

Sometimes the individual β-parameters in a model have practical significance and we want to estimate their values or test hypotheses about them. For example, if electrical usage y is related to home size x by the straight-line relationship

$$y = \beta_0 + \beta_1 x + \varepsilon$$

β_1 has a very practical interpretation. That is, you saw in Chapter 12 that β_1 is the mean increase in electrical usage, y, for a 1 unit increase in home size x.

As proposed in the preceding sections, suppose that electrical usage y is related to home size x by the quadratic model

$$y = \beta_0 + \beta_1 x + \beta_2 x^2 + \varepsilon$$

Then the mean value of y for a given value of x is

$$E(y) = \beta_0 + \beta_1 x + \beta_2 x^2$$

What is the practical interpretation of β_2? As noted earlier, the parameter β_2 measures the curvature in this response curve. That is, it would not be surprising to observe the electrical usage y rise almost proportional to home size x. Then, eventually, as the size of the home increases, the increase in electrical usage for a 1 unit increase in home size might begin to decrease. Thus, a forecaster of electrical usage would want to determine whether curvature actually was present in the response curve, or, equivalently, the forecaster would want to test the null hypothesis

$$H_0: \quad \beta_2 = 0 \quad \text{(no curvature in the response curve)}$$

against the alternative hypothesis

$$H_a: \quad \beta_2 < 0 \quad \text{(downward curvature in the response curve)}$$

A test of this hypothesis can be performed by using a Student's t-test. The t-test utilizes a test statistic analogous to that used to make inferences about the slope of the simple straight-line model (Section 12.5). The t-statistic is formed by dividing the sample estimate $\hat{\beta}_2$ of the population coefficient β_2 by the estimated standard deviation of the repeated sampling distribution of $\hat{\beta}_2$.

Test statistic: $t = \dfrac{\hat{\beta}_2}{s_{\hat{\beta}_2}}$

We use the symbol $s_{\hat{\beta}_2}$ to represent the estimated standard deviation of $\hat{\beta}_2$. The formula for computing $s_{\hat{\beta}_2}$ is very complex and its presentation is beyond the scope of this text,* but this omission will not cause difficulty. Most computer packages list the estimated standard deviation $s_{\hat{\beta}_i}$ for each of the estimated model coefficients $\hat{\beta}_i$. Moreover, they usually give the calculated t-values for each coefficient in the model.

The rejection region for the test is found in exactly the same way as the rejection regions for the t-tests in previous chapters. That is, we consult Table V in the Appendix to obtain an upper-tail value of t. This is a value t_α such that $P(t > t_\alpha) = \alpha$. We can then use this value to construct rejection regions for either one-tailed or two-tailed tests. To illustrate, in the electrical usage example the error degrees of freedom is $(n - 3) = 7$, the denominator of the estimate of σ^2. Then the rejection region (shown in Figure 13.4) for a one-tailed test with $\alpha = .05$ is

Rejection region: $t < -t_\alpha$

$t < -1.895$

Figure 13.4   Rejection Region for Test of β_2

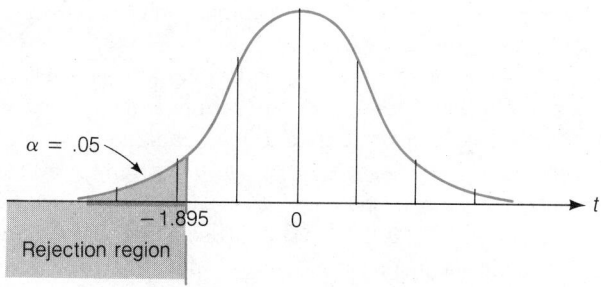

In Figure 13.5 (page 646) we again show a portion of the SAS printout for the electrical usage example. The following quantities are shaded:

1. The estimated coefficients, $\hat{\beta}_0$, $\hat{\beta}_1$, and $\hat{\beta}_2$
2. The SSE and the MSE (estimate of σ^2, variance of ε)
3. The t-statistic, observed significance level, and standard error of $\hat{\beta}_2$ for testing H_0: $\beta_2 = 0$

---

* Because most of the formulas in a multiple regression analysis are so complex, the only reasonable way to present them is by using matrix algebra. We do not assume a prerequisite of matrix algebra for this text and, in any case, we think the formulas can be omitted in an introductory course without serious loss. They are programmed into almost all standard multiple regression computer packages and are presented in some of the texts listed in the references.

Figure 13.5   Output from the SAS System

SOURCE	DF	SUM OF SQUARES	MEAN SQUARE	F-VALUE	PR > F
MODEL	2	831069.54637065	415534.77318533	189.71	0.0001
ERROR	7	15332.55362935	2190.36480419		STD DEV
CORRECTED TOTAL	9	846402.10000000		R-SQUARE	46.8013333
				0.981885	

PARAMETER	ESTIMATE	T FOR H0: PARAMETER = 0	PR > |T|	STD ERROR OF ESTIMATE
INTERCEPT	−1216.14388700	−5.01	0.0016	242.80636850
X	2.39893018	9.76	0.0001	0.24583560
X * X	−0.00045004	−7.62	0.0001	0.00005908

The estimated standard deviations for the model coefficients appear under the column labeled STD ERROR OF ESTIMATE. The t-statistics for testing the null hypothesis that the true coefficients are equal to zero appear under the column headed T FOR H0: PARAMETER = 0. The t-value corresponding to the test of the null hypothesis H_0: $\beta_2 = 0$ is the last one in the column, that is, $t = -7.62$. Since this value is less than -1.895, we conclude that the quadratic term $\beta_2 x^2$ makes a contribution to the prediction model of electrical usage.

The SAS printout shown in Figure 13.5 also lists the two-tailed significance levels for each t-value. These values appear under the column headed PR > $|T|$. The significance level .0001 corresponds to the quadratic term, and this implies that we would reject H_0: $\beta_2 = 0$ in favor of H_a: $\beta_2 \neq 0$ at any α level larger than .0001. Since our alternative was one-sided, H_a: $\beta_2 < 0$, the significance level is half that given in the printout, that is, $\frac{1}{2}(.0001) = .00005$. Thus, there is very strong evidence that the mean electrical usage increases more slowly per square foot for large houses than for small houses.

We can also form a confidence interval for the parameter β_2 as follows:

$$\hat{\beta}_2 \pm t_{\alpha/2} s_{\hat{\beta}_2} = -.000450 \pm (2.365)(.0000591)$$

or $(-.000590, -.000310)$. Note that the t-value 2.365 corresponds to $\alpha/2 = .025$ and $(n - 3) = 7$ df. This interval constitutes a 95% confidence interval for β_2 and can be used to estimate the rate of curvature in mean electrical usage as home size is increased. Note that all values in the interval are negative, reconfirming the conclusion of our test.

Testing an hypothesis about a single β-parameter that appears in any multiple regression model is accomplished in exactly the same manner as described for the quadratic electrical usage model. The t-test and a confidence interval for a β-parameter are shown in the boxes.

Test of an Individual Parameter Coefficient in the Multiple Regression Model

One-Tailed Test	**Two-Tailed Test**
H_0: $\beta_i = 0$	H_0: $\beta_i = 0$
H_a: $\beta_i < 0$	H_a: $\beta_i \neq 0$
(or H_a: $\beta_i > 0$)	

Test statistic: $t = \dfrac{\hat{\beta}_i}{s_{\hat{\beta}_i}}$ Test statistic: $t = \dfrac{\hat{\beta}_i}{s_{\hat{\beta}_i}}$

Rejection region: $t < -t_\alpha$ Rejection region: $t < -t_{\alpha/2}$
        (or $t > t_\alpha$ or $t > t_{\alpha/2}$
        when H_a: $\beta_i > 0$)

where

    n = Number of observations

  $k + 1$ = Number of β-parameters in the model

and t_α and $t_{\alpha/2}$ are based on $n - (k + 1)$ degrees of freedom.

Assumptions: See Section 13.1 for assumptions about the probability distribution for the random error component ε.

A $100(1 - \alpha)\%$ Confidence Interval for a β-Parameter

    $\hat{\beta}_i \pm t_{\alpha/2} s_{\hat{\beta}_i}$

where $t_{\alpha/2}$ is based on $n - (k + 1)$ degrees of freedom and

    n = Number of observations

  $k + 1$ = Number of β-parameters in the model

Example 13.1   A collector of antique grandfather clocks knows that the price received for the clocks increases linearly with the age of the clocks. Moreover, the collector hypothesizes that the auction price of the clocks will increase linearly as the number of bidders increases. Thus, the following model is hypothesized:

$$y = \beta_0 + \beta_1 x_1 + \beta_2 x_2 + \varepsilon$$

where

    y = Auction price

  x_1 = Age of clock (years)

  x_2 = Number of bidders

Table 13.2 Auction Price Data

AGE x_1	NUMBER OF BIDDERS x_2	AUCTION PRICE y	AGE x_1	NUMBER OF BIDDERS x_2	AUCTION PRICE y
127	13	1,235	170	14	2,131
115	12	1,080	182	8	1,550
127	7	845	162	11	1,884
150	9	1,522	184	10	2,041
156	6	1,047	143	6	854
182	11	1,979	159	9	1,483
156	12	1,822	108	14	1,055
132	10	1,253	175	8	1,545
137	9	1,297	108	6	729
113	9	946	179	9	1,792
137	15	1,713	111	15	1,175
117	11	1,024	187	8	1,593
137	8	1,147	111	7	785
153	6	1,092	115	7	744
117	13	1,152	194	5	1,356
126	10	1,336	168	7	1,262

A sample of thirty-two auction prices of grandfather clocks, along with their age and the number of bidders, is given in Table 13.2. The model $y = \beta_0 + \beta_1 x_1 + \beta_2 x_2 + \varepsilon$ is fit to the data, and a portion of the SAS printout is shown in Figure 13.6. Test the hypothesis that the mean auction price of a clock increases as the number of bidders increases when age is held constant, that is, $\beta_2 > 0$. Use $\alpha = .05$.

Figure 13.6 SAS Printout for Example 13.1

SOURCE	DF	SUM OF SQUARES	MEAN SQUARE	F VALUE	PR > F
MODEL	2	4277159.70740504	2138579.85170252	120.65	0.0001
ERROR	29	514034.51534496	17725.32811534		STD DEV
CORRECTED TOTAL	31	4791194.21875000	R-SQUARE		133.13650181
			0.892713		

PARAMETER	ESTIMATE	T FOR H0: PARAMETER = 0	PR > |T|	STD ERROR OF ESTIMATE
INTERCEPT	−1336.72205214	−7.71	0.0001	173.35612607
X1	12.73619884	14.11	0.0001	0.90238049
X2	85.81513260	9.86	0.0001	8.70575681

Solution The hypotheses of interest concern the parameter β_2. Specifically,

$$H_0: \quad \beta_2 = 0 \qquad H_a: \quad \beta_2 > 0$$

Test statistic: $\quad t = \dfrac{\hat{\beta}_2}{s_{\hat{\beta}_2}} = \dfrac{85.815}{8.706} = 9.86$

Rejection region: For $\alpha = .05$ and $n - (k + 1) = 32 - (2 + 1) = 29$ df,
reject H_0 if $t > 1.699$.

This rejection region is shown in Figure 13.7. The calculated t-value, $t = 9.86$, exceeds 1.699 and therefore falls in the rejection region. Thus, the collector can conclude that the mean auction price of a clock increases as the number of bidders increases, when age is held constant.

Figure 13.7 Rejection Region for $H_0: \beta_2 = 0$

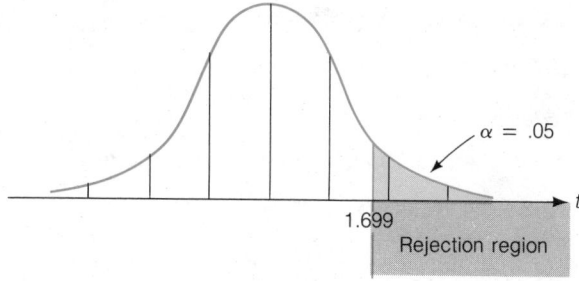

Note that the values $\hat{\beta}_1 = 12.74$ and $\hat{\beta}_2 = 85.82$ (shaded in Figure 13.6) are easily interpreted. We estimate that the mean auction price increases $12.74 per year of age of the clock (when the number of bidders is held constant), and the mean price increases by $85.82 per additional bidder (when the age of the clock is held constant.)

Be careful not to try to interpret the estimated intercept $\hat{\beta}_0 = -1,336.72$ in the same way as we interpreted $\hat{\beta}_1$ and $\hat{\beta}_2$. You might think that this implies a negative price for clocks zero years of age with zero bidders. However, these zeros are meaningless numbers in this example, since the ages range from 108 to 194 and the number of bidders ranges from 5 to 15. Interpretations of predicted y-values for values of the independent variables outside the sampled range can be very misleading.

Some computer programs use the F-statistic (SAS does not) to conduct two-tailed tests concerning the individual β-parameters, because the square of a Student's t with ν degrees of freedom is equal to an F-statistic with 1 degree of freedom in the numerator and ν degrees of freedom in the denominator. If you conduct a two-tailed t-test and reject the null hypothesis if $t > t_{\alpha/2}$ or $t < -t_{\alpha/2}$, the corresponding F-test implies rejection if the computed value of F (which is equal to the square of the computed t-statistic) is larger than F_α. Thus,

$$t_{\alpha/2}^2 = F_\alpha$$

As an example, when we tested an hypothesis about the curvature parameter β_2 in the quadratic model relating electrical usage to home size, the computed t-value was -7.62 (see Figure 13.5). The equivalent F-test yields a value $F = t^2 = (-7.62)^2 = 58.06$. And the upper-tail rejection region for a two-tailed test with $\alpha = .05$, 1 df in the numerator, and 7 df in the denominator is

$$F > F_{.05} = 5.59$$

Note that the F-value, 5.59, is equal to the square of 2.365, the value of t that corresponds to $t_{.025}$ based on 7 df. You can see that the conclusion is the same no matter which test statistic is used: There is very strong evidence that curvature is present in the model. ∎

Exercises 13.1–13.8

[Note: Starred (*) exercises require the use of a computer.]

Learning the Mechanics

13.1 Suppose you fit the multiple regression model

$$y = \beta_0 + \beta_1 x_1 + \beta_2 x_2 + \beta_3 x_3 + \varepsilon$$

to $n = 30$ data points and obtain the following result:

$$\hat{y} = 3.4 - 4.6x_1 + 2.7x_2 + .93x_3$$

The estimated standard errors of $\hat{\beta}_2$ and $\hat{\beta}_3$ are 1.86 and .29, respectively.

a. Test the null hypothesis H_0: $\beta_2 = 0$ against the alternative hypothesis H_a: $\beta_2 \neq 0$. Use $\alpha = .05$.

b. Test the null hypothesis H_0: $\beta_3 = 0$ against the alternative hypothesis H_a: $\beta_3 \neq 0$. Use $\alpha = .05$.

c. The null hypothesis H_0: $\beta_2 = 0$ is not rejected. In contrast, the null hypothesis H_0: $\beta_3 = 0$ is rejected. Explain how this can happen even though $\hat{\beta}_2 > \hat{\beta}_3$.

13.2 Suppose you fit the second-order model,

$$y = \beta_0 + \beta_1 x + \beta_2 x^2 + \varepsilon$$

to $n = 25$ data points. Your estimate of β_2 is $\hat{\beta}_2 = .47$ and the estimated standard error of the estimate is $s_{\hat{\beta}_2} = .15$.

a. Test the null hypothesis that the mean value of y is related to x by the (first-order) linear model

$$E(y) = \beta_0 + \beta_1 x$$

(H_0: $\beta_2 = 0$) against the alternative hypothesis that the true relationship is given by the quadratic model (a second-order linear model),

$$E(y) = \beta_0 + \beta_1 x + \beta_2 x^2$$

(H_a: $\beta_2 \neq 0$). Use $\alpha = .05$.

b. Suppose you want to determine only whether the quadratic curve opens upward; i.e., as x increases, the slope of the curve increases. Give the test statistic and the rejection region for the test for $\alpha = .05$. Do the data support the theory that the slope of the curve increases as x increases? Explain.

c. What is the value of the F-statistic for testing the null hypothesis H_0: $\beta_2 = 0$ against H_a: $\beta_2 \neq 0$?

d. Could the F-statistic in part c be used to conduct the test in part b? Explain.

13.3 How is the number of degrees of freedom available for estimating σ^2 (the variance of ε) related to the number of independent variables in a regression model?

*13.4 Use a computer to fit a second-order model to the following data:

x	0	1	2	3	4	5	6
y	1	2.7	3.8	4.5	5.0	5.3	5.2

a. Find SSE and s^2.
b. Do the data provide sufficient evidence to indicate that the second-order term provides information for the prediction of y? [*Hint:* Test H_0: $\beta_2 = 0$.]
c. Find the least squares prediction equation.
d. Plot the data points and graph $\hat{y}$. Does your prediction equation provide a good fit to the data?

Applying the Concepts

13.5 An employer believes that factory workers who are with the company longer tend to invest more in a company investment program per year than workers with less time with the company. The following model is believed to be adequate in modeling the relationship of annual amount invested, y, to years working for the company, x:

$$y = \beta_0 + \beta_1 x + \beta_2 x^2 + \varepsilon$$

The employer checks the records for a sample of fifty factory employees for a previous year and fits the model to get $\hat{\beta}_2 = .0015$ and $s_{\hat{\beta}_2} = .000712$.
a. What is the practical significance of a value of β_2 that is larger than zero?
b. Test to determine whether the employer can conclude that $\beta_2 > 0$. Use $\alpha = .05$.

13.6 A researcher wishes to investigate the effects of two independent variables on a teacher's attitude toward handicapped students. A study is conducted involving forty randomly selected teachers. The response y, a teacher's attitude toward handicapped students, is measured with a standardized attitude scale. Independent variables in the study are

x_1 = Average number of handicapped children taught per year
x_2 = Number of years of teaching experience

The researcher fits the model

$$y = \beta_0 + \beta_1 x_1 + \beta_2 x_2 + \beta_3 x_2^2 + \varepsilon$$

to the data with the following results:

$$\hat{y} = 50 + 1.5x_1 + 5x_2 - .1x_2^2$$
$$s_{\hat{\beta}_3} = .03$$

a. Do these data provide sufficient evidence to indicate that the quadratic term in years of experience, x_2^2, is useful for predicting attitude score? Use $\alpha = .05$.

b. Sketch the predicted attitude score $\hat{y}$ as a function of the number of years of experience x_2 for $x_1 = 5$. Repeat this for $x_1 = 10$. [*Note:* For each value of x_1 ($x_1 = 5$ and $x_1 = 10$), plot $\hat{y}$ for $x_2 = 0, 2, 4, 6, 8,$ and 10. Observe that (for this model), the vertical distance between the two prediction curves is the same for all values of x_2.]

13.7 To project personnel needs for the Christmas shopping season, a department store wants to project sales for the season. The sales for the previous Christmas season are an indication of what to expect for the current season. However, the projection should also reflect the current economic environment by taking into consideration sales for a more recent period. The following model might be appropriate:

$$y = \beta_0 + \beta_1 x_1 + \beta_2 x_2 + \varepsilon$$

where

  y = Christmas sales for a year

  x_1 = August sales for the same year

  x_2 = Christmas sales for the previous year

(All units are in thousands of dollars.) Data for 10 previous years are used to fit the prediction equation, and the following are calculated:

$$\hat{\beta}_1 = .55 \qquad s_{\hat{\beta}_1} = .10$$
$$\hat{\beta}_2 = .62 \qquad s_{\hat{\beta}_2} = .27$$

Do these data provide sufficient evidence to indicate that, using the proposed model, August sales help to predict Christmas sales? Test using $\alpha = .05$.

13.8 In a recent study (Tanner, 1983), an attempt was made to discover the factors in a person's education that determine future wages. A first-order model was fit to a set of $n = 60$ data points and the following prediction equation and t-test values were obtained:

$$y = 0 - .0945x_1 - .032x_2 + .009x_3 - .0028x_4 + .007x_5 + .105x_6 + .469x_7$$
$$(-2.61) \quad (-.96) \quad (2.74) \quad (-2.23) \quad (5.71) \quad (4.02) \quad (2.21)$$

where

  y = Future wages

  x_1 = Amount of business course work in high school

  x_2 = Amount of college prep work in high school

  x_3 = Math aptitude

  x_4 = High school grade-point average

  x_5 = Measure of socioeconomic status

  x_6 = 1 if the individual is married and 0 if not

  x_7 = Amount of on-the-job training

The t-values used to test the individual model parameters are shown in parentheses below their respective estimates.

a. What are the interpretations of the coefficients?

b. Are they statistically significant at the $\alpha = .01$ level?

c. What happens to the intercept term when x_6 is 1?

d. What happens to the intercept term when x_6 is zero?

13.5 Checking the Utility of a Model: R^2 and the Analysis of Variance F-Test

Conducting t-tests on each β-parameter in a model is *not* a good way to determine whether a model is contributing information for the prediction of y. If we were to conduct a series of t-tests to determine whether the independent variables are contributing to the predictive relationship, we would be very likely to make one or more errors in deciding which terms to retain in the model and which to exclude. For example, even if all the β-parameters (except β_0) are equal to zero, $100(\alpha)\%$ of the time you will reject the null hypothesis and conclude that some β-parameter differs from zero. Thus, in multiple regression models for which a large number of independent variables is being considered, conducting a series of t-tests may include a large number of insignificant variables and exclude some useful ones. If we want to test the utility of a multiple regression model, we will need a global test (one that encompasses all the β-parameters). We would also like to find some statistical quantity that measures how well the model fits the data.

We commence with the easier problem—finding a measure of how well a linear model fits a set of data. For this we use the multiple regression equivalent of r^2, the coefficient of determination for the straight-line model (Chapter 12). Thus, we define the ***multiple coefficient of determination, R^2,*** as

$$R^2 = 1 - \frac{\sum(y - \hat{y})^2}{\sum(y - \bar{y})^2} = 1 - \frac{SSE}{SS_{yy}}$$

where $\hat{y}$ is the predicted value of y for the model. Just as for the simple linear model, R^2 represents the fraction of the sample variation of the y-values (measured by SS_{yy}) that is explained by the least squares prediction equation. Thus, $R^2 = 0$ implies a complete lack of fit of the model to the data and $R^2 = 1$ implies a perfect fit with the model passing through every data point. In general, the larger the value of R^2, the better the model fits the data.

To illustrate, the value $R^2 = .982$ for the electrical usage example is indicated in a reprint of the SAS printout (Figure 13.8, page 654). This very high value of R^2 implies that using the independent variable home size in a quadratic model explains 98.2% of the total ***sample variation*** (measured by SS_{yy}) of electrical usage y. Thus, R^2 is a sample statistic that tells how well the model fits the data and thereby represents a measure of the utility of the entire model.

The fact that R^2 is a sample statistic implies that it can be used to make inferences about the utility of the entire model for predicting the population of y-values at each

Figure 13.8 SAS Printout for Electrical Usage Example

SOURCE	DF	SUM OF SQUARES	MEAN SQUARE	F VALUE	PR > F
MODEL	2	831069.54637065	415534.77318533	189.71	0.0001
ERROR	7	15332.55362935	2190.36480419		STD DEV
				R-SQUARE	46.8013333
CORRECTED TOTAL	9	846402.10000000		0.981885	

PARAMETER	ESTIMATE	T FOR H0: PARAMETER = 0	PR > |T|	STD ERROR OF ESTIMATE
INTERCEPT	−1216.14388700	−5.01	0.0016	242.80636850
X	2.39893018	9.76	0.0001	0.24583560
X * X	−0.00045004	−7.62	0.0001	0.00005908

setting of the independent variables. In particular, for the electrical usage data the test

H_0: $\beta_1 = \beta_2 = 0$

H_a: At least one of the coefficients is nonzero

would formally test the global utility of the model. The test statistic used to test this null hypothesis is

$$\text{Test statistic:} \quad F = \frac{R^2/k}{(1 - R^2)/[n - (k + 1)]}$$

where n is the number of data points and k is the number of parameters in the model not including β_0. The test statistic F will have the F probability distribution with k degrees of freedom in the numerator and $[n - (k + 1)]$ degrees of freedom in the denominator. The tail values of the F-distribution are given in Tables VII, VIII, IX, and X of the Appendix.

The F test statistic becomes large as the coefficient of determination R^2 becomes large. To determine how large F must be before we can conclude at a given significance level that the model is useful for predicting y, we set up the rejection region as follows:

Rejection region: $F > F_\alpha$, where F is based on k numerator and $n - (k + 1)$
                    denominator degrees of freedom

For the electrical usage example ($n = 10$, $k = 2$, $n - (k + 1) = 7$, and $\alpha = .05$), we will reject H_0: $\beta_1 = \beta_2 = 0$ if

$F > F_{.05} = 4.74$

From the computer printout (Figure 13.8), we find that the computed F is 189.71. Since this value greatly exceeds the tabulated value of 4.74, we conclude that at least one of the model coefficients β_1 and β_2 is nonzero. Therefore, this global F-test indicates that the quadratic model $y = \beta_0 + \beta_1 x + \beta_2 x^2 + \varepsilon$ is useful for predicting electrical usage.

Testing the Utility of a Multiple Regression Model: The Global F-Test

H_0: $\beta_1 = \beta_2 = \cdots = \beta_k = 0$

H_a: At least one of the β-parameters does not equal zero

Test statistic: $F = \dfrac{R^2/k}{(1 - R^2)/[n - (k + 1)]}$

Assumptions: See Section 13.1 for assumptions about the random component ε.

Rejection region: $F > F_\alpha$, based on k numerator and $n - (k + 1)$ denominator degrees of freedom, where

n = Number of data points

k = Number of β-parameters in the model, excluding β_0

Example 13.2

Refer to Example 13.1, in which an antique collector modeled the auction price y of grandfather clocks as a function of the age of the clock, x_1, and the number of bidders, x_2. The hypothesized model is

$$y = \beta_0 + \beta_1 x_1 + \beta_2 x_2 + \varepsilon$$

A sample of thirty-two observations is obtained with the results summarized in the SAS printout repeated here in Figure 13.9. Discuss the coefficient of determination

Figure 13.9 SAS Printout for Example 13.2

SOURCE	DF	SUM OF SQUARES	MEAN SQUARE	F VALUE	PR > F
MODEL	2	4277159.70740504	2138579.85170252	120.65	0.0001
ERROR	29	514034.51534496	17725.32811534		STD DEV
CORRECTED TOTAL	31	4791194.21875000	R-SQUARE		133.13650181
			0.892713		

PARAMETER	ESTIMATE	T FOR H0: PARAMETER = 0	PR > |T|	STD ERROR OF ESTIMATE
INTERCEPT	− 1336.72205214	−7.71	0.0001	173.35612607
X1	12.73619884	14.11	0.0001	0.90238049
X2	85.81513260	9.86	0.0001	8.70575681

R^2 for this example, and then conduct the global F-test of model utility at the $\alpha = .05$ level of significance.

Solution   The R^2-value is .89 (see Figure 13.9). This implies that the least squares model has explained 89% of the total variation, SS_{yy}. We now test

$$H_0: \quad \beta_1 = \beta_2 = 0 \qquad (k = 2)$$

H_a: At least one of the two model coefficients is nonzero

Test statistic: $$F = \frac{R^2/k}{(1 - R^2)/[n - (k + 1)]}$$

Rejection region: $F > F_\alpha$

For this example, $n = 32$, $k = 2$, and $n - (k + 1) = 32 - 3 = 29$. Then, for $\alpha = .05$, we will reject H_0: $\beta_1 = \beta_2 = 0$ if $F > F_{.05}$, where F is based on $k = 2$ numerator and $n - (k + 1) = 29$ denominator degrees of freedom, i.e., if $F > 3.33$. The computed value of the F-test statistic is 120.65 (see Figure 13.9). Since this value of F falls in the rejection region ($F = 120.65$ greatly exceeds $F_{.05} = 3.33$), the data provide strong evidence that at least one of the model coefficients is nonzero. The model appears to be useful for predicting auction prices.   ■

Can we be sure the best prediction model has been found if the global F-test indicates that a model is useful? Unfortunately, we cannot. There is no way of knowing (without further analysis) whether the addition of other independent variables will improve the utility of the model, as Example 13.3 indicates.

Example 13.3   Refer to Examples 13.1 and 13.2. Suppose the collector, having observed many auctions, believes that the *rate of increase* of the auction price with age will be driven upward by a large number of bidders. Thus, instead of a relationship like that shown in Figure 13.10(a), in which the rate of increase in price with age is the same for any

Figure 13.10   Examples of No Interaction and Interaction Models

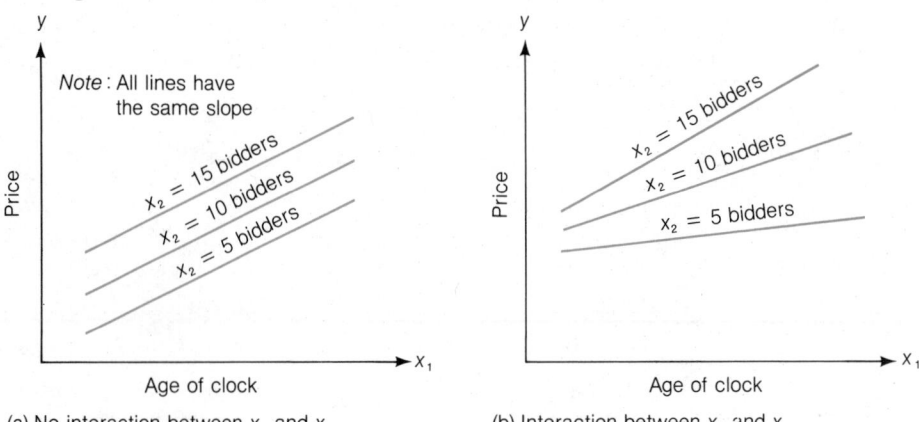

(a) No interaction between x_1 and x_2

(b) Interaction between x_1 and x_2

number of bidders, the collector believes the relationship is like that shown in Figure 13.10(b). Note that as the number of bidders increases from five to fifteen, the slope of the price versus age line increases. When the slope of the relationship between y and one independent variable (x_1) depends on the value of a second independent variable (x_2), as is the case here, we say that x_1 and x_2 **interact**.* The model that includes interaction is written

$$y = \beta_0 + \beta_1 x_1 + \beta_2 x_2 + \beta_3 x_1 x_2 + \varepsilon$$

Note that the increase in the mean price, $E(y)$, for each 1-year increase in age, x_1, is no longer given by the constant β_1 but is now $\beta_1 + \beta_3 x_2$. *That is, the amount that $E(y)$ increases for each 1 unit increase in x_1 is dependent on the number of bidders, x_2. Thus, the two variables x_1 and x_2 interact to affect y.*

The thirty-two data points listed in Table 13.2 were used to fit the model with inter-action. A portion of the SAS printout is shown in Figure 13.11. Test the hypothesis that the price–age slope increases as the number of bidders increases, i.e., that age and number of bidders, x_2, interact positively.

Figure 13.11   Portion of the SAS Printout for the Model with Interaction

SOURCE	DF	SUM OF SQUARES	MEAN SQUARE	F VALUE	PR > F
MODEL	3	4572547.98717668	1524182.66239223	195.19	0.0001
ERROR	28	218646.23157332	7808.79398476		STD DEV
CORRECTED TOTAL	31	4791194.21875000		R-SQUARE	88.36738077
				0.954365	

PARAMETER	ESTIMATE	T FOR H0: PARAMETER = 0	PR > |T|	STD ERROR OF ESTIMATE
INTERCEPT	322.75435309	1.10	0.2806	293.32514660
X1	0.87328775	0.43	0.6688	2.01965115
X2	−93.40991991	−3.14	0.0039	29.70767946
X1 * X2	1.29789828	6.15	0.0001	0.21102602

Solution   The model is

$$y = \beta_0 + \beta_1 x_1 + \beta_2 x_2 + \beta_3 x_1 x_2 + \varepsilon$$

and the hypotheses of interest to the collector concern the parameter β_3. Specifically,

$$H_0: \quad \beta_3 = 0 \qquad H_a: \quad \beta_3 > 0$$

Test statistic: $t = \dfrac{\hat{\beta}_3}{s_{\hat{\beta}_3}}$

Rejection region: For $\alpha = .05$, $t > t_{.05}$

* A detailed discussion of interaction is given in Chapter 14.

where $n = 32$, $k = 3$, and $t_{.05} = 1.701$, based on $n - (k + 1) = 28$ df. [*Remember:* $(k + 1) = 4$ is the number of parameters in the regression model.]

The *t*-value corresponding to $\hat{\beta}_3$ is indicated in Figure 13.11. The value $t = 6.15$ exceeds 1.701 and therefore falls in the rejection region. Thus, the collector can conclude that the rate of change of the mean price of the clocks with age increases as the number of bidders increases; that is, x_1 and x_2 interact. Thus, it appears that the interaction term should be included in the model.

One note of caution: Although the coefficient of x_2 is negative ($\hat{\beta}_2 = -93.41$), this does *not* imply that auction price decreases as the number of bidders increases. Since interaction is present, the rate of change (slope) of mean auction price with the number of bidders *depends on x_1*, the age of the clock. Thus, for example, the estimated rate of change of y for a unit increase in x_2 (one new bidder) for a 150-year-old clock is

$$\text{Estimated } x_2 \text{ slope} = \hat{\beta}_2 + \hat{\beta}_3 x_1 = -93.41 + 1.30(150)$$
$$= 101.59$$

In other words, we estimate that the auction price of a 150-year-old clock will *increase* by about $101.59 for every additional bidder. Although the rate of increase will vary as x_1 is changed, it will remain positive for the range of values of x_1 included in the sample. Extreme care is needed in interpreting the signs and sizes of coefficients in a multiple regression model. ∎

To summarize the discussion in this section, the value of R^2 is an indicator of how well the prediction equation fits the data. More important, it can be used (in the *F*-statistic) to determine whether the data provide sufficient evidence to indicate that the model contributes information for the prediction of y. Intuitive evaluations of the contribution of the model based on the computed value of R^2 must be examined with care. The value of R^2 increases as more and more variables are added to the model. Consequently, you could force R^2 to take a value very close to 1 even though the model contributes no information for the prediction of y. In fact, R^2 equals 1 when the number of terms in the model equals the number of data points. Therefore, you should not rely solely on the value of R^2 to tell you whether the model is useful for predicting y. Use the *F*-test.

Exercises 13.9–13.17

[*Note:* Starred (*) exercises require the use of a computer.]

Learning the Mechanics

13.9 Suppose you fit the model

$$y = \beta_0 + \beta_1 x_1 + \beta_2 x_2 + \beta_3 x_1 x_2 + \beta_4 x_1^2 + \beta_5 x_2^2 + \varepsilon$$

to $n = 30$ data points and obtain

$$\text{SSE} = .33 \qquad R^2 = .92$$

a. Do the values of SSE and R^2 suggest that the model provides a good fit to the data? Explain.

b. Is the model of any use in predicting y? Test the null hypothesis that
$$E(y) = \beta_0$$

that is,
$$H_0: \quad \beta_1 = \beta_2 = \cdots = \beta_5 = 0$$

against the alternative hypothesis

$$H_a: \quad \text{At least one of the parameters } \beta_1, \beta_2, \ldots, \beta_5 \text{ is nonzero}$$

Use $\alpha = .05$.

13.10 Suppose you fit the model
$$y = \beta_0 + \beta_1 x_1 + \beta_2 x_2 + \beta_3 x_3 + \varepsilon$$

to $n = 20$ data points and obtain

$$\sum (y_i - \hat{y}_i)^2 = 19.2 \quad \text{and} \quad \sum (y_i - \bar{y})^2 = 27.5$$

a. Find R^2. Does the value of R^2 suggest that the model provides a good fit to the data? Explain.

b. Test the null hypothesis
$$H_0: \quad \beta_1 = \beta_2 = \beta_3 = 0$$

against the alternative hypothesis

$$H_a: \quad \text{At least one of the } \beta\text{-parameters is nonzero}$$

Use $\alpha = .05$.

Applying the Concepts

13.11 In an article titled "The Time Crunch," K. D. Fox and S. Y. Nickols (1983) report on a study of the impact of a wife's employment on the number of hours available for household tasks. Some 206 families were employed in the study. One independent variable measured for each family was the total time y (in minutes per day) that the wife spent on household chores. Two independent variables were also recorded:

WEMP: $x_1 = $ Wife's hours of employment per week

YAGE: $x_2 = $ Age of the youngest child in the family

The authors explain that they first attempted to fit the model

$$E(y) = \beta_0 + \beta_1 x_1 + \beta_2 x_2 + \beta_3 x_1 x_2$$

If the regression analysis did not indicate statistical significance for the interaction term, it was eliminated from the model. Looking at the accompanying computer printout, you can see that the interaction term was eliminated from their final regression analysis.

Regression Analyses of Wives' and Husbands' Time Spent in Household Work

DEPENDENT VARIABLE: WIFE'S HOUSEHOLD WORK   (MEAN = 401.3[a])

SOURCE OF VARIATION	DF	SUMS OF SQUARES	MEAN SQUARE	F	R^2
TOTAL	205	5,432,296		62.10***	0.38
MODEL	2	2,061,918	1,030,959		
ERROR	203	3,370,378	16,603		

SOURCE WITHIN MODEL	DF	SEQUENTIAL S.S.	PARTIAL S.S.	b-ESTIMATES
WEMP	1	1,810,154***	1,110,288***	−4.08
YAGE	1	251,765***	251,765***	−7.02

DEPENDENT VARIABLE: HUSBAND'S HOUSEHOLD WORK   (MEAN = 105.2[a])

REGRESSION ANALYSIS NOT SIGNIFICANT

Note: In this particular printout "b-estimate" is our β-estimate, or parameter estimate. Also, the asterisks indicate the level of significance of test statistics (or sum of squares used to compute test statistics). Here, *** means $p < .001$.

[a] Minutes per day.

a. Based on Fox and Nickols's computer printout, does the model contribute information for the prediction of y?

b. Find R^2 and give its practical implications.

c. Do the data provide sufficient information to indicate that the model contributes information for the prediction of y? Test using $\alpha = .05$.

d. In concluding, Fox and Nickols state that "for each additional hour of employment, wives decreased household work time 4 minutes per day." Do you agree? [*Note:*   We will test H_0: $\beta_1 = 0$ in Exercise 14.29.]

13.12 In hopes of increasing the company's share of the fine food market, researchers for a meat-processing firm are working to improve the quality of its hickory-smoked hams. One of their studies concerns the effect on the flavor of the ham of time spent in the smokehouse. Hams that were in the smokehouse for varying amounts of time were each subjected to a taste test by a panel of ten food experts. The following model was thought to be appropriate by the researchers:

$$y = \beta_0 + \beta_1 t + \beta_2 t^2 + \varepsilon$$

where

y = Mean of the taste scores for the ten experts

t = Time in the smokehouse (hours)

Assume that the least squares model estimated using a sample of twenty hams is

$$\hat{y} = 20.3 + 5.2t - .0025t^2$$

and that $s_{\hat{\beta}_2} = .0011$. The coefficient of determination is $R^2 = .79$.

a. Is there evidence to indicate that the overall model is useful? Test at $\alpha = .05$.
b. Is there evidence to indicate that the quadratic term is important in this model? Test at $\alpha = .05$.

13.13 Refer to Exercise 13.6. Recall that the dependent variable y is a teacher's attitude toward handicapped students as measured by a standardized attitude scale. The independent variables are the average number of handicapped children taught per year (x_1) and the teacher's years of experience (x_2). Suppose the interaction terms ($x_1 x_2$ and $x_1 x_2^2$) are added to the model proposed in Exercise 13.6 to produce

$$y = \beta_0 + \beta_1 x_1 + \beta_2 x_2 + \beta_3 x_2^2 + \beta_4 x_1 x_2 + \beta_5 x_1 x_2^2 + \varepsilon$$

This model is fit to the same forty observations used in Exercise 13.6 with the result

$$\hat{y} = 50 + x_1 + 6x_2 - .2x_2^2 + x_1 x_2 - .5x_1 x_2^2$$

and

$$R^2 = .87.$$

a. Interpret the value of R^2.
b. Is there sufficient evidence to indicate that this model is useful for predicting attitude score? Test H_0: $\beta_1 = \beta_2 = \beta_3 = \beta_4 = \beta_5 = 0$ using $\alpha = .05$.
c. Sketch the predicted attitude score $\hat{y}$ as a function of the number of years of experience, x_2, for teachers who have taught an average of ten handicapped children per year ($x_1 = 10$). Repeat this for $x_1 = 5$. Compare these sketches with those obtained for the noninteraction model of Exercise 13.6. [*Note:* In Chapter 14 we will test to determine whether this model provides more information for the prediction of y than the noninteraction model.]

13.14 Because the coefficient of determination R^2 always increases when a new independent variable is added to the model, it is tempting to include many variables in a model to force R^2 to be near 1. However, doing so reduces the degrees of freedom available for estimating σ^2, which adversely affects our ability to make reliable inferences. Suppose you want to use eighteen economic indicators to predict next year's GNP. You fit the model

$$y = \beta_0 + \beta_1 x_1 + \beta_2 x_2 + \cdots + \beta_{17} x_{17} + \beta_{18} x_{18} + \varepsilon$$

where $y = $ GNP and $x_1, x_2, \ldots, x_{18}$ are indicators. Only 20 years of data ($n = 20$) are used to fit the model, and you obtain $R^2 = .95$. Test to see whether this impressive-looking R^2 is large enough for you to infer that this model is useful, i.e., that at least one term in the model is important for predicting GNP. Use $\alpha = .05$.

13.15 The length of a mosquito's proboscis plays a large role in determining its feeding habits. An entomologist has proposed the following model for predicting the length of a mosquito's proboscis:

$$E(y) = \beta_0 + \beta_1 x_1 + \beta_2 x_2 + \beta_3 x_3$$

where

y = Length of proboscis (millimeters)

x_1 = Dry weight (milligrams)

x_2 = Length of wing (millimeters)

x_3 = Width of wing (millimeters)

An analysis of data obtained from a sample of forty-four mosquitoes of a certain species produces the least squares model

$$\hat{y} = .968 + .292x_1 + .614x_2 - .201x_3$$

Also,

$$s_{\hat{\beta}_1} = .248 \qquad s_{\hat{\beta}_2} = .131 \qquad s_{\hat{\beta}_3} = .267 \qquad R^2 = .536$$

a. Do these statistics indicate that the model is useful in predicting proboscis length? Use $\alpha = .10$.

b. Given the first-order model specified above, do the data provide sufficient evidence to indicate that x_3 (width of wing) is an important variable for predicting y? Test using $\alpha = .05$.

*13.16 In Exercise 12.16, we found the least squares line relating the age x of a sand lance to its length y. The data are shown in the table (Winters, 1983).

MEAN LENGTH (MILLIMETERS)	176	194	212	226	236	244	254
AGE (YEARS)	2	3	4	5	6	7	8

a. Use a computer to fit a second-order model to the data.

b. Do the data provide sufficient evidence to indicate that the second-order term contributes information for the prediction of sand lance length? Test using $\alpha = .05$.

c. Find the p-value for the test in part b.

d. Find the coefficient of determination for the second-order model and interpret it.

*13.17 In Exercise 12.56, advanced ticket sales were used to predict the number of hot dogs purchased at a baseball stadium during the coming week. Another variable that might indicate attendance and help predict food needs is the visiting team's standing in their division. The data in the table at the top of the next page were collected for eight games at the stadium.

a. Fit the following model to the data:

$$y = \beta_0 + \beta_1 x_1 + \beta_2 x_2 + \beta_3 x_1 x_2 + \varepsilon$$

b. Find R^2 for the least squares equation of part a. Interpret your result.

c. Is there sufficient evidence to indicate that the overall model is useful for predicting y? Test using $\alpha = .05$.

d. Is there sufficient evidence to indicate that the interaction term should be included in the model? Test using $\alpha = .05$.

e. Interpret the values of $\hat{\beta}_0$, $\hat{\beta}_1$, $\hat{\beta}_2$, and $\hat{\beta}_3$ found in part a.

HOT DOGS PURCHASED DURING WEEK y (thousands)	ADVANCED TICKET SALES FOR WEEK x_1 (thousands)	VISITING TEAM'S STANDING x_2
54.2	55.6	1
48.9	63.5	3
50.3	60.1	2
56.1	58.9	1
20.0	45.4	6
53.8	66.6	2
46.4	59.3	4
52.7	64.8	2

13.6 Using the Model for Estimation and Prediction

In Section 12.8 we discussed the use of the least squares line for estimating the mean value of y, $E(y)$, for some value of x, say $x = x_p$. We also showed how to use the same fitted model to predict, when $x = x_p$, some value of y to be observed in the future. Recall that the least squares line yielded the same value for both the estimate of $E(y)$ and the prediction of some future value of y. That is, both are the result of substituting x_p into the prediction equation $\hat{y} = \hat{\beta}_0 + \hat{\beta}_1 x$ and calculating $\hat{y}_p$. There the equivalence ends. The confidence interval for the mean $E(y)$ was narrower than the prediction interval for y, because of the additional uncertainty attributable to the random error ε when predicting some future value of y.

These same concepts carry over to the multiple regression model. Suppose we want to estimate the mean electrical usage for a given home size, say $x_p = 1,500$ square feet. Assuming that the quadratic model represents the true relationship between electrical usage and home size, we want to estimate

$$E(y) = \beta_0 + \beta_1 x_p + \beta_2 x_p^2$$
$$= \beta_0 + \beta_1(1,500) + \beta_2(1,500)^2$$

Substituting into the least squares prediction equation, we find the estimate of $E(y)$ to be

$$\hat{y} = \hat{\beta}_0 + \hat{\beta}_1(1,500) + \hat{\beta}_2(1,500)^2$$
$$= -1,216.144 + 2.3989(1,500) - .00045004(1,500)^2$$
$$= 1,369.7$$

To form a confidence interval for the mean, we need to know the standard deviation of the sampling distribution for the estimator $\hat{y}$. For multiple regression models, the form of this standard deviation is rather complex. However, the SAS regression package allows us to obtain the confidence intervals for mean values of y for any given combination of values of the independent variables. This portion of the SAS output for the electrical usage example is shown in Figure 13.12 (page 664). The mean value and corresponding 95% confidence interval for $x_p = 1,500$ are shown in the columns labeled ESTIMATED MEAN VALUE, LOWER 95% CL FOR MEAN, and UPPER 95% CL FOR MEAN.

Figure 13.12 SAS
Printout for Estimated Mean
Value and Corresponding
Confidence Interval for
$x_p = 1,500$

X	ESTIMATED MEAN VALUE	LOWER 95% CL FOR MEAN	UPPER 95% CL FOR MEAN
1500	1369.66088739	1324.98831001	1414.33346477

Note that

$$\hat{y} = 1,369.7$$

which agrees with our earlier calculation. The 95% confidence interval for the true mean of y is shown to be 1,325.0 to 1,414.3 (see Figure 13.13).

Figure 13.13 Confidence Interval for Mean
Electrical Usage

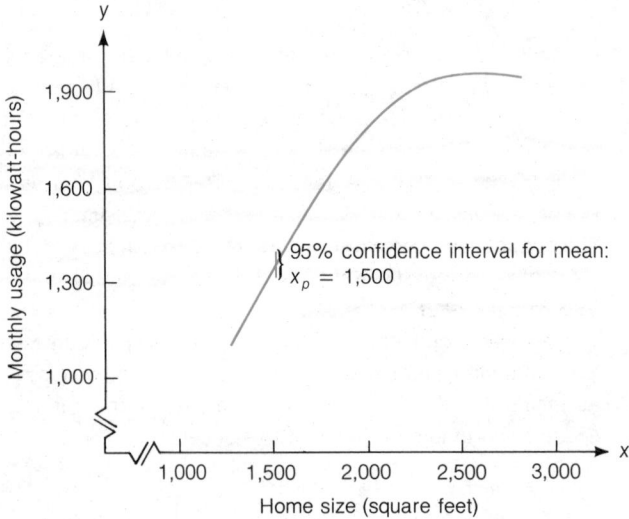

If we were interested in predicting the electrical usage for a specific 1,500-square-foot home, $\hat{y} = 1,369.7$ would be used as the predicted value. However, the prediction interval for a particular value of y is wider than the confidence interval for the mean value. This is reflected by the printout shown in Figure 13.14, which gives the predicted value of y and corresponding 95% prediction interval when $x_p = 1,500$ square feet. The prediction interval for $x_p = 1,500$ is 1,250.3 to 1,489.0 (see Figure 13.15).

Unfortunately, not all computer packages can produce confidence intervals for means and prediction intervals for specific y values. This is a rather serious oversight,

Figure 13.14 SAS
Printout for Predicted
Value and Corresponding
Prediction Interval for
$x_p = 1,500$

X	PREDICTED VALUE	LOWER 95% CL INDIVIDUAL	UPPER 95% CL INDIVIDUAL
1500	1369.66088739	1250.31627944	1489.00549533

Figure 13.15 Prediction Interval for Electrical Usage

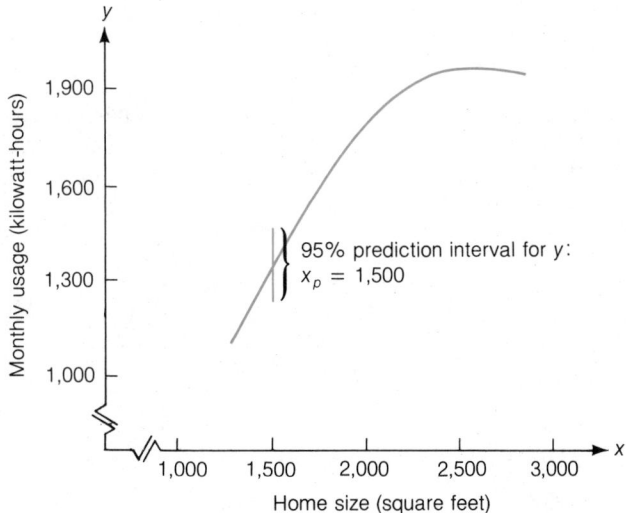

since the estimation of mean values and the prediction of specific values represent the culmination of our model-building efforts: using the model to make inferences about the dependent variable y.

Exercise 13.18

[*Note: Starred (*) exercises require the use of a computer.*]

Applying the Concepts

13.18 A physiologist wishes to investigate the relationship between the physical characteristics of preadolescent boys and their maximal oxygen uptake (measured in milliliters of oxygen per kilogram of body weight). The data shown in the table are collected on a random sample of ten preadolescent boys.

MAXIMAL OXYGEN UPTAKE y	AGE x_1 (years)	HEIGHT x_2 (centimeters)	WEIGHT x_3 (kilograms)	CHEST DEPTH x_4 (centimeters)
1.54	8.4	132.0	29.1	14.4
1.74	8.7	135.5	29.7	14.5
1.32	8.9	127.7	28.4	14.0
1.50	9.9	131.1	28.8	14.2
1.46	9.0	130.0	25.9	13.6
1.35	7.7	127.6	27.6	13.9
1.53	7.3	129.9	29.0	14.0
1.71	9.9	138.1	33.6	14.6
1.27	9.3	126.6	27.7	13.9
1.50	8.1	131.8	30.8	14.5

As a first step in the data analysis, the researcher fits the regression model

$$y = \beta_0 + \beta_1 x_1 + \beta_2 x_2 + \beta_3 x_3 + \beta_4 x_4 + \varepsilon$$

to the data. The output for a SAS regression analysis is shown here.

SOURCE	DF	SUM OF SQUARES	MEAN SQUARE	F VALUE	PR > F
MODEL	4	0.20206274	0.05051568	23.18	0.0020
ERROR	5	0.01089726	0.00217945		STD DEV
				R-SQUARE	0.04668461
CORRECTED TOTAL	9	0.21296000		0.948830	

PARAMETER	ESTIMATE	T FOR H0: PARAMETER = 0	PR > |T|	STD ERROR OF ESTIMATE
INTERCEPT	−3.42634621	−4.08	0.0096	0.84078488
AGE	−0.03999447	−1.94	0.1105	0.02065180
HEIGHT	0.04814593	6.34	0.0014	0.00759695
WEIGHT	0.00179116	0.33	0.7556	0.00544641
CHEST	−0.07709906	−0.92	0.3979	0.08343441

a. Give the value of R^2 and interpret its value.

b. It seems reasonable to assume that the greater a child's chest depth, the greater should be the maximal oxygen uptake. But note that $\hat{\beta}_4$, the estimated coefficient of chest depth, x_4, is negative. Give an explanation for this result.

c. It would seem that the weight of a child should be positively correlated to lung volume and hence to maximal oxygen uptake. Can you explain the small t-value associated with $\hat{\beta}_3$?

*d. If you have access to the appropriate computer package, find a 95% prediction interval for maximal oxygen uptake for a boy with Age = 8.8, Height = 128.1, Weight = 28.0, and Chest depth = 13.7.

13.7 Multiple Regression: An Example

In Case Study 13.1, we described an application of regression analysis—its use by Towers, Perrin, Forster & Crosby (TPF&C) to develop a prediction equation for corporate executive compensation. We will now use this interesting application of multiple regression to demonstrate all the ideas we have introduced in this chapter. Suppose the list of independent variables given in Table 13.3 is to be used to build a model for the salaries of corporate executives.

Step 1. The first step is to hypothesize a model relating executive salary to the independent variables listed in Table 13.3. TPF&C have found that executive compensation models that use the logarithm of salary as the dependent variable provide better predictive power than those using the salary as the dependent variable. This is prob-

Table 13.3

List of Independent
Variables for
Executive Compensation
Example

INDEPENDENT VARIABLE	DESCRIPTION
x_1	Years of experience
x_2	Years of education
x_3	1 if male; 0 if female
x_4	Number of employees supervised
x_5	Corporate assets (millions of dollars)
x_6	x_1^2
x_7	$x_3 x_4$

ably because salaries tend to be incremented in *percentages* rather than dollar values. When a response variable undergoes percentage changes as the independent variables are varied, the logarithm of the response variable will be more suitable as a dependent variable. The model we propose is

$$y = \beta_0 + \beta_1 x_1 + \beta_2 x_2 + \beta_3 x_3 + \beta_4 x_4 + \beta_5 x_5 + \beta_6 x_6 + \beta_7 x_7 + \varepsilon$$

where $y = \log(\text{Executive salary})$, $x_6 = x_1^2$ (quadratic term in years of experience), and $x_7 = x_3 x_4$ (cross product or interaction term between sex and number of employees supervised). The variable x_3 is a **dummy variable;** it is used to describe an independent variable that is not measured on a numerical scale but instead is qualitative (categorical) in nature. Sex is such a variable, since its values, male and female, are categories rather than numbers. Thus, we assign the value $x_3 = 1$ if the executive is male and $x_3 = 0$ if the executive is female. For more detail on the use and interpretation of dummy variables, see Chapter 14. The interaction term $x_3 x_4$ accounts for the fact that the relationship between corporate salary and the number of employees supervised, x_4, is dependent on sex, x_3. That is, as the number of supervised employees increases, a woman's salary (with all other factors being equal) might rise more (or less) rapidly than a man's. This concept (interaction) too is explained in more detail in Chapter 14.

Step 2. Now we estimate the model coefficients $\beta_0, \beta_1, \ldots, \beta_7$. Suppose that a sample of 100 executives is selected, and the variables y and $x_1, x_2, \ldots, x_7$ are recorded (or, in the case of x_6 and x_7, calculated). The sample is then used as input for the SAS regression routine; the output is shown in Figure 13.16 (page 668). The least squares model is

$$\hat{y} = 8.88 + .045x_1 + .033x_2 + .119x_3 + .00033x_4 + .0020x_5 - .00072x_6 + .00031x_7$$

Step 3. The next step is to specify the probability distribution of ε, the random error component. We assume that ε is normally distributed with a mean of zero and a constant variance σ^2. Furthermore, we assume that the errors are independent. The estimate of the variance σ^2 is given in the SAS printout as

$$s^2 = \text{MSE} = \frac{\text{SSE}}{n - (k + 1)} = \frac{\text{SSE}}{100 - (7 + 1)} = 0.0021$$

Step 4. We now want to see how well the model predicts salaries. First, note that $R^2 = .993$. This implies that 99.3% of the variation in y (the logarithm of salaries) for

Figure 13.16 SAS Printout for Executive Compensation Example

SOURCE	DF	SUM OF SQUARES	MEAN SQUARE	F VALUE	PR > F
MODEL	7	27.06425564	3.85632223	1819.30	0.0001
ERROR	92	0.19551523	0.00212517		STD DEV
CORRECTED TOTAL	99	27.25977087		R-SQUARE	0.0460995
				0.992828	

PARAMETER	ESTIMATE	T FOR H0: PARAMETER = 0	PR > |T|	STD ERROR OF ESTIMATE
INTERCEPT	8.87878688	192.49	0.0001	0.04612667
X1 (EXPERIENCE)	0.04460301	26.83	0.0001	0.00166257
X2 (EDUCATION)	0.03326230	12.31	0.0001	0.00270306
X3 (SEX)	0.11892473	6.89	0.0001	0.01724977
X4 (EMPLOYEES SUPERVISED)	0.00033216	19.97	0.0001	0.00001.664
X5 (ASSETS)	0.00201021	73.25	0.0001	0.00002744
X6 (= X1 * X1)	−0.00071702	−15.11	0.0001	0.00004746
X7 (= X3 * X4)	0.00031244	16.16	0.0001	0.00001933

these 100 sampled executives is accounted for by the model. The significance of this can be tested:

H_0: $\beta_1 = \beta_2 = \cdots = \beta_7 = 0$

H_a: At least one of the model coefficients is nonzero

Test statistic: $F = \dfrac{R^2/k}{(1 - R^2)/[n - (k + 1)]}$

Rejection region: For $\alpha = .05$, $F > F_{.05}$

where from Table VIII of the Appendix the tabulated value of F for $\alpha = .05$ and based on $k = 7$ and $n - (k + 1) = 92$ df is $F_{.05} \approx 2.1$. The test statistic is given on the SAS printout. Since $F = 1,819.3$ exceeds the tabulated value of F, we conclude that the model does contribute information for predicting executive salaries. It appears that at least one of the β-parameters in the model differs from zero.

We may be particularly interested in whether these data provide evidence that the mean salaries of executives increase as the asset value of the company increases when all other variables (experience, education, etc.) are held constant. In other words, we may want to know whether the data provide sufficient evidence to show that $\beta_5 > 0$. We use the following test:

H_0: $\beta_5 = 0$ H_a: $\beta_5 > 0$

Test statistic: $t = \dfrac{\hat{\beta}_5}{s_{\hat{\beta}_5}}$

For $\alpha = .05$, $n = 100$, $k = 7$, and $n - (k + 1) = 92$, we will reject H_0 if $t > t_{.05}$, where (because the number of degrees of freedom, 92, of t is so large) $t_{.05} \approx z_{.05} = 1.645$.

Thus, we reject H_0 if

$$t > 1.645$$

Table 13.4

Values of Independent
Variables for an Executive

$x_1 = 12$ years of
  experience
$x_2 = 16$ years of
  education
$x_3 = 0$ (female)
$x_4 = 400$ employees
  supervised
$x_5 = \$160.1$ million (the
  firm's asset value)
$x_6 = x_1^2 = 144$
$x_7 = x_3 x_4 = 0$

The t-value is indicated in Figure 13.16. The value corresponding to the independent variable x_5 is 73.25. Since this value exceeds 1.645, we find evidence that the mean salaries of executives do increase as the firm's assets increase.

Step 5. The culmination of the modeling effort is the use of the model for estimation and/or prediction. Suppose a firm is trying to determine fair compensation for an executive with the characteristics shown in Table 13.4. The least squares model can be used to obtain a predicted value for the logarithm of salary. That is,

$$\hat{y} = \hat{\beta}_0 + \hat{\beta}_1(12) + \hat{\beta}_2(16) + \hat{\beta}_3(0) + \hat{\beta}_4(400) + \hat{\beta}_5(160.1) + \hat{\beta}_6(144) + \hat{\beta}_7(0)$$

This predicted value, $\hat{y} = 10.298$, is given in Figure 13.17, a partial reproduction of the SAS regression printout for this problem. The 95% prediction interval is also given: from 10.203 to 10.392. To predict the salary of an executive with these characteristics we take the antilog of these values. That is, the predicted salary is $e^{10.298} = \$29,700$ (rounded to the nearest hundred) and the 95% prediction interval is from $e^{10.203}$ to $e^{10.392}$ (from \$27,000 to \$32,600). Thus, an executive with the characteristics in Table 13.4 should be paid between \$27,000 and \$32,600 to be consistent with the sample data.

Figure 13.17 SAS Printout for Executive Compensation Problem

X1	X2	X3	X4	X5	X6	X7	PREDICTED VALUE	LOWER 95% CL INDIVIDUAL	UPPER 95% CL INDIVIDUAL
12	16	0	400	160.1	144	0	10.29766682	10.20298295	10.39235070

13.8 Statistical Computer Programs

There are a number of different statistical program packages; some of the most popular are Biomed, Minitab, SAS, and SPSS. (See the references at the end of the chapter.) Some can be used on all large computers produced by a specific manufacturer. Consequently, you may have access to one or more of these packages at your computer center.

The multiple regression computer programs for these packages may differ in what they are programmed to do, how they do it, and the appearance of their computer printouts, but all of them print the basic outputs needed for a regression analysis. Some will compute confidence intervals for $E(y)$ and prediction intervals for y; others will not. Some test the null hypotheses that the individual β-parameters equal zero using Student's t-tests; others use F-tests.* But all give the least squares estimates, the values of SSE, s^2, etc.

To illustrate, the Minitab, SAS, and SPSS regression analysis computer printouts for Example 13.3 are shown in Figure 13.18 (page 670). For that example, we fit the model

$$y = \beta_0 + \beta_1 x_1 + \beta_2 x_2 + \beta_3 x_1 x_2 + \varepsilon$$

* A two-tailed Student's t-test based on v degrees of freedom is equivalent to an F-test where the F-statistic possesses 1 degree of freedom in the numerator and v degrees of freedom in the denominator. See Section 13.4.

Figure 13.18   Computer Printouts for Example 13.3

A. Minitab Regression Printout

THE REGRESSION EQUATION IS
Y =   323. + 0.873 X1 − 93.4 X2
    + 1.30 X3

	COLUMN	COEFFICIENT	ST. DEV. OF COEF.	T-RATIO = COEF/S.D.
	—	323.	293.	1.10
X1	C1	0.87	2.02	0.43
X2	C2	−93.4	29.7	−3.14
X3	C4	1.298	0.211	6.15

THE ST. DEV. OF Y ABOUT REGRESSION LINE IS
S =   88.4
WITH (32 − 4) = 28 DEGREES OF FREEDOM

R-SQUARED = 95.4 PERCENT
R-SQUARED = 94.9 PERCENT, ADJUSTED FOR D.F.

ANALYSIS OF VARIANCE

DUE TO	DF	SS	MS = SS/DF
REGRESSION	3	4572524.	1524174.
RESIDUAL	28	218645.	7809.
TOTAL	31	4791168.	

B. SAS Regression Printout

DEPENDENT VARIABLE: Y   AUCTION PRICE

SOURCE	DF	SUM OF SQUARES	MEAN SQUARE	F VALUE	PR > F	R-SQUARE	C.V.
MODEL	3	4572547.98717668	1524182.66239223	195.19	0.0001	0.954365	6.6584
ERROR	28	218646.23157332	7808.79398476		STD DEV		Y MEAN
CORRECTED TOTAL	31	4791194.21875000			88.36738077		1327.15625000

PARAMETER	ESTIMATE	T FOR H0: PARAMETER = 0	PR > |T|	STD ERROR OF ESTIMATE
INTERCEPT	322.75435309	1.10	0.2806	293.32514660
X1	0.87328775	0.43	0.6688	2.01965115
X2	−93.40991991	−3.14	0.0039	29.70767946
X1 * X2	1.29789828	6.15	0.0001	0.21102602

C. SPSS Regression Printout

DEPENDENT VARIABLE . .   Y   AUCTION PRICE

VARIABLE(S) ENTERED ON STEP NUMBER   1 . .   X1   AGE
                                      X2   NUMBER OF BIDDERS
                                      X3   AGE*BIDDERS

MULTIPLE R	0.97692	ANALYSIS OF VARIANCE	DF	SUM OF SQUARES	MEAN SQUARE	F
R SQUARE	0.95436	REGRESSION	3.	4572547.98718	1524182.66239	195.18797
ADJUSTED R SQUARE	0.94948	RESIDUAL	28.	218646.23157	7808.79398	
STANDARD ERROR	88.36738					

———————— VARIABLES IN THE EQUATION ————————

VARIABLE	B	BETA	STD ERROR B	F
X1	0.8732878	0.06085	2.01965	0.187
X2	−93.40992	−0.67471	29.70768	9.887
X3	1.297898	1.37032	0.21103	37.828
(CONSTANT)	322.7544			

to $n = 32$ data points. The variables in the model are

y = Auction price

x_1 = Age of clock (years)

x_2 = Number of bidders

Notice that the Minitab printout gives the prediction equation at the top of the printout. The independent variables, shown in the prediction equation and listed at the left side of the printout, are x_1, x_2, and x_3. Thus, Minitab treats the product $x_1 x_2$ as a third independent variable, x_3, which must be computed before the fitting commences. For this reason, the Minitab prediction equation always appears on the printout as first-order even though some of the independent variables shown in the prediction equation may actually be the squares or cross products of other independent variables. The inclusion of the squares or cross products of independent variables is treated in the same manner in the SPSS program shown in Figure 13.18. The SAS program is the only one of these three that can be automatically instructed to include these terms, and they appear in the printout with a star (*) that indicates multiplication. Thus, in the SAS printout $x_1 x_2$ is printed as X1*X2.

The estimates of the regression coefficients appear opposite the identifying variable in the Minitab column titled COEFFICIENT, in the SAS column titled ESTIMATE, and in the SPSS column titled B. Compare the estimates given in these three columns. Note that the Minitab printout gives the estimates with a much lesser degree of accuracy (fewer decimal places) than the SAS and SPSS. (Ignore the column titled BETA in the SPSS printout. These are standardized estimates and will not be discussed in this text.)

The standard errors of the estimates are given in the Minitab column titled ST. DEV. OF COEF., in the SAS column titled STD ERROR OF ESTIMATE, and in the SPSS column titled STD ERROR B.

The values of the test statistics for testing H_0: $\beta_i = 0$, where $i = 1, 2, 3$, are shown in the Minitab column titled T-RATIO = COEF/S.D. and in the SAS column titled T FOR H0: PARAMETER = 0. Note that the computed t-values shown in the Minitab and SAS columns are identical (except for the number of decimal places) and that Minitab does not give the observed significance level of the test. Consequently, to draw conclusions from the Minitab printout you must compare the computed values of t with the critical values given in a t-table (Table V in the Appendix). In contrast, the SAS printout gives the observed significance level for each t-test in the column titled PR > |T|. Note that these observed significance levels have been computed assuming that the tests are two-tailed. The observed significance levels for one-tailed tests would equal one-half of these values.

The SPSS conducts tests of H_0: $\beta_i = 0$, where $i = 1, 2, 3$, using the F-statistic. The computed F-values, which are appropriate only for two-tailed tests of the null hypothesis H_0: $\beta_i = 0$, are shown in the SPSS printout in the column titled F. The observed significance level is not given, so you must compare these values with those shown in the F-table for numerator degrees of freedom $v_1 = 1$ and the denominator degrees of freedom equal to the number v_2 that is attached to the SSE. For this example, $v_2 = 28$.

The Minitab printout gives the value SSE $= 218645$ under the ANALYSIS OF VARIANCE column headed SS and in the row identified as RESIDUAL. The value of $s^2 = 7,809$ is shown in the same row under the column headed MS $=$ SS/DF, and the degrees of freedom DF appears in the same row as 28. The corresponding values are shown at the top of the SAS printout in the row labeled ERROR and in the columns designated as SUM OF SQUARES, MEAN SQUARE, and DF, respectively. These quantities appear with the same headings in the SPSS printout.

The value of R^2, as defined in Section 13.5, is given in the Minitab printout as 95.4 PERCENT. (We defined this quantity as a ratio where $0 \le R^2 \le 1$.) It is given in the top right corner of the SAS printout as 0.954365, and it is shown in the left column of the SPSS printout as 0.95436. (Ignore the quantities shown in the Minitab as R^2 ADJUSTED FOR D.F. and in the SPSS printout as ADJUSTED R SQUARE. These quantities are adjusted for the degrees of freedom associated with the total SS and SSE and are not used or discussed in this text.)

The F-statistic for testing the utility of the model (Section 13.5), i.e., testing the null hypothesis that all model parameters (except β_0) equal zero, is shown under the title, F VALUE, as 195.19 in the top center of the SAS printout. Moreover, the SAS printout gives the observed significance level of this F-test under PR $>$ F as 0.0001. This F-value, 195.18797, is also printed at the right side of the SPSS printout, but no observed significance level is given. Thus, if you are using the SPSS regression analysis package you must compare the printed F-value with those tabulated in F-tables (Tables VII, VIII, IX, and X in the Appendix). The F-statistic for testing the utility of the model is not given in the Minitab printout. If you are using Minitab and wish to obtain the value of this statistic, you must compute it by using the formula given in Section 13.5:

$$F = \frac{R^2/k}{(1 - R^2)/[n - (k + 1)]}$$

(The value of R^2 is given in the Minitab printout; $n = 32$ and $k = 3$.) You can also compute the value of the F-statistic directly from the mean square entries given in the ANALYSIS OF VARIANCE table. Thus, it can be shown (proof omitted) that the F-statistic for testing

$$H_0: \quad \beta_1 = \beta_2 = \cdots = \beta_k = 0$$

is

$$F = \frac{\text{Mean square for regression}}{\text{Mean square for error (or residuals)}} = \frac{\text{Mean square for regression}}{s^2}$$

These quantities are given in the Minitab printout under the column marked MS $=$ SS/DF. Thus,

$$F = \frac{1,524,174}{7,809} = 195.18$$

a value that agrees with the values given in the SAS and SPSS printouts. The logic behind this test and other tests of hypotheses concerning sets of the β-parameters is presented in Section 14.4.

We will not comment on the merits or demerits of the various packages because you will have to use the package available at your computer center and become familiar with its output. Most of the computer printouts are similar, and it is relatively easy to learn how to read one output after you have become familiar with another. We have used different packages in the solution of the examples to help you with this problem.

13.9
Some Pitfalls: Estimability, Multi-collinearity, and Extrapolation

There are several problems you should be aware of when constructing a prediction model for some response y. A few of the most important are discussed in this section.

Problem 1
Parameter Estimability

Suppose you want to fit a model relating annual yield y to the total expenditure for fertilizer, x. We propose the first-order model

$$E(y) = \beta_0 + \beta_1 x$$

Now suppose we have 3 years of data and $1,000 is spent on fertilizer each year. The data are shown in Figure 13.19. You can see the problem: The parameters of the model cannot be estimated when all the data are concentrated at a single x-value. Recall that it takes two points (x-values) to fit a straight line. Thus, the parameters are not estimable when only one x-value is observed.

Figure 13.19 Yield and Fertilizer Expenditure Data: 3 Years

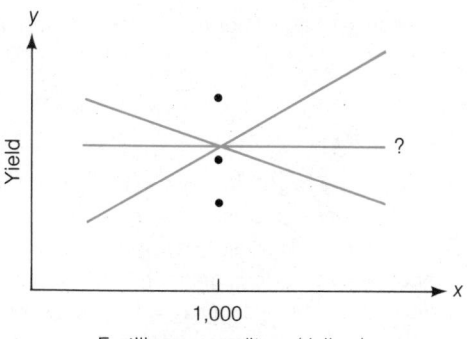

Fertilizer expenditure (dollars)

A similar problem would occur if we attempted to fit the quadratic model

$$E(y) = \beta_0 + \beta_1 x + \beta_2 x^2$$

to a set of data for which only one or two different x-values were observed (see Figure 13.20, page 674). At least three different x-values must be observed before a quadratic model can be fit to a set of data (that is, before all three parameters are estimable).

Figure 13.20 Only Two x-Values Observed— the Quadratic Model Is Not Estimable

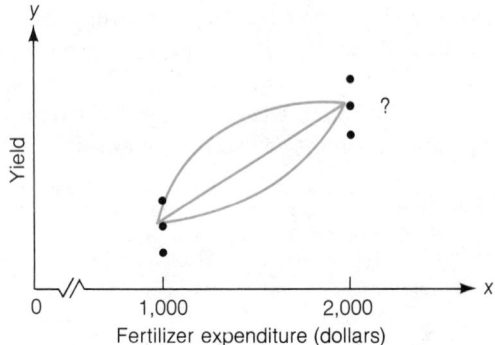

In general, the number of levels of observed x-values must be one more than the order of the polynomial in x that you want to fit.

For controlled experiments, the researcher can select experimental designs that will permit estimation of model parameters. Even when the values of the independent variables cannot be controlled by the researcher, the independent variables are almost always observed at a sufficient number of levels to permit estimation of the model parameters. When the computer program you use suddenly refuses to fit a model, however, the problem may be inestimable parameters.

Problem 2
Multicollinearity

Often, two or more of the independent variables used in the model for $E(y)$ contribute redundant information. That is, the independent variables are correlated with each other. Suppose we want to construct a model to predict the gasoline mileage rating of a truck as a function of its load, x_1, and the horsepower, x_2, of its engine. In general, you would expect heavy loads to require greater horsepower and to result in lower mileage ratings. Thus, although both x_1 and x_2 contribute information for the prediction of mileage rating, some of the information is overlapping because x_1 and x_2 are correlated.

If the model

$$E(y) = \beta_0 + \beta_1 x_1 + \beta_2 x_2$$

were fit to a set of data, we might find that the t-values for both $\hat{\beta}_1$ and $\hat{\beta}_2$ (the least squares estimates) are nonsignificant. However, the F-test for H_0: $\beta_1 = \beta_2 = 0$ would probably be highly significant. The tests may seem to produce contradictory conclusions, but really they do not. The t-tests indicate that the contribution of one variable, say x_1 = Load, is not significant after the effect of x_2 = Horsepower has been taken into account (because x_2 is also in the model). The significant F-test, on the other hand, tells us that at least one of the two variables is making a contribution to the prediction of y (i.e., either β_1, β_2, or both differ from zero). In fact, both are probably contributing, but the contribution of one overlaps with that of the other.

When highly correlated independent variables are present in a regression model, the results are confusing. The researcher may want to include only one of the variables in the final model. One way of deciding which one to include is by using stepwise regression. Generally, only one of a set of multicollinear independent variables is included in a stepwise regression model, since at each step every variable is tested in the presence of all the variables already in the model. For example, if at one step the variable truck load is included as a significant variable in the prediction of the mileage rating, the variable horsepower will probably never be added in a future step. Thus, if a set of independent variables is thought to be multicollinear, some screening by stepwise regression may be helpful.

Note that it would be fallacious to conclude that an independent variable x_1 is unimportant for predicting y *only* because it is not chosen by a stepwise regression procedure. The independent variable x_1 may be correlated with another one, x_2, that the stepwise procedure did select. The implication is that x_2 contributes *more* for predicting y (in the sample being analyzed), but it may still be true that x_1 alone contributes information for the prediction of y.

Problem 3

Prediction Outside the Experimental Region

By the late 1960s many research economists had developed highly technical models to relate the state of the economy to various economic indices and other independent variables. Many of these models were multiple regression models, where, for example, the dependent variable y might be next year's growth in GNP and the independent variables might include this year's rate of inflation, this year's CPI, etc. In other words, the model might be constructed to predict next year's economy using this year's knowledge.

Unfortunately, these models were almost all unsuccessful in predicting the recession in the early 1970s. What went wrong? One of the problems was that many of the regression models were used to predict y for values of the independent variables that were outside the region in which the model was developed. For example, the inflation rate in the late 1960s, when the models were developed, ranged from 6 to 8%. When the double-digit inflation of the early 1970s became a reality, some researchers attempted to use the same models to predict the growth in GNP 1 year hence. As you can see in Figure 13.21, the model may be very accurate for predicting y when x is in the range of experimentation, but use of the model outside that range is a dangerous practice.

Figure 13.21 Using a Regression Model Outside the Experimental Region

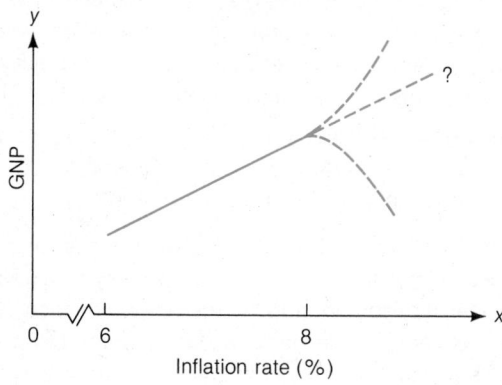

Problem 4
Correlated
Errors

Another problem associated with using a regression model to predict a variable y based on independent variables $x_1, x_2, \ldots, x_k$ arises from the fact that the data are frequently **time series.** That is, the values of both the dependent and independent variables are observed sequentially over a period of time. The observations tend to be correlated over time, which in turn often causes the prediction errors of the regression model to be correlated. Thus, the assumption of independent errors is violated, and the model tests and prediction intervals are no longer valid. The solution to this problem is to construct a **time series model;** the interested reader should consult the references for details of time series analysis.

Case Study 13.2
An Apparent
Inconsistency:
The Culprit,
Multicollinearity

Studying the relationship between worker strike activity and union membership, Neil A. Palomba (1969) encountered a common problem in modeling. Palomba attempted to relate the strike activity in a state (the percentage y of total working hours lost due to strikes) to three independent variables:

x_1 = Percentage of union members in nonagricultural establishments

x_2 = Percentage of all nonagricultural employment that is manufacturing

x_3 = Hourly earnings of workers on manufacturing payrolls

Fitting a series of first-order models involving x_1, x_2, and x_3 to y he got conflicting signals concerning the importance of these independent variables in predicting strike activity.

The first step in Palomba's modeling process, relating x_1 to y by using a simple linear regression model, produced the prediction equation

$$\hat{y} = .018201 + .005927x_1 \qquad \text{with } R^2 = .137 \qquad (1)$$
$$(.002167)$$
$$2.735$$

The standard error of $\hat{\beta}_1$, .002167, is shown in parentheses below $\hat{\beta}_1$ and the t-value for testing H_0: $\beta_1 = 0$, 2.735, is shown beneath the standard error. Since this t-value is highly significant, Palomba says it "indicates that union membership has a statistically significant positive effect on the level of strike activity."

Does Palomba's statement imply a cause and effect relationship between x_1 and y? If this is the suggestion, then a problem quickly develops when he fits a first-order model to the data using all three independent variables. The prediction equation, with the standard errors (in parentheses) and computed t-values for the parameter estimates, is

$$\hat{y} = -.040205 + .000308x_1 + .003905x_2 + .184774x_3 \qquad \text{with } R^2 = .232$$
$$(.003179) \qquad (.002260) \qquad (.081428)$$
$$.097 \qquad 1.728 \qquad 2.269$$

As Palomba observes, this equation "reveals some surprising results." He notes that R^2 increases dramatically (although it is still very small). But above all he notes that union membership, so important in prediction equation (1), is no longer statistically significant. Neither is x_2, the measure of the labor force involved in manufacturing. The statistically significant variable in the prediction equation is hourly wage x_3, "meaning that the level of wages has a positive effect upon strike activity, while the level of union membership has no effect." Since this conclusion is the reverse of the conclusion reached in analyzing model (1), we wonder how many conflicting conclusions we could create if we fit many models and, more important, we might wonder why the conflicting conclusions occur.

Based on Palomba's analysis, which variable—union membership x_1 or hourly earnings x_3—causes strike activity y to increase? A regression analysis cannot answer this question. All that a regression analysis can do is indicate whether an independent variable, as it is entered into the model, contributes information for the prediction of y. It does not indicate a cause and effect relationship. In Palomba's modeling, it seems reasonable to suspect a relationship between union membership x_1 and hourly earnings x_3. Perhaps workers who are union members receive higher hourly earnings. In any case, it appears that x_1 and x_3 are correlated and therefore contribute overlapping information for the prediction of y. This common situation is called **multicollinearity.** When multicollinearity is present, a regression analysis introduces the variables into the model in a way that ensures the best prediction equation. If changes in one of the variables actually cause the strike activity y to change, it may or may not appear to be statistically significant. Some variable highly correlated to the causative variable may replace it in the prediction equation.

Summary

We have discussed some of the methodology of multiple regression analysis, a technique for modeling a dependent variable y as a function of several independent variables $x_1, x_2, \ldots, x_k$. The steps we follow in constructing and using multiple regression models are much the same as those for the simple straight-line models:

1. The form of the probabilistic model is hypothesized.
2. The model coefficients are estimated using least squares.
3. The probability distribution of ε is specified and σ^2 is estimated.
4. The utility of the model is checked.
5. If the model is deemed useful, it may be used to make estimates of $E(y)$ and to predict values of y to be observed in the future.

We have covered steps 2 to 5 in Chapter 13, assuming that the model was specified. Model construction—step 1—is discussed in Chapter 14. Additional material on these topics can be found in the references.

Supplementary Exercises 13.19–13.42

[*Note:* Starred (*) exercises require the use of a computer.]

Learning the Mechanics

13.19 Suppose you fit the model

$$y = \beta_0 + \beta_1 x_1 + \beta_2 x_1^2 + \beta_3 x_2 + \beta_4 x_1 x_2 + \varepsilon$$

to $n = 25$ data points and find that

$$\hat{\beta}_0 = 1.26 \qquad \hat{\beta}_1 = -2.43 \qquad \hat{\beta}_2 = .05 \qquad \hat{\beta}_3 = .62 \qquad \hat{\beta}_4 = 1.81$$

$$\text{SSE} = .41 \qquad R^2 = .83$$

$$s_{\hat{\beta}_1} = 1.21 \qquad s_{\hat{\beta}_2} = .16 \qquad s_{\hat{\beta}_3} = .26 \qquad s_{\hat{\beta}_4} = 1.49$$

a. Is there sufficient evidence to conclude that at least one of the parameters, β_1, β_2, β_3, or β_4, is nonzero? Test using $\alpha = .05$.

b. Test H_0: $\beta_1 = 0$ against H_a: $\beta_1 < 0$. Use $\alpha = .05$.

c. Test H_0: $\beta_2 = 0$ against H_a: $\beta_2 > 0$. Use $\alpha = .05$.

d. Test H_0: $\beta_3 = 0$ against H_a: $\beta_3 \neq 0$. Use $\alpha = .05$.

13.20 After a regression model is fit to a set of data, a confidence interval for the mean value of y at a given setting of the independent variables is *always* narrower than the corresponding prediction interval for a particular value of y at the same setting of the independent variables. Why?

13.21 Suppose you use Minitab to fit the model

$$y = \beta_0 + \beta_1 x_1 + \beta_2 x_2 + \varepsilon$$

to $n = 15$ data points and obtain the computer printout shown here.

```
THE REGRESSION EQUATION IS
Y =    90.1 −  1.84 X1 +  .285 X2

                                                    ST. DEV.        T-RATIO =
                 COLUMN          COEFFICIENT        OF COEF.        COEF/S.D.
                   —                   90.1            23.1          · 3.90
     X1          C2                  −1.836            .367            −5.01
     X2          C3                    .285            .231             1.24

THE ST. DEV. OF Y ABOUT REGRESSION LINE IS
S =       10.7
WITH (   15 − 3) =    12 DEGREES OF FREEDOM

R-SQUARED = 91.6 PERCENT
R-SQUARED = 90.2 PERCENT, ADJUSTED FOR D.F.

ANALYSIS OF VARIANCE

DUE TO          DF       SS      MS = SS/DF
REGRESSION       2     14801.        7400.
RESIDUAL        12      1364.         114.
TOTAL           14     16165.
```

a. What is the least squares prediction equation?

b. Find R^2 and interpret its value.

c. Is there sufficient evidence to indicate that the model is useful for predicting y? Conduct an F-test using $\alpha = .05$.

d. Test H_0: $\beta_1 = 0$ against H_a: $\beta_1 \neq 0$. Use $\alpha = .05$.

y	x_1	x_2
−12	3	9
28	8	7
−15	3	1
−25	2	6
2	9	1
−11	5	1
25	7	7
−47	0	8
−28	1	6
12	5	9
−29	1	4
−3	5	5
5	7	3
7	6	4
6	5	9

*13.22 A response variable y was observed for $n = 15$ settings of two independent variables, x_1 and x_2. The data are shown in the table.

a. Fit the model $y = \beta_0 + \beta_1 x_1 + \beta_2 x_2 + \varepsilon$ to the data.

b. Fit the model $y = \beta_0 + \beta_1 x_1 + \beta_2 x_2 + \beta_3 x_1 x_2 + \varepsilon$ to the data.

c. Which of the models in parts a and b best describes the relationship between y and x_1 and x_2? Explain.

d. Repeat part a using a different statistical program package. For example, if you originally used Minitab, now use SAS, SPSS, or another available package. Compare these results with those found in part a.

13.23 Suppose you have developed a regression model to explain the relationship between y and x_1, x_2, and x_3. A set of $n = 15$ data points is used to find the least squares prediction equation. The ranges of the variables you observed were as follows: $10 \leq y \leq 100$, $5 \leq x_1 \leq 55$, $.5 \leq x_2 \leq 1$, and $1,000 \leq x_3 \leq 2,000$. Will the error of prediction be smaller when you use the least squares equation to predict y when $x_1 = 30$, $x_2 = .6$, and $x_3 = 1,300$, or when $x_1 = 60$, $x_2 = .4$, and $x_3 = 900$? Why?

Applying the Concepts

13.24 A large government agency would like to predict the number of people it will hire within the next year to fill the thirty positions that are currently open. Historically, the agency has been unable to fill all its job openings. It has been decided to model the number of positions filled in a year, y, as a function of the number of positions open, x_1, and the recruiting budget (in dollars) for the year, x_2 (e.g., for advertising the positions, paying travel expenses, etc.). A random sample of 10 years of recruiting records was drawn from the agency's 30 years of records. The model

$$E(y) = \beta_0 + \beta_1 x_1 + \beta_2 x_2$$

was fit to these data using the Minitab regression computer program package. The results shown in the computer printout on the next page were obtained.

a. Identify the least squares prediction equation.

b. Is there sufficient evidence to indicate that the model contributes information for predicting the number y of positions that will be filled? Conduct an F-test using $\alpha = .05$.

c. Test the null hypothesis H_0: $\beta_2 = 0$ against the alternative hypothesis H_a: $\beta_2 \neq 0$ using $\alpha = .05$. Interpret the results of your test in the context of the problem.

d. Use the least squares prediction equation to predict the number of the thirty positions that the agency will fill next year if its recruiting budget is $10,000.

e. Can you think of a situation for which the prediction in part d might possibly be inaccurate? Explain.

f. Which (if any) of the assumptions we make about ε in regression analysis are likely to be violated in this problem? Explain.

THE REGRESSION EQUATION IS
$Y = .0562 + .273 X1 + .0006 X2$

	COLUMN	COEFFICIENT	ST. DEV. OF COEF.	T-RATIO = COEF/S.D.
	—	.056	.902	.06
X1	C2	.2733	.0971	2.81
X2	C3	.000560	.000129	4.34

THE ST. DEV. OF Y ABOUT REGRESSION LINE IS
$S = 1.33$
WITH ($10 - 3$) = 7 DEGREES OF FREEDOM

R-SQUARED = 97.9 PERCENT
R-SQUARED = 97.3 PERCENT, ADJUSTED FOR D.F.

ANALYSIS OF VARIANCE

DUE TO	DF	SS	MS = SS/DF
REGRESSION	2	583.18	291.59
RESIDUAL	7	12.42	1.77
TOTAL	9	595.60	

*13.25   The data set shown here gives the number y of births per year (in thousands), the number x_1 of marriages per year (in thousands), the number x_2 of women in the work force (in thousands), and the availability (yes) or nonavailability (no) of the birth control pill over the years 1949–1982.

a.   Use a computer to fit the model

$$E(y) = \beta_0 + \beta_1 x_1 + \beta_2 x_2 + \beta_3 x_3$$

YEAR	BIRTHS	MARRIAGES	WOMEN WORKING	TAKING THE PILL	YEAR	BIRTHS	MARRIAGES	WOMEN WORKING	TAKING THE PILL
1949	3,560	1,580	18,030	No	1966	3,606	1,857	26,820	Yes
1950	3,632	1,667	18,680	No	1967	3,521	1,927	27,545	Yes
1951	3,750	1,595	19,309	No	1968	3,502	2,069	28,778	Yes
1952	3,847	1,539	19,559	No	1969	3,606	2,145	29,898	Yes
1953	3,902	1,546	19,668	No	1970	3,731	2,159	31,233	Yes
1954	4,017	1,490	19,970	No	1971	3,556	2,190	31,778	Yes
1955	4,097	1,518	20,842	No	1972	3,258	2,282	33,152	Yes
1956	4,168	1,569	21,808	No	1973	3,137	2,284	34,195	Yes
1957	4,255	1,518	22,097	No	1974	3,160	2,230	35,708	Yes
1958	4,204	1,451	22,482	No	1975	3,144	2,153	36,981	Yes
1959	4,245	1,494	22,865	No	1976	3,168	2,155	38,399	Yes
1960	4,258	1,523	23,619	No	1977	3,327	2,178	40,053	Yes
1961	4,268	1,548	24,199	No	1978	3,333	2,282	41,747	Yes
1962	4,167	1,580	23,978	No	1979	3,494	2,331	43,844	Yes
1963	4,098	1,654	24,675	Yes	1980	3,598	2,413	44,934	Yes
1964	4,027	1,725	25,399	Yes	1981	3,646	2,438	46,415	Yes
1965	3,760	1,800	25,952	Yes	1982	3,704	2,495	47,095	Yes

Source: Statistical Abstract of the U.S.

to the data. The independent variable x_3 is a dummy (or indicator) variable defined as follows:

$$x_3 = \begin{cases} 1 & \text{if birth control pills were available in a given year} \\ 0 & \text{if not available} \end{cases}$$

Dummy variables enable us to enter qualitative independent variables into a model. Their use and interpretation are discussed in Chapter 14.

b. Interpret the results of your regression analysis.

13.26 Most companies institute rigorous programs to assure employee safety. Suppose sixty accident reports over the last year at a company are sampled, and the number of hours the employee had worked before the accident occurred, x, and the amount of time the employee lost from work, y, are recorded. A quadratic model is proposed to investigate a fatigue hypothesis that more serious accidents occur near the end of workdays than near the beginning. Thus, the proposed model is

$$E(y) = \beta_0 + \beta_1 x + \beta_2 x^2$$

A portion of the computer printout appears as shown.

SOURCE	DF	SUM OF SQUARES	MEAN SQUARE	F VALUE
MODEL	2	112.110	56.055	1.28
ERROR	57	2496.201	43.793	R-SQUARE
TOTAL	59	2608.311		0.0430

a. Do these data support the fatigue hypothesis? Use $\alpha = .05$ to test whether the proposed model is useful in predicting the lost work time y.

b. Does the result of the test in part a necessarily mean that no fatigue factor exists? Explain.

13.27 Refer to Exercise 13.26. Suppose the company persists in using the quadratic model despite its apparent lack of utility. The fitted model is

$$\hat{y} = 12.3 + .25x - .0033x^2$$

where $\hat{y}$ is the predicted time lost (days) and x is the number of hours worked prior to an accident.

a. Use the model to predict the number of days missed by an employee who has an accident after 6 hours of work.

b. Suppose the 95% prediction interval for the predicted value in part a is (1.35, 26.01). What is the interpretation of this interval? Does this interval support your conclusion about this model in Exercise 13.26?

*13.28 Refer to Exercise 12.26. The breeder of thoroughbred horses has been advised that the prediction model could probably be improved if a quadratic term were added. The following model is therefore proposed:

$$y = \beta_0 + \beta_1 x + \beta_2 x^2 + \varepsilon$$

where, as before,

y = Life span of horse (years)

x = Gestation period of horse (days)

a. Find the least squares prediction equation and test its adequacy. (Use a computer regression routine.)

b. Has the addition of the quadratic term contributed significantly to the prediction of a thoroughbred horse's life span? Test H_0: $\beta_2 = 0$ against the alternative H_a: $\beta_2 \neq 0$ using $\alpha = .05$.

13.29 To increase the motivation and productivity of workers, an electronics manufacturer decides to experiment with a new pay incentive structure at one of two plants. The experimental plan will be tried at plant A for 6 months, while workers at plant B will remain on the original pay plan. To evaluate the effectiveness of the new plan, the average assembly time for part of an electronic system is measured for employees at both plants at the beginning and end of the 6-month period. Suppose the following model is proposed:

$$y = \beta_0 + \beta_1 x_1 + \beta_2 x_2 + \varepsilon$$

where

y = Assembly time (hours) at end of 6-month period

x_1 = Assembly time (hours) at beginning of 6-month period

$$x_2 = \begin{cases} 1 & \text{if plant A} \\ 0 & \text{if plant B} \end{cases} \quad \text{(dummy variable)}$$

A sample of $n = 42$ observations yields

$$\hat{y} = .11 + .98x_1 - .53x_2$$

where

$$s_{\hat{\beta}_1} = .231 \qquad s_{\hat{\beta}_2} = .48$$

Test to determine whether, after allowing for the effect of initial assembly time, plant A had a lower mean assembly time than plant B. Use $\alpha = .01$. [*Note:* When the $(0, 1)$ coding is used to define a dummy variable, the coefficient of the variable represents the difference between the mean response at the two levels represented by the variable. Thus, the coefficient β_2 is the difference in mean assembly time between plant A and plant B at the end of the 6-month period and $\hat{\beta}_2$ is the sample estimator of that difference.]

13.30 The EPA wants to model the gas mileage ratings, y, of automobiles as a function of their engine size, x. A quadratic (second-order) model

$$y = \beta_0 + \beta_1 x + \beta_2 x^2 + \varepsilon$$

is proposed. A sample of fifty engines of varying sizes is selected and the miles per gallon rating of each is determined. The least squares prediction equation is

$$\hat{y} = 51.3 - 10.1x + .15x^2$$

The size x of the engine is measured in hundreds of cubic inches. Moreover, $s_{\hat{\beta}_2} = .0037$ and $R^2 = .93$.

a. Sketch the prediction equation for values of x between $x = 1$ and $x = 4$.

b. Is there evidence that the quadratic term in the model is contributing to the prediction of the miles per gallon rating, y? Use $\alpha = .05$.

c. Use the model to estimate the mean miles per gallon rating for all cars with 350 cubic inch engines ($x = 3.5$).

d. Suppose a 95% confidence interval for the quantity estimated in part c is (17.2, 18.4). Interpret this interval.

e. Suppose you purchase an automobile with a 350 cubic inch engine and determine that the miles per gallon rating is 14.7. Is the fact that this value lies outside the confidence interval given in part d surprising? Explain.

13.31 To determine whether extra personnel are needed for the day, the owners of a water adventure park would like to find a model that would allow them to predict the day's attendance each morning before opening based on the day of the week and weather conditions. The model is of the form

$$E(y) = \beta_0 + \beta_1 x_1 + \beta_2 x_2 + \beta_3 x_3$$

where

y = Daily admissions

$$x_1 = \begin{cases} 1 & \text{if weekend} \\ 0 & \text{otherwise} \end{cases} \quad \text{(dummy variable)}$$

$$x_2 = \begin{cases} 1 & \text{if sunny} \\ 0 & \text{if overcast} \end{cases} \quad \text{(dummy variable)}$$

x_3 = Predicted daily high temperature (°F)

These data were recorded for a random sample of 30 days and a regression model was fit to the data. The least squares analysis produced the following results:

$$\hat{y} = -105 + 25x_1 + 100x_2 + 10x_3$$

with

$$s_{\hat{\beta}_1} = 10 \qquad s_{\hat{\beta}_2} = 30 \qquad s_{\hat{\beta}_3} = 4 \qquad R^2 = .65$$

a. Interpret the estimated model coefficients.

b. Is there sufficient evidence to conclude that this model is useful in the prediction of daily attendance? Use $\alpha = .05$.

c. Is there sufficient evidence to conclude that mean attendance increases on weekends? Use $\alpha = .10$.

d. Use the model to predict the attendance on a sunny weekday with a predicted high temperature of 95°F.

e. Suppose the 90% prediction interval for part d is (645, 1,245). Interpret this interval.

13.32 Refer to Exercise 13.31. The owners of the water adventure park are advised that the prediction model could probably be improved if interaction terms were added. In particular, it is thought that the *rate* of increase in mean attendance with increases in predicted high temperature will be greater on weekends than on weekdays. The following model is therefore proposed:

$$E(y) = \beta_0 + \beta_1 x_1 + \beta_2 x_2 + \beta_3 x_3 + \beta_4 x_1 x_3$$

The same 30 days of data used in Exercise 13.31 are again used to obtain the least squares model

$$\hat{y} = 250 - 700x_1 + 100x_2 + 5x_3 + 15x_1 x_3$$
$$s_{\hat{\beta}_4} = 3.0 \qquad R^2 = .96$$

a. Graph the predicted day's attendance, y, against the day's predicted high temperature, x_3, for a sunny weekday and for a sunny weekend day. Graph both on the same paper for x_3 between 70°F and 100°F. Note the increase in slope for the weekend day.

b. Do the data indicate that the interaction term is a useful addition to the model? Use $\alpha = .05$.

c. Use this model to predict the attendance for a sunny weekday with a predicted high temperature of 95°F.

d. Suppose the 90% prediction interval for part c is (800, 850). Compare this result with the prediction interval for the model without interaction in Exercise 13.31, part e. Do the relative widths of the confidence intervals support or refute your conclusion about the utility of the interaction term (part b)?

13.33 Refer to Exercise 13.32. The owners, noting that the coefficient $\hat{\beta}_1 = -700$, conclude the model is ridiculous because it seems to imply that the mean attendance will be 700 less on weekends than on weekdays. Explain why this is *not* the case.

13.34 Many students must work part-time to help finance their college education. A survey of 100 students was completed at a university to see if the number of hours worked per week, x, was affecting their grade-point averages, y. A quadratic model was proposed:

$$y = \beta_0 + \beta_1 x + \beta_2 x^2 + \varepsilon$$

The 100 observations yielded the least squares model

$$\hat{y} = 2.8 - .005x - .0002x^2$$

with $R^2 = .12$.

a. Do these statistics indicate that the model is useful in predicting grade-point averages? Use $\alpha = .05$.

b. Interpret the value $R^2 = .12$. Do you think the relationship between y and x is strong? Would you expect to be able to predict grade-point averages precisely (narrow prediction interval) if you knew how many hours students work per week?

13.35 Recent increases in gasoline prices have increased interest in modes of transportation other than the automobile. A metropolitan bus company wants to know if changes in numbers of bus riders are related to changes in gasoline prices. By using information from company files and gasoline price information obtained from fuel distributors, the company plans to fit the following model:

$$y = \beta_0 + \beta_1 x_1 + \beta_2 x_2 + \beta_3 x_1 x_2 + \varepsilon$$

where

x_1 = Average wholesale price for regular gas in a given month

$$x_2 = \begin{cases} 1 & \text{if the bus travels a city route only} \\ 0 & \text{if the bus travels a suburb–city route} \end{cases}$$

y = Total number of riders in a bus over the month

a. For this model, how would you test to determine whether the relationship between the mean number of riders and gasoline price is different for the two different bus routes?

b. Suppose 12 months of data are kept, and the least squares model is

$$\hat{y} = 500 + 50x_1 + 5x_2 - 10x_1 x_2$$

Graph the predicted relationship between number of riders and gas price for city buses and for suburb–city buses. Compare the slopes.

c. If $s_{\hat{\beta}_3} = 3.0$, do the data indicate that gas price affects the number of riders differently for city and suburb–city buses? Use $\alpha = .05$.

13.36 During the winter months a sample of 100 homes is taken to obtain information concerning the relationship between kilowatt usage, y, and total window and glass area, x (measured as a percentage of total wall area). The correlation coefficient for the data is equal to .24. Test to determine whether the data indicate that a linear relationship exists between y and x. Use $\alpha = .05$. [*Hint:* Use the methods of Section 13.5.]

13.37 Several states now require all high school seniors to pass a proficiency examination before they can graduate. On the test the seniors must demonstrate their familiarity with basic verbal and mathematical skills. Suppose an educational researcher wants to model the score y on such a proficiency examination as a function of the student's IQ, x_1, and socioeconomic status (SES). The SES is a categorical (or *qualitative*) variable with three levels: low, medium, and high. As we will demonstrate in Chapter 14, two dummy (indicator) variables are needed to describe a qualitative independent variable with three levels. Thus we define

$$x_2 = \begin{cases} 1 & \text{if SES is medium} \\ 0 & \text{if SES is low or high} \end{cases} \qquad x_3 = \begin{cases} 1 & \text{if SES is high} \\ 0 & \text{if SES is low or medium} \end{cases}$$

Data are collected for a random sample of sixty seniors who have taken the examination and the model

$$E(y) = \beta_0 + \beta_1 x_1 + \beta_2 x_2 + \beta_3 x_3$$

is fit to these data, with the results shown in the printout on the next page.

SOURCE	DF	SUM OF SQUARES	MEAN SQUARE	F VALUE	PR > F
MODEL	3	12268.56439492	4089.52146497	188.33	0.0001
ERROR	56	1216.01893841	21.71462390		STD DEV
CORRECTED TOTAL	59	13484.58333333		R-SQUARE	4.65989527
				0.909822	

PARAMETER	ESTIMATE	T FOR H0: PARAMETER = 0	PR > |T|	STD ERROR OF ESTIMATE
INTERCEPT	−13.06166081	−3.21	0.0022	4.07101383
X1	0.74193946	17.56	0.0001	0.04224805
X2	18.60320572	12.49	0.0001	1.48895324
X3	13.40965415	8.97	0.0001	1.49417069

a. Identify the least squares equation.

b. Interpret the value of R^2 and test to determine whether the data provide sufficient evidence to indicate that this model is useful for predicting proficiency examination scores.

c. Sketch the relationship between predicted proficiency examination score and IQ for the three levels of SES. [*Note:*   Three graphs of $\hat{y}$ versus x_1 must be drawn: the first for the low-SES model ($x_2 = x_3 = 0$), the second for the medium-SES model ($x_2 = 1$, $x_3 = 0$), and the third for the high-SES model ($x_2 = 0$, $x_3 = 1$). The increase in predicted proficiency examination score per unit increase in IQ is the same for all three levels of SES; i.e., all three lines are parallel.]

13.38 Refer to Exercise 13.37. We now use the same data to fit the model

$$E(y) = \beta_0 + \beta_1 x_1 + \beta_2 x_2 + \beta_3 x_3 + \beta_4 x_1 x_2 + \beta_5 x_1 x_3$$

Thus, we now add the interaction between IQ and SES to the model. The SAS printout for this model is shown at the top of the next page.

a. Identify the least squares prediction equation.

b. Interpret the value of R^2 and test to determine whether the data provide sufficient evidence to indicate that this model is useful for predicting proficiency examination scores.

c. Sketch the relationship between predicted proficiency examination score and IQ for the three levels of SES. [*Note:*   The interaction terms in the model allow nonparallelism among the three SES models; i.e., the mean increase in proficiency examination score per unit increase in IQ differs for the three levels of SES. To determine whether the interaction between IQ and SES is contributing to the prediction of proficiency examination score, we must test the null hypothesis H_0: $\beta_4 = \beta_5 = 0$ against the alternative that at least one of the coefficients of the interaction terms is nonzero. The method for testing portions of a regression model involving more than one β-parameter (but fewer than all of them) is discussed in Chapter 14.]

SOURCE	DF	SUM OF SQUARES	MEAN SQUARE	F VALUE	PR > F
MODEL	5	12515.10021009	2503.02004202	139.42	0.0001
ERROR	54	969.48312324	17.95339117		STD DEV
CORRECTED TOTAL	59	13484.58333333		R-SQUARE	4.23714422
				0.928104	

PARAMETER	ESTIMATE	T FOR H0: PARAMETER = 0	PR > |T|	STD ERROR OF ESTIMATE
INTERCEPT	0.60129643	0.11	0.9096	5.26818519
X1	0.59526252	10.70	0.0001	0.05563379
X2	−3.72536406	−0.37	0.7115	10.01967496
X3	−16.23196444	−1.90	0.0631	8.55429931
X1 * X2	0.23492147	2.29	0.0260	0.10263908
X1 * X3	0.30807756	3.53	0.0009	0.08739554

13.39 Plastics made under different environmental conditions are known to have differing strengths. A scientist would like to know which combination of temperature and pressure yields a plastic with a high breaking strength. A small preliminary experiment is run at two pressure levels and two temperature levels. The following model is proposed:

$$E(y) = \beta_0 + \beta_1 x_1 + \beta_2 x_2 + \beta_3 x_1 x_2$$

where

y = Breaking strength (pounds)

x_1 = Temperature (°F)

x_2 = Pressure (pounds per square inch)

A sample of $n = 16$ observations yields

$$\hat{y} = 226.8 + 4.9x_1 + 1.2x_2 - .7x_1 x_2$$

with

$$s_{\hat{\beta}_1} = 1.11 \qquad s_{\hat{\beta}_2} = .27 \qquad s_{\hat{\beta}_3} = .34$$

Do the data indicate an interaction between temperature and pressure? Test using $\alpha = .05$.

*13.40 A florist is interested in how sunlight and misting affect the thickness of the leaves on a new variety of ivy. Sixteen ivy plants of the same age and appearance are randomly divided into four groups of four plants each. One group receives no misting and indirect sunlight; one, misting and indirect sunlight; one, no misting and direct sunlight; and the last, misting and direct sunlight. After 3 months, the thickness of a center leaf from each plant is measured with the results (in millimeters) listed in the table on the next page.

		SUNLIGHT	
		Direct	*Indirect*
MISTING	*No mist*	.64, .63, .61, .62	.60, .61, .58, .59
	Mist	.65, .63, .64, .64	.62, .61, .60, .60

a. Fit the model $y = \beta_0 + \beta_1 x_1 + \beta_2 x_2 + \beta_3 x_1 x_2 + \varepsilon$ to the data, where

$$x_1 = \begin{cases} 1 & \text{if mist was present} \\ 0 & \text{if no mist} \end{cases} \qquad x_2 = \begin{cases} 1 & \text{if direct sunlight} \\ 0 & \text{if indirect sunlight} \end{cases}$$

b. Is there sufficient evidence to conclude the model is useful in predicting thickness of leaves on this variety of ivy? Use $\alpha = .05$.

c. Is there evidence of an interaction between sunlight conditions and misting conditions? Test using $\alpha = .05$.

*13.41 Many colleges and universities develop regression models for predicting the grade-point average (GPA) of incoming freshmen. This predicted GPA can then be used to make admission decisions. Although most models use many independent variables to predict GPA, we will illustrate by choosing two variables:

$x_1 = $ Verbal score on college entrance examination (percentile)

$x_2 = $ Mathematics score on college entrance examination (percentile)

The data in the table are obtained for a random sample of forty freshmen at one college.

VERBAL x_1	MATHEMATICS x_2	GPA y	VERBAL x_1	MATHEMATICS x_2	GPA y
81	87	3.49	79	75	3.45
68	99	2.89	81	62	2.76
57	86	2.73	50	69	1.90
100	49	1.54	72	70	3.01
54	83	2.56	54	52	1.48
82	86	3.43	65	79	2.98
75	74	3.59	56	78	2.58
58	98	2.86	98	67	2.73
55	54	1.46	97	80	3.27
49	81	2.11	77	90	3.47
64	76	2.69	49	54	1.30
66	59	2.16	39	81	1.22
80	61	2.60	87	69	3.23
100	85	3.30	70	95	3.82
83	76	3.75	57	89	2.93
64	66	2.70	74	67	2.83
83	72	3.15	87	93	3.84
93	54	2.28	90	65	3.01
74	59	2.92	81	76	3.33
51	75	2.48	84	69	3.06

a. Fit the first-order model (no quadratic and no interaction terms)

$$y = \beta_0 + \beta_1 x_1 + \beta_2 x_2 + \varepsilon$$

Interpret the value of R^2, and test whether these data indicate that the terms in the model are useful for predicting a freshman's GPA. Use $\alpha = .05$.

b. Sketch the relationship between predicted GPA, $\hat{y}$, and verbal score x_1 for the following mathematics scores: $x_2 = 60, 75,$ and 90.

*13.42 Refer to Exercise 13.41. Now fit the following second-order model to the data:

$$y = \beta_0 + \beta_1 x_1 + \beta_2 x_2 + \beta_3 x_1^2 + \beta_4 x_2^2 + \beta_5 x_1 x_2 + \varepsilon$$

a. Interpret the value of R^2, and test whether the data indicate that this model is useful for predicting a freshman's GPA. Use $\alpha = .05$.

b. Sketch the relationship between predicted GPA, $\hat{y}$, and the verbal score x_1 for the following mathematics scores: $x_2 = 60, 75,$ and 90. Compare these graphs with those for the first-order model in Exercise 13.41.

c. Test to see whether the interaction term, $\beta_5 x_1 x_2$, is important for the prediction of GPA. Use $\alpha = .10$. Note that this term permits the distance between three mathematics score curves for GPA versus verbal score to change as the verbal score changes.

On Your Own . . .

[*Note: The use of a computer is required for this study.*]

This is a continuation of the "On Your Own" section in Chapter 12, in which you selected three independent variables as predictors of a dependent variable of your choosing and obtained at least ten data values. Now fit the following multiple regression model (using an available computer package, if possible):

$$y = \beta_0 + \beta_1 x_1 + \beta_2 x_2 + \beta_3 x_3 + \varepsilon$$

where

$y = $ Dependent variable you chose

$x_1 = $ First independent variable you chose

$x_2 = $ Second independent variable you chose

$x_3 = $ Third independent variable you chose

a. Compare the coefficients $\hat{\beta}_1, \hat{\beta}_2,$ and $\hat{\beta}_3$ to their corresponding slope coefficients in the Chapter 12 "On Your Own," where you fit three separate straight-line models. How do you account for the differences?

b. Calculate the coefficient of determination R^2, and conduct the F-test of the null hypothesis H_0: $\beta_1 = \beta_2 = \beta_3 = 0$. What is your conclusion?

If the independent variables you chose are themselves highly correlated, you may encounter some results that are difficult to explain. For example, the coefficients $\hat{\beta}_1$, $\hat{\beta}_2$, and $\hat{\beta}_3$ may assume signs that run counter to what you expected. Or you may get a highly significant F-value in part b, but the individual t-statistics for x_1, x_2, and x_3 may all be nonsignificant. This phenomenon—a high correlation between the independent variables in a regression model—is known as ***multicollinearity.*** We will discuss it in more detail in Chapter 14.

References

Brown, M. B., ed. *Biomedical computer programs.* Berkeley: University of California Press, 1977.

Draper, N., & Smith, H. *Applied regression analysis.* New York: Wiley, 1966.

Fox, K. D., & Nickols, S. Y. "The time crunch." *Journal of Family Issues,* 1983, *4.*

Graybill, F. *Theory and application of the linear model.* North Scituate, Mass.: Duxbury, 1976. Chapters 8 and 10–13.

Kleinbaum, D., & Kupper, L. *Applied regression analysis and other multivariable methods.* North Scituate, Mass.: Duxbury, 1978. Chapters 9 and 10.

Mendenhall, W. *Introduction to linear models and the design and analysis of experiments.* Belmont, Ca.: Wadsworth, 1968.

Mendenhall, W., & McClave, J. T. *A second course in business statistics: regression analysis.* San Francisco: Dellen, 1981.

Neter, J., & Wasserman, W. *Applied linear statistical models.* Homewood, Ill.: Richard Irwin, 1974, Chapter 7.

Nie, N., Hull, C. H., Jenkins, J. G., Steinbrenner, K., & Bent, D. H. *Statistical package for the social sciences.* 2nd ed. New York: McGraw-Hill, 1975.

Palomba, N. A. "Strike activity and union membership: an empirical approach." *University of Washington Business Review,* Winter 1969.

Ray, A. A., ed. *SAS user's guide: statistics.* Cary, N.C.: SAS Institute, Inc. 1982.

Ryan, T. A., Joiner, B. L., & Ryan, B. F. *Minitab student handbook.* North Scituate, Mass.: Duxbury, 1979.

Tanner, M. B. "Vocational education and earnings for white males." *Southern Economic Journal,* 1983, *50.*

Younger, M. S. *A handbook for linear regression.* North Scituate, Mass.: Duxbury, 1979.

CHAPTER 14

Model Building

Where We've Been . . .

One of the most important topics in applied statistics, regression analysis, was presented in Chapters 12 and 13. Simple linear regression, using a straight line to model the relationship between the mean of a population and a single independent variable, was the topic of Chapter 12. Multiple regression, relating a population mean to any number of qualitative or quantitative independent variables, was the topic of Chapter 13. In both chapters we learned how to fit regression models to a set of data and how to use the model to estimate the mean value of y or to predict a future value of y for a given value of x.

Where We're Going . . .

Chapters 12 and 13 discussed, in general, how to use regression analysis to solve many practical problems. But an important problem was circumvented—the selection of a model that is appropriate for the given data. No matter how much you know about regression analysis, or how well you can fit a model to a set of data and interpret the results, the information will be of little value if you choose an ill-fitting model to relate the mean value of y to the independent variables. How to choose a reasonable model and how to use the data to modify and improve it is called model building. We will introduce you to this topic in Chapter 14.

Contents

We have indicated in Chapters 12 and 13 that the first step in the construction of a regression model is to hypothesize the form of the deterministic portion of the probabilistic model. This **model building,** or model construction, is the key to the success (or failure) of the regression analysis. If the hypothesized model does not reflect, at least approximately, the true nature of the relationship between the mean response $E(y)$ and the independent variables $x_1, x_2, \ldots, x_k$, the modeling effort will usually be unrewarded.

By model building we mean developing a model that provides a good fit to a set of data and gives good estimates of the mean value of y and good predictions of future values of y for given values of the independent variables. To illustrate, suppose that you wish to relate a student's score y on an introductory sociology examination to the amount of time x the student has spent studying for the examination and that (unknown to you) the second-order model

$$E(y) = \beta_0 + \beta_1 x + \beta_2 x^2$$

would permit you to predict y with a very small error of prediction; see Figure 14.1(a). However, suppose you erroneously choose the first-order model

$$E(y) = \beta_0 + \beta_1 x$$

to explain the relationship between y and x; see Figure 14.1(b).

The consequence of choosing the wrong model is clearly demonstrated by comparing Figures 14.1(a) and (b). The errors of prediction for the second-order model in Figure 14.1(a) are relatively small in comparison to those for the first-order model shown in Figure 14.1(b). The lesson to be learned from this simple example is very clear. Choosing a good set of independent (predictor) variables $x_1, x_2, \ldots, x_k$ does not guarantee a good multivariable prediction equation. In addition to selecting independent variables

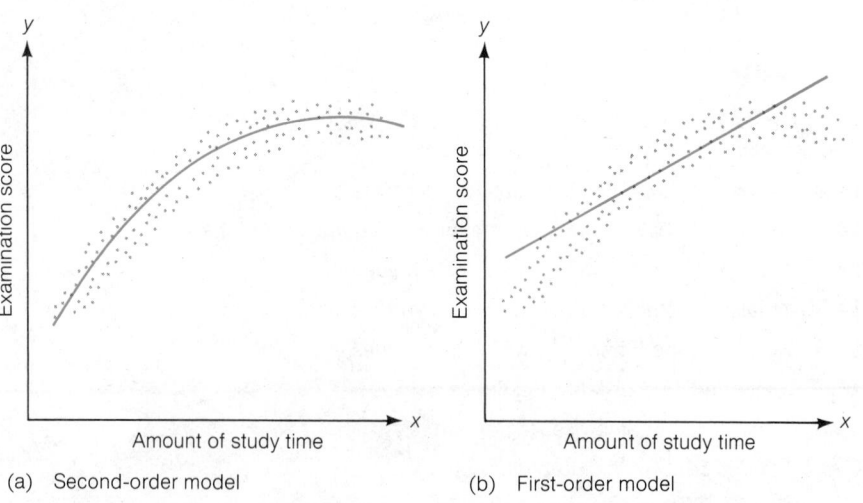

Figure 14.1 Two Models for Relating Examination Score y to Amount of Study Time x

(a) Second-order model

(b) First-order model

that contain information about y, you must also specify an equation relating y to x_1, $x_2, \ldots, x_k$ that will provide a good fit to your data.

In the following sections, we will present some useful models for relating a response y to one or more predictor variables. For a general discussion of this topic, we refer you to Mendenhall & McClave (1981, Chapter 6).

14.1
The Two Types of Independent Variables: Quantitative and Qualitative

In Chapter 2 we defined two types of data and, correspondingly, two types of variables that may be observed: quantitative and qualitative. In a regression analysis the dependent variable is always quantitative, but the independent variables may be either quantitative or qualitative. As you will see, the way an independent variable enters the model depends on its type.

Definition 14.1

A **quantitative** independent variable is one that assumes numerical values. An independent variable that is not quantitative is called **qualitative.**

The age of a person, the number of cigarettes smoked per day, the dissolved oxygen content of a river, and the amount of fertilizer applied to an experimental agricultural plot are all examples of quantitative independent variables. On the other hand, suppose three different types of fertilizer, A, B, and C, are used in an agricultural experiment. This independent variable is qualitative, since it is not measured on a numerical scale. Certainly, the type of fertilizer applied to a crop is an independent variable that may affect its yield, and we would want to include it in a model describing the crop yield y.

Definition 14.2

The **levels** of an independent variable are its different intensity settings.

For a quantitative independent variable, the levels correspond to the numerical values it assumes. For example, the number x of cigarettes smoked per day by a person may assume values 0, 1, 2, 3, 4, ...* Therefore, the independent variable x assumes levels corresponding to the numbers 0, 1, 2, 3, 4, ...

The levels of a qualitative variable are not numerical. They can be defined only by describing them. Type of fertilizer, the independent variable, was observed at three levels, one corresponding to each of the types, A, B, and C.

* There undoubtedly is some positive integer that represents the maximum number of cigarettes a person might smoke per day. Since this number is unknown, however, we give no limit to the number of values that x may assume.

Example 14.1   In Chapter 13 we considered the problem of predicting executive salary as a function of several independent variables. Consider the following four independent variables that may affect executive salaries:

a. Number of years of experience
b. Sex of the employee
c. Firm's net asset value
d. Rank of the employee

For each of these independent variables, give its type and describe the nature of the levels you would expect to observe.

Solution   a. The independent variable for the number of years of experience is quantitative since its values are numerical. We would expect to observe levels ranging from 0 to 40 (approximately) years.
b. The independent variable for sex is qualitative since its levels can be described only by the nonnumerical labels "female" and "male."
c. The independent variable for the firm's net asset value is quantitative, with a very large number of possible levels corresponding to the range of dollar values representing various firms' net asset values.
d. Suppose the independent variable for the rank of the employee is observed at three levels: supervisor, assistant vice president, and vice president. Since we cannot assign a realistic measure of relative importance to each position, rank is a qualitative independent variable.   ∎

Quantitative independent variables are treated differently than qualitative variables in regression modeling. In the next section, we will begin our discussion of how quantitative variables are used.

**Exercises
14.1–14.4**

Applying the Concepts

14.1 An experiment was conducted to investigate the effect of the following independent variables on the heart rate of humans:
a. Pounds over (or under) optimal weight
b. A subjective rating of physical condition on a scale of 1 to 10
c. Sex
d. Age
e. Race
Classify each of the variables as quantitative or qualitative, and describe the levels the variables may assume.

14.2 Companies keep personnel files that contain important information on each employee's background. These data could be used to predict an employee's score on a psychological test designed to measure the person's level of aggression. Identify the independent variables listed below as qualitative or quantitative. For qualitative variables, suggest several levels that might be observed. For quantitative variables, give a range of values (levels) for which the variable might be observed.

a. Age
b. Years of experience with the company
c. Highest educational degree
d. Job classification
e. Marital status
f. Religious preference
g. Salary
h. Sex

14.3 Which of the assumptions about ε (Section 12.3) prohibits the use of a qualitative variable as a dependent variable?

14.4 The marketing department of a large food products company conducted a study to investigate the effect of the following independent variables on the total monthly sales of one of the company's new products:

a. Advertising expenditures
b. Type of container
c. Color of container
d. Media used for advertising
e. Net weight of the product

Classify each of the variables as quantitative or qualitative, and describe the levels that the variables may assume.

14.2 Models with a Single Quantitative Independent Variable

The most common linear models relating y to a single quantitative independent variable x are those derived from a polynomial expression of the type shown in the box. Specific models, obtained by assigning values to p, are listed below.

> ## Formula for a pth-Order Polynomial with One Independent Variable
>
> $$E(y) = \beta_0 + \beta_1 x + \beta_2 x^2 + \beta_3 x^3 + \cdots + \beta_p x^p$$
>
> where p is an integer and $\beta_0, \beta_1, \ldots, \beta_p$ are unknown parameters that must be estimated.

1. First-Order Model

$E(y) = \beta_0 + \beta_1 x$ (Figure 14.2, page 696)

Comments on Model Parameters

$\beta_0 = y$-intercept

$\beta_1 = $ Slope of the line

Figure 14.2 Graph of a First-Order Model

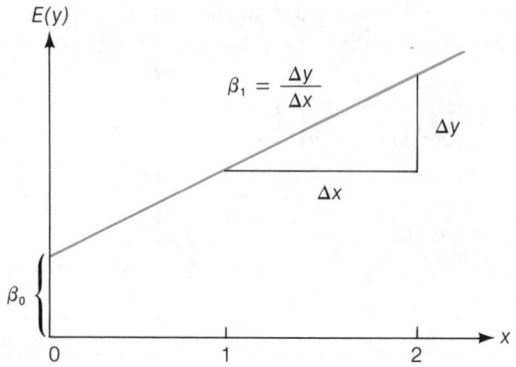

General Comments   The first-order model is used when you expect the rate of change in y, per unit change in x, to remain fairly stable over the range of values of x over which you wish to predict y.* Most relationships between $E(y)$ and x are curvilinear, but the curvature over the range of values of x for which you wish to predict y may be very slight. When this occurs, a first-order (straight-line) model should provide a good fit to your data.

2. Second-Order Model

$$E(y) = \beta_0 + \beta_1 x + \beta_2 x^2 \qquad \text{(Figure 14.3)}$$

Comments on the Parameters

$\beta_0 = y$-intercept

β_1:   Changing the value of β_1 shifts the parabola to the right or left. Increasing the value of β_1 causes the parabola to shift to the left.

β_2:   Rate of curvature

Figure 14.3 Graphs for Two Second-Order Models

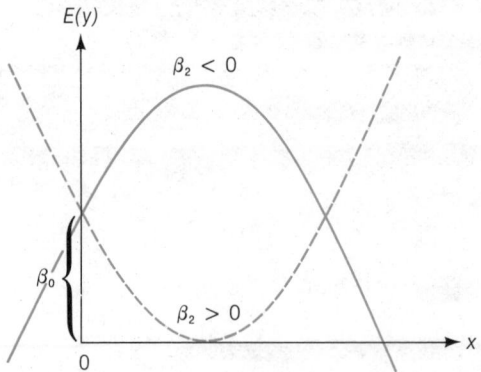

* The symbol Δy represents the change in y corresponding to a change in x equal to Δx.

General Comments A second-order model traces a parabola, one that opens either downward ($\beta_2 < 0$) or upward ($\beta_2 > 0$). Since most relationships possess some curvature, a second-order model is often a good choice to relate y to x.

3. Third-Order Model

$E(y) = \beta_0 + \beta_1 x + \beta_2 x^2 + \beta_3 x^3$ (Figure 14.4)

Comments on the Parameters

$\beta_0 = y$-intercept

β_3: The magnitude of β_3 controls the rate of reversal of curvature for the curve.

Figure 14.4 Graphs for Two Third-Order Models

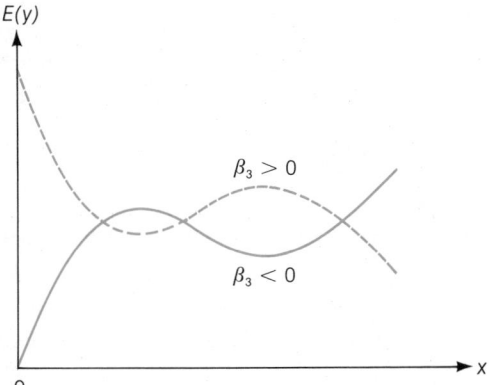

General Comments Reversals in curvature are not common, but such relationships can be modeled by third and higher-order polynomials. As can be seen in Figure 14.3, a second-order model contains no reversal in curvature. The slope either continues to increase or continues to decrease as x increases and produces either a trough or a peak. A third-order model (see Figure 14.4) contains one reversal in curvature and produces one peak and one trough; in general, the graph of a kth-order polynomial contains a total of $k - 1$ peaks and troughs.

Most functional relationships in nature seem to be smooth (except for random error), i.e., they are not subject to rapid and irregular reversals in direction. Consequently, the second-order polynomial model is perhaps the most useful of those described here. To develop a better understanding of how this model is employed, consider the following example.

Example 14.2

Power companies have to be able to predict the peak power load at their various stations in order to operate effectively. The peak power load is the maximum amount of power that must be generated each day in order to meet demand.

Suppose a power company located in the southern part of the United States decides to model daily peak power load, y, as a function of the daily high temperature, x, and the model is to be constructed for the summer months when demand is greatest.

Table 14.1
Power Load Data

TEMPERATURE ($°F$)	PEAK LOAD (*Megawatts*)	TEMPERATURE ($°F$)	PEAK LOAD (*Megawatts*)	TEMPERATURE ($°F$)	PEAK LOAD (*Megawatts*)
94	136.0	106	178.2	76	100.9
96	131.7	67	101.6	68	96.3
95	140.7	71	92.5	92	135.1
108	189.3	100	151.9	100	143.6
67	96.5	79	106.2	85	111.4
88	116.4	97	153.2	89	116.5
89	118.5	98	150.1	74	103.9
84	113.4	87	114.7	86	105.1
90	132.0				

Although we would expect the peak power load to increase as the high temperature increases, the *rate* of increase in $E(y)$ might also increase as x increases. That is, a 1 unit increase in high temperature from 100 to 101°F might result in a larger increase in power demand than would a 1 unit increase from 80 to 81°F. Therefore, we postulate the second-order model

$$E(y) = \beta_0 + \beta_1 x + \beta_2 x^2$$

and we expect β_2 to be positive.

A random sample of twenty-five summer days is selected, and the data are shown in Table 14.1. Fit a second-order model using these data, and test the hypothesis that the power load increases at an increasing *rate* with temperature, i.e., that $\beta_2 > 0$.

Solution   The SAS printout shown in Figure 14.5 gives the least squares fit of the second-order model using the data in Table 14.1. The prediction equation is

$$\hat{y} = 385.048 - 8.293x + .05982x^2$$

A plot of this equation and the observed values is given in Figure 14.6.

We now test to determine whether the sample value, $\hat{\beta}_2 = .05982$, is large enough to conclude *in general* that the power load increases at an increasing rate with

Figure 14.5   Portion of the SAS Printout for the Second-Order Model of Example 14.2

SOURCE	DF	SUM OF SQUARES	MEAN SQUARE	F VALUE	PR > F
MODEL	2	15011.77199776	7505.88599888	259.69	0.0001
ERROR	22	635.87840224	28.90356374		STD DEV
CORRECTED TOTAL	24	15647.65040000		R-SQUARE	5.37620347
				0.959363	

PARAMETER	ESTIMATE	T FOR H0: PARAMETER = 0	PR > |T|	STD ERROR OF ESTIMATE
INTERCEPT	385.04809323	6.98	0.0001	55.17243578
TEMP	−8.29252680	−6.38	0.0001	1.29904502
TEMP * TEMP	0.05982337	7.93	0.0001	0.00754855

Figure 14.6 Plot of the Observations and the Second-Order Least Squares Fit, Example 14.2

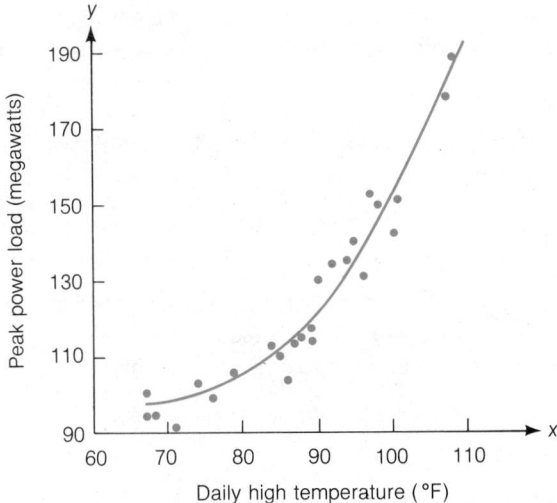

temperature:

$$H_0:\ \beta_2 = 0 \qquad H_a:\ \beta_2 > 0$$

Test statistic: $\quad t = \dfrac{\hat{\beta}_2}{s_{\hat{\beta}_2}}$

For $\alpha = .05$, $n = 25$, and $k = 2$, we will reject H_0 if

$$t > t_{.05}$$

where $t_{.05} = 1.717$ (from Table V in the Appendix) possesses $n - (k + 1) = 22$ degrees of freedom. From Figure 14.5 the calculated value of t is 7.93. Since this value exceeds $t_{.05} = 1.717$, we reject H_0 and conclude that the mean power load increases at an increasing rate with temperature. ∎

Exercises 14.5–14.16

Learning the Mechanics

14.5 Graph the following polynomials and identify the order of each on your graph:
a. $E(y) = 2 + 3x$
b. $E(y) = 2 + 3x^2$
c. $E(y) = 1 + 2x + 2x^2 + x^3$
d. $E(y) = 2x + 2x^2 + x^3$
e. $E(y) = 2 - 3x^2$
f. $E(y) = -2 + 3x$

14.6 The following graphs depict pth-order polynomials for one independent variable:

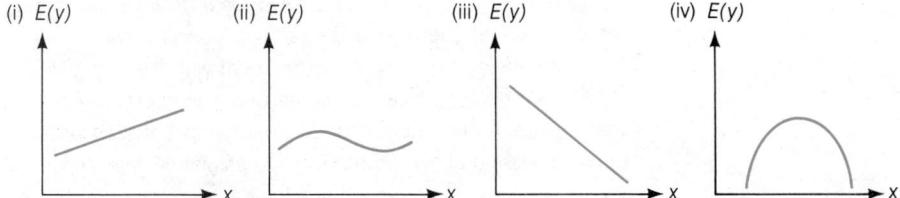

(i) $E(y)$ (ii) $E(y)$ (iii) $E(y)$ (iv) $E(y)$

a. For each graph, identify the order of the polynomial.

b. Using the parameters β_0, β_1, β_2, etc., write an appropriate model relating $E(y)$ to x for each graph.

c. By examining the graphs, you can determine the signs ($+$ or $-$) of many of the parameters in the models of part b. Give the signs of those parameters that can be determined.

14.7 Suppose $E(y)$ can best be modeled by a second-order polynomial in x, where x is a quantitative variable. Write the probabilistic model for y.

14.8 Describe the relationship between a dependent variable y and a quantitative independent variable x when the model is:

a. First order   **b.** Second order   **c.** Third order

14.9 Consider the following polynomial model:

$$E(y) = 5 - 3x + x^2$$

a. Give the order of this polynomial.

b. Sketch the curve corresponding to the equation for $E(y)$.

c. How would the graph change if the coefficient of x^2 were negative rather than positive?

14.10 Consider the following polynomial model:

$$E(y) = 2 - 4x$$

a. Give the order of this polynomial.

b. Sketch the curve corresponding to the equation for $E(y)$.

c. How would the graph change if the coefficient of x were positive instead of negative?

Applying the Concepts

[*Note: Starred (*) exercises require the use of a computer.*]

14.11 The amount of pressure used to produce a certain plastic is thought to be related to the strength of the plastic. Researchers believe that as pressure is increased, the strength of the plastic increases until, at some point, increases in pressure will have a detrimental effect on strength. Write a model to relate the strength y of the plastic to pressure x that would reflect these beliefs. Sketch the model.

14.12 A company is considering having the employees on its assembly line work 4 days for 10 hours each instead of 5 days for 8 hours. The management is concerned that the effect of fatigue due to longer afternoons of work might increase assembly times to an unsatisfactory level. An experiment with the 4-day week is planned in which time studies will be conducted on some of the workers during the afternoons. It is believed that an adequate model of the relationship between assembly time y and time since lunch, x, should allow for the average assembly time to decrease for a while after lunch (as workers get back in the groove) before it starts to increase as the workers become tired.

a. Give a model relating $E(y)$ to x that would reflect management's belief.

b. Define all terms in the model given in part a.

c. Sketch the curve described by the hypothesized model.

14.13 Underinflated or overinflated tires can increase tire wear. A new tire was tested for wear at different pressures with the results shown in the table.

a. Plot the data on a scattergram.

b. If you were given only the information for $x = 30, 31, 32, 33$, what kind of model would you suggest? For $x = 33, 34, 35, 36$? For all the data?

PRESSURE x (pounds per square inch)	MILEAGE y (thousands)
30	29
31	32
32	36
33	38
34	37
35	33
36	26

14.14 An experiment was conducted to relate the number of visits per patient per year to a doctor's office to the age x of a patient. It is suspected that young and old patients tend to seek medical assistance more frequently than those of middle age. Write an appropriate model relating mean number of visits per patient per year to the patient's age.

*__14.15__ A veterinarian who works for a large midwestern pig cooperative has developed a daily vitamin pellet that she believes will substantially increase the weight of mature pigs within 1 month. However, she is uncertain of the relationship between the daily dosage (amount of the vitamin) and the percentage gain in weight after 1 month. To understand this relationship better, she randomly selects sixteen pigs of the same age and weight and feeds them different dosage levels for a 1-month trial period. The data are given in the table.

WEIGHT GAIN (% of original weight)	DAILY PELLET DOSE
82	0
78	0
80	1
87	1
87	2
95	2
97	3
90	3
90	4
95	4
89	5
93	5
90	6
85	6
84	7
90	7

a. Plot the data on a scattergram.

b. Fit the model $E(y) = \beta_0 + \beta_1 x + \beta_2 x^2$ to the data.

c. Is there evidence to support the inclusion of the quadratic (second-order) term in the model of part b? Test using $\alpha = .05$.

d. Plot the fitted model of part b on the scattergram of part a.

e. Use the fitted model to estimate the optimum daily dosage (i.e., the dosage that encourages the greatest weight gain). [*Note:* We would want to express this estimate as a confidence interval, but its computation is beyond the scope of this text. The procedure is described in the references. You may also find that it can be obtained by using your computer program package.]

14.16 An economist has proposed the following model to describe the relationship between the number of items produced per day (output) and the number of work-hours expended per day (input) in a particular production process:

$$y = \beta_0 + \beta_1 x + \beta_2 x^2 + \varepsilon$$

where

$y =$ Number of items produced per day

$x =$ Number of work-hours per day

A portion of the Minitab computer printout that results from fitting this model to a sample of 25 weeks of production data is shown here. Test the hypothesis that as the amount of input increases, the amount of output also increases but at a decreasing rate. Do the data provide sufficient evidence to indicate that the *rate* of increase in output per unit increase of input decreases as the input increases? Test using $\alpha = .05$.

THE REGRESSION EQUATION IS
Y = −6.17 + 2.04 X1 − .0323 X2

	COLUMN	COEFFICIENT	ST. DEV. OF COEF.	T-RATIO = COEF/S.D.
	—	−6.173	1.666	−3.71
X1	C2	2.036	.185	11.02
X2	C3	−.03231	.00489	−6.60

THE ST. DEV. OF Y ABOUT REGRESSION LINE IS
S =   1.243
WITH (25 − 3) = 22 DEGREES OF FREEDOM

R-SQUARED = 95.5 PERCENT
R-SQUARED = 95.1 PERCENT, ADJUSTED FOR D.F.

ANALYSIS OF VARIANCE

DUE TO	DF	SS	MS = SS/DF
REGRESSION	2	718.168	359.084
RESIDUAL	22	33.992	1.545
TOTAL	24	752.160	

14.3 Models with Two or More Quantitative Independent Variables

The best way to understand how to write models relating a mean response $E(y)$ to a set of quantitative independent variables is to examine the different types of models when only two independent variables are involved. You can see geometrically (using three-dimensional figures) the relationships implied by the different models and will more easily understand the relationships you will encounter when the model contains three or more quantitative independent variables.

1. First-Order Model
Two Independent Variables

$$E(y) = \beta_0 + \beta_1 x_1 + \beta_2 x_2$$

Comments on the Parameters

β_0 = y-intercept, the value of $E(y)$ when $x_1 = x_2 = 0$

β_1 = Change in $E(y)$ for a 1 unit increase in x_1 when x_2 is held fixed

β_2 = Change in $E(y)$ for a 1 unit increase in x_2 when x_1 is held fixed

General Comments The graph in Figure 14.7 traces a ***response surface*** (in contrast to the response curve that is used to relate $E(y)$ to a *single* quantitative variable). Particularly, a first-order model relating $E(y)$ to two independent quantitative variables, x_1 and x_2, graphs as a plane in a three-dimensional space. The plane traces the value of $E(y)$ for every combination of values (x_1, x_2) that correspond to points in the x_1, x_2-plane. Most response surfaces in the real world are well behaved (smooth), and they possess curvature. Consequently, a first-order model is appropriate only if the response surface is fairly flat over the (x_1, x_2) region that is of interest.

The assumption that a first-order model will adequately characterize the relationship between $E(y)$ and the variables x_1 and x_2 is equivalent to assuming that x_1 and x_2 do not interact; i.e., you assume that the effect on $E(y)$ of changes in x_1 are the same,

Figure 14.7 Graph of a First-Order Model

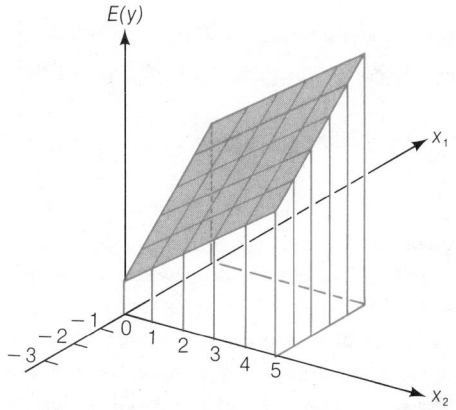

Figure 14.8 Graph Indicating No Interaction Between x_1 and x_2

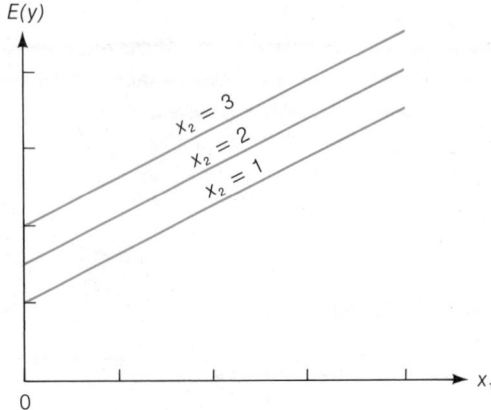

regardless of the value of x_2 (and vice versa). Thus, no interaction means that the effect of changes in one variable (say x_1) on $E(y)$ is *independent* of the value of the second variable (say x_2). For example, if we assign values to x_2 in a first-order model, the graph of $E(y)$ as a function of x_1 would produce parallel lines as shown in Figure 14.8. These lines, called **contour lines,** show the contours of the surface when it is sliced by three planes, each of which is parallel to the $E(y)$, x_1-plane, at distances $x_2 = 1, 2$, and 3 from the origin.

Definition 14.3

Two variables x_1 and x_2 are said to **interact** if the change in $E(y)$ for a 1 unit increase in x_1 (when x_2 is held fixed) is dependent on the value of x_2.

---

2. Interaction Model (Second-Order)
Two Independent Variables

$$E(y) = \beta_0 + \beta_1 x_1 + \beta_2 x_2 + \beta_3 x_1 x_2$$

Comments on the Parameters

$\beta_0 = y$-intercept, the value of $E(y)$ when x_1 and x_2 equal zero

β_1, β_2: Changing β_1 and β_2 causes the surface to shift along the x_1-axis and the x_2-axis

β_3: Controls the rate of twist in the surface (Figure 14.9)

General Comments The model described here is said to be second-order because the highest-order term ($x_1 x_2$) in x_1 and x_2 is 2; i.e., the sum of the exponents of x_1 and x_2 equals 2. This interaction model traces a twisted surface in a three-dimensional space (see Figure 14.9). A graph of $E(y)$ as a function of x_1 for given values of x_2 (say $x_2 = 1, 2$, and 3) produces nonparallel contour lines (see Figure 14.10), thus indicating that the change in $E(y)$ for a given change in x_1 is dependent on the value of x_2 and, therefore, that x_1 and x_2 interact. Interaction is an extremely important concept because it is easy to get in the habit of fitting first-order models and individually examining

Figure 14.9 Graph for an Interaction Model (Second-Order)

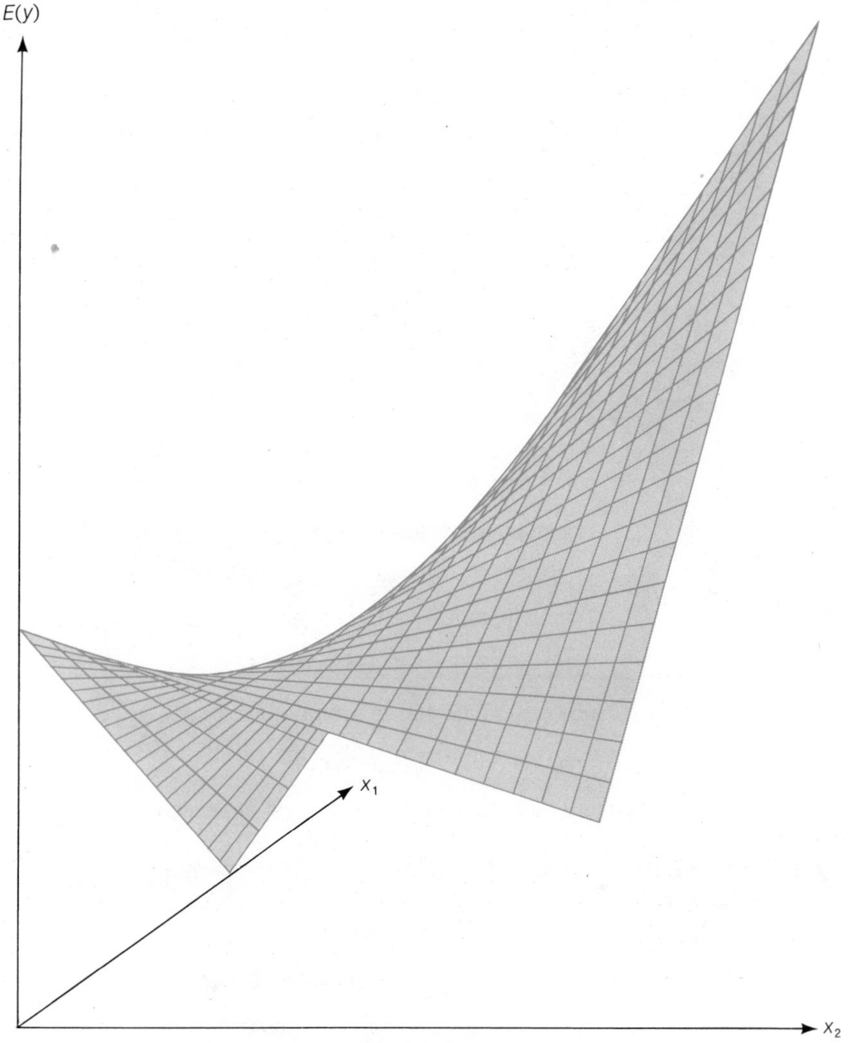

$E(y)$

x_1

x_2

Figure 14.10 Graph Indicating Interaction Between x_1 and x_2

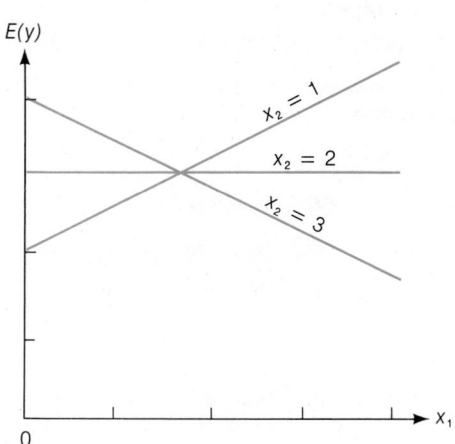

$E(y)$

$x_2 = 1$

$x_2 = 2$

$x_2 = 3$

x_1

0

the relationships between $E(y)$ and a set of independent variables, $x_1, x_2, \ldots, x_k$. Such a procedure is meaningless when interaction exists (which is, at least to some extent, almost always the case), and it can lead to gross errors in interpretation. For example, suppose the relationship between $E(y)$ and x_1 and x_2 is as shown in Figure 14.10 and you have observed y for each of the $n = 9$ combinations of values of x_1 and x_2, $x_1 = 1, 2, 3$ and $x_2 = 1, 2, 3$. If you fit a first-order model in x_1 and x_2 to the data, the fitted plane would be (except for random error) approximately parallel to the x_1, x_2-plane, thus suggesting that x_1 and x_2 contribute very little information about $E(y)$. Figure 14.10 clearly indicates that this is not the case. Fitting a first-order model to the data would not allow for the twist in the true surface and would therefore give a false impression of the relationship between $E(y)$ and x_1 and x_2. The procedure for detecting interaction between two independent variables can be seen by examining the model. The interaction model differs from the noninteraction first-order model only in the inclusion of the $\beta_3 x_1 x_2$ term.[*]

Interaction model: $E(y) = \beta_0 + \beta_1 x_1 + \beta_2 x_2 + \beta_3 x_1 x_2$

First-order model: $E(y) = \beta_0 + \beta_1 x_1 + \beta_2 x_2$

Therefore, to test for the presence of interaction we test

H_0: $\beta_3 = 0$ (No interaction)

against the alternative hypothesis

H_a: $\beta_3 \neq 0$ (Interaction)

using the familiar Student's t-test of Section 13.4.

3. Complete Second-Order Model
Two Independent Variables

$$E(y) = \beta_0 + \beta_1 x_1 + \beta_2 x_2 + \beta_3 x_1 x_2 + \beta_4 x_1^2 + \beta_5 x_2^2$$

Comments on the Parameters

$\beta_0 = y$-intercept, the value of $E(y)$ when $x_1 = x_2 = 0$

β_1, β_2: Changing β_1 and β_2 causes the surface to shift along the x_1-axis and the x_2-axis

β_3: Value of β_3 controls the rotation of the surface

β_4, β_5: Signs and values of these parameters control the type of surface and rates of curvature

The parameters β_4 and β_5 produce the following three types of surfaces:

β_4 and β_5 positive: Paraboloid that opens upward (Figure 14.11(a))

β_4 and β_5 negative: Paraboloid that opens downward (Figure 14.11(b))

β_4 and β_5 differ in sign: Saddle-shaped surface (Figure 14.11(c))

---

[*] The *order of a term* is equal to the sum of the exponents of the variables included in the term. Thus, the $x_1 x_2$ term is second-order, because the exponents of x_1 and x_2 are both equal to 1 and their sum is 2. The *order of a model* is determined by its highest-order term. Therefore, the interaction model is second-order.

Figure 14.11 Graphs for Three Second-Order Surfaces

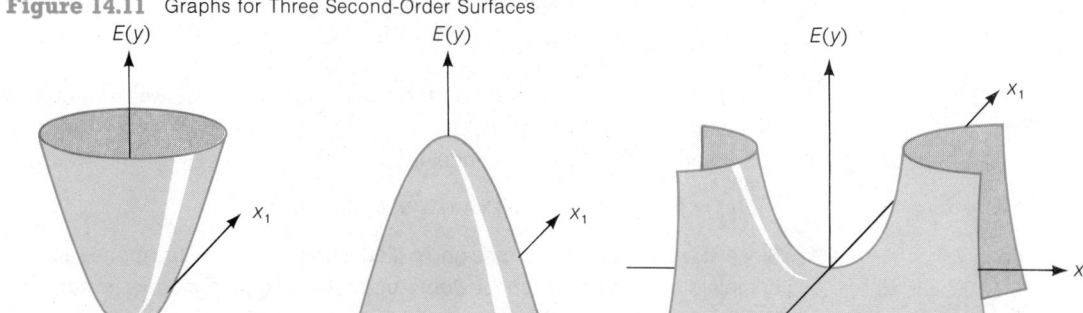

(a)   (b)   (c)

General Comments A complete second-order model is the three-dimensional equivalent of a second-order model in a single quantitative variable. Instead of tracing parabolas, it traces paraboloids and saddle surfaces. Since only a portion of the complete surface is used to fit the data, it provides a very large variety of gently curving surfaces that can be used to fit data. It is a good choice for a model if you expect curvature in the response surface relating $E(y)$ to x_1 and x_2.

Example 14.3 A social scientist would like to relate the number of hours worked per week (outside the home) by a married woman to the number of years of formal education completed by the woman and the number of children in the family.

    a. Identify the dependent variable and the independent variables.
    b. Write the first-order model for this example.
    c. Modify the model in part b to include an interaction term.
    d. Write a complete second-order model for $E(y)$.

Solution **a.** The dependent variable is

$$y = \text{Number of hours worked per week by a married woman}$$

The two independent variables, both quantitative in nature, are

$$x_1 = \text{Number of years of formal education completed by the woman}$$
$$x_2 = \text{Number of children in the family}$$

    b. The first-order model is

$$E(y) = \beta_0 + \beta_1 x_1 + \beta_2 x_2$$

This model would probably not be appropriate in this situation because it is quite possible that x_1 and x_2 may interact and that second-order terms corresponding to x_1^2 and x_2^2 may be needed to obtain a good model for $E(y)$.

c. Adding the interaction term, we obtain

$$E(y) = \beta_0 + \beta_1 x_1 + \beta_2 x_2 + \beta_3 x_1 x_2$$

This model should be better than the model in part b, since we have now allowed for interaction between x_1 and x_2.

d. The complete second-order model is

$$E(y) = \beta_0 + \beta_1 x_1 + \beta_2 x_2 + \beta_3 x_1 x_2 + \beta_4 x_1^2 + \beta_5 x_2^2$$

Since it would not be surprising to find curvature in the response surface, the complete second-order model would be preferred to the models in parts b and c. How can we tell whether the complete second-order model really does provide better predictions of hours worked than the models in parts b and c? The answers to these and similar questions are examined in Section 14.4.   ■

Models relating a mean response $E(y)$ to more than two quantitative variables are extensions of the corresponding models in two quantitative variables. For example, a first-order model for three quantitative independent variables, x_1, x_2, and x_3, is

First-order model: $E(y) = \beta_0 + \beta_1 x_1 + \beta_2 x_2 + \beta_3 x_3$

As in the case of two independent variables, this model implies that the relationship between $E(y)$ and any one variable, say x_1, is linear when x_2 and x_3 are held constant. Further, it implies that there is no interaction among x_1, x_2, and x_3; i.e., the slope of the linear relation is exactly the same, regardless of the values of x_2 and x_3. If one of the independent variables, say x_3, is held constant, the graph of $E(y)$ as a function of x_1 and x_2 is a plane. (See Figure 14.7.) Thus, for given values of β_0, β_1, β_2, and β_3, $E(y)$ will graph as a set of parallel planes, one for each different value of x_3.

Similarly, as in the case of two quantitative independent variables, we can add second-order cross-product terms, $x_1 x_2$, $x_1 x_3$, and $x_2 x_3$, to the first-order model and obtain a second-order model that allows for interaction among x_1, x_2, and x_3:

Interaction model: $E(y) = \beta_0 + \beta_1 x_1 + \beta_2 x_2 + \beta_3 x_3 + \beta_4 x_1 x_2 + \beta_5 x_1 x_3 + \beta_6 x_2 x_3$

If one of the independent variables, say x_3, is held constant, the graph of $E(y)$ as a function of x_1 and x_2 is a twisted plane of the type shown in Figure 14.9.

Finally, we could add second-order terms involving x_1^2, x_2^2, and x_3^2 to the interaction model and obtain the complete second-order model:

Complete second-order model:

$$E(y) = \beta_0 + \beta_1 x_1 + \beta_2 x_2 + \beta_3 x_3 + \beta_4 x_1 x_2 + \beta_5 x_1 x_3 + \beta_6 x_2 x_3 + \beta_7 x_1^2 + \beta_8 x_2^2 + \beta_9 x_3^2$$

If one of the independent variables, say x_3, is held constant, the graph of $E(y)$ as a function of x_1 and x_2 will trace a conic surface. (See Figure 14.11.)

Most relationships between $E(y)$ and two or more quantitative independent variables are second-order and require the use of either the interactive or the complete second-order model to obtain a good fit to a data set. As in the case of a single quantitative independent variable, however, the curvature in the response surface may be very slight over the range of values of the variables in the data set. When this happens, a first-order model may provide a good fit to the data.

How can you tell which model to use in a particular situation? We will answer this question in Section 14.4.

Case Study 14.1
Modeling Strike
Activity

Case Study 13.2 describes Neil A. Palomba's efforts to model strike activity (the percentage y of total working hours lost due to strikes) as a function of three independent variables:

x_1 = Percentage of union members in nonagricultural establishments

x_2 = Percentage of all nonagricultural employment that is manufacturing

x_3 = Hourly earnings of workers on manufacturing payrolls

The regression analysis using Palomba's most complete model, a first-order model involving all three independent variables, produced an R^2 that was only equal to .232. In other words, Palomba's best model accounted for only 23% of the variability of the y-values about their mean. How can we improve the model so that we can account for a substantial portion of the remaining 77%?

Table 14.2
Strike Activity Data,
Case Study 14.1

STATE	y	1964 x_1	1964 x_2	1964 x_3	STATE	y	1964 x_1	1964 x_2	1964 x_3
Alabama	.14	18.0	30.5	2.17	Nebraska	.05	19.3	16.6	2.36
Alaska	.11	32.2	8.9	3.54	Nevada	.36	32.8	4.6	3.16
Arizona	.09	20.08	15.3	2.72	New Hampshire	.03	20.9	40.9	2.00
Arkansas	.10	26.2	29.2	1.78	New Jersey	.27	37.7	37.1	2.67
California	.16	33.8	24.9	2.96	New Mexico	.09	13.4	6.8	2.29
Colorado	.04	21.6	15.8	2.74	New York	.11	39.4	28.2	2.60
Connecticut	.08	24.6	42.5	2.62	North Carolina	.01	6.7	41.6	1.75
Delaware	.41	21.5	36.1	2.65	North Dakota	.03	14.8	5.8	2.28
Florida	.20	13.1	15.5	2.11	Ohio	.38	35.7	39.0	2.91
Georgia	.13	12.7	31.8	1.92	Oklahoma	.01	13.7	15.5	2.35
Hawaii	.02	24.2	12.1	2.14	Oregon	.12	34.8	26.5	2.85
Idaho	.11	19.2	18.8	2.50	Pennsylvania	.14	38.4	37.8	2.55
Illinois	.18	37.9	33.5	2.76	Rhode Island	.09	29.6	38.2	2.11
Indiana	.16	34.1	40.8	2.81	South Carolina	.01	7.9	42.7	1.80
Iowa	.16	20.8	25.4	2.71	South Dakota	.16	9.5	8.8	2.34
Kansas	.11	18.8	20.6	2.65	Tennessee	.23	17.6	34.6	2.03
Kentucky	.17	25.7	26.6	2.43	Texas	.06	13.3	19.4	2.42
Louisiana	.10	17.1	17.8	2.49	Utah	.66	19.7	17.6	2.77
Maine	.15	20.3	36.6	2.00	Vermont	.26	19.3	30.9	2.08
Massachusetts	.07	29.1	33.0	2.37	Virginia	.04	15.5	26.5	2.04
Michigan	.83	38.9	40.7	3.11	Washington	.16	43.1	25.6	2.98
Minnesota	.02	33.0	24.0	2.64	West Virginia	.45	42.0	27.4	2.67
Mississippi	.14	11.6	30.5	1.76	Wisconsin	.21	31.5	37.0	2.66
Missouri	.14	39.8	28.5	2.53	Wyoming	.01	19.2	7.7	2.82
Montana	.28	36.2	12.2	2.71	Maryland				

There are two possibilities. Palomba's model may not contain the key information-contributing variables. Perhaps there are other independent variables that are highly related to strike activity but have not been included in the model. Locating these variables, if they exist, is a task for a sociologist or a specialist in industrial relations. The other possibility is that Palomba's three independent variables are the most important variables in explaining strike activity and the form of the model is wrong. Instead of being first-order, perhaps the model should be second-order or a more complex mathematical function of x_1, x_2, and x_3.

Palomba's data giving the values of y, x_1, x_2, and x_3 for forty-nine states (excluding Maryland) in 1964 are shown in Table 14.2. You will have an opportunity to improve on Palomba's model in Exercise 14.31.

**Exercises
14.17–14.25**

Learning the Mechanics

14.17 a. Write a first-order model relating $E(y)$ to two quantitative independent variables, x_1 and x_2.
b. Modify the model you constructed in part a to include an interaction term.
c. Modify the model you constructed in part b to make it a complete second-order model.

14.18 Suppose the true relationship between $E(y)$ and the quantitative independent variables x_1 and x_2 is described by the following first-order model:

$$E(y) = 3 + x_1 - 2x_2$$

a. Describe the corresponding response surface.
b. Plot the contour lines of the response surface for $x_1 = 2, 3, 4$, where $0 \le x_2 \le 5$.
c. Plot the contour lines of the response surface for $x_2 = 2, 3, 4$, where $0 \le x_1 \le 5$.
d. Use the contour lines you plotted in parts b and c to explain how changes in the settings of x_1 and x_2 affect $E(y)$.
e. Use your graph from part b to determine how much $E(y)$ changes when x_1 is changed from 4 to 2 and x_2 is simultaneously changed from 1 to 2.

14.19 Suppose the true relationship between $E(y)$ and the quantitative independent variables x_1 and x_2 is

$$E(y) = 3 + x_1 + 2x_2 - x_1 x_2$$

a. Describe the corresponding response surface.
b. Plot the contour lines of the response surface for $x_1 = 0, 1, 2$, where $0 \le x_2 \le 5$.
c. Explain why the contour lines you plotted in part b are not parallel.
d. Use the contour lines you plotted in part b to explain how changes in the settings of x_1 and x_2 affect $E(y)$.
e. Use your graph from part b to determine how much $E(y)$ changes when x_1 is changed from 2 to zero and x_2 is simultaneously changed from 4 to 5.

14.20 If two variables, x_1 and x_2, do not interact, how would you describe their effect on the mean response $E(y)$?

Applying the Concepts

14.21 Baseball fans are involved in a never-ending search for variables (and a model) that will enable them to predict a professional baseball team's success for a given baseball season. Adam Clymer in an article titled "New Statistics for Baseball" (*New York Times,* May 20, 1984) notes that John Thorn and Pete Palmer in their new book (*The Hidden Game of Baseball,* Doubleday, 1984) "measure players' offensive contributions by what they call a linear weights measure, which gives separate values to at bats, singles, doubles, . . . , home runs, . . . , stolen bases, and times caught stealing."

   a. Would a regression analysis, using a first-order model, produce Clymer's description of a "linear weights measure"? Explain.

   b. List some of the independent variables that you think might be related to a professional baseball team's percentage of games won. Construct a first-order model to relate percentage of games won to these variables.

   c. Can you propose a better model than the model in part b? Explain.

14.22 An energy conservationist wants to develop a model that will estimate the mean annual gasoline consumption y in the United States (in millions of barrels) as a function of two independent variables:

   x_1 = Number of cars (millions) in use during year

   x_2 = Number of trucks (millions) in use during year

   a. Identify the independent variables as quantitative or qualitative.

   b. Write the first-order model for $E(y)$.

   c. Write the complete second-order model for $E(y)$.

   d. With respect to the model of part c, specify the null and alternative hypotheses you would employ in testing for the presence of interaction between x_1 and x_2.

14.23 The dissolved oxygen content y in rivers and streams is related to the amount x_1 of nitrogen compounds per liter of water and the temperature x_2 of the water. Write the complete second-order model relating $E(y)$ to x_1 and x_2.

14.24 Some corporations, instead of owning a fleet of cars, rent cars from a rental agency. A corporation may do this because it is sometimes more economical to rent new cars for a year than to buy new cars each year. A major rental agency wants to develop a model that will allow it to estimate the average annual cost to the prospective customer of renting cars, y, as a function of two independent variables:

   x_1 = Number of cars rented

   x_2 = Average number of miles driven per car during year (in thousands)

   a. Identify the independent variables as quantitative or qualitative.

   b. Write the first-order model for $E(y)$.

c. Write a model for $E(y)$ that contains all first-order and interaction terms. Sketch typical response curves showing $E(y)$, the mean cost, versus x_2, the average mileage driven, for different values of x_1. (Assume that x_1 and x_2 interact.)

d. Write the complete second-order model for $E(y)$.

14.25 Refer to Exercise 14.24. Suppose the model from part c is fit with the following result:

$$\hat{y} = 1 + .05x_1 + x_2 + .05x_1x_2$$

(The units of $\hat{y}$ are thousands of dollars.) Graph the estimated cost $\hat{y}$ as a function of the average number of miles driven, x_2, over the range $x_2 = 10$ to $x_2 = 50$ (10,000 to 50,000 miles) for $x_1 = 1$, 5, and 10. Do these functions agree (approximately) with the graphs you drew for Exercise 14.24, part c?

14.4
Model Building: Testing Portions of a Model

The presentation of models with one and two quantitative independent variables raises a very general question. Do certain terms in the model contribute information for the prediction of y?

To illustrate, suppose you have collected data on a response y and two independent quantitative variables, x_1 and x_2, and you are considering the use of either a first-order or a second-order model to relate $E(y)$ to x_1 and x_2. Will the second-order model provide better predictions of y than the first-order model? To answer this question, examine the two models. Note that the second-order model contains all terms contained in the first-order model plus three additional terms, those involving β_3, β_4, and β_5:

First-order model: $E(y) = \beta_0 + \beta_1x_1 + \beta_2x_2$

$$\text{Second-order terms}$$

Second-order model: $E(y) = \beta_0 + \beta_1x_1 + \beta_2x_2 + \overbrace{\beta_3x_1x_2 + \beta_4x_1^2 + \beta_5x_2^2}$

Therefore, asking whether the second-order model contributes more information for the prediction of y than the first-order model is equivalent to asking whether $\beta_3 = \beta_4 = \beta_5 = 0$, i.e., whether the terms involving β_3, β_4, and β_5 should be retained in the model. To test whether the second-order terms should be included in the model, we test the null hypothesis

$$H_0: \quad \beta_3 = \beta_4 = \beta_5 = 0$$

(i.e., the second-order terms do not contribute information for the prediction of y) against the alternative hypothesis

$$H_a: \quad \text{At least one of the parameters, } \beta_3, \beta_4, \beta_5, \text{ differs from zero}$$

(i.e., at least one of the second-order terms contributes information for the prediction of y).

In Section 13.4 we presented the t-test for a single coefficient, and in Section 13.5 we gave the F-test for *all* the β-parameters (except β_0) in the model. We now need a test for *some* of the β-parameters in the model. The test procedure is intuitive: First, we use the method of least squares to fit the first-order model and calculate the corresponding sum of squares for error, SSE_1 (the sum of squares of the deviations between observed and predicted y-values). Next, we fit the second-order model and calculate its sum of squares for error, SSE_2. Then we compare SSE_1 to SSE_2 by calculating the difference $SSE_1 - SSE_2$. If the second-order terms contribute to the model, then SSE_2 should be much smaller than SSE_1, and the difference $SSE_1 - SSE_2$ will be large. That is, the larger the difference, the greater the weight of evidence that the second-order model provides better predictions of y than does the first-order model.

The sum of squares for error always decreases when new terms are added to the model. The question is whether this decrease is large enough to conclude that it is due to more than just an increase in the number of model terms and to chance. To test the null hypothesis that the parameters of the second-order terms, β_3, β_4, and β_5, simultaneously equal zero, we use an F-statistic calculated as follows:

$$F = \frac{(SSE_1 - SSE_2)/3}{SSE_2/[n - (5 + 1)]}$$

$$= \frac{\text{Drop in SSE/Number of } \beta\text{-parameters being tested}}{s^2 \text{ for the complete second-order model}}$$

When the assumptions listed in Section 13.1 about the error term ε are satisfied and the β-parameters for the second-order terms are all zero (H_0 is true), this F-statistic has an F-distribution with $v_1 = 3$ numerator df and $v_2 = n - 6$ denominator df. Note that v_1 is the number of β-parameters being tested and v_2 is the number of degrees of freedom associated with s^2 in the complete model.

If the second-order terms do contribute to the model (H_a is true), we expect the F-statistic to be large. Thus, we use a one-tailed test and reject H_0 when F exceeds some critical value, F_α, as shown in Figure 14.12. The steps employed in testing the null hypothesis that a set of model parameters are all equal to zero are summarized in the box on page 714.

Figure 14.12 Rejection Region for the F-Test H_0: $\beta_3 = \beta_4 = \beta_5 = 0$

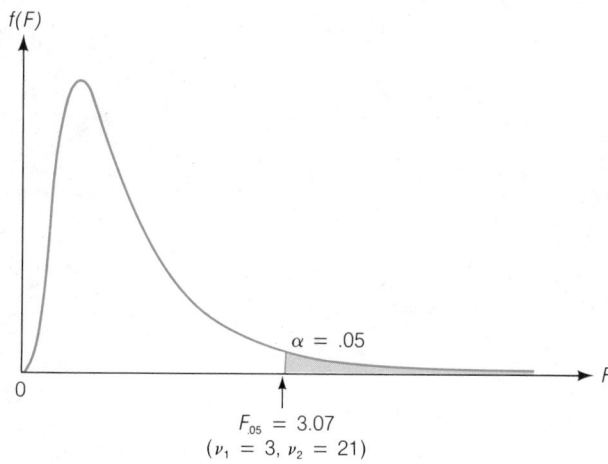

$f(F)$

$\alpha = .05$

F

0

$F_{.05} = 3.07$
$(v_1 = 3, v_2 = 21)$

F-Test for Testing the Null Hypothesis: Set of β-Parameters Equal Zero

Reduced model: $E(y) = \beta_0 + \beta_1 x_1 + \cdots + \beta_g x_g$

Complete model: $E(y) = \beta_0 + \beta_1 x_1 + \cdots + \beta_g x_g + \beta_{g+1} x_{g+1} + \cdots + \beta_k x_k$

H_0: $\beta_{g+1} = \beta_{g+2} = \cdots = \beta_k = 0$

H_a: At least one of the β-parameters being tested is nonzero

Test statistic: $F = \dfrac{(SSE_1 - SSE_2)/(k - g)}{SSE_2/[n - (k + 1)]}$

where

   SSE_1 = Sum of squared errors for the reduced model

   SSE_2 = Sum of squared errors for the complete model

   $k - g$ = Number of β-parameters specified in H_0

   $k + 1$ = Number of β-parameters in the complete model

   n = Total sample size

Rejection region: $F > F_\alpha$

where F_α is based on $v_1 = (k - g)$ numerator df and $v_2 = [n - (k + 1)]$ denominator df.

Example 14.4   Suppose you wish to study the growth of carnations as a function of the temperature x_1 in a greenhouse and the amount of fertilizer x_2 applied to the soil. A total of twenty-seven plots of equal size are treated with fertilizer in amounts varying between 50 and 60 kilograms per plot, and these plots are mechanically kept at constant temperatures between 80 and 100°F. Small carnation plants (approximately 15 centimeters in height) are planted in each plot, and their height y is measured after a 6-week growing period. The resulting data are shown in Table 14.3.

Table 14.3

Temperature, Amount of Fertilizer, and Height of Carnations

x_1 (°F)	x_2 (kilograms per plot)	y (centimeters)	x_1 (°F)	x_2 (kilograms per plot)	y (centimeters)	x_1 (°F)	x_2 (kilograms per plot)	y (centimeters)
80	50	50.8	90	50	63.4	100	50	46.6
80	50	50.7	90	50	61.6	100	50	49.1
80	50	49.4	90	50	63.4	100	50	46.4
80	55	93.7	90	55	93.8	100	55	69.8
80	55	90.9	90	55	92.1	100	55	72.5
80	55	90.9	90	55	97.4	100	55	73.2
80	60	74.5	90	60	70.9	100	60	38.7
80	60	73.0	90	60	68.8	100	60	42.5
80	60	71.2	90	60	71.3	100	60	41.4

    a. Fit a complete second-order model to the data.

    b. Sketch the response surface.

    c. Do the data provide sufficient evidence to indicate that the second-order terms contribute information for the prediction of y?

Solution   **a.** The complete second-order model is

$$E(y) = \beta_0 + \beta_1 x_1 + \beta_2 x_2 + \beta_3 x_1 x_2 + \beta_4 x_1^2 + \beta_5 x_2^2$$

The data in Table 14.3 were used to fit this model, and a portion of the SAS output is shown in Figure 14.13.

Figure 14.13 Portion of the SAS Printout for Example 14.4

SOURCE	DF	SUM OF SQUARES	MEAN SQUARE	F VALUE	PR > F
MODEL	5	8402.26453714	1680.45290743	596.32	0.0001
ERROR	21	59.17842582	2.81802028		STD DEV
CORRECTED TOTAL	26	8461.44296296		R-SQUARE . 0.993006	1.67869601

PARAMETER	ESTIMATE	T FOR H0: PARAMETER = 0	PR > |T|	STD ERROR OF ESTIMATE
INTERCEPT	−5127.89907417	−46.49	0.0001	110.29601483
X1	31.09638889	23.13	0.0001	1.34441322
X2	139.74722222	44.50	0.0001	3.14005411
X1 * X2	−0.14550000	−15.01	0.0001	0.00969196
X1 * X1	−0.13338889	−19.46	0.0001	0.00685325
X2 * X2	−1.14422222	−41.74	0.0001	0.02741299

The least squares prediction equation is

$$\hat{y} = -5{,}127.90 + 31.10x_1 + 139.75x_2 - .146x_1 x_2 - .133x_1^2 - 1.14x_2^2$$

    b. A three-dimensional graph of this prediction model is shown in Figure 14.14 (page 716). Note that the height seems to be greatest for temperatures of about 85–90°F and for applications of about 55–57 kilograms of fertilizer per plot.* Further experimentation in these ranges might lead to a more precise determination of the optimal temperature–fertilizer combination.

    c. To determine whether the data provide sufficient information to indicate that the second-order terms contribute information for the prediction of y, we will wish to test

$$H_0: \quad \beta_3 = \beta_4 = \beta_5 = 0$$

---

* Students with knowledge of the calculus should note that we can solve for the exact temperature and amount of fertilizer that maximize height in the least squares model by solving $\partial \hat{y}/\partial x_1 = 0$ and $\partial \hat{y}/\partial x_2 = 0$ for x_1 and x_2. These estimated optimal values are $x_1 = 86.25°F$ and $x_2 = 55.58$ kilograms per plot. Remember, however, that these represent only sample estimates of the optimal values.

Figure 14.14 Plot of Second-Order Least Squares Model for Example 14.4

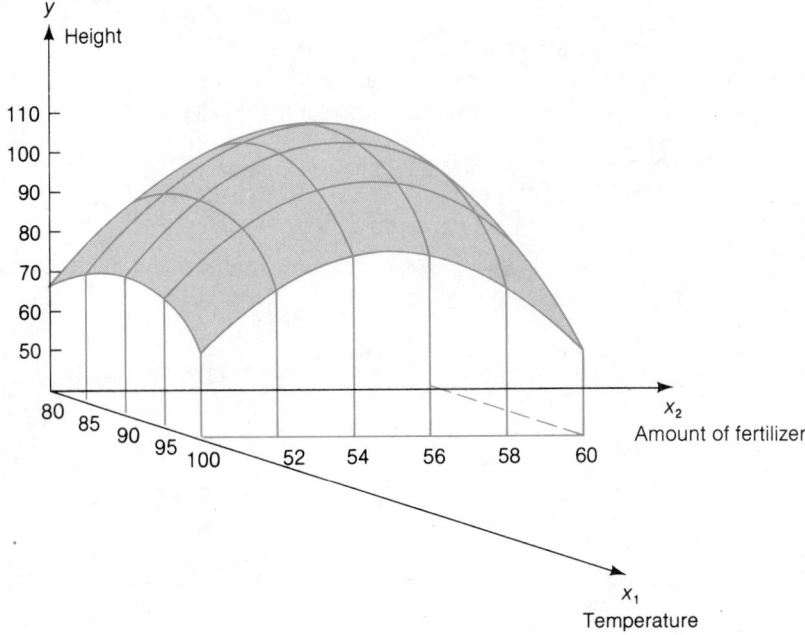

against the alternative hypothesis

H_a: At least one of the parameters β_3, β_4, β_5 differs from zero

The first step in conducting the test is to drop the second-order terms out of the complete (second-order) model and fit the reduced model

$$E(y) = \beta_0 + \beta_1 x_1 + \beta_2 x_2$$

to the data. The SAS computer printout for this procedure is shown in Figure 14.15.

Figure 14.15 SAS Computer Printout for the Reduced (First-Order) Model, Example 14.4

DEPENDENT VARIABLE: Y

SOURCE	DF	SUM OF SQUARES	MEAN SQUARE	F VALUE	PR > F	R-SQUARE
MODEL	2	1789.93444444	894.96722222	3.22	0.0577	0.211540
ERROR	24	6671.50851852	277.97952160		STD DEV	
CORRECTED TOTAL	26	8461.44296296			16.67271788	

PARAMETER	ESTIMATE	T FOR H0: PARAMETER = 0	PR > |T|	STD ERROR OF ESTIMATE
INTERCEPT	106.08518519	1.90	0.0700	55.94500427
X1	−0.91611111	−2.33	0.0285	0.39297973
X2	0.78777778	1.00	0.3262	0.78595946

You can see that the sums of squares for error, given in Figures 14.13 and 14.15 for the complete and reduced models, are

$$SSE_2 = 59.17842582$$

$$SSE_1 = 6,671.50851852$$

and that s^2 for the complete model is

$$s_2^2 = 2.81802028$$

Recall that $n = 27$, $k = 5$, and $g = 2$. Therefore, the calculated value of the F-statistic, based on $v_1 = (k - g) = 3$ numerator df and $v_2 = [n - (k + 1)] = 21$ denominator df, is

$$F = \frac{(SSE_1 - SSE_2)/(k - g)}{SSE_2/[n - (k + 1)]} = \frac{(SSE_1 - SSE_2)/(k - g)}{s_2^2}$$

where $v_1 = (k - g)$ is equal to the number of parameters involved in H_0 and s_2^2 is the value of s^2 for the complete model. Therefore,

$$F = \frac{(6,671.50851852 - 59.17842582)/3}{2.81802028}$$

$$= 782.1$$

The final step in the test is to compare this computed value of F with the tabulated value based on $v_1 = 3$ and $v_2 = 21$ df. If we choose $\alpha = .05$, then $F_{.05} = 3.07$. Since the computed value of F falls in the rejection region (see Figure 14.12), i.e., it exceeds $F_{.05} = 3.07$, we reject H_0 and conclude that at least one of the second-order terms contributes information for the prediction of y. In other words, the data support the contention that the curvature we see in the response surface is not due simply to random variation in the data. The second-order model appears to provide better predictions of y than a first-order model. ■

Example 14.4 demonstrates the motivation for testing an hypothesis that each one of a set of β-parameters equals zero; it also demonstrates the procedure. Other applications of this test will appear in the following sections.

Exercises 14.26–14.35

Learning the Mechanics

14.26 Suppose you fit the regression model

$$y = \beta_0 + \beta_1 x_1 + \beta_2 x_2 + \beta_3 x_1 x_2 + \beta_4 x_1^2 + \beta_5 x_2^2 + \varepsilon$$

to $n = 30$ data points and you wish to test

$$H_0: \quad \beta_3 = \beta_4 = \beta_5 = 0$$

a. State the alternative hypothesis, H_a.

b. Explain in detail how you would find the quantities necessary to compute the F-statistic for this test of hypothesis.

c. What are the numerator and denominator degrees of freedom associated with the F-statistic?

14.27 Suppose you fit the complete and reduced models for the hypothesis test described in Exercise 14.26 and obtain $SSE_1 = 246.1$ and $SSE_2 = 215.2$. Conduct the hypothesis test and interpret the results of your test. Test using $\alpha = .05$.

14.28 Explain why the F-test used to compare complete and reduced models is a one-tailed, upper-tail test.

Applying the Concepts

[*Note: Starred (*) exercises require the use of a computer.*]

14.29 Exercise 13.11 discussed one aspect of the analysis of working wives' "time crunch" data. In their article "The Time Crunch," K. D. Fox and S. Y. Nickols (1983) report on a study of the impact of a wife's employment on the number of hours available for household tasks. Some 206 families were employed in the study. One independent variable measured for each family was the total time y (in minutes per day) that the wife spent on household chores. Two independent variables were also recorded:

WEMP: $x_1 =$ Wife's hours of employment per week

YAGE: $x_2 =$ Age of the youngest child in the family

The computer printout for the regression analysis is repeated here. The sums of squares shown under SEQUENTIAL SS are those you would obtain by sequentially adding vari-

Regression Analyses of Wives' and Husbands' Time Spent in Household Work

DEPENDENT VARIABLE: WIFE'S HOUSEHOLD WORK		(MEAN = 401.3[a])			
SOURCE OF VARIATION	DF	SUMS OF SQUARES	MEAN SQUARE	F	R^2
TOTAL	205	5,432,296		62.10***	0.38
MODEL	2	2,061,918	1,030,959		
ERROR	203	3,370,378,	16,603		
SOURCE WITHIN MODEL	DF	SEQUENTIAL S.S.	PARTIAL S.S.	b-ESTIMATES	
WEMP	1	1,810,154***	1,110,288***	−4.08	
YAGE	1	251,765***	251,765***	−7.02	
DEPENDENT VARIABLE: HUSBAND'S HOUSEHOLD WORK		(MEAN = 105.2[a])			
REGRESSION ANALYSIS NOT SIGNIFICANT					

Note: In this particular printout "b-estimate" is our β-estimate, or parameter estimate. Also, the asterisks indicate the level of significance of test statistics (or sum of squares used to compute test statistics). Here, *** means $p < .001$.

ables to the model. For example, the sum of squares opposite WEMP is for the model

$$E(y) = \beta_0 + \beta_1 x_1$$

The sum of squares opposite YAGE is the drop in SSE for the model obtained by adding x_2, that is, for

$$E(y) = \beta_0 + \beta_1 x_1 + \beta_2 x_2$$

Therefore, this sum of squares can be used in the F-statistic to test H_0: $\beta_2 = 0$.

In contrast, the sums of squares shown under PARTIAL SS are the differences in the sums of squares between reduced and complete models. These are the quantities needed to conduct individual F-tests on both of the parameters, β_1 and β_2.

For example, 1,110,288 is the difference in the SSE values for the models

$$E(y) = \beta_0 + \beta_2 x_2 \quad \text{and} \quad E(y) = \beta_0 + \beta_1 x_1 + \beta_2 x_2$$

Therefore, it can be used to compute the F-statistic for testing H_0: $\beta_1 = 0$. [*Note:* This sum of squares differs substantially from the corresponding sum of squares, 1,810,154, shown under the SEQUENTIAL SS column.]

a. Use the information in the printout to test H_0: $\beta_1 = 0$. Test using $\alpha = .05$ and interpret your results.

b. State the alternative hypothesis implied in the test in part a.

c. Use the information in the printout to test H_0: $\beta_2 = 0$. Test using $\alpha = .05$ and interpret your results.

d. In concluding, Fox and Nickols state that "the wife's employment explained substantially more variation in the wife's housework time than age of the younger child." Do you agree? Explain.

14.30 A large hospital rates the performance of each member of its technical staff once a year. Each person is rated on a scale of 0 to 100 by his or her immediate supervisor, and this merit rating is used to determine the size of the person's pay raise for the coming year. The hospital's personnel department is interested in developing a regression model to help forecast the merit rating that an applicant for a technical position will receive after being employed 3 years. The hospital proposes to use the following model to forecast the merit ratings of applicants who have just completed their graduate studies and have no prior related job experience:

$$E(y) = \beta_0 + \beta_1 x_1 + \beta_2 x_2 + \beta_3 x_1 x_2 + \beta_4 x_1^2 + \beta_5 x_2^2$$

where

$y =$ Applicant's merit rating after 3 years

$x_1 =$ Applicant's grade-point average (GPA) in graduate school

$x_2 =$ Applicant's total score (verbal plus quantitative) on the Graduate Record Examination (GRE)

A random sample of $n = 40$ employees who have been on the technical staff of the hospital more than 3 years is selected. Each employee's merit rating after 3 years, graduate school GPA, and total score on the GRE are recorded. The model is fit to these data with the aid of a computer. A portion of the resulting computer printout is shown here:

SOURCE	DF	SUM OF SQUARES	MEAN SQUARE
MODEL	5	4911.56	982.31
ERROR	34	1830.44	53.84
TOTAL	39	6742.00	R-SQUARE
			0.73

The reduced model $E(y) = \beta_0 + \beta_1 x_1 + \beta_2 x_2$ is fit to the same data. The resulting computer printout is partially reproduced here:

SOURCE	DF	SUM OF SQUARES	MEAN SQUARE
MODEL	2	3544.84	1772.42
ERROR	37	3197.16	86.41
TOTAL	39	6742.00	R-SQUARE
			0.53

a. Identify the appropriate null and alternative hypotheses to test whether the complete (second-order) model contributes information for the prediction of y.

b. Conduct the test of hypothesis given in part a. Test using $\alpha = .05$. Interpret the results in the context of this problem.

c. Identify the appropriate null and alternative hypotheses to test whether the complete model contributes more information than the reduced (first-order) model for the prediction of y.

d. Conduct the test of hypothesis given in part b. Test using $\alpha = .05$. Interpret the results in the context of this problem.

e. Which model, if either, would you use to predict y? Explain.

*14.31 Case Study 14.1 presents Neil A. Palomba's data giving a measure of strike activity (the percentage y of total working hours lost due to strikes) and three quantitative independent variables:

x_1 = Percentage of union members in nonagricultural establishments

x_2 = Percentage of all nonagricultural employment that is manufacturing

x_3 = Hourly earnings of workers on manufacturing payrolls

The SAS computer printout for fitting a first-order model to Palomba's data is shown here. Note that $R^2 = .231906$. Can you improve on Palomba's model?

DEPENDENT VARIABLE: Y				
SOURCE	DF	SUM OF SQUARES	MEAN SQUARE	F VALUE
MODEL	3	0.28839448	0.09613149	4.53
ERROR	45	0.95518919	0.02122643	PR > F
CORRECTED TOTAL	48	1.24358367		0.0074
R-SQUARE	C.V.	ROOT MSE	Y MEAN	
0.231906	88.7929	0.14569292	0.16408163	

a. Use a computer to fit an interaction model to the data, and test to see whether this model provides more information for the prediction of y than does the first-order model. Test using $\alpha = .05$.

b. Fit a complete second-order model to the data. Test to see whether this model provides more information for the prediction of y than the interaction model. Test using $\alpha = .05$.

c. Comment on the results of parts a and b.

*14.32 The data in the table were obtained from an experiment designed to investigate the relationship between the yield y of potatoes and the levels of three minerals in the soil, x_1, x_2, and x_3. [*Note:* The mineral levels have been coded by subtracting an appropriate constant from each of the x-values.]

y	x_1	x_2	x_3	y	x_1	x_2	x_3
16.40	−1	−1	−1	2.75	1	1	1
13.51	1	−1	−1	14.33	−1.682	0	0
14.41	−1	1	−1	5.44	1.682	0	0
9.38	1	1	−1	19.80	0	−1.682	0
10.77	−1	−1	1	20.00	0	1.682	0
11.78	1	−1	1	9.37	0	0	−1.682
4.11	−1	1	1	10.03	0	0	1.682

a. Fit a second-order polynomial,

$$y = \beta_0 + \beta_1 x_1 + \beta_2 x_2 + \beta_3 x_3 + \beta_4 x_1 x_2 + \beta_5 x_1 x_3 + \beta_6 x_2 x_3 + \beta_7 x_1^2$$
$$+ \beta_8 x_2^2 + \beta_9 x_3^2 + \varepsilon$$

by least squares.

b. Test the hypothesis H_0: $\beta_4 = \beta_5 = \cdots = \beta_9 = 0$. That is, test whether the data provide sufficient evidence to indicate that a second-order model contributes more information for the prediction of yield than a first-order model. Test using $\alpha = .05$.

14.33 Refer to Exercise 14.22 in which an energy conservationist wants to develop a regression model to forecast annual gasoline consumption in the United States. The complete and reduced models for the test you described in part d of Exercise 14.22 were fit to $n = 25$ data points. The resulting values of SSE_1 and SSE_2 were 1,065.9 and 400.6, respectively.

a. Conduct the test to determine whether the data present sufficient evidence to indicate interaction between x_1 and x_2. Test using $\alpha = .05$.

b. Which model seems better for forecasting annual gasoline consumption? Why?

*14.34 A firm would like to be able to forecast its yearly sales in each of its sales regions. The firm has decided to base its forecasts on regional population size and its yearly regional advertising expenditures. The population data in the table were obtained from the U.S. Bureau of the Census, and the advertising and sales data were obtained from the firm's internal records.

SALES REGION	SALES (Thousands of units)	POPULATION OF REGION (Thousands)	ADVERTISING EXPENDITURES (Thousands of dollars)
1	65	200	8
2	80	210	10
3	85	205	9
4	100	300	8.5
5	108	320	12
6	114	290	10
7	40	90	6
8	45	85	8
9	150	450	9
10	42	87	9
11	220	480	13
12	200	500	15

a. Fit a complete second-order model to these data.

b. Is the complete second-order model useful for forecasting sales? Test using $\alpha = .05$.

c. The firm is planning to market its product in a new sales region next year. The region has a population of 400,000, and the firm plans to spend $12,000 on advertising. Use the fitted model you obtained in part a to forecast next year's sales in this new region. [*Note:* We would want to express this estimate as a prediction interval, but its computation is beyond the scope of this text. The procedure is described in the references. You may also find that it can be obtained by using your computer program package.]

14.35 Refer to Exercise 14.34 in which a firm would like to develop a regression model to forecast its yearly sales in each of its sales regions. The model under consideration is a complete second-order model:

$$E(y) = \beta_0 + \beta_1 x_1 + \beta_2 x_2 + \beta_3 x_1 x_2 + \beta_4 x_1^2 + \beta_5 x_2^2$$

where

y = Yearly regional sales

x_1 = Population of sales region

x_2 = Yearly regional advertising expenditures

Here is a portion of the computer printout that results from fitting this model to the $n = 12$ data points given in Exercise 14.34:

SOURCE	DF	SUM OF SQUARES	MEAN SQUARE
MODEL	5	38638.97	7727.79
ERROR	6	159.94	26.66
TOTAL	11	38798.91	R-SQUARE
			0.996

The reduced first-order model

$$E(y) = \beta_0 + \beta_1 x_1 + \beta_2 x_2$$

was fit to the same data and the resulting computer printout is partially reproduced here:

SOURCE	DF	SUM OF SQUARES	MEAN SQUARE
MODEL	2	36704.5	18352.2
ERROR	9	2094.4	232.7
TOTAL	11	38798.9	R-SQUARE
			0.946

Do these data present sufficient evidence to conclude that a second-order model contributes more information for the prediction of y than does a first-order model? Test using $\alpha = .05$.

14.5
Models with One Qualitative Independent Variable

Suppose we want to develop a model for the mean operating cost per mile, $E(y)$, of compact cars as a function of the car's manufacturer. (For purposes of explanation, we will ignore other independent variables that might affect the response.) Further suppose there are three manufacturers of interest, which we will identify as A, B, and C. Then the automobile manufacturer is a single qualitative variable with three levels, A, B, and C. Note that with a qualitative independent variable, we cannot attach a meaningful quantitative measure to a given level. Even if we were to call the manufacturers 1, 2, and 3, the numbers would simply be identifiers of the manufacturers and would have no meaningful quantitative interpretation.

To simplify our notation, let μ_A be the mean cost per mile for compact cars produced by manufacturer A, and let μ_B and μ_C be the corresponding mean costs per mile for those produced by B and C. Our objective is to write a single equation that will give the mean value of y (cost per mile) for the three manufacturers. This can be done as follows:

$$E(y) = \beta_0 + \beta_1 x_1 + \beta_2 x_2$$

where

$$x_1 = \begin{cases} 1 & \text{if the car is manufactured by B} \\ 0 & \text{if the car is not manufactured by B} \end{cases}$$

$$x_2 = \begin{cases} 1 & \text{if the car is manufactured by C} \\ 0 & \text{if the car is not manufactured by C} \end{cases}$$

The variables x_1 and x_2 are not meaningful independent variables as for the case of the models with quantitative independent variables. Instead, they are **dummy** (or **indicator**) **variables** that make the model function. To see how they work, let $x_1 = 0$ and $x_2 = 0$. This condition will apply when we are seeking the mean response for A. (Neither B nor C can be the manufacturer; hence, it must be A.) Then the mean cost per mile, $E(y)$, when A is the manufacturer is

$$\mu_A = E(y) = \beta_0 + \beta_1(0) + \beta_2(0) = \beta_0$$

This tells us that the mean cost per mile for A is β_0. Or, using our notation, it means that $\mu_A = \beta_0$.

Now suppose we want to represent the mean cost per mile, $E(y)$, for manufacturer B. Checking the dummy variable definitions, we see that we should let $x_1 = 1$ and $x_2 = 0$:

$$\mu_B = E(y) = \beta_0 + \beta_1 x_1 + \beta_2 x_2 = \beta_0 + \beta_1(1) + \beta_2(0) = \beta_0 + \beta_1$$

or, since $\beta_0 = \mu_A$,

$$\mu_B = \mu_A + \beta_1$$

Then it follows that the interpretation of β_1 is

$$\beta_1 = \mu_B - \mu_A$$

which is the difference between the mean costs per mile for manufacturers B and A.

Finally, if we want the mean value of y when C is the manufacturer, we let $x_1 = 0$ and $x_2 = 1$:

$$\mu_C = E(y) = \beta_0 + \beta_1(0) + \beta_2(1) = \beta_0 + \beta_2$$

or, since $\beta_0 = \mu_A$,

$$\mu_C = \mu_A + \beta_2$$

Then it follows that the interpretation of β_2 is

$$\beta_2 = \mu_C - \mu_A$$

Note that we were able to describe *three levels* of the qualitative variable with only *two dummy variables*. This is because the mean of the base level (manufacturer A, in this case) is accounted for by the intercept β_0.

Figure 14.16 Bar Chart Comparing $E(y)$ for Three Automobile Manufacturers

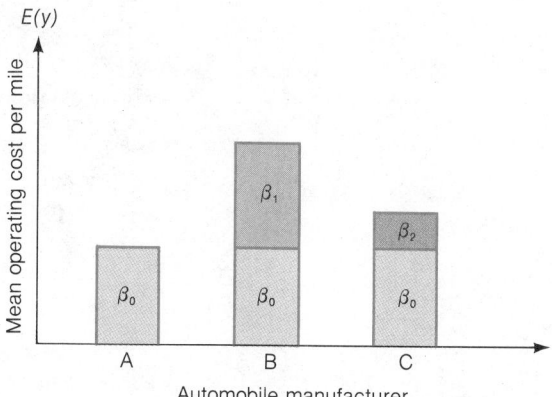

Since the automobile manufacturer is a qualitative variable, we will use a bar graph to show the value of mean operating cost per mile $E(y)$ for the three levels of automobile manufacturer (see Figure 14.16). Particularly, note that the height of the bar, $E(y)$, for each level of automobile manufacturer is equal to the sum of the model parameters shown in the preceding equations. You can see that the height of the bar corresponding to manufacturer A is β_0, that is, $E(y) = \beta_0$. Similarly, the heights of the bars corresponding to manufacturers B and C are $E(y) = \beta_0 + \beta_1$ and $E(y) = \beta_0 + \beta_2$, respectively.*

Now carefully examine the model with a single qualitative independent variable at three levels, because we will use exactly the same pattern for any number of levels. Moreover, the interpretation of the parameters will always be the same.

One level is selected as the base level. (We used manufacturer A as the base level.) Then for the 1−0 system of coding** for the dummy variables,

$$\mu_A = \beta_0$$

The coding for all dummy variables is as follows: To represent the mean value of y for a particular level, let that dummy variable equal 1; otherwise, the dummy variable is set equal to zero. Using this system of coding,

$$\mu_B = \beta_0 + \beta_1 \qquad \mu_C = \beta_0 + \beta_2$$

and so on. Because $\mu_A = \beta_0$, any other model parameter will represent the difference between means for that level and the base level:

$$\beta_1 = \mu_B - \mu_A \qquad \beta_2 = \mu_C - \mu_A$$

and so on.

---

* Note that either β_1, β_2, or both could be negative. If, for example, β_1 were negative, the height of the bar corresponding to manufacturer B would be *reduced* (rather than increased) from the height of the bar for manufacturer A by the amount β_1. Figure 14.16 is constructed assuming that β_1 and β_2 are positive quantities.

** You do not have to use a 1−0 system of coding for the dummy variables. Any two-value system will work, but the interpretation given to the model parameters will depend on the code. Using the 1−0 system makes the model parameters easier to interpret.

Procedure for Writing a Model with One Qualitative Independent Variable at k Levels

---

$$E(y) = \beta_0 + \beta_1 x_1 + \beta_2 x_2 + \cdots + \beta_{k-1} x_{k-1}$$

where x_i is the dummy variable for level i and

$$x_i = \begin{cases} 1 & \text{if } E(y) \text{ is the mean for level } i \\ 0 & \text{otherwise} \end{cases}$$

Then for this system of coding,

$$\mu_A = \beta_0 \qquad \mu_B = \beta_0 + \beta_1$$
$$\mu_C = \beta_0 + \beta_2 \qquad \mu_D = \beta_0 + \beta_3$$

and so on. Note also that

$$\beta_1 = \mu_B - \mu_A \qquad \beta_2 = \mu_C - \mu_A \qquad \beta_3 = \mu_D - \mu_A$$

and so on.

Example 14.5

Suppose a sociologist wants to compare the mean dollar amounts owed by delinquent credit card customers in three different annual income groups: under \$12,000, \$12,000–\$25,000, and over \$25,000. A sample of ten customers with delinquent accounts is selected from each group, and the amount owed by each is recorded as shown in Table 14.4. Do the data provide sufficient evidence to indicate that the mean dollar amounts owed by customers differ for the three annual income groups?

Table 14.4
Dollars Owed

GROUP 1 *Under* \$12,000	GROUP 2 \$12,000–\$25,000	GROUP 3 *Over* \$25,000
\$148	\$513	\$335
76	264	643
393	433	216
520	94	536
236	535	128
134	327	723
55	214	258
166	135	380
415	280	594
153	304	465

Solution

The model relating $E(y)$ to the single qualitative variable, annual income group, is

$$E(y) = \beta_0 + \beta_1 x_1 + \beta_2 x_2$$

where

$$x_1 = \begin{cases} 1 & \text{if group 2} \\ 0 & \text{if not} \end{cases} \qquad x_2 = \begin{cases} 1 & \text{if group 3} \\ 0 & \text{if not} \end{cases}$$

and

$$\beta_1 = \mu_2 - \mu_1 \qquad \beta_2 = \mu_3 - \mu_1$$

where μ_1, μ_2, and μ_3 are the mean responses for income groups 1, 2, and 3, respectively. Testing the null hypothesis that the means for the three groups are equal, that is, $\mu_1 = \mu_2 = \mu_3$, is equivalent to testing

$$H_0: \quad \beta_1 = \beta_2 = 0$$

because if $\beta_1 = \mu_2 - \mu_1 = 0$ and $\beta_2 = \mu_3 - \mu_1 = 0$, then μ_1, μ_2, and μ_3 must be equal. The alternative hypothesis is

$$H_a: \quad \text{At least one of the parameters, } \beta_1 \text{ or } \beta_2, \text{ differs from zero}$$

There are two ways to conduct this test. We can fit the complete model shown above and the reduced model (deleting the terms involving β_1 and β_2)

$$E(y) = \beta_0$$

and conduct the F-test described in the preceding section. (We leave this as an exercise for you.) Or we can use the F-test of the complete model (Section 13.5), which tests the null hypothesis that all parameters in the model, with the exception of β_0, equal zero. Either way you conduct the test, you will obtain the same computed value of F, the value shown on the SAS printout. The SAS printout for fitting the complete model,

$$E(y) = \beta_0 + \beta_1 x_1 + \beta_2 x_2$$

is shown in Figure 14.17; the value of the F-statistic for testing the complete model, $F = 3.48$, is shaded. We will wish to compare this value with the tabulated value of F based on $\nu_1 = 2$ numerator df and $\nu_2 = 27$ denominator df. If we choose $\alpha = .05$, we will reject $H_0: \quad \beta_1 = \beta_2 = 0$ if the computed value of F exceeds $F_{.05} = 3.35$. Since the

Figure 14.17 SAS Computer Printout for Example 14.5

DEPENDENT VARIABLE: Y

SOURCE	DF	SUM OF SQUARES	MEAN SQUARE	F VALUE	PR > F	R-SQUARE
MODEL	2	198772.46666667	99386.23333333	3.48	0.0452	0.205038
ERROR	27	770670.90000000	28543.36666667		STD DEV	
CORRECTED TOTAL	29	969443.36666667			168.94782232	

PARAMETER	ESTIMATE	T FOR H0: PARAMETER = 0	PR > |T|	STD ERROR OF ESTIMATE
INTERCEPT	229.60000000	4.30	0.0002	53.42599243
X1	80.30000000	1.06	0.2973	75.55576307
X2	198.20000000	2.62	0.0141	75.55576307

computed value of F ($F = 3.48$) exceeds $F_{.05} = 3.35$, we reject H_0 and conclude that at least one of the parameters, β_1 or β_2, differs from zero. Or, equivalently, we conclude that the data provide sufficient evidence to indicate that the mean indebtedness does vary from one income group to another. ∎

We must make two additional comments about Example 14.5. A regression analysis is not the easiest way to analyze these data (unless you have ready access to a computer and a good regression program). A simpler procedure for calculating the value of the F-statistic, known as an *analysis of variance,* was described in Chapter 9. If you choose to analyze the data by fitting complete and reduced models (Section 14.4), you will find that the least squares estimate of β_0 in the reduced model

$$E(y) = \beta_0$$

is $\bar{y}$, the mean of all $n = 30$ observations, and that the sum of squares for error for the reduced model is

$$SSE_1 = \sum(y_i - \hat{y}_i)^2 = \sum(y_i - \bar{y})^2 = 969{,}443.367$$

This value is shown in the SAS printout (Figure 14.17) as the sum of squares corresponding to CORRECTED TOTAL. We leave the remaining steps—calculating the drop in SSE and the resulting F-statistic—to you. You will find that the value you obtain is exactly the same as the value of F shown in the SAS printout.

Exercises 14.36–14.45

Learning the Mechanics

14.36 Write a regression model relating the mean value of y to a qualitative independent variable that can assume two levels. Interpret all the terms in the model.

14.37 Write a regression model relating $E(y)$ to a qualitative independent variable that can assume three levels. Interpret all the terms in the model.

14.38 The following model was used to relate $E(y)$ to a single qualitative variable with four levels:

$$E(y) = \beta_0 + \beta_1 x_1 + \beta_2 x_2 + \beta_3 x_3$$

where

$$x_1 = \begin{cases} 1 & \text{if level 2} \\ 0 & \text{if not} \end{cases} \qquad x_2 = \begin{cases} 1 & \text{if level 3} \\ 0 & \text{if not} \end{cases} \qquad x_3 = \begin{cases} 1 & \text{if level 4} \\ 0 & \text{if not} \end{cases}$$

This model was fit to $n = 30$ data points and the following result was obtained:

$$\hat{y} = 10.2 - 4x_1 + 12x_2 + 2x_3$$

Use the least squares prediction equation to find an estimate of $E(y)$ for each level of the qualitative independent variable.

14.39 Refer to Exercise 14.38. Specify the null and alternative hypotheses you would employ to test whether $E(y)$ is the same for all four levels of the independent variable.

Applying the Concepts

[*Note: Starred (*) exercises require the use of a computer.*]

14.40 The manager of a radio and television retail store would like to compare the life span in months of four different brands of color television picture tubes. Data are gathered on ten television sets selected at random from each of four brands, Sony, Zenith, Magnavox, and RCA. Write a model that will give the mean life span for the four brands, and interpret all parameters used in the model.

***14.41** Five varieties of peas are currently being tested in Ohio to determine which is best suited for production in that state. A field was divided into twenty plots, and each variety of peas was planted in four plots. The yield in bushels of peas produced from each plot was as follows:

VARIETY OF PEAS				
A	B	C	D	E
26.2	29.2	29.1	21.3	20.1
24.3	28.1	30.8	22.4	19.3
21.8	27.3	33.9	24.3	19.9
28.1	31.2	32.8	21.8	22.1

a. Write a model for the data to reflect the yield in bushels as a function of pea variety and interpret all parameters in the model.
b. Fit the proposed model to the data.
c. Do the data provide sufficient evidence to indicate the model is useful for predicting harvest yield? Use $\alpha = .05$.
d. Note that the hypothesis tested in part c is equivalent to

$$H_0: \mu_A = \mu_B = \mu_C = \mu_D = \mu_E$$

That is, the mean yield is the same for the five varieties. A test for comparing k population means using a completely randomized design was presented in Section 9.1. Repeat part c using the method given in Section 9.1 for comparing k population means, and verify that the results are identical.

14.42 The manager of a supermarket wants to model the total weekly sales of beer, y, as a function of brand. (This model will enable the manager to plan the store's inventory.) The market carries three brands, B_1, B_2, and B_3.
a. What type of independent variable is brand of beer?
b. Write the model relating mean weekly beer sales, $E(y)$, as a function of brand of beer. Be sure to explain any dummy variables you use.
c. Interpret the parameters (β's) of your model in part b.
d. In terms of the model parameters, what is the mean weekly sales for brand B_3?

14.43 Refer to Exercise 14.42. Suppose the manager uses brand B_1 as the base level and obtains the model

$$\hat{y} = 450 + 60x_1 + 30x_2$$

where

$$x_1 = \begin{cases} 1 & \text{if brand } B_2 \\ 0 & \text{otherwise} \end{cases} \qquad x_2 = \begin{cases} 1 & \text{if brand } B_3 \\ 0 & \text{otherwise} \end{cases}$$

a. What is the estimated difference between the mean* weekly sales for brands B_2 and B_1?

b. What is the estimated mean weekly sales for brand B_2?

14.44 The amount of fructose present in an athlete's bloodstream is critical to performance. A researcher ran an experiment to determine whether diet had any effect on the level of fructose in the blood after 10 minutes of running on a treadmill. Twenty-one subjects were selected, and each subject's fructose level was measured after 10 minutes on a treadmill. Each subject was randomly assigned to one of three diets. One diet was high in protein, one high in carbohydrates, and one high in fruits. After 1 month on the diet, each subject was again run on the treadmill and the fructose level in the bloodstream was measured. The dependent variable was the difference in the level of fructose in the blood between the second and the first runs on the treadmill.

a. Identify the independent variables in the experiment.

b. Write an appropriate regression model relating mean difference in fructose in the blood, $E(y)$, to the independent variables. Identify and code all dummy variables.

14.45 A fisheries researcher believes that the growth rate of fish is related to the age of the fish, its feeding location, and other factors. Suppose that the researcher wishes to collect fish of a certain species from four different locations.

a. Write a model relating the mean growth rate $E(y)$ to location, a qualitative variable.

b. Use your model to find the mean growth rate of the fish at location 1.

c. Use your model to find the mean growth rate of the fish at location 4.

d. In terms of your model, what is the difference in mean growth rates between locations 2 and 3?

14.6
Comparing the Slopes of Two or More Lines

Suppose you wish to relate the mean monthly sales $E(y)$ of a company to monthly advertising expenditure x for three different advertising media (say newspaper, radio, and television), and you wish to use first-order (straight-line) models to model the responses for all three media. Graphs of these three relationships might appear as shown in Figure 14.18.

---

* We would generally form confidence intervals to assess the reliability of these estimates. Our objective in these exercises is to develop the ability to use the models to obtain the estimates. The corresponding confidence intervals can be obtained by using the methods of Chapter 13.

Figure 14.18 Graphs of the Relationship Between Mean Sales $E(y)$ and Advertising Expenditure x

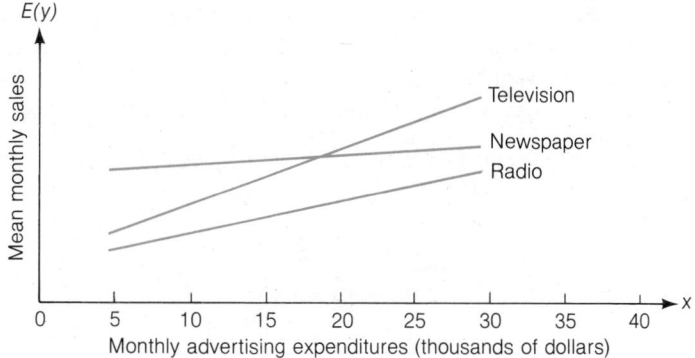

Since the lines in Figure 14.18 are hypothetical, a number of practical business questions arise. Is one advertising medium as effective as any other? That is, do the three mean sales lines differ for the three advertising media? Do the increases in mean sales per dollar input in advertising differ for the three advertising media? That is, do the slopes of the three lines differ? Note that the two practical business questions have been rephrased into questions about the parameters that define the three lines of Figure 14.18. To answer them, we must write a single linear statistical model that will characterize the three lines of Figure 14.18 and that, by testing hypotheses about the lines, will answer the practical business questions.

The response described above, monthly sales, is a function of *two* independent variables, one quantitative (advertising expenditure x_1) and one qualitative (type of medium). We will proceed, in stages, to build a model relating $E(y)$ to these variables and will show graphically the interpretation we would give to the model at each stage. This will help you to see the contributions of the various terms in the model.

1. The straight-line relationship between mean sales $E(y)$ and advertising expenditure is the same for all three media, i.e., a single line will describe the relationship between $E(y)$ and advertising expenditure x_1 for all of the media (see Figure 14.19).

$$E(y) = \beta_0 + \beta_1 x_1 \qquad \text{where } x_1 = \text{Advertising expenditure}$$

Figure 14.19 The Relationship Between $E(y)$ and x_1 Is the Same for All Media

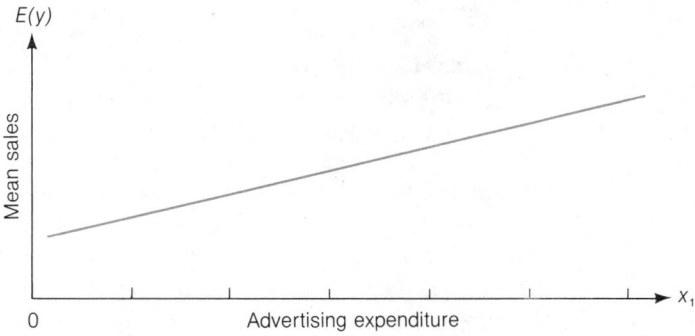

2. The straight lines relating mean sales $E(y)$ to advertising expenditure x_1 differ from one medium to another, but the rate of increase in mean sales per increase in dollar advertising expenditure x_1 is the same for all media, i.e., the lines are parallel but possess different y-intercepts (see Figure 14.20).

$$E(y) = \beta_0 + \beta_1 x_1 + \beta_2 x_2 + \beta_3 x_3$$

where

$$x_1 = \text{Advertising expenditure}$$

$$x_2 = \begin{cases} 1 & \text{if radio medium} \\ 0 & \text{if not} \end{cases}$$

$$x_3 = \begin{cases} 1 & \text{if television medium} \\ 0 & \text{if not} \end{cases}$$

Figure 14.20   Parallel Response Lines for the Three Media

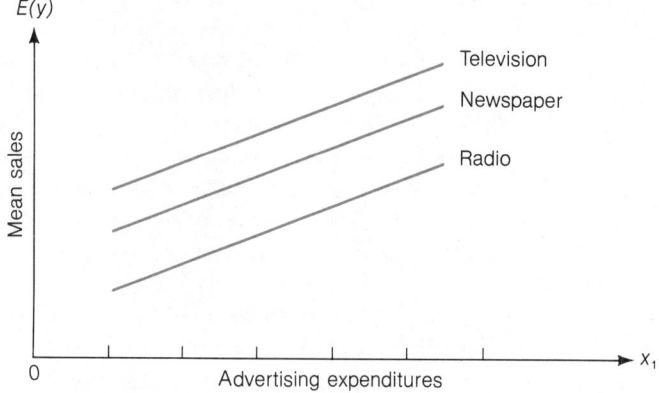

Notice that this model is essentially a combination of a first-order model with a single quantitative variable and the model with a single qualitative variable:

First-order model, single quantitative variable: $E(y) = \beta_0 + \boxed{\beta_1 x_1}$

Model with a single qualitative variable,
three levels: $E(y) = \beta_0 + \boxed{\beta_2 x_2 + \beta_3 x_3}$

where x_1, x_2, and x_3 are defined as above. The model described here implies no interaction between the two independent variables, advertising expenditure x_1, and the qualitative variable, type of advertising medium. The change in $E(y)$ for a 1 unit increase in x_1 is identical (the slopes of the lines are equal) for all three advertising media. The terms corresponding to each of the independent variables are called **main effect** terms because they imply no interaction.

3. The straight lines relating mean sales $E(y)$ to advertising expenditure x_1 differ for the three advertising media; i.e., both line intercepts and slopes differ (see Figure 14.21). As you will see, this interaction model is obtained by adding terms involving the cross-product terms, one each from each of the two independent variables:

Figure 14.21 Different Response Lines for the Three Media

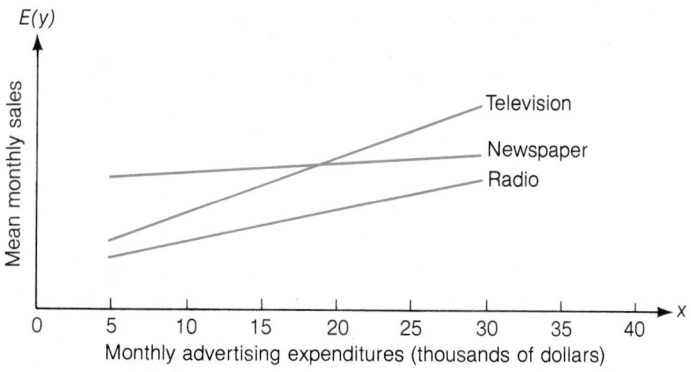

$$E(y) = \beta_0 + \overbrace{\beta_1 x_1}^{\substack{\text{Main effect,} \\ \text{Advertising} \\ \text{expenditure}}} + \overbrace{\beta_2 x_2 + \beta_3 x_3}^{\substack{\text{Main effect,} \\ \text{Type of} \\ \text{medium}}} + \overbrace{\beta_4 x_1 x_2 + \beta_5 x_1 x_3}^{\text{Interaction}}$$

Note that each of the preceding models is obtained by adding terms to model 1, the single first-order model used to model the responses for all three media. Model 2 is obtained by adding the main effect terms for type of medium, the qualitative variable. Model 3 is obtained by adding the interaction terms to model 2.

Will a single line (Figure 14.19) characterize the responses for all three media? Or do all three response lines differ as shown in Figure 14.21? A test of the null hypothesis that a single first-order model adequately describes the relationship between $E(y)$ and advertising expenditure x_1 for all three media is a test of the null hypothesis that the parameters of model 3, β_2, β_3, β_4, and β_5, equal zero, i.e.,

$$H_0: \quad \beta_2 = \beta_3 = \beta_4 = \beta_5 = 0$$

This hypothesis can be tested by fitting the complete model (model 3) and the reduced model (model 1) and conducting an F-test.

Suppose we assume that the response lines for the three media will differ but wonder whether the data present sufficient evidence to indicate a difference among the slopes of the lines. To test the null hypothesis that model 2 adequately describes the relationship between $E(y)$ and advertising expenditure x_1, we wish to test

$$H_0: \quad \beta_4 = \beta_5 = 0$$

i.e., that the two independent variables, advertising expenditure x_1 and type of medium, do not interact. This test can be conducted by fitting the complete model (model 3) and the reduced model (model 2), calculating the drop in the sum of squares for error, and conducting an F-test as described in Section 14.4.

Example 14.6 Substitute the appropriate values of the dummy variables in model 3 to obtain the equations of the three response lines shown in Figure 14.21.

Solution   The complete model that characterizes the three lines in Figure 14.21 is

$$E(y) = \beta_0 + \beta_1 x_1 + \beta_2 x_2 + \beta_3 x_3 + \beta_4 x_1 x_2 + \beta_5 x_1 x_3$$

where

$x_1 = $ Advertising expenditure

$$x_2 = \begin{cases} 1 & \text{if radio medium} \\ 0 & \text{if not} \end{cases}$$

$$x_3 = \begin{cases} 1 & \text{if television medium} \\ 0 & \text{if not} \end{cases}$$

Examining the coding, you can see that $x_2 = x_3 = 0$ when the advertising medium is newspaper. Substituting these values into the expression for $E(y)$, we obtain the newspaper medium line:

$$E(y) = \beta_0 + \beta_1 x_1 + \beta_2(0) + \beta_3(0) + \beta_4 x_1(0) + \beta_5 x_1(0)$$
$$= \beta_0 + \beta_1 x_1$$

Similarly, we substitute the appropriate values of x_2 and x_3 into the expression for $E(y)$ to obtain the radio medium line:

$$E(y) = \beta_0 + \beta_1 x_1 + \beta_2(1) + \beta_3(0) + \beta_4 x_1(1) + \beta_5 x_1(0)$$

$$= \underbrace{(\beta_0 + \beta_2)}_{\text{y-intercept}} + \underbrace{(\beta_1 + \beta_4)}_{\text{Slope}} x_1$$

and the television medium line:

$$E(y) = \beta_0 + \beta_1 x_1 + \beta_2(0) + \beta_3(1) + \beta_4 x_1(0) + \beta_5 x_1(1)$$

$$= \underbrace{(\beta_0 + \beta_3)}_{\text{y-intercept}} + \underbrace{(\beta_1 + \beta_5)}_{\text{Slope}} x_1$$

If you were to fit the general equation (model 3), obtain estimates of $\beta_0, \beta_1, \beta_2, \ldots, \beta_5$, and substitute them into the equations for the three media lines shown above, you would obtain exactly the same prediction equations as you would obtain if you were to fit three separate straight lines, one to each of the three sets of media data. You may ask why we would not fit the three lines separately. Why bother fitting a model that combines all three lines (model 3) into the same equation? The answer is that you need to use this procedure if you wish to use statistical tests to compare the three media lines. We need to be able to express a practical question about the lines in terms of an hypothesis that a set of parameters in the model equal zero. You could not do this if you performed three separate regression analyses and fit a line to each set of media data.

■

Example 14.7   An industrial psychologist conducted an experiment to investigate the relationship between worker productivity and a measure of salary incentive for two manufacturing plants; one, plant A, had union representation and the other, plant B, had nonunion representation. The productivity y per worker was measured by recording the number of machined castings that a worker could produce in a 4-week period of 40 hours per week. The incentive was the amount x_1 of bonus (in cents per casting) paid for all castings produced in excess of 1,000 per worker for the 4-week period. Nine workers were selected from each plant, and three from each group of nine were assigned to receive a 20¢ bonus per casting, three a 30¢ bonus, and three a 40¢ bonus. The productivity data for the nine workers, three for each plant type and incentive combination, are shown in the table.

| | INCENTIVE | | |
TYPE OF PLANT	20¢/casting	30¢/casting	40¢/casting
Union	1,435, 1,512, 1,491	1,583, 1,529, 1,610	1,601, 1,574, 1,636
Nonunion	1,575, 1,512, 1,488	1,635, 1,589, 1,661	1,645, 1,616, 1,689

a. Assume that the relationship between mean productivity and incentive is first-order. Plot the data points and graph the prediction equations for the two productivity lines.

b. Do the data provide sufficient evidence to indicate a difference in mean worker responses to incentives between the two plants?

Solution   If we assume that a first-order model* is adequate to detect a change in mean productivity as a function of incentive x_1, then the model that produces two productivity lines, one for each plant, is

$$E(y) = \beta_0 + \beta_1 x_1 + \beta_2 x_2 + \beta_3 x_1 x_2$$

where

$$x_1 = \text{Incentive} \qquad x_2 = \begin{cases} 1 & \text{if nonunion plant} \\ 0 & \text{if union plant} \end{cases}$$

a. The SAS computer printout for the regression analysis is shown in Figure 14.22 (page 736). Reading the parameter estimates from the printout, you can see that

$$\hat{y} = 1365.833 + 6.217x_1 + 47.778x_2 + .033x_1 x_2$$

The prediction equation for the union plant can be obtained (see the coding) by substituting $x_2 = 0$ into the general prediction equation. Then

$$\hat{y} = \hat{\beta}_0 + \hat{\beta}_1 x_1 + \hat{\beta}_2(0) + \hat{\beta}_3 x_1(0)$$
$$= \hat{\beta}_0 + \hat{\beta}_1 x_1$$
$$= 1,365.833 + 6.217x_1$$

---

* Although the model contains a term involving $x_1 x_2$, it is first-order (graphs as a straight line) in the quantitative variable x_1. The variable x_2 is a dummy variable that introduces or deletes terms in the model. The order of a model is determined only by the quantitative variables that appear in the model.

Figure 14.22 SAS Computer Printout for the Complete Model, Example 14.7

DEPENDENT VARIABLE: Y

SOURCE	DF	SUM OF SQUARES	MEAN SQUARE	F VALUE	PR > F	R-SQUARE
MODEL	3	57332.38888889	19110.79629630	11.46	0.0005	0.710600
ERROR	14	23349.22222223	1667.80158730		STD DEV	
CORRECTED TOTAL	17	80681.61111112			40.83872656	

PARAMETER	ESTIMATE	T FOR H0: PARAMETER = 0	PR > |T|	STD ERROR OF ESTIMATE
INTERCEPT	1365.83333333	26.35	0.0001	51.83641257
X1	6.21666667	3.73	0.0022	1.66723403
X2	47.77777778	0.65	0.5251	73.30775769
X1X2	0.03333333	0.01	0.9889	2.35782498

Similarly, the prediction equation for the nonunion plant is obtained by substituting $x_2 = 1$ into the general prediction equation. Then

$$\hat{y} = \hat{\beta}_0 + \hat{\beta}_1 x_1 + \hat{\beta}_2 x_2 + \hat{\beta}_3 x_1 x_2$$
$$= \hat{\beta}_0 + \hat{\beta}_1 x_1 + \hat{\beta}_2(1) + \hat{\beta}_3 x_1(1)$$

$$= \overbrace{(\hat{\beta}_0 + \hat{\beta}_2)}^{\text{y-intercept}} + \overbrace{(\hat{\beta}_1 + \hat{\beta}_3)}^{\text{Slope}} x_1$$
$$= (1{,}365.833 + 47.778) + (6.217 + .033)x_1$$
$$= 1{,}413.611 + 6.250x_1$$

A graph of these prediction equations is shown in Figure 14.23.

Figure 14.23 Graphs of the Prediction Equations for the Two Productivity Lines, Example 14.7

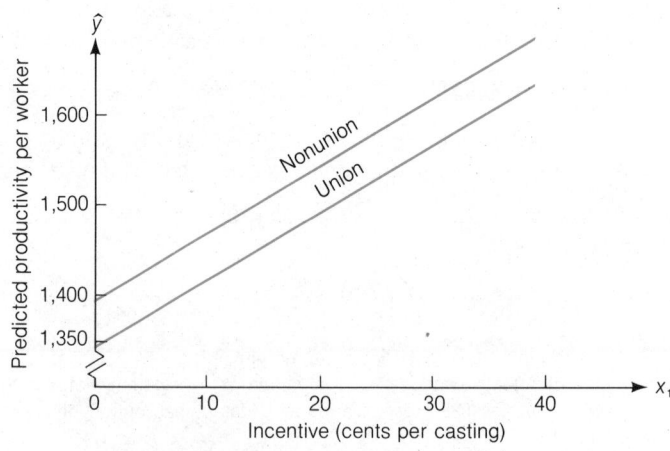

b. To determine whether the data provide sufficient evidence to indicate a difference in mean worker responses to incentives between the two plants, we test the null hypothesis that a *single* line characterizes the relationship between productivity per worker y and the amount of incentive x_1 against the alternative hypothesis that we need two separate lines to characterize the relationship—one for each plant. If there is no difference in mean response $E(y)$ to x_1 between the two plants, then we do not need the variable Type of plant in the model, i.e., we do not need the terms involving x_2. Therefore, we wish to test

$$H_0: \quad \beta_2 = \beta_3 = 0$$

against the alternative hypothesis

$H_a:$ At least one of the two parameters, β_2 or β_3, is not equal to zero

The SAS computer printout for fitting the reduced model

$$E(y) = \beta_0 + \beta_1 x_1$$

to the data is shown in Figure 14.24. Reading SSE_2 and SSE_1 from Figures 14.22 and 14.24, respectively, we obtain

Complete model: $SSE_2 = 23{,}349.22$

Reduced model: $SSE_1 = 34{,}056.28$

Drop in SSE $= SSE_1 - SSE_2 = 10{,}707.06$

The value of s^2 for the complete model (Figure 14.22) is 1,667.80. Substituting these values, along with $k = 3$ and $g = 1$, into the formula for the F-statistic yields

$$F = \frac{(SSE_1 - SSE_2)/(k - g)}{s^2} = \frac{(10{,}707.06)/2}{1{,}667.80}$$

$$= 3.21$$

Figure 14.24   SAS Computer Printout for the Reduced Model, Example 14.7

DEPENDENT VARIABLE: Y

SOURCE	DF	SUM OF SQUARES	MEAN SQUARE	F VALUE	PR > F	R-SQUARE
MODEL	1	46625.33333333	46625.33333333	21.91	0.0003	0.577893
ERROR	16	34056.27777778	2128.51736111		STD DEV	
CORRECTED TOTAL	17	80681.61111112			46.13585765	

PARAMETER	ESTIMATE	T FOR H0: PARAMETER = 0	PR > |T|	STD ERROR OF ESTIMATE
INTERCEPT	1389.72222222	33.56	0.0001	41.40819949
X1	6.23333333	4.68	0.0003	1.33182749

The numerator degrees of freedom (the number of parameters involved in H_0) is $v_1 = 2$ and the denominator degrees of freedom (the degrees of freedom associated with s^2 in the complete model) is $v_2 = 14$. If we choose $\alpha = .05$, the tabulated value of $F_{.05}$ given in Table VIII of the Appendix is 3.74. Since the computed value, $F = 3.21$, is less than the tabulated value, $F_{.05} = 3.74$, there is insufficient evidence (at the $\alpha = .05$ significance level) to indicate a difference in mean worker responses to incentives between the two plants. Therefore, there is no evidence to indicate that two different lines, one for each plant, are needed to describe the relationship between mean productivity per worker $E(y)$ and the amount of incentive x_1.   ■

Example 14.8

Refer to Example 14.7 and explain how you would determine whether the data provide sufficient evidence to indicate that the incentive x_1 affects mean productivity.

Solution

If the variable Incentive did not affect mean productivity, then we would not need terms involving x_1 in the model. Therefore, we would test the null hypothesis

$$H_0: \quad \beta_1 = \beta_3 = 0$$

against the alternative hypothesis

$$H_a: \quad \text{At least one of the parameters, } \beta_1 \text{ or } \beta_3, \text{ differs from zero}$$

We would fit the reduced model

$$E(y) = \beta_0 + \beta_2 x_2$$

to the data and find SSE_1. The values of SSE_2 and s^2 for the complete model would be the same as those used in Example 14.7. Finally, you would calculate the value of the F-statistic and compare it with a tabulated value of F based on $v_1 = 2$ numerator df and $v_2 = 14$ denominator df. If the test leads to rejection of H_0, you have evidence to indicate that the increase in mean productivity that appears to be present in the graphs shown in Figure 14.23 is not a figment of your imagination.   ■

Exercises 14.46–14.58

Learning the Mechanics

14.46 Write a first-order model that relates $E(y)$ to one quantitative independent variable.

14.47 Add the main effect terms for one qualitative variable at three levels to the model of Exercise 14.46.

14.48 Add terms to the model of Exercise 14.47 to allow for interaction between the quantitative and qualitative independent variables.

14.49 Under what circumstances will the response lines of the model in Exercise 14.48 be parallel?

14.50 Under what circumstances will the model of Exercise 14.48 have only one response line?

Applying the Concepts

[*Note:* Starred (*) exercises require the use of a computer.]

14.51 Researchers for a dog food company have developed a new puppy food that they hope will compete with the major brands. One premarketing test involved the comparison of the new food with that of two competitors in terms of weight gain. Fifteen 8-week-old German shepherd puppies, each from a different litter, were divided into three groups of five puppies each. Each group was fed one of the three brands of food.

a. Set up a model that assumes the final weight y is linearly related to initial weight x but does not allow for differences among the three brands; i.e., assume that the response curve is the same for the three brands of dog food. Sketch the response curve as it might appear.

b. Set up a model that assumes the final weight is linearly related to initial weight and allows the intercepts of the lines to differ for the three brands. In other words, assume that the initial weight and brand both affect final weight, but the two variables do not interact. Sketch typical response curves.

c. Now write the main effects with interaction model. For this model we assume that the final weight is linearly related to initial weight, but both the slopes and the intercepts of the lines depend on the brand. Sketch typical response curves.

14.52 A company is studying three different safety programs, A, B, and C, in an attempt to reduce the number of work-hours lost due to accidents. Each program is to be tried at three of the company's nine factories, and the plan is to monitor the lost work-hours, y, for a 1-year period beginning 6 months after the new safety program is instituted.

a. Write a main effects model relating $E(y)$ to the lost work-hours, x_1, the year before the plan is instituted and to the type of program that is instituted.

b. In terms of the model parameters from part a, what hypothesis would you test to determine whether the mean work-hours lost differ for the three safety programs?

14.53 Refer to Exercise 14.52. After the three safety programs have been in effect for 18 months, the complete main effects model is fit to the $n = 9$ data points. Using safety program A as the base level, the following results were obtained:

$$\hat{y} = -2.1 + .88x_1 - 150x_2 + 35x_3 \qquad \text{SSE} = 1{,}527.27$$

Then the reduced model $E(y) = \beta_0 + \beta_1 x_1$ is fit, with the result

$$\hat{y} = 15.3 + .84x_1 \qquad \text{SSE} = 3{,}113.14$$

Test to determine whether the mean work-hours lost differ for the three programs. Use $\alpha = .05$.

14.54 An insurance company is experimenting with three different training programs, A, B, and C, for its salespeople. The following main effects model is proposed:

$$E(y) = \beta_0 + \beta_1 x_1 + \beta_2 x_2 + \beta_3 x_3$$

where

y = Monthly sales (in thousands of dollars)

x_1 = Number of months experience

$x_2 = \begin{cases} 1 & \text{if training program B was used} \\ 0 & \text{otherwise} \end{cases}$

$x_3 = \begin{cases} 1 & \text{if training program C was used} \\ 0 & \text{otherwise} \end{cases}$

Training program A is the base level.

a. What hypothesis would you test to determine whether the mean monthly sales differ for salespeople trained by the three programs?

b. After experimenting with fifty salespeople over a 5-year period, the complete model is fit with the following result:

$$\hat{y} = 10 + .5x_1 + 1.2x_2 - .4x_3 \qquad \text{SSE} = 140.5$$

Then the reduced model $E(y) = \beta_0 + \beta_1 x_1$ is fit to the same data with the following result:

$$\hat{y} = 11.4 + .4x_1 \qquad \text{SSE} = 183.2$$

Test the hypothesis you formulated in part a. Use $\alpha = .05$.

*14.55 In Exercise 12.17, we fit a regression line to relate the mean gain in weight of a rainbow trout to the trout's ration level (Kovacs & Leduc, 1982). Twenty fish were fed at each of the four ration levels while living in water maintained at 6°C. We also fit regression lines to similar data for fish fed in water at 12°C and at 18°C. The table gives the means and standard deviations (in parentheses) for each sample of .twenty fish.

RATION (% body weight per day)	MEAN WET WEIGHT GAIN (%) 6°C	12°C	18°C
.0	−8.14 (3.27)	−10.33 (2.35)	−13.21 (1.96)
.8	12.31 (3.37)		
1.2		14.29 (5.32)	
1.5	28.19 (5.39)		15.51 (4.48)
2.5	29.12 (5.67)	38.05 (6.99)	
3.5			51.13 (7.14)
4.0		60.13 (12.71)	
4.5			65.49 (13.04)
Maintenance ration	.32	.48	.69

a. Write a single model to relate the mean weight gain to ration level and water temperature (°C). A graphic representation of your model should depict three straight lines with differing intercepts and equal slopes.

b. Use a computer to fit the model in part a to the sample means.

c. Write a single model to relate the mean weight gain to ration level and water temperature (°C). A graphic representation of your model should depict three straight lines with differing intercepts and slopes.

d. Use a computer to fit the model in part c to the sample means.

e. Do the data provide sufficient evidence to indicate differences in mean gain in weight for a 1% change in the ration level (i.e., the slopes of the lines) for the three levels of water temperatures? Test using $\alpha = .05$.

14.56 Many women suffer from anemia. A female physician, who is also an avid jogger, wanted to know if women who exercise regularly have a different mean red blood cell count than women who do not. She also wanted to know if the amount of a particular iron supplement a woman takes has any effect and whether the effect is the same for both groups. Write a model that will reflect the relationship between red blood cell count and the two independent variables described above.

a. Assume that the effect of the iron supplement on mean blood cell count is the same regardless of whether a woman exercises regularly.

b. Assume that the effect of the iron supplement on mean blood cell count depends on whether a woman exercises regularly.

14.57 A research physician wishes to find a model that will predict the mean time to relief after the administration of a certain drug. Two important independent variables are considered to be good predictors of relief time. They are age of the patient and method of administration. Physicians administer a standard dose of a drug to a patient in one of three different ways: orally in liquid form (method 1), orally in pill form (method 2), and intravenously (method 3). Consequently, we might use the model

$$E(y) = \beta_0 + \beta_1 x_1 + \beta_2 x_2 + \beta_3 x_3 + \beta_4 x_1 x_2 + \beta_5 x_1 x_3$$

where

$y =$ Time to relief (in minutes)

$x_1 =$ Age of patient (in years)

$x_2 = \begin{cases} 1 & \text{if drug is administered orally in pill form} \\ 0 & \text{if not} \end{cases}$

$x_3 = \begin{cases} 1 & \text{if drug is administered intravenously} \\ 0 & \text{if not} \end{cases}$

The data in the table, obtained on twelve patients, were used to fit this model.

METHOD 1		METHOD 2		METHOD 3	
Age	Time to relief	Age	Time to relief	Age	Time to relief
51	22	46	28	37	19
36	25	40	24	60	17
31	20	26	23	25	21
20	25	32	25	38	20

a A multiple regression analysis produced the results shown at the top of the next page. Test whether the model is useful in predicting mean time to relief. Use $\alpha = .05$.

b. What hypothesis would you use to test whether the age–drug method interaction terms contribute to the prediction of mean time to relief?

SOURCE	DF	SUM OF SQUARES	MEAN SQUARE	F	PR > F	R-SQUARE
MODEL	5	87.473	17.495	4.90	0.0394	0.803120
ERROR	6	21.443	3.574			
CORRECTED TOTAL	11	108.916				

c. The reduced model

$$E(y) = \beta_0 + \beta_1 x_1 + \beta_2 x_2 + \beta_3 x_3$$

was fit to the data and produced an SSE of 38.289. Does this result provide sufficient evidence at the $\alpha = .05$ level of significance to indicate that the interaction ·terms should be kept in the model?

d. Use an available computer program to check the results given in parts a and c.

14.58 An economist is interested in modeling the mean monthly demand $E(y)$ (in thousands of units) for a certain product as a function of its price (in dollars) and the season of the year. The following model has been proposed:

$$E(y) = \beta_0 + \beta_1 x_1 + \beta_2 x_2 + \beta_3 x_3 + \beta_4 x_4$$

where

$x_1 = $ Price

$$x_2 = \begin{cases} 1 & \text{if spring} \\ 0 & \text{otherwise} \end{cases} \quad x_3 = \begin{cases} 1 & \text{if summer} \\ 0 & \text{otherwise} \end{cases} \quad x_4 = \begin{cases} 1 & \text{if fall} \\ 0 & \text{otherwise} \end{cases}$$

A portion of the Minitab computer printout that results from fitting this model to a sample of 16 months of sales data selected from among the last 2 years is shown here:

```
THE REGRESSION EQUATION IS
Y =    12.1 —  1.11 X1 +  3.94 X2
      + 7.17 X3 +  3.72 X4
```

	COLUMN	COEFFICIENT	ST. DEV. OF COEF.	T-RATIO = COEF/S.D.
	—	12.067	1.473	8.19
X1	C2	−1.113	.187	−5.93
X2	C3	3.942	.858	4.59
X3	C4	7.166	.815	8.80
X4	C5	3.724	.819	4.55

```
THE ST. DEV. OF Y ABOUT REGRESSION LINE IS
S =        1.135
WITH (   16 —  5) =  11 DEGREES OF FREEDOM

R-SQUARED = 93.1 PERCENT
R-SQUARED = 90.6 PERCENT, ADJUSTED FOR D.F.

ANALYSIS OF VARIANCE
```

DUE TO	DF	SS	MS = SS/DF
REGRESSION	4	191.590	47.697
RESIDUAL	11	14.160	1.287
TOTAL	15	205.750	

The reduced model $E(y) = \beta_0 + \beta_1 x_1$ was also fit to the same data and the resulting computer printout is partially reproduced here:

```
THE REGRESSION EQUATION IS
Y =   17.4 −  1.35 X1

                                              ST. DEV.        T-RATIO =
               COLUMN          COEFFICIENT     OF COEF.        COEF/S.D.
                 —               17.394          2.833           6.08
X1             C2                −1.348           .403          −3.34

THE ST. DEV. OF Y ABOUT REGRESSION LINE IS
S =       2.859
WITH (   16 −  2) = 14 DEGREES OF FREEDOM

R-SQUARED = 44.4 PERCENT
R-SQUARED = 40.4 PERCENT, ADJUSTED FOR D.F.

ANALYSIS OF VARIANCE

DUE TO          DF          SS       MS = SS/DF
REGRESSION       1       91.345       91.345
RESIDUAL        14      114.405        8.172
TOTAL           15      205.750
```

a. Do these data present sufficient evidence to conclude that mean monthly demand depends on the season of the year? Test using $\alpha = .05$.

b. Find an estimate for $E(y)$ when the price is $8 and it is summer. Interpret your result. [*Note:* We would prefer to use a confidence interval for this estimate, but its calculation is beyond the scope of this text. The procedure can be found in the references.]

14.7 Comparing Two or More Response Curves

Suppose we think that the relationship between mean monthly sales $E(y)$ and advertising expenditure x_1 (Section 14.6) is second-order. We will construct, stage by stage, a model relating $E(y)$ to the quantitative variable x_1 and the qualitative variable, Type of medium. This will enable you to compare the procedure with the stage-by-stage construction of the first-order model (Section 14.6). The graphic interpretations will help you understand the contributions of the model terms.

1. The mean sales curves are identical for all three advertising media, i.e., a single second-order curve will suffice to describe the relationship between $E(y)$ and x_1 for all the media (see Figure 14.25, page 744):

$$E(y) = \beta_0 + \beta_1 x_1 + \beta_2 x_1^2 \qquad \text{where } x_1 = \text{Advertising expenditure}$$

2. The response curves possess the same shapes but different y-intercepts (see Figure 14.26):

$$E(y) = \beta_0 + \beta_1 x_1 + \beta_2 x_1^2 + \beta_3 x_2 + \beta_4 x_3$$

Figure 14.25 The Relationship Between $E(y)$ and x_1 Is the Same for All Media

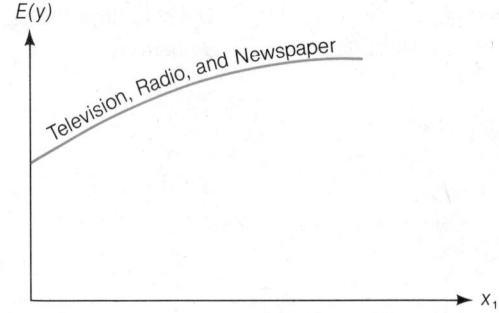

Figure 14.26 The Response Curves Have the Same Shapes But Different y-Intercepts

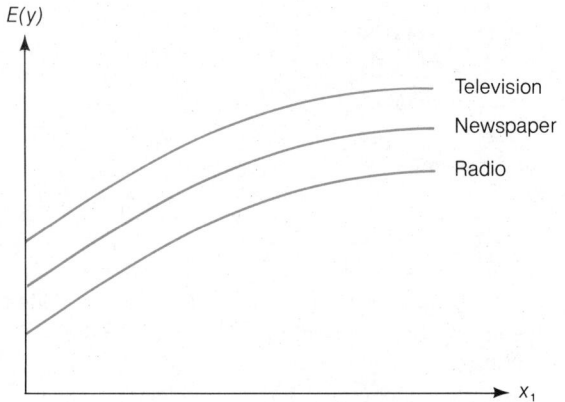

where

$x_1 =$ Advertising expenditure

$$x_2 = \begin{cases} 1 & \text{if radio medium} \\ 0 & \text{if not} \end{cases} \qquad x_3 = \begin{cases} 1 & \text{if television medium} \\ 0 & \text{if not} \end{cases}$$

3. The response curves for the three advertising media are different (i.e., Advertising expenditure and Type of medium interact), as shown in Figure 14.27:

$$E(y) = \beta_0 + \beta_1 x_1 + \beta_2 x_1^2 + \beta_3 x_2 + \beta_4 x_3 + \beta_5 x_1 x_2 + \beta_6 x_1 x_3 + \beta_7 x_1^2 x_2 + \beta_8 x_1^2 x_3$$

Now that you know how to write a model with two independent variables—one qualitative and one quantitative—we ask a question. Why do it? Why not write a separate second-order model for each type of medium where $E(y)$ is a function of only advertising expenditure? One reason we wrote the single model representing all three response curves is so that we can test to determine whether the curves are different. We will illustrate this procedure in Examples 14.10 and 14.11. A second reason for writing a single model is that we obtain a pooled estimate of σ^2, the variance of the random error component ε. If the variance of ε is truly the same for each type of medium,

Figure 14.27 The Response Curves for the Three Media Differ

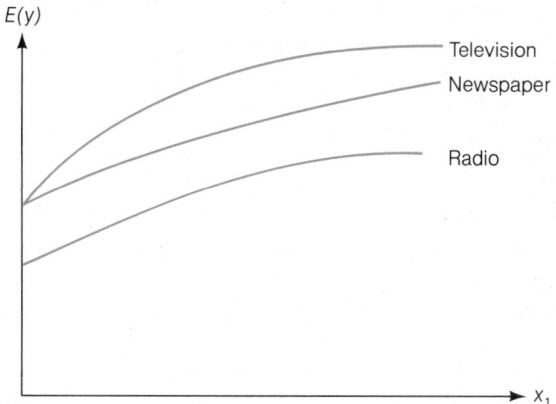

the pooled estimate is superior to calculating three separate estimates by fitting a separate model for each type of medium.

Example 14.9 Give the equation of the second-order model for the radio advertising medium.

Solution Model 3 characterizes the relationship between $E(y)$ and x_1 for the radio advertising medium (see the coding) when $x_2 = 1$ and $x_3 = 0$. Substituting these values into model 3, we obtain

$$E(y) = \beta_0 + \beta_1 x_1 + \beta_2 x_1^2 + \beta_3 x_2 + \beta_4 x_3 + \beta_5 x_1 x_2 + \beta_6 x_1 x_3 + \beta_7 x_1^2 x_2 + \beta_8 x_1^2 x_3$$
$$= \beta_0 + \beta_1 x_1 + \beta_2 x_1^2 + \beta_3(1) + \beta_4(0) + \beta_5 x_1(1) + \beta_6 x_1(0) + \beta_7 x_1^2(1) + \beta_8 x_1^2(0)$$
$$= (\beta_0 + \beta_3) + (\beta_1 + \beta_5)x_1 + (\beta_2 + \beta_7)x_1^2 \qquad \blacksquare$$

Example 14.10 What null hypothesis about the parameters of model 3 would you test if you wished to determine whether the second-order curves for the three media are identical?

Solution If the curves were identical, we would not need the independent variable Type of medium in the model; i.e., we would delete all terms involving x_2 and x_3. This would produce model 1

$$E(y) = \beta_0 + \beta_1 x_1 + \beta_2 x_1^2$$

and the null hypothesis would be

$$H_0: \quad \beta_3 = \beta_4 = \beta_5 = \beta_6 = \beta_7 = \beta_8 = 0 \qquad \blacksquare$$

Example 14.11 Suppose we assume that the response curves for the three media differ, but we want to know whether the second-order terms contribute information for the prediction of y. Or, equivalently, will a second-order model give better predictions than a first-order model?

Solution　The only difference between model 3 and a first-order model are those terms involving x_1^2. Therefore, the null hypothesis, "second-order terms contribute no information for the prediction of y," is equivalent to

$$H_0: \quad \beta_2 = \beta_7 = \beta_8 = 0$$

Examples 14.10 and 14.11 identify two tests that answer practical questions concerning a collection of second-order models. Other comparisons between the curves can be made by testing appropriate sets of model parameters (see the exercises).

The models described in the preceding sections provide only an introduction to statistical modeling. Models can be constructed to relate $E(y)$ to any number of quantitative and qualitative independent variables. You can compare response curves and surfaces for different levels of a qualitative variable or for different combinations of levels of two or more qualitative independent variables. A general explanation of how to write linear statistical models can be found in Mendenhall & McClave (1981, Chapter 6).

Exercises 14.59–14.70

Learning the Mechanics

14.59 Write a complete second-order model that relates $E(y)$ to one quantitative independent variable.

14.60 Add the main effect terms for one qualitative variable at three levels to the model of Exercise 14.59.

14.61 Add terms to the model of Exercise 14.60 to allow for interaction between the quantitative and qualitative independent variables.

14.62 Under what circumstances will the response curves of the model in Exercise 14.61 possess the same shape but have different y-intercepts?

14.63 Under what circumstances will the response curves of the model in Exercise 14.61 be parallel lines?

14.64 Under what circumstances will the response curves of the model in Exercise 14.61 be identical?

14.65 Write a model that relates $E(y)$ to two independent variables—one quantitative and one qualitative—at four levels. Construct a model that allows the associated response curves to be second-order but does not allow for interaction between the two independent variables.

Applying the Concepts

[*Note: Starred (*) exercises require the use of a computer.*]

14.66 An equal rights group has charged that women are being discriminated against in terms of the salary structure in a state university system. It is thought that a complete second-order model will be adequate to describe the relationship between salary and years of experience for both men and women. A sample is to be taken from the records for faculty members (all of equal rank) within the system and the following model is to be fit:*

$$E(y) = \beta_0 + \beta_1 x_1 + \beta_2 x_1^2 + \beta_3 x_2 + \beta_4 x_1 x_2 + \beta_5 x_1^2 x_2$$

where

y = Annual salary (in thousands of dollars)

x_1 = Experience (years)

$x_2 = \begin{cases} 1 & \text{if female} \\ 0 & \text{if male} \end{cases}$

a. What hypothesis would you test to determine whether the *rate* of increase of mean salary with experience is different for males and females?

b. What hypothesis would you test to determine whether there are differences in mean salaries that are attributable to sex?

14.67 Refer to Exercise 14.66 and the model that was proposed to describe the relationship between salary and years of experience for both men and women. Here is a portion of the computer printout that results from fitting this model to a sample of 200 faculty members in the university system:

SOURCE	DF	SUM OF SQUARES	MEAN SQUARE
MODEL	5	2351.70	470.34
ERROR	194	783.90	.04
TOTAL	199	3135.60	R-SQUARE
			0.750

The reduced model $E(y) = \beta_0 + \beta_1 x_1 + \beta_2 x_1^2$ is fit to the same data, and the resulting computer printout is partially reproduced here:

SOURCE	DF	SUM OF SQUARES	MEAN SQUARE
MODEL	2	2340.37	1170.185
ERROR	197	795.23	4.04
TOTAL	199	3135.60	R-SQUARE
			0.746

Do these data provide sufficient evidence to support the claim that the mean salary of faculty members is dependent on sex? Use $\alpha = .05$.

* In practice, we would include other variables in the model. We include only two here to simplify the exercise.

14.68 A medical researcher wants to model the extent of lung damage in emphysema patients as a function of two variables: the number of years the patient has smoked (x_1) and the sex of the patient (x_2). Fifty emphysema patients are used in the study, and the response y for each patient is a subjective score ranging from 0 to 50. (High scores indicate extensive lung damage.)

a. Identify the independent variables as qualitative or quantitative.

b. Write the main effects (first-order) model for predicting the mean lung damage score, $E(y)$, from the two variables identified in part a.

c. Using a female patient as the base level, the researcher obtained the prediction equation

$$\hat{y} = -.85 + .65x_1 + .09x_2$$

What is the estimated difference between the mean lung damage scores for a male and female who have both smoked for 20 years?

d. Find the estimated mean lung damage score for a female patient who has smoked for 15 years.

[*Note:* With the original data and an appropriate computer program package, you would be able to produce confidence intervals for the estimates in parts c and d.]

e. It would not be surprising if the functional relation between lung damage and years of smoking were second-order. Further, it is conceivable that years of smoking might affect men differently than women. Write a model that allows for these conditions.

14.69 An operations manager is interested in modeling $E(y)$, the expected length of time per month (in hours) that a machine will be shut down for repairs, as a function of the type of machine (001 or 002) and the age of the machine (in years). The manager has proposed the following model:

$$E(y) = \beta_0 + \beta_1 x_1 + \beta_2 x_1^2 + \beta_3 x_2$$

where

$$x_1 = \text{Age of machine} \qquad x_2 = \begin{cases} 1 & \text{if machine type 001} \\ 0 & \text{if machine type 002} \end{cases}$$

Data obtained on $n = 20$ machine breakdowns were used to estimate the parameters of this model. A portion of the regression analysis computer printout is shown here:

SOURCE	DF	SUM OF SQUARES	MEAN SQUARE
MODEL	3	2396.364	798.788
ERROR	16	128.586	8.037
TOTAL	19	2524.950	R-SQUARE
			0.949

The reduced model $E(y) = \beta_0 + \beta_1 x_1 + \beta_3 x_2$ was fit to the same data. The regression analysis computer printout is partially reproduced here:

SOURCE	DF	SUM OF SQUARES	MEAN SQUARE
MODEL	2	2342.42	1171.21
ERROR	17	182.53	10.74
TOTAL	19	2524.95	R-SQUARE
			0.928

Do these data provide sufficient evidence to conclude that the second-order (x_1^2) term in the model proposed by the operations manager is necessary? Test using $\alpha = .05$.

*14.70 Refer to Exercise 14.69. The data used to fit the operations manager's complete and reduced models are displayed in the table.

DOWNTIME (Hours per month)	MACHINE AGE x_1 (years)	MACHINE TYPE	x_2	DOWNTIME (Hours per month)	MACHINE AGE x_1 (years)	MACHINE TYPE	x_2
10	1.0	001	1	10	2.0	002	0
20	2.0	001	1	20	4.0	002	0
30	2.7	001	1	30	5.0	002	0
40	4.1	001	1	44	8.0	002	0
9	1.2	001	1	9	2.4	002	0
25	2.5	001	1	25	5.1	002	0
19	1.9	001	1	20	3.5	002	0
41	5.0	001	1	42	7.0	002	0
22	2.1	001	1	20	4.0	002	0
12	1.1	001	1	13	2.1	002	0

a. Use these data to test the null hypothesis that $\beta_1 = \beta_2 = 0$. Test using $\alpha = .10$.
b. Interpret the results of the test in the context of the problem.

14.8
Model Building: Stepwise Regression

The problem of predicting executive salaries was discussed in Chapter 13. Perhaps the biggest problem in building a model to describe executive salaries is choosing the important independent variables to be included in the model. The list of potentially important independent variables is extremely long, and we need some objective method of screening out those that are not important.

The problem of deciding which of a large set of independent variables to include in a model is a common one. Trying to determine which variables influence the profit of a firm, affect blood pressure of humans, or are related to a student's performance in college are only a few examples.

A systematic approach to building a model with a large number of independent variables is difficult because the interpretation of multivariable interactions and higher-order polynomials is tedious. We therefore turn to a screening procedure known as **stepwise regression.**

The most commonly used stepwise regression procedure, available in most popular computer packages, works as follows: The user first identifies the response, y, and the set of potentially important independent variables, $x_1, x_2, \ldots, x_k$, where k is generally large. [*Note:* This set of variables could include both first-order and higher-order terms. However, we may often include only the main effects of both quantitative variables (first-order terms) and qualitative variables (dummy variables), since the inclusion of second-order terms greatly increases the number of independent variables.] The response and independent variables are then entered into the computer, and the stepwise procedure begins.

Step 1. The computer fits all possible one-variable models of the form

$$E(y) = \beta_0 + \beta_1 x_i$$

to the data, where x_i is the ith independent variable, $i = 1, 2, \ldots, k$. For each model, the test of the null hypothesis

$$H_0: \quad \beta_1 = 0$$

against the alternative hypothesis

$$H_a: \quad \beta_1 \neq 0$$

is conducted using the t (or the equivalent F) test for a single β-parameter. The independent variable that produces the largest (absolute) t-value is declared the best one-variable predictor of y.* Call this independent variable x_1.

Step 2. The stepwise program now begins to search through the remaining $(k-1)$ independent variables for the best two-variable model of the form

$$E(y) = \beta_0 + \beta_1 x_1 + \beta_2 x_i$$

This is done by fitting all two-variable models containing x_1 and each of the other $(k-1)$ options for the second variable x_i. The t-values for the test $H_0: \quad \beta_2 = 0$ are computed for each of the $(k-1)$ models (corresponding to the remaining independent variables x_i, $i = 2, 3, \ldots, k$), and the variable having the largest t is retained. Call this variable x_2.

At this point, some computer packages diverge in methodology. The better packages now go back and check the t-value of $\hat{\beta}_1$ after $\hat{\beta}_2 x_2$ has been added to the model. If the t-value has become nonsignificant at some specified α level (say $\alpha = .10$), the variable x_1 is removed and a search is made for the independent variable with a β-parameter that will yield the most significant t-value in the presence of $\hat{\beta}_2 x_2$. Other packages do not recheck $\hat{\beta}_1$ but proceed directly to step 3.

---

* Note that the variable with the largest t-value is also the one with the largest (absolute) Pearson product moment correlation, r (Section 12.6), with y.

The reason the t-value for x_1 may change from step 1 to step 2 is that the meaning of the coefficient β_1 changes. In step 2, we are approximating a complex response surface in two variables with a plane. The best-fitting plane may yield a different value for β_1 than that obtained in step 1. Thus, both the value of $\hat{\beta}_1$ and, therefore, its significance usually change from step 1 to step 2. For this reason, the computer packages that recheck the t-values at each step are preferred.

Step 3. The stepwise procedure now checks for a third independent variable to include in the model with x_1 and x_2. That is, we seek the best model of the form

$$E(y) = \beta_0 + \beta_1 x_1 + \beta_2 x_2 + \beta_3 x_i$$

To do this, we fit all the $(k - 2)$ models using x_1, x_2, and each of the $(k - 2)$ remaining variables, x_i, as a possible x_3. The criterion is again to include the independent variable with the largest t-value. Call this best third variable x_3.

The better programs now recheck the t-values corresponding to the x_1 and x_2 coefficients, replacing the variables that have t-values that have become nonsignificant. This procedure is continued until no further independent variables can be found that yield significant t-values (at the specified α level) in the presence of the variables already in the model.

The result of the stepwise procedure is a model containing only those terms with t-values that are significant at the specified α level. Thus, in most practical situations only several of the large number of independent variables remain. However, it is very important *not* to jump to the conclusion that all the independent variables important for predicting y have been identified or that the unimportant independent variables have been eliminated. Remember, the stepwise procedure is using only *sample estimates* of the true model coefficients (β's) to select the important variables. An extremely large number of single β-parameter t-tests have been conducted, and the probability is very high that one or more errors have been made in including or excluding variables. That is, we have very probably included some unimportant independent variables in the model (Type I errors) and eliminated some important ones (Type II errors).

There is a second reason why we might not have arrived at a good model. When we choose the variables to be included in the stepwise regression, we may often omit high-order terms (to keep the number of variables manageable). Consequently, we may have initially omitted several important terms from the model. Thus, we should recognize stepwise regression for what it is: an objective screening procedure.

Now we will consider second-order terms (for quantitative variables) and other interactions among variables screened by the stepwise procedure. It would be best to develop this response surface model with a second set of data independent of that used for the screening, so the results of the stepwise procedure can be partially verified with new data. This is not always possible, however, because in many modeling situations only a small amount of data is available.

Do not be deceived by the impressive-looking t-values that result from the stepwise procedure—it has retained only the independent variables with the largest t-values.

Also, be certain to consider second-order terms in systematically developing the prediction model. Finally, if you have used a first-order model for your stepwise procedure, remember that it may be greatly improved by the addition of higher-order terms.

Example 14.12   In Section 13.7 we fit a multiple regression model for executive salaries as a function of experience, education, sex, etc. A preliminary step in the construction of this model was the determination of the most important independent variables. Ten independent variables were considered, as shown in Table 14.5. It would be very difficult to construct a second-order model with ten independent variables. Therefore, use the sample of 100 executives from Section 13.7 to decide which of the ten variables should be included in the construction of the final model for executive salaries.

Table 14.5

Independent Variables in the Executive Salary Example

INDEPENDENT VARIABLE	DESCRIPTION
x_1	Experience (years)—quantitative
x_2	Education (years)—quantitative
x_3	Sex (1 if male, 0 if female)—qualitative
x_4	Number of employees supervised—quantitative
x_5	Corporate assets (millions of dollars)—quantitative
x_6	Board member (1 if yes, 0 if no)—qualitative
x_7	Age (years)—quantitative
x_8	Company profits (past 12 months, millions of dollars)—quantitative
x_9	Has international responsibility (1 if yes, 0 if no)—qualitative
x_{10}	Company's total sales (past 12 months, millions of dollars)—quantitative

Solution   We will use stepwise regression with the main effects of the ten independent variables to identify the most important variables. The dependent variable y is the natural logarithm of the executive salaries. The SAS stepwise regression printout is shown in Figure 14.28.* Note that the first variable in the model is x_4, the number of employees supervised by the executive. At the second step, x_5, corporate assets, enters the model. At the sixth step, x_6, a dummy variable for the qualitative variable Board member or not, is brought into the model. However, because the significance (.2295) of the F-statistic (SAS uses the $F = t^2$ statistic rather than the t-statistic in the stepwise procedure) for x_6 is above the preassigned $\alpha = .10$, x_6 is then removed from the model. Thus, at step 7 the procedure indicates that the five-variable model including x_1, x_2, x_3, x_4, and x_5 is best. That is, none of the other independent variables can meet the $\alpha = .10$ criterion for admission to the model.

Thus, in our final modeling effort (Section 13.7) we concentrated on these five independent variables and determined that several second-order terms were important in the prediction of executive salaries.   ◼

* Note that there are several slight changes in the SAS printout labels for the stepwise procedure. For example, the $\hat{\beta}$-values, labeled ESTIMATE in the multiple regression procedure, are labeled B VALUE in the stepwise procedure.

Figure 14.28 Stepwise Regression Printout for Example 14.12

STEP 1
VARIABLE X4 ENTERED   R-SQUARE = 0.42071677

	DF	SUM OF SQUARES	MEAN SQUARE	F	PROB>F
REGRESSION	1	11.46854285	11.46854285	71.17	0.0001
ERROR	98	15.79112802	0.16113396		
TOTAL	99	27.25977087			

	B VALUE	STD ERROR		F	PROB>F
INTERCEPT	10.20077500				
X4 (EMPLOYEES SUPERVISED)	0.00057284	0.00006790		71.17	0.0001

STEP 2
VARIABLE X5 ENTERED   R-SQUARE = 0.78299675

	DF	SUM OF SQUARES	MEAN SQUARE	F	PROB>F
REGRESSION	2	21.34431198	10.67215599	175.00	0.0001
ERROR	97	5.91545889	0.06098411		
TOTAL	99	27.25977087			

	B VALUE	STD ERROR		F	PROB>F
INTERCEPT	9.87702903				
X4 (EMPLOYEES SUPERVISED)	0.00058353	0.00004178		195.06	0.0001
X5 (ASSETS)	0.00183730	0.00014438		161.94	0.0001

STEP 3
VARIABLE X1 ENTERED   R-SQUARE = 0.89667614

	DF	SUM OF SQUARES	MEAN SQUARE	F	PROB>F
REGRESSION	3	24.44318616	8.14772872	277.71	0.0001
ERROR	96	2.81658471	0.02933942		
TOTAL	99	27.25977087			

	B VALUE	STD ERROR		F	PROB>F
INTERCEPT	9.66449288				
X1 (EXPERIENCE)	0.01870784	0.00182032		105.62	0.0001
X4 (EMPLOYEES SUPERVISED)	0.00055251	0.00002914		359.59	0.0001
X5 (ASSETS)	0.00191195	0.00010041		362.60	0.0001

(continued)

Figure 14.28 Continued

STEP 4

VARIABLE X3 ENTERED   R-SQUARE = 0.94815717

	DF	SUM OF SQUARES	MEAN SQUARE	F	PROB>F
REGRESSION	4	25.84654710	6.46163678	434.37	0.0001
ERROR	95	1.41322377	0.01487604		
TOTAL	99	27.25977087			

	B VALUE	STD ERROR	F	PROB>F
INTERCEPT	9.40077349			
X1 (EXPERIENCE)	0.02074868	0.00131310	249.68	0.0001
X3 (SEX)	0.30011726	0.03089939	94.34	0.0001
X4 (EMPLOYEES SUPERVISED)	0.00055288	0.00002075	710.15	0.0001
X5 (ASSETS)	0.00190876	0.00007150	712.74	0.0001

STEP 5

VARIABLE X2 ENTERED   R-SQUARE = 0.96039323

	DF	SUM OF SQUARES	MEAN SQUARE	F	PROB>F
REGRESSION	5	26.18009940	5.23601988	455.87	0.0001
ERROR	94	1.07967147	0.01148587		
TOTAL	99	27.25977087			

	B VALUE	STD ERROR	F	PROB>F
INTERCEPT	8.85387930			
X1 (EXPERIENCE)	0.02141724	0.00116047	340.61	0.0001
X2 (EDUCATION)	0.03315807	0.00615303	29.04	0.0001
X3 (SEX)	0.31927842	0.02738298	135.95	0.0001
X4 (EMPLOYEES SUPERVISED)	0.00056061	0.00001829	939.84	0.0001
X5 (ASSETS)	0.00193684	0.00006304	943.98	0.0001

STEP 6

VARIABLE X6 ENTERED

R-SQUARE = 0.96100666

	DF	SUM OF SQUARES	MEAN SQUARE	F	PROB>F
REGRESSION	6	26.19682148	4.36613691	382.00	0.0001
ERROR	93	1.06294939	0.01142956		
TOTAL	99	27.25977087			

	B VALUE	STD ERROR	F	PROB>F
INTERCEPT	8.87509152			
X1 (EXPERIENCE)	0.02133460	0.00115963	338.48	0.0001
X2 (EDUCATION)	0.03272195	0.00614851	28.32	0.0001
X3 (SEX)	0.31093801	0.02817264	121.81	0.0001
X4 (EMPLOYEES SUPERVISED)	0.00055820	0.00001835	925.32	0.0001
X5 (ASSETS)	0.00193764	0.00006289	949.31	0.0001
X6 (BOARD)	0.03866226	0.03196369	1.46	0.2295

STEP 7

VARIABLE X6 REMOVED

R-SQUARE = 0.96039323

	DF	SUM OF SQUARES	MEAN SQUARE	F	PROB>F
REGRESSION	5	26.18009940	5.23601988	455.87	0.0001
ERROR	94	1.07967147	0.01148587		
TOTAL	99	27.25977087			

	B VALUE	STD ERROR	F	PROB>F
INTERCEPT	8.85387930			
X1 (EXPERIENCE)	0.02141724	0.00116047	340.61	0.0001
X2 (EDUCATION)	0.03315807	0.00615303	29.04	0.0001
X3 (SEX)	0.31927842	0.02738298	135.95	0.0001
X4 (EMPLOYEES SUPERVISED)	0.00056061	0.00001829	939.84	0.0001
X5 (ASSETS)	0.00193684	0.00006304	943.98	0.0001

Case Study 14.2
A Statistical Method for Land Appraisal

New factors and a lack of knowledge about the importance of factors that affect value continue to complicate the job of the rural appraiser. In order to provide knowledge on the subject, this article reports on and evaluates a study in which multiple linear regression equations were used to evaluate and quantify factors affecting value. . . . It is believed that the findings obtained with these equations, and the relationships they indicate, will be of value to the appraiser.

The authors of this statement, James O. Wise and H. Jackson Dover (1974),* use stepwise regression to identify a number of important factors (variables) that can be used to predict rural property values. They obtained their results by analyzing a sample of 105 cases from seven counties in the state of Georgia. Some of their findings are duplicated in Table 14.6. The variable names are listed in the order in which the stepwise regression procedure identified their importance, and the t-values found at each step are given for each variable. Note that both qualitative and quantitative variables have been included. Since each qualitative variable is at two levels, only one main effect term could be included in the model for each factor.

Table 14.6
Stepwise Regression Analysis of Price per Acre

VARIABLE NAME	t-VALUES
Residential land (yes–no)	10.466
Seedlings and saplings (number)	6.692
Percent ponds (%)	4.141
Distance to state park (miles)	3.985
Branches or springs (yes–no)	3.855
Site index (ratio)	3.160
Size (acres)	1.142
Farmland (yes–no)	2.288

Since there were 105 cases used in the study, a large number of degrees of freedom is associated with each t-statistic (first 103, then 102, etc.). Thus, we should compare the value of the test statistic to a corresponding z-value (1.645 for $\alpha = .10$ and the two-sided alternative hypothesis H_a: $\beta_i \neq 0$) when we judge the importance of each variable. Although Wise and Dover imply that the variable Size is important, we might not include it, since the t-value is only 1.142.

Finally, we have now only isolated a set of important variables. We still must decide exactly how each should be entered into our prediction equation, remembering that second-order terms often improve the model.

Exercises 14.71–14.74

Learning the Mechanics

14.71 There are six independent variables, x_1, x_2, x_3, x_4, x_5, and x_6, that might be useful in predicting a response y. A total of $n = 50$ observations are available, and it is decided to employ stepwise regression to help in selecting the independent variables

* The opinions and statements set forth herein do not necessarily reflect the viewpoint of the American Institute of Real Estate Appraisers or its individual members, and neither the Institute nor its editors and staff assume responsibility for such expressions of opinion or statements.

that appear to be useful. The computer fits all possible one-variable models of the form

$$E(y) = \beta_0 + \beta_1 x_i$$

where x_i is the ith independent variable, $i = 1, 2, \ldots, 6$. The information in the table is provided from the computer printout.

INDEPENDENT VARIABLE	$\hat{\beta}_1$	$s_{\hat{\beta}_1}$
x_1	1.6	.42
x_2	−.9	.01
x_3	3.4	1.14
x_4	2.5	2.06
x_5	−4.4	.73
x_6	.3	.35

a. Which independent variable is declared the best one-variable predictor of y? Explain.

b. Would this variable be included in the model at this stage? Explain.

c. Describe the next phase that a stepwise procedure would execute.

Applying the Concepts

[*Note:* *Starred (*) exercises require the use of a computer.*]

14.72 In a paper titled "The Influence of Self-Concept on Academic Success in Technological Careers," Debra A. G. Robinson and Stewart E. Cooper (1984) provide an analysis of data collected on 230 male and 72 female freshmen attending a midwestern university specializing in engineering and technology. Data collected on each student included a score on a self-concept of ability test (SCPT), scores on the SAT, ACT, and various other achievement tests, a standardized measure (called a T-score) of the student's rank in high school class, and the student's first-semester grade-point average (GPA).

One aspect of the study involved fitting by stepwise regression involving first-order terms for the test scores shown in the first column of the table. The variables are listed in the order in which they entered the model. Columns 2 and 3 give the values of R

VARIABLE	R	R^2	r	F
T-score	.37	.14	.37	18.44[a]
Mathematics test	.42	.18	.34	10.93[a]
SAT	.49	.24	.18	29.57[a]
ACT	.66	.44	.27	14.15[a]
Verbal test	.70	.50	.30	3.93
SCPT	.72	.52	.29	1.89
Trigonometry test	.72	.52	.28	.16

[a] $p < .01$.

and R^2 at each stage of the stepwise regression. Column 4 gives the simple coefficient of correlation between each variable and GPA. Column 5 gives the F-value at each stage of the stepwise regression for testing the contribution of the last variable included in the stepwise model for the prediction of GPA.

a. Write the models used in the first, second, third, and fourth stages of the fitting process.

b. Does the self-concept of ability (SCPT) score come into the model?

c. Compare the values of r between GPA and the individual variables. How do you explain your answer to part b given that the correlation coefficient for the SCPT score is larger than that for the SAT score?

d. Use the value of R^2 given in stage 4 of the stepwise regression to calculate the value of the F-statistic for testing the utility of the model.

e. Calculate the approximate p-value for the F-statistic in part d and interpret it.

14.73 Many power plants dump hot waste water into surrounding rivers, streams, and oceans, an action that may have an adverse effect on the marine life in the dumping areas. A marine biologist was hired by the EPA to determine whether the hot water runoff from a particular power plant located near a large gulf is having an adverse effect on the marine life in the area. In the initial phase of the study, the biologist's goal is to acquire a prediction equation for the number of marine animals located at certain predesignated areas, or stations, in the gulf. Based on past experience, the biologist considered the following environmental factors as predictors for the number of animals at a particular station:

x_1 = Temperature of water (TEMP)

x_2 = Salinity of water (SAL)

x_3 = Dissolved oxygen content of water (DO)

x_4 = Turbidity index, a measure of the turbidity of the water (TI)

x_5 = Depth of the water at the station (ST_DEPTH)

x_6 = Total weight of sea grasses in sampled area (TGRSWT)

As a preliminary step in the construction of this model, the biologist used a stepwise regression procedure to identify the most important of these six variables. A total of 716 samples were taken at different stations in the gulf, producing the SAS printout shown here. (The response measured was y, the log of the number of marine animals found in the sampled area.)

a. According to the SAS printout, which of the six independent variables should be used in the model? (Use $\alpha = .10$.)

b. Are we able to assume that the marine biologist has identified all the important independent variables for the prediction of y? Why?

c. Using the variables identified in part a, write the first-order model with interaction that may be used to predict y.

d. How would the marine biologist determine whether the model specified in part c is better than the first-order model?

SAS Printout for Exercise 14.73

STEP 1
VARIABLE ST_DEPTH ENTERED   R-SQUARE = 0.12227337

	DF	SUM OF SQUARES	MEAN SQUARE	F	PROB>F
REGRESSION	1	57.44041114	57.44041114	99.47	0.0001
ERROR	714	412.32998120	0.57749297		
TOTAL	715	469.77039233			

	B VALUE	STD ERROR	F	PROB>F
INTERCEPT	8.38559344			
ST_DEPTH	−0.43678519	0.04379580	99.47	0.0001

STEP 2
VARIABLE TGRSWT ENTERED   R-SQUARE = 0.18211026

	DF	SUM OF SQUARES	MEAN SQUARE	F	PROB>F
REGRESSION	2	85.55000871	42.77500435	79.38	0.0001
ERROR	713	384.22038363	0.53887852		
TOTAL	715	469.77039233			

	B VALUE	STD ERROR	F	PROB>F
INTERCEPT	8.07681529			
ST_DEPTH	−0.35355301	0.04384775	65.02	0.0001
TGRSWT	0.00271332	0.00037568	52.16	0.0001

STEP 3
VARIABLE TI ENTERED   R-SQUARE = 0.18700047

	DF	SUM OF SQUARES	MEAN SQUARE	F	PROB>F
REGRESSION	3	87.84728446	29.28242815	54.59	0.0001
ERROR	712	381.92310787	0.53640886		
TOTAL	715	469.77039233			

	B VALUE	STD ERROR	F	PROB>F
INTERCEPT	7.38863937			
TI	0.65773503	0.31782817	4.28	0.0389
ST_DEPTH	−0.31451227	0.04764144	43.58	0.0001
TGRSWT	0.00261166	0.00037802	47.73	0.0001

(continued)

SAS Printout for Exercise 14.73, Continued

STEP 4

VARIABLE DO__B ENTERED   R-SQUARE = 0.18892367

	DF	SUM OF SQUARES	MEAN SQUARE	F	PROB>F
REGRESSION	4	88.75074870	22.18768717	41.40	0.0001
ERROR	711	381.01964364	0.53589261		
TOTAL	715	469.77039233			

	B VALUE	STD ERROR	F	PROB>F
INTERCEPT	7.22576380			
DO	0.01769145	0.01362532	1.69	0.1946
TI	0.67347023	0.31790625	4.49	0.0345
ST__DEPTH	−0.30417372	0.04827962	39.69	0.0001
TGRSWT	0.00266958	0.00038047	49.23	0.0001

STEP 5

VARIABLE DO__B REMOVED   R-SQUARE = 0.18700047

	DF	SUM OF SQUARES	MEAN SQUARE	F	PROB>F
REGRESSION	3	87.84728446	29.28242815	54.59	0.0001
ERROR	712	381.92310787	0.53640886		
TOTAL	715	469.77039233			

	B VALUE	STD ERROR	F	PROB>F
INTERCEPT	7.38863937			
TI	0.65773503	0.31782817	4.28	0.0389
ST__DEPTH	−0.31451227	0.04764144	43.58	0.0001
TGRSWT	0.00261166	0.00037802	47.73	0.0001

e. Note the small value of R^2. What action might the biologist take to improve the model?

*14.74 [*Note:* This exercise is for students who have access to a stepwise regression computer program.] Literacy rate is a reflection of the educational facilities and quality of education available in a country, and mass communication plays a large part in the educational process. In an effort to relate the literacy rate of a country to various mass communication outlets, a demographer has proposed to relate the response

y = Literacy rate

to the independent variables

x_1 = Number of daily newspaper copies (per 1,000 population)

x_2 = Number of radios (per 1,000 population)

x_3 = Number of television sets (per 1,000 population)

Use stepwise regression to find a suitable model relating y to x_1, x_2, and x_3.

COUNTRY	NEWSPAPER COPIES Per 1,000	RADIOS Per 1,000	TELEVISION SETS Per 1,000	LITERACY RATE
Czechoslovakia	280	266	228	.98
Italy	142	230	201	.93
Kenya	10	114	2	.25
Norway	391	313	227	.99
Panama	86	329	82	.79
Philippines	17	42	11	.72
Tunisia	21	49	16	.32
USA	314	1,695	472	.99
USSR	333	430	185	.99
Venezuela	91	182	89	.82

Summary

Although this chapter provides only an introduction to the very important topic of *model building,* it will enable you to construct many interesting and useful models. You can build on this foundation and, with experience, develop competence in this fascinating area of statistics. Successful model building requires a delicate blend of knowledge of the process being modeled, geometry, and formal statistical testing.

The first step in model building is to *identify the response variable y and a set of independent variables.* Each independent variable is then classified as either *quantitative or qualitative,* and *dummy variables* are defined to represent the qualitative independent variables. If the total number of independent variables is large, you may want to use *stepwise regression* to screen out those that do not seem important for the prediction of *y.*

When the number of independent variables is manageable, the model builder is ready to begin a systematic effort. At least **second-order models,** those containing **two-way interactions and quadratic terms** in the quantitative variables, should be considered. Remember that a model with no interaction terms implies that each of the independent variables affects the response independently of the other independent variables. **Quadratic terms add curvature** to the contour curves when $E(y)$ is plotted as a function of the independent variable. The F-test for testing a set of β-parameters aids in deciding the final form of the prediction model.

Many problems can arise in regression modeling, and the intermediate steps are often tedious and frustrating. However, the end result of a careful and determined modeling effort is very rewarding—you will have a better understanding of the process and a predictive model for the dependent variable y.

Supplementary Exercises 14.75–14.94

[*Note:* Starred (*) exercises require the use of a computer.]

Learning the Mechanics

14.75 The values of SSE for the regression analyses in Exercises 13.6 and 13.13 are 210 and 130, respectively. Do the data provide sufficient information to indicate that the interaction terms contribute to the model? Test using $\alpha = .05$.

14.76 Why is the model-building step the key to the success or failure of a regression analysis?

14.77 Suppose you fit the regression model

$$E(y) = \beta_0 + \beta_1 x_1 + \beta_2 x_2 + \beta_3 x_2^2 + \beta_4 x_1 x_2 + \beta_5 x_1 x_2^2$$

to $n = 35$ data points and wish to test the null hypothesis

$$H_0: \quad \beta_4 = \beta_5 = 0$$

a. State the alternative hypothesis.
b. Explain in detail how to compute the F-statistic needed to test the null hypothesis.
c. What are the numerator and denominator degrees of freedom associated with the F-statistic in part b?
d. Give the rejection region for the test if $\alpha = .05$.

14.78 Graph each of the following polynomials, and give the order of each:
a. $E(y) = 6 + 3x - 2x^2$ **b.** $E(y) = 4x$
c. $E(y) = x^2$ **d.** $E(y) = 2 - 4x$
e. $E(y) = 1 + x - x^2 + x^3$ **f.** $E(y) = 1 - 4x$

14.79 Write a model relating $E(y)$ to one qualitative independent variable that is at four levels. Define all the terms in your model.

14.80 Explain the difference between qualitative and quantitative variables.

14.81 It is desired to relate $E(y)$ to a quantitative variable x_1 and a qualitative variable at three levels.
a. Write a first-order model.
b. Write a model that will graph as three different second-order curves—one for each level of the qualitative variable.

14.82 Explain why stepwise regression is used. What is its value in the model-building process?

14.83 a. Write a first-order model relating $E(y)$ to two quantitative independent variables, x_1 and x_2.
b. Write a complete second-order model.

14.84 To model the relationship between y, a dependent variable, and x, an independent variable, a researcher has taken one measurement on y at each of three different x-values. Drawing on his mathematical expertise, the researcher realizes that he can fit the second-order polynomial model

$$E(y) = \beta_0 + \beta_1 x + \beta_2 x^2$$

and it will pass exactly through all three points, yielding SSE = 0. The researcher, delighted with the "excellent" fit of the model, eagerly sets out to use it to make inferences. What problems will he encounter in attempting to make inferences?

Applying the Concepts

14.85 The number of practicing attorneys has more than doubled in the decade 1973–1983 with the percentage of female attorneys increasing from 5% to 15% of the legal profession. A random sample of 400 female attorneys and 200 male attorneys, from among the approximately 606,000 total number of attorneys in the United States, revealed that 25% of the women finished in the top 10% of their classes versus 18% of the males. The survey also found that women tended to enter law school at a later age than men. (Nearly 33% of women began practicing after age 30 compared with 14% for men.) For older women and men beginning practice, women's starting salaries tended to be higher than those of men of the same age. Finally, the median salary for different age groups increased with age, peaking for men at ages 51–55 and leveling off thereafter (*American Bar Association Journal,* October 1983).

Construct a model that estimates attorneys' salaries as a function of age, sex, years of experience, and rank in class. Then write another model and include other variables you would consider useful in estimating attorneys' salaries. Explain why they would be useful.

14.86 Due to the increase in gasoline prices, many service stations are offering self-service gasoline at reduced prices. Suppose an oil company wants to model the mean monthly gasoline sales, $E(y)$, of its affiliated stations as a function of the type of service they offer: self-service, full service, or both.

a. How many dummy variables will be needed to describe the qualitative independent variable Type of service?

b. Write the main effects model relating $E(y)$ to the type of service. Describe the coding of the dummy variables.

14.87 An experiment was conducted to compare three weight-reducing programs. Ten people were assigned to each of the three diets and their weights, at the beginning and end of a 1-month period, are recorded (in pounds) in the table.

DIET A		DIET B		DIET C	
x_1 (weight before)	y (weight loss)	x_1 (weight before)	y (weight loss)	y (weight before)	y (weight loss)
227	14	255	19	206	7
286	16	193	8	222	9
180	−2	186	4	168	2
176	8	145	15	132	0
204	15	219	16	173	−3
155	5	273	19	210	8
303	17	289	25	269	10
146	7	168	6	275	15
215	15	194	12	241	8
187	6	248	21	219	5

a. Construct a regression model to relate the weight loss y at the end of the 1-month period to

$$x_1 = \text{Weight before program} \qquad x_2 = \begin{cases} 1 & \text{if diet B} \\ 0 & \text{otherwise} \end{cases} \qquad x_3 = \begin{cases} 1 & \text{if diet C} \\ 0 & \text{otherwise} \end{cases}$$

Assume that the effect of initial weight on weight loss is identical for each of the three diets. Sketch the type of response curve you would expect to observe, showing y as a function of x_1 for each of the three diets.

b. Suppose the effect of initial weight on weight loss varies from diet to diet. Write the appropriate regression model for this case. Sketch typical response curves depicting this situation.

*c. Fit the models in parts a and b to the data given in the table.

*d. Do the data provide sufficient evidence to indicate an interaction between initial weight and type of diet? That is, is the model of part b preferable to the model of part a?

e. If you wish to test the null hypothesis of no difference among diets, which model parameters would be included in the hypothesis? [*Note:* Assume you are using the model from part b.]

*f. Conduct the test in part e using $\alpha = .05$.

*14.88 The data set shown here (also presented in Exercise 13.25) gives the number y of births per year (in thousands), the number x_1 of marriages per year (in thousands), the number x_2 of women in the work force (in thousands), and the availability (yes) or nonavailability (no) of the birth control pill over the years 1949–1982.

YEAR	BIRTHS	MARRIAGES	WOMEN WORKING	TAKING THE PILL	YEAR	BIRTHS	MARRIAGES	WOMEN WORKING	TAKING THE PILL
1949	3,560	1,580	18,030	No	1966	3,606	1,857	26,820	Yes
1950	3,632	1,667	18,680	No	1967	3,521	1,927	27,545	Yes
1951	3,750	1,595	19,309	No	1968	3,502	2,069	28,778	Yes
1952	3,847	1,539	19,559	No	1969	3,606	2,145	29,898	Yes
1953	3,902	1,546	19,668	No	1970	3,731	2,159	31,233	Yes
1954	4,017	1,490	19,970	No	1971	3,556	2,190	31,778	Yes
1955	4,097	1,518	20,842	No	1972	3,258	2,282	33,152	Yes
1956	4,168	1,569	21,808	No	1973	3,137	2,284	34,195	Yes
1957	4,255	1,518	22,097	No	1974	3,160	2,230	35,708	Yes
1958	4,204	1,451	22,482	No	1975	3,144	2,153	36,981	Yes
1959	4,245	1,494	22,865	No	1976	3,168	2,155	38,399	Yes
1960	4,258	1,523	23,619	No	1977	3,327	2,178	40,053	Yes
1961	4,268	1,548	24,199	No	1978	3,333	2,282	41,747	Yes
1962	4,167	1,580	23,978	No	1979	3,494	2,331	43,844	Yes
1963	4,098	1,654	24,675	Yes	1980	3,598	2,413	44,934	Yes
1964	4,027	1,725	25,399	Yes	1981	3,646	2,438	46,415	Yes
1965	3,760	1,800	25,952	Yes	1982	3,704	2,495	47,095	Yes

Source: Statistical Abstract of the U.S.

a. Specify a model relating the number y of annual births to one or more of the independent variables in the data set.
b. Use a computer to fit your model of part a to the data set.
c. Do the data provide sufficient evidence to indicate that your model provides information for the prediction of y? Test using $\alpha = .05$.
d. Give the p-value for the test in part c.
e. Find R^2 and interpret its value.
f. How might you improve your model of part a? Explain.

14.89 To make a product more appealing to the consumer, an automobile manufacturer is experimenting with a new type of paint that is supposed to help the car maintain its new car look. The durability of this paint depends on the length of time the car body is in the oven after it has been painted. In the initial experiment, three groups of ten car bodies each are baked for three different lengths of time—12, 24, and 36 hours—at the standard temperature setting. Then the paint finish of the thirty cars is analyzed to determine a durability rating, y.
a. Write a second-order model relating the mean durability $E(y)$ to the length of baking.
b. Could a third-order polynomial be fit to these data? Explain.

14.90 Many companies must accurately estimate their costs before a job is begun in order to acquire a contract and make a profit. For example, a heating and plumbing contractor may base cost estimates for new homes on the total area of the house and whether central air conditioning is to be installed.

a. Write a main effects model relating the mean cost of material and labor, $E(y)$, to the area and central air conditioning variables.

b. Write a complete second-order model for the mean cost as a function of the same two variables.

c. What hypothesis would you test to determine whether the second-order terms are useful for predicting mean cost?

d. Explain how you would compute the F-statistic needed to test the hypothesis of part c.

14.91 Refer to Exercise 14.90. The contractor samples twenty-five recent jobs and fits both the complete second-order model (part b) and the reduced main effects model (part a), so that a test can be conducted to determine whether the additional complexity of the second-order model is necessary. The resulting SSE and R^2 values are shown in the table.

	SSE	R^2
MAIN EFFECTS	8.548	.950
SECOND-ORDER	6.133	.964

a. Is there sufficient evidence to conclude that the second-order terms are important for predicting the mean cost? Use $\alpha = .05$.

b. Suppose the contractor decides to use the main effects model to predict costs. Use the global F-test (Section 13.5) to determine whether the main effects model is useful for predicting costs.

14.92 One factor that must be considered in developing a shipping system that is beneficial to both the customer and the seller is time of delivery. A manufacturer of farm equipment can ship its products by either rail or truck. Quadratic models are thought to be adequate in relating time of delivery to distance traveled for both modes of transportation. Consequently, it has been suggested that the following model be fit:

$$E(y) = \beta_0 + \beta_1 x_1 + \beta_2 x_1^2 + \beta_3 x_2 + \beta_4 x_1 x_2 + \beta_5 x_1^2 x_2$$

where

y = Shipping time

x_1 = Distance to be shipped $x_2 = \begin{cases} 1 & \text{if rail} \\ 0 & \text{if truck} \end{cases}$

a. What hypothesis would you test to determine whether the data indicate that the quadratic distance terms are useful in the model, i.e., whether curvature is present in the relationship between mean delivery time and distance?

b. What hypothesis would you test to determine whether there is a difference between mean delivery times by rail and truck?

14.93 Refer to Exercise 14.92. Suppose the model is fit to a total of fifty observations on delivery time. The sum of squared errors is SSE = 226.12. Then the reduced model

$$E(y) = \beta_0 + \beta_1 x_1 + \beta_2 x_1^2$$

is fit to the same data, and SSE = 259.34. Test to determine whether the data indicate that the mean delivery time differs for rail and truck deliveries. Use $\alpha = .05$.

*14.94 A firm has developed a new type of light bulb and is interested in evaluating its performance in order to decide whether to market it. It is known that the light output of the bulb depends on the cleanliness of its surface area and the length of time the bulb has been in operation. Use the data in the table and the procedures you learned in this chapter to build a regression model that relates drop in light output to bulb surface cleanliness and length of operation.

DROP IN LIGHT OUTPUT (% original output)	BULB SURFACE (C = Clean, D = Dirty)	LENGTH OF OPERATION (Hours)	DROP IN LIGHT OUTPUT (% original output)	BULB SURFACE (C = Clean, D = Dirty)	LENGTH OF OPERATION (Hours)
0	C	0	0	D	0
16	C	400	4	D	400
22	C	800	6	D	800
27	C	1,200	8	D	1,200
32	C	1,600	9	D	1,600
36	C	2,000	11	D	2,000
38	C	2,400	12	D	2,400

On Your Own . . .

We continue our "On Your Own" theme from Chapters 12 and 13. Remember that you selected three independent variables related to the annual GNP. Now increase your list of three variables to include approximately ten that you think would be useful in predicting the GNP. Obtain data for as many years as possible for the new list of variables and the GNP. With the aid of a computer analysis package, employ a stepwise regression program to choose the important variables among those you have listed. To test your intuition, list the variables in the order you think they will be selected before you conduct the analysis. How does your list compare with the stepwise regression results?

After the group of ten variables has been narrowed to a smaller group of variables by the stepwise analysis, try to improve the model by including interactions and quadratic terms. Be sure to consider the meaning of each interaction or quadratic term before adding it to the model—a quick sketch can be very helpful. See if you can systematically construct a useful model for predicting the GNP. You might want to hold out the last several years of data to test the predictive ability of your model after it is constructed. (As noted in Section 14.8, using the same data to construct *and* to evaluate predictive ability can lead to invalid statistical tests and a false sense of security.)

References

Draper, N., & Smith, H. *Applied regression analysis.* New York: Wiley, 1966.

Fox, K. D., & Nickols, S. Y. "The time crunch." *Journal of Family Issues* 1983, *4.*

Graybill, F. A. *Theory and application of the linear model,* North Scituate, Mass.: Duxbury, 1976.

Mendenhall, W. *Introduction to linear models and the design and analysis of experiments.* Belmont, Ca.: Wadsworth, 1968.

Mendenhall, W., & McClave, J. T. *A second course in business statistics: regression analysis.* San Francisco: Dellen, 1981.

Robinson, D. A. G., & Cooper, S. E. "The influence of self-concept on academic success in technological careers." *Journal of College Student Personnel,* 1984.

Wise, J. O., & Dover, H. J. "An evaluation of a statistical method of appraising rural property." *Appraisal Journal,* 1974, *42,* 103–113.

APPENDIX

Tables

Contents

Table I Random Numbers

COLUMN / ROW	1	2	3	4	5	6	7	8	9	10	11	12	13	14
1	10480	15011	01536	02011	81647	91646	69179	14194	62590	36207	20969	99570	91291	90700
2	22368	46573	25595	85393	30995	89198	27982	53402	93965	34095	52666	19174	39615	99505
3	24130	48360	22527	97265	76393	64809	15179	24830	49340	32081	30680	19655	63348	58629
4	42167	93093	06243	61680	07856	16376	39440	53537	71341	57004	00849	74917	97758	16379
5	37570	39975	81837	16656	06121	91782	60468	81305	49684	60672	14110	06927	01263	54613
6	77921	06907	11008	42751	27756	53498	18602	70659	90655	15053	21916	81825	44394	42880
7	99562	72905	56420	69994	98872	31016	71194	18738	44013	48840	63213	21069	10634	12952
8	96301	91977	05463	07972	18876	20922	94595	56869	69014	60045	18425	84903	42508	32307
9	89579	14342	63661	10281	17453	18103	57740	84378	25331	12566	58678	44947	05585	56941
10	85475	36857	53342	53988	53060	59533	38867	62300	08158	17983	16439	11458	18593	64952
11	28918	69578	88231	33276	70997	79936	56865	05859	90106	31595	01547	85590	91610	78188
12	63553	40961	48235	03427	49626	69445	18663	72695	52180	20847	12234	90511	33703	90322
13	09429	93969	52636	92737	88974	33488	36320	17617	30015	08272	84115	27156	30613	74952
14	10365	61129	87529	85689	48237	52267	67689	93394	01511	26358	85104	20285	29975	89868
15	07119	97336	71048	08178	77233	13916	47564	81056	97735	85977	29372	74461	28551	90707
16	51085	12765	51821	51259	77452	16308	60756	92144	49442	53900	70960	63990	75601	40719
17	02368	21382	52404	60268	89368	19885	55322	44819	01188	65255	64835	44919	05944	55157
18	01011	54092	33362	94904	31273	04146	18594	29852	71585	85030	51132	01915	92747	64951
19	52162	53916	46369	58586	23216	14513	83149	98736	23495	64350	94738	17752	35156	35749
20	07056	97628	33787	09998	42698	06691	76988	13602	51851	46104	88916	19509	25625	58104
21	48663	91245	85828	14346	09172	30168	90229	04734	59193	22178	30421	61666	99904	32812
22	54164	58492	22421	74103	47070	25306	76468	26384	58151	06646	21524	15227	96909	44592
23	32639	32363	05597	24200	13363	38005	94342	28728	35806	06912	17012	64161	18296	22851
24	29334	27001	87637	87308	58731	00256	45834	15398	46557	41135	10367	07684	36188	18510
25	02488	33062	28834	07351	19731	92420	60952	61280	50001	67658	32586	86679	50720	94953
26	81525	72295	04839	96423	24878	82651	66566	14778	76797	14780	13300	87074	79666	95725
27	29676	20591	68086	26432	46901	20849	89768	81536	86645	12659	92259	57102	80428	25280
28	00742	57392	39064	66432	84673	40027	32832	61362	98947	96067	64760	64584	96096	98253
29	05366	04213	25669	26422	44407	44048	37937	63904	45766	66134	75470	66520	34693	90449
30	91921	26418	64117	94305	26766	25940	39972	22209	71500	64568	91402	42416	07844	69618
31	00582	04711	87917	77341	42206	35126	74087	99547	81817	42607	43808	76655	62028	76630
32	00725	69884	62797	56170	86324	88072	76222	36086	84637	93161	76038	65855	77919	88006
33	69011	65795	95876	55293	18988	27354	26575	08625	40801	59920	29841	80150	12777	48501
34	25976	57948	29888	88604	67917	48708	18912	82271	65424	69774	33611	54262	85963	03547

(continued)

Table I Continued

ROW\COLUMN	1	2	3	4	5	6	7	8	9	10	11	12	13	14
35	09763	83473	73577	12908	30883	18317	28290	35797	05998	41688	34952	37888	38917	88050
36	91576	42595	27958	30134	04024	86385	29880	99730	55536	84855	29080	09250	79656	73211
37	17955	56349	90999	49127	20044	59931	06115	20542	18059	02008	73708	83517	36103	42791
38	46503	18584	18845	49618	02304	51038	20655	58727	28168	15475	56942	53389	20562	87338
39	92157	89634	94824	78171	84610	82834	09922	25417	44137	48413	25555	21246	35509	20468
40	14577	62765	35605	81263	39667	47358	56873	56307	61607	49518	89656	20103	77490	18062
41	98427	07523	33362	64270	01638	92477	66969	98420	04880	45585	46565	04102	46880	45709
42	34914	63976	88720	82765	34476	17032	87589	40836	32427	70002	70663	88863	77775	69348
43	70060	28277	39475	46473	23219	53416	94970	25832	69975	94884	19661	72828	00102	66794
44	53976	54914	06990	67245	68350	82948	11398	42878	80287	88267	47363	46634	06541	97809
45	76072	29515	40980	07391	58745	25774	22987	80059	39911	96189	41151	14222	60697	59583
46	90725	52210	83974	29992	65831	38857	50490	83765	55657	14361	31720	57375	56228	41546
47	64364	67412	33339	31926	14883	24413	59744	92351	97473	89286	35931	04110	23726	51900
48	08962	00358	31662	25388	61642	34072	81249	35648	56891	69352	48373	45578	78547	81788
49	95012	68379	93526	70765	10592	04542	76463	54328	02349	17247	28865	14777	62730	92277
50	15664	10493	20492	38391	91132	21999	59516	81652	27195	48223	46751	22923	32261	85653
51	16408	81899	04153	53381	79401	21438	83035	92350	36693	31238	59649	32523	72772	02338
52	18629	81953	05520	91962	04739	13092	97662	54822	94730	06496	35090	04822	86774	98289
53	73115	35101	47498	87637	99016	71060	88824	71013	35970	20286	23153	72924	35165	43040
54	57491	16703	23167	49323	45021	33132	12544	41035	80780	45393	44812	12515	98931	91202
55	30405	83946	23792	14422	15059	45799	22716	19792	09983	74353	68668	30429	70735	25499
56	16631	35006	85900	98275	32388	52390	16815	69298	82732	38480	73817	32523	41961	44437
57	96773	20206	42559	78985	05300	22164	24369	54224	35083	19687	11052	91491	60383	19746
58	38935	64202	14349	82674	66523	44133	00697	35552	35970	19124	63318	29686	03387	59846
59	31624	76384	17403	53363	44167	64486	64758	75366	76554	31601	12614	33072	60332	92325
60	78919	19474	23632	27889	47914	02584	37680	20801	72152	39339	34806	08930	85001	87820
61	03931	33309	57047	74211	63445	17361	62825	39908	05607	91284	68833	25570	38818	46920
62	74426	33278	43972	10119	89917	15665	52872	73823	73144	88662	88970	74492	51805	99378
63	09066	00903	20795	95452	92648	45454	09552	88815	16553	51125	79375	97596	16296	66092
64	42238	12426	87025	14267	20979	04508	64535	31355	86064	29472	47689	05974	52468	16834
65	16153	08002	26504	41744	81959	65642	74240	56302	00033	67107	77510	70625	28725	34191
66	21457	40742	29820	96783	29400	21840	15035	34537	33310	06116	95240	15957	16572	06004
67	21581	57802	02050	89728	17937	37621	47075	42080	97403	48626	68995	43805	33386	21597
68	55612	78095	83197	33732	05810	24813	86902	60397	16489	03264	88525	42786	05269	92532
69	44657	66999	99324	51281	84463	60563	79312	93454	68876	25471	93911	25650	12682	73572
70	91340	84979	46949	81973	37949	61023	43997	15263	80644	43942	89203	71795	99533	50501
71	91227	21199	31935	27022	84067	05462	35216	14486	29891	68607	41867	14951	91696	85065
72	50001	38140	66321	19924	72163	09538	12151	06878	91903	18749	34405	56087	82790	70925

73	65390	05224	72958	28609	81406	39147	25549	48542	42627	45233	57202	94617	23772	07896
74	27504	96131	83944	41575	10573	08619	64482	73923	36152	05184	94142	25299	84387	34925
75	37169	94851	39117	89632	00959	16487	65536	49071	39782	17095	02330	74301	00275	48280
76	11508	70225	51111	38351	19444	66499	71945	05422	13442	78675	84081	66938	93654	59894
77	37449	30362	06694	54690	04052	53115	62757	95348	78662	11163	81651	50245	34971	52924
78	46515	70331	85922	38329	57015	15765	97161	17869	45349	61796	66345	81073	49106	79860
79	30986	81223	42416	58353	21532	30502	32305	86482	05174	07901	54339	58861	74818	46942
80	63798	64995	46583	09785	44160	78128	83991	42865	92520	83531	80377	35909	81250	54238
81	82486	84846	99254	67632	43218	50076	21361	64816	51202	88124	41870	52689	51275	83556
82	21885	32906	92431	09060	64297	51674	64126	62570	26123	05155	59194	52799	28225	85762
83	60336	98782	07408	53458	13564	59089	26445	29789	85205	41001	12535	12133	14645	23541
84	43937	46891	24010	25560	86355	33941	25786	54990	71899	15475	95434	98227	21824	19585
85	97656	63175	89303	16275	07100	92063	21942	18611	47348	20203	18534	03862	78095	50136
86	03299	01221	05418	38982	55758	92237	26759	86367	21216	98442	08303	56613	91511	75928
87	79626	06486	03574	17668	07785	76020	79924	25651	83325	88428	85076	72811	22717	50585
88	85636	68335	47539	03129	65651	11977	02510	26113	99447	68645	34327	15152	55230	93448
89	18039	14367	61337	06177	12143	46609	32989	74014	64708	00533	35398	58408	13261	47908
90	08362	15656	60627	36478	65648	16764	53412	09013	07832	41574	17639	82163	60859	75567
91	79556	29068	04142	16268	15387	12856	66227	38358	22478	73373	88732	09443	82558	05250
92	92608	82674	27072	32534	17075	27698	98204	63863	11951	34648	88022	56148	34925	57031
93	23982	25835	40055	67006	12293	02753	14827	23235	35071	99704	37543	11601	35503	85171
94	09915	96306	05908	97901	28395	14186	00821	80703	70426	75647	76310	88717	37890	40129
95	59037	33300	26695	62247	69927	76123	50842	43834	86654	70959	79725	93872	28117	19233
96	42488	78077	69882	61657	34136	79180	97526	43092	04098	73571	80799	76536	71255	64239
97	46764	86273	63003	93017	31204	36692	40202	35275	57306	55543	53203	18098	47625	88684
98	03237	45430	55417	63282	90816	17349	88298	90183	36600	78406	06216	95787	42579	90730
99	86591	81482	52667	61582	14972	90053	89534	76036	49199	43716	97548	04379	46370	28672
100	38534	01715	94964	87288	65680	43772	39560	12918	86537	62738	19636	51132	25739	56947

Source: Abridged from W. H. Beyer, Ed., *CRC Standard Mathematical Tables,* 24th ed. (Cleveland: The Chemical Rubber Company), 1976. Reproduced by permission of the publisher.

Table II Binomial Probabilities

Tabulated values are $\sum_{x=0}^{k} p(x)$. (*Computations are rounded at the third decimal place.*)

a. $n = 5$

k \ p	0.01	0.05	0.10	0.20	0.30	0.40	0.50	0.60	0.70	0.80	0.90	0.95	0.99
0	.951	.774	.590	.328	.168	.078	.031	.010	.002	.000	.000	.000	.000
1	.999	.977	.919	.737	.528	.337	.188	.087	.031	.007	.000	.000	.000
2	1.000	.999	.991	.942	.837	.683	.500	.317	.163	.058	.009	.001	.000
3	1.000	1.000	1.000	.993	.969	.913	.812	.663	.472	.263	.081	.023	.001
4	1.000	1.000	1.000	1.000	.998	.990	.969	.922	.832	.672	.410	.226	.049

b. $n = 6$

k \ p	0.01	0.05	0.10	0.20	0.30	0.40	0.50	0.60	0.70	0.80	0.90	0.95	0.99
0	.941	.735	.531	.262	.118	.047	.016	.004	.001	.000	.000	.000	.000
1	.999	.967	.886	.655	.420	.233	.109	.041	.011	.002	.000	.000	.000
2	1.000	.998	.984	.901	.744	.544	.344	.179	.070	.017	.001	.000	.000
3	1.000	1.000	.999	.983	.930	.821	.656	.456	.256	.099	.016	.002	.000
4	1.000	1.000	1.000	.998	.989	.959	.891	.767	.580	.345	.114	.033	.001
5	1.000	1.000	1.000	1.000	.999	.996	.984	.953	.882	.738	.469	.265	.059

c. $n = 7$

k \ p	0.01	0.05	0.10	0.20	0.30	0.40	0.50	0.60	0.70	0.80	0.90	0.95	0.99
0	.932	.698	.478	.210	.082	.028	.008	.002	.000	.000	.000	.000	.000
1	.998	.956	.850	.577	.329	.159	.063	.019	.004	.000	.000	.000	.000
2	1.000	.996	.974	.852	.647	.420	.227	.096	.029	.005	.000	.000	.000
3	1.000	1.000	.997	.967	.874	.710	.500	.290	.126	.033	.003	.000	.000
4	1.000	1.000	1.000	.995	.971	.904	.773	.580	.353	.148	.026	.004	.000
5	1.000	1.000	1.000	1.000	.996	.981	.937	.841	.671	.423	.150	.044	.002
6	1.000	1.000	1.000	1.000	1.000	.998	.992	.972	.918	.790	.522	.302	.068

d. $n = 8$

k \ p	0.01	0.05	0.10	0.20	0.30	0.40	0.50	0.60	0.70	0.80	0.90	0.95	0.99
0	.923	.663	.430	.168	.058	.017	.004	.001	.000	.000	.000	.000	.000
1	.997	.943	.813	.503	.255	.106	.035	.009	.001	.000	.000	.000	.000
2	1.000	.994	.962	.797	.552	.315	.145	.050	.011	.001	.000	.000	.000
3	1.000	1.000	.995	.944	.806	.594	.363	.174	.058	.010	.000	.000	.000
4	1.000	1.000	1.000	.990	.942	.826	.637	.406	.194	.056	.005	.000	.000
5	1.000	1.000	1.000	.999	.989	.950	.855	.685	.448	.203	.038	.006	.000
6	1.000	1.000	1.000	1.000	.999	.991	.965	.894	.745	.497	.187	.057	.003
7	1.000	1.000	1.000	1.000	1.000	.999	.996	.983	.942	.832	.570	.337	.077

e. $n = 9$

p k	0.01	0.05	0.10	0.20	0.30	0.40	0.50	0.60	0.70	0.80	0.90	0.95	0.99
0	.914	.630	.387	.134	.040	.010	.002	.000	.000	.000	.000	.000	.000
1	.997	.929	.775	.436	.196	.071	.020	.004	.000	.000	.000	.000	.000
2	1.000	.992	.947	.738	.463	.232	.090	.025	.004	.000	.000	.000	.000
3	1.000	.999	.992	.914	.730	.483	.254	.099	.025	.003	.000	.000	.000
4	1.000	1.000	.999	.980	.901	.733	.500	.267	.099	.020	.001	.000	.000
5	1.000	1.000	1.000	.997	.975	.901	.746	.517	.270	.086	.008	.001	.000
6	1.000	1.000	1.000	1.000	.996	.975	.910	.768	.537	.262	.053	.008	.000
7	1.000	1.000	1.000	1.000	1.000	.996	.980	.929	.804	.564	.225	.071	.003
8	1.000	1.000	1.000	1.000	1.000	1.000	.998	.990	.960	.866	.613	.370	.086

f. $n = 10$

p k	0.01	0.05	0.10	0.20	0.30	0.40	0.50	0.60	0.70	0.80	0.90	0.95	0.99
0	.904	.599	.349	.107	.028	.006	.001	.000	.000	.000	.000	.000	.000
1	.996	.914	.736	.376	.149	.046	.011	.002	.000	.000	.000	.000	.000
2	1.000	.988	.930	.678	.383	.167	.055	.012	.002	.000	.000	.000	.000
3	1.000	.999	.987	.879	.650	.382	.172	.055	.011	.001	.000	.000	.000
4	1.000	1.000	.998	.967	.850	.633	.377	.166	.047	.006	.000	.000	.000
5	1.000	1.000	1.000	.994	.953	.834	.623	.367	.150	.033	.002	.000	.000
6	1.000	1.000	1.000	.999	.989	.945	.828	.618	.350	.121	.013	.001	.000
7	1.000	1.000	1.000	1.000	.998	.988	.945	.833	.617	.322	.070	.012	.000
8	1.000	1.000	1.000	1.000	1.000	.998	.989	.954	.851	.624	.264	.086	.004
9	1.000	1.000	1.000	1.000	1.000	1.000	.999	.994	.972	.893	.651	.401	.096

g. $n = 15$

p k	0.01	0.05	0.10	0.20	0.30	0.40	0.50	0.60	0.70	0.80	0.90	0.95	0.99
0	.860	.463	.206	.035	.005	.000	.000	.000	.000	.000	.000	.000	.000
1	.990	.829	.549	.167	.035	.005	.000	.000	.000	.000	.000	.000	.000
2	1.000	.964	.816	.398	.127	.027	.004	.000	.000	.000	.000	.000	.000
3	1.000	.995	.944	.648	.297	.091	.018	.002	.000	.000	.000	.000	.000
4	1.000	.999	.987	.836	.515	.217	.059	.009	.001	.000	.000	.000	.000
5	1.000	1.000	.998	.939	.722	.403	.151	.034	.004	.000	.000	.000	.000
6	1.000	1.000	1.000	.982	.869	.610	.304	.095	.015	.001	.000	.000	.000
7	1.000	1.000	1.000	.996	.950	.787	.500	.213	.050	.004	.000	.000	.000
8	1.000	1.000	1.000	.999	.985	.905	.696	.390	.131	.018	.000	.000	.000
9	1.000	1.000	1.000	1.000	.996	.966	.849	.597	.278	.061	.002	.000	.000
10	1.000	1.000	1.000	1.000	.999	.991	.941	.783	.485	.164	.013	.001	.000
11	1.000	1.000	1.000	1.000	1.000	.998	.982	.909	.703	.352	.056	.005	.000
12	1.000	1.000	1.000	1.000	1.000	1.000	.996	.973	.873	.602	.184	.036	.000
13	1.000	1.000	1.000	1.000	1.000	1.000	1.000	.995	.965	.833	.451	.171	.010
14	1.000	1.000	1.000	1.000	1.000	1.000	1.000	1.000	.995	.965	.794	.537	.140

(*continued*)

Table II   Continued

h. $n = 20$

k	0.01	0.05	0.10	0.20	0.30	0.40	0.50	0.60	0.70	0.80	0.90	0.95	0.99
0	.818	.358	.122	.012	.001	.000	.000	.000	.000	.000	.000	.000	.000
1	.983	.736	.392	.069	.008	.001	.000	.000	.000	.000	.000	.000	.000
2	.999	.925	.677	.206	.035	.004	.000	.000	.000	.000	.000	.000	.000
3	1.000	.984	.867	.411	.107	.016	.001	.000	.000	.000	.000	.000	.000
4	1.000	.997	.957	.630	.238	.051	.006	.000	.000	.000	.000	.000	.000
5	1.000	1.000	.989	.804	.416	.126	.021	.002	.000	.000	.000	.000	.000
6	1.000	1.000	.998	.913	.608	.250	.058	.006	.000	.000	.000	.000	.000
7	1.000	1.000	1.000	.968	.772	.416	.132	.021	.001	.000	.000	.000	.000
8	1.000	1.000	1.000	.990	.887	.596	.252	.057	.005	.000	.000	.000	.000
9	1.000	1.000	1.000	.997	.952	.755	.412	.128	.017	.001	.000	.000	.000
10	1.000	1.000	1.000	.999	.983	.872	.588	.245	.048	.003	.000	.000	.000
11	1.000	1.000	1.000	1.000	.995	.943	.748	.404	.113	.010	.000	.000	.000
12	1.000	1.000	1.000	1.000	.999	.979	.868	.584	.228	.032	.000	.000	.000
13	1.000	1.000	1.000	1.000	1.000	.994	.942	.750	.392	.087	.002	.000	.000
14	1.000	1.000	1.000	1.000	1.000	.998	.979	.874	.584	.196	.011	.000	.000
15	1.000	1.000	1.000	1.000	1.000	1.000	.994	.949	.762	.370	.043	.003	.000
16	1.000	1.000	1.000	1.000	1.000	1.000	.999	.984	.893	.589	.133	.016	.000
17	1.000	1.000	1.000	1.000	1.000	1.000	1.000	.996	.965	.794	.323	.075	.001
18	1.000	1.000	1.000	1.000	1.000	1.000	1.000	.999	.992	.931	.608	.264	.017
19	1.000	1.000	1.000	1.000	1.000	1.000	1.000	1.000	.999	.988	.878	.642	.182

i. $n = 25$

k \ p	0.01	0.05	0.10	0.20	0.30	0.40	0.50	0.60	0.70	0.80	0.90	0.95	0.99
0	.778	.277	.072	.004	.000	.000	.000	.000	.000	.000	.000	.000	.000
1	.974	.642	.271	.027	.002	.000	.000	.000	.000	.000	.000	.000	.000
2	.998	.873	.537	.098	.009	.000	.000	.000	.000	.000	.000	.000	.000
3	1.000	.966	.764	.234	.033	.002	.000	.000	.000	.000	.000	.000	.000
4	1.000	.993	.902	.421	.090	.009	.000	.000	.000	.000	.000	.000	.000
5	1.000	.999	.967	.617	.193	.029	.002	.000	.000	.000	.000	.000	.000
6	1.000	1.000	.991	.780	.341	.074	.007	.000	.000	.000	.000	.000	.000
7	1.000	1.000	.998	.891	.512	.154	.022	.001	.000	.000	.000	.000	.000
8	1.000	1.000	1.000	.953	.677	.274	.054	.004	.000	.000	.000	.000	.000
9	1.000	1.000	1.000	.983	.811	.425	.115	.013	.000	.000	.000	.000	.000
10	1.000	1.000	1.000	.994	.902	.586	.212	.034	.002	.000	.000	.000	.000
11	1.000	1.000	1.000	.998	.956	.732	.345	.078	.006	.000	.000	.000	.000
12	1.000	1.000	1.000	1.000	.983	.846	.500	.154	.017	.000	.000	.000	.000
13	1.000	1.000	1.000	1.000	.994	.922	.655	.268	.044	.002	.000	.000	.000
14	1.000	1.000	1.000	1.000	.998	.966	.788	.414	.098	.006	.000	.000	.000
15	1.000	1.000	1.000	1.000	1.000	.987	.885	.575	.189	.017	.000	.000	.000
16	1.000	1.000	1.000	1.000	1.000	.996	.946	.726	.323	.047	.000	.000	.000
17	1.000	1.000	1.000	1.000	1.000	.999	.978	.846	.488	.109	.002	.000	.000
18	1.000	1.000	1.000	1.000	1.000	1.000	.993	.926	.659	.220	.009	.000	.000
19	1.000	1.000	1.000	1.000	1.000	1.000	.998	.971	.807	.383	.033	.001	.000
20	1.000	1.000	1.000	1.000	1.000	1.000	1.000	.991	.910	.579	.098	.007	.000
21	1.000	1.000	1.000	1.000	1.000	1.000	1.000	.998	.967	.766	.236	.034	.000
22	1.000	1.000	1.000	1.000	1.000	1.000	1.000	1.000	.991	.902	.463	.127	.002
23	1.000	1.000	1.000	1.000	1.000	1.000	1.000	1.000	.998	.973	.729	.358	.026
24	1.000	1.000	1.000	1.000	1.000	1.000	1.000	1.000	1.000	.996	.928	.723	.222

Table III Exponentials

c	e^{-c}	c	e^{-c}	c	e^{-c}
0.00	1.000000	2.35	.095369	4.70	.009095
0.05	.951229	2.40	.090718	4.75	.008652
0.10	.904837	2.45	.086294	4.80	.008230
0.15	.860708	2.50	.082085	4.85	.007828
0.20	.818731	2.55	.078082	4.90	.007447
0.25	.778801	2.60	.074274	4.95	.007083
0.30	.740818	2.65	.070651	5.00	.006738
0.35	.704688	2.70	.067206	5.05	.006409
0.40	.670320	2.75	.063928	5.10	.006097
0.45	.637628	2.80	.060810	5.15	.005799
0.50	.606531	2.85	.057844	5.20	.005517
0.55	.576950	2.90	.055023	5.25	.005248
0.60	.548812	2.95	.052340	5.30	.004992
0.65	.522046	3.00	.049787	5.35	.004748
0.70	.496585	3.05	.047359	5.40	.004517
0.75	.472367	3.10	.045049	5.45	.004296
0.80	.449329	3.15	.042852	5.50	.004087
0.85	.427415	3.20	.040762	5.55	.003887
0.90	.406570	3.25	.038774	5.60	.003698
0.95	.386741	3.30	.036883	5.65	.003518
1.00	.367879	3.35	.035084	5.70	.003346
1.05	.349938	3.40	.033373	5.75	.003183
1.10	.332871	3.45	.031746	5.80	.003028
1.15	.316637	3.50	.030197	5.85	.002880
1.20	.301194	3.55	.028725	5.90	.002739
1.25	.286505	3.60	.027324	5.95	.002606
1.30	.272532	3.65	.025991	6.00	.002479
1.35	.259240	3.70	.024724	6.05	.002358
1.40	.246597	3.75	.023518	6.10	.002243
1.45	.234570	3.80	.022371	6.15	.002133
1.50	.223130	3.85	.021280	6.20	.002029
1.55	.212248	3.90	.020242	6.25	.001930
1.60	.201897	3.95	.019255	6.30	.001836
1.65	.192050	4.00	.018316	6.35	.001747
1.70	.182684	4.05	.017422	6.40	.001661
1.75	.173774	4.10	.016573	6.45	.001581
1.80	.165299	4.15	.015764	6.50	.001503
1.85	.157237	4.20	.014996	6.55	.001430
1.90	.149569	4.25	.014264	6.60	.001360
1.95	.142274	4.30	.013569	6.65	.001294
2.00	.135335	4.35	.012907	6.70	.001231
2.05	.128735	4.40	.012277	6.75	.001171
2.10	.122456	4.45	.011679	6.80	.001114
2.15	.116484	4.50	.011109	6.85	.001059
2.20	.110803	4.55	.010567	6.90	.001008
2.25	.105399	4.60	.010052	6.95	.000959
2.30	.100259	4.65	.009562	7.00	.000912

c	e^{-c}	c	e^{-c}	c	e^{-c}
7.05	.000867	8.05	.000319	9.05	.000117
7.10	.000825	8.10	.000304	9.10	.000112
7.15	.000785	8.15	.000289	9.15	.000106
7.20	.000747	8.20	.000275	9.20	.000101
7.25	.000710	8.25	.000261	9.25	.000096
7.30	.000676	8.30	.000249	9.30	.000091
7.35	.000643	8.35	.000236	9.35	.000087
7.40	.000611	8.40	.000225	9.40	.000083
7.45	.000581	8.45	.000214	9.45	.000079
7.50	.000553	8.50	.000204	9.50	.000075
7.55	.000526	8.55	.000194	9.55	.000071
7.60	.000501	8.60	.000184	9.60	.000068
7.65	.000476	8.65	.000175	9.65	.000064
7.70	.000453	8.70	.000167	9.70	.000061
7.75	.000431	8.75	.000158	9.75	.000058
7.80	.000410	8.80	.000151	9.80	.000056
7.85	.000390	8.85	.000143	9.85	.000053
7.90	.000371	8.90	.000136	9.90	.000050
7.95	.000353	8.95	.000130	9.95	.000048
8.00	.000336	9.00	.000123	10.00	.000045

Table IV
Normal Curve Areas

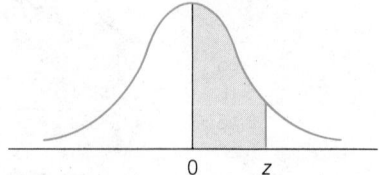

z	.00	.01	.02	.03	.04	.05	.06	.07	.08	.09
0.0	.0000	.0040	.0080	.0120	.0160	.0199	.0239	.0279	.0319	.0359
0.1	.0398	.0438	.0478	.0517	.0557	.0596	.0636	.0675	.0714	.0753
0.2	.0793	.0832	.0871	.0910	.0948	.0987	.1026	.1064	.1103	.1141
0.3	.1179	.1217	.1255	.1293	.1331	.1368	.1406	.1443	.1480	.1517
0.4	.1554	.1591	.1628	.1664	.1700	.1736	.1772	.1808	.1844	.1879
0.5	.1915	.1950	.1985	.2019	.2054	.2088	.2123	.2157	.2190	.2224
0.6	.2257	.2291	.2324	.2357	.2389	.2422	.2454	.2486	.2517	.2549
0.7	.2580	.2611	.2642	.2673	.2704	.2734	.2764	.2794	.2823.	.2852
0.8	.2881	.2910	.2939	.2967	.2995	.3023	.3051	.3078	.3106	.3133
0.9	.3159	.3186	.3212	.3238	.3264	.3289	.3315	.3340	.3365	.3389
1.0	.3413	.3438	.3461	.3485	.3508	.3531	.3554	.3577	.3599	.3621
1.1	.3643	.3665	.3686	.3708	.3729	.3749	.3770	.3790	.3810	.3830
1.2	.3849	.3869	.3888	.3907	.3925	.3944	.3962	.3980	.3997	.4015
1.3	.4032	.4049	.4066	.4082	.4099	.4115	.4131	.4147	.4162	.4177
1.4	.4192	.4207	.4222	.4236	.4251	.4265	.4279	.4292	.4306	.4319
1.5	.4332	.4345	.4357	.4370	.4382	.4394	.4406	.4418	.4429	.4441
1.6	.4452	.4463	.4474	.4484	.4495	.4505	.4515	.4525	.4535	.4545
1.7	.4554	.4564	.4573	.4582	.4591	.4599	.4608	.4616	.4625	.4633
1.8	.4641	.4649	.4656	.4664	.4671	.4678	.4686	.4693	.4699	.4706
1.9	.4713	.4719	.4726	.4732	.4738	.4744	.4750	.4756	.4761	.4767
2.0	.4772	.4778	.4783	.4788	.4793	.4798	.4803	.4808	.4812	.4817
2.1	.4821	.4826	.4830	.4834	.4838	.4842	.4846	.4850	.4854	.4857
2.2	.4861	.4864	.4868	.4871	.4875	.4878	.4881	.4884	.4887	.4890
2.3	.4893	.4896	.4898	.4901	.4904	.4906	.4909	.4911	.4913	.4916
2.4	.4918	.4920	.4922	.4925	.4927	.4929	.4931	.4932	.4934	.4936
2.5	.4938	.4940	.4941	.4943	.4945	.4946	.4948	.4949	.4951	.4952
2.6	.4953	.4955	.4956	.4957	.4959	.4960	.4961	.4962	.4963	.4964
2.7	.4965	.4966	.4967	.4968	.4969	.4970	.4971	.4972	.4973	.4974
2.8	.4974	.4975	.4976	.4977	.4977	.4978	.4979	.4979	.4980	.4981
2.9	.4981	.4982	.4982	.4983	.4984	.4984	.4985	.4985	.4986	.4986
3.0	.4987	.4987	.4987	.4988	.4988	.4989	.4989	.4989	.4990	.4990

Source: Abridged from Table I of A. Hald, *Statistical Tables and Formulas* (New York: John Wiley & Sons, Inc.), 1952. Reproduced by permission of A. Hald and the publisher.

Table V

Critical Values of t

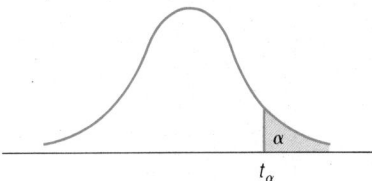

DEGREES OF FREEDOM	$t_{.100}$	$t_{.050}$	$t_{.025}$	$t_{.010}$	$t_{.005}$
1	3.078	6.314	12.706	31.821	63.657
2	1.886	2.920	4.303	6.965	9.925
3	1.638	2.353	3.182	4.541	5.841
4	1.533	2.132	2.776	3.747	4.604
5	1.476	2.015	2.571	3.365	4.032
6	1.440	1.943	2.447	3.143	3.707
7	1.415	1.895	2.365	2.998	3.499
8	1.397	1.860	2.306	2.896	3.355
9	1.383	1.833	2.262	2.821	3.250
10	1.372	1.812	2.228	2.764	3.169
11	1.363	1.796	2.201	2.718	3.106
12	1.356	1.782	2.179	2.681	3.055
13	1.350	1.771	2.160	2.650	3.012
14	1.345	1.761	2.145	2.624	2.977
15	1.341	1.753	2.131	2.602	2.947
16	1.337	1.746	2.120	2.583	2.921
17	1.333	1.740	2.110	2.567	2.898
18	1.330	1.734	2.101	2.552	2.878
19	1.328	1.729	2.093	2.539	2.861
20	1.325	1.725	2.086	2.528	2.845
21	1.323	1.721	2.080	2.518	2.831
22	1.321	1.717	2.074	2.508	2.819
23	1.319	1.714	2.069	2.500	2.807
24	1.318	1.711	2.064	2.492	2.797
25	1.316	1.708	2.060	2.485	2.787
26	1.315	1.706	2.056	2.479	2.779
27	1.314	1.703	2.052	2.473	2.771
28	1.313	1.701	2.048	2.467	2.763
29	1.311	1.699	2.045	2.462	2.756
∞	1.282	1.645	1.960	2.326	2.576

Source: From M. Merrington, "Table of Percentage Points of the t-Distribution," *Biometrika*, 1941, *32*, 300 Reproduced by permission of the *Biometrika* trustees.

Table VI

Critical Values of χ^2

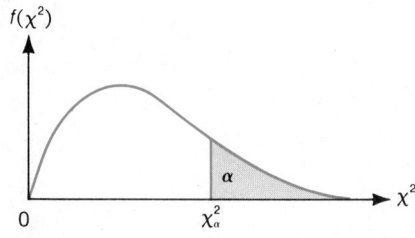

DEGREES OF FREEDOM	$\chi^2_{.995}$	$\chi^2_{.990}$	$\chi^2_{.975}$	$\chi^2_{.950}$	$\chi^2_{.900}$
1	0.0000393	0.0001571	0.0009821	0.0039321	0.0157908
2	0.0100251	0.0201007	0.0506356	0.102587	0.210720
3	0.0717212	0.114832	0.215795	0.351846	0.584375
4	0.206990	0.297110	0.484419	0.710721	1.063623
5	0.411740	0.554300	0.831211	1.145476	1.61031
6	0.675727	0.872085	1.237347	1.63539	2.20413
7	0.989265	1.239043	1.68987	2.16735	2.83311
8	1.344419	1.646482	2.17973	2.73264	3.48954
9	1.734926	2.087912	2.70039	3.32511	4.16816
10	2.15585	2.55821	3.24697	3.94030	4.86518
11	2.60321	3.05347	3.81575	4.57481	5.57779
12	3.07382	3.57056	4.40379	5.22603	6.30380
13	3.56503	4.10691	5.00874	5.89186	7.04150
14	4.07468	4.66043	5.62872	6.57063	7.78953
15	4.60094	5.22935	6.26214	7.26094	8.54675
16	5.14224	5.81221	6.90766	7.96164	9.31223
17	5.69724	6.40776	7.56418	8.67176	10.0852
18	6.26481	7.01491	8.23075	9.39046	10.8649
19	6.84398	7.63273	8.90655	10.1170	11.6509
20	7.43386	8.26040	9.59083	10.8508	12.4426
21	8.03366	8.89720	10.28293	11.5913	13.2396
22	8.64272	9.54249	10.9823	12.3380	14.0415
23	9.26042	10.19567	11.6885	13.0905	14.8479
24	9.88623	10.8564	12.4011	13.8484	15.6587
25	10.5197	11.5240	13.1197	14.6114	16.4734
26	11.1603	12.1981	13.8439	15.3791	17.2919
27	11.8076	12.8786	14.5733	16.1513	18.1138
28	12.4613	13.5648	15.3079	16.9279	18.9392
29	13.1211	14.2565	16.0471	17.7083	19.7677
30	13.7867	14.9535	16.7908	18.4926	20.5992
40	20.7065	22.1643	24.4331	26.5093	29.0505
50	27.9907	29.7067	32.3574	34.7642	37.6886
60	35.5346	37.4848	40.4817	43.1879	46.4589
70	43.2752	45.4418	48.7576	51.7393	55.3290
80	51.1720	53.5400	57.1532	60.3915	64.2778
90	59.1963	61.7541	65.6466	69.1260	73.2912
100	67.3276	70.0648	74.2219	77.9295	82.3581

DEGREES OF FREEDOM	$\chi^2_{.100}$	$\chi^2_{.050}$	$\chi^2_{.025}$	$\chi^2_{.010}$	$\chi^2_{.005}$
1	2.70554	3.84146	5.02389	6.63490	7.87944
2	4.60517	5.99147	7.37776	9.21034	10.5966
3	6.25139	7.81473	9.34840	11.3449	12.8381
4	7.77944	9.48773	11.1433	13.2767	14.8602
5	9.23635	11.0705	12.8325	15.0863	16.7496
6	10.6446	12.5916	14.4494	16.8119	18.5476
7	12.0170	14.0671	16.0128	18.4753	20.2777
8	13.3616	15.5073	17.5346	20.0902	21.9550
9	14.6837	16.9190	19.0228	21.6660	23.5893
10	15.9871	18.3070	20.4831	23.2093	25.1882
11	17.2750	19.6751	21.9200	24.7250	26.7569
12	18.5494	21.0261	23.3367	26.2170	28.2995
13	19.8119	22.3621	24.7356	27.6883	29.8194
14	21.0642	23.6848	26.1190	29.1413	31.3193
15	22.3072	24.9958	27.4884	30.5779	32.8013
16	23.5418	26.2962	28.8454	31.9999	34.2672
17	24.7690	27.5871	30.1910	33.4087	35.7185
18	25.9894	28.8693	31.5264	34.8053	37.1564
19	27.2036	30.1435	32.8523	36.1908	38.5822
20	28.4120	31.4104	34.1696	37.5662	39.9968
21	29.6151	32.6705	35.4789	38.9321	41.4010
22	30.8133	33.9244	36.7807	40.2894	42.7956
23	32.0069	35.1725	38.0757	41.6384	44.1813
24	33.1963	36.4151	39.3641	42.9798	45.5585
25	34.3816	37.6525	40.6465	44.3141	46.9278
26	35.5631	38.8852	41.9232	45.6417	48.2899
27	36.7412	40.1133	43.1944	46.9630	49.6449
28	37.9159	41.3372	44.4607	48.2782	50.9933
29	39.0875	42.5569	45.7222	49.5879	52.3356
30	40.2560	43.7729	46.9792	50.8922	53.6720
40	51.8050	55.7585	59.3417	63.6907	66.7659
50	63.1671	67.5048	71.4202	76.1539	79.4900
60	74.3970	79.0819	83.2976	88.3794	91.9517
70	85.5271	90.5312	95.0231	100.425	104.215
80	96.5782	101.879	106.629	112.329	116.321
90	107.565	113.145	118.136	124.116	128.299
100	118.498	124.342	129.561	135.807	140.169

Source: From C. M. Thompson, "Tables of the Percentage Points of the χ^2-Distribution," *Biometrika*, 1941, *32*, 188–189. Reproduced by permission of the *Biometrika* trustees.

Table VII   Percentage Points of the F-Distribution, $\alpha = .10$

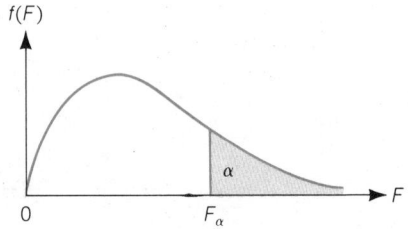

v_1				NUMERATOR DEGREES OF FREEDOM					
v_2	1	2	3	4	5	6	7	8	9
1	39.86	49.50	53.59	55.83	57.24	58.20	58.91	59.44	59.86
2	8.53	9.00	9.16	9.24	9.29	9.33	9.35	9.37	9.38
3	5.54	5.46	5.39	5.34	5.31	5.28	5.27	5.25	5.24
4	4.54	4.32	4.19	4.11	4.05	4.01	3.98	3.95	3.94
5	4.06	3.78	3.62	3.52	3.45	3.40	3.37	3.34	3.32
6	3.78	3.46	3.29	3.18	3.11	3.05	3.01	2.98	2.96
7	3.59	3.26	3.07	2.96	2.88	2.83	2.78	2.75	2.72
8	3.46	3.11	2.92	2.81	2.73	2.67	2.62	2.59	2.56
9	3.36	3.01	2.81	2.69	2.61	2.55	2.51	2.47	2.44
10	3.29	2.92	2.73	2.61	2.52	2.46	2.41	2.38	2.35
11	3.23	2.86	2.66	2.54	2.45	2.39	2.34	2.30	2.27
12	3.18	2.81	2.61	2.48	2.39	2.33	2.28	2.24	2.21
13	3.14	2.76	2.56	2.43	2.35	2.28	2.23	2.20	2.16
14	3.10	2.73	2.52	2.39	2.31	2.24	2.19	2.15	2.12
15	3.07	2.70	2.49	2.36	2.27	2.21	2.16	2.12	2.09
16	3.05	2.67	2.46	2.33	2.24	2.18	2.13	2.09	2.06
17	3.03	2.64	2.44	2.31	2.22	2.15	2.10	2.06	2.03
18	3.01	2.62	2.42	2.29	2.20	2.13	2.08	2.04	2.00
19	2.99	2.61	2.40	2.27	2.18	2.11	2.06	2.02	1.98
20	2.97	2.59	2.38	2.25	2.16	2.09	2.04	2.00	1.96
21	2.96	2.57	2.36	2.23	2.14	2.08	2.02	1.98	1.95
22	2.95	2.56	2.35	2.22	2.13	2.06	2.01	1.97	1.93
23	2.94	2.55	2.34	2.21	2.11	2.05	1.99	1.95	1.92
24	2.93	2.54	2.33	2.19	2.10	2.04	1.98	1.94	1.91
25	2.92	2.53	2.32	2.18	2.09	2.02	1.97	1.93	1.89
26	2.91	2.52	2.31	2.17	2.08	2.01	1.96	1.92	1.88
27	2.90	2.51	2.30	2.17	2.07	2.00	1.95	1.91	1.87
28	2.89	2.50	2.29	2.16	2.06	2.00	1.94	1.90	1.87
29	2.89	2.50	2.28	2.15	2.06	1.99	1.93	1.89	1.86
30	2.88	2.49	2.28	2.14	2.05	1.98	1.93	1.88	1.85
40	2.84	2.44	2.23	2.09	2.00	1.93	1.87	1.83	1.79
60	2.79	2.39	2.18	2.04	1.95	1.87	1.82	1.77	1.74
120	2.75	2.35	2.13	1.99	1.90	1.82	1.77	1.72	1.68
∞	2.71	2.30	2.08	1.94	1.85	1.77	1.72	1.67	1.63

DENOMINATOR DEGREES OF FREEDOM

v_2 \ v_1						NUMERATOR DEGREES OF FREEDOM				
						30	40	60	120	∞
1						62.26	62.53	62.79	63.06	63.33
2						9.46	9.47	9.47	9.48	9.49
3						5.17	5.16	5.15	5.14	5.13
4						3.82	3.80	3.79	3.78	3.76
5						3.17	3.16	3.14	3.12	3.10
6						2.80	2.78	2.76	2.74	2.72
7						2.56	2.54	2.51	2.49	2.47
8						2.38	2.36	2.34	2.32	2.29
9						2.25	2.23	2.21	2.18	2.16
10	2.32	2.28	2.24	2.20	2.18	2.16	2.13	2.11	2.08	2.06
11	2.25	2.21	2.17	2.12	2.10	2.08	2.05	2.03	2.00	1.97
12	2.19	2.15	2.10	2.06	2.04	2.01	1.99	1.96	1.93	1.90
13	2.14	2.10	2.05	2.01	1.98	1.96	1.93	1.90	1.88	1.85
14	2.10	2.05	2.01	1.96	1.94	1.91	1.89	1.86	1.83	1.80
15	2.06	2.02	1.97	1.92	1.90	1.87	1.85	1.82	1.79	1.76
16	2.03	1.99	1.94	1.89	1.87	1.84	1.81	1.78	1.75	1.72
17	2.00	1.96	1.91	1.86	1.84	1.81	1.78	1.75	1.72	1.69
18	1.98	1.93	1.89	1.84	1.81	1.78	1.75	1.72	1.69	1.66
19	1.96	1.91	1.86	1.81	1.79	1.76	1.73	1.70	1.67	1.63
20	1.94	1.89	1.84	1.79	1.77	1.74	1.71	1.68	1.64	1.61
21	1.92	1.87	1.83	1.78	1.75	1.72	1.69	1.66	1.62	1.59
22	1.90	1.86	1.81	1.76	1.73	1.70	1.67	1.64	1.60	1.57
23	1.89	1.84	1.80	1.74	1.72	1.69	1.66	1.62	1.59	1.55
24	1.88	1.83	1.78	1.73	1.70	1.67	1.64	1.61	1.57	1.53
25	1.87	1.82	1.77	1.72	1.69	1.66	1.63	1.59	1.56	1.52
26	1.86	1.81	1.76	1.71	1.68	1.65	1.61	1.58	1.54	1.50
27	1.85	1.80	1.75	1.70	1.67	1.64	1.60	1.57	1.53	1.49
28	1.84	1.79	1.74	1.69	1.66	1.63	1.59	1.56	1.52	1.48
29	1.83	1.78	1.73	1.68	1.65	1.62	1.58	1.55	1.51	1.47
30	1.82	1.77	1.72	1.67	1.64	1.61	1.57	1.54	1.50	1.46
40	1.76	1.71	1.66	1.61	1.57	1.54	1.51	1.47	1.42	1.38
60	1.71	1.66	1.60	1.54	1.51	1.48	1.44	1.40	1.35	1.29
120	1.65	1.60	1.55	1.48	1.45	1.41	1.37	1.32	1.26	1.19
∞	1.60	1.55	1.49	1.42	1.38	1.34	1.30	1.24	1.17	1.00

DENOMINATOR DEGREES OF FREEDOM

Source: From M. Merrington and C. M. Thompson, "Tables of Percentage Points of the Inverted Beta (F)-Distribution," Biometrika, 1943, 33, 73–88. Reproduced by permission of the Biometrika trustees.

Table VIII   Percentage Points of the F-Distribution, $\alpha = .05$

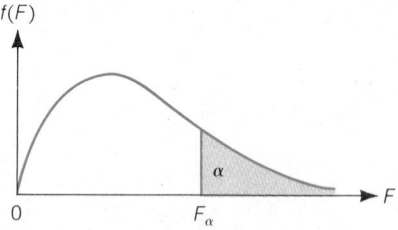

ν_2	NUMERATOR DEGREES OF FREEDOM								
ν_1	1	2	3	4	5	6	7	8	9
1	161.4	199.5	215.7	224.6	230.2	234.0	236.8	238.9	240.5
2	18.51	19.00	19.16	19.25	19.30	19.33	19.35	19.37	19.38
3	10.13	9.55	9.28	9.12	9.01	8.94	8.89	8.85	8.81
4	7.71	6.94	6.59	6.39	6.26	6.16	6.09	6.04	6.00
5	6.61	5.79	5.41	5.19	5.05	4.95	4.88	4.82	4.77
6	5.99	5.14	4.76	4.53	4.39	4.28	4.21	4.15	4.10
7	5.59	4.74	4.35	4.12	3.97	3.87	3.79	3.73	3.68
8	5.32	4.46	4.07	3.84	3.69	3.58	3.50	3.44	3.39
9	5.12	4.26	3.86	3.63	3.48	3.37	3.29	3.23	3.18
10	4.96	4.10	3.71	3.48	3.33	3.22	3.14	3.07	3.02
11	4.84	3.98	3.59	3.36	3.20	3.09	3.01	2.95	2.90
12	4.75	3.89	3.49	3.26	3.11	3.00	2.91	2.85	2.80
13	4.67	3.81	3.41	3.18	3.03	2.92	2.83	2.77	2.71
14	4.60	3.74	3.34	3.11	2.96	2.85	2.76	2.70	2.65
15	4.54	3.68	3.29	3.06	2.90	2.79	2.71	2.64	2.59
16	4.49	3.63	3.24	3.01	2.85	2.74	2.66	2.59	2.54
17	4.45	3.59	3.20	2.96	2.81	2.70	2.61	2.55	2.49
18	4.41	3.55	3.16	2.93	2.77	2.66	2.58	2.51	2.46
19	4.38	3.52	3.13	2.90	2.74	2.63	2.54	2.48	2.42
20	4.35	3.49	3.10	2.87	2.71	2.60	2.51	2.45	2.39
21	4.32	3.47	3.07	2.84	2.68	2.57	2.49	2.42	2.37
22	4.30	3.44	3.05	2.82	2.66	2.55	2.46	2.40	2.34
23	4.28	3.42	3.03	2.80	2.64	2.53	2.44	2.37	2.32
24	4.26	3.40	3.01	2.78	2.62	2.51	2.42	2.36	2.30
25	4.24	3.39	2.99	2.76	2.60	2.49	2.40	2.34	2.28
26	4.23	3.37	2.98	2.74	2.59	2.47	2.39	2.32	2.27
27	4.21	3.35	2.96	2.73	2.57	2.46	2.37	2.31	2.25
28	4.20	3.34	2.95	2.71	2.56	2.45	2.36	2.29	2.24
29	4.18	3.33	2.93	2.70	2.55	2.43	2.35	2.28	2.22
30	4.17	3.32	2.92	2.69	2.53	2.42	2.33	2.27	2.21
40	4.08	3.23	2.84	2.61	2.45	2.34	2.25	2.18	2.12
60	4.00	3.15	2.76	2.53	2.37	2.25	2.17	2.10	2.04
120	3.92	3.07	2.68	2.45	2.29	2.17	2.09	2.02	1.96
∞	3.84	3.00	2.60	2.37	2.21	2.10	2.01	1.94	1.88

DENOMINATOR DEGREES OF FREEDOM

ν_1 ν_2	NUMERATOR DEGREES OF FREEDOM									
	10	12	15	20	24	30	40	60	120	∞
1	241.9	243.9	245.9	248.0	249.1	250.1	251.1	252.2	253.3	254.3
2	19.40	19.41	19.43	19.45	19.45	19.46	19.47	19.48	19.49	19.50
3	8.79	8.74	8.70	8.66	8.64	8.62	8.59	8.57	8.55	8.53
4	5.96	5.91	5.86	5.80	5.77	5.75	5.72	5.69	5.66	5.63
5	4.74	4.68	4.62	4.56	4.53	4.50	4.46	4.43	4.40	4.36
6	4.06	4.00	3.94	3.87	3.84	3.81	3.77	3.74	3.70	3.67
7	3.64	3.57	3.51	3.44	3.41	3.38	3.34	3.30	3.27	3.23
8	3.35	3.28	3.22	3.15	3.12	3.08	3.04	3.01	2.97	2.93
9	3.14	3.07	3.01	2.94	2.90	2.86	2.83	2.79	2.75	2.71
10	2.98	2.91	2.85	2.77	2.74	2.70	2.66	2.62	2.58	2.54
11	2.85	2.79	2.72	2.65	2.61	2.57	2.53	2.49	2.45	2.40
12	2.75	2.69	2.62	2.54	2.51	2.47	2.43	2.38	2.34	2.30
13	2.67	2.60	2.53	2.46	2.42	2.38	2.34	2.30	2.25	2.21
14	2.60	2.53	2.46	2.39	2.35	2.31	2.27	2.22	2.18	2.13
15	2.54	2.48	2.40	2.33	2.29	2.25	2.20	2.16	2.11	2.07
16	2.49	2.42	2.35	2.28	2.24	2.19	2.15	2.11	2.06	2.01
17	2.45	2.38	2.31	2.23	2.19	2.15	2.10	2.06	2.01	1.96
18	2.41	2.34	2.27	2.19	2.15	2.11	2.06	2.02	1.97	1.92
19	2.38	2.31	2.23	2.16	2.11	2.07	2.03	1.98	1.93	1.88
20	2.35	2.28	2.20	2.12	2.08	2.04	1.99	1.95	1.90	1.84
21	2.32	2.25	2.18	2.10	2.05	2.01	1.96	1.92	1.87	1.81
22	2.30	2.23	2.15	2.07	2.03	1.98	1.94	1.89	1.84	1.78
23	2.27	2.20	2.13	2.05	2.01	1.96	1.91	1.86	1.81	1.76
24	2.25	2.18	2.11	2.03	1.98	1.94	1.89	1.84	1.79	1.73
25	2.24	2.16	2.09	2.01	1.96	1.92	1.87	1.82	1.77	1.71
26	2.22	2.15	2.07	1.99	1.95	1.90	1.85	1.80	1.75	1.69
27	2.20	2.13	2.06	1.97	1.93	1.88	1.84	1.79	1.73	1.67
28	2.19	2.12	2.04	1.96	1.91	1.87	1.82	1.77	1.71	1.65
29	2.18	2.10	2.03	1.94	1.90	1.85	1.81	1.75	1.70	1.64
30	2.16	2.09	2.01	1.93	1.89	1.84	1.79	1.74	1.68	1.62
40	2.08	2.00	1.92	1.84	1.79	1.74	1.69	1.64	1.58	1.51
60	1.99	1.92	1.84	1.75	1.70	1.65	1.59	1.53	1.47	1.39
120	1.91	1.83	1.75	1.66	1.61	1.55	1.50	1.43	1.35	1.25
∞	1.83	1.75	1.67	1.57	1.52	1.46	1.39	1.32	1.22	1.00

DENOMINATOR DEGREES OF FREEDOM

Source: From M. Merrington and C. M. Thompson, "Tables of Percentage Points of the Inverted Beta (*F*)-Distribution," *Biometrika*, 1943, *33*, 73–88. Reproduced by permission of the *Biometrika* trustees.

Table IX   Percentage Points of the *F*-Distribution, $\alpha = .025$

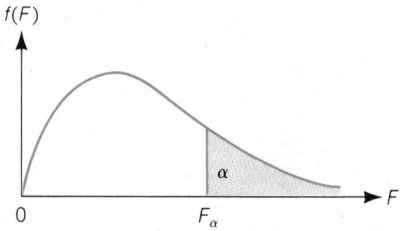

ν_1 ν_2	NUMERATOR DEGREES OF FREEDOM								
	1	2	3	4	5	6	7	8	9
1	647.8	799.5	864.2	899.6	921.8	937.1	948.2	956.7	963.3
2	38.51	39.00	39.17	39.25	39.30	39.33	39.36	39.37	39.39
3	17.44	16.04	15.44	15.10	14.88	14.73	14.62	14.54	14.47
4	12.22	10.65	9.98	9.60	9.36	9.20	9.07	8.98	8.90
5	10.01	8.43	7.76	7.39	7.15	6.98	6.85	6.76	6.68
6	8.81	7.26	6.60	6.23	5.99	5.82	5.70	5.60	5.52
7	8.07	6.54	5.89	5.52	5.29	5.12	4.99	4.90	4.82
8	7.57	6.06	5.42	5.05	4.82	4.65	4.53	4.43	4.36
9	7.21	5.71	5.08	4.72	4.48	4.32	4.20	4.10	4.03
10	6.94	5.46	4.83	4.47	4.24	4.07	3.95	3.85	3.78
11	6.72	5.26	4.63	4.28	4.04	3.88	3.76	3.66	3.59
12	6.55	5.10	4.47	4.12	3.89	3.73	3.61	3.51	3.44
13	6.41	4.97	4.35	4.00	3.77	3.60	3.48	3.39	3.31
14	6.30	4.86	4.24	3.89	3.66	3.50	3.38	3.29	3.21
15	6.20	4.77	4.15	3.80	3.58	3.41	3.29	3.20	3.12
16	6.12	4.69	4.08	3.73	3.50	3.34	3.22	3.12	3.05
17	6.04	4.62	4.01	3.66	3.44	3.28	3.16	3.06	2.98
18	5.98	4.56	3.95	3.61	3.38	3.22	3.10	3.01	2.93
19	5.92	4.51	3.90	3.56	3.33	3.17	3.05	2.96	2.88
20	5.87	4.46	3.86	3.51	3.29	3.13	3.01	2.91	2.84
21	5.83	4.42	3.82	3.48	3.25	3.09	2.97	2.87	2.80
22	5.79	4.38	3.78	3.44	3.22	3.05	2.93	2.84	2.76
23	5.75	4.35	3.75	3.41	3.18	3.02	2.90	2.81	2.73
24	5.72	4.32	3.72	3.38	3.15	2.99	2.87	2.78	2.70
25	5.69	4.29	3.69	3.35	3.13	2.97	2.85	2.75	2.68
26	5.66	4.27	3.67	3.33	3.10	2.94	2.82	2.73	2.65
27	5.63	4.24	3.65	3.31	3.08	2.92	2.80	2.71	2.63
28	5.61	4.22	3.63	3.29	3.06	2.90	2.78	2.69	2.61
29	5.59	4.20	3.61	3.27	3.04	2.88	2.76	2.67	2.59
30	5.57	4.18	3.59	3.25	3.03	2.87	2.75	2.65	2.57
40	5.42	4.05	3.46	3.13	2.90	2.74	2.62	2.53	2.45
60	5.29	3.93	3.34	3.01	2.79	2.63	2.51	2.41	2.33
120	5.15	3.80	3.23	2.89	2.67	2.52	2.39	2.30	2.22
∞	5.02	3.69	3.12	2.79	2.57	2.41	2.29	2.19	2.11

DENOMINATOR DEGREES OF FREEDOM

v_1 / v_2	NUMERATOR DEGREES OF FREEDOM									
	10	12	15	20	24	30	40	60	120	∞
1	968.6	976.7	984.9	993.1	997.2	1001	1006	1010	1014	1018
2	39.40	39.41	39.43	39.45	39.46	39.46	39.47	39.48	39.49	39.50
3	14.42	14.34	14.25	14.17	14.12	14.08	14.04	13.99	13.95	13.90
4	8.84	8.75	8.66	8.56	8.51	8.46	8.41	8.36	8.31	8.26
5	6.62	6.52	6.43	6.33	6.28	6.23	6.18	6.12	6.07	6.02
6	5.46	5.37	5.27	5.17	5.12	5.07	5.01	4.96	4.90	4.85
7	4.76	4.67	4.57	4.47	4.42	4.36	4.31	4.25	4.20	4.14
8	4.30	4.20	4.10	4.00	3.95	3.89	3.84	3.78	3.73	3.67
9	3.96	3.87	3.77	3.67	3.61	3.56	3.51	3.45	3.39	3.33
10	3.72	3.62	3.52	3.42	3.37	3.31	3.26	3.20	3.14	3.08
11	3.53	3.43	3.33	3.23	3.17	3.12	3.06	3.00	2.94	2.88
12	3.37	3.28	3.18	3.07	3.02	2.96	2.91	2.85	2.79	2.72
13	3.25	3.15	3.05	2.95	2.89	2.84	2.78	2.72	2.66	2.60
14	3.15	3.05	2.95	2.84	2.79	2.73	2.67	2.61	2.55	2.49
15	3.06	2.96	2.86	2.76	2.70	2.64	2.59	2.52	2.46	2.40
16	2.99	2.89	2.79	2.68	2.63	2.57	2.51	2.45	2.38	2.32
17	2.92	2.82	2.72	2.62	2.56	2.50	2.44	2.38	2.32	2.25
18	2.87	2.77	2.67	2.56	2.50	2.44	2.38	2.32	2.26	2.19
19	2.82	2.72	2.62	2.51	2.45	2.39	2.33	2.27	2.20	2.13
20	2.77	2.68	2.57	2.46	2.41	2.35	2.29	2.22	2.16	2.09
21	2.73	2.64	2.53	2.42	2.37	2.31	2.25	2.18	2.11	2.04
22	2.70	2.60	2.50	2.39	2.33	2.27	2.21	2.14	2.08	2.00
23	2.67	2.57	2.47	2.36	2.30	2.24	2.18	2.11	2.04	1.97
24	2.64	2.54	2.44	2.33	2.27	2.21	2.15	2.08	2.01	1.94
25	2.61	2.51	2.41	2.30	2.24	2.18	2.12	2.05	1.98	1.91
26	2.59	2.49	2.39	2.28	2.22	2.16	2.09	2.03	1.95	1.88
27	2.57	2.47	2.36	2.25	2.19	2.13	2.07	2.00	1.93	1.85
28	2.55	2.45	2.34	2.23	2.17	2.11	2.05	1.98	1.91	1.83
29	2.53	2.43	2.32	2.21	2.15	2.09	2.03	1.96	1.89	1.81
30	2.51	2.41	2.31	2.20	2.14	2.07	2.01	1.94	1.87	1.79
40	2.39	2.29	2.18	2.07	2.01	1.94	1.88	1.80	1.72	1.64
60	2.27	2.17	2.06	1.94	1.88	1.82	1.74	1.67	1.58	1.48
120	2.16	2.05	1.94	1.82	1.76	1.69	1.61	1.53	1.43	1.31
∞	2.05	1.94	1.83	1.71	1.64	1.57	1.48	1.39	1.27	1.00

DENOMINATOR DEGREES OF FREEDOM

Table X Percentage Points of the F-Distribution, $\alpha = .01$

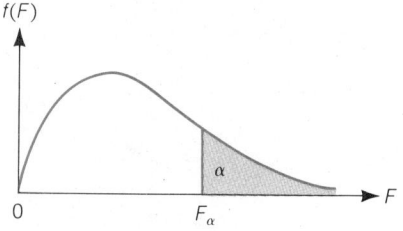

ν_2 \ ν_1	NUMERATOR DEGREES OF FREEDOM								
	1	2	3	4	5	6	7	8	9
1	4,052	4,999.5	5,403	5,625	5,764	5,859	5,928	5,982	6,022
2	98.50	99.00	99.17	99.25	99.30	99.33	99.36	99.37	99.39
3	34.12	30.82	29.46	28.71	28.24	27.91	27.67	27.49	27.35
4	21.20	18.00	16.69	15.98	15.52	15.21	14.98	14.80	14.66
5	16.26	13.27	12.06	11.39	10.97	10.67	10.46	10.29	10.16
6	13.75	10.92	9.78	9.15	8.75	8.47	8.26	8.10	7.98
7	12.25	9.55	8.45	7.85	7.46	7.19	6.99	6.84	6.72
8	11.26	8.65	7.59	7.01	6.63	6.37	6.18	6.03	5.91
9	10.56	8.02	6.99	6.42	6.06	5.80	5.61	5.47	5.35
10	10.04	7.56	6.55	5.99	5.64	5.39	5.20	5.06	4.94
11	9.65	7.21	6.22	5.67	5.32	5.07	4.89	4.74	4.63
12	9.33	6.93	5.95	5.41	5.06	4.82	4.64	4.50	4.39
13	9.07	6.70	5.74	5.21	4.86	4.62	4.44	4.30	4.19
14	8.86	6.51	5.56	5.04	4.69	4.46	4.28	4.14	4.03
15	8.68	6.36	5.42	4.89	4.56	4.32	4.14	4.00	3.89
16	8.53	6.23	5.29	4.77	4.44	4.20	4.03	3.89	3.78
17	8.40	6.11	5.18	4.67	4.34	4.10	3.93	3.79	3.68
18	8.29	6.01	5.09	4.58	4.25	4.01	3.84	3.71	3.60
19	8.18	5.93	5.01	4.50	4.17	3.94	3.77	3.63	3.52
20	8.10	5.85	4.94	4.43	4.10	3.87	3.70	3.56	3.46
21	8.02	5.78	4.87	4.37	4.04	3.81	3.64	3.51	3.40
22	7.95	5.72	4.82	4.31	3.99	3.76	3.59	3.45	3.35
23	7.88	5.66	4.76	4.26	3.94	3.71	3.54	3.41	3.30
24	7.82	5.61	4.72	4.22	3.90	3.67	3.50	3.36	3.26
25	7.77	5.57	4.68	4.18	3.85	3.63	3.46	3.32	3.22
26	7.72	5.53	4.64	4.14	3.82	3.59	3.42	3.29	3.18
27	7.68	5.49	4.60	4.11	3.78	3.56	3.39	3.26	3.15
28	7.64	5.45	4.57	4.07	3.75	3.53	3.36	3.23	3.12
29	7.60	5.42	4.54	4.04	3.73	3.50	3.33	3.20	3.09
30	7.56	5.39	4.51	4.02	3.70	3.47	3.30	3.17	3.07
40	7.31	5.18	4.31	3.83	3.51	3.29	3.12	2.99	2.89
60	7.08	4.98	4.13	3.65	3.34	3.12	2.95	2.82	2.72
120	6.85	4.79	3.95	3.48	3.17	2.96	2.79	2.66	2.56
∞	6.63	4.61	3.78	3.32	3.02	2.80	2.64	2.51	2.41

DENOMINATOR DEGREES OF FREEDOM

v_1 / v_2	NUMERATOR DEGREES OF FREEDOM									
	10	12	15	20	24	30	40	60	120	∞
1	6,056	6,106	6,157	6,209	6,235	6,261	6,287	6,313	6,339	6,366
2	99.40	99.42	99.43	99.45	99.46	99.47	99.47	99.48	99.49	99.50
3	27.23	27.05	26.87	26.69	26.60	26.50	26.41	26.32	26.22	26.13
4	14.55	14.37	14.20	14.02	13.93	13.84	13.75	13.65	13.56	13.46
5	10.05	9.89	9.72	9.55	9.47	9.38	9.29	9.20	9.11	9.02
6	7.87	7.72	7.56	7.40	7.31	7.23	7.14	7.06	6.97	6.88
7	6.62	6.47	6.31	6.16	6.07	5.99	5.91	5.82	5.74	5.65
8	5.81	5.67	5.52	5.36	5.28	5.20	5.12	5.03	4.95	4.86
9	5.26	5.11	4.96	4.81	4.73	4.65	4.57	4.48	4.40	4.31
10	4.85	4.71	4.56	4.41	4.33	4.25	4.17	4.08	4.00	3.91
11	4.54	4.40	4.25	4.10	4.02	3.94	3.86	3.78	3.69	3.60
12	4.30	4.16	4.01	3.86	3.78	3.70	3.62	3.54	3.45	3.36
13	4.10	3.96	3.82	3.66	3.59	3.51	3.43	3.34	3.25	3.17
14	3.94	3.80	3.66	3.51	3.43	3.35	3.27	3.18	3.09	3.00
15	3.80	3.67	3.52	3.37	3.29	3.21	3.13	3.05	2.96	2.87
16	3.69	3.55	3.41	3.26	3.18	3.10	3.02	2.93	2.84	2.75
17	3.59	3.46	3.31	3.16	3.08	3.00	2.92	2.83	2.75	2.65
18	3.51	3.37	3.23	3.08	3.00	2.92	2.84	2.75	2.66	2.57
19	3.43	3.30	3.15	3.00	2.92	2.84	2.76	2.67	2.58	2.49
20	3.37	3.23	3.09	2.94	2.86	2.78	2.69	2.61	2.52	2.42
21	3.31	3.17	3.03	2.88	2.80	2.72	2.64	2.55	2.46	2.36
22	3.26	3.12	2.98	2.83	2.75	2.67	2.58	2.50	2.40	2.31
23	3.21	3.07	2.93	2.78	2.70	2.62	2.54	2.45	2.35	2.26
24	3.17	3.03	2.89	2.74	2.66	2.58	2.49	2.40	2.31	2.21
25	3.13	2.99	2.85	2.70	2.62	2.54	2.45	2.36	2.27	2.17
26	3.09	2.96	2.81	2.66	2.58	2.50	2.42	2.33	2.23	2.13
27	3.06	2.93	2.78	2.63	2.55	2.47	2.38	2.29	2.20	2.10
28	3.03	2.90	2.75	2.60	2.52	2.44	2.35	2.26	2.17	2.06
29	3.00	2.87	2.73	2.57	2.49	2.41	2.33	2.23	2.14	2.03
30	2.98	2.84	2.70	2.55	2.47	2.39	2.30	2.21	2.11	2.01
40	2.80	2.66	2.52	2.37	2.29	2.20	2.11	2.02	1.92	1.80
60	2.63	2.50	2.35	2.20	2.12	2.03	1.94	1.84	1.73	1.60
120	2.47	2.34	2.19	2.03	1.95	1.86	1.76	1.66	1.53	1.38
∞	2.32	2.18	2.04	1.88	1.79	1.70	1.59	1.47	1.32	1.00

DENOMINATOR DEGREES OF FREEDOM

Source: From M. Merrington and C. M. Thompson, "Tables of Percentage Points of the Inverted Beta (*F*)-Distribution," *Biometrika,* 1943, *33,* 73–88. Reproduced by permission of the *Biometrika* trustees.

Table XI

Critical Values of T_L and T_U for the Wilcoxon Rank Sum Test: Independent Samples

Test statistic is rank sum associated with smaller sample (if equal sample sizes, either rank sum can be used).

a. $\alpha = .025$ one-tailed; $\alpha = .05$ two-tailed

n_2 \ n_1	3		4		5		6		7		8		9		10	
	T_L	T_U	T_L	T_U	T_L	T_U	T_L	T_U	T_L	T_U	T_L	T_U	T_L	T_U	T_L	T_U
3	5	16	6	18	6	21	7	23	7	26	8	28	8	31	9	33
4	6	18	11	25	12	28	12	32	13	35	14	38	15	41	16	44
5	6	21	12	28	18	37	19	41	20	45	21	49	22	53	24	56
6	7	23	12	32	19	41	26	52	28	56	29	61	31	65	32	70
7	7	26	13	35	20	45	28	56	37	68	39	73	41	78	43	83
8	8	28	14	38	21	49	29	61	39	73	49	87	51	93	54	98
9	8	31	15	41	22	53	31	65	41	78	51	93	63	108	66	114
10	9	33	16	44	24	56	32	70	43	83	54	98	66	114	79	131

b. $\alpha = .05$ one-tailed; $\alpha = .10$ two-tailed

n_2 \ n_1	3		4		5		6		7		8		9		10	
	T_L	T_U	T_L	T_U	T_L	T_U	T_L	T_U	T_L	T_U	T_L	T_U	T_L	T_U	T_L	T_U
3	6	15	7	17	7	20	8	22	9	24	9	27	10	29	11	31
4	7	17	12	24	13	27	14	30	15	33	16	36	17	39	18	42
5	7	20	13	27	19	36	20	40	22	43	24	46	25	50	26	54
6	8	22	14	30	20	40	28	50	30	54	32	58	33	63	35	67
7	9	24	15	33	22	43	30	54	39	66	41	71	43	76	46	80
8	9	27	16	36	24	46	32	58	41	71	52	84	54	90	57	95
9	10	29	17	39	25	50	33	63	43	76	54	90	66	105	69	111
10	11	31	18	42	26	54	35	67	46	80	57	95	69	111	83	127

Source: From F. Wilcoxon and R. A. Wilcox, "Some Rapid Approximate Statistical Procedures," 1964, 20–23. Reproduced with the permission of American Cyanamid Company.

Table XII

Critical Values of T_0 in the Wilcoxon Paired Difference Signed Rank Test

ONE-TAILED	TWO-TAILED	$n = 5$	$n = 6$	$n = 7$	$n = 8$	$n = 9$	$n = 10$
$\alpha = .05$	$\alpha = .10$	1	2	4	6	8	11
$\alpha = .025$	$\alpha = .05$		1	2	4	6	8
$\alpha = .01$	$\alpha = .02$			0	2	3	5
$\alpha = .005$	$\alpha = .01$				0	2	3
		$n = 11$	$n = 12$	$n = 13$	$n = 14$	$n = 15$	$n = 16$
$\alpha = .05$	$\alpha = .10$	14	17	21	26	30	36
$\alpha = .025$	$\alpha = .05$	11	14	17	21	25	30
$\alpha = .01$	$\alpha = .02$	7	10	13	16	20	24
$\alpha = .005$	$\alpha = .01$	5	7	10	13	16	19
		$n = 17$	$n = 18$	$n = 19$	$n = 20$	$n = 21$	$n = 22$
$\alpha = .05$	$\alpha = .10$	41	47	54	60	68	75
$\alpha = .025$	$\alpha = .05$	35	40	46	52	59	66
$\alpha = .01$	$\alpha = .02$	28	33	38	43	49	56
$\alpha = .005$	$\alpha = .01$	23	28	32	37	43	49
		$n = 23$	$n = 24$	$n = 25$	$n = 26$	$n = 27$	$n = 28$
$\alpha = .05$	$\alpha = .10$	83	92	101	110	120	130
$\alpha = .025$	$\alpha = .05$	73	81	90	98	107	117
$\alpha = .01$	$\alpha = .02$	62	69	77	85	93	102
$\alpha = .005$	$\alpha = .01$	55	61	68	76	84	92
		$n = 29$	$n = 30$	$n = 31$	$n = 32$	$n = 33$	$n = 34$
$\alpha = .05$	$\alpha = .10$	141	152	163	175	188	201
$\alpha = .025$	$\alpha = .05$	127	137	148	159	171	183
$\alpha = .01$	$\alpha = .02$	111	120	130	141	151	162
$\alpha = .005$	$\alpha = .01$	100	109	118	128	138	149
		$n = 35$	$n = 36$	$n = 37$	$n = 38$	$n = 39$	
$\alpha = .05$	$\alpha = .10$	214	228	242	256	271	
$\alpha = .025$	$\alpha = .05$	195	208	222	235	250	
$\alpha = .01$	$\alpha = .02$	174	186	198	211	224	
$\alpha = .005$	$\alpha = .01$	160	171	183	195	208	
		$n = 40$	$n = 41$	$n = 42$	$n = 43$	$n = 44$	$n = 45$
$\alpha = .05$	$\alpha = .10$	287	303	319	336	353	371
$\alpha = .025$	$\alpha = .05$	264	279	295	311	327	344
$\alpha = .01$	$\alpha = .02$	238	252	267	281	297	313
$\alpha = .005$	$\alpha = .01$	221	234	248	262	277	292
		$n = 46$	$n = 47$	$n = 48$	$n = 49$	$n = 50$	
$\alpha = .05$	$\alpha = .10$	389	408	427	446	466	
$\alpha = .025$	$\alpha = .05$	361	379	397	415	434	
$\alpha = .01$	$\alpha = .02$	329	345	362	380	398	
$\alpha = .005$	$\alpha = .01$	307	323	339	356	373	

Source: From F. Wilcoxon and R. A. Wilcox, "Some Rapid Approximate Statistical Procedures," 1964, 28, Reproduced with the permission of American Cyanamid Company.

Table XIII

Critical Values of Spearman's Rank Correlation Coefficient

The α-values correspond to a one-tailed test of H_0: $\rho_s = 0$. The value should be doubled for two-tailed tests.

n	$\alpha = .05$	$\alpha = .025$	$\alpha = .01$	$\alpha = .005$
5	.900	—	—	—
6	.829	.886	.943	—
7	.714	.786	.893	—
8	.643	.738	.833	.881
9	.600	.683	.783	.833
10	.564	.648	.745	.794
11	.523	.623	.736	.818
12	.497	.591	.703	.780
13	.475	.566	.673	.745
14	.457	.545	.646	.716
15	.441	.525	.623	.689
16	.425	.507	.601	.666
17	.412	.490	.582	.645
18	.399	.476	.564	.625
19	.388	.462	.549	.608
20	.377	.450	.534	.591
21	.368	.438	.521	.576
22	.359	.428	.508	.562
23	.351	.418	.496	.549
24	.343	.409	.485	.537
25	.336	.400	.475	.526
26	.329	.392	.465	.515
27	.323	.385	.456	.505
28	.317	.377	.448	.496
29	.311	.370	.440	.487
30	.305	.364	.432	.478

Source: From E. G. Olds, "Distribution of Sums of Squares of Rank Differences for Small Samples," *Annals of Mathematical Statistics*, 1938, 9. Reproduced with the permission of the editor, *Annals of Mathematical Statistics*.

Table XIV Percentage Points of the Studentized Range, $q(k, v)$, Upper 5%

v \ k	2	3	4	5	6	7	8	9	10	11	12	13	14	15	16	17	18	19	20
1	17.97	26.98	32.82	37.08	40.41	43.12	45.40	47.36	49.07	50.59	51.96	53.20	54.33	55.36	56.32	57.22	58.04	58.83	59.56
2	6.08	8.33	9.80	10.88	11.74	12.44	13.03	13.54	13.99	14.39	14.75	15.08	15.38	15.65	15.91	16.14	16.37	16.57	16.77
3	4.50	5.91	6.82	7.50	8.04	8.48	8.85	9.18	9.46	9.72	9.95	10.15	10.35	10.52	10.69	10.84	10.98	11.11	11.24
4	3.93	5.04	5.76	6.29	6.71	7.05	7.35	7.60	7.83	8.03	8.21	8.37	8.52	8.66	8.79	8.91	9.03	9.13	9.23
5	3.64	4.60	5.22	5.67	6.03	6.33	6.58	6.80	6.99	7.17	7.32	7.47	7.60	7.72	7.83	7.93	8.03	8.12	8.21
6	3.46	4.34	4.90	5.30	5.63	5.90	6.12	6.32	6.49	6.65	6.79	6.92	7.03	7.14	7.24	7.34	7.43	7.51	7.59
7	3.34	4.16	4.68	5.06	5.36	5.61	5.82	6.00	6.16	6.30	6.43	6.55	6.66	6.76	6.85	6.94	7.02	7.10	7.17
8	3.26	4.04	4.53	4.89	5.17	5.40	5.60	5.77	5.92	6.05	6.18	6.29	6.39	6.48	6.57	6.65	6.73	6.80	6.87
9	3.20	3.95	4.41	4.76	5.02	5.24	5.43	5.59	5.74	5.87	5.98	6.09	6.19	6.28	6.36	6.44	6.51	6.58	6.64
10	3.15	3.88	4.33	4.65	4.91	5.12	5.30	5.46	5.60	5.72	5.83	5.93	6.03	6.11	6.19	6.27	6.34	6.40	6.47
11	3.11	3.82	4.26	4.57	4.82	5.03	5.20	5.35	5.49	5.61	5.71	5.81	5.90	5.98	6.06	6.13	6.20	6.27	6.33
12	3.08	3.77	4.20	4.51	4.75	4.95	5.12	5.27	5.39	5.51	5.61	5.71	5.80	5.88	5.95	6.02	6.09	6.15	6.21
13	3.06	3.73	4.15	4.45	4.69	4.88	5.05	5.19	5.32	5.43	5.53	5.63	5.71	5.79	5.86	5.93	5.99	6.05	6.11
14	3.03	3.70	4.11	4.41	4.64	4.83	4.99	5.13	5.25	5.36	5.46	5.55	5.64	5.71	5.79	5.85	5.91	5.97	6.03
15	3.01	3.67	4.08	4.37	4.60	4.78	4.94	5.08	5.20	5.31	5.40	5.49	5.57	5.65	5.72	5.78	5.85	5.90	5.96
16	3.00	3.65	4.05	4.33	4.56	4.74	4.90	5.03	5.15	5.26	5.35	5.44	5.52	5.59	5.66	5.73	5.79	5.84	5.90
17	2.98	3.63	4.02	4.30	4.52	4.70	4.86	4.99	5.11	5.21	5.31	5.39	5.47	5.54	5.61	5.67	5.73	5.79	5.84
18	2.97	3.61	4.00	4.28	4.49	4.67	4.82	4.96	5.07	5.17	5.27	5.35	5.43	5.50	5.57	5.63	5.69	5.74	5.79
19	2.96	3.59	3.98	4.25	4.47	4.65	4.79	4.92	5.04	5.14	5.23	5.31	5.39	5.46	5.53	5.59	5.65	5.70	5.75
20	2.95	3.58	3.96	4.23	4.45	4.62	4.77	4.90	5.01	5.11	5.20	5.28	5.36	5.43	5.49	5.55	5.61	5.66	5.71
24	2.92	3.53	3.90	4.17	4.37	4.54	4.68	4.81	4.92	5.01	5.10	5.18	5.25	5.32	5.38	5.44	5.49	5.55	5.59
30	2.89	3.49	3.85	4.10	4.30	4.46	4.60	4.72	4.82	4.92	5.00	5.08	5.15	5.21	5.27	5.33	5.38	5.43	5.47
40	2.86	3.44	3.79	4.04	4.23	4.39	4.52	4.63	4.73	4.82	4.90	4.98	5.04	5.11	5.16	5.22	5.27	5.31	5.36
60	2.83	3.40	3.74	3.98	4.16	4.31	4.44	4.55	4.65	4.73	4.81	4.88	4.94	5.00	5.06	5.11	5.15	5.20	5.24
120	2.80	3.36	3.68	3.92	4.10	4.24	4.36	4.47	4.56	4.64	4.71	4.78	4.84	4.90	4.95	5.00	5.04	5.09	5.13
∞	2.77	3.31	3.63	3.86	4.03	4.17	4.29	4.39	4.47	4.55	4.62	4.68	4.74	4.80	4.85	4.89	4.93	4.97	5.01

Source: From E. S. Pearson and H. O. Hartley (Eds.), *Biometrika Tables for Statisticians*, vol. I, 3rd ed. (Cambridge: Cambridge University Press, 1966). Reproduced by permission of Professor E. S. Pearson and the *Biometrika* trustees.

Table XV Percentage Points of the Studentized Range, $q(k, v)$, Upper 1%

v \ k	2	3	4	5	6	7	8	9	10	11	12	13	14	15	16	17	18	19	20
1	90.03	135.0	164.3	185.6	202.2	215.8	227.2	237.0	245.6	253.2	260.0	266.2	271.8	277.0	281.8	286.3	290.0	294.3	298.0
2	14.04	19.02	22.29	24.72	26.63	28.20	29.53	30.68	31.69	32.59	33.40	34.13	34.81	35.43	36.00	36.53	37.03	37.50	37.95
3	8.26	10.62	12.17	13.33	14.24	15.00	15.64	16.20	16.69	17.13	17.53	17.89	18.22	18.52	18.81	19.07	19.32	19.55	19.77
4	6.51	8.12	9.17	9.96	10.58	11.10	11.55	11.93	12.27	12.57	12.84	13.09	13.32	13.53	13.73	13.91	14.08	14.24	14.40
5	5.70	6.98	7.80	8.42	8.91	9.32	9.67	9.97	10.24	10.48	10.70	10.89	11.08	11.24	11.40	11.55	11.68	11.81	11.93
6	5.24	6.33	7.03	7.56	7.97	8.32	8.61	8.87	9.10	9.30	9.48	9.65	9.81	9.95	10.08	10.21	10.32	10.43	10.54
7	4.95	5.92	6.54	7.01	7.37	7.68	7.94	8.17	8.37	8.55	8.71	8.86	9.00	9.12	9.24	9.35	9.46	9.55	9.65
8	4.75	5.64	6.20	6.62	6.96	7.24	7.47	7.68	7.86	8.03	8.18	8.31	8.44	8.55	8.66	8.76	8.85	8.94	9.03
9	4.60	5.43	5.96	6.35	6.66	6.91	7.13	7.33	7.49	7.65	7.78	7.91	8.03	8.13	8.23	8.33	8.41	8.49	8.57
10	4.48	5.27	5.77	6.14	6.43	6.67	6.87	7.05	7.21	7.36	7.49	7.60	7.71	7.81	7.91	7.99	8.08	8.15	8.23
11	4.39	5.15	5.62	5.97	6.25	6.48	6.67	6.84	6.99	7.13	7.25	7.36	7.46	7.56	7.65	7.73	7.81	7.88	7.95
12	4.32	5.05	5.50	5.84	6.10	6.32	6.51	6.67	6.81	6.94	7.06	7.17	7.26	7.36	7.44	7.52	7.59	7.66	7.73
13	4.26	4.96	5.40	5.73	5.98	6.19	6.37	6.53	6.67	6.79	6.90	7.01	7.10	7.19	7.27	7.35	7.42	7.48	7.55
14	4.21	4.89	5.32	5.63	5.88	6.08	6.26	6.41	6.54	6.66	6.77	6.87	6.96	7.05	7.13	7.20	7.27	7.33	7.39
15	4.17	4.84	5.25	5.56	5.80	5.99	6.16	6.31	6.44	6.55	6.66	6.76	6.84	6.93	7.00	7.07	7.14	7.20	7.26
16	4.13	4.79	5.19	5.49	5.72	5.92	6.08	6.22	6.35	6.46	6.56	6.66	6.74	6.82	6.90	6.97	7.03	7.09	7.15
17	4.10	4.74	5.14	5.43	5.66	5.85	6.01	6.15	6.27	6.38	6.48	6.57	6.66	6.73	6.81	6.87	6.94	7.00	7.05
18	4.07	4.70	5.09	5.38	5.60	5.79	5.94	6.08	6.20	6.31	6.41	6.50	6.58	6.65	6.73	6.79	6.85	6.91	6.97
19	4.05	4.67	5.05	5.33	5.55	5.73	5.89	6.02	6.14	6.25	6.34	6.43	6.51	6.58	6.65	6.72	6.78	6.84	6.89
20	4.02	4.64	5.02	5.29	5.51	5.69	5.84	5.97	6.09	6.19	6.28	6.37	6.45	6.52	6.59	6.65	6.71	6.77	6.82
24	3.96	4.55	4.91	5.17	5.37	5.54	5.69	5.81	5.92	6.02	6.11	6.19	6.26	6.33	6.39	6.45	6.51	6.56	6.61
30	3.89	4.45	4.80	5.05	5.24	5.40	5.54	5.65	5.76	5.85	5.93	6.01	6.08	6.14	6.20	6.26	6.31	6.36	6.41
40	3.82	4.37	4.70	4.93	5.11	5.26	5.39	5.50	5.60	5.69	5.76	5.83	5.90	5.96	6.02	6.07	6.12	6.16	6.21
60	3.76	4.28	4.59	4.82	4.99	5.13	5.25	5.36	5.45	5.53	5.60	5.67	5.73	5.78	5.84	5.89	5.93	5.97	6.01
120	3.70	4.20	4.50	4.71	4.87	5.01	5.12	5.21	5.30	5.37	5.44	5.50	5.56	5.61	5.66	5.71	5.75	5.79	5.83
∞	3.64	4.12	4.40	4.60	4.76	4.88	4.99	5.08	5.16	5.23	5.29	5.35	5.40	5.45	5.49	5.54	5.57	5.61	5.65

Source: From E. S. Pearson and H. O. Hartley (Eds.), *Biometrika Tables for Statisticians*, vol. I, 3rd ed. (Cambridge: Cambridge University Press, 1966). Reproduced by permission of Professor E. S. Pearson and the *Biometrika* trustees.

ANSWERS TO SELECTED EXERCISES

Chapter 2

2.1. a. Quantitative   **b.** Qualitative   **c.** Qualitative   **d.** Quantitative

2.2. a. Quantitative   **b.** Quantitative   **c.** Qualitative   **d.** Qualitative

2.14. a. 7.5 to 9.5   **b.** .15   **c.** .20   **d.** 20   **2.15. b.** .15   **2.17. b.** .65

2.21. a. 8.5   **b.** 25   **c.** .78   **d.** 13.44   **2.22. a.** 2.5, 3, 3   **b.** 3.08, 3.0, 3   **c.** 49.6, 49, 50

2.23. a. Mean is smaller   **b.** Mean is larger   **c.** They are equal

2.24. a. 1.78, 2   **b.** 3.67, 1   **c.** 0, 0   **d.** -27.7, 0   **2.25. c.** 81.15, 83, 83

2.26. a. Right   **b.** Left   **c.** Right   **d.** Symmetric   **e.** Right   **f.** Left

2.27. a. Mean 15,325; median 11,450; modes 9,000, 10,000, 12,000, 17,000

b. Mean 11,352.63; median 11,400; modes 9,000, 10,000, 12,000, 17,000; mean is most sensitive

2.28. a. 78.15, 84, 88   **b.** Median   **c.** 83.06, 86, 88; mean is most sensitive   **2.29. a.** No   **b.** 110

2.30. a. Mean   **b.** Median   **2.31.** Median 74.0; mean 73.65; no   **2.32.** Median 1,742.28; mean 2,384.84; yes

2.33. Median .13; mean .16; yes   **2.34. a.** 4, 2.30, 1.52   **b.** 6, 3.62, 1.90   **c.** 10, 7.11, 2.67   **d.** 5, 1.62, 1.27

2.35. a. 4.89, 2.21   **b.** 3.33, 1.83   **c.** .19, .43   **2.36. a.** 5, 3.70, 1.92   **b.** 99, 1,949.25, 44.15   **c.** 98, 1,307.84, 36.16

2.37. a. 3, 1.30, 1.14   **b.** 3, 1.30, 1.14   **c.** 3, 1.30, 1.14   **d.** No effect   **2.38.** First; less variability

2.39. b. 10, 10   **c.** 20.99, 5.73   **d.** Variance   **2.40. a.** Perhaps very few   **b.** At least $\frac{3}{4}$ **c.** At least $\frac{8}{9}$

2.41. a. Approx. 68%   **b.** Approx. 95%   **c.** Approx. all

2.42. a. 8.24, 3.36, 1.83   **b.** 72%, 96%, 100%   **d.** 7, $\frac{7}{4} = 1.75$

2.43. a. 73.65, 6.56   **b.** Yes   **c.** 70.1%, 98%, 100%; yes

2.44. .16, .026, .16; 87.8%, 95.9%, 95.9%

2.45. a. 2,384.84, 2,379.37   **b.** Yes; $s \approx 2,947.31$ **c.** 90%, 95%, 97.5%

2.46. Approx. 68% in (131.16, 235.70); approx. 95% in (78.89, 287.97); essentially all in (26.62, 340.24)

2.47. a. At most 25%   **b.** Approx. 2.5%   **2.48.** Do not buy. At most 450 are at least 40 feet tall.

2.49. .025; yes   **2.50. a.** At least $\frac{8}{9}$ **b.** Approx. .84   **c.** Yes

2.51. a. 9.1% are in $1 \pm .98$ **b.** 76.2% are in $1 \pm .99$ **c.** 89.3% are in $1 \pm .99$

2.52. At least $\frac{3}{4}$ had performance times in the interval (21.3, 102.5); at least $\frac{8}{9}$ in the interval (1.0, 122.8).

2.53. a. $z = 2$ **b.** $z = .5$ **c.** $z = 0$ **d.** $z = -2.5$ **e.** Sample; population; population; sample

f. Above by 2; above by .5; equal to the mean; below by 2.5

2.54. a. $z = 1.86$ **b.** $z = -2.86$; suspect outlier   **c.** $z = .25$ **d.** $z = 4$; definite outlier

2.55. a. 25% above, 75% below   **b.** 50% above, 50% below   **c.** 80% above, 20% below   **d.** 16% above, 84% below

2.56. a. $z = 2$ (most above)   **b.** $z = -3$ (most below)   **c.** $z = -2$ **d.** $z = 1.67$ **2.57.** 14.5, 17, 13, 4

2.58. Inner fence (7, 23); outer fence (1, 29); 30 is an outlier   **2.59.** 3.9, 4.0, 3.8, .2

2.60. Inner fence (3.5, 4.3); outer fence (3.2, 4.6); suspect outliers: 4.4 and 4.5; outliers: 5.2, 2.1, and .8

2.61. b. 90% of the test scores are below 660   **c.** 94% of the test scores are below your score

2.62. a. $z = -3$ **b.** Approx. 0%   **c.** 90 or above ($z = 2$)   **2.63. a.** $z = -3$ **b.** Yes   **c.** Yes, since $z = -1.5$

2.64. Yes, $z = 35$

2.65. a. 74.0, 68.8, 78.4, 9.6   **b.** Inner fence (54.4, 92.8); outer fence (40.0, 107.2); no outliers or suspect outliers

2.66. a. 1,742.28, 880.84, 2,796.42, 1,915.58

b. Inner fence $(-1,992.53, 5,669.79)$; outer fence $(-4,865.90, 8,543.16)$; outliers: 11,968.23, 9,139.21; suspect outlier: 6,815.69

2.67. a. .13, .07, .18, .11   **b.** Inner fence $(-.095, .345)$; outer fence $(-.26, .51)$; suspect outliers: .41, .45, .36, .38; outliers: .83, .66

2.68. c. 50.8, 57.06, 7.55   **d.** (43.25, 58.35), very few; (35.70, 65.90), at least 75%; (28.15, 73.45), at least 89%

e. 63.3%, 100%, 100%   **f.** Inner fence (31, 71); outer fence (16, 86); no outliers

2.70. a. $-1, 1, 2$ **b.** $-2, 2, 4$ **c.** 1, 3, 4   **d.** .1, .3, .4   **2.71.** 5

2.72. a. 2.67, 14.27, 3.78   **b.** 6, 3.33, 1.83   **c.** 12.86, 57.81, 7.60   **2.73. a.** $-1, 0, 9$ **b.** 25, 25, 16   **c.** 77.5, 80, 73

2.74. a. 2.58, .982, .991   **b.** .057, .0013, .036

2.75. a. Qualitative   **b.** Quantitative   **c.** Quantitative   **d.** Qualitative

2.76. 10, 32, 5.66   **2.77.** Decrease; smaller; smaller

2.78. d. Inner fence (23.5, 35.5); outer fence (19, 40); no outliers   **e.** 30, 4.83, 2.20

f. Perhaps very few; at least 75%; at least 89%   **g.** 70%, 96.7%, 100%

2.79. No, $z = 1.90$ **2.80. a.** 11, 2.75   **b.** 14.73, 11.21, 3.35   **c.** All

2.81. c. Inner fence (35.5, 79.5); outer fence (19, 96); no outliers   **d.** 58.24, 65.44, 8.09

e. Perhaps very few; at least 75%; at least 89%   **f.** 64%, 96%, 100%

2.82. b. Mean 16.48, median 15, mode 14   **c.** Inner fence (3.5, 31.5); outer fence $(-7, 42)$; no outliers   **d.** 22, 5.5

e. 33.93, 5.82   **f.** All

2.83. a. 68%   **b.** 47.5%   **c.** 16%   **2.84.** 52 minutes   **2.85. a.** 18.2, 13.64, 3.69   **b.** 20.2, 13.64, 3.69

2.86. a. 98.67, 330.24, 18.17   **b.** 49.33, 82.56, 9.09   **2.87. a.** Approx. none   **b.** No   **c.** Yes $(z = -1.33)$

2.88. 13.8, 11, 13.29, 3.65   **2.89. a.** At most 25%   **b.** Approx. 2.5%

2.90. 3,100 to 3,150; 3,075 to 3,175; 3,050 to 3,200; perhaps very few; at least $\frac{3}{4}$; at least $\frac{8}{9}$

2.91. a. 23.76, 6.61, 2.57   **b.** Perhaps very few; at least 75%; at least 89%   **c.** 68%, 96%, 100%

Chapter 3

3.1. a. .4   **b.** .25   **c.** .6   **3.2. a.** .25   **b.** .4   **3.3.** $P(A) = .6, P(B) = .5, P(C) = .75$

3.4. a. $\{HHH, HHT, HTH, THH, HTT, THT, TTH, TTT\}$ **b.** Each simple event has probability $\frac{1}{8}$ **c.** $\frac{7}{8}, \frac{3}{8}, \frac{1}{2}$

3.6. a. $\frac{4}{9}$ **b.** $\frac{1}{3}$ **c.** 0

3.7. a. (1, 1), (1, 2), (1, 3), (1, 4), (1, 5), (1, 6)   **b.** $\frac{1}{36}, \frac{1}{2}, \frac{1}{6}, \frac{11}{36}, \frac{1}{6}$ **3.8.** $A, \frac{1}{12}; B, \frac{1}{4}; C, \frac{1}{2}; D, \frac{1}{2}$ **3.9.** $\frac{1}{10}, \frac{6}{10}, \frac{3}{10}$

(2, 1), (2, 2), (2, 3), (2, 4), (2, 5), (2, 6)

(3, 1), (3, 2), (3, 3), (3, 4), (3, 5), (3, 6)

(4, 1), (4, 2), (4, 3), (4, 4), (4, 5), (4, 6)

(5, 1), (5, 2), (5, 3), (5, 4), (5, 5), (5, 6)

(6, 1), (6, 2), (6, 3), (6, 4), (6, 5), (6, 6)

3.11. .81, .19, .18   **3.12. a.** $\frac{1}{4}$ **b.** $\frac{1}{2}$ **c.** $\frac{1}{4}$ **3.13. a.** $\frac{1}{3}$ **b.** $\frac{1}{3}$ **c.** $\frac{1}{6}$ **d.** $\frac{1}{6}$ **3.14. a.** $\frac{1}{4}$ **b.** $\frac{1}{2}$ **c.** 0

3.15. a. $\frac{1}{6}$ **b.** $\frac{1}{6}$ **c.** $\frac{2}{3}$ **3.16.** $\frac{3}{4}$; no, the probability of being odd man out remains $\frac{1}{4}$

3.17. a. (W, R), (W, D), (W, I), (B, R), (B, D), (B, I)   **b.** Sample space   **c.** .30   **d.** .35   **e.** .55   **f.** .12

3.18. a. $\frac{1}{2}$ **b.** .225   **3.19. a.** $\frac{1}{10}$ **b.** $\frac{3}{10}$ **c.** $\frac{7}{10}$

3.20. a. (H, T, L), (H, T, 220), (H, T, 100), (H, L, 220), (H, L, 100), (H, 220, 100), (T, L, 220), (T, L, 100), (T, 220, 100), (L, 220, 100)

b. $\frac{3}{10}$

3.21. a. 18 simple events   **b.** $\frac{1}{18}$ **c.** $\frac{1}{3}$

3.22. a. A: $\{HHH, HHT, HTH, THH, HTT, THT, TTH\}$; B: $\{HHH, HTT, THT, TTH\}$; $A \cup B$: Same as A; A': $\{TTT\}$; AB: Same as B

b. $\frac{7}{8}, \frac{1}{2}, \frac{7}{8}, \frac{1}{8}, \frac{1}{2}$

3.23. a. A: $\{(1, 6), (2, 5), (3, 4), (4, 3), (5, 2), (6, 1)\}$;

B: $\{(1, 4), (2, 4), (3, 4), (4, 1), (4, 2), (4, 3), (4, 4), (4, 5), (4, 6), (5, 4), (6, 4)\}$; AB: $\{(3, 4), (4, 3)\}$;

$A \cup B$: $\{(1, 4), (1, 6), (2, 4), (2, 5), (3, 4), (4, 1), (4, 2), (4, 3), (4, 4), (4, 5), (4, 6), (5, 2), (5, 4), (6, 1), (6, 4)\}$;

A': The 30 simple events not in A **b.** $\frac{1}{6}, \frac{11}{36}, \frac{1}{18}, \frac{15}{36}, \frac{5}{6}$

3.24. a. $\frac{3}{4}$ **b.** $\frac{13}{20}$ **c.** 1   **d.** $\frac{2}{5}$ **e.** $\frac{1}{4}$ **f.** $\frac{7}{20}$ **g.** 1   **h.** $\frac{1}{4}$

3.25. a. .5   **b.** .19   **c.** .5   **d.** 1   **e.** .31   **f.** .69   **3.26.** $\frac{4}{10}, \frac{2}{10}, \frac{3}{10}, \frac{5}{10}, \frac{9}{10}, \frac{7}{10}, \frac{5}{10}, \frac{1}{10}$

3.27. a. $\{G_1P_1, G_2P_1\}$ **b.** S **c.** $\{P_1P_2\}$ **d.** $\frac{5}{6}, \frac{3}{6}, \frac{2}{6}, 1, \frac{1}{6}$ **3.28. a.** BC **b.** A' **c.** $C \cup B$ **d.** AC'

3.29. a. $\{11, 13, 15, 17, 29, 31, 33, 35\}$

b. $\{1, 2, 3, 4, 5, 6, 7, 8, 9, 10, 11, 13, 15, 17, 19, 20, 21, 22, 23, 24, 25, 26, 27, 28, 29, 31, 33, 35\}$

c. $\frac{18}{38}, \frac{18}{38}, \frac{8}{38}, \frac{28}{38}, \frac{18}{38}$ **d.** $\{11, 13, 15, 17\}$ **e.** $\frac{4}{38}$ **f.** All simple events except $\{00, 0, 30, 32, 34, 36\}$ **g.** $\frac{32}{38}$

3.30. a. Yes **b.** .26, .35, .72, .28, .05 **c.** .56, .05, .77 **d.** .74 **3.31.** .21, .5 **3.32.** $\frac{3}{4}, \frac{1}{2}, \frac{1}{2}, 1, \frac{2}{3}$

3.33. a. .55, .40, .35 **b.** .3, 0, 0 **c.** .75, 0, 0, .78 **3.34.** $\frac{1}{3}, 0, \frac{1}{14}, \frac{1}{7}, 1$ **3.35. a.** .09, .51, .09 **b.** .18, 1

3.36. a. $\frac{1}{15}$ **b.** $\frac{2}{5}$ **c.** $\frac{14}{15}$ **d.** $\frac{3}{7}$ **e.** $\frac{5}{14}$ **3.37.** .6 **3.38. a.** .51 **b.** .24 **c.** .60 **d.** 1

3.39. a. .105 **b.** .638 **c.** .517 **3.40. a.** .30 **b.** .27 **c.** .64 **d.** .79

3.41. a. .65 **b.** 0 **c.** 0 **d.** .60 **e.** No; $P(C) \neq P(C|B)$

3.42. a. .37 **b.** .68 **c.** .15 **d.** .22 **e.** 0 **f.** 0

3.43. a. $\frac{7}{8}, \frac{3}{8}, \frac{3}{8}, \frac{3}{2}, \frac{3}{8}, \frac{3}{8}, 0, 0$ **b.** $\frac{7}{8}, 1, \frac{3}{4}, \frac{3}{7}, \frac{3}{4}, 0$ **c.** No; no **d.** No; no **e.** No; yes

3.44. a. No; $P(A) \neq P(A|B)$ **b.** No; $P(AB) \neq 0$ **3.45. a.** .7 **b.** 0 **c.** 0 **d.** No; $P(A) \neq P(A|B)$

3.46. a. .08, .4, .52 **b.** .12, .3

3.47. a. $\frac{24}{50}$ **b.** $\frac{16}{50}$ **c.** $\frac{32}{50}$ **d.** $\frac{8}{50}$ **e.** $\frac{16}{50}$ **f.** $\frac{1}{2}$ **g.** No; $P(A|B) \neq P(A)$ **h.** No; AB is nonempty

3.48. a. .3 **b.** .58 **c.** .7 **d.** .3 **3.49.** .891 **3.50.** $1 - (.95)^{10} \approx .40$; independence

3.51. a. .729 **b.** .001 **c.** .027 **d.** .9 **3.52. a.** .008 **b.** .001 **c.** $\frac{1}{8}$ **3.53. a.** $\frac{2}{9}$ **b.** $\frac{1}{36}$ **c.** $\frac{11}{36}$

3.54. .04, .38 **3.55. a.** $\frac{27}{64}$ **b.** $\frac{1}{64}$ **c.** $\frac{27}{64}$ **3.56. a.** .25 **b.** .016 **c.** .422 **3.57. b.** $(.001)^3$ **c.** .997

3.58. a. .4096 **b.** .0016 **c.** .9984 **3.59. a.** 4 **b.** 8 **c.** 32 **d.** 2^n

3.60. a. 20 **b.** 20 **c.** 12 **d.** 4,950 **e.** 1 **f.** 1 **g.** 60 **h.** 1

3.61. a. 35 **b.** 15 **c.** 435 **d.** 45 **e.** $\dfrac{q!}{(q-r)!r!}$ **3.62. a.** 6 **b.** 36 **c.** 1,296 **d.** 6^n **3.63.** 6

3.64. 2,598,960 **3.65.** 2,600,000 **3.66.** No; 80 **3.67. a.** 24 **b.** 12 **3.68.** Yes

3.69. a. 17,576,000 **b.** 35,152,000 **3.70.** 100,000; 100 **3.71. a.** 56 **b.** 1,680 **c.** 6,720

3.72. 720, $\frac{1}{6}$ **3.73. a.** $\frac{21}{252}$ **b.** $\frac{21}{252}$ **c.** $\frac{105}{252}$

3.74. a. $\frac{5,148}{2,598,960} \approx .002$ **b.** $\frac{10,240}{2,598,960} = .00394$ **c.** $\frac{40}{2,598,960} = .0000154$

3.75. a. $(1, H), (1, T)$ **b.** Probability of a simple event with odd number first is $\frac{1}{12}$; otherwise, $\frac{1}{36}$

$(2, 1), (2, 2), (2, 3), (2, 4), (2, 5), (2, 6)$ **c.** $\frac{1}{4}, \frac{1}{2}$

$(3, H), (3, T)$ **d.** A' contains all simple events except $(1, H), (3, H)$, and $(5, H)$;

$(4, 1), (4, 2), (4, 3), (4, 4), (4, 5), (4, 6)$ B': $\{(2, 1), (2, 2), (2, 3), (2, 4), (2, 5), (2, 6), (4, 1), (4, 2), (4, 3), (4, 4), (4, 5),$

$(5, H), (5, T)$ $(4, 6), (6, 1), (6, 2), (6, 3), (6, 4), (6, 5), (6, 6)\}$; AB: $\{(1, H), (3, H), (5, H)\}$;

$(6, 1), (6, 2), (6, 3), (6, 4), (6, 5), (6, 6)$ $A \cup B$: $\{(1, H), (1, T), (3, H), (3, T), (5, H), (5, T)\}$

e. $\frac{3}{4}, \frac{1}{2}, \frac{1}{4}, \frac{1}{2}, \frac{1}{2}, 1$ **f.** No; no

3.76. a. 720 **b.** 10 **c.** 10 **d.** 30 **e.** 20 **f.** 1 **g.** 5,040 **h.** 2,450

3.77. a. 0, .1, .8, 1, .7, .3, $\frac{1}{3}$, 0 **b.** No; yes **c.** No; no **3.78. a.** .4 **b.** .3 **c.** .375 **d.** No

3.79. a. No; $P(AB) = .03$ **b.** .3, .1 **c.** .37 **3.81. a.** $\frac{34}{60} \approx .567$ **b.** $\frac{4}{34} \approx .118$

3.82. a. BC: $\{HHH\}$ **b.** $A \cup B$: $\{HHH, HHT, HTH, THH, THT\}$ **c.** $\frac{1}{8}$ **d.** $\frac{5}{8}$ **e.** $\frac{5}{8}$ **f.** $\frac{5}{8}$

3.83. a. .64, .32, .04 **b.** .72, .22, .06 **c.** Dependent

3.85. a. .5, .5, .3, .15, .2, .35 **b.** 1 **c.** .05 **d.** .05 **e.** 0 **f.** .45 **g.** .35

3.87. a. .15, .85, .70, .15, .05, .10 **b.** 1 **c.** .08 **d.** .85 **e.** 0 **f.** .87

3.88. $\frac{15}{36}, \left(\frac{21}{36}\right)\left(\frac{15}{36}\right), \left(\frac{21}{36}\right)^2\left(\frac{15}{36}\right), \left(\frac{21}{36}\right)^{n-1}\left(\frac{15}{36}\right)$ **3.89. a.** $\left(\frac{1}{2}\right)^5$ **b.** $5\left(\frac{1}{2}\right)^5$ **3.90. b.** 18%

3.91. a. $\frac{6}{30}$ **b.** $\frac{6}{29} \approx .207$ **c.** $\left(\frac{4}{5}\right)\left(\frac{23}{29}\right) \approx .634$ **3.92.** .79 **3.93.** $\frac{.7425}{.7475} \approx .993$

3.94. a. $\frac{48}{1,326} \approx .036$ **b.** $\left(\frac{48}{1,326}\right)\left(\frac{1,192}{1,225}\right) \approx .035$ **3.95.** .046, .747

3.96. a. There are 20 simple events **b.** $\frac{1}{20}$ **c.** $\frac{1}{20}$ **d.** $\frac{1}{2}$

3.97. a. 184,756 **b.** $\frac{1}{184,756}$ **c.** $\frac{1}{184,756}$ **d.** $\frac{1}{184,756}$

3.98. a. 38,798,760 **b.** $\frac{4,084,080}{38,798,760} \approx .105$ **c.** $\frac{9,189,180}{38,798,760} \approx .237$

3.99. a. 142,506 **b.** $\frac{33,649}{142,506} \approx .236$ **c.** $\frac{95,634}{142,506} \approx .671$

3.100. a. 68,880 **b.** $\frac{2,548}{68,880} \approx .03699$ **c.** $\frac{6,552}{68,880} \approx .095$ **3.101.** 4.4739×10^{-28}

Chapter 4

4.1. a. Discrete **b.** Continuous **c.** Continuous **d.** Discrete **e.** Continuous **f.** Continuous
4.2. a. Discrete **b.** Discrete **c.** Continuous **d.** Continuous **e.** Discrete **f.** Discrete
4.3. a. Continuous **b.** Discrete **c.** Discrete **d.** Discrete **e.** Continuous **f.** Continuous
4.4. a. $-4, 0, 1, 3$ **b.** 1 **c.** .6 **d.** 0 **4.5. a.** .30 **b.** .35 **c.** .85
4.6. a. No, $\sum p(x) \neq 1$ **b.** Yes **c.** No, $p(4) < 0$ **d.** No; $\sum p(x) \neq 1$
4.7. a. .10 **b.** 0 **c.** .25 **d.** .70 **e.** .90 **f.** .80
4.8. a. .65 **b.** .75 **c.** .80 **d.** .85 **e.** .60 **f.** .65

4.9. a.

Simple event	HHH	HHT	HTH	THH	HTT	THT	TTH	TTT
x	3	2	2	2	1	1	1	0

b.

x	0	1	2	3
$p(x)$	$\frac{1}{8}$	$\frac{3}{8}$	$\frac{3}{8}$	$\frac{1}{8}$

d. $\frac{1}{2}$

4.10. a.

x	2	3	4	5	6	7	8	9	10	11	12
$p(x)$	$\frac{1}{36}$	$\frac{2}{36}$	$\frac{3}{36}$	$\frac{4}{36}$	$\frac{5}{36}$	$\frac{6}{36}$	$\frac{5}{36}$	$\frac{4}{36}$	$\frac{3}{36}$	$\frac{2}{36}$	$\frac{1}{36}$

b. $\frac{15}{36}$ **c.** $\frac{21}{36}$ **d.** $\frac{18}{36}, \frac{18}{36}$ **e.** $\frac{6}{36}$

4.11. a.

x	0	1	2	3
$p(x)$	$\frac{27}{64}$	$\frac{27}{64}$	$\frac{9}{64}$	$\frac{1}{64}$

c. $\frac{37}{64}$ **4.12.** $\frac{7}{8}$ **4.13. a.** .118 **b.** .302 **c.** .58

4.14. a.

x	0	1	2	3
$p(x)$	.3446	.4408	.1879	.0267

c. .6554

4.15. a.

x	0	1	2	3	4
$p(x)$	.0016	.0256	.1536	.4096	.4096

c. .0272 **4.16.**

x	0	1	2
$p(x)$	$\frac{1}{6}$	$\frac{2}{3}$	$\frac{1}{6}$

4.17. a. 31, 169, 13 **c.** .95 **4.18. a.** .3, .011, .105 **c.** .85 **d.** 1 **4.19. a.** 2.7 **b.** 6.61 **c.** 2.571
4.20. a. 1.7 **b.** 1.21 **c.** 1.1 **4.21. a.** 0, 2.94, 1.715 **c.** .96
4.22. a. $E(x) = 1$ for both **b.** The first **c.** 1, .6; 1, .2 **4.23.** 1.29 **4.24. a.** $81.25 **b.** 195.3125 **c.** .90
4.25. a. $\frac{3}{4}$ **b.** $\frac{9}{16}$ **c.** .984 **4.26. a.** 4.2 **b.** 1.12 **c.** .989 **4.27.** $E(x) = 11,500$; yes
4.28. No, expected cost = $79 **4.29.** $0.25 **4.30. a.** 10 **b.** 20 **c.** 1 **d.** 1 **e.** 6
4.31. a. .36015 **b.** .375 **c.** .512 **d.** .2304 **e.** .3456 **f.** .4410

4.32. a.

x	0	1	2	3	4	5	6
$p(x)$	.016	.094	.234	.312	.234	.094	.016

b. $\mu = 3$, $\sigma^2 = 1.5$ **d.** .968

4.33. b.

x	0	1	2	3	4
$p(x)$	.4096	.4096	.1536	.0256	.0016

4.34. a. 12, 7.2, 2.683   **b.** 24, 16.8, 4.099   **c.** 30, 21, 4.583   **d.** 63, 6.3, 2.510   **e.** 30, 15, 3.873   **f.** 30, 29.1, 5.394
4.35. a. .234   **b.** .151   **c.** .263   **d.** 0   **e.** .722   **f.** .285

4.36. a.

x	0	1	2	3	4	5	6
$p(x)$	.047	.186	.311	.277	.138	.037	.004

  b. 2.4, 1.44   **d.** .959

4.37. a.

x	0	1	2	3	4	5
$p(x)$	.031	.157	.312	.312	.157	.031

  b. 2.5, 1.25   **d.** .938

4.38. a. .005   **b.** .973   **c.** .403   **d.** .966   **e.** .009   **f.** .207
4.39. e. $p = .5$, symmetric; $p < .5$, skewed to the right; $p > .5$, skewed to the left
4.40. a. No   **b.** Yes   **c.** No   **d.** For part b, $\mu = 4, \sigma^2 = 3.2, \sigma = 1.79$ **4.41. a.** 7.2   **b.** $(.4)^{12} \approx .000017$ **c.** Yes
4.42. a. .013   **b.** .783   **4.43.** .748   **4.44. a.** 0   **b.** Yes; t-PA is useful.   **4.45. a.** .590   **b.** .410   **4.46.** .098
4.47. a. 152   **b.** $\mu \pm 2\sigma$ is about 140 to 164   **4.48. a.** .009   **c.** Yes   **4.49.** $\mu = .5, \sigma = .707$, no
4.50. a. 0   **b.** .151   **4.51. a.** .2   **b.** 4   **c.** .196   **d.** No   **4.52. a.** .000   **b.** Drug seems effective
4.53. a. 0, .000, .000, .098, .873, 1   **c.** 0, .000, .000, .234, .966, 1
4.54. a. .0498   **b.** .0111   **c.** .1827   **d.** .0025   **e.** .2240   **f.** .2138   **g.** .0419
4.55. a. .9197   **b.** .0733   **c.** .8647   **d.** .0003   **4.56. b.** $\mu = 2, \sigma = 1.414$ **c.** .947
4.57. b. $\mu = 4, \sigma = 2$ **c.** .9788
4.58. Binomial: $p(0) = .277, p(1) = .365, p(2) = .231$; Poisson: $p(0) = .287, p(1) = .358, p(2) = .224$ **4.59. a.** .1255   **b.** .7769
4.60. .1088   **4.61. a.** .25   **b.** .557   **c.** .011   **4.62. a.** .67   **4.63. a.** Approx. .3012   **b.** Approx. .1205
4.64. $P(x \geq 8) \approx .00004$; yes   **4.65. a.** .193   **b.** .660   **4.66. a.** .090   **b.** .910   **4.67. a.** .080   **b.** No
4.68. a. 2   **b.** No, probability is .003
4.69. a. .000669   **b.** 7.31, 2.704   **c.** Unlikely; $x = 20$ lies 4.69 standard deviations above the mean.
4.70. a. $\frac{15}{56}$ **b.** $\frac{1}{56}$ **c.** $\frac{10}{56}$ **d.** 0   **4.71. a.** $\frac{3}{10}$ **b.** $\frac{5}{42}$ **c.** $\frac{1}{6}$ **d.** 0

4.72. a.

x	2	3	4	5
$p(x)$	$\frac{21}{252}$	$\frac{105}{252}$	$\frac{105}{252}$	$\frac{21}{252}$

  b. $\mu = 3.5, \sigma^2 = .5833$ **d.** 1

4.73. a.

x	2	3	4	5	6
$p(x)$	$\frac{15}{495}$	$\frac{120}{495}$	$\frac{225}{495}$	$\frac{120}{495}$	$\frac{15}{495}$

  b. 4, .8528   **d.** .9394

4.74. a.

x	3	4	5	6	7
$p(x)$	.071	.354	.424	.141	.010

  b. 4.67, .841   **c.** 2.99 to 6.35   **d.** .99

4.75. a. 0   **b.** .354   **c.** .424   **d.** .576   **e.** 0   **f.** 0
4.76. a. .4096   **b.** .4560   **c.** Binomial; hypergeometric   **d.** .4096, .4201   **4.77. a.** .119   **b.** .881
4.78. a. .05   **b.** .05   **c.** .50   **4.79. a.** .323   **b.** .406   **4.80.** $\frac{1}{4}$ **4.81. a.** .1   **b.** .7
4.82. .6, .3333, .1667, .0714   **4.84.** Accept for $x = 0$; probability of acceptance is lower when defectives are present.
4.85. a. 113.24, 4.194   **b.** $x = 140$ lies 6.38 standard deviations above the mean
4.86. a. .16   **b.** .128   **c.** .488   **d.** 1   **e.** .1024   **f.** .64
4.87. a. 100, 9,900, 99.50   **b.** 10, 90, 9.49   **c.** 1.25, .3125, .559   **d.** 3.33, 7.78, 2.79   **e.** 1.01, .0102, .101   **f.** 1, 0, 0
4.88. b. $\mu = 1.429; \sigma^2 = .6122$ **c.** .91   **4.89. b.** $\mu = 3.33; \sigma^2 = 7.778$ **c.** .88
4.90. a. .51   **b.** .343   **c.** 1   **d.** .8823
4.91. a. .488   **b.** 5   **c.** No; $x = 20$ lies 3.35 standard deviations above the mean.   **4.92.** .0819

4.93. a. .081 **b.** .10 **4.94. a.** .7738 **b.** 20 **c.** Yes, $x = 30$ lies only .51 standard deviation above the mean.
4.95. .343 **4.96. a.** .729 **b.** 10 **4.97. a.** .243 **b.** .131 **c.** .36 **d.** .157 **e.** .128 **f.** .121
4.98. a. Discrete **b.** Continuous **c.** Continuous **d.** Continuous
4.99. a. 16.2, 18.76, 4.33 **b.** .4 **c.** 7.54 to 24.86 **d.** 1
4.100. a. .124 **b.** .245 **c.** .755 **d.** .975 **e.** .927 **f.** 12, 4.8, 2.19 **g.** .963
4.101. a. $\frac{9}{15}$ **b.** $\frac{1}{10}$ **c.** $\frac{12}{35}$ **4.102. a.** .128 **b.** .0625 **c.** .09 **4.103. a.** .1804 **b.** .0153 **c.** .0758
4.104. a. .1494 **b.** .2240 **c.** .0498 **d.** .1008 **e.** .4232 **f.** .8008
4.105. a. .16 **b.** .3904 **c.** .36 **d.** .0819 **e.** .4096 **f.** .4419 **4.106. b.** $\mu = 2$; $\sigma = 1.414$ **c.** .9375
4.107. a. .599 **b.** .988 **4.108. a.** .027 **b.** .376 **c.** No **4.109. a.** .126 **b.** .057
4.110. a. .033 **b.** .985 **4.111.** .401 **4.112.** 22,250
4.113. a. .53 **b.** $1,240 **c.** $240 **d.** $\sigma^2 = 312,400$; $\sigma = 558.93$ **4.114. a.** .021 **b.** 292
4.115. a. .4019 **b.** .1608 **d.** $50,000 **4.116. a.** .343 **b.** .902 **c.** .004
4.117. a. A: 4.6; B: 3.7 **b.** A: $46,000; B: $55,500 **c.** A: $\sigma^2 = 1.34$, $\sigma = 1.16$; B: $\sigma^2 = 1.21$, $\sigma = 1.1$ **d.** A: .95; B: .95
4.118. a. .657 **b.** .027 **4.119.** .1715 **4.120. a.** .346 **b.** .683 **4.121. a.** 5, 4 **b.** .617 **c.** .006
4.122. a. .006 **4.123.** .059 **4.124.** 4.664% **4.125.** .512 **4.126. a.** .265 **b.** .175
4.127. a. .230 **b.** .143 **c.** .100 **4.128. a.** $\frac{11}{12}$ **b.** $\frac{1}{30}$ **4.129.** .1024 **4.130.** .042
4.131. a. .571 **b.** .786 **4.132. a.** .0315 **b.** .1067

Chapter 5

5.1. a. .4772 **b.** .3413 **c.** .4987 **d.** .2190 **e.** .4772 **f.** .3413 **g.** .4545 **h.** .2190
5.2. a. .4772 **b.** .4082 **c.** .4989 **d.** .2734
5.3. a. .6826 **b.** .9544 **c.** .6971 **d.** .5558 **e.** .9744 **f.** .9949
5.4. a. .1151 **b.** .0808 **c.** .2092 **d.** .2459 **e.** .5000 **f.** .8544
5.5. a. 0 **b.** 1.96 **c.** -1.96 **d.** 2.00 **e.** 2.06 **f.** -1.75
5.6. a. -2.07 **b.** 2.17 **c.** 1.96 **d.** 1.645 **e.** 1.00 **f.** 2.81
5.7. a. .9686 **b.** .7558 **c.** .0028 **d.** .9474 **e.** .1064 **f.** .9987 **g.** .0011 **h.** .8530
5.8. a. -2.03 **b.** 1.35 **c.** .74 **d.** 1.41 **e.** .51 **f.** 1.80 **5.9. a.** 1 **b.** -1 **c.** 0 **d.** -2.5 **e.** 3
5.10. a. 0 **b.** 1 **c.** 2.5 **d.** -3 **e.** 5 **f.** 1.4
5.11. a. -2.4 **b.** $-.4$ **c.** .6 **d.** -1.4 **e.** -4.4 **f.** 7.6
5.12. a. .3830 **b.** .3023 **c.** .1525 **d.** .7333 **e.** .1314 **f.** .9545
5.13. a. .6915 **b.** .1736 **c.** .6476 **d.** .1054 **e.** .9756 **f.** .0244
5.14. a. 30 **b.** 14.32 **c.** 40.24 **d.** 16.84 **e.** 19.76 **f.** 36.72 **g.** 48.64
5.15. a. .9544 **b.** .0228 **c.** .1587 **d.** .8185 **e.** .1498 **f.** .9974
5.16. a. .0304 **b.** .0304 **5.17. a.** 3.75% **b.** Many extremely high values
5.18. a. 6.2 **b.** 3.51% **c.** 6.27% **5.19. a.** .0150 **b.** $\mu \pm 2\sigma$ is 5.1 to 7.5 **5.20. a.** .0048 **b.** Yes
5.21. .0019 **5.22.** Yes, z-score = 27.83 **5.23. a.** .1151 **b.** .6554 **c.** 10.50 **5.24. a.** .0307 **b.** .0893
5.25. 5.068 **5.26.** 398.8 **5.27.** 473 **5.28. a.** 7,667 **b.** 45.62%
5.29. a. Interpolating; $Q_L = -.6745$; $Q_U = +.6745$ **b.** IQR = 1.349 **c.** ± 2.698 **d.** ± 4.7215 **e.** .0026 **f.** .0070 **g.** Approx. 0 **h.** No
5.32. a. .425; approx. .4207 **b.** .966; approx. .9671 **c.** .151; approx. .1498 **d.** .833; approx. .8339
5.33. a. 1 **b.** .3305 **c.** .1562 **5.34. a.** .4602 **b.** .9738 **c.** .0000 **d.** .9147 **e.** .0314
5.35. a. .5398 **b.** .7330 **c.** .9345 **5.36.** .9732 **5.37.** Approx. .0559 **5.38.** Approx. .0023
5.39. a. Approx. .7486 **b.** Approx. .2514 **5.40. a.** 6,750, 3,712.5 **b.** Approx. .0068 **c.** $P(x \geq 7,000) \approx 0$; no
5.41. a. Approx. .0011 **b.** Yes; z = 27.8 **c.** No **5.42.** Approx. 0 **5.43. a.** Approx. .9808 **b.** Approx. 0
5.44. a. Approx. .0885 **b.** Approx. .7123 **5.45. a.** Approx. .4681 **b.** Approx. .0436 **c.** Approx. .9822
5.46. a. Approx. .0559 **b.** Yes **5.47. a.** $f(x) = \frac{1}{5}$ for $10 \leq x \leq 15$ **b.** 12.5, 1.443 **c.** $\mu \pm 2\sigma$ is 9.61 to 15.39
5.48. a. .4 **b.** .36 **c.** .24 **d.** .8 **e.** 0 **f.** .4 **g.** .9

5.49. a. $f(x) = \frac{1}{2}$ for $2 \le x \le 4$ **b.** 3, .577 **c.** .577 **d.** .61 **e.** .65 **f.** 0 **5.50.** $\mu = 2$; $\sigma = 1.155$; .3125

5.51. a. 10 **b.** .05 **5.52. a.** Continuous **c.** $\mu = 7$; $\sigma = \dfrac{1}{\sqrt{12}} = .289$; 6.422 to 7.578

5.53. a. .5 **b.** 1.0 **c.** .25 **d.** .25 **e.** 0 **f.** .75 **5.54.** $(.25)^6 = .0002$
5.55. a. .667 **b.** .333 **c.** 82.5°F **5.56. b.** .5, .083 **c.** .05, .95 **d.** $f(x) = 20$ for $.90 \le x \le .95$
5.57. a. .001930 **b.** .332871 **c.** .472367 **d.** .033373 **e.** .063928 **f.** .000410
5.58. a. .3679 **b.** .9975 **c.** .0498 **d.** .99996 **5.59. a.** .8647 **b.** .7788 **c.** .6321 **d.** .2231
5.60. .9502 **5.61. a.** .35 hour **b.** .535 **5.62. a.** 13.8 minutes **b.** .82 **5.63.** .0952
5.64. a. 10 **b.** .7534 **c.** .1353 **5.65. a.** .1353 **b.** .3935 **5.66.** .095 **5.67.** .181 **5.68.** .2231
5.69. a. .2212 **b.** .0024 **c.** .0821 **5.71. a.** .4750 **b.** .9500 **c.** .8064 **d.** .1261 **e.** .2074 **f.** .8980
5.72. a. .8643 **b.** .1357 **c.** .9686 **d.** .2882 **e.** .6409 **f.** .0281
5.73. a. .85 **b.** 1.29 **c.** 0 **d.** .88 **e.** −1.04 **f.** 2.54
5.74. a. .6915 **b.** .0228 **c.** .5328 **d.** .3085 **e.** 0 **f.** .9938
5.75. a. 60 **b.** 78.96 **c.** 37.84 **d.** 76 **e.** 49.76 **f.** 53.28
5.76. a. .0094 **b.** .8524 **c.** .0262 **d.** .2469 **e.** .0733 **f.** .0256
5.77. a. $f(x) = \frac{1}{40}$ for $30 \le x \le 70$ **b.** 50; 11.547 **d.** .375 **e.** .3 **f.** 1 **g.** .577 **h.** .25
5.78. a. .3935 **b.** .6065 **c.** 0 **d.** .9502 **e.** .3611 **5.79.** .8315 **5.80. a.** .8790 **b.** .7967 **c.** About 163
5.81. a. .0456 **b.** .9082 **c.** .0023 **5.82. a.** .2033 **b.** $12,800 **5.83.** .0721 **5.84. a.** .8850 **b.** .0304
5.85. a. .0087 **b.** 22.7 **5.86. a.** .0853 **b.** .1858 **5.87. a.** .0122 **b.** .0062 **5.88. a.** .1469 **b.** .0216
5.89. a. Normal **c.** $\bar{x} = 9.9825$; $s = .1960$ **5.90. a.** .0548 **b.** .6006 **c.** .3446 **d.** $6,503.80
5.91. .1056 **5.92.** Approx. .982 **5.93.** $(.9938)^{100} = .5369$; approx. 1.0 **5.94. a.** Approx. 1.0 **b.** Approx. .8830
5.95. a. .0901 **b.** .0384 **c.** .9573 **5.96. a.** .6065 **b.** .6321
5.97. a. Approx. .9922 **b.** Approx. .6618 **5.98.** .7534 **5.99.** $P(x \ge 400 \text{ when } p = .2) \approx 0$ **5.100.** Approx. .1446
5.101. a. .2865 **b.** .2865 **5.102. a.** .4880 **b.** .1093 **c.** .8926 **5.103. a.** .1353 **b.** .5940
5.104. a. Approx. .1922 **b.** Approx. .4681 **c.** Approx. $(.4681)^3$

Chapter 6

6.1. a.-b.

Samples	(0, 0)	(0, 2)	(0, 4)	(0, 6)	(2, 2)	(2, 4)	(2, 6)	(4, 4)	(4, 6)	(6, 6)
		(2, 0)	(4, 0)	(6, 0)		(4, 2)	(6, 2)		(6, 4)	
$\bar{x}$	0	1	2	3	2	3	4	4	5	6

c. $\frac{1}{16}$ **d.**

$\bar{x}$	0	1	2	3	4	5	6
$p(\bar{x})$	$\frac{1}{16}$	$\frac{2}{16}$	$\frac{3}{16}$	$\frac{4}{16}$	$\frac{3}{16}$	$\frac{2}{16}$	$\frac{1}{16}$

6.3. a.

$\bar{x}$	1	1.5	2	2.5	3	3.5	4	4.5	5
$p(\bar{x})$	.04	.12	.17	.20	.20	.14	.08	.04	.01

c. .05 **d.** No

6.4. $\mu = 2.7$ **6.5. a.** Identical to sampling distribution of $\bar{x}$

6.6 b.

$\bar{x}$	1	$1\frac{1}{3}$	$1\frac{2}{3}$	2	$2\frac{1}{3}$	$2\frac{2}{3}$	3
$p(\bar{x})$	.216	.324	.270	.135	.045	.009	.001

6.7. a.

m	1	2	3
$p(m)$	.648	.324	.028

6.8. b.

$\bar{x}$	1	1.25	1.50	1.75	2	2.25	2.50	2.75	3
$p(\bar{x})$	.1296	.2592	.2808	.1944	.0945	.0324	.0078	.0012	.0001

6.9. b.

m	1	1.5	2	2.5	3
$p(m)$	.4752	.3240	.1701	.0270	.0037

6.12. a. $\mu = 2;\ \sigma^2 = \frac{14}{3}$ **b.**

$\bar{x}$	0	$\frac{1}{2}$	1	$\frac{5}{2}$	3	5
$p(\bar{x})$	$\frac{1}{9}$	$\frac{2}{9}$	$\frac{1}{9}$	$\frac{2}{9}$	$\frac{2}{9}$	$\frac{1}{9}$

c. $E(\bar{x}) = 2 = \mu$

d.

s^2	0	$\frac{1}{2}$	8	$\frac{25}{2}$
$p(s^2)$	$\frac{3}{9}$	$\frac{2}{9}$	$\frac{2}{9}$	$\frac{2}{9}$

e. $E(s^2) = \frac{42}{9} = \frac{14}{3} = \sigma^2$

6.13. a. $\mu = 5$ **b.**

$\bar{x}$	2	$\frac{8}{3}$	$\frac{10}{3}$	4	$\frac{13}{3}$	5	$\frac{17}{3}$	$\frac{20}{3}$	$\frac{22}{3}$	9
$p(\bar{x})$	$\frac{1}{27}$	$\frac{3}{27}$	$\frac{3}{27}$	$\frac{1}{27}$	$\frac{3}{27}$	$\frac{6}{27}$	$\frac{3}{27}$	$\frac{3}{27}$	$\frac{3}{27}$	$\frac{1}{27}$

$E(\bar{x}) = \frac{135}{27} = 5 = \mu$

c.

m	2	4	9
$p(m)$	$\frac{7}{27}$	$\frac{13}{27}$	$\frac{7}{27}$

d. The sample mean, $\bar{x}$; it is unbiased

$E(m) = \frac{129}{27} = 4.778 < \mu = 5$

6.14. a. $\mu = 1$ **b.**

$\bar{x}$	0	$\frac{1}{3}$	$\frac{2}{3}$	1	$\frac{4}{3}$	$\frac{5}{3}$	2
$p(\bar{x})$	$\frac{1}{27}$	$\frac{3}{27}$	$\frac{6}{27}$	$\frac{7}{27}$	$\frac{6}{27}$	$\frac{3}{27}$	$\frac{1}{27}$

c.

m	0	1	2
$p(m)$	$\frac{7}{27}$	$\frac{13}{27}$	$\frac{7}{27}$

d. $E(\bar{x}) = \frac{27}{27} = 1 = \mu;\ E(m) = \frac{27}{27} = 1 = \mu$ **e.** Variance of $\bar{x}$ is $\frac{2}{9} \approx .222$; variance of m is $\frac{14}{27} \approx .519$.   **f.** $\bar{x}$; smaller variance

6.16. a. $E(\bar{x}) = 1.5$ **b.** .15   **c.** .945

6.17. a.

s^2	0	.333	1	1.333
$p(s^2)$	.244	.522	.108	.126

b. .45   **c.** $E(s^2) = .45$

d.

s	0	.577	1	1.155
$p(s^2)$	.244	.522	.108	.126

e. $E(s) = .555$

6.18. $E(m) = 1.38$; m is a biased estimator of μ

6.19. a. 50, 3.536   **b.** 50, 1.414   **c.** 50, .707   **d.** 50, 1   **e.** 50, .316   **f.** 50, .224

6.20. a. 5, .6   **b.** 100, .4   **c.** 10, 8   **d.** 10, 40

6.21. a. 3.1, 5.69, 2.385   **b.** Approx. normal; $\mu_{\bar{x}} = 3.1$; $\sigma_{\bar{x}} = 1.687$

6.22. No; to be sure, sample must be large.

6.23. a. .8944   **b.** .0228   **c.** .1303 (interpolating)   **d.** .9696 (interpolating)

6.24. a. .9591   **b.** .9738   **c.** .2041   **d.** .0034

6.25. a. $\mu \pm 2\sigma_{\bar{x}}$ is 99.333 to 100.667   **b.** $3\sigma_{\bar{x}} = 1$ **c.** No

6.27. a. $\mu_{\bar{x}} = 1.3$; $\sigma_{\bar{x}} \approx .24$ **b.** Yes; large random sample   **c.** .1056   **d.** .0062

6.28. a. .0125   **b.** .0125

6.29. a. Approx. normal; $\mu_{\bar{x}} = 100$; $\sigma_{\bar{x}} = 15/\sqrt{200} \approx 1.06$ **b.** .0023

c. Claim seems incorrect, since a sample mean less than 97 is so rare.

d. $P(\bar{x} \leq 98.5) \approx .0793$; not as rare; not as much evidence to doubt claim

6.30. a. .0013   **b.** Employees are using fewer sick days (using rare event approach).

6.31. a. $\mu_{\bar{x}} = 6$; $\sigma_{\bar{x}} = 2.5/\sqrt{50} \approx .35$ **b.** .5222   **c.** .0793   **d.** Same mean; $\sigma_{\bar{x}} = 2.5/\sqrt{100} = .25$

6.32. a. $\mu = \frac{5}{3}$

b.

Sample	(0, 0, 0)	(0, 0, 2) (0, 2, 0) (2, 0, 0)	(0, 0, 3) (0, 3, 0) (3, 0, 0)	(0, 2, 2) (2, 0, 2) (2, 2, 0)	(0, 3, 3) (3, 0, 3) (3, 3, 0)	(0, 2, 3),(0, 3, 2) (2, 0, 3),(2, 3, 0) (3, 0, 2),(3, 2, 0)	(2, 2, 2)	(2, 2, 3) (2, 3, 2) (3, 2, 2)	(2, 3, 3) (3, 2, 3) (3, 3, 2)	(3, 3, 3)
Mean, $\bar{x}$	0	$\frac{2}{3}$	1	$\frac{4}{3}$	2	$\frac{5}{3}$	2	$\frac{7}{3}$	$\frac{8}{3}$	3
Median, m	0	0	0	2	3	2	2	2	3	3

c.

$\bar{x}$	0	$\frac{2}{3}$	1	$\frac{4}{3}$	$\frac{5}{3}$	2	$\frac{7}{3}$	$\frac{8}{3}$	3
$p(\bar{x})$	$\frac{1}{27}$	$\frac{3}{27}$	$\frac{3}{27}$	$\frac{3}{27}$	$\frac{6}{27}$	$\frac{4}{27}$	$\frac{3}{27}$	$\frac{3}{27}$	$\frac{1}{27}$

m	0	2	3
$p(m)$	$\frac{7}{27}$	$\frac{13}{27}$	$\frac{7}{27}$

d. $E(\bar{x}) = \frac{45}{27} = \frac{5}{3} = \mu; E(m) = \frac{47}{27} > \mu = \frac{5}{3}$

6.33. a. $\mu_{\bar{x}} = 120; \sigma_{\bar{x}} = \sqrt{\frac{410}{75}} \approx 2.34$ **b.** Approx. normal **c.** .1949 **d.** .8835 **e.** .9756 **f.** .1251

6.34. a. .5 **b.** .0418 **c.** .0749 **d.** .8729

6.35. a.

$\bar{x}$	1	1.5	2	2.5	3	3.5	4	4.5	5	5.5	6
$p(\bar{x})$	$\frac{1}{36}$	$\frac{2}{36}$	$\frac{3}{36}$	$\frac{4}{36}$	$\frac{5}{36}$	$\frac{6}{36}$	$\frac{5}{36}$	$\frac{4}{36}$	$\frac{3}{36}$	$\frac{2}{36}$	$\frac{1}{36}$

b. $\frac{33}{36}$

6.37. b. 4.68 **c.** 7.931 **d.** 4.78, 1.642 **e.** 4.68, 1.172 (means rounded to two decimals) **f.** $\bar{x}$

6.38. c.

$\bar{x}$	0	.5	1
$p(\bar{x})$	$\frac{1}{4}$	$\frac{1}{2}$	$\frac{1}{4}$

6.39. c.

$\bar{x}$	0	$\frac{1}{3}$	$\frac{2}{3}$	1
$p(\bar{x})$	$\frac{1}{8}$	$\frac{3}{8}$	$\frac{3}{8}$	$\frac{1}{8}$

6.40. b. 4.84, 1.313 **6.41. b.** 4.68, .8270

6.42.

n	1	5	10	20	30	40	50
$\sigma/\sqrt{n}$	10	4.472	3.162	2.236	1.826	1.581	1.414

6.43. a. .9887 **b.** Normal distribution for $\bar{x}$

6.44. a. Approx. normal; $\mu_{\bar{x}} = 400; \sigma_{\bar{x}} = 11.86$ **b.** .0174 **c.** .5 **6.45. a.** .2327 **b.** .2215

6.46. a. .5 **b.** .0023 **c.** Normal distribution for $\bar{x}$ **6.47. b.** $\mu_y = .5; \sigma_y = .29/\sqrt{n}$ **c.** Normal

6.48. .383 **6.49.** .9772

6.50. a. 45 samples; $\{(1, 2), (1, 3), \ldots, (9, 10)\}$ **b.** $\{(1, 4), (1, 5), \ldots, (8, 10)\}; \frac{21}{45}$

c. $p(0) = \frac{3}{45}, p(1) = \frac{21}{45}, p(2) = \frac{21}{45}$

6.51. a. Approx. normal; $\mu_{\bar{x}} = 45; \sigma_{\bar{x}} = .258$ **b.** .8508 **c.** .000 **6.52. a.** .0082 **b.** None

6.53. a. 1.342 **b.** Approx. .0026 **6.54.** .9332

6.55. a. Approx. normal; $\mu = 7,500; \sigma = 246.48$ **b.** .9576 **c.** .0075

6.56. a. Approx. normal **b.** $\mu_{\bar{x}} = \theta; \sigma_{\bar{x}} = \theta/\sqrt{n}$

Chapter 7

7.1. a. (3.82, 6.18) **7.2. a.** (3.01, 3.24) **7.3. a.** (30.0, 30.8) **b.** (30.06, 30.74) **c.** (29.87, 30.93)

7.4. a. (27.04, 28.96) **b.** (101.52, 102.48) **c.** (16.98, 17.02) **d.** (.80, 1.24) **7.5. a.** (7.97, 12.27) **b.** (8.75, 11.49)

7.6. a. (75.96, 77.84) **c.** (75.66, 78.14) **d.** Increases **7.8.** Yes, for a large random sample

7.9. a. (26.91, 27.89) **b.** (27.155, 27.645) **c.** .98, .49; width is divided by 2 **7.10.** (18.12, 25.88)

7.11. (8.9, 10.3) **7.12. a.** .65 **b.** (4.15, 4.45) **7.13.** (.67, .77) **7.14.** (18.575, 21.025) **7.15.** (31.79, 33.41)

7.16. (2.22, 2.30)　**7.19. a.** .025　**b.** .05　**c.** .005　**d.** .10　**e.** .10　**f.** .01

7.20. a. $z = 1.35$; do not reject H_0　**b.** $z = 2.52$; reject H_0　**c.** $z = -6.32$; reject H_0

7.21. a. Reject H_0 if $z > 1.5$　**b.** $\alpha = .0668$

7.22. a. $z = 1.52$; do not reject H_0　**b.** $z = 3.39$; reject H_0　**c.** $z = -1.45$; do not reject H_0

7.23. a. $z = 3.86$; reject H_0　**b.** $z = -1.08$; do not reject H_0　**c.** $z = -1.59$; reject H_0

7.24. a. $z = -1.71$; reject H_0　**b.** $z = -1.71$; do not reject H_0　**7.26.** No

7.28. a. H_0: $\mu = 43$; H_a: $\mu > 43$　**b.** $z = 2.92$; reject H_0　**7.29.** $z = -4.34$; yes

7.30. a. .166　**b.** .0853　**c.** $z = 2.36$; yes　**7.31. a.** H_0: $\mu = 8$; H_a: $\mu > 8$　**d.** $z = .87$; do not reject H_0

7.32. $z = 2.0$; yes　**7.33.** .7764　**7.34.** .4522

7.35. a. $z = 3.54$; yes　**b.** .9616 (using $.985 < \bar{x} < 1.015$ as the acceptance region)　**7.36.** .0643　**7.37.** .0930

7.38. a. .0735　**b.** .0150　**c.** .0043　**d.** .0985　**e.** .1230　**f.** .0329

7.40. a. .1052　**b.** .0050　**c.** .0614　**d.** .0444　**e.** .1738　**f.** .1212　**7.42.** .0304 (with interpolation)

7.43. a. H_0: $\mu = 4{,}627$; H_a: $\mu > 4{,}627$　**b.** .1075　**7.44. a.** $z = -1.06$; no; p-value $= .2892$　**7.45.** p-value $= .0018$

7.46. p-value $= .1230$　**7.47.** p-value $= .1922$　**7.48.** p-value $= .0228$

7.49. a. 2.306　**b.** 2.764　**c.** -2.898　**d.** -1.761

7.50. a. Reject H_0 if $t > 2.145$ or $t < -2.145$.　**b.** Reject H_0 if $t > 2.539$.　**c.** Reject H_0 if $t > 1.415$.

d. Reject H_0 if $t < -2.583$.　**e.** Reject H_0 if $t > 1.721$ or $t < -1.721$.　**f.** Reject H_0 if $t < -2.015$.

7.51. a. 1.740　**b.** 4.604　**c.** 3.012　**d.** 2.086　**e.** 1.725

7.52. a. Reject H_0 if $t < -1.476$; $t = -1.92$, so reject H_0.　**b.** Reject H_0 if $t > 2.015$ or $t < -2.015$; do not reject H_0.

c. (3.43, 6.37)　**d.** (3.75, 6.05)

7.53. a. $t = -1.63$; do not reject H_0　**b.** $t = -1.63$; do not reject H_0

c. $.05 < p$-value $< .10$ for part a; $.10 < p$-value $< .20$ for part b　**d.** (2.76, 6.84)

7.54. a. $t = -1.40$; do not reject H_0　**b.** $t = -1.40$; do not reject H_0

c. p-value $> .10$ for part a; p-value $> .20$ for part b　**d.** $(-.31, 3.97)$

7.55. a. $t = 3.56$; reject H_0　**b.** $t = 3.56$; do not reject H_0　**c.** $t = 3.56$; reject H_0　**d.** (3.12, 6.88)　**e.** (2.06, 7.94)

7.56. a. (51.13, 54.07)　**7.57. a.** $t = -2.01$; yes　**b.** Type II error　**c.** Type I error　**7.58.** (19.73, 19.99)

7.59. a. $t = -2.33$; no (reject for $t > 1.833$)　**b.** $(-30.01, -3.59)$　**7.60. a.** $t = .866$; no　**b.** p-value $> .10$

7.61. (40,314.82, 98,974.89)　**7.62.** $t = -4.19$; yes　**7.63.** (49.58, 49.82)　**7.64. a.** $t = 2.33$; yes

7.65. a. $t = -3.63$; yes　**b.** p-value $< .01$　**7.66. a.** (20.16, 56.70)　**7.67. b.** $t = -2.24$; no　**c.** $.05 < p$-value $< .10$

7.68. a. $t = -2.20$; no　**7.69.** (30,394.65, 35,298.95)

7.70. a. (.44, .68)　**b.** (.46, .66)　**c.** $z = .96$; do not reject H_0　**d.** $z = .96$; do not reject H_0

7.71. a. $z = 1.89$; do not reject H_0　**b.** $z = 1.89$; reject H_0　**c.** $z = -5.33$; reject H_0　**d.** (.65, .83)　**e.** (.63, .85)

7.72. b. $z = -1.39$; do not reject H_0　**7.74. a.** $z = 2.31$; reject H_0　**b.** p-value $= .0104$　**7.75.** (.58, .68)

7.76. a. Yes　**b.** (.09, .15)　**7.77.** (.059, .107)　**7.78. a.** $z = -1.51$; no　**b.** .0655

7.79. a. KOCO: (.69, .85); WCBS: (.44, .76); WXIA: (.38, .52)　**b.** $z = 2.91$; yes

7.80. a. $z = -8.49$; yes　**b.** p-value $< .001$　**c.** (.09, .19)

7.81. a. (.89, .95)　**b.** 1.84 million　**c.** .384 million　**d.** $\mu + 3\sigma$ is 2.99 million　**7.82.** $1.96\sqrt{\hat{p}\hat{q}/n} = .04$

7.83. a. 865　**b.** 7,780　**c.** 2,341　**d.** 423　**7.84. a.** 1,498　**b.** 375　**c.** 1,025　**d.** 1,624　**e.** 385

7.85. 577　**7.86. a.** 1,421　**b.** 1,692　**7.87.** 68　**7.88. a.** .98, .784, .56, .392, .196　**7.89. a.** 722　**b.** 174

7.90. 139　**7.91.** 62　**7.92.** 135　**7.93.** 260　**7.94.** 2,401　**7.95.** 55　**7.96.** .68

7.97. a. 14.6837　**b.** 21.0261　**c.** 12.8325

7.98. a. Reject H_0 if $\chi^2 > 30.1910$ or $\chi^2 < 7.56418$.　**b.** Reject H_0 if $\chi^2 > 36.1908$.　**c.** Reject H_0 if $\chi^2 > 21.0642$.

d. Reject H_0 if $\chi^2 < 3.05347$.　**e.** Reject H_0 if $\chi^2 > 14.0671$ or $\chi^2 < 2.16735$.　**f.** Reject H_0 if $\chi^2 < 13.8484$.

7.99. a. 26.1190, 5.62872　**b.** 23.6848, 6.57063　**c.** 31.3193, 4.07468

7.100. a. Sampled population is normal　**b.** $\chi^2 = 19.36$; reject H_0　**c.** $\chi^2 = 19.36$; reject H_0　**d.** (2.04, 27.24)

7.101. a. $\chi^2 = 479.16$; reject H_0　**c.** (3.85, 6.15)　**7.102. a.** $\chi^2 = 1.64$; no　**b.** (3.46, 46.15)

7.103. a. $\chi^2 = 47.43$; no　**b.** (3.77, 29.00)　**7.104.** $\chi^2 = 22.54$; no　**7.105.** (1.63, 7.17)　**7.106.** $\chi^2 = 4.5$; no

7.107. $\chi^2 = 54$; yes　**7.108.** $\chi^2 = 133.9$; yes　**7.109.** (.0014, .0053)

7.110. a. (70.90, 74.30)   **b.** $t = -7.51$; reject H_0 **c.** $t = -7.51$; reject H_0 **d.** (69.78, 75.42)   **e.** 75
7.111. a. $z = -1.78$; reject H_0 **b.** $z = -1.78$; do not reject H_0 **c.** (.23, .35)   **d.** (.21, .37)   **e.** 549
7.112. a. (8.08, 8.32)   **b.** $z = -1.67$; do not reject H_0 **c.** $z = -3.35$; reject H_0
7.113. a. (34.15, 71.84)   **b.** (28.52, 91.97)   **c.** $\chi^2 = 63.48$; reject H_0 **d.** $\chi^2 = 63.48$; reject H_0
7.114. a. $z = -2.99$; reject H_0 **b.** p-value $= .0028$ **7.115.** $t = 3.51$; no
7.116. a. $t = 1.56$; yes   **b.** $.05 < p$-value $< .10$ **7.117. a.** $z = 6.13$; yes   **b.** p-value ≈ 0 **7.118.** $z = 3.79$; yes
7.119. a. $t = 3.82$; yes   **7.120. a.** $z = 1.41$; no   **7.121.** Small   **7.122. a.** (10.555, 13.845)   **b.** 166
7.123. $z = 1.07$; no   **7.124.** 65   **7.125. a.** (13.72, 14.48)   **b.** (219.52, 231.68)
7.126. a. (.038, .112)   **b.** $z = 1.62$; reject H_0 **7.127.** 667   **7.128.** (3.27, 3.93)   **7.129.** $z = -1.67$; yes
7.130. (.82, .90)   **7.131. a.** (47.86, 60.14)   **b.** Maximum depth of snowfall   **7.132.** 119   **7.133.** (.52, .66)
7.134. (.030, .220)   **7.135.** (.58, .72)   **7.136.** $z = -1.45$; no   **7.137.** (.54, .58)   **7.138.** (24.86, 27.94)
7.139. (1.53, 2.47)   **7.140.** $z = -3.03$; yes   **7.141.** (.0028, .0105)   **7.142.** $\chi^2 = 3.81$; no   **7.143.** $\chi^2 = 23.66$; no

Chapter 8

8.1. a. $(-.75, -.25)$ **b.** $z = -3.96$; reject H_0 **c.** p-value $< .002$
8.2. a. $(-5.50, -1.90)$ **b.** $(-6.52, -.88)$ **c.** $z = -3.38$; reject H_0 **d.** p-value $< .001$
8.3. a. $z = 1.68$; reject H_0 **b.** $z = 1.68$; do not reject H_0 **c.** $(-.37, 4.77)$
8.4. Yes, for large, random, and independent samples   **8.5. b.** $z = -3.76$; yes   **c.** $(-4.89, -.91)$ **d.** Narrower
8.6. a. $z = 2.82$; yes   **b.** (9.21, 50.79)   **8.7.** $(-5.36, 12.16)$ **8.8.** $z = 3.43$; yes   **8.9.** $z = -2.09$; yes
8.10. a. $z = -4.22$; yes   **b.** p-value is near 0   **c.** $(-1.53, -.67)$ **8.11. a.** $z = 2.77$; yes   **c.** p-value $= .0028$
8.13. a. $(-4.91, -4.29)$ **b.** Approx. 0   **c.** Samples not random   **8.14. a.** $z = 5.68$; reject H_0 **b.** p-value $< .002$
8.15. a. $z = 5.36$; reject H_0 **b.** p-value $< .002$ **8.16. a.** $z = 10.63$; reject H_0 **b.** p-value $< .002$
8.17. $z = 1.963$; with p-value $= .05$, a significant difference exists   **8.18. a.** .5989   **b.** $t = -2.39$; yes   **c.** $(-2.22, -.26)$
8.19. a. $t = -1.65$; do not reject H_0 **b.** $(-5.62, .62)$ **8.22.** $t = -.94$; no   **8.23.** $t = -4.22$; yes
8.24. a. $t = 3.08$; yes   **b.** (4.45, 15.55)   **8.25. a.** $t = 2.76$; reject H_0 **b.** $.005 < p$-value $< .01$
8.26. a. $t = .445$; no   **b.** p-value $> .20$ **8.27. a.** AFC: 120.5, 510.4231; NFC: 115.07, 364.2253   **b.** $t = .69$; no
8.28. $t = 1.57$; no   **8.29. a.** 88.622, 68.489   **b.** Yes   **c.** $t = .61$; no   **8.30. b.** $t = 1.54$; no   **c.** $.05 < p$-value $< .10$
8.31. a. 1.67, 2.67   **b.** $\mu_D = \mu_1 - \mu_2$ **c.** $(-.04, 3.38)$ **d.** $t = 2.50$; do not reject H_0
8.32. a. $t = -5.29$; yes   **b.** $(-4.98, -2.42)$ **8.33. a.** $t = -3.08$; no   **b.** $(-63.30, 1.04)$ **8.34.** (3.04, 6.96)
8.35. a. $t = 3.71$; yes   **b.** Population of differences is normal   **8.36.** (3.60, 4.86)
8.37. a. $t = 2.98$; yes   **b.** $.01 < p$-value $< .025$ **8.38. a.** $t = .80$; no   **b.** $(-142.3, 310.64)$
8.39. a. $t = 7.68$; yes   **c.** (.28, .56)   **8.40. a.** $t = .40$; no   **b.** $(-1.88, 2.72)$ **d.** No
8.41. a. Not independent random samples   **b.** $t = 1.18$; no
8.42. a. (.003, .137)   **b.** $(-.026, .146)$ **c.** $(-.281, -.019)$
8.43. a. $z = -4.02$; reject H_0 **b.** $z = -4.02$; reject H_0 **c.** $z = -4.02$; reject H_0 **d.** $(-.141, -.059)$
8.49. $z = -1.67$; no (with $\alpha \le .05$)   **8.50. a.** $z = 4$; yes   **b.** (.12, .28)   **8.51. a.** $z = -.63$; no   **b.** $(-.206, -.134)$
8.52. a. $z = .92$; no   **b.** $z = .92$; no   **8.53. a.** $z = -6.60$; yes   **b.** p-value $< .001$ **c.** $(-.359, -.201)$
8.54. (.13, .21)   **8.56. a.** $z = .79$; do not reject H_0 **b.** p-value $= .2148$
8.57. 29 and under: $z = 4.03$ with p-value $< .002$, reject H_0; 30–49: $z = 5.92$ with p-value $< .002$, reject H_0; 50 and over: $z = 3.00$ with p-value $= .0026$, reject H_0
8.58. $(-.236, -.154)$ **8.59.** $z = 1.93$; no   **8.60. a.** Approx. 0   **b.** $z = -6.36$ with p-value $< .002$; reject H_0
8.61. a. $n_1 = n_2 = 179$ **b.** $n_1 = n_2 = 54$ **c.** $n_1 = n_2 = 121$
8.62. a. $n_1 = n_2 = 27{,}292$ **b.** $n_1 = n_2 = 542$ **c.** $n_1 = n_2 = 325$ **8.63.** 136   **8.64.** $n_1 = n_2 = 4{,}802$
8.65. $n = 34$ **8.66.** $n_1 = 260$; $n_2 = 520$ **8.67.** $n_1 = n_2 = 136$ **8.68.** $n_1 = n_2 = 35$ **8.69.** $n_1 = n_2 = 2{,}135$
8.70. a. Reject H_0 if $F > 2.16$ (if $s_1^2 > s_2^2$).   **b.** Reject H_0 if $F > 2.94$ (if $s_2^2 > s_1^2$).
c. Reject H_0 if $F > 2.11$ (if $s_1^2 > s_2^2$) or if $F > 2.29$ (if $s_2^2 > s_1^2$).
d. Reject H_0 if $F > 2.30$ (if $s_2^2 > s_1^2$).   **e.** Reject H_0 if $F > 2.78$ (if $s_1^2 > s_2^2$) or if $F > 3.95$ (if $s_2^2 > s_1^2$).

8.71. a. $F_{L,\alpha/2} = 4.82$; $F_{U,\alpha/2} = 4.20$ **b.** $F_{L,\alpha/2} = 4.20$; $F_{U,\alpha/2} = 3.51$ **c.** $F_{L,\alpha/2} = 2.18$; $F_{U,\alpha/2} = 2.59$
d. $F_{L,\alpha/2} = 3.18$; $F_{U,\alpha/2} = 2.96$
8.72. a. $F = 2.88$; reject H_0 **b.** (.142, .939) **c.** $F = 2.88$; reject H_0
8.73. a. (.02, .99) **b.** $F = 5.48$; do not reject H_0 **c.** $F = 5.48$; reject H_0 **8.75.** $F = 3.17$; yes **8.76.** (1.02, 4.41)
8.77. $F = 4.07$; reject H_0 (choose A) **8.78.** $F = 2.63$; reject H_0 (choose line 1)
8.79. a. (.01, .96) **b.** Instrument 1 is more precise.
8.80. a. $t = .78$; do not reject H_0 **b.** (−6.49, 11.49) **c.** $n_1 = n_2 = 225$
8.81. a. (.09, .69) **b.** $F = 3.84$; reject H_0 **c.** No; conclude $\sigma_1^2 \neq \sigma_2^2$.
8.82. a. (3.59, 4.21) **b.** $z = 20.60$; reject H_0 **c.** $n_1 = n_2 = 346$
8.83. a. $z = -2.04$; reject H_0 **b.** (−.196, −.004) **c.** $n_1 = n_2 = 18,248$
8.84. a. $t = 5.73$; reject H_0 **b.** (1.96, 5.64) **8.86. a.** $t = 1.064$; do not reject H_0
8.87. a. $F = 1.21$; do not reject H_0 **b.** (.38, 3.85) **8.88.** $t = -4.02$; reject H_0 **8.89.** $z = .90$; no
8.90. (−3.45, −1.95) **8.91.** $n_1 = n_2 = 113$ **8.92.** $n_1 = n_2 = 542$ **8.93.** $z = 1.67$; yes
8.94. (−.22, 1.02); yes **8.95. a.** (−.14, −.04) **b.** (.27, .35) **8.96.** $n_1 = n_2 = 195$ **8.97.** $t = -.70$; no
8.98. $z = -3.98$; yes **8.99. a.** $t = -2.08$; no **8.100.** $n_1 = n_2 = 193$ **8.101. a.** $t = 2.27$; yes
8.102. a. (1.0, 18.8) **b.** (.20, 1.76) **8.103.** $t = 6.14$; yes **8.104.** (−.003, .065) **8.105.** $z = -1.30$; do not reject H_0
8.106. (−256.02, −97.98) **8.107.** $t = -2.12$; no **8.108.** $z = -2.62$; yes **8.109.** $F = 2.79$; no
8.110. a. (−3,160.5, 877.1) **b.** No **8.111. a.** $p_1 - p_2$ **b.** (−.01, .15) **8.112.** (−.13, .05)
8.113. $n_1 = n_2 = 1,184$ **8.114. a.** $t = -2.84$; yes **b.** (−76.94, −8.66) **8.115.** $z = -2.93$; reject H_0
8.116. (−.13, .03) **8.117. a.** $t = -1.38$; yes **8.118.** $F = 1.82$; do not reject H_0 **8.119.** $n_1 = n_2 = 70$
8.120. $t = -2.0$; yes **8.121. a.** $F = 1.74$; no **b.** p-value > .10 **8.122.** (−7.06, 2.06) **8.123.** $z = -.54$; no

Chapter 9

9.1. a.

SOURCE	df	SS	MS	F
Treatments	2	5.00	2.50	1.18
Error	6	12.72	2.12	
Total	8	17.72		

b. $F = 1.18$; do not reject H_0 **c.** (−.58, 4.58) **d.** Narrower
e. (1.18, 6.22) **f.** 51

9.2. a.

SOURCE	df	SS	MS	F
Treatments	6	16.9	2.817	3.48
Error	35	28.3	.809	
Total	41	45.2		

b. 7 **c.** $F = 3.48$; yes **d.** $p < .01$ **e.** $t = -.83$; no
f. (−1.19, .39) **g.** (3.14, 4.26)

9.3. a.

SOURCE	df	SS	MS	F
Treatments	2	11.075	5.538	3.15
Error	7	12.301	1.757	
Total	9	23.376		

b. Do not reject H_0. **c.** 2.54 ± 1.92 **d.** Narrower
e. 3.975 ± 1.567 **f.** 43

9.4. a.

SOURCE	df	SS	MS	F
Treatments	4	24.7	6.18	4.90
Error	30	37.7	1.26	
Total	34	62.4		

b. 5 **c.** Yes **d.** $t = -.667$; no **e.** $-.4 \pm .987$
f. $3.7 \pm .698$

9.8. a. Completely randomized   **b.** $F = 6.79$; yes   **c.** 21.8 ± 11.01 **d.** 5.1 ± 11.68

9.9. a. $F = 8.63$; yes   **b.** p-value is smaller than .01

9.10. a. $F = 4.55$; yes   **b.** 52.83

9.11. a. $T_1 = 953.91$;   9352   **b.** SSE = 72.1479

c.

SOURCE	d				yes	**e.** (2.70, 10.54)	**f.** $t = 2.17$; yes
Treatments							
Error	107						
Total	109						

9.12. a. $F = 5.29$; yes ± 37.09

9.14. a.

SOURCE			**c.** $F = 2.21$; no **d.** $(-2.20, 2.20)$
Treatment			
Block			
Error			
Total	11		

9.15. a.

SOURCE	df		**c.** $F = 6.07$; yes **d.** $(-3.93, -.87)$
Treatment	3		
Block	5		
Error	15		
Total	23		

9.19. a. $F = 6.36$; yes   **b.**   4.10, reject H_0 **e.** p-value $< .01$

9.21. a. $F = .3381$; no   **b.** p

c.

SOURCE	df	SS	MS	F	**d.** $-.1833 \pm 1.234$
Treatment	2	.9411	.4706	.3381	
Block	5	10,240.0	2,048.0	1,471.0	
Error	10	13.92	1.392		
Total	17	10,254.9			

9.22. a. $F = 39.46$; yes   **b.** $F = 1.93$; no   **c.** -27 ± 6.85

9.23. a. Randomized block   **b.**

SOURCE	df	SS	MS	F
Year	3	433.2746	144.4249	31.72
Location	5	12,554.8238	2,510.9648	551.55
Error	15	68.2879	4.5525	
Total	23	13,056.3863		

c. $F = 31.72$; yes   **d.** -8.47 ± 2.63

9.24. a.

SOURCE	df	SS	MS	F
Year	2	280.4406	140.2203	37.39
City	11	9,759.1122	887.1920	236.57
Error	22	82.5061	3.7503	
Total	35	10,122.0589		

b. $F = 37.39$; yes   **c.** $t = 8.63$; yes

9.25. b.

SOURCE	df	SS	MS	F
Company	2	.1820	.0910	9.07
City	11	29.2598	2.6600	265.06
Error	22	.2208	.0100	
Total	35	29.6626		

c. $F = 9.07$; yes   **d.** $t = -.04$; no

9.26. a. Randomized block   **b.** No   **c.**

SOURCE	df	SS	MS	F
Treatment	3	646.6800	215.5600	.627
Error	20	6,874.6333	343.7317	
Total	23	7,521.3133		

$F = .627$; no

9.28. b.

SOURCE	df	SS	MS	F
A	2	.8	.400	.80
B	3	5.3	1.767	3.53
AB	6	9.6	1.600	3.20
Error	12	6.0	.500	
Total	23	21.7		

d. Yes; $F = 3.20$

9.29. a. 4 levels of A, 2 levels of B **b.** 3   **c.**

SOURCE	df	SS	MS	F
A	3	2.25	.75	5.00
B	1	.95	.95	6.33
AB	3	.90	.30	2.00
Error	16	2.40	.15	
Total	23	6.50		

d. No interaction; main effects both significant

9.30. a.

SOURCE	df	SS	MS	F
A	1	4.4408	4.4408	18.06
B	2	4.1267	2.0633	8.39
AB	2	18.0067	9.0033	36.62
Error	6	1.4750	.2458	
Total	11	28.0492		

b. $F = 36.62$; yes   **d.** $(-3.01, -1.09)$

9.31. a.

SOURCE	df	SS	MS	F
A	1	658.845	658.8450	46.65
B	1	255.380	255.3800	18.08
AB	1	2.000	2.0000	.14
Error	4	56.490	14.1225	
Total	7	972.715		

b. $F = .14$; no   **c.** $F = 46.65$; yes   **d.** $F = 18.08$; yes

e. $(-18.68, -3.92)$

9.32. a.

SOURCE	df	SS	MS	F
Agency	1	39.9675	39.9675	7.01
Medium	2	198.3317	99.1658	17.39
AM	2	77.3450	38.6725	6.78
Error	6	34.2050	5.7008	
Total	11	349.8492		

b. $F = 6.78$; yes

9.33. b. CM = 11,858.45   **g.**

SOURCE	df	SS	MS	F
Temp	1	227.138	227.138	9.34
Ration	1	36,671.048	36,671.048	1,508.72
TR	1	618.272	618.272	25.44
Error	76	1,847.264	24.306	
Total	79	39,363.722		

h. $F = 25.44$; yes

i. $t = -5.73$; yes

9.34. b. 6,739.605   **g.**

SOURCE	df	SS	MS	F
Sex	1	114.0050	114.0050	4.04
Weight	1	41.4050	41.4050	1.47
SW	1	18.6050	18.6050	.66
Error	28	789.9682	28.21315	
Total	31	963.9832		

h. $F = .66$; no   **i.** $t = 2.00$; no

j. $t = .85$; no

9.35. a.

SOURCE	df	SS	MS	F
Sex	1	10,686.361	10,686.361	68.74
Weight	1	538.756	538.756	3.47
SW	1	831.744	831.744	5.35
Error	36	5,596.908	155.4697	
Total	39	17,653.769		

b. $F = 5.35$; yes   **c.** $t = 7.50$; yes

d. $t = 4.23$; yes

9.36. f.

SOURCE	df	SS	MS	F
Sex	1	123.12338	123.12338	.19
Weight	1	851.64338	851.64338	1.35
SW	1	355.02337	355.02337	.56
Error	56	35,443.05700	632.91173	
Total	59	36,772.84713		

g. $F = .56$; no

9.37.

SOURCE	df	SS	MS	F
Sex	1	496.80038	496.80038	1.68
Weight	1	2,241.75940	2,241.75940	7.56
SW	1	339.15035	339.15035	1.14
Error	56	16,597.30400	296.38043	
Total	59	19,675.01413		

9.38. a. 3.95 **b.** 5.56 **c.** 6.20 **d.** 4.20

9.39. a. 15 **b.** .048 **c.** .48 **d.** $\bar{x}_4$ $\bar{x}_3$ $\bar{x}_1$ $\bar{x}_5$ $\bar{x}_2$ (Underline indicates no significant difference.)

9.40. a. 6 **b.** 5.19 **c.** 5.19 **9.41.** $\omega = 12.89$; $\bar{x}_1$ $\bar{x}_3$ $\bar{x}_2$; μ_1 and μ_2 are different

9.42. $\omega = 8.13$; $\bar{x}_5$ $\bar{x}_4$ $\bar{x}_1$ $\bar{x}_2$ $\bar{x}_3$; only μ_3 and μ_5 are different **9.43.** $\omega = 24.29$; no significant difference in means

9.44. $\omega = 16.62$; normal male differs from overweight male and overweight female means

9.45. b. 52,040.36 **c.** 464.6461 **d.** $\omega = 24.27$; no **9.46. a.** $F = 157.39$; yes **b.** $F = 111.33$; yes **c.** $-1.9 \pm .30$

9.47. a. $F = 7.70$; yes **b.** 2 ± 1.4 **c.** 11.4 ± 1.2

9.48. a. Completely randomized design **b.** $F = 6.30$; yes **c.** $4.50 \pm .23$

9.49. a. $F = 3.64$; do not reject H_0 ($\alpha = .05$) **b.** $F = 46.85$ (blocking was appropriate)

9.50. a. $F = 4.88$; yes **b.** p-value is smaller than .01 **c.** 1.66 ± 2.37 **d.** 11.99 ± 1.02

9.51. a. $F = 11.03$; yes **b.** $.319 \pm .194$ **9.52. a.** $F = 5.78$; no **b.** $-.0167 \pm .0799$ **c.** $1.6567 \pm .0565$

9.53. a. $F = 12.22$; yes **b.** 8.3 ± 2.27 **c.** $t = 1.74$; yes (at $\alpha = .05$) **d.** 16.5 ± 1.60

9.54. a. Randomized block **b.** $F = 19.48$; yes **c.** $F = 5.77$; yes

9.55. a. Randomized block **b.** $F = 3.884$; yes ($\alpha = .05$) **c.** 2.4 ± 33.48 (95% confidence)

9.56. a.

SOURCE	df	SS	MS	F
Company	1	3,237.2	3,237.2	19.62
Error	98	16,167.7	164.98	
Total	99	19,404.9		

b. $F = 19.62$; yes **c.** No, need sample means

9.57. a. $F = 22.23$; yes **b.** $F = 20.45$; yes **c.** 1.593 ± 4.088

9.58. a. $F = 2.32$; no **b.** $F = 4.68$; no **9.59. a.** $F = 5.85$; yes **b.** -130 ± 139.7

9.60. a. $F = 83.82$; yes **b.** $-1.64 \pm .4877$ **9.61. a.** $F = 4.80$; no

9.62. a.

SOURCE	df	SS	MS	F
Extractor	5	84.71	16.94	12.46
Truckload	14	159.29	11.38	8.37
Error	70	95.33	1.36	
Total	89	339.33		

b. $F = 12.46$; yes

9.63. a. $F = 10.00$; yes ($\alpha = .05$) **b.** 2.0 ± 11.5 **c.** 70.5 ± 7.04

9.64. a. $F = 7.613$; yes **b.** $8.067 \pm .4491$ **9.65.** $F = 10.66$; yes ($\alpha = .05$)

9.66. a. $F = 15.58$; yes ($\alpha = .05$) **9.67.** $F = 2.95$; no

9.68. a. Completely randomized **b.**

SOURCE	df	SS	MS	F
Treatment	2	57.6042	28.8020	.3375
Error	13	1,109.3333	85.3333	
Total	15	1,166.9375		

c. $F = .3375$; no

9.69. a. $F = 3.91$; no   **b.** p-value $= .10$ **c.** $F = 95.51$; yes   **d.** -7.84 ± 5.26

9.70. a. $F = .1925$; no   **b.**

SOURCE	df	SS	MS	F
Treatment	2	.6317	.3158	.1925
Block	3	.8558	.2853	.1739
Error	6	9.8417	1.6403	
Total	11	11.3292		

c. $F = .1739$; no   **d.** $.55 \pm 3.36$

9.71. a. Completely randomized   **b.** $F = 7.02$; yes   **c.** $8.15 \pm .449$

9.72. a. $F = 9.747$; yes   **b.** $F = 4.881$; yes   **c.**

SOURCE	df	SS	MS	F
Treatment	2	1,277.2	638.60	9.75
Block	4	1,279.1	319.78	4.88
Error	8	524.1	65.52	
Total	14	3,080.4		

d. 11.6 ± 11.805

9.73. a. Completely randomized   **b.** $F = 7.79$; yes   **c.** -5.65 ± 3.25
9.74. a. $F = 25.07$; yes   **b.** -6 ± 3.58 **9.75. a.** $F = 6.63$; yes   **b.** -2.1 ± 1.28
9.76. b. Significant interaction   **9.77. c.** Sample sizes are not equal.   **9.79.** No; sample sizes are not equal.

Chapter 10

10.1. a. Reject H_0 if $T_B \geq 58$ or $T_B \leq 32$.   **b.** Reject H_0 if $T_A \geq 40$.   **c.** Reject H_0 if $T_B \geq 98$.
10.2. a. Reject H_0 if $T_B \geq 39$.   **b.** Reject H_0 if $T_B \geq 41$ or $T_B \leq 15$.   **10.3.** $T_B = 42.5$; reject H_0
10.4. a. $T_A = 67.5$; reject H_0 **b.** $T_A = 67.5$; reject H_0 **10.6.** $z = -2.47$; yes
10.7. a. $t = -1.258$; do not reject H_0; independent samples from normally distributed populations with equal variances
b. $T_A = 37.5$; do not reject H_0 ($\alpha = .05$)
10.8. $T_A = 16$; yes   **10.9.** $T_B = 29$; do not reject H_0 **10.10. a.** $T_A = 82$; no
10.11. $T_A = 69.5$ ($T_B = 35.5$); yes   **10.12. a.** $T_A = 150.5$; reject H_0 **10.13.** $z = 3.44$; reject H_0
10.14. a. Reject H_0 if $T \leq 101$ ($T = $ smaller of T_+ and T_-).   **b.** Reject H_0 if $T_- \leq 303$.   **c.** Reject H_0 if $T_+ \leq 0$.
10.15. a. Reject H_0 if $T_- \leq 2$.   **b.** Reject H_0 if $T \leq 4$ ($T = $ smaller of T_+ and T_-).
10.16. a. $T_- = 2.5$; reject H_0 **b.** $T_- = 2.5$; reject H_0
10.17. a. Reject H_0 if $z > 1.645$.   **b.** $z = 2.97$; reject H_0 **c.** p-value $= .0015$ **10.18.** $T_- = 4.5$; reject H_0; yes
10.19. $T_- = 13$; no   **10.20.** $T_- = 3$; do not reject H_0 **10.21.** $T_- = 24.5$; no   **10.22.** $T_- = 36$; no
10.23. a. 31.4104   **b.** 27.4884   **c.** 50.8922   **d.** 96.5782   **e.** 5.99147   **f.** 25.1882
10.24. a. .995   **b.** .025   **c.** .10   **d.** .975   **e.** .025   **f.** .100
10.25. a. Completely randomized   **c.** Reject H_0 for $\chi^2 \geq 9.21$.   **d.** $H = 15.944$; reject H_0
10.26. a. $H = 8.01$; yes   **10.28.** $H = 1.57$; no   **10.29. c.** $H = 14.75$; reject H_0 **10.30.** $H = 7.15$; yes
10.31. a. $H = 6.66$; yes   **b.** $T_A = 71$ ($T_B = 34$); yes   **10.32.** $H = 2.03$; no   **10.33.** $H = 15.33$; yes
10.34. a. $F = 1.326$; do not reject H_0; lifetimes for the three brands are normally distributed with equal variances
b. $H = 1.22$; do not reject H_0
10.35. b. Reject H_0 if $\chi^2 \geq 4.605$.   **c.** $F_r = 6.93$; reject H_0 **10.36.** $F_r = 13$; reject H_0 **10.37.** $F_r = 39.41$; yes
10.38. $F_r = 15.25$; reject H_0 **10.39.** $F_r = 6.35$; yes   **10.40.** $F_r = 7.85$; reject H_0 **10.41.** $F_r = 12.3$; yes
10.42. $F_r = 1.75$; no
10.43. a. Reject H_0 if $r_s \geq .683$ or $r_s \leq -.683$.   **b.** Reject H_0 if $r_s \geq .400$.   **c.** Reject H_0 if $r_s \leq -.478$.
10.44. a. 1   **b.** $-.90$ **c.** 1   **d.** .20   **10.45. a.** $r_s = .866$; yes   **b.** $r_s = .866$; yes   **10.46.** .9879

10.47. $r_s = .657$; no ($\alpha = .05$)   **10.48.** $r_s = .881$; yes   **10.49.** $r_s = .9341$; yes   **10.50. a.** $r_s = .86$ **b.** Yes
10.51. a. $r_s = .452$ **b.** No   **10.52.** $r_s = -.804$; yes   **10.53.** $r_s = -.854$; yes   **10.54.** $H = 9.86$; reject H_0
10.55. a. $r_s = .40$; no   **b.** $T_- = 1.5$; yes   **10.56.** $T_B = 55.5$; no   **10.57.** $F_r = 14.9$; yes   **10.58.** $T_A = 139$; yes
10.59. $T_A = 21$ ($T_B = 34$); no   **10.60.** $T_{After} = 19$; yes   **10.61.** $T_- = 3$; yes   **10.62.** $T_- = 4$; yes   **10.63.** $T_+ = 6$; no
10.64. a. $T_+ = 1$; yes   **b.** $t = -2.96$; reject H_0 **10.65.** $T_+ = 19.5$; no   **10.66.** $H = 14.27$; yes   **10.67.** $H = 19.47$; yes
10.68. $H = 5.85$; no   **10.69. a.** $H = 14.61$; yes   **b.** $T_{III} = 16.5$; reject H_0 **10.70.** $F_r = 26.95$; yes
10.71. $F_r = 4.5$; no ($\alpha = .05$)   **10.72.** $F_r = 1.33$; no   **10.73.** $F_r = 6.21$; no   **10.74.** $r_s = .421$; no
10.75. $T_{After} = 94.5$; reject H_0 **10.76.** $r_s = .643$; no   **10.77.** $r_s = .364$; no   **10.78.** $r_s = -.148$; no
10.79. $T_- = 3.5$; yes   **10.80.** $T_{After} = 77.5$; yes   **10.81. a.** $F_r = 11.77$; yes   **b.** $T_+ = 2.5$; do not reject H_0
10.82. a. $H = 16.39$; yes   **b.** $T_1 = 59$; reject H_0 **10.83.** $r_s = .929$; yes   **10.84.** $F_r = 1.55$; no
10.85. a. $H = 12.79$; yes   **b.** $T_A = 28.5$; do not reject H_0 **10.86.** $r_s = .479$; no ($\alpha = .05$)   **10.87.** $r_s = .771$; no
10.88. $T_+ = 17.5$; no   **10.89. a.** $H = 3.225$; no   **b.** $T_A = 78$; do not reject H_0 **10.90.** $F_r = 6.35$; yes

Chapter 11

11.1. a. 24.9958   **b.** 70.0648   **c.** 18.5494   **d.** 10.5966   **11.2. a.** .01   **b.** .01   **c.** .90   **d.** .005
11.3. a. $X^2 > 4.60517$ **b.** $X^2 > 13.2767$ **c.** $X^2 > 7.81473$ **11.5.** $X^2 = 2.72$; no   **11.6.** $X^2 = 17.15$; yes
11.7. a. $X^2 = 6.00$; no   **b.** $(.23, .41)$ **11.8. a.** $X^2 = 12.37$; do not reject H_0 **b.** $.05 < p\text{-value} < .10$
11.9. a. $X^2 = 17.51$; yes   **b.** $.93 \pm .04$ **11.10.** $X^2 = 17.36$; yes   **11.11.** $X^2 = 4.4$; no   **11.12.** $X^2 = 9.48$; yes
11.13. $X^2 = 1.5$; no   **11.14.** $X^2 = 3.92$; no
11.15. a. Reject H_0 if $X^2 \geq 26.2962$.   **b.** Reject H_0 if $X^2 \geq 15.9871$.   **c.** Reject H_0 if $X^2 \geq 9.21034$.
11.16. $X^2 = 15.275$; reject H_0 **11.17. a.** $X^2 = 67.778$; reject H_0 **b.** $.038 \pm .014$ **11.18.** $X^2 = 19.72$; yes
11.19. a. $X^2 = 5.898$; yes   **b.** $.01 < p\text{-value} < .025$ **11.20. a.** $X^2 = 5.074$; yes   **b.** $.01 < p\text{-value} < .025$
11.21. $X^2 = 1.35$; no   **11.22.** $X^2 = 9.34$; yes   **11.23.** $X^2 = 40.70$; yes   **11.24.** $X^2 = .621$; no   **11.25.** $X^2 = 14.406$; yes
11.26. $p\text{-value} > .10$
11.27. a. Union: $X^2 = 13.367$, reject H_0; non-union: $X^2 = 9.155$, do not reject H_0
b. Union: $.025 < p\text{-value} < .05$; non-union: $p\text{-value} > .10$
11.28. a. $X^2 = 14.212$; yes   **b.** $(.304, .496)$ **c.** $(.002, .198)$
11.29. a. $X^2 = 31.259$; yes   **b.** $(.373, .519)$ **c.** $(.222, .432)$ **11.30. a.** $X^2 = 44.39$; yes   **11.31.** $X^2 = 5.57$; no
11.32. $X^2 = 1.07$; no   **11.33.** $X^2 = 2.40$; no   **11.34.** $X^2 = 93.64$; yes   **11.35. b.** $X^2 = 894.5$; yes   **c.** $z = 12.54$; yes
11.36. a. $X^2 = 54.14$; yes   **b.** No   **c.** Yes   **11.37. a.** $X^2 = 2.13$; no   **b.** $.23 \pm .06$
11.38. $X^2 = 5.506$; no   **11.39.** $X^2 = 4.977$; yes   **11.40.** $X^2 = 2.601$; no   **11.41. a.** $X^2 = 9.16$; reject H_0 **b.** $.90 \pm .06$
11.42. $X^2 = 103.1$; yes   **11.43.** $X^2 = 7.384$; no ($\alpha \leq .05$)   **11.44.** $X^2 = 180.874$; yes   **11.45.** $X^2 = 8.664$; no
11.46. a. $X^2 = 6.929$; no   **b.** $.284 \pm .0571$ **11.47.** $X^2 = 38.68$; yes   **11.48.** $X^2 = 269.91$; yes   **11.49.** $X^2 = 30.507$; yes
11.50. $X^2 = 7.4$; do not reject H_0 **11.51. a.** $X^2 = 31.86$; yes   **b.** $.12 \pm .0947$
11.52. a. $X^2 = 3.132$; no   **b.** $.0952 \pm .1280$ **11.53.** $X^2 = 42.542$; yes   **11.54.** $X^2 = 6.74$; yes
11.55. a. $X^2 = 46.25$; yes ($\alpha = .05$)   **b.** $.1429 \pm .0417$ **11.56.** $X^2 = 9.495$; yes
11.57. a. $X^2 = 9.647$ **b.** $X^2_{.05} = 7.8147$ **c.** Yes   **11.58.** $X^2 = 9.468$; no ($\alpha = .05$)

Chapter 12

12.2. a. $4, -4$ **b.** $\frac{4}{3}, 3$ **c.** $\frac{3}{4}, -\frac{5}{4}$ **d.** $-4, 11$
12.3. a. $y = -4 + 4x$ **b.** $y = 3 + \frac{4}{3}x$ **c.** $y = -\frac{5}{4} + \frac{3}{4}x$ **d.** $y = 11 - 4x$
12.5. a. $-1.5, 1$ **b.** $-1, .5$ **c.** $4, -1$ **d.** $-2, 3$ **e.** $2, 0$ **12.8. c.** $\hat{y} = 2.05 + 1.3x$
12.9. c. $\hat{y} = -1.467 + 1.371x$ **12.10. a.** $\hat{\beta}_0 = 2.6$; $\hat{\beta}_1 = -1.7$ **c.** .9   **d.** 5.15
12.11. a. $\hat{y} = 57.91 - .81x$ **c.** 9.22

12.12. a. Positive   **b.** $\hat{y} = 50.803 + .251x$ [using $x = 1,000$ (Team batting average $- .265$)]
12.13. a. Negative   **b.** $\hat{y} = 102.083 - 13.151x$ **12.14. b.** $\hat{\beta}_0 = 24.45$; $\hat{\beta}_1 = 2.38$ **c.** 33.38
12.15. a. $\hat{y} = 3.375 + 1.201x$ **12.16. a.** $\hat{y} = 156.357 + 12.786x$
12.17. a. 6°C: $\hat{y} = -2.792 + 15.135x$; 12°C: $\hat{y} = -8.368 + 17.612x$; 18°C: $\hat{y} = -12.068 + 17.599x$
12.18. a. .0613   **b.** $2s \approx .495$ **12.19. a.** .05, .025   **b.** 2.419, .6048   **c.** .3, .1
12.20. a. $\hat{y} = -45.78 + 1.562x$ **c.** 19.60, 4.90, 2.214   **d.** $(-4.43, 4.43)$
12.21. a. $\hat{y} = -.246 + 1.315x$ **c.** 7.65, 11.59   **d.** 12.44, 2.487, 1.577   **e.** $(-3.15, 3.15)$
12.22. a. 1.3, .025, 5, 2   **b.** 1.371, .6048, 17.5, 4   **c.** $-1.7, .1, 10, 3$ **12.23. a.** $t = 18.385$; yes   **b.** $t = 7.38$; yes
12.24. a. $t = -17.0$; yes   **b.** $t = -8.98$; yes   **12.25. a.** $t = -2.415$; yes   **b.** $t = 2.61$; yes
12.26. a. $\hat{y} = 18.89 + .0109x$ **b.** $t = .813$; no   **c.** $.01 \pm .027$ **12.27. a.** $\hat{y} = 70.8 - .28x$ **b.** $t = -.48$; no
12.28. a. $\hat{y} = 10.394 + 90.921x$ **b.** $H_a: \beta_1 > 0$; $t = 3.182$; reject H_0 **c.** 90.92 ± 57.58
12.29. a. 7.3243, .7324   **b.** $t = 12.81$; yes   **c.** $p < .01$ **12.30. a.** $t = 13.69$; yes   **b.** $p < .01$
12.30. a. $t = 13.69$; yes   **b.** $p < .01$ **12.31. a.** 15.676, 7.838   **b.** 17.612 ± 4.045
12.32. a. .9971, .9941   **b.** .9652, .9315   **c.** $-.9948$, .9897   **12.33. a.** Positive   **b.** .4732, .2239
12.34. a. Negative   **b.** $-.5718$, .3270   **12.35.** .9426, .9709   **12.36. a.** $-.688$, .473   **b.** $t = -2.68$; yes
12.37. $r = .9836$, $r^2 = .9675$ **12.38. a.** $-.366$, .134   **b.** $t = -.79$; no   **12.39. a.** .69   **b.** .476   **c.** $t = 3.44$; yes
12.40. a. Positive   **b.** .7418, .5503   **c.** $t = 3.498$; yes
12.41. a. Negative   **b.** $-.7624$, .5813   **c.** $\hat{y} = 47.473 - .0088x$ **d.** $t = -4.408$; yes
12.42. a. $4.3 \pm .41$ **b.** $4.3 \pm .85$
12.43. a. $\hat{y} = .25 + 1.25x$ **c.** .75   **d.** .0375   **e.** $1.5 \pm .20$ **f.** $2.125 \pm .43$ **g.** $.25 \pm .28$
12.44. a. $t = 6.96$; yes   **b.** 5.26 ± 2.10 **c.** 21.22 ± 4.01 **d.** 21.22 ± 7.90 **12.45. a.** $11.78 \pm .79$ **b.** 11.78 ± 2.06
12.46. a. $\hat{y} = 40.926 - .1343x$ **b.** $t = -4.21$; yes   **c.** $-.1343 \pm .0576$ **d.** $31.26 \pm .91$ **e.** 33.14 ± 2.21
12.47. a. $\hat{y} = 4.554 + 2.671x$ **c.** $r = .967$, $r^2 = .935$ **d.** 13.10 ± 1.19 **e.** $12.57 \pm .43$
12.48. a. 207.5 ± 5.37 **b.** 207.5 ± 13.80 **12.49.** 35.66 ± 6.46
12.50. a. $\hat{y} = 27.78 - .6x$ **c.** 7.0   **d.** .5385   **e.** $-.6 \pm .184$ **f.** $26.70 \pm .35$ **g.** 26.70 ± 1.35
12.51. b. $-.1245$, .016   **c.** $t = -.35$; no
12.52. a. $\hat{y} = 40.7842 + .7656x$ **c.** $t = 4.38$; reject H_0 **d.** $.7656 \pm .4035$ **e.** 79.06 ± 17.03 **f.** 67.58 ± 7.74
12.53. a. $\hat{y} = 22.32 + 2.650x$ **c.** $t = 6.280$; yes   **d.** .9421, .8875   **e.** 109.8 ± 2.155 **f.** 109.8 ± 6.088
12.54. a. $\hat{y} = 18.22 + 1.166x$ **c.** $t = 4.967$; yes $(\alpha = .05)$ **d.** .8044   **e.** 123.2 ± 19.09
12.55. b. $\hat{y} = 11,898.8 - 763.1x$ **d.** $t = -3.944$; reject H_0 **e.** -763.1 ± 375.9 **f.** $-.8495$ **g.** .7216
h. $5,793.9 \pm 1,609.3$ **i.** $5,793.9 \pm 677.2$
12.56. a. $\hat{y} = 2.084 + .6682x$ **c.** $t = 16.28$; yes   **d.** .9815   **e.** $35.49 \pm .979$ **f.** 38.83 ± 2.766
12.57. a. $\hat{y} = -13.4903 - .0528x$ **b.** $t = -6.84$; reject H_0 **c.** .8538   **d.** $.9320 \pm .3333$
12.58. a. $\hat{y} = 16.28 + 2.078x$ **c.** $t = 6.672$; yes $(\alpha = .05)$ **d.** 57.84 ± 6.146
12.59. a. $\hat{y} = 20.11 + .9720x$ **c.** .9815, .9634   **d.** $27.88 \pm .98$ **e.** 26.91 ± 2.57
12.60. a. $\hat{y} = 6.5143 + 10.8294x$ **b.** $t = 6.336$; reject H_0 $(\alpha = .05)$ **c.** $12.904 \pm .3594$
12.61. a. $\hat{y} = 13.48 + .0560x$ **c.** .7402, .5479   **d.** 17.40 ± 2.474
12.62. a. $\hat{y} = 2.76 + .866x$ **b.** .9853, .9707   **c.** 20.081 ± 1.003
12.63. a. 7.9416, 3.9708   **b.** 17.60 ± 2.46 **d.** 58.33 ± 5.86

Chapter 13

13.1. a. $t = 1.45$; do not reject H_0 **b.** $t = 3.21$; reject H_0
13.2. a. $t = 3.13$; reject H_0 **b.** $t = 3.13$; yes   **c.** $F = 9.82$ **d.** No
13.3. df for estimating σ^2 is $n - (k + 1)$, where $k = $ number of independent variables
13.4. a. SSE $= .0438$, $s^2 = .01095$ **b.** $t = -13.97$; yes   **c.** $\hat{y} = 1.095 + 1.636x - .1595x^2$ **13.5.** $t = 2.11$; reject H_0
13.6. a. $t = -3.33$; yes   **13.7.** $t = 5.5$; yes

13.8. b. $\beta_1, \beta_3, \beta_5, \beta_6$ are significantly different from zero   **c.** Intercept is .105   **d.** Intercept is 0
13.9. a. Yes   **b.** $F = 55.2$; yes   **13.10. a.** .3018; no   **b.** $F = 2.31$, do not reject H_0
13.11. b. .38   **c.** $F = 62.10$; yes   **d.** Yes, assuming x_2 is included in the model
13.12. a. $F = 31.98$; yes   **b.** $t = -2.27$; yes   **13.13. b.** $F = 45.51$; yes   **13.14.** $F = 1.06$; do not reject H_0
13.15. a. $F = 15.40$; yes   **b.** $t = -.75$; no
13.16. a. $\hat{y} = 131.857 + 24.452x - 1.167x^2$ **b.** $t = -7.65$; yes   **c.** $p < .01$ **d.** .9983
13.17. a. $\hat{y} = 75.079 - .290x_1 - 19.148x_2 + .268x_1x_2$ **b.** .9909   **c.** $F = 145.55$; yes   **d.** $t = 3.92$; yes   **13.18. a.** .9488
13.19. a. $F = 24.41$; yes   **b.** $t = -2.01$; reject H_0 **c.** $t = .31$; do not reject H_0 **d.** $t = 2.38$; reject H_0
13.21. a. $\hat{y} = 90.1 - 1.84x_1 + .285x_2$ **b.** $R^2 = .916$ **c.** $F = 65.43$; yes   **d.** $t = -5.01$; reject H_0
13.23. $x_1 = 30$, $x_2 = .6$, $x_3 = 1{,}300$ are in the experimental region
13.24. a. $\hat{y} = .0562 + .273x_1 + .0006x_2$ **b.** $F = 163.17$; yes   **c.** $t = 4.34$; reject H_0 **d.** $\hat{y} = 14.2462$ (about 14 positions)
e. Yes   **f.** Normality
13.25. a. $\hat{y} = 5{,}716.911 - 1.546x_1 + .033x_2 - 32.557x_3$ **13.26. a.** $F = 1.28$; do not reject H_0 **13.27. a.** 13.68
13.28. a. $\hat{y} = 20.70 - .0145x + .00005x^2$; $F = 26.25$; reject H_0 $(\alpha = .05)$ **b.** $t = .96$; no
13.29. $t = -1.104$; do not reject H_0 **13.30. b.** $t = 40.54$; yes   **c.** $\hat{y} = 17.79$ **e.** Yes
13.31. b. $F = 16.1$; yes   **c.** $t = 2.5$; yes   **d.** 945   **13.32. b.** $t = 5$; yes   **c.** 825   **d.** Support
13.33. The estimate of the difference in mean attendance between weekends and weekdays is $15x_3 - 700$.
13.34. a. $F = 6.614$; yes   **13.35. a.** Test H_0: $\beta_2 = \beta_3 = 0$.   **c.** $t = -3.333$; yes   **13.36.** $F = 5.990$; reject H_0
13.37. a. $\hat{y} = -13.062 + .742x_1 + 18.603x_2 + 13.410x_3$ **b.** $F = 188.33$; reject H_0
13.38. a. $\hat{y} = .601 + .595x_1 - 3.725x_2 - 16.232x_3 + .235x_1x_2 + .308x_1x_3$ **b.** $F = 139.42$; reject H_0 **13.39.** $t = -2.06$; no
13.40. a. $\hat{y} = .5950 + .0125x_1 + .0300x_2 + .0025x_1x_2$ **b.** $R^2 = .760$; $F = 12.67$; yes $(\alpha = .05)$ **c.** $t = .23$; no
13.41. a. $\hat{y} = -1.5705 + .02573x_1 + .033614x_2$; $F = 39.51$; reject H_0 $(\alpha = .05)$
13.42. a. $F = 100.41$; reject H_0 **c.** $t = 1.67$; do not reject H_0

Chapter 14

14.1. a. Quantitative   **b.** Quantitative   **c.** Qualitative   **d.** Quantitative   **e.** Qualitative
14.2. a. Quantitative   **b.** Quantitative   **c.** Qualitative   **d.** Qualitative   **e.** Qualitative   **f.** Qualitative
g. Quantitative   **h.** Qualitative
14.3. Normality   **14.4. a.** Quantitative   **b.** Qualitative   **c.** Qualitative   **d.** Qualitative   **e.** Quantitative
14.6. a. (i) First; (ii) Third; (iii) First; (iv) Second
b. (i) $E(y) = \beta_0 + \beta_1 x$; (ii) $E(y) = \beta_0 + \beta_1 x + \beta_2 x^2 + \beta_3 x^3$; (iii) $E(y) = \beta_0 + \beta_1 x$; (iv) $E(y) = \beta_0 + \beta_1 x + \beta_2 x^2$
c. (i) $\beta_0 > 0$, $\beta_1 > 0$; (ii) $\beta_0 > 0$, $\beta_3 > 0$; (iii) $\beta_0 > 0$, $\beta_1 < 0$; (iv) $\beta_0 < 0$, $\beta_1 > 0$, $\beta_2 < 0$
14.7. $y = \beta_0 + \beta_1 x + \beta_2 x^2 + \varepsilon$
14.8. a. $E(y) = \beta_0 + \beta_1 x$ **b.** $E(y) = \beta_0 + \beta_1 x + \beta_2 x^2$ **c.** $E(y) = \beta_0 + \beta_1 x + \beta_2 x^2 + \beta_3 x^3$
14.9. a. Second   **c.** Curve would open downward.   **14.10. a.** First   **c.** Line would slope upward.
14.11. $E(y) = \beta_0 + \beta_1 x + \beta_2 x^2$; $\beta_2 < 0$ **14.12. a.** $E(y) = \beta_0 + \beta_1 x + \beta_2 x^2$; $\beta_2 > 0$ **14.13. b.** Linear; linear; quadratic
14.14. $E(y) = \beta_0 + \beta_1 x + \beta_2 x^2$; $\beta_2 > 0$ **14.16.** $t = -6.60$; yes
14.17. a. $E(y) = \beta_0 + \beta_1 x_1 + \beta_2 x_2$ **b.** $E(y) = \beta_0 + \beta_1 x_1 + \beta_2 x_2 + \beta_3 x_1 x_2$
c. $E(y) = \beta_0 + \beta_1 x_1 + \beta_2 x_2 + \beta_3 x_1 x_2 + \beta_4 x_1^2 + \beta_5 x_2^2$
14.18. e. Decreases from 5 to 1   **14.19. e.** Increases from 5 to 13
14.22. a. Both quantitative   **b.** $E(y) = \beta_0 + \beta_1 x_1 + \beta_2 x_2$ **c.** Include $\beta_3 x_1 x_2 + \beta_4 x_1^2 + \beta_5 x_2^2$.
d. H_0: $\beta_3 = 0$ against H_a: $\beta_3 \neq 0$
14.23. $E(y) = \beta_0 + \beta_1 x_1 + \beta_2 x_2 + \beta_3 x_1 x_2 + \beta_4 x_1^2 + \beta_5 x_2^2$
14.24. a. Both quantitative   **b.** $E(y) = \beta_0 + \beta_1 x_1 + \beta_2 x_2$ **c.** Include $\beta_3 x_1 x_2$.   **d.** Include $\beta_4 x_1^2 + \beta_5 x_2^2$.
14.26. a. H_a: At least one parameter (β_3, β_4, or β_5) is not 0   **c.** 3, 24   **14.27.** $F = 1.15$; do not reject H_0
14.29. a. $F = 66.87$; reject H_0 **b.** H_a: $\beta_1 \neq 0$ **c.** $F = 15.16$; reject H_0

14.30. a. $H_0: \beta_1 = \beta_2 = \beta_3 = \beta_4 = \beta_5 = 0$; H_a: At least one β is not 0   **b.** $F = 18.39$; reject H_0
c. $H_0: \beta_3 = \beta_4 = \beta_5 = 0$; H_a: At least one β is not 0   **d.** $F = 8.46$; reject H_0 **e.** Complete model
14.31. a. $F = 1.32$; do not reject H_0 **b.** $F = .48$; do not reject H_0 **c.** No improvement in model
14.32. a. $\hat{y} = 59.0717 - 1.7003x_1 - 1.5722x_2 - 1.6971x_3 - .5638x_1x_2 + .9463x_1x_3 - 1.1963x_2x_3 - 17.3858x_1^2 - 13.8459x_2^2 - 17.4512x_3^2$
b. $F = 1.78$; do not reject H_0

14.33. a. $F = 31.55$; reject H_0 **b.** Complete model   **14.35.** $F = 24.19$; yes   **14.36.** $E(y) = \beta_0 + \beta_1 x_1$; $x_1 = \begin{cases} 1 & \text{if level 2} \\ 0 & \text{if level 1} \end{cases}$

14.37. $E(y) = \beta_0 + \beta_1 x_1 + \beta_2 x_2$, where $x_1 = \begin{cases} 1 & \text{if level 2} \\ 0 & \text{otherwise} \end{cases}$ and $x_2 = \begin{cases} 1 & \text{if level 3} \\ 0 & \text{otherwise} \end{cases}$

14.38. $\hat{y} = 10.2$ for level 1; $\hat{y} = 6.2$ for level 2; $\hat{y} = 22.2$ for level 3; $\hat{y} = 12.2$ for level 4
14.39. $H_0: \beta_1 = \beta_2 = \beta_3 = 0$ against H_a: At least one β is not 0   **14.40.** $E(y) = \beta_0 + \beta_1 x_1 + \beta_2 x_2 + \beta_3 x_3$
14.41. a. $E(y) = \beta_0 + \beta_1 x_1 + \beta_2 x_2 + \beta_3 x_3 + \beta_4 x_4$ **b.** $\hat{y} = 25.10 + 3.85x_1 + 6.22x_2 - 2.65x_3 - 4.75x_4$
c. $F = 23.97$; reject H_0 **d.** $F = 23.97$; reject H_0

14.42. a. Qualitative   **b.** $E(y) = \beta_0 + \beta_1 x_1 + \beta_2 x_2$, where $x_1 = \begin{cases} 1 & \text{if brand } B_2 \\ 0 & \text{otherwise} \end{cases}$ and $x_2 = \begin{cases} 1 & \text{if brand } B_3 \\ 0 & \text{otherwise} \end{cases}$

c. $\beta_0 =$ Mean response, beer B_1; $\beta_1 =$ (Mean response, beer B_2) $-$ (Mean response, beer B_1);   **d.** $\beta_0 + \beta_2$
$\beta_2 =$ (Mean response, beer B_3) $-$ (Mean response, beer B_1)
14.43. a. 60   **b.** 510

14.44. a. Diet   **b.** $E(y) = \beta_0 + \beta_1 x_1 + \beta_2 x_2$, where $x_1 = \begin{cases} 1 & \text{if diet is high in carbohydrates} \\ 0 & \text{otherwise} \end{cases}$

and $x_2 = \begin{cases} 1 & \text{if diet is high in fruits} \\ 0 & \text{otherwise} \end{cases}$

14.45. a. $E(y) = \beta_0 + \beta_1 x_1 + \beta_2 x_2 + \beta_3 x_3$, where $x_1 = \begin{cases} 1 & \text{if location 2} \\ 0 & \text{otherwise} \end{cases}$ $x_2 = \begin{cases} 1 & \text{if location 3} \\ 0 & \text{otherwise} \end{cases}$ $x_3 = \begin{cases} 1 & \text{if location 4} \\ 0 & \text{otherwise} \end{cases}$
b. β_0 **c.** $\beta_0 + \beta_3$ **d.** $\beta_1 - \beta_2$
14.46. $E(y) = \beta_0 + \beta_1 x_1$ **14.47.** Include $\beta_2 x_2 + \beta_3 x_3$.   **14.48.** Include $\beta_4 x_1 x_2 + \beta_5 x_1 x_3$.   **14.49.** If $\beta_4 = \beta_5 = 0$
14.50. If $\beta_2 = \beta_3 = \beta_4 = \beta_5 = 0$ **14.51. a.** $E(y) = \beta_0 + \beta_1 x_1$ **b.** Include $\beta_2 x_2 + \beta_3 x_3$.   **c.** Include $\beta_4 x_1 x_2 + \beta_5 x_1 x_3$.
14.52. a. $E(y) = \beta_0 + \beta_1 x_1 + \beta_2 x_2 + \beta_3 x_3$, where $x_2 = \begin{cases} 1 & \text{if program B} \\ 0 & \text{otherwise} \end{cases}$ and $x_3 = \begin{cases} 1 & \text{if program C} \\ 0 & \text{otherwise} \end{cases}$

b. $H_0: \beta_2 = \beta_3 = 0$ against H_a: At least one β is not 0   **14.53.** $F = 2.60$; do not reject H_0 **14.54. a.** $H_0: \beta_2 = \beta_3 = 0$ **b.** $F = 6.99$; reject H_0

14.55. a. $E(y) = \beta_0 + \beta_1 x_1 + \beta_2 x_2 + \beta_3 x_3$, where $x_1 =$ Ration level, $x_2 = \begin{cases} 1 & \text{if 12°C} \\ 0 & \text{otherwise} \end{cases}$ and $x_3 = \begin{cases} 1 & \text{if 18°C} \\ 0 & \text{otherwise} \end{cases}$

c. $E(y) = \beta_0 + \beta_1 x_1 + \beta_2 x_2 + \beta_3 x_3 + \beta_4 x_1 x_2 + \beta_5 x_1 x_3$, with x_1, x_2, and x_3 defined in part a   **e.** $F = .32$; no
14.56. a. $E(y) = \beta_0 + \beta_1 x_1 + \beta_2 x_2$ **b.** Include $\beta_3 x_1 x_2$.
14.57. a. $F = 4.90$; reject H_0 **b.** $H_0: \beta_4 = \beta_5 = 0$ **c.** $F = 2.36$; no
14.58. a. $F = 25.96$; yes   **b.** $\hat{y} = 10.39$ (10,390 units)   **14.59.** $E(y) = \beta_0 + \beta_1 x_1 + \beta_2 x_1^2$ **14.60.** Add $\beta_3 x_2 + \beta_4 x_3$.
14.61. Add $\beta_5 x_1 x_2 + \beta_6 x_1 x_3 + \beta_7 x_1^2 x_2 + \beta_8 x_1^2 x_3$.   **14.62.** If $\beta_5 = \beta_6 = \beta_7 = \beta_8 = 0$ **14.63.** If $\beta_5 = \beta_6 = \beta_7 = \beta_8 = 0$
14.64. If $\beta_3 = \beta_4 = \beta_5 = \beta_6 = \beta_7 = \beta_8 = 0$ **14.65.** $E(y) = \beta_0 + \beta_1 x_1 + \beta_2 x_1^2 + \beta_3 x_2 + \beta_4 x_3 + \beta_5 x_4$
14.66. a. $H_0: \beta_4 = \beta_5 = 0$ **b.** $H_0: \beta_3 = \beta_4 = \beta_5 = 0$ **14.67.** $F = .93$; no
14.68. a. x_1 is quantitative; x_2 is qualitative   **b.** $E(y) = \beta_0 + \beta_1 x_1 + \beta_2 x_2$ **c.** .09   **d.** $\hat{y} = 8.9$
e. Include $\beta_3 x_1^2 + \beta_4 x_1 x_2 + \beta_5 x_1^2 x_2$.
14.69. $F = 6.71$; yes   **14.71. a.** x_2 ($t = -90$ is the largest)   **b.** Yes

14.72. b. No

c. In the presence of the first 5 variables, SCPT does not contribute significantly; in their absence, SCPT has a higher value of r than others.

d. $F = 58.34$ **e.** $p < .01$

14.73. a. x_4, x_5, x_6 **b.** No, may be other important variables as yet unspecified

c. $E(y) = \beta_0 + \beta_1 x_4 + \beta_2 x_5 + \beta_3 x_6 + \beta_4 x_4 x_5 + \beta_5 x_4 x_6 + \beta_6 x_5 x_6$

14.74. $\hat{y} = .52795 + .001484 x_1$ **14.75.** $F = 10.46$; yes

14.77. a. At least one β is not 0.   **c.** 2, 29   **d.** Reject H_0 if $F > 3.33$.

14.78. a. Second   **b.** First   **c.** Second   **d.** First   **e.** Third   **f.** First   **14.79.** $E(y) = \beta_0 + \beta_1 x_1 + \beta_2 x_2 + \beta_3 x_3$

14.81. a. $E(y) = \beta_0 + \beta_1 x_1 + \beta_2 x_2 + \beta_3 x_3$ **b.** $E(y) = \beta_0 + \beta_1 x_1 + \beta_2 x_1^2 + \beta_3 x_2 + \beta_4 x_3 + \beta_5 x_1 x_2 + \beta_6 x_1 x_3 + \beta_7 x_1^2 x_2 + \beta_8 x_1^2 x_3$

14.83. a. $E(y) = \beta_0 + \beta_1 x_1 + \beta_2 x_2$ **b.** $E(y) = \beta_0 + \beta_1 x_1 + \beta_2 x_1^2 + \beta_3 x_2 + \beta_4 x_2^2 + \beta_5 x_1 x_2$ **14.84.** No estimate of σ^2

14.86. a. Two   **b.** $E(y) = \beta_0 + \beta_1 x_1 + \beta_2 x_2$, where $x_1 = \begin{cases} 1 & \text{if full-service only} \\ 0 & \text{otherwise} \end{cases}$ and $x_2 = \begin{cases} 1 & \text{if both types of service} \\ 0 & \text{otherwise} \end{cases}$

14.87. a. $E(y) = \beta_0 + \beta_1 x_1 + \beta_2 x_2 + \beta_3 x_3$ **b.** Include $\beta_4 x_1 x_2 + \beta_5 x_1 x_3$.

c. $\hat{y} = -10.8343 + .1007 x_1 + 3.4837 x_2 - 4.3625 x_3$

$\hat{y} = -8.2688 + .0884 x_1 - 1.7528 x_2 - 7.5202 x_3 + .0246 x_1 x_2 + .0151 x_1 x_3$

d. $F = .2199$; do not reject H_0 **e.** $H_0: \beta_2 = \beta_3 = \beta_4 = \beta_5 = 0$ **f.** $F = 4.989$; reject H_0 ($\alpha = .05$)

14.89. a. $E(y) = \beta_0 + \beta_1 x + \beta_2 x^2$ **b.** No, need at least four different baking times

14.90. a. $E(y) = \beta_0 + \beta_1 x_1 + \beta_2 x_2$, where $x_1 =$ Total area and $x_2 = \begin{cases} 1 & \text{if central air conditioning is installed} \\ 0 & \text{otherwise} \end{cases}$

b. $E(y) = \beta_0 + \beta_1 x_1 + \beta_2 x_1^2 + \beta_3 x_2 + \beta_4 x_1 x_2 + \beta_5 x_1^2 x_2$ **c.** $H_0: \beta_2 = \beta_4 = \beta_5 = 0$

14.91. a. $F = 2.49$; no   **b.** $F = 209$; reject H_0 **14.92. a.** $H_0: \beta_2 = \beta_5 = 0$ **b.** $H_0: \beta_3 = \beta_4 = \beta_5 = 0$

14.93. $F = 2.15$; no

Index

Cornerstones

OF
Cost Accounting

First Canadian Edition

Don R. Hansen
Oklahoma State University

Maryanne M. Mowen
Oklahoma State University

George A. Gekas
Ryerson University

David J. McConomy
Queen's University

NELSON / EDUCATION

Cornerstones of Cost Accounting, First Canadian Edition
by Don R. Hansen, Maryanne M. Mowen, George A. Gekas, and David J. McConomy

Vice President, Editorial Higher Education:
Anne Williams

Executive Editor:
Amie Plourde

Technical Reviewer:
Ross Meacher

Developmental Editor:
Tammy Scherer

Photo Researcher/Permissons Coordinator:
Sheila Hall

Senior Content Production Manager:
Imoinda Romain

Production Service:
Cenveo Publisher Services

Copy Editor:
Matthew Kudelka

Proofreader:
Lina Suresh

Indexer:
Ruti Mesker

Manufacturing Manager:
Joanne McNeil

Design Director:
Ken Phipps

Managing Designer:
Franca Amore

Cover Concept:
Jennifer Leung

Cover Design:
Johanna Liburd

Cover Image:
Jumper/Getty Images

Compositor:
Cenveo Publisher Services

Printer:
R.R. Donnelley

Library and Archives Canada Cataloguing in Publication

Cornerstones of cost accounting / Don R.
Hansen ... [et al.]. – 1st Canadian ed.

Includes bibliographical references and index.
ISBN 978-0-17-650093-1

1. Cost accounting–Textbooks.
I. Hansen, Don R.

HF5686.C8C665 2012   657'.42
C2012-900458-8

ISBN-13: 978-0-17-650093-1
ISBN-10: 0-17-650093-6

Dedications

To my wife, Jan, and my four children, Kenn, Neil, Ryan, and Makenzie.

—Don Hansen

To my husband, John, and our children, Katherine and Cara.

—Maryanne Mowen

To my parents for their disciplined upbringing, and to my sons for their protective love.

—George Gekas

This book is dedicated to my wife, Candace, without whose support and encouragement I would never have succeeded.

—David McConomy

Brief Contents

Contents

CORNERSTONES—
THE FOUNDATION
FOR SUCCESSFUL LEARNING

Carefully crafted from the ground up, the "Cornerstones" in this text will help you set up and solve fundamental calculations or procedures. And the Cornerstones go beyond simple preparation by focusing on the underlying accounting principle. There is a Cornerstone for every major concept in the book, serving as a "How To" guide for when students are struggling to complete homework assignments. By being able to master the foundations of cost accounting, it will be easier for students to understand how accounting is used for decision making in the business world, making them more marketable to future employers!

Each Cornerstone has five parts: **Information, Why, Required, What If,** and **Solution.** Through this learning system, students understand both the calculations and the conceptual meaning behind them.

The *Information* portion of each Cornerstone provides the necessary data to arrive at a solution.

The *Why* section explains the reason each exercise is important and how it fits into the big picture of cost accounting.

The *Required* section of each example provides students with each step that must be solved.

The *What If* aspect of the example asks them to consider the implications if a variable were to change. This helps them to grasp the true conceptual meaning behind the calculation.

The *Solution* ends each Cornerstone, showing the calculations for each of the required steps in the problem. This helps students understand the necessary concepts.

CORNERSTONE 2-4

The HOW and WHY of Using the Regression Results for Fixed Cost and Variable Rate to Construct and Use a Cost Formula

Information:
Anderson Company had 10 months of data on materials handling cost and number of moves, as shown in Cornerstone 2-3. Regression was run on these data and the coefficients shown by the regression program results are:

Intercept	854.4994
X variable 1	12.39153

Why:
Regression gives the best linear unbiased estimates of the intercept and slope for a set of data points. These can be used to find the fixed cost and variable rate in a cost scenario, and can be used to predict cost for a given amount of the independent variable.

Required:
1. Construct the cost formula for the materials handling activity showing the fixed cost and the variable rate.
2. If Anderson Company estimates that November will have 350 moves, what is the total estimated materials handling cost for that month?
3. **What if** Anderson wants to estimate materials handling cost for the coming year and expects 3,940 moves? What will estimated total materials handling cost be? What is the total fixed materials handling cost? Why doesn't it equal the fixed cost calculated in Requirement 2 above?

Solution:
1. Rounding the regression estimates to the nearest cent, the formula for monthly materials handling cost is:

    Total materials handling cost = $854.50 + ($12.39 × Moves)
2. Materials handling cost = $854.50 + $12.39(350) = $5,191
3. Materials handling cost for the year = 12($854.50) + $12.39(3,940)
    = $10,254 + $48,816.60 = $59,070.60

    The fixed cost for the year is 12 times the fixed cost for the month. Thus, instead of $854.50, the yearly fixed cost is $10,254.

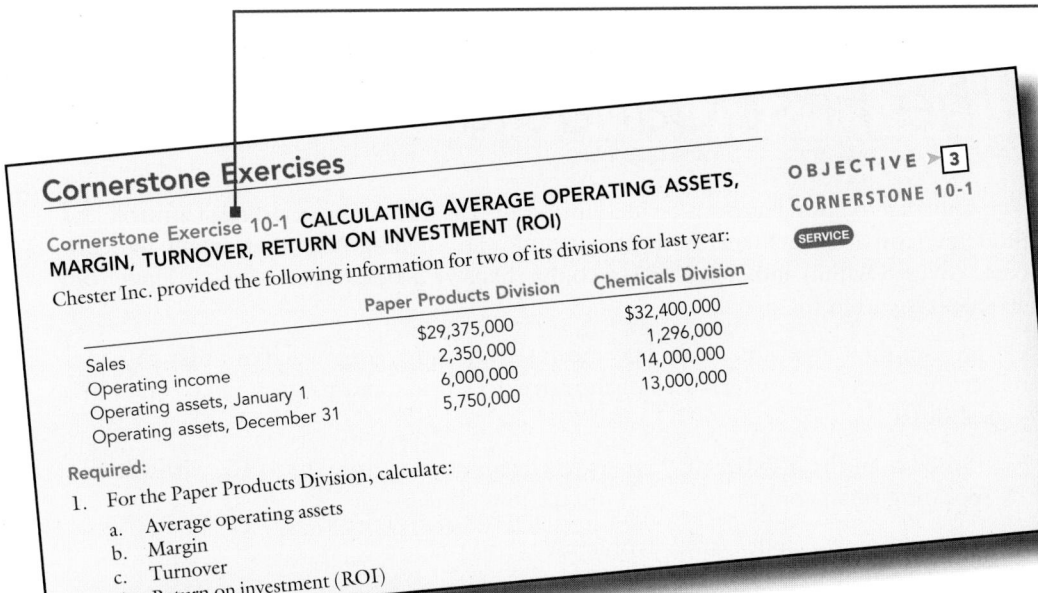

Cornerstone Exercises

Cornerstone Exercise 10-1 CALCULATING AVERAGE OPERATING ASSETS, MARGIN, TURNOVER, RETURN ON INVESTMENT (ROI)

Chester Inc. provided the following information for two of its divisions for last year:

OBJECTIVE ▸ 3
CORNERSTONE 10-1
SERVICE

	Paper Products Division	Chemicals Division
Sales	$29,375,000	$32,400,000
Operating income	2,350,000	1,296,000
Operating assets, January 1	6,000,000	14,000,000
Operating assets, December 31	5,750,000	13,000,000

Required:

1. For the Paper Products Division, calculate:
   a. Average operating assets
   b. Margin
   c. Turnover
   d. Return on investment (ROI)

End-of-Chapter Cornerstone Exercises are linked to specific Cornerstone examples in the text, providing a valuable reference as students try to complete homework on their own. This provides a model for students as they go on to complete more complex problems, helping them become independent learners!

The primary driver of success in accounting is homework. Students believe a textbook helps them succeed, but they are using books differently than in previous generations. Students use books as a source of examples and descriptions to help them complete homework. They may skim the text before or after class, but very few read the text from beginning to end. As a result of this research, *Cornerstones* was fine-tuned to provide greater efficiency and more relevance, promising better results. *Cornerstones* provides students with the confidence to be more independent, allowing them more time to learn additional concepts.

The ***Cornerstones Learning System*** is based on how students learn accounting today. This system incorporates the following key features:

- The actual Cornerstones within the chapters—unique to this family of texts!
- The Cornerstones references within the end-of-chapter Cornerstones Exercises.
- The summary of Cornerstones at the end of each chapter, with page references.
- Online reinforcement of Cornerstones concepts with video podcasts at **http://www.hansen1ce.nelson.com**.

"I like the direct ability to transfer and hence reproduce the successful mastery of a concept … [which] reinforces the material. It is simple, straightforward, and identifiable to and by the student. It would work well in a distance course too."

—Elliott Currie, University of Guelph

Features of the First Canadian Edition

- The Canadian edition has been divided into four parts: Planning, Costing, Control, and Strategic Cost Management. Each part includes an opening page that details the concepts covered within and closes with a comprehensive problem that uses concepts from all chapters within the part.

- Chapters in the Canadian edition have been significantly reorganized in order to better reflect the sequence in which Canadian instructors teach and to group like topics in a logical order.

- For this edition, the number of chapters has been reduced from 20 to 18, with Chapter 18 appearing online only.

- Each chapter now includes CMA problems.

- The number of Cornerstones boxes has been reduced to avoid redundancy or unnecessary material.

- A number of Canadian examples have been included throughout the text. Companies and institutions include Canada Post, Bombardier, RIM, Magna, Toronto Maple Leafs, Air Canada, Porter Airlines, Petro Canada, and the Canadian Institute of Chartered Accountants.

INSTRUCTOR'S RESOURCES

NETA   The Nelson Education Teaching Advantage (NETA) program is designed to deliver research-based resources that promote student engagement and higher-order thinking so as to enable the success of Canadian students and educators.

Testing Advantage Resources   Nelson Education Ltd. understands that the highest quality multiple-choice Test Bank provides the means to measure *higher-level thinking* skills as well as recall. In response to instructor concerns, and recognizing the importance of multiple-choice testing in today's classroom, we have created the assessment component of the Nelson Education Teaching Advantage (NETA) program to ensure the value of our Test Banks.

The assessment component of our NETA program was created in partnership with David DiBattista, a 3M National Teaching Fellow, professor of psychology at Brock University, and researcher in the area of multiple-choice testing. All Test Bank authors have received training by Professor DiBattista in constructing effective multiple-choice questions and creating questions that assess higher-level thinking.

All NETA Test Banks are accompanied by David DiBattista's guide for instructors, *Multiple Choice Tests: Getting Beyond Remembering*. This guide has been designed to help you use Nelson Test Banks to achieve the desired outcomes in your course. Select the "NETA Assessment" button on the Instructor's Resource CD for a digital copy of this valuable resource, as well as the *Cornerstones of Cost Accounting* Test Bank and computerized Test Bank (see "Instructor's Resource CD" below for more information).

Instructor's Resource CD Key instructor ancillaries are provided on the Instructor's Resource CD (ISBN 0-17-663345-6), offering instructors the ultimate tool for customizing lectures and presentations. The IRCD includes the following:

- *Instructor's Manual.* The Instructor's Manual for *Cornerstones of Cost Accounting*, prepared by first Canadian edition co-author George Gekas, includes learning objectives; lesson summaries and teaching strategies; information about exercises, problems, and cases; and more.
- *NETA Test Bank and ExamView® Computerized Test Bank.* Test Bank files include multiple-choice questions and problems and are provided in rich text format for easy editing and printing with all common word-processing formats. All Test Bank questions are also provided in the ExamView® computerized version. This easy-to-use software is compatible with Microsoft Windows and Mac. Create tests by selecting questions from the question bank, modifying these questions as desired, and adding new questions you write yourself. You can administer quizzes online and export tests to WebCT, Blackboard, and other formats. A copy of *Multiple Choice Tests: Getting Beyond Remembering* accompanies the testing materials for *Cornerstones of Cost Accounting*.
- *Microsoft® PowerPoint®.* Key concepts from *Cornerstones of Cost Accounting* are presented in PowerPoint format, with generous use of figures, photographs, and short tables from the text. The PowerPoint slides for the First Canadian Edition were prepared by George Gekas.
- *Solutions Manual.* This manual, prepared by the authors of the text, has been independently checked for accuracy by Ross Meacher, C.A. It contains complete solutions to each discussion question, Cornerstone exercise, problem, and CMA problem in the text. Also included are solutions to the part-ending Integrative Exercises. The Cyber Research Cases and Collaborative Learning Exercises can be found on the companion website at **http://www.hansen1ce.nelson.com**.

- *Spreadsheet Solutions.* The complete solutions to the Excel-based Spreadsheet Exercises are provided.

STUDENT RESOURCES

Companion Website Visit **http://www.hansen1ce.nelson.com** for additional resources, including spreadsheet templates, self-testing review quizzes, glossary terms, flashcards, and the *Cornerstones* video podcasts.

ACKNOWLEDGMENTS AND THANKS

We have received assistance from many people who have contributed to this book. We are grateful for the assistance and support provided by Ryerson University and the Queen's University School of Business. Our appreciation is also extended to the Canadian Institute of Chartered Accountants (CICA), the Institute of Chartered Accountants of Ontario (ICAO), the Certified Management Accountants of Ontario (CMA Ontario), and to the other sources as indicated for their generous permission to use or adapt problems from their publications.

We would like to thank the following reviewers who helped shape the first edition:

Lindsay Brock, Kwantlen Polytechnic University

Gillian Bubb, University of the Fraser Valley

Elliott Currie, University of Guelph

Elin Maher, University of New Brunswick

Andrews Oppong, Dalhousie University

Brad Witt, Humber College

Lior Yitzhaky, Ryerson University

This book was adapted from the U.S. first edition by Don R. Hansen and Maryanne M. Mowen. We appreciate the U.S. authors' willingness to share their work with us. This book has certainly benefited from their experience and contribution.

Many people at Nelson Education also earn our deepest thanks for their thoughtful contributions. Special thanks go to Anne Williams, Vice-President, Editorial—Higher Education; Amie Plourde, Executive Editor; Jenny O'Reilly and Tammy Scherer, Developmental Editors; Imoinda Romain, Senior Content Production Manager; Joanne McNeil, Manufacturing Manager; Margaret Strawbridge, Regional Sales Manager; Tiffany Reid, Sales and Editorial Representative; and Johanna Liburd, Cover Designer. We are also grateful to Matthew Kudelka, Copy Editor; Sheila Hall, Permissions Researcher; and Ross Meacher, C.A., Technical Reviewer.

Finally, special recognition is due to our families for understanding the demands of our work and supporting our efforts.

George Gekas
David J. McConomy
March 2012

About the **Authors**

George A. Gekas is an Associate Professor of Management Accounting at the Ted Rogers School of Business Management at Ryerson University in Toronto. Prior to his appointment to Ryerson, he taught for more than 25 years at several universities in Canada, the United States, and Europe. His academic career includes appointments with the University College of Cape Breton, Algoma University College, Ryerson University, the University of Maryland, the American College of Greece (Deree), and the University of Western Ontario.

Dr. Gekas received his first honours B.A. degree from the Athens University of Economics in Greece. He then earned a Master of Economics from Lakehead University, honours Bachelor of Commerce and MBA degrees from the University of Windsor, and a Ph.D. from Hull University in England. In addition, he is a Certified Management Accountant (CMA), awarded the prestigious FCMA designation, and is actively involved in the Society of Management Accountants of Ontario, for which he has held a number of positions. Currently he is Chair of its Complaints Committee. Recently, he was appointed CMA in residence at the Ted Rogers School of Business Management, the first such appointment in Canada.

Dr. Gekas has published a number of articles on a wide variety of business topics, as well as teaching aids and manuals. His expertise covers the financial management spectrum, with a focus on strategic management, operational efficiency, accountability, and corporate performance.

As a university educator and educational administrator, Dr. Gekas has been involved in a variety of capacities in a number of post-secondary educational projects and is fully immersed in the issues facing Canadian post-secondary institutions and universities. He also has a strong interest in community and social development.

David J. McConomy is an Assistant Professor at the Queen's University School of Business. He graduated with an MBA from Queen's University with a major in Finance and Accounting in 1969 after receiving a BA (Econ) from Loyola College (now part of Concordia University) in Montreal. Since qualifying as a chartered accountant with Arthur Andersen & Co. in Toronto, his 30-year career in business has focused on the finance and accounting requirements of small to medium-sized companies.

He has gained considerable experience assisting young companies, mainly in the technology field, in strategic planning and corporate finance. As Chief Financial Officer of Systemhouse Ltd. (which was later sold to MCI and subsequently to EDS), he was instrumental in taking that company public. During his tenure with Systemhouse, revenues rose from $16.5 million to in excess of $100 million.

With Antares Electronics Inc., he raised $20 million (through venture capital investments, sale-and-leaseback transactions, and the restructuring of bank lines of credit) to enable the company to become one of *Profit* magazine's fastest growing companies, as revenues increased over a three-year period from $36 million to

$100 million. Later, he helped ComnetiX Computer Systems Inc. to become a public company and to raise funds from a variety of sources to fund its growth.

Since joining the faculty of the Queen's School of Business in 2001, he has won numerous Faculty Teaching Awards. He prides himself in engaging with students both within the classroom and outside of it.

Mr. McConomy has taught accounting, financial management, and business strategy (at the undergraduate and MBA levels) on both a full-time and part-time basis at Queen's University, the University of Ottawa, and internationally with the Academy of Economic Studies in Bucharest, Romania, and in Beijing, China.

Don R. Hansen is the Head of the School of Accounting and Kerr McGee Chair at Oklahoma State University. He received his Ph.D. from the University of Arizona in 1977. He has an undergraduate degree in mathematics from Brigham Young University. His research interests include activity-based costing and mathematical modelling. He has published articles in both accounting and engineering journals including *The Accounting Review, The Journal of Management Accounting Research, Accounting Horizons,* and *Accounting Organizations and Society.* He has served on the editorial board of *The Accounting Review.* His outside interests include family, church activities, reading, movies, watching sports, and studying Spanish.

Maryanne M. Mowen is Associate Professor of Accounting at Oklahoma State University. She received her Ph.D. from Arizona State University. She brings an interdisciplinary perspective to teaching and writing in cost and management accounting, with degrees in history and economics. She also teaches classes in ethics and the impact of the Sarbanes-Oxley Act on accountants. Her scholarly research is in the areas of management accounting, behavioural decision theory, and compliance with the Sarbanes-Oxley Act. She has published articles in journals such as *Decision Science, The Journal of Economics and Psychology,* and *The Journal of Management Accounting Research.* Dr. Mowen has served as a consultant to mid-sized and Fortune 100 companies and works with corporate controllers on management accounting issues. Outside the classroom, she enjoys hiking, travelling, reading mysteries, and crossword puzzles.

PLANNING

In any organization, the management process consists of several very important steps including planning, implementing, evaluating, and taking corrective action. Managers are expected to take a forward-looking approach to running the organization and to planning the manner in which actions will be taken to make the organization a success.

The first part of this textbook deals with the elements required in the planning phase. By understanding the basic cost management concepts, we will be able to determine how the actions that managers take will influence the outcomes. By focusing on costs, managers can have a significant impact on the end result.

Fundamental to being able to predict what the outcomes of certain actions will be is the requirement to understand how costs behave. By understanding cost behaviour, a manager can predict what changes will result if one course of action is pursued rather than another.

In most organizations, the object of the exercise is to generate a profit. Companies that have the luxury of pursuing actions that will benefit the greater good are those that are producing a profit. Therefore, understanding the relationship between the activities of an organization and the profits that it can generate will form a significant part of the decision-making process. The cost-volume-profit relationships will help managers understand the probable impact of their actions.

After studying this chapter, you should be able to:

1 Describe a cost management information system, its objectives, and its major subsystems, and indicate how it relates to other operating and information systems.

2 Explain the cost assignment process.

3 Define tangible and intangible products, and explain why there are different product cost definitions.

4 Prepare income statements for manufacturing and service organizations.

5 Understand the importance of ethical behaviour for management accountants.

CHAPTER

1

Basic Cost Management Concepts

Financial Accounting versus Cost Management Accounting: A Systems Framework

A systems framework helps us understand the variety of topics that appear in the field of **cost management**. It also facilitates our ability to understand the differences between financial accounting and cost management. An **accounting information system** consists of interrelated manual and computer parts and uses processes such as collecting, recording, summarizing, analyzing, and managing data to transform inputs into information that is provided to users.

The accounting information system within an organization has two major subsystems: (1) *the financial accounting information system* and (2) *the cost management accounting information system*. One of the major differences between the two systems is the targeted user.

A Systems Framework

A **system** is a set of interrelated parts that performs one or more processes to accomplish specific objectives. Consider a home theatre system. This system has a number of interrelated parts such as the speakers, the receiver, the amplifier, the television, and the DVD player. The most obvious process (or series of actions designed to accomplish an objective) is the playing of a movie; another is the delivery of surround sound throughout the room. The primary objective of the system is to provide a theatre-quality experience while watching a movie. Notice that each part of the system is critical to achieve the overall objective. For example, if the speakers were missing, the amplifier and receiver would not be able to provide theatre-quality sound even if the other parts were present and functional.

A system works by using processes to transform inputs into outputs that satisfy the system's objectives. Consider the movie-playing process. This process requires inputs such as a movie (typically on Blu-ray or DVD), a Blu-ray or DVD player, a television set, and electricity. The inputs are transformed into the replay of the movie, an output of this process. The output of the process, delivery of surround sound, is obviously critical to achieving the overall objective of the system. The encoded sounds on the DVD become the inputs to the delivery process. This process transforms the inputs so that tracks of sound are delivered to each of the speakers throughout the room.

Accounting Information Systems

An information system is designed to provide information to people in the company who might need it. For example, the human resource (HR) information system and the materials requirements planning (MRP) system are both information systems. The HR system tracks people as they are hired. It includes data on date of hire, entry-level title and salary/wages, and any information needed for determining employee benefits. The MRP is a computerized system that keeps track of the purchase and use of raw materials used in manufacturing.

Like any system, an accounting information system has objectives, interrelated parts, processes, and outputs. The overall objective of an accounting information system is to provide information to users. The interrelated parts include order entry and sales, billing, accounts receivable and cash receipts, inventory, general ledger, and cost accounting. Each of these interrelated parts is itself a system and is therefore referred to as a *subsystem* of the accounting information system. Processes include activities such as collecting, classifying, summarizing, and managing data. Some processes may also be formal decision models—models that use inputs and provide recommended decisions as the information output. The outputs are data and reports that provide needed information for users.

Two key features of the accounting information system distinguish it from other information systems. First, an accounting information system's inputs are usually economic events. Second, the operational model of an accounting information system is critically involved with the user of information, since the output of the information system influences users and may serve as the basis for action. This is particularly true for tactical and strategic decisions but less true for day-to-day decisions. In other cases, the output may serve to confirm that the actions taken had the intended effects.[1] Another possible output is feedback, which becomes an input for subsequent

---

[1] This role of information is described in William J. Bruns, Jr., and Sharon M. McKinnon, "Information and Managers: A Field Study," *Journal of Management Accounting Research* 5 (Fall 1993): 86–108. The paper reports on a field study of how managers use accounting information. The authors point out that formal information output does not seem to be used for day-to-day decisions. Managers often use interpersonal relationships to acquire information for daily use. Support for this view can be found in David Marginson, "Information Processing and Management Control: A Note Exploring the Role Played by Information Media in Reducing Role Ambiguity," *Management Accounting Research* 17 (June 2006): 187–197.

Exhibit 1-1

Operational Model of an Accounting Information System

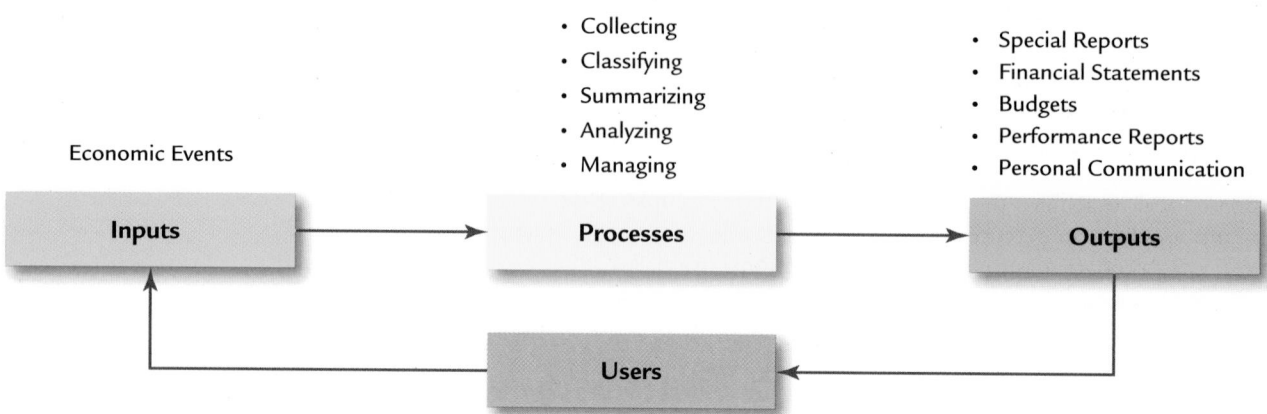

- Collecting
- Classifying
- Summarizing
- Analyzing
- Managing

- Special Reports
- Financial Statements
- Budgets
- Performance Reports
- Personal Communication

Economic Events

Inputs	→	Processes	→	Outputs

Users

operational system performance. The operational model for an accounting information system is illustrated in Exhibit 1-1. Examples of the inputs, processes, and outputs are provided in the exhibit. (The list is not intended to be exhaustive.) Notice that personal communication is an information output. Often, users may not wish to wait for formal reports and can obtain needed information on a timelier basis by communicating directly with accountants.

The accounting information system can be divided into two major subsystems: (1) the *financial accounting information system* and (2) the *cost management information system*. While we emphasize the second, it should be noted that the two systems need not be independent.[2] Ideally, the two systems should be integrated and have linked databases. Output of each of the two systems can be used as input for the other system.

The Financial Accounting Information System

The **financial accounting information system** is primarily concerned with producing outputs for *external* users. It uses well-specified economic events (e.g., payment of wages, purchases of materials) as inputs, and its processes follow certain rules and conventions. For financial accounting, the nature of the inputs and the rules and conventions governing processes, generally accepted accounting principles (GAAP), are defined by the Canadian Institute of Chartered Accountants (CICA), the provincial securities commissions, and the International Accounting Standards Board (IASB). Among its outputs are financial statements such as the balance sheet, income statement, and statement of cash flows for external users (investors, creditors, government agencies, and other outside users). Financial accounting information is used for investment decisions, stewardship evaluation, activity monitoring, and regulatory measures.

The Cost Management Information System

The **cost management information system** is primarily concerned with producing outputs for *internal* users using inputs and processes needed to satisfy management objectives. The cost management information system is not bound by externally imposed criteria that define inputs and processes. Instead, the criteria that govern the inputs and processes are set by people in the company. The cost management information system provides information for three broad objectives:

---

[2] Much of the material from this point on in this section relies on information found in the following articles: Robert S. Kaplan, "The Four-Stage Model of Cost Systems Design," *Management Accounting* (February 1990): 22–26; Steven C. Schnoebelen, "Integrating an Advanced Cost Management System into Operating Systems (Part 1)," *Journal of Cost Management* (Winter 1993): 50–54; and Steven C. Schnoebelen, "Integrating an Advanced Cost Management System into Operating Systems (Part 2)," *Journal of Cost Management* (Spring 1993): 60–67.

1. Costing services, products, and other objects of interest to management
2. Planning and control
3. Decision making

How much does a product or service cost? That depends on the reason why management wants to know the cost. For example, product costs calculated in accordance with GAAP are needed to value inventories for the balance sheet and to calculate the cost of goods sold expense on the income statement. These product costs include the cost of materials, labour, and overhead. In other cases, managers may want to know all costs that are associated with a service for purposes of tactical and strategic profitability analysis. For example, a bank might want to know the costs and revenues associated with providing small business loans. Then additional cost information may be needed concerning service provision, the cost of funds, collection costs, and so on.

Cost information is also needed for planning and control. It should help managers decide what should be done, why it should be done, how it should be done, and how well it is being done. For example, pharmaceutical companies may want to consider life cycle costing of individual drugs or drug families. The expected revenues and costs may cover the entire life of the new product. Thus, projected costs of research, development, testing, production, marketing, distribution, and servicing would be essential information. These costs form the basis of the value chain.

The **value chain** is the set of activities required to design, develop, produce, market, deliver, and provide post-sales service for the products and services sold to customers. Exhibit 1-2 illustrates the business processes of the value chain. Emphasizing customer value forces managers to determine which activities in the value chain are important to customers. The cost management information system should track information about the wide variety of activities that span the value chain. Consider, for example, the delivery segment. Timely delivery of a product or service is part of the total product and, thus, is of value to the customer. Customer value can be increased by increasing the speed of delivery and response. **Federal Express** exploited this part of the value chain and successfully developed a service that was not being offered by **Canada Post**. Today, many customers believe that delivery delayed is delivery denied. This indicates that a good cost management information system ought to develop and measure indicators of customer satisfaction.

Exhibit 1-2

The Value Chain

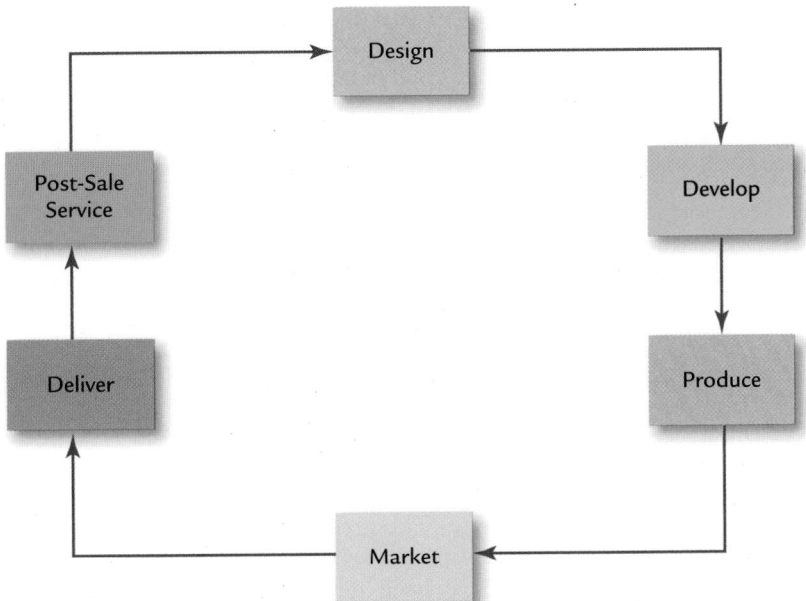

Companies have internal customers as well. For example, the procurement process acquires and delivers parts and materials to producing departments. Providing high-quality parts on a timely basis to managers of producing departments is just as vital for procurement as it is for the company as a whole to provide high-quality goods to external customers. The emphasis on managing the internal value chain and servicing internal customers has revealed the importance of a cross-functional perspective. Internal and external value chains will be discussed in more detail in Chapter 13.

Finally, cost information is important for many managerial decisions. For example, a manager may need to decide whether to continue making a component in-house or to buy it from an external supplier. In this case, the manager would need to know the cost of materials, labour, and other productive resources associated with the manufacture of the component and which of these costs would disappear if the product were no longer produced. Also needed is information about the cost of purchasing the component, including any increase in cost for internal activities such as receiving and storing goods.

Relationship of Cost Accounting to Other Operational Systems and Functions

The cost information produced by the cost management information system benefits the whole organization and should have an organization-wide perspective. Managers in many different areas of a business require cost information. For example, an engineering manager must make strategic decisions concerning product design. Later costs of production, marketing, and servicing can vary widely, depending on the design. An engineer at **Hewlett-Packard** once told us that 70 percent of eventual product costs are "locked in" during the design process. To provide accurate cost information for the different design options, the cost management system must interact not only with the design and development system but also with the production, marketing, and customer service systems. Cost information for tactical decision making is also important. For example, a sales manager needs reliable and accurate cost information when faced with a decision concerning an order that may be sold for less than the normal selling price. Such a sale may only be feasible if the production system has idle capacity. In this case, a sound decision requires interaction among the cost management system, the marketing and distribution system, and the production system. These two examples illustrate that the cost management system should have an organization-wide perspective and that it must be properly integrated with the nonfinancial functions and systems within an organization.

An integrated cost management system receives information from and provides information to all operational systems. To the extent possible, the cost management system should be integrated with the organization's operational systems. Integration reduces redundant storage and use of data, improves the timeliness of information, and increases the efficiency of producing reliable and accurate information. One way of accomplishing this is to implement an enterprise resource planning (ERP) system. ERP systems strive to input data once and make it available to people across the company for whatever purpose it may serve. For example, a sales order entered into an ERP system is used by marketing to update customer records, by production to schedule the manufacture of the goods ordered, and by accounting to record the sale.

Factors Affecting Cost Management

Worldwide competitive pressures, deregulation, growth in the service industry, and advances in information and manufacturing technology have changed the nature of our economy and caused many manufacturing and service industries to dramatically change the way in which they operate. These changes, in turn, have prompted the development of innovative and relevant cost management practices. For example, activity-based accounting systems have been developed and implemented in many organizations. Additionally, the focus of cost management accounting systems has been broadened to enable managers to better serve the needs of customers and manage the firm's business processes that are used to create customer value. A firm can establish a competitive advantage by providing more customer value for less cost than its competitors. To secure and maintain a competitive advantage, managers seek to improve time-based performance, quality, and efficiency. Accounting information must be produced to support these three fundamental organizational goals.

Global Competition

Vastly improved transportation and communication systems have led to a global market for many manufacturing and service firms. Several decades ago, firms neither knew nor cared what similar firms in Japan, France, Germany, and Singapore were producing. These foreign firms were not competitors since their markets were separated by geographical distance. Now, both small and large firms are affected by the opportunities offered by global competition. **Stillwater Designs**, a small firm that designs and markets Kicker speakers, has significant markets in Europe. The manufacture of the Kicker speakers is mostly outsourced to Asian producers. At the other end of the size scale, **Bombardier**, **Research In Motion**, and **Magna International** are developing sizable markets in China. Automobiles manufactured in Japan can be in North America in two weeks. Investment bankers and management consultants can communicate with foreign offices instantly. Improved transportation and communication in conjunction with higher quality products that carry lower prices have upped the ante for all firms. This new competitive environment has increased the demand not only for more cost information but also for more accurate cost information. Cost information plays a vital role in reducing costs, improving productivity, and assessing product line profitability.

Growth of the Service Industry

As traditional industries have declined in importance, the service sector of the economy has increased in importance. The service sector now comprises approximately three quarters of many economies and their workers. Many services—among them accounting services, transportation, and medical services—are exported. Experts predict that this sector will continue to expand in size and importance as service productivity grows. Deregulation of many services (e.g., airlines and telecommunications in the past and utilities in the present) has increased competition in the service industry. Many service organizations are scrambling to survive. The increased competition has made managers in this industry more conscious of the need to have accurate cost information for planning, controlling, continuous improvement, and decision making. Thus, the changes in the service sector add to the demand for innovative and relevant cost management information.

Advances in Information Technology and the Manufacturing Environment

Three significant advances relate to information technology. One is intimately connected with computer-integrated applications. With automated manufacturing, computers are used to monitor and control operations. Because a computer is being used, a considerable amount of useful information can be collected, and managers can be informed about what is happening within an organization almost as it happens. It is now possible to track products continuously as they move through the factory and to report (on a real-time basis) such information as units produced, material used, scrap generated, and product cost. The outcome is an operational information system that fully integrates manufacturing with marketing and accounting data.

Manufacturing management approaches such as the theory of constraints and just-in-time have allowed firms to increase quality, reduce inventories, eliminate waste, and reduce costs. Automated manufacturing has produced similar outcomes.

The impact of improved manufacturing technology and practices on cost management is significant. Product costing systems, control systems, allocation, inventory management, cost structure, capital budgeting, variable costing, and many other accounting practices are being affected.

Theory of Constraints

The **theory of constraints** is a method used to continuously improve manufacturing and nonmanufacturing activities. It is characterized as a "thinking process" that begins by recognizing that all resources are finite. Some resources, however, are more critical than others. The most critical limiting factor, called a constraint, becomes the focus of attention. By managing this constraint, performance can be improved. To manage the constraint, one must identify and exploit it (i.e., performance must be maximized subject to the constraint). All other actions are subordinate to the exploitation decision. Finally, to improve performance, the constraint must be elevated. The process is repeated until the constraint is eliminated (i.e., it is no longer the critical performance-limiting factor). The process then begins anew with the resource that has now become the critical limiting factor. Using this method, lead times and, thus, inventories can be reduced.

Just-in-Time Manufacturing

A demand-pull system, **just-in-time (JIT) manufacturing** strives to produce a product only when it is needed and only in the quantities demanded by customers. Demand, measured by customer orders, pulls products through the manufacturing process. Each operation produces only what is necessary to satisfy the demand of the succeeding operation. No production takes place until a signal from a succeeding process indicates the need to produce. Parts and materials arrive just in time to be used in production.

JIT manufacturing typically reduces inventories to much lower levels (theoretically to insignificant levels) than those found in conventional systems, increases the emphasis on quality control, and produces fundamental changes in the way production is organized and carried out. Basically, JIT manufacturing focuses on continual improvement by reducing inventory costs and dealing with other economic problems. Reducing inventories frees up capital that can be used for more productive investments. Increasing quality enhances the competitive ability of the firm. Finally, changing from a traditional manufacturing setup to JIT manufacturing allows the firm to focus more on quality and productivity and, at the same time, allows a more accurate assessment of what it costs to produce products.

Computer-Integrated Manufacturing

Automation of the manufacturing environment allows firms to reduce inventory, increase productive capacity, improve quality and service, decrease processing time, and increase output. Automation can produce a competitive advantage for a firm. The implementation of an automated manufacturing facility typically follows JIT and is a response to the increased needs for quality and shorter response times. As more firms automate, competitive pressures will force other firms to do likewise. For many manufacturing firms, automation may be equivalent to survival.

If automation is justified, it may mean installation of a computer-integrated manufacturing (CIM) system. CIM implies the following capabilities: (1) the products are designed through the use of a computer-assisted design (CAD) system; (2) a computer-assisted engineering (CAE) system is used to test the design; (3) the product is manufactured using a computer-assisted manufacturing (CAM) system (CAMs use computer-controlled machines and robots); and (4) an information system connects the various automated components.

A particular type of CAM is the flexible manufacturing system. Flexible manufacturing systems are capable of producing a family of products from start to finish using robots and other automated equipment under the control of a mainframe computer. This ability to produce a variety of products with the same set of equipment is clearly advantageous.

Cost Assignment: Direct Tracing, Driver Tracing, and Cost Allocation

To study cost accounting and operational control systems, we need to understand the meaning of cost and to become familiar with the cost terminology associated with the two systems. We must also understand the process used to assign costs. Cost assignment is one of the key processes of the cost accounting system. Improving the cost assignment process has been one of the major developments in the cost management field in the past 20 to 30 years. First, let's define cost.

Cost is the cash or cash equivalent value sacrificed for goods and services that are expected to bring a current or future benefit to the organization. We say *cash equivalent* because noncash assets can be exchanged for the desired goods or services. For example, it may be possible to trade equipment for materials used in production.

Costs are incurred to produce future benefits. In a profit-making firm, future benefits usually mean revenues. As costs are used up in the production of revenues, they are said to expire. Expired costs are called **expenses**. In each period, expenses are deducted from revenues on the income statement to determine the period's profit. A **loss** is a cost that expires without producing any revenue benefit. For example, the cost of uninsured inventory destroyed by a flood would be classified as a loss on the income statement.

Many costs do not expire in a given period. These unexpired costs are classified as **assets** and appear on the balance sheet. Computers and factory buildings are examples of assets lasting more than one period. Note that the main difference between a cost being classified as an expense or as an asset is timing. This distinction is important and will be referred to in the development of other cost concepts later in the text.

Cost Objects

Cost accounting information systems are structured to measure and assign costs to cost objects. **Cost objects** can be anything for which costs are measured and assigned; they may include products, customers, departments, projects, activities, and so on. For example, if we want to determine what it costs to produce a bicycle, then the cost object is the bicycle. If we want to determine the cost of operating a maintenance department within a plant, then the cost object is the maintenance department. If we want to determine the cost of developing a new toy, then the cost object is the new toy development project. Activities are a special kind of cost object. An **activity** is a basic unit of work performed within an organization. An activity can also be defined as an aggregation of actions within an organization useful to managers for purposes of planning, controlling, and decision making. In recent years, activities have emerged as important cost objects. Activities play a prominent role in assigning costs to other cost objects and are essential elements of an activity-based cost accounting system. Examples of activities include setting up equipment for production, moving materials and goods, purchasing parts, billing customers, paying bills, maintaining equipment, expediting orders, designing products, and inspecting products. Notice that an activity is described by an action verb (e.g., paying or designing) and an object that receives the action (e.g., bills or products).

Cost Accumulation

Costs are incurred by an organization for a variety of purposes. Once a cost object has been identified, costs are accumulated by that object to allow us to determine its cost. The accumulation by cost object can be on the basis of individual jobs, on the basis of the processes of a particular department, on activity that is being performed, or by a responsibility centre within the company. The accumulation of costs will be based on the nature of the cost object and the nature of the company itself.

Accuracy of Cost Assignments

Assigning costs *accurately* to cost objects is crucial. Accuracy is not evaluated based on knowledge of some underlying "true" cost. Rather, it is a relative concept and has to do with the reasonableness and logic of the cost assignment methods that are being used. The objective is to measure and assign as accurately as possible the cost of the resources used by a cost object. Some cost assignment methods are clearly more accurate than others. For example, suppose you want to determine the cost of lunch for Elaine Day, a student who frequents Hideaway, an off-campus pizza parlour. One cost assignment approach is to count the number of customers at the Hideaway between 12:00 P.M. and 1:00 P.M. and then divide that into the total sales receipts earned by Hideaway during this period. Suppose that this comes to $6.25 per lunchtime customer. Based on this approach we would conclude that Elaine spends $6.25 per day for lunch. Another approach is to go with Elaine and observe how much she spends. Suppose that she has a chef's salad and a medium drink each day, costing $4.50. It is easy to see which cost assignment is more accurate. The $6.25 cost assignment is distorted by the consumption patterns of other customers (cost objects). As it turns out, most lunchtime clients order the luncheon special for $5.95 (a mini-pizza, small salad, and medium drink).

Distorted cost assignments can produce poor decisions. For example, if a plant manager is trying to decide whether to continue producing power internally or to buy it from a local utility company, then an accurate assessment of how much it is costing to produce the power internally is fundamental to the analysis. If the cost of internal power production is overstated, the manager might decide to shut down the internal power department in favour of buying power from an outside company, whereas a more accurate cost assignment might suggest the opposite. It is easy to see that poor cost assignments can prove to be costly.

Traceability   Understanding the relationship of costs to cost objects can increase the accuracy of cost assignments. Costs are directly or indirectly associated with cost objects. **Indirect costs** are costs that cannot be traced easily and accurately to a cost object. **Direct costs** are those costs that can be traced easily and accurately to a cost object.[3] For costs to be traced easily means that the costs can be assigned in an economically feasible way. For costs to be traced accurately means that the costs are assigned using a *causal relationship*. Thus, **traceability** is the ability to assign a cost directly to a cost object in an economically feasible way by means of a causal relationship. The more costs that can be traced to the object, the greater the accuracy of the cost assignments. One additional point needs to be emphasized. Cost management systems typically deal with many cost objects. Thus, it is possible for a particular cost item to be classified as both a direct cost and an indirect cost. It all depends on *which* cost object is the point of reference. For example, if the plant is the cost object, then the cost of heating and cooling the plant is a direct cost; however, if the cost objects are products produced in the plant, then this utility cost is an indirect cost.

Methods of Tracing   Traceability means that costs can be assigned easily and accurately, using a causal relationship. Tracing costs to cost objects can occur in one of two ways: (1) *direct tracing* and (2) *driver tracing*. **Direct tracing** is the process of identifying and assigning costs to a cost object that are specifically or physically associated with the cost object. Direct tracing is most often accomplished by *physical observation*. For example, assume that the power department is the cost object. The salary of the power department's supervisor and the fuel used to produce power are examples of costs that can be specifically identified (by physical observation) with the cost object (the power department). As a second example, consider a pair of blue jeans. The materials (denim, zipper, buttons, and thread) and labour (to cut the denim according to the pattern and sew the pieces together) are physically observable;

[3] This definition of direct costs is based on the glossary prepared by Computer Aided Manufacturing International Inc. (CAM-I). See Norm Raffish and Peter B. B. Turney, "Glossary of Activity-Based Management," *Journal of Cost Management* (Fall 1991): 53–63. Other terms defined in this chapter and in the text also follow the CAM-I glossary.

therefore, the costs of materials and labour can be directly charged to a pair of jeans. Ideally, all costs should be charged to cost objects using direct tracing.

Unfortunately, it is often impossible to physically observe the exact amount of resources being used by a cost object. The next best approach is to use cause-and-effect reasoning to identify factors—called *drivers*—that can be observed and that measure a cost object's resource consumption. **Drivers** are factors that *cause* changes in resource usage, activity usage, costs, and revenues. **Driver tracing** is the use of *drivers* to assign costs to cost objects. Although less precise than direct tracing, driver tracing can be accurate if the cause-and-effect relationship is sound. Consider the cost of electricity for the jeans manufacturing plant. The factory manager might want to know how much electricity is used to run the sewing machines. Physically observing how much electricity is used would require a meter to measure the power consumption of the sewing machines, which may not be practical. Thus, a driver such as "machine hours" could be used to assign the cost of electricity. If the electrical cost per machine hour is $0.10 and the sewing machines use 200,000 machine hours in a year, then $20,000 of the electricity cost ($0.10 × 200,000) would be assigned to the sewing activity. The use of drivers to assign costs to activities will be explained in more detail in Chapter 6.

Assigning Indirect Costs   Indirect costs cannot be traced to cost objects. Either there is no causal relationship between the cost and the cost object, or tracing is not economically feasible. Assignment of indirect costs to cost objects is called **allocation**. Since no causal relationship exists, allocating indirect costs is based on *convenience* or some *assumed* linkage. For example, consider the cost of heating and lighting a plant that manufactures five products. Suppose that this utility cost is to be assigned to the five products. Clearly, it is difficult to see any causal relationship. A convenient way to allocate this cost is simply to assign it in proportion to the direct labour hours used by each product. Arbitrarily allocating indirect costs to cost objects reduces the overall accuracy of the cost assignments. Accordingly, the best costing policy may be that of assigning only traceable direct costs to cost objects. However, it must be admitted that allocations of indirect costs may serve other purposes besides accuracy. For example, allocating indirect costs to products may be required for external reporting. Nonetheless, most managerial uses of cost assignments are better served by accuracy. At the very least, direct and indirect cost assignments should be reported separately.

Cost Assignment Summarized   There are three methods of assigning costs to cost objects: direct tracing, driver tracing, and allocation. Of the three methods, direct tracing is the most precise since it relies on physically observable causal relationships. Driver tracing relies on causal factors called drivers to assign costs to cost objects. The precision of driver tracing depends on the strength of the causal relationship described by the driver. Identifying drivers and assessing the quality of the causal relationship is more costly than either direct tracing or allocation. Allocation, while the simplest and least expensive method, is the least accurate cost assignment method; its use should be avoided where possible. In many cases, the benefits of increased accuracy by driver tracing outweigh its additional measurement cost. This cost–benefit issue is discussed more fully later in the chapter. The process really entails choosing among competing cost management systems.

Product and Service Costs

OBJECTIVE ▶ 3
Define tangible and intangible products, and explain why there are different product cost definitions.

One of the most important cost objects is the output of organizations. The two types of output are tangible products and services. **Tangible products** are goods produced by converting raw materials into finished products through the use of labour and capital inputs such as plant, land, and machinery. Televisions, hamburgers, automobiles, computers, clothes, and furniture are examples of tangible products. **Services** are tasks or activities performed for a customer or an activity performed by a customer using an organization's products or facilities. Services are also produced using materials, labour, and capital inputs. Insurance coverage, medical care, dental care, funeral care, and accounting are examples of service activities performed for customers. Car rental, video rental, and skiing are examples of services where the customer uses an organization's products or facilities.

Services differ from tangible products on three important dimensions: intangibility, perishability, and inseparability. **Intangibility** means that buyers of services cannot see, feel, hear, or taste a service before it is bought. Thus, services are *intangible products*. **Perishability** means that services cannot be stored (there are a few unusual cases where tangible goods cannot be stored). Finally, **inseparability** means that producers of services and buyers of services must usually be in direct contact for an exchange to take place. In effect, services are often inseparable from their producers. For example, an eye examination requires both the patient and the optometrist to be present. However, producers of tangible products need not have direct contact with the buyers of their goods. Buyers of automobiles, for instance, never need to have contact with the engineers and assembly line workers who produce automobiles.

Organizations that produce tangible products are called *manufacturing* organizations. Those that produce intangible products are called *service* organizations. Managers of organizations that produce goods or services need to know how much individual products cost for a number of reasons, including profitability analysis and strategic decisions concerning product design, pricing, and product mix.

For example, **McDonald's Corporation** needed to know the cost of individual products to determine whether to keep them on the Dollar Menu. The double cheeseburger, a very popular item, rose in cost to well over $1. Many franchisees refused to sell the item as part of the Dollar Menu—some charging over $2 for it.[4] In late 2008, McDonald's exchanged the double cheeseburger for a double hamburger with just one slice of cheese. This change made it possible for its franchisees to continue to offer a well rounded Dollar Menu without taking the guaranteed loss incurred every time a double cheeseburger was sold at less than cost.

Service companies also relate cost to profit.

A number of professional sports teams, including the **Toronto Maple Leafs**, have gone the extra mile to keep season ticket holders happy. Realizing that there are numerous competing entertainment options, they have hired hospitality specialists and concierges to offer more services to season ticket holders. These additional services may include special tours of the locker room or chances to speak to upper management about their concerns. While the additional services are not cheap, they are an important part of maintaining consistent revenue even in the face of a disappointing win-loss record.[5]

Given the importance of cost to both manufacturing and service firms, when we discuss product costs, we are referring to both intangible and tangible products.

Product Costs and External Financial Reporting

An important objective of a cost management system is the calculation of product costs for external financial reporting. Externally imposed conventions require costs to be classified in terms of the special purposes, or functions, they serve. Costs are subdivided into two major functional categories: production and nonproduction. **Production (or product) costs** are those costs associated with manufacturing goods or providing services. **Nonproduction costs** are those costs associated with the functions of selling and administration. For tangible goods, production and nonproduction costs are often referred to as *manufacturing costs* and *nonmanufacturing costs*, respectively. Production costs can be further classified as *direct materials*, *direct*

---

[4] Richard Gibson, "Franchisees Balk at Dollar Menu," *The Wall Street Journal* (November 14, 2007): B3f.

[5] Adam Thompson, "The Nosebleed VIPs," *The Wall Street Journal* (March 19, 2007): B1 (adapted).

labour, and *overhead*. Only these three cost elements can be assigned to products for external financial reporting.

Direct Materials

Direct Materials   **Direct materials** are those materials traceable to the good or service being produced. The cost of these materials can be directly charged to products because physical observation can be used to measure the quantity used by each product. Materials that become part of a tangible product or those materials that are used in providing a service are usually classified as direct materials. For example, steel in an automobile, wood in furniture, alcohol in cologne, denim in jeans, braces for correcting teeth, surgical gauze and anesthesia for an operation, ribbon in a corsage, and soft drinks on an airline are all direct materials.

Direct Labour

Direct Labour   **Direct labour** is labour that is traceable to the goods or services being produced. As with direct materials, physical observation is used to measure the quantity of labour used to produce a product or service. Employees who convert raw materials into a product or who provide a service to customers are classified as direct labour. Workers on an assembly line at Research In Motion, a chef in a restaurant, a surgical nurse for an open-heart operation, and a pilot for **Air Canada** are examples of direct labour.

Overhead

Overhead   All production costs other than direct materials and direct labour are lumped into one category called **overhead**. In a manufacturing firm, overhead is also known as *factory burden* or *manufacturing overhead*. The overhead cost category contains a wide variety of items. Many inputs other than direct labour and direct materials are needed to produce products. Examples include depreciation on buildings and equipment, maintenance, supplies, supervision, materials handling, power, property taxes, landscaping of factory grounds, and plant security. **Supplies** are generally those materials necessary for production that do not become part of the finished product or are not used in providing a service. Dishwasher detergent in a fast-food restaurant and oil for production equipment are examples of supplies.

Direct materials that form an insignificant part of the final product are usually lumped into the overhead category called **indirect materials**. This treatment is justified on the basis of cost and convenience. The cost of the tracing is greater than the benefit of increased accuracy. The glue used in making furniture or toys is an example.

The cost of overtime for direct labour is usually assigned to overhead as well. The rationale is that typically no particular production run caused the overtime. Accordingly, overtime cost is common to all production runs and is therefore an indirect manufacturing cost. Note that *only* the overtime cost itself is treated this way. If workers are paid $16 per hour regular rate and a premium of $8 per overtime hour, then only the $8 overtime premium is assigned to overhead. The $16 regular rate is still regarded as a direct labour cost. In certain cases, however, overtime is associated with a particular production run, such as a special order taken when production is at 100 percent capacity. In these special cases, it is appropriate to treat overtime premiums as a direct labour cost.

Prime and Conversion Costs

Prime and Conversion Costs   The manufacturing and nonmanufacturing classifications give rise to some related cost concepts. The functional distinction between manufacturing and nonmanufacturing costs is the basis for the concepts of inventoriable costs and noninventoriable costs—at least for purposes of external reporting. Combinations of different production costs also produce the concepts of prime costs and conversion costs.

Prime cost is the sum of direct materials cost and direct labour cost. **Conversion cost** is the sum of direct labour cost and overhead cost. For a manufacturing firm, conversion cost can be interpreted as the cost of converting raw materials into a final product. Cornerstone 1-1 shows how and why to calculate prime cost, conversion cost, and product cost.

**CORNERSTONE
1-1**

The HOW and WHY of Calculating Prime Cost, Conversion Cost, Variable Product Cost, and Total Product Cost

Information:

Carreker Company manufactures cell phones. For next year, Carreker predicts that 30,000 units will be produced, with the following total costs:

Direct materials	$150,000
Direct labour	90,000
Variable overhead	30,000
Fixed overhead	450,000

Why:

Product costs are basic to management control and decision making. Managers use these costs for budgeting to check the impact of an increase or a decrease in unit sales on operating income. Since fixed costs stay the same when units change, knowledge of prime cost, conversion cost, variable product cost, and overall product cost give important information, allowing analysis of costs at differing levels of production.

Required:

1. Calculate the prime cost per unit.
2. Calculate the conversion cost per unit.
3. Calculate the total variable product cost per unit.
4. Calculate the total product (manufacturing) cost per unit.
5. **What if** 32,000 cell phones could be manufactured next year? Explain in words how that would affect the unit prime cost, the unit conversion cost, the unit variable product cost, and the unit total product cost.

Solution:

1. Unit prime cost = (Direct materials + Direct labour)/Number of units
$$= (\$150,000 + \$90,000)/30,000$$
$$= \$8$$

2. Unit conversion cost = (Direct labour + Overhead)/Number of units
$$= (\$90,000 + \$30,000 + \$450,000)/30,000$$
$$= \$19$$

3. Unit variable product cost = (Direct materials + Direct labour
$$+ \text{ Variable overhead})/\text{Number of units}$$
$$= (\$150,000 + \$90,000 + \$30,000)/30,000$$
$$= \$9$$

4. Unit product cost = (Direct materials + Direct labour +
$$\text{Variable overhead} + \text{Fixed overhead})/\text{Number of units}$$
$$= (\$150,000 + \$90,000 + \$30,000 + \$450,000)/30,000$$
$$= \$24$$

5. If the number of units produced increases, there will be no impact on any unit variable cost. Thus, unit prime cost and unit variable cost would stay the same. However, unit conversion cost and unit product cost would go down due to the presence of fixed factory overhead. Fixed overhead will remain the same in total, but decrease per unit as the number of units goes up. Conversely, if the number of units goes down, unit fixed overhead will increase.

Nonproduction Costs Nonproduction costs are divided into two categories: marketing (selling) costs and administrative costs. Marketing and administrative costs are not inventoried and are called *period* costs. **Period costs** are expensed in the period in which they are incurred. Thus, period costs are not inventoried and are not assigned to products. Period costs appear on the income statement—not the balance sheet. In a manufacturing organization, the level of these costs can be significant (often greater than 25 percent of sales revenue), and controlling them may bring greater cost savings than the same control exercised in the area of production costs.

> **Procter & Gamble** spends enormous amounts on advertising in order to develop and dominate the market for shampoo and detergent in China. P&G buys more air time each month than even the most media-conscious Chinese companies spend in a year. Couple that with the cost of free samples and salaries for the thousands of Chinese who distribute them, we see that marketing expense in China is a significant portion of P&G's budget.[6]

For service organizations, the relative importance of selling and administrative costs depends on the nature of the service being produced. Physicians and dentists, for example, generally do very little marketing and thus have very low selling costs. An airline, on the other hand, may incur substantial marketing costs.

Those costs necessary to market and distribute a product or service are **marketing (selling) costs**. They are often referred to as *order-getting* and *order-filling* costs. Examples of marketing costs include the following: salaries and commissions of sales personnel, advertising, warehousing, shipping, and customer service. The first two items are examples of order-getting costs; the last three are order-filling costs.

All costs that cannot be reasonably assigned to either marketing or production are **administrative costs**. Administration is responsible for ensuring that the various activities of the organization are properly integrated in accordance with the overall mission of the firm. The president of the firm, for example, is concerned with the efficiency of *both* marketing and production as they carry out their respective roles. Proper integration of these two functions is essential for maximizing the overall profits of a firm. Examples of administrative costs are top-executive salaries, legal fees, printing and distributing the annual report, and general accounting. Research and development is also part of administrative costs, and is usually expensed in the period incurred. Exhibit 1-3 illustrates the various types of production and nonproduction costs.

Exhibit 1-3

Production and Nonproduction Costs

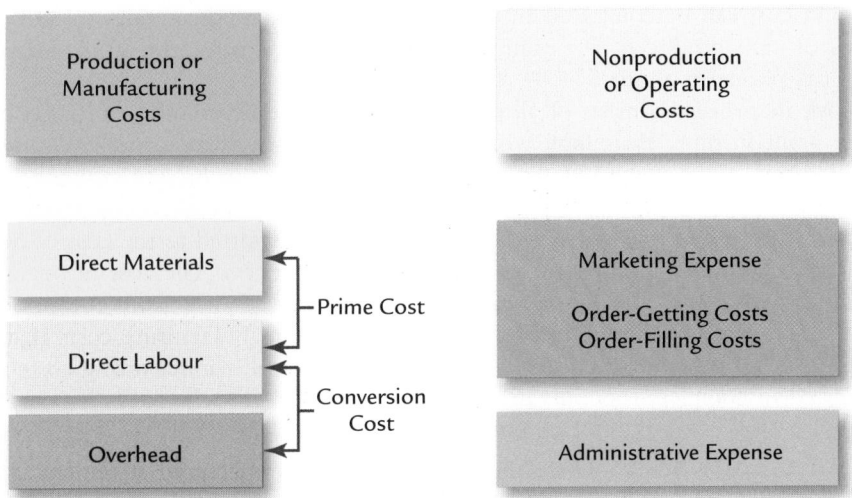

[6] Joseph Kahn, "P&G Viewed China as a National Market and Is Conquering It," *The Wall Street Journal* (September 12, 1995): A1, A6.

Financial Statements

The functional classification is the cost classification required for external reporting. In preparing an income statement, production and nonproduction costs are separated. The reason for the separation is that production costs are product costs—costs that are inventoried until the units are sold—and the nonproduction costs of marketing and administration are viewed as period costs. Thus, production costs attached to the units sold are recognized as an expense (cost of goods sold) on the income statement. Production costs attached to units that are not sold are reported as inventory on the balance sheet. Marketing and administrative expenses are viewed as costs of the period and must be deducted each and every period as expenses on the income statement. Nonproduction costs never appear on the balance sheet.

Income Statement: Manufacturing Firm

The income statement prepared for external parties follows the standard format taught in an introductory financial accounting course. This income statement is frequently referred to as **absorption-costing income** or **full-costing income** because *all* manufacturing costs (direct materials, direct labour, and overhead) are fully assigned to the product.

Under the absorption-costing approach, expenses are separated according to function and then deducted from revenues to arrive at operating income. The two major functional categories of expense are cost of goods sold and operating expenses. These categories correspond to a firm's manufacturing and nonmanufacturing (marketing and administrative) expenses. **Cost of goods sold** is the cost of direct materials, direct labour, and overhead attached to the units sold. To compute the cost of goods sold, it is first necessary to determine the cost of goods manufactured.

Cost of Goods Manufactured   The **cost of goods manufactured** represents the total manufacturing cost of goods completed during the current period. The only costs assigned to goods completed are the manufacturing costs of direct materials, direct labour, and overhead. The details of this cost assignment are given in a supporting schedule, called the *statement of cost of goods manufactured*. Cornerstone 1-2 shows how to create the statement of cost of goods manufactured.

Notice in Cornerstone 1-2 that the *total manufacturing costs* of the period are added to the manufacturing costs found in beginning work in process. The costs found in ending work in process are then subtracted to arrive at the cost of goods manufactured. If the cost of goods manufactured is for a single product, then the average unit cost can be computed by dividing the cost of goods manufactured by the number of units produced. For example, for Carreker Company, the average cost per unit of cell phones is about $24.64 ($739,300/30,000).

Work in process consists of all partially completed units found in production at a given point in time. Beginning work in process consists of the partially completed units on hand at the beginning of a period. Ending work in process consists of the incomplete units on hand at the period's end. In the statement of cost of goods manufactured, the cost of these partially completed units is reported as the cost of beginning work in process and the cost of ending work in process. The cost of beginning work in process represents the manufacturing costs carried over from the prior period; the cost of ending work in process represents the manufacturing costs that will be carried over to the next period. In both cases, additional manufacturing costs must be incurred to complete the units in work in process.

Cost of Goods Sold   Once the cost of goods manufactured statement is prepared, the cost of goods sold can be computed. The cost of goods sold is the manufacturing cost of the units that were sold during the period. It is important to remember that the cost of goods sold may or may not equal the cost of goods manufactured. In addition, we must remember that the cost of goods sold is an expense,

The HOW and WHY of Preparing the Statement of Cost of Goods Manufactured

CORNERSTONE
1-2

Information:

Carreker Company manufactures cell phones. For next year, Carreker predicts that 30,000 units will be produced, with the following total costs:

Direct materials	$154,300
Direct labour	90,000
Variable overhead	30,000
Fixed overhead	450,000

Carreker expects to purchase $147,900 of direct materials next year. Projected beginning and ending inventories for direct materials and work in process are as follows:

	Direct Materials Inventory	Work-in-Process Inventory
Beginning	$53,400	$75,000
Ending	47,000	60,000

Why:

The primary use for the statement of cost of goods manufactured is for external financial reporting. It is a crucial input to the statement of cost of goods sold and to the income statement.

Required:

1. Prepare a statement of cost of goods manufactured in good form.
2. *What if* 32,000 cell phones were to be manufactured next year? Explain which lines of the statement of cost of goods manufactured would be affected and how.

Solution:

1.

Carreker Company
Statement of Cost of Goods Manufactured
For the Coming Year

Direct materials		
Beginning inventory	$ 53,400	
Add: Purchases	147,900	
Materials available	201,300	
Less: Ending inventory	47,000	
Direct materials used in production		$154,300
Direct labour		90,000
Manufacturing (factory) overhead		480,000
Total manufacturing costs added		724,300
Add: Beginning work in process		75,000
Less: Ending work in process		60,000
Cost of goods manufactured		$739,300

CORNERSTONE 1-2
(continued)

2. If the number of units produced increases, the cost of direct materials used in production will increase. Since there are sufficient direct materials in beginning inventory, it is not clear whether purchases would increase, or instead, if ending materials inventory would go down. Direct labour would increase to reflect the additional units. Overhead would increase due to the increase in variable overhead, but the fixed overhead component would remain the same. No clear need for changes in beginning and ending WIP are required as long as the additional 2,000 units come from current production.

and it belongs on the income statement. Cornerstone 1-3 shows the cost of goods sold schedule for a manufacturing company.

Finally, we are ready to prepare an income statement for a manufacturing firm. Cornerstone 1-4 shows how the results of the statement of cost of goods sold are included with nonmanufacturing expenses to calculate operating income. Gross margin, also called gross profit, (the difference between sales and cost of goods sold) is an important number on the income statement.

Often, the income statement includes a column showing each line item as a percentage of sales. Clearly, sales is 100 percent of sales. Management can review these percentages and compare them with past history of the firm and with industry averages to see whether expenses are in line with expectations. If the industry generally spends 15 percent of sales on selling expense, then a company that spends significantly more or less than that amount may want to carefully consider whether its marketing strategy is appropriate.

Income Statement: Service Organization

The income statement for a service organization looks very similar to the one shown in Cornerstone 1-4 for a manufacturing organization. However, the cost of goods sold does differ in some key ways. For one thing, the service firm has no finished goods inventories since services cannot be stored, although it is possible to have work in process for services. For example, an architect may have drawings in process and an orthodontist may have numerous patients in various stages of processing for braces. Additionally, some service firms add order fulfillment costs to the cost of goods sold.

The Role of the Management Accountant

World-class firms are those that are at the cutting edge of customer support. They know their market and their product. They strive continually to improve product design, manufacture, and delivery. These companies can compete with the best of the best in a global environment. Accountants, too, can be termed world class. Those who merit this designation are intelligent and well prepared. They not only have the education and training to accumulate and provide financial information, but they stay up to date in their field and in business. In addition, world-class accountants must be familiar with the customs and financial accounting rules of the countries in which their firm operates.

The Controller   The **controller**, the chief accounting officer, supervises all accounting departments. Because of the critical role that management accounting plays in the operation of an organization, the controller is often viewed as a member of the top management team and encouraged to participate in planning, controlling, and decision-making activities. As the chief accounting officer, the controller has responsibility for both internal and external accounting requirements. This charge may include direct responsibility for internal auditing, cost accounting, financial accounting (including securities commission reports and financial statements), systems accounting (including analysis, design, and internal controls), budgeting support, economic

The HOW and WHY of Preparing the Statement of Cost of Goods Sold

**CORNERSTONE
1-3**

Information:
Carreker Company manufactures cell phones. For next year, Carreker predicts that 30,000 units will be produced with the following total costs:

Direct materials	$154,300
Direct labour	90,000
Variable overhead	30,000
Fixed overhead	450,000

Carreker expects to purchase $147,900 of direct materials next year. Projected beginning and ending inventories for direct materials and work in process are as follows:

	Direct Materials Inventory	Work-in-Process Inventory
Beginning	$53,400	$75,000
Ending	47,000	60,000

Carreker Company expects to sell 34,000 units. Beginning inventory of finished goods is expected to be $151,000, and ending inventory of finished goods is expected to be $45,000.

Why:
The primary use for the statement of cost of goods sold is for external financial reporting. It is a crucial input to the income statement.

Required:
1. Prepare a statement of cost of goods sold in good form.
2. **What if** only 32,000 cell phones were to be sold next year? Explain which lines of the statement of cost of goods sold would be affected and how.

Solution:
1.

Carreker Company Statement of Cost of Goods Sold For the Coming Year	
Cost of goods manufactured (Cornerstone 1-2)	$739,300
Add: Beginning finished goods	151,000
Cost of goods available for sale	890,300
Less: Ending finished goods	45,000
Cost of goods sold	845,300

2. If the number of units sold decreases, and production remains the same, then ending finished goods will be higher as the unsold units remain in inventory.

analysis, and taxes. The duties and organization of the controller's office vary from firm to firm. In some companies, the internal audit department may report directly to the financial vice president; similarly, the systems department may report directly to the financial vice president or even to another staff vice president.

**CORNERSTONE
1 - 4**

The HOW and WHY of Preparing the Income Statement for a Manufacturing Firm

Information:
Carreker Company manufactures cell phones. For next year, Carreker predicts that 30,000 units will be produced with the following total costs:

Direct materials	$154,300
Direct labour	90,000
Variable overhead	30,000
Fixed overhead	450,000

Carreker expects to purchase $147,900 of direct materials next year. Projected beginning and ending inventories for direct materials and work in process are as follows:

	Direct Materials Inventory	Work-in-Process Inventory
Beginning	$53,400	$75,000
Ending	47,000	60,000

Carreker Company expects to sell 34,000 units at a price of $35 each. Beginning inventory of finished goods is expected to be $151,000, and ending inventory of finished goods is expected to be $45,000. Total selling expense is projected at $62,000, and total administrative expense is projected at $187,000.

Why:
The primary use for the income statement is for external financial reporting. Investors and outside parties use it to determine the financial health of a firm.

Required:
1. Prepare an income statement in good form. Give percentages of sales for each major line item.
2. **What if** only 32,000 cell phones were to be sold next year? Explain which lines of the income statement would be affected and how.

Solution:
1.

	Carreker Company Income Statement For the Coming Year		
			%
Sales ($35 × 34,000)		$1,190,000	100.00
Less: Cost of goods sold (Cornerstone 1-3)		845,300	71.03
Gross margin		344,700	28.97
Less operating expenses:			
Selling expenses	$ 62,000		
Administrative expenses	187,000	249,000	20.92
Operating income		$ 95,700	8.04*

*Difference is due to rounding.

> 2. If the number of units sold decreases, both sales and cost of goods sold will decrease, as will gross margin. Since no variable elements have been noted for selling and administrative expense, it is assumed that they are fixed and will not change if sales volume changes. Operating income will decrease.

CORNERSTONE 1-4 *(continued)*

The Treasurer The **treasurer** is responsible for the finance function. Specifically, the treasurer raises capital and manages cash (banking and custody), investments, and investor relations. The treasurer may also be in charge of credit and collections as well as insurance. The treasurer reports to the financial vice president.

Information for Planning, Controlling, Continuous Improvement, and Decision Making

The cost and management accountant is responsible for generating financial information required by the firm for internal and external reporting. This involves responsibility for collecting, processing, and reporting information that will help managers in their planning, controlling, and other decision-making activities.

Planning The detailed formulation of future actions to achieve a particular end is the management activity called **planning**. Planning therefore requires setting objectives and identifying methods to achieve those objectives. A firm may have the objective of increasing its short- and long-term profitability by improving the overall quality of its products. By improving product quality, the firm should be able to reduce scrap and rework, decrease the number of customer complaints and the amount of warranty work, reduce the resources currently assigned to inspection, and so on, thus increasing profitability. This is accomplished by working with suppliers to improve the quality of incoming raw materials, establishing quality control circles, and studying defects to ascertain their cause.

Controlling The processes of monitoring a plan's implementation and taking corrective action as needed are referred to as **controlling**. Control is usually achieved with the use of **feedback**. Feedback is information that can be used to evaluate or correct the steps that are actually being taken to implement a plan. Based on the feedback, a manager may decide to let the implementation continue as is, take corrective action of some type to put the actions back in harmony with the original plan, or do some midstream replanning.

Feedback is a critical facet of the control function. It is here that accounting once again plays a vital role. Accounting reports that provide feedback by comparing planned (budgeted) data with actual data are called **performance reports**. Exhibit 1-4 shows a performance report that compares budgeted sales and cost of goods sold with the actual amounts for the month of August. Deviations from the planned amounts that increase profits are labelled "favourable," while those that decrease profits are called "unfavourable." These performance reports can have a dramatic impact on managerial actions—but they must be realistic and supportive of management plans. Revenue and spending targets must be based (as closely as possible) on actual operating conditions.

Continuous Improvement In a dynamic environment, firms must continually improve their performance to remain competitive or to establish a competitive advantage. A company pursuing continuous improvement has the goal of performing better than before and better than competitors. **Continuous improvement** has been defined as "the relentless pursuit of improvement in the delivery of value to customers."[7] In practical terms, continuous improvement means searching for ways to increase overall efficiency by reducing waste, improving quality, and reducing costs. Cost management

---

[7] W. Maguire and D. Heath, "Capacity Management for Continuous Improvement," *Journal of Cost Management* (January 1997): 26–31.

Exhibit 1-4

Performance Report Illustrated

Golding Foods Inc.
Performance Report
For the Month Ended August 31, 2011

Budget Item	Actual	Budgeted	Variance
Sales	$800,000	$900,000	$100,000 U
Cost of goods sold	600,000	650,000	50,000 F

Note: U = Unfavourable; F = Favourable.

supports continuous improvement by providing information that helps identify ways to improve and then reports on the progress of the methods that have been implemented. It also plays a critical role by developing a control system that locks in and maintains any improvements realized.

Decision Making   The process of choosing among competing alternatives is **decision making**. Decisions can be improved if information about the alternatives is gathered and made available to managers. One of the major roles of the accounting information system is to supply information that facilitates decision making. This pervasive managerial function is an important part of both planning and control. A manager cannot plan without making decisions. Managers must choose among competing objectives and methods to carry out the chosen objectives. Only one of numerous mutually exclusive plans can be chosen. Similar comments can be made concerning the control function.

Certification in Management Accounting

In 1920, the Canadian Society of Cost Accountants, the predecessor of CMA Canada, was incorporated in Hamilton, Ontario. In 1941, the Registered Industrial Accountant (RIA) program was established, which evolved into the Certified Management Accountant program in 1985. A **Certified Management Accountant (CMA)** has passed a rigorous qualifying examination, has met an experience requirement, and participates in continuing education.

One of the main purposes of creating the CMA program was to establish management accounting as a recognized, professional discipline, separate from the profession of public accounting. Since its inception, the CMA program has been very successful. Many firms now sponsor and pay for classes that prepare their management accountants for the qualifying examination, as well as provide other financial incentives to encourage acquisition of the CMA certificate.

OBJECTIVE ➤ 5
Understand the importance of ethical behaviour for management accountants.

Ethical Behaviour

Virtually all managerial accounting practices were developed to assist managers in maximizing profits. Traditionally, actions regarding the economic performance of the firm have been the overriding concern. Yet managers and managerial accountants should not become so focused on profits that they develop a belief that the only goal of a business is maximizing its net worth. The objective of profit maximization should be constrained by the requirement that profits be achieved through legal and ethical means. While this has always been an implicit assumption of managerial accounting, the assumption should be made explicit. To help achieve this objective, many of the problems in this text require explicit consideration of ethical issues.

Ethical behaviour involves choosing actions that are right, proper, and just. Behaviour can be right or wrong; it can be proper or improper; and the decisions we make can be fair or unfair. Though people often differ in their views of the meaning of the ethical terms cited, there seems to be a common principle underlying all ethical systems. This principle is expressed by the belief that each member of a group bears some responsibility for the well-being of other members. Willingness to sacrifice one's self-interest for the well-being of the group is the heart of ethical action.

This notion of sacrifice produces some core values—values that describe what is meant by right and wrong in more concrete terms. James W. Brackner, writing for the "Ethics Column" in *Management Accounting*, made the following observation:

> *For moral or ethical education to have meaning, there must be agreement on the values that are considered "right." Ten of these values are identified and described by Michael Josephson in "Teaching Ethical Decision Making and Principled Reasoning." The study of history, philosophy, and religion reveals a strong consensus as to certain universal and timeless values essential to the ethical life.*
>
> *These 10 core values yield a series of principles that delineate right and wrong in general terms. Therefore, they provide a guide to behaviour.*[8]

The 10 core values referred to in the quotation include the following:

1. Honesty
2. Integrity
3. Promise keeping
4. Fidelity
5. Fairness
6. Caring for others
7. Respect for others
8. Responsible citizenship
9. Pursuit of excellence
10. Accountability

Many of the well-known accounting scandals, such as those involving **Adelphia**, **WorldCom**, **HealthSouth**, and **Parmalat**, provide evidence of the pressures faced by top managers and accountants to produce large net income numbers, especially in the short term. Unfortunately, such individuals often give into these pressures when faced with questionable revenue- and cost-related judgments. For example, the scandal at WorldCom was committed because the CEO, Bernie Ebbers, coerced several of the top accountants at WorldCom to wrongfully record journal entries in the company's books that capitalized millions of dollars in costs as assets (i.e., on the balance sheet) rather than as expenses (i.e., on the income statement) that would have dramatically lowered current period net income. Eventually, WorldCom was forced to pay hundreds of millions of dollars to the U.S. government and to shareholders for its illegal and unethical actions. In addition, several of the top executives were sentenced to extensive prison time for their actions. The recent subprime mortgage crisis also highlights the importance of ethical considerations as some banks tried to increase their profits either by lending individuals more money than they could reasonably afford or using terms that were intentionally less clear, or transparent, than many outsiders thought they should be.[9]

As some of these examples point out, though it may seem contradictory, sacrificing self-interest for the collective good might not only be right and bring a sense of individual worth but might also make good business sense. Companies with a strong code of ethics can create strong customer and employee loyalty. While liars and cheats may win on occasion, their victories often are short-lived. Companies in business for the long term find that it pays to treat all of their constituents with honesty and loyalty.

---

[8] James W. Brackner, "Consensus Values Should Be Taught," *Management Accounting* (August 1992): 19. For a more complete discussion of the 10 core values, see also Michael Josephson, *Teaching Ethical Decision Making and Principled Reasoning, Ethics Easier Said Than Done* (The Josephson Institute, Winter Los Angeles, CA: 1988): 29–30.

[9] Jane Sasseen, "FBI Widens Net Around Subprime Industry: With 14 Companies Under Investigation, the Bureau's Scope is the Entire Securitization Process," *Business Week Online* (January 30, 2008). Taken from http://www.businessweek.com/bwdaily/dnflash/content/jan2008/db20080129_728982.htm?chan=search on February 12, 2008.

Company Codes of Ethical Conduct

To promote ethical behaviour by managers and employees, organizations commonly establish standards of conduct referred to as Company Codes of Conduct. One needs only to hear the name "Enron" to be reminded of the importance of ethical conduct.

In 2009, **Loblaws Companies** outlined its approach to social responsibility and expressed its achievements and core values in a Corporate Social Responsibility Report (CSR). That document linked the company to five core values:

- Respecting the environment
- Sourcing with integrity
- Making a positive difference in the community
- Reflecting the nation's diversity
- Being a great place to work

Reflecting those values, in 2009, Loblaws implemented the following:

- *Plastic bag diversion:* It applied a national 5-cent charge for every plastic bag provided at checkout. This led to more than 1.3 billion plastic bags being diverted from Canadian landfills by the end of 2009.
- *Improved fuel efficiency:* In its efforts to reduce its carbon footprint, it achieved a 2 percent improvement in transport fleet fuel efficiency per kilometre.
- *Sourcing with integrity:* The company committed itself to sustainably sourcing 100 percent of all seafood sold in its stores by year end 2013.

In addition, Loblaws did the following:

- Reduced its national refrigerant leak rate by 5 percent as a result of 56 corporate banner stores having alternative refrigeration systems that significantly reduced refrigerant requirements.
- Established a target to reduce nonrecyclable packaging on private label brands by 50 percent by 2013. Once this is achieved, the company's packaging will be 79 percent recyclable.
- Installed a wind turbine at the Atlantic Superstore in Porters Lake, Nova Scotia.
- Granted $8.9 million to more than 1,500 families across Canada through the President's Choice Children's Charity.
- Announced corporate donations of more than $24 million to help support local charities, programs, and organizations across Canada.
- Grew sales of Canadian produce by 16 percent during its Grown Close to Home campaign.
- Reduced sodium content in more than 50 private label products.
- Increased the number of female store managers by 53.7 per cent (since 2008).[10]

Important parts of corporate codes of conduct are integrity, performance of duties, and compliance with the rule of law. They also uniformly prohibit the acceptance of kickbacks and improper gifts, insider trading, and misappropriation of corporate information and assets.

Standards of Ethical Conduct for Managerial Accountants

Organizations commonly establish standards of conduct for their managers and employees. Professional associations also establish ethical standards. All three Canadian accounting bodies: Canadian Institute of Chartered Accountants (CICA), Certified Management Accountants (CMA), and **Certified General Accountants (CGA)**—have established ethical standards for accountants. Professional accountants are bound by these codes of conduct, which stress the importance of competence, confidentiality, integrity, and credibility or objectivity. The CMA Code of Professional Ethics is provided in Exhibit 1-5.

[10] © 2010 Loblaw Inc. Reproduced with permission. http://www.loblaw.com/Theme/Loblaw/files/en/csr_2009/targets.htm.

Exhibit 1-5

CMA Code of Professional Ethics*

All Members, Students, Firms, Public Accounting Firms and Professional Corporations will adhere to the following Code of Professional Ethics of CMA Ontario:

A Member, Student, Firm, Public Accounting Firm or Professional Corporation will act at all times with:

(a)   responsibility for and fidelity to public needs;

(b)   fairness and loyalty to such Member's, Student's, Firm's, Public Accounting Firm's or Professional Corporation's associates, clients and employers; and

(c)   competence through devotion to high ideals of personal honour and professional integrity.

A Member, Student, Firm, Public Accounting Firm or Professional Corporation will:

(a)   maintain at all times independence of thought and action;

(b)   not express an opinion on financial reports or statements without first assessing her, his or its relationship with her, his or its client to determine whether such Member, Student, Firm, Public Accounting Firm or Professional Corporation might expect her or his opinion to be considered independent, objective and unbiased by one who has knowledge of all the facts;

(c)   when preparing financial reports or statements or expressing an opinion on financial reports or statements, disclose all material facts known to such Member, Student, Firm, Public Accounting Firm or Professional Corporation in order not to make such financial reports or statements misleading, acquire sufficient information to warrant an expression of opinion and report all material misstatements or departures from generally accepted accounting principles; and

(d)   comply with the requirements of the CMA Ontario Independence Regulation for Assurance, Audit and Review Engagements.

A Member, Student, Firm, Public Accounting Firm or Professional Corporation will:

(a)   not disclose or use any confidential information concerning the affairs of such Member's, Student's, Firm's, Public Accounting Firm's or Professional Corporation's employer or client unless authorized to do so or except when such information is required to be disclosed in the course of any defence of himself, herself or itself or any associate or employee in any lawsuit or other legal proceeding or against alleged professional misconduct by order of lawful authority of the Board or any Committee of CMA Ontario in the proper exercise of their duties but only to the extent necessary for such purpose and only as permitted by law;

(b)   obtain, at the outset of an engagement, written agreement from any party or parties to whom work is contracted not to disclose or use any confidential information concerning the affairs of such Member's, Student's, Firm's, Public Accounting Firm's or Professional Corporation's employer or client unless authorized to do so or except when such information is required to be disclosed in the course of any defence of himself, herself or itself or any associate or employee in any lawsuit or other legal proceeding but only to the extent necessary for such purpose and only as permitted by law;

(c)   inform his, her or its employer or client of any business connections or interests of which such Member's, Student's, Firm's, Public Accounting Firm's or Professional Corporation's employer or client would reasonably expect to be informed;

(d)   not, in the course of exercising his, her or its duties on behalf of such Member's, Student's, Firm's, Public Accounting Firm's or Professional Corporation's employer or client, hold, receive, bargain for or acquire any fee, remuneration or benefit without such employer's or client's knowledge and consent; and

(e)   take all reasonable steps, in arranging any engagement as a consultant, to establish a clear understanding of the scope and objectives of the work before it is commenced and will furnish the client with an estimate of cost, preferably before the engagement is commenced, but in any event as soon as possible thereafter.

A Member, Student, Firm, Public Accounting Firm or Professional Corporation will:

(a)   conduct himself, herself or itself toward Members, Students, Firms, Public Accounting Firms and Professional Corporations with courtesy and good faith;

(b)   not commit an act discreditable to the profession;

(c)   not engage in or counsel any business or occupation which, in the opinion of CMA Ontario, is incompatible with the professional ethics of a management accountant or public accountant;

(d)   not accept any engagement to review the work of a Member, Student, Firm, Public Accounting Firm or Professional Corporation for the same employer except with the knowledge of that Member, Student, Firm, Public Accounting Firm or Professional Corporation, or except where the connection of that Member, Student, Firm, Public Accounting Firm or Professional Corporation with the work has been terminated, unless the Member, Student, Firm, Public Accounting Firm or Professional Corporation reviews the work of others as a normal part of his, her or its responsibilities;

(e)   not attempt to gain an advantage over Members, Students, Firms, Public Accounting Firms and Professional Corporations by paying or accepting a commission in securing management accounting or public accounting work;

(f)   uphold the principle of adequate compensation for management accounting and public accounting work; and

(g)   not act maliciously or in any other way which may adversely reflect on the public or professional reputation or business of a Member, Student, Firm, Public Accounting Firm or Professional Corporation.

A Member, Student, Firm, Public Accounting Firm or Professional Corporation will:

(a)   at all times maintain the standards of competence expressed by the Board from time to time;

(b)   disseminate the knowledge upon which the profession of management accounting is based to others within the profession and generally promote the advancement of the profession;

(c)   undertake only such work as he, she or it is competent to perform by virtue of his, her or its training and experience and will, where it would be in the best interests of an employer or client, engage, or advise the employer or client to engage, other specialists;

(d)   expose before the proper tribunals of CMA Ontario any incompetent, unethical, illegal or unfair conduct or practice of a Member, Student, Firm, Public Accounting Firm or Professional Corporation which involves the reputation, dignity or honour of CMA Ontario; and

(e)   endeavour to ensure that a professional partnership, company or individual, with which such Member, Student, Firm, Public Accounting Firm or Professional Corporation is associated as a partner, principal, director, officer, associate or employee, abides by the Code of Professional Ethics and the Rules of Professional Conduct established by CMA Ontario.

*© 2011 CMA Ontario. Reproduced with permission. http://www.cma-ontario.org/multimedia/Ontario/attachments/ProfessionalMisconductAndCodeOf ProfessionalEthics.pdf

One of the key requirements for obtaining the CMA certificate or designation is passing a qualifying examination. Four areas are emphasized: (1) business analysis; (2) management accounting and reporting; (3) strategic management; and (4) business applications. The parts to the examination reflect the needs of management accounting and underscore the earlier observation that management accounting has more of an interdisciplinary flavour than other areas of accounting.

Summary of Learning Objectives

1. **Describe a cost management information system, its objectives, and its major subsystems, and indicate how it relates to other operating and information systems.**

  - Cost management system, a subsystem of the accounting information system, designed to satisfy costing, controlling, and decision-making objectives
  - Two major subsystems: cost accounting system and the operational control system

2. **Explain the cost assignment process.**

  - Objective of the cost accounting system is assigning costs to cost objects
  - Three methods of cost assignment:

    - Direct tracing—physical observation, most accurate
    - Driver tracing—more expensive, more accurate than allocation
    - Allocation—least accurate, easiest to apply

3. **Define tangible and intangible products, and explain why there are different product cost definitions.**

  - Products are tangible.
  - Services are:

    - Intangible
    - Perishable (cannot be inventoried)
    - Inseparable (buyer and provider interact)

  - Product cost definitions:

    - Value chain includes research and development, production, marketing, and customer service. Used for pricing decisions, product mix decisions, strategic profitability analysis.
    - Operating product costs include production, marketing, and customer service. Used for strategic design decisions, tactical profitability analysis.
    - Traditional product costs include only production (direct materials, direct labour, overhead) and are used for external financial reporting.

4. **Prepare income statements for manufacturing and service organizations.**

  - Income statements rely on:

    - Cost of goods manufactured or services provided
    - Cost of goods sold or services sold (typically the same as services provided)

  - Gross margin is the difference between sales revenue and the cost of goods (or services) sold.
  - Operating income is the difference between gross margin and selling (or marketing) and administrative expense.

5. Understand the importance of ethical behaviour for management accountants.

- Management accounting aids managers in their efforts to improve the economic performance of the firm.
- Unfortunately, some managers have overemphasized the economic dimension and have engaged in unethical and illegal actions. Many of these actions have relied on the management accounting system to bring about and even support that unethical behaviour.
- To emphasize the importance of the ever present constraint of ethical behaviour, this text presents ethical issues in many of the problems appearing at the end of each chapter.

CORNERSTONE 1-1	The HOW and WHY of calculating prime cost, conversion cost, variable product cost, and total product cost, page 14
CORNERSTONE 1-2	The HOW and WHY of preparing the statement of cost of goods manufactured, page 17
CORNERSTONE 1-3	The HOW and WHY of preparing the statement of cost of goods sold, page 19
CORNERSTONE 1-4	The HOW and WHY of preparing the income statement for a manufacturing firm, page 20

CORNERSTONES FOR CHAPTER 1

Review Problems

I. Types of Costs, Cost of Goods Manufactured, Absorption-Costing Income Statement

Palmer Manufacturing produces weather vanes. For the year just ended, Palmer produced 10,000 weather vanes with the following total costs:

Direct materials	$20,000
Direct labour	35,000
Overhead	10,000
Selling expenses	6,250
Administrative expenses	14,400

During the year, Palmer sold 9,800 units for $12 each. Beginning finished goods inventory consisted of 630 units with a total cost of $4,095. There were no beginning or ending inventories of work in process.

Required:

1. Calculate the unit costs for the following: direct materials, direct labour, overhead, prime cost, and conversion cost.
2. Prepare schedules for cost of goods manufactured and cost of goods sold.
3. Prepare an absorption-costing income statement for Palmer Manufacturing.

Solution:

1. Unit direct materials = $20,000/10,000 = $2.00
   Unit direct labour = $35,000/10,000 = $3.50
   Unit overhead = $10,000/10,000 = $1.00
   Unit prime cost = $2.00 + $3.50 = $5.50
   Unit conversion cost = $3.50 + $1.00 = $4.50

2. Statement of cost of goods manufactured:

Direct materials used	$20,000
Direct labour	35,000
Overhead	10,000
Total manufacturing costs added	65,000
Add: Beginning work in process	0
Less: Ending work in process	(0)
Cost of goods manufactured	65,000

Cost of goods sold schedule:

Cost of goods manufactured	$65,000
Add: Beginning finished goods inventory	4,095
Less: Ending finished goods inventory*	(5,395)
Cost of goods sold	$63,700

*Units in ending finished goods inventory = 10,000 + 630 − 9,800 = 830; 830 ×
($2.00 + $3.50 + $1.00) = $5,395.

3. Income statement:

Sales (9,800 × $12)		$117,600
Less: Cost of goods sold		63,700
Gross margin		53,900
Less: Operating expenses:		
Selling expenses	$ 6,250	
Administrative expenses	14,400	20,650
Operating income		$ 33,250

II. Systems Concepts

Kate Myers is a student at Memorial University. Her system for tracking finances
includes the following. Kate has two credit cards; each day she places the receipts for
any items purchased on credit in a manila envelope on her desk. She checks these
receipts against the credit card bills at the end of the month. Any other financial item
that Kate thinks might be useful later is also placed into the envelope. (An example
would be a payroll stub from her job as a worker in the campus cafeteria.) Kate records
any cheque written in her chequebook register at the time she writes it. Shortly after her
bank statement arrives, she enters any cheques written and deposits made into
Quicken® (the software program she uses to balance her chequebook). She then recon-
ciles her bank statement against the Quicken account and prints a reconciliation report.
From time to time, Kate phones home to ask her mother to add more money to her
bank account. (Kate could e-mail or text her mom, but she's found that her mother
appreciates the personal touch of a phone call, and the money appears more quickly
whenever Kate phones.) Her mother, who has copies of the deposit slips for Kate's
account, mails a cheque (from her own account) with a deposit slip to Kate's account.
Whenever this occurs, Kate logs on to Bluemountain.com and e-mails her mother an
electronic thank you card.

    The following items are associated with this financial system:

a. Manila envelope
b. Chequebook
c. Cheques and deposit slips
d. Computer and printer
e. Quicken program
f. Credit cards
g. Credit card receipts

h. Payroll stubs, etc.
i. Monthly bank statements
j. Reconciliation report
k. Phone

Required:

1. What are the objectives of Kate's financial system? What processes can you identify?
2. Classify the items into one of the following categories:

    a. Interrelated parts
    b. Inputs
    c. Outputs

3. Draw an operational model for the financial system.

Solution:

1. The objectives of Kate's financial system are to keep her financially solvent and to provide a clear and accurate picture of her chequing account balance and bills incurred at any point in time. Processes include filing the credit card receipts, entering cheques written and deposits made into both the manual and computerized systems, reconciling the bank statement with the computerized system, phoning home for additional funds, and e-mailing a thank you card.
2. The items are classified as follows:
    a. Manila envelope—interrelated part
    b. Chequebook—interrelated part
    c. Cheques and deposit slips—input
    d. Computer and printer—interrelated part
    e. Quicken program—interrelated part
    f. Credit cards—interrelated part
    g. Credit card receipts—input
    h. Payroll stubs, etc.—input
    i. Monthly bank statements—interrelated part
    j. Reconciliation report—output
    k. Phone—interrelated part
3. Operational model of Kate's financial system:

Inputs	Processes	Objectives
Cheques	Filing credit card receipts	Stay financially solvent
Deposit slips	Entering cheques/deposits	Be aware of bills incurred
Credit card receipts	Reconciling statements	Know account balance
Payroll stubs, etc.	Phoning for additional funds	
	E-mailing thank you card	

Key Terms

Absorption-costing income, 16

Accounting information system, 2

Activity, 9

Administrative costs, 15

Allocation, 11

Assets, 9

Certified General Accountants (CGA), 24

Certified Management Accountant (CMA), 22

Continuous improvement, 21

Controller, 18

Discussion Questions

1. What is an accounting information system?
2. What is the difference between a financial accounting information system and a cost management information system?
3. What are the objectives of a cost management information system?
4. What is a cost object? Give some examples.
5. What is an activity? Give some examples of activities within a manufacturing firm.
6. What is a direct cost? An indirect cost?
7. What does traceability mean?
8. What is allocation?
9. Explain how driver tracing works.
10. What is a tangible product?
11. What is a service? Explain how services differ from tangible products.
12. Identify the three cost elements that determine the cost of making a product (for external reporting).
13. How do the income statements of a manufacturing firm and a service firm differ?

Cornerstone Exercises

OBJECTIVE ▶ 3
CORNERSTONE 1-1

Cornerstone Exercise 1-1 PRODUCT COSTS

Sodowsky Manufacturing Inc. produces brightly coloured clog-style shoes. For next year, Sodowsky predicts that 150,000 units will be produced, with the following total costs:

Direct materials	$300,000
Direct labour	90,000
Variable overhead	45,000
Fixed overhead	420,000

Required:

1. Calculate the prime cost per unit.
2. Calculate the conversion cost per unit.
3. Calculate the total variable cost per unit.
4. Calculate the total product (manufacturing) cost per unit.
5. *What if* the number of units increased to 165,000 and all unit variable costs stayed the same? Explain what the impact would be on the following costs: total direct materials, total direct labour, total variable overhead, total fixed overhead, unit prime cost, unit conversion cost. What would the product cost per unit be in this case?

Cornerstone Exercise 1-2 COST OF GOODS MANUFACTURED

OBJECTIVE > 4
CORNERSTONE 1-2

Refer to **Cornerstone Exercise 1-1**. For next year, Sodowsky predicts that 150,000 units will be produced, with the following total costs:

Direct materials	$300,000
Direct labour	90,000
Variable overhead	45,000
Fixed overhead	420,000

Next year, Sodowsky expects to purchase $292,400 of direct materials. Projected beginning and ending inventories for direct materials and work in process are as follows:

	Direct Materials Inventory	Work-in-Process Inventory
Beginning	$22,400	$45,000
Ending	14,800	40,000

Required:

1. Prepare a statement of cost of goods manufactured in good form.
2. *What if* the ending inventory of direct materials increased by $2,000? Which line items on the statement of cost of goods manufactured would be affected and in what direction (increase or decrease)?

Cornerstone Exercise 1-3 COST OF GOODS SOLD

OBJECTIVE > 4
CORNERSTONE 1-3

Refer to **Cornerstone Exercises 1-1 and 1-2**.

Sodowsky expects to produce 150,000 units and sell 140,000 units. Beginning inventory of finished goods is $25,000 and ending inventory of finished goods is expected to be $74,000.

Required:

1. Prepare a statement of cost of goods sold in good form.
2. *What if* the beginning inventory of finished goods decreased by $5,000? What would be the effect on the cost of goods sold?

Cornerstone Exercise 1-4 INCOME STATEMENT

OBJECTIVE > 4
CORNERSTONE 1-4

Refer to **Cornerstone Exercises 1-1, 1-2, and 1-3**. Next year, Sodowsky expects to produce 150,000 units and sell 140,000 units at a price of $7.50 each. Beginning inventory of finished goods is $25,000 and ending inventory of finished goods is expected to be $74,000. Total selling expense is projected at $33,000 and total administrative expense is projected at $145,000.

Required:

1. Prepare an income statement in good form. Be sure to include the percent of sales column.
2. *What if* the cost of goods sold percentage for the past few years was 80 percent? Explain how management might react.

OBJECTIVE ▶ 3

CORNERSTONE 1-1

Cornerstone Exercise 1-5 COSTS OF SERVICES

Jean and Tom Perritz own and manage Happy Home Helpers Inc. (HHH), a house cleaning service. Each cleaning (cleaning one house one time) takes a team of three house cleaners about 1.5 hours. On average, HHH completes about 15,000 cleanings per year. The following total costs are associated with the total cleanings:

Direct materials	$ 27,000
Direct labour	472,500
Variable overhead	15,000
Fixed overhead	18,000

Required:

1. Calculate the prime cost per cleaning.
2. Calculate the conversion cost per cleaning.
3. Calculate the total variable cost per cleaning.
4. Calculate the total service cost per cleaning.
5. *What if* rent on the office that Jean and Tom use to run HHH increased by $1,500? Explain the impact on the following:
    a. Prime cost per cleaning
    b. Conversion cost per cleaning
    c. Total variable cost per cleaning
    d. Total service cost per cleaning

OBJECTIVE ▶ 4

CORNERSTONE 1-2

SERVICE

Cornerstone Exercise 1-6 COST OF SERVICES PRODUCED

Jean and Tom Perritz own and manage Happy Home Helpers Inc. (HHH), a house cleaning service. Each cleaning (cleaning one house one time) takes a team of three house cleaners about 1.5 hours. On average, HHH completes about 15,000 cleanings per year. The following total costs are associated with the total cleanings:

Direct materials	$ 27,000
Direct labour	472,500
Variable overhead	15,000
Fixed overhead	18,000

Next year, HHH expects to purchase $25,600 of direct materials. Projected beginning and ending inventories for direct materials are as follows:

Direct Materials Inventory	
Beginning	$4,000
Ending	2,600

There is no work-in-process inventory; in other words, a cleaning is started and completed on the same day.

Required:

1. Prepare a statement of services produced in good form.
2. *What if* HHH planned to purchase $30,000 of direct materials? Assume there would be no change in beginning and ending inventories of materials. Explain which line items on the statement of services produced would be affected and how (increase or decrease).

OBJECTIVE ▶ 4

CORNERSTONE 1-3

Cornerstone Exercise 1-7 COST OF SERVICES SOLD

Jean and Tom Perritz own and manage Happy Home Helpers Inc. (HHH), a house cleaning service. Each cleaning (cleaning one house one time) takes a team of three house cleaners about 1.5 hours. On average, HHH completes about 15,000 cleanings per year. The following total costs are associated with the total cleanings:

Direct materials	$ 27,000
Direct labour	472,500
Variable overhead	15,000
Fixed overhead	18,000

Next year, HHH expects to purchase $25,600 of direct materials. Projected beginning and ending inventories for direct materials are as follows:

Direct Materials Inventory	
Beginning	$4,000
Ending	2,600

There is no work-in-process inventory and no finished goods inventory; in other words, a cleaning is started and completed on the same day.

Required:

1. Prepare a statement of cost of services sold in good form.
2. How does this cost of services sold statement differ from the cost of goods sold statement for a manufacturing firm?

Cornerstone Exercise 1-8 INCOME STATEMENT

OBJECTIVE ➤ 4
CORNERSTONE 1-4

Jean and Tom Perritz own and manage Happy Home Helpers Inc. (HHH), a house cleaning service. Each cleaning (cleaning one house one time) takes a team of three house cleaners about 1.5 hours. On average, HHH completes about 15,000 cleanings per year. The following total costs are associated with the total cleanings:

Direct materials	$ 27,000
Direct labour	472,500
Variable overhead	15,000
Fixed overhead	18,000

Next year, HHH expects to purchase $25,600 of direct materials. Projected beginning and ending inventories for direct materials are as follows:

Direct Materials Inventory	
Beginning	$4,000
Ending	2,600

There is no work-in-process inventory and no finished goods inventory; in other words, a cleaning is started and completed on the same day. HHH expects to sell 15,000 cleanings at a price of $45 each next year. Total selling expense is projected at $22,000, and total administrative expense is projected at $53,000.

Required:

1. Prepare an income statement in good form.
2. ***What if*** Jean and Tom increased the price to $50 per cleaning and no other information was affected? Explain which line items in the income statement would be affected and how.

Exercises

Exercise 1-9 SYSTEMS CONCEPTS

OBJECTIVE ➤ 1

In general, systems are described by the following pattern: (1) interrelated parts, (2) processes, and (3) objectives. Operational models of systems also identify inputs and outputs.

The dishwashing system of a university cafeteria consists of the following steps. First, students dispose of any waste paper (e.g., napkins) in a trash can, then they file by an opening to the dishwashing area and drop off their trays. Persons 1 and 2 take the trays; rinse the extra food down the disposal; and stack the dishes, glasses, and silverware in heavy-duty plastic racks. These racks slide along a conveyor into the automatic dishwasher. When the racks emerge from the other end of the dishwasher, they contain clean, germ-free items. Person 3 removes the racks and, with Person 4, empties them of clean items, stacking the dishes, silverware, glasses, and trays for future use. The empty racks are returned to the starting position in front of Persons 1 and 2. The following items are associated with this dishwashing system:

a. Automatic dishwasher
b. Racks to hold the dirty glasses, silverware, and dishes
c. Electricity
d. Water
e. Waste disposal
f. Sinks and sprayers
g. Dish detergent
h. Gas heater to heat water to 85 degrees Celsius
i. Conveyor belt
j. Persons 1, 2, 3, and 4
k. Clean, germ-free dishes
l. Dirty dishes
m. Half-eaten dinner
n. Aprons

Required:

1. What is the objective of the dishwashing system? What processes can you identify?
2. Classify the items into one of the following categories:
   a. Interrelated parts
   b. Inputs
   c. Outputs
3. Draw an operational model for the dishwashing system.
4. Discuss how a cost management information system is similar to and different from the dishwashing system.

OBJECTIVE ▶ 1   **Exercise 1-10 COST ACCOUNTING INFORMATION SYSTEM**

The following items are associated with a cost accounting information system:

a. Usage of direct materials
b. Assignment of direct materials cost to each product
c. Direct labour cost incurrence
d. Depreciation on production equipment
e. Cost accounting personnel
f. Submission of a bid, using product cost plus 25 percent
g. Power cost incurrence
h. Materials handling cost incurrence
i. Computer
j. Assignment of direct labour costs to products
k. Costing out of products
l. Decision to continue making a part rather than buying it
m. Printer
n. Report detailing individual product costs
o. Assignment of overhead costs to individual products

Required:

1. Classify the preceding items into one of the following categories:
   a. Interrelated parts
   b. Processes

c.   Objectives
d.   Inputs
e.   Outputs
f.   User actions

2. Draw an operational model that illustrates the cost accounting information system—with the preceding items used as examples for each component of the model.

3. Based on your operational model, identify which product cost definition is being used: value-chain, operating, or product (manufacturing).

Exercise 1-11 COST ASSIGNMENT METHODS

OBJECTIVE ➤ 2

Nizam Company produces speaker cabinets. Recently, Nizam switched from a traditional departmental assembly line system to a manufacturing cell in order to produce the cabinets. Suppose that the cabinet manufacturing cell is the cost object. Assume that all or a portion of the following costs must be assigned to the cell:

a.   Depreciation on electric saws, sanders, and drills used to produce the cabinets
b.   Power to heat and cool the plant in which the cell is located
c.   Salary of cell supervisor
d.   Wood used to produce the cabinet housings
e.   Maintenance for the cell's equipment (provided by the maintenance department)
f.   Labour used to cut the wood and to assemble the cabinets
g.   Replacement sanding belts
h.   Cost of janitorial services for the plant
i.   Ordering costs for materials used in production
j.   The salary of the industrial engineer (she spends about 20 percent of her time on work for the cell)
k.   Cost of maintaining plant and grounds
l.   Cost of plant's personnel office
m.   Depreciation on the plant
n.   Plant receptionist's salary and benefits

Required:

Identify which cost assignment method would likely be used to assign the costs of each of the preceding activities to the cabinet manufacturing cell: direct tracing, driver tracing, or allocation. When driver tracing is selected, identify a potential activity driver that could be used for the tracing.

Exercise 1-12 PRODUCT COST DEFINITIONS

OBJECTIVE ➤ 3

Three possible product cost definitions were introduced: (1) value-chain, (2) operating, and (3) product or manufacturing. Identify which of the three product cost definitions best fits the following situations (justify your choice):

a.   Determining which of several potential new products should be developed, produced, and sold
b.   Deciding whether to produce and sell a product whose design and development costs were higher than budgeted
c.   Setting the price for a new product
d.   Valuation of finished goods inventories for external reporting
e.   Determining whether to add a complementary product to the product line
f.   Choosing among competing product designs
g.   Calculating cost of goods sold for external reporting
h.   Deciding whether to increase the price of an existing product
i.   Deciding whether to accept or reject a special order, where the price offered is lower than the normal selling price

OBJECTIVE ▸ 3 4   ## Exercise 1-13 COST DEFINITIONS

Labrador Company provided the following information for the past calendar year:

Beginning inventory:	
Direct materials	$56,800
Work in process	34,700
Ending inventory:	
Direct materials	31,000
Work in process	29,700

During the year, direct materials purchases amounted to $160,200, direct labour cost was $225,600, and overhead cost was $308,400. There were 10,000 units produced.

Required:

1. Calculate the total cost of direct materials used in production.
2. Calculate the cost of goods manufactured. Calculate the unit manufacturing cost.
3. Of the unit manufacturing cost calculated in Requirement 2, $18.60 is direct materials and $30.85 is overhead. What is the prime cost per unit? Conversion cost per unit?

OBJECTIVE ▸ 3 4   ## Exercise 1-14 COST DEFINITIONS AND CALCULATIONS

For each of the following independent situations, calculate the missing values:

1. The Avoyelles plant purchased $143,000 of direct materials during June. Beginning direct materials inventory was $9,000, and direct materials used in production were $110,000. What is ending direct materials inventory?
2. Bienville Company produced 8,000 units at an average cost of $11.80 each. The beginning inventory of finished goods was $3,422. (The average unit cost was $11.80.) Bienville sold 8,120 units. How many units remain in ending finished goods inventory?
3. Beginning WIP was $20,000, and ending WIP was $18,750. If total manufacturing costs were $40,000, what was the cost of goods manufactured?
4. If the conversion cost is $84 per unit, the prime cost is $70, and the manufacturing cost per unit is $120, what is the direct materials cost per unit?
5. Total manufacturing costs for August were $446,900. Prime cost was $290,000, and beginning WIP was $160,000. The cost of goods manufactured was $512,000. Calculate the cost of overhead for August and the cost of ending WIP.

OBJECTIVE ▸ 4   ## Exercise 1-15 COST OF GOODS MANUFACTURED AND SOLD

Favourite Brands Company produces condensed soups at its Red Deer plant. At the beginning of June, the following information was supplied by its accountant:

Direct materials inventory	$34,000
Work-in-process inventory	24,500
Finished goods inventory	46,000

During June, direct labour cost was $78,000, direct materials purchases were $346,000, and the total overhead cost was $380,600. The inventories at the end of June were:

Direct materials inventory	$56,000
Work-in-process inventory	37,500
Finished goods inventory	56,000

Required:

1. Prepare a cost of goods manufactured statement for June.
2. Prepare a cost of goods sold schedule for June.

OBJECTIVE ▸ 3 4   ## Exercise 1-16 PRIME COST, CONVERSION COST, PREPARATION OF INCOME STATEMENT: MANUFACTURING FIRM

Roundabout Shoe Company makes walking shoes. During the past calendar year, a total of 90,000 pairs of shoes were made, and 89,000 were sold for $54.00 per pair. The actual unit cost per pair of shoes is as follows:

Direct materials	$13.20
Direct labour	5.80
Variable overhead	3.50
Fixed overhead	16.75
Total unit cost	$39.25

The selling expenses consisted of a commission of $2.70 per pair sold and advertising co-payments totalling $236,000. Administrative expenses, all fixed, equalled $183,000. There were no beginning and ending work-in-process inventories. Beginning finished goods inventory was $235,500 for 6,000 pairs of shoes.

Required:

1. Calculate the number and the dollar value of walking shoes in ending finished goods inventory.
2. Prepare a cost of goods sold statement.
3. Prepare an absorption-costing income statement.

Exercise 1-17 COST OF GOODS MANUFACTURED AND SOLD

OBJECTIVE ➤ 4

Lucero Company, a manufacturing firm, has supplied the following information from its accounting records for the past calendar year:

Direct labour cost	$206,780
Purchases of direct materials	160,400
Freight-in on materials	830
Factory supplies used	37,800
Factory utilities	46,000
Commissions paid	47,562
Factory supervision and indirect labour	190,000
Advertising	145,600
Materials handling	26,750
Work-in-process inventory, January 1	201,000
Work-in-process inventory, December 31	98,000
Direct materials inventory, January 1	47,000
Direct materials inventory, December 31	17,000
Finished goods inventory, January 1	18,000
Finished goods inventory, December 31	62,700

Required:

1. Prepare a cost of goods manufactured statement.
2. Prepare a cost of goods sold statement.

Exercise 1-18 INCOME STATEMENT, DIRECT AND INDIRECT COST CONCEPTS, SERVICE COMPANY

OBJECTIVE ➤ 3 4

Janine Wellington owns and operates a package mailing store near a university. Her store, Send 'n' Deliver, helps customers wrap items and send them via UPS, FedEx, and Canada Post. Send 'n' Deliver also rents mailboxes to customers by the month. In May, purchases of materials (stamps, cardboard boxes, tape, Styrofoam peanuts, bubble wrap, etc.) equalled $11,450; the beginning inventory of materials was $1,050, and the ending inventory of materials was $950. Payments for direct labour during the month totalled $5,570. Overhead incurred was $8,130 (including rent, utilities, and insurance, as well as payments of $4,050 to UPS and FedEx for the delivery services sold). Since Send 'n' Deliver is a franchise, Janine owes a monthly franchise fee of 5 percent of sales. She spent $750 on advertising during the month. Other administrative costs (including accounting and legal services and a trip to Calgary for training) amounted to $3,650 for the month. Revenues for May were $36,100.

Required:

1. What was the cost of materials used for packaging and mailing services during May?
2. What was the prime cost for May?

3.   What was the conversion cost for May?
4.   What was the total cost of services for May?
5.   Prepare an income statement for May.
6.   Of the overhead incurred, is any of it direct? Indirect? Explain.

OBJECTIVE ➤ 1   Exercise 1-19 **PRODUCT COST DEFINITIONS, VALUE CHAIN**

Millennium Pharmaceuticals Inc. (MPI) designs and manufactures a variety of drugs. One new drug, Glaxane, has been in development for seven years. Health Canada approval has just been received, and MPI is ready to begin production and sales.

Required:

Which costs in the value chain would be considered by each of the following managers in their decision regarding Glaxane?

1.   Shelly Roberts is plant manager of the New Glasgow, Nova Scotia plant where Glaxane will be produced. Shelly has been assured that Glaxane capsules will use well-understood processes and not require additional training or capital investment.
2.   Leslie Bothan is vice president of marketing. Leslie's job involves pricing and selling Glaxane. Because Glaxane is the first drug in its "drug family" to be commercially produced, there is no experience with potential side effects. Extensive testing did not expose any real problems (aside from occasional heartburn and insomnia), but the company cannot be sure that such side effects do not exist.
3.   Dante Fiorello is chief of research and development. His charge is to ensure that all research projects, taken as a whole, eventually produce drugs that can support the R&D labs. He is assessing the potential for further work on drugs in the Glaxane family.

OBJECTIVE ➤ 3 4   Exercise 1-20 **DIRECT MATERIALS COST, PRIME COST, CONVERSION COST, COST OF GOODS MANUFACTURED**

Tremblay Company provided the following information for the past calendar year:

Beginning inventory:	
Direct materials	$59,000
Work in process	13,000
Finished goods	34,000
Ending inventory:	
Direct materials	27,500
Work in process	14,500
Finished goods	70,100

During the year, direct materials purchases amounted to $125,000, direct labour cost was $320,000, and overhead cost was $490,000. During the year, 50,000 units were completed.

Required:

1.   Calculate the total cost of direct materials used in production.
2.   Calculate the cost of goods manufactured. Calculate the unit manufacturing cost.
3.   Of the unit manufacturing cost calculated in Requirement 2, $3.20 is direct materials and $9.80 is overhead. What is the prime cost per unit? Conversion cost per unit?

OBJECTIVE ➤ 4   Exercise 1-21 **COST OF GOODS SOLD, INCOME STATEMENT**

Refer to **Exercise 1-20**. Last calendar year, Tremblay recognized revenue of $1,320,000 and had selling and administrative expenses of $204,600.

Required:

1.   What is the cost of goods sold for last year?
2.   Prepare an income statement for Tremblay for last year.

Problems

Problem 1-22 COST ASSIGNMENT METHODS

Brody Company makes industrial cleaning solvents. Various chemicals, detergent, and water are mixed together and then bottled in 40-litre drums. Brody provided the following information for last year:

Raw materials purchases	$250,000
Direct labour	140,000
Depreciation on factory equipment	45,000
Depreciation on factory building	30,000
Depreciation on headquarters building	50,000
Factory insurance	15,000
Property taxes:	
Factory	20,000
Headquarters	18,000
Utilities for factory	34,000
Utilities for sales office	1,800
Administrative salaries	150,000
Indirect labour salaries	156,000
Sales office salaries	90,000
Beginning balance, Raw Materials	124,000
Beginning balance, Work in Process	124,000
Beginning balance, Finished Goods	84,000
Ending balance, Raw Materials	102,000
Ending balance, Work in Process	130,000
Ending balance, Finished Goods	82,000

Last year, Brody completed 100,000 units. Sales revenue equalled $1,200,000, and Brody paid a sales commission of 5 percent of sales.

Required:

1. Calculate the direct materials used in production for last year.
2. Calculate total prime cost.
3. Calculate total conversion cost.
4. Prepare a cost of goods manufactured statement for last year. Calculate the unit product cost.
5. Prepare a cost of goods sold statement for last year.
6. Prepare an income statement for last year. Show the percentage of sales that each line item represents.

Problem 1-23 INCOME STATEMENT, COST OF GOODS MANUFACTURED

Spencer Company produced 200,000 cases of sports drinks during the past calendar year. Each case of 1-litre bottles sells for $36. Spencer had 2,500 cases of sports drinks in finished goods inventory at the beginning of the year. At the end of the year, there were 11,500 cases of sports drinks in finished goods inventory. Spencer's accounting records provide the following information:

Purchases of direct materials	$2,350,000
Direct materials inventory, January 1	290,000
Direct materials inventory, December 31	112,000
Direct labour	1,100,000
Indirect labour	334,000
Depreciation, factory building	525,000
Depreciation, factory equipment	416,000
Property taxes on factory	65,000
Utilities, factory	150,000
Insurance on factory	200,000

(continued)

Salary, sales supervisor	$ 85,000
Commissions, salespersons	216,000
Advertising	500,000
General administration	390,000
Work-in-process inventory, January 1	450,000
Work-in-process inventory, December 31	750,000
Finished goods inventory, January 1	107,500
Finished goods inventory, December 31	488,750

Required:

1. Prepare a cost of goods manufactured statement.
2. Compute the cost of producing one case of sports drink last year.
3. Prepare an income statement on an absorption-costing basis. Include a column showing the percent of each line item of sales. (Round your percentage answers to two significant digits, e.g., 45.67%.)

OBJECTIVE ▶ 2 4

Problem 1-24 COST OF GOODS MANUFACTURED, COST IDENTIFICATION, SOLVING FOR UNKNOWNS

Skilz-Accountants Company creates, produces, and sells CD-ROM-based CA review courses for individual use. Jeretta Chan, head of human resources, is convinced that question development employees must have strong analytical and problem-solving skills. She has asked Terrell Slater, controller for Skilz-Accountants, to help develop problems for use in screening applicants before they are interviewed. One of the problems Terrell has developed is based on the following data for a mythical company for the previous year:

a. Conversion cost was $240,000 and was three times the prime cost.
b. Direct materials used in production equalled $45,000.
c. Cost of goods manufactured was $295,000.
d. Ending work in process is 20 percent of the cost of beginning work in process.
e. There are no beginning or ending inventories for direct materials.
f. Cost of goods sold was 80 percent of cost of goods manufactured.
g. Beginning finished goods inventory was $14,400.

Required:

1. Using the above information, prepare a cost of goods manufactured statement.
2. Using the above information, prepare a cost of goods sold statement.

OBJECTIVE ▶ 3 4

SERVICE

Problem 1-25 INCOME STATEMENT, COST OF SERVICES PROVIDED, SERVICE ATTRIBUTES

Mason, Singh, and Westbrook (MSW) is a tax services firm. The firm is located in Thunder Bay, Ontario, and employs 15 professionals and eight staff. The firm does tax work for small businesses and well-to-do individuals. The following data are provided for the past fiscal year. (The Mason, Singh, and Westbrook fiscal year runs from July 1 through June 30.)

Returns processed	3,000
Returns in process, beginning of year	$ 44,000
Returns in process, end of year	13,000
Cost of services sold	1,557,500
Beginning direct materials inventory	20,000
Purchases, direct materials	40,000
Direct labour	1,400,000
Overhead	100,000
Administrative expenses	257,000
Selling expenses	65,000

Required:

1. Prepare a statement of cost of services sold.
2. Refer to the statement prepared in Requirement 1. What is the dominant cost? Will this always be true of service organizations? If not, provide an example of an exception.
3. Assuming that the average fee for processing a return is $850, prepare an income statement for Mason, Singh, and Westbrook.
4. Discuss three differences between services and tangible products. Calculate the average cost of preparing a tax return for last year. How do the differences between services and tangible products affect the ability of MSW to use the past year's average cost of preparing a tax return in budgeting the cost of tax return services to be offered next year?

Problem 1-26 COST OF GOODS MANUFACTURED, INCOME STATEMENT

OBJECTIVE ➤ 3 4

Paulisse Company produces hand lotion for resale by discount chains. For last year, Paulisse reported the following:

Work-in-process inventory, January 1	$ 13,250
Work-in-process inventory, December 31	28,250
Finished goods inventory, January 1	113,000
Finished goods inventory, December 31	85,000
Direct materials inventory, January 1	16,200
Direct materials inventory, December 31	10,700
Direct materials used	170,200
Direct labour	72,000
Plant depreciation	9,500
Salary, production supervisor	45,000
Indirect labour	40,600
Utilities, factory	5,700
Sales commissions	40,000
Salary, sales supervisor	75,000
Depreciation, factory equipment	25,000
Administrative expenses	162,000
Supplies (40% used in the factory, 60% used in the sales office)	8,000

Last year, Paulisse produced 230,000 units and sold 250,000 units at $4 per unit.

Required:

1. Prepare a statement of cost of goods manufactured.
2. Prepare an absorption-costing income statement.

Problem 1-27 ETHICAL ISSUES

OBJECTIVE ➤ 5

John Biggs and Patty Jorgenson are both cost accounting managers for a division of a service firm. During lunch yesterday, Patty told John that she was planning on quitting her job in three months because she had accepted a position as controller of a small company in a neighbouring province. The starting date was timed to coincide with the retirement of the current controller. Patty was excited because it allowed her to live near her family. Today, the divisional controller took John to lunch and informed him that he was taking a position at headquarters and that he had recommended that Patty be promoted to his position. He indicated to John that it was a close call between him and Patty and that he wanted to let John know personally about the decision before it was announced officially.

Required:

What should John do? Describe how you would deal with his ethical dilemma (considering the CMA Code of Professional Ethics for management accountants in your response).

OBJECTIVE ➤ 5   ## Problem 1-28 ETHICAL ISSUES

Emily Thibauld, controller of an oil exploration division, has just been approached by Tim Wilson, the divisional manager. Tim told Emily that the projected quarterly profits were unacceptable and that expenses need to be reduced. He suggested that a clean and easy way to reduce expenses would be to assign the exploration and drilling costs of four dry holes to those of two successful holes. The costs could then be capitalized and not expensed, reducing the costs that needed to be recognized for the quarter. He further argued that the treatment would be reasonable because the exploration and drilling all occurred in the same field; thus, the unsuccessful efforts really were the costs of identifying the successful holes. "Besides," he argued, "even if the treatment is wrong, it can be corrected in the annual financial statements. Next quarter's revenues will be more and can absorb any reversal without causing any severe damage to that quarter's profits. It's this quarter's profits that need some help."

Emily is uncomfortable with the request because generally accepted accounting principles do not sanction the type of accounting measures proposed by Tim.

Required:

1. Using the CMA Code of Professional Ethics for management accountants, recommend the approach that Emily should take.
2. Suppose Tim insists that his suggested accounting treatment be implemented. What should Emily do?

CMA Problems

OBJECTIVE ➤ 5

CMA Problem 1-1 ETHICAL ISSUES*

Silverado Inc. is a closely held brokerage firm that has been very successful over the past five years, consistently providing most members of the top management group with 50 percent bonuses. In addition, both the chief financial officer and the chief executive officer have received 100 percent bonuses. Silverado expects this trend to continue.

Recently, the top management group of Silverado, which holds 40 percent of the outstanding shares of common stock, has learned that a major corporation is interested in acquiring Silverado. Silverado's management is concerned that this corporation may make an attractive offer to the other shareholders and that management would be unable to prevent the takeover. If the acquisition occurs, this executive group is uncertain about continued employment in the new corporate structure. As a consequence, the management group is considering changes to several accounting policies and practices that, although not in accordance with generally accepted accounting principles, would make the company a less attractive acquisition. Management has told Larry Stewart, Silverado's controller, to implement some of these changes. Stewart has also been informed that Silverado's management does not intend to disclose these changes at once to anyone outside the immediate top management group.

Required:

Using the CMA Code of Professional Ethics for management accountants, evaluate the changes that Silverado's management is considering, and discuss the specific steps that Larry Stewart should take to resolve the situation. *(CMA adapted)*

OBJECTIVE ➤ 5   ### CMA Problem 1-2 ETHICAL ISSUES*

Emery Manufacturing Company produces component parts for the farm equipment industry and has recently undergone a major computer system conversion. Jake Murray, the controller, has established a troubleshooting team to alleviate accounting problems that have occurred since the conversion. Jake has chosen Gus Swanson, assistant controller, to head the team, which will include Linda Wheeler, cost accountant; Cindy Madsen, financial analyst; Randy Leung, general accounting supervisor; and Max Crandall, financial accountant.

The team has been meeting weekly for the past month. Gus insists on being part of all the team conversations in order to gather information, to make the final decision on any ideas or actions that the team develops, and to prepare a weekly report for Jake. He has also used this team as a forum to discuss issues and disputes about him and other members of Emery's top management team. At last week's meeting, Gus told the team that he thought a competitor might purchase the common stock of Emery, because he had overheard Jake talking about this on the telephone. As a result, most of Emery's employees now informally discuss the sale of Emery's common stock and how it will affect their jobs.

Required:

Is Gus Swanson's discussion with the team about the prospective sale of Emery unethical? Discuss, citing specific standards from the CMA Code of Professional Ethics for management accountants to support your position. *(CMA adapted)*

CMA Problem 1-3 ETHICAL ISSUES*

OBJECTIVE ▶ 5

SERVICE

The external auditors for Heart Health Procedures (HHP) are currently performing the annual audit of HHP's financial statements. As part of the audit, the external auditors have prepared a representation letter to be signed by HHP's chief executive officer (CEO) and chief financial officer (CFO). The letter provides, among other items, a representation that appropriate provisions have been made for:

> *Reductions of any excess or obsolete inventories to net realizable values, and Losses from any purchase commitments for inventory quantities in excess of requirements or at prices in excess of market.*

HHP began operations by developing a unique balloon process to open obstructed arteries to the heart. In the past several years, HHP's market share has grown significantly because its major competitor was forced by Health Canada to cease its balloon operations. HHP purchases the balloon's primary and most expensive component from a sole supplier. Two years ago, HHP entered into a five-year contract with this supplier at the then current price, with inflation escalators built into each of the five years. The long-term contract was deemed necessary to ensure adequate supplies and discourage new competition. However, during the past year, HHP's major competitor developed a technically superior product, which utilizes an innovative, less costly component. This new product was recently approved by Health Canada and has been introduced to the medical community, receiving high acceptance. It is expected that HHP's market share, which has already seen softness, will experience a large decline and that the primary component used in the HHP balloon will decrease in price as a result of the competitor's use of its recently developed superior, cheaper component. The new component has been licensed by the major competitor to several outside supply sources to maintain available quantity and price competitiveness. At this time, HHP is investigating the purchase of this new component.

HHP's officers are on a bonus plan that is tied to overall corporate profits. Jim Honig, vice president of manufacturing, is responsible for both manufacturing and warehousing. During the course of the audit, he advised the CEO and CFO that he was not aware of any obsolete inventory nor any inventory or purchase commitments where current or expected prices were significantly below acquisition or commitment prices. Jim took this position even though Marian Nevins, assistant controller, had apprised him of both the existing excess inventory attributable to the declining market share and the significant loss associated with the remaining years of the five-year purchase commitment.

Marian has brought this situation to the attention of her superior, the controller, who also participates in the bonus plan and who reports directly to the CFO. Marian worked closely with the external audit staff and subsequently ascertained that the external audit manager was unaware of the inventory and purchase commitment problems. Marian is concerned about the situation and is not sure how to handle the matter.

Required:

1. Assuming that the controller did not apprise the CEO and CFO of the situation, explain the ethical considerations of the controller's apparent lack of action by discussing specific provisions of the CMA Code of Professional Ethics for management accountants.

2. Assuming Marian Nevins believes the controller has acted unethically and not apprised the CEO and CFO of the findings, describe the steps that she should take to resolve the situation. Refer to the CMA Code of Professional Ethics for management accountants in your answer.

3. Describe actions that HHP can take to improve the ethical situation within the company. *(CMA adapted)*

The Collaborative Learning Exercises can be found on the product support site at www.hansen1ce.nelson.com.

© GERENME/ISTOCK

CHAPTER

Cost Behaviour

2

Costs can display variable, fixed, or mixed behaviour. Knowing how costs change as activity changes is essential to planning, controlling, and decision making. For example, budgeting, deciding to keep or drop a product line, and evaluating the performance of a segment all depend on an understanding of cost behaviour. Not knowing and understanding cost behaviour can lead to poor—and even disastrous—decisions. This chapter discusses cost behaviour in depth so that a proper foundation is laid for its use in studying other cost management topics. Cost-volume-profit analysis (Chapter 3) and variable-costing systems (Chapter 12), for example, require that all costs be classified as fixed or variable. This chapter describes ways of separating costs into fixed and variable categories, discusses the assumptions and limitations underlying these methods, and assesses the reliability of these procedures.

OBJECTIVE ▶ 1
Define and describe fixed, variable, and mixed costs.

Basics of Cost Behaviour

Cost behaviour is the term used to describe whether a cost changes when the level of output changes. A cost that does not change as output changes is a *fixed cost*. A *variable cost*, on the other hand, increases in total with an increase in output and decreases in total with a decrease in output. While economics may *assume* that fixed and variable costs are known, in the real world, management accountants must determine them. Let's first review the basics of cost behaviour. Then, we will look at fixed, variable, and mixed costs. Finally, we will assess the impact of time horizon on cost behaviour.

Cost Objects

Recall from Chapter 1 that a cost object is the item for which managers want cost information. So the first step is to determine appropriate cost objects. This is relatively easy in a manufacturing firm; the cost object is typically the tangible product. For service firms, the logical cost object is the service. For example, hospitals may view particular services such as blood tests or radiology services as primary cost objects. There are, however, a variety of cost objects for which managers may need to know cost behaviour.

The Internet has fundamentally changed the way companies do business with their suppliers and customers. Price competition is severe so firms cannot, typically, succeed using a low-price strategy. Instead, they use a customer-service strategy. Internet-based companies strive to provide a shopping experience that is user friendly, with an abundance of information tailored to customer needs and a secure payment system. Ideally, the company provides a seamless interface for customers, taking them from information search, through product/service choice, payment, and post-sale follow-up. Software that tracks ongoing customer preferences is a large part of the enhanced customer shopping experience. **Amazon.com** is an excellent example of this, as it welcomes new and returning customers and makes the shopping experience fun and easy. As a result, "Internet-based firms rely much less on traditional infrastructure assets, such as buildings, and more on speakers, specialized software, and intellectual capital that cater to customers in cyberspace." This means that the customer is the appropriate cost object, and activities and drivers that are tied to customer service are important data to Internet-based firms.[1]

Activity Drivers and Measures of Output

The terms *fixed cost* and *variable cost* only have meaning when related to some output measure or driver. Therefore, we must first determine the underlying activities and the associated drivers that measure the output of an activity. For example, materials handling may be measured by the number of moves; shipping goods may be measured by the units sold; and laundering hospital linen may be measured by the kilograms of laundry. The choice of driver is tailored not only to the particular firm but also to the particular activity or cost being measured.

Activity drivers explain changes in activity costs by measuring changes in activity output (usage). The two general categories of activity drivers are *unit-level drivers* and *non-unit-level drivers*. Recall that drivers are factors that *cause* changes in resource usage, activity usage, costs, and revenues. **Unit-level drivers** explain changes in cost as units produced change. Kilograms of direct materials, kilowatt-hours used to run production machinery, and direct labour hours are examples of unit-based activity drivers. Each of these drivers varies proportionately with the number of units produced. **Non-unit-level drivers** explain how costs change as factors other than the number of units produced change. Examples of non-unit-based output measures

[1] Taken from Lawrence A. Gordon and Martin P. Loeb, "Distinguishing Between Direct and Indirect Costs Is Crucial for Internet Companies," *Management Accounting Quarterly* II, no. 4 (Summer 2001): 12–17.

include the number of setups, work orders, engineering change orders, inspection hours, and material moves.

In a traditional cost management system, cost behaviour is assumed to be described by unit-based drivers only. In an activity-based cost management system, both unit- and non-unit-based drivers are used. Thus, the ABC system tends to produce a much richer view of cost behaviour than would a traditional, unit-based, system. As a result, under the ABC system, cost behaviour patterns for a much broader set of activities must now be identified.

We now take a closer look at fixed, variable, and mixed costs.

Fixed Costs

Fixed costs are costs that *in total* are constant within the relevant range as the level of the activity driver varies. To illustrate fixed cost behaviour, consider a plant operated by Echo Audio Systems Inc. that produces speakers for home audio systems. One department in the plant produces a 9-centimetre voice coil and inserts it into each speaker passing through the department. The activity is voice-coil production, and the activity driver is the number of voice coils produced. The department operates two production lines, and each can make up to 100,000 voice coils per year. The production workers of each line are supervised by a production-line manager who is paid $60,000 per year. For production up to 100,000 units, only one manager is needed; for production between 100,001 and 200,000 units, the second line is activated and two managers are needed. The cost of supervision for several levels of production for the plant is given as follows:

Echo Audio Systems Inc.

Supervision Cost	Voice Coils Produced	Unit Cost
$ 60,000	40,000	$1.50
60,000	80,000	0.75
60,000	100,000	0.60
120,000	120,000	1.00
120,000	160,000	0.75
120,000	200,000	0.60

The first step in assessing cost behaviour is defining an appropriate activity driver. In this case, the activity driver is the number of voice coils produced. The second step is defining what is meant by **relevant range**, the range over which the assumed cost relationship is valid for the normal operations of a firm. Suppose that the relevant range is 120,000 to 200,000 speakers processed. Notice that the *total* cost of supervision remains constant within this range as more voice coils are produced. Echo Audio Systems pays $120,000 for supervision regardless of whether it produces 120,000, 160,000, or 200,000 voice coils.

Pay particular attention to the words *in total* in the definition of fixed costs. While the total cost of supervision remains unchanged as more voice coils are produced, the unit cost does change as the level of the activity driver changes. As the example shows, within the relevant range, the unit cost of supervision decreases from $1.00 to $0.60. Because of the behaviour of per-unit fixed costs, it is easy to get the impression that fixed costs are affected by changes in the level of the activity driver, when in reality they are not. Unit fixed costs can be misleading and may adversely affect some decisions. It is often safer to work with total fixed costs.

Exhibit 2-1 is a graph of fixed cost behaviour. For the relevant range, fixed cost behaviour is described by a horizontal line. Notice that for 120,000 voice coils produced, supervision cost is $120,000; for 160,000 voice coils produced, supervision cost is still $120,000. This line visually demonstrates that cost remains unchanged as the level of the activity driver varies. For the relevant range, total fixed costs can be represented by the following simple linear equation:

$$F = \text{Total fixed costs}$$

Exhibit 2-1

Fixed Cost Behaviour

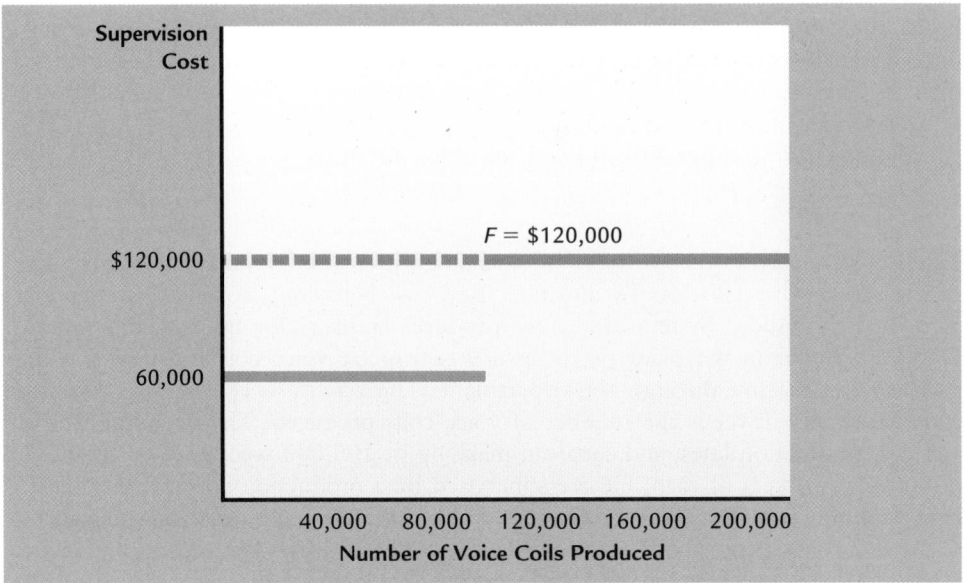

In the example for Echo Audio Systems, supervision cost amounted to $120,000 for any level of output between 100,001 and 200,000 voice coils produced. Thus, supervision is a fixed cost, and the fixed cost equation in this case is $F = \$120,000$. Strictly speaking, this equation assumes that the fixed costs are $120,000 for all levels (as if the line extends to the vertical axis as indicated by the dashed portion in Exhibit 2-1). Although this assumption is not true, it is harmless as long as the operating decisions are confined to the relevant range.

Can fixed costs change? Of course they can, but this does not make them variable. They are fixed at a new higher (or lower) level. Suppose that Echo Audio Systems gives a raise to the voice coil line supervisors. Instead of being paid $60,000 per year, they are paid $64,000 per year. Now the cost of supervision is $128,000 per year ($2 \times \$64,000$). However, supervisory costs are still fixed with respect to the number of voice coils produced. Can you draw in the new fixed cost line on Exhibit 2-1?[2]

Variable Costs

Variable costs are defined as costs that, in total, vary in direct proportion to changes in an activity driver. To illustrate, let's expand the Echo Audio Systems example to examine the direct materials cost of the voice coils. The cost is the cost of direct materials for the voice coils, and the activity driver is the number of voice coils produced. Each voice coil requires direct materials costing $3. The total direct materials cost of voice coils for various levels of production is given as follows:

Echo Audio Systems Inc.

Total Direct Materials Cost of Voice Coils	Voice Coils Produced	Unit Direct Materials Cost of Voice Coils
$120,000	40,000	$3
240,000	80,000	3
360,000	120,000	3
480,000	160,000	3
600,000	200,000	3

[2] The new line is a horizontal line that intersects the y-axis at $128,000. Note that it is drawn parallel to and above the original fixed cost line.

As more voice coils are produced, the total cost of direct materials increases in direct proportion. For example, as production doubles from 80,000 to 160,000 units, the *total* cost of voice coils doubles from $240,000 to $480,000. Notice also that the unit cost of voice coils is constant.

Variable costs can also be represented by a linear equation. Here, total variable costs depend on the level of activity driver. This relationship can be described by the following equation:

$$Y_v = VX$$

where

Y_v = Total variable costs
V = Variable cost per unit
X = Number of units of the driver

The relationship describing the cost of direct materials is $Y_p = \$3X$, where X = the number of voice coils produced. Exhibit 2-2 shows graphically that variable cost behaviour is represented by a straight line coming from the origin. At zero units processed, total variable cost is zero. However, as units produced increase, the total variable cost also increases. Note that total variable cost increases in direct proportion to increases in the number of voice coils produced (the activity driver); the rate of increase is measured by the slope of the line. At 120,000 voice coils produced, the total variable cost of direct materials is $360,000 ($3 × 120,000); at 160,000 voice coils produced, the total variable cost is $480,000 ($3 × 160,000).

Linearity Assumption

The definition of variable costs just given and the graph in Exhibit 2-2 imply a linear relationship between the cost of direct materials and the number of voice coils produced. How reasonable is the assumption that costs are linear? Do costs really increase in direct proportion to increases in the level of the activity driver? If not, then how well does this assumed linear cost function approximate the underlying cost function?

Economists usually assume that variable costs increase at a decreasing rate up to a certain volume, at which point they increase at an increasing rate. This type of

Exhibit 2-2

Variable Cost Behaviour

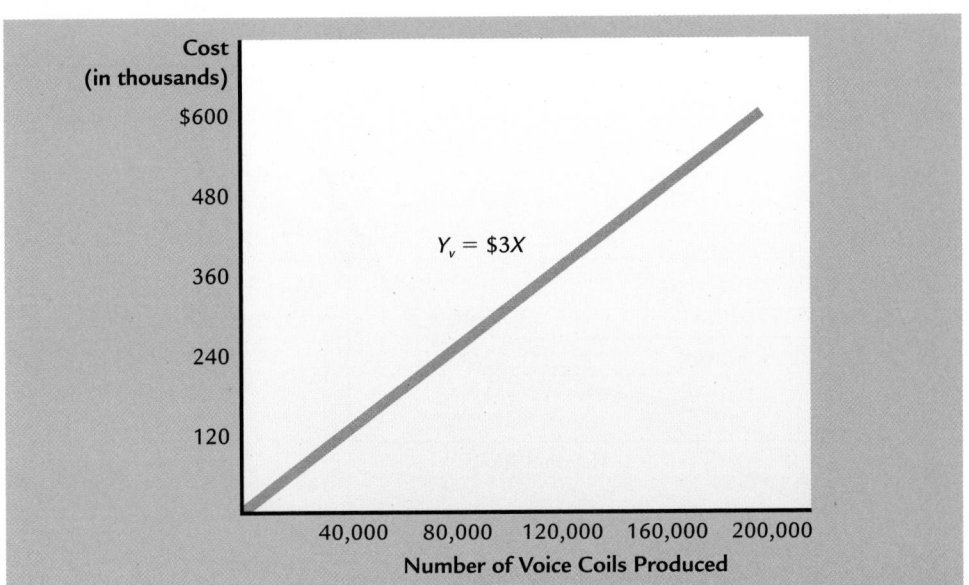

nonlinear behaviour is displayed in Exhibit 2-3. Here, variable costs increase as the number of units increases, but not in direct proportion.

If the nonlinear view more accurately portrays reality, what should we do? One possibility is to determine the actual cost function—but every activity could have a different cost function, and this approach could be very time consuming and expensive (if it can even be done). It is much simpler to assume a linear relationship.

If the linear relationship is assumed, then the main concern is how well this assumption approximates the underlying cost function. Exhibit 2-4 gives us some idea of the consequences of assuming a linear cost function. As with fixed costs, we can define the *relevant range* as the range of activity for which the assumed cost

Exhibit 2-3

Nonlinearity of Variable Costs

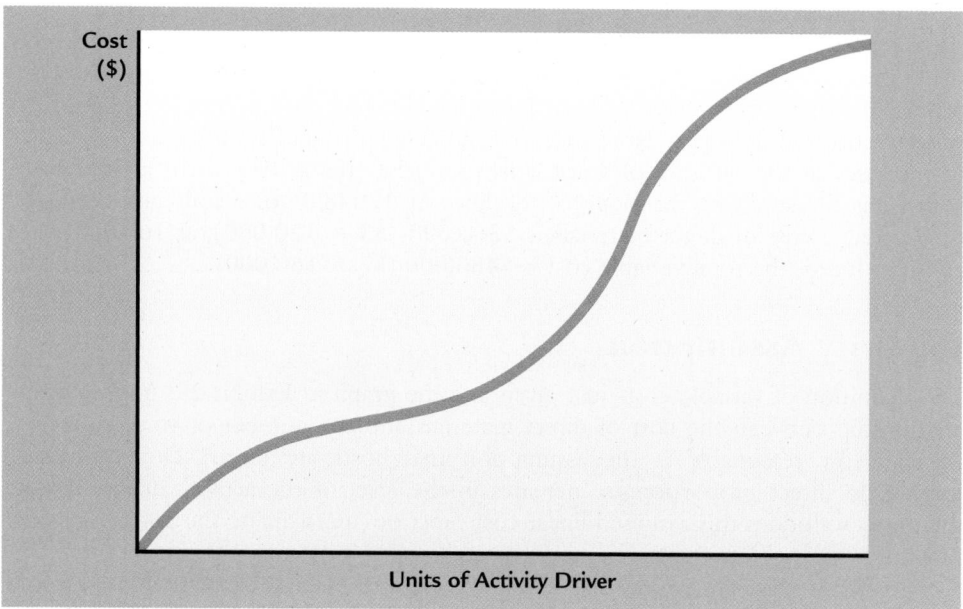

Exhibit 2-4

Relevant Range for Variable Costs

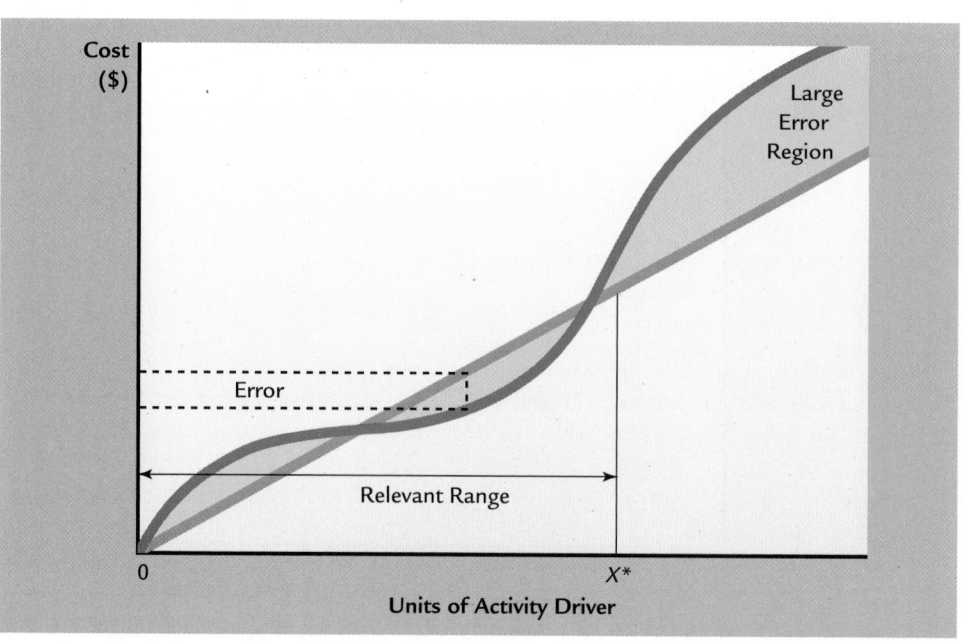

The HOW and WHY of Forming an Equation to Describe Mixed Cost

**CORNERSTONE
2 - 1**

Information:
Echo Audio Systems has 10 sales representatives, each earning a salary of $30,000 per year plus a commission of $5 per speaker sold. Last year, 100,000 speakers were sold.

Why:
As long as 100,000 speakers is in the relevant range, then a straight line depicts the cost relationship well. If the cost function is known, sensitivity analysis can be used to see what the total selling cost would be at differing levels of sales.

Required:
1. Develop a cost equation for total selling cost.
2. Compute the total variable selling cost last year.
3. Compute the total selling cost last year.
4. Compute the unit selling cost for last year.
5. *What if* 110,000 speakers had been sold last year? What would be the total selling cost and the unit selling cost? Explain why the unit selling cost decreased.

Solution:
1. Total selling cost = Fixed selling cost + (Variable rate × Units sold)

    = $300,000 + ($5 × Units sold)

2. Total variable selling cost = Variable rate × Units sold

    = $5 × 100,000

    = $500,000

3. Total selling cost = $300,000 + ($5 × Units sold)

    = $300,000 + $500,000

    = $800,000

4. Unit selling cost = Total selling cost/Units sold

    = $800,000/100,000

    = $8

5. Total selling cost = $300,000 + ($5 × 110,000) = $850,000

    Unit selling cost = $850,000/110,000 = $7.73 (rounded)

    The unit selling cost went down because the fixed cost, which stays the same, is spread out over a greater number of units.

relationships are valid. Here, validity refers to how closely the linear cost function approximates the underlying cost function. Note that for units of the activity driver beyond X^*, the approximation appears to break down.

Mixed Costs

Mixed costs are costs that have both a fixed and a variable component. For example, sales representatives are often paid a salary plus a commission on sales. Suppose that Echo Audio Systems has 10 sales representatives, each earning a salary of $30,000 per year plus commission of $5 per speaker sold. The activity is selling, and the activity driver is units sold. If 100,000 speakers are sold, then the total selling cost (associated with the sales representatives) is $800,000—the sum of the fixed salary cost of $300,000 (10 × $30,000) and the variable cost of $500,000 ($5 × 100,000). Cornerstone 2-1 shows how and why the linear equation can be used to describe a mixed cost.

Exhibit 2-5

Mixed Cost Behaviour

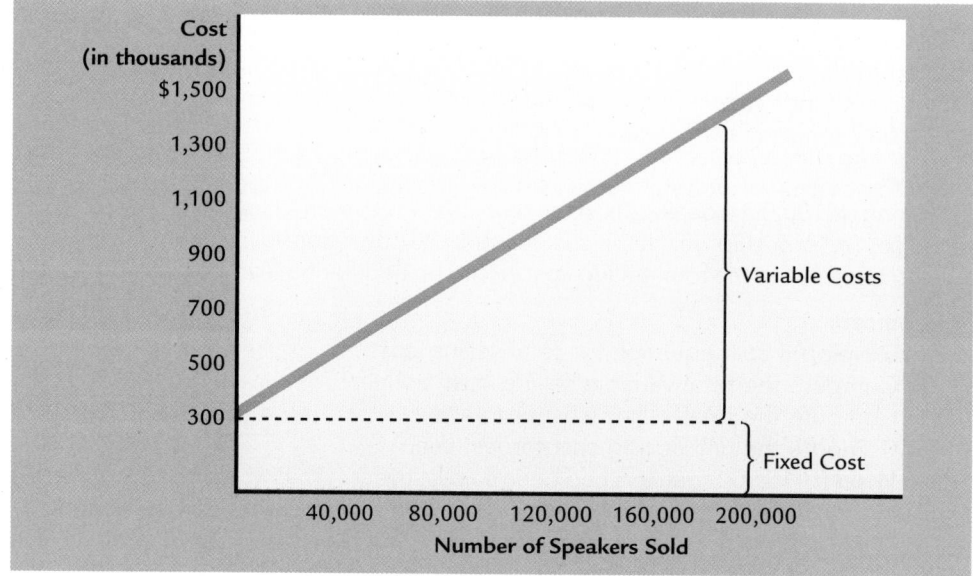

The graph for our mixed cost example is given in Exhibit 2-5. (The graph assumes that the relevant range is 0 to 200,000 units.) Mixed costs are represented by a line that intercepts the vertical axis (at $300,000, for this example). The intercept corresponds to the fixed cost component, and the slope of the line gives the variable cost per unit of activity driver (slope is $5 for the example portrayed).

Time Horizon

Determining whether a cost is fixed or variable depends on the time horizon. According to economics, in the **long run**, all costs are variable; in the **short run**, at least one cost is fixed. But how long is the short run? Different costs have short runs of different lengths. Direct materials, for example, are relatively easy to adjust. **Tim Hortons** may treat coffee beans (a direct material) as strictly variable, even though for the next few hours the amount already on hand is fixed. The lease of space for one of its coffee shops, however, is more difficult to adjust; it may run for one or more years. This cost is typically seen as fixed. The length of the short-run period depends to some extent on management judgment and the purpose for which cost behaviour is being estimated. For example, submitting a bid on a one-time, special order may span only a month—long enough to create a bid and produce the order. Other types of decisions, such as product mix decisions, will affect costs over a much longer period of time. In this case, the costs that must be considered are long-run variable costs, including product design, product development, market development, and market penetration. Short-run costs often do not adequately reflect all the costs necessary to design, produce, market, distribute, and support a product. Recently, there have been some insights that help shed light on the nature of long- and short-run cost behaviours.[3] These insights relate to activities and the resources needed to enable an activity to be performed.

OBJECTIVE ▸ 2

Explain the use of resources and activities and their relationship to cost behaviour.

Resources, Capacity and Cost Behaviour

Resources are economic elements that enable one to perform activities. Common resources of a manufacturing plant include direct materials, direct labour, electricity,

---

[3] For more on these concepts, see the following. Robert S. Kaplan and Robin Cooper, *Cost & Effect: Using Integrated Cost Systems to Drive Profitability and Performance*, 9th ed. (Cambridge, MA: Harvard Business Press, 1998); and Alfred M. King, "The Current Status of Activity-Based Costing: An Interview with Robin Cooper and Robert S. Kaplan," *Management Accounting* (September 1991): 22–26.

equipment, and so on. When a company spends money on resources, it is *acquiring* the ability or capacity to perform an activity.

Activity is a task, such as setting up equipment, purchasing materials, assembling materials, and packing completed units in boxes. When a firm acquires the resources needed to perform an activity, it obtains **activity capacity**. Usually, the amount of activity capacity needed corresponds to the level where the activity is performed efficiently. This efficient level of activity performance is called **practical capacity**.

If all of the activity capacity acquired is not used, then there is **unused capacity**, which is the difference between the acquired capacity and the actual amount of the activity used. The relationship between resource spending and resource usage can be used to define variable and fixed cost behaviour.

Flexible Resources

Resources can be categorized as (1) flexible and (2) committed. **Flexible resources** are supplied as used and needed. The organization is free to buy what it needs, when it needs it, so the quantity of the resource supplied equals the quantity demanded. There is no unused capacity for this category of resources (resources used equal resources supplied).

Since the cost of flexible resources equals the cost of resources used, the total cost of the resource increases as demand for the resource increases. The cost of a flexible resource is a variable cost. For example, in a just-in-time manufacturing environment, materials are purchased when needed and are used right away. Thus, as the units produced increase, the amount (and cost) of direct materials increases proportionately. Similarly, power is a flexible resource. Using kilowatt-hours as the driver, as the demand for power increases, the cost of power increases. Note that in each example, resource supply and usage is measured by an output measure, or driver.

Committed Resources

Committed resources are supplied in advance of usage. An explicit or implicit contract is used to obtain a given quantity of resource, regardless of whether that amount is fully used or not. Because the amount of committed resource supplied may exceed the firm's demand for it, unused capacity is possible.

Many resources are acquired before the actual demands for the resource are realized. There are two examples of this category of resource acquisition. First, organizations acquire *multi-period service capacities* by paying cash up front or by entering into an explicit contract that requires periodic cash payments. Buying or leasing buildings and equipment are examples of this form of advance resource acquisition. The annual expense associated with the multi-period category is independent of actual usage of the resource. Often, these expenses are referred to as **committed fixed expenses**. They essentially correspond to committed resources—costs incurred that provide long-term activity capacity.

Discretionary Fixed Resources Some organizations acquire resources in advance through implicit contracts—usually with their employees. These implicit contracts require an ethical focus, since they imply that the organization will maintain employment and salary levels even though there may be temporary downturns in the quantity of activity used. Hiring three engineers for $150,000 who can supply the capacity of processing 7,500 change orders (the driver) is an example of implicit contracting. Often, in response to customer feedback and competitive pressures, products need to be redesigned or modified. An engineering change order is the document that initiates this process. Certainly, none of the three engineers would expect to be laid off if only 5,000 change orders were actually processed—unless, of course, the downturn in demand is viewed as being permanent.

Companies can manage economic ups and downs with lower-level salaries and then vary the level of bonuses at the end of the year. In addition, many companies use a lower level of permanent employees and a fluctuating level of temporary, or contingent, workers. This is a growing trend that includes both manufacturing and service industries as well as unskilled (e.g., day labourers) and skilled workers (e.g., nurses and information technology specialists).[4]

> **Google** has had a policy of using temporary staff. By early 2009, the company was rumoured to have asked as many as 10,000 temporary workers to leave. This tactic was taken in response to declining growth in the amount of Internet advertising as the falling economy led many other companies to cut their advertising budgets.[5]

A key reason for the use of contingent workers is flexibility—in meeting demand fluctuations, in controlling downsizing, and in buffering core workers against job loss.[6] Resource spending for this category essentially corresponds to **discretionary fixed expenses**—costs incurred for the acquisition of short-term activity capacity.

Resource Usage Implications for Control and Decision Making

The resource usage model just described can improve both managerial control and decision making. Operational control information systems encourage managers to pay more attention to controlling resource usage and spending. A well-designed operational system would allow managers to assess the changes in resource demands that will occur from new product mix decisions. Adding new, customized products may increase the demand for various overhead activities; if sufficient unused activity capacity does not exist, then resource spending must increase.

Similarly, if resource usage can be reduced, bringing about unused capacity, managers must carefully consider what to do with the excess capacity. Eliminating the excess capacity may decrease resource spending and thus improve overall profits. Alternatively, using the excess capacity to increase output could increase revenues without a corresponding increase in resource spending.

The activity-based resource usage model also allows managers to better calculate the changes in resource supply and demand resulting from decisions such as make or buy, accept or reject special orders, and keep or drop product lines. The model increases the power of a number of traditional management accounting decision-making models. These are explored in the decision-making Chapters 15 to 17, and in Chapter 18 (online at www.hansen1ce.nelson.com).

Step-Cost Behaviour

So far, we have assumed that the cost function is continuous. In reality, some cost functions may be discontinuous. One such discontinuous function, a step function, is shown in Exhibit 2-6. A **step-cost function** displays a constant level of cost for a range of output and then jumps to a higher level of cost at some point, where it remains for a similar range of activity. In Exhibit 2-6, the cost is $100, as long as output is between 0 and 20 units. If the volume is between 20 and 40 units, the cost jumps to $200.

Step-Variable Costs   Items that display a step-cost behaviour must be purchased in chunks. The width of the step defines the range of activity output for which a particular quantity of the resource must be acquired. The width of the step in Exhibit 2-6 is 20 units of activity. If the width of the step is narrow, as in Exhibit 2-6, the cost of the resource changes in response to fairly small changes in resource usage (as measured by activity output). Costs that follow a step-cost behaviour with narrow steps are defined as **step-variable costs**. If the width of the step is narrow, step-variable costs can be approximated by a strictly variable cost.

Step-Fixed Costs   In reality, many so-called fixed costs may be best described by a step-cost function. Many committed resources—particularly those that involve implicit contracting—follow a step-cost function. Suppose, for example, that a company hires three engineers who are responsible for redesigning existing products to meet

[4] Yukako Ono, "Why Do Firms Use Temporary Workers?" *Chicago Fed Letter*, no. 260 (March 2009).

[5] Nick Farrell, "Google Cuts Staff: Doing Evil after Christmas," *The Inquirer*, January 8, 2009, http://www.theinquirer.net/inquirer/news/1050285/google-cuts-staff. (Accessed May 23, 2009.)

[6] "Contingent Employment on the Rise," *Deloitte & Touche Review* (September 4, 1995): 1–2.

Exhibit 2-6

Step-Cost Function

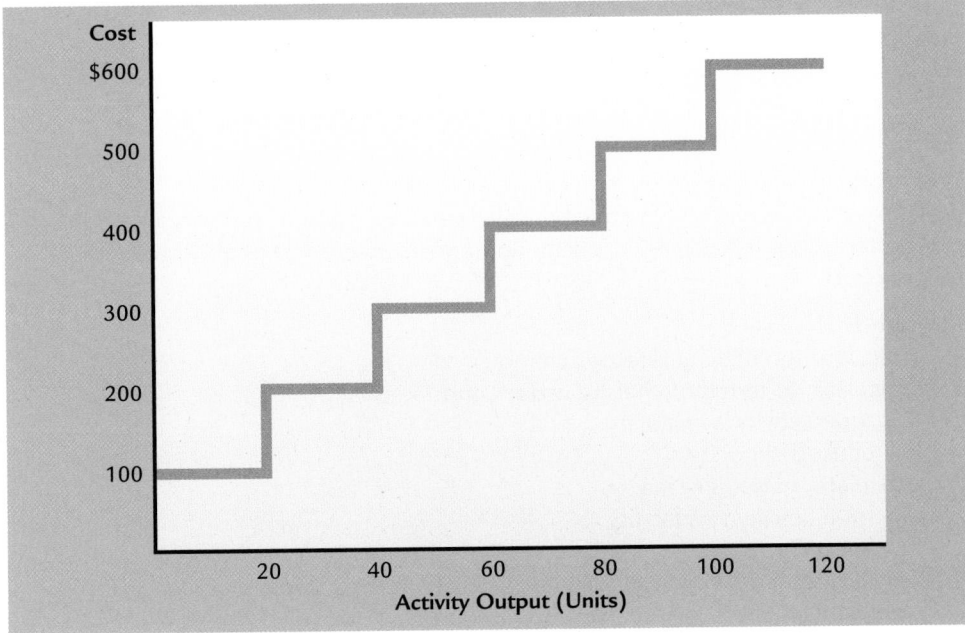

customer requirements. By hiring the engineers, the company has acquired the ability to perform an activity: engineering redesign. The salaries paid to the engineers represent the cost of acquiring the engineering redesign capacity. The number of engineering changes that can be *efficiently* processed by the three engineers is a measure of that capacity. The nature of this resource requires that the capacity be acquired in chunks (one engineer hired at a time). The cost function for this example is displayed in Exhibit 2-7. Notice that the width of the steps is 2,500 units, a much wider step than the cost function displayed in Exhibit 2-6. Costs that follow a step-cost behaviour with wide steps are defined as **step-fixed costs**. Step-fixed costs are assigned to the fixed cost category, since most are fixed over the firm's normal operating range.

When resources are acquired in advance, there may be a difference between the *resources supplied* and the *resources used (demanded)* to perform activities. This can only occur for costs that display fixed cost behaviour (resources acquired in advance

Exhibit 2-7

Step-Fixed Costs

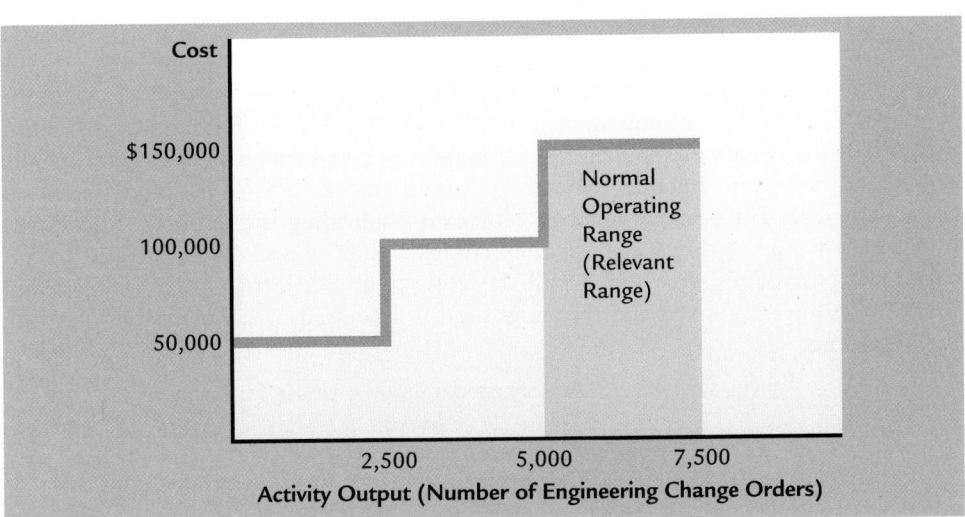

**CORNERSTONE
2-2**

The HOW and WHY of Calculating Activity Availability, Capacity Used, and Unused Capacity

Information:
Davin Company has three engineers, each of whom is paid $50,000 per year and is able to process 2,500 change orders. Last year, 6,000 change orders were processed by the three engineers.

Why:
If managers know the total capacity available as well as the capacity used, they can better utilize the activity capacity and know when additional capacity must be acquired.

Required:
1. Calculate the activity rate per change order.
2. Calculate, in terms of change orders, the:
   a. total activity availability
   b. unused capacity
3. Calculate, in dollars, the:
   a. total activity availability
   b. unused capacity
4. Express total activity availability in terms of activity capacity used and unused capacity.
5. *What if* the number of change orders processed equalled 7,500? What would unused capacity be?

Solution:
1. Activity rate = Total cost of engineers/Number of change orders

    $$= (3 \times \$50,000)/(3 \times 2,500)$$

    $$= \$20/\text{change order}$$

2. a. Total activity availability = 3 × 2,500 = 7,500 change orders
    b. Unused capacity = 7,500 − 6,000 = 1,500 change orders
3. a. Total activity availability = $20(3 × 2,500) = $150,000
    b. Unused capacity = $20(7,500 − 6,000) = $30,000
4. Total activity availability = Activity capacity used + Unused capacity

    $$7,500 = 6,000 + 1,500$$

    or

    $$\$150,000 = \$120,000 + \$30,000$$

5. If the actual change orders processed equalled 7,500, then all three engineers would be working at capacity and there would be no unused capacity.

of usage). The traditional cost management system provides information only about the cost of the resources supplied. A contemporary cost management system, on the other hand, tells how much of the activity is used and the cost of its usage, based on the activity rate. The average unit cost, obtained by dividing the resource expenditure by the activity's practical capacity, is the **activity rate**. The activity rate is used to calculate the cost of resource usage and the cost of unused activity. The relationship between resources supplied and resources used is expressed by either of the following two equations:

$$\text{Activity availability} = \text{Activity used (output)} + \text{Unused capacity} \tag{3.1}$$

$$\text{Cost of available activity} = \text{Cost of activity used (output)} + \text{Cost of unused activity} \tag{3.2}$$

Cornerstone 2-2 illustrates the way a company may determine the cost of capacity used and unused capacity.

Notice that the cost of unused capacity shown in Cornerstone 2-2 occurs because the resource (engineering redesign) must be acquired in lumpy (whole) amounts. Even if the company had anticipated the need for only 6,000 change orders, it would have been difficult to hire the equivalent of 2.4 engineers (6,000/2,500).

When activities use a mix of resources that are acquired in advance and resources that are acquired as needed, they display mixed cost behaviour. Suppose that a plant has its own Power Department; it has acquired long-term capacity for supplying power by investing in a building and equipment (resources acquired in advance). The plant also acquires fuel to produce power as needed (resources acquired as needed). The cost of the building and equipment is independent of the kilowatt-hours produced, but the cost of fuel increases as the demand for kilowatt-hours increases. The activity of supplying power has both a fixed cost component and a variable cost component, using kilowatt-hours as the output measure.

Judgment-Based Methods for Separating Mixed Costs

Sometimes it is easy to identify the variable and fixed components of a mixed cost, as in the example in Cornerstone 2-1 for the sales representatives. Many times, however, the only information available is the total cost of an activity and a measure of output (the variables Y and X). For example, the accounting system will usually record both the total cost of the maintenance activity for a given period and the number of maintenance hours provided during that period. The accounting records do not reveal the fixed and variable components of total maintenance cost.

Need for Cost Separation   Since accounting records typically show only the total cost and the associated output of a mixed cost item, it is necessary to separate the total cost into its fixed and variable components. Only through a formal effort to separate costs can they be classified into the appropriate cost behaviour categories.

Quantitative Methods for Separating Mixed Costs

OBJECTIVE ▸ 3
Separate mixed costs into their fixed and variable components using the high-low method, the scatterplot method, and the method of least squares.

The three widely used quantitative methods of separating a mixed cost into its fixed and variable components are the high-low method, the scatterplot method, and the method of least squares. Each method requires us to make the simplifying assumption of a linear cost relationship. Therefore, before we examine each of these methods more closely, let's review the expression of cost as an equation for a straight line from Cornerstone 2-1.

$$Y = F + VX$$

where

$Y =$ Total cost (the dependent variable)
$F =$ Fixed cost component (the intercept parameter)
$V =$ Variable cost per unit (the slope parameter)
$X =$ Measure of output (the independent variable)

The **dependent variable** is a variable whose value depends on the value of another variable. In the preceding equation, total activity cost is the dependent variable; it is the cost we are trying to predict. The **independent variable** is a variable that measures output and explains changes in the activity cost. It is an activity driver. A good independent variable causes or is closely associated with the dependent variable. The **intercept parameter** corresponds to fixed cost. Graphically, the intercept parameter is the point at which the mixed cost line intercepts the cost (vertical) axis. The **slope parameter** corresponds to the variable cost per unit of activity. Graphically, this represents the slope of the mixed cost line.

Since the accounting records reveal only X and Y, those values must be used to estimate the parameters F and V. With estimates of F and V, the fixed and variable components can be estimated, and the behaviour of the mixed cost can be predicted as output changes.

Three methods will be described for estimating *F* and *V*: the high-low method, the scatterplot method, and the method of least squares. The same data will be used with each method so that comparisons among them can be made. In the example, the plant manager for Anderson Company wants to determine the fixed and variable components of materials handling costs. He believes that the number of material moves is a good driver for the activity. Data for 10 months of materials handling costs and number of material moves are given in Cornerstone 2-3.

The High-Low Method

Basic geometry tells us that two points determine a line. *F*, the fixed cost component, is the intercept of the total cost line, and *V*, the variable cost per unit, is the slope of the line. Given two points, the slope and the intercept can be determined. The **high-low method** preselects the two points that are used to compute the parameters *F* and *V*. The *high point* is defined as the point with the *highest activity level*. The *low point* is defined as the point with the *lowest activity level*. Note that the high and low points are determined by the independent variable, not the dependent (typically cost) variable. Cornerstone 2-3 shows how and why the high-low method can be used to determine the fixed cost and variable rate.

Notice that the last requirement of Cornerstone 2-3 asks us to compute the materials handling cost for the year, not for a month. Since monthly data were used to determine the cost formula, fixed cost must be multiplied by 12 to get the fixed cost for the year instead of the month. If the materials handling cost for the quarter were desired, then fixed cost would be multiplied by three (the number of months in a quarter). If weekly data had been used to determine the cost formula, the fixed cost for the year would be the weekly fixed cost multiplied by 52, the number of weeks in a year.

The high-low method has two advantages. First, it is objective. That is, any two people using the high-low method on a particular data set will arrive at the same answer. Second, it is simple to calculate. The high-low method allows a manager to get a quick fix on a cost relationship using only two data points. For example, a manager may have only two years of data. Sometimes, this will be enough to get a crude approximation of the cost relationship.

The high-low method is usually not as good as the other methods for two reasons. First, the high and low points can be what are known as outliers. They may represent atypical cost-activity relationships. If so, the cost formula computed using these two points will not represent what usually takes place. The **scatterplot method** can help a manager avoid this trap by selecting two points that appear to be representative of the general cost-activity pattern. Second, even if these points are not outliers, other pairs of points may clearly be more representative. Again, the scatterplot method allows the choice of the more representative points.

An important point must be made regarding the estimates of fixed and variable costs yielded by the high-low method. These estimates should "look reasonable" to the cost analyst. For example, suppose that the high-low method returns a negative fixed cost estimate. That cannot be right; a negative fixed cost implies that a zero amount of the driver would result in revenue to the company. This is another reason that the scatterplot method can be useful. Perhaps the high or the low point is an outlier, such that the line drawn through it is very different from a line that would be drawn if the outlier were thrown out and the second-highest (or lowest) point were selected.

Scatterplot Method

The first step in applying the scatterplot method is to plot the data points so that the relationship between materials handling costs and activity output can be seen. This plot is referred to as a **scattergraph** and is shown in Exhibit 2-8, Graph A. The

The HOW and WHY of Using the High-Low Method to Determine Fixed Cost and Variable Rate

Information:
Anderson Company had the following 10 months of data on materials handling cost and number of moves:

**CORNERSTONE
2-3**

Month	Materials Handling Cost	Number of Moves
January	$2,000	100
February	3,090	125
March	2,780	175
April	1,990	200
May	7,500	500
June	5,300	300
July	4,300	250
August	6,300	400
September	5,600	475
October	6,240	425

Why:
The high-low method gives managers a quick way of estimating cost behaviour. Only two data points are needed, the high and low activity points, so this method is especially easy for companies without a long history.

Required:
1. Determine the high point and the low point.
2. Calculate the variable rate for materials handling based on the number of moves.
3. Calculate the fixed monthly cost of materials handling.
4. Write the cost formula for the materials handling activity showing the fixed cost and the variable rate.
5. If Anderson Company estimates that November will have 350 moves, what is the total estimated materials handling cost for that month?
6. *What if* Anderson wants to estimate materials handling cost for the coming year and expects 3,940 moves? What will estimated total materials handling cost be? What is the total fixed materials handling cost? Why doesn't it equal the fixed cost calculated in Requirement 3 above?

Solution:
1. The high number of moves is in May, and the low number of moves in January. (*Hint:* Did you notice that the low cost of $1,990 was for April, yet April is not the low point because its number of moves is not the lowest activity level?)

2. Variable rate = (High cost − Low cost)/(High moves − Low moves)
$$= (\$7,500 - \$2,000)/(500 - 100) = \$5,500/400$$
$$= \$13.75 \text{ per move}$$

3. Fixed cost = Total cost − (Variable rate × Moves)
   Let's choose the high point with cost of $7,500 and 500 moves.
   Fixed cost = $7,500 − ($13.75 × 500)
$$= \$625$$

   (*Hint:* Check your work by computing fixed cost using the low point.)

**CORNERSTONE
2-3**
(continued)

4. If the variable rate is $13.75 per move and fixed cost is $625 per month, then the formula for monthly materials handling cost is: Total materials handling cost = $625 + ($13.75 × Moves)
5. Materials handling cost = $625 + $13.75(350) = $5,437.50
6. Materials handling cost for the year = 12($625) + $13.75(3,940)
    = $7,500 + $54,175 = $61,675

The fixed cost for the year is 12 times the fixed cost for the month. Thus, instead of $625, the yearly fixed cost is $7,500.

vertical axis is total activity cost (materials handling cost), and the horizontal axis is the driver or output measure (number of moves). Looking at Exhibit 2-8, Graph A, we see that the relationship between materials handling costs and number of moves is reasonably linear; cost goes up as the number of moves goes up, and vice versa.

Now let's examine Exhibit 2-8, Graph B, to see if the line determined by the high and low points is representative of the overall relationship. It does look relatively representative. Does that mean that the high-low line should be chosen? Not necessarily. Suppose that management believes the variable costs of materials handling will go down in the near future. In that case, the high-low line gives a somewhat higher variable cost (slope) than desired. The scatterplot line will be chosen with a shallower slope.

Thus, one purpose of a scattergraph is to assess the validity of the assumed linear relationship. Additionally, inspecting the scattergraph may reveal several points that do not seem to fit the general pattern of behaviour. Upon investigation, it may be discovered that these points (the outliers) were due to some irregular occurrences. This knowledge can provide justification for their elimination and perhaps lead to a better estimate of the underlying cost function.

A scattergraph can provide insight concerning the relationship between cost and output by allowing one to visually fit a line to the points on the scattergraph. In doing so, the line should appear to best fit the points. In making that choice, a cost analyst is free to use past experience with the behaviour of the cost item. Experience may provide the analyst with a good intuitive sense of how materials handling costs behave; the scattergraph then becomes a useful tool to quantify this intuition. Fitting a line to the points in this way is how the scatterplot method works. Keep in mind that the scattergraph and the other statistical aids are tools that help managers improve their judgment. Using the tools does not prevent the manager from using his or her own judgment to alter any of the estimates produced by formal methods.

Examine Exhibit 2-8, Graph A, carefully. Based only on the information contained in the graph, how would you fit a line to the points in it? Of course, there are an infinite number of lines that might go through the data, but let's choose one that goes through the point for January (100, $2,000) and intersects the y-axis at $800. Now, we have the straight line shown in Exhibit 2-8, Graph C. The fixed cost is the intercept, $800. The high-low method can be used to determine the variable rate.

The two chosen points are (100, $2,000) and (0, $800). These two points are used to compute the slope:

$$V = (Y_2 - Y_1)/(X_2 - X_1)$$
$$= (\$2,000 - \$800)/(100 - 0)$$
$$= \$1,200/100$$
$$= \$12$$

Thus, the variable cost per material move is $12.

The fixed and variable components of the materials handling cost have now been identified. The cost formula for the materials handling activity can be expressed as:

$$Y = \$800 + \$12X$$

Exhibit 2-8

Scattergraph for Anderson Company's Materials Handling Costs

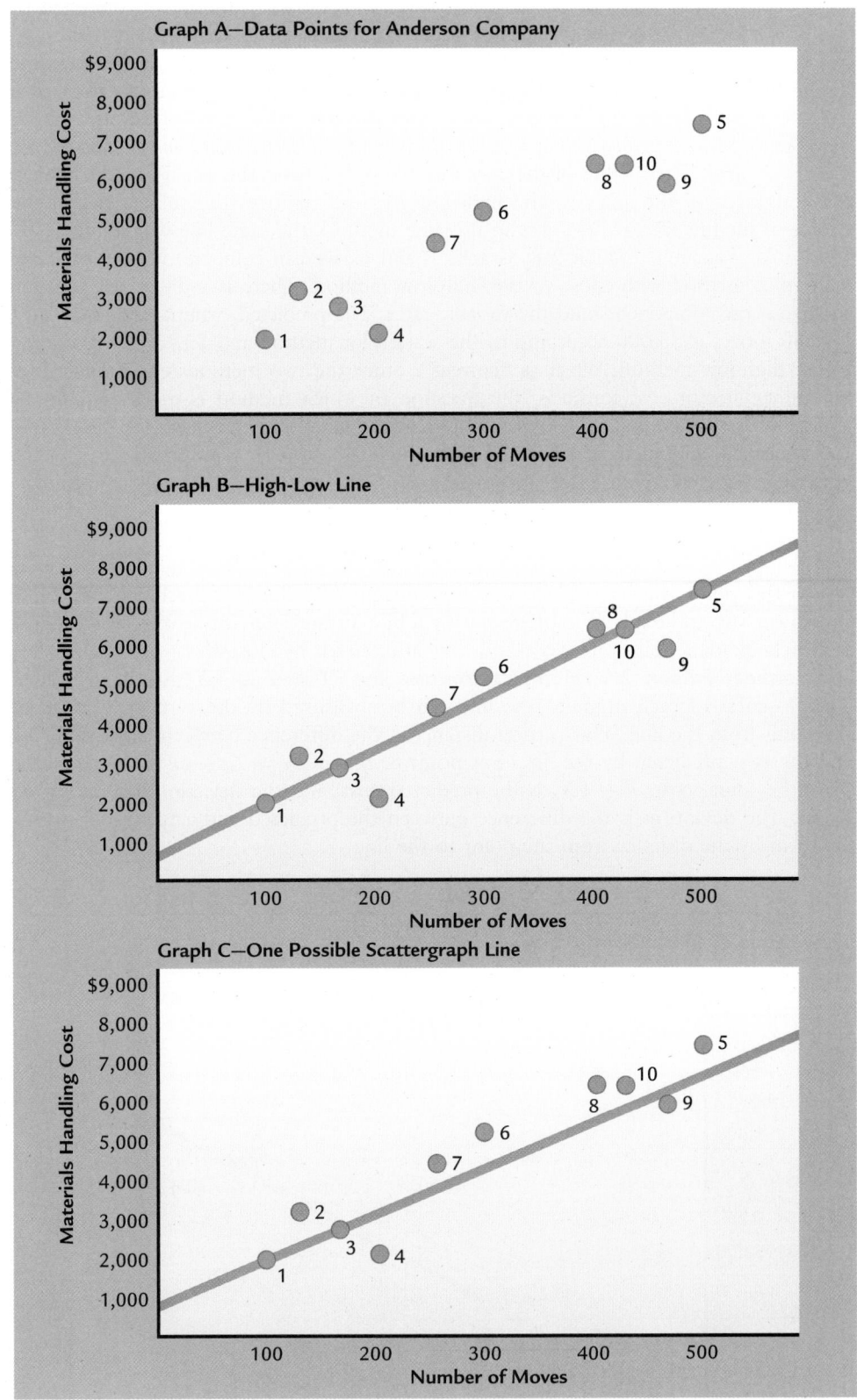

Using this formula, the total cost of materials handling for moves between 100 and 500 can be predicted and then broken down into fixed and variable components. Assume that 350 moves are planned for November. Using the cost formula, the predicted cost is $5,000 [$800 + ($12 × 350)]. Of this total cost, $800 is fixed, and $4,200 is variable.

A significant advantage of the scatterplot method is that it allows a cost analyst to inspect the data visually. The cost formula for materials handling was obtained by fitting a line to two points [(0, $800) and (100, $2,000)] in Exhibit 2-8, Graph C. We used our judgment to select the line. Whereas one person may decide that the best-fitting line is the one passing through those two points, others, using their own judgment, may decide that the best line passes through other pairs of points.

The scatterplot method suffers from the lack of any objective criterion for choosing the best-fitting line. The quality of the cost formula depends on the quality of the subjective judgment of the analyst. The high-low method removes the subjectivity in the choice of the line. Regardless of who uses the method, the same line will result.

Looking again at Exhibit 2-8, Graphs B and C, we can compare the results of the scatterplot method with those of the high-low method. There is a difference between the fixed cost components and the variable rates. The predicted materials handling cost for 350 moves is $5,000 according to the scatterplot method and $5,437.50 according to the high-low method. Which is "correct"? Since the two methods can produce significantly different cost formulas, the question of which method is the best naturally arises. Ideally, a method that is objective and, at the same time, produces the best-fitting line is needed. The **method of least squares** defines *best-fitting* and is objective in the sense that using the method for a given set of data will produce the same cost formula.

The Method of Least Squares

Up to this point, we have alluded to the concept of a line that best fits the points shown on a scattergraph. What is meant by a best-fitting line? Intuitively, it is the line to which the data points are closest. But what is meant by closest?

Consider Exhibit 2-9. Here, an arbitrary line ($Y = F + VX$) has been drawn. The closeness of each point to the line can be measured by the vertical distance of the point from the line. This vertical distance is the difference between the actual cost and the cost predicted by the line. For point 8, this is $E_8 = Y_8 - (F + VX_8)$, where Y_8 is the actual cost, $F + VX_8$ is the predicted cost, and the deviation is represented by E_8. The **deviation** is the difference between the predicted and actual costs, which is shown by the distance from the point to the line.

Exhibit 2-9

Deviations of Data from a Line

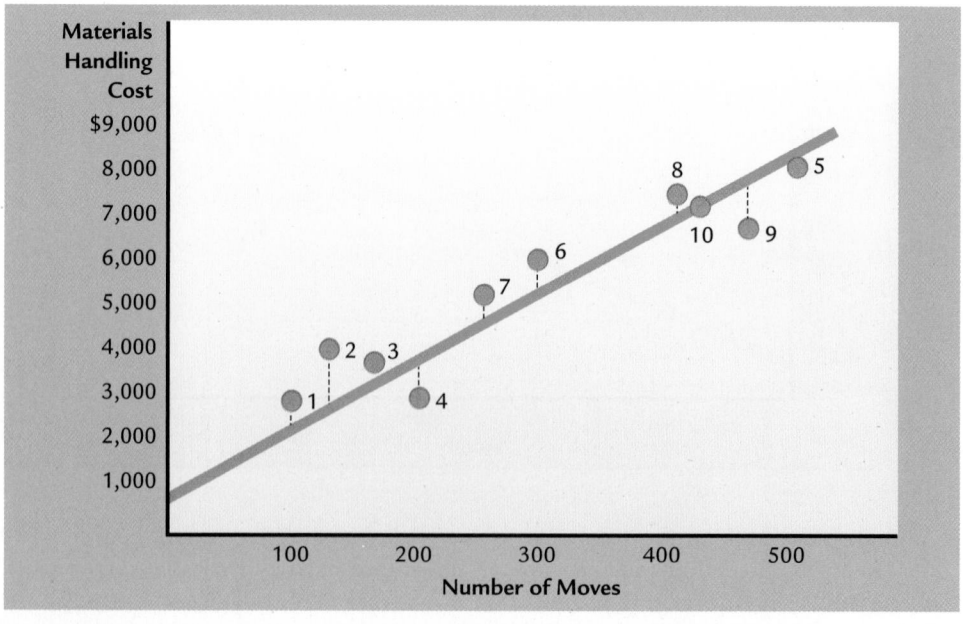

The vertical distance measures the closeness of a single point to the line, but we really need a measure of closeness of *all* points to the line. One possibility is to add all the single measures to obtain an overall measure. However, since the single measures can be positive or negative, this overall measure may not be very meaningful. For example, the sum of small positive deviations could result in an overall measure greater in magnitude than the sum of large positive deviations and large negative deviations because of the cancelling effect of positive and negative numbers. To correct this problem, each single measure of closeness is first squared, and then these squared deviations are summed as the overall measure of closeness. Squaring the deviations avoids the cancellation problem caused by a mix of positive and negative numbers.

To illustrate this concept, a measure of closeness will be calculated for the cost formula produced by the scatterplot method for Anderson Company's materials handling costs.

Actual Cost	Predicted Cost[a]	Deviation[b]	Deviation Squared
$2,000	$2,000	0	0
3,090	2,300	790	624,100
2,780	2,900	−120	14,400
1,990	3,200	−1,210	1,464,100
7,500	6,800	700	490,000
5,300	4,400	900	810,000
4,300	3,800	500	250,000
6,300	5,600	700	490,000
5,600	6,500	−900	810,000
6,240	5,900	340	115,600
Total measure of closeness			5,068,200

[a] Predicted cost = $800 + $12X$, where X is the actual measure of activity output associated with the actual activity cost and cost is rounded to the nearest dollar.
[b] Deviation = Actual cost − Predicted cost.

Since the measure of closeness is the sum of the squared deviations of the points from the line, the smaller the measure, the better the line fits the points. For example, the scatterplot method line has a closeness measure of 5,068,200. A similar calculation produces a closeness measure of 5,402,013 for the high-low line. Thus, the scatterplot line fits the points better than the high-low line. This outcome supports the earlier claim that the use of judgment in the scatterplot method is superior to the high-low method.

In principle, comparing closeness measures can produce a ranking of all lines from best to worst. The line that fits the points better than any other line is called the *best-fitting line*. It is the line with the smallest (least) sum of squared deviations. The method of least squares identifies the best-fitting line. We rely on statistical theory to obtain the formulas that produce the best-fitting line.

Using Regression Programs

Computing the regression formula manually is tedious, even with only a few data points. As the number of data points increases, manual computation becomes impractical. (When multiple regression is used, manual computation is virtually impossible.) Fortunately, spreadsheet packages such as Microsoft Excel® have regression routines that will perform the computations.[7] All you need to do is input the data. The spreadsheet regression program supplies more than the estimates of the coefficients. It also provides information that can be used to see how reliable the cost equation is, a feature that is not available for the scatterplot and high-low methods.

The first step in using the computer to calculate regression coefficients is to enter the data. The next step is to run the regression. In Excel, the regression routine is located under the "Tools" menu (toward the top left of the screen). When you pull down the "Tools" menu, you will see a number of menu possibilities. If you see "Data Analysis" just click on that and then click on "Regression." (If you don't see "Data Analysis," then choose

[7] Excel is a registered trademark of Microsoft Corporation. Any further reference to Excel refers to this footnote.

CORNERSTONE 2-4

The HOW and WHY of Using the Regression Results for Fixed Cost and Variable Rate to Construct and Use a Cost Formula

Information:

Anderson Company had 10 months of data on materials handling cost and number of moves, as shown in Cornerstone 2-3. Regression was run on these data and the coefficients shown by the regression program results are:

Intercept	854.4994
X variable 1	12.39153

Why:

Regression gives the best linear unbiased estimates of the intercept and slope for a set of data points. These can be used to find the fixed cost and variable rate in a cost scenario, and can be used to predict cost for a given amount of the independent variable.

Required:

1. Construct the cost formula for the materials handling activity showing the fixed cost and the variable rate.
2. If Anderson Company estimates that November will have 350 moves, what is the total estimated materials handling cost for that month?
3. **What if** Anderson wants to estimate materials handling cost for the coming year and expects 3,940 moves? What will estimated total materials handling cost be? What is the total fixed materials handling cost? Why doesn't it equal the fixed cost calculated in Requirement 2 above?

Solution:

1. Rounding the regression estimates to the nearest cent, the formula for monthly materials handling cost is:

   Total materials handling cost = $854.50 + ($12.39 × Moves)

2. Materials handling cost = $854.50 + $12.39(350) = $5,191

3. Materials handling cost for the year = 12($854.50) + $12.39(3,940)

   = $10,254 + $48,816.60 = $59,070.60

   The fixed cost for the year is 12 times the fixed cost for the month. Thus, instead of $854.50, the yearly fixed cost is $10,254.

"Add-Ins" and then select "Analysis ToolPak." This will add the data analysis tools. When the data analysis tools have been added, "Data Analysis" will appear at the bottom of the "Tools" menu; click on "Data Analysis," and then "Regression.")

When the "Regression" screen pops up, you can tell the program where the dependent and independent variables are located. In the box labelled "Input Y Range," place the cursor at the beginning of the rectangle, click, and then (again using the cursor) block the values under the dependent variable column. Then, move the cursor to the beginning of the box for the "Input X Range," click, and block the values in the appropriate cells. Finally, you need to tell the computer where to place the output. You can choose to put it on a separate worksheet or on the current worksheet. Let's assume you are going to save the output to the current worksheet. Click on the radio button by your choice, and then, using your cursor, block a nice-sized rectangle and click on "OK." In less than the blink of an eye, the regression output is complete.

Cornerstone 2-4 takes the results of the regression program and uses them to construct a cost formula. That cost formula can then be used to determine the predicted cost given an estimate of the independent variable.

Since the regression cost formula is the best-fitting line, it should produce better predictions of materials handling costs than either the high-low or scatterplot methods. From Cornerstone 2-4 for 350 moves, the estimate predicted by the least-squares line is $5,191 [$854.50 + ($12.39 × 350)], with a fixed component of $854.50 plus a variable component of $4,336.50. Using this prediction as a standard, the scatterplot line most closely approximates the least-squares line.

While computer output can give us the fixed and variable cost coefficients, its major usefulness lies in its ability to provide information about the reliability of the estimated cost formula.

Reliability of Cost Formulas

OBJECTIVE ▸ 4
Evaluate the reliability of the cost formula.

Regression routines provide information to help assess the reliability of the estimated cost formula. This is a feature not provided by either the scatterplot or high-low methods. There are three assessments of the cost formula's reliability: *hypothesis test of cost parameters*, *goodness of fit*, and *confidence intervals*. The **hypothesis test of cost parameters** indicates whether the parameters are different from zero. For our setting, **goodness of fit** measures the degree of association between cost and activity output. This measure is important because the method of least squares identifies the best-fitting line, but it does not reveal how good the fit is. The best-fitting line may not be a good-fitting line. It may perform miserably when it comes to predicting costs. A **confidence interval** provides a range of values for the actual cost with a prespecified degree of confidence. Confidence intervals allow managers to predict a range of values instead of a single prediction. Of course, if the degree of association is perfect, then the confidence interval will consist of a single point and the actual cost will always coincide with the predicted cost. Thus, goodness of fit and confidence intervals are related, and they provide cost analysts some idea of how reliable the resulting cost equation is.

R^2 Coefficient of Variation and Coefficient of Correlation

The coefficient of variation also known as coefficient of determination R^2 explains the percentage of variability of the dependent variable by an independent variable. For example, the materials handling cost (dependent variable) is explained by the number of material moves (independent variable). The higher the percentage of variability of materials handling cost, the better a job the independent variable (number of moves) does in explaining the dependent variable (materials handling costs).

The R^2 always varies from 0 to 1.00. If R^2 is, say, .75, it means that 75 percent of the variability in the materials handling cost is explained by the number of material moves. The rest is due to other variables, which often are unexplained. Depending on one's tolerance for error, the 75 percent may or may not be good enough. There is not an absolute cut-off point for good or bad coefficient of variation. In general, though, the closer the R^2 is to 1, the more acceptable the coefficient of variation. If an R^2 of 75 percent is considered low, one can seek another independent variable that better explains the variation in materials handling costs. For example, materials handling hours may produce, say, a .90 R^2, which clearly better explains the materials handling cost behaviour. Increasing the number of observations of the independent variable (number of material moves) makes the R^2 more reliable. That is why regression analysis takes into account the number of observations of independent variables and is adjusted accordingly.

An alternative measure to coefficient of determination is the coefficient of correlation, which is the square root of R^2. Since square roots can be negative, the value of the coefficient of correlation fluctuates between −1 and +1. If the coefficient of correlation is positive, it means there is a positive relationship between the dependent and independent variable—that is, both move in the same direction: if one increases, so does the other; as the number of materials handling moves increases, so do the material handling costs. If the correlation coefficient is negative, it suggests there is an inverse relationship between the dependent and independent variable, so that when one increases, the other decreases, and vice versa. The closer the correlation coefficient is to zero, the lower the correlation. A zero correlation coefficient value suggests no correlation between the dependent and independent variable whatsoever.

The regression analysis can be used to predict costs (materials handling cost) at different levels of activity (number of materials handling moves, materials handling labour hours). The predicted costs often differ from the actual costs. The difference is due to the fact that we use only one independent variable, whereas more variables may be shaping up the materials handling costs. Also, the difference may be caused by the inadequacy of the sample of observations, which may not be very reflective of reality.

The size of the difference between predicted and actual results is measured by the prediction standard error. The standard error measures the confidence level around the predicted cost. By adding and subtracting the standard error to predicted cost (PC +/− SE), we create a range of possible values. The width of the range diminishes the value of the coefficient of determination. A narrow width is needed for the prediction to be of good value. To reduce the range, we could use a larger sample, which may decrease the standard error. Using t statistics, a degree of confidence—that is, the likelihood that the prediction range will contain the actual cost—can be specified. A 95 percent degree of confidence means that 95 times out of 100 the actual cost will be within the prediction range.

Methods of Determining Cost Behaviour

OBJECTIVE ▸ 5
Discuss the use of managerial judgment in determining cost behaviour.

In practice, companies use a variety of methods of estimating costs. Among these are two judgment-based methods—the industrial engineering method and the account analysis method—as well as a variety of quantitative statistical methods. The best cost estimators are individuals who thoroughly understand the process, the cost drivers, and the degree of variability between the driver, the activity, and the cost.

The **industrial engineering method** is a forward-looking method of determining, through physical observation and analysis, just what activities, in what amounts, are needed to complete a process. Time and motion studies may be used in conjunction with this method. Industrial engineers may literally stand behind production workers with a stop watch to determine precisely how many minutes it takes to produce a unit of product. Once completed, the engineering studies are very precise. However, they are expensive to implement and seldom updated once they are done. This method is most frequently used for manufacturing processes where there is a direct link between materials and labour inputs with the output. An advantage of engineering methods is that they can be applied to new processes and designs. Industrial engineers determine the amount of each direct material needed and the amount of labour time each process will take. Then accountants and purchasing specialists can apply the appropriate unit costs. While this method is useful in determining the cost of manufactured items, where the process stays the same from unit to unit, it is less useful in services where different customers or circumstances may require varying amounts of time and types of service.

The **account analysis method** can be used to estimate costs by classifying accounts in the general ledger as fixed, variable, or mixed. In practice, accounts are usually put into either the fixed or variable category based on the predominant nature of the costs in the individual account. This method is often used in practice because it is simple and straightforward to apply. Accountants with a good knowledge of the cost behaviour of the various accounts can create credible cost functions using the approach. As an alternative, some accountants have set up subaccounts in the chart of accounts that are designed to separate relatively fixed from relatively variable cost categories. If costs in each account are predominantly of one of the two types, this method will give reasonable results.

To use the account analysis method, the accountant uses judgment and experience to separate the accounts into two categories—fixed and variable. Once the fixed categories are known, the average monthly cost can be computed and this is the fixed amount. The variable categories need to be further separated into categories according to the driver the accountant wishes to associate with the account. For example, accounts that are variable with respect to direct labour hours can be separated, their average costs determined, and then that total divided by the average amount of direct labour hours to obtain the variable rate per direct labour hour. Similarly, accounts driven by machine hours, purchase orders, and so on, can be averaged and then divided by their average amount of driver to obtain the rates. Cornerstone 2-5 shows how the account analysis method can be used to separate fixed and variable costs, determine a cost function, and use that cost function in budgeting.

The industrial engineering method and the account analysis method are judgment-based methods of determining cost behaviour. Quantitative methods also exist that rely on past data to generate a linear model that describes the variable and fixed portions of a cost (see Objective 3 in this chapter).

Managerial Judgment

Managerial judgment is critically important in determining cost behaviour and is by far the most widely used method in practice. Many managers simply use their experience and past observation of cost relationships to determine fixed and variable costs. This method, however, may take a number of forms. Some managers simply assign particular activity costs to the fixed category and others to the variable category. They ignore the possibility of mixed costs. Thus, a chemical firm may regard materials and utilities as strictly variable, with respect to kilograms of chemical produced, and all other costs as fixed. Even labour, the textbook example of a unit-based variable cost, may be fixed for this firm. The appeal of this method is simplicity. Before opting for this course of action, management would do well to make sure that each cost is predominantly fixed or variable and that the decisions being made are not highly sensitive to errors in classifying costs as fixed or variable.

> To illustrate the use of judgment in assessing cost behaviour, consider **Elgin Sweeper Company**, a leading manufacturer of motorized street sweepers. Using production volume as the measure of activity output, Elgin revised its chart of accounts to organize costs into fixed and variable components. Elgin's accountants used their knowledge of the company to assign expenses to either a fixed or variable category, using a decision rule that categorized an expense as fixed if it were fixed 75 percent of the time and as variable if it were variable 75 percent of the time.[8]

Management may instead identify mixed costs and divide these costs into fixed and variable components by deciding just what the fixed and variable parts are—that is, using experience to say that a certain amount of a cost is fixed and therefore that the rest must be variable. Then, the variable component can be computed using one or more cost/volume data points. This use of judgment has the advantage of accounting for mixed costs but is subject to a similar type of error as the strict fixed/variable dichotomy. That is, management may be wrong in its assessment.

> Management may use experience and judgment to refine statistical estimation results. Perhaps the experienced manager might "eyeball" the data and throw out several points as being highly unusual, or the manager might revise results of estimation to take into account projected changes in cost structure or technology. For example, **Tecnol Medical Products Inc.** radically changed its method of manufacturing medical face masks. Traditionally, face mask production was very labour intensive, requiring hand stitching. Tecnol developed its own highly automated equipment and became the industry's low-cost supplier—besting both **Johnson & Johnson** and **3M**. Tecnol's rapid expansion

[8] John P. Callan, Wesley N. Tredup, and Randy S. Wissinger, "Elgin Sweeper Company's Journey Toward Cost Management," *Management Accounting* (July 1991): 24–27.

**CORNERSTONE
2-5**

The HOW and WHY of Using Account Analysis to Determine Fixed and Variable Costs

Information:

The controller for Morrisey Company wants to determine the cost behaviour of factory overhead. Based on observation and discussions with plant workers, she feels that five accounts are most relevant. Two are fixed—supervisory salaries and depreciation—and the remaining three are variable. Indirect labour is primarily used to move materials and varies with number of moves. The largest component of utilities is electricity to run production machinery, which is driven by machine hours. Purchasing seems to be driven by the number of purchase orders. The accounts and their balances for the past six months are as follows:

	Indirect Labour Cost	Utilities	Purchasing	Supervisory Salaries	Plant & Equipment Depreciation
July	$ 14,250	$12,000	$ 38,200	$ 20,000	$ 6,500
August	15,800	10,600	35,400	23,000	6,500
September	16,800	12,500	37,600	32,000	6,500
October	20,700	12,500	40,200	27,800	6,500
November	20,000	12,500	39,900	25,400	6,500
December	17,000	12,500	39,700	17,000	6,500
Total	$104,550	$72,600	$231,000	$145,200	$39,000

Information on machine hours, number of moves, and number of purchase orders for the six-month period follows:

	Number of Moves	Machine Hours (Mhr)	Purchase Orders (PO)
July	340	5,400	300
August	380	5,200	250
September	400	5,800	380
October	500	6,200	450
November	480	6,000	340
December	420	5,600	200
Total	2,520	34,200	1,920

Why:

By separating accounts with primarily fixed costs from those with primarily variable costs, and associating the variable costs with relevant drivers, it is possible to determine cost behaviour and use it in budgeting, performance evaluation, and decision making.

Required:

1. Why did the controller decide that supervisory salaries and depreciation on the plant were fixed?
2. Calculate the average account balances for each of the five accounts. Calculate the average monthly amount of each of the three drivers.
3. Calculate the total fixed overhead for the month and the variable rates for indirect labour, utilities, and purchasing. Express the results in the form of an equation for total overhead cost.

> **CORNERSTONE 2-5**
> *(continued)*

4. In January, 490 moves, 4,375 machine hours, and 220 purchase orders are expected. What is the total overhead cost expected for the factory in January?
5. **What if** purchase orders predicted for January were 300? How would that affect the predicted overhead cost?

Solution:

1. Clearly, depreciation is fixed at $6,500 per month, and will not change unless equipment is bought or sold. While supervisory salaries did change during the six-month period, they were no doubt placed in the fixed category because they do not vary with the drivers under consideration: number of moves, machine hours, and purchase orders.

2. Average indirect labour cost = $104,550/6 = $17,425
   Average utilities = $72,600/6 = $12,100
   Average purchasing = $231,000/6 = $38,500
   Average supervisory salaries = $145,200/6 = $24,200
   Average depreciation = $39,000/6 = $6,500
   Average number of moves = 2,520/6 = 420
   Average machine hours = 34,200/6 = 5,700
   Average purchase orders = 1,920/6 = 320

3. Total fixed overhead cost = $24,200 + $6,500 = $30,700
   Variable rate for indirect labour = $17,425/420 = $41.49 per move (rounded)
   Variable rate for utilities = $12,100/5,700 = $2.12 per Mhr (rounded)
   Variable rate for purchasing = $38,500/320 = $120.31 per PO (rounded)
   Total overhead cost = $30,700 + $41.49(moves) + $2.12(machine hours) + 120.31(purchase orders)

4. Total overhead cost = $30,700 + ($41.49 × 490) + ($2.12 × 4,375) + ($120.31 × 220) = $86,773 (rounded)

5. If purchase orders increased by 80, then predicted January overhead cost would increase by $9,624.80 ($120.31 × 80) to a total of $96,398 (rounded).

into new product lines and European markets means that historical data on costs and revenues are, for the most part, irrelevant. Tecnol's management must look forward, not back, to predict the impact of changes on profit. Statistical techniques are highly accurate in depicting the past, but they cannot foresee the future, which of course is what management really wants.[9]

The advantage of using managerial judgment to separate fixed and variable costs is its simplicity. In situations in which the manager has a deep understanding of the firm and its cost patterns, this method can give good results. However, if the manager does not have good judgment, errors will occur. Therefore, it is important to consider the experience of the manager, the potential for error, and the effect that error could have on related decisions.

Multiple Regression

OBJECTIVE ▶ 6
Explain how multiple regression can be used to assess cost behaviour.

In the case of two explanatory variables (activity drivers), the linear equation is expanded to include the additional variable:

$$Y = F + V_1 X_1 + V_2 X_2$$

[9] Stephanie Anderson Forest, "Who's Afraid of J&J and 3M?" *Business Week* (December 5, 1994): 66, 68.

where

$$X_1 = \text{Number of moves}$$
$$X_2 = \text{The total distance}$$

With three variables (Y, X_1, X_2), a minimum of three points is needed to compute the parameters F, V_1, and V_2. Seeing the points becomes difficult because they must be plotted in three dimensions. Using the scatterplot method or the high-low method is not practical.

However, the extension of the method of least squares is straightforward. It is relatively simple to develop a set of equations that provides values for F, V_1, and V_2 that yield the best-fitting equation. Whenever least squares is used to fit an equation involving two or more independent variables, the method is called **multiple regression**. The computations required for multiple regression are far more complex than in simple (one independent variable) regression. In fact, any practical application of multiple regression requires use of a computer.

Let's return to the Anderson Company example. Recall that the R^2 is just 85 percent and that the fixed cost coefficient was not significant. Perhaps another variable can help explain materials handling costs. Suppose that Anderson Company's controller finds that in some months many more kilograms of materials were moved than in other months. The heavier materials required additional equipment to handle the increased load.

The controller adds the variable "kilograms moved" and gathers information on that variable for the 10 months. These data are shown below.

Month	Materials Handling Cost	Number of Moves	Kilograms Moved
January	$2,000	100	6,000
February	3,090	125	15,000
March	2,780	175	7,800
April	1,990	200	600
May	7,500	500	29,000
June	5,300	300	23,000
July	4,300	250	17,000
August	6,300	400	25,000
September	5,600	475	12,000
October	6,240	425	22,400

Multiple regression can be run using the number of moves and the number of kilograms moved as the independent variables. Running multiple regression using the Excel program is no more difficult than using it to run simple regression.

OBJECTIVE ▸ 7
Define the learning curve, and discuss its impact on cost behaviour.

The Learning Curve and Nonlinear Cost Behaviour

A number of cost behaviour patterns do not follow a linear pattern. We have already seen that total cost can increase at a decreasing rate, as is the case when there are discounts for large purchases of materials. An important type of nonlinear cost curve is the learning curve. The **learning curve** shows how the labour hours worked per unit decrease as the number of units produced increases. The basis of the learning curve is almost intuitive—as we perform an action over and over, we improve, and each additional performance takes less time than the preceding ones. We learn how to do the task, become more efficient, and smooth out the rough spots. In a manufacturing firm, learning takes place throughout the process: workers learn their tasks and managers learn to schedule production more efficiently and to arrange the flow of work. Each time cumulative volume doubles, costs fall by a constant and predictable percentage. This effect was first documented in the aircraft industry.

Managers can see that the ideas behind the learning curve can extend to the service industry as well as to manufacturing firms. Costs in marketing, distribution, and service after the sale also decrease as the number of units produced and sold increases. When used in this way, the learning curve is often called the experience curve. The **experience curve** relates cost to increased efficiency, such that the more often a task is performed, the lower will be the cost of doing it. The experience curve can be applied to any task, including production, selling, distribution, post-sales service, and so on.

The learning curve model takes two common forms: the cumulative average-time learning curve model and the incremental unit-time learning curve model. The difference between the two lies in the assumption made about the speed of learning.

Cumulative Average-Time Learning Curve

The **cumulative average-time learning curve model** states that the cumulative average time per unit decreases by a constant percentage, or learning rate, each time the cumulative quantity of units produced doubles. The **learning rate** is expressed as a percent, and it gives the percentage of time needed to make the next unit, based on the time it took to make the previous unit. The learning rate is determined through experience and must be between 50 and 100 percent. A 50 percent learning rate would eventually result in no labour time per unit—an absurd result. A 100 percent learning rate implies no learning (since the amount of decrease is zero). An 80 percent learning curve is often used to illustrate this model, possibly because the original learning curve work with the aircraft industry found an 80 percent learning curve. Cornerstone 2-6 shows how to calculate the amount of time needed for producing successive units given an 80 percent learning rate and 100 direct labour hours for the first unit.

Cornerstone 2-6 shows the cumulative average time and cumulative total time according to the doubling formula. How do we obtain these amounts for units that are not doubles of the original amount? This is done by realizing that the cumulative average-time learning model takes a logarithmic relationship.

$$Y = pX^q$$

where

Y = Cumulative average time per unit
X = Cumulative number of units produced
p = Time in labour hours required to produce the first unit
q = Rate of learning

Therefore:

$$q = \ln(\text{percent learning})/\ln 2$$

For an 80 percent learning curve:

$$q = -0.2231/0.6931 = -0.3219$$

So, when $X = 3$, $p = 100$, and $q = -0.3219$,

$$Y = 100 \times 3^{-0.3219} = 70.21 \text{ labour hours}$$

Excel can be used to calculate the number of hours required for units that are not doubles of the first. Exhibit 2-10 shows an Excel screenshot for the example from Cornerstone 2-6. The rows in bold are the cumulative number of units that obey the doubling rule. Cornerstone 2-6 tells how to calculate columns C and D for those rows. For row 7, corresponding to 3 units, follow the following steps:

Step 1: Cell F5: enter "=LN(0.8)/LN(2)". After the value "−0.32192809" appears, copy the cell and then paste it into cells F5 through F20.

Step 2: Cell A7: enter "3"

**CORNERSTONE
2-6**

The HOW and WHY of Calculating the Cumulative Average-Time Learning Curve

Information:

Lindstrom Company installs computerized patient record systems in hospitals and medical centres. Lindstrom has noticed that each general type of system is subject to an 80 percent learning curve. The installation takes a team of professionals to set up and test the system. Assume that the first installation takes 1,000 hours, and the team of professionals is paid an average of $50 per hour.

Why:

As learning occurs, workers become more familiar with the task and can complete it more quickly. The first system installed takes the longest time; by the eighth to sixteenth time the task is performed, the workers have incorporated learning effects and the task takes much less time. Managers need to know how quickly the learning will occur and what effect that will have on labour cost for budgeting, bidding, and performance evaluation.

Required:

1. Set up a table with columns showing the cumulative number of units, cumulative average time per unit in hours, and cumulative total time in hours. Show results by row for total production of one system, two systems, four systems, eight systems, sixteen systems, and thirty-two systems.
2. What is the total labour cost if Lindstrom installs the following number of systems: one; four; sixteen? What is the average cost per installed system for the following number of systems: one, four, sixteen?
3. **What if** Lindstrom is budgeting labour cost for next year based on the installation of 16 additional systems? Calculate total budgeted labour cost for a team that had previously completed 16 systems the prior year. Calculate total budgeted labour cost for a new team that had not completed any systems to date.

Solution:

1.

Cumulative Number of Systems	Cumulative Average Time per System in Hours	Cumulative Total Time: Labour Hours
(1)	(2)	(3) = (1) × (2)
1	1,000	1,000.0
2	800.0 (0.8 × 1,000)	1,600.0
4	640.0 (0.8 × 800)	2,560.0
8	512.0 (0.8 × 640)	4,096.0
16	409.6 (0.8 × 512)	6,553.6
32	327.7 (0.8 × 409.6)	10,486.4

Notice that every time the number of systems installed doubles, the cumulative average time per unit (in column 2) is just 80 percent of the previous amount.

2.   Cost for installing one system = 1,000 hours × $50 = $50,000
     Cost for installing four systems = 2,560 hours × $50 = $128,000
     Cost for installing sixteen systems = 6,553.6 hours × $50 = $327,680
     Average cost per system for one system = $50,000/1 = $50,000
     Average cost per system for four systems = $128,000/4 = $32,000
     Average cost per system for sixteen systems = $327,680/16 = $20,480

3.   Budgeted labour cost for experienced team = (10,486.4 − 6,553.6) × $50
                                               = $196,640
     Budgeted labour cost for new team = 6,553.6 × $50 = $327,680

**CORNERSTONE
2-6**
(continued)

Step 3: Cell B7: enter "=1000*POWER(A7,F7)" You are entering "1000" because that is the cumulative average time per unit for one unit. In different examples, you will enter a different cumulative average time per unit for one unit. That is, if the first unit had taken 78 hours, you would have entered "=78*POWER(A7,F7)"

Step 4: Cell C7: enter "=A7*B7"

Step 5: Cell D7: enter "=C7-C6"

You can now copy and paste cell B7 into cells B8 through B20, cell C7 into cells C8 through C20, and so on.

Let's take a closer look at the time for the last unit in Exhibit 2-10. See how the time it takes to complete the last unit drops from the first unit (1,000 hours) to the sixteenth unit (just 280.6 hours). This learning helps companies realize efficiencies as more and more units are completed. Accountants can use this information in budgeting and preparing bids, as they realize that the time for the first unit of a new type of job will not be equal to the time it takes to complete the last unit. Cost goes down. Accountants can also use this information to advise managers on the need to keep experienced employees rather than having excessive turnover. The turnover requires more training and does not give the company the benefit of the experienced employee's ability to do the job more quickly and competently.

Exhibit 2-11 shows the graph of both the cumulative average time per unit (the bottom line) and the cumulative total hours required (top line). We can see that the time per unit decreases as output increases, but that it decreases at a decreasing rate. We also see that the total labour hours increase as output increases, but they increase at a decreasing rate. Again, the implication for costing is that average cost will decrease as more experience is gained.

Incremental Unit-Time Learning Curve

The **incremental unit-time learning curve model** decreases by a constant percentage each time the cumulative quantity of units produced doubles. The same general assumptions for the learning curve hold; however, the learning rate is assumed to apply to the last unit produced, not to the cumulative average of all units to date. For an 80 percent learning rate, the cumulative average-time learning model assumes that the cumulative *average* time for every unit produced is just 80 percent of the amount for the previous output level. Thus, when we look at the time to produce two units, the average time for each of the units is assumed to be 80 percent of the time for the first unit. However, the incremental unit-time learning model assumes that only the *last* (incremental) unit experiences the decrease in time, so the second unit takes 80 hours, but the first still takes 100 hours. Thus, the total time is 180 (100 + 80) hours. Further explanation of the incremental unit-time learning curve will be left for more advanced courses.

Exhibit 2-10

Spreadsheet for Cumulative Average-Time Learning Model

	A	B	C	D	E	F
	Cumulative Average Time Learning Model.xls					
1	Cornerstone 2-6					
2	Cumulative Average-Time Learning Model					
3						
4	Cumulative Number of Units	Cumulative Average Time per Unit	Cumulative Total Time	Time for Last Unit		Value of q OR = ln(.8)/ln(2)
5	1	1000	1000	1000		−0.32192809
6	**2**	**800**	**1600**	**600**		**−0.32192809**
7	3	702.1037028	2106.31111	506.3111		−0.32192809
8	**4**	**640**	**2560**	**453.6889**		**−0.32192809**
9	5	595.6373436	2978.18672	418.1867		−0.32192809
10	6	561.6829622	3370.09777	391.9111		−0.32192809
11	7	534.4895247	3741.42667	371.3289		−0.32192809
12	**8**	**512**	**4096**	**354.5733**		**−0.32192809**
13	9	492.9496095	4436.54649	340.5465		−0.32192809
14	10	476.5098749	4765.09875	328.5523		−0.32192809
15	11	462.1111387	5083.22253	318.1238		−0.32192809
16	12	449.3463698	5392.15644	308.9339		−0.32192809
17	13	437.9155217	5692.90178	300.7453		−0.32192809
18	14	427.5916197	5986.28268	293.3809		−0.32192809
19	15	418.1991845	6272.98777	286.7051		−0.32192809
20	**16**	**409.6**	**6553.6**	**280.6122**		−0.32192809
21						
22						

Sheet1 / Sheet2 / Sheet3

Exhibit 2-11

Graph of Cumulative Total Hours Required and the Cumulative Average Time per Unit

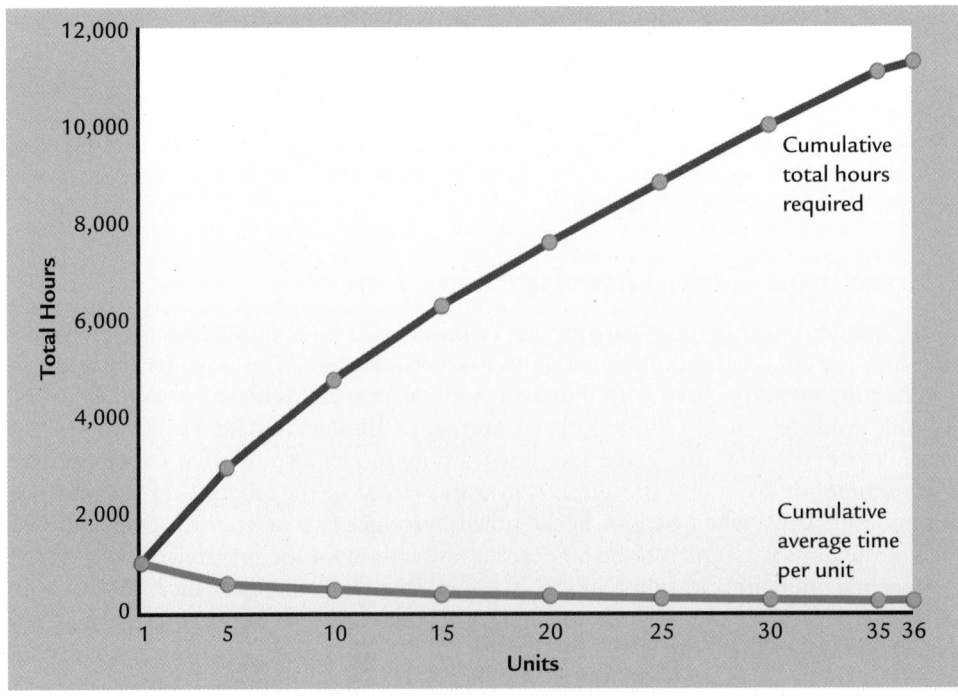

The use of the learning curve concepts permits management to be more accurate in budgeting and performance evaluation for processes in which learning occurs. While the learning curve was originally developed for manufacturing processes, it can also apply in service industries. For example, insurance companies develop new policies and new methods of selling policies. There is a learning component to each new policy as employees discover glitches that were unexpected in the development process and then learn how to fix those glitches and become more efficient.

Summary of Learning Objectives

1. **Define and describe fixed, variable, and mixed costs.**

- Variable costs change in total as activity usage changes.
- Usually, variable costs increase in direct proportion to increases in activity output.
- Fixed costs do not change in total as activity output changes.
- Mixed costs have both a variable and a fixed component.

2. **Explain the use of resources and activities and their relationship to cost behaviour.**

- Flexible resources are acquired as used and needed.

  - Flexible resources have no excess capacity for these resources.
  - They are usually considered to be variable costs.

- Committed resources are acquired in advance of usage.

  - May have excess capacity
  - Frequently considered fixed

- Step costs are acquired in lumpy amounts.

  - Narrow steps approximated by a variable cost function
  - Wide steps approximated as fixed

3. **Separate mixed costs into their fixed and variable components using the high-low method, the scatterplot method, and the method of least squares.**

- High-low method uses the high and the low data points to form a straight line.

  - Slope is variable rate.
  - Intercept is fixed cost.
  - Advantages: objective and easy
  - Disadvantage: nonrepresentative high or low point leads to misestimated cost function

- Scatterplot method plots data—two points chosen to determine a line

  - Intercept is fixed cost.
  - Slope is variable rate.
  - Advantages: identify nonlinearity, outliers, shifts in the cost relationship
  - Disadvantage: subjectivity

4. **Evaluate the reliability of the cost formula.**

- Coefficient of correlation shows degree to which two variables move together.

  - Perfect positive correlation $= 1.0$
  - Perfect negative correlation $= -1.0$

- Coefficient of determination (R^2) shows amount of cost variability explained by driver.
  - $0 \le R^2 \le 1.0$
  - Often multiplied by 100 and used as percent
- Standard error of estimate used to build a prediction interval for cost

5. **Discuss the use of managerial judgment in determining cost behaviour.**

- Used alone or in conjunction with the high-low, scatterplot, or least-squares methods.
- Experienced managers use knowledge to identify outliers, adjust parameters, and anticipate changing conditions.

6. **Explain how multiple regression can be used to assess cost behaviour.**

- Has two or more independent variables
- Useful when dependent variable is affected by more than one independent variable

7. **Define the learning curve, and discuss its impact on cost behaviour.**

- Nonlinear relationship between labour hours and output
- Doubling of output requires less than a doubling of labour time.
- Cumulative average-time learning curve assumes the cumulative average time per unit decreases by a constant percentage, or learning rate, each time the cumulative quantity of units produced doubles.
- Incremental unit-time learning curve assumes the incremental unit time decreases by a constant percentage each time the cumulative quantity of units produced doubles.

CORNERSTONES FOR CHAPTER 2

CORNERSTONE 2-1	The HOW and WHY of forming an equation to describe mixed cost, page 51
CORNERSTONE 2-2	The HOW and WHY of calculating activity availability, capacity used, and unused capacity, page 56
CORNERSTONE 2-3	The HOW and WHY of using the high-low method to determine fixed cost and variable rate, page 59
CORNERSTONE 2-4	The HOW and WHY of using the regression results for fixed cost and variable rate to construct and use a cost formula, page 64
CORNERSTONE 2-5	The HOW and WHY of using account analysis to determine fixed and variable costs, page 68
CORNERSTONE 2-6	The HOW and WHY of calculating the cumulative average-time learning curve, page 72

Review Problems

I. Resource Usage and Cost Behaviour

Thompson Manufacturing Company has three salaried clerks to process purchase orders. Each clerk is paid a salary of $28,000 and is capable of processing 5,000 purchase orders per year (working efficiently). In addition to the salaries, Thompson spends $7,500 per year for forms, postage, etc. Thompson assumes 15,000 purchase orders will be processed. During the year, 12,500 orders were processed.

Required:

1. Calculate the activity rate for the purchase order activity. Break the activity into fixed and variable components.

2. Compute the total activity availability, and break this into activity output and unused activity.
3. Calculate the total cost of the resource supplied, and break this into the cost of activity output and the cost of unused activity.

Solution:

1.
$$\text{Activity rate} = [(3 \times \$28{,}000) + \$7{,}500]/15{,}000$$
$$= \$6.10/\text{order}$$
$$\text{Fixed rate} = \$84{,}000/15{,}000$$
$$= \$5.60/\text{order}$$
$$\text{Variable rate} = \$7{,}500/15{,}000$$
$$= \$0.50/\text{order}$$

2.
$$\text{Activity availability} = \text{Activity output} + \text{Unused activity}$$
$$15{,}000 \text{ orders} = 12{,}500 \text{ orders} + 2{,}500 \text{ orders}$$

3.
$$\text{Cost of activity supplied} = \text{Cost of activity output} + \text{Cost of unused activity}$$
$$\$84{,}000 + (\$0.50 \times 12{,}500) = (\$6.10 \times 12{,}500) + (\$5.60 \times 2{,}500)$$
$$\$90{,}250 = \$76{,}250 + \$14{,}000$$

II. High-Low Method and Method of Least Squares

Linda Jones, an accountant for Golding Inc., has decided to estimate the fixed and variable components associated with the company's repair activity. She has collected the following data for the past six months:

Repair Hours	Total Repair Costs
10	$ 800
20	1,100
15	900
12	900
18	1,050
25	1,250

Required:

1. Estimate the fixed and variable components for the repair costs using the high-low method. Using the cost formula, predict the total cost of repair if 14 hours are used.
2. Estimate the fixed and variable components using the method of least squares. Translate your results into the form of a cost formula, and using that formula, predict the total cost of repairs if 14 hours are used.
3. Using the method of least squares, what are the coefficient of determination and the coefficient of correlation?

Solution:

1. The estimate of fixed and variable costs using the high-low method, where $Y =$ total cost and $X =$ number of hours, is as follows:

$$V = (Y_2 - Y_1)/(X_2 - X_1)$$
$$= (\$1{,}250 - \$800)/(25 - 10)$$
$$= \$450/15$$
$$= \$30 \text{ per hour}$$
$$F = Y_2 - VX_2$$
$$= \$1{,}250 - \$30(25)$$
$$= \$500$$
$$Y = \$500 + \$30X$$
$$= \$500 + \$30(14)$$
$$= \$920$$

2.   Regression is performed using Excel, with the results as follows:

Summary Output

Regression Statistics

Multiple R	0.984523
R^2	0.969285
Adjusted R^2	0.961607
Standard Error	32.19657
Observations	6

ANOVA

	df	SS	MS	F	Significance F
Regression	1	130853.5	130853.5	126.2311	0.000357
Residual	4	4146.476	1036.619		
Total	5	135000			

	Coefficients	Standard Error	t Stat	P-value	Lower 95%	Upper 95%	Lower 95.0%	Upper 95.0%
Intercept	509.9119	45.55789	11.19261	0.000363	383.4227	636.4011	383.4227	636.4011
X Variable 1	29.40529	2.617232	11.23526	0.000357	22.13867	36.6719	22.13867	36.6719

The calculation using the method of least squares is as follows:

$$Y = \$509.91 + \$29.41X$$
$$= \$509.91 + \$29.41(14)$$
$$= \$921.65$$

3.   The coefficient of determination (R^2) is 0.962, and the correlation coefficient (r) is 0.984 (the square root of 0.969).

Key Terms

Account analysis method, 66

Activity capacity, 53

Activity rate, 56

Committed fixed expenses, 53

Committed resources, 53

Confidence interval, 65

Cost behaviour, 46

Cumulative average-time learning curve model, 71

Dependent variable, 57

Deviation, 62

Discretionary fixed expenses, 54

Experience curve, 71

Fixed costs, 47

Flexible resources, 53

Goodness of fit, 65

High-low method, 58

Hypothesis test of cost parameters, 65

Incremental unit-time learning curve model, 73

Independent variable, 57

Industrial engineering method, 66

Intercept parameter, 57

Learning curve, 70

Learning rate, 71

Long run, 52

Method of least squares, 62

Mixed costs, 51

Multiple regression, 70

Non-unit-level drivers, 46

Practical capacity, 53

Relevant range, 47

Scattergraph, 58

Scatterplot method, 58

Short run, 52

Slope parameter, 57

Step-cost function, 54

Step-fixed costs, 55

Step-variable costs, 54

Unit-level drivers, 46

Unused capacity, 53

Variable costs, 48

Discussion Questions

1. Why is knowledge of cost behaviour important for managerial decision making? Give an example to illustrate your answer.
2. How does the length of the time horizon affect the classification of a cost as fixed or variable? What is the meaning of short run? Long run?
3. Explain the difference between resource spending and resource usage.
4. What is the relationship between flexible resources and cost behaviour?
5. What is the relationship between committed resources and cost behaviour?
6. Describe the difference between a variable cost and a step-variable cost. When is it reasonable to treat step-variable costs as if they were variable costs?
7. Why do mixed costs pose a problem when it comes to classifying costs into fixed and variable categories?
8. Why is a scattergraph a good first step in separating mixed costs into their fixed and variable components?
9. What are the advantages of the scatterplot method over the high-low method? The high-low method over the scatterplot method?
10. Describe the method of least squares. Why is this method better than either the high-low method or the scatterplot method?
11. What is meant by the best-fitting line? Is the best-fitting line necessarily a good-fitting line? Explain.
12. When is multiple regression required to explain cost behaviour?
13. Explain the meaning of the learning curve. How do managers determine the appropriate learning curve percentage to use?
14. Assume you are the manager responsible for implementing a new service. The time to perform the service is subject to the learning curve. Would you prefer that the new service have a learning rate of 85 percent or 80 percent? Why?
15. Some firms assign mixed costs to either the fixed or variable cost categories without using any formal methodology to separate them. Explain how this practice can be defended.

Cornerstone Exercises

Cornerstone Exercise 2-1 MIXED COSTS AND COST FORMULA

OBJECTIVE ➤ 1
CORNERSTONE 2-1
SERVICE

Bo's Gym is a complete fitness centre. Owner Bo Sanderson employs various fitness trainers who are expected to staff the front desk and to teach fitness classes. While on the front desk, trainers answer the phone, handle walk-ins and show them around the gym, answer member questions about the weight machines, and do light cleaning (wiping down the equipment, vacuuming the floor). The trainers also teach fitness classes (e.g., pilates, spinning, body pump) according to their own interest and training level. The cost of the fitness trainers is $500 per month and $25 per class taught. Last month, 100 classes were taught.

Required:

1. Develop a cost equation for total cost of labour.
2. What was total variable labour cost last month?
3. What was total labour cost last month?
4. What was the unit cost of labour (per class) for last month?
5. *What if* Bo increased the number of classes offered by 100 percent? What would be the total labour cost? The unit labour cost? Explain why the unit labour cost decreased.

Cornerstone Exercise 2-2 ACTIVITY AVAILABILITY, CAPACITY USED, UNUSED CAPACITY

OBJECTIVE ➤ 2
CORNERSTONE 2-2
SERVICE

Metternich Company has a Purchasing Department staffed by four purchasing agents. Each agent is paid $30,000 per year and is able to process 3,000 purchase orders. Last year, 10,000 purchase orders were processed by the four agents.

Required:

1. Calculate the activity rate per purchase order.
2. Calculate, in terms of purchase orders, the:
   a. total activity availability
   b. unused capacity

3. Calculate the dollar cost of:
   a. total activity availability
   b. unused capacity

4. Express total activity availability in terms of activity capacity used and unused capacity.
5. *What if* one of the purchasing agents agreed to work half time for $15,000? How many purchase orders could be processed by three and a half purchasing agents? What would unused capacity be in purchase orders?

OBJECTIVE ➤ 3
CORNERSTONE 2-3

Cornerstone Exercise 2-3 HIGH-LOW METHOD TO DETERMINE FIXED COST AND VARIABLE RATE

White Swan, a health care facility, had the following 12 months of data on purchasing cost and number of purchase orders.

Month	Purchasing Cost	Number of Purchase Orders
January	$20,068	330
February	20,890	370
March	20,750	410
April	22,050	400
May	21,900	450
June	21,300	460
July	23,426	560
August	21,670	440
September	22,250	500
October	21,200	470
November	21,800	480
December	20,800	370

Required:

1. Determine the high point and the low point.
2. Calculate the variable rate for purchasing cost based on the number of purchase orders.
3. Calculate the fixed monthly cost of purchasing.
4. Write the cost formula for the purchasing activity showing the fixed cost and the variable rate.
5. If White Swan estimates that next month will have 430 purchase orders, what is the total estimated purchasing cost for that month?
6. *What if* White Swan wants to estimate purchasing cost for the coming year and expects 5,340 purchase orders? What will estimated total purchasing cost be? What is the total fixed purchasing cost? Why doesn't it equal the fixed cost calculated in Requirement 3 above?

OBJECTIVE ➤ 3
CORNERSTONE 2-4

Cornerstone Exercise 2-4 USING REGRESSION RESULTS TO CONSTRUCT AND APPLY A COST FORMULA

White Swan had the following 12 months of data on purchasing cost and number of purchase orders.

Month	Purchasing Cost	Number of Purchase Orders
January	$20,068	330
February	20,890	370
March	20,750	410
April	22,050	400
May	21,900	450
June	21,300	460
July	23,426	560
August	21,670	440
September	22,250	500
October	21,200	470
November	21,800	480
December	20,800	370

The controller for White Swan ran regression on the above data, and the coefficients shown by the regression program (rounded to the nearest cent) are:

Intercept	16,403.85
X variable 1	11.69

Required:

1. Construct the cost formula for the purchasing activity showing the fixed cost and the variable rate.
2. If White Swan estimates that next month will have 430 purchase orders, what is the total estimated purchasing cost for that month? (Round your answer to the nearest dollar.)
3. *What if* White Swan wants to estimate purchasing cost for the coming year and expects 5,340 purchase orders? What will estimated total purchasing cost be? (Round your answer to the nearest dollar.) What is the total fixed purchasing cost? Why doesn't it equal the fixed cost calculated in Requirement 1 above?

Cornerstone Exercise 2-5 ACCOUNT ANALYSIS TO DETERMINE COST BEHAVIOUR

OBJECTIVE ▶ 5

CORNERSTONE 2-5

SERVICE

Lance Leffler, owner of Leffler Inc., a warehousing facility, wants to determine the cost behaviour of labour and overhead. Lance pays his workers a salary; during busy times, everyone works to get the orders out. Temps (temporary workers hired through an agency) may be hired to pack and prepare completed orders for shipment. During slower times, Lance catches up on bookkeeping and administrative tasks while the salaried workers do preventive maintenance, clean the lines and building, etc. Temps are not hired during slow times. Lance found that workers' salaries, temp agency payments, rentals, utilities, and plant and equipment depreciation are the largest dollar accounts. He believes that workers' salaries and plant and equipment depreciation are fixed, temp agency payments are associated with the number of orders (since temp workers are used to pack and prepare completed orders for shipment), and electricity is associated with the number of machine hours. When the number of different parts stored by Leffler exceeds the space in the materials storeroom, Lance rents nearby warehouse space. He can rent as much or as little space as he wants on a month-to-month basis. Therefore, he believes warehouse rental payments are variable with the number of parts purchased and stored. The account balances for the past six months as well as the six-month total are as follows:

	Workers' Salaries	Temp Agency Payments	Warehouse Rental	Electricity	Plant & Equipment Depreciation
January	$ 6,300	$ 0	$ 150	$ 300	$ 2,800
February	6,300	320	400	410	2,800
March	6,300	1,200	460	680	2,800
April	6,300	1,350	500	720	2,800
May	6,300	1,750	480	750	2,800
June	6,300	1,400	220	700	2,800
Total	$37,800	$6,020	$2,210	$3,560	$16,800

Information on number of machine hours, orders, and parts for the six-month period follows:

	Machine Hours	Number of Orders	Number of Parts
January	2,000	10	200
February	3,100	40	450
March	5,800	350	550
April	6,200	400	580
May	6,500	510	480
June	6,000	410	300
Total	29,600	1,720	2,560

Required:

1. Calculate the monthly average account balance for each account. Calculate the average monthly amount for each of the three drivers.
2. Calculate fixed monthly cost and the variable rates for temp agency payments, warehouse rent, and electricity. Express the results in the form of an equation for total cost.
3. In July, Lance predicts there will be 400 orders, 280 parts, and 5,900 machine hours. What is the total labour and overhead cost for July?
4. *What if* Lance buys a new machine in July for $18,000? The machine is expected to last 10 years and will have no salvage value at the end of that time. What part of the cost equation will be affected? How? What is the new expected cost in July?

OBJECTIVE ▶ 6
CORNERSTONE 2-4
SERVICE

Cornerstone Exercise 2-6 USING MULTIPLE REGRESSION RESULTS TO CONSTRUCT AND APPLY A COST FORMULA

The controller for White Swan Company felt that the number of purchase orders alone did not explain the monthly purchasing cost. He knew that nonstandard orders (e.g., one requiring an overseas supplier) took more time and effort. He collected data on the number of nonstandard orders for the past 12 months and added that information to the data on purchasing cost and number of purchase orders.

Month	Purchasing Cost	Number of Purchase Orders	Number of Nonstandard Orders
January	$20,068	330	35
February	20,890	370	61
March	20,750	410	14
April	22,050	400	73
May	21,900	450	55
June	21,300	460	30
July	23,426	560	80
August	21,670	440	51
September	22,250	500	50
October	21,200	470	12
November	21,800	480	27
December	20,800	370	53

Multiple regression was run on the above data; the coefficients shown by the regression program (rounded to the nearest cent) are:

Intercept	15,866.55
X variable 1	10.90
X variable 2	19.54

Required:

1. Construct the cost formula for the purchasing activity showing the fixed cost and the variable rate.

2. If White Swan estimates that next month will have 430 purchase orders and 45 non-standard orders, what is the total estimated purchasing cost for that month? (Round your answer to the nearest dollar.)

3. *What if* White Swan wants to estimate purchasing cost for the coming year and expects 5,340 purchase orders and 580 nonstandard orders? What will estimated total purchasing cost be? What is the total fixed purchasing cost? Why doesn't it equal the fixed cost calculated in Requirement 2 above? (Round your answers to the nearest dollar.)

Cornerstone Exercise 2-7 CUMULATIVE AVERAGE-TIME LEARNING CURVE

OBJECTIVE ➤ 7
CORNERSTONE 2-6

Tam Company makes aircraft engines. Tam has noticed that, in general, each new engine design is subject to an 85 percent learning rate. Assume that the first unit produced takes 400 hours, and direct labour is paid an average of $30 per hour.

Required:

1. Set up a table with columns showing the cumulative number of units, cumulative average time per unit in hours, and cumulative total time in hours. Show results by row for total production of one engine, two engines, four engines, eight engines, sixteen engines, and thirty-two engines. (Round hour answers to two significant digits.)

2. What is the total labour cost if Tam manufactures the following number of engines: one; four; sixteen? What is the average cost per engine for the following number of engines: one, four, sixteen? (Round your answers to the nearest dollar.)

3. *What if* Tam is preparing a bid to build 16 engines? Calculate budgeted labour cost for an engine design that Tam has built before (assume that 16 of these engines had been made previously and the first unit took 400 hours). Calculate budgeted labour cost for a new engine design that Tam's workers have never made before (assume the first unit will take 400 hours).

Exercises

Exercise 2-8 VARIABLE, FIXED, AND MIXED COSTS

OBJECTIVE ➤ 1

Classify the following costs of activity inputs as variable, fixed, or mixed. Identify the activity and the associated activity driver that allow you to define the cost behaviour. For example, assume that the resource input is "cloth in a shirt." The activity would be "sewing shirts," the cost behaviour "variable," and the activity driver "units produced." Prepare your answers in the following format:

Activity	Cost Behaviour	Activity Driver
Sewing shirts	Variable	Units produced

a. Flu vaccine
b. Salaries, equipment, and materials used for moving materials in a factory
c. Forms used to file insurance claims
d. Salaries, forms, and postage associated with purchasing
e. Printing and postage for advertising circulars
f. Equipment, labour, and parts used to repair and maintain production equipment
g. Power to operate sewing machines in a clothing factory
h. Wooden cabinets enclosing audio speakers
i. Advertising
j. Sales commissions
k. Fuel for a delivery van
l. Depreciation on a warehouse
m. Depreciation on a forklift used to move partially completed goods
n. X-ray film used in the Radiology Department of a hospital
o. Rental car provided for a client

OBJECTIVE ▶ 1

Exercise 2-9 COST BEHAVIOUR

Logic Inc. is a data processing firm. Based on past experience, Logic has found that its total annual overhead costs can be represented by the following formula: Overhead cost = $720,000 + $0.90X$, where X equals number of clients. Last year, Logic served 25,000 clients. Actual overhead costs for the year were as expected.

Required:

1. What is the driver for the overhead activity?
2. What is the total overhead cost incurred by Logic last year?
3. What is the total fixed overhead cost incurred by Logic last year?
4. What is the total variable overhead cost incurred by Logic last year?
5. What is the overhead cost per client served?
6. What is the fixed overhead cost per client?
7. What is the variable overhead cost per client?
8. Recalculate Requirements 5, 6, and 7 for the following levels of clientele: (a) 22,000 clients and (b) 27,000. (Round your answers to the nearest cent.) Explain this outcome.

OBJECTIVE ▶ 1

Exercise 2-10 TYPES OF COSTS

Cashion Company produces chemical mixtures for veterinary pharmaceutical companies. Its factory has four mixing lines that mix various powdered chemicals together according to specified formulas. Each line can produce up to 5,000 barrels per year. Each line has one supervisor who is paid $34,000 per year. Depreciation on equipment averages $16,000 per year. Direct materials and power cost about $4.50 per unit.

Required:

1. Prepare a graph for each of these three costs: equipment depreciation, supervisors' wages, and direct materials and power. Use the vertical axis for cost and the horizontal axis for units (barrels). Assume that sales range from 0 to 20,000 units.
2. Assume that the normal operating range for the company is 16,000 to 19,000 units per year. How would you classify each of the three types of cost?

OBJECTIVE ▶ 2

Exercise 2-11 RESOURCE USAGE MODEL AND COST BEHAVIOUR

For the following activities and their associated resources, identify the following: (1) a cost driver, (2) flexible resources, and (3) committed resources. Also, label each resource as one of the following with respect to the cost driver: (a) variable or (b) fixed.

Activity	Resource Description
Maintenance	Equipment, labour, and parts
Inspection	Test equipment, inspectors (each inspector can inspect five batches per day), and units inspected (process requires destructive sampling*)
Packing	Materials, labour (each packer places five units in a box), and conveyor belt
Payable processing	Clerks, materials, equipment, and facility
Assembly	Conveyor belt, supervision (one supervisor for every three assembly lines), direct labour, and materials

*Destructive sampling occurs whenever it is necessary to destroy a unit as inspection occurs.

OBJECTIVE ▶ 2

Exercise 2-12 RESOURCE USAGE AND SUPPLY, ACTIVITY RATES, SERVICE ORGANIZATION

EnviroLabs performs tests on water samples supplied by outside companies to ensure that their waste water meets environmental standards. Customers deliver water samples to the lab and receive the lab reports via the Internet. The EnviroLabs facility is built and staffed to handle the processing of 100,000 tests per year. The lab facility cost $250,000 to build and is expected to last 10 years and will have no salvage value. Processing equipment cost $245,500 and has a life expectancy of five years and will have no salvage value. Both facility and equipment are depreciated on a straight-line basis. EnviroLabs has eight salaried laboratory technicians, each of whom is paid $24,000. In addition to the salaries, facility, and equipment, EnviroLabs expects to spend $60,000 for chemicals and other supplies (assuming 100,000 tests are performed). Last year, 86,000 tests were performed.

Required:

1. Classify the resources associated with the water testing activity into one of the following types: (1) committed resources or (2) flexible resources.
2. Calculate the total annual activity rate for the water testing activity. Break the activity rate into fixed and variable components. (Round your answers to three significant digits.)
3. Compute the total activity availability, and break this into activity output and unused activity.
4. Calculate the total cost of resources supplied, and break this into the cost of activity used and the cost of unused activity.

Exercise 2-13 STEP COSTS, RELEVANT RANGE

OBJECTIVE ▶ 1 2

SERVICE

Eastern University has a group of tutors assisting schoolmates with their studies. Each tutor is paid $25,000 and can tutor up to 500 students per year. Eastern also hires supervisors to oversee the work of the tutors. Given the planning and supervisory work, a supervisor can oversee three tutors, at most. Eastern's accounting history reveals the following relationships between students tutored and the costs of direct labour (tutors) and supervision (measured on an annual basis):

Students served	Direct Labour	Supervision
0–500	$ 25,000	$ 40,000
501–1,000	50,000	40,000
1,001–1,500	75,000	40,000
1,501–2,000	100,000	80,000
2,001–2,500	125,000	80,000
2,501–3,000	150,000	80,000
3,001–3,500	175,000	120,000
3,501–4,000	200,000	120,000

Required:

1. Prepare two graphs: one that illustrates the relationship between direct labour cost and students tutored and one that illustrates the relationship between the cost of supervision and students tutored. Let cost be the vertical axis and students tutored the horizontal axis.
2. How would you classify each cost? Why?
3. Suppose that the normal range of activity is between 2,400 and 2,450 students and that the exact number of tutors is currently hired to support this level of activity. Further suppose that tutored students for the next year is expected to increase by an additional 400 students. How much will the cost of direct labour increase (and how will this increase be realized)? Cost of supervision?

Exercise 2-14 ACCOUNT ANALYSIS METHOD

OBJECTIVE ▶ 5

SERVICE

Penny LeClerc runs the Shear Beauty Salon near a university campus. Several months ago, Penny used some unused space at the back of the salon and bought two used tanning beds. She hired a receptionist and kept the salon open for extended hours each week so that tanning clients would be able to use the benefits of their tanning packages. After

Required:

1. Using the high-low method, calculate the variable rate per hour and the fixed cost for the nursing care activity.

2. Run a regression on the data, using hours of nursing care as the independent variable. Predict cost for the cardiac nursing care for September 2013, if 1,400 hours of nursing care are forecast.

3. Upon looking into the events that happened at the end of 2012, you find that the cardiology ward bought a cardiac-monitoring machine for the nursing station. Administrators also decided to add a new supervisory position for the evening shift. Monthly depreciation on the monitor and the salary of the new supervisor together total $10,000. Now, run two regression equations, one for the observations from 2012 and the second using only the observations for the eight months in 2013. Discuss your findings. What is your predicted cost of the cardiac nursing care activity for September 2013?

OBJECTIVE ➤ 1 3 4 6

SERVICE

Problem 2-32 COMPARISON OF REGRESSION EQUATIONS

Friendly Bank is attempting to determine the cost behaviour of its small business lending operations. One of the major activities is the application activity. Two possible activity drivers have been mentioned: application hours (number of hours to complete the application) and number of applications. The bank controller has accumulated the following data for the setup activity:

Month	Application Costs	Application Hours	Number of Applications
February	$ 7,700	2,000	70
March	7,650	2,100	50
April	10,052	3,000	50
May	9,400	2,700	60
June	9,584	3,000	20
July	8,480	2,500	40
August	8,550	2,400	60
September	9,735	2,900	50
October	10,500	3,000	90

Required:

1. Estimate a regression equation with application hours as the activity driver and the only independent variable. If the bank forecasts 2,600 application hours for the next month, what will be the budgeted application cost?

2. Estimate a regression equation with number of applications as the activity driver and the only independent variable. If the bank forecasts 80 applications for the next month, what will be the budgeted application cost?

3. Which of the two regression equations do you think does a better job of predicting application costs? Explain.

4. Run a multiple regression to determine the cost equation using both activity drivers. What are the budgeted application costs for 2,600 application hours and 80 applications?

OBJECTIVE ➤ 2 4 6

Problem 2-33 MULTIPLE REGRESSION, CONFIDENCE INTERVALS, RELIABILITY OF COST FORMULAS

Ivan Ivanovich, controller, has been given the charge to implement an advanced cost management system. As part of this process, he needs to identify activity drivers for the activities of the firm. During the past four months, Ivan has spent considerable effort identifying activities, their associated costs, and possible drivers for the activities' costs.

Initially, Ivan made his selections based on his own judgment using his experience and input from employees who perform the activities. Later, he used regression analysis to confirm his judgment. Ivan prefers to use one driver per activity, provided that an R^2 of at least 80 percent can be produced. Otherwise, multiple drivers will be used, based on evidence provided by multiple regression analysis. For example, the activity of inspecting finished goods produced an R^2 of less than 80 percent for any single activity driver. Ivan believes, however, that a satisfactory cost formula can be developed using two activity drivers: the number of batches and the number of inspection hours. Data collected for a 14-month period are as follows:

Inspection Costs	Hours of Inspection	Number of Batches
$17,689	100	10
18,350	120	20
13,125	60	15
28,000	320	30
30,560	240	25
31,755	200	40
40,750	280	35
29,500	230	22
47,570	350	50
36,740	270	45
43,500	350	38
26,780	200	18
28,500	140	28
17,000	160	14

Required:

1. Calculate the cost formula for inspection costs using the two drivers, inspection hours and number of batches. Are both activity drivers useful? What does the R^2 indicate about the formula?

2. Using the formula developed in Requirement 1, calculate the inspection cost when 300 inspection hours are used and 30 batches are produced.

Problem 2-34 LEARNING CURVE

OBJECTIVE ➤ 7

Harriman Industries manufactures engines for the aerospace industry. It has completed manufacturing the first unit of the new ZX-9 engine design. Management believes that the 1,000 labour hours required to complete this unit are reasonable and is prepared to go forward with the manufacture of additional units. An 80 percent cumulative average-time learning curve model for direct labour hours is assumed to be valid. Data on costs are as follows:

Direct materials	$10,500
Direct labour	$30 per direct labour hour
Variable manufacturing overhead	$40 per direct labour hour

Required:

1. Set up a table with columns for cumulative number of units, cumulative average time per unit in hours, cumulative total time in hours, and individual unit time for the nth unit in hours. Complete the table for 1, 2, 4, 8, 16, and 32 units. (Use the logarithmic equation to get the individual unit time.)

2. What are the total variable costs of producing 1, 2, 4, 8, 16, and 32 units? What is the variable cost per unit for 1, 2, 4, 8, 16, and 32 units?

Problem 2-35 LEARNING CURVE

Thames Assurance Company sells a variety of life and health insurance products. Recently, Thames developed a long-term care policy for sale to members of university and college alumni associations. Thames estimated that the sale and service of this type of policy would be subject to a 90 percent cumulative average-time learning curve model. Each unit consists of 350 policies sold. The first unit is estimated to take 1,000 hours to sell and service.

Required:

1. Set up a table with columns for cumulative number of units, individual unit time for the nth unit in hours, cumulative total time in hours, and cumulative average time per unit in hours. Complete the table for 1, 2, 4, 8, 16, and 32 units.
2. Suppose that Thames revises its assumption to an 80 percent learning curve. How will this affect the amount of time needed to sell and service eight units? How do you suppose that Thames estimates the percent learning rate?

CMA Problem

CMA Problem 2-1 SIMPLE AND MULTIPLE REGRESSION, EVALUATING RELIABILITY OF AN EQUATION*

The Lockit Company manufactures doorknobs for residential homes and apartments. Lockit is considering the use of simple (single-driver) and multiple regression analyses to forecast annual sales because previous forecasts have been inaccurate. The new sales forecast will be used to initiate the budgeting process and to identify more completely the underlying process that generates sales.

Larry Husky, the controller of Lockit, has considered many possible independent variables and equations to predict sales and has narrowed his choices to four equations. Husky used annual observations from 20 prior years to estimate each of the four equations.

Following are definitions of the variables used in the four equations and a statistical summary of these equations:

Statistical Summary of Four Equations

Equation	Dependent Variable	Independent Variable(s)	Intercept	Independent Variable (Rate)	Standard Error	R_2	t-Value
1	S_t	S_{t-1}	$ 500,000	$ 1.10	$500,000	0.94	5.50
2	S_t	G_t	1,000,000	0.00001	510,000	0.90	10.00
3	S_t	G_{t-1}	900,000	0.000012	520,000	0.81	5.00
4	S_t		600,000		490,000	0.96	
		N_{t-1}		10.00			4.00
		G_t		0.000002			1.50
		G_{t-1}		0.000003			3.00

S_t = Forecasted sales in dollars for Lockit in period t

S_{t-1} = Actual sales in dollars for Lockit in period $t - 1$

G_t = Forecasted Canadian gross domestic product in period t

G_{t-1} = Actual Canadian gross domestic product in period $t - 1$

N_{t-1} = Lockit's net income in period $t - 1$

*© 2011, Institute of Management Accountants, Inc. http://www.imanet.org. Reprinted with permission.

Required:

1. Write Equations 2 and 4 in the form $Y = a + bx$.
2. If actual sales are $1,500,000 in 2012, what would be the forecasted sales for Lockit in 2013?
3. Explain why Husky might prefer Equation 3 to Equation 2.
4. Explain the advantages and disadvantages of using Equation 4 to forecast sales.

(CMA adapted)

The Collaborative Learning Exercises can be found on the product support site at www.hansen1ce.nelson.com.

After studying this chapter, you should be able to:

1 Determine the number of units and amount of sales revenue needed to break even and to earn a target profit.

2 Determine the number of units and sales revenue needed to earn an after-tax target profit.

3 Apply cost-volume-profit analysis in a multiple-product setting.

4 Prepare a profit-volume graph and a cost-volume-profit graph, and explain the meaning of each.

5 Explain the impact of risk, uncertainty, and changing variables on cost-volume-profit analysis.

6 Discuss the impact of non-unit cost drivers on cost-volume-profit analysis.

CHAPTER 3

Cost-Volume-Profit Analysis

Cost-volume-profit analysis (CVP analysis) is a powerful tool for planning and decision making. Because CVP analysis emphasizes the interrelationships of costs, quantity sold, and price, it brings together all of the financial information of the firm. CVP analysis can be a valuable tool in identifying the extent and magnitude of the economic trouble a company is facing and helping pinpoint the necessary solution. The severe recession beginning in 2008 led a number of companies to concentrate on breaking even.

Real-World Example

For example, the **Mayo Clinic** announced that it had broken even for 2008 despite missing its revenue goal by $133 million.[1] The airline industry uses cost-volume-profit analysis in decisions ranging from whether to add another flight to whether to even start a new airline. Lower fuel costs and reduced

[1] Sea Stachura, "Mayo Clinic Breaks Even in 2008," Minnesota Public Radio, March 12, 2009. http://minnesota.publicradio.org/display/web/2009/03/12/mayobudget/?refid=0, accessed March 15, 2009.

capacity led **Delta Air Lines** to estimate it would break even in the first quarter of 2009.[2] Other airlines faced different market conditions and a less favourable outcome. **Air India** and **Jet Airways** were hurt by the weak rupee (Indian currency). The unfavourable currency exchange rate meant that costs the airlines must pay for fuel (denominated in dollars) would have to be offset by revenues that are denominated in rupees. The fall of the value of the rupee increased costs such that **SpiceJet** and Jet Airways faced failure to break even.[3]

CVP analysis can address many issues, such as the number of units that must be sold to break even, the impact a given reduction in fixed costs can have on the break-even point, and the impact an increase in price can have on profit. Additionally, CVP analysis allows managers to conduct sensitivity analyses by examining the impact of various price or cost levels on profit.

While this chapter deals with the mechanics and terminology of CVP analysis, your objective in studying CVP analysis is more than to learn the mechanics. CVP analysis is an integral part of financial planning and decision making. Every accountant and manager should thoroughly understand and be able to apply its concepts.

The Break-Even Point and Target Profit in Units and Sales Revenue

OBJECTIVE ▸ 1
Determine the number of units and amount of sales revenue needed to break even and to earn a target profit.

To find out how revenues, expenses, and profits behave as volume changes, it is natural to begin by finding the firm's break-even point in units sold and in sales revenue. Two frequently used approaches to finding the break-even point are the operating income approach and the contribution margin approach. We will discuss these two approaches to find the **break-even point** (the point of zero profit) and then see how each can determine the total sales revenue at break-even. The determination of units or revenue needed to achieve a target profit is a generalized case of the break-even formulas.

The first step in implementing a units-sold approach to CVP analysis is to determine just what a unit is. For manufacturing firms, the answer is obvious.

Research In Motion may define a unit as a BlackBerry device. Service firms face more varied choices. **Porter Airlines** may define a unit as a passenger mile or a one-way trip. **Canada's Wonderland** counts the number of visitor-days. The **Hospital for Sick Children** in Toronto treats many seriously ill patients from all across the province. They define services to patients in diagnostic-related groups, so that more complicated treatment is weighted more heavily than simple procedures.

The second step is to separate costs into fixed and variable components. CVP analysis focuses on the factors that change the components of profit. Because we are looking at CVP analysis in terms of units sold, we need to determine the fixed and variable components of cost and revenue with respect to units. (This assumption is relaxed when we incorporate activity-based costing into CVP analysis.) It is important to realize that CVP focuses on the firm as a whole. Therefore, *all* costs of the company—manufacturing, marketing, and administrative—are taken into account. Variable costs include *all costs* that increase as more units are sold, including direct

[2] Ann Keaton, "Delta Air Expects a Profit for '09; Keeps Order for 18 787s," *CNN Money.com*, March 10, 2009. http://money.cnn.com/news/newsfeeds/articles/djf500/200903101408DOWJONESDJONLINE000657_FORTUNE5.htm, accessed March 15, 2009.

[3] Mithun Roy, "Falling Re May Force Domestic Airlines to Revise Profit Targets," *The Economic Times*, March 3, 2009. http://economictimes.indiatimes.com/News/News-By-Industry/Transportation/Airlines–Aviation/Falling-Re-may-force-domestic-airlines-to-revise-profit-targets/articleshow/4215213.cms, accessed March 15, 2009.

materials, direct labour, variable manufacturing overhead, and variable selling and administrative costs. Similarly, fixed costs are composed of all fixed manufacturing overhead and fixed selling and administrative expenses.

Basic Concepts for CVP Analysis

The fundamental concept underlying CVP analysis is that the firm's costs can be analysed into variable and fixed costs. A useful tool for organizing the firm's costs into fixed and variable categories is the contribution-margin-based income statement. Note that **operating income** is income or profit *before* income taxes. Operating income includes only revenues and expenses from the firm's normal operations. The term **net income** is used to mean operating income minus income taxes. Cornerstone 3-1 illustrates basic CVP terms and the preparation of the contribution-margin-based income statement.

Cornerstone 3-1 shows that the contribution-margin-based operating income statement is a powerful tool for analyzing a company's projected performance. Notice that the existence of fixed costs means that sales above the estimated 10,000 units, say a 1,000-unit or 10 percent increase, would yield more than a 10 percent increase in operating income. Similarly, a 1,000-unit or 10 percent decrease would decrease operating income by more than 10 percent. This is why an understanding of fixed and variable costs is so important to managers as they examine the impact of changing sales on income.

The Equation Method for Break-Even and Target Income

Companies frequently want to know how many units must be produced and sold to break even or to earn a target income. In other words, how many units will yield the desired (at break-even, zero) profit? The basic break-even/target income equation can be easily derived from the contribution-margin-based operating income statement.

$$\text{Operating income} = \text{Sales revenues} - \text{Variable expenses} - \text{Fixed expenses}$$

This operating income equation can be expanded by expressing sales revenue and variable expenses in terms of unit dollar amounts and number of units. Thus, sales revenue equals the unit selling price times the number of units sold, and total variable costs equal the unit variable cost times the number of units sold. With these expressions, the operating income statement becomes:

$$\text{Operating income} = (\text{Price} \times \text{Number of units}) - (\text{Variable cost per unit} \times \text{Number of units}) - \text{Total fixed costs}$$

Finally, the equation for a target profit is put in terms of units:

$$\text{Units for a target profit} = (\text{Total fixed cost} + \text{Target income})/(\text{Price} - \text{Variable cost per unit})$$

For the special case when target income is zero, the break-even equation becomes:

$$\text{Break-even units} = (\text{Total fixed cost} + 0)/(\text{Price} - \text{Variable cost per unit})$$
$$= \text{Total fixed cost}/(\text{Price} - \text{Variable cost per unit})$$

An important advantage of the operating income statement is that all further CVP equations are derived from the contribution-margin-based income statement. As a result, any CVP problem can be solved by using this approach. Cornerstone 3-2 shows how and why to calculate the units needed to break even and to achieve a target profit.

The HOW and WHY of Basic Cost Calculations and the Contribution-Margin-Based Income Statement

CORNERSTONE 3-1

Information:

Blazin-Boards Company plans to sell 10,000 snowboards at $400 each in the coming year. Product costs include:

Direct materials per snowboard	$80
Direct labour per snowboard	$125
Variable overhead per snowboard	$15
Total fixed factory overhead	$800,000

Variable selling expense is a commission of 5 percent of price; fixed selling and administrative expense totals $400,000.

Why:

Since variable *product* cost per unit consists of variable production or manufacturing costs, the plant manager would use this data. The plant manager is responsible for making a quality product as inexpensively and efficiently as possible. Knowing that variable product cost is $220 per unit provides a starting place for seeing what process improvements might do to the unit cost.

The sales manager would be interested in the total variable cost per unit. Since this cost includes the sales commission, it reflects all of the variable costs. Sales managers can see the impact of the commissions (for which they are responsible) and can also see what impact a one-time discount might have on overall profitability.

Top management would use the unit contribution margin for budgeting to see what impact an increase or a decrease in unit sales would have on operating income. Since fixed costs stay the same when units change, the contribution margin gives important information.

Required:

1. Calculate the:
   a. Variable product cost per unit
   b. Variable selling expense per unit
   c. Total variable cost per unit
   d. Contribution margin per unit
   e. Contribution margin ratio
   f. Total fixed expense for the year
2. Prepare a contribution-margin-based income statement for Blazin-Boards Company for the coming year.
3. **What if** 13,000 boards could be manufactured and sold next year; how would that affect operating income? By what percent?

Solution:

1. a. Variable product cost per unit = Direct materials + Direct labour
   $$+ \text{Variable overhead}$$
   $$= \$80 + \$125 + \$15 = \$220$$

   b. Variable selling expense per unit = $400 \times 0.05 = \$20$

   c. Variable cost per unit = Direct materials + Direct labour
   $$+ \text{Variable overhead} + \text{Variable}$$
   $$\text{selling expense}$$
   $$= \$80 + \$125 + \$15 + \$20 = \$240$$

**CORNERSTONE
3-1**
(continued)

d. Contribution margin per unit = Price − Variable cost per unit

$$= \$400 − \$240 = \$160$$

e. Contribution margin ratio = (Price − Variable cost per unit)/Price

$$= (\$400 − \$240)/\$400 = 0.40 = 40\%$$

OR

$$= (\text{Sales} − \text{Total variable cost})/\text{Sales}$$
$$= (\$4,000,000 − \$2,400,000)/\$4,000,000$$
$$= 0.40 = 40\%$$

f. Total fixed expense = $800,000 + $400,000 = $1,200,000

2.

Blazin-Boards Company
Contribution-Margin-Based Operating Income Statement
For the Coming Year

	Total	Per Unit
Sales ($400 × 10,000 snowboards)	$4,000,000	$400
Total variable expense ($240 × 10,000)	2,400,000	240
Total contribution margin	1,600,000	$160
Total fixed expense	1,200,000	
Operating income	$ 400,000	

3.

Increase in sales (3,000 boards × $400)	$1,200,000
Less:	
Increase in variable cost (3,000 boards × $240)	720,000
Increase in fixed cost	0
Increase in operating income	$ 480,000

Operating income will be $480,000 higher, or $880,000 in total. This is a 120 percent ($480,000/$400,000) increase in operating income, even though the number of units sold would increase by only 30 percent. (Since fixed costs have already been covered, any increase in contribution margin goes directly to operating income.)

Contribution Margin Approach

A refinement of the equation approach is the contribution margin approach. It simply recognizes that at break-even, the total contribution margin equals the fixed expenses. The **contribution margin** is sales revenue minus total variable costs. By substituting the unit contribution margin for price minus unit variable cost in the operating income equation, and solving for the number of units, the following break-even expression is obtained:

Number of units = Fixed costs/Unit contribution margin

Recall that Cornerstone 3-1, Requirement 2, shows the income statement for the budgeted sales for the coming year for Blazin-Boards Company. The contribution margin per unit can be computed in one of two ways. One way is to divide the total

The HOW and WHY of Calculating the Units Needed to Break Even and to Achieve a Target Profit

CORNERSTONE 3-2

Information:
Blazin-Boards Company plans to sell 10,000 snowboards at $400 each in the coming year. Product costs include:

Direct materials per snowboard	$80
Direct labour per snowboard	$125
Variable overhead per snowboard	$15
Total fixed factory overhead	$800,000

Variable selling expense is a commission of 5 percent of price; fixed selling and administrative expense totals $400,000.

Why:
At the break-even point, total revenue equals total cost. Once the break-even point is reached, all fixed costs are covered and additional units add only variable costs. Thus, contribution margin earned above break-even will go toward operating profit. The target operating income is treated as fixed cost for the purpose of calculating the number of units that must be produced and sold. Knowing the break-even units gives managers an easy way to tell just when during the year the firm moves out of the red and into the black.

Required:
1. Calculate the number of units Blazin-Boards must sell to break even. Prepare a contribution-margin-based income statement for the calculated units.
2. Calculate the number of units Blazin-Boards must sell to achieve target operating income (profit) of $240,000.
3. **What if** Blazin-Boards wanted to achieve a target operating income of $300,000? Would the number of snowboards be larger or smaller than the number calculated in Requirement 2? Why?

Solution:
1. Break-even units = Total fixed costs/(Price − Unit variable cost)

$$= \$1,200,000/(\$400 - \$240)$$

$$= 7,500$$

Sales (7,500 units @ $400)	$3,000,000
Less: Variable expenses	1,800,000
Contribution margin	1,200,000
Less: Fixed expenses	1,200,000
Operating income	$ 0

Indeed, selling 7,500 units does yield a zero profit.

2. Units for $240,000 = (Total fixed costs + Target profit)/

(Price − Unit variable cost)

$$= (\$1,200,000 + \$240,000)/(\$400 - \$240)$$

$$= 9,000$$

3. For a target profit of $300,000, more than 9,000 units must be sold. In fact, 9,375 units will yield this profit.

Units for $300,000 = ($1,200,000 + $300,000)/($400 − $240) = 9,375

contribution margin by the units to be sold for a result of $160 per unit ($1,600,000/ 10,000). A second way is to compute price minus variable cost per unit. Doing so yields the same result, $160 per unit ($400 − $240). Now, we can use the contribution margin approach to calculate the break-even number of units.

$$\text{Number of units} = \$1,200,000/(\$400 - \$240)$$
$$= \$1,200,000/\$160 \text{ per unit}$$
$$= 7,500 \text{ units}$$

Of course, the answer is identical to the one computed previously using the equation approach.

Another way to check this number of units computed in Cornerstone 3-2 is to use the break-even point. As was shown in Cornerstone 3-1, Blazin-Boards must sell 10,000 snowboards, or 2,500 more than the break-even volume of 7,500 units, to earn an operating profit of $400,000. The contribution margin per snowboard is $160. Multiplying $160 by the 2,500 snowboards *above* break-even produces the operating profit of $400,000 ($160 × 2,500). This outcome demonstrates that contribution margin per unit for each unit above break-even is equivalent to the operating profit per unit. Since the break-even point had already been computed, the number of snowboards to be sold to yield a $900,000 operating income could have been calculated by dividing the unit contribution margin into the target profit and adding the resulting amount to the break-even volume.

Suppose that Blazin-Boards sells 11,000 snowboards rather than the 10,000 budgeted. What will operating income be in that case? Since fixed costs have already been covered when 10,000 units are sold, the only costs that must be covered on the additional units are the variable costs of $240 per unit. A quicker, more direct way to calculate the new, higher operating income is to take the original operating income of $400,000 for 10,000 units sold and add the contribution margin on the additional 1,000 units, or $160,000 (1,000 × $160). Thus, the total operating income for 11,000 units sold is $560,000 ($400,000 + $160,000).

Break-Even Point and Target Income in Sales Revenue   Sometimes, managers prefer to use sales revenue as the measure of sales activity instead of units sold. A units-sold measure can be converted to a sales-revenue measure simply by multiplying the unit sales price by the units sold. For example, the break-even point for Blazin-Boards Company was 7,500 snowboards. At the selling price per snowboard of $400, the break-even sales revenue is $3,000,000 ($400 × 7,500). Any answer expressed in units sold can be easily converted to an answer expressed in terms of sales revenue, if the break-even units can be easily computed. However, this is seldom the case in a multi-product firm. Fortunately, break-even revenue can be computed directly by developing a separate formula based on total fixed costs, target profit, and the contribution margin ratio. In this case, the important variable is sales revenue, so both the revenue and the variable costs must be expressed in dollars instead of units. Sales revenue is always expressed in dollars, so measuring that variable is no problem. Let's look more closely at variable costs and see how they can be expressed in terms of sales revenue.

To express variable cost in terms of sales revenue, we compute the **variable cost ratio**, which is the proportion of each sales dollar that must be used to cover variable costs. The variable cost ratio can be computed by using either total data or unit data. Of course, the percentage of sales revenue remaining after variable costs are covered is the contribution margin ratio. The **contribution margin ratio** is the proportion of each sales dollar available to cover fixed costs and provide for profit. In Exhibit 3-1, if the variable cost ratio is 60 percent of sales, then the contribution margin must be the remaining 40 percent of sales. It makes sense that the complement of the variable cost ratio is the contribution margin ratio. After all, the proportion of the sales revenue left after variable costs are covered should be the contribution margin component.

Where do fixed costs fit in? Since the contribution margin is revenue remaining after variable costs are covered, it must be the revenue available to cover fixed costs and

Exhibit 3-1

Division of Revenue into Variable Cost and Contribution Margin

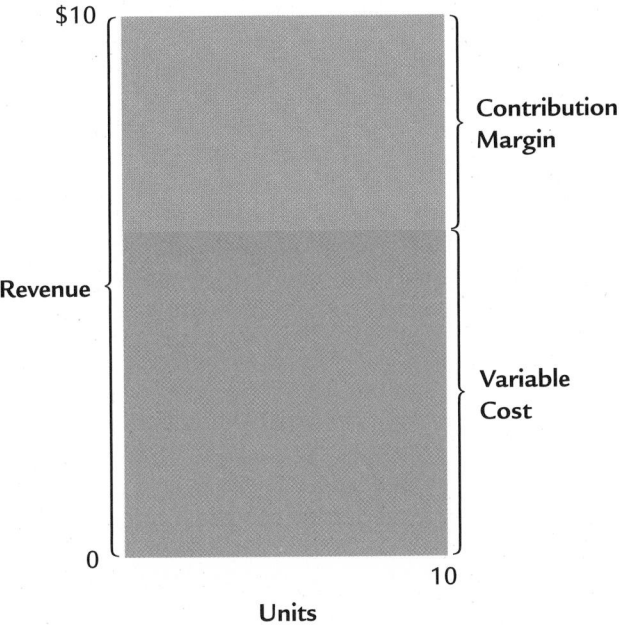

contribute to profit. In other words, we compare total fixed costs to the total contribution margin. If total fixed costs equal the contribution margin, profit is zero. (The company is at break-even.) If total fixed costs are less than the contribution margin, the company earns an operating profit equal to the excess of contribution margin over fixed costs. Finally, if total fixed costs are greater than the contribution margin, the company faces an operating loss.

Now, let's consider the **sales-revenue approach** by looking at the basic income statement.

Operating income = Sales − Variable costs − Total fixed costs

Operating income = Sales − (Variable cost ratio × Sales) − Total fixed costs

Operating income = Sales (1 − Variable cost ratio) − Total fixed costs

Operating income = Sales × Contribution margin ratio − Total fixed costs

Sales = (Total fixed costs + Operating income)/Contribution margin ratio

At break-even, operating income equals zero, so the equation becomes:

Break-even sales = Total fixed costs/Contribution margin ratio

To earn a targeted operating income, sales equal the sum of the total fixed costs and target income divided by the contribution margin ratio.

What about the equation approach used in determining the break-even point in units? We can use that approach here as well. Recall that the formula for the break-even point in units is as follows:

Break-even point in units = Total fixed costs/(Price − Unit variable cost)

If we multiply both sides of the above equation by price, the left-hand side will equal sales revenue at break-even.

Break-even units × Price = Price [Total fixed costs/(Price − Unit variable cost)]

Break-even sales = Total fixed costs × [Price/(Price − Unit variable cost)]

Break-even sales = Total fixed costs × (Price/Contribution margin)

Break-even sales = Total fixed costs/Contribution margin ratio

**CORNERSTONE
3-3**

The HOW and WHY of Calculating Revenue for Break-Even and for a Target Profit

Information:
Blazin-Boards Company plans to sell 10,000 snowboards at $400 each in the coming year. Unit variable cost equals $240. Total fixed costs equal $1,200,000.

Why:
Companies frequently prefer to express the break-even point in sales revenue. To do that, we recognize that total sales revenue must cover both total fixed costs and desired operating income. That is, the proportion of revenue left after variable costs are covered is what is left to cover fixed costs and income.

Required:
1. What is the contribution margin per unit? What is the contribution margin ratio?
2. Calculate the sales revenue needed to break even.
3. Calculate the sales revenue needed to achieve a target operating profit of $240,000.
4. **What if** Blazin-Boards had target operating income (profit) of $350,000? Would sales revenue be larger or smaller than the one calculated in Requirement 3? Why? By how much?

Solution:
1. Contribution margin per unit = Price − Unit variable cost
   $$= \$400 - \$240 = \$160$$
   Contribution margin ratio = $160/$400 = 0.40, or 40%

2. Break-even sales revenue = Total fixed cost/Contribution margin ratio
   $$= \$1,200,000/0.40 = \$3,000,000$$

3. Target sales revenue = (Total fixed cost + Target profit)/
   Contribution margin ratio
   $$= (\$1,200,000 + \$240,000)/0.40 = \$3,600,000$$

4. Target profit of $350,000 is larger than $240,000, so the sales revenue needed would be larger by $275,000.

   New target sales revenue = ($1,200,000 + $350,000)/0.40 = $3,875,000
   Increase in sales revenue = $3,875,000 − $3,600,000 = $275,000

Just as target income was added to total fixed costs in determining unit sales, target operating income is added to total fixed costs when calculating the sales revenue needed for a target operating income. Cornerstone 3-3 illustrates the calculation of break-even sales revenue and sales revenue needed to achieve a target operating profit for Blazin-Boards Company.

In general, assuming that fixed costs remain unchanged, the contribution margin ratio can be used to find the profit impact of a change in sales revenue. To obtain the total change in operating profits from a change in revenue, simply multiply the contribution margin ratio by the change in sales. For example, if sales revenue is $4,000,000 instead of $4,600,000, how will the expected profits be affected? A decrease in sales revenue of $600,000 will cause a decrease in profits of $240,000 (0.40 × $600,000).

Targeted Income as a Percent of Sales Revenue Assume that Blazin-Boards Company wants to know the number of snowboards that must be sold in order to earn a profit equal to 15 percent of sales revenue. Sales revenue is selling price multiplied by the quantity sold. Thus, the targeted operating income is 15 percent of selling price times quantity. Using the operating income approach (which is simpler in this case), we obtain the following:

$$0.15(\$400)(\text{Units}) = (\$400)(\text{Units}) - (\$240)(\text{Units}) - \$1{,}200{,}000$$
$$(\$60)(\text{Units}) = (\$160)(\text{Units}) - \$1{,}200{,}000$$
$$\$100(\text{Units}) = \$1{,}200{,}000$$
$$\text{Units} = 12{,}000$$

Does a volume of 12,000 snowboards achieve an operating profit equal to 15 percent of sales revenue? For 12,000 snowboards, the total revenue is $4,800,000 ($400 × 12,000). The operating profit can be computed without preparing a formal income statement. Remember that above break-even, the contribution margin per unit is the operating profit per unit. The break-even volume is 7,500 snowboards. If 12,000 snowboards are sold, then 4,500 (12,000 − 7,500) snowboards above the break-even point are sold. The before-tax profit, therefore, is $720,000 ($160 × 4,500), which is 15 percent of sales ($720,000/$4,800,000).

Comparison of the Break-Even Point in Units and Sales Revenue

For a single-product setting, converting the break-even point in units answer to a sales-revenue answer is simply a matter of multiplying the unit sales price by the units sold. Then why bother with a separate formula for the sales-revenue approach? For a single-product setting, neither approach has any real advantage over the other. Both offer much the same level of conceptual and computational difficulty.

However, in a multiple-product setting, CVP analysis is more complex, and the sales-revenue approach is significantly easier. This approach maintains essentially the same computational requirements found in the single-product setting, whereas the units-sold approach becomes more difficult. Even though the conceptual complexity of CVP analysis does increase with multiple products, the operation is reasonably straightforward.

After-Tax Profit Targets

OBJECTIVE ▶ 2
Determine the number of units and sales revenue needed to earn an after-tax target profit.

Income taxes are generally calculated as a percentage of income. When calculating the break-even point, income taxes play no role because the taxes paid on zero income are zero. However, when the company needs to know how many units to sell to earn a particular net income, some additional consideration is needed. Recall that net income is operating income minus income taxes and that our targeted income figure was expressed in before-tax terms. As a result, when the income target is expressed as net income, we must add back the income taxes to get operating income. Therefore, to use either the equation method or the contribution margin approach, the after-tax profit target must first be converted to a before-tax profit target.

In general, taxes are computed as a percentage of income. The after-tax profit, or net income, is computed by subtracting income taxes from the operating income (or before-tax profit).

Net income = Operating income − Income taxes

Net income = Operating income − (Tax rate × Operating income)

Net income = Operating income (1 − Tax rate)

or

Operating income = Net income/(1 − Tax rate)

Thus, to convert the after-tax profit to before-tax profit, simply divide the after-tax profit by the quantity (1 − Tax rate). Cornerstone 3-4 shows how to calculate the number of units needed to achieve an after-tax profit target.

**CORNERSTONE
3-4**

The HOW and WHY of Calculating the Number of Units to Generate an After-Tax Target Profit

Information:

Blazin-Boards Company wants to earn $390,000 in net (after-tax) income next year. Snowboards are priced at $400 each for the coming year. Product costs include:

Direct materials per snowboard	$80
Direct labour per snowboard	$125
Variable overhead per snowboard	$15
Total fixed factory overhead	$800,000

Variable selling expense is a commission of 5 percent of price; fixed selling and administrative expense totals $400,000. Blazin-Boards has a tax rate of 35 percent.

Why:

Top management may be interested in target net income, since income taxes are a legitimate expense of the business and owners are interested in after-tax income. The accountant must first convert net income into operating income since the tax rate is a variable that is not taken into account in the break-even equation. Once the conversion is made, the break-even equation can be applied.

Required:

1. Calculate the before-tax profit needed to achieve an after-tax target of $422,500.
2. Calculate the number of boards that will yield operating income calculated in Requirement 1 above.
3. Prepare an income statement for Blazin-Boards Company for the coming year based on the number of boards computed in Requirement 2.
4. *What if* Blazin-Boards had a 30 percent tax rate. Would the units sold to reach a $422,500 target net income be higher or lower than the result in Requirement 2? Calculate the number of units needed.

Solution:

1. Before-tax income = After-tax income/(1 − Tax rate)

    $$= \$422,500/(1 - 0.35)$$
    $$= \$422,500/(0.65)$$
    $$= \$650,000$$

2. Number of boards = (Total fixed cost + Target profit)/
    (Price − Variable cost per unit)

    $$= (\$1,200,000 + \$650,000)/(\$400 - \$240)$$
    $$= 11,563 \text{(rounded)}$$

3.

Blazin-Boards Company
Income Statement
For the Coming Year

	Total	Per Unit
Sales ($400 × 11,563 snowboards)	$4,625,200	$400
Total variable expense ($240 × 11,563)	2,775,120	240
Total contribution margin	1,850,080	$160

	Total	Per Unit
Total fixed expense	1,200,000	
Operating income	650,080	
Less: Income taxes ($650,080 × 0.35)	227,528	
Net income*	$ 422,552	

**CORNERSTONE
3-4**
(continued)

*Note that net income is $52 higher than the target due to rounding the units up from 11,562.5 to 11,563.

4. The units would be lower than 11,563 since the lower tax rate means that a smaller operating income would be needed to yield the same target net income.

$$\text{Before-tax income} = \text{After-tax income}/(1 - \text{Tax rate})$$
$$= \$422,500/(1 - 0.30)$$
$$= \$603,571 \text{(rounded)}$$
$$\text{Number of boards} = (\text{Total fixed cost} + \text{Target profit})/$$
$$(\text{Price} - \text{Variable cost per unit})$$
$$= (\$1,200,000 + \$603,571)/(\$400 - \$240)$$
$$= 11,272 \text{(rounded)}$$

Multiple-Product Analysis

OBJECTIVE ➤ 3
Apply cost-volume-profit analysis in a multiple-product setting.

Blazin-Boards Company has decided to offer two models of snowboards: a regular snowboard to sell for $400 and a deluxe snowboard, using graphite and designed for championship-calibre boarders, to sell for $600. The Marketing Department is convinced that 10,000 regular snowboards and 2,500 deluxe snowboards can be sold during the coming year. The controller has prepared the following projected income statement based on the sales forecast:

	Regular Snowboards	Deluxe Snowboards	Total
Sales	$4,000,000	$1,500,000	$5,500,000
Less: Variable expenses	2,400,000	750,000	3,150,000
Contribution margin	1,600,000	750,000	2,350,000
Less: Direct fixed expenses	400,000	200,000	600,000
Product margin	$1,200,000	$ 550,000	1,750,000
Less: Common fixed expenses			200,000
Operating income			$1,550,000

Note that the controller has separated direct fixed expenses from common fixed expenses. The **direct fixed expenses** are those fixed costs that can be traced to each segment and that would be avoided if the segment did not exist. Examples of direct fixed expenses include salaries of the individual segment's supervisors, any equipment that must be leased or bought just for that segment, and so on. The **common fixed expenses** are the fixed costs that are not traceable to the segments and that would remain even if one of the segments was eliminated. Corporate headquarters costs are common fixed expenses, as are the costs of the factory manager and factory landscaping.

Break-Even Point in Units and Sales Revenue for the Multiple-Product Setting

The owner of Blazin-Boards is apprehensive about adding a new product line and wants to know how many of each model must be sold to break even. If you were given the responsibility to answer this question, how would you respond?

One possible response is to use the equation we developed earlier in which fixed costs were divided by the contribution margin. However, this equation was developed for a single-product analysis. For two products, there are two unit contribution margins. The regular snowboard has a contribution margin per unit of $160 ($400 − $240), and the deluxe snowboard has one of $300 ($600 − $300). One possible solution is to apply the analysis separately to each product line. It is possible to obtain individual break-even points when income is defined as product margin. Break-even for the regular snowboard is as follows:

$$\text{Regular snowboard break-even units} = \text{Fixed costs}/(\text{Price} - \text{Unit variable cost})$$
$$= \$400,000/\$160$$
$$= 2,500 \text{ units}$$

Break-even for the deluxe snowboard can be computed as well.

$$\text{Deluxe snowboard break-even units} = \text{Fixed costs}/(\text{Price} - \text{Unit variable cost})$$
$$= \$200,000/\$300$$
$$= 667 \text{ units (rounded)}$$

Thus, 2,500 regular snowboards and 667 deluxe snowboards must be sold to achieve a break-even product margin. But a break-even product margin covers only direct fixed costs; the common fixed costs remain to be covered. Selling these numbers of snowboards would result in a loss equal to the common fixed costs. No break-even point for the firm as a whole has yet been identified. Somehow, the common fixed costs must be factored into the analysis.

Allocating the common fixed costs to each product line before computing a break-even point may resolve this difficulty. The problem with this approach is that allocation of the common fixed costs is arbitrary. Thus, no meaningful break-even volume is readily apparent.

Another possible solution is to convert the multiple-product problem into a single-product problem. If this can be done, then all of the single-product CVP methodology can be applied directly. The key to this conversion is to identify the expected sales mix, in units, of the products being sold.

Sales Mix   **Sales mix** is the relative combination of products being sold by a firm. Sales mix can be measured in units sold or in proportion of revenue. For example, if Blazin-Boards plans to sell 10,000 regular snowboards and 2,500 deluxe snowboards, then the sales mix in units is 10,000:2,500. Usually, the sales mix is reduced to the smallest possible whole numbers. Thus, the relative mix 10,000:2,500 can be reduced to 100:25 and further to 4:1. That is, for every four regular snowboards sold, one deluxe snowboard is sold.

Alternatively, the sales mix can be represented by the percent of total revenue contributed by each product. In that case, the regular snowboard revenue is $4,000,000 ($400 × 10,000), and the deluxe snowboard revenue is $1,500,000 ($600 × 2,500). The regular snowboard accounts for 70 percent of total revenue, and the deluxe snowboard accounts for the remaining 30 percent (where the percentages are rounded). It may seem as though the two sales mixes are different. The sales mix in units is 4:1; that is, of every five snowboards sold, 80 percent are regular snowboards and 20 percent are deluxe snowboards. However, the revenue-based sales mix is 70 percent for the regular snowboards. There is really no difference. The sales mix in revenue takes the sales mix in units and weights it by price. Therefore, even though the underlying proportion of snowboards sold remains 4:1, the lower priced regular snowboards are weighted less heavily when price is factored in. In the remaining discussion, we will use the sales mix expressed in units.

A number of different sales mixes can be used to define the break-even volume. For example, a sales mix of 5:1 will define a break-even point of 3,637 regular snowboards and 727 deluxe snowboards. The total contribution margin produced by this mix is $800,020 [($160 × 3,637) + ($300 × 727)]. Similarly, if 2,353 regular snowboards and 1,412 deluxe snowboards are sold (corresponding to a 5:3 sales mix), the total contribution margin is $800,080 [($160 × 2,353) + ($300 × 1,412)]. Since total fixed costs are $800,000, both sales mixes define break-even points. Fortunately, every sales mix need not be considered. Can Blazin-Boards really expect a sales mix of 5:1 or 5:3? For every two regular snowboards sold, does Blazin-Boards expect to sell a deluxe snowboard? Or for every regular snowboard, can Blazin-Boards really sell one deluxe snowboard?

According to Blazin-Boards's marketing study, a sales mix of 4:1 can be expected. This is the ratio that should be used; the others can be ignored. The sales mix that is expected to prevail should be used for CVP analysis.

Sales Mix and CVP Analysis   Defining a particular sales mix allows us to convert a multiple-product problem to a single-product CVP format. Since Blazin-Boards expects to sell four regular snowboards for every deluxe snowboard, it can define the single product it sells as a package containing four regular snowboards and one deluxe snowboard. By defining the product as a package, the multiple-product problem is converted into a single-product one. Cornerstone 3-5 illustrates the use of the package approach to calculating break-even units in the multi-product firm.

For a given sales mix, CVP analysis can be used as if the firm were selling a single product. However, actions that change the prices of individual products can affect the sales mix because consumers may buy relatively more or less of the product. Accordingly, pricing decisions may involve a new sales mix and must reflect this possibility. Keep in mind that a new sales mix will affect the units of each product that need to be sold in order to achieve a desired profit target. If the sales mix for the coming period is uncertain, it may be necessary to look at several different mixes. This is sensitivity analysis, and it gives managers insight into the possible outcomes facing the firm.

The complexity of the break-even-point-in-units approach increases dramatically as the number of products increases. Imagine performing this analysis for a firm with several hundred products. This observation seems more overwhelming than it actually is. Computers can easily handle a problem with so much data. Furthermore, many firms simplify the problem by analyzing product groups rather than individual products. Another way to handle the increased complexity is to switch from the units-sold to the sales-revenue approach. This approach can accomplish a multiple-product CVP analysis using only the summary data found in an organization's income statement. The computational requirements are much simpler.

To illustrate the break-even point in sales revenue, the same examples will be used. However, the only information needed is the projected income statement for Blazin-Boards Company as a whole.

	Total Snowboards
Sales	$5,500,000
Less: Variable expenses	3,150,000
Contribution margin	2,350,000
Less: Total fixed expenses	800,000
Operating income	$1,550,000

**CORNERSTONE
3-5**

The HOW and WHY of Calculating the Break-Even Number of Units in a Multi-Product Firm

Information:
Blazin-Boards Company plans to sell 10,000 regular snowboards and 2,500 deluxe snowboards in the coming year. Product price and cost information includes:

	Regular Snowboard	Deluxe Snowboard
Price	$ 400	$ 600
Unit variable cost	240	300
Direct fixed cost	400,000	200,000

Common fixed selling and administrative expense totals $200,000.

Why:
The break-even point in units gives managers a starting point for increasing profitability. If the company is making a loss, the break-even point tells management just what needs to be done to stop losing money. Once the break-even point is passed, the company will earn a profit. By looking at break-even points for each product, managers can see whether one product is being "carried" by other products.

Required:
1. What is the sales mix estimated for next year (calculated to the lowest whole number for each product)?
2. Using the sales mix from Requirement 1, form a package of regular and deluxe snowboards. Taking the package contribution margin to three decimal places, calculate the break-even number of regular snowboards and deluxe snowboards.
3. Prepare a contribution-margin-based income statement for Blazin-Boards Company based on the unit sales calculated in Requirement 2.
4. *What if* Blazin-Boards believed that 10,000 regular snowboards and 5,000 deluxe snowboards could be sold? What is the sales mix, and how many regular and deluxe snowboards must be produced and sold at break-even?

Solution:
1. Sales mix of regular to deluxe snowboards = 10,000:2,500 = 4:1
2.

Product	Price	Unit Variable Cost	Unit Contribution Margin	Sales Mix	Unit Contribution Margin × Sales Mix
Regular	$400	$240	$160	4	$640[a]
Deluxe	$600	$300	$300	1	300[b]
Package contribution margin					$940

[a]Found by multiplying the number of units in the package (4) by the unit contribution margin ($160).
[b]Found by multiplying the number of units in the package (1) by the unit contribution margin ($300).

Break-even packages = Total fixed cost/Package contribution margin

= ($400,000 + $200,000 + $200,000)/$940

= 851.064 packages

Break-even regular snowboards $= (4 \times 851.064) = 3,404$

Break-even deluxe snowboards $= (1 \times 851.064) = 851$

Note: Packages are not rounded off to a whole number because the number of packages is not an end in itself. The decimal amount may be important when multiplied by the sales mix. The number of snowboards is rounded to whole units, since no one will buy a fraction of a snowboard.

3.

**Blazin-Boards
Income Statement
For the Coming Year**

	Regular Snowboards	Deluxe Snowboards	Total
Sales	$1,361,600	$510,600	$1,872,200
Less: Variable expenses	816,960	255,300	1,072,260
Contribution margin	544,640	255,300	799,940
Less: Direct fixed expenses	400,000	200,000	600,000
Product margin	$ 144,640	$ 55,300	199,940
Less: Common fixed expenses			200,000
Operating income			$ (60)

4. The sales mix is 10,000:5,000, or 2:1.

Product	Price	Unit Variable Cost	Unit Contribution Margin	Sales Mix	Unit Contribution Margin × Sales Mix
Regular	$400	$240	$160	2	$320[a]
Deluxe	$600	$300	$300	1	300[b]
Package contribution margin					$620

[a]Found by multiplying the number of units in the package (2) by the unit contribution margin ($160).
[b]Found by multiplying the number of units in the package (1) by the unit contribution margin ($300).

Break-even packages = Total fixed cost/Package contribution margin

$$= (\$400,000 + \$200,000 + \$200,000)/\$620$$

$$= 1,290.323 \text{ packages}$$

Break-even regular snowboards $= (2 \times 1,290.323) = 2,581 \text{(rounded)}$

Break-even deluxe snowboards $= (1 \times 1,290.323) = 1,290 \text{(rounded)}$

Notice that this income statement corresponds to the total column of the more detailed income statement examined previously. The projected income statement rests on the assumption that 10,000 regular snowboards and 2,500 deluxe snowboards will be sold (a 4:1 sales mix). The break-even point in sales revenue also rests on the expected sales mix. (As with the units-sold approach, a different sales mix will produce different results.)

With the income statement, the usual CVP questions can be addressed. For example, how much sales revenue must be earned to break even? To answer this question, we divide the total fixed costs of $800,000 by the contribution margin ratio of 0.4273 ($2,350,000/$5,500,000).

$$\text{Break-even sales} = \text{Fixed costs/Contribution margin ratio}$$
$$= \$800,000/0.4273$$
$$= \$1,872,221$$

The break-even point in sales revenue implicitly uses the assumed sales mix but avoids the requirement of building a package contribution margin. No knowledge of individual product data is needed. The computational effort is similar to that used in the single-product setting. Moreover, the answer is still expressed in sales revenue. Unlike the break-even point in units, the answer to CVP questions using sales revenue is still expressed in a single summary measure. The sales-revenue approach, however, does sacrifice information concerning individual product performance.

Graphical Representation of CVP Relationships

OBJECTIVE ▶ 4
Prepare a profit-volume graph and a cost-volume-profit graph, and explain the meaning of each.

A graphical representation can help managers see the difference between variable cost and revenue and deepens their understanding of CVP relationships. It may also help managers understand quickly what impact an increase or decrease in sales will have on the break-even point. Two basic graphs, the profit-volume graph and the cost-volume-profit graph, are presented here.

The Profit-Volume Graph

A **profit-volume graph** portrays the relationship between profits and sales volume. The profit-volume graph is the graph of the operating income equation [Operating income = (Price × Units) − (Unit variable cost × Units) − Fixed costs]. In this graph, operating income (profit) is the dependent variable, and number of units is the independent variable. Usually, values of the independent variable are measured along the horizontal axis and values of the dependent variable along the vertical axis.

To make this discussion more concrete, a simple set of data will be used. Assume that Gordon Company produces a single product with the following cost and price data:

Total fixed costs	$100
Variable cost per unit	$ 5
Selling price per unit	$ 10

Using these data, operating income can be expressed as follows:

$$\text{Operating income} = (\$10 \times \text{Units}) - (\$5 \times \text{Units}) - \$100$$
$$= (\$5 \times \text{Units}) - \$100$$

This relationship is graphed by plotting units along the horizontal axis and operating income (or loss) along the vertical axis. Two points are needed to graph a linear equation. While any two points will do, the two points often chosen are those that correspond to zero sales volume and zero profits. When units sold are zero, Gordon experiences an operating loss of $100 (or a profit of −$100). The point corresponding to zero sales volume, therefore, is (0, −$100). In other words, when no sales take place, the company suffers a loss equal to its total fixed costs. When operating income is zero, the units sold are equal to 20. The point corresponding to zero profits (break-even) is (20, $0). These two points, plotted in Exhibit 3-2, define the profit graph shown in the same figure.

Exhibit 3-2

Profit-Volume Graph

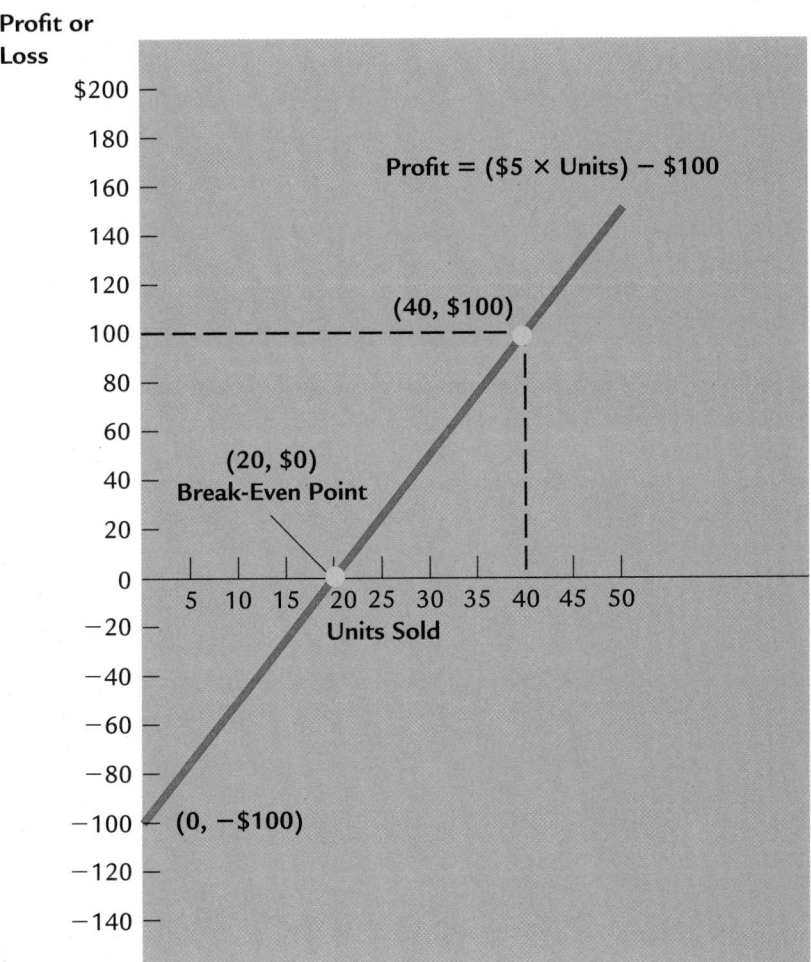

The graph in Exhibit 3-2 can be used to assess Gordon's profit (or loss) at any level of sales activity. For example, the profit associated with the sale of 40 units can be read from the graph by (1) drawing a vertical line from the horizontal axis to the profit line and (2) drawing a horizontal line from the profit line to the vertical axis. As we can see, the profit associated with sales of 40 units is $100. The profit-volume graph, while easy to interpret, fails to reveal how costs change as sales volume changes. A more comprehensive graph provides this detail.

The Cost-Volume-Profit Graph

The **cost-volume-profit graph** depicts the relationships among cost, volume, and profits. To obtain the more detailed relationships, it is necessary to graph two separate lines: the total revenue line and the total cost line. These lines are represented, respectively, by the following two equations:

$$\text{Revenue} = \text{Price} \times \text{Units}$$
$$\text{Total cost} = (\text{Unit variable cost} \times \text{Units}) + \text{Fixed costs}$$

Using the Gordon Company example, the revenue and cost equations are as follows:

$$\text{Revenue} = \$10 \times \text{Units}$$
$$\text{Total cost} = (\$5 \times \text{Units}) + \$100$$

To portray both equations in the same graph, the vertical axis is measured in dollars and the horizontal axis in units sold.

Two points are needed to graph each equation. We will use the same x-coordinates used for the profit-volume graph. For the revenue equation, setting number of units equal to zero results in revenue of $0; setting number of units equal to 20 results in revenue of $200. Therefore, the two points for the revenue equation are (0, $0) and (20, $200). For the cost equation, units sold of zero and units sold equal to 20 produce the points (0, $100) and (20, $200). The graphs of both equations appear in Exhibit 3-3.

Notice that the total revenue line begins at the origin and rises with a slope equal to the selling price per unit (a slope of 10). The total cost line intercepts the vertical axis at a point equal to total fixed costs and rises with a slope equal to the variable cost per unit (a slope of 5). When the total revenue line lies below the total cost line, a loss region is defined. Similarly, when the total revenue line lies above the total cost line, a profit region is defined. The point where the total revenue line and the total cost line intersect is the break-even point. To break even, Gordon Company must sell 20 units and thus receive $200 in total revenues.

Now, let's compare the information available from the CVP graph to that available from the profit-volume graph. To do so, consider the sale of 40 units. Recall that the profit-volume graph revealed that selling 40 units produced profit of $100. Examine Exhibit 3-3 again. The CVP graph also shows profits of $100, but it reveals more than that. The CVP graph discloses that total revenues of $400 and total costs of $300 are associated with the sale of 40 units. Furthermore, the total costs can be broken down into fixed costs of $100 and variable costs of $200. The CVP graph provides revenue and cost information not provided by the profit-volume graph. Unlike the profit-volume graph, some computation is needed to determine the profit associated with a given sales volume. Nonetheless, because of the greater information content, managers are likely to find the CVP graph a more useful tool.

Exhibit 3-3

Cost-Volume-Profit Graph

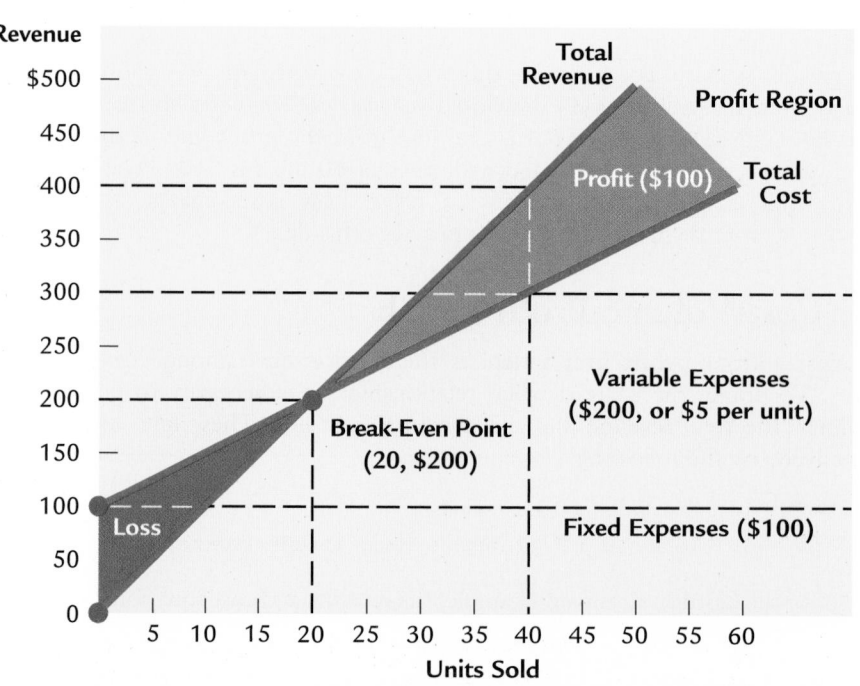

Assumptions of Cost-Volume-Profit Analysis

The profit-volume and cost-volume-profit graphs just illustrated rely on some important assumptions. Some of these assumptions are as follows:

1. The analysis assumes a linear revenue function and a linear cost function.
2. The analysis assumes that price, total fixed costs, and unit variable costs can be accurately identified and remain constant over the relevant range.
3. The analysis assumes that what is produced is sold.
4. For multiple-product analysis, the sales mix is assumed to be known.
5. The selling prices and costs are assumed to be known with certainty.

The first assumption, linear cost and revenue functions, deserves additional consideration. Let's take a look at the underlying revenue and total cost functions identified in economics. Exhibit 3-4, Panel A, portrays the curvilinear revenue and cost functions. We see that as quantity sold increases, revenue also increases, but eventually revenue begins to rise less steeply than before. This is explained quite simply by the need to decrease price as many more units are sold. The total cost function is more complicated, rising steeply at first, then levelling off somewhat (as increasing returns to scale develop), and then rising steeply again (as decreasing returns to scale develop). How can we deal with these complicated relationships?

Relevant Range   Fortunately, we do not need to consider all possible ranges of production and sales for a firm. Remember that CVP analysis is a short-run decision-making tool. (We know that it is short run in orientation because some costs are fixed.) It is only necessary for us to determine the current operating range, or **relevant range**, for which the linear cost and revenue relationships are valid. Exhibit 3-4, Panel B, illustrates a relevant range from 5,000 to 15,000 units. Note that the cost and revenue relationships are roughly linear in this range, allowing us to use our linear CVP equations. Of course, if the relevant range changes, different fixed and variable costs and different prices must be used.

The second assumption is linked to the definition of relevant range. Once a relevant range has been identified, then the cost and price relationships are assumed to be known and constant.

Production Equal to Sales   The third assumption is that what is produced is sold. There is no change in inventory over the period. The fact that inventory has no

---

> **Exhibit 3-4**

Cost and Revenue Relationships

Panel A: Curvilinear CVP Relationships

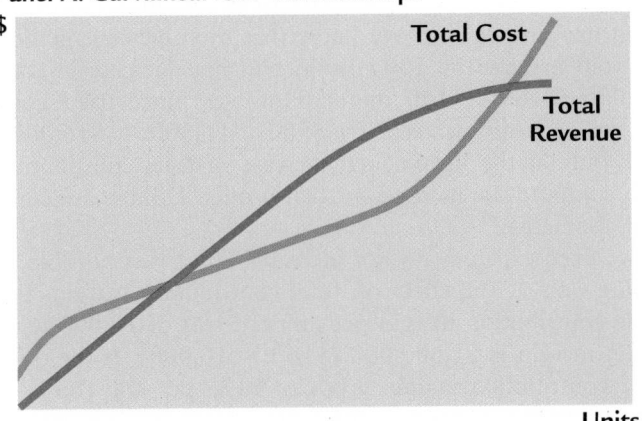

Panel B: Relevant Range and Linear CVP Relationships

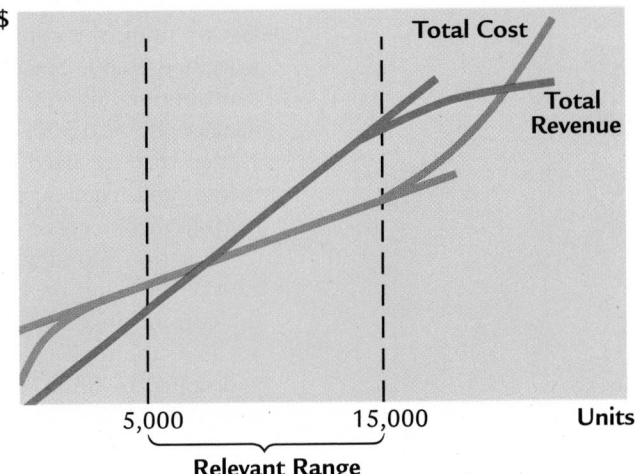

impact on break-even analysis makes sense. Break-even analysis is a short-run decision-making technique, so we are looking to cover all costs of a particular period of time. Inventory embodies costs of a previous period and is not considered.

Constant Sales Mix   The fourth assumption is a constant sales mix. In single-product analysis, the sales mix is obviously constant—100 percent of sales is applied to one product. Multiple-product break-even analysis requires a constant sales mix. However, it is virtually impossible to predict the sales mix with certainty. Typically, this constraint is handled in practice through sensitivity analysis. By using the capabilities of spreadsheet analysis, the sensitivity of variables to a variety of sales mixes can be readily assessed.

Prices and Costs Known with Certainty   Finally, the fifth assumption is that prices and costs are known. In actuality, firms seldom know variable costs and fixed costs with certainty. A change in one variable usually affects the value of others. Often, there is a probability distribution with which to contend. There are formal ways of explicitly building uncertainty into the CVP model. Exploration of these issues is introduced in the next section.

Changes in the CVP Variables

OBJECTIVE ▶ 5

Explain the impact of risk, uncertainty, and changing variables on cost-volume-profit analysis.

Because firms operate in a dynamic world, they must be aware of changes in prices, variable costs, and fixed costs. They must also account for the effects of risk and uncertainty. We will take a look at the effects on the break-even point of changes in price, unit variable cost, and fixed costs. We will also look at ways managers can handle risk and uncertainty within the CVP framework.

Let's return to the Blazin-Boards Company example before the deluxe snowboard was introduced (when only the regular snowboard is produced). Suppose that the Sales Department recently conducted a market study that revealed three different alternatives.

Alternative 1: If advertising expenditures increase by $16,500, sales will increase from 10,000 units to 10,100 units.

Alternative 2: A price decrease from $400 per snowboard to $375 per snowboard would increase sales from 10,000 units to 12,000 units.

Alternative 3: Decreasing prices to $375 and increasing advertising expenditures by $16,500 will increase sales from 10,000 units to 12,000 units.

Should Blazin-Boards maintain its current price and advertising policies, or should it select one of the three alternatives described by the marketing study?

Consider the first alternative. What is the effect on profits if advertising costs increase by $16,500 and sales increase by 100 units? This question can be answered just by using the contribution margin per unit. We know that the unit contribution margin is $160. Since units sold increase by 100, the incremental increase in total contribution margin is $16,000 ($160 × 100 units). However, since fixed costs increase by $16,500, profits will actually decrease by $500 ($16,500 − $16,000). Notice that we need to look only at the incremental increase in total contribution margin and fixed expenses to compute the increase in total profits. Exhibit 3-5 summarizes the effects of the first alternative.

For the second alternative, fixed expenses do not increase. Thus, it is possible to answer the question by looking only at the effect on total contribution margin. For the current price of $400, the contribution margin per unit is $160. If 10,000 units are sold, the total contribution margin is $1,600,000 ($160 × 10,000). If the price is dropped to $375, then the contribution margin drops to $135 per unit ($375 − $240). If 12,000 units are sold at the new price, then the new total contribution margin is $1,620,000 ($135 × 12,000). As shown in Exhibit 3-6, dropping the price results in a profit increase of $20,000 ($1,620,000 − $1,600,000).

Exhibit 3-5

Summary of the Effects of Alternative 1

	Before the Increased Advertising	With the Increased Advertising
Units sold	10,000	10,100
Unit contribution margin	× $160	× $160
Total contribution margin	$1,600,000	$1,616,000
Less: Fixed expenses	1,200,000	1,216,500
Profit	$ 400,000	$ 399,500

	Difference in Profit
Change in sales volume	100
Unit contribution margin	× $160
Change in contribution margin	$16,000
Less: Increase in fixed expenses	16,500
Decrease in profit	$ (500)

Exhibit 3-6

Summary of the Effects of Alternative 2

New contribution margin ($135 × 12,000 units)	$1,620,000
Old contribution margin ($160 × 10,000 units)	1,600,000
Increased contribution margin	$ 20,000

The third alternative calls for a decrease in the unit selling price and an increase in advertising costs. Like the first alternative, the profit impact can be assessed by looking at the incremental effects on contribution margin and fixed expenses. The incremental profit change can be found by (1) computing the incremental change in total contribution margin, (2) computing the incremental change in fixed expenses, and (3) adding the two results.

The current total contribution margin for 10,000 units sold is $1,600,000. Since the new unit contribution margin is $135, and units sold increase to 12,000, the new total contribution margin is $1,620,000 ($135 × 12,000 units). Thus, the incremental increase in total contribution margin is $20,000 ($1,620,000 − $1,600,000). However, to achieve this incremental increase in contribution margin, an incremental increase of $16,500 in fixed costs is needed. The net effect is an incremental increase in profits of $3,500. The effects of the third alternative are summarized in Exhibit 3-7.

Of the three alternatives identified by the marketing study, both the second and third alternatives promise a benefit. They increase total profits by $20,000 (alternative 2) and $3,500 (alternative 3). Clearly, alternative 2 has a higher profit potential.

These examples are all based on a units-sold approach. However, we could just as easily have applied a sales-revenue approach. The answers would be the same.

Introducing Risk and Uncertainty

An important assumption of CVP analysis is that prices and costs are known with certainty. This is seldom the case. Risk and uncertainty are a part of business decision making and must be considered. Formally, risk differs from uncertainty in that with risk, the probability distributions of the variables are known. With uncertainty, the probability distributions are not known. For our purposes, however, the terms will be used interchangeably.

Exhibit 3-7

Summary of the Effects of Alternative 3

	Before Changes	With the Increased Advertising and Decreased Price
Units sold	10,000	12,000
Unit contribution margin	× $160	× $135
Total contribution margin	$1,600,000	$1,620,000
Less: Fixed expenses	1,200,000	1,216,500
Profit	$ 400,000	$ 403,500

	Difference in Profit
Decrease in contribution margin on 10,000 units	$(250,000)
Increase in contribution margin on 2,000 units	270,000
Change in contribution margin	$ 20,000
Less: Increase in fixed expenses	16,500
Increase in profit	$ 3,500

Managers deal with risk and uncertainty in a variety of ways. First, of course, management must realize the uncertain nature of future prices, costs, and quantities. Next, managers move from consideration of a break-even point to what might be called a break-even band. In other words, given the uncertain nature of the data, perhaps a firm might break even when 1,800 to 2,000 units are sold—instead of the point estimate of 1,900 units. Furthermore, managers may engage in sensitivity or what-if analyses. Here, a computer spreadsheet is helpful, as managers set up the break-even (or targeted profit) relationships and then check to see the impact that varying costs and prices have on quantity sold. Two concepts useful to management are *margin of safety* and *operating leverage*. Both of these may be considered measures of risk. Each requires knowledge of fixed and variable costs.

Margin of Safety   The **margin of safety** is the units sold or expected to be sold or the revenue earned or expected to be earned above the break-even volume. Cornerstone 3-6 illustrates the margin of safety.

The margin of safety can be viewed as a crude measure of risk. There are always events, unknown when plans are made, that can lower sales below the original expected level. If a firm's margin of safety is large given the expected sales for the coming year, the risk of suffering losses should sales take a downward turn is less than if the margin of safety is small. Managers who face a low margin of safety may wish to consider actions to increase sales or decrease costs.

For example, **Walt Disney Company** faced lower theme park earnings in the last quarter of 2004 due to the unprecedented number of hurricanes that hit Florida during August. Disney's CFO explained that "near-term local attendance could be impacted as people put their lives together" after the disasters. He also noted that the company would focus on "increasing occupancy at theme park hotels, per capita spending by visitors to the theme parks, and managing costs." The objective is to reach an operating margin of at least 20 percent over the next three to four years.[4] A more robust operating margin at all theme parks would cushion Disney in the event of unforeseen events.

[4] Dwight Oestricher, "Disney CFO Staggs Sees Theme Park 1Q Hurt by Storms," *The Wall Street Journal* (September 30, 2004): B1 and B2.

The HOW and WHY of Calculating Margin of Safety

**CORNERSTONE
3-6**

Information:
Blazin-Boards Company plans to sell 10,000 snowboards at $400 each in the coming year. Product costs include:

Direct materials per snowboard	$80
Direct labour per snowboard	$125
Variable overhead per snowboard	$15
Total fixed factory overhead	$800,000

Variable selling expense is a commission of 5 percent of price; fixed selling and administrative expense totals $400,000.

Why:
Margin of safety is a crude measure of risk. The further above the break-even point, the larger the margin of safety and the further the company is away from break-even and a loss.

Required:
1. Calculate the margin of safety in units for the coming year. (Recall that the break-even point in units was calculated in Cornerstone 3-2.)
2. Calculate the break-even sales and the margin of safety in sales for the coming year.
3. **What if** Blazin-Boards Company actually sells 9,800 snowboards in the coming year? Calculate the margin of safety in units and sales revenue.

Solution:
1. Margin of safety = 10,000 units − 7,500 units = 2,500 units
2. Break-even sales = 7,500 units × $400 = $3,000,000
   Margin of safety = (10,000 units × $400) − $3,000,000 = $1,000,000
3. Margin of safety = 9,800 units − 7,500 units = 2,300 units
   Margin of safety = (9,800 units × $400) − $3,000,000 = $920,000

Operating Leverage   In physics, a lever is a simple machine used to multiply force. Basically, the lever magnifies the amount of effort applied to create a greater effect. The larger the load moved by a given amount of effort, the greater the mechanical advantage. In financial terms, operating leverage is concerned with the relative mix of fixed costs and variable costs in an organization. It is sometimes possible to trade off fixed costs for variable costs. As variable costs decrease, the unit contribution margin increases, making the contribution of each unit sold that much greater. In such a case, the effect of fluctuations in sales on profitability increases. Thus, firms that have lowered variable costs by increasing the proportion of fixed costs will benefit with greater increases in profits as sales increase than will firms with a lower proportion of fixed costs. Fixed costs are being used as leverage to increase profits. Unfortunately, it is also true that firms with a higher operating leverage will also experience greater reductions in profits as sales decrease. Therefore, **operating leverage** is the use of fixed costs to extract higher percentage changes in profits as sales activity changes.

The greater the degree of operating leverage, the more that changes in sales activity will affect profits. Because of this phenomenon, the mix of costs that an organization chooses can have a considerable influence on its operating risk and profit level.

The **degree of operating leverage** can be measured for a given level of sales by taking the ratio of total contribution margin to profit, as follows:

$$\text{Degree of operating leverage} = \text{Total contribution margin}/\text{Profit}$$

If fixed costs are used to lower variable costs such that contribution margin increases and profit decreases, then the degree of operating leverage increases—signalling an increase in risk. Cornerstone 3-7 illustrates the degree of operating leverage and the way it can be used to calculate the change in profit given a percentage change in sales.

As Cornerstone 3-7 shows, the degree of operating leverage is a valuable piece of information. It can be used to quickly determine the impact of a percentage change in sales on operating income. We see that a 40 percent increase in sales can bring a significant benefit to the firm. However, the effect is a two-edged sword. As sales decrease, the automated system will also show much higher percentage profit decreases. Moreover, the increased operating leverage is available under the automated system because of the presence of increased fixed costs. The break-even point for the automated system is 7,500 units ($375,000/$50), whereas the break-even point for the manual system is 5,000 units ($100,000/$20). Clearly, at a lower level of sales the manual system is better and at a higher level of sales the automated system is better. What is the sales level at which the manual and the automated systems are equally profitable? We can compute that by setting the operating income equations for each system equal to each other.

$$(\$100 \times \text{units}) - (\$50 \times \text{units}) - \$375,000 = (\$100 \times \text{units}) - (\$80 \times \text{units}) - \$100,000$$
$$(\$50 \times \text{units}) - \$375,000 = (\$20 \times \text{units}) - \$100,000$$
$$\text{Units} = 9,167 \text{(rounded)}$$

In choosing between the automated and manual systems, the manager must assess the likelihood that sales will exceed 9,167 units. If, after careful study, there is a strong belief that sales will easily exceed this level, the choice is obvious: the automated system. On the other hand, if sales are unlikely to exceed 9,167 units, the manual system is preferable. Exhibit 3-8 summarizes the relative difference between the manual and automated systems in terms of some of the CVP concepts.

Sensitivity Analysis and CVP

The pervasiveness of personal computers and spreadsheets has made cost analysis within reach of most managers. An important tool is **sensitivity analysis**, a what-if technique that examines the impact of changes in underlying assumptions on an

---

Exhibit 3-8

Differences between Manual and Automated Systems

	Manual System	Automated System
Price	Same	Same
Variable costs	Relatively higher	Relatively lower
Fixed costs	Relatively lower	Relatively higher
Contribution margin	Relatively lower	Relatively higher
Break-even point	Relatively lower	Relatively higher
Margin of safety	Relatively higher	Relatively lower
Degree of operating leverage	Relatively lower	Relatively higher
Downside risk	Relatively lower	Relatively higher
Upside potential	Relatively lower	Relatively higher

The HOW and WHY of Calculating Degree of Operating Leverage and Percent Change in Profit

CORNERSTONE
3-7

Information:

Sharda Company is planning to add a new product line. To do so, the firm can choose to rely heavily on automation or on labour. Relevant data for a sales level of 10,000 units follow:

	Automated System	Manual System
Sales	$1,000,000	$1,000,000
Variable expenses	500,000	800,000
Contribution margin	500,000	200,000
Less total fixed expenses	375,000	100,000
Operating income	$ 125,000	$ 100,000
Unit selling price	$ 100	$ 100
Unit variable cost	50	80
Unit contribution margin	50	20

Why:

The automated system has higher fixed costs, lower variable costs, and a higher contribution margin per unit. The higher fixed costs are used to extract more contribution margin from each unit sold, and this system will pay off nicely—if unit sales are high enough. The manual system will be less risky if unit sales are lower. The degree of operating leverage can help a firm determine how much riskier the automated system is.

Required:

1. Compute the degree of operating leverage for each system.
2. Suppose that sales are 40 percent higher than budgeted. By what percentage will operating income increase for each system? What will be the *increase* in operating income for each system?
3. **What if** unit sales are 30 percent lower than budgeted? By what percentage will operating income decrease for each system? What will be the *total* operating income for each system?

Solution:

1. Degree of operating leverage = Contribution margin/Profit

    Automated system degree of operating leverage = $500,000/$125,000 = 4.0

    Manual system degree of operating leverage = $200,000/$100,000 = 2.0

2. Automated system increase in profit percentage = 4.0 × 40% = 160%

    Manual system increase in profit percentage = 2.0 × 40% = 80%

    Automated system increase in profit = 1.6 × $125,000 = $200,000

    Manual system increase in profit = 0.8 × $100,000 = $80,000

    Automated system new profit = $125,000 + $200,000 = $325,000

    Manual system new profit = $100,000 + $80,000 = $180,000

3. Automated system decrease in profit percentage = 4.0 × 30% = 120%

    Manual system decrease in profit percentage = 2.0 × 30% = 60%

    Automated system decrease in profit = 1.2 × $125,000 = $150,000

    Manual system decrease in profit = 0.6 × $100,000 = $60,000

    Automated system new profit = $125,000 − $150,000 = $(25,000)

    Manual system new profit = $100,000 − $60,000 = $40,000

answer. It is relatively simple to input data on prices, variable costs, fixed costs, and sales mix and set up formulas to calculate break-even points and expected profits. Then, the data can be varied as desired to see what impact changes have on the expected profit.

In Cornerstone 3-7, a company analyzed the impact on profit of using an automated versus a manual system. The computations were essentially done by hand, and too much variation was cumbersome. Using the power of a computer, it would be an easy matter to change the sales price in $1 increments between $75 and $125, with related assumptions about quantity sold. At the same time, variable and fixed costs could be adjusted. For example, suppose that the automated system has fixed costs of $375,000, but that those costs could easily range up to twice as much in the first year and come back down in the second and third years as bugs are worked out of the system and workers learn to use it. Again, the spreadsheet can effortlessly handle the many computations.

We must note that the spreadsheet, while wonderful for cranking out numerical answers, cannot do the most difficult job in CVP analysis. That job is determining the data to be entered in the first place. The accountant must be familiar with the cost and price distributions of the firm, as well as with the impact of changing economic conditions on these variables. The fact that variables are seldom known with certainty is no excuse for ignoring the impact of uncertainty on CVP analysis. Fortunately, sensitivity analysis can also give managers a feel for the degree to which a poorly forecast variable will affect an answer. That is also an advantage.

ETHICS Finally, it is important to note that the CVP results are only one input into business decisions. There are many other factors that may bear on decisions to choose one type of process over another, for example, or whether to delete certain costs. Businesses and nonprofit entities often face trade-offs involving safety. Ethical concerns also have an important place in CVP analysis. One possibility is that the cost of potential problems can be estimated and included in the CVP results. Often, however, the costs and probabilities are not known with sufficient certainty. In that case, these factors are included as qualitative factors in the ultimate decision-making process. Chapter 11, on short-run decision making, covers this topic in more detail.◆

OBJECTIVE ➤ 6

Discuss the impact of non-unit cost drivers on cost-volume-profit analysis.

CVP Analysis and Non-Unit Cost Drivers

Conventional CVP analysis assumes that all costs of the firm can be divided into two categories: those that vary with sales volume (variable costs) and those that do not (fixed costs). Furthermore, costs are assumed to be a linear function of sales volume. Frequently, however, there are costs that vary with non-unit cost drivers. An activity-based costing (ABC) system, in which costs are divided into unit- and non-unit-based categories, is a good example of this. Some costs, such as setting up production equipment, vary with the number of batches; other costs, such as purchasing and receiving costs, may vary with the number of different products. In conventional CVP, those non-unit variable costs are assumed to be fixed. However, CVP can be modified to take account of this richer set of variable costs. This type of modification can make CVP even more useful, since it provides more accurate insights concerning cost behaviour. These insights produce better decisions.

To illustrate, assume that a company's costs can be explained by three variables: a unit-level cost driver, units sold; a batch-level cost driver, number of setups; and a product-level cost driver, engineering hours. The cost equation can then be expressed as follows:

$$\text{Total cost} = \text{Fixed costs} + (\text{Unit variable cost} \times \text{Number of units}) + (\text{Setup cost} \times \text{Number of setups}) + (\text{Engineering cost} \times \text{Number of engineering hours})$$

Operating income, as before, is total revenue minus total cost. This is expressed as follows:

$$\text{Operating income} = \text{Total revenue} - [\text{Fixed costs} + (\text{Unit variable cost} \times \text{Number of units}) + (\text{Setup cost} \times \text{Number of setups}) + (\text{Engineering cost} \times \text{Number of engineering hours})]$$

Let's use the contribution margin approach to calculate the break-even point in units. At break-even, operating income is zero, and the number of units that must be sold to achieve break-even is as follows:

Break-even units = [Fixed costs + (Setup cost × Number of setups) + (Engineering cost ×
Number of engineering hours)]/(Price − Unit variable cost)

A comparison of the ABC break-even point with the conventional break-even point reveals two significant differences. First, the fixed costs differ. Some costs previously identified as being fixed may actually vary with non-unit cost drivers, in this case setups and engineering hours. Second, the numerator of the ABC break-even equation has two non-unit-variable cost terms: one for batch-related activities and one for product-sustaining activities.

Example Comparing Conventional and ABC Analysis

To make the previous discussion more concrete, a comparison of conventional cost-volume-profit analysis with activity-based costing is useful. Let's assume that a company wants to compute the units that must be sold to earn a before-tax profit of $20,000. The analysis is based on the following data:

Data about Variables

Cost Driver	Unit Variable Cost	Level of Cost Driver
Units sold	$10	—
Setups	1,000	20
Engineering hours	30	1,000

Other data:

Total fixed costs (conventional)	$100,000
Total fixed costs (ABC)	50,000
Unit selling price	20

The units that must be sold to earn a before-tax profit of $20,000 are computed as follows:

Units = (Targeted income + Fixed costs)/(Price − Unit variable cost)
= ($20,000 + $100,000)/($20 − $10)
= $120,000/$10
= 12,000

Using the ABC equation, the units that must be sold to earn an operating income of $20,000 are computed as follows:

Units = [Targeted income + Fixed costs + (Setup cost × Setups) +
(Engineering rate × Engineering hours)]/(Price − Unit variable cost)
= ($20,000 + $50,000 + $20,000 + $30,000)/($20 − $10)
= $120,000/$10
= 12,000

The number of units that must be sold is identical under both approaches. The reason is simple. The total fixed cost pool under conventional costing consists of non-unit-based variable costs plus costs that are fixed regardless of the cost driver. ABC breaks out the non-unit-based variable costs. These costs are associated with certain levels of each cost driver. For the batch-level cost driver, the level is 20 setups; for the product-level variable, the level is 1,000 engineering hours. As long as the levels of activity for the non-unit-based cost drivers remain the same, the results for the conventional and ABC computations will also be the same. But these levels can change, and because of this, the information provided by the two approaches can be significantly different. The ABC equation for CVP analysis is a richer representation of the underlying cost behaviour and can provide important strategic insights. To see this, let's use the same data provided previously and look at a different application.

Strategic Implications: Conventional CVP Analysis versus ABC Analysis

Suppose that after the conventional CVP analysis, marketing indicates that only 10,000 units can be sold, not the 12,000 anticipated earlier. The president of the company directs the product design engineers to find a way to reduce the cost of making the product. The engineers also have been told that the conventional cost equation, with fixed costs of $100,000 and a unit variable cost of $10, is accurate. The variable cost of $10 per unit consists of the following: direct labour, $4; direct materials, $5; and variable overhead, $1. To comply with the request to reduce the break-even point, engineering produces a new design that requires less labour, thereby reducing the direct labour cost by $2 per unit. The design would not affect direct materials or variable overhead. The new variable cost is $8 per unit, and the break-even point is calculated as follows:

$$\text{Units} = \text{Fixed costs}/(\text{Price} - \text{Unit variable cost})$$
$$= \$100{,}000/(\$20 - \$8)$$
$$= 8{,}333$$

The projected income if 10,000 units are sold is computed as follows:

Sales ($20 × 10,000)	$200,000
Less: Variable expenses ($8 × 10,000)	80,000
Contribution margin	120,000
Less: Fixed expenses	100,000
Operating income	$ 20,000

Excited, the president approves the new design. A year later, the president discovers that the expected increase in income did not materialize. In fact, there was a loss. Why? The answer is provided by an ABC approach to CVP analysis.

The original ABC cost relationship for the example is as follows:

$$\text{Total cost} = \$50{,}000 + (\$10 \times \text{Units}) + (\$1{,}000 \times \text{Setups}) + (\$30 \times \text{Engineering hours})$$

Suppose that the new design requires a more complex setup, increasing the cost per setup from $1,000 to $1,600. Also, suppose that the new design, because of increased technical content, requires a 40 percent increase in engineering support (from 1,000 hours to 1,400 hours). The new cost equation, including the reduction in unit-level variable costs, is as follows:

$$\text{Total cost} = \$50{,}000 + (\$8 \times \text{Units}) + (\$1{,}600 \times \text{Setups}) + (\$30 \times \text{Engineering hours})$$

The break-even point, setting operating income equal to zero and using the ABC equation, is calculated as follows (assume that 20 setups are still performed):

$$\text{Units} = [\$50{,}000 + (\$1{,}600 \times 20) + (\$30 \times 1{,}400)]/(\$20 - \$8)$$
$$= \$124{,}000/\$12$$
$$= 10{,}333$$

And the income for 10,000 units is (recall that a maximum of 10,000 can be sold) as follows:

Sales ($20 × 10,000)		$200,000
Less: Unit-based variable expenses ($8 × 10,000)		80,000
Contribution margin		120,000
Less non-unit-based variable expenses:		
Setups ($1,600 × 20)	$32,000	
Engineering support ($30 × 1,400)	42,000	74,000
Traceable margin		46,000
Less: Fixed expenses		50,000
Operating income (loss)		$ (4,000)

How could the engineers have been off by so much? Didn't they know that the new design would increase setup cost and engineering support? Yes and no. They were probably aware of the increases in these two variables, but the conventional cost equation diverted attention from figuring just how much impact changes in those variables would have on units needed to break even. The information conveyed to the engineers by the conventional equation gave the impression that any reduction in labour cost—not affecting direct materials or variable overhead—would reduce total costs, since changes in the level of labour activity would not affect the fixed costs. The ABC equation, however, indicates that a reduction in labour input that adversely affects setup activity or engineering support might be undesirable. When more insight is provided, better design decisions can be made. Providing ABC cost information to the design engineers would probably have led them down a different—and better—path for the company.

CVP Analysis and JIT

If a firm has adopted JIT, the variable cost per unit sold is reduced, and fixed costs are increased. Direct labour, for example, is now viewed as fixed instead of variable. Direct materials, on the other hand, is still a unit-based variable cost. In fact, the emphasis on total quality and long-term purchasing makes the assumption even more true that direct materials cost is strictly proportional to units produced (because waste, scrap, and quantity discounts are eliminated). Other unit-based variable costs such as power and sales commissions also persist. Additionally, the batch-level variable is gone (in JIT, the batch is one unit). Thus, the cost equation for JIT can be expressed as follows:

$$\text{Total cost} = \text{Fixed costs} + (\text{Unit variable cost} \times \text{Units}) +$$
$$(\text{Engineering cost} \times \text{Number of engineering hours})$$

Since its application is a special case of the ABC equation, no example will be given.

CVP Analysis, Multiple Drivers, and Nonprofit Entities

Clearly, cost-volume-profit analysis is helpful for manufacturing and service firms. It is also useful for not-for-profit entities. In the case of a nonprofit organization, there may be a variety of ways of gaining revenue and a variety of programs that generate costs. As a result, the straightforward application of cost-volume-profit equations may not be possible. Then, the managers will need to become aware of the different types of costs, the different drivers, and the underlying economic conditions that affect them.

Canadians are passionate about hockey, but it is difficult for teams below the **National Hockey League** level to make a profit. The teams that compete for the Allan Cup (viewed by some to be second only to the Stanley Cup in terms of prestige) often come from smaller communities and must rely on various sources of revenue in order to break even. Variable costs include "game day" costs for labour and supplies; fixed costs relate to the cost of the venue, the players and coaches, stipends, and equipment. Revenues come from many sources, including gate receipts, concession and merchandise sales, and other community support, which often takes the form of fundraising efforts. All sources of revenue and expenses must be considered in order to determine the revenue level that will allow a team to break even.

Knowing the cost structure makes it easier for the team to see just where cost cuts are possible and where revenue increases must be found. In cases like this, it might be best to use a spreadsheet to calculate costs and revenues under various scenarios.

Summary of Learning Objectives

1. **Determine the number of units and amount of sales revenue needed to break even and to earn a target profit.**

- At break-even, total costs (variable and fixed) equal total sales revenue.
- Break-even units equal total fixed costs divided by the contribution margin (price minus variable cost per unit).
- Break-even revenue equals total fixed costs divided by the contribution margin ratio.
- To earn a target (desired) profit, total costs (variable and fixed) plus the amount of target profit must equal total sales revenue.
- Units to earn target profit equal total fixed costs plus target profit divided by the contribution margin.
- Sales revenue to earn target profit equals total fixed costs plus target profit divided by the contribution margin ratio.

2. **Determine the number of units and sales revenue needed to earn an after-tax target profit.**

- Desired after-tax profit must be converted into before-tax profit to calculate units or revenue needed.
- To find the operating income implied by a certain after-tax profit, divide the after-tax profit by (1 − tax rate).
- Apply the break-even equations as before to the newly calculated before-tax profit target.

3. **Apply cost-volume-profit analysis in a multiple-product setting.**

- Multiple-product analysis requires the expected sales mix.
- Break-even units for each product will change as the sales mix changes.
- Increased sales of high contribution margin products decrease the break-even point.
- Increased sales of low contribution margin products increase the break-even point.

4. **Prepare a profit-volume graph and a cost-volume-profit graph, and explain the meaning of each.**

- CVP assumes linear revenue and cost functions, no finished goods ending inventories, constant sales mix, and that selling prices and fixed and variable costs are known with certainty.
- Profit-volume graphs plot the relationship between profit (operating income) and units sold. Break-even units are shown where the profit line crosses the horizontal axis.
- CVP graphs plot a line for total costs and a line for total sales revenue. The intersection of these two lines is the break-even point in units.

5. **Explain the impact of risk, uncertainty, and changing variables on cost-volume-profit analysis.**

- Uncertainty regarding costs, prices, and sales mix affect the break-even point.
- Sensitivity analysis allows managers to vary costs, prices, and sales mix to show various possible break-even points.
- Margin of safety shows how far the company's actual sales and/or units are above or below the break-even point.
- Operating leverage is the use of fixed costs to increase the percentage changes in profits as sales activity changes.

6. **Discuss the impact of non-unit cost drivers on cost-volume-profit analysis.**

- Under ABC, cost drivers are separated into unit-based and non-unit-based drivers.
- Variable rates for the non-unit-based drivers are multiplied by the estimated level of the drivers and added to total fixed costs.
- The standard CVP models still hold under ABC.

Summary of Important Equations

The subject of cost-volume-profit analysis naturally lends itself to the use of numerous equations. Some of the more common equations used in this chapter are summarized in Exhibit 3-9.

Exhibit 3-9

Summary of Important Equations

1. Sales revenue = Price × Units sold
2. Operating income = (Price × Units) − (Unit variable cost × Units) − Fixed cost
3. Break-even point in units = Fixed cost/(Price − Unit variable cost)
4. Contribution margin ratio = Contribution margin/Sales

   or   = (Price − Unit variable cost)/Price
5. Variable cost ratio = Total variable cost/Sales

   or   = Unit variable cost/Price
6. Break-even point in sales revenue = Fixed cost/Contribution margin ratio

   or   = Fixed cost/(1 − Variable cost ratio)
7. Margin of safety = Sales − Break-even sales

   or   = Units sold − Break-even units
8. Degree of operating leverage = Total contribution margin/Operating income
9. Percentage change in operating income = Degree of operating leverage × Percent change in sales

CORNERSTONE 3-1	The HOW and WHY of basic cost calculations and the contribution-margin-based income statement, page 103
CORNERSTONE 3-2	The HOW and WHY of calculating the units needed to break even and to achieve a target profit, page 105
CORNERSTONE 3-3	The HOW and WHY of calculating revenue for break-even and for a target profit, page 108
CORNERSTONE 3-4	The HOW and WHY of calculating the number of units to generate an after-tax target profit, page 110
CORNERSTONE 3-5	The HOW and WHY of calculating the break-even number of units in a multi-product firm, page 114
CORNERSTONE 3-6	The HOW and WHY of calculating margin of safety, page 123
CORNERSTONE 3-7	The HOW and WHY of calculating degree of operating leverage and percent change in profit, page 125

CORNERSTONES FOR CHAPTER 3

Review Problems

I. Break-Even Point, Targeted Profit, Margin of Safety

Cutlass Company's projected profit for the coming year is as follows:

	Total	Per Unit
Sales	$200,000	$20
Less: Variable expenses	120,000	12
Contribution margin	80,000	$ 8
Less: Fixed expenses	64,000	
Operating income	$ 16,000	

Required:

1. Compute the break-even point in units.
2. How many units must be sold to earn a profit of $30,000?
3. Compute the contribution margin ratio. Using that ratio, compute the additional profit that Cutlass would earn if sales were $25,000 more than expected.
4. Suppose Cutlass would like to earn operating income equal to 20 percent of sales revenue. How many units must be sold for this goal to be realized? Prepare an income statement to prove your answer.
5. For the projected level of sales, compute the margin of safety.
6. For the projected level of sales, compute the degree of operating leverage. What is the percent change in profit if sales increase by 15 percent?

Solution:

1. The break-even point is as follows:

$$\text{Units} = \text{Fixed costs}/(\text{Price} - \text{Unit variable cost})$$
$$= \$64,000/(\$20 - \$12)$$
$$= \$64,000/\$8$$
$$= 8,000$$

2. The units sold to earn a profit of $30,000 is as follows::

$$\text{Units} = (\$64,000 + \$30,000)/\$8$$
$$= \$94,000/\$8$$
$$= 11,750$$

3. The contribution margin ratio is $8/$20 = 0.40. With additional sales of $25,000, the additional profit would be 0.40 × $25,000 = $10,000.
4. To find the number of units sold for a profit equal to 20 percent of sales, let target income equal (0.20)(Price × Units) and solve for units.

$$\text{Operating income} = (\text{Price} \times \text{Units}) - (\text{Unit variable cost} \times \text{Units}) - \text{Fixed costs}$$
$$(0.2)(\$20)\text{Units} = (\$20 \times \text{Units}) - (\$12 \times \text{Units}) - \$64,000$$
$$\$4 \times \text{Units} = \$64,000$$
$$\text{Units} = 16,000$$

The income statement is as follows:

Sales (16,000 × $20)	$320,000
Less: Variable expenses (16,000 × $12)	192,000
Contribution margin	128,000
Less: Fixed expenses	64,000
Operating income	$ 64,000

Operating income/Sales = $64,000/$320,000 = 0.20, or 20%

5. The margin of safety is 10,000 − 8,000 = 2,000 units, or $40,000 in sales revenues.
6.   Degree of operating leverage = $80,000/$16,000 = 5.0
      Percent change in profit = 15% × 5.0 = 75%

II. CVP with Activity-Based Costing

Dory Manufacturing Company produces T-shirts that are screen-printed with the logos of various sports teams. Each shirt is priced at $10. Costs are as follows:

Cost Driver	Unit Variable Cost	Level of Cost Driver
Units sold	$ 5	—
Setups	$450	80
Engineering hours	$ 20	500

Other data:

Total fixed costs (conventional)	$96,000
Total fixed costs (ABC)	50,000

Required:

1. Compute the break-even point in units using conventional analysis.
2. Compute the break-even point in units using activity-based analysis.
3. Suppose that Dory could reduce the setup cost by $150 per setup and could reduce the number of engineering hours needed to 425. How many units must be sold to break even in this case?

Solution:

1. Break-even units = Fixed costs/(Price − Unit variable cost) = $96,000/($10 − $5)

   = 19,200 units

2. Break-even units = [Fixed costs + (Setups × Setup cost) +

   (Engineering hours × Engineering cost)]/

   (Price − Unit variable cost)

   = [$50,000 + ($450 × 80) + ($20 × 500)]/($10 − $5)

   = $96,000/$5

   = 19,200 units

3. Break-even units = [$50,000 + ($300 × 80) + ($20 × 425)]/($10 − $5)

   = $82,500/$5

   = 16,500 units

Key Terms

Break-even point, 101

Common fixed expenses, 111

Contribution margin, 104

Contribution margin ratio, 106

Cost-volume-profit graph, 117

Degree of operating leverage, 124

Direct fixed expenses, 111

Margin of safety, 122

Net income, 102

Operating income, 102

Operating leverage, 123

Profit-volume graph, 116

Relevant range, 119

Sales mix, 112

Sales-revenue approach, 107

Sensitivity analysis, 124

Variable cost ratio, 106

Discussion Questions

1. Explain how CVP analysis can be used for managerial planning.
2. Describe the difference between the units-sold approach to CVP analysis and the sales-revenue approach.
3. Define the term *break-even point*.
4. Explain why contribution margin per unit becomes profit per unit above the break-even point.
5. A restaurant owner who had yet to earn a monthly profit said, "The busier we are, the more we lose." What do you think is happening in terms of contribution margin?
6. What is the variable cost ratio? The contribution margin ratio? How are the two ratios related?
7. If the contribution margin increases from 30 to 35 percent of sales, what will happen to the break-even point, and why will this occur?
8. Suppose a firm with a contribution margin ratio of 0.3 increased its advertising expenses by $10,000 and found that sales increased by $30,000. Was it a good decision to increase advertising expenses? Why is this simple problem an important one for businesspeople to understand?
9. Define the term *sales mix*, and give an example to support your definition.

10. Explain how CVP analysis developed for single products can be used in a multiple-product setting.
11. Why might a multiple-product firm choose to calculate just overall break-even revenue rather than the break-even quantity by product?
12. How do income taxes affect the break-even point and CVP analysis?
13. Explain how a change in sales mix can change a company's break-even point.
14. Define the term *margin of safety*. Explain what is meant by the term *operating leverage*. What impact does an increase in the margin of safety have on risk? What impact does an increase in leverage have on risk?
15. Why does the activity-based costing approach to CVP analysis offer more insight than the conventional approach does?

Cornerstone Exercises

OBJECTIVE ▶ 1
CORNERSTONE 3-1

Cornerstone Exercise 3-1 VARIABLE COSTS, CONTRIBUTION MARGIN, CONTRIBUTION MARGIN RATIO

Custom Screenprinting Company plans to sell 8,000 T-shirts at $15 each in the coming year. Product costs include:

Direct materials per T-shirt	$6.00
Direct labour per T-shirt	$1.25
Variable overhead per T-shirt	$0.75
Total fixed factory overhead	$32,000

Variable selling expense is the redemption of a coupon, which averages $1.00 per T-shirt; fixed selling and administrative expense totals $10,000.

Required:

1. Calculate the:
   a. Variable product cost per unit
   b. Total variable cost per unit
   c. Contribution margin per unit
   d. Contribution margin ratio
   e. Total fixed expense for the year

2. Prepare a contribution-margin-based income statement for Custom Screenprinting Company for the coming year.

3. *What if* the per unit selling expense increased from $1.00 to $1.75? Calculate the new values for the following:
   a. Variable product cost per unit
   b. Total variable cost per unit
   c. Contribution margin per unit
   d. Contribution margin ratio
   e. Total fixed expense for the year

OBJECTIVE ▶ 1
CORNERSTONE 3-2

Cornerstone Exercise 3-2 BREAK-EVEN UNITS, UNITS FOR TARGET PROFIT

Dorion Company makes an in-car navigation system. Next year, Dorion plans to sell 6,000 units at a price of $300 each. Product costs include:

Direct materials	$75
Direct labour	$50
Variable overhead	$12
Total fixed factory overhead	$500,000

Variable selling expense is a commission of 6 percent of price; fixed selling and administrative expense totals $109,000.

Required:

1. Calculate the sales commission per unit sold. Calculate the contribution margin per unit.
2. How many units must Dorion Company sell to break even? Prepare an income statement for the calculated number of units.
3. Calculate the number of units Dorion Company must sell to achieve target operating income (profit) of $333,500.
4. *What if* Dorion Company wanted to achieve a target operating income of $314,650? Would the number of units needed increase or decrease compared to your answer in Requirement 3? Compute the number of units needed for the new target operating income.

Cornerstone Exercise 3-3 BREAK-EVEN SALES, SALES FOR TARGET PROFIT

OBJECTIVE ▶ 1

CORNERSTONE 3-3

SERVICE

Kalpna Company is a placement agency for temporary nurses. It serves hospitals and clinics throughout the metropolitan area. Kalpna Company believes it will place temporary nurses for a total of 25,000 hours next year. Kalpna charges the hospitals and clinics $75 per hour and has variable costs of $60 per hour (this includes the payment to the nurse). Total fixed costs equal $321,000.

Required:

1. Calculate the contribution margin per unit and the contribution margin ratio.
2. Calculate the sales revenue needed to break even.
3. Calculate the sales revenue needed to achieve a target profit of $100,000.
4. *What if* Kalpna had target operating income (profit) of $90,000? Would sales revenue be larger or smaller than the one calculated in Requirement 3. Why? By how much?

Cornerstone Exercise 3-4 AFTER-TAX PROFIT TARGETS

OBJECTIVE ▶ 2

CORNERSTONE 3-4

LaFramboise Company wants to earn $420,000 in net (after-tax) income next year. Its product is priced at $480 per unit. Product costs include:

Direct materials	$150
Direct labour	$100
Variable overhead	$25
Total fixed factory overhead	$352,000

Variable selling expense is $45 per unit; fixed selling and administrative expense totals $340,000. LaFramboise has a tax rate of 30 percent.

Required:

1. Calculate the before-tax profit needed to achieve an after-tax target of $420,000.
2. Calculate the number of units that will yield operating income calculated in Requirement 1 above.
3. Prepare an income statement for LaFramboise Company for the coming year based on the number of units computed in Requirement 2.
4. *What if* LaFramboise had a 35 percent tax rate? Would the units sold to reach a $420,000 target net income be higher or lower than the units calculated in Requirement 3? Calculate the number of units needed at the new tax rate.

Cornerstone Exercise 3-5 MULTIPLE-PRODUCT BREAK-EVEN AND TARGET PROFIT

OBJECTIVE ▶ 1 3

CORNERSTONE 3-5

Sandman Enterprises produces and sells two products: a bedside lamp decorated with comic book characters, and a baby mobile that hangs above a crib and can play lullabies. Sandman plans to sell 30,000 bedside lamps and 20,000 lullaby mobiles in the coming year. Product price and cost information includes:

	Bedside Lamp	Lullaby Mobile
Price	$24	$40
Unit variable cost	$12	$10
Direct fixed cost	$23,600	$45,000

Common fixed selling and administrative expense totals $85,000.

Required:

1. What is the sales mix estimated for next year (calculated to the lowest whole number for each product)?
2. Using the sales mix from Requirement 1, form a package of bedside lamps and lullaby mobiles. How many bedside lamps and lullaby mobiles are sold at break-even?
3. Prepare a contribution-margin-based income statement for Sandman Enterprises based on the unit sales calculated in Requirement 2.
4. *What if* Sandman Enterprises wanted to earn operating income equal to $14,400? Calculate the number of bedside lamps and lullaby mobiles that must be sold to earn this level of operating income. (*Hint:* Remember to form a package of bedside lamps and lullaby mobiles based on the sales mix and to first calculate the number of packages to earn operating income of $14,400.)

OBJECTIVE ▶ 1 5
CORNERSTONE 3-6

Cornerstone Exercise 3-6 BREAK-EVEN UNITS AND SALES REVENUE, MARGIN OF SAFETY

Dunst and Dafoe Copy Shop (D&D) provides photocopying service. Next year, D&D estimates it will copy 2,400,000 pages at a price of $0.08 each in the coming year. Product costs include:

Direct materials	$0.020
Direct labour	$0.004
Variable overhead	$0.001
Total fixed overhead	$50,000

There is no variable selling expense; fixed selling and administrative expense totals $21,500.

Required:

1. Calculate the break-even point in units.
2. Calculate the break-even point in sales revenue.
3. Calculate the margin of safety in units for the coming year.
4. Calculate the margin of safety in sales revenue for the coming year.
5. What if the total fixed overhead increases to $53,300? Recalculate:
   a. Break-even point in units
   b. Break-even point in sales revenue
   c. Margin of safety in units for the coming year
   d. Margin of safety in sales revenue for the coming year

OBJECTIVE ▶ 1 5
CORNERSTONE 3-7

Cornerstone Exercise 3-7 DEGREE OF OPERATING LEVERAGE, PERCENT CHANGE IN PROFIT

Dorval Company is considering two different processes to make its product—process 1 and process 2. Process 1 requires Dorval to manufacture subcomponents of the product in-house. As a result, materials are less expensive, but fixed overhead is higher. Process 2 involves purchasing all subcomponents from outside suppliers. The direct materials costs are higher, but fixed factory overhead is considerably lower. Relevant data for a sales level of 30,000 units follow:

	Process 1	Process 2
Sales	$6,000,000	$6,000,000
Variable expenses	2,700,000	4,200,000
Contribution margin	3,300,000	1,800,000
Less total fixed expenses	1,925,000	600,000
Operating income	$1,375,000	$1,200,000
Unit selling price	$200	$200
Unit variable cost	$90	$140
Unit contribution margin	$110	$60

Required:

1. Compute the degree of operating leverage for each process.
2. Suppose that sales are 30 percent higher than budgeted. By what percentage will operating income increase for each process? What will be the *increase* in operating income for each system? What will be the *total operating income* for each process?
3. *What if* unit sales are 10 percent lower than budgeted? By what percentage will operating income decrease for each process? What will be the *total operating income* for each process?

Exercises

Exercise 3-8 CONTRIBUTION MARGIN, BREAK-EVEN UNITS, CONTRIBUTION MARGIN INCOME STATEMENT, MARGIN OF SAFETY

OBJECTIVE ➤ 1

Kool-skinz Company manufactures custom-designed skins (covers) for iPods® and other portable MP3 devices. Variable costs are $10.80 per custom skin, the price is $18, and fixed costs are $66,960.

Required:

1. What is the contribution margin for one custom skin?
2. How many custom skins must Kool-skinz Company sell to break even?
3. If Kool-skinz Company sells 12,000 custom skins, what is the operating income?
4. Calculate the margin of safety in units and in sales revenue if 12,000 custom skins are sold.

Exercise 3-9 BREAK-EVEN IN UNITS

OBJECTIVE ➤ 1

Shellenberger Company manufactures high-end gas grills. Fixed costs amount to $19,980,000 per year. Variable costs per gas grill are $395, and the average price per gas grill is $950.

Required:

1. How many gas grills must Shellenberger Company sell to break even?
2. If Shellenberger Company sells 41,000 gas grills in a year, what is the operating income?
3. If Shellenberger Company's variable costs increase to $420 per grill while the price and fixed costs remain unchanged, what is the new break-even point? (Round up to the next higher whole unit.)

Exercise 3-10 CONTRIBUTION MARGIN RATIO, BREAK-EVEN SALES REVENUE, SALES REVENUE FOR TARGET PROFIT

OBJECTIVE ➤ 1

Parker Pharmaceuticals Inc. plans to sell 500,000 units of anti-venom at an average price of $6 each in the coming year. Total variable costs equal $600,000. Total fixed costs equal $8,000,000.

Required:

1. What is the contribution margin per unit? What is the contribution margin ratio?
2. Calculate the sales revenue needed to break even.
3. Calculate the sales revenue needed to achieve a target profit of $650,000.
4. *What if* the average price per unit increased to $7? Recalculate:
   a. Contribution margin per unit
   b. Contribution margin ratio (rounded to four decimal places)
   c. Sales revenue needed to break even
   d. Sales revenue needed to achieve a target profit of $650,000

OBJECTIVE ➤ 1 5   **Exercise 3-11 BREAK-EVEN IN UNITS, TARGET INCOME, NEW UNIT VARIABLE COST, DEGREE OF OPERATING LEVERAGE, PERCENT CHANGE IN OPERATING INCOME**

McDuffy's Inc. has developed a chew-proof dog bed—the McTuffie. Fixed costs are $144,000 per year. The average price for the McTuffie is $32, and the average variable cost is $24 per unit. Currently, McDuffy's produces and sells 20,000 McTuffies.

Required:

1. How many McTuffies must be sold to break even?
2. If McDuffy's wants to earn $46,000 in profit, how many McTuffies must be sold? Prepare a variable-costing income statement to verify your answer.
3. Suppose that McDuffy's would like to lower the break-even units to 12,000. The company does not believe that the price or fixed cost can be changed. Calculate the new unit variable cost that would result in break-even units of 12,000.
4. What is McDuffy's current contribution margin and operating income? Calculate the degree of operating leverage (round your answer to two decimal places). If sales increased by 10 percent next year, what would the percent change in operating income be? What would the new total operating income for next year be?

OBJECTIVE ➤ 1 2

Exercise 3-12 BREAK-EVEN FOR A SERVICE FIRM

Sasha Melton owns and operates The Green Belt Company (GBC), which provides live plants and flower arrangements to professional offices. Sasha has fixed costs of $2,380 per month for office/greenhouse rent, advertising, and a delivery van. Variable costs for the plants, fertilizer, pots, and other supplies average $25 per job. GBC charges $60 per month for the average job.

Required:

1. How many jobs must GBC average each month to break even?
2. What is the operating income for GBC in a month with 65 jobs? With 90 jobs?
3. Sasha faces a tax rate equal to 30 percent. How many jobs must Sasha have per month to earn an after-tax income of $980?
4. Suppose that Sasha's fixed costs increase to $2,500 per month and she decides to increase the price to $75 per job. What is the new break-even point in number of jobs per month?

OBJECTIVE ➤ 1 5

Exercise 3-13 BREAK-EVEN IN SALES REVENUE, MARGIN OF SAFETY

StarSports Inc. represents professional athletes and movie and television stars. The agency had revenue of $10,780,000 last year, with total variable costs of $5,066,600 and fixed costs of $2,194,200.

Required:

1. What is the contribution margin ratio for StarSports based on last year's data? What is the break-even point in sales revenue?
2. What was the margin of safety for StarSports last year?
3. One of StarSports's agents proposed that the firm begin cultivating high school sports stars around the nation. This proposal is expected to increase revenue by $150,000 per year, with increased fixed costs of $140,000. Is this proposal a good idea? Explain.

OBJECTIVE ➤ 1 2   **Exercise 3-14 CVP, BEFORE- AND AFTER-TAX TARGETED INCOME**

Head-Gear Company produces helmets for bicycle racing. Currently, Head-Gear charges $230 per helmet. Variable costs are $80.50 per helmet, and fixed costs are $1,255,800. The tax rate is 25 percent. Last year, 14,000 helmets were sold.

Required:

1. What is Head-Gear's net income for last year?
2. What is Head-Gear's break-even revenue? (Round to the nearest dollar.)

3. Suppose Head-Gear wants to earn before-tax operating income of $900,000. How many units must be sold? (Round to the nearest unit.)
4. Suppose Head-Gear wants to earn after-tax net income of $650,000. How many units must be sold? (Round to the nearest unit.)
5. Suppose the income tax rate rises to 35 percent. How many units must be sold for Head-Gear to earn after-tax income of $650,000? (Round to the nearest unit.)

Exercise 3-15 CVP, BEFORE- AND AFTER-TAX TARGETED INCOME

OBJECTIVE ➤ 1 4

Sara Pacheco is a sophomore in university and earns a little extra money by making beaded key ring accessories. She sells them on Saturday mornings at the local flea market. Sara charges $5 per unit and has unit variable costs (beads, wire rings, etc.) of $2. Her fixed costs consist of small pliers, a glue gun, etc., which cost her $90.

Required:

1. Calculate Sara's break-even units.
2. Prepare a profit-volume graph for Sara.
3. Prepare a cost-volume-profit graph for Sara.

Exercise 3-16 ASSUMPTIONS AND USE OF VARIABLES

OBJECTIVE ➤ 1 5

Choose the *best* answer for each of the following multiple-choice questions.

1. Cost-volume-profit analysis includes some simplifying assumptions. Which of the following is **not** one of these assumptions?
   a. Cost and revenues are predictable.
   b. Cost and revenues are linear over the relevant range.
   c. Changes in beginning and ending inventory levels are insignificant in amount.
   d. Sales mix changes are irrelevant.
2. The term *relevant range*, as used in cost accounting, means the range
   a. over which costs may fluctuate
   b. over which cost relationships are valid
   c. of probable production
   d. over which production has occurred in the past 10 years
3. How would the following be used in calculating the number of units that must be sold to earn a targeted operating income?

	Price per Unit	Targeted Operating Income
a.	Denominator	Numerator
b.	Numerator	Numerator
c.	Not used	Denominator
d.	Numerator	Denominator

4. Information concerning Korian Corporation's product is as follows:

Sales	$300,000
Variable costs	240,000
Fixed costs	40,000

   Assuming that Korian increased sales of the product by 20 percent, what should the operating income be?
   a. $20,000
   b. $24,000
   c. $32,000
   d. $80,000
5. The following data apply to Chan Company for last year:

Total variable costs per unit	$3.50
Contribution margin/Sales	30%
Break-even sales (present volume)	$1,000,000

Chan wants to sell an additional 50,000 units at the same selling price and contribution margin. By how much can fixed costs increase to generate additional profit equal to 10 percent of the sales value of the additional 50,000 units to be sold?

a. $50,000

b. $57,500

c. $67,500

d. $125,000

6. Bryan Company's break-even point is 8,500 units. Variable cost per unit is $140, and total fixed costs are $297,500 per year. What price does Bryan charge?

a. $140

b. $35

c. $175

d. cannot be determined from the above data

OBJECTIVE ▶ 1 5

Exercise 3-17 CONTRIBUTION MARGIN, CVP, NET INCOME, MARGIN OF SAFETY

Tintique Inc. produces novelty nail polishes. Each bottle sells for $3.84. Variable unit costs are as follows:

Acrylic base	$0.75	Bottle, packing material	$1.15
Pigments	0.38	Selling commission	0.25
Other ingredients	0.35		

Fixed overhead costs are $12,000 per year. Fixed selling and administrative costs are $6,720 per year. Tintique sold 35,000 bottles last year.

Required:

1. What is the contribution margin per unit for a bottle of nail polish? What is the contribution margin ratio?
2. How many bottles must be sold to break even? What is the break-even sales revenue?
3. What was Tintique's operating income last year?
4. What was the margin of safety in revenue?
5. Suppose that Tintique raises the price to $4.00 per bottle, but anticipated sales will drop to 29,800 bottles. What will the new break-even point in units be? Should Tintique raise the price? Explain.

OBJECTIVE ▶ 1 5

Exercise 3-18 OPERATING LEVERAGE

Income statements for two different companies in the same industry are as follows:

	Trimax Inc.	Quintex Inc.
Sales	$500,000	$500,000
Less: Variable costs	250,000	100,000
Contribution margin	250,000	400,000
Less: Fixed costs	200,000	350,000
Operating income	$ 50,000	$ 50,000

Required:

1. Compute the degree of operating leverage for each company.
2. Compute the break-even point for each company. Explain why the break-even point for Quintex is higher.
3. Suppose that both companies experience a 50 percent increase in revenues. Compute the percentage change in profits for each company. Explain why the percentage increase in Quintex's profits is so much greater than that of Trimax.

OBJECTIVE ▶ 3

Exercise 3-19 CVP ANALYSIS OF MULTIPLE PRODUCTS

Rossi Company produces GPS devices. One model is the GPS-auto, a basic model that is designed to attach to the windshield of a car. Another model, the GPS-marine, has extensive charts of oceans around the world. For the coming year, Rossi expects to sell

80,000 GPS-autos and 5,000 GPS-marines. A segmented income statement for the two products is as follows:

	GPS-Auto	GPS-Marine	Total
Sales	$16,000,000	$3,000,000	$19,000,000
Less: Variable costs	12,000,000	2,000,000	14,000,000
Contribution margin	4,000,000	1,000,000	5,000,000
Less: Direct fixed costs	1,200,000	960,000	2,160,000
Segment margin	$ 2,800,000	$ 40,000	2,840,000
Less: Common fixed costs			1,280,000
Operating income			$ 1,560,000

Required:

1. Compute the number of GPS-autos and GPS-marines that must be sold to break even.

2. Using information only from the total column of the income statement, compute the sales revenue that must be generated for the company to break even. (Round the contribution margin ratio to five significant digits and the sales revenue to the nearest dollar.)

Exercise 3-20 CVP WITH ACTIVITY-BASED COSTING

OBJECTIVE ➤ 1 6

Busy-Bee Baking Company produces a variety of breads. The average price of a loaf of bread is $1. Costs are as follows:

Cost Driver	Unit Variable Cost	Level of Cost Driver
Units sold	$0.65	—
Setups	$300	150
Maintenance hours	$15	2,500

Other data:

Total fixed costs (traditional)	$140,000
Total fixed costs (ABC)	57,500

Required:

1. Compute the break-even point in units using conventional analysis.
2. Compute the break-even point in units using activity-based analysis.
3. Suppose that Busy-Bee could reduce the setup cost by $100 per setup and could reduce the number of maintenance hours needed to 1,000. How many units must be sold to break even in this case? (Round answer up to whole units.)

Exercise 3-21 CVP WITH ACTIVITY-BASED COSTING AND MULTIPLE PRODUCTS

OBJECTIVE ➤ 1 3 6

Busy-Bee Baking Company produces a variety of breads. The plant manager would like to expand production into sweet rolls as well. The average price of a loaf of bread is $1. Anticipated price for a package of sweet rolls is $1.50. Costs for the new level of production are as follows:

Cost Driver	Unit Variable Cost	Level of Cost Driver
Loaf of bread	$0.65	—
Package of sweet rolls	$0.93	—
Setups	$300	250
Maintenance hours	$15	3,500

Other data:

Total fixed costs (traditional)	$185,000
Total fixed costs (ABC)	57,500

Busy-Bee believes it can sell 600,000 loaves of bread and 200,000 packages of sweet rolls in the coming year.

Required:

1. Prepare a contribution-margin-based income statement for next year. Be sure to show sales and variable costs by product and in total.
2. Compute the break-even sales for the company as a whole using conventional analysis.
3. Compute the break-even sales for the company as a whole using activity-based analysis.
4. Compute the break-even units of each product in units. Does it matter whether you use conventional analysis or activity-based analysis? Why or why not?
5. Suppose that Busy-Bee could reduce the setup cost by $100 per setup and could reduce the number of maintenance hours needed to 1,000. How many units of each product must be sold to break even in this case? (Round answers up to whole units.)

Problems

OBJECTIVE ➤ 3

Problem 3-22 USING A COMPUTER SPREADSHEET TO SOLVE MULTIPLE-PRODUCT BREAK-EVEN, VARYING SALES MIX

More-Power Company has projected sales of 75,000 regular sanders and 30,000 mini-sanders for next year. The projected income statement is as follows:

	Regular Sander	Mini-Sander	Total
Sales	$3,000,000	$1,800,000	$4,800,000
Less: Variable expenses	1,800,000	900,000	2,700,000
Contribution margin	1,200,000	900,000	2,100,000
Less: Direct fixed expenses	250,000	450,000	700,000
Product margin	$ 950,000	$ 450,000	1,400,000
Less: Common fixed expenses			600,000
Operating income			$ 800,000

Required:

1. Set up the given income statement on a spreadsheet (e.g., Excel™). Then substitute the following sales mixes and calculate operating income. Be sure to print the results for each sales mix (a through d).

	Regular Sander	Mini-Sander
a.	75,000	37,500
b.	60,000	60,000
c.	30,000	90,000
d.	30,000	60,000

2. Calculate the break-even units for each product for each of the preceding sales mixes.

OBJECTIVE ➤ 1

Problem 3-23 CONTRIBUTION MARGIN, UNIT AMOUNTS

Consider the following information on four independent companies.

	A	B	C	D
Sales	$10,000	$?	$?	$9,000
Less: Variable costs	8,000	11,700	9,750	?
Contribution margin	2,000	7,800	?	?
Less: Fixed costs	?	4,500	?	900
Operating income	$ 1,000	$?	$8,000	$2,850

	A	B	C	D
Units sold	?	1,300	300	500
Price/Unit	$4	$?	$130	$?
Variable cost/Unit	$?	$9	$?	$?
Contribution margin/Unit	$?	$6	$?	$?
Contribution margin ratio	?	?	75%	?
Break-Even in units	?	?	?	?

Required:

Calculate the correct amount for each question mark. Be sure to round any fractional break-even units *up* to the next whole number.

Problem 3-24 **BREAK-EVEN IN SALES REVENUE, VARIABLE-COSTING RATIO, CONTRIBUTION MARGIN RATIO, MARGIN OF SAFETY**

OBJECTIVE ➤ 1 2 5

SERVICE

Furyk Company runs a driving range and golf shop. The budgeted income statement for the coming year is as follows.

Sales	$800,000
Less: Variable expenses	344,000
Contribution margin	456,000
Less: Fixed expenses	310,000
Income before taxes	146,000
Less: Income taxes	51,100
Net income	$ 94,900

Required:

1. What is Furyk's variable cost ratio? Its contribution margin ratio?
2. Suppose Furyk's actual revenues are $150,000 greater than budgeted. By how much will before-tax profits increase? Give the answer without preparing a new income statement.
3. How much sales revenue must Furyk earn in order to break even? What is the expected margin of safety? (Round your answers to the nearest dollar.)
4. How much sales revenue must Furyk generate to earn a before-tax profit of $120,000? An after-tax profit of $120,000? (Round your answers to the nearest dollar.) Prepare a contribution margin income statement to verify the accuracy of your last answer.

Problem 3-25 **CHANGES IN BREAK-EVEN POINTS WITH CHANGES IN UNIT PRICES**

OBJECTIVE ➤ 1 5

Cabrera Inc. produces and sells bobblehead dolls. Last year, Cabrera sold 156,250 units. The income statement for Cabrera Inc. for last year is as follows:

Sales	$625,000
Less: Variable expenses	343,750
Contribution margin	281,250
Less: Fixed expenses	180,000
Operating income	$101,250

Required:

1. Compute the break-even point in units and in revenues. Compute the margin of safety in sales revenue for last year.
2. Suppose that the selling price decreases by 10 percent. Will the break-even point increase or decrease? Recompute the break-even point in units. (Round up to the next whole unit.)
3. Suppose that the variable cost per unit decreases by $0.25. Will the break-even point increase or decrease? Recompute the break-even point in units. (Round up to the next whole unit.)
4. Can you predict whether the break-even point increases or decreases if both the selling price and the unit variable cost decrease? Recompute the break-even point in

units incorporating both of the changes in Requirements 2 and 3. (Round up to the next whole unit.)

5. Assume that total fixed costs increase by $50,000. (Assume no other changes from the original data.) Will the break-even point increase or decrease? Recompute it. (Round up to the next whole unit.)

OBJECTIVE ▶ 1 2 5   Problem 3-26 BREAK-EVEN, AFTER-TAX TARGET INCOME, MARGIN OF SAFETY, OPERATING LEVERAGE

Faldo Company produces a single product. The projected income statement for the coming year, based on sales of 200,000 units, is as follows:

Sales	$2,000,000
Less: Variable costs	1,400,000
Contribution margin	600,000
Less: Fixed costs	450,000
Operating income	$ 150,000

Required:

1. Compute the unit contribution margin and the units that must be sold to break even. Suppose that 30,000 units are sold above the break-even point. What is the profit?
2. Compute the contribution margin ratio and the break-even point in dollars. Suppose that revenues are $200,000 greater than expected. What would the total profit be?
3. Compute the margin of safety in sales revenue.
4. Compute the operating leverage. Compute the new profit level if sales are 20 percent higher than expected.
5. How many units must be sold to earn a profit equal to 10 percent of sales?
6. Assume the income tax rate is 40 percent. How many units must be sold to earn an after-tax profit of $180,000?

OBJECTIVE ▶ 1 2 5   Problem 3-27 BASIC CVP CONCEPTS

Katayama Company produces a variety of products. One division makes neoprene wet-suits. The division's projected income statement for the coming year is as follows:

Sales (65,000 units)	$15,600,000
Less: Variable expenses	8,736,000
Contribution margin	6,864,000
Less: Fixed expenses	4,012,000
Operating income	$ 2,852,000

Required:

1. Compute the contribution margin per unit, and calculate the break-even point in units. Repeat, using the contribution margin ratio.
2. The divisional manager has decided to increase the advertising budget by $140,000 and cut the average selling price to $200. These actions will increase sales revenues by $1 million. Will this improve the division's financial situation? Prepare a new income statement to support your answer.
3. Suppose sales revenues exceed the estimated amount on the income statement by $612,000. Without preparing a new income statement, determine by how much profits are underestimated.
4. How many units must be sold to earn an after-tax profit of $1.254 million? Assume a tax rate of 34 percent. (Round your answer up to the next whole unit.)
5. Compute the margin of safety in dollars based on the given income statement.
6. Compute the operating leverage based on the given income statement. (Round to three significant digits.) If sales revenues are 20 percent greater than expected, what is the percentage increase in profits?

Problem 3-28 CVP ANALYSIS: SALES-REVENUE APPROACH, PRICING, AFTER-TAX TARGET INCOME

OBJECTIVE ➤ 2 5

SERVICE

Mahan Consulting is a service organization that specializes in the design, installation, and servicing of mechanical, hydraulic, and pneumatic systems. For example, some manufacturing firms, with machinery that cannot be turned off for servicing, need some type of system to lubricate the machinery during use. To deal with this type of problem for a client, Mahan designed a central lubricating system that pumps lubricants intermittently to bearings and other moving parts.

The operating results for the firm for the previous year are as follows:

Sales	$974,880
Less: Variable expenses	534,234
Contribution margin	440,646
Less: Fixed expenses	264,300
Operating income	$176,346

In the coming year, Mahan expects variable costs to increase by 4 percent and fixed costs by 3 percent.

Required:

1. What is the contribution margin ratio (rounded to three significant digits) for the previous year?
2. Compute Mahan's break-even point for the previous year in dollars.
3. Suppose that Mahan would like to see a 6 percent increase in operating income in the coming year. What percent (on average) must Mahan raise its bids to cover the expected cost increases and obtain the desired operating income? Assume that Mahan expects the same mix and volume of services in both years.
4. In the coming year, how much revenue must be earned for Mahan to earn an after-tax profit of $175,000? Assume a tax rate of 40 percent.

Problem 3-29 MULTIPLE PRODUCTS, BREAK-EVEN ANALYSIS, OPERATING LEVERAGE, SEGMENTED INCOME STATEMENTS

OBJECTIVE ➤ 3 5

Ironjay Inc. produces two types of weight-training equipment: the Jay-flex (a weight machine that allows the user to perform a number of different exercises) and a set of free weights. Ironjay sells the Jay-flex to sporting goods stores for $200. The free weights sell for $75 per set. The projected income statement for the coming year follows:

Sales	$600,000
Less: Variable expenses	390,000
Contribution margin	210,000
Less: Fixed expenses	157,500
Operating income	$ 52,500

The owner of Ironjay estimates that 40 percent of the sales revenues will be produced by sales of the Jay-flex, with the remaining 60 percent by free weights. The Jay-flex is also responsible for 40 percent of the variable expenses. Of the fixed expenses, one-third are common to both products, and one-half are directly traceable to the Jay-flex line.

Required:

1. Compute the sales revenue that must be earned for Ironjay to break even.
2. Compute the number of Jay-flex machines and free weight sets that must be sold for Ironjay to break even.
3. Compute the degree of operating leverage for Ironjay. Now, assume that the actual revenues will be 40 percent higher than the projected revenues. By what percentage will profits increase with this change in sales volume?
4. Ironjay is considering adding a new product—the Jay-rider. The Jay-rider is a cross between a rowing machine and a stationary bicycle. For the first year, Ironjay estimates that the Jay-rider will cannibalize 600 units of sales from the Jay-flex. Sales of free-weight sets will remain unchanged. The Jay-rider will sell for $180 and have

variable costs of $140. The increase in fixed costs to support manufacture of this product is $5,700. Compute the number of Jay-flex machines, free-weight sets, and Jay-riders that must be sold for Ironjay to break even. For the coming year, is the addition of the Jay-rider a good idea? Why or why not? Why might Ironjay choose to add the Jay-rider anyway?

OBJECTIVE ▶ 1 5

Problem 3-30 BREAK-EVEN IN UNITS AND SALES DOLLARS, MARGIN OF SAFETY

Dragon Company produces a single product. Last year's income statement is as follows:

Sales (20,000 units)	$1,218,000
Less: Variable costs	812,000
Contribution margin	406,000
Less: Fixed costs	300,000
Operating income	$ 106,000

Required:

1. Compute the break-even point in units and sales revenue.
2. What was the margin of safety for Dragon Company last year?
3. Suppose that Dragon Company is considering an investment in new technology that will increase fixed costs by $250,000 per year, but will lower variable costs to 45 percent of sales. Units sold will remain unchanged. Prepare a budgeted income statement assuming Dragon makes this investment. What is the new break-even point in units, assuming the investment is made?

OBJECTIVE ▶ 1 6

Problem 3-31 CVP ANALYSIS, IMPACT OF ACTIVITY-BASED COSTING

Salem Electronics currently produces two products: a programmable calculator and a tape recorder. A recent marketing study indicated that consumers would react favourably to a radio with the Salem brand name. Owner Kenneth Booth was interested in the possibility. Before any commitment was made, however, Kenneth wanted to know what the incremental fixed costs would be and how many radios must be sold to cover these costs.

In response, Betty Johnson, the marketing manager, gathered data for the current products to help in projecting overhead costs for the new product. The overhead costs based on 30,000 direct labour hours follow. (The high-low method using direct labour hours as the independent variable was used to determine the fixed and variable costs.)

	Fixed	Variable
Materials handling	$ —	$18,000
Power	—	22,000
Engineering	100,000	—
Machine costs	30,000*	80,000
Inspection	40,000	—
Setups	60,000	—

*All depreciation.

The following activity data were also gathered:

	Calculators	Recorders
Units produced	20,000	20,000
Direct labour hours	10,000	20,000
Machine hours	10,000	10,000
Material moves	120	120
Kilowatt-hours	1,000	1,000
Engineering hours	4,000	1,000
Hours of inspection	700	1,400
Number of setups	20	40

Betty was told that a plantwide overhead rate was used to assign overhead costs based on direct labour hours. She was also informed by engineering that if 20,000 radios were produced and sold (her projection based on her marketing study), they would have the same activity data as the recorders (use the same direct labour hours, machine hours, set-ups, and so on).

Engineering also provided the following additional estimates for the proposed product line:

Prime costs per unit	$ 18
Depreciation on new equipment	18,000

Upon receiving these estimates, Betty did some quick calculations and became quite excited. With a selling price of $26 and just $18,000 of additional fixed costs, only 4,500 units would have to be sold to break even. Since Betty was confident that 20,000 units could be sold, she was prepared to strongly recommend the new product line.

Required:

1. Reproduce Betty's break-even calculation using conventional cost assignments. How much additional profit would be expected under this scenario, assuming that 20,000 radios were sold?
2. Using an activity-based costing approach, calculate the break-even point and the incremental profit that would be earned on sales of 20,000 units.
3. Explain why the CVP analysis done in Requirement 2 is more accurate than the analysis done in Requirement 1. What recommendation would you make?

Problem 3-32 ABC AND CVP ANALYSIS: MULTIPLE PRODUCTS

OBJECTIVE ▶ 3 6

Good Scent Inc. produces two colognes: Rose and Violet. Of the two, Rose is more popular. Data concerning the two products follow:

	Rose	Violet
Expected sales (in cases)	50,000	10,000
Selling price per case	$100	$80
Direct labour hours	36,000	6,000
Machine hours	10,000	3,000
Receiving orders	50	25
Packing orders	100	50
Material cost per case	$50	$43
Direct labour cost per case	$10	$7

The company uses a conventional costing system and assigns overhead costs to products using direct labour hours. Annual overhead costs follow. They are classified as fixed or variable with respect to direct labour hours.

	Fixed	Variable
Direct labour benefits	$ —	$200,000
Machine costs	200,000*	262,000
Receiving department	225,000	—
Packing department	125,000	—
Total costs	$550,000	$462,000

*All depreciation.

Required:

1. Using the conventional approach, compute the number of cases of Rose and the number of cases of Violet that must be sold for the company to break even.
2. Using an activity-based approach, compute the number of cases of each product that must be sold for the company to break even.

CMA Problems

CMA Problem 3-1 BREAK-EVEN IN SALES REVENUE*

Big Blue Motors Inc. employs 24 sales personnel to market its line of luxury automobiles. The average car sells for $85,000, and a 6 percent commission is paid to the salesperson. Big Blue Motors is considering a change to the commission arrangement where the company would pay each salesperson a salary of $1,600 per month plus a commission of 2 percent of the sales made by that salesperson. What is the amount of total monthly car sales at which Big Blue Motors would be indifferent as to which plan to select? *(CMA adapted)*

CMA Problem 3-2 BREAK-EVEN IN UNITS, AFTER-TAX TARGET INCOME, CVP ASSUMPTIONS*

Shapiro Company manufactures and sells adjustable canopies that attach to motor homes and trailers. The market covers both new unit purchases as well as replacement canopies. Shapiro developed its 2013 business plan based on the assumption that canopies would sell at a price of $400 each. The variable costs for each canopy were projected at $200, and the annual fixed costs were budgeted at $120,000. Shapiro's after-tax profit objective was $225,000; the company's effective tax rate is 40 percent.

While Shapiro's sales usually rise during the second quarter, the May financial statements reported that sales were not meeting expectations. For the first five months of the year, only 350 units had been sold at the established price, with variable costs as planned, and it was clear that the 2013 after-tax profit projection would not be reached unless some actions were taken. Shapiro's president assigned a management committee to analyze the situation and develop several alternative courses of action. The following mutually exclusive alternatives, labelled A, B, and C, were presented to the president:

A. Lower the variable costs per unit by $25 through the use of less expensive materials and slightly modified manufacturing techniques. The sales price will also be reduced by $30, and sales of 2,200 units for the remainder of the year are forecast.

B. Reduce the sales price by $40. The sales organization forecasts that with the significantly reduced sales price, 2,700 units can be sold during the remainder of the year. Total fixed and variable unit costs will stay as budgeted.

C. Cut fixed costs by $10,000, and lower the sales price by 5 percent. Variable costs per unit will be unchanged. Sales of 2,000 units are expected for the remainder of the year.

Required:

1. Determine the number of units that Shapiro Company must sell in order to break even assuming no changes are made to the selling price and cost structure.
2. Determine the number of units that Shapiro Company must sell in order to achieve its after-tax profit objective.
3. Determine which one of the alternatives Shapiro Company should select to achieve its annual after-tax profit objective. Be sure to support your selection with appropriate calculations.
4. The precision and reliability of CVP analysis are limited by several underlying assumptions. Identify at least four of these assumptions. *(CMA adapted)*

CMA Problem 3-3 BREAK-EVEN IN SALES REVENUE, CHANGES IN VARIABLES*

Carmichael Corporation is in the process of preparing next year's budget. The pro forma income statement for the current year is as follows:

Sales		$1,800,000
Cost of sales:		
Direct materials	$250,000	
Direct labour	180,000	
Variable overhead	106,000	
Fixed overhead	100,000	636,000
Gross profit		1,164,000

Selling and administrative expenses:		
Variable	$400,000	
Fixed	350,000	750,000
Operating income		$ 414,000

Required:

1. What is the break-even sales revenue (rounded to the nearest dollar) for Carmichael Corporation for the current year?
2. For the coming year, the management of Carmichael Corporation anticipates an 8 percent increase in variable costs and a $60,000 increase in fixed expenses. What is the break-even point in dollars for next year? *(CMA adapted)*

CMA Problem 3-4 AFTER-TAX TARGET INCOME, PROFIT ANALYSIS*

OBJECTIVE ▸ 2 3 5

X-Cee-Ski Company recently expanded its manufacturing capacity, which will allow it to produce up to 21,000 pairs of cross-country skis of the mountaineering model or the touring model. The Sales Department assures management that it can sell between 9,000 and 14,000 pairs of either product this year. Because the models are very similar, X-Cee-Ski will produce only one of the two models.

The following information was compiled by the Accounting Department:

Per-Unit (Pair) Data

	Mountaineering	Touring
Selling price	$180	$120
Variable costs	130	90

Fixed costs will total $320,000 if the mountaineering model is produced but will be only $220,000 if the touring model is produced. X-Cee-Ski is subject to a 40 percent income tax rate.

Required:

1. If X-Cee-Ski Company desires an after-tax net income of $48,000, how many pairs of touring model skis will the company have to sell?
2. Suppose that X-Cee-Ski Company decided to produce only one model of skis. What is the total sales revenue at which X-Cee-Ski Company would make the same profit or loss regardless of the ski model it decided to produce?
3. If the Sales Department could guarantee the annual sale of 12,000 pairs of either model, which model would the company produce, and why? *(CMA adapted)*

CMA Problem 3-5 BREAK-EVEN IN UNITS*

OBJECTIVE ▸ 1

SERVICE

Don Masters and two of his colleagues are considering opening a law office in a large metropolitan area that would make inexpensive legal services available to those who could not otherwise afford these services. The intent is to provide easy access for their clients by having the office open 360 days per year, 16 hours each day from 7:00 a.m. to 11:00 p.m. The office would be staffed by a lawyer, paralegal, legal secretary, and clerk-receptionist for each of the two 8-hour shifts.

In order to determine the feasibility of the project, Don hired a marketing consultant to assist with market projections. The results of this study show that if the firm spends $500,000 on advertising the first year, the number of new clients expected each day would have the following probability distribution:

Number of New Clients per Day	Probability
20	0.10
30	0.30
55	0.40
85	0.20

Don and his associates believe these numbers are reasonable and are prepared to spend the $500,000 on advertising. Other pertinent information about the operation of the office is as follows.

The only charge to each new client would be $30 for the initial consultation. All cases that warranted further legal work would be accepted on a contingency basis with the firm earning 30 percent of any favourable settlements or judgments. Don estimates that 20 percent of new client consultations will result in favourable settlements or judgments averaging $2,000 each. Repeat clients are not expected during the first year of operations.

The hourly wages of the staff are projected to be $25 for the lawyer, $20 for the paralegal, $15 for the legal secretary, and $10 for the clerk-receptionist. Fringe benefit expenses will be 40 percent of the wages paid. A total of 400 hours of overtime is expected for the year; this will be divided equally between the legal secretary and the clerk-receptionist positions. Overtime will be paid at one and one-half times the regular wage, and the fringe benefit expense will apply to the full wages.

Don has located 6,000 square metres of suitable office space, which rents for $28 per square metre annually. Associated expenses will be $22,000 for property insurance and $32,000 for utilities.

It will be necessary for the group to purchase malpractice insurance, which is expected to cost $180,000 annually. The initial investment in office equipment will be $60,000; this equipment has an estimated useful life of four years. The cost of office supplies has been estimated to be $4 per expected new client consultation.

Required:

1. Determine how many new clients must visit the law office being considered by Don Masters and his colleagues in order for the venture to break even during its first year of operations.

2. Using the information provided by the marketing consultant, determine if it is feasible for the law office to achieve break-even operations. *(CMA adapted)*

The Collaborative Learning Exercises can be found on the product support site at www.hansen1ce.nelson.com.

After studying this chapter, you should be able to:

▶1 Differentiate the cost accounting systems of service and manufacturing firms and of unique and standardized products.

▶2 Discuss the interrelationship of cost accumulation, cost measurement, and cost assignment.

▶3 Identify the source documents used in job-order costing.

▶4 Describe the cost flows associated with job-order costing, and prepare the journal entries.

▶5 Explain how activity-based costing is applied to job-order costing.

▶6 Explain how spoiled units are accounted for in a job-order costing system.

CHAPTER

4

Job-Order Costing Systems

Now that we have an understanding of basic cost terminology, we need to look more closely at the system that the firm sets up to account for costs. In other words, we need to determine how we accumulate costs, how to measure/classify them, and then how to assign them to units manufactured or units of service delivered.

Manufacturing Firms versus Service Firms

In general, a firm's cost management system mirrors the production process. A cost management system modelled after the production process allows managers to better monitor the economic performance of the firm. A production process may yield a tangible product or a service. Those products or services may be similar in nature or unique. These characteristics of the production process determine the best approach for developing a cost management system.

Manufacturing involves combining direct materials, direct labour, and overhead to produce a new product. The good produced is tangible and can be inventoried and

OBJECTIVE ▶1
Differentiate the cost accounting systems of service and manufacturing firms and of unique and standardized products.

transported from the plant to the customer. A service is characterized by its intangible nature. It is not separable from the customer and cannot be inventoried. Traditional cost accounting has emphasized manufacturing and virtually ignored services. Now, more than ever, that approach will not do. Our economy has become increasingly service oriented. Managers must be able to track the costs of services rendered just as precisely as they track the costs of goods manufactured. In fact, a company's controller may find it necessary to cost both goods and services as managers take an internal customer approach.

The range of manufacturing and service firms can be represented by a continuum as shown in Exhibit 4-1. The pure service, shown at the left, involves no raw materials and no tangible item for the customer. There are few pure services. Perhaps an example would be an Internet cafe. In the middle of the continuum, and still very much a service, is a beauty salon, which uses direct materials such as hair spray and styling gel. At the other end of the continuum is the manufactured product. Examples include automobiles, cereals, cosmetics, and drugs. Even these, however, often have a service component. For example, a prescription drug must be prescribed by a physician and dispensed by a licensed pharmacist. Automobile dealers stress the continuing service associated with their cars. And what about fast food? Does **Second Cup** provide a product or a service? There are elements of both.

Four areas in which services differ from products are intangibility, inseparability, heterogeneity, and perishability. **Intangibility** refers to the nonphysical nature of services as opposed to products. **Inseparability** means that production and consumption are inseparable for services. **Heterogeneity** refers to the greater chances for variation in the performance of services than in the production of products.

Perishability means that services cannot be inventoried but must be consumed when performed. These differences affect the types of information needed for planning, control, and decision making in the production of services. Exhibit 4-2 illustrates the features associated with the production of services and their interface with the cost management system.

Intangibility

Intangibility of services leads to a major difference in the accounting for services as opposed to products. A service company cannot inventory the service and therefore has a minimal to moderate inventory of supplies. A manufacturing company has inventories of raw materials, supplies, work in process, and finished goods. Because of the significance and complexity of inventories in manufacturing, we will spend more time on manufacturing companies in accounting for the cost of inventories.

Service companies typically rank lower than manufacturing companies in ratings of customer satisfaction.[1] An important reason for this is that service firms have a greater degree of heterogeneity of labour. Service firms are keenly aware of the importance of human resources; the service is provided by people. A key assumption of microeconomics is the homogeneity of labour. That is, one direct labourer is assumed to be identical to another. This assumption is the basis of labour standards in standard costing. Service companies know that one worker is not identical to another.

Exhibit 4-1

Continuum of Services and Manufactured Products

Pure Service			Manufactured Product
Internet cafe	Beauty salon	Restaurant Software	Automobiles Cereals

---

[1] Jaclyn Fierman, "Americans Can't Get No Satisfaction," *Fortune* (December 11, 1995): 186–194.

Exhibit 4-2

Features of Service Firms and Their Interface with the Cost Management System

Feature*	Relationship to Business	Impact on Cost Management System
Intangibility	Services cannot be stored.	There are no inventory accounts.
	Services cannot be protected through patents.	There is a stronger need for an ethical code of conduct.
	Services cannot readily be displayed or communicated.	
	It is more difficult to set prices.	Costs must be related to entire organization.
Inseparability	Consumer is involved in production.	Costs are accounted for by customer type.
	Other consumers are involved in production.	
	Centralized mass production of services is difficult.	System must be generated to encourage consistent quality.
Heterogeneity	Standardization and quality control are difficult.	Productivity measurement is ongoing. TQM is critical.
Perishability	Service benefits expire quickly.	There are no inventories.
	Service may be repeated often for one customer.	There needs to be a standardized system to handle repeat customers.

*First two columns adapted from Valarie Zeithaml, A. Parasuraman, and Leonard L. Berry, "Problems and Strategies in Services Marketing," *Journal of Marketing* 49 (Spring 1985): 34–46.

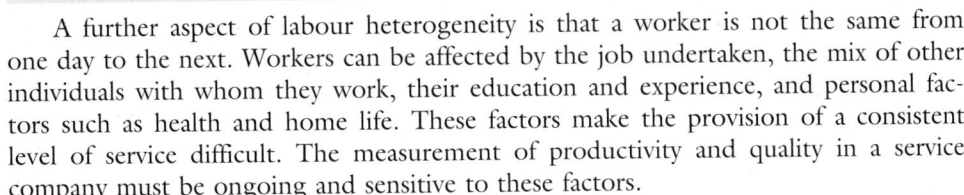

For example, **Canada's Wonderland** hires "backstage employees" and "on-stage employees." The backstage employees may do maintenance, sew costumes, and work in personnel, but they do not work with the paying public (guests). On-stage employees, hired both for their particular skills and for their ability to interact well with people, work directly with the guests.

A further aspect of labour heterogeneity is that a worker is not the same from one day to the next. Workers can be affected by the job undertaken, the mix of other individuals with whom they work, their education and experience, and personal factors such as health and home life. These factors make the provision of a consistent level of service difficult. The measurement of productivity and quality in a service company must be ongoing and sensitive to these factors.

Inseparability

Inseparability means that differences in customers affect the service firm more than the manufacturing firm. When **Proctor-Silex®** sells a toaster, the mood and personal qualities of the customer are irrelevant. When **Toronto-Holland Orthopedic and Arthritic Centre** sells a service to a customer, however, the disposition of the customer may affect the amount of service required as well as the quality of the service rendered. Inseparability also means that customers evaluate services differently from products. As a result, service companies may need to spend more money on some resources and less on others than would be necessary in a manufacturing plant. For example, consumers may use price and physical facilities as the major cues to service quality. Service firms, then, tend to incur higher costs for attractive places of business than do manufacturing firms. Your initial impression of a manufacturing plant may be how large, noisy, and dingy it is. Floors are concrete; the ceiling is typically unfinished. In short, it is not a pretty sight. However, as long as a high-quality product is made, the consumer does not care. This is very different from most consumers' attitudes toward the service environment. Banks, doctors' offices, and restaurants are pleasant places, tastefully decorated, and filled with plants. This is cost effective to the extent that customers are drawn to such an environment to conduct business. In addition, the environment may allow the service firm to charge a higher price—signalling its higher quality.

Perishability of services is very similar to intangibility. For example, there are no work-in-process or finished goods inventories of services. However, there is a subtle distinction between intangibility and perishability that merits discussion. A service is perishable if the effects are short term. Not all services fall into this category. Plastic surgery is not perishable, but haircuts are. The impact on cost management is that perishable services require systems to easily handle repeat customers. The repetitive nature of the service also leads us to the use of standardized processes and costing. Examples are financial services (e.g., cheque clearing by banks), janitorial services, and beauty and barber shops.

ETHICS Customers may perceive greater risk when buying services than when buying products. Ethics are important here. The internal accountant who is responsible for gathering data on service quality must accurately report the bad news as well as the good. A customer who has been stung once by misleading advertising or by a firm's failure to deliver the promised performance will be loathe to try that firm again. A manufacturer can offer a warranty or product replacement. But the service firm must consider the customer's wasted time. Therefore, the service firm must be especially careful to avoid promising more than can or will be delivered.◆

Consider the example of **Lexus** which discovered a defect shortly after introducing the car into the United States. Lexus dealers contacted each buyer personally and arranged for loaner cars while the defect was being fixed. In the case of buyers who lived far from a dealership, Lexus brought the repair people to the buyers. More recently, Lexus discovered a larger transmission problem and immediately swapped out the affected cars for new cars.[2] Contrast this experience with service issues experienced by many GM buyers, who must deal with several layers of automotive hierarchy in order to get a defect repaired. Clearly, Lexus understood the value of customers' time in arranging the service.

Service companies are particularly interested in planning and control techniques that apply to their special types of firms. Productivity measurement and quality control are very important. Pricing may involve different considerations for the service firm.

The important point is that service and manufacturing companies may have different needs for accounting data and techniques. It is important for the accountant to be aware of relevant differences in order to provide appropriate support, and to be cross-functionally trained.

McDonald's is an example of both a manufacturing and a service company. In the kitchen, McDonald's runs a production line. The product is rigidly consistent. Each hamburger contains the same amount of meat, mustard, ketchup, and pickles. The buns are identical. The burgers are cooked the prescribed amount of time to a set temperature. They are wrapped in a methodical manner and join other burgers in the warming bin. Standard cost accounting techniques work well for this phase, and McDonald's uses them. At the counter, however, the company becomes a service organization. Customers want their orders taken and filled quickly and correctly. In addition, they want pleasant service and maybe some help finding certain items on the menu. Clean restrooms are critical. McDonald's emphasizes nonfinancial measures of performance for service areas: counter customers are to be served within 60 seconds; drive-through customers are to be served within 90 seconds; restrooms are to be checked and cleaned at least once an hour.

[2] Bill Taylor, "More Lessons from Lexus—Why It Pays to Do the Right Thing." Harvard Business Publishing, December 12, 2007. http://blogs.harvardbusiness.org/taylor/2007/12/more_lessons_from_lexuswhy_it.html (Accessed May 28, 2009).

Heterogeneity versus Standardized Products and Services

A second way of characterizing products and services is according to the degree of uniqueness. If a firm produces unique products in small batches, and if those products incur different costs, then the firm must keep track of the costs of each product or batch. This is referred to as a job-order costing system, the focus of this chapter. At the other extreme, the company may make many identical units of the same product. Since the units are the same, the costs of each unit are also the same. Accounting for the costs of the identical units is relatively simple and is referred to as process costing, examined in Chapter 5.

It is important to note that the uniqueness of the products (or units) for cost accounting purposes relates to unique costs. Consider a large construction company that builds houses in developments across southern Ontario. While the houses are based on several standard models, buyers can customize their houses by selecting different types of brick, tile, carpet, and so on. However, these selections are taken from a set menu of choices. While one house is painted white and its neighbour house is painted green, the cost is the same. However, if different selections have different costs, then those costs must be accounted for separately. Thus, if one home buyer selects a whirlpool tub while another selects a standard model, the different cost of the two tubs must be tracked to the correct house. A builder has to offer choices to clients but also keep track of the costs of each choice. A production process that appears to produce similar products may incur different costs for each product. In this type of situation, the firm should track costs using a job-order costing system.

Both service and manufacturing firms use the job-order costing approach. Custom cabinet makers and home builders manufacture unique products, which must be accounted for using a job-order costing approach. Dental and medical services also use job-order costing. The costs associated with a simple dental filling clearly differ from those associated with a root canal. Printing, automotive repair, and appliance repair are also services using job-order costing.

Firms in process industries mass-produce large quantities of similar, or homogeneous, products. Each product is essentially indistinguishable from its companion product. Examples of process manufacturers include food, cement, petroleum, and chemical firms. The important point here is that the cost of one unit of product is identical to the cost of another. Therefore, service firms can also use a process-costing approach. Discount stockbrokers, for example, incur much the same cost to execute a customer order for one stock as for another; cheque-clearing departments of banks incur a uniform cost to clear a cheque, no matter the value of the cheque or to whom it is written.

A third type of costing system is operation costing. **Operation costing** is a hybrid of job-order and process costing. Units within a batch are the same and can be accounted for using a process approach. However, each batch is different from other batches and the costs of the batches are handled separately in a job-order costing manner. Some clothing and electronics firms use operation costing.

Interestingly, companies are gravitating toward job-order costing because of the increased variety of products and increased demand for small orders and prototypes.

An excellent example is Etobicoke-based apparel company **Fairweather Ltd**. The company manufactures women's and girls' nightware, foundation garments, lingerie, and loungewear. Fairweather can handle small and bigger orders. It emphasizes on-time delivery and fills orders quickly. The actual products that customers receive are the same as the samples the company uses for marketing purposes. Flexible manufacturing of specialized products leads Fairweather into an approximate job-order costing environment.

Perishability

When a client misses an appointment with a doctor, dentist, hairdresser, or lawyer, that time can never be recaptured. When a commercial airliner flies with empty seats, or hotel rooms are empty, or theatre tickets go unsold, the potential for sales/profits vanishes. The perishability of these services creates difficulties in balancing supply and demand. Demand may be seasonal; for example, Christmas may be the peak season for retailers, or summer the peak season for hotels and airlines (i.e., when most people take vacation). When demand fluctuates and capacity is fixed, it can be a challenge to maintain high performance levels. During tax season, for example, tax preparers may not have the time to provide clients with the same personalized attention as at other times of the year. Supply/demand issues for goods can usually be solved through production scheduling and inventory management; this is not the case with services. Because of perishability, inventory is nil and creative thinking is necessary in order to better utilize available capacity.

Canada's largest theatre chain is **Cineplex Entertainment**. The movie theatre industry today is facing several challenges, the main one being that, like many other service-based businesses, Cineplex must deal with the fact that its services are both intangible and perishable. Cineplex has tackled the first challenge by making the theatre experience more tangible. The company is remodelling its complexes, upgrading their projection and sound systems, and installing more comfortable seating. It is also developing nonfilm activities to turn its venues into entertainment destinations. A new 4,180-square-metre cinema complex in Oakville, for example, offers a six-lane bowling alley, billiards, party rooms, a Kids Club, babysitting services, and private VIP rooms. All of this is in addition to 12 movie screens. In partnership with Scotiabank, Cineplex has also launched a loyalty program. Members of that program earn points, which they can redeem for tangible rewards such as concession-stand items and movie tickets. This program, named Scene, has been a hit, having reached 90 percent of its membership target in its first six months. In just one year, more than 600,000 people joined the program.

Perishability is another big issue for Cineplex. A theatre that is only half full on weeknights means permanently lost revenue. Cineplex addresses this problem through alternative program offerings that bring in new audiences, smooth out demand, and boost revenues. Instead of movies, Cineplex has broadcast live concerts by popular artists, as well as opera performances on a pay-per-view basis. It also offers various sporting events (professional wrestling, World Cup soccer, NHL games), which customers can see at a fraction of the cost of seeing them live. These offerings have helped neutralize perishability effects. More important, these add-on events to movies have allowed Cineplex to reach audiences—demographically speaking—that seldom if ever go to the movies.

Courtesy of Magic Marketing, www.magicmarketing.ca

OBJECTIVE ▶2
Discuss the interrelationship of cost accumulation, cost measurement, and cost assignment.

Setting Up the Cost Accounting System

Once the characteristics of a firm's production process are understood, the accountant can set up a system for generating appropriate cost information. A good cost accounting information system is flexible and reliable. It provides information for a variety of purposes and can be used to answer different types of questions. In general, the system is used to satisfy the needs for cost accumulation, cost measurement, and cost assignment. **Cost accumulation** is the recognition and recording of costs. **Cost measurement** involves determining the dollar amounts of direct materials, direct labour, and overhead used in production. **Cost assignment** is the association of production costs with the units produced. Exhibit 4-3 illustrates the relationship of cost accumulation, cost measurement, and cost assignment.

Cost Accumulation

Cost accumulation refers to the recognition and recording of costs. The cost accountant needs to develop source documents that keep track of costs as they occur. A **source document** describes a transaction. Data from these source documents can then be recorded in a database. The recording of data in a database allows

Exhibit 4-3

Relationship of Cost Accumulation, Cost Measurement, and Cost Assignment

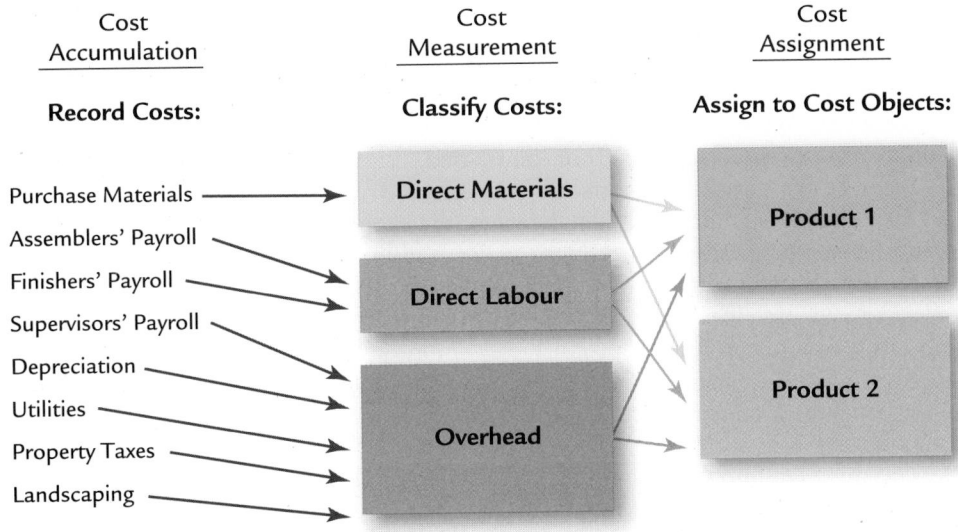

accountants and managers the flexibility to analyze subsets of the data as needed to aid in management decision making. The cost accountant can also use the database to see that the relevant costs are recorded in the general ledger and posted to appropriate accounts for purposes of external financial reporting.

Well-designed source documents can supply information in a flexible way. In other words, the information can be used for multiple purposes. For example, the sales receipt written up or input by a clerk when a customer buys merchandise lists the date, the items purchased, the quantities, the prices, the sales tax paid, and the total dollar amount received. Just this one source document can be used in determining sales revenue for the month, the sales by each product, the tax owed to the government, and the cash received or the accounts receivable recorded. Similarly, employees often fill in labour time tickets, indicating which jobs they worked on, on what date, and for how long. Data from the labour time ticket can be used in determining direct labour cost used in production, the amount to pay the worker, the degree of productivity improvement achieved over time, and the amount to budget for direct labour for an upcoming job.

Cost Measurement

Once costs are accumulated (recorded), they can be classified or organized in a meaningful way. Cost measurement refers to classifying the costs. For example, in manufacturing it may consist of determining the dollar amounts of direct materials, direct labour, and overhead used in production. The dollar amounts may be the actual amounts expended for the manufacturing inputs or they may be estimated amounts. Often, bills for overhead items arrive after the unit cost must be calculated; therefore, estimated amounts are used to ensure timeliness of cost information and to control costs.

There are two commonly used ways to *measure* the costs associated with production: *actual costing* and *normal costing*. Actual costing requires the firm to use the actual cost of all resources used in production to determine unit cost. The second method, normal costing, requires the firm to apply actual costs of direct materials and direct labour to units produced. However, overhead is applied based on a predetermined estimate. Normal costing is more widely used in practice.

Actual versus Normal Costing   An **actual cost system** uses actual costs for direct materials, direct labour, and overhead to determine unit cost. In practice, strict actual cost systems are rarely used because they cannot provide accurate unit cost information on a timely basis. Per-unit computation of the direct materials and direct labour costs is not the problem. Direct materials and direct labour can be traced to units produced. The main problem with using actual costs for calculation of unit cost is with manufacturing overhead. There are three reasons why this is so.

First, a traditional system applies overhead using unit-based drivers. However, many overhead items cannot be traced to units of production. Depreciation on plant and equipment, purchasing, and receiving are costs that are not associated with unit-based drivers. Activity-based costing is a way of overcoming this difficulty by using multiple drivers—both unit- and non-unit-based.

Second, many overhead costs are not incurred uniformly throughout the year; they can change significantly from one month to the next. For example, a factory located in northern Alberta may incur higher utilities costs in the winter as it heats the factory. Even if the factory always produces 10,000 units a month, the per-unit overhead cost in December will be higher than the per-unit overhead cost in June. As a result, one unit of product costs more in one month than another, even though the units are identical, and the production process is the same. The difference in the per-unit overhead cost is due to actual overhead costs that were incurred nonuniformly.

The third reason is that per-unit overhead costs fluctuate dramatically because of nonuniform production levels. Suppose a factory has seasonal production; it may produce 10,000 units in March, but 30,000 units in September as it gears up for the Christmas buying season. Then, if all other costs remain the same, month to month, the per-unit overhead of the product will be approximately three times as high in March as in September. Again, the units are identical, and the production process is the same.

The problem of fluctuating per-unit overhead costs can be avoided if the firm waits until the end of the year to assign the overhead costs. Unfortunately, waiting until the end of the year to determine overhead costs per unit is unacceptable. A company needs timely unit cost information throughout the year, both for interim financial statements and to help managers make decisions such as pricing. Most decisions requiring unit cost information simply cannot wait until the end of the year. Managers must react to day-to-day conditions in the marketplace in order to maintain a sound competitive position.

Normal costing solves the problems associated with actual costing. A cost system that measures overhead costs on a predetermined basis and uses actual costs for direct materials and direct labour is called a **normal costing system**. Predetermined overhead or activity rates are calculated at the beginning of the year and are used to apply overhead to production as the year goes on. Any difference between actual and applied overhead is handled as an overhead variance (overapplied–underapplied overhead).

Virtually all firms assign overhead to production on a predetermined basis. This fact seems to suggest that most firms successfully approximate the end-of-the-year overhead rate. Thus, the measurement problems associated with the use of actual overhead costs are solved by the use of estimated overhead costs. A job-order costing system that uses actual costs for direct materials and direct labour and estimated costs for overhead is called a *normal job-order costing system*.

Cost Assignment

Once costs have been accumulated and measured, they are assigned to units of product manufactured or units of service delivered. Unit costs are important for a wide variety of purposes. For example, bidding is a common requirement in markets for custom homes and industrial buildings. It is virtually impossible to submit a meaningful bid without knowing the costs associated with the units to be produced. Product cost information is vital in a number of other areas as well. Decisions concerning product design and introduction of new products are affected by expected unit costs. Decisions to make or buy a product, to accept or reject a special order, or to keep or drop a product line require unit cost information.

In its simplest form, computing the unit manufacturing or service cost is easy. The unit cost is the total product cost associated with the units produced divided by the number of units produced. For example, if a toy company manufactures 100,000 tricycles and the total cost of direct materials, direct labour, and overhead for these tricycles is $1,500,000, then the cost per tricycle is $15 ($1,500,000/100,000). Although the concept is simple, the practical reality of the computation is more complex and breaks down when there are products that differ from one another or when the company needs to know the cost of the product before all of the actual costs associated with its production are known.

Importance of Unit Costs to Manufacturing Firms   Unit cost is a critical piece of information for a manufacturer. Unit costs are essential for valuing inventory, determining income, and making a number of important decisions.

Disclosing the cost of inventories and determining income are financial reporting requirements that a firm faces at the end of each period. In order to report the cost of its inventories, a firm must know the number of units on hand and the unit cost. The cost of goods sold, used to determine income, also requires knowledge of the units sold and their unit cost.

Whether or not the unit cost information should include all manufacturing costs depends on the purpose for which the information is going to be used. For financial reporting, full or absorption unit cost information is required. If a firm is operating below its production capacity, however, variable cost information may be much more useful in a decision to accept or reject a special order. Thus, unit cost information needed for external reporting may not supply the information necessary for a number of internal decisions, especially those decisions that are short run in nature. Different costs are needed for different purposes.

Full cost information is useful as an input for a number of important internal decisions as well as for financial reporting. In the long run, for any product to be viable, its price must cover its full cost. Decisions to introduce a new product, to continue a current product, and to analyze long-run prices are examples of important internal decisions that rely on full unit cost information.

Importance of Unit Costs to Nonmanufacturing Firms   Service and nonprofit firms also require unit cost information. Conceptually, the way companies accumulate and assign costs is the same whether or not the firm is a manufacturing firm. The service firm must first identify the service "unit" being provided. In an auto repair shop, the service unit would be the work performed on an individual customer's car. Because each car is different in terms of the work required (an oil change versus a transmission overhaul, for example), the costs must be assigned individually to each job. A hospital would accumulate costs by patient, patient day, and type of procedure (e.g., X-ray, complete blood count test). A governmental agency must also identify the service provided. For example, city government might provide household trash collection and calculate the cost by truck run or by collection per house.

Service firms use cost data in much the same way that manufacturing firms do. They use costs to determine profitability, the feasibility of introducing new services, and so on. However, because service firms do not produce physical products, they do not need to value work-in-process and finished goods inventories. Of course, they may have supplies, and the inventory of supplies is simply valued at historical cost.

Nonprofit firms must track costs to be sure that they provide their services in a cost-efficient way. Governmental agencies have a fiduciary responsibility to taxpayers to use funds wisely. This requires accurate accounting for costs.

Direct materials and direct labour costs are traced to units of production. There is a clear relationship between the amount of materials and labour used and the level of production. Actual costs can be used because the actual cost of materials and labour are known reasonably well at any point in time.

Overhead is applied using a predetermined rate based on budgeted overhead costs and budgeted amount of driver. Two considerations arise. One is the choice of the activity base or driver. The other is the choice of activity base or driver for plant-wide, departmental, and activity-based overhead rates.

Choosing the Activity Level

Once the measure(s) of activity are chosen, we still need to predict the level of activity usage that applies to the coming year. Although any reasonable level of activity could be chosen, the two leading candidates are expected actual activity and normal activity. **Expected activity level** is simply the production level the firm expects to attain for the coming year. **Normal activity level** is the average activity usage that a firm experiences in the long term (normal volume is computed over more than one year).

For example, assume that Paulos Manufacturing expects to produce 18,000 units next year and has budgeted overhead for the year at $216,000. Over the past four years, Paulos Manufacturing produced the following number of units:

Year 1	22,000
Year 2	17,000
Year 3	21,000
Year 4	20,000

If expected actual capacity is used, Paulos Manufacturing will apply overhead using a predetermined rate of $12 ($216,000/18,000). However, if normal capacity is used, then the denominator of the equation for predetermined overhead is the average of the past four years of activity, or 20,000 units [(22,000 + 17,000 + 21,000 + 20,000)/4]. Then the predetermined overhead rate to be used for the coming year is $10.80 ($216,000/20,000).

Which choice is better? Of the two, normal activity has the advantage of using much the same activity level year after year. As a result, it produces less fluctuation from year to year in the assignment of per-unit overhead cost. Of course, if activity stays fairly stable, then the normal capacity level is roughly equal to the expected actual capacity level.

Other activity levels used for computing **predetermined overhead rates** are those corresponding to the theoretical and practical levels. **Theoretical activity level** is the absolute maximum production activity of a manufacturing firm. It is the output that can be realized if everything operates perfectly. **Practical activity level** is the maximum output that can be realized if everything operates efficiently. Efficient operation allows for some imperfections such as normal equipment breakdowns, some shortages, and workers operating at less than peak capability. Normal and expected actual activities tend to reflect consumer demand, while theoretical and practical activities reflect a firm's production capabilities.

OBJECTIVE ▶ 3

Identify the source documents used in job-order costing.

The Job-Order Costing Overview

As we have seen, manufacturing and service firms can be divided into two major industrial types based on the uniqueness of their product. The degree of product or service heterogeneity affects how we track costs. As a result, three different cost assignment systems have been developed: job-order costing, operation costing, and process costing.

Firms operating in job-order industries produce a wide variety of products or jobs that are usually quite distinct from one another. Customized or built-to-order products fit into this category, as do services that vary from customer to customer. Examples of job-order processes include printing, construction, furniture making, automobile repair, and beautician services. In manufacturing, a job may be a single unit such as a house, or it may be a batch of units such as eight tables. Job-order systems may be used to produce goods for inventory that are subsequently sold in the general market. Often, however, a job is associated with a particular customer order.

The key feature of job-order costing is that the cost of one job differs from that of another job and must be monitored separately.

For job-order production systems, costs are accumulated by *job*. This approach to assigning costs is called a **job-order costing system**. In a job-order firm, collecting costs by job provides vital information for management. Once a job is completed, the unit cost can be obtained by dividing the total manufacturing costs by the number of units produced. For example, if the production costs for printing 100 wedding announcements total $350, then the unit cost for this job is $3.50. The manager of the printing firm can compare the unit cost information with the prevailing market price to see if there is a reasonable profit margin. If there is not, then this may signal that the costs are out of line with other printing firms, and the manager may work to reduce costs, or alternatively, seek to emphasize other types of jobs for which the firm can earn a reasonable profit margin. In fact, the profit contributions of different printing jobs offered by the firm can be computed, and this information can then be used to select the most profitable mix of printing services to offer.

In illustrating job-order costing, we will assume a normal costing measurement approach. The actual costs of direct materials and direct labour are assigned to jobs along with overhead applied using a predetermined overhead rate. *How* these costs are actually assigned to the various jobs, however, is the central issue. In order to assign these costs, we must identify each job and the direct materials and direct labour associated with it. Additionally, some mechanism must exist to allocate overhead costs to each job.

The document that identifies each job and accumulates its manufacturing costs is the **job-order cost sheet**. An example is shown in Exhibit 4-4. The cost accounting department creates such a cost sheet upon receipt of a production order. Orders are written up in response to a specific customer order or in conjunction with a

Exhibit 4-4

The Job-Order Cost Sheet

Job Number		16	
For	Benson Company	Date Ordered	April 2, 2013
Item Description	Valves	Date Completed	April 24, 2013
Quantity Completed	100	Date Shipped	April 25, 2013

Direct Materials		Direct Labour				Overhead		
Requisition Number	Amount	Ticket Number	Hours	Rate	Amount	Hours	Rate	Amount
12	$300	68	8	$6	$ 48	8	$10	$ 80
18	450	72	10	7	70	10	10	100
	$750				$118			$180

Cost Summary

Direct materials $ 750

Direct labour   118

Overhead   180

Total cost $1,048

Unit cost $10.48

beginning of February, Bob purchases, on account, $2,500 of direct materials. This purchase is recorded as follows:

1. Materials Inventory   2,500
        Accounts Payable   2,500

Materials Inventory is an inventory account. It also is the controlling account for all raw materials. When materials are purchased, the cost of these materials "flows" into the materials inventory account.

From January 2 to January 19, the production supervisor used three requisition forms to remove $1,000 of direct materials from the storeroom. From January 20 to January 31, two additional requisition forms for $500 of direct materials were used. The first three forms revealed that the direct materials were used for Job 101; the last two requisitions were for Job 102. Thus, for January, the cost sheet for Job 101 would have a total of $1,000 in direct materials posted, and the cost sheet for Job 102 would have a total of $500 in direct materials posted. In addition, the following entry would be made:

2. Work-in-Process Inventory   1,500
        Materials Inventory   1,500

This second entry captures the flow of direct materials flowing from the storeroom to work in process. All such flows are summarized in the work-in-process inventory account and are posted individually to the respective jobs. Work-in-Process Inventory is a controlling account, and the job cost sheets are the subsidiary accounts. Exhibit 4-7 summarizes the direct materials cost flows. Notice that the source document that drives the direct materials cost flows is the materials requisition form.

Accounting for Direct Labour Cost

Since two jobs were in progress during January, time tickets filled out by direct labourers must be sorted by each job. Once the sorting is completed, the hours worked and the wage rate of each employee are used to assign the direct labour cost to each job. For Job 101, the time tickets showed 60 hours at an average wage rate of $10 per hour, for a total direct labour cost of $600. For Job 102, the total was $250, based on 25 hours at an average hourly wage of $10. In addition to the postings to each job's cost sheet, the following summary entry would be made:

3. Work-in-Process Inventory   850
        Wages Payable   850

OBJECTIVE ▶[
Describe the cost flows
job-order costing, and
journal entries.

Exhibit 4-7

Summary of Direct Materials Cost Flows

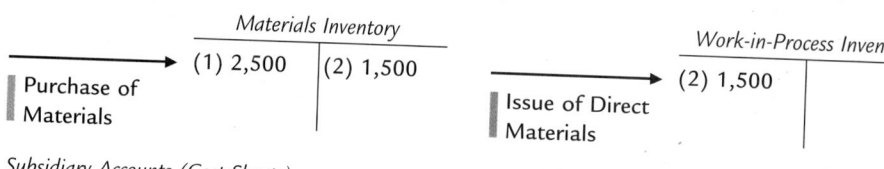

	Materials Inventory				Work-in-Process Inventory
	(1) 2,500	(2) 1,500			(2) 1,500
Purchase of Materials				Issue of Direct Materials	

Subsidiary Accounts (Cost Sheets)

Job 101 Direct Materials	
Req. No.	Amount
1	$ 300
2	200
3	500
	$1,000

Job 102 Direct Materials	
Req. No.	Amount
4	$250
5	250
	$500

Source Documents: Materials Requisitions Forms

Exhibit 4-8

Summary of Direct Labour Cost Flows

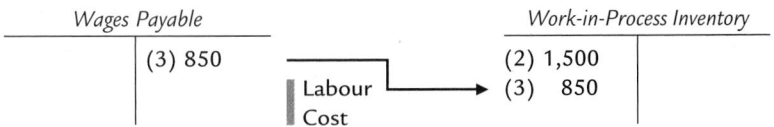

Work-in-Process Inventory Subsidiary
Accounts (Cost Sheets)

Job 101 Labour			
Ticket	Hours	Rate	Amount
1	15	$10	$150
2	20	10	200
3	25	10	250
	60		$600

Job 102 Labour			
Ticket	Hours	Rate	Amount
4	15	$10	$150
5	10	10	100
	25		$250

Source Documents: Time Tickets

The summary of the direct labour cost flows is given in Exhibit 4-8. Notice that the direct labour costs assigned to the two jobs exactly equal the total assigned to Work-in-Process Inventory. Note also that the time tickets filled out by the individual labourers are the source of information for posting the labour cost flows. Remember that the labour cost flows reflect only direct labour cost. Indirect labour is assigned as part of overhead.

Accounting for Overhead

Under a normal costing approach, actual overhead costs are *never* assigned to jobs. Overhead is applied to each individual job using a predetermined overhead rate. Recall, however, that a company must still account for *actual overhead costs* incurred. We will first describe how to account for applied overhead and then discuss accounting for actual overhead.

Accounting for Overhead Application   Assume that Bob has estimated overhead costs for the year at $9,600. Additionally, since he expects business to increase throughout the year as he becomes established, he estimates 2,400 total direct labour hours. Accordingly, the predetermined overhead rate is as follows:

$$\text{Overhead rate} = \$9,600/2,400 = \$4 \text{ per direct labour hour}$$

Overhead costs flow into Work-in-Process Inventory via the predetermined rate. Since direct labour hours are used to assign overhead into production, the time tickets serve as the source documents for assigning overhead to individual jobs and to the controlling work-in-process inventory account.

For Job 101, with a total of 60 hours worked, the amount of overhead cost posted is $240 ($4 × 60). For Job 102, the overhead cost is $100 ($4 × 25). A summary entry reflects a total of $340 (i.e., all overhead applied to jobs worked on during January) in applied overhead.

4.	Work-in-Process Inventory	340	
	Overhead Control		340

The credit balance in the overhead control account equals the total applied overhead at a given point in time. In normal costing, only applied overhead ever enters the work-in-process inventory account.

Accounting for Actual Overhead Costs

Accounting for Actual Overhead Costs   To illustrate how actual overhead costs are recorded, assume that All Signs Company incurred the following indirect costs for January:

Lease payment	$200
Utilities	50
Equipment depreciation	100
Indirect labour	65
Total overhead costs	$415

As indicated earlier, actual overhead costs never enter the work-in-process inventory account. The usual procedure is to record actual overhead costs on the debit side of the overhead control account. For example, the actual overhead costs would be recorded as follows:

5.	Overhead Control	415
	Lease Payable	200
	Utilities Payable	50
	Accumulated Depreciation—Equipment	100
	Wages Payable	65

Thus, the debit balance in Overhead Control gives the total actual overhead costs at a given point in time. Since actual overhead costs are on the debit side of this account and applied overhead costs are on the credit side, the balance in Overhead Control is the overhead variance at a given point in time. For All Signs Company at the end of January, the actual overhead of $415 and applied overhead of $340 produce underapplied overhead of $75 ($415 − $340).

The flow of overhead costs is summarized in Exhibit 4-9. To apply overhead to work-in-process inventory, a company needs information from the time tickets and a predetermined overhead rate based on direct labour hours.

Exhibit 4-9

Summary of Overhead Cost Flows

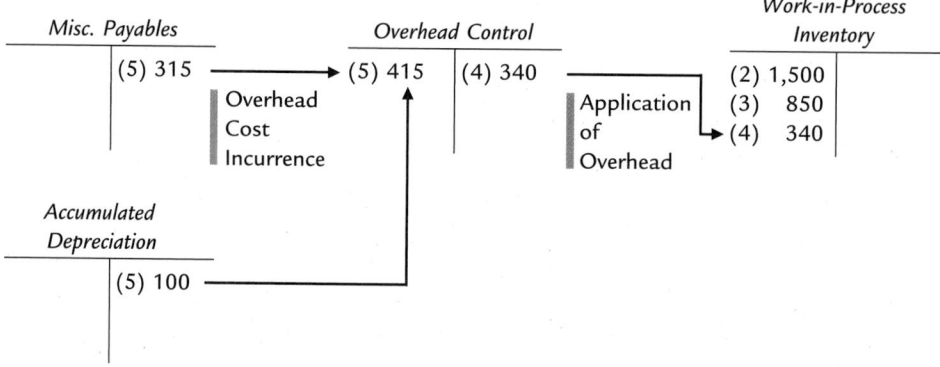

Work-in-Process Inventory Subsidiary Accounts (Cost Sheets)

Job 101
Applied Overhead

Hours	Rate	Amount
60	$4	$240

Job 102
Applied Overhead

Hours	Rate	Amount
25	$4	$100

Source Documents: Time Ticket
Other Source: Predetermined Rate

Accounting for Finished Goods Inventory

We have already seen what takes place when a job is completed. The columns for direct materials, direct labour, and applied overhead are totalled. These totals are then transferred to another section of the cost sheet, where they are summed to yield the manufacturing cost of the job. This job cost sheet is then transferred to a finished goods inventory file. Simultaneously, the costs of the completed job are transferred from the work-in-process inventory account to the finished goods inventory account.

For example, assume that Job 101 was completed in January with the completed job-order cost sheet shown in Exhibit 4-10. Since Job 101 is completed, the total manufacturing costs of $1,840 must be transferred from the work-in-process inventory account to the finished goods inventory account. This transfer is described by the following entry:

6. Finished Goods Inventory   1,840
       Work-in-Process Inventory   1,840

Exhibit 4-11 shows a summary of the cost flows that occur when a job is finished.

Completion of goods in a manufacturing process represents an important step in the flow of manufacturing costs. Because of the importance of this stage in a manufacturing operation, a schedule of the cost of goods manufactured is prepared periodically to summarize the cost flows of all production activity. This report is an important input for a firm's income statement and can be used to evaluate a firm's manufacturing effort. The statement of cost of goods manufactured was first introduced in Chapter 1. However, in a normal costing system, the report is somewhat different from the actual cost report presented in that chapter.

Exhibit 4-10

Completed Job-Order Cost Sheet

Job Number	101

For	Housing Development	Date Ordered	Jan. 1, 2013
Item Description	Street Signs	Date Started	Jan. 2, 2013
Quantity Completed	20	Date Finished	Jan. 15, 2013

Direct Materials		Direct Labour				Applied Overhead		
Requisition Number	Amount	Ticket Number	Hours	Rate	Amount	Hours	Rate	Amount
1	$ 300	1	15	$10	$150	15	$4	$ 60
2	200	2	20	10	200	20	4	80
3	500	3	25	10	250	25	4	100
	$1,000				$600			$240

Cost Summary

Direct materials	$1,000
Direct labour	600
Overhead	240
Total cost	$1,840
Unit cost	$92

Exhibit 4-11

Summary of Finished Goods Cost Flow

Work-in-Process Inventory		Finished Goods Inventory	
(2) 1,500	(6) 1,840	(6) 1,840	
(3) 850			
(4) 340			

Transfer of Finished Goods →

The statement of cost of goods manufactured presented in Exhibit 4-12 summarizes the production activity of All Signs Company for January. The key difference between this report and the one appearing in Chapter 1 is the use of applied overhead to arrive at the cost of goods manufactured. Finished goods inventories are carried at *normal cost* rather than the *actual cost*.

Notice that ending work-in-process inventory is $850. Where did we obtain this figure? Of the two jobs, Job 101 was finished and transferred to Finished Goods Inventory at a cost of $1,840. This amount is credited to Work-in-Process Inventory, leaving an ending balance of $850. Alternatively, we can add up the amounts debited to Work-in-Process Inventory for all remaining unfinished jobs. Job 102 is the only job still in process. The manufacturing costs assigned thus far are direct materials, $500; direct labour, $250; and overhead applied, $100. The total of these costs gives the cost of ending work-in-process inventory.

Accounting for Cost of Goods Sold

In a job-order firm, units can be produced for a particular customer or they can be produced with the expectation of selling the units as market conditions warrant.

Exhibit 4-12

Statement of Cost of Goods Manufactured

All Signs Company
Statement of Cost of Goods Manufactured
For the Month Ended January 31, 2013

Direct materials		
Beginning direct materials inventory	$ 0	
Add: Purchases of direct materials	2,500	
Total direct materials available	2,500	
Less: Ending direct materials	1,000	
Direct materials used		$1,500
Direct labour		850
Manufacturing overhead		
Lease	200	
Utilities	50	
Depreciation	100	
Indirect labour	65	
	415	
Less: Underapplied overhead	75	
Overhead applied		340
Current manufacturing costs		2,690
Add: Beginning work-in-process inventory		0
Less: Ending work-in-process inventory		(850)
Cost of goods manufactured		$1,840

When the job is shipped to the customer, the cost of the finished job becomes the cost of the goods sold. When Job 101 is shipped, the following entries will made. (Recall that the selling price is 150 percent of manufacturing cost.)

7a.	Cost of Goods Sold	1,840	
	Finished Goods Inventory		1,840

7b.	Accounts Receivable	2,760	
	Sales Revenue		2,760

In addition to these entries, a statement of cost of goods sold usually is prepared at the end of each reporting period (e.g., monthly and quarterly). Exhibit 4-13 presents such a statement for All Signs Company for January. Typically, the overhead variance is not material and is therefore closed to the cost of goods sold account. Cost of goods sold *before* adjustment for an overhead variance is called **normal cost of goods sold**. After adjustment for the period's overhead variance takes place, the result is called the **adjusted cost of goods sold**. It is this latter figure that appears as an expense on the income statement. Cornerstone 4-2 shows how and why the job-order cost sheet can be used in determining the ending balances of Work in Process, Finished Goods, and Cost of Goods Sold.

Closing the overhead variance to the cost of goods sold account is done once, at the end of the year. Variances are expected each month because of nonuniform production and nonuniform actual overhead costs. As the year unfolds, these monthly variances should, by and large, offset one another so that the year-end variance is small. Nonetheless, to illustrate how the year-end overhead variance would be treated, we will close out the overhead variance experienced by All Signs Company in January.

Closing the underapplied overhead to cost of goods sold requires the following entry:

8.	Cost of Goods Sold	75	
	Overhead Control		75

Notice that debiting Cost of Goods Sold is equivalent to adding the underapplied amount to the normal cost of goods sold figure. If the overhead variance is overapplied, then the entry will reverse, and Cost of Goods Sold will be credited.

If Job 101 had not been ordered by a customer but had been produced with the expectation that the signs could be sold to various other developers, then all 20 units may not be sold at the same time. Assume that on January 31, 15 signs were sold. In this case, the cost of goods sold figure is the unit cost times the number of units sold ($92 × 15, or $1,380).

Exhibit 4-13

Statement of Cost of Goods Sold

All Signs Company
Statement of Cost of Goods Sold
For the Month Ended January 31, 2013

Beginning finished goods inventory	$ 0
Cost of goods manufactured	1,840
Goods available for sale	1,840
Less: Ending finished goods inventory	0
Normal cost of goods sold	1,840
Add: Underapplied overhead	75
Adjusted cost of goods sold	$1,915

CORNERSTONE 4-2

The HOW and WHY of Using a Job-Order Cost Sheet to Determine the Balances of Work in Process, Finished Goods, and Cost of Goods Sold

Information:

(We use the same data as Cornerstone 4-1.) All-Round Fence Company installs fences for homeowners and small commercial firms. During March, All-Round worked on three jobs. Completed job-order cost sheets for March (from Cornerstone 4-1) follow:

	Job 62	Job 63	Job 64
Beginning balance	$ 620	$ 0	$ 0
Materials requisitioned	4,900	4,600	3,000
Direct labour cost	2,500	1,740	1,600
Applied overhead	1,500	1,044	960
Total cost	$9,520	$7,384	$5,560

During March, Job 62 was completed and sold at 125 percent of cost. Jobs 63 and 64 remain unfinished at the end of the month.

Why:

Since all costs are tracked by job, the balance in Work in Process can be computed by summing the costs of all completed jobs. The amount added to Finished Goods is the sum of jobs completed but not sold. Cost of Goods Sold must be the total cost of all jobs sold during the month.

Required:

1. What is the ending balance of Work in Process for March?
2. Assume that the March 1 balance of Finished Goods was zero. What is the ending balance of Finished Goods for March?
3. What is the cost of goods sold for March?
4. **What if** the March 1 balance of Finished Goods was $4,560 (consisting of Job 61)? What is the ending balance of Finished Goods in March?

Solution:

1. Since Jobs 63 and 64 are unfinished by March 31, their total cost must be the balance in Work in Process.

    March 31 Work in Process = $7,384 + $5,560 = $12,944

2. Since no jobs were completed but not sold, nothing is added to Finished Goods at the end of March. Since the beginning balance in Finished Goods was zero, then the ending balance must also be zero.

3. Cost of goods sold = Job 62 = $9,520

4. If the beginning balance of Finished Goods was $4,560, any newly completed jobs were sold by the end of March, and Job 61 was not sold during March, the ending balance would remain at $4,560.

Closing out the overhead variance to Cost of Goods Sold completes the description of manufacturing cost flows. To facilitate a review of these important concepts, Exhibit 4-14 shows a complete summary of the manufacturing cost flows for All Signs Company. Notice that these entries summarize information from the underlying job-order cost sheets. Although the description in this exhibit is specific to the

Exhibit 4-14

All Signs Company Summary of Manufacturing Cost Flows

Materials Inventory			
(1)	2,500	(2)	1,500

Wages Payable	
(3)	850

Overhead Control			
(5)	415	(4)	340
		(8)	75

Work-in-Process Inventory			
(2)	1,500	(6)	1,840
(3)	850		
(4)	340		

Finished Goods Inventory			
(6)	1,840	(7a)	1,840

Cost of Goods Sold	
(7a)	1,840
(8)	75

(1)	Purchase of direct materials	$2,500
(2)	Issue of direct materials	1,500
(3)	Incurrence of direct labour cost	850
(4)	Application of overhead	340
(5)	Incurrence of actual overhead cost	415
(6)	Transfer of Job 101 to finished goods	1,840
(7a)	Cost of goods sold of Job 101	1,840
(8)	Closing out underapplied overhead	75

example, the pattern of cost flows shown would be found in any manufacturing firm that uses a normal job-order costing system.

Manufacturing cost flows, however, are not the only cost flows experienced by a firm. Nonmanufacturing costs are also incurred. A description of how we account for these costs follows.

Accounting for Nonmanufacturing Costs

Recall that costs associated with selling and general administrative activities are classified as nonmanufacturing costs. These costs are period costs and are never assigned to the product in a traditional costing system. They are not part of the manufacturing cost flows. They do not belong to the overhead category and are treated as a totally separate category.

To illustrate how these costs are accounted for, assume that All Signs Company had the following additional transactions in January:

Advertising circulars	$ 75
Sales commission	125
Office salaries	500
Depreciation, office equipment	50

The following compound entry could be used to record the preceding costs:

Selling Expense Control	200	
Administrative Expense Control	550	
Accounts Payable		75
Wages Payable		625
Accumulated Depreciation—Office Equipment		50

Controlling accounts accumulate all of the selling and administrative expenses for a period. At the end of the period, all of these costs flow to the period's income statement. An income statement for All Signs Company is shown in Exhibit 4-15.

With the description of the accounting procedures for selling and administrative expenses completed, the basic essentials of a normal job-order costing system are also complete. This description has assumed that a single plantwide overhead rate was being used.

Exhibit 4-15

Income Statement

All Signs Company
Income Statement
For the Month Ended January 31, 2013

Sales		$2,760
Less: Cost of goods sold		1,915
Gross margin		845
Less selling and administrative expenses:		
Selling expenses	$200	
Administrative expenses	550	750
Operating income		$ 95

OBJECTIVE ▶ 5
Explain how activity-based costing is applied to job-order costing.

Job-Order Costing with Activity-Based Costing

Using a single rate based on direct labour hours to assign overhead may result in inaccurate cost assignments, in that too much or too little overhead is assigned. Departmental overhead rates and activity-based costing are suggested as ways of solving this problem. In job-order costing, departmental overhead rates and activity-based costing affect only the application of overhead. Thus, the job-order costing sheet has additional lines for overhead application, and the source documents must include all drivers for which overhead is applied. Cornerstone 4-3 shows how and why to set up a job-order cost sheet for a company using activity-based costing.

OBJECTIVE ▶ 6
Explain how spoiled units are accounted for in a job-order costing system.

Accounting for Spoiled Units in a Traditional Job-Order Costing System

Throughout this chapter, we have assumed that the units produced are good units. However, on occasion, mistakes are made; defective units are produced and are either thrown away or reworked and sold. How do we account for those costs?

Traditional job-order costing makes a distinction between normal and abnormal spoilage. **Normal spoilage** is expected due to the nature of the typical production process. This spoilage may require extra work to make the units saleable, or may result in the units being discarded. For example, from time to time maintenance workers oil the sewing machines in a jeans factory. The next pair of jeans to be sewn may pick up some drops of the machine oil. The jeans are spoiled and discarded. This is normal spoilage and the cost is included in overhead, which is then applied to all units produced. **Abnormal spoilage** is due to the exacting nature of a particular job. This type of spoilage is charged to the job that caused it. Cornerstone 4-4 tells how to treat spoilage and why the distinction between normal and abnormal spoilage is made.

The treatment of spoilage in a job-order environment is to determine whether the spoilage is normal or abnormal and charge the job if it is abnormal. Normal spoilage is considered a cost of doing business. It is subsumed in the overhead rate and spread across all jobs through applied overhead.

The HOW and WHY of Using Activity-Based Costing in Job-Order Costing

**CORNERSTONE
4-3**

Information:

Chow Company is a job-order costing firm that uses activity-based costing to apply overhead to jobs. Chow identified three overhead activities and related drivers. Budgeted information for the year is as follows:

Activity	Cost	Driver	Amount of Driver
Engineering design	$120,000	Engineering hours	3,000
Purchasing	80,000	Number of parts	10,000
Other overhead	250,000	Direct labour hours	40,000

Chow worked on four jobs in July. Data are as follows:

	Job 60	Job 61	Job 62	Job 63
Balance, July 1	$32,450	$40,770	$29,090	$ 0
Direct materials	$26,000	$37,900	$25,350	$11,000
Direct labour	$40,000	$38,500	$43,000	$20,900
Engineering hours	20	10	15	100
Number of parts	150	180	200	500
Direct labour hours	2,500	2,400	2,600	1,200

By July 31, Jobs 60 and 62 were completed and sold. The remaining jobs were in process.

Why:

ABC requires data to be collected on each activity cost and driver. Then the activity cost is assigned to each job since it is unique in its use of activity drivers.

Required:

1. Calculate the activity rates for each of the three overhead activities.
2. Prepare job-order cost sheets for each job showing all costs through July 31.
3. Calculate the balance in Work in Process on July 31.
4. Calculate cost of goods sold for July.
5. *What if* Job 61 required no engineering hours? What is the new cost of Job 61? How would the cost of the other jobs be affected?

Solution:

1. Engineering design rate = $120,000/3,000 = $40 per engineering hour

    Purchasing rate = $80,000/10,000 = $8 per part

    Other overhead = $250,000/40,000 = $6.25 per direct labour hour

2.

	Job 60	Job 61	Job 62	Job 63
Balance, July 1	$ 32,450	$ 40,770	$ 29,090	$ 0
Direct materials	26,000	37,900	25,350	11,000
Direct labour	40,000	38,500	43,000	20,900
Engineering design	800	400	600	4,000
Purchasing	1,200	1,440	1,600	4,000
Other overhead	15,625	15,000	16,250	7,500
Total cost	$116,075	$134,010	$115,890	$47,400

<table>
<tr><td>**CORNERSTONE
4-3**
(continued)</td><td>3. Work in Process = Job 61 + Job 63 = $134,010 + $47,400 = $181,410

4. Cost of goods sold = Job 60 + Job 62 = $116,075 + $115,890 = $231,965

5. If Job 61 required no engineering time, then the engineering applied to Job 61 would be zero and the cost of Job 61 would decrease by $400. The new Job 61 cost would be $133,610. The cost of the other three jobs would not be affected.</td></tr>
</table>

**CORNERSTONE
4-4**

The HOW and WHY of Accounting for Normal and Abnormal Spoilage in a Job-Order Environment

Information:
Petris Inc. manufactures cabinets on a job-order basis. Job 98-12 calls for 100 units with direct materials of $2,000 and direct labour of $1,000 ($10 per hour times 100 hours). Overhead is applied at the rate of 150 percent of direct labour dollars. At the end of the job, 100 units are produced; however, three of the cabinets required rework due to improper installation of shelving. The rework involved six extra direct labour hours and an additional $50 of material.

Why:
Normal spoilage requires that the additional cost of any rework be charged to Overhead Control. The cost of the job is not assigned any rework cost. If the spoilage is abnormal, any additional cost is assigned to the job requiring the rework.

Required:
1. Assume that the spoilage was due to assigning new, untrained workers to the job and, therefore, is normal spoilage.
   a. Calculate the cost of Job 98-12.
   b. Make any needed journal entry to the overhead control account.
2. Assume that the spoilage is abnormal and is a result of exacting specifications for this job.
   a. Calculate the cost of Job 98-12.
   b. Make any needed journal entry to the overhead control account.
3. *What if* three of the cabinets in Job 98-12 were not completely up to specifications due to unevenly applied stain? The stain could not be reworked, but the customer was willing to accept those three for a $20 per cabinet discount in the price. What is the total cost of the job? Would there be any additional entries to Overhead Control?

Solution:

1.

   a.

Job 98-12		**Overhead Control**	
Direct materials	$2,000	Direct materials	$ 50
Direct labour	1,000	Direct labour (6 × $10)	60
Overhead ($1,000 × 150%)	1,500	Overhead ($60 × 150%)	90
Total job cost	$4,500	Total	$200
÷ Units	÷ 100		
Unit cost	$ 45		

b. Since the spoilage is normal, none of the rework cost can be charged to the job, and instead must be charged (debited) to Overhead Control.

Overhead Control	110	
Materials		50
Payroll		60

CORNERSTONE 4-4
(continued)

2.

a.

Direct materials ($2,000 + $50)	$2,050
Direct labour [$1,000 + (6 × $10)]	1,060
Overhead ($1,060 × 150%)	1,590
Total job cost	$4,700
÷ Units	÷ 100
Unit cost	$ 47

b. No additional entry is needed to Overhead Control since all costs of the job are added to the job-order cost sheet and flow through Work in Process.

3. If no rework is done, then the job-order cost sheet will look like the one in Requirement 1. Total cost is $4,500, and no additional entries are made to Overhead Control. The price discount will affect the price charged; it will be lower than it otherwise would be.

Summary of Learning Objectives

1. **Differentiate the cost accounting systems of service and manufacturing firms and of unique and standardized products.**

- Manufacturing firms produce tangible products and need costs for:

  - Inventory measurement on the balance sheet (Materials, Work in Process, Finished Goods)
  - Cost of Goods Sold on the income statement

- Service firms produce intangible products with the following characteristics:

  - Intangibility
  - Inseparability
  - Heterogeneity
  - Perishability

- Uniqueness of units of service or production affect costing method.

  - Job-order costing is used for unique units with unique costs of production.
  - Operation costing is a hybrid of job-order and process costing. Batches consist of unique units.
  - Process costing is used when units are homogeneous.

2. **Discuss the interrelationship of cost accumulation, cost measurement, and cost assignment.**

- Cost accumulation is the recording of costs in the general ledger.
- Cost measurement refers to the classification and organization of costs.
- Cost assignment determines the cost of particular cost objects (such as units).

3. **Identify the source documents used in job-order costing.**

- Job-order cost sheets summarize all costs assigned to a job.
  - Subsidiary to work-in-process account
  - Includes direct materials, direct labour, and applied overhead

- Materials requisition forms record materials signed out for use on a job.
- Time tickets are used to keep track of direct labour time used on each job.
- Other source documents track the amount of activity drivers used by each job.

4. **Describe the cost flows associated with job-order costing, and prepare the journal entries.**

- Costs flow into Work in Process as debits for:
  - Direct materials (credit the materials account)
  - Direct labour (credit the payroll account)
  - Applied overhead (credit the overhead control account)

- Cost of completed jobs is:
  - Debited to Finished Goods (if inventoried) or Cost of Goods Sold (if sold immediately upon completion)
  - Credited to Work in Process

- Cost of Goods Sold is:
  - Debited to Cost of Goods Sold
  - Credited to Finished Goods (if removed from inventory) or Work in Process (if sold immediately upon completion)

- Jobs sold are:
  - Debited to Accounts Receivable or Cash
  - Credited to Sales Revenue

5. **Explain how activity-based costing is applied to job-order costing.**

- Use of activity drivers must be tracked by job.
- Activity cost is applied to each job by multiplying the activity rate by the job's use of the associated driver.

6. **Explain how spoiled units are accounted for in a job-order costing system.**

- Normal spoilage is expected.
  - Cost of normal spoilage is included in the overhead rate.
  - All units pick up the cost of normal spoilage.

- Abnormal spoilage is caused by a particular job and is charged directly to that job.

CORNERSTONES FOR CHAPTER 4

CORNERSTONE 4-1   The HOW and WHY of setting up a simplified job-order cost sheet, page 165

CORNERSTONE 4-2   The HOW and WHY of using a job-order cost sheet to determine the balances of work in process, finished goods, and cost of goods sold, page 172

CORNERSTONE 4-3   The HOW and WHY of using activity-based costing in job-order costing, page 175

CORNERSTONE 4-4   The HOW and WHY of accounting for normal and abnormal spoilage in a job-order environment, page 176

Discussion Quest

Cornerstone Exe

Cornerstone Exercise 4-1

Oliphant Company is an R
overhead for the year was
The average wage rate for di
phant Company worked on

Beginning balance
Materials requisitioned
Direct labour cost

Overhead is assigned as a pe
were completed; Job 39 wa
oped Job 40 to order for a
the chance of Oliphant beii

Review Problem

Job Cost, Applied Overhead, Unit Cost

Burnaby Company uses a normal job-order costing system. It processes most jobs through two departments. Selected budgeted and actual data for the past year follow. Data for one of several jobs completed during the year also follow.

	Department A	Department B
Budgeted overhead	$100,000	$500,000
Actual overhead	$110,000	$520,000
Expected activity (direct labour hours)	50,000	10,000
Expected machine hours	10,000	50,000

	Job 10
Direct materials	$20,000
Direct labour cost:	
Department A (5,000 hrs @ $6 per hr)	$30,000
Department B (1,000 hrs @ $6 per hr)	$6,000
Machine hours used:	
Department A	100
Department B	1,200
Units produced	10,000

Burnaby Company uses a plantwide, predetermined overhead rate to assign overhead (OH) to jobs. Direct labour hours (DLH) are used to compute the predetermined overhead rate. Burnaby prices its jobs at cost plus 30 percent.

Required:

1. Compute the predetermined overhead rate.
2. Using the predetermined rate, compute the per-unit manufacturing cost for Job 10.
3. Assume that Job 10 was completed in May and sold in September. Prepare journal entries for the completion and sale of Job 10.
4. Recalculate the unit manufacturing cost for Job 10 using departmental overhead rates. Use direct labour hours for Department A and machine hours for Department B. Does this approach provide a more accurate unit cost? Explain.
5. Assume that Job 10 was completed in May and sold in September. Using your work from Requirement 4, prepare journal entries for the completion and sale of Job 10.

Solution:

1. Predetermined overhead rate = $600,000/60,000 = $10 per DLH. Add the budgeted overhead for the two departments, and divide by the total expected direct labour hours (DLH = 50,000 + 10,000).

2.

Direct materials	$ 20,000
Direct labour	36,000
Overhead ($10 × 6,000 DLH)	60,000
Total manufacturing costs	$116,000
Unit cost ($116,000/10,000)	$ 11.60

Integrative Exercise 1

Part 1
Chapters 1–4

CableTech Bell Corporation (CTB) operates in the telecommunications industry. CTB has two divisions: the Phone Division and the Cable Service Division. The Phone Division manufactures telephones in several plants located in western Canada. The product lines run from relatively inexpensive touch-tone wall and desk phones to expensive, high-quality cellular phones. CTB also operates a cable TV service in Manitoba. The Cable Service Division offers three products: a basic package with 25 channels; an enhanced package, which is the basic package plus 15 additional channels and two movie channels; and a premium package, which is the basic package plus 25 additional channels and three movie channels.

The Cable Service Division reported the following activity for the month of March:

	Basic	Enhanced	Premium
Sales (units)	50,000	500,000	300,000
Price per unit	$16	$30	$40
Unit costs:			
Directly traced	$ 3	$ 5	$ 7
Driver traced	$ 2	$ 4	$ 6
Allocated	$10	$13	$15

The unit costs are divided as follows: 70 percent production and 30 percent marketing and customer service. Direct labour cost is the only driver used for tracing. Typically, the division uses only production costs to define unit costs. The preceding unit product cost information was provided at the request of the marketing manager and was the result of a special study.

Bryce Yee, the president of CTB, is reasonably satisfied with the performance of the Cable Service Division. March's performance is fairly typical of what has been happening over the past two years. The Phone Division, however, is another matter. Its overall profit performance has been declining. Two years ago, income before income taxes had been about 25 percent of sales. March's dismal performance was also typical for what has been happening this year and is expected to continue—unless some action by management is taken to reverse the trend. During March, the Phone Division reported the following results:

Inventories:	
Materials, March 1	$ 23,000
Materials, March 31	40,000
Work in process, March 1	130,000
Work in process, March 31	45,000
Finished goods, March 1	480,000
Finished goods, March 31	375,000
Costs:	
Direct labour	117,000
Plant and equipment depreciation	50,000
Materials handling	85,000
Inspections	60,000
Scheduling	30,000
Power	30,000
Plant supervision	12,000
Manufacturing engineering	21,000
Sales commissions	120,000
Salary, sales supervisor	10,000
Supplies	17,000
Warranty work	40,000
Rework	30,000

During March, the Phone Division purchased materials totalling $312,000. There are no significant inventories of supplies (beginning or ending). Supplies are accounted for separately from materials. CTB's Phone Division had sales totalling $1,170,000 for March.

Based on March's results, Bryce decided to meet with three of the Phone Division's managers; Violette Leduc, divisional manager; Rajnit Surachi, divisional controller; and Larry Hartley, sales manager. A transcript of their recorded conversation is given next:

Bryce: March's profit performance is down once again, and I think we need to see if we can identify the problem and correct it—before it's too late. Violette, what's your assessment of the situation?

Violette: Foreign competition is eating us alive. They are coming in with lower-priced phones of comparable or higher quality than our own. I've talked with several of the retailers that carry our lines, and they say the same. They are convinced that we can sell more if we lower our prices.

Larry: They're right. If we could lower our prices by 10 to 15 percent, I think that we'd regain most of our lost market share. But we also need to make sure that the quality of our products meets that of our competitors. As you know, we are spending a lot of money each month on rework and warranties. That worries me. I'd like to see that warranty cost cut by 70 to 80 percent. If we could do that, then customers would be more satisfied with our products, and I bet that we would not only regain our market share but increase it.

Rajnit: Lowering prices without lowering per-unit costs will not help us increase our profitability. I think we need to improve our cost accounting system. I am not confident that we really know how much each of our product lines is costing us. It may be that we are overpricing some of our units because we are overcosting them. We may be underpricing other units.

Larry: This sounds promising—especially if the overcosting is for some of our high-volume lines. A price decrease for these products would make the biggest difference—and if we knew they were overcosted, then we could offer immediate price reductions.

Bryce: Rajnit, I need more explanation. We have been using the same cost accounting system for the past 10 years. Why would it be a problem?

Rajnit: I think that our manufacturing environment has changed. Over the years, we have added a lot of different product lines. Some of these products make very different demands on our manufacturing overhead resources. We trace—or attempt to trace—overhead costs to the different products using direct labour cost, a unit-based cost driver. We may be doing more allocation than tracing. If so, then we probably don't have a very good idea of our actual product costs. Also, as you know, with the way computer technology has changed over time, it is easier and cheaper to collect and use detailed information—information that will allow us to assign costs more accurately.

Bryce: This may be something we should explore. Rajnit, what do you suggest?

Rajnit: If we want more accurate product costs and if we really want to get in the cost reduction business, then we need to understand how costs behave. In particular, we need to understand activity cost behaviour. Knowing what activities we perform, why we perform them, and how well we perform them will help us identify areas for improvement. We also need to know how the different products consume activity resources. What this boils down to is the need to use an activity-based management system. But before we jump into this, we need some idea of whether non-unit-based drivers add anything. Activity-based management is not an inexpensive undertaking. So I suggest that we do a preliminary study to see if direct labour cost is adequate for tracing. If not, then maybe some non-unit-drivers might be needed. In fact, if you would like, I can gather some data that will provide some evidence on the usefulness of the activity-based approach.

Bryce: What do you think, Violette? It's your division.

Violette: What Rajnit has said sounds promising. I think he should pursue it and do so quickly. I also think that we need to look at improving our quality. It sounds like we have a problem there. If quality could be improved, then our costs will drop. I'll talk to our quality people. Rajnit, in the meantime, find out for us if moving to an activity-based system is the way to go. How much time do you need?

Rajnit: I have already been gathering data. I could probably have a report within two weeks.

MEMO

TO: Violette Leduc
FROM: Rajnit Surachi
SUBJECT: Preliminary Analysis

Based on my initial analysis, I am confident that an ABC system will offer significant improvement. For one of our conventional phone plants, I regressed total monthly overhead cost on monthly direct labour cost using the following 15 months of data:

Overhead	Direct Labour Cost
$360,000	$110,000
300,000	100,000
350,000	90,000
400,000	100,000
320,000	90,000
380,000	100,000
300,000	90,000
280,000	90,000
340,000	95,000
410,000	115,000
375,000	100,000
360,000	85,000
340,000	85,000
330,000	90,000
300,000	80,000

The results were revealing. Although direct labour cost appears to be a driver of overhead cost, it really doesn't explain a lot of the variation. I then searched for other drivers—particularly non-unit drivers—that might offer more insight into overhead cost behaviour. Every time a batch is produced, material movement occurs, regardless of the size of the batch. The number of moves seemed like a more logical driver. I was able to gather only 10 months of data for this. (Our information system doesn't provide the number of moves, so I had to build the data set by interviewing production personnel.) This information is provided next:

Materials Handling Cost	Number of Moves
$80,000	1,500
60,000	1,000
70,000	1,250
72,000	1,300
65,000	1,100
85,000	1,700
67,000	1,200
73,500	1,350
83,000	1,400
84,000	1,700

The regression results were impressive. There is no question in my mind that the number of moves is a good driver of materials handling costs. Using the number of moves to assign materials handling costs to products would likely be better than the cost assignment using direct labour cost. Furthermore, since small batches use the same number of moves as large batches, we have some evidence that we may be overcosting our high-volume products.

I looked at one more overhead activity: inspecting products. We have 15 inspectors who are paid an average of $4,000 per month. Each inspector offers about 160 hours of inspection capacity per month. However, it appears that they actually work only about 80 percent of those hours. The drop in demand we have experienced explains this idle time. I see no evidence of variable cost behaviour here. I'm not exactly sure how to treat inspection cost, but I think that it is more related to inspection hours than direct labour cost. Some of the other overhead activities seem to be non-unit-level, as well—enough, in fact, to be concerned about how we assign costs.

After receiving the memo, Violette was intrigued. She then asked Rajnit to use the same phone plant as a pilot for a preliminary ABC analysis. She instructed him to assign all overhead costs to the plant's two products (Regular and Deluxe models), using only four activities. The four activities were rework, moving materials, inspecting products, and a general catch-all activity labelled "other manufacturing activities." From the special study already performed, she knew that materials handling and inspecting involved significant cost; from production reports, she also knew that the rework activity involved significant cost. If the ABC and unit-based cost assignments did not differ by breaking out these three major activities, then ABC may not matter.

Pursuant to the request, Rajnit produced the following cost and driver information:

Activity	Expected Cost	Driver	Activity Capacity
Other activities	$2,000,000	Direct labour dollars	$1,250,000
Moving materials	900,000	Number of moves	18,000
Inspecting	720,000	Inspection hours	24,000
Reworking	380,000	Rework hours	3,800
Total overhead cost	$4,000,000		

Expected activity demands:

	Regular Model	Deluxe Model
Units completed	100,000	40,000
Direct labour dollars	$875,000	$375,000
Number of moves	7,200	10,800
Inspection hours	6,000	18,000
Rework hours	1,900	1,900

Required:

1. Compute two different unit costs for each of the Cable Service Division's products. What managerial objectives are being served by these unit cost computations?

2. Three different cost categories are provided by the Cable Service Division: direct tracing, driver tracing, and allocation. Discuss the meaning of each. Based on how costs are assigned, do you think that the Cable Service Division is using a functional-based or an activity-based cost accounting system? What other differences exist between functional-based and activity-based cost accounting systems?

3. Discuss the differences between the Cable Service Division's products and the Phone Division's products.

4. Prepare an income statement for the Cable Service Division for March.

5. Prepare an income statement for the Phone Division for March. Include a supporting cost of goods manufactured statement.

6. The Phone Division has been using the same cost accounting system for over 10 years. Explain why its cost accounting system may be outmoded. What factors determine when a new cost accounting system is warranted?

7. Using the method of least squares, calculate two cost formulas: one for overhead using direct labour cost as the driver, and one for materials handling cost using number of moves as the driver. Comment on Rajnit's observations concerning the outcomes.

8. How would you describe the cost behaviour of the inspection activity? Assume that the quality control manager implements a program that reduces the number of defective units by 50 percent. Because of the improved quality, the demand for inspection hours will also drop by 50 percent. What is the potential monthly reduction in inspection costs? How did knowledge of inspection's cost behaviour help?

9. Calulate the overhead cost per unit for each phone model using direct labour cost to assign all overhead costs to products.

COSTING

In order for a manager to make the right decisions in a variety of circumstances, it is essential for that manager to use the proper information. If the costs of a product or service are not calculated properly, the manager can be influenced to make the wrong choices. By making the wrong choices, a manager may be putting his organization at a disadvantage relative to the competition, with very serious consequences.

The cost of a product typically consists of direct material, direct labour, and manufacturing overhead. The definitions of these elements are relatively straightforward; however, in the real world, there can be many choices in terms of how to assign these costs to the product or service being offered. As with choices in any other context, different companies will compute their costs differently, and the impact of the choices used to compute the costs will influence many aspects of the operations of a business.

The process of allocating costs will influence manpower planning decisions, pricing decisions, marketing decisions, expansion or contraction decisions, and many other decisions on a day-to-day basis. Since every organization is different and every management team has to decide how to manage the business, there are many different approaches to assigning costs to a product or a service. Part 2 of this text examines a variety of ways to determine cost, depending on the nature of the company's operations and the purpose of assigning the costs.

After studying this chapter,
you should be able to:

➤ **1** Describe the basic characteristics of process costing, including cost flows, journal entries, and the cost of production report.

➤ **2** Describe process costing for settings without work-in-process inventories.

➤ **3** Describe process costing for situations with ending work-in-process inventories.

➤ **4** Prepare a departmental production report using the FIFO method.

➤ **5** Prepare a departmental production report using the weighted average method.

➤ **6** Prepare a departmental production report with transferred-in goods and changes in output measures.

➤ **7** Describe the basic features of operation costing.

➤ **8** Explain how spoilage is treated in a process-costing system.

CHAPTER

5   Process Costing

Basic Operational and Cost Concepts

To understand a process-costing system, it is necessary to understand the underlying operational system. An operational process system is characterized by a large number of homogeneous products passing through a series of *processes*, where each process is responsible for one or more operations that bring a product one step closer to completion. Thus, a **process** is a series of activities (operations) that are linked to perform a specific objective. Valeant Pharmaceuticals International Inc. a Mississauga, Ontario-based manufacturer of a wide variety of medications for pain management, cardiovascular disease, neurology, and dermatology, uses process costing in all its plants. For example, its Montreal plant produces a pain management medication using three processes: blending, encapsulating, and bottling. Typically, the blending process consists of four linked activities: selecting, sifting, measuring, and mixing. Direct labourers select the appropriate chemicals (active and inert ingredients) and sift the materials to remove any foreign substances. Then the materials are *measured* and *combined* in a mixer to blend them thoroughly in the prescribed proportions.

In each process, materials, labour, and overhead inputs may be needed (typically in equal amounts for each unit of product). Upon completion of a particular process, the partially completed goods are transferred to another process. For example, when the mix prepared by the Blending Department is finished, the resulting mixture is sent to the encapsulating process. The encapsulating process consists of four linked activities: loading, filling, sealing, and drying. Initially, the blend and a gelatin mass are loaded into a machine. Two thin ribbons of gel are formed, one on each side of the machine. The mix is fed to a positive displacement pump, which inserts an accurate dose between the two ribbons of gel. The two ribbons are then sealed together using heat and pressure. Finally, the capsules are placed in tumble dryers and then conveyed to a drying room. Once sufficiently dry, they can be sent to bottling. The final process is bottling. It has four linked activities: loading, counting, capping, and packing. Capsules are transferred to this department, loaded into a hopper, and automatically counted into bottles. Filled bottles are mechanically capped, and direct labour then manually packs the correct number of bottles into boxes that are transferred to the warehouse. Exhibit 5-1 summarizes the operational process system for pain management manufacturing.

Exhibit 5-1

An Operational Process System: Pain Management Manufacturing

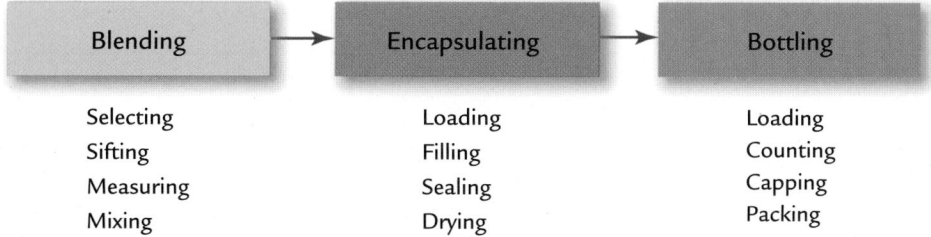

Blending	Encapsulating	Bottling
Selecting	Loading	Loading
Sifting	Filling	Counting
Measuring	Sealing	Capping
Mixing	Drying	Packing

Cost Flows

The cost flows for a process-costing system are basically similar to those of a job-order costing system. There are two key differences. First, a job-order costing system accumulates production costs by job, and a process-costing system accumulates production costs by process. Second, for manufacturing firms, the job-order costing system uses a single work-in-process (WIP) account, while the process-costing system has a WIP account for every process. Exhibit 5-2 illustrates the first key difference: the different approaches to cost accumulation. Notice that job systems assign manufacturing costs to jobs (which act as subsidiary work-in-process accounts) and transfer these costs directly to the finished goods account when the job is completed. When units are finished for a process, manufacturing costs are transferred from one process department's account to the next. A cost transferred from a prior process to a subsequent process is referred to as a **transferred-in cost**. The last process transfers the costs to Finished Goods. Cornerstone 5-1 reviews the rationale for process cost flows, shows how the cost flows are calculated (without WIP inventories), and shows how journal entries are made.

Cornerstone 5-1 illustrates that when goods are completed in one process, they are transferred with their costs to the subsequent process. Exhibit 5-3 illustrates this transfer of costs using T-accounts. For example, Blending transferred $15,000 of its costs to Encapsulating, and Encapsulating (after further processing) transferred $22,500 of costs to Bottling. These transferred-in costs are (from the viewpoint of the process receiving them) a type of direct materials cost. This is true because the subsequent process receives a partially completed unit that must be subjected to additional manufacturing activity, which includes more direct labour, more overhead, and, in some cases, additional direct materials. For example, the second journal entry for the Encapsulating Department reveals that $7,500 of additional manufacturing costs were added after

Exhibit 5-2

Comparison of Cost Accumulation Methods

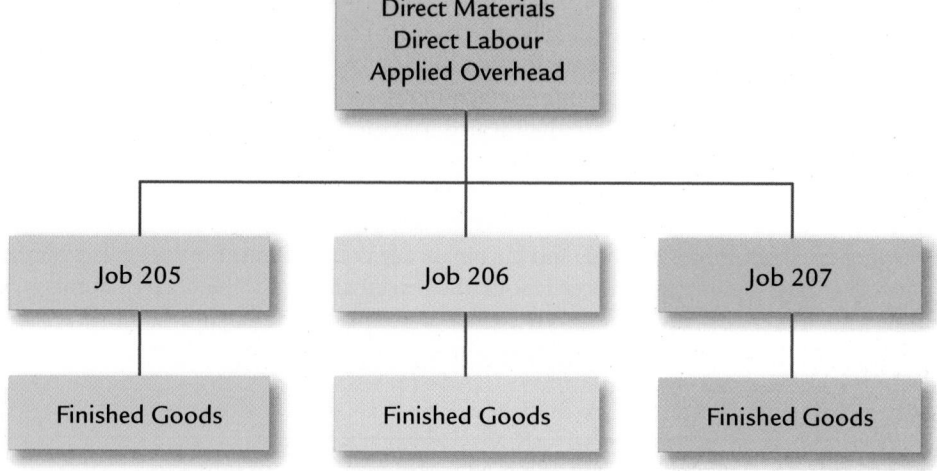

JOB-ORDER COSTING   Manufacturing Costs

PROCESS COSTING   Manufacturing Costs

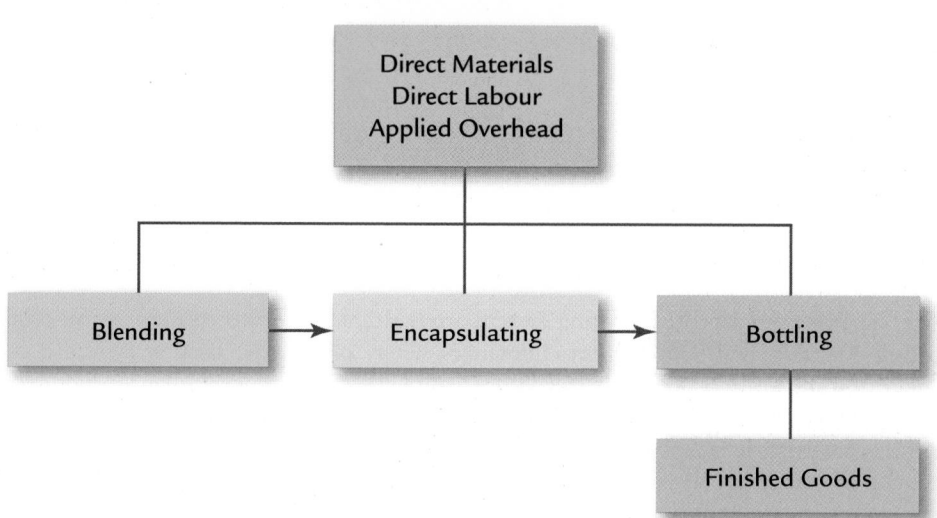

Exhibit 5-3

Process Cost Flows Illustrated Using T-Accounts: No Ending WIP

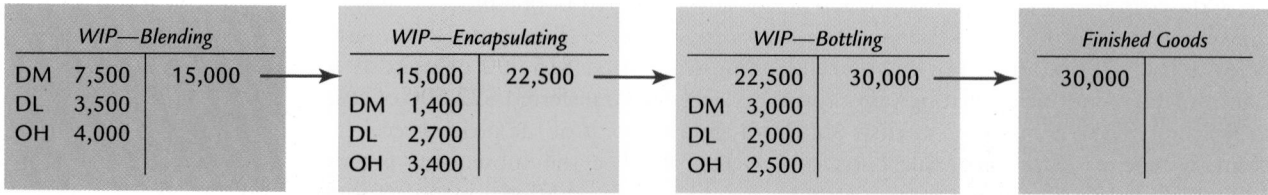

Note: DM = Direct Materials; DL = Direct Labour; OH = Overhead

The HOW and WHY of Cost Flows: Process Costing

**CORNERSTONE
5 - 1**

Information:

Rite-Way Pharmaceutical's Lethbridge plant produced 10,000 bottles of pain management medication with the following costs:

	Blending Process	Encapsulating Process	Bottling Process
Direct materials	$7,500	$1,400	$3,000
Direct labour	3,500	2,700	2,000
Applied overhead	4,000	3,400	2,500

Why:

In process costing, each department (process) accumulates its costs in a WIP account. As the work is finished in a process, the partially completed units and all their associated costs are transferred to the next process. Costs are transferred by debiting the WIP account of the process receiving the units while the WIP account of the transferring department is credited.

Required:

1. Calculate the costs transferred out of each department. Assume no WIP inventories.
2. Prepare the journal entries corresponding to these transfers. Also, prepare the journal entry for Encapsulating that reflects the costs added to the transferred-in goods received from Blending.
3. **What if** the Blending Department had an ending WIP of $5,000? Calculate the cost transferred out and provide the journal entry that would reflect this transfer. What is the effect on finished goods calculated in the first requirement, assuming the other two departments have no ending WIP?

Solution:

1.

	Blending	Encapsulating	Bottling
Direct materials	$ 7,500	$ 1,400	$ 3,000
Direct labour	3,500	2,700	2,000
Applied overhead	4,000	3,400	2,500
Costs added	15,000	7,500	7,500
Costs transferred in	0	15,000	22,500
Costs transferred out	$15,000	$22,500	$30,000

2. Transfer entries:

Work in Process—Encapsulating	15,000	
Work in Process—Blending		15,000
Work in Process—Bottling	22,500	
Work in Process—Encapsulating		22,500
Finished Goods	30,000	
Work in Process—Bottling		30,000

CORNERSTONE 5-1
(continued)

Cost-added entry (Encapsulating only):

Work in Process—Encapsulating	7,500	
Materials		1,400
Payroll		2,700
Overhead Control		3,400

3. The cost transferred out would be $10,000 ($15,000 − $5,000). The journal entry is:

Work in Process—Encapsulating	10,000	
Work in Process—Blending		10,000

Finished goods is reduced by $5,000.

receiving the transferred-in goods from Blending. Thus, while Blending sees the active and inert powders as a combination of direct materials, direct labour, and overhead costs, Encapsulating sees only the powder—a direct material, costing $15,000.

Although a process-costing system has more work-in-process accounts than a job-order costing system, it is a simpler and less expensive system to operate. In a process-costing system, there are no individual jobs, no job-order cost sheets, and no need to track materials to individual jobs. Materials are tracked to processes, but there are far fewer processes than jobs. Furthermore, there is no need to use time tickets for assigning labour costs to processes. Since labourers typically work their entire shift within a particular process, no detailed tracking of labour is needed. In fact, in many firms, labour costs are such a small percentage of total process costs that they are simply combined with overhead costs, creating a conversion cost category.

The Production Report

In process-costing systems, costs are accumulated by department for a period of time. The **production report** is the document that summarizes the manufacturing activity that takes place in a process department for a given period of time. The production report also serves as a source document for transferring costs from the work-in-process account of a prior department to the work-in-process account of a subsequent department. In the department that handles the final stage of processing, it serves as a source document for transferring costs from the work-in-process account to the finished goods account.

A production report provides information about the physical units processed in a department and also about the manufacturing costs associated with them. Thus, a production report is divided into a unit information section and a cost information section. The unit information section has two major subdivisions: (1) units to account for and (2) units accounted for. Similarly, the cost information section has two major subdivisions: (1) costs to account for and (2) costs accounted for. In summary, a production report traces the flow of units through a department, identifies the costs charged to the department, shows the computation of unit costs, and reveals the disposition of the department's costs for the reporting period. One must remember that all costs *must* be accounted for. We cannot omit any costs during the process.

Unit Costs

A key input to the cost of production report is unit costs. In principle, calculating unit costs in a process-costing system is very simple. First, measure the manufacturing costs for a process department for a given period of time. Second, measure the

Exhibit 5-4

Basic Features of a Process-Costing System

1. Homogeneous units pass through a series of similar processes.
2. Each unit in each process receives a similar dose of manufacturing costs.
3. Manufacturing costs are accumulated by a process for a given period of time.
4. There is a work-in-process account for each process.
5. Manufacturing cost flows and the associated journal entries are generally similar to job-order costing.
6. The departmental production report is the key document for tracking manufacturing activity and costs.
7. Unit costs are computed by dividing the departmental costs of the period by the output of the period.

output of the process department for the same period of time. Finally, the unit cost for a process is computed by dividing the costs of the period by the output of the period. The unit cost for the final process is the cost of the fully completed product. Exhibit 5-4 summarizes the basic features of a process-costing system.

While the basic features seem relatively simple, the actual details of process-costing systems are somewhat more complicated. A major source of difficulty is dealing with how costs and output of the period are defined when calculating the unit cost of each process. The presence of significant work-in-process inventories complicates the cost and output definitions needed for the unit cost calculation. For example, partially finished units in the beginning work-in-process inventory carry with them work and costs associated with a prior period. Yet, these units must be finished this period, so they will also have current-period costs and work associated with them. A fundamental question is how to deal with the prior-period costs and work. Another important and related complicating factor is nonuniform application of production costs, i.e., units half completed may not have half of each input needed. Much of our discussion of process-costing systems will deal with the approaches taken to deal with these complicating factors.

Process Costing with No Work-In-Process Inventories

OBJECTIVE ▶ 2
Describe process costing for settings without work-in-process inventories.

Perhaps it is best to begin with a discussion of process costing in settings where there are no work-in-process inventories. Seeing how process costing works without work-in-process inventories makes it easier to understand the procedures that are needed to deal with work-in-process inventories. Study of the no-inventory setting is also justified because many service organizations and just-in-time (JIT) manufacturing firms operate in such a setting.

Service Organizations

Services that are basically homogeneous and repetitively produced can take advantage of a process-costing approach. Processing tax returns, sorting mail by postal code, cheque processing in a bank, changing oil, air travel between Toronto and Vancouver, checking baggage, and laundering and pressing shirts are all examples of homogeneous services that are repetitively produced. Although many services consist of a single process, some services require a sequence of processes. Air travel between Toronto and Vancouver, for example, involves the following sequence of services: reservation, ticketing, baggage checking and seat confirmation, flight, and baggage delivery and pickup. Although services cannot be stored, it is possible for firms engaged in service production to have work-in-process inventories. For example, a batch of tax returns can be partially completed at the end of a period. However, many services are provided

The two methods use different total costs and different measures of output. The FIFO method is the more theoretically appealing because it divides the cost of the period by the output of the period. The weighted average method, however, merges costs in beginning work in process with current-period costs and merges the output found in beginning work in process with current-period output. This creates the possibility for errors—particularly if the weighted average method is used for situations where input costs are changing significantly from one period to the next.

In the Blending Department example, the FIFO method unit cost and the weighted average method unit cost for conversion costs are the same; evidently, the cost of this input remained the same for the two periods being considered. The unit direct materials cost for the FIFO method, however, is $0.18 versus $0.17 for the weighted average method. Apparently, the cost of direct materials has increased, and merging the lower direct materials cost of the prior period with that of the current period creates a weighted average direct materials cost that underestimates the current-period direct materials cost. The resulting difference in the cost of a fully completed unit is only $0.01 ($0.23 − $0.22). On the surface, this seems harmless.

The difference in the costs reported under each method for goods transferred out and the ending work-in-process inventories is only $300 (see Exhibits 5-5 and 5-6). This is a less than 2 percent difference for goods transferred out and only about a 5 percent difference for ending work in process. The $0.01 unit cost difference does not appear to be material. Yet, if the final product is considered, even a $0.01 difference may be significant. Recall that Rite-Way passes the powder from the Blending Department to the Encapsulating Department, where the powder is converted to capsules. Next, the capsules are sent to the Bottling Department where eight capsules are placed in small bottles. The output of the Blending Department is measured in units of 10 grams each. Suppose that 40 grams of powder convert to eight capsules. The difference in the cost of the final product would be understated by $0.04—not $0.01. Using this unit cost information may produce erroneous decisions such as under- or overpricing. Furthermore, if the other two departments also use the weighted average method, the costs in those departments could also be understated. The cumulative effect could produce a significant distortion in cost for the final product—magnifying the effect.

A second disadvantage of weighted average costing should be mentioned as well. The weighted average method also combines the performance of the current period with that of a prior period. Often, it is desirable to exercise control by comparing the actual costs of the current period with the budgeted or standard costs for the period. The weighted average method makes this comparison suspect because the performance of the current period is not independent of the prior period.

The major benefit of the weighted average method is simplicity. When all units in beginning work in process are treated as belonging to the current period, all equivalent units belong to the same time period when it comes to calculating unit costs. As a consequence, the requirements for computing unit cost are greatly simplified. Yet, as has been discussed, accuracy and performance measurement are impaired. The FIFO method overcomes both of these disadvantages. It should be mentioned, however, that both methods are widely used. Perhaps we can conclude that there are many settings in which the distortions caused by the weighted average method are not serious enough to be of concern.

OBJECTIVE ▶ 6
Prepare a departmental production report with transferred-in goods and changes in output measures.

Treatment of Transferred-In Goods

In process manufacturing, some departments invariably receive partially completed goods from prior departments. For example, under the FIFO method, the transfer of goods from Blending to Encapsulating is valued at $19,500. These transferred-in goods are a type of direct material for the subsequent process—materials that are added at the beginning of the subsequent process. The usual approach is to treat transferred-in goods as a separate material category when calculating equivalent units (the what-if question of Cornerstone 5-6 illustrates the possibility of multiple material categories). Thus, we now have three categories of manufacturing inputs: transferred-in materials, direct materials added, and conversion costs. For the Rite-Way Pharmaceuticals example, Encapsulating receives transferred-in materials, a powdered mixture, from Blending, loads the powder into gelatin capsules (a material added), seals the capsules, and dries them. The process uses labour and overhead to convert the powder into capsules.

In dealing with transferred-in goods, three important points should be remembered. First, the cost of this material is the cost of the goods transferred out computed in the prior department. Second, the units started in the subsequent department correspond to the units transferred out from the prior department, assuming that there is a one-to-one relationship between the output measures of both departments. Third, the units of the transferring department may be measured differently than the units of the receiving department. If this is the case, then the goods transferred in must be converted to the units of measure used by the second department.

To illustrate how process costing works for a department that receives transferred-in work, we will use the Encapsulating Department of Rite-Way's Lethbridge plant. The Encapsulating Department receives a powder from Blending and fills capsules with the powder. The units of the Blending Department are measured in 100 gram units, and the units of the Encapsulating Department are measured in capsules. To convert grams to capsules, we need to know the relationship between grams and capsules. Every 100 grams of the transferred-in mix converts to 4.4 capsules. Thus, to convert the transferred-in materials to the new output measure, multiply the transferred-in units by 4.4.

Now let's consider the month of May for the Lethbridge plant and focus our attention on the Encapsulating Department. We will assume that the Lethbridge plant uses the weighted average method. May's cost and production data for the Encapsulating Department are given in Exhibit 5-7. Notice that the transferred-in cost for May is the Mixing Department's transferred-out cost. (Exhibit 5-6 shows that the Mixing Department transferred out 90,000 100 gram units of powder, costing $19,800.) Also notice that output for the Encapsulating Department is measured in capsules. Given the data in Exhibit 5-7, the five steps of process costing can be illustrated for the Encapsulating Department.

Exhibit 5-7

Production and Cost Data: Encapsulating Department

Rite-Way Pharmaceuticals, Lethbridge Plant Encapsulating Department Production and Cost Data for May	
Production:	
Units in process, May 1, 80% complete[a]	24,000 (capsules)
Units completed and transferred out	375,000
Units in process, May 31, 30% complete*	45,000
Costs:	
Work in process, May 1:	
Transferred-in costs	$1,200
Direct materials (gelatin capsules)	450
Conversion costs	270
Total work in process	$1,920
Current costs:	
Transferred-in costs	$19,800
Direct materials (gelatin capsules)[b]	3,750
Conversion costs	7,500
Total current costs	$31,050

[a] With respect to conversion costs. Direct materials are 100 percent complete because they are added at the beginning of the process.
[b] The cost of capsule coating materials is insignificant and therefore added to the conversion costs category.
* With respect to conversion costs.

Step 1: Physical Flow Schedule   In constructing a physical flow schedule for the Encapsulating Department, its dependence on the Blending Department must be considered:

Units to account for:		
Units, beginning work in process		24,000
Units transferred in during May		396,000*
Total units to account for		420,000
Units accounted for:		
Units completed and transferred out:		
Started and completed	351,000	
From beginning work in process	24,000	375,000
Units, ending work in process		45,000
Total units accounted for		420,000

*90,000 × 4.4 (converts transferred-in units from 100 grams to capsules)

Step 2: Calculation of Equivalent Units   The calculation of equivalent units of production using the weighted average method is shown in Exhibit 5-8. Notice that the transferred-in goods from Blending are treated as materials added at the beginning of the process. Transferred-in materials are always 100 percent complete, since they are added at the beginning of the process.

Step 3: Computation of Unit Costs   The unit cost is computed by calculating the unit cost for each input category:

$$\text{Unit transferred-in cost} = (\$1,200 + \$19,800)/420,000 = \$0.05$$
$$\text{Unit direct materials cost} = (\$450 + \$3,750)/420,000 = \$0.01$$
$$\text{Unit conversion costs} = (\$270 + \$7,500)/388,500 = \$0.02$$
$$\text{Total unit cost} = \$0.05 + \$0.01 + \$0.02$$
$$= \$0.08$$

Step 4: Valuation of Inventories   The cost of goods transferred out is simply the total unit cost multiplied by the goods completed:

$$\text{Cost of goods transferred out} = \$0.08 \times 375,000 = \$30,000$$

Costing out ending work in process is done by computing the cost of each input and then adding to obtain the total:

Transferred-in materials: $0.05 × 45,000	$2,250
Direct materials added: $0.01 × 45,000	450
Conversion costs: $0.02 × 13,500	270
Total	$2,970

Exhibit 5-8

Equivalent Units of Production: Weighted Average Method

	Transferred-In Materials	Direct Materials Added	Conversion Costs
Units completed	375,000	375,000	375,000
Add: Units in ending work in process × Percentage complete:			
45,000 × 100%	45,000	—	—
45,000 × 100%	—	45,000	—
45,000 × 30%	—	—	13,500
Equivalent units of output	420,000	420,000	388,500

Exhibit 5-9

Production Report: Encapsulating Department

Rite-Way Pharmaceuticals, Lethbridge Plant
Encapsulating Department
Production Report for May
(Weighted Average Method)

UNIT INFORMATION

Units to account for:		Units accounted for:	
Units, beginning work in process	24,000	Units completed	375,000
Units started	396,000	Units, ending work in process	45,000
Total units to account for	420,000	Total units accounted for	420,000

	Equivalent Units		
	Transferred-In Materials	Direct Materials	Conversion Costs
Units completed	375,000	375,000	375,000
Units, ending work in process	45,000	45,000	13,500
Equivalent units of output	420,000	420,000	388,500

COST INFORMATION

	Transferred-In Materials	Direct Materials	Conversion Costs	Total
Costs to account for:				
Beginning work in process	$ 1,200	$ 450	$ 270	$ 1,920
Incurred during the period	19,800	3,750	7,500	31,050
Total costs to account for	$ 21,000	$ 4,200	$ 7,770	$32,970
Divided by equivalent units	÷420,000	÷420,000	÷388,500	
Cost per equivalent unit	$ 0.05	$ 0.01	$ 0.02	$ 0.08

Costs accounted for:			
Units transferred out ($0.08 × 375,000)			$30,000
Ending work in process:			
Transferred-in materials ($0.05 × 45,000)		$2,250	
Direct materials ($0.01 × 45,000)		450	
Conversion costs ($0.02 × 13,500)		270	2,970
Total costs accounted for			$32,970

The cost of production report for the Encapsualting Department for the month of May, including Step 5 (which was skipped), is shown in Exhibit 5-9.

The only additional complication introduced in the analysis for a subsequent department is the presence of the transferred-in category. As we have just shown, dealing with this category is similar to handling any other category. However, remember that the current cost of this special type of material is the cost of the units transferred in from the prior process and that the units transferred in are the units started (adjusted for any differences in output measurement).

Operation Costing

OBJECTIVE ▶ 7
Describe the basic features of operation costing.

Not all manufacturing firms have a pure job production environment or a pure process production environment. Some manufacturing firms have characteristics of both job and process environments. Firms in these *hybrid* settings often use *batch production processes*. **Batch production processes** produce batches of different products that are identical in many ways but differ in others. In particular, many firms produce products that make virtually the same demands on conversion inputs but different demands on direct materials inputs. Thus, the conversion activities are similar or identical, but the direct materials used are significantly different. For example, the conversion activities required to produce cans of pie filling are essentially identical for apple or cherry pie filling, but the cost of the direct materials can differ significantly. Similarly, the conversion activities for women's skirts may be identical, but the cost of direct materials can differ dramatically, depending on the nature of the fabric used (wool versus polyester, for example). Clothes, textiles, shoes, and food industries are examples where batch production may take place. For these firms, a costing system known as *operation costing* is often adopted.

Basics of Operation Costing

Operation costing is a blend of job-order and process-costing procedures applied to batches of homogeneous products. This costing system uses *job-order procedures* to assign direct materials costs to batches and *process procedures* to assign conversion costs. A hybrid costing approach is used because each batch uses different doses of direct materials but makes the same demands on the conversion resources of individual processes (usually called operations). Although different batches may pass through different operations, the demands for conversion activities for the *same* process do not differ among batches.

Work orders are used to collect production costs for each batch. Work orders also are used to initiate production. Using work orders to initiate and track costs to each batch is a job-costing characteristic. However, since individual products of different batches consume the same conversion resources as they pass through the same operation, then each product (regardless of batch membership) can be treated as a single homogeneous unit. This last trait is a process-costing characteristic and can be exploited to simplify the assignment of conversion costs.

Materials requisition forms are used to identify the direct materials, quantity and prices, and work order number. Using the materials requisition form as the source document, the cost of direct materials is posted to the work order sheet. Conversion costs are collected by *process* and assigned to products using a *predetermined conversion rate* (identical in concept to predetermined overhead rates). Conversion costs are budgeted for each department, and a single conversion rate is computed for each department (process) using a unit-based activity driver such as direct labour hours or machine hours. For example, assume that the budgeted conversion costs for a sewing operation are $100,000 (consisting of items such as direct labour, depreciation, supplies, and power), and the practical capacity of the operation is 10,000 machine hours. The conversion rate is computed as follows:

$$\text{Conversion rate} = \$100,000/10,000 \text{ machine hours}$$
$$= \$10 \text{ per machine hour}$$

Now consider two batches of shoes that pass through the sewing operation: one batch consists of 50 pairs of men's leather boots, and the second batch consists of 50 pairs of women's leather sandals. First, it should be clear that the batches have different direct material requirements so the cost of direct materials should be tracked separately (job-costing feature). Second, it should also be obvious that the sewing activity is the same for each in the sense that one hour of sewing time should consume the same resources regardless of whether the product is boots or sandals (the process-costing feature). If the batch of boots takes 25 machine hours, the batch will be assigned $250 of conversion costs ($10 × 25 hours). If the batch of sandals takes 12 machine hours, it will be assigned $120 of conversion costs ($10 × 12). Again, even though the products consume the same resources per machine hour, the batches can differ in total amount of resources consumed in an operation. So it is necessary to use a work order for each batch to collect costs.

Exhibit 5-10 illustrates the physical flow and cost flow features of operation costing. The illustration is for two batches and three processes. Panel A illustrates the physical flows, and Panel B shows the cost flows. The letters *a* and *f* represent the assignment of direct materials cost to the two batches. This example assumes that all direct materials are issued at the very beginning. Thus, direct materials cost would be assigned to the

Exhibit 5-10

Basic Features of Operation Costing

Panel A: Physical Flows

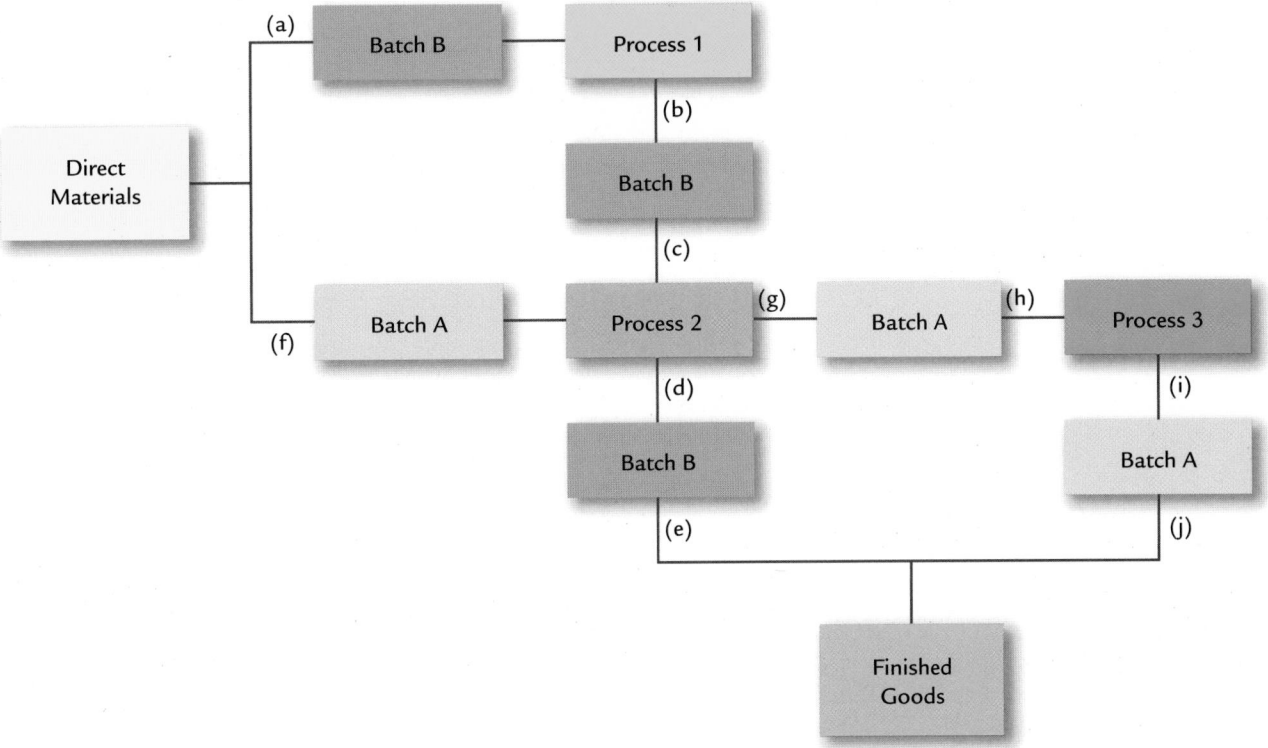

Panel B: Cost Flows (shown by letter in Panel A and in dollars below)

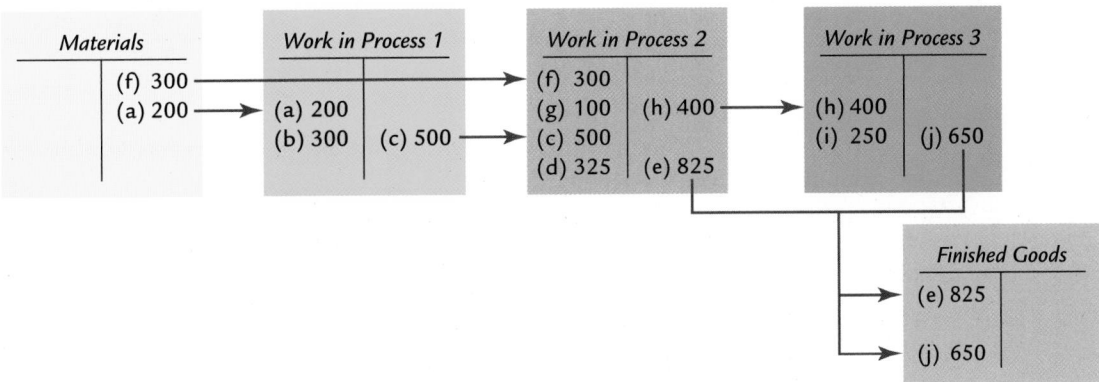

work-in-process account for the beginning process for each batch. The example also illustrates that batches do not have to participate in every process. Batch A uses Processes 2 and 3, while Batch B uses Processes 1 and 2. The letters immediately following the process represent the application of conversion costs to the respective batches.

Operation Costing Example

To illustrate operation costing, consider the Gimli plant of Rite-Way Pharmaceuticals. The Gimli plant produces a variety of vitamin and mineral products. The company produces a multivitamin and mineral product as well as single vitamin and mineral products—for example, bottles of vitamins C and E, calcium, and so on. Assume that the company also produces different strengths of vitamins (for example, 200 mg and 1,000 mg doses of vitamin C). The company also uses different sizes of bottles (for example, 60 and 120 capsules). There are four operations: blending, encapsulating, tableting, and bottling. Consider the following two work orders:

	Work Order 100	Work Order 101
Direct materials	Ascorbic acid	Vitamin E
	Capsules	Vitamin C
	Bottle (100 capsules)	Vitamin B-1
	Cap and labels	Vitamin B-2
		Vitamin B-4
		Vitamin B-12
		Biotin
		Zinc
		Bottle (60 tablets)
		Cap and labels
Operations	Blending	Blending
	Encapsulating	Tableting
	Bottling	Bottling
Number in batch	5,000 bottles	10,000 bottles

Notice how the work order specifies the direct materials needed, the operation required, and the size of the batch. Assume that the following costs are collected by work order:

	Work Order 100	Work Order 101
Direct materials	$4,000	$15,000
Conversion costs:		
Blending	1,000	3,000
Encapsulating	3,000	—
Tableting	—	4,000
Bottling	1,500	2,000
Total production costs	$9,500	$24,000

The journal entries associated with Work Order 100 are illustrated below. The first entry assumes that all materials needed for the batch are requisitioned at the start. Another possibility is to requisition the materials needed for the batch in each process as the batch enters that process.

1.	Work in Process—Blending	4,000	
	Materials		4,000
2.	Work in Process—Blending	1,000	
	Conversion Costs Applied		1,000
3.	Work in Process—Encapsulating	5,000	
	Work in Process—Blending		5,000
4.	Work in Process—Encapsulating	3,000	
	Conversion Costs Applied		3,000
5.	Work in Process—Bottling	8,000	
	Work in Process—Encapsulating		8,000
6.	Work in Process—Bottling	1,500	
	Conversion Costs Applied		1,500
7.	Finished Goods	9,500	
	Work in Process—Bottling		9,500

The journal entries for the other work order are not shown but would follow a similar pattern.

Summary of Learning Objectives

1. **Describe the basic characteristics of process costing, including cost flows, journal entries, and the cost of production report.**

- Process systems are characterized by a larger number of homogeneous products passing through a series of processes.
- Materials, labour, and overhead are applied in each process.
- Costs are accumulated by process and are transferred from one process to another by debiting the WIP of the receiving process and crediting the WIP of the transferring process.
- The production report summarizes manufacturing activity and costs for a process for a given period of time.

2. **Describe process costing for settings without work-in-process inventories.**

- No WIP inventories can occur in service organizations and JIT manufacturing firms.
- The unit cost is the costs of the period divided by the output of the period.

3. **Describe process costing for situations with ending work-in-process inventories.**

- With EWIP, output is measured using equivalent units.
- Equivalent units are the complete units that could have been produced given the total amount of effort expended.
- Five steps are followed to prepare a cost of production report:
    - Physical flow analysis
    - Equivalent unit calculation
    - Calculation of unit cost
    - Valuation of inventories
    - Cost reconciliation
- If materials are not added uniformly, multiple calculations of equivalent units are needed, one for each type of input.

4. **Prepare a departmental production report using the FIFO method.**

- The FIFO method excludes the equivalent output and costs in BWIP when the current-period unit cost is calculated.
- FIFO follows the process-costing principle.
- When calculating costs of goods transferred out, two categories of completed units are needed:
    - Units completed from BWIP
    - Units started and completed
- For the BWIP category, the cost is the sum of the prior-period cost and the current cost to complete the BWIP.

5. **Prepare a departmental production report using the weighted average method.**

- The weighted average method treats the equivalent output and costs in BWIP as if they belong to the current period when calculating unit cost.
- The costing of goods transferred out is simplified as there is only one category of completed units.

6. **Prepare a departmental production report with transferred-in goods and changes in output measures.**

- Partially completed goods (transferred-out goods) received from a prior department are transferred-in goods.

- For the receiving department, transferred-in goods are materials that are added at the beginning of the process.
- Transferred-in goods may need to be remeasured to reflect the output measure of the receiving department.

7. **Describe the basic features of operation costing.**

- Operation costing is a blend of job-order and process-costing procedures and can be used whenever batches of homogeneous products are produced.
- Job-order procedures are used to assign direct materials costs.
- Process procedures are used to assign conversion costs.

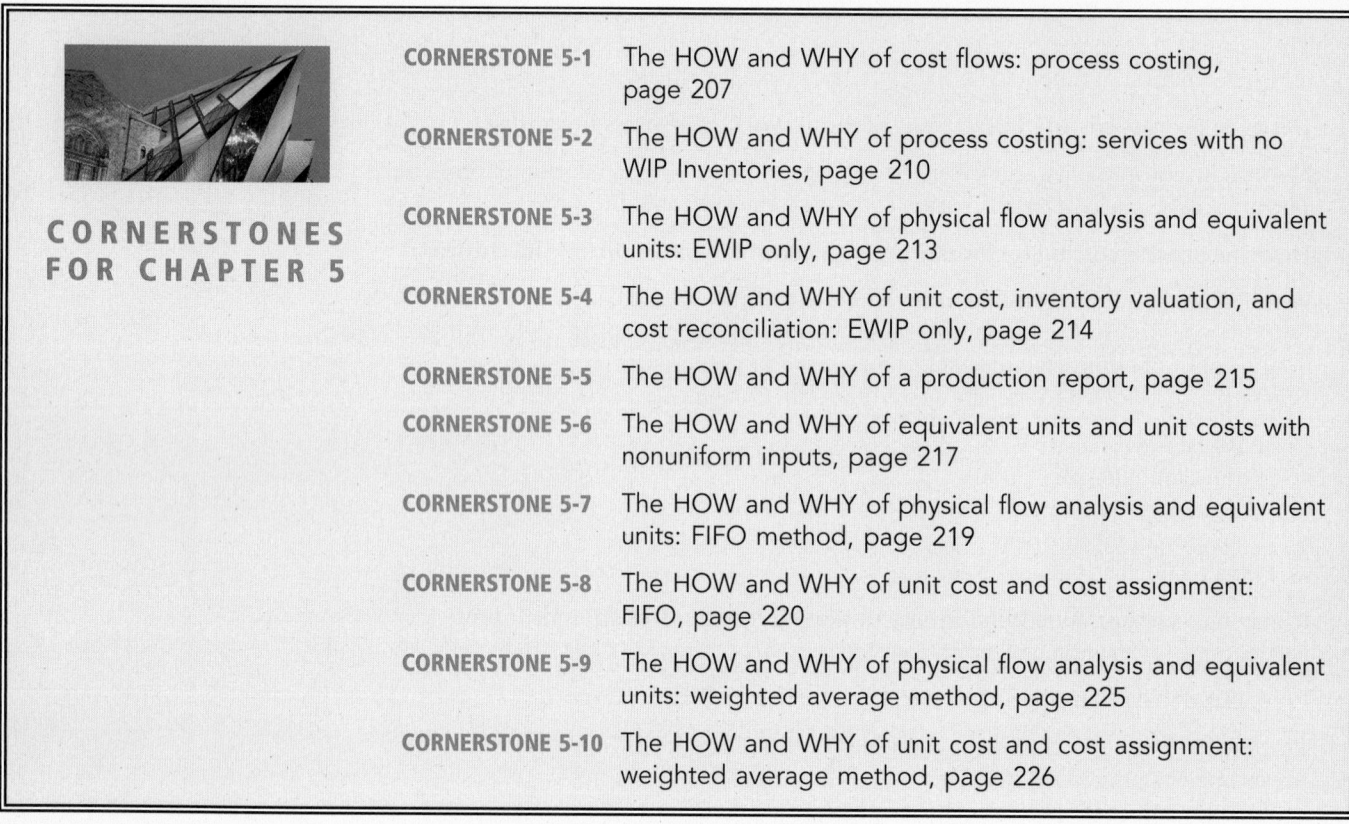

CORNERSTONES FOR CHAPTER 5

OBJECTIVE ▶ 8
Explain how spoilage is treated in a process-costing system.

Appendix: Spoiled Units

When spoilage takes place in a process-costing situation, its effects ripple through the cost of production report. Let's take Payson Company as an example. Payson Company produces a product that passes through two departments: Mixing and Cooking. In the Mixing Department, all direct materials are added at the beginning of the process. All other manufacturing inputs are added uniformly. The following information pertains to the Mixing Department for February:

a. Beginning work in process (BWIP), February 1: 100,000 kilograms, 40 percent complete with respect to conversion costs. The costs assigned to this work are as follows:

Direct materials	$20,000
Direct labour	10,000
Overhead	30,000

b. Ending work in process (EWIP), February 28: 50,000 kilograms, 60 percent complete with respect to conversion costs.

c. Units completed and transferred out: 360,000 kilograms. The following costs were added during the month:

Direct materials	$211,000
Direct labour	100,000
Overhead	270,000

d. All units are inspected at the 80 percent point of completion, and any spoiled units identified are discarded. During February, 10,000 kilograms were spoiled.

We can look at the five steps of the cost of production report. First, we must create a physical flow schedule.

Units to account for:	
Units, beginning work in process	100,000
Units started	320,000
Total units to account for	420,000
Units accounted for:	
Units transferred out	360,000
Units spoiled	10,000
Units, ending work in process	50,000
Total units accounted for	420,000

The second step is the creation of a schedule of equivalent units, shown below.

	Direct Materials	Conversion Costs
Units completed	360,000	360,000
Units spoiled × Percentage complete:		
Direct materials (10,000 × 100%)	10,000	
Conversion costs (10,000 × 80%)		8,000
Units in ending work in process		
× Percentage complete:		
Direct materials (50,000 × 100%)	50,000	—
Conversion costs (50,000 × 60%)	—	30,000
Equivalent units of output	420,000	398,000

The cost per equivalent unit is as follows:

DM unit cost ($20,000 + $211,000)/420,000	$0.55
CC unit cost ($40,000 + $370,000)/398,000	1.03*
Total cost per equivalent unit	$1.58

*Rounded.

Now we must calculate the cost of goods transferred out and the cost of ending work in process. If the spoilage is normal (expected), the cost of spoiled units is added to the cost of the good units. In this case, the inspection occurred at the 80 percent point of completion. Therefore, none of the spoiled units are from ending work in process (as these units are only 60 percent complete and have not yet been inspected). Thus, all spoilage cost is assigned to the good units transferred out.

Cost of goods transferred out:	
Good units ($1.58 × 360,000)	$568,800
Spoiled units ($0.55 × 10,000) + ($1.03 × 8,000)	13,740
	$582,540

Cost of ending work in process = ($0.55 × 50,000) + ($1.03 × 30,000)
= $58,400

Costs are reconciled as follows:

Costs to account for:	
Beginning work in process	$ 60,000
Costs added	581,000
Total costs to account for	$641,000
Costs accounted for:	
Goods transferred out	$582,540
Ending work in process	58,400
Total costs accounted for	$640,940*

*$60 difference is due to rounding.

Suppose that the spoilage was abnormal. Then the spoilage cost is assigned to a spoilage loss account. The costs are accounted for as follows:

$$\text{Cost of good units transferred out} = \$1.58 \times 360,000 = \$568,800$$
$$\text{Spoiled units} = (\$0.55 \times 10,000) + (\$1.03 \times 8,000)$$
$$= \$13,740$$
$$\text{Cost of ending work in process} = (\$0.55 \times 50,000) + (\$1.03 \times 30,000)$$
$$= \$58,400$$

Costs are reconciled as follows:

Costs to account for:	
Beginning work in process	$ 60,000
Costs added	581,000
Total costs to account for	$641,000
Costs accounted for:	
Goods transferred out	$568,800
Loss from abnormal spoilage	13,740
Ending work in process	58,400
Total costs accounted for	$640,940*

*$60 difference is due to rounding.

Notice the difference between the treatment of normal and abnormal spoilage. When spoilage is assumed to be normal, it is not tracked separately but is embedded in the total cost of good units. As a result, no one knows precisely how much spoilage adds to total manufacturing costs and whether an effort should be made to reduce it. The treatment of spoilage as abnormal is more in keeping with an emphasis on total quality management, where there is no tolerance allowed for waste. At least the product cost of spoiled goods is tracked in a separate account. Of course, a factory engaged in total quality management would not stop at classifying spoilage as abnormal. It would also identify the activities that are associated with these spoiled goods in an effort to discover the root causes of poor quality.

Review Problem

Weighted Average Method, Single Department; Equivalent Units, FIFO Method

Payson Company produces a product that passes through two departments: Mixing and Cooking. Both departments use the weighted average method. In the Mixing Department, all direct materials are added at the beginning of the process. All other manufacturing inputs are added uniformly. Payson uses the weighted average method. The following information pertains to the Mixing Department for February:

a. Beginning work in process (BWIP), February 1: 100,000 kilograms, 100 percent complete with respect to direct materials and 40 percent complete with respect to conversion costs. The costs assigned to this work are as follows:

Direct materials	$20,000
Direct labour	10,000
Overhead	30,000

b. Ending work in process (EWIP), February 28: 50,000 kilograms, 100 percent complete with respect to direct materials and 60 percent complete with respect to conversion costs.

c. Units completed and transferred out: 370,000 kilograms. The following costs were added during the month:

Direct materials	$211,000
Direct labour	100,000
Overhead	270,000

Required:

1. Prepare a physical flow schedule.
2. Prepare a schedule of equivalent units.
3. Compute the cost per equivalent unit.
4. Compute the cost of goods transferred out and the cost of ending work in process.
5. Prepare a cost reconciliation.
6. Repeat Requirements 2–4 using the FIFO method.

Solution:

1. Physical flow schedule:

Units to account for:		
Units, BWIP		100,000
Units started		320,000
Total units to account for		420,000
Units accounted for:		
Units completed and transferred out:		
Started and completed	270,000	
From BWIP	100,000	370,000
Units, EWIP		50,000
Total units accounted for		420,000

2. Schedule of equivalent units:

	Direct Materials	Conversion Costs
Units completed	370,000	370,000
Units, EWIP × Percentage complete:		
Direct materials (50,000 × 100%)	50,000	—
Conversion costs (50,000 × 60%)	—	30,000
Equivalent units of output	420,000	400,000

3. Cost per equivalent unit:

DM unit cost ($20,000 + $211,000)/420,000	$0.550
CC unit cost ($40,000 + $370,000)/400,000	1.025
Total cost per equivalent unit	$1.575

4. Cost of goods transferred out and cost of ending work in process:

Cost of goods transferred out = $1.575 × 370,000 = $582,750

Cost of EWIP = ($0.55 × 50,000) + ($1.025 × 30,000) = $58,250

5. Cost reconciliation:

Costs to account for:	
BWIP	$ 60,000
Costs added	581,000
Total costs to account for	$641,000
Costs accounted for:	
Goods transferred out	$582,750
EWIP	58,250
Total costs accounted for	$641,000

6. FIFO results:

Schedule of equivalent units:

	Direct Materials	Conversion Costs
Units started and completed	270,000	270,000
Units, BWIP × Percentage to complete:	—	60,000
Units, EWIP × Percentage complete:		
Direct materials (50,000 × 100%)	50,000	—
Conversion costs (50,000 × 60%)	—	30,000
Equivalent units of output	320,000	360,000

Cost per equivalent unit:

DM unit cost ($211,000/320,000)	$0.659*
CC unit cost ($370,000/360,000)	1.028*
Total cost per equivalent unit	$1.687

*Rounded.

Cost of goods transferred out and cost of ending work in process:

$$\text{Cost of goods transferred out} = (\$1.687 \times 270{,}000) + (\$1.028 \times 60{,}000)$$
$$+ \$60{,}000$$
$$= \$577{,}170^*$$
$$\text{Cost of EWIP} = (\$0.659 \times 50{,}000) + (\$1.028 \times 30{,}000) = \$63{,}790^*$$

*Difference of $40 in total costs due to rounding

Key Terms

Batch production processes, 231

Cost reconciliation, 212

Equivalent units of output, 211

FIFO costing method, 216

Operation costing, 232

Physical flow schedule, 212

Process, 204

Process-costing principle, 210

Production report, 208

Transferred-in cost, 205

Weighted average costing method, 222

Work orders, 232

Discussion Questions

1. What is a process? Provide an example that illustrates the definition.
2. Describe the differences between process costing and job-order costing.
3. What journal entry would be made as goods are transferred out from one department to another department? From the final department to the warehouse?
4. What are transferred-in costs?
5. Explain why transferred-in costs are a special type of material for the receiving department.
6. What is a production report? What purpose does this report serve?
7. Can process costing be used for a service organization? Explain. Describe how process costing can be used for JIT manufacturing firms.
8. What are equivalent units? Why are they needed in a process-costing system?
9. How is the equivalent unit calculation affected when direct materials are added at the beginning or end of the process rather than uniformly throughout the process?
10. Describe the five steps in accounting for the manufacturing activity of a processing department, and indicate how they interrelate.
11. Under the weighted average method, how are prior-period costs and output treated? How are they treated under the FIFO method?
12. Under what conditions will the weighted average and FIFO methods give essentially the same results?
13. In assigning costs to goods transferred out, how do the weighted average and FIFO methods differ?
14. How are transferred-in costs treated in the calculation of equivalent units?
15. What is operation costing? When is it used?

Cornerstone Exercises

Cornerstone Exercise 5-1 COST FLOWS

OBJECTIVE ▶ 1
CORNERSTONE 5-1

Beauchemin Company produced 50,000 metal components for tractors. There were no beginning or ending work-in-process inventories in any department. Beauchemin incurred the following costs for October:

	Moulding Department	Grinding Department	Finishing Department
Direct materials	$6,500	$2,600	$4,000
Direct labour	5,000	4,400	6,000
Applied overhead	8,500	7,000	5,500

Required:

1. Calculate the costs transferred out of each department.
2. Prepare the journal entries corresponding to these transfers. Also, prepare the journal entry for Grinding that reflects the costs added to the transferred-in goods received from Moulding.
3. What if the Grinding Department had an ending WIP of $6,000? Calculate the cost transferred out and provide the journal entry that would reflect this transfer. What is the effect on finished goods calculated in Requirement 1, assuming the other two departments have no ending WIP?

Cornerstone Exercise 5-2 UNIT COST, NO WORK-IN-PROCESS INVENTORIES

OBJECTIVE ▶ 2
CORNERSTONE 5-2
SERVICE

Goldman Dentistry has a hygienist that performs cleanings for its patients. During May, Goldman had the following cost and output information:

Direct materials	$800
Hygienist's salary	$4,450
Overhead	$6,750
Number of cleanings	600

Required:

1. Calculate the cost per cleaning for May.
2. Calculate the cost of services sold for May.
3. What if Goldman found a way to reduce overhead costs by 40 percent? How would this affect the profit per cleaning?

OBJECTIVE ➤ 3

CORNERSTONE 5-3

SERVICE

Cornerstone Exercise 5-3 PHYSICAL FLOW AND EQUIVALENT UNITS WITH EWIP

Fleming, Fleming, and Chan, a local CA firm, provided the following data for individual tax returns processed for March (output is measured in number of returns):

Units, beginning work in process	—
Units started	6,000
Units completed	5,000
Units, ending work in process (50% complete)	1,000
Total production costs	$5,500

Required:

1. Prepare a physical flow schedule
2. Prepare an equivalent units schedule. Explain why output is measured in equivalent units.
3. *What if* EWIP is 80 percent complete? How would this change affect the physical flow schedule? The equivalent units schedule?

OBJECTIVE ➤ 3

CORNERSTONE 5-4

SERVICE

Cornerstone Exercise 5-4 COST INFORMATION

During June, Herring Associates incurred total production costs of $40,000 for copy editing manuscripts and had the following equivalent units schedule:

Units completed	190
Units in EWIP × Fraction complete:	
25 × 0.40	10
Equivalent units	200

Required:

1. Calculate the cost of copy editing one manuscript for June.
2. Assign costs to manuscripts completed and to EWIP and then do a cost reconciliation.
3. *What if* the costs assigned to manuscripts completed and EWIP were calculated using a unit cost of $250? What is the discrepancy between the costs assigned and the costs to account for? What could have caused an incorrect unit cost?

OBJECTIVE ➤ 3

CORNERSTONE 5-5

Cornerstone Exercise 5-5 PRODUCTION REPORT

Edge Company produces power drinks. The Mixing Department, the first process department, mixes the ingredients required for the drinks. The following data are for April:

Work in process, April 1	—
Litres started	30,000
Litres transferred out	25,000
Litres in EWIP	5,000
Direct materials cost	$28,000
Direct labour cost	$56,000
Overhead applied	$112,000

Direct materials are added at the beginning of the process. Ending inventory is 60 percent complete with respect to direct labour and overhead.

Required:

1. Why would a manager want a production report?
2. Prepare a production report for the Mixing Department for April.

Cornerstone Exercise 5-6 NONUNIFORM INPUTS

OBJECTIVE ➤ 3
CORNERSTONE 5-6

Dulce Company produces premium chocolate candy bars. Conversion costs are added uniformly. For March, EWIP is 40 percent complete with respect to conversion costs. The following information is provided for March:

Physical flow schedule:

Units to account for:		
Units in BWIP		0
Units started		50,000
Total units to account for		50,000
Units accounted for:		
Units completed:		
From BWIP	0	
Started and completed	40,000	40,000
Units in EWIP		10,000
Total units accounted for		50,000

Inputs	
Direct Materials	**Conversion Costs**
$25,000	$44,000

Required:

1. Calculate the equivalent units for each input category.
2. Calculate the unit cost for each category and in total.
3. *What if* a different type of material is *also* added at the end of the process (a candy wrapper), costing $4,000? Calculate the new unit cost.

Cornerstone Exercise 5-7 UNIT INFORMATION WITH BWIP, FIFO METHOD

OBJECTIVE ➤ 4
CORNERSTONE 5-7

Fedoruk Products produces a barbeque sauce using three departments: Cooking, Mixing, and Bottling. In the Cooking Department, all materials are added at the *beginning* of the process. Output is measured in 100-gram units. The production data for July are as follows:

Production:	
Units in process, July 1, 60% complete*	10,000
Units completed and transferred out	80,000
Units in process, July 31, 80% complete*	15,000

*With respect to conversion costs.

Required:

1. Prepare a physical flow schedule for July.
2. Prepare an equivalent units schedule for July using the FIFO method.
3. *What if* 60 percent of the materials were added at the beginning of the process and 40 percent were added at the end of the process (all ingredients used are treated as the same type or category of materials)? How many equivalent units of materials would there be?

Cornerstone Exercise 5-8 COST INFORMATION AND FIFO

OBJECTIVE ➤ 4
CORNERSTONE 5-8

Hassan Company had the equivalent units schedule and cost information for its Sewing Department for the month of December shown below.

Required:

1. Calculate the unit cost for December, using the FIFO method.
2. Calculate the cost of goods transferred out, calculate the cost of EWIP, and reconcile the costs assigned with the costs to account for.
3. *What if* you were asked for the unit cost from the month of November? Calculate November's unit cost and explain why this might be of interest to management.

	Direct Materials	Conversion Costs
Units started and completed	40,000	40,000
Add: Units in beginning work in process ×		
Percentage to complete:		
5,000 × 0% direct materials	—	—
5,000 × 50% conversion costs	—	2,500
Add: Units in ending work in process ×		
Percentage complete:		
10,000 × 100% direct materials	10,000	—
10,000 × 35% conversion costs	—	3,500
Equivalent units of output	50,000	46,000
Costs:		
Work in process, December 1:		
Direct materials		$ 35,000
Conversion costs		10,000
Total work in process		$ 45,000
Current costs:		
Direct materials		$400,000
Conversion costs		184,000
Total current costs		$584,000

OBJECTIVE ▶ 5
CORNERSTONE 5-9

Cornerstone Exercise 5-9 UNIT INFORMATION WITH BWIP, WEIGHTED AVERAGE METHOD

Fedoruk Products produces a barbeque sauce using three departments: Cooking, Mixing, and Bottling. In the Cooking Department, all materials are added at the *beginning* of the process. Output is measured in 100-gram units. The production data for July are as follows:

Production:	
Units in process, July 1, 60% complete*	10,000
Units completed and transferred out	80,000
Units in process, July 31, 80% complete*	15,000

*With respect to conversion costs.

Required:

1. Prepare a physical flow schedule for July.
2. Prepare an equivalent units schedule for July using the weighted average method.
3. *What if* you were asked to calculate the FIFO equivalent units beginning with the weighted average equivalent units? Calculate the FIFO equivalent units by subtracting out the prior-period output found in BWIP.

OBJECTIVE ▶ 5
CORNERSTONE 5-10

Cornerstone Exercise 5-10 COST INFORMATION AND THE WEIGHTED AVERAGE METHOD

Morrison Company had the equivalent units schedule and cost information for its Sewing Department for the month of December shown below.

Required:

1. Calculate the unit cost for December, using the weighted average method.
2. Calculate the cost of goods transferred out, calculate the cost of EWIP, and reconcile the costs assigned with the costs to account for.
3. *What if* you were asked to show that the weighted average unit cost for materials is the blend of the November unit materials cost and the December unit materials cost? The November unit materials cost is $6.60 ($66,000/10,000), and the December unit materials cost is $12.22 ($550,000/45,000). The equivalent units in BWIP are 10,000, and the FIFO equivalent units are 45,000. Calculate the weighted average unit materials cost using weights defined as the proportion of total units completed from each source (BWIP output and current output).

	Direct Materials	Conversion Costs
Units completed	45,000	45,000
Add: Units in ending work in process ×		
Percentage complete:		
10,000 × 100% direct materials	10,000	—
10,000 × 45% conversion costs	—	4,500
Equivalent units of output	55,000	49,500
Costs:		
Work in process, December 1:		
Direct materials		$ 66,000
Conversion costs		14,000
Total work in process		$ 80,000
Current costs:		
Direct materials		$550,000
Conversion costs		184,000
Total current costs		$734,000

Exercises

Exercise 5-11 JOURNAL ENTRIES

K-Briggs Company has three process departments: Mixing, Encapsulating, and Bottling. At the beginning of the year, there were no work-in-process or finished goods inventories. The following data are available for the month of July:

Department	Manufacturing Costs Added*	Ending Work in Process
Mixing	$216,000	$54,000
Encapsulating	198,000	45,000
Bottling	180,000	9,000

*Includes only the direct materials, direct labour, and the overhead used to process the partially finished goods received from the prior department. The transferred-in cost is not included.

Required:

1. Prepare journal entries that show the transfer of costs from one department to the next (including the entry to transfer the costs of the final department).
2. Prepare T-accounts for the entries made in Requirement 1. Use arrows to show the flow of costs.

Exercise 5-12 PROCESS COSTING, SERVICE ORGANIZATION

A local barbershop cuts the hair of 1,000 customers per month. The clients are men, and the barbers offer no special styling. During the month of March, 1,000 customers were serviced. The cost of haircuts includes the following:

Direct labour	$ 7,000
Direct materials	1,000
Overhead	2,000
Total	$10,000

Required:

1. Explain why process costing is appropriate for this haircutting operation.
2. Calculate the cost per haircut.
3. Can you identify some possible direct materials used for this haircutting service? Is the usage of direct materials typical of services? If so, provide examples of services that use direct materials. Can you think of some services that would not use direct materials?

OBJECTIVE ➤ 1 2   Exercise 5-13 **JIT MANUFACTURING AND PROCESS COSTING**

Romano Company uses JIT manufacturing. There are several manufacturing cells set up within one of its factories. One of the cells makes stands for flat-screen televisons. The cost of production for the month of April is given below.

Cell labour	$ 40,000
Direct materials	100,000
Overhead	80,000
Total	$220,000

During May, 20,000 stands were produced and sold.

Required:

1. Explain why process costing can be used for computing the cost of production for the stands.
2. Calculate the cost per unit for a stand.
3. Explain how activity-based costing can be used to determine the overhead assigned to the cell.

OBJECTIVE ➤ 2 3   Exercise 5-14 **PHYSICAL FLOW, EQUIVALENT UNITS, UNIT COSTS, NO BEGINNING WIP INVENTORY, ACTIVITY-BASED COSTING**

Arnez Inc. produces a subassembly used in the production of hydraulic cylinders. The subassemblies are produced in three departments: Plate Cutting, Rod Cutting, and Welding. Overhead is applied using the following drivers and activity rates:

Driver	Rate	Actual Usage (by Plate Cutting)
Direct labour cost	75% of direct labour	$732,000
Inspection hours	$20 per hour	7,450 hours
Purchase orders	$500 per order	800 orders

Other data for the Plate Cutting Department are as follows:

Beginning work in process	—
Units started	740,000
Direct materials cost	$3,700,000
Units, ending work in process (100% materials; 80% conversion)	40,000

Required:

1. Prepare a physical flow schedule.
2. Calculate equivalent units of production for:
   a. Direct materials
   b. Conversion costs
3. Calculate unit costs for:
   a. Direct materials
   b. Conversion costs
   c. Total manufacturing
4. Provide the following information:
   a. The total cost of units transferred out
   b. The journal entry for transferring costs from Plate Cutting to Welding
   c. The cost assigned to units in ending inventory

OBJECTIVE ➤ 1 3   Exercise 5-15 **PRODUCTION REPORT, NO BEGINNING INVENTORY**

Piel Suave Company manufactures sun protection lotion. The Mixing Department, the first process department, mixes the chemicals required for the lotion. The following data are for 2013:

Required:

1. Prepare a physical flo~
2. Compute the cost pe~
3. Determine the cost o~
4. Prepare the journal e~

Exercise 5-22 WEIGHT~
COST, MULTIPLE DEP~

Layton Company has a~
Polishing. During Decer~
the Polishing Departmen~
was $40,000. Direct ma~
measured the same way in~

The second departm~
December:

Units to acco~	
Units, beg~	
Units start~	
Total units~	
Units accoun~	
Units, end~	
Units com~	
Units acco~	

Costs in beginning work~
$5,000; conversion costs~
month: direct materials, $~

Required:

1. Assuming the use of t~
2. Compute the unit co~

Exercise 5-23 FIFO M~
DEPARTMENTS

Using the same data foun~

Required:

Prepare a schedule of e~
December.

Exercise 5-24 JOURN~

Baxter Company has two~
mined overhead rate of~
company experienced th~

a. Materials issued to A~
b. Direct labour cost: A~
   $8 per hour
c. Overhead applied to~
d. Goods transferred t~
e. Goods transferred t~
f. Actual overhead inc~

Required:

1. Prepare the required~
2. Assuming Assembly~
   determine the cost c~

Work in process, January 1, 2013	—
Units started	900,000
Units transferred out	756,000
Direct materials cost	$900,000
Direct labour cost	$1,785,600
Overhead applied	$2,678,400

Direct materials are added at the beginning of the process. Ending inventory is 95 per-cent complete with respect to direct labour and overhead.

Required:

Prepare a production report for the Mixing Department for 2013.

Exercise 5-16 WEIGHTED AVERAGE METHOD, FIFO METHOD, PHYSICAL FLOW, EQUIVALENT UNITS

 OBJECTIVE ➤ 3 4 5

Darim Company manufactures a product that passes through two processes: Fabrication and Assembly. The following information was obtained for the Fabrication Department for June:

a. All materials are added at the beginning of the process.
b. Beginning work in process had 60,000 units, 30 percent complete with respect to conversion costs.
c. Ending work in process had 12,000 units, 25 percent complete with respect to conversion costs.
d. Started in process, 75,000 units.

Required:

1. Prepare a physical flow schedule.
2. Compute equivalent units using the weighted average method.
3. Compute equivalent units using the FIFO method.

Exercise 5-17 FIFO METHOD, VALUATION OF GOODS TRANSFERRED OUT AND ENDING WORK IN PROCESS

OBJECTIVE ➤ 4

Alden Company uses the FIFO method to account for the costs of production. For Crushing, the first processing department, the following equivalent units schedule has been prepared:

	Direct Materials	Conversion Costs
Units started and completed	22,000	22,000
Units, beginning work in process:		
10,000 × 0%	—	—
10,000 × 40%	—	4,000
Units, ending work in process:		
6,000 × 100%	6,000	—
6,000 × 75%	—	4,500
Equivalent units of output	28,000	30,500

The cost per equivalent unit for the period was as follows:

Direct materials	$3.00
Conversion costs	5.00
Total	$8.00

The cost of beginning work in process was direct materials, $30,000; conversion costs, $25,000.

Required:

1. Determine the cost of ending work in process and the cost of goods transferred out.
2. Prepare a physical flow schedule.

OBJECT

OBJECT

OBJECT

OBJECT

After studying this chapter, you should be able to:

➤ **1** Describe the basics of plantwide and departmental overhead costing.

➤ **2** Explain why plantwide and departmental overhead costing may not be accurate.

➤ **3** Provide a detailed description of activity-based product costing.

➤ **4** Explain how ABC can be simplified.

CHAPTER

6 Activity-Based Costing

In Chapter 1, we mentioned that cost management information systems can be divided into two types: unit-based and activity-based. The unit-based costing systems use only unit-based activity drivers to assign overhead to products. This chapter begins by describing how unit-based costing is used for computing traditional product costs. This enables us to compare and contrast unit-based and activity-based costing approaches. An activity-based cost accounting system offers greater product costing accuracy but at an increased cost. The justification for adopting an activity-based costing approach must rely on the benefits of improved decision making. It is important to understand that a necessary condition for improved decisions is that the accounting numbers produced by an activity-based costing system be significantly different from those produced by a unit-based costing system. When will this be the case? Are there any signals that indicate unit-based costing is no longer working? Finally, assuming that an activity-based cost accounting system is called for, how does it work? What are its features? What steps must be followed for successful implementation of an ABC system? This chapter addresses these questions and other related issues.

Unit-Based Product Costing

Unit-based product costing assigns only manufacturing costs to products. Assigning the cost of direct materials and direct labour to products poses no particular challenge. These costs can be assigned to products using direct tracing, and most unit-based costing systems are designed to ensure that this tracing takes place. Overhead costs, on the other hand, pose a different problem. The physically observable input–output relationship that exists between direct labour, direct materials, and products is simply not available for overhead. Thus, assignment of overhead must rely on a predetermined overhead rate based on unit-based drivers.

A **predetermined overhead rate** is calculated at the beginning of the year using the following formula:

Overhead rate = Budgeted annual overhead/Budgeted annual driver units

Predetermined overhead rates are used because overhead and production often are incurred nonuniformly throughout the year, and it is not possible to wait until the end of the year to calculate the actual overhead cost assignments (managers need unit product cost information throughout the year). A cost system that uses predetermined overhead rates for overhead and actual costs for direct materials and direct labour is referred to as a **normal costing system**. Budgeted overhead is simply the firm's best estimate of the amount of overhead (utilities, indirect labour, depreciation, etc.) to be incurred in the coming year. The estimate is often based on last year's figures, adjusted for anticipated changes in the coming year. Budgeted annual driver units represent the predicted activity driver units. Assignment of overhead costs should follow, as nearly as possible, a cause-and-effect relationship, and the *unit-based drivers* used simply measure the consumption of overhead by products. The five most commonly used unit-level drivers are:

1. Units produced
2. Direct labour hours
3. Direct labour dollars
4. Direct machine hours
5. Direct material dollars

The use of only unit-based drivers to assign overhead costs to products assumes that all overhead consumed by products is highly correlated with the number of units produced. Thus, the higher the units of product produced, the higher the unit-level drivers consumed. To the extent that this assumption is true, unit-based costing can produce accurate cost assignments.

Plantwide or departmental predetermined overhead rates are used to assign or apply overhead costs to production according to the actual production activity. The total overhead assigned to actual production at any point in time is called **applied overhead**. Applied overhead is computed using the following formula:

Applied overhead = Overhead rate × Actual driver usage

Once the applied overhead is assigned, the unit cost is calculated by dividing the total applied overhead by the units produced.

Overhead Application: Plantwide Rates

For plantwide rates, all budgeted overhead costs are assigned to a single plantwide pool. Next, a plantwide rate is computed using a single unit-level driver, which is usually direct labour hours. Finally, overhead costs are assigned to products by multiplying the rate by the actual total direct labour hours used by each product (second-stage assignment). The corresponding calculations and their rationales are illustrated in Cornerstone 6-1.

**CORNERSTONE
6-1**

The HOW and WHY of Applied Overhead and Unit Overhead Cost: Plantwide Rates

Information:

PlayFun Inc. produces two types of battery-operated toys: robots and race cars. Its plant uses a plantwide rate based on direct labour hours to assign its overhead costs. The company has the following estimated and actual data for the coming year:

Estimated overhead	$350,000
Expected activity	50,000
Actual activity (direct labour hours):	
Robots	10,000
Race cars	40,000
Units produced:	
Robots	50,000
Race cars	250,000

Why:

Product cost information is needed for such things as financial statement preparation, pricing decisions, and keep-or-drop decisions. Predetermined overhead rates (based on expected overhead and expected activity) are used because overhead and production are incurred nonuniformly and managers cannot wait until the end of the year to obtain product cost information. A plantwide rate is used under the assumption that all overhead costs are largely caused by a single, unit-level cost driver such as direct labour hours or machine hours.

Required:

1. Calculate the predetermined plantwide overhead rate and the applied overhead for each product, using direct labour hours.
2. Calculate the overhead cost per unit for each product.
3. *What if* robots used 5,000 hours (to produce 50,000 units) instead of 10,000 hours? Calculate the effect on the profitability of this product line if all 50,000 units are sold, and then discuss the implications of this outcome.

Solution:

1. Plantwide rate = $350,000/50,000 = $7.00 per hour

   Applied overhead:

	Robots	Race Cars
$7.00 × 10,000	$70,000	
$7.00 × 40,000		$280,000

2. Overhead per unit (robots) = $70,000/50,000 = $1.40
   Overhead per unit (race cars) = $280,000/250,000 = $1.12

3. There would be a reduction of $35,000 ($7.00 × 5,000) of overhead assigned to the robots and so profitability for this product line would increase by this amount. Overhead assignments affect product cost and profitability and thus can affect many decisions (e.g., pricing). This conclusion, in turn, implies that the way overhead is assigned is important.

Calculation and Disposition of Overhead Variances

From Cornerstone 6-1, the initial calculation of applied overhead is $350,000. It is possible (and likely) that the applied amount in a period differs from the actual overhead incurred for the period. Since the predetermined overhead rate is based on estimated data, applied overhead will rarely equal actual overhead. The difference between actual overhead and applied overhead is an **overhead variance**. If actual overhead is greater than applied overhead, then the variance is called **underapplied overhead**. If applied overhead is greater than actual overhead, then the variance is called **overapplied overhead**.

Overhead variances occur because it is impossible to perfectly estimate future overhead costs and production activity. Accordingly, at the end of a reporting period, procedures must exist to dispose of any overhead variance. An overhead variance is disposed of in one of two ways:

1. If immaterial, it is assigned to cost of goods sold.
2. If material, it is allocated among work-in-process inventory, finished goods inventory, and cost of goods sold.

The most common practice is simply to assign the entire overhead variance to cost of goods sold. This practice is justified on the basis of materiality, the same principle used to justify expensing the entire cost of a pencil sharpener in the period acquired rather than allocating (through depreciation) its cost over the life of the sharpener. Thus, the overhead variance is added to cost of goods sold if underapplied and subtracted from cost of goods sold if overapplied. A journal entry is the mechanism for adding or subtracting the overhead variance. Cost of Goods Sold would be debited (credited) if under- (over-) applied.

If the overhead variance is material, it should be allocated to the period's production. Conceptually, the overhead costs of a period belong to goods started but not completed (work-in-process inventory), goods finished but not sold (finished goods inventory), and goods finished and sold (cost of goods sold). The recommended way to achieve this allocation is to *prorate the overhead variance based on the ending applied overhead balances in each account*. Using applied overhead captures the original cause-and-effect relationships used to assign overhead. Using another balance to prorate, such as total manufacturing costs, may result in an unfair assignment of the additional overhead. For example, two products identical on all dimensions except for the cost of direct material inputs should receive the same overhead assignment. Yet if total manufacturing costs were used to allocate an overhead variance, then the product with the more expensive direct materials would receive a higher overhead assignment. The prorating adds the amount to each account if underapplied and subtracts an amount from each account if overapplied. Again, a journal entry is the mechanism used. Cornerstone 6-2 illustrates the calculation and disposal of overhead variances.

Overhead Application: Departmental Rates

Under departmental rates, overhead costs are assigned to individual production departments, creating departmental overhead cost pools. Budgeted overhead costs are assigned using direct tracing, driver tracing, and allocation. Once costs are assigned to individual production departments, then unit-level drivers such as direct labour hours (for labour-intensive departments) and machine hours (for machine-intensive departments) are used to compute predetermined overhead rates for each department. Products passing through the departments are assumed to consume overhead resources in proportion to the departments' unit-based drivers (machine hours or direct labour hours used). Overhead is assigned to products by multiplying the departmental rates by the amount of the driver used in the respective departments. The total overhead assigned to each department is equal to the total overhead assigned to products. Increased accuracy is the usual justification offered for the use of departmental rates.

The PlayFun example will again be used to illustrate departmental rates. Assume that PlayFun has two producing departments: Moulding and Assembly. Machine

CORNERSTONE 6-2

The HOW and WHY of Overhead Variances and Their Disposal

Information:

PlayFun Inc.'s plant produces two types of battery-operated toys: robots and race cars. The company has the following data for the past year:

		Prorate Percentage
Actual overhead	$380,000	
Applied overhead:		
Work-in-process inventory	$ 70,000	20% ($70,000/$350,000)
Finished goods inventory	105,000	30% ($105,000/$350,000)
Cost of goods sold	175,000	50% ($175,000/$350,000)
Total	$350,000	100%

The PlayFun plant uses the overhead control account to accumulate both actual and applied overhead.

Why:

At the end of the period, the total actual amount of overhead incurred must be reported as a product cost. Financial reports use actual production costs and, thus, applied and actual overhead must be reconciled. First, the difference is calculated: Actual overhead − Applied overhead (called an overhead variance). Next, the variance balance, which is either under- or overapplied overhead, must be removed through an adjustment at the end of the period. If the amount of the overhead variance is not material, then it is typically closed out to cost of goods sold. If material, the variance is prorated among Work in Process, Finished Goods, and Cost of Goods Sold.

Required:

1. Calculate the overhead variance for the year and close it to Cost of Goods Sold.
2. Assume the variance calculated is material. After prorating, close the variances to the appropriate accounts and provide the final ending balances of these accounts.
3. **What if** the variance is overapplied instead of underapplied? Provide the appropriate adjusting journal entries (if immaterial and then if material).

Solution:

1. Overhead variance = $380,000 − $350,000 = $30,000 underapplied

Cost of Goods Sold	30,000	
Overhead Control		30,000

2. Proration: (0.20 × $30,000; 0.30 × $30,000; 0.50 × $30,000)

Work-in-Process Inventory	6,000	
Finished Goods Inventory	9,000	
Cost of Goods Sold	15,000	
Overhead Control		30,000

	Unadjusted Balance	Prorated Underapplied Overhead	Adjusted Balance
Work-in-Process Inventory	$ 70,000	$ 6,000	$ 76,000
Finished Goods Inventory	105,000	9,000	114,000
Cost of Goods Sold	175,000	15,000	190,000

3.

Overhead Control	30,000	
Cost of Goods Sold		30,000

Overhead Control	30,000	
Work-in-Process Inventory		6,000
Finished Goods Inventory		9,000
Cost of Goods Sold		15,000

hours are used to assign the overhead of moulding, and direct labour hours are used to assign the overhead of assembly. Cornerstone 6-3 illustrates the calculations and summarizes their rationale.

Limitations of Plantwide and Departmental Rates

OBJECTIVE ▸ 2
Explain why plantwide and departmental overhead costing may not be accurate.

Plantwide and departmental rates have been used for a long time by many organizations. In some settings, however, they do not work well and may actually cause severe product cost distortions. Of course, to cause a significant cost distortion, overhead costs must be a significant percentage of total manufacturing costs. For some manufacturers, overhead costs are a small percentage (e.g., 5 percent or less), and therefore no matter what method is used to apply overhead, there is little difference in the per-unit costs. In this case, using a very simple, uncomplicated approach such as plantwide rates is appropriate. Assuming, however, that the overhead costs are a significant percentage of total manufacturing costs, at least two major factors can impair the ability of the unit-based plantwide and departmental rates to assign overhead costs accurately: (1) the proportion of non-unit-related overhead costs to total overhead costs is large, and (2) the degree of product diversity produced and the amount of overhead each product consumes are great.

1. Non-Unit-Related Overhead Costs

The use of either plantwide rates or departmental rates assumes that a product's consumption of overhead resources is related strictly to the units produced. But what if there are overhead activities that are unrelated to the number of units produced? Setup costs, for example, are incurred each time a batch of products is produced. A batch may consist of 1,000 or 10,000 units, and the cost of setup is the same. Yet as more setups are done, setup costs increase. The number of setups, not the number of units produced, is the cause of setup costs. Furthermore, product engineering costs may depend on the number of different engineering work orders rather than the units produced of any given product. Both these examples illustrate the existence of non-unit-based drivers. **Non-unit-based drivers** are factors, other than the number of units produced, that measure the demands that cost objects place on activities.

**CORNERSTONE
6-3**

The HOW and WHY of Departmental Overhead Rates

Information:

The data for the two producing departments of the PlayFun Inc. plant are given below.

	Moulding	Assembly	Total
Estimated overhead	$250,000	$100,000	$350,000
Direct labour hours (expected and actual):			
Robots	5,000	5,000	10,000
Race cars	5,000	35,000	40,000
Total	10,000	40,000	50,000
Machine hours:			
Robots	17,000	3,000	20,000
Race cars	3,000	7,000	10,000
Total	20,000	10,000	30,000

Machine hours are used to assign the overhead of the Moulding Department, and direct labour hours are used to assign the overhead of the Assembly Department. There are 50,000 robots produced and sold and 250,000 race cars.

Why:

Product costs that reflect the consumption of resources actually used are relatively more accurate and improve decision making and control. Overhead intensity and patterns of consumption by products can differ from department to department. The argument is that departmental overhead rates will better reflect each product's use of resources and thus will be more accurate than a single plantwide rate.

Required:

1. Calculate the overhead rates for each department.
2. Assign overhead to the two products and calculate the overhead cost per unit. How does this compare with the plantwide rate unit cost of Cornerstone 6-1?
3. **What if** the machine hours in Moulding were 5,000 for robots and 15,000 for race cars and the direct labour hours used in Assembly were 4,000 and 36,000, respectively? Calculate the overhead cost per unit for each product, and compare with the plantwide rate unit cost of Cornerstone 6-1. What can you conclude from this outcome?

Solution:

1. Moulding: $250,000/20,000 = $12.50 per machine hour
   Assembly: $100,000/40,000 = $2.50 per direct labour hour
2. Overhead assignment:

	Robots	Race Cars
($12.50 × 17,000) + ($2.50 × 5,000)	$225,000	
($12.50 × 3,000) + ($2.50 × 35,000)	_____	$125,000

	Robots	Race Cars
Total applied overhead	$225,000	$125,000
Units of production	÷50,000	÷250,000
Unit overhead cost	$ 4.50	$ 0.50

CORNERSTONE
6-3
(continued)

The cost increased dramatically for robots (from $1.40 to $4.50) and decreased significantly for race cars (from $1.12 to $0.50).

3. Overhead assignment:

	Robots	Race Cars
($12.50 × 5,000) + ($2.50 × 4,000)	$72,500	
($12.50 × 15,000) + ($2.50 × 36,000)		$277,500
Total applied overhead	$72,500	$277,500
Units of production	÷50,000	÷250,000
Unit overhead cost	$ 1.45	$ 1.11

Compared to the plantwide unit overhead costs, the cost is $0.05 more for robots and $0.01 less for racing cars. The message is that departmental rates may not necessarily cause a significant change in the assignments. It depends on the complexity of each product and how the resource demands are made in each department. However, implementation of departmental rates would probably be done based on the observation that significant differences in resource consumption do exist, justifying the decision.

Thus, unit-level drivers cannot assign these costs accurately to products. In fact, using only unit-level drivers to assign non-unit-related overhead costs can create distorted product costs. The severity of this distortion depends on what proportion of total overhead costs these non-unit-based costs represent. For many companies, this percentage can be significant—reaching more than 40 or 50 percent of the total. Clearly, as this percentage decreases, the acceptability of using unit-based drivers for assigning costs increases.

2. Product Diversity

Significant non-unit overhead costs will not cause product cost distortions provided that products consume the non-unit overhead activities in the same proportion as the unit-level overhead activities. Product diversity, on the other hand, can cause product cost distortion. **Product diversity** simply means that different products consume overhead activities in different proportions. Product diversity is caused by such things as differences in product size, product complexity, setup time, and size of batches. The proportion of each activity consumed by a product is referred to as the **consumption ratio**. The way that non-unit overhead costs and product diversity can produce distorted product costs (when only unit-level drivers are used to assign overhead costs) will be illustrated by providing detailed data for PlayFun Inc.

  To illustrate the failure of plantwide and departmental rates, let's once again consider PlayFun's plant, which produces battery-operated toy robots and race cars. The two producing departments are Moulding and Assembly. Moulding is responsible for shaping the plastic components of each product, and Assembly is responsible for assembling the internally produced plastic components with outside purchased

Exhibit 6-1

Product Costing Data

I. Activity Usage Measures (expected and actual)

	Robots	Race Cars	Total
Units produced	50,000	250,000	—
Prime costs	$200,000	$750,000	$950,000
Direct labour hours	10,000	40,000	50,000
Machine hours	20,000	10,000	30,000
Number of setups	25	75	100
Inspection hours	1,200	2,800	4,000
Number of moves	140	210	350

II. Departmental Data (expected and actual)

	Moulding	Assembly	Total
Direct labour hours:			
Robots	5,000	5,000	10,000
Race cars	5,000	35,000	40,000
Total	10,000	40,000	50,000
Machine hours:			
Robots	17,000	3,000	20,000
Race cars	3,000	7,000	10,000
Total	20,000	10,000	30,000
Overhead costs:			
Machining	$120,000	$ 30,000	$150,000
Moving materials	40,000	30,000	70,000
Setting up	70,000	10,000	80,000
Inspecting products	20,000	30,000	50,000
Total	$250,000	$100,000	$350,000

electronic parts. Expected product costing data are given in Exhibit 6-1. Because the quantity of race cars produced is five times greater than that of robots, we can label the race cars a high-volume product and robots a low-volume product. Because different moulds are needed, the products are produced in batches. The moulds for robots are larger and more varied than those for race cars; thus, batches for robots tend to be smaller and take longer to process.

For ease of presentation, only four types of overhead activities, performed by four distinct support departments, are assumed: setting up the equipment for each batch, machining, inspecting, and moving a batch. Each batch of products is inspected after each department's operations. After moulding, a sample of the components is inspected to ensure correct size and shape. After assembly, a sample is also tested to ensure that each unit works as expected. Overhead costs are assigned to the two production departments using the direct method. Effectively, costs are assigned using direct and driver tracing.

Plantwide Departmental Overhead and Activity Rates

The traditional unit product cost is the unit prime cost plus unit overhead cost. Prime costs are assigned to each of the products using direct tracing. From Exhibit 6-2, the unit prime cost for robots is $4.00 ($200,000/50,000), and the unit prime cost for race cars is $3.00 ($750,000/250,000). Cornerstones 6-1 and 6-3 provide the unit overhead cost calculations for plantwide and overhead rates. Adding the unit

Exhibit 6-2

Unit Product Cost: Plantwide and Departmental Rates

I. Plantwide

	Robots	Race Cars
Prime cost[a]	$4.00	$3.00
Overhead cost[b]	1.40	1.12
Unit cost	$5.40	$4.12

II. Departmental

	Robots	Race Cars
Prime cost[a]	$4.00	$3.00
Overhead cost[c]	4.50	0.50
Unit cost	$8.50	$3.50

[a] $200,000/50,000; $750,000/250,000
[b] From Cornerstone 6-1
[c] From Cornerstone 6-3

prime costs to the unit overhead costs produces the desired unit product cost. Exhibit 6-2 summarizes and provides the details of these calculations.

Problems with Costing Accuracy   The accuracy of the overhead cost assignment can be challenged regardless of whether plantwide or departmental rates are used. The main problem with either procedure is the assumption that machine hours and/or direct labour hours drive or cause all overhead costs.

From Exhibit 6-1, we know that race cars, the high-volume product, use four times the direct labour hours used by robots, the low-volume product (40,000 hours versus 10,000 hours). Thus, if a plantwide rate is used, the race cars will receive four times more overhead cost than will the robots. But is this reasonable? Do unit-based activity drivers explain the consumption of all overhead activities? In particular, can we reasonably assume that each product's consumption of overhead increases in direct proportion to the direct labour hours used? Let's look at the four overhead activities and see if unit-based drivers accurately reflect the demands of the two products for overhead resources.

Of the four activities, only machining appears to be a unit-level cost, since machining will occur each time a unit is produced. Thus, using direct labour hours or machine hours on the surface appears reasonable. However, the data in Exhibit 6-1 suggest that a significant portion of overhead costs is not driven or caused by the units produced (measured by direct labour hours). For example, each product's demands for the setup, material moving, and inspection activities are more logically related to the number of setups, number of moves, and inspection hours, respectively. These non-unit-level activities represent more than 50 percent ($200,000/$350,000) of the total overhead costs—a significant percentage. Notice that the high-volume product, race cars, uses three times the number of setups of robots, about 2.33 times as many inspection hours, and only one and one-half times as many moves. However, use of direct labour hours, a unit-based activity driver, and a plantwide rate assigns four times more setup, inspection, and materials handling costs to the race cars than to the robots. Thus, we have product diversity, and we should expect product cost distortion because the quantity of unit-based overhead that each product consumes does not vary in direct proportion to the quantity consumed of non-unit-based overhead. How to calculate the consumption ratios for the various activities is shown in Cornerstone 6-4. Consumption ratios are simply the proportion of each activity consumed by a product. The *assumed* consumption ratios can also be calculated for the plantwide and overhead rates. Comparing the

**CORNERSTONE
6-4**

The HOW and WHY of Consumption Ratios

Information:
Product costing data from Exhibit 6-1.

Why:
Consumption ratios reflect the proportion of an activity consumed by the individual products. They are especially useful to assign costs of a shared resource. For example, two individuals sharing the cost of a pizza would logically do so in proportion to the amount of the pizza consumed. In a multiple-product firm, there are many shared resources and it is reasonable to assign the costs of shared resources in proportion to the resource consumed. Activity drivers are a measure of activity output and thus can be used as measures of activity consumption.

Required:
1. Calculate the activity consumption ratios for each product.
2. Calculate the *assumed* consumption ratios for plantwide (direct labour hours) and departmental rates.
3. ***What if*** the activity consumption ratios were approximately equal to the consumption ratio associated with direct labour hours? What does this tell you?

Solution:
1.

Overhead Activity	Consumption Ratios		Activity Driver
	Robots	**Race Cars**	
Machining	0.67[a]	0.33[a]	Machine hours
Setups	0.25[b]	0.75[b]	Number of setups
Inspecting products	0.30[c]	0.70[c]	Inspection hours
Moving materials	0.40[d]	0.60[d]	Number of moves

[a] 20,000/30,000 (robots) and 10,000/30,000 (race cars)
[b] 25/100 (robots) and 75/100 (race cars)
[c] 1,200/4,000 (robots) and 2,800/4,000 (race cars)
[d] 140/350 (robots) and 210/350 (race cars)

2.

Overhead Activity	Consumption Ratios		Activity Driver
	Robots	**Race Cars**	
Plantwide:			
Manufacturing	0.20[a]	0.80[a]	Direct labour hours
Departmental:			
Moulding	0.85[b]	0.15[b]	Machine hours
Assembly	0.13[c]	0.87[c]	Direct labour hours

[a] 10,000/50,000 (robots) and 40,000/50,000 (race cars)
[b] 17,000/20,000 (robots) and 3,000/20,000 (race cars)
[c] 5,000/40,000 (robots) and 35,000/40,000 (race cars)

3. If the activity ratios were approximately the same (all about 0.20 and 0.80 for each product, respectively), it would indicate that there is little product diversity—that the products are consuming all activities in the same ratio as direct labour hours. This outcome would signal that a plantwide rate is functioning quite well in assigning overhead costs to products. There would be no need to use either departmental or activity rates.

consumption ratios with the assumed consumption pattern of a plantwide rate suggests that using only direct labour hours to assign costs will overcost the race cars and undercost the robots. Comparing the departmental consumption ratios with the plantwide ratios (in Cornerstone 6-4) and the product costs illustrated in Exhibit 6-2 indicates that the departmental rates are likely making a correction in the right direction (more overhead is being assigned to the robots and less to the race cars), but whether the correction is about right, too little, or too much can be assessed by calculating activity-based costs.

The most direct method of overcoming the distortions caused by the unit-level rates is to expand the number of rates used so that the rates reflect the actual consumption of overhead costs by the various products. Thus, instead of pooling the overhead costs in plant or departmental pools, rates are calculated for each individual overhead activity. The rates are based on causal factors that measure consumption (unit- and non-unit-level activity drivers). Costs are assigned to each product by multiplying the activity rates by the amount consumed by each activity (as measured by the activity driver). Cornerstone 6-5 illustrates the calculations and summarizes the rationale for activity-based costing.

Comparison of Different Product Costing Methods   In Exhibit 6-3, the unit costs and unit overhead costs from activity-based costing are compared with the unit costs produced by unit-based costing using either a plantwide or departmental rate. This comparison clearly illustrates the effects of using only unit-based activity drivers to assign overhead costs. The activity-based cost assignment follows a cause-and-effect pattern of overhead consumption and is therefore the most accurate of the three costs shown in Exhibit 6-3. Using a plantwide overhead rate undercosts the robots and overcosts the race cars. In fact, relative to the ABC cost, the plantwide assignment decreases the total unit cost of the robots by at least 25 percent [($7.26 − $5.40)/$7.26] and increases the unit cost of the race cars by about 10 percent [($4.12 − $3.75)/$3.75]. The effect is even more dramatic when comparing only unit overhead costs. Departmental overhead rates overcorrect and produce distortions as well, although, in this example, the distortion is reduced (about a 17 percent error for robots and 7 percent for race cars, relative to ABC assignments). Thus, in the presence of significant non-unit overhead costs and product diversity, using only unit-based activity drivers can lead to one product subsidizing another (for plantwide rates, the race cars subsidize the robots). This subsidy could create the appearance that one group of products is highly profitable and can adversely impact the pricing and competitiveness of another group of products. In a highly competitive environment, the more accurate the cost information, the better the planning and decision making.

The PlayFun example also helps us understand when ABC may be useful for a firm. First, ABC offers no increase in product costing accuracy for a single-product setting; there must be product diversity. Second, if products consume non-unit-level activities in the same proportion as unit-level activities, then ABC assignments will be the same as unit-based assignments. Third, non-unit-level overhead must be a significant percentage of production cost. If it is not, then it hardly matters how it is assigned. Thus, firms that have plants with multiple products, high product diversity, and significant non-unit-level overhead are candidates for an ABC system.

Exhibit 6-3

Comparison of Unit Costs

	Total Unit Cost			Unit Overhead Cost		
	Robots	**Race Cars**		**Robots**	**Race Cars**	
Activity-based cost	$7.26	$3.75	Cornerstone 6-5	$3.26	$0.75	Cornerstone 6-5
Unit-based cost:						
Plantwide rate	5.40	4.12	Exhibit 6-2	1.40	1.12	Cornerstone 6-1
Departmental rates	8.50	3.50	Exhibit 6-2	4.50	0.50	Cornerstone 6-3

CORNERSTONE 6-5

The HOW and WHY of Activity-Based Costing

Information:

Activity usage and costs from Exhibit 6-1:

	Robots	Race Cars	Total
Units produced	50,000	250,000	—
Prime costs	$200,000	$750,000	$950,000
Machine hours	20,000	10,000	30,000
Number of setups	25	75	100
Number of moves	140	210	350
Inspection hours	1,200	2,800	4,000

Overhead costs:	
Machining	$150,000
Setting up	80,000
Moving materials	70,000
Inspecting products	50,000

Why:

An activity rate is calculated for each activity and the activity cost is assigned to products based on how much they use of each activity. The assignment is done using cause-and-effect relationships. Causal factors, called activity drivers, measure the amount of activity consumed by a product. The activity rate multiplied by the amount used of the activity determines the amount of activity cost assigned to a particular product. The total of all the assigned activity costs is the amount of overhead consumed by a product. Because the assignment uses causal factors, it tends to be *relatively* more accurate than assignments that use only unit-level drivers.

Required:

1. Calculate the four activity rates.
2. Calculate the unit costs using activity rates. Also, calculate the overhead cost per unit (see Exhibit 6-2 for unit prime costs).
3. **What if** consumption ratios were used to assign costs instead of activity rates? Show the cost assignment for moving materials.

Solution:

1.   Machining rate: $150,000/30,000 = $5.00 per machine hour
            Setup rate: $80,000/100 = $800 per setup
      Moving materials rate: $70,000/350 = $200 per move
          Inspecting rate: $50,000/4,000 = $12.50 per hour

2.

	Robots	Race Cars
Prime costs	$200,000	$750,000
Overhead costs:		
Machining:		
$5 × 20,000	100,000	
$5 × 10,000		50,000

CORNERSTONE
6-5
(continued)

	Robots	Race Cars
Setting up:		
$800 × 25	$ 20,000	
$800 × 75		$ 60,000
Moving materials:		
$200 × 140	28,000	
$200 × 210		42,000
Inspecting products:		
$12.50 × 1,200	15,000	
$12.50 × 2,800		35,000
Total manufacturing costs	$363,000	$937,000
Units of production	÷ 50,000	÷ 250,000
Unit cost	$ 7.26	$ 3.75 (rounded)

Overhead cost per unit: Robots: $7.26 − $4.00* = $3.26

Cars: $3.75 − $3.00* = $0.75

*Prime cost per unit:

Robots = $200,000 ÷ 50,000 = $4.00

Cars = $750,000 ÷ 250,000 = $3.00

3. Using consumption ratios will yield exactly the same overhead assignments as activity rates, if the actual activity usage is the same as the expected usage (assuming no rounding error for the ratios). For moving materials, the consumption ratio is 0.40 for robots and 0.60 for race cars. Thus, the assignment is 0.40 × $70,000 = $28,000 (robots) and 0.60 × $70,000 = $42,000 (race cars), which is the same assignment obtained using activity rates.

One survey studied this concept.[1] Of those firms surveyed, 49 percent had adopted ABC. When compared with non-adopting firms, it was found that adopting firms reported a higher potential for distorted costs and a higher level of overhead when expressed as a percentage of total production costs. Adopting firms also reported a greater need or utility for accurate cost information for decision making.

Activity-Based Costing System

OBJECTIVE ▶ 3
Provide a detailed description of activity-based product costing.

The PlayFun example shows quite clearly that prime costs are assigned in the same way for functional as for activity-based costing. The example also demonstrates that the total amount of overhead costs is assigned under either approach. The amount assigned to each product, though, can differ significantly, depending on which method is used. The theoretical premise of activity-based costing is that it assigns costs according to the resource consumption of each product. If this is true, then activity-based costing should produce more accurate product costs if there is product diversity simply because unit-based drivers cannot capture the full consumption pattern of products. The PlayFun example suggests that we simply need to choose among a plantwide cost pool, departmental cost pools, or activity cost pools. If there is no product diversity and a plantwide cost pool is chosen, all we need is the cost of overhead resources: depreciation, salaries, utilities, rent, and so forth. On the other hand, departmental cost pools require more detail because costs must be assigned to every producing

[1] Kip Krumwiede, "ABC: Why It's Tried and How It Succeeds," *Management Accounting* (April 1998): 32–38.

Exhibit 6-4

Activity-Based Costing Model

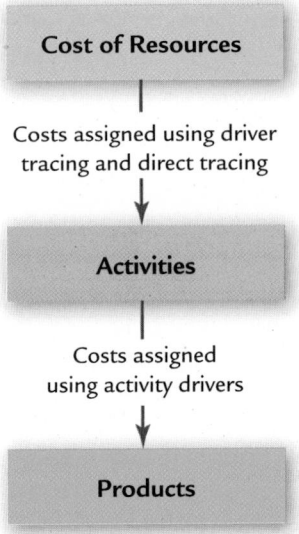

Cost of Resources

Costs assigned using driver
tracing and direct tracing

Activities

Costs assigned
using activity drivers

Products

department. Finally, activity-based costing requires the most detail because each activity performed and its associated costs must be identified.

As Exhibit 6-4 illustrates, an **activity-based costing (ABC) system** first traces costs to activities and then to products and other cost objects. The underlying assumption is that activities consume resources, and that products and other cost objects consume activities. In designing an ABC system, there are six essential steps, as listed in Exhibit 6-5.

Step 1: Identify, Define, and Classify Activities

Identifying activities is a logical first step in designing an activity-based costing system. Activities represent actions taken or work performed. Identifying an activity is equivalent to describing action taken—usually by using an action verb and an object that receives the action. A simple list of the activities identified is called an **activity inventory**. A sample activity inventory for an electronics manufacturer is listed in Exhibit 6-6. Of course, the actual inventory of activities for most organizations would list more than 12 activities (220 to 300 are not uncommon).

Exhibit 6-5

Design Steps for an ABC System

Stage I (identify, classify, and determine cost of each activity)

1. Identify, define, and classify activities.
2. Assign the cost of resources to each activity.
3. Assign the cost of secondary activities to primary activities.

Stage II (determine rates for each activity, consumption of each activity, and assign activity costs to products/services)

4. Identify cost objects and specify the amount of each activity consumed by each cost object.
5. Calculate primary activity rates.
6. Assign activity costs to cost objects.

Exhibit 6-6

Sample Activity Inventory

1. Providing space
2. Providing utilities
3. Purchasing materials
4. Receiving materials
5. Paying for materials
6. Collecting engineering data
7. Developing test programs
8. Testing products
9. Setting up lots
10. Handling wafer lots

Activity Definition   Once an inventory of activities exists, then activity attributes are used to define activities. **Activity attributes** are nonfinancial and financial information items that describe individual activities. An **activity dictionary** lists the activities in an organization along with desired attributes. The attributes selected depend on the purpose being served. Examples of activity attributes with a product costing objective include tasks that describe the activity, types of resources consumed by the activity, amount (percentage) of time spent on an activity by workers, cost objects that consume the activity, and a measure of activity consumption (activity driver). Activities are the building blocks for both product costing and continuous improvement.

Activity Classification   Attributes define and describe activities and, at the same time, become the basis for activity classification. Activity classification facilitates the achievement of key managerial objectives such as product or customer costing, continuous improvement, total quality management, and environmental cost management. For example, for costing purposes, activities can be classified as primary or secondary. A **primary activity** is an activity that is consumed by a final cost object such as a product or customer. A **secondary activity** is one that is consumed by intermediate cost objects such as primary activities, materials, or other secondary activities. Recognizing the difference between the two types of activities facilitates product costing. Exhibit 6-4 indicates that activities consume resources. Thus, in the first stage of activity-based costing, the cost of resources is assigned to activities. Exhibit 6-4 also reveals that products consume activities—but only primary activities. Thus, before assigning the costs of primary activities to products, the costs of the secondary activities consumed by primary activities must be assigned to the primary activities. Many other useful activity classifications exist. For example, activities can be classified as *value-added* or *non-value-added* (defined and discussed in detail in Chapter 14), as *quality-related* or as *environmental* (discussed in Chapter 16). In designing an activity costing system, the desired attributes and essential classifications need to be characterized up front so that the necessary data can be collected for the activity dictionary.

Gathering the Necessary Data   Interviews, questionnaires, surveys, and observation are means of gathering data for an ABC system. Interviews with managers or other knowledgeable representatives of functional departments are perhaps the most common approach for gathering the needed information. Interview questions can be used to identify activities and activity attributes needed for costing or other managerial purposes. The information derived from interview questions serves as the basis for constructing an activity dictionary and provides data helpful for assigning resource costs to individual activities. In structuring an interview, the questions should reveal certain key attributes. Interview questions should be structured to provide answers that allow the desired attributes to be identified and measured. An example is perhaps the best way to show how an interview can be used to collect the data for an activity dictionary.

Illustrative Example   Suppose that a hospital is carrying out an ABC pilot study to determine the nursing cost for different types of cardiology patients. The cardiology unit is located on one floor of the hospital. The interview with the unit's nursing supervisor is provided below. Questions are given along with their intended purposes and the supervisor's responses. The interview is not intended to be viewed as an exhaustive analysis but rather represents a sample of what could occur.

Question 1 (Activity Identification): Can you describe what your nurses do for patients in the cardiology unit? (Activities are people doing things for other people.)

Response: There are four major activities: treating patients (administering medicine and changing dressings), monitoring patients (checking vital signs and posting patient information), providing hygienic and physical care for patients (bathing, changing bedding and clothes, walking the patient, etc.), and responding to patient requests (counselling, providing snacks, and answering calls).

Question 2 (Activity Identification): Do any patients make use of any equipment? (Activities also can be equipment doing work for other people.)

Response: Yes. In the cardiology unit, monitors are used extensively. Monitoring is an important activity for this type of patient.

Question 3 (Activity Identification): What role do you have in the cardiology unit? (Activities are people doing things for other people.)

Response: I have no direct contact with the patients. I am responsible for scheduling, evaluations, and resolving problems with the ward's nurses.

Question 4 (Resource Identification): What resources are used by your nursing care activities (equipment, materials, energy)? (Activities consume resources in addition to labour.)

Response: Uniforms (which are paid for by the hospital), computers, nursing supplies such as scissors and instruments (supplies traceable to a patient are charged to the patient), and monitoring equipment at the nursing station.

Question 5 (Resource Driver Identification): How much time do nurses spend on each activity? How much equipment time is spent on each activity? (Information is needed to assign the cost of labour and equipment to activities.)

Response: We recently completed a work survey. About 25 percent of a nurse's time is spent treating patients, 20 percent providing hygienic care, 40 percent responding to patient requests, and 15 percent on monitoring patients. My time is 100 percent supervision. The monitoring equipment is used 100 percent for monitoring activity. Use of the computer is divided between 40 percent for supervisory work and 60 percent for monitoring. (Posting readings to patient records is viewed as a monitoring task.)

Question 6 (Potential Activity Drivers): What are the outputs of each activity? That is, how would you measure the demands for each activity? (This question helps identify activity drivers.)

Response: Treating patients: number of treatments; providing hygienic care: hours of care; responding to patient requests: number of requests; and monitoring patients: monitoring hours.

Question 7 (Potential Cost Objects Identified): Who or what uses the activity output? (Identifies the cost object: products, other activities, customers, etc.)

Response: Well, for supervising, I schedule, evaluate performance, and try to ensure that the nurses carry out their activities efficiently. Nurses benefit from what I do. Patients receive the benefits of the nursing care activities. We have three types of cardiology patients: intensive care, intermediate care, and normal care. These patients make quite different demands on the nursing activities. For example, intensive care patients rarely have walking time but use a lot of treatments and need more monitoring time.

Exhibit 6-7

Activity Dictionary: Cardiology Unit

Activity Name	Activity Description	Activity Type	Cost Object(s)	Activity Driver
Supervising nurses	Scheduling, coordinating, and performance evaluation	Secondary	Activities within department	Percentage of time nurses spend on each activity
Treating patients	Administering medicine and changing dressings	Primary	Patient types	Number of treatments
Providing hygienic care	Bathing, changing bedding and clothes, walking patients	Primary	Patient types	Labour hours
Responding to patient requests	Answering calls, counselling, providing snacks, etc.	Primary	Patient types	Number of requests
Monitoring patients	Checking vital signs and posting patient information	Primary	Patient types	Monitoring hours

Activity Dictionary   Based on the answers to the interview, an activity dictionary can now be prepared. Exhibit 6-7 illustrates the dictionary for the cardiology unit. The activity dictionary names the activity (typically by using an action verb and an object that receives the action), describes the tasks that make up the activity, classifies the activity as primary or secondary, lists the users (cost objects), and identifies a measure of activity output (activity driver). For example, the supervising activity is consumed by the following primary activities: treating patients, providing hygienic care, responding to patient requests, and monitoring patients. The three products—intensive care patients, intermediate care patients, and normal care patients—in turn, consume the primary activities.

Steps 2 and 3: Assign Costs to Activities

After identifying and describing activities, the next task is determining how much it costs to perform each activity. The cost of an activity is simply the cost of the resources consumed by each activity. Activities consume resources such as labour, materials, energy, and capital. The cost of these resources is found in the general ledger, but how much is spent on each activity is not revealed. Resource costs must be assigned to activities using direct and driver tracing. For example, consider the labour resource. The time spent on each activity is the driver used to assign the labour costs to the activity. If the time spent is 100 percent, then labour is exclusive to the activity, and direct tracing is the cost assignment method (such as the labour cost of nursing supervision). On the other hand, if the nursing resource is shared by several activities, then driver tracing is used for the cost assignment. These drivers are called resource drivers. **Resource drivers** are factors that measure the consumption of resources by activities. For labour resources, a *work distribution matrix* is often used. A work distribution matrix simply identifies the amount of labour consumed by each activity and is derived from the interview process (or a written survey). Interviews, survey forms, questionnaires, and timekeeping systems are examples of tools that can be used to collect data on resource drivers. Notice that tracking the effort spent on different activities is similar to tracking the time that labourers spend on different jobs. However, there is one critical difference. The percent of effort spent on various activities is usually fairly constant and may only need to be measured periodically (perhaps annually). In effect, the labour time is a standard used to assign the cost of resources.

Labour is only one of many resources consumed by activities. Activities also consume materials, capital, and energy. The interview, for example, reveals that cardiology care activities also include the use of monitors (capital), a computer (capital), uniforms (materials), and supplies (materials). The cost of these other resources is also assigned to activities using direct tracing and resource drivers. Assigning costs to activities completes the first stage of activity-based costing. In this first stage, activities are classified as primary and secondary. If there are secondary activities, then intermediate stages exist. In an intermediate stage, the cost of secondary activities is assigned to those activities (or other intermediate cost objects) that consume their output. These calculations and concepts are illustrated in Cornerstone 6-6.

The assignment of resource costs to activities requires that the resource costs described in the general ledger be unbundled and reassigned. In a traditional accounting system, the general ledger reports costs by department and by spending account (based on a chart of accounts). The $340,000 of nursing salaries, for example, would be recorded as part of the total salaries of the cardiology unit. The general ledger indicates what is spent, but it does not reveal how the resources are spent. In an activity-based cost system, costs must be reported by activity. Thus, an ABC system must restate the general ledger costs so that the new system reveals how the resources are being consumed. Exhibit 6-8 illustrates the unbundling concept for nursing care activities in the cardiology unit. As the exhibit indicates, the reassignment of resource costs to individual activities contributes to the creation of an ABC database for the organization.

Step 4: Identify Cost Objects

Once the costs of primary activities are determined, these costs can then be assigned to products or other cost objects in proportion to their usage of the activity, as measured by activity drivers. However, before any assignment is made, the cost objects must be identified and the demands these objects place on the activities must be measured. Many different cost objects are possible: products, materials, customers, distribution channels, suppliers, and geographical regions are some examples. For our example, the cost objects are products (services): intensive cardiology care, intermediate cardiology care, and normal cardiology care. How to deal with cost assignment for other cost objects is discussed in a later section. **Activity drivers** measure the demands that cost objects place on activities. Most ABC system designs choose between one of two types of activity drivers: transaction drivers and duration drivers. **Transaction drivers** measure the number of times an activity is performed, such as the number of treatments and the number of requests. **Duration drivers** measure the demands in terms of the time it takes to perform an activity, such as hours of hygienic care and monitoring hours.

Exhibit 6-8

Unbundling of General Ledger Costs

General Ledger		→ ABC Database	
Cardiology Unit			
Chart of Accounts View		**ABC View**	
Supervision	$ 50,000	Supervising nurses	$ 60,000
Supplies and uniforms	60,000	Treating patients	98,500
Salaries	340,000	Providing hygienic care	78,800
Computer	10,000	Responding to requests	98,500
Monitor	26,000	Monitoring patients	150,200
Total	$486,000	Total	$486,000

The HOW and WHY of Assigning Resource Costs to Activities

Information:

Resources		Activities	Nursing Hours	
Supervision	$ 50,000	Supervising nurses	2,000	10.0%
Supplies and uniforms	60,000	Treating patients	4,500	22.5
Salaries	340,000	Providing hygienic care	3,600	18.0
Computer	10,000	Responding to requests	4,500	22.5
Monitor	26,000	Monitoring patients	5,400	27.0
Total	$486,000	Total	20,000	100.0%
		Total without supervising	18,000	

CORNERSTONE 6-6

- Monitors are used only by the monitoring activity.
- The one computer is used 800 hours for supervisory work (40 percent) and 1,200 hours for monitoring work (60 percent).
- The nursing resources (supplies, uniforms, and labour) are assigned to activities using nursing hours. The supervisor spends 100 percent of her time on supervision.

Why:

Activities consume resources, and other cost objects consume activities. The cost of each activity must therefore be determined. The cost of resources is assigned to activities using direct tracing and driver tracing. Resource drivers are used to assign shared resources. After this initial assignment, the costs of secondary activities are assigned to primary activities.

Required:

1. Prepare a work distribution matrix for the five activities.
2. Calculate the cost of each activity.
3. What if the cost of the supervising activity is assigned to the other four activities? Why would this be done? If it is done, what is the final cost of these four primary activities?

Solution:

1.

Percentage of Time on Each Activity

Activity	Supervisor	Nurses	Supporting Calculation
Supervising nurses	100%	0%	(2,000/2,000)
Treating patients	0	25	(4,500/18,000)
Providing hygienic care	0	20	(3,600/18,000)
Responding to requests	0	25	(4,500/18,000)
Monitoring patients	0	30	(5,400/18,000)

CORNERSTONE 6-6 (continued)

2.

Activities	Monitor[a]	Computer[b]	Nursing Resources[c]	Total
Supervising nurses		$4,000	$ 56,000	$ 60,000
Treating patients			98,500	98,500
Providing hygienic care			78,800	78,800
Responding to requests			98,500	98,500
Monitoring patients	$26,000	6,000	118,200	150,200

[a] Exclusive use by monitoring (100% × $26,000)
[b] 0.40 × $10,000; 0.60 × $10,000
[c] $50,000 + (0.10 × $60,000); [(0.25 × $340,000) + (0.225 × $60,000)]; [(0.20 × $340,000) + (0.18 × $60,000)]; [(0.25 × $340,000) + (0.225 × $60,000)]; [(0.30 × $340,000) + (0.27 × $60,000)]

3. Supervising is a secondary activity, and its costs are consumed by primary activities (assigned in proportion to the labour content of each activity).

Treating patients	$113,500[a]
Providing hygienic care	90,800[b]
Responding to requests	113,500[c]
Monitoring patients	168,200[d]

[a] $98,500 + (0.25 × $60,000)
[b] $78,800 + (0.20 × $60,000)
[c] $98,500 + (0.25 × $60,000)
[d] $150,200 + (0.30 × $60,000)

Duration drivers should be used when the time required to perform an activity varies from transaction to transaction. If, for example, treatments for normal care patients average 10 minutes but for intensive care patients average 45 minutes, then treatment hours may be a much better measure of the demands placed on the activity of treating patients than the number of treatments.

With the drivers defined, a bill of activities can be created. A **bill of activities** specifies the product, expected product quantity, activities, and amount of each activity expected to be consumed by each product. Exhibit 6-9 presents a bill of activities for the cardiology care example.

Exhibit 6-9

Bill of Activities: Cardiology Unit

Activity	Driver	Normal	Intermediate	Intensive	Total
Production (output)	Patient days	10,000	5,000	3,000	
Treating patients	Treatments	500	1,000	1,500	3,000
Providing hygienic care	Hygienic hours	1,125	562	1,913	3,600
Responding to requests	Requests	3,000	4,000	1,000	8,000
Monitoring patients	Monitoring hours	540	1,620	3,240	5,400

Steps 5 and 6: Calculate Primary Activity Rates and Assign Them to Cost Objects

Primary activity rates are computed by dividing the budgeted activity costs by practical activity capacity, where activity capacity is the amount of activity output (as measured by the activity driver). Practical capacity is the activity output that can be produced if the activity is performed efficiently. Using data from Cornerstone 6-6 and Exhibit 6-9, the activity rates for the cardiology unit nursing care example can now be calculated:

Rate Calculations:

Treating patients:	$113,500/3,000 = $37.83 per treatment
Providing hygienic care:	$90,800/3,600 = $25.22 per hour of care
Responding to requests:	$113,500/8,000 = $14.19 per request
Monitoring patients:	$168,200/5,400 = $31.15 per monitoring hour

Note: Rates are rounded to the nearest cent.

These rates provide the price charged for activity usage. Using these rates, costs are assigned as shown in Exhibit 6-10. As should be evident, the assignment process is the same as that for the PlayFun example illustrated earlier in Cornerstone 6-5.

Unit-, Batch-, Product-, and Facility-Level Activities   To help identify activity drivers and enhance the management of activities, activities are often classified into one of the following four general activity categories: (1) unit-level, (2) batch-level, (3) product-level, and (4) facility-level. **Unit-level activities** are those that are performed each time a unit is produced. Grinding, polishing, and assembly are examples of unit-level activities. **Batch-level activities** are those that are performed each time a batch is produced. The costs of batch-level activities vary with the number of batches but are fixed (and, therefore, independent) with respect to the number of units in each batch. Setups, inspections (if done by sampling units from a batch), purchasing, and materials handling are examples of batch-level activities. **Product-level activities** are those activities performed that enable the various products of a company to be produced. These activities and their costs tend to increase as the number of different products increases. Engineering changes (to products), developing product-testing procedures, introducing new

Exhibit 6-10

Assigning Costs: Final Cost Objects

	Normal	Intermediate	Intensive
Treating patients:			
$37.83 × 500	$ 18,915		
$37.83 × 1,000		$ 37,830	
$37.83 × 1,500			$ 56,745
Providing hygienic care:			
$25.22 × 1,125	28,373		
$25.22 × 562		14,174	
$25.22 × 1,913			48,246
Responding to requests:			
$14.19 × 3,000	42,570		
$14.19 × 4,000		56,760	
$14.19 × 1,000			14,190
Monitoring patients:			
$31.15 × 540	16,821		
$31.15 × 1,620		50,463	
$31.15 × 3,240			100,926
Total costs	$106,679	$159,227	$220,107
Units	÷ 10,000	÷ 5,000	÷ 3,000
Nursing cost per patient day*	$ 10.67	$ 31.85	$ 73.37

*Rounded to nearest cent.

products, and expediting goods are examples of product-level activities. **Facility-level activities** are those that sustain a factory's general manufacturing processes. Providing facilities, maintaining grounds, and providing plant security are examples.

Classifying activities into these general categories facilitates product costing because the costs of activities associated with the different levels respond to different types of activity drivers. (Cost behaviour differs by level.) Knowing the activity level is important because it helps management identify the activity drivers that measure the amount of each activity output being consumed by individual products. Activity-based costing systems improve product costing accuracy by recognizing that many of the so-called fixed overhead costs vary in proportion to changes other than production volume. Level classification also provides insights concerning the root causes of activities and thus can help managers in their efforts to improve activity performance.

By understanding what causes these costs to increase or decrease, they can be traced to individual products. This cause-and-effect relationship allows managers to improve product costing accuracy, which can significantly improve decision making. Additionally, this large pool of fixed overhead costs is no longer so mysterious. Knowing the underlying behaviour of many of these costs allows managers to exert more control over the activities that cause the costs. It also allows managers to identify which of the activities add value and which do not. Value analysis is the heart of activity-based management and is the basis for continuous improvement. Activity-based management and continuous improvement are explored in later chapters.

OBJECTIVE ▶ 4
Explain how ABC can be simplified.

Reducing the Size and Complexity of an ABC System

As should be evident from the discussion up to this point, ABC systems are expensive to create and implement, complex to operate, and difficult to modify or update. You may wish to review Exhibit 6-5, which listed the six steps for the process of creating, implementing, and operating an ABC system. The first three steps correspond to the first stage of ABC (Stage 1), and the last three steps correspond to the second stage of ABC (Stage 2). Stage 1 requires time-consuming and costly interviewing and surveying with the objective of identifying and classifying activities and then determining the cost of each activity. This Stage 1 process produces results that are subjective and difficult to validate. Stage 2 requires an activity rate for each activity. An organization may have hundreds of different activities and, thus, hundreds of activity rates. Activity rates require the identification of activity drivers that measure the consumption of activities by cost objects. Both Stage 1 and Stage 2 are complex and costly. Efforts to simplify ABC have been proposed that involve either before-the-fact simplification or after-the-fact simplification. One prominent before-the-fact simplification approach is *Time-Driven ABC (TDABC)* and is concerned with simplifying Stage 1. Two after-the-fact simplification approaches that simplify Stage 2 are the *Approximately Relevant ABC System* and the *Equally Accurate ABC System*.

Before-the-Fact Simplification: TDABC

Time-Driven Activity-Based Costing (TDABC) is a before-the-fact simplification method that simplifies Stage 1 by eliminating the need for detailed interviewing and surveying to determine resource drivers.[2] Activities still must be identified. However, TDABC assigns resource costs to activities in a very simple and straightforward way. First, it calculates the total operating cost of a department or process for supplying resource capacity (cost of all resources such as equipment, personnel, materials, etc.). Second, it calculates a capacity cost rate by dividing the total resource cost by the practical capacity (as measured by resource time used in the department) of the resources supplied:

Capacity cost rate = Cost of resources supplied/Practical capacity of resources supplied

---

[2] TDABC is described in Robert S. Kaplan and Steven R. Anderson, "Time-Driven Activity-Based Costing," *Harvard Business Review* (November 2004): 131–138; "The Innovation of Time-Driven Activity-Based Costing," *Cost Management* 21, 2 (March/April): 5–15.

Third, it estimates the time to perform one unit of activity. One unit of activity is one unit of an activity driver; thus, multiplying the capacity cost rate by the time to perform one unit of activity and then by the total activity output (as measured by the activity driver) yields the activity cost:

$$\text{Activity cost} = \text{Capacity cost rate} \times \text{Time to perform one unit of activity}$$
$$\times \text{Total activity output}$$
$$= \text{Activity rate} \times \text{Total activity output}$$

In practical terms, the resource cost can be driven directly to products without formally calculating the activity cost. Since by multiplying the capacity cost rate by the time it takes to perform one unit of activity yields the activity rate, resource costs can be assigned to individual products by simply multiplying the activity rate by the amount of activity consumed by each product. Cornerstone 6-7 illustrates the basic concepts of TDABC.

TDABC Features

Easy-to-Update Requirements Cornerstone 6-7 illustrates that the detailed requirements typically found in Stage 2 are not needed. Cornerstone 6-7 also shows that TDABC has a significant advantage when it comes to updating requirements. If new activities are added or identified, there is no need to engage in detailed interviews as with traditional ABC. Instead, all that is needed is observation to determine how long it takes to produce one unit of output for each new activity. Other changes in operations such as changes in resource costs or time (e.g., resource price increases, acquisition of new equipment, process improvements, increase in activity efficiency, etc.) are easily updated by adjusting the capacity cost rate. This then produces new activity rates. Updates are easily obtained as changes occur.

Cost of Unused Capacity Although not discussed in Cornerstone 6-7, another feature of TDABC is its ability to calculate the cost of unused capacity. The unit time multiplied by the activity output is the total time used by an activity. If the actual activity quantities differ from the practical capacity quantities (practical capacity ranges from 80–90 percent of theoretical capacity), then the cost assigned to products will be less than the cost of the total resources. The difference is the cost of unused capacity.

$$\text{Cost of unused capacity} = \text{Total cost of resources} - \text{Total resource cost assigned}$$
$$\text{to products}$$

For example, if the total cost of resources is $486,000 and the total cost of resources assigned to products is $476,000, then the cost of unused capacity is $10,000 ($486,000 − $476,000).

Response Time Occasionally, the time to process a transaction driver, such as response to requests, may differ depending on the category of the patient treated. For example, suppose that it takes 0.4 hour to respond to normal-care patient requests; however, an additional 0.3 hour is needed to respond to critical-care patient requests and an additional 0.8 hour is needed to respond to emergency-state patient requests. In a traditional ABC system, this complexity can be handled by creating three different activities. TDABC, on the other hand, handles this increased complexity in a very simple, straightforward way. TDABC estimates the resource demand using a time equation:

$$\text{Response time} = 0.40 + 0.30 \text{ (if critical patient)} +$$
$$0.80 \text{ (if emergency-state patient)}$$

Suppose that the capacity cost rate is $30 per hour. The cost per response for a normal patient is $30 \times 0.4 = 12. For a critical care patient, the cost per response is $30(0.40 + 0.30) = 21 and for an emergency-state patient it is $30(0.40 + 0.80) = 36. The time equation allows different response times to be calculated depending on the category of patient and thus different costs can be assigned to the different request categories.

**CORNERSTONE
6-7**

The HOW and WHY of TDABC

Information:

Exhibit 6-9 and the following information on a cardiology unit:

Resources		Activities	Time/Unit of Activity
Supervision	$ 50,000	Treating patients	1.40 hrs
Supplies and uniforms	60,000	Providing hygienic care	1.00 hr
Salaries	340,000	Responding to requests	0.60 hr
Computer	10,000	Monitoring patients	1.00 hr
Monitor	26,000		
Total	$486,000		
Total nursing hours	18,000	(practical capacity)	

Why:

TDABC avoids detailed interviewing, surveying, and timekeeping systems required for assessing resource drivers to assign resource costs to activities. All that is needed is the total labour time in a department or process (measured at practical capacity), the total resource costs, and the time required to perform one activity. The first two data items are readily obtained through objective estimates. The time to perform one unit of activity is simply the unit of time for a duration driver. For transaction drivers, the amount of time required to perform one transaction (such as a setup) is obtained by observation or interview. The capacity cost rate is total resource cost/total time at practical capacity. This rate is multiplied by the time per unit of an activity to obtain the activity rate. This rate multiplied by the total activity output provides the activity cost.

Required:

1. Calculate the capacity cost rate for the cardiology unit.
2. Calculate the activity rate for each activity and the cost of the two activities: Treating patients and Monitoring patients.
3. *What if* at mid-year, the nursing supervisor resigns and a new supervisor is hired for a salary of $63,200 and the cardiology nurses also receive a 12 percent increase in salaries? Update the activity rates calculated in Requirement 2.

Solution:

1. Capacity cost rate = $486,000/18,000 = $27 per hour

2.   Treating patients: $27 × 1.40 = $37.80 per patient

      Providing hygienic care: $27 × 1 = $27 per care hour

      Responding to requests: $27 × 0.60 = $16.20 per request

      Monitoring patients: $27 × 1 = $27 per monitoring hour

      Treating patients: $37.80 × 3,000 = $113,400

      Monitoring patients: $27 × 5,400 = $145,800

3. New capacity cost rate = $540,000/18,000 = $30 (Resource costs increase by $54,000 due to salary changes of the supervisor and nurses.) New activity rates are therefore:

> ### CORNERSTONE
> ### 6-7
> *(continued)*

Treating patients: $30 × 1.40 = $42 per patient

Providing hygienic care: $30 × 1 = $30 per care hour

Responding to requests: $30 × 0.60 = $18 per request

Monitoring patients: $30 × 1 = $30 per monitoring hour

Real-World Example

An interesting application of time equations is reported for the inter-library loan service of the **KULeuven Arenberg Library**.[3] The initial activity for a requested book or article from another library is defined as *processing the request*. The transaction driver is defined as the number of requests processed. Thus, TDABC must estimate the time required to process one request. The estimated time required to process one request is 6.8 minutes and is the sum of the following tasks: receive the request, select the library that has the requested book or article, print a hard copy of the request, enter data in an Excel™ file, and classify all printouts. The time for processing a request can be increased or decreased depending on two additional complexities. First, if the library patron asks for feedback, then an additional 6.3 minutes are required to provide the feedback via e-mail, telephone, or personal contact at the library desk. Second, the lending library may respond negatively and indicate that the book or article is not available. In this case, the process needs to be repeated with a new potential lending library, which adds an additional 6.6 minutes. The resulting time equation for this initial activity is given as:

Process time = 6.8 + 6.3 (if feedback is requested) + 6.6 (if negative response)

After-the-Fact Simplification

Although TDABC simplifies Stage I, Stage 2 still has to deal with hundreds of different activity rates. While information technology is capable of handling this volume, there is merit to reducing the number of rates if it can be done without suffering a significant decrease in the accuracy of the cost assignments. Fewer activity rates may produce more readable and manageable product cost reports, reducing the perceived complexity of an activity-based costing system and increasing its likelihood of managerial acceptance. For example, if there are a large number of activities on a bill of activities, managers are likely to find it too complex to read, interpret, and use. In this case, the more complex ABC or TDABC system may not be sustained. One of the oft-cited reasons for refusing to implement an ABC system, or for abandoning it once implemented, is the perceived complexity of the system. Fewer rates may also reduce the ongoing cost of operating an ABC system. Predetermined rates require that actual activity data be collected so that overhead can be applied. Fewer rates thus reduce the ongoing data collection activity required. In practical terms, a complex ABC system may not be sustainable simply because there is too much actual driver data to collect effectively.

Consider the data presented in Exhibit 6-11 for Patna Company, a manufacturer of wafers for integrated circuits. Patna produces two types of wafers: Wafer A and Wafer B. A wafer is a thin slice of silicon used as a base for integrated circuits or other electronic components. The dies on each wafer represent a particular configuration—a configuration designed for use by a particular end product. Patna produces wafers in batches, where each batch corresponds to a particular type of wafer (A or B). In the wafer inserting and sorting process, dies are inserted, and the wafers are tested to ensure that the dies are not defective. From Exhibit 6-11, we see that the activity-based costs for Wafer A and Wafer B are $800,000 and $1,200,000, respectively. These activity-based costs are calculated using the 12 drivers. A key question is whether or not the benefits of an ABC system can essentially

[3] Eli Pernot, Filip Roodhooft, and Alexandria Van den Abbeele, "Time-Driven Activity-Based Costing for Inter-Library Services: A Case Study in a University," *The Journal of Academic Libraryship* 33, 5 (September 2007): 551–560.

Exhibit 6-11

Data for Patna Company

Activity	Budgeted Activity Cost	Driver	Quantity[a]	Expected Consumption Ratios	
				Wafer A	Wafer B
Inserting and sorting process:					
1. Developing test programs	$ 400,000	Engineering hours	10,000	0.25	0.75
2. Making probe cards	58,750	Development hours	4,000	0.10	0.90
3. Testing products	300,000	Test hours	20,000	0.60	0.40
4. Setting up batches	40,000	Number of batches	100	0.55	0.45
5. Engineering design	80,000	Number of change orders	50	0.15	0.85
6. Handling wafer lots	90,000	Number of moves	200	0.45	0.55
7. Inserting dies	350,000	Number of dies	2,000,000	0.70	0.30
Procurement process:					
8. Purchasing materials	450,000	Number of purchase orders	2,500	0.20	0.80
9. Unloading materials	60,000	Number of receiving orders	3,000	0.35	0.65
10. Inspecting materials	75,000	Inspection hours	5,000	0.65	0.35
11. Moving materials	30,000	Distance moved	3,000	0.50	0.50
12. Paying suppliers	66,250	Number of invoices	3,500	0.30	0.70
Total activity cost	$2,000,000				
Unit-level (plantwide) cost assignment[b]				$1,400,000	$600,000
Activity cost assignment[c]				$800,000	$1,200,000

[a] Total amount of the activity expected to be used by both products
[b] Calculated using *number of dies* as the single unit-level driver:
   Wafer A = 0.7 × $2,000,000; Wafer B = 0.3 × $2,000,000
[c] Calculated using *each* activity cost and either the associated consumption ratios or activity rates. For example, the cost assigned to Wafer A using the consumption ratio for *developing testing programs* is 0.25 × $400,000 = $100,000. Repeating this for each activity and summing yields a total of $800,000 assigned to Wafer A.

be captured with a system using a significantly reduced number of drivers. We will consider two approaches for simplification: (1) Approximately Relevant ABC Systems and (2) Equally Accurate Reduced ABC Systems.

Approximately Relevant ABC Systems   It is possible that an organization is better off having an approximately relevant ABC system rather than a precisely useless one.[4] One intriguing suggestion for obtaining an approximately relevant ABC system is to do an analysis of the activity accounting system and to use only the most expensive activities for ABC assignment.[5] The costs of all other activities can be added to the cost pools of the expensive activities. For example, the costs of the less expensive activities could be allocated in proportion to the costs in each of the expensive activities. In this way, most costs will be assigned to the products accurately. The costs of the most expensive activities will still be assigned using appropriate cause-and-effect drivers, while the added costs will be assigned somewhat arbitrarily. The advantages of this approach are that it is simple, easy to understand, and easy to implement. It also often provides a good approximation of the ABC costs. Cornerstone 6-8 illustrates this approach.

Cornerstone 6-8 illustrates that the ABC costs are approximated quite well by the reduced system of four drivers. Furthermore, it seems that the cost is much better than the plantwide rate, even when the system has significant error relative to the ABC assignments. If activity costs roughly follow the Pareto principle or 80/20 rule (80 percent of the overhead costs are caused by 20 percent of the activities), then this approach for

[4] Tom Pryor, "Simplify Your ABC," *Cost Management Newsletter* 15 (June 2004): accessed online at http://www.icms .net/news-21.htm.
[5] Ibid.

The HOW and WHY of Approximately Relevant ABC Systems

Information:
Exhibit 6-11.

Why:
The number of drivers used to assign costs can be reduced by using only the drivers associated with the most expensive activities. Costs of the less expensive activities are allocated to the more expensive activities in proportion to their original cost. This provides a cost system that assigns most of the costs using causal relationships and yet is simple to understand and easy to use. For this method to be of value, a high percentage of the overhead costs must be attributable to a relatively small number of activities.

**CORNERSTONE
6-8**

Required:
1. Using the four most expensive activities, calculate the overhead cost assigned to each product.
2. Calculate the error relative to the fully specified ABC product cost and comment on the outcome.
3. **What if** activities 1, 5, 8, and 12 each had a cost of $400,000 and the remaining activities had a cost of $50,000? Calculate the cost assigned to Wafer A by a fully specified ABC system and then by an approximately relevant ABC approach. Comment on the implications for the approximately relevant approach.

Solution:
1.

Activity	Budgeted Activity Cost[a]	Driver	Quantity	Expected Consumption Ratios Wafer A	Wafer B
1. Developing test programs	$ 533,333	Engineering hours	10,000	0.25	0.75
3. Testing products	400,000	Test hours	20,000	0.60	0.40
7. Inserting dies	466,667	Number of dies	2,000,000	0.70	0.30
8. Purchasing materials	600,000	Purchase orders	2,500	0.20	0.80
Total activity cost	$2,000,000				
Approximate ABC cost[b]				$820,000	$1,180,000

[a] Original activity cost plus share of the costs of the remaining "inexpensive" activities (allocated in proportion to the original costs of the expensive activities (as shown in Exhibit 6-11): For example, the cost pool for purchasing materials is $450,000 + [($450,000/$1,500,000) × $500,000] = $600,000.
[b] Reduced system ABC assignment (using consumption ratios): Wafer A: [(0.25 × $533,333) + (0.60 × $400,000) + (0.70 × $466,667) + (0.20 × $600,000)]; Wafer B: [(0.75 × $533,333) + (0.40 × $400,000) + (0.30 × $466,667) + (0.80 × $600,000)]

2. Relative error, Wafer A: ($820,000 − $800,000)/$800,000 = 0.025 (2.5%)
Relative error, Wafer B: ($1,180,000 − $1,200,000)/$1,200,000 = −0.017 (−1.7%)

The maximum error is a 2.5 percent overstatement of the ABC cost of Exhibit 6-11, when 12 drivers are used. This is a very good approximation indicating that the approach has merit.

3. Using consumption ratios, the ABC cost of Wafer A is $400,000(0.25 + 0.15 + 0.20 + 0.30) + $50,000(0.10 + 0.60 + 0.55 + 0.45 + 0.70 + 0.35 + 0.65 + 0.50) = $555,000. Since the cost is the same for each of the four

**CORNERSTONE
6-8
(continued)**

most expensive activities, the reassigned cost for each of the four activities is $500,000 (each receives the same amount of the less expensive activities). Thus, using consumption ratios, the approximately relevant cost is $500,000(0.25 + 0.15 + 0.20 + 0.30) = $450,000. The difference between the ABC cost and the approximately relevant cost is −$105,000 ($450,000 − $555,000) or a relative error of about −19 percent. It appears that a significant error can occur even when the expensive activities account for about 80 percent of the total overhead. However, this is still a vast improvement over the plantwide rate assignment (which is $1,400,000 vs. $555,000).

reducing the size of the system has considerable promise. For example, if a system has 100 activities, then the top 20 activities (as measured by their cost) need to account for a very high percentage of the total costs. In those cases where this holds, a reduced system may work reasonably well because *most* of the costs are assigned using cause-and-effect relationships. Even so, there may be some who would balk at the notion of using 15 to 20 drivers. The approach also loses its usefulness for those companies where a small number of activities do not account for a large share of the overhead costs.

Equally Accurate Reduced ABC Systems   Another approach is to use expected consumption ratios to reduce the number of drivers. Although the theoretical motivation for this approach is beyond the scope of the text, the methodology is straightforward. Consider again the 12 activities of Exhibit 6-11. The product costs assigned to Wafer A and Wafer B were $800,000 and $1,200,000, respectively. Thus, Wafer A is expected to consume 40 percent ($800,000/$2,000,000) of the total cost being assigned, and Wafer B is expected to consume 60 percent ($1,200,000/ $2,000,000) of the total cost being assigned. Wafer A has an *expected global consumption ratio* of 0.40, and Wafer B has an *expected global consumption ratio* of 0.60. The **expected global consumption ratio** is the proportion of the total activity costs consumed by a given product (cost object). The expected global consumption ratio pattern for Patna Company is (0.40, 0.60). Each activity also has a consumption ratio pattern.

For a two-product firm, the activity consumption ratio patterns are always described by an array (vector) of two components. For the Patna Company example, the first ratio in the array is the proportion of the activity consumed by Wafer A, and the second ratio is the proportion consumed by Wafer B. For example, the activity, developing test programs, has a consumption pattern of (0.25, 0.75), where Wafer A consumes 25 percent of the activity cost and Wafer B consumes 75 percent of the activity cost. Similarly, the activity, inserting dies, has a consumption pattern of (0.70, 0.30), where Wafer A consumes 70 percent of the activity cost and Wafer B consumes 30 percent. As the number of products increases, the number of consumption ratio components also increases. The dimension of the consumption ratio pattern array corresponds to the number of products. When the number of activities is more than the number of products, it is always possible to find a reduced system that *duplicates* the cost assignments of the larger system. To achieve this duplication, the number of drivers needed is *at most* equal to the number of products (two drivers for our example). Thus, two drivers can be used to match the larger 12-driver system cost assignments. A key step in the reduction process is expressing each global consumption ratio as a weighted combination of the consumption ratios for each product. For example, using the activities, developing test programs and inserting dies, the weighted combination for Wafer A is $0.25w_1 + 0.70w_2 = 0.40$. A similar equation can be developed for Wafer B: $0.75w_1 + 0.30w_2 = 0.60$. Solving these two equations yields values for w_1 and w_2. These values are *allocation ratios* and when multiplied by the total overhead costs define two cost pools (one for the first activity and one for the second activity). Using the consumption ratios or drivers for each activity then assigns the appropriate amount of cost to each product. How this is achieved and the motivation are summarized in Cornerstone 6-9.

The HOW and WHY of Equally Accurate Reduced ABC Systems

CORNERSTONE
6-9

Information:

From Exhibit 6-11, the following data are extracted:

Activity	Driver	Quantity	Expected Consumption Ratios	
			Wafer A	Wafer B
3. Testing products	Test hours	20,000	0.60	0.40
8. Purchasing materials	Purchase orders	2,500	0.20	0.80
1. Developing test programs	Engineering hours	10,000	0.25	0.75
7. Inserting dies	Number of dies	2,000,000	0.70	0.30
ABC assignment			$800,000	$1,200,000
Total overhead cost			$2,000,000	

Why:

It is always possible to find a reduced system that matches the accuracy of the larger ABC system. Using fewer drivers facilitates acceptance and use of an ABC system. The steps that should be followed to achieve the desired simplification are: (1) Calculate the expected global consumption ratio (ABC product cost/total overhead cost); (2) Select the needed number of activities (equal to the number of products); (3) Form equations for each product by multiplying the consumption ratios of each product by the allocation weights and setting the result equal to the product's global consumption ratio; (4) Solve the simultaneous set of equations; (5) Use the weights to form the cost pools that will duplicate the larger ABC system cost assignments; and (6) Use the consumption ratios (or drivers) to assign the cost pools to individual products.

Required:

1. Form reduced system cost pools for activities 3 and 8.
2. Assign the costs of the reduced system cost pools to Wafer A and Wafer B.
3. **What if** the two activities were 1 and 7? Repeat Requirements 1 and 2. What does this imply?

Solution:

1. Global ratios = 0.40 ($800,000/$2,000,000) for Wafer A and 0.60 ($1,200,000/$2,000,000) for Wafer B.
   Equations:

   $0.60w_1 + 0.20w_2 = 0.40$ (Wafer A)

   $0.40w_1 + 0.80w_2 = 0.60$ (Wafer B)

   Multiplying both sides of the first equation by 4, subtracting the second from the first, and solving, we obtain:

   Solving: $w_1 = 1/2$ and $w_2 = 1/2$

   Testing products cost pool: $0.5 \times \$2,000,000 = \$1,000,000$

   Purchasing cost pool: $0.5 \times \$2,000,000 = \$1,000,000$

**CORNERSTONE
6-9
*(continued)***

2. Using the consumption ratios, the same cost assignment is realized with two drivers:
   Wafer A: $(0.60 \times \$1,000,000) + (0.20 \times \$1,000,000) = \$800,000$
   Wafer B: $(0.40 \times \$1,000,000) + (0.80 \times \$1,000,000) = \$1,200,000$

3. Equations:

   $0.25w_1 + 0.70w_2 = 0.40$ (Wafer A)

   $0.75w_1 + 0.30w_2 = 0.60$ (Wafer B)

   Solving: $w_1 = 2/3$ and $w_2 = 1/3$

   Cost pool (test programs): $(2/3) \times \$2,000,000 = \$1,333,333$
   Cost pool (inserting dies): $(1/3) \times \$2,000,000 = \$666,667$

   Wafer A: $(0.25 \times \$1,333,333) + (0.70 \times \$666,667) = \$800,000$
                                                        (rounded)

   Wafer B: $(0.75 \times \$1,333,333) + (0.30 \times \$666,667) = \$1,200,000$
                                                        (rounded)

   The implication is that any two activities will work—but negative allocations may occur if the global ratio on the right-hand side does not lie between the coefficients of the two allocation weights.

Cornerstone 6-9 shows that an equally accurate simplified system can be derived from the more complex ABC system. Instead of using 12 drivers, it is possible to use only two drivers and achieve the same cost assignment of the more complex system. This reduced system represents an *after-the-fact* simplification. The reduced system is derived from an *existing* complex ABC data set. Of course, the same is true for the approximately relevant reduced system that uses the Pareto principle to achieve the reduction. The value of after-the-fact simplification is based on two key justifications. First, the reduced system eliminates the perceived complexity of the system. For example, it is much easier for nonfinancial users to read, interpret, and use a two-driver system compared to a 12-driver system. Second, the reduced ABC system needs to collect actual driver data only for the drivers being used to assign the costs to products. For example, in the case of Patna Company, only actual data for testing hours and number of purchase orders need to be collected so that overhead costs can be assigned (applied) to the two products. This is much less costly than collecting actual data for 12 drivers. Finally, it should also be pointed out that the two drivers in Exhibit 6-11 are only one of many two-driver combinations that can be used to reduce the ABC system without sacrificing the assignment accuracy of the more complex system.

Summary of Learning Objectives

1. **Describe the basics of plantwide and departmental overhead costing.**

- Budgeted overhead costs are accumulated into plantwide or departmental pools and predetermined overhead rates are calculated.
- Predetermined rates use unit-level drivers such as direct labour hours and machine hours.
- Overhead is assigned by multiplying the rate by the actual total amount of unit-level driver (e.g., direct labour hours).

- The difference between the actual overhead and applied overhead is an overhead variance and is either under- or overapplied. If the variance is immaterial, it is closed to Cost of Goods Sold; otherwise, it is allocated among work-in-process inventory, finished goods, and cost of good sold.

2. **Explain why plantwide and departmental overhead costing may not be accurate.**

- Overhead assignments should reflect the amount of overhead demanded (consumed) by each product.
- Many overhead activities are unrelated to the units produced, and assigning overhead using unit-level drivers may distort product costs.
- If overhead is a significant proportion of total manufacturing costs, this distortion can be serious.
- Activity-based costing uses both unit-level and non-unit-level drivers and thus reflects a more accurate picture of the actual overhead consumed by products.

3. **Provide a detailed description of activity-based product costing.**

- Identify, define, and classify activities.
- Assign the cost of resources to each activity.
- Assign the cost of secondary activities to primary activities.
- Identify cost objects and specify the amount of each activity consumed by each cost object.
- Calculate primary activity rates.
- Assign activity costs to cost objects.

4. **Explain how ABC can be simplified.**

- TDABC, a before-the-fact simplification approach, eliminates the need to identify resource drivers to assign resource costs to activities, eliminating the need for much of the detailed implementation interviews.
- TDABC also makes it easier to update ABC when changes occur.
- Simplified ABC systems can be derived from complex ABC systems.
- Simplified systems facilitate the presentation and use of ABC information and reduce the cost of collecting actual driver data.
- Two after-the-fact approaches were discussed: the approximately relevant ABC system and the equally accurate reduced ABC system. The first approach may be useful for those firms where a few activities account for most of the overhead costs. The second system is useful whenever the number of activities is greater than the number of products (which is usually the case).

CORNERSTONES FOR CHAPTER 6

Review Problem

Unit-Based Costing versus Activity-Based Costing

Abat-Jour Lamp Company is noted for its full line of quality lamps. The company operates one of its plants in Montreal, Quebec. That plant produces two types of lamps: classical and modern. Jane Martinez, president of the company, recently decided to change from a unit-based, traditional costing system to an activity-based costing system. Before making the change companywide, she wanted to assess the effect on the product costs of the Montreal plant. This plant was chosen because it produces only two types of lamps; most other plants produce at least a dozen.

To assess the effect of the change, the following data have been gathered (for simplicity, assume one process):

Lamp	Quantity	Prime Costs	Machine Hours	Material Moves	Setups
Classical	400,000	$800,000	81,250	300,000	100
Modern	100,000	$150,000	43,750	100,000	50
Dollar amount	—	$950,000	$500,000*	$900,000	$600,000

*The cost of operating the production equipment.

Under the current system, the costs of operating equipment, materials handling, and setups are assigned to the lamps on the basis of machine hours. Lamps are produced and moved in batches.

Required:

1. Compute the unit cost of each lamp using the current unit-based approach.
2. Compute the unit cost of each lamp using an activity-based costing approach.
3. Show how a reduced system using two cost pools and two drivers, moves and setups, can be used to achieve the same cost assignments obtained in Requirement 2.

Solution:

1. Total overhead is $2,000,000. The plantwide rate is $16 per machine hour ($2,000,000/125,000). Overhead is assigned as follows:

    Classical lamps: $16 × 81,250 = $1,300,000
    Modern lamps: $16 × 43,750 = $700,000

    The unit costs for the two products are as follows:

    Classical lamps: ($800,000 + $1,300,000)/400,000 = $5.25
    Modern lamps: ($150,000 + $700,000)/100,000 = $8.50

2. In the activity-based approach, a rate is calculated for each activity:

    Machining: $500,000/125,000 = $4.00 per machine hour
    Moving materials: $900,000/400,000 = $2.25 per move
    Setting up: $600,000/150 = $4,000 per setup

    Overhead is assigned as follows:

Classical lamps:	
$4 × 81,250	$ 325,000
$2.25 × 300,000	675,000
$4,000 × 100	400,000
Total	$1,400,000

Modern lamps:

$4 × 43,750	$ 175,000
$2.25 × 100,000	225,000
$4,000 × 50	200,000
Total	$ 600,000

This produces the following unit costs:

Classical lamps:

Prime costs	$ 800,000
Overhead costs	1,400,000
Total costs	$2,200,000
Units produced	÷ 400,000
Unit cost	$ 5.50

Modern lamps:

Prime costs	$ 150,000
Overhead costs	600,000
Total costs	$ 750,000
Units produced	÷ 100,000
Unit cost	$ 7.50

3. First, calculate the activity consumption ratios:

	Moving	Setups
Classical	300,000/400,000 = 3/4	100/150 = 2/3
Modern	100,000/400,000 = 1/4	50/150 = 1/3

Second, calculate the global consumption ratios (information from Requirement 2 is needed):

ABC Assignments		Global Ratios
Overhead assigned to classical:	$1,400,000	$1,400,000/$2,000,000 = 0.70
Overhead assigned to modern:	600,000	$600,000/$2,000,000 = 0.30
Total	$2,000,000	

Third, set up and solve the consumption ratio equations:

$$(3/4)w_1 + (2/3)w_2 = 0.70$$
$$(1/4)w_1 + (1/3)w_2 = 0.30$$

Solving, we have the allocation ratios: $w_1 = 0.40$ and $w_2 = 0.60$. Thus, the cost pools for the two activities are:

Moving: $0.40 × $2,000,000 = $800,000$

Setups: $0.60 × $2,000,000 = $1,200,000$

The activity rates for the reduced system would be:

Moving: $800,000/400,000 = $2.00 per move

Setups: $1,200,000/150 = $8,000 per setup

Overhead cost assignments:

Classical lamps:

$2.00 × 300,000	$ 600,000
$8,000 × 100	800,000
Total	$1,400,000

Modern lamps:

$2.00 × 100,000	$ 200,000
$8,000 × 50	400,000
Total	$ 600,000

Key Terms

Activity attributes, 273

Activity dictionary, 273

Activity drivers, 276

Activity inventory, 272

Activity-based costing (ABC) system, 272

Applied overhead, 259

Batch-level activities, 279

Bill of activities, 278

Consumption ratio, 265

Duration drivers, 276

Expected global consumption ratio, 286

Facility-level activities, 280

Non-unit-based drivers, 263

Normal costing system, 259

Overapplied overhead, 261

Overhead variance, 261

Predetermined overhead rate, 259

Primary activity, 273

Product diversity, 265

Product-level activities, 279

Resource drivers, 275

Secondary activity, 273

Time-Driven Activity-Based Costing (TDABC), 280

Transaction drivers, 276

Underapplied overhead, 261

Unit-based drivers, 259

Unit-level activities, 279

Discussion Questions

1. What is a predetermined overhead rate? Explain why it is used.
2. Describe what is meant by under- and overapplied overhead.
3. Explain how a plantwide overhead rate, using a unit-based driver, can produce distorted product costs. In your answer, identify two major factors that impair the ability of plantwide rates to assign cost accurately.
4. What are non-unit-related overhead activities? Non-unit-based cost drivers? Give some examples.
5. What is an overhead consumption ratio?
6. Overhead costs are the source of product cost distortions. Do you agree or disagree? Explain.
7. What is activity-based product costing?
8. What are the six steps that define the design of an activity-based costing system?
9. Explain how the cost of resources is assigned to activities. What is meant by the phrase "unbundling the general ledger accounts"?
10. What is a bill of activities?
11. Identify and define two types of activity drivers.
12. What are unit-level activities? Batch-level activities? Product-level activities? Facility-level activities?
13. How does TDABC simplify ABC?
14. Explain why it is easy to update a TDABC model.
15. Describe two ways to reduce a complex ABC system. Of the two ways, which has the most merit?

Cornerstone Exercises

OBJECTIVE ▶ 1

CORNERSTONE 6-1

Cornerstone Exercise 6-1 APPLIED OVERHEAD AND UNIT OVERHEAD COST: PLANTWIDE RATES

Liang Inc. produces two types of speakers: deluxe and regular. Liang uses a plantwide rate based on direct labour hours to assign its overhead costs. The company has the following estimated and actual data for the coming year:

Estimated overhead	$750,000
Expected activity	25,000
Actual activity (direct labour hours):	
Deluxe speaker	5,000
Regular speaker	20,000
Units produced:	
Deluxe speaker	10,000
Regular speaker	100,000

Required:

1. Calculate the predetermined plantwide overhead rate and the applied overhead for each product, using direct labour hours.
2. Calculate the overhead cost per unit for each product.
3. *What if* the deluxe product used 10,000 hours (to produce 10,000 units) instead of 5,000 hours (total expected hours remain the same)? Calculate the effect on the profitability of this product line if all 10,000 units are sold, and then discuss the implications of this outcome.

Cornerstone Exercise 6-2 OVERHEAD VARIANCES AND THEIR DISPOSAL

Aphrodite Company has the following data for the past year:

Actual overhead	$ 960,000
Applied overhead:	
Work-in-process inventory	$ 150,000
Finished goods inventory	250,000
Cost of goods sold	600,000
Total	$1,000,000

Aphrodite uses the overhead control account to accumulate both actual and applied overhead.

Required:

1. Calculate the overhead variance for the year and close it to Cost of Goods Sold.
2. Assume the variance calculated is material. After prorating, close the variances to the appropriate accounts and provide the final ending balances of these accounts.
3. *What if* the variance is of the opposite sign calculated in Requirement 1? Provide the appropriate adjusting journal entries for Requirements 1 and 2.

Cornerstone Exercise 6-3 DEPARTMENTAL OVERHEAD RATES

Fleming Inc. provided the following data for its two producing departments:

	Polishing	Painting	Total
Estimated overhead	$500,000	$100,000	$600,000
Direct labour hours (expected and actual):			
Frame A	2,000	5,000	7,000
Frame B	3,000	15,000	18,000
Total	5,000	20,000	25,000
Machine hours:			
Frame A	7,000	3,000	10,000
Frame B	3,000	2,000	5,000
Total	10,000	5,000	15,000

Machine hours are used to assign the overhead of the Polishing Department, and direct labour hours are used to assign the overhead of the Painting Department. There are 60,000 units of Frame A produced and sold and 100,000 of Frame B.

Required:

1. Calculate the overhead rates for each department.
2. Using departmental rates, assign overhead to the two products and calculate the overhead cost per unit. How does this compare with the plantwide rate unit cost, using direct labour hours?
3. *What if* the machine hours in Polishing were 2,000 for Frame A and 8,000 for Frame B and the direct labour hours used in Painting were 14,000 and 6,000, respectively? Calculate the overhead cost per unit for each product using departmental rates, and compare with the plantwide rate unit costs calculated in Requirement 2. What can you conclude from this outcome?

OBJECTIVE ▶ 2

CORNERSTONE 6-4

Cornerstone Exercise 6-4 CONSUMPTION RATIOS

Larissa Inc. produces two types of electronic parts and has provided the following data:

	Part X12	Part YK7	Total
Units produced	100,000	600,000	—
Direct labour hours	30,000	70,000	100,000
Machine hours	50,000	300,000	350,000
Number of setups	40	80	120
Testing hours	1,000	9,000	10,000
Number of purchase orders	500	3,500	4,000

There are four activities: machining, setting up, testing, and purchasing.

Required:

1. Calculate the activity consumption ratios for each product.
2. Calculate the consumption ratios for a plantwide (direct labour hours) rate. When compared with the activity ratios, what can you say about the relative accuracy of a plantwide rate? Which product is undercosted?
3. *What if* the machine hours were used for the plantwide rate? Would this remove the cost distortion of a plantwide rate?

OBJECTIVE ▶ 2

CORNERSTONE 6-5

Cornerstone Exercise 6-5 ACTIVITY-BASED PRODUCT COSTING

Kim Company produces two lawn mowers: basic and self-propelled. The company has four activities: machining, engineering, receiving, and packing. Information on these activities and their drivers is given below.

	Basic	Self-Propelled	Total
Units produced	250,000	750,000	—
Prime costs	$20,000,000	$75,000,000	$95,000,000
Machine hours	250,000	1,250,000	1,500,000
Engineering hours	1,000	9,000	10,000
Receiving orders	1,000	3,000	4,000
Inspection hours	2,000	4,000	6,000

Overhead costs:

Machining	$15,000,000
Engineering	5,000,000
Receiving	1,400,000
Inspecting products	900,000

Required:

1. Calculate the four activity rates.
2. Calculate the unit costs using activity rates. Also, calculate the overhead cost per unit.
3. *What if* consumption ratios were used to assign costs instead of activity rates? Show the cost assignment for the inspection activity.

Cornerstone Exercise 6-6 ASSIGNING COST OF RESOURCES TO ACTIVITIES, UNBUNDLING THE GENERAL LEDGER

OBJECTIVE ▶ 3

CORNERSTONE 6-6

SERVICE

Perlman Bank provided the following data about its resources and activities for its chequing account process:

Resources		Activities	Clerical Hours
Supervision	$ 70,000	Processing accounts	10,000
Phone and supplies	90,000	Issuing statements	5,000
Salaries	275,000	Processing transactions	7,000
Computer	25,000	Answering customer inquiries	3,000
Total	$460,000	Total	25,000

- Computers are used only by the issuing (30 percent) and processing transaction (70 percent) activities.
- Phone and supplies are 60 percent customer inquiries with the other 40 percent divided equally among the remaining activities, including supervising the chequing operation.
- The supervisor spends 100 percent of her time on supervision. In addition to the 25,000 clerical hours, there are 2,000 hours of supervision used (the hours used by the supervising clerks activity, which is not listed above).

Required:

1. Prepare a work distribution matrix for the five primary activities.
2. Calculate the cost of each activity.
3. *What if* the cost of the supervising activity is assigned to the other four activities? Why would this be done? If it is done, what is the final cost of these four primary activities?

Cornerstone Exercise 6-7 SIMPLIFYING THE ABC SYSTEM: TDABC

OBJECTIVE ▶ 4

CORNERSTONE 6-7

SERVICE

Perlman Bank provided the following data about its resources and activities for its chequing account process:

Resources		Activities	Time per Unit	Activity Driver
Supervision	$ 70,000	Processing accounts	0.20 hr	No. of accounts
Phone and supplies	90,000	Issuing statements	0.10 hr	No. of statements
Salaries	275,000	Processing transactions	0.05 hr	No. of transactions
Computer	25,000	Answering customer inquiries	0.15 hr	No. of inquiries
Total	$460,000			
Total cheque processing hours	25,000 (practical capacity)			

Required:

1. Calculate the capacity cost rate for the chequing account process.
2. Calculate the activity rates for the four activities. If the total number of statements issued was 20,000, calculate the cost of the issuing statements activity.
3. *What if* process improvements decreased the number of customer inquiries, leading to a 10 percent reduction in cheque processing hours and a $10,000 reduction in total resource costs? Update all the activity rates for these changes in operating conditions.

Cornerstone Exercise 6-8 SIMPLIFYING THE ABC SYSTEM: APPROXIMATELY RELEVANT ABC SYSTEMS

OBJECTIVE ▶ 4

CORNERSTONE 6-8

Patra Company produces wafers for integrated circuits. Data for the most recent year are provided:

			Expected Consumption Ratios	
Activity		Driver	Wafer A	Wafer B
Inserting and sorting process activities:				
1. Developing test programs	$ 50,000	Engineering hours	0.25	0.75
2. Making probe cards	60,000	Development hours	0.10	0.90
3. Testing products	600,000	Test hours	0.60	0.40
4. Setting up batches	135,000	Number of batches	0.55	0.45
5. Engineering design	90,000	Number of change orders	0.15	0.85
6. Handling wafer lots	300,000	Number of moves	0.45	0.55
7. Inserting dies	700,000	Number of dies	0.70	0.30
Procurement process activities:				
8. Purchasing materials	400,000	Number of purchase orders	0.20	0.80
9. Unloading materials	60,000	Number of receiving orders	0.35	0.65
10. Inspecting materials	75,000	Inspection hours	0.65	0.35
11. Moving materials	500,000	Distance moved	0.50	0.50
12. Paying suppliers	30,000	Number of invoices	0.30	0.70
Total activity cost	$3,000,000			

		Wafer A	Wafer B
Unit-level (plantwide) cost assignment[a]		$2,100,000	$900,000
Activity cost assignment[b]		$1,500,000	$1,500,000

[a] Calculated using *number of dies* as the single unit-level driver
[b] Calculated by multiplying the consumption ratio of each product by the cost of each activity

Required:

1. Using the five most expensive activities, calculate the overhead cost assigned to each product. Assume that the costs of the other activities are assigned in proportion to the cost of the five activities.
2. Calculate the error relative to the fully specified ABC product cost and comment on the outcome.
3. *What if* activities 1, 2, 5, and 8 each had a cost of $650,000 and the remaining activities had a cost of $50,000? Calculate the cost assigned to Wafer A by a fully specified ABC system and then by an approximately relevant ABC approach. Comment on the implications for the approximately relevant approach.

OBJECTIVE ▶ 4
CORNERSTONE 6-9

Cornerstone Exercise 6-9 SIMPLIFYING THE ABC SYSTEM: EQUALLY ACCURATE REDUCED ABC SYSTEMS

Selected activities and other information are provided for Patra Company for its most recent year of operations.

			Expected Consumption Ratios	
Activity	Driver	Quantity	Wafer A	Wafer B
7. Inserting dies	Number of dies	2,000,000	0.70	0.30
8. Purchasing materials	Number of pur- chase orders	2,500	0.20	0.80
1. Developing test programs	Engineering hours	10,000	0.25	0.75
3. Testing products	Test hours	20,000	0.60	0.40
ABC assignment			$1,500,000	$1,500,000
Total overhead cost				$3,000,000

Required:

1. Form reduced system cost pools for activities 7 and 8.
2. Assign the costs of the reduced system cost pools to Wafer A and Wafer B.
3. *What if* the two activities were 1 and 3? Repeat Requirements 1 and 2. What does this imply?

Exercises

Exercise 6-10 **PREDETERMINED OVERHEAD RATE, APPLIED OVERHEAD, UNIT COST**

OBJECTIVE ▶ 1

Glencoe Inc. costs products using a normal costing system. The following data are available for last year:

Budgeted:		
Overhead	$476,000	
Machine hours	140,000	
Direct labour hours	17,000	
Actual:		
Overhead	$475,000	
Machine hours	137,000	
Direct labour hours	16,550	
Prime cost	$1,750,000	
Number of units	250,000	

Overhead is applied on the basis of direct labour hours.

Required:

1. What was the predetermined overhead rate?
2. What was the applied overhead for last year?
3. Was overhead over- or underapplied, and by how much?
4. What was the total cost per unit produced (carry your answer to four significant digits)?

Exercise 6-11 **PREDETERMINED OVERHEAD RATE, APPLICATION OF OVERHEAD**

OBJECTIVE ▶ 1

Jackson Company and Jalil Company both use predetermined overhead rates to apply manufacturing overhead to production. Jackson's is based on machine hours, and Jalil's is based on materials cost. Budgeted production and cost data for Jackson and Jalil are as follows:

	Jackson	Jalil
Manufacturing overhead	$608,000	$660,000
Units	20,000	60,000
Machine hours	32,000	22,500
Materials cost	$300,000	$1,200,000

At the end of the year, Jackson Company had incurred overhead of $610,000 and had produced 19,600 units using 31,980 machine hours and materials costing $294,000. Jalil Company had incurred overhead of $648,000 and had produced 61,500 units using 22,650 machine hours and materials costing $1,185,000.

Required:

1. Compute the predetermined overhead rates for Jackson Company and Jalil Company.
2. Was overhead over- or underapplied for each company, and by how much?

Exercise 6-12 **PREDETERMINED OVERHEAD RATE, OVERHEAD VARIANCES, JOURNAL ENTRIES**

OBJECTIVE ▶ 1

Menotti Company uses a predetermined overhead rate to assign overhead to jobs. Because Menotti's production is machine intensive, overhead is applied on the basis of machine hours. The expected overhead for the year was $3.8 million, and the practical level of activity is 250,000 machine hours.

During the year, Menotti used 255,000 machine hours and incurred actual overhead costs of $3.82 million. Menotti also had the following balances of applied overhead in its accounts:

Work-in-process inventory	$ 384,000
Finished goods inventory	416,000
Cost of goods sold	1,200,000

Required:

1. Compute a predetermined overhead rate for Menotti.
2. Compute the overhead variance, and label it as under- or overapplied.
3. Assuming the overhead variance is immaterial, prepare the journal entry to dispose of the variance at the end of the year.
4. Assuming the overhead variance is material, prepare the journal entry that appropriately disposes of the variance at the end of the year.

OBJECTIVE ▶ 1

Exercise 6-13 DEPARTMENTAL OVERHEAD RATES

Narpet Company produces machine tools and currently uses a plantwide overhead rate, based on machine hours. Ray Johnson, the plant manager, has heard that departmental overhead rates can offer significantly better cost assignments than can a plantwide rate. Narpet has the following data for its two departments for the coming year:

	Department A	Department B
Overhead costs (expected)	$240,000	$60,000
Normal activity (machine hours)	40,000	20,000

Required:

1. Compute a predetermined overhead rate for the plant as a whole based on machine hours.
2. Compute predetermined overhead rates for each department using machine hours.
3. Suppose that a machine tool (Product #12X75) used 20 machine hours from Department A and 50 machine hours from Department B. A second machine tool (Product #32Y15) used 50 machine hours from Department A and 20 machine hours from Department B. Compute the overhead cost assigned to each product using the plantwide rate computed in Requirement 1. Repeat the computation using the departmental rates found in Requirement 2. Which of the two approaches gives the fairest assignment? Why?
4. Repeat Requirement 3 assuming the expected overhead cost for Department B is $120,000. Now would you recommend departmental rates over a plantwide rate?

OBJECTIVE ▶ 2

CORNERSTONE 6-3

Exercise 6-14 DRIVERS AND PRODUCT COSTING ACCURACY

Balint Company produces two types of leather purses: standard and handcrafted. Both purses use equipment for cutting and stitching. The equipment also has the capability of creating standard designs. The standard purses use only these standard designs. They are all of the same size to accommodate the design features of the equipment. The handcrafted purses can be cut to any size because the designs are created manually. Many of the manually produced designs are in response to specific requests of retailers. The equipment must be specially configured to accommodate the production of a batch of purses that will receive a handcrafted design. Balint Company assigns overhead using direct labour dollars. Merle Jones, sales manager, is convinced that the purses are not being costed correctly.

To illustrate his point, he decided to focus on the expected annual setup and machine-related costs, which are as follows:

Setup equipment	$18,000
Depreciation	20,000*
Operating costs	22,000

*Computed on a straight-line basis, book value at the beginning of the year was $100,000.

The machine has the capability of supplying 100,000 machine hours over its remaining life.

Merle also collected the expected annual prime costs for each purse, the machine hours, and the expected production (which is the normal output for the company).

	Standard Purse	Handcrafted Purse
Direct labour	$12,000	$36,000
Direct materials	$12,000	$12,000
Units	3,000	3,000
Machine hours	18,000	2,000
Number of setups	40	40
Setup time	400 hrs	200 hrs

Required:

1. Do you think that the direct labour costs and direct materials costs are accurately traced to each type of purse? Explain.
2. The controller has suggested that overhead costs be assigned to each product using a plantwide rate based on direct labour dollars. Machine costs and setup costs are overhead costs. Assume that these are the only overhead costs. For each type of purse, calculate the overhead per unit that would be assigned using a direct labour dollars overhead rate. Do you think that these costs are traced accurately to each purse? Explain.
3. Now calculate the overhead cost per unit per purse using two overhead rates: one for the setup activity and one for the machining activity. In choosing a driver to assign the setup costs, did you use number of setups or setup hours? Why? As part of your explanation, define transaction and duration drivers. Do you think machine costs are traced accurately to each type of purse? Explain.

Exercise 6-15 MULTIPLE VERSUS SINGLE OVERHEAD RATES, ACTIVITY DRIVERS

OBJECTIVE ▶ 3 4

Tardif Company has identified the following overhead activities, costs, and activity drivers for the coming year:

Activity	Expected Cost	Activity Driver	Activity Capacity
Setting up equipment	$120,000	Number of setups	300
Ordering costs	90,000	Number of orders	9,000
Machine costs	210,000	Machine hours	21,000
Receiving	100,000	Receiving hours	5,000

Tardif produces two models of dishwashers with the following expected prime costs and activity demands:

	Model A	Model B
Direct materials	$150,000	$200,000
Direct labour	$120,000	$120,000
Units completed	8,000	4,000
Direct labour hours	3,000	1,000
Number of setups	200	100
Number of orders	3,000	6,000
Machine hours	12,000	9,000
Receiving hours	1,500	3,500

The company's normal activity is 4,000 direct labour hours.

Required:

1. Determine the unit cost for each model using direct labour hours to apply overhead.
2. Determine the unit cost for each model using the four activity drivers.
3. Which method produces the more accurate cost assignment? Why?

Exercise 6-16 ACTIVITY-BASED COSTING, ACTIVITY IDENTIFICATION, ACTIVITY DICTIONARY

OBJECTIVE ▶ 3

Perlman Bank is in the process of implementing an activity-based costing system. A copy of an interview with the manager of Perlman's Credit Card Department follows.

QUESTION 1: How many employees are in your department?

RESPONSE: There are eight employees, including me.

QUESTION 2: What do they do (please describe)?

RESPONSE: There are four major activities: supervising employees, processing credit card transactions, issuing customer statements, and answering customer questions.

QUESTION 3: Do customers outside your department use any equipment?

RESPONSE: Yes. Automatic bank tellers service customers who require cash advances.

QUESTION 4: What resources are used by each activity (equipment, materials, energy)?

RESPONSE: We each have our own computer, printer, and desk. Paper and other supplies are needed to operate the printers. Of course, we each have a telephone as well.

QUESTION 5: What are the outputs of each activity?

RESPONSE: Well, for supervising, I manage employees' needs and try to ensure that they carry out their activities efficiently. Processing transactions produces a posting for each transaction in our computer system and serves as a source for preparing the monthly statements. The number of monthly customer statements has to be the product for the issuing activity, and I suppose that the number of customers served is the output for the answering activity. And I guess that the number of cash advances would measure the product of the automatic teller activity, although the teller really generates more transactions for other products such as chequing and savings accounts. So, perhaps the number of teller transactions is the real output.

QUESTION 6: Who or what uses the activity output?

RESPONSE: We have three products: classic, gold, and platinum credit cards. Transactions are processed for these three types of cards, and statements are sent to clients holding these cards. Similarly, answers to questions are all directed to clients who hold these cards. As far as supervising, I spend time ensuring the proper coordination and execution of all activities except for the automatic teller. I really have no role in managing that particular activity.

QUESTION 7: How much time do workers spend on each activity? By equipment?

RESPONSE: I just completed a work survey and have the percentage of time calculated for each worker. All seven clerks work on each of the three departmental activities. About 40 percent of their time is spent processing transactions, with the rest of their time split evenly between issuing statements and answering questions. Phone time for all seven workers is used only for answering client questions. Computer time is 70 percent transaction processing, 20 percent statement preparation, and 10 percent question answering. Furthermore, my own time and that of my computer and telephone are 100 percent administrative. Credit card transactions represent about 20 percent of the total automatic teller transactions.

Required:

Prepare an activity dictionary using five columns: activity name, activity description, activity type (primary or secondary), cost object(s), and activity driver.

OBJECTIVE ▶ 3

SERVICE

Exercise 6-17 **ASSIGNING RESOURCE COSTS TO ACTIVITIES, RESOURCE DRIVERS, PRIMARY AND SECONDARY ACTIVITIES**

Refer to the interview in **Exercise 6-16** (especially to Questions 4 and 7). The general ledger reveals the following annual costs:

Supervisor's salary	$ 64,600
Clerical salaries	210,000
Computers, desks, and printers	32,000
Computer supplies	7,200
Telephone expenses	4,000
ATM	1,250,000

All nonlabour resources, other than the ATM, are spread evenly among the eight credit department employees (in terms of assignment and usage). Credit department employees have no contact with ATMs. Printers and desks are used in the same ratio as computers by the various activities.

Required:

1. Determine the cost of all primary and secondary activities.
2. Assign the cost of secondary activities to the primary activities.

Exercise 6-18 ASSIGNING RESOURCE COSTS TO ACTIVITIES, RESOURCE DRIVERS, PRIMARY AND SECONDARY ACTIVITIES

OBJECTIVE ▶ 3

Bob Randall, cost accounting manager for Hemple Products, was asked to determine the costs of the activities performed within the company's Manufacturing Engineering Department. The department has the following activities: creating bills of materials (BOMs), studying manufacturing capabilities, improving manufacturing processes, training employees, and designing tools. The general ledger accounts reveal the following expenditures for Manufacturing Engineering:

Salaries	$500,000
Equipment	100,000
Supplies	30,000
Total	$630,000

The equipment is used for two activities: improving processes and designing tools. The equipment's time is divided by two activities: 40 percent for improving processes and 60 percent for designing tools. The salaries are for nine engineers, one who earns $100,000 and eight who earn $50,000 each. The $100,000 engineer spends 40 percent of her time training employees in new processes and 60 percent of her time on improving processes. One engineer spends 100 percent of her time on designing tools, and another engineer spends 100 percent of his time on improving processes. The remaining six engineers spend equal time on all activities. Supplies are consumed in the following proportions:

Creating BOMs	10%
Studying capabilities	5
Improving processes	35
Training employees	20
Designing tools	30

After determining the costs of the engineering activities, Bob was then asked to describe how these costs would be assigned to jobs produced within the factory. (The company manufactures machine parts on a job-order basis.) Bob responded by indicating that creating BOMs and designing tools were the only primary activities. The remaining were secondary activities. After some analysis, Bob concluded that studying manufacturing capabilities was an activity that enabled the other four activities to be realized. He also noted that all of the employees being trained are manufacturing workers—employees who work directly on the products. The major manufacturing activities are cutting, drilling, lathing, welding, and assembly. The costs of these activities are assigned to the various products using hours of usage (grinding hours, drilling hours, etc.). Furthermore, tools were designed to enable the production of specific jobs. Finally, the process improvement activity focused only on the five major manufacturing activities.

Required:

1. What is meant by unbundling general ledger costs? Why is it necessary?
2. What is the difference between a general ledger database system and an activity-based database system?
3. Using the resource drivers and direct tracing, calculate the costs of each manufacturing engineering activity. What are the resource drivers?
4. Describe in detail how the costs of the engineering activities would be assigned to jobs using activity-based costing. Include a description of the activity drivers that might be used. Where appropriate, identify both a possible transaction driver and a possible duration driver.

OBJECTIVE ▶ 3   Exercise 6-19 **PROCESS IDENTIFICATION AND ACTIVITY CLASSIFICATION**

Calzado Company produces leather shoes in batches. The shoes are produced in one plant located on 20 acres. The plant operates two shifts, five days per week. Each time a batch is produced, just-in-time suppliers deliver materials to the plant. When the materials arrive, a worker checks the quantity and type of materials with the bill of materials for the batch. The worker then makes an entry at a PC terminal near the point of delivery acknowledging receipt of the material. An accounts payable clerk reviews all deliveries at the end of each day and then prints and mails cheques the same day materials are received. Prior to producing a batch, the equipment must be configured to reflect style and size features. Once configured, the batch is produced passing through three operations: cutting, sewing, and attaching buckles and other related parts such as heels. At the end of the production process, a sample of shoes is inspected to ensure the right level of quality.

After inspection, the batch is divided into lots based on the customer orders for the shoes. The lots are packaged in boxes and then transferred to a staging area to await shipment. After a short wait (usually within two hours), the lots are loaded onto trucks and delivered to customers (retailers).

Within the same plant, the company also has a team of design engineers who respond to customer feedback on style and comfort issues. This department modifies existing designs, develops new shoe designs, builds prototypes, and test markets the prototypes before releasing the designs for full-scale production.

Required:

1. Identify Calzado's processes and their associated activities.
2. Classify each activity within each process as unit-level, batch-level, product-level, or facility-level.

OBJECTIVE ▶ 4   Exercise 6-20 **TDABC**

Bob Randall, cost accounting manager for Hemple Products, was asked to determine the costs of the activities performed within the company's Manufacturing Engineering Department. The department has the following activities: creating bills of materials (BOMs), studying manufacturing capabilities, improving manufacturing processes, training employees, and designing tools. The resource costs (from the general ledger) and the times to perform one unit of each activity are provided below.

Resource Costs		Activities	Unit Time	Driver
Salaries	$500,000	Creating BOMs	0.5 hr	No. of BOMs
Equipment	100,000	Designing tools	5.4 hrs	No. of tool designs
Supplies	30,000	Improving processes	1.0 hr	Process improvement hrs
Total	$630,000	Training employees	2.0 hrs	No. of training sessions

Total machine and labour hours (at practical capacity):

Machine hours	2,000
Engineering hours	18,000
Total hours	20,000

The activity, designing tools, uses number of tools designed as the activity driver. Using a traditional approach, the cost of the designing tools activity was determined to be $179,000 (see Exercise 6-18) with an expected activity output of 1,000 for the coming year. During the first week of the year, two jobs (Job 150 and Job 151) had a demand for 10 and 20 new tools, respectively.

Required:

1. Calculate the capacity cost rate for the Manufacturing Engineering Department.
2. Using the capacity cost rate, determine the activity rates for each activity.
3. Calculate the cost of designing tools that would be assigned to each job using the TDABC derived activity rate and then repeat using the traditional ABC rate. What might be the cause or causes that would explain the differences in the two approaches?
4. Now suppose that time for creating BOMs is 0.5 for a standard product but that creating a BOM for a custom product adds an additional 0.3 hour. Express the time equation for this added complexity and then calculate the activity rate for the activity of creating a BOM for custom products.

Exercise 6-21 APPROXIMATELY RELEVANT ABC

OBJECTIVE ▶ 4

Silva Company has identified the following overhead activities, costs, and activity drivers for the coming year:

Activity	Expected Cost	Activity Driver	Activity Capacity
Setting up equipment	$126,000	Number of setups	150
Ordering materials	18,000	Number of orders	900
Machining	126,000	Machine hours	10,500
Receiving	30,000	Receiving hours	1,250

Silva produces two models of cell phones with the following expected activity demands:

	Model X	Model Y
Units completed	5,000	10,000
Number of setups	100	50
Number of orders	300	600
Machine hours	6,000	4,500
Receiving hours	375	875

Required:

1. Determine the total overhead assigned to each product using the four activity drivers.
2. Determine the total overhead assigned to each model using the two most expensive activities. The costs of the two relatively inexpensive activities are allocated to the two expensive activities in proportion to their costs.
3. Using ABC as the benchmark, calculate the percentage error and comment on the accuracy of the reduced system. Explain why this approach may be desirable.

Exercise 6-22 EQUALLY ACCURATE REDUCED ABC SYSTEM

OBJECTIVE ▶ 4

Refer to **Exercise 6-21**.

Required:

1. Calculate the global consumption ratios for the two products.
2. Using the activity consumption ratios for number of orders and number of setups, show that the same cost assignment can be achieved using these two drivers as that of the complete, four-driver ABC system.

Problems

OBJECTIVE ➤ 1

Problem 6-23 PREDETERMINED OVERHEAD RATES, OVERHEAD VARIANCES, UNIT COSTS

Ultima Company produces two products and uses a predetermined overhead rate to apply overhead. Ultima currently applies overhead using a plantwide rate based on direct labour hours. Consideration is being given to the use of departmental overhead rates where overhead would be applied on the basis of direct labour hours in Department 1 and on the basis of machine hours in Department 2. At the beginning of the year, the following estimates are provided:

	Department 1	Department 2
Direct labour hours	800,000	160,000
Machine hours	20,000	240,000
Overhead cost	$480,000	$1,440,000

Actual results reported by department and product during the year are as follows:

	Department 1	Department 2
Direct labour hours	784,000	168,000
Machine hours	22,000	256,000
Overhead cost	$500,000	$1,540,000

	Product 1	Product 2
Direct labour hours:		
Department 1	600,000	184,000
Department 2	120,000	48,000
Machine hours:		
Department 1	10,000	12,000
Department 2	56,000	200,000

Required:

1. Compute the plantwide predetermined overhead rate and calculate the overhead assigned to each product.
2. Calculate the predetermined departmental overhead rates and calculate the overhead assigned to each product.
3. Using departmental rates, compute the applied overhead for the year. What is the under- or overapplied overhead for the firm?
4. Prepare the journal entry that disposes of the overhead variance calculated in Requirement 3, assuming it is not material in amount. What additional information would you need if the variance is material to make the appropriate journal entry?

OBJECTIVE ➤ 2

Problem 6-24 UNIT-BASED VERSUS ACTIVITY-BASED COSTING

Apollo Company produces exercise bikes. One of its plants produces two versions: a standard model and a deluxe model. The deluxe model has a wider and sturdier base and a variety of electronic gadgets to help the exerciser monitor heartbeat, calories burned, distance travelled, and so on. At the beginning of the year, the following data were prepared for this plant:

	Standard Model	Deluxe Model
Expected quantity	30,000	15,000
Selling price	$370	$700
Prime costs	$4.5 million	$5.25 million
Machine hours	37,500	37,500
Direct labour hours	75,000	75,000
Engineering support (hours)	13,500	31,500
Receiving (orders processed)	3,000	4,500

	Standard Model	Deluxe Model
Materials handling (number of moves)	15,000	45,000
Purchasing (number of requisitions)	750	1,500
Maintenance (hours used)	6,000	24,000
Paying suppliers (invoices processed)	3,750	3,750
Setting up batches (number of setups)	60	540

Additionally, the following overhead activity costs are reported:

Maintenance	$ 600,000
Engineering support	900,000
Materials handling	1,200,000
Setups	750,000
Purchasing	450,000
Receiving	300,000
Paying suppliers	300,000
	$4,500,000

Required:

1. Calculate the cost per unit for each product using direct labour hours to assign all overhead costs.
2. Calculate activity rates and determine the overhead cost per unit. Compare these costs with those calculated using the unit-based method. Which cost is the most accurate? Explain.

Problem 6-25 ABC, RESOURCE DRIVERS, SERVICE INDUSTRY

Cushing Medical Clinic operates a cardiology care unit and a maternity care unit. Ned Carson, the clinic's administrator, is investigating the charges assigned to cardiology patients. Currently, all cardiology patients are charged the same rate per patient day for daily care services. Daily care services are broadly defined as occupancy, feeding, and nursing care. A recent study, however, has revealed several interesting outcomes. First, the demands patients place on daily care services vary with the severity of the case being treated. Second, the occupancy activity is a combination of two activities: lodging and use of monitoring equipment. Since some patients require more monitoring than others, these activities should be separated. Third, the daily rate should reflect the differences in demands resulting from differences in patient type. Separating the occupancy activity into two separate activities would also require the determination of the cost of each activity. Determining the costs of the monitoring activity will be fairly easy because its costs are directly traceable. Lodging costs, however, are shared by two activities: lodging cardiology patients and lodging maternity care patients. The total lodging costs for the two activities are $3,800,000 per year and consist of such items as building depreciation, building maintenance, and building utilities. The cardiology floor and the maternity floor each occupy 20,000 square metres. Carson has determined that lodging costs will be assigned to each unit based on square metres.

To compute a daily rate that reflects the difference in demands, patients are placed in three categories according to illness severity, and the following annual data have been collected:

Activity	Cost of Activity	Activity Driver	Quantity
Lodging	$1,900,000	Patient days	15,000
Monitoring	1,400,000	Monitoring hours used	20,000
Feeding	300,000	Patient days	15,000
Nursing care	3,000,000	Nursing hours	150,000
Total	$6,600,000		

The demands associated with patient severity are also provided:

Severity	Patient Days	Monitoring Hours	Nursing Hours
High	5,000	10,000	90,000
Medium	7,500	8,000	50,000
Low	2,500	2,000	10,000

Required:

1. Suppose that the costs of daily care are assigned using only patient days as the activity driver (which is also the measure of output). Compute the daily rate using this unit-based approach of cost assignment.
2. Compute activity rates using the given activity drivers (combine activities with the same driver).
3. Compute the charge per patient day for each patient type using the activity rates from Requirement 2 and the demands on each activity.
4. Suppose that the product is defined as "stay and treatment" where the treatment is bypass surgery. What additional information would you need to cost out this newly defined product?
5. Comment on the value of activity-based costing in service industries.

OBJECTIVE ▶ 2 3

SERVICE

Problem 6-26 ACTIVITY-BASED COSTING, SERVICE FIRM

The First National Bank operated for years under the assumption that profitability can be increased by increasing dollar volumes. Historically, First National's efforts were directed toward increasing total dollars of sales and total dollars of account balances. In recent years, however, First National's profits have been eroding. Increased competition, particularly from credit unions, was the cause of the difficulties. As key managers discussed the bank's problems, it became apparent that they had no idea what their products were costing. Upon reflection, they realized that they had often made decisions to offer a new product that promised to increase dollar balances without any consideration of what it cost to provide the service.

After some discussion, the bank decided to hire a consultant to compute the costs of three products: chequing accounts, personal loans, and the gold VISA. The consultant identified the following activities, costs, and activity drivers (annual data):

Activity	Activity Cost	Activity Driver	Activity Capacity
Providing ATM service	$ 100,000	No. of transactions	200,000
Computer processing	1,000,000	No. of transactions	2,500,000
Issuing statements	800,000	No. of statements	500,000
Customer inquiries	360,000	Telephone minutes	600,000

The following annual information on the three products was also made available:

	Chequing Accounts	Personal Loans	Gold VISA
Units of product	30,000	5,000	10,000
ATM transactions	180,000	0	20,000
Computer transactions	2,000,000	200,000	300,000
Number of statements	300,000	50,000	150,000
Telephone minutes	350,000	90,000	160,000

In light of the new cost information, Larry Roberts, the bank president, wanted to know whether a decision made two years ago to modify the bank's chequing account product was sound. At that time, the service charge was eliminated on accounts with an average annual balance greater than $1,000. Based on increases in the total dollars in chequing, Larry was pleased with the new product. The chequing account product is described as follows: (1) chequing account balances greater than $500 earn interest of 2 percent per year, and (2) a service charge of $5 per month is charged for balances less than $1,000. The bank earns 4 percent on chequing account deposits. Fifty percent of the accounts are less than $500 and have an average balance of $400 per account. Ten percent of the accounts are between $500 and $1,000 and average $750 per account. Twenty-five percent of the accounts are between $1,000 and $2,767; the average balance is $2,000. The remaining accounts carry a balance greater than $2,767. The average balance for these accounts is $5,000. Research indicates that the $2,000 category was by far the greatest contributor to the increase in dollar volume when the chequing account product was modified two years ago.

Required:

1. Calculate rates for each activity.
2. Using the rates computed in Requirement 1, calculate the cost of each product.
3. Evaluate the chequing account product. Are all accounts profitable? Compute the average annual profitability per account for the four categories of accounts described in the problem. What recommendations would you make to increase the profitability of the chequing account product? (Break-even analysis for the unprofitable categories may be helpful).

Problem 6-27 PRODUCT COSTING ACCURACY, CORPORATE STRATEGY, ABC

OBJECTIVE ▸ 2 3

Autotech Manufacturing is engaged in the production of replacement parts for automobiles. One plant specializes in the production of two parts: Part #127 and Part #234. Part #127 produced the highest volume of activity, and for many years it was the only part produced by the plant. Five years ago, Part #234 was added. Part #234 was more difficult to manufacture and required special tooling and setups. Profits increased for the first three years after the addition of the new product. In the last two years, however, the plant has faced intense competition, and its sales of Part #127 have dropped. In fact, the plant showed a small loss in the most recent reporting period. Much of the competition was from foreign sources, and the plant manager was convinced that the foreign producers were guilty of selling the part below the cost of producing it. The following conversation between Patty Burt, plant manager, and Joseph Fielding, divisional marketing manager, reflects the concerns of the division about the future of the plant and its products.

JOSEPH: You know, Patty, the divisional manager is real concerned about the plant's trend. He indicated that in this budgetary environment, we can't afford to carry plants that don't show a profit. We shut one down just last month because it couldn't handle the competition.

PATTY: Joe, you and I both know that Part #127 has a reputation for quality and value. It has been a mainstay for years. I don't understand what's happening.

JOSEPH: I just received a call from one of our major customers concerning Part #127. He said that a sales representative from another firm offered the part at $20 per unit— $11 less than what we charge. It's hard to compete with a price like that. Perhaps the plant is simply obsolete.

PATTY: No. I don't buy that. From my sources, I know we have good technology. We are efficient. And it's costing a little more than $21 to produce that part. I don't see how these companies can afford to sell it so cheaply. I'm not convinced that we should meet the price. Perhaps a better strategy is to emphasize producing and selling more of Part #234. Our margin is high on this product, and we have virtually no competition for it.

JOSEPH: You may be right. I think we can increase the price significantly and not lose business. I called a few customers to see how they would react to a 25 percent increase in price, and they all said that they would still purchase the same quantity as before.

PATTY: It sounds promising. However, before we make a major commitment to Part #234, I think we had better explore other possible explanations. I want to know how our production costs compare to those of our competitors. Perhaps we could be more efficient and find a way to earn our normal return on Part #127. The market is so much bigger for this part. I'm not sure we can survive with only Part #234. Besides, my production people hate that part. It's very difficult to produce.

After her meeting with Joseph, Patty requested an investigation of the production costs and comparative efficiency. She received approval to hire a consulting group to make an independent investigation. After a three-month assessment, the consulting group provided the following information on the plant's production activities and costs associated with the two products:

	Part #127	Part #234
Production	500,000	100,000
Selling price	$31.86	$24.00
Overhead per unit*	$12.83	$5.77
Prime cost per unit	$8.53	$6.26
Number of production runs	100	200
Receiving orders	400	1,000
Machine hours	125,000	60,000
Direct labour hours	250,000	22,500
Engineering hours	5,000	5,000
Material moves	500	400

*Calculated using a plantwide rate based on direct labour hours. This is the current way of assigning the plant's overhead to its products.

The consulting group recommended switching the overhead assignment to an activity-based approach. It maintained that activity-based cost assignment is more accurate and will provide better information for decision making. To facilitate this recommendation, it grouped the plant's activities into homogeneous sets with the following costs:

Overhead:	
Setup costs	$ 240,000
Machine costs	1,750,000
Receiving costs	2,100,000
Engineering costs	2,000,000
Materials handling costs	900,000
Total	$6,990,000

Required:

1. Verify the overhead cost per unit reported by the consulting group using direct labour hours to assign overhead. Compute the per-unit gross margin for each product.
2. After learning of activity-based costing, Patty asked the controller to compute the product cost using this approach. Recompute the unit cost of each product using activity-based costing. Compute the per-unit gross margin for each product.
3. Should the company switch its emphasis from the high-volume product to the low-volume product? Comment on the validity of the plant manager's concern that competitors are selling below the cost of making Part #127.
4. Explain the apparent lack of competition for Part #234. Comment also on the willingness of customers to accept a 25 percent increase in price for Part #234.
5. Assume that you are the manager of the plant. Describe what actions you would take based on the information provided by the activity-based unit costs.

OBJECTIVE ▶ 4

SERVICE

Problem 6-28 TIME-DRIVEN ACTIVITY-BASED COSTING COMPARED TO ABC, STAGE 1

The Bienestar Cardiology Clinic has two major activities: diagnostic and treatment. The two activities use four resources: nursing, medical technicians, cardiologists, and equipment. Detailed interviews have provided the following work distribution matrix:

	Resources				
Activity	Nursing	Technicians	Cardiologists	Equipment	Total Activity Time
Diagnosing patients	0.70	0.80	0.40	0.60	12,000 hrs
Treating patients	0.30	0.20	0.60	0.40	8,000 hrs
Total time (hrs)	4,000	4,000	6,000	6,000	20,000
Cost	$80,000	$80,000	$320,000	$320,000	

The total time estimated corresponds to practical capacity (interviewers adjusted the total time to about 80 percent of the available time). The equipment time is measured in machine hours. Thus, the total time (at practical capacity) in the system is 20,000 hours. In considering the implementation of a TDABC model, the following unit times and transaction information are also provided:

	Unit Time	Driver	Expected Activity Driver Quantity
Diagnosing patients	3 hrs	No. of patients	4,000
Treating patients	0.8 hr	No. of treatments	10,000

Required:

1. Calculate the cost of each activity using the indicated values of the resource drivers.
2. Calculate the capacity cost rate for TDABC. Using the capacity cost rate, calculate the cost of each activity under TDABC. Compare these values with those obtained in Requirement 1 and discuss possible reasons for any differences.
3. Suppose that the actual activity driver quantities are 3,500 and 9,000. Calculate the cost of unused capacity.
4. Suppose that the clinic acquires new equipment that reduces the total time required for the two activities from 20,000 to 18,000 hours. The equipment cost remains the same. Explain how the ABC system would be updated and then describe how TDABC would provide updates.
5. Suppose that diagnosing patients without any cardiac disease takes two hours while diagnosing patients with mildly diseased hearts takes an additional 1.5 hours and those with more severe problems takes four hours. Prepare a time equation and using the capacity cost rate from Requirement 2, calculate the activity rate for each of the three types of patients.

Problem 6-29 **ACTIVITY-BASED COSTING, REDUCING THE NUMBER OF DRIVERS WITH EQUAL ACCURACY**

OBJECTIVE ▶ 2 4

Reducir Inc. produces two different types of hydraulic cylinders. Reducir produces a major subassembly for the cylinders in the Cutting and Welding Department. Other parts and the subassembly are then assembled in the Assembly Department. The activities, expected costs, and drivers associated with these two manufacturing processes are given below.

Process	Activity	Cost	Activity Driver	Expected Quantity
Cutting and Welding	Welding	$ 776,000	Welding hours	4,000
	Machining	450,000	Machine hours	10,000
	Inspecting	448,250	No. of inspections	1,000
	Materials handling	300,000	No. of moves	12,000
	Setups	240,000	No. of setups	100
		$2,214,250		
Assembly	Changeover	$ 180,000	Changeover hours	1,000
	Rework	61,750	Rework orders	50
	Testing	300,000	No. of tests	750
	Materials handling	380,000	No. of parts	50,000
	Engineering support	130,000	Engineering hours	2,000
		$1,051,750		

Note: In the assembly process, the materials handling activity is a function of product characteristics rather than batch activity.

Other overhead activities, their costs, and drivers are listed below.

Activity	Cost	Activity Driver	Quantity
Purchasing	$135,000	Purchase requisitions	500
Receiving	274,000	Receiving orders	2,000
Paying suppliers	225,000	No. of invoices	1,000
Providing space and utilities	100,000	Machine hours	10,000
	$734,000		

Other production information concerning the two hydraulic cylinders is also provided:

	Cylinder A	Cylinder B
Units produced	1,500	3,000
Welding hours	1,600	2,400
Machine hours	3,000	7,000
Inspections	500	500
Moves	7,200	4,800
Setups	45	55
Changeover hours	540	460
Rework orders	5	45
No. of tests	500	250
Parts	40,000	10,000
Engineering hours	1,500	500
Requisitions	425	75
Receiving orders	1,800	200
Invoices	650	350

Required:

1. Using a plantwide rate based on machine hours, calculate the total overhead cost assigned to each product and the unit overhead cost.
2. Using activity rates, calculate the total overhead cost assigned to each product and the unit overhead cost. Comment on the accuracy of the plantwide rate.
3. Calculate the global consumption ratios.
4. Calculate the consumption ratios for welding and materials handling (Assembly) and show that two drivers, welding hours and number of parts, can be used to achieve the same ABC product costs calculated in Requirement 2. Explain the value of this simplification.
5. Calculate the consumption ratios for inspection and engineering and show that the drivers for these two activities also duplicate the ABC product costs calculated in Requirement 2.

OBJECTIVE ▸ 2   **Problem 6-30 APPROXIMATELY RELEVANT ABC**

Refer to the data given in **Problem 6-29** and suppose that the expected activity costs are reported as follows (all other data remain the same):

Process	Activity	Cost
Cutting and Welding	Welding	$2,000,000
	Machining	1,000,000
	Inspecting	50,000
	Materials handling	72,000
	Setups	400,000
		$3,522,000
Assembly	Changeover	$ 28,000
	Rework	50,000
	Testing	40,000
	Materials handling	60,000
	Engineering support	70,000
		$ 248,000

Other overhead activities:

Activity	Cost
Purchasing	$ 50,000
Receiving	70,000
Paying suppliers	80,000
Providing space and utilities	30,000
	$230,000

The per unit overhead cost using the 14 activity-based drivers is $1,108 and $779 for Cylinder A and Cylinder B, respectively.

Required:

1. Determine the percentage of total costs represented by the three most expensive activities.
2. Allocate the costs of all other activities to the three activities identified in Requirement 1. Allocate the other activity costs to the three activities in proportion to their individual activity costs. Now assign these total costs to the products using the drivers of the three chosen activities.
3. Using the costs assigned in Requirement 2, calculate the percentage error using the ABC costs as a benchmark. Comment on the value and advantages of this ABC simplification.

Problem 6-31 PRODUCT COSTING ACCURACY, PLANTWIDE AND DEPARTMENTAL RATES, ABC

OBJECTIVE ➤ 1 2 4

Waterloo Company produces two type of calculators: scientific and business. Both products pass through two producing departments. The business calculator is by far the most popular. The following data have been gathered for these two products:

Product-Related Data	Scientific	Business
Units produced per year	75,000	750,000
Prime costs	$250,000	$2,500,000
Direct labour hours	100,000	1,000,000
Machine hours	50,000	500,000
Production runs	100	150
Inspection hours	2,000	3,000
Maintenance hours	2,250	9,000

Department Data	Department 1	Department 2
Direct labour hours:		
Scientific calculator	75,000	25,000
Business calculator	112,500	887,500
Total	187,500	912,500
Machine hours:		
Scientific calculator	25,000	25,000
Business calculator	400,000	100,000
Total	425,000	125,000
Overhead costs:		
Setup costs	$225,000	$225,000
Inspection costs	175,000	175,000
Power	250,000	150,000
Maintenance	200,000	250,000
Total	$850,000	$800,000

Required:

1. Compute the overhead cost per unit for each product using a plantwide, unit-based rate using direct labour hours.
2. Compute the overhead cost per unit for each product using departmental rates. In calculating departmental rates, use machine hours for Department 1 and direct labour hours for Department 2. Repeat using direct labour hours for Department 1 and machine hours for Department 2.
3. Compute the overhead cost per unit for each product using activity-based costing.
4. Comment on the ability of departmental rates to improve the accuracy of product costing.

CMA Problem

CMA Problem 6-1 ACTIVITY-BASED COSTING*

Oineon Corporation manufactures several types of printed circuit boards; however, two of the boards account for most of the company's sales. The first board, a television circuit board, has been a standard in the industry for several years. The market for this type of board is competitive and, therefore, price sensitive. Oineon plans to sell 65,000 of the TV boards in 2013 at a price of $150 per unit. The second high-volume product, a personal computer (PC) circuit board, is a recent addition to Oineon's product line. Because the PC board incorporates the latest technology, it can be sold at a premium price. The 2013 plans include the sale of 40,000 PC boards at $300 per unit.

Oineon's management group is meeting to discuss strategies for 2013. The current topic of conversation is how to spend the sales and promotion dollars for next year. The sales manager believes that the market share for the TV board could be expanded by concentrating Oineon's promotional efforts in this area. In response to this suggestion, the production manager said, "Why don't you go after a bigger market for the PC board? The cost sheets I get show the contribution from the PC board is more than double the contribution from the TV board. I know we get a premium price for the PC board, so selling it should help overall profitability."

Oineon uses a standard cost system. The following data apply to the TV and PC boards:

	TV Board	PC Board
Direct material	$80	$140
Direct labour	1.5 hours	4 hours
Machine time	.5 hours	1.5 hours

Variable factory overhead is applied on the basis of direct labour hours. For 2013, variable factory overhead is budgeted at $1,120,000 and direct labour hours are estimated at $280,000. The hourly rates for machine time and direct labour are $10 and $14, respectively. Oineon applies a materials handling charge at 10 percent of material cost. This materials handling charge is not included in variable factory overhead. Total 2013 expenditures for material are budgeted at $10,600,000.

Ed Welch, Oineon's controller, believes that before management proceeds with the discussion about allocating sales and promotional dollars to individual products, it might be worthwhile to look at the products on the basis of the activities involved in their production. As Welch explained to the group, "Activity-based costing integrates the cost of all activities, known as cost drivers, into individual product costs rather than including these costs in overhead pools." Welch has prepared the schedule shown below to help management understand this concept.

*© 2011, CMA Ontario. http://www.cma-ontario.org. Reprinted with permission.

	Budgeted Cost	Cost Driver	Annual Activity for Cost Driver
Material overhead:			
Procurement	$ 400,000	Number of parts	4,000,000 parts
Production scheduling	220,000	Number of boards	110,000 boards
Packaging and shipping	440,000	Number of boards	110,000 boards
	$1,060,000		
Variable overhead:			
Machine setup	$ 446,000	Number of setups	278,750 setups
Hazardous waste disposal	48,000	Kilograms of waste	16,000 kilograms
Quality control	560,000	Number of inspections	160,000 inspections
General supplies	66,000	Number of boards	110,000 boards
	$1,120,000		
Manufacturing:			
Machine insertion	$1,200,000	Number of parts	3,000,000 parts
Manual insertion	4,000,000	Number of parts	1,000,000 parts
Wave soldering	132,000	Number of boards	110,000 boards
	$5,332,000		

Required Per Unit	TV Board	PC Board
Parts	25	55
Machine insertions	24	35
Manual insertions	1	20
Machine setups	2	3
Hazardous waste	.02 kg	.35 kg
Inspections	1	2

"Using this information," Welch explained, "we can calculate an activity-based cost for each TV and PC board and then compare it to the standard cost we have been using. The only cost that remains the same for both cost methods is the cost of direct material. The cost drivers will replace the direct labour, machine time, and overhead costs in the standard cost."

Required:

1. Prepare a gross profit report based on standard costs for each of the products.
2. Prepare a gross profit report based on activity-based costing for each of the products.
3. Comment on the likely actions that management would take if they used the report in Requirement 1 and if they used the report in Requirement 2.
   (Adapted from CMA Ontario)

> *The Collaborative Learning Exercises can be found on the product support site at www.hansen1ce.nelson.com.*

After studying this chapter, you should be able to:

1 Describe the difference between support departments and producing departments.

2 Calculate charging rates, and distinguish between single and dual charging rates.

3 Allocate support centre costs to producing departments using the direct method, the sequential method, and the reciprocal method.

4 Calculate departmental overhead rates.

5 Identify the characteristics of the joint production process, and allocate joint costs to products.

CHAPTER

7

Allocating Costs of Support Departments and Joint Products

Mutually beneficial costs, which occur when the same resource is used in the output of two or more services or products, are known as **common costs**. These common costs may pertain to periods of time, individual responsibilities, sales territories, and classes of customers. A special case of common costs is that of the joint production process. This chapter will first focus on the costs common to departments and to products, and then on the common costs of the joint production process.

OBJECTIVE ▶ 1
Describe the difference between support departments and producing departments.

An Overview of Cost Allocation

The complexity of many modern firms leads the accountant to allocate costs of support departments to producing departments and individual product lines. Allocation is simply a means of dividing a pool of costs and assigning those costs to various subunits. It is important to realize that allocation does not affect the total cost. Total cost is neither reduced nor increased by allocation. However, the amounts of cost assigned to the subunits can be affected by the allocation procedure chosen. Because cost allocation can affect bid prices, the perceived profitability of individual products, and the behaviour of managers, it is an important topic.

Producing and Support Departments

There are two categories of departments: producing departments and support departments. **Producing departments** are directly responsible for creating the products or services sold to customers. In a large public accounting firm, examples of producing departments are auditing, tax, and management advisory services (computer systems services). In a manufacturing setting such as **Volkswagen (VW)**, producing departments are those that work directly on the products being manufactured (e.g., assembly and painting). **Support departments** provide essential services for producing departments. These departments are indirectly connected with an organization's services or products. At VW, those departments might include engineering, maintenance, personnel, and building and grounds.

Imperial Parking of Vancouver, B.C., with over 2,000 locations and 3,700 employees, has gathered together a number of its support departments to form a shared service centre (SSC). The SSC performs activities that are used across a wide array of the company's departments. Payroll, customer billing, and accounts receivable processing are examples of SCC "components". The company reaps the savings from economies of scale and standardized process design. Tools to measure performance are also incorporated into the SSC design. The SSC faces three important cost questions:

1. What causes costs in our operation?
2. How much should be charged back to the customers/producing departments?
3. How do our costs compare with those of outsourcing firms that perform the same service?

The drivers used to develop charging rates are seldom unit-based drivers (based on production). Instead, they might include the number of transactions processed and the percentage of errors in customer-provided information. Because activity-based costing (ABC) provides a better understanding of costs and their related drivers, it provides a better framework for managing SSC costs than traditional cost accounting systems.[1]

Once the producing and support departments have been identified, the overhead costs traceable to each department can be determined. A factory cafeteria, for example, would have food costs, wages of cooks and servers, depreciation on dishwashers and stoves, and supplies (e.g., napkins and plastic forks). Overhead directly associated with a producing department, such as assembly in a furniture-making plant, would include supplies used by that department, supervisory salaries, and depreciation on departmental equipment. Overhead that cannot be easily assigned to a producing or support department is assigned to a catchall department such as general factory. General factory might include depreciation on the factory building, rental of a Santa Claus suit for the factory Christmas party, the cost of restriping the parking lot, the plant manager's salary, and telephone service. In this way, all costs are assigned to a department.

Exhibit 7-1 shows how a manufacturing firm and a service firm can be divided into producing and support departments. The manufacturing plant, which makes furniture, may be departmentalized into two producing departments (Assembly and Finishing) and four support departments (Materials Storeroom, Cafeteria, Maintenance, and General Factory). The service firm, a bank, might be departmentalized into three producing departments (Auto Loans, Commercial Lending, and Personal Banking) and three support departments (Drive Through, Data Processing, and Bank

[1] Ann Triplett and Jon Scheumann, "Managing Shared Services with ABM," *Strategic Finance* (February 2000): 40–45. (adapted)

Exhibit 7-1

Examples of Departmentalization for a Manufacturing Firm and a Service Firm

Manufacturing Firm: Furniture Maker	
Producing Departments	**Support Departments**
Assembly: 　Supervisors' salaries 　Small tools 　Indirect materials 　Depreciation on machinery Finishing: 　Sandpaper 　Depreciation on sanders and buffers	Materials Storeroom: 　Clerk's salary 　Depreciation on forklift Cafeteria: 　Food 　Cooks' salaries 　Depreciation on stoves Maintenance: 　Janitors' salaries 　Cleaning supplies 　Machine oil and lubricants General Factory: 　Depreciation on building 　Security 　Utilities

Service Firm: Bank	
Producing Departments	**Support Departments**
Auto Loans: 　Loan processors' salaries 　Forms and supplies Commercial Lending: 　Lending officers' salaries 　Depreciation on office equipment 　Bankruptcy prediction software Personal Banking: 　Supplies and postage for statements	Drive Through: 　Tellers' salaries 　Depreciation on equipment Data Processing: 　Personnel salaries 　Software 　Depreciation on hardware Bank Administration: 　Salary of CEO· 　Receptionist's salary 　Telephone costs 　Depreciation on bank and vault

Administration). Overhead costs are traced to each department. Note that each factory or service company overhead cost must be assigned to one, and only one, department.

Once the company is departmentalized and all overhead costs are traced to the individual departments, support department costs are assigned to producing departments, and overhead rates for the producing departments are developed to cost products. Although support departments do not work directly on the products or services that are sold, the costs of providing these support services are part of the total product cost and must be assigned to the products. This assignment of costs consists of a two-stage allocation: (1) allocation of support department costs to producing departments and (2) assignment of these allocated costs to individual products. The second-stage allocation, achieved through the use of departmental overhead rates, is necessary because there are multiple products being worked on in each producing department. If there were only one product within a producing department, all the support costs allocated to that department would belong to that product. Recall that a predetermined overhead rate is computed by taking total estimated overhead for a department and dividing it by an estimate of an appropriate allocation base. Now we see that a producing department's overhead consists of two parts: overhead directly associated with a producing department and overhead allocated to the producing department from the support departments. A support department cannot have an overhead rate that assigns overhead

Exhibit 7-2

Steps in Allocating Support Department Costs to Producing Departments

1. Departmentalize the firm.
2. Classify each department as a support department or a producing department.
3. Trace all overhead costs in the firm to a support or producing department.
4. Allocate support department costs to the producing departments.
5. Calculate predetermined overhead rates for producing departments.
6. Allocate overhead costs to the units of individual product through the predetermined overhead rates.

costs to units produced, because products are not produced in support departments. The nature of support departments is to service producing departments, not the products that pass through the producing departments. For example, maintenance personnel repair and maintain the equipment in the Assembly Department, not the furniture that is assembled in that department. Exhibit 7-2 summarizes the steps involved.

Types of Allocation Bases

In effect, producing departments *cause* support activities. **Causal factors** are variables or activities within a producing department that provoke the incurrence of support costs. In choosing a basis for allocating support department costs, appropriate causal factors (activity drivers) should be identified. Using causal factors results in more accurate product costs. Furthermore, if the causal factors are known, managers are more able to control the consumption of services.

To illustrate the types of causal factors, or activity drivers, that can be used, consider the following three support departments: Power, Personnel, and Materials Handling. For power costs, a logical allocation base is kilowatt-hours, which can be measured by separate metres for each department. If separate metres do not exist, perhaps machine hours used by each department would be a good proxy, or a means of approximating power usage. For personnel costs, both the number of producing department employees and the labour turnover (e.g., number of new hires) are possible activity drivers. For materials handling, the number of material moves, the hours of materials handling used, and the quantity of material moved are all possible activity drivers. Exhibit 7-3 lists some possible activity drivers for allocating support department costs. When competing activity drivers exist, managers choose the factor that is most easily measured and that provides the most convincing relationship.

Exhibit 7-3

Examples of Possible Activity Drivers for Support Departments

Accounting:
   Number of transactions
Cafeteria:
   Number of employees
Data Processing:
   Number of lines entered
   Number of hours of service
Engineering:
   Number of change orders
   Number of hours
Maintenance:
   Machine hours
   Maintenance hours
Materials Storeroom:
   Number of material moves
   Kilograms of material moved
   Number of different parts

Payroll:
   Number of employees
Personnel:
   Number of employees
   Number of firings or layoffs
   Number of new hires
   Direct labour cost
Power:
   Kilowatt-hours
   Machine hours
Purchasing:
   Number of orders
   Cost of orders
Shipping:
   Number of orders

While the use of a causal factor to allocate common cost is the best solution, sometimes an easily measured causal factor cannot be found. In that case, the accountant looks for a good proxy. For example, the common cost of plant depreciation may be allocated to producing departments on the basis of square metres. Though square metres do not cause depreciation, it can be argued that the number of square metres a department occupies is a good proxy for the services provided to it by the factory building. The choice of a good proxy to guide allocation is dependent upon the company's objectives for cost allocation.

Objectives of Cost Allocation

A number of important objectives are associated with the allocation of support department costs to producing departments and ultimately to specific products. The following major objectives have been identified by the IMA:[2]

1. To obtain a mutually agreeable price
2. To compute product-line profitability
3. To predict the economic effects of planning and control
4. To value inventory
5. To motivate managers

Competitive pricing requires a good understanding of costs. If costs are not accurately allocated, some costs could be overstated, resulting in prices, or bids, that are too high and a loss of potential business. Alternatively, if the costs are understated, bids could be too low, producing losses on these products.

Good estimates of individual product costs also allow managers to assess the profitability of individual products and services. Multiproduct companies need to be sure that all products are profitable and that the overall profitability of the firm is not disguising the poor performance of individual products. This meets the profitability objective identified by the IMA.

By assessing the profitability of various support services, a manager may evaluate the mix of support services offered by the firm. From this evaluation, executives may decide to drop some support services, reallocate resources from one to another, reprice certain support services, or exercise greater cost control in some areas. These steps would meet the IMA's planning and control objective.

For a service organization such as a law firm, the IMA objective of inventory valuation is not relevant. For manufacturing organizations, however, this objective requires special attention. Rules of financial reporting (GAAP) require that all direct and indirect manufacturing costs be assigned to the products produced. Since support department costs are indirect manufacturing costs, they must be assigned to products. This is accomplished through support department cost allocation. Inventories and cost of goods sold, then, include direct materials, direct labour, and all manufacturing overhead, including the cost of support departments.

Allocations can be used to motivate managers. If support department costs are not allocated to producing departments, managers tend to overuse these services. Consumption of a support service may continue until the marginal benefit of the service equals zero. Of course, the marginal cost of a service is greater than zero. By allocating the costs and holding managers of producing departments responsible for the economic performance of their units, the organization ensures that managers will use a support service until the marginal benefit of that service equals its marginal cost. Thus, allocation of support department costs helps each producing department select the correct level of support service use.

There are other behavioural benefits. Allocation of support department costs to producing departments encourages managers of those departments to monitor the performance of support departments. Since support department costs affect the economic performance of their own departments, those managers have an incentive to influence the

---

[2] *Statements of Management Accounting (Statement 4B),* "Allocation of Service and Administrative Costs" (Montvale, NJ: NAA, 1985). The NAA is now known as the Institute of Management Accountants (IMA).

control over these costs through means other than simple usage of the support service. For instance, the managers can compare the internal costs of the support service with the costs of acquiring it externally. If a support department is not as cost effective as an outside source, perhaps the company should not continue to supply the service internally.

Many university libraries, for example, are moving toward the use of outside contractors for photocopying services. They have found that these contractors are more cost efficient and provide a higher level of service to library users than did the previous method of using professional librarians to make change, keep the copy machines supplied with paper, fix paper jams, etc.

Monitoring by managers of producing departments will also encourage managers of support departments to be more sensitive to the needs of the producing departments.

Clearly, there are good reasons for allocating support department costs. The validity of these reasons depends, however, on the accuracy and fairness of the cost assignments made. Although it may not be possible to identify a single method of allocation that simultaneously satisfies all of these objectives, several guidelines have been developed to assist in determining the best allocation method. These guidelines are cause and effect, benefits received, fairness, and ability to bear. Another guideline to be used in conjunction with any of the others is cost–benefit. That is, the method used must provide sufficient benefits to justify any effort involved.

Cause and effect requires the determination of causal factors to guide allocation. For example, a corporate legal department may track the number of hours spent on legal work for its various divisions (e.g., handling patent applications, lawsuits, etc.). The number of hours worked by lawyers and paralegals has a clear cause-and-effect relationship with the overall cost of the Legal Department and may be used to allocate these costs to the various company divisions.

The benefits-received guideline associates the cost with perceived benefits. Research and development (R&D) costs, for example, may be allocated on the basis of the sales of each division. While some R&D efforts may be unsuccessful and the successful efforts may happen to benefit one division in one year, all divisions have a stake in corporate R&D and will at some point have increased sales because of it.

Fairness or equity is a guideline often mentioned in government contracting. In the case of cost allocation methods, fairness usually means that the government contract should be costed in a method similar to that for nongovernmental contracts. For example, an airplane engine manufacturer may allocate a portion of corporate Legal Department costs to the government contract if these costs are usually allocated to private contracts.

Ability to bear is the least desirable guideline. It tends to "penalize" the most profitable division by allocating to it the largest proportion of a support department cost—regardless of whether the profitable division receives any services from the allocated department. This is why no motivational benefits of allocation are realized.

In determining how to allocate support department costs, the guideline of cost-benefit must be considered. In other words, the costs of implementing a particular allocation scheme must be compared to the benefits expected to be derived. This is why companies try to use easily measured and understood bases for allocation.

Allocating One Department's Costs to Other Departments

OBJECTIVE ▶ 2
Calculate charging rates, and distinguish between single and dual charging rates.

Frequently, the costs of a support department are allocated to other departments through the use of a charging rate. For example, a company's Data Processing Department may serve various other departments. The cost of operating the Data Processing Department is then allocated to the user departments. While this seems simple and straightforward, a number of considerations go into determining an appropriate charging rate. The two major factors are (1) the choice of a single or a dual charging rate and (2) the use of budgeted versus actual support department costs.

A Single Charging Rate

Some companies prefer to develop a single charging rate. This method is similar in concept to a plantwide overhead rate in that all support department costs are accumulated in the numerator and some measure of usage is in the denominator. There is only one rate, and it is relatively simple to apply. Suppose, for example, that Parminder and Murphy, a large regional public accounting firm, develops an in-house Photocopying Department to serve its three producing departments (Audit, Tax, and Management Advisory Services, or MAS). The firm wants to charge the using departments for their use of the photocopying service. Cornerstone 7-1 illustrates the calculation and use of the single charging rate when it is applied to budgeted amounts.

Under the single charging rate, the amount charged to the producing departments is based solely on the number of pages copied. Cornerstone 7-1 shows that the amount charged to the three using departments for their actual usage—272,000 pages copied—is $32,640. A single rate treats the fixed cost as if it were variable. In fact, to the producing departments, photocopying is strictly variable. Did the Photocopying Department need $32,640 to copy 272,000 pages? No, it needed only $32,446 [$26,190 + (272,000 × $0.023)]. The extra amount charged is due to the treatment of a fixed cost in a variable manner. Similarly, if the total number of pages is less than the amount budgeted, the single rate has the effect of undercompensating the Photocopying Department. Cornerstone 7-1 shows that if the actual pages copied had been 268,000, then $32,160 would be charged. However, the Photocopying Department needed $32,354 [$26,190 + (268,000 × $0.023)]. Again, the culprit is the treatment of fixed costs as if they were variable.

Multiple Charging Rates

Sometimes a single charging rate masks the variety of causal factors that lead to a support department's total costs. The Parminder and Murphy Photocopying Department is a good example. We saw that a single charging rate was based on the number of pages copied. Then, it looked like every page copied cost $0.12. But this is not true. A large portion of the costs of the Photocopying Department are fixed; they are not affected by the number of pages copied. Recall that $26,190 per year is spent on wages and rental of the photocopier. Why is this cost incurred? A talk with the photocopying company representative quickly yields the information that the size of the machine rented depends not on the number of pages copied per year, but on monthly peak usage. When Parminder and Murphy established the Photocopying Department, it surveyed the Audit, Tax, and MAS departments to determine each one's highest monthly usage. The Audit and MAS departments have fairly even copying needs throughout the year. The Tax Department, however, expects to need one-third of its yearly estimate in the month of April. Given this information, it appears that two charging rates are needed—one for variable costs based on the number of pages copied, and one for fixed costs based on estimated peak usage.

Developing a Variable Rate   The variable costs of the Photocopying Department are for paper and toner; these equal $0.023 per page. That is the variable rate to be used.

Developing a Fixed Rate   Fixed service costs are incurred to provide the capacity needed to deliver the service required by the producing departments. When the support department was established, its capacity was designed to serve the long-term needs of the producing departments. Since the original support needs caused the creation of the support service capacity, it seems reasonable to allocate fixed costs based on those needs.

The HOW and WHY of Calculating and Using a Single Charging Rate

**CORNERSTONE
7-1**

Information:
The expected (budgeted) cost of Parminder and Murphy's Photocopying Department for the coming year include:

Fixed costs (machine rental, salaries): $26,190 per year
Variable costs (paper and toner): $0.023 per page copied

Estimated (budgeted) usage by:

Audit Department	94,500
Tax Department	67,500
MAS Department	108,000
Total pages	270,000

Actual usage by:

Audit Department	92,000
Tax Department	65,000
MAS Department	115,000
Total pages	272,000

Why:
Many companies want to charge the costs of using support departments to the using departments. This makes the using departments responsible for their usage and helps prevent overuse of resources.

Required:
1. Calculate a single charging rate for the Photocopying Department.
2. Use this rate to assign the costs of the Photocopying Department to the user departments based on actual usage. Calculate the total amount charged for photocopying for the year.
3. **What if** the Audit and Tax departments used 92,000 and 65,000 pages, respectively, but the MAS Department only used 111,000 pages? How much would have been charged out to the three departments?

Solution:
1. Total expected costs of the Photocopying Department:

Fixed costs	$26,190
Variable costs ($0.023 × 270,000 pages)	6,210
Total costs	$32,400

Single charging rate = $32,400/270,000 = $0.12 per page

2. Charge based on actual usage = Charging rate × Actual pages
   Audit Department charge = $0.12 × 92,000 = $11,040
   Tax Department charge = $0.12 × 65,000 = $7,800
   MAS Department charge = $0.12 × 115,000 = $13,800
   Total amount charged = $11,040 + $7,800 + $13,800
   = $32,640

**CORNERSTONE
7-1
*(continued)***

3. Audit Department charge = $0.12 × 92,000 = $11,040
    Tax Department charge = $0.12 × 65,000 = $7,800
    MAS Department charge = $0.12 × 111,000 = $13,320
    Total amount charged = $11,040 + $7,800 + $13,320
    = $32,160

Either the normal or peak activity of the producing departments provides a reasonable measure of original support service needs. Normal capacity is the average capacity achieved over more than one fiscal period. If service is required uniformly over the time period, normal capacity is a good measure of activity. Peak capacity allows for variation in the need for the support department, and the size of the department is structured to allow for maximum need. In our example, the Tax Department may need much more photocopying during the first four months of the year, and its usage may be based on that need. The choice of normal or peak capacity in allocating budgeted fixed service costs depends on the needs of the individual firm. Budgeted fixed costs are allocated in this way regardless of whether the purpose is product costing or performance evaluation.

The allocation of fixed costs follows a three-step procedure:

1. *Determination of budgeted fixed support service costs.* The fixed support service costs that should be incurred for a period need to be identified.
2. *Computation of the allocation ratio.* The practical or normal capacity of each producing department is used to compute an allocation ratio. The allocation ratio gives a producing department's share or percentage of the total capacity of all producing departments.

    Allocation ratio = Producing department capacity/Total capacity

3. *Allocation.* The fixed support service costs are allocated in proportion to each producing department's original support service needs.

    Allocation = Allocation ratio × Budgeted fixed support service costs

Cornerstone 7-2 shows how and why to calculate two charging rates—one for the variable costs of the support department, and the other rate for the fixed costs.

Total Allocation   Under the dual charging rates, the fixed photocopying rates are charged to the departments in accordance with their original capacity needs. Especially in a case like the Photocopying Department example, in which fixed costs are such a high proportion of total costs, the additional effort needed to develop the dual rates may be worthwhile.

Comparing Cornerstone 7-1 results with those of Cornerstone 7-2, we see that the allocation of Photocopying Department costs is very different when the two charging rates are used. In this case, the Tax Department absorbs a larger proportion of the cost, because its peak usage is responsible for the size of the department. Notice, too, that the total amount charged of $32,446 is very close to the actual cost of running the department. With the two charging rates, each based on a strong causal factor, the allocation of cost to the using departments is close to the amount of cost that they actually cause the support department. The development of dual charging rates (which are used as the basis for pricing) is particularly important in companies such as public utilities.

The dual-rate method has the benefit of sending the correct signal regarding increased usage of the support department. Suppose that the Tax Department wants to have several research articles on tax law changes photocopied for clients. Should this be done "in house" by the Photocopying Department or sent to a private photocopying firm

The HOW and WHY of Calculating and Using Multiple Charging Rates

CORNERSTONE 7-2

Information:

The expected (budgeted) cost of Parminder and Murphy's Photocopying Department for the coming year include:

Fixed costs (machine rental, salaries): $26,190 per year
Variable costs (paper and toner): $0.023 per page copied

The Audit and MAS departments expect to use photocopying services evenly throughout the year. The Tax Department expects that one-third of its annual usage will occur in April.

Estimated (budgeted) usage:

	Yearly Pages	Monthly Peak Pages
Audit Department	94,500	7,875
Tax Department	67,500	22,500
MAS Department	108,000	9,000
Total pages	270,000	39,375

Actual usage in year:

Audit Department	92,000
Tax Department	65,000
MAS Department	115,000
Total pages	272,000

Why:

Two rates are calculated; the variable rate is based on number of pages and the fixed rate is based on peak usage. These rates more accurately assign support department costs to the using departments.

Required:

1. Calculate a variable rate for the Photocopying Department. Calculate the allocated fixed cost for each using department based on its budgeted monthly peak usage in pages.
2. Use the two rates to assign the costs of the Photocopying Department to the user departments based on actual usage. Calculate the total amount charged for photocopying for the year.
3. **What if** the Audit and Tax departments actually used 92,000 and 65,000 pages, respectively, but the MAS Department only used 111,000 pages? How much would have been charged out to the three departments?

Solution:

1. Variable rate = $0.023 per page

   The fixed allocation is calculated for each department based on budgeted monthly peak usage. Monthly peak usage for Audit and MAS is one-twelfth of the yearly amount. The monthly peak usage for Tax is one-third of the yearly amount (the amount for April). The allocation is given in the following table:

**CORNERSTONE
7-2
(continued)**

Department	Peak Number of Copies	Percent*	Budgeted Fixed Cost	Allocated Fixed Cost
Audit	7,875	20%	$26,190	$ 5,238
Tax	22,500	57	26,190	14,928
MAS	9,000	23	26,190	6,024
Total	39,375	100%		$26,190

*Percent for Audit = $7,875/$39,375 = 0.20, or 20%
Percent for Tax = $22,500/$39,375 = 0.57, or 57% (rounded)
Percent for MAS = $9,000/$39,375 = 0.23, or 23% (rounded)

2.

Department	Actual Number of Copies	Variable Rate	Variable Amount	Fixed Amount	Total Charge
Audit	92,000	$0.023	$2,116	$ 5,238	$ 7,354
Tax	65,000	0.023	1,495	14,928	16,423
MAS	115,000	0.023	2,645	6,024	8,669
Total	272,000		$6,256	$26,190	$32,446

3.

Department	Actual Number of Copies	Variable Rate	Variable Amount	Fixed Amount	Total Charge
Audit	92,000	$0.023	$2,116	$ 5,238	$ 7,354
Tax	65,000	0.023	1,495	14,928	16,423
MAS	111,000	0.023	2,553	6,024	8,577
Total	268,000		$6,164	$26,190	$32,354

that charges $0.06 per page? Under the single-rate method, the in-house charge would be too high because it wrongly assumes that fixed cost will increase as pages copied increase. However, under the dual-rate method, the additional cost would be only $0.023 per page, which correctly approximates the additional cost of the job.

Could there be more than two charging rates? Definitely. However, as a company breaks down support department resources and causal factors more finely, it may be approaching activity-based costing. The extra precision of charging rates must be balanced against the cost of determining and applying those rates. As always, the company must consider costs and benefits.

Budgeted versus Actual Usage

In Cornerstones 7-1 and 7-2, the allocation bases for determining the charging rates were based on budgeted amounts, not actual amounts. This is valuable for two reasons. First, the use of budgeted data permits producing departments to use the support department allocations in developing overhead rates that are used for product or service costing. Recall that the overhead rate is calculated at the beginning of the period, when actual costs are unknown. Thus, budgeted costs must be used. The second usage of allocated support department costs is for performance evaluation. In this case, too, budgeted support department costs are allocated to producing departments.

Managers of support and producing departments usually are held accountable for the performance of their departments. Their ability to control costs is an important factor in their performance evaluations. This ability is usually measured by comparing actual costs with planned or budgeted costs. If actual costs exceed budgeted costs, the department may be operating inefficiently, with the difference between the two costs serving as the measure of that inefficiency. Similarly, if actual costs are less than budgeted costs, the department may be operating efficiently.

A general principle of performance evaluation is that managers should be held responsible for costs or activities over which they have control. Since managers of producing departments have significant input regarding the level of support service consumed, they should be held responsible for their share of support service costs. This statement, however, has an important qualification: A department's evaluation should not be affected by the degree of efficiency achieved by another department.

This qualifying statement has an important implication for the allocation of support department costs. *Actual* costs of a support department should not be allocated to producing departments because they may include efficiencies or inefficiencies achieved by the support department. Managers of producing departments do not control the degree of efficiency achieved by a support department manager. When *budgeted* costs are allocated instead of actual costs, no inefficiencies or efficiencies are transferred from one department to another.

Whether budgeted usage or actual usage is employed depends on the purpose of the allocation. For *product costing*, the allocation is done at the beginning of the year on the basis of budgeted usage so that a predetermined overhead rate can be computed. If the purpose is *performance evaluation*, however, the allocation is done at the end of the period and is based on actual usage. The use of cost information for performance evaluation is covered in more detail in Chapter 9.

Let's return to our photocopying example. Recall that annual budgeted fixed costs were $26,190 and the budgeted variable cost per page was $0.023. The three producing departments—Audit, Tax, and MAS—estimated usage at 94,500 copies, 67,500 copies, and 108,000 copies, respectively. Given these data, the costs allocated to each department at the *beginning* of the year are shown in Exhibit 7-4.

⟨ **Exhibit 7-4** ⟩

Use of Budgeted Data for Product Costing: Comparison of Single- and Dual-Rate Methods

	Single-Rate Method				
	Number of Copies	×	Total Rate	=	Allocated Cost
Audit	94,500		$0.12		$11,340
Tax	67,500		0.12		8,100
MAS	108,000		0.12		12,960
Total	270,000				$32,400

	Dual-Rate Method					
	Number of Copies	×	Variable Rate	+	Fixed Allocation	=	Allocated Cost
Audit	94,500		$0.023		$ 5,238		$ 7,411*
Tax	67,500		0.023		14,928		16,481*
MAS	108,000		0.023		6,024		8,508
Total	270,000						$32,400

*Rounded.

When the allocation is done for the purpose of budgeting the producing departments' costs, then, of course, the budgeted support department costs are used. The photocopying costs allocated to each department would be added to other producing department costs—including those directly traceable to each department plus other support department allocations—to compute each department's anticipated spending. In a manufacturing plant, the allocation of budgeted support department costs to the producing departments would precede the calculation of the predetermined overhead rate.

During the year, each producing department would also be responsible for actual charges incurred based on the actual number of pages copied. Going back to the actual usage assumed previously, a second allocation is now made to measure the actual performance of each department against its budget. The actual photocopying costs allocated to each department for performance evaluation purposes are shown in Exhibit 7-5.

Fixed versus Variable Bases: A Note of Caution

Using normal or practical capacity to allocate fixed support service costs provides a *fixed* base. As long as the capacities of the producing departments remain at the original level, there is no reason to change the allocation ratios. Thus, each year, the Audit Department receives 35 percent of the budgeted fixed photocopying costs, the Tax Department 25 percent, and the MAS Department 40 percent, no matter what their actual usage is. If the capacities of the departments change, the ratios should be recalculated.

In practice, some companies choose to allocate fixed costs in proportion to actual usage or expected actual usage. Since usage may vary from year to year, allocation of fixed costs would then use a variable base. Variable bases, however, have a significant drawback: they allow the actions of one department to affect the amount of cost allocated to another department.

To see how this is demonstrated, let's return to Parminder and Murphy's Photocopying Department and assume that fixed costs are allocated on the basis of anticipated usage for the coming year. The Audit and Tax departments budget the same number of copies as before. However, the MAS Department anticipates much less activity due to a regional recession, which will cut down the number of new clients served; the anticipated number of photocopies for this department falls to 68,000. The adjusted fixed cost allocation ratios and allocated fixed cost based on the newly budgeted usage are as follows:

Department	Number of Copies	Percent	Allocated Fixed Cost
Audit	94,500	41.1%	$10,764
Tax	67,500	29.3	7,674
MAS	68,000	29.6	7,752
Total	230,000	100.0%	$26,190

Notice that both the Audit and Tax departments' allocation of fixed costs increased even though the fixed costs of the Photocopying Department remained unchanged. This increase was caused by a decrease in the MAS Department's use of photocopying. In effect, the Audit and Tax departments have been penalized by MAS's decision to reduce the number of pages copied for its department. Imagine the feelings of the first two managers when they realize that their copying charges have increased due to the increase in allocated fixed costs! The penalty occurs because a variable base has been used to allocate fixed support service costs; this can be avoided by using a fixed base.

Exhibit 7-5

Use of Actual Data for Performance Evaluation Purposes: Comparison of Single- and Dual-Rate Methods

Single-Rate Method				
	Number of Copies	×	Total Rate	=	Allocated Cost
Audit	92,000		$0.12		$11,040
Tax	65,000		0.12		7,800
MAS	115,000		0.12		13,800
Total	272,000				$32,640

Dual-Rate Method					
	Number of Copies	×	Variable Rate	+	Fixed Allocation	=	Allocated Cost
Audit	92,000		$0.023		$ 5,238		$ 7,354
Tax	65,000		0.023		14,928		16,423
MAS	115,000		0.023		6,024		8,669
Total	272,000						$32,446

Allocating Multiple Support Department Costs to Other Departments

OBJECTIVE ▶ 3

Allocate support centre costs to producing departments using the direct method, the sequential method, and the reciprocal method.

So far, we have considered cost allocation from a single support department to several producing departments. We used the direct method of support department cost allocation, in which support department costs are allocated only to producing departments. This was appropriate in the earlier example because no other support departments existed. This would also be appropriate when there is no possibility of interaction among support departments. Many companies do have multiple support departments and they frequently interact. For example, in a factory, Personnel and Cafeteria serve each other, other support departments, and the producing departments.

Ignoring these interactions and allocating support costs directly to producing departments may produce unfair and inaccurate cost assignments. For example, Power, although a support department, may use 30 percent of the services of the Maintenance Department. The maintenance costs caused by the Power Department belong to the Power Department. When these costs are not assigned to the Power Department, its costs are understated. In effect, some of the costs caused by Power are "hidden" in the Maintenance Department because maintenance costs would be lower if the Power Department did not exist. As a result, a producing department that is a heavy user of power and an average or below-average user of maintenance may then receive, under the direct method, a cost allocation that is understated.

In determining which support department cost allocation method to use, companies must determine the extent of support department interaction. In addition, they must weigh the costs and benefits associated with the three methods described and illustrated in the following sections: the direct, sequential, and reciprocal methods. Exhibit 7-6 presents data for a factory with two support departments—Power and Maintenance, and two producing departments—Grinding and Assembly. The activity drivers for the support departments are kilowatt-hours (for Power) and maintenance hours (for Maintenance). The direct overhead costs for each department are listed first. For the support departments, direct overhead costs include all costs of running the support departments. For the producing departments, the direct overhead costs are those overhead costs that are traced directly to those departments such as supervisory salaries and equipment depreciation. A final point should be

Exhibit 7-6

Data for Support and Producing Departments

	Support Departments		Producing Departments	
	Power	**Maintenance**	**Grinding**	**Assembly**
Direct costs	$250,000	$160,000	$100,000	$ 60,000
Normal activity:				
Kilowatt-hours	—	200,000	600,000	200,000
Maintenance hours	1,000	—	4,500	4,500

made regarding the dashes for kilowatt-hours for Power and for maintenance hours for Maintenance. Doesn't the Power Department use power? Absolutely, however, it does not matter how many kilowatt-hours are used by Power for the purposes of allocating Power Department cost. Similarly for Maintenance, for allocation purposes, it does not matter how many hours the Maintenance Department spends on maintaining its own department.

Direct Method of Allocation

When companies allocate support department costs only to the producing departments, they are using the **direct method** of allocation. The direct method is the simplest and most straightforward way to allocate support department costs. All costs of the support departments are allocated directly to producing departments in proportion to each producing department's usage of the service. This method does not allocate any support department costs to another support department, even if other support departments use the services of a support department. This usage of one support department by another is called support department reciprocity. Under the direct method, no support department reciprocity or interaction is recognized.

Cornerstone 7-3 shows how and why to allocate support department costs to producing departments using the direct method.

Examine Cornerstone 7-3 carefully. Notice that Requirement 1 shows how to calculate the allocation ratios. Since no support department cost is allocated to another support department, there are no percentages shown for Power or Maintenance. All of the support department output is assigned to the producing departments. In Requirement 2, the costs of Power and Maintenance are allocated to the producing departments. We see that all cost in each support department is divided up between the producing departments. Once those costs are allocated, there is zero cost remaining in the support departments.

Finally, it is a good idea to check the pre-and post-allocation totals. Before allocation, the total overhead in the factory is $570,000 ($250,000 + $160,000 + $100,000 + $60,000). After allocation is complete, total factory overhead is still $570,000 ($367,500 + $202,500). These totals will always be the same (except for rounding error). Allocation does not increase or decrease total overhead, it just redistributes it to the producing departments.

Sequential Method of Allocation

The **sequential (or step) method** of allocation recognizes that interactions among the support departments do occur; however, the sequential method takes only partial account of this interaction. Cost allocations are performed in step-down fashion, following a predetermined ranking procedure. This ranking can be performed in various ways. One possibility is to rank the support departments in order of the percentage of service provided to other support departments. Another possibility is to rank the support departments in order of their total cost, from highest cost department to lowest.

Once the support departments have been ranked, the top ranking department is allocated to lower ranking support departments and the producing departments. It is then closed out (has a total cost remaining of zero) and the remaining support departments cannot allocate cost back to it. Then, the costs of the support

The HOW and WHY of Allocating Support Department Costs to Producing Departments Using the Direct Method

**CORNERSTONE
7-3**

Information:
Refer to Exhibit 7-6 for data on the two support and two producing departments. The costs of the Power Department are allocated on the basis of kilowatt-hours, and the costs of the Maintenance Department are allocated on the basis of maintenance hours. The factory uses the direct method of support department cost allocation.

Why:
Support department costs must be allocated to the producing departments so that the producing departments can calculate their overhead rates. The direct method is simple and easy to use. If there is relatively little support department reciprocity, it does a fairly good job.

Required:
1. Calculate the allocation ratios for the four departments using the direct method.
2. Using the direct method, allocate the costs of the Power and Maintenance departments to the Grinding and Assembly departments.
3. **What if** the Maintenance Department used only 100,000 kilowatt-hours? How would that affect the allocation of Power Department costs to the Grinding and Assembly departments?

Solution:
1. Allocation ratios:

	Proportion of Driver Used by			
	Power	**Maintenance**	**Grinding**	**Assembly**
Power	—	—	0.75[1]	0.25[2]
Maintenance	—	—	0.50[3]	0.50[4]

[1] Proportion of kilowatt-hours used by Grinding = 600,000/(600,000 + 200,000) = 0.75
[2] Proportion of kilowatt-hours used by Assembly = 200,000/(600,000 + 200,000) = 0.25
[3] Proportion of maintenance hours used by Grinding = 4,500/(4,500 + 4,500) = 0.50
[4] Proportion of maintenance hours used by Assembly = 4,500/(4,500 + 4,500) = 0.50

2.

	Support Departments		Producing Departments	
	Power	**Maintenance**	**Grinding**	**Assembly**
Direct costs	$ 250,000	$ 160,000	$100,000	$ 60,000
Allocate:				
Power[1]	(250,000)	—	187,500	62,500
Maintenance[2]	—	(160,000)	80,000	80,000
Total after allocation	$ 0	$ 0	$367,500	$202,500

[1] Grinding = 0.75 × $250,000 = $187,500; Assembly = 0.25 × $250,000 = $62,500
[2] Grinding = 0.50 × $160,000 = $80,000; Assembly = 0.50 × $160,000 = $80,000

3. Since none of the Power cost is allocated to Maintenance, it does not matter how many kilowatt-hours are used by Maintenance.

department next in sequence are similarly allocated, and so on. In the sequential method, once a support department's costs are allocated, it never receives a subsequent allocation from another support department. In other words, costs of a support department are never allocated to support departments *above* it in the sequence. Also note that the costs allocated from a support department are its direct costs *plus* any costs it receives in allocations from other support departments. The direct costs of a department are those that are directly traceable to the department.

Cornerstone 7-4 shows how and why to use the sequential method. The data originally given in Exhibit 7-6 are used. First, the support departments are ranked. Power provides relatively more service to Maintenance than Maintenance provides to Power. In addition, the cost of Power is higher than the cost of Maintenance. So, no matter which ranking system is used, Power is allocated first, then Maintenance.

As before, it is a good idea to check the pre- and post-allocation totals. Before allocation, the total overhead in the factory is $570,000 ($250,000 + $160,000 + $100,000 + $60,000). After allocation is complete, total factory overhead is $570,000 ($355,000 + $215,000). These totals will always be the same (except for rounding error). Allocation does not increase or decrease total overhead, it just redistributes it to the producing departments.

The sequential method may be more accurate than the direct method because it recognizes some interactions among the support departments. It does not recognize all interactions, however; no maintenance costs were assigned to the Power Department even though it used 10 percent of the Maintenance Department's output. The reciprocal method corrects this deficiency, as we will see later.

Reciprocal Method of Allocation

The **reciprocal method** of allocation recognizes all interactions of support departments. Under the reciprocal method, the usage of one support department by another is used to determine the total cost of each support department, where the total cost reflects interactions among the support departments. Then, the new total of support department costs is allocated to the producing departments. This method fully accounts for support department interaction.

Total Cost of Support Departments

To determine the total cost of a support department so that this total cost reflects interactions with other support departments, a system of simultaneous linear equations must be solved. Each equation, which is a cost equation for a support department, is the sum of the department's direct costs plus the proportion of service received from other support departments.

$$\text{Total cost} = \text{Direct costs} + \text{Allocated costs}$$

The same data set contained in Exhibit 7-6 used to illustrate the direct and sequential methods will be used to illustrate the reciprocal method. Cornerstone 7-5 shows how and why the reciprocal method is used to allocate support department costs.

As Cornerstone 7-5 shows, the steps for the reciprocal method are:

1. Compute the allocation ratios for all support and producing departments.
2. Form a simultaneous equations system with one equation for each support department. The interpretation of each equation is that the total reciprocated cost of the support department equals its original cost plus any cost that it imposes on any other support departments.
3. Solve the simultaneous equations system for each unknown to obtain the total reciprocated cost of each support department.
4. Allocate the total reciprocated costs of the support departments to other support departments and to the producing departments based on the allocation ratios developed in Step 1.

The HOW and WHY of Allocating Support Department Costs to Producing Departments Using the Sequential (Step) Method

**CORNERSTONE
7-4**

Information:
Refer to Exhibit 7-6 for data on the two support and two producing departments. The costs of the Power Department are allocated on the basis of kilowatt-hours, and the costs of the Maintenance Department are allocated on the basis of maintenance hours. The factory uses the sequential method of support department cost allocation.

Why:
Support department costs must be allocated to the producing departments so that the producing departments can calculate their overhead rates. The sequential method takes some account of support department reciprocity and is, therefore, somewhat better than the direct method.

Required:
1. Calculate the allocation ratios for the four departments using the sequential method.
2. Using the sequential method, allocate the costs of the Power and Maintenance departments to the Grinding and Assembly departments.
3. **What if** the Maintenance Department used only 100,000 kilowatt-hours? How would that affect the allocation of Power Department costs to the Grinding and Assembly departments?

Solution:
1. Power is allocated first because 20 percent of its service [200,000/(200,000 + 600,000 + 200,000) = 20%] is used for other support departments (in this case, Maintenance). Only 10 percent [1,000/(1,000 + 4,500 + 4,500) = 10%] of Maintenance services are used by other support departments (i.e., Power).
   Allocation ratios:

	Proportion of Driver Used by			
	Power	**Maintenance**	**Grinding**	**Assembly**
Power	—	0.20^1	0.60^2	0.20^3
Maintenance	—	—	0.50^4	0.50^5

[1] Proportion of kilowatt-hours used by Maintenance = 200,000/(200,000 + 600,000 + 200,000)
   = 0.20
[2] Proportion of kilowatt-hours used by Grinding = 600,000/(200,000 + 600,000 + 200,000) = 0.60
[3] Proportion of kilowatt-hours used by Assembly = 200,000/(200,000 + 600,000 + 200,000) = 0.20
[4] Proportion of maintenance hours used by Grinding = 4,500/(4,500 + 4,500) = 0.50
[5] Proportion of maintenance hours used by Assembly = 4,500/(4,500 + 4,500) = 0.50

2.

	Support Departments		**Producing Departments**	
	Power	**Maintenance**	**Grinding**	**Assembly**
Direct costs	$ 250,000	$ 160,000	$100,000	$ 60,000
Allocate:				
Power[1]	(250,000)	50,000	150,000	50,000
Maintenance[2]	—	(210,000)	105,000	105,000
Total after allocation	$ 0	$ 0	$355,000	$215,000

[1] Maintenance = 0.2 × $250,000 = $50,000; Grinding = 0.60 × $250,000 = $150,000; Assembly = 0.20 × $250,000 = $50,000
[2] Grinding = 0.50 × ($160,000 + $50,000) = $105,000; Assembly = 0.50 × ($160,000 + $50,000) = $105,000

CORNERSTONE 7-4
(continued)

3. If Maintenance used only 100,000 kilowatt-hours, then the proportion of service it uses would drop to 11.11% [100,000/(100,000 + 600,000 + 200,000)]. Power would still be allocated first; however, the allocation ratios for Power would change to: Maintenance, 11.11%; Grinding, 66.67%; and Assembly, 22.22% (rounded). Thus, relatively fewer dollars would be allocated to Maintenance, and relatively more to Grinding and Assembly. The new allocations would be as follows:

| | Support Departments | | Producing Departments | |
	Power	Maintenance	Grinding	Assembly
Direct costs	$ 250,000	$ 160,000	$100,000	$ 60,000
Allocate:				
Power[1]	(250,000)	27,775	166,675	55,550
Maintenance[2]	—	(187,775)	93,888	93,887*
Total after allocation	$ 0	$ 0	$360,563	$209,437

[1] Maintenance = 0.1111 × $250,000 = $27,775; Grinding = 0.6667 × $250,000 = $166,675; Assembly = 0.2222 × $250,000 = $55,550
[2] Grinding = 0.50 × ($160,000 + $27,775) = $93,887; Assembly = 0.50 × ($160,000 + $27,775) = $93,887*
* rounded

After the equations are solved, the total costs of each support department are known. These total costs, unlike the direct or sequential methods, reflect all interactions between support departments. As a result, the reciprocal method is the best method in terms of accounting for all interactions among the support departments.

Comparison of the Three Cost Allocation Methods

Exhibit 7-7 gives the cost allocations from the Power and Maintenance departments to the Grinding and Assembly departments using the three support department cost allocation methods. How different are the results? Does it really matter which method is used? Depending on the degree of support department interaction, the three allocation methods can give quite different results. In this particular example, the direct method (as compared to the sequential method) allocated $12,500 more to the Grinding Department (and $12,500 less to the Assembly Department). Surely, the manager of the Assembly Department would prefer the direct method, and the manager of the Grinding Department would prefer the sequential method. Because allocation methods do affect the cost responsibilities of managers, it is important for

Exhibit 7-7

Comparison of Support Department Cost Allocations Using the Direct, Sequential, and Reciprocal Methods

| | Direct Method | | Sequential Method | | Reciprocal Method | |
	Grinding	Assembly	Grinding	Assembly	Grinding	Assembly
Direct costs	$100,000	$ 60,000	$100,000	$ 60,000	$100,000	$ 60,000
Allocated from power	187,500	62,500	150,000	50,000	162,857	54,285
Allocated from maintenance	80,000	80,000	105,000	105,000	96,429	96,429
Total cost	$367,500	$202,500	$355,000	$215,000	$359,286	$210,714

The HOW and WHY of Allocating Support Department Costs to Producing Departments Using the Reciprocal Method

CORNERSTONE 7-5

Information:

Refer to Exhibit 7-6 for data on the two support and two producing departments. The costs of the Power Department are allocated on the basis of kilowatt-hours, and the costs of the Maintenance Department are allocated on the basis of maintenance hours. The factory uses the reciprocal method of support department cost allocation.

Why:

Support department costs must be allocated to the producing departments so that the producing departments can calculate their overhead rates. The reciprocal method takes full account of support department reciprocity and is, therefore, the theoretically best method.

Required:

1. Calculate the allocation ratios for the four departments in preparation for the reciprocal method.
2. Develop a simultaneous equations system of total costs for the support departments. Solve for the total reciprocated costs of each support department.
3. Using the reciprocal method, allocate the fully reciprocated costs of the Power and Maintenance departments to the Grinding and Assembly Departments.
4. **What if** the Maintenance Department used only 100,000 kilowatt-hours? How would that affect the allocation of Power Department costs to the Grinding and Assembly departments?

Solution:

1. Allocation ratios:

	Proportion of Driver Used by			
	Power	**Maintenance**	**Grinding**	**Assembly**
Power	—	0.20[1]	0.60[2]	0.20[3]
Maintenance	0.10[4]	—	0.45[5]	0.45[6]

[1] Proportion of kilowatt-hours used by Maintenance = 200,000/(200,000 + 600,000 + 200,000)
= 0.20
[2] Proportion of kilowatt-hours used by Grinding = 600,000/(200,000 + 600,000 + 200,000) = 0.60
[3] Proportion of kilowatt-hours used by Assembly = 200,000/(200,000 + 600,000 + 200,000) = 0.20
[4] Proportion of maintenance hours used by Power = 1,000/(1,000 + 4,500 + 4,500) = 0.10
[5] Proportion of maintenance hours used by Grinding = 4,500/(1,000 + 4,500 + 4,500) = 0.45
[6] Proportion of maintenance hours used by Assembly = 4,500/(1,000 + 4,500 + 4,500) = 0.45

2. Let P = Fully reciprocated costs for Power; and

M = Fully reciprocated costs for Maintenance

$P = \$250,000 + 0.1M$

$M = \$160,000 + 0.2P$

Solve for P by substituting ($\$160,000 + 0.2P$) for M:

$P = \$250,000 + 0.1(\$160,000 + 0.2P)$

$P - 0.02P = \$250,000 + \$16,000$

$0.98P = \$266,000$

$P = \$271,429$ (rounded)

**CORNERSTONE
7-5**
(continued)

Solve for M:

$M = \$160,000 + 0.2(\$271,429) = \$214,286$ (rounded)

3.

	Support Departments		Producing Departments	
	Power	**Maintenance**	**Grinding**	**Assembly**
Direct costs	$ 250,000	$ 160,000	$100,000	$ 60,000
Allocate:				
Power[1]	(271,429)	54,286	162,857	54,286*
Maintenance[2]	21,429	(214,286)	96,429	96,429
Total after allocation	$ 0	$ 0	$359,286	$210,714

[1] Maintenance $= 0.20 \times \$271,429 = \$54,286$; Grinding $= 0.60 \times \$271,429 = \$162,857$;
Assembly $= 0.20 \times \$271,429 = \$54,286$

[2] Power $= 0.10 \times \$214,286 = \$21,429$; Grinding $= 0.45 \times \$214,286 = \$96,429$; Assembly $= 0.45 \times$
$\$214,286 = \$96,429$

* rounded

4. If Maintenance used only 100,000 kilowatt-hours, then the proportion of service it uses would drop to 11.11% [100,000/(100,000 + 600,000 + 200,000)]. The allocation ratios for Power would change to: Maintenance, 11.11%; Grinding, 66.67%; and Assembly, 22.22% (rounded). This would affect the simultaneous equations and the subsequent allocation.

$$P = \$250,000 + 0.1M$$
$$M = \$160,000 + 0.1111P$$

Solve for P by substituting ($\$160,000 + 0.1111P$) for M:

$$P = \$250,000 + 0.1(\$160,000 + 0.1111P)$$
$$P - 0.01111P = \$250,000 + \$16,000$$
$$0.98889P = \$266,000$$
$$P = \$268,989 \text{ (rounded)}$$

Solve for M:

$$M = \$160,000 + 0.1111(\$268,989) = \$189,885 \text{ (rounded)}$$

The new allocations would be as follows:

	Support Departments		Producing Departments	
	Power	**Maintenance**	**Grinding**	**Assembly**
Direct costs	$ 250,000	$ 160,000	$100,000	$ 60,000
Allocate:				
Power[1]	(268,989)*	29,885	179,335	59,769
Maintenance[2]	18,989	(189,885)	85,448	85,448
Total after allocation	$ 0	$ 0	$364,783	$205,217

[1] Maintenance $= 0.1111 \times \$268,989 = \$29,885$; Grinding $= 0.6667 \times \$268,989 = \$179,335$; Assembly $= 0.2222$
$\times \$268,989 = \$59,769$

[2] Power $= 0.10 \times \$189,885 = \$18,989$; Grinding $= 0.45 \times \$189,885 = \$85,448$; Assembly $= 0.45 \times \$189,885 =$
$\$85,448$

* rounded

the accountant to understand the consequences of the different methods and to have good reasons for the eventual choice.

It is important to keep a cost–benefit perspective in choosing an allocation method. The accountant must weigh the advantages of better allocation against the increased cost using a more theoretically preferred method, such as the reciprocal method. For example, about 30 years ago, the controller for the **IBM** Poughkeepsie plant decided that the reciprocal method of cost allocation would do a better job of allocating support department costs. He identified over 700 support departments and solved the system of equations using a computer. Computationally, he had no problems. However, the producing department managers did not understand the reciprocal method. They were sure that extra cost was being allocated to their departments, but they were not sure just how. After months of meetings with the line managers, the controller threw in the towel and returned to the sequential method—which everyone did understand.[3]

Another factor to be considered in allocating support department cost is the rapid change in technology. Many firms currently find that support department cost allocation is useful for them. However, the move toward activity-based costing and just-in-time manufacturing can virtually eliminate the need for support department cost allocation. In the case of the JIT factory with manufacturing cells, much of the service (e.g., maintenance, materials handling, and setups) is performed by cell workers. Allocation is not necessary.

Departmental Overhead Rates and Product Costing

OBJECTIVE ▶ 4
Calculate departmental overhead rates.

Upon allocating all support service costs to producing departments, an overhead rate can be computed for each department. This rate is computed by adding the allocated service costs to the overhead costs that are directly traceable to the producing department and dividing this total by some measure of activity, such as direct labour hours or machine hours. Cornerstone 7-6 shows how and why to use the allocated support department costs to develop departmental overhead rates.

One might wonder, however, just how accurate are the job costs calculated in Cornerstone 7-6? Is this amount the true cost of the product in question? Since materials and labour are directly traceable to products, the accuracy of product costs depends largely on the accuracy of the assignment of overhead costs. This in turn depends on the degree of correlation between the factors used to allocate support service costs to departments and on the factors used to allocate the department's overhead costs to the products. For example, if power costs are highly correlated with kilowatt-hours and machine hours are highly correlated with a product's consumption of the Grinding Department's overhead costs, then we can have some confidence that the $5 overhead rate accurately assigns costs to individual products. However, if the allocation of support service costs to the Grinding Department or the use of machine hours is faulty—or both—then product costs will be distorted. The same reasoning can be applied to the Assembly Department. To ensure accurate product costs, great care should be used in identifying and using causal factors for both stages of overhead assignment. Activity-based costing, explained in Chapter 6, can be used to develop more accurate product costs.

Outsourcing Support Department Costs

Today many companies are examining the cost savings that might be obtained by outsourcing certain of their nonessential activities. Payroll services, computer systems operations, routine accounting functions, call centre operations, customer service functions, and other routine activities lend themselves to being provided by outside companies.

[3] This is based on conversations between the author and the IBM controller.

**CORNERSTONE
7-6**

The HOW and WHY of Using Allocated Support Department Costs to Calculate Departmental Overhead Rates

Information:

Assume that the factory in our example uses the sequential method to allocate support department costs. The cost allocation is shown in Cornerstone 7-4. The Grinding Department overhead rate is based on normal activity of 71,000 machine hours. The Assembly Department overhead rate is based on normal activity of 107,500 direct labour hours.

Job 189 required 20 machine hours in Grinding and five direct labour hours in Assembly. Total direct materials cost was $465, and total direct labour cost was $370.

Why:

One reason for support department cost allocation is to allow producing departments to calculate overhead rates. The overhead rates are then used to cost product.

Required:

1. Calculate the overhead rate for Grinding based on machine hours and the overhead rate for Assembly based on direct labour hours.
2. Using the overhead rates calculated in Requirement 1, calculate the cost of Job 189.
3. **What if** Job 189 had required five machine hours in Grinding and 20 direct labour hours in Assembly? Direct labour and direct materials costs remained the same. Calculate the new cost of Job 189.

Solution:

1. Grinding Department overhead rate = $355,000/71,000

    = $5 per machine hour

    Assembly Department overhead rate = $215,000/107,500

    = $2 per direct labour hour

2. Cost of Job 189:

Direct materials	$465
Direct labour cost	370
Applied overhead:	
Grinding (20 × $5)	100
Assembly (5 × $2)	10
Total cost	$945

3. New Cost of Job 189:

Direct materials	$465
Direct labour cost	370
Applied overhead:	
Grinding (5 × $5)	25
Assembly (20 × $2)	40
Total cost	$900

Many firms have developed a business by offering specialized expertise to other companies on the basis that they can provide those services on a cost-effective basis. The reasoning is that a company that is providing a single service to a variety of clients can develop expertise in that function that a company whose main focus is in another area, such as manufacturing or service provision, cannot develop itself.

So it is increasingly important that managers understand the true cost of providing such services and which parts of the organization are using the services. When considering outsourcing, companies must evaluate the qualitative as well as the quantitative factors to ensure that they are not entrusting essential activities to others in ways that might work against them in the marketplace.

Accounting for Joint Production Processes

O B J E C T I V E ➤ 5

Identify the characteristics of the joint production process, and allocate joint costs to products.

Joint products are two or more products produced simultaneously by the same process up to a "split-off" point. The **split-off point** is the point at which the joint products become separate and identifiable. For example, oil and natural gas are joint products. When a company drills for oil, it gets natural gas as well. As a result, the costs of exploration, acquisition of mineral rights, and drilling are incurred to the initial split-off point. Such costs are necessary to bring crude oil and natural gas out of the ground, and they are common costs to both products. Of course, some joint products may require processing beyond the split-off point. For example, crude oil can be processed further into aviation fuel, gasoline, kerosene, naptha, and other petrochemicals. The key point, however, is that the direct materials, direct labour, and overhead costs incurred up to the initial split-off point are joint costs that can only be allocated to the final product in some arbitrary manner. Joint products are so enmeshed that once the decision to produce has been made, management decision has little effect on the output, at least to the initial split-off point. Exhibit 7-8 depicts the joint production process. Joint products are related to one another such that an increase in the output of one increases the output of the others, although not necessarily in the same ratio. Up to the split-off point, you cannot get more of one product without getting more of the other(s).

Costs are either separable or not. **Separable costs** are easily traced to individual products and offer no particular problem. If not separable, they are allocated to various products for various reasons. Cost allocations are arbitrary. That is, there is no well-accepted theoretical way to determine which product incurs what part of the joint cost. In reality, all joint products benefit from the entire joint cost. The objective in joint cost allocation is to determine the most appropriate way to allocate a cost

Exhibit 7-8

Joint Production Process

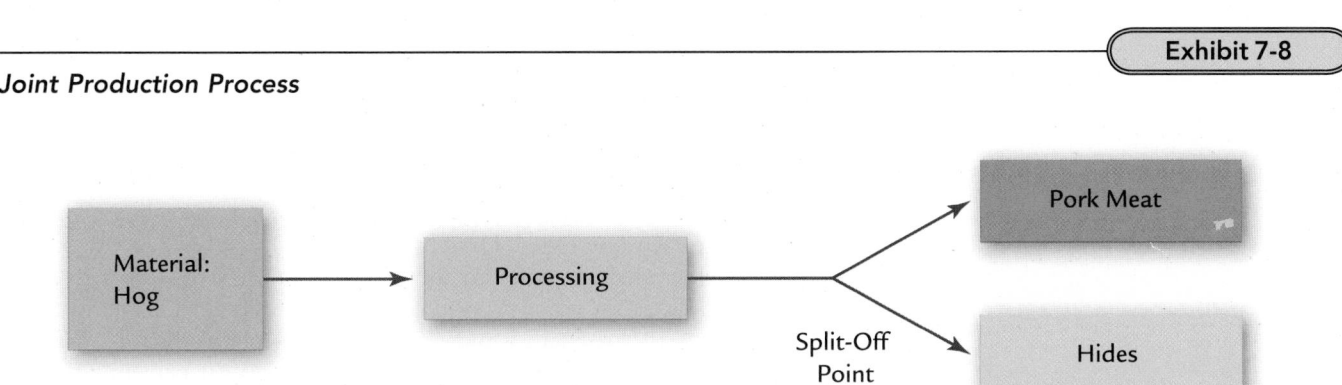

that is not really separable. The primary reason for joint cost allocation is that financial reporting (GAAP) and federal income tax law require it. In addition, these product costs are somewhat useful in calculating the cost of special lots or orders including government cost-type contracts and in justifying prices for legislative or administrative regulations. It is important to note that the allocation of joint costs is not appropriate for certain types of management decisions. The impact of joint costs on decision making is reserved for Chapter 11.

There are two important differences between costs incurred up to the split-off point in joint product situations and those indirect costs incurred for products that are produced independently. First, certain costs such as direct materials and direct labour, which are directly traceable to products when two or more products are separately produced, become indirect and indivisible when used prior to the split-off point. For example, if ore contains both iron and zinc, the direct material itself is a joint product. Since neither zinc nor iron can be produced alone prior to the split-off point, the related processing costs of mining, crushing, and splitting the ore are also joint costs. Second, manufacturing overhead becomes even more indirect in joint product situations. Consider the purchase of pineapples. A pineapple, in and of itself, is not a joint product. However, when pineapples are purchased for canning, the initial processing or trimming of the fruit results in a variety of products (skin for animal feed, trimmed core for further slicing and dicing, and juice). The processing costs to the point of split-off, as well as the cost of the original pineapples, are mutually beneficial to all products produced to that point. Both of these phenomena are caused either because the material itself is a joint product or because processing results in the simultaneous output of more than one product.

Accounting for Joint Product Costs

The accounting for overall joint costs of production (direct materials, direct labour, and overhead) is no different from the accounting for product costs in general. It is the *allocation* of joint costs to the individual products that is the source of difficulty. Still, the allocation must be done for financial reporting purposes—to value inventory carried on the balance sheet and to determine income. Thus, an allocation method must be found that, though arbitrary, allocates the costs on as reasonable a basis as possible. Because judgment is involved, equally competent accountants can arrive at different costs for the same product. There are a variety of methods for allocating joint costs. These methods include the physical units method, the weighted average method, the sales-value-at-split-off method, the net realizable value method, and the constant gross margin percentage method. These are covered in the following sections.

Physical Units Method   Under the **physical units method**, joint costs are distributed to products on the basis of some physical measure. These physical measures may be expressed in units such as kilograms, tonnes, litres, board feet, atomic weight, or heat units. If the joint products do not share the same physical measure (e.g., one product is measured in litres, another in kilograms), some common denominator may be used. For example, a producer of fuels may take litres, barrels, and tonnes and convert each one into BTUs (British thermal units) of energy.

Computationally, the physical units method allocates the same proportion of joint cost to each product as the underlying proportion of units. So, if a joint process yields 300 kilograms of Product A and 700 kilograms of Product B, Product A receives 30 percent of the joint cost and Product B receives 70 percent. Alternatively, one can divide total joint costs by total output to find an average unit cost. The average unit cost is then multiplied by the number of units of each product. Cornerstone 7-7 shows how and why the physical units method can be used to allocate joint cost.

The HOW and WHY of Using the Physical Units Method to Allocate Joint Product Costs

**CORNERSTONE
7-7**

Information:
A sawmill processes logs into four grades of lumber totalling 3,000,000 board feet as follows:

Grades	Board Feet
First and second	450,000
No. 1 common	1,200,000
No. 2 common	600,000
No. 3 common	750,000
Total	3,000,000

Total joint cost is $186,000.

Why:
The joint cost must be allocated to the various grades of lumber in order to cost product and value inventory. Physical units allocate the cost in proportion to the number of units and is useful when the value of one product (here, grade) is close to the value of another product.

Required:
1. Allocate the joint cost to the four grades of lumber using the physical units method.
2. Allocate the joint cost to the four grades of lumber by finding the average joint cost per board foot and multiplying it by the number of board feet in the grade.
3. *What if* First and second and No. 1 common each had 825,000 board feet? How would that affect the allocation of cost to these two grades? How would it affect the allocation of cost to the No. 2 and No. 3 common grades?

Solution:
1.

Grades	Board Feet	Percent of Units*	Joint Cost Allocation
	(2)	(3)	(3) × $186,000
First and second	450,000	15%	$ 27,900
No. 1 common	1,200,000	40	74,400
No. 2 common	600,000	20	37,200
No. 3 common	750,000	25	46,500
Total	3,000,000	100%	$186,000

* Percent for First and second = 450,000/3,000,000 = 0.15, or 15%
  Percent for No. 1 common = 1,200,000/3,000,000 = 0.40, or 40%
  Percent for No. 2 common = 600,000/3,000,000 = 0.20, or 20%
  Percent for No. 3 common = 750,000/3,000,000 = 0.25, or 25%

2. Average joint cost = $186,000/3,000,000 board feet = $0.062

   First and second joint cost allocation = $0.062 × 450,000 = $27,900

   No. 1 common joint cost allocation = $0.062 × 1,200,000 = $74,400

> **CORNERSTONE 7-7**
> *(continued)*
>
> No. 2 common joint cost allocation = $0.062 \times 600,000 = \$37,200$
>
> No. 3 common joint cost allocation = $0.062 \times 750,000 = \$46,500$
>
> (*Note:* Either method gives the same allocation results.)
>
> 3. If First and second and No. 1 common each had 825,000 board feet, then each would receive 27.5 percent (825,000/3,000,000) of the joint cost, or $51,150 (27.5% \times \$186,000)$. There would be no impact on the allocation to No. 2 common and No. 3 common since their proportion of total board feet did not change.

Although the physical units method is not wholly satisfactory, it has a measure of logic behind it. Since all products are manufactured by the same process, it is impossible to say that one costs more per unit to produce than the other. For example, manufacturers of forest products may add the average cost of logs entering the mill to the average conversion cost to arrive at an average finished product cost. This cost is applied to all finished products, no matter their type, grade, or market value. This method serves the purpose of product costing.

The physical units method may be used in any industry that processes joint products of differing grades (e.g., flour milling, tobacco, and lumber). However, a disadvantage of the physical units method is that high profits may be reflected from the sale of the high grades, with low profits or losses reflected on the sale of lower grades. This may result in incorrect managerial decisions if the data are not properly interpreted.

The physical units method presumes that each unit of material in the final product costs just as much to produce as any other. This is especially true where the dominant element can be traced to the product. Many feel this method often is unsatisfactory because it ignores the fact that not all costs are directly related to physical quantities. Also, the product might not have been produced at all if it had been physically separable before the split-off point from the part desired.

Weighted Average Method Some shortcomings encountered under the physical units method can be overcome by using weight factors. These weight factors may include such diverse elements as amount of material used, difficulty to manufacture, time consumed, difference in type of labour used, and size of unit. These factors and their relative weights are usually combined in a single value, called the **weight factor**.

An example of the use of weight factors is found in the canning industry.[4] One type of weight factor is used to convert different-size cases of peaches into a uniform size for purposes of allocating joint costs to each case. Thus, if a basic case contains 24 cans of peaches in size $2\frac{1}{2}$ cans, that case is assigned a weight factor of 1.0. A case with 24 cans in size 303 (a can roughly half the size of the $2\frac{1}{2}$ can) receives a weight of 0.57, and so on. Once all types of cases have been converted into basic cases using the weight factors, joint costs can be allocated according to the physical units method. Peaches can also be assigned weight factors according to grade (e.g., fancy, choice, standard, and pie). If the standard grade is weighted at 1.00, then the better grades are weighted more heavily and the pie grade less heavily. Cornerstone 7-8 shows how and why the weighted average method can be used to allocate joint costs to different products.

As Cornerstone 7-8 shows, once the weight factors are applied, the physical units can be applied to obtain the percentage of weighted cases for each grade. These

---

[4] The peach-canning example is adapted from K. E. Jankowski, "Cost and Sales Control in the Canning Industry," *N.A.C.A. Bulletin* 36 (November 1954): 376.

The HOW and WHY of Using the Weighted Average Method to Allocate Joint Product Costs

**CORNERSTONE
7-8**

Information:

A peach-canning factory purchases $5,000 of peaches; grades them into fancy, choice, standard, and pie quality; and then cans each grade. The following data on grade, number of cases, and weight factor follow:

	Number of Cases	Weight Factor
Fancy	100	1.30
Choice	120	1.10
Standard	303	1.00
Pie	70	0.50
Total	593	

Why:

The joint cost must be allocated to the various grades of peaches in order to cost the product and value inventory. The weighted average method allows firms to place relatively more value on certain types or grades of units than on others.

Required:

1. Allocate the joint cost to the four grades of peaches using the weighted average method.
2. **What if** the factory found that peaches for pie were being valued more by customers and decided to increase the weight factor for pie peaches to 1.00? How would that affect the allocation of cost to pie peaches? How would it affect the allocation of cost to the remaining grades?

Solution:

1.

Grades	Number of Cases	Weight Factor	Weighted Number of Cases	Percent	Allocated Joint Cost
Fancy	100	1.30	130	0.21667	$1,083
Choice	120	1.10	132	0.22000	1,100
Standard	303	1.00	303	0.50500	2,525
Pie	70	0.50	35	0.05833	292
Total			600		$5,000

2. If the pie grade weight factor is increased to 1.00, then the weighted number of cases would double and pie peaches would receive a relatively larger amount of joint cost. However, the allocation of cost to all other grades will decrease since the increased weighted cases for pie will impact all percentages. The following table shows what would happen:

Grades	Number of Cases	Weight Factor	Weighted Number of Cases	Percent	Allocated Joint Cost
Fancy	100	1.30	130	0.2047	$1,023*
Choice	120	1.10	132	0.2079	1,040
Standard	303	1.00	303	0.4772	2,386
Pie	70	1.00	70	0.1102	551
Total			635		$5,000

* rounded

percentages are then multiplied by the joint cost to yield the allocated joint cost. The effect is to allocate relatively more of the joint cost to the fancy and choice grades because they represent more desirable peaches. The pie grade peaches, the good bits and pieces from bruised peaches, are relatively less desirable and are assigned a lower weight.

Frequently, weight factors are predetermined and set up as part of either an estimated cost or a standard cost system. The use of carefully constructed weight factors enables the cost accountant to give more attention to several influences and, therefore, results in more reasonable allocations. The real danger, of course, is that weights may be used that are either inappropriate in the first place or become so through the passage of time. Obviously, if arbitrary rates are used, the resulting costs of individual products will be arbitrary.

Allocation Based on Relative Market Value

Many accountants believe that joint costs should be allocated to individual products according to their ability to absorb joint costs. The advantage of this approach is that joint cost allocation will not produce consistently profitable or unprofitable items. The rationale for using ability to bear is the assumption that costs would not be incurred unless the jointly produced products together would yield enough revenue to cover all costs plus a reasonable return. On the other hand, fluctuations in the market value of any one or more of the end products automatically change the apportionment of the joint costs, although they actually cost no more or no less to produce than before.

The relative market value approach to joint cost allocation is better than the physical units approach provided that two conditions hold: (1) the physical mix of output can be altered by incurring more (less) total joint costs and (2) this alteration produces more (less) total market value.[5] Several variants of the relative market value method are found in practice.

Allocation Based on Sales Value at Split-Off Point

The **sales-value-at-split-off method** allocates joint cost based on each product's proportionate share of market or sales value at the split-off point. Under this method, the higher the market value, the greater the share of joint cost charged against the product. As long as the prices at split-off are stable, or the fluctuations in prices of the various products are synchronized (not necessarily in amount, but in the rate of change), their respective allocated costs remain constant. Cornerstone 7-9 shows how and why to allocate joint costs using the sales-value-at-split-off method.

The sales-value-at-split-off method can be approximated through the use of weighting factors based on price. The advantage is that the price-based weights do not change as market prices do. An example of this method is found in the glue industry. Material is put into process in the Cooking Department. The products resulting from the cooking operations are the several "runs of glue." The first run is of the highest grade, has the highest market value, and costs the least. Successive runs require higher temperatures, cost more, and produce lower grades of products. Glue factories do not attempt to determine the actual cost of each skimming because the effect would be to show the lowest cost on the first grade of product and the highest cost on the lowest grade. Instead, the cost of all glue produced is determined, and this total cost is spread over the various grades on the basis of their respective tests of purity. The relative degree of purity is an indicator of the quality and, therefore, of the market value of each run or grade produced. Hence,

[5] William Cats-Baril, James F. Gatti, and D. Jacque Grinnell, "Joint Product Costing in the Semiconductor Industry," *Management Accounting* (February 1986): 29.

The HOW and WHY of Using the Sales-Value-at-Split-Off Method to Allocate Joint Product Costs

**CORNERSTONE
7-9**

Information:

A sawmill processes logs into four grades of lumber totalling 3,000,000 board feet as follows:

Grades	Board Feet	Price at Split-Off
First and second	450,000	$0.300
No. 1 common	1,200,000	0.200
No. 2 common	600,000	0.121
No. 3 common	750,000	0.070
Total	3,000,000	

Total joint cost is $186,000.

Why:

The joint cost must be allocated to the various grades of lumber in order to cost product and value inventory. The sales-value-at-split-off method allocates the joint cost in proportion to each product's sales value at the split-off point.

Required:

1. Allocate the joint cost to the four grades of lumber using the sales-value-at-split-off method.
2. **What if** First and second and No. 1 common each had 825,000 board feet? How would that affect the allocation of cost to these two grades? How would it affect the allocation of cost to the No. 2 and No. 3 common grades?

Solution:

1.

Grades	Board Feet Produced	Price at Split-Off	Sales Value at Split-Off	Percent of Total Market Value	Allocated Joint Cost
First and second	450,000	$0.300	$135,000	0.2699	$ 50,202*
No. 1 common	1,200,000	0.200	240,000	0.4799	89,261
No. 2 common	600,000	0.121	72,600	0.1452	27,007
No. 3 common	750,000	0.070	52,500	0.1050	19,530
Total	3,000,000		$500,100		$186,000

* rounded

Sales value at split-off for First and second = 450,000 × $0.300 = $135,000

Sales value at split-off for No. 1 common = 1,200,000 × $0.200 = $240,000

Sales value at split-off for No. 2 common = 600,000 × $0.121 = $72,600

Sales value at split-off for No. 3 common = 750,000 × $0.070 = $52,500

Percent for First and second = $135,000/$500,100 = 0.2699, or 26.99%

Percent for No. 1 common = $240,000/$500,100 = 0.4799, or 47.99%

Percent for No. 2 common = $72,600/$500,100 = 0.1452, or 14.52%

Percent for No. 3 common = $52,500/$500,100 = 0.1050, or 10.50%

**CORNERSTONE
7-9
(continued)**

First and second joint cost allocation = 0.2699 × $186,000 = $50,202*

No. 1 common joint cost allocation = 0.4799 × $186,000 = $89,261

No. 2 common joint cost allocation = 0.1452 × $186,000 = $27,007

No. 3 common joint cost allocation = 0.1050 × $186,000 = $19,530

2. If First and second and No. 1 common each had 825,000 board feet, then First and second would have a much higher sales value at split-off and would receive a higher percentage of joint cost. No. 1 common would have a lower sales value at split-off and receive a lower joint cost allocation. While the sales value at split-off of No. 2 common and No. 3 common would not be affected, their sales value as a percent of the total would go down since the increased value for First and second went up. Results of this change follow:

Grades	Board Feet Produced	Price at Split-Off	Sales Value at Split-Off	Percent of Total Market Value	Allocated Joint Cost
First and second	825,000	$0.300	$247,500	0.4604	$ 85,635*
No. 1 common	825,000	0.200	165,000	0.3069	57,083
No. 2 common	600,000	0.121	72,600	0.1350	25,110
No. 3 common	750,000	0.070	52,500	0.0977	18,172
Total	3,000,000		$537,600		$186,000

* rounded

multiplying the yield for each run by its relative purity is equivalent to multiplying it by the market value. The amounts weighted by purity are used to allocate the joint costs to each run. Additional runs would be undertaken, of course, only as long as the incremental revenue of the additional run is equal to or exceeds the incremental costs incurred.

The weighting factor based on market value at split-off is conceptually the same as the weighting factor method under physical units. However, in this case, the weighting factor is based on sales value, while the weighting factor described in the physical units section could be based on other considerations such as processing difficulty, size, and so on, that may or may not be related to market value.

Net Realizable Value Method   When market value is used to allocate joint costs, we are talking about market value *at the split-off point*. However, on occasion, there is no ready market price for the individual products at the split-off point. In this case, the net realizable value method can be used. First, we obtain a **hypothetical sales value** for each joint product by subtracting all separable (or further) processing costs from the eventual market value. This approximates the sales value at split-off. Then, the **net realizable value method** can be used to prorate the joint costs based on each product's share of hypothetical sales value. Cornerstone 7-10 shows how and why to use the net realizable value method to allocate joint costs.

The net realizable value method is particularly useful when one or more products cannot be sold at the split-off point but must be processed further.

Constant Gross Margin Percentage Method   The net realizable value method is easy to apply. However, it assigns all profit to the hypothetical market value. In other words, the further processing costs are assumed to have no profit

The HOW and WHY of Using the Net Realizable Value Method to Allocate Joint Product Costs

**CORNERSTONE
7-10**

Information:
A company manufactures two products, Alpha and Beta, from a joint process. Each production run costs $5,750 and results in 1,000 litres of Alpha and 3,000 litres of Beta. Neither product is salable at split-off, but must be further processed such that the separable cost for Alpha is $1 per litre and for Beta is $2 per litre. The eventual market price for Alpha is $5 and for Beta, $4.

Why:
The net realizable value method is used when one or more of the joint products cannot be sold at split-off. In this case, a hypothetical market value is constructed so that joint cost allocation can be done as close to the split-off point as possible.

Required:
1. Allocate the joint cost to Alpha and Beta using the net realizable value method.
2. *What if* it cost $2 to process each litre of Alpha beyond the split-off point? How would that affect the allocation of joint cost to these two products?

Solution:
1.

Product	Market Price (1)	−	Further Processing Cost (2)	=	Hypothetical Market Price (3)	×	Number of Units (4)	=	Hypothetical Market Value (5)	Percent*	Allocated Joint Cost**
Alpha	$5.00		$1.00		$4.00		1,000		$ 4,000	0.40	$2,300
Beta	4.00		2.00		2.00		3,000		6,000	0.60	3,450
Total									$10,000		$5,750

* Percent for Alpha = $4,000/$10,000 = 0.40, or 40%
  Percent for Beta = $6,000/$10,000 = 0.60, or 60%
** Alpha joint cost allocation = 0.40 × $5,750 = $2,300
  Beta joint cost allocation = 0.60 × $5,750 = $3,450

2. If it cost $2 to process each litre of Alpha, the hypothetical market price would be less, the hypothetical market value would be less, and Alpha would receive a smaller allocation of joint cost. The following table shows the results:

Product	Market Price (1)	−	Further Processing Cost (2)	=	Hypothetical Market Price (3)	×	Number of Units (4)	=	Hypothetical Market Value (5)	Percent*	Allocated Joint Cost**
Alpha	$5.00		$2.00		$3.00		1,000		$3,000	0.3333	$1,916
Beta	4.00		2.00		2.00		3,000		6,000	0.6667	3,834
Total									$9,000		$5,750

* Percent for Alpha = $3,000/$9,000 = 0.3333, or 33.33% (rounded)
  Percent for Beta = $6,000/$9,000 = 0.6667, or 66.67% (rounded)
** Alpha joint cost allocation = 0.3333 × $5,750 = $1,916 (rounded)
  Beta joint cost allocation = 0.6667 × $5,750 = $3,834 (rounded)

value even though they are critical to selling the products. The **constant gross margin percentage method** corrects for this by recognizing that costs incurred after the split-off point are part of the cost total on which profit is expected to be earned, and it allocates joint cost such that the gross margin percentage is the same for each product. Cornerstone 7-11 shows how and why to apply the constant gross margin percentage method in joint cost allocation.

Notice that the constant gross margin percentage method allocates more joint cost to Alpha than did the net realizable value method. This is due to the assumption of a relationship between cost and the cost-created value. That is, the net realizable value assumed no gross margin attributable to further processing costs, while the constant gross margin percentage method assumed not only that further processing yields profit but also that it yields an identical profit percentage across products. Which assumption is correct? There are two important questions: first, whether there is a "direct relationship" between cost and value and, second, whether the relationship is necessarily the same for all products jointly produced before and after the split-off point. The practice of product-line pricing to meet competition tends to make such assumptions invalid. Although exceptions exist, many companies do not try to maintain more-or-less equal margins between prices and full costs on their various products.

Accounting for By-Products

The distinction between joint products and **by-products** rests solely on the relative importance of their sales value. A by-product is a secondary product recovered in the course of manufacturing a primary product. It is a product whose total sales value is relatively minor in comparison with the sales value of the main product(s). This is not a sharp distinction, but rather one of degree. The first distinction is whether the operation is characterized by joint production. Then any by-products must be distinguished from main or joint products. By-products can be characterized by their relationship to the main products in the following manner:

1. By-product resulting from scrap, trimmings, and so forth, of the main products in essentially non-joint product types of undertakings (e.g., fabric trimmings from clothing pieces)
2. Scrap and other residue from essentially joint product types of processes (e.g., fat trimmed from beef carcasses)
3. A minor joint product situation (fruit skins and trimmings used as animal feed)

Relationships between joint products and by-products change, as do the classes of products within each of these classifications. When the relative importance of the individual products changes, the products need to be reclassified and the costing procedures changed. In fact, many by-products begin as waste materials, become economically significant (and thus become by-products), and grow in importance to finally become full-fledged joint products. For example, sawdust and wood chips in sawmill operations were originally waste, but over the years, they have gained value as a major component of particle board. The various methods of accounting for by-products reflect this development. Generally, accounting for by-products began as an extension of accounting for waste material. Revenue from the sale of the by-products is recorded as separate income, when the amount of income is so small that it has little impact on either overall cost or sales. As the value of by-product revenues becomes more significant, the cost of the main product is reduced by recoveries, and finally the by-products achieve near main product status and are allocated a share of the joint cost incurred prior to split-off.

Pork production offers examples of different types of by-products. Of course, the joint (main) products include pork roasts, bacon, ribs, sausage, and so on. Many different by-products are also produced during the meat-packing process.

The HOW and WHY of Using the Constant Gross Margin Percentage Method to Allocate Joint Product Costs

**CORNERSTONE
7-11**

Information:

A company manufactures two products, Alpha and Beta, from a joint process. Each production run costs $5,750 and results in 1,000 litres of Alpha and 3,000 litres of Beta. Neither product is salable at split-off, but must be further processed such that the separable cost for Alpha is $1 per litre and for Beta is $2 per litre. The eventual market price for Alpha is $5 and for Beta, $4.

Why:

The constant gross margin percentage method is used to avoid assuming that all profit occurs at the split-off point. It allocates joint cost to ensure that the same gross profit is applicable to all products.

Required:

1. Calculate the total revenue, total costs, and total gross profit the company will earn on the sale of Alpha and Beta.
2. Allocate the joint cost to Alpha and Beta using the constant gross margin percentage method.
3. *What if* it cost $2 to process each litre of Alpha beyond the split-off point? How would that affect the allocation of joint cost to these two products?

Solution:

1.

Total revenue [($5 × 1,000) + ($4 × 3,000)]		$17,000
Further processing costs [($1 × 1,000) + ($2 × 3,000)]	$7,000	
Joint processing costs	5,750	12,750
Total gross margin		$ 4,250

2. Gross margin percentage = Gross margin/Total revenue
$$= \$4,250/\$17,000 = 0.25, \text{ or } 25\%$$

	Alpha	Beta
Eventual market value	$5,000	$12,000
Less: Gross margin at 25% of market value	1,250	3,000
Cost of goods sold	3,750	9,000
Less separable costs:		
Alpha = $1 × 1,000 units	1,000	
Beta = $2 × 3,000 units		6,000
Allocated joint cost	$2,750	$ 3,000

3. An increase in the further processing cost of Alpha will reduce the gross margin percentage and will decrease the joint cost allocated to Alpha.

Total revenue [($5 × 1,000) + ($4 × 3,000)]		$17,000
Further processing costs [($2 × 1,000) + ($2 × 3,000)]	$8,000	
Joint processing costs	5,750	13,750
Total gross margin		$ 3,250

**CORNERSTONE
7-11**
(continued)

Gross margin percentage = Gross margin/Total revenue
= $3,250/$17,000
= 0.1912, or 19.12% (rounded)

	Alpha	Beta
Eventual market value	$5,000	$12,000
Less: Gross margin at 19.12% of market value	956	2,294
Cost of goods sold	4,044	9,706
Less separable costs:		
Alpha = $2 × 1,000 units	2,000	
Beta = $2 × 3,000 units		6,000
Allocated joint cost	$2,044	$ 3,706

Seaboard Foods, for example, thoroughly washes its production facility at the end of each day's shift. The waste water, which contains blood and small trimmings, sluices down drains in the floor. These lead to pipes, which channel the waste water into covered containment ponds, where anaerobic bacteria get to work breaking down the proteins and producing methane. Seaboard then recovers the methane for use in utility production for the plant. There is no accounting needed for the use of the methane. It is used solely within the plant and is not resold to outside users.

Another pork by-product is heart valves for use in transplantation. These valves are sold to heart valve manufacturers, who take up to four weeks to process the bovine or porcine valves into medical grade valves. Since no further processing occurs in the packing plant, this use can be accounted for as revenue from the sale of by-products, or as an offset against the cost of the main product(s).

Treatment of the By-Product as Other Revenue   If the by-product can be sold, the company can choose to credit the sale to "Other Income" or to set up an account for "Sale of By-Product." Then, the revenue from the sale of the by-product would be credited to that account. Under this method, no cost is assigned to the by-product. All joint cost is allocated to the main products. Suppose that Edwards Company manufactures several main products and one by-product from a joint production process. One production run has the following costs:

Direct materials	$15,000
Direct labour	6,500
Applied overhead	4,550
Total joint production cost	$26,050

From each production run, Edwards obtains 1,600 kilograms of Product A, 400 kilograms of Product B, and 30 kilograms of a by-product. The by-product can be sold for $5 per kilogram. When the 30 kilograms of by-product are sold, the following journal entry will be made:

Accounts Receivable	150	
Sale of By-Product		150

Notice that under this method, no cost is assigned to the by-product and it is not carried in inventory. All joint production cost ($26,050 per batch) is allocated to the main products.

Treatment of the By-Product as a Reduction in the Cost of the Main Products   An alternative method is to account for any revenue received from sale of the by-product as a reduction in the joint costs of the main products. In the Edwards example, the joint cost of $26,050 would be reduced by $150 from the sale of the by-product. Then, $25,900 would be the joint cost allocated to the main products, Product A and Product B. If Edwards used the physical units method of joint cost allocation, then the following allocations would be made:

	Units	Percent	Joint Cost Allocation
Product A	1,600	80%	$20,720
Product B	400	20	5,180
Total	2,000		$25,900

In summary, there are a number of ways to account for by-products. The treatments of by-product revenue as other income or as a deduction in the cost of the main products are the most commonly used accounting methods. By definition, by-product is immaterial. Thus, the accounting treatment focuses on methods that are relatively quick and simple.

Ethical Implications of Cost Allocation

ETHICS   This chapter has dealt with the subject of cost allocation, that is, moving cost from one department or product to another department or product. There are good reasons for reallocating costs and many widely accepted and used ways of doing this. However, the ability to allocate costs among various cost objects gives management a fair amount of discretion as to how the allocation is done. We have seen that some ways of allocating costs assign relatively more costs to a particular support department or joint product than other ways of allocating costs. The question arises, is this cost allocation ethical? As always, we return to the fundamentals of business. The business is ethical if it treats all parties fairly and does not attempt to mislead or misstate results.

If managers are not careful in the method they choose to allocate costs, one department may be unduly penalized while another may get an advantage. The issue then becomes one of favouring one department (and its managers) over another. Ethically, this is wrong.◆

Summary of Learning Objectives

1.   **Describe the difference between support departments and producing departments.**

- Producing departments create the products or services that the firm is in business to make and sell.
- Support departments serve producing departments but do not create a salable product.
- The costs of the support departments must be allocated to producing departments for:

  - Inventory valuation
  - Product-line profitability
  - Pricing
  - Planning and control. Allocation can also be used to encourage favourable managerial behaviour.

2.   **Calculate charging rates, and distinguish between single and dual charging rates.**

- Single charging rate combines variable and fixed costs of the support department.

  - Budgeted fixed and variable costs are in the numerator and budgeted usage is in the denominator.
  - Actual usage by using departments is multiplied by the charging rate to get the amount charged.

      b.   Physical units method at split-off

      c.   Estimated net realizable value method

2.   Prepare an analysis for Sonimad Sawmill Inc. to compare processing the decorative pieces further as it presently does, with selling the rough-cut product immediately at split-off. Be sure to provide all calculations.

3.   Assume Sonimad Sawmill Inc. announced that in six months it will sell the rough-cut product at split-off due to increasing competitive pressure. Identify at least three types of likely behaviour that will be demonstrated by the skilled labour in the planing and sizing process as a result of this announcement. Explain how this behaviour could be improved by management. *(CMA adapted)*

The Collabourative Learning Exercises can be found on the product support site at www.hansen1ce.nelson.com.

After studying this chapter, you should be able to:

➤ **1** Define budgeting, and discuss its role in planning, controlling, and decision making.

➤ **2** Prepare the operating budget, identify its major components, and explain the interrelationships of the various components.

➤ **3** Identify the components of the financial budget, and prepare a cash budget.

➤ **4** Define flexible budgeting, and discuss its role in planning, control, and decision making.

➤ **5** Define activity-based budgeting, and discuss its role in planning, control, and decision making.

➤ **6** Identify and discuss the key features that a budgetary system should have to encourage managers to engage in goal-congruent behaviour.

Budgeting for Planning and Control

CHAPTER

8

Careful planning, whether formal or informal, is vital to the health of any organization. Business managers must know their resource capabilities and have a plan that shows how those resources will be used. In this chapter, the basics of budgeting are discussed, and traditional master budgets are developed. Flexible and activity-based budgeting are also presented, along with extensive discussion of the behavioural aspects of budgeting and its use in control.

The Role of Budgeting in Planning and Control

OBJECTIVE ➤ 1
Define budgeting, and discuss its role in planning, controlling, and decision making.

Budgets are quantitative plans for the future, stated in either physical or financial terms or both. Budgeting is critically important to both planning and control. When used for planning, a budget is a method for translating the goals and strategies of an organization into operational terms. **Control** is the process of setting standards, receiving feedback on actual performance, and taking corrective action

Exhibit 8-1

The Master Budget and Its Interrelationships

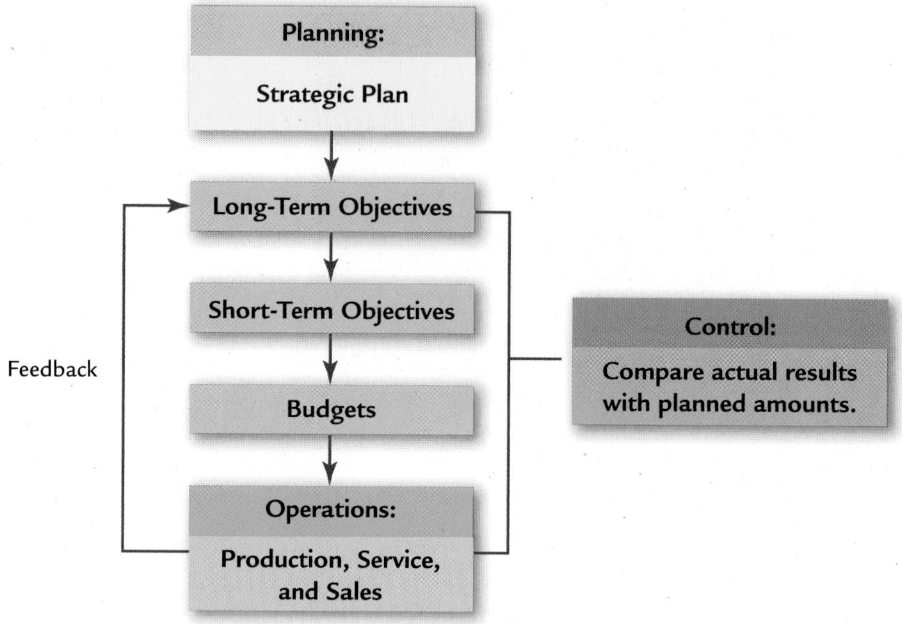

whenever actual performance deviates significantly from planned performance. Thus, when budgets are used for controlling purposes, a budget is used to compare actual outcomes with budgeted outcomes, and can steer operations back on course, if necessary. This comparison provides feedback both for operations and for future budgets.

Exhibit 8-1 illustrates the relationship of budgets to planning, operations, and control. Budgets evolve from the long-run objectives of the firm; they form the basis for operations.

Purposes of Budgeting

Budgets are usually prepared for organizational units (departments, plants, divisions, and so on). Also, budgets are prepared for activities, (sales, production, research, and so on). This system of budgets serves as the comprehensive financial plan for the entire organization and gives an organization several advantages:

1. It forces managers to plan.
2. It improves communication and coordination among different organizational units and activities.
3. It provides resource information that can be used to improve decision making.
4. It aids in resource allocation and sets benchmarks that can be used for the subsequent evaluation of performance.

Budgeting forces management to plan for the future—to develop an overall direction for the organization, foresee problems, and develop future policies. When managers plan, they grow to understand the capabilities of their businesses and where the resources of the business should be used. All businesses and not-for-profit entities should budget. All large businesses do budget. In fact, the budgeting activity of a company such as **IBM** takes significant amounts of time and involves many managers at a variety of levels. Some small businesses do not budget, and many of those go out of business in short order.

Budgets help managers make better decisions. For example, a cash budget points out potential shortfalls. If a company foresees a cash deficiency, it may want to improve accounts receivable collection or postpone plans to purchase new assets.

Budgets set standards for the use of a company's resources and help control and motivate employees. Businesses with successful budgets ensure that steps are taken to achieve the objectives outlined in an organization's master plan.

Budgets are also used for communication and coordination of employee efforts, so that all employees can be aware of their role in achieving the organization's objectives. This is why explicitly linking the budget to the long-run plans of the organization is so important. The budget is not a series of vague, rosy scenarios, but a set of specific plans to achieve those objectives. Budgets encourage coordination because the various areas and activities of the organization must all work together to achieve the stated objectives. The role of communication and coordination becomes more important as an organization grows larger.

The budgeting process can range from the fairly informal process undergone by a small firm, to an elaborately detailed, several-month procedure employed by large firms. Key features of the process include directing and coordinating the overall budget. Every organization must have someone responsible for directing and coordinating the overall budgeting process. This **budget director** works under the direction of the budget committee and is usually the controller or someone who reports to the controller. The **budget committee** is responsible for reviewing the budget, providing policy guidelines and budgetary goals, resolving differences that may arise as the budget is prepared, approving the final budget, and monitoring the actual performance of the organization as the year unfolds. The budget committee ensures that the budget is linked to the strategic plan of the organization. The president of the organization appoints the members of the committee, who are usually the president, vice presidents, and the controller.

The Budgeting Process

Types of Budgets

The **master budget** is a comprehensive financial plan made up of various individual departmental and activity annual budgets. A master budget can be divided into *operating* and *financial* budgets. **Operating budgets** are concerned with the income-generating activities of a firm: sales, production, and finished goods inventories. The ultimate outcome of the operating budgets is a pro forma or budgeted income statement. Note that "pro forma" is synonymous with "budgeted" and "estimated." In effect, the pro forma income statement is done "according to form" but with estimated, not historical, data. **Financial budgets** are concerned with the inflows and outflows of cash and with financial position. Planned cash inflows and outflows are detailed in a cash budget, and expected financial position at the end of the budget period is shown in a budgeted, or pro forma, balance sheet. Exhibit 8-2 illustrates the components of the master budget.

The master budget is usually prepared for a one-year period corresponding to the company's fiscal year. The yearly budgets are broken down into quarterly and monthly budgets. Using shorter time periods helps managers compare actual data with budgeted data as the year unfolds. Because progress can be checked more frequently, problems can be identified and handled before they become serious.

Most organizations prepare the budget for the coming year during the last four or five months of the current year. However, some organizations have developed a continuous budgeting philosophy. A **continuous (or rolling) budget** is a moving 12-month budget. As a month expires in the budget, an additional month in the future is added so that the company always has a 12-month plan on hand. Proponents of continuous budgeting maintain that it forces managers to plan ahead constantly. The majority of CFOs believe that rolling forecasts are very valuable, and

Exhibit 8-2

Components of the Master Budget

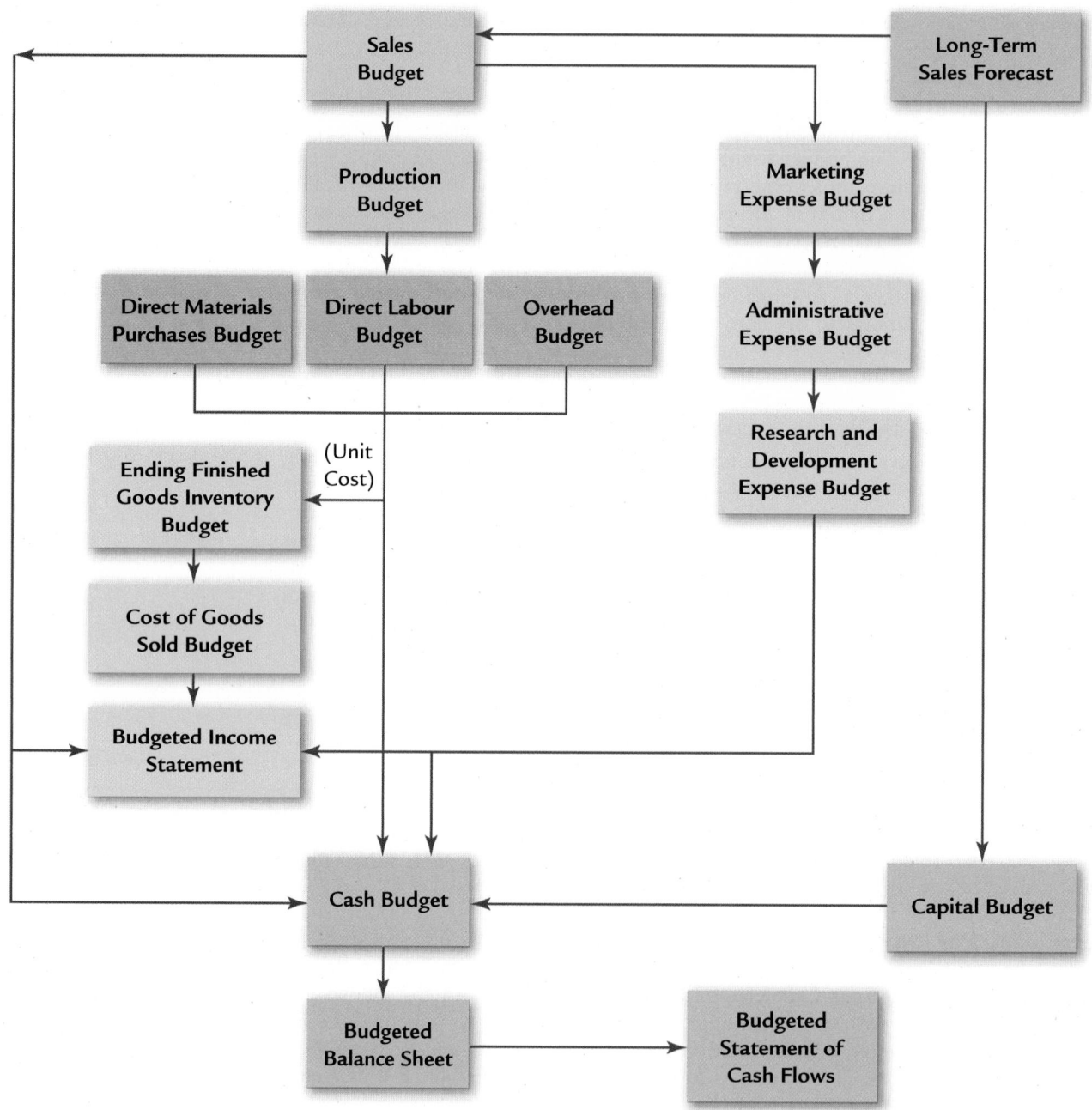

companies that do use them typically roll the forecasts out for five or six quarters rather than four.[1]

Similar to a continuous budget is *a continuously updated budget.* The objective of this budget is not to have 12 months of budgeted information at all times, but instead to update the master budget each month as new information becomes available. For example, every autumn, Chandler Engineering prepares a budget for the coming year. Then at the end of each month of the year, the budget is transformed into a rolling forecast by recording year-to-date results and the forecast for the

---

[1] Omar Aguilar, "How Strategic Performance Management Is Helping Companies Create Business Value," *Strategic Finance* (January 2003): 44–49.

remainder of the year. In essence, the budget is continually updated throughout the year.

Technological advances permit companies to keep much closer track of sales and production needs throughout the year. **Revlon** has adopted a new computer system that manages sales data for each item in each store. It can track sales as specifically as by colour of nail polish. This faster, better information allows Revlon to adjust budgets continually throughout the year. As a result, the company can manage operations by cutting the production and shipment of slow-selling cosmetics and ramping up production of the hot sellers.[2]

Gathering Information for Budgeting

At the beginning of the master budgeting process, the budget director alerts all segments of the company to begin gathering budget information. The data used to create the budget come from many sources. Historical data are one possibility. For example, last year's direct materials costs may give the production manager a good feel for potential materials costs for next year. Still, historical data alone cannot tell a company what to expect in the future.

Forecasting Sales   The sales forecast is the basis for the sales budget, which, in turn, is the basis for all of the other operating budgets and most of the financial budgets. Accordingly, the accuracy of the sales forecast strongly affects the soundness of the entire master budget.

    Creating the sales forecast is usually the responsibility of the Marketing Department. One approach is for the chief sales executive to have individual salespeople submit sales predictions, which are aggregated to form a total sales forecast. The accuracy of this sales forecast may be improved by considering other factors such as the general economic climate, competition, advertising, pricing policies, and so on. Some companies supplement the Marketing Department forecast with more formal approaches, such as time-series analysis, correlation analysis, econometric modelling, and industry analysis.

    To illustrate an actual sales forecasting approach, consider the practices of a company that manufactures oil field equipment on a job-order basis. Each month, the Finance and Sales departments' heads meet to construct a sales forecast based on bookings. A booking is a probable sales order submitted by sales personnel in the field; it is meant to alert the Engineering and Manufacturing departments to a potential job. Past experience has shown that bookings are generally followed by sales/shipments within 30 to 45 days. Exhibit 8-3 shows the short-term bookings forecast for the company. Notice that the dollar amount of each booking is multiplied by its probability of occurrence to obtain a weighted dollar amount. The sum of weighted amounts is the forecast for sales for the month. The probability estimate is determined jointly by the salesperson and the controller. Each probability is initially set at 50 percent. Then, it is adjusted upward or downward based on any additional information about the sale. The probability is really a prediction of a compound event: the prediction of both getting the order and determining the month in which it will happen. The Sales Department tends toward overconfidence—both in terms of getting the order and in landing it sooner rather than later. Thus, the controller takes a more pessimistic view and modifies the forecast. The end result is the form shown in the exhibit.

Forecasting Other Variables   Sales are not the only concern in budgeting. Costs and cash-related items are critical. Many of the same factors considered in sales forecasting apply to cost forecasting. Here, historical amounts can be of real value. Managers can adjust past figures based on their knowledge of coming events. For example, a three-year union contract takes much of the uncertainty out of wage prediction.

[2] Emily Nelson, "Revlon Chief Banks on Risky Strategy as He Seeks New Image for Ailing Firm," *The Wall Street Journal* (November 21, 2000): B1.

Exhibit 8-3

Short-Term Bookings Forecast for Oil Field Equipment Company

Quote #	Region/Country	Customer	Product	Dollar Amount	Probability	Weighted Month Total
March 2013						
1194-17	Spain	Valencia	repair 3224	$ 37,500	100%	$ 37,500
1294-03	Bulgaria	Luecim	1256, 7188	74,145	80	59,316
0195-55	USA	Exxon	4498	25,000	95	23,750
0295-19	USA	BP/TX	6766, 1267	150,442	100	150,442
0295-23	China	China Res	7541, 8875	55,900	75	41,925
0295-45	China	China Res	8879, 0944	34,500	80	27,600
0395-36	Abu Dhabi	ADES	7400, 6751, 5669 & spares	30,000	50	15,000
March Total						$355,533
April 2013						
1294-14	China	Jiang Han	6524, 5523, 0412, 4578, 3340	$234,000	80%	$187,200
0295-43	Russia	Geoserv	3356	76,800	60	46,080
0295-10	Venezuela	Petrolina	4450, 6713, 7122	112,500	90	101,250
0395-37	Indonesia	Chevron	8890, 0933	98,000	65	63,700
0395-71	Italy	CV International	7815	16,000	70	11,200
April Total						$409,430
May 2013						
0295-21	Mexico	Instituto Mexicana	8900 & spares	$ 34,000	40%	$ 13,600
0395-29	Venezuela	Petrolina	8416, 8832	165,000	50	82,500
0495-11	USA	Branchwater Inc.	9043, 8891	335,000	60	201,000
0495-68	Saudi Arabia	Aramco	0453	3,500	50	1,750
May Total						$298,850

(Of course, if the contract is expiring, the uncertainty returns.) Alert purchasing agents will have an idea of changing materials prices. In fact, large companies such as **Nestlé** and the **Coca-Cola Company** have entire departments devoted to the forecasting of commodity prices and supplies. They invest in commodity futures to smooth out price fluctuations, an action that facilitates budgeting. Overhead is broken down into its component costs; these can be predicted using past data and relevant inflation figures.

The cash budget is a critically important part of the master budget, and some of its components, especially payment of accounts receivable, also require forecasting. This is discussed in more detail in the section on cash budgeting.

OBJECTIVE ▶ 2

Prepare the operating budget, identify its major components, and explain the interrelationships of the various components.

Preparing the Operating Budget

The first section of the master budget is the operating budget. It consists of a series of schedules for all phases of operations, culminating in a pro forma income statement. The following are the components of the operating budget.

1. Sales budget
2. Production budget
3. Direct materials purchases budget
4. Direct labour budget
5. Overhead budget
6. Ending finished goods inventory budget
7. Cost of goods sold budget
8. Marketing expense budget
9. Research and development expense budget
10. Administrative expense budget
11. Pro forma income statement

You may want to refer back to Exhibit 8-2 to see how these components of the operating budget fit into the master budget.

The example used to illustrate the components of the operating budget is based on ABT Inc. a manufacturer of concrete block and pipe for the construction industry. For simplicity, we will prepare the operating budget for ABT's concrete block line. (The budget for the pipe product line is prepared in the same way and merged into the overall company budget.)

Sales Budget

The **sales budget** is the projection approved by the budget committee that describes expected sales for each product in units and dollars. The sales budget must be constructed first, before other budgets can be constructed.

Cornerstone 8-1 illustrates the sales budget for ABT's concrete block line. The sales budget reveals that ABT's sales fluctuate seasonally. Most sales (75 percent) take place in the spring and summer. Also, note that ABT expects price to increase from $0.70 to $0.80 in the summer quarter. Because of the price change within the year, an average price must be used for the column that describes the total year's activities ($0.75 = $12,000,000/16,000,000 units). If ABT has two types of concrete blocks, a separate sales budget would be prepared for each type. This is shown in Requirement 2 of Cornerstone 8-1.

Production Budget

The **production budget** describes how many units must be produced in order to meet sales needs and satisfy ending inventory requirements. The production budget depends on the unit sales shown in the sales budget.

A separate production budget is constructed for each product manufactured (or service provided). Both unit sales and unit finished goods inventories desired are required for the production budget. The basic equation for the production budget is:

$$\text{Units to be produced} = \text{Desired units in ending inventory} + \text{Unit sales} - \text{Units in beginning inventory}$$

Of course, if there were no inventories, the number of units to be produced would equal the number of units to be sold. In service firms, units of service provided equal the units of service sold since services are not inventories. Similarly, in the JIT firm, units sold equal units produced, since a customer order triggers production. The production budget must, however, consider the existence of beginning and ending inventories. Notice that the production budget is expressed in terms of units; we do not yet know how much they will cost. Cornerstone 8-2 shows how and why a production budget is constructed.

Direct Materials Purchases Budget

After the production budget is completed, budgets for direct materials, direct labour, and overhead can be prepared. The **direct materials purchases budget** is similar in format to the production budget; it is based on the amount of materials needed for production and the inventories of direct materials.

Expected direct materials usage is determined by the input-output relationship (the technical relationship existing between direct materials and output). This relationship is often determined by the Engineering Department or the industrial designer. For example, one lightweight concrete block requires approximately 2.6 kilograms of materials (cement, sand, gravel, shale, pumice, and water). The relative mix of these ingredients is fixed for a specific kind of concrete block. Thus, it is fairly easy to determine expected usage for each material from the production budget by multiplying the amount of material needed per unit of output times the number of units of output.

**CORNERSTONE
8-1**

The HOW and WHY of Constructing a Sales Budget

Information:
ABT Inc. manufactures and sells concrete block for residential and commercial building. ABT expects to sell the following next year, in 2013:

	Quarter 1	Quarter 2	Quarter 3	Quarter 4
Units	2,000,000	6,000,000	6,000,000	2,000,000
Unit selling price	$0.70	$0.70	$0.80	$0.80

Why:
The sales budget is the foundation for the master budget; all other budgets are based in part on the units sold or revenue given in the sales budget.

Required:
1. Construct a sales budget for the ABT concrete block line for the coming year. Show total sales by quarter and in total for the year.
2. *What if* there were two types of concrete block (type 1 and type 2) and that 60 percent of the sales in each quarter were for type 1? Assume the selling price for type 1 is $0.60 in the first quarter and $0.70 for the rest of the year. The selling price of type 2 is $0.80 in Quarters 1 and 2 and is $0.90 per unit for the rest of the year. Construct a sales budget for ABT showing sales for both types and in total.

Solution:
1.

Sales Budget
For the Year Ended December 31, 2013

	Quarter 1	Quarter 2	Quarter 3	Quarter 4	Total
Units	2,000,000	6,000,000	6,000,000	2,000,000	16,000,000
Unit selling price	× $0.70	× $0.70	× $0.80	× $0.80	× $0.75
Sales	$1,400,000	$4,200,000	$4,800,000	$1,600,000	$12,000,000

2.

Sales Budget
For the Year Ended December 31, 2013

	Quarter 1	Quarter 2	Quarter 3	Quarter 4	Total
Type 1 block[1]					
Units	1,200,000	3,600,000	3,600,000	1,200,000	9,600,000
Unit selling price	× $0.60	× $0.70	× $0.70	× $0.70	× $0.6875[2]
Sales	$ 720,000	$2,520,000	$2,520,000	$ 840,000	$ 6,600,000
Type 2 block[3]					
Units	800,000	2,400,000	2,400,000	800,000	6,400,000
Unit selling price	× $0.80	× $0.80	× $0.90	× $0.90	× $0.85[4]
Sales	$ 640,000	$1,920,000	$2,160,000	$ 720,000	$ 5,440,000
Total sales	$1,360,000	$4,440,000	$4,680,000	$1,560,000	$12,040,000

[1] Units in each quarter equal 60% of the unit sales in Requirement 1.
[2] Average price for the year = $6,600,000/9,600,000 = $0.6875
[3] Units in each quarter equal 40% of the unit sales in Requirement 1.
[4] Average price for the year = $5,440,000/6,400,000 = $0.85

The HOW and WHY of Constructing a Production Budget

Information:

ABT expects the following unit sales and desired ending inventory next year, in 2013:

Quarter	Unit Sales	Ending Inventory
1	2,000,000	500,000
2	6,000,000	500,000
3	6,000,000	100,000
4	2,000,000	100,000

**CORNERSTONE
8-2**

Inventory on both January 1, 2013, and January 1, 2014, is expected to be 100,000 blocks.

Why:

The production budget is needed to tell production how much to produce in the coming year. The number of units produced will be used to determine budgeted costs for direct materials, direct labour, and overhead.

Required:

1. Construct a production budget for the ABT concrete block line for the coming year. Show total units produced by quarter and in total for the year.
2. **What if** ABT did not provide the desired ending inventory in units, but instead relied on an inventory rule—that the desired ending inventory of blocks was equal to 5 percent of the next period's sales? Further, assume that the budgeted unit sales in Quarter 1, 2014, equalled 2,500,000, and that the beginning inventory for Quarter 1, 2013, met the inventory rule. Construct a production budget for ABT showing units produced by quarter and in total for the year.

Solution:

1.

Production Budget
For the Year Ended December 31, 2013

	Quarter 1	Quarter 2	Quarter 3	Quarter 4	Total
Unit sales	2,000,000	6,000,000	6,000,000	2,000,000	16,000,000
Desired ending inventory	500,000	500,000	100,000	100,000	100,000
Total needed	2,500,000	6,500,000	6,100,000	2,100,000	16,100,000
Less: Beginning inventory*	100,000	500,000	500,000	100,000	100,000
Units produced	2,400,000	6,000,000	5,600,000	2,000,000	16,000,000

*Beginning inventory for Quarter 1 is given. Beginning inventory for the succeeding quarters is equal to the ending inventory of the previous quarter. That is, beginning inventory for Quarter 2 is equal to desired ending inventory for Quarter 1.

Notice that desired ending inventory for the year equals the desired ending inventory for Quarter 4. Beginning inventory for the year equals the beginning inventory for Quarter 1.

2. If desired ending inventory of blocks equals 5 percent of the next quarter's sales, then the desired ending inventory for each quarter is as follows:

Quarter 1 ending inventory = 0.05 × 6,000,000 = 300,000

Quarter 2 ending inventory = 0.05 × 6,000,000 = 300,000

Quarter 3 ending inventory = 0.05 × 2,000,000 = 100,000

Quarter 4 ending inventory = 0.05 × 2,500,000 = 125,000

For example, the **Toronto Raptors** basketball team budgets the number of seats it expects to fill at each game and the price per ticket. Other revenues (such as television royalties and concession sales) are also budgeted.

In a not-for-profit service firm, the sales budget is replaced by a budget that identifies the levels of the various services that will be offered for the coming year and the sources of funds to pay for producing those services. The source of the funds may be tax revenues, contributions, payments by users of the services, or some combination. For example, a local **United Way**'s board of directors will budget the campaign target (dollars of contributions) for the coming year and then distribute the total funds among the qualifying agencies according to three possible levels of contribution—pessimistic, expected, and optimistic.

Both for-profit and not-for-profit service organizations lack finished goods inventory budgets. However, all the remaining operating budgets found in a manufacturing organization have counterparts in service organizations. A not-for-profit service organization's income statement is replaced by a statement of sources and uses of funds.

We have seen how the firm developed a master budget and used it to plan for the coming year. Once the plan is developed, the budget can be used for control and decision making. For meaningful comparisons, however, it may be necessary to recalculate some measures of output. Flexible budgeting can be used to create plans for various levels of activity. Furthermore, the company that uses activity-based costing may find activity-based budgeting (ABB) to be more valuable than traditional budgeting. Activity-based budgets can be more accurate in planning and are more useful for control.

Preparing the Financial Budget

OBJECTIVE ▶ 3

Identify the components of the financial budget, and prepare a cash budget.

The remaining budgets found in the master budget are the financial budgets. Typical financial budgets include the budget for capital expenditures, the cash budget, the budgeted balance sheet, and the budgeted statement of cash flows.

While the master budget is a plan for one year, the **capital expenditures budget** is a financial plan outlining the expected acquisition of long-term assets and typically covers a number of years. Details on the budgeted statement of cash flows are appropriately reserved for another course. Accordingly, only the cash budget and the budgeted balance sheet will be illustrated here.

The Cash Budget

Understanding cash flow is critical to managing a business. Often, a business is successful in producing and selling a product but fails because of timing problems associated with cash inflows and outflows. By knowing when cash deficiencies and surpluses are likely to occur, a manager can plan to borrow cash when needed and to repay the loans during periods of excess cash. Bank loan officers use a company's cash budget to document the need for cash, as well as the company's ability to repay. Because cash flow is the lifeblood of an organization, the cash budget is one of the most important budgets in the master budget.

Components of the Cash Budget   The **cash budget** is the detailed plan that shows all expected sources and uses of cash. The cash budget, illustrated in Exhibit 8-4A, has the following five main sections:

1. Total cash available
2. Cash disbursements
3. Cash excess or deficiency
4. Financing
5. Cash balance

The *total cash available* section consists of the beginning cash balance and the expected cash receipts. Expected cash receipts include all sources of cash for the

Exhibit 8-4

Cash Budget and Balance Sheet

A. The Cash Budget

```
      Beginning cash balance
    + Cash receipts
      Cash available
    − Cash disbursements
    − Minimum cash balance
      Excess or deficiency of cash
    − Repayments
    + Loans
    + Minimum cash balance
      Ending cash balance
```

B. Balance Sheet for ABT Inc.

ABT Inc.
Balance Sheet
December 31, 2012

Assets

Current assets:		
Cash	$ 120,000	
Accounts receivable	300,000	
Materials inventory	50,000	
Finished goods inventory	55,000	
Total current assets		$ 525,000
Property, plant, and equipment (PP&E):		
Land	2,500,000	
Buildings and equipment	9,000,000	
Accumulated depreciation	(4,500,000)	
Total PP&E		7,000,000
Total assets		$7,525,000

Liabilities and Shareholders' Equity

Current liabilities:		
Accounts payable		$ 100,000
Shareholders' equity:		
Common stock	$ 600,000	
Retained earnings	6,825,000	
Total shareholders' equity		7,425,000
Total liabilities and shareholders' equity		$7,525,000

period being considered. One source of cash is cash sales. However, often a significant proportion of sales is on account; thus, a major task of an organization is to determine the pattern of collection for its accounts receivable.

If a company has been in business for a while, it can use past experience to create an accounts receivable aging schedule. In other words, the company can determine, on average, what percentages of its accounts receivable are paid in the months following the sales. Cornerstone 8-11 shows how and why to prepare a cash receipts budget, including an accounts receivable aging schedule.

The *cash disbursements* section lists all planned cash outlays for the period except for interest payments on short-term loans (these payments appear in the financing section). All expenses not resulting in a cash outlay are excluded from the list. (Depreciation, for example, is never included in the disbursements section.)

**CORNERSTONE
8-11**

The HOW and WHY of Constructing a Cash Receipts Budget with an Accounts Receivable Aging Schedule

Information:

Recall from Cornerstone 8-1 that in 2013 ABT sales are Quarter 1, $1,400,000; Quarter 2, $4,200,000; Quarter 3, $4,800,000; and Quarter 4, $1,600,000. In ABT's experience, 50 percent of sales are paid in cash. Of the sales on account, 70 percent are collected in the quarter of sale; the remaining 30 percent are collected in the quarter following the sale. Total sales for the fourth quarter of 2012 totalled $2,000,000.

Why:

The cash receipts budget shows sources of cash for the period. Some sales may be cash; some are on account and may be received later. The embedded accounts receivable aging schedule helps managers determine how much of a period's sales on account will actually be received in cash. Cash receipts are a critical part of the cash budget.

Required:

1. Calculate cash sales expected in each quarter of 2013.
2. Construct a cash receipts budget including an accounts receivable aging schedule for ABT Inc. for each quarter of the coming year.
3. *What if* ABT determined that the percentage received in the quarter after the quarter of sale was 25 percent and that the remaining 5 percent was never collected? How would that affect cash received in each quarter?

Solution:

1. Quarter 1, 2013, Cash sales = 0.50 × $1,400,000 = $700,000

    Quarter 2, 2013, Cash sales = 0.50 × $4,200,000 = $2,100,000

    Quarter 3, 2013, Cash sales = 0.50 × $4,800,000 = $2,400,000

    Quarter 4, 2013, Cash sales = 0.50 × $1,600,000 = $800,000

2.

	Quarter 1	Quarter 2	Quarter 3	Quarter 4
Cash sales	$ 700,000	$2,100,000	$2,400,000	$ 800,000
Received on account from:				
Quarter 4, 2012[a]	300,000			
Quarter 1, 2013[b]	490,000	210,000		
Quarter 2, 2013[c]		1,470,000	630,000	
Quarter 3, 2013[d]			1,680,000	720,000
Quarter 4, 2013[e]				560,000
Total cash receipts	$1,490,000	$3,780,000	$4,710,000	$2,080,000

[a] $1,000,000 × 0.30 = $300,000
[b] $700,000 × 0.70 = $490,000; $700,000 × 0.30 = $210,000
[c] $2,100,000 × 0.70 = $1,470,000; $2,100,000 × 0.30 = $630,000
[d] $2,400,000 × 0.70 = $1,680,000; $2,400,000 × 0.30 = $720,000
[e] $800,000 × 0.70 = $560,000

3. The 5 percent of accounts receivable that ABT never collects is bad debts expense. It never appears on the cash budget since it is never collected in cash. The cash received in the quarter after the quarter of sale would be reduced and total cash would be reduced. The following is the cash

receipts budget with the updated accounts receivable aging schedule
using the new assumption.

CORNERSTONE
8-11
(continued)

	Quarter 1	Quarter 2	Quarter 3	Quarter 4
Cash sales	$ 700,000	$2,100,000	$2,400,000	$ 800,000
Received on account from:				
Quarter 4, 2012	250,000			
Quarter 1, 2013	490,000	175,000		
Quarter 2, 2013		1,470,000	525,000	
Quarter 3, 2013			1,680,000	600,000
Quarter 4, 2013				560,000
Total cash receipts	$1,440,000	$3,745,000	$4,605,000	$1,960,000

The *cash excess or deficiency* section compares the cash available with the cash needed. Cash needed includes the total cash disbursements plus the minimum cash balance required by company policy. The minimum cash balance is simply the lowest amount of cash on hand that the firm finds acceptable. Consider your own chequing account. You probably try to keep at least some cash in the account, perhaps because a minimum balance avoids service charges or because it allows you to make an unplanned purchase. Similarly, companies also require minimum cash balances. The amount varies from firm to firm and is determined by each company's particular needs and policies. If the total cash available is less than the cash needs, a deficiency exists. In such a case, a short-term loan will be needed. On the other hand, with a cash excess (cash available is greater than the firm's cash needs), the firm can repay loans and perhaps make some temporary investments.

The *financing* section of the cash budget consists of borrowings and repayments. If there is a deficiency, the financing section shows the necessary amount to be borrowed. When excess cash is available, the financing section shows planned repayments, including interest.

The final section of the cash budget is the planned ending cash balance. Remember that the minimum cash balance was subtracted to find the cash excess or deficiency. However, the minimum cash balance is not a disbursement, so it must be added back to yield the planned ending balance.

Once all sections of the cash budget are understood, it is time to construct one. Cornerstone 8-12 shows how and why to prepare the cash budget.

The cash budget shown in Cornerstone 8-12 underscores the importance of breaking down the annual budget into smaller time periods. The cash budget for the year implies that there is enough cash from operations to buy the new equipment. Quarterly information, however, shows that short-term borrowing is needed to buy the new equipment earlier in the year rather than later. Breaking down the annual cash budget into quarterly or monthly time periods conveys more information. Most firms prepare monthly cash budgets; some even prepare weekly and daily cash budgets.

ABT's cash budget shows another important piece of information. By the end of the fourth quarter, the firm holds a considerable amount of cash ($488,300). ABT should consider investing this cash in an interest-earning account or short-term marketable securities rather than allow it to sit idly in a bank account. The management of ABT could also consider making additional long-term investments. Once plans are finalized for use of the excess cash, the cash budget should be revised to reflect those plans. Budgeting is a dynamic process. As the budget is developed, new information becomes available and is incorporated in the budgetary plans.

**CORNERSTONE
8-12**

The HOW and WHY of Constructing a Cash Budget

Information:

The information needed to prepare the cash budget comes from Cornerstones 8-1 through 8-11 and from the following information.

a. ABT requires a $100,000 minimum cash balance for the end of each quarter. On December 31, 2012, the cash balance was $120,000.

b. Money can be borrowed and repaid in multiples of $100,000. Interest is 12 percent per year. Interest payments are made only for the amount of the principal being repaid. All borrowing and repayment take place at the end of a quarter.

c. All materials are purchased on account; 80 percent of purchases are paid for in the quarter of purchase. The remaining 20 percent are paid in the following quarter. The purchases for the fourth quarter of 2012 were $500,000.

d. Budgeted depreciation is $200,000 per quarter for overhead; $5,000 for marketing expense; and $12,000 for administrative expense. (Remember that depreciation is not a cash expense and must be deleted from total expenses before the cash budget is prepared.)

e. The capital budget for 2013 revealed plans to purchase additional equipment for $600,000 in the first quarter. The acquisition will be financed with operating cash, supplementing it with short-term loans as necessary.

f. Corporate income taxes of $20,700 will be paid at the end of the fourth quarter.

Why:

The cash budget is critical to managers' planning. It shows how much cash will be available each time period. Companies without sufficient cash may go under even if their net income is positive.

Required:

1. Calculate cash payments for purchases expected in each quarter of 2013. (*Hint*: Use Cornerstone 8-3.)

2. Prepare a cash budget for ABT Inc. for each quarter of the coming year.

3. ***What if*** ABT did not have access to short-term financing? How would that affect the cash budget? ABT's ability to stay in business?

Solution:

1. Payments in current quarter $= 0.8$(current quarter purchases)
$$+ \; 0.2(\text{prior quarter purchases})$$

Payments Quarter 1 $= 0.8(\$654,000) + 0.2(\$500,000) = \$623,200$

Payments Quarter 2 $= 0.8(\$1,560,000) + 0.2(\$654,000) = \$1,378,800$

Payments Quarter 3 $= 0.8(\$1,426,000) + 0.2(\$1,560,000)$
$$= \$1,452,800$$

Payments Quarter 4 $= 0.8(\$520,000) + 0.2(\$1,426,000) = \$701,200$

2.

	Quarter 1	Quarter 2	Quarter 3	Quarter 4	Year
Beginning balance (a)*	$ 120,000	$ 100,800	$ 123,000	$ 190,200	$ 120,000
Collections (C8-11):					
Cash sales	700,000	2,100,000	2,400,000	800,000	6,000,000
Received on account:					
Current quarter sales	490,000	1,470,000	1,680,000	560,000	4,200,000
Prior quarter sales	300,000	210,000	630,000	720,000	1,860,000
Total cash available	$1,610,000	$3,880,800	$4,833,000	$2,270,200	$12,180,000

CORNERSTONE
8-12
(continued)

	Quarter 1	Quarter 2	Quarter 3	Quarter 4	Year
Disbursements:					
Payments for purchases:					
Current quarter purchases	$ 523,200	$1,248,000	$1,140,800	$ 416,000	$ 3,328,000
Prior quarter purchases	100,000	130,800	312,000	285,200	828,000
Direct labour (C8-4)	504,000	1,260,000	1,176,000	420,000	3,360,000
Overhead (C8-5, d)	408,000	840,000	792,000	360,000	2,400,000
Marketing expense (C8-8, d)	133,000	338,000	333,000	133,000	937,000
Administrative expense (C8-9, d)	41,000	41,000	41,000	41,000	164,000
Income tax (f)				20,700	20,700
Equipment (e)	600,000				600,000
Total disbursements	2,309,200	3,857,800	3,794,800	1,675,900	11,637,700
Minimum cash balance (a)	100,000	100,000	100,000	100,000	100,000
Total cash needs	2,409,200	3,957,800	3,894,800	1,775,900	11,737,700
Excess (deficiency)	(799,200)	(77,000)	938,200	494,300	442,300
Financing (b):					
Borrowings	800,000	100,000			900,000
Repayments			800,000	(100,000)	(900,000)
Interest			(48,000)	(6,000)	(54,000)
Total financing	800,000	100,000	(848,000)	(106,000)	(54,000)
Add: Minimum cash balance	100,000	100,000	100,000	100,000	100,000
Ending cash balance	$ 100,800	$ 123,000	$ 190,200	$ 488,300	$ 488,300

*Parenthetical references refer to Cornerstones, for example, C8-11 is Cornerstone 8-11, or to the information stated above.

3. If ABT had no access to short-term financing, then the company would be in severe trouble by the end of Quarter 1. By the end of Quarter 1, ABT has a shortfall of $699,200. It will not be able to make the cash payments it plans to make and will very possibly be forced out of business. The seasonal nature of ABT's sales make it imperative for the company to use borrowed money early in the year and to make up for it later in the year.

Budgeted Balance Sheet

The budgeted balance sheet for the coming year develops from information contained in the balance sheet for the previous year and in the various budgets in the master budget. It represents the culmination of the financial events expected of the coming year and shows management where the company is expected to be at the end of the year. The balance sheet for the beginning of the year is given in Exhibit 8-4B. This balance sheet is necessary in preparing the end-of-the-year budgeted balance sheet that is shown in Exhibit 8-5.

As we have described the individual budgets that make up the master budget, the interdependencies of the component budgets have become apparent. You may want to refer back to Exhibit 8-2 to review these interrelationships.

Exhibit 8-5

Budgeted Balance Sheet for ABT Inc.

ABT Inc.
Budgeted Balance Sheet
December 31, 2013

Assets

Current assets:		
Cash[a]	$ 488,300	
Accounts receivable[b]	240,000	
Materials inventory[c]	50,000	
Finished goods inventory[d]	67,000	
Total current assets		$ 845,300
Property, plant, and equipment (PP&E):		
Land[e]	$ 2,500,000	
Buildings and equipment[f]	9,600,000	
Accumulated depreciation[g]	(5,368,000)	
Total PP&E		6,732,000
Total assets		$7,577,300
Liabilities and Shareholders' Equity		
Current liabilities:		
Accounts payable[h]		$ 104,000
Shareholders' equity:		
Common stock[i]	$ 600,000	
Retained earnings[j]	6,873,300	
Total shareholders' equity		7,473,300
Total liabilities and shareholders' equity		$7,577,300

[a] Ending cash balance for the year from Cornerstone 8-12.
[b] From Cornerstone 8-1 and Cornerstone 8-11, fourth quarter credit sales times 0.3 (percentage to be collected in the following quarter).
[c] From Cornerstone 8-3, fourth quarter desired ending inventory of 5,000,000 kg times $0.01 (cost per kilogram).
[d] From Cornerstone 8-6.
[e] From Exhibit 8-4B, December 31, 2012, balance sheet, Land account.
[f] From Exhibit 8-4B, December 31, 2012, balance sheet, Buildings and Equipment account plus $600,000 for new equipment purchase.
[g] From Exhibit 8-4B, December 31, 2012, balance sheet, accumulated depreciation balance plus depreciation balances from Cornerstone 8-12 ($4,500,000 + $800,000 + $20,000 + $48,000).
[h] Equals 20 percent of fourth-quarter purchases of direct materials, see Cornerstones 8-3 and 8-12.
[i] From Exhibit 8-4B, December 31, 2012, balance sheet, Common Stock account.
[j] From Exhibit 8-4B, December 31, 2012, balance sheet, retained earnings balance plus net income from Cornerstone 8-10.

Shortcomings of the Traditional Master Budget Process

Shortcomings of the master budget can be classified into several categories. The traditional master budget is:

1. department oriented and does not recognize the interdependencies among departments.
2. static, not dynamic.
3. results, not process, oriented.

Let's look more closely at each of these.

Departmental Orientation   In traditional budgeting, each department develops its own budget. These budgets are then aggregated to form the overall company budget. A department may start by determining what resources (i.e., labour, supplies, etc.) it currently has and then adjust those levels for the potential level of output. That is, in the traditional budget, departments plan from resources to outputs. The activity-based budgeting (ABB) approach is the opposite. ABB starts by asking what level of output is desired and then works backward to see what resources are necessary to achieve that level of output. We might ask, what difference does it make? Couldn't you achieve the same effect whether you go backward or forward? The answer, rooted in human behaviour, is no. By concentrating on last year's costs and going forward, a department locks in past ways of doing things. Companies that use ABB, however, start first with the desired output and then figure out what resources are needed. That level of resources may or may not be the same as last year's level.

As a result, traditional budgeting may have managers feeling embattled. There is a sense of "every department for itself." Managers feel encouraged to use every cent of budgeted resources, whether or not those resources are needed. Indeed, if the department did not use the full level of budgeted resources, it would have a hard time making a case for increased—or even the same level of—resources in the coming year.

Static Rather than Dynamic Budgets   A **static budget** is one developed for a single level of activity. Recall that the master budget is based on budgeted sales for the coming year. Once that amount is determined, production, marketing, and administrative budgets are built around it. An adjunct to the static nature of the budget is the use of last year's budget to create this year's budget. Often, the current budget is based on last year's amounts as adjusted for inflation. This approach to budgeting, called the **incremental approach**, can incorporate last year's inefficiencies into the current budget. Under the incremental approach, heads of budgeting units often strive to spend all of the year's budget so that no surplus exists at the end of the year. (This is particularly true for government agencies.) This action is taken to maintain the current level of the budget and enable the head of the unit to request additional funds.

For example, a military base was faced with the possibility of a surplus at the end of the fiscal year. The base commander, however, found ways to spend the extra money before the year ended. Officers and other personnel residing around the base were given several bags of lawn fertilizer. Also, new furniture was acquired for the officer quarters.

The waste and inefficiency portrayed in this example is often perpetuated and encouraged by incremental budgeting.

Zero-base budgeting is an alternative approach.[3] Unlike with incremental budgeting, the prior year's budgeted level is not taken for granted. Existing operations are analyzed, and continuance of the activity or operation must be justified on the basis of its need or usefulness to the organization. The burden of proof is on each manager to justify why any money should be spent at all. Zero-base budgeting requires extensive, in-depth analysis. Although this approach has been used successfully in industry and government, it is time consuming and costly. Advocates of the incremental approach argue that incremental budgeting also uses extensive, in-depth reviews but not as frequently because they are not justified on a cost-benefit basis. A reasonable compromise may be to use zero-base budgeting every three to five years in order to weed out waste and inefficiency. Especially in a period of intense competition and re-engineering, zero-base budgeting can force managers to "break set" and see their units in a different perspective.

Results Orientation   This shortcoming is closely allied to the static nature of the master budget. By focusing on results instead of process, managers, in effect, disconnect the process from its output. When budgets are resource driven rather than output driven, then managers concentrate on resources and may fail to see the link between resources and output. Then, when the need for cost cutting arises, they make across-the-board cuts, slicing every department's budget by the same percentage. This has the superficial appearance of fairness—in that every department "shares the pain." Unfortunately, some departments have more fat than others, and some may be downright unneeded. Across-the-board cuts do not cut true waste and inefficiency; that is not their point.

[3] Zero-base budgeting was developed by Peter Pyhrr of Texas Instruments. For a detailed discussion of the approach, see Peter Pyhrr, *Zero-Base Budgeting* (New York: Wiley, 1973).

It is important to realize that the master budget is not inherently flawed. That is why the traditional approach to budgeting has been used for so long. In fact, it has been very useful over the decades and many managers strongly agree that "budgets are indispensable and companies couldn't manage without them."[4] However, the past 30 or so years have been characterized by rapid change. In a period of change, managers may not realize that previously acceptable ways of doing things no longer work. This is the case for the master budget. For example, consider its static nature. If sales are much the same from year to year, if the production process does not change, and if the firm's product mix is fairly simple and stable, then a static budget based in large part on last year's numbers makes sense. However, this is not the situation for the vast majority of businesses today. Flexible budgets can give managers some feel for the impact of fixed and variable costs. Activity-based budgets go further, by recognizing the numerous drivers for variable costs and by starting with outputs and working backwards to resources.

Flexible Budgets versus Static Budgets

OBJECTIVE ➤ 4
Define flexible budgeting, and discuss its role in planning, control, and decision making.

Budgets are useful control measures. To be used in performance evaluation, however, it is necessary to determine how budgeted amounts should be compared with actual results. Master budget amounts, while vital for planning, are less useful for control. The reason is that the anticipated level of activity rarely equals the actual level of activity. Therefore, the costs and revenues associated with the anticipated level of activity cannot be readily compared with actual costs and revenues for a different level of activity.

Static Budgets   Budgets that are developed around a single expected level of activity (expected sales for the year) are static budgets. Because the revenues and costs prepared for static budgets depend on a level of activity that rarely equals actual activity, they are not very useful when it comes to preparing performance reports.

To illustrate, let's return to the ABT Inc. example used in developing the master budget. Suppose that ABT provides quarterly performance reports. Recall that ABT anticipated sales of 2 million units in the first quarter and had budgeted production of 2.4 million units to support that level of sales (Cornerstone 8-2). Suppose instead that sales activity was greater than expected in the first quarter; 2.6 million concrete blocks were sold instead of the 2 million budgeted in the sales budget; and, because of increased sales activity, production was increased over the planned level. Instead of producing 2.4 million units, ABT produced 3 million units. A performance report comparing the actual production costs for the first quarter with the original planned production costs is given in Exhibit 8-6A.

According to the report, unfavourable variances occurred for direct materials, direct labour, supplies, indirect labour, and rent. However, something is fundamentally wrong with the report. Actual costs for production of *3 million concrete blocks* are being compared with planned costs for production of *2.4 million*. Because direct materials, direct labour, and variable overhead are variable costs, we expect them to be greater as more is produced. Thus, even if cost control were perfect for the production of 3 million units, unfavourable variances would be shown for all variable costs.

To create a meaningful performance report, actual costs and expected costs must be compared at the *same* level of activity. Since actual output often differs from planned output, some method is needed to compute what the costs should have been for the actual output level.

Flexible Budgets   The budget that provides expected costs for a variety of activity levels is called a **flexible budget**. Flexible budgeting can be used in planning by showing what costs will be at various levels of activity. When used this way,

---

[4] This quote is taken from an article discussing the results of a survey of IMA members on budgeting. Theresa Libby and R. Murray Lindsay, "Beyond Budgeting or Better Budgeting? IMA Members Express Their Views," *Strategic Finance* (August 2007): 46–51. (Quote taken from pp. 48–49.)

Exhibit 8-6

Performance Reports

A. ABT Performance Report for Quarter 1: Comparison of Actual with Static (Master) Budget Amounts

	Actual	Budgeted	Variance	
Units produced	3,000,000	2,400,000	600,000	F[a]
Direct materials cost	$ 927,300	$ 624,000[b]	$303,300	U[c]
Direct labour cost	630,000	504,000[d]	126,000	U
Overhead:[e]				
Variable:				
Supplies	80,000	72,000	8,000	U
Indirect labour	220,000	168,000	52,000	U
Power	40,000	48,000	(8,000)	F
Fixed:				
Supervision	90,000	100,000	(10,000)	F
Depreciation	200,000	200,000	0	
Rent	30,000	20,000	10,000	U
Total	$2,217,300	$1,736,000	$481,300	U

[a] F means the variance is favourable.
[b] 2,400,000 units × $0.26 (Cornerstone 8-6 gives unit costs for direct materials and direct labour).
[c] U means the variance is unfavourable.
[d] 2,400,000 units × $0.21 (Cornerstone 8-6 gives unit costs for direct materials and direct labour).
[e] Variable overhead equals 2,400,000 units times: $0.03 for supplies; $0.07 for indirect labour; and $0.02 for power. Budgeted fixed overhead per quarter is given in Cornerstone 8-5.

B. Managerial Performance Report: Quarterly Production (in thousands)

	Actual Results (1)	Flexible Budget (2)	Flexible Budget Variances (3) = (1) − (2)	Static Budget (4)	Volume Variances (5) = (2) − (4)	
Units produced	3,000,000	3,000,000	0	2,400,000	600,000	F
Direct materials cost	$ 927,300	$ 780,000	$147,300 U	$ 624,000	$156,000	U
Direct labour cost	630,000	630,000	0	504,000	126,000	U
Overhead:						
Variable:						
Supplies	80,000	90,000	(10,000) F	72,000	18,000	U
Indirect labour	220,000	210,000	10,000 U	168,000	42,000	U
Power	40,000	60,000	(20,000) F	48,000	12,000	U
Fixed:						
Supervision	90,000	100,000	(10,000) F	100,000	0	
Depreciation	200,000	200,000	0	200,000	0	
Rent	30,000	20,000	10,000 U	20,000	0	
Total	$2,217,300	$2,090,000	$127,300 U	$1,736,000	$354,000	U

managers can deal with uncertainty by examining the expected financial results for a number of plausible scenarios. Spreadsheets are particularly useful in developing this type of flexible budget.

The budget that provides budgeted costs for the actual level of activity is also a flexible budget. The flexible budget can be used after the fact, for control purposes to compute what costs should have been for the actual level of activity. Once expected costs are known for the actual level of activity, a performance report that compares those expected costs to actual costs can be prepared. When used for control, flexible budgets help managers compare "apples to apples" in assessing performance. Cornerstone 8-13 shows how and why to prepare a flexible budget for varying levels of activity.

week to week. The receptionist opened all the mail and sorted it into folders by client. It took approximately two hours a day to perform this task.

- The second activity is "paying bills." There were approximately 1,000 bills per month, or 12,000 per year. The number of bills varied widely from client to client. The administrative assistants performed this activity, using computer software to enter and pay bills. Based on the amount of time this took and the cost of supplies, software, and postage, the average cost of paying one bill was $1.75.
- The third activity is "reconciling accounts." The administrative assistants performed this activity, and it took about 30 minutes per account each month. There were 350 accounts. This averaged out to one administrative assistant working full time on reconciling accounts. Related supplies and the use of a computer and software added another $4,900 to the total.
- The firm advertised for and interviewed caregivers for their clients as needed. The driver for this activity is number of new hires. The yearly cost, including newspaper advertising and the time of the administrative assistants, totalled $7,200 per year. On average, there were estimated to be 60 new hires in a year.
- A private investigator was retained to perform thorough background checks of prospective caregivers. Each background check cost $25, and an average of four prospective caregivers were checked for every successful new hire.
- Every month, the administrative assistants made personal visits to each client. The number of clients was a good driver for this activity, and the total cost was about $650 per client, per year.
- Each month, Brad or one of the administrative assistants prepared a monthly report for every client. The report detailed the financial activity and included the notes taken from the home visits. Prospective issues and problems were raised. These reports were sent to the clients as well as to interested adult children. The cost of time, supplies, and postage averaged $175 per client, per year.
- The final activity is managing the department and signing up new clients. Brad is responsible for the bulk of this activity. The activity does not have a driver, but instead, consists of the remaining costs of the department.

The Secure-Care Department's activity-based budget is shown in Exhibit 8-7C. Notice that the department has identified eight activities and four drivers. This level of detail is much richer than that for the flexible budget presented in Exhibit 8-7B, where there was only one driver, the number of clients. With an activity-based budget, we get a feel for the diversity among the clients. Some have more accounts, and some more bills to pay. In other words, "clients" are not all the same. There is considerable product diversity, and this diversity is not captured in either the traditional or the simple flexible budget.

The traditional, flexible, and activity-based budgets for the Secure-Care Department all total $273,800. But notice the richness of detail in the activity-based budget. Here we can see the relationship between output and resource usage. Also, the manager's attention is focused on the most costly activities: paying bills, reconciling accounts, and visiting homes. Brad may want to use this information in pricing the various parts of the secure-care service.

Earlier, we noted that both the traditional and flexible budgeting approaches worked well for particular sets of circumstances. Recall that a key feature is that the environment of the company remains stable. When that is the case, one year is much like the next. The technology is the same, and there is little product diversity. A single volume-based driver works well to account for any changes. However, many companies now face an environment that is changing rapidly in many ways. These companies are ill served by budgets that are founded on the notion that everything remains the same. Companies in a changing environment, whether it relates to changing technology, competition, or customer base, need a much more flexible technique for planning and control. The activity-based budget can be extended to include feature costing. This provides an even more powerful tool for planning and control.

Feature costing assigns costs to activities and products or services based on the product's or service's features.[6] In the Secure-Care Department, we could see that one client was not necessarily the same as another. In other words, different clients had different features that required the department to use different sets of activities to handle them. A client with only one chequing account and a few repetitive bills took little time. Other clients had numerous accounts and bills. Some clients may be difficult to get along with, leading to rapid turnover of their caregivers and necessitating additional interviewing and background investigation. If the company wanted to extend the ABB process, it could add feature costing. That is, it could determine what features of clients differentiate them into groups that require different sets of activities. We can easily imagine that the company might delve further into the various features, asking what leads to the different features (root cause analysis) and what could be done to modify the more costly features. For example, perhaps the monthly reports could be posted, using appropriate security, on the Internet. The reports could be updated relatively easily, and postage and printing costs could be minimized.

The Behavioural Dimension of Budgeting

OBJECTIVE ▶ 6
Identify and discuss the key features that a budgetary system should have to encourage managers to engage in goal-congruent behaviour.

Budgets are often used to judge the actual performance of managers. Bonuses, salary increases, and promotions are all affected by a manager's ability to achieve or beat budgeted goals. Since a manager's financial status and career can be affected, budgets can have significant behavioural effects. Whether those effects are positive or negative depends to a large extent on how budgets are used.

Positive behaviour occurs when the goals of individual managers are aligned with the goals of the organization and the manager has the drive to achieve them. The alignment of managerial and organizational goals is often referred to as **goal congruence**. In addition to goal congruence, however, a manager must also exert effort to achieve the goals of the organization.

If the budget is improperly administered, the reaction of subordinate managers may be negative. This negative behaviour can be manifested in numerous ways, but the overall effect is subversion of the organization's goals. **Dysfunctional behaviour** involves individual behaviour that is in basic conflict with the goals of the organization.

ETHICS A theme underlying the behavioural dimension of budgeting is ethics. The importance of budgets in performance evaluation and managers' pay raises and promotions leads to the possibility of unethical action. All of the dysfunctional actions regarding budgets that a manager may choose to take can have an unethical aspect. For example, a manager who deliberately underestimates sales and overestimates costs for the purpose of making the budget easier to achieve is engaging in unethical behaviour. It is the responsibility of the company to create budgetary incentives that do not encourage unethical behaviour. It is the responsibility of the manager to avoid engaging in such behaviour.◆

Characteristics of a Good Budgetary System

An ideal budgetary system is one that achieves complete goal congruence and simultaneously creates a drive in managers to achieve the organization's goals in an ethical manner. While an ideal budgetary system probably does not exist, research and practice have identified some key features that promote a reasonable degree of positive behaviour. These features include the following:

- frequent feedback on performance,
- monetary and nonmonetary incentives,
- participation,
- realistic standards,
- controllability of costs, and
- multiple measures of performance.

[6] J. A. Brimson, "Feature Costing: Beyond ABC," *Journal of Cost Management* (January/February 1998): 6–12.

Frequent Feedback on Performance Managers need to know how they are doing as the year unfolds. Providing them with frequent, timely performance reports allows them to know how successful their efforts have been and gives managers time to take corrective actions and change plans as necessary. Frequent performance reports can reinforce positive behaviour and give managers the time and opportunity to adapt to changing conditions. Continuous monitoring to see if actual costs and revenues are in accord with budgeted amounts and selective investigation of significant variances allows managers to focus only on areas that need attention. This process is called *management by exception.*

Monetary and Nonmonetary Incentives A sound budgetary system encourages goal-congruent behaviour. **Incentives** are the means that are used to encourage managers to work toward achieving the organization's goals. Incentives can be either negative or positive. Negative incentives use fear of punishment to motivate; positive incentives use rewards. What incentives should be tied to an organization's budgetary system?

The most successful companies view people as their most important asset. Their budgets reflect their underlying philosophy by including significant expenditures on recruiting and career development in good times. Even in difficult economic times, employees are protected to the extent possible.

For example, in 2008–2009, **FedEx** worked to keep costs under control and help save jobs. Pay was cut 5 percent across the board, and the CEO took a 20 percent pay cut. Similarly, **General Electric** CEO Jeff Immelt took a 28 percent pay cut and asked for no bonus for 2008.[7]

Of course, negative incentives can be used as well. The most serious negative incentive is the threat of dismissal. Other negative incentives include loss of bonuses, promotions, or raises.

Participative Budgeting Rather than imposing budgets on subordinate managers, **participative budgeting** allows subordinate managers considerable say in how the budgets are established. Typically, overall objectives are communicated to the manager, who helps develop a budget that will accomplish these objectives. In participative budgeting, the emphasis is on the accomplishment of the broad objectives, not on individual budget items.

The budget process described earlier for ABT uses participative budgeting. The company provides the sales forecast to its profit centres and requests a budget that shows planned expenditures and expected profits given that specific level of sales. The managers of the profit centres are fully responsible for preparing the budgets by which they will later be evaluated. Although the budgets must be approved by the president, disapproval is uncommon; the budgets are usually in line with the sales forecast and last year's operating results adjusted for expected changes in revenues and costs.

Participative budgeting communicates a sense of responsibility to subordinate managers and fosters creativity. Since the subordinate manager creates the budget, it is more likely that the budget's goals will become the manager's personal goals, resulting in greater goal congruence. Advocates of participative budgeting claim that the increased responsibility and challenge inherent in the process provide nonmonetary incentives that lead to a higher level of performance. They argue that individuals involved in setting their own standards will work harder to achieve them. In addition to the behavioural benefits, participative budgeting has the advantage of involving individuals whose knowledge of local conditions may enhance the entire planning process.

[7] "World's Most Admired Companies 2009," *Fortune* (March 16, 2009), http://money.cnn.com/magazines/fortune/mostadmired/2009/index.html.

Participative budgeting has three potential problems that should be mentioned:

1. Setting standards that are either too high or too low
2. Building slack into the budget (often referred to as *padding the budget*)
3. Pseudoparticipation

Some managers may tend to set the budget either too loose or too tight. Since budgeted goals tend to become the manager's goals when participation is allowed, making this mistake in setting the budget can result in decreased performance levels. If goals are too easily achieved, a manager may lose interest, and performance may actually drop. Challenge is important to aggressive and creative individuals. Similarly, setting the budget too tight ensures failure to achieve the standards and frustrates the manager. This frustration, too, can lead to poor performance. The trick is to get managers in a participative organization to set high but achievable goals.

The second problem with participative budgeting is the opportunity for managers to build slack into the budget. **Budgetary slack** exists when a manager deliberately underestimates revenues or overestimates costs. Either approach increases the likelihood that the manager will achieve the budget and consequently reduces the risk that the manager faces. Padding the budget also unnecessarily ties up resources that might be used more productively elsewhere.

Slack in budgets can be virtually eliminated if top management dictates lower expense budgets. However, the benefits to be gained from participation may far exceed the costs associated with padding the budget. Even so, top management should carefully review budgets proposed by subordinate managers and provide input, where needed, in order to decrease the effects of building slack into the budget.

The third problem with participation occurs when top management assumes total control of the budgeting process, seeking only superficial participation from lower-level managers. This practice is termed **pseudoparticipation**. Top management is simply obtaining formal acceptance of the budget from subordinate managers, not seeking real input. Accordingly, none of the behavioural benefits of participation will be realized.

Realistic Standards Budgeted objectives are used to gauge performance; accordingly, they should be based on realistic conditions and expectations. Budgets should reflect operating realities such as actual levels of activity, seasonal variations, efficiencies, and general economic trends. Flexible budgets, for example, are used to ensure that the budgeted costs provide standards that are compatible with the actual activity level. Another factor to consider is that of seasonality. Some businesses receive revenues and incur costs uniformly throughout the year; thus, spreading the annual revenues and costs evenly over quarters and months is reasonable for interim performance reports. However, for businesses with seasonal variations, this practice would result in distorted performance reports.

Factors such as efficiency and general economic conditions are also important. Occasionally, top management makes arbitrary cuts in prior-year budgets with the belief that the cuts will reduce fat or inefficiencies that allegedly exist. In reality, some units may be operating efficiently and others inefficiently. An across-the-board cut without any formal evaluation may impair the ability of some units to carry out their missions. General economic conditions also need to be considered. Budgeting for a significant increase in sales when a recession is projected is not only foolish but also potentially harmful.

For example, for years, **Kodak** confidently predicted that its film business would grow by 8 percent, when the industry was growing by only 4 percent.[8] The predicted growth did not occur. This type of unfounded optimism did nothing to improve sales and only hurt stock analysts' perception of the company.

---

[8] Peter Nulty, "Digital Imaging Had Better Boom Before Kodak Film Busts," *Fortune* (May 1, 1995): 80–83.

Controllability of Costs   Conventional thought maintains that managers should be held accountable only for costs over which they have control. **Controllable costs** are costs whose level a manager can influence. In this view, a manager who has no responsibility for a cost should not be held accountable for it. For example, divisional managers have no power to authorize such corporate-level costs as research and development and salaries of top managers. Therefore, they should not be held accountable for the incurrence of those costs.

Many firms, however, do put noncontrollable costs in the budgets of subordinate managers. Making managers aware of the need to cover all costs is one rationale for this practice. If noncontrollable costs are included in a budget, they should be separated from controllable costs and labelled as *noncontrollable*.

Multiple Measures of Performance   Often, organizations make the mistake of using budgets as their only measure of managerial performance. Overemphasis on this measure can lead to a form of dysfunctional behaviour called *milking the firm* or *myopia*. **Myopic behaviour** occurs when a manager takes actions that improve budgetary performance in the short run but bring long-run harm to the firm.

There are numerous examples of myopic behaviour. To meet budgeted cost objectives or profits, managers can reduce expenditures for preventive maintenance, advertising, and new product development. Managers can also fail to promote deserving employees to keep the cost of labour low and can choose to use lower-quality materials to reduce the cost of materials. In the short run, these actions will lead to improved budgetary performance, but in the long run, productivity will fall, market share will decline, and capable employees will leave for more attractive opportunities.

Managers who engage in this kind of behaviour often have a short tenure. That is, they spend three to five years before being promoted or moving to a new area of responsibility. Their successors are the ones who pay the price for their myopic behaviour. The best way to prevent myopic behaviour is to measure the performance of managers on several dimensions, including some long-run attributes. Productivity, quality, and personnel development are examples of other areas of performance that could be evaluated. Financial measures of performance are important, but overemphasis on them can be counterproductive.

Summary of Learning Objectives

1. **Define budgeting, and discuss its role in planning, controlling, and decision making.**

- A budget is a financial plan for the future.
- Budgeting is important for planning, control, and decision making.
- The master budget is the comprehensive plan for the coming year. It consists of:

  - The operating budget
  - The financial budget

2. **Prepare the operating budget, identify its major components, and explain the interrelationships of the various components.**

- The sales budget shows the expected sales quantity and price of each product or service.
- The production budget shows the budgeted units to be produced in each period to meet sales and desired ending inventory needs. It includes:

  - The direct materials purchases budget
  - The direct labour budget
  - The overhead budget

- Ending finished goods inventory and cost of goods sold budgets are used in the budgeted income statement.
- Operating expense budgets include:

    - The marketing expense budget
    - The administrative expense budget
    - Any other needed budgets for operating departments (e.g., Research and Development)

- The budgeted income statement is the culmination of the operating budget.

3. Identify the components of the financial budget, and prepare a cash budget.

- The cash budget shows the sources and disbursements of cash by period for the coming year.

    - Only cash items are shown in the cash budget.
    - The accounts receivable aging schedule helps companies determine the timing of cash receipts.

- The cash budget is critically important to the ability of a company to meet its obligations.
- The budgeted balance sheet shows the expected assets, liabilities, and owners' equity for the end of the coming year.

4. Define flexible budgeting, and discuss its role in planning, control, and decision making.

- A flexible budget shows costs for varying levels of activity.

    - Useful for planning
    - Useful for sensitivity analysis

- A flexible budget can be constructed for the actual level of activity.

    - Useful for control
    - Compares actual costs to budgeted amounts for actual level of activity

5. Define activity-based budgeting, and discuss its role in planning, control, and decision making.

- Activity-based budgeting recognizes interdependencies among departments.
- It also focuses on business processes.

6. Identify and discuss the key features that a budgetary system should have to encourage managers to engage in goal-congruent behaviour.

- Dysfunctional behaviour can occur when budgets are overemphasized as a control mechanism.
- Budgets are better performance measures when used with:

    - Participative budgeting
    - Other nonmonetary incentives
    - Frequent feedback on performance
    - Ensuring that the budgetary objectives reflect reality
    - Holding managers accountable for only controllable costs

---

CORNERSTONE 8-1	The HOW and WHY of constructing a sales budget, page 378	
CORNERSTONE 8-2	The HOW and WHY of constructing a production budget, page 379	
CORNERSTONE 8-3	The HOW and WHY of constructing a direct materials purchases budget, page 381	**CORNERSTONES FOR CHAPTER 8**

Review Problems

I. Sales, Production, Direct Materials, and Direct Labour Budgets

Young Products produces coat racks. The projected sales for the first quarter of the coming year and the beginning and ending inventory data are as follows:

Sales	100,000 units
Unit price	$15
Beginning inventory	8,000 units
Targeted ending inventory	12,000 units

The coat racks are moulded and then painted. Each rack requires four kilograms of metal, which cost $2.50 per kilogram. The beginning inventory of materials is 4,000 kilograms. Young Products wants to have 6,000 kilograms of metal in inventory at the end of the quarter. Each rack produced requires 30 minutes of direct labour time, which is billed at $9 per hour.

Required:

1. Prepare a sales budget for the first quarter.
2. Prepare a production budget for the first quarter.
3. Prepare a direct materials purchases budget for the first quarter.
4. Prepare a direct labour budget for the first quarter.

Solution:

1.

Young Products
Sales Budget
For the First Quarter

Units	100,000
Unit selling price	× $15
Sales	$1,500,000

2.

Young Products
Production Budget
For the First Quarter

Sales (in units)	100,000
Desired ending inventory	12,000
Total needs	112,000
Less: Beginning inventory	8,000
Units to be produced	104,000

3.

Young Products
Direct Materials Purchases Budget
For the First Quarter

Units to be produced	104,000
Direct materials per unit (kg)	× 4
Production needs (kg)	416,000
Desired ending inventory (kg)	6,000
Total needed (kg)	422,000
Less: Beginning inventory (kg)	4,000
Materials to be purchased (kg)	418,000
Cost per kilogram	× $2.50
Total purchase cost	$1,045,000

4.

Young Products
Direct Labour Budget
For the First Quarter

Units to be produced	104,000
Labour time per unit	× 0.5
Total hours needed	52,000
Wage per hour	× $9
Total direct labour cost	$468,000

II. Flexible Budgeting

Archambault Company manufactures backpacks, messenger bags, and rolling duffel bags. Archambault's accountant has estimated the following cost formulas for overhead:

Indirect labour cost = $90,000 + $0.50 per direct labour hour

Maintenance = $45,000 + $0.40 per machine hour

Power = $0.15 per machine hour

Depreciation = $150,000

Other = $63,000 + $1.30 per direct labour hour

In the coming year, Archambault is considering three budgeting scenarios: conservative (assumes increased competition from other companies), expected, and optimistic (assumes a particularly robust economy). Anticipated quantities sold of each type of product appear in the following table:

Product	Conservative	Expected	Optimistic
Backpacks	50,000	100,000	150,000
Messenger bags	20,000	40,000	80,000
Rolling duffel bags	15,000	25,000	50,000

The standard amounts for one unit of each type of product are as follows:

	Backpacks	Messenger Bags	Rolling Duffel Bags
Direct materials	$5.00	$4.00	$8.00
Direct labour hours	1.2 hours	1.0 hour	2.5 hours
Machine hours	1.0 hour	0.75 hour	2.0 hours

Direct labour costs $8 per hour.

Required:

1. Prepare an overhead budget for the three potential scenarios.
2. Now, suppose that the actual level of activity for the year was 120,000 backpacks, 45,000 messenger bags, and 40,000 rolling duffel bags. Actual overhead costs were as follows:

Indirect labour	$230,400
Maintenance	145,500
Power	38,000
Depreciation	150,000
Other	435,350

Prepare a performance report for overhead costs.

Solution:

1.

Direct Labour Hours	Conservative	Expected	Optimistic
Backpacks (@ 1.2 DLH)	60,000	120,000	180,000
Messenger bags (@ 1.0 DLH)	20,000	40,000	80,000
Rolling duffel bags (@ 2.5 DLH)	37,500	62,500	125,000
Total direct labour hours	117,500	222,500	385,000

Machine Hours	Conservative	Expected	Optimistic
Backpacks (@ 1.0 MHr.)	50,000	100,000	150,000
Messenger bags (@ 0.75 MHr.)	15,000	30,000	60,000
Rolling duffel bags (@ 2.0 MHr.)	30,000	50,000	100,000
Total machine hours	95,000	180,000	310,000

Flexible Overhead Budget	Conservative	Expected	Optimistic
Variable overhead:			
Indirect labour ($0.50 × DLH)	$ 58,750	$111,250	$ 192,500
Maintenance ($0.40 × MHr.)	38,000	72,000	124,000
Power ($0.15 × MHr.)	14,250	27,000	46,500
Other ($1.30 × DLH)	152,750	289,250	500,500
Total variable overhead	$263,750	$499,500	$ 863,500
Fixed overhead:			
Indirect labour	$ 90,000	$ 90,000	$ 90,000
Maintenance	45,000	45,000	45,000
Depreciation	150,000	150,000	150,000
Other	63,000	63,000	63,000
Total fixed overhead	348,000	348,000	348,000
Total overhead	$611,750	$847,500	$1,211,500

2. Flexible budget based on actual output:

	Direct Labour Hours	Machine Hours
Backpacks:		
(1.2 × 120,000)	144,000	
(1.0 × 120,000)		120,000
Messenger bags:		
(1.0 × 45,000)	45,000	
(0.75 × 45,000)		33,750
Rolling duffel bags:		
(2.5 × 40,000)	100,000	
(2.0 × 40,000)		80,000
Total	289,000	233,750

	Flexible Budget Amount*	Actual	Variance
Indirect labour	$234,500	$230,400	$4,100 F
Maintenance	138,500	145,500	7,000 U
Power	35,063	38,000	2,937 U
Depreciation	150,000	150,000	—
Other	438,700	435,350	3,350 F
Total overhead	$996,763	$999,250	$2,487 U

*Indirect labour = $90,000 + ($0.50 × 289,000)
 Maintenance = $45,000 + ($0.40 × 233,750)
 Power = $0.15 × 233,750
 Other = $63,000 + ($1.30 × 289,000)

Key Terms

Administrative expense budget, 384

Budget committee, 373

Budget director, 373

Budgetary slack, 409

Budgets, 371

Capital expenditures budget, 390

Cash budget, 390

Continuous (rolling) budget, 373

Control, 371

Controllable costs, 410

Direct labour budget, 380

Direct materials purchases budget, 377

Dysfunctional behaviour, 407

Effectiveness, 401

Efficiency, 401

Ending finished goods inventory budget, 382

Feature costing, 407

Financial budgets, 373

Flexible budget, 398

Flexible budget variances, 401

Goal congruence, 407

Incentives, 408

Incremental approach, 397

Marketing expense budget, 384

Master budget, 373

Myopic behaviour, 410

Operating budgets, 373

Overhead budget, 380

Participative budgeting, 408

Production budget, 377

Pseudoparticipation, 409

Research and development expense budget, 385

Sales budget, 377

Static budget, 397

Variable budget, 401

Zero-base budgeting, 397

Discussion Questions

1. Define *budget*. How are budgets used in planning?
2. Define *control*. How are budgets used to control?
3. Discuss some of the reasons for budgeting.
4. What is the master budget? An operating budget? A financial budget?
5. Explain the role of a sales forecast in budgeting. What is the difference between a sales forecast and a sales budget?
6. All budgets depend on the sales budget. Is this true? Explain.
7. What is an accounts receivable aging schedule? Why is it important?
8. Suppose that the vice president of sales is a particularly pessimistic individual. If you were in charge of developing the master budget, how, if at all, would you be influenced by this knowledge?
9. Suppose that the controller of your company's largest factory is a particularly optimistic individual. If you were in charge of developing the master budget, how, if at all, would you be influenced by this knowledge?
10. What impact does the learning curve have on budgeting? What specific budgets might be affected? (*Hint:* Refer to Chapter 2 for material on the learning curve.)
11. While many small firms do not put together a complete master budget, nearly every firm creates a cash budget. Why do you think that is so?
12. Discuss the shortcomings of the traditional master budget. In what situations would the master budget perform well?
13. Define *static budget*. Give an example that shows how reliance on a static budget could mislead management.
14. What are the two meanings of a flexible budget? How is the first type of flexible budget used? The second type?
15. What are the steps involved in building an activity-based budget? How do these steps differentiate the ABB from the master budget?

Cornerstone Exercises

OBJECTIVE ▶ 2
CORNERSTONE 8-1
SERVICE

Cornerstone Exercise 8-1 SALES BUDGET

StrikeSmart Company manufactures and sells soccer balls for teams of children in elementary and high school. StrikeSmart's best-selling lines are the training ball line—durable soccer balls for training and practice; and the match ball line—high performance soccer balls used in games. In the first four months of next year, StrikeSmart expects to sell the following:

	Training Balls		Match Balls	
	Units	Selling Price	Units	Selling Price
January	70,000	$9.50	10,000	$18.00
February	65,000	$9.50	8,000	$18.00
March	100,000	$9.50	15,000	$18.00
April	120,000	$9.50	25,000	$18.00

Required:

1. Construct a sales budget for StrikeSmart for the first three months of the coming year. Show total sales for each product line by month and in total for the first quarter.
2. **What if** StrikeSmart added a third line—tournament quality soccer balls that were expected to take 40 percent of the units sold of the match balls and would have a selling price of $40 each? Prepare a sales budget for StrikeSmart for the first three months of the coming year. Show total sales for each product line by month and in total for the first quarter.

Cornerstone Exercise 8-2 PRODUCTION BUDGET

OBJECTIVE ➤ 2
CORNERSTONE 8-2

Refer to **Cornerstone Exercise 8-1**, through Requirement 1. StrikeSmart requires ending inventory of product to equal 20 percent of the next month's unit sales. Beginning inventory in January was 10,000 training balls and 1,500 match balls.

Required:

1. Construct a production budget for each of the two product lines for StrikeSmart Company for the first three months of the coming year.
2. *What if* StrikeSmart wanted a production budget for the two product lines for the month of April? What additional information would you need to prepare this budget?

Cornerstone Exercise 8-3 DIRECT MATERIALS PURCHASES BUDGET

OBJECTIVE ➤ 2
CORNERSTONE 8-3

Refer to **Cornerstone Exercise 8**-2 for the production budgets for training balls and match balls. Every training ball requires 0.7 square metre of polyvinyl chloride panels, one bladder with valve (to fill with air), and 3 grams of glue. StrikeSmart's policy is that 10 percent of the following month's production needs for raw materials be ending inventory. Beginning inventory in January for all raw materials met this requirement.

Required:

1. Construct a direct materials purchases budget for each type of raw materials for the training ball line for January and February of the coming year.
2. *What if* StrikeSmart increased the ending inventory percentage to 20 percent of the next month's production needs? What impact would that have on the direct materials purchases budgets prepared in Requirement 1?

Cornerstone Exercise 8-4 DIRECT LABOUR BUDGET FOR SERVICE

OBJECTIVE ➤ 2
CORNERSTONE 8-4

The School of Accountancy and Finance of Ryerson University is planning its annual fundraising campaign. This year, the school is planning a call-a-thon and will ask the Accounting and Finance Society (AFS) members to volunteer to make phone calls to a list of 5,000 alumni. The dean's office has agreed to let the AFS use their offices from 6 p.m. to 9 p.m. each weekday so that they will have access to phones. Each volunteer will be provided with a phone and a script with an introduction and suggested responses to various questions that had been asked in the past. Carol Johnson, AFS faculty advisor, estimates the following:

1. Of the 5,000 phone numbers, roughly 10 percent will be wrong numbers (because alumni change addresses/phone numbers without updating the university records). In that case, the student is instructed to apologize to the answering party, hang up, and move on to the next phone number. Each of these calls takes about three minutes.
2. Another 15 percent will be correct numbers, but no one is home and Call Answer picks up. In that case, the student is instructed to simply hang up and move on to the next phone number. Each of these calls takes about two minutes.
3. Each time an alumnus answers the phone, the student is instructed to introduce him or herself and read the scripted introduction. The student is encouraged to engage the alumnus in conversation and reminiscences about Ryerson University, and bring the alum up to date on the wonderful things that are happening in the AFS. Some calls are longer, some shorter, but the average call length is 10 minutes.

Required:

1. Prepare a direct labour budget, in hours, for the fundraising call-a-thon. If 15 students volunteer, how many evenings will the call-a-thon take? (Round to two significant digits.)
2. *What if* the call-a-thon can be moved to the Ryerson University Foundation phone bank? That facility has an automated calling system that automatically dials the phone numbers and routes all answered calls directly to students. As a result, no time is spent dialling and listening to Call Answer. The time savings due to having the numbers automatically dialled and routed mean that the average length of a wrong

2. Communications plays an important role in the budgetary process whether a participative or an imposed budgetary approach is used.

   a. Discuss the differences between communication flows in these two budgetary approaches.

   b. Discuss the behavioural implications associated with the communication process for each of the budgetary approaches. (CMA adapted)

OBJECTIVE > 1 6

CMA Problem 8-6 INFORMATION FOR BUDGETING, ETHICS*

Norton Company, a manufacturer of infant furniture and carriages, is in the initial stages of preparing the annual budget for the coming year. Scott Ford has recently joined Norton's accounting staff and is interested in learning as much as possible about the company's budgeting process. During a recent lunch with Thai Wong, sales manager, and Abdul Saif, production manager, Ford initiated the following conversation.

FORD: Since I'm new around here and am going to be involved with the preparation of the annual budget, I'd be interested in learning how the two of you estimate sales and production numbers.

WONG: We start out very methodically by looking at recent history, discussing what we know about current accounts, potential customers, and the general state of consumer spending. Then, we add that usual dose of intuition to come up with the best forecast we can.

SAIF: I usually take the sales projections as the basis for my projections. Of course, we have to make an estimate of what this year's closing inventories will be, which is sometimes difficult.

FORD: Why does that present a problem? There must have been an estimate of closing inventories in the budget for the current year.

SAIF: Those numbers aren't always reliable since Thai makes some adjustments to the sales numbers before passing them on to me.

FORD: What kind of adjustments?

WONG: Well, we don't want to fall short of the sales projections so we generally give ourselves a little breathing room by lowering the initial sales projection anywhere from 5 to 10 percent.

SAIF: So, you can see why this year's budget is not a very reliable starting point. We always have to adjust the projected production rates as the year progresses, and of course, this changes the ending inventory estimates. By the way, we make similar adjustments to expenses by adding at least 10 percent to the estimates; I think everyone around here does the same thing.

Required:

1. Thai Wong and Abdul Saif have described the use of budgetary slack.

   a. Explain why Wong and Saif behave in this manner, and describe the benefits they expect to realize from the use of budgetary slack.

   b. Explain how the use of budgetary slack can adversely affect Wong and Saif.

2. As a management accountant, Scott Ford believes that the behaviour described by Thai Wong and Abdul Saif may be unethical and that he may have an obligation not to support this behaviour. Explain why the use of budgetary slack may be unethical in terms of the general standards of competency, integrity, and credibility as you understand them. (CMA adapted)

> The Collaborative Learning Exercises can be found on the product support site at www.hansen1ce.nelson.com.

After studying this chapter, you should be able to:

1 Describe how unit input standards are developed, and explain why standard costing systems are adopted.

2 Explain the purpose of a standard cost sheet.

3 Compute and journalize the direct materials and direct labour variances, and explain how they are used for control.

4 Compute overhead variances three different ways, and explain overhead accounting.

5 Calculate mix and yield variances for direct materials and direct labour.

CHAPTER 9

Standard Costing: A Functional-Based Control Approach

Budgets help managers in planning and also set standards that are used to control and evaluate managerial performance. In Chapter 8, we saw that budgets can be classified as static or flexible. Static budgets are not very useful for assessing efficiency; their main value is assessing whether or not the targeted level of activity is achieved and, thus, provide some insight concerning managerial effectiveness. Flexible budgets evaluate efficiency by comparing the actual costs and actual revenues with the corresponding budgeted amounts for the *same* level of activity. These flexible budget variances generate important feedback for managers but fail to reveal whether the sources of the variances are attributable to input prices, input quantities, or both.

Developing Unit Input Standards

Although flexible budget variances provide significant information for control, developing standards for input prices and input quantities allows a more detailed understanding of the sources of these variances. By developing standards for the products and services

OBJECTIVE ▶ 1
Describe how unit input standards are developed, and explain why standard costing systems are adopted.

being offered, managers can compare performance to them. **Price standards** specify how much should be paid for the quantity of the input to be used. **Quantity standards** specify how much of the input should be used per unit of output. The **unit standard cost** is defined as the product of these two standards: Standard price × Standard quantity ($SP \times SQ$).

For example, Peterson Company, a manufacturer of specialty ice creams and frozen yogurts, may decide that 750 grams of yogurt should be used for every litre of frozen yogurt produced (the quantity standard) and that the price to be paid for the yogurt should be $0.12 per 100 grams (the price standard). The standard cost of the yogurt per litre of frozen yogurt is then $0.90 ($0.12 × 7.5). The standard cost of yogurt per litre can be used to predict what the total cost of yogurt should be as the activity level varies; it thus becomes a flexible budget formula. If 20,000 litres of frozen yogurt are produced, the total expected cost of yogurt is $18,000 ($0.90 × 20,000); if 30,000 litres are produced, the total expected cost of yogurt is $27,000 ($0.90 × 30,000). Standard costs, therefore, facilitate budgeting, but the input price and quantity standards will also allow us to obtain a more detailed analysis of the flexible budget variance.

Establishing Standards

Developing standards requires significant input from a variety of sources. Historical experience, engineering studies, and input from operating personnel are three potential sources of quantitative standards. Historical experience should be used with caution because relying on input-output relationships from the past may perpetuate operating inefficiencies. Engineers and operating personnel can provide valuable insights concerning efficient levels of input quantities. Similar comments can be made about input price standards. Price standards are the joint responsibility of operations, purchasing, personnel, and accounting. Operations determine the quality of the inputs required; personnel and purchasing are responsible for acquiring the input quality requested at the lowest price. Market forces, trade unions, and other external forces limit the range of choices for price standards. In setting price standards, purchasing must consider discounts, freight, and quality; personnel must consider payroll taxes, fringe benefits, and qualifications. Accounting is responsible for recording price standards and for preparing reports that compare actual performance to the standard.

Standards are often classified as either *ideal* or *currently attainable*. **Ideal standards** demand maximum efficiency and can be achieved only if everything operates perfectly. No machine breakdowns, slack, or lack of skill (even momentarily) are allowed. **Currently attainable standards** can be achieved under efficient operating conditions. Allowance is made for normal breakdowns, interruptions, less than perfect skill, and so on. These standards are demanding but achievable. One cautionary observation about standards should be made. If standards are too tight and never achievable, workers become frustrated, and performance levels decline. However, challenging but achievable standards can lead to higher performance levels—particularly when the individuals subject to the standards have participated in their creation.

Kaizen Standards   **Kaizen standards** are continuous improvement standards. They reflect planned improvement and are a type of currently attainable standard. Kaizen standards by their very nature have a cost reduction focus and because of their emphasis on continuous improvement are constantly changing. (They are dynamic standards.) Kaizen standards are discussed in detail in Chapter 14. This chapter focuses on the more traditional standard cost system.

Standards and Activity-Based Costing   Standards also play an important role in activity-based systems. An activity's cost is determined by the amount of resources consumed by each activity. To avoid measuring the amount of resource consumption on an ongoing basis for literally hundreds of activities, standard consumption patterns are identified based on historical experience. The purpose of standards in this case is to facilitate cost assignments. Control is not an issue. Standards used in this sense were discussed in Chapter 6. Activity-based systems also use standards for control, where control is specifically defined as cost reduction. Activities are classified as either value-added or non-value-added. For each activity, the ideal output is identified and then efforts are made to reduce activity production to this ideal level. This activity-based approach to control is described in Chapter 14.

Usage of Standard Costing Systems

Standard costing systems are widely used. For example, according to one survey, 74 percent of the respondents were using a standard costing system, with the usage emphasis being placed on planning and control.[1] Several reasons for adopting a standard costing system include managing costs, improving planning and control, facilitating decision making, and facilitating product costing.

Cost Management   Standard costing allows managers to manage costs by establishing standards that reflect efficient operating conditions. Standards also help managers understand what needs to be done to improve current and future performance. Furthermore, for firms concerned with continuous improvement, kaizen standards are useful aids in achieving significant cost reductions.

Planning and Control   Standard costing systems enhance planning and control and improve performance measurement. Unit standards are a fundamental requirement for a flexible budgeting system, which is a key feature of a meaningful planning and control system. Budgetary control systems compare actual costs with budgeted costs by computing variances, the difference between the actual and planned costs for the actual level of activity. By developing unit price and quantity standards, an overall variance can be decomposed into a *price variance* and a *usage* or *efficiency variance*. By performing this decomposition, a manager has more and better information. For example, a manager can tell whether the variance is due to differences between planned prices and actual prices, to differences between planned usage and actual usage, or to both. Thus, in principle, the use of efficiency variances enhances operational control. Additionally, by breaking out the price variance, over which managers have little control, the system provides an improved measure of managerial efficiency.

Decision Making and Product Costing   Standard costing systems are useful for decision making and product costing. For example, standard costing systems provide readily available unit cost information that can be used for pricing decisions. This is particularly useful for companies that engage in extensive bidding and for companies that are paid on a cost-plus basis. Standard product costs are determined using quantity and price standards for direct materials, direct labour, and overhead. In contrast, a normal costing system predetermines overhead costs for the purpose of product costing but assigns direct materials and direct labour to products by using actual costs. An actual costing system assigns the actual costs of all three manufacturing inputs to products. Exhibit 9-1 summarizes these three cost assignment approaches.

Exhibit 9-1

Cost Assignment Approaches

	Manufacturing Costs		
	Direct Materials	**Direct Labour**	**Overhead**
Actual costing system	Actual	Actual	Actual
Normal costing system	Actual	Actual	Budgeted
Standard costing system	Standard	Standard	Standard

[1] Norwood Whittle, "Older and Wiser," *Management Accounting* (July/August 2000): 34–36.

Standard costing also simplifies product costing for firms in process industries. For example, if a process-costing system uses standard costing to assign product costs, there is no need to compute a unit cost for each equivalent unit-cost category. A standard unit cost would exist for direct materials, transferred-in materials, and conversion costs categories.[2] Usually, a standard process-costing system will follow the equivalent-unit calculation of the FIFO approach. That is, *current* equivalent units of work are calculated. By calculating current equivalent units of work, current actual production costs can be compared with standard costs (costs allowed for current production) for control purposes.

OBJECTIVE ▶ 2

Explain the purpose of a standard cost sheet.

Standard Cost Sheets

Standard costing systems can be used in both manufacturing and service organizations. Both products and services use inputs such as direct materials, direct labour, and overhead. Standard costing simply establishes price and quantity standards for these inputs regardless of whether the inputs are associated with tangible or intangible products. To illustrate standard costing for a service setting, consider a hospital. Hospital costing systems often use a homogeneous work unit called a relative value unit (*RVU*). An *RVU* measures the relative amount of time required to perform a procedure. Although the exact time to perform a particular test is not revealed, the relative time for performing two or more distinct tests has been computed. Thus, a test with an *RVU* of three will take three times as long to perform as a test with an *RVU* of one. Historical standards can be computed by dividing the variable direct labour costs of a hospital department by the number of *RVU*s performed by that department. This standard direct labour cost per *RVU* can then be multiplied by the *RVU*s of a given procedure to obtain the standard direct labour cost for that procedure.[3]

As indicated, standard costs are developed for direct materials, direct labour, and overhead used in producing a product or service. Using these costs, the **standard cost per unit** is computed. The **standard cost sheet** provides the detail underlying the standard unit cost. To illustrate, let us develop a standard cost sheet for a litre of deluxe strawberry frozen yogurt, produced by Peterson Company. The production of the strawberry frozen yogurt begins by creating two different mixtures. The first mixture consists of milk and gelatin. These two ingredients are mixed, heated, and then cooled. The second mixture consists of yogurt, cream, and crushed strawberries. The two mixtures are blended and mixed well. This final mixture is then poured into a one-litre container and frozen. The process is automated. Direct labour is used to operate the equipment and inspect the product for consistency and flavour. The standard cost sheet is given in Exhibit 9-2.

Five materials are used to produce the deluxe strawberry frozen yogurt: yogurt, strawberries, milk, cream, and gelatin. The container in which the yogurt is placed is also classified as a direct material. Direct labour consists of machine operators (who also inspect). Variable overhead, applied using direct labour hours, is made up of three costs: gas (used in cooking), electricity (used to operate the equipment), and water (used for cleaning). Fixed overhead, also applied using direct labour hours, consists of salaries, depreciation, taxes, and insurance. Notice that 1,150 grams of liquids (yogurt, milk, and cream) are used to produce a litre of frozen yogurt. This extra input is needed for two reasons. First, some liquid is lost through evaporation. Second, Peterson wants slightly more than 1,000 grams of frozen yogurt placed in each container to ensure customer satisfaction and to meet government requirements for the government's office of weights and measures.

Exhibit 9-2 reveals other important insights. The standard usage for variable and fixed overhead is tied to the direct labour standards. For variable overhead, the rate is

---

[2] If you have not read the chapter on process costing (Chapter 5), the discussion on the merits of standard costing will not be as meaningful. However, the point being made is still relevant. Standard costing can produce useful computational savings.

[3] For an interesting description of how historical labour standards can be developed in a hospital setting, see Richard D. McDermott, Kevin D. Stocks, and Joan Ogden, *Code Blue* (Syracuse, Utah: Traemus Books, 2000): 212–221.

Exhibit 9-2

Standard Cost Sheet for Deluxe Strawberry Frozen Yogurt

Description	Standard Price (100 grams)		Standard Usage		Standard Cost	Subtotal
Direct materials:						
Yogurt	$ 0.12	×	750 g	=	$0.90	
Strawberries	0.05	×	400 g	=	0.20	
Milk	0.10	×	300 g	=	0.30	
Cream	0.24	×	100 g	=	0.24	
Gelatin	0.04	×	50 g	=	0.02	
Container	0.06	×	1	=	0.06	
Total direct materials						$1.72
Direct labour:						
Machine operators	16.00	×	0.01 hr	=	$0.16	
Total direct labour						0.16
Overhead:						
Variable overhead	12.00	×	0.01 hr	=	$0.12	
Fixed overhead	40.00	×	0.01 hr	=	0.40	
Total overhead						0.52
Total standard unit cost						$2.40

$12.00 per direct labour hour. Since one litre of frozen yogurt uses 0.01 direct labour hour, the variable overhead cost assigned to a litre is $0.12 ($12.00 × 0.01). For fixed overhead, the rate is $40 per direct labour hour, making the fixed overhead cost per litre $0.40 ($40 × 0.01). Using direct labour hours as the only driver to assign overhead reveals that Peterson uses a traditional, volume-based cost accounting system.

The standard cost sheet reveals the quantity of each input that should be used to produce one unit of output. The unit quantity standards can be used to compute the total amount of inputs allowed for the actual output. This computation is an essential component in computing efficiency variances. A manager should be able to compute the **standard quantity of materials allowed** (SQ) and the **standard hours allowed** (SH) for the actual output. This computation must be done for every class of direct material and for every class of direct labour. Cornerstone 9-1 shows how and why the standard amounts for actual production are computed.

Variance Analysis for Direct Materials and Direct Labour

OBJECTIVE ▶ 3
Compute and journalize the direct materials and direct labour variances, and explain how they are used for control.

A flexible budget can be used to identify the direct material or direct labour input costs that should have been incurred for the actual level of activity. This planned cost is obtained by multiplying the amount of input allowed for the actual output by the standard unit price. Letting SP be the standard unit price of an input and SQ the standard quantity of inputs allowed for the actual output, the planned or budgeted input cost is $SP \times SQ$. The actual input cost is $AP \times AQ$, where AP is the actual price per unit of the input, and AQ is the actual quantity of input used. The **total budget variance** is the difference between the actual cost of the input and its standard cost:

$$\text{Total budget variance} = (AP \times AQ) - (SP \times SQ)$$

The total budget variance measures the difference between the actual cost of direct materials and direct labour and their budgeted costs for the actual level of activity. While it is interesting to know whether or not the actual costs were as planned, the detail of the standard cost card allows managers to determine what aspects of

**CORNERSTONE
9-1**

The HOW and WHY of Computing Standard Quantities Allowed (*SQ* and *SH*)

Information:
During the first week of April, Peterson Company produced 20,000 litres of deluxe strawberry frozen yogurt. Exhibit 9-2 shows that the unit quantity standard is 750 grams of yogurt per litre and that the unit standard is 0.01 direct labour hour per litre.

Why:
Unit standards must be converted to the standard quantities of inputs allowed for actual production in order to determine how much of each resource is expected to be used. Managers can use the standard quantities allowed in planning (to estimate how much will be required for planned production) or in control (to compare with the actual quantities used).

Required:
1. Calculate the grams of yogurt that should have been used (*SQ*) for the production of 20,000 litres of frozen yogurt.
2. Calculate the hours of direct labour that should have been used (*SH*) for the production of 20,000 litres of frozen yogurt.
3. **What if** 22,000 litres of frozen yogurt had actually been produced in the first week of April? Would the standard quantities of yogurt (in grams) and of direct labour hours be higher or lower than the amounts calculated in Requirements 1 and 2? What would the new standard quantities be?

Solution:
1. Yogurt allowed:

    SQ = Unit quantity standard $\times$ Actual output

    = 750 $\times$ 20,000

    = 15,000,000 grams

2. Operator hours allowed:

    SH = Unit labour standard $\times$ Actual output

    = 0.01 $\times$ 20,000

    = 200 direct labour hours

3. If 22,000 litres were produced instead of 20,000, the standard quantities allowed would be higher, since the production of more frozen yogurt takes more yogurt and more direct labour hours. The *SQ* for yogurt would be 16,500,000 grams (750 $\times$ 22,000), and the *SH* for direct labour hours would be 220 hours (0.01 $\times$ 22,000).

total cost were different than planned. The next sections discuss the way that the total variances can be decomposed into the price and usage variances for direct materials and the rate and efficiency variances for direct labour.

Calculating the Direct Materials Price Variance and Direct Materials Usage Variance

The total budget variance can be broken down into price and usage variances. **Price (rate) variance** is the difference between the actual and standard unit prices of an input multiplied by the actual quantity of inputs. **Usage (efficiency) variance** is the difference between the actual and standard quantity of inputs multiplied by the standard unit price of the input. An **unfavourable (U) variance** occurs whenever actual prices or usage of inputs are greater than standard prices or usage. When the opposite occurs, a **favourable (F) variance** is obtained. Every nonzero variance must be tagged as favourable or unfavourable. This lets the manager know the direction of the deviation from standard.

The price and usage variances can be computed using formulas or a graphical, three-pronged approach. The choice is up to the individual; some people find the formulas more meaningful, others appreciate the graphical approach. Both approaches will be illustrated in the Cornerstones. First, we will set up the formulas for the direct materials price and usage variances.

Let:

AP = Actual price per unit
SP = Standard price per unit
AQ = Actual quantity of direct material used in production
SQ = Standard quantity
MPV = Materials price variance
MUV = Materials usage variance

The **direct materials price variance (MPV)** is the difference between what was actually paid for direct materials and what would have been paid for the actual quantity bought if it had been bought at the standard price. Thus, the materials price variance is:

$$MPV = (AP \times AQ) - (SP \times AQ)$$

or, factoring, we have:

$$MPV = (AP - SP)AQ$$

The MPV is calculated as the difference between actual and standard prices multiplied by the actual quantity. If the actual price is greater than standard, the MPV is U (unfavourable). If the actual price is less than the standard price, the MPV is F (favourable).

The **direct materials usage variance (MUV)** is the difference between the amount of materials actually used and what should have been used for the actual quantity of units produced multiplied by the standard price. Thus, the materials usage variance is:

$$MUV = (SP \times AQ) - (SP \times SQ)$$

or, factoring, we have:

$$MUV = (AQ - SQ)SP$$

The MUV is quickly calculated as the difference between actual and standard amounts of direct materials multiplied by the standard price. If the actual quantity is greater than standard, the MUV is U (unfavourable). If the actual quantity is less than the standard quantity, the MUV is F (favourable). Cornerstone 9-2 shows the how and why of calculating the direct materials price and usage variances.

Cornerstone 9-2 calculated only the MPV and MUV for one input, yogurt, in order to simplify the example. In actuality, Peterson would calculate these two variances for each type of direct material used.

Timing of the Price Variance Computation

The direct materials price variance can be computed at one of two points: (1) when the direct materials are issued for use in production or (2) when they are purchased. Computing the price variance at the point of purchase is preferable. It is better to have information on variances earlier rather than later. The more timely the information, the more likely proper managerial action can be taken. Old information is often useless information. Direct materials may sit in inventory for weeks or months before they are needed in production. By the time the direct materials price variance is computed, signalling a problem, it may be too late to take corrective action. Or, even if corrective action is still possible, the delay may cost the company thousands of dollars.

If the direct materials price variance is computed at the point of purchase, then AQ needs to be redefined as the actual quantity of direct materials *purchased*, rather

The **variable overhead efficiency variance** measures the change in variable overhead consumption that occurs because of efficient (or inefficient) use of direct labour. The efficiency variance is computed using the following formula:

$$\text{Variable overhead efficiency variance} = (SVOR \times AH) - (SVOR \times SH)$$
$$= (AH - SH)SVOR$$

Cornerstone 9-7 shows the how and why of calculating the variable overhead spending and efficiency variances. It uses both the formula approach explained above and the three-pronged graphical approach.

**CORNERSTONE
9-7**

The HOW and WHY of Computing the Variable Overhead Spending Variance and the Variable Overhead Efficiency Variance

Information:
Peterson Company provided the following information for the month of May:

Variable overhead rate (standard)	$12.00 per direct labour hour[a]
Actual variable overhead costs	$16,120
Actual hours worked	1,300
Litres of deluxe strawberry frozen yogurt produced	120,000
Hours allowed for actual production	1,200[b]
Applied variable overhead	$14,400[c]

[a] See Exhibit 9-2 for the standard cost card.
[b] 0.01 direct labour hour at standard × 120,000 actual litres produced (See Exhibit 9-2 for unit standards and prices.)
[c] $12.00 × 1,200 (Overhead is applied using standard hours allowed.)

Why:
The total variable overhead variance is broken into the variable overhead spending and efficiency variances. The spending variance shows the difference between the actual variable overhead rate and the standard variable overhead rate. The variable overhead efficiency variance shows the impact of a difference between actual hours worked and the standard hours that should have been worked on the variable overhead. These variances indicate where managers could begin to investigate overhead variances.

Required:
1. Calculate the variable overhead spending variance using the formula approach.
2. Calculate the variable overhead efficiency variance using the formula approach.
3. Calculate the variable overhead spending variance and variable overhead efficiency variance using the three-pronged graphical approach.
4. **What if** only 1,190 direct labour hours were actually worked in May? What impact would that have had on the variable overhead spending variance? On the variable overhead efficiency variance?

Solution:
1. Variable overhead spending variance $= (AVOR - SVOR)AH$
$$= [(\$16,120/1,300) - \$12.00]1,300$$
$$= (\$12.40 - \$12.00) \times 1,300 = \$520\ U$$

2. Variable overhead efficiency variance $= (AH - SH)SVOR$

    $$= (1{,}300 - 1{,}200)\$12 = \$1{,}200 \text{ U}$$

3.

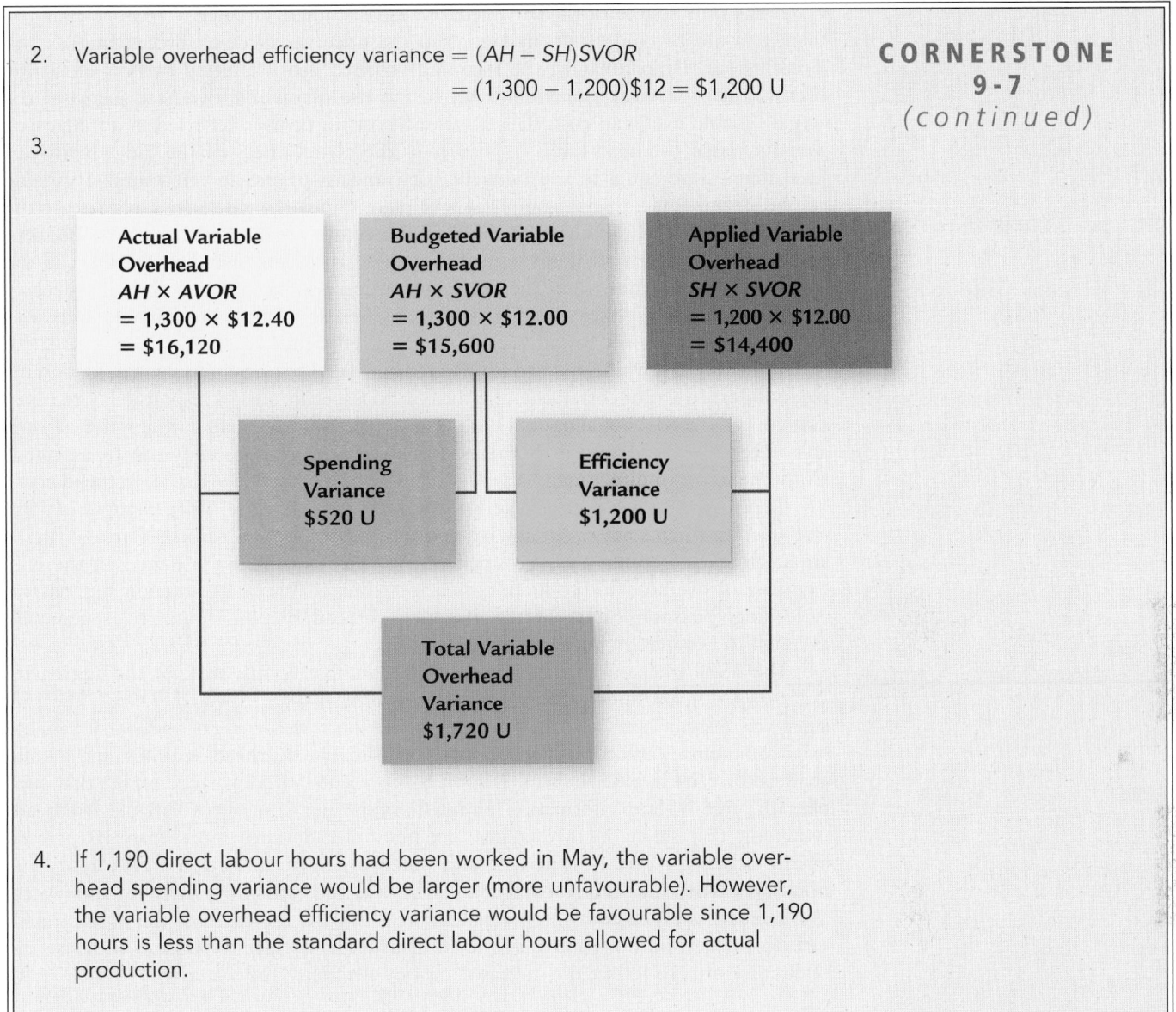

4. If 1,190 direct labour hours had been worked in May, the variable overhead spending variance would be larger (more unfavourable). However, the variable overhead efficiency variance would be favourable since 1,190 hours is less than the standard direct labour hours allowed for actual production.

Interpreting the Variable Overhead Variances

The variable overhead spending variance and the variable overhead efficiency variance give managers information they can use in controlling costs.

Interpreting the Variable Overhead Spending Variance

The variable overhead spending variance is similar to the price variances of direct materials and direct labour, although there are some conceptual differences. Variable overhead is not a homogeneous input—it is made up of a large number of individual items such as indirect materials, indirect labour, electricity, maintenance, and so on. The standard variable overhead rate represents the weighted cost per direct labour hour that should be incurred for all variable overhead items. The difference between what should have been spent per hour and what actually was spent per hour is a type of price variance.

A variable overhead spending variance can arise because prices for individual variable overhead items have increased or decreased. Assume, for the moment, that the price changes of individual overhead items are the only cause of the spending variance. If the spending variance is unfavourable, then price increases for individual variable overhead items are the cause; if the spending variance is favourable, then price decreases dominate.

If the only source of the variable overhead spending variance were price changes, then it would be completely analogous to the price variances of direct materials and direct labour. Unfortunately, the spending variance also is affected by how efficiently overhead is used. Waste or inefficiency in the use of variable overhead increases the actual variable overhead cost. This increased cost, in turn, is reflected in an increased actual variable overhead rate. Thus, even if the actual prices of the individual overhead items were equal to the budgeted or standard prices, an unfavourable variable overhead spending variance could still take place. Similarly, efficiency can decrease the actual variable overhead cost and decrease the actual variable overhead rate. Efficient use of variable overhead items contributes to a favourable spending variance. If the waste effect dominates, then the net contribution will be unfavourable; if efficiency dominates, then the net contribution is favourable. Thus, the variable overhead spending variance is the result of both price and efficiency.

Many variable overhead items are affected by several responsibility centres. For example, utilities are a joint cost. Assigning the cost to a specific area of responsibility requires that cost be traced—not allocated—to the area. To the extent that consumption of variable overhead can be traced to a responsibility centre, responsibility can be assigned. Consumption of indirect materials is an example of a traceable variable overhead cost.

Controllability is a prerequisite for assigning responsibility. Price changes of variable overhead items are essentially beyond the control of supervisors. If price changes are small (as they often are), the spending variance is primarily a matter of the efficient use of overhead in production, which is controllable by production supervisors. Accordingly, responsibility for the variable overhead spending variance is generally assigned to production departments.

The $520 unfavourable spending variance simply reveals that, in the aggregate, Peterson Company spent more on variable overhead than expected. Even if the variance was insignificant, it reveals nothing about how well costs of individual variable overhead items were controlled. Control of variable overhead requires line-by-line analysis for each individual item. Exhibit 9-4 presents a performance report that supplies the line-by-line information essential for proper control of variable overhead. Assuming that Peterson investigates any item that deviates more than 10 percent from budget, the cost of electricity and water would be investigated. The investigation reveals that the utility companies increased the rates for electricity and water. The increase is expected to be permanent. In this case, the cause of the unfavourable variances is beyond the control of the company. The correct response is to revise the budget formula to reflect the increased cost of electricity and water.

Interpreting the Variable Overhead Efficiency Variance   The variable overhead efficiency variance is directly related to the direct labour efficiency or usage variance. If variable overhead is truly driven by direct labour hours, then like the direct labour usage variance, the variable overhead efficiency variance is caused by efficient or

Exhibit 9-4

Variable Overhead Spending Variance by Item

		Peterson Company Performance Report For the Month Ended May 31, 2013		
	Cost Formula[a]	Actual Costs	Budget[b]	Spending Variance
Natural gas	$ 7.60	$ 9,640	$ 9,880	$240 F
Electricity	4.00	5,850	5,200	650 U
Water	0.40	630	520	110 U
Total	$12.00	$16,120	$15,600	$520 U

[a] Per direct labour hour.
[b] The budget allowance is computed using the cost formula and 1,300 actual direct labour hours.

inefficient use of direct labour. If more (or fewer) direct labour hours are used than the standard calls for, then the total variable overhead cost will increase (or decrease). The validity of the measure depends on the validity of the relationship between variable overhead costs and direct labour hours. In other words, do variable overhead costs *really* change in proportion to changes in direct labour hours? If so, responsibility for the variable overhead efficiency variance should be assigned to the individual who has responsibility for the use of direct labour: the production manager.

The reasons for the unfavourable variable overhead efficiency variance are generally the same as those offered for the unfavourable labour usage variance. For example, some of the variance can be explained by the fact that overtime hours were used during the first week to make up for a bad batch of yogurt. The remaining deficiency was caused by the use of new employees who took longer to carry out tasks because of their lack of experience.

More information concerning the effect of direct labour usage on variable overhead is available in a line-by-line analysis of individual variable overhead items. This can be accomplished by comparing the budget allowance for the actual hours used with the budget allowance for the standard hours allowed for each item. A performance report that makes this comparison for all variable overhead costs is shown in Exhibit 9-5. From Exhibit 9-5, we can see that the cost of natural gas is affected most by inefficient use of direct labour. For example, inexperienced labourers may heat the mix of gelatin and milk longer than is really needed, thus using more gas.

The column labelled *Budget for Standard Hours* gives the amount that should have been spent on variable overhead for the actual output. The total of all items in this column is the applied variable overhead, the amount assigned to production in a standard costing system. Note that in a standard costing system, variable overhead is applied using the hours allowed for the actual output (*SH*), while in normal costing, variable overhead is applied using actual hours. Although not shown in Exhibit 9-5, the difference between actual costs and this column is the total variable overhead variance (underapplied by $1,720). Thus, the underapplied variable overhead variance is the sum of the spending and efficiency variances.

Four-Variance Analysis: The Two Fixed Overhead Variances

The total fixed overhead variance is the difference between the actual fixed overhead and the applied fixed overhead. To help managers understand why fixed overhead may differ from applied fixed overhead, the total variance can be broken down into two variances: the fixed overhead spending variance and the fixed overhead volume variance.

Exhibit 9-5

Variable Overhead Spending and Efficiency Variances by Item

Peterson Company
Performance Report
For the Month Ended May 31, 2013

	Cost Formula[a]	Actual Costs	Budget[b]	Spending Variance	Budget for Standard Hours[c]	Efficiency Variance
Natural gas	$ 7.60	$ 9,640	$ 9,880	$240 F	$ 9,120	$ 760 U
Electricity	4.00	5,850	5,200	650 U	4,800	400 U
Water	0.40	630	520	110 U	480	40 U
Total	$12.00	$16,120	$15,600	$520 U	$14,400	$1,200 U

[a] Per direct labour hour.
[b] The budget allowance is computed using the cost formula and 1,300 actual direct labour hours.
[c] Standard hours for actual production equal 1,200 (0.01 hours × 120,000 litres).

Calculating the Fixed Overhead Spending Variance and Fixed Overhead Volume Variance

The **fixed overhead spending variance** is defined as the difference between the actual fixed overhead and the budgeted fixed overhead. If less is spent on fixed overhead items than was budgeted, the spending variance is favourable, and vice versa. The formula for computing the fixed overhead variance follows ($AFOH$ = Actual fixed overhead and $BFOH$ = Budgeted fixed overhead):

Fixed overhead spending variance = AFOH − BFOH

Any difference between actual fixed overhead and budgeted fixed overhead must be due to a change in the amount of fixed overhead—some item has increased or decreased vis-à-vis what was expected. This difference is called a spending variance.

The **fixed overhead volume variance** is the difference between budgeted fixed overhead and applied fixed overhead.

Fixed overhead volume variance = Budgeted fixed overhead − Applied fixed overhead

Keep in mind that the budgeted fixed overhead was determined in advance of the year, and that the fixed overhead rate, used to apply fixed overhead to production, was calculated then as well. Thus, the fixed overhead rate is the rate that it would take to apply fixed overhead to production *assuming that the actual production equals the budgeted production.* For example, if 2,000 units are budgeted, each unit taking three direct labour hours, then 6,000 direct labour hours are budgeted. If the budgeted fixed overhead is $24,000, then the fixed overhead rate would be $4. If 2,000 units are actually produced, then $24,000 will be applied to production ($4 × 2,000 units × 3 direct labour hours). There is no volume variance. Suppose instead that 2,100 units are actually produced. Then $25,200 ($4 × 2,100 units × 3 direct labour hours) is applied, an amount that is $1,200 higher than the budgeted fixed overhead. This difference is solely due to the increased production. We refer to this variance as "favourable" and a variance in which the actual production is less than budgeted as "unfavourable." As a rule, if actual production is less than budgeted production, the volume variance will be unfavourable; if actual production is more than budgeted production, the volume variance will be favourable.

Cornerstone 9-8 shows the how and why of calculating the fixed overhead spending and volume variances.

Interpreting the Fixed Overhead Variances

As was the case with the variable overhead variances, managers can gain useful information from the fixed overhead variances. However, due to the fixed nature of the costs involved, managers find that it is useful to spend time looking at fixed overhead on an item-by-item basis.

Interpreting the Fixed Overhead Spending Variance   Fixed overhead is made up of a number of individual items such as salaries, depreciation, taxes, and insurance. Many fixed overhead items—long-run investments, for instance—are not subject to change in the short run; consequently, fixed overhead costs are often beyond the immediate control of management. Since many fixed overhead costs are affected primarily by long-run decisions, not by changes in production levels, the budget variance is usually small. For example, depreciation, salaries, taxes, and insurance costs are not likely to be much different than planned.

Because fixed overhead is made up of many individual items, a line-by-line comparison of budgeted costs with actual costs provides more information concerning the causes of the spending variance. Exhibit 9-6 provides such a report. The report reveals that the fixed overhead spending variance is essentially in line with expectations. The fixed overhead spending variances, both on a line-item basis and in the aggregate, are relatively small (all less than 10 percent of the budgeted costs).

The HOW and WHY of Computing the Fixed Overhead Spending Variance and the Fixed Overhead Volume Variance

CORNERSTONE 9-8

Information:
Peterson Company provided the following information for the month of May:

Budgeted/planned items for May:	
Budgeted fixed overhead	$40,000
Expected production in litres of frozen yogurt	100,000
Expected activity in direct labour hours (0.01 × 100,000)	1,000 direct labour hours
Standard fixed overhead rate ($40,000/1,000)	$40 per direct labour hour
Actual results for May:	
Actual production of yogurt in litres	120,000 litres
Actual fixed overhead cost	$40,500
Standard hours allowed for actual production (0.01 × 120,000)	1,200 direct labour hours

Why:
The total fixed overhead variance is broken into the fixed overhead spending and volume variances. The spending variance shows the difference between the actual fixed overhead and the budgeted fixed overhead. The volume variance shows the impact of a difference between actual units produced and the budgeted units. These variances indicate where managers could begin to investigate overhead variances.

Required:
1. Calculate the fixed overhead spending variance using the formula approach.
2. Calculate the fixed overhead volume variance using the formula approach.
3. Calculate the fixed overhead spending variance and fixed overhead volume variance using the three-pronged graphical approach.
4. **What if** only 95,000 litres of frozen yogurt had actually been produced in May? What impact would that have had on the fixed overhead spending variance? On the fixed overhead volume variance?

Solution:
1. Fixed overhead spending variance = Actual fixed overhead
    − Budgeted fixed overhead
    = $40,500 − $40,000
    = $500 U

2. Fixed overhead volume variance = Budgeted fixed overhead − Applied fixed overhead
    = Budgeted fixed overhead − (Fixed overhead rate × *SH*)
    = $40,000 − ($40 × 1,200) = $8,000 F

CORNERSTONE 9-8 *(continued)*

3.

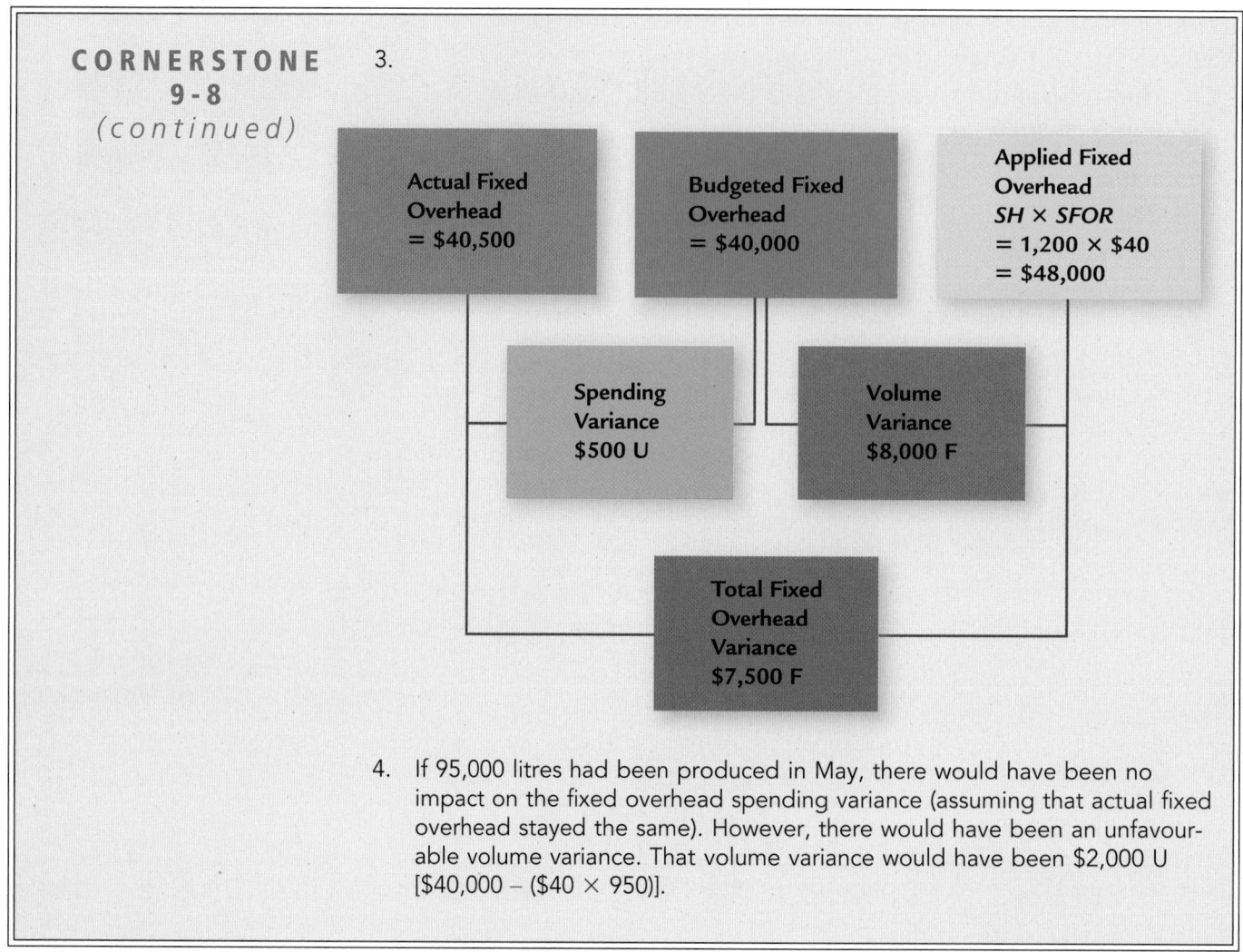

4. If 95,000 litres had been produced in May, there would have been no impact on the fixed overhead spending variance (assuming that actual fixed overhead stayed the same). However, there would have been an unfavourable volume variance. That volume variance would have been $2,000 U [$40,000 − ($40 × 950)].

Exhibit 9-6

Fixed Overhead Spending Variance by Item

	Peterson Company Performance Report For the Month Ended May 31, 2013		
	Actual Costs	**Budgeted Cost**	**Spending Variance**
Depreciation	$10,000	$10,000	$ 0
Salaries	26,300	25,800	500 U
Taxes	2,200	1,200	1,000 U
Insurance	2,000	3,000	1,000 F
Total	$40,500	$40,000	$ 500 U

Interpreting the Fixed Overhead Volume Variance   The volume variance occurs because the actual output differs from the budgeted output or volume. At the beginning of the month, if management had expected 120,000 litres with 1,200 standard hours, the volume variance would not have existed. In this view, the volume variance is seen as prediction error—a measure of the inability of management to select the correct volume over which to spread fixed overhead.

If, however, the budgeted volume represented the amount that management believed *could* be produced and sold, the volume variance conveys more significant information. If the actual volume is more than the budgeted volume, the volume variance signals that a gain has occurred (relative to expectations). That gain is not equivalent, however, to the dollar value of the volume variance. The gain is equal to the increase in contribution margin on the extra units produced and sold. However, the volume variance is positively correlated with the gain. Suppose that the contribution margin per standard direct labour hour is $100. By producing 120,000 litres of frozen yogurt instead of 100,000 litres, the company gained sales of 20,000 litres. This is equivalent to 200 hours (0.01 × 20,000). At $100 per hour, the gain is $20,000 ($100 × 200). The favourable volume variance of $4,000 signals this gain but understates it. In this sense, the volume variance is a measure of this year's *planned* utilization of capacity.

On the other hand, if *practical capacity* is used as the budgeted volume, then the volume variance is a direct measure of capacity utilization. Practical capacity measures the most that can be produced under efficient operating conditions (and, thus, represents the productive capacity the firm has acquired). The difference between available hours of production and actual hours is a measure of underutilization, and when multiplied by the standard fixed overhead rate, the volume variance becomes a measure of the cost of underutilization of capacity. This is similar in concept to the activity capacity utilization measure described in Chapter 2. The principal difference is that the fixed overhead rate used to measure the cost of unused capacity contains more than the cost of acquiring the productive capacity. Fixed overhead is made up of many costs incurred for reasons other than obtaining productive capacity (e.g., the salaries of the plant supervisor, janitors, and industrial engineers).

Assuming that volume variance measures capacity utilization implies that the general responsibility for this variance should be assigned to the production department. At times, however, investigation into the reasons for a significant volume variance may reveal the cause to be factors beyond the control of production. Then, specific responsibility may be assigned elsewhere. For example, if purchasing acquires direct materials of lower quality than usual, significant rework time may result, causing lower production and an unfavourable volume variance. In this case, responsibility for the variance rests with purchasing, not production.

Accounting for Overhead Variances

Overhead is applied to production by debiting Work in Process and crediting variable and fixed overhead control accounts. The amount assigned is simply the respective overhead rates multiplied by the standard hours allowed for actual production. The actual overhead is accumulated on the debit side of the overhead control accounts. Periodically (e.g., monthly), overhead variance reports are prepared. At the end of the year, the applied variable and fixed overhead costs and the actual fixed overhead costs are closed out and the variances isolated. The overhead variances are then disposed of by closing them to Cost of Goods Sold if they are not material or by prorating them among Work in Process, Finished Goods, and Cost of Goods Sold if they are material. We will use the May transactions for Peterson Company to illustrate the process that would occur at the end of the year. Essentially, we are assuming that the May transactions reflect an entire year for illustrative purposes.

To recognize the incurrence of actual overhead, the following entry is needed:

Variable Overhead Control	16,120	
Fixed Overhead Control	40,500	
Various Accounts		56,620

To assign overhead to production, we have the following entry:

Work in Process	62,400	
Variable Overhead Control		14,400
Fixed Overhead Control		48,000

To recognize the variances, the following entry is needed:

Fixed Overhead Control	7,500	
Variable Overhead Spending Variance	520	
Variable Overhead Efficiency Variance	1,200	
Fixed Overhead Spending Variance	500	
Variable Overhead Control		1,720
Fixed Overhead Volume Variance		8,000

Finally, to close out the variances to Cost of Goods Sold, we would have the following entries. (Entries assume that variances are immaterial.)

Fixed Overhead Volume Variance	8,000	
Cost of Goods Sold		8,000
Cost of Goods Sold	2,220	
Variable Overhead Spending Variance		520
Variable Overhead Efficiency Variance		1,200
Fixed Overhead Spending Variance		500

Some students may find it easier to follow the flow of these transactions through the accounts by using T-accounts. In the end, all overhead and overhead variance accounts carry a zero balance.

Two- and Three-Variance Analysis Methods

One drawback of the four-variance method is that it requires a company to identify the actual variable and fixed costs as well as budgeted rates and costs. For companies that wish to avoid the need to track actual variable and fixed costs, the two- and three-variance methods can be used.

The two- and three-variance analyses do not require knowledge of actual variable and actual fixed overhead. These methods provide less detail and, thus, less information. We will simply present the method of computation for the two forms of analysis. The four-variance method is recommended over these two approaches. The May data for Peterson Company will be used to illustrate the two methods with the assumption that only the total actual overhead is known: $56,620.

Two-Variance Analysis
The two-variance analysis is shown in Exhibit 9-7. (*SVOR* designates the standard variable overhead rate.) Several points should be made relative to the four-variance analysis shown in Cornerstones 9-7 and 9-8. First, the total variance is the sum of the total fixed and variable overhead variances. Second, the volume variance is the same as that of the four-variance method. Notice that in the computation of the volume variance, the applied variable overhead term, *SVOR* × *SH*, is common to the middle and right prongs of the diagram. Thus, when the right number is subtracted from the left number, we are left with budgeted fixed overhead minus applied fixed overhead, which is the fixed overhead volume variance. Third, the budget variance is the sum of the spending and efficiency variances of the four-variance method ($520 U + $500 U + $1,200 U = $2,220 U). As indicated, the two-variance method sacrifices a lot of information.

Three-Variance Analysis
The three-variance analysis is shown in Exhibit 9-8. Again, some observations can be made about this method relative to the four-variance method. First, the total variance is again the sum of the total variable and fixed overhead variances. Second, the spending variance is the sum of the variable and fixed overhead spending variances. The variable overhead efficiency and the fixed overhead volume variances are the same. The three-variance method also illustrates that the budget variance of the two-variance method breaks down into spending and efficiency variances.

Exhibit 9-7

Two-Variance Analysis: Peterson Company

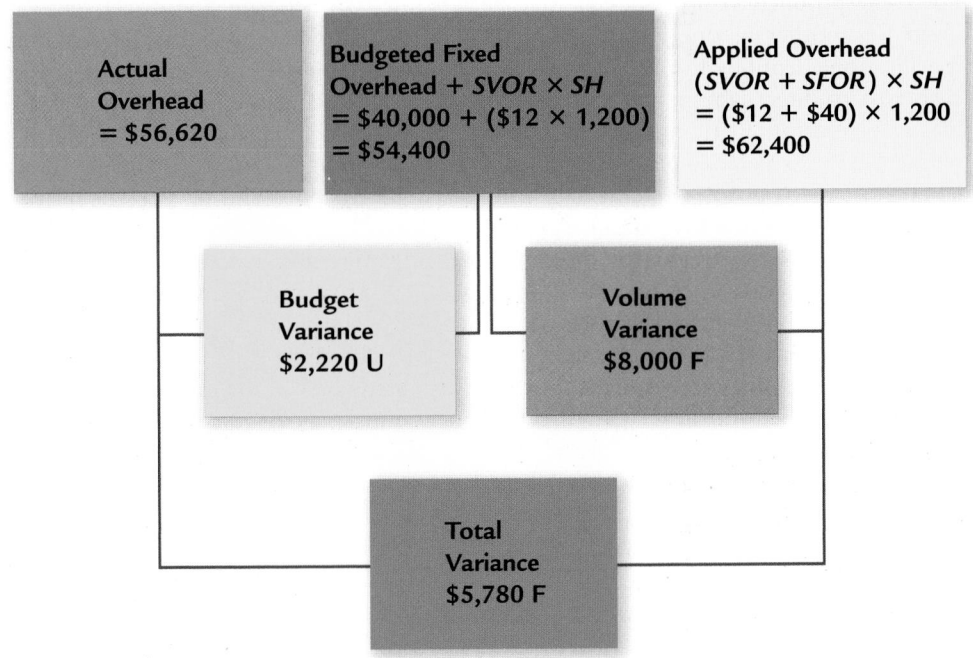

Three-Variance Analysis: Peterson Company

Exhibit 9-8

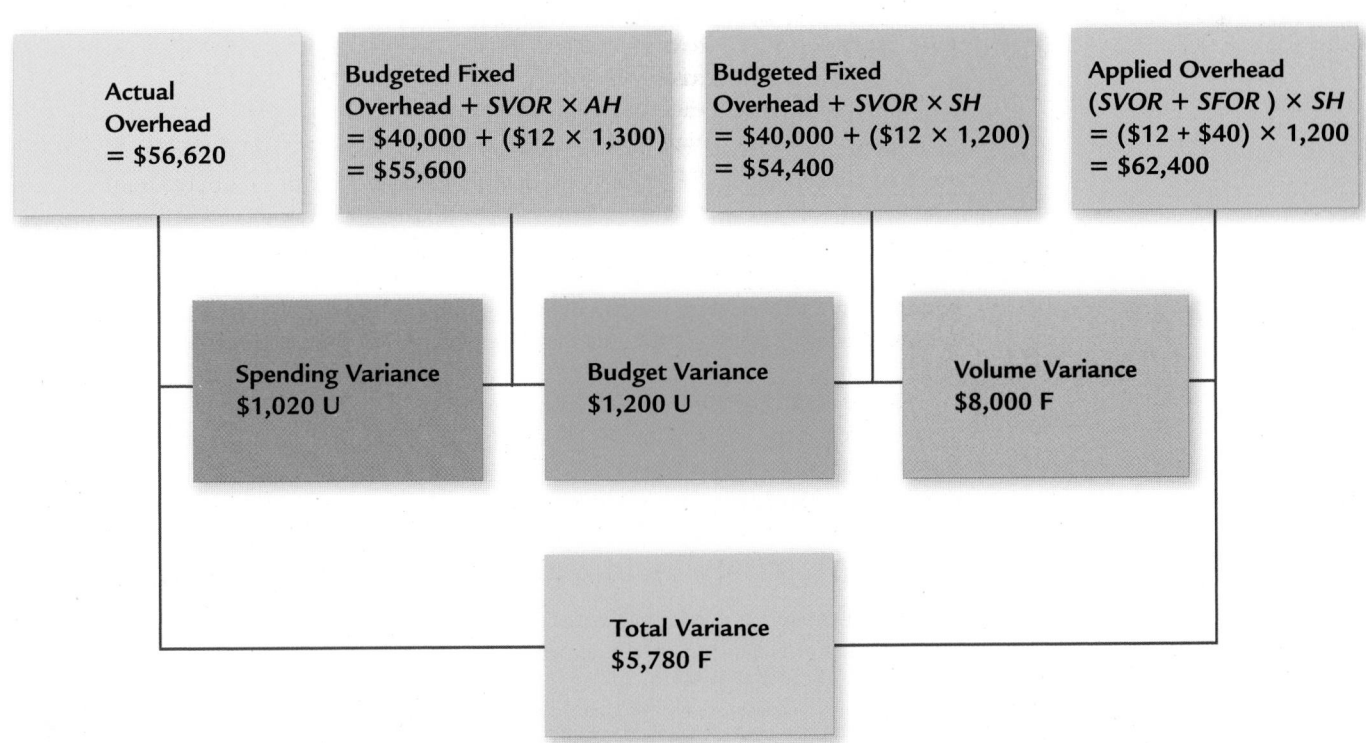

Mix and Yield Variances: Materials and Labour

For some production processes, it may be possible to substitute one direct material input for another or one type of direct labour for another. Usually, a standard mix specification identifies the proportion of each direct material and the proportion of each type of direct labour that should be used for producing the product. For example, in producing an orange-pineapple fruit drink, the standard direct materials mix may call for 30 percent pineapple and 70 percent orange, and the standard direct labour mix may call for 33 percent of fruit preparation labour and 67 percent of fruit processing labour. Clearly, within reason, it is possible to make input substitutions. Substituting direct materials or direct labour, however, may produce *mix* and *yield* variances. A **mix variance** is created whenever the actual mix of inputs differs from the standard mix. A **yield variance** occurs whenever the actual yield (output) differs from the standard yield. For example, a basic recipe for chocolate chip cookies says it will make three dozen two-inch cookies. But many of us who have baked these cookies know that you never get three dozen two-inch cookies because some of the cookie dough "disappears" before it ever gets to the baking sheet. This difference between the number of cookies you should get and the number you actually do get is the yield variance. For direct materials, the sum of the mix and yield variances equals the direct materials usage variance; for direct labour, the sum is the direct labour efficiency variance.

Direct Materials Mix Variance

The mix variance is the difference in the standard cost of the actual mix of inputs used and the standard cost of the mix of inputs that should have been used. Let *SM* be the quantity of each input that should have been used given the total actual input quantity. This quantity is computed as follows for each direct material input:

$$SM = \text{Standard mix proportion} \times \text{Total actual input quantity}$$

The standard mix quantity is computed for each input.[6] The total actual input quantity is the sum of the quantities of all inputs put into production.

Given *SM*, the mix variance is computed as follows:

$$\text{Mix variance} = \Sigma(AQi - SMi)SPi \qquad (9.1)$$

Basically, the mix variance is the sum of the differences between the actual amount of each input and its standard mix amount, multiplied by the standard price. If relatively more of a more expensive input is used, the mix variance will be unfavourable. If relatively more of a less expensive input is used, the mix variance will be favourable. Cornerstone 9-9 shows the how and why of calculating the mix variance.

Notice in Cornerstone 9-9 that the mix variance is unfavourable. This occurs because more almonds are used than are called for in the standard mix, and almonds are a more expensive input. If the mix variance is material, then an investigation should be undertaken to determine the cause of the variance so that corrective action can be taken.

Notice that Cornerstone 9-9 can also be used to calculate the mix variance for inputs other than direct materials. For example, there may be different types of direct labour needed to make a product. It is possible to substitute relatively more of a more or less expensive type of direct labour and obtain a direct labour mix variance.

Direct Materials Yield Variance

The direct materials yield variance is designed to show the extent to which the amount of input resulted in the expected amount of output. Using the standard mix

[6] The standard mix amounts are not the standard quantities allowed for actual output. The total standard quantity allowed is computed by dividing the actual yield by the standard yield ratio. The total standard input allowed is then multiplied by the standard mix ratios to compute the quantity of each direct material input that should have been used of the actual output. Alternatively, the unit direct material standards can be developed by dividing the standard input mix quantity by the standard yield. Multiplying the unit standards by the actual yield will also produce SQ for each input.

The HOW and WHY of Computing the Mix Variance

CORNERSTONE
9 - 9

Information:

Malcom Nut Company produces mixed nuts using peanuts and almonds. Malcom developed the following standard mix for producing 120 kilograms of mixed nuts. (Almonds and peanuts are purchased in the shell and processed.)

Direct Material	Mix	Mix Proportion	SP	Standard Cost
Peanuts	128 kg	0.80	$0.50	$64
Almonds	32	0.20	1.00	32
Total	160 kg			$96

Malcom put a batch of 1,600 kilograms of nuts into process. Of the total, 1,120 kilograms were peanuts, and the remaining 480 kilograms were almonds. The actual yield was 1,300 kilograms.

Why:

The materials usage variance tells managers whether total materials are in accordance with standards. The mix variance gives further information about materials usage since different materials have different standard prices.

Required:

1. Calculate the standard mix (*SM*) in kilograms for peanuts and for almonds.
2. Calculate the mix variance.
3. Calculate the actual proportion used of peanuts and almonds. Use these results to explain why the mix variance is unfavourable.
4. *What if* of the total 1,600 kilograms of nuts put into process, 1,360 kilograms were peanuts and 240 kilograms were almonds? How would that affect the mix variance?

Solution:

1. *SM* = Standard mix proportion × Total actual input quantity

   SM peanuts = 0.80 × 1,600 = 1,280 kilograms

   SM almonds = 0.20 × 1,600 = 320 kilograms

2. The formula can be applied most easily using the following approach:

Direct Material	AQ	SM	AQ − SM	SP	(AQ − SM)SP
Peanuts	1,120	1,280	(160)	$0.50	$ (80)
Almonds	480	320	160	1.00	160
Mix variance					$ 80 U

3. Actual mix proportion peanuts = 1,120/1,600 = 0.70, or 70%

   Actual mix proportion almonds = 480/1,600 = 0.30, or 30%

   The mix variance is unfavourable because a larger percentage of the relatively more expensive input, almonds, was used.

4. Since peanuts now account for 85 percent (1,360/1,600) of the total, almonds account for only 15 percent. The mix variance will be favourable since relatively more peanuts, the cheaper input, are used.

**CORNERSTONE
9-10**

The HOW and WHY of Computing the Yield Variance

Information:
Malcom Nut Company produces mixed nuts using peanuts and almonds. Malcom developed the following standard mix for producing 120 kilograms of mixed nuts. (Almonds and peanuts are purchased in the shell and processed.)

Direct Material	Mix	Mix Proportion	SP	Standard Cost
Peanuts	128 kg	0.80	$0.50	$64
Almonds	32	0.20	1.00	32
Total	160 kg			$96

Malcom put a batch of 1,600 kilograms of nuts into process. Of the total, 1,120 kilograms were peanuts, and the remaining 480 kilograms were almonds. The actual yield was 1,300 kilograms.

Why:
The yield variance tells managers whether total inputs resulted in the amount of output expected.

Required:
1. Calculate the yield ratio based on the standard amounts given.
2. Calculate the standard cost per kilogram of the yield.
3. Calculate the standard yield for actual input of 1,600 kilograms of nuts.
4. Calculate the yield variance.
5. **What if** the total 1,600 kilograms of nuts put into process resulted in a yield of 1,190? How would that affect the yield variance?

Solution:
1. Using the standard mix for 120 kilograms of mixed nuts:
   Yield ratio = 120 kilograms of output/160 kilograms of input = 0.75

2. Standard cost of the yield $(SPy) = \$96/120$ kilograms of yield
   $$= \$0.80 \text{ per kilogram}$$

3. Standard yield = Yield ratio × Actual amount of inputs
   $$= 0.75 \times 1,600 \text{ kilograms} = 1,200 \text{ kilograms}$$

4. Yield variance = (Standard yield − Actual yield)SPy
   $$= (1,200 - 1,300)\$0.80 = \$80 \text{ F}$$

5. If the 1,600 kilograms of nuts put into process resulted in only 1,190 kilograms of mixed nuts, then the yield variance would be unfavourable. That is, the actual yield of 1,190 is less than the standard yield of 1,200.

information and the actual results, the yield variance is computed by the following formula:

$$\text{Yield variance} = (\text{Standard yield} - \text{Actual yield})SPy \qquad (9.2)$$
where

Standard yield = Yield ratio × Total actual inputs
Yield ratio = Total output/Total input
SPy = Standard cost of the yield (equal to total cost of a
standard batch divided by the amount of the yield)

Cornerstone 9-10 shows the how and why of calculating the yield variance.

The yield variance in Cornerstone 9-10 is favourable because the actual yield is greater than the standard yield. Direct material yield variance should be investigated to find the root causes. Corrective action to restore the process to the standards may be required or it may lead to a change in standards if the joint effect of the mix and yield variances is favourable.

Direct Labour Mix and Yield Variances

The direct labour mix and yield variances are computed in the same way as the direct materials mix and yield variances. Specifically, Equations 9.1 and 9.2 apply to direct labour in the same way with the notation defined appropriately for direct labour. For example, AQ, in Equation 9.1, is interpreted as AH, the actual hours used, and SP as the standard price of labour. With this understanding, the computation of mix and yield variances will be illustrated using the Malcom Nut Company example. Suppose that Malcom has two types of direct labour, shelling labour and mixing labour. Malcom has developed the following standard mix for direct labour. (Yield, of course, is measured in kilograms of output and corresponds to the same batch size used for the direct materials standards.)

Standard Mix Information: Direct Labour

Direct Labour Type	Mix	Mix Proportion	SP	Standard Cost
Shelling	3 hrs	0.60	$ 8.00	$24
Mixing	2	0.40	15.00	30
Total	5 hrs			$54
Yield	120 kg			

Yield ratio: $24 = (120/5)$, or 2,400%

Standard cost of the yield (SPy) : $0.45 per kilogram (54/120 kilograms of yield)

Suppose that Malcom processes 1,600 kilograms of nuts and produces the following actual results:

Direct Labour Type	Actual	Mix Percentages*
Shelling	20 hrs	40.0%
Mixing	30	60.0
Total	50 hrs	100.0%
Yield	1,300 kg	2,600.0%

*Uses 50 hours as the base.

Direct Labour Mix Variance   The standard mix proportion for shelling labour is 0.60. Thus, if 50 hours of actual input is used, then the mix standard calls for the following amount of shelling labour:

$$SM(\text{shelling}) = 0.60 \times 50$$
$$= 30 \text{ hours}$$

A similar computation produces $SM = 20$ hours for mixing labour (0.40×50).

Given SM, the direct labour mix variance is computed as follows (using Equation 9.1):

Direct Labour Type	AH	SM	AH − SM	SP	(AH − SM)SP
Shelling	20	30	(10)	$ 8.00	$ (80)
Mixing	30	20	10	15.00	150
Direct labour mix variance					$ 70 U

Notice that the direct labour mix variance is unfavourable. This occurs because more mixing labour was used than was called for in the standard mix, and mixing labour is more expensive than shelling labour.

Direct Labour Yield Variance   Using the standard mix information and the actual results, the direct labour yield variance is computed as follows:

$$\text{Direct labour yield variance} = (\text{Standard yield} - \text{Actual yield})SP_y$$
$$= [(24 \times 50) - 1{,}300]\$0.45$$
$$= (1{,}200 - 1{,}300)\$0.45$$
$$= \$45\ F$$

The direct labour yield variance is favourable because the actual yield is greater than the standard yield.

Summary of Learning Objectives

1. **Describe how unit input standards are developed, and explain why standard costing systems are adopted.**

- A standard costing system budgets quantities and costs on a unit basis for direct labour, direct materials, and overhead.
- Standard costs are the amount that should be expended to produce a product or service. They are set using:

  - Historical experience,
  - Engineering studies, and
  - Input from operating personnel, marketing, and accounting.

- Currently attainable standards are those that can be achieved under efficient operating conditions.
- Ideal standards are those achievable under maximum efficiency or ideal operating conditions.
- Standard costing systems are used for:

  - Planning
  - Operating
  - Control
  - Decision making

2. **Explain the purpose of a standard cost sheet.**

- The standard cost sheet shows the amount and cost of direct materials, direct labour, and overhead needed to make one unit of output.
- Using these unit quantity standards, the standard quantity of direct materials allowed and the standard hours allowed can be computed for the actual output.
- These computations play an important role in variance analysis.

3. **Compute and journalize the direct materials and direct labour variances, and explain how they are used for control.**

- The direct materials price variance compares the actual price of materials with the standard price. This difference is multiplied by the actual amount purchased.
- The direct materials usage variance compares the actual amount of materials used with the standard amount of materials for actual production. This difference is multiplied by the standard price.
- The direct labour rate variance is the difference between actual wage and standard wage multiplied by the actual number of direct labour hours.
- The direct labour efficiency variance is the difference between the actual hours worked and the standard hours for actual production multiplied by the standard wage.
- All variances are closed out at the end of the year.

  - Immaterial variances are closed to Cost of Goods Sold.
  - Material variances are prorated among Work in Process, Finished Goods, and Cost of Goods Sold.

4. **Compute overhead variances three different ways, and explain overhead accounting.**

- The four-variance method is the most detailed. It includes the following variances:
  - Variable overhead spending variance
  - Variable overhead efficiency variance
  - Fixed overhead spending variance
  - Fixed overhead volume variance

- The three-variance method does not require dividing costs into fixed and variable amounts. It includes the following variances:
  - Spending variance
  - Efficiency variance
  - Volume variance

- The two-variance method does not require dividing costs into fixed and variable amounts. It includes the following variances:
  - Budget variance
  - Volume variance

5. **Calculate mix and yield variances for direct materials and direct labour.**

- The mix variance shows the impact of different input proportions on the cost of the output.
- The yield variance shows the difference between the amount of output that was produced versus the expected output for a given amount of input.

CORNERSTONE 9-1	The HOW and WHY of computing standard quantities allowed (*SQ* and *SH*), page 440
CORNERSTONE 9-2	The HOW and WHY of computing the direct materials price variance (*MPV*) and direct materials usage variance (*MUV*), page 442
CORNERSTONE 9-3	The HOW and WHY of computing the direct labour rate variance (*LRV*) and direct labour efficiency variance (*LEV*), page 445
CORNERSTONE 9-4	The HOW and WHY of using control limits to determine when to investigate a variance, page 449
CORNERSTONE 9-5	The HOW and WHY of closing the balances in the variance accounts at the end of the year, page 451
CORNERSTONE 9-6	The HOW and WHY of calculating the total variable overhead variance, page 453
CORNERSTONE 9-7	The HOW and WHY of computing the variable overhead spending variance and the variable overhead efficiency variance, page 454
CORNERSTONE 9-8	The HOW and WHY of computing the fixed overhead spending variance and the fixed overhead volume variance, page 459
CORNERSTONE 9-9	The HOW and WHY of computing the mix variance, page 465
CORNERSTONE 9-10	The HOW and WHY of computing the yield variance, page 466

CORNERSTONES FOR CHAPTER 9

Responsibility accounting is a system that measures the results of each responsibility centre and compares those results with some expected or budgeted outcome. The four major types of responsibility centres are as follows:

1. **Cost centre:** A responsibility centre in which a manager is responsible only for costs.
2. **Revenue centre:** A responsibility centre in which a manager is responsible only for revenues.
3. **Profit centre:** A responsibility centre in which a manager is responsible for both revenues and costs.
4. **Investment centre:** A responsibility centre in which a manager is responsible for revenues, costs, and investments.

A production department within the factory, such as Assembly or Finishing, is an example of a cost centre. The supervisor of a production department does not set price or make marketing decisions, but can control manufacturing costs. Therefore, the production department supervisor is evaluated on the basis of how well costs are controlled.

The Marketing Department manager sets price and projected sales. Therefore, the Marketing Department may be evaluated as a revenue centre. Direct costs of the Marketing Department and overall sales are the responsibility of the sales manager.

In some companies, plant managers are responsible for pricing and selling products they manufacture. These plant managers control both costs and revenues, putting them in control of a profit centre. Operating income is an important performance measure for profit centre managers.

Finally, divisions are often cited as examples of investment centres. In addition to having control over cost and pricing decisions, divisional managers can make investment decisions, such as plant closings and openings, and decisions to keep or drop a product line. As a result, both operating income and some type of return on investment are important performance measures for investment centre managers.

It is important to realize that while the responsibility centre manager has responsibility for only the activities of that centre, decisions made by that manager can affect other responsibility centres. For example, the sales force at a floor care products firm routinely offers customers price discounts at the end of the month. Sales increase dramatically, but the factory is forced to put in overtime shifts to keep up with demand.

The Role of Information and Accountability

Information is the key to appropriately holding managers responsible for outcomes. For example, a production department manager is responsible for departmental costs but not for sales. This is because the production department manager not only controls some of these costs but also is best informed regarding them. Any deviation between actual and expected costs can best be explained at this level. Sales are the responsibility of the sales manager, because this manager understands and can explain price and quantity sold.

The management accountant has an expanded role in the development of a responsibility accounting system in the global business environment. Business looks to the accountant for financial and business expertise. The accountant's job is not cut and dried. Knowledge, creativity, and flexibility are needed to help managers make decisions. Good training, education, and staying up to date with one's field are important to any accountant. However, the job of the accountant in the international firm is made more challenging by the ambiguous and ever-changing nature of global business. Since much of the accountant's job is to provide relevant information to management, staying up to date requires reading in a variety of business areas, including information systems, marketing, management, politics, and economics. In addition, the accountant must be familiar with the financial accounting rules of the countries in which the firm operates.

Responsibility also entails accountability. Accountability implies performance measurement, which means that actual outcomes are compared with expected or budgeted outcomes. This system of responsibility, accountability, and performance evaluation is often referred to as *responsibility accounting* because of the key role that accounting measures and reports play in the process.

Decentralization

OBJECTIVE ▸ 2
Explain why firms choose to decentralize.

Firms with multiple responsibility centres choose one of two approaches to manage their diverse and complex activities: centralized decision making or decentralized decision making. In **centralized decision making**, decisions are made at the very top level, and lower-level managers are charged with implementing these decisions. On the other hand, **decentralized decision making** allows managers at lower levels to make and implement key decisions pertaining to their areas of responsibility. **Decentralization** is the practice of delegating decision-making authority to the lower levels.

Organizations range from highly centralized to strongly decentralized. Although some firms lie at either end of the continuum, most fall somewhere between the two extremes, with the majority of these tending toward a decentralized approach. A special case of the decentralized firm is the **multinational corporation (MNC)**. The MNC is a corporation that "does business in more than one country in such a volume that its well-being and growth rest in more than one country."[1]

Reasons for Decentralization

There are seven main reasons for delegating decision-making authority to lower levels of management.

1. Better Access to Local Information   Decision quality is affected by the quality of information available. Lower-level managers who are in contact with immediate operating conditions (e.g., the strength and nature of local competition, the nature of the local labour force, and so on) have better access to local information. As a result, local managers are often in a position to make better decisions. This advantage of decentralization is particularly applicable to multinational corporations, where far-flung divisions may be operating in a number of different countries, subject to various legal systems and customs.

2. Cognitive Limitations   Even if central management had good local information, those managers would face another problem. In a large, complex organization that operates in diverse markets with hundreds or thousands of different products, no one person has all of the expertise and training needed to process and use the information. Cognitive limitation means that individuals with specialized skills would still be needed. Rather than having different individuals at headquarters for every specialized area, why not let these individuals have direct responsibility in the field? In this way, the firm can avoid the cost and bother of collecting and transmitting local information to headquarters. The structure of Canadian business is changing. No longer are middle managers individuals with "people skills" and organization skills only. They must have specific fields of expertise in addition to managerial talent. For example, a middle manager in a bank may refer to herself as a financial specialist even though she manages 20 people. The capability to add skilled expertise is seen as crucial in today's downsized environment.

---

[1] Yair Aharoni, "On the Definition of a Multinational Corporation," in A. Kapoor and Phillip D. Grub, eds., *The Multinational Enterprise in Transition* (Princeton, NJ: Darwin Press, 1972): 4.

3. More Timely Response   In a centralized setting, it takes time to transmit local information to headquarters and then transmit the decision back to the local unit. These two transmissions cause delay and increase the potential for miscommunication, decreasing the effectiveness of the response. In a decentralized organization, where the local manager both makes and implements the decision, this problem does not arise.

Local managers in the MNC are able to respond quickly to customer discount demands, local government demands, and changes in the political climate. The different languages native to managers of divisions in the MNC make miscommunication an even greater problem. MNCs address this problem in two ways. First, a decentralized structure pushes decision making down to the local manager level, eliminating the need to interpret instructions from above. Second, MNCs are learning to incorporate technology that overrides the language barrier and eases cross-border data transfer. Technology is of great help in smoothing communication difficulties between parent and subsidiary and between one subsidiary and another.

4. Focusing of Central Management   Managers at higher levels of the hierarchical pyramid have broader responsibilities and powers. By decentralizing the operating decisions, central management is free to focus on strategic planning and decision making. Central management can concentrate on the long-run survival of the organization rather than day-to-day operations.

5. Training and Evaluation of Segment Managers   An organization always needs well-trained managers to replace higher-level managers who retire or move on to other opportunities. By decentralizing, lower-level managers are given opportunities to make and implement decisions. What better way to prepare a future generation of higher-level managers than by giving them the chance to make significant decisions? This also enables top managers to evaluate the local manager's capabilities, so that those who make the best decisions can be promoted to central management.

Just as decentralization gives the lower-level managers in the home country a chance to develop managerial skills, foreign subsidiary managers also gain valuable experience. Additionally, home country managers gain broader experience by interacting with managers of foreign divisions. The chance to learn from one another is much greater in a decentralized MNC. Off and on throughout the past 50 years, a tour of duty at a foreign subsidiary has been a part of the manager's climb to the top. Now, foreign subsidiary managers may expect to spend some time at headquarters in the home office, as well.

6. Motivation of Segment Managers   By giving local managers freedom to make decisions, some of their higher-level needs (self-esteem and self-actualization) are being met. Greater responsibility can produce more job satisfaction and motivate the local manager to work harder. Initiative and creativity are encouraged. Of course, the extent to which the motivational benefits can be realized depends to a large degree on how managers are evaluated and rewarded for their performance.

7. Enhanced Competition   In a highly centralized company, large overall profit margins can mask inefficiencies within the various subdivisions. A decentralized approach allows the company to determine each division's contribution to profit and to expose each division to market forces.

The Units of Decentralization

Decentralization is usually achieved by segmenting the company into *divisions*. One way in which divisions are differentiated is by the types of goods or services produced.

For example, **Armstrong World Industries, Inc.** has four product divisions: floor coverings (resilient sheet and tile); building products (acoustical ceilings and wall panels); industry products (insulation for heating, cooling, plumbing, and refrigeration systems); and ceramic tile. **PepsiCo** divisions include the **PepsiCo Americas**

Beverages (including **SoBe, Tropicana, Gatorade,** and **Aquafina Water**, as well as its flagship soft drink division), **PepsiCo Americas Foods** (including **Frito-Lay, Quaker Foods & Snacks, Sabritas, Gamesa,** and **Latin America Foods**), and **PepsiCo International.** Some divisions depend on other divisions. For example, PepsiCo spun off **KFC, Taco Bell,** and **Pizza Hut** into **Yum Brands.** In these restaurants, the cola you purchase will be Pepsi—not Coke.

In a decentralized setting, some interdependencies usually exist; otherwise, a company would merely be a collection of totally separate entities. The presence of these interdependencies creates the need for transfer pricing, which is discussed later in this chapter.

Similarly, companies create divisions according to the type of customer served.

Wal-Mart has five retail divisions. The Wal-Mart stores division targets discount store customers. The supercentre division targets customers of Wal-Mart's supercentre stores, which sell a variety of food, drug, and household items. **Sam's Club** focuses on buyers for small business. **Wal-Mart Neighborhood Markets** offer smaller convenience stores. Finally, the international division concentrates on global opportunities.

Organizing divisions as responsibility centres differentiates them on the degree of decentralization and creates opportunities to control them through responsibility accounting. Control of cost centres is achieved by evaluating the efficiency and the effectiveness of divisional managers. **Efficiency** means how well activities are performed; it might be measured by the number of units produced per hour or by the cost of those units. **Effectiveness** can be defined as whether the manager has performed the right activities. Measures of effectiveness might focus on value-added versus non-value-added activities.

Measuring Divisional Performance of Investment Centres

OBJECTIVE ▶ 3

Compute and explain return on investment (ROI), residual income (RI), and economic value added (EVA).

Companies maintain control of responsibility centres by developing performance measures for each centre and basing rewards on a manager's ability to control the responsibility centre.

Performance measures are developed to provide some direction for managers of decentralized divisions and to evaluate their performance. The development of performance measures and the specification of a reward structure are major issues for a decentralized organization. Because performance measures can affect the behaviour of managers, the measures chosen should encourage goal congruence. **Goal congruence** means that the goals of the manager are closely aligned with the goals of the firm. Well-chosen performance measures influence managers to pursue the company's objectives. Three performance evaluation measures for investment centres are return on investment, residual income, and economic value added.

Return on Investment

While divisional net income could be used to rank the divisions of a company, it may provide misleading information about segment performance. For example, suppose that two divisions report profits of $100,000 and $200,000, respectively. Is the second division performing better than the first? What if the first division used an investment of $500,000 to produce the contribution of $100,000, while the second used an investment of $2 million to produce the $200,000 contribution? Clearly, relating the reported operating profits to the assets used to produce them is a more meaningful measure of performance.

One way to relate operating profits to assets employed is to compute the profit earned per dollar of investment. For example, the first division earned $0.20 per dollar invested ($100,000/$500,000); the second division earned only $0.10 per dollar invested ($200,000/$2,000,000). In percentage terms, the first division provides a 20 percent rate of return and the second division, 10 percent. This method of computing the relative profitability of investments is known as the return on investment.

Return on investment (ROI) is the most common measure of performance for an investment centre. It is useful both externally and internally. Externally, ROI is used by shareholders to indicate the health of a company. Internally, ROI is used to measure the relative performance of divisions.

ROI can be defined in the following three ways:

ROI = Operating income/Average operating assets

= Operating income/Sales × (Sales/Average operating assets)

= Operating income margin × Operating asset turnover

Operating income refers to earnings before interest and income taxes and is typically used for divisions. Net income is used in the calculation of ROI for the company as a whole. **Operating assets** include all assets used to generate operating income. They usually include cash, receivables, inventories, land, buildings, and equipment. Average operating assets is computed as follows:

Average operating assets = (Beginning net book value + Ending net book value)/2

Opinions vary regarding how long-term assets (plant and equipment) should be valued (e.g., gross book value versus net book value or historical cost versus current cost). Most firms use historical cost net book value.[2] Cornerstone 10-1 shows the how and why of calculating average operating assets, margin, turnover, and return on investment.

Margin and Turnover   The ROI formula is broken into two component ratios: *margin* and *turnover*. **Margin** is the ratio of operating income to sales. It shows the portion of sales that is available for interest, income taxes, and profit. **Turnover** is a different measure; it is found by dividing sales by average operating assets. The result shows how productively assets are being used to generate sales.

Both measures can affect ROI. Let's examine the relationship of margin, turnover, and ROI more closely by considering Cornerstone 10-1. Both divisions have the same return on investment, 18 percent. However, the Snack Foods Division has a margin of 6 percent versus the Appliance Division margin of 3 percent. This tells us that the Snack Foods Division earns twice as much per dollar of sales than the Appliance Division. However, the Appliance Division has higher turnover, indicating that it is using its operating assets more effectively than the Snack Foods Division. That is, it takes fewer dollars of assets to support every dollar of income earned.

Consider a second year of data for each of the two divisions:

	Snack Foods Division	Appliance Division
	Year 2	Year 2
Sales	$40,000,000	$117,000,000
Operating income	$2,000,000	$2,925,000
Average operating assets	$10,000,000	$19,500,000
Margin	5%	2.5%
Turnover	4.0	6.0
ROI	20%	15%

The Snack Foods Division improved its ROI from 18 percent to 20 percent from Year 1 to Year 2, while the Appliance Division's ROI dropped from 18 percent to 15 percent. Notice that the margins for both divisions dropped from Year 1 to Year 2. A declining margin could be explained by increasing expenses, by competitive pressures (forcing a decrease in selling prices), or both.

---

[2] For a discussion of the relative merits of gross book value, see James S. Reese and William R. Cool, "Measuring Investment Centre Performance," *Harvard Business Review* (May–June 1978): 28–46, 174–176.

The HOW and WHY of Calculating Average Operating Assets, Margin, Turnover, and Return on Investment (ROI)

**CORNERSTONE
10-1**

Information:

Multidiv Inc. provided the following information for two of its divisions for last year:

	Snack Foods Division	Appliance Division
Sales	$30,000,000	$117,000,000
Operating income	1,800,000	3,510,000
Operating assets, January 1	9,600,000	17,500,000
Operating assets, December 31	10,400,000	21,500,000

Why:

Return on investment is a key measure of performance. It relates the income earned to the investment needed to produce that income. It is appropriate for companies and for investment centres.

Required:

1. For the Snack Foods Division, calculate:
   a. Average operating assets
   b. Margin
   c. Turnover
   d. Return on investment (ROI)
2. For the Appliance Division, calculate:
   a. Average operating assets
   b. Margin
   c. Turnover
   d. Return on investment (ROI)
3. **What if** ending assets for the Snack Foods Division were $14,400,000? How would that affect average operating assets? Margin? Turnover? ROI?

Solution:

1. a. Average operating assets = (Beginning assets + Ending assets)/2

$$= (\$9,600,000 + \$10,400,000)/2$$

$$= \$10,000,000$$

   b. Margin = Operating income/Sales

$$= \$1,800,000/\$30,000,000$$

$$= 0.06, \text{ or } 6\%$$

   c. Turnover = Sales/Average operating assets

$$= \$30,000,000/\$10,000,000$$

$$= 3.0$$

   d. ROI = Margin × Turnover

$$= 0.06 \times 3.0$$

$$= 0.18, \text{ or } 18\%$$

   OR

   ROI = Operating income/Average operating assets

$$= \$1,800,000/\$10,000,000$$

$$= 0.18, \text{ or } 18\%$$

CORNERSTONE 10-1 *(continued)*

2. a. Average operating assets = (Beginning assets + Ending assets)/2
                                 = ($17,500,000 + $21,500,000)/2
                                 = $19,500,000

    b. Margin = Operating income/Sales
               = $3,510,000/$117,000,000
               = 0.03, or 3%

    c. Turnover = Sales/Average operating assets
                 = $117,000,000/$19,500,000
                 = 6.0

    d. ROI = Margin × Turnover
            = 0.03 × 6.0
            = 0.18, or 18%

        OR

        ROI = Operating income/Average operating assets
            = $3,510,000/$19,500,000
            = 0.18, or 18%

3. If ending operating assets for the Snack Foods Division were $14,400,000, then the average operating assets would be higher. Higher average operating assets leads to lower turnover and lower ROI. Margin would not be affected. New amounts would be:

    Average operating assets = ($9,600,000 + $14,400,000)/2 = $12,000,000
    Turnover = $30,000,000/$12,000,000 = 2.5
    ROI = 0.06 × 2.5 = 0.15, or 15%

Despite the declining margin, the Snack Foods Division increased its rate of return. This increase resulted from an increase in the turnover rate that more than compensated for the decline in margin. The increase in turnover could be explained by a deliberate policy to reduce inventories (the average assets remained the same for the Snack Foods Division even though sales increased by $10 million).

The Appliance Division, on the other hand, had lower ROI because margin declined and the turnover rate stayed constant. Although more information is needed before any definitive conclusion is reached, the different responses to similar difficulties may say something about the relative skills of the two managers.

Advantages of the ROI Measure   When ROI is used to evaluate performance, division managers naturally try to increase it. This can be accomplished by increasing sales, decreasing costs, and/or decreasing investment. Three advantages of using ROI are as follows:

1. It encourages investment centre managers to pay careful attention to the relationships among sales, expenses, and investment.
2. It encourages cost efficiency.
3. It discourages excessive investment in operating assets.

Each of these three advantages is discussed in turn.

The first advantage is that ROI encourages managers to consider the interrelationship of income and investment. Suppose that a division manager is faced with the suggestion from her marketing vice president that the advertising budget be increased by $100,000. The marketing vice president is confident that this will boost sales by $200,000 and raise the contribution margin by $110,000. If the division were evaluated on the basis of operating income, this information is enough. However, if the division is evaluated on the basis of ROI, the manager will want to know how much additional investment is required to support the increased production and sales. Suppose that an additional $50,000 of operating assets will be needed. Currently, the division has sales of $2 million, operating income of $150,000, and operating assets of $1 million. Current ROI is 15 percent ($150,000/$1,000,000).

If advertising increased by $100,000 and the contribution margin by $110,000, operating income would increase by $10,000 ($110,000 − $100,000). Investment in operating assets must also increase by $50,000. With the additional advertising, the ROI is 15.24 percent ($160,000/$1,050,000). Since the ROI is increased by the proposal, the divisional manager should increase advertising.

The second advantage is that ROI encourages cost efficiency. The manager of an investment centre always has control over costs. Therefore, increasing efficiency through judicious cost reduction is a common method of increasing ROI. For example, decreasing non-value-added activities is a good way to decrease cost without decreasing production, sales, or quality. (Chapter 14 explains this in more detail.) There are ways to decrease costs in the short run that have a harmful effect on the business. This possibility is discussed in the section on disadvantages of ROI.

The third advantage is that ROI encourages efficient investment. Divisions that have cut costs to the extent possible must focus on investment reduction. For example, operating assets can be trimmed through the reduction of materials inventory and work-in-process inventory, perhaps by installing just-in-time purchasing and manufacturing systems. New, more productive machinery can be installed, inefficient plants can be closed, and so on. Companies are taking a hard look at their level of investment and acting to reduce it. This is a positive result of ROI-based evaluation.

Disadvantages of the ROI Measure   The use of ROI to evaluate performance also has disadvantages. Two negative aspects associated with ROI are frequently mentioned.

1. It discourages managers from investing in projects that would decrease the divisional ROI but would increase the profitability of the company as a whole. (Generally, projects with an ROI less than a division's current ROI would be rejected.)
2. It can encourage myopic behaviour, in that managers may focus on the short run at the expense of the long run.

The first disadvantage can be illustrated by an example. Suppose that the Snack Foods Division has the opportunity to invest in two projects for the coming year. The first project is a new cheese-coated corn chip that requires additional factory space and special coating machinery. The second project is star-shaped corn chips. That project will require special extruding machinery to create the desired shapes. The outlay required for each investment, the dollar returns, and the ROI are as follows:

	Project I	Project II
Investment	$10,000,000	$4,000,000
Operating income	$1,500,000	$760,000
ROI	15%	19%

The division is currently earning an ROI of 18 percent, using operating assets of $10 million to generate operating income of $1.8 million. Corporate headquarters will approve up to $15 million in new investment capital and requires that all investments earn at least 12 percent. Any capital not used by a division is invested by headquarters so that it earns exactly 12 percent.

The divisional manager has four alternatives: (a) add Project I, (b) add Project II, (c) add both Projects I and II, and (d) maintain the status quo (invest in neither project). The divisional ROI has been computed for each alternative.

	Add Project I	Add Project II	Add Both Projects	Maintain Status Quo
Operating income	$3,300,000	$2,560,000	$4,060,000	$1,800,000
Operating assets	$20,000,000	$14,000,000	$24,000,000	$10,000,000
ROI	16.50%	18.29%	16.92%	18.00%

The divisional manager chooses to invest only in Project II, since it will have a favourable effect on the division's ROI (18.29 percent is greater than 18.00 percent).

Assuming that any capital not used by the division is invested at 12 percent, the manager's choice will produce a lower profit for the company than could have been realized. If Project I had been selected, the company would have earned $1.5 million. By not selecting Project I, the $10 million in capital is invested at 12 percent, earning only $1.2 million (0.12 × $10,000,000). By maximizing the division's ROI, then, the divisional manager cost the company $300,000 in profits ($1,500,000 − $1,200,000).

The second disadvantage of using ROI to evaluate performance is that it can encourage myopic behaviour. An advantage of ROI is that it encourages cost reduction. However, while cost reduction can result in more efficiency, it can also result in lower efficiency in the long run. The emphasis on short-run results at the expense of the long run is **myopic behaviour**. Examples are laying off more highly paid employees, cutting the advertising budget, delaying promotions and employee training, reducing preventive maintenance, and using cheaper materials.

Each of these steps reduces expenses, increases income, and raises ROI. While these actions increase profits and ROI in the short run, they have some long-run negative consequences. Laying off more highly paid salespeople may hurt the division's future sales. For example, it has been estimated that the average monthly cost of replacing a sales representative with five to eight years' experience with a representative with less than one year of experience is $36,000 of lost sales. Low employee turnover has been linked to high customer satisfaction.[3] Future sales could also be harmed by cutting back on advertising and using cheaper materials. By delaying promotions, employee morale will be affected, which can, in turn, lower productivity and future sales. Finally, reducing preventive maintenance will likely cut into the productive capability of the division by increasing downtime and decreasing the life of the productive equipment. While these actions raise current ROI, they lead to lower future ROI.

Residual Income

To avoid managers using ROI to turn down investments that are profitable for the company but that lower a division's ROI, some companies have adopted an alternative performance measure known as *residual income*. **Residual income** is the difference between operating income and the minimum dollar return required on a company's operating assets:

Residual income = Operating income − (Minimum rate of return × Operating assets)

Cornerstone 10-2 shows the how and why of calculating residual income.

Cornerstone 10-2 shows that the residual incomes of the two divisions are different, even though their ROIs are the same. Clearly, Multidiv earns more from the larger Appliance Division than it does from the Snack Foods Division.

[3] James L. Heskett, Thomas O. Jones, Gary W. Loveman, W. Earl Sasser, Jr., and Leonard A. Schlesinger, "Putting the Service-Profit Chain to Work," *Harvard Business Review* 74, 2 (March/April 1994): 164–174.

The HOW and WHY of Calculating Residual Income

**CORNERSTONE
10-2**

Information:
Multidiv Inc. provided the following information for two of its divisions for last year:

	Snack Foods Division	Appliance Division
Sales	$30,000,000	$117,000,000
Operating income	1,800,000	3,510,000
Average operating assets	10,000,000	19,500,000

Multidiv Inc. requires a 12 percent minimum rate of return.

Why:
Residual income is measured in dollar amounts rather than percentages. It relates the income earned to the minimum required return on investment and overcomes the tendency for managers to turn down profitable projects that might lower divisional ROI.

Required:
1. Calculate residual income for the Snack Foods Division.
2. Calculate residual income for the Appliance Division.
3. **What if** the minimum required rate of return was 16 percent? How would that affect the residual income of the two divisions?

Solution:
1. Residual income = Operating income − (Minimum rate of return × Operating assets)
$$= \$1,800,000 - (0.12 \times \$10,000,000)$$
$$= \$600,000$$

2. Residual income = Operating income − (Minimum rate of return
$$\times \text{ Operating assets})$$
$$= \$3,510,000 - (0.12 \times \$19,500,000)$$
$$= \$1,170,000$$

3. If the minimum rate of return was 16 percent, the residual income of both divisions would be lower.

Snack Foods residual income = $1,800,000 − (0.16 × $10,000,000) = $200,000

Appliance residual income = $3,510,000 − (0.16 × $19,500,000) = $390,000

Advantages of Residual Income   Residual income is a dollar measure of performance. Even though the percentage rate of return is a familiar format for managers, and takes away the impact of size from the measure, at the end of the day, the dollar income does count. A manager can become so focused on the return on investment that profitable projects that return more than their cost of capital may be rejected. Residual income refocuses the manager on the profit.

To illustrate the use of residual income, consider the Snack Foods Division example again. Recall that the division manager rejected Project I because it would have reduced divisional ROI, which cost the company $300,000 in profits. The use of residual income as the performance measure would have prevented this loss. The residual income for each project is computed below.

Project I:

Residual income = Operating income − (Minimum rate of return × Operating assets)

= $1,500,000 − (0.12 × $10,000,000)

= $300,000

Project II:

Residual income = Operating income − (Minimum rate of return × Operating assets)

= $760,000 − (0.12 × $4,000,000)

= $280,000

Notice that both projects increase residual income; in fact, Project I increases the division's residual income more than Project II does. Thus, both would be selected by the divisional manager. For comparative purposes, the divisional residual income for each of the four alternatives identified earlier follows:

	Add Project I	Add Project II	Add Both Projects	Maintain Status Quo
Operating assets	$20,000,000	$14,000,000	$24,000,000	$10,000,000
Operating income	$ 3,300,000	$ 2,560,000	$ 4,060,000	$ 1,800,000
Minimum return*	2,400,000	1,680,000	2,880,000	1,200,000
Residual income	$ 900,000	$ 880,000	$ 1,180,000	$ 600,000

*Minimum return = 0.12 × Operating assets.

When residual income is used as the performance measure, both projects are clearly profitable and will be chosen. Managers are encouraged to move beyond a focus on the percentage return on investment to look at the absolute dollar value of the additional profit.

Disadvantages of Residual Income   Two disadvantages of residual income are that it is an absolute measure of return and that it does not discourage myopic behaviour. Absolute measures of return make it difficult to directly compare the performance of divisions. For example, consider the residual income computations for Division A and Division B, where the minimum required rate of return is 8 percent.

	Division A	Division B
Average operating assets	$15,000,000	$2,500,000
Operating income	$ 1,500,000	$ 300,000
Minimum return[a]	1,200,000	200,000
Residual income	$ 300,000	$ 100,000
Residual return[b]	2%	4%

[a] 0.08 × Operating assets.
[b] Residual income divided by operating assets.

At first, it is tempting to claim that Division A outperforms Division B, since its residual income is three times higher. Notice, however, that Division A used six times as many assets to produce this difference. If anything, Division B is more efficient.

One possible way to correct this disadvantage is to compute a residual return on investment by dividing residual income by average operating assets. This measure indicates that Division B earned 4 percent while Division A earned only 2 percent. Another possibility is to compute both return on investment and residual income and use both measures for performance evaluation. ROI could then be used for interdivisional comparisons.[4]

---

[4] In their study, Reese and Cool found that only 2 percent of the companies surveyed used residual income by itself, whereas 28 percent used both residual income and return on investment. See Reese and Cool, "Measuring Investment Centre Performance."

The second disadvantage of residual income is that it can encourage a short-run orientation. Just as a manager can choose to cut maintenance, training, and sales force expenses when being evaluated under ROI, the manager being evaluated on the basis of residual income can take the same actions. The problem of myopic behaviour is not solved by switching to this measure. A preferable method of reducing the myopic behaviour problem of residual income is the economic value added method, discussed next.

Economic Value Added

Another measure of profitability for performance evaluation of investment centres is *economic value added*.[5] **Economic value added (EVA)** is after-tax operating income minus the total annual cost of capital. If EVA is positive, the company is creating wealth. If it is negative, then the company is destroying wealth. Over the long term, only those companies creating capital, or wealth, can survive. Many companies today use EVA to adjust management compensation; EVA encourages managers to use existing and new capital for maximum gain. **The Coca-Cola Company**, **General Electric**, and **Intel**, a few of the companies that have seen increasing EVA during the past 15 years.[6]

EVA is a dollar figure, not a percentage rate of return. However, it does bear a resemblance to rates of return such as ROI because it links net income (return) to capital employed. The key feature of EVA is its emphasis on *after-tax* operating income and the *actual* cost of capital. Other return measures may use accounting book value numbers which may or may not represent the true cost of capital. Residual income, for example, typically uses a minimum expected rate of return. Investors like EVA because it relates profit to the amount of resources needed to achieve it.

Calculating EVA   EVA is after-tax operating income minus the dollar cost of capital employed. The equation for EVA is expressed as follows:

$$\text{EVA} = \text{After-tax operating income} - (\text{Weighted average cost of capital} \times \text{Total capital employed})$$

The difficulty faced by most companies is computing the cost of capital employed. Two steps are involved: (1) determine the **weighted average cost of capital** (a percentage figure) and (2) determine the total dollar amount of capital employed.

To calculate the weighted average cost of capital, the company must identify all sources of invested funds. Typical sources are borrowing and equity (stock issued). Typically, borrowed money has an interest rate attached, and that rate is adjusted for its tax deductibility. For example, if a company issued 10-year bonds at an annual interest rate of 8 percent and the tax rate is 40 percent, then the after-tax cost of the bonds is 4.8 percent [0.08 − (0.4 × 0.08)]. Equity is handled differently. The cost of equity financing is the opportunity cost to investors. Over time, shareholders have received an average return that is six percentage points higher than the return on long-term government bonds. If these bond rates are about 6 percent, then the average cost of equity is 12 percent (6% + 6%). Riskier stocks command a higher return; more stable and less risky stocks offer a somewhat lower return. Finally, the proportionate share of each method of financing is multiplied by its percentage cost and summed to yield the total dollar amount of capital employed.

Suppose that a company has two sources of financing: $2 million of long-term bonds paying 9 percent interest and $6 million of common stock, which is considered to be of average risk. If the company's tax rate is 35 percent and the rate of interest on long-term government bonds is 3 percent, the company's weighted average cost of capital is computed as follows:

---

[5] EVA® is a registered trademark of Stern Stewart & Co.

[6] Richard Teitelbaum, "America's Greatest Wealth Creators," *Fortune* (November 10, 1997): 265–276; and Tad Leahy, "Measures of the Future," *Business Finance*, (February 1999), http://businessfinancemag.com/article/measures-future-editorial-supplement-appraising-value-finance-0201.

	Amount	Percent	×	After-Tax Cost	=	Weighted Cost
Bonds	$2,000,000	0.25		0.09(1 − 0.35) = 0.0585		0.0146
Equity	6,000,000	0.75		0.06 + 0.03 = 0.090		0.0675
Total	$8,000,000					0.0821

Thus, the company's weighted average cost of capital is 8.21 percent.

Next we need to know the amount of capital employed. Clearly, the amount paid for buildings, land, and machinery must be included. However, other expenditures meant to have a long-term payoff, such as research and development, employee training, and so on, should also be included. Despite the fact that the latter are classified by GAAP as expenses, EVA is an internal management accounting measure, and therefore, these expenses can be thought of as the investments that they truly are. Cornerstone 10-3 shows the how and why of calculating the weighted average cost of capital, the total dollar amount of capital employed, and EVA.

Behavioural Aspects of EVA   Some companies have found that EVA helps to encourage the right kind of behaviour from their divisions in a way that emphasis on operating income alone cannot. The underlying reason is EVA's reliance on the true cost of capital. In many companies, the responsibility for investment decisions rests with corporate management. As a result, the cost of capital is considered a corporate expense. If a division builds inventories and investment, the cost of financing that investment is passed on to the overall income statement and does not reduce the division's operating income. Investment seems free to the divisions, and of course, they want more. As a result, EVA should be measured for subsets of the company.

Suppose that Supertech Inc. has two divisions, the Hardware Division and the Software Division. Operating income statements for the divisions are as follows:

	Hardware Division	Software Division
Sales	$5,000,000	$2,000,000
Cost of goods sold	2,000,000	1,100,000
Gross profit	3,000,000	900,000
Divisional selling and administrative expenses	2,000,000	400,000
Operating income	$1,000,000	$ 500,000

It looks as if the Hardware Division is doing a good job, and so is Software. Now, consider each division's use of capital. Suppose that Supertech's weighted average cost of capital is 11 percent. Hardware, by increasing inventories of components and finished goods, use of warehouses, and so on, uses capital amounting to $10 million, so its dollar cost of capital is $1,100,000 (0.11 × $10,000,000). Software does not need large materials inventories, but it does invest heavily in training and research and development. Its capital usage is $2 million, and its total dollar cost of capital is $220,000 (0.11 × $2,000,000). The EVA for each division can be calculated as follows:

	Hardware Division	Software Division
Operating income	$1,000,000	$500,000
Less: Cost of capital	1,100,000	220,000
EVA	$ (100,000)	$280,000

Now, it is clear that the Hardware Division is actually losing money by using too much capital. The Software Division, on the other hand, has created wealth for Supertech. By using EVA, the Hardware Division's manager will no longer

The HOW and WHY of Calculating the Weighted Average Cost of Capital and EVA

**CORNERSTONE
10-3**

Information:
Furman Inc. had after-tax operating income last year of $1,583,000. Three sources of financing were used by the company: $2 million of mortgage bonds paying 8 percent interest, $3 million of unsecured bonds paying 10 percent interest, and $10 million in common stock, which was considered to be no more or less risky than other stocks. (Over time, shareholders have received an average return that is six percentage points higher than the return on long-term government bonds.) The rate of return on long-term treasury bonds is 6 percent. Furman Inc. pays a marginal tax rate of 40 percent.

Why:
Economic value added adjusts earnings by the true cost of capital employed. As a result, it is a measure of wealth created or destroyed by a company.

Required:
1. Calculate the after-tax cost of each method of financing.
2. Calculate the weighted average cost of capital for Furman Inc. Calculate the total dollar amount of capital employed for Furman Inc.
3. Calculate economic value added (EVA) for Furman Inc. for last year. Is Furman Inc. creating or destroying wealth?
4. **What if** Furman Inc. had $15 million in common stock and no mortgage bonds or unsecured bonds? How would that affect the weighted average cost of capital? How would it affect EVA?

Solution:
1. After-tax cost of mortgage bonds = Interest rate − (Tax rate × Interest rate)
$$= [0.08 − (0.4 × 0.08)] = 0.048$$

After-tax cost of unsecured bonds = Interest rate − (Tax rate × Interest rate)
$$= [0.10 − (0.4 × 0.10)] = 0.06$$

Cost of common stock = Return on long-term treasury bonds
$$+ \text{Average premium}$$
$$= 0.06 + 0.06 = 0.12$$

2.

	Amount	Percent ×	After-Tax Cost =	Weighted Cost
Mortgage bonds	$ 2,000,000	0.1333	0.048	0.0064
Unsecured bonds	3,000,000	0.2000	0.060	0.0120
Common stock	10,000,000	0.6667	0.120	0.0800
Total	$15,000,000			0.0984

Weighted average percentage cost of capital = 0.0984, or 9.84%
Total dollar amount of capital employed = 0.0984 × $15,000,000
= $1,476,000

3.

After-tax operating income	$1,583,000
Less: Total dollar amount of capital employed	1,476,000
EVA	$ 107,000

**CORNERSTONE
10-3**
(continued)

Furman Inc. is creating capital because EVA is positive (the after-tax earnings are greater than the after-tax cost of capital).

4. If all $15 million of financing were in common stock, the weighted average percentage cost of capital would be 12 percent and the total dollar amount of capital employed would be $1,800,000 (0.12 × $15,000,000). EVA would be negative, and Furman Inc. would be destroying wealth, not creating it.

$$EVA = \$1,583,000 - \$1,800,000 = \$(217,000)$$

consider inventories and warehouses to be "free" goods. Instead, the manager will strive to reduce capital usage and increase EVA. A reduction of capital usage to $8 million, for example, would boost EVA to $120,000 [$1,000,000 − (0.11 × $8,000,000)].

Quaker Oats faced a similar situation. Prior to 1991, Quaker Oats evaluated its business segments on the basis of quarterly profits. To keep quarterly earnings on an upward march, segment managers sharply discounted products at the end of each quarter. This resulted in huge orders from retailers and surges in production at Quaker's plants at the end of each three-month period. This practice is called trade loading because it "loads up the trade" (retail stores) with product. However, trade loading is expensive because it requires massive amounts of capital (e.g., working capital, inventories, and warehouses to store the quarterly spikes in output). Before EVA, a Quaker plant could run well below capacity early in the quarter. Purchasing, however, would buy huge quantities of boxes, plastic wrappers, granola, and chocolate chips, in anticipation of the production surge of the last six weeks of the quarter. As the products were finished, Quaker packed warehouses with finished goods. All costs associated with inventories were absorbed by corporate headquarters. Thus, they appeared to be free to the plant managers, who were encouraged to build ever higher inventories. The advent of EVA and the cancellation of trade loading led to a smoothing of production throughout the quarter, higher overall production (and sales), and lower inventories.

EVA can be used in the public sector, as well.

Multiple Measures of Performance

ROI, residual income, and EVA are important measures of managerial performance. However, they are financial measures, and may tempt managers to focus only on dollar figures. This focus may not tell the whole story for the company. In addition, lower-level managers and employees may feel helpless to affect net income or investment. To counter this, nonfinancial operating measures have been developed. For example, top management could look at such factors as market share, customer complaints, personnel turnover ratios, and personnel development. By letting lower-level managers know that attention to long-run factors is also vital, the tendency to overemphasize financial measures is reduced.

Modern managers are especially likely to use multiple measures of performance and to include nonfinancial as well as financial measures.

The Balanced Scorecard (discussed in Chapter 15) was developed to measure a firm's performance in multiple areas.

Measuring and Rewarding the Performance of Managers

While some companies consider the performance of the division to be equivalent to the performance of the manager, there is a compelling reason to separate the two. Often, the performance of the division is subject to factors beyond the manager's control. It is particularly important, then, to take a responsibility accounting approach and evaluate managers on the basis of factors under their control. A serious concern is the creation of a compensation plan that is closely tied to the performance of the division.

Incentive Pay for Managers—Encouraging Goal Congruence

Managerial evaluation and incentive pay would be of little concern if all managers were equally likely to perform up to the best of their abilities, and if those abilities were known in advance. In the case of a small company, owned and managed by the same person, there is no problem. The owner puts in as much effort as she or he wishes and receives all of the income as a reward for performance. However, in most companies, the owner hires managers to operate the company on a day-to-day basis and delegates decision-making authority to them. The shareholders of a company hire the CEO through the board of directors, and division managers are hired by the CEO to operate their divisions on behalf of the owners. Then, the owners must ensure that the managers are providing good service.

Why wouldn't managers provide good service? There are three reasons: (1) they may be unable to perform the job, (2) they may prefer not to work hard, and (3) they may prefer to spend company resources on perquisites. The first reason requires owners to discover information about the manager before hiring him. Think back to the reasons for decentralization—one was that it provided training for future managers. The training process provides signals about the managerial ability of division managers. The second and third reasons require the owner to monitor the manager or to arrange an incentive scheme that will more closely ally the manager's goals with those of the owner. Some managers may not want to do hard or routine work. Some may be risk-averse and not take actions that expose them, and the company, to risky situations. Thus, it is necessary to compensate them for undertaking risk and hard work. Closely related to the desire of some managers to shirk responsibility is the tendency of managers to overuse perquisites. **Perquisites** are a type of fringe benefit received over and above salary. Some examples are a nice office, use of a company car or jet, expense accounts, and company-paid country club memberships. While some perquisites are legitimate uses of company resources, they can be abused. A well-structured incentive pay plan can help to encourage goal congruence between managers and owners.

Managerial Rewards

Managerial rewards frequently include incentives tied to performance. The objective is to encourage goal congruence so that managers will act in the best interests of the firm. Managerial rewards may include the following:

1. Cash Compensation Cash compensation includes salaries and bonuses. Raises are one way for a company to reward good managerial performance. However, once the raise takes effect, it is usually permanent. Bonuses give a company more flexibility. Many companies use a combination of salary and bonus to reward performance by keeping salaries fairly level and allowing bonuses to fluctuate with reported income.

Managers may find their bonuses tied to divisional net income or to targeted increases in net income. For example, a division manager may receive an annual salary of $75,000 and a yearly bonus of 5 percent of the increase in reported net income. If net income does not rise, the manager's bonus is zero. This incentive pay scheme makes increasing net income, an objective of the owner, important to the manager as well.

Profit-sharing plans make employees partial owners in the sense that they receive a share of the profits. They are not owners in the sense of decision making or downside risk sharing. This is a form of risk sharing, in particular, sharing of upside risk. Typically, employees are paid a flat rate, and then, any profits to be shared are over and above wages. The objective is to provide an incentive for employees to work harder and smarter.

Income-based compensation can encourage dysfunctional behaviour. The manager may engage in unethical practices, such as postponing needed maintenance. If the bonus is capped at a certain amount (say the bonus is equal to 1 percent of net income but cannot exceed $50,000), managers may postpone revenue recognition from the end of the year in which the maximum has already been achieved to the next year. Those who structure the reward systems need to understand both the positive incentives built into the system as well as the potential for negative behaviour.

2. Stock Options (Noncash Compensation) Stock is a share in the company, and theoretically, it should increase in value as the company does well and decrease in value as the company does poorly. Thus, the issuance of stock to managers makes them part owners of the company and should encourage goal congruence. Many companies encourage employees to purchase shares of stock, or they grant shares as a bonus. A disadvantage of stock as compensation is that share price can fall for reasons beyond the control of managers.

Companies frequently offer stock options to managers. A **stock option** is the right to buy a certain number of shares of the company's stock, at a particular price and after a set length of time. The objective of awarding stock options is to encourage managers to focus on the longer term. The price of the option shares is usually set at market price at the time of issue. Then, if the stock price rises in the future, the manager may exercise the option, thus purchasing stock at a below-market price and realizing an immediate gain.

Assume than an executive was granted an option to purchase 100,000 shares of company stock at the current market price of $20 per share. The option was granted in August 2011 and could be exercised after two years. If, by August 2013, the stock has risen to $23 per share, the executive can purchase all 100,000 shares for $2,000,000 (100,000 × $20 option price) and immediately sell them for $2,300,000 (100,000 × $23) for a profit of $300,000. Of course, if the stock price drops below $20, the executive will not exercise the option. Typically, however, stock prices rise along with the market, and the executive can safely bet on a future profit.

Companies are becoming more aware of the impact on options of the overall movement of the stock market. If the market moves strongly higher, there is the potential for windfall profits. That is, any profit realized from selling stock based on low cost options may be more closely related to the overall rise in the stock market and less related to outstanding performance by top management. In addition, top executives with a number of options may focus on the short-term movements of the stock price rather than on the long-term indicators of company performance. In essence, they may trade long-term returns for short-term returns.

Typically, there are constraints on the exercise of the options. For example, the stock purchased with options may not be sold for a certain period of time. A disadvantage of stock options is that the price of the stock is based on many factors and is not completely within the manager's control.

Issues to Consider in Structuring Cash-Based Compensation Single measures of performance, which are often the basis of bonuses, are subject to gaming behaviour in that managers may increase short-term measures at the expense of long-term measures. For example, a manager may keep net income high by refusing to invest in more modern and efficient equipment. Depreciation expense remains low, but so do productivity and quality. Clearly, the manager has an incentive to understand the computation of the accounting numbers used in performance evaluation. An accounting change in inventory valuation or in the method of depreciation, for example, will change net income even though sales and costs remain unchanged. Frequently, we see that a new CEO of a troubled corporation will take a number of losses (e.g., inventory write-downs) all at once. This is referred to as the "big bath" and usually results in very low (or negative) net income in that year. Then, the books are cleared for a good increase in net income, and a correspondingly large bonus, for the next year.

Both cash bonuses and stock options can encourage a short-term orientation. To encourage a longer-term orientation, some companies require top executives to purchase and hold a certain amount of company stock to retain employment.

Another issue to be considered in structuring management compensation plans is that owners and managers may be affected differently by risk. When managers have so much of their own capital—both financial and human—invested in the company, they may be less apt to take risks. Owners, because of their ability to diversify away some of the risk, may prefer a more risk-taking attitude. As a result, managers must be somewhat insulated from catastrophic downside risk in order to encourage them to make entrepreneurial decisions.

Noncash Compensation   We often see managers who trade off increased salary for improvements in title, office location and trappings, use of expense accounts, and so on. Perquisites can be well used to make the manager more efficient. For example, a busy manager may be able to effectively employ several assistants and may find that use of a corporate jet allows him or her to more efficiently schedule travel in overseeing far-flung divisions. However, perquisites may be abused as well. One wonders how the shareholders of a corporation can benefit from extravagant parties organized by the CEO in exotic, expensive locations. These parties are intended to promote the corporation's image. They sometimes also satisfy the vanity and ego of some executives.

Measuring Performance in the Multinational Firm

It is important for the MNC to separate the evaluation of the *manager* of a division from the evaluation of the *division*. The manager's evaluation should not include factors over which he exercises no control, such as currency fluctuations, income taxes, and so on. It is particularly difficult to compare the performance of a manager of a division (or subsidiary) in one country with the performance of a manager of a division in another country. Even divisions that appear to be similar in terms of production may face very different economic, social, or political forces. Instead, managers should be evaluated on the basis of revenues and costs incurred. Once a manager is evaluated, then the subsidiary financial statements can be restated to the home currency and uncontrollable costs can be allocated.[7]

International environmental conditions may be very different from, and more complex than, domestic conditions. Environmental variables facing local managers of divisions include economic, legal, political, social, and educational factors. Some important economic variables are inflation, foreign currency exchange rates, income taxes, and transfer prices.

Legal and political factors are important. For example, a country may not allow cash outflows or may forbid the import of certain items, such as guns.

Educational, infrastructure, and cultural variables affect how the multinational firm is treated by the subsidiary's country. Many clothing distributors depend on factories in developing countries to do the manufacturing. However, first, those companies had to develop the area, putting in roads and communication equipment and providing training for workers.

---

[7] Helen Gernon and Gary Meek, *Accounting: An International Perspective* (Homewood, IL: Richard D. Irwin-McGraw-Hill, 2001).

Comparison of Divisional ROI   The existence of differing environmental factors makes interdivisional comparison of ROI potentially misleading. For example, the lack of consistency in internal reporting may obscure interdivisional comparison. A minimum wage law in one country may restrict the manager's ability to affect labour costs. Another country may prevent the export of cash. Still others may have a well-educated workforce but poor infrastructure (transportation and communication facilities). Therefore, the corporation must be aware of and control these differing environmental factors when assessing managerial performance.

The modern accountant who works in a global environment must be aware of more than business and finance. Political and legal systems have important implications for the company. Sometimes, the political system changes quickly, throwing the company into crisis mode. Other times, the situation evolves more slowly.

On occasion, the political structure may mean that North American standards of control may not "work" in foreign countries. The business objective for a socialist country may not be efficiency or effectiveness, but compliance with the central plan. This culture of altering the plan to match the actual results continues to exist in some countries.

Multiple Measures of Performance   Rigid evaluation of the performance of foreign divisions of the MNC ignores the overarching strategic importance of developing a global presence. The interconnectedness of the global company weakens the independence or stand-alone nature of any one segment. As a result, residual income and ROI are less important measures of managerial performance for divisions of the MNC. MNCs must use additional measures of performance that relate more closely to the long-run health of the company. In addition to ROI and residual income, top management looks at such factors as market potential and market share.

Additionally, the use of ROI and RI in the evaluation of managerial performance in divisions of an MNC is subject to problems beyond those faced by a decentralized company that operates in only one country. It is particularly important, then, to take a responsibility accounting approach and evaluate managers on the basis of factors under their control. Multiple measures of performance, keyed to local operating conditions, can spotlight managers' responses to different and difficult operating conditions.

OBJECTIVE ▸ 5

Explain the role of transfer pricing in a decentralized firm.

Transfer Pricing

Often, the output of one division can be used as input for another division. For example, integrated circuits produced by one division can be used by a second division to make video recorders. **Transfer prices** are the prices charged for goods produced by one division and transferred to another. The price charged affects the revenues of the transferring division and the costs of the receiving division. As a result, the profitability, return on investment, and managerial performance evaluation of both divisions are affected.

The Impact of Transfer Pricing on Income

Exhibit 10-1 illustrates the effect of the transfer price on two divisions of ABC Inc. Division A produces a component and sells it to another division of the same company, Division C. The $30 transfer price is revenue to Division A and increases division income; clearly, Division A wants the price to be as high as possible. Conversely, the $30 transfer price is cost to Division C and decreases division income, just like the cost of any materials. Division C prefers a lower transfer price. For the company as a whole, A's revenue minus C's cost equals zero.

While the actual transfer price nets out for the company as a whole, transfer pricing can affect the level of profits earned by the company as a whole if it affects divisional behaviour. Divisions, acting independently, may set transfer prices that maximize divisional profits but adversely affect firmwide profits. For example, suppose that Division A in Exhibit 10-1 sets a transfer price of $30 for a component that

Exhibit 10-1

Impact of Transfer Price on Transferring Divisions and the Company as a Whole

ABC Inc.	
Division A	**Division C**
Produces component and transfers it to C for transfer price of $30 per unit	Purchases component from A at transfer price of $30 per unit and uses it in production of final product
Transfer price = $30 per unit	Transfer price = $30 per unit
Revenue to A	Cost to C
Increases net income	Decreases net income
Increases ROI	Decreases ROI

Transfer price revenue = Transfer price cost
Zero impact on ABC Inc.

costs $24 to produce. If Division C can obtain the component from an outside supplier for $28, it will refuse to buy from Division A. Division C will realize a savings of $2 per component ($30 internal transfer price − $28 external price). However, if Division A cannot replace the internal sales with external sales, the company as a whole will be worse off by $4 per component ($28 external cost − $24 internal cost). This outcome would increase the total cost to the firm as a whole. Thus, how transfer prices are set can be critical for profits of the business as a whole.

Setting Transfer Prices

OBJECTIVE ▶ 6
Discuss the methods of setting transfer prices.

A transfer pricing system should satisfy three objectives: accurate performance evaluation, goal congruence, and preservation of divisional autonomy.[8] Accurate performance evaluation means that no one divisional manager should benefit at the expense of another (in the sense that one division is made better off while the other is made worse off). Goal congruence means that divisional managers select actions that maximize firmwide profits. Autonomy means that central management should not interfere with the decision-making freedom of divisional managers. The **transfer pricing problem** concerns finding a system that simultaneously satisfies all three objectives.

We can evaluate the degree to which a transfer price satisfies the objectives of a transfer pricing system by considering the opportunity cost of the goods transferred. The *opportunity cost approach* can be used to describe a wide variety of transfer pricing practices. Under certain conditions, this approach is compatible with the objectives of performance evaluation, goal congruence, and autonomy.

The **opportunity cost approach** identifies the minimum price that a selling division would be willing to accept and the maximum price that the buying division would be willing to pay. These minimum and maximum prices correspond to the opportunity costs of transferring internally and they define a *bargaining range*. They are defined for each division as follows:

1. The **minimum transfer price**, or floor, is the transfer price that would leave the selling division no worse off if the good is sold to an internal division. Note that the selling division would prefer a higher price; however, the minimum transfer price is the absolute lowest that could be accepted.
2. The **maximum transfer price**, or ceiling, is the transfer price that would leave the buying division no worse off if an input is purchased from an internal

[8] Joshua Ronen and George McKinney, "Transfer Pricing for Divisional Autonomy," *Journal of Accounting Research* (Spring 1970): 100–101.

division. Note that the buying division would prefer a lower price; however, the maximum transfer price is the absolute highest that could be accepted.

The opportunity cost approach tells us that a good should be transferred internally whenever the opportunity cost (minimum price) of the selling division is less than the opportunity cost (maximum price) of the buying division. By definition, this approach ensures that neither divisional manager is made worse off by transferring internally. This means that total divisional profits are not decreased by the internal transfer.

Central management rarely sets specific transfer prices. Instead, most companies develop some general policies that divisions must follow. Three commonly used policies are market-based transfer pricing, negotiated transfer pricing, and cost-based transfer pricing. Each of these can be evaluated according to the opportunity cost approach.

Market Price

If there is an outside market for the good to be transferred and that outside market is perfectly competitive, the correct transfer price is the market price.[9] In such a case, divisional managers' actions will simultaneously optimize divisional profits and firm-wide profits. No division can benefit at the expense of another division and central management will not be tempted to intervene.

The opportunity cost approach also signals that the correct transfer price is the market price. Since the selling division can sell all that it produces at the market price, transferring internally at a lower price would make that division worse off. Similarly, the buying division can always acquire the intermediate good at the market price, so it would be unwilling to pay more for an internally transferred good. Since the minimum transfer price for the selling division is the market price and since the maximum price for the buying division is also the market price, the only possible transfer price is the market price.

In fact, moving away from the market price will decrease the overall profitability of the firm. This principle can be used to resolve divisional conflicts that may occur, as the following example illustrates.

Yarrow Company is a decentralized manufacturer of small appliances. The Parts Division, which is at capacity, produces parts that are used by the Motor Division. The parts can also be sold to other manufacturers and to wholesalers at a market price of $8. For all practical purposes, the market for the parts is perfectly competitive.

Suppose that the Motor Division, operating at 70 percent capacity, receives a special order for 100,000 motors at a price of $30. Full manufacturing cost of the motors is $31, broken down as follows:

Direct materials	$10
Transferred-in part	8
Direct labour	2
Variable overhead	1
Fixed overhead	10
Total cost	$31

Notice that the motor includes a part transferred in from the Parts Division at a market-based transfer price of $8. Should the Parts Division lower the transfer price to allow the Motor Division to accept the special order? The opportunity cost approach helps us to answer this question.

Since the Parts Division can sell all that it produces, the minimum transfer price is the market price of $8. Any lower price would make the Parts Division worse off.

---

[9] A perfectly competitive market for the intermediate product requires four conditions: (1) the division producing the intermediate product is small relative to the market as a whole and cannot influence the price of the product; (2) the intermediate product is indistinguishable from the same product of other sellers; (3) firms can easily enter and exit the market; and (4) consumers, producers, and resource owners have perfect knowledge of the market.

For the Motor Division, identifying the maximum transfer price that can be paid so that it is no worse off is a bit more complex.

Since the Motor Division is under capacity, the fixed overhead portion of the motor's cost is not relevant. The relevant costs are those additional costs that will be incurred if the order is accepted. These costs, excluding for the moment the cost of the transferred-in component, equal $13 ($10 + $2 + $1). Thus, the contribution to profits before considering the cost of the transferred-in component is $17 ($30 − $13). The division could pay as much as $17 for the component and still break even on the special order. However, since the component can always be purchased from an outside supplier for $8, the maximum price that the division should pay internally is $8. Thus, the market price is the best transfer price.

Negotiated Transfer Prices

Perfectly competitive markets rarely exist. In most cases, producers *can* influence price (e.g., by being large enough to influence demand by dropping the price of the product or by selling closely related but differentiated products). When imperfections exist in the market for the intermediate product, market price may no longer be suitable. In this case, negotiated transfer prices may be a practical alternative. Opportunity costs help define the boundaries of the negotiation set.

Example 1: Avoidable Distribution Costs Assume that a division produces a circuit board. Currently, the division sells 1,000 units per day, with variable manufacturing costs of $12 per unit. The division can sell all that it produces to the outside market at $22. Any outside sales incur a distribution cost of $2 per unit. Alternatively, the board can be sold internally to the company's recently acquired Electronic Games Division. There is no distribution cost if the board is sold internally.

The Electronic Games Division, also at capacity, produces and sells 350 games per day. These games sell for $45 per unit and have variable manufacturing costs of $32 per unit. Variable selling expenses of $3 per unit are also incurred. Sales and production data for each division are summarized in Exhibit 10-2.

How could the Games Division and the Circuit Board Division set a transfer price? If the Games Division currently pays $22 per circuit board, it would refuse to pay more than $22; thus, the maximum transfer price is $22. The minimum transfer price is set by the Circuit Board Division. While this division prices its circuit boards at $22, it will avoid $2 of distribution cost if it sells internally. Therefore, the minimum transfer price is $20 ($22 − $2). The bargaining range for the transfer price is between $20 and $22.

Suppose that the Games Division manager offered a transfer price of $20. That division would be better off by $2 per circuit board, since it had previously paid $22

Exhibit 10-2

Summary of Sales and Production Data

	Circuit Board Division	Games Division
Units sold:		
Per day	1,000	350
Per year*	260,000	91,000
Unit data:		
Selling price	$22	$45
Variable costs:		
Manufacturing	$12	$32
Selling	$2	$3
Annual fixed costs	$1,480,000	$610,000

*There are 260 selling days in a year.

Exhibit 10-3

Comparative Income Statements

Before Negotiation: All Sales External			
	Circuit Board Division	Games Division	Total
Sales	$ 5,720,000	$ 4,095,000	$ 9,815,000
Less variable expenses:			
Cost of goods sold	(3,120,000)	(2,912,000)	(6,032,000)
Variable selling	(520,000)	(273,000)	(793,000)
Contribution margin	2,080,000	910,000	2,990,000
Less: Fixed expenses	1,480,000	610,000	2,090,000
Operating income	$ 600,000	$ 300,000	$ 900,000

After Negotiation: Internal Transfers @ $21.10			
	Circuit Board Division	Games Division	Total
Sales	$ 5,638,100	$ 4,095,000	$ 9,733,100
Less variable expenses:			
Cost of goods sold	(3,120,000)	(2,830,100)	(5,950,100)
Variable selling	(338,000)	(273,000)	(611,000)
Contribution margin	2,180,100	991,900	3,172,000
Less: Fixed expenses	1,480,000	610,000	2,090,000
Operating income	$ 700,100	$ 381,900	$ 1,082,000
Change in operating income	$ 100,100	$ 81,900	$ 182,000

per board. Its profits would increase by $700 per day ($2 × 350 units per day). The Circuit Board Division, on the other hand, would be no better, or worse, off than before and would earn no additional profit. While a transfer price of $20 per circuit board is possible, it is unlikely that the Circuit Board manager would agree to it.

Now suppose that the Circuit Board Division counters with an offer of $21.10 per board. That transfer price allows the Circuit Board Division to increase its profits by $385 per day [($21.10 − $20) × 350 units]. The Games Division would increase its profits by $315 per day [($22 − $21.10) × 350 units].

While we cannot tell exactly where the Circuit Board Division and the Games Division would set a transfer price, we can see that it will be somewhere within the bargaining range. [The minimum transfer price ($20) and the maximum transfer price ($22) set the limits of the bargaining range.] Exhibit 10-3 provides income statements for each division before and after the agreement. Notice how the profit increase is split between the two divisions.

Example 2: Excess Capacity   In perfectly competitive markets, the selling division can sell all that it wishes at the prevailing market price and would produce at capacity. In a less ideal setting, a selling division may be unable to sell all that it produces; accordingly, the division may have excess capacity.[10]

To illustrate the role of transfer pricing and negotiation in this setting, consider the dialogue between Sharena Casper, manager of a Plastics Division, and Manny Rogers, manager of a Pharmaceutical Division:

MANNY: Sharena, my division has shown a loss for the past three years. When I took over the division at the beginning of the year, I set a goal with headquarters to break

---

[10]Output can be increased by decreasing selling price. Of course, decreasing selling price to increase sales volume may not increase profits—in fact, profits could easily decline. We assume in this example that the divisional manager has chosen the most advantageous selling price and that the division is still left with excess capacity.

even. At this point, projections show a loss of $5,000—but I think I have a way to reach my goal, if I can get your cooperation.

SHARENA: If I can help, I will. What do you have in mind?

MANNY: I need a special deal on your plastic bottle Model 3. A large West Coast retail chain wants to buy 250,000 bottles. But we have to give them a real break on price. They have offered $0.85 per unit. My variable cost per unit is $0.60, not including the cost of the plastic bottle. I normally pay $0.40 for your bottle, but if I do that, the order will lose me $37,500. I can't afford that kind of loss. I know that you have excess capacity. If you can make 250,000 bottles, I'll pay your variable cost per unit, provided it is no more than $0.25. Are you interested? Can you handle that size order?

SHARENA: I have enough excess capacity to handle the order easily and my variable cost is $0.15. However, I want part of the profit. I'll let you have the order for $0.20. That way, we both make $0.05 per bottle, for a total contribution of $12,500. That'll put you in the black and help me get closer to my budgeted profit goal.

MANNY: Great! Thanks so much. If this West Coast chain provides more orders in the future—as I expect it will—and at better prices, I'll make sure you get our business.

Notice the role that opportunity costs play in the negotiation. In this case, the minimum transfer price is the Plastic Division's variable cost ($0.15), representing the incremental outlay if the order is accepted. Since the division has excess capacity, only variable costs are relevant to the decision. By covering the variable costs, the order does not affect the division's total profits. For the buying division, the maximum transfer price is the purchase price that would allow the division to cover its incremental costs on the special order ($0.25). Adding the $0.25 to the other costs of processing ($0.60), the total incremental costs incurred are $0.85 per unit. Since the selling price is also $0.85 per unit, the division is made no worse off. Both divisions, however, can be better off if the transfer price is between the minimum price of $0.15 and the maximum price of $0.25.

Comparative statements showing the contribution margin earned by each division and the firm as a whole are shown in Exhibit 10-4 for each of the four transfer prices discussed. These statements show that the firm earns the same profit for all four transfer prices; however, different prices do affect the individual divisions' profits differently. Because of the autonomy of each division, there is no guarantee that the firm will earn the maximum profit. For example, if Sharena had insisted on maintaining the price of $0.40, no transfer would have taken place, and the overall $25,000 increase in profits would have been lost.

Disadvantages of Negotiated Transfer Prices   Negotiated transfer prices have three disadvantages that are commonly mentioned.

1. One divisional manager with private information may take advantage of another divisional manager.
2. Performance measures may be distorted by the negotiating skills of managers.
3. Negotiation can consume considerable time and resources.

ETHICS It is interesting that Manny, the manager of the Pharmaceutical Division, did not know the variable cost of producing the plastic bottle. Yet that cost was a key to the negotiation. Clearly, he had not done his homework before starting the negotiation. This lack of knowledge gave Sharena, the other divisional manager, the opportunity to exploit the situation. For example, she could have claimed that the variable cost was $0.27 and offered to sell for $0.25 per unit as a favour to Manny, saying that she would absorb a $5,000 loss in exchange for a promise of future business. In this case, she would capture the full $25,000 benefit of the transfer. Alternatively, she could have misrepresented the figure and used it to turn down the request, thus preventing Manny from achieving his budgetary goal; after all, she may be competing with Manny for promotions, bonuses, salary increases, and so on.

Exhibit 10-4

Comparative Statements

Transfer Price of $0.40			
	Pharmaceutical	Plastics	Total
Sales	$212,500	$100,000	$312,500
Less: Variable expenses	250,000	37,500	287,500
Contribution margin	$ (37,500)	$ 62,500	$ 25,000
Transfer Price of $0.25			
Sales	$212,500	$62,500	$275,000
Less: Variable expenses	212,500	37,500	250,000
Contribution margin	$ 0	$25,000	$ 25,000
Transfer Price of $0.20			
Sales	$212,500	$50,000	$262,500
Less: Variable expenses	200,000	37,500	237,500
Contribution margin	$ 12,500	$12,500	$ 25,000
Transfer Price of $0.15			
Sales	$212,500	$37,500	$250,000
Less: Variable expenses	187,500	37,500	225,000
Contribution margin	$ 25,000	$ 0	$ 25,000

Fortunately, Sharena displayed sound judgment and acted with integrity.[11] For negotiation to work, managers must be willing to share relevant information. How can this requirement be satisfied? Perhaps the best course of action is to hire managers with integrity—managers who have a commitment to ethical behaviour. Additionally, top management can take other actions to discourage the use of private information for exploitive purposes. For example, corporate headquarters could base some part of the management reward structure on overall profitability to encourage actions that are in the best interests of the company as a whole.◆

The second disadvantage of negotiated transfer prices is that the practice distorts the measurement of managerial performance. According to this view, divisional profitability may be affected too strongly by the negotiating skills of managers, masking the actual management of resources entrusted to each manager. Although this argument may have some merit, it ignores the fact that negotiating is a desirable skill. Perhaps divisional profitability should reflect differences in negotiating skills.

The third criticism of this technique is that negotiating is time consuming. The time spent in negotiation could be spent managing other activities necessary to the success of the division. Sometimes, negotiations reach an impasse, forcing top management to spend time mediating the process.[12] Although negotiating takes time, a mutually satisfactory outcome can increase profits for the divisions and the firm. Furthermore, negotiation does not have to be repeated each time for similar transactions.

Advantages of Negotiated Transfer Prices
Although time consuming, negotiated transfer prices offer some hope of complying with the three criteria of goal

---

[11] Because of the excess capacity, her agreement to work with Manny was beneficial for both of their divisions. Note that if she had not had excess capacity, it would have been better for her division and for the company as a whole to refuse the special offer to Manny's division and to sell for full price to outsiders.

[12] The involvement of top management may be very cursory, however. In the case of a very large oil company that negotiates virtually all transfer prices, two divisional managers could not come to an agreement after several weeks of effort and appealed to their superior. His response: "Either come to an agreement within 24 hours, or you are both fired." Needless to say, an agreement was reached within the allotted time.

congruence, autonomy, and accurate performance evaluation. Just as important, however, is the process of making sure that actions of the different divisions mesh together so that the company's overall goals are attained. If negotiation helps ensure goal congruence, there is no need for central management to intervene. Finally, if negotiating skills of divisional managers are comparable or if the firm views these skills as an important managerial skill, concerns about motivation and accurate performance measures are avoided. Cornerstone 10-4 shows the how and why of calculating market-based and negotiated transfer prices.

Cost-Based Transfer Prices

Three forms of cost-based transfer pricing will be considered: full cost, full cost plus markup, and variable cost plus fixed fee. In all cases, standard costs should be used to avoid passing on the inefficiencies of one division to another. A more important issue, however, is the propriety of cost-based transfer prices. Under what circumstances, if any, should they be used?

Full-Cost Transfer Pricing Perhaps the least desirable type of transfer pricing approach is full cost. Its only real virtue is simplicity. Full-cost transfer pricing can provide perverse incentives and distort performance measures. As we have seen, the opportunity costs of both the buying and selling divisions are essential for determining the propriety of internal transfers. At the same time, they provide useful reference points for determining a mutually satisfactory transfer price. Only rarely will full cost provide accurate information about opportunity costs.

A full-cost transfer price would have shut down the negotiations described earlier. In the first example, the manager would never have considered transferring internally if the price had to be full cost. Yet by transferring at selling price less some distribution expenses, both divisions—and the firm as a whole—were better off. In the second example, the manager of the Pharmaceutical Division could never have accepted the special order with the West Coast chain. Both divisions and the company would have been worse off.

Full Cost Plus Markup Full cost plus markup suffers from virtually the same problems as full cost. It is somewhat less perverse, however, if the markup can be negotiated. For example, a full-cost-plus-markup formula could have been used to represent the negotiated transfer price of the first example. In some cases, a full-cost-plus-markup formula may be the outcome of negotiation; if so, it is simply another example of negotiated transfer pricing. In these cases, the use of this method is fully justified. Using full cost plus markup to represent all negotiated prices, however, is not possible (e.g., it could not be used to represent the negotiated price of the second example). The superior approach is negotiation, since more cases can be represented, and full consideration of opportunity costs is possible.

Variable Cost Plus Fixed Fee Like full cost plus markup, variable cost plus fixed fee can be a useful transfer pricing approach provided that the fixed fee is negotiable. This method has one advantage over full cost plus markup: if the selling division is operating below capacity, variable cost is its opportunity cost. Assuming that the fixed fee is negotiable, the variable cost approach can be equivalent to negotiated transfer pricing. Negotiation with full consideration of opportunity costs is preferred.

Propriety of Use Despite the disadvantages of cost-based transfer prices, many companies use these methods, especially full cost and full cost plus markup. These methods are simple and objective. In addition, often transfers between divisions have a small impact on the profitability of either division. Thus, it may be cost effective to use an easy-to-identify, cost-based formula rather than spending valuable time and resources on negotiation.

**CORNERSTONE
10-4**

The HOW and WHY of Calculating Market-Based and Negotiated Transfer Prices

Information:

Omni Inc. has a number of divisions, including Alpha Division, producer of circuit boards, and Delta Division, a heating and air conditioning manufacturer.

Alpha Division produces the cb-117 model that can be used by Delta Division in the production of thermostats that regulate the heating and air conditioning systems. The market price of the cb-117 is $14. Cost information for the cb-117 model is:

Variable product cost	$2.50
Fixed cost	6.50
Total product cost	$9.00

Delta needs 30,000 units of model cb-117 per year. Alpha Division is at full capacity (100,000 units of cb-117).

Required:

1. If Omni Inc. has a transfer pricing policy that requires transfer at market price, what would the transfer price be? Do you suppose that Alpha and Delta divisions would choose to transfer at that price?

2. Now suppose that Omni Inc. allows negotiated transfer pricing and that Alpha Division can avoid $3 of selling and distribution expense by selling to Delta Division. Which division sets the minimum transfer price, and what is it? Which division sets the maximum transfer price, and what is it? Do you suppose that Alpha and Delta divisions would choose to transfer somewhere in the bargaining range?

3. *What if* Alpha Division plans to produce and sell only 65,000 units of cb-117 next year (excess capacity exists)? Which division sets the minimum transfer price, and what is it? Which division sets the maximum transfer price, and what is it? Do you suppose that Alpha and Delta divisions would choose to transfer somewhere in the bargaining range?

Solution:

1. The market price is $14. Both Delta and Alpha divisions would be willing to transfer at that price (since neither division would be worse off than if it bought/sold in the outside market).

2. Minimum transfer price = $14 − $3 = $11. It is set by Alpha, the selling division. Maximum transfer price = $14. It is the market price and is set by Delta, the buying division.

   Yes, both divisions would be willing to accept a transfer price within the bargaining range. The actual transfer price set depends on the negotiating skills of the Alpha and Delta division managers.

3. Minimum transfer price = $2.50 (the variable cost of production). This price is set by Alpha, the selling division. Maximum transfer price = $14. This is the market price and is set by Delta, the buying division.

   Both divisions would be willing to accept a transfer price within the bargaining range. The actual transfer price depends on the negotiating skills of the Alpha and Delta division managers. (Notice that the fixed product costs are not included in the minimum transfer price because Alpha will have to pay total fixed cost no matter how many units are produced.)

In other cases, the use of full cost plus markup may simply be the formula agreed upon in negotiations. That is, the full-cost-plus-markup formula is the outcome of negotiation, but the transfer pricing method being used is reported as full cost plus markup. Once established, this formula could be used until the original conditions change to the point where renegotiation is necessary. In this way, the time and resources of negotiation can be minimized. For example, the goods transferred may be custom-made, and the managers may have little ability to identify an outside market price. In this case, reimbursement of full costs plus a reasonable rate of return may be a good surrogate for the transferring division's opportunity costs. Cornerstone 10-5 shows how and why cost-based transfer prices are calculated.

Transfer Pricing and the Multinational Firm

For the multinational firm, transfer pricing must accomplish two objectives: performance evaluation and optimal determination of income taxes. If all countries had the same tax structure, then transfer prices would be set independently of income taxes. However, there are high-tax countries (like Canada) and low-tax countries (such as the Cayman Islands). As a result, MNCs may use transfer pricing to shift costs to high-tax countries and shift revenues to low-tax countries.

Exhibit 10-5 illustrates this concept, as two transfer prices are set. The first transfer price is $100 as title for the goods passes from the Belgian subsidiary to the reinvoicing centre in Puerto Rico. Because the first transfer price is equal to full cost, profit is zero, and income taxes on zero profit also equal zero. The second transfer price is set at $200 by the reinvoicing centre in Puerto Rico. The transfer from Puerto Rico to Canada does result in profit, but this profit does not result in any income tax because Puerto Rico has no corporate income taxes. Finally, the Canadian subsidiary sells the product to an external party at the $200 transfer price. Again, price equals cost, so there is no profit on which to pay income taxes. Consider what would have happened without the reinvoicing centre. The goods would have gone directly from Belgium to Canada. If the transfer price was set at $200, the profit in Belgium would have been $100, subject to the 42 percent tax rate. Alternatively, if the transfer price set was $100, no Belgian income tax would have been paid, but the Canadian subsidiary would have realized a profit of $100, and that would have been subject to the Canadian corporate income tax rate of 35 percent.

Canadian-based multinationals are subject to CRA regulations on the pricing of intercompany transactions. CRA has the authority to reallocate income and deductions among divisions if it believes that such reallocation will reduce potential tax evasion. Transfer prices and sales among companies should be made at arm's length. That is, the transfer price set should match the price that would be set if the transfer were being made by unrelated parties, adjusted for differences that have a measurable effect on the price. Differences include landing costs and marketing costs. Landing costs (e.g., freight, insurance, customs duties, and special taxes) can increase the allowable transfer price. Marketing costs are usually avoided for internal transfers and reduce the transfer price. The CRA allows three pricing methods that approximate arm's-length pricing. In order of preference, these are the comparable uncontrolled price method, the resale price method, and the cost-plus method. The **comparable uncontrolled price method** is essentially market price. The **resale price method** is equal to the sales price received by the reseller less an appropriate markup. That is, the subsidiary purchasing a good for resale sets a transfer price equal to the resale price less a gross profit percentage. The **cost-plus method** is simply the cost-based transfer price. Cornerstone 10-6 shows the how and why of using the comparable uncontrolled price method and the resale price method.

The determination of an arm's-length price is a difficult one. Many times, the transfer pricing situation facing a company does not "fit" any of the three preferred methods just outlined. At such times, there may be room for a negotiated transfer price agreed upon between the CRA and the company. The CRA, taxpayers, and the Tax Court have struggled with negotiated transfer prices for years. However, this type

**CORNERSTONE
10-5**

The HOW and WHY of Calculating Cost-Based Transfer Prices

Information:

Omni Inc. has a number of divisions, including Alpha Division, producer of circuit boards, and Delta Division, a heating and air conditioning manufacturer.

Alpha Division produces the cb-117 model that can be used by Delta Division in the production of thermostats that regulate the heating and air conditioning systems. The market price of the cb-117 is $14. Cost information for the cb-117 model is:

Variable product cost	$2.50
Fixed cost	6.50
Total product cost	$9.00

Delta needs 30,000 units of model cb-117 per year. Alpha Division is at full capacity (100,000 units of cb-117).

Required:

1. If Omni Inc. has a transfer pricing policy that requires transfer at full product cost, what would the transfer price be? Do you suppose that Alpha and Delta divisions would choose to transfer at that price?
2. If Omni Inc. has a transfer pricing policy that requires transfer at full cost plus 25 percent, what would the transfer price be? Do you suppose that Alpha and Delta divisions would choose to transfer at that price?
3. If Omni Inc. has a transfer pricing policy that requires transfer at variable product cost plus a fixed fee of $12.00 per unit, what would the transfer price be? Do you suppose that Alpha and Delta divisions would choose to transfer at that price?
4. ***What if*** Alpha Division plans to produce and sell only 65,000 units of cb-117 next year? The Omni Inc. policy is that all transfers be at full cost. Which division sets the minimum transfer price, and what is it? Which division sets the maximum transfer price, and what is it? Do you suppose that Alpha and Delta divisions would choose to transfer?

Solution:

1. The full cost transfer price is $9.00. Delta Division would be delighted with that price, but Alpha Division would refuse to transfer since $14 could be earned in the outside market.
2. The cost-plus transfer price is $11.25 ($9.00 + $2.25). Again, Delta Division would be delighted with that price, but Alpha Division would refuse to transfer since $14 could be earned in the outside market.
3. The variable product cost plus fixed fee is $14.50 ($2.50 + $12). In this case, Alpha would be delighted, but Delta would refuse, since it can buy all it needs on the outside market for $14.
4. Minimum transfer price = $9.00 (the full cost of production). This price is set by Alpha, the selling division. Maximum transfer price = $14. This is the market price and is set by Delta, the buying division.

   Yes, both divisions would be willing to accept the transfer price of $9.00 per unit.

Exhibit 10-5

Use of Transfer Pricing to Affect Income Taxes Paid

Action	Tax Impact
Belgian subsidiary of Parent Company produces a component at a cost of $100 per unit. Title to the component is transferred to a reinvoicing centre* in Puerto Rico at a transfer price of $100/unit.	42% tax rate $100 revenue − $100 cost = $0 Taxes paid = $0
Reinvoicing centre in Puerto Rico, also a subsidiary of Parent Company, transfers title of component to Canadian subsidiary of Parent Company at a transfer price of $200/unit.	0% tax rate $200 revenue − $100 cost = $100 Taxes paid = $0
Canadian subsidiary sells component to external company at $200 each.	35% tax rate $200 revenue − $200 cost = $0 Taxes paid = $0

*A reinvoicing centre takes title to the goods but does not physically receive them. The primary objective of a reinvoicing centre is to shift profits to divisions in low-tax countries.

of negotiation occurs after the fact—that is, after income tax returns have been submitted and the company is being audited. **The CRA may or may not accept a transfer price as a transfer of intangibles (such as royalties on licences), sales of property, provision of services, and other items**.

Transfer pricing abuses are illegal—if they can be proved to be abuses. For example in some celebrated cases, toothbrushes were priced at more than $5,600 each for import from the United Kingdom into the United States, car seats were exported to Belgium for $1.66 each, and missile and rocket launchers were exported to Israel for just $52 each.

Of course, MNCs are also subject to taxation by other countries as well as Canada. Since income taxes are virtually universal, consideration of income tax effects pervades management decision making. Canada, the United States, Japan, the European Union, and South Korea have all issued transfer pricing regulations within the past 20 years. This increased emphasis on transfer price justification may account for the increased use of market prices as the transfer price by MNCs. However, research has shown that even market-based transfer prices can vary significantly from market prices set for a virtually identical arm's-length transaction.[13] It is thought that a highly important environmental variable considered by MNCs in setting a transfer pricing policy is overall profit to the company—with overall profit including the income tax impact of intra-company transfers.

Managers may legally avoid income taxes; they may not evade them. The distinction is important. Unfortunately, the difference between avoidance and evasion is less a line than a blurry gray area. While the situation depicted in Exhibit 10-5 is clearly abusive, other tax-motivated actions are not. For example, an MNC may decide to establish a needed research and development centre within an existing subsidiary in a high-tax country, since the costs are deductible. MNCs may have income tax-planning information systems that attempt to accomplish global income tax minimization. This is not an easy task.

[13] Andrew B. Bernard, J. Bradford Jensen, and Peter K. Schotts, "Transfer Pricing by U.S.-Based Multinational Firms," (September 2008). Tuck School of Business Working Paper No. 2006-33; U.S. Census Bureau Centre for Economic Studies Paper No. CES-WP-08-29. Available at SSRN: http://papers.ssrn.com/sol3/papers.cfm?abstract_id=924573.

**CORNERSTONE
10-6**

The HOW and WHY of Using the Comparable Uncontrolled Price Method and the Resale Price Method in Calculating Transfer Prices

Information:
ABC Inc. has a number of divisions around the world. Division B (in the United States) purchases a component from Division C (in Canada). The component can be purchased externally for $38 each. The freight and insurance on the item amount to $5; however, commissions of $3.80 need not be paid.

Required:
1. Calculate the transfer price using the comparable uncontrolled price method.
2. Suppose that there is no outside market for the component that Division C transfers to Division B. Further assume that Division B sells the component for $42 and normally receives a 40 percent markup on cost of goods sold. Calculate the transfer price using the resale price method.
3. Now assume that there is no external market for the component transferred from Division C to Division B, and that the component is used in the manufacture of another product (i.e., it is not resold). Calculate the transfer price using the cost-plus method. Further assume that Division C's manufacturing cost for the component is $20.
4. **What if** freight and insurance were $4 per unit? How would that affect the comparable uncontrolled price? The cost-plus price?

Solution:
1. The comparable uncontrolled price is calculated as follows:

Market price	$38.00
Plus: Freight and insurance	5.00
Less: Commissions	(3.80)
Transfer price	$39.20

2. With no outside market for Division C, and a resale price for Division B, the transfer price is calculated as follows:

$$\text{Resale price} = \text{Transfer price} + (\text{Markup percentage} \times \text{Transfer price})$$
$$\$42 = 1.40 \times \text{Transfer price}$$
$$\text{Transfer price} = \$42/1.40$$
$$= \$30$$

3. $\text{Cost-plus transfer price} = \text{Manufacturing cost} + \text{Freight and insurance}$
$$= \$20 + \$5$$
$$= \$25$$

4. If freight and insurance decreased by $1, the comparable uncontrolled price and the cost-plus price would be $1 lower.

Summary of Learning Objectives

1. **Define responsibility accounting, and describe the four types of responsibility centres.**

- Responsibility accounting is a system that measures the results of each responsibility centre and compares those results with some expected or budgeted outcome.
- In a decentralized organization, lower-level managers make and implement decisions; in a centralized organization, lower-level managers are responsible only for implementing decisions.
- Four types of responsibility centres are:

  - Cost centres—manager is responsible for costs.
  - Revenue centres—manager is responsible for price and quantity sold.
  - Profit centres—manager is responsible for costs and revenues.
  - Investment centres—manager is responsible for costs, revenues, and investment.

2. **Explain why firms choose to decentralize.**

- Local managers can make better decisions using local information.
- Local managers can also provide a more timely response to changing conditions.
- Cognitive limitations make it difficult for any one central manager to be fully knowledgeable about all products and markets.
- Decentralization permits training and motivating local managers.
- Top management is free to spend time on longer-range activities, such as strategic planning.
- Decentralizations enhance competition among the divisions.

3. **Compute and explain return on investment (ROI), residual income (RI), and economic value added (EVA).**

- ROI is the ratio of operating income to average operating assets.
- Margin is operating income divided by sales *or* margin times turnover.
- Turnover is sales divided by average operating assets.
- Advantage: ROI encourages managers to focus on improving sales, controlling costs, and using assets efficiently.
- Disadvantage: ROI can encourage managers to sacrifice long-run benefits for the short run.
- Residual income (RI) is operating income minus a minimum percentage cost of capital times capital employed.

  - If RI > 0, then the division is earning more than the minimum cost of capital.
  - If RI < 0, then the division is earning less than the minimum cost of capital.
  - If RI = 0, then the division is earning just the minimum cost of capital.

- Economic value added is *after-tax* operating profit minus the *actual* total annual cost of capital.

  - If EVA > 0, then the company is creating wealth.
  - If EVA < 0, then the company is destroying capital.

4. **Discuss methods of evaluating and rewarding managerial performance.**

- Goal congruence means that the goals of the manager are aligned with the goals of the company.
- Firms encourage goal congruence by constructing management compensation programs that reward managers for taking actions that benefit the firm. These programs can include:

  - Salary
  - Bonuses

- Stock options
- Noncash benefits (perquisites)

- The accountant in the international firm faces the ambiguous and ever-changing nature of global business.
- The accountant in the multinational company (MNC) must stay up to date on numerous business areas including:

  - Information systems
  - Marketing
  - Management
  - Political and legal factors
  - Economics
  - The financial accounting rules of the countries in which his or her firm operates

5. Explain the role of transfer pricing in a decentralized firm.

- When one division of a company produces a product that can be used in production by another division, transfer pricing exists.
- A transfer price is the price charged by one division of a company to another division of the same company.
- The transfer price is revenue to the selling division and cost to the buying division.

6. Discuss the methods of setting transfer prices.

- Three methods are commonly used to set transfer prices:

  - Market-based price
  - Cost-based price
  - Negotiated price

- The buying division sets the maximum transfer price.
- The selling division sets the minimum transfer price.
- MNCs with subsidiaries in both high- and low-tax countries may use transfer prices to minimize income taxes.
- The CRA accepts three transfer pricing policies:

  - Comparable uncontrolled price method
  - Resale price method
  - Cost-plus method

CORNERSTONES FOR CHAPTER 10

Review Problems

I. Transfer Pricing

The Components Division produces a part that is used by the Goods Division. The cost of manufacturing the part is as follows:

Direct materials	$10
Direct labour	2
Variable overhead	3
Fixed overhead*	5
Total cost	$20

*Based on a practical volume of 200,000 parts.

Other costs incurred by the Components Division are as follows:

Fixed selling and administrative expense	$500,000
Variable selling expense	$1 per unit

The part usually sells for between $28 and $30 in the external market. Currently, the Components Division is selling it to external customers for $29. The division is capable of producing 200,000 units of the part per year; however, because of a weak economy, only 150,000 parts are expected to be sold during the coming year. The variable selling expenses are avoidable if the part is sold internally.

The Goods Division has been buying the same part from an external supplier for $28. It expects to use 50,000 units of the part during the coming year. The manager of the Goods Division has offered to buy 50,000 units from the Components Division for $18 per unit.

Required:

1. Determine the minimum transfer price that the Components Division would accept.
2. Determine the maximum transfer price that the manager of the Goods Division would pay.
3. Should an internal transfer take place? Why or why not? If you were the manager of the Components Division, would you sell the 50,000 components for $18 each? Explain.
4. Suppose that the average operating assets of the Components Division total $10 million. Compute the ROI for the coming year, assuming that the 50,000 units are transferred to the Goods Division for $21 each.

Solution:

1. The minimum transfer price is $15. The Components Division has idle capacity and so must cover only its incremental costs, which are the variable manufacturing costs. (Fixed costs are the same whether or not the internal transfer occurs; the variable selling expenses are avoidable.)
2. The maximum transfer price is $28. The Goods Division would not pay more for the part than the price it would have to pay an external supplier.
3. Yes, an internal transfer ought to occur; the opportunity cost of the selling division is less than the opportunity cost of the buying division. The Components Division would earn an additional $150,000 profit ($3 × 50,000). The total joint benefit, however, is $650,000 ($13 × 50,000). The manager of the Components Division should attempt to negotiate a more favourable outcome for that division.
4. Income statement:

Sales [($29 × 150,000) + ($21 × 50,000)]	$ 5,400,000
Less: Variable cost of goods sold ($15 × 200,000)	(3,000,000)
Variable selling expenses ($1 × 150,000)	(150,000)
Contribution margin	2,250,000
Less: Fixed overhead ($5 × 200,000)	(1,000,000)
Fixed selling and administrative	(500,000)
Operating income	$ 750,000

$$\text{ROI} = \text{Operating income/Average operating assets}$$
$$= \$750{,}000/\$10{,}000{,}000$$
$$= 0.075$$

II. EVA

Surfit Company, which manufactures surfboards, has been in business for six years. Sam Foster, owner of Surfit, is pleased with the firm's profit picture and is considering taking the company public (i.e., selling stock in Surfit on the Toronto exchange). Data for the past year are as follows:

After-tax operating income	$ 250,000
Total capital employed	1,060,000
Long-term debt (interest at 9%)	100,000
Owner's equity	900,000

Surfit Company pays taxes at the rate of 35 percent.

Required:

1. Calculate the weighted average cost of capital, assuming that owner's equity is valued at the average cost of common stock of 12 percent. Calculate the total cost of capital for Surfit Company last year.
2. Calculate EVA for Surfit Company.

Solution:

1.

	Amount	Percent	×	After-Tax Cost	=	Weighted Cost
Long-term debt	$ 100,000	0.10		0.0585*		0.0059
Owner's equity	900,000	0.90		0.1200		0.1080
Total	$1,000,000					0.1139

*0.09 × (1 − 0.035) = 0.0585

The weighted average cost of capital is 11.39 percent.

The cost of capital last year = 0.1139 × $1,060,000 = $120,734.

2. EVA = $250,000 − $120,734 = $129,266

Key Terms

Centralized decision making, 491

Comparable uncontrolled price method, 517

Cost centre, 490

Cost-plus method, 517

Decentralization, 491

Decentralized decision making, 491

Economic value added (EVA), 501

Effectiveness, 493

Efficiency, 493

Goal congruence, 493

Investment centre, 490

Margin, 494

Maximum transfer price, 509

Minimum transfer price, 509

Multinational corporation (MNC), 491

Myopic behaviour, 498

Operating assets, 494

Operating income, 494

Opportunity cost approach, 509

Perquisites, 505

Profit centre, 490

Resale price method, 517

Residual income, 498

Responsibility accounting, 490

Discussion Questions

1. What is decentralization? Discuss the differences between centralized and decentralized decision making.
2. Explain why firms choose to decentralize.
3. Explain how access to local information can improve decision making.
4. What are margin and turnover? Explain how these concepts can improve the evaluation of an investment centre.
5. What are the three benefits of ROI? Explain how each can lead to improved profitability.
6. What are two disadvantages of ROI? Explain how each can lead to decreased profitability.
7. What is residual income? Explain how residual income overcomes one of ROI's disadvantages.
8. What is EVA? How does it differ from ROI and residual income?
9. What is a stock option? How can it encourage goal congruence?
10. What is a transfer price?
11. What is the transfer pricing problem?
12. If the minimum transfer price of the selling division is less than the maximum transfer price of the buying division, the intermediate product should be transferred internally. Do you agree or disagree? Why?
13. If an outside, perfectly competitive market exists for the intermediate product, what should the transfer price be? Why?
14. Identify three cost-based transfer prices. What are the disadvantages of cost-based transfer prices? When might it be appropriate to use cost-based transfer prices?

Cornerstone Exercises

Cornerstone Exercise 10-1 CALCULATING AVERAGE OPERATING ASSETS, MARGIN, TURNOVER, RETURN ON INVESTMENT (ROI)

OBJECTIVE ▶ 3
CORNERSTONE 10-1
SERVICE

Chester Inc. provided the following information for two of its divisions for last year:

	Paper Products Division	Chemicals Division
Sales	$29,375,000	$32,400,000
Operating income	2,350,000	1,296,000
Operating assets, January 1	6,000,000	14,000,000
Operating assets, December 31	5,750,000	13,000,000

Required:

1. For the Paper Products Division, calculate:
   a. Average operating assets
   b. Margin
   c. Turnover
   d. Return on investment (ROI)

2. For the Chemicals Division, calculate:
   a. Average operating assets
   b. Margin
   c. Turnover
   d. Return on investment (ROI)

3. **What if** operating income for the Paper Products Division was $2,000,000? How would that affect average operating assets? Margin? Turnover? ROI? Calculate any changed ratios (round to four significant digits).

OBJECTIVE ➤ 3

CORNERSTONE 10-2

SERVICE

Cornerstone Exercise 10-2 CALCULATING RESIDUAL INCOME

Refer to **Cornerstone Exercise 10-1**. Chester Inc. requires a 10 percent minimum rate of return.

Required:

1. Calculate residual income for the Paper Products Division.
2. Calculate residual income for the Chemicals Division.
3. **What if** the minimum required rate of return was 8 percent? How would that affect the residual income of the two divisions?

OBJECTIVE ➤ 3

CORNERSTONE 10-3

Cornerstone Exercise 10-3 CALCULATING WEIGHTED AVERAGE COST OF CAPITAL AND ECONOMIC VALUE ADDED (EVA)

Duhamel Industries Inc. had after-tax operating income last year of $1,996,500. Three sources of financing were used by the company: $4 million of mortgage bonds paying 8 percent interest, $5 million of unsecured bonds paying 10 percent interest, and $11 million in common stock, which was considered to be relatively risky (with a risk premium of 8 percent). The rate on long-term treasuries is 6 percent. Duhamel Industries Inc. pays a marginal tax rate of 40 percent.

Required:

1. Calculate the after-tax cost of each method of financing.
2. Calculate the weighted average cost of capital for Duhamel Industries Inc. Calculate the total dollar amount of capital employed for Duhamel Industries Inc.
3. Calculate economic value added (EVA) for Duhamel Industries Inc. for last year. Is the company creating or destroying wealth?
4. **What if** Duhamel Industries Inc. had common stock that was less risky than other stocks and commanded a risk premium of 5 percent? How would that affect the weighted average cost of capital? How would it affect EVA?

OBJECTIVE ➤ 6

CORNERSTONE 10-4

Cornerstone Exercise 10-4 DETERMINING MARKET-BASED AND NEGOTIATED TRANSFER PRICES

Deng Inc. has a number of divisions, including Aberdeen Division, producer of surgical blades, and Fairfield Division, a manufacturer of medical instruments.

Aberdeen Division produces a 2.6 cm steel blade that can be used by Fairfield Division in the production of scalpels. The market price of the blade is $22.60. Cost information for the blade is:

Variable product cost	$12.40
Fixed cost	7.00
Total product cost	$19.40

Fairfield needs 15,000 units of the 2.6 cm blade per year. Aberdeen Division is at full capacity (90,000 units of the blade).

Required:

1. If Deng Inc. has a transfer pricing policy that requires transfer at market price, what would the transfer price be? Do you suppose that Aberdeen and Fairfield divisions would choose to transfer at that price?
2. Now suppose that Deng Inc. allows negotiated transfer pricing and that Aberdeen Division can avoid $1.75 of selling and distribution expense by selling to Fairfield Division. Which division sets the minimum transfer price, and what is it? Which division sets the maximum transfer price, and what is it? Do you suppose that Aberdeen and Fairfield divisions would choose to transfer somewhere in the bargaining range?

3. ***What if*** Aberdeen Division plans to produce and sell only 65,000 units of the 2.6 cm blade next year? Which division sets the minimum transfer price, and what is it? Which division sets the maximum transfer price, and what is it? Do you suppose that Aberdeen and Fairfield divisions would choose to transfer somewhere in the bargaining range?

Cornerstone Exercise 10-5 DETERMINING MARKET-BASED AND NEGOTIATED TRANSFER PRICES

OBJECTIVE ➤ 6
CORNERSTONE 10-5

Refer to **Cornerstone Exercise 10-4**.

Required:

1. If Deng Inc. has a transfer pricing policy that requires transfer at full product cost, what would the transfer price be? Do you suppose that Aberdeen and Fairfield divisions would choose to transfer at that price?
2. If Deng Inc. has a transfer pricing policy that requires transfer at full cost plus 25 percent, what would the transfer price be? Do you suppose that Aberdeen and Fairfield divisions would choose to transfer at that price?
3. If Deng Inc. has a transfer pricing policy that requires transfer at variable product cost plus a fixed fee of $2.00 per unit, what would the transfer price be? Do you suppose that Aberdeen and Fairfield divisions would choose to transfer at that price?
4. ***What if*** Aberdeen Division plans to produce and sell only 65,000 units of the 2.6 cm blade next year? The Deng Inc. policy is that all transfers be at full cost. Which division sets the minimum transfer price, and what is it? Which division sets the maximum transfer price, and what is it? Do you suppose that Aberdeen and Fairfield divisions would choose to transfer?

Cornerstone Exercise 10-6 DETERMINING MARKET-BASED AND NEGOTIATED TRANSFER PRICES

OBJECTIVE ➤ 6
CORNERSTONE 10-6

Dalloway Inc. has a number of divisions around the world. Division CA (in Canada) purchases a component from Division ND (in the Netherlands). The component can be purchased externally for $18.50 each. The freight and insurance on the item amount to $1.90; however, commissions of $0.95 need not be paid.

Required:

1. Calculate the transfer price using the comparable uncontrolled price method.
2. Suppose that there is no outside market for the component that Division ND transfers to Division CA. Further assume that Division CA sells the component for $22.10 and normally receives a 30 percent markup on cost of goods sold. Calculate the transfer price using the resale price method.
3. Now assume that there is no external market for the component transferred from Division ND to Division CA, and that the component is used in the manufacture of another product (i.e., it is not resold). Calculate the transfer price using the cost-plus method. Further assume that Division ND's manufacturing cost for the component is $16.30.
4. ***What if*** commissions avoided were $2 per unit? How would that affect the comparable uncontrolled price? The resale price? The cost-plus price?

Exercises

Exercise 10-7 ROI, MARGIN, TURNOVER

OBJECTIVE ➤ 3

Pirelli Inc. presented two years of data for its Clothing Division and its Camping Division.

Clothing Division:	Year 1	Year 2
Sales	$3,400,000	$3,750,000
Operating income	190,400	191,250
Average operating assets	1,062,500	1,062,500

(continued)

Camping Division:

	Year 1	Year 2
Sales	$2,600,000	$2,680,000
Operating income	119,600	107,200
Average operating assets	1,000,000	1,000,000

Required:

1. Compute the ROI and the margin and turnover ratios for each year for the Clothing Division. (Round your answers to four significant digits.)
2. Compute the ROI and the margin and turnover ratios for each year for the Camping Division. (Round your answers to four significant digits.)
3. Explain the change in ROI from Year 1 to Year 2 for each division.

OBJECTIVE ▶ 3

Exercise 10-8 ROI AND INVESTMENT DECISIONS

Refer to **Exercise 10-7** for data. At the end of Year 2, the manager of the Camping Division is concerned about the division's performance. As a result, he is considering the opportunity to invest in two independent projects. The first is called the "Ever-Tent"; it is a small two-person tent capable of withstanding the high winds at the top of Mt. Everest. While the market for actual Everest climbers is small, the manager expects that well-to-do weekend campers will buy it due to the cachet of the name and its light weight. The second is a "KiddieKamp" kit which includes a child-sized sleeping bag and a colourful pup tent that can be set up easily in one's backyard. Without the investments, the division expects that Year 2 data will remain unchanged. The expected operating incomes and the outlay required for each investment are as follows:

	Ever-Tent	KiddieKamp
Operating income	$ 5,500	$ 3,800
Outlay	50,000	40,000

Pirelli's corporate headquarters has made available up to $100,000 of capital for this division. Any funds not invested by the division will be retained by headquarters and invested to earn the company's minimum required rate of return, 9 percent.

Required:

1. Compute the ROI for each investment.
2. Compute the divisional ROI (rounded to four significant digits) for each of the following four alternatives:
   a. The Ever-Tent is added.
   b. The KiddieKamp is added.
   c. Both investments are added.
   d. Neither investment is made; the status quo is maintained.
   Assuming that divisional managers are evaluated and rewarded on the basis of ROI performance, which alternative do you think the divisional manager will choose?

OBJECTIVE ▶ 3

Exercise 10-9 RESIDUAL INCOME AND INVESTMENT DECISIONS

Refer to the data given in **Exercise 10-8.**

Required:

1. Compute the residual income for each of the opportunities. (Round to the nearest dollar.)
2. Compute the divisional residual income (rounded to the nearest dollar) for each of the following four alternatives:
   a. The Ever-Tent is added.
   b. The KiddieKamp is added.
   c. Both investments are added.
   d. Neither investment is made; the status quo is maintained.

Assuming that divisional managers are evaluated and rewarded on the basis of residual income, which alternative do you think the divisional manager will choose?

3.   Based on your answer in Requirement 2, compute the profit or loss from the divisional manager's investment decision. Was the correct decision made?

Exercise 10-10   CALCULATING EVA

OBJECTIVE ▶ 3

Mortimer Company manufactures elderberry wine. Last year, Mortimer earned operating income of $206,000 after income taxes. Capital employed equalled $2 million. Mortimer is 60 percent equity and 40 percent 10-year bonds paying 5 percent interest. Mortimer's marginal tax rate is 40 percent. The company is considered a fairly risky investment and probably commands a 12-point premium above the 6 percent rate on long-term T-bills.

Jonathan Mortimer's aunts, Abby and Martha, have just retired, and Mortimer is the new CEO of Mortimer Company. He would like to improve EVA for the company. Compute EVA under each of the following independent scenarios that Mortimer is considering. (Use a spreadsheet to perform your calculations and round all percentage figures to four significant digits.)

Required:

1.   No changes are made; calculate EVA using the original data.
2.   Sugar will be used to replace another natural ingredient (atomic number 33) in the elderberry wine. This should not affect costs but will begin to affect the market assessment of Mortimer Company, bringing the premium above long-term T-bills to 10 percent the first year and 7 percent the second year. Calculate revised EVA for both years.
3.   Mortimer is considering expanding but needs additional capital. The company could borrow money, but it is considering selling more common stock, which would increase equity to 80 percent of total financing. Total capital employed would be $3,000,000. The new after-tax operating income would be $450,000. Using the original data, calculate EVA. Then, recalculate EVA assuming the materials substitution described in Requirement 2. New after-tax income will be $450,000, and in Year 1, the premium will be 10 percent above the long-term rate. In Year 2, it will be 7 percent above the long-term rate. (*Hint:* You will calculate three EVAs for this requirement.)

Exercise 10-11   OPERATING INCOME FOR SEGMENTS

OBJECTIVE ▶ 3

Venpool Inc. manufactures and sells cooktops and ovens through three divisions: Home, Restaurant, and Specialty. Each division is evaluated as a profit centre. Data for each division for last year are as follows (numbers in thousands):

	Home	Restaurant	Specialty
Sales	$3,450	$3,000	$2,100
Cost of goods sold	2,400	2,200	1,400
Selling and administrative expenses	840	370	250

The income tax rate for Venpool Inc. is 30 percent. Venpool Inc. has two sources of financing: bonds paying 6 percent interest, which account for 30 percent of total investment, and equity accounting for the remaining 70 percent of total investment. Venpool Inc. has been in business for over 15 years and is considered a relatively stable stock, despite its link to the cyclical construction industry. As a result, Venpool stock has an opportunity cost of 5 percent over the 6 percent long-term government bond rate. Venpool's total capital employed is $3.34 million ($2,100,000 for the Home Division, $700,000 for the Restaurant Division, and the remainder for the Specialty Division).

Required:

1.   Prepare a segmented income statement for Venpool Inc. for last year.
2.   Calculate Venpool's weighted average cost of capital. (Round to four significant digits.)

3. Calculate EVA for each division and for Venpool Inc.
4. Comment on the performance of each of the divisions.

OBJECTIVE ▶ 5 6

Exercise 10-12 TRANSFER PRICING, IDLE CAPACITY

Mouton & Perrier Inc. has a number of divisions that produce liquors, bottled water, and glassware. The Glassware Division manufactures a variety of bottles which can be sold externally (to soft-drink and juice bottlers) or internally to Mouton & Perrier's Bottled Water Division. Sales and cost data on a case of 24 basic 1 litre bottles are as follows:

Unit selling price	$2.95
Unit variable cost	$1.25
Unit product fixed cost*	$0.70
Practical capacity in cases	500,000

*$350,000/500,000

During the coming year, the Glassware Division expects to sell 390,000 cases of this bottle. The Bottled Water Division currently plans to buy 100,000 cases on the outside market for $2.95 each. Ellyn Burridge, manager of the Glassware Division, approached Justin Thomas, manager of the Bottled Water Division, and offered to sell the 100,000 cases for $2.89 each. Ellyn explained to Justin that she can avoid selling costs of $0.12 per case by selling internally and that she would split the savings by offering a $0.06 discount on the usual price.

Required:

1. What is the minimum transfer price that the Glassware Division would be willing to accept? What is the maximum transfer price that the Bottled Water Division would be willing to pay? Should an internal transfer take place? What would be the benefit (or loss) to the firm as a whole if the internal transfer takes place?
2. Suppose Justin knows that the Glassware Division has idle capacity. Do you think that he would agree to the transfer price of $2.89? Suppose he counters with an offer to pay $2.40. If you were Ellyn, would you be interested in this price? Explain with supporting computations.
3. Suppose that Mouton & Perrier's policy is that all internal transfers take place at full manufacturing cost. What would the transfer price be? Would the transfer take place?

OBJECTIVE ▶ 6

Exercise 10-13 TRANSFER PRICING

Comfort Furniture Manufacturing Inc. has a division in Canada that produces and sells furniture for discount furniture stores. One type of sofa is made in the International Division in China. The sofas are sold externally in Canada for $250 each. It costs $7.50 per sofa for shipping and $10 per sofa for import duties. When the sofas are sold externally, Comfort Furniture Manufacturing spends $25 each for commissions and an average of $1.30 per sofa for advertising.

Required:

1. Which CRA method should be used to calculate the allowable transfer price?
2. Using the appropriate CRA method, calculate the transfer price.

OBJECTIVE ▶ 6

Exercise 10-14 TRANSFER PRICING

Desant Inc. has a division in Indonesia that makes dyestuff in a variety of colours used to dye denim for jeans, and another division in Canada that manufactures denim clothing. The Dyestuff Division incurs manufacturing costs of $3.76 for one kilogram of powdered dye.

The Clothing Division currently buys its dye powder from an outside supplier for $4.10 per kilogram. If the Clothing Division purchases the powder from the Indonesian division, the shipping costs will be $0.16 per kilogram, but sales commissions of $0.04 per kilogram will be avoided with an internal transfer.

Required:

1. Which Section 482 method should be used to calculate the allowable transfer price? Calculate the appropriate transfer price per kilogram.
2. Assume that the Clothing Division cannot buy this type of powder externally since it has an unusual formula that results in a colour particular to Desant's jeans. Which CRA method should be used to calculate the allowable transfer price? Calculate the appropriate transfer price per kilogram.

Exercise 10-15 TRANSFER PRICING

OBJECTIVE ➤ 6

Rao Inc. has a division in Canada that makes paint. Rao has another U.S. division, the Retail Division, that operates a chain of home improvement stores. The Retail Division would like to buy the unique, long-lasting paint from the Canadian division, since this type of paint is not currently available. The Paint Division incurs manufacturing costs of $4.12 for 1 litre of paint.

If the Retail Division purchases the paint from the Canadian division, the shipping costs will be $0.40 per litre, but sales commissions of $0.60 per litre will be avoided with an internal transfer. The Retail Division plans to sell the paint for $11.68 per litre. Normally, the Retail Division earns a gross margin of 60 percent above cost of goods sold.

Required:

1. Which CRA method should be used to calculate the allowable transfer price?
2. Calculate the appropriate transfer price per litre.

Exercise 10-16 ROI AND RESIDUAL INCOME

OBJECTIVE ➤ 3

A multinational corporation has a number of divisions, two of which are the North American Division and the South American Division. Data on the two divisions are as follows:

	North American	South American
Average operating assets	5,000,000	3,900,000
Operating income	618,000	397,500
Minimum required return	10%	10%

Round all rates of return to four significant digits.

Required:

1. Compute residual income for each division. By comparing residual income, is it possible to make a useful comparison of divisional performance? Explain.
2. Compute the residual rate of return by dividing the residual income by the average operating assets. Is it possible now to say that one division outperformed the other? Explain.
3. Compute the return on investment for each division. Can we make meaningful comparisons of divisional performance? Explain.
4. Add the residual rate of return computed in Requirement 2 to the required rate of return. Compare these rates with the ROI computed in Requirement 3. Will this relationship always be the same?

Exercise 10-17 MARGIN, TURNOVER, ROI

OBJECTIVE ➤ 3

Consider the data for each of the following four independent companies:

	A	B	C	D
Revenue	$10,000	$48,000	$96,000	?
Expenses	$8,000	?	$90,000	?
Operating income	$2,000	$12,000	?	?
Assets	$40,000	?	$48,000	$9,600
Margin	?	25%	?	6.25%
Turnover	?	0.50	?	2.00
ROI	?	?	?	?

Required:

1. Calculate the missing values in the above table.
2. Assume that the cost of capital is 9 percent for each of the four firms. Compute the residual income for each of the four firms.

OBJECTIVE ▶ 3

Exercise 10-18 **ROI, RESIDUAL INCOME**

The following selected data pertain to the Argent Division for last year:

Sales	$1,000,000
Variable costs	$624,000
Traceable fixed costs	$100,000
Average invested capital	$1,500,000
Imputed interest rate	15%

Required:

1. How much is the residual income?
2. How much is the return on investment? (Rounded to four significant digits.)

OBJECTIVE ▶ 4

Exercise 10-19 **STOCK OPTIONS**

Fermat Inc. has acquired two new companies, one in consumer products and the other in financial services. Fermat's top management believes that the executives of the two newly acquired companies can be most quickly assimilated into the parent company if they own shares of Fermat stock. Accordingly, on April 1, Fermat approved a stock option plan whereby each of the top four executives of the new companies could purchase up to 20,000 shares of Fermat stock at $15 per share. The option will expire in five years.

Required:

1. If Fermat stock rises to $26.50 per share by December 1, what is the value of the option to each executive?
2. Discuss some of the advantages and disadvantages of the Fermat stock option plan.

Problems

OBJECTIVE ▶ 4

SERVICE

Problem 10-20 **BONUSES AND STOCK OPTIONS**

Lawanna Davis graduated from Eastern University with a major in accounting five years ago. She obtained a position with a well-known professional services firm upon graduation and has become one of their outstanding performers. In the course of her work, she has developed numerous contacts with business firms in the area. One of them, Bumaby Inc. recently offered her a position as head of their Financial Services Division. The offer includes a salary of $50,000 per year, annual bonuses of 1 percent of divisional operating income, and a stock option for 10,000 shares of Bumaby stock to be exercised at $15 per share in two years. Last year, the Financial Services Division earned $1,110,000. This year, it is budgeted to earn $1,600,000. Bumaby stock has increased in value at the rate of 16 percent per year over the past five years. Lawanna currently earns $65,000.

Required:

Advise Lawanna on the relative merits of the Bumaby offer.

OBJECTIVE ▶ 5 6

Problem 10-21 **SETTING TRANSFER PRICES—MARKET PRICE VERSUS FULL COST**

Omicron Inc. manufactures heating and air conditioning units in its six divisions. One division, the Components Division, produces electronic components that can be used by the other five. All the components produced by this division can be sold to outside customers; however, from the beginning, about 70 percent of its output has been used internally. The current policy requires that all internal transfers of components be transferred at full cost.

Recently, Cynthia Busby, the new chief executive officer of Omicron, decided to investigate the transfer pricing policy. She was concerned that the current method of pricing internal transfers might force decisions by divisional managers that would be suboptimal for the firm. As part of her inquiry, she gathered some information concerning Part 4CM, used by the Small AC Division in its production of a window air conditioner, Model 7AC.

The Small AC Division sells 100,000 units of Model 7AC each year at a unit price of $58. Given current market conditions, this is the maximum price that the division can charge for Model 7AC. The cost of manufacturing the air conditioner is computed as follows:

Part 4CM	$ 6.45
Direct materials	23.00
Direct labour	15.00
Variable overhead	3.50
Fixed overhead	6.50
Total unit cost	$54.45

The window unit is produced efficiently, and no further reduction in manufacturing costs is possible.

The manager of the Components Division indicated that he could sell 10,000 units (the division's capacity for this part) of Part 4CM to outside buyers at $12 per unit. The Small AC Division could also buy the part for $12 from external suppliers. The following detail on the manufacturing cost of the component was provided:

Direct materials	$2.75
Direct labour	0.80
Variable overhead	1.10
Fixed overhead	1.80
Total unit cost	$6.45

Required:

1. Compute the firmwide contribution margin associated with Part 4CM and Model 7AC. Also, compute the contribution margin earned by each division.
2. Suppose that Cynthia Busby abolishes the current transfer pricing policy and gives divisions autonomy in setting transfer prices. Can you predict what transfer price the manager of the Components Division will set? What should be the minimum transfer price for this part? The maximum transfer price?
3. Given the new transfer pricing policy, predict how this will affect the production decision for Model 7AC of the manager of the Small AC Division. How many units of Part 4CM will the manager of the Small AC Division purchase, either internally or externally?
4. Given the new transfer price set by the Components Division and your answer to Requirement 3, how many units of Part 4CM will be sold externally?
5. Given your answers to Requirements 3 and 4, compute the firmwide contribution margin. What has happened? Was Cynthia's decision to grant additional decentralization good or bad?

Problem 10-22 TRANSFER PRICING WITH IDLE CAPACITY

OBJECTIVE ▸ 3 5 6

Oriole Inc. owns a number of food service companies. Two divisions are the Coffee Division and the Donut Shop Division. The Coffee Division purchases and roasts coffee beans for sale to supermarkets and specialty shops. The Donut Shop Division operates a chain of donut shops where the donuts are made on the premises. Coffee is an important item for sale along with the donuts and, to date, has been purchased from the Coffee Division. Company policy permits each manager the freedom to decide whether or not to buy or sell internally. Each divisional manager is evaluated on the basis of return on investment and residual income.

Recently, an outside supplier has offered to sell coffee beans, roasted and ground, to the Donut Shop Division for $4.30 per kilogram. Since the current price paid to the

	JSC	RLI
Current liabilities	$ 1,400,000	$ 850,000
Long-term liabilities	3,800,000	1,200,000
Shareholders' equity	2,800,000	950,000
Total liabilities and shareholders' equity	$ 8,000,000	$ 3,000,000

Required:

1. If Mason Industries continues to use return on investment as the sole measure of division performance, explain why JSC would be reluctant to acquire RLI. Be sure to support your answer with appropriate calculations.

2. If Mason Industries could be persuaded to use residual income to measure the performance of JSC, explain why JSC would be more willing to acquire RLI. Be sure to support your answer with appropriate calculations.

3. Discuss how the behaviour of division managers is likely to be affected by the use of:

   a. Return on investment as a performance measure
   b. Residual income as a performance measure *(CMA adapted)*

The Collaborative Learning Exercises can be found on the product support site at www.hansen1ce.nelson.com.

Part 2
Chapters 5–10

Beauville Furniture Corporation produces sofas, recliners, and lounge chairs. Beauville is located in a medium-sized community in southeastern Quebec. It is a major employer in the community. In fact, the economic well-being of the community is tied very strongly to Beauville. Beauville operates a sawmill, a fabric plant, and a furniture plant in the same community.

The sawmill buys logs from independent producers. The sawmill then processes the logs into four grades of lumber: firsts and seconds, No. 1 common, No. 2 common, and No. 3 common. All costs incurred in the mill are common to the four grades of lumber. All four grades of lumber are used by the furniture plant. The mill transfers everything it produces to the furniture plant, and the grades are transferred at cost. Trucks are used to move the lumber from the mill to the furniture plant. Although no outside sales exist, the mill could sell to external customers, and the selling prices of the four grades are known.

The fabric plant is responsible for producing the fabric that is used by the furniture plant. To produce three totally different fabrics (identified by fabric ID codes: FB60, FB70, and FB80, respectively), the plant has three separate production operations—one for each fabric. Thus, production of all three fabrics occurs at the same time in different locations in the plant. Each fabric's production operation has two processes: the weaving and pattern process, and the colouring and bolting process. In the weaving and pattern process, yarn is used to create yards of fabric with different designs. In the next process, the fabric is dyed, cut into 25-metre sections, and wrapped around cardboard rods to form 25-metre bolts. The bolts are transported by forklift to the furniture plant's Receiving Department. All of the output of the fabric plant is used by the furniture plant (to produce the sofas and chairs). For accounting purposes, the fabric is transferred at cost to the furniture plant.

The furniture plant produces orders for customers on a special-order basis. The customers specify the quantity, style, fabric, lumber grade, and pattern. Typically, jobs are large (involving at least 500 units). The plant has two production departments: Cutting and Assembly. In the Cutting Department, the fabric and wooden frame components are sized and cut. Other components are purchased from external suppliers and are removed from stores as needed for assembly. After the fabric and wooden components are finished for the entire job, they are moved to the Assembly Department. The Assembly Department takes the individual components and assembles the sofas (or chairs).

Beauville Furniture has been in business for over two decades and has a good reputation. However, during the past five years, Beauville experienced eroding profits and declining sales. Bids were increasingly lost (even aggressive bids) on the more popular models. Yet, the company was winning bids on some of the more-difficult-to-produce items. Louis Renaud, the owner and manager, was frustrated. He simply couldn't understand how some of his competitors could sell for such low prices. On a common sofa job involving 500 units, Beauville's bids were running $25 per unit, or $12,500 per job more than the winning bids (on average). Yet, on the more difficult items, Beauville's bids were running about $60 per unit less than the next closest bid. Giselle Boucher, vice president of finance, was assigned the task of preparing a cost analysis of the company's product lines. Louis wanted to know if the company's costs were excessive. Perhaps the company was being wasteful, and it was simply costing more to produce furniture than it was costing its competitors.

Giselle prepared herself by reading recent literature on cost management and product costing and attending several conferences that explored the same issues. She then reviewed the costing procedures of the company's mill and two plants and did a preliminary assessment of their soundness. The production costs of the mill were common to all lumber grades and were assigned using the physical units method. Since the output and production costs were fairly uniform throughout the year, the mill used an actual costing system. Although Giselle had no difficulty with actual costing, she decided to explore the effects of using the sales-value-at-split-off method. Thus, cost and production data for the mill were gathered so that an analysis could be conducted. The two plants used normal costing systems. The fabric plant used process costing, and the furniture plant used job-order costing. Both plants used plantwide overhead rates based on direct labour hours. Based on her initial reviews, she concluded that the costing procedures for the fabric plant were satisfactory. Essentially, there was no evidence of product diversity. A statistical analysis revealed that about 90 percent of the variability in the plant's overhead cost could be explained by direct labour hours. Thus, the use of a plantwide overhead rate based on direct labour hours seemed justified. What did concern her, though, was the material waste that she observed in the plant. Maybe a standard cost system would be useful for increasing the overall cost efficiency of the plant. Consequently, as part of her report to Louis, she decided to include a description of the fabric plant's costing procedures—at least for one of the fabric types. She also decided to develop a standard cost sheet for the chosen fabric. The furniture plant, however, was a more difficult matter. Product diversity was present and could be causing some distortions in product costs. Furthermore, statistical analysis revealed that only about 40 percent of the variability in overhead cost was explained by the direct labour hours. She decided that additional analysis was needed so that a sound product costing method could be recommended. One possibility would be to increase the number of overhead rates. Thus, she decided to include departmental data so that the effect of moving to departmental rates could be assessed. Finally, she also wanted to explore the possibility of converting the sawmill and fabric plant into profit centres and changing the existing transfer pricing policy.

With the cooperation of the cost accounting manager for the mill and each plant's controller, she gathered the following data for last year:

Sawmill:

Joint manufacturing costs: $900,000

Grade	Quantity Produced (board feet)	Price at Split-Off (per 1,000 board foot)
Firsts and seconds	1,500,000	$300
No. 1 common	3,000,000	225
No. 2 common	1,875,000	140
No. 3 common	1,125,000	100
Total	7,500,000	

Fabric Plant:

Budgeted overhead: $1,200,000 (50 percent fixed)
Practical volume (direct labour hours): 120,000 hours
Actual overhead: $1,150,000 (50 percent fixed)
Actual hours worked:

	Weaving and Pattern	Colouring and Bolting	Total
Fabric FB60	20,000	12,000	32,000
Fabric FB70	28,000	14,000	42,000
Fabric FB80	26,000	18,000	44,000
Total	74,000	44,000	118,000

Departmental data on Fabric FB70 (actual costs and actual outcomes):

	Weaving and Pattern	Colouring and Bolting
Beginning inventories:		
Units*	20,000	400
Costs:		
Transferred in	$ 0	$100,000
Materials	$80,000	$8,000
Labour	$18,000	$6,600
Overhead	$22,000	$9,000

	Weaving and Pattern	Colouring and Bolting
Current production:		
Units started	80,000	?
Units transferred out	80,000	3,200
Costs:		
Transferred in	$ 0	?
Materials	$320,000	$82,000
Labour	$208,000	$99,400
Overhead	?	?
Percentage completion:		
Beginning inventory	30%	40%
Ending inventory	40%	50%

* Units are measured in metres for the Weaving and Pattern Department and in bolts for the Colouring and Bolting Department. *Note:* With the exception of the cardboard bolt rods, materials are added at the beginning of each process. The cost of the rods is relatively insignificant and is included in overhead.

Proposed standard cost sheet for Fabric FB70 (for the Colouring and Bolting Department only):

Transferred-in materials (25 metres @ $10)	$250.00
Other materials (2.8 metres @ $7.15)	20.00 (rounded)
Labour (3.1 hours @ $8)	24.80
Fixed overhead (3.1 hours @ $5)	15.50
Variable overhead (3.1 hours @ $5)	15.50
Standard cost per unit	$325.80

Furniture Plant:

Departmental data (budgeted):

	Service Departments				Producing Departments	
	Receiving	Power	Maintenance	General Factory	Cutting	Assembly
Overhead	$450,000	$600,000	$300,000	$525,000	$750,000	$375,000
Machine hours	—	—	—	—	60,000	15,000
Receiving orders	—	—	—	—	13,500	9,000
Square metres	1,000	5,000	4,000	—	15,000	10,000
Direct labour hours	—	—	—	—	50,000	200,000

After some discussion with the furniture plant controller, Giselle decided to use machine hours to calculate the overhead rate for the Cutting Department and direct labour hours for the Assembly Department rate (the Cutting Department was more automated than the Assembly Department). As part of her report, she wanted to compare the effects of plantwide rates and departmental rates on the cost of jobs. She wanted to know if overhead costing could be the source of the pricing problems the company was experiencing.

To assess the effect of the different overhead assignment procedures, Giselle decided to examine two prospective jobs. One job, Job A500, could produce 500 sofas, using a frequently requested style and Fabric FB70. Bids on this type of job were being lost more frequently to competitors. The second job, Job B75, would produce 75 specially designed recliners. This job involved a new design and was more difficult for the workers to build. It involved some special cutting requirements and an unfamiliar assembly. Recently, the company seemed to be winning more bids on jobs of this type. To compute the costs of the two jobs, Giselle assembled the following information on the two jobs:

Job A500:

Direct materials:	
Fabric FB70	180 bolts @ $350
Lumber (No. 1 common)	20,000 board feet @ $0.12
Other components	$26,600
Direct labour:	
Cutting Department	400 hours @ $10
Assembly Department	1,600 hours @ $8.75
Machine time:	
Cutting Department	350 machine hours
Assembly Department	50 machine hours

Job B75:

Direct materials:	
Fabric FB70	26 bolts @ $350
Lumber (first and seconds)	2,200 board feet @ $0.12
Other components	$3,236
Direct labour:	
Cutting Department	70 hours @ $10
Assembly Department	240 hours @ $8.75
Machine time:	
Cutting Department	90 machine hours
Assembly Department	15 machine hours

Required:

1. Allocate the joint manufacturing costs to each grade, and calculate the cost per board foot for each grade: (a) using the physical units method of allocation and (b) using the sales-value-at-split-off method. Which method should the mill use? Explain. What is the effect on the cost of each proposed job if the mill switches to the sales-value-at-split-off method?

2. Calculate the plantwide overhead rate for the fabric plant.

3. Calculate the amount of under- or overapplied overhead for the fabric plant.

4. Using the weighted average method, calculate the cost per bolt for Fabric FB70.

5. Assume that the weaving and pattern process is not a separate process for each fabric. Also, assume that the yarn used for each fabric differs significantly in cost. In this case, would process costing be appropriate for the weaving and pattern process? What costing approach would you recommend? Describe your approach in detail.

6. In the Colouring and Bolting Department, 3,200 tonnes of other materials were used to produce the output of the period. Using the proposed standard cost sheet, calculate the following variances for the Colouring and Bolting Department:
   a. Materials price variance (for other materials only)
   b. Materials usage variance (for other materials only)
   c. Labour rate variance
   d. Labour efficiency variance

   In calculating the variances, which method did you use to compute the actual output of the period—FIFO or weighted average? Explain.

7. Assume that the standard hours allowed for the actual total output of the fabric plant are 115,000. Calculate the following variances:
   a. Fixed overhead spending variance
   b. Fixed overhead volume variance
   c. Variable overhead spending variance
   d. Variable overhead efficiency variance

8. Suppose that the fabric plant has 500 bolts of FB70 in beginning finished goods inventory. The current-year plan is to have 1,000 bolts of FB70 in finished goods inventory at the end of the year. This fabric has an external market price of $400 per bolt. If the fabric plant is set up as a profit centre, it could sell 3,000 bolts per year to outside customers and supply 2,000 bolts per year internally to Beauville's furniture plant. If the fabric plant were designated as a profit centre, the plant would transfer all goods internally at market price. Using the proposed standard cost sheet (as needed) and any other relevant data, prepare the following for Fabric FB70:
   a. Sales budget
   b. Production budget
   c. Direct labour budget
   d. Cost of goods sold budget

9. Calculate the following overhead rates for the furniture plant: (1) plantwide rate and (2) departmental rates. Use the direct method for assigning service costs to producing departments.

10. For each of the overhead rates computed in Requirement 9, calculate unit bid prices for Jobs A500 and B75. Assume that the company's aggressive bidding policy is unit cost plus 50 percent. Did departmental overhead rates have any effect on Beauville's winning or losing bids? What recommendation would you make? Explain. Now, adjust the costs and bids for departmental rate bids using the proposed standard costs for the Colouring and Bolting Department. Did this make a difference? What does this tell you?

11. Suppose that the fabric plant is set up as a profit centre. Bolts of Fabric FB70 sell for $400 (or can be bought for $400 from outside suppliers). The fabric plant and the furniture plant both have excess capacity. Assume that Job A500 is a special order. The fabric and furniture plants have sufficient excess capacity to satisfy the demands of Job A500. What is the minimum transfer price for a bolt of FB70? If the maximum transfer price is $400, by how much do the fabric plant's profits increase if the two profit centres negotiate a transfer price that splits the joint benefit?

After studying this chapter, you should be able to:

➤ **1** Describe the tactical decision-making model.

➤ **2** Define the concept of relevant costs and revenues.

➤ **3** Explain how the activity resource usage model is used in assessing relevancy.

➤ **4** Apply the tactical decision-making concepts in a variety of business situations.

CHAPTER

11 Tactical Decision Making

A major role of the cost management information system is supplying cost and revenue data that are useful in tactical decision making. This chapter focuses on the use of cost and revenue data in tactical decision making. To make sound decisions, the user of the cost information must be able to decide what is relevant to the decision and what is not relevant.

OBJECTIVE ➤ 1

Describe the tactical decision-making model.

Tactical Decision Making

Tactical decision making consists of choosing among alternatives with an immediate or limited end in view. Accepting a special order for less than the normal selling price to use idle capacity and increase this year's profits is an example. The immediate objective is to exploit idle productive capacity so that short-run profits can be increased. Thus, some tactical decisions tend to be *short-run* in nature; however, it should be emphasized that short-run decisions often have long-run consequences. Consider a second example. Suppose that a company is considering producing a component instead of buying it from suppliers. The immediate objective may be to lower the cost of making the main product. Yet, this tactical decision may be a small part of the overall strategy of establishing a cost leadership position for the firm. Thus, tactical decisions are often *small-scale actions* that serve a larger purpose. Recall that the overall objective of strategic decision making is to establish a long-term competitive advantage. Tactical decision making should support this overall objective, even if the immediate objective is short run (accepting a one-time special order) or small scale (making instead of buying a component). Thus, *sound* tactical decisions achieve not only the limited objective but also serve a larger purpose. In fact, all tactical decisions should serve the overall strategic goals of an organization.

The Tactical Decision-Making Process

A general tactical decision-making model is outlined here. The six steps describing the process are listed below.

1. Recognize and define the problem.
2. Identify alternatives as possible solutions to the problem, and eliminate any unfeasible alternatives.
3. Identify the costs and benefits associated with each feasible alternative. Eliminate the costs and benefits that are not relevant to the decision.
4. Compare the *relevant* costs and benefits for each alternative.
5. Assess qualitative factors.
6. Select the alternative with the greatest overall benefit.

Step 1: Define the Problem   To illustrate the steps of the process, consider an apple producer. Each year, 25 percent of the apples harvested are small and odd-shaped. These apples cannot be sold in the normal distribution channels and have simply been dumped in the orchards for fertilizer. The owner is not satisfied with this approach and wants to determine the best way to handle these apples.

Step 2: Identify Feasible Alternatives   Several alternatives are being considered:

1. Sell the apples to pig farmers.
2. Bag the apples (two-kilogram bags) and sell them to local supermarkets as seconds.
3. Rent a local canning facility and convert the apples to applesauce.
4. Rent a local canning facility and convert the apples to pie filling.
5. Continue with the current dumping practice.

Of the five alternatives, alternative one was eliminated because there were not enough local pig farmers interested in the apples; alternative five represented the status quo and was eliminated at the request of the owner; alternative four was also eliminated because the local canning facility did not have the equipment needed to produce pie filling. However, the local canning facility's equipment could be used for producing applesauce. Thus, alternative three was possible. Furthermore, since local supermarkets agreed to buy two-kilogram bags of irregular apples and bagging could be done at the warehouse, alternative two was also a possibility. Thus, two alternatives were deemed feasible.

Step 3: Predict Costs and Benefits and Eliminate Irrelevant Costs   Suppose that the apple producer predicts that labour and materials (bags and ties) for the bagging option would cost $0.10 per kilogram. The two-kilogram bags of apples could be sold for $1.10 per bag to the local supermarkets. Making applesauce would cost $0.80 per kilogram for facility rental, labour, apples, cans, and other materials. It takes three kilograms of apples to produce five, 500-millilitre cans of applesauce. Each 500-millilitre can will sell for $0.78. The cost of growing and harvesting the apples is not relevant to choosing between the bagging alternative and the applesauce alternative.

Step 4: Compare Relevant Costs and Benefits   The bagging alternative costs \$0.20 to produce a two-kilogram bag (\$0.10 × 2 kilograms), and the revenue is \$1.10 per bag, or \$0.55 per kilogram. Thus, the net benefit is \$0.45 per kilogram (\$0.55 − \$0.10). For the applesauce alternative, three kilograms of apples produce five 500-millilitre cans of applesauce. The revenue for five cans is \$3.90 (5 × \$0.78), which converts to \$1.30 per kilogram (\$3.90/3). Thus, the net benefit is \$0.50 per kilogram (\$1.30 − \$0.80). Of the two alternatives, the applesauce option offers \$0.05 more per kilogram than the bagging option.

Step 5: Assess the Qualitative Factors   Qualitative factors are those that are very difficult to translate into dollars. For the apple example, the producer currently is not involved in producing any apple consumer products and is reluctant to move into applesauce production. He has no experience in this part of the industrial value chain and knows little about the channels of distribution for applesauce. An outside expert would need to be hired. Additionally, the rental opportunity is a year-to-year issue. In the long term, a major capital commitment might be needed. Bagging the small apples, on the other hand, is a product differentiation strategy that allows the producer to operate within familiar territory.

Step 6: Select the Best Alternative   While the applesauce option is somewhat more lucrative, the qualitative factors argue against it. Therefore, the bagging alternative should be chosen. This alternative maintains the current position in the industrial value chain and strengthens the producer's competitive position by following a differentiation strategy for the small, odd-shaped apples.

Summary of Decision-Making Process   The six steps define a simple decision model. A **decision model** is a set of procedures that, if followed, will lead to a decision. **Tactical cost analysis** is the use of relevant cost data to identify the alternative that provides the greatest benefit to the organization. Thus, tactical cost analysis includes predicting costs, identifying relevant costs, and comparing relevant costs. As we have seen, however, tactical cost analysis is only part of the overall decision process. Qualitative factors deserve more discussion.

Qualitative Factors

Step five of the decision-making model is critically important. While cost and revenue information is important, other information, often qualitative in nature, is needed to make an informed decision. For example, the relationship of the alternatives being considered to the organization's strategic objectives is essentially a qualitative assessment.

Other qualitative factors are also important. For example, the spike in gasoline prices in 2008–2009 led to renewed interest in hybrid vehicles. While **Toyota's** Prius, **Nissan's** Altima, and **Chevrolet's** two-mode Tahoe SUV had payback periods shorter than five years, the payback on many of these vehicles is quite long. Some take over 10 years to pay back the difference in cost between a hybrid version and nonhybrid version of a vehicle, and a few take close to 100 years![1] Given the fact that the economics are frequently stacked against buying the hybrid version, why do so many people opt for it? Qualitative reasons are often the answer. Individuals like the idea of reducing their carbon footprint and "going green." There is satisfaction in doing something good for the environment. Bradley Berman, editor of Hybridcars.com, mentions the "tech appeal" of the hybrid vehicles. "Hybrids definitely appeal to people who are into 'fun technology,'" he says. "If you were one of the folks who went out and

---

[1] John O'Dell, "Payback for Many Hybrids Grows as Gas Costs Rise," *Edmunds.com* (June 11, 2008), http://blogs. edmunds.com/greencaradvisor/2008/06/payback-for-many-hybrids-grows-as-gas-costs-rise.html.

> got an iPod or iPhone as soon as they came out, and if you use a TiVO instead
> of a VCR, then you'll probably like the fact that today's hybrids are the most
> advanced vehicles out there today in terms of electronics."[2]

How should qualitative factors be handled in the decision-making process? First of all, they must be identified. Second, the decision maker should try to quantify them. Often, qualitative factors are simply more difficult to quantify, but not impossible. For example, possible unreliability of an outside supplier might be quantified as the probable number of days late multiplied by the labour cost of downtime in the plant. Finally, truly qualitative factors, such as the impact of late orders on customer relations, or the apple producer's discomfort with the canning option, must be taken into consideration in the final step of the decision-making model—the selection of the alternative with the greatest overall benefit.

Relevant Costs and Revenues

OBJECTIVE ▶ 2
Define the concept of relevant costs and revenues.

In choosing between two alternatives, only the costs and revenues relevant to the decision should be considered. Identifying and comparing relevant costs and revenues is the heart of the tactical decision model. **Relevant costs (revenues)** are future costs (revenues) that differ across alternatives. Since relevant revenues are treated in the same way as relevant costs, we will simplify the discussion by concentrating on costs. All decisions relate to the future; accordingly, only future costs can be relevant. In addition, the cost also must differ from one alternative to another. If a future cost is the same for more than one alternative, it has no effect on the decision. Such a cost is an *irrelevant* cost. The ability to identify relevant and irrelevant costs is an important decision-making skill.

Relevant Costs Illustrated

To illustrate the concept of relevant costs, consider Avicom Inc., a company that makes jet engines for commercial aircraft. A supplier has approached the company and offered to sell one component, nacelles (enclosures for jet engines), for what appears to be an attractive price. Avicom faces a make-or-buy decision. Assume that the cost of direct materials used to produce the nacelles is $270,000 per year (based on normal volume). Is this cost relevant? It is certainly a future cost. To produce the component for another year requires that materials be purchased. In addition, it differs across the two alternatives. If the component is purchased from an external supplier, no internal production is needed, and no direct materials need be purchased, reducing the materials cost to zero. Since the cost of direct materials differs across alternatives ($270,000 for the make alternative and $0 for the buy alternative), it is a relevant cost.

Irrelevant Cost Illustrated

Avicom uses machinery to manufacture nacelles. This machinery was purchased five years ago and has an annual depreciation cost of $50,000. Is depreciation a future cost that differs across the two alternatives?

Past Costs Depreciation, in this case, represents an allocation of a cost already incurred. It is a **sunk cost**, because no future decision can alter the original cost of the machinery; the original cost is the same for both alternatives. Although we allocate this sunk cost to future periods and call it *depreciation*, none of the original cost is avoidable. Sunk costs are past costs and are always irrelevant. Thus, the acquisition cost of the machinery and its associated depreciation should not be a factor in the make-or-buy decision.

---

[2] Kevin Ransom, "Reasons to Buy a Hybrid—or Not," *CNN.com* (January 28, 2008), http://www.cnn.com/2008/LIVING/wayoflife/01/28/buy.hybrid/index.html.

Future Costs   Assume that the cost to heat and cool the plant—$40,000 per year—is allocated to different production departments, including the department that produces nacelles, which receives $4,000 of the cost. Is this $4,000 cost relevant to the make-or-buy decision facing Avicom?

The cost of providing plant utilities is a future cost, since it must be paid in future years. But does the cost differ across the make-and-buy alternatives? It is unlikely that the cost of heating and cooling the plant will change whether nacelles are produced or not. Thus, the cost is the same across both alternatives. The amount of the utility payment allocated to the remaining departments may change if production of nacelles is stopped, but the level of the total payment is unaffected by the decision. It is therefore an irrelevant cost.

Relevant Costs and Benefits in International Trade

Relevant costs and benefits are useful in decision making in the international trade arena. For example, a company may import materials for use in production. While this transaction may look identical to the purchase of materials from domestic suppliers, import tariffs add complexity and cost. A **tariff** is a tax on imports levied by the federal government. Any cost associated with the purchase of materials, such as freight-in or a tariff, is a materials cost. Companies search for ways to reduce tariffs. They may restrict the amount of imported materials, alter the materials by adding Canadian resources (to increase the domestic content and gain more favourable tariff status), or utilize foreign trade zones.

Some Canadian companies set up manufacturing plants within **foreign trade zones** (which are physically in Canada but considered to be outside of Canada for tariff purposes). Since tariffs are not paid until the imported materials leave the zone, as part of a finished product, the company can postpone payment of duty and the associated loss of working capital. Additionally, the company does not pay duty on defective materials or inventory that has not yet been included in finished products.

An example may help to illustrate the potential cost advantages. Suppose that Roadrunner Inc. operates a petrochemical plant located in a foreign trade zone. The plant imports volatile materials (i.e., chemicals that experience substantial evaporation loss during processing) for use in production. Wilycoyote Inc. operates an identical plant just outside the foreign trade zone. Consider the impact on duty and related expenditures for the two plants for the purchase of $400,000 of crude oil imported from Venezuela. Both Roadrunner and Wilycoyote use the oil in chemical production. Each purchases the oil about three months before use in production, and the finished chemicals remain in inventory about five months before sale and shipment to the customer. About 30 percent of the oil is lost through evaporation during production. Duty is assessed at 6 percent of cost. Each company faces a 12 percent carrying cost.

Wilycoyote pays duty, at the point of purchase, of $24,000 (0.06 × $400,000). In addition, Wilycoyote has carrying cost associated with the duty payment of 12 percent per year times the portion of the year that the oil is in materials or finished goods inventory. In this case, the months in inventory equal 8 (3 + 5). Total duty-related carrying cost is $1,920 (0.12 × 8/12 × $24,000). Together, duty and duty-related carrying cost totals $25,920. Roadrunner, on the other hand, pays duty at the time of sale because it is in a foreign trade zone. Imported goods do not incur duty until (unless) they are moved out of the zone. Since 70 percent of the original imported oil remains in the final product, duty equals $16,800 (0.7 × $400,000 × 0.06). There is no carrying cost associated with the duty. The duty-related costs for the two companies are summarized below.

	Roadrunner	Wilycoyote
Duty paid at purchase	$ 0	$24,000
Carrying cost of duty	0	1,920
Duty paid at sale	16,800	0
Total duty and duty-related cost	$16,800	$25,920

Clearly, Roadrunner's foreign trade zone location has saved $9,120 ($25,920 − $16,800) on just one purchase of imported materials.

In the above example, the underlying business decision involves whether or not to locate in a foreign trade zone. Relevant costs include the cost of duty and the carrying cost of duty for plants located inside and outside the zone. Additional potential for cost reduction inside the zone occurs when goods that do not meet Canadian health, safety, and pollution control regulations are subject to fine. Noncomplying foreign goods can be imported into foreign trade zones and modified to comply with the law without being subject to the fine. Another example of the efficient use of foreign trade zones is the assembly of high-tariff component parts into a lower-tariff finished product. In this case, the addition of domestic labour raises the domestic content of the finished product and makes the embedded foreign parts eligible for more favourable tariff treatment.[3] A qualitative factor is that logistics may be streamlined by using foreign trade zones, leading to quicker and more efficient clearance of customs.

The Activity Resource Usage Model

OBJECTIVE ▶ 3
Explain how the activity resource usage model is used in assessing relevancy.

Understanding cost behaviour is basic in determining relevancy. When costs were primarily unit-based, a simple distinction between fixed and variable costs could be made. Now, however, the ABC model has us focus on unit-level, batch-level, product-level, and facility-level costs. The first three are variable, but with respect to different types of activity drivers. The activity resource usage model can help us sort out the behaviour of various activity costs and assess their relevance.

The **activity resource usage model** focuses on the use of resources and has two categories: (1) flexible resources and (2) committed resources. Recall from Chapter 2 that flexible resources are those that are acquired as used and needed. Committed resources are acquired in advance of usage. These categories and their usefulness in relevant costing are described in the following sections.

Flexible Resources

Resource spending is the cost of acquiring activity capacity. The amount paid for the supply of an activity is the activity cost. For flexible resources, the resources demanded (used) equal the resources supplied. Thus, for this category, *if the demand for an activity changes across alternatives*, then resource spending will change and the cost of the activity is relevant to the decision. For example, electricity supplied internally uses fuel for the generator. Fuel is a flexible resource. Consider the following two alternatives: (1) accept a special, one-time order and (2) reject the special order. If accepting the order increases the demand for kilowatt-hours (power's activity driver), then the cost of power will differ across alternatives by the increase in fuel consumption. Thus, power cost is relevant to the decision.

Committed Resources

Committed resources are acquired in advance of usage through implicit contracting, and they are usually acquired in lumpy amounts. Consider an organization's employees. The implicit understanding may be that the organization will maintain employment levels even though there may be temporary downturns in the amount of an activity used, meaning that an activity may have unused capacity. Increased demand for an activity across alternatives may not mean increased cost—if there is sufficient unused capacity. For example, assume a company has five manufacturing engineers who can each work 2,000 hours and earn $50,000 per year; total engineering

---

[3] These examples are taken and adapted from James E. Groff and John P. McCray, "Foreign-Trade Zones: Opportunity for Strategic Development in the Southwest," *Journal of Business Strategies* (Spring 1992): 14–26.

capacity is 10,000 (5 × 2,000 hours) engineering hours. Suppose that this year the company expects to use only 9,000 engineering hours for its normal business. This means that the engineering activity has 1,000 hours of unused capacity. If there is a special order that requires 500 engineering hours, the cost of engineering would be irrelevant. The order can be filled using unused engineering capacity, and the resource spending is the same for each alternative ($250,000 will be spent whether or not the order is accepted).

However, *if a change in demand for the activity requires a change in resource supply*, then the activity cost will be relevant to the decision. This change in cost can occur in one of two ways: (1) the demand for the resource exceeds the supply (increases resource spending) or (2) the demand for the resource drops permanently and supply exceeds demand enough so that activity capacity can be reduced (decreases resource spending).

To illustrate the first change, suppose that the special order requires 1,500 engineering hours. This exceeds the unused capacity of 1,000 hours. To meet the demand, the organization would need to hire a sixth engineer or perhaps use a consulting engineer. Either way, spending on engineering increases if the order is accepted; the cost of engineering is now a relevant cost.

To illustrate the second type of change, suppose that the company could purchase a component used for production instead of making it in house. Recall that 10,000 engineering hours are available and that 9,000 are used. If the component is purchased, then the demand for engineering hours will drop from 9,000 to 7,000. This is a permanent reduction because engineering support will no longer be needed for manufacturing the component. Because unused capacity is now 3,000 hours, and engineering capacity is acquired in chunks of 2,000 hours, the company can reduce capacity and resource spending by laying off one engineer or reassigning the engineer to another plant where the services are in demand. Either way, the resource supply is reduced to 8,000 hours. Engineering cost would differ by $50,000 (the salary for one engineer) across the make-or-buy alternatives. This cost is then relevant to the decision.

Often, committed resources are acquired in advance for multiple periods—before resource demands are known. Leasing or buying a building is an example. Buying multiperiod activity capacity is often done by paying cash up front. In this case, an annual expense may be recognized, but no additional resource spending is needed. Up-front resource spending is a sunk cost and never relevant. Periodic resource spending, such as leasing, is essentially independent of resource usage. Even if a permanent reduction of activity usage is experienced, it is difficult to reduce resource spending because of formal contractual commitments.

For example, assume a company leases a plant for $100,000 per year for 10 years. The plant is capable of producing 20,000 units of a product—the level expected when the plant was leased. After five years, the demand for the product drops and the plant needs to produce only 15,000 units each year. The annual lease payment of $100,000 must still be paid even though units produced have decreased. Suppose instead that demand increases beyond the 20,000-unit capability. The company may consider buying or leasing an additional plant. Here, resource spending could change across alternatives. Exhibit 11-1 summarizes the activity resource usage model's role in assessing relevancy.

OBJECTIVE ▶4
Apply the tactical decision-making concepts in a variety of business situations.

Illustrative Examples of Tactical Decision Making

The activity resource usage model and the concept of relevancy are valuable tools in making tactical decisions. It is important to see how they are used to solve a variety of problems. Applications include decisions to make or buy a component, to keep or drop a segment or product line, to accept or reject a special order at less than the usual price, and to process a joint product further or sell it at the split-off point. Of course, this is not an exhaustive list. The same decision-making principles can be applied to other settings. Once you see how they are used, it is relatively easy to apply them in any appropriate setting. In illustrating the applications, we assume that the

Exhibit 11-1

Resource Demand and Supply

Category	Relationships	Relevancy
Flexible	Supply = Demand a. Demand changes b. Demand constant	 a. Relevant b. Not relevant
Committed	Supply − Demand = Unused capacity a. Demand increase < Unused capacity b. Demand increase > Unused capacity c. Demand decrease (permanent) 　　1. Activity capacity reduced 　　2. Activity capacity unchanged	 a. Not relevant b. Relevant 1. Relevant 2. Not relevant

first two steps of the tactical decision-making model have already been done. Thus, the emphasis is on tactical cost analysis.

Make-or-Buy Decisions

Organizations are often faced with a **make-or-buy decision**—a decision of whether to make or to buy components or services used in making a product or providing a service. For example, a physician can buy laboratory tests from external suppliers (hospitals or for-profit laboratories), or these lab tests can be done internally. Similarly, a PC computer manufacturer can make its own disk drives, or they can be bought from external suppliers.

Outsourcing of technical and professional jobs is becoming an important make-or-buy issue. **Outsourcing** refers to the move of a business function to another company, either inside or outside Canada.

For example, some newspapers are closing down their foreign news bureaus and outsourcing the jobs to other companies.[4] CAs find that income tax preparation can be outsourced to lower-cost providers in India.[5]

Qualitative considerations are important in the outsourcing decision. Time is a valuable resource, and many companies have found that a global presence leads to time and quality enhancement. For example, software companies have found that call centres located in Ireland and the United States provide better customer service. At 8 a.m., a customer in Toronto who needs an answer to a question may not get help from a California-based call centre, but will get help from a Dublin-based centre. On the negative side, the political ramifications of outsourcing, with its overtones of "exporting jobs," have led companies to weigh the decision more carefully.

Make-or-buy decisions are not short run in nature but fall into the small-scale tactical decision category. For example, the decision to make or buy may be motivated by cost leadership and/or differentiation strategies. Making instead of buying (or vice versa) may be one way to reduce the cost of producing the main product. Alternatively, choosing to make or buy may be a way of increasing the quality of the component and thus increasing the overall quality of the final product (differentiating on the basis of quality). Cornerstone 11-1 shows the how and why of structuring the make-or-buy decision.

Cost Analysis: Activity-Based Cost Management System   The make-or-buy problem can also be illustrated in an activity-based costing format. The

[4] Russell Adams and Shira Ovide, "Newspapers Move to Outsource Foreign Coverage," *The Wall Street Journal* (January 15, 2009): B4.

[5] Gary S. Shamis, M. Cathryn Green, Susan M. Sorensen, and Donald L. Kyle, "Outsourcing, Offshoring, Nearshoring: What to Do?" *Journal of Accountancy* (June 2005), http://www.journalofaccountancy.com/Issues/2005/Jun/Outsourcing OffshoringNearshoringWhatToDo.htm.

**CORNERSTONE
11-1**

The HOW and WHY of Structuring a Make-or-Buy Decision

Information:

Ling Company produces 100,000 units of Part 34B, used in one of its snow-blower engines, each year. An outside supplier has offered to supply the part for $4.75. The unit cost is:

Direct materials	$0.50
Direct labour	2.40
Variable overhead (power)	0.90
Fixed overhead	1.05
Total unit cost	$4.85

Overhead is applied on the basis of machine hours; Part 34B requires 30,000 machine hours per year.

Why:

The make-or-buy situation requires the company to focus on relevant costs and benefits. The problem is set up with relevant costs and benefits organized under column headings for each alternative. The difference between the alternatives gives the quantitative advantage/disadvantage for each alternative.

Required:

1. What are the alternatives for Ling Company?
2. Assume that none of the fixed cost is avoidable. List the relevant cost(s) of internal production and of external purchase.
3. Which alternative is more cost effective and by how much?
4. **What if** $60,000 of fixed overhead is supervision for Part 34B that is avoided if the part is purchased? Which alternative is more cost effective and by how much?

Solution:

1. The alternatives are to make the part in house or buy the part externally.
2. The relevant costs of making the part are direct materials, direct labour, and variable factory overhead. The relevant cost of buying the part is the purchase price.
3.

	Make	Buy	Difference
Direct materials	$ 50,000	$ 0	$ 50,000
Direct labour	240,000	0	240,000
Variable overhead	90,000	0	90,000
Purchase price	0	475,000	(475,000)
Totals	$380,000	$475,000	$ (95,000)

Because the fixed overhead is not relevant, the analysis shows a $95,000 advantage in favour of making the part in house.

4.

	Make	Buy	Difference
Direct materials	$ 50,000	$ 0	$ 50,000
Direct labour	240,000	0	240,000
Variable overhead	90,000	0	90,000
Supervision	60,000	0	60,000
Purchase price	0	475,000	(475,000)
Totals	$440,000	$475,000	$ (35,000)

Now, supervision (part of fixed overhead) is relevant; the analysis shows a $35,000 advantage in favour of making the part in house.

Exhibit 11-2

Activity and Cost Information

	Activity Cost Formula:				
Activity	**Fixed Cost**		**Variable Rate**		**Amount of Driver**
Providing power	=	$0	+	$3	×	Machine hours
Providing supervision	=	$0	+	$20,000	×	Lines
Moving materials	=	$250,000	+	$0.60	×	Number of moves
Inspecting product	=	$280,000	+	$1.50	×	Inspection hours
Setting up equipment	=	$0	+	$10	×	Setup hours
Providing space	=	$971,000				Square metres
Depreciation	=	$120,000				Units

Activity Driver	Total Capacity	Expected Usage	Part 34B Usage	Units of Purchase
Machine hours	As needed	750,000	30,000	1
Supervisory lines	15	15	3	3
Moves	250,000	240,000	40,000	25,000
Inspection hours	16,000	14,000	2,000	2,000
Setup hours	60,000	58,000	6,000	2,000
Providing space	971,000	971,000	5,000	50,000
Depreciation	620,000	100,000	100,000	15,000

structure of the problem is the same as that shown in Cornerstone 11-1; however, typically the relevant costs are more extensive and care must be taken to determine which activities are relevant, and by how much. To illustrate the ABC analysis for Ling Company's Part 34B, we will use the data in Exhibit 11-2 along with the data from Cornerstone 11-1. All activity capacities are annual capacity measures. The cost of providing space includes annual plant depreciation, property taxes, and annual maintenance. This cost is allocated to the products based on the square metres of space occupied by the product's production equipment. The variable component of each activity represents the cost of flexible resources. The fixed cost component represents the cost of committed resources acquired in advance of usage. Units of purchase indicate how many units of the activity (as measured by its driver) must be acquired at a time. For example, if more capacity for moving materials is needed, it must be bought in lump sums of 25,000 moves at a time.

To determine whether Ling should continue to make Part 34B or buy it from an external supplier depends on how much *resource spending* can be reduced because of the ability to reduce resource usage (by buying instead of making). As is done in Cornerstone 11-1, the problem is structured as follows:

	Make	Buy	Difference
Direct materials	$ 50,000	$ 0	$ 50,000
Direct labour	240,000	0	240,000
Providing power	90,000	0	90,000
Providing supervision	60,000	0	60,000
Moving materials	49,000	0	49,000
Inspecting product	38,000	0	38,000
Setting up equipment	60,000	0	60,000
Purchase price	0	475,000	(475,000)
Totals	$587,000	$475,000	$ 112,000

If Ling buys Part 34B instead of making it, *resource usage* decreases for each of the seven activities. Let's review them.

Direct materials, direct labour, and power are strictly variable and their amounts are identical to those calculated in Cornerstone 11-1.

$$\text{Direct materials} = \$0.50 \times 100,000 \text{ units} = \$50,000$$
$$\text{Direct labour} = \$2.40 \times 100,000 \text{ units} = \$240,000$$
$$\text{Power} = \$3.00 \times 30,000 \text{ units} = \$90,000$$

Supervision is a lumpy resource and must be acquired in units of three lines. Since the making of Part 34B requires exactly three lines, the amount of supervision needed is $60,000 ($20,000 $\times$ 3 lines).

Moving materials and inspecting product are a bit more complicated since there is a variable and a fixed amount.

$$\text{Moving materials} = (\$250,000/250,000 \text{ moves at capacity})(25,000)$$
$$+ (\$0.60 \times 40,000 \text{ moves})$$
$$= \$49,000$$
$$\text{Inspecting product} = (\$280,000/16,000 \text{ inspection hours at capacity})(2,000)$$
$$+ (\$1.50 \times 2,000 \text{ inspection hours})$$
$$= \$38,000$$

Notice that the fixed amount associated with Part 34B is the amount by which fixed resource spending can be reduced, so it depends on the total capacity—250,000 moves for moving materials and 16,000 inspection hours for inspecting product. The fixed rate is multiplied by the lumpy amount or the units that must be purchased at once. The variable amount of resource spending associated with Part 34B is the amount of driver actually used in producing the part times the variable rate.

Setting up equipment is strictly variable with respect to the number of setup hours. Since Part 34B uses 6,000 setup hours, and the rate is $10 per setup hour, the production of the part requires $60,000.

Notice that providing space and equipment depreciation are ignored since they are irrelevant costs. They will remain the same in total no matter whether the part is made internally or purchased externally.

As we can see, the additional information provided by activity-based costing changed the analysis so that purchasing the part is better. The company will save $112,000 per year if the part is bought externally. Of course, this is just the quantitative analysis. There may be compelling qualitative factors that Ling should consider. For example, will the outside supplier maintain the quality needed by Ling? Will the supplier be able to meet delivery requirements? Only a full analysis that considers both quantitative and qualitative factors will give management the support to make a good decision.

Keep-or-Drop Decisions

Often, a manager needs to determine whether a segment, such as a product line, should be kept or dropped. **General Motors**, for example, decided to drop a number of car lines, including Oldsmobile, Hummer, Saab, and Saturn.[6] A **keep-or-drop decision** uses relevant cost analysis to determine whether a segment or line of business should be kept or dropped. In a traditional cost management system, segmented income statements, using unit-based fixed or variable costs, improve the ability to make keep-or-drop decisions. Cornerstone 11-2 shows the how and why of structuring the keep-or-drop decision analysis.

As Cornerstone 11-2 shows, revenues and costs that are directly attributable to a segment must be identified. If the segment is dropped, then only the traceable revenues and costs should vanish. Furthermore, the traceable income (loss) determines whether a segment should be dropped or kept. If the product (or segment) margin is

---

[6]Todd Lassa, "GM's Survival Strategy: Divisions, Nameplates to Disappear," *Motor Trend* (March 23, 2009), http://www.motortrend.com/features/auto_news/2009/112_0903_gm_survival_strategy/index.html.

The HOW and WHY of Structuring a Keep-or-Drop Product Line Decision

CORNERSTONE 11-2

Information:

Dexter Company makes three types of GPS devices. The Basic GPS model is an entry-level automotive GPS device; it is sold through discounters and Amazon.com. The Runner's GPS is a miniaturized model that allows the runner to track mileage, steps, and heart rate while running; it is sold through athletic stores and on sports gear websites. The Chart Plotter is a specialized GPS device for sailors; it can be customized with maps of the sea floor and specific geographic areas of coast line and deep water. It is sold via the Web on dedicated GPS sites. Dexter Company is considering dropping the Basic GPS line and keeping the Runner's GPS and Chart Plotter. The segmented income statement is presented on the following page.

	Basic GPS	Runner's GPS	Chart Plotter	Total
Sales	$ 450,000	$ 980,000	$1,670,000	$ 3,100,000
Less variable costs	(324,000)	(372,000)	(601,600)	(1,297,600)
Contribution margin	126,000	608,000	1,068,400	1,802,400
Less direct fixed costs:				
Advertising	(85,000)	(124,000)	(130,000)	(339,000)
Supervision	(60,000)	(115,000)	(135,000)	(310,000)
Product margin	$ (19,000)	$ 369,000	$ 803,400	1,153,400
Less common fixed expenses				915,000
Operating income				$ 238,400

Why:

Companies need to consider whether a segment or product line should remain. This problem requires a look at the relevant costs and benefits of dropping the segment.

Required:

1. List the alternatives being considered.
2. List the relevant benefits and costs for each alternative.
3. Which alternative is more cost effective and by how much?
4. **What if** dropping the Basic GPS line would mean a 10 percent loss of volume for the Runner's GPS device and a 2 percent loss in volume for the Chart Plotter? Which alternative would be more cost effective and by how much?

Solution:

1. The two alternatives are to keep the Basic GPS line or to drop it.
2. The relevant benefits and costs of keeping the Basic GPS line include sales of $450,000, variable costs of $324,000, advertising cost of $85,000, and supervision cost of $60,000. All common fixed costs are irrelevant. None of the relevant benefits and costs of keeping the Basic GPS line would occur under the drop alternative.

CORNERSTONE 11-2
(continued)

3.

	Keep	Drop	Differential Amount to Keep
Sales	$450,000	$0	$ 450,000
Less variable costs	324,000	0	(324,000)
Contribution margin	126,000	0	126,000
Less direct fixed costs:			
Advertising	(85,000)	0	(85,000)
Supervision	(60,000)	0	(60,000)
Product margin	$ (19,000)	0	$ (19,000)

There is a $19,000 loss if the Basic GPS line is kept.

4.

	Basic GPS	Runner's GPS	Chart Plotter	Total
Sales	$0	$ 882,000	$1,636,600	$2,518,600
Less variable costs	0	(334,800)	(589,568)	(924,368)
Contribution margin	0	547,200	1,047,032	1,594,232
Less direct fixed costs:				
Advertising	0	(124,000)	(130,000)	(254,000)
Supervision	0	(115,000)	(135,000)	(250,000)
Product margin	$0	$ 308,200	$ 782,032	1,090,232
Less common fixed expenses				915,000
Operating income				$ 175,232

Difference in income = Income with all three lines − Income with only two lines

= $238,400 − $175,232 = $63,168

Because of the impact that dropping the Basic GPS line has on the sales of the other two lines, the analysis shows that dropping the line will actually decrease income by $63,168. Therefore, the Basic GPS line should be kept. However, it would be a good idea to consider ways to make production more efficient.

positive, then the segment is kept; if negative, then the segment may be dropped. Cornerstone 11-2 shows a traditional segmented income statement, where products are defined as segments. The statement indicates that both Runner's GPS and Chart Plotter models provide positive product margins and the Basic GPS model has a negative product margin. Thus, management would likely consider dropping the Basic GPS model. However, when the analysis considers potential complementary effects—the impact of the dropped product line on sales of the other two product lines—the decision likely would change. In the latter case, it is clear that customers prefer a full line of products, and that the Basic GPS somehow adds to the profitability of the other two GPS models.

A company can improve the differential analysis by looking beyond the traditional product costing model. That is, managers can consider more than the unit-based variable versus fixed cost categories by looking at the impact of non-unit costs.

For example, convenience stores constantly balance the need to offer a wide selection of products with the need to streamline offerings so that they can fit into the small-store format. In the past, the stores determined which products to stock based on each one's profitability. Profit was calculated as the difference between wholesale and retail prices. While this sounds reasonable, it completely ignores the additional costs associated with carrying and stocking each product line. In early 2001, the **American Wholesale Marketers Association** and the **National Association of Convenience Stores** presented the results of a study of new software designed to "assess each item's profitability by factoring in the operating, labour, inventory, and overhead costs of each item." In the past, the cost of handling a product was not considered when determining per-product costs. However, handling costs are a significant part of the total cost structure.

One owner of a chain of convenience stores tested the software and learned that every auto fuse and bulb sold resulted in a loss of 50 cents. He surveyed customers and found that they were willing to pay a higher price. As a result, he raised the price by $1. This achieved two goals. The bulbs and fuses now make money, and customers still appreciate the opportunity to pop into the convenience store for suddenly needed products. The same chain determined that three kinds of laundry detergent were two too many. It pared its offering to one brand and displayed it more prominently. Sales increased by 20 percent, while costs fell because the sole brand could be ordered by the case.[7]

Let's continue the Dexter Company example using activity data. Suppose that Dexter Company finds that the common fixed expenses actually include some traceable fixed expenses that can be assigned to the product lines based on non-unit driver usage. In particular, three such expenses and their activity data are shown in the following table:

Activity	Driver	Total Capacity	Unused Capacity	Basic GPS Usage	Units of Purchase
Inspecting products	Number of batches	200	15	80	40
Customer service	Number of calls	30,000	900	12,000	1,000
Material handling	Number of moves	2,800	400	1,400	350

Now we can reanalyze the keep-or-drop decision with the additional activity information.

	Keep	Drop	Differential Amount to Keep
Sales	$ 450,000	$0	$ 450,000
Less variable costs	324,000	0	(324,000)
Contribution margin	126,000	0	126,000
Less direct fixed costs:			
Advertising	(85,000)	0	(85,000)
Supervision	(60,000)	0	(60,000)
Inspection[a]	(56,000)	0	(56,000)
Customer service[b]	(60,000)	0	(60,000)
Material handling[c]	(70,000)	0	(70,000)
Product margin	$(205,000)	0	$(205,000)

[a] Inspection rate = $140,000/200 = $700; $700 × 80 batches = $56,000
[b] Customer service rate = $150,000/30,000 = $5; $5 × 12,000 calls = $60,000
[c] Material handing rate = $140,000/2,800 = $50; $50 × 1,400 moves = $70,000

---

[7] Ann Zimmerman, "Convenience Stores Create Software to Boost Profitability and Cut Costs," *The Wall Street Journal Interactive Edition* (February 15, 2001).

Notice that the amount of each traceable fixed activity used by the Basic GPS can be eliminated. There is a $205,000 loss if the Basic GPS line is kept. While the use of ABC does not change the structure or conceptual basis of the keep-or-drop decision, it does give managers a better idea of just which costs will be affected by the analysis.

As always, qualitative factors are considered in the keep-or-drop decision. If a line is being dropped, how will it affect customer loyalty? Will employees need to be laid off, or can the excess labour be absorbed into other lines? Some of these factors can be quantified, or probabilities can be assigned so that managers can use sensitivity analysis. Others are truly qualitative and must be considered subjectively.

Special-Order Decisions

In general, price discrimination laws require that firms engaged in interprovincial commerce sell identical products at the same price to competing customers in the same market. These restrictions do not apply to competitive bids or to noncompeting customers. Bid prices can vary to customers in the same market, and firms often have the opportunity to consider one-time special orders from potential customers in markets not ordinarily served. A **special-order decision** focuses on whether a specially priced order should be accepted or rejected. Special-order decisions are examples of tactical decisions with a short-term focus. Increasing short-term profits is the limited objective represented by this type of decision. Care should be taken so that acceptance of special orders does not jeopardize normal distribution channels or adversely affect other strategic elements. With this qualification, it should be noted that special orders often can be attractive, especially when the firm has unused capacity. For this situation, the company can focus its analysis on resources acquired as needed—because this will be the source of any increase in resource spending attributable to the order.

Suppose, for example, that Polarcreme Inc., an ice-cream company, is operating at 80 percent of its productive capacity, 10 million litre units. An ice-cream distributor from a geographic region not normally served by the company has offered to buy 2 million units of premium ice cream at $1.75 per unit, provided its own label can be attached to the product. Normal selling price is $2.50 per unit. Cornerstone 11-3 shows how and why the special-order decision should be structured.

Notice that the special order in Cornerstone 11-3 has a price of $1.75 per unit, well below the normal selling price of $2.50; in fact, it is even below the total unit cost. Even so, accepting the order was profitable for the company. The company has sufficient idle capacity, and the order will not displace other units being produced to sell at the normal price. Additionally, some of the costs are not relevant, such as the commissions, distribution, and fixed cost. The added cost attributable to the special order is, of course, included in the analysis. Requirement 4 of Cornerstone 11-3 asks us to consider the potential impact of the special order on existing customers. In this case, it was easy to quantify the impact and see that the special order should be rejected. In other cases, managers may have to consider the impact on other customers as part of the qualitative factors impinging on the order.

Decisions to Sell or Process Further

As discussed in Chapter 7, joint products have common processes and costs of production up to a split-off point. At that point, they become distinguishable. For example, certain minerals such as copper and gold may both be found in a given ore. The ore must be mined, crushed, and treated before the copper and gold are separated. The point of separation is called the split-off point. The costs of mining, crushing, and treatment are common to both products.

Often, joint products are sold at the split-off point. But sometimes it is more profitable to process a joint product further, beyond the split-off point, prior to selling it. Determining whether to **sell or process further** is an important decision that a manager must make. The key point in this decision is that all of the joint

The HOW and WHY of Structuring a Special-Order Decision

**CORNERSTONE
11-3**

Information:

Polarcreme Inc., an ice-cream company, is operating at 80 percent of its productive capacity, 10 million litre units. An ice-cream distributor from a different geographic region has offered to buy 2 million units of premium ice cream at $1.75 per unit, provided its own label can be attached to the product. Normal selling price is $2.50 per unit. Cost information for the premium ice cream follows:

	Total of 8,000,000 Units	Unit Cost
Variable costs:		
Direct materials	$ 7,600,000	$0.95
Direct labour	2,000,000	0.25
Packaging	1,600,000	0.20
Commissions	160,000	0.02
Distribution	240,000	0.03
Other variable costs	400,000	0.05
Non-unit-level costs:		
Purchasing ($8 × 40,000 purchase orders)	320,000	0.04
Receiving ($6 × 80,000 receiving orders)	480,000	0.06
Setting up ($8,000 × 50 setups)	400,000	0.05
Fixed costs	1,600,000	0.20
Total costs	$14,800,000	$1.85

The special order will not require commissions or distribution (the buyer will pick up the order at Polarcreme's factory). The order will require 10,000 purchase orders, 20,000 receiving orders, and 13 setups. In addition, a one-time cost for the special order's label template will be required at $24,500.

Why:

A special order is "special" because the price is lower than normal. Companies need to consider all relevant costs and benefits when considering a special order.

Required:

1. List the alternatives being considered.
2. List the relevant benefits and costs for each alternative.
3. Which alternative is more cost effective and by how much?
4. *What if* accepting the special order upset a regular customer who was considering expanding into the new geographical region and decided, then, to take their regular annual order of 2 million units of premium ice cream to another company? Which alternative would be better?

Solution:

1. The two alternatives are to accept or reject the special order.
2. The relevant benefits and costs of accepting the order include revenue, direct materials, direct labour, packaging, other variable costs, purchasing, receiving, setting up, and the cost of the label template. No fixed costs will be affected. If the order is rejected, the net benefit is zero.

CORNERSTONE 11-3
(continued)

3.

	Accept	Reject	Differential Amount to Accept
Sales	$ 3,500,000	$0	$ 3,500,000
Direct materials	(1,900,000)	0	(1,900,000)
Direct labour	(500,000)	0	(500,000)
Packaging	(400,000)	0	(400,000)
Other variable costs	(100,000)	0	(100,000)
Purchasing ($8 × 10,000 purchase orders)	(80,000)	0	(80,000)
Receiving ($6 × 20,000 receiving orders)	(120,000)	0	(120,000)
Setting up ($8,000 × 13 setups)	(104,000)	0	(104,000)
Label template	(24,500)	0	(24,500)
Net benefit	$ 271,500	0	$ 271,500

There is a $271,500 increase in operating income if the special order is accepted.

4. In this case, the regular order, at $2.50 per unit, would be better than the special order at $1.75 per unit and the company would be better off rejecting the special order. Even though the special order avoids the commission and distribution charge, those total only $0.05 per unit, and the company would be better off making the additional $0.75 in price with the regular customer, not to mention avoiding the $24,500 for the special label template.

production costs are irrelevant to the sell or process further decision. By the time the split-off point is reached, all joint costs are sunk and therefore irrelevant.

To illustrate, consider Lafleur Corporation. Lafleur is an agricultural corporation that produces and sells fresh produce and canned food products. The Leamington Division of Lafleur specializes in tomato products. Leamington has a large tomato farm that produces all the tomatoes used in its products. The farm is divided into manageable plots. Each plot produces approximately 1,500 kilograms of tomatoes; this defines a load. Each plot must be cultivated, fertilized, sprayed, watered, and harvested. When the tomatoes have ripened, they are harvested. The tomatoes are then transported to a warehouse, where they are washed and sorted. The approximate cost of all these activities is $200 per load.

Tomatoes are sorted into two grades (A and B). Grade A tomatoes are larger and better shaped than Grade B. Grade A tomatoes are sold to large supermarkets. Grade B tomatoes are sent to the canning plant, where they are processed into catsup, tomato sauce, and tomato paste. Each load produces about 1,000 kilograms of Grade A tomatoes and 500 kilograms of Grade B tomatoes. Recently, the manager of the canning plant requested that the Grade A tomatoes be used for a Lafleur hot sauce. Studies have indicated that the Grade A tomatoes provided a better flavour and consistency for the sauce than did Grade B tomatoes. Furthermore, Grade B tomatoes are fully utilized for other products. Cornerstone 11-4 shows the how and why of structuring a sell at split-off or process further decision.

Cornerstone 11-4 shows that the joint cost of production, the $200 per load to grow and harvest the tomatoes, is irrelevant and can be ignored. It should be reiterated that the allocation of joint cost to the various joint products is done solely for the purposes of costing product and valuing inventories. It is not a part of the sell or process further decision. There is one other situation in which the joint cost is

The HOW and WHY of Structuring a Sell at Split-Off or Process Further Decision

Information:

Lafleur Company grows and sells fresh and canned food products. The Leamington farm grows and harvests tomatoes. Each plot yields 1,500 kilograms of tomatoes, referred to as a load; of the 1,500 kilograms, 1,000 kilograms are Grade A tomatoes and 500 are Grade B. The cost of growing and harvesting the tomatoes is $200 per load. Lafleur can sell the 1,000 kilograms of Grade A tomatoes in a load to grocers for $0.40 per kilogram. Alternatively, the tomatoes could be processed into hot sauce. Each bottle of hot sauce sells for $1.50 and requires one kilogram of tomatoes. The cost of additional processing averages $1 per bottle; this amount includes the remaining ingredients, bottles, labour, and needed processing activities.

CORNERSTONE 11-4

Why:

Because joint costs are incurred prior to the split-off point, they are sunk costs in determining whether to sell a product at split off or process it further. Only the sales value at split-off, the further processing costs, and the eventual sales value are relevant to this decision.

Required:

1. List the alternatives being considered.
2. List the relevant benefits and costs for each alternative.
3. Which alternative is more cost effective and by how much?
4. *What if* the best of the Grade A tomatoes, Premium A's, could be sold to grocers for $0.80 per kilogram? Of the 1,000 kilograms of Grade A tomatoes in a load, about 30 percent are Premium A's. However, the grocers will not buy the Premium A's unless they are also sold the regular Grade A tomatoes. (They will deal with another supplier instead.) It will cost an additional $50 per load to separate the Premium A's from the regular Grade A's. Which alternative would be better?

Solution:

1. The two alternatives are to sell the Grade A tomatoes at split-off or process them further.
2. The relevant benefits and costs of selling at split-off versus processing the tomatoes further include revenue from sale to grocers and revenue from selling the hot sauce less the additional (further) processing costs. The $200 per load cost of growing and harvesting the tomatoes is sunk and need not be considered.
3.

	Sell at Split-Off	Process Further	Differential Amount to Process Further
Sales	$400	$ 1,500	$ 1,100
Further processing cost	0	(1,000)	(1,000)
Total	$400	$ 500	$ 100

There is a $100 per load advantage to processing the Grade A tomatoes into hot sauce.

4. In this case, 300 kilograms of the Grade A tomatoes (Premium A's) are sold for $0.80 and the remaining 700 kilograms are sold for $0.40. The total revenue at split-off would be $520 ($240 + $280). (You might think that the original alternative still exists—sell all of the Grade A tomatoes at

CORNERSTONE 11-4 *(continued)*

split-off for $0.40 per kilogram. While that alternative does exist, it is so clearly dominated by the new alternative with the higher-priced Premium A's that it can be safely ignored. The firm will no longer consider it.)

	Sell at Split-Off	Process Further	Differential Amount to Process Further
Sales	$520	$ 1,500	$ 980
Further processing cost	(50)	(1,000)	(950)
Total	$470	$ 500	$ 30

There is a $30 per load advantage to processing the Grade A tomatoes into hot sauce.

considered, and that would be the management decision to engage in the joint production process at all. If the total revenues do not cover all costs (both joint and further processing), then the company may want to reconsider being in that line of business.

Relevant Costing and Ethical Behaviour

ETHICS Relevant costs are used in making tactical decisions—decisions that have an immediate view or limited objective in mind. In making these decisions, however, decision makers should always keep the decisions within an ethical framework. Reaching objectives is important, but how you get there is perhaps even more important. Unfortunately, many managers have the opposite view. Part of the reason for the problem is the extreme pressure to perform that many managers feel. Often, the individual who is not a top performer may be laid off or demoted. Under such conditions, the temptation is often great to engage in questionable behaviour.◆

Real-World Example

For example, the price of cashmere decreased greatly during the 1990s. The lower price of cashmere fibre meant that sweaters and coats became much more affordable, and imports from China and Hong Kong more than doubled. Unfortunately, the cashmere content of the clothing was uneven, and, on occasion, misrepresented to the eventual seller. In the fall of 2000, **Lands' End** found that one of its blazers, advertised as a blend of lambswool and 30 percent cashmere, tested in the range of 10 to 30 percent cashmere. The company advised its operators to tell prospective purchasers of the variability and to offer $20 off the price to those who still wanted the jackets. Other sellers chose to take the "low road" and continued to advertise and sell their variable mix fibre sweaters and blazers at the higher percentage of cashmere.

There can be endless debates about what is right and what is wrong. Chapter 1 discusses some ethical standards that have been developed to provide guidance for individuals. Additionally, many companies are hiring full-time ethics officers. Often, these officers set up hotlines so that employees can call and register complaints or ask about the propriety of certain actions. However, as pointed out in an article in *Fortune*: "The old advice is still the best: Don't do anything on the job you wouldn't want your mother to read about with her morning coffee."[8]

[8] Kenneth Labich, "The New Crisis in Business Ethics," *Fortune* (April 20, 1992): 172.

Relevant Cost Analysis in Personal Decision Making

Finally, it is useful to note that relevant costing analysis is important in personal decision making. Nearly any short-term decision can be improved by following the decision model outlined in this chapter.

For many parents of young skiers, an important decision is whether to buy the child a new pair of skis or go with the seasonal rental approach. The problem, of course, is that children are growing, and a pair of skis that works this year may well not be the right size next year. Children also change their minds. The child who can't wait to get onto the slopes may be sick and tired of the sport after an unsuccessful morning. Or the child may decide to switch to snowboarding. Further complicating the problem is that so many alternatives are available—daily ski rental, seasonal ski rental, lease-to-own, purchasing new, or purchasing used. Computer scientists have actually developed a ski-rental algorithm to help people decide when skis should be bought versus rented. In the final analysis, parents use a combination of their budgets, their assessment of the probabilities that their children will enjoy skiing and will go frequently, and the importance of the "coolness factor" of snazzy graphics applied to the new skis.[9]

Throughout the years, many of our students have found it enlightening to use the model to consider their decisions to keep or buy a car, get a pet, choose a university, and so on. They can see that they have already implicitly used the decision-making model, and how explicit use can improve their decision making. They can also see how important qualitative factors are in those decisions, and in many cases they become more comfortable with the ultimate decision once they can see the legitimacy of relying on qualitative factors.

Summary of Learning Objectives

1. **Describe the tactical decision-making model.**

- Tactical decisions consist of choosing among alternatives with an immediate end in view.

    - Short term.
    - Larger strategic objectives are served.

- Six steps of the decision-making model are:

    - Recognize and define the problem.
    - Identify feasible alternatives.
    - Identify costs and benefits for each feasible alternative.
    - Total relevant costs and benefits for each alternative.
    - Assess qualitative factors.
    - Select best alternative.

2. **Define the concept of relevant costs and revenues.**

- Relevant costs are:

    - Future costs that differ across alternatives
    - Frequently variable costs—called flexible resources

[9] Nancy Keates, "The Cost of Taking Half Pints on the Half-Pipe," *The Wall Street Journal* (November 10, 2007): W1.

- Past costs:

  - Are sunk and never relevant
  - May be used to predict future costs

3. **Explain how the activity resource usage model is used in assessing relevancy.**

- Resources can be classified as flexible resources and committed resources.

  - Flexible resources are acquired as needed.
  - Committed resources are acquired in advance of usage.

- The cost of flexible resources is relevant.
- The cost of committed resources is relevant if demand changes across alternatives lead to a change in capacity.
- Changes in activity capacity cause resource spending to change.

4. **Apply the tactical decision-making concepts in a variety of business situations.**

- Make-or-buy decision
- Keep-or-drop decision
- Special-order decision
- Further processing of joint products

CORNERSTONES FOR CHAPTER 11

CORNERSTONE 11-1 The HOW and WHY of structuring a make-or-buy decision, page 556

CORNERSTONE 11-2 The HOW and WHY of structuring a keep-or-drop product line decision, page 559

CORNERSTONE 11-3 The HOW and WHY of structuring a special-order decision, page 563

CORNERSTONE 11-4 The HOW and WHY of structuring a sell at split-off or process further decision, page 565

Review Problem

Activity Resource Usage Model, Strategic Elements, and Relevant Costing

Perkins Company has idle capacity. Recently, Perkins received an offer to sell 2,000 units of one of its products to a new customer in a geographic region not normally serviced. The offering price is $10 per unit. The product normally sells for $14. The activity-based accounting system provides the following information:

	Cost Driver	Unused Capacity	Quantity Demanded*	Activity Rate** Fixed	Variable
Direct materials	Units	0	2,000	—	$3.00
Direct labour	Direct labour hours	0	400	—	7.00
Setups	Setup hours	0	25	$50.00	8.00
Machining	Machine hours	6,000	4,000	4.00	1.00

*This represents only the amount of resources demanded by the special order being considered.
**Fixed activity rate is the price that must be paid per unit of activity capacity. The variable activity rate is the price per unit of resource for resources acquired as needed.

Although the fixed activity rate for setups is $50 per hour, any expansion of this resource must be acquired in blocks. The unit of purchase for setups is 100 hours of setup servicing. Thus, any expansion of setup activity must be done 100 hours at a time. The price per hour is the fixed activity rate.

Required:

1. Compute the change in income for Perkins Company if the order is accepted. Comment on whether or not the order should be accepted. (In particular, discuss the strategic issues.)
2. Suppose that the setup activity had 50 hours of unused capacity. How does this affect the analysis?

Solution:

1. The relevant costs are those that change if the order is accepted. These costs would consist of the variable activity costs (flexible resources) plus any cost of acquiring additional activity capacity (committed resources). The income will change by the following amount:

Revenues ($10 × 2,000 units)	$20,000
Less increase in resource spending:	
Direct materials ($3 × 2,000 units)	(6,000)
Direct labour ($7 × 400 direct labour hours)	(2,800)
Setups [($50 × 100 hours) + ($8 × 25 hours)]	(5,200)
Machining ($1 × 4,000 machine hours)	(4,000)
Income change	$ 2,000

Special orders should be examined carefully before acceptance. This order offers an increase in income of $2,000, but it does require expansion of the setup activity capacity. If this expansion is short run in nature, then it may be worth it. If it entails a long-term commitment, then the company would be exchanging a one-year benefit of $2,000 for an annual commitment of $5,000. In this case, the order should be rejected. Even if the commitment is short term, other strategic factors need to be considered. Will this order affect any regular sales? Is the company looking for a permanent solution to its idle capacity, or are special orders becoming a habit (a response pattern that may eventually prove disastrous)? Will acceptance adversely affect the company's normal distribution channels? Acceptance of the order should be consistent with the company's strategic position.

2. If 50 hours of excess setup capacity exist, then the setup activity can absorb the special order's activity demands with no additional resource spending required for additional capacity. Thus, the profitability of the special order would be increased by $5,000 (the increase in resource spending that would have been required). Thus, total income would increase by $7,000 if the order is accepted.

Key Terms

Discussion Questions

1. What is tactical decision making?
2. "Tactical decisions are often small-scale decisions that serve a larger purpose." Explain what this means.
3. What is tactical cost analysis? What steps in the tactical decision-making model correspond to tactical cost analysis?

4. Describe a tactical decision you personally have had to make. Apply the tactical decision-making model to your decision. How did it turn out? (*Hint:* You could discuss buying a car, choosing a university, buying a puppy, etc.)

5. What is a relevant cost? Explain why depreciation on an existing asset is always irrelevant.

6. Give an example of a future cost that is not relevant.

7. Relevant costs always determine which alternative should be chosen. Do you agree or disagree? Explain.

8. Can direct materials ever be irrelevant in a make-or-buy decision? Explain. Give an example of a fixed cost that is relevant.

9. What role do past costs play in tactical cost analysis?

10. When will flexible resources be relevant to a decision?

11. When will the cost of committed resources be relevant to a decision?

12. What are the main differences between a traditional and an activity-based make-or-buy analysis?

13. Explain why activity-based segmented reporting provides more insight concerning keep-or-drop decisions.

14. Should joint costs be considered in a sell-or-process-further decision? Explain.

15. Why would a firm ever offer a price on a product that is below its full cost?

Cornerstone Exercises

OBJECTIVE ▶ 2 4
CORNERSTONE 11-1

Cornerstone Exercise 11-1 MAKE-OR-BUY DECISION, ALTERNATIVES, RELEVANT COSTS

Each year, Subramanian Company produces 20,000 units of a component used in radar detectors. An outside supplier has offered to supply the part for $2.36. The unit cost is:

Direct materials	$0.50
Direct labour	0.62
Variable overhead	0.24
Fixed overhead	2.75
Total unit cost	$4.11

Overhead is applied on the basis of machine hours; the component requires 10,000 machine hours per year.

Required:

1. What are the alternatives for Subramanian Company?

2. Assume that none of the fixed cost is avoidable. List the relevant cost(s) of internal production and of external purchase.

3. Which alternative is more cost effective and by how much?

4. *What if* $37,000 of fixed overhead is rental of equipment used only in production of the component that can be avoided if the component is purchased? Which alternative is more cost effective and by how much?

OBJECTIVE ▶ 2 4
CORNERSTONE 11-2

Cornerstone Exercise 11-2 KEEP-OR-DROP DECISION, ALTERNATIVES, RELEVANT COSTS

Ambi Company makes three types of rug shampooers. Model 1 is the basic model rented through hardware stores and supermarkets. Model 2 is a more advanced model with both dry and wet vacuuming capabilities. Model 3 is the heavy duty riding shampooer sold to hotels and convention centres. A segmented income statement is shown below.

	Model 1	Model 2	Model 3	Total
Sales	$ 345,000	$ 618,000	$ 575,000	$1,538,000
Less variable costs of goods sold	(125,000)	(183,160)	(386,250)	(694,410)
Less commissions	(6,900)	(37,080)	(28,750)	(72,730)
Contribution margin	$ 213,100	$ 397,760	$ 160,000	770,860

	Model 1	Model 2	Model 3	Total
Less common fixed expenses:				
Fixed factory overhead				(398,000)
Fixed selling and administrative				(290,000)
Operating income				$ 82,860

While all models have positive contribution margins, Ambi Company is concerned because operating income is less than 10 percent of sales and is low for this type of company. The company's controller gathered additional information on fixed costs to see why they were so high. The following information on activities and drivers was gathered:

	Activity		Driver Usage by Model		
Activity	Cost	Activity Driver	Model 1	Model 2	Model 3
Engineering	$ 30,000	Engineering hours	50	75	875
Setting up	180,000	Setup hours	2,500	7,500	20,000
Customer service	110,000	Service calls	1,000	1,500	17,500

In addition, Model 3 requires the rental of specialized equipment costing $20,000 per year.

Required:

1. Reformulate the segmented income statement using the additional information on activities.
2. Using your answer to Requirement 1, assume that Ambi Company is considering dropping any model with a negative product margin. What are the alternatives? Which alternative is more cost effective and by how much? (Assume that any traceable fixed costs can be avoided.)
3. *What if* Ambi Company can only avoid 375 hours of engineering time and 10,000 hours of setup time that are attributable to Model 3? How does that affect the alternatives presented in Requirement 2? Which alternative is more cost effective and by how much?

Cornerstone Exercise 11-3 SPECIAL-ORDER DECISION, ALTERNATIVES, RELEVANT COSTS

OBJECTIVE ➤ 2 4
CORNERSTONE 11-3

Zhang Leaf Paper Products Inc. manufactures boxed stationery for sale to specialty shops. Currently, the company is operating at 90 percent of capacity. A chain of drugstores has offered to buy 50,000 boxes of Zhang Leaf's blue-bordered thank you notes as long as the box can be customized with the drugstore chain's logo. While the normal selling price is $3.00 per box, the chain has offered just $2.45 per box. Zhang Leaf Paper Products can accommodate the special order without affecting current sales. Unit cost information for a box of thank you notes follows:

Direct materials	$1.22
Direct labour	0.25
Variable overhead	0.18
Fixed overhead	1.10
Total cost per box	$2.75

Fixed overhead is $420,000 per year and will not be affected by the special order. Normally, there is a commission of 10 percent of price; this will not be paid on the special order since the drugstore chain is dealing directly with the company. The special order will require additional fixed costs of $16,000 for the design and setup of the machinery to stamp the drugstore chain's logo on each box.

Required:

1. List the alternatives being considered. List the relevant benefits and costs for each alternative.
2. Which alternative is more cost effective and by how much?

3. *What if* Zhang Leaf Paper Products was operating at capacity and accepting the special order would require rejecting an equivalent number of boxes sold to existing customers? Which alternative would be better?

OBJECTIVE ➤ 2 4
CORNERSTONE 11-4

Cornerstone Exercise 11-4 SELL AT SPLIT-OFF OR PROCESS FURTHER DECISION, ALTERNATIVES, RELEVANT COSTS

Avalon Chemicals Company processes a number of chemical compounds for use in veterinary medicine. One compound is decomposed into two chemicals: avacol and selgene. The cost of processing one batch of compound is $74,000, and the result is 4,000 litres of avacol and 6,000 litres of selgene. Avalon Chemicals can sell the avacol at split-off for $13 per litre and the selgene for $4.50 per litre. Alternatively, the avacol can be processed further at a cost of $9 per litre (of avacol) into flourcine. It takes 2 litres of avacol for every litre of flourcine. A litre of flourcine sells for $46.

Required:

1. List the alternatives being considered.
2. List the relevant benefits and costs for each alternative.
3. Which alternative is more cost effective and by how much?
4. *What if* the production of avacol into flourcine required additional purchasing and quality inspection activity? Every 500 litres of avacol that undergo further processing require 20 more purchase orders at $10 each and 15 more quality inspection hours at $25 each. Which alternative would be better and by how much?

Exercises

OBJECTIVE ➤ 2

Exercise 11-5 DETERMINING RELEVANT COSTS

Six months ago, Lee Anna Martelli purchased a fire-engine red, used LeBaron convertible for $10,000. Lee Anna was looking forward to the feel of the sun on her shoulders and the wind whipping through her hair as she zipped along the highways of life. Unfortunately, the wind turned her hair into straw, and she didn't do much zipping along since the car spent so much of its time in the shop. So far, she has spent $1,200 on repairs, and she's afraid there is no end in sight. In fact, Lee Anna anticipates the following costs of restoration:

Rebuilt engine	$1,250
New paint job	560
Tires	460
New interior	500
Miscellaneous maintenance	340
Total	$3,110

On a visit to a used car dealer, Lee Anna found a five-year-old Honda CR-V in excellent condition for $9,100—Lee Anna thinks she might really be more the sport-utility type anyway. Lee Anna checked the Blue Book values and found that she can sell the LeBaron for only $3,600. If she buys the CR-V, she will pay cash but would need to sell the LeBaron.

Required:

1. In trying to decide whether to restore the LeBaron or buy the CR-V, Lee Anna is distressed because she has already spent $11,200 on the LeBaron. The investment seems too much to give up. How would you react to her concern?
2. List all costs that are relevant to Lee Anna's decision. What advice would you give her?

OBJECTIVE ➤ 2 3 4
SERVICE

Exercise 11-6 RESOURCE SUPPLY AND USAGE, SPECIAL ORDER, RELEVANCY

Barker Inc. has six salaried clerks to process purchase orders. Each clerk is paid a salary of $26,300 and is capable of processing as many as 6,500 purchase orders per year. Each clerk uses a PC and laser printer in processing orders. Time available on each PC system is sufficient to process 6,500 orders per year. The depreciation on each PC system is $1,100 per year. In addition to the salaries, Barker spends $27,300 for forms, postage, and other supplies (assuming 39,000 purchase orders are processed). During the year, 38,200 orders were processed.

Required:

1. Classify the resources associated with purchasing as (1) flexible or (2) committed.
2. Compute the total activity availability, and break this into activity usage and unused activity.
3. Calculate the total cost of resources supplied (activity cost), and break this into the cost of activity used and the cost of unused activity.
4. (a) Suppose that a large special order will cause an additional 500 purchase orders. What purchasing costs are relevant? By how much will purchasing costs increase if the order is accepted? (b) Suppose that the special order causes 1,000 additional purchase orders. How will your answer to (a) change?

Exercise 11-7 RESOURCE SUPPLY AND USAGE, ADDING A SERVICE LINE, RELEVANCY

OBJECTIVE ▶ [2] [3] [4]

SERVICE

Roxanne Sawchuk owns a beauty shop with eight hair and nail professionals. Her shop is relatively large and has four rooms at the back, three of which are currently empty (the fourth is used to store supplies). Roxanne's is popular and typically busy Monday through Friday all day and Saturday until noon. She has been thinking about adding a tanning salon in the back of the shop. She figures that she can buy two tanning beds for $10,000 each. The necessary supplies (cleaning materials, complimentary skin lotion in each room) will run about $450 per month. The extra electricity will cost about $100 per month. Currently, Roxanne pays part-time help to staff the front desk, make appointments, check clients in, and so on. For the hours that the beauty salon is open, the current staff can easily handle the additional tanning salon duties. However, Roxanne knows that she'll have to stay open an additional four hours each weeknight (from 5 p.m. until 9 p.m) as well as six hours on both Saturday and Sunday afternoons. Hiring additional staff for the tanning salon during those hours will cost about $8 per hour.

Required:

1. Classify the resources associated with the tanning salon as (1) flexible or (2) committed.
2. (a) Suppose that the tanning salon does very well and Roxanne decides to add one more tanning bed, using the third room at the back of the shop. What additional costs are relevant? (b) Suppose that Roxanne decides to add two more tanning beds? How will your answer to (a) change?

Exercise 11-8 SPECIAL-ORDER DECISION, TRADITIONAL ANALYSIS, QUALITATIVE ASPECTS

OBJECTIVE ▶ [4]

Sportz-a-Lot Inc. manufactures toys and sporting equipment, including golf kits for pre-schoolers. A national sporting goods chain recently submitted a special order for 7,600 golf kits. Sportz-a-Lot was not operating at capacity and could use the extra business. Unfortunately, the order's offering price of $16.50 per golf kit was below the cost to produce the sets. The controller was opposed to taking a loss on the deal. However, the personnel manager argued in favour of accepting the order even though a loss would be incurred; it would avoid the problem of layoffs and would help maintain the community image of the company. The full cost to produce a golf kit is presented below.

Direct materials	$ 7.90
Direct labour	5.40
Variable overhead	4.75
Fixed overhead	3.10
Total	$21.15

No variable selling or administrative expenses would be associated with the order. Non-unit-level activity costs are a small percentage of total costs and are therefore not considered.

Required:

1. Assume that the company would accept the order only if it increased total profits. Should the company accept or reject the order? Provide supporting computations.
2. Suppose that Sportz-a-Lot has negotiated with the potential customer, and has determined that it can substitute cheaper materials, reducing direct materials cost by $1.09 per unit. In addition, the company's engineers have found a way to reduce direct labour cost by $2.40 per unit. Should the company accept or reject the order? Provide supporting computations.
3. Consider the personnel manager's concerns. Discuss the merits of accepting the order even if it decreases total profits.

OBJECTIVE ➤ 2 4

Exercise 11-9 MAKE-OR-BUY, TRADITIONAL ANALYSIS

Savard Company is currently manufacturing Part KAV-71, producing 35,000 units annually. The part is used in the production of several products made by Savard. The cost per unit for KAV-71 is as follows:

Direct materials	$53.80
Direct labour	12.00
Variable overhead	2.75
Fixed overhead	1.30
Total	$69.85

Of the total fixed overhead assigned to KAV-71, $12,950 is direct fixed overhead (the annual lease cost of machinery used to manufacture Part KAV-71), and the remainder is common fixed overhead. An outside supplier has offered to sell the part to Savard for $64. There is no alternative use for the facilities currently used to produce the part. No significant non-unit-based overhead costs are incurred.

Required:

1. Should Savard Company make or buy Part KAV-71?
2. What is the maximum amount per unit that Savard would be willing to pay to an outside supplier?

OBJECTIVE ➤ 3 4

Exercise 11-10 MAKE-OR-BUY, TRADITIONAL AND ABC ANALYSIS

Venable Inc., a manufacturer of snowmobiles, has just received an offer from a supplier to provide 3,000 units of a component used in its main product. The component is a track assembly that is currently produced internally. The supplier has offered to sell the track assembly for $64 per unit. Venable is currently using a traditional, unit-based costing system that assigns overhead to jobs on the basis of direct labour hours. The estimated traditional full cost of producing the track assembly is as follows:

Direct materials	$40.00
Direct labour	12.50
Variable overhead	6.00
Fixed overhead	40.00

Prior to making a decision, the company's CEO commissioned a special study to see whether there would be any decrease in the fixed overhead costs. The results of the study revealed the following:

3 setups—$1,350 each.

One half-time inspector is needed. The company already uses part-time inspectors hired through a temporary employment agency. The yearly cost of the part-time inspectors for the track assembly operation is $12,300 and could be totally avoided if the part were purchased.

Engineering work: 515 hours, $30/hour. (Although the work decreases by 515 hours, the engineer assigned to the track assembly line also spends time on other products, and there would be no reduction in his salary.)

260 fewer material moves at $35 per move.

Required:

1. Ignore the special study, and determine whether the track assembly should be produced internally or purchased from the supplier.
2. Now, using the special study data, repeat the analysis.
3. Discuss the qualitative factors that would affect the decision, including strategic implications.
4. After reviewing the special study, the controller made the following remark: "This study ignores the additional activity demands that purchasing would cause. For example, although the demand for inspecting the part on the production floor decreases, we may need to inspect the incoming parts in the receiving area. Will we actually save any inspection costs?" Is the controller right?

Exercise 11-11 RESOURCE USAGE MODEL, SPECIAL ORDER

OBJECTIVE ▶ 3 4

Ehrling Inc. manufactures metal racks for hanging clothing in retail stores. Ehrling was approached by the CEO of Carly's Corner, a regional nonprofit food bank, with an offer to buy 350 heavy-duty metal racks for storing canned goods and dry food products. While racks normally sell for $245 each, Carly's Corner offered $75 per rack. The CEO explained that the number of families they served had grown significantly over the past two years, and that the charity needed additional storage for the donated food items. Since Ehrling is operating at 80 percent of capacity, and Ehrling employees have "adopted" Carly's Corner as their annual charity, the company wants to make the special order work. Ehrling's controller looked into the cost of the storage racks using the following information from the activity-based accounting system:

	Activity Driver	Unused Capacity	Quantity Demanded*	Activity Rate** Fixed	Activity Rate** Variable
Direct materials	Number of racks	0	350	—	$82
Direct labour	Direct labour hours	0	525	—	15
Setups	Setup hours	60	1	$150	5
Inspection	Inspection hours	800	20	10	5
Machining	Machine hours	6,000	175	40	3

*This represents only the amount of resources demanded by the special order being considered.
**This is expected activity cost divided by activity capacity.

Expansion of activity capacity for setups, inspection, and machining must be done in steps. For setups, each step provides an additional 20 hours of setup activity and costs $3,000. For inspection, activity capacity is expanded by 2,000 hours per year, and the cost is $20,000 per year (the salary for an additional inspector). Machine capacity can be leased for a year at a rate of $40 per machine hour. Machine capacity must be acquired, however, in steps of 1,500 machine hours.

Required:

1. Compute the change in income for Ehrling Inc. if the order is accepted.
2. Does the order require any change in capacity for setups, inspection, or machining?
3. Suppose that the inspection activity can be eliminated for this order since the customer is in town and does not need to have the racks boxed and shipped. Because of this, direct materials can be reduced by $13 per unit, and direct labour can be reduced by 0.5 hour per unit. How is the analysis affected?
4. Ehrling can find no other cost saving measures for this special order. Why might the company decide to accept it even if it shows a loss?

Exercise 11-12 KEEP-OR-DROP: TRADITIONAL VERSUS ACTIVITY-BASED ANALYSIS

Harding Ltd. produces two types of cough syrup: Basic and Multi-Symptom. Of the two, Basic is the more popular. Data concerning the two products follow:

	Basic	Multi-Symptom	Unused Capacity[a]	Units of Purchase[b]
Expected sales (in cases)	46,000	12,000	—	—
Selling price per case	$72	$192	—	—
Direct labour hours	38,000	12,000	—	As needed
Machine hours	11,500	3,000	—	2,500
Receiving orders	250	500	250	500
Packing orders	700	1,200	100	250
Material cost per case	$50	$80	—	—
Direct labour cost per case	$12	$15	—	—
Advertising costs	$100,000	$160,000	—	—

[a]Practical capacity less expected usage (all unused capacity is permanent).
[b]In some cases, activity capacity must be purchased in steps (whole units). These steps are provided as necessary. The cost per step is the fixed activity rate multiplied by the step units. The fixed activity rate is the expected fixed activity costs divided by practical activity capacity.

Annual overhead costs are listed below. These costs are classified as fixed or variable with respect to the appropriate activity driver.

Activity	Fixed[a]	Variable[b]
Direct labour benefits	$ 0	$200,000
Machine	200,000	250,000
Receiving	200,000	22,500
Packing	80,000	43,700
Total costs	$480,000	$516,200

[a]Costs associated with practical activity capacity. The machine fixed costs are all depreciation.
[b]These costs are for the actual levels of the cost driver.

Required:

1. Prepare a traditional segmented income statement, using a unit-level overhead rate based on direct labour hours. Using this approach, determine whether the Basic cough syrup product line should be kept or dropped.
2. Prepare an activity-based segmented income statement. Repeat the keep-or-drop analysis using an ABC approach.

Exercise 11-13 SELL OR PROCESS FURTHER, BASIC ANALYSIS

Schulzer Ltd. is a pork processor. Its plants, located on the Prairies, produce several products from a common process: sirloin roasts, chops, spare ribs, and the residual. The roasts, chops, and spare ribs are packaged, branded, and sold to supermarkets. The residual consists of organ meats and leftover pieces that are sold to sausage and hot dog processors. The joint costs for a typical week are as follows:

Direct materials	$84,000
Direct labour	30,000
Overhead	18,000

The revenues from each product are as follows: sirloin roasts, $65,000; chops, $70,000; spare ribs, $33,000; and residual, $10,000.

Schulzer's management has learned that certain organ meats are a prized delicacy in Asia. They are considering separating those from the residual and selling them abroad for $56,000. This would bring the value of the residual down to $3,750. In addition, the organ meats would need to be packaged and then air freighted to Asia. Further processing cost per week is estimated to be $27,500 (the cost of renting additional packaging equipment, purchasing materials, and hiring additional direct labour). Transportation cost would be $11,900 per week. Finally, resource spending would need to be expanded for other activities as well (purchasing, receiving, and internal shipping). The increase in resource spending for these activities is estimated to be $3,120 per week.

Required:

1. What is the gross profit earned by the original mix of products for one week?
2. Should the company separate the organ meats for shipment overseas or continue to sell them at split-off? What is the effect of the decision on weekly gross profit?

Exercise 11-14 RELEVANT COSTS, FOREIGN TRADE ZONES

OBJECTIVE ▶ 2

SERVICE

Grassley Ltd. is considering opening a new warehouse to serve the Maritime region. Dante Mauro, controller for Grassley, has been reading about the advantages of foreign trade zones. He wonders if locating in one would be of benefit to his company, which imports about 90 percent of its merchandise (chess sets from the Philippines, jewellery from Thailand, pottery from Mexico). Dante estimates that the new warehouse will store imported merchandise costing about $23.7 million per year. Inventory shrinkage at the warehouse (due to breakage and mishandling) is about 7 percent of the total. The average tariff rate on these imports is 5.5 percent.

Required:

1. If Grassley locates the warehouse in a foreign trade zone, how much will be saved in tariffs? Why?
2. Suppose that, on average, the merchandise stays in a Grassley warehouse for nine months before shipment to retailers. Carrying cost for Grassley is 8 percent per year. If Grassley locates the warehouse in a foreign trade zone, how much will be saved in carrying costs? What will the total tariff-related savings be? (Round your answers to the nearest dollar.)
3. Suppose that the shifting economic situation leads to a new tariff rate of 15 percent and a new carrying cost of 10 percent per year. To combat these increases, Grassley has instituted a total quality program emphasizing reducing shrinkage. The new shrinkage rate is 5 percent. Given this new information, if Grassley locates the warehouse in a foreign trade zone, how much will be saved in carrying costs? What will the total tariff-related savings be? (Round your answers to the nearest dollar.)

Exercise 11-15 PROVIDE IN-HOUSE OR OUTSOURCE DECISION, SERVICES, QUALITATIVE ASPECTS

OBJECTIVE ▶ 4

SERVICE

Tony and Tina Roselli own and run TNT's Pizza Restaurant. Tony is responsible for managing the day-to-day aspects, hiring workers, overseeing the kitchen, building, and grounds. He is the chief cook and handles all purchasing. Tina is the hostess and manages the front of the house (restaurant talk for the dining area). She schedules the wait staff, ensures that customers are well taken care of, and pitches in to bus tables and refill drinks as needed. Tina also handles the financial aspects of the business and is responsible for bookkeeping and tax compliance. Two years ago, Tony and Tina became parents of a baby boy, Joseph, nicknamed "LJ" for Little Joe. Tina brings LJ to work each day, and both Rosellis as well as the restaurant staff help out watching him. Recently, the restaurant has grown busier, so Tony and Tina expanded the hours of operation. As a result, the staff rarely has any free time and Tina feels she has too much to handle. Tony and Tina are considering outsourcing their bookkeeping and tax filing needs to a local accountant.

Typically, Tina spends 15 hours per month on bookkeeping and taxes. This increases to 40 hours in April. She uses a room off the kitchen as her office (a room that is sorely needed for additional food storage given the expansion). If Tina continues to do the financial work, the restaurant will need to make up for 75 percent of her time by hiring additional help at $10 per hour (hourly wage plus the restaurant's cost of employee benefits). The local accountant will charge $25 per hour for bookkeeping services; he expects this service to average eight hours per month. Taxes are filed quarterly for labour as well as GST and local taxes. These tax forms should cost about $75 per quarter. The annual income tax filing is estimated to cost $350, payable at the time of filing in April.

Required:

1. Given the information, determine whether Tina should do the bookkeeping and tax work in house or outsource it to the accountant.
2. Discuss the qualitative factors that would affect the decision, including strategic implications.

OBJECTIVE ▶ 4

Exercise 11-16 SPECIAL-ORDER DECISION, SERVICES, QUALITATIVE ASPECTS

Jason Rogers works full time for UPS and runs a lawn mowing service part time after work during the months of April through October. Jason has three men working with him, each of whom is paid $6 per lawn mowing. Jason has 30 residential customers who contract with him for once-weekly lawn mowing during the months of May through September, and twice per month mowings during April and October. On average, Jason charges $40 per lawn mowed. Recently, LStar Property Management Services asked Jason to mow the lawn at each of its 20 rental houses twice per month during the months of May through September. LStar has offered to pay $20 per lawn mowing, and would forgo the lawn edging that normally takes Jason's team about half of its regular mowing time. If Jason accepts the job, he can assign a two-man team to mow the rental house yards, and will have to buy an additional power lawn mower for about $350 used. Fuel to run the additional mower will be about $0.50 per mowing.

Required:

1. If Jason accepts the special order, by how much will his income increase or decrease?
2. What are some of the qualitative reasons why Jason might want to accept or decline the special order?

OBJECTIVE ▶ 4

Exercise 11-17 KEEP-OR-DROP, SERVICES, QUALITATIVE ASPECTS

Jem Farber owns Jem's Special Event Planning Service, a full-service event planner. Jem does much of the work herself and hires additional help as needed. She plans corporate events, weddings, and special occasion parties. Each of these is considered a separate line of business due to the specialized aspects of each type of event. Last year, Jem's accountant provided the following segmented income statement:

	Corporate	Wedding	Special Occasion	Total
Revenue	$ 55,300	$195,000	$168,000	$ 418,300
Less variable costs	(22,120)	(97,500)	(50,400)	(170,020)
Contribution margin	$ 33,180	$ 97,500	$117,600	248,280
Less common fixed expenses:				
Fixed operating expense				(175,000)
Fixed selling				(55,000)
Operating income				$ 18,280

Jem was not pleased with last year's results; corporate events were down considerably from the previous few years. In addition, she thinks that dealing with the corporate party-throwers may be more work than it is worth. Two important aspects of event planning are negotiating with vendors (e.g., caterers, florists, bands and orchestras, venues) on price and setting up for and being present at the event itself. The corporate negotiating seemed to consume extra time and their restrictions on the price they would pay made the negotiations particularly difficult. She decided to gather some data on the negotiation and setting-up activities:

	Corporate	Wedding	Special Occasion
Negotiating hours	400	1,200	400
Setting-up hours	100	400	500
Total cost of negotiating	$40,000		
Total cost of setting up	$60,000		

Required:

1. Prepare a segmented income statement using the activity data for negotiating and setting up. The total cost of these two activities can be subtracted from the fixed operating expense. The remaining fixed operating expense will be the common fixed operating expense. What does this income statement suggest about the relative profitability of the three product lines?

2. Jem believes that next year will be even worse. Her hunch is that corporate business will be down and that these clients will be especially intent on saving money by reducing the rate paid to Jem. She believes total corporate revenue may decrease by 25 percent overall, while the variable costs associated with those events will only decrease by 20 percent. Weddings, on the other hand, Jem expects to increase. Her reputation is growing and she thinks she can raise her revenues in this area by 15 percent even if the number of weddings does not increase. As a result, she expects variable costs of weddings to remain static. The special occasions (wedding anniversary parties, bar and bat mitzvahs, and so on) line is also expected to increase—with revenue and variable costs expected to increase by 10 percent. Jem does not know quite what to expect with respect to the negotiating and setting-up activities, so she thinks she'll just keep those constant for planning purposes. Prepare a segmented income statement using the activity data and these assumptions. What does this income statement suggest about dropping the corporate segment?

Problems

Problem 11-18 IDENTIFYING PROBLEMS AND ALTERNATIVES, RELEVANT COSTS

OBJECTIVE ▸ 1 2

Norton Products Inc. manufactures potentiometers. (A potentiometer is a device that adjusts electrical resistance.) Currently, all parts necessary for the assembly of products are produced internally. Norton has a single plant located in The Pas, Manitoba. The facilities for the manufacture of potentiometers are leased, with five years remaining on the lease. All equipment is owned by the company. Because of increases in demand, production has been expanded significantly over the five years of operation, straining the capacity of the leased facilities. Currently, the company needs more warehousing and office space, as well as more space for the production of plastic mouldings. The current output of these mouldings, used to make potentiometers, needs to be expanded to accommodate the increased demand for the main product.

Leo Tidwell, owner and president of Norton Products, has asked his vice president of marketing, John Tidwell, and his vice president of finance, Linda Thayn, to meet and discuss the problem of limited capacity. This is the second meeting the three have had concerning the problem. In the first meeting, Leo rejected Linda's proposal to build the company's own plant. He believed it was too risky to invest the capital necessary to build a plant at this stage of the company's development. The combination of leasing a larger facility and subleasing the current plant was also considered but was rejected; subleasing would be difficult, if not impossible. At the end of the first meeting, Leo asked John to explore the possibility of leasing another facility comparable to the current one. He also assigned Linda the task of identifying other possible solutions. As the second meeting began, Leo asked John to give a report on the leasing alternative.

JOHN: After some careful research, I'm afraid that the idea of leasing an additional plant is not a very good one. Although we have some space problems, our current level of production doesn't justify another plant. In fact, I expect it will be at least five years before we need to be concerned about expanding into another facility like the one we have now. My market studies reveal a modest growth in sales over the next five years. All this growth can be absorbed by our current production capacity. The large increases in demand that we experienced the past five years are not likely to be repeated. Leasing another plant would be an overkill solution.

LEO: Even modest growth will aggravate our current space problems. As you both know, we are already operating three production shifts. But, John, you are right—except for plastic mouldings, we could expand production, particularly during the graveyard shift. Linda, I hope that you have been successful in identifying some other possible solutions. Some fairly quick action is needed.

LINDA: Fortunately, I believe that I have two feasible alternatives. One is to rent an additional building to be used for warehousing. By transferring our warehousing needs to the new building, we will free up internal space for offices and for expanding the production of plastic mouldings. I have located a building within two kilometres of our plant that we could use. It has the capacity to handle our current needs and the modest growth that John mentioned. The second alternative may be even more attractive. We currently produce all the parts that we use to manufacture potentiometers, including shafts and bushings. In the past several months, the market has been flooded with these two parts. Prices have tumbled as a result. It might be better to buy shafts and bushings instead of making them. If we stop internal production of shafts and bushings, this would free up the space we need. Well, Leo, what do you think? Are these alternatives feasible? Or should I continue my search for additional solutions?

LEO: I like both alternatives. In fact, they are exactly the types of solutions we need to consider. All we have to do now is choose the one best for our company.

Required:

1. Define the problem facing Norton Products.
2. Identify all the alternatives that were considered by Norton Products. Which ones were classified as not feasible? Why? Now identify the feasible alternatives.
3. For the feasible alternatives, what are some potential costs and benefits associated with each alternative? Of the costs that you have identified, which do you think are relevant to the decision?

OBJECTIVE ▶ 2 4

SERVICE

Problem 11-19 KEEP-OR-DROP FOR SERVICE FIRM, COMPLEMENTARY EFFECTS, TRADITIONAL ANALYSIS

Devern Assurance Company provides both property and automobile insurance. The projected income statements for the two products are as follows:

	Property Insurance	Automobile Insurance
Sales	$4,200,000	$12,000,000
Less variable expenses	3,830,000	9,600,000
Contribution margin	370,000	2,400,000
Less direct fixed advertising expenses	400,000	500,000
Segment margin	(30,000)	1,900,000
Less common fixed expenses (allocated)	100,000	200,000
Operating income (loss)	$ (130,000)	$ 1,700,000

The president of the company is considering dropping the property insurance. However, some policyholders prefer having their property and automobile insurance with the same company, so if property insurance is dropped, sales of automobile insurance will drop by 12 percent. No significant non-unit-level activity costs are incurred.

Required:

1. If Devern Assurance Company drops property insurance, by how much will income increase or decrease? Provide supporting computations.
2. Assume that dropping all advertising for the property insurance line and increasing the corporate advertising budget by $450,000 will increase sales of property insurance by 10 percent and automobile insurance by 8 percent. Prepare a segmented income statement that reflects the effect of increased advertising. Should advertising be increased?

Problem 11-20 RESOURCE USAGE, SPECIAL ORDER

OBJECTIVE ▶ 3 4

SERVICE

St. John's Medical Centre (SJMC) has five medical technicians who are responsible for conducting cardiac catheterization testing in SJMC's Cath Lab. Each technician is paid a salary of $36,000 and is capable of conducting 1,000 procedures per year. The cardiac catheterization equipment is one year old and was purchased for $250,000. It is expected to last five years. The equipment's capacity is 25,000 procedures over its life. Depreciation is computed on a straight-line basis, with no salvage value expected. The reading of the catheterization results is conducted by an outside physician whose fee is $120 per test. The technician's report with the outside physician's note of results is sent to the referring physician. In addition to the salaries and equipment, SJMC spends $50,000 for supplies and other costs needed to operate the equipment (assuming 5,000 procedures are conducted). When SJMC purchased the equipment, it fully expected to perform 5,000 procedures per year. In fact, during its first year of operation, 5,000 procedures were run. However, a larger hospital has established a clinic in the city and will siphon off some of SJMC's business. During the coming years, SJMC expects to run only 4,200 cath procedures yearly. SJMC has been charging $850 for the procedure—enough to cover the direct costs of the procedure plus an assignment of general overhead (e.g., depreciation on the hospital building, lighting and heating, and janitorial services).

At the beginning of the second year, a hospital from a neighbouring community approached SJMC and offered to send its clients to SJMC for cardiac catheterization provided that the charge per procedure would be $550. The hospital estimates that it can provide about 500 patients per year. The hospital has indicated that the arrangement is temporary—for one year only. The hospital expects to have its own testing capabilities within one year.

Required:

1. Classify the resources associated with the cardiac catheterization activity into one of the following: (1) committed resources or (2) flexible resources.

2. Calculate the activity rate for the cardiac catheterization activity. Break the activity rate into fixed and variable components. Now, classify each activity resource as relevant or irrelevant with respect to the following alternatives: (1) accept the hospital's offer or (2) reject the hospital's offer. Explain your reasoning.

3. Assume that SJMC will accept the hospital's offer if it reduces the hospital's operating costs. Should the hospital's offer be accepted?

4. Jerold Bosserman, SJMC's controller, argued against accepting the hospital's offer. Instead, he argued that SJMC should be increasing the charge per procedure rather than accepting business that doesn't even cover full costs. He also was concerned about local physician reaction if word got out that SJMC was performing procedures for $550. Discuss the merits of Jerold's position. Include in your discussion an assessment of the price increase that would be needed if the objective is to maintain total revenues from cardiac catheterizations experienced in the first year of operation.

5. Chandra Danton, SJMC's administrator, has been informed that one of the Cath Lab technicians is leaving for an opportunity at a larger hospital. She met with the other technicians, and they agreed to increase their hours to pick up the slack so that SJMC won't need to hire another technician. By working a couple hours extra every week, each remaining technician can perform 1,050 procedures per year. They agreed to do this for an increase in salary of $2,000 per year. How does this outcome affect the analysis of the hospital's offer?

6. Assuming that SJMC wants to bring in the same revenues earned in the cardiac catheterization activity's first year less the reduction in resource spending attributable to using only four technicians, how much must SJMC charge for a procedure?

Problem 11-21 ACTIVITY-BASED RESOURCE USAGE MODEL, MAKE-OR-BUY

OBJECTIVE ▶ 3 4

Brandy Dees recently bought Blade Enterprises, a company that manufactures ice skates. Brandy decided to assume management responsibilities for the company and appointed herself president shortly after the purchase was completed. When she bought the company, Brandy's investigation revealed that with the exception of the blades, all parts of the skates are produced internally. The investigation also revealed that Blade once produced the blades internally and still owned the equipment. The equipment was in good condition and was stored in a local warehouse. Blade's former owner had decided three years earlier to purchase the blades from external suppliers.

Brandy Dees is seriously considering making the blades instead of buying them from external suppliers. The blades are purchased in sets of two and cost $8 per set. Currently, 100,000 sets of blades are purchased annually.

Skates are produced in batches, according to shoe size. Production equipment must be reconfigured for each batch. The blades could be produced using an available area within the plant. Prime costs will average $5.00 per set. There is enough equipment to set up three lines of production, each capable of producing 80,000 sets of blades. A supervisor would need to be hired for each line. Each supervisor would be paid a salary of $40,000. Additionally, it would cost $1.50 per machine hour for power, oil, and other operating expenses. Since three types of blades would be produced, additional demands would be made on the setup activity. Other overhead activities affected include purchasing, inspection, and materials handling. The company's ABC system provides the following information about the current status of the overhead activities that would be affected. (The lumpy quantity indicates how much capacity must be purchased should any expansion of activity supply be needed—the units of purchase. The purchase cost per unit is the fixed activity rate. The variable rate is the cost per unit of resources acquired as needed for each activity.)

Activity	Cost Driver	Current Activity Capacity	Activity Usage	Lumpy Quantity	Fixed Activity Rate	Variable Activity Rate
Setups	Number of setups	1,000	800	100	$200	$500
Purchasing	Number of orders	50,000	47,000	5,000	10	0.50
Inspecting	Inspection hours	20,000	18,000	2,000	15	none
Materials handling	Number of moves	9,000	8,700	500	30	1.50

The demands that *production* of blades place on the overhead activities are as follows:

Activity	Resource Demands
Machining	50,000 machine hours
Setups	250 setups
Purchasing	4,000 purchase orders (associated with materials)
Inspection	1,500 inspection hours
Materials handling	650 moves

If the blades are made, the purchase of the blades from outside suppliers will cease. Therefore, purchase orders will decrease by 6,500 (the number associated with their purchase). Similarly, the moves for the handling of incoming blades will decrease by 400. Any unused activity capacity is viewed as permanent.

Required:

1. Should Blade make or buy the blades?
2. Explain how the ABC resource usage model helped in the analysis. Also, comment on how a conventional approach would have differed.

OBJECTIVE ▶ 2 4

Problem 11-22 MAKE-OR-BUY, TRADITIONAL ANALYSIS, QUALITATIVE CONSIDERATIONS

Beliveau Dental Services is part of a private clinic that operates in a large metropolitan area. Currently, Beliveau has its own dental laboratory to produce two varieties of porcelain crowns—all porcelain and porcelain fused to metal (PFM). The unit costs to produce the crowns are as follows:

	All Porcelain	PFM
Direct materials	$190	$ 80
Direct labour	50	20
Variable overhead	25	5
Fixed overhead	60	40
Total	$325	$145

Fixed overhead is detailed as follows:

Salary (supervisor)	$30,000
Depreciation	8,000
Rent (lab facility)	22,000

Overhead is applied on the basis of direct labour hours. The rates above were computed using 8,000 direct labour hours. No significant non-unit-level overhead costs are incurred.

A local dental laboratory has offered to supply Beliveau all the crowns it needs. Its price is $265 for all-porcelain crowns and $145 for porcelain-fused-to-metal crowns; however, the offer is conditional on supplying both types of crowns—it will not supply just one type for the price indicated. If the offer is accepted, the equipment used by Beliveau's laboratory would be scrapped (it is old and has no market value), and the lab facility would be closed. Beliveau uses 2,500 all-porcelain crowns and 1,000 porcelain-fused-to-metal crowns per year.

Required:

1. Should Beliveau continue to make its own crowns, or should they be purchased from the external supplier? What is the dollar effect of purchasing?
2. What qualitative factors should Beliveau consider in making this decision?
3. Suppose that the lab facility is owned rather than rented and that the $22,000 is depreciation rather than rent. What effect does this have on the analysis in Requirement 1?
4. Refer to the original data. Assume that the volume of crowns is 5,000 all porcelain and 2,000 porcelain fused to metal. Should Beliveau make or buy the crowns? Explain the outcome.

Problem 11-23 **SELL OR PROCESS FURTHER**

OBJECTIVE ▶ 4

Pharmaco Corporation buys three chemicals that are processed to produce two popular ingredients for liquid pain relievers. The three chemicals are in liquid form. The purchased chemicals are blended for two to three hours and then heated for 15 minutes. The results of the process are two separate ingredients, PR1 and PR2. For every 4,300 litres of chemicals used, 2,000 litres of each pain reliever are produced. The pain relievers are sold to companies that process them into their final form. The selling prices are $34 per litre for PR1 and $45 per litre for PR2. The costs to produce one batch (containing 2,000 litres of each chemical) are as follows:

Chemicals	$23,400
Direct labour	9,000
Catalyst	3,600
Overhead	8,000

The pain relievers are bottled in five-litre plastic containers and shipped. The cost of each container is $2.10. The costs of shipping are $0.50 per container.

Pharmaco Corporation could process PR1 further by mixing it with inert powders and flavouring to form tablets. The tablets can be sold directly to retail drug stores as a generic brand. If this route is taken, the revenue received per case of tablets would be $13.50, with eight cases produced by every litre of PR1. The costs of processing into tablets total $11.00 per litre of PR1. Packaging costs $5.16 per case. Shipping costs are $1.68 per case.

Required:

1. Should Pharmaco sell PR1 at split-off, or should PR1 be processed and sold as tablets?
2. If Pharmaco normally sells 26,000 litres of PR1 per year, what will be the difference in profits if PR1 is processed further?

OBJECTIVE ▶ 1 2 4

Problem 11-24 MAKE-OR-BUY, TRADITIONAL ANALYSIS

Morrill Company produces two different types of gauges: a density gauge and a thickness gauge. The segmented income statement for a typical quarter follows.

	Density Gauge	Thickness Gauge	Total
Sales	$150,000	$80,000	$230,000
Less variable expenses	80,000	46,000	126,000
Contribution margin	70,000	34,000	104,000
Less direct fixed expenses*	20,000	38,000	58,000
Segment margin	$ 50,000	$ (4,000)	46,000
Less common fixed expenses			30,000
Operating income			$ 16,000

*Includes depreciation.

The density gauge uses a subassembly that is purchased from an external supplier for $25 per unit. Each quarter, 2,000 subassemblies are purchased. All units produced are sold, and there are no ending inventories of subassemblies. Morrill is considering making the subassembly rather than buying it. Unit-level variable manufacturing costs are as follows:

Direct materials	$2
Direct labour	3
Variable overhead	2

No significant non-unit-level costs are incurred.

Morrill is considering two alternatives to supply the productive capacity for the subassembly.

1. Lease the needed space and equipment at a cost of $27,000 per quarter for the space and $10,000 per quarter for a supervisor. There are no other fixed expenses.
2. Drop the thickness gauge. The equipment could be adapted with virtually no cost and the existing space utilized to produce the subassembly. The direct fixed expenses, including supervision, would be $38,000, $8,000 of which is depreciation on equipment. If the thickness gauge is dropped, sales of the density gauge will not be affected.

Required:

1. Should Morrill Company make or buy the subassembly? If it makes the subassembly, which alternative should be chosen? Explain and provide supporting computations.
2. Suppose that dropping the thickness gauge will decrease sales of the density gauge by 10 percent. What effect does this have on the decision?
3. Assume that dropping the thickness gauge decreases sales of the density gauge by 10 percent and that 2,800 subassemblies are required per quarter. As before, assume that there are no ending inventories of subassemblies and that all units produced are sold. Assume also that the per-unit sales price and variable costs are the same as in Requirement 1. Include the leasing alternative in your consideration. Now, what is the correct decision?

OBJECTIVE ▶ 2

Problem 11-25 EXPORTING, FOREIGN TRADE ZONES

Qatar Company manufactures plain-paper fax machines in a small factory in Barrie, Ontario. Sales have increased by 50 percent in each of the past three years, as Qatar has expanded its market from Canada to the United States and Mexico. As a result, the

Barrie factory is at capacity. Beryl Adams, president of Qatar, has examined the situation and developed the following alternatives.

1. Add a permanent second shift at the plant. However, the semiskilled workers who assemble the fax machines are in short supply, and the wage rate of $15 per hour would probably have to be increased across the board to $18 per hour in order to attract sufficient workers from out of town. The total wage increase (including fringe benefits) would amount to $125,000. The heavier use of plant facilities would lead to increased plant maintenance and small tool cost.
2. Open a new plant and locate it in Mexico. Wages (including fringe benefits) would average $3.50 per hour. Investment in plant and equipment would amount to $300,000.
3. Open a new plant and locate it in a foreign trade zone, possibly in Quebec City. Wages would be somewhat lower than in Barrie, but higher than in Mexico. The advantages of postponing tariff payments on parts imported from Asia could amount to $50,000 per year.

Required:

Advise Beryl of the advantages and disadvantages of each of her alternatives.

CMA Problems

CMA Problem 11-1 SPECIAL ORDER, TRADITIONAL ANALYSIS*

OBJECTIVE ▶ 2 4

Caron Company manufactures two types of cold-pressed olive oil, Refined Oil and Top Quality Oil, out of a joint process. The joint (common) costs incurred are $84,000 for a standard production run that generates 40,000 litres of Refined Oil and 20,000 litres of Top Quality Oil. Additional processing costs beyond the split-off point are $2.25 per litre for Refined Oil and $1.80 per litre for Top Quality Oil. Refined Oil sells for $3.75 per litre, while Top Quality Oil sells for $6.80 per litre.

Marche LcBeau, a supermarket chain, has asked Caron to supply it with 40,000 litres of Top Quality Oil at a price of $6.30 per litre. Marche LcBeau plans to have the oil bottled with its own label.

If Caron accepts the order, it will save $0.10 per litre in packaging of Top Quality Oil. There is sufficient excess capacity for the order. However, the market for Refined Oil is saturated, and any additional sales of Refined Oil would take place at a price of $2.15 per litre. Assume that no significant non-unit-level activity costs are incurred.

Required:

1. What is the profit normally earned on one production run of Refined Oil and Top Quality Oil?
2. Should Caron accept the special order? Explain. *(CMA adapted)*

CMA Problem 11-2 PLANT SHUTDOWN OR CONTINUE OPERATIONS, QUALITATIVE CONSIDERATIONS, TRADITIONAL ANALYSIS*

OBJECTIVE ▶ 2 4

KarlAuto Corporation manufactures automobiles, vans, and trucks. Among the various KarlAuto plants throughout Canada is the Regina plant, where vinyl covers and upholstery fabric are sewn. These are used to cover interior seating and other surfaces of KarlAuto products.

Pam Teegin is the plant manager for the Regina cover plant—the first KarlAuto plant in the region. As other area plants were opened, Teegin, in recognition of her management ability, was given the responsibility to manage them. Teegin functions as a regional manager, although the budget for her and her staff is charged to the Regina plant.

Teegin has just received a report indicating that KarlAuto could purchase the entire annual output of the Regina cover plant from outside suppliers for $32 million. Teegin was astonished at the low outside price, because the budget for the Regina plant's operating costs was set at $56.45 million. Teegin believes that the Regina plant will have to close down operations in order to realize the $24.45 million in annual cost savings.

The budget (in thousands) for the Regina plant's operating costs for the coming year follows:

Materials		$12,000
Labour:		
Direct	$13,800	
Supervision	3,750	
Indirect plant	4,300	21,850
Overhead:		
Depreciation—Equipment	5,000	
Depreciation—Building	3,000	
Pension expense	5,600	
Plant manager and staff	3,000	
Corporate allocation	6,000	22,600
Total budgeted costs		$56,450

Additional facts regarding the plant's operations are as follows:

Due to the Regina plant's commitment to use high-quality fabrics in all of its products, the Purchasing Department was instructed to place blanket orders with major suppliers to ensure the receipt of sufficient materials for the coming year. If these orders are cancelled as a consequence of the plant closing, termination charges would amount to 18 percent of the cost of direct materials.

Approximately 600 plant employees will lose their jobs if the plant is closed. This includes all direct labourers and supervisors as well as the plumbers, electricians, and other skilled workers classified as indirect plant workers. Some would be able to find new jobs, but many others would have difficulty. All employees would have difficulty matching the Regina plant's base pay of $29.40 per hour, the highest in the area. A clause in the Regina plant's contract with the union may help some employees; the company must provide employment assistance to its former employees for 12 months after a plant closing. The estimated cost to administer this service would be $1 million for the year.

Some employees would probably elect early retirement because the company has an excellent pension plan. In fact, $4.6 million of next year's pension expense would continue whether or not the plant is open.

Teegin and her staff would not be affected by the closing of the Regina plant. They would still be responsible for administering three other area plants.

Equipment depreciation for the plant is considered to be a variable cost and the units-of-production method is used to depreciate equipment; the Regina plant is the only KarlAuto plant to use this depreciation method. However, it uses the customary straight-line method to depreciate its building.

Required:

1. Prepare a quantitative analysis to help in deciding whether or not to close the Regina plant. Explain how you treated the nonrecurring relevant costs.
2. Consider the analysis in Requirement 1, and add to it the qualitative factors that you believe are important to the decision. What is your decision? Would you close the plant? Explain. *(CMA adapted)*

The Collabourative Learning Exercises can be found on the product support site at www.hansen1ce.nelson.com.

After studying this chapter, you should be able to:

➤ **1** Discuss basic pricing concepts.

➤ **2** Calculate a markup on cost and a target cost.

➤ **3** Discuss the impact of the legal system and ethics on pricing.

➤ **4** Discuss the variations in price, cost, and profit over the product life cycle.

➤ **5** Explain why firms measure profit, and calculate measures of profit using absorption and variable costing.

➤ **6** Compute the sales price, price volume, contribution margin, contribution margin volume, sales mix, market share, and market size variances.

➤ **7** Describe some of the limitations of profit measurement.

CHAPTER

Pricing and Profitability Analysis 12

Henry Ford said, "A business that does not make a profit for the buyer of a commodity, as well as for the seller, is not a good business. Buyer and seller must both be wealthier in some way as a result of a transaction, else the balance is broken."[1] Ford reminds us that the relationship between buyer and seller is an exchange relationship. Both expect to profit from it. Typically, we measure profit as the difference between revenues and costs. Price and revenue will be discussed first. Then, we will look at profit—the interplay of price and cost.

Basic Pricing Concepts

OBJECTIVE ➤ **1**
Discuss basic pricing concepts.

One of the more difficult decisions facing a company is pricing. The accountant is the primary resource the firm turns to when financial data are needed, whether that information relates to cost or to price. Therefore, accountants must be familiar with

---

[1] Henry Ford, *Today and Tomorrow* (Portland, OR: Productivity Press, 1926, reprinted in 1988).

sources of revenue data as well as the economic and marketing concepts needed to interpret those data.

Demand and Supply

Customers want high-quality goods and services at a low price. Although customer demand is studied in detail in marketing classes and demand and supply in economics courses, accountants need to be aware of the way demand interacts with supply.

With all else equal, customers will buy more at lower prices and less at higher prices. Producers, on the other hand, are able to supply more at higher prices than they can at lower prices. The market-clearing or equilibrium price is located at the intersection of the supply and demand curves. At this price, the amount that producers supply just equals the amount that consumers demand. If firms charge a price that is higher than the market-clearing price, demand falls short of supply. Producers see inventories pile up as consumers buy other goods. If the price is lower than the market-clearing price, everything that is produced is bought. Shortages and backlogs occur, signalling the need to increase production and/or to raise prices.

Factors other than price that influence demand include consumer income, quality of goods offered for sale, availability of substitutes, demand for complementary goods, whether the good is a necessity or a luxury, and so on. However, the basic demand-supply relationship remains, and producers know that raising prices nearly inevitably results in less sold. Price elasticity and market structure are two factors that influence companies' ability to adjust price.

Price Elasticity of Demand

Since price affects quantity sold, producers want to know just how much a price change will change quantity demanded. **Price elasticity of demand** is measured as the percentage change in quantity divided by the percentage change in price. If demand is relatively elastic, a small percent change in price will lead to a greater percent change in quantity demanded. The opposite is true for inelastic demand.

Goods that are price elastic tend to have many substitutes, are not necessities, and take a relatively large amount of consumer income. The demand for movie tickets, restaurant meals, and automobiles is relatively elastic.

Price-inelastic goods have few substitutes, are necessities, or constitute a relatively small percentage of consumer income. Prescription drugs, electricity, and toothpicks are examples of price-inelastic goods. While price elasticity of demand is difficult to compute in real-world situations, it is possible to see its effects at work.

For example, **Unilever**, maker of Dove soap, Lipton teas, and Hellmann's mayonnaise, found its profit margins slipping in 2008 after it raised prices on many products. Demand fell precipitously, leading to falling profit margins. The new CEO, Paul Polman, quickly reversed that strategy, lowering prices and increasing quantity sold. Apparently, many of Unilever's products face elastic demand. The various soaps, teas, and so on, have numerous competitors. While a consumer may like Dove soap, for example, a price increase may send him or her to another brand.[2]

Other companies may have products with inelastic demand. For example, airlines define their core market as business travellers, who have inelastic demand for air travel. They need the flexibility to purchase tickets at the last minute, to change reservations, and to fly during the work week. Prices for tickets bought under these circumstances stay relatively high.

[2] Aaron O. Patrick, "Unilever CEO's Push to Cut Prices Drives Increase in Sales," *The Wall Street Journal* (August 7, 2009): B1.

Types of Market Structure and Price

Market structure affects price, as well as the costs necessary to support that price. In general, there are four types of market structure: perfect competition, monopolistic competition, oligopoly, and monopoly. These markets differ according to the number of buyers and sellers, the uniqueness of the product, and the relative ease of entry by firms into and out of the market (i.e., barriers to entry).

The **perfectly competitive market** has many buyers and sellers—no one of which is large enough to influence the market—a homogeneous product, and easy entry into and exit from the industry. Firms in a perfectly competitive market cannot charge a higher price than the market price because no one would buy their product, and they will not set a lower price because they can sell all they can produce at the market price.

At the opposite extreme is a monopoly. In a **monopoly**, barriers to entry are so high that there is only one firm in the market and the product is unique. The monopolistic firm is a price setter. While the monopolist sets the price, that does not mean it can force consumers to buy. It does mean that a somewhat higher price (with a lower quantity sold) can be set than would be set in a competitive market. Some monopolies have legally enforced barriers to entry (e.g., **Canada Post**). Other firms are monopolies because of patent protection, specialized knowledge, or exceptionally high-cost production equipment. Pharmaceutical companies have a monopoly on new drugs due to patent protection. When the patent expires, generic drug companies can produce it, and the price of the drug plummets.

Monopolistic competition has characteristics of both monopoly and perfect competition, but it is much closer to the competitive situation. There are many sellers and buyers, but the products are differentiated on some basis. Restaurants are good examples of monopolistic competitors. Each restaurant serves food but attempts to differentiate itself in some way—ethnic style of food, closeness to work or schools, availability of a party room, gourmet versus casual atmosphere, and so on. The end result is to slightly raise prices above the perfectly competitive price, as customers agree to pay a little more for the unique feature that appeals to them.

An **oligopoly** is characterized by a few sellers. Typically, barriers to entry are high, and they are usually cost related. For example, the cereal industry is dominated by **Kellogg's**, **General Mills**, and **Quaker Oats**. The reason is not the high cost of manufacturing corn flakes. Instead, the huge selling expenditures (e.g., advertising and shelf space fees) of the big three effectively prevent smaller companies from entering the market.

The various types of market structure and their characteristics are summarized in Exhibit 12-1. Companies must be aware of the market structure in which they operate in order to understand their pricing options. Note that these market structures also have implications for the supply or cost side. The firm in the perfectly competitive industry has lower marketing costs (advertising, positioning, discounting, coupons) than the firm in the monopolistically competitive industry, which must constantly reinforce the consumer's perception of its product's uniqueness. The monopolist typically incurs expenses to protect its monopoly position, often through legal fees and lobbying (included in administrative expenses).

Exhibit 12-1

Characteristics of the Four Basic Types of Market Structure

Market Structure Type	Number of Firms in Industry	Barriers to Entry	Uniqueness of Product	Expenses Related to Structure Type
Perfect competition	Many	Very low	Not unique	No special expenses
Monopolistic competition	Many	Low	Some unique features	Advertising, coupons, costs of differentiation
Oligopoly	Few	High	Fairly unique	Costs of differentiation, advertising, rebates, coupons
Monopoly	One	Very high	Very unique	Legal and lobbying expenditures

Cost and Pricing Policies

Companies use various strategies to set price. Since cost is an important determinant of supply and known to the producer, many companies base price on cost. Still other companies use a target-costing strategy, or strategies based on the initial conditions in the market.

Cost-Based Pricing

Demand is one side of the pricing equation; supply is the other side. Since revenue must cover cost for the firm to make a profit, many companies start with cost to determine price. That is, they calculate product cost and add the desired profit. The mechanics of this approach are straightforward. Usually, there is some cost base and a markup. The **markup** is a percentage applied to base cost; it includes desired profit and any costs not included in the base cost. Companies that bid for jobs routinely base bid price on cost. Cornerstone 12-1 shows the how and why of calculating a markup on cost.

As can be seen in Cornerstone 12-1, the markup on cost of goods sold is 43 percent. Notice that the 43 percent markup covers both profit and selling and administrative expenses. The markup percentage of 186 percent of direct materials cost would yield the same amount of profit, assuming the level of operations and other expenses remained stable. The markup percentage on direct materials covers direct labour, overhead, selling and administrative expenses, and profit. The choice of base and markup percentage generally rests on convenience.

When the markup percentages calculated in Cornerstone 12-1 were used in determining bid price, they were initial prices. Chris can adjust the price based on her knowledge of competition for this type of job and other factors. The markup is a guideline, not an absolute rule.

If a company actually sets its prices based on markup percentages, is it guaranteed to make a profit? Not at all. If very few jobs are won, the entire markup will go toward selling and administrative expenses, the costs not explicitly included in the pricing calculations.

Markup pricing is often used by retail stores, and their typical markup is 100 percent of cost. If a sweater is purchased by Graham Department Store for $24, the retail price marked is $48 [$24 + (1.00 × $24)]. That 100 percent markup is meant to cover the salaries of the clerks, payment for space and equipment (cash registers, etc.), utilities, advertising, and so on, as well as profit. A major advantage of markup pricing is that standard markups are easy to apply. Consider the difficulty of setting a price for every piece of merchandise in a store. For example, **The Bay** department store stocks a wide variety of goods, from glassware and pottery to furniture and textiles. Assessing the supply and demand characteristics of each item is time consuming. It is much simpler to apply a uniform markup to cost and then adjust prices as needed if less is demanded than anticipated.

Target Costing and Pricing

Most North American companies, and nearly all European firms, set the price of a new product as the sum of the costs and the desired profit. The rationale is that the company must earn sufficient revenues to cover all costs and yield a profit. Peter Drucker writes, "This is true but irrelevant: Customers do not see it as their job to ensure manufacturers a profit. The only sound way to price is to start out with what the market is willing to pay."[3]

---

[3] Peter Drucker, "The Five Deadly Business Sins," *The Wall Street Journal* (October 21, 1993): A22.

The HOW and WHY of Calculating a Markup on Cost

Information:
AudioPro Company, owned and operated by Chris Brown, sells and installs audio equipment in homes and vehicles. Direct materials and direct labour costs are easy to trace to the jobs. Assemblers receive, on average, $12 per hour. AudioPro's income statement for last year is as follows.

CORNERSTONE 12-1

Revenues		$350,350
Cost of goods sold:		
Direct materials	$122,500	
Direct labour	73,500	
Overhead	49,000	245,000
Gross profit		105,350
Selling and administrative expenses		25,000
Operating income		$ 80,350

Chris wants to find a markup on cost of goods sold that will allow her to earn about the same amount of profit on each job as was earned last year.

Why:
Firms use a markup on cost as an easy way to price items so that, in general, all other costs and profit are included in the price. The cost is a known quantity and must be covered by price in order for the firm to earn a profit.

Required:
1. What is the markup on cost of goods sold (COGS) that will maintain the same profit as last year?
2. Suppose that Chris wants to expand her company's product line to include automobile alarm systems and electronic remote car door openers. She estimates the following costs for the sale and installation of one electronic remote car door opener.

Direct materials	$ 80.60
Direct labour (3 hours × $12)	36.00
Applied overhead	23.40
Total cost	$140.00

What is the price Chris will use for this new product given the markup percentage calculated in Requirement 1?
3. **What if** Chris wants to calculate a markup on direct materials cost, since it is the largest cost of doing business? What is the markup on direct materials cost that will maintain the same profit as last year? What is the bid price Chris will use for the job given in Requirement 2 if the markup percentage is calculated on the basis of direct materials cost?

Solution:
1. The markup percentage must include all costs that are not a part of cost of goods sold plus desired profit.

$$\text{Markup on COGS} = (\text{Selling and administrative expenses}$$
$$+ \text{Operating income})/\text{COGS}$$
$$= (\$25,000 + \$80,350)/\$245,000$$
$$= 0.43, \text{ or } 43\% \text{ of cost of goods sold}$$

CORNERSTONE 12-1
(continued)

2. Price for new product = $140 + (0.43 × $140) = $140 + $60.20 = $200.20
   = $140 × 1.43 = $200.20

3. Markup on direct materials = (Direct labour + Overhead + Selling and administrative expenses + Operating income)/ Direct materials
   = ($73,500 + $49,000 + $25,000 + $80,350)/ $122,500
   = 1.86, or 186% of direct materials cost
   Bid price = $80.60 + (1.86 × $80.60) = $80.60 + $149.92
   = $230.52 (rounded)

Target costing sets the cost of a product or service based on the price (target price) that customers are willing to pay. The Marketing Department determines what characteristics and price for a product are most acceptable to consumers. Then, it is the job of the company's engineers to design and develop the product such that cost and profit can be covered by that price. Japanese firms have been doing this for years; North American companies are beginning to use target costing.

Retail stores engage in a form of target costing when they look for goods that can be priced at a particular level to appeal to customers.

For example, many department stores work with clothing companies to develop house labels. The house label goods are typically good quality items that cost less and are priced lower than comparable name brand items. The house label gives the store flexibility. The store is not in the business of manufacturing sweaters, for example, but can find a source that will deliver sweaters of particular quality for the cost that will allow the store to achieve a target price and profit. Kenmore and Craftsman are house brands of **Sears**, and MasterCraft is a house brand of **Canadian Tire**.

Let's return to the AudioPro Company example in Cornerstone 12-1. Suppose Chris finds that other aftermarket audio installers price the remote car door opener at $155, while her initial price was $200.20. Should she drop her plans to expand into this product line? No, not if she can tailor her price to the market price. Recall that the original price called for $80 of direct materials and $36 of direct labour. Perhaps Chris could offer one remote device instead of two, saving $15 in cost. In addition, she might be able to shave some time off the direct labour, once the workers are trained and able to work more efficiently. This would result in $16 of savings. Prime cost would be $85.60 ($80.60 − $15 + $36 − $16) instead of the original $116.60.

AudioPro Company applies overhead at the rate of 65 percent of direct labour cost. However, Chris must think carefully about this job. Perhaps somewhat less overhead will be incurred because purchasing is reduced. (Only one reliable supplier is needed, and the tools and facilities can be shared with the audio installation.) Perhaps overhead for this job will amount to $10 (50 percent of direct labour). That would make the cost of one job $95.60 ($65.60 + $20 + $10).

Now, if the standard markup of 43 percent is applied, the price would be $137, well within the other firms' price of $155. As you can see, target costing is an iterative process. Chris will go through the cycle until she either achieves the target cost or determines that she cannot. Note, however, that target costing has given Chris a chance to develop a profitable market—a chance she might not have had if the original cost-based price had been set.

Target costing involves much more upfront work than cost-based pricing. However, let's not forget the additional work that must be done if the cost-based price turns out to be higher than what customers will accept. Then, the arduous task of bringing costs into line to support a lower price, or the opportunity cost of missing the market altogether, begins.

Other Pricing Policies

Penetration pricing is the pricing of a new product at a low initial price, perhaps even lower than cost, to build market share quickly. This is useful when the product or service is new and customers have great uncertainty as to its value. Penetration pricing is not predatory pricing; the important difference is the intent. The penetration price is not meant to destroy competition. Accountants, lawyers, and other professionals with new practices often use penetration pricing to establish a customer base.

Price skimming means that a higher price is charged when a product or service is first introduced. In essence, the company skims the cream off the market. It is used most effectively when the product is new, a small group of consumers values it, and the company enjoys a monopolistic advantage. Companies that engage in price skimming are hoping to recoup the expenses of research and development through high initial pricing. A cost consideration is that, in the start-up phase of production, economies of scale and learning effects have not occurred.

For example, in the late 1960s, **Hewlett-Packard** produced hand-held calculators. These were truly novel and very expensive. Priced at over $400, only scientists and engineers, who used the calculators in their work, felt the need for this product. As the market for hand-held calculators grew and technology improved, economies of scale kicked in, and the cost and price dropped dramatically. By the 1980s, tiny solar calculators were being given away as enticements to new subscribers of magazines.

Closely related to skimming is price gouging. **Price gouging** is said to occur when firms with market power price products "too high." How high is too high? Surely, cost is a consideration. Any time price just covers cost, gouging does not occur. This is why many firms go to considerable trouble to explain their cost structure and point out costs that consumers may not realize exist. Pharmaceutical companies, for example, emphasize the research and development costs associated with new drugs. When a high price is not clearly supported by cost, buyers take offence.

The Legal System and Pricing

OBJECTIVE ▶ 3
Discuss the impact of the legal system and ethics on pricing.

Government also plays an important role in pricing. Over time, many laws have been passed regulating how firms can set prices. The basic principle behind much pricing regulation is that competition is good and should be encouraged. Therefore, collusion by companies to set prices and deliberate attempts to drive competitors out of business are prohibited.

Predatory Pricing

Predatory pricing is the practice of setting prices below cost for the purpose of injuring competitors and eliminating competition. It is important to note that pricing below cost is not necessarily predatory pricing. Companies frequently price an item below cost, by running weekly specials in a grocery store, or practising penetration pricing, for example. Twenty-two U.S. states have laws against predatory pricing, each differing somewhat in definition and rules.

In 2001, **Air Canada**'s dominance of the Canadian airline sector was challenged in court by the country's antitrust watchdog, which charged it with predatory pricing against two smaller rivals. On March 8, 2001, Canada's Competition Bureau announced that it would ask the Competition Tribunal, a specialized court that rules on antitrust matters, to issue an order prohibiting Air Canada from pricing its fares below cost in eastern Canada.

Air Canada's pricing changes were squeezing low-cost carriers **CanJet** (based in Halifax) and **WestJet** (based in Calgary) out of the market, the bureau said. The bureau wanted Air Canada to stop operating flights in eastern Canada at fares that do not cover its "avoidable cost" of providing the service.

"The bureau believes Air Canada's pricing and capacity management would result in WestJet and CanJet abandoning these routes," said Conrad von Finckenstein, Canada's commissioner of competition. "It is concerned their exit will result in higher prices in the long term. With the ongoing restructuring of the airline industry, the bureau is determined to ensure that new entrants have a fair opportunity to compete."

Montreal-based Air Canada, which controlled some 80 percent of Canada's domestic airline market and was the world's eleventh-largest carrier at the time, said it would "vigorously challenge" the Competition Bureau's allegations.

Air Canada took issue with the bureau's interpretation of avoidable costs, which included fuel, labour, and aircraft expenses that would have been avoided if the service or flight had not been provided. It also asked for an expedited hearing before the Competition Tribunal to clarify the rules on pricing in the Canadian airline industry.

Predatory pricing on the international market is called **dumping**, which occurs when companies sell below cost in other countries, and domestic industry is injured. The defence against a charge of dumping is demonstrating that the price is indeed above or equal to costs, or that domestic industry is unhurt.

Price Discrimination

Price discrimination refers to the charging of different prices to different customers for essentially the same product. In the United States, price discrimination was covered by the landmark Robinson-Patman Act of 1936. In Canada, such matters are covered by the Competition Act, the oldest antitrust statute in the Western world. Enacted in 1889, the Competition Act makes it an offence to adopt a practice of granting a discount, rebate, price concession, allowance, or any other price-related advantage to one customer—that is, to not make the same advantage available to competing customers who purchase like quantity and quality.

Besides price discrimination, the Competition Act covers price fixing, bid rigging, exclusive dealing, refusal to deal, promotional allowance, predatory pricing, tied selling (requiring a customer to buy a product as a condition of supplying the customer with another product), market restrictions (requiring a customer to sell a product only in a defined market as a condition of supplying that product), and so on. Price discrimination under certain specified conditions may be allowed: (1) if the competitive situation demands it and (2) if costs (including costs of manufacture, sale, or delivery) can justify the lower price. Clearly, this second condition is important for the accountant, as a lower price offered to one customer must be justified by identifiable cost savings. Additionally, the amount of the discount must be at least equalled by the amount of cost saved.

The burden of proof for firms accused of violating the Competition Act is on the firms. The cost justification argument must be buttressed by substantial cost data. Proving a cost justification is an absolute defence. The availability of large databases, the development of activity-based costing, and powerful computing make it easier to justify costs. Still, problems remain. Cost allocations make such determinations particularly thorny. In justifying quantity discounts to larger companies, a company might keep track of sales calls, differences in time and labour required to make small and large deliveries, and so on.

In computing a cost differential, the company must create classes of customers based on the average costs of selling to those customers and then charge all customers in each group a cost-justifiable price. Cornerstone 12-2 shows the how and why of calculating cost and profit by customer segment.

The HOW and WHY of Calculating Cost and Profit by Customer Class

**CORNERSTONE
12-2**

Information:

Cobalt Inc. manufactures vitamin supplements with an average manufacturing cost of $163 per case (a case contains 100 bottles of vitamins). Cobalt Inc. sold 250,000 cases last year to the following three classes of customer.

Customer	Price per Case	Cases Sold
Large drugstore chain	$200	125,000
Small local pharmacies	232	100,000
Individual health clubs	250	25,000

The large drugstore chain special labelling costs $0.03 per bottle. The chain orders through electronic data interchange (EDI), which costs Cobalt about $50,000 annually in operating expenses and depreciation. Cobalt pays all shipping costs, which amounted to $1.5 million last year.

The small local pharmacies order in smaller lots that require special picking and packing in the factory; the special handling adds $20 to the cost of each case sold. Sales commissions to the independent jobbers who sell Cobalt products to the pharmacies average 10 percent of sales. Bad debts expense amounts to 1 percent of sales.

Individual health clubs purchase vitamins in even smaller lots; the special picking and packaging costs average $30 per case. There are no sales commissions for the health clubs. Instead, Cobalt advertises in health club management magazines, accepts orders by phone, and supplies point-of-sale posters and displays for the clubs. These marketing costs are $100,000 per year. Bad debts expense for this class of customer averages 10 percent.

Why:

Firms covered by price discrimination laws must be sure that price differentials are supported by cost differentials. On average, profit for each customer type is about the same.

Required:

1. Calculate the total cost per case for each of the three customer classes.
2. Using the costs from Requirement 1, calculate the profit per case per customer class. Does the cost analysis support the charging of different prices? Why or why not?
3. **What if** Cobalt charged the average price per case to all customer classes? How would that affect the profit percentages?

Solution:

1. _____

Chain store:

Manufacturing cost per case	$163.00
Special labelling cost ($0.03 × 100)	3.00
EDI ($50,000/125,000 cases)	0.40
Shipping ($1,500,000/125,000 cases)	12.00
Total cost per case	$178.40

Small pharmacies:

Manufacturing cost per case	$163.00
Special handling per case	20.00

**CORNERSTONE
12-2**
(continued)

Sales commission ($232 × 0.10)	$ 23.20
Bad debts expense ($232 × 0.01)	2.32
Total cost per case	$208.52

Health clubs:

Manufacturing cost per case	$163.00
Special handling per case	30.00
Selling expense ($100,000/25,000 cases)	4.00
Bad debts expense ($250 × 0.10)	25.00
Total cost per case	$222.00

2.

	Chain Store	Small Pharmacies	Health Clubs
Price per case	$200.00	$232.00	$250.00
Less: Cost per case	178.40	208.52	222.00
Profit per case	$ 21.60	$ 23.48	$ 28.00
Profit percent per case	10.80%	10.12%	11.20%

The profit percentages range from 10.12 percent to 11.20 percent. There appears to be cost justification for the price differentials among the three customer classes.

3. The average price per case is $227.33. If this price were charged to all three customers, the profit percentage for the chain store would increase and the profit percentages to the small pharmacies and health clubs would decrease. While Cobalt would earn the same overall profit percentage, this assumes that the chain store would continue to purchase the vitamin supplements from Cobalt at the new higher price. This assumption may be wrong. The chain store may well refuse to buy any product from Cobalt, leaving Cobalt with fewer units sold overall and a lower profit from the remaining customers.

Cornerstone 12-2 shows that price differences must be linked to cost differences. When this is done, the company's contention that higher prices are related to higher costs may shield it from charges of price discrimination and may also act as a behavioural prod to more expensive customers to change their way of doing business to qualify for price breaks.

ETHICS Just as a company can practise unethical behaviour in applying costs, it can mislead in pricing. A good example is the practice some airlines have of adding on fees outside of advertised prices. According to an article from May 31, 2011, in *The Wall Street Journal*, airline revenue from add-ons to ticket sales jumped to almost $22 billion last year and continues to soar as more carriers chase extra sources of income. A growing number of carriers worldwide are charging passengers for services once included in ticket prices, such as baggage and meals. Carriers are also finding new revenue sources, such as in-flight Internet connections. Forty-seven of the world's largest airlines, which together account for almost half of all airline revenue, last year reported ancillary sales of 15.11 billion euros, up 38 percent from 2009. Other carriers surveyed did not specify how much passenger revenue they receive from sources other than fares.

No-frills budget carriers began charging for extras more than a decade ago, when the rise of Internet ticket sales allowed them to split out elements more easily and charge passengers directly. They attract fliers with bargain ticket prices and earn more than 20 percent of their total revenue from ancillary sources. But more traditional carriers are expanding quickly, especially because fuel prices have risen significantly over recent years and airlines have not been able to raise fares.

In 2007, only 23 airlines reported ancillary revenues and the total was less than $2.5 billion, according to the article.[4] For example, reserving seats prior to 24 hours before flight time costs up to $15 extra and must be paid with a second credit-card charge. This is not illegal since passengers are not required to reserve seats. However, some customers have found the practice misleading.[5]◆

The Product Life Cycle

OBJECTIVE ➤ 4
Discuss the variations in price, cost, and profit over the product life cycle.

There are a number of views of the product life cycle. Many products have a predictable profit or product life cycle. From the perspective of marketing, the **product life cycle** describes the profit history of the product according to four stages: introduction, growth, maturity, and decline. In the introductory phase, profits are low for two reasons. First, revenues are low as the product gains market acceptance. Second, investment and learning may be high, leading to higher expenses. The growth stage is characterized by increasing market acceptance and sales, as well as economies of scale, which bring down expenses. The product breaks even, and profit rises. In the maturity phase, profits stabilize. The product has found its market, and revenues are relatively stable. Investment is down, and all learning effects in production are realized, leading to stable costs. Finally, in the decline phase, the product reaches the end of its cycle, and revenues and profits decline. Costs may still be low, but not enough to slip in below sales. Exhibit 12-2 illustrates the interaction of profit and the product life cycle with its four stages.

Exhibit 12-2

Product Life Cycle and Profitability

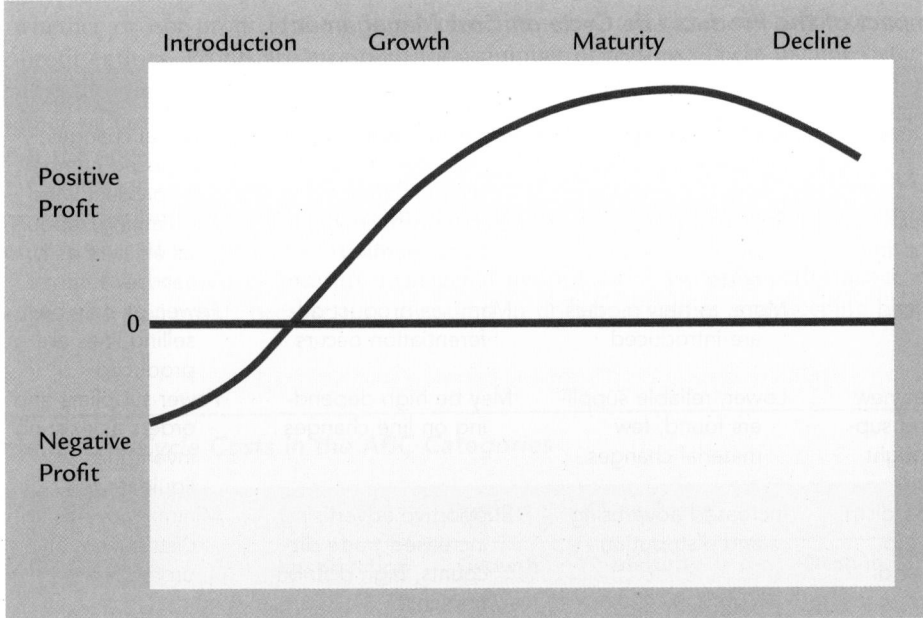

[4] Daniel Michaels, "Extra Airline Fees & Growth Market," *The Wall Street Journal* (May 31, 2011).
[5] Scott McCartney, "The Next Airline Fee: Buying Tickets?" *The Wall Street Journal* (March 3, 2009): D4.

operating income is equal to variable-costing income. In September, inventory increased, and absorption-costing operating income is higher than variable-costing operating income. The difference of $4,000 ($7,750 − $3,750), is just equal to the fixed overhead per unit multiplied by the increase in inventory ($16 × 250 units).

What happens when inventory decreases? Again, there is an effect on operating income under absorption costing but not under variable costing. Let's take Lasersave into the month of October, when production is 1,250 units (just like September), but 1,300 units are sold.

In this case, when inventory decreases (or production is less than sales), variable-costing operating income is greater than absorption-costing operating income. The difference of $800 ($14,475 − $13,675) is equal to the 50 units that, under absorption costing, came from inventory with $16 of the previous month's fixed manufacturing overhead attached. Exhibit 12-5 summarizes the impact of changes in inventory on operating income under absorption costing and variable costing.

To summarize, when inventories change from the beginning to the end of the period, the two costing approaches will give different operating incomes. The reason for this is that absorption costing assigns fixed manufacturing overhead to units produced. If those units are sold, the fixed overhead appears on the income statement under cost of goods sold. If the units are not sold, the fixed overhead goes into inventory. Under variable costing, however, all fixed overhead for the period is expensed. As a result, absorption costing allows managers to manipulate operating income by producing for inventory.

The variable-costing income statement has an advantage in addition to providing better signals regarding performance. It also provides more useful information for management decision making. For example, how much more will Lasersave earn if it sells one more unit? Cornerstone 12-3 indicates that $17 ($60 − $43) is the per-unit gross profit. However, that figure includes some fixed overhead, and fixed overhead will not change if another unit is produced and sold. The variable-costing income statement in Cornerstone 12-4 gives more useful information. Additional contribution margin of the extra unit is $35.75 ($60 − $23 − $1.25). The key insight of variable costing is that fixed expenses do not change as units produced and sold change. Therefore, while the variable-costing income statement cannot be used for external reporting, it is a valuable tool for some management decisions.

Profitability of Segments and Divisions

Companies often want to know the profitability of a segment of the business. That segment could be a product, division, sales territory, or customer group. Determining the profit attributable to subdivisions of the company is harder than determining overall profit because of the need to allocate expenses. Segmented income statements using variable, absorption, and activity-based costing have been covered in previous chapters. For example, segmented income statements in the keep-or-drop decision were covered in Chapter 11. Activity-based segmented income statements by product line or customer class are covered in Chapters 6 and 14. As a result, we will not go into the computations in depth here. Instead, we will focus on the managerial use of variable-costing segmented income statements.

Exhibit 12-5

Changes in Inventory under Absorption and Variable Costing

If	Then
1. Production > Sales	Absorption-costing income > Variable-costing income
2. Production < Sales	Absorption-costing income < Variable-costing income
3. Production = Sales	Absorption-costing income = Variable-costing income

Profit by Product Line   It is easy to understand why a firm would like to know whether or not a particular product is profitable. A product that consistently loses money and has no potential to become profitable could be dropped. This would free up resources for a product with higher potential. On the other hand, a profitable product may merit additional time and attention.

Product line profitability would be easy to compute if all costs and revenues were easily traceable to each product. This is seldom the case. Therefore, companies must first determine how profit will be computed. Let's examine Chang Company, which manufactures two products: basic fax machines and multi-function fax machines. The basic fax machine has telephone and fax capability. This type of machine is less expensive and easier to produce. The multi-function fax machine is the high-end machine. It is a combination of two-line telephone, fax, computer printer, and copier. The multi-function fax machine uses more advanced technology and is more difficult to produce. Data on each product follow.

	Basic	Multi-Function
Number of units	20,000	10,000
Direct labour hours	40,000	15,000
Price	$200	$350
Prime cost per unit	$55	$95
Overhead per unit*	$30	$22.50

*Annual overhead is $825,000, and overhead is applied on the basis of direct labour hours.

Marketing expenses, all variable, amount to 10 percent of sales. Administrative expenses of $2 million, all fixed, are allocated to the products in accordance with revenue. Absorption-costing income by product line is shown in Exhibit 12-6.

Clearly, the multi-function fax machine is more profitable. But what does this tell us? Can we conclude that each basic fax machine sold adds $41.65 ($833,000/ 20,000 units) to profit? Does each multi-function fax machine sold add $104.20 ($1,042,000/10,000) to profit? No, Chang Company has intermingled variable and fixed costs and has allocated administrative expenses on the basis of revenue, when there is no reason to believe that revenue drives administrative expenses. Additionally, overhead has been assigned to the products on a per-unit basis, but we do not know just what it includes. Is $22.50 an accurate representation of the overhead resources required to produce one multi-function fax machine? A variable-costing segmented income statement will give better information.

Using Variable Costing to Measure Segment Profit   Chang Company could use variable costing and segregate direct fixed and common fixed expenses as well. To apply variable costing to Chang Company, we need additional information on fixed and variable costs of overhead. Suppose that total variable overhead is $360,000 and total fixed overhead is $465,000. Since overhead is applied on the basis of direct labour hours, the variable overhead assigned to basic fax machines is $261,818 [$360,000 × (40,000/55,000)]. The variable overhead assigned to multi-function

---

Exhibit 12-6

Chang Company Absorption-Costing Income Statement
(In thousands of dollars)

	Basic	Multi-Function	Total
Sales	$ 4,000	$3,500	$ 7,500
Less: Cost of goods sold	1,700	1,175	2,875
Gross profit	2,300	2,325	4,625
Less:			
Marketing expenses	(400)	(350)	(750)
Administrative expenses	(1,067)	(933)	(2,000)
Operating income	$ 833	$1,042	$ 1,875

Exhibit 12-7

Chang Company Variable-Costing Income Statement (In thousands of dollars)

	Basic	Multi-Function	Total
Sales	$ 4,000	$ 3,500	$ 7,500
Less:			
Variable cost of goods sold	(1,362)	(1,048)	(2,410)
Sales commissions	(400)	(350)	(750)
Contribution margin	$ 2,238	$ 2,102	4,340
Less:			
Fixed overhead			(465)
Administrative expenses			(2,000)
Operating income			$ 1,875

fax machines is $98,182 [$360,000 × (15,000/55,000)]. The variable-costing income statement is given in Exhibit 12-7. Notice that all fixed expenses that are not attributable to either of the product lines are subtracted from the total column.

Divisional Profit Just as companies want to know the relative profitability of different products, they may want to assess the relative profitability of different divisions of the company. Divisional profit is often used in evaluating the performance of managers. Failure to earn a profit can lead to the division's closing. For example, **General Motors** decided to drop the Oldsmobile line due to its continued unprofitability.

Divisional profit may be calculated using any of three approaches described in the preceding section. Usually, the absorption-based approach is used, and a share of corporate expense is allocated to each division to remind them that all expenses of the company must be covered. Suppose that Polyglyph Inc. is a conglomerate with four divisions: Alpha, Beta, Gamma, and Delta. Corporate expenses of $10 million are allocated to each division on the basis of sales. The divisional income statements are as follows:

	Alpha	Beta	Gamma	Delta	Total
Sales	$ 90	$ 60	$ 30	$120	$300
Less: Cost of goods sold	35	20	11	98	164
Gross profit	55	40	19	22	136
Less:					
Division expenses	(20)	(10)	(15)	(20)	(65)
Corporate expenses	(3)	(2)	(1)	(4)	(10)
Operating income (loss)	$ 32	$ 28	$ 3	$ (2)	$ 61

How might Polyglyph view these results? Clearly, Delta has an operating loss. Corporate management would raise questions about Delta's continuing viability. However, notice that Delta's operating loss would be eliminated if allocated corporate expenses were not included. In fact, all divisions would look more profitable if corporate expenses were not allocated to the divisions. As a result, management might concentrate on Delta's potential for an improved profit picture. Delta's divisional expenses are relatively high. Perhaps this is due to an ambitious research and development program. If payoffs from this program can be anticipated, corporate management will be much less concerned than if the divisional expenses do not have potential. Corporate management will also be concerned with trends over time and the immediate and long-term prospects for each division. Even a seemingly profitable division, like Alpha, may need attention if it is in a declining industry or if it uses significantly more resources than indicated by the corporate expense allocation. Additional material on divisional profitability and responsibility accounting is covered in Chapter 10.

Overall Profit The computation of segmented profit is clearly useful in many management decisions. However, the allocation problems inherent in computing profit on divisions, segments, and product lines may mean that overall profit is most useful in some contexts. It is certainly easiest to compute, and it does have meaning. If the overall profit is consistently positive, the company remains in business, even if one or more segments is losing money.

Analysis of Profit-Related Variances

OBJECTIVE ▶ 6
Compute the sales price, price volume, contribution margin, contribution margin volume, sales mix, market share, and market size variances.

Managers frequently want to compare actual profit earned with expected profit. This leads naturally to variance analysis, in which actual and budgeted amounts are compared. Profit variances centre on the difference between budgeted and actual prices, volumes, and contribution margin.

Sales Price and Price Volume Variances

Actual revenue may differ from expected revenue because actual price differs from expected price or because quantity sold differs from expected quantity sold, or both. The **sales price variance** is the difference between actual price and expected price multiplied by the actual quantity or volume sold. In equation form, it is the following:

$$\text{Sales price variance} = (\text{Actual price} - \text{Expected price}) \times \text{Quantity sold}$$

The **price volume variance** is the difference between actual volume sold and expected volume sold multiplied by the expected price. It can be expressed in the following equation:

$$\text{Price volume variance} = (\text{Actual volume} - \text{Expected volume}) \times \text{Expected price}$$

The overall sales variance is the sum of the sales price variance and the price volume variance.

$$\text{Overall sales variance} = \text{Sales price variance} + \text{Price volume variance}$$

As is the case with all variances, the sales price and price volume variances are labelled favourable (F) if the variance increases profit above the amount expected. They are labelled unfavourable (U) if the variance decreases profit below the amount expected. Cornerstone 12-5 shows the how and why of calculating the sales price, price volume, and overall sales variances.

As is shown in Cornerstone 12-5, the sum of the sales price and price volume variances is the **total (overall) sales variance**. Of course, this is simply the difference between actual and expected revenue. Breaking the overall sales variance into price and volume components gives managers a better feel for why actual revenue may differ from budgeted revenue.

It is important to note that these variances just begin to alert managers to problems in pricing and sales. As is the case with all variances, significant variances are investigated to discover the underlying reasons for the difference between expected and actual results. In the case of an unfavourable sales price variance, the reason may be the giving of unanticipated price discounts, perhaps to meet competitors' prices. The sales price and price volume variances interact. For example, an unfavourable sales price variance may be paired with a favourable price volume variance because the lower price raised quantity sold.

Contribution Margin Variance

We have just looked at the price and sales variances. The cost variances were covered in Chapter 9. Now it is time to put sales and cost together and calculate any variances between actual and expected contribution margin. The **contribution margin variance** is the difference between actual and budgeted contribution margin.

$$\text{Contribution margin variance} = \text{Actual contribution margin} - \text{Budgeted contribution margin}$$

**CORNERSTONE
12-5**

The HOW and WHY of Calculating the Sales Price Variance, the Price Volume Variance, and the Overall Sales Variance

Information:

Armada Company distributes produce. In May, Armada Company expects to sell 20,000 kilograms of produce at an average price of $0.20 per kilogram. Actual results are 23,000 kilograms sold at an average price of $0.19 per kilogram.

Why:

The sales price variance tells managers what impact a difference between actual and expected sales price has on revenue. The price volume variance tells managers what impact a difference between actual and expected units sold has on revenue.

Required:

1. Calculate the sales price variance for May.
2. Calculate the price volume variance for May.
3. Calculate the overall sales variance for May. Explain why it is favourable or unfavourable.
4. **What if** May sales were actually 19,000 kilograms? How would that affect the sales price variance? The price volume variance? The overall sales variance?

Solution:

1. Sales price variance = (Actual price − Expected price) × Quantity sold
   $$= [(\$0.19 - \$0.20) \times 23{,}000] = \$230 \text{ U}$$

2. Price volume variance = (Actual volume − Expected volume)
   $$\times \text{ Expected price}$$
   $$= [(23{,}000 - 20{,}000) \times \$0.20] = \$600 \text{ F}$$

3. Overall sales variance = Sales price variance + Price volume variance
   $$= \$230 \text{ U} + \$600 \text{ F} = \$370 \text{ F}$$

   The overall sales variance is favourable because the favourable price volume variance is larger than the unfavourable sales price variance. That is, the lower than expected sales price did reduce revenue; however, the greater than expected volume overcame that effect and raised revenue overall.

4. If May sales in kilograms were 19,000, there would be a decrease in the sales price variance, since the actual number of kilograms sold decreased. There would be an unfavourable price volume variance, and the overall sales variance would be unfavourable because both the sales price variance and the price volume variance are unfavourable.

This variance is favourable if the actual contribution margin earned is higher than the budgeted amount. Cornerstone 12-6 shows the how and why of calculating the contribution margin variance.

The contribution margin variance is an overall variance. It can be broken into the contribution margin volume variance and the sales mix variance.

Contribution Margin Volume Variance The **contribution margin volume variance** is the difference between the actual quantity sold and the budgeted quantity sold multiplied by the budgeted average unit contribution margin. Note the difference between the contribution margin volume variance and the price volume variance. Both look at the difference between actual and budgeted volume sold. However, the price volume variance multiplies that difference by sales price, while the

The HOW and WHY of Calculating the Contribution Margin Variance

CORNERSTONE 12-6

Information:

Birdwell Inc. produces and sells two types of bird feeders. The regular type is a simple plastic and wood model, which can be hung from a tree branch. The deluxe model is a larger, stand-alone model, which includes a post and a round squirrel shield to prevent squirrels from eating the bird seed. Budgeted and actual data for the two models are shown below.

Budgeted Amounts:

	Regular Model	Deluxe Model	Total
Sales:			
($10 × 1,500)	$15,000		
($50 × 500)		$25,000	$40,000
Variable expenses	9,000	17,500	26,500
Contribution margin	$ 6,000	$ 7,500	$13,500

Actual Amounts:

	Regular Model	Deluxe Model	Total
Sales:			
($10 × 1,250)	$12,500		
($50 × 625)		$31,250	$43,750
Variable expenses	7,500	21,875	29,375
Contribution margin	$ 5,000	$ 9,375	$14,375

Why:

The contribution margin variance tells managers the difference between actual and expected contribution margin. This is a starting point for analyzing the factors that led to any difference between actual and expected profit.

Required:

1. Calculate the contribution margin variance.
2. **What if** actual units sold of the deluxe bird feeder decreased? How would that affect the contribution margin variance? What if actual units sold of the deluxe bird feeder increased? How would that affect the contribution margin variance?

Solution:

1. Contribution margin variance = Actual contribution margin

                      − Expected contribution margin

               = $14,375 − $13,500 = $875 F

2. If units sold of the deluxe bird feeder decreased while everything else stayed the same, the contribution margin variance would decrease. Whether it turned unfavourable would depend on the amount of decrease in deluxe sales. On the other hand, if units sold of the deluxe bird feeder increased while everything else stayed the same, the contribution margin variance would become larger and still be favourable.

**CORNERSTONE
12-7**

The HOW and WHY of Calculating the Contribution Margin Volume Variance

Information:

Recall from Cornerstone 12-6 that Birdwell Inc. provided the following information:

	Budgeted	**Actual**
Sales in units, regular model	1,500	1,250
Sales in units, deluxe model	500	625
Total contribution margin	$13,500	$14,375

Why:

The contribution margin volume variance tells managers what impact a difference between actual and expected sales volume has on contribution margin. Unlike the price volume variance, the contribution margin variance weights the difference between actual and expected volume by the contribution margin, which includes both price and variable cost. Thus, it is more closely related to profit.

Required:

1. Calculate the budgeted average unit contribution margin.
2. Calculate the contribution margin volume variance.
3. **What if** actual units sold of the deluxe bird feeder decreased? How would that affect the contribution margin volume variance? What if actual units sold of the deluxe bird feeder increased? How would that affect the contribution margin volume variance?

Solution:

1. Budgeted average unit contribution margin

   = Budgeted total contribution margin/Budgeted total units

   = $13,500/(1,500 + 500) = $6.75

2. Contribution margin volume variance

   = (Actual quantity sold − Budgeted quantity sold)

   × Budgeted average unit contribution margin

   = [(1,250 + 625) − (1,500 + 500)] × $6.75 = $843.75 U

3. If actual units sold of the deluxe bird feeder decreased while everything else stayed the same, the contribution margin volume variance would decrease and become even more unfavourable. If actual units sold of the deluxe bird feeder increased while everything else stayed the same, the contribution margin volume variance would increase and become less unfavourable. Whether it turned favourable would depend on the amount of the increase in deluxe bird feeder sales.

contribution margin volume variance multiplies that difference by contribution margin. Therefore, the contribution margin volume variance gives management information about gained or lost profit due to changes in the quantity of sales.

Contribution margin volume variance = (Actual quantity sold − Budgeted quantity sold)
× Budgeted average unit contribution margin

The budgeted average unit contribution margin is the total budgeted contribution margin divided by the budgeted total number of units of all products to be sold. Cornerstone 12-7 shows the how and why of calculating the contribution margin volume variance.

As Cornerstone 12-7 shows, the unfavourable contribution margin volume variance is the result of selling fewer units, in total, than budgeted. Still, we can see that Birdwell Inc. actually had a higher contribution margin than expected. The shift in the sales mix explains why.

Sales Mix Variance The sales mix represents the proportion of total sales yielded by each product. A company that produces only one product obviously has a sales mix of 100 percent for that product, and there is no effect of changing sales mix on profit. Multiproduct firms, however, do experience shifts in their sales mix. If relatively more of the high-profit product is sold, profit will be higher than expected. If the sales mix shifts toward the low-profit product, profit will be lower than expected. We can define the **sales mix variance** as the sum of the change in units for each product multiplied by the difference between the budgeted contribution margin and the budgeted average unit contribution margin.

Sales mix variance = [(Product 1 actual units − Product 1 budgeted units)
× (Product 1 budgeted contribution margin
− Budgeted average unit contribution margin)]
+ [(Product 2 actual units − Product 2 budgeted units)
× (Product 2 budgeted contribution margin
− Budgeted average unit contribution margin)]

The preceding sales mix variance equation, as detailed in Cornerstone 12-8, is for two products. If three products were produced, we would simply keep adding the change in units times the change in contribution margin for every additional product.

Now, we can see that the favourable sales mix variance of $1,718.75, combined with the unfavourable contribution margin volume variance of $843.75, explains the overall favourable contribution margin variance of $875.

Market Share and Market Size Variances

Managers not only want to look inward at contribution margin through the volume and sales mix variances, but they also want to look outward to see how their company is doing compared with the rest of their industry. **Market share** gives the proportion of industry sales accounted for by a company. **Market size** is the total revenue for the industry. Clearly, both market share and market size have an impact on a company's profits.

The **market share variance** is the difference between the actual market share percentage and the budgeted market share percentage multiplied by actual industry sales in units times budgeted average unit contribution margin. The **market size variance** is the difference between actual and budgeted industry sales in units multiplied by the budgeted market share percentage times the budgeted average unit contribution margin.

Market share variance = [(Actual market share percentage − Budgeted market share
percentage) × Actual industry sales in units] × Budgeted
average unit contribution margin
Market size variance = [(Actual industry sales in units − Budgeted industry sales
in units) × Budgeted market share percentage] × Budgeted
average unit contribution margin

Cornerstone 12-9 shows the how and why of calculating the market share variance and the market size variance.

As Cornerstone 12-9 shows, the market share variance for Birdwell is $2,869 unfavourable. In other words, Birdwell's reduction in market share from 10 percent to 8.152 percent cost the company $2,869 in contribution margin.

CORNERSTONE 12-8

The HOW and WHY of Calculating the Sales Mix Variance

Information:

Recall from Cornerstone 12-6 that Birdwell Inc. provided the following information:

	Budgeted	Actual
Sales in units, regular model	1,500	1,250
Sales in units, deluxe model	500	625
Unit contribution margin, regular model	$4.00	
Unit contribution margin, deluxe model	$15.00	
Total contribution margin	$13,500	$14,375

Why:

The sales mix variance tells managers what impact a difference between actual and expected percentages of products sold has on contribution margin. Only budgeted contribution margins are used to weight the differences in the sales mix.

Required:

1. Calculate the sales mix variance.
2. **What if** actual units sold of the deluxe bird feeder decreased? How would that affect the sales mix variance? What if actual units sold of the deluxe bird feeder increased? How would that affect the sales mix variance?

Solution:

1. Sales mix variance = [(Product 1 actual units – Product 1 budgeted units) × (Product 1 budgeted contribution margin – Budgeted average unit contribution margin)] + [(Product 2 actual units – Product 2 budgeted units) × (Product 2 budgeted contribution margin – Budgeted average unit contribution margin)]

$$= [(1,250 - 1,500) \times (\$4.00 - \$6.75)] + [(625 - 500) \times (\$15.00 - \$6.75)]$$
$$= \$1,718.75 \text{ F}$$

2. If actual units sold of the deluxe bird feeder (the high contribution margin product) decreased while everything else stayed the same, the sales mix variance would decrease and become less favourable. Depending on the amount of decrease, the sales mix variance could become unfavourable. If, on the other hand, actual units sold of the deluxe bird feeder increased, then the sales mix variance would increase and become more favourable.

The impact of changing market size on Birdwell's profits can be assessed through the market size variance. It is $2,025 favourable. This means that the company's contribution margin would have increased by this amount had the actual market share percentage equalled the budgeted market share percentage. Unfortunately for Birdwell, the market share percentage slipped. Still, Birdwell is better off due to increasing market size, since a market share of 8.2 percent would yield even smaller profits from a smaller market.

While the contribution margin variances and the market share and market size variances yield important insights into profitability, companies may want to analyze profit further.

The HOW and WHY of Calculating the Market Share Variance and the Market Size Variance

**CORNERSTONE
12-9**

Information:
Budgeted unit sales for the entire bird feeder industry were 20,000 (of all model types), and actual unit sales for the industry were 23,000. Recall from Cornerstone 12-6 that Birdwell Inc. provided the following information:

	Budgeted	Actual
Sales in units, regular model	1,500	1,250
Sales in units, deluxe model	500	625
Total contribution margin	$13,500	
Budgeted average unit contribution margin	$6.75	

Why:
The market share and market size variances allow firms to compare their performance with the market as a whole. This gives managers a chance to look outside their own companies and to see what the possibilities are in the market for their products.

Required:
1. Calculate the market share variance.
2. Calculate the market size variance.
3. **What if** Birdwell actually sold a total of 2,300 units (in total of the two models)? How would that affect the market share variance? The market size variance?

Solution:
1. Market share variance = [(Actual market share percentage − Budgeted market share percentage) × Actual industry sales in units] × Budgeted average unit contribution margin
   Actual market share percentage = 1,875/23,000 = 0.08152, or 8.152% (rounded)

   Budgeted market share percentage = 2,000/20,000 = 0.10, or 10%

   Market share variance = [(0.08152 − 0.10) × 23,000] × $6.75

   = $2,869 U (rounded to the nearest dollar)

   Note that the market share variance is unfavourable because Birdwell's actual share of the market is less than the budgeted share of the market.
2. Market size variance = [(Actual industry sales in units − Budgeted industry sales in units) × Budgeted market share percentage] × Budgeted average unit contribution margin

   = [(23,000 − 20,000) × 0.10] × $6.75
   = $2,025 F

   Note that the market size variance is favourable because the actual units sold in the market is larger than the number of units expected to be sold in the market.
3. If Birdwell actually sold a total of 2,300 units, then the actual market share percentage would be 10 percent, exactly equal to the budgeted market share percentage. The market share variance would be zero. There would be no impact on the market size percentage.

OBJECTIVE ▸ 7
Describe some of the limitations of
profit measurement.

Limitations of Profit Measurement

Profit measurement is important, but there is more to life and business than monetary profit measurement.

One limitation to profitability analysis is its focus on past, not future, performance. The economic environment is unpredictable, and consistent profitability—brought about by great management, productive employees, and a high-quality product—does not guarantee success when economic conditions change. At that point, shifts in strategy may prove crucial. For example, the shift from payment for costs incurred to payment by diagnosis code has changed life considerably in the health-care industry. Previously, insurance companies and the government paid doctors and hospitals for all costs incurred. Clearly, cost cutting was not important. Now, the emphasis on efficiency and cost control has had a significant impact on all participants in the medical field.

> **Johnson & Johnson**, for example, worked hard to change the rate of reimbursement for stents used in angioplasty. The J&J stent was technically superior to others on the market and cost more. However, Medicare in the United States paid hospitals the same amount no matter which stent was used. J&J was able to show, using data on 200,000 Medicare patients, that patients using the J&J stent were able to avoid a second and third angioplasty. Stent reimbursement increased.[8]

The point is that companies must remain flexible and be aware of changing business conditions. The savvy cost manager is aware of economic and environmental trends outside the company. These can determine the success of management plans. They also help provide a reference point for management in determining whether profits are good or bad. A small increase in profit during a recession may signal outstanding performance. The same increase during economic expansion raises doubts about management's ability.

Another limitation is profit's emphasis on quantifiable measures. Henry Ford said that both buyer and seller must be wealthier in some way as a result of a transaction. But must wealth always be measured in money? Some aspects of profit are, no doubt, qualitative. Start-up companies may be thrilled to have made it past the one-year mark. The confidence that comes with being able to successfully start and continue a business is part of their wealth. Many companies give back a portion of their profits to their communities; this, too, is a form of wealth.

Finally, we must remember that profit has a strong impact on people's behaviour. Predictably, individuals prefer profit to loss. Their jobs, promotions, and bonuses may depend on the annual profit, and this dependence can affect their behaviour in expected and unexpected ways. As accountants, it is important to realize that profit measurement can lead to different incentives for individuals to work harder and to act ethically.

ETHICS People's desire to avoid losses and their inclination to take a short-run perspective can affect the potential for unethical conduct. Unethical conduct can take any number of forms, but basically it comes down to lying. Companies may try to pass off inferior work or materials as high-quality work—worthy of a higher price. Companies may keep two sets of books—for the purpose of cheating on income and other taxes. They may overstate the value of inventory in order to understate the cost of goods sold and thereby overstate net income.

Companies that value numerical profit above all else should not be surprised if employees act accordingly and do what is in their power to increase the numbers. Not only does this overreliance on numerical profit lead to unethical behaviour, but

---

[8] Ron Winslow, "Johnson & Johnson Misses Beat with Device for Cardiac Surgery," *The Wall Street Journal* (September 18, 1998): A1.

it also provides incentives to ignore the less measurable outcomes that might benefit the company. Workers basically look for companies to "put their money where their mouth is." If raises, promotions, and bonuses are awarded only on the basis of profit, employees will work to increase profits. Even if the company says other factors are important (e.g., good corporate citizenship, innovation, and high-quality products), this will be seen as mere lip service.

The ever-present salience of monthly, quarterly, and annual profit and loss statements may cause companies to emphasize short-run results. Too much emphasis on short-run optimization can lead to ethical problems. A solution is to focus on the long run. Companies that take a long-run orientation know that they cannot cheat customers and expect to retain their business. Eventually, shoddy materials and workmanship will be realized by the customer. The customer will go elsewhere, and regaining trust once lost is an agonizingly slow process. As a result, ethical people and companies often emphasize the long run as the best basis for behaviour.◆

Summary of Learning Objectives

1. **Discuss basic pricing concepts.**

- The basic economic interplay between demand and supply helps to set price.

  - Customers buy less at a high price than they do at a low price.
  - Producers (suppliers) are able to supply more at a high price than at a low price.
  - Equilibrium price is set where quantity demanded equals quantity supplied.

- Price elasticity of demand is the percent change in quantity demanded for a given percent change in price.

  - Products with elastic demand tend to:

    - Have many substitutes
    - Not be necessities
    - Take a relatively large amount of consumer income

  - Products with inelastic demand tend to:

- Have few substitutes
- Be necessities
- Take a relatively small amount of consumer income

- Market structure affects the relationship between the company and other companies in its industry.

  - Perfectly competitive markets have many buyers and sellers. Price is set in the market.
  - Monopolistic competition is characterized by some ability to differentiate one's product from that of other firms.
  - Monopoly is characterized by one seller with the ability to increase price somewhat above that of a competitive market. There may be legal reasons for the monopoly.
  - Oligopoly is characterized by few sellers and many buyers. There are frequently high barriers to entry in this market.

2. **Calculate a markup on cost and a target cost.**

- Many firms use cost-based pricing.

  - Price is based on cost plus desired profit.
  - The markup is NOT pure profit—it also includes all costs not included in the base cost.
  - Strategy is supply based—it does not take demand into account until late in the process.

- Target cost-based pricing strategy begins with price and subtracts desired profit to determine allowable cost.

3. **Discuss the impact of the legal system and ethics on pricing.**

- The legal system supports competition and outlaws certain business practices.

  - Predatory pricing
  - Some forms of price discrimination

- Fairness and ethical conduct may prevent the exploitation of market power.

  - Price gouging
  - Dumping

4. **Discuss the variations in price, cost, and profit over the product life cycle.**

- The product life cycle has an important impact on price.

  - Introduction phase usually has negative profit.
  - Growth phase shows increasing profit.
  - Maturity phase is accompanied by a levelling off of profit.
  - Decline phase is the end of the product life cycle.

- Learning effects and increasing efficiency help costs decrease as the product life cycle changes from introduction through growth and maturity.

5. **Explain why firms measure profit, and calculate measures of profit using absorption and variable costing.**

- Profit is measured to assess performance.
- Absorption-costing income measurement is required for external financial reporting.

  - All manufacturing costs are attached to units of product including:

    - Direct materials
    - Direct labour
    - Variable factory overhead
    - A portion of fixed factory overhead

  - Product costs for the period are assigned to units sold or units put into inventory.

- Variable costing is useful for management decision making.

  - All variable manufacturing costs are attached to units of product including:

    - Direct materials
    - Direct labour
    - Variable overhead

  - All fixed costs (including fixed factory overhead and fixed selling and administrative expense) are treated as period expenses on the income statement.

- Variable costing and ABC give better signals regarding performance and incremental costs.

- Profitability analysis can be accomplished for individual segments, including:
    - Product lines
    - Divisions
    - Customer groups

6. **Compute the sales price, price volume, contribution margin, contribution margin volume, sales mix, market share, and market size variances.**

- Profit-related variances are used to analyze the changes in profit from one time period to another.
    - Sales price variance compares expected price with actual price and multiplies by actual volume.
    - Price volume variance compares actual volume with expected volume and multiplies by expected price.
- Contribution margin variance considers interplay of price and variable cost.
    - Contribution margin volume variance shows the impact of a difference between expected and actual sales volume on contribution margin.
    - Sales mix variance shows the impact on contribution margin of changes in the actual versus expected sales mix.
- Market share and size variances allow a firm to compare its performance against competing firms.
    - Market share variance shows the impact of a difference between actual and expected percentage of market volume multiplied by budgeted average contribution margin.
    - Market size variance shows the impact on profit of a difference between actual volume sold in the market and expected volume.

7. **Describe some of the limitations of profit measurement.**

- Limitations of profit include:
    - Focus on past performance
    - Uncertain economic conditions
    - Difficulty of capturing all important factors in financial measures
- Successful firms measure far more than accounting profit.
    - Impact on the community
    - Employees
- Ethical behaviour is fostered by appropriate emphasis on profit.

CORNERSTONE 12-1   The HOW and WHY of calculating a markup on cost, page 591

CORNERSTONE 12-2   The HOW and WHY of calculating cost and profit by customer class, page 595

CORNERSTONE 12-3   The HOW and WHY of calculating inventory cost and preparing the income statement using absorption costing, page 601

CORNERSTONE 12-4   The HOW and WHY of calculating inventory cost and preparing the income statement using variable costing, page 604

CORNERSTONES FOR CHAPTER 12

CORNERSTONE 12-5   The HOW and WHY of calculating the sales price variance, the price volume variance, and the overall sales variance, page 610

CORNERSTONE 12-6   The HOW and WHY of calculating the contribution margin variance, page 611

CORNERSTONE 12-7   The HOW and WHY of calculating the contribution margin volume variance, page 612

CORNERSTONE 12-8   The HOW and WHY of calculating the sales mix variance, page 614

CORNERSTONE 12-9   The HOW and WHY of calculating the market share variance and the market size variance, page 615

Review Problems

I. Pricing

Melcher Company produces and sells small household appliances. A few years ago, it designed and developed a new hand-held mixer, named the "Mixalot." The Mixalot can be used to mix milkshakes and light batter. With the mincer attachment, it can mince up to a cup of vegetables or fruits. The Mixalot was very different from the standard table model Melcher mixer. Because of this, over $250,000 was spent on design and development. Another $50,000 was spent on consumer focus groups, in which prototypes of the Mixalot were kitchen tested by consumers. It was in those groups that safety problems surfaced. For example, one of the testers sliced his hand. This necessitated adding a plastic guard around the blade. Moulding and attaching the blade would add $1.50 to prime costs of the Mixalot, which had originally been estimated to cost $3.50 to produce. Information regarding the first five years of operations is as follows:

	Year 1	Year 2	Year 3	Year 4	Year 5
Unit sales	25,000	150,000	400,000	400,000	135,000
Price	$15	$20	$20	$18	$15
Prime cost	$125,000	$600,000	$1,640,000	$1,640,000	$526,500
Setup cost	5,000	9,600	80,000	80,000	12,000
Purchase of special equipment	65,000	—	—	—	—
Expediting	—	15,000	40,000	35,000	—
Rework	12,500	45,000	60,000	60,000	6,750
Other overhead	50,000	300,000	800,000	800,000	270,000
Warranty repair	6,250	7,500	10,000	10,000	3,375
Commissions (5%)	18,750	150,000	400,000	360,000	101,250
Advertising	250,000	150,000	100,000	100,000	25,000

During the first year, Melcher's prime costs included the safety guard. The special equipment was for moulding and attaching the guard. The equipment had a life of five years with no salvage value.

Required:

1. What is the cost of goods sold per unit for the Mixalot in each of the five years?
2. What marketing expenses were associated with the Mixalot in each of the five years? Calculate them on a per-unit basis.

3. Calculate operating income for the Mixalot in each of the five years. Then, compare all costs to revenues for the Mixalot over the entire product life cycle. Was the Mixalot profitable?

4. Discuss the pricing strategy of Melcher Company for the Mixalot, initially and over the product life cycle.

Solution:

1.

	Year 1	Year 2	Year 3	Year 4	Year 5
Prime cost	$ 125,000	$ 600,000	$1,640,000	$1,640,000	$ 526,500
Setup cost	5,000	9,600	80,000	80,000	12,000
Depreciation on special equipment	13,000	13,000	13,000	13,000	13,000
Expediting	—	15,000	40,000	35,000	—
Rework	12,500	45,000	60,000	60,000	6,750
Other overhead	50,000	300,000	800,000	800,000	270,000
Total COGS	$ 205,500	$ 982,600	$2,633,000	$2,628,000	$ 828,250
Divided by units	÷ 25,000	÷ 150,000	÷ 400,000	÷ 400,000	÷ 135,000
Unit COGS	$ 8.22	$ 6.55	$ 6.58	$ 6.57	$ 6.14

2.

	Year 1	Year 2	Year 3	Year 4	Year 5
Warranty repair	$ 6,250	$ 7,500	$ 10,000	$ 10,000	$ 3,375
Commissions (5%)	18,750	150,000	400,000	360,000	101,250
Advertising	250,000	150,000	100,000	100,000	25,000
Total marketing expenses	$275,000	$307,500	$510,000	$470,000	$ 129,625
Divided by units	÷ 25,000	÷ 150,000	÷ 400,000	÷ 400,000	÷ 135,000
Unit marketing expense	$ 11.00	$ 2.05	$ 1.28	$ 1.18	$ 0.96

3.

	Year 1	Year 2	Year 3	Year 4	Year 5
Sales	$ 375,000	$3,000,000	$8,000,000	$7,200,000	$2,025,000
Less: COGS	205,500	982,600	2,633,000	2,628,000	828,250
Gross profit	169,500	2,017,400	5,367,000	4,572,000	1,196,750
Less: Marketing expenses	275,000	307,500	510,000	470,000	129,625
Operating income (loss)	$(105,500)	$1,709,900	$4,857,000	$4,102,000	$1,067,125

Five-year operating income	$11,630,525
Less: Design and development expenses	300,000
Excess of revenue over all costs	$11,330,525

Yes, the Mixalot was profitable over the five-year cycle, even after the design and development expenses were subtracted. Note that these expenses do not appear on the operating income statement required for external reporting.

4. The initial price set for the Mixalot was $15. This is the lowest price of those charged during the five-year period. It appears that Melcher Company was using a penetration pricing strategy for the Mixalot. This makes sense given that the Mixalot was not a radically new product (i.e., there were other appliances on the market that could do what the Mixalot could do). There were blenders to mix milkshakes, knives and chopping boards to cut up vegetables, and food processors to mix and chop. Melcher Company needed to get the Mixalot out into actual kitchens to build demand. Notice, too, the large marketing expenditures in the first year to create awareness. This also helps to support price increases down the line. Finally, by the

fifth year, the Mixalot is in the declining stage of the product life cycle. Probably other companies have begun producing competing products, and the number of new Mixalots demanded has declined.

II. Absorption and Variable Costing

Acme Novelty Company produces coin purses and key chains. Selected data for the past year are as follows:

	Coin Purse	Key Chain
Production (units)	100,000	200,000
Sales (units)	90,000	210,000
Selling price	$5.50	$4.50
Direct labour hours	50,000	80,000
Manufacturing costs:		
Direct materials	$ 75,000	$100,000
Direct labour	250,000	400,000
Variable overhead	20,000	24,000
Fixed overhead	50,000	80,000
Nonmanufacturing costs:		
Variable selling	30,000	60,000
Direct fixed selling	35,000	40,000
Common fixed selling*	25,000	25,000

*Common fixed selling cost totals $50,000 and is divided equally between the two products.

Budgeted fixed overhead for the year, $130,000, equalled the actual fixed overhead. Fixed overhead is assigned to products using a plantwide rate based on expected direct labour hours, which were 130,000. The company had 10,000 key chains in inventory at the beginning of the year. These key chains had the same unit cost as the key chains produced during the year.

Required:

1. Compute the unit cost for the coin purses and key chains using the variable-costing method. Compute the unit cost using absorption costing.
2. Prepare an income statement using absorption costing.
3. Prepare an income statement using variable costing.
4. Explain the reason for any difference between absorption- and variable-costing operating incomes.
5. Prepare a segmented income statement using products as segments.

Solution:

1. The unit cost for the coin purse is as follows:

Direct materials ($75,000/100,000)	$0.75
Direct labour ($250,000/100,000)	2.50
Variable overhead ($20,000/100,000)	0.20
Variable cost per unit	$3.45
Fixed overhead [(50,000 × $1.00)/100,000]	0.50
Absorption cost per unit	$3.95

The unit cost for the key chain is as follows:

Direct materials ($100,000/200,000)	$0.50
Direct labour ($400,000/200,000)	2.00
Variable overhead ($24,000/200,000)	0.12
Variable cost per unit	2.62
Fixed overhead [80,000 × $1.00)/200,000]	0.40
Absorption cost per unit	$3.02

Notice that the only difference between the two unit costs is the assignment of the fixed overhead cost. Notice also that the fixed overhead unit cost is assigned using the predetermined fixed overhead rate ($130,000/130,000 DLHrs = $1 per DLH). For example, the coin purses used 50,000 direct labour hours and so receive $1 × 50,000, or $50,000, of fixed overhead. This total, when divided by the units produced, gives the $0.50 per-unit fixed overhead cost. Finally, observe that variable nonmanufacturing costs are not part of the unit cost under variable costing. For both approaches, only manufacturing costs are used to compute the unit costs.

2. The income statement under absorption costing is as follows:

Sales [($5.50 × 90,000) + ($4.50 × 210,000)]	$1,440,000
Less: Cost of goods sold [($3.95 × 90,000) +	
($3.02 × 210,000)]	989,700
Gross profit	450,300
Less: Selling expenses*	215,000
Operating income	$ 235,300

*The sum of selling expenses for both products.

3. The income statement under variable costing is as follows:

Sales [($5.50 × 90,000) + ($4.50 × 210,000)]	$1,440,000
Less:	
Variable cost of goods sold [($3.45 ×	
90,000) + ($2.62 × 210,000)]	(860,700)
Variable selling expenses	(90,000)
Contribution margin	489,300
Less:	
Fixed overhead	(130,000)
Fixed selling expenses	(125,000)
Operating income	$ 234,300

4. Variable-costing income is $1,000 less ($235,300 − $234,300) than absorption-costing income. This difference can be explained by the net change of fixed overhead found in inventory under absorption costing.

Coin purses:	
Units produced	100,000
Units sold	90,000
Increase in inventory	10,000
Unit fixed overhead	× $0.50
Increase in fixed overhead	$ 5,000
Key chains:	
Units produced	200,000
Units sold	210,000
Decrease in inventory	(10,000)
Unit fixed overhead	× $0.40
Decrease in fixed overhead	$ (4,000)

The net change is a $1,000 ($5,000 − $4,000) increase in fixed overhead in inventories. Thus, under absorption costing, there is a net flow of $1,000 of the current period's fixed overhead into inventory. Since variable costing recognized all of the current period's fixed overhead as an expense, variable-costing income should be $1,000 lower than absorption-costing income, as it is.

5.   Segmented income statement:

	Coin Purses	Key Chains	Total
Sales	$ 495,000	$ 945,000	$1,440,000
Less variable expenses:			
Variable cost of goods sold	(310,500)	(550,200)	(860,700)
Variable selling expenses	(30,000)	(60,000)	(90,000)
Contribution margin	154,500	334,800	489,300
Less direct fixed expenses:			
Fixed overhead	(50,000)	(80,000)	(130,000)
Direct selling expenses	(35,000)	(40,000)	(75,000)
Product margin	$ 69,500	$ 214,800	284,300
Less common fixed expenses:			
Common selling expenses			(50,000)
Operating income			$ 234,300

Key Terms

Absorption costing, 600

Contribution margin variance, 609

Contribution margin volume variance, 610

Dumping, 594

Market share, 613

Market share variance, 613

Market size, 613

Market size variance, 613

Markup, 590

Monopolistic competition, 589

Monopoly, 589

Oligopoly, 589

Penetration pricing, 593

Perfectly competitive market, 589

Predatory pricing, 593

Price discrimination, 594

Price elasticity of demand, 588

Price gouging, 593

Price skimming, 593

Price volume variance, 609

Product life cycle, 597

Sales mix variance, 613

Sales price variance, 609

Target costing, 592

Total (overall) sales variance, 609

Variable costing, 603

Discussion Questions

1.   Define *price elasticity of demand*. Give an example of a product with relatively elastic demand and an example of a product with relatively inelastic demand. (Give examples not given in the text.)

2.   What are the features of a perfectly competitive market? Give two examples of competitive markets. How could a firm in such a market move to a less competitive market?

3.   How do you calculate the markup on cost of goods sold? Is the markup pure profit? Explain.

4.   How does target costing differ from traditional costing? How does a target cost relate to price?

5.   What is the difference between penetration pricing and price skimming?

6.   Why do gas stations in the middle of town typically charge a little less for gasoline than do gas stations located on highway turnoffs?

7.   What is price discrimination? Is it legal?

8.   Describe the product life cycle. How do unit-level costs behave in relation to the product life cycle? Batch-level costs? Product-level costs? Facility-level costs?

9. Why do firms measure profit? Why do regulated firms care about the level of profit?
10. What is a segment, and why would a company want to measure profits of segments?
11. Suppose that Alpha Company has four product lines, three of which are profitable and one (let's call it "Loser") that generally incurs a loss. Give several reasons why Alpha Company may choose not to drop the Loser product line.
12. How does absorption costing differ from variable costing? When will absorption-costing operating income exceed variable-costing operating income?
13. What are some advantages and disadvantages of using net income as a measure of profitability?
14. Why do some firms measure customer profitability? In what situation(s) would a firm not want to measure customer profitability?
15. What variances do managers use in trying to understand the difference between actual and planned revenue?

Cornerstone Exercises

Cornerstone Exercise 12-1 MARKUP ON COST, JOB PRICING

OBJECTIVE ▸ 2

CORNERSTONE 12-1

SERVICE

Cliff Meyers owns and operates Cliff's Car Repair Company. Cliff maintains and repairs automobiles and trucks. Direct materials and direct labour costs are easy to trace to the jobs. Cliff's income statement for last year is as follows:

Revenues		$273,920
Cost of goods sold:		
Direct materials	$124,000	
Direct labour	45,000	
Overhead	45,000	214,000
Gross profit		59,920
Selling and administrative expenses		19,400
Operating income		$ 40,520

Cliff wants to find a markup on cost of goods sold that will allow him to earn about the same amount of profit on each job as was earned last year.

Required:

1. What is the markup on cost of goods sold (COGS) that will maintain the same profit as last year?
2. A customer brings in a car that needs a water pump replacement and some general maintenance. The job will have the following costs:

Direct materials	$230
Direct labour	50
Applied overhead	20
Total cost	$300

What is the price that Cliff will quote given the markup percentage calculated in Requirement 1?
3. ***What if*** Cliff wants to calculate a markup on direct materials cost, since it is the largest cost of doing business? What is the markup on direct materials cost that will maintain the same profit as last year? What is the bid price Cliff will use for the job given in Requirement 2 if the markup percentage is calculated on the basis of direct materials cost?

Cornerstone Exercise 12-2 COSTS OF DIFFERENT CUSTOMER CLASSES

OBJECTIVE ▸ 3

CORNERSTONE 12-2

Giroux Food Products Company manufactures canned mixed nuts with an average manufacturing cost of $48 per case (a case contains 24 cans of nuts). Garrity sold 150,000 cases last year to the following three classes of customer:

Customer	Price per Case	Cases Sold
Supermarkets	$55	80,000
Small grocers	90	40,000
Convenience stores	88	30,000

The supermarkets require special labelling on each can costing $0.03 per can. They order through electronic data interchange (EDI), which costs Giroux about $50,000 annually in operating expenses and depreciation. Giroux delivers the nuts to the stores and stocks them on the shelves. This distribution costs $50,000 per year.

The small grocers order in smaller lots that require special picking and packing in the factory; the special handling adds $20 to the cost of each case sold. Sales commissions to the independent jobbers who sell Giroux products to the grocers average 10 percent of sales. Bad debts expense amounts to 8 percent of sales.

Convenience stores also require special handling that costs $30 per case. In addition, Giroux is required to co-pay advertising costs with the convenience stores at a cost of $15,000 per year. Frequent stops are made to each convenience store by Giroux delivery trucks at a cost of $30,000 per year.

Required:

1. Calculate the total cost per case for each of the three customer classes.
2. Using the costs from Requirement 1, calculate the profit per case per customer class. Does the cost analysis support the charging of different prices? Why or why not?
3. *What if* Giroux charged the average price per case to all customer classes? How would that affect the profit percentages?

OBJECTIVE ➤ 5
CORNERSTONE 12-3

Cornerstone Exercise 12-3 ABSORPTION COSTING, VALUE OF ENDING INVENTORY, OPERATING INCOME

Habib Products Inc. began operations in October and manufactured 30,000 units during the month with the following unit costs:

Direct materials	$ 6.00
Direct labour	3.00
Variable overhead	2.00
Fixed overhead*	10.00
Variable marketing cost	2.50

*Fixed overhead per unit = $300,000/30,000 units produced = $10

Total fixed factory overhead is $300,000 per month. During October, 28,000 units were sold at a price of $35, and fixed marketing and administrative expenses were $130,500.

Required:

1. Calculate the cost of each unit using absorption costing.
2. How many units remain in ending inventory? What is the cost of ending inventory using absorption costing?
3. Prepare an absorption-costing income statement for Habib Products Inc. for the month of October.
4. *What if* November production was 30,000 units, costs were stable, and sales were 31,000 units? What is the cost of ending inventory? What is operating income for November?

OBJECTIVE ➤ 5
CORNERSTONE 12-4

Cornerstone Exercise 12-4 VARIABLE COSTING, VALUE OF ENDING INVENTORY, OPERATING INCOME

Refer to **Cornerstone Exercise 12-3**.

Required:

1. Calculate the cost of each unit using variable costing.
2. How many units remain in ending inventory? What is the cost of ending inventory using variable costing?

3. Prepare a variable-costing income statement for Habib Products Inc. for the month of October.
4. **What if** November production was 30,000 units, costs were stable, and sales were 31,000 units? What is the cost of ending inventory? What is operating income for November?

Cornerstone Exercise 12-5 SALES PRICE VARIANCE, PRICE VOLUME VARIANCE, OVERALL SALES VARIANCE

OBJECTIVE ▶ 6
CORNERSTONE 12-5
SERVICE

Plenty Company is a pet food wholesale firm. In December, Plenty Company expects to sell 20,000 bags of pet food at an average price of $2.20 per bag. Actual results are 18,500 bags sold at an average price of $2.25 per bag.

Required:

1. Calculate the sales price variance for December.
2. Calculate the price volume variance for December.
3. Calculate the overall sales variance for December. Explain why it is favourable or unfavourable.
4. **What if** December sales were actually 22,000 bags? How would that affect the sales price variance? The price volume variance? The overall sales variance?

Cornerstone Exercise 12-6 CONTRIBUTION MARGIN VARIANCE

OBJECTIVE ▶ 6
CORNERSTONE 12-6

Park Inc. produces and sells two types of power lawn mowers—the basic mower and the self-propelled mower. Budgeted and actual data for the two models are shown below.

Budgeted Amounts:

	Basic Mower	Self-Propelled Mower	Total
Sales:			
($250 × 15,000)	$3,750,000		
($300 × 45,000)		$13,500,000	$17,250,000
Variable expenses	1,500,000	9,000,000	10,500,000
Contribution margin	$2,250,000	$ 4,500,000	$ 6,750,000

Actual Amounts:

	Basic Mower	Self-Propelled Mower	Total
Sales:			
($238 × 14,800)	$3,522,400		
($310 × 44,000)		$13,640,000	$17,162,400
Variable expenses	1,628,000	7,920,000	9,548,000
Contribution margin	$1,894,400	$ 5,720,000	$ 7,614,400

Required:

1. Calculate the contribution margin variance.
2. **What if** actual units sold of the self-propelled mower increased? How would that affect the contribution margin variance? What if actual units sold of the self-propelled mower decreased? How would that affect the contribution margin variance?

OBJECTIVE ▶ 6

CORNERSTONE 12-7

Cornerstone Exercise 12-7 CONTRIBUTION MARGIN VOLUME VARIANCE

Refer to **Cornerstone Exercise 12-6**.

Required:

1. Calculate the budgeted average unit contribution margin.
2. Calculate the contribution margin volume variance.
3. *What if* actual units sold of the self-propelled mower decreased? How would that affect the contribution margin volume variance? What if actual units sold of the self-propelled mower increased? How would that affect the contribution margin volume variance?

OBJECTIVE ▶ 6

CORNERSTONE 12-8

Cornerstone Exercise 12-8 SALES MIX VARIANCE

Refer to **Cornerstone Exercise 12-6**.

Required:

1. Calculate the sales mix variance.
2. *What if* actual units sold of the basic mower increased? How would that affect the sales mix variance? What if actual units sold of the self-propelled mower increased? How would that affect the sales mix variance?

OBJECTIVE ▶ 6

CORNERSTONE 12-9

Cornerstone Exercise 12-9 MARKET SHARE VARIANCE, MARKET SIZE VARIANCE

Budgeted unit sales for the entire lawn mower industry were 1,200,000 (of all model types), and actual unit sales for the industry were 1,190,000. Recall from **Cornerstone Exercise 12-6** that Park Inc. provided the following information:

	Budgeted	Actual
Sales in units, basic mower	15,000	14,800
Sales in units, self-propelled mower	45,000	44,000
Total contribution margin	$6,750,000	
Budgeted average unit contribution margin	$112.50	

Required:

1. Calculate the market share variance (take percentages out to five significant digits).
2. Calculate the market size variance.
3. *What if* Park actually sold a total of 61,000 units (in total of the two models)? How would that affect the market share variance? The market size variance?

Exercises

OBJECTIVE ▶ 1

Exercise 12-10 ELASTICITY OF DEMAND AND MARKET STRUCTURE

Janet and Phil Hopkins graduated several years ago with M.S. degrees in accounting and set up a full-service accounting firm. Janet and Phil have many small business clients and have noticed some pricing trends while compiling annual financial statements. The following data are for five of the pizza parlours that are Janet and Phil's clients.

	Quantity Sold	Average Price
Mom's	18,000	$10.00
Dad's	21,000	7.90
Auntie's	22,000	8.00
Uncle's	30,000	7.00
Wally's	24,000	7.50

Required:

1. Is the demand for pizza relatively more elastic or inelastic?
2. What type of market structure characterizes the pizza industry? How do you suppose that Mom's can charge so much more per pizza than Uncle's does?

Exercise 12-11 DEMAND CURVE AND CHARACTERISTICS OF MARKET STRUCTURE

OBJECTIVE ▶ 1

SERVICE

Amy Chang wants to start a business supplying florists with field-grown flowers. She has located an appropriate plot of land and believes she can grow daisies, asters, chrysanthemums, carnations, and other assorted types during a nine-month growing period. By growing the flowers in a field as opposed to a greenhouse, Amy expects to save a considerable amount on herbicide and pesticide. She is considering passing the savings along to her customers by charging $1.25 per standard bunch versus the prevailing price of $1.50 per standard bunch.

Amy has turned to her neighbour, Bob Winters, for help. Bob is an accountant in town who is familiar with general business conditions. Bob gathered the following information for Amy:

a. There are 50 growers within a one-hour drive of Amy's land.
b. In general, there is little variability in price. Flowers are treated as commodities, and one aster is considered to be pretty much like any other aster.
c. There are numerous florists in the city, and the amount that Amy would supply could be easily absorbed by the florists at the prevailing price.

Required:

1. What type of market structure characterizes the flower-growing industry in Amy's region? Explain.
2. Given your answer to Requirement 1, what price should Amy charge per standard bunch? Why?

Exercise 12-12 BASICS OF DEMAND, LIFE-CYCLE PRICING

OBJECTIVE ▶ 1 4

SERVICE

Paul Bourdain is an accountant just ready to open an accounting firm in his hometown. He has heard that established accountants in town charge $65 per hour. That sounds good to Paul. In fact, he believes that he should be able to charge $75 an hour given his high GPA and the fact that he is up to date on current accounting issues.

Required:

Should Paul charge $75 per hour? What would you advise him to do?

Exercise 12-13 MARKUP ON COST, COST-BASED PRICING

OBJECTIVE ▶ 2

SERVICE

Kapoor Designs Company is a general contractor that specializes in custom residential housing. Each job requires a bid that includes Kapoor's direct costs and subcontractor costs as well as an amount referred to as "overhead and profit." Kapoor's bidding policy is to estimate the costs of direct materials, direct labour, and subcontractors' costs. These are totalled, and a markup is applied to cover overhead and profit. In the coming year, the company believes it will be the successful bidder on 20 jobs with the following total revenues and costs:

Revenue		$16,240,000
Direct materials	$3,450,000	
Direct labour	4,100,000	
Subcontractors	6,450,000	14,000,000
Overhead and profit		$ 2,240,000

Required:

1. Given the preceding information, what is the markup percentage on total direct costs?
2. Suppose Kapoor Designs is asked to bid on a job with estimated direct costs of $465,000. What is the bid? If the customer complains that the profit seems pretty high, how might Kapoor counter that accusation?

OBJECTIVE ▶ 2   **Exercise 12-14 MARKUP ON COST**

Many different businesses employ markup on cost to arrive at a price. For each of the following situations, explain what the markup covers and why it is the amount that it is.

a. Department stores have a markup of 100 percent of purchase cost.
b. Jewellery stores charge anywhere from 100 percent to 300 percent of the cost of the jewellery. (The 300 percent markup is referred to as "keystone.")
c. Johnson Construction Company charges 12 percent on direct materials, direct labour, and subcontracting costs.
d. Hamilton Auto Repair charges customers for direct materials and direct labour. Customers are charged $45 per direct labour hour worked on their job; however, the employees actually cost Hamilton $15 per hour.

OBJECTIVE ▶ 5   **Exercise 12-15 ABSORPTION AND VARIABLE COSTING WITH OVER- AND UNDERAPPLIED OVERHEAD**

Egnatia Inc. has just completed its first year of operations. The unit costs on a normal costing basis are as follows:

Manufacturing costs (per unit):	
Direct materials (3 kg @ $1.50)	$ 4.50
Direct labour (0.5 hr @ $14)	7.00
Variable overhead (0.5 hr @ $6)	3.00
Fixed overhead (0.5 hr @ $9)	4.50
Total	$19.00
Selling and administrative costs:	
Variable	$2 per unit
Fixed	$138,000

During the year, the company had the following activity:

Units produced	24,000
Units sold	21,300
Unit selling price	$34
Direct labour hours worked	12,000

Actual fixed overhead was $12,000 less than budgeted fixed overhead. Budgeted variable overhead was $5,000 less than the actual variable overhead. The company used an expected actual activity level of 12,000 direct labour hours to compute the predetermined overhead rates. Any overhead variances are closed to Cost of Goods Sold.

Required:

1. Compute the unit cost using (a) absorption costing and (b) variable costing.
2. Prepare an absorption-costing income statement.
3. Prepare a variable-costing income statement.
4. Reconcile the difference between the two income statements.

OBJECTIVE ▶ 5   **Exercise 12-16 VARIABLE COSTING, ABSORPTION COSTING**

During its first year of operations, Arkady Inc. produced 40,000 plastic snow scoops. Snow scoops are oversized shovel-type scoops that are used to push snow away. Unit sales were 38,200 scoops. Fixed overhead was applied at $0.75 per unit produced. Fixed overhead was underapplied by $2,900. This fixed overhead variance was closed to Cost of Goods Sold. There was no variable overhead variance. The results of the year's operations are as follows (on an absorption-costing basis):

Sales (38,200 units @ $20)	$764,000
Less: Cost of goods sold	546,260
Gross margin	217,740
Less: Selling and administrative expenses (all fixed)	184,500
Operating income	$ 33,240

Required:

1. Calculate the cost of the firm's ending inventory under absorption costing. What is the cost of the ending inventory under variable costing?
2. Prepare a variable-costing income statement. Reconcile the difference between the two income figures.

Exercise 12-17 COST-BASED PRICING, TARGET PRICING

Carina Franks operates a catering company in Toronto. She provides food and servers for parties. She also rents tables, chairs, dinnerware, glassware, and linens. Estefan and Maria Montero have contacted Carina about plans for their daughter's graduation party. The Monteros would like a catered affair on the lawn of a rural church. They have requested an open bar, sit-down dinner for 350 people, a large tent, and a dance floor. Of course, they expect Carina to supply serving staff, tables with linens, dinnerware, and glassware. They will handle the flowers, decorations, and hiring the band on their own. Carina put together this bid:

Food (350 × $25)	$ 8,750
Beverages (350 × $15)	5,250
Servers (6 × 4 hours × $10)	240
Bartenders (2 × 4 hours × $10)	80
Clean-up staff (3 × 3 hours × $10)	90
Rental of:	
Dance floor	300
Linens	80
Tables	200
Dinnerware	120
Glassware	150
Total	$15,260

Required:

1. Explain where costs for Carina's services and profit are calculated in the preceding bid.
2. Suppose that the Monteros blanch when they see the preceding bid. One of them suggests that they had hoped to spend no more than $10,000 or so on the party. How could Carina work with the Monteros to achieve a target cost of that amount?
3. Estefan Montero protests the cost of dance floor rental. He said, "I've seen those for rent at U-Rent-It for $75." How would you respond to this remark if you were Carina? (*Hint:* You want this job and so telling him "Go ahead and do it yourself, Cheapskate!" is not an option.)

Exercise 12-18 LIFE-CYCLE PRICING, SALES PRICE, AND PRICE VOLUME VARIANCES

Data for Thermo Company are as follows:

Budgeted price	$14.30
Actual price	$15.00
Budgeted quantity	1,450
Actual quantity sold	1,500

Required:

1. Calculate the sales price variance.
2. Calculate the price volume variance.
3. Suppose that the product is in the introductory stage of the product life cycle. What information do these two variances provide to Thermo's managers?

Exercise 12-19 PRICING STRATEGY, SALES VARIANCES

Hanadarko Inc. manufactures and sells three products: K, M, and P. In January, Hanadarko Inc. budgeted sales of the following.

	Budgeted Volume	Budgeted Price
Product K	110,000	$45.00
Product M	165,000	21.50
Product P	20,000	20.00

At the end of the year, actual sales revenues for Product K and Product M were $5,600,000 and $3,270,000, respectively. The actual price charged for Product K was $50 and for Product M was $20. Only $10 was charged for Product P to encourage more consumers to buy it, and actual sales revenue equalled $600,000 for this product.

Required:

1. Calculate the sales price and price volume variances for each of the three products based on the original budget.
2. Suppose that Product P is a new product just introduced during the year. What pricing strategy is Hanadarko Inc. following for this product?

Problems

OBJECTIVE ▶ 3

Problem 12-20 PRICE DISCRIMINATION, CUSTOMER COSTS

Jorell Inc. manufactures and distributes a variety of labellers. Annual production of labellers averages 340,000 units. A large chain store purchases about 30 percent of Jorell's production. Several thousand independent retail office supply stores purchase the other 70 percent. Jorell incurs the following costs of production per labeller:

Direct materials	$ 8.90
Direct labour	2.40
Overhead	3.20
Total	$14.50

Jorell has two salespeople assigned to the chain store account at a cost of $55,000 each per year. Delivery is made in 1,500 unit batches at a delivery cost of $750 per batch. Eight salespeople service the remaining accounts. They call on the stores and incur salary and mileage expenses of approximately $41,000 each. Delivery costs vary from store to store, averaging $0.60 per unit.

Jorell charges the chain store $16.50 per labeller and the independent office supply stores $20 per labeller.

Required:

Is Jorell's pricing policy supported by cost differences in serving the two different classes of customer? Support your answer with relevant calculations.

OBJECTIVE ▶ 5

Problem 12-21 UNIT COSTS, INVENTORY VALUATION, VARIABLE AND ABSORPTION COSTING

Liebman Company produced 90,000 units during its first year of operations and sold 87,000 at $21.80 per unit. The company chose practical activity—at 90,000 units—to compute its predetermined overhead rate. Manufacturing costs are as follows:

Direct materials	$540,000
Direct labour	99,000
Expected and actual variable overhead	369,000
Expected and actual fixed overhead	468,000

Required:

1. Calculate the unit cost and the cost of finished goods inventory under absorption costing.
2. Calculate the unit cost and the cost of finished goods inventory under variable costing.

3. What is the dollar amount that would be used to report the cost of finished goods inventory to external parties. Why?

Problem 12-22 INCOME STATEMENTS, VARIABLE AND ABSORPTION COSTING

OBJECTIVE ➤ 5

The following information pertains to Petruchio Inc. for last year:

Beginning inventory, units	1,400
Units produced	120,000
Units sold	118,000
Variable costs per unit:	
Direct materials	$7.00
Direct labour	$10.50
Variable overhead	$3.60
Variable selling expenses	$2.10
Fixed costs per year:	
Fixed overhead	$234,000
Fixed selling and administrative expenses	$236,000

There are no work-in-process inventories. Normal activity is 120,000 units. Expected and actual overhead costs are the same. Costs have not changed from one year to the next.

Required:

1. How many units are in ending inventory?
2. Without preparing an income statement, indicate what the difference will be between variable-costing income and absorption-costing income.
3. Assume the selling price per unit is $29. Prepare an income statement using (a) variable costing and (b) absorption costing.

Problem 12-23 INCOME STATEMENTS AND FIRM PERFORMANCE: VARIABLE AND ABSORPTION COSTING

OBJECTIVE ➤ 5

Veracruz Company had the following operating data for its first two years of operations:

Variable costs per unit:	
Direct materials	$ 8.00
Direct labour	4.00
Variable overhead	1.50
Fixed costs per year:	
Overhead	90,000
Selling and administrative	23,450

Veracruz produced 30,000 units in the first year and sold 25,000. In the second year, it produced 25,000 units and sold 30,000 units. The selling price per unit each year was $21. Veracruz uses an actual costing system for product costing.

Required:

1. Prepare income statements for both years using absorption costing. Has firm performance, as measured by income, improved or declined from Year 1 to Year 2?
2. Prepare income statements for both years using variable costing. Has firm performance, as measured by income, improved or declined from Year 1 to Year 2?
3. Which method do you think most accurately measures firm performance? Why?

Problem 12-24 CONTRIBUTION MARGIN VARIANCE, CONTRIBUTION MARGIN VOLUME VARIANCE, SALES MIX VARIANCE

OBJECTIVE ➤ 6

Elburty Company provides management services for apartments and rental units. In general, Elburty packages its services into two groups: basic and complete. The basic package includes advertising vacant units, showing potential renters through them, and collecting monthly rent and remitting it to the owner. The complete package adds maintenance of units and bookkeeping to the basic package. Packages are priced on a per-rental unit basis. Actual results from last year are as follows:

	Basic	Complete
Sales (rental units)	1,700	300
Selling price	$130	$280
Variable expenses	$90	$230

Elburty had budgeted the following amounts:

	Basic	Complete
Sales (units)	1,715	285
Selling price	$120	$290
Variable expenses	$90	$240

Required:

1. Calculate the contribution margin variance.
2. Calculate the contribution margin volume variance.
3. Calculate the sales mix variance.

OBJECTIVE ➤ 6   **Problem 12-25 CONTRIBUTION MARGIN VARIANCE, CONTRIBUTION MARGIN VOLUME VARIANCE, SALES MIX VARIANCE**

Laconia Company produces three models of a product. Actual results from last year are as follows:

	Model 1	Model 2	Model 3
Unit sales	2,725	1,310	965
Selling price	$52	$68	$34
Variable expenses	$18	$34	$14

Laconia had budgeted the following amounts:

	Model 1	Model 2	Model 3
Unit sales	2,700	1,300	1,000
Selling price	$50	$70	$30
Variable expenses	$20	$30	$10

Required:

1. Calculate the contribution margin variance.
2. Calculate the contribution margin volume variance.
3. Calculate the sales mix variance.

OBJECTIVE ➤ 5   **Problem 12-26 IMPACT OF INVENTORY CHANGES ON ABSORPTION-COSTING INCOME, DIVISIONAL PROFITABILITY**

Dana Baird was manager of a new Medical Supplies Division. She had just finished her second year and had been visiting with the company's vice president of operations. In the first year, the operating income for the division had shown a substantial increase over the prior year. Her second year saw an even greater increase. The vice president was extremely pleased and promised Dana a $5,000 bonus if the division showed a similar increase in profits for the upcoming year. Dana was elated. She was completely confident that the goal could be met. Sales contracts were already well ahead of last year's performance, and she knew that there would be no increases in costs.

At the end of the third year, Dana received the following data regarding operations for the first three years:

	Year 1	Year 2	Year 3
Production	10,000	11,000	9,000
Sales (in units)	8,000	10,000	12,000
Unit selling price	$10	$10	$10
Unit costs:			
Fixed overhead*	$2.90	$3.00	$3.00
Variable overhead	$1.00	$1.00	$1.00
Direct materials	$1.90	$2.00	$2.00

	Year 1	Year 2	Year 3
Direct labour	$1.00	$1.00	$1.00
Variable selling	$0.40	$0.50	$0.50
Actual fixed overhead	$29,000	$30,000	$30,000
Other fixed costs	$9,000	$10,000	$10,000

*The predetermined fixed overhead rate is based on expected actual units of production and expected fixed overhead. Expected production each year was 10,000 units. Any under- or overapplied fixed overhead is closed to Cost of Goods Sold.

Yearly Income Statements

	Year 1	Year 2	Year 3
Sales revenue	$80,000	$100,000	$120,000
Less: Cost of goods sold*	54,400	67,000	86,600
Gross margin	25,600	33,000	33,400
Less: Selling and administrative expenses	12,200	15,000	16,000
Operating income	$13,400	$ 18,000	$ 17,400

*Assumes a LIFO inventory flow, as allowed in the United States.

Upon examining the operating data, Dana was pleased. Sales had increased by 20 percent over the previous year, and costs had remained stable. However, when she saw the yearly income statements, she was dismayed and perplexed. Instead of seeing a significant increase in income for the third year, she saw a small decrease. Surely, the Accounting Department had made an error.

Required:

1. Explain to Dana why she lost her $5,000 bonus.
2. Prepare variable-costing income statements for each of the three years. Reconcile the differences between the absorption-costing and variable-costing incomes.
3. If you were the vice president of Dana's company, which income statement (variable-costing or absorption-costing) would you prefer to use for evaluating Dana's performance? Why?

Problem 12-27 ETHICAL ISSUES, ABSORPTION COSTING, PERFORMANCE MEASUREMENT

OBJECTIVE ▸ 3 5 7

Bill Fremont, division controller and CMA, was upset by a recent memo he received from the divisional manager, Steve Preston. Bill was scheduled to present the division's financial performance at headquarters in one week. In the memo, Steve had given Bill some instructions for this upcoming report. In particular, Bill had been told to emphasize the significant improvement in the division's profits over last year. Bill, however, didn't believe that there was any real underlying improvement in the division's performance and was reluctant to say otherwise. He knew that the increase in profits was because of Steve's conscious decision to produce more inventory.

In an earlier meeting, Steve had convinced his plant managers to produce more than they knew they could sell. He argued that by deferring some of this period's fixed costs, reported profits would jump. He pointed out two significant benefits. First, by increasing profits, the division could exceed the minimum level needed so that all the managers would qualify for the annual bonus. Second, by meeting the budgeted profit level, the division would be better able to compete for much-needed capital. Bill objected but had been overruled. The most persuasive counterargument was that the increase in inventory could be liquidated in the coming year as the economy improved. Bill, however, considered this event unlikely. From past experience, he knew that it would take at least two years of improved market demand before the productive capacity of the division was exceeded.

Required:

1. Discuss the behaviour of Steve Preston, the divisional manager. Was the decision to produce for inventory an ethical one?
2. What should Bill Fremont do? Should he comply with the directive to emphasize the increase in profits? If not, what options does he have?
3. Identify any ethical standards that may apply to this situation.

OBJECTIVE ▶ 5   Problem 12-28 **SEGMENTED INCOME STATEMENTS, ADDING AND DROPPING PRODUCT LINES**

Dan Petrov has just been appointed manager of Kirchner Glass Products Division. He has two years to make the division profitable. If the division is still showing a loss after two years, it will be eliminated, and Dan will be reassigned as an assistant divisional manager in another division. The divisional income statement for the most recent year is as follows:

Sales	$4,590,000
Less: Variable expenses	3,953,450
Contribution margin	636,550
Less: Direct fixed expenses	675,000
Divisional margin	(38,450)
Less: Common fixed expenses (allocated)	200,000
Divisional profit (loss)	$ (238,450)

Upon arriving at the division, Dan requested the following data on the division's three products:

	Product A	Product B	Product C
Sales (units)	12,000	14,500	15,000
Unit selling price	$150.00	$120.00	$70.00
Unit variable cost	$100.00	$83.00	$103.33
Direct fixed costs	$100,000.00	$425,000.00	$150,000.00

He also gathered data on a proposed new product (Product D). If this product is added, it will displace one of the current products; the quantity that can be produced and sold will equal the quantity sold of the product it is displacing, although demand limits the maximum quantity that can be sold to 20,000 units. Because of specialized production equipment, it is not possible for the new product to displace part of the production of a second product. The information on Product D is as follows:

Unit selling price	$ 80
Unit variable cost	30
Direct fixed costs	240,000

Required:

1. Prepare segmented income statements for Products A, B, and C.
2. Determine the products that Dan should produce for the coming year. Prepare segmented income statements that prove your combination is the best for the division. By how much will profits improve given the combination that you selected? (*Hint:* Your combination may include one, two, or three products.)

OBJECTIVE ▶ 5   Problem 12-29 **OPERATING INCOME FOR SEGMENTS**

Symetria Inc. manufactures and sells automotive tools through three divisions: Eastern, Southern, and International. Each division is evaluated as a profit centre. Data for each division for last year are as follows:

	Eastern	Southern	International
Sales	$3,150,000	$987,000	$6,500,000
Cost of goods sold	1,580,000	680,000	4,100,000
Selling and administrative expenses	337,000	280,000	620,000

Symetria Inc. had corporate administrative expenses equal to $585,000; these were not allocated to the divisions.

Required:

1. Prepare a segmented income statement for Symetria Inc. for last year.
2. Comment on the performance of each of the divisions.

Problem 12-30 **PRODUCT PROFITABILITY**

OBJECTIVE ▶ 4 5

SERVICE

Porter Insurance Company has three lines of insurance: automobile, property, and life. The life insurance segment has been losing money for the past five quarters, and Leah Harper, Porter's controller, has done an analysis of that segment. She has discovered that the commission paid to the agent for the first year the policy is in place is 55 percent of the first-year premium. The second-year commission is 20 percent, and all succeeding years a commission equal to 5 percent of premiums is paid. No salaries are paid to agents; however, Porter does advertise on television and in magazines. Last year, the advertising expense was $500,000. The loss rate (payout on claims) averages 50 percent. Administrative expenses equal $450,000 per year. Revenue last year was $10,000,000 (premiums). The percentage of policies of various lengths is as follows:

First year in force	65%
Second year	25
More than two years in force	10

Experience has shown that if a policy remains in effect for more than two years, it is rarely cancelled.

Leah is considering two alternative plans to turn this segment around. Plan 1 requires spending $250,000 on improved customer claim service in hopes that the percentage of policies in effect will take on the following distribution:

First year in force	50%
Second year	15
More than two years in force	35

Total premiums would remain constant at $10,000,000, and there are no other changes in fixed or variable cost behaviour.

Plan 2 involves dropping the independent agent and commission system and having potential policyholders phone in requests for coverage. Leah estimates that revenue would drop to $7,000,000. Commissions would be zero, but administrative expenses would rise by $1,200,000, and advertising (including direct mail solicitation) would increase by $1,000,000.

Required:

1. Prepare a variable-costing income statement for last year for the life insurance segment of Porter Insurance Company.
2. What impact would Plan 1 have on income?
3. What impact would Plan 2 have on income?

Problem 12-31 **CUSTOMER PROFITABILITY, LIFE-CYCLE REVENUE**

OBJECTIVE ▶ 4 5

SERVICE

Refer to the original data in **Problem 12-30**. Fred Morton has just purchased a life insurance policy from Porter with premiums equal to $1,500 per year.

Required:

1. Assume Fred holds the policy for one year and then drops it. What is his contribution to Porter's operating income?
2. Assuming Fred holds the policy for three years, what is his contribution to Porter's operating income in the second and third years? Over a three-year period? What implications does this hold for Porter's efforts to retain policyholders?

Problem 12-32 **CUSTOMER PROFITABILITY**

OBJECTIVE ▶ 5

Olin Company manufactures and distributes carpentry tools. Production of the tools is in the mature portion of the product life cycle. Olin has a sales force of 20. Salespeople are paid a commission of 7 percent of sales, plus expenses of $35 per day for days spent on the road away from home, plus $0.30 per kilometre. They deliver products in addition to making the sales, and each salesperson is required to own a truck suitable for making deliveries.

For the coming quarter, Olin estimates the following:

Sales	$1,300,000
Cost of goods sold	450,000

On average, a salesperson travels 6,000 kilometres per quarter and spends 38 days on the road. The fixed marketing and administrative expenses total $400,000 per quarter.

Required:

1. Prepare an income statement for Olin Company for the next quarter.
2. Suppose that a large hardware chain, MegaHardware Inc. wants Olin Company to produce its new SuperTool line. This would require Olin Company to sell 80 percent of total output to the chain. The tools will be imprinted with the SuperTool brand, requiring Olin to purchase new equipment, use somewhat different materials, and reconfigure the production line. Olin's industrial engineers estimate that cost of goods sold for the SuperTool line would increase by 15 percent. No sales commission would be incurred, and MegaHardware would link Olin to its EDI system. This would require an annual cost of $100,000 on the part of Olin. Mega-Hardware would pay shipping. As a result, the sales force would shrink by 80 percent. Should Olin accept MegaHardware's offer? Support your answer with appropriate calculations.

OBJECTIVE ▶ 5

Problem 12-33 SEGMENTED INCOME STATEMENTS, ANALYSIS OF PROPOSALS TO IMPROVE PROFITS

Shannon Inc. has two divisions. One produces and sells paper party supplies (napkins, paper plates, invitations); the other produces and sells cookware. A segmented income statement for the most recent quarter is as follows:

	Party Supplies Division	Cookware Division	Total
Sales	$500,000	$750,000	$1,250,000
Less: Variable expenses	425,000	460,000	885,000
Contribution margin	75,000	290,000	365,000
Less: Direct fixed expenses	85,000	110,000	195,000
Segment margin	$ (10,000)	$180,000	170,000
Less: Common fixed expenses			130,000
Operating income			$ 40,000

On seeing the quarterly statement, Madge Shatsky, president of Shatsky Inc., was distressed and discussed her disappointment with Abdel Sharif, the company's vice president of finance.

MADGE: The Party Supplies Division is killing us. It's not even covering its own fixed costs. I'm beginning to believe that we should shut down that division. This is the seventh consecutive quarter it has failed to provide a positive segment margin. I was certain that Paula Kelly could turn it around. But this is her third quarter, and she hasn't done much better than the previous divisional manager.

ABDEL: Well, before you get too excited about the situation, perhaps you should evaluate Paula's most recent proposals. She wants to spend $10,000 per quarter for the right to use familiar cartoon figures on a new series of invitations, plates, and napkins and at the same time increase the advertising budget by $25,000 per quarter to let the public know about them. According to her marketing people, sales should increase by 10 percent if the right advertising is done—and done quickly. In addition, Paula wants to lease some new production machinery that will increase the rate of production, lower labour costs, and result in less waste of materials. Paula claims that the variable cost ratio will be reduced by 30 percent. The cost of the lease is $95,000 per quarter.

Upon hearing this news, Madge calmed considerably, and, in fact, was somewhat pleased. After all, she was the one who had selected Paula, and she had a great deal of confidence in Paula's judgment and abilities.

Required:

1. Assuming that Paula's proposals are sound, should Madge be pleased with the prospects for the Party Supplies Division? Prepare a segmented income statement for the next quarter that reflects the implementation of Paula's proposals. Assume that the Cookware Division's sales increase by 5 percent for the next quarter and that the same cost relationships hold.

2. Suppose that everything materializes as Paula projected except for both increases in sales—no change in sales revenues takes place. Are the proposals still sound? What if the variable costs are reduced by 40 percent instead of 30 percent with no change in sales?

CMA Problems

CMA Problem 12-1   COST-BASED PRICING*

OBJECTIVE ▸ 2

Otero Fibres Inc. specializes in the manufacture of synthetic fibres, which the company uses in many products such as blankets, coats, and uniforms for police and firefighters. Otero has been in business since 1985 and has been profitable every year since 1993. The company uses a standard cost system and applies overhead on the basis of direct labour hours.

Otero has recently received a request to bid on the manufacture of 800,000 blankets scheduled for delivery to the Canadian military. The bid must be stated at full cost per unit plus a return on full cost of no more than 10 percent after income taxes. Full cost has been defined as including all variable costs of manufacturing the product, a reasonable amount of fixed overhead, and reasonable incremental administrative costs associated with the manufacture and sale of the product. The contractor has indicated that bids in excess of $30 per blanket are not likely to be considered.

In order to prepare the bid for the 800,000 blankets, Andrea Lightner, cost accountant, has gathered the following information about the costs associated with the production of the blankets.

Direct material	$1.70 per kilogram of fibres
Direct labour	$6.50 per hour
Direct machine costs*	$10.00 per blanket
Variable overhead	$3.00 per direct labour hour
Fixed overhead	$8.00 per direct labour hour
Incremental administrative costs	$2,450 per 1,000 blankets
Special fee**	$0.50 per blanket
Material usage	3 kilograms per blanket
Production rate	4 blankets per direct labour hour
Effective tax rate	35%

*Direct machine costs consist of items such as special lubricants, replacement of needles used in stitching, and maintenance costs. These costs are not included in the normal overhead rates.

**Otero recently developed a new blanket fibre at a cost of $750,000. In an effort to recover this cost, Otero has instituted a policy of adding a $0.50 fee to the cost of each blanket using the new fibre. To date, the company has recovered $125,000. Lightner knows that this fee does not fit within the definition of full cost, as it is not a cost of manufacturing the product.

Required:

1. Calculate the minimum price per blanket that Otero Fibres could bid without reducing the company's operating income.

2. Using the full-cost criteria and the maximum allowable return specified, calculate Otero Fibres' bid price per blanket.

3. Without prejudice to your answer to Requirement 2, assume that the price per blanket that Otero Fibres calculated using the cost-plus criteria specified is greater than the maximum bid of $30 per blanket allowed. Discuss the factors that Otero Fibres should consider before deciding whether or not to submit a bid at the maximum acceptable price of $30 per blanket. *(CMA adapted)*

OBJECTIVE ▶ 5   CMA Problem 12-2 **ABSORPTION- AND VARIABLE-COSTING INCOME STATEMENTS***

Leroux Optics Inc. specializes in manufacturing lenses for large telescopes and cameras used in space exploration. As the specifications for the lenses are determined by the customer and vary considerably, the company uses a job-order costing system.

Manufacturing overhead is applied to jobs on the basis of direct labour hours, utilizing the absorption- or full-costing method. Leroux's predetermined overhead rates for 2012 and 2013 were based on the following estimates.

	2012	2013
Direct labour hours	32,500	44,000
Direct labour cost	$325,000	$462,000
Fixed manufacturing overhead	$130,000	$176,000
Variable manufacturing overhead	$162,500	$198,000

Jim Bao, Leroux's controller, would like to use variable (direct) costing for internal reporting purposes as he believes statements prepared using variable costing are more appropriate for making product decisions. In order to explain the benefits of variable costing to the other members of Leroux's management team, Bao plans to convert the company's income statement from absorption costing to variable costing. He has gathered the following information for this purpose, along with a copy of Leroux's 2012 and 2013 comparative income statement.

Leroux Optics Inc.
Comparative Income Statement
For the Years 2012 and 2013

	2012	2013
Net sales	$1,140,000	$1,520,000
Cost of goods sold:		
Finished goods at January 1	16,000	25,000
Cost of goods manufactured	720,000	976,000
Total available	736,000	1,001,000
Less: Finished goods at December 31	25,000	14,000
Unadjusted cost of goods sold	711,000	987,000
Overhead adjustment	12,000	7,000
Cost of goods sold	723,000	994,000
Gross profit	417,000	526,000
Selling expenses	(150,000)	(190,000)
Administrative expenses	(160,000)	(187,000)
Operating income	$ 107,000	$ 149,000

Leroux's actual manufacturing data for the two years are as follows:

	2012	2013
Direct labour hours	30,000	42,000
Direct labour cost	$300,000	$435,000
Direct materials used	$140,000	$210,000
Manufacturing overhead	$132,000	$175,000

The company's actual inventory balances were as follows:

	December 31, 2011	December 31, 2012	December 31, 2013
Direct materials	$32,000	$36,000	$18,000
Work in process:			
Costs	$44,000	$34,000	$60,000
Direct labour hours	1,800	1,400	2,500

*© 2011, Institute of Management Accountants, Inc. http://www.imanet.org. Reprinted with permission.

	December 31, 2011	December 31, 2012	December 31, 2013
Finished goods:			
Costs	$16,000	$25,000	$14,000
Direct labour hours	700	1,080	550

For both years, all administrative expenses were fixed, while a portion of the selling expenses resulting from an 8 percent commission on net sales was variable. Leroux reports any over- or underapplied overhead as an adjustment to the cost of goods sold.

Required:

1. For the year ended December 31, 2013, prepare the revised income statement for Leroux Optics Inc. utilizing the variable-costing method. Be sure to include the contribution margin on the revised income statement.

2. Describe two advantages of using variable costing rather than absorption costing. *(CMA adapted)*

CMA Problem 12-3 CONTRIBUTION MARGIN VARIANCE, CONTRIBUTION MARGIN VOLUME VARIANCE, MARKET SHARE VARIANCE, MARKET SIZE VARIANCE*

OBJECTIVE ➤ 6

SERVICE

Saini Inc. produces and sells gel-filled ice packs. Saini's performance report for April follows:

	Actual	Budgeted
Units sold	100,000	90,000
Sales	$410,000	$360,000
Variable costs	375,000	315,000
Contribution margin	$ 35,000	$ 45,000
Market size (in units)	1,250,000	1,200,000

Required:

1. Calculate the contribution margin variance and the contribution margin volume variance.
2. Calculate the market share variance and the market size variance. *(CMA adapted)*

CMA Problem 12-4 SEGMENTED REPORTING AND VARIANCES*

OBJECTIVE ➤ 5 6

Pitts-Walsh Company (PWC) is a manufacturing company whose product line consists of lighting fixtures and electronic timing devices. The Lighting Fixtures Division assembles units for the upscale and mid-range markets. The Electronic Timing Devices Division manufactures instrument panels that allow electronic systems to be activated and deactivated at scheduled times for both efficiency and safety purposes. Both divisions operate out of the same manufacturing facilities and share production equipment.

PWC's budget for the year ending December 31, 2013, follows and was prepared on a business segment basis under the following guidelines:

a. Variable expenses are directly assigned to the incurring division.
b. Fixed overhead expenses are directly assigned to the incurring division.
c. The production plan is for 8,000 upscale fixtures, 22,000 mid-range fixtures, and 20,000 electronic timing devices. Production equals sales.

PWC established a bonus plan for division management that required meeting the budget's planned operating income by product line, with a bonus increment if the division exceeds the planned product line operating income by 10 percent or more.

PWC Budget
For the Year Ending December 31, 2013
(In thousands of dollars)

	Lighting Fixtures		Electronic Timing Devices	Total
	Upscale	Mid-Range		
Sales	$1,440	$ 770	$ 800	$ 3,010
Variable expenses:				
Cost of goods sold	(720)	(439)	(320)	(1,479)
				(continued)

| | Lighting Fixtures | | Electronic | |
	Upscale	Mid-Range	Timing Devices	Total
Selling and administrative	(170)	(60)	(60)	(290)
Contribution margin	550	271	420	1,241
Fixed overhead expenses	140	80	80	300
Segment margin	$ 410	$ 191	$ 340	$ 941

Shortly before the year began, the CEO, Jack Parkow, suffered a heart attack and retired. After reviewing the 2013 budget, the new CEO, Joe Kudla, decided to close the lighting fixtures mid-range product line by the end of the first quarter and use the available production capacity to grow the remaining two product lines. The marketing staff advised that electronic timing devices could grow by 40 percent with increased direct sales support. Increases above that level and increasing sales of upscale lighting fixtures would require expanded advertising expenditures to increase consumer awareness of PWC as an electronics and upscale lighting fixtures company. Kudla approved the increased sales support and advertising expenditures to achieve the revised plan. Kudla advised the divisions that for bonus purposes the original product-line operating income objectives must be met, but he did allow the Lighting Fixtures Division to combine the operating income objectives for both product lines for bonus purposes.

Prior to the close of the fiscal year, the division controllers were furnished with preliminary actual data for review and adjustment, as appropriate. These preliminary year-end data reflect the revised units of production amounting to 12,000 upscale fixtures, 4,000 mid-range fixtures, and 30,000 electronic timing devices and are presented as follows:

PWC Preliminary Actuals
For the Year Ending December 31, 2013
(In thousands of dollars)

| | Lighting Fixtures | | Electronic | |
	Upscale	Mid-Range	Timing Devices	Total
Sales	$ 2,160	$140	$1,200	$ 3,500
Variable expenses:				
Cost of goods sold	(1,080)	(80)	(480)	(1,640)
Selling and administrative	(260)	(11)	(96)	(367)
Contribution margin	820	49	624	1,493
Fixed overhead expenses	140	14	80	234
Segment margin	$ 680	$ 35	$ 544	$ 1,259

The controller of the Lighting Fixtures Division, anticipating a similar bonus plan for 2014, is contemplating deferring some revenues to the next year on the pretext that the sales are not yet final and accruing in the current year expenditures that will be applicable to the first quarter of 2014. The corporation would meet its annual plan, and the division would exceed the 10 percent incremental bonus plateau in 2013 despite the deferred revenues and accrued expenses contemplated.

Required:

1. Outline the benefits that an organization realizes from segment reporting. Evaluate segment reporting on a variable-costing basis versus an absorption-costing basis.
2. Calculate the contribution margin, contribution margin volume, and sales mix variances.
3. Explain why the variances occurred. (CMA adapted)

The Collabourative Learning Exercises can be found on the product support site at www.hansen1ce.nelson.com.

After studying this chapter, you should be able to:

1 Explain what strategic cost management is and how it can be used to help a firm create a competitive advantage.

2 Discuss value-chain analysis and the strategic role of activity-based customer and supplier costing.

3 Tell what life-cycle cost management is and how it can be used to maximize profits over a product's life cycle.

4 Identify the basic features of JIT purchasing and manufacturing.

5 Describe the effect JIT has on cost traceability and product costing.

CHAPTER

Strategic Cost Management

13

Why is one brand of ice cream viewed as better than another brand? It may reflect a deliberate decision by an ice cream producer to design and make an ice cream product that uses special ingredients and flavours rather than simply the ordinary. It is a means of differentiating the product and making it unlike those of competitors. It also may mean a conscious decision has been made to target certain types of consumers—consumers who are willing to pay for a higher-quality, specialized ice cream. Whether this is a good strategy or not depends on its profitability. Cost management plays a vital role in strategic decision making. Cost information is critical in formulating and choosing strategies as well as in evaluating the continued viability of existing strategic positions.

In Chapter 6, the basic concepts of activity-based costing were introduced. These concepts were illustrated using the traditional product cost definition. Activity-based product costing can significantly improve the accuracy of traditional product costs. Thus, inventory valuation is improved, and managers (and other information users) have better information concerning the costs of products, leading to more informed decision making. Yet the value of the traditional product cost definition is limited and may not be very useful in certain decision contexts. For example, corporations engage in decision making that affects their long-run competitive position and profitability. Strategic planning and decision making require a much broader set of cost information than that provided by product costs. Cost information about customers, suppliers, and different product designs is also needed to support strategic management objectives.

This broader set of information should satisfy two requirements. First, it should include information about the firm's environment and internal workings. Second, it must be prospective and thus should provide insight about future periods and activities. A value-chain framework with cost data to support a **value-chain analysis** satisfies the first requirement. Cost information to support product life-cycle analysis is needed to satisfy the second requirement. Value-chain analysis can produce organizational changes that fundamentally alter the nature and demand for cost information. Just-in-time (JIT) manufacturing is an example of a strategic approach that alters the nature of the cost accounting information system. In this chapter, we introduce strategic cost management, life-cycle cost management, and JIT manufacturing. The JIT approach is used to illustrate the value-chain concepts. However, given the breadth of its application and its effect on cost accounting, JIT is a topic that by itself merits study. Furthermore, JIT's linkages to strategic cost management justify this topic's inclusion in the same chapter with strategic cost management.

Strategic Cost Management: Basic Concepts

OBJECTIVE ▶ 1
Explain what strategic cost management is and how it can be used to help a firm create a competitive advantage.

Decision making that affects the long-term competitive position of a firm must explicitly consider the strategic elements of a decision. The most important strategic elements for a firm are its long-term growth and survival. Thus, **strategic decision making** involves choosing among alternative strategies with the goal of selecting a strategy, or strategies, that provide a company with reasonable assurance of long-term growth and survival. The key to achieving this goal is to gain a *competitive advantage*. **Strategic cost management**[1] is the use of cost data to develop and identify superior strategies that will produce a sustainable competitive advantage.

Strategic Positioning: The Key to Creating and Sustaining a Competitive Advantage

Competitive advantage is creating better customer value for the same or lower cost than offered by competitors or creating equivalent value for lower cost than offered by competitors. **Customer value** is the difference between what a customer receives (customer realization) and what the customer gives up (customer sacrifice). What a customer receives is more than simply the basic level of performance provided by a product.[2] What is received is called the *total product*. The **total product** is the complete range of tangible and intangible benefits that a customer receives from a purchased product. Thus, customer realization includes basic and special product features, service, quality, instructions for use, reputation, brand name, and any other factors deemed important by customers. Customer sacrifice includes the cost of purchasing the product, the time and effort spent acquiring and learning to use the product, and **post-purchase costs**, which are the costs of using, maintaining, and disposing of the product.

Increasing customer value to achieve a competitive advantage is tied closely to judicious strategy selection. Three general strategies have been identified: *cost leadership*, *product differentiation*, and *focusing*.[3]

---

[1] The idea of strategic cost management was introduced by John K. Shank and Vijay Govindarajan in their book *Strategic Cost Management* (The Free Press, 1993).

[2] Keep in mind that our definition of *product* includes services. Services are intangible products.

[3] See M. E. Porter, *Competitive Advantage: Creating and Sustaining Superior Performance* (New York: Free Press, 1985) for a more complete discussion of the three strategic positions.

Cost Leadership   The objective of a **cost leadership strategy** is to provide the same or better value to customers at a *lower cost* than offered by competitors. Essentially, if customer value is defined as the difference between realization and sacrifice, a low-cost strategy increases customer value by minimizing customer sacrifice. In this case, cost leadership is the goal of the organization. For example, a company might redesign a product so that fewer parts are needed, lowering production costs and the costs of maintaining the product after purchase.

Differentiation   A **differentiation strategy**, on the other hand, strives to increase customer value by increasing what the customer receives (customer realization). A competitive advantage is created by providing something to customers that is not provided by competitors. Therefore, product characteristics must be created that set the product apart from its competitors. This differentiation can occur by adjusting the product so that it is different from the norm or by promoting some of the product's tangible or intangible attributes. Differences can be functional, aesthetic, or stylistic. For example, a retailer of computers might offer on-site repair service, a feature not offered by other rivals in the local market. Or a producer of crackers may offer animal-shaped crackers, as **Nabisco** did with Teddy Grahams®, to differentiate its product from other brands with more conventional shapes. To be of value, however, customers must see the variations as important. Furthermore, the value added to the customer by differentiation must exceed the firm's costs of providing the differentiation. If customers see the variations as important and if the value added to the customer exceeds the cost of providing the differentiation, then a competitive advantage has been established.

Focusing   A **focusing strategy** is selecting or emphasizing a market or customer segment in which to compete. One possibility is to select the markets and customers that appear attractive and then develop the capabilities to serve these targeted segments. Another possibility is to select specific segments where the firm's core competencies in the segments are superior to those of competitors. A focusing strategy recognizes that not all segments (e.g., customers and geographic regions) are the same. Given the capabilities and potential capabilities of the organization, some segments are more attractive than others.

Strategic Positioning   In reality, many firms will choose not just one general strategy, but a combination of the three general strategies. **Strategic positioning** is the process of selecting the optimal mix of these three general strategic approaches. The mix is selected with the objective of creating a sustainable competitive advantage. A **strategy**, reflecting combinations of the three general strategies, can be defined as:

> ... *choosing the market and customer segments the business unit intends to serve, identifying the critical internal business processes that the unit must excel at to deliver the value propositions to customers in the targeted market segments, and selecting the individual and organizational capabilities required for the internal, customer, and financial objectives.*[4]

As used in the definition, "choosing the market and customer segments" is actually focusing; "deliver[ing] the value propositions" is choosing to increase customer realization and/or decrease sacrifice and, therefore, entails cost leadership and/or differentiation strategies, or a combination of the two. Developing the necessary capabilities to serve the segments is related to all three general strategies.

What is the role of cost management in strategic positioning? The *objective* of strategic cost management is to *reduce* costs while simultaneously *strengthening* the chosen strategic position. Remember that a competitive advantage is tied to costs. For example, suppose that an organization is providing the same customer value at a higher cost than its competitors. By increasing customer value for specific customer segments (e.g., differentiation and focusing are used to strengthen the strategic position) and, at the same time, *decreasing* costs, the organization might reach a state

[4] Robert S. Kaplan and David P. Norton, *The Balanced Scorecard* (Boston: Harvard Business School Press, 1996): 37.

where it is providing greater value at the same or less cost than its competitors, thus creating a competitive advantage.

Industrial Value-Chain Framework, Linkages, and Activities

Choosing an optimal (or most advantageous) strategic position requires managers to understand the activities that contribute to its achievement. Successful pursuit of a sound strategic position mandates an understanding of the *industrial value chain*. The **industrial value chain** is the linked set of value-creating activities from basic raw materials to the disposal of the finished product by end-use customers. Exhibit 13-1 illustrates a possible industrial value chain for the petroleum industry. A given firm operating in the oil industry may not—and likely will not—span the entire value chain. The exhibit illustrates that different firms participate in different portions of the value chain. Most large oil firms such as **ExxonMobil** and **Petro-Canada** (now part of **Suncor**) are involved in the value chain from exploration to service stations (like Firm A in Exhibit 13-1). Yet even these oil giants purchase oil from other producers and also supply gasoline to service station outlets that are owned by others. Furthermore, there are many oil firms that engage exclusively in smaller segments of the chain such as exploration and production or refining and distribution (like Firms B and C in Exhibit 13-1). Regardless of its position in the value chain, to create and

---

Exhibit 13-1

Value Chain for the Petroleum Industry

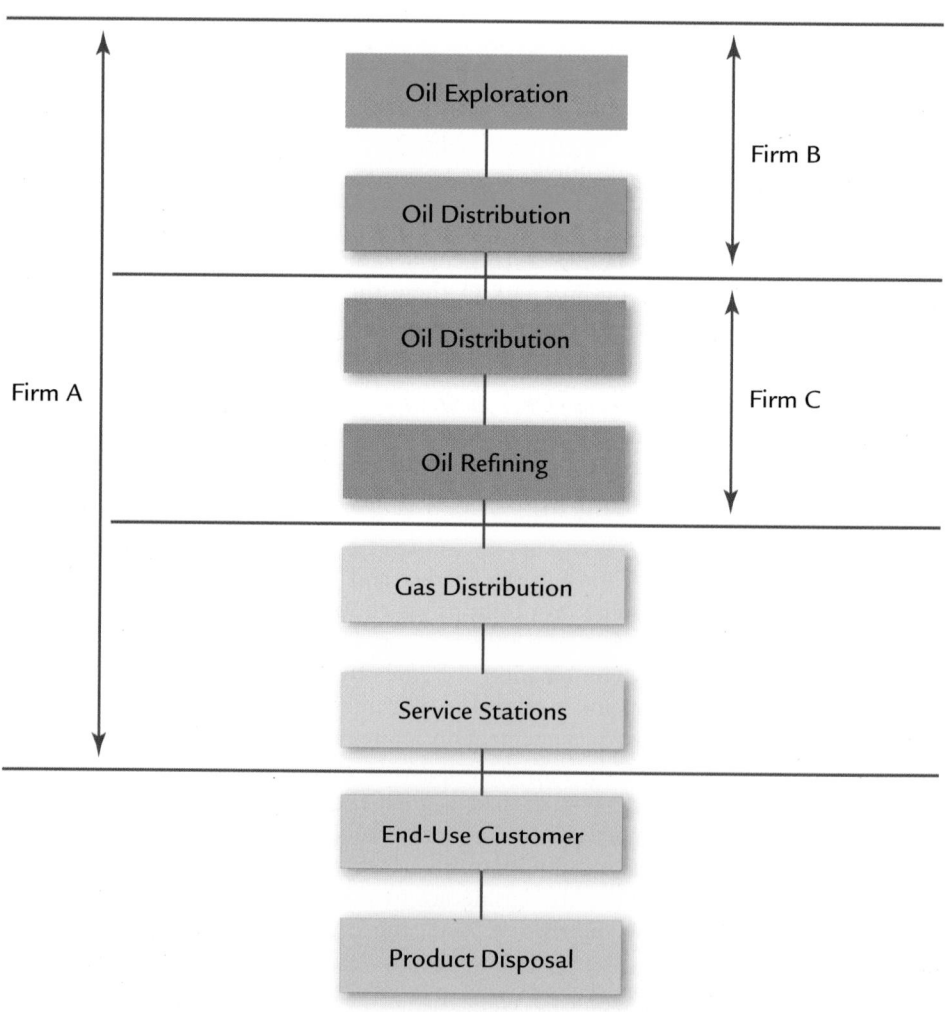

sustain a competitive advantage, a firm must understand the entire value chain and not just the portion in which it operates.

Thus, breaking down the value chain into its strategically relevant activities is basic to successful implementation of cost leadership and differentiation strategies. A value-chain framework is a compelling approach to understanding a firm's strategically important activities. Fundamental to a value-chain framework is the recognition that there exist complex linkages and interrelationships among activities both within and beyond the firm. Two types of linkages must be analyzed and understood: *internal linkages* and *external linkages*. **Internal linkages** are relationships among activities that are performed within a firm's portion of the value chain. **External linkages**, on the other hand, describe the relationship of a firm's value-chain activities that are performed with its suppliers and customers. External linkages, therefore, are of two types: *supplier linkages* and *customer linkages*.

External linkages emphasize the fact that a company must understand the entire value chain and not just the portion of the chain in which it participates. An *external* focus is needed for effective strategic cost management. A company cannot ignore supplier and customer linkages and expect to establish a sustainable competitive advantage. A company needs to understand its relative position in the industrial value chain. An assessment of the economic strength and relationships of each stage in the entire value-chain system can provide a company with several significant strategic insights. For example, knowing the revenues and costs of the different stages may reveal the need to forward or backward integrate to increase overall economic performance. Alternatively, it may reveal that divestiture and a narrowing of participation in the industrial value chain is a good strategy. Finally, knowing the supplier power and buyer power can have a significant effect on how external linkages are exploited. Supplier and buyer power can be assessed for a company by comparing the percentage of profits earned in the industrial value chain with the percentages earned by suppliers and by customers. For example, suppose that the profit earned per litre of gasoline by an independent refiner and producer is $0.15 and that the profit earned by a network of service stations that buy the gasoline (not owned by the independent) is $0.05 per litre. The percentage of profit earned in this segment of the value chain by the downstream stage is 25 percent ($0.05/$0.20), while the independent earns 75 percent of the profit. Buyer power is weak relative to the refiner and producer. If, in addition, the return on assets being earned by the service station segment is high, this may reveal that integrating forward is both desirable and possible.

To exploit a firm's internal and external linkages, we must identify the firm's activities and select those that can be used to produce (or sustain) a competitive advantage. This selection process requires knowledge of the cost and value of each activity. For strategic analysis, activities are classified as *organizational activities* and *operational activities*; the costs of these activities, in turn, are determined by *organizational* and *operational cost drivers*.

Organizational Activities and Cost Drivers

Organizational activities are of two types: *structural* and *executional*. **Structural activities** are activities that determine the underlying economic structure of the organization. **Executional activities** are activities that define the processes and capabilities of an organization and thus are directly related to the ability of an organization to execute successfully. **Organizational cost drivers** are structural and executional factors that determine the long-term cost structure of an organization. Thus, there are two types of organizational drivers: *structural cost drivers* and *executional cost drivers*. Possible structural and executional activities with their cost drivers are listed by category in Exhibit 13-2.

As the exhibit shows, it is possible (and perhaps common) that a given organizational activity can be driven by more than one driver. For example, the cost of building plants is affected by number of plants, scale, and degree of centralization. Firms that have a commitment to a high degree of centralization may build larger plants so that there can be more

Exhibit 13-2

Organizational Activities and Drivers

Structural Activities	Structural Cost Drivers
Building plants	Number of plants, scale, degree of centralization
Management structuring	Management style and philosophy
Grouping employees	Number and type of work units
Complexity	Number of product lines, number of unique processes, number of unique parts, degree of complexity
Vertically integrating	Scope, buying power, selling power
Selecting and using process technologies	Types of process technologies, experience

Executional Activities	Executional Cost Drivers
Using employees	Degree of involvement
Providing quality	Quality management approach
Providing plant layout	Plant layout efficiency
Designing and producing products	Product configuration
Providing capacity	Capacity utilization

geographic concentration and greater control. Similarly, complexity may be driven by number of different products, number of unique processes, and number of unique parts.

Organizational drivers are factors that affect an organization's long-term cost structure. This is readily understood by simply considering the various drivers shown in Exhibit 13-2. Among the structural drivers are the familiar drivers of scale, scope, experience, technology, and complexity. For example, economies and diseconomies of scale are well-known economic phenomena, and the learning curve effect (experience) is also well documented. An interesting property of structural cost drivers is that more is not always better. Moreover, the efficiency level of a structural driver can change. For example, changes in technology can affect the scale driver by changing the optimal size of a plant. In the steel industry, minimill technology has eliminated scale economies (in the form of megamills) as a competitive advantage. Plants of much smaller scale can now achieve the same level of efficiency once produced only by larger steel plants.

Of more recent interest and emphasis are executional drivers. Considerable managerial effort is being expended to improve how things are done in an organization. Continuous improvement and its many faces (employee empowerment, total quality management, process value analysis, life-cycle assessment, etc.) are what executional efficiency is all about. Consider employee involvement and empowerment. The cost of using employees decreases as the degree of involvement increases. Employee or worker involvement refers to the culture, degree of participation, and commitment to the objective of continuous improvement.

Operational Activities and Drivers

Operational activities are day-to-day activities performed as a result of the structure and processes selected by the organization. Examples include receiving and inspecting incoming parts, moving materials, shipping products, testing new products, servicing products, and setting up equipment. **Operational cost drivers** (activity drivers) are those factors that drive the cost of operational activities. They include such factors as number of parts, number of moves, number of products, number of customer orders, and number of returned products. As should be evident, operational activities and drivers are the focus of activity-based costing. Possible operational activities and their drivers are listed in Exhibit 13-3.

Exhibit 13-3

Operational Activities and Drivers

Unit-Level Activities	Unit-Level Drivers
Grinding parts	Grinding machine hours
Assembling parts	Assembly labour hours
Drilling holes	Drilling machine hours
Using materials	Kilograms of material
Using power	Number of kilowatt-hours

Batch-Level Activities	Batch-Level Drivers
Setting up equipment	Number of setups
Moving batches	Number of moves
Inspecting batches	Inspection hours
Reworking products	Number of defective units

Product-Level Activities	Product-Level Drivers
Redesigning products	Number of change orders
Expediting	Number of late orders
Scheduling	Number of different products
Testing products	Testing hours

The structural and executional activities define the number and nature of the day-to-day activities performed within the organization. For example, if an organization decides to produce more than one product at a facility, then this structural choice produces a need for scheduling, a product-level activity. Similarly, providing a plant layout defines the nature and extent of the materials handling activity (usually a batch-level activity). Furthermore, although organizational activities define operational activities, analysis of operational activities and drivers can be used to suggest strategic choices of organizational activities and drivers. For example, knowing that the number of moves is a measure of consumption of the materials handling activity by individual products may suggest that resource spending can be reduced if the plant layout is redesigned to reduce the number of moves needed. Operational and organizational activities and their associated drivers are strongly interrelated.

Internal Value Chain

OBJECTIVE ▸ 2
Discuss value-chain analysis and the strategic role of activity-based customer and supplier costing.

Sound strategic cost management mandates the consideration of that portion of the value chain in which a firm participates (called the *internal value chain*). Exhibit 13-4 reviews the internal value-chain activities for an organization. Activities before and after production must be identified and their linkages recognized and exploited. Exploiting internal linkages means that relationships between activities are assessed and used to reduce costs and increase value. For example, product design and development activities occur before production and are linked to production activities. The way the product is designed affects the costs of production. How production costs are affected requires a knowledge of cost drivers. Thus, knowing the cost drivers of activities is crucial for understanding and exploiting linkages. If design engineers know that the number of parts is a cost driver for various production activities (material usage, direct labour usage, assembly, inspection, materials handling, and purchasing are examples of activities where costs could be affected by number of parts), then redesigning the product so that it has standard parts, multiple sources, short lead times, and high quality can significantly reduce the overall cost of the product.

Exhibit 13-4

Internal Value Chain

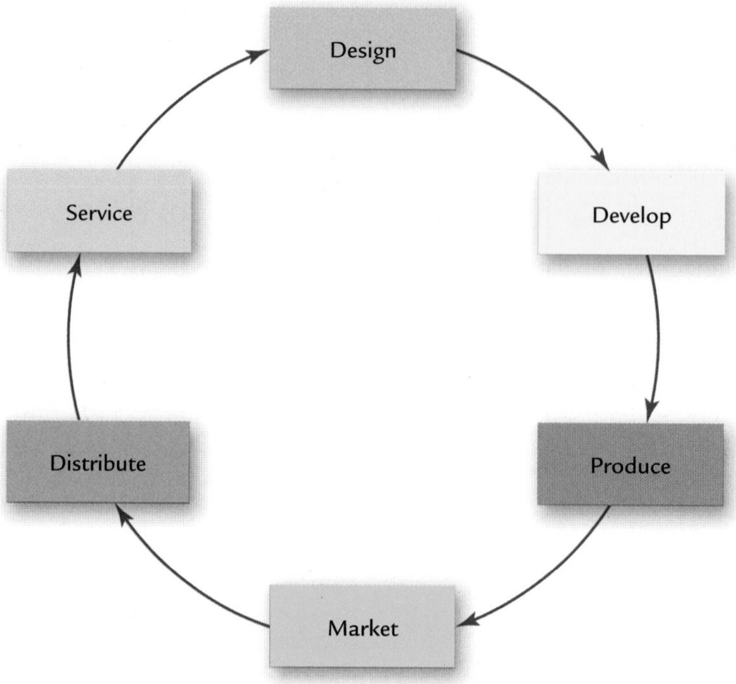

Cornerstone 13-1 illustrates how internal linkages can be exploited to reduce costs in the internal value chain.

Cornerstone 13-1 underscores the importance of individual activities for assessing the impact of the new design. Knowing the cost of different design strategies is made possible by assessing the linkages of activities and the effects of changes in demand for the activities. Notice the key role that the resource usage model plays in this analysis.[5] The purchasing activity currently supplies 15,000 units of activity capacity, acquired in steps of 5,000 units. Capacity is measured in the number of purchase orders. Unused activity for the current product configuration is 2,500 units (15,000 − 12,500). Reconfiguring the product reduces the demand from 12,500 orders to 6,500 orders. This increases the unused activity capacity to 8,500 units (15,000 − 6,500). At this point, management has the capability of reducing resource spending on the resources acquired in advance of usage. Since activity capacity is acquired in chunks of 5,000 units, resource spending can be reduced by $30,000 (the price of one purchasing clerk). Furthermore, since demand decreases, resource spending for the resources acquired as needed is also reduced $3,000 by the variable component ($0.50 × 6,000). The activity-based costing model and knowledge of activity cost behaviour are powerful and integral components of strategic cost management.

In Cornerstone 13-1, the analysis implicitly assumes that resource spending on the engineering design activity would remain unchanged. Therefore, there was no cost to exploiting the linkage. Suppose, however, that an increase in resource spending of $50,000 is needed to exploit the linkages between engineering design and activities downstream in the firm's value chain. Spending $50,000 to save $453,000 is certainly sound. Spending on one activity to save on the cost of other activities is a fundamental principle of strategic cost analysis.

---

[5] The resource usage model was introduced in Chapter 2.

The HOW and WHY of Exploiting Internal Linkages to Reduce Costs and Increase Value

CORNERSTONE 13-1

Information:
A firm currently produces a high-tech medical product with 20 parts. Design engineering has produced a new configuration for the product that requires only eight parts. Current activity capacity and demand (20-part configuration) and expected activity demand (8-part configuration) are provided.

Activities	Activity Driver	Activity Capacity	Current Activity Demand	Expected Activity Demand
Material usage	Number of parts	200,000	200,000	80,000
Assembling parts	Direct labour hours	10,000	10,000	5,000
Purchasing parts	Number of orders	15,000	12,500	6,500

Additionally, the following activity cost data are provided:

Material usage: $3 per part used; no fixed activity cost.

Assembling parts: $12 per direct labour hour; no fixed activity cost.

Purchasing parts: Three salaried clerks, each earning a $30,000 annual salary; each clerk is capable of processing 5,000 purchase orders. Variable activity costs: $0.50 per purchase order processed for forms, postage, etc.

Why:
Exploiting internal linkages means that relationships between activities in the internal value chain are assessed and used to reduce costs and increase value.

Required:
1. Calculate the cost reduction produced by the new design.
2. Suppose that 10,000 units are being produced and sold for $400 per unit and that the price per unit will be reduced by the per-unit savings. What is the new price for the eight-part configured product?
3. **What if** the expected activity demand for purchase orders was 4,500? How would this affect the answers to Requirements 1 and 2?

Solution:
1.

Material usage cost reduction [(200,000 − 80,000)$3]	$360,000
Labour usage cost reduction [(10,000 − 5,000)$12]	60,000
Purchasing cost reduction* [$30,000 + $0.50(12,500 − 6,500)]	33,000
Total savings	$453,000

*Based on the new demand, the number of purchasing agents can be reduced by one, saving $30,000.

2. New price = $400 − ($453,000/10,000) = $354.70

3. Since each purchasing agent can process only 5,000 orders, one agent is needed, saving an additional $30,000 of salary costs. Variable purchasing costs would also drop by an additional $1,000 [$0.50 × (6,500 − 4,500)]. Thus, total savings would increase by $31,000, and the new price would decrease by an additional $3.10 ($31,000/10,000) to $351.60 ($354.70 − $3.10).

Exploiting Supplier Linkages

Although each firm has its own value chain, as was shown in Exhibit 13-1 on page 646, each firm also belongs to a broader value chain—the *industrial value chain*. The value-chain system also includes value-chain activities that are performed by suppliers and buyers. A firm cannot ignore the interaction between its own value-chain activities and those of its suppliers and buyers. Linkages with activities external to the firm can also be exploited. Exploiting external linkages means managing these linkages so that both the company and the external parties receive an increase in benefits.

Suppliers provide inputs and, as a consequence, can have a significant effect on a user's strategic positioning. For example, assume that a company adopts a *total quality management* approach to differentiate and reduce overall quality costs. **Total quality management** is an approach to managing quality that demands the production of defect-free products. Reducing defects, in turn, reduces the total costs spent on quality activities. Yet if the components are delivered late and are of low quality, then there is no way the buying company can produce high-quality products and deliver them on time to its customers. To achieve a defect-free state, a company is strongly dependent on its suppliers' ability to provide defect-free parts. Once this linkage is understood, then a company can work closely with its suppliers so that the product being purchased meets its needs.

Honeywell understands this linkage and has established a supplier review board with the objective of improving business relationships and material quality. Its evaluation and selection of suppliers is based on factors such as product quality, delivery, reliability, continuous improvement, product price, and overall relations. Suppliers are expected to meet certain quality and delivery standards such as 500 parts per million (defect rate), 99 percent on-time delivery, and a 99 percent lot acceptance rate.[6]

Managing Procurement Costs Using Activity-Based Costing   To encourage purchasing managers to choose suppliers whose quality, reliability, and delivery performance are acceptable, two essential requirements have been identified.[7] First, a broader view of component costs is needed. Unit-based costing systems typically reward purchasing managers solely on purchase price (e.g., materials price variances). A broader view means that the costs associated with quality, reliability, and late deliveries are added to the purchase costs. Purchasing managers are then required to evaluate suppliers based on total cost, not just purchase price. Second, supplier costs are assigned to products using causal relationships.

Activity-based costing is the key to satisfying both requirements. To satisfy the first requirement, suppliers are defined as a cost object and costs relating to purchase, quality, reliability, and delivery performance are traced to suppliers. In the second case, products are the cost objects, and supplier costs are traced to specific products. By tracing supplier costs to products—rather than averaging them over all products as unit-based costing does—managers can see the effect of large numbers of unique components requiring specialty suppliers versus products with only standard components. Knowing the costs of more complex products helps product designers better evaluate the trade-offs between functionality and cost as they design new products. Additional functions should provide more benefits (by an increased selling price) than costs. By accurately tracing supplier costs to products, a better understanding of product profitability is produced, and product designers are more capable of choosing among competing product designs. Cornerstone 13-2 illustrates the concepts and calculations associated with activity-based supplier costing.

---

[6] As reported at http://www.honeywell.com on May 15, 2009.

[7] These requirements are discussed in Robin Cooper and Regine Slagmulder, "The Scope of Strategic Cost Management," *Management Accounting* (February 1998): 16–18. Much of the discussion in this section is based on this article.

The HOW and WHY of Activity-Based Supplier Costing

**CORNERSTONE
13-2**

Information:
A purchasing manager uses two suppliers for the source of two electronic components, X1Z and Y2Z. Data associated with these two components are supplied below.

I. Activity Costs (component failure and late delivery are attributable to suppliers; process failure is caused by internal processes):

Activity	Component Failure/ Late Delivery	Process Failure
Reworking products	$200,000	$40,000
Expediting products	50,000	10,000

II. Supplier Data

	Fielding Electronics		Oro Limited	
	X1Z	Y2Z	X1Z	Y2Z
Unit purchase price	$10	$26	$12	$28
Units purchased	40,000	20,000	5,000	5,000
Failed units	800	190	5	5
Late shipments	30	20	0	0

Why:
Activity-based supplier costing uses drivers to trace costs associated with quality, reliability, and late deliveries to individual suppliers and adds these costs to the direct purchase costs. This enables managers to improve their evaluation and selection of suppliers, with the objective of reducing total supplier costs.

Required:
1. Calculate the activity rates for assigning costs to suppliers.
2. Calculate the total unit purchasing cost for each component for each supplier.
3. **What if** the quantity of X1Z that can be purchased is limited to 50,000 units from Fielding and 30,000 units from Oro Limited? There is no limit from either source for Y2Z. Based on cost, what purchasing mix should be chosen?

Solution:
1. Reworking rate = $200,000/1,000* = $200 per failed component
   *(800 + 190 + 5 + 5)

   Expediting rate = $50,000/50* = $1,000 per late delivery
   *(30 + 20)

2.

	Fielding Electronics		Oro Limited	
	X1Z	Y2Z	X1Z	Y2Z
Reworking products:				
$200 × 800	$160,000			
$200 × 190		$38,000		
$200 × 5			$1,000	
$200 × 5				$1,000

CORNERSTONE 13-2 (continued)		Fielding Electronics		Oro Limited	
		X1Z	Y2Z	X1Z	Y2Z
Expediting products:					
$1,000 × 30		30,000			
$1,000 × 20			20,000		
Total costs		$190,000	$ 58,000	$ 1,000	$ 1,000
Units		÷ 40,000	÷ 20,000	÷ 5,000	÷ 5,000
Unit cost		4.75	2.90	0.20	0.20
Unit purchase cost		10.00	26.00	12.00	28.00
Total unit supplier cost		$ 14.75	$ 28.90	$ 12.20	$ 28.20

3. Based on lowest cost: X1Z: 15,000 units from Fielding and 30,000 units from Oro; Y2Z: 0 from Fielding and 25,000 from Oro.

The results of Cornerstone 13-2 show that the "low-cost" supplier actually costs more when the linkages with the internal activities of reworking and expediting are considered. If the purchasing manager is provided with all costs, then the choice becomes clear: Oro Limited is the better supplier. It provides a higher-quality product on a timely basis and at a lower overall cost per unit.

Exploiting Customer Linkages

Customers can also have a significant influence on a firm's strategic position. Choosing marketing segments, of course, is one of the principal elements that define strategic position. For example, selling a medium-level quality product to low-end dealers for a special, low price because of idle capacity could threaten the main channels of distribution for the product. This is true even if the dealers apply their own private labels to the product. Why? Because selling the product to low-end dealers creates a direct competitor for its regular, medium-level dealers. Potential customers of the regular retail outlets could switch to the lower-end outlets because they can buy the same quality for a lower price. And what if the regular outlets deduce what has happened? What effect would this have on the company's medium-level differentiation strategy? The long-term damage to the company's profitability may be much greater than any short-run benefit from selling the special order.

Managing Customer Costs   A key objective for strategic costing is the identification of a firm's sources of profitability. In a unit-based costing system, selling and general and administrative costs are usually treated as period costs and, if assigned to customers, are typically assigned in proportion to the revenues generated. Thus, the message of unit-based costing is that servicing customers either costs nothing or they all appear to cost the same percentage of their sales revenue. If customer-servicing costs are significant, then failure to assign them at all or to assign them accurately will prevent sales representatives from managing the customer mix effectively. Why? Because sales representatives will not be able to distinguish between customers who place significant demands on servicing resources and those who place virtually no demand on these resources. This lack of knowledge can lead to actions that will weaken a firm's strategic position. To avoid this outcome and encourage actions that strengthen strategic position, customer-related costs should be assigned to customers using activity-based costing. Accurate assignment of customer-related costs allows the firm to classify customers as profitable or unprofitable. For example, using activity-based customer costing, a small Polish company found that only 400 out of almost 1,400 customers were profitable.[8] Some of the most regular customers were actually in the unprofitable category. Analysis revealed that the most profitable customers were those who placed large orders, paid on time, received moderate volume discounts, ordered standard products, and required standard delivery conditions. Analysis of customer profitability also revealed that the most significant problem causing unprofitability was small orders.

Once customers are identified as profitable or unprofitable, actions can be taken to strengthen the strategic position of the firm. For profitable customers, an organization can undertake efforts to increase satisfaction by offering higher levels of service, lower prices, new services, or some combination of the three. For unprofitable customers, an organization can attempt to deliver the customer services more efficiently (thus, decreasing service costs), increase prices to reflect the cost of the resources being consumed, encourage unprofitable customers to leave (by reducing selling efforts to this segment), or some combination of the three actions. Cornerstone 13-3 illustrates the power and utility of activity-based customer costing.

Cornerstone 13-3 reveals some interesting insights concerning the benefits of activity-based customer costing. First, some customers may benefit by price corrections. The large customer, for example, could be granted an immediate price decrease. This price decrease would also benefit Thompson, because the price correction is needed to maintain half of its current business. A company, however, such as Thompson may also face the difficult task of announcing a price increase for some of its customers (such is the prospect regarding the 10 smaller customers). However, activity-based customer analysis should go much deeper than accurate cost assignment and fair pricing. For Thompson, identifying the right cost driver (number of orders processed) revealed a linkage between the order-filling activity and customer behaviour. Smaller, frequent orders were imposing costs on Thompson, which were then passed on to all customers through the use of the sales volume allocation. Since the total cost is marked up 20 percent, the price charged was even higher. Furthermore, decreasing the number of orders can decrease order-filling costs. Knowing this, Thompson could offer price discounts for larger orders. For example, providing an incentive (quantity discounts) to increase the size of the orders of the small customers can create sufficient savings to make it unnecessary to increase the selling price to the smaller customers. But there are other possible linkages as well. Larger and less frequent orders will also decrease the demand on other internal activities, such as setting up equipment and materials handling. Reduction in other activity demands could produce further cost reductions and additional price cuts, making companies like Thompson more competitive. Ultimately, exploiting customer linkages can make both the seller and the buyer better off.

Life-Cycle Cost Management

OBJECTIVE ▶ 3
Tell what life-cycle cost management is and how it can be used to maximize profits over a product's life cycle.

Strategic cost management emphasizes the importance of an external focus and the need to recognize and exploit both internal and external linkages. Life-cycle cost management is a related approach that builds a conceptual framework which facilitates management's ability to exploit internal and external linkages. To understand what is meant by life-cycle cost management, we first need to understand basic product life-cycle concepts.

Product Life-Cycle Viewpoints

Product life cycle is simply the time a product exists—from conception to abandonment. Usually, product life cycle refers to a product class as a whole—such as automobiles—but it can also refer to specific forms (such as station wagons) and to

[8] Dorota Kuchta and Michal Troska, "Activity-Based Costing and Customer Profitability," *Cost Management* (May/June 2007): 18–25.

**CORNERSTONE
13-3**

The HOW and WHY of Activity-Based Customer Costing

Information:

Thompson Company produces precision parts for 11 major buyers. Of the 11 customers, one accounts for 50 percent of sales and the other 10 account for the remainder of sales, who purchase parts in roughly equal quantities. Orders are priced by adding manufacturing cost to ordering costs and then adding a 20 percent markup. Under this pricing structure, the large customer approaches Thompson and reveals a bid from a Thompson competitor that is $0.50 per part less than Thompson charges and threatens to take its business elsewhere without a price concession.

	One Large Customer	Ten Smaller Customers
Units purchased	500,000	500,000
Orders placed	2	200
Manufacturing cost	$3,000,000	$3,000,000
Order-filling cost allocated*	$303,000	$303,000
Order cost per unit	$0.606	$0.606

*Order-filling capacity is purchased in blocks (steps) of 45, each step costing $40,400; variable order-filling activity costs are $2,000 per order. The activity capacity is 225 orders; thus, the total order-filling cost is $606,000 [(5 × $40,400) + ($2,000 × 202)]. Current practice allocates ordering cost in proportion to the units purchased; therefore, the large customer receives half the total ordering cost.

Why:

Activity-based customer costing assigns the costs of customer-caused activities to individual customers or customer types. Customers can then be classified as profitable or unprofitable (or as causing or not causing inefficiencies), and actions can be taken to improve efficiency and profitability.

Required:

1. Calculate the unit price offered to Thompson's customers using the current order-filling cost allocation.
2. Assume that a newly implemented ABC system concludes that the number of orders placed is the best cost driver for the order-filling activity. Assign order-filling costs using this driver to each customer type and then calculate the new unit price for each customer type. Can Thompson beat the bid of its competitor?
3. **What if** Thompson offers a discount for orders of 10,000 units or more to the smaller customers? Assume that all the small customers can and do take advantage of this offer at the minimum level possible. Can Thompson offer the original price of $7.93 (from Requirement 1) to the small customers and not decrease its profitability?

Solution:

1. Unit price for each customer type = [($3,000,000 + $303,000) × 1.20]/500,000 = $7.93 per unit (rounded to the nearest cent)
2. Order-filling rate = $606,000/202 = $3,000 per order. Large customer ordering cost = $3,000 × 2 = $6,000; Small customer ordering cost = $3,000 × 200 = $600,000. Large customer unit price = [($3,000,000 + $6,000) × 1.20]/500,000 = $7.21 (rounded to the nearest cent); Small customer unit price = [($3,000,000 + $600,000) × 1.20]/500,000 = $8.64. The new large customer price is $0.72 ($7.93 − $7.21) less and easily beats the competitor's price.
3. The number of orders for the 10 smaller customers would decrease to 50 (500,000/10,000). This means that the total order-filling cost would decrease to $184,800 [(2 × $40,400) + ($2,000 × 52)]. Thus, the new order-filling rate would be $184,800/52 = $3,554 (rounded to the nearest dollar); therefore, the new small customer ordering cost = $3,554 × 50 = $177,700. Finally, the new small customer unit price = [($3,000,000 + $177,700) × 1.20]/500,000 = $7.63 (rounded to the nearest dollar). This price would be less than the original price.

specific brands or models (such as a **Toyota** Camry). Also, by replacing "conception" with "purchase," we obtain a customer-oriented definition of product life cycle. The producer-oriented definition refers to the life of classes, forms, or brands, whereas the customer-oriented definition refers to the life of a specific unit of product. These producer and customer orientations can be refined by looking at the concepts of revenue-producing life and consumable life. **Revenue-producing life** is the time a product generates revenue for a company. A product begins its revenue-producing life with the sale of the first product. **Consumable life**, on the other hand, is the length of time that a product serves the needs of a customer. Revenue-producing life is clearly of most interest to the producer, while consumable life is of most interest to the customer. Consumable life, however, is also of interest to the producer because it can be used as a competitive tool.

Marketing Viewpoint   The producer of goods or services has two viewpoints concerning product life cycle: the marketing viewpoint and the production viewpoint. The marketing viewpoint describes the general sales pattern of a product as it passes through distinct life-cycle stages. Exhibit 13-5 illustrates the general pattern of the marketing view of product life cycle. The distinct stages identified by the exhibit are introduction, growth, maturity, and decline. The **introduction stage** is characterized by preproduction and startup activities, where the focus is on obtaining a foothold in the market. As the graph indicates, there are no sales for a period of time (the preproduction period) and then slow sales growth as the product is introduced. The **growth stage** is a period of time when sales increase more quickly. The **maturity stage** is a period of time when sales increase more slowly. Eventually, the slope (of the sales curve) in the maturity stage becomes neutral and then turns negative. This **decline stage** is when the product loses market acceptance and sales begin to decrease.

Production Viewpoint   The production viewpoint of the product life cycle defines stages of the life cycle by changes in the type of activities performed: research and development activities, production activities, and logistical activities. The production viewpoint emphasizes life-cycle costs, whereas the market viewpoint emphasizes sales revenue behaviour. **Life-cycle costs** are all costs associated with the product for its

---

Exhibit 13-5

General Pattern of Product Life Cycle: Marketing Viewpoint

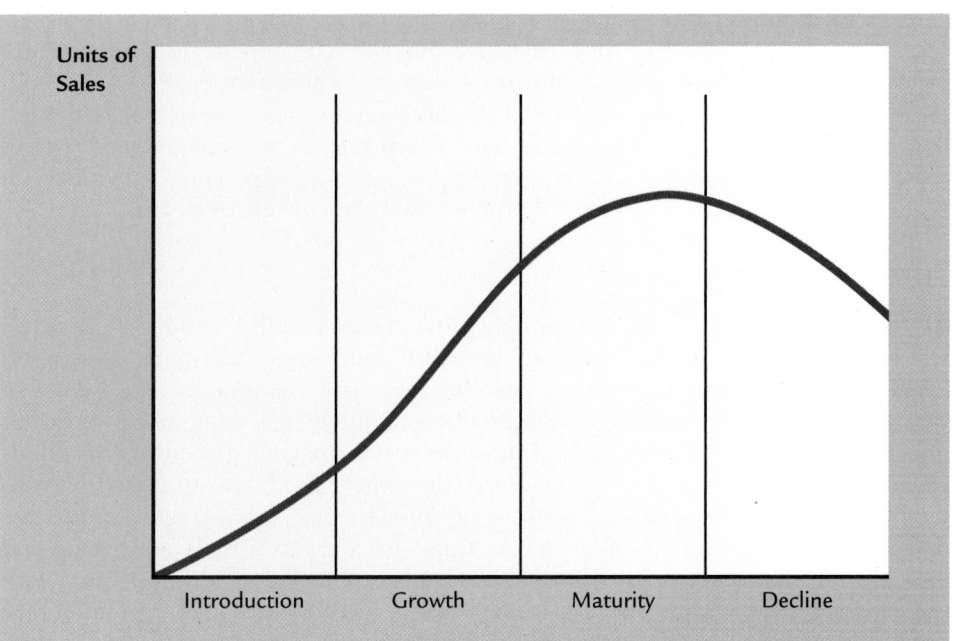

Exhibit 13-6

Product Life Cycle: Production Viewpoint

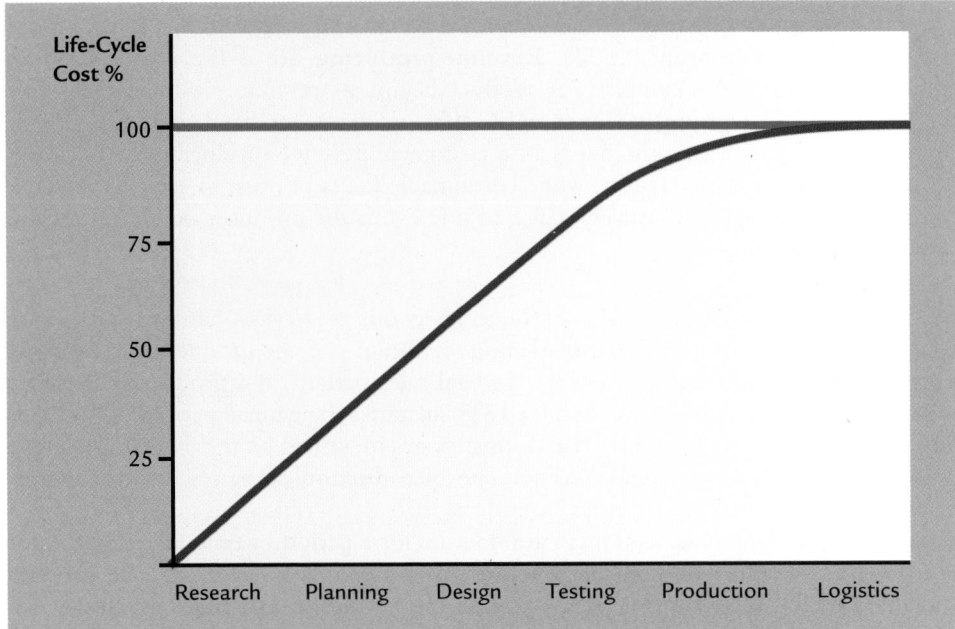

entire life cycle. These costs include research (product conception), development (planning, design, and testing), production (conversion activities), and logistics support (advertising, distribution, warranty, customer service, product servicing, and so on). The product life cycle and the associated cost commitment curve are illustrated in Exhibit 13-6. Notice that 90 percent or more of the costs associated with a product are *committed* during the development stage of the product's life cycle. Committed means that most of the costs that will be incurred are predetermined—set by the nature of the product design and the processes needed to produce the design.

Consumable Life-Cycle Viewpoint　Like the production life cycle, the consumption life cycle's stages are related to activities. These activities define four stages: purchasing, operating, maintaining, and disposal. The consumable life-cycle viewpoint emphasizes product performance for a given price. Price refers to the costs of ownership, which include the following elements: purchase cost, operating costs, maintenance costs, and disposal costs. Thus, total customer satisfaction is affected by both the purchase price and post-purchase costs. Because customer satisfaction is affected by post-purchase costs, producers also have a vital interest in managing the level of these costs. How producers can exploit the linkage of post-purchase activities with producer activities is a key element of product life-cycle cost management.

Interactive Viewpoint

All three life-cycle viewpoints offer insights that can be useful to producers of goods and services. In fact, producers cannot afford to ignore any of the three. A comprehensive life-cycle cost management program must pay attention to the variety of viewpoints that exist. This observation produces an integrated, comprehensive definition of life-cycle cost management. **Life-cycle cost management** consists of actions taken that cause a product to be designed, developed, produced, marketed, distributed, operated, maintained, serviced, and disposed of so that life-cycle profits are maximized. Maximizing life-cycle profits means that producers must understand and capitalize on the relationships that exist among the three life-cycle viewpoints. Once these relationships are understood, actions can be implemented that take advantage of revenue enhancement and cost reduction opportunities.

Relationships among Life-Cycle Viewpoints The marketing viewpoint is concerned with the nature of the sales pattern over the life cycle of the product; it is a *revenue-oriented viewpoint*. The production viewpoint, however, emphasizes the internal activities needed to develop, produce, market, and service products. The production stages exist to support the sales objectives of the marketing stages. This sales support requires resource expenditure; thus, the production life cycle can be described as a *cost-oriented viewpoint*. The consumption life cycle is concerned with product performance and price (including post-purchase costs). The ability to generate revenues and the level of resource expenditure are both related to product performance and price. The producer must be concerned with what the customer receives and what the customer gives up. Thus, the consumption life cycle can be described as a *customer-value oriented viewpoint*. Exhibit 13-7 illustrates the relationships among the stages of the three viewpoints. The stages of the marketing viewpoint are listed as columns; production and consumable life-cycle viewpoints appear as rows. These last two viewpoints are identified by the nature of their attributes: expenses for the production life cycle and customer value for the consumable life cycle. Competition and customer type are included under customer value because they affect the producer's approach to providing customer value.

The relationships described in Exhibit 13-7 are typical but can vary depending on the nature of the product and the industry in which a producer operates. Some explanation of the relationships should reveal the potential for producers to exploit them. Relationships can be viewed vertically or horizontally. Consider, for example, the introduction stage, and examine the vertical relationships. In this stage, we would expect losses or negligible profits because of high levels of expenditure in research and

Exhibit 13-7

Typical Relationships of Product Life-Cycle Viewpoints

Marketing Product Life Cycle:				
Attributes	**Introduction**	**Growth**	**Maturity**	**Decline**
Sales	Low	Rapid growth	Slow growth, peak sales	Declining

Production Life Cycle:				
Attributes	**Introduction**	**Growth**	**Maturity**	**Decline**
Expenses:				
Product research	High	Moderate	Moderate	Low
Product development	Moderate	High	Moderate	Low
Plant and equipment	Low to moderate	High	Moderate	Low
Advertising	Moderate to high	High	Moderate	Low
Service	Low	Moderate	High	Low

Consumable Life Cycle:				
Attributes	**Introduction**	**Growth**	**Maturity**	**Decline**
Customer value:				
Customer type	Innovators	Mass market	Mass market, differentiated	Laggards
Performance sensitivity	High	High	High	Moderate
Price sensitivity	Low	Moderate	High	Moderate
Competition	None	Growing	High	Low

Attributes	**Introduction**	**Growth**	**Maturity**	**Decline**
Profits	Negligible to loss	Peak levels	Moderate to high	Low

development and marketing. Customers at this stage are described as innovators. These are simply the first customers to buy the product. Innovators are venturesome, willing to try something new. They are usually more concerned with the performance of the new product than with its price. This fact, coupled with the lack of competitors, may allow a high price to be charged for the new product. If the barriers to entry in the marketplace are high, then a high price may continue to be charged for some time. However, if competition grows as indicated by the horizontal dimension of the table, and if price sensitivity increases, then the producer will need to rely on further research and development and differentiation to maintain a competitive advantage.

Revenue Enhancement   Revenue-generating approaches depend on marketing life-cycle stages and on customer value effect. Pricing strategy, for example, varies with stages. In the introductory stage, as mentioned earlier, higher prices can be charged because customers are less price sensitive and more interested in performance.

In the maturity stage, customers are highly sensitive to both price and performance. This suggests that adding features, increasing durability, improving maintainability, and offering customized products may all be good strategies to follow. In this stage, differentiation is important. For revenue enhancement to be viable, however, the customer must be willing to pay a premium for any improvement in product performance. Furthermore, this premium must exceed the cost the producer incurs in providing the new product attribute. In the decline stage, revenues may be enhanced by finding new uses and new customers for the product. A good example is the use of **Arm & Hammer's**® baking soda to absorb refrigerator odors in addition to its normal role in baking goods.[9]

Cost Reduction   Cost reduction, not cost control, is the emphasis of life-cycle cost management. Cost reduction strategies should explicitly recognize that actions taken in the early stages of the production life cycle can lower costs for later production and consumption stages. Since 90 percent or more of a product's life-cycle costs are determined during the development stage, it makes sense to emphasize management of activities during this phase of a product's existence. Studies have shown that every dollar spent on preproduction activities saves $8–$10 on production and postproduction activities, including customer maintenance, repair, and disposal costs.[10] Apparently, many opportunities for cost reduction occur before production begins. Managers need to invest more in preproduction assets and dedicate more resources to activities in the early phases of the product life cycle to reduce production, marketing, and post-purchase costs.

Product design and process design afford multiple opportunities for cost reduction by designing to reduce: (1) manufacturing costs, (2) logistical support costs, and (3) post-purchase costs, which include customer time involved in maintenance, repair, and disposal. For these approaches to be successful, managers of producing companies must have a good understanding of activities and cost drivers and know how the activities interact. Manufacturing, logistical, and post-purchase activities are not independent. Some designs may reduce post-purchase costs and increase manufacturing costs. Others may simultaneously reduce production, logistical, and post-purchase costs.

A unit-based costing system usually will not supply the information needed to support life-cycle cost management. Unit-based costing systems emphasize the use of unit-based cost drivers to describe cost behaviour, focus on production activities, ignore logistical and post-purchase activities, and expense research and development and other nonmanufacturing costs as they are incurred. Unit-based costing systems rarely, if ever, collect a complete history of a product's costs over its life cycle. An activity-based costing system, however, produces information about activities, including both preproduction and postproduction activities, and cost drivers. Activity-based costing information is critical for life-cycle cost reduction decisions as is shown by Cornerstone 13-4.

---

[9] Sak Onkvisit and John J. Shaw, "Competition and Product Management: Can the Product Life Cycle Help?" *Business Horizons* (July–August 1986): 51–52.

[10] Mark D. Shields and S. Mark Young, "Managing Product Life Cycle Costs: An Organizational Model" and R. L. Engwall, "Cost Management for Defense Contractors," *Cost Accounting for the 90's, Responding to Technological Change* (Montvale, NJ: National Association of Accountants, 1988).

The HOW and WHY of Activity-Based Life-Cycle Cost Reduction

**CORNERSTONE
13-4**

Information:

Design engineers are considering two new product designs that reduce direct materials and direct labour content. Data for both unit-based and ABC systems are provided below.

Unit-based system:

Variable conversion activity rate: $40 per direct labour hour

Material usage rate: $8 per part

ABC system:

Labour usage: $10 per direct labour hour

Material usage (direct materials): $8 per part

Machining: $28 per machine hour

Purchasing activity: $60 per purchase order

Setup activity: $1,000 per setup hour

Warranty activity: $200 per returned unit (usually requires extensive rework)

Customer repair cost: $10 per repair hour

Activity and Resource Information (annual estimates)

	Design A	Design B
Units produced	10,000	10,000
Direct material usage	100,000 parts	60,000 parts
Labour usage	50,000 hours	80,000 hours
Machine hours	25,000	20,000
Purchase orders	300	200
Setup hours	200	100
Returned units	400	75
Repair time (customer)	800 hours	150 hours

Why:

ABC produces better and more detailed information for cost reduction decisions concerning process and product designs by recognizing that manufacturing, logistical, and post-purchase activities are not independent.

Required:

1. Select the lower-cost design using unit-based costing. Are logistical and post-purchase activities considered in this analysis?
2. Select the lower-cost design using ABC analysis. Explain why the analysis differs from the unit-based analysis.
3. ***What if*** the customer repair cost were $10 *per unit* for Design A and $50 per unit for Design B? Assume that every unit must face repair by the consumer during the consumable life cycle. Now which is the better design?

CORNERSTONE 13-4 (continued)

Solution:

1.

	Design A	Design B
Direct materials[a]	$ 800,000	$ 480,000
Conversion cost[b]	2,000,000	3,200,000
Total manufacturing costs	$2,800,000	$3,680,000
Units produced	÷ 10,000	÷ 10,000
Unit cost	$ 280	$ 368

[a]$8 × 100,000; $8 × 60,000
[b]$40 × 50,000; $40 × 80,000

Logistical and post-purchase costs are not considered.

2.

	Design A	Design B
Direct materials	$ 800,000	$ 480,000
Direct labour[a]	500,000	800,000
Machining[a]	700,000	560,000
Purchasing[b]	18,000	12,000
Setups[b]	200,000	100,000
Warranty[b]	80,000	15,000
Total product costs	$2,298,000	$1,967,000
Units produced	÷ 10,000	÷ 10,000
Unit cost	$ 230*	$ 197*
Post-purchase costs[c]	$ 8,000	$ 1,500

[a] $10 × 50,000; $10 × 80,000; $28 × 25,000; $28 × 20,000
[b] $60 × 300; $60 × 200; $1,000 × 200; $1,000 × 100; $200 × 400; $200 × 75
[c] $10 × 800; $10 × 150
*Rounded to the nearest dollar.

ABC assigns manufacturing costs using both unit and non-unit drivers. It also considers the effects of manufacturing, logistical, and post-purchase activities (unit-based uses only manufacturing activities).

3. The post-purchase costs for Design A would be $240 ($230 + $10) and for Design B would be $247 ($197 + $50). Design A is the cheaper of the two designs when post-purchase costs are considered.

Role of Target Costing

Life-cycle cost management emphasizes cost reduction, not cost control. Target costing becomes a particularly useful tool for establishing cost reduction goals during the design stage. A **target cost** is the difference between the sales price needed to capture a predetermined market share and the desired per-unit profit. The sales price reflects the product specifications or functions valued by the customer (referred to as *product functionality*). If the target cost is less than what is currently achievable, then management must find cost reductions that move the actual cost toward the target cost. Finding those cost reductions is the principal challenge of target costing.

Three cost reduction methods are typically used: (1) reverse engineering, (2) value analysis, and (3) process improvement. In reverse engineering, the competitors' products are closely analyzed (a "tear down" analysis) in an attempt to discover more design features that create cost reductions. Value analysis attempts to assess the value placed on various product functions by customers. If the price customers are willing to pay for a particular function is less than its cost, the function is a candidate for elimination. Another possibility is to find ways to reduce the cost of providing the function (e.g., using common components). Both reverse engineering and value analysis focus on product design to achieve cost reductions. The processes used to produce and market the product are also sources of potential cost reductions. Thus, redesigning processes to improve their efficiency can also contribute to achieving the needed cost reductions. The target-costing model is summarized in Exhibit 13-8.

A simple example can be used to illustrate the concepts described by Exhibit 13-8. Assume that a company is considering the production of a new trencher. Current product specifications and the targeted market share call for a sales price of $250,000. The required profit is $50,000 per unit. The target cost is computed as follows:

$$\text{Target cost} = \$250,000 - \$50,000$$
$$= \$200,000$$

It is estimated that the current product and process designs will produce a cost of $225,000 per unit. Thus, the cost reduction needed to achieve the target cost and desired profit is $25,000 ($225,000 − $200,000). A tear-down analysis of a competitor's trencher revealed a design improvement that promised to save $5,000 per unit. When compared with the $25,000 reduction needed, additional effort was still

---

	Exhibit 13-8

Target-Costing Model

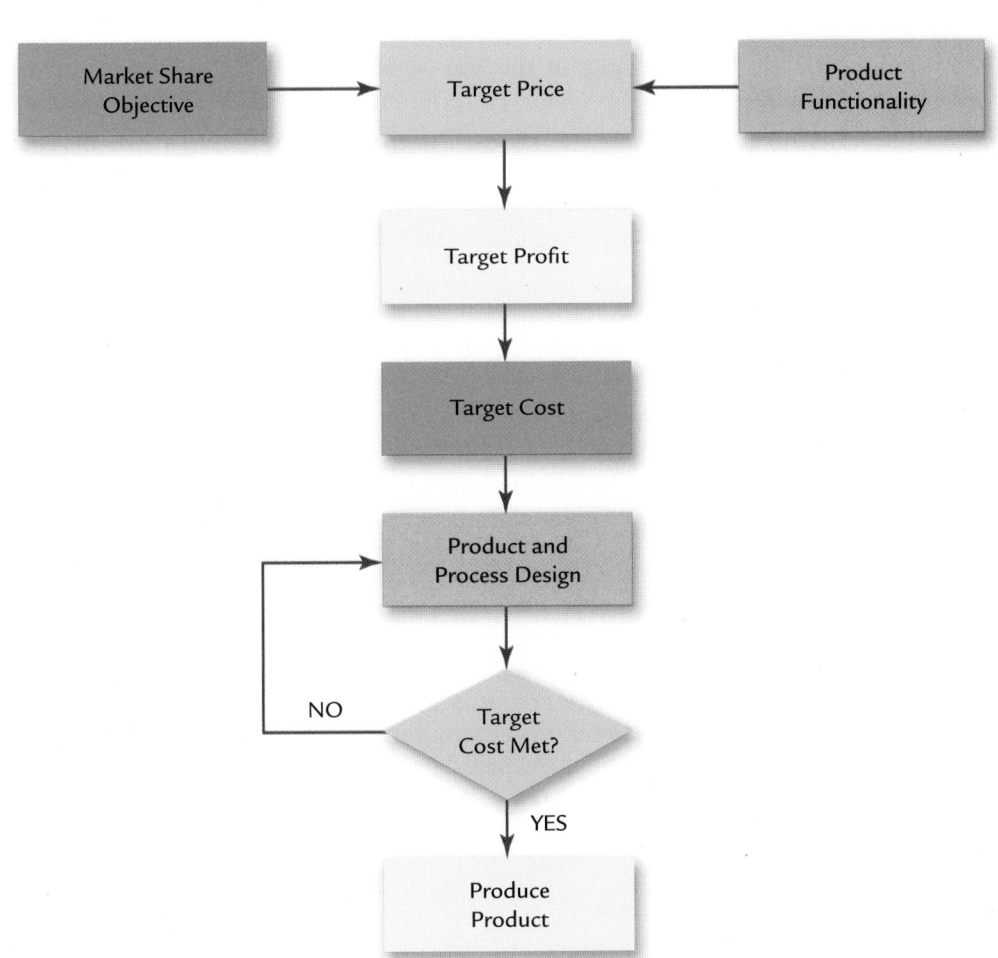

necessary. A marketing study of customer reactions to product functions revealed that the extra trenching speed in the new design was relatively unimportant. Changing the design to reflect a lower trenching speed saved $10,000. The company's supplier also proposed the use of a standardized component, reducing costs by another $5,000. Finally, the design team was able to change the process design and reduce the test time by 50 percent. This saved $6,000 per unit. The last change reached the threshold value, and production for the new model was approved.

Target costs are a type of currently attainable standard. But they are conceptually different from traditional standards. What sets them apart is the motivating force. Traditional standards are internally motivated and set, based on concepts of efficiency developed by industrial engineers and production managers. Target costs, on the other hand, are externally driven, generated by an analysis of markets and competitors.

Supplier and Firm Interaction   The example just given indicated that one source of cost reduction came from a supplier suggestion. During the design stage, target costing requires a close interaction between the firm and its suppliers. This interaction should produce lower cost solutions than would be possible if the design teams acted in isolation.[11] Joint design efforts require cooperative relationships. Incentives for such relationships come from a willingness to search for mutually beneficial solutions.

Short Life Cycles   Although life-cycle cost management is important for all manufacturing firms, it is particularly important for firms that have products with short life cycles. Products must recover all life-cycle costs and provide an acceptable profit. If a firm's products have long life cycles, profit performance can be increased by such actions as redesigning, changing prices, reducing costs, and altering the product mix. In contrast, firms that have products with short life cycles usually do not have time to react in this way so their approach must be proactive. Thus, for short life cycles, good life-cycle planning is critical, and prices must be set properly to recover all the life-cycle costs and provide a good return. Activity-based costing can be used to encourage good life-cycle planning. By careful selection of cost drivers, design engineers can be motivated to choose cost-minimizing designs.

OBJECTIVE ▶ 4
Identify the basic features of JIT purchasing and manufacturing.

Just-in-Time (JIT) Manufacturing and Purchasing

JIT manufacturing and purchasing systems offer a prominent example of how managers can use the strategic concepts discussed earlier in the chapter to bring about significant changes within an organization. Firms that implement JIT are pursuing a cost reduction strategy by redefining the structural and procedural activities performed within an organization. Cost reduction is supportive of either a cost leadership or differentiation strategy. Cost reduction is directly related to cost leadership. Successful differentiation depends on offering greater value; yet this value added must be more than the cost of providing it. JIT can help add value by reducing waste. Successful implementation of JIT has brought about significant improvements, such as better quality, increased productivity, reduced lead times, major reductions in inventories, reduced setup times, lower manufacturing costs, and increased production rates. JIT techniques have been implemented by the following companies with meaningful results:

---

[11] Robin Cooper and Regine Slagmulder, "Cost Management Beyond the Boundaries of the Firm," *Management Accounting* (March 1998): 18–20.

AT&T	General Electric	Motorola
Black & Decker	Harley-Davidson	Toys "R" Us
Canada Safeway	Hewlett-Packard	Wal-Mart
Chrysler	Intel	Westinghouse
Ford	John Deere	Xerox

Adopting a JIT manufacturing system has a significant effect on the nature of the cost management accounting system. Installing a JIT system affects the traceability of costs, enhances product costing accuracy, diminishes the need for allocation of service-centre costs, changes the behaviour and relative importance of direct labour costs, impacts job-order and process-costing systems, decreases the reliance on standards and variance analysis, and decreases the importance of inventory tracking systems. To understand and appreciate these effects, we need a fundamental understanding of what JIT manufacturing is and how it differs from traditional manufacturing.

JIT manufacturing is a demand-pull system. The objective of **JIT manufacturing** is to eliminate waste by producing a product only when it is needed and only in the quantities demanded by customers. Demand pulls products through the manufacturing process. Each operation produces only what is necessary to satisfy the demand of the succeeding operation. No production takes place until a signal from a succeeding process indicates a need to produce. Parts and materials arrive just in time to be used in production. JIT assumes that all costs other than direct materials are driven by time and space drivers. JIT then focuses on eliminating waste by compressing time and space.

Inventory Effects

Usually, the push-through system produces significantly higher levels of finished goods inventory than does a JIT system. JIT manufacturing relies on the exploitation of a customer linkage. Specifically, production is tied to customer demand. This linkage extends back through the value chain and also affects how a manufacturer deals with suppliers. **JIT purchasing** requires suppliers to deliver parts and materials just in time to be used in production. Thus, supplier linkages are also vital. Supply of parts must be linked to production, which is linked to demand. One effect of successful exploitation of these linkages is to reduce all inventories to much lower levels. Since 1980, inventories have dropped dramatically in both Canada and the United States relative to gross domestic product.

Traditionally, inventories of raw materials and parts are carried so that a firm can take advantage of quantity discounts and hedge against future price increases of the items purchased. The objective is to lower the cost of inventory. JIT achieves the same objective without carrying inventories. The JIT solution is to exploit supplier linkages by negotiating long-term contracts with a few chosen suppliers located as close to the production facility as possible and by establishing more extensive supplier involvement. Suppliers are not selected on the basis of price alone.

Performance—the quality of the component and the ability to deliver as needed—and commitment to JIT purchasing are vital considerations. Every effort is made to establish a partners-in-profits relationship with suppliers. Suppliers need to be convinced that their well-being is intimately tied to the well-being of the buyer.

To help reduce the uncertainty in demand for the supplier and establish the mutual confidence and trust needed in such a relationship, JIT manufacturers emphasize long-term contracts. Other benefits of long-term contracts exist. They stipulate prices and acceptable quality levels. Long-term contracts also reduce dramatically the number of orders placed, which helps to drive down the ordering and receiving costs. Another effect of long-term contracting is a reduction in the cost of parts and materials—usually in the range of 5 percent to 20 percent less than what was paid in a traditional setting.

The need to develop close supplier relationships often drives the supplier base down dramatically.

For example, **Mercedes-Benz U.S. International**'s factory in Vance, Alabama, saved time and money by streamlining its supplier list from 1,000 to 100 primary suppliers. In exchange for annual 5 percent price cuts, the chosen suppliers have multiyear contracts (as opposed to the yearly bidding process practised at other Mercedes' plants) and can adapt off-the-shelf parts to Mercedes' needs. The end result is lower costs for both Mercedes and its suppliers.[12]

Suppliers also benefit. The long-term contract ensures a reasonably stable demand for their products. A smaller supplier base typically means increased sales for the selected suppliers. Thus, both buyers and suppliers benefit—a common outcome when external linkages are recognized and exploited.

By reducing the number of suppliers and working closely with those that remain, the quality of the incoming materials can be improved significantly—a crucial outcome for the success of JIT. As the quality of incoming materials increases, some quality-related costs can be avoided or reduced. For example, the need to inspect incoming materials disappears, and rework requirements decline.

Plant Layout

The type and efficiency of plant layout is another executional cost driver that is managed differently under JIT manufacturing. In traditional job and batch manufacturing, products are moved from one group of identical machines to another. Typically, machines with identical functions are located together in an area referred to as a *department* or *process*. Workers who specialize in the operation of a specific machine are located in each department. Thus, the executional cost driver for a traditional setting is departmental structure. JIT replaces this traditional plant layout with a pattern of manufacturing cells. The executional cost driver for a JIT setting is cell structure. Cell structure is chosen over departmental structure because it increases the ability of the organization to "execute" successfully. The cellular manufacturing design can also affect structural activities, such as plant size and number of plants, because it typically requires less space. Space savings can reduce the demand to build new plants and will affect the size of new plants when they are needed.

Manufacturing cells contain machines that are grouped into families, usually in a semicircle. The machines are arranged so that they can be used to perform a variety of operations in sequence. Each cell is set up to produce a particular product or product family. Products move from one machine to another from start to finish. Workers are assigned to cells and are trained to operate all machines within the cell. In other words, labour in a JIT environment is multiskilled, not specialized. Each manufacturing cell is essentially a minifactory; in fact, a cell is often referred to as a *factory within a factory*.

Grouping of Employees

Another major structural difference between JIT and traditional organizations relates to how employees are grouped. As just indicated, each cell is viewed as a mini-factory; thus, each cell requires easy and quick access to support services, which means that centralized service departments must be scaled down and their personnel reassigned to work directly with manufacturing cells. For example, with respect to raw materials, JIT calls for multiple stock points, each one located near where the material will be used. There is no need for a central store location—in fact, such an arrangement actually hinders efficient production. A purchasing agent can be assigned to each cell to handle material requirements. Similarly, other service personnel, such as manufacturing and quality engineers, can be assigned to cells.

Other support services may be relocated to the cell by training cell workers to perform the services. For example, in addition to direct production work, cell workers may perform setup duties, move partially completed goods from station to station

---

[12] David Woodruff and Karen Lowry Miller, "Mercedes' Maverick in Alabama," *BusinessWeek* (September 11, 1995): 64–65.

within the cell, perform preventive maintenance and minor repairs, conduct quality inspections, and perform janitorial tasks. This multiple task capability is directly related to the pull-through production approach. Producing on demand means that production workers (formerly direct labourers) may often have "free" time. This non-production time can be used to perform some of the other support activities.

Employee Empowerment

A major procedural difference between traditional and JIT environments is the degree of participation allowed workers in the management of the organization. According to the JIT view, increasing the degree of participation (the executional cost driver) increases productivity and overall cost efficiency. Workers are allowed a say in how the plant operates. For example, workers are allowed to shut down production to identify and correct problems. Managers seek workers' input and use their suggestions to improve production processes. Workers are often involved in interviewing and hiring other employees, sometimes even prospective bosses. The reason? If the "chemistry is right," then the workforce will be more efficient, and they will work together better.

Employee empowerment, a procedural activity, also affects other structural and procedural activities. The management structure must change in response to greater employee involvement. Because workers assume greater responsibilities, fewer managers are needed, and the organizational structure becomes flatter. Flatter structures speed up and increase the quality of information exchange. The style of management needed in the JIT firm also changes. Managers in the JIT environment need to act as facilitators more than as supervisors. Their role is to develop people and their skills so that they can make value-adding contributions.

Total Quality Management

JIT necessarily carries with it a much stronger emphasis on managing quality. A defective part brings production to a grinding halt. Poor quality simply cannot be tolerated in a manufacturing environment that operates without inventories. Simply put, JIT cannot be implemented without a commitment to total quality management (TQM). TQM is essentially a never-ending quest for perfect quality: the striving for a defect-free product design and manufacturing process. This approach to managing quality is diametrically opposed to the traditional doctrine, called **acceptable quality level (AQL)**. AQL permits or allows defects to occur provided they do not exceed a predetermined level.

The major differences between JIT manufacturing and traditional manufacturing are summarized in Exhibit 13-9. These differences will be referred to and discussed in greater detail as the implications of JIT manufacturing for cost management are examined.

Exhibit 13-9

Comparison of JIT Approaches with Traditional Manufacturing and Purchasing

JIT	Traditional
1. Pull-through system	1. Push-through system
2. Insignificant inventories	2. Significant inventories
3. Small supplier base	3. Large supplier base
4. Long-term supplier contracts	4. Short-term supplier contracts
5. Cellular structure	5. Departmental structure
6. Multiskilled labour	6. Specialized labour
7. Decentralized services	7. Centralized services
8. High employee involvement	8. Low employee involvement
9. Facilitating management style	9. Supervisory management style
10. Total quality management	10. Acceptable quality level
11. Buyers' market	11. Sellers' market
12. Value-chain focus	12. Value-added focus

OBJECTIVE ➤ 5

Describe the effect JIT has on cost traceability and product costing.

JIT and Its Effect on the Cost Management System

The numerous changes in structural and procedural activities that we have described for a JIT system also change traditional cost management practices. Both the cost accounting and operational control systems are affected. In general, the organizational changes simplify the cost management accounting system and simultaneously increase the accuracy of the cost information being produced.

Traceability of Overhead Costs

Costing systems use three methods to assign costs to individual products: direct tracing, driver tracing, and allocation. Of the three methods, the most accurate is direct tracing; for this reason, it is preferred over the other two methods. In a JIT environment, many overhead costs assigned to products using either driver tracing or allocation are now directly traceable to products. Cellular manufacturing, multiskilled labour, and decentralized service activities are the major features of JIT responsible for this change in traceability.

In a departmental structure, many different products may be subjected to a process located in a single department (e.g., Grinding). After completion of the process, the products are then transferred to other processes located in different departments (e.g., Assembly, Painting, and so on). Although a different set of processes is usually required for each product, most processes are applicable to more than one product. For example, 30 different products may need grinding. Because more than one product is processed in a department, the costs of that department are common to all products passing through it, and therefore the costs must be assigned to products using activity drivers or allocation. In a manufacturing-cell structure, however, all processes necessary for the production of each product or major subassembly are collected in one area called a cell. Thus, the costs of operating that cell can be assigned to the cell's product or subassembly using direct tracing. (However, if a family of products uses a cell, then we must resort to drivers and allocation to assign costs.)

Equipment formerly located in other departments, for example, is now reassigned to cells, where it may be dedicated to the production of a single product or subassembly. In this case, depreciation is now a directly attributable product cost. Multiskilled workers and decentralized services add to the effect. Workers in the cell are trained to set up the equipment in the cell, maintain it, and operate it. Additionally, cell workers may also be used to move a partially finished part from one machine to the next or to perform maintenance, setups, and materials handling. These support functions were previously done by a different set of labourers for all product lines. Additionally, people with specialized skills (e.g., industrial engineers and production schedulers) are assigned directly to manufacturing cells. Because of multitask assignments and redeployment of other support personnel, many support costs can now be assigned to a product using direct tracing. Exhibit 13-10 compares the traceability of some selected costs in a traditional manufacturing environment with their traceability in the JIT environment (assuming single-product cells). Comparisons are based on the three cost assignment methods.

Product Costing

One consequence of increasing directly attributable costs is to increase the accuracy of product costing. Directly traceable costs are associated (usually by physical observation) with the product and can safely be said to belong to it. Other costs, however, are common to several products and must be assigned to these products using activity drivers and allocation. Because of cost and convenience, activity drivers that are less than perfectly correlated with the consumption of overhead activities may be chosen. JIT manufacturing reduces the need for this difficult assessment by converting many common costs to directly attributable costs. Note, however, that the driving force behind these changes is not the cost management system itself but the changes in the structural and procedural activities brought about by implementing a JIT system.

Exhibit 13-10

Product Cost Assignment: Traditional versus JIT Manufacturing

Manufacturing Cost	Traditional Environment	JIT Environment
Direct labour	Direct tracing	Direct tracing
Direct materials	Direct tracing	Direct tracing
Materials handling	Driver tracing	Direct tracing
Repairs and maintenance	Driver tracing	Direct tracing
Energy	Driver tracing	Direct tracing
Operating supplies	Driver tracing	Direct tracing
Supervision (department)	Allocation	Direct tracing
Insurance and taxes	Allocation	Allocation
Plant depreciation	Allocation	Allocation
Equipment depreciation	Driver tracing	Direct tracing
Custodial services	Allocation	Direct tracing
Cafeteria services	Driver tracing	Driver tracing

While activity-based costing offers significant improvement in product costing accuracy, focusing on structural and procedural activities offers even more potential improvement.

Exhibit 13-10 illustrates that JIT does not convert all costs into directly traceable costs. Even with JIT in place, some overhead activities remain common to the manufacturing cells. These remaining support activities are mostly facility-level activities. In a JIT system, the batch size is one unit of product. Thus, all batch-level activities convert into unit-level activities. Additionally, many of the batch-level activities are reduced or eliminated. For example, materials handling may be significantly reduced because of the reorganization from a departmental structure to a cellular structure. Similarly, for single-product cells, there is no setup activity. Even for cells that produce a family of products, setup times would be minimal. Furthermore, it is likely that the need to use activity drivers for the cost of product-level activities is significantly diminished because of decentralizing these support activities to the cell level. Is there, then, a role for ABC in a JIT firm?

Although JIT diminishes the value of ABC for tracing manufacturing costs to individual products, an activity-based costing system has much broader application than just tracing manufacturing costs to products. For many strategic and tactical decisions, the product cost definition needs to include nonmanufacturing costs. For example, value-line and operational product costing is an invaluable tool for strategic costing analysis and for life-cycle cost management. Also, including post-purchase costs as part of the product cost definition provides valuable insights. Thus, knowing and understanding general and administrative, research, development, marketing, customer service, and post-purchase activities and their cost drivers is essential for sound cost analysis. Furthermore, as we have already seen, using ABC to assign costs accurately to suppliers and customers is an essential part of strategic cost management.

JIT's Effect on Job-Order and Process-Costing Systems

In implementing JIT in a job-order setting, the firm should first separate its repetitive business from its unique orders. Manufacturing cells can then be established to deal with the repetitive business. For those products where demand is insufficient to justify their own manufacturing cells, groups of dissimilar machines can be set up in a cell to make families of products or parts that require the same manufacturing sequence.

With this reorganization of the manufacturing layout, job orders are no longer needed to accumulate product costs. Instead, costs can be accumulated at the cellular level. Additionally, because lot sizes are now too small (as a result of reducing work-in-process and finished goods inventories), it is impractical to have job orders for each job. Add to this the short lead time of products because of the time and space compression features of JIT (virtually no setup time and cellular structures), and it becomes difficult to track each piece moving through the cell. In effect, the job environment has taken on the nature of a process-costing system.

JIT simplifies process costing. A key feature of JIT is lower inventories. Assuming that JIT is successful in reducing work in process, the need to compute equivalent units vanishes. Calculating product costs follows the simple pattern of collecting costs for a cell for a period of time and dividing the costs by the units produced for that period.

Backflush Costing

The JIT system also offers the opportunity to simplify the accounting for manufacturing cost flows. Given low inventories, it may not be desirable to spend resources tracking the cost flows through all the inventory accounts. In a traditional system, there was a work-in-process account for each department so that manufacturing costs could be traced as work proceeded through the factory. Under JIT, there are no departments, a 14-day lead time (for example) has been decreased to four hours, and it would be absurd to trace costs from station to station within a cell. After all, if production cycle time is in minutes or hours, and goods are shipped immediately upon completion, then all of each day's manufacturing costs flow to Cost of Goods Sold. Recognizing this outcome leads to a simplified approach to accounting for manufacturing cost flows. This simplified approach, called **backflush costing**, uses trigger points to determine when manufacturing costs are assigned to key inventory and temporary accounts.

Varying the number and location of trigger points creates several types of backflush costing. Trigger points are simply events that prompt ("trigger") the accounting recognition of certain manufacturing costs. There are four variations, depending on the definition of the trigger points (which, in turn, depends on how fully the firm has implemented JIT):

1. The purchase of raw materials (trigger point 1) and the completion of goods (trigger point 2).
2. The purchase of raw materials (trigger point 1) and the sale of goods (trigger point 2).
3. The completion of goods (only trigger point).
4. The sale of goods (only trigger point).

Variations 1 and 2   For Variations 1 and 2, the first trigger point is the purchase of raw materials. When materials are purchased in a JIT system, they are immediately placed into process. Raw Materials and In Process Inventory (WIP) is debited, and Accounts Payable is credited. The inventory account is used only for tracking the cost of raw materials. There is no separate materials inventory account and no work-in-process inventory account. Combining direct labour and overhead into one category is a second feature of backflush costing. As firms implement JIT and become automated, the traditional direct labour cost category disappears. Multiskilled workers perform setup activities, machine-loading activities, maintenance, materials handling, and so on. As labour becomes multifunctional, the ability to track and report direct labour separately becomes impossible. Consequently, backflush costing usually combines direct labour costs with overhead costs in a temporary account called *Conversion Cost Control*. This account accumulates the *actual* conversion costs on the debit side and the applied conversion costs on the credit side. Any difference between the actual conversion costs and the applied conversion costs is closed to Cost of Goods Sold.

In the first variant of backflush costing, the completion of goods triggers the recognition of the manufacturing costs used to produce the goods (the second trigger point). At this point, conversion cost application is recognized by debiting Finished Goods Inventory and crediting Conversion Cost Control; the cost of direct materials is recognized by debiting Finished Goods Inventory and crediting the WIP inventory account. Therefore, the costs of manufacturing are "flushed" out of the system after the goods are completed.

In the second variant of backflush costing, the second trigger point is defined by the point when goods are sold rather than when they are completed. For this variant, the costs of manufacturing are flushed out of the system *after* the goods are sold. Thus, the application of conversion cost and the transfer of direct materials cost are accomplished by debiting Cost of Goods Sold and crediting Conversion Cost Control and WIP Inventory, respectively. Other entries are the same as Variation 1.

Variations 3 and 4   Under Variations 3 and 4, there is only one trigger point. Both variations recognize actual conversion costs by debiting Conversion Cost Control and crediting various accounts (such as Accumulated Depreciation). Neither variation makes any entry for the purchase of raw materials. For Variation 3, when the goods are completed, all costs, including direct materials cost, are flushed out of the system. This is done by debiting Finished Goods Inventory for the cost of all manufacturing inputs and crediting Accounts Payable for the cost of direct materials and Conversion Cost Control for the application of conversion costs. For Variation 4, the costs are flushed out of the system when the goods are sold. Thus, Cost of Goods Sold is debited, and Accounts Payable and Conversion Cost Control are credited. Of the four variations, only Variation 4 avoids all inventory accounts and, thus, would be the approach used for a pure JIT firm. Cornerstone 13-5 illustrates backflush costing.

The HOW and WHY of Backflush Costing

CORNERSTONE 13-5

Information:
A JIT company had the following transactions during June:

1. Purchased raw materials on account for $160,000.
2. Placed all materials received into production.
3. Incurred actual direct labour costs of $25,000.
4. Incurred actual overhead costs of $225,000.
5. Applied conversion costs of $235,000 ($25,000 of direct labour + $210,000 of applied overhead).
6. Completed all work for the month.
7. Sold all completed work.
8. Computed the difference between actual and applied costs.

Why:
Reduced cycle time and immediate shipping of goods simplifies accounting for manufacturing cost flows. How simplified depends on the completeness of the JIT system (measured by "trigger points").

Required:
1. Prepare the journal entries for traditional and backflush costing. For backflush costing, assume there are two trigger points: (1) the purchase of raw materials and (2) the completion of the goods.
2. Assume the second trigger point in Requirement 1 is the sale of goods. What would change for the backflush costing journal entries?
3. **What if** there is only one trigger point and it is (a) completion of the goods or (b) sale of goods? How would the backflush costing journal entries differ from Requirement 1 for (a) and (b)?

CORNERSTONE 13-5 (continued)

Solution:

1.

Transaction	Traditional Journal Entries		Backflush Journal Entries: Variation 1		
1. Purchase of raw materials	Materials Inventory 160,000		Raw Materials and In		
	Accounts Payable	160,000	Process Inventory 160,000		
			Accounts Payable		160,000
2. Materials issued to production	Work-in-Process		No entry		
	Inventory 160,000				
	Materials				
	Inventory	160,000			
3. Direct labour cost incurred	Work-in-Process		Combined with		
	Inventory 25,000		overhead: See		
	Wages Payable	25,000	next entry.		
4. Overhead cost incurred	Overhead Control 225,000		Conversion Cost		
	Accounts Payable	225,000	Control 250,000		
			Wages Payable		25,000
			Accounts Payable		225,000
5. Application of overhead	Work-in-Process		No entry		
	Inventory 210,000				
	Overhead Control	210,000			
6. Completion of goods	Finished Goods		Finished Goods		
	Inventory 395,000		Inventory 395,000		
	Work-in-Process		Raw Materials		
	Inventory	395,000	and In Process		
			Inventory		160,000
			Conversion Cost		
			Control		235,000
7. Goods are sold	Cost of Goods Sold 395,000		Cost of Goods Sold 395,000		
	Finished Goods		Finished Goods		
	Inventory	395,000	Inventory		395,000
8. Variance is recognized	Cost of Goods Sold 15,000		Cost of Goods Sold 15,000		
	Overhead		Conversion Cost		
	Control	15,000	Control		15,000

2. The entries for Transactions 6 and 7 in Requirement 1 are replaced with the following entry:

Cost of Goods Sold	395,000	
Raw Materials and In Process Inventory		160,000
Conversion Cost Control		235,000

All other entries follow those in Requirement 1.

3. (a) There is no entry for Transaction 1. Transaction 6 is replaced with the following entry:

CORNERSTONE
13-5
(continued)

Finished Goods Inventory	395,000	
Accounts Payable		160,000
Conversion Cost Control		235,000

(b) There is no entry for Transaction 1. Transactions 6 and 7 are replaced with the following entry:

Cost of Goods Sold	395,000	
Accounts Payable		160,000
Conversion Cost Control		235,000

Summary of Learning Objectives

1. **Explain what strategic cost management is and how it can be used to help a firm create a competitive advantage.**

- Obtaining a competitive advantage so that long-term survival is ensured is the goal of strategic cost management.
- Different strategies create different bundles of activities. By assigning costs to activities, the costs of different strategies can be assessed.
- There are three generic or general strategies: cost leadership, differentiation, and focusing. The particular mix and relative emphasis of these three strategies define a firm's strategic position.
- The objective of strategic cost management is to reduce costs while simultaneously strengthening a firm's strategic position.

2. **Discuss value-chain analysis and the strategic role of activity-based customer and supplier costing.**

- Knowledge of organizational and operational activities and their associated cost drivers is fundamental to strategic cost analysis. Knowledge of the firm's value chain and the industrial value chain is also critical.
- Value-chain analysis relies on identifying and exploiting internal and external linkages.
- Good cost management of supplier and customer linkages requires an understanding of what suppliers cost and how much it costs to service customers.
- Activity-based assignments to suppliers and customers provide the accurate cost information needed.

3. **Tell what life-cycle cost management is and how it can be used to maximize profits over a product's life cycle.**

- Life-cycle cost management is related to strategic cost analysis and, in fact, could be called a type of strategic cost analysis.
- Life-cycle cost management requires an understanding of the three types of life-cycle viewpoints: the marketing viewpoint, the production viewpoint, and the consumable life viewpoint.

- Target costing plays an essential role in life-cycle cost management by providing a methodology for reducing costs in the design stage by considering and exploiting both customer and supplier linkages.

4. Identify the basic features of JIT purchasing and manufacturing.

- JIT purchasing and manufacturing offer a totally different set of structural and procedural activities from those of the traditional organization.
- In JIT purchasing, parts and materials arrive just in time to be used in production. JIT assumes that all costs other than direct materials are driven by time and space drivers.
- Understanding supplier and customer linkages is vital for a successful JIT system.

5. Describe the effect JIT has on cost traceability and product costing.

- In a JIT environment, many overhead costs assigned to products using either driver tracing or allocation are now directly traceable to products.
- A vastly simplified process costing system is the usual structure for a JIT environment.
- Product costing is more accurate because of increased traceability of costs.
- Accounting for the cost accounting cycle is simplified using backflush costing.

CORNERSTONES FOR CHAPTER 13

CORNERSTONE 13-1 The HOW and WHY of exploiting internal linkages to reduce costs and increase value, page 651

CORNERSTONE 13-2 The HOW and WHY of activity-based supplier costing, page 653

CORNERSTONE 13-3 The HOW and WHY of activity-based customer costing, page 656

CORNERSTONE 13-4 The HOW and WHY of activity-based life-cycle cost reduction, page 661

CORNERSTONE 13-5 The HOW and WHY of backflush costing, page 671

Review Problems

I. Strategic Cost Management, Target Costing

Assume that a firm has the following activities and associated cost behaviours:

Activities	Cost Behaviour
Assembling components	$10 per direct labour hour
Setting up equipment	Variable: $100 per setup
	Step-fixed: $30,000 per step, 1 step = 10 setups
Receiving goods	Step-fixed: $40,000 per step, 1 step = 2,000 hours

Activities with step-cost behaviour are being fully utilized by existing products. Thus, any new product demands will increase resource spending on these activities.

Two designs are being considered for a new product: Design I and Design II. The following information is provided about each design (1,000 units of the product will be produced):

Activity Driver	Design I	Design II
Direct labour hours	3,000	2,000
Number of setups	10	20
Receiving hours	2,000	4,000

The company has recently developed a cost equation for manufacturing costs using direct labour hours as the driver. The equation has $R^2 = 0.60$ and is as follows:

$$Y = \$150{,}000 + \$20X$$

Required:

1. Suppose that the firm's design engineers are told that only direct labour hours drive manufacturing costs (based on the direct labour cost equation). Compute the cost of each design. Which design would be chosen based on this unit-based cost assumption?
2. Now compute the cost of each design using all driver and activity information. Which design will now be chosen? Are there any other implications associated with the use of the more complete activity information set?
3. Consider the following statement: "Strategic cost analysis should exploit internal linkages." What does this mean? Explain, using the results of Requirements 1 and 2.
4. An outside consultant indicated that target costing ought to be used in the design stage. Explain what target costing is, and describe how it requires an understanding of both supplier and customer linkages.
5. What other information would be useful to have concerning the two designs? Explain.

Solution:

1. Design I: $\$20 \times 3{,}000 = \$60{,}000 + \$150{,}000 = \$210{,}000$
   Design II: $\$20 \times 2{,}000 = \$40{,}000 + \$150{,}000 = \$190{,}000$
   The unit-based analysis would lead to the selection of Design II.
2. Design I:

Assembling components ($10 × 3,000)	$ 30,000
Setting up equipment [(10 × $100) + (1 × $30,000)]	31,000
Receiving goods (1 × $40,000)	40,000
Total	$101,000

   Design II:

Assembling components ($10 × 2,000)	$ 20,000
Setting up equipment [(20 × $100) + (2 × $30,000)]	62,000
Receiving goods (2 × $40,000)	80,000
Total	$162,000

   Design I has the lowest total cost. Notice also the difference in expected total manufacturing costs. The direct labour driver approach produces a much higher cost for both designs. This difference in cost could produce significant differences in pricing strategies.
3. Exploiting internal linkages means taking advantage of the relationships among the activities that exist within a firm's segment of the value chain. To do this, we must know what the activities are and how they are related. Activity costs and drivers are an essential part of this analysis. Using only unit-based drivers for design decisions, as in Requirement 1, ignores the effect that different designs have on non-unit-based activities. The results of Requirement 2 illustrate a significant difference between two designs—relative to the unit-based analysis. The traditional costing system simply is not rich enough to supply the information needed for a thorough analysis of linkages.
4. Target costing specifies the unit cost required to achieve a given share of the market for a product with certain functional specifications. This target cost is then compared with the expected unit cost. If the expected unit cost is greater than the target cost, then actions are taken to reduce the costs to the desired level. Three general methods of cost reduction are used: (1) tear-down engineering, (2) value analysis, and (3) process improvement. Tear-down engineering dismantles competitors' products to search for more efficient product designs. Value engineering evaluates customer reactions to proposed functions and determines whether they are worth

the cost to produce. Process improvement seeks to improve the efficiency of the process that will be used to produce the new product. The first two methods are concerned with improving product design, while the third is concerned with improving process design. Involving both customers and suppliers in the process has the objective of producing lower costs than would be obtained if the design team worked in isolation. Suppliers, for example, may suggest alternative designs that will reduce the cost of the components that go into the product. Customers, of course, can indicate whether they value a particular design feature and, if so, how much they would be willing to pay for it.

5. Linkages also extend to the rest of the firm's internal value-chain activities. It would be useful to know how design choices affect, and are affected by, logistical activities. External linkages would also help. For example, it would be interesting to know how post-purchase activities and costs are affected by the two designs.

II. Backflush Costing

Foster Company has implemented a JIT system and is considering the use of backflush costing. Foster had the following transactions for the first quarter of the current fiscal year. (Conversion cost variances are recognized quarterly.)

1. Purchased raw materials on account for $400,000.
2. Placed all materials received into production.
3. Incurred actual direct labour costs of $60,000.
4. Incurred actual overhead costs of $400,000.
5. Applied conversion costs of $470,000.
6. Completed all work for the month.
7. Sold all completed work.
8. Computed the difference between actual and applied costs.

Required:

Prepare journal entries for Variations 2 and 4 of backflush costing.

Solution:

	Transaction	Backflush Journal Entries: Variation 2		
1.	Purchase of raw materials	Raw Materials and In Process Inventory	400,000	
		Accounts Payable		400,000
4.	Overhead cost incurred	Conversion Cost Control	460,000	
		Wages Payable		60,000
		Accounts Payable		400,000
7.	Goods are sold	Cost of Goods Sold	870,000	
		Raw Materials and In Process Inventory		400,000
		Conversion Cost Control		470,000
8.	Variance is recognized	Conversion Cost Control	10,000	
		Cost of Goods Sold		10,000

	Transaction	Backflush Journal Entries: Variation 4		
4.	Overhead cost incurred	Conversion Cost Control	460,000	
		Wages Payable		60,000
		Accounts Payable		400,000

Transaction	Backflush Journal Entries: Variation 4		
7. Goods are sold	Cost of Goods Sold	870,000	
	Accounts Payable		400,000
	Conversion Cost Control		470,000
8. Variance is recognized	Conversion Cost Control	10,000	
	Cost of Goods Sold		10,000

Key Terms

Acceptable quality level (AQL), 667

Backflush costing, 670

Competitive advantage, 644

Consumable life, 657

Cost leadership strategy, 645

Customer value, 644

Decline stage, 657

Differentiation strategy, 645

Executional activities, 647

External linkages, 647

Focusing strategy, 645

Growth stage, 657

Industrial value chain, 646

Internal linkages, 647

Introduction stage, 657

JIT manufacturing, 665

JIT purchasing, 665

Life-cycle cost management, 658

Life-cycle costs, 657

Manufacturing cells, 666

Maturity stage, 657

Operational activities, 648

Operational cost drivers, 648

Organizational cost drivers, 647

Post-purchase costs, 644

Product life cycle, 655

Revenue-producing life, 657

Strategic cost management, 644

Strategic decision making, 644

Strategic positioning, 645

Strategy, 645

Structural activities, 647

Target cost, 662

Total product, 644

Total quality management (TQM), 652

Value-chain analysis, 644

Discussion Questions

1. What does it mean to obtain a competitive advantage? What role does the cost management system play in helping to achieve this goal?
2. What is customer value? How is customer value related to a cost leadership strategy? To a differentiation strategy? To strategic positioning?
3. Explain what internal and external linkages are.
4. What are organizational and operational activities? Organizational cost drivers? Operational cost drivers?
5. What is the difference between a structural cost driver and an executional cost driver? Provide examples of each.
6. What is value-chain analysis? What role does it play in strategic cost analysis?
7. What is an industrial value chain? Explain why a firm's strategies are tied to what happens in the rest of the value chain. Using total quality management as an example, explain how the success of this quality management approach is dependent on supplier linkages.
8. What are the three viewpoints of product life cycle? How do they differ?
9. What are the four stages of the marketing life cycle?
10. What are life-cycle costs? How do these costs relate to the production life cycle?
11. What are the four stages of the consumption life cycle? What are post-purchase costs? Explain why a producer may want to know post-purchase costs.
12. "Life-cycle cost reduction is best achieved during the development stage of the production life cycle." Do you agree or disagree? Explain.

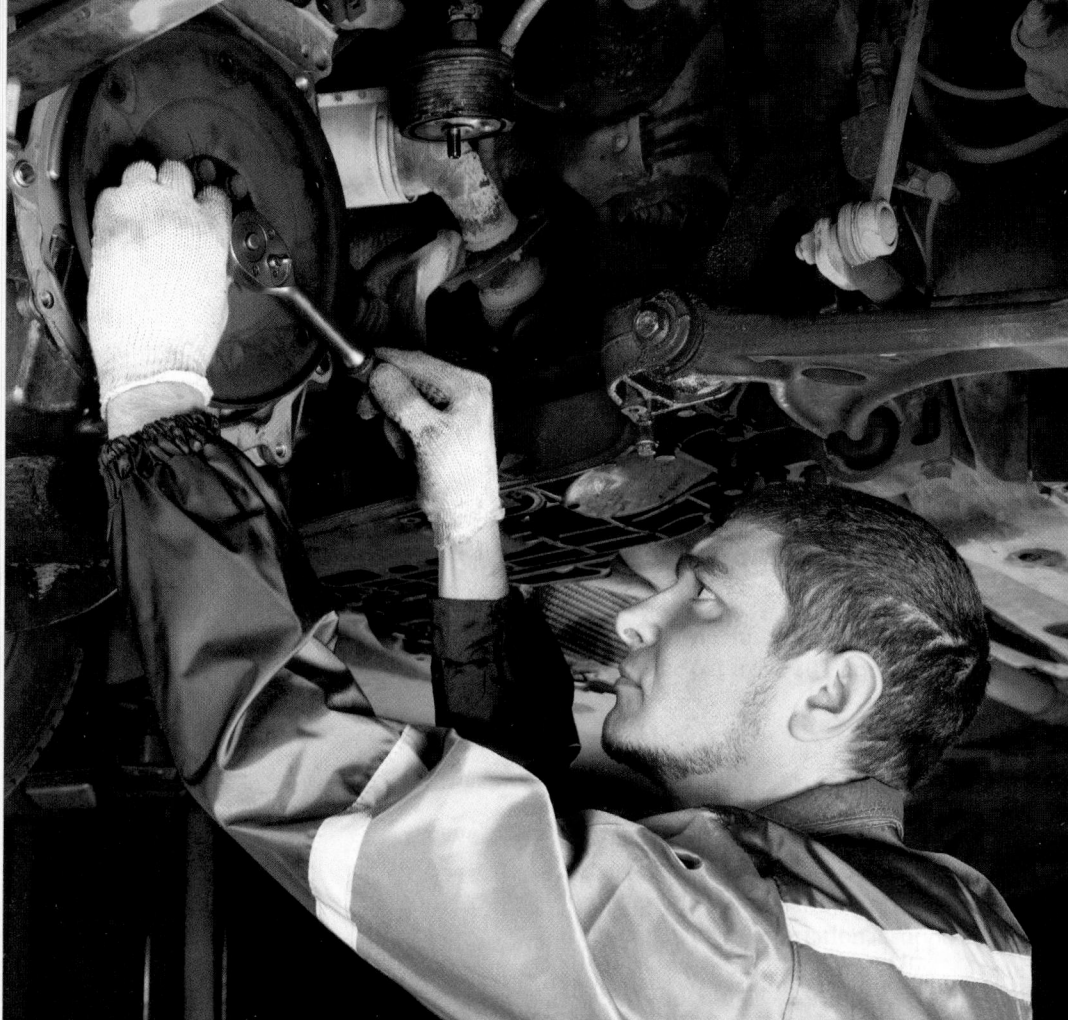

After studying this chapter, you should be able to:

➤ **1** Describe how activity-based management and activity-based costing differ.

➤ **2** Define process value analysis.

➤ **3** Describe activity-based financial performance measurement.

➤ **4** Discuss the implementation issues associated with an activity-based management system.

➤ **5** Explain how activity-based management is a form of responsibility accounting, and tell how it differs from financial-based responsibility accounting.

© DMITRY MORDVINTSEV/ISTOCK

CHAPTER

14 Activity-Based Management

Many firms operate in rapidly changing environments. Typically, these firms face stiff national and international competition. This stringent competitive environment demands that firms offer customized products and services to diverse customer segments. This, in turn, means that firms must find cost-efficient ways of producing high-variety, low-volume products. To find ways to improve performance, firms operating in this kind of environment not only must know what it currently *costs* to do things, but must also evaluate *why* and *how* they do things. Improving performance translates into constantly searching for ways to eliminate waste—a process known as **continuous improvement**. Activity-based costing and activity-based management are important tools in this ongoing improvement effort.

The Relationship of Activity-Based Costing and Activity-Based Management

OBJECTIVE ➤ 1
Describe how activity-based management and activity-based costing differ.

Processes are the source of many of the improvement opportunities that exist within an organization. Processes are made up of activities that are linked to perform a specific objective. Improving processes means improving the way activities are performed. Thus, management of activities, not costs, is the key to successful control for firms operating in continuous improvement environments. The realization that activities are crucial to both improved product costing and effective control has led to a new view of business processes called activity-based management.

Activity-based management (ABM) is a systemwide, integrated approach that focuses management's attention on activities with the objective of improving customer value as well as the profit achieved by providing this value. ABC is the major source of information for activity-based management. Thus, the activity-based management model has two dimensions: a cost dimension and a process dimension. This two-dimensional model is presented in Exhibit 14-1.

The cost dimension provides cost information about resources, activities, and cost objects of interest such as products, customers, suppliers, and distribution channels. *The objective of the cost dimension is improving the accuracy of cost assignments.* As the model suggests, the cost of resources is traced to activities, and then the cost of activities is assigned to cost objects. This activity-based costing dimension is useful for product costing, strategic cost management, and tactical analysis.

The process dimension provides information about what activities are performed, why they are performed, and how well they are performed. *This dimension's objective is cost reduction.* It is this dimension that provides the ability to engage in and measure continuous improvement. To understand how the process view connects with continuous improvement, a more explicit understanding of process value analysis is needed.

Exhibit 14-1

The Two-Dimensional Activity-Based Management Model

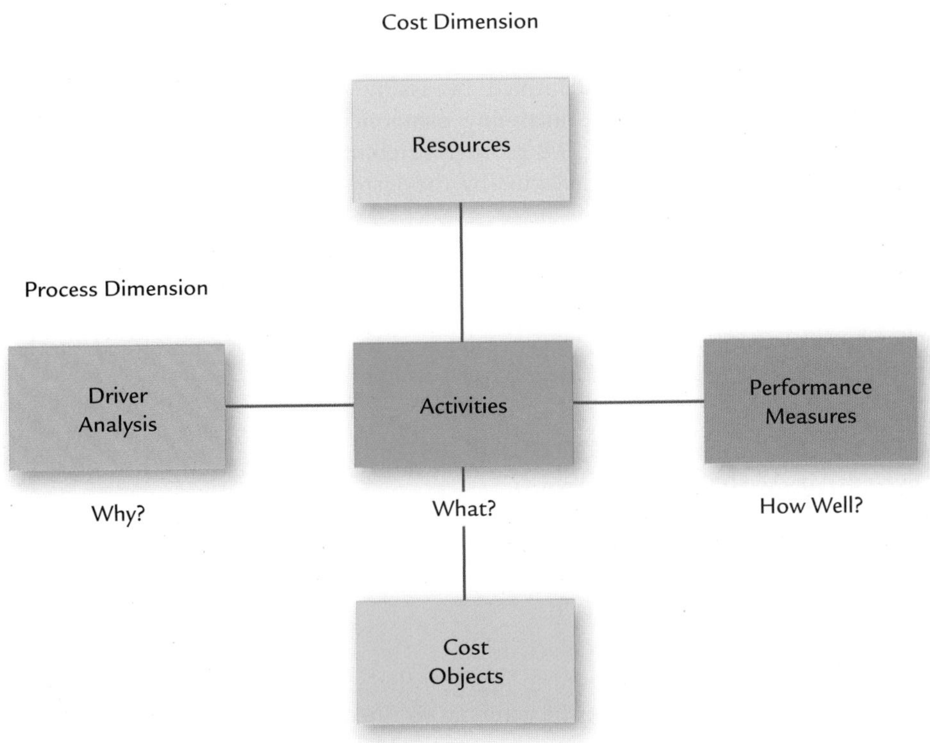

Process Dimension: Process Value Analysis

Process value analysis (PVA) is fundamental to activity-based Management. It focuses on accountability for activities rather than costs, and emphasizes the maximization of systemwide performance instead of individual performance. Process value analysis moves activity management from a conceptual basis to an operational basis. As the model in Exhibit 14-1 illustrates, process value analysis is concerned with (1) *driver analysis*, (2) *activity analysis*, and (3) *performance measurement*.

Driver Analysis: Defining Root Causes

Managing activities requires an understanding of what factors cause activities to be performed and what causes activity costs to change. Activities consume inputs (resources) and produce outputs. For example, if the activity is maintaining the payroll master file, the resources used would be such things as a payroll clerk, a computer, a printer, computer paper, and disks. The output would be an updated employee file. An **activity output measure** is the number of times the activity is performed. It is the quantifiable measure of the output. For example, the number of employee files maintained is a possible output measure for maintaining the payroll master file.

The activity output measure calculates the demands placed on an activity and is an *activity driver*. As the demands for an activity change, the cost of the activity can change. For example, as the number of employee files maintained increases, the activity of maintaining the master payroll may need to consume more inputs (labour, disks, paper, and so on). However, output measures (activity drivers), such as the number of files maintained, may not and usually do not correspond to the *root causes* of activity costs; rather, they are the consequences of the activity being performed. The purpose of **driver analysis** is to reveal the root causes of activity costs. For example, an analysis may reveal that the root cause of treating and disposing of toxic waste is product design. Once the root cause is known, action can be taken to improve the activity. Specifically, creating a new product design may reduce or eliminate the cost of treating and disposing of toxic waste.

Often, several activities may have the same root cause. For example, the costs of inspecting incoming components (output measure: number of inspection hours) and reordering (output measure: number of reorders) may both be caused by poor quality of purchased components. By working with carefully selected suppliers to help them improve their product quality, both activities may be improved.

Typically, root causes are identified by asking one or more "why" questions. Example: Why are we inspecting incoming components? Answer: Because some may be defective. Question: Why are we reordering components? Answer: Because some components are judged to be defective by the inspection. Question: Why are some purchased components defective? Answer: Because our suppliers are not providing reliable components. Once the answers to the why questions are obtained, the answers to "how" questions are possible. Example: How do we improve the quality of incoming components? Answer: By selecting (or developing) suppliers that provide higher-quality components. The why questions identify the root causes, and the how questions enable management to identify ways to improve.

Activity Analysis: Identifying and Assessing Value Content

Activity analysis is the process of identifying, describing, and evaluating the activities an organization performs. Activity analysis should produce four outcomes:

1. what activities are performed,
2. how many people perform the activities,
3. the time and resources required to perform the activities, and
4. an assessment of the value of the activities to the organization, including a recommendation to select and keep only those that add value.

Steps 1–3 have been described in Chapter 6. Those steps were critical for assigning costs. Step 4, determining the value-added content of activities, is concerned with cost reduction rather than cost assignment. Thus, this may be considered the most important part of activity analysis. Activities can be classified as *value-added* or *non-value-added*.

Value-Added Activities

Value-added activities are those activities necessary to remain in business. Value-added activities contribute to customer value and/or help meet an organization's needs. Activities that comply with legal mandates are value-added because they exist to meet organizational needs. Moreover, they add to customer value by allowing the business to continue operating so that the products and services desired by the customer can be obtained. Even though mandated activities are necessary, customers should insist that they be performed as efficiently as possible to reduce the cost impact on goods and services. Examples of mandated activities include those needed to comply with the reporting requirements of the CICA and the filing requirements of the CRA. The remaining activities in the firm are *discretionary*. Classifying discretionary activities as value-added is more of an art than a science and depends heavily on subjective judgment. However, it is possible to identify three conditions, which if simultaneously met, are sufficient to classify a discretionary activity as value-added. These conditions are as follows: (1) the activity produces a change of state, (2) the change of state was not achievable by preceding activities, and (3) the activity enables other activities to be performed.

For example, consider the production of metal components used in medical equipment. The first activity, gating, creates a wax mould replica of the final product. The next activity, shelling, creates a ceramic shell around the wax mould. After removing the wax, molten metal is poured into the resulting cavity. The shell is then broken to reveal the desired metal component. The gating activity is value-added because (1) it causes a change of state—unformed wax is transformed into a wax mould, (2) no prior activity was supposed to create this change of state, and (3) it enables the shelling activity to be performed. Similar comments hold for the shelling and pouring activities. The value-added properties are easy to see for operational activities like gating and shelling, but what about a more general activity like supervising production workers?

A managerial activity is specifically designed to manage other value-added activities—to ensure that they are performed in an efficient and timely manner. Supervision certainly satisfies the enabling condition. Is there a change in state? There are two ways of answering in the affirmative. First, supervising can be viewed as an enabling resource that is consumed by the operational activities that do produce a change of state. Thus, supervising is a secondary activity that serves as an input needed to help bring about the change of state expected for value-added primary activities. Second, it could be argued that the supervision brings order by changing the state from uncoordinated activities to coordinated activities.

Once value-added activities are identified, we can define value-added costs. **Value-added costs** are the costs to perform value-added activities with perfect efficiency. Implicit in this definition is the notion that value-added activities may contain nonessential actions that create unnecessary cost.

Non-Value-Added Activities

Non-value-added activities are unnecessary and are not valued by internal or external customers. Non-value-added activities often are those that fail to produce a change in state or those that replicate work because it wasn't done correctly the first time. Inspecting wax moulds, for example, is a non-value-added activity. Inspection is a *state-detection activity*, not a state-changing activity. (It tells us the state of the mould—whether or not it is of the right shape.) As a general rule, state-detection activities are not value-added. Now, consider the activity of recasting moulds that fail inspection. This recasting is designed to bring the mould from a nonconforming state to a conforming state. Thus, a change of state occurs. Yet, the activity is non-value-added because it *repeats* work; it is doing something that should have been done by preceding activities, the first time the wax mould was cast. Thus, it is a *state-correction activity*. **Non-value-added costs** are costs that are caused either by non-value-added activities or by the inefficient performance of value-added activities. Because of increased competition, many firms are attempting to eliminate non-value-added activities and nonessential portions of value-added activities because they add unnecessary cost and impede performance. Therefore, activity analysis attempts to identify and eventually eliminate all unnecessary activities and, simultaneously, increase the efficiency of necessary activities.

Assessing the value content of activities enables managers to eliminate waste. As waste is eliminated, costs are reduced. Cost reduction *follows* the elimination of waste. Note the value of managing the *causes* of the costs rather than the costs themselves. Increasing the efficiency of a non-value-added activity is not a good long-term strategy. For example, training inspectors in sampling procedures may increase the efficiency of the activity of inspecting incoming components, but it is better to implement a supplier evaluation program that leads to suppliers that provide defect-free components, thus eliminating the need for inspection.

For example, **US Airways** implemented an activity-based cost management (ABCM) system to manage its in-house engine maintenance business unit. First, ABCM helped determine the cost of engine maintenance with increased accuracy. Second, ABCM provided operational and financial information that allowed work teams to identify opportunities for improvement. Thus, ABCM provided accurate cost information and simultaneously revealed opportunities for improvement. ABCM identified 410 activities—activities such as tear down, welding, waiting for tooling, and rework. Of the 410 activities, 47 were identified as non-value-added. The non-value-added activities were rank-ordered on the basis of activity cost, providing information about where the most significant process improvement opportunities were located. Root cause analysis was undertaken by the various work teams to determine the causes for the efforts being expended on the non-value-added activities. Once the root causes were identified, the teams took action to reduce or eliminate the non-value-added activities. The net effect was to produce $4.3 million in process savings per year.[1]

Examples of Non-Value-Added Activities   Reordering parts, expediting production, and rework due to defective parts are examples of non-value-added activities. Other examples include warranty work, handling customer complaints, and reporting defects. Non-value-added activities can exist anywhere in the organization. In the manufacturing operation, five major activities are often cited as wasteful and unnecessary:

1. *Scheduling.* An activity that uses time and resources to determine when different products have access to processes (or when and how many setups must be done) and how much will be produced.
2. *Moving.* An activity that uses time and resources to move materials, work in process, and finished goods from one department to another.
3. *Waiting.* An activity in which materials or work in process use time and resources by waiting on the next process.
4. *Inspecting.* An activity in which time and resources are spent ensuring that the product meets specifications.
5. *Storing.* An activity that uses time and resources while a good or material is held in inventory.

None of these activities adds any value for the customer. Scheduling, for example, is not necessary if the company has learned how to produce on demand. Similarly, inspecting will not be necessary if the product is produced correctly the first time. The challenge of activity analysis is to find ways to produce the good without using any of these activities.

[1] Joe Donnelly and Dave Buchanan, "Implementation Lands $4.3 Million in Process Improvement Savings," *Better Management* (May 26, 2009) http://www.bettermanagement.com/library/library.aspx?l=536.

Ways of Cost Reduction through Activity Management   Competitive conditions dictate that companies must deliver products the customers want, on time, and at the lowest possible cost. This means that an organization must continually strive for cost improvement. **Kaizen costing** is characterized by constant, incremental improvements to existing processes and products. Activity management is a fundamental part of kaizen costing. Activity management can reduce costs in four ways:[2]

1. Activity elimination
2. Activity selection
3. Activity reduction
4. Activity sharing

Activity elimination focuses on eliminating non-value-added activities. For example, the activity of expediting production seems necessary at times to ensure that customers' needs are met. Yet this activity is necessary only because of the company's failure to produce efficiently. By improving cycle time, a company may eventually eliminate the need for expediting. Cost reduction then follows.

Activity selection involves choosing among various sets of activities that are caused by competing strategies. Different strategies cause different activities. Different product design strategies, for example, can require significantly different activities. Activities, in turn, cause costs. Each product design strategy has its own set of activities and associated costs. All other things being equal, the lowest cost design strategy should be chosen. In a kaizen cost framework, *redesign* of existing products and processes can lead to a different, lower cost set of activities. Thus, activity selection can have a significant effect on cost reduction.

Activity reduction decreases the time and resources required by an activity. This approach to cost reduction should be aimed primarily at improving the efficiency of necessary activities or act as a short-term strategy for moving non-value-added activities toward the point of elimination. For example, by improving product quality, customer complaints should decrease and, consequently, the demand for handling customer complaints should decrease.

Activity sharing increases the efficiency of necessary activities by using economies of scale. Specifically, the quantity of the cost driver is increased without increasing the total cost of the activity itself. This lowers the per-unit cost of the cost driver and the amount of cost traceable to the products that consume the activity. For example, a new product can be designed to use components already being used by other products. By using existing components, the activities associated with these components already occur, and the company avoids the creation of a whole new set of activities.

Performance Measurement Analysis

Activity performance measurement is designed to assess how well an activity was performed and the results achieved. Measures of activity performance are both financial and nonfinancial and centre on three major dimensions: (1) efficiency, (2) quality, and (3) time. *Efficiency* is concerned with the relationship of activity outputs to activity inputs. For example, activity efficiency is improved by producing the same activity output with less inputs. Costs trending downward is evidence that activity efficiency is improving. *Quality* is concerned with doing the activity right the first time it is performed. If the activity output is defective, then the activity may need to be repeated, causing unnecessary cost and reduction in efficiency. The *time* required to perform an activity is also critical. Longer times usually mean more resource consumption and less ability to respond to customer demands. Time measures of performance tend to be nonfinancial, whereas efficiency and quality measures are both financial and nonfinancial.

---

[2] Peter B.B. Turney, "How Activity-Based Costing Helps Reduce Cost," *Journal of Cost Management* (Winter 1991): 29–35.

Cost Dimension: Financial Measures of Activity Efficiency

Assessing activity performance should reveal the current level of efficiency and the potential for increased efficiency. Both financial and nonfinancial measures are used to reveal past performance and signal future potential gains in efficiency. Financial measures of activity performance are emphasized in this chapter, and nonfinancial measures are discussed in Chapter 15. **Financial measures** of performance should provide specific information about the dollar effects of activity performance changes. Thus, financial measures should indicate both potential and actual savings. Financial measures of activity efficiency include (1) value-and non-value-added activity costs, (2) trends in activity costs, (3) kaizen standard setting, (4) benchmarking, (5) activity flexible budgeting, and (6) activity capacity management.

Reporting Value- and Non-Value-Added Costs

Reducing non-value-added costs is one way to increase activity efficiency. A company's accounting system should distinguish between value-added costs and non-value-added costs because improving activity performance requires eliminating non-value-added activities and optimizing value-added activities. A firm should identify and formally report the value- and non-value-added costs of each activity. Highlighting non-value-added costs reveals the magnitude of the waste the company is currently experiencing, thus providing some information about the potential for improvement. This encourages managers to place more emphasis on controlling non-value-added activities. Progress can then be assessed by preparing trend and cost reduction reports. Tracking these costs over time permits managers to assess the effectiveness of their activity management programs.

Knowing the amount of costs saved is important for strategic purposes. For example, if an activity is eliminated, then the costs saved should be traceable to individual products. These savings can produce price reductions for customers, making the firm more competitive. Changing the pricing strategy, however, requires knowledge of the cost reductions realized by activity analysis. A cost-reporting system, therefore, is an important ingredient in an activity-based responsibility accounting system.

Value-added costs are the only costs that an organization should incur. The *value-added standard* calls for the complete elimination of non-value-added activities; for these activities, the optimal output is zero, with zero cost. The value-added standard also calls for the complete elimination of the inefficiency of activities that are necessary but inefficiently carried out. Hence, value-added activities also have an optimal output level. A **value-added standard**, therefore, identifies the optimal activity output. Identifying the optimal activity output requires activity output measurement.

Setting value-added standards does not mean that they will be (or should be) achieved immediately. The idea of continuous improvement is to move toward the ideal. Workers (teams) can be rewarded for improvement. Moreover, nonfinancial activity performance measures can be used to supplement and support the goal of eliminating non-value-added costs (these are discussed later in the chapter). Finally, measuring the efficiency of individual workers and supervisors is not the way to eliminate non-value-added activities. Remember, activities cut across departmental boundaries and are part of processes. Focusing on activities and providing incentives to improve processes is a more productive approach. Improving the process should lead to improved results.

By comparing actual activity costs with value-added activity costs, management can assess the level of activity inefficiency and the potential for improvement. To identify and calculate value- and non-value-added costs, output measures for each activity must be defined. Once output measures are defined, then value-added standard quantities (SQ) for each activity can be defined. Value-added costs can be computed by multiplying the value-added standard quantities by the price standard (SP).

Formulas for Value- and Non-Value-Added Costs

$$\text{Value-added costs} = SQ \times SP$$
$$\text{Non-value-added costs} = (AQ - SQ)SP$$

(This is in essence a quantity variance.)

where

SQ = The value-added output level for an activity

SP = The standard price per unit of activity output measure

AQ = The actual quantity used of flexible resources or the practical activity capacity acquired for committed resources

Non-value-added costs can be calculated as the difference between the actual level of the activity's output (AQ) and the value-added level (SQ), multiplied by the standard price. These formulas are presented in Exhibit 14-2. Some further explanation is needed.

For flexible resources (resources acquired as needed), AQ is the actual quantity of activity used. For committed resources (resources acquired in advance of usage), AQ represents the actual quantity of activity capacity acquired, as measured by the activity's practical capacity. This definition of AQ allows the computation of non-value-added costs for both variable and fixed activity costs. For fixed activity costs, SP is the budgeted activity costs divided by AQ, where AQ is practical activity capacity. Cornerstone 14-1 illustrates the power of these concepts.

Notice from the information in Cornerstone 14-1 that the value-added standards (SQ) for inspection and grinding call for their elimination. Ideally, there should be no defective moulds; by improving quality, changing production processes, and so on, inspection and grinding can eventually be eliminated. The cost report of Cornerstone 14-1 allows managers to see the non-value-added costs; as a consequence, it emphasizes the opportunity for improvement. By redesigning the products and reducing the number of parts required, purchase time can be reduced. By improving the moulding process and labour skill, management can reduce the demands for moulding time, inspection, and grinding. Thus, reporting value- and non-value-added costs at a point in time may trigger actions to manage activities more effectively. Once they see the amount of waste, managers may be induced to search for ways to improve activities and bring about cost reductions. Reporting these costs may also help managers improve planning, budgeting, and pricing decisions. For example, a manager might consider it possible to lower a selling price to meet a competitor's price if that manager can see the potential for reducing non-value-added costs to absorb the effect of the price reduction.

The Trend Report: Does a Decline of Non-Value-Added Costs Result in Cost Reduction?

As managers take actions to improve activities, do the cost reductions follow as expected? One way to answer this question is to compare the costs for each activity over time. The goal is activity improvement as measured by cost reduction. We should see a decline in non-value-added costs from one period to the next—provided the activity improvement initiatives are effective. The trend report will also reveal the amount of cost reduction still available. Cost reduction for value-added activities focuses on increasing the efficiency of these activities while the cost reduction goal for non-value-added activities is their eventual elimination. Cornerstone 14-2 provides an illustration of trend reporting of non-value-added costs.

CORNERSTONE
14-1

The HOW and WHY of Value- and Non-Value-Added Cost Reporting

Information:

A manufacturing firm has four activities: purchasing materials, moulding, inspecting moulds, and grinding imperfect moulds. Purchasing and moulding are necessary activities; inspection and grinding are unnecessary. The following data pertain to the four activities for the year ending 2013 (actual price per unit of the activity driver is assumed to be equal to the standard price):

Activity	Activity Driver	SQ	AQ	SP
Purchasing	Purchasing hours	20,000	24,000	$20
Moulding	Moulding hours	30,000	34,000	12
Inspecting	Inspection hours	0	6,000	15
Grinding	Number of units	0	5,000	6

Why:

A cost report that shows value and non-value-added costs allows managers to see the amount of waste, assess its materiality, and identify opportunities for improvement.

Required:

1. Prepare a cost report for the year ending 2013 that shows value-added costs, non-value-added costs, and total costs for each activity.
2. Explain why inspection and grinding are non-value-added activities.
3. *What if* purchasing cost is a step-fixed cost with each step being 2,000 hours whereas moulding cost is a variable cost? What is the implication for reducing the cost of waste for each activity?

Solution:

1.

Value- and Non-Value-Added Cost Report for the Year Ended 2013			
Activity	Value-Added Costs	Non-Value-Added Costs	Total Costs
Purchasing	$400,000	$ 80,000	$ 480,000
Moulding	360,000	48,000	408,000
Inspecting	0	90,000	90,000
Grinding	0	30,000	30,000
Total	$760,000	$248,000	$1,008,000

2. Inspection is a state-detection activity, and grinding is a state-correction activity.

3. For purchasing, cost reduction occurs only when the actual demand for purchasing hours is reduced by each block of 2,000 hours. For moulding, each hour saved produces a savings of $12. Accordingly, cost savings will likely materialize more quickly for moulding than for purchasing.

The trend report in Cornerstone 14-2 reveals that more than half of the non-value-added costs have been eliminated. It also reveals that there is still ample room for improvement, but activity improvement so far has been successful. Reporting non-value-added costs, however, not only reveals cost reduction but also indicates

The HOW and WHY of Non-Value-Added Cost Trend Reporting

Information:
See the information for Cornerstone 14-1. Assume that at the beginning of 2013, the moulding process was redesigned and the employees in Moulding were trained in a new work technique. By reducing the number of bad moulds, the firm hoped to significantly reduce waste for all four activities. Purchasing and Inspecting resources are purchased in steps of 2,000 hours. The other two activities are acquired as used and needed. At the end of 2013, the following results were reported for the four activities:

Activity	Activity Driver	SQ	AQ	SP
Purchasing	Purchasing hours	20,000	22,000	$20
Moulding	Moulding hours	30,000	32,000	12
Inspecting	Inspection hours	0	2,000	15
Grinding	Number of units	0	2,500	6

Why:
Comparing changes in non-value-added costs over time reveals where cost reductions have been achieved, allows managers to assess the effectiveness of improvement measures undertaken, and shows how much improvement potential remains.

Required:
1. Prepare a trend report that shows the non-value-added costs for each activity for 2012 and 2013 and the change in costs for the two periods. Discuss the report's implications.
2. Explain the role of activity reduction for both value-added activities and non-value-added activities.
3. *What if* at the end of 2013, the selling price of a competing product is reduced by $10 per unit? Assume that the firm produces and sells 10,000 units of its product and that its product is associated only with the four activities being considered. By virtue of the waste-reduction savings, can the competitor's price reduction be matched without reducing the unit profit margin of the product that prevailed at the beginning of the year?

Solution:
1.

Trend Report: Non-Value-Added Costs			
Activity	2012	2013*	Change
Purchasing	$ 80,000	$ 40,000	$ 40,000
Moulding	48,000	24,000	24,000
Inspecting	90,000	30,000	60,000
Grinding	30,000	15,000	15,000
Total	$248,000	$109,000	$139,000

*Since the reduction for the purchasing and inspection were in multiples of 2,000, the cost savings is simply SP multiplied by the reduction in AQ.

The trend report shows a significant reduction in non-value-added costs, validating the improvement actions taken.

2. For value-added activities, the non-value-added component is usually the result of using more of the activity than should be used; thus, activity

CORNERSTONE 14-2 (*continued*)

reduction is the objective for improving activity efficiency. For non-value-added activities, activity reduction is an intermediate step that ultimately will lead to activity elimination. Depending on the nature of the resources consumed by the activity, activity reduction can also lead to cost reductions.

3. From Requirement 1, the savings per unit of product are $13.90 ($139,000/10,000), indicating that the competitor's price reduction can be matched (or beat) without changing the unit profit margin that existed at the beginning of the year.

where the reduction occurred. It provides managers with information on how much potential for cost reduction remains, assuming that the value-added standards remain the same. Value-added standards, however, like other standards, are not cast in stone. New technology, new designs, and other innovations can change the nature of activities performed. As new ways for improvement surface, value-added standards can change. Managers should not become content but should continually seek higher levels of efficiency.

Drivers and Behavioural Effects

Activity output measures are needed to compute and track non-value-added costs. Reducing a non-value-added activity should produce a reduction in the demand for the activity and, therefore, a reduction in the activity output measures. If a team's performance is affected by its ability to reduce non-value-added costs, then the selection of activity drivers (as output measures) and the way the drivers are used can affect behaviour. For example, if the output measure for setup costs is chosen as setup time, an incentive is created for workers to reduce setup time. Since the value-added standard for setup costs calls for their complete elimination, then the incentive to drive setup time to zero is compatible with the company's objectives, and the induced behaviour is beneficial.

Suppose, however, that the objective is to reduce the number of unique parts a company processes, thus reducing the demand for activities such as purchasing and incoming inspection. If the costs of these activities are assigned to products based on the number of parts, the incentive created is to reduce the number of parts in a product. Yet if too many parts are eliminated, the functionality of the product may be reduced to a point where its marketability is adversely affected. Identifying the value-added standard number of parts for each product through the use of functional analysis can discourage this type of behaviour.[3] Designers can then be encouraged to reduce the non-value-added costs by designing to reach the value-added standard number of parts. The standard has provided a concrete objective and defined the kind of behaviour that the incentive allows.

Kaizen Costing

Kaizen costing is concerned with reducing the costs of *existing* products and processes. In operational terms, this translates into reducing non-value-added costs. Controlling this cost reduction process is accomplished through the repetitive use of two major subcycles: (1) the kaizen or continuous improvement cycle and (2) the maintenance cycle. The kaizen subcycle is defined by a Plan-Do-Check-Act sequence. If a company is emphasizing the reduction of non-value-added costs, the amount of improvement planned for the coming period (month, quarter, etc.) is set (the *Plan* step). A **kaizen standard** reflects the planned improvement for the upcoming period. The planned improvement is assumed to be attainable, and kaizen standards are a type of currently attainable standard. Actions are taken to implement the planned improvements (the *Do* step). Next, actual results (e.g., costs) are compared with the

---

[3] Functional analysis compares the price customers are willing to pay for a particular product function with the cost of providing that function.

Exhibit 14-3

Kaizen Cost Reduction Process

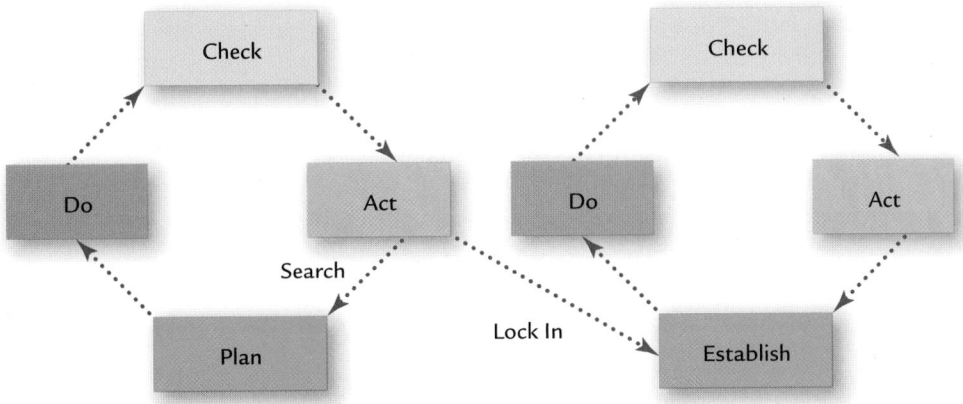

kaizen standard to provide a measure of the level of improvement attained (the *Check* step). Setting this new level as a minimum standard for future performance locks in the realized improvements and simultaneously initiates the maintenance cycle and a search for additional improvement opportunities (the *Act* step). The maintenance cycle follows a traditional Establish-Do-Check-Act sequence. A standard is set based on prior improvements (locking in these improvements). Next, actions are taken (the *Do* step) and the results checked to ensure that performance conforms to this new level (the *Check* step). If not, then corrective actions are taken to restore performance (the *Act* step). The kaizen cost reduction process is summarized in Exhibit 14-3. Cornerstone 14-3 demonstrates an application of kaizen costing.

In some cases, companies have formalized the process of revising standards. For example, **Shionogi Pharmaceuticals** first assesses whether the improvements are attributable to kaizen activities or to random fluctuations. If kaizen activities are the source, Shionogi then evaluates the *sustainability* of the kaizen improvements. Improvements are locked in through revision of standards only if the improvements are judged to be sustainable.[4]

Benchmarking

Benchmarking is complementary to kaizen costing and activity-based management, and it can be used as a search mechanism to identify opportunities for improvement. **Benchmarking** uses best practices found within and outside the organization as the standard for evaluating and improving activity performance. The objective of benchmarking is to become the best at performing activities and processes (thus, benchmarking represents an important activity management methodology). The approach certainly seems to have considerable merit. A recent APQC study revealed that benchmarking returns ranged from $1.5 million to $189.4 million.[5] Interestingly, there was a direct correlation between the level of return and the degree of senior management support.

Internal Benchmarking Benchmarking against internal operations is called *internal benchmarking*. Within an organization, different units (e.g., different plant sites) that perform the same activities are compared. The unit with the best performance for a given activity sets the standard. Other units then have a target to meet or exceed. Furthermore, the best practices unit can share information with other units on how it has achieved its superior results. Internal benchmarking has several advantages. First, a significant amount of information is often readily available that can be shared throughout the

[4] Robin Cooper, *When Lean Enterprises Collide* (Boston: Harvard Business School Press, 1995).

[5] Kate Vitasek and Karl Mandrodt, *Benchmarking: Prerequisite for Best-In-Class Supply Chains*, an APQC white paper (see knowledge base section)(February 5, 2007), http://www.apqc.org/portal/apqc/ksn?paf_gear_id=contentgearhome& paf_dm=full&pageselect=detail&docid=129520, accessed May 27, 2009.

CORNERSTONE
14-3

The HOW and WHY of Kaizen Costing

Information:

An automotive parts division has a grinding activity for the subassemblies that it produces. Activity output is measured using grinding hours. The value-added standard (SQ) for this activity is zero grinding hours. On January 1, at the beginning of the fiscal year, eight grinding hours were allowed per batch (which almost always corresponded to the actual grinding hours used). The standard wage rate is $18 per grinding hour. During January, a new procedure for production of the subassemblies was developed with the expectation that the demand for grinding would be reduced by 25 percent. The new procedure was implemented in February and expectations concerning the effect on the grinding activity were met.

Why:

Kaizen costing has the objective of continuously improving the efficiency of activities and processes. It can be characterized as a *dynamic* standard costing system. Maintenance standards are revised based on achieved, sustainable improvements produced by the kaizen subcycle.

Required:

1. What are the maintenance standard for grinding hours and the associated expected cost at the beginning of February? The kaizen standard and expected associated cost?
2. What are the maintenance standard for grinding hours and the associated cost at the end of February? Explain. What is the next step in the kaizen cost reduction process?
3. **What if** the new procedure implemented in February produced a 20 percent reduction instead of a 25 percent reduction? What would the new maintenance standard and cost be?

Solution:

1. Maintenance standard: 8 hours per batch; Expected cost per batch: $144 ($8 \times \$18$); Kaizen standard: 6 hours per batch (0.75×8); Expected cost per batch: $108 ($6 \times \$18$).

2. Maintenance standard: 6 hours per batch; Expected cost per batch: $108 ($6 \times \$18$). After determining that the suggested improvement works and is sustainable, the new level of performance is locked in by revising the maintenance standard from eight hours to six hours. The next step is to search for another improvement opportunity that will then produce a new kaizen standard and expected batch cost. The ultimate objective is to eliminate all the non-value-added cost through a series of kaizen improvements.

3. You lock in the level actually achieved by the suggested improvement approach. In this case, the maintenance standard would be 6.4 hours (0.80×8) and the standard batch cost $115.20 ($6.4 \times \$18$).

organization. Second, immediate cost reductions are often realized. Third, the best internal standards that spread throughout the organization become the benchmark for comparison against external benchmarking partners. This last advantage also suggests the major disadvantage of internal benchmarking. Specifically, the best internal performance may fall short of what others are doing, particularly direct competitors.

There are numerous examples of the benefits of internal benchmarking.[6] **Thomson Corporation** collected and broadcast best practices through internal benchmarking

[6] Frank Jossi, "Take a Peek Inside," *HRMagazine* (June 2002): 46–52.

throughout the company and saved $200 million in one year. **Chevron** saved $150 million by transferring energy use management techniques throughout the company. **Public Service Enterprise Group** used internal benchmarking to improve the process for ripping up a street, repairing a line, backfilling the hole, and repaving the area. The improvement dropped costs from an average of $2,200 to just $200 per incident.

External Benchmarking   Benchmarking that involves comparison with others outside the organization is called *external benchmarking*. The three types of external benchmarking are competitive benchmarking, functional benchmarking, and generic benchmarking. Competitive benchmarking is a comparison of activity performance with direct competitors. The main problem with competitive benchmarking is that it is very difficult to obtain information beyond that found in the public domain. At times, however, it is possible. **The Ritz-Carlton**, for example, dramatically improved its housekeeping process by studying the best practices of a competitor.[7] Functional benchmarking is a comparison with firms that are in the same industry but do not compete in the same markets. For example, a Japanese communications firm might be able to compare its customer service process with that of **Bell Canada**. Generic benchmarking studies the best practices of noncompetitors outside a firm's industry. Certain activities and processes are common to all organizations. If superior external best practices can be identified, then they can be used as standards to motivate internal improvements. For example, **Verizon** improved its field service process by studying the field service process of an elevator company.[8]

Activity Flexible Budgeting

Activity flexible budgeting is the prediction of what activity costs will be as activity output changes. Variance analysis within an activity framework makes it possible to improve traditional budgetary performance reporting. It also enhances the ability to manage activities.

In a unit-based approach, budgeted costs for the actual level of activity are obtained by assuming that a single unit-based driver (units of product or direct labour hours) drives all costs. A cost formula is developed for each cost item as a function of units produced or direct labour hours. Exhibit 14-4 presents a unit-based flexible budget based on direct labour hours. If, however, costs vary with respect to more than one driver and the drivers are not highly correlated with direct labour hours, then the predicted costs can be misleading.

Exhibit 14-4

Flexible Budget: Direct Labour Hours

	Cost Formula		Direct Labour Hours	
	Fixed	Variable	10,000	20,000
Direct materials	—	$10	$100,000	$200,000
Direct labour	—	8	80,000	160,000
Maintenance	$ 20,000	3	50,000	80,000
Machining	15,000	1	25,000	35,000
Inspections	120,000	—	120,000	120,000
Setups	50,000	—	50,000	50,000
Purchasing	220,000	—	220,000	220,000
Total	$425,000	$22	$645,000	$865,000

[7] Robert C. Camp, *Business Process Benchmarking* (Milwaukee, WI: ASQC Quality Press, 1995): 273.
[8] Ibid.

The solution, of course, is to build flexible budget formulas for more than one driver. Cost estimation procedures (high-low method, the method of least squares, and so on) can be used to estimate and validate the cost formulas for each activity. This multiple cost-formula approach allows managers to predict more accurately what costs should be for different levels of activity usage, as measured by the activity output measure. These costs can then be compared with the actual costs to help assess budgetary performance. Exhibit 14-5 provides an example of an activity flexible budget. Notice that the budgeted amounts for direct materials and direct labour are the same as those reported in Exhibit 14-4; they use the same activity output measure. The budgeted amounts for the other items differ significantly from the traditional amounts because the activity output measures differ.

Assume that the first activity level for each driver in Exhibit 14-5 corresponds to the actual activity usage levels. Exhibit 14-6 compares the budgeted costs for the actual activity usage levels with the actual costs. One item is on target, and the other six items are mixed. The net outcome is a favourable variance of $21,500.

The performance report in Exhibit 14-6 compares total budgeted costs for the actual level of activity with the total actual costs for each activity. It is also possible to compare the actual fixed activity costs with the budgeted fixed activity costs, and the actual variable activity costs with the budgeted variable costs. Moreover, Exhibit 14-5 presents the budget formulas for each activity without any indication of how these formulas can be derived. Cornerstone 14-4 demonstrates how an activity-based flexible budget formula can be derived and then used for performance reporting with a detailed breakdown of fixed and variable activity costs.

Exhibit 14-5

Activity Flexible Budget

DRIVER: DIRECT LABOUR HOURS

	Formula		Level of Activity	
	Fixed	Variable	10,000	20,000
Direct materials	—	$10	$100,000	$200,000
Direct labour	—	8	80,000	160,000
Subtotal	—	$18	$180,000	$360,000

DRIVER: MACHINE HOURS

	Fixed	Variable	8,000	16,000
Maintenance	$20,000	$5.50	$64,000	$108,000
Machining	15,000	2.00	31,000	47,000
Subtotal	$35,000	$7.50	$95,000	$155,000

DRIVER: NUMBER OF SETUPS

	Fixed	Variable	25	30
Inspections	$80,000	$2,100	$132,500	$143,000
Setups	—	1,800	45,000	54,000
Subtotal	$80,000	$3,900	$177,500	$197,000

DRIVER: NUMBER OF ORDERS

	Fixed	Variable	15,000	25,000
Purchasing	$211,000	$1	$226,000	$236,000
Total			$678,500	$948,000

Exhibit 14-6

Activity-Based Performance Report*

	Actual Costs	Budgeted Costs	Budget Variance
Direct materials	$101,000	$100,000	$ 1,000 U
Direct labour	80,000	80,000	—
Maintenance	55,000	64,000	9,000 F
Machining	29,000	31,000	2,000 F
Inspections	125,500	132,500	7,000 F
Setups	46,500	45,000	1,500 U
Purchasing	220,000	226,000	6,000 F
Total	$657,000	$678,500	$21,500 F

*Activity levels of drivers: 10,000 direct labour hours, 8,000 machine hours, 25 setups, and 15,000 orders.

As Cornerstone 14-4 shows, breaking each variance into fixed and variable components provides more insight into the source of the variation in planned and actual expenditures. Activity budgets also provide valuable information about capacity usage.

Activity Capacity Management

Activity capacity is the number of times an activity can be performed. For example, consider inspecting finished goods as the activity. A sample from each batch is taken to determine the batch's overall quality. The demand for the inspection activity determines the amount of activity capacity. For instance, suppose that the number of batches inspected measures activity output. Now, suppose that 60 batches are scheduled to be produced. Then, the activity capacity is 60 batches. Finally, assume that a single inspector can inspect 20 batches per year. Thus, three inspectors must be hired to provide the necessary capacity. If each inspector is paid a salary of $40,000, the budgeted cost of the activity capacity is $120,000. This is the cost of the resources (labour) acquired in advance of usage. The budgeted activity rate is $2,000 per batch ($120,000/60).

Several questions relate to activity capacity and its cost. First, what *should* the activity capacity be? The answer to this question provides the ability to measure the amount of improvement possible. Second, how much of the capacity acquired was actually used? The answer to this question signals a nonproductive cost and, at the same time, an opportunity for capacity reduction and cost savings.

Capacity Variances There are two capacity variances: the *activity volume variance* and the *unused capacity variance*. The **activity volume variance** is the difference between the actual quantity capacity, (AQ) and the value-added standard quantity of activity that should be used (SQ), multiplied by the budgeted activity rate (SP):

$$\text{Activity volume variance} = (AQ - SQ)SP$$

The volume variance in this framework represents the non-value-added cost of the inspection activity. It measures the amount of improvement that is possible through analysis and management of activities. However, since the supply of the activity must be acquired in advance of usage (usually in blocks or steps, for example, one inspector at a time), it is also important to measure the current demand for the activity (actual usage). If AQ is more than SQ ($AQ > SQ$), then the variance is unfavourable (indicating that non-value-added cost is present).

The **unused capacity variance** is defined as the difference between actual quantity capacity available (AQ) and activity usage (AU), multiplied by the budgeted activity rate (SP):

$$\text{Unused capacity variance} = (AU - AQ)SP$$

**CORNERSTONE
14-4**

The HOW and WHY of Activity-Based Flexible Budgeting

Information:

Thomas Company has a "maintaining equipment" activity and wants to develop a flexible budget formula for the activity. The following resources are used by the activity:

- Three portable diagnostic units, with a lease cost of $8,000 per year per unit
- Three maintenance personnel each paid a salary of $45,000 per year (A total of 6,000 maintenance hours are supplied by the three workers.)
- Parts and supplies: $100 per diagnosis
- Maintenance hours: Four hours used per diagnosis

During the year, the activity operated at 80 percent of capacity and incurred the following actual activity and resource costs:

- Lease cost: $24,000
- Salaries: $145,000
- Parts and supplies: $135,000

Why:

The variable cost component for each activity corresponds to resources acquired as needed (flexible resources), and the fixed cost component corresponds to resources acquired in advance of usage (committed resources). Performance reporting compares the actual activity costs with the costs budgeted for the actual activity level (for a given time period).

Required:

1. Prepare a flexible budget formula for the maintenance activity using maintenance hours as the driver.
2. Prepare a performance report for the maintenance activity.
3. **What if** maintenance workers were hired through outsourcing and paid $20 per hour (the diagnostic units are still leased by Thomas)? Repeat Requirement 1 for the outsourcing case.

Solution:

1. Acquired in advance of usage:

Diagnostic equipment	$ 24,000 (3 × $8,000)
Maintenance workers	135,000 (3 × $45,000)
Total fixed costs	$159,000

Acquired as needed:
   Parts and supplies: $100/4 = $25 per maintenance hour (X)
Formula: Maintenance cost = $159,000 + $25X$

2.

Activity-Based Performance Report

Activity	Actual Cost	Budgeted Cost (80% level)*	Budget Variance
Maintenance:			
Fixed cost	$169,000	$159,000	$10,000 U
Variable cost	135,000	120,000	15,000 U

*$159,000 (fixed); $25 × 0.80 × 6,000 (variable)

3. Maintenance cost = $24,000 + $45X$ (The cost of diagnostic equipment is fixed; the variable cost is the $20 per hour of contract labour plus the $25 per hour for parts and supplies.)

The unused capacity goal is to reduce the demand for the activity until such time as the unused capacity variance equals the volume variance. When capacity exceeds demand, management can take steps to reduce it. Why? Because the activity volume variance is a non-value-added cost and the unused activity variance measures the progress made in reducing this non-value-added cost. Thus, the variance is labelled as favourable. Cornerstone 14-5 shows the calculation and usage of the capacity variances.

In Cornerstone 14-5, we know that the supply of inspection resources is greater than its usage. Assume that this unused capacity exists because management has been engaged in a quality-improvement program that has reduced the need to inspect certain batches of products. When the cost of unused capacity reaches $40,000, this difference between the supply of the inspection resources and their usage should impact future spending plans. Furthermore, because of the quality-improvement program, we can expect this difference to persist and even become greater (with the ultimate goal of reducing the cost of inspection activity to zero). Management now must be willing to exploit the unused capacity it has created. Essentially, when the savings reach the price of one inspector, activity availability can be reduced; thus, the spending on inspection can be decreased. A manager can use several options to achieve this outcome. When the inspection demand has been reduced to at most 4,000 hours, the company needs only two full-time inspectors. The extra inspector could be permanently reassigned to an activity where resources are in short supply. If reassignment is not feasible, the company should lay off the extra inspector.

This example illustrates an important feature of activity capacity management. Activity improvement can create unused capacity, but managers must be willing and able to make the tough decisions to reduce resource spending on the redundant resources to gain the potential profit increase. Profits can be increased by reducing resource spending or by transferring the resources to other activities that will generate more revenues.

Implementing Activity-Based Management

OBJECTIVE ▸ 4
Discuss the implementation issues associated with an activity-based management system.

Activity-based management (ABM) is a more comprehensive system than an activity-based costing (ABC) system. ABM adds a process view to the cost view of ABC. ABM encompasses ABC and uses it as a major source of information. ABM can be viewed as an information system that has the broad objectives of (1) improving decision making by providing accurate cost information and (2) reducing costs by encouraging and supporting process value analysis (PVA) and continuous improvement efforts. The first objective is the domain of ABC, while the second objective belongs to ABM. The second objective requires more detailed data than ABC's objective of improving the accuracy of costing assignments. If a company intends to use both ABC and PVA, then its approach to implementation must be carefully conceived. For example, if ABC creates aggregate cost pools based on homogeneity, much of the detailed activity information may not be needed. Yet for PVA, this detail must be retained. Clearly, how to implement an ABM system is a major consideration. Exhibit 14-7 provides a representation of an ABM implementation model.

Discussion of the ABM Implementation Model

The model in Exhibit 14-7 shows that the overall objective of ABM is to improve a firm's profitability, an objective achieved by identifying and selecting opportunities for improvement and using more accurate information to make better decisions. Root cause analysis, for example, reveals opportunities for improvement. By identifying non-value-added costs, priorities can be established based on the initiatives that offer the most cost reduction. Furthermore, the potential cost reduction itself is measured by ABC calculations.

Exhibit 14-7 also reveals that 10 steps define an ABM implementation: two common steps and four that are associated with either ABC or PVA. The PVA steps have been discussed extensively in this chapter, whereas the ABC steps were discussed in

**CORNERSTONE
14-5**

The HOW and WHY of Activity Capacity Management

Information:
Inspecting finished goods is the activity. Activity output is measured by inspection hours. The following data pertain to the activity for the most recent year:

Activity supply: 6,000 hours (three inspectors @ 2,000 hours per year)
Inspector cost (salary): $40,000 per year
Actual usage: 4,500 inspection hours

Why:
The volume variance measures the non-value-added cost of the inspection activity and the unused capacity variance measures the progress toward reducing the activity waste. Knowing these two variances is valuable information for managing activity capacity.

Required:
1. Calculate the volume variance and explain its significance.
2. Calculate the unused capacity variance and explain its use.
3. **What if** the actual usage is 3,500 hours? What effect will this have on capacity management?

Solution:
1. Inspection generally is classified as a non-value-added activity. Thus,

$$\text{Volume variance} = (AQ - SQ)SP$$
$$= (6,000 - 0)\$20^*$$
$$= \$120,000 \text{ U}$$

*Activity rate = ($40,000 × 3)/6,000

The volume variance is a measure of the non-value-added cost. In this case, the entire cost of the activity is non-value-added. Management should strive to find ways to reduce and eventually eliminate the activity.

2. $$\text{Unused capacity variance} = (AU - AQ)SP$$
$$= (4,500 - 6,000)\$20$$
$$= \$30,000 \text{ F}$$

The demand for the activity has been reduced; however, the reduction is not sufficient to produce a reduction in activity spending.

3. Recalculating the unused capacity variance:

$$\text{Unused capacity variance} = (AU - AQ)SP$$
$$= (3,500 - 6,000)\$20$$
$$= \$50,000 \text{ F}$$

At this level of demand, only two inspectors are needed to meet the demand; thus, resource spending can be reduced by $40,000.

Chapter 6. The two common steps are (1) systems planning and (2) activity identification, definition, and classification.

Systems Planning
Systems planning provides the justification for implementing ABM and addresses the following issues:

1. The purpose and objectives of the ABM system
2. The organization's current and desired competitive position

Exhibit 14-7

ABM Implementation Model

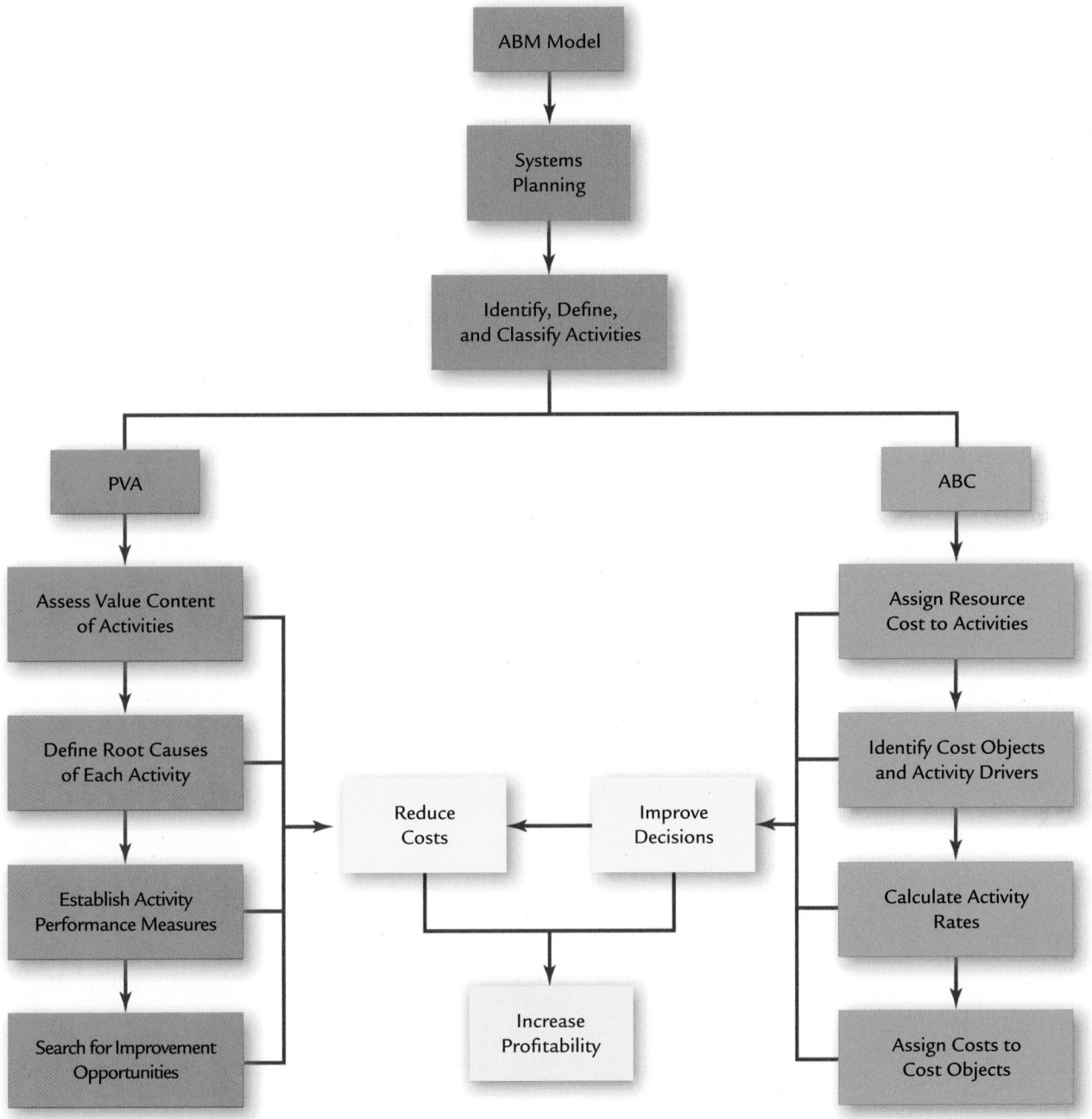

3. The organization's business processes and product mix
4. The timeline, assigned responsibilities, and resources required for implementation
5. The ability of the organization to implement, learn, and use new information

To convince operating personnel that ABM is of great value, the objectives of an ABM system must be carefully identified and related to the firm's desired competitive position, business processes, and product mix. The broad objectives have already been mentioned (improving accuracy and continuous improvement); however, it is also necessary to develop specific desired outcomes associated with each of these two objectives. For example, one specific outcome is that of changing the product mix based on more accurate costs (with the expectation that profits will increase). Another specific outcome is that of improving the firm's competitive position by increasing process efficiency through elimination of non-value-added activities. Planning also entails establishing a timeline for the implementation project, assigning specific responsibilities to individuals or teams, and developing a detailed budget. Although all five issues listed are important, the information usage issue deserves special attention. Successful implementation is strongly dependent on the organization's ability to learn how to use the new information provided by ABM. Users must be convinced that this new information can solve specific problems. They also need to be trained to use activity-based costing information to produce better decisions, and they need to understand how ABM drives and supports continuous improvement.

Identifying, defining, and classifying activities require more attention for ABM than for ABC. The activity dictionary should include a detailed listing of the tasks that define each activity. Knowing the tasks that define an activity can be very helpful for improving the efficiency of value-added activities. Classification of activities also allows ABM to connect with other continuous improvement initiatives such as JIT, total quality management, and total environmental quality cost management. For example, identifying quality-related and environmental activities enables management to focus attention on the non-value-added activities of the quality and environmental categories. ABC also provides a more complete understanding of the effect that quality and environmental costs have on products, processes, and customers. It is important to realize that successful implementation requires time and patience. This is especially true when it comes to using the new information provided by an ABM system. For example, one survey revealed that it takes an average of 3.1 years for non-accounting personnel to grow accustomed to using ABC information.[9]

Why ABM Implementations Fail

ABM can fail as a system for a variety of reasons. One of the major reasons is the lack of support of higher-level management. Not only must this support be obtained before undertaking an implementation project, but it must also be maintained. Loss of support can occur if the implementation takes too long or the expected results do not materialize.

Results may not occur as expected because operating and sales managers do not have the expertise to use the new activity information. Thus, significant efforts to train and educate need to be undertaken. Advantages of the new data need to be spelled out carefully, and managers must be taught how these data can be used to increase efficiency and productivity.

Resistance to change should be expected; it is not unusual for managers to receive the new cost information with skepticism. Showing how this information can enable them to be better managers should help to overcome this resistance. Involving nonfinancial managers in the planning and implementation stages may also reduce resistance and secure the required support.

Failure to integrate the new system is another major reason for an ABM system breakdown. The probability of success is increased if the ABM system is not in competition with other improvement programs or the official accounting system. It is important to communicate the concept that ABM complements and enhances other improvement programs. Moreover, it is important that ABM be integrated to the point that activity costing outcomes are not in direct competition with the traditional accounting numbers. Managers may be tempted to continue using the traditional accounting numbers in lieu of the new data.

---

[9] Kip R. Krumwiede, "ABC: Why It's Tried and How It Succeeds," *Management Accounting* (April 1998): 32–38.

Activity-Based Responsibility Accounting

OBJECTIVE ▶ 5
Explain how activity-based management is a form of responsibility accounting, and tell how it differs from financial-based responsibility accounting.

Responsibility accounting is a fundamental tool of managerial control and is defined by four essential elements: (1) assigning responsibility, (2) establishing performance measures or benchmarks, (3) evaluating performance, and (4) assigning rewards. The objective of responsibility accounting is to influence behaviour in such a way that individual and organizational initiatives are aligned to achieve a common goal or goals. Exhibit 14-8 illustrates the responsibility accounting model.

A particular responsibility accounting system is defined by how the four elements in Exhibit 14-8 are defined. Three types of responsibility accounting systems have evolved over time: *financial-based*, *activity-based*, and *strategic-based*. All three are found in practice today. Essentially, firms choose the responsibility accounting system that is compatible with the requirements and economics of their particular operating environment. Firms that operate in a stable environment with standardized products and processes and low competitive pressures will likely find the less complex, financial-based responsibility accounting systems to be quite adequate. As organizational complexity increases and the competitive environment becomes much more dynamic, activity-based and strategic-based systems are likely to be more suitable. Strategic-based responsibility accounting systems are discussed in Chapter 15.

The responsibility accounting system for a stable environment is referred to as *financial-based responsibility accounting*. A **financial-based responsibility accounting system** assigns responsibility to organizational units and expresses performance measures in financial terms. It emphasizes a financial perspective. *Activity-based responsibility accounting*, on the other hand, is the responsibility accounting system developed for those firms operating in continuous improvement environments. **Activity-based responsibility accounting** assigns responsibility to processes and uses both financial and nonfinancial measures of performance, thus emphasizing both financial and process perspectives. A comparison of each of the four elements of the responsibility accounting model for each responsibility system reveals the key differences between the two approaches.

Exhibit 14-8

The Responsibility Accounting Model

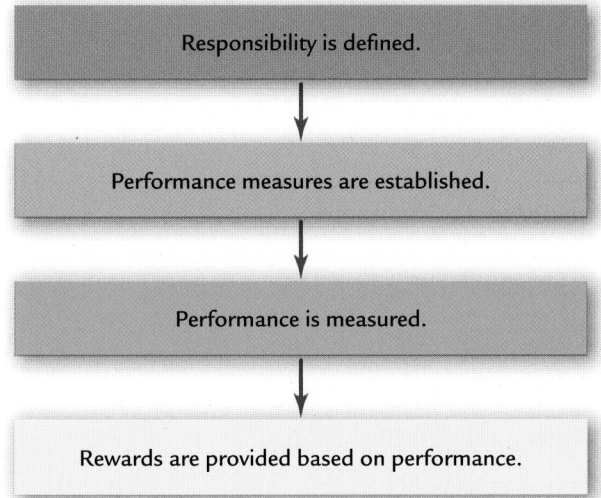

Responsibility is defined.

↓

Performance measures are established.

↓

Performance is measured.

↓

Rewards are provided based on performance.

Assigning Responsibility

Exhibit 14-9 lists the differences in responsibility assignments between the two systems. Financial-based responsibility accounting focuses on *functional* organizational units and individuals. First, a responsibility centre is identified. This centre is typically an organizational unit such as a plant, department, or production line. Whatever the functional unit is, responsibility is assigned to the individual in charge. Responsibility is defined in financial terms (e.g., costs). Emphasis is on achieving optimal financial results at the local level (i.e., organizational unit level). Exhibit 14-9 reveals that in an activity or process-based responsibility system, the focal point changes from units and individuals to processes and teams. Systemwide optimization is the emphasis. Also, financial responsibility continues to be vital. The reasons for the change in focus are simple. In a continuous improvement environment, the financial perspective translates into continuously *enhancing revenues*, *reducing costs*, and *improving asset utilization*. Creating this continuous growth and improvement requires an organization to constantly improve its capabilities of delivering value to customers and shareholders. A process perspective is chosen instead of an organizational-unit perspective because processes are the *sources* of value for customers and shareholders and because they are the key to achieving an organization's financial objectives. The customer can be internal or external to the organization. Procurement, new product development, manufacturing, and customer service are examples of processes.

Since processes are the way things are done, changing the way things are done means changing processes. Three methods can change the way things are done: *process improvement*, *process innovation*, and *process creation*.

Process improvement refers to incremental and constant increases in the efficiency of an existing process. For example, **Medtronic Xomed**, a manufacturer of surgical products (for ears, nose, and throat specialists), improved its processes by providing written instructions telling workers the best way to do their jobs. Over a three-year period, the company reduced rework by 57 percent, reduced scrap by 85 percent, and experienced a 38 percent reduction in the cost of its shipped products.[10] Activity-based management is particularly useful for bringing about process improvements. Processes are made up of activities that are linked by a common objective. Listing these activities and classifying them as value-added or non-value-added immediately suggests a way to make the process better: eliminate the non-value-added activities.

Process innovation (business re-engineering) refers to the performance of a process in a radically new way with the objective of achieving dramatic improvements in response time, quality, and efficiency. **IBM Credit**, for example, radically redesigned its credit approval process and reduced its time for preparing a quote from seven days to one; similarly, **Federal-Mogul**, a parts manufacturer, used process innovation to reduce development time for part prototypes from 20 weeks to 20 days.[11]

Exhibit 14-9

Responsibility Assignments Compared

Financial-Based Responsibility	Activity-Based Responsibility
1. Organizational units	1. Processes
2. Local operating efficiency	2. Systemwide efficiency
3. Individual accountability	3. Team accountability
4. Financial outcomes	4. Financial outcomes

---

[10] William Leventon, "Manufacturers Get Lean to Trim Waste," *Medical Device & Diagnostic Industry* (September 2004), http://www.devicelink.com/mddi/archive/04/09/016.html.

[11] Thomas H. Davenport, *Process Innovation* (Boston: Harvard Business School Press, 1993): 2.

Process creation refers to the installation of an entirely new process with the objective of meeting customer and financial objectives. **Chemical Bank**, for example, identified three *new* internal processes: understanding customer segments, developing new products, and cross-selling the product line.[12] These new internal processes were viewed as critical by the bank's management for improving the customer and profit mix and creating an enabled organization. It should be mentioned that process creation does not mean that the process has to be *original* to the organization. It means that it is *new* to the organization. For example, developing new products is a process common to many organizations but evidently was new to Chemical Bank.

Many processes cut across functional boundaries. This facilitates an integrated approach that emphasizes the firm's value-chain activities. It also means that cross-functional skills are needed for effective process management. Teams are the natural outcome of this process management requirement. Teams also improve the quality of work life by fostering friendships and a sense of belonging. Process improvement, innovation, and creation require significant group activity (and support) and cannot be carried out effectively by individuals. **General Electric**, **Xerox**, **Martin Marietta Materials**, and **Aetna** have all begun to use teams as their basic work unit.[13]

Establishing Performance Measures

Once responsibility is defined, performance measures must be identified and standards set to serve as benchmarks for performance measurement. Exhibit 14-10 provides a comparison of the two systems' approach to the task of defining performance measures. According to Exhibit 14-10, budgeting and standard costing are the cornerstones of the benchmark activity for a financial-based system. This, of course, implies that performance measures are objective and financial in nature. Furthermore, they tend to support the status quo and are relatively stable over time. Exhibit 14-10 reveals some striking differences for firms operating in a continuous improvement environment.

First, performance measures are process-oriented and, thus, must be concerned with process attributes such as process time, quality, and efficiency. Second, performance measurement standards are structured to support change. Therefore, standards are dynamic in nature. They change to reflect new conditions and new goals and to help maintain any progress that has been realized. For example, standards can be set that reflect some desired level of improvement for a process. Once the desired level is achieved, the standard is changed to encourage an additional increment of improvement. In an environment where constant improvement is sought, standards cannot be static. Third, optimal standards assume a vital role. They set the ultimate achievement target and, thus, identify the potential for improvement. Finally, standards should reflect the value added by individual activities and processes. Identifying a value-added standard for each activity is much more ambitious than the traditional financial responsibility system. It expands control to include the entire organization.

Exhibit 14-10

Performance Measures Compared

Financial-Based Measures	Activity-Based Measures
1. Organizational unit budgets	1. Process-oriented standards
2. Standard costing	2. Value-added standards
3. Static standards	3. Dynamic standards
4. Currently attainable standards	4. Optimal standards

---

[12] Norman Klein and Robert Kaplan, *Chemical Bank: Implementing the Balanced Scorecard* (Harvard Business School, Case 125–210, 1995): 5–6.

[13] Davenport, *Process Innovation*, 97.

Performance Evaluation Compared

Financial-Based Performance Evaluation	Activity-Based Performance Evaluation
1. Financial efficiency	1. Time reductions
2. Controllable costs	2. Quality improvements
3. Actual versus standard	3. Cost reductions
4. Financial measures	4. Trend measurement

Evaluating Performance

Exhibit 14-11 compares performance evaluation under financial- and activity-based responsibility accounting systems. In a financial-based framework, performance is measured by comparing actual outcomes with budgeted outcomes. In principle, individuals are held accountable only for those items over which they have control. Financial performance, as measured by the ability to meet or beat a stable financial standard, is strongly emphasized. In the activity-based framework, performance is concerned with more than just the financial perspective. The process perspective adds time, quality, and efficiency as critical dimensions of performance. Decreasing the time a process takes to deliver its output to customers is viewed as a vital objective. Thus, nonfinancial, process-oriented measures such as cycle-time and on-time deliveries become important. Performance is evaluated by gauging whether these measures are improving over time. The same is true for measures relating to quality and efficiency. Improving a process should translate into better financial results. Hence, measures of cost reductions achieved, trends in cost, and cost per unit of output are all useful indicators of whether a process has improved. Progress toward achieving optimal standards and interim standards needs to be measured. The objective is to provide low-cost, high-quality products, delivered on a timely basis.

Assigning Rewards

In both systems, individuals are rewarded or penalized according to the policies and discretion of higher management. As Exhibit 14-12 shows, many of the same financial instruments (e.g., salary increases, bonuses, profit sharing, and promotions) are used to provide rewards for good performance. Of course, the nature of the incentive structure differs in each system. For example, the reward system in a financial-based responsibility accounting system is designed to encourage individuals to achieve or beat budgetary standards. Furthermore, for the activity-based responsibility system, rewarding individuals is more complicated than it is in a unit-based setting. Individuals simultaneously have accountability for team and individual performance. Since process-related improvements are mostly achieved through team

Rewards Compared

Financial-Based Rewards	Activity-Based Rewards
1. Financial performance basis	1. Multidimensional performance basis
2. Individual rewards	2. Group rewards
3. Salary increases	3. Salary increases
4. Promotions	4. Promotions
5. Bonuses and profit sharing	5. Bonuses, profit sharing, and gainsharing

efforts, group-based rewards are more suitable than individual rewards. In one company (a producer of electronic components), for example, optimal standards have been set for unit costs, on-time delivery, quality, inventory turns, scrap, and cycle time.[14] Bonuses are awarded to the team whenever performance is maintained on all measures and improves on at least one measure. Notice the multidimensional nature of this measurement and reward system. Another difference concerns the notion of gainsharing versus profit sharing. Profit sharing is a global incentive designed to encourage employees to contribute to the overall financial well-being of the organization. Gainsharing is more specific. Employees are allowed to share in gains related to specific improvement projects. Gainsharing helps obtain the necessary buy-in for specific improvement projects inherent to activity-based management.

Summary of Learning Objectives

1. **Describe how activity-based management and activity-based costing differ.**

- Activity-based management encompasses both activity-based costing and process value analysis.
- Activity-based costing is concerned with accurate assignment of costs to cost objects and is an important source of information for managing activities. ABC, however, is not concerned with the issue or presence of waste in activities.
- Identifying waste and its causes and eliminating it fall within the domain of process value analysis.

2. **Define process value analysis.**

- Process value analysis emphasizes activity management with the intent of maximizing systemwide performance. It consists of three elements: driver analysis, activity analysis, and performance measurement.
- Driver analysis is also referred to as root cause analysis. It seeks to identify why activities are performed.
- Activity analysis identifies all activities and the resources they consume and classifies activities as value-added or non-value-added.
- Performance measurement is concerned with how well activities are performed.

3. **Describe activity-based financial performance measurement.**

- Reporting value- and non-value-added costs is an integral part of a sound activity-based management system. Tracking trends in these costs over time is an effective control measure.
- Once management determines the source of non-value-added costs, a focused program of continuous improvement can be implemented.
- Kaizen costing is a well-accepted approach for reducing costs by eliminating waste.
- Activity flexible budgeting and activity capacity management offer additional control capabilities.
- Activity flexible budgeting differs from the traditional approach by using more than unit-level drivers to predict what costs will be at different levels of activity output.
- Activity capacity management involves identification of the volume variance (non-value-added cost) and the unused capacity variance (progress toward reducing non-value-added cost).

---

[14] C. J. McNair, "Responsibility Accounting and Controllability Networks," *Handbook of Cost Management* (Boston: Warren Gorham Lamont, 1993): E41–E43.

4. **Discuss the implementation issues associated with an activity-based management system.**

- Implementing an activity-based management system requires careful planning and execution.
- The objectives of the system must be identified and explained.
- The benefits of the system and the anticipated effects should also be noted.
- A key issue is assessing and managing the ability of the organization to implement, learn, and use the new activity information. Strong support from higher management is critical for this process.

5. **Explain how activity-based management is a form of responsibility accounting, and tell how it differs from financial-based responsibility accounting.**

- A firm can adopt one of three responsibility accounting systems.
- Two are discussed in this chapter: financial-based responsibility accounting and activity-based responsibility accounting.
- Financial-based responsibility accounting focuses on organizational units such as departments and plants; uses financial outcome measures, static standards, and benchmarks to evaluate performance; and emphasizes status quo and organizational stability.
- Activity-based responsibility accounting focuses on processes, uses both operational and financial measures, employs dynamic standards, and emphasizes and supports continuous improvement.

**CORNERSTONES
FOR CHAPTER 14**

CORNERSTONE 14-1   The HOW and WHY of value- and non-value-added cost reporting, page 710

CORNERSTONE 14-2   The HOW and WHY of non-value-added cost trend reporting, page 711

CORNERSTONE 14-3   The HOW and WHY of kaizen costing, page 714

CORNERSTONE 14-4   The HOW and WHY of activity-based flexible budgeting, page 718

CORNERSTONE 14-5   The HOW and WHY of activity capacity management, page 720

Review Problems

I. Financial-Based Responsibility Accounting versus Activity-Based Responsibility Accounting

The labour standard for a company is two hours per unit produced, which includes setup time. At the beginning of the last quarter, 20,000 units had been produced and 44,000 hours used. The production manager was concerned about the prospect of reporting an unfavourable labour efficiency variance at the end of the year. Any unfavourable variance over 9 to 10 percent of the standard usually meant a negative performance rating. Bonuses were adversely affected by negative ratings. Accordingly, for the last quarter, the production manager decided to reduce the number of setups and use longer production runs. He knew that his production workers usually were within 5 percent of the standard. The real problem was with setup times. By reducing the setups, the actual hours used would be within 7 to 8 percent of the standard hours allowed.

Required:

1. Explain why the behaviour of the production manager is unacceptable for a continuous improvement environment.
2. Explain how an activity-based responsibility accounting approach would discourage the kind of behaviour described.

Solution:

1. In a continuous improvement environment, efforts are made to reduce inventories and eliminate non-value-added costs. The production manager is focusing on meeting the labour usage standard and is ignoring the impact on inventories that longer production runs may have.
2. Activity-based responsibility accounting focuses on activities and activity performance. For the setup activity, the value-added standard would be zero setup time and zero setup costs. Thus, avoiding setups would neither save labour time nor affect the labour variance. Of course, labour variances themselves would not be computed—at least not at the operational level.

II. Activity Volume Variance, Unused Activity Capacity, Value- and Non-Value-Added Cost Reports, Kaizen Standards

Pollard Manufacturing has developed value-added standards for its activities including material usage, purchasing, and inspecting. The value-added output levels for each of the activities, their actual levels achieved, and the standard prices are as follows:

Activity	Activity Driver	SQ	AQ	SP
Using lumber	Board feet	24,000	30,000	$10
Purchasing	Purchase orders	800	1,000	50
Inspecting	Inspection hours	0	4,000	12

Assume that material usage and purchasing costs correspond to flexible resources (acquired as needed) and that inspection uses resources that are acquired in blocks or steps of 2,000 hours. The actual prices paid for the inputs equal the standard prices.

Required:

1. Assume that continuous improvement efforts reduce the demand for inspection by 30 percent during the year (actual activity usage drops by 30 percent). Calculate the volume and unused capacity variances for the inspection activity. Explain their meaning. Also, explain why there is no volume or unused capacity variance for the other two activities.
2. Prepare a cost report that details value- and non-value-added costs.
3. Suppose that the company wants to reduce all non-value-added costs by 30 percent in the coming year. Prepare kaizen standards that can be used to evaluate the company's progress toward this goal. How much will these measures save in resource spending?

Solution:

1.

$SP \times SQ$	$SP \times AQ$	$SP \times AU$
12×0	$12 \times 4,000$	$12 \times 2,800$
$0	$48,000	$33,600

Volume Variance	Unused Capacity Variance
$48,000 U	$14,400 F

The activity volume variance is the non-value-added cost. The unused capacity variance measures the cost of the unused activity capacity. The other two activities have no volume variance or capacity variance because they use only flexible resources. No activity capacity is acquired in advance of usage; thus, there cannot be an unused capacity variance or a volume variance.

2.

	Costs		
	Value-Added	Non-Value-Added	Total
Using lumber	$240,000	$ 60,000	$300,000
Purchasing	40,000	10,000	50,000
Inspecting	0	48,000	48,000
Total	$280,000	$118,000	$398,000

3.

	Kaizen Standards	
	Quantity	Cost
Using lumber	28,200	$282,000
Purchasing	940	47,000
Inspecting	2,800	33,600

If the standards are met, then the savings are as follows:

Using lumber: $10 × 1,800 = $18,000
Purchasing: $50 × 60   =   3,000
Savings $21,000

There is no reduction in resource spending for inspecting because it must be purchased in increments of 2,000 and only 1,200 hours were saved—another 800 hours must be reduced before any reduction in resource spending is possible. The unused capacity variance must reach $24,000 before resource spending can be reduced.

Key Terms

Activity analysis, 704

Activity capacity, 717

Activity elimination, 707

Activity flexible budgeting, 715

Activity output measure, 704

Activity reduction, 707

Activity selection, 707

Activity sharing, 707

Activity volume variance, 717

Activity-based management (ABM), 703

Activity-based responsibility accounting, 723

Benchmarking, 713

Continuous improvement, 702

Driver analysis, 704

Financial measures, 708

Financial-based responsibility accounting system, 723

Kaizen costing, 707

Kaizen standard, 712

Non-value-added activities, 705

Non-value-added costs, 706

Process creation, 725

Process improvement, 724

Process innovation (business re-engineering), 724

Process value analysis (PVA), 704

Responsibility accounting, 723

Unused capacity variance, 717

Value-added activities, 705

Value-added costs, 705

Value-added standard, 708

Discussion Questions

1. What are the two dimensions of the activity-based management model? How do they differ?
2. What is driver analysis? What role does it play in process value analysis?
3. What is activity analysis? Why is this approach compatible with the goal of continuous improvement?
4. What are value-added activities? Value-added costs?
5. What are non-value-added activities? Non-value-added costs? Give an example of each.
6. Identify and define four different ways to manage activities so that costs can be reduced.
7. What is a kaizen standard? Describe the kaizen and maintenance subcycles.
8. Explain how benchmarking can be used to improve activity performance.
9. Explain how activity flexible budgeting differs from unit-based flexible budgeting.

10. In implementing an ABM system, what are some of the planning considerations?
11. Explain why a detailed task description is needed for ABM and not for ABC.
12. What are some of the reasons that ABM implementation may lose the support of higher management?
13. Explain how lack of integration of an ABM system may cause its failure.
14. Describe a financial-based responsibility accounting system.
15. Describe an activity-based responsibility accounting system. How does it differ from financial-based responsibility accounting?

Cornerstone Exercises

Cornerstone Exercise 14-1 VALUE- AND NON-VALUE-ADDED COST REPORTING

OBJECTIVE ▶ 3

CORNERSTONE 14-1

SERVICE

Espera Distribution Centre has four activities: receiving materials, assembly, expediting products, and storing goods. Receiving and assembly are necessary activities; expediting and storing goods are unnecessary. The following data pertain to the four activities for the year ending 2012 (actual price per unit of the activity driver is assumed to be equal to the standard price):

Activity	Activity Driver	SQ	AQ	SP
Receiving	Receiving orders	8,000	12,000	$21
Assembly	Labour hours	50,000	60,000	15
Expediting	Orders expedited	0	4,000	50
Storing	Number of units	0	8,000	7

Required:

1. Prepare a cost report for the year ending 2012 that shows value-added costs, non-value-added costs, and total costs for each activity.
2. Explain why expediting products and storing goods are non-value-added activities.
3. *What if* receiving cost is a step-fixed cost with each step being 1,000 orders whereas assembly cost is a variable cost? What is the implication for reducing the cost of waste for each activity?

Cornerstone Exercise 14-2 TREND REPORTING FOR NON-VALUE-ADDED COSTS

OBJECTIVE ▶ 3

CORNERSTONE 14-2

SERVICE

Refer to **Cornerstone Exercise 14-1**. Assume that at the beginning of 2013, Espera trained the assembly workers in a new approach that had the objective of increasing the efficiency of the assembly process. Espera also began moving toward a JIT purchasing system. When JIT is fully implemented, the demand for expediting is expected to be virtually eliminated. It is expected to take two to three years for full implementation. Assume that receiving cost is a step-fixed cost with steps of 1,000 orders. The other three activities employ resources that are acquired as used and needed. At the end of 2013, the following results were reported for the four activities:

Activity	Activity Driver	SQ	AQ	SP
Receiving	Receiving orders	8,000	8,000	$21
Assembly	Labour hours	50,000	52,000	15
Expediting	Orders expedited	0	2,000	50
Storing	Number of units	0	4,000	7

Required:

1. Prepare a trend report that shows the non-value-added costs for each activity for 2012 and 2013 and the change in costs for the two periods. Discuss the report's implications.
2. Explain the role of activity reduction for receiving and for expediting. What is the expected value of SQ for each activity after JIT is fully implemented?

3. **What if** at the end of 2013, the selling price of a competing product is reduced by $18 per unit? Assume that the firm produces and sells 20,000 units of its product and that its product is associated only with the four activities being considered. By virtue of the waste-reduction savings, can the competitor's price reduction be matched without reducing the unit profit margin of the product that prevailed at the beginning of the year? If not, how much more waste reduction is needed to achieve this outcome? In this case, what price decision would you recommend?

OBJECTIVE ▶ 3
CORNERSTONE 14-3

Cornerstone Exercise 14-3 KAIZEN COSTING

Lansky Inc. produces custom-made machine parts. A setup activity is required for the batches of parts that it produces. Activity output is measured using setup hours. The value-added standard (*SQ*) for this activity is zero. On July 1, at the beginning of the fiscal year, five setup hours were allowed and used per batch. The standard wage rate for setup labour is $15 per setup hour. During the first quarter of the new fiscal year, the company is planning to implement a new setup method developed by Lansky's industrial engineers that is expected to reduce setup time by 40 percent. The new procedure was implemented during the first quarter and the improvement expected was realized.

Required:

1. What is the setup standard for setup hours and the associated expected cost at the beginning of the first quarter? The kaizen standard and expected associated cost?
2. What is the setup standard for setup hours and the associated cost at the end of the first quarter? Explain. What is the next step in the kaizen cost reduction process?
3. **What if** the new procedure implemented in the first quarter only produced a 30 percent reduction in setup time instead of the expected 40 percent reduction? What would the new maintenance standard and cost be? What criteria would you logically expect to be met before maintenance standards and costs are modified?

OBJECTIVE ▶ 3
CORNERSTONE 14-4

Cornerstone Exercise 14-4 ACTIVITY-BASED FLEXIBLE BUDGETING

Balzac Company sells iron railings for indoor and outdoor use. In order to assemble the railings to custom size, Balzac welds them together. Balzac wants to develop a flexible budget formula for the welding. The following resources are used by the activity:

* Four welding units, with a lease cost of $12,000 per year per unit
* Six welding employees each paid a salary of $50,000 per year (A total of 9,000 welding hours are supplied by the six workers.)
* Welding supplies: $300 per job
* Welding hours: Three hours used per job

During the year, the activity operated at 90 percent of capacity and incurred the following actual activity and resource costs:

* Lease cost: $48,000
* Salaries: $315,000
* Parts and supplies: $805,000

Required:

1. Prepare a flexible budget formula for the welding activity using welding hours as the driver.
2. Prepare a performance report for the welding activity.
3. **What if** welders were hired through outsourcing and paid $30 per hour (the welding equipment is provided by Balzac)? Repeat Requirement 1 for the outsourcing case.

OBJECTIVE ▶ 3
CORNERSTONE 14-5

Cornerstone Exercise 14-5 ACTIVITY CAPACITY MANAGEMENT

Uchdorf Manufacturing just completed a study of its purchasing activity with the objective of improving its efficiency. The driver for the activity is number of purchase orders. The following data pertain to the activity for the most recent year:

Activity supply: five purchasing agents capable of processing 2,400 orders per year (12,000 orders)

Purchasing agent cost (salary): $45,600 per year

Actual usage: 10,600 orders per year

Value-added quantity: 7,000 orders per year

Required:

1. Calculate the volume variance and explain its significance.
2. Calculate the unused capacity variance and explain its use.
3. *What if* the actual usage drops to 9,000 orders? What effect will this have on capacity management? What will be the level of spending reduction if the value-added standard is met?

Exercises

Exercise 14-6 ABC VERSUS ABM

OBJECTIVE ➤ 1 2

Fresco Inc. produces elite juicers (priced at $300) as well as supreme juicers (priced at $180). Recently, Fresco has been losing market share with its supreme juicers because of competitors offering juicers with the same quality and features but at a lower price. A careful market study revealed that if Fresco could reduce the price of its supreme juicer to $170, it would regain its former share of the market. Management, however, is convinced that any price reduction must be accompanied by a cost reduction of the same amount so that per-unit profitability is not affected. Charles McManus, company controller, has indicated that poor overhead costing assignments may be distorting management's view of each product's cost and, therefore, the ability to know how to set selling prices. Charles has identified the following overhead activities: assembly, inspection, and rework. The three activities, their costs, and practical capacities are as follows:

Activity	Cost	Practical Capacity
Assembly	$2,700,000	90,000 assembly hours
Inspection	1,800,000	45,000 inspection hours
Rework	900,000	45,000 rework hours

The consumption patterns of the two products are as follows:

	Supreme	Elite
Units	100,000	30,000
Assembly hours	50,000	40,000
Inspection hours	10,000	35,000
Rework hours	7,500	37,500

Fresco assigns overhead costs to the two products using a plantwide rate based on direct labour (assembly) hours.

Required:

1. Calculate the unit overhead cost of the supreme juicer using assembly hours to assign overhead costs. Now, repeat the calculation using ABC to assign overhead costs. Did improving the accuracy of cost assignments solve Fresco's competitive problem? What did it reveal?
2. Now, assume that *in addition* to improving the accuracy of cost assignments, Charles observes that defective supplier components are the root cause of both the inspection and rework activities. Suppose further that Fresco has found a new supplier that provides higher-quality components such that inspection and rework costs are reduced by 50 percent. Now, calculate the cost of each product (assuming that inspection and rework time are also reduced by 50 percent) using ABC. The relative consumption patterns also remain the same. Comment on the difference between ABC and ABM.

OBJECTIVE ➤

Exercise 14-7 ROOT CAUSE (DRIVER ANALYSIS)

For the following two activities, ask a series of "why" questions (with your answers) that reveal the root cause. Once the root cause is identified, use a "how" question to reveal how the activity can be improved (with your answer).

Activity 1: Daily cleaning of a puddle of oil near production machinery.
Activity 2: Providing customers with sales allowances.

OBJECTIVE ➤

Exercise 14-8 NON-VALUE-ADDED ACTIVITIES: NON-VALUE-ADDED COST

Epsilon Company has 20 clerks who work in its Accounts Payable Department. A study revealed the following activities and the relative time demanded by each activity:

Activities	Percentage of Clerical Time
Comparing purchase orders and receiving orders and invoices	15%
Resolving discrepancies among the three documents	70
Preparing cheques for suppliers	10
Making journal entries and mailing cheques	5

The average salary of a clerk is $30,000.

Required:

Classify the four activities as value-added or non-value-added, and calculate the clerical cost of each activity. For non-value-added activities, indicate why they are non-value-added.

OBJECTIVE ➤

Exercise 14-9 ROOT CAUSE (DRIVER ANALYSIS)

Refer to **Exercise 14-8.**

Required:

Suppose that clerical error—either Epsilon's or the supplier's—is the common root cause of the non-value-added activities. For each non-value-added activity, ask a series of "why" questions that identify clerical error as the activity's root cause.

OBJECTIVE ➤ 2 5

Exercise 14-10 PROCESS IMPROVEMENT/INNOVATION

Refer to **Exercise 14-8.** Suppose that clerical error is the common root cause of the non-value-added activities. Paying bills is a subprocess that belongs to the procurement process. The procurement process is made up of three subprocesses: purchasing, receiving, and paying bills.

Required:

1. What is the definition of a process? Identify the common objective for the procurement process. Repeat for each subprocess.
2. Now, suppose that Epsilon decides to attack the root cause of the non-value-added activities of the bill-paying process by improving the skills of its purchasing and receiving clerks. As a result, the number of discrepancies found drops by 30 percent. Discuss the potential effect this initiative might have on the bill-paying process. Does this initiative represent process improvement or process innovation? Explain.

OBJECTIVE ➤ 2 5

Exercise 14-11 PROCESS IMPROVEMENT/INNOVATION

Refer to **Exercise 14-10.** Suppose that Epsilon attacks the root cause of the non-value-added activities by establishing a totally different approach to procurement called electronic data interchange (EDI). EDI gives suppliers access to Epsilon's online database that reveals Epsilon's production schedule. By knowing Epsilon's production schedule, suppliers can deliver the parts and supplies needed just in time for their use. When the parts are shipped, an electronic message is sent from the supplier to Epsilon that the shipment is en route. When the order arrives, a bar code is scanned with an electronic wand initiating payment for the goods. EDI involves no paper—no purchase orders—no receiving orders—and no invoices.

Required:

Discuss the potential effects of this solution on Epsilon's bill-paying process. Is this process innovation or process improvement? Explain.

Exercise 14-12 VALUE- AND NON-VALUE-ADDED COSTS, UNUSED CAPACITY

OBJECTIVE ➤ 2 3

For Situations 1 through 6, provide the following information:

a. An estimate of the non-value-added cost caused by each activity.
b. The root causes of the activity cost (such as plant layout, process design, and product design).
c. The appropriate cost reduction measure: activity elimination, activity reduction, activity sharing, or activity selection.

1. It takes 45 minutes and six kilograms of material to produce a product using a traditional manufacturing process. A process re-engineering study provided a new manufacturing process design (using existing technology) that would take 15 minutes and four kilograms of material. The cost per labour hour is $12, and the cost per kilogram of material is $8.

2. With its original design, a product requires 15 hours of setup time. Redesigning the product could reduce the setup time to an absolute minimum of 30 minutes. The cost per hour of setup time is $200.

3. A product currently requires eight moves. By redesigning the manufacturing layout, the number of moves can be reduced from eight to zero. The cost per move is $10.

4. Inspection time for a plant is 8,000 hours per year. The cost of inspection consists of salaries of four inspectors, totalling $120,000. Inspection also uses supplies costing $2 per inspection hour. A supplier evaluation program, product redesign, and process redesign reduced the need for inspection by creating a zero-defect environment.

5. Each unit of a product requires five components. The average number of components is 5.3 due to component failure, requiring rework and extra components. By developing relations with the right suppliers and increasing the quality of the purchased component, the average number of components can be reduced to five components per unit. The cost per component is $600.

6. A plant produces 100 different electronic products. Each product requires an average of eight components that are purchased externally. The components are different for each part. By redesigning the products, it is possible to produce the 100 products so that they all have four components in common. This will reduce the demand for purchasing, receiving, and paying bills. Estimated savings from the reduced demand are $900,000 per year.

Exercise 14-13 CALCULATION OF VALUE- AND NON-VALUE-ADDED COSTS, ACTIVITY VOLUME AND UNUSED CAPACITY VARIANCES

OBJECTIVE ➤ 2 3 4

Maquina Company produces custom-made machine parts. Maquina recently has implemented an activity-based management (ABM) system with the objective of reducing costs. Maquina has begun analyzing each activity to determine ways to increase its efficiency. Setting up equipment was among the first group of activities to be carefully studied. The study revealed that setup hours was a good driver for the activity. During the last year, the company incurred fixed setup costs of $504,000 (salaries of 14 employees). The fixed costs provide a capacity of 28,000 hours (2,000 per employee at practical capacity). The setup activity was viewed as necessary, and the value-added standard was set at 2,000 hours. Actual setup hours used in the most recent period were 26,200.

Required:

1. Calculate the volume and unused capacity variances for the setup activity. Explain what each variance means.

2. Prepare a report that presents value-added, non-value-added, and actual costs for setup. Explain why highlighting the non-value-added costs is important.

3. Assume that management is able to reduce the demand for the setup activity so that the actual hours needed drop from 26,200 to 4,000. What actions should now be taken regarding activity capacity management?

4. Another activity studied was inspection of supplier materials/components. Explain why inspecting incoming goods should be viewed as a non-value-added activity. In providing your explanation, consider the following counterargument: "Inspecting incoming goods adds value because it reduces the demand for other unnecessary activities such as rework, reordering, and warranty work."

OBJECTIVE **Exercise 14-14 COST REPORT, VALUE-ADDED AND NON-VALUE-ADDED COSTS**

Vecchio Company has developed value-added standards for four activities: purchasing parts, receiving parts, moving parts, and setting up equipment. The activities, the activity drivers, the standard and actual quantities, and the price standards for 2012 are as follows:

Activities	Activity Driver	SQ	AQ	SP
Purchasing parts	Purchase orders	1,300	1,820	$200
Receiving parts	Receiving orders	2,600	3,900	130
Moving parts	Number of moves	0	1,300	260
Setting up equipment	Setup hours	0	5,200	78

The actual prices paid per unit of each activity driver were equal to the standard prices.

Required:

1. Prepare a cost report that lists the value-added, non-value-added, and actual costs for each activity.

2. Which activities are non-value-added? Explain why. Also, explain why value-added activities can have non-value-added costs.

OBJECTIVE **Exercise 14-15 TREND REPORT, NON-VALUE-ADDED COSTS**

Refer to **Exercise 14-14.** Suppose that for 2013, Vecchio Company has chosen suppliers that provide higher-quality parts and redesigned its plant layout to reduce material movement. Additionally, Vecchio implemented a new setup procedure and provided training for its purchasing agents. As a consequence, less setup time is required and fewer purchasing mistakes are made. At the end of 2013, the following information is provided:

Activities	Activity Driver	SQ	AQ	SP
Purchasing parts	Purchase orders	1,300	1,560	$200
Receiving parts	Receiving orders	3,380	4,056	130
Moving parts	Number of moves	0	420	260
Setting up equipment	Setup hours	0	1,300	78

Required:

1. Prepare a report that compares the non-value-added costs for 2013 with those of 2012.

2. What is the role of activity reduction for non-value-added activities? For value-added activities?

3. Comment on the value of a trend report.

OBJECTIVE **Exercise 14-16 IMPLEMENTATION OF ACTIVITY-BASED MANAGEMENT**

Asma Haji, manager of an electronics division, was not pleased with the results that had recently been reported concerning the division's activity-based management implementation project. For one thing, the project had taken eight months longer than projected and had exceeded the budget by nearly 35 percent. But even more vexatious was the fact that after all was said and done, about three-fourths of the plants were reporting that the activity-based product costs were not much different for most of the products than those of the old costing system. Plant managers were indicating that they were continuing to use the old costs as they were easier to compute and understand. Yet, at the same time, they were complaining that they were having a hard time meeting the bids of competitors. Reliable sources were also revealing that the division's product costs were higher than many competitors'. This outcome perplexed plant managers because their control system continued to report favourable materials and labour efficiency variances. They complained that ABM had failed to produce any significant improvement in cost performance.

Asma decided to tour several of the plants and talk with the plant managers. After the tour, she realized that her managers did not understand the concept of non-value-added costs nor did they have a good grasp of the concept of kaizen costing. No efforts were being made to carefully consider the activity information that had been produced. One typical plant manager threw up his hands and said: "This is too much data. Why should I care about all this detail? I do not see how this can help me improve my plant's performance. They tell me that inspection is not a necessary activity and does not add value. I simply can't believe that inspecting isn't value-added and necessary. If we did not inspect, we would be making and sending more bad products to customers."

Required:

Explain why Asma's division is having problems with its ABM implementation.

Exercise 14-17 FINANCIAL-BASED VERSUS ACTIVITY-BASED RESPONSIBILITY ACCOUNTING

OBJECTIVE ▶ 5

For each of the following situations, two scenarios are described, labelled A and B. Choose which scenario is descriptive of a setting corresponding to activity-based responsibility accounting and which is descriptive of financial-based responsibility accounting. Provide a brief commentary on the differences between the two systems for each situation, addressing the possible advantages of the activity-based view over the financial-based view.

Situation 1

A: The purchasing manager, receiving manager, and accounts payable manager are given joint responsibility for procurement. The instructions given to the group of managers are to reduce costs of acquiring materials, decrease the time required to obtain materials from outside suppliers, and reduce the number of purchasing mistakes (e.g., wrong type of materials or the wrong quantities ordered).

B: The plant manager commended the manager of the Grinding Department for increasing his department's machine utilization rates—and doing so without exceeding the department's budget. The plant manager then asked other department managers to make an effort to obtain similar efficiency improvements.

Situation 2

A: Delivery mistakes had been reduced by 70 percent, saving over $40,000 per year. Furthermore, delivery time to customers had been cut by two days. According to company policy, the team responsible for the savings was given a bonus equal to 25 percent of the savings attributable to improving delivery quality. Company policy also provided a salary increase of 1 percent for every day saved in delivery time.

B: Bill Johnson, manager of the Product Development Department, was pleased with his department's performance on the last quarter's projects. They had managed to complete all projects under budget, virtually assuring Bill of a fat bonus, just in time to help with this year's Christmas purchases.

Situation 3

A: "Harvey, don't worry about the fact that your department is producing at only 70 percent capacity. Increasing your output would simply pile up inventory in front of the next production department. That would be costly for the organization as a whole. Sometimes, one department must reduce its performance so that the performance of the entire organization can improve."

B: "Susan, I am concerned about the fact that your department's performance measures have really dropped over the past quarter. Labour usage variances are unfavourable, and I also see that your machine utilization rates are down. Now, I know you are not a bottle-neck department, but I get a lot of flack when my managers' efficiency ratings drop."

Situation 4

A: Vasily was muttering to himself. He had just received last quarter's budgetary performance report. Once again, he had managed to spend more than budgeted for both materials and labour. The real question now was how to improve his performance for the next quarter.

B: Great! Cycle time had been reduced and, at the same time, the number of defective products had been cut by 35 percent. Cutting the number of defects reduced production costs by more than planned. Trends were favourable for all three performance measures.

Situation 5

A: Cambry was furious. An across-the-board budget cut! "How can they expect me to provide the computer services required on less money? Management is convinced that costs are out of control, but I would like to know where—at least in my department!"

B: After a careful study of the Accounts Payable Department, it was discovered that 80 percent of an accounts payable clerk's time was spent resolving discrepancies between the purchase order, the receiving document, and the supplier's invoice. Other activities such as recording and preparing cheques consumed only 20 percent of a clerk's time. A redesign of the procurement process eliminated virtually all discrepancies and produced significant cost savings.

Situation 6

A: Five years ago, the management of Breeann Products commissioned an outside engineering consulting firm to conduct a time-and-motion study so that labour efficiency standards could be developed and used in production. These labour efficiency standards are still in use today and are viewed by management as an important indicator of productive efficiency.

B: Janet was quite satisfied with this quarter's labour performance. When compared with the same quarter of last year, labour productivity had increased by 23 percent. Most of the increase was due to a new assembly approach suggested by production line workers. She was also pleased to see that materials productivity had increased. The increase in materials productivity was attributed to reducing scrap because of improved quality.

Situation 7

A: "The system converts materials into products, not people at work stations. Therefore, process efficiency is more important than labour efficiency—but we also must pay particular attention to those who use the products we produce, whether inside or outside the firm."

B: "I was quite happy to see a revenue increase of 15 percent over last year, especially when the budget called for a 10 percent increase. However, after reading the recent copy of our trade journal, I now wonder whether we are doing so well. I found out that the market expanded by 30 percent, and our leading competitor increased its sales by 40 percent."

Problems

OBJECTIVE ▶ 1 2 4

Problem 14-18 ABM IMPLEMENTATION, ACTIVITY ANALYSIS, ACTIVITY DRIVERS, DRIVER ANALYSIS, BEHAVIOURAL EFFECTS

Joseph Lee, controller of Thorpe Company, has been in charge of a project to install an activity-based cost management system. This new system is designed to support the company's efforts to become more competitive. For the past six weeks, he and the project committee members have been identifying and defining activities, associating workers with activities, and assessing the time and resources consumed by individual activities. Now, he and the project committee are focusing on three additional implementation issues: (1) identifying activity drivers, (2) assessing value content, and (3) identifying cost drivers (root causes). Joseph has assigned a committee member the responsibilities of assessing the value content of five activities, choosing a suitable activity driver for each activity, and identifying the possible root causes of the activities. Following are the five activities with possible activity drivers:

Activity	Possible Activity Drivers
Setting up equipment	Setup time, number of setups
Performing warranty work	Warranty hours, number of defective units
Welding subassemblies	Welding hours, subassemblies welded
Moving materials	Number of moves, distance moved
Inspecting components	Hours of inspection, number of defective components

The committee member ran a regression analysis for each potential activity driver, using the method of least squares to estimate the variable and fixed cost components. In all five cases, costs were highly correlated with the potential drivers. Thus, all drivers appeared to be good candidates for assigning costs to products. The company plans to reward production managers for reducing product costs.

Required:

1. What is the difference between an activity driver and a cost driver? In answering the question, describe the purpose of each type of driver.
2. For each activity, assess the value content and classify each activity as value-added or non-value-added (justify the classification). Identify some possible root causes of each activity, and describe how this knowledge can be used to improve activity performance. For purposes of discussion, assume that the value-added activities are not performed with perfect efficiency.
3. Describe the behaviour that each activity driver will encourage, and evaluate the suitability of that behaviour for the company's objective of becoming more competitive.

Problem 14-19 ABM, KAIZEN COSTING

OBJECTIVE ▶ 2 3 5

Cycleta Inc. supplies small motors for a large appliance manufacturing company. The appliance company has recently requested that Cycleta decrease its delivery time. Cycleta made a commitment to reduce the lead time for delivery from seven days to one day. To help achieve this goal, engineering and production workers had made the commitment to reduce time for the setup activity (other activities such as moving materials and rework were also being examined simultaneously). Current setup times were 18 hours. Setup cost was $400 per setup hour. For the first quarter, engineering developed a new process design that it believed would reduce the setup time from 18 hours to nine hours. After implementing the design, the actual setup time dropped from 18 hours to eight hours. Engineering believed the actual reduction was sustainable. In the second quarter, production workers suggested a new setup procedure. Engineering gave the suggestion a positive evaluation, and they projected that the new approach would save an additional six hours of setup time. Setup labour was trained to perform the new setup procedures. The actual reduction in setup time based on the suggested changes was five hours.

Required:

1. What kaizen setup standard would be used at the beginning of each quarter?
2. Describe the kaizen subcycle using the two quarters of data provided by Cycleta.
3. Describe the setup subcycle using the two quarters of data provided by Cycleta.
4. How much non-value-added cost was eliminated by the end of two quarters? Discuss the role of kaizen costing in activity-based management.
5. Explain why kaizen costing is compatible with activity-based responsibility accounting while standard costing is compatible with financial-based responsibility accounting.

OBJECTIVE ▶ 3

Problem 14-20 ACTIVITY FLEXIBLE BUDGETING, PERFORMANCE REPORT, VOLUME VARIANCE

Innovator Inc. wants to develop an activity flexible budget for the activity of moving materials. Innovator uses eight forklifts to move materials from receiving to stores. The forklifts are also used to move materials from stores to the production area. The forklifts are obtained through an operating lease that costs $12,000 per year per forklift. Innovator employs 25 forklift operators who receive an average salary of $45,000 per year, including benefits. Each move requires the use of a crate. The crates are used to store the parts and are emptied only when used in production. Crates are disposed of after one cycle (two moves), where a cycle is defined as a move from receiving to stores to production. Each crate costs $1.20. Special fuel for a forklift costs $1.80 per litre. A litre of fuel is used every 20 moves. Forklifts can make three moves per hour and are available for 280 days per year, 24 hours per day (the remaining time is downtime for various reasons). Each operator works 40 hours per week and 50 weeks per year.

Required:

1. Prepare a flexible budget for the activity of moving materials, using the number of cycles as the activity driver.
2. Calculate the activity capacity for moving materials. Suppose Innovator works at 90 percent of activity capacity and incurs the following costs:

Salaries	$1,170,000
Leases	96,000
Crates	91,200
Fuel	14,450

   Prepare the budget for the 90 percent level and then prepare a performance report for the moving materials activity.
3. Calculate and interpret the volume variance for moving materials.
4. Suppose that a redesign of the plant layout reduces the demand for moving materials to one-third of the original capacity. What would be the budget formula for this new activity level? What is the budgeted cost for this new activity level? Has activity performance improved? How does this activity performance evaluation differ from that described in Requirement 2? Explain.

OBJECTIVE ▶ 2 3

Problem 14-21 ACTIVITY-BASED MANAGEMENT, NON-VALUE-ADDED COSTS, TARGET COSTS, KAIZEN COSTING

Hassan Khalil, president of Harmony Electronics, was concerned about the end-of-the-year marketing report that he had just received. According to Emily Hagood, marketing manager, a price decrease for the coming year was again needed to maintain the company's annual sales volume of integrated circuit boards (CBs). This would make a bad situation worse. The current selling price of $18 per unit was producing a $2-per-unit profit—half the customary $4-per-unit profit. Foreign competitors keep reducing their prices. To match the latest reduction would reduce the price from $18 to $14. This would put the price below the cost to produce and sell it. How could the foreign firms sell for such a low price? Determined to find out if there were problems with the company's operations, Hassan decided to hire Jan Booth, a well-known consultant who specializes in methods of continuous improvement. Jan indicated that she felt that an activity-based management system needed to be implemented. After three weeks, Jan had identified the following activities and costs:

Batch-level activities:	
Setting up equipment	$ 125,000
Materials handling	180,000
Inspecting products	122,000
Product-sustaining activities:	
Engineering support	120,000
Handling customer complaints	100,000

Filling warranties	$ 170,000
Storing goods	80,000
Expediting goods	75,000
Unit-level activities:	
Using materials	500,000
Using power	48,000
Manual insertion labour[a]	250,000
Other direct labour	150,000
Total costs	$1,920,000[b]

[a]Diodes, resistors, and integrated circuits are inserted manually into the circuit board.
[b]This total cost produces a unit cost of $16 for last year's sales volume.

Jan indicated that some preliminary activity analysis showed that per-unit costs could be reduced by at least $7. Since Emily had indicated that the market share (sales volume) for the boards could be increased by 50 percent if the price could be reduced to $12, Hassan became quite excited.

Required:

1. What is activity-based management? What connection does it have to continuous improvement?

2. Identify as many non-value-added costs as possible. Compute the cost savings per unit that would be realized if these costs were eliminated. Was Jan correct in her preliminary cost reduction assessment? Discuss actions that the company can take to reduce or eliminate the non-value-added activities.

3. Compute the target cost required to maintain current market share, while earning a profit of $4 per unit. Now, compute the target cost required to expand sales by 50 percent. How much cost reduction would be required to achieve each target?

4. Assume that Jan suggested that kaizen costing be used to help reduce costs. The first suggested kaizen initiative is described by the following: switching to automated insertion would save $60,000 of engineering support and $90,000 of direct labour. Now, what is the total potential cost reduction per unit available? With these additional reductions, can Harmony achieve the target cost to maintain current sales? To increase it by 50 percent? What form of activity analysis is this kaizen initiative: reduction, sharing, elimination, or selection?

5. Calculate income based on current sales, prices, and costs. Now, calculate the income using a $14 price and a $12 price, assuming that the maximum cost reduction possible is achieved (including Requirement 4's kaizen reduction). What price should be selected?

Problem 14-22 VALUE-ADDED AND KAIZEN STANDARDS, NON-VALUE-ADDED COSTS, VOLUME VARIANCE, UNUSED CAPACITY

OBJECTIVE ▶ 3

Tom Zhang, vice president of EPIC Company (a producer of plastic products), has been supervising the implementation of an activity-based cost management system. One of Tom's objectives is to improve process efficiency by improving the activities that define the processes. To illustrate the potential of the new system to the president, Tom has decided to focus on two processes: production and customer service.

Within each process, one activity will be selected for improvement: moulding for production and sustaining engineering for customer service. (Sustaining engineers are responsible for redesigning products based on customer needs and feedback.) Value-added standards are identified for each activity. For moulding, the value-added standard calls for nine kilograms per mould. (Although the products differ in shape and function, their size, as measured by weight, is uniform.) The value-added standard is based on the elimination of all waste due to defective moulds (materials is by far the major cost for the moulding activity). The standard price for moulding is $15 per kilogram. For sustaining engineering, the standard is 60 percent of current practical activity capacity. This standard is based on the fact that about 40 percent of the complaints have to do with design features that could have been avoided or anticipated by the company.

Current practical capacity (at the end of 2013) is defined by the following requirements: 18,000 engineering hours for each product group that has been on the market or in development for five years or less, and 7,200 hours per product group of more than five years. Four product groups have less than five years' experience, and 10 product groups have more. There are 72 engineers, each paid a salary of $70,000. Each engineer can provide 2,000 hours of service per year. There are no other significant costs for the engineering activity.

For 2013, actual kilograms used for moulding were 25 percent above the level called for by the value-added standard; engineering usage was 138,000 hours. There were 240,000 units of output produced. Tom and the operational managers have selected some improvement measures that promise to reduce non-value-added activity usage by 30 percent in 2014. Selected actual results achieved for 2014 are as follows:

Units produced	240,000
Kilograms of material	2,600,000
Engineering hours	120,000

The actual prices paid per kilogram and per engineering hour are identical to the standard or budgeted prices.

Required:

1. For 2013, calculate the non-value-added usage and costs for moulding and sustaining engineering. Also, calculate the cost of unused capacity for the engineering activity.
2. Using the targeted reduction, establish kaizen standards for moulding and engineering (for 2014).
3. Using the kaizen standards prepared in Requirement 2, compute the 2014 usage variances, expressed in both physical and financial measures, for moulding and engineering. (For engineering, explain why it is necessary to compare actual resource usage with the kaizen standard.) Comment on the company's ability to achieve its targeted reductions. In particular, discuss what measures the company must take to capture any realized reductions in resource usage.

OBJECTIVE ▶ 2 3 ## Problem 14-23 BENCHMARKING AND NON-VALUE-ADDED COSTS, TARGET COSTING

Bienestar Inc. has two plants that manufacture a line of wheelchairs. One is located in Quebec City and the other in Winnipeg. Each plant is set up as a profit centre. During the past year, both plants sold their tilt wheelchair model for $1,620. Sales volume averages 20,000 units per year in each plant. Recently, the Winnipeg plant reduced the price of the tilt model to $1,440. Discussion with the Winnipeg manager revealed that the price reduction was possible because the plant had reduced its manufacturing and selling costs by reducing what was called "non-value-added costs." The Winnipeg manufacturing and selling costs for the tilt model were $1,260 per unit. The Winnipeg manager offered to loan the Quebec City plant his cost accounting manager to help it achieve similar results. The Quebec City plant manager readily agreed, knowing that his plant must keep pace—not only with the Winnipeg plant but also with competitors. A local competitor had also reduced its price on a similar model, and Quebec City's marketing manager had indicated that the price must be matched or sales would drop dramatically. In fact, the marketing manager suggested that if the price were dropped to $1,404 by the end of the year, the plant could expand its share of the market by 20 percent. The plant manager agreed but insisted that the current profit per unit must be maintained. He also wants to know if the plant can at least match the $1,260 per-unit cost of the Winnipeg plant and if the plant can achieve the cost reduction using the approach of the Winnipeg plant.

The plant controller and the Winnipeg cost accounting manager have assembled the following data for the most recent year. The actual cost of inputs, their value-added (ideal) quantity levels, and the actual quantity levels are provided (for production of 20,000 units). Assume there is no difference between actual prices of activity units and standard prices.

	SQ	AQ	Actual Cost
Materials (kg)	855,000	900,000	$18,900,000
Labour (hrs)	205,200	216,000	2,700,000
Setups (hrs)	—	14,400	1,080,000
Materials handling (moves)	—	36,000	2,520,000
Warranties (no. repaired)	—	36,000	3,600,000
Total			$28,800,000

Required:

1. Calculate the target cost for expanding the Quebec City plant's market share by 20 percent, assuming that the per-unit profitability is maintained as requested by the plant manager.
2. Calculate the non-value-added cost per unit. Assuming that non-value-added costs can be reduced to zero, can the Quebec City plant match the Winnipeg per-unit cost? Can the target cost for expanding market share be achieved? What actions would you take if you were the plant manager?
3. Describe the role benchmarking played in the effort of the Quebec City plant to protect and improve its competitive position.

Problem 14-24 FINANCIAL VERSUS ACTIVITY FLEXIBLE BUDGETING

OBJECTIVE ▶ 2 3 5

Kelly Gray, production manager, was upset with the latest performance report, which indicated that she was $100,000 over budget. Given the efforts that she and her workers had made, she was confident that they had met or beat the budget. Now, she was not only upset but also genuinely puzzled over the results. Three items—direct labour, power, and setups—were over budget. The actual costs for these three items follow:

	Actual Costs
Direct labour	$210,000
Power	135,000
Setups	140,000
Total	$485,000

Kelly knew that her operation had produced more units than originally had been budgeted, so more power and labour had naturally been used. She also knew that the uncertainty in scheduling had led to more setups than planned. When she pointed this out to John Huang, the controller, he assured her that the budgeted costs had been adjusted for the increase in productive activity. Curious, Kelly questioned John about the methods used to make the adjustment.

JOHN: If the actual level of activity differs from the original planned level, we adjust the budget by using budget formulas—formulas that allow us to predict what the costs will be for different levels of activity.

KELLY: The approach sounds reasonable. However, I'm sure something is wrong here. Tell me exactly how you adjusted the costs of labour, power, and setups.

JOHN: First, we obtain formulas for the individual items in the budget by using the method of least squares. We assume that cost variations can be explained by variations in productive activity where activity is measured by direct labour hours. Here is a list of the cost formulas for the three items you mentioned. The variable X is the number of direct labour hours:

$$\text{Labour cost} = \$10X$$
$$\text{Power cost} = \$5,000 + \$4X$$
$$\text{Setup cost} = \$100,000$$

KELLY: I think I see the problem. Power costs don't have a lot to do with direct labour hours. They have more to do with machine hours. As production increases, machine hours increase more rapidly than direct labour hours. Also, . . .

JOHN: You know, you have a point. The coefficient of determination for power cost is only about 50 percent. That leaves a lot of unexplained cost variation. The coefficient for labour, however, is much better—it explains about 96 percent of the cost variation. Setup costs, of course, are fixed.

KELLY: Well, as I was about to say, setup costs also have very little to do with direct labour hours. And I might add that they certainly are not fixed—at least not all of them. We had to do more setups than our original plan called for because of the scheduling changes. And we have to pay our people when they work extra hours. It seems as if we are always paying overtime. I wonder if we simply do not have enough people for the setup activity. Supplies are used for each setup, and these are not cheap. Did you build these extra costs of increased setup activity into your budget?

JOHN: No, we assumed that setup costs were fixed. I see now that some of them could vary as the number of setups increases. Kelly, let me see if I can develop some cost formulas based on better explanatory variables. I'll get back with you in a few days.

Assume that after a few days' work, John developed the following cost formulas, all with a coefficient of determination greater than 90 percent:

$$\text{Labour cost} = \$10X, \text{ where } X = \text{Direct labour hours}$$
$$\text{Power cost} = \$68,000 + 0.9Y, \text{ where } Y = \text{Machine hours}$$
$$\text{Setup cost} = \$98,000 + \$400Z, \text{ where } Z = \text{Number of setups}$$

The actual measures of each of the activity drivers are as follows:

Direct labour hours	20,000
Machine hours	90,000
Number of setups	110

Required:

1. Prepare a performance report for direct labour, power, and setups using the direct-labour-based formulas.
2. Prepare a performance report for direct labour, power, and setups using the multiple cost driver formulas that John developed.
3. Of the two approaches, which provides the most accurate picture of Kelly's performance? Why?
4. After reviewing the approach to performance measurement, a consultant remarked that non-value-added cost trend reports would be a much better performance measurement approach than comparing actual costs with budgeted costs—even if activity flexible budgets were used. Do you agree or disagree? Explain.

CMA Problem

OBJECTIVE ▸ 2 3 5   CMA Problem 14-1 **ACTIVITY FLEXIBLE BUDGETING, NON-VALUE-ADDED COSTS***

Wendy Li, controller for Marston Inc., prepared the following budget for manufacturing costs at two different levels of activity for 2013:

	Level of Activity	
Driver: Direct Labour Hours	*50,000*	*100,000*
Direct materials	$ 300,000	$ 600,000
Direct labour	200,000	400,000
Depreciation (plant)	100,000	100,000
Subtotal	$ 600,000	$1,100,000

*© 2011, Institute of Management Accountants, Inc. http://www.imanet.org. Reprinted with permission.

	Level of Activity	
Driver: Machine Hours	200,000	300,000
Maintaining equipment	$ 360,000	$ 510,000
Machining	112,000	162,000
Subtotal	$ 472,000	$ 672,000
Driver: Material Moves	20,000	40,000
Moving materials	$ 165,000	$ 290,000
Driver: Number of Batches Inspected	100	200
Inspecting products	$ 125,000	$ 225,000
Total	$1,362,000	$2,287,000

During 2013, Marston worked a total of 80,000 direct labour hours, used 250,000 machine hours, made 32,000 moves, and performed 120 batch inspections. The following actual costs were incurred:

Direct materials	$440,000
Direct labour	355,000
Depreciation	100,000
Maintaining equipment	425,000
Machining	142,000
Moving materials	232,500
Inspecting products	160,000

Marston applies overhead using rates based on direct labour hours, machine hours, number of moves, and number of batches. The second level of activity (the right column in the preceding table) is the practical level of activity (the available activity for resources acquired in advance of usage) and is used to compute predetermined overhead pool rates.

Required:

1. Prepare a performance report for Marston's manufacturing costs in 2013.
2. Assume that one of the products produced by Marston is budgeted to use 10,000 direct labour hours, 15,000 machine hours, and 500 moves and will be produced in five batches. A total of 10,000 units will be produced during the year. Calculate the budgeted unit manufacturing cost.
3. One of Marston's managers said the following: "Budgeting at the activity level makes a lot of sense. It really helps us manage costs better. But the previous budget really needs to provide more detailed information. For example, I know that the moving materials activity involves the use of forklifts and operators, and this information is lost when only the total cost of the activity for various levels of output is reported. We have four forklifts, each capable of providing 10,000 moves per year. We lease these forklifts for five years, at $10,000 per year. Furthermore, for our two shifts, we need up to eight operators if we run all four forklifts. Each operator is paid a salary of $30,000 per year. Also, I know that fuel costs about $0.25 per move."

   Assuming that these are the only three items, expand the detail of the flexible budget for moving materials to reveal the cost of these three resource items for 20,000 moves and 40,000 moves, respectively. Based on these comments, explain how this additional information can help Marston better manage its costs. (Especially consider how activity-based budgeting may provide useful information for non-value-added activities.) *(CMA adapted)*

The Collaborative Learning Exercises can be found on the product support site at www.hansen1ce.nelson.com.

Part 3
Chapters 11–14

Sabrina Hoffman is founder and CEO of Golden Care Inc., which owns and operates several assisted-living facilities. The facilities are apartment-style buildings with 25 to 30 one- or two-bedroom apartments. While each apartment has its own complete kitchen, in every building Golden Care offers communal dining options and an on-site nurse who is available 24 hours a day. Residents can choose monthly meal options that include one or two meals per day in the dining room. Residents who require nursing services (e.g., blood pressure monitoring and injections) can receive those services from the nurse. However, Golden Care facilities are not nursing homes, all residents are ambulatory, and custodial care is not an option. In the five years it has been in operation, the company has expanded from one facility to five, located in Maritime cities. The income statement for last year follows.

Golden Care Inc.
Income Statement for Last Year

Revenue	$2,880,000
Cost of services	2,016,000
Gross profit	864,000
Marketing and administrative expenses	500,000
Operating income	$ 364,000

Sabrina originally got into the business because she had trouble finding adequate facilities for her mother. The concept worked well, and income over the past five years had grown nicely at 20 percent per year. However, Sabrina sensed clouds on the horizon. She knew that the population was aging and that her current clients would be moving to more traditional forms of nursing care. As a result, Sabrina wanted to consider adding one or more nursing homes to Golden Care. These nursing homes would be staffed around the clock with RNs and LPNs. The residents would likely have more severe medical problems and would be confined to beds or wheelchairs. Sabrina knew that quality care of this type was needed. So, she contacted Peter Verdon, her marketing manager, and Bernadette Zhang, her accountant, for a brainstorming session.

Peter: Sabrina, I really like the concept. As you know, several of our facilities have faced seeing their long-term residents move out to local nursing homes. Not only are these homes of lower quality than what we could provide, but losing a resident is heartrending for the staff, as well as for the remaining residents. I like the idea of providing a transition from less care to more.

Bernadette: I agree with you, Peter. But let's not forget the differences between assisted-living and full-time, nursing-home-type care. Our expenses will really increase.

Sabrina: That's why I wanted to talk with both of you. As you know, Golden Care's mission statement emphasizes the need to make a profit. We can't continue to serve our residents and provide high-quality care if we don't make enough money to pay our staff a living wage and earn enough of a profit to smooth over the rough patches and continue to improve our business. Could the two of you look into this idea, and get back to me in a week or so?

Throughout the following week, the three communicated by e-mail. By the end of the week, a number of possibilities had surfaced, and these were summarized in a message from Bernadette to the others.

TO: sabrina.hoffman@goldencare.com, peter.verdon@goldencare.com
FROM: bernadette.zhang@goldencare.com
MESSAGE:
I've compiled the ideas from all of our e-mails into the following list. This may be a good starting point for our meeting tomorrow.

1. Buy an existing nursing home in one of Golden Care's current locations.
2. Buy an existing nursing home in another city.

3. Build a new nursing home facility in one of Golden Care's current locations.
4. Build a new nursing home facility in another city.
5. Build a wing on to an existing Golden Care facility. The Huron Junction facility has sufficient open land for an addition.

The next day, Sabrina, Peter, and Bernadette met again in Sabrina's office.

Sabrina: I didn't realize there were so many possibilities. Are we going to have to work up numbers on each of them?

Bernadette: No, I think we can eliminate a few of them pretty quickly. For example, building a new facility would cost more than the other options, and it would involve the most risk.

Peter: I agree, and I also think we might eliminate the purchase of an existing nursing home for the same reasons. Also, existing homes would not give us the option of building a facility that is state of the art and meets our needs, and it would lock us into a preexisting patient mix.

Sabrina: I like that thinking. Let's restrict our attention to Option 5.

Bernadette: I thought you might like that option, so Peter and I sketched out two alternatives for an extension of the Huron Junction building. We call the alternatives Basic Care and Lifestyle Care.

Peter: There are different markets for each type of care. If we want to concentrate on government-supported patients, the reimbursement is lower, and we would want to offer the Basic Care option. Private insurance and private-pay patients could afford more services; if we are marketing to these patients, we could offer the Lifestyle Care option. Both alternatives provide high-quality nursing care. Basic Care concentrates on the quality nursing and maintenance activities. For example, the addition would have 25 double rooms, two nursing stations, two recreation rooms, a treatment room, and an office. The Lifestyle Care option adds physical and recreational therapy with a specially equipped gym and pool. That addition would have 30 single rooms, two nursing stations, a recreation room, a swimming pool, a hydrotherapy spa and gym, a treatment room, and an office. In each case, there would be cable TV and telephone hookups in each room and a buffer area between the nursing home and the apartments.

Sabrina: Why the buffer area? Won't that add unnecessary cost?

Peter: It adds cost, but it will be well worth it. Sabrina, you must remember that the nursing home patients are different from the apartment residents. Some of the patients will have advanced dementia. We'll lose apartment residents in a hurry if they have to be reminded every day of what might be in store for them later on.

Sabrina: I see your point. Bernadette, what will these two plans cost? I'll tell you right now that I like the Lifestyle Care option better. It fits with our history of doing whatever we can to make life better for our residents.

Bernadette: I've checked into the costs of putting on a new wing and operating both alternatives. Here's a listing.

Basic Care		**Lifestyle Care**	
Construction	$1,500,000	Construction	$2,000,000
Annual operating expenses:		Annual operating expenses:	
Staff:		Staff:	
RNs (3 × $30,000)	90,000	RNs (3 × $30,000)	90,000
LPNs (6 × $22,000)	132,000	LPNs (6 × $22,000)	132,000
Aides (6 × $20,000)	120,000	Aides (6 × $20,000)	120,000
Cooks (2 × $15,000)	30,000	Physical and recreational	
Janitors (2 × $18,000)	36,000	therapists (2 × $25,000)	50,000
Other* (60% variable)	300,000	Cooks (1.5 × $15,000)	22,500
Debt service	150,000	Janitors (2 × $18,000)	36,000
Depreciation (over 20 years)	75,000	Other (60% variable)	360,000
		Debt service	200,000
		Depreciation (over 20 years)	100,000

* Other includes supplies, utilities, food, and so on.

In both cases, total administrative costs for Golden Care would increase by $30,000 per year. This seems high, but the increased legal and insurance requirements will add significantly more paperwork and accounting.

Sabrina: All this sounds reasonable, but why is reimbursement such an important factor?

Peter: Well, if you admit government-supported patients, the province will reimburse at most $30,000 per year. Private insurance policies will pay roughly $46,000 per year. We can charge up to about $65,000 for private patients, but this type of care is so expensive that many of these patients exhaust their own funds. The nice aspect of government-supported patients is that we can be virtually assured that we will operate at capacity.

Sabrina: Can we cross that bridge when we come to it?

Peter: No, not really. Once the patient is a resident of our facility, it is hard to evict him or her. Also, while it is legal to force patients out before they go on government support and to refuse to accept government-supported patients, once we do accept government-supported patients, we are prevented by law from evicting them—no matter how high our costs go.

Sabrina: OK, it looks as if we have some hard work ahead of us to decide whether or not to get into this line of business.

Required:

1. How did Sabrina, Bernadette, and Peter use the tactical decision-making model of Chapter 11?

2. Categorize each of the expenses for the Basic Care and Lifestyle Care options as flexible or committed. Further categorize the committed expenses as committed fixed or committed step costs.

3. Calculate the break-even number of patients (in total and for each type of reimbursement) for each of the following scenarios:
   a. Basic Care option, 20 percent private insurance and 80 percent government supported
   b. Basic Care option, no government support
   c. Lifestyle Care option, no government support, 75 percent private insurance, 25 percent private pay
   d. Lifestyle Care option, all insurance reimbursement

4. What is the markup percent of cost of services charged on the assisted-living expenses? What would the price per month for a Basic Care patient be if the same markup were used? For a Lifestyle Care patient? (Assume in both cases that occupancy is at 80 percent of capacity.)

5. What is the payback period for the new addition?

6. Research Assignment: What is the relevant law restricting the ability of nursing homes to evict government-supported patients? Why would nursing homes accept government-supported patients and later evict them? Is eviction of government-supported patients still a problem? Discuss the legal and ethical issues in a nursing home's decision on whether to accept government-supported patients.

STRATEGIC COST MANAGEMENT

The first three parts of this textbook have focused on the nature of costs and on how the organization can be influenced to perform in a certain manner. However, once a manager has learned the tools and techniques to understand and control costs within an organization, the question that must be asked is "How does this fit with the organization's strategy?"

Management must establish the strategic direction that an organization will pursue before entering the planning and management cycle. Once this strategic direction is established, everything else the organization does must be consistent with the strategy set forth. The tools and techniques discussed earlier must enable the organization to pursue not only a strong cost management program but also its strategic objectives.

There are a number of approaches that allow management to ensure that the organization's strategic objectives are not being sacrificed in the name of "cost control." In recent years, there has been increased emphasis on nonfinancial measures and structural changes to ensure that the organization as a whole is operating at the most efficient level possible.

Part 4 of this text introduces a number of the approaches to preserving the strategic direction while still allowing the organization to achieve the best possible results for its shareholders, managers, employees, and all other stakeholders.

After studying this chapter, you should be able to:

➤ **1** Compare and contrast activity-based and strategic-based responsibility accounting systems.

➤ **2** Discuss the basic features of the Balanced Scorecard.

➤ **3** Explain how the Balanced Scorecard links measures to strategy.

➤ **4** Describe how an organization can achieve strategic alignment.

CHAPTER

15

The Balanced Scorecard: Strategic-Based Control

Many firms operate in an environment where change is rapid. Products and processes are constantly being redesigned and improved, and stiff national and international competitors are always present. The competitive environment demands that firms offer customized products and services to diverse customer segments. This, in turn, means that firms must find cost-efficient ways of producing high-variety, low-volume products. This usually means that more attention is paid to linkages between the firm and its suppliers and customers with the goal of improving cost, quality, and response times for all parties in the value chain. Furthermore, for many industries, product life cycles are shrinking, placing greater demands on the need for innovation. Thus, organizations operating in a dynamic, rapidly changing environment are finding that adaptation and change are essential to survival. In Chapter 6, we learned that activity-based management describes the fundamental economics that drive a firm and thus allows managers to have a better understanding of the causes of cost. In turn, understanding the root causes of costs enables managers to more effectively improve performance by continuously improving processes.

Activity-based management also produced a new form of responsibility accounting, one that better fits environments that demand continuous improvement because of keen competitive conditions and dynamic change. Recall that the responsibility accounting model is defined by four essential elements: (1) assigning responsibility, (2) establishing performance measures or benchmarks, (3) evaluating performance, and (4) assigning rewards. The traditional or financial-based responsibility accounting model emphasizes financial performance of organizational units and evaluates and rewards performance using static financial-oriented standards (e.g., budgets and standard costing). While this model is useful for firms operating in a stable environment that wish to emphasize maintaining the status quo, it is certainly not suitable for firms operating in a dynamic environment that requires continuous improvement. For this reason, activity-based responsibility accounting was developed. (Chapter 14 detailed the differences between the two models.) However, while the activity-based responsibility accounting model was a significant improvement, it soon became apparent that it suffered from some limitations. This then led to the development of *strategic-based responsibility accounting*, the topic of this chapter.

Activity-Based versus Strategic-Based Responsibility Accounting

OBJECTIVE ▸ 1

Compare and contrast activity-based and strategic-based responsibility accounting systems.

Activity-based responsibility accounting represents a significant change in how responsibility is assigned, measured, and evaluated. Effectively, the activity-based system added a process perspective to the financial perspective of the functional-based responsibility accounting system. Processes represent how things are done within an organization; therefore, any effort to improve organizational performance had to involve improving processes. It also altered the financial perspective by changing the point of view from that of cost control to maintain the status quo to that of cost reduction by continuous learning and change. Thus, responsibility accounting changed from a one-dimensional system to a two-dimensional system, and from a control system to a *performance management system*. Although these changes were dramatic and in the right direction, it was soon discovered that the new approach also had some limitations. The most significant shortcoming was the fact that the continuous improvement efforts were often fragmented, and they failed to connect with an organization's overall mission and strategy. Lacking was a navigational system, and the result was undirected and rudderless continuous improvement. Consequently, at times, the expected competitive successes did not materialize.

What was needed was *directed continuous improvement*. Providing direction meant that managers needed to carefully specify a mission and strategy for their organization and identify the objectives, performance measures, and initiatives necessary to accomplish this overall mission and strategy. In other words, a strategic-based responsibility accounting system was the next step in the evolution of responsibility accounting. A **strategic-based responsibility accounting system (strategic-based performance management system)** translates the strategy of an organization into operational objectives and measures. A strategic performance management system can assume different forms, the most common being that of the Balanced Scorecard. The **Balanced Scorecard** is a strategic-based performance management system that typically identifies objectives and measures for four different perspectives: the financial perspective, the customer perspective, the process perspective, and the learning and growth perspective.[1]

The Balanced Scorecard converts a company's strategy into executable actions that are deployed throughout the organization. The Balanced Scorecard approach has spread rapidly in the United States and throughout the world. A **Bain & Company** survey of a broad range of international executives revealed that in 2008, 53 percent of the companies surveyed were using the Balanced Scorecard.[2] The usage rate

---

[1] Robert S. Kaplan and David P. Norton, *The Balanced Scorecard* (Boston: Harvard Business School Press, 1996).

[2] Darrell Rigby and Barbara Bilodeau, "Management Tools and Trends 2009," Bain & Company, http://www.bain.com/management_tools/home.asp (accessed June 1, 2009).

ranged from 49 percent in North America to 56 percent in Latin America. Global usage was expected to expand to 63 percent within a year. Because of its widespread use and popularity, we will focus our discussion of performance management on the Balanced Scorecard. A general overview of the Balanced Scorecard will first be provided by comparing the specific responsibility elements of activity-based responsibility accounting with those of the Balanced Scorecard. In the remainder of the chapter, more specific details of the Balanced Scorecard will be provided.

Assigning Responsibility

Exhibit 15-1A reveals that the strategic-based responsibility accounting system adds direction to improvement efforts by tying responsibility to the firm's strategy. It also maintains the process and financial perspectives of the activity-based approach but adds a customer and a learning and growth (infrastructure) perspective, increasing the number of responsibility dimensions to four. Although more perspectives could be added, these four perspectives are essential for creating a competitive advantage and allowing managers to articulate and communicate the organization's mission and strategy. Only perspectives that serve as a potential source for a competitive advantage should be included (e.g., an environmental perspective). This leaves open the possibility of expanding the number of perspectives. Notice that the two additional perspectives consider the interests of customers and employees, interests that were not fully considered by the activity-based responsibility system. Another difference is that the Balanced Scorecard diffuses responsibility for the perspectives throughout the entire organization. Ideally, all individuals in the organization should understand the organization's strategy and know how their specific responsibilities support achievement of the strategy. The key to this diffusion is proper and careful definition of performance measures.

Establishing Performance Measures

Exhibit 15-1B reveals that the strategic-based approach carries over the financial and process-oriented standards of the activity-based system, including the concepts of value-added and dynamic standards. None of the advances developed in an activity approach are thrown out, but the strategic-based approach adds some important refinements. In a strategic-based responsibility accounting system, performance measures must be integrated so that they are mutually consistent and reinforcing. In effect, performance measures should be designed so that they are derived from and communicate an organization's strategy and objectives. By translating the organization's strategy into objectives and measures that can be understood, communicated, and acted upon, it is possible to more completely align individual and organizational goals and initiatives. Thus, the measures must be balanced and linked to the organization's strategy.

When a firm has balanced measures, the measures selected are balanced between *lag measures* and *lead measures*, between *objective measures* and *subjective measures*, between *financial measures* and *nonfinancial measures*, and between *external measures* and *internal measures*. **Lag measures** are outcome measures—measures of results from past efforts (e.g., customer profitability). **Lead measures (performance drivers)** are factors that drive future performance (e.g., hours of employee training). **Objective measures** are those that can be readily quantified and verified (e.g., market share), whereas **subjective measures** are less quantifiable and more judgmental in nature (e.g., employee capabilities). **Financial measures** are those expressed in monetary terms, whereas **nonfinancial measures** use nonmonetary units (e.g., number of dissatisfied customers). **External measures** are those that relate to *customers* versus *shareholders* (e.g., customer satisfaction and return on investment). **Internal measures** are those measures that relate to the *processes* and *capabilities* that create value for customers and shareholders (e.g., process efficiency and employee satisfaction).

A strategic performance management system uses many different kinds of measures because of the need to build a closer link to strategy. In the traditional, financial-based responsibility model, performance measures are almost always financial and,

Exhibit 15-1

Responsibility, Performance, and Reward

A. Responsibility Assignments Compared

Activity-Based Responsibility	Strategic-Based Responsibility
1. No tie to strategy	1. Linked to strategy
2. Systemwide efficiency	2. Systemwide efficiency
3. Team accountability	3. Team accountability
4. Financial perspective	4. Financial perspective
5. Process perspective	5. Process perspective
	6. Customer perspective
	7. Learning and growth perspective

B. Performance Measures Compared

Activity-Based Measures	Strategic-Based Measures
1. Process-oriented and financial standards	1. Standards for all four perspectives
2. Value-added standards	2. Used to communicate strategy
3. Dynamic standards	3. Used to help align objectives
4. Optimal standards	4. Linked to strategy and objectives
	5. Balanced measures

C. Performance Evaluation Compared: Activity-Based versus Strategic-Based

Activity-Based Performance Evaluation	Strategic-Based Performance Evaluation
1. Time reductions	1. Time reductions
2. Quality improvements	2. Quality improvements
3. Cost reductions	3. Cost reductions
4. Trend measurements	4. Trend measurements
	5. Expanded set of metrics
	6. Stretch targets for all four perspectives

D. Rewards Compared

Activity-Based Rewards	Strategic-Based Rewards
1. Performance evaluated on two or more dimensions	1. Performance evaluated on four or more dimensions
2. Group rewards	2. Group rewards
3. Salary increases	3. Salary increases
4. Promotions	4. Promotions
5. Bonuses, profit sharing, and gainsharing	5. Bonuses, profit sharing, and gainsharing

therefore, almost always lag measures. Financial and lag measures are not sufficient to link with strategy. Many strategic objectives are nonfinancial in nature and require the use of nonfinancial measures to promote and measure progress. For example, increasing customer loyalty may be a key strategic objective that will lead to increased revenues and profits. Yet how is customer loyalty measured? The number of repeat orders is a good possible measure, and it is a nonfinancial measure. And what are some of the drivers of customer loyalty? Increasing product quality? Increasing on-time deliveries? Or both? And how are these critical success factors measured? Percentage of defective units and percentage of on-time deliveries are good possibilities. Clearly, to express the desired linkages among strategic objectives, nonfinancial measures are needed.

The concept of lead measures is also critical. A lead measure, by definition, is one that has a causal linkage with the strategy. For example, if the number of defective units decreases, will customer loyalty actually increase? If the number of repeat orders increases, will revenues and profits actually increase? Assuming a causal relationship exists, when in reality it does not, can be quite costly. For example, **Xerox** assumed that increasing customer satisfaction would lead to increased financial performance. It then spent millions on surveying and measuring customer satisfaction only to discover that increasing customer satisfaction did not increase financial performance. As it turned out, a customer loyalty measure was the correct lead measure for improving financial performance.[3]

Finally, it should be noted that to communicate an organization's strategy through the language of measurement requires both scope and flexibility. Scope implies that both internal and external measures are needed. Flexibility requires subjective and objective measurement as well as nonfinancial measures. In effect, a Balanced Scorecard expresses the complete story of a company's strategy through an integrated set of financial and nonfinancial measures that are both predictive and historical and that may be measured subjectively or objectively.

Performance Measurement and Evaluation

In an activity-based responsibility system, performance measures are process oriented. Thus, performance evaluation focuses on improvement of process characteristics, such as time, quality, and efficiency. Financial consequences of improving processes are also measured, usually by cost reductions achieved. Therefore, a financial perspective is included. A strategic performance management system expands these evaluations to include the customer and learning and growth perspectives as well as a more comprehensive financial view. The organization must also deal with performance evaluation of things, such as customer satisfaction, customer retention, employee capabilities, and revenue growth from new customers and new products. However, the difference is more profound than simply expanding the number and type of measures being evaluated. Exhibit 15-1C summarizes the comparison of performance evaluation for the activity and strategic-based approaches.

Performance evaluation in a Balanced Scorecard framework is deeply concerned with the effectiveness and viability of the organization's strategy. Furthermore, the Balanced Scorecard approach is used to drive organizational change, and much of this change emphasis is expressed through performance evaluation. This is communicated by establishing *stretch* targets for the individual performance measures of the various perspectives. **Stretch targets** are targets that are set at levels that, if achieved, will transform the organization within a period of three to five years. Performance for a given period is evaluated by comparing the actual values of the various measures with the targeted values. Two key features make stretch targets feasible: (1) the measures are linked by causal relationships and (2) because of the linkages, the targets are not set in isolation but rather through a consensus of all those in the organization. Exhibit 15-1D reveals that the reward method of the two systems are strikingly similar and differ only on the number of dimensions being evaluated.

Assigning Rewards

For any performance management system to be successful, the reward system must be linked to the performance measures. The activity- and strategic-based systems both

---

[3] Christopher Ittner and David Larcker, "Coming Up Short on Nonfinancial Performance Measurement," *Harvard Business Review* (November 2003): 88–95.

use the same financial instruments to provide compensation to those who achieve targeted performance goals. A key difference for both systems from the traditional control system is the fact that rewards are based on much more than financial measures. In the case of the Balanced Scorecard, four dimensions of performance must be considered instead of the two in an activity-based performance system. It is very unlikely that an organization can secure the needed support for a Balanced Scorecard of measures unless compensation is tied to the scorecard measures. Both systems must also face the thorny problem of team-based rewards.

Basic Concepts of the Balanced Scorecard

OBJECTIVE ➤ 2
Discuss the basic features of the Balanced Scorecard.

The Balanced Scorecard permits an organization to create a strategic focus by *translating* an organization's strategy into operational objectives and performance measures for four different perspectives: the financial perspective, the customer perspective, the internal business process perspective, and the learning and growth (infrastructure) perspective. The Balanced Scorecard is an effective way of implementing and managing a company's strategy. A number of companies attribute their recent financial success to this strategic performance management system.

Strategy Translation

Strategy, according to the creators of the Balanced Scorecard framework, is defined as:[4]

> *choosing the market and customer segments the business unit intends to serve, identifying the critical internal and business processes that the unit must excel at to deliver the value propositions to customers in the targeted market segments, and selecting the individual and organizational capabilities required for the internal, customer, and financial objectives.*

Strategy, then, is identifying and defining management's desired relationships among the four perspectives. *Strategy translation*, on the other hand, means specifying objectives, measures, targets, and initiatives for each perspective. The strategy translation process is illustrated in Exhibit 15-2. Consider, for example, a company that wishes to pursue a revenue growth strategy. For the financial perspective, the company may specify an *objective* of growing revenues by introducing new products. The *performance measure* may be the percentage of revenues from the sale of new products. The *target* or *standard* for the coming year for the measure may be 20 percent. (That is, 20 percent of the total revenues for the coming year must be from the sale of new products.) The *initiative* describes *how* this is to be accomplished. The "how," of course, involves the other three perspectives. The customer segments, internal processes, and individual and organizational capabilities that will permit the realization of the revenue growth objective must now be identified. This illustrates the fact that the financial objectives serve as the focus for the objectives, measures, and initiatives of the other three perspectives. It also illustrates the need to carefully define the relationships among the four perspectives so that strategy becomes visible and operational. However, before examining how these causal relationships define and operationalize the strategy, we first need a better understanding of the four perspectives, their objectives, and their measures.

1. The Financial Perspective, Objectives, and Measures

The **financial perspective** establishes the long- and short-term financial performance objectives expected from the organization's strategy and simultaneously describes the

[4] Kaplan and Norton, *The Balanced Scorecard*, 37.

Exhibit 15-2

Strategy Translation Process

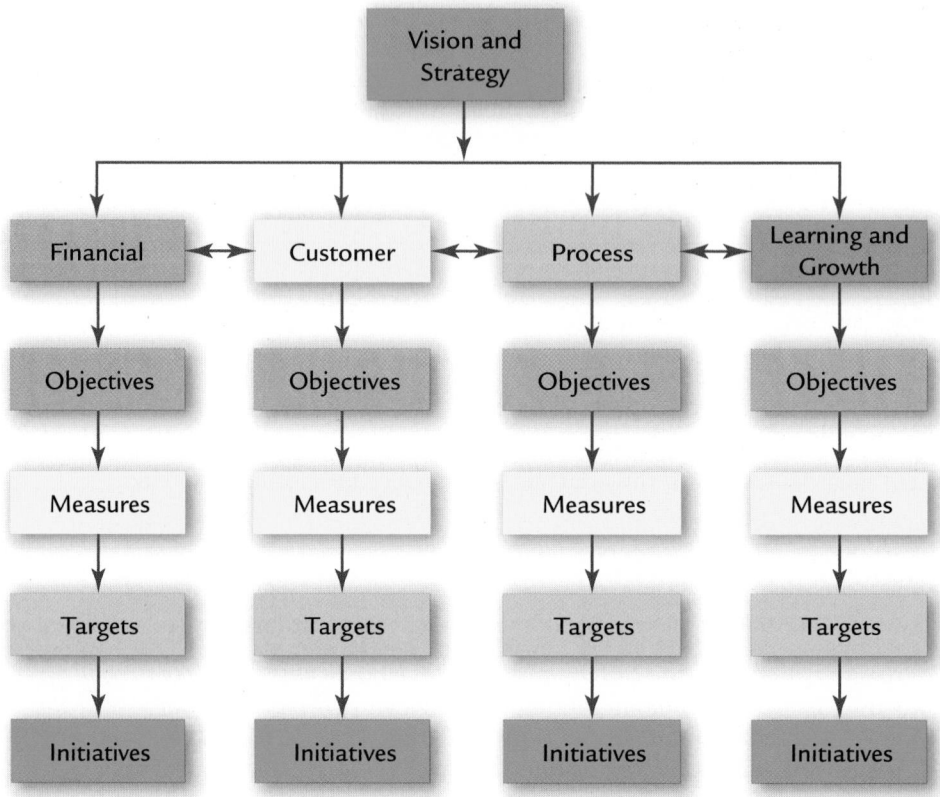

economic consequences of actions taken in the other three perspectives. This implies that the objectives and measures of the other perspectives should be chosen so that they cause or bring about the desired financial outcomes. The financial perspective has three strategic themes: revenue growth, cost reduction, and asset utilization. These themes serve as the building blocks for the development of specific operational objectives and measures. Of course, the three themes are constrained by the need for managers to manage risk.

Revenue Growth   Increasing revenues can be achieved in a variety of ways, and the potential strategic objectives reflect these possibilities. Among these possibilities are the following objectives: increase the number of new products, create new applications for existing products, develop new customers and markets, and adopt a new pricing strategy. Once operational objectives are known, performance measures can be designed. Possible measures for the preceding list of objectives (in the order given) are percentage of revenue from new products, percentage of revenue from new applications, percentage of revenues from new customers and market segments, and profitability by product or customer.

Cost Reduction   Reducing the cost per unit of product, per customer, or per distribution channel are examples of cost reduction objectives. The appropriate measures are obvious: costs per unit of the particular cost objects. Trends in these measures will tell whether or not the costs are being reduced. For these objectives, the accuracy of cost assignments is especially important. Activity-based costing can play an essential measurement role, especially for selling and administrative costs—costs not usually assigned to cost objects like customers and distribution channels.

Asset Utilization   Improving asset utilization is the principal objective. Financial measures such as return on investment and economic value added are used. Since return on investment and economic value-added measures were discussed in detail in Chapter 10, they will not be discussed here. The objectives and measures for the financial perspective are summarized in Exhibit 15-3A.

Risk Management   Managing the risk associated with the adopted strategy is another critical strategic theme—one that is common to the three strategic financial themes already discussed. Diversification of customer types, product lines, and suppliers are common means of lowering risk. Sourcing materials from only one supplier may lower costs, but it may also jeopardize the firm's throughput if something happens to the supplier (e.g., a labour strike). Similarly, revenues may be increased by relying on one very large customer—but what happens if the customer decides to buy elsewhere? Thus, any strategic initiative must be balanced with careful consideration of the risk involved.

2. Customer Perspective, Objectives, and Measures

The **customer perspective** defines the customer and market segments in which the business unit will compete and describes the way that value is created for customers. The customer perspective is the source of the revenue component for the financial objectives. Failure to deliver the right kinds of products and services to the targeted customers means revenue will not be generated.

Once the customers and segments are defined, *core objectives* and *measures* are developed. **Core objectives and measures** are those that are common across all organizations. There are five key core objectives: increase market share, increase customer retention, increase customer acquisition, increase customer satisfaction, and increase customer profitability. Possible core measures for these objectives, respectively, are market share (percentage of the market), percentage growth of business from existing customers and percentage of repeating customers, number of new customers, ratings from customer satisfaction surveys, and individual and segment profitability. Activity-based costing is a key tool in assessing customer profitability (see Chapter 13). Notice that customer profitability is the only financial measure among the core measures. This measure, however, is critical because it emphasizes the importance of the *right* kind of customers. What good is it to have customers if they are not profitable? The obvious answer spells out the difference between being customer focused and customer obsessed.

Customer Value   In addition to the core measures and objectives, measures are needed that drive the creation of *customer value* and, thus, drive the core outcomes. For example, increasing customer value builds customer loyalty (increases retention) and increases customer satisfaction. **Customer value** is the difference between realization and sacrifice, where realization is what the customer receives and sacrifice is what is given up. Realization includes such attributes as product functionality (features), product quality, reliability of delivery, delivery response time, image, and reputation. Sacrifice includes attributes such as product price, time required to learn to use the product, operating cost, maintenance cost, and disposal cost. The costs incurred by the customer *after* purchase are called **post-purchase costs**.

The attributes associated with realization and sacrifice provide the basis for the objectives and measures that will lead to improving the core outcomes. The objectives for the sacrifice side of the value equation are the simplest: decrease price and decrease post-purchase costs. Selling price and post-purchase costs are important measures of value creation. Decreasing these costs decreases customer sacrifice, and, thus, increases customer value. Increasing customer value should impact favourably on most of the core objectives. Similar favourable effects can be obtained by increasing realization. Realization objectives, for example, would include the following: improve product functionality, improve product quality, increase delivery reliability, and improve product image and reputation. Possible measures for these objectives include, respectively, feature satisfaction ratings, percentage of returns, on-time delivery percentage, and

Exhibit 15-3

Objectives and Measures: Financial, Customers, Process, and Learning/Growth Perspectives

A. Summary of Objectives and Measures: Financial Perspective

Objectives	Measures
Revenue Growth:	
Increase the number of new products	Percentage of revenues from new products
Create new applications	Percentage of revenues from new applications
Develop new customers and markets	Percentage of revenues from new sources
Adopt a new pricing strategy	Product and customer profitability
Cost Reduction:	
Reduce unit product cost	Unit product cost
Reduce unit customer cost	Unit customer cost
Reduce distribution channel cost	Cost per distribution channel
Asset Utilization:	
Improve asset utilization	Return on investment
	Economic value added

B. Summary of Objectives and Measures: Customer Perspective

Objectives	Measures
Core:	
Increase market share	Market share (percentage of market)
Increase customer retention	Percentage growth, existing customers
	Percentage of repeating customers
Increase customer acquisition	Number of new customers
Increase customer satisfaction	Ratings from customer surveys
Increase customer profitability	Customer profitability
Performance Value:	
Decrease price	Price
Decrease post-purchase costs	Post-purchase costs
Improve product functionality	Ratings from customer surveys
Improve product quality	Percentage of returns
Increase delivery reliability	On-time delivery percentage
	Aging schedule
Improve product image and reputation	Ratings from customer surveys

product recognition rating. Of these objectives and measures, delivery reliability will be used to illustrate how measures can affect managerial behaviour, indicating the need to be careful in the choice and use of performance measures.

Delivery reliability means that output is delivered on time. On-time delivery is a commonly used operational measure of reliability. To measure on-time delivery, a firm sets delivery dates and then finds on-time delivery performance by dividing the orders delivered on time by the total number of orders delivered. The goal, of course, is to achieve a ratio of 100 percent. However, this measure used by itself may produce undesirable behavioural consequences.[5] For example, plant managers may give priority to filling orders not yet late over orders that are already late. The performance measure here is encouraging managers to have one very late shipment rather than several moderately late shipments! A chart measuring the age of late deliveries could help mitigate this problem. Exhibit 15-3B summarizes the objectives and measures for the customer perspective.

[5] Joseph Fisher, "Nonfinancial Performance Measures," *Journal of Cost Management* (Spring 1992): 31–38.

C. Summary of Objectives and Measures: Process Perspective	
Objectives	**Measures**
Innovation:	
Increase the number of new products	Number of new products/total products; R&D expenses
Increase proprietary products	Percentage revenue from proprietary products
	Number of patents pending
Decrease product development cycle time	Time to market (from start to finish)
Operations:	
Increase process quality	Quality costs
	Output yields
	Percentage of defective units
Increase process efficiency	Unit cost trends
	Output/input(s)
Decrease process time	Cycle time and velocity
	MCE
Post-Sales Service:	
Increase service quality	First-pass yields
Increase service efficiency	Cost trends
	Output/input(s)
Decrease service time	Cycle time

D. Summary of Objectives and Measures: Learning and Growth Perspective	
Objectives	**Measures**
Increase employee capabilities	Employee satisfaction ratings
	Employee turnover percentages
	Employee productivity (revenue/employee)
	Hours of training
	Strategic job coverage ratio (percentage of critical job requirements filled)
Increase motivation and alignment	Suggestions per employee
	Suggestions implemented per employee
Increase information systems capabilities	Percentage of processes with real-time feedback capabilities
	Percentage of customer-facing employees with online access to customer and product information

3. Process Perspective, Objectives, and Measures

The **internal business process perspective** describes the internal processes needed to provide value for customers and owners. Processes are the means by which strategies are executed. Thus, the process perspective entails the identification of the critical processes needed that affect customer and shareholder satisfaction. To provide the framework needed for this perspective, a *process value chain* is defined. The **process value chain** is made up of three processes: the *innovation process*, the *operations process*, and the *post-sales service process*.[6] The **innovation process** anticipates the emerging and potential needs of customers and creates new products and services to satisfy those needs. It represents what is called the *long wave* of value creation. The **operations process** produces and delivers *existing* products and services to customers. It begins

[6] Kaplan and Norton, *The Balanced Scorecard*, 96.

with a customer order and ends with the delivery of the product or service. It is the *short wave* of value creation. The **post-sales service process** provides critical and responsive services to customers after the product or service has been delivered.

Innovation Process: Objectives and Measures   Objectives for the innovation process include the following: increase the number of new products, increase percentage of revenue from proprietary products, and decrease the time to develop new products. Associated measures are actual new products developed versus planned products, percentage of total revenues from new products, percentage of revenues from proprietary products, and development cycle time (time to market).

Operations Process: Objectives and Measures   Three operations process objectives are almost always mentioned and emphasized: increase process quality, increase process efficiency, and decrease process time. Examples of process quality measures are quality costs, output yields (good output/good input), and percentage of defective units (good output/total output). Quality costing and control are discussed extensively in Chapter 16. Measures of process efficiency are concerned mainly with process cost and process productivity. Measuring and tracking process costs is facilitated by activity-based costing and process value analysis. These issues were explored in depth in the activity-based management chapter (Chapter 14). Productivity measurement is explored in Chapter 17. Common process time measures are cycle time, velocity, and manufacturing cycle efficiency (MCE).

Cycle Time and Velocity   The time it takes a company to respond to a customer order is referred to as *responsiveness*. *Cycle time* and *velocity* are two operational measures of responsiveness. **Cycle time (manufacturing)** is the length of time it takes to produce a unit of output from the time materials are received (starting point of the cycle) until the good is delivered to finished goods inventory (finishing point of the cycle).[7] Thus, cycle time is the time required to produce a product (time/units produced). **Velocity** is the number of units of output that can be produced in a given period of time (units produced/time). Although cycle time has been defined for the operations process, it is defined in a similar way for innovation and post-sales service processes. For example, how long does it take to create a new product and introduce it to the market? Or, how long does it take to resolve a customer complaint (from start to finish)?

Incentives can be used to encourage operational managers to reduce manufacturing cycle time or to increase velocity, thus improving delivery performance. A natural way to accomplish this objective is to tie product costs to cycle time and reward operational managers for reducing product costs. For example, in a JIT firm, cell conversion costs can be assigned to products on the basis of the time that it takes a product to move through the cell. Using the theoretical productive time available for a period (in minutes), a value-added standard cost per minute can be computed.

Standard cost per minute = Cell conversion costs/Minutes available

To obtain the conversion cost per unit, this standard cost per minute is multiplied by the actual cycle time used to produce the units during the period. By comparing the unit cost computed using the actual cycle time with the unit cost possible using the theoretical or optimal cycle time, a manager can assess the potential for improvement. Note that the more time it takes a product to move through the cell, the greater the unit product cost. With incentives to reduce product cost, this approach to product costing encourages operational managers and cell workers to find ways to decrease cycle time or increase velocity. Cornerstone 15-1 illustrates the concepts of cycle time and velocity.

---

[7] Other definitions of cycles are possible (e.g., a cycle's starting point could begin when the customer order is received and the finishing point when the goods are delivered to the customer). For a JIT firm, delivery to the customer is a reasonable finishing point. Another possibility for the finishing point is when the customer receives the goods. Cycle time measures the time elapsed from start to finish, regardless of how the starting and finishing points are defined.

The HOW and WHY of Calculating Cycle Time and Velocity

**CORNERSTONE
15-1**

Information:
Assume that a company has the following data for one of its manufacturing cells:

> Theoretical velocity: 40 units per hour
> Productive minutes available (per year): 1,200,000
> Annual conversion costs: $4,800,000
> Actual velocity: 30 units per hour

Why:
Cycle time (time/units produced) and velocity (units produced/time) measure the time it takes for a firm to respond to such things as customer orders, customer complaints, and the development of new products.

Required:
1. Calculate the actual conversion cost per unit using actual cycle time and the standard cost per minute.
2. Calculate the ideal conversion cost per unit using theoretical cycle time and the standard cost per minute. What incentive exists for managers when cycle time costing is used?
3. **What if** the actual velocity is 36 units per hour? What is the conversion cost per unit? What effect will this improvement have on delivery performance?

Solution:
1. Actual cycle time = 60 minutes/30 units = 2 minutes per unit
   (Notice that cycle time is the reciprocal of velocity.)
   Standard cost per minute = $4,800,000/1,200,000 = $4 per minute
   Conversion cost per unit = $4 × 2 = $8 per unit

2. Theoretical cycle time = 60 minutes/40 units = 1.5 minutes per unit
   Conversion cost per unit = $4 × 1.5 = $6 per unit
   The incentive is to reduce cycle time because it reduces the cost per unit.

3. Actual cycle time = 60 minutes/36 units = 1.67 minutes
   Conversion cost per unit = $4 × 1.67 = $6.68 per unit
   The company should be able to deliver orders more quickly and performance should improve.

Manufacturing Cycle Efficiency (MCE)

Another time-based operational measure calculates manufacturing cycle efficiency (MCE) as follows:

MCE = Processing time/(Processing time + Move time + Inspection time + Waiting time + Other non-value-added time)

where processing time is the efficient or ideal time it takes to convert materials into a finished good. The other activities and their times are viewed as wasteful, and the goal is to reduce those times to zero. If this is accomplished, the value of MCE would be 1.0. Many manufacturing companies have MCEs less than 0.05.[8] As MCE improves (moves toward 1.0), cycle time decreases. Furthermore, since the only way MCE can improve is by decreasing waste, cost reduction must also follow. Cornerstone 15-2 provides a detailed illustration of MCE.

---

[8] Kaplan and Norton, *The Balanced Scorecard*, 117.

**CORNERSTONE
15-2**

The HOW and WHY of Calculating Manufacturing Cycle Efficiency (MCE)

Information:

A company has provided the following information for one of its products for each hour of production:

> Actual velocity: 100 units (per hour)
> Move time: 20 minutes
> Inspection time: 15 minutes
> Rework time: 10 minutes

Why:

MCE measures the proportion of manufacturing cycle time attributable to value-added processing. Without waste (non-value added time), the ratio should be equal to 1.0.

Required:

1. Calculate MCE. Comment on its significance.
2. What is the theoretical cycle time? Calculate MCE using actual and theoretical cycle times.
3. **What if** waste is reduced by one-third? What is the new MCE? New cycle time?

Solution:

1. Process time = 60 minutes − 20 minutes −15 minutes − 10 minutes
   $$= 15 \text{ minutes}$$

   MCE = Process time/(Process time + Move time + Inspection time + Rework time)
   $$= 15/(15 + 20 + 15 + 10)$$
   $$= 0.25$$

   A value of 0.25 indicates that 75 percent of the manufacturing cycle is attributable to waste.

2. Theoretical cycle time = 15 minutes/100 units = 0.15 minute
   Actual cycle time = 60 minutes/100 units = 0.60 (includes theoretical cycle time plus the waste)

   MCE = Theoretical cycle time/Actual cycle time
   $$= 0.15/0.60 = 0.25$$

3. New waste = (2/3)(20 minutes + 15 minutes + 10 minutes)
   $$= 30 \text{ minutes}$$
   MCE = 15/(15 + 30) = 0.33
   (It now takes 45 minutes to produce units.)

Post-Sales Service Part of Operations Process Increasing quality, increasing efficiency, and decreasing process time are also objectives that apply to the post-sales service process. Service quality, for example, can be measured by first-pass yields where first-pass yields are defined as the percentage of customer requests resolved with a single service call. Efficiency can be measured by cost trends and

[8] Kaplan and Norton, *The Balanced Scorecard*, 117.

productivity measures. Process time can be measured by cycle time where the starting point of the cycle is defined as the receipt of a customer request and the finishing point is when the customer's problem is solved. The objectives and measures for the process perspective are summarized in Exhibit 15-3C.

4. Learning and Growth Perspective

The **learning and growth (infrastructure) perspective** defines the capabilities that an organization needs to create long-term growth and improvement. This last perspective is concerned with three major *enabling factors*: employee capabilities, information systems capabilities, and employee attitudes (motivation, empowerment, and alignment). These factors enable processes to be executed efficiently. The learning and growth perspective is the source of the capabilities that enable the accomplishment of the other three perspectives' objectives. This perspective has three major objectives: increase employee capabilities; increase motivation, empowerment, and alignment; and increase information systems capabilities.

Employee Capabilities   Three core *outcome* measurements for employee capabilities are employee satisfaction ratings, employee turnover percentages, and employee productivity (e.g., revenue per employee). Examples of lead measures or performance drivers for employee capabilities include hours of training and strategic job coverage ratios (percentage of critical job requirements filled). As new processes are created, new skills are often demanded. Training and hiring are sources of these new skills. Furthermore, the percentage of the employees needed in certain key areas with the requisite skills signals the capability of the organization to meet the objectives of the other three perspectives.

> **Mackay Memorial Hospital** in Taiwan, for example, had a specific learning and growth objective of promoting employees' ability of performing research, teaching, and innovation. Two specific performance measures for this objective were the *number of science citation index (SCI) papers* and *the number of research projects*. Thus, the more specific objectives were to increase the number of SCI papers and the number of research projects. From 2003 to 2005, the number of SCI papers increased from 132 to 1,945, and the number of research projects increased from 46 to 61.[9]

Motivation, Empowerment, and Alignment   Employees must not only have the necessary skills but must also have the freedom, motivation, and initiative to use those skills effectively. The number of suggestions per employee and the number of suggestions implemented per employee are possible measures of motivation and empowerment. Suggestions per employee provide a measure of the degree of employee involvement, whereas suggestions implemented per employee signal the quality of the employee participation. The second measure also signals to employees whether or not their suggestions are being taken seriously.

Information Systems Capabilities   Increasing information system capabilities means providing more accurate and timely information to employees so that they can improve processes and effectively execute new processes. Measures should be concerned with the *strategic information availability*. For example, possible measures include percentage of processes with real-time feedback capabilities and percentage of customer-facing employees with online access to customer and product information. Exhibit 15-3D summarizes the objectives and measures for the learning and growth perspective.

[9] Wen-Cheng Chang, Yu-Chi Tung, Chun-Hsiung Huang, and Ming-Chin Yang, "Performance Improvement After Implementing the Balanced Scorecard: A Large Hospital's Experience in Taiwan," *Total Quality Management* 19, no. 11 (November 2008): 1143–1154.

OBJECTIVE ➤ 3
Explain how the Balanced Scorecard links measures to strategy.

Linking Measures to Strategy

The Balanced Scorecard is a collection of critical performance measures that have some special properties. First, the performance measures are derived from a company's vision, strategy, and objectives. To link measures to a strategy, they must be derived from strategy. Second, performance measures should be chosen so that they are *balanced* between outcome and lead measures. Outcome measures such as profitability, return on investment, and market share tend to be generic and, therefore, common to most strategies and organizations. Performance drivers make things happen; consequently, lead measures are indicators of how the outcomes are going to be realized. Lead measures usually distinguish one strategy from another. Thus, lead measures are often unique to a strategy and because of this uniqueness support the objective of linking measures to strategy. Third, all scorecard measures should be linked by cause-and-effect relationships.

The Concept of a Testable Strategy with Strategic Feedback

This last requirement—that of linking through the use of cause-and-effect relationships—is the most important requirement. Cause-and-effect relationships are the means by which lead and lag measures are integrated and simultaneously serve as the mechanism for expressing and revealing the firm's strategy. Outcome measures are important because they reveal whether the strategy is being implemented successfully with the desired economic consequences. Lead measures supposedly cause the outcome. For example, if the number of defective products is decreased (a lead measure), does this result in a greater market share (an outcome or lag measure)? Does a greater market share (acting now as a lead measure), in turn, result in more revenues and profits (lag measures)? These questions reveal the vital role of cause-and-effect relationships in expressing an operational model of a strategy—a strategy that can be expressed in a testable format. In fact, a **testable strategy** can be defined as a set of linked objectives aimed at an overall goal. The testability of the strategy is achieved by restating the strategy into a set of cause-and-effect hypotheses that are expressed by a sequence of if-then statements.[10]

Perhaps the most important message associated with the cause-and-effect structure is that the viability of the strategy is testable. Strategic feedback is available that allows managers to test the reasonableness of the strategy. For example, if the number of defective products decrease, we would expect to see an increase in market share. If not, it could be due to one of two causes: (1) implementation problems or (2) an invalid strategy. First, it is possible that a *key performance indicator* such as the number of defective units did not achieve its targeted level (i.e., the reduction in the number of defective units was less than planned). In this case, the failure to produce the expected *outcomes* for other objectives (e.g., market share and revenue) could be merely an implementation problem. On the other hand, if the targeted levels of performance drivers were achieved and the expected outcomes did not materialize, then the problem could very well lie with the strategy itself. This is an example of *double-loop feedback*. **Double-loop feedback** occurs whenever managers receive information about both the *effectiveness* of strategy implementation as well as the *validity* of the assumptions underlying the strategy. In a traditional performance management system, typically, only *single-loop feedback* is provided. **Single-loop feedback** emphasizes only effectiveness of implementation. In single-loop feedback, actual results deviating from planned results are a signal to take corrective action so that the plan (strategy) can be executed as intended. The validity of the assumptions underlying the plan is usually not questioned.

Double-loop feedback is the foundation for strategic learning. In the Balanced Scorecard framework, strategic planning is dynamic—not static. Hypothesis testing makes it possible to change and adapt once it becomes clear that some parts of the

---

[10] Kaplan and Norton, *The Balanced Scorecard*, 149. (Kaplan and Norton describe the sequence of if-then statements only as a strategy. Calling it a testable strategy distinguishes it from the earlier, more general definition offered.)

strategy may not be viable. For example, it may be that improving quality by reducing the number of defects may not increase market share. If all other competitors are also improving quality, then the correct view may be that improving quality is needed to *maintain* market share. Increasing market share may require the company to search for some other value proposition that will be unique and innovative (e.g., offering a new product).

The **strategy map** is a useful tool that graphically illustrates the cause-and-effect relationships and connects the Balanced Scorecard strategy with an organization's operating activities. The strategy map provides a concise and pictorial representation of the firm's strategy. The linkages portrayed are for each of the firm's objectives and show how these objectives are linked for each of the four perspectives.

Strategic Alignment

O B J E C T I V E ➤ 4
Describe how an organization can
achieve strategic alignment.

Creating a strategy is one thing. Implementing the strategy successfully is another. For the Balanced Scorecard to be successful, the entire organization must be committed to its achievement. The Balanced Scorecard is designed to bring about organizational change. For this change to take place, employees must be fully informed of the strategy; they must share ownership for the objectives, measures, targets, and initiatives; incentives must be structured to support the strategy; and resources must be allocated to support the strategy.

Communicating the Strategy

The scorecard objectives and measures, once developed, become the means for articulating and communicating the strategy of the organization to its employees and managers. The objectives and measures also serve the purpose of aligning individual objectives and actions with organizational objectives and initiatives. Videos, newsletters, brochures, and the company's computer network are examples of media that can be used to inform employees of the strategy, objectives, and measures associated with the Balanced Scorecard. How much specific detail to communicate is certainly a relevant question. Communicating too much detail may create a potential problem with competitors. The Balanced Scorecard is a very explicit representation of the company's targeted markets and the means required for obtaining gains in these markets. This can be very sensitive information; the more employees who are aware of it, the more likely it may end up in the hands of competitors. Yet it is important that employees have a sufficient understanding of what is happening that they will accept and agree to the strategic efforts of the organization. Articulation of the Balanced Scorecard should be clear enough that individuals can see the linkage between what they do and the organization's long-term objectives. Seeing this linkage increases the likelihood that personal goals and actions are congruent with organizational goals.

Targets and Incentives

Once objectives and measures have been defined and communicated, performance expectations must be established. Performance expectations are communicated by setting targeted values for the measures associated with each objective. Managers are held accountable for the assigned responsibility by comparing the actual values of the measures with the targeted values. Finally, compensation is linked to achievement of the scorecard objectives. It is vital that the reward system be tied to all the scorecard objectives and not just to traditional financial measures. Failure to change the compensation system will encourage managers to continue their focus on short-term financial performance with little reason to pay attention to the strategic objectives of the scorecard.

Exhibit 15-4 provides an example of targets using a set of objectives and measures for a typical strategy. The relative importance that management has assigned to each perspective and objective is revealed by weights expressed as percentages. Targets are set for

Exhibit 15-4

Targets and Weighting Scheme Illustrated

Perspectives	Objectives	Measures	Targets
Financial (25%)	Increase shareholder value (25%) Increase profits (25%) Increase revenues (25%) Decrease process costs (25%)	Share price Profits Revenues Costs	50% increase 100% increase 30% increase 20% decrease
Customer (25%)	Increase market share (20%) Increase customer retention (30%) Improve delivery reliability (50%)	Market share Repeat orders On-time percentage	25% 70% 100%
Internal Process (25%)	Improve cycle time (60%) Redesign process (40%)	Cycle time Yes or No	2 days Yes
Learning & Growth (25%)	Improve employee skills (100%)	Hours of training	30 hours per employee

both the long term and the short term (e.g., a three- to five-year horizon and a one-year horizon) and should be backed up with initiatives that can be undertaken to achieve them. For example, is it really possible to increase share prices by 50 percent over a three-year span? And how much increase will be targeted for the coming year? The increase is dependent on increasing revenues by 30 percent and decreasing costs by 20 percent. These changes are, in turn, dependent on other events in other perspectives. Can cycle time be reduced to two days (say, from a current level of five days)?

Structuring incentive compensation with multiple dimensions is a challenging task. Typically, weights that reflect the relative importance of the perspectives are used to determine the percentage of the bonus pool that will be assigned to each perspective. Thus, from Exhibit 15-4, we see that for this example each perspective would be assigned 25 percent of the total bonus pool. But within each category, there are usually multiple objectives and multiple measures. For example, within the customer category, there are three performance measures. How much of the 25 percent bonus pool should be assigned to each measure? Again, weights that reflect the relative importance of each objective within its category are used to make this determination. Exhibit 15-4, for example, reveals that management has decided to assign 50 percent of the customer category bonus to the on-time delivery objective, 30 percent to the customer retention objective, and 20 percent to the market share objective. Thus, of the original bonus pool, 12.5 percent is assigned to the delivery objective (0.50×0.25).

Distributing potential bonus money to the various perspectives and measures is one thing, but payment of incentive compensation is dependent on *performance*. The actual values of the measures are compared to the targeted values for a given time period. Compensation is then paid, based on the percentage achievement of each objective. However, there is one major qualification for the Balanced Scorecard framework. To ensure that proper (balanced) attention is given to all measures, no incentive compensation is paid unless each strategic measure exceeds a prespecified minimum threshold value.[11]

Firms adopting the Balanced Scorecard seem to realize the necessity of connecting their reward system to the objectives and measures of the new performance management system. A Mercer study in 1999 found that 88 percent of the responding companies reported that linking the reward system to the Balanced Scorecard was effective.[12] **Mobil**, for example, reported that it would not have had the same focus on the scorecard if there was not a link to compensation.[13] The CEO of **Cigna Property & Casualty** observed that linking compensation to the new measurement system was key to gaining acceptance of the new measurement approach.[14] In another survey by the

[11] Ibid., 219–220.

[12] William Mercer and Company, *Rewarding Employees: Balanced Scorecard Fax-Back Survey Results* (London UK, May 20, 1999).

[13] Robert S. Kaplan and David P. Norton, "Transforming the Balanced Scorecard from Performance Measurement to Strategic Management: Part II," *Accounting Horizons* (June 2001): 147–160.

[14] Ibid.

Hay Group, it was found that 13 of 15 firms studied linked compensation to the scorecard. Specifically, about 25 to 33 percent of the total compensation is affected by the Balanced Scorecard, with about 40 percent focused on the financial perspective and 20 percent assigned to each of the three remaining perspectives.[15]

Resource Allocation

Achieving strategic targets such as those envisioned in Exhibit 15-4 requires that resources be allocated to the corresponding strategic initiatives. This requires two major changes. First, an organization must decide how much of the strategic targets will be achieved for the coming year. Second, the operational budgetary process must be structured to provide the resources necessary for achievement of these short-term advances along the strategic path. If these changes are not incorporated, then it is difficult to imagine that the strategy will truly become actionable.

Summary of Learning Objectives

1. **Compare and contrast activity-based and strategic-based responsibility accounting systems.**

- Activity-based responsibility accounting focuses on processes, uses both operational and financial measures, employs dynamic standards, and emphasizes and supports continuous improvement.
- Strategic-based responsibility accounting expands the number of responsibility dimensions from two to four. Customer and learning and growth perspectives are added.
- Strategic-based performance measures become an integrated set of measures, linked to an organization's mission and strategy.
- Activity- and strategic-based responsibility accounting systems work best for firms operating in dynamic environments.

2. **Discuss the basic features of the Balanced Scorecard.**

- The Balanced Scorecard is a strategic performance management system that translates the vision and strategy of an organization into operational objectives and measures.
- Objectives and measures are developed for each of four perspectives: the financial perspective, the customer perspective, the process perspective, and the learning and growth perspective.

3. **Explain how the Balanced Scorecard links measures to strategy.**

- Performance measures are derived from a company's vision, strategy, and objectives.
- Performance measures are balanced between outcome and lead measures.
- All scorecard measures are linked by cause-and-effect relationships.
- The cause-and-effect relationships produce a set of testable hypotheses expressed by a sequence of if-then statements.

4. **Describe how an organization can achieve strategic alignment.**

- The entire organization must be committed to the Balanced Scorecard.
- Employees must be fully informed of the strategy and share ownership for the objectives, measures, targets, and initiatives.
- Incentives (e.g., compensation) must be structured to support the strategy, and resources must be allocated to support the strategy.
- Thus, alignment with the strategy expressed by the Balanced Scorecard is achieved by communication, incentives, and allocation of resources to support the strategic initiatives.

[15] Todd Manas, "Making the Balanced Scorecard Approach Payoff," *ACA Journal* 8, no. 2 (Second Quarter, 1999).

CORNERSTONE 15-1   The HOW and WHY of calculating cycle time and velocity, page 761

CORNERSTONE 15-2   The HOW and WHY of calculating manufacturing cycle efficiency (MCE), page 762

CORNERSTONES FOR CHAPTER 15

Review Problems

I. Perspectives, Measures, and Strategic Objectives

The following measures belong to one of four perspectives: financial, customer, process, or learning and growth.

a.   Revenues from new products
b.   On-time delivery percentage
c.   Economic value added
d.   Employee satisfaction
e.   Cycle time
f.   First-pass yields
g.   Strategic job coverage ratio
h.   Number of new customers
i.   Unit product cost
j.   Customer profitability

Required:

Classify each measure by perspective, and suggest a possible strategic objective that might be associated with the measure.

Solution:

	Perspective	Objective
a.	Financial	Increase number of new products
b.	Customer	Increase delivery reliability
c.	Financial	Improve asset utilization
d.	Learning and Growth	Increase motivation and alignment
e.	Process	Decrease process time
f.	Process	Increase service quality
g.	Learning and Growth	Increase employee capabilities
h.	Customer	Increase customer acquisition
i.	Financial	Decrease product cost
j.	Customer	Increase customer profitability

II. Cycle Time and Velocity, MCE

Currently, a company can produce 60 units per hour of a particular product. During this hour, move time and wait time take 30 minutes, while actual processing time is 30 minutes.

Required:

1.   Calculate the current MCE.
2.   Calculate the current cycle time.
3.   Suppose that move time and wait time are reduced by 50 percent. What is the new velocity? The new cycle time? The new MCE?

Solution:

1. MCE = Process time/(Process time + Move time + Wait time)

    = 30 minutes/60 minutes

    = 0.50

2. Cycle time = 1/Velocity = 1/60 hour, or 1 minute

3. The time now required to produce 60 units is 45 minutes (30 minutes process time; move and wait time of 15 minutes). Thus, velocity = 60/(3/4 hour) = 80 units per hour; cycle time = 1/80 hour, or 0.75 minute. Finally, MCE = 30/(30 + 15) = 0.67.

Key Terms

Balanced Scorecard, 751

Core objectives and measures, 757

Customer perspective, 757

Customer value, 757

Cycle time (manufacturing), 760

Double-loop feedback, 764

External measures, 752

Financial measures, 752

Financial perspective, 755

Innovation process, 759

Internal business process perspective, 759

Internal measures, 752

Lag measures, 752

Lead measures (performance drivers), 752

Learning and growth (infrastructure) perspective, 763

Nonfinancial measures, 752

Objective measures, 752

Operations process, 759

Post-purchase costs, 757

Post-sales service process, 760

Process value chain, 759

Single-loop feedback, 764

Strategic-based responsibility accounting system (strategic-based performance management system), 751

Strategy, 755

Strategy map, 765

Stretch targets, 754

Subjective measures, 752

Testable strategy, 764

Velocity, 760

Discussion Questions

1. Describe a strategic-based responsibility accounting system. How does it differ from activity-based responsibility accounting?
2. What is a Balanced Scorecard?
3. What is meant by balanced measures?
4. What is a lag measure? A lead measure?
5. What is the difference between an objective measure and a subjective measure?
6. What are stretch targets? What is their strategic purpose?
7. How does the reward system for a strategic-based system differ from the traditional approach?
8. What are the three strategic themes of the financial perspective?
9. Identify the five core objectives of the customer perspective.
10. Explain what is meant by the long wave and the short wave of value creation.
11. Define the three processes of the process value chain.
12. Identify three objectives of the learning and growth perspective.
13. What is a testable strategy?
14. What is meant by double-loop feedback?
15. Identify and explain three methods for achieving strategic alignment.

Cornerstone Exercises

OBJECTIVE ▶ 2
CORNERSTONE 15-1

Cornerstone Exercise 15-1 CYCLE TIME AND VELOCITY

Blackburn Manufacturing has the following data for one of its production departments:

> Theoretical velocity: 75 units per hour
> Productive minutes available per year: 2,500,000
> Annual conversion costs: $15,000,000
> Actual velocity: 40 units per hour

Required:

1. Calculate the actual conversion cost per unit using actual cycle time and the standard cost per minute.
2. Calculate the ideal conversion cost per unit using theoretical cycle time and the standard cost per minute. What incentive exists for managers when cycle time costing is used?
3. *What if* the actual velocity is 55 units per hour? What is the conversion cost per unit? What effect will this improvement have on delivery performance?

OBJECTIVE ▶ 2
CORNERSTONE 15-2

Cornerstone Exercise 15-2 MCE

Parron Company has provided the following information for one of its products for each hour of production:

> Actual velocity: 80 units (per hour)
> Move time: 18 minutes
> Inspection time: 24 minutes
> Rework time: 8 minutes

Required:

1. Calculate MCE. Comment on its significance.
2. What is the theoretical cycle time? Calculate MCE using actual and theoretical cycle times.
3. *What if* non-value added time is reduced by one-half? What is the new MCE? New cycle time?

Exercises

OBJECTIVE ▶ 1

Exercise 15-3 ACTIVITY-BASED RESPONSIBILITY ACCOUNTING VERSUS STRATEGIC-BASED RESPONSIBILITY ACCOUNTING

The following comment was made by the CEO of a company that recently implemented the Balanced Scorecard: "Responsibility in a strategic-based performance management system differs on the three D's: Direction, Dimension, and Diffusion."

Required:

Explain how this comment describes differences in responsibility between an activity-based and a strategic-based performance management system.

OBJECTIVE ▶ 1

Exercise 15-4 ACTIVITY-BASED RESPONSIBILITY ACCOUNTING VERSUS STRATEGIC-BASED RESPONSIBILITY ACCOUNTING

"A Balanced Scorecard expresses the complete story of a company's strategy through an integrated set of financial and nonfinancial measures that are both predictive and historical and which may be measured subjectively or objectively."

Required:

1. Using the above statement about scorecard measures, explain how scorecard measurement differs from that of an activity-based management system.

2. Explain what is meant by historical and predictive measures. Why are both types important for describing a company's strategy?

Exercise 15-5 ACTIVITY-BASED RESPONSIBILITY ACCOUNTING VERSUS STRATEGIC-BASED RESPONSIBILITY ACCOUNTING

OBJECTIVE ➤ 1 3

The Balanced Scorecard is an approach that has the objective of driving change. Performance evaluation is an integral part of this effort. Performance evaluation within the Balanced Scorecard framework is also concerned with the effectiveness and viability of the organization's strategy.

Required:

1. Describe how the Balanced Scorecard is used to drive organizational change.
2. Explain how performance evaluation is used to assess the effectiveness and viability of an organization's strategy.

Exercise 15-6 BALANCED SCORECARD, PERSPECTIVES, CLASSIFICATION OF PERFORMANCE MEASURES

OBJECTIVE ➤ 1 2

Consider the following list of scorecard measures:

a. Product profitability
b. Ratings from customer surveys
c. Number of patents pending
d. Strategic job coverage ratio
e. Revenue per employee
f. Quality costs
g. Percentage of market
h. Employee turnover percentages
i. First-pass yields
j. On-time delivery percentage
k. Percentage of revenues from new sources
l. Economic value added

Required:

Classify each measure according to the following: perspective, financial or nonfinancial, subjective or objective, and external or internal. When the perspective is process, identify which type of process: innovation, operations, or post-sales service.

Exercise 15-7 CYCLE TIME AND CONVERSION COST PER UNIT

OBJECTIVE ➤ 2

The theoretical cycle time for a product is 72 minutes per unit. The budgeted conversion costs for the manufacturing cell dedicated to the product are $6,480,000 per year. The total labour minutes available are 1,440,000. During the year, the cell was able to produce 0.5 unit of the product per hour. Suppose also that production incentives exist to minimize unit product costs.

Required:

1. Compute the theoretical conversion cost per unit.
2. Compute the applied conversion cost per minute (the amount of conversion cost actually assigned to the product).
3. Discuss how this approach to assigning conversion cost can improve delivery time performance. Explain how conversion cost acts as a performance driver for on-time deliveries.

Exercise 15-8 CYCLE TIME AND VELOCITY, MCE

OBJECTIVE ➤ 2

A manufacturing plant has the theoretical capability to produce 162,000 laptops per quarter but currently produces 60,750 units. The conversion cost per quarter is $7,290,000. There are 40,500 production hours available within the plant per quarter. In addition to the processing minutes per unit used, the production of the laptops uses 8 minutes of move time, 12 minutes of wait time, and 5 minutes of rework time. (All work is done by cell workers.)

Required:

1. Compute the theoretical and actual velocities (per hour) and the theoretical and actual cycle times (minutes per unit produced).
2. Compute the ideal and actual amounts of conversion cost assigned per laptop.
3. Calculate MCE. How does MCE relate to the conversion cost per laptop?

OBJECTIVE ➤ 2 3

Exercise 15-9 CYCLE TIME AND VELOCITY, MCE

Refer to **Exercise 15-8**. Assume that the company identifies poor plant layout as the root cause of wait time and move time.

Required:

1. Express an improvement strategy as a series of if-then statements that will reduce the conversion cost per laptop.
2. Assume that you set an MCE target of 75 percent, based on the improvement strategy described in Requirement 1. What is the expected conversion cost per unit? Explain how you can use these targets to test the viability of your quality improvement strategy.

OBJECTIVE ➤ 1 2 3

Exercise 15-10 BALANCED SCORECARD, LEAD AND LAG VARIABLES, DOUBLE-LOOP FEEDBACK

The following if-then statements were taken from a Balanced Scorecard:

a. If employee productivity increases, then process efficiency will increase.
b. If process efficiency increases, then product price can be decreased.

Required:

1. Identify the lead and lag variables, and explain your reasoning.
2. Discuss the implications of Requirement 1 for the financial and learning and growth perspectives.
3. Using the first if-then statement, explain the concept of double-loop feedback.

OBJECTIVE ➤ 3

Exercise 15-11 BALANCED SCORECARD

Halifax Civic Hospital developed the following series of if-then statements for its Balanced Scorecard strategy:

- If employee turnover rate decreases and employee satisfaction increases, then the quality of health care service will improve.
- If the quality of health care service improves, then operating efficiency will increase and patient satisfaction will increase.
- If operating efficiency increases, then operating costs will decrease.
- If patient satisfaction increases, then market share will increase.
- If market share increases, then revenues will increase.
- If revenues increase and costs decrease, then we will balance the budget.

Required:

1. Explain how a performance measure can act both as a lag variable and a lead indicator.
2. *What if* budget savings did not increase to the targeted level? Explain how this result could be attributable to either an implementation problem or an invalid strategy. What actions would likely be taken for each case?

OBJECTIVE ➤ 3

Exercise 15-12 TESTABLE STRATEGY

Consider the following quality improvement strategy as expressed by a series of if-then statements:

- If design engineers receive quality training, then they can redesign products to reduce the number of defective units.
- If the number of defective units is reduced, then customer satisfaction will increase.
- If customer satisfaction increases, then market share will increase.

- If market share increases, then sales will increase.
- If sales increase, then profits will increase.

Required:

Explain how the quality improvement strategy can be tested.

Exercise 15-13 BALANCED SCORECARD, STRATEGY TRANSLATION, DOUBLE-LOOP FEEDBACK

OBJECTIVE ➤ 2 3

Mumbai Company, an electronics firm, buys circuit boards and manually inserts various electronic devices into the printed circuit board. Mumbai sells its products to original equipment manufacturers. Profits for the past two years have been less than expected. Jocelyn Dubois, owner of Mumbai, was convinced that her firm needed to adopt a revenue growth and cost reduction strategy to increase overall profits.

After a careful review of her firm's condition, Jocelyn realized that the main obstacle for increasing revenues and reducing costs was the high defect rate of her products (a 6 percent reject rate). She was certain that revenues would grow if the defect rate was reduced dramatically. Costs would also decline as there would be fewer rejects and less rework. By decreasing the defect rate, customer satisfaction would increase, causing, in turn, an increase in market share. Jocelyn also felt that the following actions were needed to help ensure the success of the revenue growth and cost reduction strategy:

a. Improve the soldering capabilities by sending employees to an outside course.
b. Redesign the insertion process to eliminate some of the common mistakes.
c. Improve the procurement process by selecting suppliers that provide higher-quality circuit boards.

Required:

1. State the revenue growth and cost reduction strategy using a series of cause-and-effect relationships expressed as if-then statements.
2. Explain how the revenue growth strategy can be tested. In your explanation, discuss the role of lead and lag measures, targets, and double-loop feedback.

Exercise 15-14 BALANCED SCORECARD, STRATEGIC ALIGNMENT

OBJECTIVE ➤ 4

Refer to **Exercise 15-13**. Suppose that Jocelyn communicates the following weights to her CEO:

Perspective: Financial, 40%; Customer, 20%; Process, 20%; Learning and growth, 20%
Financial objectives: Profits, 50%; Revenues, 25%; Costs, 25%
Customer objectives: Customer satisfaction, 60%; Market share, 40%
Process objectives: Defects decrease, 40%; Supplier selection, 30%; Redesign process, 30%
Learning and growth objective: Training, 100%

Jocelyn next sets up a bonus pool of $100,000 and indicates that the weighting scheme just described will be used to determine the amount of potential bonus for each perspective and each objective.

Required:

1. Calculate the potential bonus for each perspective and objective.
2. Describe how Jocelyn might award actual bonuses so that her managers will be encouraged to implement the Balanced Scorecard.
3. What are some other ways that Jocelyn can encourage alignment with the company's strategic objectives (other than incentive compensation)?

Problems

Problem 15-15 ACTIVITY-BASED RESPONSIBILITY ACCOUNTING VERSUS STRATEGIC-BASED RESPONSIBILITY ACCOUNTING

Carson Wei, president of Mallory Plastics, was considering a report sent to him by Emily Sorensen, vice president of operations. The report was a summary of the progress made by an activity-based management system that was implemented three years ago. Significant progress had indeed been realized. At the conclusion of the report, Emily urged Carson to consider the adoption of the Balanced Scorecard as a logical next step in the company's efforts to establish itself as a leader in its industry. Emily clearly was impressed by the Balanced Scorecard and intrigued by the possibility that the change would enhance the overall competitiveness of Mallory. She requested a meeting of the executive committee to explain the similarities and differences between the two approaches. Carson agreed to schedule the meeting but asked Emily to prepare a memo in advance, listing the most important similarities and differences between the two approaches to responsibility accounting.

Required:

Prepare the memo requested by Carson.

Problem 15-16 SCORECARD MEASURES, STRATEGY TRANSLATION

At the end of 2011, Activo Company implemented a low-cost strategy to improve its competitive position. Its objective was to become the low-cost producer in its industry. A Balanced Scorecard was developed to guide the company toward this objective. To lower costs, Activo undertook a number of improvement activities such as JIT production, total quality management, and activity-based management. Now, after two years of operation, the president of Activo wants some assessment of the achievements. To help provide this assessment, the following information on one product has been gathered:

	2011	2013
Theoretical annual capacity*	124,800	124,800
Actual production**	104,000	117,000
Market size (in units sold)	650,000	650,000
Production hours available (20 workers)	52,000	52,000
Very satisfied customers	41,600	70,200
Actual cost per unit	$162.50	$130
Days of inventory	7.8	3.9
Number of defective units	6,500	2,600
Total worker suggestions	52	156
Hours of training	130	520
Selling price per unit	$195	$195
Number of new customers	2,600	13,000

*Amount that could be produced given the available production hours; everything produced is sold.
**Amount that was produced given the available production hours.

Required:

1. Compute the following measures for 2011 and 2013:
   a. Actual velocity and cycle time
   b. Percentage of total revenue from new customers (assume one unit per customer)
   c. Percentage of very satisfied customers (assume each customer purchases one unit)
   d. Market share
   e. Percentage change in actual product cost (for 2013 only)

f.   Percentage change in days of inventory (for 2013 only)
g.   Defective units as a percentage of total units produced
h.   Total hours of training
i.   Suggestions per production worker
j.   Total revenue
k.   Number of new customers

2.   For the measures listed in Requirement 1, list likely strategic objectives, classified according to the four Balanced Scorecard perspectives. Assume there is one measure per objective.

Problem 15-17 IF-THEN STATEMENTS, BALANCED SCORECARD

OBJECTIVE ➤ 2 3

Refer to the data in **Problem 15-16**.

Express Activo's strategy as a series of if-then statements. What does this tell you about Balanced Scorecard measures?

Problem 15-18 STRATEGIC OBJECTIVES, SCORECARD MEASURES

OBJECTIVE ➤ 2 3

The following strategic objectives have been derived from a strategy that seeks to improve asset utilization by more careful development and use of its human assets and internal processes:

a.   Increase revenue from new products.
b.   Increase implementation of employee suggestions.
c.   Decrease operating expenses.
d.   Decrease cycle time for the development of new products.
e.   Decrease rework.
f.   Increase employee morale.
g.   Increase customer satisfaction.
h.   Increase access of key employees to customer and product information.
i.   Increase customer acquisition.
j.   Increase return on investment (ROI).
k.   Increase employee productivity.
l.   Decrease the collection period for accounts receivable.
m.   Increase employee skills.

The heart of the strategy is developing the company's human resources. Management is convinced that empowering employees will lead to an increase in economic returns. Studies have shown that there is a positive relationship between employee morale and customer satisfaction. Furthermore, the more satisfied customers pay their bills more quickly. It was hypothesized that as employees became more involved and more productive their morale would improve. Thus, the strategy incorporated key objectives that would lead to an increase in productivity and involvement.

Required:

1.   Classify the objectives by perspective, and suggest a measure for each objective.
2.   Describe the likely causal relationships among the strategic objectives.

Problem 15-19 CYCLE TIME, CONVERSION COST PER UNIT, MCE

OBJECTIVE ➤ 2

A manufacturing cell has the theoretical capability to produce 375,000 carburetors per quarter. The conversion cost per quarter is $3,750,000. There are 125,000 production hours available within the cell per quarter.

Required:

1.   Compute the theoretical velocity (per hour) and the theoretical cycle time (minutes per unit produced).
2.   Compute the ideal amount of conversion cost that will be assigned per subassembly.

3. Suppose the actual time required to produce a carburetor is 35 minutes. Compute the amount of conversion cost actually assigned to each unit produced. What happens to product cost if the time to produce a unit is decreased to 25 minutes? How can a firm encourage managers to reduce cycle time? Finally, discuss how this approach to assigning conversion cost can improve delivery time.

4. Calculate MCE. How much non-value-added time is being used? How much is it costing per unit?

5. Cycle time, velocity, MCE, conversion cost per unit (theoretical conversion rate × actual conversion time), and non-value-added costs are all measures of performance for the cell process. Discuss the incentives provided by these measures.

OBJECTIVE ➤ 2 3

Problem 15-20 MCE, TESTABLE STRATEGY

Auflegger Inc. manufactures a product that experiences the following activities (and times):

	Hours
Processing (two departments)	42.0
Inspecting	2.8
Rework	7.0
Moving (three moves)	11.2
Waiting (for the second process)	33.6
Storage (before delivery to customer)	43.4

Required:

1. Compute the MCE for this product.

2. A study lists the following root causes of the inefficiencies: poor quality components from suppliers, lack of skilled workers, and plant layout. Suggest a possible cost reduction strategy, expressed as a series of if-then statements, that will reduce MCE and lower costs.

3. Is MCE a lag or a lead measure? If and when MCE acts as a lag measure, what lead measures would affect it?

OBJECTIVE ➤ 3

Problem 15-21 CYCLE TIME, VELOCITY, PRODUCT COSTING

Morrison Inc. has a JIT system in place. Each manufacturing cell is dedicated to the production of a single product or major subassembly. One cell, dedicated to the production of small four wheelers, has four operations: machining, finishing, assembly, and qualifying (testing). The machining process is automated, using computers. In this process, the model's frame and engine are constructed. In finishing, the frame is sandblasted, buffed, and painted. In assembly, the frame and engine are assembled. Finally, each model is tested to ensure operational capability.

For the coming year, the four-wheeler cell has the following budgeted costs and cell time (both at theoretical capacity):

Budgeted conversion costs	$15,500,000
Budgeted materials	$18,600,000
Cell time	24,800 hours
Theoretical output	18,600 models

During the year, the following actual results were obtained:

Actual conversion costs	$15,500,000
Actual materials	$16,120,000
Actual cell time	24,800 hours
Actual output	15,500 models

Required:

1. Compute the velocity (number of models per hour) that the cell can theoretically achieve. Now, compute the theoretical cycle time (number of hours or minutes per model) that it takes to produce one model.

2. Compute the actual velocity and the actual cycle time.

3. Compute MCE. Comment on the efficiency of the operation.

4. Compute the budgeted conversion cost per minute. Using this rate, compute the conversion cost per model if theoretical output is achieved. Using this measure, compute the conversion cost per model for actual output. Does this product costing approach provide an incentive for the cell manager to reduce cycle time? Explain.

Problem 15-22 BALANCED SCORECARD, NON-VALUE-ADDED ACTIVITIES, STRATEGY TRANSLATION, KAIZEN COSTING

OBJECTIVE ▶ 1 2 3 4

At the beginning of the last quarter of 2011, Blind River Ltd., a consumer products firm, hired Brittiny (Brit) Compton to take over one of its divisions. The division manufactured small home appliances and was struggling to survive in a very competitive market. Brit immediately requested a projected income statement for 2011. In response, the controller provided the following statement:

Sales	$25,000,000
Variable expenses	20,000,000
Contribution margin	5,000,000
Fixed expenses	6,000,000
Projected loss	$ (1,000,000)

After some investigation, Brit soon realized that the products being produced had a serious problem with quality. She once again requested a special study by the controller's office to supply a report on the level of quality costs. By the middle of November, Brit received the following report from the controller:

Inspection costs, finished product	$ 400,000
Rework costs	2,000,000
Scrapped units	600,000
Warranty costs	3,000,000
Sales returns (quality-related)	1,000,000
Customer complaint department	500,000
Total estimated quality costs	$7,500,000

Brit was surprised at the level of quality costs. They represented 30 percent of sales, which was certainly excessive. She knew that the division had to produce high-quality products to survive. The number of defective units produced needed to be reduced dramatically. Thus, Brit decided to pursue a quality-driven turnaround strategy. Revenue growth and cost reduction could both be achieved if quality could be improved. By growing revenues and decreasing costs, profitability could be increased.

After meeting with the managers of production, marketing, purchasing, and human resources, Brit made the following decisions, effective immediately (end of November 2011):

a. More will be invested in employee training. Workers will be trained to detect quality problems and empowered to make improvements. Workers will be allowed a bonus of 10 percent of any cost savings produced by their suggested improvements.

b. Two design engineers will be hired immediately, with expectations of hiring one or two more within a year. These engineers will be in charge of redesigning processes and products with the objective of improving quality. They will also be given the responsibility of working with selected suppliers to help improve the quality of their products and processes. Design engineers are considered a strategic necessity.

c. Implement a new process: evaluation and selection of suppliers. This new process has the objective of selecting a group of suppliers that are willing and capable of providing nondefective components.

d. Effective immediately, the division will begin inspecting purchased components. According to production, many of the quality problems are caused by defective components purchased from outside suppliers. Incoming inspection is viewed as a transitional activity. Once the division has developed a group of suppliers capable of delivering nondefective components, this activity will be eliminated.

e. The goal is to produce products with a defect rate less than 0.10 percent within three years. By reducing the defect rate to this level, marketing is confident that market share will increase by at least 50 percent (as a consequence of increased customer satisfaction). Products with better quality will help establish an improved product image and reputation, allowing the division to capture new customers and increase market share.

f. Accounting will be given the task of installing a quality information reporting system. Daily reports on operational quality data (e.g., percentage of defective units), weekly updates of trend graphs (posted throughout the division), and quarterly cost reports are the types of information required.

g. To help direct the improvements in quality activities, kaizen costing is to be implemented. For example, for the year 2012, a kaizen standard of 6 percent of the selling price per unit will be set for rework costs, a 25 percent reduction from the current actual cost.

To ensure that the quality improvements are directed and translated into concrete financial outcomes, Brit has also begun to implement a Balanced Scorecard for the division. By the end of 2012, progress is being made. Sales have increased to $26,000,000, and the kaizen improvements are meeting or beating expectations. For example, rework costs have dropped to $1,500,000.

At the end of 2013, two years after the turnaround quality strategy was implemented, Brit receives the following quality cost report:

Quality training	$ 500,000
Supplier evaluation	230,000
Incoming inspection costs	400,000
Inspection costs, finished product	300,000
Rework costs	1,000,000
Scrapped units	200,000
Warranty costs	750,000
Sales returns (quality-related)	435,000
Customer complaint department	325,000
Total estimated quality costs	$4,140,000

Brit also receives an income statement for 2013:

Sales	$30,000,000
Variable expenses	22,000,000
Contribution margin	8,000,000
Fixed expenses	5,800,000
Income from operations	$ 2,200,000

Brit is pleased with the outcomes. Revenues have grown, and costs have been reduced by at least as much as she had projected for the two-year period. Growth next year should be even greater as she is beginning to observe a favourable effect from the higher-quality products. Also, further quality cost reductions should materialize as incoming inspections are showing much higher-quality purchased components.

Required:

1. Identify the strategic objectives, classified by the Balanced Scorecard perspectives. Next, suggest measures for each objective.
2. Using the results from Requirement 1, describe Brit's strategy using a series of if-then statements.
3. Explain how you would evaluate the success of the quality-driven turnaround strategy. What additional information would you like to have for this evaluation?
4. Explain why Brit felt that the Balanced Scorecard would increase the likelihood that the turnaround strategy would actually produce good financial outcomes.
5. Advise Brit on how to encourage her employees to align their actions and behaviour with the turnaround strategy.

CMA Problem

CMA Problem 15-1 **THE BALANCED SCORECARD***

Aegean Inc. is pursuing a number of strategic objectives under an overall strategy to become a low-cost producer. To this end, it applies a balanced scorecard perspective. To succeed in this strategy, Aegean has undertaken a number of initiatives including total quality management, activity-based costing, and JIT. By the end of the year 2013, the new strategy had been in place for two full years, and it was felt an assessment was necessary to find out how well the new strategy had worked.

The data below were gathered to facilitate this assessment.

	2012	2013
Theoretical maximum capacity (units)	96,000	96,000
Actual production (units)	76,000	88,000
Production hours available (20 employees)	40,000	40,000
Post-purchase costs per unit	$ 20	$ 10
Kilograms of materials used	100,000	100,000
Kilograms of scrap	10,000	8,000
Selling price per unit	$ 150	$ 140
Actual cost per unit	$ 125	$ 100
Days in inventory	6	3
Number of new customers	2,000	8,000
Number of defective units	4,500	2,000
Suggestions per employee	2	6
Hours of training	100	400

Required:

1. Compute the following for both years 2012 and 2013:
   a. Theoretical velocity and cycle time
   b. Actual velocity and cycle time
   c. Labour productivity
   d. Scrap as a percentage of materials used
   e. Percentage change in post-purchase costs for 2013
   f. Percentage change in actual product costs for 2013
   g. Percentage change in days of inventory for 2013
   h. Percentage of defective units in terms of total units produced
   i. New customers metric
   j. Hours of training metric
   k. Total employee suggestions

*© 2011, CMA Ontario. http://www.cma-ontario.org. Reprinted with permission.

2. Classify the above data according to balanced scorecard perspectives (Financial, Customer, Process, Learning and Growth) indicating each metric as lead or lag variable. Then evaluate the overall strategy success of Aegean. *(Adapted from CMA Ontario)*

The Collaborative Learning Exercises can be found on the product support site at www.hansen1ce.nelson.com.

After studying this chapter, you should be able to:

1 Define quality, describe the four types of quality costs, and discuss the approaches used for quality cost measurement.

2 Prepare a quality cost report, and explain its use.

3 Explain why quality cost information is needed and how it is used.

4 Describe and prepare three different types of quality performance reports.

5 Discuss how environmental costs can be measured, reported, and reduced.

6 Show how environmental costs can be assigned to products and processes.

CHAPTER

16

Quality and Environmental Cost Management

There are numerous quality- and environmental-related activities, all of which consume resources that determine the level of quality and environmental costs incurred by a firm. Inspecting or testing parts, for example, is a quality appraisal activity that has the objective of detecting bad products, whereas contamination tests are designed to measure the level of pollution. Detecting bad products and correcting them before they are sent to customers is usually less expensive than letting them be acquired by customers. Similarly, preventing contamination and waste from entering the environment is also usually less expensive. The objective of quality and environmental cost management is to find ways to minimize total quality and environmental costs. Interestingly, there are remarkable similarities between the two approaches. Quality cost management will first be explored, followed by environmental cost management.

Competitive forces are requiring firms to pay increasing attention to quality. Customers are demanding higher-quality products and services. Improving quality may actually be the key to survival for many firms. Improving process quality and the quality of products and services is a fundamental strategic objective that is part of any well-designed Balanced Scorecard. If quality is improved, then customer satisfaction increases; if customer satisfaction increases, then market share will increase; and if market share increases, then revenues will increase; moreover, if quality improves, then operating costs will also decrease. Thus, improving quality can increase market share and sales, while simultaneously decreasing costs. The overall effect enhances a firm's financial and competitive position.

Improving quality can increase firm value because it increases a firm's profitability. Improving quality can increase profitability in at least two ways: (1) by increasing customer demand and (2) by decreasing the costs of providing goods and services.

Over the past few years, major companies but also other organizations have made significant strides in improving quality. Even so, much remains to be done. The costs of quality can be substantial and a source of significant savings. According to some experts, most companies, if they properly evaluate their costs of quality, will find that they are between 15 and 25 percent of sales.[1] Most experts tend to agree that quality costs range between 5 and 30 percent of sales.[2] Yet, quality experts indicate that the optimal quality level should be about 2 to 4 percent of sales. This difference between actual and optimal figures represents a veritable gold mine of opportunity. Improving quality can produce significant improvements in profitability.

In 2003, **Caterpillar Financial Services Corporation U.S. (CFSC)** won a Malcolm Baldrige National Quality Award. The Malcolm Baldrige Quality award recognizes companies that excel in quality management and achievement. CFSC improved its quality and increased its contributions to Caterpillar Inc.'s total earnings from 5.6 percent to more than 25 percent.[3] CFSC's efforts to improve quality produced significant after-tax savings, increased customer satisfaction levels (exceeding industry benchmarks), and improved the overall work environment (80 percent of employees indicated that they would recommend CFSC as a place to work, exceeding the national norm of 55 percent).

As companies implement quality improvement programs, a need arises to monitor and report on the progress of these programs. Managers need to know what quality costs are and how they are changing over time. Reporting and measuring quality performance is absolutely essential to the success of an ongoing quality improvement program. A fundamental prerequisite for this reporting is measuring the costs of quality. But to measure those costs, an operational definition of quality is needed.

OBJECTIVE ▶ 1
Define quality, describe the four types of quality costs, and discuss the approaches used for quality cost measurement.

Quality Defined

Operationally, a **quality product or service** is one that meets or exceeds customer expectations. In effect, quality is customer satisfaction. But what is meant by "customer expectations"? Customers can be concerned with such product attributes as reliability, durability, fitness for use, and conformance to specifications. Although many important attributes can affect customer satisfaction, the quality attributes that are measurable tend to receive more emphasis. Conformance, in particular, is strongly emphasized. In fact, many quality experts believe that **"quality of conformance"** is the best operational definition. There is some logic to this position. Product specifications should explicitly consider such things as reliability, durability, and fitness for use. Implicitly, a conforming product is reliable, durable, fit for use, and performs well. The product should be produced as the design specifies it; specifications should

[1] M. J. Harry and R. Schroeder, *Six Sigma: The Breakthrough Management Strategy Revolutionizing the World's Top Corporations* (New York: Doubleday, Random House, 2000).

[2] G. Giakatis, T. Enkawa, and K. Washitani, "Hidden Quality Costs and the Distinction between Quality Cost and Quality Loss," *Total Quality Management*, 12, 2 (2001): 179–180.

[3] S. M. Paton, "Quality Conversation with James S. Beard," *Quality Digest*, (September 2004), http://www.qualitydigest.com/sept04/articles/07_article.shtml.

be met. Conformance is the basis for defining what is meant by a nonconforming, or *defective*, product.

A **defective product** is one that does not conform to specifications. **Zero defects** means that all products conform to specifications. But what is meant by "conforming to specifications"? Traditional conformance defines an acceptable range of values for each specification or quality characteristic. A target value is defined, and upper and lower limits are set that describe acceptable product variation for a given quality characteristic. Any unit that falls within the limits is deemed nondefective. For example, the targeted specification for a machined part may be a drilled hole that is two cm in diameter, and any part that is within 3 percent of the target is acceptable. On the other hand, the *robust quality view* of conformance emphasizes exactness of conformance. **Robustness** means exact conformance to the target value (no tolerance allowed). There is no range in which variation is acceptable. A nondefective machine part in the robust setting would be one that has a drilled hole that measures exactly two cm. Since evidence exists that product variation can be costly, the robust quality definition of conformance is superior to the traditional definition.

Costs of Quality

Quality-linked activities are those activities performed because poor quality may or does exist. The costs of performing these activities are referred to as costs of quality. There are four categories of quality costs: (1) prevention costs, (2) appraisal costs, (3) internal failure costs, and (4) external failure costs.

Prevention costs are incurred to prevent poor quality in the products or services being produced. As prevention costs increase, we would expect the costs of failure to decrease. Examples of prevention costs are quality engineering, quality training programs, quality planning, quality reporting, supplier evaluation and selection, quality audits, quality circles, field trials, and design reviews.

Appraisal costs are incurred to determine whether products and services are conforming to their requirements or customer needs. Examples include inspecting and testing materials, packaging inspection, supervising appraisal activities, product acceptance, process acceptance, measurement (inspection and test) equipment, and outside endorsements. Two of these terms require further explanation.

Product acceptance involves sampling from batches of finished goods to determine whether they meet an acceptable quality level; if so, the goods are accepted. *Process acceptance* involves sampling goods while in process to see if the process is in control and producing nondefective goods; if not, the process is shut down until corrective action can be taken. The main objective of the appraisal function is to prevent nonconforming goods from being shipped to customers.

Internal failure costs are incurred because products and services do not conform to specifications or customer needs. This nonconformance is detected prior to the product being shipped or the service being delivered to outside parties. These are the failures detected by appraisal activities. Examples of internal failure costs are scrap, rework, downtime (due to defects), reinspection, retesting, and design changes. These costs disappear if no defects exist.

External failure costs are incurred because products and services fail to conform to requirements or satisfy customer needs after being delivered to customers. Of all the costs of quality, this category can be the most devastating. Costs of recalls, for example, can run into the hundreds of millions. Other examples include lost sales because of poor product performance, returns and allowances because of poor quality, warranties, repair, product liability, customer dissatisfaction, lost market share, and complaint adjustment. External failure costs, like internal failure costs, disappear if no defects exist.

Costs of quality are the costs that exist because poor quality may or does exist. This definition implies that quality costs are associated with two subcategories of quality-related activities: *control activities* and *failure activities*. **Control activities** are performed by an organization to prevent or detect poor quality (because poor quality may exist). Thus, control activities are made up of prevention and appraisal activities. **Control costs** are the costs of performing control activities. **Failure activities** are performed by an organization or its customers in response to poor quality (poor quality does exist). If the response to poor quality occurs before delivery of a bad (nonconforming, unreliable, not durable, and so on) product to a customer, the activities are classified as internal failure activities; otherwise, they are classified as external failure activities. **Failure costs** are the costs incurred by an organization because failure activities are performed. Notice that the definitions of failure activities and failure costs imply that customer response to poor quality can impose costs on an organization.

Exhibit 16-1 summarizes the four quality cost categories and lists specific examples of costs. Each of the costs could have been expressed as the cost of quality-related activities such as the cost of certifying vendors, inspecting incoming materials, adjusting complaints, and so on.

Quality Cost Measurement

Quality costs can also be classified as *observable* or *hidden*. **Observable quality costs** are those that are available from an organization's accounting records. **Hidden quality costs** are opportunity costs resulting from poor quality. (Opportunity costs are not usually recognized in accounting records.) Consider, for example, all the examples of quality costs listed in Exhibit 16-1. With the exception of lost sales, customer dissatisfaction, and lost market share, all the quality costs are observable and should be available from the accounting records. Note also that the hidden costs are all in the external failure category. These hidden quality costs can be significant and should be estimated. Although estimating hidden quality costs is not easy, three methods have been suggested: (1) the multiplier method, (2) the market research method, and (3) the Taguchi quality loss function.

Exhibit 16-1

Examples of Quality Costs by Category

Prevention Costs	Appraisal (Detection) Costs
Quality engineering	Inspection of materials
Quality training	Packaging inspection
Recruiting	Product acceptance
Quality audits	Process acceptance
Design reviews	Field testing
Quality circles	Continuing supplier verification
Marketing research	
Prototype inspection	
Vendor certification	

Internal Failure Costs	External Failure Costs
Scrap	Lost sales (performance-related)
Rework	Lost market share
Downtime (defect-related)	Customer dissatisfaction
Reinspection	Ill will
Retesting	Returns/allowances
Design changes	Recalls
Repairs	Warranties
	Discounts due to defects
	Product liability
	Complaint adjustment

The Multiplier Method   The multiplier method assumes that the total failure cost is simply some multiple of measured failure costs:

$$\text{Total external failure cost} = k(\text{Measured external failure costs})$$

where k is the multiplier effect. The value of k is based on experience. For example, **Westinghouse Electric** reports a value of k between 3 and 4.[4] Thus, if the measured external failure costs are \$3 million, the actual external failure costs are between \$9 million and \$12 million. Sampling and surveying are common methods used by companies to determine the value of the multiplier.[5] Including hidden costs in assessing the amount of external failure costs allows management to more accurately determine the level of resource spending for prevention and appraisal activities. Specifically, with an increase in failure costs, we would expect management to increase its investment in control costs.

The Market Research Method   Formal market research methods are used to assess the effect of poor quality on sales and market share. Customer surveys and interviews with members of a company's sales force can provide significant insights into the magnitude of a company's hidden costs. Market research results can be used to project future profit losses attributable to poor quality.

The Taguchi Quality Loss Function   The traditional zero defects definition assumes that hidden quality costs exist only for units that fall outside the upper and lower specification limits. The **Taguchi loss function** assumes that any variation from the target value of a quality characteristic causes hidden quality costs. Furthermore, the hidden quality costs increase quadratically as the actual value deviates from the target value. The Taguchi quality loss function, illustrated in Exhibit 16-2, can be described by the following equation:

$$L(A) = k(A - T)^2$$

where

> $k = $ A proportionality constant dependent upon the organization's external failure
> cost structure
> $A = $ Actual value of quality characteristic
> $T = $ Target value of quality characteristic
> $L = $ Quality loss

Exhibit 16-2 demonstrates that the quality cost is zero at the target value and increases symmetrically, at an increasing rate, as the actual value varies from the target value. Assume, for example, that a company produces watches and the quality characteristic is accuracy (as measured by how much time is gained or lost in three months). Assume $K = \$2$ and $T = 0$ minutes. Exhibit 16-3 illustrates the computation of the quality loss for four units. Notice that the cost quadruples when the deviation from target doubles (Units 2 and 3). Notice also that the average deviation squared and the average loss per unit can be computed. These averages can be used to compute the total expected hidden quality costs for a product. If, for example, the total units produced are 5,000 and the average squared deviation is 7.5, then the expected cost per unit is \$15 ($7.5 \times \$2$) and the total expected loss for the 5,000 units would be \$75,000 ($\$15 \times 5,000$).

---

[4]T. L. Albright and P. R. Roth, "The Measurement of Quality Costs: An Alternative Paradigm," *Accounting Horizons* (June 1992): 15–27.

[5]V. Sower, "Estimating External Failure Costs: A Key Difficulty in COQ Systems," *Quality Congress. ASQ's Annual Quality Congress Proceedings*, 58 (2004): 547–552.

Exhibit 16-2

The Taguchi Quality Loss Function

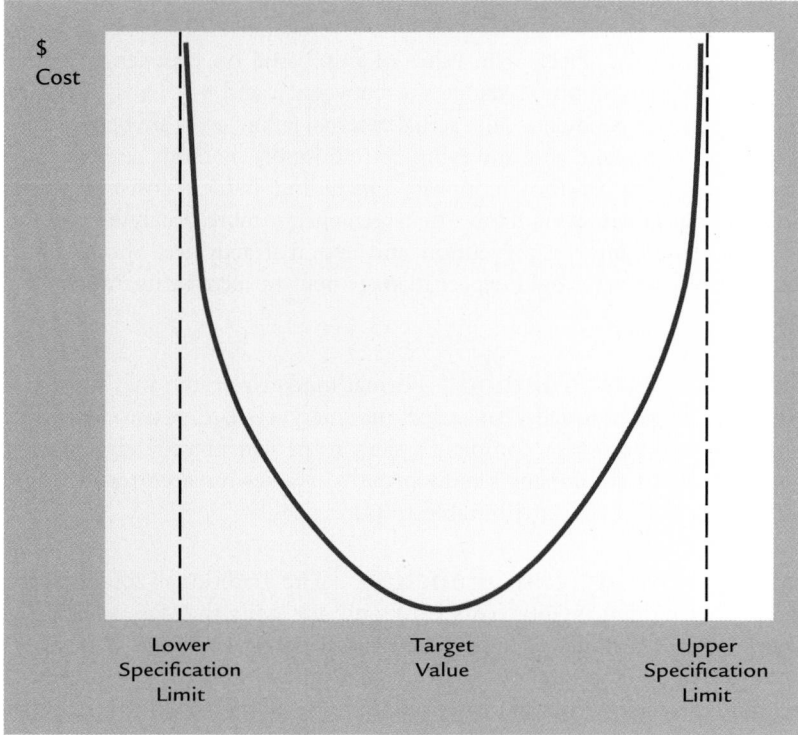

To apply the Taguchi loss function, *k* must be estimated. The value for *k* is computed by dividing the estimated cost at one of the specification limits by the squared deviation of the limit from the target value:

$$k = c/d^2$$

where

c = Loss at the lower or upper specification limit

d = Distance of limit from target value

This means that we still must estimate the loss for a given deviation from the target value. The first two methods, the multiplier method or the market research method, may be used to help in this estimation (a one-time assessment need). Once *k* is known, the hidden quality costs can be estimated for any level of variation from the target value.

Exhibit 16-3

Quality Loss Computation Illustrated

Unit No.	Time Gained (Lost) (A)	A − T	$(A − T)^2$	$k(A − T)^2$
1	−1	−1	1	$ 2.00
2	2	2	4	8.00
3	4	4	16	32.00
4	−3	−3	9	18.00
			30	$60.00
Units			÷ 4	÷ 4
Average			7.5	$15.00

Reporting Quality Costs

OBJECTIVE ▶ 2

Prepare a quality cost report, and explain its use.

A quality cost reporting system is essential to an organization serious about improving and controlling quality costs. The first and simplest step in creating such a system is assessing current actual quality costs. A detailed listing of actual quality costs by category can provide two important insights. First, it reveals the magnitude of the quality costs in each category, allowing managers to assess their financial impact. Second, it shows the distribution of quality costs by category, allowing managers to assess the relative importance of each category. Cornerstone 16-1 illustrates a quality cost report for Chesser Company.

The financial significance of quality costs can be assessed more easily by expressing these costs as a percentage of actual sales. The quality cost report in Cornerstone 16-1, for example, reports Chesser Company's quality costs as representing 20 percent of sales for fiscal 2013. Given the rule of thumb that quality costs should be no more than about 2.5 percent, Chesser Company has ample opportunity to improve profits by decreasing quality costs by improving quality.

Cornerstone 16-1 suggests that Chesser Company needs to embark on a serious quality improvement program to reduce its quality costs. But by how much should quality costs be reduced? Is there an optimal level of costs that a manager should be striving to achieve?

Zero-Defects Model and Robust Quality View

The original or traditional zero-defects model makes the claim that it is cost beneficial to reduce *nonconforming units* to zero. In the mid-1980s, the zero-defects model was taken one step further by the robust quality model, which made the definition of a defective or nonconforming unit much tighter. According to the robust view, a loss is experienced from producing products that vary from a target value; the greater the distance from the target value, the greater the loss. In other words, variation from the ideal is costly, and specification limits serve no useful purpose and, in fact, may be deceptive. The zero-defects model understates the quality costs and, thus, the potential for savings from even greater efforts to improve quality. Therefore, the robust quality model tightened the definition of a defective unit, refined our view of quality costs, and intensified the quality race.

For firms operating in an intensely competitive environment, improving quality is a competitive necessity. If the robust quality view is correct, then firms can capitalize on it, decreasing the number of defective units (robustly defined as zero tolerance) while simultaneously decreasing their total quality costs. Essentially, as firms increase their prevention and appraisal costs and reduce their failure costs, they discover that they can then cut back on their prevention and appraisal costs. Notice that failure costs can be reduced to zero according to this model and that control costs are finite at the zero-defect point.

The strategy to reduce quality costs recommended by the American Society for Quality Control, in part, states:

> The strategy for reducing quality costs is quite simple: (1) take direct attack on failure costs in an attempt to drive them to zero; (2) invest in the "right" prevention activities to bring about improvement; (3) reduce appraisal costs according to results achieved; and, (4) continuously evaluate and redirect prevention efforts to gain further improvement. This strategy is based on the premise that:
>
> • For each failure there is a root cause.
> • Causes are preventable.
> • Prevention is always cheaper.[6]

This ability to reduce total quality costs dramatically in all categories is borne out by real-world experiences. Westinghouse Electric, for example, found that its profits

[6] Jack Campanella, ed., *Principles of Quality Costs* (Milwaukee: ASQC Quality Press, 1990): 12.

**CORNERSTONE
16-1**

The HOW and WHY of Preparing a Quality Cost Report

Information:

Chesser Company had total sales of $5,000,000 for fiscal year ended March 31, 2013. Chesser's costs of quality-related activities are as follows:

Warranty	$250,000
Scrap	150,000
Reliability engineering	65,000
Rework	100,000
Quality training	10,000
Process acceptance	70,000
Materials inspection	30,000
Customer complaints	325,000

Why:

A quality cost report reveals the magnitude of the quality costs by category, and it also shows the relative distribution of these costs. The relative distribution allows the manager to assess the importance of the various categories and to determine where quality improvement emphasis is needed.

Required:

1. Prepare a quality cost report, classifying costs by category and expressing each category as a percentage of sales. What message does the cost report provide?
2. Prepare a bar graph and pie chart that illustrates each category's contribution to total quality costs. Comment on the significance of the distribution.
3. *What if* five years from now, quality costs are 2.5 percent of sales, with control costs being 80 percent of the total quality costs? What would your conclusion be?

Solution:

1.

Quality Cost Report Chesser Company For the Year Ended March 31, 2013			
		Quality Costs	**Percentage of Sales**[a]
Prevention costs:			
Quality training	$ 10,000		
Reliability engineering	65,000	$ 75,000	1.50%
Appraisal costs:			
Materials inspection	30,000		
Process acceptance	70,000	100,000	2.00
Internal failure costs:			
Scrap	150,000		
Rework	100,000	250,000	5.00
External failure costs:			
Warranty	250,000		
Customer complaints	325,000	575,000	11.50
Total quality costs		$1,000,000	20.00%[b]

[a] Actual sales of $5,000,000.
[b] $1,000,000/$5,000,000 = 20 percent

CORNERSTONE
16-1
(continued)

The report clearly indicates that quality costs are too high as 20 percent of sales are much greater than the desired 2 to 4 percent of sales that prevails for companies with good quality performance.

2. See Exhibit 16-4. The graphs reveal that failure costs are approximately 82 percent of the total quality costs, suggesting that Chesser needs to invest more in control activities to drive down failure costs.

3. First, assuming that the reduction in quality costs is due to quality improvements, the 2.5 percent level reveals that the company is producing at a very high quality level. In practical terms, if quality costs are in the 2 to 4 percent range with virtually no failure costs (0.5 percent of sales in this case), then the company has effectively and practically achieved a zero-defects state.

continued to improve until its control costs accounted for about 70 to 80 percent of total quality costs.[7] Based on this experience, we know that it is possible to reduce total quality costs significantly—in all categories—and that the process radically alters the relative distribution of the quality cost categories.

The Role of Activity-Based Cost Management

Activity-based costing (ABC) can be used to calculate the quality costs per unit of a firm's products. Once an ABC system is in place, the only requirement is to identify those activities that are quality related, such as inspection, rework, and warranty work. Assume, for example, that the cost of the rework activity is $250,000. Now, assume that a company produces 10,000 units each of two products: a regular model and a deluxe model. The number of units reworked is 1,000 for the regular model and 4,000 for the deluxe model (units reworked is the activity driver). The activity rate is $50 per reworked unit ($250,000/5,000), and the rework costs (an internal failure cost) assigned to each product are $50,000 and $200,000 for the regular model and the deluxe model, respectively. This provides a signal that the deluxe model is of lower quality than the regular model. Thus, ABC can be used as a means to identify

Exhibit 16-4

Quality Cost Categories: Relative Contribution Graphs

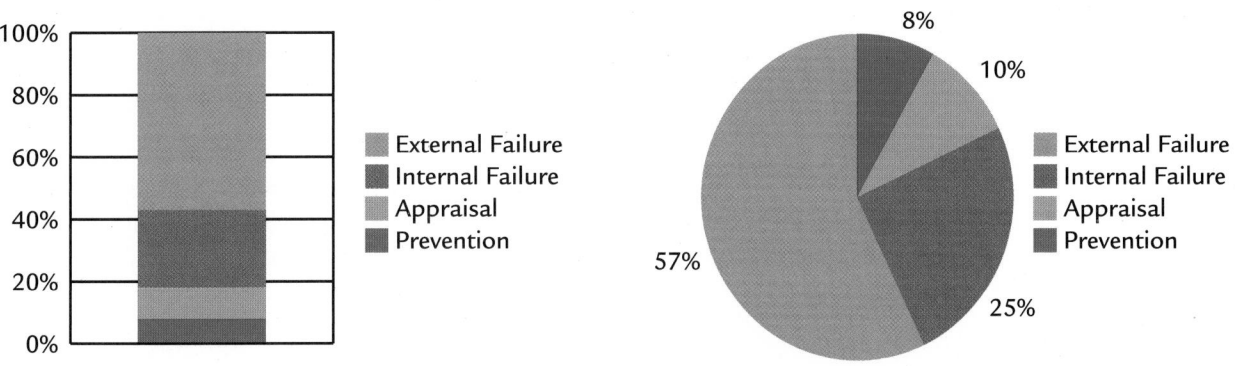

[7] These factual observations are based on those reported by Lawrence Carr and Thomas Tyson, "Planning Quality Cost Expenditures," *Management Accounting* (August 1995).

cost objects with quality problems, such as low-quality products, low-quality processes, and low-quality suppliers. This can then allow more focused management of quality costs.

Activity-based management (ABM) is also useful. ABM classifies activities as value-added and non-value-added and keeps only those that add value. This principle can be applied to quality-related activities. Appraisal and failure activities and their associated costs are non-value-added and should be eliminated (eventually). Prevention activities—performed efficiently—can be classified as value-added and should be retained. A company may be tracking all four categories of quality costs, but choose not to report prevention costs as part of the final cost of quality figures. This may be done to prevent managers from reducing quality costs by cutting prevention activities. Prevention activities pay off as they typically reduce external defects in multiples of the prevention costs.

Root causes (cost drivers) can also be identified, especially for failure activities, and used to help managers understand what is causing the costs of the activities. This information can then be used to select ways of reducing quality costs to an appropriate level. In effect, activity-based management supports the robust zero-defect view of quality costs. There is no optimal trade-off between control and failure costs; the latter are non-value-added costs and should be reduced to zero. Some control activities are non-value-added and should be eliminated. Other control activities are value-added but may be performed inefficiently, and the costs caused by the inefficiency are non-value-added. Thus, costs for these categories may also be reduced to lower levels.

OBJECTIVE ➤ 3

Explain why quality cost information is needed and how it is used.

Quality Cost Information and Decision Making

Reporting quality costs can improve managerial planning, control, and decision making. For example, if a company wants to implement a process re-engineering program to improve the quality of its products, it will need to assess the following: current quality costs by item and by category, the additional costs associated with the program, and the projected savings by item and by category. *When* the costs and savings will occur must also be projected. Then, a capital budgeting analysis can be produced to determine the merits of the proposed program. If the outcome is favourable and the program is initiated, then it becomes important to monitor the program through performance reporting.

Using quality cost information to implement and monitor the effectiveness of quality programs is only one use of a quality cost system. Other important uses can also be identified. Quality cost information is an important input to management decision making. It is also important to outside parties as they assess the quality of the company, through programs such as ISO 9000.

Managers need quality cost information in a number of decision-making contexts. Two of these contexts are strategic pricing and cost-volume-profit analysis.

Strategic Pricing

Consider AMD Inc., which produces electronic measurement devices. Market share for the company's low-level electronic measurement instruments had been steadily dropping. Linda Werther, marketing manager, identified price as the major problem. She knew that Japanese firms produced and sold the low-level instruments for less than AMD could. If AMD reduced its price to that of the competition, the new price would be below cost. Yet if something were not done, the Japanese firms would continue to expand their market share. One possibility was simply to drop the low-level line and concentrate on instruments in the medium and high-level categories. Linda knew, however, that this was a short-term solution, since soon the same Japanese firms would be competing at the higher levels. A brief income statement for the low-level instruments is as follows:

Revenues (1,000,000 @ $20)	$ 20,000,000
Cost of goods sold	(15,000,000)
Operating expenses	(3,000,000)
Product-line income	$ 2,000,000

Linda strongly believed that a 15 percent price decrease would restore the instrument line's market share and profitability to its former levels. One possibility was the implementation of total quality management. Her first action was to request information on the quality costs for the lower-level instruments. AMD's controller, Ahmet Sayed, admitted that the costs were not tracked separately. For example, the cost of scrap was buried in the work-in-process inventory account. He did promise, however, to estimate some of the costs. Data from his report for the low-level instruments are as follows:

Quality costs (estimated):	
Inspection of materials	$ 200,000
Scrap	800,000
Rejects	500,000
Rework	400,000
Product inspection	300,000
Warranty work	1,000,000
Total estimate	$3,200,000

Upon receiving the report, Linda, Ahmet, and Joe Luccheti, manager of the Quality Control Department, met to determine possible ways of reducing quality costs for the low-level line. Joe was confident that the quality costs could be reduced by 50 percent within 18 months. He had already begun planning the implementation of a new quality program. Linda calculated that a 50 percent reduction in the quality costs associated with the low-level instruments would reduce costs by about $1.60 per unit ($1,600,000/1,000,000)—which would make up slightly more than half of the $3 reduction in selling price that would be needed (the reduction is 15 percent of $20). Based on this outcome, Linda decided to implement the price reduction in three phases: a $1 reduction immediately, a $1 reduction in six months, and the final reduction of $1 in 12 months. This phased reduction would likely prevent any further erosion of market share and would start increasing market share sometime in the second phase. By phasing in the price reductions, the Quality Control Department would have time to reduce costs so that any big losses could be avoided.

The AMD Inc. example illustrates that both quality cost information and the implementation of a total quality control program contributed to a significant strategic decision. It also illustrates that improving quality was not a panacea. The reductions were not as large as needed to bear the full price reduction. Other productivity gains will be needed to ensure the long-range viability of the product line. Implementing JIT manufacturing, for example, might reduce inventories and decrease costs of materials handling and maintenance.

Cost-Volume-Profit Analysis and Strategic Design Decisions

Traditionally, cost-volume-profit analysis relies on the analysis of fixed and variable costs in conjunction with cost. Terry Foster, the marketing manager, and Sharon Fox, the design engineer, discovered shortcomings in the traditional analysis when they proposed a new product. They had been certain that a proposal for the new product was going to be approved. Instead, they received the following report from the controller's office:

Report: New Product Analysis, Project 675

Projected sales potential: 44,000 units
Production capacity: 45,000 units
Unit selling price: $60
Unit variable costs: $40

Report: New Product Analysis, Project 675

Fixed costs:

Product development	$ 500,000
Manufacturing	200,000
Selling	300,000
Total	$1,000,000

Projected break-even: 50,000 units

Decision: Reject

Reason(s): The break-even point is greater than the production capacity as well as the projected sales volume.

In an effort to discover just why the cost figures came out so poorly for a project that both individuals felt strongly would be profitable, the two met with Bob Bronstein, the assistant controller. The following conversation took place:

SHARON: Bob, I would like to know why there is a $3-per-unit scrap cost. Can you explain it?

BOB: Sure. It's based on the scrap cost that we track for existing, similar products.

SHARON: Well, I think you have overlooked the new design features of this new product. Its design virtually eliminates any waste—especially when you consider that the product will be made on a numerically controlled machine.

TERRY: Also, this $2-per-unit charge for repair work should be eliminated. The new design that Sharon is proposing solves the failure problems we have had with related products. It also means that the $100,000 of fixed costs associated with the Repair Centre can be eliminated.

BOB: Sharon, how certain are you that this new design will eliminate some of these quality problems?

SHARON: I'm absolutely positive. The early prototypes did exactly as we expected. The results of those tests are included in the proposal.

BOB: Right. Reducing the variable cost by $5 per unit and the fixed costs by $100,000 produces a break-even point of 36,000 units. These changes alone make the project viable. I'll change the report to reflect a positive recommendation.

The above scenario illustrates the importance of further classifying quality costs by behaviour. Although only unit-based behaviour is assumed, activity-based classification is also possible and could enhance the decision usefulness of quality costs. The scenario also reinforces the importance of identifying and reporting quality costs separately. The new product was designed to reduce its quality costs, and only by knowing the quality costs assigned could Sharon and Terry have discovered the error in the break-even analysis. Finally, notice the effect total quality management has on design decisions. By being aware of the quality costs and their causes, the new product's design was structured to avoid many of the existing quality problems.

Certifying Quality through ISO 9000

Just as a company assesses the quality of its suppliers, that same company may supply other companies that require vendor certification of quality. ISO (pronounced ICE-OH) 9000 is a family of international quality standards developed by the International Organization for Standardization in Geneva, Switzerland, that address quality management. These standards centre on the concept of documentation and control of nonconformance and change. ISO 9000 has been a success in Europe, and U.S. companies doing business in Europe were the first to board the ISO 9000 bandwagon, simply because it was a requirement of doing business there. A program called ISO 9001:2008 has evolved in response to the need for a standardized set of procedures for supplier quality verification.

Companies that attain ISO 9000 certification have been audited by an independent test company, which certifies that the company meets certain quality standards. The standards on which certification is currently based are ISO 9001:2008 standards.

The ISO 9001:2008 standards deal with quality systems and specifically with quality assurance models relating to quality systems that are concerned with such things as design/development, production, installation, final inspection, and testing. Many companies are certified based on 9001:2000 standards, which are essentially the same as those of 2008. The 2008 update is mostly concerned with clarification of the 2000 version of ISO 9001. It is important to understand that these standards do not apply to the production of a particular product or service. Instead, they apply to the way in which a company ensures quality, for example, by testing products, training employees, keeping records, and fixing defects.

Thus, ISO 9001:2008 certification does not certify either the quality of the product itself or the commitment of the company to continuous improvement. As a result, companies that require ISO 9001 certification do not stop auditing their suppliers. Requiring certification is just a first step.

On the plus side, many companies have found that the process of applying for ISO 9001 certification, while lengthy and expensive (it can take many months and cost $1,000,000 or more for larger companies), yields important benefits in terms of self-knowledge and improved financial performance. There are several innovative ways for a company to turn its workforce to ISO and continuous improvement. Posting large placards with simplified instructions and pictures at every workstation ensures that employees know exactly what needs to be done. In addition, replacing paper manuals that are difficult to use with an electronic system easily accessible from each employee's personal computer provides instant guidance and direction.

ISO 9000 is a first step in supplier certification. In 2009, 30,675 ISO 9001 certifications had been awarded in the United States.[8] Many large ISO 9001-certified companies are also urging their suppliers to obtain certificates.

Controlling Quality Costs

OBJECTIVE ▶ 4
Describe and prepare three different types of quality performance reports.

Good quality cost management requires that quality costs be reported and controlled (control having a cost reduction emphasis). Control enables managers to compare actual outcomes with standard outcomes to gauge performance and take any necessary corrective actions. Quality cost performance reports have two essential elements: actual outcomes and standard or expected outcomes. Deviations of actual outcomes from the expected outcomes are used to evaluate managerial performance and provide signals concerning possible problems.

Performance reports are essential to quality improvement programs. A report like the one shown in Cornerstone 16-1 forces managers to identify the various costs that should appear in a performance report, to identify the current quality performance level of the organization, and to begin thinking about the level of quality performance that should be achieved. Identifying the quality standard is a key element in a quality performance report. The standard should emphasize cost reduction opportunities.

Choosing the Quality Standard

The Total Quality Approach   The total quality management standard that will be used is referred to as the *robust zero-defects standard*. This standard calls for products and services to be produced and delivered that meet the targeted value. The need for total quality control is inherent in JIT and lean manufacturing approaches. However, JIT or lean manufacturing is not a prerequisite for moving toward total quality control. This approach can stand by itself.

Admittedly, the total quality standard is one that may not be completely attainable; however, evidence exists that it can be closely approximated. Defects are caused either by lack of knowledge or by lack of attention. Lack of knowledge can be corrected by proper training and lack of attention by effective leadership. Note also that total quality control implies the ultimate elimination of failure costs. Those who

---

[8] See www.anab.org, certification totals, accessed September 24, 2009.

believe that no defects should be permitted will continue to search for new ways to improve quality costs.

Consider the following case. A firm engaged in a significant volume of business through mailings. On average, 15 percent of the mailings were sent to the wrong address. Returned merchandise, late payments, and lost sales all resulted from this error rate. In one case, a tax payment was sent to the wrong address. By the time the payment arrived, it was late, causing a penalty of $300,000. Why not spend the resources (surely less than $300,000) to get the mailing list right and have no errors? Is a mailing list that is 100 percent accurate really impossible to achieve? Why not do it right the first time?

Quantifying the Quality Standard   Quality can be measured by its costs; as the costs of quality decrease, higher quality results—at least up to a point. Even if the standard of zero defects is achieved, a company must still have prevention and appraisal costs. A company with a well-run quality management program can get by with quality costs of about 2.5 percent of sales. (If zero defects are achieved, this cost is for prevention and appraisal.) This 2.5 percent standard is accepted by many quality control experts and many firms that are adopting aggressive quality improvement programs.

The 2.5 percent standard is for total costs of quality. Costs of individual quality factors, such as quality training or materials inspection, will be less. Each organization must determine the appropriate standard for each individual factor. Budgets can be used to set spending for each standard so that the total budgeted cost meets the 2.5 percent goal.

Physical Standards   For line managers and operating personnel, physical measures of quality—such as number of defects per unit, the percentage of external failures, billing errors, contract errors, and other physical measures—may be more meaningful. For physical measures, the quality standard is zero defects or errors. The objective is to get everyone to do it right the first time.

Use of Interim Standards   For most firms, the standard of zero defects is a long-range goal. The ability to achieve this standard is strongly tied to supplier quality. For most companies, materials and services purchased from outside parties make up a significant part of a product's cost. To achieve the desired quality level, a major campaign to involve suppliers in similar quality improvement programs may be needed. Developing the relationships and securing the needed cooperation from suppliers takes time—in fact, it takes years. Similarly, getting people within the company itself to understand the need for quality improvement and to have confidence in the program can take several years.

Because improving quality to the zero-defects level can take years, yearly quality improvement standards should be developed so that managers can use performance reports to assess the progress made on an interim basis. These **interim quality standards** express quality goals for the year. Progress should be reported to managers and employees in order to gain the confidence needed to achieve the ultimate standard of zero defects. Even though reaching the zero-defects level is a long-range project, management should expect significant progress on a yearly basis.

Types of Quality Performance Reports

Quality performance reports measure the progress realized by an organization's quality improvement program. Three types of progress can be measured and reported:

1. Progress with respect to a current-period standard or goal (an interim standard report)
2. The progress trend since the inception of the quality improvement program (a multiple-period trend report)
3. Progress with respect to the long-range standard or goal (a long-range report)

Interim Standard Report The organization must establish an interim quality standard each year and make plans to achieve this targeted level. Since quality costs are a measure of quality, the targeted level can be expressed in dollars budgeted for each category of quality costs and for each cost item within the category. Often, the interim quality standard is simply the quality costs incurred in the previous year, adjusted for management's desired reduction. At the end of the period, the **interim quality performance report** compares the actual quality costs for the period with the budgeted costs. This report measures the progress achieved within the period relative to the planned level of progress for that period. Cornerstone 16-2 illustrates such a report.

Multiple-Period Trend Report The interim quality report provides management with information concerning the within-period progress measured relative to specific goals. Also useful is a picture of how the quality improvement program has been doing since its inception. Is the multiple-period trend—the overall change in quality costs—moving in the right direction? Are significant quality gains being made each period? Answers to these questions can be given by providing a chart or graph that tracks the change in quality from the beginning of the program to the present. Such a graph is called a **multiple-period quality trend report**. By plotting quality costs as a percentage of sales against time, the overall trend in the quality program can be assessed. The first year plotted is the year prior to the implementation of the quality improvement program. Cornerstone 16-3 provides a detailed example of multiple-period trend reporting.

Long-Range Report At the end of each period, a report that compares the period's actual quality costs with the costs that the firm eventually hopes to achieve should be prepared. This report forces management to keep the ultimate quality goal in mind, reveals the room left for improvement, and facilitates planning for the coming period. Under a zero-defects philosophy, the costs of failure should be virtually nonexistent. (They are non-value-added costs.) Reducing the costs of failure increases a firm's competitive ability. It may also result in lowering external failure rates.

Remember that achieving higher quality will not totally eliminate prevention and appraisal costs. (In fact, increased emphasis on zero defects may actually increase the cost of prevention, depending on the type and level of prevention activities initially present.) Generally, we would expect appraisal costs to decrease significantly. Product acceptance, for example, may be phased out entirely as product quality increases; however, increased emphasis on process acceptance is likely. The firm must have assurance that the process is operating in a zero-defects mode. A **long-range quality performance report** compares the current actual costs with the costs that would be allowed if the zero-defects standard were being met (assuming a sales level equal to that of the current period). The target costs are, if chosen properly, value-added costs. The variances are non-value-added costs. Thus, the long-range performance report is simply a variation of the value- and non-value-added cost report. Cornerstone 16-4 illustrates this long-range report.

Incentives for Quality Improvement Most organizations provide both monetary and nonmonetary recognition for significant contributions to quality improvement. Of the two types of incentives, many quality experts believe that the nonmonetary are more useful.

Nonmonetary Incentives As with budgets, participation helps employees internalize quality improvement goals as their own. One approach used by many companies in their efforts to involve employees is the use of error cause identification forms. **Error cause identification** is a program in which employees describe problems that interfere with their ability to do the job right the first time. The error-cause-removal approach is one of the 14 steps in Philip Crosby's quality improvement program.[9]

[9] Philip Crosby, *Quality Is Free* (New York: New American Library, 1980).

CORNERSTONE 16-4

The HOW and WHY of Long-Range Quality Performance Reporting

Information:

The actual quality costs for the year ended June 30, 2013, for AMD Inc. are given below:

Prevention costs:	
Quality training	$ 80,000
Reliability engineering	160,000
Total prevention costs	240,000
Appraisal costs:	
Materials inspection	84,000
Process acceptance	96,000
Total appraisal costs	180,000
Internal failure costs:	
Scrap	50,000
Rework	100,000
Total internal failure costs	150,000
External failure costs:	
Customer complaints	65,000
Warranty	165,000
Total external failure costs	230,000
Total quality costs	$800,000

At the zero-defect state, AMD expects to spend $50,000 on quality training, $100,000 on reliability engineering, and $25,000 on process acceptance. Assume sales of $8,000,000.

Why:

The long-range performance report compares the current actual costs with the costs that would be present if no poor quality existed. It is, in effect, a listing of the non-value-added costs and reflects the potential for further savings by improving quality.

Required:

1. Prepare a long-range performance report for 2013. What does this report tell the management of AMD?
2. Explain why quality costs still are present for the zero-defect state.
3. **What if** AMD achieves the zero-defect state reflected in the report? What are some of the implications of this achievement?

Solution:

1.

AMD Inc.
Long-Range Performance Report
For the Year Ended June 30, 2013

	Actual Costs	Target Costs	Variance
Prevention costs:			
Quality training	$ 80,000	$ 50,000	$ 30,000 U
Reliability engineering	160,000	100,000	60,000 U
Total prevention costs	$240,000	$150,000	$ 90,000 U

CORNERSTONE
16-4
(continued)

	Actual Costs	Target Costs	Variance
Appraisal costs:			
Materials inspection	$ 84,000	$ 0	$ 84,000 U
Process acceptance	96,000	25,000	71,000 U
Total appraisal costs	180,000	25,000	155,000 U
Internal failure costs:			
Scrap	50,000	0	50,000 U
Rework	100,000	0	100,000 U
Total internal failure costs	150,000	0	150,000 U
External failure costs:			
Customer complaints	65,000	0	65,000 U
Warranty	165,000	0	165,000 U
Total external failure costs	230,000	0	230,000 U
Total quality costs	$800,000	$175,000	$625,000 U
Percentage of sales	10.0%	2.2%	7.81% U

AMD is spending too much money on failure activities. More effort at improving quality is still needed.

2. Prevention costs are value-added costs and would be necessary to maintain the quality gains. The presence of appraisal costs may not be necessary in a strictly theoretical sense (if there are no defective units, then there is no need to engage in detection activities).

3. By spending less money on defects, AMD can use the savings to expand and to employ additional people to support this expansion. Improved quality may naturally cause expansion by enhancing its competitive position. Thus, although improved quality may mean fewer jobs in some areas (such as inspection and warranty service), it also means that additional jobs will be created through expanded business activity.

reducing useful goods and services while simultaneously increasing profits. If true, then a more proactive approach is both needed and appropriate. Moreover, proactive environmental decisions require information about environmental costs and benefits—information that has not existed as a separate and well-defined category. According to a concept known as *ecoefficiency,* meeting sound business objectives and resolving environmental concerns are not mutually exclusive.

The Ecoefficiency Paradigm

Ecoefficiency is defined as the ability to produce competitively priced goods and services that satisfy customer needs while *simultaneously* reducing negative environmental impacts, resource consumption, and costs. Ecoefficiency means producing more goods and services using less materials, energy, water, and land, while, at the same time, minimizing air emissions, water discharges, waste disposal, and the dispersion of toxic substances. However, perhaps the most important claim of the ecoefficiency paradigm is that preventing pollution and avoiding waste is economically beneficial—that it is possible to do more with less. Moreover, it is complementary to and supportive of *sustainable development.* **Sustainable development** is defined as development that meets the needs of the present without compromising the ability of future generations to meet their own needs. Although absolute sustainability may not be attainable, progress toward its achievement certainly seems to have some merit.

Exhibit 16-7

Multiple-Period Trend Graph: Relative Quality Costs

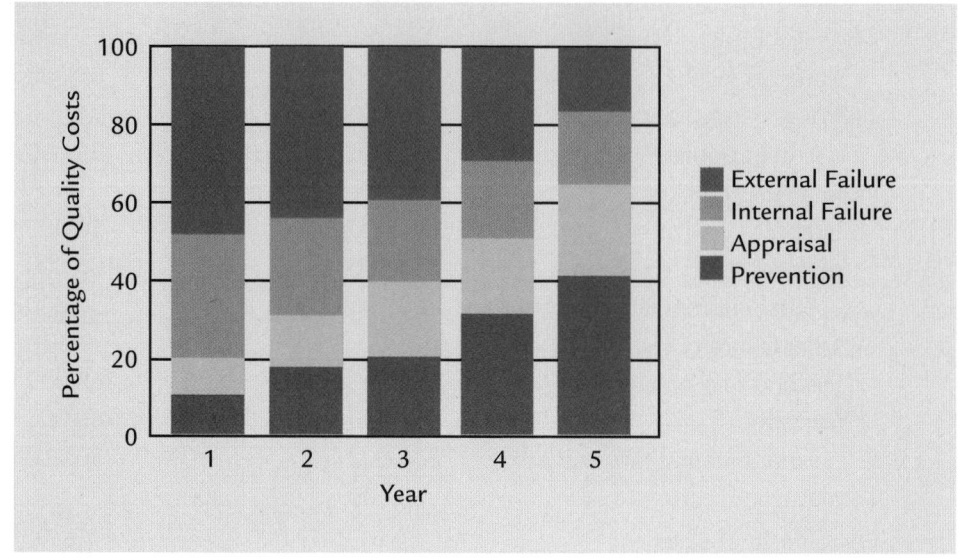

Ecoefficiency implies a positive relationship between environmental and economic performance. Exhibit 16-8 illustrates the objectives, opportunities, and outcomes that define the relationships envisioned by ecoefficiency.[11] Four broad objectives are revealed: (1) reduce the consumption of resources, (2) reduce the environmental impact, (3) increase product value, and (4) reduce environmental liability. Reducing the consumption of resources entails such things as reducing the use of energy, materials, water, and land. It also includes increasing product durability and enhancing product recyclability. Reducing environmental impact is primarily concerned with minimizing releases of pollutants into the environment and encouraging the sustainable use of renewable resources. Increasing product value means that products are produced that provide the functionality that customers need but with fewer materials and less resources. It also means that products are produced without degrading the environment, and their use and disposal are environmentally friendly. Reducing environmental liability requires that a company identify and efficiently manage the risks and opportunities relating to the environment. Achievement of the objectives requires a firm to seek opportunities to improve ecoefficiency, which brings us to the second level of Exhibit 16-8.

Process improvement and innovation are familiar methods for increasing efficiency. In this case, however, the objective is to increase ecoefficiency, which means that process changes must focus simultaneously on reducing costs and improving environmental performance. Process improvement is most useful for improving relative environmental performance, but process re-engineering is probably more suitable for major advances in ecoefficiency. Revalorizing by-products describes the search for ways to convert waste materials into useful products or useful inputs for other companies' products. For example, the sludge from wastewater treatment facilities can be converted into commercial compost. Product redesign is another key method for improving ecoefficiency. Products can be redesigned so that they use fewer materials, a smaller variety of materials, and less toxic materials and are easier to take apart for recycling while simultaneously providing a high degree of functionality for users. Finally, ecoefficiency can be improved by finding different and better ways of satisfying customer needs. This may entail redefining markets and reshaping supply and demand. For example, providing a service instead of selling a product has the potential of creating higher resource efficiency and less pollution. Car sharing is an example of this last approach.

---

[11] The objectives and opportunities are those identified by the World Business Council for Sustainable Development (WBCSD). See WBCSD, "Eco-efficiency: Creating More Value with Less Impact," (October 1, 2000), http://www.wbcsd.org/web/publications/eco_efficiency_creating_more_value.pdf.

Exhibit 16-8

Ecoefficiency Relationships

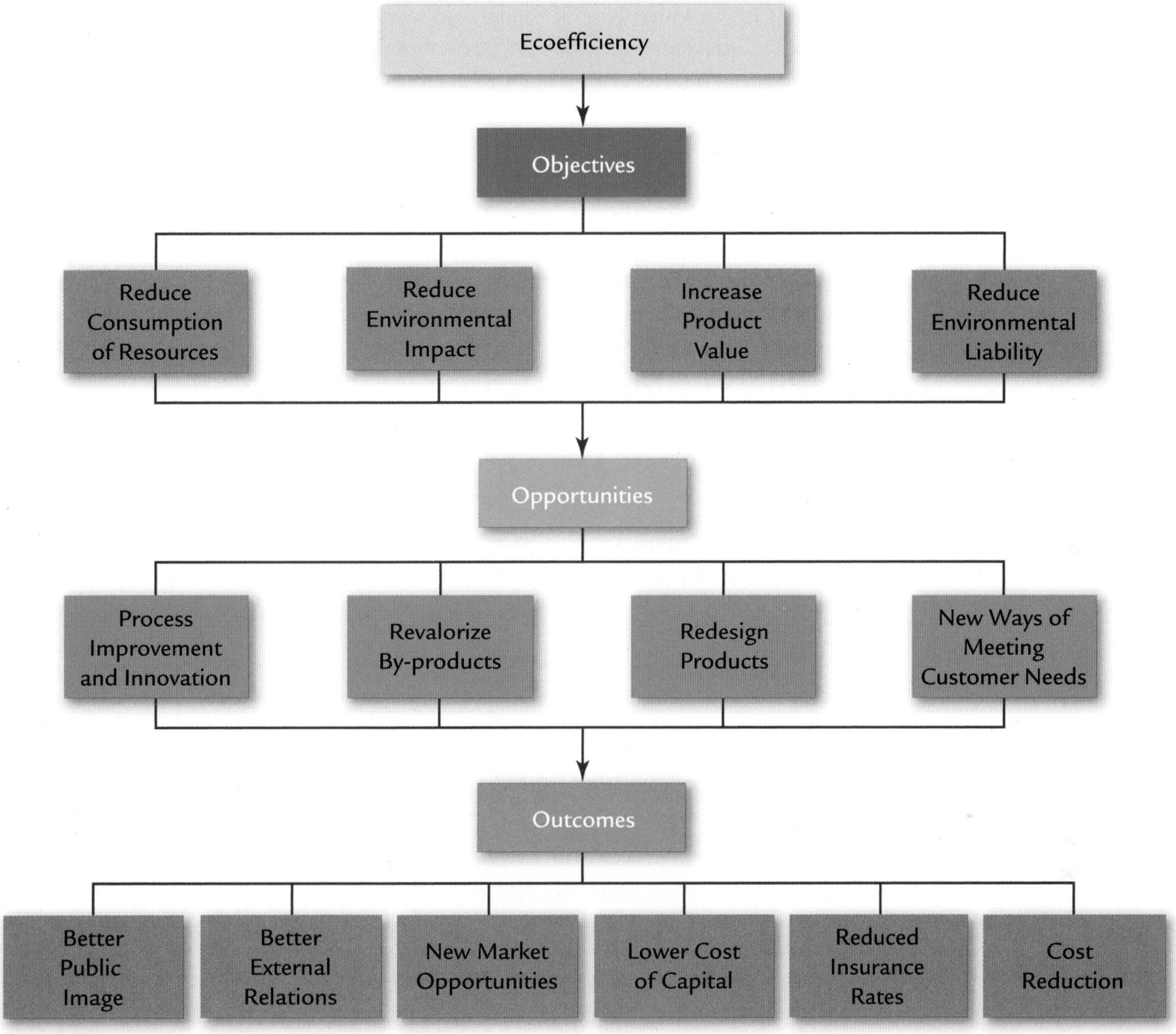

```
                         ┌─────────────────────┐
                         │    Ecoefficiency     │
                         └──────────┬──────────┘
                                    ↓
                         ┌─────────────────────┐
                         │     Objectives      │
                         └──────────┬──────────┘
        ┌───────────────┬──────────┴──────────┬───────────────┐
┌───────────────┐ ┌───────────────┐ ┌───────────────┐ ┌───────────────┐
│    Reduce     │ │    Reduce     │ │   Increase    │ │    Reduce     │
│  Consumption  │ │ Environmental │ │   Product     │ │ Environmental │
│ of Resources  │ │    Impact     │ │    Value      │ │   Liability   │
└───────────────┘ └───────────────┘ └───────────────┘ └───────────────┘
        └───────────────┴──────────┬──────────┴───────────────┘
                                    ↓
                         ┌─────────────────────┐
                         │    Opportunities    │
                         └──────────┬──────────┘
        ┌───────────────┬──────────┴──────────┬───────────────┐
┌───────────────┐ ┌───────────────┐ ┌───────────────┐ ┌───────────────┐
│    Process    │ │   Revalorize  │ │   Redesign    │ │  New Ways of  │
│  Improvement  │ │  By-products  │ │   Products    │ │   Meeting     │
│ and Innovation│ │               │ │               │ │Customer Needs │
└───────────────┘ └───────────────┘ └───────────────┘ └───────────────┘
        └───────────────┴──────────┬──────────┴───────────────┘
                                    ↓
                         ┌─────────────────────┐
                         │      Outcomes       │
                         └──────────┬──────────┘
```

Better Public Image	Better External Relations	New Market Opportunities	Lower Cost of Capital	Reduced Insurance Rates	Cost Reduction

Zipcar is a membership-based auto-sharing company that provides car reservations to its members. Its cars are billable by the hour or the day. The company parks cars at convenient locations around major centres. Members can reserve Zipcars online or by phone 24/7. They have automated access to Zipcars using an access card, which unlocks the door. The customer then finds the car keys inside. Membership is about $65 per year. Rates vary by city, day of the week, and vehicle make and model. Gas, parking, insurance, and maintenance are included in the price. Zipcar was founded in 2000 in Cambridge, Massachusetts, and today operates in major U.S. cities, in Toronto and Vancouver, and in the U.K. It has some 700,000 members.[12]

The third and final level of Exhibit 16-8 illustrates the payoffs of ecoefficiency. Pursuing the opportunities just discussed can produce a number of beneficial outcomes. Reduced environmental impacts can create social benefits like a better public image and better relations in the community and with regulators. This, in turn,

[12] Ibid.

improves the company's image and enhances its ability to sell products and services. Efforts to improve ecoefficiency also may increase revenues by creating new markets (e.g., creating outputs that were formerly classified as useless residues). Ecoefficient firms tend to reduce their environmental risks and, consequently, capture external benefits such as a lower cost of capital and lower insurance rates. Finally, cost reductions follow improvements in environmental performance.

The cost reduction outcome is particularly important. Environmental costs can be a significant percentage of total operating costs; many of these costs can be reduced or eliminated through effective management. For example, knowledge of environmental costs and their causes may lead to redesign of a process that, as a consequence, reduces the materials used and the pollutants emitted to the environment (an interaction between the innovation and cost reduction incentives). Thus, current and future environmental costs are reduced, and the firm becomes more competitive. To provide this financial information, it is necessary to define, measure, classify, and assign environmental costs to processes, products, and other cost objects of interest.

Environmental Costs Defined

Before environmental cost information can be provided to management, environmental costs must be defined. Various possibilities exist; however, an appealing approach is to adopt a definition consistent with a total environmental quality model. In the total environmental quality model, the ideal state is that of zero damage to the environment (analogous to the zero-defects state of total quality management). *Damage* is defined as either direct degradation of the environment such as the emission of solid, liquid, or gaseous residues into the environment (e.g., water contamination and air pollution) or indirect degradation such as *unnecessary* usage of materials and energy. Accordingly, environmental costs can be referred to as *environmental quality costs*. In a similar sense to quality costs, **environmental costs** are costs that are incurred because poor environmental quality exists or *may* exist. Thus, environmental costs are associated with the creation, detection, remediation, and prevention of environmental degradation. With this definition, environmental costs can be classified into four categories:

1. **Environmental prevention costs** are the costs of activities carried out to prevent the production of contaminants and/or waste that could cause damage to the environment.

2. **Environmental detection costs** are the costs of activities executed to determine if products, processes, and other activities within the firm are in compliance with appropriate environmental standards. The environmental standards and procedures that a firm seeks to follow are defined in three ways: (1) regulatory laws of governments, (2) voluntary standards (ISO 14000) developed by the International Standards Organization, and (3) environmental policies developed by management.

3. **Environmental internal failure costs** are costs of activities performed because contaminants and waste have been produced but not discharged into the environment. Thus, internal failure costs are incurred to eliminate and manage contaminants or waste once produced. Internal failure activities have one of two goals: (1) to ensure that the contaminants and waste produced are not released to the environment or (2) to reduce the level of contaminants released to an amount that complies with environmental standards.

4. **Environmental external failure costs** are the costs of activities performed *after* discharging contaminants and waste into the environment. **Realized external failure costs** are those incurred and paid for by the firm. **Unrealized external failure (societal) costs** are caused by the firm but are incurred and paid for by parties outside the firm. Societal costs can be further classified as (1) those resulting from environmental degradation and (2) those associated with an adverse impact on the property or welfare of individuals. In either case, the costs are borne by others and not by the firm even though the firm causes them.

Exhibit 16-9

Classification of Environmental Costs by Activity Type

Prevention Activities	Internal Failure Activities
Evaluating and selecting suppliers	Operating pollution control equipment
Evaluating and selecting pollution control equipment	Treating and disposing of toxic waste
Designing processes	Maintaining pollution equipment
Designing products	Licensing facilities for producing contaminants
Carrying out environmental studies	Recycling scrap
Auditing environmental risks	
Developing environmental management systems	
Recycling products	
Obtaining ISO 14001 certification	

Detection Activities	External Failure Activities
Auditing environmental activities	Cleaning up a polluted lake
Inspecting products and processes	Cleaning up oil spills
Developing environmental performance measures	Cleaning up contaminated soil
Testing for contamination	Settling personal injury claims (environmentally related)
Verifying supplier environmental performance	Restoring land to natural state
Measuring contamination levels	Losing sales due to poor environmental reputation
	Using materials and energy inefficiently
	Receiving medical care due to polluted air (S)
	Losing employment because of contamination (S)
	Losing a lake for recreational use (S)
	Damaging ecosystems from solid waste disposal (S)

Note: "S" = societal costs.

Exhibit 16-9 summarizes the four environmental cost categories and lists specific activities for each category. Within the external failure cost category, societal costs are labelled with an "S." The costs for which the firm is financially responsible are called **private costs**. All costs without the S label are private costs. Of the four categories of environmental activities, the external failure cost category is the one that causes the most economic hardship for an organization.

Environmental Cost Report

Environmental cost reporting is essential if an organization is serious about improving its environmental performance and controlling environmental costs. Reporting environmental costs by category reveals two important outcomes: (1) the impact of environmental costs on firm profitability and (2) the relative amounts expended in each category. Cornerstone 16-5 provides an example of a simple environmental cost report.

Environmental Cost Reduction

Investing more in prevention and detection activities can bring about a significant reduction in environmental failure costs. Environmental costs appear to behave in much the same way as quality costs. The lowest environmental costs are attainable at the *zero-damage point* much like the zero-defects point of the total quality cost model. Thus, an ecoefficient solution would focus on prevention with the usual justification that *prevention is cheaper than the cure*. Analogous to the total quality management model, zero damage is the lowest cost point for environmental costs.

CORNERSTONE 16-5

The HOW and WHY of an Environmental Cost Report

Information:
Operating costs for Verde Corporation as of December 31, 2013, are $30,000,000. Environmental costs are as follows:

Maintaining pollution equipment	$ 400,000
Developing measures	240,000
Operating pollution equipment	1,200,000
Designing products	600,000
Training employees	240,000
Restoring land	2,100,000
Inspecting processes	720,000
Cleaning up lake	3,300,000

Required:
1. Prepare an environmental cost report, classifying costs by quality category and expressing each as a percentage of total operating costs. What is the message of this report?
2. Prepare a pie chart that shows the relative distribution of environmental costs by category. What does this report tell you?
3. ***What if*** Verde deliberately did not include the cost of polluting a lake in the report? Offer possible reasons for this decision.

Solution:
1.

Verde Corporation
Environmental Cost Report
For the Year Ended December 31, 2013

	Environmental Costs		Percentage of Operating Costs[a]
Prevention costs:			
Training employees	$ 240,000		
Designing products	600,000	$ 840,000	2.80%
Detection costs:			
Inspecting processes	720,000		
Developing measures	240,000	960,000	3.20
Internal failure costs:			
Operating pollution equipment	1,200,000		
Maintaining pollution equipment	400,000	1,600,000	5.33
External failure costs:			
Cleaning up lake	3,300,000		
Restoring land	2,100,000	5,400,000	18.00
Total quality costs		$8,800,000	29.33%[b]

[a] Actual opening costs of $30,000,000.
[b] $8,800,000/$30,000,000 = 29.33%

CORNERSTONE
16-5
(continued)

Environmental costs are 29.33 percent of total operating costs, seemingly a significant amount. Reducing environmental costs by improving environmental performance can significantly increase a firm's profitability.

2. See Exhibit 16-10. Of the total environmental costs, only 21 percent are from the prevention and detection categories, and 79 percent of the environmental costs are failure costs. Thus, increasing prevention activities should drive down the costs of failure activities in a way that is cost beneficial.

3. The most likely reason is that the cost is a social cost and not paid for by the company and thus not of direct interest to Verde. In fact, such formal recognition may create a potential liability for the company (Could this be an ethical issue?).

An Environmental Financial Report

Ecoefficiency suggests a possible modification to environmental cost reporting. Specifically, in addition to reporting environmental costs, why not report *environmental benefits?* In a given period, there are three types of benefits:

1. Additional revenues
2. Current savings
3. Cost avoidance (ongoing savings)

Additional revenues are revenues that flow into the organization due to environmental actions such as recycling paper, finding new applications for nonhazardous waste (e.g., using wood scraps to make wood chess pieces and boards), and increased sales due to an enhanced environmental image. Current savings refer to reductions in environmental costs achieved in the current year. Cost avoidance refers to ongoing savings of costs that had been paid in prior years. By comparing benefits produced with environmental costs incurred in a given period, a type of environmental financial statement is produced. Managers can use this statement to assess progress (benefits produced) and potential for progress (environmental costs). The environmental financial statement could also form part of an environmental progress report that is provided to shareholders on an annual basis. Exhibit 16-11 provides an example of an

Exhibit 16-10

Relative Distribution: Environmental Costs

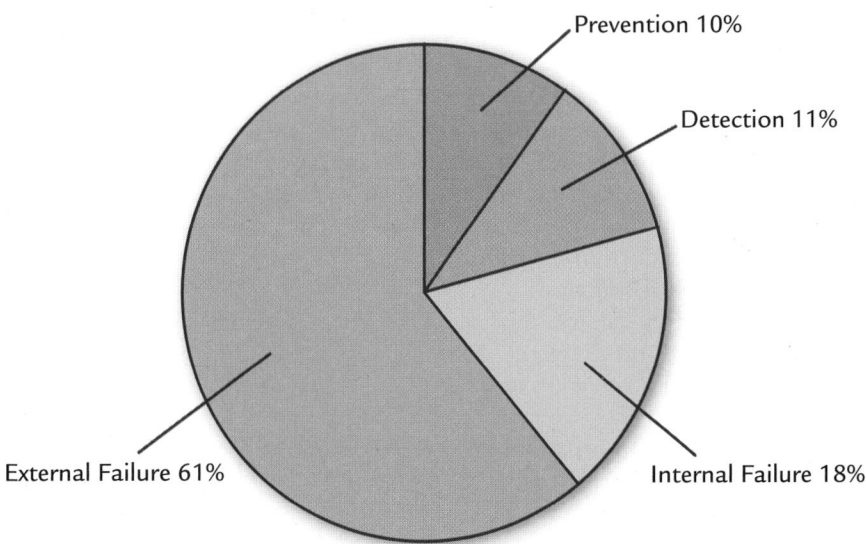

Prevention 10%
Detection 11%
Internal Failure 18%
External Failure 61%

Exhibit 16-11

Environmental Financial Statement

Verde Corporation
Environmental Financial Statement
For the Year Ended December 31, 2013

Environmental benefits:	
Income sources:	
Recycling income	$ 600,000
Revenues from waste-derived products	150,000
Ongoing savings:	
Cost reductions, contaminants	900,000
Cost reductions, hazardous waste disposal	1,200,000
Current savings:	
Energy conservation cost savings	300,000
Packaging cost reductions	450,000
Total environmental benefits	$3,600,000
Environmental costs:	
Prevention costs:	
Designing processes for the environment	$ 640,000
Supplier evaluation and selection	200,000
Detection costs:	
Testing for contamination	560,000
Measuring contamination levels	400,000
Internal failure costs:	
Waste treatment, transport, and disposal	1,500,000
Operating pollution control equipment	300,000
External failure costs:	
Inefficient materials usage	1,400,000
Cleaning up soil	4,000,000
Total environmental costs	$9,000,000

environmental financial statement. The benefits reported reveal good progress, but the costs are still two and one-half times the benefits, indicating that more improvements are clearly needed.

Assigning Environmental Costs to Products and Processes

OBJECTIVE ▶ 6
Show how environmental costs can be assigned to products and processes.

Both products and processes are sources of environmental costs. Processes that *produce* products can create solid, liquid, and gaseous residues that are subsequently introduced into the environment. These residues have the potential of degrading the environment. Residues, then, are the causes of both internal and external environmental failure costs (e.g., investing in equipment to prevent the introduction of the residues into the environment and cleaning up residues after they are allowed into the environment). Production processes are not the only source of environmental costs. Packaging is also a source.

Products themselves can be the source of environmental costs. After selling a product, its use and disposal by the customer can produce environmental degradation. These are examples of *environmental post-purchase costs*. Most of the time, environmental post-purchase costs are borne by society and not by the company and, thus, are societal costs. On occasion, however, environmental post-purchase costs are converted into realized external costs.

The environmental costs of processes that produce, market, and deliver products and the environmental post-purchase costs caused by the use and disposal of the products are examples of *environmental product costs*. **Full environmental costing** is the assignment of all environmental costs, both private and societal, to products. **Full private costing** is the assignment of only private costs to individual products. Private costing, then, would assign the environmental costs to products caused by the internal processes of the organization. Private costing is probably a good starting point for many firms. Private costs can be assigned using data created *inside* the firm. Full costs require gathering of data that are produced outside the firm by third parties.

Assigning environmental costs to products can produce valuable managerial information. For example, it may reveal that a particular product is responsible for much more toxic waste than other products. This information may lead to an alternative design for the product or its associated processes that is more efficient and environmentally friendly. It could also reveal that with the environmental costs correctly assigned, the product is not profitable. This could mean something as simple as dropping the product to achieve significant improvement in environmental performance and economic efficiency. Many opportunities for improvement may exist, but knowledge of the environmental product costs is the key. Moreover, environmental costs must be assigned accurately.

Activity-Based Environmental Cost Assignments

The environmental costs of processes that produce, market, and deliver products and the environmental post-purchase costs caused by the use and disposal of the products are examples of *environmental product costs*.

Activity-based costing facilitates environmental costing. Tracing the environmental costs to the products responsible for those costs is a fundamental requirement of a sound environmental accounting system. Each environmental activity is assigned costs, activity rates are computed, and the rates are then used to assign environmental costs to products based on usage of the activity. Cornerstone 16-6 shows how to assign environmental costs to two different types of industrial cleaners.

The HOW and WHY of Activity-Based Environmental Cost Assignments

**CORNERSTONE
16-6**

Information:
Pelideaen Company reported the following:

1. Environmental activity costs

Activity	Costs
Design processes (to reduce pollution)	$ 45,000
Inspect processes (for pollution problems)	80,000
Maintain environmental equipment	125,000
Toxic waste disposal	200,000

2. Driver data

	Cleanser A	Cleanser B
Design hours	2,000	1,000
Inspection hours	1,750	2,250
Maintenance hours	200	4,800
Kilograms of waste	1,000	19,000

After studying this chapter, you should be able to:

➤ **1** Describe the basic features of lean manufacturing.

➤ **2** Describe lean accounting.

➤ **3** Discuss and define productive efficiency and partial productivity measurement.

➤ **4** Explain what total productivity measurement is, and describe its advantages.

CHAPTER

17

Lean Accounting and Productivity Measurement

Consider a hypothetical company, Maple Autoparts Inc., that produces four major product lines: shock absorbers, aluminum alloy and steel wheels, brake systems, and aluminum radiators. Maple is contemplating expansion into new international markets and is facing such competitors as **DENSO** (Japanese), **Bosch** (German), and **Delphi** (American). To achieve success in this endeavour, Maple needs to be more efficient by streamlining operating processes, eliminating waste, and improving quality and delivery performance. Clearly, an organization must be as good as or better than its competitors at taking materials, labour, machines, power, and other inputs and turning out high-quality goods and services. A company can create a competitive advantage by using fewer inputs to produce a given output or by producing more output for a given set of inputs. Management needs to assess the potential and actual effectiveness of decisions that are geared to improve efficiency. Management also needs to monitor and control efficiency changes. Measures of productive efficiency satisfy these performance and control objectives.

Lean manufacturing is concerned with eliminating waste in manufacturing processes. Promised benefits include such outcomes as reduced lead times, improved quality, improved on-time deliveries, less inventory, less space, less human effort, lower costs, and increased profitability. *Lean accounting* is a simplified approach to costing that supports lean manufacturing with both financial and nonfinancial measures. One key area that supports efficiency improvement is *productivity measurement*, which is concerned with the relationship between outputs and inputs. As waste decreases through lean manufacturing practices, productive efficiency should increase.

Lean Manufacturing

OBJECTIVE ▶ 1
Describe the basic features of lean manufacturing.

Maple Autoparts is typical of many companies that operate in an environment where change is rapid. Products and processes are constantly being redesigned and improved, and stiff national and international competitors are always present. The competitive environment demands that firms offer customized products and services to diverse customer segments. This, in turn, means that firms must find cost-efficient ways of producing high-variety, low-volume products and pay more attention to linkages between the firm, its suppliers and customers. Furthermore, for many industries, product life cycles are shrinking, creating a greater need for innovation. Thus, organizations operating in a dynamic, rapidly changing environment are finding that adaptation and change are essential to survival. To find ways to improve performance, firms operating in this kind of environment are forced to re-evaluate how they do things. Improving performance translates into constantly searching for ways to eliminate waste (both time and money) and to undertake only those actions that bring value to the customer. This philosophical approach to manufacturing is often referred to as *lean manufacturing*. **Lean manufacturing** is thus an approach designed to eliminate waste and maximize customer value. It is characterized by delivering the right product, in the right quantity, with the right quality (zero-defect), at the exact time the customer needs it and at the lowest possible cost.

Lean manufacturing systems allow managers to eliminate waste, reduce costs, and become more efficient. Firms that implement lean manufacturing are pursuing a cost reduction strategy by redefining the activities performed within an organization. Cost reduction is directly related to cost leadership. Lean manufacturing adds value by reducing waste. Successful implementation of lean manufacturing has brought about significant improvements, such as better quality, increased productivity, reduced lead times, major reductions in inventories, reduced setup times, lower manufacturing costs, and increased production rates.

In substance, lean manufacturing is the same as the *Toyota Production System* developed by Shigeo Shingo, Taaichi Ohno, and Eiji Toyoda. *World-class manufacturing* and *just-in-time (JIT) manufacturing and purchasing* are terms that encompass many of the same ideas. Lean manufacturing is also similar in concept to **Ford**'s lean enterprise system. However, the contributions of Shingo, Ohno, and Toyoda overcame some of the major shortcomings and flaws of the Ford system. Specifically, the Ford system did not properly value employees and also was not structured to deal with product variety. High-variety, low-volume products were not compatible with the Ford production system. Employee empowerment, team structures, cellular manufacturing, reduced setup times, and small batches all came into being in the Toyota Production System and are integral parts of a lean manufacturing system.

Becoming lean requires lean thinking. Lean manufacturing is distinguished by the following five principles of lean thinking:[1]

- Precisely specify value by each particular product.
- Identify the "value stream" for each.
- Make value flow without interruption.
- Let the customer pull value from the producer.
- Pursue perfection.

Value by Product

Value is determined by the customer—at the very least, it is an item or feature for which the customer is willing to pay. Customer value is the difference between realization and sacrifice. Realization is what a customer receives. Sacrifice is what a customer

[1] James Womack and Daniel Jones, *Lean Thinking* (New York: Free Press, 2003).

gives up, including what they are willing to pay for the basic and special product features, quality, brand name, and reputation. Value thus relates to a specific product and to specific features of the product. Adding features and functions that are not wanted by the customers is a waste of time and resources. Furthermore, attempting to market features and products that customers don't want is a waste of time and resources. Assessing value is externally oriented and not internally generated. Only value-added features should be produced; non-value-added activities should be eliminated.

Value Stream

The **value stream** is made up of all activities, both value-added and non-value-added, required to bring a product group or service from its starting point (e.g., customer order or concept for a new product) to a finished product in the hands of the customer. There are several types of value streams, the most common being the *order fulfillment value stream*. The order fulfillment value stream focuses on providing current products to current customers.[2] A second type of value stream is the *new product value stream*, which focuses on developing new products for new or existing customers. A value stream reflects all that is done—both good and bad—to bring the product to a customer. Thus, analyzing the value stream allows management to identify waste. Activities within the value stream are value-added or non-value-added. Non-value-added activities are the source of waste. They are of two types: (1) activities avoidable in the short run and (2) activities unavoidable in the short run due to current technology or production methods. The first type is most quickly eliminated, while the second type requires more time and effort. Exhibit 17-1 visually portrays an order fulfillment value stream for one of Maple Autoparts' family of aluminum wheels. This particular value stream only has one manufacturing cell; other value streams may have several cells.

A value stream may be created for every product; however, it is more common to group products that use common processes into the same value stream. One way to identify the value streams is to use a simple two-dimensional matrix, where the activities/processes are listed on one dimension and the products on a second dimension. Exhibit 17-2 provides a simple matrix for the four-wheel models: aluminum Model A, aluminum Model B, steel Model C, and steel Model D. In this case,

Exhibit 17-1

Order Fulfillment Value Stream

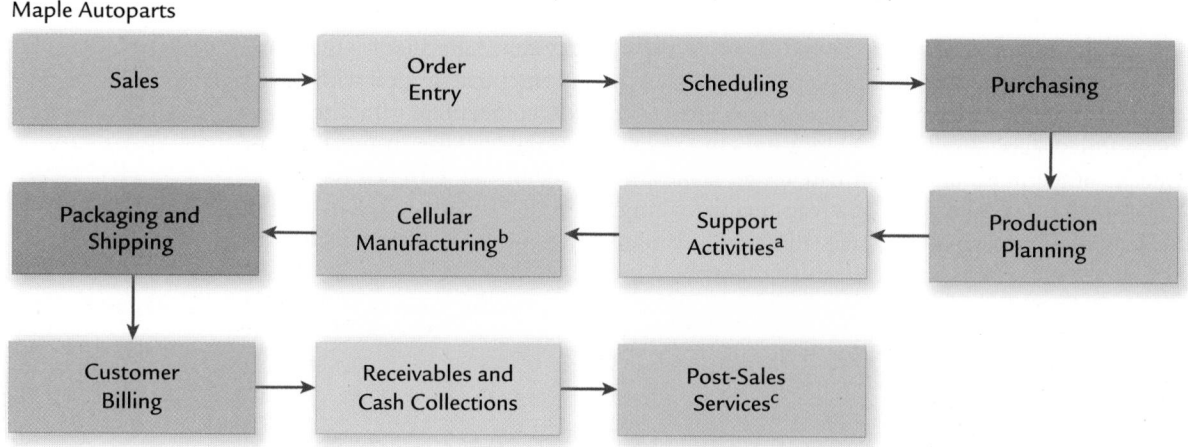

Maple Autoparts

[a]Moving materials, quality management, engineering, setting up equipment, maintenance, etc.
[b]Cutting, drilling and insertion, assembly, and finishing.
[c]Customer complaints, field repairs, warranty services, etc.

[2]For a more complete description of the different types of value streams, see Brian Maskell and Bruce Baggaley, *Practical Lean Accounting* (New York: Productivity Press, 2004), and Francis A. Kennedy and Jim Huntzinger, "Lean Accounting: Measuring and Managing the Value Stream," *Cost Management* (September/October 2005): 31–38. These two sources also recommend the matrix approach for identifying value streams illustrated in Exhibit 17-2.

Exhibit 17-2

Matrix Approach to Identifying Value Streams

| | Production Activities: Order Fulfillment Value Stream | | | | | | | |
Wheel Model	Order Entry	Production Planning	Purchasing	Aluminum Cell[a]	Steel Cell[b]	Stress Testing[c]	Packaging and Shipping	Invoicing
A	x	x	x	x			x	x
B	x	x	x	x			x	x
C	x	x	x		x	x	x	x
D	x	x	x		x	x	x	x

[a] Casting, machining, painting, and finishing.
[b] Stamping, welding, and cladding (attaching stainless steel or painted plastic components to approximate the look of chromed aluminum).
[c] To ensure that the steel wheels have the same fatigue strength as aluminum, they go through a stress test.

Models A and B would be placed in one value stream.
Models C and D would define a second value stream.

two value streams are indicated, where each is made up of two product models (notice that the steel models have two major processes different from the aluminum models, thus the need for two value streams).

Once value streams are identified, then the next step is to assign people and resources to the value streams. As a rule of thumb, each value stream should have between 25 and 150 people.[3] As much as possible, the people, the machines, the manufacturing processes, and the support activities need to be dedicated to the value streams. This allows a sense of ownership and provides a means of direct accountability. It also simplifies and facilitates product costing. In a sense, the value stream is its own independent company, and the value-stream team is responsible for its improvement, growth, and profitability.

Value Flow

Value flow is made up of all move and wait time necessary to move resources and product batches before and after the production process, to product completion. In a traditional manufacturing setup, production is organized by function into departments and products are produced in large batches, moving from department to department. This approach requires significant move time and wait time as each batch *moves* from one department to another and *waits* for its turn if there is a batch in-process in front of it. Often, lengthy changeovers are needed to prepare the equipment to produce the next batch of goods that may have some very different characteristics. Traditional batch production is not equipped to deal with product variety; furthermore, move and wait time are sources of waste. Batches must wait for a preceding batch and a subsequent setup *before* beginning a process. Once a batch starts a process, units are processed sequentially; as units are finished, they must wait for other units in the batch to be finished before the entire batch moves to the next process. For example, if a department can process one unit every five minutes, then the first unit of a batch of 10 will be completed after five minutes but must then wait an additional 45 minutes for the remaining units to be completed before moving to the next process. Thus, there is pre-process waiting and post-process waiting. Exhibit 17-3 illustrates Maple's current department layout for production of Model A aluminum wheels. The exhibit illustrates the presence of both wait and move times.

Reduced Setup/Changeover Times   With large batches, setups are infrequent, and the fixed cost of a setup is spread out over many units. Typical results produce complexity in scheduling and large work-in-process and finished goods inventories. Lean manufacturing reduces wait and move times dramatically and allows the production of small batches (low volume) of differing products (high variety). The key factors in achieving these outcomes are lower setup times and cellular manufacturing. Reducing

[3] Ibid.

Exhibit 17-3

Maple's Current Departmental Layout: Model A Aluminum Wheel Production

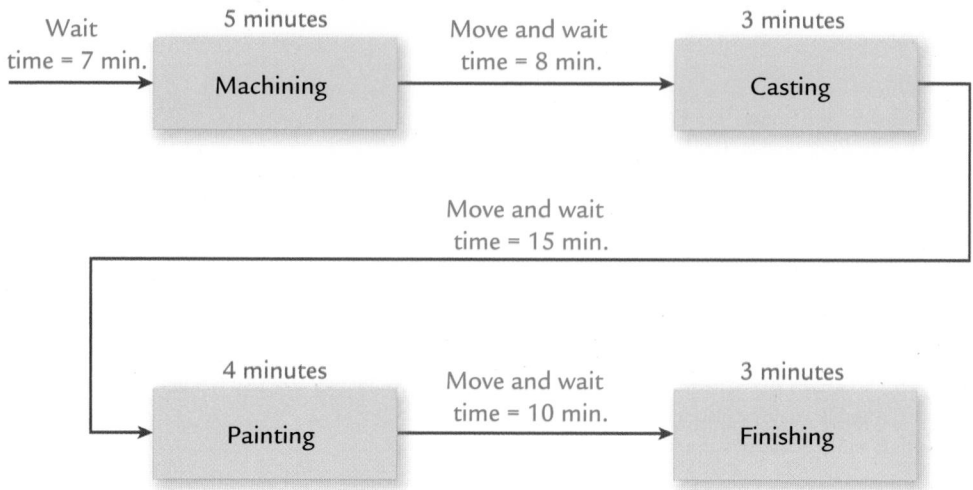

Colour Code:
Blue = Value-added process time
Red = Non-value-added move and pre-process wait time

the time to configure equipment to produce a different type of product enables *smaller batches in greater variety* to be produced. It also decreases the time it takes to produce a unit of output, thus increasing the ability to respond to customer demand. Customers do not value changeover and therefore it represents waste. While reducing setup times is important, even more critical is the use of cellular or continuous flow manufacturing.

Cellular Manufacturing   Lean manufacturing uses a series of cells to produce families of similar products. A lean manufacturing system replaces the traditional plant layout with a pattern of manufacturing cells. Cell structure is chosen over departmental structure because it reduces lead time, decreases product cost, improves quality, and increases on-time delivery. **Manufacturing cells** contain all the operations in close proximity that are needed to produce a family of products. The machines used are typically grouped in a semicircle. The reason for locating processes close to one another is to minimize move time and to keep a continuous flow between operations while maintaining zero inventory between any two operations. The cell is usually dedicated to producing products that require similar operations. Exhibit 17-4 shows a proposed cellular manufacturing structure for Model A aluminum wheels. Notice that by grouping processes closely together and dedicating the cell to a family of products, the move and wait times are essentially eliminated. Cornerstone 17-1 illustrates the value of cellular manufacturing relative to the traditional departmental approach.

Exhibit 17-4

Maple's Proposed Manufacturing Cell (Model A)

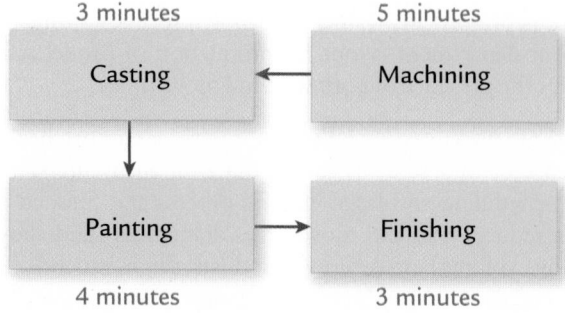

Blue = Value-added process time

The HOW and WHY of Cellular Manufacturing

**CORNERSTONE
17-1**

Information:
See Exhibits 17-3 and 17-4.

Why:
Cellular manufacturing groups process closely together, and this act effectively eliminates wait and move times. For a given batch of units, total production time is reduced with subsequent decreases in lead time and cost and improved on-time delivery.

Required:
1. Calculate the total time it takes to produce a batch of 10 units using Maple's traditional departmental structure.
2. Using cellular manufacturing, how much time is saved producing the same batch of 10 units? Assuming the cell operates continuously, what is the production rate? Which process controls this production rate?
3. **What if** the processing time of machining is reduced from five to four minutes? What is the production rate now, and how long will it take to produce a batch of 10 units?

Solution:
1. Total lead time for a batch of 10 units:

Processing time	
Machining	50 minutes
Casting	30 minutes
Painting	40 minutes
Finishing	30 minutes
Total processing	150 minutes
Move and wait times	40 minutes
Total batch time	190 minutes

2.

Processing time (10 units):	Elapsed time
First unit	15 minutes
Second unit	20 minutes (processing begins five minutes after the first)
Tenth unit	60 minutes (total processing time)

Time saved over traditional manufacturing: 190 minutes − 60 minutes = 130 minutes

If the cell is processing continuously, then a unit is produced every five minutes after the start-up unit. Thus, the production rate is 12 units per hour (60/5). The *bottleneck* process (the one with the longest per-unit processing time) controls the production rate.

3. Four minutes is now the longest per-unit processing time, and so the production rate is 60/4 = 15 units per hour. Producing 10 units will take 40 minutes [(10/15) × 60].

Pull Value

Pull value refers to a firm's ability to use inventories and produce products in the quantities needed, when needed, driven by a demand pull system. Many firms produce for inventory and then try to sell the excess goods they have produced. Efforts are made to create demand for the excess goods—goods that customers probably do not even want. Lean manufacturing uses a *demand-pull* system. The objective of lean manufacturing is to eliminate waste by producing a product only when it is needed and only in the quantities demanded by customers. Demand pulls products through the manufacturing process. Each operation produces only what is necessary to satisfy the demand of the succeeding operation. No production takes place until a signal from a succeeding process indicates a need to produce. Parts and materials arrive just in time to be used in production. Low setup times and cellular manufacturing are the major enabling factors for producing on demand. The Kanban system described in Chapter 18 [online at www.hansen1ce.nelson.com] is one way to ensure that materials and products flow according to demand.

Customer demand extends back through the value chain and affects how a manufacturer deals with suppliers. Materials inventories also represent waste. Thus, managing supplier linkages is also vital to lean manufacturing. **JIT purchasing** requires suppliers to deliver parts and materials just in time to be used in production. Supply of parts must be linked to production, which is linked to demand. One effect of successful management of customer and supplier linkages is to reduce all inventories to much lower levels.

Traditionally, inventories of raw materials and parts are carried so that a firm can take advantage of quantity discounts and hedge against future price increases of the items purchased. The objective is to lower the cost of inventory. JIT purchasing achieves the same objective without carrying inventories. The JIT solution is to exploit supplier linkages by negotiating long-term contracts with a few chosen suppliers located as close to the production facility as possible and by establishing more extensive supplier involvement. Suppliers are not selected on the basis of price alone. Performance—the quality of the component and the ability to deliver as needed—and commitment to JIT purchasing are vital considerations. Every effort is made to establish a partners-in-profits relationship with suppliers. Suppliers need to be convinced that their well-being is intimately tied to the well-being of the buyer.

To help reduce the uncertainty in demand for the supplier and establish the mutual confidence and trust needed in such a relationship, lean manufacturers emphasize long-term contracts that stipulate prices and acceptable quality levels. Long-term contracts also reduce dramatically the number of orders placed, which helps drive down the ordering costs. Another effect of long-term contracting is a reduction in the cost of parts and materials—usually in the range of 5 percent to 20 percent less than what was paid in a traditional setting. The need to develop close supplier relationships often drives the supplier base down dramatically. Suppliers also benefit, as the long-term contract ensures a reasonably stable demand for their products. A smaller supplier base typically means increased sales for the selected suppliers. Thus, both buyers and suppliers benefit, a common outcome when customer and supplier linkages are recognized and managed well. By reducing the number of suppliers and working closely with those that remain, the quality of the incoming materials can be improved significantly—a crucial outcome for the success of lean manufacturing. As the quality of incoming materials increases, some quality-related costs can be avoided or reduced. For example, the need to inspect incoming materials may disappear, and rework requirements decline.

Pursue Perfection

Zero setup times, zero defects, zero inventories, zero waste, producing on demand, increasing a cell's production rates, minimizing cost, and maximizing customer value represent ideal outcomes that a lean manufacturer seeks. As the process of becoming lean begins to unfold and improvements are realized, the possibility of achieving perfection becomes more believable. The relentless and continuous pursuit of these ideals is fundamental to lean manufacturing. As the flow increases and processes begin to improve, more hidden waste tends to be exposed. The objective is to produce the highest-quality, lowest-cost products in the least amount of time. To achieve this objective, a lean manufacturer must identify and eliminate the various forms of waste.

Sources of Waste   **Waste** consumes resources without adding value. Waste is anything customers do not value. Elimination of waste requires that its various forms be identified. The major sources of waste are listed below.

- Defective products
- Overproduction of goods not needed
- Inventories of goods awaiting further processing or consumption
- Unnecessary processing
- Unnecessary movement of people
- Unnecessary transport of goods
- Waiting
- The design of goods and services that do not meet the needs of the customer

Employee Empowerment   Employee involvement is vital for identifying and eliminating all forms of waste. A major procedural difference between traditional and lean environments is the degree of participation allowed workers in the management of the organization. In a lean environment, increasing the degree of participation increases productivity and overall cost efficiency. Managers seek workers' input and use their suggestions to improve production processes. The management structure must change in response to greater employee involvement. Because workers assume greater responsibilities, fewer managers are needed, and the organizational structure becomes flatter. Flatter structures speed up and increase the quality of information exchange. The style of management needed in a lean firm also changes. Managers in a lean environment act as facilitators more than as supervisors. Their role is to develop people and their skills so that they can make value-adding contributions.

Total Quality Control   Lean manufacturing necessarily carries with it a much stronger emphasis on managing quality. A defective part brings production to a grinding halt. Poor quality simply cannot be tolerated in a manufacturing environment that operates without inventories. Simply put, lean manufacturing cannot be implemented without a commitment to total quality control (*TQC*). *TQC* is essentially a never-ending quest for perfect quality: the striving for a defect-free product design and manufacturing process. Quality cost management is discussed extensively in Chapter 16.

Inventories   Overproduction of goods is controlled by letting customers pull goods through the system. Inventories are lowered by cellular manufacturing, low setup times, JIT purchasing, and a demand-pull system. Inventory management is of such importance that its treatment is covered in a separate chapter, Chapter 18.

Activity-Based Management   Process value analysis is the methodology for identifying and eliminating non-value-added activities. Non-value-added activities are unnecessary activities, including waiting, and thus much of the waste in a lean system is attacked using process value analysis. Process value analysis searches for the root causes of the wasteful activities and then, over time, eliminates these activities. See Chapter 14 for a detailed discussion of process value analysis.

Lean Accounting

OBJECTIVE ▶ 2
Describe lean accounting.

The numerous changes in structural and procedural activities that we have described for a lean firm also change traditional cost management practices. The traditional cost management system may not work well in the lean environment. In fact, the traditional costing and operational control approaches may actually work against lean manufacturing. Standard costing variances and departmental budgetary variances will likely encourage overproduction and work against the demand-pull system needed in lean manufacturing. For example, emphasis on labour efficiency by comparing actual hours used with hours allowed for production encourages production to keep labour occupied and productive. Similarly, emphasis on departmental efficiency (e.g., machine utilization rates) will cause non-bottleneck departments to overproduce and build work-in-process inventory. Furthermore, we already know from our study of activity-based costing (ABC) that in a multiple-product plant, the use of a plant-wide over-head rate can produce distorted product costs relative to focused manufacturing assignments or activity-based assignments. Distorted product costs can signal failure for lean manufacturing even when significant improvements may be occurring. To avoid obstacles and false signals, changes in both product-costing and operational control approaches are needed when moving to a value-stream-based lean manufacturing system.[4]

Value Streams and Traceability of Overhead Costs

Costing systems use three methods to assign costs to individual products: direct tracing, driver tracing, and allocation. Of the three methods, the most accurate is direct tracing; thus, it is preferred over the other two methods. Assume initially that a value stream is created for each product within a plant. In a lean environment, many overhead costs assigned to products using either driver tracing or allocation are now directly traceable to products. Equipment formerly located in other departments, for example, is now reassigned to value streams, and, under the single-product value-stream structure, is dedicated to the production of a single product. In this case, depreciation is now a directly traceable product cost. Multi-skilled workers and decentralized services add to the effect. Workers are assigned to the value stream and are trained to set up the equipment in the cells within the stream, maintain them, and operate them. These support functions were previously handled by a different set of labourers for all product lines. Additionally, people with specialized skills (e.g., industrial engineers and production schedulers) are assigned directly to value streams. The labour cost of these employees is now directly assigned to each value stream. Typically, implementing the value-stream structure does not require an increase in the number of people needed. Lean manufacturing eliminates wasteful activities, reducing the demand for people. For example, when production planning is reduced significantly because of an efficiently functioning demand-pull system, some of those working in production planning can be cross-trained to perform value-added activities within the value stream such as purchasing and quality control.

Exhibit 17-5 is a visual summary of value-stream cost assignments. Most costs are assigned directly to the value stream; however, some costs such as facility costs are assigned to each value stream using cost drivers. Facility costs are assigned using a cost per square metre (total cost/total square metres). If a value stream uses less square metres, it receives less cost. Thus, the purpose of this assignment is to motivate value-stream managers to find ways to occupy less space. As space is made available, it can be used for new product lines or to accommodate increased sales. For example, suppose that the facility costs are $200,000 per year for a plant occupying 20,000 square metres. The cost per square metre is $10. If a value stream occupies 5,000 square metres, it is assigned a cost of $50,000. Should the value stream figure out how to do the same tasks with 4,000 square metres, the cost would be reduced to $40,000. Any unabsorbed facility cost would be deducted from revenue as a separate item.

---

[4] Much of the material on lean accounting is based on two sources: Frances A. Kennedy and Jim Huntzinger, "Lean Accounting: Measuring and Managing the Value Stream," *Cost Management* (September/October 2005): 31–38, and Brian Maskell and Bruce Baggaley, *Practical Lean Accounting* (New York: Productivity Press, 2004).

Exhibit 17-5

Value-Stream Costs

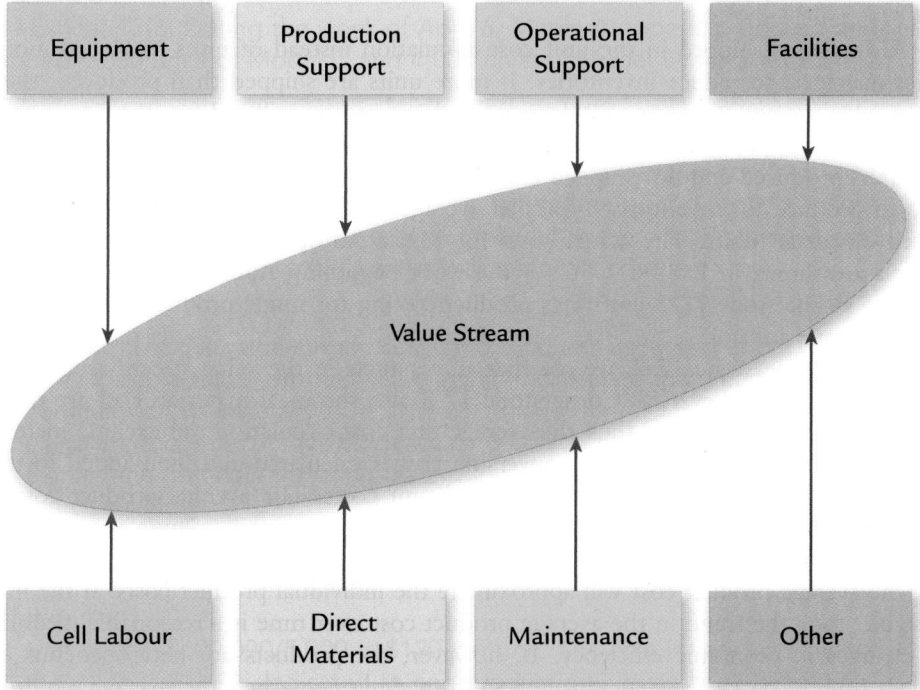

Limitations and Problems   Initially, it may not be possible to assign all the people needed exclusively to a value stream. There may be some individuals working in more than one value stream. The cost of these shared workers can be assigned to individual value streams in proportion to the time spent in each stream. It is also true that even in the most ideal of circumstances, there will be some individuals who will remain outside any particular value stream (the plant manager, for example). However, with multiple value streams, the unassigned costs are likely to be a very small percentage of the total costs. Finally, in reality, having a value stream for every product is not practical. The usual practice is to organize value streams around a family of products.

Value-Stream Costing

Product Costing: Single-Product (Focused) Value Stream   Because of multitask assignments, cross-training, and redeployment of other support personnel, most support costs are exclusive to a focused value stream and are thus assigned to a product using direct tracing. One consequence of increasing directly traceable costs is to increase the accuracy of product costing. Directly traceable costs are exclusively associated with the product and can safely be said to belong to it. Product cost is calculated by taking the costs of the period and dividing by the output. Focused value streams provide simple and accurate product costing.

Product Costing: Multiple-Product Value Stream   Value streams are formed around products with common processes (see Exhibit 17-2). Manufacturing cells within a value stream are thus structured to make a family of products or parts that require the same manufacturing sequence. The costs are assigned in the same

Exhibit 17-7

Maple Autoparts Profit and Loss Statement

	Week Ending April 13			
	Aluminum Stream	Steel Stream	Sustaining Costs	Plant Totals
Revenues	$ 700,000	$1,500,000		$2,200,000
Material costs	(280,000)	(410,000)		(690,000)
Conversion costs	(70,000)	(190,000)		(260,000)
Value-stream profit	$ 350,000	$ 900,000		1,250,000
Value-stream ROS*	50%	60%		
Employee costs			$(40,000)	(40,000)
Other expenses			(30,000)	(30,000)
Change in inventory:				
Current less prior period				(500,000)
Plant gross profit				$ 680,000
Plant ROS				31%

*ROS = Return on sales = Profit/Sales.

wheels (Models A and B) and (2) steel wheels (Models C and D). Exhibit 17-7 shows a profit and loss statement for the plant, for the week ending April 13. (The plant had significantly increased its sales of steel wheels to auto manufacturers that were replacing low-end aluminum wheels with steel units on new models.) Costs outside the value stream (sustaining costs) are reported in a separate column. The revenues and costs reported are the actual revenues and costs for the week. To avoid distorting the current week's performance, inventory reductions are reported separately from the value-stream contributions. Adding the inventory changes also allows the income to be stated correctly for external reporting.

Decision Making

Using the average product cost for a value stream means that the individual product costs are not known. In reality, a fully specified and accurate product cost is not needed for many decisions. Waste can be eliminated at the activity and process levels without knowing product costs. We do not need detailed variances by product to signal sources of waste and potential for improvement. In fact, as already noted, standard costing variances may actually impede improvement decisions. For other decisions, the effect of the decision on the profitability of value stream may be the only information needed for certain decisions. For example, special order and make-or-buy decisions can be made at the value-stream level.

Consider a make-or-buy decision. Suppose that Maple Autoparts is currently purchasing a component used in making its wheel products and is considering making the component. The decision can be made by comparing the profitability of the value stream under the buy scenario with the profitability under the make scenario. A typical analysis would be as follows for Maple's ABS value stream:

	Buy	Make
Revenue	$1,500,000	$1,500,000
Material costs	(410,000)	(380,000)
Conversion costs	(190,000)	(200,000)
Value-stream profit	$ 900,000	$ 920,000

The profitability of the value stream increases under the make alternative, and so the decision would be to make the component rather than buy it.

While analysis of the effect on value-stream profitability has its merits, it also has its perils. Many of the decisions are short term in nature and do not reflect the long-term consequences. For example, acceptance of a special order below the full cost of a product (unknown with average cost) may increase value-stream profitability because of existing unused value-stream capacity, but continued acceptance of such orders may not earn the return necessary to replace capacity that is eventually exhausted through use. Thus, other very important decisions may need individual product cost information, and a lean accounting system must provide this information.[5]

Performance Measurement

Abandoning a standard cost system also removes a major operational control system, and it must be replaced. The lean control system uses a Box Scorecard that compares operational, capacity, and financial metrics with prior week performances and with a future desired state. Trends over time and the expectation of achieving some desired state in the near future are the means used to motivate constant performance improvement. Thus, the lean control approach uses a mixture of financial and nonfinancial measures for the value stream. The future desired state reflects targets for the various measures. Operational, nonfinancial measures are also used at the cell level. A typical value-stream Box Scorecard is shown in Exhibit 17-8 (metrics and format can vary). Only a brief introduction to the Box Scorecard is made because the Balanced Scorecard is a more thorough and integrated approach that encompasses the concepts of a Box Scorecard.

For the operational measures, units sold per person is a partial labour productivity measure and is therefore a measure of labour *efficiency*. Productivity measures are

Exhibit 17-8

ABS Value-Stream Box Scorecard

For 4/6/2013			
	Last Week	**This Week (4/6/13)**	**Planned Future State (6/30/13)**
Operational			
Units sold per person	250	270	280
On-time delivery	90%	92%	97%
Dock-to-dock days	18.5	18	16
First-time through	56%	58%	65%
Average product cost	$128	$120	$115
Accounts receivable days	31	30	28
Capacity			
Productive	21%	20%	25%
Nonproductive	45%	46%	30%
Available	34%	34%	45%
Financial			
Weekly sales	$1,800,000	$1,500,000	$2,000,000
Weekly material cost	$800,000	$600,000	$600,000
Weekly conversion cost	$400,000	$300,000	$400,000
Weekly value-stream profit	$600,000	$600,000	$1,000,000
ROS	33%	40%	50%

[5] Ibid.

discussed more completely later in this chapter. Dock-to-dock is the *time* it takes for a product to be manufactured from the moment the materials arrive at the receiving dock until the finished product is shipped from the shipping dock. Dock-to-dock is a cycle time measure, a concept that was studied in Chapter 15. First-time through is a measure of *quality* and is simply the percentage of product that made it through production without being defective and thus needing to be rejected or reworked. Capacity is labelled as *productive* (value-added), *nonproductive* (non-value-added—used but wasteful), and *available* (unused) capacity. The scorecard measures are expected to improve over time and to be helpful in managing and bringing about improvement. For example, from the Box Scorecard in Exhibit 17-8, we see that the nonproductive capacity is targeted to go from 46 percent (current state) to 30 percent (future state), with productive capacity increasing from 20 percent to 25 percent and available capacity increasing from 34 percent to 45 percent. As waste is eliminated, the nonproductive capacity converts into available capacity. The machines, people, and other resources used for wasteful activities are now available for more productive work. For financial performance to improve, some decisions must be made with respect to the increase in available capacity. The most sensible and practical approach is to commit to use the freed-up resources to expand the business. One possibility is to add new product lines. Another possibility is to transfer the resources to other value streams that are in a high-growth state with increasing resource demands. Another is to realize cost reductions by reducing headcount and eliminating resources. This latter approach is the least desirable. It makes it hard to gain the cooperation and involvement of employees with the transformation into a lean workforce if their suggestions and actions are going to lead to the loss of their jobs or the jobs of their friends and coworkers.

OBJECTIVE ▶ 3
Discuss and define productive efficiency and partial productivity measurement.

Productivity

A key objective of lean manufacturing and accounting is that of increasing overall productive efficiency. **Productivity** is concerned with producing output efficiently, and it specifically addresses the relationship of output and the inputs used to produce the output. Usually, different combinations or mixes of inputs can be used to produce a given level of output. **Total productive efficiency** is the point at which two conditions are satisfied: (1) for any mix of inputs that will produce a given output, no more of any one input is used than necessary to produce the output and (2) given the mixes that satisfy the first condition, the least costly mix is chosen. The first condition is driven by technical relationships and, therefore, is referred to as **technical efficiency**. Technical improvements in productivity can be achieved by using fewer inputs to produce the same output, by producing more output using the same inputs, or by producing more output with relatively fewer inputs. The second condition is driven by relative input price relationships and, therefore, is referred to as **allocative efficiency**. Input prices determine the *relative proportions* of each input that should be used. Choosing the right combination of inputs can also produce significant improvements in economic efficiency. Exhibits 17-9 and 17-10 illustrate technical and allocative efficiency improvements. The output in the exhibits is vehicles, and the inputs are labour (number of workers) and capital (dollars invested in automated equipment).

Partial Productivity Measurement Defined

Productivity measurement is simply a quantitative assessment of productivity changes. The objective is to assess whether productive efficiency has increased or decreased. Productivity measurement can be actual or prospective. Actual productivity measurement allows managers to assess, monitor, and control changes. Prospective measurement is forward looking, and it serves as input for strategic decision making. Specifically, prospective measurement allows managers to compare relative benefits of

Exhibit 17-9

Improving Technical Efficiency

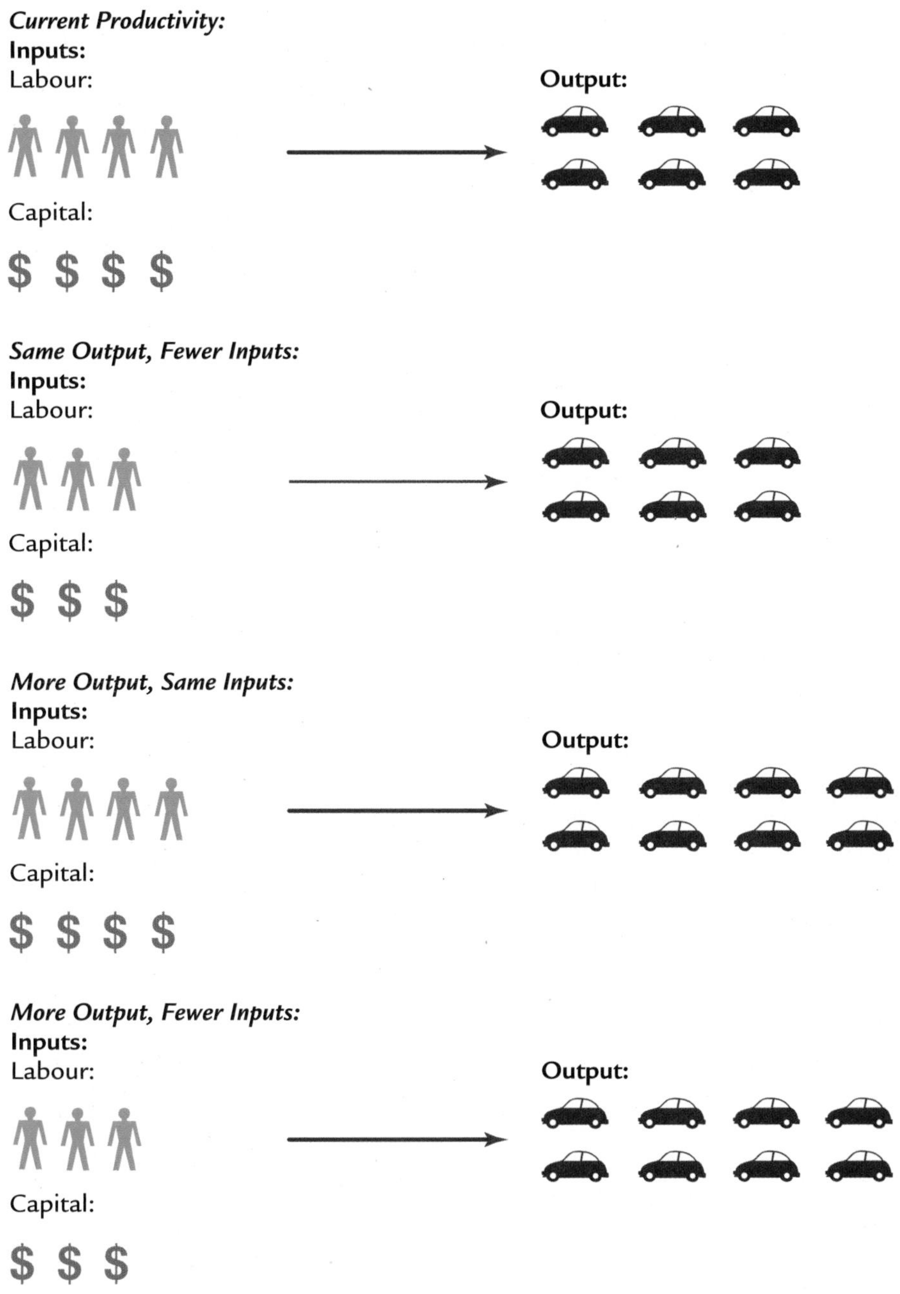

Current Productivity:
Inputs:
Labour:

Output:

Capital:

$ $ $ $

Same Output, Fewer Inputs:
Inputs:
Labour:

Output:

Capital:

$ $ $

More Output, Same Inputs:
Inputs:
Labour:

Output:

Capital:

$ $ $ $

More Output, Fewer Inputs:
Inputs:
Labour:

Output:

Capital:

$ $ $

different input combinations, choosing the inputs and input mix that provide the greatest benefit. Productivity measures can be developed for each input separately or for all inputs jointly. Measuring productivity for one input at a time is called **partial productivity measurement**.

Productivity of a single input is typically measured by calculating the ratio of the output to the input as follows:

$$\text{Productivity ratio} = \text{Output/Input}$$

Because the productivity of only one input is being measured, the measure is called a *partial productivity measure*. If both output and input are measured in physical quantities, then we have an **operational productivity measure**. If output

Exhibit 17-10

Improving Allocative Efficiency

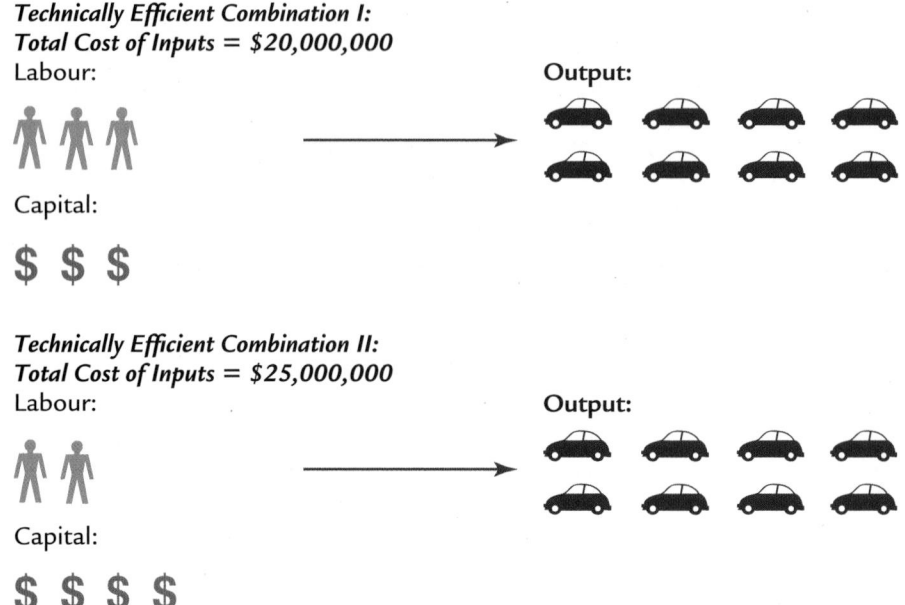

Technically Efficient Combination I:
Total Cost of Inputs = $20,000,000
Labour:

Capital:

Technically Efficient Combination II:
Total Cost of Inputs = $25,000,000
Labour:

Capital:

or input is expressed in dollars, then we have a **financial productivity measure**. Cornerstone 17-3 illustrates partial productivity measurement.

Assume, for example, that in 2012, Tydek Company produced 240,000 frames for snowmobiles and used 60,000 hours of labour. The labour productivity ratio is four frames per hour (240,000/60,000). This is an operational measure, since the units are expressed in physical terms. If the selling price of each frame is $30 and the cost of labour is $15 per hour, then output and input can be expressed in dollars. The labour productivity ratio, expressed in financial terms, is $8 of revenue per dollar of labour cost ($7,200,000/$900,000).

Measuring Changes in Productive Efficiency   The labour productivity ratio of four frames per hour measures the 2012 productivity experience of Tydek. By itself, the ratio conveys little information about productive efficiency or whether the company has improving or declining productivity. It is possible, however, to make a statement about increasing or decreasing productivity efficiency by measuring *changes* in productivity. To do so, the actual current productivity measure is compared with the productivity measure of a prior period. This prior period is referred to as the **base period** and serves to set the benchmark or standard for measuring changes in productive efficiency. The prior period can be any period desired. It could, for example, be the preceding year, the preceding week, or even the period during which the last batch of products was produced. For strategic evaluations, the base period is usually chosen as an earlier year. For operational control, the base period tends to be close to the current period—such as the preceding batch of products or the preceding week.

To illustrate, assume that 2012 is the base period and that the labour productivity standard, therefore, is four frames per hour. Further assume that late in 2012, Tydek decided to try a new procedure for producing and assembling the frames with the expectation that the new procedure would use less labour. In 2013, 250,000 frames were produced, using 50,000 hours of labour. The labour productivity ratio for 2013 is five frames per hour (250,000/50,000). The *change* in productivity is a one-unit-per-hour *increase* in productivity (from four units per hour in 2012 to five units per hour in 2013). The change is a significant improvement in labour productivity and provides evidence supporting the efficacy of the new process.

Advantages of Partial Measures   Partial measures allow managers to focus on the use of a particular input. Operating partial measures have the advantage of being easily interpreted by everyone within the organization. Consequently, partial operational measures are easy to use for assessing productivity performance of operating personnel. Labourers, for instance, can relate to units produced per hour or units produced per kilogram of material. Thus, partial operational measures provide feedback that operating personnel can relate to and understand—measures that deal with the specific inputs over which they have control. The ability of operating personnel to understand and relate to the measures increases the likelihood that the measures will be accepted. Furthermore, for operational control, the standards for performance are often very short run in nature. For example, standards can be the productivity ratios of prior batches of goods. Using this standard, productivity trends within the year itself can be tracked.

Disadvantages of Partial Measures   Partial measures, used in isolation, can be misleading. A decline in the productivity of one input may be necessary to increase the productivity of another. Such a trade-off is desirable if overall costs decline, but the effect would be missed by using either partial measure. For example, changing a process so that direct labourers take less time to assemble a product may increase scrap and waste while leaving total output unchanged. Labour productivity has increased, but productive use of materials has declined. If the increase in the cost of waste and scrap outweighs the savings of the decreased labour, then overall productivity has declined.

Two important conclusions can be drawn from this example. First, the possible existence of trade-offs mandates a total measure of productivity for assessing the merits of productivity decisions. Only by looking at the total productivity effect of all inputs can managers accurately draw any conclusions about overall productivity performance. Second, because of the possibility of trade-offs, a total measure of productivity must assess the aggregate financial consequences and, therefore, should be a financial measure.

Total Productivity Measurement

Measuring productivity for all inputs at once is called **total productivity measurement**. In practice, it may not be necessary to measure the effect of all inputs. Many firms measure the productivity of only those factors that are thought to be relevant indicators of organizational performance and success. Thus, in practical terms, total productivity measurement can be defined as focusing on a limited number of inputs, which, in total, indicates organizational success. In either case, total productivity measurement requires the development of a multifactor measurement approach. A common multifactor approach suggested in the productivity literature (but rarely found in practice) is the use of aggregate productivity indexes. Aggregate indexes are complex and difficult to interpret and have not been generally accepted. Two approaches that have gained some acceptance are *profile measurement* and *profit-linked productivity measurement*.

OBJECTIVE ▶ 4
Explain what total productivity measurement is, and describe its advantages.

Profile Productivity Measurement   Producing a product involves numerous critical inputs such as labour, materials, capital, and energy. **Profile measurement** provides a series or vector of separate and distinct partial operational measures. Profiles (vectors or series of measures) can be compared over time to provide information about productivity changes. When the partial productivity ratios move in the same direction when compared with the base period ratios, some definitive statements about productivity changes can be made. However, if the ratios move in opposite directions, a trade-off exists and the comparison of profiles provides a mixed signal about productivity changes. Furthermore, while a profile analysis reveals if a trade-off exists, it does not reveal whether the trade-off is good or bad. If the economic effect of the productivity changes is positive, then the trade-off is good; otherwise, it must be viewed as bad. Cornerstone 17-3 illustrates profile productivity measurement and reveals its limitations.

As Cornerstone 17-3 shows, profile analysis can provide managers with useful insights about changes in productivity. However, comparing productivity profiles will not always reveal the nature of the overall change in productive efficiency. Often, it may be necessary to *value* input productivity trade-offs to assess the nature of *overall* productivity change.

Profit-Linked Productivity Measurement   Assessing the effects of productivity changes on current profits is one way to value productivity changes. Profits change from the base period to the current period. Some of that profit change is attributable to productivity changes. Measuring the amount of profit change attributable to productivity change is defined as **profit-linked productivity measurement**.

Assessing the effect of productivity changes on current-period profits will help managers understand the economic importance of productivity changes. Linking productivity changes to profits is described by the following rule:

> **Profit-Linkage Rule.** *For the current period, calculate the cost of the inputs that would have been used in the absence of any productivity change and compare this cost with the cost of the inputs actually used. The difference in costs is the amount by which profits changed because of productivity changes.*

The formula corresponding to the linkage rule is given below.

$$\text{Profit-linked productivity change} = \sum PQ_i P_i - \sum AQ_i P_i$$

where

PQ_i = The amount of input i that would have been used for the current period in the absence of a productivity change

P_i = Current-period price of input i

AQ_i = Actual amount of input i used in the current period

$\sum$ represents the sum of the items that follow it.

To apply the linkage rule formula, the inputs that would have been used for the current period in the absence of a productivity change must be calculated. To determine PQ_i, divide the current-period output by the input's base-period productivity ratio:

$$PQ_i = \text{Current-period output/Base-period productivity ratio for input } i$$

The profit-linked measure computes the amount of profit change from the base period to the current period attributable to productivity changes. Generally, this will not be equal to the total profit change between the two periods. The difference between the total profit change and the profit-linked productivity change is called the **price-recovery component**. This component is the change in revenue less a change in the cost of inputs, *assuming no productivity changes*. It, therefore, measures the ability of revenue changes to cover changes in the cost of inputs, assuming no productivity change, and is calculated as follows:

$$\text{Price recovery} = \text{Total profit change} - \text{Profit-linked productivity change}$$

Cornerstone 17-4 illustrates the application of the profit-linked rule.

Cornerstone 17-4 reveals that the net effect of the process change implemented by Tydek was favourable, increasing profits by $12,500. Profit-linked productivity effects can be assigned to individual inputs. The increase in labour productivity creates a $187,500 increase in profits; however, the drop in materials productivity caused a $175,000 decrease in profits. Most of the profit decrease came from an increase in materials usage—apparently, waste, scrap, and spoiled units are much greater with the new process. Thus, the profit-linked measure provides partial measurement effects as

The HOW and WHY of Profile Productivity Measurement

**CORNERSTONE
17-3**

Information:

In 2013, Tydek Company implements a new production and assembly process affecting labour and materials with the following reported data:

	2012	2013
Number of frames produced	240,000	250,000
Labour hours used	60,000	50,000
Materials used (kg)	1,200,000	1,150,000

Why:

Profiles (vectors) of productivity measures can be compared over time to assess productivity changes. If the changes are in the same direction, then a definitive statement about productivity can be made; if a trade-off exists, valuing the individual input productivity changes is needed to assess the nature of the overall productivity change.

Required:

1. Calculate the productivity profile for 2012.
2. Calculate the productivity profile for 2013, and comment on the effect of the new production and assembly process.
3. **What if** the materials used in 2013 were 1,300,000 kilograms? What does comparison of the 2012 and 2013 profiles now communicate?

Solution:

1.

| Partial Operational
Productivity Ratios	2012 Profile*
Labour productivity ratio	4.000
Material productivity ratio	0.200

*Labour: 240,000/60,000; Materials: 240,000/1,200,000

2.

| Partial Operational
Productivity Ratios	2013 Profile*
Labour productivity ratio	5.000
Material productivity ratio	0.217

*Labour: 250,000/50,000; Materials: 250,000/1,150,000

Comparing the 2012 profile (4, 0.200) with the 2013 profile (5, 0.217), productivity increased for each input; thus, the new process has improved overall productivity.

3.

| Partial Operational
Productivity Ratios	2012 Profile[a]	2013 Profile[b]
Labour productivity ratio	4.000	5.000
Material productivity ratio	0.200	0.192

[a]Labour: 240,000/60,000; Materials: 240,000/1,200,000
[b]Labour: 250,000/50,000; Materials: 250,000/1,300,000

Labour productivity has increased, and materials productivity has decreased. A trade-off between the two inputs exists and must be valued to assess the nature of the overall productivity change.

**CORNERSTONE
17-4**

The HOW and WHY of Profit-Linked Productivity Measurement

Information:

In 2013, Tydek Company implements a new process affecting labour and materials. The following two years of expanded data are provided:

	2012	2013
Number of frames produced	240,000	250,000
Labour hours used	60,000	50,000
Materials used (kg)	1,200,000	1,300,000
Unit selling price (frames)	$30	$30
Wages per labour hour	$15	$15
Cost per kilogram of material	$3	$3.50

Why:

The productivity effect on current-period profit is the difference in the cost of the inputs that would have been used and the cost of the actual inputs used. Price recovery is the difference in the actual profit change and the profit-linked productivity change.

Required:

1. Calculate the cost of inputs in 2013, assuming no productivity change from 2012 to 2013.
2. Calculate the actual cost of inputs for 2013. What is the net value of the productivity changes? How much profit change is attributable to each input's productivity change?
3. *What if* a manager wants to know how much of the total profit change from 2012 to 2013 is attributable to price recovery? Calculate the price-recovery component and comment on its meaning.

Solution:

1. Base-period productivity ratios: 4 (labour) and 0.200 (materials). Thus, we have:

$$PQ \text{ (labour)} = 250,000/4 = 62,500 \text{ hrs.}$$
$$PQ \text{ (materials)} = 250,000/0.200 = 1,250,000 \text{ kg}$$

Cost of labour ($PQ \times P = 62,500 \times \$15$)	$ 937,500
Cost of materials ($PQ \times P = 1,250,000 \times \$3.50$)	4,375,000
Total PQ cost	$5,312,500

2.

Cost of labour ($AQ \times P = 50,000 \times \$15$)	$ 750,000
Cost of materials ($AQ \times P = 1,300,000 \times \$3.50$)	4,550,000
Total current cost	$5,300,000

Profit-linked productivity measure:

CORNERSTONE
17-4
(continued)

Input	(1) PQ	(2) PQ × P	(3) AQ	(4) AQ × P	(2) – (4) (PQ –AQ) × P
Labour	62,500	$ 937,500	50,000	$ 750,000	$ 187,500
Materials	1,250,000	4,375,000	1,300,000	4,550,000	(175,000)
		$5,312,500		$5,300,000	$ 12,500

Net productivity change = $12,500. Labour productivity change = $187,500. Materials productivity change = $(175,000).

3.

	2012	2013	2013–2012
Revenues	$7,200,000	$7,500,000	$ 300,000
Cost of inputs	4,500,000	5,300,000	(800,000)
Profit	$2,700,000	$2,200,000	$(500,000)

Price recovery = Total profit change – Profit-linked productivity change
= $(500,000) – $12,500
= $(512,500)

The increase in revenues would not have been sufficient to recover the increase in the cost of the inputs. The increase in productivity provided some relief for the price-recovery problem.

well as a total measurement effect. The total profit-linked productivity measure is the sum of the individual partial measures. This property makes the profit-linked measure ideal for assessing trade-offs. A much clearer picture of the effects of the changes in productivity emerges. Unless waste and scrap can be brought under better control, the company ought to return to the old assembly process. Of course, it is possible that the learning effects of the new process are not yet fully captured and that further improvements in labour productivity might be observed. As labour becomes more proficient at the new process, it is possible that the materials usage could also decrease.

Summary of Learning Objectives

1. **Describe the basic features of lean manufacturing.**

- Lean manufacturing has two principal objectives: eliminating waste and creating value for the customer.
- It is characterized by lean thinking—focusing on customer value, value streams, production flow, demand-pull, and perfection.
- Value streams are made up of all activities, both value-added and non-value-added, required to bring a product group or service from its starting point (e.g., customer order or concept for a new product) to a finished product in the hands of the customer.
- Value-stream analysis allows waste to be identified and eliminated.

2. **Describe lean accounting.**

- Lean accounting is an approach designed to support and encourage lean manufacturing.
- Average costing, value-stream cost reporting, and the heavy use of nonfinancial measures for operational control are typical lean accounting approaches.
- The average product cost is the total value-stream cost of the period divided by the units shipped in the period.
- Value-stream costing reports the actual revenues and actual costs on a weekly basis (for each value stream).
- The lean control system uses a Box Scorecard that compares operational, capacity, and financial metrics with prior-week performances and with a future desired state.
- Simplicity and compatibility are major characteristics of lean accounting.

3. **Discuss and define productive efficiency and partial productivity measurement.**

- Productivity deals with how efficiently inputs are used to produce the output.
- Technical efficiency is concerned with producing a given output using no more than necessary of any input.
- Allocative efficiency is concerned with choosing the least costly technically efficient combination of inputs.
- Partial measures of productivity evaluate the efficient use of single inputs.

4. **Explain what total productivity measurement is, and describe its advantages.**

- Total measures of productivity assess efficiency for all inputs.
- Profile measures are vectors of series of partial measures but provide mixed signals if the productivity changes for inputs are in opposite directions.
- Profit-linked measures value trade-offs in input productivity changes.

CORNERSTONES FOR CHAPTER 17

Review Problems

I. MCE, Lean Measures, and the Balanced Scorecard

Numark Inc. manufactures a product that experiences the following activities and times (the production processes are listed in sequential order):

	Minutes
Cutting	20
Welding	15
Assembly	7
Polishing	3
Moving (three moves)	12
Waiting	18

Required:

1. Compute the time required to produce one unit of product under the current production layout.
2. Assume that Numark creates a manufacturing cell that eliminates move and wait times. What is the production rate assuming continuous production?
3. If the time for the cutting operation is cut in half, what effect will this have on the production rate?

Solution:

1. Production time for one unit is 75 minutes $(20 + 15 + 7 + 3 + 12 + 18)$.
2. Production rate $= 60/20 = 3$ units per hour.
3. Production rate $= 60/15 = 4$ units per hour. The cycle time of the slowest operation is now welding (15 minutes).

II. Productivity

At the end of 2012, Alma Company implemented a new labour process and redesigned its product with the expectation that input usage efficiency would increase. Now, at the end of 2013, the president of the company wants an assessment of the changes in the company's productivity. The data needed for the assessment are as follows:

	2012	2013
Output	10,000	12,000
Output prices	$20	$20
Materials (kg)	8,000	8,400
Materials unit price	$6	$8
Labour (hrs)	5,000	4,800
Labour rate per hour	$10	$10
Power (kwh)	2,000	3,000
Price per kwh	$2	$3

Required:

1. Compute the partial operational measures for each input for both 2012 and 2013. What can be said about productivity improvement?
2. Prepare a partial income statement for each year, and calculate the total change in profits.
3. Calculate the profit-linked productivity measure for 2013. What can be said about the productivity program?
4. Calculate the price-recovery component. What does this tell you?

Solution:

1. Partial measures:

	2012	2013
Materials	10,000/8,000 = 1.25	12,000/8,400 = 1.43
Labour	10,000/5,000 = 2.00	12,000/4,800 = 2.50
Power	10,000/2,000 = 5.00	12,000/3,000 = 4.00

Profile analysis indicates that productive efficiency has increased for materials and labour and decreased for power. The outcome is mixed, and no statement about overall productivity improvement can be made without valuing the trade-off.

2. Income statements:

	2012	2013
Sales	$200,000	$240,000
Cost of inputs	102,000	124,200
Gross profit	$ 98,000	$115,800

Total change in profits: $115,800 - $98,000 = $17,800$ increase

3. Profit-linked measurement:

Input	(1) PQ*	(2) PQ × P	(3) AQ	(4) AQ × P	(2) – (4) (PQ × P) – (AQ × P)
Materials	9,600	$ 76,800	8,400	$ 67,200	$ 9,600
Labour	6,000	60,000	4,800	48,000	12,000
Power	2,400	7,200	3,000	9,000	(1,800)
		$144,000		$124,200	$19,800

*Materials: 12,000/1.25; Labour: 12,000/2; Power: 12,000/5

The value of the increases in efficiency for materials and labour more than offsets the increased usage of power. Thus, the productivity improvement program should be labelled as successful.

4. Price recovery:

Price-recovery component = Total profit change – Profit-linked productivity change
Price-recovery component = $17,800 – $19,800
= $(2,000)

This says that without the productivity improvement, profits would have declined by $2,000. The $40,000 increase in revenues would not have offset the increase in the cost of inputs. From the solution to Requirement 3, the cost of inputs without a productivity increase would have been $144,000 (column 2). The increase in the input cost without a productivity change would have been $144,000 – $102,000 = $42,000. This is $2,000 more than the increase in revenues. Only because of the productivity increase did the firm show an increase in profitability.

Key Terms

Allocative efficiency, 852

Base period, 854

Financial productivity measure, 854

JIT purchasing, 844

Lean manufacturing, 839

Manufacturing cells, 842

Operational productivity measure, 853

Partial productivity measurement, 853

Price-recovery component, 856

Productivity, 852

Productivity measurement, 852

Profile measurement, 855

Profit-Linkage Rule, 856

Profit-linked productivity
measurement, 856

Technical efficiency, 852

Total productive efficiency, 852

Total productivity measurement, 855

Value stream, 840

Waste, 845

Discussion Questions

1. What is lean manufacturing?
2. What are the five principles of lean thinking?
3. Identify two types of value streams and explain how they differ.
4. How are value streams identified and created?
5. Explain how lean manufacturing is able to produce small batches (low-volume products) of differing products (high variety).
6. What role does a demand-pull system have on lean manufacturing?
7. Identify eight forms and sources of waste.
8. What is a focused value stream?
9. What is the purpose of assigning facility costs to value streams, using a fixed price?
10. Why are units shipped used to calculate the value-stream product cost?

11. When will the average unit cost be useful for value streams?
12. Explain why changes in value-stream profitability may be better information than individual product cost for certain decisions.
13. Define total productive efficiency.
14. Explain the difference between technical and allocative efficiency.
15. What is productivity measurement?
16. Explain the difference between partial and total measures of productivity.
17. What is an operational productivity measure? A financial measure?
18. Discuss the advantages and disadvantages of partial measures of productivity.
19. What is the purpose of a base period?
20. What is profile measurement and analysis? What are the limitations of this approach?
21. What is profit-linked productivity measurement and analysis?
22. Explain why profit-linked productivity measurement is important.
23. What is the price-recovery component?

Cornerstone Exercises

Cornerstone Exercise 17-1 **CONTINUOUS FLOW VS. DEPARTMENTAL FLOW MANUFACTURING**

OBJECTIVE ➤ 1

CORNERSTONE 17-1

Barker Company has the following departmental manufacturing structure for one of its products:

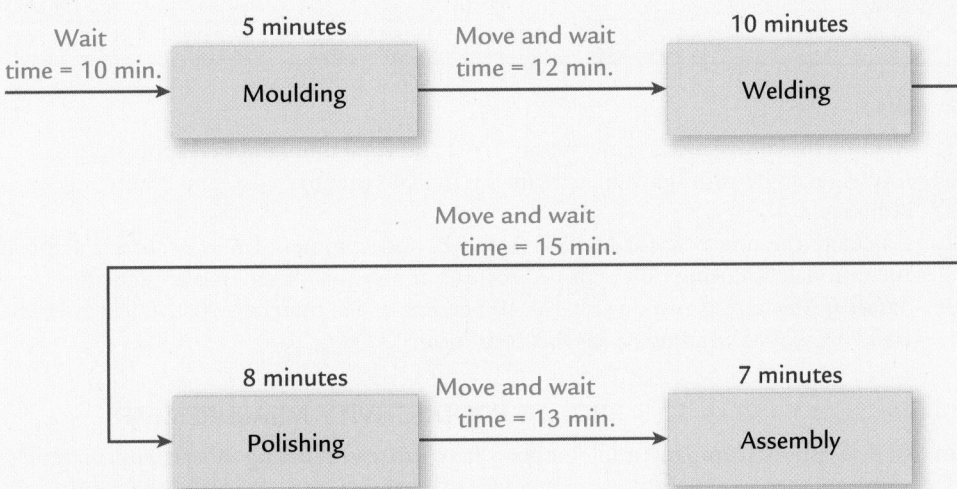

After some study, the production manager of Barker recommended the following revised cellular manufacturing approach:

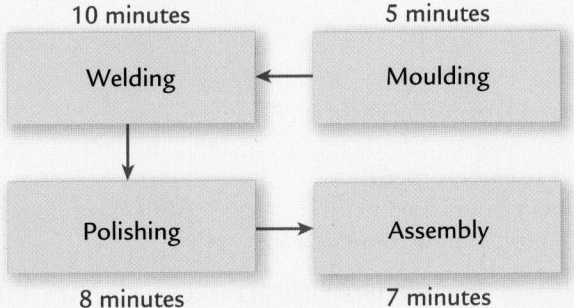

Required:

1. Calculate the total time it takes to produce a batch of 20 units using Barker's traditional departmental structure.

2. Using cellular manufacturing, how much time is saved producing the same batch of 20 units? Assuming the cell operates continuously, what is the production rate? Which process controls this production rate?

3. *What if* the processing times of moulding, welding, and assembly are all reduced to four minutes each? What is the production rate now, and how long will it take to produce a batch of 20 units?

OBJECTIVE ▶ 2

CORNERSTONE 17-2

Cornerstone Exercise 17-2 VALUE-STREAM COSTING

During the week of June 12, Leonard Manufacturing produced and shipped 7,500 units of its aluminum wheels: 1,500 units of Model A and 6,000 units of Model B. The following costs were incurred:

	Materials	Salaries/ Wages	Machining	Other	Total Cost
Order processing		$ 36,000			$ 36,000
Production planning		72,000			72,000
Purchasing		54,000			54,000
Stamping	$ 750,000	75,000	$ 72,000	$36,000	933,000
Welding	300,000	84,000	84,000	24,000	492,000
Cladding	150,000				150,000
Testing		21,000			21,000
Packaging and shipping		18,000			18,000
Invoicing		24,000			24,000
Totals	$1,200,000	$384,000	$156,000	$60,000	$1,800,000

Required:

1. Assume initially that the value-stream costs and total units shipped apply only to one model (a single-product value stream). Calculate the unit cost, and comment on its accuracy.

2. Calculate the unit cost for Models A and B, and comment on its accuracy. Explain the rationale for using units shipped instead of units produced in the calculation.

3. *What if* Model A is responsible for 40 percent of the materials cost? Show how the unit cost would be adjusted for this condition.

OBJECTIVE ▶ 3

CORNERSTONE 17-3

Cornerstone Exercise 17-3 PROFILE PRODUCTIVITY MEASUREMENT

In 2013, Brisbois Company implements a new process affecting labour and materials. The following reported data are provided to evaluate the effect on the company's productivity:

	2012	2013
Number of units produced	180,000	150,000
Labour hours used	36,000	25,000
Materials used (kg)	720,000	500,000

Required:

1. Calculate the productivity profile for 2012.

2. Calculate the productivity profile for 2013, and comment on the effect of the new production and assembly process.

3. *What if* the labour hours used in 2013 were 37,500? What does comparison of the 2012 and 2013 profiles now communicate?

OBJECTIVE ▶ 4

CORNERSTONE 17-4

Cornerstone Exercise 17-4 PROFIT-LINKED PRODUCTIVITY MEASUREMENT

Refer to **Cornerstone Exercise 17-3**. Brisbois Company provides the following additional information so that total productivity can be valued:

	2012	2013
Number of units produced	180,000	150,000
Labour hours used	36,000	37,500
Materials used (kg)	720,000	500,000
Unit selling price	$20	$22
Wages per labour hour	$12	$14
Cost per kilogram of material	$3.40	$3.50

Required:

1. Calculate the cost of inputs in 2013, assuming no productivity change from 2012 to 2013.
2. Calculate the actual cost of inputs for 2013. What is the net value of the productivity changes? How much profit change is attributable to each input's productivity change?
3. *What if* a manager wants to know how much of the total profit change from 2012 to 2013 is attributable to price recovery? Calculate the price-recovery component, and comment on its meaning.

Exercises

Exercise 17-5 **VALUE-STREAM IDENTIFICATION**

OBJECTIVE ➤ 1

Helix Inc. formed the following matrix for its five products:

Production Activities/Processes

Product Model	Order Entry	Production Planning	Subassembly 47A Cell	Basic Cell	Assembly Cell	Inspecting	Packaging and Shipping	Warranty
A	x	x		x	x	x	x	x
B	x	x	x		x	x	x	
C	x	x		x			x	x
D	x	x		x	x	x	x	x
E	x	x	x		x	x	x	

Required:

Using the information in the matrix, identify the value streams.

Exercise 17-6 **CONTINUOUS FLOW VERSUS DEPARTMENTAL FLOW MANUFACTURING**

OBJECTIVE ➤ 1

Vitacom Inc. has the following departmental structure for producing a well-known multi-vitamin:

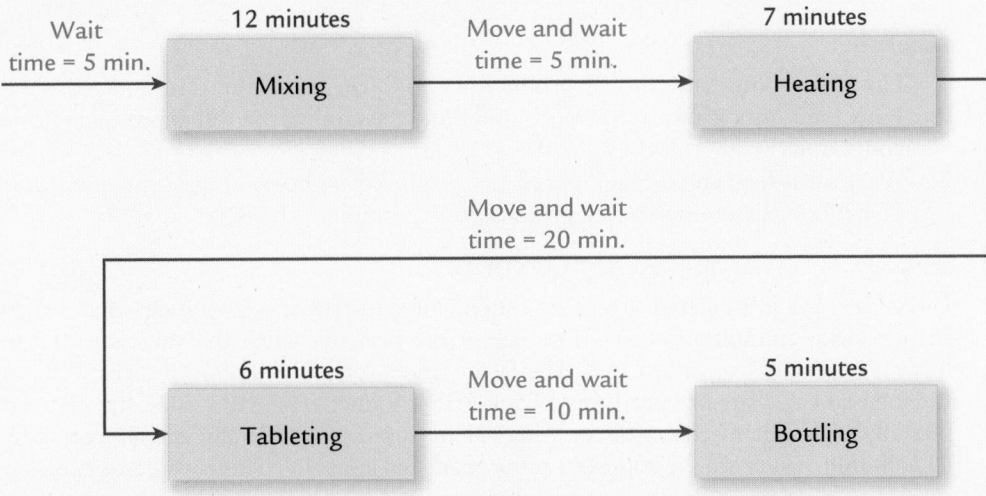

A consultant designed the following cellular manufacturing structure for the same product:

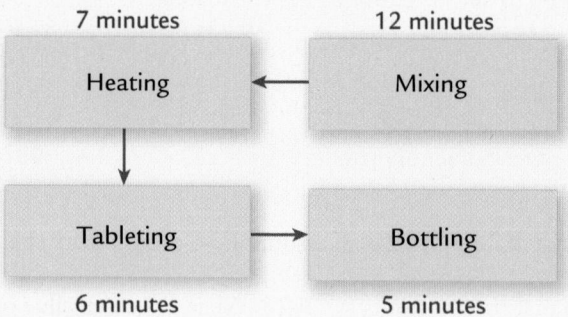

The times shown represent the time required to process one unit of product.

Required:

1. Calculate the time required to produce a batch of 12 bottles using a batch processing departmental structure.
2. Calculate the time to process 12 bottles using cellular manufacturing.
3. How much manufacturing time will the cellular manufacturing structure save for a batch of 12 bottles?

OBJECTIVE ≫ 1   Exercise 17-7   **BOTTLENECK OPERATION, IMPROVING PRODUCTION FLOW**

Vitacom Inc. implemented cellular manufacturing as recommended by a consultant. The production flow improved dramatically. However, the company was still faced with the competitive need to improve its cycle time so that it could produce one bottle every four minutes (15 bottles per hour). The cell structure is shown below; the times above the process represent the time required to process one unit.

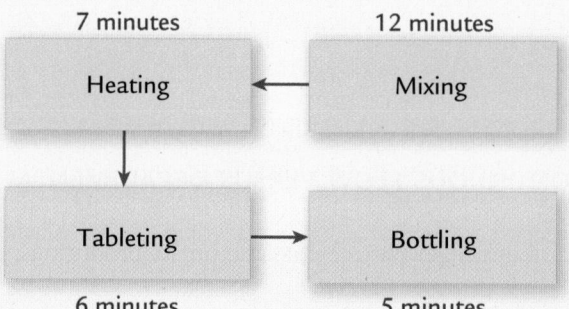

Required:

1. How many bottles can the cell produce per hour (on a continuous running basis)?
2. How long does it take to produce one bottle, assuming the cell is producing on a continuous basis?
3. What must happen so that the cell can produce one bottle every four minutes or 15 per hour, assuming the cell produces on a continuous basis?

OBJECTIVE ≫ 2   Exercise 17-8   **VALUE-STREAM COSTING**

Farber Inc. has just created five order fulfillment value streams, two focused and three that produce multiple products. The size of the plant in which the value streams are located is 100,000 square metres. The facility costs total $1,000,000 per year. One of the focused value streams produces a basic MP3 product. The MP3 value stream occupies 20,000 square metres. Not counting facility costs, the MP3 value-stream costs total $1,800,000. There are 25,000 MP3 units produced annually. There were not sufficient quality personnel for each value stream; thus, the MP3 stream had to share a quality engineer who spends 40 percent of his time with the MP3 value stream and the other

60 percent with two other value streams. While 40 percent of the time is not sufficient time for the value streams, the contribution will be workable until other arrangements can be made. His salary is $75,000 per year. Vivian Olsen, an industrial engineer, is one of two employees assigned completely to the value stream from production planning. Vivian has not been with the company as long as the other production engineer. Because of the demand-pull nature of the new value stream, only one production planner is needed.

Required:

1. Explain how the value-stream costs of $1,800,000 were most likely assigned to the MP3 value stream. Explain how facility costs will be treated and why.
2. How many employees are likely to be located within the MP3 value stream?
3. Given that only one production planner is needed, what should the company do with its extra engineer (Vivian Olsen)?
4. Calculate the unit product cost for the MP3 value stream. Comment on the accuracy of this cost and its value for monitoring value-stream performance.

Exercise 17-9 VALUE-STREAM AVERAGE COSTING, ABC COSTS AS BENCHMARKS

OBJECTIVE ▶ 2

A value stream has three activities and two products. The units produced and shipped per week are 100 of the deluxe model (Model A) and 300 of the basic model (Model B). The resource consumption patterns are shown as follows:

	Model A	Model B	Costs of Value-Stream Activities
Cell manufacturing	1,200 minutes	3,600 minutes	$38,400
Engineering	30 hours	130 hours	13,600
Testing	50 hours	110 hours	12,000
Total			$64,000

Required:

1. Calculate the ABC product cost for Models A and B.
2. Calculate the value-stream average product cost. Assuming reasonable stability in the consumption patterns of the products and product mix, assess how well the products are grouped based on similarity.

Exercise 17-10 VALUE-STREAM REPORTING WITH INVENTORY DECREASE

OBJECTIVE ▶ 2

Shorts Manufacturing Inc. has implemented lean manufacturing in its Quebec City plant as a pilot program. One of its value streams produces a family of small electric tools. The value-stream team managers were quite excited about the results, as some of their efforts to eliminate waste were proving to be effective. During the most recent three weeks, the following data pertaining to the electric tool value stream were collected:

Week 1:

Demand = 90 units @ $40

Beginning inventory = 10 units @ $20 ($5 materials and $15 conversion)

Production = 90 units using $450 of material and $1,350 of conversion cost

Week 2:

Demand = 100 units @ $40

Beginning inventory = 10 units @ $20 ($5 materials and $15 conversion)

Production = 90 units using $450 of material and $1,350 of conversion cost

Week 3:

Demand = 90 units @ $40

Beginning inventory = 0

Production = 100 units using $500 of material and $1,500 of conversion cost

Required:

1. Prepare a traditional income statement for each week.
2. Calculate the average value-stream product cost for each week. What does this cost signal, if anything?
3. Prepare a value-stream income statement for each week. Assume that any increase in inventory is valued at average cost. Comment on the financial performance of the value stream and its relationship to traditional income measurement.

OBJECTIVE ▶ 2   **Exercise 17-11 BOX SCORECARD**

The following Box Scorecard was prepared for a value stream:

	Last Week	This Week (6/30/13)	Planned Future State (12/31/13)
Operational			
Units sold per person	100	108	115
On-time delivery	85%	90%	95%
Dock-to-dock days	12	11	9
First-time through	60%	62%	70%
Average product cost	$75	$74	$70
Capacity			
Productive	25%	26%	27%
Nonproductive	65%	62%	40%
Available	10%	12%	33%
Financial			
Weekly sales	$800,000	$825,000	$1,000,000
Weekly material cost	$320,000	$330,000	$380,000
Weekly conversion cost	$280,000	$280,240	$320,000
Weekly value-stream profit	$200,000	$214,760	$300,000
ROS	25%	26%	30%

Required:

1. How many nonfinancial measures are used to evaluate performance? Why are nonfinancial measures used?
2. Classify the operational measures as time-based, quality-based, or efficiency-based. Discuss the significance of each category for lean manufacturing.
3. What is the role of the Planned Future State column?
4. Discuss the capacity category and explain the meaning of each measure and its significance.
5. Discuss the relationship between the financial measures and the measures in the operational and capacity categories.

OBJECTIVE ▶ 3   **Exercise 17-12 TECHNICAL AND PRICE EFFICIENCY**

Listed below are several possible input combinations for producing 7,500 units of a pocket PC. Two of the input combinations are technically efficient.

	Materials	Labour	Energy
Unit input prices	$100	$ 60	$ 25
Input combinations:			
A	100	192	720
B	110	180	540
C	150	200	600
D	92	190	570

Required:

1. Identify the technically efficient input combinations. Explain your choices.
2. Which of the two technically efficient input combinations should be used? Explain.

Exercise 17-13 PRODUCTIVITY MEASUREMENT, TECHNICAL AND ALLOCATIVE EFFICIENCY, PARTIAL MEASURES

OBJECTIVE ➤ 3

Corelli Company produces hand-crafted pottery that uses two inputs: materials and labour. During the past quarter, 24,000 units were produced, requiring 96,000 kilograms of materials and 48,000 hours of labour. An engineering efficiency study commissioned by the local university revealed that Carsen can produce the same 24,000 units of output using either of the following two combinations of inputs:

	Materials	Labour
Combinations:		
F1	72,000	36,000
F2	79,200	33,600

The cost of materials is $8 per kilogram; the cost of labour is $12 per hour.

Required:

1. Compute the output-input ratio for each input of Combination F1. Does this represent a productivity improvement over the current use of inputs? What is the total dollar value of the improvement? Classify this as a technical or an allocative efficiency improvement.
2. Compute the output-input ratio for each input of Combination F2. Does this represent a productivity improvement over the current use of inputs? Now, compare these ratios to those of Combination F1. What has happened?
3. Compute the cost of producing 24,000 units of output using Combination F1. Compare this cost to the cost using Combination F2. Does moving from Combination F1 to Combination F2 represent a productivity improvement? Explain.

Exercise 17-14 INTERPERIOD MEASUREMENT OF PRODUCTIVITY PROFILES

OBJECTIVE ➤ 4

Helena Company needs to increase its profits and so has embarked on a program to increase its overall productivity. After one year of operation, Kent Olson, manager of the Halifax plant, reported the following results for the base period and its most recent year of operations:

	2012	2013
Output	307,200	360,000
Power (quantity used)	38,400	18,000
Materials (quantity used)	76,800	81,000

Required:

Compute the productivity profiles for each year. Did productivity improve? Explain.

Exercise 17-15 INTERPERIOD MEASUREMENT OF PRODUCTIVITY, PROFIT-LINKED MEASUREMENT

OBJECTIVE ➤ 4

Refer to **Exercise 17-14**. Suppose the following input prices are provided for each year:

	2012	2013
Unit price (power)	$ 2	$ 3
Unit price (materials)	16	15
Unit selling price	6	8

Required:

1. Compute the profit-linked productivity measure. By how much did profits increase due to productivity?
2. Calculate the price-recovery component for 2013. Explain its meaning.

OBJECTIVE ▸ 3 4

Exercise 17-16 BASICS OF PRODUCTIVITY MEASUREMENT

Holbrook Company gathered the following data for the past two years:

	Base Year	Current Year
Output	900,000	1,080,000
Output prices	$15	$15
Input quantities:		
Materials (kg)	1,200,000	1,080,000
Labour (hrs)	300,000	540,000
Input prices:		
Materials	$5	$6
Labour	$8	$8

Required:

1. Prepare a productivity profile for each year.
2. Prepare partial income statements for each year. Calculate the total change in income.
3. Calculate the change in profits attributable to productivity changes.
4. Calculate the price-recovery component. Explain its meaning.

Problems

OBJECTIVE ▸ 3 4

Problem 17-17 FOCUSED VALUE STREAMS, PRODUCT COSTING

Sixty employees (all CAs) of a local public accounting firm eat lunch at least twice weekly at a very popular pizza restaurant. The pizza restaurant recently began offering discounts for groups of 15 or more. Groups would be seated in a separate room, served individual bowls of salad costing $2 each, pitchers of root beer costing $3 each (each pitcher has a five-glass capacity), and medium, two-topping pizzas for $10 (10 slices each). The food would have to be ordered in advance.

Thirty of the CAs commit to eating three slices of pizza, three glasses of root beer, and one bowl of salad [a consumption pattern of (3,3,1)]. The other 30 are more hearty eaters and commit to seven slices of pizza, two glasses of root beer, and one bowl of salad [a consumption pattern of (7,2,1)]. Each member of the group must pay an assessed amount for the lunch.

Required:

1. Determine the total number of pizzas, pitchers of root beer, and salads that must be ordered for the 60 employees.
2. One of the CAs offered to determine the amount that each should pay. He suggested that the easiest way is to assign the average cost to each person eating in the group. Based on this suggestion, how much would each CA pay for lunch?
3. One CA objected to using average cost, noting that half of the CAs are much lighter eaters than the other half. Based on the large differences in consumption behaviours, he suggested forming two groups: one for the light eaters and one for the heavier eaters. Calculate the lunch cost for each CA for each group. Discuss the analogy to formation of focused value streams in a manufacturing environment. Calculate the cost that would be assigned using ABC. What does this tell you?

OBJECTIVE ▸ 3 4

Problem 17-18 MULTIPLE-PRODUCT VALUE STREAMS, PRODUCT COSTING, CREATING AVAILABLE CAPACITY

Refer to **Problem 17-17**.

After some detailed polling among the 60, four types of eaters were identified: two types of light eaters and two types of heavy eaters. The consumption patterns for each group are given (slices of pizza, glasses of root beer, and bowls of salad): Light Eaters (Group A): A1 = (2,2,1) and A2 = (3,3,1); Heavy Eaters (Group B): B1 = (6,3,1) and B2 = (7,2,1). There are an equal number of CAs in each of the four groups.

Required:

1. Calculate the average lunch cost for each CA in each of the two groups, A and B. Compare this to the ABC cost assignments. Discuss the merits of grouping based on similarity. Discuss the analogy to multiple-product value streams.

2. Suppose that members of the heavy-eating group (Group B) decided that they were eating more than necessary for their health and well-being and decided to reduce their total calories. They therefore agreed to reduce consumption of pizza by one slice and consumption of root beer by one glass for each member of the group. Relative to the original order, how much extra capacity exists? If the excess capacity is eliminated by reducing the order, what is the new average cost? Suppose that the decision is to use the extra capacity to invite four guests (two of Type B1 and two of Type B2) to lunch (at the cost of the CAs). If the original order is used as the benchmark cost, what is the extra cost of the guest program? Comment on the conceptual significance of this for manufacturing firms.

Problem 17-19 BOX SCORECARD, SPECIAL ORDER DECISION

OBJECTIVE ➤ 2

Jingbao Company, a manufacturer of small tools, implemented lean manufacturing at the end of 2012. The company's goal for the year was to increase the ROS to 40 percent of sales. A value-stream team was established and began to work on lean improvements. During the year, the team was able to achieve significant results on several fronts. The Box Scorecard below reflects the performance measures at the beginning of the year, midyear, and end of year. Although the team members were pleased with their progress, they were disappointed in the financial results. They were still far from the targeted ROS of 40 percent. They were also puzzled as to why the improvements made did not translate into significantly improved financial performance.

	January 1, 2013	June 30, 2013	December 31, 2013
Operational			
Revenue per person	$15,000	$15,000	$15,000
On-time delivery	70%	90%	95%
Dock-to-dock days	15	6	5
First-time through	60%	60%	90%
Average product cost	$60	$60	$59
Capacity			
Productive	40%	40%	40%
Nonproductive	50%	30%	10%
Available	10%	30%	50%
Financial			
Weekly sales	$800,000	$800,000	$800,000
Weekly material cost	$260,000	$260,000	$240,000
Weekly conversion cost	$300,000	$300,000	$300,000
Weekly value-stream profit	$240,000	$240,000	$260,000
ROS	30%	30%	32.5%

Required:

1. From the scorecard, what was the focus of the value-stream team for the first six months? The second six months? What are the implications of these changes?

2. Using information from the scorecard, offer an explanation for why the financial results were not as good as expected.

3. Suppose that on December 31, 2013, a potential customer offered to purchase an order of goods that would increase weekly revenues in January by $100,000 and material cost by $30,000. Using the old standard cost system, the projected conversion cost of the order would be $60,000. Would you recommend that the order be accepted or rejected? Explain.

OBJECTIVE ▶ 1 2   Problem 17-20 **LEAN VERSUS STANDARD-COSTING-BASED MEASURES**

Continuous improvement is the governing principle of a lean accounting system. Following are several performance measures. Some of these measures would be associated with a traditional standard-costing accounting system, and some would be associated with a lean accounting system.

a. Materials price variances
b. Cycle time
c. Comparison of actual product costs with target costs
d. Materials quantity or efficiency variances
e. Comparison of actual product costs over time (trend reports)
f. Comparison of actual overhead costs, item by item, with the corresponding budgeted costs
g. Comparison of product costs with competitors' product costs
h. Percentage of on-time deliveries
i. First-time through
j. Reports of value- and non-value-added costs
k. Labour efficiency variances
l. Days of inventory
m. Downtime
n. Manufacturing cycle efficiency (*MCE*)
o. Unused (available) capacity variance
p. Labour rate variance
q. Using a sister plant's best practices as a performance standard

Required:

1. Classify each measure as lean or traditional (standard costing). If traditional, discuss the measure's limitations for a lean environment. If it is a lean measure, describe how the measure supports the objectives of lean manufacturing.
2. Classify the measures into operational (nonfinancial) and financial categories. Explain why operational measures are better for control at the shop level (production floor) than financial measures. Should any financial measures be used at the operational level?
3. Suggest some additional measures that you would like to see added to the list that would be supportive of lean objectives.

OBJECTIVE ▶ 3 4   Problem 17-21 **PRODUCTIVITY AND QUALITY, PROSPECTIVE ANALYSIS**

Walnut Company is considering the acquisition of a computerized manufacturing system. The new system has a built-in quality function that increases the control over product specifications. An alarm sounds whenever the product falls outside the programmed specifications. An operator can then make some adjustments on the spot to restore the desired product quality. The system is expected to decrease the number of units scrapped because of poor quality. The system is also expected to decrease the amount of labour inputs needed. The production manager is pushing for the acquisition because he believes that productivity will be greatly enhanced—particularly when it comes to labour and material inputs. Output and input data follow. The data for the computerized system are projections.

	Current System	Computerized System
Output (units)	20,000	20,000
Output selling price	$40	$40
Input quantities:		
Materials	80,000	70,000
Labour	40,000	30,000
Capital (dollars)	$40,000	$200,000
Energy	20,000	50,000
Input prices:		
Materials	$4.00	$4.00
Labour	$9.00	$9.00
Capital (percent)	10.00%	10.00%
Energy	$2.00	$2.50

Required:

1. Compute the partial operational ratios for materials and labour under each alternative. Is the production manager right in thinking that materials and labour productivity increase with the automated system?
2. Compute the productivity profiles for each system. Does the computerized system improve productivity?
3. Determine the amount by which profits will change if the computerized system is adopted. Are the trade-offs among the inputs favourable? Comment on the system's ability to improve productivity.

Problem 17-22 PRODUCTIVITY MEASUREMENT, BASICS

OBJECTIVE ➤ 4

Fowler Company produces handcrafted leather purses. Virtually all of the manufacturing cost consists of materials and labour. Over the past several years, profits have been declining because the cost of the two major inputs has been increasing. Wilma Fowler, the president of the company, has indicated that the price of the purses cannot be increased; thus, the only way to improve or at least stabilize profits is to increase overall productivity. At the beginning of 2013, Wilma implemented a new cutting and assembly process that promised less materials waste and a faster production time. At the end of 2013, Wilma wants to know how much profits have changed from the prior year because of the new process. In order to provide this information to Wilma, the controller of the company gathered the following data:

	2012	2013
Unit selling price	$16	$16
Purses produced and sold	18,000	24,000
Materials used	36,000	40,000
Labour used	9,000	10,000
Unit price of materials	$4	$4.50
Unit price of labour	$9	$10

Required:

1. Compute the productivity profile for each year. Comment on the effectiveness of the new production process.
2. Compute the increase in profits attributable to increased productivity.
3. Calculate the price-recovery component, and comment on its meaning.

Problem 17-23 PRODUCTIVITY MEASUREMENT, TECHNICAL AND PRICE EFFICIENCY

OBJECTIVE ➤ 3 4

In 2012, Farouk Chemicals used the following input combination to produce 55,000 litres of an industrial solvent:

Materials	33,000 kg
Labour	66,000 hrs

In 2013, Farouk again planned to produce 55,000 litres of solvent and was considering two different changes in process, both of which would be able to produce the desired output. The following input combinations are associated with each process change:

	Change I	Change II
Materials	38,500 kg	27,500 kg
Labour	44,000 hrs.	55,000 hrs.

The following combination is optimal for an output of 55,000 units. However, this optimal input combination is unknown to Farouk:

Materials	22,000 kg
Labour	44,000 hrs

The cost of materials is $60 per kilogram, and the cost of labour is $15 per hour. These input prices hold for 2012 and 2013.

Required:

1. Compute the productivity profiles for each of the following:

    a. The actual inputs used in 2012
    b. The inputs for each proposed 2013 process change
    c. The optimal input combination. Will productivity increase in 2013, regardless of which change is used? Which process change would you recommend based on the prospective productivity profiles?

2. Compute the cost of 2012's productive inefficiency relative to the optimal input combination. Repeat for 2013 proposed input changes. Will productivity improve from 2012 to 2013 for each process change? If so, by how much? Explain. Include in your explanation a discussion of changes in technical and allocative efficiency.

3. Since the optimal input combination is not known by Farouk, suggest a way to measure productivity improvement. Use this method to measure the productivity improvement achieved from 2012 to 2013. How does this measure compare with the productivity improvement measure computed using the optimal input combination?

CMA Problem

CMA Problem 17-1 LEAN ACCOUNTING AND PRODUCTIVITY*

Kastoria Inc. produces handcrafted leather purses. Virtually all manufacturing costs consist of materials and labour. Both material and labour costs have been increasing the past few years, and as a result the profits follow suit. Angie Diamantatos, the company's president, is aware that the increased costs cannot be passed on to the consumer in higher prices, as cheaper imports from the Far East sell at very competitive prices. She is instead focusing on productivity measures to improve profitability. To stimulate employees in 2013, she created a bonus pool over and above wages equal to 10 percent of productivity gains. She now wants to assess how the bonus pool has worked to improve productivity. The following data has been gathered:

	2012	2013
Selling price per unit	$32	$32
Unit price of materials	$8	$9
Unit price of labour	$9	$10
Sales in units	100,000	120,000
Kilograms of material used	200,000	200,000
Hours of labour used	50,000	50,000

Required:

1. What are the partial productivity ratios for each year?
2. Comment on the effectiveness of the productivity improvement program.
3. Compute the increase in profits due to improved productivity after the bonus is paid.
4. Calculate the price recovery component ignoring the bonus. What do you think? (Adapted from CMA Ontario)

> The Collaborative Learning Exercises can be found on the product support site at www.hansen1ce.nelson.com.

Integrative Exercise 4

Part 4
Chapters 15–17

Zando Pharmaceuticals is an affiliate of the German-based Heisenberg Corporation, which employs 40,000 worldwide. Zando's Vancouver facility houses the Canadian corporate headquarters and Research and Development. It produces 30 products, using 28 different batch processes. The facility has 2,000 employees on-site. In recent years, Zando's profitability has suffered, which can be attributed to increased competition, customer dissatisfaction, and regulatory pressures. Tony Brown, president of Zando, called a meeting to consider ways to improve profitability. He labelled the meeting a strategic planning session and invited the following officers: Kathy Shorts, environmental manager, Troy Lewis, head of R&D, Johnny Mizukawa, vice president of production and quality, Larry Sower, vice president of finance, and Doreen Dineen, marketing vice president.

Tony: You all have received the quarterly financial reports for the past two years. The trends are negative. We are losing market share, profits are decreasing, and our costs seem to be increasing. We need to take actions to increase sales and reduce costs, and we need to do so as quickly as possible. Given our research strengths, it seems to me that our best bet is to grow revenues by introducing new products with proprietary rights. As far as costs are concerned, we need to improve our performance on that dimension as well. Lower per-unit costs for new and existing products are needed. Any suggestions?

Troy: For our products, our ability to control costs resides in development—my area—rather than manufacturing. We probably need to pay more attention to product and process design issues to ensure a reasonably level per-unit cost. Revenues are also affected in this stage. Once we patent a drug, the clock begins to tick, and we need to reduce time to market. Significantly reducing time to market will allow us to generate revenues for a longer period of time than we are currently experiencing. It would also be helpful if we could reduce the cycle time for product development. Both actions would increase revenues. Finally, we can increase revenues by increasing the volume of new products.

Johnny: There is a lot of merit to the observation that cost reduction opportunities reside mostly in product development. Once a drug is approved, its approval includes the manufacturing process. Any future changes in the manufacturing process require approval from Health Canada. Because of this, we have been reluctant, historically, to engage in process improvement or re-engineering. However, I wonder if we shouldn't reconsider this longstanding policy. Some of the quality problems we have could be corrected by changing some of our existing processes, and the costs saved may easily exceed any cost incurred from seeking Health Canada approval. I think our quality costs are at least 15 percent of sales. That's a lot of opportunity for improvement.

Kathy: I agree that cost reduction—both in the product development stage and the manufacturing stage—should be a key strategic theme. The environmental area also offers some very good opportunities. A recent pollution prevention act passed by the legislature requires that we calculate the costs of generating hazardous substances for each process. This act was the incentive we needed to begin developing an environmental cost management system. The results so far indicate that environmental costs are much more than we realized. They are estimated to be in the range of 20 to 30 percent of total operating costs. Environmental costs can be reduced by such things as computerizing chemical inventory, eliminating the use of chlorinated solvents and other hazardous materials, reducing our use of virgin feedstocks, and redesigning processes and products so that we can reduce toxic residue release. We can really have a positive environmental impact while simultaneously reducing costs if more attention is paid to environmental issues during product development.

Doreen: I like what I am hearing because I think that it also affects our ability to increase market share and revenues. For example, environmental impact is one of our major concerns. Some retail pharmacy chains pay particular attention to green products, and right now we are not competing well. Our environmental image is negative and needs to be improved. I am convinced that doing so will allow us to increase market share. Quality is another important matter. We have had to recall two batches of products during the past two years due to poor quality, and this has hurt our image more than the environmental issue. Improving the processes to avoid these kinds of problems will save us a lot of grief. Product image and reputation are essential to increasing customer satisfaction and market share.

Tony: We started with the need to improve financial performance by increasing revenues and reducing costs. So far, we have some very good suggestions to help achieve these two objectives, but I have some concerns. First, do we have the talent and capabilities to improve quality and environmental performance? Troy, do your professionals really understand what they need to do to improve process and product designs so that we can see the desired quality and environmental improvements? Also, how can we reduce the cycle time for products and the time to market once patented?

Troy: Let me answer those question in order. First, we probably are lacking the understanding on the design issues. We will need to do some training to help our research scientists and chemical engineers understand the consequences. We may need to hire a couple of professionals who have experience in dealing with these issues. Second, we may need to make cycle time and time to market significant performance measures and reward our people for actions that reduce those measures. Our employees need to align their interests with those of the company. If we can achieve this, we should see more revenue produced per employee.

Tony: Good. Now, Johnny, tell us about production and quality. Do our manufacturing engineers and production workers need help with environmental and quality issues?

Johnny: Without question, training will be needed. Moreover, I really need to hire a couple of quality engineers.

Kathy: I also think that we need an environmental engineer with experience in pharmaceutical manufacturing processes.

Tony: Good. We certainly shouldn't ignore the necessary infrastructure to bring about the needed changes. Larry, you have been quiet, what do you think about all this? Do you have any suggestions?

Larry: Infrastructure is important. If this is all going to work, timely and accurate information will be needed. It is hard to design products and processes with cost being a significant issue without providing the right kind of cost information. We are in the process of revamping the cost management information system so that it is activity based and so that we can provide quality and environmental cost information. After listening to the comments made here, I might also suggest that we need a strategic measurement system that can be used to align the interests of our employees with our improvement strategy. People need to know what is important, that the important factors are being measured, and that they are going to be evaluated and rewarded based on these factors. Finally, I would encourage the use of target costing to help manage costs during product development. To help you all understand the importance of good information, I have assembled some activity data relating to two new products currently under development. These two products will use the same process, using different setups. The data are organized into resource, activity, and cost object modules with an accompanying list of activity drivers to facilitate the use of an ABC software package we recently acquired.

Resource Module (Projected Costs of Manufacturing Process Associated with the Two Products)

Materials	$2,000,000
Salaries and wages	1,000,000
Energy	500,000
Licence fee (environmental)	200,000
Environmental fines	400,000
Depreciation, pollution control equipment	100,000
	$4,200,000

Activity Module

	Resource Driver (Percentage Usage)				
	Materials	**Labour**	**Energy**	**Fees**	**Fines**
Supervising process*	0%	10%	0%	0%	0%
Setting up	3	20	14	0	0
Blending chemicals	80	40	30	0	0
Producing waste	10	8	10	0	0
Disposing of hazardous waste	6	12	15	40	70
Inspecting products	0	7	6	0	0
Releasing air contaminants	0	0	0	60	30
Operating pollution control equipment	1	3	25	0	0
	100%	100%	100%	100%	100%

* Secondary activity whose costs are assigned to primary activities in proportion to the labour time used.

Cost Object Module (Products and Projected Activity Usage)

Cost Objects	Antibiotic XK1	Antibiotic XK5
Expected output (kilograms)	50,000	50,000
Setup hours	12,000	7,000
Direct labour hours (blending)	24,000	16,000
Kilograms of waste	8,000	2,000
Kilograms of hazardous waste	5,000	1,000
Hours of inspection	3,000	500
Tonnes of air contaminants	4.5	0.5
Machine hours (pollution control)	2,000	500

List of Activity Drivers

Activity Drivers	Activity Capacity
Setup hours	20,000
Direct labour hours (blending)	40,000*
Kilograms of waste	10,000*
Kilograms of hazardous waste	8,000
Hours of inspection	4,000
Tonnes of air contaminants	5*
Machine hours (pollution control)	3,000

* Capacity is flexible (i.e., acquired as needed, and always matches usage). Capacity for other activities is acquired in advance of usage. For example, setups are acquired in units (steps) of 950 hours. Projected usage for setups equals practical capacity.

Required:

1. Use the comments from the executive meeting to identify strategic objectives and possible performance measures for each of five perspectives: financial, customer, environmental, process, and learning and growth. Would you recommend the Balanced Scorecard for Zando? Why or why not?

2. Suppose that Doreen suggested gainsharing in response to Troy's suggestion to reward product development employees for improving cycle time and time to market. What is gainsharing? How could it be used in the product development setting?

3. Determine the cost of all activities for the proposed new process. Now, assign the cost of the secondary activity to the primary activities.

4. Classify the primary activities into three categories: environmental, quality, and other (neither quality nor environmental). Did some activities end up in more than one category? Explain.

5. Calculate the cost per unit for each of the proposed products using primary activity rates. Now, calculate the *environmental* cost per unit and the *quality* cost per unit. What does this tell you about the relative desirability of the two products?

6. Following Larry's suggestion, Tony decided to use target costing to help improve new product profitability. Based on analyses by Tony and Doreen, the target prices for XK1 and XK5 are $50 per kilogram and $35 per kilogram, respectively. Tony has indicated that any new product should earn a gross profit equal to 20 percent of sales. Based on this information, answer the following:
   a. What is the target cost for each product? Given this information, what should be done?
   b. Suppose Doreen indicates that sales for each product can be increased by 50 percent if the selling price is lowered by 10 percent. Assuming the same target profit (Tony wants the original target profit per kilogram maintained), calculate the new target costs. If all non-value-added costs were eliminated, could the target be met? (Calculate the unit cost at the 50,000-unit level.) Now, calculate the effect on total profits under a scenario where non-value-added costs are not eliminated versus a scenario where all non-value-added costs are eliminated. (Include in this analysis any possible increase in sales volume.)

Glossary

Page references beginning with "18-" refer to Chapter 18, located at **www.hansen1ce.nelson.com.**

A

abnormal spoilage spoilage that exceeds the amount expected under normal efficient operating conditions. (p. 174)

absorption costing a costing method that assigns all manufacturing costs, including direct materials, direct labour, variable overhead, and a share of fixed overhead, to each unit of product. (p. 600)

absorption-costing income income computed by following a functional classification. (p. 16)

acceptable quality level (AQL) a predetermined level of defective products that a company permits to be sold. (p. 667)

account analysis method a method used to estimate costs by classifying accounts in the general ledger as fixed, variable, or mixed. (p. 66)

accounting information system a system consisting of interrelated manual and computer parts that uses processes such as collecting, recording, summarizing, analyzing (using decision models), and managing data to provide output information to users. (p. 2)

activity a basic unit of work performed within an organization. It also can be defined as an aggregation of actions within an organization useful to managers for purposes of planning, controlling, and decision making. (p. 9)

activity analysis the process of identifying, describing, and evaluating the activities an organization performs. (p. 704)

activity attributes financial and nonfinancial information items that provide descriptive labels for individual activities. (p. 273)

activity capacity the ability to perform activities or the number of times an activity can be performed. (p. 53, 717)

activity dictionary lists the activities in an organization along with desired attributes. (p. 273)

activity drivers measure the demands that cost objects place on activities. (p. 276)

activity elimination the process of eliminating non-value-added activities. (p. 707)

activity flexible budgeting the prediction of what activity costs will be as activity output changes. (p. 715)

activity inventory a listing of the activities performed within an organization. (p. 272)

activity output measure assesses the number of times the activity is performed. It is the quantifiable measure of the output. (p. 704)

activity rate the average unit cost, obtained by dividing the resource expenditure by the activity's practical capacity. (p. 56)

activity reduction decreasing the time and resources required by an activity. (p. 707)

activity resource usage model a model that classifies resources according to their nature, which allows the assessment of changes in resource supply (and thus resource spending) as activity demand for the resource changes. (p. 553)

activity selection the process of choosing among sets of activities caused by competing strategies. (p. 707)

activity sharing increasing the efficiency of necessary activities by using economies of scale. (p. 707)

activity volume variance the cost difference of the actual activity capacity acquired and the capacity that should be used. (p. 717)

activity-based costing (ABC) system a cost accounting system that uses both unit- and non-unit-based cost drivers to assign costs to cost objects by first tracing costs to activities and then tracing costs from activities to products. (p. 272)

activity-based management (ABM) an advanced control system that focuses management's attention on activities with the objective of improving the value received by the customer and the profit received by providing this value. It includes driver analysis, activity analysis, and performance evaluation and draws on activity-based costing as a major source of information. (p. 703)

activity-based responsibility accounting assigns responsibility to processes and uses both financial and nonfinancial measures of performance. (p. 723)

actual cost system a cost measurement system in which actual manufacturing costs are assigned to products. (p. 157)

adjusted cost of goods sold normal cost of goods sold adjusted to include overhead variance. (p. 171)

administrative costs all costs associated with the general administration of the organization that cannot be reasonably assigned to either marketing or production. (p. 15)

administrative expense budget a budget consisting of estimated expenditures for the overall organization and operation of the company. (p. 384)

allocation assignment of indirect costs to cost objects. (p. 11)

allocative efficiency the point at which given the mixes that satisfy the condition of technical efficiency, the least costly mix is chosen. (p. 852)

applied overhead the overhead assigned to production using a predetermined overhead rate. (p. 259)

appraisal costs costs incurred to determine whether or not products and services are conforming to requirements. (p. 783)

assets unexpired costs. (p. 9)

B

backflush costing a simplified approach for cost flow accounting that uses trigger points to determine when manufacturing costs are assigned to key inventory and temporary accounts. (p. 670)

Balanced Scorecard a strategic-based performance management system that typically identifies objectives and measures for four different perspectives: the financial perspective, the customer perspective, the process perspective, and the learning and growth perspective. (p. 751)

base period a prior period used to set the benchmark for measuring productivity changes. (p. 854)

batch production processes a process that produces batches of different products that are identical in many ways but differ in others. (p. 231)

batch-level activities activities performed each time a batch is produced. (p. 279)

benchmarking uses best practices as the standard for evaluating activity performance. (p. 713)

bill of activities specifies the product, product quantity, activity, and amount of each activity expected to be consumed by each product. (p. 278)

binding constraint constraints whose limited resources are fully used by a product mix. (p. 18-15)

break-even point the point where total sales revenue equals total costs (i.e., the point of zero profits). (p. 101)

budget a plan of action expressed in financial terms. (p. 371)

budget committee a committee responsible for setting budgetary policies and goals, reviewing and approving the budget, and resolving any differences that may arise in the budgetary process. (p. 373)

budget director the individual responsible for coordinating and directing the overall budgeting process. (p. 373)

budgetary slack the process of padding the budget by overestimating costs and underestimating revenues. (p. 409)

by-product a secondary product recovered in the course of manufacturing a primary product during a joint process. (p. 346)

C

capital expenditures budget a financial plan outlining the acquisition of long-term assets. (p. 390)

carrying costs the costs of holding inventory. (p. 18-2)

cash budget a detailed plan that outlines all sources and uses of cash. (p. 390)

causal factors activities or variables that invoke service costs. Generally, it is desirable to use causal factors as the basis for allocating service costs. (p. 317)

centralized decision making a system in which decisions are made at the top level of an organization and local managers are given the task of implementing them. (p. 491)

Certified General Accountant (CGA) a certified accountant who is permitted (by law) to serve as an external auditor. CGAs

must pass a national examination and be licensed by the province in which they practice. (p. 24)

Certified Management Accountant (CMA) an accountant who has satisfied the requirements to hold a certificate in management accounting. (p. 22)

committed fixed expenses costs incurred for the acquisition of long-term activity capacity, usually as the result of strategic planning. (p. 53)

committed resources acquired as used and needed, these are a strictly variable cost. The quantity supplied equals quantity demanded, so there is no excess capacity. (p. 53)

common cost the cost of a resource used in the output of two or more services or products. (p. 314)

common fixed expenses fixed costs that are not traceable to the segments and that would remain even if one of the segments were eliminated. (p. 111)

comparable uncontrolled price method the transfer price essentially equal to the market price. (p. 517)

competitive advantage creating better customer value for the same or lower cost than can competitors or equivalent value for lower cost than can competitors. (p. 644)

confidence interval prediction interval that provides a range of values for the actual cost with a prespecified degree of confidence. (p. 65)

constant gross margin percentage method a joint cost allocation method that maintains the same gross margin percentage for each product. (p. 346)

constrained optimization choosing the optimal mix given the constraints faced by the firm. (p. 18-15)

constraint set the collection of all constraints that pertain to a particular optimization problem. (p. 18-16)

constraints a mathematical expression that expresses a resource limitation. (p. 18-15)

consumable life the length of time that a product serves the needs of a customer. (p. 657)

consumption ratio the proportion of an overhead activity consumed by a product. (p. 265)

continuous (or rolling) budget a moving 12-month budget with a future month added as the current month expires. (p. 373)

continuous improvement the relentless pursuit of improvement in the delivery of value to customers; searching for ways to increase overall efficiency by reducing waste, improving quality, and reducing costs. (p. 21, 702)

continuous replenishment when a manufacturer assumes the inventory management function for the retailer. (p. 18-9)

contribution margin the difference between revenue and all variable expenses. (p. 104)

contribution margin ratio contribution margin divided by sales revenue. It is the proportion of each sales dollar available to cover fixed costs and provide for profit. (p. 106)

contribution margin variance the difference between actual and budgeted contribution margin. (p. 609)

contribution margin volume variance the difference between the actual quantity sold and the budgeted quantity sold multiplied by the budgeted average unit contribution margin. (p. 610)

control the process of setting standards, receiving feedback on actual performance, and taking corrective action whenever actual performance deviates significantly from planned performance. (p. 371)

control activities activities performed by an organization to prevent or detect poor quality (because poor quality may exist). (p. 783)

control costs costs incurred from performing control activities. (p. 784)

control limits the maximum allowable deviation from a standard. (p. 448)

controllable costs costs that managers have the power to influence. (p. 410)

controller the chief accountant of an organization. (p. 18)

controlling the monitoring of a plan through the use of feedback to ensure that the plan is being implemented as expected. (p. 21)

conversion cost the sum of direct labour cost and overhead cost. (p. 13)

core objectives and measures those objectives and measures common to most organizations. (p. 757)

cost the cash or cash equivalent value sacrificed for goods and services that are expected to bring a current or future benefit to the organization. (p. 9)

cost accounting information system a cost management subsystem designed to assign costs to individual products and services and other objects as specified by management. (p. 9)

cost accumulation the recognition and recording of costs. (p. 156)

cost assignment the process of associating manufacturing costs with the units produced. (p. 156)

cost behaviour the way in which a cost changes in relation to changes in activity usage. (p. 46)

cost centre a responsibility centre in which a manager is responsible for cost. (p. 490)

cost leadership strategy providing the same or better value to customers at a lower cost than offered by competitors. (p. 645)

cost management identifies, collects, measures, classifies, and reports information that is useful to managers in costing (determining what something costs), planning, controlling, and decision making. (p. 2)

cost management information system an accounting information subsystem that is primarily concerned with producing outputs for internal users using inputs and processes needed to satisfy management objectives. (p. 4)

cost measurement the process of assigning dollar values to cost items. (p. 156)

cost object any item such as products, departments, projects, activities, and so on, for which costs are measured and assigned. (p. 9)

cost of goods manufactured the total cost of goods completed during the current period. (p. 16)

cost of goods sold the cost of direct materials, direct labour, and overhead attached to the units sold. (p. 16)

cost reconciliation determining whether the costs assigned to units transferred out and to units in ending work in process are equal to the costs in beginning work in process plus the manufacturing costs incurred in the current period. (p. 212)

cost-plus method a transfer price that is simply a cost-based transfer price. (p. 517)

costs of quality costs incurred because poor quality may exist or because poor quality does exist. (p. 783)

cost-volume-profit graph a graph that depicts the relationships among costs, volume, and profits. It consists of a total revenue line and a total cost line. (p. 117)

cumulative average-time learning curve model the model stating that the cumulative average time per unit decreases by a constant percentage, or learning rate, each time the cumulative quantity of units produced doubles. (p. 71)

currently attainable standard a standard that reflects an efficient operating state; it is rigorous but achievable. (p. 436)

customer perspective a Balanced Scorecard viewpoint that defines the customer and market segments in which the business will compete. (p. 757)

customer value the difference between what a customer receives (customer realization) and what the customer gives up (customer sacrifice). (p. 644, 757)

cycle time (manufacturing) the length of time required to produce one unit of a product. (p. 760)

D

decentralization the granting of decision-making freedom to lower operating levels. (p. 491)

decentralized decision making a system in which decisions are made and implemented by lower-level managers. (p. 491)

decision making the process of choosing among competing alternatives. (p. 22)

decision model a set of procedures that, if followed, will lead to a decision. (p. 550)

decline stage the stage in a product's life cycle when the product loses market acceptance and sales begin to decrease. (p. 657)

defective product a product or service that does not conform to specifications. (p. 783)

degree of operating leverage a measure of the sensitivity of profit changes to changes in sales volume. It measures the percentage change in profits resulting from a percentage change in sales. (p. 124)

dependent variable a variable whose value depends on the value of another variable. For example, Y in the cost formula $Y = F + VX$ depends on the value of X. (p. 57)

deviation the difference between the cost predicted by a cost formula and the actual cost. It measures the distance of a data point from the cost line. (p. 62)

differentiation strategy an approach that strives to increase customer value by increasing what the customer receives. (p. 645)

direct costs costs that can be easily and accurately traced to a cost object. (p. 10)

direct fixed expenses fixed costs that can be traced to each segment and would be avoided if the segment did not exist. (p. 111)

direct labour labour that is traceable to the goods or services being produced. (p. 13)

direct labour budget a budget showing the total direct labour hours needed and the associated cost for the number of units in the production budget. (p. 380)

direct labour efficiency variance (*LEV*) the difference between the actual direct labour hours used and the standard direct labour hours allowed multiplied by the standard hourly wage rate. (p. 447)

direct labour rate variance (*LRV*) the difference between the actual hourly rate paid and the standard hourly rate multiplied by the actual hours worked. (p. 447)

direct materials those materials that are traceable to the good or service being produced. (p. 13)

direct materials price variance (*MPV*) the difference between the actual price paid per unit of materials and the standard price allowed per unit multiplied by the actual quantity of materials purchased. (p. 441)

direct materials purchases budget a budget that outlines the expected usage of materials production and purchases of the direct materials required. (p. 377)

direct materials usage variance (*MUV*) the difference between the direct materials actually used and the direct materials allowed for the actual output multiplied by the standard price. (p. 441)

direct method a method that allocates service costs directly to producing departments. This method ignores any interactions that may exist among service departments. (p. 328)

direct tracing the process of identifying costs that are specifically or physically associated with a cost object. (p. 10)

discretionary fixed expenses costs incurred for the acquisition of short-term capacity or services, usually as the result of yearly planning. (p. 54)

double-loop feedback information about both the effectiveness of strategy implementation and the validity of assumptions underlying the strategy. (p. 764)

driver analysis the effort expended to identify those factors that are the root causes of activity costs. (p. 704)

driver tracing the use of drivers to assign costs to cost objects. (p. 11)

drivers factors that cause changes in resource usage, activity usage, costs, and revenues. (p. 11)

drum-buffer-rope (DBR) system the TOC inventory management system that relies on the drum beat of the major constrained resource, time buffers, and ropes to determine inventory levels. (p. 18-22)

drummer the major binding constraint. (p. 18-21)

dumping predatory pricing on the international market. (p. 594)

duration drivers measure the demands in terms of the time it takes to perform an activity, such as hours of hygienic care and monitoring hours. (p. 276)

dysfunctional behaviour individual behaviour that conflicts with the goals of the organization. (p. 407)

E

ecoefficiency a concept of sustainable development where the aim is to create more goods and services while lowering costs, using fewer resources, and creating less waste and pollution. (p. 801)

economic order quantity (EOQ) the amount that should be ordered (or produced) to minimize the total ordering (or setup) and carrying costs. (p. 18-4)

economic value added (EVA) the after-tax operating profit minus the total annual cost of capital. (p. 501)

effectiveness the manager's performance of the right activities. Measures might focus on value-added versus non-value-added activities. (p. 401, 493)

efficiency the performance of activities. May be measured by the number of units produced per hour or by the cost of those units. (p. 401, 493)

electronic data interchange (EDI) an inventory management method that allows suppliers access to a buyer's online database. (p. 18-10)

ending finished goods inventory budget a budget that describes planned ending inventory of finished goods in units and dollars. (p. 382)

environmental costs costs that are incurred because poor environmental quality exists or may exist. (p. 804)

environmental detection costs costs incurred to detect poor environmental performance. (p. 804)

environmental external failure costs costs incurred after contaminants are introduced into the environment. (p. 804)

environmental internal failure costs costs incurred after contaminants are produced but before they are introduced into the environment. (p. 804)

environmental prevention costs costs incurred to prevent damage to the environment. (p. 804)

equivalent units of output the complete units that could have been produced given the total amount of productive effort expended for the period under consideration. (p. 211)

error cause identification a program in which employees describe problems that prevent them from doing their jobs right the first time. (p. 795)

ethical behaviour choosing actions that are right, proper, and just. (p. 22)

executional activities activities that define the processes of an organization. (p. 647)

expected activity level the level of production activity expected for the coming period. (p. 160)

expected global consumption ratio the proportion of the total activity costs consumed by a given product or cost object. (p. 286)

expenses expired costs. (p. 9)

experience curve relates cost to increased efficiency, such that the more often a task is performed, the lower will be the cost of doing it. (p. 71)

external constraints limiting factors imposed on the firm from external sources. (p. 18-15)

external failure costs costs incurred because products fail to conform to requirements after being sold to outside parties. (p. 783)

external linkages the relationship of a firm's activities within its segment of the value chain with those activities of its suppliers and customers. (p. 647)

external measures measures that relate to customer and shareholder objectives. (p. 752)

F

facility-level activities activities that sustain a factory's general manufacturing processes. (p. 280)

failure activities activities performed by an organization or its customers in response to poor quality. (p. 784)

failure costs the costs incurred by an organization because failure activities are performed. (p. 784)

favourable (F) variance a variance produced whenever the actual amounts are less than the budgeted or standard allowances. (p. 441)

feasible set of solutions the collection of all feasible solutions. (p. 18-16)

feasible solution a product mix that satisfies all constraints. (p. 18-16)

feature costing assigns costs to activities and products or services based on the product's or service's features. (p. 407)

feedback information that can be used to evaluate or correct steps being taken to implement a plan. (p. 21)

FIFO costing method a unit-costing method that excludes prior-period work and costs in computing current-period unit work and costs. (p. 216)

financial accounting information system an accounting information subsystem that is primarily concerned with producing outputs for external users and uses well-specified economic events as inputs and processes that meet certain rules and conventions. (p. 4)

financial budgets that portion of the master budget that includes the cash budget, the budgeted balance sheet, the budgeted statement of cash flows, and the capital budget. (p. 373)

financial measures measures expressed in dollar terms. (p. 708, 752)

financial perspective a Balanced Scorecard viewpoint that describes the financial consequences of actions taken in the other three perspectives. (p. 755)

financial productivity measure a productivity measure in which inputs and outputs are expressed in dollars. (p. 854)

financial-based responsibility accounting system a system that assigns responsibility to organizational units and typically measures performance using only financial metrics. (p. 723)

fixed costs costs that in total are constant within the relevant range as the level of the cost driver varies. (p. 47)

fixed overhead spending variance the difference between actual fixed overhead and applied fixed overhead. (p. 458)

fixed overhead volume variance the difference between budgeted fixed overhead and applied fixed overhead; it is a measure of capacity utilization. (p. 458)

flexible budget a budget that can specify costs for a range of activity. (p. 398)

flexible budget variances the difference between actual costs and expected costs given by a flexible budget. (p. 401)

flexible resources acquired as used and needed, these are a strictly variable cost. The quantity supplied equals quantity demanded, so there is no excess capacity. (p. 53)

focusing strategy selecting or emphasizing a market or customer segment in which to compete. (p. 645)

foreign trade zones areas physically on a country's soil but considered to be outside that country's commerce. Goods imported into a foreign trade zone are duty-free until they leave the zone. (p. 552)

full environmental costing the assignment of all environmental costs, both private and societal, to products. (p. 809)

full private costing the assignment of only private costs to individual products. (p. 809)

full-costing income see **absorption-costing income**. (p. 16)

G

gainsharing an incentive plan used to enhance productivity by linking compensation bonuses directly to a team's performance. (p. 797)

goal congruence the alignment of a manager's personal goals with those of the organization. (p. 407, 493)

goodness of fit the degree of association between Y and X (cost and activity). It is measured by how much of the total variability in Y is explained by X. (p. 65)

growth stage the stage in a product's life cycle when sales increase at an increasing rate. (p. 657)

H

heterogeneity refers to the greater chances for variation in the performance of services than in the production of products. (p. 152)

hidden quality costs opportunity costs resulting from poor quality. (p. 784)

high-low method a method for fitting a line to a set of data points using the high and low points in the data set. For a cost formula, the high and low points represent the high and low activity levels. It is used to break out the fixed and variable components of a mixed cost. (p. 58)

hypothesis test of cost parameters a statistical assessment of a cost formula's reliability that indicates whether the parameters are different from zero. (p. 65)

hypothetical sales value an approximation of the sales value of a joint product at split-off. It is found by subtracting all separable (or further) processing costs from the eventual market value. (p. 344)

I

ideal standards standards that reflect perfect operating conditions. (p. 436)

incentives the positive or negative measures taken by an organization to induce a manager to exert effort toward achieving the organization's goals. (p. 408)

incremental approach the practice of taking the prior year's budget and adjusting it upward or downward to determine next year's budget. (p. 397)

incremental unit-time learning curve model decreases by a constant percentage each time the cumulative quantity of units produced doubles. (p. 73)

independent variable a variable whose value does not depend on the value of another variable. For example, in the cost formula $Y = F + VX$, the variable X is an independent variable. (p. 57)

indirect costs costs that cannot be traced to a cost object. (p. 10)

indirect materials direct materials that form an insignificant part of the final product. (p. 13)

industrial engineering method a forward-looking method of determining through physical observation and analysis, just what activities, in what amounts, are needed to complete a process. (p. 66)

industrial value chain the linked set of value-creating activities from basic raw materials to end-use customers. (p. 646)

innovation process a process that anticipates the emerging and potential needs of customers and creates new products and services to satisfy those needs. (p. 759)

inseparability an attribute of services that means that production and consumption are inseparable. (p. 12, 152)

intangibility refers to the nonphysical nature of services as opposed to products. (p. 12, 152)

intercept parameter the fixed cost, representing the point where the cost formula intercepts the vertical axis. In the cost formula $Y = F + VX$, F is the intercept parameter. (p. 57)

interim quality performance report a comparison of current actual quality costs with short-term budgeted quality targets. (p. 795)

interim quality standards a standard based on short-run quality goals. (p. 794)

internal business process perspective a Balanced Scorecard viewpoint that describes the internal processes needed to provide value for customers and owners. (p. 759)

internal constraints limiting factors found within the firm. (p. 18-15)

internal failure costs costs incurred because products and services fail to conform to requirements where lack of conformity is discovered prior to external sale. (p. 783)

internal linkages relationships among activities within a firm's value chain. (p. 647)

internal measures measures that relate to the processes and capabilities that create value for customers and shareholders. (p. 752)

introduction stage a product life-cycle stage characterized by preproduction and startup activities, where the focus is on obtaining a foothold in the market. (p. 657)

inventory the money an organization spends in turning raw materials into throughput. (p. 18-20)

investment centre a responsibility centre in which a manager is responsible for revenues, costs, and investments. (p. 490)

J

JIT purchasing a system that requires suppliers to deliver parts and materials just in time to be used in production. (p. 665, 844)

job-order cost sheet a document or record used to accumulate manufacturing costs for a job. (p. 161)

job-order costing system a cost accumulation method that accumulates manufacturing costs by job. (p. 161)

joint products two or more products, each having relatively substantial value, that are produced simultaneously by the same process up to a "split-off" point. (p. 337)

just-in-case inventory management a traditional inventory model based on anticipated demand. (p. 18-1)

just-in-time inventory management the continual pursuit of productivity through the elimination of waste. (p. 18-8)

just-in-time (JIT) manufacturing a demand-pull system that strives to produce a product only when it is needed and only in the quantities demanded by customers. (p. 8, 665)

K

kaizen costing efforts to reduce the costs of existing products and processes. (p. 707)

kaizen standard an interim standard that reflects the planned improvement for a coming period. (p. 436, 712)

Kanban system an information system that controls production on a demand-pull basis through the use of cards or markers. (p. 18-11)

keep-or-drop decision a relevant costing analysis that focuses on keeping or dropping a segment of a business. (p. 558)

L

lag measures outcome measures or measures of results from past efforts. (p. 752)

lead measures (performance drivers) factors that drive future performance. (p. 752)

lead time for purchasing, the time to receive an order after it is placed. For manufacturing, the time to produce a product from start to finish. (p. 18-4)

lean manufacturing an approach designed to eliminate waste and maximize customer value; characterized by delivering the right product, in the right quantity, with the right quality (zero-defect), at the exact time the customer needs it and at the lowest possible cost. (p. 839)

learning and growth (infrastructure) perspective a Balanced Scorecard viewpoint that defines the capabilities that an organization needs to create long-term growth and improvement. (p. 763)

learning curve an important type of nonlinear cost curve that shows how the labour hours worked per unit decrease as the volume produced increases. (p. 70)

learning rate expressed as a percent, it gives the percentage of time needed to make the next unit, based on the time it took to make the previous unit. (p. 71)

life-cycle cost assessment assigning costs and benefits to environmental consequences and improvements. (p. 657)

life-cycle cost management actions taken that cause a product to be designed, developed, produced, marketed, distributed, operated, maintained, serviced, and disposed of so that life-cycle profits are maximized. (p. 658)

linear programming a method that searches among possible solutions until it finds the optimal solution. (p. 18-16)

linear programming model expresses a constrained optimization problem as a linear objective function subject to a set of linear constraints. (p. 18-15)

long run period of time for which all costs are variable (i.e., there are no fixed costs). (p. 52)

long-range quality performance report a performance report that compares current actual quality costs with long-range targeted quality costs (usually in the 2%–3% range). (p. 795)

loose constraints constraints whose limited resources are not fully used by a product mix. (p. 18-15)

loss a cost that expires without producing any revenue benefit; a negative profit. (p. 9)

M

make-or-buy decision a decision that focuses on whether a component (service) should be made (provided) internally or purchased externally. (p. 555)

manufacturing cells a plant layout containing machines grouped in families, usually in a semicircle. (p. 666, 842)

margin the ratio of net operating income to sales. (p. 494)

margin of safety the units sold or expected to be sold or sales revenue earned or expected to be earned above the break-even volume. (p. 122)

market share the proportion of industry sales accounted for by a company. (p. 613)

market share variance the difference between the actual market share percentage and the budgeted market share percentage multiplied by actual industry sales in units times budgeted average unit contribution margin. (p. 613)

market size the total revenue for the industry. (p. 613)

market size variance the difference between actual and budgeted industry sales in units multiplied by the budgeted market share percentage times the budgeted average unit contribution margin. (p. 613)

marketing expense budget a budget that outlines planned expenditures for selling and distribution activities. (p. 384)

marketing (selling) costs those costs necessary to market and distribute a product or service. (p. 15)

markup a percentage applied to base cost for the purpose of calculating price; the markup includes desired profit and any costs not included in the base. (p. 590)

master budget the collection of all area and activity budgets representing a firm's comprehensive plan of action. (p. 373)

materials requisition form a document used to identify the cost of raw materials assigned to each job. (p. 162)

maturity stage the stage in a product's life cycle when sales increase at a decreasing rate. (p. 657)

maximum transfer price the transfer price that will make the buying division no worse off if an input is acquired internally. (p. 509)

method of least squares a statistical method to find a line that best fits a set of data. It is used to break out the fixed and variable components of a mixed cost. (p. 62)

minimum transfer price the transfer price that will make the selling division no worse off if the intermediate product is sold internally. (p. 509)

mix variance the difference in the standard cost of the mix of actual material inputs and the standard cost of the material input mix that should have been used. (p. 464)

mixed costs costs that have both a fixed and a variable component. (p. 51)

monopolistic competition a market that is close to the competitive market. There are many sellers and buyers, low barriers to entry, but the products are differentiated on some basis. (p. 589)

monopoly a market in which barriers to entry are so high that there is only one firm selling a unique product. (p. 589)

multinational corporation (MNC) a corporation for which a significant amount of business is done in more than one country. (p. 491)

multiple regression the use of least-squares analysis to determine the parameters in a linear equation involving two or more explanatory variables. (p. 70)

multiple-period quality trend report a graph that plots quality costs (as a percentage of sales) against time. (p. 795)

myopic behaviour managerial actions that improve budgetary performance in the short run at the expense of the long-run welfare of the organization. (p. 410, 498)

N

net income operating income less taxes, interest expense, and research and development expense. (p. 102)

net realizable value method a method of allocating joint production costs to the joint products based on their proportionate share of eventual revenue less further processing costs. (p. 344)

nonfinancial measures measures expressed in nonmonetary units. (p. 752)

nonproduction costs those costs associated with the functions of selling and administration. (p. 12)

non-unit-based drivers factors, other than the number of units produced, that measure the demands that cost objects place on activities. (p. 263)

non-unit-level drivers explain the changes in cost as factors other than units change. (p. 46)

non-value-added activities activities either unnecessary or necessary but inefficient and improvable. (p. 705)

non-value-added costs costs that are caused either by non-value-added activities or by the inefficient performance of value-added activities. (p. 706)

normal activity level the average activity level that a firm experiences over more than one fiscal period. (p. 160)

normal cost of goods sold the cost of goods sold figure obtained when the per-unit normal cost is used. (p. 171)

normal costing system a cost measurement system in which the actual costs of direct materials and direct labour are assigned to production and a predetermined rate is used to assign overhead costs to production. (p. 158, 259)

normal spoilage spoilage that is expected with an efficient production process and that may require extra work to make the units saleable, or may result in the units being discarded. (p. 174)

O

objective function the function to be optimized, usually a profit function; thus, optimization usually means maximizing profits. (p. 18-15)

objective measures measures that can be readily quantified and verified. (p. 752)

observable quality costs those quality costs that are available from an organization's accounting records. (p. 784)

oligopoly a market structure characterized by a few sellers and high barriers to entry. (p. 589)

operating assets those assets used to generate operating income, consisting usually of cash, inventories, receivables, property, plant, and equipment. (p. 494)

operating budgets budgets associated with the income-producing activities of an organization. (p. 373)

operating expenses the money an organization spends in turning inventories into throughput. (p. 18-20)

operating income revenues minus expenses from the firm's normal operations. Income taxes are excluded. (p. 102, 494)

operating leverage the use of fixed costs to extract higher percentage changes in profits as sales activity changes. Leverage is achieved by increasing fixed costs while lowering variable costs. (p. 123)

operation costing a costing system that uses job-order costing to assign materials costs and process costing to assign conversion costs. (p. 155, 232)

operational activities day-to-day activities performed as a result of the structure and processes selected by an organization. (p. 648)

operational cost drivers those factors that drive the cost of operational activities. (p. 648)

operational productivity measure measures that are expressed in physical terms. (p. 853)

operations process a process that produces and delivers existing products and services to customers. (p. 759)

opportunity cost approach a transfer pricing system that identifies the minimum price that a selling division would be willing to accept and the maximum price that a buying division would be willing to pay. (p. 509)

ordering costs the costs of placing and receiving an order. (p. 18-2)

organizational cost drivers structural and procedural factors that determine the long-term cost structure of an organization. (p. 647)

outsourcing the payment by a company for a business function that was formerly done in-house. (p. 555)

overapplied overhead the overhead variance resulting when applied overhead is greater than the actual overhead cost incurred. (p. 261)

overhead all production costs other than direct materials and direct labour. (p. 13)

overhead budget a budget that reveals the planned expenditures for all indirect manufacturing items. (p. 380)

overhead variance the difference between the actual overhead and the applied overhead. (p. 261)

P

partial productivity measurement a ratio that measures productive efficiency for one input. (p. 853)

participative budgeting an approach to budgeting that allows managers who will be held accountable for budgetary performance to participate in the budget's development. (p. 408)

penetration pricing the pricing of a new product at a low initial price, perhaps even lower than cost, to build market share quickly. (p. 593)

perfectly competitive market a market (or industry) characterized by many buyers and sellers—no one of which is large enough to influence the market—a homogeneous product, and easy entry into and exit from the industry. (p. 589)

performance reports accounting reports that provide feedback to managers by comparing planned outcomes with actual outcomes. (p. 21)

period costs costs such as marketing and administrative costs that are expensed in the period in which they are incurred. (p. 15)

perishability an attribute of services that means that they cannot be inventoried but must be consumed when performed. (p. 12, 152)

perquisites a type of fringe benefit over and above salary that is received by managers. (p. 505)

physical flow schedule a schedule that accounts for all units flowing through a department during a period. (p. 212)

physical units method a method of allocating joint production costs based on each product's share of total units. (p. 338)

planning setting objectives and identifying methods to achieve those objectives. (p. 21)

post-purchase costs the costs of using, maintaining, and disposing of a product incurred by the customer after purchasing a product. (p. 644, 757)

post-sales service process a process that provides critical and responsive service to customers after the product or service has been delivered. (p. 760)

practical activity level the output a firm can achieve if it is operating efficiently. (p. 160)

practical capacity the efficient level of activity performance. (p. 53)

predatory pricing the practice of setting prices below cost for the purpose of injuring competitors and eliminating competition. (p. 593)

predetermined overhead rate estimated overhead divided by the estimated level of production activity. It is used to assign overhead to production. (p. 160, 259)

prevention costs costs incurred to prevent defects in products or services being produced. (p. 783)

price discrimination charging different prices to different customers for essentially the same commodity. (p. 594)

price elasticity of demand measured as the percentage change in quantity divided by the percentage change in price. (p. 588)

price gouging when firms with market power (i.e., little or no competition) price products "too high." (p. 593)

price skimming a pricing strategy in which a higher price is charged at the beginning of a product's life cycle, then lowered at later phases of the life cycle. (p. 593)

price standards the price that should be paid per unit of input. (p. 436)

price (rate) variance the difference between standard price and actual price multiplied by the actual quantity of inputs used. (p. 440)

price volume variance the difference between actual volume sold and expected volume sold multiplied by the expected price. (p. 609)

price-recovery component the difference between the total profit change and the profit-linked productivity change. (p. 856)

primary activity an activity that is consumed by a product or customer (i.e., a final cost object). (p. 273)

prime cost the sum of direct materials cost and direct labour cost. (p. 13)

private costs environmental costs that an organization has to pay. (p. 805)

process a series of activities (operations) that are linked to perform a specific objective. (p. 204)

process creation installing an entirely new process to meet customer and financial objectives. (p. 725)

process improvement incremental and constant increases in the efficiency of an existing process. (p. 724)

process innovation (business re-engineering) the performance of a process in a radically new way with the objective of achieving dramatic improvements in response time, cost, quality, and other important competitive factors. (p. 724)

process value analysis (PVA) an analysis that defines activity-based responsibility accounting, focuses on accountability for activities rather than costs, and emphasizes the maximization of system-wide performance instead of individual performance. (p. 704)

process value chain the innovation, operations, and post-sales service processes. (p. 759)

process-costing principle the period's unit cost is computed by dividing the costs of the period by the output of the period. (p. 210)

producing departments a unit within an organization responsible for producing the products or services that are sold to customers. (p. 315)

product diversity the situation present when products consume overhead in different proportions. (p. 265)

product life cycle the time a product exists—from conception to abandonment; the profit history of the product according to four stages: introduction, growth, maturity, and decline. (p. 597, 655)

product-level activities activities performed that enable the various products of a company to be produced. (p. 279)

production budget a budget that shows how many units must be produced to meet sales needs and satisfy ending inventory requirements. (p. 377)

production (or product) costs those costs associated with the manufacture of goods or the provision of services. (p. 12)

production Kanban a card or marker that specifies the quantity the preceding process should produce. (p. 18-11)

production report a report that summarizes the manufacturing activity for a department during a period and discloses physical flow, equivalent units, total costs to account for, unit cost computation, and costs assigned to goods transferred out and to units in ending work in process. (p. 208)

productivity producing output efficiently, using the least quantity of inputs possible. (p. 852)

productivity measurement assessment of productivity changes. (p. 852)

profile measurement a series or vector of separate and distinct partial operational measures. (p. 855)

profit centre a responsibility centre in which a manager is responsible for both revenues and costs. (p. 490)

Profit-Linkage Rule for the current period, calculate the cost of the inputs that would have been used in the absence of any productivity change and compare this cost with the cost of the inputs actually used. The difference in costs is the amount by which profits changed because of productivity changes. (p. 856)

profit-linked productivity measurement an assessment of the amount of profit change—from the base period to the current period—attributable to productivity changes. (p. 856)

profit-volume graph a graphical portrayal of the relationship between profits and sales activity. (p. 116)

pseudoparticipation a budgetary system in which top management solicits inputs from lower-level managers and then ignores those inputs. Thus, in reality, budgets are dictated from above. (p. 409)

Q

quality of conformance conforming to the design requirements of the product. (p. 782)

quality product or service a product that meets or exceeds customer expectations. (p. 782)

quantity standards the quantity of input allowed per unit of output. (p. 436)

R

realized external failure costs the environmental costs caused by environmental degradation and paid for by the responsible organization. (p. 804)

reciprocal method a method that simultaneously allocates service costs to all user departments. It gives full consideration to interactions among service departments. (p. 330)

relevant costs (revenues) future costs (revenues) that differ across alternatives. (p. 551)

relevant range the range over which an assumed cost relationship is valid for the normal operations of a firm. (p. 47, 119)

reorder point the point in time at which a new order (or setup) should be initiated. (p. 18-4)

resale price method computes a transfer price equal to the sales price received by the reseller less an appropriate markup. (p. 517)

research and development expense budget a budget that outlines planned expenditures for research and development. (p. 385)

residual income the difference between operating income and the minimum required dollar return on a company's operating assets. (p. 498)

resource drivers factors that measure the demands placed on resources by activities and are used to assign the cost of resources to activities. (p. 275)

responsibility accounting a system that measures the results of each responsibility centre and compares those results with some measure of expected or budgeted outcome. (p. 490, 723)

responsibility centre a segment of the business whose manager is accountable for specified sets of activities. (p. 489)

return on investment (ROI) the ratio of operating income to average operating assets. (p. 494)

revenue centre a responsibility centre in which a manager is responsible only for sales. (p. 490)

revenue-producing life the time a product generates revenue for a company. (p. 657)

robustness exact conformance to the target value (no tolerance allowed). (p. 783)

ropes actions taken to tie the rate at which raw material is released into the plant (at the first operation) to the production rate of the constrained resource. (p. 18-22)

S

safety stock extra inventory carried to serve as insurance against fluctuations in demand. (p. 18-4)

sales budget a budget that describes expected sales in units and dollars for the coming period. (p. 377)

sales mix the relative combination of products (or services) being sold by an organization. (p. 112)

sales mix variance the sum of the change in units for each product multiplied by the difference between the budgeted contribution margin and the budgeted average unit contribution margin. (p. 613)

sales price variance the difference between actual price and expected price multiplied by the actual quantity or volume sold. (p. 609)

sales-revenue approach an approach to CVP analysis that uses sales revenue to measure sales activity. Variable costs and contribution margin are expressed as percentages of sales revenue. (p. 107)

sales-value-at-split-off method a method of allocating joint production costs based on each product's share of revenue realized at the split-off point. (p. 342)

scattergraph a plot of (X, Y) data points. For cost analysis, X is activity usage and Y is the associated cost at that activity level. (p. 58)

scatterplot method a method to fit a line to a set of data using two points that are selected by judgment. It is used to break out the fixed and variable components of a mixed cost. (p. 58)

secondary activity an activity that is consumed by intermediate cost objects such as materials and primary activities. (p. 273)

sell or process further relevant costing analysis that focuses on whether or not a product should be processed beyond the split-off point. (p. 562)

sensitivity analysis a "what-if" technique that examines altering certain key variables to assess the effect on the original outcome. (p. 124)

separable costs costs incurred beyond the split-off point that can be assigned to specific products identified at the split-off point. (p. 337)

sequential (or step) method a method that allocates service costs to user departments in a sequential manner. It gives partial consideration to interactions among service departments. (p. 328)

services a task or activity performed for a customer or an activity performed by a customer using an organization's products or facilities. (p. 11)

setup costs the costs of preparing equipment and facilities so that they can be used for production. (p. 18-2)

shadow price the amount by which throughput will increase for one additional unit of scarce resource. (p. 18-17)

short run period of time in which at least one cost is fixed. (p. 52)

simplex method an algorithm that identifies the optimal solution for a linear programming problem. (p. 18-17)

single-loop feedback information about the effectiveness of strategy implementation. (p. 764)

slope parameter the variable cost per unit of activity usage, represented by V in the cost formula $Y = F + VX$. (p. 57)

source document a document that describes a transaction and is used to keep track of costs as they occur. (p. 156)

special-order decisions decisions that focus on whether a specially priced order should be accepted or rejected. (p. 562)

split-off point the point at which the joint products become separate and identifiable. (p. 337)

standard bill of materials a listing of the type and quantity of materials allowed for a given level of output. (p. 444)

standard cost per unit the per-unit cost that should be achieved given materials, labour, and overhead standards. (p. 438)

standard cost sheet a listing of the standard costs and standard quantities of direct materials, direct labour, and overhead that should apply to a single product. (p. 438)

standard hours allowed the direct labour hours that should have been used to produce the actual output (Unit labour standard × Actual output). (p. 439)

standard quantity of materials allowed the quantity of materials that should have been used to produce the actual output (Unit materials standard × Actual output). (p. 439)

static budget a budget for a particular level of activity. (p. 397)

step-cost function a cost function in which cost is defined for ranges of activity usage rather than point values. The function has the property of displaying constant cost over a range of activity usage and then changing to a different cost level as a new range of activity usage is encountered. (p. 54)

step-fixed costs a step-cost function in which cost remains constant over wide ranges of activity usage. (p. 55)

step-variable costs a step-cost function in which cost remains constant over relatively narrow ranges of activity. (p. 54)

stock option the right to purchase a certain amount of stock at a fixed price. (p. 506)

stock-out costs the costs of insufficient inventory. (p. 18-2)

strategic cost management the use of cost data to develop and identify superior strategies that will produce a sustainable competitive advantage. (p. 644)

strategic decision making choosing among alternative strategies with the goal of selecting a strategy or strategies that provide a company with reasonable assurance of long-term growth and survival. (p. 644)

strategic positioning the process of selecting the optimal mix of cost leadership, differentiation, and focusing strategies. (p. 645)

strategic-based responsibility accounting system (strategic-based performance management system) a responsibility accounting system that translates an organization's mission and strategy into operational objectives and measures for four different perspectives: the financial perspective, the customer perspective, the process perspective, and the learning and growth (infrastructure) perspective. (p. 751)

strategy choosing the market and customer segments, identifying critical internal business processes at which the firm must excel to increase customer value, and selecting the individual and organizational capabilities required to achieve the firm's internal, customer, and financial objectives. (p. 645, 755)

strategy map a detailed graphical representation of an organization's strategic objectives and the cause-and-effect relationships that exist among them. (p. 765)

stretch targets targets that are set at levels that, if achieved, will transform the organization within a period of three to five years. (p. 754)

structural activities activities that determine the underlying economic structure of the organization. (p. 647)

subjective measures measures that are nonquantifiable whose values are judgmental in nature. (p. 752)

sunk cost a past cost—a cost already incurred. (p. 551)

supplies materials necessary for production but that do not become part of the finished product or are not used in providing a service. (p. 13)

support departments a unit within an organization that provides essential support services for producing departments. (p. 315)

sustainable development development that meets the needs of the present without compromising the ability of future generations to meet their own needs. (p. 801)

system a set of interrelated parts that performs one or more processes to accomplish specific objectives. (p. 3)

T

tactical cost analysis the use of relevant cost data to identify the alternative that provides the greatest benefit to the organization. (p. 550)

tactical decision making choosing among alternatives with only an immediate or limited end in view. (p. 548)

Taguchi loss function a function that assumes any variation from the target value of a quality characteristic causes hidden quality costs. (p. 785)

tangible products goods produced by converting raw materials through the use of labour and capital inputs such as plant, land, and machinery. (p. 11)

target cost the difference between the sales price needed to achieve a projected market share and the desired per-unit profit. (p. 662)

target costing a method of determining the cost of a product or service based on the price that customers are willing to pay. Also referred to as price-driven costing. (p. 592)

tariff the tax on imports levied by the federal government. (p. 552)

technical efficiency point at which for any mix of inputs that will produce a given output, no more of any one input is used than is absolutely necessary. (p. 852)

testable strategy set of linked objectives aimed at an overall goal that can be restated into a sequence of cause-and-effect hypotheses. (p. 764)

theoretical activity level the maximum output possible for a firm under perfect operating conditions. (p. 160)

theory of constraints method used to continuously improve manufacturing activities and nonmanufacturing activities. (p. 8, 18-17)

throughput the rate at which an organization generates money through sales. (p. 18-19)

time buffer the inventory needed to keep the constrained resource busy for a specified time interval. (p. 18-21)

time ticket a document used to identify the cost of direct labour for a job. (p. 163)

Time-Driven Activity-Based Costing (TDABC) a before-the-fact simplification method that simplifies Stage 1 by eliminating the need for detailed interviewing and surveying to determine resource drivers. (p. 280)

total budget variance the difference between the actual cost of an input and its planned cost. (p. 439)

total (overall) sales variance the sum of the sales price and sales volume variances. (p. 609)

total preventive maintenance a program of preventive maintenance that has zero machine failures as its standard. (p. 18-11)

total product the complete range of tangible and intangible benefits a customer receives from a product. (p. 644)

total productive efficiency the point at which technical and price efficiency are achieved. (p. 852)

total productivity measurement an assessment of productive efficiency for all inputs combined. (p. 855)

total quality management (TQM) a philosophy that requires managers to strive to create an environment that will enable workers to manufacture perfect (zero defects) products. (p. 652)

traceability the ability to assign a cost directly to a cost object in an economically feasible way using a causal relationship. (p. 10)

transaction drivers measure the number of times an activity is performed, such as the number of treatments and the number of requests. (p. 276)

transfer prices the price charged for goods transferred from one division to another. (p. 508)

transfer pricing problem the problem of finding a transfer pricing system that simultaneously satisfies the three objectives of accurate performance evaluation, goal congruence, and autonomy. (p. 509)

transferred-in cost the cost of goods transferred in from a prior process. (p. 205)

treasurer the financial officer responsible for the management of cash and investment capital. (p. 21)

turnover a measure that is found by dividing sales by average operating assets to show how productively assets are being used to generate sales. (p. 494)

U

underapplied overhead the overhead variance resulting when the actual overhead cost incurred is greater than the applied overhead. (p. 261)

unfavourable (U) variance a variance produced whenever the actual input amounts are greater than the budgeted or standard allowances. (p. 441)

unit standard cost the product of these two standards: Standard price $\times$ Standard quantity ($SP \times SQ$). (p. 436)

unit-based drivers explain changes in cost as units produced change. (p. 259)

unit-level activities activities that are performed each time a unit is produced. (p. 279)

unit-level drivers activity drivers that explain changes in cost as units produced change. (p. 46)

unrealized external failure (societal) costs environmental costs caused by an organization but paid for by society. (p. 804)

unused capacity the difference between the acquired activity capacity and the actual activity usage. (p. 53)

unused capacity variance the difference between acquired capacity (practical capacity) and actual capacity. (p. 717)

usage (efficiency) variance the difference between standard quantities and actual quantities multiplied by standard price. (p. 440)

V

value chain the set of activities required to design, develop, produce, market, distribute, and service a product (the product can be a service). (p. 5)

value stream made up of value-added and non-value-added activities required to bring a product group or service from its starting point to a finished product in the hands of the customer. (p. 840)

value-added activities activities that are necessary to achieve corporate objectives and remain in business. (p. 705)

value-added costs costs caused by value-added activities. (p. 705)

value-added standard the optimal output level for an activity. (p. 708)

value-chain analysis identifying and exploiting internal and external linkages with the objective of strengthening a firm's strategic position. (p. 644)

variable budget (p. 401) see **flexible budget**.

variable cost ratio variable costs divided by sales revenue. It is the proportion of each sales dollar needed to cover variable costs. (p. 106)

variable costing a costing method that assigns only variable manufacturing costs to the product; these costs include direct materials, direct labour, and variable overhead. Fixed overhead is treated as a period cost and is expensed in the period incurred. (p. 603)

variable costs costs that in total vary in direct proportion to changes in a cost driver. (p. 48)

variable overhead efficiency variance the difference between the actual direct labour hours used and the standard hours allowed multiplied by the standard variable overhead rate. (p. 454)

variable overhead spending variance the difference between the actual variable overhead and the budgeted variable overhead based on actual hours used to produce the actual output. (p. 453)

velocity the number of units that can be produced in a given period of time (e.g., output per hour). (p. 760)

vendor Kanban a card or marker that signals to a supplier the quantity of materials that need to be delivered and the time of delivery. (p. 18-11)

W

waste anything customers do not value. (p. 845)

weight factor a value used to assign weights to various joint products in accordance with their relative size, difficulty to produce, etc. (p. 340)

weighted average cost of capital the proportionate share of each method of financing is multiplied by its percentage cost and summed. (p. 501)

weighted average costing method a unit-costing method that merges prior-period work and costs with current-period work and costs. (p. 222)

withdrawal Kanban a marker or card that specifies the quantity that a subsequent process should withdraw from a preceding process. (p. 18-11)

work in process consists of all partially completed units found in production at a given point in time. (p. 16)

work orders used to collect production costs for product batches and to initiate production. (p. 232)

work-in-process inventory file the collection of all job cost sheets. (p. 162)

Y

yield variance the difference in the standard material cost of the standard yield and the standard material cost of the actual yield. (p. 464)

Z

zero defects a quality performance standard that requires all products and services to be produced and delivered according to specifications. (p. 783)

zero-base budgeting a method of budgeting in which the prior year's budgeted level is not taken for granted. Existing operations are analyzed, and continuance of the activity or operation must be justified on the basis of its need or usefulness to the organization. (p. 397)

Check Figures

Check figures are given for selected exercises and problems.

Chapter 1

1-13	2. Cost of goods manufactured = $725,000
	3. Conversion cost = $53.90
1-14	1. Ending inventory = $42,000
	3. Cost of goods manufactured = $41,250
	5. Overhead = $156,900
1-15	1. Cost of goods manufactured = $769,600
	2. Cost of goods sold = $759,600
1-16	1. Finished goods ending inventory = $274,750
	2. Cost of goods sold = $3,493,250
	3. Operating income = $653,450
1-17	1. Cost of goods manufactured = $801,560
	2. Cost of goods sold = $756,860
1-18	5. Operating income = $4,645
1-20	2. Cost of goods manufactured = $965,000
	3. Conversion cost = $16.10
1-21	2. Operating income = $ 186,500
1-22	3. Total overhead = $300,000
	4. Cost of goods manufactured = $706,000
	5. Cost of goods sold = $708,000
	6. Operating income = $122,200, 10.18%
1-23	1. Cost of goods manufactured = $5,018,000
	3. Operating income = $1,048,250, 15.25%
1-24	1. Cost of goods manufactured = $295,000
	2. Cost of goods sold = $236,000
1-25	1. Cost of services sold = $1,577,500
	3. Operating income = $650,500
1-26	1. Cost of goods manufactured = $356,200
	2. Operating income = $334,000

Chapter 2

2-9	3. Total fixed overhead cost = $720,000
	6. Unit fixed cost = $28.80 per client
2-12	2. Fixed activity rate = $2.661 per test
2-14	3. April cost = $4,214
2-17	2. $Y = \$174,768.82$
2-18	3. $Y = \$94,140$
2-19	1. $Y = \$9,344 + \$8.30X$
2-20	2. $Y = \$5,791$

2-21	2. $Y = \$379,842$
2-23	2. Total labour cost for 16 sets = $202,502
2-26	4. Charge/day = $124.80
2-28	1. Total unit variable cost = $210
	3. Total unit variable cost = $222
2-30	3. $Y = \$91,815$
	4. $Y = \$87,195$
2-31	2. $Y = \$63,696$
	3. $Y = \$69,101$
2-32	1. $Y = \$9,025$
	2. $Y = \$9,227$
	4. $Y = \$9,375.80$
2-33	2. $Y = \$34,895.70$

Chapter 3

3-8	1. Contribution margin = $7.20
	3. Operating income = $19,440
3-9	1. Break-even in units = 36,000
	2. Operating income = $2,775,000
3-10	1. Contribution margin ratio = 80%
	3. Sales revenue for target profit = $10,812,500
3-11	1. Break-even in units = 18,000
	2. Operating income = $46,000
3-12	1. Break-even in units = 68 jobs per month
	4. Break-even in units = 50 jobs per month
3-13	1. Break-even sales revenue = $4,140,000
	2. Margin of safety = $6,640,000
3-14	1. Net income after taxes = $627,900
3-15	1. Break-even in units = 30
3-17	1. Contribution margin ratio = 0.25
	3. Operating income = $14,880
	4. Margin of safety = $59,520
3-19	2. Revenue = $13,071,895
3-20	1. Break-even in units = 400,000
	3. Break-even in units = 292,858
3-21	1. Operating income = $139,000
	3. Break-even sales = $513,889
3-22	1. (a) Operating income = $1,025,000
3-24	3. Margin of safety = $256,140
	4. Net income = $120,000
3-25	3. New break-even in units = 87,805
	5. New break-even in units = 127,778
3-26	3. Margin of safety = $500,000

3-27 2. Operating income = $1,292,800
5. Margin of safety = $6,481,818

3-28 2. Revenue = $584,735
4. Revenue = $1,311,386

3-29 3. Percentage change in net income = 160%

3-30 3. Net income = $119,900

3-31 2. Total unit variable cost = $21

Chapter 4

4-7 2. Total cost = $4,130

4-8 2. Total unit cost: June = $776, July = $701, August = $607

4-11 1. Job 78: Total job cost = $6,408
2. Job 80: Ending Work in Process = $9,060

4-12 3. Ending Work in Process = $52,075
4. Cost of goods sold = Job 115 = $24,180
5. Price of Job 115 = $29,016

4-14 3. Work in process, August 31 = $12,345
4. Cost of goods sold = $19,324
5. August sales revenue = $28,986

4-15 Operating income = $7,802

4-16 2. (d) Finished Goods Inventory = $1,630

4-17 2. Ending Work in Process = $2,500

4-18 2. Ending Work in Process = $12,890

4-20 3. Ending Work in Process = $72,750
4. Cost of goods sold = $258,988

4-21 3. Cost of goods manufactured = $160,000

4-22 1. Bid price: Job 1 = $8,890, Job 2 = $16,926
2. Bid price: Job 1 = $9,821, Job 2 = $17,136

4-24 1. Unit bid price: Job 97-28 = $18.75, Job 97-35 = $60.00
2. Unit bid price: Job 97-28 = $14.67, Job 97-35 = $101.01

4-25 1. Total cost = $47.50
4. Price = $641.25

4-26 1. Total cost = $45.00
2. Total cost = $51.00

4-27 4. Profit = $18,951

4-28 1. Gross profit = $19

4-29 2. Gross profit = $644.46
4. Total price = $2,980.25

Chapter 5

5-12 2. Cost per haircut = $10

5-13 2. Unit cost = $11 per unit

5-14 4. (c) Cost of EWIP = $280,000

5-15 Total costs accounted for = $5,364,000

5-16 1. Total units accounted for = 135,000

5-17 1. Total cost of goods transferred out = $251,000
2. Total units accounted for = 38,000

5-20 2. Cost of goods transferred out = $68,750

5-21 1. Total units accounted for = 70,000
2. Unit cost = $10.40 per equivalent unit

5-26 1. Total units accounted for = 135,000
3. Conversion cost = $5,871,150

5-27 3. Total unit cost = $77.420

5-28 Total costs accounted for = $246,240

5-29 Total costs accounted for = $246,000

5-35 2. Unit cost: Regular strength = $2.17, Extra strength = $2.17

5-37 2. Total unit cost = $8.50
3. Cost of units transferred out = $2,975,000

5-38 3. Total unit cost = $0.39
4. Loss due to spoilage = $390

Chapter 6

6-10 3. Underapplied overhead = $11,600
4. Unit cost = $8.8536

6-11 2. Jackson: Underapplied overhead = $2,380
Jalil: Overapplied overhead = $3,750

6-12 2. Overapplied overhead = $56,000

6-21 1. Total OH assigned: Model X = $171,000, Model Y = $129,000
2. Total OH assigned: Model X = $185,740, Model Y = $114,305

6-22 2. Total OH assigned: Model X = $171,001, Model Y = $129,002

6-24 2. Unit cost (ABC): Standard = $189.50, Deluxe = $571.00

6-27 1. Gross margin: Part 127 = $10.50, Part 234 = $11.97
2. Gross margin (loss): Part 127 = $16.60, Part 234 = $(18.54)

6-28 3. Unused capacity cost = $92,000

6-29 2. Overhead per unit: Cylinder A = $1,467, Cylinder B = $600
4. Overhead per unit: Cylinder A = $1,467, Cylinder B = $600
5. Overhead per unit: Cylinder A = $1,467, Cylinder B = $600

6-30 1. Percentage of total activity costs = 85%
2. Overhead per unit: Cylinder A = $1,004.38, Cylinder B = $832.01

6-31 3. Overhead per unit: Scientific = $5.95, Business = $1.61

Chapter 7

7-15 2. Total cost = $1.18

7-16 2. Total = $431.67

7-19 1. Total cost: Department A = $55,000, Department B = $55,000

7-20 2. Toothpaste = $56.01 per MHr, Tooth Whitener = $23.27 per MHr

7-21 2. Toothpaste = $52.60 per MHr, Tooth Whitener = $24.41 per MHr

7-22 2. Assembly = $14.00 per DLHr, Finishing = $8.18 per DLHr

7-23 2. Assembly = $13.70 per DLHr, Finishing = $8.36 per DLHr

7-24 2. Assembly = $14.09 per DLHr, Finishing = $8.12 per DLHr

7-27 2. Incremental value of further processing = $10,000

7-34 2. In-house members: Total = $233,584, Out-of-house members: Total = $41,293

Chapter 8

8-18 Total = $11,605,000

8-22 1. Ending cash balance = $2,783

8-23   Total cash receipts: August = $46,623, September = $58,105

8-24   1. Total: August = $22,350, September = $27,814
        2. Total: July = $23,683, August = $25,644

8-25   1. Total overhead costs = $188,100

8-28   1. Total = $ 3,026,000

8-29   7. Total unit cost = $76.96
        8. Budgeted cost of goods sold = $4,155,302
        9. Income before income taxes = $42,298

8-31   1. Total predicted overhead = $423,145
        2. Total predicted overhead = $423,201

8-32   1. Total predicted overhead = $423,184
        2. Total predicted overhead = $423,109

Chapter 9

9-12   1. SQ = 1,456,000 kilograms
        2. SH = 210,000 hours

9-13   1. MPV = $6,240 U, MUV = $1,040 U
        2. LRV = $0, LEV = $900 U

9-16   1. Yield ratio = 0.90
        2. SPy = $450.01
        4. Direct material mix variance = $39,973.65 U

9-18   1. Yield ratio = 5
        2. Standard cost = $3.04 per unit of yield

9-19   1. MPV = $2,714 F, MUV = $280 U
        2. LRV = $9,100 U, LEV = $12,000 U

9-22   1. Direct materials = $24,480, Direct labour = $81,600
        3. MPV = $1,075 F, MUV = $1,320 U
        4. LRV = $1,712.50 U, LEV = $600 U

9-23   1. Standard variable overhead rate = $2.22 per DLHr
        2. Total variable overhead variance = $4,920 U

9-24   1. MPV = $27,400 F, MUV = $13,780 U
        2. LRV = $2,005 F, LEV = $6,300 U

9-26   1. MPV = $5,416 F, MUV = $6,480 U
        2. LRV = $730 F, LEV = $8,100 U

9-27   2. MPV = $2,800 F, MUV (Regular) = $1,200 F
        3. LRV = $3,700 F, LEV (Regular) = $5,600 U

9-28   1. Yield variance = $42,750 F

9-29   1. Yield variance = $21,000 F

9-30   4. LEV = $7,600 F

9-32   2. Cost of goods transferred out = $1,995,000

Chapter 10

10-9   1. KiddieKamp residual income = $200

10-10   1. EVA = $(34,000)

10-12   1. Increased profit = $182,000

10-13   2. Transfer price = $241.20

10-14   1. Transfer price = $4.22
         2. Transfer price = $3.92

10-15   2. Transfer price = $7.30

10-16   1. South American = $7,500
         2. South American = 0.19%
         3. South American = 10.19%

10-17   2. D's residual income = $336

10-18   1. Residual income = $51,000
         2. ROI = 18.40%

10-19   1. Value of option = $230,000

10-21   5. Contribution margin = $73,500

10-23   3. Total addition to profits = $768,000

Chapter 11

11-8   1. Incremental loss per golf kit = $(1.55)

11-9   2. Maximum price = $68.92

11-11   1. Loss from accepting order = $(10,955)
         3. Loss from accepting order = $(3,680)

11-12   1. Operating income = $367,800
         2. Operating income = $367,800

11-13   1. Gross profit = $46,000

11-14   2. Total tariff-related savings = $96,720

11-15   1. Total annual cost = $3,050

11-17   1. Operating income = $18,280
         2. Operating income = $49,889

11-19   2. Operating income = $1,749,000

11-20   2. Activity rate per procedure = $176

11-24   1. Total relevant costs: Lease and make = $51,000, Buy = $50,000
         2. Total relevant cost: Make = $58,600, Buy = $50,000

Chapter 12

12-13   1. Markup percentage = 16%
         2. Bid = $539,400

12-15   2. Operating income = $145,900
         3. Operating income = $133,750

12-16   2. Operating income = $31,890

12-18   1. Sales price variance = $1,050 F
         2. Price volume variance = $715 F

12-19   1. Product P: Price volume variance = $800,000 F

12-21   1. Total cost = $1,476,000
         2. Total cost = $1,008,000

12-22   3. (a) Operating income = $214,400

12-23   1. Cost of goods sold: Year 1 = $412,500, Year 2 = $510,000
         2. Cost of goods sold: Year 1 = $337,500, Year 2 = $405,000

12-24   1. Contribution margin variance = $17,300 F
         3. Sales mix variance = $300 F

12-25   1. Contribution margin variance = $3,490 F

12-26   1. Net change in income = $(600)

12-28   1. Operating income (loss) = $(238.45)
         2. Operating income = $785

12-29   1. Operating income = $2,455,000

12-30   1. Operating income = $(75,000)

12-31   1. Profit on first year = $675

12-32   1. Operating income = $296,400

12-33   1. Operating income = $72,250

Chapter 13

13-10   1. Unit cost: Wood = $200.05, Gardner = $208.24

13-11   1. Customer profitability = $5,474,000

13-19   1. Cost per unit of average product = $62.50
         2. Cost per unit of average product = $63

13-20   1. Unit cost: Bach = $404.60, Rivera = $310.98

13-22   1. Total savings = $2,675,000

13-23   4. Income before taxes = $191,250

13-24   1. Target cost = $115 per unit
         3. Design A: Profit per unit = $15.43
         4. Increase in benefits = $800,000

13-26   1. After JIT unit cost = $32.54

Chapter 14

14-20   1. Total: Fixed = $1,221,000, Variable = $1.38
        4. Total: Fixed = $441,000, Variable = $1.38
14-21   2. Total = $852,000
        3. Cost reduction to maintain = $6, Cost reduction to expand = $8
        4. Unit savings = $8.35
14-22   2. Moulding: SQ = 2,538,000 kg, Engineering: SQ = 126,720 eng. hrs
14-23   1. Target cost = $1,224
        2. Unit non-value-added cost = $414

Chapter 15

15-7   1. Theoretical conversion cost per unit = $324
        2. Applied conversion cost per unit = $540
15-8   1. Cycle time (actual) = 40 minutes per laptop
        2. Assignment per unit (actual) = $120
15-16   1. (d) 2011: 16%, 2013: 18%
           (g) 2011: 6.25%, 2013: 2.22%
           (k) 2011: 2,600, 2013: 13,000
15-19   1. Cycle time (theoretical) = 20 minutes per unit
        2. Assignment per unit (theoretically) = $10.00
        4. Cost = $7.50
15-20   1. MCE = 0.30
15-21   1. Theoretical cycle time = 80 minutes per model
        2. Actual cycle time = 96 minutes per model

Chapter 16

16-9   2. Hidden cost = $225,000
16-12   1. Total quality costs = $1,350,000
16-13   2. Profit potential = $900,000
16-15   2. (c) Bonus pool = $73,600
16-21   1. Total environmental costs = $9,855,000
16-23   1. Unit cost per kilogram: Org AB = $0.542, Org XY = $0.0912
16-24   1. Treatment rate = $0.10 per kilogram
16-25   1. Total quality costs = $8,000,000
16-29   1. Total quality costs = $1,396,500
16-30   1. Total quality costs: January = $86,000, February = $100,500
16-31   1. External failure costs: 2012 = 33.2%, 2013 = 36.1%
16-38   1. Total benefits = $1,131,000, Total costs = $1,179,000

Chapter 17

17-6   1. Total time = 400 minutes
        3. Time saved = 19.83 minutes per bottle (21.33 for continuous)
17-9   1. Unit cost: Model A = $159, Model B = $160.33
17-10   2. Week 3: Average cost = $22.22
17-13   1. Value of productivity = $336,000
17-15   2. Price recovery = $857,400
17-16   2. Change in income = $300,000
        4. Price-recovery component = $(420,000)
17-17   2. Average lunch cost = $8.50
        3. Group A: Average lunch cost = $6.80
17-18   1. Group A: Average lunch cost = $6.17
17-21   1. Current system labour = 0.50
17-22   2. Increase in profits due to productivity = $56,000
17-23   3. Change I = $0, Change II = $495,000

Chapter 18 (www.hansen1ce.nelson.com)

18-5   1. Annual ordering cost = $2,400
        3. Cost of current inventory policy = $17,400
18-6   1. EOQ = 1,600
        2. Total cost = $12,000
18-7   1. EOQ = 2,500
        2. Ordering cost = $562.50
18-8   1. Reorder point = 840 units
18-9   1. EOQ = 60,000
        2. Total cost = $180,000
18-10   1. EOQ = 20,000
        2. Total cost = $180,000
18-11   1. ROP (small casings) = 26,400
18-12   Safety stock = 150
18-15   2. Total contribution margin = $7,560,000
18-17   1. EOQ = 12,000